D0759207
WITHDRAWN

The Western North Atlantic Region

Frontispiece 1. Examples of principal lithofacies in the western North Atlantic Ocean basin. All are core samples from the Deep Sea Drilling Project. Each is identified in parentheses below according to cruise leg, site number (with hole number, A., B., etc., if applicable), core number, core-section number, and depth interval in core section (in centimeters).

1. Un-named formation, Blake-Bahama Basin. Upper Callovian to lower Oxfordian dark claystone with coarser radiolarian siltstone lenses, capped at top by lighter pelmicritic limestone that is graded and laminated (76, 534A, 120, 1, 48-65 cm). These and 63 m of underlying Callovian sedimentary rocks at Site 534A represent the oldest strata cored to date in the western North Atlantic basin.

2. Cat Gap Formation, lower continental rise off New Jersey. Upper Oxfordian to lower Kimmeridgian clayey limestone, well laminated to burrowed (11, 105, 37, 5, 82-113 cm). Colors which reflect the oxidation state of iron in the sediment (reddish = oxidized, grayish = reduced) are separated by diffuse boundaries.

3. Blake-Bahama Formation, lower continental rise off New Jersey. Upper Berriasian to lower Valanginian limestone and chalky limestone (11, 105, 28, 1, 92-141 cm). Limestone is well laminated to burrowed and locally contains flow structures. Darker layers contain clay and nannofossils, and light layers are principally recrystallized nannofossil calcite.

4. Hatteras Formation, western Bermuda Rise. Upper Hauterivian-Barremian greenish gray, laminated to burrowed claystone and laminated, black radiolarian mudstone with pyrite crystals and nodules (43, 387, 37, 4, 2-48 cm). Deposition was under low-oxygen (green-gray) to anoxic conditions (black).

5. Plantagenet Formation, central Bermuda Rise. Upper Cretaceous dusky yellow brown and moderate brown zeolitic claystone (43, 386, 38, 4, 37-70 cm). Darker brown bands are zeolite-rich (~60%) compared to lighter brown bands (~30%). Plantagenet clays accumulated at extremely low rates (~1-3 m/m.y.) beneath well oxygenated waters in the late Cretaceous basin.

6. Plantagenet Formation, lower continental rise off New Jersey. Upper Cretaceous multicolored silty clay with rare sphalerite and zeolites (11, 105, 9, 2, 108-123 cm).

7. Volcaniclastic breccia, flank of Nashville Seamount (43, 382, 25, 2, 110-133 cm). Basalt clasts show variable degrees of alteration and are contained within a matrix of gray-white calcite. Nashville Seamount was the easternmost volcanic edifice formed in the New England Seamount Chain as the North American plate migrated across the New England hotspot in the Late Cretaceous.

8. Volcaniclastic breccia, flank of Vogel Seamount, New England Seamount Chain (43, 385, 23, 2, 99-123 cm). Variably altered, rounded to angular basalt clasts in this Upper Cretaceous breccia are set in a matrix of calcite cement.

9. Volcaniclastic turbidites, flank of Nashville Seamount, New England Seamount Chain (43, 382, 17, 1, 67-114 cm). These lower Campanian beds are laminated to cross-laminated and consist of volcanogenic clay and silty clay with interbeds of marly nannofossil ooze and zeolitic feldspathic silt. They form the uppermost part of the volcaniclastic apron of Nashville Seamount, above the breccias shown in panel 7.

10. Plantagenet Formation, lower continental rise off New Jersey. Upper Cretaceous or Paleogene silty clay with thin clayey quartz-sand stringers, heavy minerals, and palagonite grains (11, 105, 6, 2, 36-61 cm). Sharp color break in lower part of panel is a photographic artifact.

11. Bermuda Rise Formation, western Bermuda Rise. Lower Eocene radiolarian mudstone, more carbonate rich ($\gtrsim$25%) in lower part (43, 387, 22, 3, 116-138 cm). Fine-grained, distal turbidites were episodically deposited in the central western North Atlantic basin at this time; the panel shows the upper part of one turbidite: a thick, homogeneous unit capped by burrow-mottled low-carbonate mud at the top of the turbidite sequence. The mudstones commonly are silicified and contain porcelanitic chert, particularly at the tops and bases of turbidites.

12. Blake Ridge Formation, lower continental rise terrace off New Jersey. The middle Miocene, burrow-mottled, gray-brown hemipelagic mudstone is characteristic of the Blake Ridge Formation (11, 106B, 5, 3, 76-88 cm).

13. Blake Ridge Formation, central Bermuda Rise. Middle Eocene burrowed-mottled, marly biosiliceous ooze (top of a turbidite) capped by laminated biosiliceous ooze and biosiliceous mud (base of a turbidite) (43, 386, 17, 3, 76-98 cm). Biosiliceous component is primarily sponge spicules and radiolarians.

14. Great Abaco Member, Blake Ridge Formation, Blake-Bahama Basin. Lower Miocene silty, calcareous, biosiliceous claystone capped by laminated marly chalk forming the base of a turbidite (76, 534A, 14, 5, 50-65 cm). The Great Abaco Member is characterized by intraclastic chalks and calcareous turbidites deposited within the Blake-Bahama Basin.

2
1
3
4
6
5
8
7
9
11
10
14
13
12

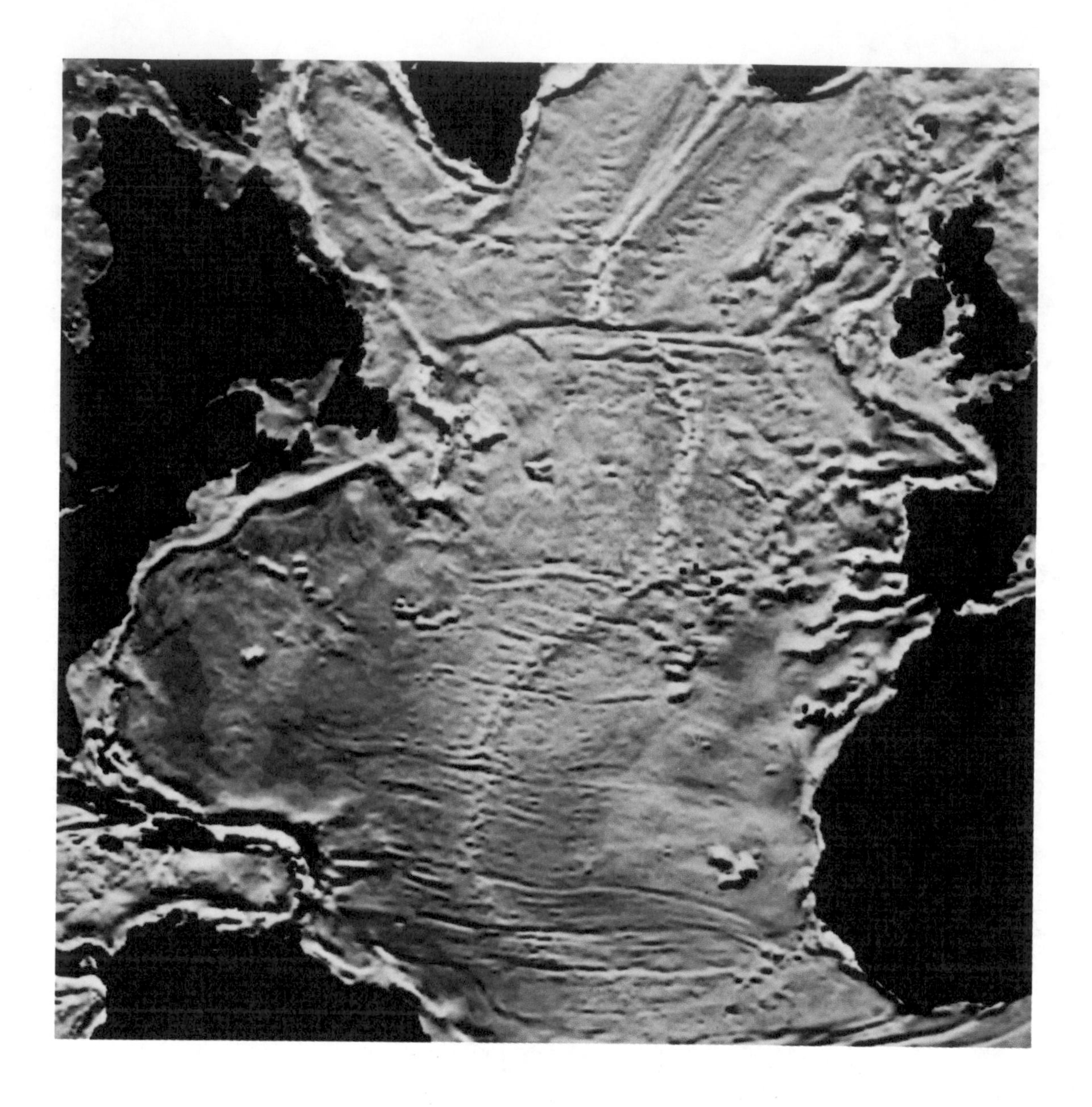

Frontispiece 2. Portion of computer-generated image "Gravity Field of the World's Oceans," derived from SEASAT altimetry by W. F. Haxby. Published by U.S. Navy Office of Naval Research.

Cover photo. Portion of computer-generated image, "Relief of the Surface of the Earth," edited by J. R. Heirtzler, Report MGG-2, National Geophysical Data Center. Submarine topography based on DBDB-5 (Digital Bathymetric Data Base) of U.S. Naval Oceanographic Office.

The Geology of North America
Volume M

The Western North Atlantic Region

Edited by

Peter R. Vogt
Naval Research Laboratory
Washington, D.C. 20375-5000

and

Brian E. Tucholke
Woods Hole Oceanographic Institution
Woods Hole, Massachusetts 02543

1986

Acknowledgment

Publication of this volume, one of the synthesis volumes of *The Decade of North American Geology Project* series, has been made possible by members and friends of the Geological Society of America, corporations, and government agencies through contributions to the Decade of North American Geology fund of the Geological Society of America Foundation.

Following is a list of individuals, corporations, and government agencies giving and/or pledging more than $50,000 in support of the DNAG Project:

ARCO Exploration Company
Chevron Corporation
Cities Service Company
Conoco, Inc.
Diamond Shamrock Exploration Corporation
Exxon Production Research Company
Getty Oil Company
Gulf Oil Exploration and Production Company
Paul V. Hoovler
Kennecott Minerals Company
Kerr McGee Corporation
Marathon Oil Company
McMoRan Oil and Gas Company
Mobil Oil Corporation
Pennzoil Exploration and Production Company
Phillips Petroleum Company
Shell Oil Company
Caswell Silver
Sohio Petroleum Corporation
Standard Oil Company of Indiana
Sun Exploration and Production Company
Superior Oil Company
Tenneco Oil Company
Texaco, Inc.
Union Oil Company of California
Union Pacific Corporation and its operating companies:
- Champlin Petroleum Company
- Missouri Pacific Railroad Companies
- Rocky Mountain Energy Company
- Union Pacific Railroad Companies
- Upland Industries Corporation

U.S. Department of Energy

Published by the Geological Society of America, Inc.
3300 Penrose Place, P.O. Box 9140, Boulder, Colorado 80301

Printed in U.S.A.

Library of Congress Cataloging-in-Publication Data

The Western North Atlantic region.

(The Geology of North America ; v. M)
Bibliography: p.
Includes index.
"One of the synthesis volumes of the Decade of North American Geology Project series"—P.
1. Geology—North Atlantic Ocean. 2. Geophysics—North Atlantic Ocean. 3. Paleoceanography—North Atlantic Ocean. 4. Mines and mineral Resources—North Atlantic Ocean. I. Vogt, Peter R. (Peter Richard), 1939– . II. Tucholke, Brian E. III. Geological Society of America. IV. Decade of North American Geology Project. V. Series.
QE71.G48 1986 vol. M 557 s 86-19550
[QE350.2] [551.46'08'0931]
ISBN 0-8137-5202-7

Contents

SURFICIAL SEDIMENTATION

BIOFACIES

PALEOCEANOGRAPHY

Plates
(in accompanying slipcase)

Preface

The Geology of North America series has been prepared to mark the Centennial of The Geological Society of America. It represents the cooperative efforts of more than 1,000 individuals from academia, state and federal agencies of many countries, and industry to prepare syntheses that are as current and authoritative as possible about the geology of the North American continent and adjacent oceanic regions.

This 29-volume series is part of the Decade of North American Geology (DNAG) Project which also includes eight wall maps at a scale of 1:5,000,000 that summarize the geology, tectonics, magnetic and gravity anomaly patterns, regional stress fields, thermal aspects, seismicity, and neotectonics of North America and its surroundings. Together, the synthesis volumes and maps are the first coordinated effort to integrate all available knowledge about the geology and geophysics of a crustal plate on a regional scale.

The products of the DNAG Project present the state of knowledge of the geology and geophysics of North America in the 1980s, and they point the way toward work to be done in the decades ahead.

In addition to the contributions from organizations and individuals acknowledged at the front of this book, major support for this volume has been provided by Woods Hole Oceanographic Institution and the U.S. Naval Research Laboratory.

A. R. Palmer
General Editor for the volumes
published by the Geological
Society of America

J. O. Wheeler
General Editor for the volumes
published by the Geological
Survey of Canada

Foreword

This volume is the first to be published in the DNAG series on *The Geology of North America.* As volume editors, we consequently have had the opportunity to break new ground in formulating how best to synthesize and present the broad spectrum of geoscience studies applicable to the North Atlantic. By the same token, we have not benefited from prior experience to guide us in avoiding the pitfalls inherent in such a large undertaking. This circumstance slowed completion but in many ways stimulated production of a better volume.

Our guiding philosophy, articulated in the 1982 "Perspectives" volume, has been to produce a synthesis of the geology and geophysics of the western North Atlantic that is useful to the widest possible spectrum of earth scientists. It should be comprehensible to earth-science graduate students and yet sophisticated enough to give a meaningful overview to the professional. It should integrate data and concepts into the kind of framework that will be useful to both Atlantic and non-Atlantic geologists. In fact, it is our hope that portions of this book will reach beyond geologists to physical oceanographers, ocean engineers, science historians, and lawyers dealing with the sea. We intended this volume to offer a set of perspectives on the status of knowledge during this Centennial decade. Individual chapters convey much of this information, and we summarize in the first chapter some of our own, broader perspectives on the past and future of geologic research in the western North Atlantic Ocean. We can look forward, in our dotage, to endless hours of evaluating the quality of our prescience and whether our efforts have been of lasting value.

Unlike many books compiled around geologic themes or subject areas from collections of contributed papers, we designed this volume carefully from its inception to give complete coverage of the geology and geophysics of the western North Atlantic Ocean basin. We realized early that most of the volume could be effectively organized under the paradigm of plate tectonics, but only if the scope was broadened beyond the physiographic "western North Atlantic" as was initially conceptualized. It made little sense to synthesize only half of a rifted ocean basin and to overlook the present Mid-Atlantic Ridge axis as the factory where crustal generation could be studied in "geological real time." Therefore, a series of chapters was organized around the "present plate boundary." Similarly, an eastward extension of the U.S.-Canadian border was a poor basis for delimiting the geographic scope of a synthesis volume. Iceland has often been used as a boundary, but from the point of view of mantle dynamics, geochemistry, depths, and gravity anomalies, Iceland is the center of a vast regional anomaly, not the edge. Some of our chapters, therefore, synthesize the North Atlantic *sensu lato* from near the equator north through the Eurasia Basin in the Arctic. Thus the entire accreting plate boundary forming the eastern edge of the North American Plate could be treated in space and time as a single system, and the entire ocean basin could be dealt with as a single paleoceanographic system. Nonetheless, discussions emphasize the western North

Atlantic, and coordination with other volumes—particularly the Arctic—has ensured complementary rather than repetitive treatments in areas of geographic overlap.

Once we established the volume outline, we identified potential authors whom we considered among the world-class authorities. Of those invited to contribute chapters, only one turned us down. Another withdrew midway through the process, allowing one of us time to write the chapter (Present Plate Motions). A third chapter on hotspots and absolute plate motion was promised but never submitted, leaving the only gap in our original outline.

The authors gave generously of their time to produce chapters through a multi-stage process of formulating chapter outlines, participating in workshops, and responding to editorial direction and peer review of manuscripts. Because of length limits on the volume, we restricted authors to pre-agreed page limits and to citing only the most relevant literature. We also attempted to see that the philosophy outlined earlier was followed, that chapters neither repeated nor missed important subject material, and that chapters effectively utilized the set of plates in the slipcase accompanying this volume. For us, this has been an extremely time-consuming but rewarding, once-in-a-lifetime task, and we trust the quality of the final product will justify the effort. Neither of us could justify claiming the lion's share of the editorial duties, so we settled the order of editorship on the volume in a logical, scientific manner—by the flip of a coin.

Brian E. Tucholke
Woods Hole, Massachusetts

Peter R. Vogt
Washington, D.C.

July, 1986

The Geology of North America
Vol. M, The Western North Atlantic Region
The Geological Society of America, 1986

Chapter 1

Perspectives on the geology of the North Atlantic Ocean

Brian E. Tucholke
Woods Hole Oceanographic Institution, Woods Hole, Massachusetts 02543
Peter R. Vogt
Naval Research Laboratory, Washington, D.C. 20375-5000

INTRODUCTION

This volume deals with the geology and geophysics of the western North Atlantic, the basin between the Mid-Atlantic Ridge and the eastern margin of North America. Although the book's focus is on the "North American Basin" (Fig. 1), individual chapters and charts extend the bounds beyond the "western North Atlantic." These extensions were dictated by the geology, which cannot be synthesized meaningfully if constrained by artificial geographic or political boundaries. Generally speaking, the content of this volume is contained within the region surrounded by the continental margins of the Antilles, eastern North America, Greenland, and Iceland in the west and north, the Mid-Atlantic Ridge crest on the east, and approximately 10°–15° N latitude to the south. However, in dealing with the Mid-Atlantic Ridge, plate kinematic models, and Atlantic paleogeography, we have extended our scope north into the Arctic region and east into the eastern Atlantic in order to present a complete synthesis of this major plate boundary and its evolution through time. The significance of Iceland for oceanic geoscience often has been overlooked or underrated; to help remedy this, two chapters deal mainly with this remarkable subaerial exposure of the Mid-Atlantic Ridge. One topic, mid-plate seismicity and stress, also demanded a plate-wide treatment covering the entire "stable" lithosphere from the Mid-Atlantic Ridge to the western Cordillera. Placing this treatment in the North Atlantic synthesis presumes that mid-plate stress is closely related to absolute plate motion (which is calculated largely from oceanic data) and/or to "ridge-push" forces from the Mid-Atlantic Ridge.

Having let plate tectonics extend the geographical scope of our volume, we were confronted with the problem of an appropriate projection for the accompanying maps. Standard Mercator projections have long been popular in marine science because of the twin conveniences of easy data plotting in the rectilinear grid, and ease of use in navigation at sea. Unfortunately, a standard Mercator projection is unsuitable for the entire Atlantic Ocean because of area distortion in high latitudes. From several possible projections we selected a Transverse Mercator with a 40°W center meridian because it runs midway down the North Atlantic and Arctic. In this projection, Florida is modestly exaggerated with respect to an area on the 40°W meridian by about the same amount as it is with respect to the Equator on a standard Mercator projection.

The area addressed in this volume is immense (up to 4×10^7 km^2), comparable in area to the North American continent. Consequently, features as large as the Mid-Atlantic Ridge can be dealt with in only part of our volume. In contrast, the comparably sized western North American Cordillera has four volumes of *The Geology of North America* devoted to it. However, given the size of present data bases, the state of our understanding in each of these areas, and their relative complexity, such apportionment is not unreasonable.

Because this book deals almost exclusively with oceanic geology, the part of the geological record that is represented (0 to ~175 Ma) is only about 5% of the age range represented by North America's known rocks. We also are synthesizing geology essentially unseen by human eyes except for oceanic islands and very limited submersible observations. Thus marine geology depends much more on geophysical techniques and on interpretation of a comparatively small collection of samples from dredging, coring, and deep-sea drilling.

This synthesis volume perhaps best approaches the uniformitarian ideal. The birth of oceanic crust can be studied along the crest of the Mid-Atlantic Ridge and compared to undisturbed older crust of various ages; the ca. 175 m.y. old ocean crust adjacent to the eastern United States is as old as any known in the world's ocean basins. Sedimentation processes occurring across the present ocean floor are mostly the same as those interpreted from older lithofacies, even though the Atlantic oceanographic environment has varied greatly in the past. Both sediments and upper igneous crust are rarely deformed by mid-plate or metamorphic igneous/tectonic processes, and particular lithofacies or

Tucholke, B. E., and Vogt, P. R., 1986, Perspectives on the geology of the North Atlantic Ocean, *in* Vogt, P. R., and Tucholke, B. E., eds., The Geology of North America, Volume M, The Western North Atlantic Region: Geological Society of America

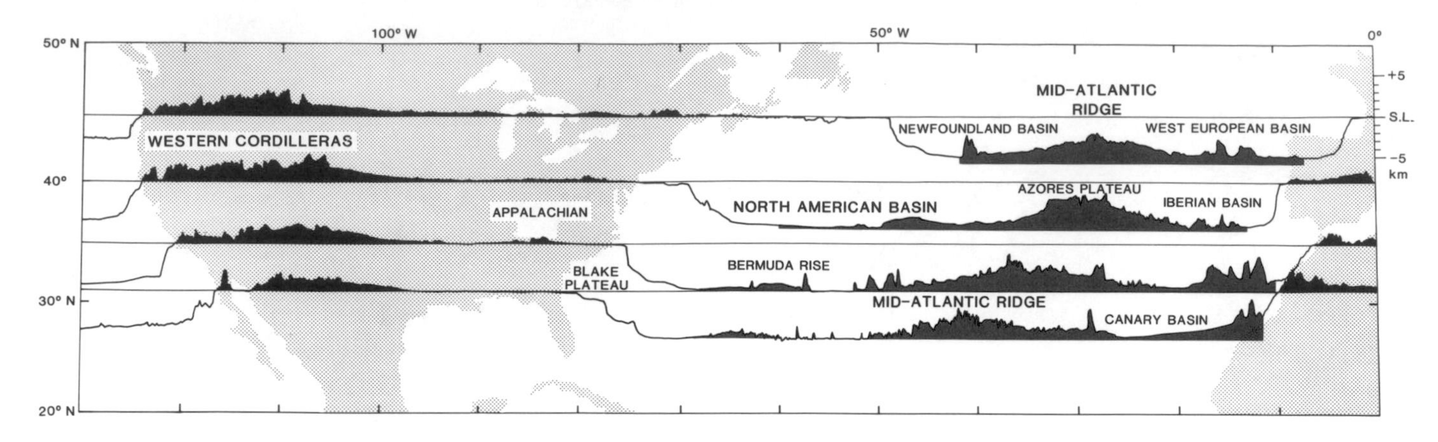

Figure 1. Topographic profiles constructed from a 5′ × 5′ digital grid (similar to that used by Heirtzler, 1985) from the Pacific across the United States and the central North Atlantic. Topography is black above sea level and red above –5,500 m in the Atlantic seaward of the continental rises. Note that the central Atlantic is about the same size as the entire U.S., the Mid-Atlantic Ridge is comparable in width and relief to the western Cordillera, and the Bermuda Rise is comparable to the Appalachians. Redrawn from cover of EOS, v. 63, September 14, 1982.

average basement characteristics are more homogeneous over areas of 10^5 and 10^6 km^2 than even continental platform sediments, let alone Phanerozoic orogens or the Precambrian shield.

Despite the differences, there also are close links between continental North America and the deep Atlantic basins. For example: (1) The Mid-Atlantic Ridge has generated a geophysical record of North American plate motion, and it even may help drive that motion. (2) Mid-plate seismicity in the western Atlantic and eastern North America may reflect much the same compressional stress field, possibly because of viscous resistance to the westward movement of the plate. (3) A much different paleostress field, recorded in Triassic-Jurassic igneous activity and normal faults along the eastern continental margin, prefigured the birth of the modern Atlantic. (4) Subsidence of the Atlantic Coastal Plain is a phenomenon that echoes the subsidence of oceanic crust and results from similar thermo-isostatic processes in the mantle. (5) Possible ophiolites in orogens like the Appalachians represent remnants of Paleozoic oceanic crust and invite comparisons with modern Atlantic crust. (6) The Atlantic offers at least three types of geologic provinces that may become accreted terranes far in the geologic future: (a) Jan Mayen Ridge and Rockall Plateau, both continental fragments; (b) Bermuda, a large volcanic edifice; and (c) Iceland, an oceanic plateau, which also could be an example of a primordial continent. (7) The sediments of the deep western Atlantic contain a 175 m.y. record of continental detritus derived under varying climatic regimes and topographic gradients from North America east of the drainage divide. Assessed together with marine biogenic debris, this sediment probably contains a better environmental history of eastern North America than can be found ashore. (8) Many historical features of important younger (<175 Ma) marine sequences that are preserved on continental North America can be understood best in relation to their deep-sea chronostratigraphic equivalents; the numerous transgressions and regressions so sensitively recorded in shallow-water strata may have oceanic causes including eustatic sea-level changes related to changing sea-floor spreading rate, mid-plate volcanism and thermal bulges, and desiccation of isolated basins. Milankovitch-type evaporation/rainfall cycles in shallow continental seas may even have modulated the thermohaline circulation that supplied oxygen to the deep Atlantic (e.g., Barron and others, 1985).

Four years ago, we published our thoughts on the probable format and content of the western North Atlantic geologic synthesis that is contained in this book (Vogt and Tucholke, 1982). With few exceptions we have been able to follow that plan. Because most geological studies in the oceans depend heavily on topographic information, the book is organized to deal first with the bathymetry of the North Atlantic Ocean floor and the methods used to image it. Equally important, the introductory part of the book presents an updated geochronologic and geomagnetic time scale, providing a foundation to assess geologic rates and times for this and succeeding volumes of *The Geology of North America.* The main part of the volume begins with the tectonics and crustal properties of the North Atlantic, with special attention to the crustal factory at the axis of the Mid-Atlantic Ridge. This is proper because, excluding "cosmogenic" effects such as solar insolation (e.g., Milankovitch cycles) and the impacts of large bolides, it is "geogenic" processes (tectonics, volcanism, erosion, sedimentation, etc.) that control the form and composition of the geologic record (see Tucholke and McCoy, this volume, Fig. 1). Succeeding chapters deal with plate kinematics, sedimentary processes, and the biofacies record. All of these subjects lead to a synthesis of the North Atlantic ocean paleoenvironment. Final chapters briefly explore satellite sensing of the ocean environment, seabed resources, and the legal problems involved in managing the seabed and its resources.

FROM PAST TO PRESENT

To appreciate how our present understanding of Atlantic geology evolved, we identify here some of the main historical trends in marine geologic research and the forces behind those trends. Although man's knowledge and exploration of the shallow seabed extends back to classical times and beyond (Vogt and Tucholke, this volume), we begin here with an eighteenth century (1775) view of the North Atlantic (Fig. 2); this was the North Atlantic (also known as the Western or Virginian Ocean) as perceived in James Hutton's time. In that era, sea monsters had vanished, to be replaced by numerous sightings of banks and rocks, mainly along the sea routes from the U.S. to Europe; almost all of these sightings now are known to be artifacts. Some reports ("Mayda Island, white and flat") were evidently icebergs far beyond their normal range because this was the time of Europe's "Little Ice Age." The depth of the blue-water ocean was still unknown. Observational ocean science thus lagged far behind theory: Newton (1642–1727) had already laid the groundwork for modern calculus, optics (e.g. Vogt and Tucholke, this volume), gravity-anomaly modeling (e.g., Rabinowitz and Jung, this volume), and satellite-orbit calculations (Anderle, this volume). Euler (1707–1783), among many other contributions, had created the mathematical basis later used to rotate lithospheric plates about "Euler poles" (e.g., Vogt, this volume, ch. 24).

Deep-sea wire-sounding began in 1840, and by 1854, the broad shape of the Atlantic sea floor was known (Schilling, this volume). With the invention of the "Deep-Sea Sounding Apparatus" by Midshipman J. M. Brooke in 1852, a sea-floor sediment sample could be returned with each depth sounding (a feat not achieved by modern sounding methods). The excellent preservation of delicate microfossil tests and the scarcity of terrigenous detritus in bottom samples from the "Telegraphic Plateau" or "Middle Ground" (Mid-Atlantic Ridge) amazed the scientific community, as much as did the recovery a few decades later of shallow-water and terrestrial debris in the deep basins. The potential of ocean-floor sediments as a biogenic paleoenvironmental record book seemed to dawn on Maury (1855) because he wrote of the ocean's surface as a nursery and its "bottom as one vast burial place." A grand cycling of elements was implied, in which the ocean is "a vast chemical bath in which the solid parts of the earth are washed, filtered, and precipitated again as solid matter." Noting the similarity of microfossils from the Atlantic sea floor to those in strata exposed on land, Maury anticipated aspects of plate tectonics by contemplating how living foraminifera "may have been preparing the ingredients for the fruitful soil of a land that some upheaval ages far away in the future may cast up from the bottom of the sea. . ."

A debate ensued over the presence or absence of benthic life. Maury came down on the wrong (absence) side of this argument. Had he been right, the un-bioturbated sea floor would have provided biostratigraphic resolution much better than the 10^3–10^4 years set by benthic sediment mixing. Before the Geological Society of America's founding in 1888, the H.M.S. *Challenger* in 1872–76 had explored the ocean basins on a mission of marine biology, physical oceanography, and sea-floor geography/geology. Benthic life was demonstrated. Phosphate and manganese modules were dredged by the *Challenger,* and the main sediment types were described (foraminiferal, radiolarian, and diatom oozes; terrigenous muds, red clay). Where hard rocks or igneous minerals were recovered, they were almost invariably volcanogenic, particularly basaltic. Taking little note of the oceanic crust's apparent volcanic character (Schilling, this volume), land geologists continued to write of sunken land bridges and even sunken continents.

In many other ways the seeds of modern marine geoscience—and most other sciences as well—were sown in the nineteenth century. In the all-important area of seismology/acoustics, major milestones include the measurement of sound speed in water by Colladon and Sturm in 1827, the development of the theory of seismic-wave propagation by Poisson (1829–1831), and the discoveries of magnetostriction by Joule in the 1840's and of piezoelectricity by Jacques and Pierre Curie in 1880. Geomagnetic field strength was first measured by Gauss in 1832; the theoretical foundation for electromagnetism was laid by Maxwell in 1873. Scientific studies of evolution began with Lamarck in 1809, and natural selection was proposed by Wallace and Darwin a half century later. Darwin's "Origin of the Species" in 1859 has often been compared, in its impact on biology/paleontology, to the discovery of plate tectonics in its impact on the earth sciences.

The age of nuclear physics dawned in the closing years of the nineteenth century with the discovery of spontaneous radiation from uranium salts by Henri Becquerel in 1896. Concurrent research by Marie and Pierre Curie led in 1898 to the discovery of radium and polonium, and the recognition of three fundamental types of "radiation" (later called alpha, beta, and gamma). In the following 75 years, outgrowths of these discoveries penetrated every area of marine geoscience. A few examples include: Proton-precession and optically-pumped helium or cesium-vapor magnetometers that now routinely measure magnetic anomalies in the ocean basins. Deep-sea boreholes and the cores extracted therefrom are scanned with gamma and neutron logging tools. Elemental composition of the oceanic crust is assayed with electron microprobes (e.g., Melson and O'Hearn, this volume), by x-ray fluorescence and, for trace elements, by neutron activation (e.g., Schilling, this volume). Isotope ratios (e.g. $^3He/^4He$, $^{87}Sr/^{86}Sr$, $^{143}Nd/^{144}Nd$, $^{206}Pb/^{204}Pb$), so crucial in determining the mantle history of oceanic basalts, are measured by mass spectometry, as is $^{18}O/^{16}O$ in biogenic precipitates used to determine paleotemperatures and glacier ice volumes. Accurate age dating of solid-earth material by measuring the products of natural radioactive decay was perhaps the single greatest gift of modern physics to the geosciences. Without it, the geologic time-scale could not have been constructed. However, the direct application of radiometric age dating to ocean-floor materials has been disappointing because oceanic basalts generally are depleted in radio-

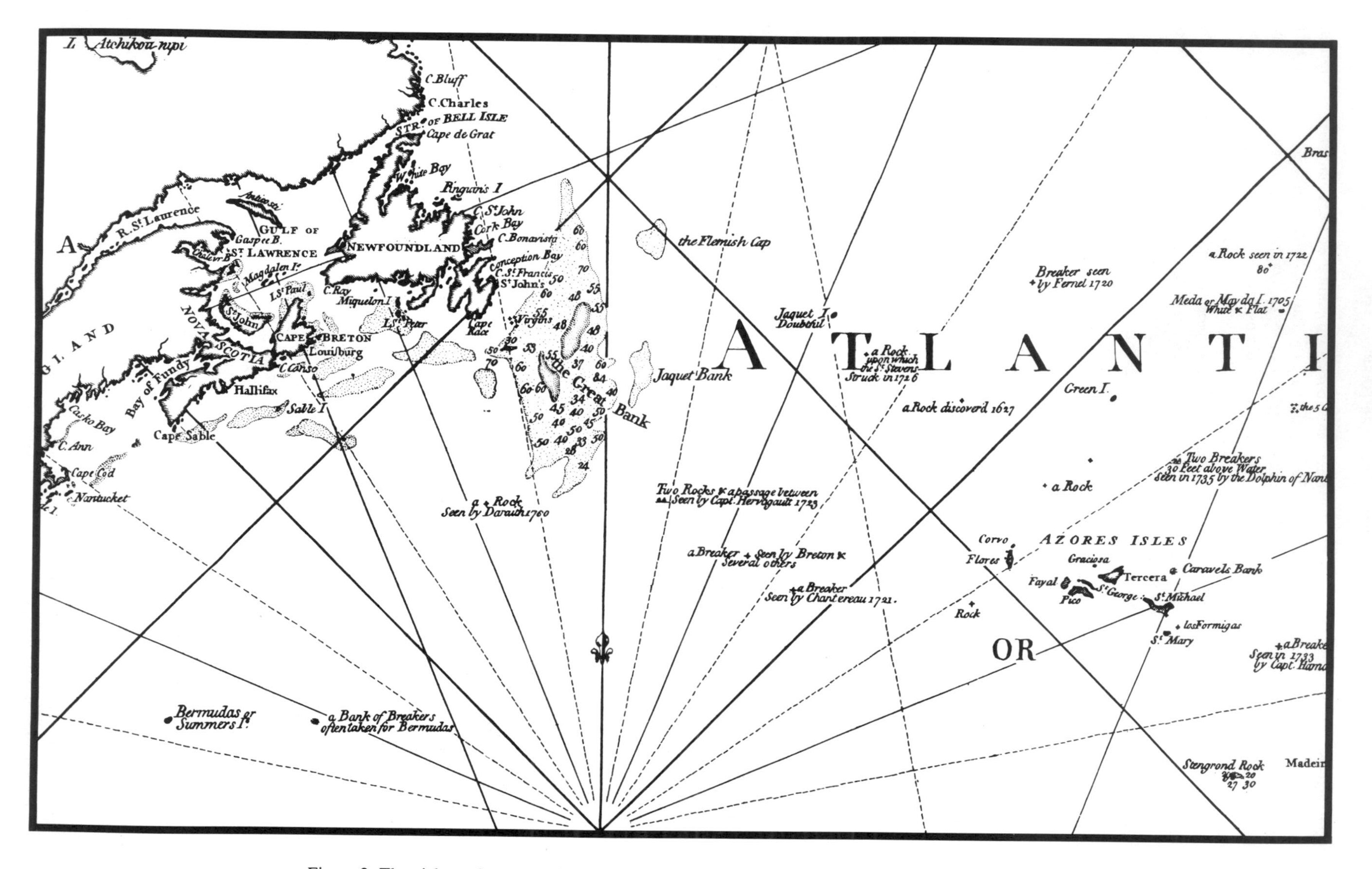

Figure 2. The eighteenth century Atlantic: Spot soundings (fathoms), banks (stippled), and numerous observations misinterpreted as rocks or even islands. The diagonal lines were used as aids for constant-course navigation on Mercator charts. Retouched from part of a chart by Sayer (1775).

genic elements such as potassium and uranium. Exceptions are volcanic islands and seamount chains (Hawaii, New England Seamounts, etc.), and age dating of these features has enabled measurement of absolute plate motion (Vogt, this volume, Ch. 24). Finally, the discovery of natural radioactivity came to explain how a 4,600 Ma planet can still be geodynamically young. Plate tectonics, the surface expression of mantle convection, could not occur without this radiogenic energy, and without knowledge of radioactivity we would be hard-pressed to explain the measured heat flow emanating from the ocean floors (e.g., Sclater and Wixon, this volume).

Although the nineteenth century produced numerous wire soundings and bottom samples of the seabed, these data yielded only low-resolution information about the form and structure of the ocean floor. A few higher-resolution discoveries included the first recognition of a continental-margin canyon (Hudson Canyon) by Dana in 1863, and Dana's earlier conclusion in 1840 that volcanos are piles of rock erupted from a vent, not flat-lying lavas punched up by doming. Darwin's hypothesis of volcanic-island erosion, subsidence, and coral-reef formation in 1842 was inspired by observations in the Pacific, but the same subsidence operates in the Atlantic. Seamount chains were unknown, but Dana noted the linearity, parallelism, and age progressions of the latest volcanism along mid-Pacific island/atoll chains. However, it was Dana's colleague, J. P. Couthouy, in the 1840s, who first suggested that the islands were "forced up in regular succession by the subterranean fire," that is, that volcanism not only ended but also began earlier in the west. This basic tenet of the hotspot model is only now coming into general acceptance for Atlantic seamount chains such as the New England Seamounts. Definitive proof still awaits deep drilling, however. Louis Agassiz in 1840 awakened the world to the glacial ages, but the importance of the sea-floor record in understanding glacial cycles was not realized for another century, even though the "astronomical" theory for the ice ages had been suggested already by Adhemar in 1842.

Real progress in understanding the subsea geomorphology of the Atlantic began in the 1920s with the advent of echo sounding. Trans-Atlantic lines obtained in 1925–27 on the *Meteor* cruises in the equatorial and South Atlantic (Stocks and Wüst, 1935) clearly demonstrated the continuity and roughness of the Mid-Atlantic Ridge that were hinted at so strongly in earlier charts. Cores obtained by *Meteor* showed how Quaternary ocean climates could be reconstructed from microfossil assemblages (Schott, 1935). Although Alfred Wegener lived until 1930, he apparently never saw the *Meteor* bathymetry, and one wonders how these data might have reshaped his ideas on continental drift, even though the data set was woefully inadequate to "illuminate" the fundamental plate tectonic grain contained in fracture zones, the median rift valley, and seamount chains. This fabric was not identified until the late 1950s and 1960s. Anticipating the work of Sclater and his colleagues many years later (e.g., Sclater and Wixon, this volume), Wegener noticed that younger ocean basins tend to be shallower, and he attributed this correctly to greater thermal expansion: "Two ocean floors 100° different in temperature as far down as the 60-km depth line and in isostatic equilibrium with each other, must show a depth difference of 160 m, the warmer floor at the greater height" (Wegener, 1929, p. 208). Had Wegener extended this thinking to the Mid-Atlantic Ridge, would sea-floor spreading have been the almost inevitable consequence? Certainly he was troubled by his own explanation of the Mid-Atlantic Ridge as a "strip which crumbled during the separation process," because the width of the ridge, which he knew as up to 1,300 km, seemed to him "still excessive, since the congruence of the present-day block margins of South America and Africa is so striking here and seems to indicate that these margins were in fairly direct connections."

The first half of the twentieth century brought many other technological achievements that laid the foundations for today's understanding of ocean-floor structure; some of the most important included: measurement of gravity at sea beginning in 1923 (Vening Meinesz, 1932); application of seismic refraction techniques in the ocean (e.g., Ewing and others, 1937); and the development of the seismic-reflection profiler (Ewing and Tirey, 1961; Hersey, 1963). Continuous measurement of marine magnetic anomalies along the mid-ocean ridges (Vogt, this volume, Ch. 15), their axial symmetry about the ridge, and their correlation to the developing geomagnetic time scale led Vine and Matthews (1963) to verify sea-floor spreading and measure its rate. The 1960s added deep-tow side-scan sonars, multibeam bathymetry, and accurate satellite navigation to the marine geologist's tool box.

Discovery of the semiconductor by W. Shockley in 1948 led to explosive development in many areas of electronics, particularly digital computers which had just been developed. Every area of marine geoscience—and every other science—has been penetrated by solid-state electronics and digital computers, from the shipboard and shore-lab storage and processing of marine geological/geophysical data to the testing of physical/numerical models, the production of graphics (including the cover of this volume), and the preparation and editing of manuscripts. Although magnetic and gravity data could have been (and were) collected and interpreted without computers, multichannel seismic profiling, climate/ocean modeling, multibeam bathymetry, seismic waveform analysis, and satellite remote sensing and navigation would be virtually impossible. Only supercomputers can invert the colossal matrices required, for example, to calculate three-dimensional velocity structure (tomography) of the ocean or of the earth's crust and mantle from a set of acoustic or seismic observations. Adverse effects of the "computer revolution" are that computer models are run off so easily that users may not have a good grasp of the underlying physics and mathematics. Furthermore, an increase in precision does not necessarily imply an increase in accuracy; no amount of computing power can overcome the "basic" geophysical problem of non-uniqueness, e.g. in modeling potential field data (Rabinowitz and Jung, this volume). There are an infinity of density distributions that will fit a marine-gravity profile, and an infinity of magnetization distributions that will fit a magnetic profile. Bounds can be placed on possible distributions, however, and as "ground-truthing" by actual sampling increases, the bounds become tighter and make geophysical data ever more valuable predictors of crustal structure, composition, and physical state.

Direct sea-floor sampling, particularly by drilling, has provided a quantum leap in our understanding of marine geology. The Deep Sea Drilling Project (DSDP) and its present-day successor, the Ocean Drilling Program (ODP), have drilled more than 170 sites in the North Atlantic from 1968 through 1986 (Plate 2). The scientific returns from this effort (65.5 km drilled, of which 31.7 km was cored) have been enormous, as shown by the large number of drilling-related publications (see Fig. 5). However, these holes have barely entered the crust; the deepest subbottom penetration has been a mere 1,740 m at DSDP Site 398 in the eastern Atlantic. This contrasts with the deepest commercial well, the Bertha Rogers gas well in Oklahoma, at 9,647 m; the super-deep hole on Kola Peninsula, at 12,063 m as of 10 August, 1984; and the deepest commercial offshore well drilled along eastern North American margin, the SW Banquereau F-34 well on the Nova Scotia Shelf, at 6,185 m subbottom. Commercial drilling recently has moved into deeper water, the present maximum water depth being 2,106 m at Shell 372-1 on the lower Continental Slope at the edge of the Baltimore Canyon Trough. On Iceland, holes have been drilled to 3,085 m depth in the Neo-Volcanic zone and to 2,820 m in older crust (Laugaland site). These sites returned only cuttings; the only deep hole in Iceland that was continuously cored was the Iceland Research Drilling Program (ERDP) site, that reached 1,919 m depth.

DSDP and ODP drilling in the ocean basins can exploit submarine erosion of sediments to reach deep stratigraphic levels, but there are no comparable areas of erosion on igneous crust except for a few sites drilled on guyots and aseismic ridges. In contrast, because of glacial and fluvial erosion on Iceland, the Laugaland and IRDP boreholes actually penetrated 4,300 m and 3,500 m, respectively, below the original land surface; they therefore represent "oceanic" basement penetration that is much greater than the 1,076 m record for ocean drilling at DSDP Site 504B in the Pacific Ocean; the Atlantic record is only 544 m at Site 418A. In the oceans, the drillship *Joides Resolution* can presently support a 9,150 m drill string. Barring safety problems, this will allow oceanic basement to be reached as far landward as the middle of the Inner Magnetic Quiet Zone east of the U.S. margin (Plates 3, 5), but the normal, 6.5 km-thick oceanic crust can not yet be entirely penetrated except at anomalously shallow sites, or near the accreting plate boundary where temperature gradients may be excessive.

A more quantitative look at the scope of recent marine-geological research is presented in Figures 3 through 6. The distribution of key marine geological/geophysical data available today is shown in Figure 3 (ship tracks) and Figure 4 (sea-floor samples) as reported to the National Geophysical Data Center. These maps do not show all existing data; even Stocks and Wüst (1935) were unable to track down all the soundings available in their time. In particular Figure 3 does not show classified or proprietary data, airborne magnetics, and satellite altimetry. Although there is now a large data base, it is evident that data density is highly uneven. This inhomogeneity has many causes, but several are obvious. Data are concentrated in (1) areas of national (mainly U.S.) economic or military (Navy) interests; (2) proximity to coasts, particularly major refueling or home ports for research vessels (note the number of tracks converging on Honolulu and ports along the U.S. east coast); and (3) regions of high scientific interest—as illustrated by data concentration at intervals along the Mid-Atlantic Ridge crest and around DSDP drill sites. By contrast, numerous areas of 10^3–10^4 km^2 have never been sampled directly by core or dredge (Fig. 4), and present sampling density can be compared to that of bathymetry in 1920, just before echo-sounding began. Of the tracks in Figure 3, very few have multi-beam echo-sounding or side-scan sonar data, and the "information width" of the single-beam tracks is very limited. Thus the present state of ocean-floor mapping and sampling is not only inhomogeneous but generally poor (see also Vogt and Tucholke, this volume). In terms of average horizontal resolution, the ocean floors are for the most part less well known than the surfaces of the Moon, Mars, and even the Uranian moon Miranda, where optical images of 1 km resolution were obtained by the Voyager spacecraft in 1986. The North Atlantic is better known than other large oceans, however (Fig. 3).

A close look at marine geological (particularly Atlantic) research activity in the last quarter century is shown in Figure 5, based on annual totals of GEOREF key-word citations since 1960 and thus including the 1966–1968 plate-tectonic revolution. GEOREF is a computer bibliographic file produced from literature searches by the American Geological Institute. The entries in GEOREF represent most but not all published scientific literature including abstracts; the major, most widely read journals are completely covered. The down-turn in numbers of citations for the last few years on most plots of Figure 5 is partly an artifact that reflects the time lag between publication date and entry in GEOREF; the sharp rise in "total references" from 1977 to 1978 reflects an increased GEOREF budget. The entire earth sciences experienced a literature explosion about 1965–1969, followed by large increases in the Icelandic literature (1971–74), in articles dealing with the Mid-Atlantic Ridge (1973–1977), and in papers covering the Atlantic as a whole (1976–1981). The Appalachian literature explosion, reflecting the impact of plate tectonics concepts on terrestrial geology, appeared about the same time as that of the Atlantic, whose late 1970s peak largely reflects publications resulting from the Deep Sea Drilling Project (DSDP volume 44 appeared in 1978; volumes 43, 45, 46, 48, and 49 in 1979; and volume 50 in 1980).

Searches done by topic (middle column, Fig. 5) generally show activity beginning in the early 1960s and extending to a peak in the late 1970s. This was followed by a decline in magnetics and Mid-Atlantic Ridge basalt studies, fairly steady activity in hydrothermal, gravity, seismicity, and bathymetry studies, and an ongoing increase in crustal-structure research. Topics in the hydrothermal and crustal-structure categories are probably those where the greatest progress can be expected in the next few years.

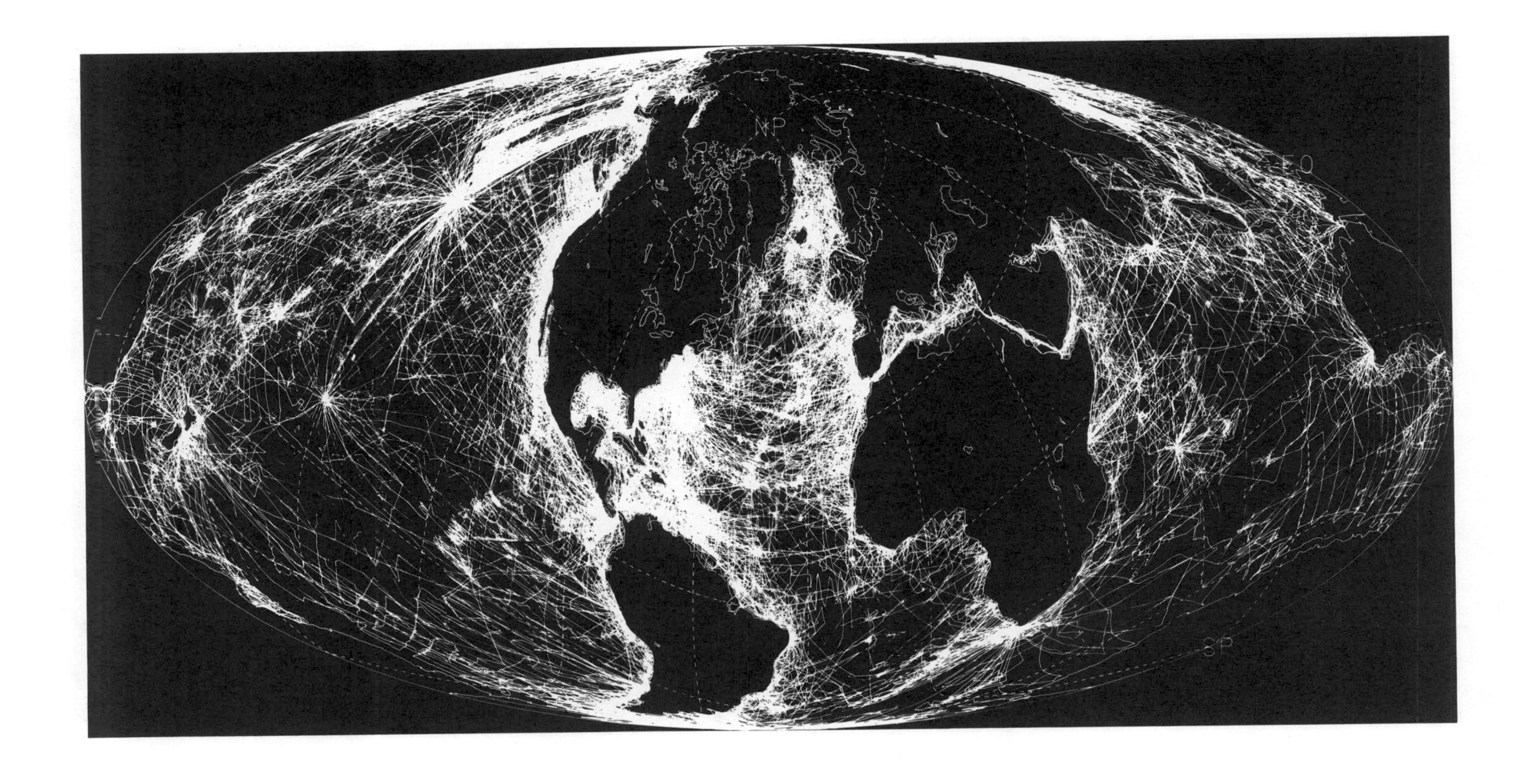

Figure 3. Ship tracks along which at least one type of marine-geophysical data (bathymetry, magnetics, or seismics) has been reported to the National Geophysical Data Center (NGDC), as of December 1984. Mollweide Projection centered on 40°N, 40°W, specifically made for this volume by NGDC, courtesy of T. Holcombe and P. Sloss.

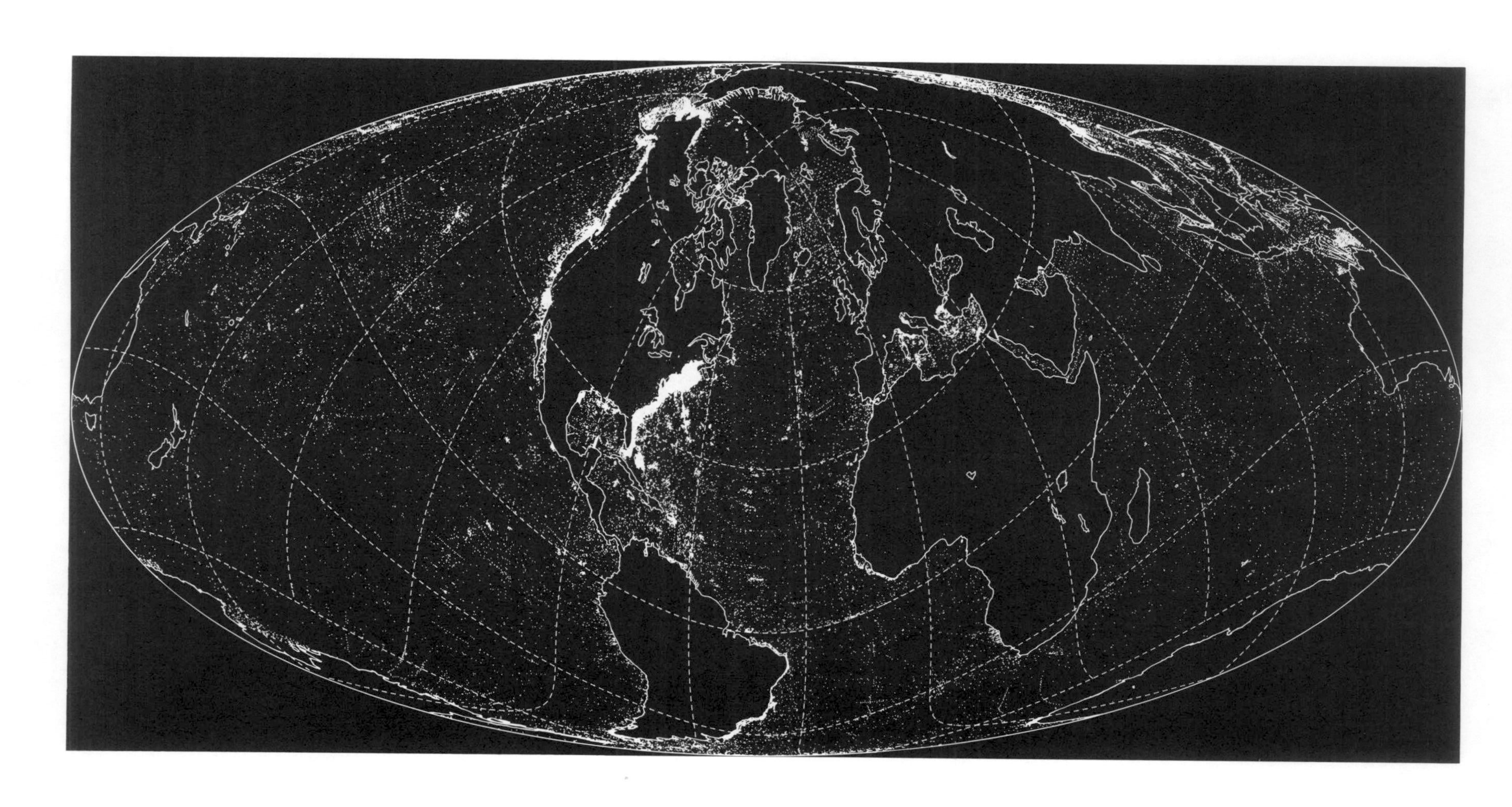

Figure 4. Distribution of $0.1° \times 0.1°$ areas for which geologic (bottom or subbottom sample) data have been reported to the National Geophysical Data Center. Courtesy of T. Holcome and C. Moore.

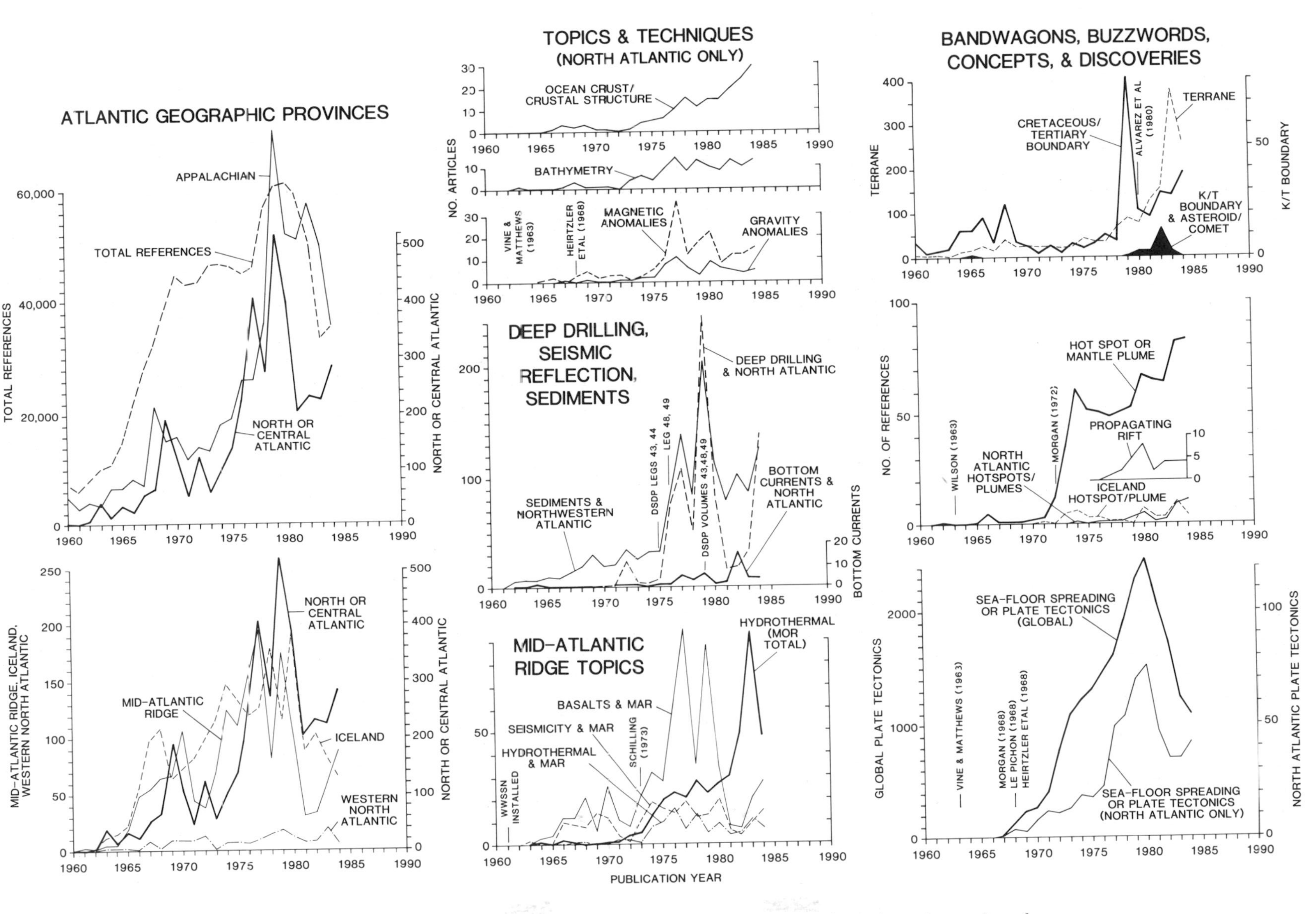

Figure 5. Research activity in different marine-geological and Atlantic subjects, shown by number of publications each year and listed under the key words indicated. Derived from GEOREF computer search (American Geological Institute). Most of the searches indicate holdings as of September 1985. Number of items (particularly for 1983–1985) is underestimated due to time lag between publication and data-file entry. Publication dates of important papers or volumes are indicated. See text for further discussion.

With increased use of multi-beam echo-sounding (e.g., SEA-BEAM) and side-looking sonars (e.g., Sea MARC), bathymetric and other imaging studies (Vogt and Tucholke, this volume) will also experience an upswing.

The most dramatic biostratigraphic event during the evolution of the North Atlantic was the Cretaceous/Tertiary (K/T) boundary extinction. It experienced an early popularity wave in the mid-to-late 1960s, followed by a spike in 1979 (Fig. 5); the spike was created by publication of a special volume that had been planned, according to its editor T. Birkelund, prior to the Alvarez (asteroid impact) hypothesis. Although an early version of the hypothesis was first published in that volume, it did not in itself cause the "publication spike." The discovery of iridium enrichment at the K/T boundary, however, stimulated a high and increasing publication rate on this subject after 1980.

Searches by concept also show a variety of patterns. Obviously the publication levels are zero until a concept is discovered and identified by a key word. Sea-floor spreading and plate tectonics, introduced in 1962–68 (LePichon, 1968; Morgan, 1968), started a long-term increase in publication rate until 1981. The subsequent decrease may partly reflect a tendency to take the plate-tectonic relevance of a paper so much for granted that "plate tectonics" is no longer entered as a key word. Development of the hotspot/mantle-plume concept (Morgan, 1972) resulted in a rapid literature explosion (possibly scientists caught with their pants down in 1967–1968 did not want to be left behind this time around). The more recent "propagating rift" concept introduced in the late 1970s (Hey and others, 1980) has not yet had a large impact in terms of publication numbers.

Another way to assess the development of the geosciences is to plot successive estimates of given geophysical quantities against publication date. Many geoscience topics, such as submarine geomorphology, would be difficult to quantify in this manner, but others adapt readily to such treatment. To illustrate the idea, we chose absolute age estimates of several geomagnetic reversals and epoch boundaries in the Mesozoic-Cenozoic time scale (Fig. 6); these quantitative values underlie estimates of many geologic parameters such as sea-floor spreading rates, sedimentation rates, crustal subsidence rates, and times of important geologic events and their correlation. The plots (Fig. 6) show that age determinations for late Tertiary reversals (particularly post-10 Ma) and for selected epoch boundaries (including the Cretaceous/Tertiary boundary) have changed very little (less than 1–2 m.y.) in the last one to two decades, and future refinements are likely to be small and slow. On the other hand, age estimates for magnetic reversals in the Late Jurassic–Early Cretaceous (M-Series) anomalies and in the Late Cretaceous–Early Tertiary have changed (and fluctuated) significantly; this partly reflects the difficulty of obtaining appropriate Cretaceous and Jurassic rocks for absolute dating. An extreme example is the age estimate for the Triassic/Jurassic boundary, which has become progressively older in the last half century, from 145 Ma (Holmes, 1937) to about 208 Ma today.

In the North Atlantic, the estimated age for initiation of sea-floor spreading (Fig. 6, bottom center), moved progressively towards younger dates through the 1960s and 1970s, principally as a result of sea-floor age calibrations by deep-sea drilling. Today, based on drilling results and multichannel-seismic studies of deep continental-margin stratigraphy, the time at which central North Atlantic sea-floor spreading began seems to be stabilized in the interval Bajocian-Bathonian, roughly 175–180 Ma on the time scale used here (Kent and Gradstein, this volume; Plate 1). Little further progress can be expected in estimating the age of this important event until additional holes are drilled through the sedimentary "drift sequence" overlying the first-formed oceanic crust.

FRONTIERS OF NORTH ATLANTIC STUDY

Introduction

The past 25 years of marine geological and geophysical studies in the North Atlantic and elsewhere were unprecedented in development of data-acquisition technology and in explanation of observations via th paradigm of plate tectonics. It is fair to say that without the technology, and its application both in detailed surveys and in routine data acquisition on regional to global scales, the plate-tectonics revolution would have been much longer in coming, and might still remain today in the realm of speculation. Technology development will similarly lead the way to advances in the future. Although we cannot predict when or if major "breakthroughs" will occur, we can identify some important problems and make reasonable estimates of impending trends that are likely to help solve them.

The common theme in this discussion is that we understand, however imperfectly, much of the gross workings of the geological processes that shape the sea floor and its sedimentary record. In many aspects of the science we also have high-resolution spatial data in very small areas, or detailed temporal records over short observation intervals. The technology is now becoming available to extend greatly these precise, but limited, records at relatively low cost. Perhaps more important is the fact that ship/instrument positioning and digital-imaging technology have reached the point where photographs, sidescan-sonar records, and multibeam bathymetric data can be melded and manipulated in an accurate frame of reference. This "upscale integration" will allow (1) much better understanding of geologic relations and the processes that engendered them, and (2) greater predictive capability when using the fragmentary data sets with which we usually have to deal.

The future of geological and geophysical research in the oceans is likely to incorporate the "systems approach" more extensively than it does now. Two systems are involved. One is the technological system, including upscale integration, as discussed above. The second is the geological system in the form of a "type area" or "natural laboratory." By concentrating the technological system on discrete geologic systems that contain important or representative tectonic, structural, or stratigraphic elements of the geologic record, we will advance our understanding of fundamental geologic processes much more rapidly. This philosophy com-

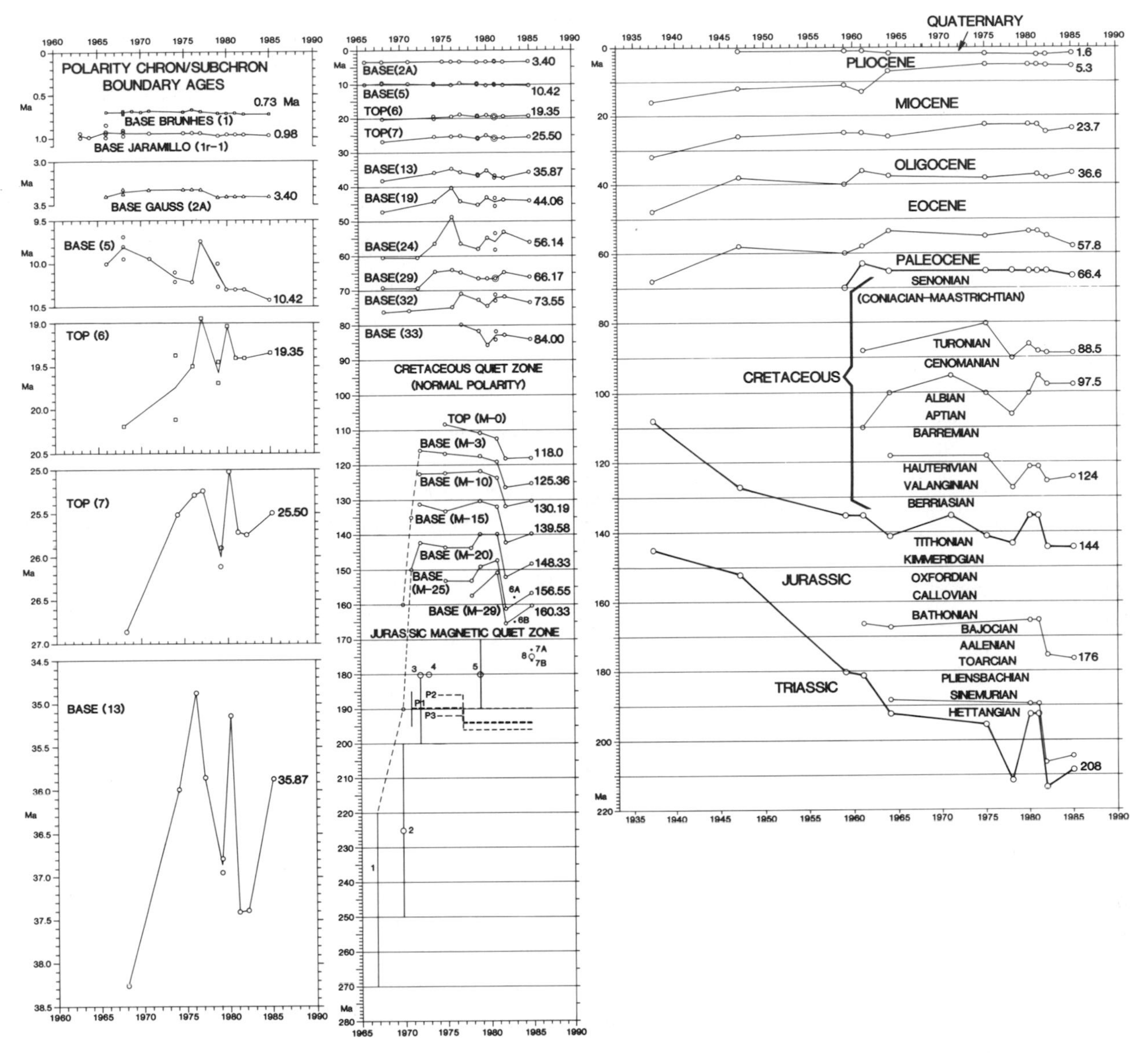

Figure 6. Age-dating the Earth. Estimates for ages of key geomagnetic reversals (reversal number in parentheses; left and center parts show time scale values used in this volume annotated at right side of each); age estimates of events related to the initial opening of the central North Atlantic (numbered at the bottom of the center plot); age estimates of period and epoch boundaries (right). Age estimates are plotted according to year of publication. Lines connect successive data points only to aid seeing the points. Where more than one age estimate is published in the same year, lines pass through numerical averages of estimates. Bottom of center plot: 1, proposed "Kiaman" age of the magnetic smooth zone bordering eastern North American and northwest Africa (Heirtzler and Hayes, 1967; Emery and others, 1970); 2, break-up age extrapolated landward from oceanic magnetic anomalies according to Vogt and others (1970; dashed line and small circle at 190 Ma shows age of magnetic smooth/rough transition); 3, extrapolated break-up age according to Pitman and Talwani (1972); 4, estimated break-up age according to Vogt (1973); 5, extrapolated breakup age according to Vogt and Einwich (1979); 6A, B, breakup age extrapolated from oceanic magnetic anomalies (A, R. Sheridan, personal communication, 1986; and B, from "breakup unconformity" drilled at COST G2 well (Plate 2), on time scale of Van Hinte [1976]; 7A, B, same as 6 but based on the time scale used in this volume (Plate 1); 8, breakup age estimate from Klitgord and Schouten, this volume; P1, K-Ar age of Palisades Sill according to Lambert (1971); P2 and P3 are $^{40}A/^{39}A$ ages of Palisades Sill and related volcanics (Dallmeyer, 1975). Steps shown in P1–P3 at 1977 reflect the revision of radioactive decay constants (Steiger and Jäger, 1977).

plements the geographic approach (e.g., regional mapping) or topical perspective (e.g., plate tectonics) so common in the geosciences to date. Each philosophy, of course, has its place, but the systems approach (e.g., the FAMOUS project) may well drive the real geoscience advances in the future. In the following pages we consider a few (but hardly all!) future directions in North Atlantic research that are likely to be productive, and we consider the role that the systems approach may play. Similar problem definition on specific subjects is contained within most other chapters of this book.

Accretion Axis

The Mid-Atlantic "crustal factory" (Macdonald, this volume) has been closely examined in only a few areas, such as Iceland (which is very anomalous) and the FAMOUS-AMAR area (which is high on the flank of the Azores regional hotspot and is therefore also atypical). Other representative parts of the Mid-Atlantic Ridge axis must be examined in equal or greater detail. The examination should cover the spectrum from typical ridge to hotspot-dominated, the range of opening rates, and such poorly understood accretion variables as asymmetric spreading, which seems to be regionally persistent only north of Iceland (Vogt and others, 1982). Plate-boundary lengths examined should be large enough (at least 50–60 km) to cover a complete fracture-zone/ridge/fracture-zone system, perhaps equivalent to the bounds of a single magmatic system (Schouten and others, 1985). The study areas should extend in swaths from the ridge axis outward onto the flanks in order to establish temporal patterns of crustal evolution. An important question to answer is whether the volcanotectonic episodes are related to plate-kinematic forcing (changes in opening rate or pole of relative plate motion) and thus synchronized along much of the boundary, or whether they are determined by episodes within the magmatic system on scales of 50–100 km. Detailed sea-floor imaging alone is insufficient to answer such questions, and a systematic program of seismic experiments and close-spaced deep drilling will be essential to resolve the various scales of crustal genesis in space and time.

Fracture Zones

The gross morphology of the North Atlantic ocean crust is well defined, and it includes fracture zones that appear to parallel the past directions of plate motion at length scales of hundreds of kilometers or less (Plates 2, 5). The average spacing between fracture-zone valleys in the North Atlantic is on the order of 50 km. This segmentation of mid-ocean ridges is thought to be a fundamental property of the plate boundary that may reflect the spacing of zones of magma genesis in the mantle (Schouten and Klitgord, 1982; Schouten and others, 1985) or thermal contraction (Sandwell, 1986). We currently have a poor understanding of how faithfully the fracture zone traces record the instantaneous directions of relative plate motion or at what scale the mechanical properties of the lithosphere will dominate over this record. Recent study of existing data on the Kane Fracture Zone (Tucholke and Schouten, 1985) suggests that the fracture zone has recorded changes in relative plate motion at intervals as short as 2–3 m.y., corresponding to bathymetric trends at least as small as 20–30 km in length scale. Thus fracture-zone structure probably is a more sensitive indicator of plate motions than magnetic anomaly patterns, which in the North Atlantic generally are poorly defined (Klitgord and Schouten, this volume). The reality of such fine-scale changes in plate motion can be tested by SEA-BEAM bathymetric swath-mapping to look for coherent structural changes in conjugate limbs of several fracture zones. If the rapid changes prove to be real and coherent, then it will be possible to document plate motions in detail and to begin investigating correlations between changes in these motions and other geologic phenomena that occur on similar time scales (e.g., sea-level changes, episodic volcanism, faunal/floral extinctions, geomagnetic reversals).

Where there are adequate data to control the observations, it is apparent that many fracture-zone traces do not strictly parallel one another. This is observed, for example, in Plate 2, although the lack of parallelism in many cases there is because the isobaths represent an averaged version of data contoured imperfectly by hand. There are several possible explanations for the lack of parallelism. In the case of fracture zones formed at a large (>20 km) transform offset, any significant change in relative plate motion could result in growth of additional transform offsets (and fracture valleys), or destruction of existing transforms/valleys (e.g., Menard and Atwater, 1969). In either case, the fracture valleys would be oblique and merge with one another.

For small-offset (<20 km) and "zero-offset" transforms, it is commonly observed that the expression of the fracture valley is less well defined and the trend of the fracture valleys can be oblique to that of larger-offset fracture zones (Johnson and Vogt, 1973; Vogt, this volume, Ch. 24). The oblique trends could represent slow rift propagation of the type described by Hey and others (1980). Johnson and Vogt (1973) proposed that such V-shaped trends may reflect relative motion between the plate boundary and magma chambers in the underlying asthenosphere. Schouten and others (1986) suggest that these magma chambers may be fixed in the asthenosphere. The oblique trends are blocked at large-offset transforms, but dominate the structure of small- and zero-offset transforms (e.g., the Kurchatov FZ; Vogt and Tucholke, this volume, Fig. 8). With multibeam swath-mapping, the capability exists to test the coherency of these oblique trends and to relate changes in their orientation to times of change in relative plate motion as recorded in the larger-offset fracture zones. If such possible correlations are confirmed, they will have fundamental consequences for understanding the history of plate motions and for predicting the structure and composition of oceanic crust.

Seamounts and Hotspots

We now know the location, height, and general configuration of most large seamounts and volcanic islands in the North Atlantic (e.g., Plate 2). Most of these volcanic edifices appear to be arranged along hotspot traces that describe the history of motion of lithospheric plates over hotspots (plumes?) in the subjacent mantle. On the basis of seamount/island locations and meager data on their age, as well as data from other plates, consistent kinematic models have been developed that describe motions of the Atlantic plates over a set of fixed hotspots (e.g., Morgan, 1981, 1983; Duncan, 1984). Many more age dates are needed along the hotspot traces, particularly at bends that mark changes in plate motion (for instance the seamounts at 33°N, 54°–55°W at the New England Seamounts/Corner Rise bend).

The geochemical/geophysical imprints of near-ridge hotspots on the oceanic crust have been studied extensively in the last 15 years (see Melson and O'Hearn, and Schilling, this volume), but many questions remain unanswered. The rare-earth and radiogenic-isotope anomalies associated with Iceland are generally interpreted in terms of anomalous mantle-hotspot sources, Iceland being "primordial" and the Azores perhaps being recycled ancient subducted continental material (Schilling, this volume). However, Oskarsson and others (1985) attribute much of Iceland's geochemical anomaly to remelting of Eurasian-plate crust as the accreting Mid-Atlantic Ridge plate boundary has migrated westward across the fixed hotspot. Postulated sub-axial asthenosphere flow (e.g., from Iceland southwest beneath Reykjanes Ridge as far as the Charlie-Gibbs FZ "dam") is supported by a few geochemical indices (e.g. $^3He/^4He$), but most geophysical/ geochemical anomalies terminate much closer to Iceland (Plate 8B; Schilling, 1973; see Vogt, 1983, for a review). V-shaped basement structures on the Reykjanes Ridge have been interpreted as traces of along-axis flow at 5–20 cm/yr (Vogt, 1971) but problems remain with this interpretation (Johansen and others, 1984). It may be possible to demonstrate "paleo-asthenosphere flow" by observation of upper-mantle seismic anisotropy; initial results from the flanks of the Reykjanes Ridge indicate ridge-parallel P-wave velocities as high as 8.6–8.9 km in the upper mantle, suggesting flow-orientation of olivine crystals (Ritzert and Jacoby, 1985). Perhaps the postulated flow of mantle asthenosphere from an off-axis ridge plume toward the accretion axis (e.g., Vink, 1984) could be tested in the same way. Certainly observed geochemical anomalies along the present Mid-Atlantic Ridge axis support such flow (Schilling and others, 1985).

Sedimentary Processes

A great deal of detailed information has been gathered about dynamic sedimentary processes in the past 25 years. In current-controlled sedimentary regimes we know the form and internal structure of bedforms ranging in size from ripples (e.g., Nowell and Hollister, 1985) to kilometer-scale mud waves. Where gravity flows such as turbidity currents and sediment slumps occur, studies have in many cases defined the general form of the transport pathways; cores and submersible or photographic observations have revealed details about the sedimentary structures and bedforms. However, in only a few instances and over relatively small areas have data been acquired and integrated across the spectrum of sensor scales available (e.g., from multi-beam bathymetry down to photography). Where gravity flows reoccur at reasonably high frequencies (e.g., on active submarine fans) it may be possible to monitor such events by deploying instrument arrays in likely areas.

The greatest advances in understanding dynamic sedimentary processes may require a systems approach to the problems. For example, a submarine canyon system including its tributaries could be surveyed from source to mouth, thus crossing the spectrum of sea-floor gradients and flow regimes. Use of appropriate sensor technology and "upscale integration" would allow each observation to be placed in proper context with respect to the development of the sedimentary system. Thus we would better understand, for example, the positioning and localization of braided channels, channel meanders, sediment waves on levees, scour zones, debris-flow fill, and numerous other geomorphic/ geodynamic features that constitute a submarine canyon system.

Stratigraphy

Up to this point, our discussion has been principally from the geomorphic viewpoint. That is proper because we usually rely on our knowledge of present geomorphology and what it tells us about geologic processes in order to decipher the stratigraphic record ("the present is the key to the past"). There is no reason to expect this philosophy to change, and our increasing knowledge of geomorphology and related processes will impact directly on our ability to interpret the stratigraphic record. In addition, there is a continuing trend toward higher resolution in all aspects of stratigraphic studies. In this area, the greater resolution will be accomplished not only by technological improvement (e.g., broad-band, and/or deeply-towed seismic reflection systems, and borehole logging) but also by changing philosophies about where sampling should be conducted. A good example of the latter is the Deep Sea Drilling Project and its successor, the Ocean Drilling Project. In the early years of ocean drilling, and continuing to a lesser degree until the present, drill holes typically were sited in "geologically unusual" areas. These included basement highs (where basement age could be determined with minimal subseabed penetration), areas of rough topography, crests and flanks of seamounts, and sedimentary sections abbreviated by large unconformities (i.e., where old sediments could be cored). While these sites yielded a wealth of important data, they all suffered from the fact that much of the sedimentary record was missing because of erosion by currents or gravity flows. This kind of siting still persists, but there is a growing trend toward drilling sites that minimize the likelihood that hiatuses are present, and toward logging the holes to optimize correlations to the seismic stratigraphic record. This new philosophy, essentially independent of the systems approach, will greatly expand our ability to resolve geologic events more precisely, and thereby to understand better their interrelationships.

Ocean Paleoenvironments and Enigmatic Events: Cosmogenic vs. Geogenic Causes

Astronomical (Milankovitch) forcing dominates the $10^4 < T < 10^6$yr part of the paleoenvironmental spectrum, at least since the mid-Pliocene (e.g., Berger, 1980) and possibly during earlier times (e.g., Barron and others, 1985). At longer time scales (10^6 to 10^8yr), the ocean paleoenvironment also shows trends (e.g., temperature and sea-level falls throughout the Tertiary) that are punctuated by short "events"; most of the events, by virtue of clustered extinctions, correspond to boundaries of the geologic time scale. The events are global (e.g. the terminal Eocene event) and cannot be understood solely by studying a particular ocean basin. However, they are best recorded in sea-floor sediments. Extracting the causes for these events and modeling the associated response of the ocean/ice/ atmosphere system, are among the most exciting and interdisciplinary geological frontiers.

The Cretaceous/Tertiary extinction event is currently the most popular, and it has been widely analyzed following the discovery of the "iridium spike" by Alvarez and others (1980). Indeed, this discovery sparked a series of "neo-catastrophist" hypotheses (e.g., Whitmire and Matese, 1985; Whitmire and Jackson, 1984) as well as other investigations up and down the geologic column, with emphasis on extraterrestrial impacts (Silver and Schultz, 1982). A cosmogenic explanation of the extinction event had been proposed by Urey (1973), but this did not stimulate careful examination of K/T boundary-layer composition. The geochemical case for asteroid impact is now very strong, yet there is still legitimate dissent (e.g., Officer and Drake, 1985), and the different environmental consequences of impact still have to be sorted out in terms of their effects on biota (heat?, cold?, fires?, etc.). Not much physical evidence for the "event" has been found aside from a moderate clustering of hiatuses in the deep-sea sediment record and a regression (Vail and others, 1977) which may or may not be precisely coeval with the extinction. For example, sediment slumping specifically triggered by the seismic shock or by the massive tsunamis that would be expected has yet to be identified in any ocean.

Vogt (1972) and Morgan (1986) have noted the apparent coincidence of the K/T boundary with the Deccan Traps in India, the largest "flood basalt" event in the last 200 m.y. The best age dates for the Deccan Traps cluster around 65 Ma (Baksi, 1986). Climatic effects of flood basalts could include heating due to CO_2 and other gases released (Fischer, in Berger and others, 1982), or cooling due to sulfate aerosols (DeVine and others, 1984). Therefore biologic effects, even extinctions, caused by the volcanism are not implausible. This leaves three almost equally unacceptable choices: (1) an exceptionally great flood basalt event accidentally coincided with an exceptionally great extraterrestrial impact; the likelihood of chance coincidence (in the same 1 m.y.) of two events with 10^8 yr recurrence times is only 1 in 10,000; (2) the Deccan Traps were caused by asteroid impact; or (3) the K/T boundary layer is actually geogenic, perhaps associated with violent exhalation from fluidization pipes developed with the Deccan Traps when a great mantle blob reached the upper mantle.

Other epoch boundaries in the geologic time scale are coincident with similar, although less severe, geologic and oceanographic events and in some instances evidence for impacts (e.g. near the Eocene/Oligocene boundary). The events are mostly explicable by geogenic "triggering" when tectonic and oceanographic thresholds were reached (Tucholke and Mountain, and Berggren and Olsson, this volume). However, the detailed "causality chains," e.g. from mantle dynamics to tectonic events to paleoceanographic events and hence to extinctions are still very poorly understood. Improved methods of detecting these events and especially of dating the time of their occurrence for correlation purposes is needed. Such studies will greatly improve our understanding of ocean paleoenvironments.

MARINE GEOSCIENCE: THE DRIVING FORCES

Whereas pure curiosity about the natural world is a primary factor in motivating scientists to perform "basic research," marine geoscience, compared to its terrestrial counterpart, has depended more heavily on other "driving forces" (military, economic, etc.) to create the economic and technical foundation for that research. This difference is easily understood when one compares the logistics and cost of most terrestrial field work (sometimes accomplished by a single researcher, a few students, and modest technology) with marine geological field work that involves daily ship costs of $\$10^4$ or more ($\$10^5$ for the *Joides Resolution*), teams of researchers, and tens of millions of dollars invested in research vessels and equipment (space research is still more extreme in this respect). However, the ocean/continent contrast in research costs is less dramatic when investigations of deep-earth structure by geophysical techniques or drilling are considered. In some instances, for example deep multichannel reflection profiling, airborne gravity, and especially satellite altimetry, it is actually cheaper and easier to gather the desired data on or above the sea.

There has never been a sharp boundary between basic and applied research in the marine area. For example, although most U.S. basic research money in marine geoscience now originates from the National Science Foundation, the U.S. Office of Naval Research (ONR) has long been a sponsor of both kinds of research (in fact, both authors were in large measure funded by ONR during the preparation of this volume). The petroleum industry has indirectly supported deep-ocean drilling and, more directly, individual research projects. While both the Navy and industry insist on the best possible basic science, they also expect to realize benefits sooner or later. Research monies are sown in fields that might be expected to bear the fruit of application. In this respect, the national decision to establish an EEZ will focus research in that 200 mile-wide band, and funding of "basic research" into hydrothermal processes are more than coincidentally related to possible mining of metallic ores or exploitation of geothermal energy. Thus, marine geoscience has been "nudged" in certain directions by its sponsors. It is not clear whether this "nudge factor" has played a significant role in the evolution of knowledge. Certainly it has figured in the global data distribution (Figs. 3, 4).

Bluewater geoscience had scarcely begun in the 1840s when the value of Lieutenant M. F. Maury's Atlantic wire soundings was challenged. What possible good was it to anyone to know the depths of the deep sea? Maury answered this with another question, a quote from Benjamin Franklin: "Of what worth is a new-born babe?" In fact, as early as the 1850's the laying of undersea telegraph cables required a knowledge of Atlantic Ocean floor topography. Although many subsequent examples can be cited (for example, sea-floor bathymetry is or may be used for purposes as disparate as waste disposal and navigation), the greatest gifts of ocean floor geology/geophysics have been a demonstration and record in the igneous crust of crustal accretion and plate motions, and a more faithful record of global paleoenvironments in the sediment cargo (submarine and even subaerial) than anything deposited on land; indeed, most of the important strata on land were deposited on past sea floors and can best be understood in terms of total oceanic paleoenvironments. Ocean-floor geology has revolutionized man's knowledge of his home planet. Although the rather modest cost of acquiring this new knowledge has undoubtedly been redeemed many times over in applications, we share the widely held view that basic research belongs to the list of important national activities.

Marine geoscience has enjoyed a unique partnership with naval, commercial, and other ocean "users," who by virtue of their priorities and much larger budgets could bear the development costs for technologies subsequently made available to geoscientists. Furthermore, these major "users" have collected an enormous volume of data (e.g., magnetic and bathymetry data by the U.S. Navy, Plates 2 and 3) and made it available. This "indirect funding" of marine geoscience is often overlooked. Among the U.S. Navy's direct and semi-direct technological contributions to basic marine geology and geophysics are: (1) the magnetometer, developed starting early in World War II to detect submarines (Spiess, 1980) and subsequently used to map sea-floor magnetic lineations that unlocked the history of plate motions; (2) the 3.5-kHz active sonar, also an anti-submarine warfare device, and now a valuable tool to map shallow-subbottom structure and echo character (the historical importance of naval sonar research to sea-floor imaging is symbolized by the "A" in GLORIA, the long-range side-scan sonar: "A" stands for asdic, the British term for sonar, and asdic in turn stands for "Allied Submarine Detection Investigation Committee"); (3) the TRANSIT and GPS satellite navigation systems; and (4) multibeam bathymetry, originally the SASS system but now widespread in the commercial form called SEABEAM. Satellite altimetry (e.g., SEASAT, see frontispiece) has been developed with the help of Department of Defense funding because of the importance of accurate gravity data for missile-guidance purposes. In the area of non-naval national security, the world-wide seismography station network (WWSSN) went into operation in 1961 to monitor Soviet compliance with the 1958 Limited Test-Ban Treaty. The WWSSN accurately located earthquakes as well, thereby delineating active plate boundaries (Plate 11), and data from the network confirmed the transform fault model of Wilson (1965) among many other contributions (Einarsson, this volume).

In the commercial arena, offshore drilling and downhole-logging technology developed for hydrocarbon extraction made deep-sea drilling possible via the *Glomar Challenger* and its successor, the *Joides Resolution.* The geological impact of deep drilling in the ocean basins can scarcely be overstated. Most of the drilling on Iceland has been in quest of geothermal energy (Edmond, this volume). Mineral prospecting contributed to magnetometer development, and in fact Vacquier had begun work on his fluxgate instrument at the Gulf Oil Company before World War II (Spiess, 1980). Multichannel seismic profiling, including many of the processing techniques, also migrated to the deep ocean from the petroleum industry. Although there still has not been much progress in deep seabed mining (for example, manganese nodules, Dillon and others, this volume), the acoustic sea-floor imaging system first developed by International Nickel (INCO) to prospect for nodules found its way as an orphan into the basic-research realm as the Sea MARC side-scan sonar system. INCO had terminated the Sea MARC program because Law of the Sea negotiations (Knauss, this volume) reduced the prospect for reasonable profit, and because of a global drop in nickel prices (J. Kosalos, personal communication, 1986). Basic science in this instance was a beneficiary of unwelcome economic development, but more often science has been the first to suffer from economic downturns and budget-cutting.

Because so many of the most productive "toys" used in marine geoscience were originally developed for other purposes, it cannot be claimed that "science drives technology." For the most part, new technologies were made available, not built to test hypotheses as in the classic scientific method. Marine scientists adapted the technologies to deep-ocean settings, and new phenomena appeared in the data, demanding scientific hypotheses to explain them. Sometimes the phenomena were not explained correctly for years. For example, in the late 1950s, magnetic anomalies were mapped by Raff and Mason (1961) off the U.S. west coast just to see what was there. The magnetometer was added to a bathymetric survey commissioned by the Navy for underwater acoustic purposes (Spiess, 1980). The lineations mapped by the surveys were not correctly explained until Vine and Matthews' (1963) work, and more subtle diagonal disruptions were not recognized as propagating rifts until the late 1970s (Hey and others, 1980). It is almost certain that some of the tomorrow's breakthroughs are already recorded in data presently available.

THE BICENTENNIAL: SOME PREDICTIONS

What will be known about the Atlantic at the GSA's

bicentennial in 2088? It is easier to make geological predictions for the geological future (m.y. time scale) than socio-economic or technological predictions for the historical future (century time scale). It is clear, however, that future progress in ocean-floor research will depend increasingly on technically complex, expensive programs requiring teams of scientists and institutions. The expense will need to be justified, if not for science, then for economic or military reasons. We cannot, of course, predict the unpredictable discoveries such as lasers or semiconductors, or hydrothermal vent communities. In fact, most of the progress to date has resulted not from application of the scientific method (designing experiments to test hypotheses) but by accidental discoveries. Raff and Mason did not map magnetic anomalies to test the Vine-Matthews hypothesis, and hydrothermal vent communities were stumbled upon, not predicted even by "exobiologists." It is easy to predict that geological parameters will be known more accurately in 2088 than in 1988, but for many such quantities the principle of "diminishing marginal returns" will rule in the end. For example, the age estimates of magnetic reversals or geologic stage boundaries can already be seen to be approaching their "final" values (never to be known exactly) by a process of iteration (Fig. 3). Successive revisions of the geological time scale will become progressively less frequent, more demanding in terms of new data, and smaller in terms of marginal improvement. As another example, consider the problem of establishing the existence of a 5–10mW/m^2 positive heat-flow anomaly in the greater Iceland area in order to prove that the underlying Iceland hotspot mantle is actually anomalously hot. In the last 20 years, about $N=10^2$ measurements have been made, giving a regional heat flow of the order 100 ± 20 mW/m^2 in round numbers. To drive the error down to the required ± 5 mW/m^2 would require 16 times the present data set (assuming a random error varying as $\sqrt{N}$), and this would require 300 years at the present average rate of five new measurements per year! As for geological understanding of ocean-floor features, we expect that the 100th fracture zone or the 100th seamount to be studied in detail will not be as interesting as the tenth. Mapping and sampling the 100th fracture zone or seamount will probably become a job, similar to land cartography, performed by a government agency rather than a university.

If the ocean floors are to be mapped at high resolution before the GSA bicentennial, manned surface ships or manned submersibles are not a practical means to do it (Vogt and Tucholke, this volume). More likely, this job will be accomplished, if at all, under a systematic, probably multinational program utilizing a small fleet of untethered, remotely operated vehicles (ROVs) including both swimmers and crawlers. In addition to routine mapping, the swimmers will patrol active plate boundaries, and other dynamic environments in a systematic search for (or deployment to) new volcanic eruptions, faults and slumps, and hydrothermal vents. Systematic sea-floor mapping will uncover numerous shipwrecks, including many that because of their age, excellent degree of preservation, or historical value should be considered a patrimony belonging to all mankind. These wrecks will have to be marked or protected using satellite-telemetering buoys to preserve them from destruction by curio-seekers. In addition to submersibles used for imaging, crawling ROVs analogous to planetary roving vehicles will track slowly back and forth on the abyssal sea floor, taking and analyzing cores of sediment and rock, measuring heat flow, and storing data for later discharge via satellite.

Before 2088, a small grid of moholes may have been drilled, along with a large number of additional holes penetrating deep into oceanic crust and continental margins. Numerous sediment cores and borehole logs will probably result in "pieced together" but complete records of magnetic reversals, microfossil zonations, and Milankovitch signatures from the present back to the oldest (~175 Ma) submarine sediments. The active plate boundaries such as the MAR may by 2088 be densely sprinkled with acoustic and optic (laser) seafloor geodetic networks measuring crustal strain, including plate motion. These data will be merged with terrestrial inter- and intra-plate strain data (from satellites such as GPS) into global geodynamic models. At the same time, the ocean basins may be "wired" acoustically for continuous ocean "tomography" to monitor changing ocean climates. Permanent seismometers will be installed in remotely operated sea-floor laboratories and in boreholes. Together with better-distributed permanent and improved mobile networks on land, deep-ocean seismometers might locate teleseisms to 1 km accuracy. In addition, tomography derived from travel-time and wave-form inversion will finally prove or disprove the existence of deep-mantle plumes.

All of the above advances presuppose a dedicated national or international ocean-exploration program, perhaps patterned after the present Ocean Drilling Project. Although economic and technical resources already exist to embark on such a long-term program, the political will seems lacking and in fact large political obstacles exist; for example, the reluctance of coastal states to permit international exploration of the EEZs. However, should this planet become significantly more peaceful in the twenty-first century, some of the staggering sums now spent by nations on defense could be reprogrammed into systematic oceanic and planetary exploration. Even if not, the world's growing appetite for energy and other resources (e.g., Dillon and others, this volume) will motivate accelerated ocean floor mapping. In either case, the bicentennial DNAG project may well include an image of the entire Atlantic sea floor comparable in resolution to that presently available for the Moon. Atlantic heat-flow contour maps with SEASAT-like resolution might be available (the work of crawling ROVs), and detailed maps of basement composition and structure may exist based on close-spaced drilling grids. Ocean-wide contour maps of depth to Moho and other sub-basement surfaces may be available from systematic multichannel survey grids.

It is easy to predict an increased understanding, by 2088, of the ocean paleoenvironment. Numerical ocean/atmosphere models will be vastly more sophisticated and will show the system response to paleoenvironmental change (such as a change of sea-floor topography). However, whether, for example, the terminal Cretaceous and terminal Eocene events will ever be truly understood as to their environmental and biological system responses will depend in the final analysis on the number and quality of data actually recorded in the Earth. No computer is powerful enough or model sophisticated enough to replace the lithosphere subducted, the sediments eroded, or the microfossils that have been dissolved.

REFERENCES

Alvarez, W., Alvarez, L. W., Asaro, F., and Michel, H. V.
1980 : Extraterrestrial cause for the Cretaceous-Tertiary extinction: Science, v. 208, p. 1095–1108.

Baksi, A. J.
1986 : A review of the geochronological data available on flood basalt provinces in India and the Columbia River basalts, USA: EOS, Transactions of the American Geophysical Union, v. 67, p. 391.

Barron, E. J., Arthur, M. A., and Kauffman, E. G.
1985 : Cretaceous rhythmic bedding sequence; A plausible link between orbital variations and climate: Earth and Planetary Science Letters, v. 72, p. 327–340.

Berger, A.
1980 : The Milankovitch astronomical theory of paleoclimates; A modern view: Vistas in Astronomy, v. 24, p. 103–122.

Berger, W. H., and others (panel)
1982 : Climate in Earth History: Studies in Geophysics, National Academy Press, 198 p.

Dallmeyer, R. D.
1975 : The Palisades Sill; A Jurassic intrusion? Evidence from $^{49}Ar/^{39}Ar$ incremental release ages: Geology, v. 3, p. 243–245.

Devine, J. D., Sigurdsson, H., and Davis, A. N.
1984 : Estimates of sulfur and chlorine yield to the atmosphere from volcanic eruptions and potential climate effects: Journal of Geophysical Research, v. 89, p. 6309–6325.

Duncan, R. A.
1984 : Age progressive volcanism in the New England Seamounts and the opening of the central Atlantic Ocean: Journal of Geophysical Research, v. 89, p. 9980–9990.

Emery, K. O., Uchupi, E., Phillips, J. D., Bowin, C. O., Bunce, E. T., and Knott, S. T.
1970 : Continental rise off eastern North American: American Association of Petroleum Geologists Bulletin, v. 54, p. 44–108.

Ewing, M., Crary, A. P., and Rutherford, H. M.
1937 : Geophysical investigations in the emerged and submerged Atlantic coastal plain: Part I, Methods and results: Geological Society of America Bulletin, v. 48, p. 753–801.

Ewing, J. I., and Tirey, G. B.
1961 : Seismic profiler: Journal of Geophysical Research, v. 66, p. 2917–2927.

Heirtzler, J. R., and Hayes, D. E.
1967 : Magnetic boundaries in the North Atlantic Ocean: Science, v. 157, p. 185–187.

Heirtzler, J. R., Dickson, G. O., Herron, E. M., Pitman, W. C., and LePichon, X.
1968 : Marine magnetic anomalies, geomagnetic field reversals, and motions of ocean floor and continents: Journal of Geophysical Research, v. 73, p. 2119–2136.

Hersey, J. B.
1963 : Continuous reflection profiling: in The Sea, Volume 3, ed. M. N. Hill: New York, Wiley, p. 47–72.

Hey, R. N., Duennebier, F. K., and Morgan, W. J.
1980 : Propagating rifts on mid-ocean ridges: Journal of Geophysical Research, v. 85, p. 2647–2658.

Holmes, A.
1937 : The Age of the Earth: London, Nelson, 263 p.

Johnson, G. L., and Vogt, P. R.
1973 : Mid-Atlantic Ridge from 47° to 51° North: Geological Society of America Bulletin, v. 84, p. 3443–3462.

Johansen, B., Vogt, P. R., and Eldholm, O.
1984 : Reykjanes Ridge; Further analysis of crutal subsidence and time-transgressive basement topography: Earth and Planetary Science Letters, v. 68, p. 249–258.

Lambert, R.S.J.
1971 : The pre-Pleistocene Phanerozoic time-scale; A supplement: Geological Society of London Special Publication 5, Part 1, p. 9–31.

LePichon, S.
1968 : Sea floor spreading and continental drift: Tectonophysics, v. 7, p. 3661–3697.

Maury, M. F.
1855 : The Physical Geography of the Sea: New York, Harper and Brothers, 287 p.

Menard, H. W., and Atwater, T.
1969 : Origin of fracture zone topography: Nature, v. 222, p. 1037–1040.

Morgan, W. J.
1968 : Rises, trenches, great faults, and crustal blocks: Journal of Geophysical Research, v. 73, p. 1959–1982.
1972 : Deep mantle convection plumes and plate motions: American Association of Petroleum Geologists Bulletin, v. 56, p. 203–213.
1981 : Hotspot tracks and the opening of the Atlantic and Indian Oceans: in The Sea Volume 7, ed. C. Emiliani: New York, Wiley, p. 443–488.
1983 : Hotspot tracks and the early rifting of the Atlantic: Tectonophysics, v. 94, p. 123–139.
1986 : Flood basalts and mass extinctions: EOS Transactions of the American Geophysical Union, v. 67, p. 391.

Nowell, A.R.M., and Hollister, C. D., eds.
1985 : Deep Ocean Sediment Transport; Preliminary Results of the High Energy Benthic Boundary Layer Experiment: Amsterdam, Elsevier, 420 p.

Officer, C. B., and Drake, C. L.
1985 : Terminal Cretaceous environmental effects: Science, v. 227, p. 1161–1167.

Oskarsson, N., Steinthorsson, S., and Sigvaldason, G. E.
1985 : Iceland geochemical anomaly; Origin, volcanotectonics, chemical fractionation, and isotope evolution of the crust: Journal of Geophysical Research, v. 90, p. 10011–10025.

Pitman, W. C., and Talwani, M.
1972 : Sea-floor spreading in the North Atlantic: Geological Society of America Bulletin, v. 83, p. 619–645.

Raff, A. D., and Mason, R. G.
1961 : Magnetic survey off the west coast of North America, 40° N. latitude to 52° N. latitude: Geological Society of America Bulletin, v. 72, p. 1267–1270.

Ritzert, M., and Jacoby, W. R.
1985 : On the lithospheric seismic structure of Reykjanes Ridge at 62.5 °N: Journal of Geophysical Research, v. 90, p. 10117–10128.

Sandwell, D.
1986 : Thermal stress and the spacings of transform faults: Journal of Geophysical Research (in press).

Sayer, R.
1775 : A chart of the Atlantic Ocean, 1 sheet, printed for R. Sayer, Map and Printseller, 53 Fleet Street, London.

Schilling, J.-G.
1973 : Iceland mantle plume; Geochemical study of Reykjanes Ridge: Nature, v. 242, p. 565–571.

Schilling, J. G., Thompson, G., Kingsley, R., and Humphris, S.
1985 : Hotspot-migrating ridge interaction in the South Atlantic: Nature,

v. 313, p. 187–191.
Schott, W.
1935 : Die Foraminiferen in dem Äquatorialen Teil des Atlantischen Ozeans; Wissenschaftliche Ergebnisse der Deutschen Atlantischen Expedition, Forschungs-Schiff Meteor, 1925–1927, v. 111, p. 43–134.
Schouten, H., and Klitgord, K. D.
1982 : The memory of the accreting plate boundary and the continuity of fracture zones: Earth and Planetary Science Letters, v. 59, p. 255–266.
Schouten, H., Klitgord, K. D., and Whitehead, J. A.
1985 : Segmentation of mid-ocean ridges: Nature, v. 317, p. 225–229.
Schouten, H., Dick, H.J.B., and Klitgord, K. D.
1986 : Migration of mid-ocean ridge volcanism: Nature (submitted).
Silver, L. T., and Schultz, P. H., eds.
1982 : Geological implications of impacts of large asteroids and comets on the Earth: Geological Society of America Special Paper 190, 528 p.
Spiess, F. N.
1980 : Some origins and perspectives in deep-ocean instrumentation development: in Oceanography: The Past, eds. M. Sears, and D. Merriman: New York, Springer, p. 226–239.
Steiger, R. H., and Jäger, E.
1977 : Subcommission of Geochronology; Convention on the use of decay constants in geo- and cosmo-chronology: Earth and Planetary Science Letters, v. 36, p. 359–362.
Stocks, T., and Wüst, G.
1935 : Die Tiefenverhältnisse des offenen Atlantischen Ozeans: Deutsche Atlantische Expedition METEOR, 1925–1927: Wissenschaftliche Ergebnisse, v. 3, Teil 1, 1. Lieferung, 31 p.
Tucholke, B. E., and Schouten, H.
1985 : Global plate motion changes recorded in the Kane Fracture Zone: Geological Society of America Abstracts with Programs, v. 17, p. 737.
Urey, H. C.
1973 : Cometary collisions and geological periods: Nature, v. 242, p. 32–33.
Vail, P. R., and Mitchum, R. M., Jr., and Thompson, S., III
1977 : Global cycles of relative changes of sea level: in Seismic stratigraphy; Applications to hydrocarbon exploration, ed. C.E. Payton: American Association of Petroleum Geologists Memoir 26, p. 83–97.
Van Hinte, J. E.
1976 : A Jurassic time scale: American Association of Petroleum Geologists Bulletin, v. 60, p. 489–497.
Vening Meinesz, F. A .
1932 : Gravity expeditions at sea, 1923–1930; Volume I: Delft, Holland, Publication of the Netherlands Geodetic Commission, p. 1–110.
Vine, F. J., and Mathews, D. H.
1963 : Magnetic anomalies over oceanic ridges: Nature, v. 199, p. 947–949.
Vink, G. E.
1984 : A hotspot model for Iceland and the Voring Plateau: Journal of Geophysical Research, v. 89, p. 9949–9959.
Vogt, P. R.
1971 : Asthenosphere motion recorded by the ocean floor south of Iceland: Earth and Planetary Science Letters, v. 13, p. 153–160.
1972 : Evidence for global synchronism in mantle plume convection, and possible significance for geology: Nature, v. 240, p. 338–342.
1973 : Early events in the opening of the North Atlantic: in Implications of Continental Drift to the Earth Sciences, eds. D. H. Tarling, and S. K. Runcorn: London, Academic Press, v. 2, p. 693–712.
1983 : The Iceland mantle plume; Status of the hypothesis after a decade of new work: in Structure and Development of the Greenland-Scotland Ridges, eds. M.H.P. Bott, S. Saxov, M. Talwani, and J. Thiede: New York, Plenum, p. 191–216.
Vogt, P. R., and Einwich, A. M.
1979 : Magnetic anomalies and sea-floor spreading in the western North Atlantic, and a revised calibration of the Keathley (M) geomagnetic reversal chronology: in Initial Reports of the Deep Sea Drilling Project, Volume 43, eds. B. Tucholke, and P. R. Vogt: Washington, D.C., U.S. Government Printing Office, p. 857–876.
Vogt, P. R., and Tucholke, B. E.
1982 : The western North Atlantic: in Perspectives in Regional Geological Synthesis, ed. A. R. Palmer: Geological Society of America, DNAG Special Publication 1, p. 117–132.
Vogt, P. R., Anderson, C. N., Bracey, D. R., and Schneider, E. D.
1970 : North Atlantic magnetic smooth zones: Journal of Geophysical Research, v. 75, p. 2955–2968.
Vogt, P. R., Kovacs, L. C., Bernero, C., and Srivastava, S. P.
1982 : Asymmetrical geophysical signatures in the Greenland-Norwegian and southern Labrador Seas and the Eurasia Basin: Tectonophysics, v. 89, p. 95–160.
Wegener, A.
1929 : The Origin of Continents and Oceans (translated from the Fourth Revised 1929 German edition by J. Biram): New York, Dover, 246 p.
Whitmire, D. P., and Jackson, A. A.
1984 : Are periodic mass extinctions driven by a distant solar companion?: Nature, v. 308, p. 713–715.
Whitmire, D. P., and Matese, J. J.
1985 : Periodic comet showers and Planet X: Nature, v. 313, p. 36–38 and 744.
Wilson, J. T.
1963 : Hypothesis of earth's behavior: Nature, v. 198, p. 925–929.
Wilson, J. T.
1965 : A new class of faults and their bearing on continental drift: Nature, v. 207, p. 343–347.

MANUSCRIPT ACCEPTED BY THE SOCIETY JULY 14, 1986

ACKNOWLEDGMENTS

We thank A. R. Palmer, Centennial Science Program Coordinator, for the opportunities to design and develop this volume and to comment, scientifically and philosophically, on the past, present, and future of marine geoscience. We also thank G. Rassam of GEOREF for assisting in the acquisition and interpretation of the bibliographic data contained in Figure 5 and Eileen Pickenpaugh for compiling the data. T. Holcombe, P. Sloss, and C. Moore kindly provided information from the National Geophysical Data Center data base in Figures 3 and 4. The opinions expressed in this chapter are those of the authors, and do not necessarily reflect the opinions of agencies funding marine geoscience research.

Printed in U.S.A.

The Geology of North America
Vol. M, The Western North Atlantic Region
The Geological Society of America, 1986

Chapter 2

Imaging the ocean floor: History and state of the art

Peter R. Vogt
Naval Research Laboratory, Washington, D.C. 20375-5000
Brian E. Tucholke
Woods Hole Oceanographic Institution, Woods Hole, Massachusetts, 02543

INTRODUCTION

Bathymetry is the science and technology of ocean depth measurement. In the narrow and traditional sense, the objective of bathymetry is to define the seafloor topographic surface, $h(x, y)$. Most bathymetric studies also interpret the depth data—usually displayed in contour charts or profiles—in terms of undersea geomorphology and, in concert with other geophysical data, as a constraint on sub-seafloor geologic structure. In many parts of the ocean, bathymetry is about the only type of geological information available. Thus, compared to land, seafloor topography necessarily plays a larger role in geologic interpretations.

The North Atlantic Ocean covers an enormous area ($4 \times 10^7 km^2$) with diverse and numerous geomorphic features at many scales. Bathymetric control is very uneven in distribution and quality. Some areas have been surveyed by overlapping swaths of multibeam bathymetry with reliable mapping of features less than 1 km apart, while other areas of 10^3 or $10^4 km^2$ have no soundings, particularly in the equatorial latitudes. Unlike many previous studies, this chapter does not offer a systematic guided tour of the submarine topography. (See Holcombe, 1977 for a glossary of ocean-bottom land forms.) Recent syntheses of Atlantic bathymetry include Vogt and Perry (1982) and Emery and Uchupi (1984). Instead, the potential of seafloor "imaging" is evaluated over a large range of spatial scales, approximately ten orders of magnitude from the form of the entire Atlantic basin down to the size of grains in seafloor sediments (Fig. 1). Under "submarine imaging" are included 3-D (stereoscopic) descriptions of bottom topography (ordinary bathymetry) as well as 2-D maps recorded by traditional sidescan sonar and photographic systems. "Subaerial imaging" of exposed oceanic crust (e.g., Iceland) includes aircraft and satellite instrumentation that derive images from radar, images illuminated (visual) or energized (near infrared) by solar radiation, and images internally energized by geothermal processes (broadband infrared). All these techniques have contributed to one total "image" of the shape, texture, and contrast in basic physical properties across a sharp physical boundary, the outer edge of the solid earth.

The main part of this chapter is organized according to the physical principles used for imaging; the principal modern types are acoustic (at roughness scales >1 m) and electromagnetic (principally in the visible spectrum for scales of $\leqslant 10$ m). The imaging is discussed briefly in terms of its historical and technical evolution, impact on geological thought, and resolution. Some techniques have been supplanted by others or are only of historical or local interest. Atlantic examples of each important image type are presented.

The term "resolution," used frequently in this chapter, means minimum detectable horizontal separation between two adjacent features. The concept of resolution is more objectively embodied in the two-dimensional transfer function relating an object to its image. Some images, notably acoustic, also have vertical resolution. Resolution cannot be specified precisely because it depends on differences in feature contrast and elongation. For example, a linear feature such as a fault scarp may be resolved even though it is narrower than the resolution limit for equidimensional features such as volcanic cones. The practical resolution of various imaging systems, together with their areal coverage, is shown in Figure 1A.

The least well known seafloor features occur at horizontal scales from about 10 m to several kilometers. Such features are now being studied by side-scan sonar, but deeply towed systems of this type can survey at ship speeds only up to about 3 km/hr. Until recently, such systems also have not provided accurate 3-D information compared to multi-beam bathymetric surveying, but the backscatter images often reveal subtle differences in fine-scale "texture" (e.g., basalt versus debris flows) that cannot be resolved in three dimensions. The ultimate seafloor images probably will combine stereoscopic (depth contour) control with various backscatter imagery (Ryan, 1986), large-area mosaic photography, and other data (color, composition, texture).

Our bathymetric contour chart of the North Atlantic (Plate 2) is one of an evolving series of "traditional" bathymetric images, which are gradually becoming more accurate as data bases become larger and navigation accuracies improve. Such regional charts do not (and cannot) show the detailed form of contours

Vogt, P. R., and Tucholke, B. E., 1986, Imaging the ocean floor: History and state of the art, *in* Vogt, P. R., and Tucholke, B. E., eds., The Geology of North America, Volume M, The Western North Atlantic Region: Geological Society of America.

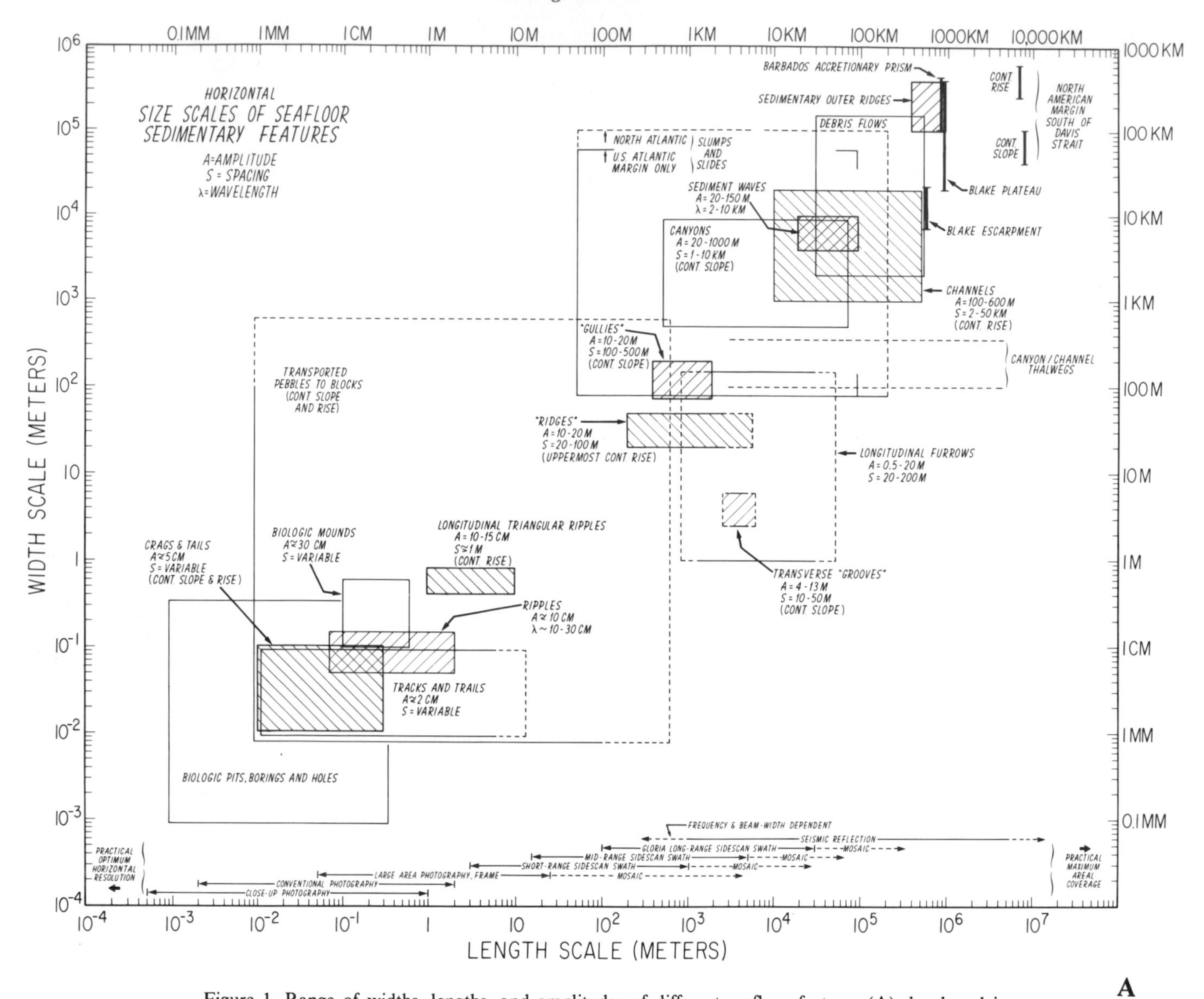

Figure 1. Range of widths, lengths, and amplitudes of different seafloor features (A) developed in seafloor sediments, and (B, facing page) of volcanic or tectonic origin, such as those to be found along the Mid-Atlantic Ridge axis. Ranges for different kinds of submarine imaging techniques are indicated at bottom of (A). Shading or hatching serves only to discriminate among overlapping boxes. (Note that by definition, parts of boxes extending above the diagonal lines are empty because objects are not wider than their length.)

because of deliberate low-pass filtering, hand-smoothing, or data file decimation. Depth measurements incorporated in most bathymetric charts are related to the instantaneous local sea level at the time of measurement. Actually ocean and atmospheric dynamics cause spatial and temporal sea-level variations. These rarely exceed ±1 m and are usually neglected, although they can cause difficulty in contouring very flat areas such as abyssal plains where gradients can be less than 1:5,000. Vertical ship motions caused by ocean waves may reach ±10 m but the resulting depth variations are easily recognized as high frequency noise in echograms. Geoid undulations also warp the surface of the North Atlantic (Vogt, this volume, Ch. 14), producing stationary sea-level anomalies up to a few meters over large seamounts and tens of meters over longer wavelength features. These amplitudes are small enough—compared to seafloor relief at comparable wavelengths—that the question of whether to reduce bathymetric data to a reference spheroid (as done for the ocean surface) is academic. Most seafloor geological processes are gravity controlled: for example lava, debris flows, and turbidites all attempt (with varying degrees of success) to distribute matter along local geopotentials. Thus, measured water depth as a departure from the geoid is physically most appropriate.

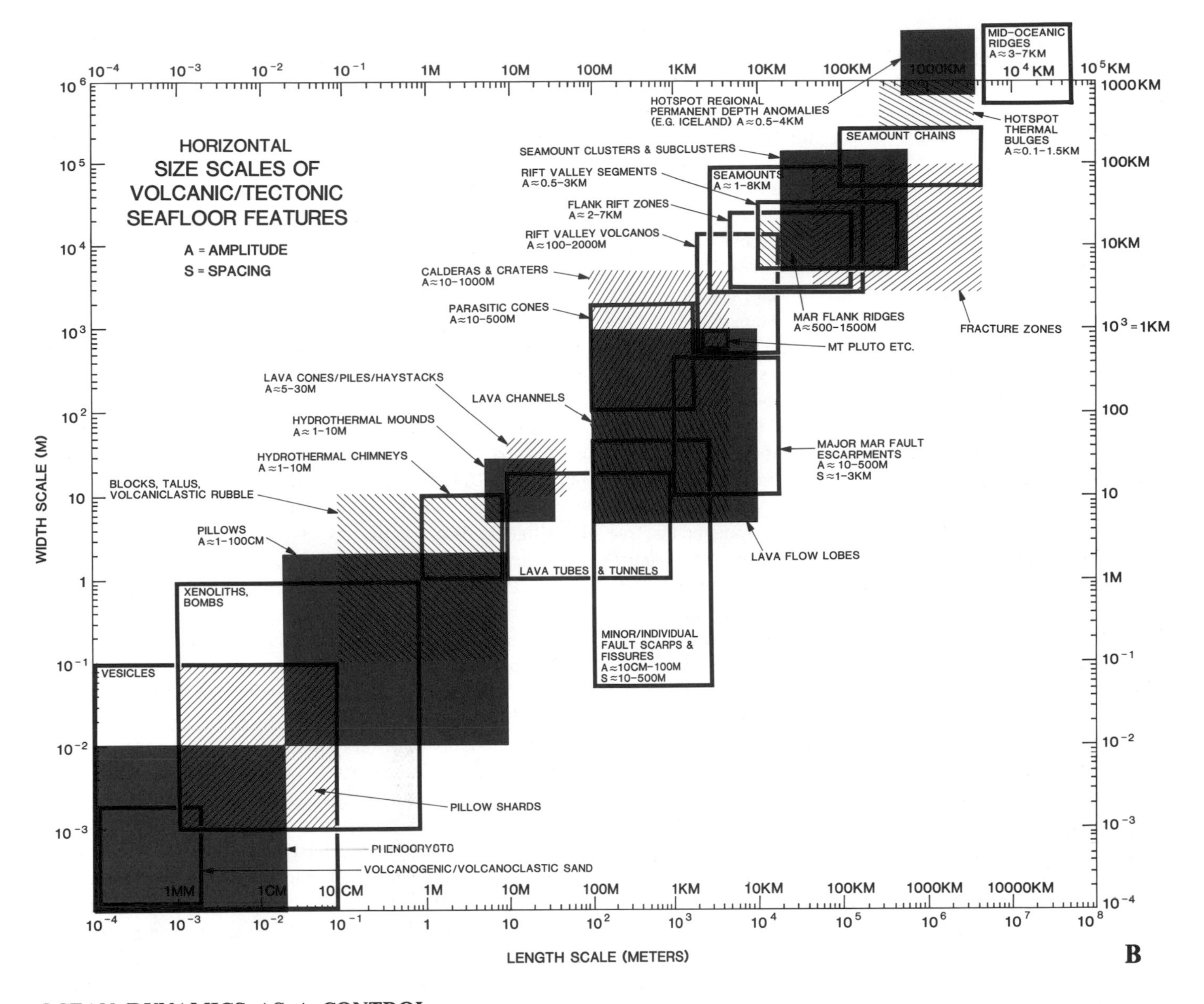

OCEAN DYNAMICS AS A CONTROL ON BATHYMETRY

Breakers mark shallow water and are one of the most obvious indicators of shallow seafloor. Breakers were used early as shoal detectors, for example, by Sayer in 1775 (Fig. 2 of Tucholke and Vogt, this volume), even though numerous observations were misinterpreted as rock or even islands (probably errant icebergs). We have learned only in the last few decades that the majority of reported breakers could not have been caused by shallows in central Atlantic waters.

However, waves do not have to break to reveal seafloor depth. G. B. Airy showed in 1845 that a sinusoidal gravity wave of wavelength L travels at a speed of $[(gL/2\pi)\ tanh\ (2\pi d/L)]^{1/2}$, where d is water depth. "Shallow-water waves" are those with $d/L < 1/20$, for which wave speed reduces to $(gd)^{1/2}$. This relation can be used to derive depth both in shallow water (for ordinary gravity waves, seiches, and bores) and in deep ocean basins (for tsunamis and tides, which have wavelengths much longer than the ocean is deep). The first reasonably correct figure (4 km) for mean ocean depth was derived in 1856 by A. D. Bache from the travel time of earthquake-generated tsunamis, in effect using tide gauges as "bathymeters" (Sverdrup and others, 1942). This replaced an earlier estimate using tides (18 km, by Laplace in 1775), which, although incorrect, was the first quantitative estimate of ocean depth.

Now the tables are turned; bathymetry is an important input to modern tide- and tsunami-prediction models, but these long waves have little new to tell us about the shape of the ocean floors. However, gravity wave patterns as mapped by airborne/-satellite photography or radar still have value for shallow-water bathymetry in remote areas, especially in water depths <50 m (Kasischke and others, 1983, 1984). In addition to the classical Airy-type wave refraction patterns, SAR (Synthetic Aperture Radar) imagery patterns reveal other ocean surface effects related to bathymetry; these include nonlinear gravity wave interactions (caused by gravity waves propagating into shallow water) and changes in current velocity, upwelling, and generation of internal waves (caused by ocean currents flowing over seafloor topographic features; Kasischke and others, 1983; Baines, 1979; Bell, 1975a; Lyzenga, 1981; Alpers and Hennings, 1984).

Besides measurements on the ocean surface, the movement of deep water masses identified by their temperature/salinity (T/S) also constrains our understanding of deep-sea topography. The T/S differences between deep waters east and west of the Mid-Atlantic Ridge showed scientists on the 1872–76 *Challenger* expedition (Thomson, 1878) that the ridge was a continuous feature long before this was documented by soundings. Oceanographic studies on the 1925–27 *Meteor* expedition (Wüst, 1936), plus scattered earlier soundings, pointed to a gap traversing the equatorial Mid-Atlantic Ridge. This gap, first postulated by Drygalski (1904) and now known to be the Romanche Fracture Zone valley, allows some Antarctic bottom water to spill eastward into the Guinea Basin. The sill depth is estimated to be 3,750 m from oceanographic data, and 4,000–4,800 m from known bathymetry (Metcalf and others, 1964; Vogt and Perry, 1982; Emery and Uchupi, 1984). In another instance, Worthington (1953) used deep water-mass properties to predict a continuous ridge separating the Eurasia and Amerasia basins in the Arctic. This prediction proved to be true, as the Lomonosov Ridge later was documented from sounding data.

BATHYMETRIC RELATIONS TO STATIC POTENTIAL FIELDS (GRAVITY AND MAGNETICS)

Because ocean water is much less dense than rock, it was early supposed that ocean basins should be associated with a strong disturbance of the earth's gravity field, in which case gravity could have been used as a direct measure of depth (Siemens, 1876). Pratt (1859) derived a formula relating anomalous vertical deflection and sea level height on the coast of India to depth of the adjacent Indian Ocean. His formula made no allowance for isostasy, and it predicted deflections of 5 sec to 20 sec for a mean depth of 4.8 km (3 mi). Because such an effect was not discernible, he later inferred (Pratt, 1871) that there must be an excess of matter beneath the seafloor.

With the advent of accurate marine gravimetry (e.g., Vening Meinesz, 1937) the ocean basins away from island arcs were found to be in approximate isostatic equilibrium. The density deficit caused by the water mass was found to be compensated by a rise in the earth's mantle, compared to the continents. Thus average gravity anomalies are near zero over the ocean basins rather than several hundred milligals negative (Rabinowitz and Jung, this volume). Isostasy notwithstanding, there still exists a positive correlation between seafloor topography and free-air gravity anomaly (McKenzie and Bowin, 1976) and geoid undulation (Vogt, this volume, Ch. 14). For broad features such as the Mid-Atlantic Ridge and hotspot-generated swells, this correlation reflects the greater distance to the low-density root, with a possible dynamic contribution by mantle upwelling. For topographic features of horizontal scale less than about 300–400 km, the correlation is largely due to regional flexural compensation of the topographic load; the strength of the lithospheric plates causes a seamount to be compensated by a broad downwarp, leaving a pronounced positive gravity anomaly even over a compensated seamount. Correlation between topography and gravity is generally highest at wavelengths of 50 to 100 km. It disappears below 20–30 km, due to the combined effects of gravity (or geoid) measurement noise and attenuation of short wavelengths in oceanic water depths.

Because survey ships carrying gravity meters usually have echo sounders, gravity data had no real value for determining bathymetry. This changed dramatically in 1978 with the launching of SEASAT, an accurate (±10 cm sea-surface height) spaceborne altimeter (Vogt, this volume, Ch. 14). The processed satellite altimetry data, whether displayed as gravity anomaly (Haxby and others, 1983; Haxby, 1985) or geoid undulation (Parke and Dixon, 1982; Haxby, 1985), provided spectacular remote-sensed "proxy" views of seafloor topography (see color frontispiece). In remote and poorly surveyed areas the SEASAT data provided a better "bathymetric" map than was previously available from accurate but widely spaced sounding lines. In the relatively well surveyed central and North Atlantic, SEASAT revealed no major surprises, although previously unknown fracture zones were delineated (Vogt and others, 1984; Vogt, this volume, Ch. 14). It is possible to "invert" altimetry or gravity to seafloor topography, using techniques such as iterative forward modelling or transfer functions, in one or both horizontal dimensions. One-dimensional (profile) inversion of SEASAT altimetry to seafloor topography has been demonstrated in the Pacific, where the bottom can be reproduced to within several hundred meters error (Dixon and others, 1983).

Where shipboard soundings are sparse, magnetic anomalies measured by aircraft can also be used to obtain generalized bathymetry. Sediments are very weakly magnetized, so magnetic anomalies relate to depth of igneous basement rather than seafloor. However, over much of the ocean basins the sediment cover is so thin (Plate 6) that a "magnetic basement depth" (e.g., Kovacs and Vogt, 1982) essentially equals the water depth. Seafloor-spreading magnetic lineations of course do not correspond to seafloor topography, but other anomalies such as those over seamounts correlate strongly (e.g., Plate 3; Vogt, this volume, Ch. 15). In the latter case, bathymetric predictions could be computed from aeromagnetic data using iterative forward modelling or transfer functions in one or two horizontal dimensions; this has not yet been attempted. The simplest and to date most useful application of aeromagnetic data to bathymetry is to define the locations of seamounts, fracture zones, and other features, and to use this information as control in contouring widely-spaced soundings or as a basis for designing bathymetric surveys.

ELECTROMAGNETIC IMAGING

Background

The electromagnetic spectrum extends from gamma rays (some with wavelength less than 10^{-12} m and frequencies above 10^{20}Hz) to the longest radio waves. The spectrum includes x-rays, ultraviolet, and the visible band that extends from 3.7 to 7.5×10^{14}Hz; optical imaging of the solid earth is considered in a subsequent section. The propagation of electromagnetic energy—whether through sea water, the atmosphere, or the vacuum of space—is basically predicted from equations derived by Maxwell in 1873 (Liebermann, 1962). In practical application, attenuation in seawater is given by the "skin depth" $(2/\omega\mu\alpha)^{1/2}$, the distance at which field intensity is diminished by 8.6dB (ω = frequency, μ = magnetic permeability, α = electrical conductivity). For seawater (α = 2.4 to 4 Siemens/m) the skin depth is about $250\omega^{1/2}$ m, or only 25 cm for 10^6 Hz (microwaves) and 25 m for 10^2 Hz (radio waves); thus the high conductivity of seawater mandates very low frequencies for "electromagnetic bathymetry." Unlike acoustic waves, the rate of propagation of electromagnetic waves is frequency dependent. Although Maxwell's equations predict that seawater also should be penetrable at high frequencies, the actual transmission window is at much higher frequencies (in the visible light range) than predicted. This occurs because conductivity increases at high frequencies and because of an additional effect due to the polarization of water molecules (Liebermann, 1962). Penetration of light through the sea is also limited by scattering from suspended matter and water molecules (even in the clearest sea water), and by absorption. These combined effects make clear seawater most transparent in the blue and blue-green (0.46–0.50 μ, or 6 to 6.5×10^{14} Hz), whereas turbid water is more transparent in the green (0.54 μ) or even yellow green (0.56 μ). Total attenuation ranges from as little as 2.5%/m in the clearest ocean at 0.46 μ to more than 30%/m in turbid coastal water at 0.55 μ (see Tyler and Preisendorfer, 1962; Clarke and Denton, 1962; Tyler, 1977).

In the following sections we briefly review subaerial radar and infrared imaging, as well as very low frequency electromagnetic sounding and "optical bathymetry" based on ocean color or laser-echo-ranging in shallow water.

Radar

Radar (radio detection and ranging) includes active or passive remote sensing generally in the microwave part of the spectrum (wavelengths of one mm to one m [3×10^{11} to 3×10^8 Hz]). As already noted, the high conductivity of seawater prevents radar echosounding of the ocean floors (i.e., the skin depth is minimal), but radar provides a "bathymetric proxy" from the shape of the sea surface. In addition, a small portion of the North Atlantic Ocean crust is exposed in Iceland and other islands (ca. 1.2×10^5 km^2, or 0.3%) and can be mapped topographically by aircraft or satellite radar (Fig. 2C) even through clouds or at night. A smaller portion (1.3×10^4 km^2) of this area is covered by glacier ice (Iceland and Jan Mayen); here the crustal topography can be mapped directly by airborne or surface radio-echosounding (RES; Bogorodskiy and others, 1985), and undulations in the ice surface as mapped by radar also can provide clues to the bedrock topography (Williams, 1983a).

Radars used in SEASAT altimetry and synthetic aperture radar (SAR) are typically in the range $1–30 \times 10^9$ Hz (30 cm to 1 cm wavelength, i.e., "microwave"). SEASAT altimetry resolves features down to 5–10 cm amplitude and 30 km wavelength, while SAR has a 25 m horizontal resolution (Fig. 2C; Ford and others, 1980; McDonough and Martin-Kaye, 1984). SAR's 100 km swath width is similar to 50–150 km widths achievable by side-looking airborne radars (SLARs), which may be either synthetic-aperture or real-aperture. SLAR images can be combined in stereo-pairs for better recognition of topography, but they generally have been used as variable-density images like those of most side-looking sonars (e.g., GLORIA, Sea MARC; Figs. 2A, 3). Multispectral radar imaging is also being developed. Radio-echosounding of glaciers (RES) operates at somewhat lower frequencies (0.005 to 1×10^9 Hz) and achieves accuracies and resolutions of a few meters in mapping subglacial topography, besides providing a wealth of glaciological data.

Infrared Imagery

Broadband infrared imagery ("thermography") of active subaerial volcanic and geothermal phenomena (Friedman and others, 1969; Williams, 1980) began in the 1960s and so far has been confined mainly to airborne sensing because spatial resolution of satellite systems is too coarse. However, the 1966 Surtsey eruption was detected on a *Nimbus* pass in the 3.4–4.1 μ band ($0.88–0.73 \times 10^{14}$ Hz; Williams, 1983b) even with a 62 km^2 pixel area. The 3.0–5.5 μ and 8–14 μ infrared bands are of greatest value for monitoring dynamic geologic processes, including terrestrial sediment plumes intruding ocean surface waters. The 3.0–5.5 μ band is most suited for studying volcanism.

Submarine Electromagnetic Imagery

For the submarine realm the electromagnetic spectrum appears devoid of bathymetric applications except in the visible-light spectrum (3.7 to 7.5×10^{14} Hz, discussed later) and at wavelengths of about 10^3 km (300 Hz) and up where skin depth finally becomes comparable to water depth. Airborne electromagnetic bathymetry (AEM) is one experimental technique using the longer wavelengths (Zollinger and others, 1985). This aircraft system uses a high-power 2 msec EM pulse to induce transient eddy currents in the ocean, and it measures the depth-dependent decay of the secondary magnetic field generated by the eddy currents. AEM bathymetry works only in shallow water, with depths of 40 m measurable to ±3% and 60 m depths measurable to ±15% at 144 Hz fundamental frequency. Better depth penetration may be obtained at 90 Hz (Zollinger and others,

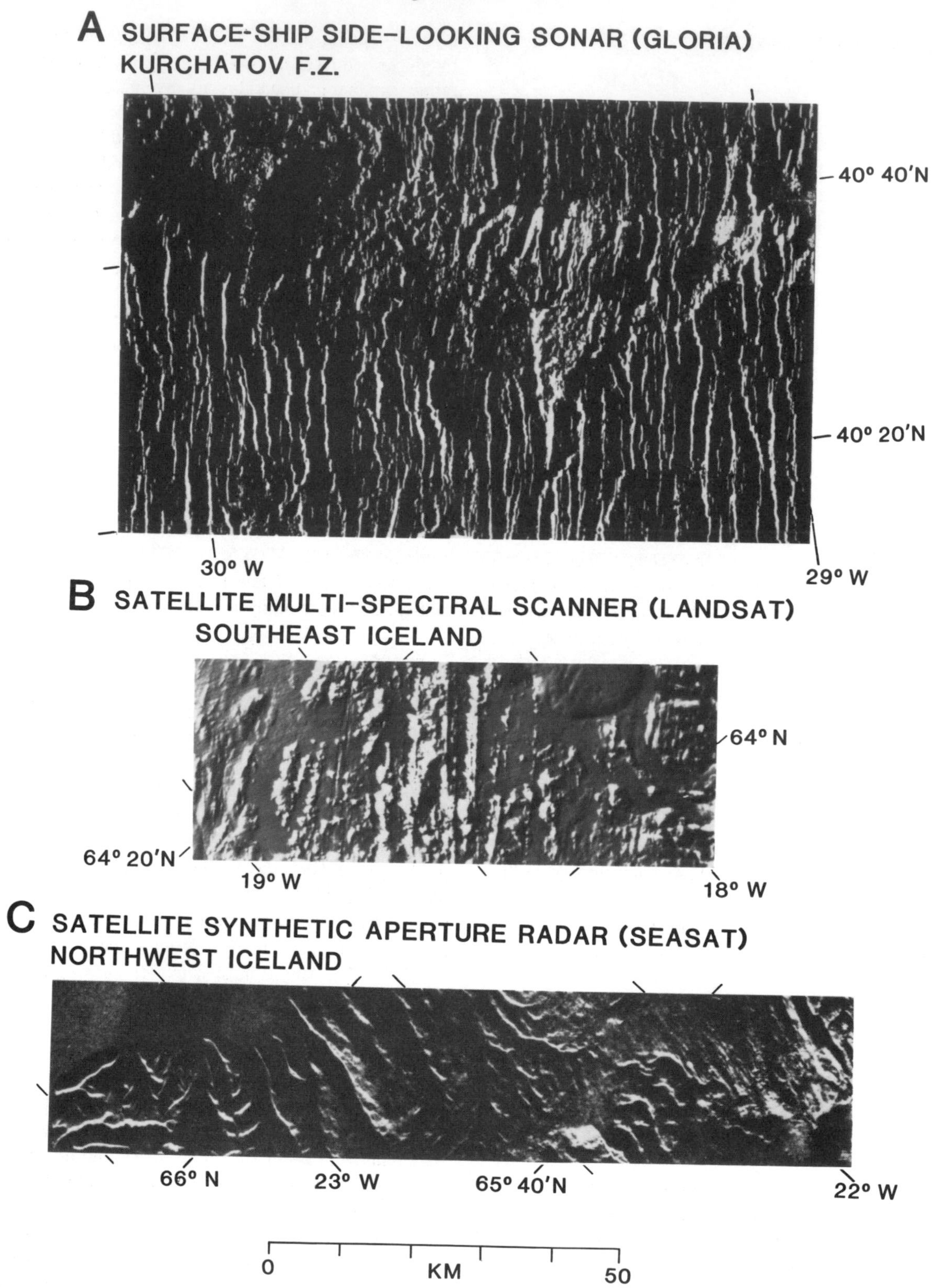

Figure 2. A, GLORIA side-scan sonar mosaic (strong backscatter is white) for Kurchatov Fracture Zone (Searle and Laughton, 1977; see also Fig. 8). B, portion of Landsat MSS image over southwest Iceland; shadows are white (Williams, 1976). C, Seasat synthetic aperture radar (SAR) mosaic for northwestern Iceland (Ford and others, 1980). Note prevalence of linear structures typically 3–6 km apart in all images. In A these features are probably normal faults and volcanic edifices, in B they are crater rows and hyaloclastite ridges, and in C they reflect structural control on erosion of 10 Ma crust.

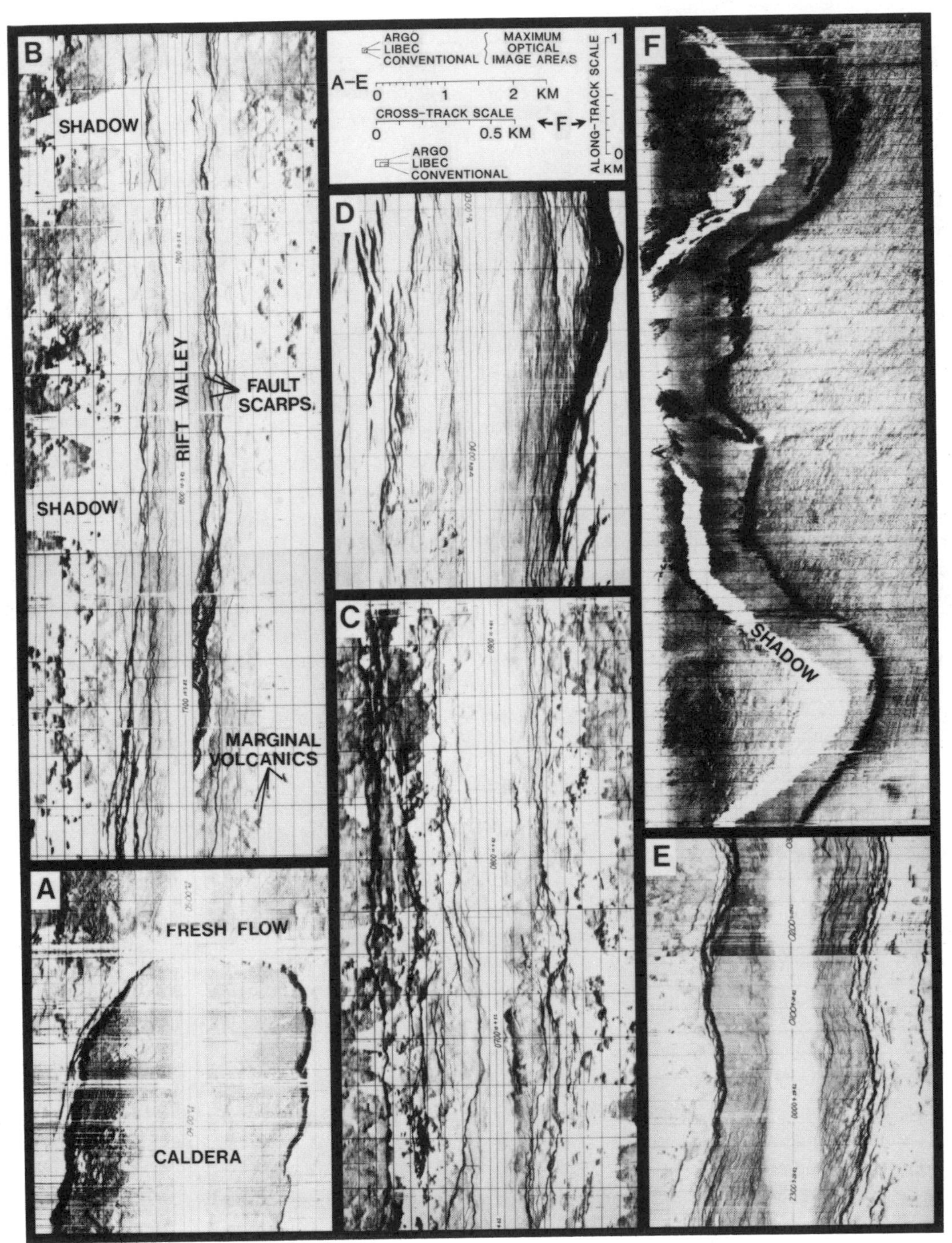

Figure 3. Examples of Sea MARC I side-scan sonar images. A to E are sonographs about 5 km wide, showing geothermal portions of axis of Juan de Fuca Ridge in the Pacific Ocean at water depths of 1,600 to 2,600 m (A, axial volcano caldera near 46°N, 130°W; B, rift valley at 46.8°N, 129.4°W; C, necking of rift valley at 47.1°N, 129.1°W; D, deep rift valley near tip of propagating rift at 47.6°N, 129°W; E, sinuous rift valley at 44.7°N, 130.3°W. Images (discussed in Crane and others, 1985) are courtesy of Steven Hammond of PMEL (NOAA). Similar features are expected to be present along parts of the Mid-Atlantic Ridge axis; F, Titanic Submarine Canyon near 41°N, 50°W on southern Grand Banks continental margin. Sonograph is compressed 2:1 in the along-track direction; water depth is about 3,800 m. The canyon is seen entrenched into a grooved erosion surface with the grooves parallel to the surface outcrop of bedding horizons. Canyon relief is 40 m from rim to floor. This is one of the first images made by Sea MARC I (August, 1980), with funding from "Titanic '80." Image courtesy of W.B.F. Ryan and F. N. Spiess. Note comparative sizes of typical optically imaged areas at same scale (top).

1985). Average water depth can also be measured on the polar sea ice, using magnetotellurics (De Laurier, 1978).

The dependence of darkness and color on depth in clear water was shown dramatically by Skylab photography of areas such as the Tongue of the Ocean area of the Bahama Banks. LANDSAT provided an opportunity to exploit this phenomenon quantitatively (Hammack, 1977). Arthur and others (1981) used LANDSAT MSS data to extract bathymetric data to ±1–1.5 m in water depths up to 20 m. Lasers, invented in 1960, extended optical bathymetry to somewhat greater depths (Link and others, 1983). For example, the airborne system described by Guenther and Thomas (1983) used a pulsed blue-green laser flown at 300 m elevation; the laser penetrates about 10 times farther than photography for any water clarity and it can map seafloor bathymetry to at least 50 m depth in very clear water, 20–40 m in typical coastal waters. The system maps a swath about 200 m wide. Under development for the Naval Oceanographic Office is an airborne system utilizing both a laser and a passive multispectral scanner (MSS) of the LANDSAT IV type (see below). The laser is used to calibrate the scanner, allowing mapping of swaths up to 550 m wide at water depths up to at least 50 m to a mean depth error of ±0.5 m (R. H. Higgs, personal communication, 1986). Deep-tow laser and MSS imaging has not yet been developed, but for a laser signal emitted underwater, 1,000 m is probably an upper limit for one-way detection range in the clearest waters (Spiess, 1985a, 1985b).

OPTICAL IMAGING

Imagery of Subaerial Oceanic Crust

Where the Atlantic seafloor emerges to form oceanic islands, a great diversity of optical imagery is now possible (Williams, 1983b; Hayden, 1985). Iceland has received the greatest attention, partly because of its large area, frequent volcanic activity, and diverse volcanic landforms (Williams and others, 1983). Systematic airborne photography was begun in Iceland during World War II. Airborne stereo-photogrammetry is now a routine method of topographic mapping, and in volcanic/tectonic terrains like Iceland, sequential imaging can be used to derive terrain changes such as volumes of erupted lava.

Beginning in 1972 the LANDSAT MSS (Multispectral Scanner System) and wideband RBV (Return Beam Vidicon) systems revolutionized geologic mapping from space, although no accurate 3-D information is obtained from such data. MSS data (Fig. 2B) are acquired in four spectral bands: green (0.5–0.6 μ), red (0.6–0.7 μ), and two bands in the near-infrared (0.7–0.8 and 0.8–1.1 μ). The green band is close to the spectral region where clear ocean water is most transparent (about 0.47–0.52 μ) and brackets the transparency window for coastal waters (0.53–0.56 μ). Images from each MSS band can be studied separately or combined in different ways to form effective "false-color" displays. Pixel sizes (to be multiplied by 2.8 to give resolution) are 79 m for LANDSAT MSS, and 30 m for the later LANDSAT RBV and LANDSAT 3 and 4 TM (Thematic Mapper). A comparison of SEASAT SAR and LANDSAT MSS data was presented by Bodechtel and others (1979). The latest improvement in visible-spectrum imagery is the Large Format Camera (LFC) photographs taken from the Space Shuttle. These stereo photographs achieve spatial resolutions of 8–10 m per line pair (Williams, 1985). Similar resolution is expected from the French SPOT-1 satellite. These improvements bring to satellite imaging of the solid earth the resolution reached on the ocean floor only at laboriously slow rates by side-scan sonars such as Sea MARC (Fig. 3).

Submarine Optical Imagery

Underwater photography was in use before the end of the nineteenth century (Boutan, 1893). Development of unmanned, automatic seafloor photography had its beginnings in 1939 (Ewing and others, 1946), but the war delayed the first successful deep-ocean photography until 1947–48 when photographs were taken in depths up to 5,534 m in the North Atlantic (Ewing and others, 1967). Stereo-pairs, image mosaics, and color photographs were obtained in subsequent years, although most images were single, oblique, close-up views in black-and-white taken 2 to 5 m above the bottom (Fig. 4A–D). The value of submarine color photography is diminished compared to land because only blue to blue-green light is transmitted any considerable distance through seawater. Large collections of bottom photographs are reproduced in Hersey (1967) and Heezen and Hollister (1971), and color photographs taken from a manned submersible are shown in Ballard and Moore (1977). Macdonald and others (1986) describe recent bottom photography of the Vema FZ using the Scripps deep-tow instrument package.

Optical images of larger areas of the seafloor have been obtained by large-area camera systems and/or photomosaics (Fig. 4 E–G). The mosaics provide an important overlap with side-scan sonar images (Figs. 1, 3). Areas 15 × 20 m could be imaged at 12 m range, but at relatively low resolution and contrast, by the LIBEC (Light Behind the Camera) system developed at the Naval Research Laboratory (Brundage and Patterson, 1976). Further advances in areas optically imaged in a single frame are possible (Brundage and Patterson, 1976) but no quantum improvements can be expected. Backscatter from within the water column limits subject-lens ranges to about 2 or 3 attenuation lengths (30–50 m) even in the clearest waters, much less in turbid coastal seas. Another large-area camera system currently in use (ANGUS) operates to 6,000 m depth, 10 m above the bottom at tow speeds of 1.3–2 km/hr. Up to 3,000 35-mm frames may be exposed (at intervals of 15 seconds to 1.5 minutes) on a single traverse of up to 30 km. Results of recent deployments of ANGUS on the Mid-Atlantic Ridge are described by Karson and Dick (1983) and OTTER (1984). "Real-time" viewing of the seafloor in television images transmitted up a cable from a towed camera is another available tool, first used in 1963 (Brundage and others, 1967). One such system (*Argo*), currently under development at Woods Hole Oceanographic Institution (Harris and

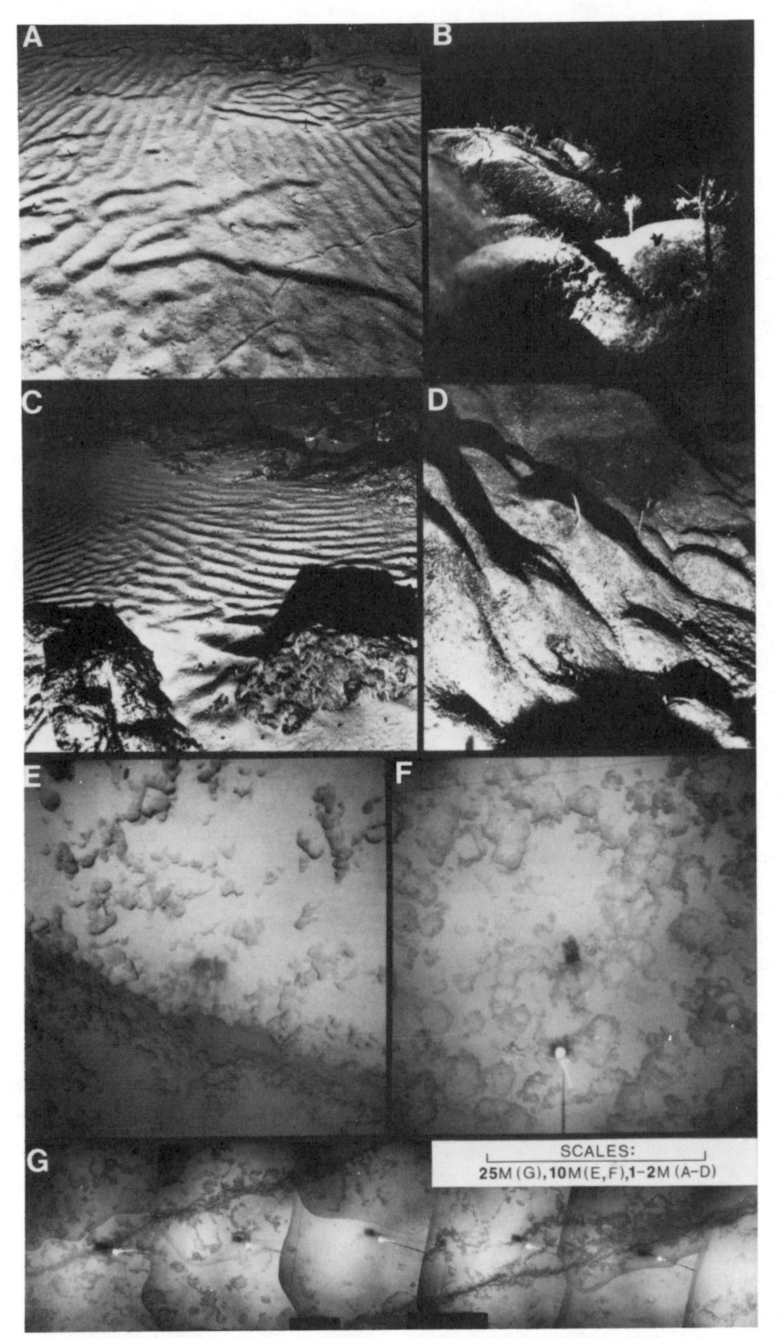

Figure 4. The Atlantic seafloor at close range: Top four images (A–D) are oblique-view bottom photographs taken about 35°31.8′N, 58°38.0′W on Gilliss Seamount in the New England Seamounts (USNS *Lynch* cruise 33B-1; Taylor and others, 1975). The areas imaged are about 2 × 3 m. Photographs by courtesy of W. H. Jahn (NORDA). A, ripple marks intersected by tracks of benthic animals, 2,944 m; B, pillow basalts and stalked, sessile animals, 3,127 m; C, basalt outcrops and intersecting ripple patterns in calcareous ooze, indicating local interaction of currents with the seafloor roughness, 2,962 m; D, pillow basalt encrusted with Fe-Mn oxides, 3,182 m. Lower images (E–G) are large-area vertical views taken from USNS *Mizar* in the FAMOUS area (Mid-Atlantic Ridge crest) with the LIBEC system (Brundage and Patterson, 1976). Square imaged areas are about 200 m^2 each. Note basalt pillows, fissure with vertical offset indicated by intensity of reflected light (E), calcareous sediment fill between pillows, a cirrate octopod (F, lower left center) and camera shadows. Photograph mosaics of such images (e.g., G) were used to plan 1974 submersible dives in the FAMOUS area (Ballard and van Andel, 1977). The mosaic (36°49.0′N, 33°15.9′W) shows two fissures 20 m apart. E, 10 October, 1973, 36°50.3′N, 33°17.0′W; F: 15 October 1973, 36°49.3′N, 33°15.8′W. Photographs courtesy of W. Brundage (NRL).

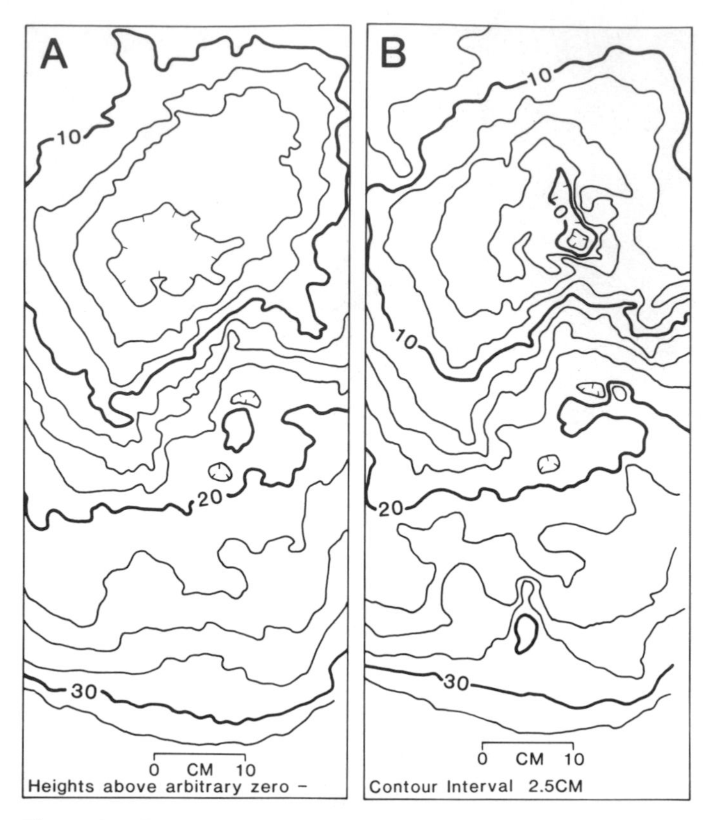

Figure 5. Microtopography of 0.35 × 0.8 m portion of sea-bed at 36°33.3′N, 72°14.2′W on central Continental Rise off Virginia. Contours were derived by stereophotogrammetry from images obtained during engineering tests for the High Energy Benthic Boundary Layer Experiment (HEBBLE). A) seafloor at 0330GMT, 20 July 1981; B) same area at 0330 GMT 16 August 1981. Contour differences show net accumulation and erosion caused by seafloor organisms and abyssal currents. Redrawn from diagrams kindly provided by M. Wimbush, URI.

Albers, 1985), employs three cameras with an effective ASA exceeding 200,000 and will be able to image areas up to 4,000 m^2 per "frame" (swath widths of 56 m at 35 m tow height). Eventually the system will include a tethered, remotely controlled vehicle (*Jason*) that can precisely navigate and sample within *Argo's* imaging area (R. D. Ballard, personal communication, 1985). Stereo color television designed for *Jason* will require transmission rates of more than 6×10^6 bits/sec that will require fiber optic cables not yet in service.

An optical-imaging technique of great importance for bathymetry is stereo-photogrammetry (Schuldt and others, 1967). Stereo photographs taken from towed vehicles show geologic relations that would easily be missed or misinterpreted in conventional 2-D images. Recently, time-lapse, close-up stereo photography has been used to study changes in sedimentary microtopography caused by the activity of abyssal currents and benthic fauna (Fig. 5). Other possible three-dimensional quantitative optical imaging techniques are discussed by Strand (1985).

The Descent of Man

Direct visual imaging of the seafloor by "free" divers entered history in the pre-Christian eastern Mediterranean (Beach, 1969), but for millenia, dives were limited to depths of 30–50 m (rarely 100 m) and times of five minutes or less. Although the use of helmets or pouches as additional air supplies was common in classical times (Geyer, 1977), the first underwater breathing apparatus using compressed air was developed by T. C. McKeen in 1863; it was soon buried and forgotten. A usable apparatus, Ohgushi's "Peerless Respirator" was developed in Japan in 1918 but went unnoticed in the West. Finally, building on innovations by Y. LePrieur in the 1920s–1930s, J. Cousteau and E. Gagnan developed the "Aqua-Lung" (SCUBA) in 1943. Using this device in its modern forms, the seafloor geologist can work safely and practically for several hours to 50 m depth using compressed air, and to 100 m depth using mixed gases. Maximum depths of 500 m have been reached by swimming divers, but the safety risk and long decompression periods leave little marine geological application. Since even 500 m is barely off the Continental Shelf, the Atlantic Basin remains basically off limits to unprotected geological aquanauts.

Tethered diving bells, diving suits, and free-swimming submarines were developed (in that order) to extend the time, comfort, and ultimately the depth at which man could operate on the seafloor (Peterson, 1969). Legend has it that Alexander the Great used a diving bell, but no contemporary drawings exist. Diving bells were in use in Europe by the sixteenth and seventeenth centuries, primarily for salvage work. In 1717, Edmund Halley's treatise solved the air renewal problem by bleeding up air from auxiliary bells at deeper levels, a system replaced in the nineteenth century with pumps. Successful diving suits date from the eighteenth century; the first modern "closed dress" suit, entirely covering the diver, was perfected by Augustus Siebe in 1837.

The first workable free-swimming submarine was David Bushnell's one-man, hand-powered *Turtle* (1776). Manpower remained the only practical submarine propulsion until the invention of the electric motor and storage battery in the late nineteenth century. The latter are still used in all modern research submersibles except for the U.S. Navy research submarine NR-1, which is nuclear. The "true submarine" (Beach, 1969), which is nuclear powered and thus freed of the need to surface for air and battery recharge, dates from the mid-1950s. Although military interest has spurred development of most submarines, it also has provided impetus to research all facets of the sea, including submarine geology in a variety of submersibles. Geological use of military submarines has been limited to gravity (Vening Meinesz, 1937) and echo-sounding. Nuclear submarines have played an important role in mapping the bathymetry below the ice-covered Arctic (Perry and others, 1985).

The development of suitable life-support systems in 1930 allowed zoologist William Beebe and Otis Barton to be lowered into the Atlantic off Bermuda in a spherical, tethered diving bell ("bathysphere"). Their record depth of 923 m set in 1934 survived until a post-war lowering of 1,360 m off southern California in 1948. In that year the first free-swimming submersible ("bathyscaphe," the FNRS II, designed by A. Picard) was lowered to 759 m in an unmanned test off Dakar. The French Navy then built FNRS III, which in 1954 made the first manned descent into the abyss (4,050 m, 300 km SW of Dakar; Houot, 1955; Houot and Willm, 1955). The similar *Trieste* (Dietz, 1968) later bought by the U.S. Navy, reached the floor of the Marianas Trench (10,810 m) in 1960. The greatest North Atlantic depths (Puerto Rico Trench) were explored with the bathyscaphe *Archimede* in 1964 (Péres, 1965).

Many types of research submersibles were built in the 1960s, and many were subsequently moth-balled due to lack of funding. Important research results through 1969 were reviewed by Ballard and Emery (1970) and more recent syntheses are in Geyer (1977). A new breed of submersible, epitomized by the *Alvin* (first used in 1966) incorporated greater mobility and research capabilities than earlier versions such as the *Trieste*. Scientist observers still look out of ports, but modern submersibles like *Alvin* are able to take sediment cores and other samples and make a variety of other measurements. Ninety-six manned submersibles are able to operate at 150 m or more today (Busby Associates, 1985), but only five are capable of operating a 3,000 m or more (*Cyana, Turtle, Alvin, Nautile,* and *Sea Cliff*). *Alvin's* original 1,830 m limit was doubled to 3,650 m when the steel pressure hull was replaced with a titanium alloy hull in 1975; subsequent changes increased the depth limit to 4,000 m. The Ti-hulled *Nautile* and *Sea Cliff* can operate down to 6,000 m, putting all but the deepest fracture zone valleys of the North Atlantic and the Puerto Rico Trench within reach.

Present knowledge of seafloor features such as hydrothermal vents can be credited partly to submersibles, although the same features could and have been investigated remotely. A common submersible observation in canyons (e.g., Trumbull and McCamis, 1967), on carbonate escarpments, on seamounts, and in the Mid-Atlantic Ridge rift valley is the occurrence of locally steep (45° to 90°) slopes with relief up to 100 m. Those who were surprised at such small-scale roughness (in view of an invariably smoother seafloor image derived from surface-ship bathymetry) were unaware of the limitations of echo-sounding. However, an earlier generation of marine scientists (Stocks and Wüst, 1935) was similarly surprised when echo-sounders revealed meso-scale (10–100 km) roughness not evident in wire-line soundings on the floor of the Atlantic.

The pros and cons of manned submersibles are discussed at length in Geyer (1977). Weighed against the cost of building and operating manned submersibles, the geological returns have been modest (compared, say, to the Deep Sea Drilling Project). It is possible to argue that the same research could have been performed remotely, if less romantically, by systems such as LIBEC, ANGUS, and *Argo,* although this overlooks the submersible's other uses, particularly in sampling and manipulation. On the other hand, these functions may soon be accomplished just as effectively, more continuously, and probably much more cheaply by remotely operated vehicles such as *Jason,* in conjunction with the *Argo* system.

"DIRECT-CONTACT" BATHYMETRY

"Line-and-sinker" was the dominant method of bathymetric sounding for millenia, and in its final form of wire sounding it survived until the advent of echo sounders in the early twentieth century. Most such sounding was on continental shelves and in coastal waters, and was obtained to avoid grounding ships on submarine shoals. Bathymetry has also been exploited as a navigational aid, particularly when bottom constitution (e.g., sand, shells, mud, or rocks) was known (Edvalson and Miscoski, 1969). Bathymetry is still used for navigation, notably by submersibles that cannot avail themselves of electromagnetic navigation aids when diving (Cohen, 1970).

Perhaps the earliest record of ocean depth measurement beyond the shelves was by the Greek Posidonius about 85 B.C., who reported a sounding of 1,000 fm off Sardinia (1,830 m assuming a modern fathom; Emery and Uchupi, 1984); this record also illustrates the antiquity of the fathom as a measure of depth. The first "modern" attempt with hand lines was by Ferdinand Magellan in 1521, whose men lowered 750 m of line in the Pacific without reaching bottom (Pettersson, 1954). Magellan is said to have proclaimed this site in the Tuamotu Archipelago, actually 3,700 m deep, the "deepest place in the ocean." Three centuries later the depths of the open ocean were still a matter of speculation. Captain C. J. Phipps of the Royal Navy obtained the first modern "deep sea" sounding (683 fm) in the Norway Basin in 1773. Sabine (1823) achieved a 1,000 fm sounding in the Caribbean, measuring at the same time the low (8°C) temperature of the deep tropical ocean. Apparent deep-sea soundings of 9,100 and 15,200 m (30,000 and even 50,000 ft) or more early in the nineteenth century resulted because sounding lines continued to pay out after the plummet had hit bottom. Credit for the first successful abyssal sounding (2,425 fm at 27°26′S, 17°29′W on the west flank of the Mid-Atlantic Ridge) goes to Sir James Ross on HMS *Erebus* in 1840 (Dietz and Knebel, 1968).

With the development of the undersea telegraph cable, deep-sea soundings acquired a practical value. In the U.S. it was the Navy's responsibility to collect such data. Initially using disposable 29 kg (64 lb) cannon shots attached to a ball of disposable twine, Lieutenant Matthew Fontaine Maury first measured depths in the North Atlantic; his bathymetric chart of the central Atlantic (Maury, 1855) showed the first evidence for the Mid-Atlantic Ridge (called "Middle Ground," Fig. 6). Maury and his co-workers at the U.S. Naval Observatory and Hydrographical Office (forerunner of the present Naval Oceanographic Office) improved sounding techniques by using detachable weights, fine-wire lines, and steam-powered winches. More sophisticated was the "Kelvin sounding machine," actually a pressure gauge from which depth of water was measured by the encroachment of seawater up the inside of a "chemical tube" coated to react with seawater.

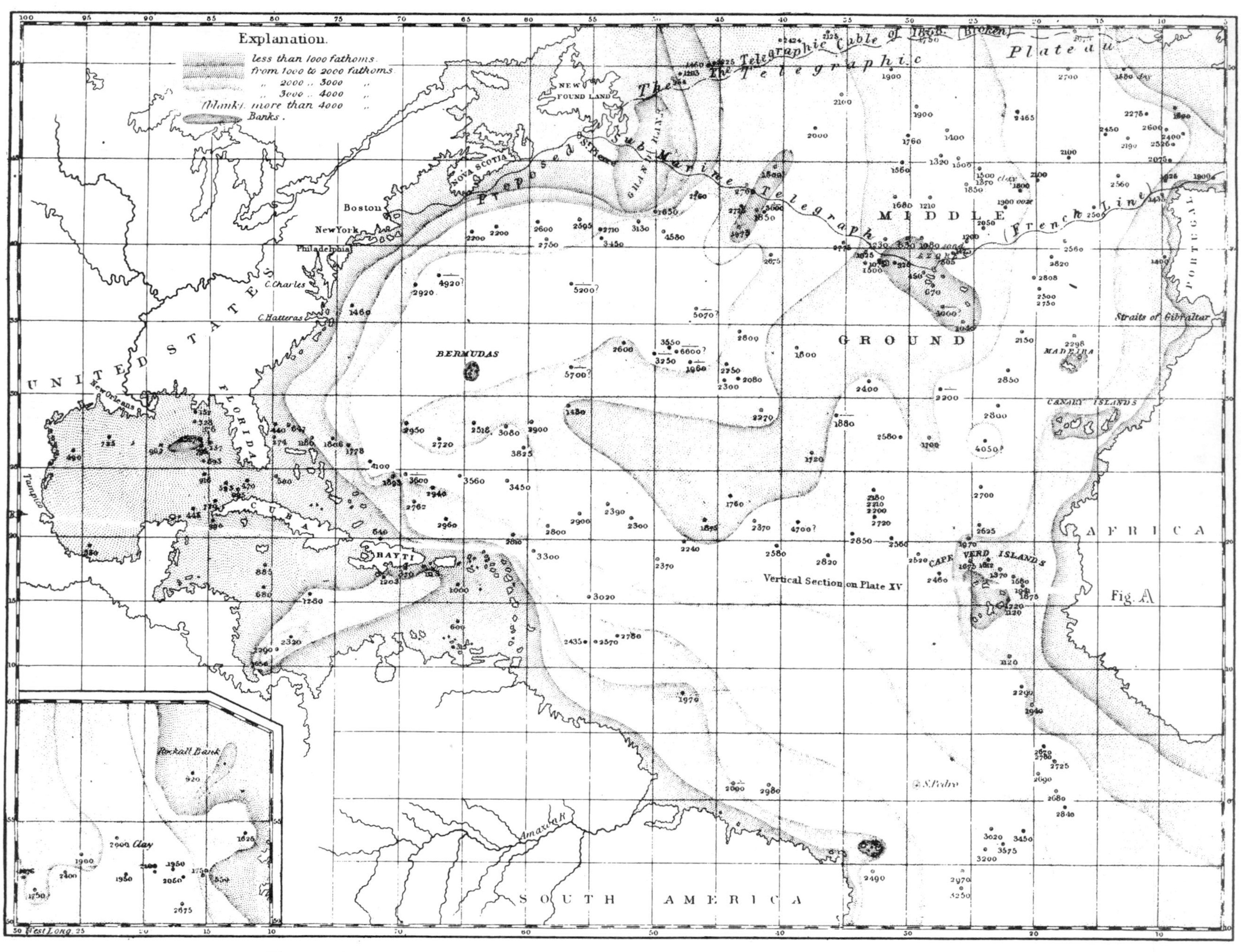

Figure 6. The nineteenth-century Atlantic: Maury's (1855) bathymetric chart of the North Atlantic, showing the first evidence of the Mid-Atlantic Ridge and major basins interpreted from a small number of wire soundings. Note Maury's question marks following suspiciously great depths.

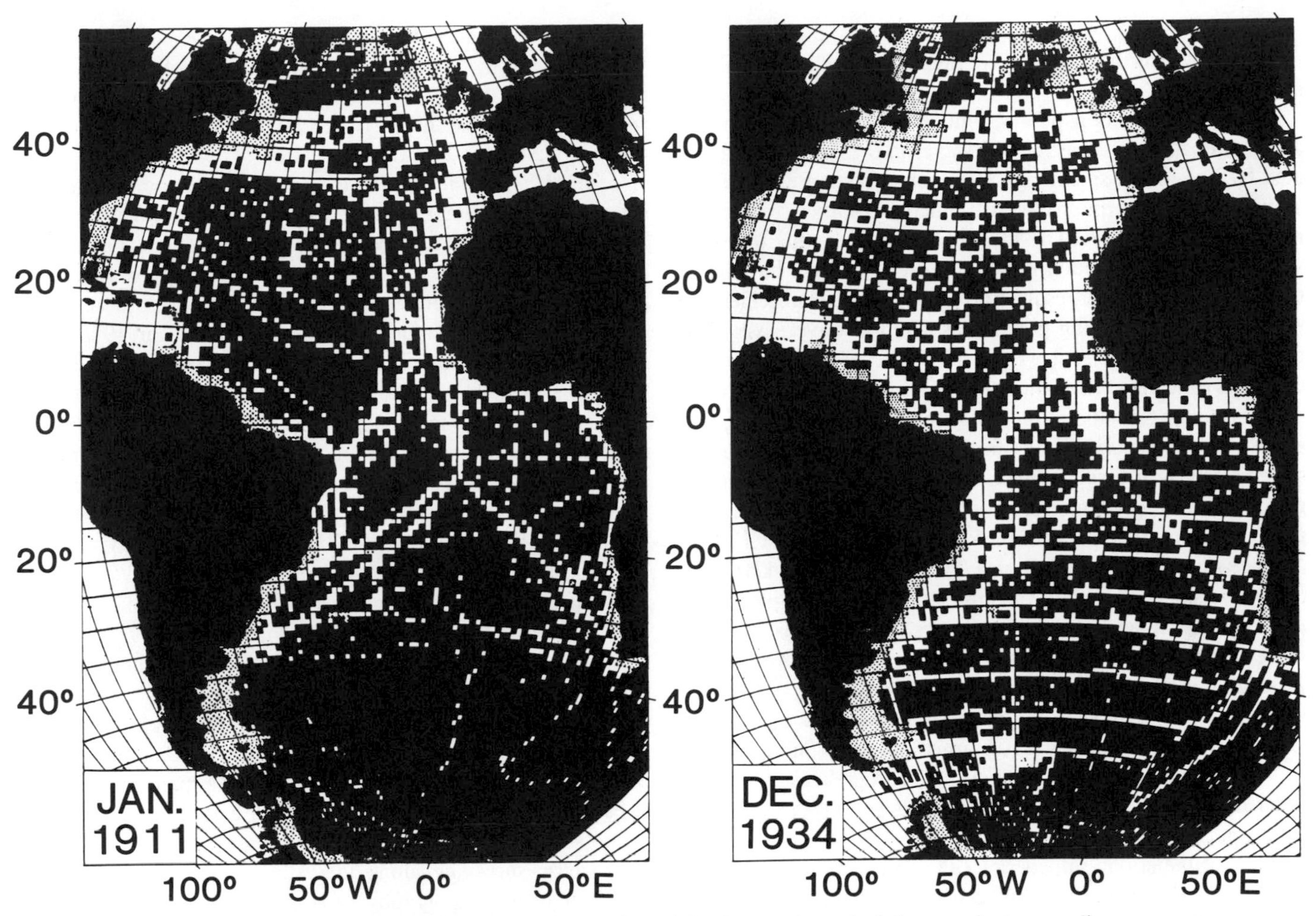

Figure 7. Regional bathymetric knowledge of the Atlantic near the end of the age of wire sounding (1911), and a quarter century later, after only 14 years of echo sounding (1934). 1° × 1° squares containing at least one depth sounding are white; squares entirely or predominantly <2,000 m deep are stippled. Adapted from Stocks and Wüst (1935).

Some 15,000 world-wide soundings deeper than 550 fm were available by the time a working echo sounder was invented in 1921 (e.g., Fig. 7). The wire-sounding data could resolve first-order features such as the Mid-Oceanic Ridge (e.g., Murray and Renard, 1891; Murray and Hjort, 1912). The addition of a massive echo-sounding data base yielded hypsometric curves (Menard and Smith, 1966) only slightly different from those computed from wire-sounding data by Kossinna (1921). In areas not subsequently resurveyed, nineteenth century wire-sounding depths are still embedded in modern data bases.

ACOUSTIC IMAGING

Background

Comprehensive discussions of acoustic-wave transmission and interaction with the seafloor are available in the literature (e.g., Vigoureux and Hersey, 1962; Urick, 1967; Caruthers, 1977; Clay and Medwin, 1977). For our present purpose (imaging the seafloor) we need only consider the transmission and reflection of longitudinal (*P*, or compressional) elastic waves.

Sound speed in water (compressional velocity $V_\rho = [\rho K]^{-1/2}$, where ρ is density and K is adiabatic compressibility) was first measured on Lake Geneva by Colladon and Sturm (1827) and the existence of sound "waves" was predicted from the laws of physics by Poisson in 1829. V_ρ averages about 1.5 km/s in the oceans but is not constant. It increases with increasing water temperature, salinity, and pressure, and therefore changes with season, time of day, depth, geographical position (Matthews, 1939; Carter, 1980), and proximity to rivers and melting ice (Clay and Medwin, 1977; Hurdle, 1986). Most sound-speed variability occurs in the upper few hundred meters. Sound speed can be calculated from empirical formulas as a function of temperature, salinity, and depth, or it can be measured by a velocimeter to ±0.1 m/s. In the Atlantic, V_ρ ranges from less than 1.44 km/s in some high latitude surface waters to more than 1.54 km/s in the deepest basins. Sound speed is essentially independent of frequency within the 1 to 10^6 Hz range of possible bathymetric interest.

Signal attenuation puts resolution and range limits on all acoustic imaging systems. Sound waves in the ocean are attenuated by wave-front spreading, reflection/refraction at interfaces, scattering, and absorption. Acoustic energy absorbed by seawater is converted into heat. The absorption is partly due to "shear viscosity" (relative motion between adjacent layers), but mainly to "bulk viscosity" that results from molecular rearrangements in response to the changing pressure as a sound wave passes. The time required for molecular reordering in response to a pressure change is called the "relaxation time." The closer the sound wave period is to the relaxation time, the greater the loss per cycle. This effect is of primary significance for high frequencies: such acoustic systems can most easily provide narrow beam width and hence good resolution but can be used only at short ranges. Absorption increases from 10^{-4} dB/m at 3 kHz to 10^{-3} dB/m at 12 kHz (surface-ship echo sounding) and 0.04 dB/m at 100 kHz (e.g., deep-tow sonars). Much of the absorption and its frequency dependence is caused by water molecules, but sound is absorbed more effectively in seawater than in freshwater (25 times more at 20 kHz); this effect was correctly attributed only after World War II to the presence of dissolved magnesium sulfate (and to a small extent boric acid), whereas NaCl turned out to be unimportant.

Strong reflection of sound waves, comprehensible to first order from simple concepts of wave fronts and rays carried over from optics, occur at both the sea surface and seafloor surfaces because of the strong impedance (ρV_ρ) mismatch across these boundaries. However, the actual echo formation process is complex and the "acoustic bottom" depends on the system used, particularly its frequency (Clarke and others, 1985). In typical seafloor topography most of the returns are back-scattered rather than reflected in th specular sense. The term scattering refers to the much more complex process by which sound waves are deflected in various directions (including forward- and back-scattering) by seafloor roughness elements having length scales smaller than the wavelength of sound. In addition some of the sound is reflected from subbottom interfaces or volume-scattered from small-scale inhomogeneities within the subbottom. (Scattering and reflection from below the seafloor becomes more important at lower frequencies, e.g., 3.5 kHz; Damuth, 1975). The sum total of such scattered and reflected sound arriving at a hydrophone from the seafloor is called bottom reverberation; there is also scattering of sound from the sea surface and from within the sea from organisms, particles, bubbles, and water-mass inhomogeneities. Back-scattering is most important for side-scan sonar imagery (also for photography and radar imagery) because energy source and receiver are close to each other. Many kinds of imaging would be impossible if the seafloor was not rough at small scales.

Conventional Echo Sounding

Echo sounding was tried in 1807 in the Atlantic off French Guiana (Wertenbaker, 1974); the early unsuccessful experiments were described by Maury (1855). Had these investigators listened underwater or through a pipe, rather than from above the sea surface, it is possible that abyssal echo sounding might have succeeded long before the dawn of electronics.

The 1912 *Titanic* disaster inspired the development of crude echosounding systems (von Arx, 1962; Dietz, 1969) that were the ancestors of various downward-looking echosounders and side-looking "sonars" (Sound Navigation and Ranging) in use today. Seafloor sounding using small explosives was first successfully demonstrated by Alexander Behm in the North Sea in 1920, with Harvey Hayes (later the first director of the Acoustics Division at the U.S. Naval Research Laboratory) developing an operational "sonic depth finder" in 1921. The main problem was not detecting the echo of a sound pulse, but accurately measuring its travel time. Assorted echosounding systems of the 1920s and 1930s used electronic oscillators and piezoelectric transducers both for transmitting and receiving sound pulses (Wood, 1969), and we use basically the same systems today.

Our knowledge of deep-ocean bathymetry began to expand dramatically with the advent of echo sounding (Fig. 7). The first trans-Atlantic echosounding profile, some 900 soundings between Newport, Rhode Island, and Gibraltar, was obtained by the destroyer USS *Stewart* in 1922 (Edvalson and Miscoski, 1969). In 1923 detailed surveying at track separations of only 10–20 km was carried out along the California margin in support of earthquake investigations, showing an early appreciation for the possible tectonic significance of seafloor topography. Similar bathymetric surveying along the Atlantic margin showed the continental-slope topography to be incised by numerous deep canyons that initially were thought to have formed subaerially (Veatch and Smith, 1939). No comparable systematic work was done in the Atlantic seaward of the continental slopes until the 1950s and 1960s. A benchmark prewar bathymetric description of the Atlantic was that of Stocks and Wüst (1935), based largely on the wide-spaced east-west *Meteor* traverses across the South Atlantic and Equatorial North Atlantic (Fig. 7). They noted the rough topography on the Mid-Atlantic Ridge, delineated its crest, plotted the regional variation in crestal depth, and wrestled with the problem of extrapolating contours between widely-spaced tracks, as we still do in poorly sounded areas.

By the start of World War II, continuous graphic recorders were in use, and systems capable of sounding at all ocean depths were developed during the war. However, the bathymetric data base was slow to expand and Atlantic contour charts remained primitive (e.g., Tolstoy and Ewing, 1949; Tolstoy, 1951). Precision echosounders with a timing accuracy better than 1 part in 3,000 were developed in the early 1950s (Luskin and others, 1954; Knott and Hersey, 1956). In the following 10 to 15 years, precision echosounding profiles were acquired by ships of numerous institutions. Along-track analysis of slope, relief, and texture got underway (Holcombe and Heezen, 1970). Depths in most of these profiles could be determined to a precision of about 1 m in ideal circumstances of flat or gently sloping bottom. Beam width typically was 30° and greater; hence, in areas of irregular topography, side echoes usually preceded and thereby masked the true bottom echo from directly beneath the ship. This limitation still exists for conventional echosounding data; it provides serious limitations for investigation of rough seafloor topography at large working scales, although for small-scale summaries (e.g., >1:1,000,000) it is not a serious problem.

High-Resolution, Narrow-Beam, and Multi-Narrow-Beam Echo Sounding

Starting in the 1960s, data obtained by more precise narrow-beam and multi-narrow-beam ("multibeam") bathymetric survey systems began to be collected in the North Atlantic. Over most of the North Atlantic the data base still consists predominately of older data, but narrow-beam and multibeam data coverage is expanding rapidly. One system used to obtain more accurate bathymetric data is the narrow-beam echosounder (total beam width ~14°). Narrow-beam data are particularly valuable in areas of irregular topography such as the Continental Slope and the Mid-Atlantic Ridge. Unfortunately even such narrow-beam soundings insonify an area with a diameter up to 0.25 times water depth, so that side echoes remain a problem on steep slopes. Detailed narrow-beam bathymetric surveys have been made over several areas, for example the Continental Slope (e.g., Bennett and others, 1978) and the Mid-Atlantic Ridge crest in the FAMOUS area (Phillips and Fleming, 1978).

Multibeam echo sounding presently is one of the best techniques for rapid, accurate mapping of seafloor bathymetry. The first deep-water system of this type (SASS) was designed for and used by the U.S. Naval Oceanographic Office in the mid-1960s (Glenn, 1970). A commercial variant (Sea Beam) became available to the academic community a decade later (Glenn, 1976). In 1977 the first such system was installed on the R/V *Charcot* (CNEXO, France) and at present more than 10 multibeam systems are in use on ships of Australia, France, Germany, Japan, the U.S., and the USSR. Several shallow-water systems (BO'SUN, now called variously Hydrochart, Bathymetric Swath Survey System, or BS^3) have also been built (Glenn, 1976; Farr, 1980). Hydrochart offers an angular coverage of 105°, or 2.6 times the water depth, and operates in up to 600 m water depth. A newer system (Hydrochart II) uses 17 beams and operates to 1,000 m water depth. Hull or tow-fish mounted multibeam bathymetric systems are also being developed in West Germany (Hydrosweep and Superchart) and Norway (Benigraph and EM 100). In the deep water Sea Beam system, linear orthogonal projecting and receiving arrays are mounted on a ship's hull; beamforming is accomplished by computer software such that the seafloor effectively is insonified by 16 adjacent beams, each 2.67° square. A seafloor swath equal in width to 0.8 times water depth can be mapped continuously, and with proper track spacing 100% seafloor coverage or overlapping coverage can be obtained. The SASS system employs up to 61 beams, each approximately 1.5° wide (total swath width up to 1.2 times water depth; Glenn, 1970; Czarnecki and Bergin, 1985). Examples of SASS-surveyed North Atlantic seafloor areas are shown as inserts A–D on Plate 2 and in Figure 8.

Side-Scan (Backscatter Strength) Sonars

Modern side-scan sonars (Somers and Stubbs, 1984) usually employ a towed "fish," and measure backscattered echo intensity on each of a number of narrow beams; together, like a multibeam bathymetry system, these map a swath of seafloor (Fig. 1A, bottom, and Figs. 2A and 3). Sidescan sonars were first used extensively in shallow water (Belderson and others, 1972). For both multibeam and sidescan, the returned sound is a combination of back-scattered and specularly reflected sound. At near-vertical incidence the specular component so dominates the total return that no useful back-scatter information is obtained from a swath directly below the towfish, hence the blank stripes that appear along the center or sides of individual side-scan swaths (e.g., Fig. 3). Upon correction for slant-range of the beam, side-scan sonars provide images similar to photographs and synthetic-aperture radars (i.e., shades-of-gray diagrams high in horizontal resolution but with little 3-D information). The angular (relative) resolution of such systems is rather constant, with absolute resolution related directly to height above the bottom. The horizontal (range) resolution is approximated by $V_\rho \tau/2$, where τ is pulse length. The narrower the pulse, the broader the spectrum, even for a sinusoidal carrier frequency, and therefore resolution is related to the bandwidth that varies as $1/\tau$. Side-scan sonar systems can be grouped roughly into three categories: (1) long-range (10–60 km total swath), (2) mid-range (2–10 km swath), and (3) short-range (<2 km swath). Ultra-long range (basin-wide) reconnaissance reverberation is a form of side-scan imaging that also has been demonstrated.

Ultra-long range back-scatter mapping was first reported by Spindel and Heirtzler (1972) from an experiment northeast of the Blake and Bahama escarpments. An advanced experiment in the eastern North Atlantic (Erskine and Franchi, 1984; and Erskine and others, 1986) imaged a seafloor area of 4×10^6 km^2 in a week-long experiment, and echoes were returned from all major topographic features (Fig. 9).

The GLORIA II system (Laughton, 1981; Teleki and others, 1981) of the Institute of Oceanographic Sciences (IOS) in the United Kingdom has been the most widely used long-range side-scan system. GLORIA II operates at about 6.5 kHz, is towed near the surface, and uses sidebeams 2.5° (horizontal) by 30° (vertical) to insonify a seafloor swath 15 to 30 km wide on each side of the ship track (Fig. 2A). An improved version, GLORIA Mark III, was acquired by the USGS for surveying the U.S. EEZ, including the Atlantic Margin, in the next several years (Hill, 1986). The GLORIA systems probably resolve coherent features with horizontal separations as little as 100 m and amplitudes down to a few meters. GLORIA imaging in the North Atlantic has been mainly on the continental margins and the Mid-Atlantic

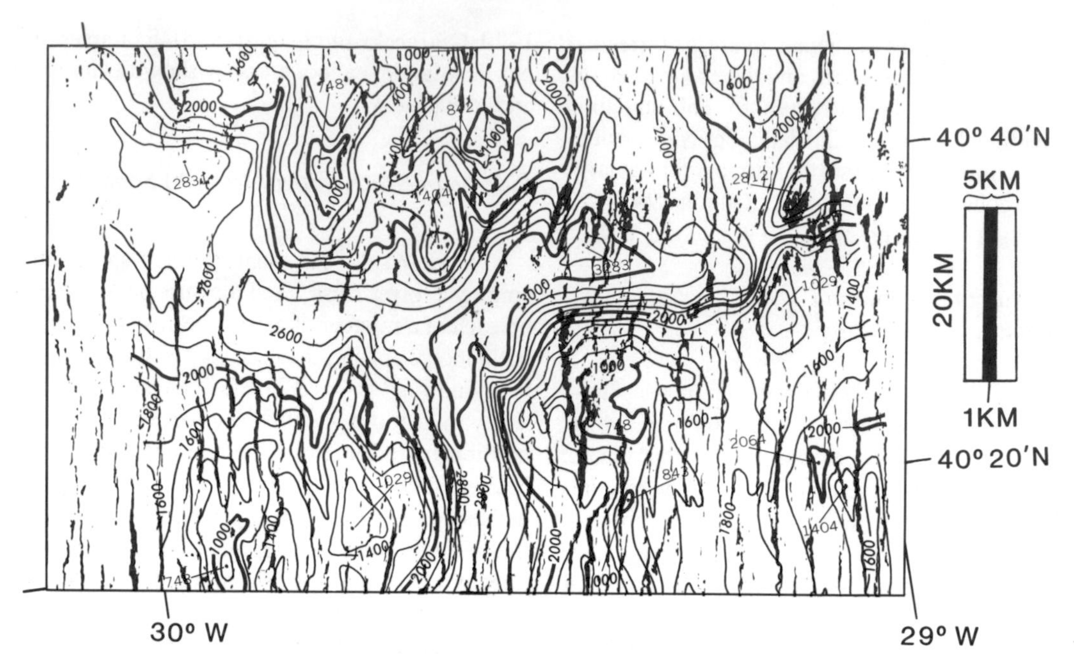

Figure 8. Example of GLORIA side-scan sonar image (strong acoustic backscatter in red) covering the Kurchatov Fracture Zone and adjacent segments of the Mid-Atlantic Rift Valley, superimposed on multibeam bathymetric contours (black; 200 m contour interval). Note the striking difference in information revealed by these two acoustic images of the same area. GLORIA data from Searle and Laughton (1977); multibeam contours courtesy R. Higgs, U.S. Naval Oceanographic Office. Typical Sea MARC surface-tow and deep-tow (black) image swath widths shown at right for scale.

Ridge (Laughton and Searle, 1979; Searle and Laughton 1981; Belderson and others, 1984). A comparison between multibeam depth contours and GLORIA backscatter imagery on the MAR (Fig. 8) shows surprising differences in information content. The GLORIA data emphasize north-south linear fault escarpments typically 5 km apart, with little information on depth, whereas the multibeam depth contours depict the rift valley, Kurchatov FZ, and rift mountain summits but do not clearly delineate the strong north-south crustal fabric. Obviously the geologically most useful image is a combination of both types. Ideally, overlapping side-scan swaths should be obtained in differing orientations because side-scan sonar imaging is azimuth dependent. Had the GLORIA system been towed east-west instead of north-south as in Figure 8, a different image would have resulted, and it would have highlighted any east-west trending structures.

Mid-range side-scan sonar work has been done principally with the Sea MARC I system (Sea Mapping and Remote Characterization; Fig. 3). This 30-kHz system (Chayes, 1983) surveys a swath up to 2.5 km on either side of the fish, which is towed 300 m above bottom at about 3 km/hr. Seafloor objects less than a meter in amplitude and spaced <20 m apart should be resolvable, and features a few meters in height and 6–12 m wide are known to have been resolved. Nearly photographic-quality images of submarine canyons have been obtained along the U.S. Atlantic Continental Slope (e.g., McGregor and others, 1982; Ryan, 1982; Farre and Ryan, 1985). Some models of Sea MARC also measure the angle of the back-scattered sound; thus they are able to generate both backscatter-strength imagery and bathymetric contours. (Sea MARC II is presently able to measure depth to within 3%, not yet up to the International Hydrographic Bureau standard of 1%.) At the same time, developments are underway to use Sea Beam returns to measure scattering strength (de Moustier, 1986), so the distinctions between Sea Beam and Sea MARC are decreasing. Deriving backscatter-strength information from multibeam bathymetric systems is difficult because for near-vertical incidence the returned sound is dominated by specular versus backscattered energy. Since back-scatter echo character depends on wavelength, multi-spectral side-scan (Ryan, 1986) can provide still more geological information, and would be an "active" sonar analogue of the Landsat MSS passive scanners and active multispectral radars described earlier. Modern geologic seafloor studies utilize both multi-beam bathymetry and side-scan sonar (e.g., Fornari and others, 1984).

Short-range side-scan sonar systems usually operate at 100–200 kHz or even higher (e.g., 1,000 kHz; Hammerstad and others, 1985). These systems can resolve seafloor features less than 1 m in amplitude and separated by a distance of a few meters or less. DEEPTOW is one such deep-towed sidescan sonar system. Developed by the Scripps Institution of Oceanography starting in 1960, it has been used in the North Atlantic to study such features as sediment bedforms (e.g., Hollister and others, 1974), acoustic properties of the bottom (Lonsdale and others, 1983), and fine-scale crustal morphology on the Mid-Atlantic Ridge crest (Luyendyk and Macdonald, 1977; Shih and others,

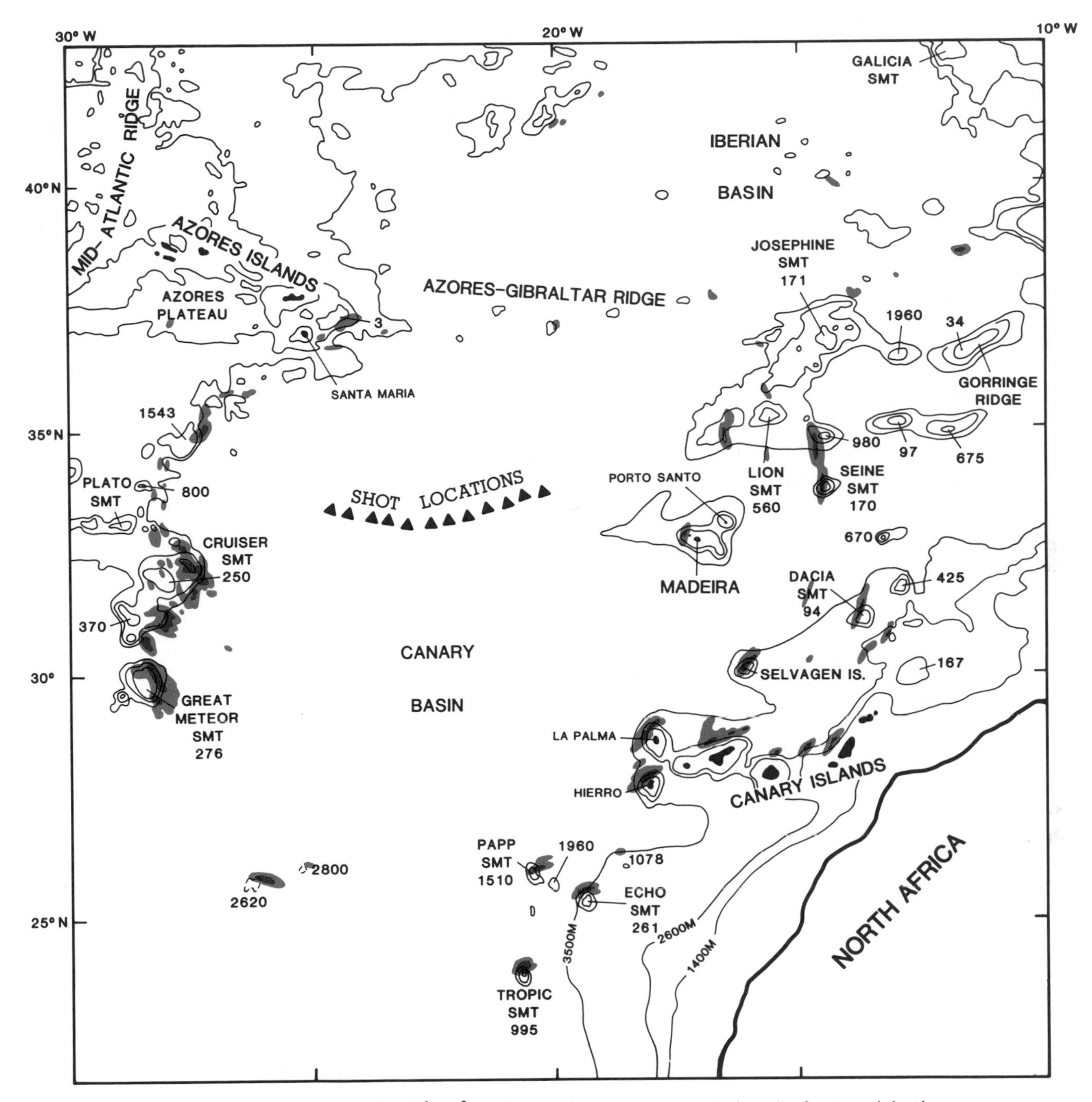

Figure 9. Very large area (4×10^6 km^2) backscatter image (strong returns in red) of eastern Atlantic, produced by Bathymetric Hazard Survey Test (BHST) conducted by NRL in September 1980 (Erskine and others, 1986; Erskine and Franchi, 1984). Simplified from Figure 2 of Erskine and others, 1985 and diagrams supplied courtesy of F. Erskine, NRL. Bathymetry (black) is in meters, based on SYNBAPS-I model of Van Wyckhouse (1973). The experiment used a horizontal towed receiving array and vertical source array of small explosive charges. Box shows location of GLORIA image, Fig. 8.

1978; Macdonald and others, 1986). A new 150 kHz deep-tow system, Sea MARC CL/B, will cover swaths 50 m to 1 km wide and provide both bathymetric contours and backscatter-strength imagery in real time (J. Kosalos, personal communication, 1986). Patterned after Sea MARC is the French *SAR* system, designed to map a 1 km wide swath from 60 m above the bottom. *SAR* is reportedly capable of distinguishing objects "as small as 30 × 76 cm" (Ryan and Rabushka, 1985). Even side-scan seismic systems are possible (Slootweg, 1986), providing images, for example, of basement topography buried under sediments.

ATLANTIC MAPS AND IMAGE DISPLAY

Image Deceptions: Bathymetric Artifacts

Artifacts (non-real features) appearing in any kind of seafloor topographic or other image can be broadly divided into *instrumental artifacts* associated with the measurement technique or system (including noise) and *interpretation artifacts* due to incorrect contouring of correct data. We here restrict ourselves to artifacts in bathymetric contours. One instrumental artifact is deep "meteorite-impact craters" (Pfannenstiel, 1960) or "Gulf Stream holes." Such isolated deeps, still locally pockmarking maps of poorly surveyed areas, were due to excessive wiresounding depths caused by ship drift (e.g., Fig. 7). Another artifact caused by instrumentation is demonstrated in the echosounder used on the *Meteor* expedition (Maurer and Stocks, 1933), which produced artificial steps of 50–150 m height due to power changes in the ship's generator. These steps are small compared to topography on the Mid-Atlantic Ridge, but they are large enough that Stocks and Wüst (1935) failed to recognize abyssal plains. "Great circle faults" seen on the 1961 world chart (U.S. Navy Hydrographic Office, 1961) reflect uncalibrated echosounders on ships traveling great circle routes (Emery and Uchupi, 1984). Some banks or seamounts disappeared from charts only after lengthy controversy showed the echoes were returned from a dense horizon of biological scatterers in the water column. A famous example is the "American Scout Seamount" (Backus and Worthington, 1965), which dominated the Newfoundland Basin in a 3-D relief chart (Anonymous, 1963). With a "summit depth" of 40–100 m it was the shallowest seamount summit in the entire western Atlantic. Such biologic artifacts were no longer commonly reported after the introduction of gated recorders, but reproduction of questionable but probably erroneous older soundings on more modern charts has continued in areas not subsequently resurveyed. Side echos are a different kind of artifact that can be easily recognized but not readily corrected for; these occur in wide-beam echosounding data over steep terrain and they result in underestimation of depth.

Some instrument artifacts have resulted in geologic misinterpretation, such as Pfannenstiel's "meteor craters." A more subtle example is Colorado Seamount, a spurious 900 m echo sounding obtained over the Mid-Atlantic Ridge crest at 33°N by the *U.S.S. Colorado* in 1924. This not unreasonable sounding was carried over from one chart to the next and became the site of the "Colorado hotspot" of Burke and Wilson (1976). The actual topography, although only 0.5–1 km deeper, is typical rift-mountain topography (Plate 2; U.S. Naval Oceanographic Office, 1977), albeit regionally elevated in the Azores area. A mere 500 m depth error (vs. 5,000 m for American Scout Seamount) was enough in this instance to produce a probably spurious mantle hotspot!

Bathymetric measurement artifacts are still produced today, particularly by operator error or instrument failure. Even modern multibeam systems generate certain artifacts (such as the "omega" and "tunnel" effects of Sea Beam systems; de Moustier and Kleinrock, 1986). However, with our now better knowledge of ocean-basin topography, contemporary artifacts are generally recognized as such and corrected for.

Navigation errors introduce a special family of artifacts because image errors result even from correct measurements. The horizontal displacement of an entire image is not as serious as displacement of parts of an image due to variable navigation errors, a typical problem faced in contouring random bathymetric tracks collected over a long period of time. Soundings collected prior to the introduction of satellite navigation ("Transit") in the mid-1960s (with typical 0.5–3 km errors) may be mispositioned by tens of kilometers except in nearshore areas (relative errors would of course be less). In view of the detailed mapping capability of modern multibeam and side-scan sonar systems, it is fortunate that navigation accuracy, already impressive (e.g., Dunham and Shostack, 1980) presently is experiencing another quantum improvement with the new GPS (Global Positioning System) satellites; those will bring absolute navigation accuracy down into the 1–10 m range (e.g., Ashjaee, 1985).

The mere choice of a contour interval also introduces interpretation artifacts. Topography with amplitudes comparable to or less than the contour interval is "highlighted" wherever the regional depth is close to one of the chosen contour levels, and it is suppressed when the regional depth is midway between two contours. This effect is more pronounced at larger contour intervals such as 1 km (Perry and others, 1981) and in areas of steep but rather low-amplitude relief such as the lower flanks of the Mid-Atlantic Ridge.

Other interpretation artifacts simply reflect data paucity (e.g., the apparent smoothness of Mid-Atlantic Ridge topography between the 15°20′N and Kane fracture zones, Plate 2), or to the nature (or lack) of interpretation methods used in contouring of the typical ridge/valley and fracture zone topography on the Mid-Atlantic Ridge.

Image Display

The topographic and textural information about the ocean floor comes in many different forms, depending on the sensing system used and the nature of the image processing and interpretation. The display of bathymetric information on charts does not seem to have an ancient history, with Juan de la Cosa first drawing spot soundings on a map in 1504 and Pierre Ancelin, a Rotterdam cartographer, first using contours in 1697 (Edvalson and Miscoski, 1969). Shallow-water bathymetry had advanced enough by 1752 that Phillippe Buache was able to demonstrate continuity between topographic features on land and seafloor (von Arx, 1962). Early bathymetric charts of the Atlantic showed spot soundings only, with stippling for banks (see Fig. 2, Tucholke and Vogt, this volume). Maury's (1855) Atlantic chart combined shading with contours and spot soundings, similar to displays still used today (Fig. 6). In modern data sets with widely separated tracks, the contouring of a large area is a matter of geological judgment, a problem geomorphologists have grappled with since Stocks and Wüst (1935). Development of digital contouring only created a new debate—whether mindless but objective and repeatable computer contouring (generally preceded by low-pass filtering, for example Slootweg, 1978) is better than hand contouring, which embodies important but sometimes poorly quantified geological prejudices. Obviously the greater the data density, the less need there is for such judgments. On the other hand, geologic advances in the last two decades have quantified geological structures at least to first order (Emery and Uchupi, 1984); for example, allowing prediction of fracture zone trends from plate-kinematic models.

A large set of echo-sounding data in the western North Atlantic was summarized on contour charts in the late 1960s by Uchupi (1965, 1968) and Pratt (1968). Pratt's map (scale 1:4 million; contour interval 100 m), especially in areas of lower data density, was a rather mechanical summary but even today provides a useful overview of bathymetry. Uchupi's maps were of larger scale (scale 1:1 million; contour interval 200 m or less) and provide considerably more insight into the geomorphology of the continental margin and the distribution of the geologic processes responsible. Belding and Holland (1970) also published a bathymetric map of the U.S. Atlantic margin. This compilation includes some industry data not available to Uchupi for his earlier maps. More recent, generally available bathymetric maps (all on a standard Mercator projection, scaled to the Equator) that include part or all of the western North Atlantic Ocean basin are listed below.

1. Uchupi (1971). The scale of these maps is about 1:5.5 million. The maps are more accurate than those previously published, but the 400 m contour interval provides much less detail. Updated versions were published by Emery and Uchupi (1984) at a scale of about 1:18.5 million.

2. National Ocean Survey (1975 to 1981). This bathymetric map series covers the U.S. Atlantic continental margin at a scale of 1:250,000 and is based on sounding data in the NOS files.

3. Perry and others (1981). The scale is 1:8.7 million, and the map extends from 4°S to 59°N. The large contour interval (1,000 m) makes this map useful for only very general purposes. Earthquake epicenters, magnetic lineations, DSDP drill sites, paleobathymetric chartlets, and other information is included.

4. General Bathymetric Chart of the Oceans, Searle and others, (1982). The scale is 1:10 million and the contour interval is variable (100 to 500 m). The GEBCO maps are compilations of previously available maps together with new data, and they are the best currently available summary of bathymetry for the North Atlantic.

5. Ocean-Margin Drilling (OMD) Synthesis—1984. A very useful overall compilation of bathymetry of the U.S. Atlantic continental margin was completed by Shor (1984a, 1984b), Shor and Flood (1984), and Flood and Shor (1984) for the Ocean Margin Drilling (OMD) program. The maps are at a scale of about 1:2.2 million, and the contour interval is 100 m.

6. U.S. Naval Oceanographic Office (various years). The scale is about 1:4.4 million, and the contour interval varies from 200 to 1,000 m (see locator insert map in Plate 2 for chart areas and years last updated). The maps include degraded Navy multibeam echosounding data.

The 1969 version of the Navoceano maps was used to create SYNBAPS (Synthetic Bathymetric Profiling System), a digital data base constructed by digitizing depth values from the contour charts at 5-minute grid intersections (Van Wyckhouse, 1973). This digital data base can be used in a variety of ways, including digital contouring and graphic displays of seafloor morphology. An updated version, called DBDB-5 (Digital Bathymetric Data Base), was made available in late 1984. The global, depth-shaded physiographic maps of Heirtzler (1985) are based on DBDB-5 (see cover, this volume).

An edited, machine-contoured map based on DBDB-5 (except in the Arctic) is shown in Plate 2. It would be wrong to conclude that simply because this DNAG bathymetric chart of the North Atlantic (Plate 2) was produced by computer contouring of a digital data base, it has fewer artifacts. DBDB-5 was derived from hand-contoured bathymetric charts (principally number 6 above), and Plate 2 is therefore no more accurate than those charts. In fact, its accuracy is reduced because of software-dependent smoothing and interpolation in the process of machine contouring. As one example of a "deliberate" artifact, the use of a 5 min by 5 min data grid represents a low-pass filter that removes or distorts features of horizontal scales less than about 10–20 km. This grid spacing and the use of digital contouring tends to suppress long, narrow ridges or troughs such as those associated with fracture zones (e.g., Smoot and Sharman, 1985) thus producing the "bubble-like" shape of the contours (Plate 2). The insets in Plate 2 show how much more irregular the ocean floor is when observed in detail by multibeam echo sounding.

As an alternative to contours for showing general topographic texture in the North Atlantic, Heezen and others (1959) and Heezen and Tharp (1968) introduced hand-sketched (hachured) physiographic diagrams based on echosounding profiles. Their first diagram (1959) was the first comprehensive summary of the principal physiographic provinces of the North Atlantic and it was an amazing interpretation of the North Atlantic considering the relatively small data set available at that time. Many of their definitions remain the standard descriptions of physiographic provinces in the North Atlantic, although some, such as the "Upper," "Middle," and "Lower Steps" of the Mid-Atlantic Ridge were probably a misinterpretation of profiles crossing fracture zones, which were then unknown. Heezen and others (1959) showed the Mid-Atlantic Ridge without fracture zones, whereas later renditions (e.g., Heezen and Tharp, 1968) were dominated by numerous transverse fracture zones. Both a larger data base and the introduction of the new plate tectonic concepts caused this change in image character.

Three-dimensional (perspective) realizations of seafloor topography are desirable for many scientific, educational, and practical purposes and have been created in several ways. Two plastic 3-D relief maps of the entire North Atlantic have been made (Anonymous, 1963; 1979), as have plaster models of smaller areas, for example a segment of the Mid-Atlantic Ridge (Collette and others, 1980). Varying the illumination angle and direction helps in detecting subtle geological trends in 3-D relief maps. Nowadays such variation is easily done by computer processing and display of digital topographic images, (e.g., Edwards and others, 1984) often with auxiliary data such as "texture" or composition. However, in many instances image quality depends on the illumination (or insonification) geometry at the time it was collected. The LANDSAT image of Iceland (Fig. 2B), although digital, was illuminated at a fortuitously low angle by the winter sun, so that it revealed subtle undulations in land and glacier surfaces, the latter in turn related to bedrock topography.

Next to relief models of the seafloor, stereo images provide the most realistic 3-D view. Stereo-pair bottom photography most closely replicates a normal view, and it also yields quantitative microtopographic contours (Fig. 5). A regular contour chart can also be converted to stereo ("anaglyph") form for viewing through red/green glasses. However, a practical problem for all kinds of 3-D seafloor displays (versus equivalent land displays) is the extreme unevenness of data coverage. Whereas contour charts can show unsurveyed areas as blank or by dashed contours, 3-D images tend to portray the entire surface and require interpolation by man or machine. Consequently they may give false impressions of structural trends, roughness, and texture. This problem will be resolved only with better data coverage.

A wholly different class of seafloor topographic images is obtained by transforming bathymetric profiles or two-dimensional gridded data from the spatial domain to the frequency domain. Power spectra in one or two dimensions (Akal and Hovem, 1978; Bell, 1975b, 1979) are the most common such image transformations. Spectra provide a way to generalize the statistical character of bottom topography, although much of the information contained in the bathymetric data is thereby lost. Power spectra of bathymetric profiles can generally be approximated by an expression of the form $P(k) = Ck^b$, where $P(k)$ is power spectral density as a function of wave number k, and C is a constant describing the overall degree of bottom roughness and is, for example, higher for the Mid-Atlantic Ridge than the East Pacific Rise (Berkson and Matthews, 1984). The constant b describing the average slope of the spectrum is typically –2 to –3 but may vary from near zero to –5. Sediment covered areas such as the U.S. Atlantic Continental Rise are characterized by b values typically –1.5 to –2 for wavelengths >3 km, whereas spectral slopes are lower ($b \approx -0.5$) at shorter wavelengths (Fox and Hayes, 1985). The steep-slope segment may reflect volcanic-tectonic (basement topographic) control on seafloor topography, whereas the shallow slopes are wholly due to sedimentary processes. Two-dimensional spectra, resembling contours of a seamount, can be computed from two-dimensional data sets such as derived from multibeam bathymetry or stereophotogrammetry (e.g., Fox and Hayes, 1985; Czarnecki and Bergin, 1985, 1986). 2-D spectra provide a means to describe topographic anisotropy at different wavelengths. Spectral "redness" (negative b-values) is a fundamental property of topographic power spectra for the seafloor, land surface, and the surfaces of terrestrial planets and moons. Topography exhibits "red" spectra because the topography-generating and modifying processes (e.g., volcanism, slumping, erosion by bottom currents) are constrained by the force of gravity, and because the materials are weak except at small scales. The existence of an "angle of repose" for unconsolidated sediments, fluid lavas, and other such materials precludes formation or preservation of short wavelength high-amplitude "pinnacles" that would be needed for "white" or even "blue" spectra.

CONCLUSIONS

Although man's image of the near-shore seafloor began to evolve millenia ago, the abyss remained unfathomed until the mid-nineteenth century. The 15,000 wire soundings collected in the following 80 years were enough to resolve the Mid-Atlantic Ridge and major basins, but on average there was still only one measurement per 2×10^4 km^2. Echosounding starting in 1921 brought a quantum improvement in resolution, but principally only in one (along-track) dimension; detailed grid-surveys were unpopular and difficult due to navigation uncertainties on the order of 10 km. The 1960s brought deep-towed side-scan sonar, multibeam bathymetry, and satellite navigation accurate to <1 km. Now, during the 1980s, implementation of GPS navigation to <10 m accuracy is eliminating the age-old problem of locating oneself on the earth's surface. Side-looking sonars (e.g., the Sea MARC series) towed at any desired depth can now produce digital images, some multispectral, of seafloor backscatter strength and bathymetry. Digitally acquired and mosaicked large-area seafloor photographs are being made, and backscatter strength information is being extracted from multibeam bathymetric data. Future studies will use integrated data from these digital systems to (1) quantify seafloor textures in different geomorphologic provinces; (2) detect and classify specific kinds of objects (fault scarps, lava fronts, pit craters, erratics, hydrothermal mounds, etc.); (3) estimate acoustic properties and structures in the shallow subbottom; and (4) even search for time changes in bottom topography or texture as a result of bottom currents, slumping, or volcano-tectonic processes in dynamic areas such as plate boundaries and shelves.

Acoustic systems continue to be the only effective way to map ocean floor texture and topography in the very wide band between what the ocean surface reveals to radar altimetry (30 km limiting resolution) down to 0.1–1 m resolution where optical methods (LIBEC, ANGUS, *Argo*) take over. By contrast, imaging of most terrestrial planets and moons—as well as exposed oceanic crust such as Iceland—is done best in the microwave, infrared, and visible parts of the electromagnetic spectrum. In both acoustic and electromagnetic regimes the trend is toward multispectral images produced at several discrete frequencies (e.g., LANDSAT, Sea MARC IV) and there is considerable overlap in processing/interpretation techniques (such as image enhancement) for subaerial and seafloor images. The similarity in microwave land-surface images and acoustic seafloor images (e.g., Fig. 2A, C) owes much to the almost identical wavelength (despite the 2×10^5 difference in wavespeed). The wavelengths used in SAR are 1 to 30 cm, almost identical to the range from the Sea MARC CL (150 kHz, 1 cm wavelength) to the 3.5 kHz profiler system.

Aircraft, still used extensively on the earth for visible, infrared and radar imaging and stereophotogrammetry, may diminish in importance as spaceborne sensing further improves. However, aircraft will continue to be the best way to map magnetic anomalies over land or sea. GPS navigation has now made aerogravity surveys with fixed-wing aircraft feasible. Over the oceans such data, compared to satellite altimetry, will be of somewhat higher resolution, will be essentially unaffected by ocean dynamics, and will therefore complement satellite altimetry (SEASAT, GEOSAT, TOPEX). Aircraft sensing of magnetic and gravity anomalies originating at the seafloor will be further enhanced when gradiometry becomes routine.

Electro-magnetics (AEM), lasers, and multispectral scanners deployed on aircraft, as well as water-wave patterns mapped by satellite or aircraft probably will continue to play a minor role in certain shallow-water (<100 m) bathymetric applications. In the abyss, lasers have been proposed for precise seafloor geodesy, for example, to measure strains along plate boundaries (Spiess, 1985a, 1985b). It may also be possible to build deep-towed laser "multibeam" swath-mapping systems, providing imagery and depth contours that are of higher resolution along a 100 m wide swath than is possible even with a deep-towed sonar (e.g., Sea MARC CL).

The ultimate image of the planetary surface will be a superposition of images made with different systems, each of different absolute resolution. In this composite image, low spatial frequencies would be sampled at a much lower rate than high ones (J. Kosalos, personal communication, 1986). For example, the marine geoid as mapped by satellite radar altimetry (Haxby, 1985) approximates a low-resolution image of the ocean floor (because the response function of gravity to seafloor topography peaks in the 50–500 km wavelength band). Over the period of a few years a satellite altimeter sampling the entire ocean surface (3.6×10^8 km^2, of which about 10% or 4×10^7 km^2 is the North Atlantic) gives an image of described by about 10^7 pixels, somewhat better than shown in the frontispiece. Resolution can be improved further (perhaps up to 2 times) using fixed-wing aircraft (e.g., Brozena, 1984), which can also—unlike the satellite—sample magnetic anomalies related to bathymetric features and plate tectonics. Assuming an aircraft can survey 0.5×10^6 km per year, a global oceanic image of about 10^8 pixels (10 km rectangular grid survey) could be produced in 150 aircraft-years. Another improvement could be gained with gravity gradiometry. Significant improvement in topographic resolution, however, requires surface ships whose slow speed (ca. 20 km/hr) greatly increases survey time. High resolution images require towing optical or acoustic systems near the bottom at rates of only a few km/hr. (Note that for acoustic systems along-track and cross-track resolution may not be the same, and they vary as a function of band width, beam width, frequency, ship's speed, and slant range.)

Sea MARC II towed at 10 km/hr at 20 km swath width (200 km^2/hr) could image the world ocean to 25 m horizontal resolution in about 200 ship years, yielding a 4×10^{12} pixel image at a cost of about 10^9 U.S. dollars, at current rates. Sea MARC I towed at 2 km/hr at 2 km swath width (4 km^2/hr) would take some 10^3 ship years, improving the resolution to 3 m and generating a 4×10^{14} pixel image. Sea MARC CL towed close to the bottom and generating a swath width of only 50 m at 2 km/hr (0.1 km^2/hr) would take some 5×10^5 ship years to produce an image with 5 cm resolution and 10^{18} pixels. Finally, an optical system of 3 mm resolution, with a 10 m swath width and towed at 2 km/hr (.02 km^2/hr) would take 2.5×10^6 ship years and yield an image of 3.6×10^{20} pixels to describe the world's ocean floors down to the scale of a small pebble. Systematic imaging of the ocean floors at high resolution is clearly a job for untethered robot submersibles, not surface ships.

At present rates of data collection—probably less than one ship-year per year total for deep-tow systems—only a few percent of the ocean floors will have been imaged in detail even by the GSA bicentennial in 2088. However, complete coverage by aircraft and surface-towed shipborne systems such as SEABEAM or Sea MARC II would be attainable if a dozen vessels and two airplanes were dedicated to the job. This would also require a systematic surveying strategy (tracks on a regular grid), such as that adopted in the 1960s by the Naval Oceanographic Office. Aside from multibeam and geophysical surveying by Navoceano, most shipborne and aircraft data have been collected in non-standardized patterns in support of special investigations in areas of current scientific or other interest. As a result our image of the ocean floors continues to evolve in a highly irregular fashion, with detailed patches here and there and transit lines radiating from major ports (see Fig. 3, Tucholke and Vogt, this volume). This is a highly inefficient way to image the ocean floors, and it will become steadily more inefficient in the future as larger and larger fractions of each cruise are spent crossing areas previously imaged. A possible remedy is international adoption of a standardized survey grid (for example, east-west lines exactly on even minutes of latitude, north-south lines on regular longitudes and becoming systematically decimated in higher latitudes, and NW-NE cross-lines 5 n.mi. apart). Survey and research vessels would be requested to follow any grid-lines not previously surveyed, and so in time a systematic and homogeneous seafloor image could be developed, similar to those describing the surfaces of other terrestrial planets and moons. Because extra transit time would be required for research ships to travel from point A to point B along gridlines, rather than on great-circle courses, funding priorities would have to be rearranged. However, the long-term result of such a "SEAGRID" program would be to have imaged the ocean floors more cheaply and efficiently and to have created more homogeneous data bases. With the introduction of GPS navigation, which will allow data collection precisely along predetermined tracks, it is not unrealistic to begin considering the design and implementation of such a program.

REFERENCES

Akal, T., and Hovem, J.
1978 : Two-dimensional space series analysis for sea-floor roughness: Marine Geotechnology, v. 3, p. 171–182.

Alpers, W., and Hennings, I.
1984 : A theory of the imaging mechanism of underwater bottom topography by real and synthetic aperture radar: Journal of Geophysical Research, v. 89, p 10529–10546.

Anonymous
1963 : Basin relief model, North Atlantic Ocean: U.S. Naval Photographic Interpretation Center, USNPIC 910/6U (three-dimensional chart), scale 1:5,800,000 at 30°N.

Anonymous
1979 : World scientific ocean floor relief model: Defense Mapping Agency, Washington, D.C., (three-dimensional chart), scale 1:12,300,000 at Equator.

Arnone, R. A., and Arthur, B. E., Jr.
1981 : Interpretation of hydrographic features using Landsat images: NORDA Report 39, Naval Ocean Research and Development Activity, NSTL Station, Mississippi, 159 p.

Ashjaee, J.
1985 : GPS Doppler processing for precise positioning in dynamic applications, *in* Ocean Engineering and the Environment, Conference Record: Marine Technology Society Institute of Electrical and Electronic Engineers Ocean Engineering Society, p. 195–203.

Backus, R. H., and Worthington, L. V.
1965 : On the existence of the seamount known as "American Scout": Deep-sea Research, v. 12, p. 457–460.

Baines, P. G.
1979 : Observations of stratified flow over two-dimensional obstacles in fluid of finite depth: Tellus, v. 31, p. 351–371.

Ballard, R. D., and Emery, K. O.
1970 : Research Submersibles in Oceanography: Marine Technology Society, Washington, D.C., 70 p.

Ballard, R. D., and Moore, J. G.
1977 : Photographic Atlas of the Mid-Atlantic Ridge Rift Valley: Springer Verlag, New York, 114 p.

Ballard, R. D., and van Andel, Tj.H.
1977 : Morphology and tectonics of the inner rift valley at lat. 36°50′N on the Mid-Atlantic Ridge: Geological Society of America Bulletin, v. 88, p.507–530.

Beach, E. L.
1969 : Man Beneath the Sea; in Exploring the ocean world, ed. C. P. Idyll; T.Y. Crowell, New York, p. 232–261.

Belderson, R. H., Kenyon, N. H., Stride, A. H., and Stubbs, A. R.
1972 : Sonographs of the Sea Floor: Elsevier Publishing Company, Amsterdam, 185 p.

Belderson, R. H., Jones, E.J.W., Gorini, M. A., and Kenyon, N. H.
1984 : A long-range side-scan sonar (Gloria) survey of the Romanche active transform in the equational Atlantic: Marine Geology, v. 56, p. 65–78.

Belding, H. F., and W. C. Holland
1970 : Bathymetric maps; Eastern continental margin, U.S.A.: American Association of Petroleum Geologists, scale 1:1,000,000.

Bell, T. H., Jr.
1975a: Topographically generated internal waves in the open ocean: Journal of Geophysical Research, v. 80, p. 320–327.
1975b: Statistical features of sea-floor topography: Deep-Sea Research, v. 22, p. 883–892.
1979 : Mesoscale sea floor roughness: Deep-Sea Research, v. 26A, p. 65–76.

Bennett, R. H., Lambert, D. N., McGregor, B. A., Forde, E. B., and Merrill, G. F.
1978 : Slope map depicting a major submarine slide on the U.S. Atlantic continental slope east of Cape May: U.S. Department of Commerce Map and Text, A-5787 USCOM-NOAA-DC.

Berkson, J. M., and Matthews, J. E.
1984 : Statistical characterization of seafloor roughness: Institute of Electrical and Electronic Engineers Journal of Oceanic Engineering, v. OE-9, p. 48–52.

Bodechtel, J., Hiller, K., and Munzer, U.
1979 : Comparison of SEASAT and LANDSAT data of Iceland for qualitative geologic applications, *in* Proceedings of the SEASAT-SAR Processor Workshop, Frascati, Italy; Paris, European Space Agency, p. 61–67.

Bogorodskiy, V. V., Bentley, C. R., and Gudmandsen, P. E.
1985 : Radioglaciology: Kluwer Academic Publishers, Hingham, Massachusetts, 254 p.

Boutan, L.
1893 : Mémoire sur la photographie sous-marine: Archives de Zoologie Expérimentale et Generale, v. 21, p. 281–324.

Brozena, J. M.
1984 : A preliminary analysis of the NRL airborne gravimetry system; Geophysics; v. 49, p. 1060–1069.

Brundage, W. L., and Patterson, R. B.
1976 : LIBEC photography as a sea floor mapping tool: OCEANS '76, MTS-IEEE paper 8B, p. 1–11.

Brundage, W. L., Jr., Buchanan, C. L., and Patterson, R. B.
1967 : Search and serendipity, *in* Deep-Sea Photography: Johns Hopkins Press, Baltimore, p. 75–87.

Burke, K., and Wilson, J. T.
1976 : Hotspots on the earth's surface: Scientific American, v. 235, p. 46–57.

Busby Associates, Incorporated
1985 : Undersea Vehicle Directory—1985: Busby Associates, Incorporated, Arlington, Virginia, 430 p., plus appendices.

Carter, D.J.T.
1980 : Echo-sounding correction tables (third edition): Hydrographic Department, Ministry of Defence, Taunton, Somerset, 150 p.

Caruthers, J. W.
1977 : Fundamentals of marine acoustics: Elsevier, Oceanography Series, no. 18, 137 p.

Chayes, D. N.
1983 : Evolution of Sea MARC I: Institute of Electrical and Electronic Engineers: Proceedings, Third Working Symposium on Oceanographic Data Systems, p. 103–108.

Clarke, G. L., and Denton, E. J.
1962 : Light and animal life; in The sea, Volume I, ed. M. N. Hill, Wiley and Sons, New York, p. 456–468.

Clarke, T. L., Proni, J., Alper, S., and Huff, L.
1985 : Definition of "Ocean Bottom" and "Ocean Bottom Depth," *in* Ocean Engineering and the Environment, Conference Record: Marine Technology Society/ Institute of Electrical and Electronic Engineers Ocean Engineering Society, p. 1212–1216.

Clay, C. S., and Medwin, H.
1977 : Acoustical Oceanography: Wiley and Sons, New York, 544 p.

Cohen, P. M.
1970 : Bathymetric navigation and charting: U.S. Naval Institute, Annapolis, Maryland, 138 p.
1986 : The coming of age of bathymetric maps: Bulletin of the American Congress on Surveying and Mapping, no. 101, p. 25–27.

Colladon, J. D., and Sturm, J.K.F.
1827 : The compression of liquids (in French), *in* Speed of sound in liquids, Part IV: Annales de Chimie et de Physique, Series 2, v. 36, p. 236–257.

Collette, B. J., Verhoef, J., and de Mulder, A.F.J.
1980 : Gravity and a model of the median valley: Journal of Geophysics, v. 24, p. 33–40.

Crane, K., Aikman, F., Embley, R., Hammond, S., Malahoff, A., and Lupton, J.
1985 : The distribution of geothermal fields on the Juan de Fuca Ridge: Journal of Geophysical Research, v. 90, p. 727–744.

Czarnecki, M., and Bergin, J. M.

1985 : Characteristics of ocean bottom roughness for several seamounts derived from multibeam bathymetric data, *in* Ocean Engineering and the Environment, Conference Record: Marine Technology Society/Institute of Electrical and Electronic Engineers Ocean Engineering Society, p. 663–672.

1986 : Statistical characterization of small scale bottom topography as derived from multibeam sonar data, *in* Proceedings of the Fourth Working Symposium on Oceanographic Data Systems: Los Angeles, Institute of Electrical and Electronic Engineers Computer Society Press, p. 15–24.

Damuth, J. E.

1975 : Echo character of the western equatorial Atlantic and its relationship to the dispersal and distribution of terrigenous sediment: Marine Geology, v. 18, p. 17–45.

De Laurier, J. M.

1978 : Arctic Ocean sediment thicknesses and upper mantle temperatures from magnetotelluric soundings: Arctic Geophysical Review, v. 45, p. 35–50.

de Moustier, C.

1986 : Beyond bathymetry: mapping acoustic backscattering from the deep seafloor with Sea Beam: Journal of the Acoustical Society of America, v. 79, p. 316–331.

de Moustier, C., and Kleinrock, M. C.

1986 : Bathymetric artifacts in SEA BEAM data: how to recognize them, what causes them: Journal of Geophysical Research, v. 91, p. 3407–3424.

Dietz, R. S.

1969 : The underwater landscape; Ch. 2; in Exploring the Ocean World, ed. C. P. Idyll; T. Y. Crowell, New York, p. 22–41.

Dietz, R. S., and Knebel, H. J.

1968 : Survey of Ross's original deep sea sounding site: Nature, v. 220, p. 751–753.

Dixon, T. H., Naraghi, M., McNutt, M. K., and Smith, S. M.

1983 : Bathymetric prediction from SEASAT altimeter data: Journal of Geophysical Research, v. 88, p. 1563–1571.

Drygalski, E. von

1904 : Zum Kontinent des eisigen Südens: Deutsche Südpolar-Expedition; Fahrten and Forschungen des Gauss 1901–1903: Georg Reimer, Berlin, p. 109 and 267.

Dunham, S. J., and Shostack, B. H.

1980 : Precise reconstruction of ship's position, *in* Institute of Electrical and Electronis Engineers 80 Position Location and Navigation Symposium Record: Institute of Electrical and Electronics Engineers, New York, p. 430–435.

Edvalson, F. M., and Miscoski, V. T.

1969 : Bathymetric Charts; Their Development and Use: U.S. Naval Oceanographic Office, Washington, D.C., 15 p.

Edwards, M. H., Arvidson, R. E., and Guiness, E. A.

1984 : Digital image processing of SEA BEAM bathymetric data for structural studies of seamounts near the East Pacific Rise: Journal of Geophysical Research, v. 89, p. 11108–11116.

Emery, K. O., and Uchupi, E.

1984 : The geology of the Atlantic ocean: Springer Verlag, New York, 1050 p. and 23 charts.

Erskine, F. T., and Franchi, E. R.

1984 : Rapid, basin-wide surveying of large undersea topography using scattered acoustic energy: Marine Technology Society and The Ocean Engineering Society of the Institute of Electrical and Electronics Engineers, Proceedings Oceans, San Francisco, 1984, p. 990–995.

Erskine, F. T., Bernstein, G. M., Franchi, E. R., and Adams, B. B.

1986 : Imaging of long range reverberation from ocean basin topography: Marine Geodesy, in press.

Ewing, M., Vine, A. C., and Worzel, J. L.

1946 : Photography of the ocean bottom: Journal of the Optical Society of America, v. 36, p. 307–321.

Ewing, M., Worzel, J. L., and Vine, A. C.

1967 : Early development of ocean-bottom photography at Woods Hole Oceanographic Institution and Lamont Geological Observatory, *in* Hersey, J. B., ed., Deep-Sea Photography: Johns Hopkins Press, Baltimore, Maryland, p. 13–39.

Farr, H. K.

1980 : Multibeam bathymetric sonar; SEA BEAM and Hydrochart: Marine Geodesy, v. 4, p. 77–93.

Farre, J. A., and Ryan, W.B.F.

1985 : A 3-D view of erosional scars on the U.S. mid-Atlantic continental margin: American Association of Petroleum Geologists Bulletin, v. 69, p. 923–932.

Flood, R. D., and A. N. Shor

1984 : Bathymetry, in Ocean margin drilling program regional atlas series, Atlas 5, eds. G. M. Bryan and J. Heirtzler; Woods Hole, Massachusetts, Marine Science International, p. 1.

Ford, J. P., Blom, R. G., Bryan, M. L., Daily, M. I., Dixon, T. H., Elachi, C., and Xenos, E. C.

1980 : Seasat views North America, the Caribbean, and western Europe with imaging radar: California Institute of Technology, Jet Propulsion Laboratory, Pasadena, 141 p.

Fornari, D. J., Ryan, W.B.F., and Fox, P. J.

1984 : The evolution of craters and calderas on young seamounts; Insights from Sea MARC I and Sea Beam sonar surveys of a small seamount group near the axis of the East Pacific Rise at 10°N: Journal of Geophysical Research, v. 89, p. 11069–11083.

Fox, C. G., and Hayes, D. E.

1985 : Quantitative methods for analyzing the roughness of the seafloor: Reviews of Geophysics, v. 23, p. 1–48.

Friedman, J. D., Williams, R. S., Pálmason, G., and Miller, C. D.

1969 : Infrared surveys in Iceland: U.S. Geological Survey Professional Paper 650-C, p. C89–C105.

Geyer, R. A., ed.

1977 : Submersibles and their use in oceanography and ocean engineering: Elsevier Oceanography Series, Amsterdam, v. 17, 383 p.

Glenn, M. F.

1970 : Introducing an operational multi-beam-array sonar: International Hydrographic Review, v. 47, p. 35–40.

1976 : Multi-narrow beam sonar systems; The Marine Technology Society and the Oceanic Engineering Society of the Institute of Electrical and Electronic Engineers Proceedings Oceans, 1976, San Francisco, p. 8D-1–8D-2.

Guenther, G. C., and Thomas, R.W.L.

1983 : System design and performance factors for airborne laser hydrography: The Marine Technology Society and the Oceanic Engineering Society of the Institute of Electrical and Electronics Engineers, Proceedings Oceans, 1983, San Francisco, p. 425–430.

Hammack, J. C.

1977 : Landsat goes to sea: Photogrammetric Engineering and Remote Sensing, v. 43, p. 386–391.

Hammerstad, E., Løvik, A., Krane, L., Steinseth, M., Halvorsen, S., and Minde, S.

1985 : The Benigraph surveying system; Field trial results: Bentech A/S, Tromas ø, Norway, 20 p.

Harris, S. E., and Albers, K.

1985 : *Argo;* Capabilities for deep ocean exploration: Oceanus, v. 28, p. 100.

Haxby, W. F.

1985 : Gravity Field of the World's Oceans: U.S. Navy Office of Naval Research (chart), scale 1:51,400,000 at Equator.

Haxby, W. F., Karner, G. D., LaBrecque, J. L., and Weissel, J. K.

1983 : Digital images of combined oceanic and continental data sets and their use in tectonic studies: EOS Transactions of the American Geophysical Union, v. 64, p. 995–1004.

Heezen, B. C., and Hollister, C. D.
1971 : The Face of the Deep: Oxford University Press, New York, 659 p.
Heezen, B. C., and Tharp, M.
1968 : Physiographic diagram of the North Atlantic Ocean (revised): Geological Society of America, Boulder, Colorado, map and chart series (chart), scale 1:5,700,000 at Equator.
Heezen, B. C., Tharp, M., and Ewing, M.
1959 : The floors of the oceans, Part I; The North Atlantic: Geological Society of America Special Paper 65, 122 p., includes physiographic diagram.
Heirtzler, J. R., ed.
1985 : Relief of the surface of the Earth (3 charts): National Geophysical Data Center, Report MGG-2, Boulder, Colorado, scale 1:38,400,000 at Equator.
Hersey, J. B., ed.
1967 : Deep-Sea Photography: Johns Hopkins Press, Baltimore, Maryland, 310 p.
Hill, G. W.
1986 : U.S. Geological Survey plans for mapping the Exclusive Economic Zone using GLORIA, *in* Proceedings; The Exclusive Economic Zone Symposium; Exploring the New Ocean Frontier, Smithsonian Institution, October 2–3, 1985: U.S. Department of Commerce, Rockville, Maryland, p. 69–78.
Holcombe, T. L.
1977 : Ocean bottom features; terminology and nomenclature: GeoJournal, v. 1, p. 25–48.
Holcombe, T. L., and Heezen, B. C.
1970 : Patterns of relative relief, slope, and topographic texture in the North Atlantic Ocean: Technical Report 8 (CU-8-70), Naval Ships Command N00024-67-1186, Lamont-Doherty Geological Observatory, Palisades, N.Y., 115 p.
Hollister, C. D., Flood, R. D., Johnson, D. A., Lonsdale, P. F., and Southard, J. B.
1974 : Abyssal furrows and hyperbolic echo traces on the Bahama Outer Ridge: Geology, v. 2, p. 395–400.
Houot, G. S.
1955 : Le bathyscaphe F.N.R.S. III au service de lèxploration des grandes profondeurs: Deep-Sea Research, v. 2, p. 247–249.
Houot, G., and Willm, P.
1955 : 2000 Fathoms Down: Dutton, New York, 249 p.
Hurdle, B. G.
1986 : The sound-speed structure, *in* Hurdle, B. G., ed., The Nordic seas: Springer Verlag, New York, p. 155–181.
Karson, J. A., and Dick, H.J.B.
1983 : Tectonics of ridge-transform intersections at the Kane fracture zone: Marine Geophysical Researches, v. 6, p. 51–98.
Kasischke, E. S., Shuchman, R. A., and Lyzenga, D. R.
1983 : Detection of bottom features on synthetic aperture radar imagery: Photogrammetric Engineering and Remote Sensing, v. 49, p. 1341–1353.
Kasischke, E., Meadows, G., and Jackson, P.
1984 : The use of SAR to detect hazards to navigation: Environmental Research Institute of Michigan Report 169-200-2-F, 194 p.
Kastens, K. A., Macdonald, K. C., Miller, S. P., and Fox, P. J.
1986 : Deep-tow studies of the Vema Fracture Zone; 2. Evidence for tectonism and currents in the sediments of the transform valley floor: Journal of Geophysical Research, v. 91, p. 3355–3367.
Knott, S. T., and Hersey, J. B.
1956 : Interpretation of high-resolution echo-sounding techniques and their use in bathymetry, marine geophysics, and biology: Deep-Sea Research, v. 4, p. 27–30.
Kossinna, E.
1921 : Die Tiefen des Weltmeeres: Institut für Meereskunde, Veröffentlichungen, Geographie-Naturwissenschaften, v. 9, 70 p.
Kovacs, L. C., and Vogt, P. R.
1982 : Depth-to-magnetic source analysis of the Arctic Ocean region: Tectonophysics, v. 89, p. 255–294.
Laughton, A. S.
1981 : The first decade of GLORIA: Journal of Geophysical Research, v. 86, p. 11511–11534.
Laughton, A. S., and Searle, R. C.
1979 : Tectonic processes on slow spreading ridges; in Deep drilling results in the Atlantic Ocean; Ocean Crust, eds. M. Talwani, C.G.A. Harrison and D. E. Hayes; American Geophysical Union, Maurice Ewing Series, p. 15–32.
Liebermann, L. N.
1962 : Other electromagnetic radiation, in The sea, Volume I, ed. M. N. Hill; Wiley and Sons, New York, p. 469–475.
Link, L. E., Krabill, W. B., and Swift, R. N.
1983 : A prospectus on airborne laser mapping systems: Advances in Space Research, v. 3, p. 309–322.
Lonsdale, P. F., Tyce, R. C., and Spiess, F. N.
1983 : Near-bottom acoustic observations of abyssal topography and reflectivity, *in* Physics of Sound in Marine Sediments: Marine Science, v. 1, p. 293–317.
Luyendyk, B. P., and Macdonald, K. C.
1977 : Physiography and structure of the inner floor of the FAMOUS rift valley: Observations with a deep-towed instrument package; Geological Society of America Bulletin, v. 88, p. 648–663.
Luskin, B., Heezen, B. C., Ewing, M., and Landisman, M.
1954 : Precision measurement of ocean depth: Deep-Sea Research, v. 1, p. 131–140.
Lyzenga, D. R.
1981 : Remote bathymetry using active and passive techniques: 1981 International Geoscience and Remote Sensing Symposium Digest, Washington, D.C., p. 777–786.
Macdonald, K. C., Castillo, D. A., Miller, S. P., Fox, P. J., Kastens, K. A., and Bonatti, E.
1986 : Deep-Tow studies of the Vema Fracture Zone, 1. Tectonics of a major flow-slipping transform fault and its intersection with the Mid-Atlantic Ridge: Journal of Geophysical Research, v. 91, p. 334–3354.
Matthews, D. J.
1939 : Tables of the velocity of sound in pure water and sea water for use in echo sounding and echo ranging: Admiralty Hydrographic Department, London, 52 p.
Maurer, H., and Stocks, T.
1933 : Die Echolotungen des "Meteor"; Deutsche Atlantische Expedition "Meteor," 1925–1927: Wissenschaftliche Ergebnisse, v. 2, 309 p.
Maury, M. F.
1855 : The Physical Geography of the Sea: Harper and Brothers, New York, 287 p.
McDonough, M., and Martin-Kaye, P.H.A.
1984 : Radargeologic interpretation of Seasat imagery of Iceland: International Journal of Remote Sensing, v. 5, p. 433–450.
McGregor, B. A., Stubblefield, W. L., Ryan, W.B.F., and Twichell, D. C.
1982 : Willmington submarine canyon; A marine fluvial-like system: Geology, v. 10, p. 27–30.
McKenzie, D., and Bowin, C.
1976 : The relationship between bathymetry and gravity in the Atlantic Ocean: Journal of Geophysical Research, v. 81, p. 1903–1915.
Menard, H. W., and Smith, S. M.
1966 : Hypsometry of the ocean basin provinces: Journal of Geophysical Research, v. 71, p. 4305–4325.
Metcalf, W. G., Heezen, B. C., and Stalcup, M. C.
1964 : The sill depth of the Mid-Atlantic Ridge in the equatorial region: Deep-Sea Research, v. 11, p. 1–10.
Murray, J., and Hjort, J.
1912 : The depths of the ocean, A general account of the modern science of oceanography based largely on the scientific researches of the Norwegian

steamer *Michael Sars* in the North Atlantic: Macmillan Publishing Company, London, 821 p.

Murray, J., and Renard, A. F.
1891 : Report on the deep-sea deposits based on the specimens collected during the voyage of H.M.S. *Challenger* in the years 1872 to 1876: Her Majesty's Stationery Office, London, 525 p.

National Ocean Survey
1975 *to* 1980: Bathymetric maps of the U.S. Atlantic margin (Numbers NH16-12, NH17-3, NH17-6, NH18-7, NI17-9, NI17-12, NI18-7, NI18-10, NJ18-1, NJ18-2, NJ18-3, NJ18-4, NJ18-5, NJ18-6, NJ18-8, NJ18-9, NJ18-11, NJ18-12, NJ19-1, NK18-4, NK19-9, NK19-10, NK18-9, NJ18-11, NJ18-12, NJ19-1, NK 19-4, NK19-9, NK19-10, NK19-11, and NK19-12): National Oceanic and Atmospheric Administration, U.S. Department of Commerce, Washington, D.C., scale 1:250,000.

OTTER
1984 : The geology of the Oceanographer Transform; The ridge-transform intersection: Marine Geophysical Researches, v. 6, p. 109–141.

Parke, M.E., and Dixon, T. H.
1982 : Topographic relief from Seasat altimeter mean sea-surface, July 7–December 10, 1978: Nature, v. 300, p. 317.

Pérès, J. M.
1965 : Apercù sur les resùltats de deux plongees èffectuèes dans le ravin de Puerto-Rico par le bathyscaphe ARCHIMEDE: Deep-Sea Research, v. 12, p. 883–891.

Perry, R. K., Fleming, H. S., Vogt, P. R., Cherkis, N. Z., Feden, R. H., Thiede, J., and Strand, J. E.
1981 : North Atlantic Ocean; Bathymetry and plate tectonic evolution: Geological Society of America Map and Chart Series MC-35, Boulder, Colorado, scale 1:8,700,000.

Perry, R. K., Fleming, H. S., Weber, J. R., Kristoffersen, Y., Hall, J. K., Grantz, A., and Johnson, G. L.
1985 : Bathymetry of the Arctic Ocean: Naval Research Laboratory, Washington, D.C., scale 1:6,000,000.

Peterson, M.
1969 : Underwater Archaeology, in Exploring the Ocean World, ed. C. P. Idyll; T. Crowell Company, New York, p. 196–231.

Pettersson, H.
1954 : The ocean floor: Yale University Press, New Haven, 181 p.

Pfannenstiel, M.
1960 : Erlaüterungen zu den bathymetrischen Karten des östlichen Mittelmeeres: Institute of Oceanography of Monaco Bulletin, v. 1192, 60 p.

Phillips, J. D., and Fleming, H. S.
1978 : Multi-beam sonar study of the Mid-Atlantic Ridge rift valley, 36°–37°N; Map and Chart Series MC-19, Geological Society of America, Boulder, CO. 5 p. plus charts at 1:36,457 scale.

Pratt, J. H.
1859 : On the influence of the ocean on the plumb line in India: Philosophical Transactions of the Royal Society of London, v. 149, p. 779–796.
1871 : On the constitution of the solid crust of the earth: Philosophical Transactions of the Royal Society of London, v. 161, part 2, p. 335–347.

Pratt, R. M.
1968 : Atlantic continental shelf and slope; Physiography and sediments of the deep sea: U.S. Geological Survey Professional Paper 529-B, 44 p.

Ryan, W.B.F.
1982 : Imaging of submarine landslides with wide-swath sonar, in Marine slides and other mass movements, eds. S. Saxov and J. K. Nieuwenhuyis; Marine Sciences, v. 6, p. 175–187.
1986 : Perspective 3-dimensional viewing of side-looking sonar images and the prospects of multispectral stereo images: EOS Transactions of the American Geophysical Union, v. 66, p. 1262–1263.

Ryan, W.B.F., and Rabushka, A.
1985 : The discovery of the *Titanic* by the U.S. and French Expedition: Oceanus, v. 28, no. 4, p. 19.

Sabine, E.
1823 : On the temperature at considerable depths of the Caribbean Sea: Philosophical Transactions of the Royal Society of London, part I, p. 206–210.

Schuldt, M. D., Cook, C. E., and Hale, B. W.
1967 : Photogrammetry applied to photography at the bottom, in Deep-sea Photography, ed. J. B. Hersey; Johns Hopkins Press, Baltimore, Maryland, p. 69–73.

Searle, R. C., and Laughton, A. S.
1977 : Sonar studies of the Mid-Atlantic Ridge and Kurchatov Fracture Zone; Journal of Geophysical Research, v. 82, p. 5313–5328.
1981 : Fine-scale sonar study of tectonics and volcanism on the Reykjanes Ridge: Institute of Oceanographic Sciences, p. 5–13.

Searle, R. C., Johnson, G. L., and Monahan, D.
1982 : General bathymetric chart of the oceans (GEBCO), (chart 5.08, North Atlantic, fifth edition, Canadian Hydrographic Survey, Ottawa, Canada, scale 1:10,000,000, 1 sheet.

Shih, J.S.F., Atwater, T., and McNutt, M.
1978 : A near-bottom geophysical traverse of the Reykjanes Ridge: Earth and Planetary Science Letters, v. 39, p. 75–83.

Shor, A. N.
1984a: Bathymetry, *in* Uchupi, E., and Shor, A. N., eds., Ocean margin drilling program regional atlas series, Atlas 3: Marine Science International, Woods Hole, Massachusetts, p. 1.
1984b: Bathymetry, *in* Uchupi, E., and Shor, A. N., eds., Ocean margin drilling program regional atlas series, Atlas 2: Marine Science International, Woods Hole, Massachusetts, p. 1.

Shor, A. N., and Flood, R. D.
1984 : Bathymetry, *in* Ewing, J. I., and Rabinowitz, P. D., eds., Ocean margin drilling program regional atlas series, Atlas 4: Marine Science International, Woods Hole, Massachusetts, p. 1.

Siemens, C. W.
1876 : On determining the depth of the sea without the use of a sounding line: Philosophical Transactions of the Royal Society of London, v. 166, p. 671–692.

Slootweg, A. P.
1978 : Computer contouring with a digital filter: Marine Geophysical Researches, v. 3, p. 401–405.
1986 : Basement imaging with side-looking seismics; Marine Geophysical Researches, v. 8, p. 149–174.

Somers, M. L., and Stubbs, A. R.
1984 : Sidescan sonar: Institute of Electrical and Electronic Engineers, Proceedings, v. 131, part F, p. 243–256.

Smoot, N. C., and Sharman, G. F.
1985 : Charlie-Gibbs; A fracture zone ridge: Tectonophysics, v. 116, p. 137–142.

Spiess, F. N.
1985a: Analysis of a possible seafloor strain measurement system: Marine Geodesy, v. 9, no. 4, p. 385–398.
1985b: Sub-oceanic geodetic measurements, in Special issue on satellite geodynamics, ed. L. S. Water; Institute of Electrical and Electronic Engineers Transactions on Geoscience and Remote Sensing, GE-23: no. 4, p. 502–510.

Spindel, R. C., and Heirtzler, J. R.
1972 : Long-range echo ranging: Journal of Geophysical Research, v. 77, p. 7073–7088.

Stocks, T., and Wüst, G.
1935 : Die Tiefenverhältnisse des offenen Atlantischen Ozeans; Deutsche Atlantische Expedition *Meteor,* 1925–27: Wissenschaftliche Ergebnisse, v. 3, Teil 1, 1. Lieferung, 31 p.

Strand, T. C.
1985 : Optical three-dimensional sensing for machine vision: Optical Engineering, v. 24, p. 33–40.

Sverdrup, H. U., Johnson, M. W., and Fleming, R. H.
1942 : The Oceans: Prentice-Hall, New York, 1060 p.

Taylor, P. T., Stanley, D. J., Simkin, T., and Jahn, W.
1975 : Gilliss seamount; Detailed bathymetry and modification by bottom currents: Marine Geology, v. 19, p. 139–157.
Teleki, P. G., Roberts, D. G., Chavez, P. S., Somers, M. L., and Twichell, D. C.
1981 : Sonar characteristics of the U.S. Atlantic continental slope; acoustic characteristics and image processing techniques: Proceedings, Offshore Technology Conference 13, v. 2, p. 91–102.
Thomson, C. W.
1878 : The voyage of the Challenger, The Atlantic, Volume 2: Harper and Brothers, New York, p. 246–326.
Tolstoy, I.
1951 : Submarine topography in the North Atlantic: Geological Society of America Bulletin, v. 62, p. 441–450.
Tolstoy, I., and Ewing, M.
1949 : North Atlantic hydrography and the Mid-Atlantic Ridge: Geological Society of America Bulletin, v. 60, p. 1527–1540.
Trumbull, J.V.A., and McCamis, J.V.A.
1967 : Geological exploration of an East Coast submarine canyon from a research submersible: Science, v. 158, p. 370–372.
Tyler, J. E., ed.
1977 : Light in the Sea; Benchmark papers in optics, Volume 3: Dowden, Hutchinson, and Ross, Stroudsburg, Pennsylvania, 384 p.
Tyler, J. E., and Preisendorfer, R. W.
1962 : Light; in The sea, Volume I, ed. M. N. Hill; Wiley and Sons, New York, p. 397–451.
Uchupi, E.
1965 : Maps showing relation of land and submarine topography, Nova Scotia to Florida: U.S. Geological Survey Miscellaneous Geological Investigations Map I-451, scale 1:1,000,000, 3 sheets.
1968 : Atlantic continental shelf and slope; Physiography: U.S. Geological Survey Professional Paper 529-C, 30 p.
1971 : Bathymetric atlas of the Atlantic, Caribbean, and Gulf of Mexico: Woods Hole Oceanographic Institution, ref. no. 71-72 (unpublished manuscript), scale 1:5,000,000, 10 sheets.
Urick, R. J.
1967 : Principles of Underwater Sound for Engineers: McGraw-Hill, New York, 342 p.
U.S. Naval Hydrographic Office
1961 : The World: Miscellaneous Publication 15,254.12, 12 sheets, Washington, D.C., scale 1:12,233,000, 500 fm contour interval.
1977 : Navy ocean area world relief maps (North Atlantic Ocean—Sheets NA6, NA7, NA9 and 9A, NA10): World Bathymetric Unit, U.S. Naval Oceanographic Office, Bay Saint Louis, Mississippi, Mercator projection, scale 1″ = 1° (unpublished).
Van Wyckhouse, R. J.
1973 : Synthetic bathymetric profiling system (SYNBAPS); U.S. Naval Oceanographic Office, NAVOCEANO, Bay Saint Louis, Mississippi, Technical Report TR-233, 138 p.
Veatch, A. C., and Smith, P. A.
1939 : Atlantic submarine valleys of the United States and the Congo Submarine Valley: Geological Society of America Special Paper 7, 101 p., 17 plates.
Vening Meinsz, F. A.
1937 : Results of maritime gravity research, 1923–1932, *in* International aspects of oceanography: National Academy of Science, Washington, D.C., p. 61–69.
Vigoureux, P., and Hersey, J. B.
1962 : Sound in the sea, in The sea, Volume 1, ed. M. N. Hill; Wiley and Sons, New York, p. 476–497.
Vogt, P. R., and Perry, R. K.
1982 : North Atlantic Ocean; Bathymetry and plate tectonics evolution: Text to accompany Geological Society of America Map and Chart Series MC-35, 21 p.
Vogt, P. R., Zondek, B., Fell, P. W., Cherkis, N. Z., and Perry, R. K.
1984 : SEASAT altimetry, the North Atlantic geoid, and evaluation of shipborne subsatellite profiles: Journal of Geophysical Research, v. 89, p. 9885–9903.
von Arx, W. S.
1962 : Introduction to Physical Oceanography: Addison-Wesley Publishing Company, Reading, Massachusetts, 422 p.
Wertenbaker, W.
1974 : The Floor of the Sea: Little, Brown and Company, Boston, 275 p.
Williams, R. S., Jr.
1976 : Vatnajökull icecap, Iceland, *in* Williams, R. S., and Carder, W. D., eds., ERTS-1, A new window on our planet: U.S. Geological Survey Professional Paper 929, p. 188–193.
1980 : The use of broadband thermal infrared images to monitor and study dynamic geological phenomena, *in* Settle, M., ed., Workshop on geological applications of thermal infrared remote sensing techniques: Lunar and Planetary Institute Report 81-06, p. 98–106.
1983a: Satellite glaciology of Iceland: Jökull, v. 33, p. 3–12.
(ed.), 1983b: Geological applications: Ch. 31 of Manual of Remote Sensing, v. II, ed. J. E. Estes: Falls Church, Virginia, American Society of Photogrammetry, 1,167 p.
1985 : Quantitative geomorphologic studies from spaceborne platforms; *in* Hayden, R. S., ed., Global Mega-Geomorphology, NASA Conference publication 2312, p. 94–97.
Williams, R. S., Jr., Thorarinsson, S., and Morris, E. C.
1983 : Geomorphic classification of Icelandic volcanoes: Jökull, v. 33, p. 19–24.
Wood, A. B.
1960 : A textbook of sound: G. Bell and Sons, London, 610 p.
Worthington, L. V.
1953 : Oceanographic results of Project Skijump I and Skijump II in the Polar Sea, 1951–1952: EOS, Transactions of the American Geophysical Union, v. 34, p. 543–551.
Wüst, G.
1936 : Schichtung und Zirkulation des Atlantischen Ozeans: Das Bodenwasser und die Stratosphäre: Wissenschaftliche Ergebnisse der Deutschen Atlantischen Expedition Meteor 1925–1927, v. 6, p. 1–288.
Zollinger, R., Becker, A., and Morrison, F.
1985 : Data analysis of airborne electromagnetic bathymetry: Naval Ocean Research and Development Activity Report 93, 16 p.

MANUSCRIPT ACCEPTED BY THE SOCIETY JUNE 23, 1986

ACKNOWLEDGMENTS

Preparation of this paper was supported in part by a Mellon Senior Study Award to B. Tucholke at Woods Hole Oceanographic Institution. We thank J. R. Heirtzler, J. Jaffe, E. Uchupi, R. Higgs, W. Brundage, and J. Bergin for comments on an early draft of the manuscript, J. Kosalos for insightful perspectives on acoustic imaging, and N. C. Smoot for providing information about the SASS multibeam bathymetric system and for the bathymetric insets on Plate 2. M. Whitney and J. Perrotta prepared the manuscript and I. Jewett the illustrations. Contribution number 6226 of Woods Hole Oceanographic Institution.

Printed in U.S.A.

The Geology of North America
Vol. M, The Western North Atlantic Region
The Geological Society of America, 1986

Chapter 3

A Jurassic to recent chronology

Dennis V. Kent
Lamont-Doherty Geological Observatory and Department of Geological Sciences, Columbia University, Palisades, New York 10964
Felix M. Gradstein
Geological Survey of Canada, Bedford Institute of Oceanography, Dartmouth, Nova Scotia B2Y 4A2, Canada

INTRODUCTION

We present an integrated geomagnetic polarity and geologic time scale for the Jurassic to Recent interval, encompassing the age range of the modern ocean floor. The time scale is based on the most recent bio-, magneto-, and radiochronologic data available.

The biostratigraphic bases for Jurassic, Cretaceous, and Cenozoic time-scales are discussed extensively elsewhere (e.g., Gradstein, this volume; Van Hinte, 1976b; Hardenbol and Berggren, 1978; Berggren and Van Couvering, 1974; Van Couvering and Berggren, 1977). Emphasis is placed here on magnetochronology and its integration with biochronology in the derivation of an internally consistent geologic time scale. The binary signal of normal and reversed geomagnetic polarity has little intrinsic absolute time value (ordinal scale), but it can be used to measure time according to its radiochronologic calibration (cardinal scale). The standard magnetic reversal sequence has a correlatable, characteristic pattern and is demonstrated to be continuous from numerous marine magnetic anomaly profiles from the world ocean. The reversal sequence is recorded by lateral accretion in sea-floor spreading and vertical accumulation in sedimentary or lava sections, allowing independent checks on the completeness and relative spacing of the reversal sequence as well as the opportunity to apply an assortment of geochronologic data for calibration. Although different phenomena and assumptions are invoked in their derivation, both magneto- and biochronologic time estimates involve indirect assessment according to calculated rates of sea-floor spreading, sedimentation, and biotic evolution. These extend the application of the relatively few reliable radiometric dates available, so that a continuous geologic time scale can be inferred.

The degree of magnetochronologic resolution possible depends on the frequency of geomagnetic reversals and the availability of a well-defined record of the polarity sequence best developed in marine magnetic lineations (Vogt, this volume, Ch. 15). Thus the Campanian to Recent and the latest Oxfordian to early Aptian intervals of frequent reversals, corresponding to the mid-ocean ridge and the M-sequence magnetic lineations, respectively, allow the construction of a precise magnetochronologic framework. In contrast, the early Aptian to Santonian, and the Callovian to late Oxfordian intervals are of predominantly constant geomagnetic polarity. They correspond to the oceanic Cretaceous and Jurassic Quiet Zones, respectively, and magnetochronologic resolution is poor. A Sinemurian to Bathonian interval of frequent reversals has been documented in magnetostratigraphic land sections, primarily from the Mediterranean region. However, oceanic crust that might carry a magnetic anomaly signature of these early and middle Jurassic reversals is apparently not present. Consequently, the detailed sequence of reversals is poorly known for this time interval.

LATE CRETACEOUS AND CENOZOIC

The chronology and chronostratigraphy of this time interval is drawn directly from Berggren and others (1984a, b). In their work, bio- and magnetostratigraphy in some European Paleogene and Neogene stratotype sections are integrated and an assessment of some 200 Cenozoic and Late Cretaceous calcareous plankton datum events are directly correlated with magnetic polarity stratigraphy in deep-sea sediment cores and land sections. The data provide improved identification of the boundaries and durations of chronostratigraphic units in terms of planktic biostratigraphy and geomagnetic polarity chrons.

The geomagnetic polarity time scale is based on the radiometric dates and magnetic polarities on lavas for 0 to 4 Ma (Mankinen and Dalrymple, 1979) and is extended in time by age calibration of the polarity sequence inferred from marine magnetic anomalies. The polarity sequence compiled by LaBrecque and others (1977) is taken as representative of the sea-floor spreading record for the Late Cretaceous and Cenozoic. Six selected high-temperature radiometric ages are used for age calibration in such a way as to minimize apparent accelerations in sea-floor spreading history. These key ages are for Anomalies 2A (3.40 Ma), 5 (8.87 Ma), 12 (32.4 Ma), 13 (34.6 Ma), 21 (49.5 Ma) and 34 (84.0 Ma) (see Berggren and others, 1984a). Calculated ages for magnetic polarity intervals are shown in Table 1. Relative precision of boundaries in the reversal sequence depends

Kent, D. V., and Gradstein, F. M., 1986, A Jurassic to recent chronology; in Vogt, P. R., and Tucholke, B. E., eds., The Geology of North America, Volume M, The Western North Atlantic Region: Geological Society of America.

TABLE 1. REVISED GEOMAGNETIC POLARITY TIME-SCALE FOR OXFORDIAN TO RECENT TIME

Normal Polarity Interval (Ma)	Anomaly	Normal Polarity Interval (Ma)	Anomaly	Anomaly (Reversed)	Normal Polarity Interval (Ma)	Anomaly (Normal)
					- 118.00	Cretaceous Quiet Zone
0.00 - 0.73	1	24.04 - 24.21	6C	M0	118.70 - 121.81	
0.91 - 0.98		25.50 - 25.60	7	M1	122.25 - 123.03	M2
1.66 - 1.88	2	25.67 - 25.97	7	M3	125.36 - 126.46	M4
2.47 - 2.92	2A	26.38 - 26.56	7A	M5	127.05 - 127.21	
2.99 - 3.08	2A	26.86 - 26.93	8	M6	127.34 - 127.52	
3.18 - 3.40	2A	27.01 - 27.74	8	M7	127.97 - 128.33	
3.88 - 3.97	3	28.15 - 28.74	9	M8	128.60 - 128.91	
4.10 - 4.24	3	28.80 - 29.21	9	M9	129.43 - 129.82	
4.40 - 4.47	3	29.73 - 30.03	10	M10	130.19 - 130.57	
4.57 - 4.77	3	30.09 - 30.33	10		130.63 - 131.00	
5.35 - 5.53	3A	31.23 - 31.58	11		131.02 - 131.36	
5.68 - 5.89	3A	31.64 - 32.06	11	M10N	131.65 - 132.53	
6.37 - 6.50		32.46 - 32.90	12	M11	133.03 - 133.08	
6.70 - 6.78	4	35.29 - 35.47	13	M11	133.50 - 134.31	
6.85 - 7.28	4	35.54 - 35.87	13		134.42 - 134.75	
7.35 - 7.41	4	37.24 - 37.46	15	M12	135.56 - 135.66	
7.90 - 8.21	4A	37.48 - 37.68	15		135.88 - 136.24	
8.41 - 8.50	4A	38.10 - 38.34	16		136.37 - 136.64	
8.71 - 8.80		38.50 - 38.79	16	M13	137.10 - 137.39	
8.92 - 10.42	5	38.83 - 39.24	16	M14	138.30 - 139.01	
10.54 - 10.59		39.53 - 40.43	17	M15	139.58 - 141.20	
11.03 - 11.09		40.50 - 40.70	17	M16	141.85 - 142.27	
11.55 - 11.73	5A	40.77 - 41.11	17	M17	143.76 - 144.33	
11.86 - 12.12	5A	41.29 - 41.73	18	M18	144.75 - 144.88	
12.46 - 12.49		41.80 - 42.23	18		144.96 - 145.98	
12.58 - 12.62		42.30 - 42.73	18	M19	146.44 - 146.75	
12.83 - 13.01	5AA	43.60 - 44.06	19		146.81 - 147.47	
13.20 - 13.46	5AB	44.66 - 46.17	20	M20	148.33 - 149.42	
13.69 - 14.08	5AC	48.75 - 50.34	21	M21	149.89 - 151.46	
14.20 - 14.66	5AD	51.95 - 52.62	22		151.51 - 151.56	
14.87 - 14.96	5B	53.88 - 54.03	23		151.61 - 151.69	
15.13 - 15.27	5B	54.09 - 54.70	23	M22	152.53 - 152.66	
16.22 - 16.52	5C	55.14 - 55.37	24		152.84 - 153.21	
16.56 - 16.73	5C	55.66 - 56.14	24		153.49 - 153.52	
16.80 - 16.98	5C	58.64 - 59.24	25	M23	154.15 - 154.48	
17.57 - 17.90	5D	60.21 - 60.75	26		154.85 - 154.88	
18.12 - 18.14	5D	63.03 - 63.54	27	M24	155.08 - 155.21	
18.56 - 19.09	5E	64.29 - 65.12	28		155.48 - 155.84	
19.35 - 20.45	6	65.50 - 66.17	29		156.00 - 156.29	
20.88 - 21.16	6A	66.74 - 68.42	30	M25	156.55 - 156.70	
21.38 - 21.71	6A	68.52 - 69.40	31		156.78 - 156.88	
21.90 - 22.06	6AA	71.37 - 71.65	32		156.96 - 157.10	
22.25 - 22.35	6AA	71.91 - 73.55	32		157.20 - 157.30	
22.57 - 22.97	6B	73.96 - 74.01			157.38 - 157.46	
23.27 - 23.44	6C	74.30 - 80.17	33		157.53 - 157.61	
23.55 - 23.79	6C	84.00 -118.00	34		157.66 - 157.85	
				PM26	158.01 - 158.21	
				PM27	158.37 - 158.66	
				PM28	158.87 - 159.80	
				PM29	160.33 -(169.00)	Jurassic Quiet Zone

on the spatial resolution of the magnetic anomaly data and on the assumption that the compiled reversal sequence represents a linear and continuous record over time intervals of at least tens of million years (but see Vogt, this volume, Ch. 24). The accuracy of the reversal chronology ultimately depends on the radiometric age data set used for calibration. Remarkably, the first extended magnetochronology proposed by Heirtzler and others (1968), based on a simple extrapolation from Anomaly 2A (Gauss/Gilbert boundary), gives age estimates for magnetochrons that are within 10 percent of the absolute age estimates summarized here and based on more extensive magnetobiostratigraphic correlations and radiometric date calibration data. This agreement indicates that the constant-spreading rate-assumption applied to selected areas of the world ocean is a very good first-order approximation in the derivation of a geomagnetic reversal chronology.

The age-calibrated magnetic reversal sequence can then be used as a vernier (analogous to use of age-calibrated stratigraphic thickness) to obtain precise age estimates for various boundaries in accordance with magnetobiostratigraphic correlations. Numerical ages on the Late Cretaceous and Cenozoic geologic time-scale (Fig. 1; Plate 1, in pocket inside back cover) are therefore based on the revised magnetochronology summarized above.

MID-CRETACEOUS

The stratigraphic interval from the lower Aptian to the top of the Santonian records predominantly normal geomagnetic polarity; this nicely accounts for the Cretaceous Quiet Zones in the oceans (Lowrie and others, 1980). Consequently, there are no well-documented magnetozones or anomalies that can be correlated (Vogt, this volume, Ch. 15). Fortunately, abundant radiometric dates are available for this interval; numerical ages for stage boundaries (Fig. 1; Plate 1B) are therefore taken directly from the chronometric estimates of Harland and others (1982).

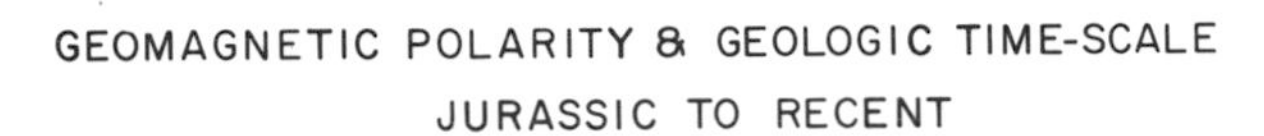

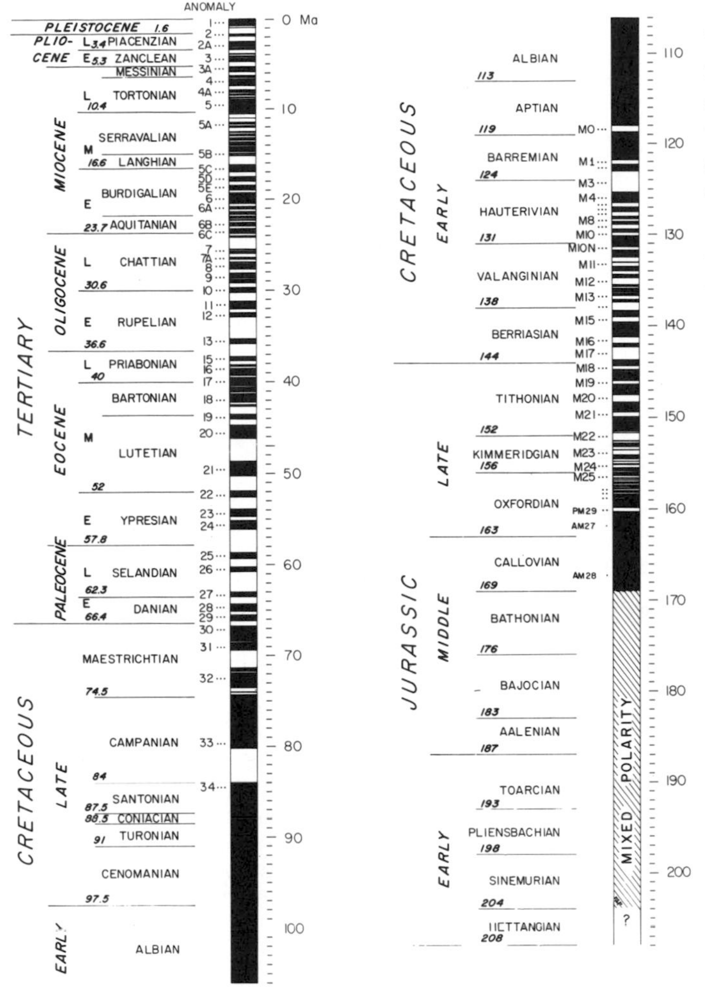

Figure 1. Geomagnetic polarity and geologic time scale for the Jurassic to recent. Methods of construction and sources of information are discussed in the text. Numerical ages of stage boundaries are indicated in millions of years (Ma).

LATE JURASSIC AND EARLY CRETACEOUS

The older end of the Cretaceous Quiet Zones in the oceans is bounded typically by lineated magnetic anomalies referred to as the M-sequence. These anomalies are best defined over the higher-spreading-rate systems in the Pacific but they are correlatable to the Keathley sequence of the North Atlantic (Larson and Chase, 1972; Larson and Pitman 1972). The standard magnetic reversal model for the M-sequence (M0 to M25 from youngest to oldest, designating key anomalies that are interpreted to correspond usually to reversed polarity) was derived from the Hawaiian lineations that are assumed to have formed at a constant rate of sea-floor spreading (Larson and Hilde, 1975). The M-sequence has been extended beyond M25 in the Pacific (Cande and others, 1978; PM26 to PM29 in Fig. 1 and Plate 1B) and in the Atlantic (Bryan and others, 1980; AM26 to AM28); the anomaly designations are not the same in both papers.

The chronologic control used by Larson and Hilde (1975) was based on biostratigraphic assignments of basal sediments of five DSDP holes drilled over identified M-anomalies; numerical ages were referred to the Geological Society of London (1964) time-scale. More exact magnetobiostratigraphic correlations have since become available (Heller, 1977; Lowrie and others, 1980; Channell and others, 1982; Ogg, 1983; Gradstein and Sheridan, 1983; Lowrie and Channell, 1984; Gradstein, this volume) and they provide a basis for a more refined chronology. The correlation of Late Jurassic and Early Cretaceous stage boundaries with the M-sequence geomagnetic reversal record (Figs. 1 and 2; Plate 1B) is based on the work cited above and is discussed in Kent and Gradstein (submitted).

Despite the improved magnetobiostratigraphic correlations, numerical age estimates for Late Jurassic and Early Cretaceous stages are still poor due to a lack of reliable radiometric dates. Harland and others (1982) adopted the "equal duration of stages concept" to interpolate between chronometrically determined age calibration tie-points at the Aptian/Albian boundary (113 Ma) and the Ladinian/Anisian boundary (238 Ma). To be consistent with our use of the Harland and others (1982) chronology for the Santonian to Albian, and to avoid an artificial discontinuity, we use their age estimates of 119 Ma for the Barremian/Aptian boundary and 156 Ma for the Oxfordian/Kimmeridgian boundary to calibrate the M-sequence.

The Barremian/Aptian falls within the M-sequence; it is the next (older) boundary from the Aptian/Albian which is considered by Harland and others (1982) to be the only chronometrically well-constrained tie-point (113 Ma) in the Early Cretaceous and the Jurassic. An isochron age of 120 Ma was determined for basalt overlain by lower Aptian sediment at DSDP Hole 417D, which was drilled on anomaly MO in the western North Atlantic (Ozima and others, 1979). This date is admittedly poor, but it nevertheless is consistent with the 119 Ma age estimate of Harland and others (1982) for the base of the Aptian. Armstrong (1978) interpolates whole-rock K/Ar dates to arrive at an age of approximately 156 Ma for the Oxfordian/Kimmeridgian boundary. This is identical to the broadly interpolated age derived by Harland and others (1982). Additional evidence for a 154–158 Ma age range for this boundary is found in the Sierra Nevada (California) (Schweickert and others, 1984).

For these reasons, we have accepted the age interpolations of 119 Ma (Barremian/Aptian) and 156 Ma (Oxfordian/Kimmeridgian) as reasonably well-defined tie-points for calibration of the M-sequence. We note, however, that these age estimates based largely on high-temperature mineral dates are older, by as much as 14–16 Ma in the case of the Oxfordian/Kimmeridgian boundary, than age estimates in Van Hinte (1976a, b) and Odin and others (1982); the discrepancy is principally in the fact that their age estimates are much more controlled by glauconite dates.

We used the magnetostratigraphic correlations, the assumption of a constant spreading rate on the Hawaiian lineations

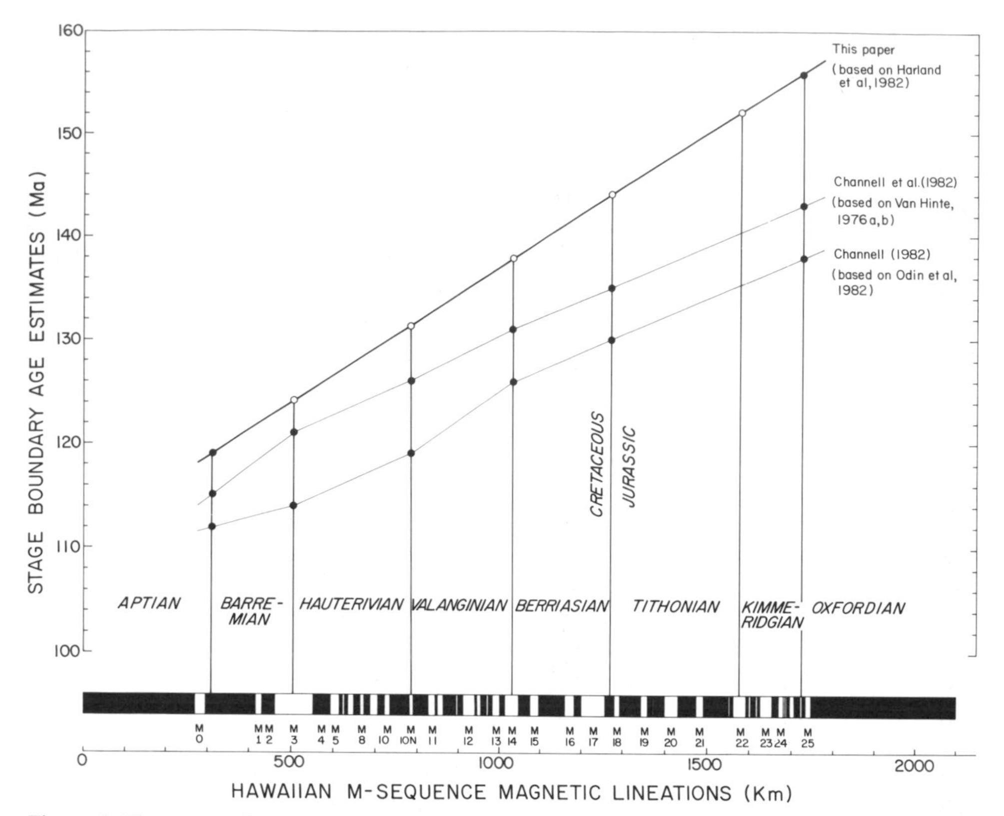

Figure 2. Three sets of age estimates for late Jurassic and early Cretaceous stages that have been similarly correlated to standard magnetic M-reversal sequence as derived from Hawaiian lineations (Larson and Hilde 1975). A constant spreading rate of 3.836 cm/yr is assumed here for the Hawaiian lineations based on age estimates (from Harland and others 1982) of 119 Ma and 156 Ma for the Barremian/Aptian and Oxfordian/Kimmeridgian boundaries, respectively; age estimates of intervening stage boundaries (open circles) are based on interpolation. In comparison, age calibration data used by Channell and others (1982) are from Van Hinte (1976a, b), and those used by Channel (1982) are from Odin and others (1982).

(Larson and Hilde, 1975), and the above calibration tie-points to derive age estimates for Kimmeridgian to Barremian stage boundaries (Fig. 2; Plate 1B) and for the M-sequence geomagnetic reversals (Table 1). This is simply another case of using the magnetic reversal sequence as a vernier to estimate ages of correlated boundaries between points of known or assumed age. In comparison to the equal-stage-duration model of Harland and others (1982), the magnetochronological method also gives approximately equal (ca. 6 m.y.) durations for the oldest four stages of the Early Cretaceous; however, it results in an apparently longer Tithonian (8 m.y.) and a shorter Kimmeridgian (4 m.y.). This factor of two ratio in relative duration of the Tithonian and Kimmeridgian stages is in good agreement with the relative duration inferred from assuming equal duration of ammonite zones (see below). Differences with previously published magnetochronologies for the M-sequence vary according to the degree of magnetobiostratigraphic data incorporated (e.g. Cox, 1982) and the geologic time-scale used in calibration (Fig. 2).

JURASSIC (PRE-KIMMERIDGIAN)

The older end of the M-sequence of anomalies is bounded by the Jurassic Quiet Zone that is represented by sea-floor areas of subdued magnetic signature. The boundary of the M-sequence with the Jurassic Quiet Zone is typically gradational or indistinct, characterized by a decreasing anomaly amplitude envelope from at least M22 to M25 and complicated by small-scale lineated anomalies extending the sequence to PM29 (Cande and others, 1978) or AM28 (Bryan and others, 1980).

Magnetostratigraphic studies on land-sections (summarized by Channel and others, 1982) indicate an interval of predominantly normal geomagnetic polarity for the Callovian and most of the Oxfordian. Thus reversed polarity magnetozones that might correlate to anomalies older than M25 have not yet been confidently identified in magnetostratigraphic sections. Whether this reflects inadequate magnetostratigraphic sampling or means that the small-amplitude, pre-M25 anomalies are not due to geomagnetic reversals is not yet clear.

Preliminary magnetostratigraphic work on Middle and Lower Jurassic land sections, primarily from the Mediterranean area, suggests that the Sinemurian to Bathonian stages are characterized by frequent geomagnetic reversals even though correlation of magnetozones between sections is difficult (Channell and others, 1982; Steiner and Ogg, 1983). No marine magnetic anomalies record this interval of frequent reversals at the oldest end of the Jurassic Quiet Zones, suggesting that the present oceanic crust is of post-Bathonian age (ca <169 Ma). However, at least in the Atlantic, this crust is deeply buried by sediment and any magnetic lineations would most likely already be low-amplitude and poorly developed.

For pre-Kimmeridgian Jurassic geochronology, the general lack of correlatable, lineated magnetic anomalies means that it is not possible to effectively use magnetochronology to calibrate the time-scale as we did for the Kimmeridgian-Aptian interval. We therefore apply bio-and radiochronologic methods for Early and Middle Jurassic stage age estimation, according to the following arguments.

We proceed from the age estimate of 156 Ma for the Oxfordian/Kimmeridgian boundary. The chronograms of Harland and others (1982) show that few radiometric dates exist to allow radiochronologic estimates of stage boundaries between the base of the Kimmeridgian and the base of the Jurassic. However, according to Armstrong (1982), at least part of the Sinemurian should be older than 203 Ma and some part of the Toarcian should fall in the age bracket of 185–189 Ma. The Triassic/Jurassic boundary age is inferred from another series of whole-rock cooling ages from Triassic and Jurassic beds in volcanogenic and sedimentary complexes in British Columbia (Canada) (Armstrong, 1982). Armstrong, taking into account all world data, suggests that the best available evidence places the base of the Jurassic at 208 Ma. This figure is just within the 200–208 Ma range of Odin and Letolle (1982) that is also based on high-temperature mineral radiometric dates. Harland and others (1982) estimate an age of 213 Ma for the Triassic/Jurassic boundary based on the interpolation between Middle Triassic and Middle Cretaceous using equal duration of stages; however, the 208 Ma estimate that we prefer is within the chronometric uncertainty they calculate. We therefore accept 208 Ma as a reasonable estimate for the Triassic/Jurassic boundary.

The Jurassic chronology is then built on the radiometric age constraints of 208 Ma for the Jurassic/Triassic boundary and an Oxfordian/Kimmeridgian boundary age of 156 Ma. We use an interpolation mechanism that also was adopted by Van Hinte (1976b), the equal duration of zones, but we use the updated ammonite zonation advocated by Hallam (1975). There are 50 zones between the base of the Hettangian (208 Ma) and the top of the Oxfordian (156 Ma), which is 1.04 m.y. per zone. Skeptics will point to the uncertainties in this number, which is only an average, but we believe it is less crude a means for apportioning time than the equal duration of stages assumed by Harland and others (1982). In fact, Harland and others (1982) justify their equal-duration criterion by evolutionary turnover; thus it may be better to assume equal duration of zones, which are the shortest (bio) stratigraphic building blocks.

Above the Oxfordian the ammonite zonation is less well established, particularly because latitudinal provincialism creates more of a problem in correlation. However, the Kimmeridgian may have four to six zones and the Tithonian seven to nine zones (Hallam, 1975); this ostensibly requires a Tithonian stage twice as long as the Kimmeridgian and is consistent with the magnetochronological estimates given above.

From this information, we derive boundary age estimates for Jurassic stages as shown in Figure 1 and Plate 1B. The older age limit of the Sinemurian (204 Ma) falls just within the age constraint of Armstrong (1982), noted earlier, and at least part of the Toarcian (193–187 Ma) is in Armstrong's 185–189 Ma range. According to our equal-zone-duration method of interpolation, Jurassic stages vary in duration by a factor of 2, but of course the average duration of the 11 Jurassic stages (5.7 m.y.) is still very near the average (ca. 6 m.y.) assumed by Harland and others (1982) in their interpolation. Our Jurassic stage-boundary age estimates therefore tend to differ at most by 5 m.y. (at the Triassic/Jurassic boundary) from Harland and others (1982) and usually they fall within 2 m.y. The question of what accuracy in time the new scale achieves will be answered only when many more well spaced and stratigraphically meaningful radiometric dates become available. At the moment time resolution, although not accuracy, is going to be far better when a biochronologic and magnetochronologic framework is used rather than radiochronology alone.

As a final note, we place age constraints with our time-scale on the pre-M25 small-amplitude magnetic anomalies. PM29 (Cande and others, 1978) has an age estimated by extrapolation from the M-sequence of ca. 160 Ma (Table 1) which should place it within the mid-Oxfordian (Fig. 1; Plate 1B). AM26 (Bryan and others, 1980) probably corresponds to PM29 (Cande and others, 1978). A minimum age for AM27 (Bryan and others, 1980) may be derived from the pinch-out of seismic reflector D on basement at AM27. Drilling at DSDP Site 534 (Sheridan and others 1983) shows D to be approximately early Oxfordian in age. AM27 is thus no younger than early Oxfordian. As also shown by Site 534 drilling on crust at anomaly AM28, this anomaly is likely to be of early or middle Callovian age.

REFERENCES

Armstrong, R. L.

1978 : Pre-Cenozoic Phanerozoic time scale—computer file of critical dates and consequences of new and in-progress decay-constant revisions; in Contributions to the geologic time scale, eds., G. Cohee, M. Glaessner, and H. Hedberg; AAPG, Tulsa, Okla, p. 73–91.

1982 : Late Triassic–Early Jurassic time-scale calibration in British Columbia, Canada; in Numerical dating in stratigraphy, ed. G.S. Odin; New York: John Wiley & Sons Ltd., p. 509–513.

Berggren, W.A. and Van Couvering, J. A.

1974 : The Late Neogene: Biostratigraphy, geochronology and paleoclimatology of the last 15 million years in Marine and Continental Sequences: Paleogeography, Paleoclimatology, and Paleoecology, v. 16, p. 1–216.

Berggren, W. A., Kent, D. V., and Flynn, J. J.

1984a: Paleogene chronology and chronostratigraphy: Geological Society of London, in press.
Berggren, W. A., Kent, D. V., and Van Couvering, J. A.
1984b: Neogene chronology and chronostratigraphy: Geological Society of London, in press.
Bryan, G. M., Markl, R. G., and Sheridan, R. E.
1980 : IPOD site surveys in the Blake-Bahama Basin: Marine Geology, v. 35, p. 43–63.
Cande, S. C., Larson, R. L., and LaBrecque, J. L.
1978 : Magnetic lineations in the Pacific Jurassic quiet zone: Earth and Planetary Science Letters, v. 41, p. 434–440.
Channell, J. E. T., Ogg, J. G., and Lowrie, W.
1982 : Geomagnetic polarity in the early Cretaceous and Jurassic: Philosophical Transactions of the Royal Society of London, v. A306, p. 137–146.
Cox, A. V.
1982 : Magnetic Reversal Time-Scale; in A geological time scale, W. R. Harland and others; Cambridge, Eng.: Cambridge University Press, 128 p.
Geological Society of London
1964 : Geological Society Phanerozoic time scale 1964: Geological Society of London Quarterly Journal, v. 120, p. 260–262.
Gradstein, F. M., and Sheridan, R. E.
1983 : On the Jurassic Atlantic Ocean and a synthesis of results of deep sea drilling project leg 76; in Initial reports of the deep sea drilling project, Volume 76, eds., R. E. Sheridan and F. M. Gradstein; Washington, D.C., U.S. Government Printing Office, p. 913–943.
Hardenbol, J., and Berggren, W. A.
1978 : A new Paleogene numerical time scale; in Contributions to the geologic time scale, eds. G. Cohee, M. Glaessner, and H. Hedberg; American Association of Petroleum Geologists, Tulsa, Oklahoma, p. 213–234.
Hallam, A.
1975 : Jurassic environments: London Cambridge University Press, Cambridge, Eng., 269 p.
Harland, W. B., Cox, A. V., Llewellyn, P. G., Pickton, C. A. G., Smith, A. G., and Walters, R.
1982 : A geologic time scale; Cambridge, Eng., Cambridge University Press, 128 p.
Heirtzler, J. R., Dickson, G. O., Herron, E. M., Pitman, W. C. III. and Le Pichon, X
1968 : Marine magnetic anomalies, geomagnetic field reversals, and motions of the ocean floor and continent: Journal of Geophysical Research, v. 73, p. 2119–2136.
Heller, F.
1977 : Palaeomagnetism of Upper Jurassic limestones from Southern Germany: Journal of Geophysics, v. 42, p. 475–488.
Kent, D. V. and Gradstein, F.
Submitted: A Cretaceous and Jurassic geochronology: Geological Society of America Bulletin.
LaBrecque, J. L., Kent, D. V., and Cande, S. C.
1977 : Revised magnetic polarity time scale for the Late Cretaceous and Cenozoic Time: Geology, v. 5, p. 330–335.
Larson, R. L., and Chase, C. G.
1972 : Late Mesozoic evolution of the western Pacific: Geological Society of America Bulletin, v. 83, p. 3627–3644.
Larson, R. L., and Hilde, T. W. C.
1975 : A revised time scale of magnetic anomalies for the Early Cretaceous and Late Jurassic: Journal of Geophysical Research, v. 80, p. 2586–2594.
Larson, R. L., and Pitman, W. C. III
1972 : Worldwide correlation of Mesozoic magnetic anomalies, and its implications: Geological Society of America Bulletin, v. 83, p. 3645–3662.
Lowrie, W., and Channell, J. E. T.
1984 : Magnetostratigraphy of the Jurassic-Cretaceous boundary in the Maiolica Limestone (Umbria, Italy): Geology, v. 12, p. 44–47.
Lowrie, W., Channell, J. E. T., and Alvarez, W.
1980 : A review of magnetic stratigraphic investigations in Cretaceous pelagic carbonate rocks: Journal of Geophysical Research, v. 89, p. 3597–3605.
Mankinen, E. A., and Dalrymple, G. B.
1979 : Revised geomagnetic polarity time scale for the interval 0–5 m.y. B.P.: Journal of Geophysical Research, v. 84, p. 615–626.
Odin, G. S., and Letolle, R.
1982 : The Triassic time scale in 1981; in Numerical dating in stratigraphy, ed. G. S. Odin; New York, John Wiley & Sons, Ltd., p. 523–533.
Odin, G. S., Curry, D., Gale, N. H., and Kennedy, W. J.
1982 : The Phanerozoic time scale in 1981: in Numerical Dating in Stratigraphy, ed. G. S. Odin; New York, John Wiley & Sons, Ltd., p. 957–960.
Ogg, J. G.
1983 : Magnetostratigraphy of Upper Jurassic and Lower Cretaceous sediments. DSDP Site 534, Western North Atlantic; in Initial Reports of the deep sea drilling project, v. 76, Sheridan R. E., and others; Washington, D.C., U.S. Government Printing Office, p. 685–697.
Ozima, M., Kaneoka, I., and Yanaqisawa, M.
1979 : 40 AR=40 AR=39 AR Geochronological studies of drilled basalts from leg 51 and 52; in Initial reports of the deep sea drilling project, v. 51, 52, 53, eds. T. Donnelly, J. Francheteau, P. Robinson, M. Flower and M. Salisbury; Washington, D.C., U.S. Government Printing Office, p. 1127–1128.
Schweickert, R. A., Bogen, N. L., Girty, G. H., Hanson, R. E., and Merquerian, C.
1984 : Contrasting styles of deformation during the nevadan Orogeny: Geological Society of America Bulletin, v. 95, p. 967–979.
Sheridan, R. E., Gradstein, F. M., et al.
1983 : in Initial reports of the deep sea drilling project, v. 76, U.S. Government Printing Office, Washington, D.C., 949 p.
Steiner, M. B. and Ogg, J. G.
1983 : Jurassic magnetic polarity time scale: EOS, v. 45, p. 689.
Van Couvering, J. A., and Berggren, W. A.
1977 : Biostratigraphical basis of the Neogene time scale; in Concepts and methods in biostratigraphy, eds. E. G. Kauffman and J. E. Hazel: Stroudsburg, Pa., Dowden, Hutchinson and Ross, Inc., p. 283–306.
Van Hinte, J. E.
1976a: A Cretaceous time scale: American Association of Petroleum Geologists Bulletin, v. 60, p. 498–516.
1976b: A Jurassic time scale: American Association of Petroleum Geologists Bulletin, v. 60, p. 489–497.

MANUSCRIPT ACCEPTED BY THE SOCIETY DECEMBER 3, 1984
LAMONT-DOHERTY GEOLOGICAL OBSERVATORY CONTRIBUTION NO. 3522

ACKNOWLEDGMENTS

We thank J. Ogg and G. Westermann for valuable stratigraphic advice, J. Channell and J. LaBrecque for constructive criticism of the manuscript, and especially B. Tucholke for careful editorial review. Financial support was provided by the U.S. National Science Foundation, Submarine Geology and Geophysics (DVK) and the Geological Survey of Canada (FMG).

The Geology of North America
Vol. M, The Western North Atlantic Region
The Geological Society of America, 1986

Chapter 4

The crest of the Mid-Atlantic Ridge: Models for crustal generation processes and tectonics

Ken C. Macdonald
Department of Geological Sciences and Marine Science Institute, University of California at Santa Barbara, Santa Barbara, California 93106

HISTORICAL INTRODUCTION

By 1854, there were enough deep ocean soundings to allow M. F. Maury to draw the first bathymetric chart of the North Atlantic Ocean in 1855. He defined the great shoaling "middle ground" of the Atlantic basin, later to be known as the Mid-Atlantic Ridge (MAR). It is interesting to note that during the Challenger expedition (1872–1876), Sir Wyville Thomson predicted, on the basis of water temperatures alone, that the MAR is a nearly continuous topographic barrier dividing the Atlantic Ocean. In the 1950s, continuous echogram profiles revealed the presence of a deep median valley along the axis of the MAR (Heezen and others, 1959; Hill 1960). Seismicity (Gutenburg and Richter, 1954) and the occurrence of youthful basaltic lavas on the MAR (Shand, 1949) indicated that the median valley was a geologically active rift zone. Heezen (1960) first suggested that extension across the rift valley might be responsible for accretion of basaltic oceanic crust and the drift of the continents, but his contention that this rifting resulted from expansion of the earth was incorrect. High heat flow measured on the MAR (Bullard and Day, 1961) and large axial magnetic anomalies (Heezen and others, 1959) were added to seismicity and recent volcanism as enigmas associated with the rift valley.

In his seafloor spreading hypothesis Hess (1962) elegantly explained these puzzling observations in a unified model, and the bilateral symmetry of magnetic anomalies found on other mid-ocean ridges (Vine and Matthews, 1963; Vine, 1966) elevated seafloor spreading to a widely accepted theory. However, there were early indications that the MAR from 10 to 50°N was problematical. During the flurry of magnetic anomaly studies in the 1960s that demonstrated seafloor spreading in one ocean basin after another, magnetic anomalies over the "type" mid-ocean ridge in the northern Atlantic eluded a clear simple seafloor spreading interpretation, except over the Reykjanes Ridge.

There was little doubt, however, that the MAR had to be a major oceanic spreading center and that the median valley must mark the axis of spreading. To test this hypothesis, the first extensive study of the MAR median valley was launched by Canadian investigators at 45°N based on Hill's (1960) earlier work in the area (Fig. 1). In the next several years, the ridge at 45°N was studied in unprecedented detail, including high density magnetic, gravity, and bathymetric surveys (e.g., Loncarevic and others, 1966; Loncarevic and Parker, 1971), seismic refraction (e.g., Keen and Tramontini, 1970), and extensive sampling and experimental work (e.g., Aumento, 1968; Barrett and Aumento, 1970; Irving, 1970). Numerous petrologic and geochemical studies yielded the first testable models for oceanic crustal formation (e.g., Cann, 1970; Aumento and others, 1971). At about the same time, significant but less intensive studies were carried out at 23°N (van Andel and Bowin, 1968) and at 43°N (Phillips and others, 1969).

In spite of these detailed and carefully designed studies, the geophysical structure, fine-scale tectonics and petrogenesis of the MAR remained obscure. For example, correlatable magnetic anomalies, the fundamental measure of seafloor spreading, were still poorly resolved, and even the most sophisticated techniques for analyzing magnetic data yielded only marginal results (Loncarevic and Parker, 1971). Many of the unanswered questions were addressed by more advanced technology that became available in the late 1960s and early 1970s. In the Trans-Atlantic geotraverse program (TAG), new efforts included long, high-resolution geophysical profiles followed by detailed narrow-beam mapping and sampling of the MAR from 24 to 27°N (Fig. 1). A dive program followed in 1982 using the research submersible ALVIN (Rona and others, 1982).

The most ambitious and comprehensive MAR program was project FAMOUS at 37°N (Heirtzler and Le Pichon, 1974). This program centered around the first use of manned submersibles in mid-ocean ridge exploration, but many other technologies were also used and tested during FAMOUS (French-American Mid-Ocean Undersea Study). These included deep-tow, ANGUS (Acoustically Navigated Geologic Underwater System), ocean bottom seismometer refraction, transponder navigated dredging, GLORIA long-range side-scan sonar, SASS multi-beam echo-sounding, LIBEC large area photography, and near-bottom mag-

Macdonald, K. C., 1986, The crest of the Mid-Atlantic Ridge: Models for crustal generation processes and tectonics; in Vogt, P. R., and Tucholke, B. E., eds., The Geology of North America, Volume M, The Western North Atlantic Region: Geological Society of America.

netic and thermal measurements. Although the new techniques elucidated and more carefully defined many of the vexing problems concerning MAR tectonics, some of the old problems remained and many new ones arose as we investigated crustal generation processes on a finer scale. The MAR seemed fairly simple in the mid to late 1960s once it was recognized as a major plate boundary where the oceanic crust is accreted within and beneath the rift valley. Resolving the tectonics on the scale of terrestrial field investigations, however, proved to be a difficult and rewarding challenge.

In this chapter I shall discuss the detailed investigations of the MAR from 36 to 37°N carried out during the FAMOUS (1973-74) and AMAR (ALVIN Mid-Atlantic Ridge) (1978) ALVIN expeditions. In addition, reference will be made to work from 45°N, 26°N and 23°N, and to recent studies near the Oceanographer, Kane, and Vema fracture zones. The following topics will be discussed: (1) gross structure of the MAR, (2) tectonics and volcanism within the rift valley, (3) models for the axial magma chamber, (4) disruption of spreading center processes by closely spaced transform faults.

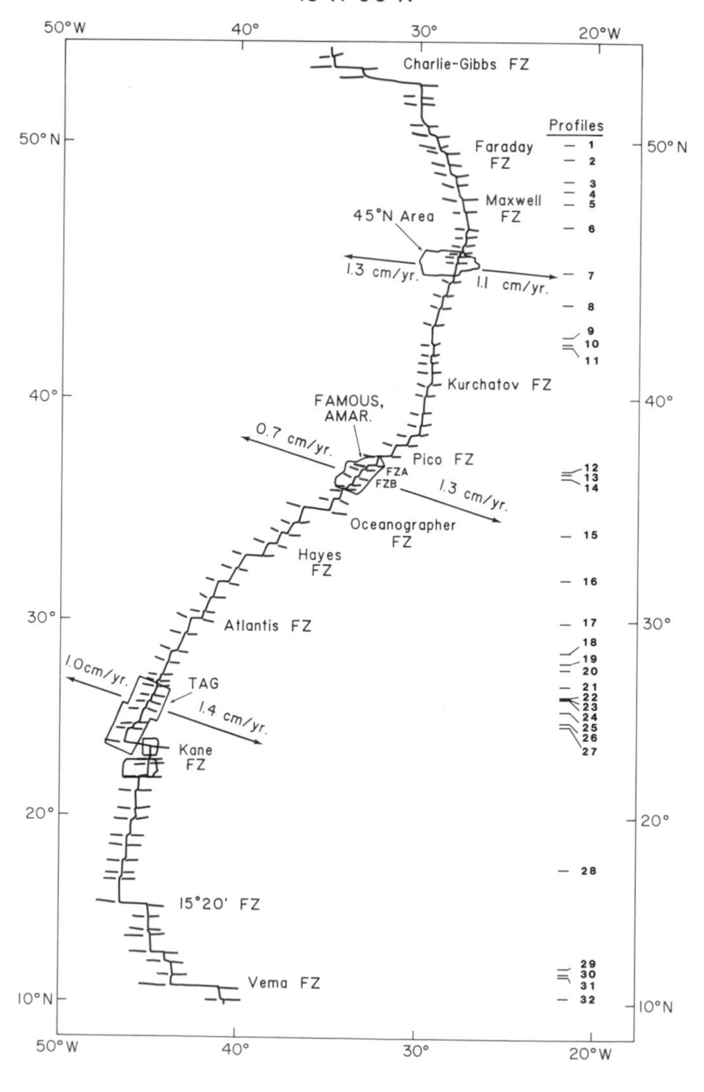

Figure 1. Detailed information of the Mid-Atlantic Ridge plate boundary based on detailed surveys (areas enclosed in boxes) and on GEBCO charts. Detailed study areas discussed in the text are shown with local spreading rates indicated. Fracture zones are shown in red. The latitudes of numbered median rift profiles in Figure 3 are indicated on the right. Notice how frequently the MAR is interrupted by transform faults. Many of the transform faults offset the ridge only a short distance (<30 km) and appear as bends in the overall trend of the MAR. Only the first 100–200 km of the fracture zone traces are shown in red. Many of the traces are not colinear with the associated transform faults suggesting minor shifts in the locations of the transform faults as shown in Figure 10D (see section 4 for discussion.)

1. GROSS STRUCTURE OF THE MID-ATLANTIC RIDGE

The large-scale geomorphology of oceanic spreading centers seems to be controlled primarily by total opening rate, which varies from about 1 to 16 cm/yr. At the slow total opening rates of 1 to 3 cm/yr characteristic of the MAR in the north Atlantic, a deep rift valley marks the spreading center (Figs. 2, 3). At spreading rates of 3 to 5 cm/yr (slow-intermediate transition), characteristic of the south Atlantic and parts of the Indian Ocean, an axial rift valley is well developed only near transform faults, shoaling significantly and almost disappearing between transform faults. The median valley is a "nested" rift, with an inner floor bounded by block-faulted inner walls (Needham and Francheteau, 1974). Beyond are faulted but relatively horizontal terraces bounded by the outer walls (Macdonald, 1975) (Fig. 4). The extremely rough and faulted topography created in the rift valley is largely preserved in the ocean basin, diminished only by sediment burial. At intermediate opening rates of 5 to 9 cm/yr, the rift valley, if present, is only 50 to 200 m deep. This shallow rift is

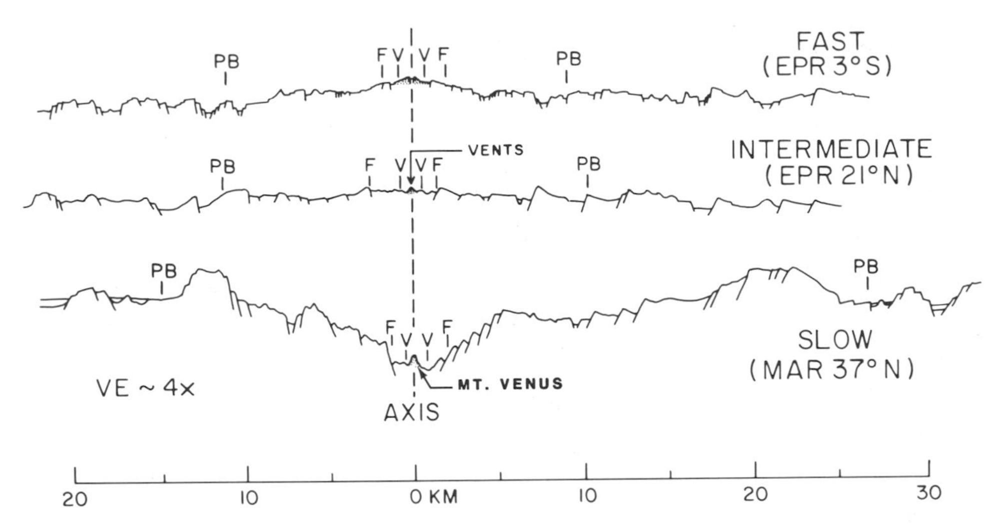

Figure 2. High resolution deep-tow profiles of the Mid-Atlantic Ridge rift valley in the FAMOUS area (bottom) with profiles of the intermediate and fast spreading parts of the East Pacific Rise for comparison. Neovolcanic zone bracketed by Vs, zone of fissuring by Fs, plate boundary zone (width of active fault zone) by PBs. Data from Macdonald and Luyendyk (1977), Lonsdale (1977), Normark (1976), and Shih (1980).

superposed on a broad axial high and the flanking topography is relatively smooth (Fig. 2). At fast spreading rates (greater than 9 cm/yr) there is no rift valley: instead a triangular, semi-circular or rectangular-shaped axial high is observed (e.g., EPR south of 15°N; Rea, 1978; Macdonald and others, 1984). The topography is relatively smooth with a fine-scale horst and graben structure.

Careful inspection of the regional bathymetry of the MAR (Plate 2) reveals that the rift valley is continuous except near major hot spots. The ubiquity of a rift valley on all non-hot spot slow spreading ridges suggests that it is a steady state structure (Fig. 5) (Deffeyes, 1970; but see Vogt, this volume, Ch. 24). If this is true, a problem arises in explaining the disappearance of the axially dipping regional slope of the rift valley as it is transformed into the horizontal undulating relief of the rift mountains (Harrison and Watkins, 1977; Verosub and Moores, 1981). One model is that the rift valley walls are tilted back to approximately horizontal in the rift mountains. The rift valley/rift mountain transition may be accomplished by a modest rotation (5–9°) of the entire rift valley half section as it passes into the rift mountains. A second model is that the rift valley is undone by "unfaulting" along pre-existing inward-facing faults (Harrison, 1974). Thus, as new normal faults are created near the center of the valley, relict normal faults are collapsed by reverse faulting at the valley edges. A third hypothesis is that the rift valley staircase is effectively over-printed by normal faults that dip away from the valley axis (Macdonald and Atwater, 1978a). These may be new fault planes or reactivated faults that previously had small offsets. All three processes must be acting in the rift mountains to maintain a steady state rift valley (Macdonald and Atwater, 1978a; Harrison and Stieltjes, 1977). These processes are related to a significant tilting of fault blocks away from the rift axis by at least 10–30° (Macdonald and Atwater, 1978b). This tilting is either manifested by flexure and resulting failure of the crust by outward-dipping normal faulting, or by simple tilting of crustal blocks (Macdonald and Atwater, 1978b). These processes occur as far as 30 kilometers off axis, which accounts for the greater width of the plate boundary zone on slow spreading ridges (approximately 60 kilometers), compared with fast spreading centers (about 8–20 kilometers) (Shih, 1980).

There is still considerable debate over, the size, occurrence and mechanisms of Atlantic oceanic crust rotations. One hypothesis is that the rotations may be surprisingly large, and that significant outward tilting may be accomplished by rotation along listric normal faults that occur in the rift valley and rift mountains. A listric normal fault is one that is curved and concave upward, such that the fault dip is high or medium angle (greater than 45°) near the surface, decreasing to near horizontal at depth. However, it should be noted that there is little evidence from ophiolites for crustal tilts exceeding 60° as suggested by Verosub and Moores (1981).

Bathymetric profiles of the rift valley reveal that, while the valley is present nearly everywhere, its shape, depth and width may vary considerably (Fig. 3). Depth of the valley floor relative to the rift mountains may vary in the extreme, from 1.0 to 2.8 km, with a mean value of 1.4 km (Fig. 5). The width between the shallowest points of the bounding scarps generally varies from 20 to 40 km. The variation in depth to the floor of the rift valley is related to proximity to transform faults.

Any model for the creation of the MAR rift valley must account for its extraordinary depth and its disappearance at intermediate spreading rates. I have compiled data from the MAR, the East Pacific Rise, (EPR), the Indian Ocean Ridge, and the Pacific Antarctica Ridge to investigate the disappearance of the rift valley as a function of spreading rate (Fig. 5). It is almost always present (except for major hot spots) at spreading half-rates lower than 1.7 cm/yr. At spreading half-rates above 4.5 cm/yr,

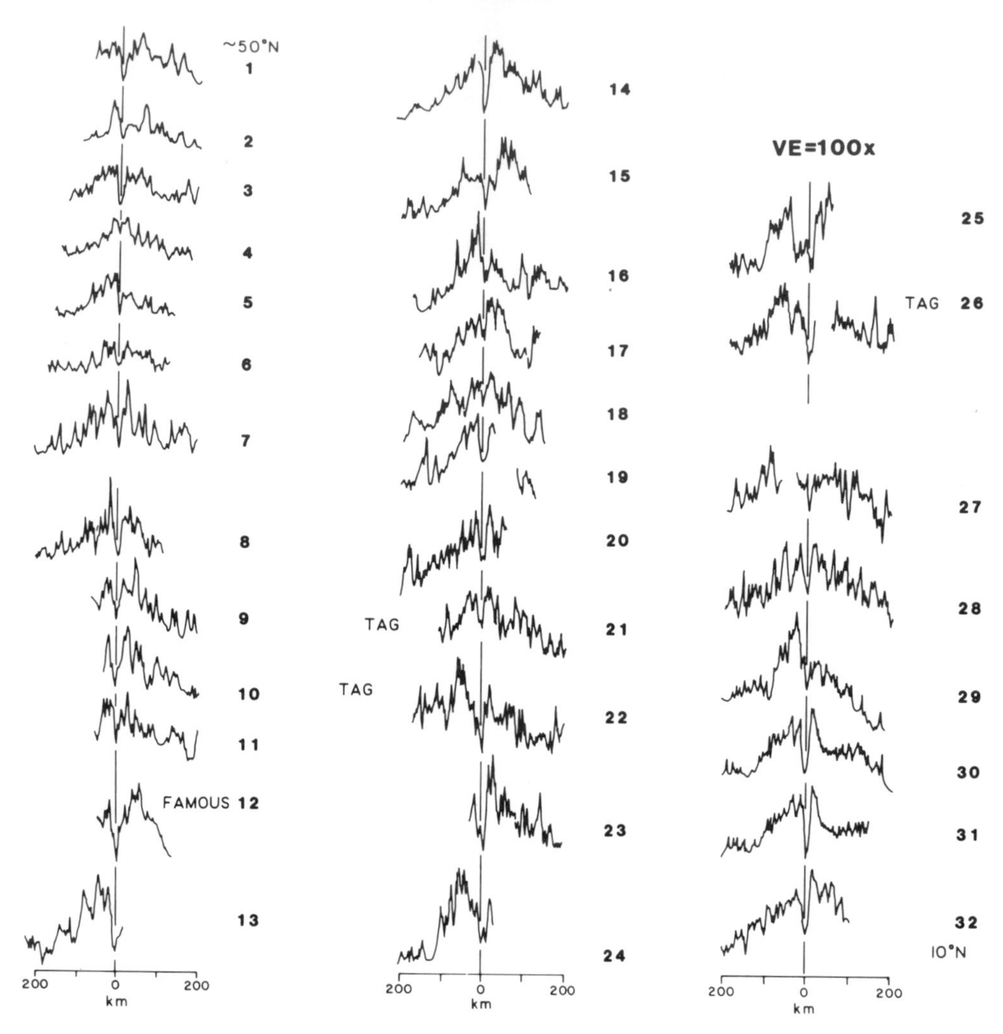

Figure 3. Profiles of the MAR rift valley from 10° to 50°N. All are surface vessel echo-sounding profiles except for profile 12 in the FAMOUS area (which is the same as the deep-tow MAR profile in Figure 2). Notice how the median rift valley is ubiquitous, its depth varying from 1.0 to 2.8 km. On profile 4 the valley is partially filled in by a seamount. The shape of the valley varies from a V-shaped to a U-shaped structure as discussed in the text.

an axial high is always present, while for half-rates between 1.7 and 4.5 cm/yr the axis may have either a rift valley or a gentle axial high with a very shallow summit rift zone (Figs. 2 and 5).

The transition from rift valley to axial high as a function of spreading rate is more gradual than one might conclude from the data in Figure 5. For spreading half-rates exceeding 1.7 cm/yr, a rift valley greater than 300 m deep occurs only within 20 km of a triple junction or transform fault intersection. A good example is the EPR near 23°N where a 600 m deep rift valley occurs 12 km away from the intersection of the EPR with the Tamayo Fracture zone. It is indistinguishable from the inner rift of the FAMOUS area (see Fig. 6 in Macdonald and others, 1979 and compare with Figures 2 and 3). Yet, only 30 km from the intersection, the EPR is marked by a clear axial high. In addition, many of the profiles in Figure 5 occur near ridge-transform intersections. Thus a model for rift valley dynamics must account for a gradual disappearance of the rift valley at spreading half-rates exceeding 1.7 cm per year, and the persistence of a rift valley at rates up to 9.0 cm per year near transform fault or triple junction intersections.

There is a plethora of rift valley models. Of the kinematic models, there are several that explain rift valleys and axial peaks as caused by imbalances in the width of the crustal accretion zones versus the width of the lithospheric acceleration zones (Deffeyes, 1970; Anderson and Noltimier, 1973; Nelson, 1981; Reid and Jackson, 1981). These models suggest that creation of the oceanic crust takes place over a wider region than the active tectonic zone, that is, an area wider than 60 km. This contradicts *Alvin* observations and deep-tow magnetic inversion studies that place upper bounds on the width of the zone crustal formation (Ballard and van Andel, 1977; Macdonald, 1977). Pálmason's (1980; this volume) kinematic model does not encounter this contradiction, but does not address the problem of creating a deep rift valley since one does not exist in Iceland.

The two most applicable models are the *viscous head loss*

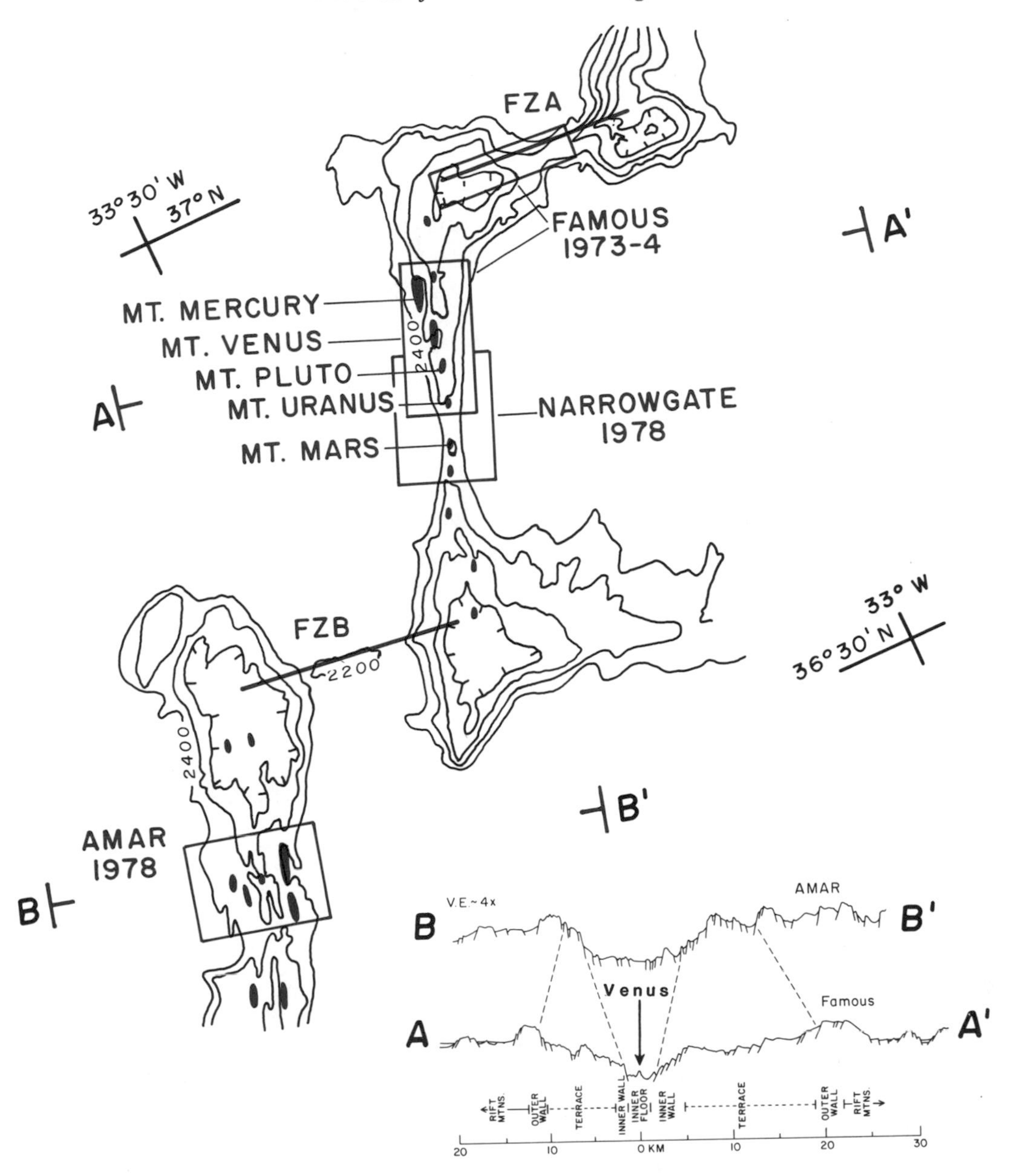

Figure 4. Charts of the FAMOUS and AMAR rifts, contour interval 200 m (see also Vogt, this volume Ch. 24). The rift segment from FZA to FZB is referred to in the literature as the *FAMOUS* rift or *rift valley 2,* while the segment south of FZB is known as the *AMAR* rift or *rift valley 3.* Areas of intensive study using submersibles are shown in red boxes. *Alvin* studies were also carried out in the rift mountains and on FZB, while other instruments such as GLORIA, Deep-Tow, and ANGUS were used over a much broader area. Inner floor volcanoes are shown as red patches and the inferred active transform fault traces are shown as red lines. Many of the volcanoes in the AMAR rift are relatively old and fault-bounded. Profile A–A′ of the FAMOUS rift is typical of a V-shaped valley cross-section with a narrow inner floor and wide terraces, while deep-tow profile B–B′ of the AMAR rift is typical of a U-shaped valley, a cross-section with a wide inner floor and narrow terraces. An evolutionary sequence may exist with the U- and V-shaped rift valleys as end-members in which the walls bound the width of the neovolcanic zone and influence the fidelity of the recording of magnetic anomalies. Note that the FAMOUS and AMAR rifts are 20° oblique to FZA and FZB, however they are perpendicular to the Minster and Jordan (1978) pole of opening for the North Atlantic. Hence oblique spreading here is only apparent. The short offset transform faults have a component of extension across them and do not trace small circles about the pole of opening. The resulting fracture zones, however do trace small circle paths. Note also that the AMAR and FAMOUS rifts appear to overlap at FZB. However, the neovolcanic zones have only a minimal overlap, and the large apparent overlap may be due to north-south lateral shifts of FZB as shown in Figure 9D.

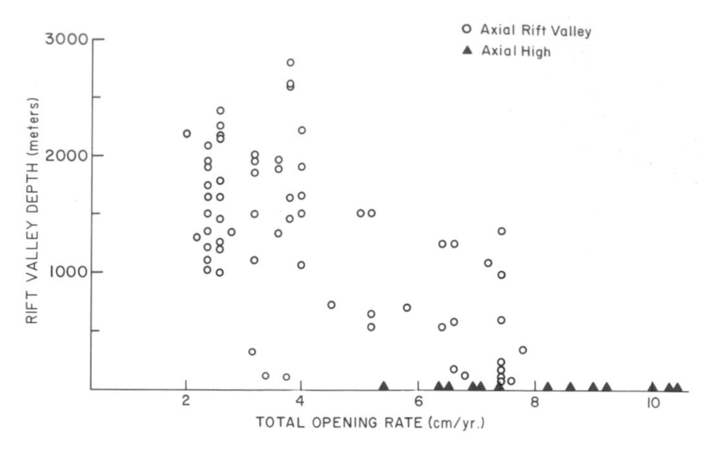

Figure 5. Depth of the median valley as a function of spreading rate. A 1.0–2.8 km deep valley is ubiquitous for spreading half-rates less than 1.7 cm/yr and an axial high is always present at half-rates above 4.5 cm/yr (solid red symbols). Between 1.7 and 4.5 cm/yr a rift valley may occur, but generally near transform fault intersections or triple junction intersections. Thus the disappearance of the rift valley as a function of spreading rate (away from transform or triple junction intersections) is gradual, occurring between half-rates of 1.7 and 4.5 cm/yr.

model of Sleep and others (Sleep, 1969; Lachenbruch, 1973, 1976; Sleep and Rosendahl, 1979), and the *steady-state necking* model (Tapponier and Francheteau, 1978). According to Sleep's model, along the sides of an idealized conduit tapping upwelling limbs of asthenospheric material, viscous forces are sufficient to cause a significant loss of hydraulic head, resulting in a topographic depression over the spreading center. To conserve energy, this head loss is regained by uplift of the rift valley walls relative to the valley floor. The head loss is directly proportional to the upwelling velocity and inversely proportional to the cube of the conduit width. A deep rift valley is present at slow spreading rates because material upwells through a relatively narrow conduit formed by the frozen lithosphere. At faster spreading rates, there is far less head loss because a wider conduit overwhelms the dependence on flow rate. The rift valley is replaced by flat topography or a crestal peak that may be caused by the presence of a low density crustal magma chamber. Thus, the cross section of fast spreading ridges closely approaches that expected from buoyancy in a relaxed state while slow spreading ridges exhibit significant dynamic effects.

In the "steady-state necking" model the rift valley is caused by necking or thinning in a ductile layer beneath the rift valley. The analogy is that of a beam plastically necking-out under tensional stress. The layer does not actually break like a necking beam because new material is added constantly from below (maintaining steady-state), while the entire region is continually uplifted by buoyancy forces. At slow spreading rates the strength of the crust and the axial zone is presumed great enough for necking to be significant, creating a rift valley. At faster rates, the young crust is too hot and weak at shallow levels for this process to be significant. There is still considerable controversy over which models best describe rift valley dynamics.

While the large scale bilateral symmetry of the MAR suggests a symmetric spreading history, high resolution magnetic studies reveal that spreading is generally asymmetric. For example, in the FAMOUS area, the spreading rate has been grossly symmetrical for 10 m.y. (Bird and Phillips, 1975). However, deep-tow profiles reveal that the grossly symmetrical spreading is composed of episodes of highly asymmetric spreading that reverse in sense. At 37°N the spreading rates for 0–2 Ma are 0.7 cm/yr to the west and 1.3 cm/yr to the east, while the rates for 2–4 Ma are 1.3 cm/yr to the west and 1.0 cm/yr to the east (Macdonald, 1977). Asymmetric spreading that reverses in sense through time results in large scale symmetry when averaged over 10^7 years. Asymmetric spreading for crust 0–5 m.y. old in the Atlantic has been reported at 45°N (Loncarevic and Parker, 1971), 26°N (McGregor and Rona, 1975) and 6°S (van Andel and Heath, 1970). Adjacent segments of the MAR often have asymmetric spreading in opposite senses requiring the intervening transform fault to lengthen or shorten. Long term persistent asymmetry in spreading rate and topography has not been reported for the MAR except north of Iceland (Vogt and others, 1982).

2. TECTONICS AND VOLCANISM WITHIN THE RIFT VALLEY

The Neovolcanic Zone

The neovolcanic zone is the area straddling the spreading axis in which most of the recent and ongoing surficial volcanism occurs. On the MAR it is characterized by a highly discontinuous chain of volcanoes that are elongate parallel to the spreading axis (Needham and Francheteau, 1974; Macdonald and others, 1975). Mount Venus, in the FAMOUS area, is a typical example with dimensions of 1 × 4 km and a height of 250 m (ARCYANA, 1975). While the axial volcanoes are primarily composed of fresh sediment-free pillow basalts, thin sheet flow basalts (approximately 10–50 cm thick) are also observed, especially on flatter flanking areas and near volcanic collapse pits (Atwater, 1979). The boundaries of the neovolcanic zone are difficult to define precisely because of the lack of reliable radioactive age dating tools for young basaltic rocks. Instead we must use fine-scale sediment cover and the fresh appearance of basaltic lava morphologies. The roughness of the pillow lava surfaces shield local sediment pockets less than 50 cm deep from bottom current erosion, so that sediment cover can be a fairly reliable measure of age. Submarine lavas also exhibit a wide range of fragile ornamentation and glassy surfaces that show visible degradation in only 10^2–10^3 years. Other useful criteria for age are the progressive inward palagonitization of pillow rims and a progressive coating of manganese that changes the surface appearance of pillows from glassy to stony (Bryan and Moore, 1977). Based on submersible and deep-tow camera studies using these criteria, it has been estimated that the neovolcanic zone is only 1 to 3 km wide in the FAMOUS area of the MAR (Ballard and van Andel, 1977; Luyendyk and Macdonald, 1977; ARCYANA, 1975). Credence is lent to these qualitative observations by determinations of other workers at other spreading centers showing that the neovolcanic zone is rarely wider than 3 km regardless of spreading rate (RISE, 1980; Corliss and others, 1979; Rona and Grey, 1980). The neovolcanic zone at various spreading centers is bracketed by V's in Figure 2.

Observations of fresh lavas and sparse sediment cover give only an instantaneous view of the neovolcanic zone. One way to determine temporal variations in the width and location of the neovolcanic zone is through high-resolution deep-tow studies of magnetic anomaly transitions. The magnetic anomaly polarity transition width is a measure of the crustal accretion zone width, including the neovolcanic zone and underlying magnetized plutonics (Harrison, 1968). On the MAR, the transition zone appears to vary between 1 and 8 km in width (Macdonald, 1977). Thus, while the neovolcanic zone is generally only 1 to 3 km wide over short periods of time, there may be a tendency for the crustal accretion zone to wander laterally or periodically widen up to 5 or even 10 km within the rift valley inner floor.

Another way in which magnetic anomalies may be used to define the neovolcanic zone is by mapping maxima in the axial magnetization. The zones of most recent volcanism tend to be marked by very high magnetizations because the basaltic rocks have not yet been exposed to extensive low-temperature oxidation that tends to reduce magnetization (Klitgord and others, 1975). In the FAMOUS and AMAR study areas, inversion of near-bottom magnetic anomalies and direct paleomagnetic sampling show that the most recent lavas have magnetizations of approximately 20 to 40 amps per meter, relative to magnetizations of only 2 to 10 amps per meter for rocks older than 0.5 Ma. While caution must be exercised in using this approach because of other petrologic and geochemical variations that may affect the magnetization (Prévot and others, 1979), this method has proven very useful for mapping zones of recent volcanism on many spreading centers (Klitgord and others, 1975; Macdonald, 1977; Van Wagonner and Johnson, 1983). The width of the active volcanic zone inferred from crustal magnetization in the FAMOUS area is very narrow, less than 2 to 3 km.

Off-axis volcanism does occur on the Mid-Atlantic Ridge (Heirtzler and Ballard, 1977). In the FAMOUS area, deep-tow magnetic data indicate that up to 10 percent of the volcanism occurs outside the main neovolcanic zone in crust 0.5 to 2.0 m.y. old, that is, 5 to 20 km off axis (Macdonald, 1977; Atwater, 1979).

How often do major volcanic eruptions occur along the MAR neovolcanic zone? A combination of magnetic and physical property studies coupled with seismic refraction, deep-sea drilling analyses, and ophiolite field studies suggest that the thickness of the volcanic section is approximately 1 ± 0.5 km (Hall and Robinson, 1979; Fox and others, 1973; Christensen and Salisbury, 1975; Pallister and Hopson, 1981). This information can be combined with the precise knowledge of the dimensions and spacing of volcanoes (based on multi-beam charts in the FAMOUS area) to estimate the spatial density of volcanoes in the crustal section. When combined with the spreading rate, the frequency of major eruptive cycles large enough to construct a volcano the size of Mt. Venus can be estimated to be once every 5000 to 10,000 years (Bryan and Moore, 1977; Atwater, 1979). Do these axial eruptions occur rapidly with long periods of intervening quiescence, or is actively fairly continuous at a slow rate? Analogy with terrestrial eruptions suggests the former. Perhaps the best evidence is from deep-sea drilling project (DSDP) holes. For holes greater than 500 m deep, thick crustal units occur in which magnetic, petrologic, and geochemical properties are nearly uniform and are significantly different from crustal sections above or below. In the presence of secular variation of the magnetic field, this suggests eruption episodes of short duration, 1 to 100 yrs, separated by long periods of quiescence (Hall 1976). Eruptive episodes may be far more frequent in Iceland since it is the epicenter of a major hotspot (Saemundsson, this volume).

Another indication of shifting or widening of the neovolcanic zone comes from evidence that reversely magnetized rocks exist within the inner floor of the MAR (Ade-Hall and others, 1973; Johnson and Atwater, 1977). This is quite startling because the crust here should all be of positive polarity. Small outcrops of reversely magnetized rocks sampled from *Alvin* are most likely caused by eruption during a brief reversal within the Brunhes epoch (Van Waggoner and Johnson, 1983). However, the large negative anomalies measured by surface as well as deep-tow magnetometers may be caused by blocks of older crust (>0.7 m.y.) left behind by small jumps of the neovolcanic zone and buried by subsequent lava flows (Macdonald, 1977). There is a 14 percent probability of older negative crust becoming temporarily stranded near the axis due to small lateral shifts (1–5 km) of the neovolcanic zone within the inner floor (Macdonald, 1977).

A difficult problem is how to accumulate a volcanic section approximately 1 km thick when the tallest axial volcanoes are only 250 m high. Occasionally the 1.5 km deep rift valley may be filled with a seamount (Johnson and Vogt, 1973; Vogt, this volume, Ch. 24), but this is observed at only two places along the entire length of the MAR at this time. In general the volcanic section must be approximately 3 to 6 volcanoes thick. The volcanoes must be either down-faulted (Atwater, 1979) or tilted steeply toward the axis (Moore and others, 1974) to allow for vertical accumulation of successive flows. Modelling of magnetic fields (McGregor and Rona, 1975) suggest that down-faulting must occur with, at most, only slight rotation of the volcanoes. Whether the axial volcanoes are actually split in half or transported laterally as intact units is yet another question and is still a point of debate (Atwater, 1979; Ballard and van Andel, 1977).

Observations outside the inner floor of the rift valley suggest that at least some of the axial volcanoes have survived the precarious trip into the rift mountains and onto the terraces. Small volcanoes that are elongate parallel to the sreading axis are superposed on many of the fault blocks outside the inner floor. Curiously, most of the volcanic highs are not randomly situated, but form lips at the edges of fault blocks (Fig. 6). Of 107 volcanoes mapped outside the inner floor of the FAMOUS area rift valley, 72 (70%) were lips. If volcanoes averaging 500 m wide were randomly distributed in the rift mountains, only 30% would be lips. It is probable that these volcanic lips are created within the inner floor of the rift valley like Mount Venus. If the axial volcano has been dormant for some time, block faulting is likely to be concentrated along its edges where the crust is thinnest, resulting in a volcanic lip perched at the edge of the block fault. However, if the volcano has been recently active and the isotherms are still peaked beneath the edifice, the crust may be thinnest and weakest along its axis and it may be split in two by faulting and rifting. Both examples are common within the FAMOUS and AMAR study areas of the Mid-Atlantic Ridge.

Crustal Fissuring

Before the seafloor is 100,000 years old the crust becomes intensely fissured. These fissures closely resemble the *gjar* in Iceland and are typically 1 to 3 m wide, extending 10 m to 2 km along strike (Luyendyk and Macdonald, 1977; Ballard and van Andel, 1977). The most intense fissuring occurs between the distal edges of the neovolcanic zone and the inner walls delimiting the inner floor of the rift valley (the area inside the F's in Fig. 2). Their azimuth closely parallels the strike of the ridge, suggesting that they are caused by tensional failure of the crust during spreading. The tensional stresses producing pervasive fissuring are caused by horizontal acceleration of crust from zero at the idealized center of the neovolcanic zone to the full spreading rate at the edge of the plate boundary zone.

It is likely that these fissure fields provide the principal access for cold seawater to penetrate the young, hot oceanic crust (Lister, 1972; Lowell, 1975). Detailed field observations in the FAMOUS and AMAR areas, as well as recent observations in the TAG area suggest that fissuring is particularly intense on the Mid-Atlantic Ridge with average spacings of approximately 10 to 50 m between major fissures (Ballard and van Andel, 1977; Rona and others, 1982). This intense fissuring may lead to extremely rapid and efficient cooling of the oceanic crust and, as a result, high-temperature hydrothermal activity at the seafloor-ocean interface may be extremely rare. Earthquake focal depths of up to 8–10 km beneath the axis suggest that the fissures may extend to significant depths (Lilwall and others, 1978; Toomey and others, 1982). Instead, most high temperature hydrothermal activity and deposition on the Mid-Atlantic Ridge may be confined to depth with considerable subsurface mixing of fresh seawater with hydrothermal fluids.

Faulting

At a distance of 1 to 5 km from the spreading axis, some fissures develop large vertical offsets by normal faulting, with most of the faults dipping toward the spreading center. For the profiles shown in Figure 2, significant faulting begins near points

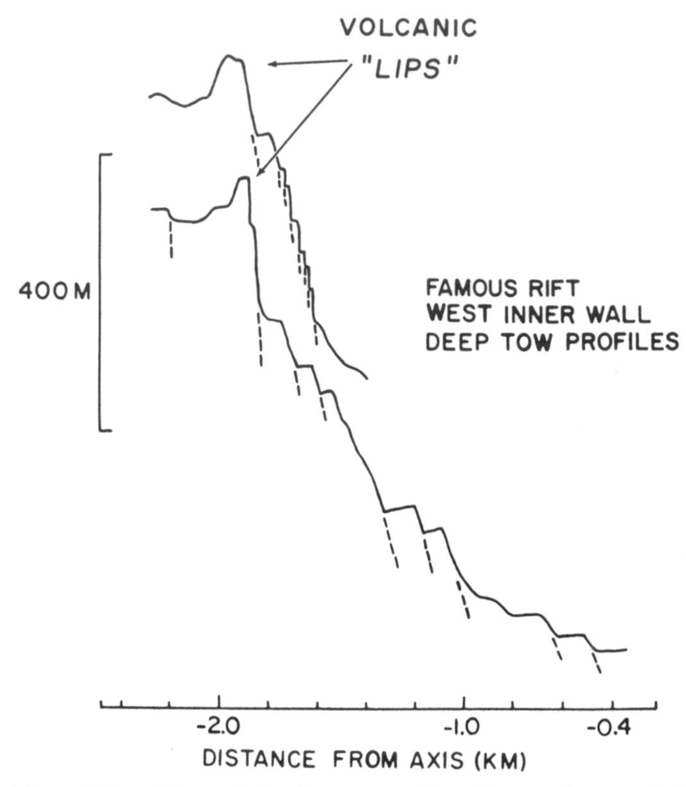

Figure 6. Two high resolution deep-tow profiles of the west inner wall of the FAMOUS rift near Mt. Venus. The character of the inner wall varies considerably even though these profiles are only 1.2 km apart. Note that the inner wall does not provide a deep "road cut" into the oceanic crust since the escarpment consists of at least nine faulted slivers. Also note the volcanic "lip" perched at the edge of the escarpment on both profiles. These volcanoes have been transported out of the inner floor by faulting. Vertical exaggeration = 2.5x.

F. The initiation of normal faulting creates the inner walls of the rift valley that bound the inner floor. Beyond the inner walls are the relatively horizontal, undulating terraces, followed by the outer walls marking the boundary between the rift valley and the rift mountains (Fig. 4, bottom).

While major escarpments on the MAR, such as the inner and outer valley walls, may reach heights of 500 to 1000 m, deep windows are not provided by these scarps because they actually consist of many closely spaced smaller normal faults. The west inner wall of the FAMOUS rift is a typical example (see Fig. 6) in which a single, narrow escarpment 500 to 700 m high actually consists of at least eight normally faulted, 20 to 100 m wide slivers with individual throws of 15 m to a maximum of 200 m. *Alvin* dives reveal that the fault scarps are further obscured by a layer of welded fault breccia.

Long range GLORIA side-scan sonar records indicate that the average spacing of major faulted escarpments on the MAR near 45°N, 40°N, and 36°N average 2 km, both inside and outside the rift valley (Laughton and Searle, 1979). The spacing is a matter of scale and resolution as *Alvin* and deep-tow records show much closer spacings (Macdonald and Luyendyk, 1977;

Ballard and van Andel, 1977). The question of average fault length and continuity encounters the same problem. Major escarpments such as the inner walls of the rift valley are continuous between transform faults and reach 40 to 60 km in length. However, as shown in Figure 6, high resolution profiles only 1.2 km apart show significant changes in fault spacing and throw. Rapid appearance and disappearance of individual faults along strike has been documented during the AMAR *Alvin* dives as well, with faults as large as 200 m in throw persisting only 1 to 2 km along strike as distinct structures.

Both the faults and the fissures result in a significant horizontal extension of the crust. In the FAMOUS area, the combined extension is 11 percent to the west and 18 percent to the east, the difference being in the same sense as the spreading rate asymmetry of 0.7 cm/yr to the west and 1.3 cm/yr to the east (Macdonald and Luyendyk, 1977). This suggests that asymmetric spreading results in asymmetric crustal extension as well as asymmetric crustal accretion.

The depth of faulting along the MAR is a critical unknown. Spectral modeling of surface waves suggest that normal faulting is shallow, only 3 to 5 km deep on the MAR (Weidner and Aki, 1973). Ocean bottom seismometer (OBS) studies on the MAR at 45°N suggest focal depths as great as 10 km (Lilwall and others, 1978), but the results are highly questionable since most of the events occur slightly outside of an array of only three OBSs. However, a recent study on the MAR near 23°N with a larger number of instruments and much better hypocentral control documents focal depths of 8 ± 1 km (Toomey and others, 1982). In addition, refraction work reveals a normal 5–6 km thick oceanic crustal section beneath the inner floor of the rift (Purdy and others, 1982). Thus, normal faulting to depths of 5 to 10 km along the axis is well documented. Francis (1981) has suggested that earthquakes up to 10 kilometers deep imply that normal faulting penetrates through to the mantle near the axis of the MAR. If so, extensive serpentinization of ultramafic mantle material may occur and vertical intrusion of serpentinite may invade much of the base of the crust (Bonatti and Honnorez, 1976; OTTER Scientific team, in press).

Episodicity and Variations Along Strike

The commonly-used two-dimensional approximation of MAR structure and tectonics just discussed is a useful starting point for analysis, but recent high-resolution studies reveal important variations along-strike (Plate 8A) as well as episodicity in time. Detailed studies using deeply-towed cameras and sonars, submersibles, and multi-beam bathymetry, suggest that volcanism within the inner floor is highly episodic and may follow cycles (Francheteau and others, 1979). Mapping of various lava types and analogues with terrestrial volcanic cycles suggest that pillow lavas and sheet flows are the submarine equivalents of tube-fed and surface-fed pahoehoe, respectively (Ballard and others, 1979; Crane and Ballard, 1980). The analogy suggests that sheet flows erupt from a new volcanic vent at very high effusion rates during the initial stage of the volcanic cycle. As the volcanic edifice builds, the lava starts to flow through a volcanic catacomb of tunnels and tubes rather than erupting directly from fissures. Channeling of lava through the volcanic plumbing system and diminishing effusion rates produce pillow lavas rather than sheet flows. This sequence is supported by detailed petrologic studies in the FAMOUS and AMAR areas (Stakes and others, 1984).

Very detailed paleomagnetic studies using samples collected from *Alvin* and by precise transponder navigated dredging have helped to unravel some of the fine-scale volcano-tectonic development within the inner floor of the FAMOUS and AMAR rifts (Van Waggoner and Johnson, 1983). Magnetization of the collected samples decays by a factor of 5 in only 10^6 years (Vogt, this volume, Ch. 15). This result agrees well with magnetization determinations made by modeling of deep-tow magnetic fields (Macdonald, 1977). Low magnetizations correlate with high Curie temperatures, suggesting that low temperature oxidation of the magnetic minerals takes place quickly after the basalts are erupted. Assuming other petrologic effects to be negligible, Van Waggoner and Johnson use the rapid decay of magnetic intensity to establish an approximate age determination technique. They find that surficial rocks representing the most recent volcanic phase are 10,000 years old in the FAMOUS area, 30,000 years old in the AMAR area, and 40,000 in the Narrowgate area. The highest magnetizations are found in the center of the Narrowgate area (84 amps/m) suggesting that this area is experiencing a renewed phase of volcanism after a long period of quiescence.

From the FAMOUS-Narrowgate area to the AMAR valley, the period of time since the last phase of volcanism increases and the terrain becomes increasingly controlled by tectonic rather than volcanic processes. During the small volume volcanic episodes in AMAR, the rate of extension and normal faulting exceeds the rate of volcanic extrusion so that no significant axial volcanoes have developed in the neovolcanic zone. While the most recent volcanism in the FAMOUS-Narrowgate area is confined to the narrow inner floor, the broad AMAR inner floor has resulted from a long phase of tectonic extension and normal faulting, combined with a long period of volcanic quiescence.

Tectonism may be quite continuous relative to the periods of quiescence (approximately 10,000 years long) between major episodes of volcanism. The entire length of the MAR under consideration here is seismically active with essentially no earthquake gaps even for short spreading center segments and for time periods as short as ten years (Einarsson, 1979). The continuity of tectonic activity extends even to a daily time span with a steady pulse of microearthquake activity of about 5 to 20 events of magnitude 0 to 1 per day (Reid and Macdonald, 1973; Spindel and others, 1974; Francis and others, 1977, 1978). Most of the microearthquake seismicity is associated with normal faulting in the inner walls of the rift.

The FAMOUS and AMAR projects (Fig. 4) produced the first detailed geologic map of a spreading center on a scale comparable to detailed studies on land. Based on these studies, volcanic and tectonic episodes of activity within the inner floor are being described in detail. The earlier FAMOUS work is summarized in the April and May 1977 issues of *Geological Society of America Bulletin.* The very fruitful AMAR program is summarized by Atwater (1979), Crane and Ballard (1981) Stakes and others (1984) and Van Wagonner and Johnson (1983). Much of the detailed geological work is still in progress at this writing.

3. MODELS FOR THE AXIAL MAGMA CHAMBER

It has been proposed that crustal volcanic and plutonic rocks are formed by differentiation of mantle-derived parent magmas in a shallow axial magma chamber (AMC), rather than by injection and eruption directly from the mantle (Melson and O'Hearn, this volume). The AMC model has been strongly supported by analysis of ophiolites (e.g., Cann, 1974; Dewey and Kidd, 1976; Pallister and Hopson, 1981), petrologic studies of oceanic rocks (e.g., Natland, 1980; Bryan and Moore, 1977; Rhodes and Dungan, 1979), thermal models (Sleep, 1975, 1978; Sleep and Rosendahl, 1979), and seismic experiments (Orcutt and others, 1976). The principal evidence comes from seismic refraction, where an AMC is inferred to exist beneath the axis of the fast spreading East Pacific Rise from travel time delays, and shadow zones for P waves at shot/receiver ranges of 15–35 km (Orcutt and others, 1976; Reid and others, 1977; see Macdonald, 1982, for summary). Such data require a low velocity zone interpreted as a shallow crustal AMC beneath the spreading axis. Refraction studies have indicated AMCs at depths of 2 to 6 km beneath the EPR near 22°N (McClain and Lewis, 1980), 21°N (Reid and others, 1977), 15°N (Trehu, 1982), 13°N (Orcutt and others, in press), 12°N (Lewis and Garmany, 1982), 10°S (Bibee, 1979), and the Galapagos spreading center near 86°W (Bibee, 1979). The original refraction results at the East Pacific Rise at 9°N are substantiated by multichannel seismic reflection work (Herron and others, 1980; Hale and others, 1982). An AMC is not detectable on the EPR near 11°20′N (Bratt and Solomon, 1984) or near 12°N (Lewis, 1983). However, both of these sites are near overlapping spreading centers.

In contrast, numerous seismic refraction studies over the slow spreading MAR have not provided clear evidence for an AMC (Poehls, 1974; Whitmarsh, 1975; Fowler, 1976; Nisbet and Fowler, 1978; Fowler and Keen, 1979; Toomey and others, 1982; Bunch and Kennett, 1980). These experiments include four attempts in the FAMOUS area at 37°N, two at 45°N, one at 23°N and one on the Reykjanes Ridge. On slow spreading centers, however, it is important to keep in mind that severe limitations are imposed by closely-spaced offsets of the spreading center by transform faults. In addition, rift valley topography is large in amplitude and makes travel time corrections difficult. However, the carefully executed refraction and microearthquake experiment at 23°N clearly precludes a shallow crustal magma chamber of any significant size. Normal oceanic crustal structure with a mature thickness of 5 to 6 km occurs beneath the axis, and microearthquakes occur to depths of 8 ± 1 km beneath the axis (Toomey and others, 1982; Purdy and others, 1982). While it is possible for a seismically undetectable AMC to exist beneath the MAR, it would have to be narrower than one seismic wavelength (approximately 1–2 km).

There is little doubt that AMCs have existed beneath all spreading centers at one time or another, but are they steady state features of the mid-ocean ridge system? Are they steady state at fast but not slow spreading centers? The debate continues on this topic. Unless this is a particularly unusual and unique period in the history of mid-ocean ridges, the evidence for AMCs beneath eight out of 10 ridge segments spreading faster than 6 cm/yr suggests that magma chambers are at least quasi-steady state features beneath intermediate and fast spreading centers. By quasi-steady state, I mean that they are permanent, but may vary with time in size and shape.

Numerous thermal models have been proposed for spreading centers, but unfortunately the results are somewhat ambiguous. Sleep's model (1975, 1978) addresses the fine scale thermal structure beneath the spreading center and predicts a steady state magma chamber at intermediate to fast spreading rates. More recent models that account for the importance of hydrothermal heat loss still allow for maintenance of a steady state magma chamber at intermediate to fast spreading rates but not at slow spreading rates (Sleep and Rosendahl, 1979; Sleep, 1983). Kuznir (1980) however, allows for hydrothermal cooling and still obtains a steady state magma chamber approximately 1 to 3 km wide for a slow half spreading rate of 1 cm/yr. At the other extreme, Lister (1983) maintains that hydrothermal circulation precludes the maintenance of a steady state magma chamber at even the fastest known spreading rates. The conflict in thermal models stems from poorly constrained estimates for the magnitude and areal distribution of hydrothermal heat loss at spreading centers, and ignorance of the depth and rate of crustal fissuring. Another important unknown concerns the rate at which cracks propagate downward in the axial zone versus the importance of closing of the cracks by hydrothermal precipitation.

While seismic evidence for AMCs beneath the MAR is negative or at best unresolved, petrologic evidence (Melson and O'Hearn, this volume) suggests that the AMC is quasi-steady state even at slow spreading rates (Bryan and others, 1979; Bryan and Moore, 1977; Stakes and others, 1984). The very narrow range of petrologic composition of basalt samples from the MAR suggests that AMCs are steady state even at spreading rates of 2 cm/yr. If magma chambers were non-steady state (i.e. transient), very primitive or highly fractionated basalts should erupt during the formative and waning stages of the magma chamber, respectively.

In response to this controversy, a number of AMC models have been proposed (Fig. 7). The *infinite onion* model features a steady state AMC in which the edges successively freeze (Cann, 1974). This brought tears to the eyes of seismologists who had found no evidence of an AMC anywhere along the MAR. An *infinite leek* model has been proposed which features an ephemeral AMC and creates the gabbro layer by repeated freezing and coalescing of transient magma chambers (Nisbet and Fowler,

MAR MAGMA CHAMBER MODELS

A. FUNNEL–SHAPED SEMAIL OPHIOLITE MODEL

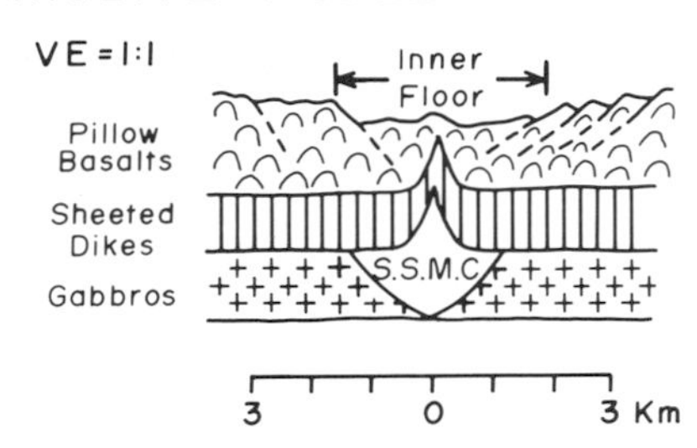

C. "INFINITE LEEK" MODEL

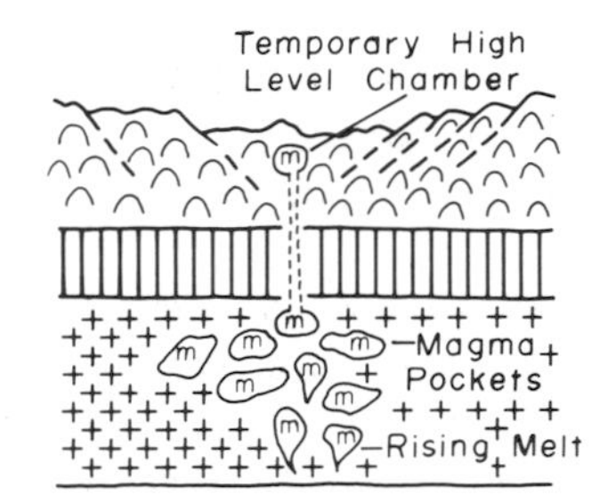

B. "INFINITE ONION" MODEL

D. ALONG STRIKE VIEW OF MAGMA CHAMBER FOR MODELS A,B

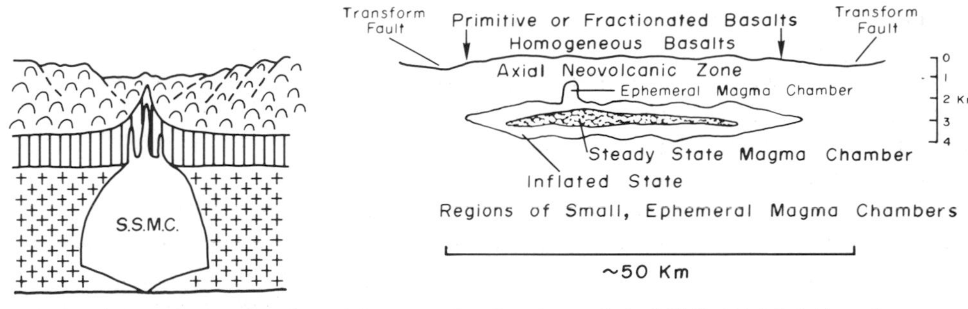

Figure 7. Three of the possible models for axial magma chambers beneath the MAR. (A) Model based on observations in the Semail ophiolite in Oman after Pallister and Hopson (1981), (B) the *infinite onion* model after Cann (1974) and Bryan and Moore (1977), (C) the *infinite leek* model after Nisbet and Fowler (1978). The Semail and Infinite onion models predict a steady state magma chamber (S.S.M.C.). As drawn, both chamber models should be seismically detectable, but would not be if they were less than 1–2 km wide. Model (C) predicts a non-steady state magma chamber. See text for discussion. For simplicity, possible subsurface extensions of normal faults into the dike and gabbro layers are not shown. (D) shows a possible along-strike section for the steady state magma chamber models. The chamber may exist in the steady state along only 30 percent of the length of the MAR, confined to the elevated regions of the rift valley away from transform fault intersections.

1978). However, each *leek* is a closed system intrusion that would produce highly fractionated lavas upon freezing. Unless these highly fractionated magmas fail to erupt for some reason, this model directly conflicts with the detailed petrologic sampling of the FAMOUS and AMAR programs (Bryan and Moore, 1977; Stakes and others, in press).

Stakes and others (1984) have proposed an AMC model similar to that for Oman (Pallister and Hopson, 1981) for the MAR based on extensive sampling of the FAMOUS and AMAR rifts. The AMC is funnel-shaped with the narrow end down (Fig. 7). In this *funnel* model, steady state replenishing and continued magma mixing account for the uniform composition of the basalts collected from *Alvin*. The chamber is postulated to be only 1 to 2 km wide, sidestepping the negative seismic evidence by making it too narrow to be detected. In addition, the AMC exists as a steady state feature only near the elevated midpoints of the inner floor of the rift valley, away from the cooling edge effects of transform faults. Note that Mt. Venus, which is 15 km away from FZA, is characterizead by very primitive basalts that may have erupted directly from the mantle rather than fractionating in a magma chamber. Thus, given that the entire length of the FAMOUS rift is only 43 km, only the central 10 to 15 km of the rift valley or only 1/3 of its total length would retain a steady state AMC in this model. In the past I have advocated a non-steady state AMC at slow spreading rates (Macdonald, 1982), but since the petrologists have retreated to such a tiny $1 \times 2 \times 10$ km AMC, perhaps the geophysicists should let them keep it! In any case, even if a small AMC such as this persists in a steady state on the MAR, the lack of significant along-strike continuity of the AMC could explain many of the spreading rate-related structural, tectonic, and volcanic differences between the slow spreading MAR and the fast spreading East Pacific Rise (Macdonald, 1983a, b).

4. DISRUPTION OF SPREADING CENTER PROCESSES BY CLOSELY SPACED TRANSFORM FAULTS

Transform faults are very closely and uniformly spaced on the MAR (Fig. 8). Given that the thermal and structural effects of transform faults extend 15 to 30 km away from their intersections with spreading centers (Fox and Gallo, 1984; and this volume) and that the inter-transform spacing on the MAR is 55 km (Fig. 8), there are essentially no segments of the MAR that are normal or free of transform fault edge effects. In this light it is folly to discuss the tectonics of the MAR without assessing the influences of the transform faults. I shall restrict myself to ways in which the MAR is modified near transform fault intersections. There are four processes that occur near ridge transform (RT) intersections discussed here: (1) deepening and widening of the rift valley due to increased hydraulic head loss; (2) thinning of the oceanic crust due to thermal edge effects; (3) transmission of shear stresses associated with the transform fault into the spreading center domain with resultant rotation of fault trajectories; and (4) a phenomenon I shall call apparent oblique spreading, involving small offset transform faults that do not follow small circles about the pole of opening.

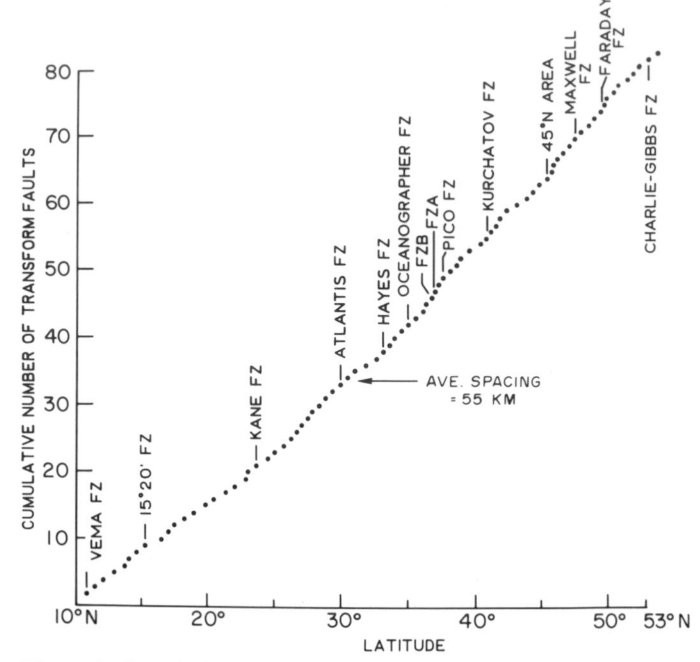

Figure 8. Cumulative number of transform faults on the MAR from 10°N–53°N. The average spacing is 55 km, so close that essentially every segment of the MAR is altered along most of its length by transform fault edge effects. See also Vogt, this volume, Ch. 12.

Valley Deepening and Crustal Thinning at RT Intersections

The imposition of a third rigid boundary at RT intersections may increase the hydraulic head loss by 30 to 40 percent (Sleep and Biehler, 1970). This closely matches the observed increase in depth at RT intersections. For example, the intersection deep at FZA is 600 m or 40 percent deeper than "Narrowgate," which is the inner floor midway between FZA and FZB (Fig. 4). The structural effects on the rift valley are considerable. Both the inner floor and the major normal faults creating the inner and outer walls of the valley plunge towards the intersections with topographic gradients of 15 to 30 m per km (see Fig. 16 in Macdonald and Luyendyk, 1977). The inner floor also widens by as much as a factor of two. Dynamic effects associated with migration of magma along strike beneath the spreading axis may also contribute to the deepening of the valley floor (Parmentier and Forsyth, 1984). Indeed, many of the transform faults are too short (and the juxtaposed lithosphere too thin) to significantly affect the adjoining spreading segment.

In addition to the viscous effects pointed out above, transform faults have profound thermal effects on ridge structure and crustal accretion. The degree of partial melting that may occur in the rising asthenosphere is reduced and the degree of fractional crystallization is enhanced near a RT intersection because the cold bounding edge of the transform is a significant heat sink (Fox and others, 1980; Langmuir and Bender, 1984; Fox and Gallo, this volume). How significant a heat sink depends on the age contrast of the lithosphere near the intersection. As a result, the flux of basaltic melt is reduced and the crust is thinner. Seismic refraction results suggest that the oceanic crust may be thinned by as much as a factor of 2 near the Kane Fracture Zone (Detrick and Purdy, 1980). At large offset transforms on the MAR such as the Vema, lithosphere 30 to 50 km thick may be juxtaposed against the accreting plate boundary "all but nullifying the processes that lead to the emplacement of normal oceanic lithosphere (Fox and Gallo, 1984)." These effects are far greater on the slow spreading MAR than on the EPR because the age contrast for a given offset transform fault is 3 to 9 times greater.

The Transmission of Shear Stresses Into the Spreading Center Domain

Faults oblique to both spreading center and transform fault trends are common near RT intersections, and are hypothesized to be oblique trending normal faults (Crane, 1976; Lonsdale, 1977; Macdonald and others, 1979). Recent *Alvin* studies on the Oceanographer and Tamayo transform faults confirm that these faults are spreading center-related normal faults that are 30 to 45° oblique to the spreading axis near the RT intersection (OTTER Scientific Team, in press; Tamayo Tectonic Team, 1984). Deep-tow studies near the Vema Fracture Zone show that the normal faults follow intermediate (σ_2) stress trajectories that rotate smoothly and continuously from spreading center-parallel

INFLUENCES OF TRANSFORM FAULTS ON AXIAL MAR TECTONICS

A. SHORT OFFSET, 5-20KM. : A "BEND" IN THE RIFT VALLEY (e.g. TAG area transforms, Kurchatov F.Z.)

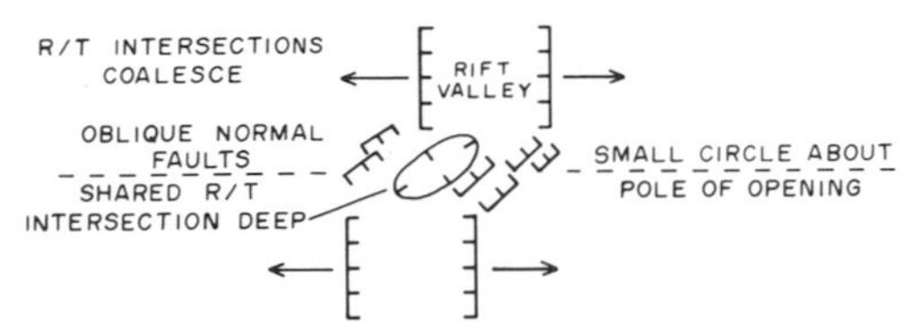

B. INTERMEDIATE OFFSETS, 20-30KM. : TRANSFORM VALLEY DOES NOT FOLLOW SMALL CIRCLE ABOUT POLE OF OPENING, "APPARENT OBLIQUE SPREADING" (e.g. F.Z.A.)

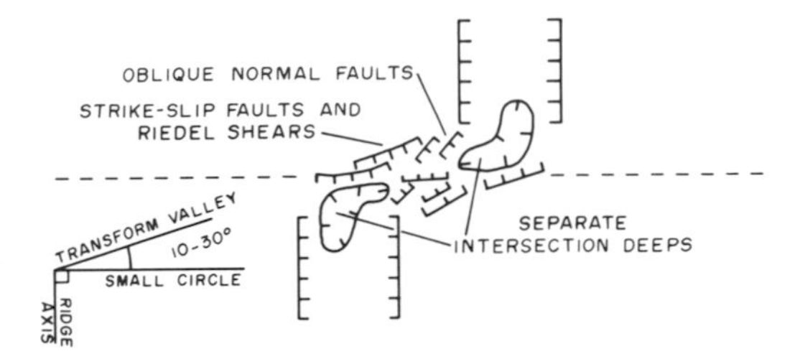

C. LARGE OFFSET, >40KM. ROTATED NEOVOLCANIC ZONE AND NORMAL FAULTS NEAR R/T INTERSECTION (e.g. Vema)

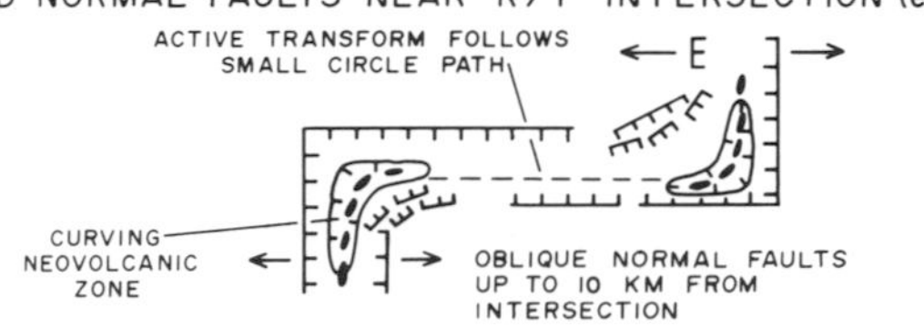

D. EFFECTS OF LATERAL SHIFTS OF TRANSFORM FAULTS (e.g. FZB)

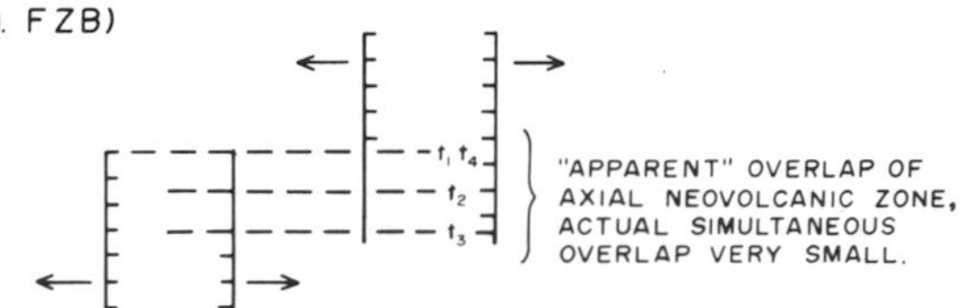

Figure 9. Various ways in which MAR structure is modified near transform fault intersections depending on the length of the offset. Small transform faults (A, B) result in a "bend" in the rift valley trend that may not follow a small circle about the pole of opening. "Apparent" oblique spreading results. Large offset transform faults (C) create a significant rotation of the neovolcanic zone and rift valley normal faults within 5–10 km of ridge/transform intersections. Another common phenomenon is the "apparent" overlapping of spreading centers caused by north-south lateral shifts of the transform fault (D). These shifts cause en echelon overlap of intersection deeps but not a large overlap of the neovolcanic zones.

to 45° oblique to the spreading axis (Macdonald and others, submitted). Rotation of stress trajectories is caused by transmission of transform-related shear stresses into the spreading center domain. Near the Vema transform fault these shear stresses extend into the MAR rift valley at least 7 km along strike (Macdonald and others, submitted).

"Apparent" Oblique Spreading

A fundamental premise of plate tectonics is that transform faults trace out small circles about the pole of opening (McKenzie and Parker, 1967). This may not be true for small offset transform faults on the MAR. Notice the enormous number of very small offset (<20 km) transform faults on the MAR in Figure 1 and how they are not sharp perpendicular offsets of the MAR, but appear as gentle bends in the rift valley (Vogt, this volume, Ch. 12). The Kurchatov FZ and the TAG area transform faults are well studied examples of these phenomena.

Detailed GLORIA side-scan sonar studies of the Kurchatov FZ (Searle and Laughton, 1977) do not reveal any prominent structures perpendicular to the MAR. Instead the normal faults bounding the rift valley veer approximately 45° clockwise for 18 km of offset, then back to a spreading center trend (Fig. 9). In the TAG area, where the transform offsets are even less (approximately 10 to 12 km) there is only a single intersection depression shared between the two RT intersections creating a broad 45° bend in the trend of the rift valley (Rona and Gray, 1980). Similar structures have been mapped by Johnson and Vogt (1973) on the MAR from 47°N to 51°N. On transform faults whose offset is less than 15 to 20 km, the fault zone is one continuous RT intersection dominated by oblique normal faults.

Detailed studies of FZA and FZB, however, do reveal clear east-west strike slip faults that are punctuated by microearthquake activity along their entire length between RT intersections (Detrick and others, 1973; Reid and Macdonald, 1973). However, FZA and FZB are not perpendicular to the adjoining spreading centers, but intersect at angles of 70°–75°. the propensity of oblique intersections between spreading centers and their adjoining transform faults in the Atlantic led us to propose that oblique spreading is the rule, not the exception, at slow spreading centers (Atwater and Macdonald, 1977). However, it may be the transform faults rather than the spreading centers that are oblique (Collette and others, 1979; Searle and Laughton, 1977). Comparing the strikes of FZA and FZB to Minster and Jordan's predicted trends (1978), we find that spreading centers away from RT intersections are orthogonal to small circles about the pole of opening, and the transform faults are oblique relative to small circle trends.

Thus, there are essentially three types of transform faults (Fig. 9): (1) large offset (>30 km) such as the Oceanographer, Kane, and Vema Fracture Zones that are parallel to small circles about the pole of opening and orthogonal to their spreading centers; (2) small offset transforms in which the entire transform valley is simply two coalesced RT intersections, and structures within the valley strike 45° relative to small circle trends; and (3) transitional transform faults such as FZA and FZB that have clear strike slip faults but are 10 to 20° oblique to small circle trends due to intersection effects.

Short offset transform faults (types 2 and 3) give rise to *apparent oblique spreading* because the angle between the gross morphology of the transform valley and the MAR rift valley is not 90°, but 80° to 45° (Vogt, this volume, Ch. 12). It is not true oblique spreading because away from the intersections the MAR is orthogonal to small circles about the pole of opening. This is not the same as postulated leaky transform faulting caused by a change in the pole of opening (Menard and Atwater, 1968).

Transform fault types 2 and 3 may be the present day equivalents to Schouten's hypothetical "zero offset transform faults" (Schouten and White, 1980). If so, then the long-term time-averaged fracture zone trends of these small offset transform faults may indeed trace small circles about the pole of opening, while the short, equidimensional transform valleys do not.

Small lateral offsets of the East Pacific Rise are manifested by "overlapping spreading centers" (OSCs) that are structurally very different from small transform faults in the Atlantic. Whether or not they have a common origin is a point of some controversy (Klitgord and Schouten, 1983; Macdonald and Fox, 1983b; Macdonald and others, 1984).

Heterogeneity of Crust Accreted at Slow-spreading Centers

From the discussion above it is clear that transform faults disrupt the two-dimensionality of spreading centers and contribute significantly to heterogeneity in the physical and geochemical properties of the crust. We have reviewed evidence that the oceanic crust thins significantly near the ridge/transform intersections, that transform fault related shear stresses affect spreading center processes within approximately 10 km of the ridge/transform intersections, and that the dynamic forces responsible for the maintenance of the axial rift valley may be enhanced or otherwise perturbed at the ridge transform intersection and up to several tens of kilometers along the strike of the axial rift valley. In addition, the AMC may be a steady state feature along a short length of the ridge segment removed from the RT intersection, but it is probably transient near these intersections. Considering that the MAR is disrupted every 55 km by a transform fault (Figs. 1 and 9), it is unlikely that any of the crust created along the MAR is free of the disrupting edge effects of transform faults. The result is a highly heterogeneous oceanic crust.

Rock magnetization is one physical property that can be inferred indirectly on a global scale via the magnetic field generated by crustal rocks, and the degree of heterogeneity of the crust should directly affect the clarity of magnetic anomalies (Vogt, this volume, Ch. 15). It has been known for some time that regional magnetic anomalies measured at the sea surface are far clearer and easier to decipher in the Pacific than in the Atlantic. For example, compare the careful magnetic anomaly study on the MAR at 45°N (Loncarevic and Parker, 1971) with magnetic anomalies on the East Pacific Rise at 21°N (Larson 1971) or 20°S (Rea and Blakely, 1975) or the Galapagos spreading center (Hey, 1977; Klitgord and others, 1975). To explain the difference in magnetic anomaly clarity we have suggested that crustal accretion processes and resulting magnetic structures vary significantly with spreading rate (Macdonald and others, 1983). Geologic processes that may have a spreading rate dependence and affect crustal heterogeneity include the frequency of volcanic eruptions, the stability of the neovolcanic zone, and the extent of tectonic disruption of the crust (Macdonald, 1982). An important factor to add to that list is the close spacing of transform faults in the Atlantic. If one considers transform faults whose offsets represent 1 m.y. age contrasts or greater (that is, the age of the lithosphere juxtaposed to the spreading center by the transform fault) then the average spacing is 70 km in the slow-spreading north Atlantic and approximately 500 km in the fast-spreading Pacific. The swath of crust disrupted by a transform fault is approximately 30 km, so roughly 50 percent of the Atlantic basin is significantly affected by transform fault edge effects while only 5 percent of the Pacific crust is so affected. Thus the close spacing of transform faults may be a major cause of the high degree of crustal heterogeneity observed in the Atlantic, and this factor must be considered in models of crustal accretion processes in the Atlantic.

REFERENCES

Ade-Hall, J. M., Aumento, F., Ryall, P.J.C., Gerstein, R. E., Brooke, J., and McKeown, D. L.
1973 : The Mid-Atlantic ridge near 45°N, 21, Magnetic results from basalt drill cores from the median valley: Canadian Journal of Earth Sciences, v. 10, p. 676–696.

ARCYANA
1975 : Transform fault and rift valley from bathyscaph and diving saucer: Science, v. 190, p. 108–116.

Anderson, R. N., and Noltimier, H. C.
1973 : A model for the horst and graben structure of mid-ocean ridge crests based upon spreading velocity and basalt delivery to the oceanic crust: Geophysical Journal of the Royal Astronomical Society, v. 34:137–147.

Atwater, T. M.
1979 : Constraints from the FAMOUS area concerning the structure of the oceanic section; in Deep drilling results in the Atlantic Ocean: Ocean crust, eds. M. Talwani, C. G., Harrison, D. E. Hayes, 2, 33–42.

Atwater, T. M., and Macdonald, K. C.
1977 : Slowly spreading ridge crests; are they perpendicular to their transform faults?: Nature, v. 270, 715–719.

Atwater, T., Ballard, R., Crane, K., Grover, N., Hopson, C., Johnson, H. P.,

Luyendyke, B., Macdonald, K., Peirce, J., Shih, J., Shure, L., Stakes, D., Walker, N., van Andel, Tj. H.
1978 : AMAR 78: A coordinated submersible, photographic mapping and dredging program in the Mid-Atlantic Rift valleys near 36.5°N: Expedition Cruise Report, unpublished manuscript.
Aumento, F.
1968 : The Mid-Atlantic Ridge near 45°N, 2, basalts from the area of Confederation Peak: Canadian Journal of Earth Sciences, v. 5, p. 1–21.
Aumento, F., Loncarevic, B. D., and Ross, D. I.
1971 : Hudson geotraverse: geology of the Mid-Atlantic ridge at 45°N: Philosophical Transactions of Royal Society, London, ser. A, v. 268, p. 623–650.
Ballard, R. D., Holcomb, R. T., and van Andel, Tj. H.
1979 : The Galapagos Rift at 86°W: 3. Sheet flows, collapse pits, and lava lakes of the rift valley: Journal of Geophysical Research, v. 84, p. 5407–5422.
Ballard, R. D., and van Andel, Tj. H.
1977 : Morphology and tectonics of the inner rift valley at lat. 36°50′N on the Mid-Atlantic Ridge: Geological Society of America Bulletin, v. 88, p. 507–530.
Barrett, D. L., and Aumento, F.
1970 : The Mid-Atlantic ridge near 45°N, 11, Seismic velocity, density and layering of the crust: Canadian Journal of Earth Sciences, v. 7, no. 4, p. 1117–1124.
Bibee, L. D.
1979 : Crustal structure in areas of active crustal accretion: [Ph.D. Thesis], University of California, San Diego, p. 155.
Bird, P., and Phillips, J. D.
1975 : Oblique spreading near the Oceanographer fracture zone: Journal of Geophysical Research, v. 80, p. 4021–4027.
Bonatti, E., and J. Honnorez
1976 : Sections of the earth crust in the equatorial Atlantic: Journal of Geophysical Research, v. 81, p. 4104–4117.
Bratt, S. R., and Solomon, S. C.
1984 : Compressional and shear wave structure of the East Pacific Rise at 11°20′N: Constraints from three-component Ocean Bottom Seismometer data: Journal of Geophysical Research, v. 89, p. 6095–6110.
Bryan, W. B., and Moore, J. G.
1977 : Compositional variations of young basalts in the Mid-Atlantic ridge rift valley near lat. 36°49′N: Geological Society of America Bulletin, v. 88, p. 556–570.
Bryan, W. B., Thompson, G., and Michael, P. J.
1979 : Compositional variation in a steady-state zoned magma chamber: Mid-Atlantic Ridge at 36°50′N: Tectonophysics, v. 55, p. 63–85.
Bullard, E. C., and Day, A.
1961 : The flow of heat through the floor of the Atlantic Ocean: Geophysical Journal, vol. 4, p. 282–292.
Bunch, A.W.H., and Kennett, B.L.N.
1980 : The crustal structure of the Reykjanes ridge at 59° 30′N: Geophysical Journal of Royal Astronomical Society, v. 61, p. 141–166.
Cann, J. R.
1970 : New model for the structure of the oceanic crust: Nature, vol. 226, p. 928–930.
1974 : A model for oceanic crustal structure developed: Geophysical Journal the Royal Astronomical Society, v. 39, p. 169–187.
Christiansen, N. I., and Salisbury, M. H.
1975 : Structure and constitution of the lower oceanic crust. Reviews of Geophysics and Space Physics, v. 13, p. 57–86.
Colette, B. J., Slootweg, A. P., and Twigt, W.
1979 : Mid-Atlantic ridge crest topography between 12° and 15°N: Earth and Planetary Sciences Letters, v. 42, p. 103–108.
Corliss, J. B., Dymond, J., Gordon, L. I., Edmond, J. M., Von Herzen, R. P., Ballard, R. D., Green, K., Williams, D., Bainbridge, A., Crane, K., and van Andel, Tj. H.
1979 : Submarine thermal springs on the Galapagos Rift: Science, v. 203, p. 1073–1083.
Crane, K.
1976 : The Intersection of the Siqueiros transform fault and the East Pacific Rise: Marine Geology, v. 21, p. 25–46.
Crane, K., and Ballard, R. D.
1980 : The Galapagos rift at 86°W: 4, Structure and morphology of hydrothermal fields and their relationship to the volcanic and tectonic processes of the rift valley: Journal of Geophysical Research, v. 85, p. 1443–1454.
1981 : Volcanics and structure of the famous Narrowgate rift: evidence for cyclic evolution, AMAR1: Journal of Geophysical Research, v. 86, p. 5112–5124.
Deffeyes, K. S.
1970 : The axial valley: a steady rate feature in the terrain; in Megatectonics of continents and oceans, eds. H. Johnson and B. C. Smith; Rutgers University Press, Brunswick, N.J., p. 194–222.
Detrick, R. S., Mudie, J. D., Luyendyk, B. P., and Macdonald, K. C.
1973 : Near-bottom observations of an active transform fault (Mid-Atlantic ridge at 37°N): Nature, v. 246, no. 152, p. 59–61.
Detrick, R. S., and Purdy, G. M.
1980 : The crustal structure of the Kane Fracture Zone from seismic refraction studies: Journal of Geophysical Research, v. 85, p. 3759–3777.
Dewey, J. F., and Kidd, W.S.F.
1977 : Geometry of plate accretion: Geological Society of America Bulletin, v. 88, p. 960–968.
Einarsson, P.
1979 : Seismicity and earthquake focal mechanisms along the Mid-Atlantic plate boundary between Iceland and the Azores: Tectonophysics, v. 55, p. 127–153.
Fowler, C.M.R.
1976 : Crustal structure of the Mid-Atlantic ridge crest at 37°N: Geophysical Journal of the Royal Astronomical Society, v. 47, p. 459–491.
Fowler, C.M.R., and Keen, C. E.
1979 : Oceanic crustal structure, Mid-Atlantic ridge at 45°N: Geophysical Journal, v. 56, p. 219–226.
Fox, P. J., Detrick, R. S., and Purdy, G. M.
1980 : Evidence for crustal thinning near fracture zones: Implications for ophiolites, in Ophiolites, Panayiotou, A.; Proceedings of International Ophiolite Symposium, Cyprus: Geological Survey Department of Cyprus, Nicosia.
Fox, P. J., and Gallo, D. G.
1984 : A tectonic model for ridge-transform-ridge plate boundaries: Implications for the structure of oceanic lithosphere: Tectonophysics, v. 104, p. 205–242.
Fox, P. J., and Schreiber, E.J.J.
1973 : The geology of the oceanic crust: Compressional wave velocities of oceanic rocks: Journal of Geophysical Research, v. 78, p. 5155–5172.
Francheteau, J., Needham, H. D., Choukroune, R., Juteau, T., Seguret, M., Ballard, R. D., Fox, P. J., Normark, W., Carranza, A., Cordoba, D., Guerrero, J., Rangan, C., Bargault, H., Cambon, P., and Hekinian, R.
1979 : Massive deep-sea sulfide ore deposits discovered on the East Pacific Rise: Nature, v. 277, p. 523–528.
Francis, T.J.G.
1981 : Serpentinization faults and their role in the tectonics of slow spreading ridges: Journal of Geophysical Research, v. 86, p. 11, 616–11, 622.
Francis, T.J.G., Porter, I. T., and Lilwall, R. C.
1978 : Microearthquakes near the eastern end of St. Paul's Fracture Zone: Geophysical Journal of Royal Astronomical Society, v. 53, p. 201–217.
Francis, T.J.G., Porter, I. T., and McGrath, J. R.
1977 : Ocean bottom seismograph observations on the Mid-Atlantic ridge near 37°N: Geological Society of America Bulletin, v. 88, p. 664–677.
Gutenberg, B., and Richter, C. F.
1954 : Seismicity of the earth and associated phenomena: Princeton Univer-

sity Press, Princeton, New Jersey, p. 309.
Hale, L. D., Morton, C. J., and Sleep, N. H.
1982 : Reinterpretation of seismic reflection data over the East Pacific Rise: Journal of Geophysical Research, v. 87, p. 7707–7719.
Hall, J. M.
1976 : Major problems regarding the magnetization of oceanic crustal layer 2: Journal of Geophysical Research, v. 81, p. 4223–4230.
Hall, J. M., and Robinson, P. T.
1979 : Deep crustal drilling in the North Atlantic Ocean: Science, v. 204, p. 573–586.
Harrison, C.G.A
1968 : Formation of magnetic anomaly patterns by dyke injection: Journal of Geophysical Research, v. 73, p. 2137–2142.
1974 : Tectonics of mid-ocean ridges: Tectonophysics, v. 22, p. 301–310.
Harrison, C.G.A., and Stieltjes, L.
1977 : Faulting within the median valley: Tectonophysics, v. 38, p. 137–144.
Harrison, C.G.A., and Watkins, N. D.
1977 : Shallow inclinations of remanent magnetism in deep-sea drilling project igneous cores: Geomagnetic field behavior or postemplacement effects?: Journal of Geophysical Research, v. 82, p. 4869–4877.
Heezen, B. C.
1960 : The rift in the ocean floor: Scientific American, v. 203, no. 4, p. 99–110.
Heezen, B. C., Tharp, M., and Ewing, M.
1959 : The floors of the oceans, I, the North Atlantic Ocean: Geological Society of America Special Paper 65, 122 p.
Heirtzler, J. R., and Ballard, R. D.
1977 : Submersible observations at the Hole 332B area; in Initial reports of the deep sea drilling project, Volume 37, eds. F. Aumento, W. G. Melson, and others; p. 363–366.
Heirtzler, J. R., and Le Pichon, X.
1974 : FAMOUS: A plate tectonics study of the genesis of the lithosphere: Geology, v. 2, no. 6, p. 273–278.
Herron, T. J., Stoffa, P. L., and Ball, P.
1980 : Magma chamber and mantle reflection—East Pacific Rise: Geophysical Research Letters, v. 7, p. 989–992.
Hess, H. H.
1962 : History of the ocean basins, Petrologic Studies: A volume to honor A. F. Buddington: Geological Society of America, p. 599–620.
Hey, R. N.
1977 : A new class of pseudo-faults and their bearing on plate tectonics: A propagating rift model: Earth and Planetary Science Letters, v. 37, p. 321–325.
Hill, M. N.
1960 : A median valley of the Mid-Atlantic ridge: Deep-Sea Research, v. 6, no. 3, p. 193–205.
Irving, E.
1970 : The Mid-Atlantic ridge at 45°N, 14, oxidation and magnetic properties of basalt, review and discussion: Canadian Journal of Earth Sciences, v. 7, p. 1528–1538.
Johnson, H. P., and Atwater, T. M.
1977 : Magnetic study of basalts from the Mid-Atlantic ridge: Geological Society of America Bulletin, v. 88, p. 637–647.
Johnson, G. L., and Vogt, P. R.
1973 : Mid-Altantic ridge from 47° to 51° north: Geological Society of America Bulletin, v. 84, p. 3443–3462.
Keen, C. E., and Tramontini, C.
1970 : A seismic refraction survey on the Mid-Atlantic ridge: Geophysical Journal of the Royal Astronomical Society, v. 20, p. 473–491.
Klitgord, K. D., Huestis, S. P., Parker, R. L., and Mudie, J. D.
1975 : An analysis of near-bottom magnetic anomalies: Sea floor spreading, the magnetized layer, and the geomagnetic time scale: Geophysical Journal of the Royal Astronomical Society, v. 43, p. 387–424.
Kusznir, N. J.
1980 : Thermal evolution of the oceanic crust; its dependence on spreading rate and effect on crustal structure: Geophysical Journal of Royal Astronomical Society, v. 61, p. 167–181.
Lachenbruch, A. H.
1973 : A simple mechanical model for oceanic spreading centers: Journal of Geophysical Research, v. 78, p. 3395–3417.
1976 : Dynamics of a passive spreading center: Journal of Geophysical Research, v. 81, p. 1883–1902.
Langmuir, C. H. and Bender, J. F.
1984 : The geochemistry of oceanic basalts in the vicinity of transform faults: observations and implications: Earth and Planetary Science Letters, v. 69, p. 107–127.
Larson, R. L.
1971 : Near-bottom geophysical studies of the East Pacific Rise crest: Geological Society of America Bulletin, v. 82, p. 823–842.
Laughton, A. S., and Searle, R. C.
1979 : Tectonic processes on slow-spreading ridges; in Implications of deep drilling results in the Atlantic Ocean: Ocean crust, eds., M. Talwani and others; Maurice Ewing Series, American Geophysical Union, v. 2, p. 15–32.
Lewis, B.R.T.
1983 : The process of formation of ocean crust: Science, v. 220, p. 151–156.
Lewis, B.T.R., and Garmany, J. D.
1982 : Constraints on the structure of the East Pacific Rise from seismic refraction data: Journal of Geophysical Research, v. 87, p. 8417–8425.
Lilwall, R. C., Francis, T.J.G., and Porter, I. T.
1978 : Ocean bottom seismograph observations on the Mid-Atlantic ridge near 45°N—further results: Geophysical Journal of Royal Astronomical Society, v. 55, p. 255–262.
Lister, C.R.B.
1972 : On the thermal balance of a mid-ocean ridge: Geophysical Journal of Royal Astronomical Society, v. 26, p. 515–535.
1983 : On the intermittency and crystallization mechanisms of sub-seafloor magma chambers: Geophysical Journal of Royal Astronomical Society, v. 73, p. 351–365.
Loncarevic, B. D., Mason, C. S., and Matthews, D. H.
1966 : The Mid-Atlantic ridge near 45°N, I, The median valley: Canadian Journal of Earth Sciences, v. 3, p. 327–349.
Loncarevic, B. D., and Parker, R. L.
1971 : The Mid-Atlantic ridge near 45°N, 17, Magnetic anomalies and seafloor spreading: Canadian Journal of Earth Sciences, v. 8, p. 883–898.
Lonsdale, P.
1977 : Structural geomorphology of a fast-spreading rise crest: The East Pacific Rise near 3°25′S: Marine Geophysical Research, v. 3, p. 251–293.
Lowell, R. P.
1975 : Circulation in fractures, hot springs and convective heat transport on mid-ocean ridge crests: Geophysical Journal of Royal Astronomical Society, v. 40, p. 351–365.
Luyendyk, B. P., and Macdonald, K. C.
1977 : Physiography and structure of the inner floor of the Famous rift valley: observations with a deeply towed instrument package: Geological Society of America Bulletin, v. 88, p. 648–663.
MacDonald, K. C.
1976 : Geomagnetic reversals and the deep drill hole at DSDP site 32: Journal of Geophysical Research, v. 81, p. 4163–4165.
1977 : Near-bottom, magnetic anomalies, asymmetric spreading, oblique spreading, and tectonics of the Mid-Atlantic ridge near 37°N: Geological Society of America Bulletin, v. 88, p. 541–555.
1982 : Mid-ocean ridges: Fine-scale tectonic, volcanic and hydrothermal processes within the plate boundary zone: Annual Review of Earth and Planetary Sciences, v. 10, p. 155–190.
1983a: Crustal processes at spreading centers: Reviews of Geophysics and Space Physics, v. 21, p. 1441–1453.

1983b: A geophysical comparison between fast and slow spreading centers: constraints on magma chamber formation and hydrothermal activity; in Hydrothermal processes at seafloor spreading centers, eds. P. Rona and others; p. 27–51, NATO Conference series IV: 12, Plenum, N.Y. 796 pp.

MacDonald, K. C., and Atwater, T. M.
1978a: AMAR78, Preliminary results, I, Evolution of the median rift (abstract): EOS, v. 59, p. 1198.
1978b: Evolution of rifted ocean ridges: Earth and Planetary Sciences Letters, v. 39, p. 319–327.

MacDonald, Ken C., Castillo, D., Miller, S., Fox, P. J., Kastens, K. A., and Bonatti, E.
submitted: Deep tow studies of the Vema fracture zone: I. The tectonics of a major slow-slipping transform fault and its intersection with the Mid-Atlantic Ridge; Journal of Geophysical Research.

MacDonald, K. C., and Fox, P. J.
1983a: Overlapping spreading centres: new accretion geometry on the East Pacific Rise: Nature, v. 302, p. 55–58.
1983b: Overlapping spreading centers on the East Pacific Rise—discussion and reply: Nature, v. 303, p. 549–550.

MacDonald, K. C., Kastens, K., Spiess, F. N., and Miller, S. P.
1979 : Deep tow studies in the Tamayo Transform Fault: Marine Geophysical Research, v. 4, p. 37–70.

MacDonald, K. C., and Luyendyk, B. P.
1977 : Deep-tow studies of the structure of the Mid-Atlantic ridge crest near 37°N (Famous): Geological Society of America Bulletin, v. 88, p. 621–636.

MacDonald, K. C., Luyendyk, B. P., Mudie, J. D., and Spiess, F. N.
1975 : Near-bottom geophysical study of the Mid-Atlantic Ridge median valley near lat. 37 N: Preliminary observations: Geology, v. 3, p. 211–215.

MacDonald, K. C., Miller, S. P., Luyendyk, B. P., Atwater, T. M., and Shure, L.
1983 : Investigation of a Vine-Matthews magnetic lineation from a submersible: The source and character of marine magnetic anomalies: Journal of Geophysical Research, v. 88, p. 3403–3418.

MacDonald, K. C., Sempere, J. C., and Fox, P. J.
1984 : The East Pacific Rise from the Siqueiros to the Orozco fracture zones: along-strike continuity of the neovolcanic zone and the structure and evolution of overlapping spreading center: Journal of Geophysical Research, v. 89, no. B7, p. 6049–6306.

McClain, J. S., and Lewis, B.T.R.
1980 : A seismic experiment of the axis of the East Pacific Rise: Marine Geology, v. 35, p. 147–169.

McGregor, B. A., and Rona, P. A.
1975 : Crest of the Mid-Atlantic ridge at 26°N: Journal of Geophysical Research, v. 80, p. 3307–3314.

McKenzie, D. P., and Parker, R. L.
1967 : The North Pacific: An example of tectonics on a sphere: Nature, v. 216, p. 1276–1280.

Maury, M. F.
1963 : The physical geography of the sea: Cambridge, Mass., Harvard University Press, (from eighth edition, published 1861).

Menard, H. W., and Atwater, T.
1968 : Changes in the direction of seafloor spreading: Nature, v. 219, p. 463–467.

Minster, J. B., and Jordan, T. H.
1978 : Present-day plate motions: Journal of Geophysical Research, v. 83, p. 5331–5354.

Moore, J. G., Fleming, H. S., and Phillips, J. D.
1974 : Preliminary model for extrusion and rifting of the axis of the Mid-Atlantic ridge, 36°48′N: Geology, v. 2, p. 437–440.

Natland, J. H.
1980 : Effect of axial magma chambers beneath spreading centers on the compositions of Basaltic Rocks: Initial Report Deep Sea Drilling Project LIV, p. 833–850.

Needham, H. D., and Francheteau, J.
1974 : Some characteristics of the rift valley in the Atlantic Ocean near 36°48′N: Earth and Planetary Sciences Letters, v. 22, p. 29–43.

Nelson, K. D.
1981 : A simple thermal-mechanical model for mid-ocean ridge topographic variation: Geophysical Journal of Royal Astronomical Society, v. 65, p. 19–30.

Nisbet, E. G., and Fowler, C.M.R.
1978 : The Mid-Atlantic ridge at 37 and 45°N: some geophysical and petrologic constraints: Geophysical Journal Royal Astronomical Society, v. 54, p. 631–660.

Normark, W. R.
1976 : Delineation of the main extrusion zone of the East Pacific Rise at Lat. 21°N: Geology, v. 4, p. 681–685.

Orcutt, J. A., Kennett, B.L.N., and Dorman, L. M.
1976 : Structure of the East Pacific Rise from an ocean bottom seismometer array: Geophysical Journal of the Royal Astronomical Society, v. 45, p. 305–320.

Orcutt, J. A., McClain, J. S., and Burnett, M.
in press: Seismic constraints on the generation, evolution and structure of the ocean crust: Geological Society of London Special Publication.

Osmaston, M. F.
1971 : Genesis of ocean ridge median valleys and continental rift valleys: Tectonophysics, v. 11, p. 387–405.

OTTER Scientific Team
in press : The geology of the Oceanographer Transform: The eastern ridge–transform intersection: Marine Geophysical Research.

Pallister, J. S., and Hopson, C. A.
1981 : Semail ophiolite plutonic suite: Field relations, phase variation, cryptic variation and layering, and a model of a spreading ridge magma chamber: Journal of Geophysical Research, v. 86, p. 2593–2644.

Pálmason, G.
1980 : A continuum model of crustal generation in Iceland kinematic aspects: Journal of Geophysics, v. 47, p. 7–18.

Phillips, J. P., Thompson, G., Von Herzen, R. P., and Bowen, V. T.
1969 : Mid-Atlantic ridge near 43°N latitude: Journal of Geophysical Research, v. 74, p. 3069–3081.

Poehls, K.
1974 : Seismic refraction on the Mid-Atlantic ridge at 37°N: Journal of Geophysical Research, v. 79, p. 3370–3373.

Prevot, M., Lecaille, A., Hekinian, R.
1979 : Magnetism of the Mid-Atlantic ridge crest near 37°N from FAMOUS and DSDP results: A review; in Deep sea drilling results in the Atlantic ocean: Ocean crust, eds. M. Talwani, C. G. Harrison, D. E. Hayes; Maurice Ewing Series 2, American Geophysical Union, Washington, D.C., 431 p.

Parmentier, E. M. and Forsyth, D. W.
1985 : Three-dimensional flow beneath a slow-spreading ridge axis: a dynamic contribution to the deepening of the median valley toward fracture zones: Journal of Geophysical Research, v. 90, p. 678–684.

Purdy, G. M., Detrick, R. S., and Cormier, M.
1982 : Seismic constraints on the crustal structure at a ridge-fracture zone intersection: EOS, v. 63, p. 1100.

Ramberg, I. B., Gray, D. F., and Raynolds, R.G.H.
1977 : Tectonic evolution of the FAMOUS area of the Mid-Atlantic Ridge, lat. 35°50′ to 37°20′N: Geological Society of America Bulletin, v. 88, p. 609–620.

Rea, D. K.
1978 : Evolution of the East Pacific Rise between 3°S and 13°S since the middle miocene: Geophysical Research Letters, v. 5, p. 561–564.

Rea, D. K., Blakely, R. J.
1975 : Short-wavelength magnetic anomalies in a region of rapid seafloor spreading: Nature, v. 255, p. 126–128.

Reid, I. D., and Jackson, H. R.

1981 : Oceanic spreading rate and crustal thickness: Marine Geophysical Research, v. 5, p. 165–172.
Reid, I. D., and MacDonald, K. C.
1973 : Microearthquake study of the Mid-Atlantic Ridge near 37°N using sonobuoys: Nature, v. 246, p. 88–90.
Reid, I. D., Orcutt, J. A., and Prothero, W. A.
1977 : Seismic evidence for a narrow zone of partial melting underlying the East Pacific Rise at 21°N: Geological Society of American Bulletin, v. 88, p. 678–682.
Rhodes, J. M., and Dungan, M. A.
1979 : The evolution of ocean-floor basaltic magmas, in Deep drilling results in the Atlantic Ocean: Ocean Crust; eds. M. Talwani and others; v. 2, p. 262–272.
RISE Team
1980 : East Pacific Rise: Hot springs and geophysical experiments: Science, v. 207, p. 1421–1433.
Rona, P. A., Thompson, G., Mottl, M. J., Karson, J., Jenkins, W. J., Graham, D., Von Damm, K., and Edmond, J. M.
1982 : Direct observations of hydrothermal mineralization at the TAG hydrothermal field, Mid-Atlantic ridge 26°N: EOS, v. 63, p. 1014.
Rona, P. A., and Gray, D. F.
1980 : Structural behavior of fracture zones symmetric and asymmetric about a spreading axis: Mid-Atlantic ridge (latitude 23°N to 27°N): Geological Society of America Bulletin, v. 91, p. 485–494.
Schouten, H., and Klitgord, K. D.
1982 : The memory of the accreting plate boundary and the continuity of fracture zones: Earth and Planetary Sciences Letters, v. 59, p. 255–266.
1983 : Overlapping spreading centers on the East Pacific Rise—comment: Nature, v. 303, p. 549–550.
Schouten, H., and White, R. S.
1980 : Zero-offset fracture zones: Geology, v. 8, p. 175–179.
Searle, R. C.
1979 : Side-scan sonar studies of North Atlantic fracture zones: Journal of the Geological Society, v. 136, part 3, p. 283–292.
Searle, R. C., and Laughton, A. S.
1977 : Sonar studies of the Mid-Atlantic ridge and Kurchatov Fracture Zone: Journal Geophysical Research, v. 82, p. 5313–5328.
Shand, S. J.
1949 : Rocks of the Mid-Atlantic ridge: Journal Geology, v. 57, p. 89–91.
Shih, J.S.F.
1980 : The nature and origin of fine-scale seafloor relief; [Ph.D. thesis] Massachusetts Institute of Technology, Cambridge, Mass., 222 pp.
Sleep, N. H.
1969 : Sensitivity of heat flow and gravity to the mechanism of seafloor spreading: Journal of Geophysical Research, v. 74, p. 542–549.
1975 : Formation of ocean crust: Some thermal constraints: Journal of Geophysical Research, v. 80, p. 4037–4042.
1978 : Thermal structure of mid-oceanic ridge axes, some implications to basaltic volcanism: Geophysical Research Letters, v. 5, p. 426–428.
Sleep, N. H., and Biehler, S.
1970 : Topography and tectonics at the intersections of fracture zones with central rifts: Journal of Geophysical Research, v. 75, p. 2748–2752.
Sleep, N. H., and Rosendahl, B. R.
1979 : Topography and tectonics of mid-ocean ridge axes: Journal of Geophysical Research, v. 84, p. 6831–6839.
1983 : Hydrothermal convection at ridge axes: in Hydrothermal processes at seafloor spreading centers, ed. P. A. Rona: NATO conference series, series 4, v. 12, Plenum Press, New York, p. 71–82.
Spindel, R. C., Davis, S. B., Macdonald, K. C., Porter, R. P., and Phillips, J. D.
1974 : Microearthquake survey of median valley of the Mid-Atlantic ridge at 36°30′N: Nature, v. 248, p. 577–579.
Stakes, D., Shervals, J. W., and Hopson, C. A.
1984 : The volcano-tectonic cycle of the FAMOUS and AMAR valleys, Mid-Atlantic Ridge (36°47′N): Evidence from basalt glass and basalt phenocryst compositional variations for a steady-state magma chamber beneath the valley midsections: Journal of Geophysical Research, v. 89, p. 6995–7028.
Tamayo Tectonic Team
1984 : Tectonics at the Intersection of the East Pacific Rise with the Tamayo Transform Fault: Marine Geophysical Research, v. 6, p. 159–185.
Tapponnier, P., and Francheteau, J.
1978 : Necking of the lithosphere and the mechanics of slowly accreting plate boundaries: Journal of Geophysical Research, v. 83, p. 3955–3970.
Toomey, D. R., Murray, M. H., Purdy, G. M., and Murray, M. H.
1982 : Microearthquakes on the Mid-Atlantic ridge near 23°N: new observations with a large network: EOS, v. 63, p. 1103.
Trenu, A. M.
1982 : Seismicity and structure of the Orozco transform from ocean bottom seismic observation, WHOl-reference-82-13; [Ph.D. dissertation] Woods Hole Oceanographic Institution/Massachusetts Institute of Technology.
van Andel, Tj. H., and Ballard, R. D.
1979 : The Galapagos Rift at 86°W, 2, Volcanism, structure and evolution of the rift valley: Journal of Geophysical Research, v. 84, p. 5390–5406.
van Andel, Tj. H., and Bowin, C. O.
1968 : Mid-Atlantic ridge between 22° and 23°N latitude and the tectonics of mid-ocean rises: Journal of Geophysical Research, v. 73, p. 1279–1298.
van Andel, Tj. H., and Heath, C. R.
1970 : Tectonics of the Mid-Atlantic ridge 6-8° south latitude: Marine Geophysical Research, v. 1, p. 5–36.
Van Wagoner, N. A., and Johnson, H. P.
1983 : Magnetic properties of three segments of the Mid-Atlantic ridge at 37°N: FAMOUS, Narrowgate and AMAR: AMAR2: Journal of Geophysical Research, v. 88, p. 5065–5083.
Verosub, K. L., and Moores, E. M.
1981 : Tectonic rotations in extensional regimes and their paleomagnetic consequences for oceanic basalts: Journal of Geophysical Research, v. 86, p. 6335–6350.
Vine, F. J.
1966 : Ocean floor spreading: new evidence: Science, v. 154, no. 3755, p. 1405–1415.
Vine, F. J., and Matthews, D. H.
1963 : Magnetic anomalies over oceanic ridges: Nature, v. 199, p. 947–949.
Weidner, D. J. and Aki, K.
1973 : Focal depth and mechanism of mid-ocean ridge earthquakes: Journal of Geophysical Research, v. 78, p. 1818.
Vogt, P. R., Kovacs, L. C., Bernero, C., and Srivastava, S. P.
1982 : Asymmetric geophysical signatures in the Greenland-Norwegian and south Labrador seas and the Eurasian basin, Tectonophysics, v. 89, p. 95–160.
Whitmarsh, R. B.
1975 : Axial intrusion zone beneath the median valley of the Mid-Atlantic ridge at 37°N detected by explosion seismology: Geophysical Journal of Royal Astronomical Society, v. 42, p. 189–215.

MANUSCRIPT ACCEPTED BY THE SOCIETY JULY 1, 1984

Acknowledgments

I would like to thank C. Hopson whose review was so thorough he should probably be a co-author, P. Vogt who cajoled me into writing this paper, and P. J. Fox, E. Bonatti, A. Palmer, D. Stakes, R. Haymon, J. Bicknell, and P. Vogt who improved the paper with their careful reviews. My efforts in the research presented in this paper have been supported by the National Science Foundation and the Office of Naval Research.

Printed in U.S.A.

The Geology of North America
Vol. M, The Western North Atlantic Region
The Geological Society of America, 1986

Chapter 5

Subaerial volcanism in the western North Atlantic

Kristján Saemundsson
Orkustofnun, Grensasv. 9, 108 Reykjavík, Iceland

INTRODUCTION

Three regions of subaerial volcanism occur near the eastern edge of the North America plate: Iceland, Jan Mayen and the Azores. In Iceland the ridge axis emerges above sea level, most of the volcanic pile was generated subaerially along its trace, and superimposed flank zones are situated off the accretion zone. The Azores lie on both sides of the oceanic ridge axis topping an extensive platform adjoining it. The platform was generated by excess volcanic production on the transecting ridge crest similar to the ridge formation in Iceland. The islands are stratovolcanoes formed off the ridge crest by flank volcanism. The volcanism and tectonic setting of Jan Mayen and the Azores is compared to the flank zones of Iceland, of which they appear to be close analogs.

Direct access to an active oceanic spreading axis makes Iceland particularly significant. The crust generated there suffers erosion which exposes the uppermost 1-2 km for three dimensional view.

Subaerial volcanism in Iceland's accretion zone produces a layered igneous sequence due to a wider rift zone and extensive lateral spread of lava flows as opposed to the deep ocean ridge with a much narrower rift and outflow zone of lava. Subaerial spreading may also have been important in early rifting of the North Atlantic margins (Hinz, 1981). As a result, submerged layered extrusives with units thickening downdip towards the spreading axis and offlapping in the direction of spreading may exist near some of the North Atlantic continental margins (Tucholke and Ludwig, 1982).

Submarine plateaus and chains or clusters of seamounts on the floor of the NW-Atlantic may have been generated partly by subaerial volcanism. These are briefly discussed and compared to possible present day analogs. In conclusion, certain aspects of the geology of the three subaerial regions are discussed with regard to the hotspot model.

ICELAND

Tectonic Frame and Morphology

The geological structure of Iceland is controlled by its position on the Mid-Atlantic Ridge, with extensional features predominating. Active deformation occurs in axial rift zones, non-rifting flank zones and along oblique or transverse fracture zones that connect offset segments of the axial rift zones either internally or to the submarine mid-ocean ridge axis (Fig. 1).

Iceland is the longest ridge segment (400 km) and the largest landmass (103,000 km^2) anywhere exposed along the Mid-Ocean Ridge System. Morphologically there are two major units: 1) **The active volcanic areas** with central volcanoes, and their associated fissure swarms, the loci of fissure eruptions, fault scarps and open cracks. The larger central volcanoes of the interior are glacier covered, and subglacial volcanic products predominate. The subglacial eruptive environment modifies the volcanic products to impressive piles of pillow lava and hyaloclastite that dominate in the landscape. Erosion in the active volcanic areas is outweighed by volcanism. However, sheet floods connected with subglacial eruptions convey and deposit large quantities of volcanic material which forms extensive outwash plains and debris cones off the coasts. 2) Two thirds of the area of Iceland forms **marginal blocks** flanking the active volcanic zones. The morphology of these is characterized by stream and ice erosion which has stripped off the topmost part of the original lava pile and carved valleys and hummocky plains into it. The term lava pile refers to the monotonous sequence of primarily lava flows exposed in cliff or valley sections sometimes over 1 km high. 3) A third morphological unit is the **shelf around Iceland,** essentially an abrasion platform overlain by volcanogenic sediment and extended oceanward by prograding sediment deposition.

Early History of Iceland Block

Little is known about exactly when the Icelandic insular basement (Iceland Block, Fig 1) began to form. Marine magnetic anomalies largely fail due to extensive outflow of lavas in the subaerial environment. There are clues, however. Nunns, (1983) reconstruction for the NE-Atlantic and Norwegian Greenland Sea suggests that the Iceland Block began to form between 44 and 26 m.y. ago. Vogt and others (1980) on the basis of mag-

Saemundsson, K., 1986, Subaerial volcanism in the western North Atlantic; *in* Vogt, P. R., and Tucholke, B. E., eds., The Geology of North America, Volume M, The Western North Atlantic Region: Geological Society of America.

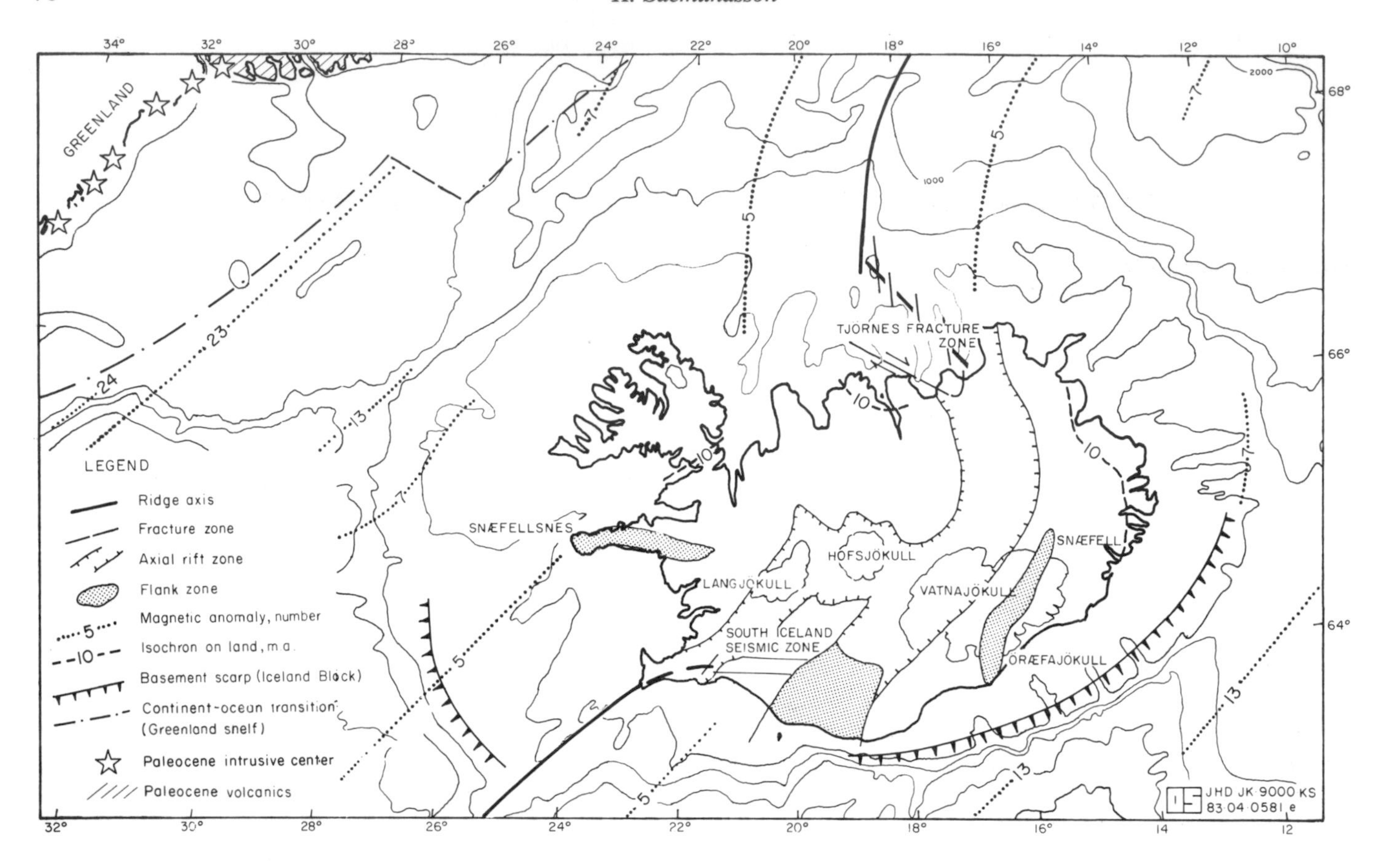

Figure 1. Simplified morphotectonic map of the Iceland Block and the hotspot track towards Greenland. Compiled from Larsen (1980), Talwani and others (1971), Talwani and Eldholm (1977), Vogt and others (1980), Saemundsson (1979), Fleischer (1974), Kristjánsson (1976), Egloff and Johnson (1979) and Nunns (1983).

netic anomaly identification, and Sigurdsson and Loebner (1981) on the basis of increased subaerial explosive volcanism recorded in deep sea sediments, relate formation of the Iceland Block to increased hotspot activity around 25 Ma. From the vesicularity of basalt drilled on the west flank of Reykjanes Ridge at 63-64° north, Duffield (1978) concluded that water depth at the axis had diminished from about 2500 m to about 700 m between eruption of the oldest basalt, at 35.5 Ma to the youngest basalt, at 2.7 Ma, with an intermediate datum of about 1400 m at 19.5 Ma. The evidence suggests that excessive volcanism on the Iceland Block began sometime between 19.5 and 35.5 Ma. There is no firm evidence yet to indicate that Iceland developed from submarine to subaerial growth. However, the observed shoaling of the Reykjanes Ridge makes this probable.

Geology of the Lava Pile Marginal to Axial Rift Zones

Age of Lava Pile. Two stratigraphic groups make up the marginal blocks. These are an Upper Tertiary series and a Plio-Pleistocene series. A distinction is made on the basis of widespread and frequent intercalations of glacial debris and subglacially extruded volcanics in the Plio-Pleistocene series, combined with magnetic reversal patterns and radiometric age dating (McDougall and Wensink, 1966; Saemundsson, 1979). The Plio-Pleistocene series defined on this basis includes the Matuyama epoch and the Gauss epoch above the Mammoth event, and ranges in age from 3.1 to 0.7 Ma. Superpositioning and the present configuration of axial rift zones predict that the oldest exposed rocks in Iceland should occur in the farthest northwest, north and east. Radiometric dating indicates that the oldest exposed rocks in eastern Iceland are 13.6 Ma (Watkins and Walker, 1977), in the Northwest about 15 Ma (Moorbath and others, 1968, McDougall and others, 1984) and about 12 Ma in the North (Saemundsson and others, 1980). These ages appear unexpectedly low compared to the age of the adjoining sea bottom from magnetic anomalies (Nunns, 1983) if extrapolated over Iceland. The explanation may be extensive outflow of subaerial lavas, these ages being derived from the uppermost 1000 m of the pile below which another 2-3 km of lavas have been drilled (Pálmason and others, 1978). An even greater thickness is suggested by model calculations (Pálmason, this volume). Southeastward displacement of axial rift zones with time combined with simultaneous crustal spreading on two parallel axes is also important (Saemundsson, 1979).

Rocks of Lava Pile and Environment During Build Up. The marginal areas consist of dominantly tholeiitic basalts (90%)

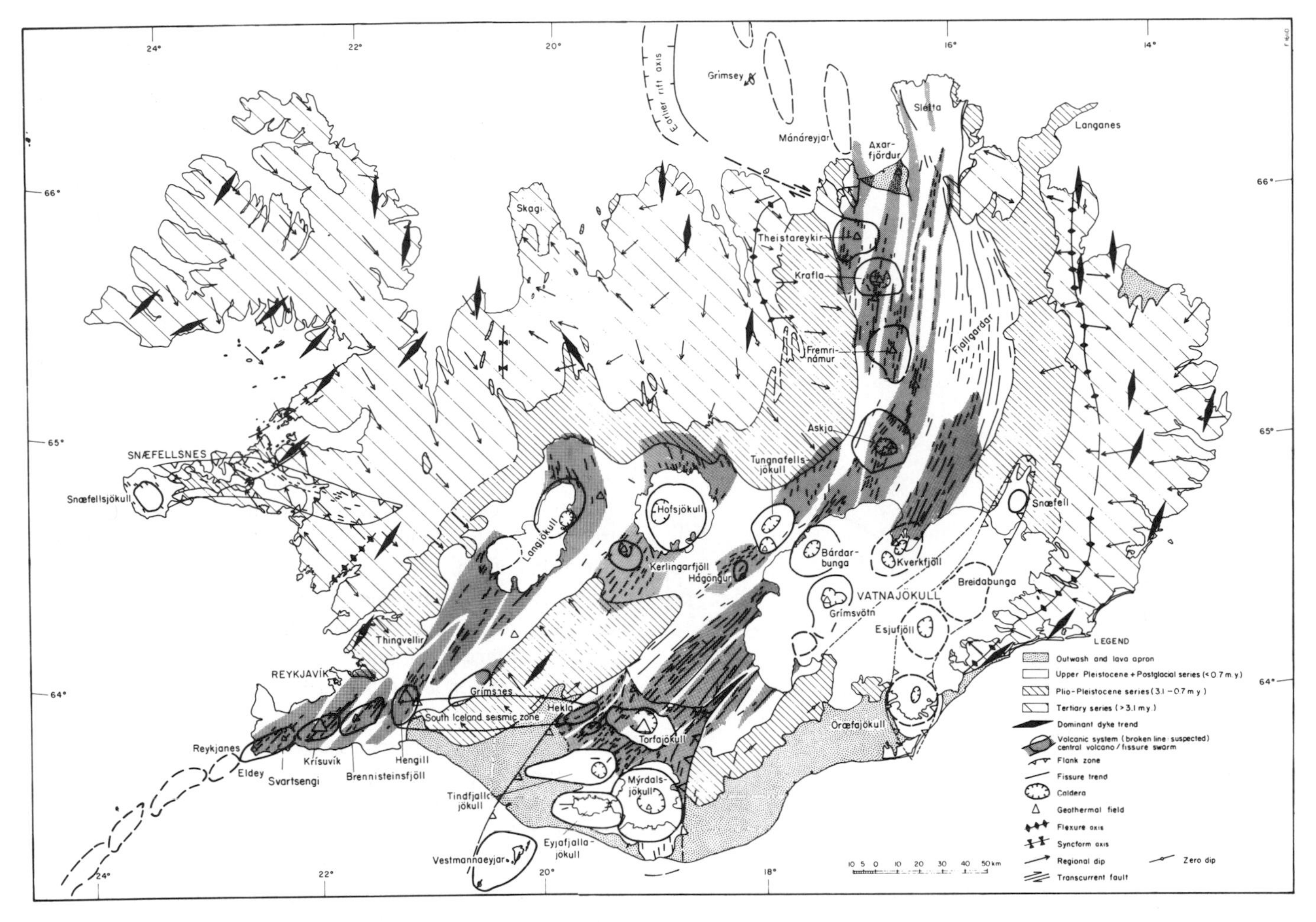

Figure 2. Structural outline map of Iceland. From Saemundsson (1979) slightly modified. Offshore structures in Tjörnes Fracture Zone after McMaster and others (1977).

with acid and intermediate rocks constituting about 5%. Tuffs and volcanogenic sediments amount to some 5% in a lava pile not influenced by subglacial volcanism (Walker, 1959) but increase considerably in the Plio-Pleistocene series. The volcaniclastic material is found as thin partings between the lavas, characteristically red colored, windblown basaltic tuff. Thicker clastic beds occur also which are composed of tuff, reworked debris and sometimes lignite. They are traceable for tens of km along strike (Fig. 3). In the Plio-Pleistocene series subglacial volcanics and glacially derived detrital beds gain in volume and may exceed 50% of the lava pile on a regional scale in South Iceland. In southeastern Iceland ice sheets developed locally in the Pliocene probably under conditions similar to the present day Vatnajökull, i.e., high ground and high precipitation. Deposits of glacial origin are found there interspersed within the lava pile back to about 5 Ma (Torfason, 1979). In eastern and western Iceland the glacial intercalations begin around 3 Ma (McDougall and Wensink, 1966; Saemundsson and Noll, 1974) whereas in northeastern Iceland they first occur around 2 Ma (Albertsson, 1978). This clearly reflects the influence of orographic conditions and precipitation. Sections of the lower part of the Plio-Pleistocene strata in western Iceland (Saemundsson and Noll, 1974, Kristjansson and others, 1980) from the Mammoth event of the Gauss epoch up to the Gilsá event of the Matuyama (3.1 up to 1.8 Ma) indicate that glaciations recurred every 100,000-120,000 years. Information about the frequency of glaciations for the upper part of the Plio-Pleistocene, is available only from the Tjörnes sequence on Iceland's north coast, where 6 glaciations are reported for the period from the Gilsá event (1.8 Ma) up to the end of the Matuyama epoch (0.7 Ma) (Albertsson, 1978).

Structure of Lava Pile. Structural relationships indicate that the lava pile grew as lenticular units from elongate volcanic systems which included swarms of dykes and fissures intersecting central volcanoes (Walker, 1963; Gibson and Piper, 1972). The width of dykes most commonly is in the range 1-5 m. The dykes tend to thicken with depth and towards the central volcanoes (Helgason and Zentilli, 1985). The majority of faults and dykes stand nearly normal to the stratification, implying a subvertical attitude at the time of formation. Normal faults with a hade of 60-70° also occur. The faults appear contemporaneous with dyking and may represent the near surface expression of dyke swarms (Walker 1965, Helgason and Zentilli, 1985). The throw of

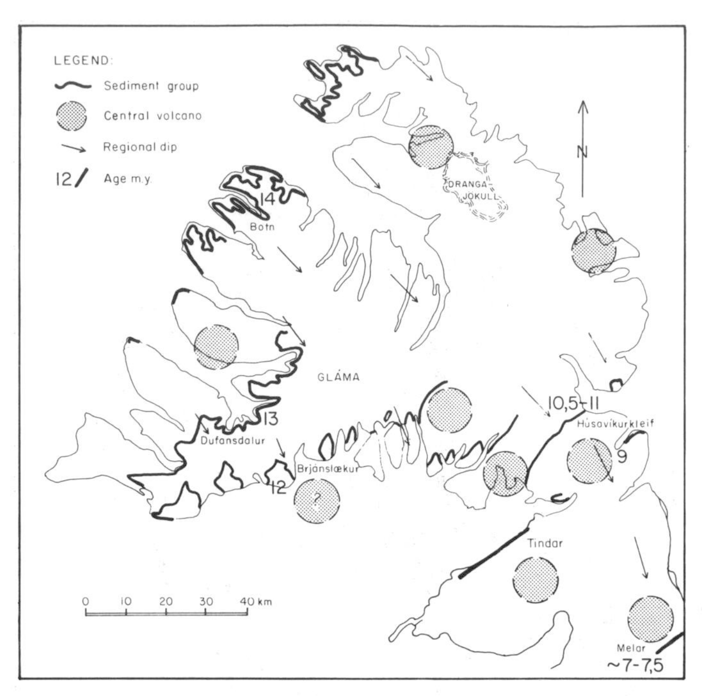

Figure 3. Major clastic cycles in NW-Iceland. Most of them include several layers intercalated with basalt flows. Preferred interpretation is that they separate between volcanic systems that migrated eastward with time. Ages are from McDougall and others (1984).

individual faults varies from a few meters up to several hundred m. Detailed unpublished maps show a range of spacing of less than 1 to rarely over 5 faults per kilometer.

The lenticular units attain their greatest thickness in the area of the central volcanoes which are also marked by the occurrence of acid and intermediate rocks among the copious basalt outpourings. They commonly develop calderas. Intrusive bodies of dolerite, gabbro and/or granophyre up to 10 km in area, are exposed in many eroded central volcanoes. They are also pervaded by intrusive sheets (Walker, 1975) which sometimes form discrete cone sheet swarms of up to 15 km in diameter. The proportion of intrusive rocks may locally exceed 50% in the more deeply eroded central volcanoes (Walker, 1963) whereas the dyke intensity outside them rarely exceeds 10% in surface exposure. The great proportion of intrusive rocks at shallow depth gave rise to temporary hot water convection cells which caused propylitization of the core of the central volcanoes. The hydrothermal alteration of the core areas is accordingly more intense than the zeolite grade metamorphism of the surrounding volcanic pile with an almost horizontal zonation (Walker, 1960). The deepest zeolite zone with laumontite, is rarely reached outside areas of dense sheet or dyke swarms (Walker, 1975). Boreholes outside intrusive centers (Pálmason and others, 1978; Mehegan and others, 1982) show the laumontite zone, between 1800 and 2800 m below the original surface of the lava pile, followed downcore by an epidote chlorite zone. The fossil thermal gradients at the time of metamorphic overprinting accordingly ranged from 70° to 90°C/km (Kristmannsdottir, 1982).

The life spans of the volcanic systems have been found to vary from 300,000 to over 1 million years. They are preserved as entities in the volcanic pile, indicating that they grew, drifted off towards the margin of the current volcanic zone and then became extinct. New ones replaced them over the more or less stationary deep-seated zone of magma generation. The central volcanoes during their active periods stand apart topographically, later to become buried by onlapping lavas from younger volcanic systems. Prominent and widespread clastic beds in western Iceland, traceable for tens of kilometers along strike (Fig. 3) may relate to transition periods when new volcanic systems took over (McDougall and others, 1977; Saemundsson, 1979), or to larger scale rift jumping (Saemundsson, 1974; Johannesson, 1980). Possibly some of them represent drastic climatic fluctuations (Mudie and Helgason, 1983).

Thickness, Tilt and Accumulation Rate of Lava Pile. Vertical sections of the volcanic sequence in Iceland expose up to 1500 m of rocks below which lie at least another 2-5 km of extrusives. At this depth seismic layer 3 (Vp = 6.5 km/s) is reached (Pálmason, 1971), which is interpreted by most workers as a metamorphic transition (Christensen and Wilkens, 1982) and might also constitute the base of extrusive rocks. Deep drilling has confirmed a minimum 4.5 km vertical thickness of the lava pile (Pálmason and others, 1978). The lavas of the pile dip gently on a regional scale, generally toward the central part of Iceland (Fig. 2). The dips increase gradually from near zero at the highest exposed levels to about 5-10° at sea level. The increase in dip is matched by individual lava groups thickening down the direction of dip (Walker, 1959). The regional tilt thus must have been imparted to the pile during its growth. Extension and subsidence was matched by dyke injections and lava pouring out at the surface. Lateral transport due to crustal spreading would gradually remove the growing pile away from the zone of accretion and no more lavas would add to it. Systematic mapping on a regional scale has been carried out in most parts of the Tertiary and Plio-Pleistocene areas. Well documented accumulation rates for composite but continuous sections are listed in Table 1. These sequences do not generally involve central volcanoes, with the few exceptions where short segments were measured across their flanks. High extrusion rate usually correlates with a deep level of exposure as is evident for the IRDP-borehole which correlates with the East Iceland sequence in Table 1. High extrusion rates in East, North and NW-Iceland are as yet inconclusive as regards magmatic pulses starting at about 12.5 and 7 Ma proposed on the basis of ridge crest morphology north and south of Iceland (Vogt and others, 1980).

Geology of the Axial Rift Zones

Definition and Rock Types. The axial rift zones are defined as comprising rocks formed during the Brunhes magnetic epoch (last 0.7 m.y.). Two series are distinguished: an Upper Pleistocene series and a Postglacial series. Axial rift zones so defined cover about 1/3 of the area of Iceland. Usually an unconformity is found between rocks of the Plio-Pleistocene and the

TABLE 1. ACCUMULATION RATES FOR LONG CONTINUOUS LAVA SEQUENCES IN ICELAND

Area/thickness of section (km)	Period Ma	Accumulation rate (m/m.y.)	Depth below orig. surface (km)	Reference
East Iceland (9 km)	13.4- 7.6 7.6- 6.5 6.5- 2.3	720 2600 360	0.4-1.3 0.8-1.7 0.2-0.8	Watkins and Walker (1977)
East Iceland IRDP borehole (1.9 km)	9.5-11	1300	1.5-3.4	Helgason and Zentilli (1982)
West Iceland (3.5 km)	7.0- 2.0	730	0.3-1.0	McDougall et al. (1977)
West Iceland (2.1 km)	4.2- 1.8	875	0.3-1.1	Kristjansson et al. (1980)
North Iceland (5 km)	12.0- 9.5 9.5- 9.0	1000 4000	0.3-1.2 0.2-1.0	Saemundsson et al. (1980)
NW Iceland (6.9 km)	14 -11.9 12.7- 8.0	1820 670	0.3-0.9 0.3-1.2	McDougall et al. (1984)

Upper-Pleistocene due to volcanic products of the axial rift zones forming a transgressive apron of lava flows. The volcanic rocks of the Upper Pleistocene fall into two types with regard to structure and morphology: subaerial lava flows and subglacial piles of pillow lava and hyaloclastite. Rocks belonging to the Upper Pleistocene series are to a large extent related to active volcanic systems (Fig. 2). The sequence of glacials and interglacials during the Upper Pleistocene is incompletely known in terms of Icelandic lava pile stratigraphy. However, groups of lava flows alternating with tillite beds and hyaloclastite suggest several interglacial/glacial cycles.

Active Volcanic Systems. Deformation of the axial rift zones is concentrated into fissure swarms commonly 5-10 km broad and up to 100 km long (see satellite image, Vogt and Tucholke, this volume). Their trend is variable but uniform within each branch of the axial rift zones. They typically form en echelon arrays which may be dextral or sinistral depending on the trend of the individual branches with respect to the direction of relative plate separation, which is near N 100°E (Minster and Jordan, 1978). The dominant elements of the fissure swarms are the volcanic fissures, lava shields and non-eruptive open cracks and faults. Most of the fissure swarms pass through central volcanoes, these together constituting a volcanic system. The boundaries of individual volcanic systems may be diffuse in areas of interfingering; also, lava shields sometimes occupy a marginal position which makes their alliance problematic. Over 20 volcanic systems have been defined within the axial rift zones of Iceland (Saemundsson, 1978). Of the active central volcanoes, 10 have developed calderas and a related circular fracture system. Despite being the locus of maximum lava production on the volcanic systems, the central volcanoes of the axial rift zones have low relief relative to their surroundings except where they grew under glaciers. Most central volcanoes in the axial rift zone can be classified as composite shield volcanoes on the basis of their shape and the rock types erupted. The total discharge of lavas in the axial rift zones along Iceland's 400 km MAR segment is about 40 $km^3/1000$ years, with a maximum in central Iceland (Jakobsson, 1972).

The faults have nearly vertical hades and are spaced a few tens of meters up to a few kilometers apart. They are typically parallel stepped and branching. When traced along their trend across rocks of variable age, the faults emerge as growth faults smoothed by lava from time to time. Cross faults and volcanic fissures striking normal to the main trend are sometimes seen but they are rare. Fissures trending normal to the spreading direction, as described by Searle and Laughton (1981) are present in oblique axial rift zones such as the Reykjanes Peninsula, but are subordinate. The faults and fissures with their vertical hade and horizontal extension are probably associated with dyke injection (Walker, 1965). The active fissure swarms have prominent graben structures with most of the throw, up to several hundred m, occurring towards their margins. Listric fault planes would be expected at depth (Tryggvason, 1970, Nakamura, 1970).

Mechanism of Crustal Growth. Volcanic eruptions in Iceland average one every 5 years over the last few centuries (Thorarinsson, 1966). Only a few of those are associated with major ground fissuring. Three events of this kind have occurred in northern Iceland since 1724. The last which affected the Krafla volcanic system has been in progress since 1975 and is still being monitored. Throughout the active period tiltmeters at Krafla showed continuous inflation of the central volcano. Inflation was interrupted by recurring short deflation events (Björnsson et al., 1979). Deflation was accompanied by ground fissuring and E-W widening of limited segments of the fissure swarm, and often by extrusion of lava mainly during 1980 to 1984.

The kinematics of rifting can be explained by gradual buildup of tensional stress across the axial rift zone in response to plate divergence at an average rate of 2 cm/y for the last few m.y.

(Vogt, 1983). The intervals of tensional stress buildup last 100 years or more and alternate with episodes of active rifting that last only a few years. During an active episode, magma ascends below the central volcano and is stored in shallow magma chambers (Einarsson, 1978). At Krafla the inflow of magma is on the order of 5 m^3/s, a figure based on the rate and area of inflation (Björnsson and others, 1979). During the short deflation/rifting events magma is expelled from the shallow magma chambers into the fissure swarm along laterally injecting dykes (Brandsdóttir and Einarsson, 1979, Einarsson and Brandsdóttir, 1980). As a result the tensional stress is gradually released and the accompanying magmatism changes from primarily dyke injection in the early stage of the rifting episode to lava outflow at a later stage. The extension across the Krafla fissure swarm along its 80 km length varies from less than 1 m up to 8 m in the middle part (Tryggvason, 1983). Besides rifting in the central part of the fissure swarm, geodetic measurements have shown that the stretched flanks contract due to elastic rebound (Björnsson and others, 1979). Long term gravity changes, tilt variations and geodetic measurements across the axial zones in NE-Iceland are consistent with crustal separation as slow continuous stretching during quiet periods and episodic rifting with up to several m of extension occurring within a few years period. This kind of stress release presumably holds for other parts of the axial rift zone as well, although the time scale and magnitude of rifting may be different. Evidence from stress measurements in the Tertiary and Plio-Pleistocene areas (Haimson and Rummel, 1982) is conjectural as to whether the lithospheric plates are being forced apart into compression or dragged apart into tension.

Rift Jumping. The regional dips within the lava pile define shallow synforms and low antiforms (Fig. 2). The synforms may have formed in a similar way as the presently active axial rift zones from crustal extension and downsagging, thus indicating the position of extinct rift zones (Saemundsson, 1967). There are two such structures in western Iceland, one trending SW-NE in the Snaefellsnes area and the other trending N-S in the area west of Skagi (Fig. 2). Volcanism came to extinction in these synform areas about 6-7 m.y. ago (Saemundsson, 1979; Johannesson, 1980). At the same time, new rift zones were initiated or intensified, out of which grew the present day Reykjanes-Langjökull zone and the axial rift zone in northern Iceland. In northern Iceland an age and structural discontinuity is present within the Tertiary basalts on both sides of the axial rift zone (Saemundsson, 1974). The older sequence of basalts is downwarped and commonly dips 20-30° below the edge of the younger unit. In western Iceland the shift of the axial rift zone from Snaefellsnes to Reykjanes - Langjökull led to an antiform structure of the Tertiary lavas in between (Saemundsson, 1977; Johannesson, 1980). The western flank was formed prior to 7 m.y. ago within the now extinct NE-SW trending Snaefellsnes axis, and the eastern flank formed from about 7 m.y. ago up to Recent within the Reykjanes-Langjökull zone.

Further evidence of the transient character of the axial rift zones is found off Iceland's north coast (Fig. 2). The extinct southern continuation of Kolbeinsey Ridge, now marked by a deep graben structure west of Grimsey Island, once connected to the northern axial rift zone along a narrow transform fault zone (Húsavík faults). A set of active en echelon fissure swarms east of this island now connects the Kolbeinsey Ridge with the axial rift zone in Northern Iceland (McMaster and others, 1977). An age of about 1 m.y. has been suggested for this rift jump (Saemundsson, 1978).

The two segments of the axial rift zones north of the South Iceland seismic zone appear complementary as regards their volcanic products. The western segment erupts olivine tholeiite forming lava shields, whereas the eastern segment erupts more evolved quartz tholeiite and lava shields are absent. The western segment is currently at a low in volcanic production, a part of it even lies below sea level in Lake Thingvallavatn. The situation is comparable to the graben west of Grímsey referred to above. The tectonics and volcanism suggest an ongoing rift jump from the western zone to the eastern zone, concurrent with an eastward propagation of the oblique Reykjanes rift zone along the South Iceland seismic zone (Saemundsson, 1978).

Flank Zones

Three branches of active volcanic zones with poorly developed extensional features are located outside the main accretion axis. Those are the Snaefellsnes volcanic zone, the southern part of the eastern volcanic zone and the Öraefajökull-Snaefell zone in SE-Iceland (Fig. 2). Their volcanic products lie unconformably upon older piles of volcanics that at least partially suffered erosion before the flank volcanism started. The flank zones erupt mostly transitional to alkalic lava types and the volume production decreases markedly with increasing alkalic affinities (Jakobsson, 1972). Most central volcanoes of the flank zones form large stratovolcanoes, cone shaped or elongated parallel to the fissure trend. Many of them have a large summit crater and sometimes a caldera.

The Snaefellsnes volcanic zone has an E-W trend and consists of three WNW-ESE trending volcanic arrays and normal faults (Sigurdsson, 1970). The total volcanic discharge amounts to only 0.6 $km^3/1000$ years averaged over the Holocene (Jakobsson, 1972). The zone crosses the age discontinuity of the antiform structure in western Iceland. Recent volcanism is confined to the older western flank (pre 7 Ma), but faults continue eastward into the younger sequence. They diminish, however, rather abruptly and become dispersed. The fracture pattern favours dextral shear with the volcanic arrays developing parallel with the direction of maximum horizontal compression. E-W and north-southerly fracture trends are both present but less conspicuous and may represent the two conjugate strike-slip fracture trends. The zone has been interpreted also in terms of horizontal north-southerly extension based on focal mechanism solutions in the eastern part of the zone (Einarsson and others, 1977) or dextral shear resulting from supposed different spreading rates north and south of the zone (Sigurdsson, 1970).

In southern Iceland the flank zone forms the southern propagation of the spreading eastern volcanic zone. From an analysis of the tectonic fracture pattern associated with the Surtsey, 1963–1967, and Heimaey, 1973, eruptions, within the Vestmannaeyjar volcanic field, a conjugate shear set with faults trending NNE-SSW and ENE-WSW has been inferred (Brander and Wadge, 1973). The orientation of the volcanic axes through the stratovolcanoes on land, including Hekla, follows the same general trends (Saemundsson, 1978). The southern part of the eastern volcanic zone, here referred to as a flank zone, may thus be structurally controlled by sinistral shear like the adjoining South Iceland seismic zone.

The Öraefajökull-Snaefell NE-SW chain of stratovolcanoes is located in the eastern Vatnajökull region some 50 km east of and parallel to the main rift axis (Fig. 2). Snaefell, north of Vatnajökull, and Öraefajökull, (2119 m) south of the glacier and the only one active in Postglacial time, are both slightly N-S elongated. Both lack the fissure swarms characteristic of the axial rift zones. The rocks of Öraefajökull are tholeiitic, however, with alkalic trends (Prestvik, 1979). It is possible that similar volcanoes existed in this part of Iceland in Tertiary time, also offset eastward relative to the main spreading axis (Torfason, 1979). Evidence of this is seen in the large gabbro and granophyre intrusions, up to 20 km^2 in area, of southeastern Iceland (Fig. 4). They bear no obvious relation to the build up of the surrounding lava pile having been intruded during a later magmatic period, and have no elongated dyke swarms localized about them. They most likely formed in the roots of stratovolcanoes that grew on top of the lava pile, removed from the zone of active spreading. This flank zone closely parallels the axial rift zone in the area of maximum volcanic discharge across the Icelandic hotspot (Tryggvason and others, 1983) suggesting a causative connection.

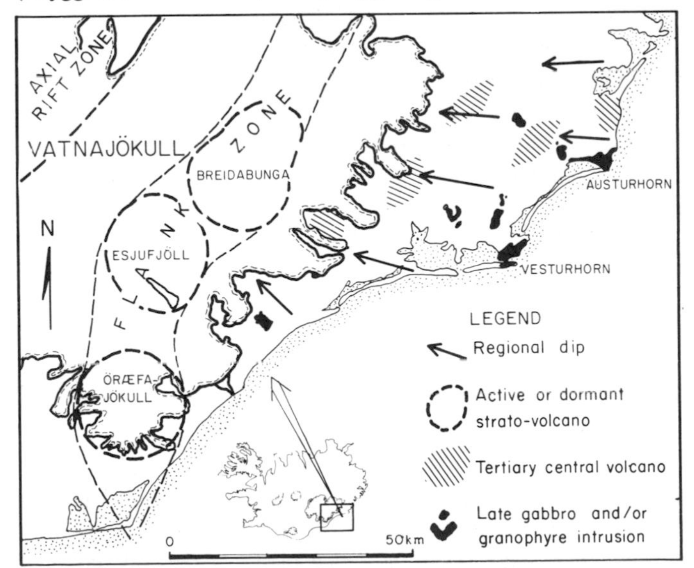

Figure 4. Major intrusions in SE-Iceland unrelated to central volcanoes of the lava pile may have formed in the roots of later volcanoes superimposed on the lava pile. Their relation to the axial rift zone of their day was probably similar as that of the Öraefajökull-Snaefell flank zone which parallels the currently active axial rift zone. Based on information from Helgi Torfason (1979).

Fracture Zones

The axial rift zones in Iceland connect with the submerged ocean ridges along transverse fault zones which were defined by Sykes (1967) and Ward (1971) as fracture zones from earthquake distribution patterns. They display a variety of tectonic features along their strike, including oblique segments of the axial rift zones with pronounced en echelon features, as well as older crust broken up by active faulting. One such fracture zone, the South-Iceland seismic zone (Fig. 2), connects the Reykjanes Peninsula and Iceland's Eastern volcanic zone (Björnsson and Einarsson, 1974). It lies across eroded Plio-Pleistocene rocks dipping to the northwest and north, and is broken up by a system of young faults trending N-S and ENE-WSW. Together they form a conjugate shear set indicating sinistral shear. Only minor lateral offsets of the strata have been observed.

The axial rift zone in northern Iceland connects with the Kolbeinsey Ridge along the Tjörnes Fracture Zone. It involves both NW-SE trending transcurrent faults and N-S grabens and volcanic fissure swarms arranged in a left stepping en echelon pattern (McMaster and others, 1977). The total width of the zone from north to south, as defined from earthquake distribution, is about 70 km. Geological evidence points to a narrow zone of true transform character along the NW-SE trending Húsavík fault zone (Saemundsson, 1974) that preceded the N-S en echelon fissure swarms (Fig. 1). The Húsavík faults are traceable for 25 km on land from the edge of the axial rift zone. They juxtapose rocks of contrasting dips and ages. The faults can be followed from Húsavík on the sea floor for 50 km towards NW where they merge with a deep graben structure extending northwards up to Kolbeinsey Ridge. Gravity measurements have revealed a very marked low coinciding with the trace of the fault zone in this area indicating a sediment filled graben (Pálmason, 1974; Gunnarsson, 1980). The rocks exposed on land nearest to the supposed transcurrent Húsavík faults are tectonically crushed and mineralized to a degree unknown elsewhere in surface exposure in Iceland and show a 90° dextral bend indicating considerable right lateral offset (Voight and others, 1983). They formed part of an active transform fault zone during a time interval when a spreading axis lay west of the Grímsey shoal (Fig. 1). The transform motion was greatly reduced about a million years ago when this axis jumped to a new position east of the shoal, joining Iceland's northern axial rift zone by way of the presently active N-S fissure swarms (Saemundsson, 1978).

The transverse E-W zone across central Iceland extending from Langjökull to Vatnajökull has also been referred to as having transform character (Sigurdsson, 1970, Schaefer, 1972). However, supporting evidence from fracture patterns, topography or seismicity is rather vague.

The Iceland Crust and Upper Mantle

The velocity structure of the Iceland crust from refraction data (Pálmason, 1971) remodelled by Flóvenz (1980) shows a rapid P-wave velocity increase with depth down to the 6.5 km/s isovelocity level, below which it becomes nearly constant down to layer 4. The depth to the 6.5 km/s isovelocity level, layer 3 of Pálmason's (1971) model, is in the range 3-6 km but decreases to approximately 1-3 km beneath eroded central volcanoes.

Pálmason's (1971) calculated velocities from 7.2 to 7.4 km/s for layer 4 indicate an anomalous mantle. Pálmason concluded from temperature gradients in boreholes that temperatures close to the melting range of basalt were reached at the base of the crust at about 8 km depth in SW Iceland.

Observations of teleseismic traveltime residuals (Tryggvason, 1964; Long and Mitchell, 1970) suggest an anomalous low-velocity mantle beneath Iceland down to some 250 km. S-wave travel-time residuals (Francis, 1969; Girardin and Poupinet, 1974) indicate a more pronounced S- than P-wave velocity anomaly in the upper mantle beneath Iceland. A generalized crustal and upper mantle cross section from the flank of the Reykjanes Ridge into Iceland shows a sharp transition from a normal oceanic to the anomalous Iceland seismic structure (RRISP Working Group, 1980). Gebrande and others (1980) found that S-waves travelling through layer 4 were delayed and attenuated, and no S-waves were recorded beyond 250 km from the shot points. The disappearance of S-waves was not found to be associated with the axial rift zones only, but seemed to be a general feature of the anomalous mantle under Iceland. Magnetotelluric measurements across the axial rift zone in northern Iceland and the marginal Tertiary areas (Beblo and Björnsson, 1980) have disclosed a 5-10 km thick low resistivity layer (15 ohm-m) in the uppermost mantle beneath Iceland. The depth to this layer is 10 km below the rift zone. It increases to 20 km below the 10 m.y. old basalt pile east and west of it. Earlier magnetotelluric measurements gave similar results in other parts of Iceland (Hermance, 1973; Hermance and others, 1976). Comparing their results with temperature-resistivity data of basaltic and ultramafic rocks, Beblo and Björnsson concluded that the low-resistivity layer most likely consists of partially molten basalt at a temperature of 1000-1100°C. This conclusion is supported by observed temperature gradients of 40-60°C/km in the marginal Tertiary rocks and around 100°C/km closest to the axial rift zone (Pálmason and Saemundsson, 1974; Pálmason et al., 1978). S-wave attenuation, low resistivity and the extent of the Bouguer gravity low of Iceland (Einarsson, 1954; Bott, 1965) suggest that the anomalous mantle and partial melting are not confined to the axial rift zones only but may exist below the whole island, and be responsible for the high mobility of the Iceland crust and the occurrence of volcanic activity in flank zones outside the accretion axis (Sanford and Einarsson, 1982).

Modification by Erosional Agencies

Iceland, being an Arctic oceanic island, is heavily sculptured by marine and glacial erosion. However, modification of the crust by erosion takes place almost wholly outside the zones of crustal growth. The marginal blocks flanking these zones are primarily modified by glacial erosion, whereas marine erosion has cut extensive abrasion platforms into their edges. For postglacial time an erosion rate of about 1 cm/y has been calculated for the east and northwest coasts (H. Tómasson, unpublished). In this way the uppermost 1-1.5 km of the lava pile have been levelled off and the debris transported oceanward. The deep erosion is compensated by isostatic uplift of the marginal blocks. Another explanation of their uplift which reaches up to 1 km relative to the paralleling axial rift zone, involves late intrusion of inclined sheets into their base (Piper and Gibson, 1972). However, evidence points to compensation of the sheet intrusions by subsidence (Walker, 1975).

The shelf around Iceland extends from 15 up to more than 100 km from the coast with depths mostly between 100 and 300 m. Progradation of the shelf by sediment deposition is characteristic outside the adjoining mid-ocean ridge axes, amounting to some 12-14 km off SE-Iceland (Kristjánsson, 1976) and 10-30 km off SW-Iceland (Egloff and Johnson, 1979). Sediments on the inner part of the shelf are thin, except for local accumulations in subsiding troughs. One such graben structure in the Tjörnes Fracture Zone contains up to 4 km of sediment (Gunnarsson, 1980). Great sediment thicknesses occur also in a graben continuing southward from Kolbeinsey Ridge (McMaster and others, 1977). A basin fill is partly exposed on land on the Tjörnes peninsula north of the Húsavík faults within the Tjörnes Fracture Zone. The sequence was formed in a near-shore environment and marine strata alternate with subaerial lava and lignite seams. The total aggregate thickness is about 1 km, of which ½ is marine. The age of these deposits is Pliocene and Plio-Pleistocene (Albertsson, 1978).

The shelf edge is indented by submarine valleys which are the continuation of Iceland's major rivers or fjord systems (Fig. 1). These are pathways of suspension flows of volcanogenic and erosionally drived debris which is deposited as fans in the adjoining deeper ocean basins (Ruddiman, 1972). The active volcanoes contribute material to the surrounding oceanic areas both by sheet floods carrying enormous quantities of hyaloclastite tuffs resulting from subglacial volcanism (Johnson and Pálmason, 1980) and also in lesser amounts, as wind borne ash produced by explosive eruptions (Sigurdsson and Loebner, 1981).

AZORES

Tectonic Frame and Morphology

The Azores Archipelago (surface area 2300 km^2) lies astride the MAR with 2 islands west of the ridge axis and 7 islands stretching up to 550 km east of it. Flank volcanism apparently created much of the extensive Azores platform, the bulk of which also lies east of the ridge axis (Fig. 5). Flank volcanism approaches the ridge axis to within 100 km according to magnetic data (Searle, 1980). The Azores are located at a triple plate junction where the MAR changes its regional trend SW-NE south of the junction to almost due N-S to the north. In a general

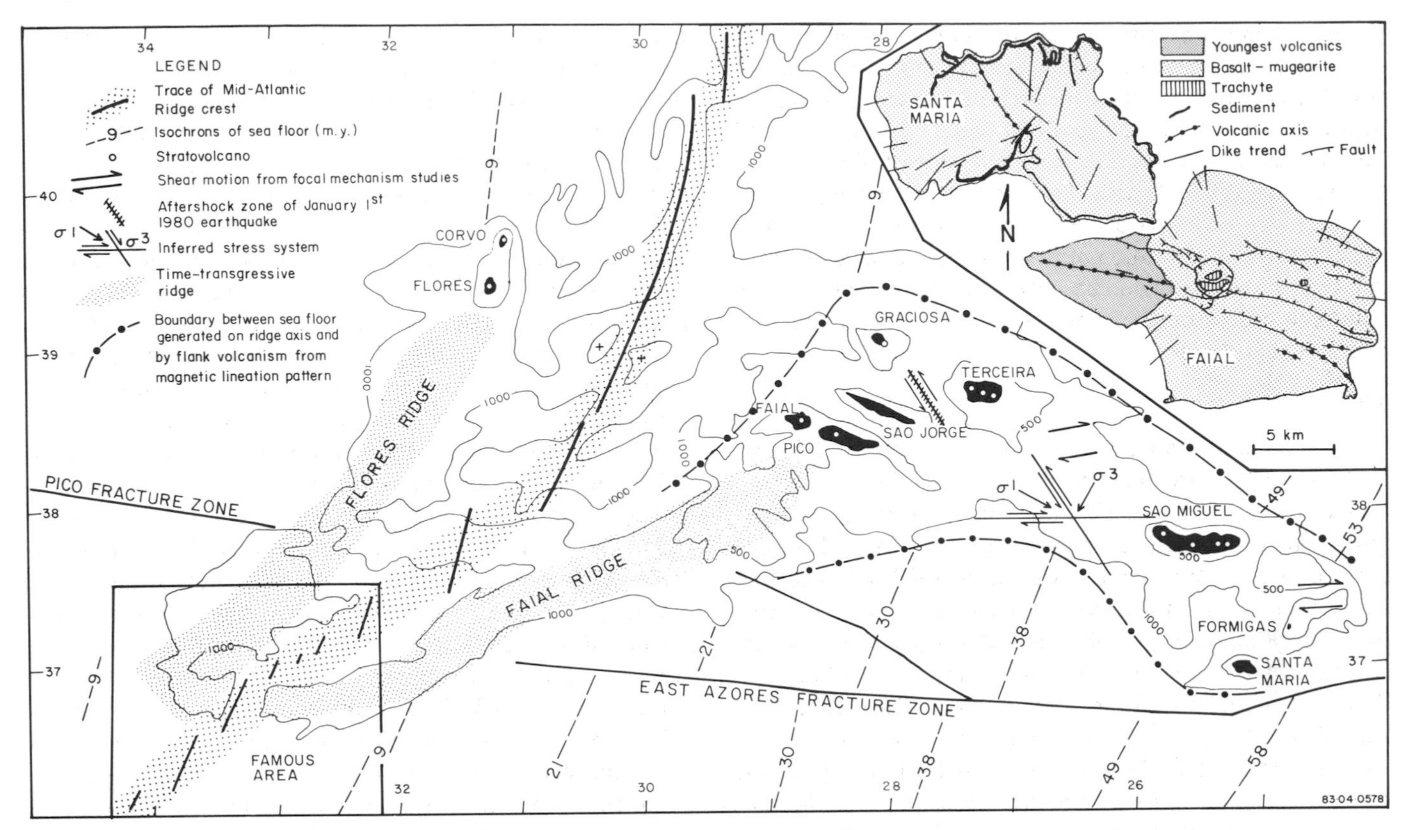

Figure 5. Simplified morphotectonic map of Azores region. Bathymetry from Vogt (1979). Isochrons from Laughton and Whitmarsh (1974). Focal mechanism solutions from Udias and others (1976) and Hirn and others (1980). The inferred stress pattern derived from the focal mechanism solutions appears consistent with Feraud's and others (1980) definition of σ_1. Boundary between ridge and flank volcanism is drawn on the basis of magnetics (Searle 1980). Sea floor ages from magnetic anomaly identification (Searle 1980) is in good agreement with Laughton and Whitmarsh (1974). Geology from Zbyszewski and others (1959, 1961).

way the Azores Islands east of the MAR lie along the supposed boundary between the African and Eurasian plates. The character of this boundary is probably a combination of slow oblique spreading and dextral shearing. The plate boundary is represented by the East Azores Fracture Zone, which is seismically active to the east of the Azores but apparently dead to the south. The boundary now passes through the Azores platform following the NW-SE array of islands and basement features which are seismically active (Laughton and Whitmarsh, 1974; Udias and others, 1976) and merge with the active East Azores Fracture Zone in the SE.

The Azores Islands are the upper parts of large stratovolcanoes which coalesce to form the largest islands. They rise from depths of locally 2000 m below, to over 2000 m above sea level (Pico 2320 m). They are all active or dormant except Santa Maria and the Formigas in the extreme SE. The shape of the islands has been modified by marine and fluvial erosion to a degree corresponding to their age. Caldera collapses, slumping and faulting also have played an important role and interacted with erosion to shape the islands. Volcanic debris supplied by fluvial erosion, slumping and marine erosion together with primary volcaniclastic rocks resulting from explosive eruptions (Walker and Croasdale, 1971, Huang and others, 1979) may have considerable thickness in the vicinity of the Azores. Seismic reflection profiles indicate sediment thicknesses of up to 500 m south of the Azores but less elsewhere over the platform (Searle, 1976; 1980).

The Azores Platform

The Azores platform is a topographic bulge adjoining the Mid Atlantic Ridge. The ridge axis transects the western part of this platform with an overall NE to NNE trend. A median valley is absent on the platform, apart from a shallow trough of 200 m depth or less (Searle, 1980), but present on the adjoining ridge segments (Vogt and Perry, 1982). The platform crust is oceanic in character, about 8 km thick with most of the thickening taking place in layer 2 (Searle, 1976) probably as a result of flank volcanism. It is underlain by anomalous mantle of 7.3-7.7 km/s P-wave velocity (Sapin, 1975). The Azores crust, although thinner than the crust under Iceland, has a similar velocity structure and the sub-Moho velocities are the same. Basalts occurring along the crest of the platform are tholeiite with typical hotspot characteristics (White and others, 1979, Schilling, 1975, Schilling and others, 1983; Schilling, this volume). The platform seafloor obviously can not be dated on the basis of linear magnetic anomalies where modified by constructional flank volcanism. Vogt (1979) has drawn attention to isochron-transgressive basement ridges that are symmetrical about the MAR—axis and trend SW from Flores, west of the Ridge, and Fayal, east of it; they point towards the FAMOUS area. These suggest a magma pulse moving southwest under the accretion axis starting about 10 Ma, an age given by the separation of its ends and the spreading rate of 2.6 m.y. (Minster and Jordan, 1978). Similar time-transgressive basement ridges occur on the Reykjanes Ridge southwest of Iceland (Vogt, 1974).

The Azores Islands

Deep-sea tephra indicates that Azores flank volcanism goes back to mid-Miocene time (Huang and others, 1979). The islands are made up almost exclusively of volcanic material, locally intercalated with calcareous marine sediment. The oldest basalts dated as 4.2-5.5 Ma (Feraud and others, 1980) occur on Santa Maria. This is a submarine sequence (Ribeiro and others, 1979) overlain by limestone described as mid-Miocene from macrofossil evidence (da Veiga Ferreira, 1961) but as Upper Miocene to Pliocene on the basis of foraminiferal assemblages (Krejci-Graf and others, 1958). The submarine units are overlain by a subaerial volcanic sequence. Erosion has exposed dykes (Fig. 5) radiating from the high ground on the central part of the island (Zbyscewski and others, 1961). Formigas islets northeast of Santa Maria, which are remnants truncated to wave base, consist of similarly old basalts cut by dykes and also Miocene to Pliocene sediment (Ribeiro and others, 1979, Abdel–Monem and others, 1975). Up to 4 m.y. old rocks have been reported from the eastern part of Sao Miguel (Abdel-Monem and others, 1975). Drilling in the central part of this island has disclosed a subaerial sequence of lava flows and pyroclastics down to 700 m below sea level, followed downcore by a transition sequence below which occur subaqueous volcanics down to 900 m (Muecke and others, 1974). This sequence was erupted between 0.1 and 0.3 m.y. ago according to K/Ar age dating and normal polarity of the core (Muecke and others, 1974). The easternmost islands give evidence of considerable vertical crustal movement. Santa Maria and Formigas are uplifted, but San Miguel has subsided. The rather fast vertical displacements indicated may relate to graben formation as on Terceira and Graciosa. The middle and western parts of Sao Miguel and the other seven islands are young volcanoes of Brunhes epoch age, less than 0.7 Ma. The volcanic edifices vary from slightly elongated to ridge shaped, parallel with their respective volcanic axis. According to geological maps each island represents one volcanic system, except Sao Miguel which consists of four systems, and Terceira with two or three. Corvo and Flores, west of the MAR-crest, were built up from N-S elongated volcanic axes whereas the islands east of the ridge crest were constructed on WNW-ESE striking axes. Three volcanic systems on Sao Miguel, two on Terceira, and the ones on Fayal, Pico and Sao Jorge have erupted during the 550 years of human habitation (Mitchell-Thomé, 1981). Reported submarine eruptions between Sao Miguel and Terceira and south of Sao Miguel indicate that the seamounts distributed over the floor of the Azores platform are volcanoes. The Azores volcanic systems produce alkalic rock suites ranging in composition from alkali basalts to trachytes and peralkaline rhyolites. Some have developed calderas that formed consequently upon eruption of large pumice or ash flow sheets (Walker and Croasdale, 1971; Ribeiro and others, 1979). The volume production of volcanism on the Azores is low. For San Miguel about 1 km^3/thousand years is believed to be a close average for the past 50,000 years (Booth and others, 1978).

The fracture pattern of the Azores is manifested in the alignment of vents marking individual volcanic systems and in the occurrence of normal faults that are most pronounced in the Fayal and Pico systems and also on eastern Terceira. Mapping of submarine topography using long-range side-scan sonar has revealed the major tectonic trends of the sea floor in the Azores region (Searle, 1980). Earthquake distribution (Udias and others, 1976; Hirn and others, 1980) yields supplementary evidence. The sonograph data (Searle, 1980) show a dominating N-S Mid-Atlantic Ridge trend on the Azores platform out to about 100 km from the ridge axis. East of this three structural trends occur: On the islands, a WNW-ESE trend dominates, whereas E-W and NNW-SSE trends are prominent as fracture valleys on the platform but locally also as fault zones on land (Sao Jorge, Sao Miguel). The degree of faulting in the Azores is far less than that of any typical axial rift zone but is comparable to fracturing observed in Iceland's flank zones. The tectonic pattern of the Azores, east of the MAR-axis, indicates dextral shear which is consistent with the inferred movement between the Africa and Eurasia plates. This interpretation is in accord with sinistral shear on a NNW-SSE fault-plane found for aftershocks of the January 1st 1980 earthquake (Hirn and others, 1980) and dextral shear on conjugate E-W fault planes (Udias, 1976). Radial volcanic fissure patterns elongated in a preferred WNW-ESE direction are conspicuous among the Azores volcanoes (Fig. 5). Following Nakamura (1977) this is also the direction of maximum horizontal compression of the regional stress which bisects the angle between the two conjugate trends of strike-slip faulting (Fig. 5). The Terceira Rift, proposed as a secondary spreading axis by Krause and Watkins (1970), appears as one array within a transform fault zone which is split into en echelon segments. Some N-S extension and subsidence is likely to occur along the WNW-ESE en echelon arrays as they trend obliquely to the E-W spreading direction. Troughs along the Terceira trend may have formed by sagging and extension as proposed by Krause and Watkins (1970) or alternatively are holes left vacant during volcano growth around them. The seismically inactive fracture zone increments south of the Azores probably preceded the en echelon arrays farther north. Timing of this shift is indicated in Laughton and Whitmarsh (1974) by the dying out of the fracture zone in 9-21 m.y. old sea floor. The structural trends and location of the western islands opposite the culmination of the MAR-crest on the Azores platform is strikingly similar to the previously described Öraefajökull-Snaefell flank zone in Iceland, suggesting a common origin.

JAN MAYEN

Tectonic Frame and Morphology

Jan Mayen is located in an area of shallow banks to the south of the Jan Mayen Fracture Zone along which the spreading axis is offset some 200 km from Kolbeinsey Ridge to Mohns Ridge (Fig. 6). The fracture zone is marked by a valley over 2000 m deep immediately north of the island. Opposite Jan Mayen across the fracture zone lies the spreading axis of Mohns Ridge. The position of Jan Mayen is further borne out by the proximity of the Jan Mayen Ridge, a continental sliver trending south from the island. It is not known whether this sliver extends up to or even underlies the Jan Mayen volcanoes.

Morphologically, Jan Mayen Island, 380 km^2, is dominated by glacier crowned Beerenberg volcano (2277 m) in the northeast, and an icefree volcanic range just over 700 m high in the southwest. A low isthmus connects the two volcanic complexes. The island is little eroded except by marine erosion and by outlet glaciers on Beerenberg volcano. More than half the icefree area is covered by fresh scoria and lava (Imsland, 1978).

The Shoals Around Jan Mayen

Jan Mayen Island is located on the northern end of the Jan Mayen Ridge. Seismic refraction shooting has revealed about 18 km thick crust of oceanic character below the island and a sub-Moho P-wave velocity of 8.27 km/s (Sörnes and Narvestad, 1976). It is contended that the Jan Mayen Ridge was split from the Greenland continental margin following a jump of the oceanic accretion axis from Aegir Ridge in the Norwegian Sea to a new position out of which Kolbeinsey Ridge on the Iceland Plateau began to develop between 44 to 26 Ma (Nunns, 1983). The Kolbeinsey Ridge axis ends about 150 km west of Jan Mayen in another shoal area reaching within 50 m of the surface on the Eggvin shoal. From the dispersion of seismic surface waves a crustal thickness of 12-15 km has been found near the axis of Kolbeinsey Ridge increasing to about 20 km elsewhere on the Iceland Plateau (Evans and Sacks, 1979). Tholeiites from Eggvin indicate a hotspot origin whereas tholeiites elsewhere along Kolbeinsey Ridge are of normal oceanic chemistry (Dittmer and others, 1975; Schilling, this volume).

Geology of Jan Mayen

Jan Mayen Island is just over 50 km long, with its long axis trending NE-SW. It apparently consists of two volcanic systems (Dollar, 1966) each about 25 km long (Fig. 6). A third volcanic system may lie just west of Sör Jan in the area of Stimen. This is represented by a narrow submarine ridge 15-20 km long

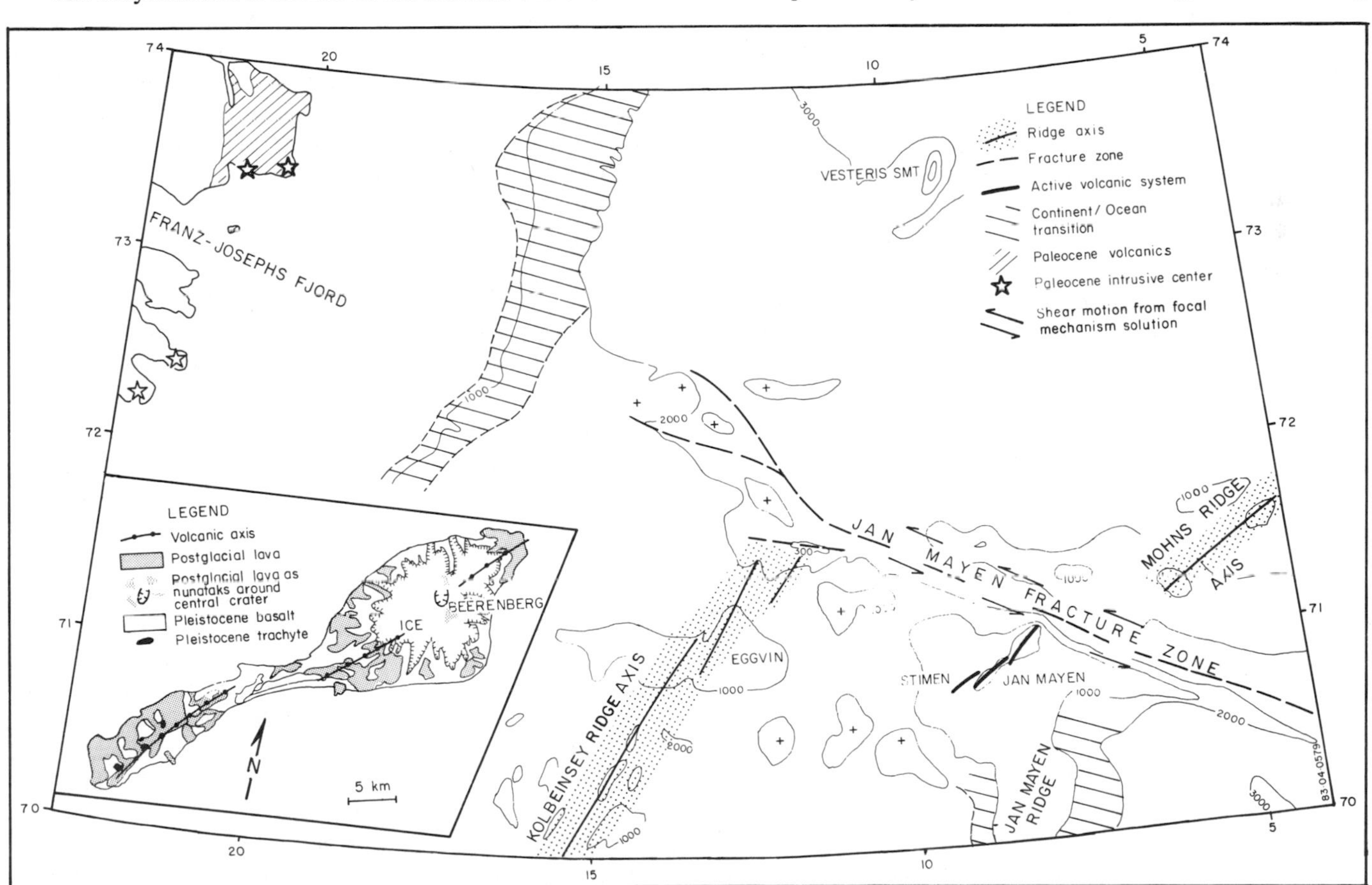

Figure 6. Simplified morphotectonic map of Jan Mayen region. Bathymetry from Johnson and Campsie (1976). Focal mechanism solutions from Bungum and Husebye (1977). Geology from Imsland (1978).

trending WSW-ENE, and reaching within 100 m of the surface. Beerenberg, a stratovolcano elongated NE-SW constitutes the eastern system. This volcano is entirely built up of basaltic lava and scoria and subordinate hyaloclastite tuff of submarine origin (Imsland, 1978). The alignment of craters around the lower slopes of Beerenberg volcano suggests radial dyking superimposed on the prominent NE-SW volcanic fissure trend (Fitch, 1964). Sör Jan constitutes the western volcanic system, a mountainous range dotted with numerous fissure-aligned craters and trachyte domes, but without a central cone. The rocks range from ankaramite to trachyte, and were erupted subaerially except the oldest unit, a submarine hyaloclastite (Imsland, 1978). Although fissure alignment of craters is conspicuous on both systems, faults are indistinct. Both volcanic systems have been active in the recent geologic past. Beerenberg is the most active and has had a few recorded eruptions. The rocks of Jan Mayen are normally magnetized (Fitch and others, 1965a) and are assigned to the Brunhes magnetic epoch on the basis of the obvious youth of the volcanoes and radiometric ages which yield about 0.5 Ma for the oldest dated sample (Fitch and others, 1965b). Calculations of average volcanic productivity using 5.35 km^3 of lava erupted during the Holocene as reference, give an age of about 0.4 Ma for the island (Imsland, 1978). A late Pleistocene uplift of Jan Mayen by at least 170 m is borne out by the submarine hyaloclastite and marine fossils (Imsland, 1978).

The deep zone of melting indicated by the alkalic rocks of Jan Mayen may be related to propagation efforts of the Mohns Ridge axis beyond the Jan Mayen Fracture Zone into an area affected by the melting zone of a hotspot located in the general Eggvin-Jan Mayen area. It is possible also that the presence of continental crust in the Jan Mayen ridge provides a further lithospheric weakness (Vink and others, 1984) to guide volcanism. The closest analog to Jan Mayen may exist in the propagating flank zone of South-Iceland which lies in the prolongation of a spreading axis intercepted by the South-Iceland seismic zone (Fig. 2).

ANCIENT SUBAERIAL COMPLEXES

Several shallow platforms and seamounts are dispersed over the floor of the western North-Atlantic. A number of these were probably emergent with subaerial volcanism from their summits. They have been interpreted as hotspot tracks (Morgan, 1981), but simultaneous or progressive volcanism along fractures is a viable alternative in the case of mid-plate seamounts. Most of the seamounts are probably younger than the oceanic crust upon which they are built, but platforms may have formed by axial accretion in the Icelandic manner.

Platforms

The hotspot track from Greenland to Iceland (Fig. 1) may have been generated largely by subaerial spreading beginning immediately prior to anomaly 24 (Brooks and Nielsen, 1982). Larsen (1980) located the ocean-continent transition on the Greenland-Iceland ridge. The sediment thickness on the oceanic side is given as 1.2 km on the ridge, which contrasts with sediment thicknesses of 6-8 km to the north and 3-4 km to the south of the buried ridge. The waist in the physiography of the Iceland-Greenland ridge can be taken as a measure of volcanic production which appears to have passed through a low in Mid-Tertiary. Increased activity beginning about 25 Ma (Vogt and others, 1980) gave rise to Iceland and the shallow platform around it (Fig. 1). A low in subaerial explosive volcanism is indicated during the Oligocene by scarcity of acid tephra in sedimentary cores of the North-Atlantic Basin (Sigurdsson and Loebner, 1981). Another trace of this hotspot or another one similar, with possibly subaerial volcanism occurs offshore of West Greenland (Keen and Clarke, 1974).

A topographically less obvious hotspot track trends from Jan Mayen towards the basalt area around Franz-Josephs Fjord, East Greenland (Fig. 6) (Johnson and Campsie, 1976, Upton and others, 1980). The basalts are of Paleocene age, as are the ones at the western end of the Iceland hotspot track (Beckinsdale and others, 1970). Information on basement structure offshore East-Greenland between 72°N and 74°N has not been acquired. Larsen (1980) marks the ocean-continent transition in the region (Fig. 6) with the uneven physiography of the general Jan Mayen transverse area closely approaching the transition zone. The Vöring Plateau which is supposedly of subaerial origin (Mutter and others, 1982) occupies a conjugate position in the NE-Atlantic. Therefore subaerial volcanics may well exist offshore on the East-Greenland side beneath the sediment. Johnson and Campsie (1976) have suggested an intermittently active hotspot in the Jan Mayen area to account for the discontinuous volcanic peaks and shoal areas. The hotspot activity in the Franz-Josephs Fjord–Vöring plateau area possibly predated that of the Greenland-Faeroe plateau. The two plateaus may be genetically related to the same hotspot as envisaged by Vink (1984).

A third major igneous complex of probably subaerial origin is the area of the Southeast Newfoundland Ridge (SENR) and the J-Anomaly Ridge south of Grand Banks (Fig. 7; Plate 3). The SENR trends SE from the Grand Banks for over 300 km, broadening towards the continental shelf and the junction with the J-Anomaly Ridge. It follows the trace of the Newfoundland Fracture Zone. Sullivan (1983) proposed a seafloor spreading model for the Newfoundland Basin. According to her model the SENR was formed between 102 and 80 Ma during a period of oblique transform faulting. The resultant leaky transform fault produced the volcanic edifice of the SENR of which particularly the northwestern part is likely to have formed subaerially like the contiguous J-Anomaly Ridge. Tucholke and Ludwig (1982) have studied the J-Anomaly Ridge by seismic refraction and multichannel reflection profiles. It is about 800 km long and 40-60 km wide with a NE-SW strike subparallel with the regional seafloor isochrons. It is visible in seismic profiles as a basement high or step covered by thick sediment towards SW. Only the northern end is visible in bottom morphology. Deep sea drilling has shown that this part of it was above sea level in

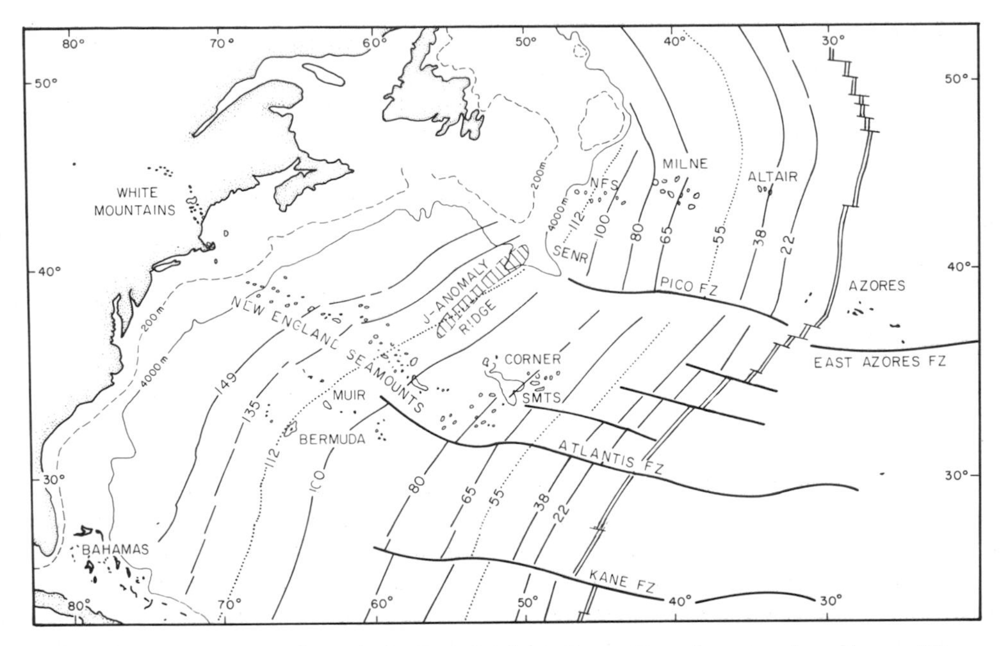

Figure 7. Seamounts and platforms in the Central Atlantic Ocean. Crustal ages are from Vogt and Perry (1982). J-Anomaly Ridge after Tucholke and Ludwig (1982). NFS = Newfoundland Seamounts. SENR = Southeast Newfoundland Ridge.

Mid-Cretaceous time (Peterson and others, 1979). The cored, once subaerially exposed basement, now at 4.2 km depth, is of normal tholeiitic composition, however, slightly enriched in light rare-earths indicative of hotspot origin (Tucholke, Vogt and others, 1979). Seismic velocity layering indicates anomalously thick layer 2 (Tucholke and Ludwig, 1982). Reflectors within the basement show offlapping units in the direction of spreading which have been interpreted by Tucholke and Ludwig as formed during emergence of the spreading axis under conditions analogous to those which produced the lenticular units of the lava pile in Iceland.

Seamounts

The seamount groups are located primarily on Jurassic to Cretaceous oceanic crust in the large bight between the Grand Banks and Bahama-Blake plateau. The most prominent of these is an igneous belt that stretches from the White Mountains along the **New England Seamounts** to the Corner Seamounts (Fig. 7; Plates 2, 3) over a distance of more than 2000 km. Formation of the chain falls into an age range between 124 Ma in the White Mountains and 70-75 Ma at Corner Seamounts (Duncan, 1984) in the extreme east. The age gradient and breaks in the chain favour migratory or episodic volcanism and a hotspot origin. Subaerial volcanics have not been conclusively shown to exist on any of the New England Seamounts so far, although vesicular volcaniclastic debris on the flanks must have been erupted in shallow water (Vogt and Tucholke, 1979). However, their relief of 6-6.5 km above the oceanic basement (Vogt and Perry, 1982), and the flat-topped summits on some of them favour subaerial growth. The seamounts grew on the flank of the MAR (Vogt and Tucholke, 1979). The Corner Seamounts were formed closest to the ridge axis and are nearly synchronous with the adjacent sea floor (Vogt and Perry, 1982).

The **Newfoundland Seamounts** constitute a major group of seamounts with an overall E-W distributional trend. Two other seamount groups cluster farther west in the area of Milne and Altair Seamounts (Fig. 7); the youngest is on about 38 Ma crust. Age data from the Newfoundland Seamounts indicate that their formation was completed during the same period as the Southeast Newfoundland Ridge (Sullivan, 1983). The position of the Newfoundland Seamounts has been related to fracture zone trends of the region (Sullivan, 1983). However, they also may be looked upon as part of a hotspot track extending to Altair Seamount (Duncan, 1984).

Bermuda is the only seamount presently above sea level in the western North-Atlantic. These islands are built of arenaceous limestone which overlies a volcanic wave cut platform at an average depth of only 80 m below sealevel (Gees and Medioli, 1970). It is the largest of three seamounts crowning the NE-SW trending Bermuda Rise. Radiometric data indicate that the Bermuda Seamount formed about 33 Ma on oceanic crust now about 112 m.y. old (Reynolds and Aumento, 1974). Similar conclusions were reached from DSDP drilling southeast of Bermuda (Vogt and Tucholke, 1979). Drilling on Bermuda penetrated subaqueous tholeiitic basalt intruded by lamprophyric sheets from 26-800 m below sealevel. Subaerial volcanics if ever present appear to have been removed by erosion. A mechanism involving intrusion of lamprophyric sheets into oceanic crust causing local upswelling has been invoked to explain the formation of the Bermuda Seamount (Aumento and Sullivan, 1974).

The ancient submarine complexes have all been related to hotspot activity. Many of them have conjugate submarine platforms or seamount chains developed in the eastern Atlantic suggesting location of a hotspot near the accretion axis (Tucholke and Ludwig, 1982). Others, like Bermuda or the New England Seamounts, would have formed in the track of an intraplate hotspot of episodic character. Obviously much more data is needed particularly on the time scales involved in seamount formation, on the associated structural patterns and on their geochemistry in order to clarify the picture.

CONCLUDING REMARKS ON CAUSATIVE MECHANISMS

The present-day subaerial volcanic complexes in the Western North-Atlantic have developed under the influence of primarily two tectonic domains. One is the accretion axis of the MAR presently emergent in Iceland only. The other domain includes the Azores, Jan Mayen and the flank zones of Iceland where older crust is broken up, allowing ascent of magma. Ancient, now submerged, platforms and seamounts apparently were generated under conditions similar to either of these.

In plate tectonics theory Iceland, the Azores and Jan Mayen have been referred to as hotspots (Wilson, 1965; Morgan, 1972) as have indeed most or all seamount groups in the NW-Atlantic (Duncan, 1984). Under this term, a plume originating deep within the mantle is generally surmised. The nature of plumes is still being debated. From variations in the topographic expression of their tracks, plumes in the North Atlantic must have been of fluctuating intensity—even isolated diapiric mantle blobs have been suggested (Schilling, 1975). Alternatively, a very large plume has been suggested (Talwani and Eldholm, 1977) to account for the elevated ocean floor adjoining the Iceland area. Upwelling plumes would generate radial stress fields (Burke and Dewey, 1973) that also may be a controlling factor in the tectonics of the three regions. The plumes are of lower viscosity and higher temperature than the surrounding mantle. They give rise to excessive volcanism and an abnormally thick oceanic crust. Geophysical experiments in the area of the Iceland hotspot indicate that temperatures in the melting range of basalts are reached at the base of the crust underneath much of the island, which measures about 500 km diameter. This has not been demonstrated in the Azores and Jan Mayen areas. In the Azores the E-W extension of active volcanism, however, is of similar dimension as Iceland. This anomalous upper mantle may be regarded as a yielding cushion of a wide circumference below the crust which is easily broken up by fracture zones, propagating rift axes or even by thermal contraction (Pálmason, 1981) to allow ascent of magma in flank zones. Evidence from Iceland and the Azores with their scattered flank zone volcanism indicates that the hotspot centers are not necessarily below these, but are located beneath the spreading axis itself in the area where its topography is highest, i.e., below Vatnajökull in Iceland's Eastern axial rift zone, and below the MAR-axis on the Azores platform. In these areas tholeiites with most pronounced hotspot affinities are erupted (Sigvaldason and Steinthorsson, 1974; Schilling, 1975). Similar reasoning for the Jan Mayen area would place the hotspot near Eggvin at the northern end of Kolbeinsey Ridge. Jan Mayen Island would be located marginal to the hotspot in an area still influenced by high temperature at the base of a crust strained by the Jan Mayen Fracture Zone and spreading Mohns Ridge axis.

Certain aspects of the geology of Iceland and the Azores region could be interpreted as a direct expression of hotspot activity. This particularly applies to lava production passing through a maximum along the accretion axis, a 120° angle in the trend of the accretion axis, geochemical variation along the accretion axis, and v-shaped time-transgressive ridges symmetric about the accretion axis.

The increase in lava production correlates with an increasingly higher topography of the axial rift zone towards central Iceland. This is likely to hold for the shoaling of the MAR axis across the Azores platform as well.

Burke and Dewey (1973) conjectured that upwelling plumes will cause a concentric uplift above them. Tangential stresses will break up the crust in three radial rifts oriented approximately 120° to each other. If spreading is involved, two rifts will develop into active spreading axes; the third arm will die away or become suppressed. This concept may apply to the change in trend of extensional features across central Iceland (Tryggvason, 1973; Saemundsson, 1978; Wyss, 1980) and along the MAR crest on the Azores platform (Searle, 1980).

Tholeiites erupted in a hotspot environment are enriched in $^{87}Sr/^{86}Sr$ and large ion lithophile elements as well as in light rare earths relative to normal oceanic tholeiites. Tholeiites from Iceland (Schilling, 1973; Sigvaldason and Steinthorsson, 1974), the Azores (White and others, 1979; Schilling, 1975) and the Eggvin Bank west of Jan Mayen (Dittmer and others, 1975) are all of this type showing geochemical gradients from a presumed hotspot center.

V-shaped time-transgressive ridges observed on the Reykjanes Ridge (Vogt, 1971) south of the Azores (Vogt, 1979) as well as on the J-Anomaly Ridge (Tucholke and Ludwig, 1982) have been explained as the result of SW-ward channeling of asthenosphere from plumes along the spreading axis. These ridges are indicative of magmatic pulses moving away from the plume center. High extrusion rates occurring in Eastern Iceland at 6-7 Ma and on Iceland's NW-Peninsula between 14 and 12 Ma correlate with Vogt's (1979) magmatic pulses and plate acceleration documented in marine magnetics north of Iceland (Vogt and others, 1980). Also, the rift jumps occurring in Iceland at about 6-7 Ma are coincident with the proposed increase in volcanic discharge migrating SW from Iceland about 7 Ma. These have been interpreted as a result of refocussing of the axial rift zone above an eastward migrating plume center relative to the MAR spreading axis (Tryggvason, 1973, Saemundsson, 1974). Increased plume flow may have strained the crust and uppermost mantle sufficiently to bring about a new rift opening.

REFERENCES

Abdel-Monem, A. A., Fernandes, L. A., and Bonne G. M.
1975 : K-Ar ages from the eastern Azores Group (Santa Maria, Sao Miguel and the Formigas Islands): Lithos, v. 8, p. 247–254.

Albertsson, K. J.
1978 : Um aldur jardlaga á Tjörnesi (Some notes on the age of the Tjörnes strata sequence): Natturufraedingurinn, v. 48, p. 1–8.

Aumento, F. and Sullivan, K. D.
1974 : Deep drill investigations of the oceanic crust in the North Atlantic; in Geodynamics of Iceland and the North Atlantic area, ed. L. Kristjansson; Reidel, p. 83–103.

Beblo, M. and Björnsson, A.
1980 : A model of electrical resistivity beneath NE-Iceland, correlation with temperature: Journal of Geophysics, v. 47, p. 184–190.

Beckinsdale, R. D., Brooks, C. K., and Rex, D. C.
1970 : K-Ar ages for the Tertiary of East-Greenland: Bulletin Geological Society of Denmark, v. 20, p. 27–37.

Björnsson, A., Johnsen, G., Sigurdsson, S., and Thorbergsson, G.
1979 : Rifting of the plate boundary in North Iceland 1975–1978: Journal of Geophysical Research, v. 84, no. B6, p. 3029–3038.

Björnsson, S. and Einarsson, P.
1974 : Seismicity of Iceland; in Geodynamics of Iceland and the North Atlantic area, ed. L. Kristjánsson; Dordrecht, D. Reidel, p. 225–239.

Booth, B., Croasdale, R., and Walker, G.P.L.
1978 : A quantitative study of five thousand years of volcanism on Sao Miguel, Azores: Philosophical Transactions, Royal Society of London, ser. A, v. 288, p. 271–319.

Bott, M.H.P.
1965 : The upper mantle beneath Iceland: Geophysical Journal, Royal Astronomical Society, v. 9, p. 275–277.

Brander, J. and Wadge, G.
1973 : Distance measurements across the Heimaey eruptive fissue: Nature, v. 244, p. 496–498.

Brandsdóttir, B. and Einarsson, P.
1979 : Seismic activity associated with the Sept. 1977 deflation of the Krafla central volcano, NE-Iceland: Journal of Volcanology and Geothermal Research, v. 6, p. 197–212.

Brooks, C. K. and Nielsen, T.F.D.
1982 : The East Greenland continental margin: a transition between oceanic and continental magmatism: Journal, Geological Society of London, v. 139, p. 265–275.

Bungum, H. and Husebye, E. S.
1977 : Seismicity of the Norwegian Sea: The Jan Mayen Fracture Zone: Tectonophysics, v. 40, p. 351–360.

Burke, K. and Dewey, J. F.
1973 : Plume generated triple junctions: Key indicators in applying plate tectonics in old rocks: Journal of Geology, v. 81, p. 406–433.

Christensen, N. I. and Wilkens, R. H.
1982 : Seismic properties, density and composition of the Icelandic crust near Reydarfjördur: Journal of Geophysical Research, v. 87, no. B8, p. 6389–6395.

da Veiga Ferreira, O.
1961 : Afloramentos de calcario miocenico da Ilha de Santa Maria (Azores): Comunicaiones de Servicios Geologicas Portugal, v. 45, p. 493–501.

Dittmer, F., Fine, S., Rasmussesn, M., Bailey, J. C., Campsie, J.
1975 : Dredged basalts from the mid-oceanic ridge north of Iceland: Nature, v. 254, p. 298–301.

Dollar, A.T.J.
1966 : Genetic aspects of the Jan Mayen fissure volcano group on the Mid-Oceanic submarine Mohns Ridge, Norwegian Sea: Abstract, Bulletin Volcanologique, v. 29, p. 25–26.

Duffield, W. A.
1978 : Vesicularity of basalt erupted at Reykjanes Ridge crest: Nature, v. 274, p. 217–220.

Duncan, R. A.
1984 : Age progressive volcanism in the New England Seamounts and the opening of the Central Atlantic Ocean: Journal of Geophysical Research, v. 89, p. 9980–9990.

Egloff, J. and Johnson, G. L.
1979 : Erosional and depositional structures of the Southwest Iceland insular margin: Thirteen geophysical profiles: Geological and Geophysical Investigations of Continental Margins, American Association of Petroleum Geologists Memoir 29, p. 43–63.

Einarsson, P.
1978 : S-wave shadows in the Krafla Caldera in NE-Iceland, evidence for a magma chamber in the crust: Bulletin Volcanologique, v. 41, p. 187–195.

Einarsson, P. and Brandsdóttir, B.
1980 : Seismological evidence for lateral magma intrusion during the July 1978 deflation of the Krafla volcano in NE-Iceland: Journal of Geophysics, v. 47, p. 160–165.

Einarsson, P., Klein, F. W., and Björnsson, S.
1977 : The Borgarfjördur earthquakes of 1974 in West Iceland: Seismological Society of America Bulletin, v. 67, p. 187–208.

Einarsson, T.
1954 : A survey of gravity in Iceland: Societas Scientiarium Islandica, Rit 30, 22 p.

Evans, J. R. and Sacks, I. S.
1979 : Deep structure of the Iceland Plateau: Journal of Geophysical Research, v. 84, no. B12, p. 6859–6866.

Feraud, G., Kaneoka, I., and Allegre, C. L.
1980: : K/Ar ages and stress pattern in the Azores: Geodynamic implications: Earth and Planetary Science Letters, v. 46, p. 275–286.

Fitch, F. J.
1964 : The development of the Beerenberg volcano Jan Mayen: Proceedings Geologist's Association, v. 75, p. 133–165.

Fitch, F. J., Grasty, R. J., and Miller, J. A.
1965b: Potassium-Argon ages of rocks from Jan Mayen and an outline of its volcanic history: Nature, v. 207, p. 1349–1351.

Fitch, F. J., Nairn, A.E.M., and Talbot, C. J.
1965a: Paleomagnetic studies on rocks from North Jan Mayen: Norsk Polar-institut Arbok 1963, p. 49–60.

Fleischer, U.
1974 : The Reykjanes Ridge—A summary of geophysical data; in Geodynamics of Iceland and the North Atlantic area, ed. L. Kristjansson; Dordrecht, D. Reidel, p. 17–31.

Flóvenz, O. G.
1980 : Seismic structure of the Icelandic crust above layer 3 and the relation between body wave velocity and the alteration of the basaltic crust: Journal of Geophysics, v. 47, p. 211–220.

Francis, T.J.G.
1969 : Upper Mantle structure along the axis of the mid-Atlantic ridge near Iceland: Geophysical Journal, Royal Astronomical Society, v. 17, p. 507–520.

Gebrande, H., Miller, H., and Einarsson, P.
1980 : Seismic structure of Iceland along RRISP-Profile I: Journal of Geophysics, v. 47, p. 239–249.

Gees, R. A. and Medioli, F.
1970 : A continuous seismic survey of the Bermuda platform; part I, Castle Harbour: Marine Sediments, v. 6, p. 21–25.

Gibson, I. L. and Piper, J.D.A.
1972 : Structure of the Icelandic basalt plateau and the process of drift: Philosophical Transactions Royal Society of London, ser. A, v. 271, p. 141–150.

Girardin, N. and Poupinet, G.
1974 : Teleseismic S travel time delay for Mid-Atlantic Ridge earthquakes: Physics of Earth and Planetary Interiors, v. 9, p. 306–313.

Gunnarsson, K.

1980 : Seabottom around Iceland, History of evolution and sediment deposition: Icelandic Energy Authority, Professional Paper OS80025/JHD14, 79 p (in Icelandic).

Haimson, B. C. and Rummel, F.
1982 : Hydrofracturing stress measurements in the IRDP drillhole at Reydarfjordur, Iceland: Journal of Geophysical Research, v. 87, no. B8, p. 6631–6649.

Helgason, J. and Zentilli, M.
1982 : Stratigraphy and correlation of the region surrounding the IRDP drillhole 1978, Reydarfjordur, eastern Iceland: Journal of Geophysical Research, v. 87, no. B8, p. 6405–6417.
1985 : Field characteristics of laterally emplaced dikes: anatomy of an exhumed Miocene dike swarm in Reydarfjordur, eastern Iceland: Tectonophysics, v. 115, p. 247–274.

Hermance, J. F.
1973 : An electrical model for the sub-Icelandic crust: Geophysics, v. 38, p. 3–13.
1981 : Crustal genesis in Iceland. Geophysical constraints on crustal thickening with age: Geophysical Research Letters, v. 8, p. 203–206.

Hermance, J. F., Thayer, R. E., and Björnsson, A.
1976 : The telluric-magnetotelluric method in the regional assessment of geothermal potential: Proceedings Second U.N. Symposium on the Development and use of Geothermal Resources, v. 2, p. 1037–1048.

Hinz, K.
1981 : A hypothesis on terrestrial catastrophes. Wedges of very thick oceanward dipping layers beneath passive continental margins.—Their origin and paleoenvironmental significance: Geologisches Jahrbuch Reihe E, Geophysik, v. 22, p. 3–28.

Hirn, A., Haessler, H., Hoang Trong, P., Wittlinger, G., and Mendes Victor, L. A.
1980 : Aftershock sequence of the January 1st, 1980, earthquake and present-day tectonics in the Azores: Geophysical Research Letters, v. 7, p. 501–504.

Huang, T. C., Watkins, N. D., and Wilson, L.
1979 : Deep-sea tephra from the Azores during the past 300,000 years: Eruptive cloud height and ash volume estimates: Summary: Geological Society of America Bulletin, Part I, v. 90, p. 131–133.

Imsland, P.
1978 : The geology of the volcanic island Jan Mayen, Arctic Ocean: Nordic Volcanological Institute Report, v. 78-13, 74 p.

Jakobsson, S. P.
1972 : Chemistry and distribution pattern of Recent basaltic rocks in Iceland: Lithos, v. 5, p. 365–386.

Johannesson, H.
1980 : Jardlagaskipan og thróun rekbelta á Vesturlandi (Evolution of rift zones in western Iceland): Natturufraedingurinn, v. 50, p. 13–31.

Johnson, G. L. and Campsie, J.
1976 : Morphology and structure of the western Jan Mayen Fracture Zone: Norsk Polarinstitut Arbok 1974, p. 69–81.

Johnson, G. L. and Pálmason, G.
1980 : Observations of the morphology and structure of the sea floor south and west of Iceland: Journal of Geophysics, v. 47, p. 23–30.

Keen, M. J. and Clarke, D. B.
1974 : Tertiary basalts of Baffin Bay: Geochemical evidence for a fossil hotspot: in Geodynamics of Iceland and the North Atlantic area, ed. L. Kristjansson; Dordrecht, D. Reidel, p. 127–132.

Krause, D. C. and Watkins, N. D.
1970 : North Atlantic crustal genesis in the vicinity of the Azores: Geophysical Journal, Royal Astronomical Society, v. 19, p. 261–283.

Krejci–Graf, K., Frechen, J., Wetzel, W., and Colom, G.
1958 : Gesteine und Fossilen von den Azoren: Senckenbergiana, v. 39, p. 303–351.

Kristjánsson, L.
1976 : A marine magnetic survey off Southern Iceland: Marine Geophysical Research, v. 2, p. 315–326.

Kristjánsson, L., Fridleifsson, I. B., and Watkins, N. D.
1980 : Stratigraphy and paleomagnetism of the Esja, Eyrarfjall and Akrafjall mountains, SW-Iceland: Journal of Geophysics, v. 47, p. 31–42.

Kristmannsdottir, H.
1982 : Alteration in the IRDP drillhole compared with other drillholes in Iceland: Journal of Geophysical Research, v. 87, no. B8, p. 6525–6531.

Larsen, H. C.
1980 : Geological perspectives of the East Greenland continental margin: Bulletin Geological Society of Denmark, v. 29, p. 77–101.

Laughton, A. S. and Whitmarsh, R. B.
1974 : The Azores-Gibraltar plate boundary; in Geodynamics of Iceland and the North-Atlantic Area, ed. L. Kristjansson; Dordrecht, D. Reidel, p. 63–81.

Long, R. E. and M. G. Mitchell
1970 : Teleseismic P-wave delay time in Iceland: Geophysical Journal, Royal Astronomical Society, v. 20, p. 41–48.

McDougall, I., Kristjansson, L., and Saemundsson, K.
1984 : Magnetostratigraphy and geochronology of NW-Iceland: Journal of Geophysical Research, v. 89, p. 7029–7060.

McDougall, I., Saemundsson, K., Johannesson, H., Watkins, N. D., and Kristjansson, L.
1977 : Extension of the geomagnetic polarity time scale to 6.5 m.y.: K-Ar dating, geological and paleomagnetic study of a 3,500 m lava succession in western Iceland: Geological Society of America Bulletin, v. 88, p. 1–15.

McDougall, I. and Wensink, H.
1966 : Paleomagnetism and geochronology of the Pliocene—Pleistocene lavas in Iceland: Earth and Planetary Science Letters, v. 1, p. 232–236.

McMaster, R. L., Schilling, J. G., and Pinet, P.
1977 : Plate boundary within Tjörnes Fracture Zone on northern Iceland's insular margin: Nature, v. 269, p. 663–668.

Mehegan, J. M., Robinson, P. T., and Delaney, J. R.
1982 : Secondary mineralization and hydrothermal alteration in the Reydarfjordur drill core, eastern Iceland: Journal of Geophysical Research, v. 87, no. B8, p. 6511–6524.

Minster, J. B. and Jordan, T. H.
1978 : Present day plate motions: Journal of Geophysical Research, v. 83, no. B11, p. 5331–5354.

Mitchell-Thomé, R. C.
1981 : Vulcanicity of historic times in the Middle Atlantic islands: Bulletin volcanologique, v. 44, p. 57–69.

Moorbath, S., Sigurdsson, H., and Goodwin, R.
1968 : K-Ar ages of oldest exposed rocks in Iceland: Earth and Planetary Science Letters, v. 4, p. 197–205.

Morgan, W. J.
1972 : Plate motions and deep mantle convection: Geological Society of America, Memoir 132, p. 7–22.
1981 : Hotspot tracks and the opening of the Atlantic and Indian Oceans: The Sea, v. 7, p. 443–487.

Mudie, P. J. and Helgason, J.
1983 : Palynological evidence for Miocene climatic cooling in eastern Iceland about 9.8 Ma: Nature, v. 303, p. 689–692.

Muecke, G. K., Ade-Hall, J. M., Aumento, F., Macdonald, A., Reynolds, P. H., Hyndman, R. D., Quintino, J., Opdyke, N., and Lowrie, W.
1974 : Deep drilling in an active geothermal area in the Azores: Nature, v. 252, p. 181–185.

Mutter, J. C., Talwani, M., and Stoffa, P. L.
1982 : Origin of seaward-dipping reflectors in oceanic crust off the Norwegian margin by "subaerial sea-floor spreading": Geology, v. 10, p. 353–357.

Nakamura, K.
1970 : En echelon features of icelandic ground fissures: Acta Naturalia Islandica, v. 2, no. 8, 15 p.
1977 : Volcanoes as possible indicators of tectonic stress orientation—

principle and proposal: Journal of Volcanology and Geothermal Research, v. 2, p. 1–16.

Nunns, A. G.

1983 : Platetectonic evolution of the Greenland-Scotland Ridge and surrounding regions; in "Structure and development of the Greenland-Scotland Ridge: New methods and concepts, eds. M.H.P. Bott, S. Saxov, M. Talwani and J. Thiede; Plenum, p. 11–30.

Pálmason, G.

1971 : Crustal structure of Iceland from explosion seismology: Socetas Scientiarium Islandica, Rit 40, 187 p.

1974 : Insular margins of Iceland; in Geology of Continental Margins, eds. C. A. Burke and C. L. Drake; New York, Springer, p. 375–379.

1981 : Continuum model of crustal generation in Iceland, kinematics aspects: Journal of Geophysics, v. 47, p. 7–18.

Pálmason, G., Arnorsson, S., Fridleifsson, I. B., Kristmannsdóttir, H., Saemundsson, K., Stefansson, V., Steingrímsson, B., Tomasson, J., and Kristjansson, L.

1978 : The Iceland crust: Evidence from drillhole data on structure and processes: American Geophysical Union, Ewing Series, v. 3, p. 43–65.

Pálmason, G. and Saemundsson, K.

1974 : Iceland in relation to the Mid-Atlantic Ridge: Annual Reviews Earth and Planetary Sciences, v. 2, p. 25–50.

Peterson, N., Bleil, V., and Eisenack, P.

1979 : Rock and paleomagnetism of leg 43 basalts: Initial Reports Deep Sea Drilling Program, v. 43, p. 773–780.

Piper, J.D.A. and Gibson, I. L.

1972 : Stress control of processes at extensional plate margins: Nature Physical Sciences v. 238, no. 84, p. 83–86.

Prestvik, T.

1979 : Petrology of hybrid intermediate and silicic rocks from Öraefajökull, Southeast Iceland: Geologiska Föreningen Forhandlingar v. 101, p. 299–308.

Reynolds, P. H. and Aumento, F.

1974 : Deep Drill, 1972. Potassium/Argon dating of the Bermuda drill core: Canadian Journal of Earth Sciences, v. 11, p. 1269–1273.

Ribeiro, A., Antunes, M. T., Ferreira, M. P., Rocha, R. B., Soares, A. F., Zbyszewski, G., Moitinho de Almeida, F., de Carvalho, D., and Monteiro, J. H.

1979 : Introduction a la geologie generale du Portugal: Servicos Geologicos de Portugal, 114 p.

RRISP Working Group (14 authors)

1980 : Reykjanes Ridge Iceland Seismic Experiment (RRISP 1977): Journal of Geophysics, v. 47, p. 228–238.

Ruddiman, W. F.

1972 : Sediment redistribution on the Reykjanes Ridge: Seismic evidence: Geological Society of America Bulletin, v. 83, p. 2039–2062.

Saemundsson, K.

1967 : Outline of the structure of SW-Iceland: Societas Scientiarium Islandica, Rit 38, p. 151–159.

1974 : Evolution of the axial rifting zone in northern Iceland and the Tjörnes Fracture Zone: Geological Society of America Bulletin, v. 85, p. 495–504.

1977 : Origin of anticlinal structures in Iceland; in Some problems of riftogenesis, ed. N. A. Logatchev; Novosibirsk, p. 175–181. (in russian)

1978 : Fissure swarms and central volcanoes of the neovolcanic zones of Iceland: Geological Journal Special Issue 10, p. 415–432.

1979 : Outline of the geology of Iceland: Jökull, v. 29, p. 7–28.

Saemundsson, K., Kristjansson, L., McDougall, I., and Watkins, N. D.

1980 : K-Ar dating, geological and paleomagnetic study of a 5 km lava succession in northern Iceland: Journal of Geophysical Research, v. 85, no. B7, p. 3628–3646.

Saemundsson, K. and Noll, H.

1974 : K/Ar ages of rocks from Húsafell, western Iceland and the development of the Húsafell central volcano: Jökull, v. 24, p. 40–59.

Sanford, A. R. and Einarsson, P.

1982 : Magma chambers in rifts; in Continental and oceanic rifts, ed. G. Pálmason; Geodynamics Series, v. 8, p. 147–168.

Sapin, M.

1975 : Premiére campagne de sismologie experimentale aux Acores: 3éme Réunion Annuelle de Science de la Terre April 1975, Montpellier (Abstract).

Schaefer, K.

1972 : Transform faults in Island: Geologische Rundschau, v. 61, p. 942–960.

Schilling, J.-G.

1973 : Iceland mantle plume: geochemical evidence along Reykjanes Ridge: Nature, v. 242, p. 565–571.

1975 : Azores mantle blob: rare earth evidence: Earth and Planetary Science Letters, v. 25, p. 103–115.

Schilling, J.-G., Zajac, M., Evans, R., Johnston, T., White, W., Devine, J. D., and Kingsley, R.

1983 : Petrologic and Geochemical Variations along the Mid-Atlantic Ridge from 29° N to 73° N: American Journal of Science, v. 283, no. 6, p. 510–586.

Searle, R. C.

1976 : Lithosperic structure of the Azores Plateau from Rayleigh-wave dispersion: Geophysical Journal, Royal Astronomical Society, v. 44, p. 537–546.

1980 : Tectonic pattern of the Azores spreading centre and and triple junction: Earth and Planetary Science Letters, v. 51, p. 415–434.

Searle, R. C. and Laughton, A. S.

1981 : Fine-scale sonar study of tectonics and volcanism on the Reykjanes Ridge: Oceanologica Acta, no. SP, p. 5–13.

Sigurdsson, H.

1970 : Structural origin and plate tectonics of the Snaefellsnes volcanic zone, western Iceland: Earth and Planetary Science Letters, v. 10, p. 129–135.

Sigurdsson, H. and Loebner, B.

1981 : Deep-sea record of Cenozoic explosive volcanism in the North Atlantic; in Tephra studies, ed. S. Self and R.S.J. Sparks; Dordrecht, D. Reidel, p. 289–316.

Sigvaldason, G. E. and Steinthorsson, S.

1974 : Chemistry of tholeiitic basalts and their relation to the Kverkfjöll hotspot; in Geodynamics of Iceland and the North-Atlantic area, ed. L. Kristjansson; Dordrecht, D. Reidel, p. 155–164.

Sullivan, K. D.

1983 : The Newfoundland Basin: ocean-continent boundary and Mesozoic seafloor spreading history: Earth and Planetary Science Letters, v. 62, p. 321–339.

Sörnes, A. and Narvestad, T.

1976 : Seismic Survey of Jan Mayen; Norsk Polarinstitut Arbok 1975, p. 37–52.

Sykes, L. R.

1967 : Mechanism of earthquakes and nature of faulting on the mid ocean ridges: Journal of Geophysical Research, v. 72, p. 2131–2153.

Talwani, M., Windisch, C. C., and Langseth, M. G.

1971 : Reykjanes Ridge crest: A detailed geophysical study: Journal of Geophysical Research, v. 76, no. 2, p. 473–517.

Talwani, M. and Eldholm, O.

1977 : Evolution of the Norvegian-Greenland Sea: Geological Society of America Bulletin, v. 88, p. 969–999.

Thorarinsson, S.

1966 : Surtsey: the new island in the North Atlantic: Almenna bokafelagid, Reykjavík, 47 p.

Torfason, H.

1979 : Investigations into the structure of Southeastern Iceland: University of Liverpool, Ph.D. Thesis, 587 p. (unpubl.).

Tryggvason, E.

1964 : Arrival times of P-waves and upper mantle structure. Seismological Society of America Bulletin, v. 54, p. 727–736.

1970 : Surface deformation and fault displacement associated with an earthquake swarm in Iceland: Journal of Geophysical Research, v. 75, P. 4407–4422.
1973 : Seismicity, earthquake swarms and plate boundaries in the Iceland region: Seismological Society of America Bulletin, v. 63, p. 1327–1348.
1983 : The widening of the Krafla fissure swarm during the 1975-1981 volcanotectonic episoce: Nordic Volcanological Institute, Publ. 8304, 48 p.

Tryggvason, K., Husebye, E. S., and Stefansson, R.
1983 : Seismic image of the hypothesized Icelandic hotspot: Tectonophysics, v. 100, p. 97–118.

Tucholke, B. E., and Ludwig, W. J.
1982 : Structure and origin of the J-Anomaly Ridge, Western North Atlantic Ocean: Journal of Geophysical Research, v. 87, no. B11, p. 9389–9407.

Tucholke, B. E., and others
1979 : Site 384: The Cretaceous/Tertiary boundary, Aptian reefs and the J-Anomaly Ridge: in B. E. Tucholke and P. R. Vogt, eds., Initial Reports Deep Sea Drilling Project, v. 43, Washington, D.C., U.S. Government Printing Office, p. 107–154.

Udías, A., López-Arroyo, A., and Mezcua, J.
1976 : Seismotectonics of the Azores-Alboran Region: Tectonophysics, v. 31, p. 259–289.

Upton, B.G.J., Emeleus, C. H., and Hald, N.
1980 : Tertiary volcanism in northern E. Greenland: Gauss Halvö and Hold with Hope: Journal Geological Society of London, v. 137, p. 491–508.

Vink, G. E.
1984 : A hotspot model for Iceland and the Vöring Plateau: Journal of Geophysical Research, v. 89, p. 9949–9959.

Vink, G. E., and Morgan, W. J. and Zhao, W. L.
1984 : Preferential rifting of continents: A source of displaced terranes: Journal of Geophysical Research, v. 89, p. 10072–10076.

Vogt, P. R.
1971 : Asthenophere motion recorded by the ocean floor south of Iceland: Earth and Planetary Science Letters, v. 13, p. 153–160.
1974 : The Iceland Phenomenon: Imprints of a hotspot on the ocean crust and implications for flow below the plates; in Geodynamics of Iceland and the North Atlantic area, ed. L. Kristjansson; Dordrecht, D. Reidel, p. 105–126.
1979 : Global magmatic episodes: New evidence and implications for the steady-state mid-oceanic ridge: Geology, v. 7, p. 93–98.
1983 : The Iceland mantle plume: Status of the hypothesis after a decade of new work; in "Structure and development of the Greenland-Scotland Ridge, eds. M.H.P. Bott, S. Saxov, M. Talwani and J. Thiede; Plenum, p. 191–213.

Vogt, P. R., Johnson, G. L., and Kristjansson, L.
1980 : Morphology and magnetic anomalies north of Iceland: Journal of Geophysics, v. 47, p. 67–80.

Vogt, P. R. and Perry, R. K.
1982 : North Atlantic Ocean: Bathymetry and plate tectonic evolution, Geological Society of America, Map and Chart Series, MC-35.

Vogt, P. R. and Tucholke, B. E.
1979 : The New England Seamounts: Testing origin: Initial Reports Deep Sea Drilling Project, v. 43, p. 847–856.

Voight, B., Young, K., Jancin, M., Orkan, N., Aronson, J., and Saemundsson, K.
1983 : Húsavík fault system, Tjörnes Fracture Zone, Iceland: International Union of Geology and Geophysics 18th General Assembly Hamburg. (Abstract).

Walker, G.P.L.
1959 : Geology of the Reydarfjordur area, eastern Iceland: Quarterly Journal Geological Society of London, v. 114, p. 367–393.
1960 : Zeolite zones and dike distribution in relation to the structure of the basalts of eastern Iceland: Journal of Geology, v. 68, p. 515–528.
1963 : The Breiddalur central volcano, eastern Iceland: Quarterly Journal Geological Society of London, v. 119, p. 29–63.
1965 : Evidence of crustal drift from Icelandic Geology: Philosophical Transactions Royal Society (London), v. 258, p. 199–204.
1975: : Intrusive sheet swarms and the identity of crustal layer 3 in Iceland: Quarterly Journal Geological Society London, v. 131, p. 143–161.

Walker, G.P.L. and Croasdale, R.
1971 : Two plinian-type eruptions in the Azores: Quarterly Journal Geological Society London, v. 127, p. 17–55.

Ward, P. L.
1971 : New interpretation of the geology of Iceland: Geological Society of America Bulletin, v. 82, p. 2991–3012.

Watkins, N. D. and Walker, G.P.L.
1977 : Magnetostratigraphy of eastern Iceland: American Journal of Science, v. 277, p. 513–584.

White, W. M., Tapia, M.D.M., and Schilling, J.-G.
1979 : The petrology and chemistry of the Azores Islands: Contributions to Mineralogy and Petrology, v. 69, p. 201–213.

Wilson, J. T.
1965 : Evidence from ocean islands suggesting movement in the Earth: Philosophical Transactions Royal Society (London), v. 258, p. 145–167.

Wyss, M.
1980 : Hawaiian rifts and recent Icelandic volcanism: Expressions of plume generated radial stress fields: Journal of Geophysics, v. 47, p. 19–22.

Zbyszewski, G., da Veiga Ferreira, O., and Torre de Assuncao, C.
1961 : Carta geologica de Portugal 1:25,000. Noticia explicativa da folha de Ilha de Santa Maria (Acores): Servicos Geologicos de Portugal.

Zbyszewski, G., Moitinho de Alineida, F., da Veiga Ferreira, O., and Torre de Assuncao, C.
1959 : Carta geologica de Portugal 1:25,000. Noticia explicativa da folha Faial (Acores): Servicos Geologicos de Portugal.

Manuscript Accepted by the Society April 12, 1985

Printed in U.S.A.

The Geology of North America
Vol. M. The Western North Atlantic Region
The Geological Society of America, 1986

Chapter 6

Model of crustal formation in Iceland, and application to submarine mid-ocean ridges

Gudmundur Pálmason
Orkustofnun, Grensásvegur 9, 108 Reykjavík, Iceland

INTRODUCTION

During the early 1960s, when the idea of seafloor spreading was crystallizing out of geophysical data gathered over oceanic areas, similar ideas were beginning to take shape in Icelandic geology, but on the basis of a different kind of data. Geological studies of the structure of the eastern Iceland Tertiary lava pile led Walker (1959, 1960) to suggest that regional dips of the lavas were due to sagging accompanying lava deposition in the active volcanic zone. These ideas were further elaborated by Bodvarsson and Walker (1964), who tried to estimate the amount of crustal drift that might have accompanied the injection of dikes as observed in the eastern Iceland lava pile, taking into account the increase in dike fraction with depth in the crust. Subsequent studies of the Tertiary lava pile in other parts of Iceland have essentially confirmed a structure analogous to that given by Walker for eastern Iceland (Saemundsson, 1978).

Observations from two distinct geological regions can be brought to bear on the process of crustal accretion in Iceland. One region is the present active zone of rifting and volcanism, the axial rift zone. The other region is the flanking Tertiary lava pile where the internal structure of the uppermost well-developed crust can be observed to a depth of 1–1.5 km in valleys that have been carved into the crust by glacier action.

In the axial rift zones the constructional processes of extrusive volcanism, faulting, and fissuring take place as discrete but closely associated events. The frequent en echelon arrangement of volcanic-tectonic swarms within the axial zones has been pointed out by Saemundsson (1978). The Krafla rifting episode 1975–1984 has thrown light on the accretion process on one such swarm, the 80 km long and 5 km wide Krafla swarm (Björnsson and others, 1977, 1979; Pálmason, 1981). Evidence presented by Laughton and Searle (1979) shows that an en echelon structure of the axial rift zone also occurs on the submarine Reykjanes Ridge southwest of Iceland.

In the terminology of Saemundsson (1978), the neovolcanic zones in Iceland comprise both the axial rift zones where most of the crustal separation takes place and the lateral rift zones where rifting is inconspicuous. Fracture zones complicate the pattern of the Mid-Atlantic Ridge through Iceland. Only within the axial rift zones are all the aspects of the accretion process displayed. Here the extrusive volcanism and the subsidence balance to keep the surface at a level determined by isostatic conditions. The evidence for slow subsidence in the axial zone (Pálmason and Saemundsson, 1974) comes from geodetic measurements and drillhole logs within the active zone as well as from observations of the structure of the older crust flanking the axial zone.

A quantitative model of crustal accretion in the axial zone was developed by Pálmason (1973, 1980, 1981). It has been tested primarily by comparison with data from Iceland, but is not limited in its scope to that segment of the mid-ocean ridges. It contains parameters that can be varied to simulate probable conditions on other segments of the mid-ocean ridges.

One of the main objectives of the model has been to calculate in a realistic way the thermal state of the crust in the accretion zone. When combined with the kinematic aspects of the accretion process this has provided insight into the relationship between processes in the axial zone and the structure of the oceanic crust. Several models of crustal accretion at the mid-ocean ridges have been suggested in the literature, in most cases to explain a limited set of data. For a review of these models the reader is referred to Macdonald (1982; this volume).

In recent years the oceanic crust has been studied in more detail than before, by submersible diving and other means, in particular in the FAMOUS (French American Mid-Ocean Undersea Study) area of the Mid-Atlantic Ridge at 36–37 °N. Drilling to more than 500 m into the oceanic crustal basement has also been achieved in several areas. These studies have provided a wealth of new information and have shown that some of the early models of the crust (constructed, for example, to explain the magnetic anomalies observed over the ocean floor) are inaccurate and contradicted by some of the new data.

The present paper has two main objectives. One is to review the model developed for crustal accretion in Iceland and its application to that area. The other objective is to investigate to what extent the model may, with appropriate parameter values, fit the

Pálmason, G., 1986, Model of crustal formation in Iceland, and application to submarine mid-ocean ridges; *in* Vogt, P. R., and Tucholke, B. E., eds., The geology of North America, Volume M, The Western North Atlantic Region: Geological Society of America.

new data from ocean crustal basement drillings and from studies in the slow spreading FAMOUS area. Some preliminary results for fast spreading ridges will also be mentioned briefly.

KINEMATIC ASPECTS OF THE MODEL

The assumptions involved in the kinematics of the model have been described by Pálmason (1973, 1980, 1981). The model attempts to describe the overall time-averaged movement of solid crustal elements during the accretion process, taking into account certain material balance conditions. The actual accretion process consists of discrete episodes, involving extrusive volcanism, dike formation, formation of open fissures, normal faulting, and so on. These accretion episodes occur at intervals of a few hundred years in the same segment of the accretion zone. The large scale features of the crust, as described by the model, are formed in a time interval on the order of one million years, requiring a few thousand accretion episodes. This time perspective has to be kept in mind when comparing the model with observations, for example, from the deeply eroded valleys in the Tertiary crust of Iceland, or from drillholes into the oceanic crustal basement.

The two-dimensional model relates constructional processes in the accretion zone in a quantitative way to the structure of the crust. Most other models that have been proposed (for example, to explain the magnetic anomaly pattern, or the seismic structure of the crust) are essentially one-dimensional models, and do not include in a quantitative way, a prediction of the thermal state of the crust resulting from the model. The Iceland model is not necessarily limited in its application to conditions in the Iceland segment of the mid-ocean ridges, as it includes variable constructional properties such as spreading rate, width of accretion zone, rate of lava production, and other parameters from which a resulting crustal structure may be calculated, and compared with data from drillholes into the oceanic crustal basement. The term lavas is here used to include all extrusive rocks, although some part of these may be in the form of pyroclastic deposits.

The kinematic aspects of the model are characterized by two main assumptions. One is that lava is deposited at a certain rate with a certain distribution across the accretion zone. For a steady-state process, conservation of material requires that the deposition of lava at the surface is compensated by an equivalent amount of subsidence of the solid crust in the axial zone. The other assumption is that the horizontal extension (strain) by intrusive activity (dikes) and normal faulting, which accompanies the acceleration of the crust from rest at the axis to a final spreading velocity away from the accretion zone, is distributed in a certain way across the accretion zone. Both these distributions are assumed to be Gaussian, but with different standard deviations, σ_1 for the extension rate and σ_2 for the lava deposition rate. An additional parameter ϵ gives the fraction of the total horizontal extension rate that is due to normal faulting.

THERMAL ASPECTS OF THE MODEL

As the heat sources associated with extension by dike formation are directly related to the kinematics of the model, it is possible to calculate the thermal state of the model crust, taking into account both the heating effect of the intrusions and the cooling effect of crustal subsidence. These two processes control the thermal state of the crust in the accretion zone. They in turn depend on the spreading rate, the width of the accretion zone, the lava production rate, and the relative importance of dikes and normal faults in the extensional process.

As the method of calculating the thermal state of the model crust in the accretion zone has not been given elsewhere, it will be outlined briefly below. The appropriate equation describing the thermal state is the heat transport equation for a moving medium with heat sources:

$$\alpha \nabla^2 T + \frac{A}{c\rho} - \bar{v} \cdot \nabla T = \frac{\partial T}{\partial t}$$

where α = thermal diffusivity (= $k/\rho \times c$), T = temperature, A = heat source function, $\bar{v}$ = velocity of lava elements, t = time, k = thermal conductivity, ρ = density and c = specific heat. Only steady state solutions are considered here (i.e. $\partial T/\partial t = 0$).

The heat source function can be written as follows:

$$A = n c\rho \operatorname{div} \bar{v} \, (T_k - T)$$

Here $T_k = T_l + L/c$ where T_l is the liquidus temperature of basalt and L is the latent heat of fusion. The divergence of the flow field of lava elements, div $\bar{v}$, represents the intensity of intrusives formation. The factor n takes into account that during ascent of magma some heat is conducted from the magma channel (dike) to the surrounding rock. This represents a heat source to be added to the heat of the magma remaining in the channel at the end of the magmatic episode. A value of n = 2 has been used in the calculation of the temperature field (Pálmason, 1973).

For a given flow field $\bar{v}$ of lava elements, and for given boundary conditions, the heat transport equation has been solved numerically by a finite difference method.

The following numerical values for the material properties have been used in the thermal calculations: k = 1.7 W/m°C; ρ = 2900 kg/m^3; c = 1.05 kJ/kg°C; T_l = 1200 °C; L = 420 kJ/kg.

BOUNDARY CONDITIONS FOR THE TEMPERATURE FIELD

The calculation of the crustal temperature is carried out over a rectangular slab of constant thickness. The boundary conditions at the axis and at the surface are

$\frac{\partial T}{\partial x} = 0$ at x = 0 (x = horizontal coordinate)

T = 0 at z = 0 (z = vertical coordinate)

At the bottom of the slab and at its distant vertical boundary the asymptotic solution given by Parker and Oldenburg (1973) (see also Oldenburg, 1975) is used as a boundary condition. This solution has the form:

$$T(x,z) = T_1 \times \frac{\operatorname{erf}\left(\frac{z\sqrt{v_d}}{2\sqrt{\alpha \times x}}\right)}{\operatorname{erf}(\beta)}$$

where v_d is the spreading half rate, and β is determined from the boundary condition at the bottom of the lithosphere, that is,

$$k\frac{\partial T}{\partial z} = \frac{\partial z}{\partial t}\rho L$$

where $\partial z/\partial t$ is the rate of growth of the lithosphere at its bottom. The equation determining β is

$$\beta e^{\beta^2} \times \operatorname{erf}(\beta) = \frac{c \times T_1}{\sqrt{\pi} \times L}$$

Using the previously given numerical values, one obtains β = 0.914. The asymptotic solution above becomes

$$T(x,z) = 1493 \times \operatorname{erf}\left(\frac{z\sqrt{v_d}}{2\sqrt{\alpha \times x}}\right) {}^\circ C$$

This asymptotic solution is strictly valid only on the slab boundary where $T < T_1$. Formally, it is used here also for the slab bottom below the axial zone, because it can be shown by repeated calculations for different slab thicknesses that the crustal temperature near the axis is not sensitive to the choice of slab thickness, as long as the thickness is greater than the depth to the partial melt region beneath the axis ($T > \sim 1000°C$). The temperature field in the axial zone is controlled primarily by the surface boundary condition and by the processes of magma intrusion and the downward movement of cold lava from the surface. In the final solution the calculated temperature field above a certain maximum temperature is disregarded, as it belongs to the regime of partial melting that the model does not treat in its present form. The maximum temperature is here taken as 1000°C.

HYDROTHERMAL CIRCULATION

It is well known from observations in Iceland and on submarine mid-ocean ridges that hydrothermal activity is a significant process of heat discharge in the axial zone. Drillholes in Iceland have shown that hydrothermal circulation takes place to depths of at least 3–4 km (Pálmason and others, 1979). The most intensive hydrothermal fields, the so-called high-temperature fields, are often associated with centers of volcanic activity within the axial zone, and probably derive their heat from intrusive sources beneath the volcanic centers. These volcanic centers have a limited life span of 3–5 × 10^5 years (Saemundsson, 1978), and similarly, the hydrothermal activity associated with a particular center may be expected to have a limited life span. Few reliable estimates of the natural heat discharge from hydrothermal fields in the axial zone have been made, which makes it difficult to assess their long-term average contribution to the total heat discharged in the axial zone.

The hydrothermal circulation is not explicitly taken into account in the model in its present form. In nature it modifies the crustal temperature field, elevating the isotherms in some parts of the accretion zone and depressing them in other parts. The intrusive activity providing the heat source for both the hydrothermal and conductive heat transport in the axial zone is taken into account in the model. Therefore the total calculated heat flux in the model may be viewed as a first approximation to the combined heat transport by conduction and water convection. With better knowledge of the long-term average effect of the hydrothermal activity it may be possible to modify the heat transport equation (for example by a heat sink distribution) to take into account explicitly the effect of the hydrothermal circulation in the averaging sense of the model.

LITHOSPHERIC MODEL CROSS SECTION FOR ICELAND

The choice of model parameters to use for the Iceland case was discussed at some length by Pálmason (1980). A structural and thermal cross section calculated with parameters believed to be representative for Iceland was given by Pálmason (1981), and is reproduced in Figure 1. Some of the more important properties of the model lithosphere can be discussed on the basis of this section. These include the structure of the uppermost crust (lavas, dikes), the low-temperature alteration, the seismic velocity structure, and the magnetization structure. Furthermore, the composition of erupted magma may be affected by crustal processes in the axial zone.

The structure of the uppermost model crust with downdip thickening of lava units and an increase of dike fraction with depth, fits well the regional structure of the eastern Iceland Tertiary lava pile as described by Walker (1960). Furthermore the thermal state of the model crust is consistent with predicted temperatures from the northeastern Iceland axial zone based on magnetotelluric measurements (Beblo and Björnsson, 1978, 1980) and surface thermal gradients (Pálmason and others, 1979).

The structure of the deeper crust can be discussed on the basis of the model section in Figure 1. It is clear from the form of the lava trajectories and the isotherms that a certain zonation of the crust with depth takes place in the outer part of the accretion zone. Cold lava elements originating at the surface at various distances from the axis go through a thermal history of heating to a maximum temperature and then cooling again as they subside and move sideways along their trajectories. Lavas originating close to the axis reach the greatest depths in the crust and the highest temperatures on their way. Properties of the crust that are influenced by the maximum temperature reached by the lavas on their way into the frozen-in crust will show a certain zonation.

Low-temperature alteration zoning of the upper crust has been studied in many parts of Iceland (Walker, 1960; Kristmannsdóttir, 1975, 1979; Kristmannsdóttir and Tómasson, 1978; Pálmason and others, 1979; Mehegan and others, 1982). This

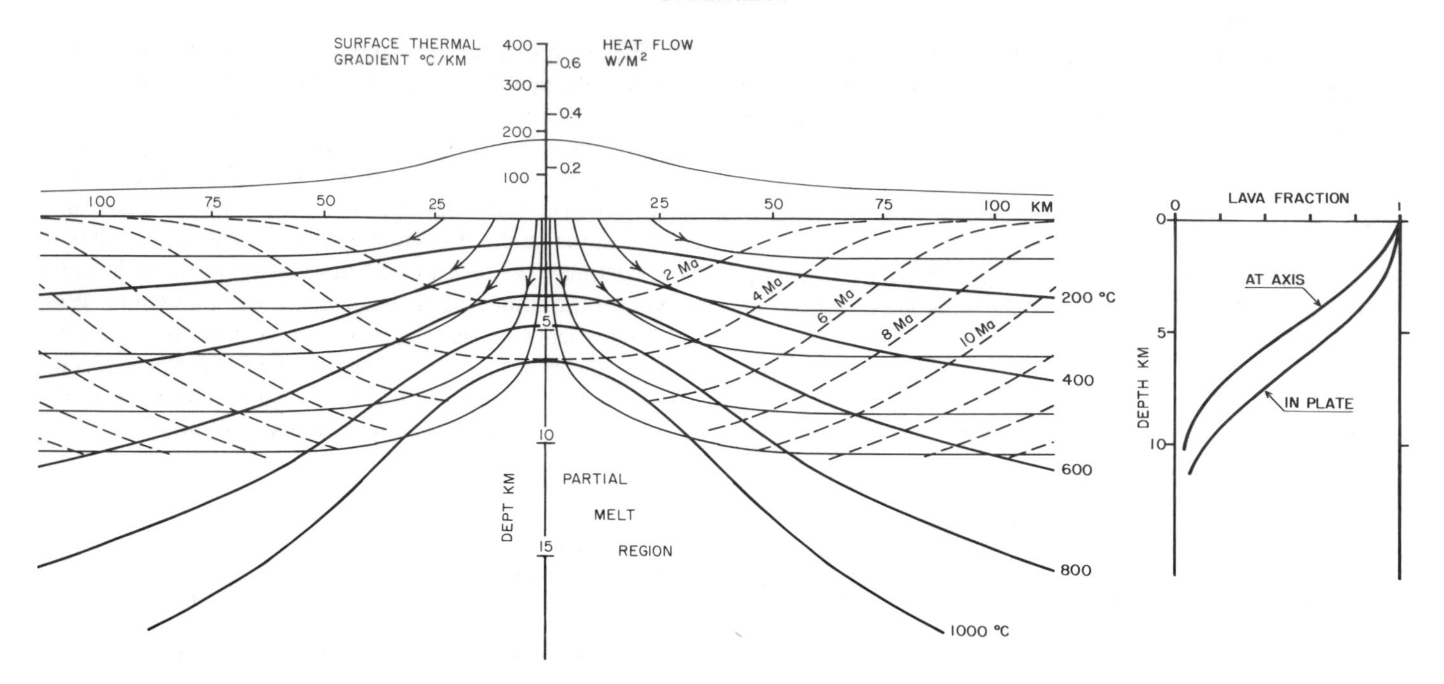

Figure 1. A structural and thermal model cross section for an axial zone, with parameters appropriate for Iceland (Pálmason, 1981). Trajectories (arrows) and isochrons (dashed lines) of lava elements are shown, as well as calculated isotherms and surface gradient and heat flow. The lava fraction (extrusives fraction) at the axis and in the distant lithosphere is shown on the right. The model parameters used are: spreading half rate 1 cm/yr, lava production rate 1.33×10^{-4} km^2/yr, crustal strain rate standard deviation 15 km, lava deposition rate standard deviation 20 km, and normal fault fraction of crustal strain 0.75.

process is presently occurring in active geothermal fields, and on the basis of data from such areas a relationship between alteration zones and temperature has been deduced (e.g. Kristmannsdóttir, 1982). An example of this process is the zeolite-greenschist facies transition which, according to the above studies, takes place at about 250°C. The model section indicates that lavas reheated to 250°C during the accretion process reach a final depth in the model crust of about 3 km, in good agreement with the depth at which this transition has been found in drillholes in the Tertiary areas of northern and eastern Iceland (Kristmannsdóttir, 1982).

The seismic velocity zoning of the crust is probably caused largely by metamorphic processes as suggested by many workers (Einarsson, 1965; Pálmason, 1971; Flóvenz 1980). The uppermost few km of the crust are characterized by a downward gradient in seismic velocity, considered to be due to an increasing degree of metamorphism (Flóvenz, 1980). The lower part of the crust, Layer 3, has an almost constant P-wave velocity of 6.5 km/s (Pálmason, 1971). The transition zone between the upper and lower parts of the crust is at 5–6 km depth. The model section in Figure 1 indicates that lavas reaching the depth of this transition zone have been reheated to about 600 °C before cooling down again in the distant crust. A temperature interval around 600 °C is thus predicted by the model for the metamorphic reactions responsible for the transition from the upper to the lower crust, be this the greenschist-amphibolite facies transition (Pálmason, 1971) or simply an increase in epidote content (Christiansen and Wilkens, 1982).

As suggested by the model, the magnetization structure of the crust is rather complex, and depends at least partly on the thermal history of lava elements reaching various depths in the crust. In the upper part of the crust where the reheating of lavas has been only moderate, the original magnetization of the lavas, perhaps somewhat reduced by low-temperature oxidation, would persist and lead to alternating normally and reversely magnetized zones bounded by isochrons of rather gentle dips. The lavas are intruded by younger dikes of either polarity, increasing in relative abundance with depth. Below a certain depth the lavas would have been reheated to temperatures high enough to destroy their magnetization. Subsequently they would cool again upon moving laterally through sloping isotherms, and thereby presumably acquire a new magnetization. The boundaries between zones of alternating magnetizations in this lower crust would have the slope of the isotherm at which remagnetization takes place. The rocks in this depth range would mostly be intrusive, probably gabbros, with a decreasing lava component with depth (see Fig. 1, right side).

The crust would thus consist of two magnetic layers with different magnetization structures. The magnetic zones in the lower layer are displaced laterally away from the axis relative to the corresponding zones in the upper layer. The boundary between the layers would form at a depth determined by the isotherm above which the lavas lose their magnetization. For Icelandic basalts, it has been argued by Kristjánsson and Watkins (1977) that the Curie point of magnetite, 580 °C, is the temperature where this happens. With this assumption, the depth to the boundary between the two magnetic layers will be between 5 and 6 km, according to Figure 1, that is the same as the depth to the boundary between the upper and lower seismic layers of the crust.

The two-layered magnetization structure of the crust resulting from the kinematic and thermal characteristics of the model is in some aspects similar to the structure proposed by Cande and Kent (1976) to explain the skewness of magnetic anomalies observed in the Pacific (see also Vogt, this volume, Ch. 15). The model relates the two-layered structure to processes in the accretion zone. It remains to be seen whether or not this structure can generate the magnetic field as observed over the Iceland area.

The model suggests that crustal processes in the axial zone have a significant effect on the composition of extrusive rocks. The subsidence in the axial rift zone implies a countercurrent flow of lava originating at the surface and magma originating in the mantle. Interaction or mixing of magma occurs especially in the depth range where partial remelting (anatexis) of the downmoving lavas takes place. This mixing changes the composition of the upwards moving magma. According to Figure 1, about 30% of the downward moving crustal material at 6 km depth, where the temperature is about 1000°C, is lava erupted at the surface about four million years ago. This material will undergo partial remelting.

Some of the petrogenetic consequences of the model for the Iceland area have been discussed by Óskarsson and others (1982), who find that the observed petrochemical range of Icelandic rocks can arise from the interaction of mantle-derived olivine tholeiite magma with crust-derived magma. Furthermore, the low $^{18}O/^{16}O$ ratios reported for Icelandic rocks relative to oceanic ridges and islands (Muehlenbachs and others, 1974) can be explained by a mixing of mantle magma with crust-derived silicic magma depleted in ^{18}O by interaction with low ^{18}O meteoric water.

DOES THE ICELAND ACCRETION MODEL APPLY TO SUBMARINE RIDGE SEGMENTS?

It is natural to ask whether the Iceland accretion model, which seems capable of correlating various processes and structures in the Icelandic crust, can also contribute to an understanding of the accretion process in the submarine environment. It is often considered that the accretion processes are very different in the two environments, perhaps because of the diffuse magnetic pattern over Iceland. The evidence suggests a much narrower axial zone in the submarine environment (Macdonald, this volume). A narrower width, however, does not prevent the model from being used elsewhere, as width is one of the parameters that can be varied to fit available observations. The lava production rate in the Iceland zone is probably greater than on normal oceanic segments with comparable spreading rates. This may be due to hot spot activity in the mantle beneath Iceland. The lava production rate is poorly known for oceanic areas, but it is another model parameter that can be varied.

In the following, an attempt will be made to look at some of the evidence that has emerged from studies of submarine rifts, in particular from submersible dives in the FAMOUS area of the Mid-Atlantic Ridge at about 37°N, and from DSDP Leg 37 drillings into the basaltic basement of the oceanic crust, in the light of the Iceland model. A useful summary of some of the DSDP drilling results is given by Hall and Robinson (1979). The main source of information on the FAMOUS area is a series of papers in the 1977 volume of the Geological Society of America Bulletin. A summary of the present state of knowledge of mid-ocean ridges, excluding Iceland, was given by Macdonald (1982).

In order to apply the model to a new area it is necessary to assume values for some of the model parameters. The average spreading half rate v_d in the FAMOUS area will be assumed to be 1 cm/yr (Phillips and others, 1975). The lava production rate q (per unit length of the axis) is not known, but reasonable limits can be put on likely values. An equivalent thickness H of a crustal lava layer, that is, the thickness of a layer consisting entirely of lavas (i.e. extrusive rocks), is related to q by $q = 2 \times v_d \times H$. Leg 37 drillings suggest that H is well over 500 m. The average production rate of lavas in Iceland corresponds to an equivalent layer thickness of some 6.7 km. An intermediate value of 2.5 km for H will be chosen here for the FAMOUS area, giving $q = 5 \times 10^{-5}$ km^2/yr. This is almost six times higher than the value suggested by Moore and others (1974), but less than one half the average value for Iceland. The standard deviations for the extension rate and the lava deposition rate will be assumed to be equal, that is, $\sigma_1 = \sigma_2 = \sigma$ as there is little observational evidence to support a significant difference. The actual value of σ will be left open to discussion in the light of available data. In the absence of better evidence it is assumed here that $\epsilon = 0.5$, independent of depth, that is, normal faults and dikes contribute equal amounts to the horizontal crustal extension.

EVIDENCE FROM DSDP DRILLHOLES INTO CRUSTAL BASEMENT OUTSIDE THE AXIAL ZONE

Some of the new data from DSDP drillings relevant to the present discussion are enumerated below, and discussed in the light of the Iceland model.

1. Most of the basalt penetrated by the drillholes is of *extrusive* origin with only a small fraction of intrusives. The model predicts that at 500 m depth the intrusive fraction will be 18 percent, and less at shallower depths. This result is dependent on assumptions regarding the values of v_d and q. Large lateral variations may be expected in the intrusive fraction, and results from drillholes are therefore highly uncertain as far as the regional behavior of this property is concerned.

2. *Sediments* are commonly interlayered with the extrusive basalts, particularly in the upper 200–300 m of basement (Hall and Robinson, 1979). This is difficult to reconcile with the assumption of a very narrow extrusion zone. The effect of the width

of the extrusion zone can be studied with the help of the Iceland model, if it is modified to include a process of constant rate sedimentation superimposed on the Gaussian distribution of lava deposition. The sedimentation rate will be assumed to be 3 cm/1000 yrs (on the basis of Leg 37 drillholes). The details of the calculations will not be given here, but the results are shown in Figure 2, calculated at a distance of 100 km from the axis for four different widths of the accretion zone.

A transition zone is obtained where the structure changes from 100 percent sediments to 100 percent basalts (excluding the small effect of intrusions in this depth range). If the sediment-basement transition zone thickness H_{sb} is defined as the depth interval over which the structure changes from 95 percent sediments to 95 percent lavas, the following relation is obtained between H_{sb} and σ:

σ km	H_{sb} m
0	0
2	55
5	150
10	310

Thus the thickness of the transition zone is strongly dependent on the width of the lava deposition zone. A value of 5–10 km for σ seems more likely on the basis of the observed sediment structure than a smaller value of 1–2 km. It should be noted that changing the distance from the axis merely shifts the vertical axis in Figure 2, but does not change the structure of the transition zone.

3. Distinct *lithologic units* and *magnetic reversals* are observed in the drillhole sections, and have been interpreted to reveal a certain episodicity in the volcanic activity on a time scale of less than 10^5 years (Hall and Robinson, 1979). These observations are not unexpected on the basis of the Iceland model, according to which the uppermost basement crust is formed not in the highly active median valley close to the axis, but in the outer region of the accretion zone where volcanic activity is sporadic. This has been discussed in some detail for the Iceland crust exposed in eroded valleys in eastern Iceland (Pálmason, 1980). The uppermost 500 m of the oceanic crustal basement as penetrated by the DSDP drillholes is due to lava deposition that has taken place at distances greater than 1.34 σ from the axis of the accretion zone. For σ equal to 2 or 10 km this means outside a zone of a total width of 5.4 or 27 km, respectively.

In order to estimate the frequency of eruptions producing lava flows in this outer zone, as exposed in the drillhole sections, it is useful to look at the time variable (age of lavas) in a vertical profile through the uppermost model crust away from the accretion zone. This is calculated from equation (9) of Pálmason (1980), and is shown in Figure 3. The horizontal line $t = t_o$ corresponds to the classical model with age given by $t_o = x/v_d$. The lower curve $t = t(z)$ gives the age variation of lavas with depth, assuming a Gaussian distribution of lava deposition in the accretion zone. Near the surface the deposition rate is very low, but increases with depth. At 500 m depth it is 890 m/m.y. (for σ = 10 km); an average value for the depth interval 0–500 meters may be 320 m/m.y. If an average thickness of a lava flow is 5 m one finds an average time interval of 15,600 years between volcanic events producing lava flows. The lithologic units, some 50 meters thick (Hall and Robinson, 1979), if formed over a relatively short time, would be separated by time intervals of some 150,000 years. The corresponding time interval for σ = 2 km would be 30,000 years.

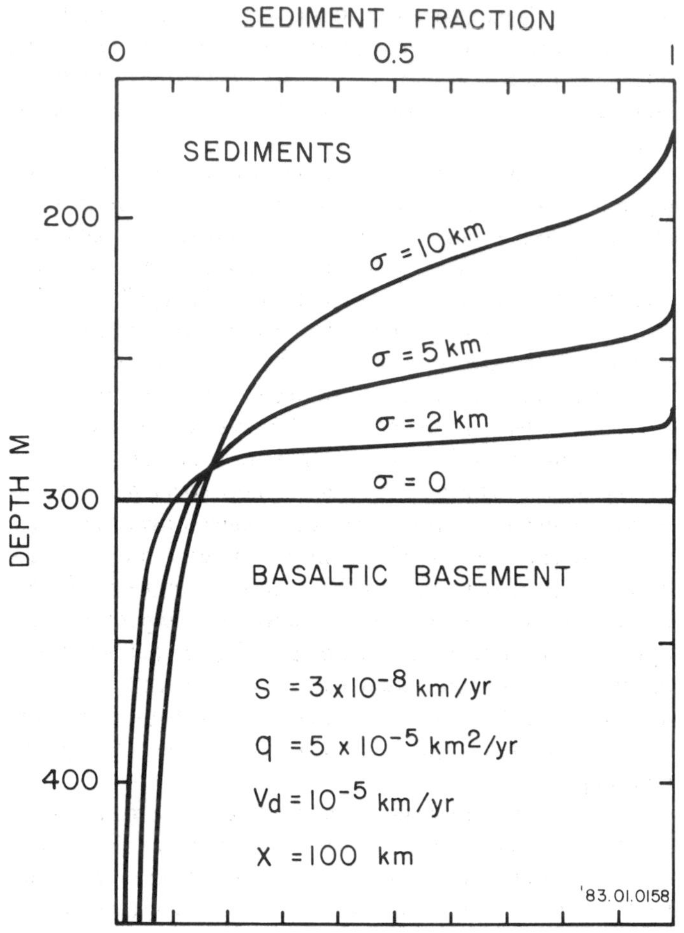

Figure 2. The structure of the sediment-basement transition in the oceanic crust, calculated at a distance of 100 km from the axis, and for four different widths of the accretion zone. At other distances the depth scale will merely be shifted. S = sedimentation rate.

According to the model, the oceanic crustal basement section 0–500 m covers a time interval proportional to the standard deviation of the lava deposition rate in the accretion zone. With v_d = 1 cm/yr and q = 5 × 10^{-5} km^2/yr, a standard deviation of 10 km gives a time interval of about 1.7 m.y. for the section, whereas a standard deviation of 1 km gives an interval of 170,000 yrs. The probability of finding magnetic reversals in such a section thus increases with an increased σ. From the number of reversals observed in the holes (Hall and Robinson, 1979) and the average interval between reversals for the past 10 m.y. (Lowrie and Kent, 1983) one obtains a rough estimate of about 3 km for σ.

The above discussion shows that the apparent episodicity in the volcanic processes, deduced from the distinct lithologic units found in the drillholes, may be explained in terms of the Iceland model by the sporadic occurrence of extrusive events in the outer part of the accretion zone, from where the drilled section originates. No conclusions regarding the volume-time relation of extrusive events within the central median valley can be drawn from the drilled sequence.

4. *Hydrothermal alteration* is generally absent in the drilled sections (Hall and Robinson, 1979). This may seem surprising because of known frequent occurrences of strong hydrothermal activity along rifts. Again, the Iceland model explains this to be due to the origin of the drilled sections in the outer part of the accretion zone. The most intense hydrothermal circulation takes place in the innermost part of the accretion zone, where permeability by active fractures is large and the heat source is at the shallowest level. The thermally altered rocks in this central area will, however, sink to depths in the crust well below the depth reached by the holes. To study these thermally altered rocks, deeper holes into the oceanic crustal basement are needed.

EVIDENCE FROM THE FAMOUS AREA

"On a gross scale, one is impressed by the symmetry of the Mid-Atlantic Ridge. . . . However, on a scale of kilometres or hundred thousands of years, it is difficult to find any parameter that is symmetrical." (Macdonald and Luyendyk, 1977). This quotation describes in a nutshell the failure of the very detailed studies in the FAMOUS area to confirm the existence of a very narrow zone of crustal construction that would have remained stable for long periods of time to produce the symmetrical, classical magnetic structure of the oceanic crust.

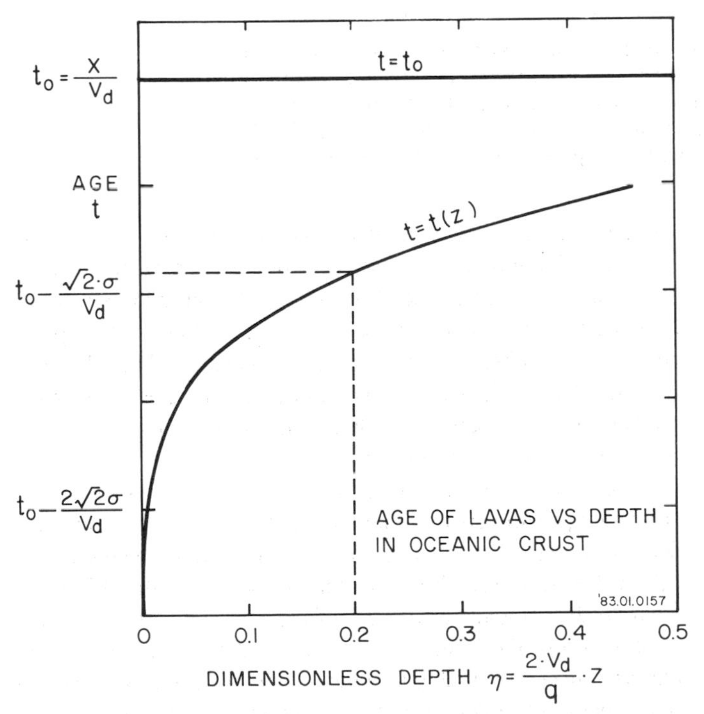

Figure 3. Age of lavas in a vertical profile through the uppermost model ocean crust. The horizontal line corresponds to the classical model with age given by $t = x/v_d$, while the curve results from a Gaussian distribution of the lava deposition rate. The dashed lines correspond to the bottom of a 500 m deep hole, for the accretion parameters used.

Although the 2–3 km wide inner floor of the median valley is clearly the site of very recent volcanic and tectonic activity (Ballard and Van Andel, 1977; Macdonald, 1977; Bryan and Moore, 1977; Luyendyk and Macdonald, 1977), there is considerable evidence to show that over longer periods of time extrusive volcanism takes place over a wider area, with diminishing intensity outwards. The highly asymmetric spreading rates deduced about a narrow zone of crustal accretion (Needham and Francheteau, 1974; Macdonald, 1977) contrasts sharply with the symmetrical overall pattern of magnetic anomalies for the past 10 m.y. in the FAMOUS area (Phillips and others, 1975). The asymmetry changes sense at some distance from the axis to finally produce an approximately symmetrical pattern (Macdonald, 1977). On other segments of the rift the asymmetry is smaller. The asymmetry indicates a certain instability in the location of the narrow extrusion zone that seems to wander about relative to the crustal plates away from the accretion zone, which presumably move with a uniform velocity. The amount of asymmetry deduced in the Rift Valley 2 is sufficient to move the axis laterally over a distance of some 5–6 km in about 1.7 m.y.

Although it is difficult to establish ages of the rocks in the median valley, there is some evidence to indicate that very young rocks occur over a somewhat wider area than the inner floor. Studies of the thickness of palagonite and manganese crusts indicate that the flank lava flows are considerably younger than those corresponding to the inferred spreading age of the crust on which they occur (Bryan and Moore, 1977). Magnetization studies from direct modeling (Macdonald, 1977) indicate that strong magnetization corresponding to relatively fresh basalts occurs sporadically to distances of some 10–15 km from the axis. Both these indications tend to support the possibility of a somewhat wider accretion zone than corresponds to the inner floor of the median valley. In addition, the finding of reversely magnetized rocks from oriented samples on the median valley floor (Johnson and Atwater, 1977) is puzzling in the light of the commonly accepted model of a very narrow zone (1–2 km) of crustal construction at the ridge axis. Evidence for reversely magnetized rocks in the median valley comes not only from studies in the FAMOUS area, but has also been reported at 23°N (Johnson and Atwater, 1977) and at 45°N (Ade-Hall and others, 1973). There is also additional evidence from inversion of deep-tow magnetic anomaly data for the occurrence of reversely magnetized rocks in the FAMOUS median valley (Macdonald, 1977).

Not all the rift valley segments in the FAMOUS area exhibit the same regular morphology as Rift Valley 2. The inner floor of Rift Valley 3 south of Rift Valley 2 broadens towards its southern end to about 18 km, where it is associated with an en echelon arrangement of elongate hills and depressions (Ramberg and van

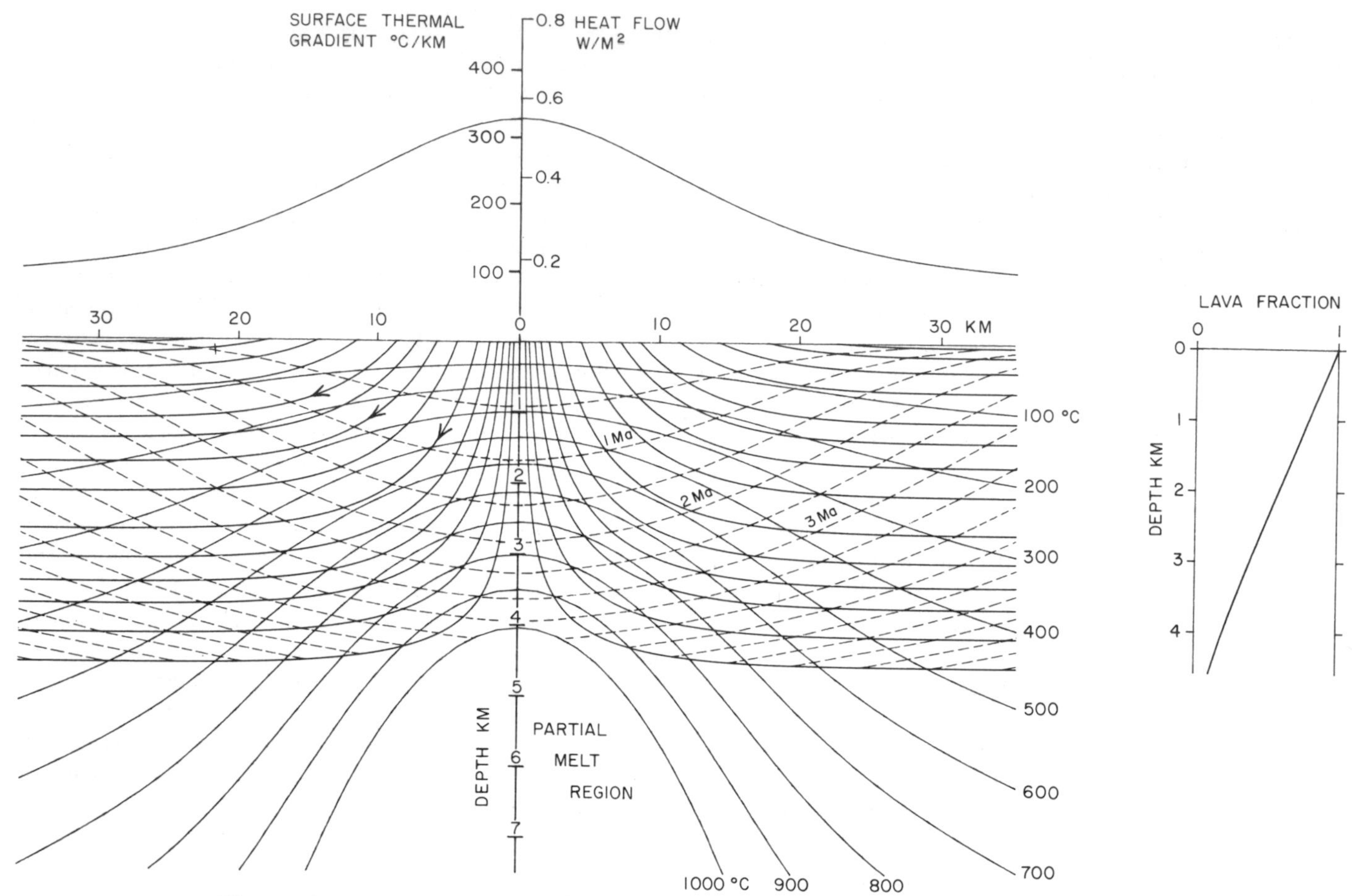

Figure 4. A structural and thermal model cross section for the FAMOUS area, showing lava trajectories (arrows), lava isochrons (dashed), calculated isotherms as well as surface gradient and heat flow. The lava fraction (extrusives fraction) is shown at the right. Here it is independent of the distance from the axis. The model parameters used are: spreading half rate, 1 cm/yr; lava production rate, 5×10^{-5} km^2/yr; standard deviations of crustal strain rate and lava deposition rate, 10 km; and normal fault fraction of crustal strain, 0.5.

Andel, 1977). Although the nature of the volcanism in Rift Valley 3 is not known in as much detail as in Rift Valley 2, a wider zone of active volcanism than in Rift Valley 2 is suggested by the available data.

The evidence discussed above shows that, when viewed on a time scale of 0.5–1.0 m.y., the time needed to form the large scale features of the crust in the FAMOUS area, the accretion zone is probably wider than the inner floor of the median valley. When interpreted in terms of the Iceland model the data from the Leg 37 crustal basement drillings suggest that the standard deviation of the accretion process is in the range of 3–10 km. This conclusion has still to be tested by calculation of the magnetic field resulting from the two-layered magnetization structure indicated by the model.

GENERAL DISCUSSION AND POSSIBLE MODEL CROSS SECTIONS FOR A SLOW SPREADING RIDGE, EXEMPLIFIED BY THE FAMOUS AREA

When applying the model to processes in the accretion zone and the structure of the oceanic crust, it is necessary to keep in mind that the uppermost crust does not derive from processes in the central part of the accretion zone, but from processes in its outer parts. When this is taken into account, many of the apparently puzzling observations from the basement drillings become obvious. This includes volcanic episodicity, magnetic reversals, absence of hydrothermal alteration, and structure of the sediment-basement transition zone.

Several crustal basement model sections have been calculated with parameters that might apply to a slow spreading and a fast spreading ridge. Two of these, which might represent the FAMOUS area, are shown in Figures 4 and 5, the difference being in the standard deviations of the accretion process (2 and 10 km). From these sections one can predict the depth to lavas that have been reheated to a certain temperature during the accretion process and may correspond to certain seismic or magnetic horizons in the oceanic crust. In Table 1 the depth to lavas that have been heated to about 600 °C (Z (600 °C)) is given as well as the axial depth to the 1000°C isotherm (Z_0 (1000°C)) for several sets of model parameters, both for slow spreading and fast spread-

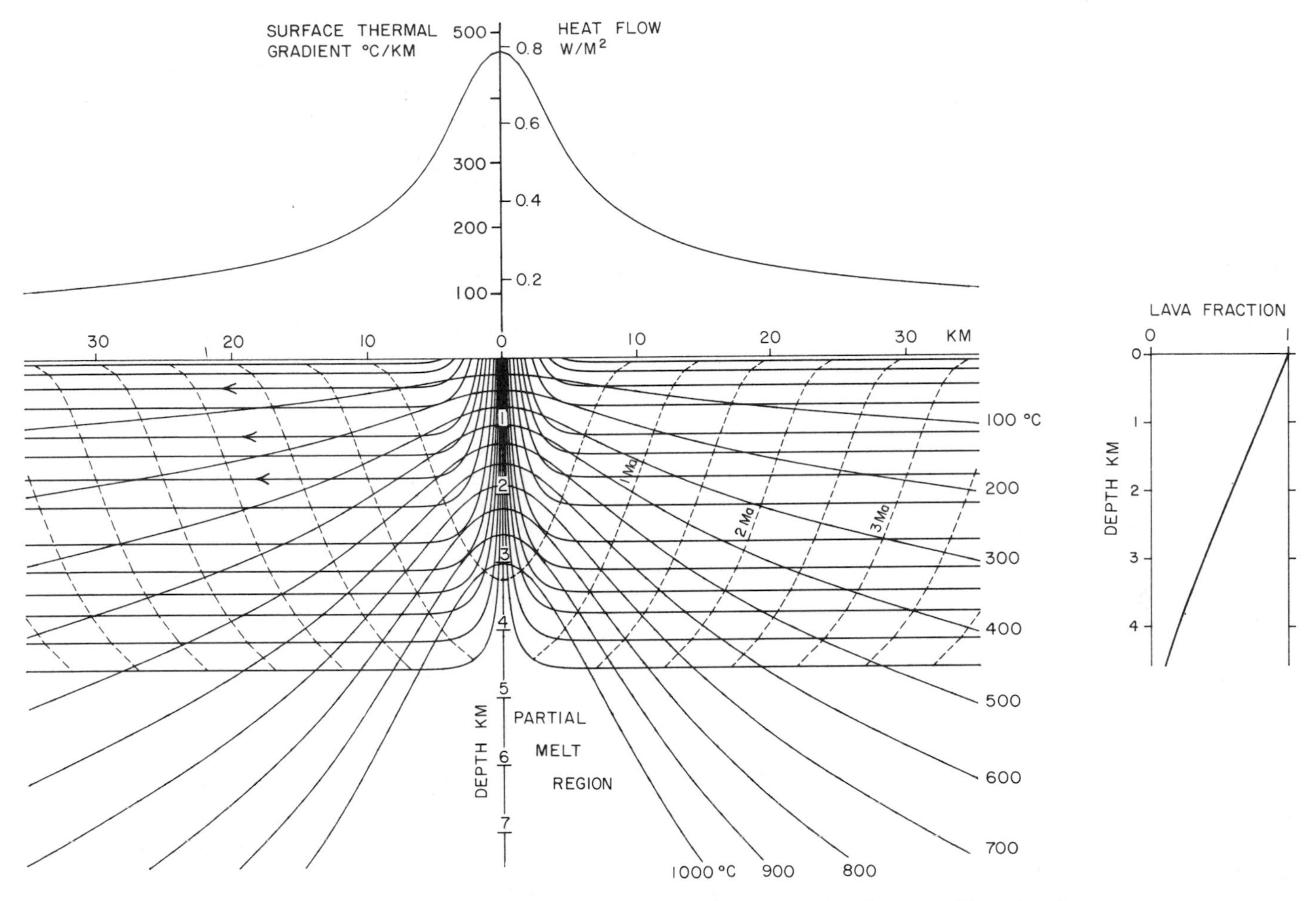

Figure 5. Same as Figure 4, but with the standard deviations of crustal strain rate and lava deposition rate of 2 km.

ing conditions. The earlier discussion of the Iceland section indicates that temperatures of about 600 °C and higher may be associated with the formation of Layer 3, found at 5–6 km depth in the Iceland crust. It is therefore of interest to see what will be predicted on that basis for the depth to Layer 3 in the oceanic crust. For the slow spreading parameters in Table 1 a depth of 1.4 to 3.1 km is obtained, but for the fast spreading parameters a range of 0.9 to 1.8 km is obtained. These depths are in reasonable agreement with seismic studies of the oceanic crust (Ewing and Houtz, 1979), although a more detailed comparison is necessary for individual areas. These calculations may also suggest likely depths to the boundary between an upper and a lower layer of magnetization in the crust, if this is associated with the Curie point of magnetite, 580°C.

The surface thermal gradients calculated from the model can be converted to heat flow by multiplying with the thermal conductivity k. A value of k = 1.7 W/m °C has been used in the present work. The effect of hydrothermal circulation on the temperature distribution in the crust has not been taken into account, but the intrusive activity providing the heat source for the surface heat flux is included in the model. Therefore, the calculated model heat flow can be viewed as a first approximation to the total heat flux by conduction and convection. The third component of heat transport to the surface, by advection of magma, is directly related to the assumed value of q.

Difficulty arises in calculating the total heat flux for a unit length of the axial zone, as its width is not well defined. For a total width of 50 km, the estimated heat flux by conduction (Q_c) and convection (Q_g) for the calculated sections in Figures 1, 4 and 5 are shown in Table 2. Also shown is the estimated heat transport by advection of magma (Q_v). The total heat transport

TABLE 1. MODEL CALCULATIONS OF DEPTH TO THE 1000°C ISOTHERM ($Z_o(1000°C)$) BENEATH AN AXIAL ZONE, AND THE DEPTH TO LAYER 3 ($Z(600°C)$) IN THE OCEANIC CRUST, FOR VARIOUS VALUES OF THE MODEL PARAMETERS v_d, q and σ

v_d cm/yr	q km²/yr	σ km	Z_o(1000°C) km	Z (600°C) km
1	5x10⁻⁵	2	2.95	2.0
"	"	5	3.6	2.6
"	"	10	4.0	3.1
"	2x10⁻⁵	2	1.85	1.4
6	15x10⁻⁵	2	1.25	0.85
"	"	5	1.5	1.05
"	"	10	1.95	1.35
"	30x10⁻⁵	2	2.0	1.3
"	"	5	2.25	1.5
"	"	10	2.6	1.8

TABLE 2. THE THERMAL BALANCE OF THE ICELAND AND FAMOUS SEGMENTS OF THE MAR (IN MW/KM)

	$Q_c + Q_g$	Q_v	Total
Iceland	13.7	21	34.7
FAMOUS (σ = 2 km)	18.6	7.8	26.4
" (σ = 10 km)	19.4	7.8	27.2

to the surface over a 5 km-wide axial zone is thus estimated on the basis of the model to be 35 MW/km for Iceland and 27 MW/km for the FAMOUS area. More detailed calculations are needed with variable model parameters (q and σ) for the FAMOUS area in order to assess how significant the difference between these two values is. Perhaps the similarity is primarily due to the similar spreading rate in the two areas.

A considerably higher estimate of 28 MW/km has been given by Bodvarsson (1982) for the hydrothermal component in Iceland, in part on the basis of a very high heat release of 5000 MW from the subglacial Grímsvötn geothermal area in south-eastern Iceland (Björnsson and others, 1982). Similarly, a very high hydrothermal heat flux has been reported for the black smoker chimneys of the RISE hydrothermal field on the East Pacific Rise (Macdonald, 1982). These high heat fluxes are difficult to reconcile with steady state models. Most likely they represent transient heat releases from cooling bodies of lavas erupted subglacially or in a submarine environment.

REFERENCES

Ade-Hall, J. M., Aumento, F., Ryall, P., Gerstein, R., Brooke, J., and Keown, D. Mc.
1973 : The Mid-Atlantic Ridge near 45°N, XXI, Magnetic results from basalt drill cores from the median valley: Canadian Journal of Earth Sciences, v. 10, p. 679–696.

Ballard, R. D., and van Andel, T. H.
1977 : Morphology and tectonics of the inner rift valley at lat 36°50′N on the Mid-Atlantic Ridge: Geological Society of America Bulletin, v. 88, p. 507–530.

Beblo, M., and Björnsson, A.
1978 : Magnetotelluric investigations of the lower crust and upper mantle beneath Iceland: Journal of Geophysics, v. 45, p. 1–16.
1980 : A model of electrical resistivity beneath NE-Iceland, correlation with temperature: Journal of Geophysics, v. 47, p. 184–190.

Björnsson, A., Saemundsson, K., Einarsson, P., Tryggvason, E., and Grönvold, K.
1977 : Current rifting episode in North Iceland: Nature, v. 266, p. 318–323.

Björnsson, A., Johnsen, G., Sigurdsson, S., Thorbergsson, G., and Tryggvason, E.
1979 : Rifting of the plate boundary in North Iceland 1975–1978: Journal of Geophysical Research, v. 84, p. 3029–3038.

Björnsson, H., Björnsson, S., and Sigurgeirsson, Th.
1982 : Penetration of water into hot rock boundaries of magma at Grímsvötn: Nature, v. 295, p. 580–581.

Bodvarsson, G.
1982 : Terrestrial energy currents and transfer in Iceland; in Continental and Oceanic Rifts, ed. G. Pálmason; Geodynamics Series v. 8, American Geophysical Union and Geological Society of America, p. 271–282.

Bodvarsson, G., and Walker, G.P.L.
1964 : Crustal drift in Iceland, Geophysical Journal of the Royal Astronomical Society, v. 8, p. 285–300.

Bryan, W., and Moore, J. G.
1977 : Compositional variations of young basalts in the Mid-Atlantic Ridge rift valley near lat. 36°49′N. Geological Society of America Bulletin, v. 88, p. 556–570.

Cande, S. C., and Kent, D. V.
1976 : Constraints imposed by the shape of marine magnetic anomalies on the magnetic source: Journal of Gcophysical Rcscarch, v. 81, p. 4157–4162.

Christiansen, N. I., and Wilkens, R. H.
1982 : Seismic properties, density, and composition of the Icelandic crust near Reydarfjördur: Journal of Geophysical Research, v. 87, p. 6389–6395.

Einarsson, T.
1965 : Remarks on crustal structure in Iceland: Geophysical Journal of the Royal Astronomical Society, v. 10, p. 283–288.

Ewing, J., and Houtz, R.
1979 : Acoustic stratigraphy and structure of the oceanic crust; in Deep Drilling Results in the Atlantic Ocean: Ocean Crust, ed. M. Talwani, C. G. Harrison and D. E. Hayes, Maurice Ewing Series, v. 2, American Geophysical Union, Washington, D.C., p. 1–14.

Flóvenz, Ó. G.
1980 : Seismic structure of the Icelandic crust above layer three and the relation between body wave velocity and alteration of the basaltic crust: Journal of Geophysics, v. 47, p. 211–220.

Hall, J. M., and Robinson, P. T.
1979 : Deep crustal drilling in the North Atlantic Ocean: Science, v. 204, p. 573–586.

Johnson, H. P., and Atwater, T.
1977 : Magnetic study of basalts from the Mid-Atlantic Ridge, lat 37°N. Geological Society of America Bulletin, v. 88, p. 637–647.

Kristjánsson, L., and Watkins, N. D.
1977 : Magnetic study of basaltic fragments recovered by deep drilling in Iceland, and the "magnetic layer" concept: Earth and Planetary Sciences Letters, v. 34, p. 365–374.

Kristmannsdóttir, H.
1975 : Hydrothermal alteration of basaltic rocks in Icelandic geothermal areas: Proceedings of the Second U.N. Symposium on the Development and Use of Geothermal Resources, San Francisco, p. 441–445.
1979 : Alteration of basaltic rocks by hydrothermal activity at 100–300°C: in International Clay Conference 1978, ed. M. M. Mortland and V. Cc Farmer; Amsterdam, Elsevier, p. 359–367.
1982 : Alteration in the IRDP drill hole compared with other drill holes in Iceland: Journal of Geophysical Research, v. 87, p. 6525–6531.

Kristmannsdóttir, H. and Tómasson, J.
1978 : Zeolite zones in geothermal areas in Iceland; in Natural Zeolites: Occurrence, Properties, Use, ed. L. B. Sand and F. A. Mumpton; Oxford, Pergamon Press, p. 277–248.

Laughton, A. S., and Searle, R. C.
1979 : Tectonic processes on slow spreading ridges; in Deep Drilling Results in the Atlantic Ocean: Ocean Crust, ed. M. Talwani, C. G. Harrison and D. E. Hayes; Maurice Ewing Series, American Geophysical Union, v. 2, p. 15–32.

Lowric, W., and Kcnt, D. V.
1983 : Geomagnetic reversal frequency since the Late Cretaceous: Earth and Planetary Science Letters, v. 62, p. 305–313.

Luyendyk, B. P., and Macdonald, K. C.
1977 : Physiography and structure of the inner floor of the FAMOUS rift valley: Observation with a deep-towed instrument package: Geological Society of America Bulletin, v. 88, p. 648–663.

Macdonald, K. C.
1977 : Near-bottom magnetic anomalies, assymmetric spreading, oblique spreading, and tectonics of the Mid-Atlantic Ridge near lat 37°N: Geological Society of America Bulletin, v. 88, p. 541–555.

1982 : Mid-ocean ridges: Fine scale tectonic, volcanic and hydrothermal processes within the plate boundary zone: Annual Review of Earth and Planetary Sciences, v. 10, p. 155–190.

Macdonald, K. C., and Luyendyk, B. P.
1977 : Deep-tow studies of the structure of the Mid-Atlantic Ridge crest near lat. 37°N: Geological Society of America Bulletin, v. 88, p. 621–636.

Mehegan, J. M., Robinson, P. T., and Delaney, J. R.
1982 : Secondary mineralization and hydrothermal alteration in the Reydarfjordur drill core, eastern Iceland: Journal of Geophysical Research, v. 87, p. 6511–6524.

Moore, J. G., Fleming, H. S., and Phillips, J. D.
1974 : Preliminary model for extrusion and rifting at the axis of the Mid-Atlantic Ridge, 36°48′ North: Geology, Sept. 1974, p. 437–440.

Muehlenbachs, K., Anderson, Jr., A. T., and Sigvaldason, G. E.
1974 : Low 0 basalts from Iceland: Geochimica et Cosmochimica Acta, v. 38, p. 577–588.

Needham, H. D., and Francheteau, J.
1974 : Some characteristics of the rift valley in the Atlantic Ocean near 36°48′ north: Earth and Planetary Science Letters, v. 22, p. 29–43.

Oldenburg, D. W.
1975 : A physical model for the creation of the lithosphere: Geophysical Journal of the Royal Astronomical Society, v. 43, p. 425–451.

Óskarsson, N., Sigvaldason, G. E., and Steinthórsson, S.
1982 : A dynamic model of rift zone petrogenesis and the regional petrology of Iceland: Journal of Petrology, v. 23, p. 28–74.

Pálmason, G.
1971 : Crustal Structure of Iceland from Explosion Seismology: Publication 40, Societas Scientiarum Islandica, Reykjavík, 187 pp.
1973 : Kinematics and heat flow in a volcanic rift zone, with application to Iceland: Geophysical Journal of the Royal Astronomical Society, v. 33, p. 451–481.
1980 : A continuum model of crustal generation in Iceland: Kinematic aspects: Journal of Geophysics, v. 47, p. 7–18.
1981 : Crustal rifting, and related thermo-mechanical processes in the lithosphere beneath Iceland: Geologische Rundschau, v. 70, p. 244–260.

Pálmason, G., and Saemundsson, K.
1974 : Iceland in relation to the Mid-Atlantic Ridge: Annual Review of Earth and Planetary Science, v. 2, p. 25–50.

Pálmason, G., Arnórsson, S., Fridleifsson, I. B., Kristmannsdóttir, H., Saemundsson, K., Stefánsson, V., Steingrímsson, B., Tómasson, J., and Kristjánsson, L.
1979 : The Iceland crust: Evidence from drillhole data on structure and processes; in Deep Drilling Results in the Atlantic Ocean: Ocean Crust, ed. M. Talwani, C. G. Harrison, and D. E. Hayes; Maurice Ewing Series 2, American Geophysical Union, Washington, D.C., p. 43–65.

Parker, R. L., and Oldenburg, D. W.
1973 : Thermal model of ocean ridges: Nature Physical Sciences, v. 242, p. 137–139.

Phillips, J. D., Fleming, N. S., Feden, R., King, W. E., and Perry, R.
1975 : Aeromagnetic study of the Mid-Atlantic Ridge near the Oceanographer Fracture Zone: Geological Society of America Bulletin, v. 86, p. 1348–1357.

Ramberg, I. B., and van Andel, T. H.
1977 : Morphology and tectonic evolution of the rift valley at lat 36°30′N, Mid-Atlantic Ridge: Geological Society of American Bulletin, v. 88, p. 577–586.

Saemundsson, K.
1978 : Fissure swarms and Central volcanoes of the neovolcanic zones of Iceland; in Crustal Evolution in Northwestern Britain and Adjacent Regions, ed. D. R. Bowes and B. E. Leake: Geological Journal, Special Issue No. 10, p. 415–432.

Walker, G.P.L.
1959 : Geology of the Reydarfjördur area, eastern Iceland: Quarterly Journal of the Geological Society, London, v. 114, p. 367–391.
1960 : Zeolite zones and dyke distribution in relation to the structure of the basalts in eastern Iceland: Journal of Geology, v. 68, p. 515–528.

Manuscript Accepted by the Society October 15, 1984

ACKNOWLEDGMENTS

Asmundur Jakobsson has helped greatly in the computer calculations used in the paper. Norman Sleep, Peter Vogt and an anonymous reviewer made valuable critical comments on the manuscript. All these contributions are gratefully acknowledged.

Printed in U.S.A.

The Geology of North America
Vol. M, The Western North Atlantic Region
The Geological Society of America, 1986

Chapter 7

Seismicity along the eastern margin of the North American Plate

Páll Einarsson
Science Institute, University of Iceland, Dunhaga 3, 107 Reykjavik, Iceland

INTRODUCTION

The eastern boundary of the North American Plate is marked by a narrow and continuous belt of seismic activity. The earthquake zone follows the crest of the Mid-Atlantic Ridge and its associated fracture zones (F.Z.) (Fig. 1; Plate 8A; Vogt, this volume, Ch. 12.) and is intimately related to the process of plate separation, transform faulting, and generation of oceanic crust. This part of the Mid-Atlantic plate boundary has the advantage of being relatively close to the dense seismograph networks of North America and Europe, and has therefore been the subject of more study than most other parts of the oceanic rift system. For this reason it has sometimes been taken as the type example of an oceanic rift, which can be misleading. Other segments of the ridge system, such as parts of the East Pacific Rise, have been shown to possess a distinctly different seismicity pattern (Stover, 1973), with most earthquakes occurring along the fracture zones and practically no seismicity along the ridge axes.

In this paper we will describe the main features of the seismicity of the mid-oceanic ridge system in the North Atlantic and Arctic Oceans and how it relates to the physical state and processes near the plate boundary. Intraplate earthquakes, that is events within the North American Plate, are dealt with by Zoback and others (this volume).

Because of the enormous amount of literature on the subject no attempt has been made to give complete reference to every paper. For more complete references the reader is referred to recent review papers on different aspects of the subject, for example those by Savostin and Karasik (1981) and Husebye and others (1975) on Arctic seismicity; Einarsson (1979) and Einarsson and Björnsson (1979) on the seismicity of Iceland and the ridge to the south; Duschenes and others (1983) and Whitmarsh and Lilwall (1983) on ocean bottom seismograph studies, Sanford and Einarsson (1982) on the detection of magma chambers in rifts; and Lilwall (1982) on the seismicity of oceanic rifts. Studies of focal mechanisms are summarized by Einarsson (1985).

Early studies of the world's seismicity identified the Atlantic Ocean as a seismically active area even though the distribution of known earthquakes was understandably limited to inhabited areas along the coast and on Atlantic islands. The identification of a continuous seismic zone in the middle of the ocean was first possible with instrumental observation, first by Tams (1922, 1927a, b) and later by Gutenberg and Richter (1941, 1949).

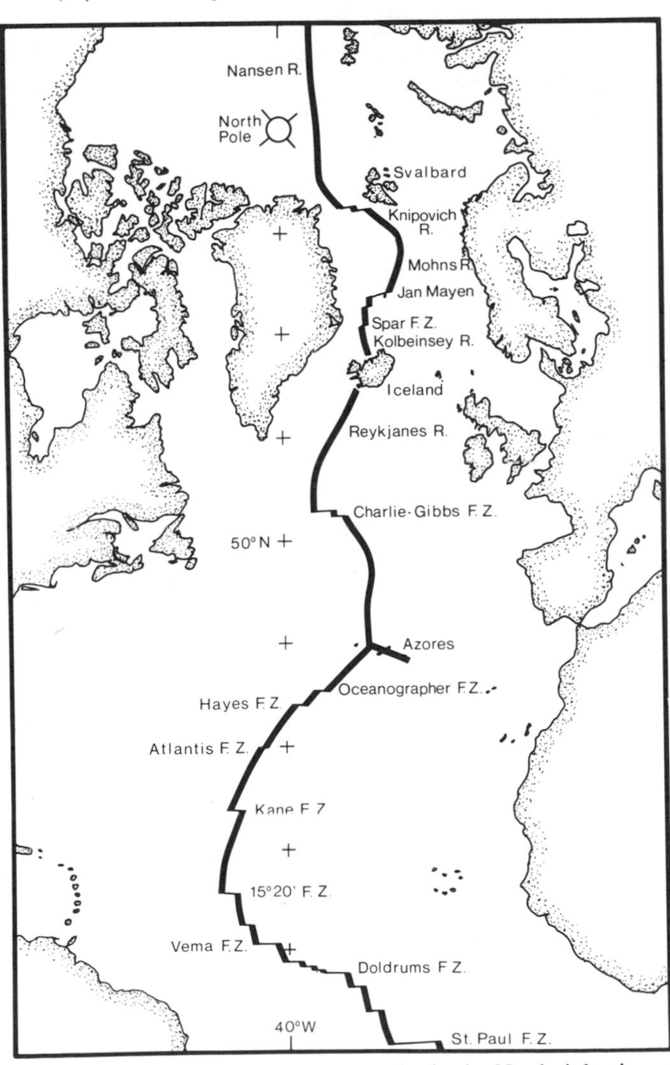

Figure 1. Index map of the plate boundaries in the North Atlantic and Arctic Oceans.

Einarsson, P., 1986 Seismicity along the eastern margin of the North American Plate; *in* Vogt, P. R., and Tucholke, B. E., eds., The Geology of North America, Volume M, The Western North Atlantic Region: Geological Society of America.

A major step in the attempt to define the seismic zones followed the implementation in the early sixties of the World Wide Standardized Seismograph Network (WWSSN), and the use of electronic computers and improved Earth models to locate earthquakes. Sykes (1965) published a map of relocated epicenters of the period 1955–1964 in the Arctic region, which revealed more details than possible before. The continuity and narrowness of the seismic zone was further established, as well as its coincidence with rifted mid-oceanic ridges, an extension of the Mid-Atlantic Ridge into the Arctic.

The increased resolution of the seismic networks coincided with a rapid increase in knowledge of sea floor morphology. It now became possible to correlate the seismic zones with specific topographic features such as fracture zones and rifts along ridge crests. New seismicity maps of the world, such as the one by Barazangi and Dorman (1969) showed that seismic zones form a continuous net encircling the Earth. The global network of long period seismographs offered new opportunities to obtain reliable fault plane solutions for earthquakes. Hypotheses on sea floor spreading and plate tectonics provided a new framework of thought, and were strongly supported by improved seismological data (Isacks and others, 1968). Seismic zones delineate lithospheric plates and fault plane solutions along plate boundaries indicate the relative plate movements. For oceanic plate boundaries the landmark paper was written by Sykes (1967), who demonstrated the main features of mid-oceanic ridge seismicity. Earthquakes along fracture zones were shown to be limited to the section between the adjacent ridge axes, and were accompanied by strike-slip faulting along the fracture zone, in an opposite sense to that expected for a simple offset of the ridge crest. This means that the offset of ridge segments across the fracture zones is not caused by transcurrent motions along the fracture zones. This was an important support to the transform fault hypothesis of Wilson (1965). Earthquakes along the ridge crests were found to be associated with normal faulting, indicating crustal extension in the axial zone. Further differences between ridge crest and fracture zone earthquakes were demonstrated by Francis (1968a,b), who found significantly different magnitude-frequency relationships for the two classes of events. Earthquake swarms were found to be frequent along the ridge axes (Sykes, 1970), whereas earthquakes along the fracture zones tended to occur in mainshock-aftershock sequences. The studies of Sykes and Francis showed that the more or less orthogonal system of ridges and fracture zones is not the expression of horizontal shear along conjugate fault planes, as had been suggested by some earlier authors (e.g. Van Bemmelen, 1964; Tr. Einarsson, 1968). In fact, the theory of sea-floor spreading provided the only consistent explanation for the distribution of earthquake foci and focal mechanisms along the Mid-Atlantic Ridge.

DATA

For obvious reasons one must rely on instrumental data when studying Mid-Atlantic Ridge seismicity, except in Iceland, where considerable information on historical seismicity is available. Instrumental data for this region come from three types of sources.

1. Teleseismic data give relatively homogeneous information over the North-Atlantic area. These data give epicentral location for all earthquakes larger than magnitude 4.5, and fault plane solutions can be obtained for earthquakes of body wave magnitude 5.5 and larger. In time the data are relatively homogeneous since 1963, that is, after the implementation of the WWSSN. Epicentral locations and fault plane solutions of earthquakes since 1963 are shown on Plate 11 (see also Plate 8A).

2. Local seismograph networks in Iceland give detailed information on the seismicity of that part of the ridge system. Four stations were in operation in the late sixties, but the number of short period, permanent stations was greatly increased in the early seventies and has since varied between 20 and 40 stations. In addition, dense, multielement networks have been in operation in different areas over short periods of time. These studies have yielded accurate hypocentral depths and fault plane solutions for a large number of small events.

3. Ocean-bottom seismographs have been deployed in a few areas on the Mid-Atlantic Ridge, notably near 1°N, 23°N, 37°N, and 45°N. For practical reasons, these studies are limited in space and time and their results must be interpreted with these limitations in mind.

FROM THE EQUATOR TO THE AZORES

This part of the Mid-Atlantic Ridge is cut by an unusually large number of fracture zones with large offsets (Vogt, this volume, Ch. 12). South of 25°N, almost all the fracture zones offset the ridge crest to the left, but to the north all offsets are to the right. This shapes the ridge system into an arc-like structure, concave towards east, reflecting the original shape of the continents at the time of break-up. The present plate boundaries between the North American and African Plates and the South American and African Plates are faithfully traced by the seismicity, which shows a narrow zone of brittle deformation. In most places the zone is 20 km wide or less, which is just about the resolution of the teleseismic locations. The boundary between the North and South American Plates, on the other hand, is not delineated by a well-defined seismic zone. The intraplate seismicity west of the Mid-Atlantic Ridge at latitudes 10–20°N appears to be slightly higher than normal, indicating plate deformation in a broad zone separating the North and South American Plates.

All major transform faults can be identified on the seismicity map by one or more of their seismic characteristics, that is east-west alignment of epicenters, offsets in the ridge crest seismic zone, and fault plane solutions indicating strike-slip faulting in the transform sense. All these characteristics have been found in the Vema, 15°20′, Kane, and the Doldrums fracture zones. Other major fracture zones, such as the Atlantis F. Z., are only identifiable on the seismicity map as an offset in the ridge crest seismicity. Most fracture zones have a clear E-W seismicity belt, up to 610 km in length. Examples of right-lateral and left-lateral transform faulting are shown in Figure 2.

All fault plane solutions obtained for earthquakes at or near the ridge crest in this sector show normal faulting. Although the orientation of the fault planes and stress axes cannot be determined with confidence, the stress conditions in the axial regions appear to be non-uniform and may change considerably over short distances and with time.

Two seismic peculiarities found near the Oceanographer and Hayes fracture zones are worth mentioning. A prominent cluster of epicenters is found west of the plate boundary, where the Oceanographer Fracture Zone joins the ridge axis (Fig. 2). This area has shown persistent activity over the last 20 years with several events of magnitude 5 or larger. The events occur as far as

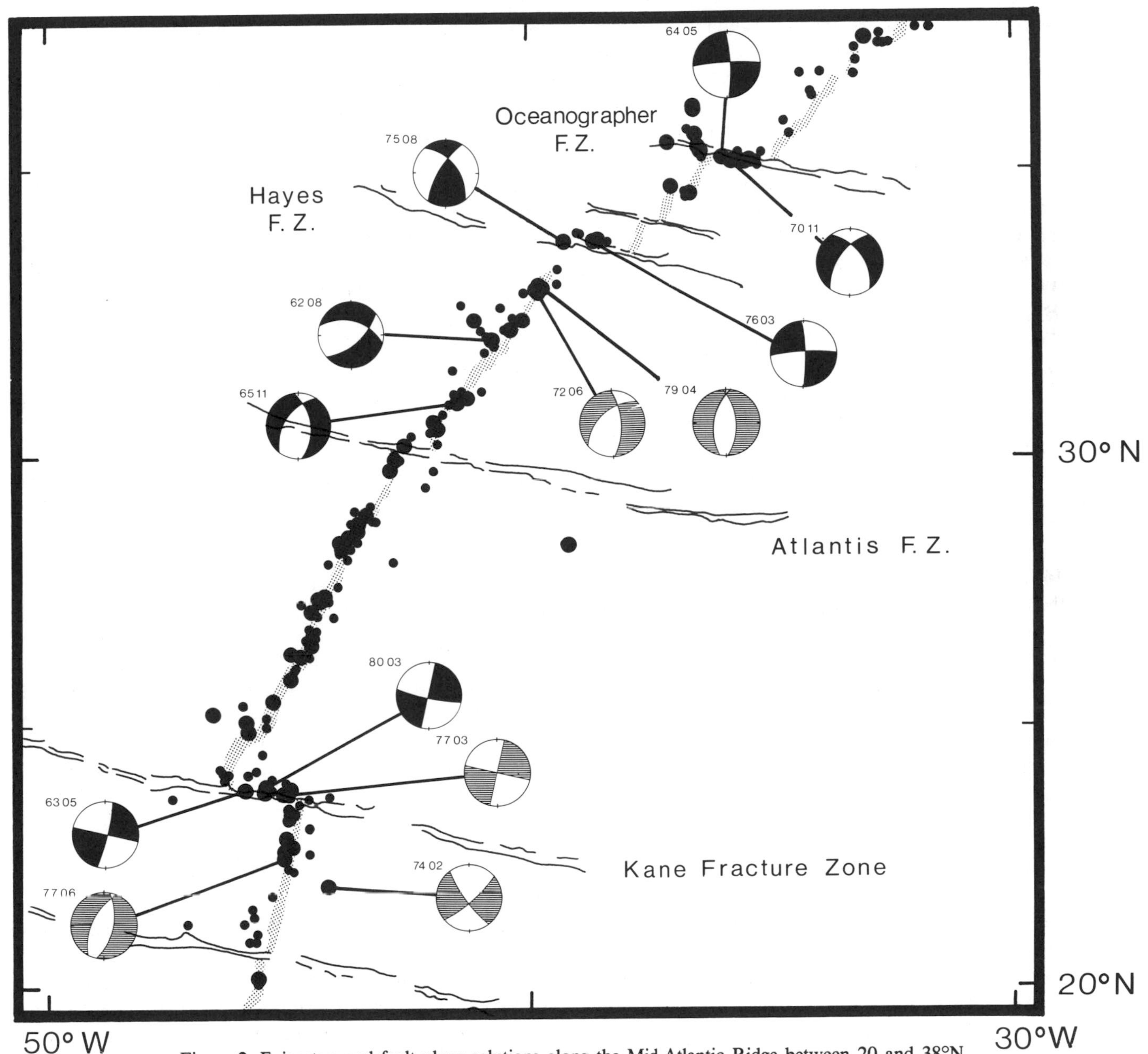

Figure 2. Epicenters and fault plane solutions along the Mid-Atlantic Ridge between 20 and 38°N, showing transform faulting in the Kane, Hayes, and Oceanographer fracture zones, and normal faulting along the ridge crest. Epicenters are from the PDE lists of the U.S. Geological Survey for the period 1963–1981, only epicenters determined with 10 or more stations are included. Fault plane solutions are shown schematically on lower hemisphere stereographic projection of the focal sphere, compressional quadrants are black. Numbers denote year and month of the event. Solutions are from Sykes (1967), Udías and others (1976), and Einarsson (1985).

100 km off the plate boundary, which is much more than the expected uncertainty in the locations. No unusual topographic features are known in this area, and the nature of this seismic activity is unclear. The other case is a fault plane solution of one event near the western end of the Hayes transform fault that shows a significant but unusual component of thrust faulting. Large scale vertical movements are indicated by the presence of transverse ridges in some of the major fracture zones (Bonatti, 1978; Bonatti and Chermak, 1981; Bonatti and others, 1983) and seem to be an integral part of the tectonic regime of a transform fault. Occasional occurrence of thrust faulting events in fracture zones should therefore not be too surprising.

This sector of the ridge system has been the object of several microearthquake surveys using both ocean bottom and free floating instruments. Duschenes and others (1983), studying the hypocentral resolution of microearthquake surveys carried out at sea, concluded that networks consisting of sonobuoys alone (Reid and MacDonald, 1973, Spindel and others, 1974) are only of marginal use in work of this kind and warned against overinterpretation of these data.

Earthquakes were recorded by ocean bottom instruments in the FAMOUS area (Francis and others, 1977). These originated both in the median valley and in the adjacent fracture zone, but since only two instruments were operational, too much significance should not be attached to the locations. Rowlett (1981) studied seismicity at the eastern ridge junction of the Oceanographer Fracture Zone, using two ocean bottom seismographs for about five days. Hypocenters could, of course, not be determined with any accuracy, but making some assumptions it was concluded, that the recorded seismicity was distributed across the corner between the ridge axis and the transform fault. Similar results were obtained by Rowlett and Forsyth (1984) for the western ridge intersection of the Vema Fracture Zone and by Francis and others (1978) for the eastern ridge intersection of the St. Paul Fracture Zone. In the latter study hypocentral depths could be resolved. Earthquakes were found to originate at 0–8 km depths.

To date, the most extensive ocean bottom seismic experiment on the Mid-Atlantic Ridge was conducted at the ridge axis near 23°N, slightly south of the Kane Fracture Zone (Toomey and others, 1985). Ten instruments were in operation for a period of three weeks. Earthquakes were located beneath the central valley and the eastern valley wall, with hypocentral depths between 5 and 8 km. Composite fault plane solutions show normal faulting. These results indicate that this section is undergoing active extension and has cooled to temperatures within the brittle field of behavior to a depth of at least 7–8 km, even in the center of the median valley.

THE AZORES-GIBRALTAR ZONE

In the Atlantic, the Eurasian and African Plates are separated by a seismic zone that extends westward from the Straits of Gibraltar to form a triple junction with the Mid-Atlantic Ridge within the Azores archipelago. The nature and evolution of this plate boundary and the triple junction has been the subject of considerable debate. See for example Krause and Watkins (1970), McKenzie (1972), Laughton and Whitmarsh (1974), Udias and others (1976), Hirn and others (1980), Searle (1980), Udias (1982), and Moreira (1982).

The zone can be divided into three sections according to its seismic characteristics. The westernmost section extends from the triple junction, through the Azores, to the eastern end of the archipelago. On the seismicity map the zone appears relatively narrow, comparable to the zone on the Mid-Atlantic Ridge. Fault plane solutions show that strike-slip is the principal mode of faulting, one nodal plane with an easterly strike and the other striking northerly. It appears likely that the northerly striking nodal planes of the fault plane solutions are the fault planes, which are therefore not parallel to the main seismic zone or the trend of the archipelago (Hirn and others, 1980). The earthquakes here seem to occur in response to a stress field set up and maintained by relative movement of the two plates, but the seismic zone has not developed into a steady state feature. Several authors have concluded that the plate boundary in this region has a complex history (e.g. McKenzie, 1972; Laughton and Whitmarsh, 1974). The situation here may be somewhat similar to the South Iceland Seismic Zone where faults active in individual earthquakes strike transversely to the overall trend of the zone (Einarsson and Eiríkson, 1982).

The middle section of the Azores-Gibraltar zone has been seismically quiet for several decades. It is either locked or temporarily inactive.

East of 18°W the plate boundary is no longer defined by a narrow seismic belt. The high, but diffuse seismicity in this region shows that plate deformation occurs within a 400–500 km wide zone that extends into the Gulf of Cadiz and is connected to the seismic belt of Morocco and Algiers. Earthquakes in this zone may reach magnitude 8 (as did the shocks of February 1969 and May 1975) or even more (as must have been the case in 1755 when Lisbon was destroyed). Focal mechanisms are characterized by thrust faulting; occasionally strike-slip mechanisms are seen. A common feature is the maximum compressional axis trending N to NW, reflecting convergence between the Eurasian and African Plates.

FROM THE AZORES TO ICELAND

This section of the Mid-Atlantic plate boundary consists of two gently arcuate ridges, offset near 52°N by the Charlie-Gibbs Fracture Zone, a major transform fault. All other fracture zones, including the Kurchatov Fracture Zone near 40.5°N, have too small an offset to be resolved on the seismicity map. The ridge axis immediately to the north of the Azores is relatively straight and plate separation occurs at a right angle to the plate boundary. Seismic activity is fairly uniform both in space and time and earthquakes larger than magnitude 5 are rare (Einarsson, 1979). Two fault plane solutions show normal faulting at the ridge axis. Using three instruments, an ocean bottom seismograph study was conducted in the axial region near 45°N (Lilwall and others, 1977, 1978). All the recorded activity, including a swarm, was located in a narrow, elongate zone under the median valley.

The ridge section between 48 and 51°N appears to have some peculiar features. The seismic zone has a general NNW trend. Oblique spreading must therefore occur along this plate boundary, if the spreading direction is assumed to be parallel to the Charlie-Gibbs Fracture Zone immediately to the north. The structure of this part of the ridge is characterized by alternating N-S trending and oblique spreading axes. The N-S axes are associated with transverse basement ridges that trend slightly north of the spreading direction on both sides of the plate boundary. This "herringbone" pattern in the topography was interpreted by Johnson and Vogt (1973) to result from asthenospheric flow southward from the Iceland hot spot. One must then assume that the intersection of the transverse ridge with the plate boundary is the locus of unusually high production of eruptive material that slowly migrates southward along the plate boundary.

Focal mechanism studies in this region show some unexpected results. Fault plane solutions for two events at or near the ridge axis, near 49.5°N and 51°N, have a significant component of thrust faulting. The axis of maximum compression is horizontal, trending NE nearly perpendicular to the basement ridges. Several possible explanations for these unusual fault plane solutions were given by Einarsson (1979), including magmatic activity within a central volcano complex. Forcible intrusion of viscous magma at shallow depth can cause thrust faulting in the adjacent region, and the deflation of a magma chamber will cause reverse faulting in the chamber roof. Explanations of this kind are favoured here, especially since the only other known examples of reverse faulting at a divergent plate boundary are found under the Bárdarbunga central volcano in the eastern rift zone of Iceland (see later in this paper). It is therefore suggested that the transverse ridges described by Johnson and Vogt (1973) are the traces of central volcano complexes.

The Charlie-Gibbs Fracture Zone, between 52 and 53°N, offsets the ridge crest about 350 km to the left. The structure of this zone is described in considerable detail in the literature, most recently by Searle (1981). The structure is dominated by two parallel troughs, 45 km apart, that are separated by an E-W ridge. In the western part of the transform section the northern trough is more pronounced, but in the eastern part the southern trough is better developed. Vogt and Avery (1974) and later Searle (1981) concluded from topography, sonographs, and magnetic and seismic data that the two troughs were joined by a spreading center cutting across the transverse ridge at 31°45′W. This is seen clearly in the seismicity map in the back of this volume. During the last two decades, at least, practically all the recorded seismicity has been limited to the northern trough west of the central spreading axis. Transform faulting is shown by five fault plane solutions. The southern trough has been seismically inactive, except possibly at the eastern junction with the ridge axis. One fault plane solution indicates oblique faulting at this junction.

Kanamori and Stewart (1976) studied the source parameters of earthquakes and the mode of strain release along the Charlie-Gibbs Fracture Zone. They found that earthquakes of M_S between 6 and 7 occurred with an average repeat time of 13 years, the latest in 1967 ($M_S = 6.5$) and 1974 ($M_S = 6.9$). Seismic moment (M_O) was found to be anomalously large for the respective M_S, and similarly M_S was anomalously large if compared to m_b. These disparities were explained by large fault length (60–70 km) and anomalously low dislocation particle velocity (20 cm/s). Okal and Stewart (1982), who studied this phenomenon for a number of fracture zones, concluded that "slow" earthquakes tended to occur in fracture zones in the neighborhood of hot spot volcanism.

The seismic zone north of the Charlie-Gibbs Fracture Zone follows the crest of the Reykjanes Ridge to Iceland. A systematic variation along the plate boundary is seen in several of the characteristics of the ridge. Near its southern end the ridge trends N-S, has a rough topography, a well developed rift valley and high rate of seismicity. These features are rather typical for the Mid-Atlantic Ridge, and here the spreading occurs at a right angle to the plate boundary. Near 56°N the ridge bends to a N35°E direction. All available fault plane solutions in this area show normal faulting (Einarsson, 1979; Tréhu and others, 1981). The spreading north of this point appears to be oblique to the plate boundary. Near 58½°N another change occurs. The topography becomes smooth and the central valley gives way to a central horst (Fleischer, 1974). The seismicity, which is relatively high to the south, is low north of 58½°N. This seismicity pattern has persisted for the past half century at least (Francis, 1973). The northern part of the Reykjanes Ridge thus seems to have some of the characteristics of a fast spreading ridge, in spite of its low spreading rate of 1 cm/yr (Talwani and others, 1971). There seems to be a general consensus in the literature to ascribe this to the proximity to the Iceland hot spot (see e.g. Vogt, 1978).

ICELAND

The plate boundary in Iceland is displaced to the east by two major fracture zones, the South Iceland Seismic Zone in the south and the Tjörnes Fracture Zone in the north. Both zones have rather complex structures and lack the clear topographic expression typical of oceanic fracture zones. They are defined primarily by their high seismicity, earthquake focal mechanisms, and configuration with respect to the spreading axes (Ward, 1971; Tryggvason, 1973). The largest earthquakes in Iceland occur within these zones and may exceed magnitude 7 (M_S). The divergent plate boundary between the fracture zones is expressed by volcanic rift zones, with one branch in northern Iceland and two parallel branches in southern Iceland. Earthquakes occur along the rift zones, but they rarely exceed magnitude 5 (m_b). This activity is highly clustered both in time and in space, and a large part of it appears to be related to central volcanoes.

Teleseismic locations of earthquakes of the period 1963–1981 are shown in Figure 3. This map shows many of the characteristics of Icelandic seismicity, even though some of the

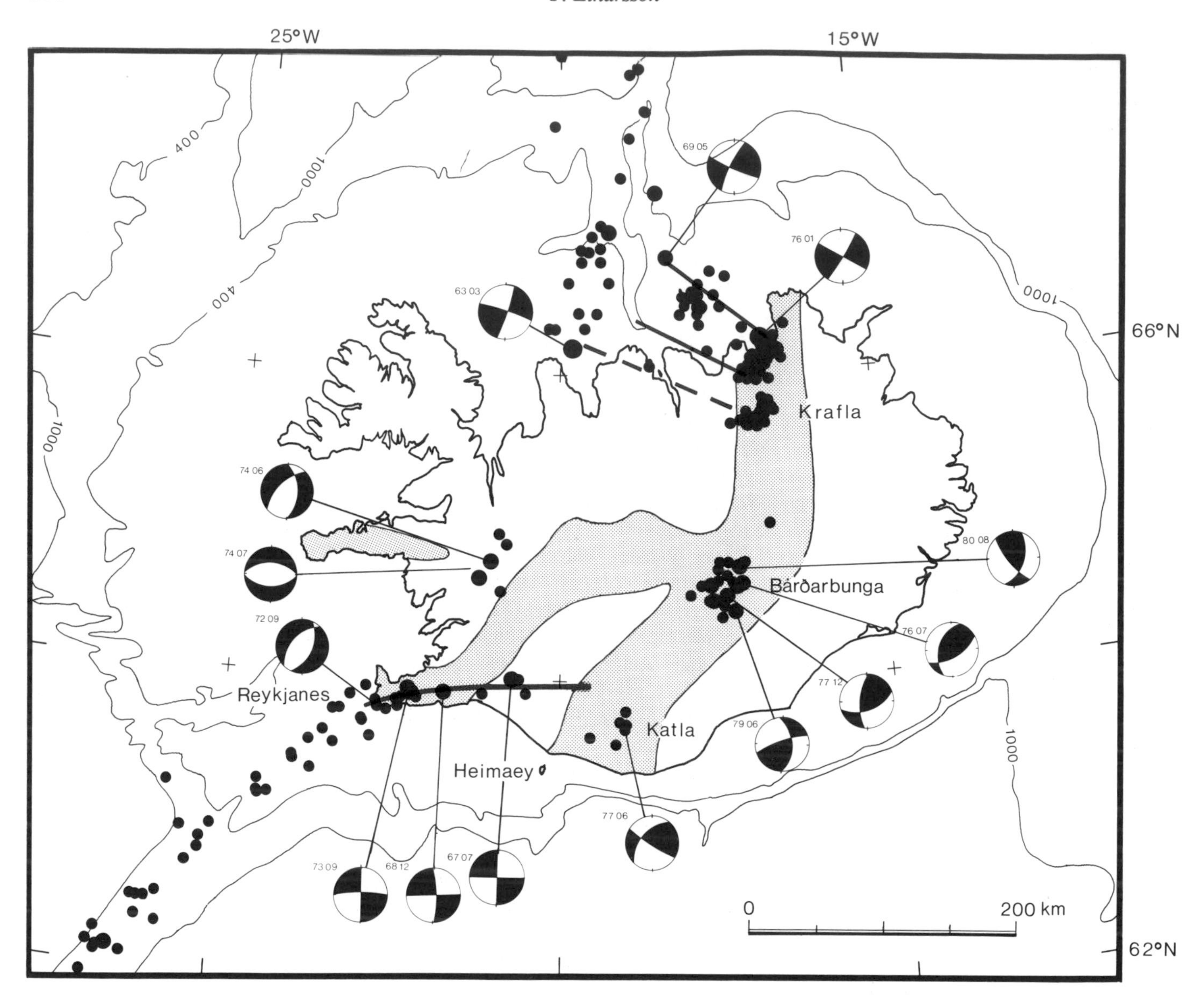

Figure 3. Epicenters and single event focal mechanism solutions in the Iceland area. Epicenters are taken from the PDE lists of the U.S. Geological Survey for the period 1963–1981, only epicenters determined with 10 or more stations are included. Larger dots are events of m_b = 5 and larger. Focal mechanism solutions are shown schematically on lower hemisphere stereographic projection of the focal sphere, compressional quadrants are black. For further references on the solutions see Einarsson (1979, 1985). The volcanic zones are stippled red and red lines show the seismic belts in the fracture zones.

locations may be in error by as much as 40 km. Concentration of activity is seen in the Tjörnes Fracture Zone near the coast of northern Iceland, in southwestern Iceland on the Reykjanes Peninsula, and in the South Iceland Seismic Zone. The focal mechanisms indicate strike-slip faulting. If the easterly striking nodal planes are taken as fault planes, the sense of motion is right-lateral in northern Iceland and left-lateral in southwestern Iceland, which is consistent with a transform fault interpretation of these zones. Outside of the fracture zones, clusters of activity are seen in the Borgarfjördur area in western Iceland, in the volcanic zone in Central Iceland, and near the volcanoes Katla in southern Iceland and Krafla in northern Iceland. More detailed information reveals that each of these zones has distinct seismological characteristics, which will be described below.

Historical seismicity of the three last centuries is shown in Figure 4. Except for the Borgarfjördur earthquakes of 1974, all major events are within the two transform zones. In southern Iceland the events are concentrated in a narrow, E-W belt, but in northern Iceland the estimated epicenters are distributed in a broader zone. The same pattern is seen in the instrumentally determined epicenters shown in Figure 3. It is clear that the large-scale seismotectonic pattern of the two zones is different, even though there are similarities in some of the details.

The Reykjanes Peninsula is an area of high seismicity and

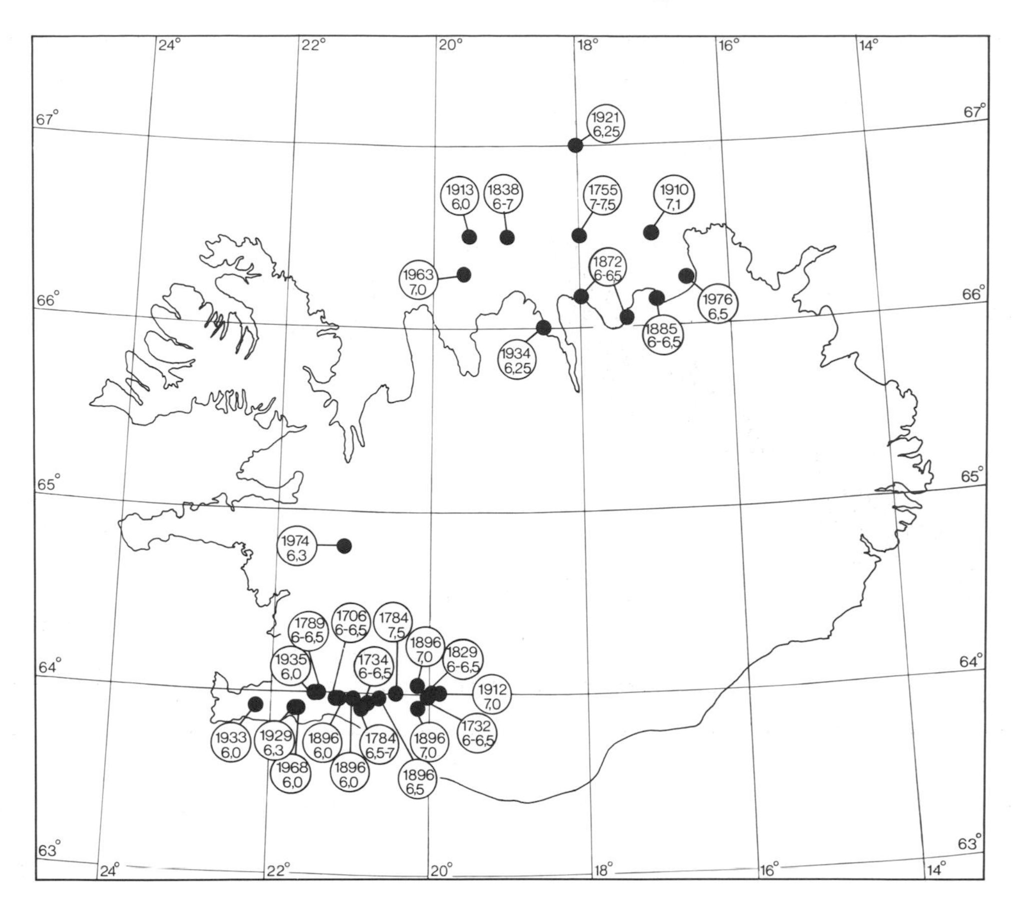

Figure 4. Estimated epicenters of historical, destructive earthquakes in Iceland after 1700. Year of occurrence and magnitude are given in the circles. Magnitudes of events prior to 1900 are estimated from the destruction areas. From Björnsson and Einarsson (1981).

recent volcanism that forms a transition between the Reykjanes Ridge to the west, and the western volcanic zone and the South Iceland seismic zone to the east (Fig. 3). The Mid-Atlantic plate boundary as defined by the seismicity enters Iceland near the tip of Reykjanes and then runs along the peninsula in an easterly direction. Detailed studies by Klein and others (1973, 1977) show that the seismic zone is less than 2 km wide in most places (Fig. 5). The earthquakes are mostly at a depth of 1–5 km and are not located on a single fault. The seismicity seems to be caused by deformation of the brittle crust above a deeper seated, aseismic deformation zone. Small scale structures can be resolved in the seismicity within the zone. Several seismic lineations or faults can be identified, striking obliquely or even transversely to the main zone.

Focal mechanisms have been determined for a large number of small earthquakes, using data from dense, local networks, and for two earthquakes larger than m_b 5 using teleseismic data. The minimum compressive stress is consistently oriented in a horizontal, NW direction. The maximum compressive stress rotates between the vertical direction, causing normal faulting on NE-striking faults, and the horizontal NE direction, causing strike-slip faulting on N or E striking faults. Thus the stress regime is characterized by the NW-trending minimum stress. The other principal stresses are probably nearly equal and may change directions according to local, or time dependent, conditions. Dykes open up against the minimum stress and strike NE, as shown by the eruptive fissures observed on the surface.

The mode of strain release changes systematically along the peninsula. Near the tip of Reykjanes, earthquakes occur in swarms; that is, in sequences where no single event is much larger than the others. Normal faulting is the most common faulting mechanism. Toward the east earthquakes tend to occur more in mainshock-aftershock sequences and strike-slip faulting becomes more prominent. The plate boundary in the eastern part of the peninsula has not been mapped in detail because of the low activity there in recent years.

The South Iceland Seismic Zone bridges the gap between the two branches of the volcanic rift zones in South Iceland. Few instrumentally determined epicenters are available, but the epicenters of historical events determined from the destruction areas (Fig. 4) define a narrow, E-W trending zone. In spite of this E-W orientation there is no indication on the surface of a major E-W fault in this area. However, many of the earthquakes have been associated with surface faulting; but the faults or fault systems have a N-S orientation (Einarsson and Eiríksson, 1982). The earthquakes appear to be caused by right-

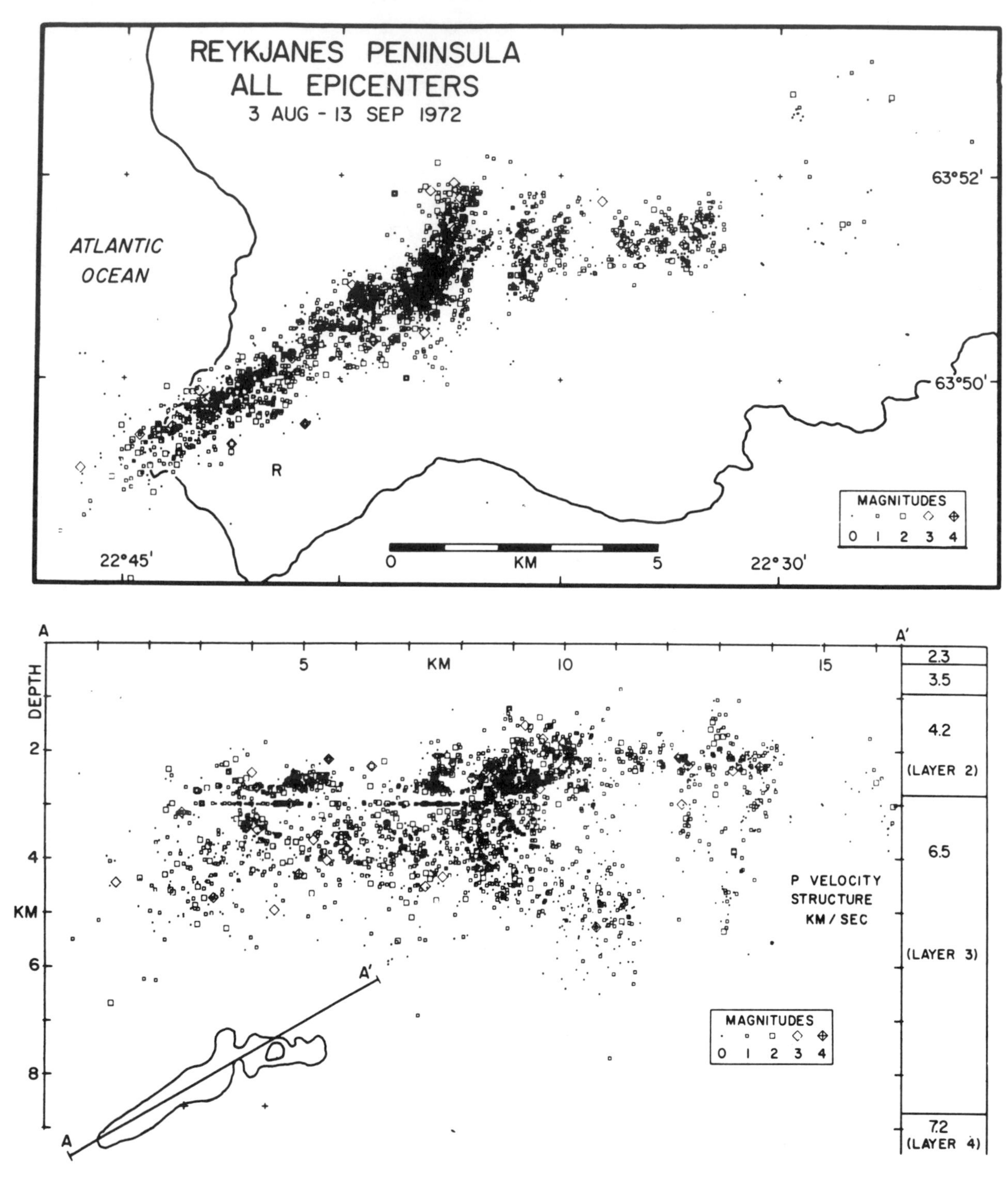

Figure 5. Epicenters and a hypocentral cross section of the earthquake swarm of Aug.–Sept. 1972, on the Reykjanes Peninsula in Iceland, from Klein and others (1977). Horizontal and vertical error of most hypocenters is less than 0.5 km, velocity model used in the locations is shown at the right. The position of the cross section with respect to the epicentral zone is shown in the inset.

lateral strike-slip on northerly striking faults, arranged side by side within the E-W zone. This is supported by the N-S elongate destruction areas (Einarsson and others, 1981). The reason for this unusual configuration is poorly understood, but a tentative interpretation (Einarsson and Eiríksson, 1982) is that the transform zone is migrating sideways in response to the southward propagation of the eastern rift zone.

The strain in the South Iceland Seismic Zone is released in sequences of large earthquakes with a recurrence time of 50–120 years. The sequence often starts with a magnitude 7–7½ shock in the eastern part of the zone followed by slightly smaller shocks in the western part.

The Tjörnes Fracture Zone also has a complicated structure, but in a different way. It is a broad zone of faulting and seismicity that connects the southern end of the submarine Kolbeinsey Ridge to the volcanic rift zone in North Iceland. The seismicity is too diffuse to be associated with a single fault or a simple plate boundary. Instead, the transform motion appears to be taken up by a series of parallel NW-striking faults or seismic zones (Einarsson, 1976; Einarsson and Björnsson, 1979). Two such zones have been identified; the Grímsey zone marking the northern boundary of the Tjörnes Fracture Zone off shore and the Húsavík zone, which can be traced on land from the rift zone through the town of Húsavík, and then offshore as a bathymetric feature (Saemundsson, 1974) and a strong gravity anomaly (Pálmason, 1974). A third zone has been tentatively identified near the southern boundary of the Tjörnes F. Z. Three fault plane solutions confirm transform faulting.

Earthquakes in the volcanic zones are generally smaller than in the fracture zones. Volcanic eruptions are usually accompanied by earthquakes, but between eruptions most parts of the volcanic zones are seismically quiet. A few areas of persistent seismic activity are found, the most prominent ones in Central Iceland and near the subglacial volcano Katla in South Iceland (Fig. 3).

The seismic area in Central Iceland, near the center of the Iceland hot spot, is largely covered by the ice cap Vatnajökull so its tectonic structure is poorly known. Recent studies of ERTS images of this area (Thorarinsson and others, 1973) seem to indicate that the structure is dominated by a group of central volcanoes, and it is tempting to relate the earthquakes to volcanic processes. An unusual sequence of earthquakes occurred in this area during the period 1974 to 1980. Seven earthquakes of m_b 5 and larger occurred in this period, but events of that size were unknown there before. Fault plane solutions of four of these events (Fig. 3) all show that the sequence is associated with reverse faulting in the caldera region of the subglacial Bárdarbunga volcano. It is difficult to see how this type of faulting could be related to plate movements. Brittle failure of the crust above a deflating magma chamber, however, could produce reverse faulting earthquakes, and this is considered to be the most likely explanation for the Bárdarbunga events.

The subglacial Katla volcano is located near the southern end of the eastern volcanic zone, south of its junction with the South Iceland Seismic Zone. The structure of this part of the zone is characterized by several central volcanoes; rifting structures are less significant. Historic eruptions of Katla have been preceded by felt earthquakes and, because of the potential danger of future eruptions, the seismicity at Katla is monitored by a relatively dense seismograph network. The epicenters located so far delineate two active areas 15 km apart. One poorly constrained fault plane solution indicates strike-slip with a significant component of reverse faulting. As in the case of Bárdarbunga, a deflating magma chamber may offer an explanation for this type of faulting. The seismic activity in the Mýrdalsjökull area shows a pronounced annual cycle. The probability of an earthquake occurring within a given time interval is several times higher in the second half of the year than in the first half. This annual cycle was first noted by Tryggvason (1973) for the years 1952–1958 and has been confirmed by later data.

The Heimaey eruption in 1973 was preceded by an intensive swarm of small earthquakes that started 30 hours before the eruption. Earthquakes also accompanied the eruption, but seismicity declined as lava production diminished. No shock reached local magnitude 4. The earthquakes during the eruption originated at the depth of 15–25 km and occupied a spherical volume centered under Heimaey. It seems likely that the erupted magma was either stored or formed within this volume.

The depth of the Heimaey earthquakes is much larger than observed elsewhere in Iceland. In this area the upper boundary of the anomalous mantle underlying Iceland is at a depth of 12–15 km. Earthquakes at the depth of 15–25 km may be taken to imply brittle failure in the Icelandic, anomalous mantle where creep or ductile behaviour is normally assumed. In this volcanic region it is possible, however, that high strain rates associated with magmatic processes may cause brittle failure in material that would be ductile at lower strain rates.

A major rifting episode has been in progress since 1975 in the volcanic rift zone in northeastern Iceland. The activity has been confined to the Krafla central volcano and its associated fissure swarm (Björnsson and others, 1977, 1979) and provides a demonstration of a process that seems to play an important role in Icelandic tectonics. The activity is characterized by repeated cycles of relatively slow inflation and rapid deflation of the volcano. Magma apparently accumulates at a constant rate under the volcano during the inflation periods and during the deflation events the magma escapes from the reservoir area. Each cycle of activity is accompanied by a characteristic pattern of seismic activity as described by Brandsdóttir and Einarsson (1979) and Einarsson and Brandsdóttir (1980). Continuous volcanic tremor starts in the caldera region at about the same time as the deflation. Small earthquakes also occur in the caldera, but the epicentral area is soon extended along the Krafla fissure swarm to the north or to the south, as shown by the example in Figure 6, the rifting event of July 1978. The rate of propagation of the seismic activity is highest during the first few hours, typically 0.5 m/sec., but the speed decreases as the deflation rate decreases and the epicentral zone is extended. The earthquake activity culminates after the maximum in tremor and deflation rate is reached. The largest earthquakes are located within a well defined, but each time different, section of the fissure swarm. Local magnitude only rarely exceeds 4.5. The depth of hypocenters is in the range 0–9 km. Extensive fault movements, both normal faulting and fissuring, occur in the area of maximum earthquake activity. The propagating seismic activity suggests that the magma escaping from the Krafla reservoir is injected laterally into the fissure swarm to form a dyke. The dykes may be as long as 40–60 km.

The first and the most violent deflation event started on Dec.

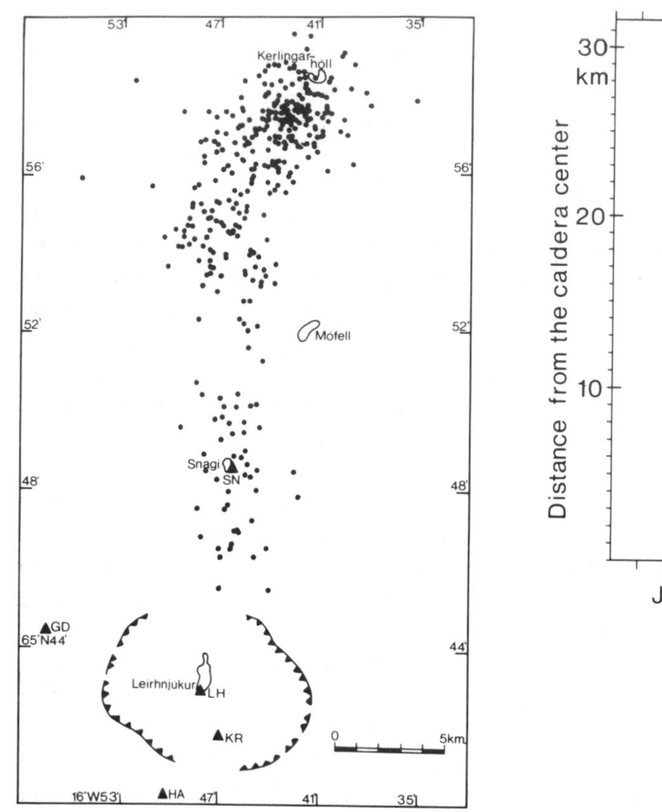

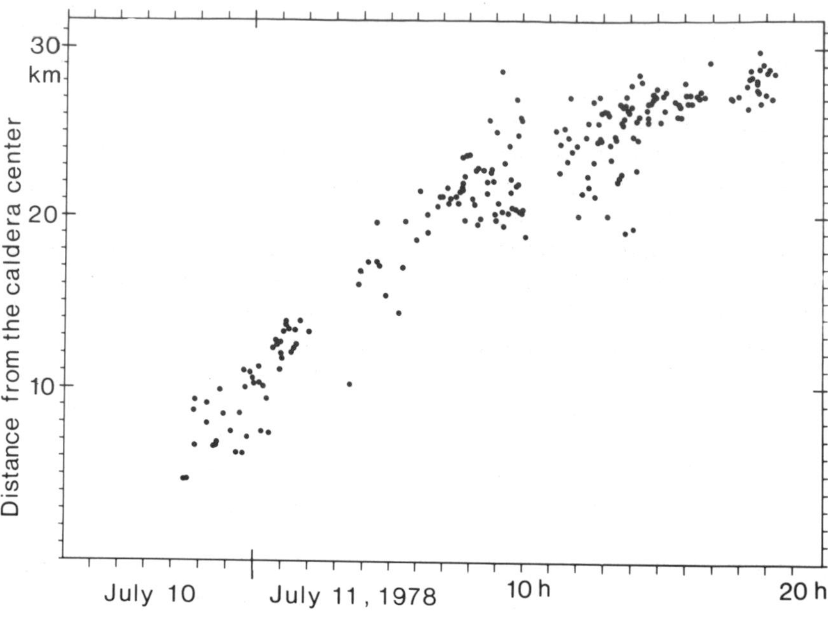

Figure 6. Epicentral map and migration of epicenters during the intrusion of July 1978 in the Krafla fissure swarm in North Iceland. The origin in the time-distance plot corresponds to the center of the caldera and the time when deflation of the caldera region and volcanic tremor began. After Einarsson and Brandsdóttir (1980).

20, 1975. The deflation of the caldera exceeded 2 m and the accompanying earthquake swarm lasted about eight weeks. Most of the epicenters that appear in the northern part of the volcanic zone in Figure 3 belong to this swarm. The largest earthquakes were confined to two separate areas. One area was within the caldera where earthquakes were apparently associated with faulting above the deflating magma reservoir. Depth of most hypocenters was 0–4 km. The largest earthquakes reached magnitude 5. The other epicentral area was near the junction between the Krafla fissure swarm and the Grímsey seismic lineament. The largest earthquake was of m_b 6 and the focal mechanism shows right-lateral strike-slip along the Grímsey zone. This earthquake sequence demonstrates well the relationship between rifting along the diverging plate boundary and transform faulting in the fracture zone. The present Krafla events are assumed to be the result of interaction between magma pressure under the Krafla volcano and rifting of the plate boundary. The rifting is triggered by increasing magma pressure in the reservoir when a fluid-filled extensional crack propagates horizontally along the Krafla fault swarm. The driving force of this process is the tectonic stress at the plate boundary, but the mode of strain release is modified by the presence of fluid.

Prior to July 1980 the deflation events were associated with mostly rifting and subsurface transport of magma. Eruptions to the surface were only minor. But character of the deflation events changed in July 1980. In most events since then a large proportion of the mobilized magma has reached the surface and rifting has played a secondary role.

Earthquakes are rare outside of the volcanic zones and the transform zones in southern and northern Iceland. However, intraplate earthquakes are known in the Iceland region, for example, near the insular shelf margin east of Iceland and in Borgarfjördur in western Iceland where a significant earthquake sequence was recorded in 1974. This sequence was studied with portable instruments, giving detailed locations and fault plane solutions (Einarsson and others, 1977). The depth of hypocenters was 0–8 km and the area was shown to be undergoing horizontal extension.

THE PLATE BOUNDARY IN THE ARCTIC

North of Iceland the plate boundary follows the Kolbeinsey Ridge to the Jan Mayen Fracture Zone. This part of the plate boundary is anomalous in several ways, because of its relatively high elevation, low seismicity, and its asymmetric position with respect to the adjacent continents. All these features may be related in some way to the existence of the Iceland hot spot to the south. The topographic anomaly is part of a much larger regional bulge centered on Iceland. The low seismicity is comparable to that of the northern part of the Reykjanes Ridge and thus displays a certain symmetry with respect to Iceland. The off-center position of the Kolbeinsey Ridge is the result of one or more ridge jumps to the west that may be related to drifting of the plate boundary off the Iceland mantle plume. The axis of the Kolbeinsey Ridge is offset by two minor transform faults, the Spar Fracture Zone near 69°N and another small fracture zone, probably located near 71°N, immediately south of the Eggvin Bank. Neither zone is clearly expressed on the seismicity map. The plate boundary in the Eggvin Bank region is poorly defined and may be complicated by the topographic high suggested by Saemundsson (this volume) to be the trace of a hot spot in this area.

The Jan Mayen Fracture Zone is the most significant fracture zone in the Arctic region, displacing the ridge axis 210 km to the right. The transform section of the fracture zone is a pronounced but highly asymmetric trough, because of the high topography connected with the Eggvin Bank and the Jan Mayen continental sliver to the south. The volcanic island of Jan Mayen protrudes into the fracture zone about 55 km west of the eastern ridge intersection. The overall trend of the transform section is 110°, and all fault plane solutions are consistent with left-lateral strike-slip along a plane with strike varying between 120° in the east to 102° in the west.

During the last two decades there has been a marked difference in the strain release pattern between the western and the eastern parts of the Jan Mayen Fracture Zone. Few, but large earthquakes have occurred in the western part. In the eastern part, on the other hand, the seismicity is characterized by relatively frequent, but smaller, earthquakes. Much of this activity is concentrated where Jan Mayen protrudes into the fracture zone. Soernes and Fjeldskaar (1980) concluded that the activity had increased in this area, notably after the eruption of Beerenberg on Jan Mayen in 1970. Earthquake swarms are found to be associated with Beerenberg, but most of the larger earthquakes seem to be related to slip along the transform fault.

In most respects the Mohns Ridge is a typical mid-oceanic ridge, centrally located in the Greenland-Norwegian basin and uninterrupted by fracture zones. The spreading is slightly oblique, especially at its western end, where it joins with the Jan Mayen Fracture Zone at an angle of 120°. The seismic zone is well defined, narrow and continuous along the rifted crest. The seismicity is higher than that of the Kolbeinsey Ridge and the northern Reykjanes Ridge, but comparable to other parts of the Mid-Atlantic Ridge. Fault plane solutions by Conant (1972) and Savostin and Karasik (1981) show normal faulting at the ridge axis.

The pattern of seismicity changes abruptly from the Mohns to the Knipovich Ridge (Vogt, this volume, Ch. 12). Spreading along the Knipovich Ridge is highly oblique and the seismicity is diffuse. This diffuse distribution of epicenters is not an artifact of the location accuracy as one can see by comparison with the tightly concentrated seismicity of the Mohns Ridge immediately to the south. Vogt and others (1982) note that the change in the seismicity pattern coincides with a change in the ridge crest topography. The rift valley widens where the seismicity drops off. The scattered epicenters imply that deformation takes place within a wide zone, a feature possibly inherited from the time when this part of the plate boundary changed from being primarily of strike-slip character to being a divergent boundary. Subsequent jumping of the rift axis may also contribute to the complexity.

The intraplate seismicity is considerably higher east than west of the Knipovich Ridge. This asymmetry was noted by Vogt and others (1982) in connection with asymmetry of other features of the Arctic ridge system such as topography, gravity, and spreading rate. A good part of the intraplate events east of the plate boundary appears to be related to a seismicity anomaly in Svalbard.

The axis of the Knipovich Ridge approaches the continental margin west of Svalbard and is there displaced to the left by two 100 km long transform faults connected by a short ridge axis. The southern fracture zone has been relatively quiet in the last two decades, whereas several earthquakes larger than $m_b = 5$ have occurred in the northern one. Fault plane solutions of five events show transform faulting (Horsfield and Maton, 1970; Conant, 1972; Savostin and Karasik, 1981).

The Svalbard Archipelago stands out on most seismicity maps as an area of high seismicity. Mitchell and others (1979) and Bungum and others (1982) operated temporary local networks in 1976, 1977, and 1979, and found that most of the activity was concentrated in two zones, although scattered activity was also recorded in other parts of the archipelago. Teleseismic maps show activity on the continental shelf west of Svalbard, possibly connecting it to the Knipovich plate boundary. The concentrated earthquake zones are elongate in a WNW-ESE direction and fault plane solutions show left-lateral strike-slip. This seismicity was attributed by Bungum and others (1982) to an interaction between the present tectonic stress field and older zones of weakness. Savostin and Karasik (1981), on the other hand, felt that this seismicity implied the existence of a separate plate, the Spitsbergen microplate, even though its eastern boundary could not be clearly delineated.

Northwest of Svalbard, near 83°N, the plate boundary turns to a NE trend and follows the Nansen (Gakkel) Ridge across the Arctic Basin. The Nansen Ridge is uninterrupted by large offset transform faults over a distance of more than 2000 km. The seismicity is moderately high and all available fault plane solutions show normal faulting at the ridge crest (Savostin and Karasik, 1981; Conant, 1972; Sykes, 1967). The seismicity of this part of the plate boundary is elaborated on further by Fujita (1985).

EARTHQUAKE SEQUENCES IN TIME AND SPACE

Earthquake sequences are often classified into three principal types (Mogi, 1963): mainshock-aftershock sequences, foreshock-mainshock-aftershock sequences, and earthquake swarms, that is, sequences without one event that is distinctly larger than the others. Examples of different types of sequences are shown in

Figure 7. The sequence of September 1969 in the 15°20′ Fracture Zone is a typical mainshock-aftershock sequence, with four recorded aftershocks following a strike-slip earthquake ($m_b = 5.7$) within a few hours. The sequence of October 1974 in the Charlie-Gibbs Fracture Zone is similar except that a foreshock preceded the mainshock by nine minutes. The Reykjanes Ridge sequence of March 1967 is a typical swarm that terminated a very active period in this part of the plate boundary. Furthermore, sequences are found that are difficult to classify according to this scheme, for example, the complex sequence of December 1975 to February 1976 in northern Iceland, involving deflation of the Krafla volcano, rifting and intrusion in the rift zone, and transform faulting. It may be described as two earthquake swarms and a mainshock-aftershock sequence superimposed on one another. Sequences that are best described as a series of mainshocks also exist, for example, in the Charlie-Gibbs Fracture Zone in July 1965 (Sykes, 1970) and in the South Iceland Seismic Zone in 1896 (Einarsson and others, 1981). In these sequences each mainshock is outside the source area of the previous one.

Soon after the installation of networks of long period seismographs it became apparent that earthquake swarms were common along the world rift system. Sykes (1970) studied the spatial distribution of swarms and found that most of them originated in the crestal area of the ridges. Fracture zone seismicity, on the other hand, is characterized by mainshock-aftershock sequences, although swarms occasionally occur there. Clustering of events, both in time and space, is thus found to be more common along the rift zones than along fracture zones (Francis and Porter, 1971).

In addition to clustering of epicenters within distances of ten kilometers, several examples can be found of apparent temporal correlation of seismicity over distances of hundreds of kilometers along the plate boundary. These include active periods in the vicinity of the Vema Fracture Zone in 1979, the Charlie-Gibbs Fracture Zone in 1965–1967 (Einarsson, 1979), and the South Iceland Seismic Zone (Einarsson and others, 1981). These examples demonstrate temporal correlation of activity over large areas, and yet the source volume of each earthquake is much too small for one event to directly influence the occurrence of the others. It seems likely that the events are triggered by a regional strain pulse that affects a large part of the plate boundary, possibly related to large scale plate motions.

Several cases of migration of seismicity along the plate boundary have been documented, ranging over distances of kilometers to thousands of kilometers, and at rates from 1 to 100 km per day.

Limond and Recq (1981) observed an apparent migration of seismic events along the western boundary of the Eurasian Plate, from Jan Mayen southward to the Azores and then east toward Gibraltar. By correlating the seismicity of different subsections of this plate boundary, a migration velocity of 3 to 10 km per day was found. The significance of the correlation is critically dependent on the occurrence of three earthquake sequences; a large swarm in March 1967 on the Reykjanes Ridge near 56°N, a

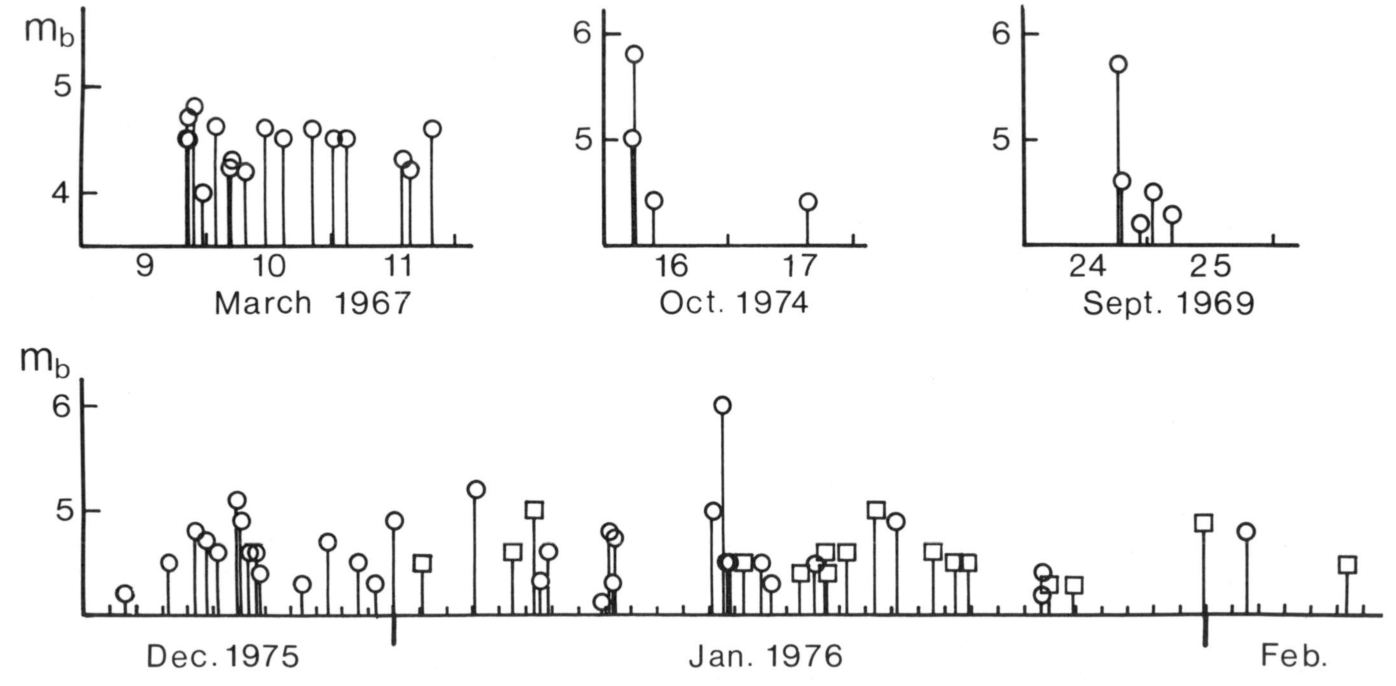

Figure 7. Magnitude as a function of time for four earthquake sequences, a swarm on the Reykjanes Ridge near 56°N, a foreshock-mainshock-aftershock sequence in the Charlie-Gibbs Fracture Zone, a mainshock-aftershock sequence in the 15°20′ Fracture Zone, and a mixed sequence in northern Iceland associated with deflation of the Krafla Volcano (squares), and rifting and transform faulting near the junction of the Krafla fissure swarm and the Grimsey seismic zone (circles). Data from the PDE listings of the U.S. Geological Survey. Note the different time scales.

swarm north of the Azores near 41°N in January 1968, and a large earthquake on February 28, 1969 in the Goringe Bank area. Removing these three peaks from the time series destroys the correlation.

Migration of epicenters was already mentioned in conjunction with lateral injection of magma from the Krafla Volcano in northern Iceland (see Fig. 6), but a direct connection between rifting of a divergent plate boundary and strike-slip motion along an adjacent transform fault was also found during the Krafla rifting episode. The first magmatic event of the Krafla Volcano that occurred between December 1975 and February 1976, began with an eruption and deflation of the caldera region, followed by migration of epicenters along the rift zone towards the transform fault where strike-slip faulting subsequently occurred (Björnsson and others, 1977). An 80 km long section of the plate boundary was involved during this two month event. In a typical Krafla rifting event the hypocenters probably mark the tip of a propagating, fluid-filled crack. The rate of propagation is governed by the rate at which magma can be fed to the crack tip, which again is a function of the magma viscosity, dike width, dike length, and the pressure gradient (Einarsson and Brandsdóttir, 1980). Migration speeds in the range of 0.5–4 km per hour have been observed.

Migration of hypocenters is well documented for the Reykjanes Peninsula earthquake swarm of September 1972 in Iceland (Klein and others, 1977). The activity started with small earthquakes the central part of the epicentral zone, and then spread laterally away from this nucleation point at a rate of 1–2 km per day. Individual subswarms also showed similar behaviour, usually spreading bilaterally from the center of each subswarm. Klein and others (1977) suggested that the migration was governed by a propagating dislocation in a viscoelastic fault zone or by a fluid diffusion process.

Major earthquake sequences in the South Iceland Seismic Zone follow a pattern of migration, beginning with a large (M_s about 7) event in the eastern part of the zone followed by smaller shocks farther west (Einarsson and others, 1981). Even though the first event is the largest one of the sequence, the later events are not aftershocks. They occur outside the source area of the first event but are probably triggered by the change in the regional strain field associated with it. The time delay between events may be due to viscoelastic response of the crust or coupling between the elastic lithosphere and the viscous asthenosphere.

THE FREQUENCY-MAGNITUDE DISTRIBUTION AND B-VALUE

Earthquakes in general are found to follow a relation of the form: $\log N = A - bM$ where N is the number of events of magnitude larger or equal to M. A and b are constants, describing the overall level of seismicity and the relative importance of large versus small earthquakes. Thus a low b-value means that large earthquakes are relatively frequent, whereas high b-values are found where the seismicity is characterized by small events. The value of b depends on the type of magnitude scale used, but has been found to be remarkably similar between different seismic areas. A significant difference was, however, demonstrated between the b-value of earthquakes along the fracture zones and the axial areas of the Mid-Atlantic Ridge (Francis, 1968 a, b). These two populations of events had b-values of 0.99 and 1.72 respectively, if the m_b-scale was used, and 0.65 and 1.33 if surface wave magnitudes (Ms) were used. Francis used earthquakes of the period 1963 to 1967 in his study. A larger data set, for the period 1963–1981, reveals an even greater difference between ridge and fracture zone earthquakes. Magnitude distribution plots are shown in Figure 8 for events along the ridge axis 40–51.9°N and 53.1–60°N (dots), and for four major fracture zones, the Charlie-Gibbs, Kane, 15°20′, and Vema zones (crosses). The magnitude detection limit is around $m_b = 4.5$ and is not significantly variable between areas. The plots are reasonably linear and the difference in the slopes is evident. Maximum likelihood estimates of the respective b-values give 2.3 ± 0.3 and 1.2 ± 0.3 (95% confidence limits).

The b-values of ridge axes and fracture zone earthquakes in some way reflect the difference between these two tectonic regimes, although the exact physical mechanism responsible for the different b-values is unclear. The b-value in fracture zones is close to that of most other seismic zones, so it seems to be the high b-value at ridge crests that needs explaining. In laboratory tests on

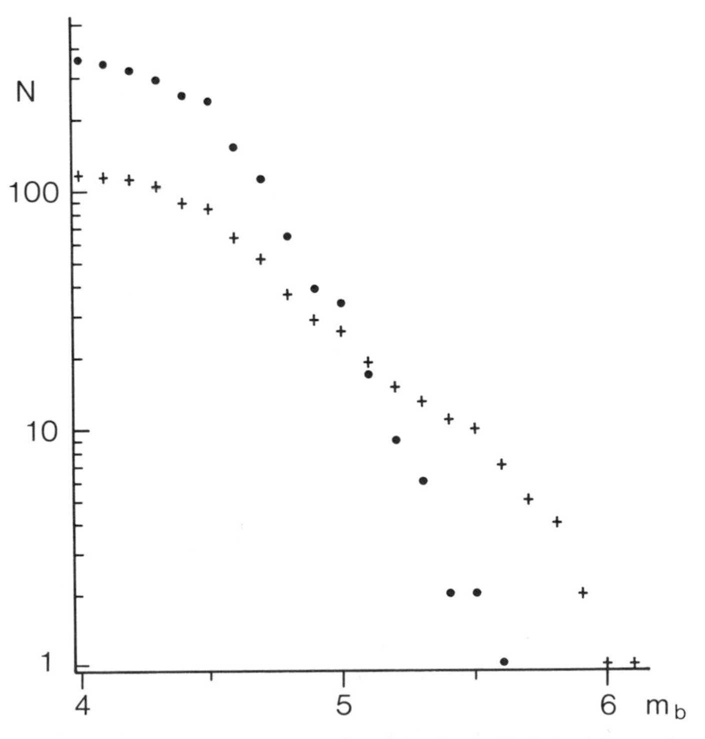

Figure 8. Cumulative number as a function of magnitude (m_b) for earthquakes along ridge axes (dots) and fracture zones (crosses). Data are taken from the PDE lists of the U.S. Geological Survey for the period 1963–1981.

microcracking, high b-values have been found to be associated with heterogeneous material and concentrated sources of stress (Mogi, 1963) or fracturing at low stress (Scholz, 1968). Cataclastic deformation has also been found to be accompanied by microcracking with high b-values (Scholz, 1968). Although a direct correspondence between microfracturing in laboratory specimens and large scale fracturing of the crust during earthquakes has not been experimentally verified, it seems plausible that some of these physical factors may contribute to the high b-values observed at the crest of the Mid-Atlantic Ridge. The crust in the axial zone is known to be extensively fractured by normal faulting and fissuring. Intrusions further add to the heterogeneity of the crust. Uniform loading of this heterogeneous zone by the relative motion of the adjacent plates leads to a heterogeneous stress field, which is further complicated by the local stress field around new magmatic intrusions, both of mechanical and thermal origin. Hydrothermal circulation has an effect by lowering the shear strength of the rock and by changing the thermal conditions. When this zone is loaded towards failure, the areas of high stress and/or low strength will fail first, even at relatively low average stress. Conditions for a large earthquake, high, uniform stress over a wide area, are not likely to develop. Thus the scale of the inhomogeneities is directly related to the source dimensions of the earthquakes. Inhomogeneities with linear dimensions of 1 to 10 km may be expected to significantly affect the observed b-value of earthquakes in the magnitude range 4 to 6, if one assumes that the relationships between magnitude, seismic moment, and source dimensions given by Kanamori (1977) apply for ridge crest earthquakes.

For further discussion of high b-values at the axis of mid-oceanic ridge the reader is referred to Francis (1968), Sykes (1970) and Francis and Porter (1971).

DEPTH OF HYPOCENTERS AND THE QUESTION OF MAGMA CHAMBERS

The source depth of earthquakes gives important information on the physical state of the crust and constraints for models of crustal growth. Thus the deepest hypocenters in an area mark the transition from brittle to totally ductile behavior of crustal material. The depth of this transition depends on temperature, creep rate, and water content (Meissner and Strehlau, 1982).

Brittle failure does not take place within magma bodies, but earthquakes may occur in regions surrounding and even beneath magma pockets and intrusions in response to pressure changes in the magma. The depth distribution of hypocenters therefore constrains the possible locations of magma chambers.

Routine teleseismic locations of Mid-Atlantic Ridge earthquakes give little information about their source depth, except that they all fall into the category of shallow events, that is, depth less than 70 km. Reliable determination of source depth along the oceanic part of the ridge system is therefore restricted to a few teleseismic studies of individual earthquakes and to ocean bottom surveys where three or more instruments have been deployed temporarily. In Iceland, on the other hand, dense temporary networks have been operated in several areas, and in some parts of the active zones the permanent seismograph network gives reliable depth determinations on a routine basis.

Teleseismic studies of earthquake sources have been made for several Mid-Atlantic events, yielding focal depth as a resulting parameter (Weidner and Aki, 1973; Duschenes and Solomon, 1977; Hart, 1978; Tréhu and others, 1981). Ridge crest earthquakes are generally found to be shallower than 5 km, and fracture zone events possibly slightly deeper. Although the few events studied by teleseismic data represent only a small fraction of the total number of events recorded from the Mid-Atlantic Ridge system, the results deserve serious attention. These events are among the largest and most significant earthquakes in this area during the period of instrumental observation.

Results of studies where local seismograph arrays have been used to determine focal depths are compared in Figure 9. Most of the results are from Iceland, where 5–23 stations have been used. Only two ocean bottom investigations are included, from the ridge crest near 45°N where three instruments were used, and near the eastern end of the St. Paul's transform fault where four instruments were deployed.

Three of the examples in Figure 9 are from the Krafla area in the axial rift zone in northern Iceland. These are only representative samples of seismicity during the magmatic and tectonic activity that began in this region in 1975. The first example shows the depth distribution of hypocenters within the caldera of the Krafla Volcano during one inflation period of the volcano between the deflation and rifting events of January 20 and April 27, 1977. Earthquakes in the caldera region are correlated with the level of inflation, and may be regarded as the response of the magma chamber roof to the increasing magma pressure. Seismicity is highest between the depth of 1 and 4 km, above and surrounding the top of the magma chamber or chambers, as determined from surface deformation (Björnsson and others, 1979) and attenuation of S-waves (Einarsson, 1978). Below is an earthquake-free zone at 5–6 km depth, but earthquakes again occur at about 7 km depth. S-waves also propagate without abnormal attenuation at 7 km depth under the volcano, presumably marking the bottom of the magma chamber.

In contrast to the inflation periods, characterized by relatively continuous seismicity extending over weeks or months, the rifting events are accompanied by intense earthquake swarms lasting hours or days. Epicenters migrate along the rift zone, out of the caldera and away from the magma chamber, either to the north or to the south (Brandsdóttir and Einarsson, 1979; Einarsson and Brandsdóttir, 1980). Depth distribution of two such swarms is shown in Figure 9. It is remarkable that the depth distribution is significantly different during these two intrusion events, and yet the epicentral zones are nearly identical. The event of September 1977 was associated with extensive faulting and rifting at the surface within a 16 km long segment of the rift zone, extending from the northern caldera rim southward across the caldera and along the southern rift zone. A small lava eruption

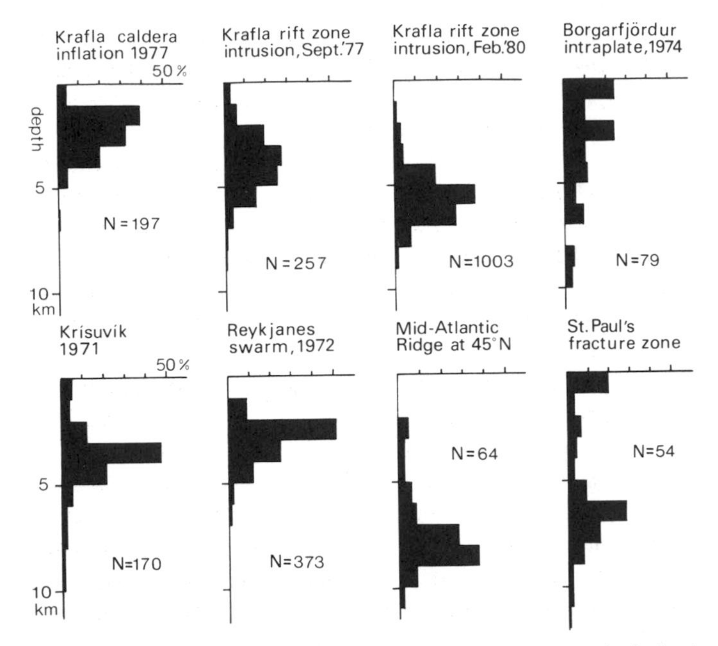

Figure 9. Histograms for depth distribution of hypocenters in Iceland (Krafla, Borgarfjördur, Krísuvík, Reykjanes) and on the submarine part of the ridge system (45°N, St. Paul's Fracture Zone). N is the total number of events used in the plot.

occurred near the northern caldera rim and a small amount of basaltic pumice was erupted through a drill hole in the southern rift. The hypocentral distribution suggests a dike intrusion mainly at the depth of 3–4 km (Brandsdóttir and Einarsson, 1979) in the southern rift. The February 1980 event was in many respects different. Only minor rifting and faulting was observed at the surface, no eruption occurred, and the maximum in the depth distribution was at 5–7 km. This suggests that the dike of February 1980 was injected below the September 1977 dike. The lesson to be learned here is that the frequency of earthquakes versus depth during a short recording period is not only dependent on material properties and physical state of the crust, it also depends on the previous course of events in the area. It is significant, however, that earthquakes occur down to the depth of 9 km, even under the most active part of the rift zone.

The depth distribution of the intraplate earthquakes in Borgarfjördur in 1974 (Einarsson and others, 1977) is shown in Figure 9 for comparison. These earthquakes were associated with normal faulting of a 3–7 m.y. old crust within the North American Plate. The hypocenters reach a depth of 9–10 km, not significantly deeper than the rift zone earthquakes.

The results of two studies on the Reykjanes Peninsula are shown in Figure 9, one in Krísuvík area in the central part of the peninsula (Klein and others, 1973), and the other near the tip of Reykjanes (Klein and others, 1977). Dense, multi-element arrays were used in both studies and most of the hypocenters have vertical errors considerably less than 1 km. In Krísuvík most hypocenters are between 2 and 5 km depth, but earthquakes also occur close to the surface and as deep as 9 km. Most of the Reykjanes earthquakes belong to a large swarm of more than 17000 recorded events (M $\leq$4.4) that occurred during a six week recording period in 1972. Earthquakes occur at 1–7 km depth, but the distribution is sharply peaked at 2–3 km depth. The peak is not as sharp if the distribution of energy release is considered. The energy distribution is essentially flat between 2 and 5 km depth (Klein and others, 1977).

The activity recorded by Lilwall and others (1977, 1978) at the ridge crest near 45°N is located between 2 and 11 km depth, with a peak near 8 km. The data of Francis and others (1978) from the eastern junction of the St. Paul's Fracture Zone with the ridge axis show a similar peak at 6–8 km depth, but in addition there is a peak at 0–1 km depth. Duschenes and others (1983) suggest that the shallow peak may be artifacts of the location technique used. The experience from Iceland warns against assigning too much significance to the peaks in the distribution, especially since both data sets were obtained during short recording periods. It is of considerable interest to note that earthquakes occur down to 11–12 km depth under the rift valley of the Mid-Atlantic Ridge. The results of Toomey and others (1985) show brittle behavior down to at least 7–8 km depth beneath the central valley at 23°N.

Several models of crustal generation by sea-floor spreading assume a magma chamber at small but usually unspecified depth beneath and extending along the spreading center (e.g. Cann, 1974; Sleep, 1975, 1978; Rosendahl, 1976; Hall and Robinson, 1979). Evidence for and against the existence of such a magma chamber was recently reviewed by Sanford and Einarsson (1982) and Lewis (1983), who concluded that in the case of the Mid-Atlantic Ridge no evidence could be found for an extensive magma chamber at crustal levels (see also Macdonald, this volume). Moreover, depth of hypocenters and the efficient propagation of crustal S-waves across the crestal zone of the ridge seem to preclude anything but small, isolated pockets of magma in the crust. Evidence for such small magma chambers is found under the Krafla Volcano (Einarsson, 1978) and possibly on the Reykjanes Peninsula in Iceland (Klein and others, 1977). Evidence is also found for a zone of melting in the mantle under the Mid-Atlantic Ridge. Molnar and Oliver (1969), for example, found a narrow zone of inefficient propagation of the Sn phase centered on the ridge crest. Sn is a shear wave that normally propagates over long distances in the lithosphere at subcrustal levels. This zone of partial melting appears to be particularly well developed in the mantle under Iceland, where it is characterised by low density, low P-velocity, high P- to S-velocity ratio, high S-wave attenuation and a layer of low electrical resistivity near the crust-mantle boundary (Sanford and Einarsson, 1982).

CONCLUSIONS

1. The eastern boundary of the North American Plate, which follows the Mid-Atlantic Ridge system and its continuation into the Arctic, is marked by a continuous belt of seismicity, that is less than 30 km wide in most places.

2. Earthquakes in fracture zones occur almost exclusively

within its transform section, connecting the ridge crests on either side.

3. A distinct difference in seismicity pattern is found between the ridge crests and fracture zones. Mainshock-aftershock sequences and normal b-values are found along the fracture zones, whereas the seismicity along ridge crests is characterized by earthquake swarms and high b-values. The difference is probably related to structural heterogeneity in the zone of volcanism and crustal accretion along the ridge crest.

4. Almost all available fault plane solutions in fracture zones show strike-slip faulting in the transform sense, sometimes with a small component of reverse faulting.

5. A large majority of fault plane solutions along ridge crests indicates normal faulting with tension axes perpendicular to the trend of the ridge.

6. Hypocenters reach depths of 8 km under at least some parts of the Mid-Atlantic Ridge, and seem to rule out the existence of a large, continuous magma chamber at crustal levels. If shallow magma chambers play an important role in the process of crustal genesis at the Mid-Atlantic Ridge, they must be intermittent features and of limited extent, possibly similar to the magma chamber at 3-7 km depth under the Krafla Volcano in the axial rift zone in northeastern Iceland.

7. Focal mechanism solutions with large components of reverse faulting have been found at the ridge axis in two areas, near 50°N and in the Bárdarbunga central volcano near the center of the Iceland hot spot. Several mechanisms may be found to explain reverse faulting at a divergent plate boundary, but the one favoured here is brittle failure above a deflating, localized magma chamber.

8. Several cases of earthquake migration or correlation over considerable distances are found. In the case of the Krafla Volcano epicentral migration along the rift occurs during deflation of the caldera region, indicating lateral injection of magma. In other cases the correlation of seismic activity occurs over too great distances for each earthquake to directly influence the occurrence of the other. In these instances regional strain pulses associated with the large scale plate motions must be invoked to explain the correlations.

REFERENCES

Barazangi, M., and Dorman, J.
1969 : World seismicity of E.S.S.A. Coast and Geodetic Survey epicenter data for 1961–1967: Bulletin of the Seismological Society of America, v. 59, p. 369–380.

Björnsson, A., Saemundsson, K., Einarsson, P., Tryggvason, E. and Grönvold, K.
1977 : Current rifting episode in north Iceland: Nature, v. 266, p. 318–323.

Björnsson, A., Johnsen, G., Sigurdsson, S., Thorbergsson, G. and Tryggvason, E.
1979 : Rifting of the plate boundary in north Iceland 1975–1978: Journal of Geophysical Research, v. 84, p. 3029–3038.

Björnsson, S. and Einarsson, P.
1981 : Jardskjálftar (Earthquakes); in Náttúra Íslands (in Icelandic), ed. H. Thórarinsdóttir, S. Steinthórsson: Almenna Bókafélagid, Reykjavík, p. 121–155.

Bonatti, E.
1978 : Vertical tectonism in oceanic fracture zones: Earth and Planetary Science Letters, v. 37, p. 369–379.

Bonatti, E., and Chermak, A.
1981 : Formerly emerging crustal blocks in the equatorial Atlantic: Tectonophysics, v. 72, p. 165–180.

Bonatti, E., Sartori, R., and Boersma, A.
1983 : Vertical crustal movements at the Vema Fracture Zone in the Atlantic: Evidence from dredged limestones: Tectonophysics, v. 91, p. 213–232.

Brandsdóttir, B., and Einarsson, P.
1979 : Seismic activity associated with the September 1977 deflation of the Krafla central volcano in NE Iceland: Journal of Volcanology and Geothermal Research, v. 6, p. 197–212.

Bungum, H., Mitchell, B. J., and Kristoffersen, Y.
1982 : Concentrated earthquake zones in Svalbard: Tectonophysics, v. 82, p. 175–188.

Cann, J. R.
1974 : A model for oceanic crustal structure developed: Geophysical Journal of the Royal Astronomical Society, v. 39, p. 169–187.

Conant, D. A.
1972 : Six new focal mechanism solutions for the Arctic and a center of rotation for plate movements; [M.A. Thesis] Columbia University, New York.

Duschenes, J., Lilwall, R. C., and Francis, T.J.G.
1983 : The hypocentral resolution of microearthquake surveys carried out at sea: Geophysical Journal of the Royal Astronomical Society, v. 72, p. 435–451.

Duschenes, J. D., and Solomon, S. C.
1977 : Shear wave travel time residuals from oceanic earthquakes and the evolution of oceanic lithosphere: Journal of Geophysical Research, v. 82, p. 1985–2000.

Einarsson, P.
1976 : Relative location of earthquakes within the Tjörnes Fracture Zone: Societas Scientiarum Islandica, Greinar V, p. 45–60.
1978 : S-wave shadows in the Krafla caldera in NE-Iceland, evidence for a magma chamber in the crust: Bulletin Volcanologique, v. 41, p. 1–9.
1979 : Seismicity and earthquake focal mechanisms along the mid-Atlantic plate boundary between Iceland and the Azores: Tectonophysics, v. 55, p. 127–153.
1985 : Compilation of earthquake fault plane solutions in the North Atlantic and Arctic Oceans: Manuscript, 27 pp.

Einarsson, P. and Björnsson, S.
1979 : Earthquakes in Iceland: Jökull, v. 29, p. 37–43.

Einarsson, P., Björnsson, S., Foulger, G., Stefánsson, R., and Skaftadóttir, T.
1981 : Seismicity pattern in the South Iceland seismic zone: in Earthquake Prediction—An International Review, Maurice Ewing Series 4, American Geophysical Union, p. 141–151.

Einarsson, P. and Brandsdóttir, B.
1980 : Seismological evidence for lateral magma intrusion during the July 1978 deflation of the Krafla volcano in NE-Iceland: Journal of Geophysics, v. 47, p. 160–165.

Einarsson, P. and Eiríksson, J.
1982 : Earthquake fractures in the districts Land and Rangárvellir in the South Iceland Seismic Zone: Jökull, v. 32, p. 113–120.

Einarsson, P., Klein, F. W., and Björnsson, S.
1977 : The Borgarfjördur earthquakes of 1974 in West Iceland: Bulletin of the Seismological Society of America, v. 67, p. 187–208.

Einarsson, T.
1968 : Submarine ridges as an effect of stress fields: Journal of Geophysical Research, v. 73, p. 7561–7575.

Fleischer, U.
1974 : The Reykjanes Ridge—A summary of geophysical data; in Geodynamics of Iceland and the North Atlantic Area, ed. L. Kristjánsson: Dordrecht, Netherlands, Boston, U.S.A., D. Reidel Publishing Company, p. 17–31.

Francis, T.J.G.
1968a: The detailed seismicity of mid-ocean ridges: Earth and Planetary Science Letters, v. 4, p. 39–46.
1968b: Seismicity of mid-oceanic ridges and its relation to properties of the upper mantle and crust: Nature, v. 220, p. 899–901.
1973 : The seismicity of the Reykjanes Ridge: Earth and Planetary Science Letters, v. 18, p. 119–124.

Francis, T.J.G., and Porter, I. T.
1971 : A statistical Study of mid-Atlantic Ridge earthquakes: Geophysical Journal of the Royal Astronomical Society, v. 24, p. 31–50.

Francis, T.J.G., Porter, I. T., and Lilwall, R. C.
1978 : Microearthquakes near the eastern end of St. Paul's Fracture Zone: Geophysical Journal of the Royal Astronomical Society, v. 53, p. 201–217.

Francis, T.J.G., Porter, I. T., and McGrath, J. R.
1977 : Ocean-bottom seismograph observations on the Mid-Atlantic Ridge near 37°N: Geological Society of America Bulletin, v. 88, p. 664–677.

Fujita, K., Hasegawa, H., Wetmiller, R., and Forsyth, D.
1985 : Seismicity and focal mechanisms in the Arctic region: in press

Gutenberg, B., and C. F. Richter
1941 : Seismicity of the Earth: Geological Society of America, Special Papers No. 34, 125 pp.
1949 : Seismicity of the Earth and Associated Phenomena: Princeton University Press, 273 p.

Hall, J. M., and Robinson, P. T.
1979 : Deep crustal drilling in the North Atlantic Ocean: Science, v. 204, p. 573–586.

Hart, R. S.
1978 : Body wave observations from the September, 1969, North Atlantic Ridge earthquake (Abstract): EOS, Transactions of the American Geophysical Union, v. 59, p. 326.

Hirn, A., Haessler, A., Hoang Trong, P., Wittlinger, G., and Mendes Victor, L. A.
1980 : Aftershock sequence of the January 1st, 1980, earthquake and present-day tectonics in the Azores: Geophysical Research Letters, v. 7, p. 501–504.

Horsfield, W. T., and Maton, P. I.
1970 : Transform faulting along the De Geer line: Nature, v. 226, p. 256–257.

Husebye, E. S., Gjoystdal, H., Bungum, H., and Eldholm, O.
1975 : The seismicity of the Norwegian and Greenland Seas and adjacent continental shelf areas: Tectonophysics, v. 26, p. 55–70.

Isacks, B., Oliver, J., and Sykes, L. R.
1968 : Seismicity and the new global tectonics: Journal of Geophysical Research, v. 73, p. 5855–5899.

Johnson, G. L., and Vogt, P. R.
1973 : Mid-Atlantic ridge from 47° to 51°N: Geological Society of America Bulletin, v. 84, p. 3443–3462.

Kanamori, H.
1977 : The energy release in great earthquakes: Journal of Geophysical Research, v. 82, p. 2981–2987.

Kanamori, H., and Stewart, G. S.
1976 : Mode of strain release along the Gibbs fracture zone, mid-Atlantic ridge: Physics of Earth and Planetary Interiors, v. 11, p. 312–332.

Klein, F. W., Einarsson, P., and Wyss, M.
1973 : Microearthquakes on the mid-Atlantic plate boundary on the Reykjanes Peninsula in Iceland: Journal of Geophysical Research, v. 78, p. 5084–5099.
1977 : The Reykjanes Peninsula, Iceland, earthquake swarm of September 1972 and its tectonic significance: Journal of Geophysical Research, v. 82, p. 865–888.

Krause, D. C., and Watkins, N. D.
1970 : North Atlantic crustal genesis in the vicinity of the Azores: Geophysical Journal of the Royal Astronomical Society, v. 19, p. 261–283.

Laughton, A. S., and Whitmarsh, R. B.
1974 : The Azores-Gibraltar plate boundary; in Geodynamics of Iceland and the North Atlantic Area, ed. L. Kristjánsson: Dordrecht, Netherlands, D., Reidel Publishing Company, p. 63–81.

Lewis, B.T.R.
1983 : The process of formation of oceanic crust: Science, v. 220, p. 151–157.

Lilwall, R. C.
1982 : Seismicity of the oceanic rifts; in Continental and Oceanic Rifts, ed. G. Pálmason: Geodynamics Series, v. 8, American Geophysical Union, p. 63–80.

Lilwall, R. C., Francis, T.J.G., and Porter, I. T.
1977 : Ocean-bottom seismograph observations on the Mid-Atlantic Ridge near 45°N.: Geophysical Journal of the Royal Astronomical Society, v. 51, p. 357–370.
1978 : Ocean-bottom seismograph observations on the Mid-Atlantic Ridge near 45°N—further results: Geophysical Journal of the Royal Astronomical Society, v. 55, p. 255–262.

Limond, W. Q., and Recq, M.
1981 : Possible earthquake migration along the Western European plate boundary: Journal of Geophysical Research, v. 86, p. 11623–11630.

McKenzie, D.
1972 : Active tectonics of the Mediterranean region: Geophysical Journal of the Royal Astronomical Society, v. 30, p. 109–185.

Meissner, R., and Strehlau, J.
1982 : Limits of stresses in continental crust and their relation to the depth-frequency distribution of shallow earthquakes: Tectonics, v. 1, p. 73–89.

Mitchell, B. J., Zollweg, J. E., Kohsmann, J. J., Cheng, C.-C., and Hang, E. J.
1979 : Intraplate earthquakes in the Svalbard archipelago: Journal of Geophysical Research, v. 84, p. 5620–5626.

Mogi, K.
1963 : Some discussion on aftershocks, foreshocks and earthquake swarms—The fracture of a semi-infinite body caused by an inner stress origin and its relation to the earthquake phenomena, 3: Bulletin of the Earthquake Research Institute, Tokyo University, v. 41, p. 615–658.

Molnar, P., and Oliver, J.
1969 : Lateral variations of attenuation in the upper mantle and discontinuities in the lithosphere: Journal of Geophysical Research, v. 74, p. 2648–2682.

Moreira, V.
1982 : Seismotectonics of mainland Portugal and its adjacent region in the Atlantic (in Portuguese, with English abstract), Lisbon, 29 pp.

Okal, E. A., and Stewart, L. M.
1982 : Slow earthquakes along oceanic fracture zones: Evidence for asthenospheric flow away from hotspots? Earth and Planetary Science Letters, v. 57, p. 75–87.

Pálmason, G.
1974 : The insular margins of Iceland; in The geology of continental margins, ed. C. A. Burk and C. L. Drake: Berlin, Springer Verlag, p. 375–379.

Reid, I., and Macdonald, K.
1973 : Microearthquake study of the Mid-Atlantic Ridge near 37°N, using sonobuoys: Nature, v. 246, p. 88–90.

Rosendahl, B. R.
1976 : Evolution of oceanic crust, 2. Constraints, implications, and inferences: Journal of Geophysical Research, v. 81, p. 5305–5314.

Rowlett, H.
1981 : Seismicity at intersections of spreading centers and transform faults: Journal of Geophysical Research, v. 86, p. 3815–3820.

Rowlett, H., and Forsyth, D. W.
1984 : Recent faulting and microearthquakes at the intersection of the Vema Fracture Zone and the Mid-Atlantic Ridge: Journal of Geophysical Re-

search, v. 89, p. 6079–6094.

Saemundsson, K.
1974 : Evolution of the axial rifting zone in Northern Iceland and the Tjörnes fracture zone: Geological Society of America Bulletin, v. 85, p. 495–504.

Sanford, A. R., and Einarsson, P.
1982 : Magma chambers in rifts; in Continental and oceanic rifts, ed. G. Pálmason: Geodynamics Series, v. 8, American Geophysical Union, p. 147–168.

Savostin, L. A., and Karasik, A. M.
1981 : Recent plate tectonics of the Arctic Basin and of northeastern Asia: Tectonophysics, v. 74, p. 111–145.

Scholz, C. H.
1968 : The frequency-magnitude relation of micro-fracturing in rock and its relation to earthquakes: Bulletin of the Seismological Society of America, v. 58, p. 399–415.

Searle, R.
1980 : Tectonic pattern of the Azores spreading centre and triple junction: Earth and Planetary Science Letters, v. 51, p. 415–434.
1981 : The active part of Charlie-Gibbs Fracture Zone: A study using sonar and other geophysical techniques: Journal of Geophysical Research, v. 86, p. 243–262.

Sleep, N. H.
1975 : Formation of oceanic crust: Some thermal constraints: Journal of Geophysical Research, v. 80, p. 4037–4042.
1978 : Thermal structure and kinematics of mid-oceanic ridge axis, some implications to basaltic volcanism: Geophysical Research Letters, v. 5, p. 426–428.

Soernes, A., and Fjeldskaar, W.
1980 : The local seismicity in the Jan Mayen area: Norsk Polarinstitutt, Skrifter Nr. 172, p. 21–32.

Spindel, R. C., Davis, S. B., Macdonald, K. C., Porter, R. P., and Philips, J. D.
1974 : Microearthquake survey of median valley of the Mid-Atlantic Ridge at 36°30′N: Nature, v. 248, p. 577–579.

Stover, C. W.
1973 : Seismicity and tectonics of the East Pacific Ocean: Journal of Geophysical Research, v. 78, p. 5209–5220.

Sykes, L. R.
1965 : The seismicity of the Arctic: Bulletin of the Seismological Society of America, v. 55, p. 519–536.
1967 : Mechanism of earthquakes and nature of faulting on the mid-ocean ridges: Journal of Geophysical Research, v. 72, p. 2131–2153.
1970 : Earthquake swarms and sea-floor spreading: Journal of Geophysical Research, v. 75, p. 6598–6611.

Talwani, M., Windisch, C. C., and Langseth, M. G.
1971 : Reykjanes Ridge crest: A detailed geophysical study: Journal of Geophysical Research, v. 76, p. 473–517.

Tams, E.
1922 : Die seismischen Verhältnisse des Europäischen Nordmeeres: Centralblatt für Mineralogie, Geologie und Palaeontologie, Jahrgang 1922, p. 385–397.
1927a: Die seismischen Verhältnisse des offenen Atlantischen Ozeans: Zeitschrift für Geophysik, v. 3, p. 361–363.
1927b: Die seismischen Verhältnisse des offenen Atlantischen Ozeans: Gerlands Beiträge zur Geophysik, v. 18, p. 319–353.

Thorarinsson, S., Saemundsson, K., and Williams, R. S.
1973 : ERTS—1 Image of Vatnajökull: Analysis of glaciological, structural and volcanic features: Jökull, v. 23, p. 7–17.

Toomey, D. R., Solomon, S. C., Purdy, G. M., and Murray, M. H.
1985 : Microearthquakes beneath the median valley of the Mid-Atlantic Ridge near 23°N: Hypocenters and focal mechanisms: Journal of Geophysical Research, in press.

Tréhu, A. M., Nábělek, J. L., and Solomon, S. C.
1981 : Source characterization of two Reykjanes Ridge earthquakes: Surface waves and moment tensors; P waveforms and non-orthogonal nodal planes: Journal of Geophysical Research, v. 86, p. 1701–1724.

Tryggvason, E.
1973 : Seismicity, earthquake swarms and plate boundaries in the Iceland region: Bulletin of the Seismological Society of America, v. 63, p. 1327–1348.

Udias, A.
1982 : Seismicity and seismotectonic stress field in the Alpine-Mediterranean region: Alpine Mediterranean Geodynamics. American Geophysical Union, Geodynamics Series, v. 7, p. 75–82.

Udias, A., Lopez Arroyo, A., and Mezcua, J.
1976 : Seismotectonics of the Azores-Alboran region: Tectonophysics, v. 31, p. 259–289.

Van Bemmelen, R. W.
1964 : The evolution of the Atlantic mega-undation: Tectonophysics, v. 1, p. 385–430.

Vogt, P. R.
1978 : Long wavelength gravity anomalies and intraplate seismicity: Earth and Planetary Science Letters, v. 37, p. 465–475.

Vogt, P. R., and Avery, O. E.
1974 : Detailed magnetic surveys in the northeast Atlantic and Labrador Sea: Journal of Geophysical Research, v. 79, p. 363–389.

Vogt, P. R., Kovacs, L. C., Bernero, C., and Srivastava, S. P.
1982 : Asymmetric geophysical signatures in the Greenland-Norwegian and Southern Labrador Seas and the Eurasian Basin: Tectonophysics, v. 89, p. 95–160.

Ward, P. L.
1971 : New interpretation of the geology of Iceland: Geological Society of America Bulletin, v. 82, p. 2991–3012.

Weidner, D. J., and Aki, K.
1973 : Focal depth and mechanism of mid-ocean ridge earthquakes: Journal of Geophysical Research, v. 78, p. 1818.

Whitmarsh, R. B., and Lilwall, R. C.
1983 : Ocean-bottom seismographs; in Structure and Development of the Greenland-Scotland Ridge, ed. M.H.P. Bott, S. Saxov, M. Talwani and J. Thiede. Plenum Press, New York and London, p. 257–286.

Wilson, J. T.
1965 : A new class of faults and their bearing on continental drift: Nature, v. 207, p. 343–347.

MANUSCRIPT ACCEPTED BY THE SOCIETY FEBRUARY 25, 1985

ACKNOWLEDGMENTS:

The author received help from various institutions and individuals during the preparation of this paper. Financial contribution was obtained from the Icelandic Science Fund. Extensive use was made of the WWSSN film chips library at Lamont-Doherty Geological Observatory, the help and hospitality of Drs. R. Bilham and D. Simpson is gratefully acknowledged. Sigfús Johnsen, Sigurdur E. Pálsson, and Bryndís Brandsdóttir helped with computing and plotting, Kristín Pálsdóttir and Ágústa Thorláksdóttir typed the manuscript.

Printed in U.S.A.

The Geology of North America
Vol. M, The Western North Atlantic Region
The Geological Society of America, 1986

Chapter 8

"Zero-age" variations in the composition of abyssal volcanic rocks along the axial zone of the Mid-Atlantic Ridge

William G. Melson
Tim O'Hearn
Smithsonian Institution, Washington DC 20560

INTRODUCTION

The mid-ocean ridge spreading centers mark the zones where the asthenosphere comes closest to the earth's surface. Such centers are termed accreting plate margins because the oceanic lithosphere forms along these centers. Magmas that have erupted along spreading centers are minimally modified or not modified at all by contamination with older lithosphere, compared to magmas erupted through the thicker lithosphere beneath continents and islands. Thus, their compositions presumably provide us with least-modified mantle-derived liquids. Controversy remains as to the extent that these liquids—which are overwhelmingly basaltic—represent "primitive" liquids (O'Hara, 1982).

One of the most incisive models of accreting margin magma chambers is Cann's "cocktail-onion" model (1970, 1974). This model (Macdonald, this volume) envisions a permanent magma chamber beneath fast-spreading centers and periodic magma chambers beneath slow, rifted spreading centers. The Mid-Atlantic Ridge (MAR) belongs to the latter, rifted spreading centers.

Initial work on abyssal basalt from spreading centers revealed a remarkable uniformity in composition and their "depleted" character compared to oceanic island basalts (Engel and others, 1965) in regard to elements such as K and the light rare earth elements (LREE). Since that time, a wealth of new analytical data has revealed much more diversity and certain systematic patterns related to such factors as distance from on-ridge volcanic islands, spreading rate, water depth, and broad regional groupings. A breakthrough in this area included the discovery of systematic variations in Sr-isotopic ratios moving south from Iceland along the Reykjanes Ridge, revealing decreasingly less radiogenic ratios southward, with values approaching "normal" abyssal basalt values away from Iceland (Schilling, 1973; and this volume; Hart and others, 1973).

Basalts away from the regional highs and islands show LREE depletions and have low Sr^{87}/Sr^{86} ratios (Schilling, 1971; Hart, 1971). Tarney and others (1980) have termed these normal, or N-type Mid-Ocean Ridge Basalt (MORB), and the "enriched," platform and island-associated basalt, E-type basalt. Transitional types have been referred to as T-types. This, admittedly subjective, nomenclature is used throughout this paper.

This paper focuses on what is known and inferred about young, median-valley, abyssal volcanic rocks along the Mid-Atlantic Ridge from the Jan Mayen Fracture Zone to just south of the equator. We deal mainly with the major and minor element composition of MAR abyssal volcanic rocks and intend to complement the papers on the trace and isotopic analyses by Schilling (this volume) and by Bryan and Frey (this volume) on older, now off-ridge abyssal volcanic rocks.

The plan of the paper involves (1) a brief review of the tectonic setting of the northern Mid-Atlantic Ridge median zone, (2) a segment-by-segment review of abyssal volcanic rock compositions, and (3) a brief model that accounts for some of the observed variations.

A number of important subjects relevant to the MAR-axial zone are not covered here because they have been reviewed recently. Fox and Stroup (1981) have given an extensive review of analogs between ophiolites and oceanic crust and Fox and Gallo (this volume) look specifically at some North Atlantic fracture zones and some of their relations to ophiolite models. Pálmason (this volume) has proposed thermal models for the axial zone involving hydrothermal cooling. Elthon (1981) gives an extensive review of metamorphism in oceanic crust.

MAJOR AND MINOR ELEMENT DATA SETS

The main data sets of major element analyses of glasses and bulk rocks used here are from the median zone of the Mid-Atlantic Ridge and are those of the Smithsonian Volcanic Glass Project data, referred to herein as SIVGP data, consisting of 1732 individual analyses and 481 group analyses. The literature data files used here consist of glass and bulk rock analyses of the University of Rhode Island Data Set, referred to herein as the URI data, and containing 174 bulk rock analyses and 101 glass analyses (Schilling and others, 1983 and Sigurdsson, 1981) as well as other literature sources. The URI and SIVGP data have focused

Melson, W. G., and O'Hearn, T., 1986, "Zero-age" variations in the composition of abyssal volcanic rocks along the axial zone of the Mid-Atlantic Ridge; *in* Vogt, P. R., and Tucholke, B. E., eds., The Geology of North America, Volume M, The Western North Atlantic Region: Geological Society of America.

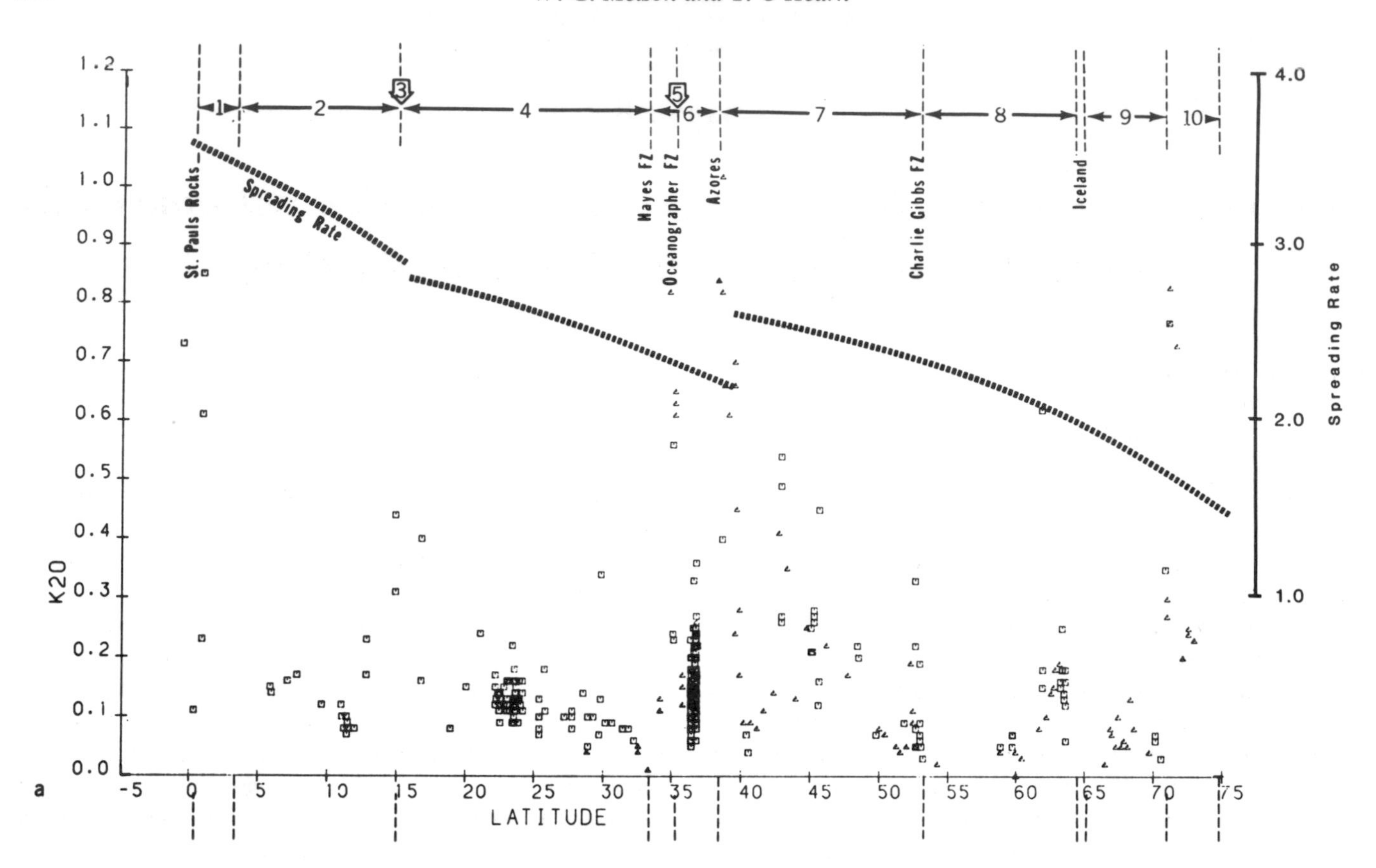

Figure 1 (this and following pages). (a) K_2O content of abyssal volcanic glasses from the northern Mid-Atlantic Ridge. Data from SIVGP and literature. Boundaries between the ten compositional-geographic groups defined in the text are delineated. (b) Al_2O_3, (c) FeO* (total iron as FeO), (d) MgO, (e) CaO, (f) Na_2O, (g) TiO_2. Squares are data from SIVGP and triangles are data from Sigurdsson (1981).

on the problem of broad regional variations in abyssal volcanic rocks and have been done in each case by the same analytical methods. The SIVGP set and the URI glass analyses give similar results on an interlaboratory glass standard (Sigurdsson, 1981). Numerous other reliable analyses that have been published from different labs are not compiled here.

The distribution of analysed samples in the above mentioned data sets is uneven (Fig. 1). Note, for example, the strong concentration of data from the FAMOUS area (around 37°N). Because of this uneven distribution, an unweighted overall average for the zero-age samples does not accurately represent the average or typical Mid-Atlantic Ridge. The sample distribution will be repeatedly referred to in the region-by-region overviews.

Much of the data of concern here comes from analyses of the glassy rinds of abyssal volcanic rocks. The paper includes previously unpublished glass analyses from the SIVGP (Table 1), many of them from north of the Azores. New analyses from south of the Azores extending previously published coverage are also included (Melson and O'Hearn, 1979). Relationships between these compositional data, spreading rate, bathymetry, and proposed hotspots are the principal foci of this paper (see also Plates 8A, 8B).

THERMAL PLUMES AND HOTSPOTS

The centers of excess volcanism, the oceanic islands, have been termed hotspots (e.g. Morgan 1973; 1981). These islands and their submarine extensions are composed of volcanic rocks that have distinctive trace and radiogenic isotopic characteristics compared to deeper MAR-axial zone lavas (e.g. Schilling, 1973; 1975; this volume).

The crust beneath Iceland, the Azores, and possibly St. Paul's Rocks is thicker than elsewhere along the ridge, reflecting a greater contribution of ultimately mantle-derived liquids compared to the topographically deeper axial zones. Magmas rising beneath these islands and their shallow submarine axial-zone extensions travel through a greater crustal thickness, increasing the likelihood of crystal-liquid fractionation and wall-rock interactions. These effects must contribute to some extent to the much greater diversity of lavas from Iceland and the Azores compared to lavas from deep axial zones with presumably thinner lithosphere. Also, the islands can be much more effectively sampled

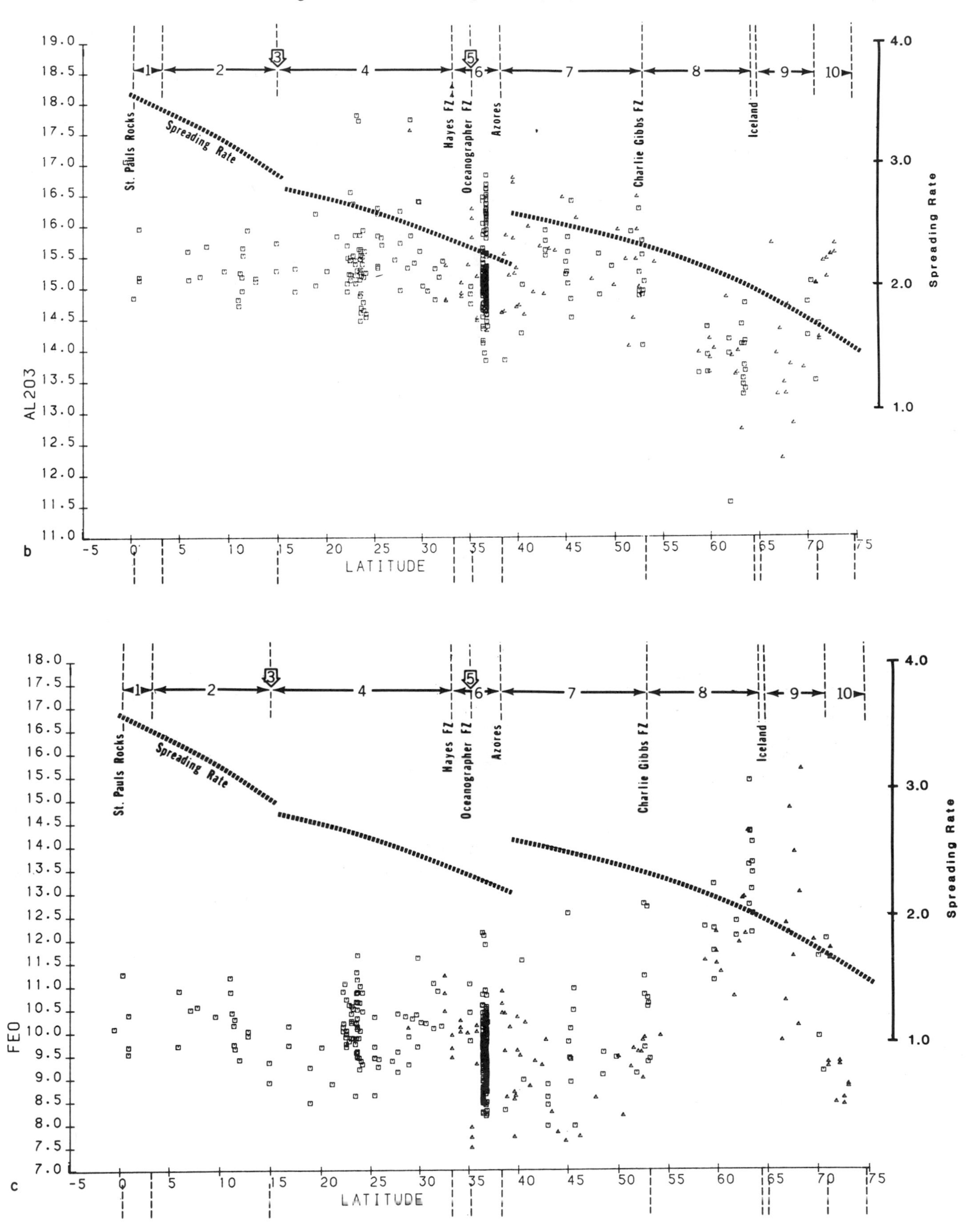

AL203
LATITUDE
Spreading Rate
St. Pauls Rocks
Hayes FZ
Oceanographer FZ
Azores
Charlie Gibbs FZ
Iceland
b
FEO
c

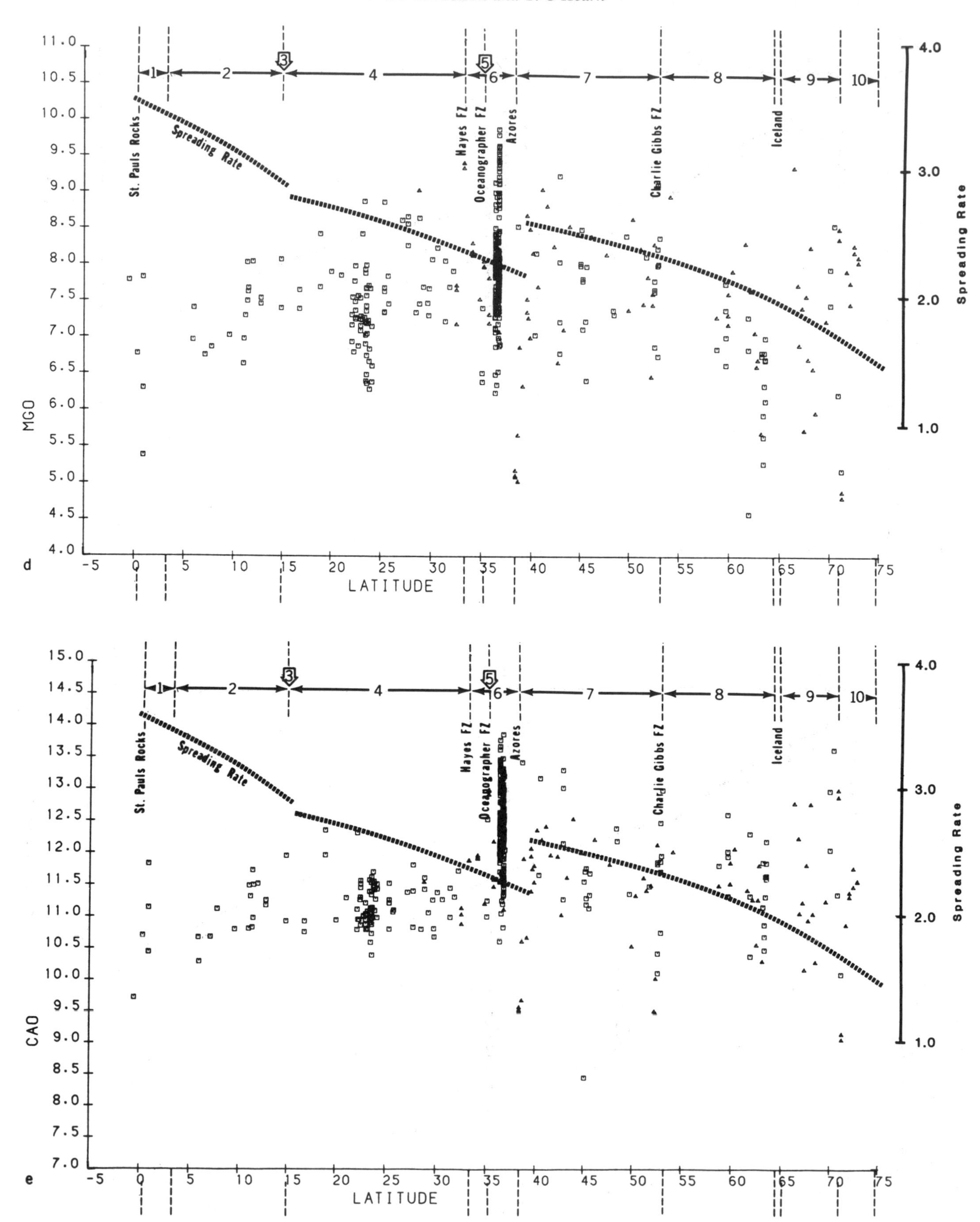

11.0
10.5
10.0
9.5
9.0
8.5
8.0
7.5
7.0
6.5
6.0
5.5
5.0
4.5
4.0
MGO
d
-5 0 5 10 15 20 25 30 35 40 45 50 55 60 65 70 75
LATITUDE
1
2
3
4
5
6
7
8
9
10
St. Pauls Rocks
Spreading Rate
Hayes FZ
Oceanographer FZ
Azores
Charlie Gibbs FZ
Iceland
4.0
3.0
2.0
1.0
Spreading Rate
15.0
14.5
14.0
13.5
13.0
12.5
12.0
11.5
11.0
10.5
10.0
9.5
9.0
8.5
8.0
7.5
7.0
CAO
e

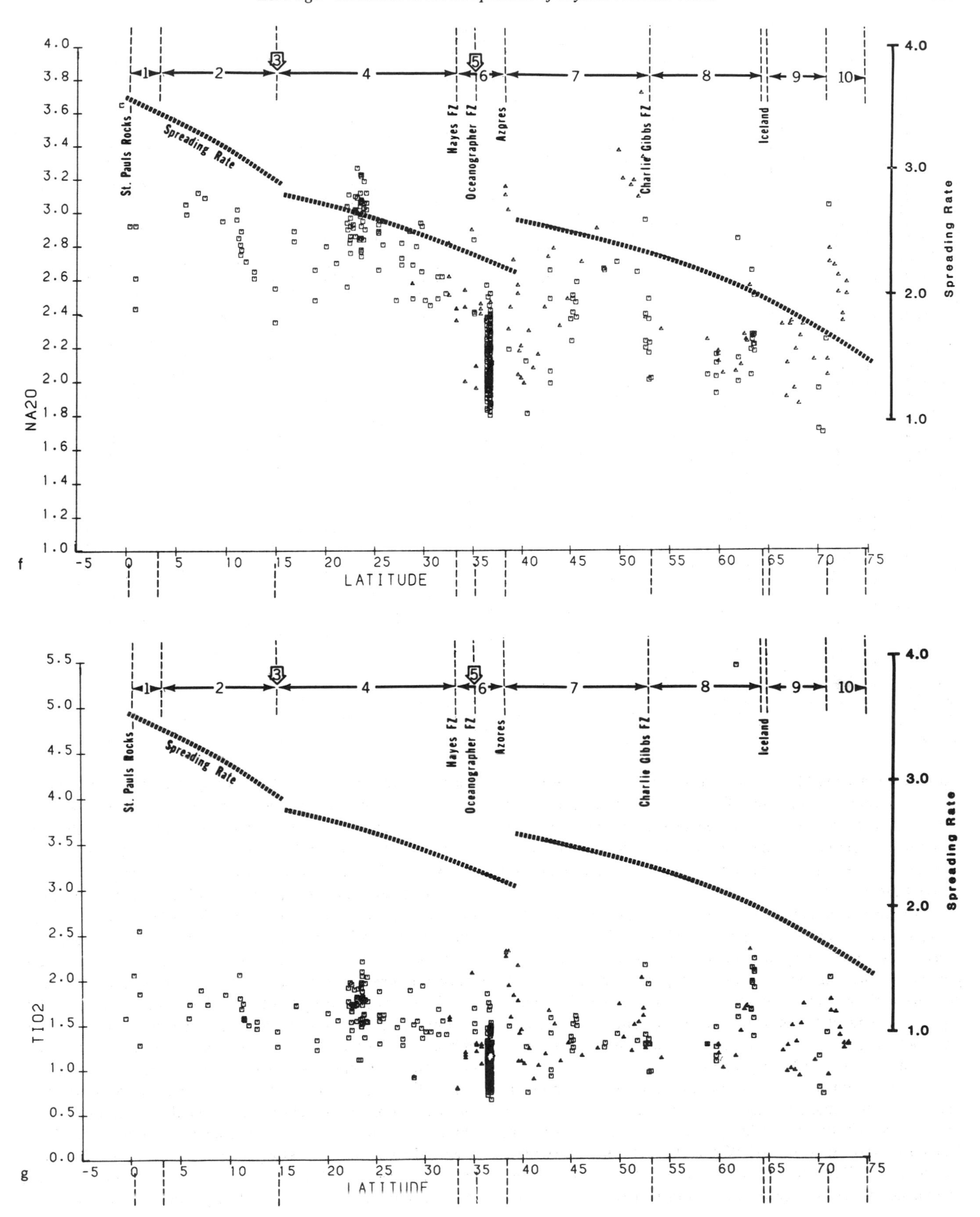
NA2O
LATITUDE
Spreading Rate
St. Pauls Rocks
Hayes FZ
Oceanographer FZ
Azores
Charlie Gibbs FZ
Iceland
f
TIO2
g

TABLE 1. CHEMICAL TYPES OF DEEP-SEA VOLCANIC GLASSES FROM THE OCEANIC BASEMENT OF THE MID-ATLANTIC RIDGE BETWEEN 38N AND 71N.
Smithsonian glass analyses by T. O'Hearn

Group	Long	N	R	SiO_2	Al_2O_3	Feo*	MgO	CaO	Na_2O	K_2O	TiO_2	P_2O_5	Sum	S	A	E	Dep	So	Ship
A 0.53S1	20.52W	8	0	48.38	17.06	10.09	7.78	9.72	3.65	0.73	1.58	0.21	99.20	FZ	Q	E	5760	W	AII 20-D21
A 0.33N1	16.95W	1	0	50.88	14.85	11.27	6.77	10.70	2.92	0.11	2.06	0.18	99.74	FZ	Q	E	5175	W	AII 20-D13
A 0.91N1	30.28W	7	0	50.96	15.96	9.54	7.82	11.83	2.43	0.23	1.28	0.12	100.17	FZ	Q	E	2815	W	AII 20-D34
A 0.93N1	29.37W	8	0	50.76	15.18	9.69	6.30	11.14	2.61	0.61	1.85	0.21	98.35	FZ	Q	E	2304	W	AII 20-D36
A 0.93N2	29.37W	1	1	50.61	15.13	10.39	5.38	10.44	2.92	0.85	2.55	0.27	98.54	FZ	Q	E	2304	W	AII 20-D36
A 5.92N1	33.25W	1	0	51.45	15.60	9.71	6.96	10.67	3.05	0.15	1.58	0.17	99.34	R	<1	E	4660	M	P 6903-26
A 6.01N1	33.28W	3	0	51.69	15.14	10.91	7.40	10.29	2.99	0.14	1.73	0.16	100.45	R	<1	E	1750	M	P 69-28
A 7.15N1	33.97W	1	0	52.19	15.19	10.50	6.75	10.68	3.12	0.16	1.89	0.21	100.69	R	<1	E	4450	M	P 69-34
A 7.82N1	38.04W	1	0	51.22	15.68	10.56	6.86	11.12	3.09	0.17	1.73	0.20	100.63	R	<1	E	2880	M	P 69-40
A 9.60N1	40.65W	27	0	51.16	15.28	10.36	7.02	10.80	2.95	0.12	1.84	0.15	99.68	FZ	<1	E	3340	W	AII 20-D58
A11.02N1	43.67W	8	1	50.65	14.82	11.19	6.63	10.81	2.96	0.12	2.06	0.16	99.40	FZ	<1	E	4750	W	AII 20-D 3
A11.07N1	44.22W	2	0	51.28	14.72	10.88	6.97	11.49	3.02	0.10	1.80	0.16	100.42	FZ	<1	E	4370	M	P 70-11
A11.22N1	43.60W	7	0	51.27	15.25	10.43	7.29	11.32	2.85	0.08	1.68	0.13	100.30	FZ	<1	E	4460	W	AII 20-D 6
A11.40N1	43.50W	1	0	50.50	15.18	10.16	8.02	10.83	2.81	0.10	1.74	0.17	99.51	R	<1	E	4100	M	P 70-28
A11.42N1	43.58W	1	0	51.36	14.96	9.74	7.49	11.73	2.75	0.07	1.57	0.14	99.81	R	<1	E	2650	M	P 70-32
A11.48N1	43.63W	5	0	50.47	15.65	10.29	7.62	10.98	2.89	0.09	1.58	0.14	99.71	R	<1	E	3400	M	P 70-30
A11.52N1	43.60W	3	0	50.80	15.53	9.65	7.67	11.50	2.78	0.08	1.56	0.13	99.70	R	<1	E	3000	M	P 70-31
A11.97N1	43.70W	5	0	51.19	15.94	9.41	8.03	11.52	2.71	0.08	1.50	0.13	100.51	R	<1	E	4600	M	P 70-12
A12.80N1	44.79W	1	0	51.84	15.11	9.93	7.45	11.18	2.61	0.17	1.46	0.17	99.92	R	<1	E	2440	N	MM81-D13
A12.81N1	44.79W	1	0	51.36	15.16	10.03	7.53	11.26	2.65	0.23	1.54	0.17	99.93	R	<1	E	1725	N	MM81-D14
A14.89N1	44.94W	1	0	51.54	15.28	9.35	8.06	10.93	2.55	0.44	1.43	0.19	99.77	R	<1	E	3480	N	MM81-D11
A14.90N1	44.90W	1	0	50.87	15.73	8.91	7.39	11.96	2.35	0.31	1.26	0.14	98.92	R	<1	E	3300	N	MM81-D10
A16.79N1	46.40W	1	0	50.92	14.95	10.14	7.38	10.76	2.89	0.16	1.71	0.17	99.08	R	<1	E	3575	N	MM81-D 8
A16.80N1	46.38W	1	0	50.37	15.32	9.72	7.64	10.93	2.83	0.40	1.72	0.20	99.49	R	<1	E	3300	N	MM81-D 9
A18.92N1	46.02W	1	0	50.98	15.05	9.24	7.68	12.36	2.66	0.08	1.33	0.10	99.48	R	<1	E	3100	N	M7841
A18.92N1	46.09W	2	0	50.64	16.21	8.47	8.41	11.97	2.48	0.08	1.22	0.12	99.60	R	<1	E	3100	N	MM81-D 7
A20.07N1	45.65W	1	0	50.09	15.28	9.68	7.89	10.95	2.80	0.15	1.63	0.15	98.62	R	<1	E	3791	N	M7831
A21.09N1	45.74W	1	0	50.49	15.84	8.88	7.84	11.30	2.70	0.24	1.55	0.15	98.99	R	<1	E	3334	N	M7822
A22.17N1	45.25W	10	1	50.55	15.69	10.12	7.30	11.12	2.94	0.15	1.76	0.15	99.78	R	<1	E	3100	V	TW65-D10
A22.17N2	45.25W	2	0	51.27	14.95	10.03	7.16	12.32	2.56	0.12	1.36	0.10	99.87	R	<1	E	3100	V	TW65-D10
A22.17N3	45.25W	2	0	50.78	15.08	10.89	6.93	10.80	3.04	0.17	1.91	0.14	99.74	R	<1	E	3100	V	TW65-D10
A22.24N1	45.02W	18	3	50.53	15.49	10.20	7.54	10.96	2.90	0.13	1.71	0.14	99.60	R	<1	E	3150	V	TW65-D 9
A22.36N1	45.20W	4	0	50.76	15.23	11.06	6.79	10.96	3.11	0.14	1.97	0.16	100.18	R	<1	E	3525	V	TW65-D 5
A22.45N1	45.02W	3	0	50.63	15.45	9.90	7.76	11.27	2.82	0.11	1.53	0.13	99.60	R	<1	E	2676	N	M7813
A22.50N1	45.00W	2	0	50.43	15.46	9.75	7.48	11.31	2.84	0.14	1.65	0.14	99.20	R	<1	E	2820	N	MM81-D 5
A22.52N1	45.05W	1	0	50.76	15.22	9.97	7.24	11.57	2.92	0.12	1.65	0.16	99.61	R	<1	E	3200	N	MM81-D 3
A22.52N1	45.05W	24	0	50.82	15.51	10.04	7.36	11.46	2.86	0.13	1.73	0.14	100.05	R	<1	E	1985	V	TW65-D 4
A22.52N2	45.05W	4	0	50.49	15.13	10.72	7.26	10.87	2.97	0.14	1.95	0.15	99.68	R	<1	E	1985	V	TW65-D 4
A22.52N3	45.05W	1	0	50.61	16.56	9.69	7.97	11.52	2.76	0.09	1.44	0.12	100.76	R	<1	E	1985	V	TW65-D 4
A22.80N1	45.20W	2	0	50.00	16.37	9.81	7.56	11.01	2.91	0.12	1.71	0.11	99.60	R	<1	E	2600	V	TW65-D 8
A22.80N2	45.20W	2	0	50.38	15.53	10.60	6.87	11.02	2.93	0.15	1.89	0.15	99.52	R	<1	E	2600	V	TW65-D 8
A23.02N1	44.96W	4	0	50.59	15.86	9.88	7.53	11.22	2.86	0.11	1.54	0.16	99.75	R	<1	E		W	AII 92-D31
A23.02N2	44.96W	19	1	50.77	15.45	10.41	7.08	10.96	3.01	0.16	1.72	0.17	99.73	R	<1	E		W	AII 92-D31
A23.03N1	44.91W	22	1	50.92	15.21	10.21	7.15	11.05	3.10	0.13	1.73	0.17	99.67	R	<1	E		W	AII 92-D30
A23.03N1	45.03W	13	3	50.48	15.39	10.25	7.24	10.88	3.02	0.12	1.79	0.20	99.37	R	<1	E	2889	W	AII 92-D 5
A23.07N1	45.17W	2	0	51.74	15.08	10.57	7.28	10.80	3.00	0.16	1.81	0.16	100.60	R	<1	E	3255	W	AII 78-D 6
A23.22N1	44.92W	3	0	48.53	17.82	9.87	8.41	10.82	3.09	0.11	1.11	0.09	99.85	R	<1	E	3990	W	AII 78-D 3
A23.22N2	44.92W	2	0	50.25	15.29	10.52	7.23	11.09	3.27	0.16	1.90	0.17	99.88	R	<1	E	3990	W	AII 78-D 3

than the submarine MAR, also contributing to their greater apparent compositional diversity. These two factors can account for the fundamental major and minor element abundance differences between the islands and deep abyssal axial MAR but cannot account for the different radiogenic isotope and certain trace element differences.

Radiogenic isotopic ratios of Sr, Nd, Pb, and He and certain trace element abundance patterns as well as absolute abundances reveal that these emergent hotspot magmas were derived ultimately from less "depleted" mantle compared to normal, abyssal erupted magmas. This observation has been interpreted as reflecting (1) tapping of "primitive," undepleted mantle (Wasserburg and DePaolo, 1979), (2) "recycled" oceanic lithosphere (Hofmann and White, 1982), or (3) both.

At least four hotspots are situated on or near the northern half of the Mid-Atlantic Ridge: (1) the Sierra Leone hotspot (ca. 2°N, 31°W), (2) the Azores hotspot (ca. 38.5°N, 28°W), (3) the Iceland hotspot (64.6°N, 16.7°W), and (4) the Jan Mayen hotspot at about 72°N. Morgan's (1981) view is an end-member one in that he envisions outward flow from thermal plumes as the main if not the only source of spreading-center magmas. Schilling (1973, 1975, this volume) has explained systematic trace and isotopic variations along the Reykjanes Ridge and south of the Azores as reflecting binary mixing of mantle plume (hotspot or E-type) and depleted asthenospheric or N-type magmas.

Three high-K_2O groups occur where there are no current islands along the axial zone of the Mid-Atlantic Ridge: the regions around 15°N, 35°N, and 45°N. These, however, are associated with axial valley highs at 15°N and 45°N and off-axis transverse ridges or seamounts that are probably hotspot traces.

Some igneous petrologists who have worked on the Mid-Atlantic Ridge have not examined petrogenetic models in terms of hotspot influences. This tendency ignores the now well-known, systematic, regionally well-defined compositional groups and trends. Often, sub-horizontal crustal and upper asthenospheric magma motions are simply not considered, although some geophysicists commonly give a major role to such flow (Vogt, 1971; Vogt and Johnson, 1975; Morgan, 1973). Schilling (1973; Schilling and others, 1983) has been a staunch advocate of hotspot influences and his ideas are frequently referred to in this paper.

TABLE 1. (CONTINUED)

Group	Long	N	R	SiO_2	Al_2O_3	FeO*	MgO	CaO	Na_2O	K_2O	TiO_2	P_2O_5	Sum	S	A	E	Dep	So	Ship
A23.52N1	44.82W	2	0	50.18	15.86	9.56	7.92	11.12	3.04	0.10	1.48	0.14	99.40	FZ	<1	E	3686	W	AII 96-D18
A23.52N1	44.91W	2	0	50.27	15.14	10.67	6.90	10.91	3.06	0.13	1.91	0.18	99.17	R	<1	E	4650	W	KN79-D34
A23.52N1	44.98W	1	0	50.65	15.27	10.20	7.20	11.12	2.98	0.13	1.79	0.17	99.51	R	<1	E	4248	W	GS104-D19
A23.53N1	44.96W	7	0	50.18	14.88	10.86	7.02	10.60	3.12	0.12	1.95	0.20	98.93	R	<1	E	4560	W	GS104-D25
A23.53N2	44.96W	1	0	49.83	15.47	9.87	7.87	11.14	2.85	0.09	1.48	0.14	98.74	R	<1	E	4560	W	GS104-D25
A23.56N1	44.93W	2	0	51.27	14.48	11.31	6.39	10.98	3.23	0.18	1.97	0.18	99.99	R	<1	E	4900	W	KN79-D21
A23.60N1	45.03W	3	0	50.73	15.60	9.49	7.66	11.54	2.77	0.11	1.52	0.14	99.56	FZ	<1	E	3611	W	AII 96-D14
A23.61N1	44.78W	1	0	51.19	15.63	10.31	7.35	11.35	2.99	0.12	1.80	0.17	100.91	R	<1	E	3570	W	KN79-D26
A23.62N1	44.74W	1	0	50.63	14.61	11.15	6.48	10.98	3.22	0.14	2.09	0.19	99.49	FZ	<1	E	4100	W	KN79-D10
A23.62N1	44.77W	5	0	50.52	15.24	10.50	7.42	10.74	3.07	0.12	1.87	0.17	99.65	R	<1	E	3700	W	KN79-D27
A23.62N1	45.04W	1	0	50.94	15.30	10.13	7.18	11.06	2.92	0.16	1.78	0.19	99.66	FZ	<1	E	4820	W	GS104-D17
A23.62N2	45.04W	4	0	50.58	15.47	9.59	7.56	11.40	2.78	0.13	1.57	0.15	99.23	FZ	<1	E	4820	W	GS104-D17
A23.63N1	44.81W	2	0	50.60	14.92	10.88	6.74	10.88	3.08	0.14	2.04	0.18	99.46	R	<1	E	3870	W	KN79-D28
A23.63N1	45.05W	5	0	50.65	15.54	9.46	7.67	11.63	2.74	0.11	1.51	0.15	99.46	F	<1	E	4380	W	GS104-D20
A23.63N2	45.05W	2	0	50.88	15.53	10.08	7.98	11.01	2.85	0.16	1.72	0.17	100.38	FZ	<1	E	4380	W	GS104-D20
A23.65N1	44.80W	1	0	50.84	14.69	11.68	6.36	10.40	3.23	0.12	2.20	0.22	99.74	FZ	<1	E	3880	W	GS104-D29
A23.67N1	44.78W	2	0	50.50	15.20	10.66	7.04	11.15	3.07	0.15	1.81	0.16	99.74	R	<1	E	3800	W	KN79-D25
A23.78N1	46.30W	1	0	52.11	15.39	9.88	7.18	11.71	2.95	0.13	1.61	0.16	101.12	R	<1	E	4580	W	GS104-D47
A23.80N1	46.28W	1	0	51.59	15.51	9.45	7.22	11.49	3.00	0.13	1.62	0.16	100.17	FZ	<1	E	5220	W	GS104-D40
A23.85N1	46.22W	2	0	50.92	15.94	9.20	7.70	11.56	2.84	0.09	1.36	0.14	99.75	FZ	<1	E	4590	W	GS104-D35
A23.88N1	46.26W	5	0	50.90	14.78	10.65	6.66	11.49	3.03	0.13	1.78	0.16	99.58	FZ	<1	E	5505	W	AII 96-D 6
A23.92N1	46.24W	1	0	51.04	15.18	10.04	6.83	11.43	3.06	0.16	1.78	0.16	99.68	FZ	<1	E	4259	W	AII 96-D 7
A23.92N2	46.24W	1	0	50.89	14.65	11.01	6.28	11.10	3.19	0.13	1.96	0.16	99.37	FZ	<1	E	4259	W	AII 96-D 7
A24.16N1	46.27W	5	0	50.82	15.25	9.31	7.14	11.52	3.02	0.16	1.53	0.14	98.89	R	<1	E	2970	W	GS104-D2
A24.16N2	46.27W	1	0	51.01	14.54	10.87	6.38	10.99	3.12	0.14	2.03	0.18	99.26	R	<1	E	3970	W	GS104-D2
A24.18N1	46.28W	3	0	51.15	14.59	10.46	6.60	11.44	3.06	0.11	1.76	0.15	99.32	R	<1	E	4090	W	GS104-D1
A25.40N1	45.30W	8	0	51.05	15.85	9.45	7.62	11.22	2.93	0.10	1.60	0.13	99.95	R	<1	E	3274	L	V25-D1
A25.40N2	45.03W	9	0	50.55	16.30	9.44	7.67	11.22	2.88	0.10	1.61	0.14	99.91	R	<1	E	3274	L	V25-D1
A25.40N3	45.30W	6	1	51.32	15.36	9.69	7.34	11.27	2.89	0.08	1.54	0.13	99.62	R	<1	E	3274	L	V25-D1
A25.40N4	45.30W	3	0	50.36	16.20	8.63	8.85	11.53	2.66	0.07	1.29	0.11	99.70	R	<1	E	3274	L	V25-D1
A25.40N5	45.30W	2	0	49.75	15.33	10.34	7.33	10.81	2.88	0.13	1.87	0.17	98.61	R	<1	E		L	V25-D1
A25.77N1	45.16W	1	0	51.32	15.82	9.25	7.83	11.08	2.95	0.18	1.57	0.14	100.14	R	<1	E	4304	T	TAG3-ST12
A25.84N1	45.04W	1	0	50.87	15.70	9.42	7.45	11.11	2.81	0.11	1.61	0.16	99.24	R	<1	E	3200	N	MM81-D 2
A27.20N1	43.55W	1	0	50.94	15.46	9.38	8.60	11.39	2.48	0.10	1.47	0.10	99.92	R	<1	E	3670	W	AII 59-D5
A27.77N1	44.08W	1	0	50.39	16.25	9.14	8.65	11.82	2.73	0.08	1.27	0.10	100.43	R	<1	E		H	JCH98-D15
A27.77N2	44.08W	1	0	50.36	14.97	10.41	8.25	10.84	2.82	0.11	1.55	0.13	99.44	R	<1	E		H	JCH98-D15
A28.56N1	44.83W	2	0	49.75	15.33	10.35	7.33	10.81	2.89	0.14	1.88	0.18	98.66	R	<1	E	4815	L	V25-D4
A28.90N1	43.32W	22	0	48.80	17.75	9.30	8.63	11.62	2.49	0.05	0.91	0.08	99.63	R	<1	E	3701	W	CHN 21-D2
A28.90N2	43.32W	10	0	51.08	15.85	9.91	7.45	11.44	2.69	0.10	1.50	0.14	100.16	R	<1	E	3701	W	CHN 21-D2
A29.28N1	43.08W	1	0	49.95	15.41	10.30	7.68	11.07	2.80	0.10	1.54	0.16	99.01	R	<1	E		H	JCH98-D14
A29.71N1	42.65W	1	0	50.21	16.41	10.39	7.46	11.27	2.94	0.07	1.46	0.15	100.36	R	<1	E	2430	S	
A29.81N1	42.73W	1	0	51.04	16.40	9.69	7.29	10.82	2.65	0.13	1.35	0.11	99.48	R	<1	E	1420	S	
A29.86N1	42.77W	1	0	50.08	15.60	11.62	7.66	10.69	2.92	0.34	1.93	0.23	101.07	R	<1	E	3200	S	
A30.17N1	41.92W	4	0	50.49	15.04	10.22	8.06	11.39	2.48	0.09	1.42	0.14	99.33	R	<1	E		H	JCH98-D12
A30.68N1	41.82W	4	0	50.43	14.96	10.20	8.22	11.28	2.45	0.09	1.42	0.12	99.17	R	<1	E		H	JCH98-D11
A31.47N1	40.95W	4	0	50.76	15.32	10.09	8.04	11.46	2.49	0.08	1.39	0.12	99.75	R	<1	E		H	JCH98-D 9
A31.47N2	40.95W	1	0	50.68	14.82	11.07	7.21	11.00	2.62	0.08	1.67	0.16	99.31	R	<1	E		H	JCH98-D 9
A31.90N1	40.97W	2	0	50.76	15.18	10.90	7.68	11.32	2.62	0.08	1.54	0.14	100.22	R	<1	E		H	JCH98-D 8
A32.28N1	40.35W	4	0	51.20	15.44	10.14	7.90	11.73	2.52	0.06	1.39	0.11	100.49	R	<1	E		H	JCH98-D 7
A35.13N1	35.73W	1	0	50.69	14.91	11.06	6.50	11.01	2.84	0.24	1.68	0.22	99.15	FZ	<1	E	2150	L	V30-D 7
A35.15N1	34.90W	4	0	51.41	14.75	10.44	6.38	11.26	2.41	0.56	1.51	0.25	98.97	FZ	<1	E	3350	L	V30-D10
A35.18N1	36.23W	1	0	51.03	15.03	9.82	7.39	12.54	2.40	0.23	1.42	0.17	100.03	R	<1	E		H	JCH97-D 2

Increasingly detailed coverage of the composition of "zero-age" magmas along the Mid-Atlantic Ridge provides the opportunity to re-examine the relative roles of vertical and horizontal magma or mantle-diapir flow. The principal geochemical data examined here, the major and minor element contents, are of limited yet critical value in such an assessment. The radiogenic isotopes of Rb, Nd, U, and Th, along with certain trace elements, especially the rare earth elements, are even more critical and are dealt with in the companion papers in this volume by Schilling and by Bryan and Frey. There is, however, a good correlation between K and both Sr^{87}/Sr^{86} and La/Sm ratios (Fig. 2); thus, K is used here as a pointer to distinguishing E-, T-, and N-type basalts.

SPREADING RATES

The Mid-Atlantic Ridge is often cited as a slowly-accreting plate margin with a typical well-developed rift valley. Recent detailed examination of near-axis magnetic anomalies has led to the conclusion that the spreading rate has ranged considerably over the past few million years (Le Douran and others, 1982). Spreading rates, for example, have been considerably higher over the past 0.8 m.y. (3 cm/yr full rate) compared to 1.25 cm/yr full rate for the previous 1.2 m.y. in the 20°N to 37°N axial zone. A number of authors have argued that rapid spreading rates favor a steady-state magma chamber (Stakes and others, 1984). Such steady-state magma chambers can produce magmas that are more fractionated than magmas derived more directly from mantle sources. The most extreme fractionated magmas are those that are produced during the final stage of magma chamber crystallization, regardless of whether it was a single cycle or multiple cycle (repeated influxes of primitive liquids) magma chamber.

Vogt (Plate 8A) gives spreading rates as a function of latitude and distance from poles of opening along the Mid-Atlantic Ridge. These rates, derived from the model of Minster and Jordan (1978), represent an average for the last few million years, and they smooth out fluctuations found by more detailed studies (Le Douran and others, 1982). The general trend is that of decreasing spreading rates moving north from the equator. Maximum total opening rates of 4.0 cm/yr occur at about 1°N. At the Azores Triple Junction, at about 39.4°N, rates are down to about 2.1 cm/yr. A slight increase in spreading rate occurs here, where

TABLE 1. (CONTINUED)

Group	Long	N	R	SiO_2	Al_2O_3	FeO*	MgO	CaO	Na_2O	K_2O	TiO_2	P_2O_5	Sum	S	A	E	Dep	So	Ship
A36.40N1	33.77W	1	0	51.51	14.68	10.59	7.26	12.00	2.36	0.23	1.30	0.16	100.09	R	<1	E	2671	W	AII 77-D 4
A36.42N1	33.65W	9	0	50.91	15.08	9.67	7.98	12.40	2.04	0.10	1.08	0.11	99.37	R	<1	E		W	GS103-38
A36.42N1	33.66W	1	0	51.49	15.06	9.75	7.93	12.31	2.00	0.14	1.19	0.13	100.00	R	<1	E		W	GS103-7
A36.42N1	33.65W	4	0	51.39	15.06	9.94	7.47	12.01	2.25	0.15	1.31	0.15	99.73	R	<1	E	2616	W	AII 77-D14
A36.43N1	33.68W	1	0	51.66	14.54	9.76	7.10	12.14	1.97	0.18	1.13	0.14	98.62	R	<1	E	2507	W	AII 77-P10
A36.43N1	33.65W	6	0	51.38	15.10	10.12	7.64	12.11	2.12	0.11	1.18	0.11	99.87	R	<1	E		W	GS103-5
A36.43N2	33.65W	1	0	51.46	15.16	9.46	8.05	12.65	2.06	0.08	1.02	0.10	100.04	R	<1	E		W	GS103-5
A36.43N1	33.66W	1	0	51.31	14.95	9.51	7.84	12.44	2.21	0.12	1.12	0.12	99.62	R	<1	E	2560	W	AII 77-P9
A36.44N1	33.64W	3	0	51.35	14.39	10.24	7.27	12.17	2.18	0.14	1.13	0.13	99.00	R	<1	E		W	ALV 829
A36.44N1	33.65W	2	0	51.29	14.70	10.60	7.06	11.72	2.33	0.16	1.46	0.18	99.50	R	<1	E		W	ALV 829
A36.44N1	33.65W	2	0	51.22	14.75	10.43	7.12	11.86	2.36	0.15	1.40	0.16	99.45	R	<1	E		W	ALV 829
A36.44N1	33.65W	1	0	51.56	14.95	9.30	7.84	12.63	2.16	0.14	1.16	0.13	99.87	R	<1	E		W	ALV 829
A36.44N1	33.64W	6	0	52.19	14.12	10.83	6.48	11.56	2.19	0.15	1.32	0.15	98.99	R	<1	E	2518	W	AII 77-D16
A36.44N2	33.64W	7	0	51.31	14.35	12.17	6.23	10.62	2.57	0.10	1.84	0.20	99.49	R	<1	E	1518	W	AII 77-D16
A36.44N1	33.66W	1	0	51.29	14.16	10.19	7.50	12.39	2.16	0.14	1.28	0.15	99.26	R	<1	E		W	ALV 825-
A36.44N2	33.66W	1	0	51.04	15.65	8.54	8.32	13.64	2.03	0.06	0.81	0.08	100.17	R	<1	E		W	ALV 825-
A36.44N1	33.66W	1	0	51.01	15.46	8.56	8.39	13.46	2.03	0.06	0.81	0.10	99.88	R	<1	E		W	ALV 825-
A36.44N1	33.67W	1	0	50.98	15.21	8.60	8.41	13.48	1.92	0.10	0.93	0.10	99.73	R	<1	E		W	ALV 825
A36.44N2	33.67W	1	0	51.39	14.70	9.77	7.34	12.09	2.21	0.16	1.11	0.15	98.92	R	<1	E		W	ALV 825
A36.44N1	33.68W	2	0	51.44	14.88	9.12	7.56	12.55	1.90	0.13	0.96	0.10	98.64	R	<1	E	2437	W	AII 77-D15
A36.45N1	33.62W	4	0	51.25	15.26	8.74	8.26	13.24	1.96	0.12	0.90	0.10	99.83	R	<1	E		W	GS102-4
A36.45N2	33.62W	2	0	51.40	14.92	9.70	7.42	11.90	2.18	0.18	1.12	0.14	98.96	R	<1	E		W	GS103-4
A36.45N1	33.65W	1	0	51.53	15.34	9.42	7.84	12.39	2.18	0.15	1.21	0.12	100.18	R	<1	E		W	ALV 829
A36.45N1	33.66W	2	0	51.82	15.16	8.58	8.23	12.96	1.99	0.10	0.88	0.12	99.84	R	<1	E		W	GS103-D41
A36.45N1	33.66W	2	0	50.64	15.20	8.94	8.26	12.98	2.02	0.08	0.92	0.10	99.14	R	<1	E		W	ALV 831
A36.45N1	33.66W	1	0	50.88	15.12	8.68	8.28	13.36	1.95	0.10	0.96	0.11	99.44	R	<1	E		W	ALV 831
A36.45N1	33.69W	4	0	51.17	15.11	9.34	7.86	12.49	2.13	0.14	1.08	0.12	99.44	R	<1	E		W	ALV 828
A36.45N1	33.70W	1	0	51.06	15.10	9.34	8.05	12.47	2.06	0.12	1.14	0.13	99.47	R	<1	E		W	ALV 828
A36.45N1	33.65W	1	0	51.76	15.28	9.66	7.82	12.20	2.25	0.16	1.24	0.13	100.50	R	<1	E		W	ALV 829
A36.45N1	33.69W	1	0	51.20	15.12	9.52	7.81	12.52	2.13	0.13	1.10	0.13	99.66	R	<1	E		W	ALV 828
A36.45N1	33.70W	1	0	49.20	16.05	8.46	9.12	13.48	1.83	0.05	0.74	0.07	99.00	R	<1	E		W	ALV 828
A36.45N1	33.66W	1	0	52.19	14.98	10.26	7.39	12.00	2.22	0.14	1.15	0.11	100.44	R	<1	E		W	ALV 831
A36.45N2	33.66W	1	0	50.81	15.60	8.52	8.49	13.45	2.09	0.08	0.80	0.08	99.92	R	<1	E		W	ALV 831
A36.45N1	33.66W	1	0	50.90	15.29	8.74	8.40	13.32	1.95	0.09	0.97	0.09	99.75	R	<1	E		W	ALV 831
A36.45N1	33.68W	1	0	51.45	14.65	9.89	7.48	11.94	2.22	0.15	1.24	0.14	99.16	R	<1	E		W	ALV 828
A36.46N1	33.65W	1	0	51.39	14.96	10.12	7.50	12.31	2.18	0.20	1.13	0.12	99.91	R	<1	E		W	ALV 829
A36.46N1	33.65W	3	0	51.59	14.69	10.04	7.34	12.06	2.16	0.18	1.12	0.13	99.31	R	<1	E		W	GS103-37
A36.46N1	33.66W	1	0	51.63	15.22	9.53	7.92	12.54	2.17	0.12	1.08	0.10	100.31	R	<1	E		W	ALV 831
A36.46N2	33.66W	1	0	51.19	15.54	8.70	8.10	13.69	2.07	0.06	0.80	0.07	100.22	R	<1	E		W	ALV 831
A36.46N1	33.69W	1	0	50.10	16.42	8.23	8.82	13.38	1.84	0.07	0.78	0.06	99.70	R	<1	E		W	ALV 826
A36.46N1	33.58W	4	0	50.82	15.41	8.81	8.02	12.92	2.00	0.14	1.01	0.12	99.25	R	<1	E	2231	W	AII 77-D11
A36.46N1	33.66W	2	0	50.87	15.14	9.00	8.10	13.10	2.14	0.08	0.93	0.08	99.44	R	<1	E		W	ALV 831
A36.46N1	33.66W	1	0	51.35	15.14	8.78	8.12	13.43	2.01	0.09	0.97	0.10	99.99	R	<1	E		W	ALV 831
A36.46N1	33.68W	2	0	50.98	15.32	8.78	8.12	13.16	2.12	0.08	0.90	0.09	99.55	R	<1	E		W	ALV 828
A36.46N1	33.62W	1	0	51.97	15.18	9.21	7.82	12.33	2.31	0.12	1.06	0.12	100.12	R	<1	E		W	ALV 827
A36.46N1	33.62W	1	0	51.18	15.23	9.22	8.05	12.82	2.01	0.12	0.97	0.13	99.73	R	<1	E		W	ALV 827
A36.46N1	33.63W	1	0	51.42	15.08	8.54	8.16	12.81	1.93	0.15	1.03	0.15	99.27	R	<1	E		W	ALV 827
A36.46N2	33.63W	1	0	51.56	15.04	9.78	7.50	12.07	2.37	0.12	1.23	0.13	99.80	R	<1	E		W	ALV 827
A36.46N1	33.70W	1	0	49.69	16.48	8.26	8.98	13.79	1.88	0.05	0.71	0.10	99.94	R	<1	E		W	ALV 826
A36.46N1	33.63W	2	0	51.02	15.32	8.70	8.24	13.02	1.98	0.13	0.96	0.13	99.50	R	<1	E		W	ALV 827
A36.46N1	33.63W	11	0	51.88	14.94	9.82	7.53	12.16	2.19	0.16	1.10	0.12	99.90	R	<1	E		W	GS103-9
A36.46N1	33.66W	1	0	51.12	14.93	9.16	7.89	13.08	2.18	0.09	0.98	0.09	99.52	R	<1	E		W	ALV 831
A36.47N1	33.63W	4	0	51.83	14.88	9.84	7.45	12.18	2.19	0.16	1.15	0.13	99.81	R	<1	E		W	ALB 827

the new pole of opening (Plate 8A, this volume) takes over. North of this jump, spreading rates decrease from 2.6 to 2.3 cm/yr up to the Charlie Gibbs Fracture Zone at 52°N. Spreading rates continue to decrease along the Reykjanes Ridge from 2.3 to 2.0 cm/yr, and from 2.0 to 1.9 cm/yr over Iceland. The decrease along the Kolbeinsey Ridge is from 1.9 to 1.7 cm/yr and from 1.7 to 1.5 cm/yr over the Mohns Ridge. Thus, over the North Atlantic from near the equator to about 74°N, total opening rates decrease mainly progressively from about 4.0 to 1.5 cm/yr.

REVIEW OF MID-ATLANTIC RIDGE "ZERO-AGE" BASALT VARIATIONS

Mid-Atlantic Ridge abyssal basalt has been classified using a variety of major and minor element parameters. These include (1) Na_2O and TiO_2 abundances (Shido and Miyashiro, 1973), (2) Na_2O and TiO_2 abundances, among other parameters (Melson and O'Hearn, 1979), (3) FeO enrichment and Al_2O_3 "depletions" (Sigurdsson, 1981), (4) degree of silica saturation (Morel and Hekinian, 1980), and (5) liquidus trends in the normative ternary plagioclase-olivine-pyroxene systems (Bryan and Dick, 1982).

A number of papers have pointed out significant patterns in the distribution of major and minor element basalt compositions along major segments of the Mid-Atlantic Ridge. The following summary highlights the results of this work. Many more studies have focused on an understanding of specific segments of the Mid-Atlantic Ridge. The work of Aumento (1968, 1969) at 45°N and of numerous researchers on the FAMOUS and 22°–24°N regions have been particularly informative.

Shido and Miyashiro (1973) noted that the Mid-Atlantic Ridge can be divided into two provinces based on the whole rock compositional data available to them at that time. North of the Azores, CaO was found to be higher and Na_2O lower than in basalt south of the Azores. The southern group was found to typically contain phenocrysts of plagioclase and olivine, whereas those north of the Azores contained these phases and commonly augite phenocrysts and microphenocrysts as well. They found no regional difference between FeO*/MgO, an expected conse-

TABLE 1. (CONTINUED)

Group	Long	N	R	SiO_2	Al_2O_3	FeO*	MgO	CaO	Na_2O	K_2O	TiO_2	P_2O_5	Sum	S	A	E	Dep	So	Ship
A36.47N1	33.63W	1	0	51.77	14.65	10.34	6.86	12.07	2.38	0.18	1.25	0.16	99.66	R	<1	E		W	ALV 827
A36.47N2	33.63W	1	0	51.56	15.29	9.87	7.59	12.40	2.26	0.12	1.30	0.17	100.56	R	<1	E		W	ALV 827
A36.47N1	33.62W	1	0	51.74	14.70	9.77	7.36	12.00	2.18	0.17	1.20	0.14	99.26	R	<1	E		W	ALV 827-14
A36.48N1	33.71W	1	0	51.16	15.16	9.59	7.63	12.26	2.28	0.10	1.06	0.12	99.36	R	<1	E		W	ALV 826-6
A36.48N1	33.66W	1	0	50.31	15.87	8.95	8.74	12.64	2.29	0.12	1.12	0.10	100.14	R	<1	E		W	GS103-8
A36.49N1	33.65W	1	0	51.17	15.24	9.22	8.40	12.98	2.20	0.08	0.96	0.10	100.35	R	<1	E	2589	W	AII 73-D13
A36.49N2	33.65W	3	0	51.64	14.99	9.92	7.69	12.18	2.22	0.20	1.12	0.12	100.08	R	<1	E	2589	W	AII 73-D13
A36.54N1	33.52W	1	0	51.62	14.77	9.68	7.84	12.59	2.01	0.14	1.11	0.12	99.88	R	<1	E	2108	W	KN42-ST146-
A36.58N1	33.60W	1	0	51.99	13.96	12.11	6.34	11.06	2.50	0.18	1.74	0.22	100.10	R	<1	E	2913	W	AII77-ST23-
A36.58N1	33.52W	1	0	50.02	15.93	9.55	8.92	12.42	2.00	0.13	0.85	0.11	99.93	R	<1	E	2437	W	AII77-ST52-
A36.60N1	33.47W	5	0	51.58	15.21	8.71	8.31	12.52	1.95	0.25	1.10	0.13	99.76	FZ	<1	E	2118	W	KN42-D24
A36.62N1	33.69W	2	0	50.96	15.32	9.45	7.96	12.42	2.00	0.12	1.09	0.12	99.44	R	<1	E		W	GS103-27
A36.62N2	33.69W	4	1	51.50	14.96	10.31	7.33	11.69	2.32	0.14	1.28	0.13	99.66	R	<1	E		W	GS103-27
A36.62N3	33.69W	4	0	51.22	15.27	8.45	8.14	13.13	1.97	0.15	0.96	0.11	99.40	R	<1	E		W	GS103-27
A36.64N1	33.47W	1	0	51.14	14.64	10.44	7.45	11.26	2.30	0.33	1.66	0.19	99.41	R	<1	E	1828	W	KN42-ST121-
A36.64N1	33.46W	5	0	51.19	15.29	8.84	8.06	12.46	2.04	0.21	1.07	0.13	99.29	R	<1	E	2137	W	KN42-D11
A36.66N1	33.34W	5	1	51.32	14.95	9.46	7.87	12.79	2.05	0.17	1.16	0.12	99.89	R	<1	E	2901	W	KN42-D12
A36.66N2	33.34W	8	0	51.36	14.63	10.18	7.31	12.05	2.14	0.22	1.46	0.16	99.51	R	<1	E	2901	W	KN42-D12
A36.70N1	33.29W	1	0	51.01	14.87	9.32	8.02	12.80	1.97	0.13	1.06	0.13	99.31	R	<1	E	2560	W	AII 73-D 5
A36.70N2	33.29W	6	0	50.75	14.86	10.27	7.72	11.92	2.14	0.17	1.28	0.14	99.25	R	<1	E	2560	W	AII 73-D 5
A36.70N3	33.29W	1	0	51.01	14.66	10.92	7.17	11.63	2.21	0.20	1.46	0.19	99.45	R	<1	E	2560	W	AII 73-D 5
A36.72N1	33.33W	4	0	51.01	15.07	9.24	8.31	13.03	2.04	0.11	1.02	0.11	99.94	R	<1	E	2437	W	AII 77-D 6
A36.72N2	33.33W	1	0	52.17	13.84	11.91	6.52	11.21	2.45	0.20	1.70	0.18	100.18	R	<1	E	2437	W	AII 77-D 6
A36.72N1	33.34W	5	0	50.61	14.99	9.58	7.71	12.81	2.11	0.19	1.27	0.17	99.44	R	<1	E	2443	W	AII 77-D 7
A36.72N2	33.34W	2	0	51.09	14.96	9.33	7.96	13.30	2.03	0.12	1.07	0.12	99.98	R	<1	E	2443	W	AII 77-D 7
A36.72N1	33.30W	3	1	50.95	15.16	9.40	8.17	12.51	2.00	0.16	1.12	0.13	99.60	R	<1	E		W	ALV 830
A36.72N1	33.31W	2	0	51.00	15.18	9.42	7.99	12.58	2.02	0.16	1.16	0.12	99.63	R	<1	E		W	ALV 830-4
A36.73N1	33.30W	1	0	50.51	15.25	8.66	8.35	13.19	1.95	0.19	1.06	0.12	99.28	R	<1	E		W	ALV 832
A36.73N1	33.31W	2	0	51.17	15.30	8.82	8.29	13.28	1.90	0.15	1.02	0.10	100.03	R	<1	E		W	ALV 830
A36.73N1	33.30W	1	0	50.84	15.02	9.95	7.45	12.32	2.34	0.17	1.37	0.15	99.61	R	<1	E		W	ALV 832
A36.73N1	33.29W	2	0	51.18	15.09	8.77	8.14	13.50	2.02	0.14	0.92	0.10	99.86	R	<1	E		W	GS103-D32
A36.73N2	33.29W	1	0	51.15	14.92	9.71	7.54	12.71	2.30	0.17	1.18	0.13	99.81	R	<1	E		W	GS103-D32
A36.73N1	33.31W	1	0	50.88	15.06	9.16	7.96	12.97	2.14	0.15	1.10	0.11	99.53	R	<1	E		W	ALV 832
A36.73N1	33.32W	2	0	51.42	15.02	10.05	7.38	12.18	2.22	0.19	1.28	0.14	99.88	R	<1	E		W	ALV 830
A36.73N1	33.31W	13	0	50.94	15.02	9.56	7.83	12.56	2.09	0.14	1.18	0.13	99.45	R	<1	E	2519	W	AII 73-D 4
A36.73N2	33.31W	3	0	50.57	14.89	10.16	7.92	12.14	2.29	0.16	1.24	0.14	99.51	R	<1	E	2519	W	AII 73-D 4
A36.74N1	33.29W	2	0	51.62	14.78	10.00	6.98	12.19	2.30	0.20	1.46	0.15	99.68	R	<1	E		W	ALV 824
A36.75N1	33.29W	1	0	51.16	15.10	9.26	7.82	12.90	2.00	0.14	1.14	0.11	99.63	R	<1	E	2476	W	AII 73-D 3
A36.75N1	33.28W	1	0	51.30	15.20	9.00	7.93	12.72	2.07	0.17	1.06	0.12	99.57	R	<1	E		W	ALV 821
A36.75N1	33.30W	1	0	51.07	15.05	9.36	7.92	12.67	2.06	0.16	1.14	0.11	99.54	R	<1	E		W	ALV 824
A36.75N1	33.29W	2	0	51.02	14.79	9.74	7.52	12.39	1.90	0.18	1.20	0.13	98.87	R	<1	E		W	ALV 820
A36.75N1	33.31W	2	0	51.10	15.08	9.17	7.72	12.92	1.96	0.20	1.03	0.12	99.30	R	<1	E		W	ALV 824
A36.75N1	33.28W	1	0	50.77	14.92	9.70	7.96	12.18	1.97	0.25	1.24	0.17	99.16	R	<1	E		W	ALV 821
A36.75N1	33.30W	1	0	51.00	14.95	8.79	8.18	13.09	1.86	0.14	1.04	0.10	99.15	R	<1	E		W	ALV 821
A36.75N1	33.27W	1	0	49.99	16.25	8.42	8.97	13.34	2.01	0.10	0.82	0.09	99.99	R	<1	E		W	ALV 822-8
A36.75N1	33.29W	1	0	51.46	14.82	9.81	7.64	12.64	2.03	0.16	1.22	0.12	99.90	R	<1	E		W	ALV 820
A36.75N1	33.29W	2	0	50.68	14.91	9.70	7.71	12.37	2.02	0.16	1.24	0.13	98.92	R	<1	E		W	ALV 820
A36.75N1	33.30W	1	0	50.99	15.02	9.60	7.64	12.43	2.04	0.17	1.22	0.15	99.26	R	<1	E		W	ALV 820-4
A36.75N1	33.30W	2	0	51.15	15.10	9.38	7.84	12.40	1.96	0.17	1.14	0.14	99.28	R	<1	E		W	ALV 824
A36.76N1	33.29W	4	0	51.31	15.17	9.44	7.83	12.48	1.98	0.16	1.13	0.14	99.64	R	<1	E		W	GS103-D46
A36.76N1	33.28W	1	0	51.17	15.17	9.00	8.07	13.03	2.09	0.13	1.02	0.10	99.78	R	<1	E		W	ALV 822
A36.76N1	33.28W	1	0	50.42	15.21	9.40	7.84	12.25	2.18	0.23	1.18	0.14	98.85	R	<1	E		W	ALV 821
A36.76N1	33.29W	2	0	50.78	15.06	9.48	7.97	12.34	2.14	0.14	1.18	0.13	99.32	R	<1	E		W	ALV 821
A36.76N1	33.29W	1	0	50.83	15.12	9.00	7.97	12.81	1.90	0.13	1.04	0.13	98.93	R	<1	E		W	ALV 821

quence of fractional crystallization that would produce lower CaO and higher Na_2O, and thus ascribed these differences to differences in their primary magma parents. Shido and Miyashiro (1973) also note that the Azores mark the boundary between the opening of the Eurasian and American plates to the north, and of the African and South American plates to the south, and suggest this tectonic difference in some way affects magma compositions. However, no specific model was offered to explain why this might be so. The large amount of data now available from the FAMOUS area show that low Na_2O, high CaO basalts are a characteristic of that region, and thus indicate that if there is a boundary between these possible two groupings, it is south of the Azores. It may be the same boundary pointed to by Bougault and Treuil (1980) and Melson and O'Hearn (1979).

Sigurdsson (1981) has recognized two chemical groups in a suite of 101 basaltic glasses from the Mid-Atlantic Ridge between 29° and 73°N. He uses FeO abundances and Fe-enrichment trends, Al_2O_3 "depletion," lower SiO_2 and the alkalies, and higher CaO/Al_2O_3, to define a northern group (Group A) located from the Charlie Gibbs Fracture Zone along the Reykjanes Ridge on the north to at least 70°N. He notes a southern subgroup like these from 29° to 34°N. These are characterized by plagioclase and olivine phenocrysts, with clinopyroxene in only the more evolved samples. The more fractionated samples in this group involve about 60 percent crystallization from the most primitive member in the ratio of olivine: plagioclase: clinopyroxene of 1:3.3:1.5.

His second group involves the region from 35°N to the Charlie Gibbs Fraction Zone (53°N), and the Mohns Ridge north of the Jan Mayen Fracture Zone at 71°N to at least 73°N. This group, termed Group B, has higher SiO_2 and alkalies, low Fe, and lacks Al_2O_3 depletion and iron enrichment trends, and contains calcic augite throughout the composition range, whereas sub-calcic augite is absent. The most fractionated members require subtraction of olivine: plagioclase: clinopyroxene in the ratio of 1:5:3.9 from the more primitive compositions. Augite fractionation is required throughout the trend. It appears that most of the glasses examined by Melson and O'Hearn (1979) and

TABLE 1. (CONTINUED)

Group	Long	N	R	SiO_2	Al_2O_3	FeO*	MgO	CaO	Na_2O	K_2O	TiO_2	P_2O_5	Sum	S	A	E	Dep	So	Ship
A36.76N1	33.29W	1	0	50.23	15.17	9.60	7.77	12.34	1.93	0.15	1.20	0.12	98.51	R	<1	E		W	ALV 821
A36.76N1	33.30W	1	0	51.06	14.83	9.75	7.60	12.54	2.06	0.17	1.24	0.13	99.38	R	<1	E		W	ALV 820-5
A36.76N1	33.31W	5	0	50.82	14.90	10.07	8.00	12.34	2.22	0.16	1.18	0.14	99.83	R	<1	E	2589	W	AII 73-D 6
A36.76N1	33.28W	2	0	50.94	15.15	8.78	8.31	13.22	1.87	0.16	0.98	0.14	99.55	R	<1	E		W	ALV 822
A36.76N1	33.34W	10	0	51.46	15.02	9.42	7.77	12.65	2.04	0.15	1.09	0.13	99.73	R	<1	E	2391	W	AII 73-D 2
A36.76N1	33.27W	2	0	50.74	15.18	8.18	8.42	13.88	1.82	0.16	0.86	0.11	99.35	R	<1	E		W	GS103-D30
A36.76N1	33.30W	1	1	51.15	15.19	9.35	8.14	12.34	1.98	0.14	1.11	0.14	99.54	R	<1	E		W	GS103-D23
A36.78N1	33.29W	2	0	51.18	14.90	9.76	7.71	12.35	2.10	0.16	1.16	0.11	99.43	R	<1	E		W	ALV 818-4
A36.78N1	33.30W	1	0	50.98	15.01	8.95	8.18	13.25	1.99	0.22	1.04	0.13	99.75	R	<1	E		W	ALV 818
A36.78N1	33.30W	1	0	51.31	14.87	9.48	7.59	12.94	2.25	0.22	1.06	0.11	99.83	R	<1	E		W	ALV 818
A36.78N1	33.30W	1	0	50.47	14.96	8.87	8.14	13.28	2.02	0.22	1.05	0.14	99.15	R	<1	E		W	ALV 818
A36.78N1	33.26W	4	0	50.86	15.34	8.93	8.64	12.72	1.80	0.18	1.01	0.12	99.60	R	<1	E		W	GS103-D33
A36.78N2	33.26W	1	0	51.02	15.28	9.40	7.87	12.52	2.08	0.21	1.18	0.16	99.72	R	<1	E		W	GS103-D33
A36.78N1	33.28W	10	0	51.41	14.71	10.19	7.34	10.84	2.16	0.18	1.39	0.16	98.38	R	<1	E		W	GS103-20
A36.79N1	33.27W	2	0	50.95	14.78	9.52	7.50	12.57	2.05	0.18	1.16	0.12	98.83	R	<1	E	2642	W	ALV 534-3
A36.80N1	33.27W	2	0	49.24	16.21	9.05	9.32	12.39	1.98	0.12	0.95	0.12	99.38	R	<1	E	2679	W	ALV 534-2
A36.80N1	33.29W	4	0	51.12	14.88	10.00	7.56	12.13	2.10	0.16	1.24	0.15	99.34	R	<1	E		W	GS103-35
A36.80N1	33.27W	1	0	50.04	15.29	10.39	7.88	11.34	2.33	0.18	1.30	0.14	98.89	R	<1	E	2643	W	ALV 534-1
A36.80N1	33.26W	1	0	49.54	16.62	8.74	9.43	12.84	2.08	0.09	0.74	0.08	100.16	R	<1	E	2526	W	ALV 528-4
A36.81N1	33.27W	1	0	49.00	16.37	8.88	9.55	12.38	1.96	0.10	0.86	0.09	99.19	R	<1	E	2646	W	ALV 530-3
A36.81N1	33.27W	2	0	51.32	14.93	9.89	7.63	12.21	2.17	0.18	1.30	0.14	99.77	R	<1	E	1663	W	ALV 530-2
A36.81N1	33.26W	1	0	48.68	15.89	9.16	9.46	12.38	2.13	0.11	0.88	0.09	98.78	R	<1	E	2526	W	ALV 522-2
A36.81N1	33.26W	2	0	49.45	16.30	9.00	9.28	12.49	2.11	0.12	0.85	0.10	99.70	R	<1	E	2505	W	ALV 528-3
A36.81N1	33.25W	1	0	50.17	15.32	10.56	7.91	11.60	2.38	0.24	1.40	0.16	99.74	R	<1	E	2512	W	ALV 527-5
A36.81N1	33.25W	2	0	50.84	15.06	10.21	7.80	11.60	2.28	0.25	1.42	0.16	99.62	R	<1	E	2565	W	ALV 527-6
A36.81N1	33.25W	1	0	50.11	15.16	10.37	8.02	11.54	2.40	0.23	1.45	0.18	99.46	R	<1	E	2584	W	ALV 527-4
A36.81N1	33.27W	1	0	49.77	15.95	9.08	8.82	12.32	1.98	0.12	0.90	0.08	99.02	R	<1	E	2750	W	ALV 525-4
A36.81N1	33.26W	4	0	48.79	16.84	8.99	9.41	12.70	2.05	0.06	0.66	0.07	99.57	R	<1	E	2707	W	ALV 527-1
A36.81N1	33.26W	4	0	49.18	15.81	9.10	9.78	12.22	2.01	0.12	0.88	0.10	99.20	R	<1	E	2625	W	ALV 525-5
A36.81N1	33.26W	1	0	49.52	16.04	8.96	9.86	12.39	2.07	0.17	0.85	0.09	99.95	R	<1	E	2625	W	ALV 528-2
A36.81N1	33.28W	8	0	51.17	14.83	9.95	7.80	12.53	2.05	0.18	1.07	0.12	99.70	FZ	<1	E	2250	W	KN42-D23
A36.81N2	33.28W	1	0	51.49	14.57	10.51	6.95	11.52	2.29	0.23	1.48	0.15	99.19	FZ	<1	E	2250	W	KN42-D23
A36.81N1	33.27W	3	0	48.36	16.07	9.33	9.61	12.50	2.05	0.11	0.77	0.08	98.88	R	<1	E	2669	W	ALV 522-1
A36.81N1	33.27W	1	0	49.71	16.20	9.10	8.94	12.53	1.92	0.11	0.81	0.11	99.43	R	<1	E	2680	W	ALV 528-1
A36.81N1	33.27W	1	0	50.74	14.68	10.29	7.40	12.27	2.21	0.23	1.20	0.16	99.18	R	<1	E	2715	W	ALV 525-2
A36.81N1	33.26W	1	0	49.08	16.48	9.08	9.20	12.61	2.01	0.09	0.77	0.09	99.41	R	<1	E	2700	W	ALV 519-5
A36.81N1	33.26W	2	0	48.81	16.44	9.13	9.52	12.51	2.07	0.11	0.78	0.09	99.46	R	<1	E	2690	W	ALV 519-4
A36.82N1	33.26W	3	0	49.45	16.52	9.17	9.15	12.53	2.00	0.09	0.77	0.10	99.78	R	<1	E	2709	W	ALV 519-3
A36.82N1	33.26W	2	0	49.05	16.20	9.25	9.63	12.53	2.03	0.11	0.78	0.08	99.66	R	<1	E	2689	W	ALV 520-1
A36.82N1	33.27W	3	0	49.14	16.37	9.16	9.37	12.56	2.01	0.11	0.76	0.08	99.56	R	<1	E	2728	W	ALV 520-2
A36.82N1	33.26W	3	0	49.90	16.69	9.13	9.10	12.71	1.92	0.08	0.76	0.08	100.37	R	<1	E	2735	W	ALV 519-2
A36.82N1	33.24W	1	0	51.56	14.64	9.17	7.93	13.01	1.96	0.14	1.02	0.11	99.54	R	<1	E	2604	W	ALV 526-6
A36.82N1	33.25W	1	0	50.72	14.72	10.37	7.76	11.36	2.33	0.24	1.52	0.16	99.18	R	<1	E	2550	W	ALV 526-5
A36.82N1	33.25W	1	0	50.16	15.55	9.88	8.45	11.92	2.20	0.16	1.12	0.11	99.55	R	<1	E	2663	W	ALV 526-3
A36.82N1	33.25W	1	0	50.40	14.79	10.36	8.06	11.61	2.30	0.22	1.45	0.16	99.35	R	<1	E	2585	W	ALV 526-4
A36.82N1	33.25W	3	0	49.82	16.11	8.48	9.11	13.00	1.85	0.13	0.85	0.10	99.45	R	<1	E	2694	W	ALV 526-1
A36.82N1	33.25W	1	0	49.97	15.52	9.38	8.81	11.77	2.19	0.14	1.13	0.13	99.04	R	<1	E	2692	W	ALV 526-2
A36.82N1	33.26W	1	0	50.67	16.16	9.37	8.77	11.99	2.23	0.10	1.04	0.10	100.43	R	<1	E	2690	W	ALV 518-1
A36.82N1	33.26W	1	0	50.17	15.27	9.80	8.88	11.59	2.39	0.16	1.11	0.14	99.51	R	<1	E	2649	W	ALV 529-4
A36.82N1	33.27W	2	0	51.14	16.28	8.22	8.88	12.37	2.25	0.20	0.95	0.11	100.40	R	<1	E	2705	W	ALV 529-3
A36.82N1	33.26W	3	0	50.34	16.61	9.05	8.98	11.99	2.19	0.08	0.85	0.10	100.19	R	<1	E	2685	W	ALV 518-2
A36.82N1	33.27W	1	0	51.91	14.58	10.07	7.26	12.34	2.08	0.22	1.14	0.13	99.73	R	<1	E	2603	W	ALV 521-1
A36.82N1	33.28W	1	0	51.06	14.59	10.81	6.87	11.40	2.52	0.27	1.47	0.16	99.15	R	<1	E	2449	W	ALV 524-2
A36.82N1	33.26W	4	0	51.75	15.15	9.96	7.35	11.82	2.28	0.15	1.28	0.13	99.87	R	<1	E	2640	W	ALV 518-3

the clearly distinguishable two groups (0° to 29°N and FAMOUS groups) belong to Sigurdsson's (1981) Group B glasses. There is a clear need to cross-correlate the groupings, as different authors are examining different data sets with different regional distributions, and choosing different parameters for discrimination.

Bryan and Dick (1982) examine three regional groups: the FAMOUS, 0° to 29°N, and Cayman Trough glasses. They point out that each has a distinct liquidus trend in the normative ternary plagioclase-pyroxene-olivine system that probably reflects major element mantle heterogeneity as well as varying degrees of partial melting in the mantle source. The FAMOUS group, 29°N and Sites 395 and 396B group, and Cayman Trough glasses have parallel liquidus trends but are successively displaced toward plagioclase and are from successively deeper (water depths) spreading centers. In simplest terms, these might reflect successively smaller extents of partial melting. If isostatic compensation is assumed, the probable, successively thinner axial asthenosphere would suggest that less extensive melting is in fact one control in generating these three groups.

Morel and Hekinian (1980) have divided the Mid-Atlantic Ridge into two basalt provinces: (1) a southern province from 25°S to 33°N and (2) a northern province from 33°N to 53°N. The first province has the characteristics previously described for the 0° to 29°N group of Melson and O'Hearn (1979). Morel and Hekinian (1980) point out in addition that these rocks are typically saturated with regard to SiO_2 whereas their northern group, from 33°N to 53°N, is typically oversaturated with regard to SiO_2. It is significant that so many papers on regional variations in basalt chemistry along the axis of the MAR have pointed to a boundary somewhere near 33°–34°N.

SEGMENT BY SEGMENT UPDATE OF "ZERO-AGE" VARIATIONS

The various approaches and data outlined above need to be

TABLE 1. (CONTINUED)

Group	Long	N	R	SiO_2	Al_2O_3	FeO*	MgO	CaO	Na_2O	K_2O	TiO_2	P_2O_5	Sum	S	A	E	Dep	So	Ship
A36.82N1	33.24W	1	0	49.67	14.80	10.40	7.94	11.64	2.36	0.24	1.43	0.15	98.63	R	<1	E	2540	W	ALV 523-3
A36.82N1	33.24W	1	0	49.44	14.99	10.44	7.74	11.73	2.42	0.24	1.45	0.15	98.60	R	<1	E	2532	W	ALV 523-4
A36.83N1	33.25W	1	0	49.72	14.47	10.37	7.71	11.45	2.37	0.36	1.48	0.18	98.11	R	<1	E	2743	W	ALV 523-2
A36.83N1	33.25W	1	0	50.24	15.57	9.30	8.45	12.24	2.29	0.24	1.16	0.14	99.63	R	<1	E	2726	W	ALV 523-1
A36.83N1	33.27W	1	0	50.89	16.21	8.29	9.40	12.26	2.06	0.16	0.89	0.12	100.28	R	<1	E	2800	W	JCH31-D1
A36.85N1	33.32W	1	0	51.11	14.70	9.99	7.83	12.60	2.20	0.17	1.18	0.12	99.90	R	<1	E	2271	W	KN42-St88
A36.85N1	32.97W	1	0	51.05	15.36	8.71	8.08	13.22	1.87	0.14	0.95	0.11	99.49	R	<1	E	2000	O	BA75-D14
A36.92N1	33.13W	3	0	51.39	14.39	10.23	6.89	12.06	2.11	0.22	1.30	0.16	98.75	R	<1	E	2889	W	KN42-D 4
A38.70N1	29.25W	1	0	51.10	13.85	8.31	8.51	13.44	2.19	0.40	1.48	0.18	99.46	R	<1	E		H	JCH97-D 5
A40.42N1	29.54W	1	0	51.97	14.28	11.57	7.02	11.67	2.12	0.07	1.25	0.09	100.04	R	<1	E	2780	S	
A40.55N1	29.21W	1	0	52.29	15.06	8.97	8.14	13.19	1.81	0.04	0.74	0.07	100.31	R	<1	E	2780	S	
A42.95N1	29.33W	2	0	50.57	15.62	7.96	9.21	13.32	1.99	0.26	0.92	0.10	99.95	R	<1	E	2168	W	AII 32-D11
A42.95N2	29.33W	1	0	51.25	15.94	8.59	8.02	13.04	2.06	0.27	0.99	0.11	100.27	R	<1	E	2168	W	AII 32-D11
A42.96N1	29.20W	2	0	51.09	15.77	8.42	7.20	12.17	2.49	0.49	1.40	0.16	99.19	R	<1	E	1816	W	AII 32-D12
A42.96N2	29.20W	1	0	51.77	15.53	8.87	6.77	11.29	2.66	0.54	1.57	0.18	99.18	R	<1	E	1816	W	AII 32-D12
A45.12N1	28.14W	1	0	51.23	15.43	9.80	7.97	11.57	2.37	0.25	1.36	0.11	100.09	R	<1	E		W	CH43-D 28
A45.12N1	28.33W	1	0	51.13	15.23	12.59	8.00	8.46	2.37	0.21	1.31	0.12	99.42	R	<1	E		W	CH43-D103
A45.22N1	28.00W	1	0	51.53	15.27	9.45	8.47	11.73	2.24	0.21	1.20	0.12	100.22	R	<1	E	3076	A	HUD66-ST56
A45.33N1	28.03W	2	0	51.27	15.83	8.92	7.79	11.78	2.41	0.28	1.24	0.09	99.61	R	<1	E	1792	A	HUD66-ST 9
A45.33N2	28.03W	1	0	51.63	15.08	10.09	7.10	11.21	2.49	0.26	1.50	0.14	99.50	R	<1	E	1792	A	HUD66-ST 9
A45.33N3	28.03W	1	0	51.11	15.59	9.42	7.76	11.31	2.51	0.27	1.32	0.12	99.41	R	<1	E	1792	A	HUD66-ST 9
A45.62N1	27.72W	1	0	51.18	14.83	10.49	7.21	11.15	2.47	0.12	1.59	0.10	99.14	R	<1	E	2377	A	HUD68-ST173
A45.68N1	27.83W	1	0	52.50	14.53	10.96	6.40	11.36	2.38	0.16	1.55	0.12	99.96	R	<1	E	3486	A	HUD68-ST197
A45.73N1	27.72W	1	0	49.70	16.42	7.96	7.96	11.70	2.59	0.45	1.48	0.17	98.43	R	<1	E	3370	C	DISC-60-D1
A48.46N1	28.01W	8	0	50.49	15.56	9.08	7.36	12.41	2.67	0.22	1.24	0.15	99.18	R	<1	E	2030	N	LYN73-D58
A48.54N1	28.04W	4	0	51.07	14.90	9.57	7.30	12.21	2.66	0.20	1.29	0.13	99.33	R	<1	E	2432	N	LYN73-D57
A49.81N1	28.65W	1	0	51.21	15.36	9.46	8.38	11.38	2.71	0.07	1.39	0.09	100.05	R	<1	E	3402	N	LYN73-D56
A51.84N1	30.07W	11	0	51.13	15.92	9.11	8.09	11.50	2.65	0.09	1.31	0.12	99.92	R	<1	E	2780	N	LYN73-D53
A52.67N1	34.94W	53	1	50.84	14.95	10.83	7.61	11.85	2.24	0.08	1.28	0.12	99.80	FZ	<1	E	3240	N	LYN71-D 7
A52.67N2	34.94W	3	0	49.49	15.26	11.23	6.86	10.44	2.96	0.33	2.15	0.23	98.95	FZ	<1	E	3240	N	LYN71-D 7
A52.67N3	34.94W	2	1	51.04	14.88	12.80	7.64	10.13	2.20	0.05	1.27	0.07	100.08	FZ	<1	E	3240	N	LYN71-D 7
A52.67N4	34.94W	1	0	50.90	16.29	9.68	7.76	11.88	2.40	0.22	1.38	0.13	100.64	FZ	<1	E	3240	N	LYN71-D 7
A52.99N1	34.99W	1	0	49.85	15.77	9.36	9.08	12.50	2.01	0.09	0.96	0.09	99.71	R	<1	E	1975	H	JCH4-D1
A52.99N2	34.99W	1	0	50.82	15.54	10.57	7.97	11.89	2.37	0.07	1.33	0.13	100.69	R	<1	E	1975	H	JCH4-D1
A52.99N3	34.99W	7	1	50.76	14.90	10.76	8.20	11.73	2.17	0.06	1.28	0.11	99.97	R	<1	E	1975	H	JCH4-D1
A52.99N4	34.99W	1	0	51.55	14.08	12.73	6.73	10.77	2.49	0.19	1.94	0.17	100.65	R	<1	E	1975	H	JCH4-D1
A53.07N1	35.03W	7	0	51.06	14.96	10.64	7.99	11.98	2.23	0.05	1.27	0.12	100.30	R	<1	E	940	N	LYN73-D50
A53.19N1	35.09W	3	0	51.21	15.11	9.43	8.35	13.01	2.02	0.03	0.97	0.09	100.22	R	<1	E	2707	N	LYN73-D49
A58.87N1	30.95W	4	0	51.07	13.64	12.31	6.83	11.83	2.04	0.05	1.27	0.10	99.14	R	<1	E	1380	N	LYN71-D10
A59.75N1	29.87W	1	0	51.17	14.38	11.14	7.72	12.63	1.93	0.07	1.08	0.07	100.19	R	<1	E	1116	N	BA52-D7
A59.75N2	29.87W	37	0	50.78	14.38	11.78	7.36	12.05	2.03	0.05	1.14	0.09	99.66	R	<1	E	1116	N	BA52-D7
A59.77N1	29.80W	37	0	50.20	13.93	12.27	6.99	11.97	2.11	0.07	1.24	0.09	98.87	R	<1	E	997	N	BA52-D6
A59.77N2	29.80W	1	0	51.75	13.65	13.23	6.61	11.35	2.16	0.07	1.46	0.13	100.41	R	<1	E	997	N	BA52-D6
A61.97N1	26.59W	1	0	49.18	13.95	12.11	7.26	12.33	2.00	0.15	1.69	0.08	98.75	R	<1	E	878	N	BA52-D3
A61.99N1	26.59W	63	0	50.74	14.18	12.43	6.82	11.35	2.14	0.18	1.57	0.16	99.57	R	<1	E	658	N	BA52-D5
A61.99N2	26.59W	1	0	45.30	11.55	17.37	4.58	10.40	2.85	0.62	5.43	0.42	98.52	R	<1	E	658	N	BA52-D5
A63.29N1	24.23W	16	2	50.27	14.10	12.78	6.78	11.88	2.04	0.15	1.65	0.18	99.83	R	<1	E	99	N	LYN71-D20
A63.19N2	24.23W	1	0	50.51	14.42	13.65	6.74	11.18	2.19	0.16	1.66	0.22	100.73	R	<1	E	99	N	LYN71-D20
A63.42N1	23.87W	1	0	51.76	13.45	15.48	5.27	10.50	2.28	0.18	2.12	0.18	101.22	R	<1	E	68	N	LYN71-D19
A63.42N1	23.87W	1	0	50.89	13.55	14.36	5.63	10.71	2.66	0.25	1.97	0.20	100.22	R	<1	E	68	N	LYN71-D19
A63.42N3	23.87W	13	1	51.15	13.29	14.38	5.93	10.90	2.25	0.18	1.95	0.20	100.23	R	<1	E	68	N	LYN71-D19
A63.54N1	23.67W	1	0	51.15	14.10	12.18	6.78	11.66	2.27	0.14	1.36	0.14	99.78	R	<1	E	112	N	LYN71-D18
A63.54N2	23.67W	2	0	51.32	13.76	13.12	6.33	11.34	2.22	0.13	1.56	0.14	99.92	R	<1	E	112	N	LYN71-D18
A63.64N1	23.41W	1	0	49.25	14.76	12.59	6.98	12.21	2.22	0.12	1.90	0.25	100.28	R	<1	E	67	N	LYN71-D17

integrated into a single whole. What follows is one such approach to this longer term goal. Table 1 gives new glass analyses from various regions of the Mid-Atlantic Ridge, much of it from north of the Azores. The summary begins in the Equatorial Atlantic and moves northward. In all, ten compositional-topographic elevation zones are defined.

(1) Equatorial North Atlantic: Sierra Leone hotspot

Abyssal basaltic glasses from the region near St. Paul's Rocks commonly have high K_2O and high P_2O_5, presumably due to the effects of ongoing activity of the Sierra Leone hotspot. This is the area of maximum spreading rates in the North Atlantic (ca. 4.0 cm/yr total opening rate) and involves opening about a pole near 62.98°N and 39.14°W (Africa-South America Pole). The axial zone becomes deeper north of St. Paul's Rocks. Several major transform fault (fracture) zones displace the ridge westward: (see also Fox and Gallo, this volume; Edmond, this volume). (1) the Chain Fracture Zone around 4°S, (2) the Romanche Fracture Zone near the equator, (3) the St. Paul's Fracture Zone near 1°N, (4) the 4°N Fracture Zone, (5) the Doldrums Fracture Zone at about 7°N, (6) the Vema Fracture Zone at about 11°N, and (7) the 14°N Fracture Zone. The Romanche Fracture Zone is by far the largest of these, marking a left-lateral offset of the axial zone of about 700 kilometers. These large equatorial fracture zones appear to involve some opening perpendicular to their axes (Thompson and Melson, 1972).

Although no volcanic islands occur near the postulated Sierra Leone hotspot, the unique ultrabasic to alkaline ultrabasic rocks of St. Paul's Rocks are exposed near the ridge axis at 1°N. These and some associated alkaline basaltic rocks dredged on the north flank of the islets have typical LREE enriched abundances and major element characteristics like those of hotspots (Frey, 1970 and Melson and others, 1972). This alkaline basalt contains ultramafic xenoliths that appear in some cases to be the residue of the partial melting that produced the alkaline basalt, while others

TABLE 1. (CONTINUED)

Group	Long	N	R	SiO_2	Al_2O_3	FeO*	MgO	CaO	Na_2O	K_2O	TiO_2	P_2O_5	Sum	S	A	E	Dep	So	Ship
A63.64N2	23.41W	1	0	49.62	14.15	13.70	6.69	11.87	2.51	0.06	2.09	0.19	100.88	R	<1	E	67	N	LYN71-D17
A63.64N3	23.41W	13	1	49.22	13.67	13.48	6.68	11.67	2.18	0.16	2.07	0.19	99.32	R	<1	E	67	N	LYN71-D17
A63.64N4	23.41W	2	1	49.70	13.38	14.14	6.12	11.62	2.28	0.18	2.22	0.24	99.88	R	<1	E	67	N	LYN71-D17
A70.17N1	15.28W	4	0	51.04	14.78	9.92	7.92	13.01	1.72	0.06	0.79	0.08	99.32	R	<1	E	1280	N	LYN73-D 7
A70.17N2	15.28W	3	0	50.46	14.24	11.66	7.43	12.08	1.96	0.07	1.14	0.11	99.15	R	<1	E	1280	N	LYN73-D 7
A70.57N1	14.96W	2	0	50.39	15.11	9.16	8.52	13.66	1.70	0.03	0.72	0.07	99.36	R	<1	E	1170	N	LYN73-D30
A70.93N1	12.99W	1	0	52.09	13.51	12.05	6.21	11.38	2.25	0.35	1.40	0.18	99.42	R	<1	E	110	N	LYN73-D29
A71.28N1	5.75W	3	0	51.14	14.43	11.63	5.17	10.12	3.05	0.77	2.01	0.30	98.62	R	<1	E	1463	N	LYN73-D18

Explanation of column headings: Group consists of latitude collected (A = Atlantic) and chemical type designation, e.g. A71.28N is the first chemical type from this locality. Long. = longitude of the type; N - number of samples which were averaged for the chemical type; R = number of analyses very close to but outside the chemical discriminant limits used in defining the chemical type. These are rejected in defining the type. S = probable geologic setting; R = present or ancient Mid-Atlantic Ridge rift valley, FZ = fracture zone. A = probable age of extrusive (E) from magnetic anomaly age, in millions of years. E = mode of emplacement as eruptive (E) or intrusive (I), DEP = depth, in meters, SO = institution or individuals who contributed samples - see acknowledgments for source codes. Ship = ship code, cruise, and dredge or dive number.

appear associated with cumulates from tholeiitic magmas (Sinton, 1979). Basaltic glasses near St. Paul's Rocks include both low and anomalously high K_2O (to .85 wgt %) varieties. To the north, such high-K_2O glasses do not occur.

Morgan (1981) has postulated that the Verde hotspot lies beneath the St. Paul's Rocks region, yet his reconstructions show clearly that it lies beneath the Verde Islands at present. The Sierra Leone Rise to the east and the Amazon Ridge to the west of St. Paul's Rocks may well have been generated by the ongoing activity of the Sierra Leone hotspot and appear totally unrelated to the Verde hotspot. The slightly northward trends of these ridges are consistent with northward motion of the North American and African plates.

(2) 3°N to 14°N

Although sampling density is very poor in this region, only low-K_2O, N-type glasses have been found north of St. Paul's Rocks to 14°N (Fig. 2). Basaltic glasses from and near the Vema Fracture Zone appear to be normal depleted MORB, with K_2O ranging from .13 to .17. In earlier work, the basalts of the 2°S to 15°N region were grouped with a high TiO_2, high Na_2O group (the 0° to 29°N Group, Melson and O'Hearn, 1979). It seems more likely that these grade northward into that group, but that those found near St. Paul's Rocks represent a separate group, related to magma influxes from the Sierra Leone thermal plume.

Off-axis, normal to the 3° to 14°N region, deep Atlantic sea-floor occurs, separated to the north and south by less deep zones. This is in accord with the model of the generation of thinner crust and hence deeper sea floor along N-type spreading segments.

(3) 15°N hotspot

High-K_2O basaltic glasses occur near 15°N (Fig. 1). H. Bougault (unpublished data) reports glasses with trace element characteristics of E-type basalt but, to our knowledge, radiogenic isotopic ratios have not been determined. The 15°N axial zone is a topographic high compared to the axial zones both north and south of it. Also to the west, the well-developed Researcher Ridges (Peter and Westbrook, 1976) may reflect past activity of this possible mantle plume. A northward-opening V-shaped topographic high in the Atlantic sea floor moves outward from the 15°N axial zone, also possibly reflecting past activity of this plume.

(4) 16°N to 34°N (Hayes Fracture Zone)

North of about 16°N is a major, well-studied group of N-type basalts. This is a deep region along the MAR axial zone and connects with deeps in off-axis Atlantic sea floor. A number of publications indicate that a northern, major petrological break in regard to glass compositions occurs near the Hayes Fracture Zone (34°N, Bougault and Treuil, 1980), based largely on minor and trace elements. This boundary was located near 29°N based on less closely spaced sample data using major and minor element glass compositions (Melson and O'Hearn, 1979). Here, it is assumed that this break is near the Hayes Fracture Zone. Compared to the northern, or FAMOUS group, the basaltic glasses between 16°N and the Hayes Fracture Zone have higher TiO_2, Na_2O, and Hf/La and lower Sr^{87}/Sr^{86}, and Pb^{206}/Pb^{204} among other differences. The samples studied by Schilling and others (1983) and by Sigurdsson (1981) included only a few south of the Hayes Fracture Zone and thus did not reveal this rather important boundary. Bougault and Treuil (1980) attribute this break to mantle heterogeneity because it appears to involve elemental ratios and Pb and Sr isotopic ratios that are not affected by fractional crystallization or partial melting.

Dmitriev and others (1984) have found that the 16°N to 34°N glasses have a higher Fe_2O_3/FeO ratio than those from the FAMOUS region (37°N). They attribute this to a source mantle that is grossly layered, with a shallower, higher Fe_2O_3/FeO zone. The major element compositional differences between these two regions is attributed to partial melting and equilibration of the 16°N to 34°N glasses within the lower pressure (shallower depth) plagioclase-peridotite facies. The 35°N to 40°N glasses are attributed to tapping of a deeper, lower Fe_2O_3/FeO zone of melting within the spinel-peridotite facies.

Bryan and others (1981) have studied extensively the basalts from just north of, within, and just south of the Kane Fracture Zone (24°N). They find all are typical large-ion lithophile element depleted N-type lavas. They find no compositional breaks across the fracture zone, implying that if there is a component of magma flow along the ridge axis, that flow is not disrupted by this major fracture zone. This region falls well within the 0° to 29°N group of Melson and O'Hearn (1979).

On the other hand, Miyashiro and others (1970) found a highly diverse suite of rocks within the Kane Fracture Zone, ranging from N-type basalt, N-type gabbros, titanian ferrogabbro, and aplite. This remarkable diversity includes particularly extreme fractionation in the presence of the titanian ferrogabbro (TiO_2 = 7.05, FeO* = 18.39). These intrusive samples are related to exposures in part within the fracture zone and compositionally equivalent extrusives have yet to be recovered in this region.

The region at 23°N and 28°N has produced the highest Al_2O_3 contents of any abyssal North Atlantic basaltic glasses. These have been interpreted to prove the existence of high-alumina basaltic liquids (Nicholls and others, 1964), as most high-alumina basalts owe their unusual compositions to the presence of plagioclase phenocrysts, which may or may not have crystallized from a magma with the bulk composition of the rock in which they are found, that is, the basalts may be cumulates. Flower (1982) proposes that such high-Al_2O_3 basalts form by resorption of accumulated plagioclase.

The depth to the median valley decreases southward from the Azores to about 10°N. Thus, the southern group of lavas in general was erupted at greater water depth (thinner lithosphere) than the northern group, and erupted at a somewhat higher spreading rate.

(5) 34°N to the Azores

One of the most significant concepts to arise from studies of the FAMOUS area at 37°N is that of a steady-state zoned magma chamber to account for excess enrichments of incompatible elements over that expected from closed system fractional crystallization (Bryan and others, 1979). Stakes and others (1984) have developed a similar model for a steady-state magma chamber in the central regions between fracture zones in the FAMOUS and AMAR valley to the south. They have put forth the notions of volcano-tectonic cycles involving periodic replenishment of long-term magma chambers with "primitive" liquids.

We concur with Schilling (1975) that the Azores mantle plume is influencing the compositions of the northern group (Hayes Fracture Zone to the Azores). The Azores hotspot appears to be a young but highly active one, beginning around 20 m.y. ago (Morgan, 1981). One of the characteristics of the FAMOUS region glass suite is the occurrence of the most MgO-rich glass yet found in the mid-ocean ridge system among the approximately 6000 glass analyses from the SIVGP data set. In spite of this high MgO, this particular sample has an elevated K_2O content compared to the next highest MgO glasses in the SIVGP data set:

Ocean	A	I	P
Lat	36.91N	9.83N	2.64N
Lon	33.26W	57.96E	9.55W
SiO_2	49.52	50.50	50.81
Al_2O_3	16.04	16.75	15.61
FeO*	8.96	7.71	8.64
MgO	9.86	9.80	9.71
CaO	12.39	13.15	13.29
Na_2O	2.07	1.71	1.84
K_2O	0.17	0.04	0.08
TiO_2	0.85	0.70	0.95
P_2O_5	0.09	0.08	0.09

(6) 34°N E-type basalt

These samples lie within the latitudinal range of the 34°N to Azores group but have such distinctive compositions that we have broken them into a separate group. Around 34°N, basalts with clear-cut "plume-type" compositions have been noted (Schilling and others, 1983) even though the region is not topographically high and is bordered to the south by LILE-depleted glasses and to the north by glasses but slightly enriched in LIL elements, probably due to southward flow from the Azores hotspot. New glass analyses of samples provided by P. J. Fox and H. Bougault from around 35°N (Table 1, analyses A35.13N1, A35.15N1, and A35.18N1) show abnormally high K_2O and P_2O_5, in accord with the notion of "plume-type" (E-type) magmas in that region.

The seamounts east of this region (Atlantis, Plato, Cruiser, and Meteor) and to the west (Corner Seamounts) may in some way be connected to the E-type basalt from 35°N. Thus, although there is no pronounced topographic axial high around 35°N, there is need for a close look at the location, if any, of a currently active hotspot in or near this region.

(7) Azores to the Charlie Gibbs Fracture Zone (53°N)

This region is a complex one in which no simple chemical groupings or gradients have been defined to our knowledge. Part of that complexity is brought out by the work of Shibata and others (1979) who found both tholeiitic and alkali basalts at 43°N.

At 45°N, Aumento (1969) records some of the most acidic rocks—diorites—yet recovered from the Mid-Atlantic Ridge. This single occurrence has figured prominently in many comparisons with the volumetrically small but commonplace acidic rocks associated with ophiolites. The exposure of such rocks in a regime of non-erosion points to the structural complexity of the 45°N region. Petrologically, Pallister and Hopson (1981) point out that such acidic differentiates can occur toward the sides (wings) of a magma chamber away from the source of replenishment where closed-system fractionation takes over as spreading carries the two sides of the solidifying chamber apart. To date, acidic volcanic rocks or glasses, like those reported from the Galapagos Rift (Byerly, 1980), have yet to be recovered from the North Atlantic seafloor, although they do occur rarely on Iceland and other Atlantic volcanic islands.

In the 45°N region Aumento (1968) found alkali olivine basalts distributed away from the median zone and tholeiitic basalts along the axial zone. This was one of the first, if not the first, discoveries of lateral zoning in the median valley. Zonation was later found to characterize the FAMOUS and AMAR zones just south of the Azores (Bryan and others, 1979 and Stakes and others, 1984), but the compositional differences are less dramatic. The region is characterized by the Group B (see above) petrographic province of Sigurdsson (1981) and many of the characteristics of this group are based on his glass analyses from this region.

The SIVGP and URI glass analyses data sets used here show peaks in K_2O abundances in the region from 41° to 46°N (Fig. 1). The existence of N-type basalts along the axis of the MAR just west of the Azores is evidenced by low K_2O basalt around 40°N (note A40.55N1, Table 1). Just to the north and on up to 46°N, K_2O reaches over 0.5 percent (e.g. 45.62N1, Table 1).

Elevations from 40° to 46°N remain high along the axial zone moving north from the Azores, and there are no major fracture zones. The simplest interpretation for the abundance of the E-type basalts is thus northward flow from the Azores thermal plume. However, the lack, so far, of E-type basalt just north of the Azores argues against this model. Instead, we may be seeing a non-localized along-axis rise of plume-type and depleted-type source materials. As in the case of Iceland, both E- and N-type basalts occur.

(8) Reykjanes Ridge

This section of the MAR is bounded on the south by the Charlie Gibbs Fracture Zone and on the north by Iceland. The Reykjanes Ridge has an axial valley in the south and to the north the valley tapers off to an axial high. The Reykjanes Ridge is dominated by linear volcanic ridges trending oblique to the axial zone (Searle and Laughton, 1981).

The segment of the North Atlantic north of the Charlie Gibbs Fracture Zone began opening about 85 million years ago, while that to the south, between Iberia and the Grand Banks, began opening about 40 million years before that (Srivastava, this volume). The Charlie Gibbs Fracture Zone was in existence by 85 Ma and has remained a major transform fault as well as a petrologic boundary.

Campsie and others (1973) noted that the Reykjanes Ridge is characterized by low Al_2O_3 and Na_2O and higher FeO* relative to the rest of the MAR. Schilling (1973) pointed out a trend of decreasing K_2O, TiO_2 and P_2O_5 away from Iceland. Schilling invoked magma mixing of a normal MORB (N-type) reservoir with Icelandic mantle plume magmas (E-type) to account for these trends. Our glass data show the same sorts of trends documented by Campsie and others (1973) and Schilling (1973). However, data from Leg 49, Site 409, shows as much variation in K_2O, P_2O_5 and TiO_2 as the entire Reykjanes Ridge (Wood and others, 1979). These findings indicate a heterogenous mantle source over space and time for this segment of the North Atlantic.

Remarkably, the most extreme titanium ferrobasalt (FETI basalt) yet recovered from the world-ocean spreading centers is from the Reykjanes Ridge. This sample is one of 64 basaltic glass samples analyzed from the same dredge haul just south of Iceland (see groups A61.99N, Table 1):

	Reykjanes Ridge	Galapagos Rift	Juan de Fuca Ridge-Blanco Trough Intersection
Ocean	A	P	P
Lat	61.99N2	0.41	44.44N
Lon	26.59W	85.23W	130.30W
SiO_2	45.30	51.02	50.84
Al_2O_3	11.55	11.46	11.46
FeO*	17.37	17.75	15.17
MgO	4.58	3.48	5.00
CaO	10.40	8.68	10.15
Na_2O	2.85	2.80	2.72
K_2O	0.62	0.22	0.21
TiO_2	5.43	3.70	2.87
P_2O_5	0.42	0.46	0.26

Note that among these three extreme FETI basalts, the Reykjanes Ridge glass has by far the highest K_2O content. This may reflect extreme crustal fractionation from a parental magma already high in K_2O and derived from the Iceland thermal plume.

(9) Iceland to 71°N: Kolbeinsey Ridge

Basalts from the median zone of this region have the lowest average K_2O of any in the data sets used here (Fig. 2). Their composition appears to rule out any northward flow of E-type magmas or deeper plume-type peridotite from the Iceland thermal plume. Additional basalt analyses from the Kolbeinsey Ridge show similarly low K_2O contents (Dittmer, 1975).

(10) 71° to 75°N: Mohns Ridge

The few analyses available from this region are from the URI data set (Schilling and others, 1983 and Sigurdsson, 1981). From 71° to 72°N, near Eggvin Bank, E-type basalt occurs, presumably related to the proposed Jan Mayen thermal plume. North of 72°N, the compositions are again N-type basalt, and Sigurdsson (1981) groups them with his Group B samples.

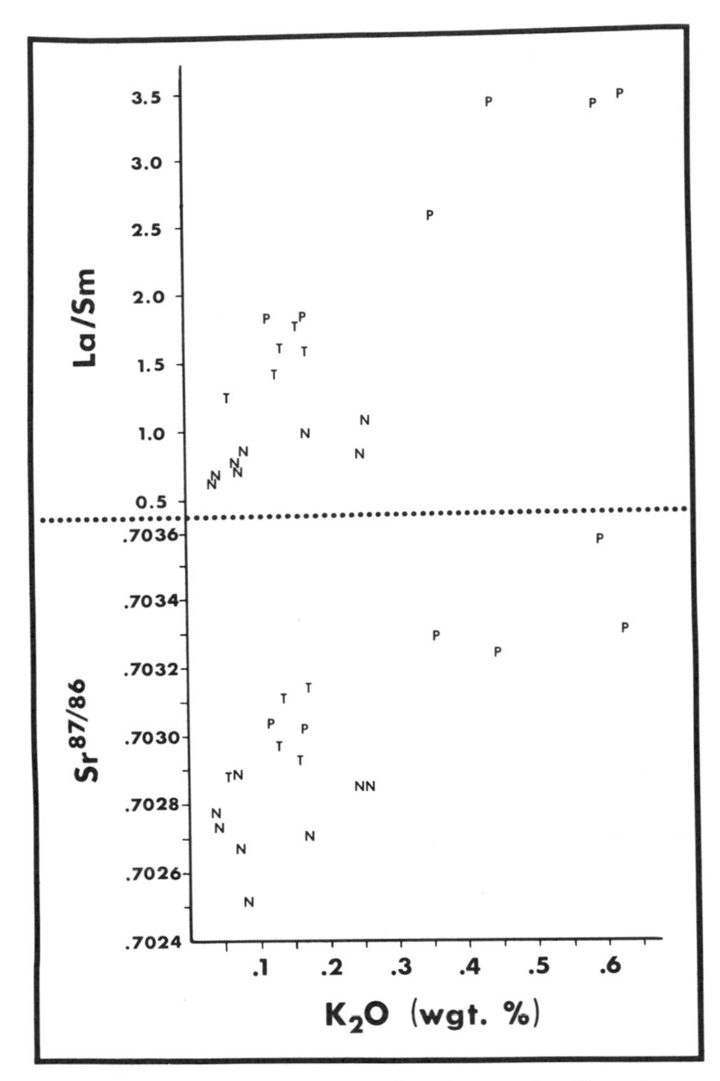

Figure 2. Positive correlation between K and average Lm/Sm and average Sr^{87}/Sr^{86} for plume (P), transitional (T) and normal (N) abyssal basalt along the Mid-Atlantic Ridge. Data from Schilling and others (1983).

PETROLOGIC MODELS OF THE MID-ATLANTIC RIDGE

The above summary outlines ten compositional-bathymetric groups along the MAR axial zone. What processes account for these groups? Numerous models have been developed for the processes involved in the formation of new lithosphere and crust along accreting plate margins. The compositional controls on the MAR abyssal volcanic rocks involve one or more of the following processes:

(1) Subhorizontal flow beneath the ridge axis of magma or partially melted mantle peridotite below the crust, sometimes outward from thermal plumes (e.g. northern Reykjanes Ridge; southward and northward from the Azores). Compositional breaks at some fracture zones (e.g. the Charlie Gibbs Fracture Zone at 55°N and the Hayes Fracture Zone at 34°N) would reflect damming of this subhorizontal flow.

(2) The stage of evolution of the magma chamber (steady state, nearly solidified, and so on) and the extent of plagioclase, olivine, and clinopyroxene fractionation.

(3) High pressure fractionation processes during or after separation of liquids from the mantle.

(4) Parent mantle composition, extent of partial melting, and pressure and temperature regime of the partial melting.

All of these processes have been documented as affecting the compositions of MAR abyssal volcanic rocks but it has not always been possible to produce models that separate these effects from one another (Bryan and others, 1981). Stakes and others (1984) have provided a model of the stages involved in an ephemeral magma chamber which, in "mid-life" is a steady-state magma chamber involving an approximate balance between the volume of magma influx and outflow based on data around 37°N. Ranging from magmas derived directly from the mantle to those from magma chambers in advance stages of solidification, erupted lavas become increasingly fractionated. These stages in magma chamber evolution must be imposed on "primitive" magmas with a certain initial range in major and minor element composition and a larger range in trace element abundances and in radiogenic isotopic ratios.

The major and minor element data sets used in this review include analyses that can be placed into the various stages in magma chamber evolution defined by Stakes and others (1984). Table 2 presents a sampling of such compositions into such a model. The first stage is envisioned as involving the most primitive, possibly totally undifferentiated mantle liquids. These are postulated to have the highest MgO contents. Olivine settling or other olivine-phenocryst and microlite enrichment mechanisms complicate interpretations of bulk rock analyses as having ever represented liquids. Porphyritic rocks from the Mid-Atlantic Ridge can have quite high MgO contents. The two highest in the URI bulk rock data have FeO* and MgO contents of 10.07 and 12.05 at 33.37°N 39.08°W, and 8.23 and 16.00 at 45.18°N and 27.90°W, respectively. The highest MgO recorded for any North Atlantic "zero-age" basaltic glasses is 9.86 with FeO* = 8.96 at 36.81°N and 33.26°W, within the FAMOUS area. O'Hara (1982) argues that even these most basic liquids have undergone extensive fractionation, involving loss of much olivine, prior to eruption.

The extreme FETI basalt (titanian ferrobasalt) from the Reykjanes Ridge (Table 2) may represent extreme fractionation from a nearly solidified magma chamber. Eruptions of similar FETI basalt has been attributed in the case of the Juan de Fuca and Galapagos spreading centers to propagating rifts (e.g. Sinton and others, 1983). They interpret such extreme fractionation products as closed-system fractionation of the newly formed magma chambers just "upstream" from the propagating rift.

Experimental phase equilibria studies are a critical part of interpreting the petrogenesis of MORB. Considerable data of this nature has been gathered, some of it specifically from samples taken along the northern MAR. The generation of MORB by partial melting and subsequent evolution by fractional crystalliza-

TABLE 2. SELECTED MAJOR AND MINOR ELEMENT ANALYSES FROM THE SMITHSONIAN VOLCANIC GLASS DATA SET PLACED IN THE CONTEXT OF THE STAGES OF EVOLUTION OF AN EPHERMIRAL MAGMA CHAMBER MODEL AS OUTLINED BY STAKES AND OTHERS (1984)

SiO_2	Al_2O_3	FeO*	MgO	CaO	K_2O	Na_2O	TiO_2	P_2O_5	Lat	Long	Ref
(1) Melts derived directly from the mantle are erupted on the sea floor.											
49.52	16.04	8.96	9.86	12.39	0.17	2.07	0.85	0.09	36.81N	33.26W	1
(2) Growth of new magma chamber, minor fractionation; rapid replenishment. Such magmas would not be particularly distinctive and therefore are not included.											
(3) Steady-state magma chamber.											
49.67	14.80	10.40	7.94	11.64	0.24	2.36	1.43	0.09	36.82N	33.24W	1
(4) Early stage of magma chamber solidification.											
45.30	11.55	17.37	4.58	10.40	0.62	2.85	5.43	0.42	61.99N	26.59W	1
(5) Advanced stage of magma chamber solidification. Intermediate and acidic rocks. No glasses noted from the MAR from this stage. Diorites from 45°N (Aumento, 1969) and aplite from the Kane Fracture zone (Miyashiro and others, 1970) may be the only representatives so far found that are derived from this stage.											

tion and, in some cases, by magma mixing has been a specific focus of much of this work. A central question centers on the extent to which the most magnesian MORBs are direct products of partial fusion of the upper mantle. There is no disagreement that most MORBs show some evidence of modification by crystal fractionation. Walker and others (1979) have developed cogent arguments for the important role of magma mixing in the generation of some MORBs from the Oceanographer Fracture Zone. Rhodes and Dungan (1979) and Rhodes and others (1979) also point out evidence of magma mixing from melt inclusions in phenocrysts and bulk compositional data.

The proposed partial fusion models fall most simply into two types (Fugii and Bougault, 1983). The first model involves generation of picritic liquids reaching the seafloor only after extensive fractionation (e.g., O'Hara, 1982). Such liquids were envisioned as forming by partial fusion at 25–35 kbar. Elthon and Scarfe (1980) also propose a primary MORB picritic magma, forming at 25 kbar by partial melting of a garnet peridotite. Stolper (1980) also proposes picritic liquids as parents to MORBs, with segregation from parent mantle at 15–20 kbar based on experimental equilibration between harzburgite and basaltic liquid. Elthon (1979) also proposes the existence of extreme picritic magmas in producing ophiolitic assemblages of probable mid-ocean ridge type.

The second group of partial melting models for the origins of MORB envisions the most magnesian MORBs as direct partial melts of the upper mantle (e.g. Kushiro, 1973; Presnall and others, 1979; Fugii and Bougault, 1983). These models point to partial melting at an invariant point involving liquid + spinel + olivine + orthopyroxene + clinopyroxene + plagioclase, that is, at the invariant point along the boundary between plagioclase and spinel peridotites. This is at about 10 kbar and 1300°C. (Presnall, 1979; Fugii and Bougault, 1983). O'Hara (1982) argued that no MORBs defined at that time had orthopyroxene as a stable phase at any pressure, and thus that none were primary liquids. Fugii and Bougault (1983) found that orthopyroxene became a stable liquidus phase in a magnesian basalt from the FAMOUS area above around 10 kbar. Thus, they conclude that such magnesian MORBs may well be primary liquids. Nonetheless, such magnesian MORBs are quite rare.

We conclude that magnesian MORBs may in some cases be primary mantle liquids and that the existence of picritic primary parental liquids for some MORBs is likely based on the abundance of olivine-rich cumulates in ophiolitic analogs of modern seafloor spreading centers. Over 6000 analyses from the world seafloor spreading centers include no picritic liquids. Among these analyses, the maximum MgO content remains slightly less than 10 weight percent, with MgO contents of greater than 9.5 percent composing but 1 percent of the 963 chemical groups now represented in the SIVGP data set. Presumably the scarcity of high-MgO liquids reflects extensive crustal and possibly upper mantle fractionation producing olivine-rich cumulates.

Walker and others (1979) give compelling data and arguments for the important role of magma mixing in generating many of the basalts from the Oceanographer Fracture Zone. They envision repeated episodes of mixing of newly injected primitive magmas with magmas previously evolved by crystal fractionation. The arguments center on the observation that somewhat evolved, clinopyroxene-component-rich magmas could not have reached their compositions by fractionation of a more magnesian parent through any feasible pressure or temperature change scenario. This view of an important role for magma mixing seems in accord with the views of Stakes and others (1984) and Bryan and others (1979) for the existence of magma chambers, which, although not necessarily long-lived, typically involve periodic recharging with primitive liquids.

The total suite of SIVGP glasses of "zero-age" fall well within the previously defined field for MORBs in the diopside-olivine-silica ternary (Walker and others, 1979). The basaltic glasses from around 37°N show a large grouping of compositions in the diopside field, that is, they are perched relative to the one atmosphere liquid line of descent of this ternary (Fig. 3b), and

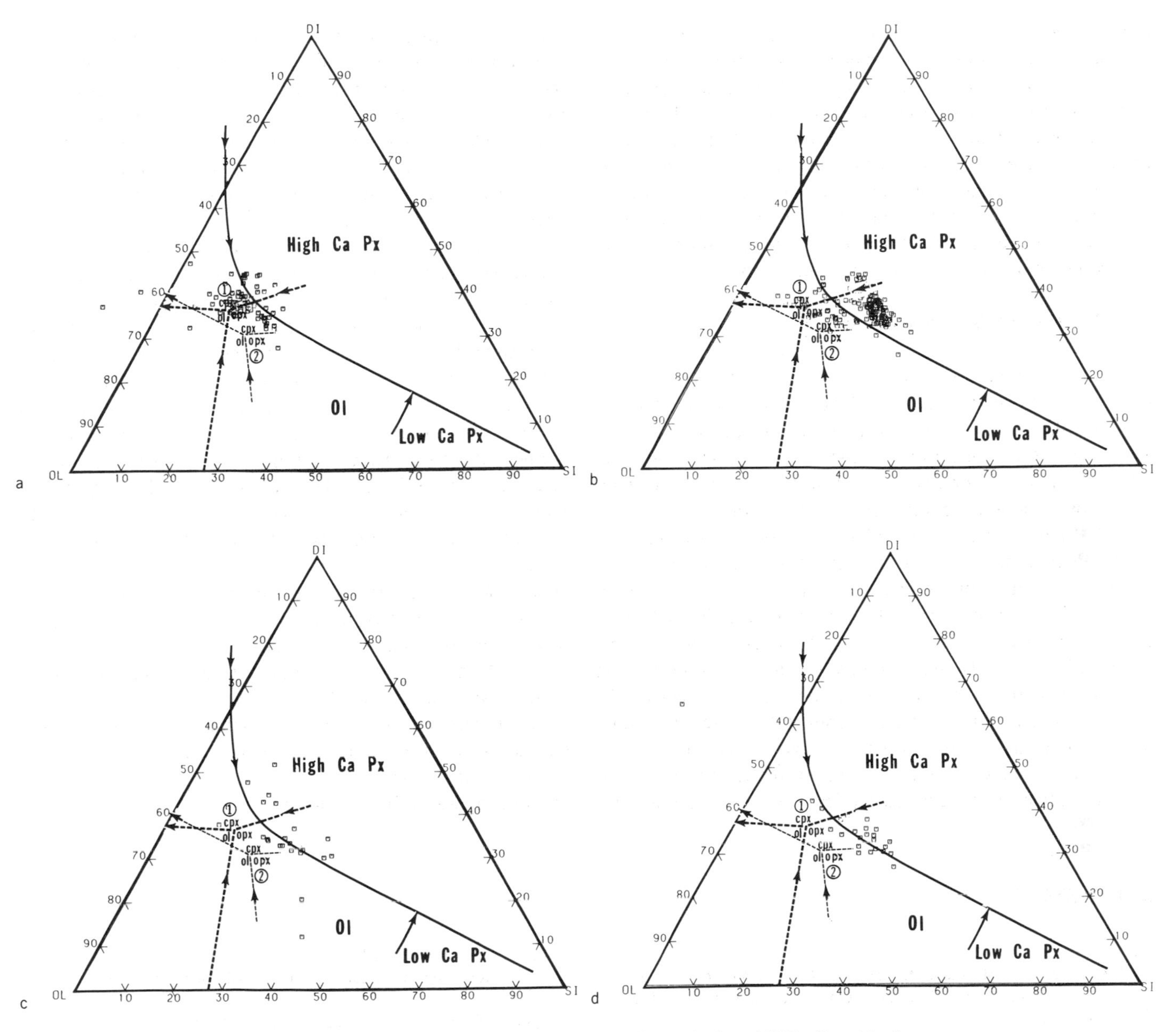

Figure 3. Projections into Diopside-Olivine-Silica (see Walker and others 1979) of basaltic glass groups from (a) 14°–34°N, (b) 34°N–Azores, (c) Azores to Charlie Gibbs Fracture Zone and (d) the Reykjanes Ridge. Ternary "eutectics" at 10 Kilobars are from Fugii and Bougault (1983), labeled 1, and from Stolper (1980), labeled 2.

may well be a result of the kind of magma mixing invoked by Walker (1979). Basaltic glasses from 15°N to the Hayes Fracture Zone, on the other hand, have very few compositions that fall in that field (Fig. 3a) and presumably in this deeper axial zone have undergone rare magma mixing of the Walker type. The 15°N–34°N glasses tend to define a trend which falls near the diopside-olivine cotectic at 10 kbar (Fugii and Bougault, 1983) in a Walker-type diagram (Fig. 3b).

The glass samples from the Azores north to the Charlie Gibbs Fracture Zone resemble the 34°–39°N group in diopside-olivine-silica space, with a large number of "perched" compositions probably reflecting magma mixing (Fig. 3c). The highly evolved basaltic glasses that characterize the Reykjanes Ridge define a trend that falls very close to the nearly linear, more evolved end of the liquid line of descent. Mixing among the various members in this group would produce compositions therefore that would also fall on the liquid line of descent (Fig. 3d). In interpretations like the above, it must be kept in mind that the Walker-type diagram is quite sensitive to silica concentrations and any attendant uncertainties and also does not reflect faithfully Mg/Fe variations.

In summary, most of the northern MAR basalts show evidence of extensive crustal magma chamber fractionation, the Reykjanes Ridge samples in particular. Also, there appears to be a tendency for the shallowest group of samples, the Reykjanes Ridge samples, to show the closest proximity to evolved compositions along the liquid line of descent in Walker-type ternary projections (Di-O1-Si). Samples from deeper zones depart from this liquid line of descent, and commonly show evidence of magma mixing, and, in the high-MgO members, probable direct magma derivation. Depth of final mantle equilibration is probably around 30 km (10 kbar) for most such primitive liquids.

SUMMARY AND CONCLUSIONS

(1) The dominantly basaltic extrusives along the axial zone of the Mid-Atlantic Ridge can be divided into ten compositional-water depth zones. Such classifications remain in a state of flux, as many regions of the northern MAR are still poorly sampled. Also, major and minor element classifications differ in the literature, depending on the parameters used as discriminants.

(2) Shallow axial zones tend to have more evolved compositions and are typically enriched (E-type) or transitional (T-type) basalts. The notion of thermal plume models, based largely on trace and isotopic analyses, finds support in the major and minor element compositions. Imposed on such probable thermal plume compositional effects are imprints of the stages of magma chamber evolution at the time of eruption. The tendency for the basalt erupted at shallower axial zones to be more highly fractionated than those from deeper axial zones probably reflects longer residence times in one or more magma chambers of originally primitive mantle-derived liquids. For axial-zone or near-axial-zone volcanic islands, compositional diversity is much greater than that found in submarine extensions. This reflects (a) the ease of spotting and sampling such diversity on land compared to beneath the sea, (b) longer paths and hence longer transit times and longer pre-eruption cooling histories enhancing opportunities for fractionation, (c) a possible greater diversity of primary mantle compositions, and (d) more opportunity for interactions with wall rocks.

(3) Good correlations between K and trace element and radiogenic isotopic ratios in basalts allow determinations of K alone as a pointer to trace and radiogenic isotopic values.

(4) High K_2O, E-type basalt, mostly connected with axial topographic highs, occurs near (a) St. Paul's Rocks (Sierra Leone "thermal plume"), (b) 15°N (unnamed "thermal plume"), (c) 35°N, (d) the Azores (Azores "thermal plume"), (e) erratically between the Azores and Charlie Gibbs Fracture Zone, (f) the northern Reykjanes Ridge, (Iceland "thermal plume"), and (g) along the Mohns Ridge (Jan Mayen "thermal plume"). The probable "plume" at 15°N is just now coming to light through the work of a number of researchers. Although this paper clearly has a bias toward thermal plume models to explain E-type compositional distributions, other models involving extensive wall-rock assimilation, particularly of extensively hydrothermally altered magma chamber roof rocks, may also be valid.

(5) Magma mixing is a common feature of northern MAR basalts, particularly from the 37°N region. This is probably a result of renewed injections of primitive liquids into evolved magma chambers. The length of time such magma chambers exist probably differs from region to region. For 37°N, such magma chambers appear to be ephemeral.

REFERENCES

Aumento, F.
1968 : The Mid-Atlantic Ridge near 45°N. II. Basalts from the area of Confederation Peak: Canadian Journal of Earth Sciences, v. 5, no. 1, p. 1–21.
1969 : Diorites from the Mid-Atlantic Ridge at 45°N: Science, v. 165, p. 1112–1113.

Bougault, H., and Treuil, M.
1980 : Mid-Atlantic Ridge: Zero-age geochemical variations between Azores and 22°N: Nature, v. 286, p. 209–212.

Bryan, W. B., and Dick, H.J.B.
1982 : Contrasted abyssal liquidus trends: Evidence for mantle major element heterogeneity: Earth and Planetary Science Letters, v. 58, p. 15–26.

Bryan, W. B., Thompson, G., and Ludden, J. N.
1981 : Compositional variation in normal MORB from 22°–25°N: Mid-Atlantic Ridge and Kane Fracture Zone: Journal of Geophysical Research, v. 86, p. 11815–11836.

Bryan, W. B., Thompson, G., and Michael, P. J.
1979 : Compositional variation in a steady-state zoned magma chamber: Mid-Atlantic Ridge at 36°50′N: Tectonophysics, v. 55, p. 63–85.

Byerly, G.
1980 : The nature of differentiation trends in some volcanic rocks from the Galapagos spreading center: Journal of Geophysical Research, v. 85, p. 3797–3810.

Burke, K., Kidd, W.S.F., and Wilson, J. T.
1973 : Relative and latitudinal motion of Atlantic hotspots: Nature, v. 245, p. 133–137.

Campsie, J., Bailey, J. C., Rasmussen, M., and Dittmer, F.
1973 : Chemistry of tholeiites from the Reykjanes Ridge and Charlie Gibbs Fracture Zone: Nature Physical Sciences, v. 244, p. 71–73.

Cann, J. R.
1970 : New model for the structure of ocean crust: Nature, v. 226, p. 928–930.
1974 : A model for the oceanic crustal structure developed: Geophysical Journal of the Royal Astronomical Society, v. 39, p. 169–197.

Dittmer, F., Fine, S., Rasmussen, M., Bailey, J. C., and Campsie, J.
1975 : Dredged basalts from the mid-oceanic ridge north of Iceland: Nature, v. 254, p. 298–301.

Dmitriev, L. V., Sobolev, A. V., Uchanov, A. V., Malysheva, T. V., and Melson, W. G.
1984 : Primary differences in oxygen fugacity and depth of melting in the mantle source regions for oceanic basalts: Earth and Planetary Science Letters, v. 70, p. 303–310.

Elthon, D.
1979 : High magnesia basalts as the parental magma for ocean floor basalts: Nature, v. 278, p. 514–518.
1981 : Metamorphism in oceanic spreading centers; in The Oceanic Litho-

sphere, in The Sea, Volume 7, ed. C. Emiliani; New York: Wiley-Interscience, p. 285–304.

Elthon, D., and Scarfe, C. M.
1980 : High-pressure phase equilibria of a high-magnesia basalt: Implications for the origin of mid-ocean ridge basalts: Carnegie Institution of Washington Yearbook 79, p. 277–281.

Engel, A.E.J., Engel, C., and Havens, R. G.
1965 : Chemical characteristics of oceanic basalts and the upper mantle: Geological Society of America Bulletin, v. 76, p. 719–734.

Flower, M.F.J.
1982 : Cryptocumulate tholeiite magma as evidence for magma-mixing at an intermediate rate spreading axis: Nature, v. 299, p. 542.

Fox, P. J., and Stroup, J. B.
1981 : The plutonic foundation of the oceanic crust; in The Oceanic Lithosphere, in The Sea, Volume 7, ed. C. Emiliani; New York: Wiley-Interscience, p. 119–218.

Frey, F. A.
1970 : Rare earth and potassium abundances in St. Paul's Rocks: Earth and Planetary Science Letters, v. 7, p. 351–360.

Fugii, T., and Bougault, H.
1983 : Melting relations of a magnesian abyssal tholeiite and the origin of MORBs: Earth and Planetary Science Letters, v. 62, p. 283–295.

Hart, S. R.
1971 : K, Rb, Cs, Sr, and Ba and Sr isotope ratios of ocean-floor basalts: Philosophical Transactions of the Royal Society of London, v. A268, p. 573–588.

Hart, S. R., Schilling, J.-G., and Powell, J. L.
1973 : Basalts from Iceland and along the Reykjanes Ridge: Sr-isotope geochemistry: Nature Physical Sciences, v. 268, p. 707–725.

Hofmann, A. W., and White, W. M.
1982 : Mantle plumes from ancient oceanic crust: Earth and Planetary Science Letters, v. 57, p. 421–436.

Kushiro, I.
1973 : Origin of some magmas in oceanic and circumoceanic regions: Tectonophysics, v. 17, p. 211–222.

Le Douran, S., Needham, H. D., and Francheteau, J.
1982 : Pattern of opening rates along the axis of the Mid-Atlantic Ridge: Nature, v. 300, p. 254–257.

Melson, W. G., Hart, S. R., and Thompson, G.
1972 : St. Paul's Rocks, equatorial Atlantic: Petrogenesis, radiometric ages, and implications on sea-floor spreading; in Studies in Earth and Space Sciences, ed. R. Shagam: Geological Society of America Memoir, no. 132, p. 241–272.

Melson, W. G., and O'Hearn, T.
1979 : Basaltic glass erupted along the Mid-Atlantic Ridge between 0–37°N: relationships between composition and latitude; in Deep Drilling Results in the Atlantic: Ocean Crust, ed. M. Talwani, C. G. Harrison, and D. E. Hayes; American Geophysical Union, Washington, D.C., p. 249–261.

Minster, J. B., and Jordan, T. H.
1978 : Present-day plate motions: Journal of Geophysical Research, v. 83, p. 5331–5354.

Miyashiro, A., Shido, F., and Ewing, M.
1970 : Crystallization and differentiation in abyssal tholeiites and gabbros from the mid-oceanic ridges: Earth and Planetary Science Letters, v. 7, p. 361–365.

Morel, J. M., and Hekinian, R.
1980 : Compositional variations of volcanics along recent spreading ridges: Contributions to Mineralogy and Petrology, v. 72, p. 425–436.

Morgan, W. J.
1973 : Plate motions and deep mantle convection; in Studies in Earth and Space Sciences, ed. R. Shagam; Geological Society of America Memoir, no. 132, p. 7–22.
1981 : Hotspot traces and the opening of the Atlantic and Indian Oceans; in The Oceanic Lithosphere, The Sea, Volume 7, ed. C. Emiliani, New York, Wiley-Interscience, p. 442–488.

Nicholls, G. D., Nalwalk, A. J., and Hayes, E. E.
1964 : The nature and composition of rock samples dredged from the Mid-Atlantic Ridge between 22°N and 52°N: Marine Geology, v. 1, p. 333–343.

O'Hara, M. J.
1982 : MORB—a mohole misbegotten?: Transactions of the American Geophysical Union, EOS, June 15, p. 537–538.

Pallister, J. S., and Hopson, C. A.
1981 : Samail ophiolite plutonic suite: field relations, phase variation, cryptic variation and layering, and a model of a spreading ridge magma chamber: Journal of Geophysical Research, v. 86, p. 2593–2645.

Peter, G. P., and Westbrook, G. K.
1976 : Tectonics of Southwestern North Atlantic and Barbados Ridge Complex: American Association of Petroleum Geologists Bulletin, v. 60, p. 1078–1106.

Presnall, D. C., Dixon, J. R., O'Donnell, T. H., and Dixon, S. A.
1979 : Generation of mid-ocean ridge tholeiites: Journal of Petrology, v. 20, p. 3–35.

Rhodes, J. M., and Dungan, M. A.
1979 : The evolution of ocean-floor basalts: in Deep Drilling Results in the Atlantic Ocean: Ocean Crust, ed. M. Talwani, C. G. Harrison, and D. E. Hayes; American Geophysical Union, Washington, D.C., p. 262–272.

Rhodes, J. M., Dungan, M. A., Blanchard, D. P., and Long, P. E.
1979 : Magma mixing at mid-ocean ridges: Evidence from basalt drilled near 22°N on the Mid-Atlantic Ridge: Tectonophysics, v. 55, p. 35–61.

Schilling, J.-G.
1971 : Sea-floor evolution: Rare-earth evidence: Philosophical Transactions of the Royal Society of London, v. 268, p. 663–706.
1973 : Iceland mantle plume: Geochemical study of Reykjanes Ridge: Nature, v. 242, p. 565–571.
1975 : Azores mantle blob: Rare earth evidence: Earth and Planetary Science Letters, v. 25, p. 103–115.

Schilling, J.-G., Zajac, M., Evans, R., Johnston, T., White, W., Devine, J. D., and Kingsley, R.
1983 : Petrologic and geochemical variations along the Mid-Atlantic Ridge from 29°N to 73°N: American Journal of Science, v. 283, p. 510–586.

Searle, R. C., and Laughton, A. S.
1981 : Fine-scale sonar study of the tectonics and volcanism on the Reykjanes Ridge: Oceanologica Acta, v. 4 (supplement), p. 5–13.

Shibata, T., Thompson, G., and Frey, F. A.
1979 : Tholeiitic and alkali basalts from the Mid-Atlantic Ridge at 43°N: Contributions to Mineralogy and Petrology, v. 70, p. 127–141.

Shido, F., and Miyashiro, A.
1973 : Compositional difference between abyssal tholeiites from north and south of the Azores on the Mid-Atlantic Ridge: Nature, v. 143, p. 59–60.

Sigurdsson, H.
1981 : First-order major element variation in basalt glasses from the Mid-Atlantic Ridge: 29°N to 73°N: Journal of Geophysical Research, v. 86, p. 9483–9502.

Sinton, J. M.
1979 : Ultramafic xenoliths and high-pressure xenocrysts in submarine basanitoid, Equatorial Mid-Atlantic Ridge: Contributions to Mineralogy and Petrology, v. 70, p. 49–57.

Sinton, J. M., Wilson, D. S., Christie, D. M., Hey, R. N., and Delaney, J. R.
1983 : Petrologic consequences of rift propagation on oceanic spreading ridges: Earth and Planetary Science Letters, v. 62, p. 193–207.

Stakes, D., Shervais, J. W., and Hopson, C. A.
1984 : The volcano-tectonic cycle of the FAMOUS and AMAR valleys, Mid-Atlantic Ridge (36°47′N): Evidence from basalt glass and phenocryst compositional variations for a steady-state magma chamber beneath valley midsections: Journal of Geophysical Research, v. 89, p. 6995–7028.

Stolper, E.

1980: A phase diagram for mid-ocean ridge basalts: Preliminary results and implications for petrogenesis: Contributions to Mineralogy and Petrology, v. 74, p. 13–27.

Tarney, J., Wood, D. A., Saunders, A. D., Cann, J. R., and Varet, J.
1980 : Nature of mantle heterogeneity in the North Atlantic: Evidence from Deep Sea Drilling: Philosophical Transactions of the Royal Society of London, v. A297, p. 179–202.

Thompson, G., and Melson, W. G.
1972 : Petrology of oceanic crust across fracture zones: Evidence for a new kind of sea-floor spreading: Journal of Geology, v. 80, p. 526–538.

Vogt, P. R.
1971 : Asthenosphere motion recorded by the ocean floor south of Iceland: Earth and Planetary Science Letters, v. 13, p. 153–160.

Vogt, P. R., and Johnson, L.
1975 : Transform faults and longitudinal flow beneath the midoceanic ridge: Journal of Geophysical Research, v. 80, p. 1399–1428.

Wasserberg, G. J., and DePaolo, D. J.
1979 : Models of earth structure inferred from neodymium and strontium isotopic abundances: Proceedings of the National Academy of Sciences, v. 76, p. 3594–3598.

Walker, D., Shibata, T., and DeLong, S. E.
1979 : Abyssal tholeiites from the Oceanographer Fracture Zone: Contributions to Mineralogy and Petrology, v. 70, p. 111–125.

Wood, D. A., Varet, J., Bougault, H., Corre, O., Joron, J. L., Treuil, M., Bizouard, H., Norry, M. J., Hawkesworth, C. J., and Roddick, C. J.
1979 : The petrology, geochemistry, and mineralogy of North Atlantic basalts: A discussion based on IPOD Leg 49; in Initial Reports of the Deep Sea Drilling Project, v. 49, ed. B. P. Luyendyk, J. R. Cann, and others; Washington, D.C., U.S. Government Printing Office, p. 595–655.

Manuscript Accepted by the Society September 5, 1984

ACKNOWLEDGMENTS

We gratefully acknowledge the donation of samples from the following institutions and individuals (codes are from Table 1): A—F. Aumento; C—J. R. Cann; H—R. Hekinian and H. Bougault; L—Lamont-Doherty Geological Observatory; N—NOAA, through P. Rona; O—U.S. Naval Oceanographic Office, through P. Vogt and G. L. Johnson; S—A. Sharaskin; W—Woods Hole Oceanographic Institution, through W. B. Bryan, V. Bowen, and G. Thompson; 1—DSDP; and D. Stakes. At the U.S. National Museum, E. Jarosewich, J. Nelen, C. Obermeyer, and J. Collins assisted with the electron microprobe analyses and wet chemical analyses of standards. All this assistance is gratefully acknowledged. R. Johnson and F. Walkup prepared the polished grain mounts. Finally, we gratefully acknowledge the thoughtful reviews provided by C. A. Hopson, D. Clague, S. Jakobsson, and P. Vogt on an earlier version of this paper.

Printed in U.S.A.

The Geology of North America
Vol. M, The Western North Atlantic Region
The Geological Society of America, 1986

Chapter 9

Geochemical and isotopic variation along the Mid-Atlantic Ridge axis from 79°N to 0°N

Jean-Guy Schilling
Graduate School of Oceanography, University of Rhode Island, Kingston, Rhode Island 02881

INTRODUCTION

This chapter describes our current knowledge of the isotopic and trace element content of lavas erupted along the present axis of the Mid-Atlantic Ridge (MAR). We will emphasize how their geographical distribution relates to changes in the morphology, tectonics and elevation of the ridge crest, and spreading rates. This approach provides constraints on processes of formation of ocean crust by seafloor spreading and on mantle processes including the scale, nature and dynamics of mantle heterogeneities.

SOME HISTORICAL PERSPECTIVES

The first hint of the existence of the Mid-Atlantic Ridge occurred in the period of 1849–1853 after soundings were made by US Navy Ships under the direction of USN Lieut. M. F. Maury. Congress had authorized the effort to assist Maury in his study of winds and currents and apparently also in the preparation of laying the first Trans-Atlantic cable. It was first called the Telegraphic Plateau. Murray and Peake (1904) gave a detailed account and colorful anecdotes surrounding the initial discovery of the Mid-Atlantic Ridge, and how gradually its topographic definition and latitudinal extent were made known in the 1850 to 1900 period. In the summary volumes of the "*Challenger* Reports," the MAR is shown to be continuous from 50°N southward throughout the South Atlantic. The widest part of the Mid-Atlantic Ridge, between 30°N and 50°N, was referred to as the Dolphin Plateau. The narrow portion south of 30°N was named the Connecting Plateau, as it was thought to connect the Dolphin Plateau to the Challenger Plateau running down the center of the South Atlantic; but Murray and Peake (1904) pointed out that this connection is broken for a short interval at the Equator, because of deep soundings found there (evidently a first hint of the Equatorial Fracture Zones). The name of the Mid-Atlantic Ridge seems to first appear in Murray and Hjort's (1912) classic book, *The Depths of the Ocean,* and was probably coined in the 1904–1912 period.

Only in the 1950s was it discovered that the crest of the MAR was rifted along a great part of its length (e.g., Heezen, 1960), that it coincided with a relatively narrow zone of earthquake epicenters (Rothé, 1954), and it was also characterized by high heat-flow (e.g., Bullard and others, 1956).

Realization that the crest of the MAR is volcanic in origin came even more slowly. For example Fisher (1881 and 1889) points out that the MAR (midocean platform), from which volcanic islands rise, cannot be an anticline resulting from folding of the continental crust, apparently the prevailing view at the time. He speculated that it may be partly volcanic from emplacement and congelation (accretion) of basaltic magma rising into fissures opened from below. His book, entitled *Physics of the Earth's Crust,* is an entire thesis set up to demonstrate quantitatively that the earth's crust must be underlain by a thin layer of basaltic liquid, or at least plastic material of basic composition, capable of flowing and laterally transporting the crust above and generating mountain ranges by compression. He showed from consideration on density and isostasy that the oceanic crust must be denser and thinner than the continental crust (2.955 vs. 2.68 gm/cm^3 and 20 km rather than 25 km thick respectively!). Fisher's model was soon forgotten as a result of objections raised by the influential W. Thomson (later Lord Kelvin) who believed that the Earth's interior was as rigid as steel. Even in Holmes' (1928) model of continental drift, the Mid-Ocean Ridge is considered to be a sialic splinter of the original continents (old sial overlying the sima) left behind over the stagnant zone of rising currents of a convection cell. This idea still prevailed in Daly's 1942 book, *The Floor of the Ocean—New Light on Old Mysteries,* though Daly noted that the continental splinter may not be continuous along the 7000-km-long MAR. These ideas were influenced by occurrences of sialic inclusions in lavas from Iceland, the Azores and Ascension Island, and Wilson's (1940) evidence of intermediate surface wave velocities along a north-south path in the Atlantic relative to continental and other oceanic paths. Daly's model for the oceanic crust was essentially opposite to ours. He shows a thicker oceanic than continental crust, and the gabbro layer on the top of the vitreous sima layer.

Post-war seismic work established the oceanic crust above the Moho over the deep basin to be about 5–6 km thick and

Schilling, J.-G., 1986, Geochemical and isotopic variation along the Mid-Atlantic Ridge axis from 79°N to 0°N; *in* Vogt, P. R., and Tucholke, B. E., eds., The Geology of North America, Volume M, The Western North Atlantic Region: Geological Society of America.

composed of three layers; but even Hess, one of the fathers of the concept of seafloor spreading (Hess, 1962; Dietz, 1961) hesitated for ten years before concluding that the crest of the MAR is primarily basaltic (Hess, 1965). To account for the elevation of the ridge crest he first thought that it was made of peridotite-basalt breccias (Hess, 1955), and later of a highly serpentinized outcropping zone, being impressed by the uniform thickness of Layer 3 on the flanks, the occurrence of serpentinites dredged in the Atlantis F. Z. (later described by Shand, 1949; Quon and Ehler, 1963), and the occurrence of ultramafics on St. Paul's Rocks near the Equator (first described by Darwin, 1876; Renard, 1882). Until 1965, Hess attached little importance to Layer 2 which, with the seismic resolution available, appeared irregular in thickness and velocities. He believed Layer 2 was composed mostly of sediments and some lava flows. Dietz (1961), who coined the name "sea-floor spreading," also expressed doubts and suggested that the topographic roughness of the crest of the Mid-Atlantic Ridge, although suggestive of youth, "resembles neither volcanic flows nor incipient volcanoes." But similar topography and magnetic anomaly strips over and parallel to the Carlsberg Ridge axis were later interpreted by Vine and Mathews (1963) as basic extrusives, such as volcanoes and fissure eruptions. By 1966, Vine considered the Mid-Atlantic Ridge crest to be volcanic in origin.

The first hint from direct sampling that the MAR might be volcanic occurred in 1874 when the steamship *Faraday,* while repairing a telegraph cable, brought up by accident a 21-pound piece of basalt that had been ripped from an outcrop after a 27.5 ton pull on a rope. Hall (1876), who was just returning from the 4-year-long voyage of the *Challenger,* noted that "any but a rope of marvelously perfect manufacture would have yielded." The location was approximately 200 miles east of the MAR-axis. This black vesicular basalt, partly filled with zeolites (probably derived from a seamount), was dismissed by Hall as not being in situ but rather ice-rafted in origin, although the petrologist who examined and described the sample thought otherwise.

Apparently, the very first chips of fresh basaltic glass from the Mid-Atlantic Ridge were recovered in 1898, again by accident, while a French company was repairing some submarine cables. They were recovered at 3100 meters depth with the teeth of a grapnel during several lowerings at 47°0′N and 29°40′W relative to the Paris Zero Meridian (which once corrected relative to the Greenwich Zero Meridian plots right on the crest of the MAR). A note on these glasses prepared by Termier (1899) was presented by the renowned mineralogist M. Michel-Lévy at the next spring meeting of the French Academy. The terrain was described as mountainous, with high peaks, steep slopes and deep valleys. The summits were rocky and no mud was present in the bottom of these valleys. Termier seemed convinced that the chips had been freshly broken off of outcrops on the seafloor. He reported a density of these glasses of 2.784 g/cm^3 at 20°C, and expressed regrets that the chips were too small to conduct a chemical analysis, but identified the glass to be basaltic on the basis of its density. He also noted the presence of strange aureoles (varioles), and of some gas inclusions. Termier concluded that the fact that the seafloor 500 km north of the Azores, "on a line passing through the Azores and Iceland," is composed of eruptive rocks, was interesting in itself. He thought it curious that volcanic glass would be found at such great depth and noted that to explain this anomaly, "of course, one could invoke subsidence of the Atlantic seafloor after eruption." The prevailing views in France at the time were the oceans represented depressions formed by folding and occasional foundering of the earth's crust resulting from the cooling of an igneous molten core (e.g., De Lapparent, 1900). One cannot help but detect in Termier's remark some greater vision on what this discovery meant to his scientific mind, but which was too speculative to be presented at the Academy. Termier (1899b) reported further on the mineralogic nature of the brown varioles, and concluded that these glass chips were not only a geological but also a mineralogical curiosity.

In the report of the voyage of the Danish Ingolf-expedition (1895–1896), Wandel (1899) suggested that the Reykjanes Ridge (which was discovered and named during the expedition) must be young and volcanic in origin. In support, he mentioned dredging and trawling evidence indicating an abundance of glacial erratics on both flanks of the ridge, their absence on the axis itself, and two incidental records of earthquake activities felt previously from ships crossing the region. However, Wandel thought that these observations corroborated an older idea that the region may represent the locus of a sunken volcanic island (Busse Island) previously suggested in an 1897 report of another Danish Arctic Expedition (1605–1620) published by the Hakluyt Society. Subsequently, Boeggild (1900) made a very careful study of the nature and geographical distribution of 91 sediment samples and additional dredged boulders (volcanic and glacial erratics) obtained during the same Ingolf expedition. The study included among other things the sorting of minute glasses of basaltic and acidic (obsidian) composition and determination of their vesicle content. Boeggild could not readily explain how a large amount of volcanic debris of various sizes seemed to have spread from Iceland northward (toward Jan Mayen) and southwestward (along the Reykjanes Ridge), both in a direction opposite to existing predominant oceanic currents in the two regions. It is quite clear in reading these two reports that Wandel and Boeggild, although coming close, never suspected that submarine volcanic activity may have been taking place directly on the seafloor along the ridge axis. A brief account of the subsequent evolution of these early ideas pursued by Icelandic scientists can be found in Drake and Girdler (1982).

The first chemical analysis of a MAR pillow basalt with a glassy rind was reported by Correns (1930) from a dredged haul recovered just south of the Equator during the *Meteor* expedition. He pointed out its low K_2O content and similarity with other glassy submarine basalts collected in the Pacific during the *Challenger* expedition, and suggested that this might be more characteristic of submarine eruptions than previously believed. He also provided a lengthy chemical account of alteration of the glass rind into palagonite by seawater weathering and formation of manganese crust and its possible origin.

The first extensive rock-dredging of the crest of the MAR in the North Atlantic was done in 1947 by M. Ewing with R/V *Atlantis* (Cruise 150) around 30°N. The most abundant rock types recovered were pillow basalts with glassy rims, followed by serpentinites and some gabbros (including anorthosite-rich gabbros) and consolidated sediments (Shand, 1949). An He-age from one of these pillows provided evidence of the youth of the ridge (Carr and Kulp, 1953). Another description of these rocks and another dredge haul at 43°N appeared in Quon and Ehler (1963), but one must await Muir and Tilley (1966) and later Miyashiro and others (1969) for chemical analyses and a detailed petrologic account of these rocks. Efforts to dredge the seafloor intensified in the late 1950s using improved techniques including camera and pinger attachments (Nalwalk and others, 1961). Three papers on the chemistry and petrology of the MAR basalts from 50°N, 45°N, 29°N, and 22°N appeared simultaneously in 1964 (Nicholls and others, 1964; Engel and Engel, 1964; Muir and Tilley, 1964). Both Engel and Engel and Nicholls pointed out the low K_2O of the MAR basalts at 50°N relative to island or plateau basalts. This had also been previously emphasized by Wiseman (1937) for basalts dredged on the Carlsberg Ridge in the Indian Ocean and by Engel and Engel (1961, 1963) for basalts from the Northeastern Pacific. In contrast, Muir and Tilley (1964) noted that pillow basalts from 45°N were not low in K_2O but rather transitional or slightly alkalic. They were struck by the youth of the ridge crest from K/Ar dating of the pillows and bottom photographs, and by the high heat flow (Bullard and Day, 1961). They ventured that the ridge could be the result of "very recent igneous intrusions, perhaps in the form of dyke swarms that accompany the extrusion of pillow lavas." They also predicted that "well-developed layered gabbroic intrusions" should occur at quite moderate depths below the basalts forming the crust of the rift. The presence of xenocrysts partly reabsorbed also lead them to suggest the possible presence and frequent replenishment of a magma chamber beneath the ridge axis, associated with processes of assimilation. Engel and Engel (1963) suggested that the degree of alkalinity correlates positively with elevation of the seafloor and that islands are derived from the parent tholeiitic magma by differentiation in shallow reservoirs. These three studies established clearly that the MAR axis is very young, probably mostly volcanic in origin, and that ocean island volcanism could no longer be taken as chemically representative of mid-ocean ridge volcanism. In the same period, interpretation of trace elements data was significantly enhanced by (1) the use of relative abundances rather than absolute concentrations including normalization to chondrites (Coryell and others, 1963) and (2) quantitative modeling of trace element behavior during fractional crystallization (McIntyre, 1963) and partial melting (Schilling, 1966; Schilling and Winchester, 1967). The later model was first presented by Schilling in 1964 at a conference on andesite at Scripps Institute of Oceanography.

At the same time Frey and Haskin (1964) further showed that MORB from 30°N, 32°N, and 34°N had relative rare-earth abundance patterns nearly undisturbed relative to chondrites and apparently very primitive, except perhaps for the light-REE that seemed to be progressively depleted with decreasing atomic numbers or increasing ionic size. Gast (1960 and 1965) emphasized the high K/Rb ratio of Mid-Ocean Ridge Basalts (MORB) (>1500), and its similarity with achondrites and he suggested that the mantle source of MAR basalts and achondrites may have a similar overall composition—a suggestion that would not stand the test of time. Tatsumoto and others (1965) pointed out the very low Th, U, Rb, K, $^{87}Sr/^{86}Sr$, Th/U, and Rb/K (abundance ratios of large over smaller ionic size) of such MORB relative to island basalts such as those from Hawaii. More significantly, they recognized that either the Rb/Sr ratios in MAR basalts were too low to have produced in a closed system the $^{87}Sr/^{86}Sr$ of these MORBs, or the Rb/Sr of these basalts were not representative of their source in the mantle but could have been affected by partial melting. They opted for the former and suggested that the mantle source of MORB may represent a depleted basic residual layer in the upper mantle, from which most radioactive elements (and large ions) had been previously transported to the continental crust. The depletion was dated as 1 to 1.5 Ga assuming a two-step differentiation model. At the same time Gast and others (1964) also reported very distinct Pb and Sr isotopic compositions for the South Atlantic islands, Ascension, St. Helena, and Gough, and attributed them to long-lived heterogeneities in the parent/daughter U/Pb and Rb/Sr ratios of the sources of these lavas in the mantle (ca. 1.5 Ga). These and other trace element and isotopic studies conducted in the same period firmly established that mid-ocean ridges and ocean island basalts were in fact derived from distinct mantle sources (e.g., Schilling, 1966; Bence, 1966; Schilling and Winchester, 1967).

Subsequent petrological studies aimed at testing the concept of seafloor spreading were undertaken across the MAR at 22°N (van Andel and others, 1965; Melson and van Andel, 1966; van Andel and Bowin, 1968), at 45°N (Aumento, 1967, 1968; Aumento and others, 1971) and at 60°N (Krause and Schilling, 1969; DeBoer and others, 1969). These studies revealed the volcanic nature of the ridge crest, its youth, the limited zone width of extrusion taking place along the rift, the apparent increase in age and degree of low temperature alteration the basalts were suffering while being transported to the flanks of the ridge, and the extent of symmetry of such processes about the ridge axis. An indication that hydrothermal circulation may be significant was provided by the study of greenstones found around 22°N by Melson and van Andel (1966). Increasing alkalinity with increasing distance from the ridge axis in lava extruded as seamounts across the ridgecrest at 45°N led Aumento (1967) to suggest decreasing geothermal gradients on the flanks of the MAR and derivation of basalts from increasingly greater depth. All these problems and others were fruitfully debated at a symposium on "Ocean Floor Rocks" at the Royal Society of London in the fall of 1969 (Bullard, 1971). It was clear by this time that most MORB were derived from a mantle source rather uniformly

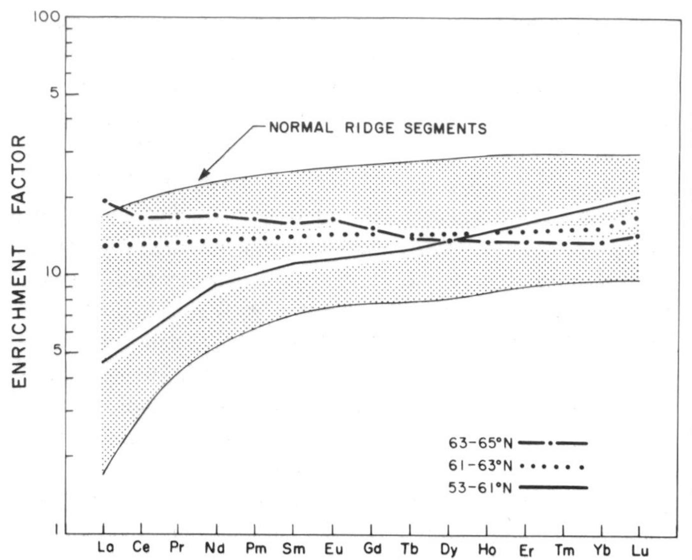

Figure 1. Variation of rare earth abundance patterns in basalts along the Reykjanes Ridge and the Southwest Rift Zone on Iceland, averaged out per 2 or more degrees of latitude. The enrichment factors are relative to chondrites. The envelopes show the range observed along normal MAR segments between ridge-centered hotspots, where basalts are derived from the LILE-depleted asthenosphere (see Figure 3 for locations of these segments).

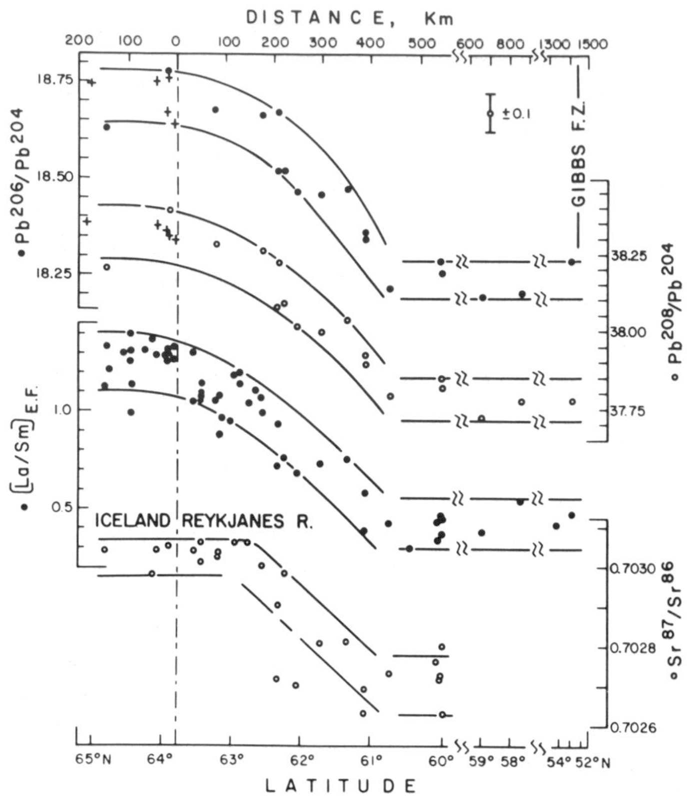

Figure 2. Variation of $^{206}Pb/^{204}Pb$, $^{208}Pb/^{204}Pb$, $^{87}Sr/^{86}Sr$ and $(La/Sm)_{E.F.}$ in basalts erupted along the Reykjanes Ridge (taken from Sun and others 1975).

depleted in large ion lithophile elements (LILE, e.g., the light REE) and world-wide in extent (e.g., Frey and others, 1968; Kay and others, 1970; Schilling, 1971; Fig. 1). However, in view of the great mobility imparted to the mantle from the theory of seafloor spreading and plate tectonics, it was difficult to visualize how the mantle sources of island and MOR basalts would retain their isotope integrity and identity for billions of years without, somehow, physically interacting with each other!

An obvious test was to sample the Mid-Atlantic Ridge along its axis up and over Iceland to study a transition between such mantle source domains. In 1971 we undertook a cruise along the Reykjanes Ridge with R/V *Trident,* thinking that a sharp transition in chemistry might occur where the ridge axis sharply rises at the edge of the Iceland insular platform. Instead, we found a transition spreading regularly along 400 km (Fig. 2). Establishment of such a long gradient in LILE by diffusion between two juxtaposed partially molten mantle domains of different concentrations would require at least 10^{12} years (i.e., 1000 times greater than the age of the Earth)! Clearly either a mechanism considering mantle convective flow outward from Iceland involving successive melt extractions, or a mechanism of mixing between material with two distinct REE, LILE and isotopic signatures had to be invoked (Schilling, 1973a).

Independently, Morgan (1971) introduced the mantle plume model for oceanic island hotspots and Vogt (1971) suggested, from V-shaped basement features branching symmetrically about the ridge axis, that the discharge of plume material may take place from Iceland southward along the Reykjanes Ridge at a rate of some 20 cm/yr. This evidence, and indications that Pb and Sr isotope ratios were also gradational (Hart and others, 1973; Sun and others, 1975) gave further credence to Schilling's (1973a) mantle plume–LILE–depleted asthenosphere binary mixing model.

The model stimulated controversies (e.g., O'Hara, 1973, 1975, 1977; Schilling, 1973b; O'Nions and Pankhurst, 1974; Sigvaldason and others, 1974; Flower and others, 1975; Langmuir and others, 1978); but whatever the real cause of the gradients, the approach of using the Mid-Atlantic Ridge axis as a window into the upper mantle to probe and constrain probable mantle flow patterns with trace element or isotope tracers was established and opened a new era in the study of the Earth's mantle.

Geological studies of the crest of the MAR in the later parts of the 1970s concentrated on two main approaches, namely: (1) detailed studies and sampling of small segments of the rift valley or transform fault intersections using a variety of modern exploration techniques, including direct observation from manned submersibles (e.g., FAMOUS) and (2) large-scale reconnaissance dredging along the ridge. The first type of study was aimed at understanding in detail the tectonic and magmatic processes associated with the formation of new oceanic crust. The second type of study was directed towards revealing the scale, geochemical nature, and architecture (forms and shapes) of mantle heterogeneities beneath the ridge and possibly mantle dynamics.

TABLE 1. SOURCE OF DATA AND SYMBOLS USED FOR CONSTRUCTING GEOCHEMICAL VARIATION DIAGRAMS ALONG THE MAR AXIS SHOWN IN FIGURES 1-12 AND PLATE 8B

Reference	Symbol	$^{87}Sr/^{86}Sr$	$^{143}Nd/^{144}Nd$	$^{206}Pb/^{204}Pb$	REE	Other tr.elem.	K&Ti
Aumento, 1968	A						x
Basaltic Volc.Std.Proj,1981	B				x	x	x
Bougault, 1977	C				x	x	x
Bougault and Treuil, 1980	D				x	x	x
Bryan and others, 1981	E				x	x	x
Carlson and others, 1978	G		x				
Cohen and others, 1980	H	x	x	x			
Cohen and O'Nions, 1982	I	x	x	x			
Corliss, 1970	J				x		
De Paolo and Wasserburg,1976	K		x				
Dupré and Allegre, 1980	L	x	x	x			
Engel and Engel, 1964	M						x
Erlank and Kable, 1976	N					x	x
Melson and O'Hearn, 1979	O						x
Frey and others, 1968	P				x		
Graham and Nicholls, 1969	Q				x		
Hart and others, 1973	R	x					
Ito and others, 1980	S	x	x				
Kay and others, 1970	T				x		x
Machado and others, 1982	U	x	x				
Melson and Van Andel, 1966	V						x
Frey and Haskin, 1964	W				x		
Muir and others, 1964	X						x
Nicholls and ohthers, 1964	Y						x
O'Nions and Gronvold, 1973	Z	x			x		
O'Nions and Pankhurst, 1974	Z	x	x				
O'Nions and others, 1977	Z	x	x				
Richard and others, 1976	1		x				
Scott, per. com.	2				x		
Sigvaldson, 1974	3				x		
Sun, 1980	4			x			
Sun and Jahn, 1975	4	x		x			
Sun and others, 1975	4			x			
Sun and others, 1979	4	x			x	x	x
Tatsumoto, 1978	5			x			
White and Hofmann, 1982	6	x	x				
Wood and others, 1979	7	x					
Zindler and others, 1979	8	x	x				
Shibata and others, 1979	9				x	x	x
Schilling and others, 1983, and G. Waggoner, pers. com.	+	x	x		x	x	x
FAMOUS and AMAR							
Bryan and Moore, 1977	F						x
Dupre and others, 1981	F	x		x			
Frey and Stakes, per. com.	F				x		x
Langmuir and others, 1977	F				x	x	x
Le Roex and others, 1981	F					x	x
White, 1979	F	x	x				
White and Bryan, 1977	F	x			x		

MANTLE HETEROGENEITIES BENEATH THE MAR

For this chapter we have compiled all the data we could find in the literature on critical LILE contents and isotopic ratios of dredged basalts from the MAR axis. Basalts outside of the rift valley were generally not included, nor those from fracture zones except within their active "transform" section. The source of these geochemical data and their nature are listed in Table 1. The locations of the samples are shown in Plate 8B. Concentration variation of key LILE and their ratios, and radiogenic isotope ratios for Sr, Pb, Nd, and He are shown in Figure 3 and Plate 8B. The La/Sm and La/Ti ratios (highly incompatible to incompatible ratios, that is the larger of the two ions in the numerator) were selected on the basis that they are least affected by fractionation processes during magma formation and emplacement into the oceanic crust, that is, partial melting and fractional crystallization as well as seawater alteration (hydrothermal activity or simple weathering). These ratios are thus fairly representative of their source derivation in the mantle. La/Sm and La/Ti also provide the maximum geographical coverage. For comparison, we also show the elevation of the MAR axis based on the depth of recovery of the dredged samples, complemented with the depth profile of Le Douaran and Francheteau (1981) using a method of moving average.

These profiles show that north of 34°N, Layer 2A is dominated by geochemical anomaly highs over the Azores, Iceland, and Jan Mayen hotspots, which are themselves anomalously elevated. This is also the case for the gravity field (Rabinowitz and Jung, this volume; Vogt, this volume, Ch. 14). Clearly the anomaly highs in gravity, ridge elevation, and the geochemical enrichment are paired in the sense that their maxima tend to be geographically in phase, but their wavelengths are not the same. The gravity anomalies have the longest wavelengths, the geochemical anomalies the shortest, and elevation anomalies are intermediate. Also superimposed over the Iceland and Azores suprastructures are shorter wavelength geochemical and ridge elevation anomalies centered over the Jan Mayen, 45°N, 35°N and 14°N, and 0–1°N regions. These do not seem to be accompanied by distinct anomalies in the gravity field. The short wavelength geochemical anomalies are characterized by large local geochemical dispersion. This is evident within single stations where the ratios range from highly enriched to nearly as low as normal MORBs. Similar spike-highs have been observed in the South Atlantic at the latitude of Ascension, St. Helena, and Tristan-Gough off-ridge or near-ridge hotspots (Schilling and others, 1984). These short anomalies are in marked contrast to the long wavelength geochemical anomalies that show smaller local dispersion about the regular gradients, as for example indi-

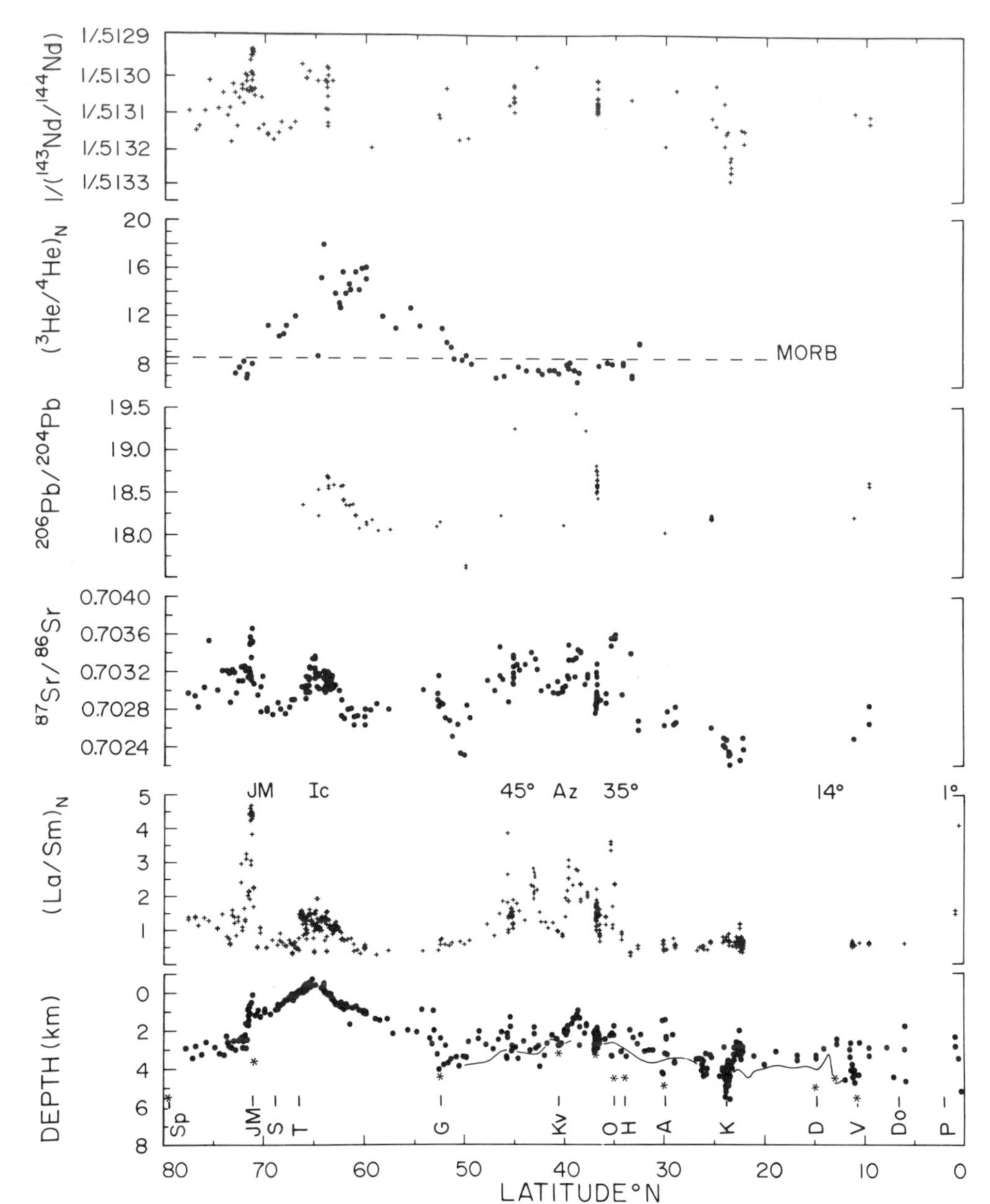

Figure 3. Variation of radiogenic bearing isotope ratios and La/Sm (normalized to chondrites) in Layer 2A along the axis of the MAR based on pillow basalts mostly dredged or collected by manned submersible (FAMOUS-AMAR). Variations over Iceland are confined to the SW and NE neovolcanic rift zones only. $^{3}He/^{4}He$ is relative to the atmospheric ratio. Depth of MAR axis is estimated from depth of sample recovery, and the continuous line is taken from Le Douaran and Francheteau (1981). Stars are the maximum depth of ridge-transform modal deeps documented by Fox and Gallo (1984). For source of data and sample locations with respect to bathymetry see Table 1 and Plate 8B. Fracture Zone abbreviations are : Sp–Spitzbergen; JM–Jan Mayen; S–Spar; T–Tjornes; G–Charlie-Gibbs; Ku–Kurchatov; O–Oceanographer; H–Hayes; A–Atlantis; K–Kane; D–Desirade (Fifteen-Twenty); V–Vema; Do–Doldrum; and P–St. Paul. Note anomaly-highs over Jan Mayen (JM), Iceland (Ic), 45°N, Azores (Az), 35°N, 14°N, and 1°N.

cated within single dredge hauls, or short ridge segments densely sampled such as the northern part of the Reykjanes Ridge and FAMOUS (White and Bryan, 1977; Dupre and others, 1981) (Fig. 3).

The long wavelength geochemical anomalies are asymmetrical about the ridge-centered hotspot. One side is characterized by a long (>500 km) regular gradient such as south of the Azores, north of 45°N, south of Iceland on the Reykjanes Ridge, and northeast of Jan Mayen on the Mohns Ridge. The latter is not readily apparent on the latitudinal plots because it is highly compressed since it is projected from the 30° strike of the Mohns Ridge, and also because the Kolbeinsey Ridge and the Mohns

Ridge latitudinally overlap between 71° and 72°N. There is no correlation between the presence of such gradients and the morphology of the ridge axis. The Reykjanes Ridge is featured by an elevated horst, whereas south of the Azores or northeast of Jan Mayen there is a rift. Most of the Kolbeinsey Ridge axis is elevated where no gradient is observed. However, V-shaped, time-transgressive ridges are apparent north and south of Iceland and possibly south of the Azores (e.g., Plate 8A; Saemundsson, this volume). The other side of these long wavelength anomalies is represented by a steep gradient or a discontinuity, but the sampling intervals are inadequate to decide. These discontinuities either tend to coincide with a major fracture zone (e.g., on the Kolbeinsey Ridge just south of Jan Mayen F.Z., or across the Tjörnes F.Z.), or occur where relatively recent rift jumps seem to have taken place (e.g., Tjörnes F.Z. or Kurchatov F.Z.).

Small discontinuities along normal ridge segments have also been observed across two major fracture zones, the Gibbs F.Z. where the La/Sm (but not $^{87}Sr/^{86}Sr$) steps up from north to south (Schilling and others, 1983; White and Schilling, 1978), and the Kane F.Z. where the $^{87}Sr/^{86}Sr$ steps down from north to south (and the converse for $^{143}Nd/^{144}Nd$).

There also seem to be important differences between the base lines drawn through the anomaly minima, once these short and long wavelength geochemical and elevation anomalies have been filtered out. The La/Sm ratio stays nearly constant at a level typical of normal-MORB from 10°N to 60°N, then increases northward, whereas the $^{87}Sr/^{86}Sr$ and the elevation of the ridge-crest increase more progressively northward (shoal). These trends correlate with decreasing age of the opening of the North Atlantic and the decreasing spreading rate. However, the general shoaling may be an artifact of fracture zones that are more numerous near the equator. Because of this uncertainty, we have also plotted the maximum depths of ridge-transform modal deeps so far documented by Fox and Gallo (1984). The northward shoaling tendency remains with the exception of the Spitsbergen F.Z. Similarly, the possible increase in La/Sm along normal ridge segments (not influenced by hotspots) may be biased by the high La/Sm of the Knipovich Ridge; and the $^{87}Sr/^{86}Sr$ also by high values in the Knipovich Ridge and unusually low values just south of the Gibbs F.Z., Kane F.Z. and 22°N.

The probable causes of the anomalies in LILE and isotopic ratio variations remain subject to interpretation despite the fact that these parameters are the most direct we currently have available to probe the mantle. It is clear that spreading rate variation along the MAR cannot be the cause since it increases regularly southward as shown in Plate 8A. Other possibilities are:

(1) The variation observed in Layer 2A along the zero-age profile reflects similar heterogeneities in the underlying upper mantle. In this model the enriched mantle domains are embedded randomly within the depleted asthenosphere and passively transported and deformed as the asthenosphere convects and flows toward the spreading ridge axis (e.g., Davies, 1981; or Richter and Ribe, 1979).

(2) The variation may also reflect vertical zoning of the mantle if lavas are derived from progressively greater depth as the center of the anomalously elevated regions are reached, a supposition that has received some support from petrological considerations (Schilling and others, 1983b).

(3) Perhaps a more popular idea is that the enriched mantle domains are buoyant and rise as plumes or blobs that interact and mix while penetrating the LILE-depleted asthenosphere. The plumes remain essentially fixed with respect to each other and to the lower mantle (Morgan, 1981, 1983), and have variable lifetimes with some extending up to several tens to even hundreds of millions of years (for example, Iceland, 60 m.y. and still active from geochemical considerations, though perhaps in decline, Schilling and others, 1982). The ascending plume or blob begins to melt at a greater depth than the LILE asthenosphere ascending at ridge axis (Schilling, 1973a). In the case of a ridge-centered hotspot, the flux of plume material in excess of what is required by accretion per unit ridge-length of new crust by seafloor spreading is taken up in two ways: (a) thick lava flows may pile up and produce thicker crust over the hotspot, and (b) the excess plume material tends to spread laterally along the spreading axis which acts as a mass sink (Fig. 4). The subcrustal flow of plume material is progressively used up and diluted with LILE-depleted asthenospheric material, thus generating a compositionally gradational Layer 2A.

In the case of a ridge migrating away from the fixed hotspot, Morgan (1978) has suggested that the spreading ridge axis continues to act as a sink for the plume. A sublithospheric channel may develop and extend with time between the hotspot and the migrating ridge thus producing a geochemical anomaly on the ridgecrest nearest the hotspot. The model has received support in the case of the Galapagos hotspot and even in the South Atlantic where the hotspots are now located as far as 700 km away from the MAR (Schilling and others, 1985). It has also been shown that the width of the geochemical and residual ridge elevation anomalies progressively decreases as the distance between the hotspot and the ridge increases, and mixing becomes more and more irregular until the sublithospheric flow stops and the hotspot finally becomes totally intraplate (Schilling, 1985). The residual depth anomalies disappear first. The 35°N has been explained this way; that the Great Meteor hotspot prior to 80 Ma formed the New England Seamounts, the MAR overrode it, and then the hotspot-migrating ridge connection was established. During the past 57 m.y., the African plate overrode this hotspot and produced the Great Meteor, Hyere, and Atlantis Seamount Chain (Morgan, 1981; 1983). Results from DSDP Leg 82 are supportive (Rideout and Schilling, 1983; Duncan, 1984). A similar explanation could be advanced for the 0–1°N and 15°N hint of anomalies, but density of sampling must be increased relative to existing off-ridge hotspots in the region to further test this working hypothesis (Morgan, 1981; 1983; Duncan, 1984).

(4) Alternatively, Chase (1979) has suggested that the asymmetric geochemical gradients skewed southward around Iceland and the Azores are the results of an asthenospheric return flow that would smear southward along the MAR axis the

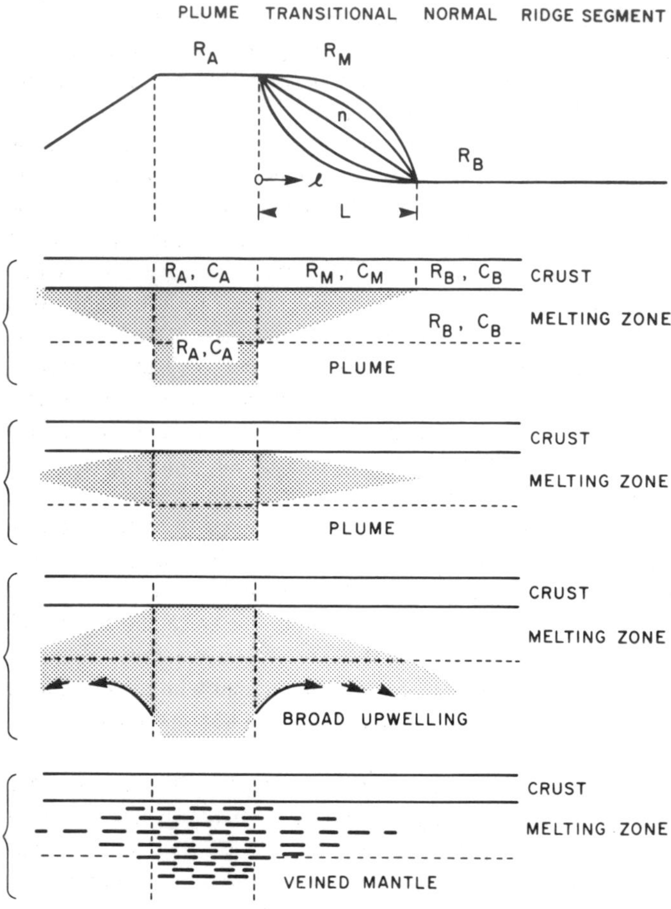

Figure 4. Some possible mantle topologies that could satisfy the isotope and La/Sm gradients observed along the Reykjanes Ridge shown in Figure 2 by mixing of the LILE-enriched plume and the LILE-depleted asthenosphere source within the depth zone of partial melting. R_A, R_M, R_B are for LILE or isotope ratios in the plume, the mixture and the depleted asthenosphere and "n" which defines the curvature, is the concentration ratio between sources A and B of the element or isotope at the denominator of the R's ratio. Curves concave down are for $n > 1$ and concave up for $n < 1$, whereas the straight line is for $n = 1$. The fraction of plume in the mix is assumed to decrease linearly with distance ℓ from Iceland.

point source anomalous plume material rising from below. Chase's model shows that the return asthenospheric flow in the North Atlantic is essentially parallel to the strike of the MAR. This explanation cannot account for the gradient running northeast of Jan Mayen along the Mohns Ridge.

Geochemical discontinuities noted at fracture zones have been explained in several ways. In the shallower depth subaxial pipe flow model of Vogt (1971), the discontinuity at the Tjörnes F.Z. and near the Kurchatov F.Z. would have resulted from damming by the thick lithosphere plate opposite the fracture zone, since plates grow at a rate proportional to the square root of time (Vogt and Johnson, 1975). Schilling and others (1983) noted that in these two cases the MAR may have jumped to its present location in relatively recent time (e.g., Talwani and Eldholm, 1977; Quintino and Machado, 1977), and the excess plume flow along the ridge axis may not have had time to readjust to the new configuration after the rift jump. However, another explanation must be provided for the Nd-Sr isotopic discontinuity at the Kane F.Z. since it is remote from any hotspot influence. Machado and others (1982) have suggested that fracture zones may confine (broad) mantle domains with distinct identities and evolution histories coinciding or predating the opening of the North Atlantic. The discontinuity precludes the presence of a long magma chamber beneath the ridge, but rather suggests magma segregation from discrete pods.

Finally, these models must be tempered by the fact that our ability to detect heterogeneities in the mantle is partly biased by some sampling problems inherent in using lava as probes of the upper mantle. The heterogeneities we detect in lavas erupted along the ridgecrest depend in part on the lengthscale and geometry of the melting zone relative to that of the actual mantle heterogeneities we like to detect, and thus also operative degrees of melting. The exact mode of melt coalescence, migration, and residence time in shallow magma chambers underlying the ridgecrest are also influential, since through mixing these processes tend to rehomogenize and obliterate any initial mantle heterogeneities sought in the first place (Cohen and O'Nions, 1982). The lengthscale of the magma chamber along-ridge relative to the sampling interval is also a factor that needs to be considered in interpreting these spatial variations.

Sleep (1984) has also pointed out that some spatial separation and bias can be caused from the fact that domains richer in incompatible elements and H_2O (and other volatiles) may melt and segregate sooner than the surrounding asthenosphere during advection toward the ridge. His analysis of the stress pattern suggests that magmas derived from these rich domains may follow pathways that lead toward flank eruptions. This mechanism does not explain the zero age latitudinal variation discussed here, but could explain some variation noted across the ridge axis (e.g., Bryan and Frey, this volume).

GEOCHEMICAL NATURE OF THE NORTH ATLANTIC HOTSPOTS

The degree of LILE enrichment and radiogenic Sr and Pb contents varies from one hotspot to another, thus suggesting different evolutionary histories relative to the LILE-depleted source and relative to each other. The Azores, Jan Mayen, 45°N, and 35°N mantle sources appear the most enriched, whereas the enrichment beneath Iceland is more subdued. The MAR basalts inversely correlate in terms of $^{87}Sr/^{86}Sr$ versus $^{143}Nd/^{144}Nd$ variation and form a continuum. The trend overlaps the so-called "mantle array" (O'Nions and Pankhurst, 1973), but with one notable exception on the Reykjanes Peninsula whose significance is discussed by Zindler and others (1979). The difference between hotspots is best revealed by the $^3He/^4He$ variation in MAR glasses along the ridge and with respect to $^{87}Sr/^{86}Sr$ (Figure 3). Iceland is a high $^3He/^4He$ province relative to the LILE-depleted source and shows a positive correlation with $^{87}Sr/^{86}Sr$, whereas

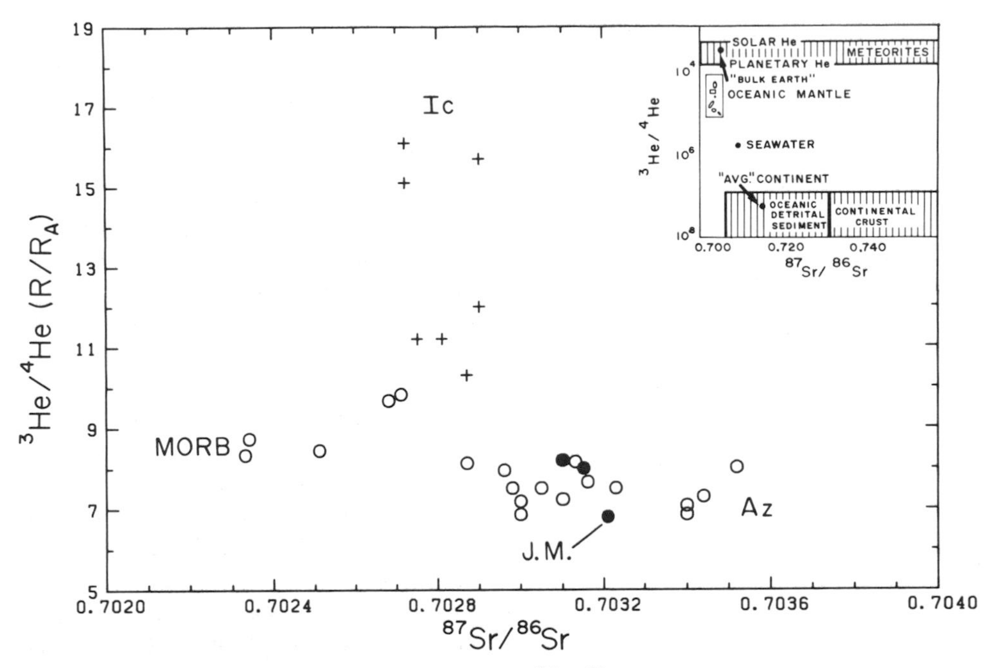

Figure 5. $^3He/^4He$ (relative to atmospheric ratio) vs. $^{87}Sr/^{86}Sr$ variation for MAR basalts from the Iceland Province (52–71°N; crosses), the Azores Province (52–28°N; open circles) and the Jan Mayen Province (71–74°N; filled circles). Upper corner insert shows MAR variation with respect to other important reservoirs (He isotopic data taken from Kurz and others, 1982b; and Poreda and others, 1983).

the Azores, 45°N, 35°N, and Jan Mayen are mantle source domains lower in $^3He/^4He$ that show a negative correlation with $^{87}Sr/^{86}Sr$ (Poreda and others, 1980; Kurz and others, 1982a and b). These trends suggest mixing with the Iceland mantle source containing a component of primordial He gas, whereas, the Azores, Jan Mayen, 45°N, and 35°N mantle anomalies may be related to some recycling of continental or hydrospheric materials (Kurz and others, 1982b).

The fundamental geochemical distinctions among hotspots are also apparent in the bulk composition of the basalts. Figure 6 shows that at any fixed index of fractionation measured by the Mg-value, the MAR basalts erupted over Iceland are significantly richer in Fe-Mg-Ca components, whereas basalts over the Azores and Jan Mayen Platforms (and 45°N and 35°N) are significantly richer in Si-Al-Alkali components, the very same components with which the continental crust is also enriched.

It has also been suggested that the geochemical differences in major and minor elements may also influence the partitioning of elements such as Fe, Mg, Mn, or Ni in olivine crystallizing from these magmas. For example, Figure 7 shows that at a fixed forsterite content the olivine/melt partition coefficients of Fe or Mg, are significantly lower over the Iceland province than the Azores (Jan Mayen, 45°, and 35°N) and thus can produce very distinct evolutionary trends during magma emplacement and formation of the oceanic crust over these provinces (e.g., Sigurdsson, 1981; Hermes and Schilling, 1976).

Finally, MAR basalts over the Azores Platform and Iceland are rich in clinopyroxene phenocrysts, whereas this mineral is essentially absent along normal ridge segments (Schilling and

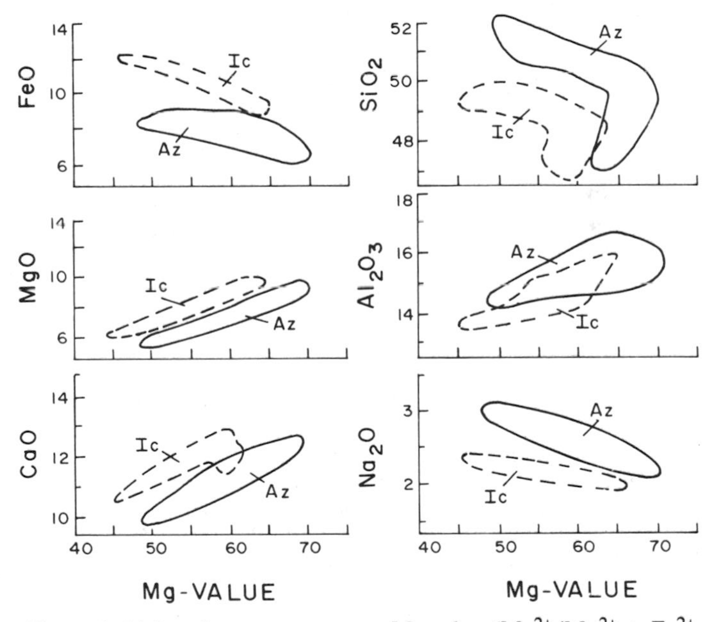

Figure 6. Major element contents vs Mg-value ($Mg^{2+}/Mg^{2+} + Fe^{2+}$, atomic) of MAR basalts from the center of the Azores Province (40°N–38°N, solid line) and Iceland (Southwest Rift Zone; dashed line). Data are from Schilling and others (1983).

others, 1983). The clinopyroxene phenocrysts are the results of more extensive fractional crystallization, since basalts from Iceland seem to have segregated over a wider range of depths and may be emplaced through a thicker crust than along normal ridge segments (see Saemundsson, this volume). The clinopyroxene saturation may also be the result of more extensive melting taking

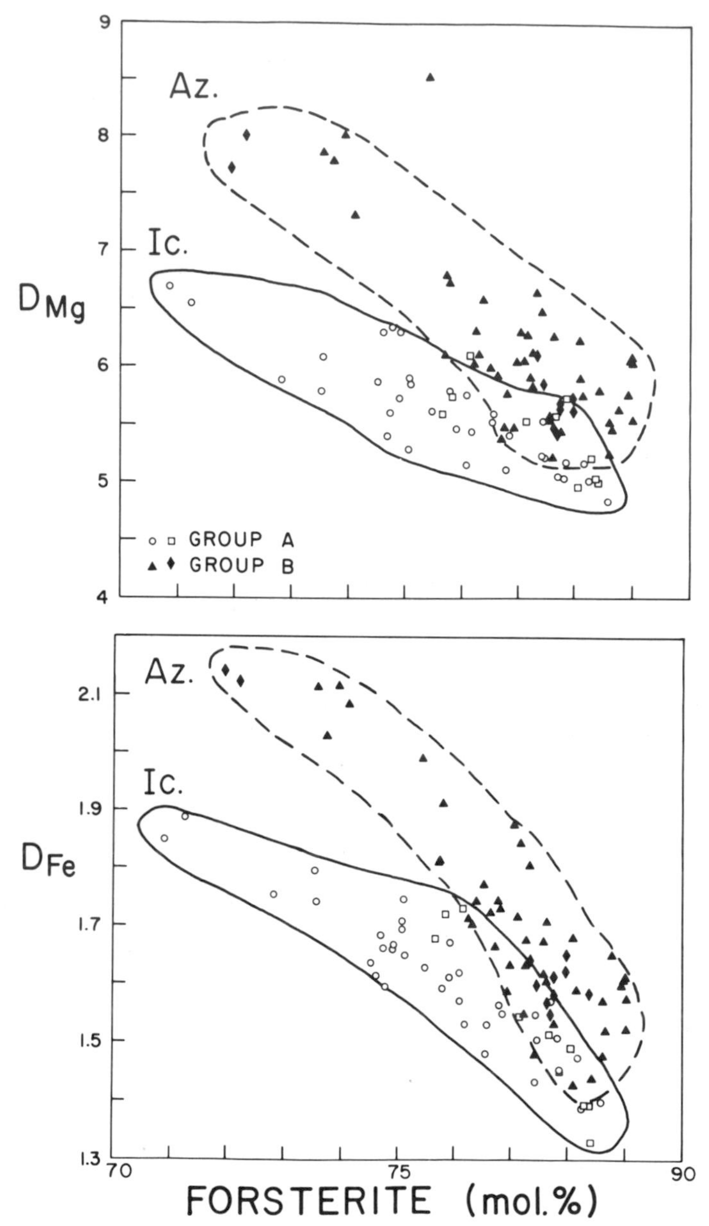

Figure 7. Olivine/melt partition coefficient variation for Fe (D_{Fe}) and Mg (D_{Mg}) versus forsterite content of the olivine in MAR basaltic glasses. Note the systematically higher D_{Mg} and D_{Fe} values for the Azores Province (Az, 28°–52°N) relative to the Iceland Province (Ic, 53°–71°N). Schilling previously unpublished data.

place beneath the two platforms. This is suggested by the absence of clinopyroxene and the more residual nature of peridotite exposed along fracture zones in the vicinity of the Azores compared to peridotite sampled south along the MAR (Michael and Bonatti, 1983). Finally, saturation of clinopyroxene on the Azores transect seems to have taken place over a wider range of pressure than for Iceland, and this may also be attributed to the higher content of basic relative to acid components of the magmas from the Azores Province (Schilling and others, 1983). See also Melson and O'Hearn (this volume) for further petrological information.

FUNDAMENTAL GEOCHEMISTRY

Important insights into the fundamentals of trace element geochemistry are gained by considering the magnitude of relative trace element enrichment within a single province. It is now well known that the smooth rare earth fractionation patterns observed in basalts (e.g., Figure 1) are directly related to the size of these trivalent ions, since the reciprocal of their ionic radius covaries linearly with their atomic number. Goldschmit (1958) had in fact predicted the effect from the lanthanide contraction. For example, Y^{3+} would behave very similarly to the heavy rare earth and element pairs such as Hf/Zr or Nb/Ta would have very similar and highly coherent behaviors despite their large difference in atomic mass number (>30). Early rare-earth work on basalts, including MAR basalts, fully confirmed the prediction by demonstrating that the Y^{3+} enrichment factor relative to chondrites was identical to Er^{3+} (Schilling, 1972). These two elements are hardly fractionated during magmatic processes involving silicate crystals and melts. The approach was extended by Bougault (1980) for other high field-strength cations (large valences, small radii) such as Nb and Ta, which are essentially analogues of La, whereas Zr, Hf, and Ti are close analogues of the middle REE and straddle Sm behavior, and V tends to simulate Lu behavior. Thus, chondrite normalized enrichment patterns similar to those for the REE can be obtained with Nb, Zr, Ti, Y, and V. By comparing the enrichment factors of these elements with those for the REE for which there is considerable crystal/melt partition coefficient data available, Bougault proposed the following scale of relative increase in incompatibility:

$$D_{Sc} > D_V > D_Y \simeq D_{Tb} > D_{Ti} > D_{Zr} = D_{Hf} > D_{Nb} = D_{Ta} = D_{La} > D_{Th}$$

Bougault and others (1979) were also the first to point out that refractory element pairs with similar geochemical behavior, such as Y/Tb, Zr/Hf, and Nb/Ta, remain essentially constant along the MAR and in DSDP holes in the North Atlantic, despite the fact that the absolute concentration of these elements varies over a wide range and the lavas are erupted over or far away from hotspots. They also emphasized that these ratios are similar to those in chondrites, thus suggesting a homogeneous primordial mantle source of chondritic composition. For example, Figure 8 further demonstrates the constancy and the chondritic nature of such ratios along the MAR, based on available data for two highly incompatible elements such as Nb/La, and pairs of lesser incompatibility such as Sm/Zr and Sm/Ti. All these incompatible elements are also refractory. The approach was extended to more volatile trace elements, including the alkalis and the halogens (Schilling and others, 1980). Figure 9 illustrates that the enrichment in the basalts over the Azores platform relative to adjacent normal MORB segments increases either (1) as the ionic radius increases for ions of valence 3 or less, positive or negative, or (2) as the degree of incompatibility based on known crystal/melt partition coefficient data increases, that is, the crystal/melt partition coefficients decrease. Ratios of element pairs as dispar-

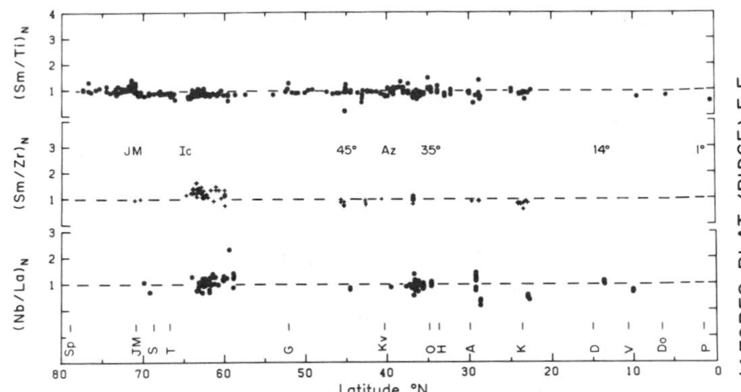

Figure 8. Chondrite normalized variation of trace element ratios for pairs with similar degrees of incompatibility. Note the constancy and essentially chondritic nature (value = 1) of these ratios. For source of data see Table 1 and Plate 8B. See Figure 3 for meaning of abbreviations.

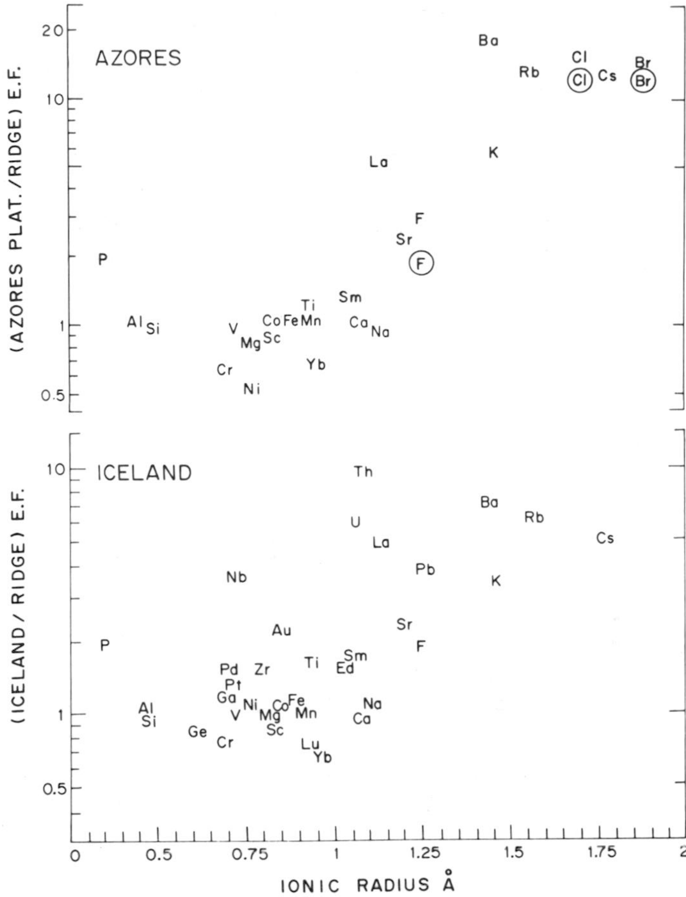

Figure 9. Covariation of ionic radius with relative enrichment contents of MAR basalts from the Azores Platform (37–40°N) and Iceland (63–65°N) relative to adjacent normal ridge segments (data from Schilling and others 1982, 1983, and unpublished.) Note the broad inverse correlation for incompatible elements (ionic radius >0.8A). All analyses were made on the interior of pillow basalts. F, Cl, and Br analyses on glasses are circled.

ate as Sr^{2+}/F^-, Rb^+/Ba^{2+}, and Ba^{2+}/Cl^- remain nearly constant along the entire profile studied in the vicinity of the Azores (52–30°N), and provide a clue to the composition of a precursor mantle prior to the fractionation(s) that produced the LILE-depleted mantle source and the mantle heterogeneities associated with the North Atlantic hotspots. However, in the case of ratios involving a volatile trace element (e.g., alkalis or halogens), the value is no longer the same as for chondrites but tends to approach more closely values independently estimated for the bulk earth (e.g. Ganapathy and Anders, 1974). The constancy of element ratios with similar degrees of incompatibility, such as Ba/Rb, Cs/Rb, and Ba/Cs, has been further confirmed by Hofmann and White (1983) by using a larger set of mid-ocean ridge basalts from the Atlantic and the Pacific. They suggested that these ratios can indeed be taken as representative of bulk earth values since they do not fractionate. Relative to a content of 17 ppm for Sc and chondritic ratios for other refractory elements, they calculated a Rb/Sr ratio of 0.29 for the bulk earth, which is slightly lower than the independent estimate based on the $^{143}Nd/^{144}Nd$ versus $^{87}Sr/^{86}Sr$ covariation for oceanic basalts. Thus a certain degree of consistency is emerging, which makes the use of MAR basalt's ratio of elements with similar incompatibility very helpful in estimating more primordial mantle source compositions.

VOLATILES ALONG THE MAR

There is considerable interest in knowing the content and isotopic composition of volatiles in basalt glasses quenched at various depths (pressures) below sea level in various volcanic and tectonic environments. These studies provide clues to the content of volatiles in the mantle and their possible origin (primordial, recycled, etc.), the mechanism of outgassing of the earth, the role they play in controlling melting conditions, magma compositions, and rheology of the mantle. These studies also provide important limits on budget calculations on rate of growth and compositional evolution of the hydrosphere and atmosphere through geologic time. Finally, the extent of volatiles outgassing from magmas can also influence the mode of magmatic eruptions and the morphology of their products and constructional volcanism on the seafloor, as well as the magnetic properties of the oceanic crust (e.g., ash deposits versus pillow basalts). Moore (1965) pioneered such studies by measuring the H_2O content of glass rims of pillow basalts as a function of depth below sea level along the extension of the Kilauea-Iki Rift on the flank of the island of Hawaii. The MAR has subsequently played an important role since it offers a natural laboratory to study empirically relationships with respect to (1) depth (or pressure), (2) the transition from submarine to subaerial volcanic regime (e.g., the Reykjanes Ridge–Iceland), and (3) distinct mantle sources.

For example, the amount of vesicles in glass rims of MAR basalts and their size generally increase with decreasing depth (Fig. 10), partly as a result of solubility of gases decreasing with pressure and the volume of gases expanding with decreasing pressure (e.g., Moore and others, 1977). However, the degree of

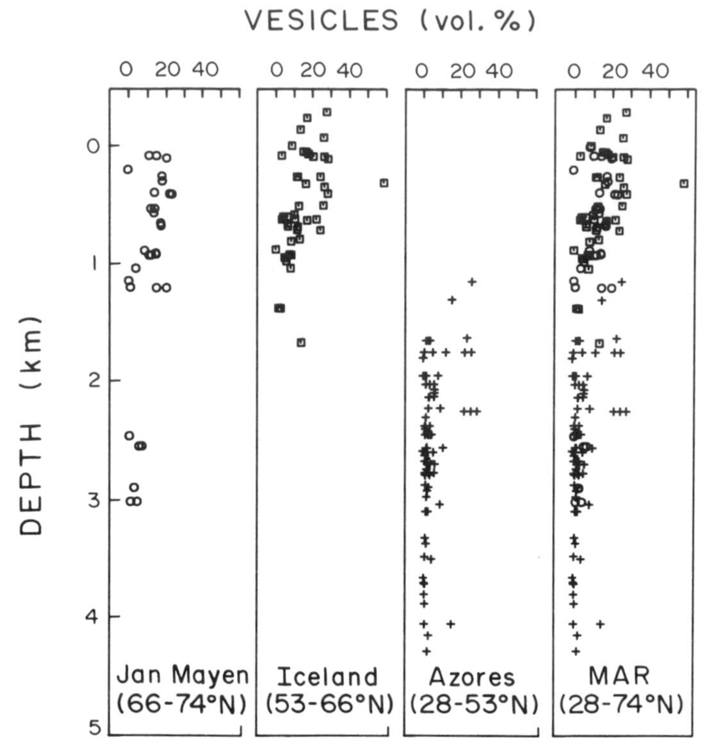

Figure 10. Volume (%) of vesicles contained in the glassy rim of basalts as a function of depth of eruption below sea-level. Note increase below critical depth of outgassing (<700 m). Data from Schilling and others (1983).

vesicularity may also be influenced by magma bulk compositions, as well as perhaps by the nature of mantle sources (Moore, 1979). At equal depth of eruption below sealevel, alkali basalts appear more vesicular than tholeiites (Moore, 1979). It is generally agreed that the composition of gases contained in vesicles trapped in the glassy margin of pillow basalts is dominated by CO_2. This is probably the result of the low solubility of carbon compounds in basaltic melts under existing P, T conditions of melting and emplacement beneath mid-ocean ridges, and possibly also due to the relatively high abundance of CO_2 in the mantle. H_2O is the next subordinate volatile found in vesicles, followed by sulfur species (SO_2 dominant), HCl, N_2, methane (and other organic gaseous compounds), H_2, He, and other rare gases. The fact that some vesicles form as deep as 5000 m indicates that the volatile content of glasses, vesicle-free or not, is not necessarily representative of primary melts at the time of their formation by partial melting; or in the mantle source, assuming that the degree of partial melting were known. This has prompted Anderson (1974) or Delaney and others (1978) to consider the content and composition of volatiles in fluids or melt inclusions in phenocrysts contained in MORBs. CO_2 is also the dominant gas in these inclusions. In contrast H_2O appears undersaturated in MAR basalts erupted deeper than 500–600 m depth (or 60 bars). The 500–600 m critical depth of outgassing of H_2O was theoretically predicted by McBirney (1963) and empirally confirmed on the flank of Hawaii (Moore, 1965) and along the Reykjanes Ridge up to Iceland (Moore and Schilling, 1973).

This part of the MAR has served as a model for the study of outgassing of lesser abundant volatiles (Unni and Schilling, 1978). Above the critical depth of H_2O-outgassing, vesicularity increases rapidly, and not only H_2O decreases in the quenched melts but also volatiles such as S (probably SO_2) and some of the halogens, which, in the absence of an H_2O-vapor phase would normally be undersaturated in the melt (Figure 11) (Unni and Schilling, 1978; Rowe and Schilling, 1979). The H_2O-phase exsolving above 500–600 m water depth seems to act as a carrier of other volatiles, because these volatiles are more soluble and partition preferentially into the H_2O-rich phase than into the basaltic melt under such P, T conditions. Thus, exolving water from melts acts as a fluxing agent for outgassing other minor volatiles. The mechanism can strongly fractionate volatiles even among a group as homologous as the halogens (Rowe and Schilling, 1979). While Cl and Br are outgassed above 500 m, F remains in the melt and the F/Cl ratio in the quenched glasses changes by an order of magnitude due to H_2O vesiculation (Fig. 11). Experimental partition coefficient data for Cl and F between silicate melts and H_2O-vapor are fully supportive of this interpretation; however, Rayleigh distillation calculations show also that the amount of Cl loss exceeds that predicted from the amount of H_2O apparently lost, again suggesting that another major volatile phase is also outgassing deeper than 500 m. It seems that CO_2 must be the other phase. Mysen and others (1975) have shown that CO_2 solubility in basaltic melts increases in the presence of H_2O. Loss of H_2O may further enhance CO_2 exsolution and may be responsible for a renewed outburst of CO_2 from magmas decompressing through 50–60 bars pressure, thus contributing further to the rapid increase in vesicularity of magma passing through this critical threshold. Thus differences in the gas content of vesicles trapped in basaltic melts quenched above or below the critical depth of H_2O outgassing may be observed in the future.

The fact that H_2O and the halogens are apparently not significantly outgassed at depths greater than 500–600 m of water has prompted Schilling and others (1980) to study possible mantle source heterogeneities in these volatiles. Cl, Br, F, and H_2O increase toward the Azores (and Iceland) as incompatible elements do (e.g., K_2O, Plate 8B), as long as the depth of the ridge axis is deeper than 500–600 m. The enrichment of Cl and Br over the Azores transect is similar to that of Ba, Cs, and Rb, whereas F enrichment is comparable to P or Sr, and again the Azores is significantly more enriched in these volatiles than Iceland. H_2O is at least two to three times more abundant in MAR basalts erupted over the Azores platform than adjacent normal ridge segments. This is a minimum for the mantle source underlying the Azores, since residual peridotites from fracture zones in the vicinity are more depleted in basaltic components than usual, suggesting a higher operative degree of melting beneath the platform (Michael and Bonatti, 1983). These observations have led Schilling and others (1983) to suggest that hotspot regions may also be *wetspots,* and the enhanced content of H_2O and other volatiles

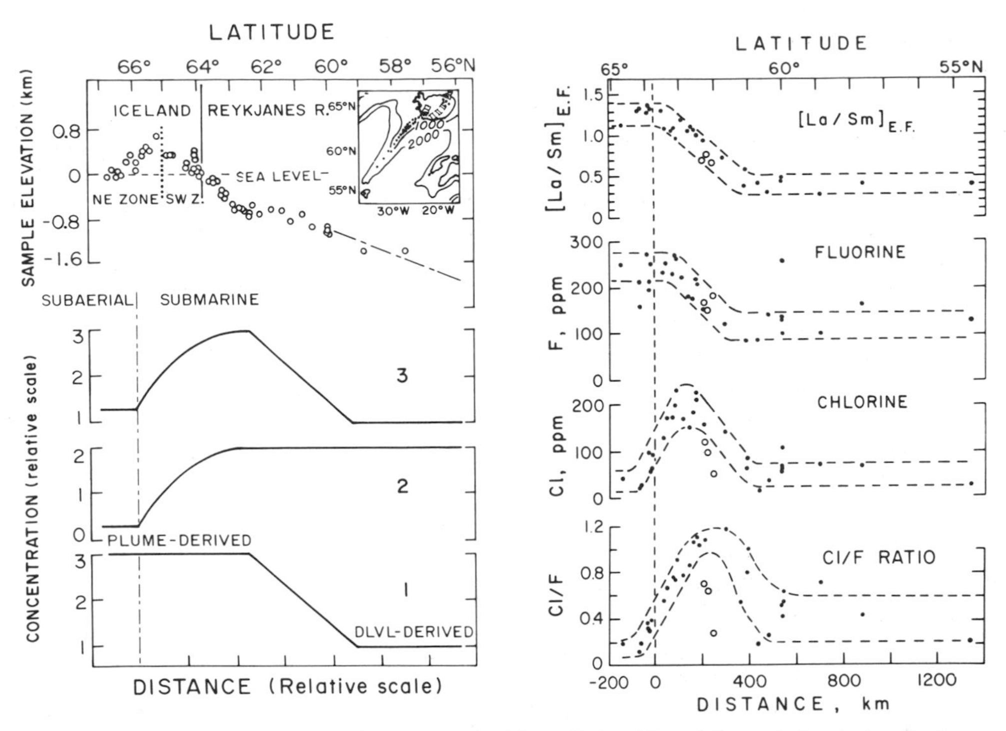

Figure 11. Left: Curves 1, 2, and 3 represent model predicting C1 and Br variation in basalt along Reykjanes Ridge resulting from mantle source mixing and outgassing at shallow depth.

Right: Actual variations observed. Note the maximum for Cl and Br due to outgassing as predicted by the model (adapted from Unni and Schilling, 1978; and Rowe and Schilling, 1979.) Black dots are for pillow basalts; open circles are for highly vesicular outgassed subaerial lavas subsequently rounded by wave action during subsidence (or submergence). This is corroborated by the low Cl and Br content of these subaerial looking lavas, and in contrast lack any depletion in refractory elements such as La, Sm, and F.

may be partly responsible for the unusually large rate of magma outpouring and buildup of these shallow volcanic platforms. The mantle plume upwelling beneath these platforms may have been triggered by a flux of volatiles that would have swept out trace elements and metasomatized the mantle beneath hotspots (Schilling and others, 1980). Sr and Nd isotopic systematics suggest that such a metasomatic event would have occurred 400–600 m.y. ago at the earliest, that is, prior to the last opening of the Atlantic.

Outgassing of sulfur species appears even more complex than H_2O or CO_2. On the one hand, Wendlandt (1982), among others, has shown experimentally that S species may be saturated with respect to H_2S during partial melting and may have resulted in the separation of an immiscible Fe-sulfide rich melt. The globules would scavenge elements such as Ni, Co, Cu, and noble metals. As pressure decreases, sulfur solubility decreases and may be essentially controlled by FeO activity in the melt (and perhaps Ti as well) as evident from the strong correlation of S content in submarine basaltic glasses with FeO (and Ti) (e.g., Mathez, 1976). Finally, some sulfur must also be lost even below the critical depth of outgassing, since vesicles are often lined up with tiny sulfide globules that apparently precipitated during rapid cooling above the rigid temperature of 800–1000°C (e.g., Moore and Calk, 1971; Moore, 1979).

It is clear that volatiles play important roles in magma genesis. It may also affect some of the morphology of the ridge, hydrothermal activity, and ore formation on or below the seafloor. The facts that CO_2 may be saturated even at or prior to the depth of partial melting and that a flux of CO_2 bubbles may exist separately from ascent of magma, have important implications for understanding the outgassing of other trace volatiles, the fractionation of certain trace element pairs, and perhaps stable isotope ratios or even $^3He/^4He$ ratios. For example, Pineau and Javoy (1983) have shown that the unusually low ^{13}C isotopic composition of carbon species remaining dissolved in quenched MAR and EPR basaltic glasses relative to CO_2 contained in trapped vesicles may be the result of such CO_2 flux. The isotopic fractionation of C in turn has led these authors to estimate a probable minimum content of 0.2 to 1 wt% carbon in primary basaltic melts erupted along the Mid-Atlantic Ridge. In view of the complexities that may have taken place during processes of outgassing, the identification of possible mantle sources using H/D, C, O, and S isotope ratios must be done with caution as work progresses.

THE FAMOUS PROJECT

The French American Mid-Ocean Undersea Study (FAMOUS Project) was designed to study in detail the magmatic and tectonic processes associated with the formation of new oceanic crust by seafloor spreading at a ridge axis. Its location is around 36°45′N just south of the Azores (Heirtzler and van Andel, 1977). The axis is typically rifted and bounded on the north and south by two fracture zones with only a small offset. Geochemically, FAMOUS lies within a transition approximately mid-way along the gradient between the Azores platform and a so-called normal ridge segment which starts at 34°N, just south of the Hayes F.Z. Detailed geophysical surveys revealed that the rift valley floor is relatively broad and asymmetrically bounded by normal faults and that spreading has also been asymmetric (0.7 cm/yr westward and 1.5 cm/yr eastward) (Phillips and Fleming, 1978; Needham and Francheteau, 1974). Small elongated central volcanoes (Mt. Venus, Mt. Pluto) with 100–200 m relief mark the "center" of the rift valley floor where the most recent spreading and volcanic activity has taken place.

Detailed precision dredging and sampling with the manned submersibles *Archimede, Cyana,* and *Alvin* and petrologic studies were also conducted (Ballard and others, 1975; ARCYANA, 1975, 1977). These have revealed a valley floor petrographically and compositionally zoned, again somewhat asymmetrically.

Petrographically, the center of the valley is composed of olivine (Fo_{84-87}) basalts (Mt. Pluto and Mt. Venus) or picritic basalts, which in turn are surrounded by aphyric or plagioclase-clinopyroxene-olivine (Fo_{73-80}) phenocryst-bearing basalts. A field of basalts containing plagioclase-rich phenocrysts is found near the base of the rift wall west of Mt. Venus (Hekinian and others, 1976; Bryan and Moore, 1977). These basalts contain notably large, rounded, and partly resorbed xenocrysts of plagioclase, olivine of intermediate compositions and microphenocrysts of chrome spinel embedded in glass matrices of composition similar to the plag-cpx-ol-basalt (Fig. 12). Age-wise, the olivine basalts and picrites are the youngest, the plag-rich basalts the oldest (<100,000 y.) and the plag-cpx-ol-basalts intermediate. However, the flank lavas are always younger than the age of the seafloor inferred from magnetic anomalies, thus suggesting that flank eruptions have taken place particularly at the intersection of the rift walls and the floor of the valley, a region which is intensely fractured and more seismically active as a result of tectonic adjustment.

Compositionally, the FeO/MgO ratio and incompatible element contents (e.g., Ti, K_2O, La, Rb, Ba, Sm, including H_2O) of these lavas increase asymmetrically outward from the rift center, whereas compatible elements (e.g., Ni, Cr) decrease in the glass or whole rock as well (Hekinian and others, 1976; Bryan and Moore, 1977; White and Bryan, 1977). The picrites and olivine basalts have the lowest incompatible element contents (e.g., TiO_2 = 0.6 wt%) and the plag-cpx-ol-basalts the highest (e.g., TiO_2 up to 2 wt%). The range of composition and petrographic types observed in the limited area of FAMOUS (8 × 8 km) approaches that observed along the entire Mid-Atlantic Ridge. In contrast, Sr, Pb, and Nd isotope ratios are remarkably uniform (White and Bryan, 1977; White, 1979; Dupre and others, 1981). The ranges are .70286 to .70299 for $^{87}Sr/^{86}Sr$, .513150 to .513192 for $^{143}Nd/^{144}Nd$, and 18.33 to 18.89 for $^{206}Pb/^{204}Pb$, and approach the analytical uncertainties, thus suggesting a well-mixed single uniform mantle source of intermediate composition between the LILE-depleted source present south of the Hayes FZ and the mantle plume source underlying the center of the Azores Platform.

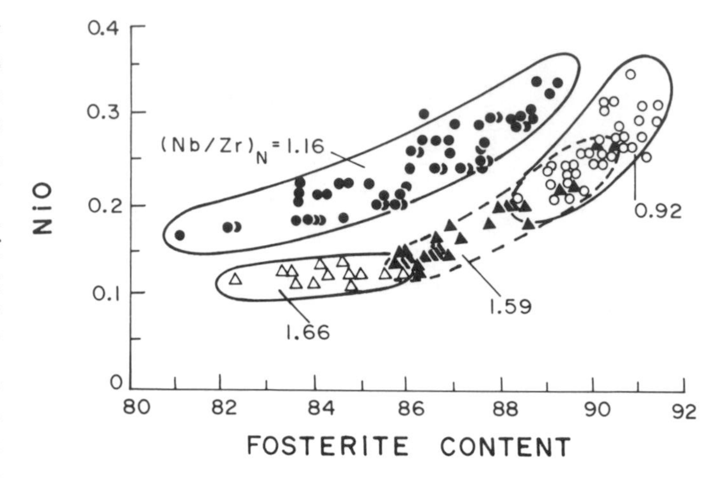

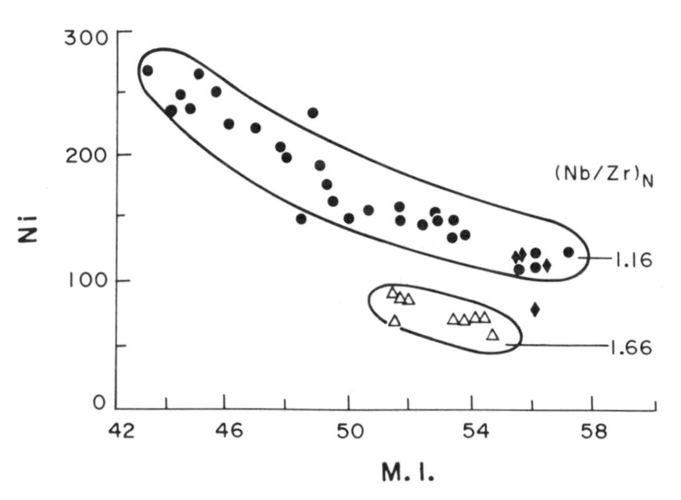

Figure 12. Top: Variation in NiO content with respect to forsterite content of olivine phenocrysts in FAMOUS basalts. Closed circle–olivine basalts; open triangle–plag-cpx basalts; filled triangles–plag-phyric basalts; and open circles–picritic basalts. Note also the different Nb/Zr ratios (normalized to chondrites) and NiO content of these different basalt petrographic types. Bottom: Variation in Ni content with respect to MI = (100 (FeO = Fe_2O_3)/(MgO + Fe_2O_3, in wt %) in phyric to sparsely phyric FAMOUS basalts. Symbols: dots–olivine basalt; diamond–plag-ol basalts; and triangles–plag-cpx basalts. Note also different Nb/Zr ratios and Ni content of these different basalt petrographic types. Adapted from Le Roex and others (1981).

Finally, concentration ratios of large to smaller incompatible ions are more variable than the above isotopic ratios and show an asymmetrical zonation across the FAMOUS valley floor. However, their ranges do not exceed those found along the gradient from the Azores Platform down to the Hayes FZ, possibly suggesting local variations in partial melting conditions.

These early petrological results have led Bryan and Moore

(1977) to suggest that FAMOUS is underlain by a small shallow magma chamber that is compositionally zoned asymmetrically outward and perpendicular to the ridge axis, with gradients and bounding walls steeper on the west side. Furthermore, they noted that enrichments in incompatible elements on the flank of this magma chamber appear in excess of that predicted by fractional crystallization models. This led them to suggest that the trace and minor elements could have diffused outward (perhaps as volatile complexes) from the center of the magma chamber along a downward temperature gradient, since associated geothermometric studies (Hekinian and others, 1976) have also shown that the olivine and picritic basalts may have temperatures of 40–60°C in excess of the plag-cpx-ol-basalts present on the flanks. However, even among the picrite and olivine basalts, rare earth patterns vary significantly and some crossing of the patterns exists. This has led Langmuir and others (1977) to rule out lava derivation from a single magma chamber where mixing would probably take place. Instead they proposed a direct derivation from the mantle of distinct magma batches produced by a complex process of partial melting and melt segregation variable in time and place. They named the process dynamical melting.

Independently, Bougault and Hekinian (1974) and Le Roex and others (1981) have shown that at least two, and possibly more, independent magma types and evolution series, derived from distinct mantle regions (depth), were required for the variation of Ni with respect to FeO/MgO in the glass, olivine, or spinel phenocrysts present. The recognized petrographic lava types shown in Figure 12 also have distinct ranges of $(Nb/Zr)_N$ ratio (an analogue of $(La/Sm)_N$). These differences have been taken to indicate that the different magma groups observed in FAMOUS could not have been processed through the same magma chamber and would preclude any direct interrelationships with open or closed system fractional crystallization (Le Roex and others, 1981).

Subsequent *Alvin* dives just south of FAMOUS, in the so-called Narrowgate and AMAR regions south of Fracture Zone B have revealed a picture quite different from FAMOUS: possibly cyclic conditions or alternation between periods of more intense tectonic activity followed by periods of intense volcanic activity on a time scale of perhaps 0.2 to 0.5 m.y. (Stakes and others, 1984). The AMAR lavas show a more limited range of compositions. The variations are more stratigraphic than lateral as in the FAMOUS region. The FAMOUS and AMAR results and the fact that seismic and thermal studies in the region preclude the size of a present-day magma chamber beneath the FAMOUS valley from being more than about 2 km^2 have led Stakes and others (1984) to suggest the presence of a shallow lenticular magma chamber beneath this segment of the MAR axis, which would fluctuate in length (expand and contract) with time. The chamber would currently be in a contraction period, so that primitive melts from the mantle are leaking almost straight through to the surface at Mt. Venus and Mt. Pluto, whereas mixed magmas from a steady-state reservoir would be reaching the surface further south (e.g., Mt. Mars and AMAR). Trace element and isotopic data are currently being gathered to test this model.

In a similar vein, White and Bryan (1977) have proposed a transitory magma chamber beneath FAMOUS that is periodically renewed by episodes of increased melting in the mantle. Finally White (1979), after restudying the FAMOUS problem and establishing a zonation in ratios as close in incompatibility as Sm/Nd, modified his model to include separate small magma chambers or pockets. He suggested that there is either (1) a broad zonation in degree of melting that may be a common feature of Mid-Ocean Ridges or (2) a gradual temporal increase in degrees of melting, causing a progressive LILE-depletion of primary magmas separating and ascending from the mantle. However, in both cases central lavas experienced less fractional crystallization because the residence time of magma through the crust or in separate small magma chambers (or pockets) increases with distance from the central axis.

In conclusion, the presence of a continuous magma chamber at a steady state of magma supply and eruption beneath the MAR (e.g., Walker and others, 1979; O'Hara, 1977) is clearly ruled out by the observations. The presence of small, discontinuous magma chambers at variable transient stages may be more appropriate for the slow spreading MAR, as well as from a petrological viewpoint. However, the magma chambers have yet to be identified geophysically.

CONCLUDING REMARKS

The North Atlantic morphology, gravity field, crustal thickness, lava production rates, and geochemical diversity of eruptive products strongly suggest that unusual processes, superimposed over those associated with accretion of new crust by seafloor spreading, are taking place over and beneath the Jan Mayen, Iceland, and Azores Plateaus. Most intense tectonic and magmatic activities over these regions do not always coincide with the MAR axis and appear variable in time and space (e.g., Saemundsson, this volume; and Bryan and Frey, this volume). The propagation of new secondary rifts, parallel to the main MAR axis, apparently ephemeral in time, may contribute to the unusual conditions over these hotspots, such as the tectonic complexities and the more extreme variability of their magmatic products (see Pálmason, this volume). The frequencies of such variations remain to be established, partly because of the poor resolution of our geochronologic methods, insufficient sampling density for detecting high-frequency variation of relatively young events, or simply because our observations provide only an instantaneous picture and have not yet spanned one magmatic cycle or episode for time scales of the order of a few tens of years to a few hundred thousands of years.

It is clear that these *wet hotspots* are underlain by mantle domains significantly richer in volatiles as well as in elements that are heat and radiogenic isotope producing (K, U, Th, and Sm). These enrichments may be at the source of such unusually intense magmatic and tectonic activities over these platforms. However, detailed scrutiny and comparison of isotopic covariation and systematics between these regions suggest that these mantle domains have been isolated from each other over a time scale greater than that of the opening of the North Atlantic, and their origins in the mantle must be several: some perhaps reflecting recycling into the mantle of older earth's crust (e.g., Azores, Jan Mayen); others, more primary and possibly related to heterogeneities resulting from the mode of formation of the earth by accretion (e.g., Iceland).

The study of secular variation in geochemical enrichments is in an infantile stage but needs to be pursued actively since it provides a means to study the scale of these mantle heterogeneities, their dynamics, and probable origin. Where sufficiently dense sampling over long enough time spans (e.g., opening of the North Atlantic) has been possible, such as at the latitude of Iceland (Schilling and Noe Nygaard, 1974; Schilling and others, 1982), such studies have revealed very regular secular decreases in geochemical enrichments over a period of several million years, followed by periods of great variability in amplitude and frequency as observed over the last 700,000 years of Iceland history, or on the Faeroes Islands just at the onset of the Atlantic Opening. It is this kind of evidence and variable plume flow implied by topography (Plate 8A; Vogt, 1971) which has led to the suggestion that such heterogeneities are more blob-like, a few hundred kilometers in size, rather than represented by continuous upwelling plumes as initially suggested by Morgan (1971). Mantle heterogeneities of a scale of a few meters to several kilometers, veined- or schlieren-like, have also been suggested (e.g., Wood, 1979). The discrepancy in these interpretations, we feel, is partly a problem of (1) irregular sampling density, (2) where the sampling is done with respect to the large-scale gradients observed along the MAR axis and (3) the lack of sufficient attention paid to the apparent nature of secular variations over these platforms (i.e., frequency of the geochemical anomalies we have just seen appear to be time-dependent as well). For example, we have seen that if we were to limit sampling to Iceland terrain formed during the last 700,000 years, it would suggest small-scale mantle heterogeneities, whereas sampling of similar density of Tertiary terrain so far would suggest large-scale uniformity.

Another possibility of studying the scales and origins of mantle heterogeneities beneath the North Atlantic has been to look at variations of only young volcanic products along the MAR axis. Large-scale gradients of the order of either 500–1000 km or 100–200 km scale have been noted, as well as discontinuities at some fracture zones. Interpretation of their origin is nonunique. But some regularities are also emerging if the width of the geochemical anomalies is considered within the reference of fixed hotspots and the migrating MAR axis.

Better and more systematic sampling in space and time is clearly required in the future to choose between these various models and asserting further emerging regularities. However, we need to remember that our ability to detect mantle heterogeneities may still be partly biased by the sampling problem inherent in our indirect way of probing the mantle by means of lavas, namely: (1) the relative lengthscale and geometry of the melting zone relative to that of the mantle heterogeneities sought and (2) the presence or absence of well-mixed magma chambers and their individual lengthscales relative to that of the sampling interval.

Even in such areas as Iceland (e.g., Einarsson, this volume; Saemundsson, this volume) or the FAMOUS-AMAR region on the flank of the Azores anomaly, two regions, which primarily for logistic reasons, have been studied in much greater detail using various modern detection techniques, it has not yet been possible to decide with certainty whether magma chambers are indeed present beneath the active ridge axis and if so, what may be their size, lifetime, and so on. A detailed study, similar to FAMOUS, along a so-called *normal* ridge segment where geochemical data suggest greater simplicity and uniformity remains to be undertaken and should be among our future priorities.

REFERENCES

Anderson, A. T.
1974 : Chlorine, sulfur and water in magmas and oceans: Geological Society of America Bulletin, v. 85, p. 1485–1492.

ARCYANA.
1975 : Transform fault and rift valley from bathyscaphe and diving saucer: Science, v. 190, p. 108–116.
1977 : Rocks collected in the FAMOUS area by bathyscaphe and diving saucer from rift valleys of the Mid-Atlantic Ridge: Petrological diversity and structural settling: Deep Sea Research,v . 24, p. 565–569.

Aumento, F.
1967 : Magmatic evolution on the Mid-Atlantic Ridge: Earth and Planetary Science Letters, v. 2, p. 225–230.
1968 : The Mid-Atlantic Ridge near 45°N II. Basalts from the area of Confederation Peak: Canadian Journal of Earth Sciences, v. 5, p. 1–21.

Aumento, F.; Loncarevic, B. D., and Ross, D. I.
1971 : Hudson geotraverse: geology of the Mid-Atlantic ridge at 45°N: Philosophical Transactions of the Royal Society of London, v. A268, p. 623–650.

Ballard, R. D., Bryan, W. B., Heirtzler, J. R., Keller, G., Moore, J. G., and van Andel, Tj. H.
1975 : Manned submersible observations in the FAMOUS area, Mid-Atlantic Ridge: Science, v. 190, p. 103–108.

Basaltic Volcanism Study Project
1981 : Basalt Volcanism on the Terrestrial Planets: New York, Pergamon Press, Inc., 1286 p.

Bence, A. E.
1966 : Rubidium-strontium isotopic relationships in oceanic basalts: [Ph.D thesis]: Massachusetts Institute of Technology, 199 p.

Boeggild, O. B.
1900 : The deposits of the sea-bottom, in the Danish Ingolf-expedition, v. 1, Bianco Luno (F. Dreyer), Printer to the Court, Copenhagen, v. 1, pt. 3, p. 1–87.

Bougault, H.
1980 : Contribution des éléments de transition à la compréhension de la genèse des basalts océaniques: [Thesis], Université de Paris VII, 221 p.

Bougault, H., and Hekinian, R.
1974 : Rift valley in the Atlantic Ocean near 36°50′N: Petrology and Geochemistry of Basaltic Rocks: Earth and Planetary Science Letters, v. 24, p. 249–261.

Bougault, H., and Treuil, M.

1980 : Mid-Atlantic Ridge: Zero-age geochemical variations between Azores and 22°N: Nature, v. 286, p. 209–212.

Bougault, H., Joron, J-L., and Treuil, M.
1979 : Alteration, fractional crystallization, partial melting, mantle properties from trace elements in basalts recovered in the North Atlantic; in Deep Drilling Results in the Atlantic Ocean: Ocean Crust, eds. M. Talwani, D. G. Harrison and D. E. Hayes; American Geophysical Union Maurice Ewing Series, v. 2, p. 352–368.

Bryan, W. B.
1979 : Regional variation and petrogenesis of basalt glasses from the FAMOUS area, Mid-Atlantic Ridge: Journal of Petrology, v. 20, p. 293–325.

Bryan, W. B., and Moore, J. G.
1977 : Compositional variations of young basalts in the Mid-Atlantic Ridge rift valley near latitude 36°49′N: Geological Society of America Bulletin, v. 88, p. 556–570.

Bryan, W. B., Thompson, G., and Ludden, J. N.
1981 : Compositional variation in normal MORB from 22°–25°N: Mid-Atlantic Ridge and Kane Fracture Zone: Journal of Geophysical Research, v. 86, p. 11815–11836.

Bryan, W. B., Thompson, G., and Michael, P. J.
1979 : Compositional variation in a steady-state zoned magma chamber: Mid-Atlantic Ridge at 36°50′N: Tectonophysics, v. 55, p. 63–85.

Bullard, E. C.
1971 : A discussion on the petrology of igneous and metamorphic rocks from the ocean floor: Philosophical Transactions of the Royal Society of London, v. A268, p. 383.

Bullard, E. C., and Day, A.
1961 : The flow of heat through the floor of the Atlantic Ocean: Geophysical Journal of the Royal Astronomical Society, v. 4, p. 282–292.

Bullard, E. C., Maxwell, A. E., and Revelle, R.
1956 : Heat flow through the deep sea floor: Advances in Geophysics, v. 3, p. 153–181.

Carlson, R. W., Macdougall, J. D., and Lugmair, G. W.
1978 : Differential Sm/Nd evolution in oceanic basalts: Geophysical Research Letters, v. 5, p. 229–232.

Carr, D. R., and Kulp, J. L.
1953 : Age of a Mid-Atlantic Ridge basalt boulder: Geological Society of America Bulletin, v. 64, p. 253–254.

Chase, C.
1979 : Asthenospheric counterflow: a kinematic model: Geophysical Journal of the Royal Astronomical Society, v. 56, p. 1–18.

Cohen, R. S., and O'Nions, R. K.
1982 : The lead, neodymium and strontium isotopic structure of ocean ridge basalts: Journal of Petrology, v. 25, p. 299–324.

Cohen, R. S., Evansen, N. M., Hamilton, P. J., and O'Nions, R. K.
1980 : U-Pb, Sm-Nd and Rb-Sr systematics of mid-ocean ridge basalt glasses: Nature, v. 283, p. 149–153.

Corliss, J. B.
1970 : Mid-Ocean Ridge Basalts. I. The origin of submarine hydrothermal solutions: II. Regional diversity along the Mid-Atlantic Ridge [Ph.D. diss.], University of California at San Diego, 148 p.

Correns, C. W.
1930 : Uber einen Basalt vom Boden des atlantischen Ozeans and Seine Zersetzungsrinde: Chemie der Erde, v. 5, p. 76–86.

Coryell, C. D., Chase, J. W., and Winchester, J. W.
1963 : A procedure for geochemical interpretation of terrestrial rare-earth abundance patterns: Journal of Geophysical Research, v. 68, p. 559–566.

Craig, H., and Lupton, J. E.
1981 : Helium-3 and mantle volatiles in the ocean and the oceanic crust: in The Sea, v. 7, New York, Wiley-Interscience, p. 391–428.

Daly, R. A.
1942 : The Floor of the Ocean, New Light on Old Mysteries: University of North Carolina Press, Chapel Hill, 177 p.

Darwin, C.
1876 : Geological Observations on the Volcanic Islands and Parts of South America during the Voyage of the H.M.S. Beagle: London, Smith Elder and Company, 647 p.

Davies, G. F.
1981 : Earth's neodymium budget and structure and evolution of the mantle: Nature, v. 290, p. 208–213.

DeBoer, J., Schilling, J-G., and Krause, D. C.
1969 : Magnetic polarity of pillow basalts from the Reykjanes Ridge: Science, v. 166, p. 996–998.

Delaney, J. R., Muenow, D. W., and Graham, D. G.
1978 : Abundance and distribution of water, carbon and sulfur in the glassy rims of submarine pillow basalts: Geochimica et Cosmochimica Acta, v. 42, p. 581–594.

De Lapparent, A.
1900 : Traité de Géologie, Volumes 1 to 3, Fourth Edition: Paris, France, Masson, 1912 p.

De Paolo, D. J., and Wasserburg, G. J.
1976 : Inferences about magma sources and mantle structure from variations of $^{143}Nd/^{144}Nd$: Geophysical Research Letters, v. 3, p. 743–746.

Dietz, R. S.
1961 : Continent and ocean basin evolution by spreading of the sea floor: Nature, v. 190, p. 854–857.

Drake, C. L., and Girdler, R. W.
1982 : History of rift studies; in Continental and Oceanic Rifts, ed. G. Palmason; American Geophysical Union Dynamics, Series 8, p. 5–15.

Duncan, R. A.
1984 : Age progressive volcanism in the New England Seamounts and the opening of the Central Atlantic Ocean; Journal of Geophysical Research, v. 89, p. 9980–9990.

Dupré, B., and Allegre, C. J.
1980 : Pb-Sr-Nd isotopic correlation and the chemistry of the N. Atlantic mantle: Nature, v. 286, p. 17–22.

Dupré, B., Lambret, B., Rousseau, D., and Allegre, C. J.
1981 : Limitations on the scale of mantle heterogeneities under oceanic ridges, lead and strontium isotope variations in the FAMOUS and CYAMEX areas: Nature, v. 294, p. 552–554.

Engel, C. G., and Engel, A. E.
1961 : Composition of basalt cored in Mohole Project: American Association of Petroleum Geologists Bulletin, v. 45, p. 1799.
1963 : Basalts dredged from the north-eastern Pacific Ocean floor: Science, v. 140, p. 1321–1325.

Engel, A.E.J., and Engel, C. G.
1964 : Composition of basalts from the Mid-Atlantic Ridge: Science, v. 144, p. 1330–1333.

Erlank, A. J., and Kable, E.J.D.
1976 : The significance of incompatible elements in Mid-Atlantic Ridge basalts from 45°N with particular reference to Zr/Nb: Contributions to Mineralogy and Petrology, v. 54, p. 281–291.

Fisher, O.
1881 : Physics of the Earth's Crust: London, MacMillan and Company, 299 p.
1889 : Physics of the Earth's Crust, Second Edition: London, MacMillan and Company, 391 p.

Flower, M.F.J., Schmincke, H-U., and Thompson, R. N.
1975 : Phlogopite stability and the $^{87}Sr/^{86}Sr$ step in basalts along the Reykjanes Ridge: Nature, v. 254, p. 404–405.

Fox, P. J., and Gallo, D. G.
1984 : A tectonic model for ridge-transform-ridge-plate boundaries: implications for the structure of oceanic lithosphere: Tectonophysics, v. 104, p. 205–242.

Frey, F. A., and Haskin, L.
1964 : Rare earths in Oceanic Basalts: Journal of Geophysical Research, v. 69, p. 775–780.

Frey, F. A., Haskin, M. A., Poetz, J. A., and Haskin, L. A.

1968 : Rare Earth Abundances in some basic rocks: Journal of Geophysical Research, v. 73, p. 6085–6098.
Ganapathy, R., and Anders, E.
1974 : Bulk compositions of the moon and earth, estimated from meteorites: Geochimica et Cosmochimica Acta Supplement 5, p. 1181–1206.
Gast, P. W.
1960 : Limitations on the composition of the upper mantle: Journal of Geophysical Research, v. 65, p. 1287–1297.
1965 : Terrestrial ratio of potassium to rubidium and the composition of the Earth's mantle: Science, v. 147, p. 858–860.
Gast, P. W., Tilton, G. R., and Hedge, C.
1964 : Isotopic composition of lead and strontium from Ascension and Gough Islands: Science, v. 145, p. 1181–1185.
Goldschmidt, V. M.
1958 : Geochemistry: Oxford, Clarendon Press, 730 p.
Graham, A. L., and Nicholls, G. D.
1969 : Mass spectrographic determinations of lanthanide element contents in basalts: Geochimica et Cosmochimica Acta, v. 33, p. 555–568.
Hall, M.
1876 : Note upon a portion of basalt from Mid-Atlantic: Mineralogical Magazine, v. 1, p. 1–13.
Hart, S. R., Schilling, J-G., and Powell, J. L.
1973 : Basalts from Iceland and along the Reykjanes Ridge: Sr Isotope Geochemistry: Nature: Physical Science, v. 246, p. 104–107.
Heezen, B. C.
1960 : The rift in the ocean floor: Scientific American, v. 203, p. 98–110.
Heirtzler, J. R., and van Andel, Tj. H.
1977 : Project FAMOUS: Its origin, programs and setting: Geological Society of America Bulletin, v. 88, p. 481–487.
Hekinian, R., Moore, J. G., and Bryan, W. B.
1976 : Volcanic rocks and processes of the Mid-Atlantic Ridge rift valley Near 36°49′N: Contribution to Mineralogy and Petrology, v. 58, p. 83–110.
Hermes, O. D., and Schilling, J-G.
1976 : Olivine from Reykjanes Ridge and Iceland tholeiites, and its significance to the two-mantle source model: Earth and Planetary Science Letters, v. 29, p. 7–20.
Hess, H. H.
1954 : Geological hypotheses and the earth's crust under the oceans: Proceedings of the Royal Society of London, v. A 222, p. 341–348.
1955 : The oceanic crust: Journal of Marine Research, v. 14, p. 423–439.
1962 : History of Ocean Basins; in Petrologic Studies: in A Volume to Honor A. F. Buddington, eds. A. E. Engel, J. James and B. F. Leonard; Geological Society of America, p. 599–620.
1965 : Mid-oceanic ridges and tectonics of the sea-floor; in Submarine Geology and Geophysics, eds. W. F. Whittad and R. Bradshaw; Butterworth, London, p. 317–332.
Hofmann, A. W., and White, W. M.
1983 : Ba, Rb, and Cs in the Earth's mantle: Zeitschrift fur Naturforschung, v. 38a, p. 256–266.
Holmes, A.
1928 : Radioactivity and Earth movements: Transactions of the Geological Society of Glasgow, v. 18, p. 579–606.
Ito, E., White, W. M., von Drach, V., Hofmann, A. W., and James, D. E.
1980 : Isotopic studies of ocean ridge basalts: Annual Report of the Director, Department of Terrestrial Magnetism, Carnegie Institution, p. 465–471.
Jagoutz, E., Palme, H., Baddenhausen, H., Blum, K., Cendales, M., Dreibus, G., Spettel, B., Lorenz, V., and Wanke, H.
1979 : The abundances of major, minor and trace elements in the earth's mantle as derived from primitive ultramafic nodules: Geochimica et Cosmochimica Acta, Supplement 11, v. 2, p. 2031–2050.
Javoy, M., Pineau, F., and Iiyama, I.
1978 : Experimental determination of the isotopic fractionation between gaseous CO_2 and carbon dissolved in tholeiitic magma: Contributions to Mineralogy and Petrology, v. 67, p. 35–39.
Kay, R., Hubbard, N. J., and Gast, P. W.
1970 : Chemical characteristics and origin of oceanic ridge volcanic rocks: Journal of Geophysical Research, v. 75, p. 1585–1613.
Krause, D., and Schilling, J-G.
1969 : Dredged basalts from the Reykjanes Ridge, North Atlantic: Nature, v. 224, p. 791–793.
Kurz, M. D., Jenkins, W. J., and Hart, S. R.
1982a: Helium-isotopic systematics of oceanic islands and mantle heterogeneity: Nature, v. 297, p. 43–47.
Kurz, M. D., Jenkins, W. J., Schilling, J-G., and Hart, S. R.
1982b: Helium isotopic variations in the mantle beneath the North Atlantic Ocean: Earth and Planetary Science Letters, v. 58, p. 1–14.
Langmuir, C. H., Bender, J. F., Bence, A. E., Hanson, G. N., and Taylor, S. R.
1977 : Petrogenesis of basalts from the FAMOUS area: Mid-Atlantic Ridge: Earth and Planetary Science Letters, v. 36, p. 133–156.
Langmuir, C. H., Vocke, R. D., Jr., Hanson, G. N., and Hart, S.
1978 : A general mixing equation with applications to Icelandic basalts: Earth and Planetary Science Letters, v. 37, p. 380–392.
Le Douaran, S., and Francheteau, J.
1981 : Axial depth anomalies from 10 to 50° north along the Mid-Atlantic Ridge: correlation with other mantle properties: Earth and Planetary Science Letters, v. 54, p. 29–47.
Le Roex, A. P., Erlank, A. J., and Needham, H. D.
1981 : Geochemical and mineralogical evidence for the occurrence of at least three distinct magma types in the FAMOUS region: Contributions to Mineralogy and Petrology, v. 77, p. 24–37.
Machado, N., Brooks, J. N., and Thompson, G.
1982 : Fine-scale isotopic heterogeneity in the sub-Atlantic mantle: Nature, v. 295, p. 226–228.
Mathez, E. A.
1976 : Sulfur solubility and magmatic sulfides in submarine basalt glass: Journal of Geophysical Research, v. 81, p. 4269–4276.
McBirney, A. R.
1963 : Factors governing the nature of submarine volcanism: Bulletin Volcanologique, v. 26, p. 455–469.
McIntire, W. L.
1963 : Trace element partition coefficients—a review of theory and applications to geology: Geochimica et Cosmochimica Acta, v. 27, p. 1209–1264.
Melson, W. G., and O'Hearn, T.
1979 : Basaltic glass erupted along the Mid-Atlantic Ridge between 0–37° N: Relationships between composition and latitude: American Geophysical Union, Maurice Ewing Series, v. 2, p. 249–261.
Melson, W. G., and van Andel, Tj. H.
1966 : Metamorphism in the Mid-Atlantic Ridge, 22°N latitude: Marine Geology, v. 4, p. 165–186.
Michael, P. J., and Bonatti, E.
1983 : Peridotites from DSDP Leg 82 and Oceanographer F.Z.: EOS, v. 64, p. 345.
Miyashiro, A., Shido, F., and Ewing, M.
1969 : Diversity and origin of abyssal tholeiite from the Mid-Atlantic Ridge near 24° and 30° north latitude: Contributions to Mineralogy and Petrology, v. 23, p. 38–52.
Moore, J. G.
1965 : Petrology of deep-sea basalt near Hawaii: American Journal of Science, v. 263, p. 40–52.
1979 : Vesicularity and CO_2 in mid-ocean ridge basalt: Nature, v. 282, p. 250–253.
Moore, J. G., and Calk, L.
1971 : Sulfide spherules in vesicles of dredged pillow basalt: American Mineralogist, v. 56, p. 476–488.
Moore, J. G., and Schilling, J-G.
1973 : Vesicles, water and sulphur in Reykjanes Ridge basalts: Contributions to Mineralogy and Petrology, v. 41, p. 105–118.
Moore, J. G., Batchelder, J. N., and Cunningham, C. G.

1977 : CO_2-filled vesicles in mid-ocean basalt: Journal of Volcanology and Geothermal Research, v. 2, p. 309–327.

Morgan, W. J.
1971 : Convection plumes in the lower mantle: Nature, v. 230, p. 42–43.
1978 : Rodriguez, Darwin, Amsterdam, A second type of hotspot island: Journal of Geophysical Research, v. 83, p. 5355–5360.
1981: Hotspot tracks and the opening of the Atlantic and Indian Oceans; in The Sea, v. 7, The Oceanic Lithosphere, ed. C. Emiliani; New York, Wiley-Interscience, p. 443–487.
1983 : Hotspot tracks and the early rifting of the Atlantic; in Processes of continental rifting, ed. Paul Morgan; Tectonophysics, v. 94, p. 123–139.

Muir, I. D., and Tilley, C. E.
1964 : Basalts from the northern part of the rift zone of the Mid-Atlantic Ridge: Journal of Petrology, v. 5, p. 409–434.
1966 : Basalts from the northern part of the Mid-Atlantic Ridge II, the Atlantis Collection near 30°N: Journal of Petrology, v. 7, p. 193–201.

Muir, I. D., Tilley, C. E., and Scoon, J. H.
1964 : Basalts from the northern part of the rift zone of the Mid-Atlantic Ridge: Journal of Petrology, v. 5, p. 409–434.

Murray, J., and Hjort, J.
1912 : The Depths of the Ocean: London, MacMillan, 821 p.

Murray, J., and Peake, R. E.
1904 : On recent contributions to our knowledge of the floor of the North Atlantic Ocean: Royal Geographical Society, v. 1, p. 1–35.

Mysen, B. O., Arculus, R. J., and Egger, D.
1975 : Solubility of carbon dioxide in melts of andesite, tholeiite and olivine nephelinite composition to 30 kbar pressure: Contributions to Mineralogy and Petrology, v. 53, p. 227–239.

Nalwalk, A. J., Hersey, J. B., Reitzel, J. S., and Edgerton, H. E.
1961 : Improved techniques of deep-sea rock dredging: Deep Sea Research, v. 3–4, p. 301–302.

Needham, H. D., and Francheteau, J.
1974 : Some characteristics of the rift valley in the Atlantic Ocean near 36°48′North: Earth and Planetary Science Letters, v. 22, p. 29–43.

Nicholls, G. D., Nalwalk, A. J., and Hays, E. E.
1964 : The nature and composition of rock samples dredged from the Mid-Atlantic Ridge between 22°N and 52°N: Marine Geology, v. 1, p. 333–343.

O'Hara, M. J.
1973 : Non-primary magmas and dubious mantle plume beneath Iceland: Nature, v. 243, p. 507–508.
1975 : Is there an Icelandic mantle plume?: Nature, v. 253, p. 708–710.
1977 : Geochemical evolution during fractional crystallization of a periodically refilled magma chamber: Nature, v. 266, p. 503–507.

O'Nions, R. K., and Grönvold, K.
1973 : Petrogenetic relationships of acid and basic rocks in Iceland: Sr-isotopes and rare-earth elements in late and post-glacial volcanics: Earth and Planetary Science Letters, v. 19, p. 397–409.

O'Nions, R. K., Hamilton, P. J., and Evansen, N. M.
1977 : Variations in $^{143}Nd/^{144}Nd$ and $^{87}Sr/^{86}Sr$ ratios in oceanic basalts: Earth and Planetary Science Letters, v. 34, p. 13–22.

O'Nions, R. K., and Pankhurst, R. J.
1974 : Petrogenetic significance of isotope and trace element variations in volcanic rocks from the Mid-Atlantic: Journal of Petrology, v. 15, p. 603–634.

Peter, G., and Westbrook, G. K.
1976 : Tectonics of Southwestern North Atlantic and Barbados Ridge Complex: American Association of Petroleum Geologists Bulletin, v. 60, p. 1078–1106.

Phillips, J. D., and Fleming, H. S.
1978 : Multi-beam sonar study of the Mid-Atlantic Ridge rift valley, 36°–37°N: Geological Society of America Map and Chart Series, MC-19.

Pineau, F., and Javoy, M.
1983 : Carbon isotopes and concentrations in mid-oceanic ridge basalts: Earth and Planetary Science Letters, v. 62, p. 239–257.

Poreda, R., Craig, H., and Schilling, J-G.
1980 : $^3He/^4He$ variations along the Reykjanes Ridge: EOS, v. 61, p. 1158.

Quintino, J., and Machado, F.
1977 : Heat flow and the Mid-Atlantic rift volcanism of San Miguel Island, Azores: Tectonophysics, v. 41, p. 173–179.

Quon, S. H., and Ehler, E. G.
1963 : Rocks of Northern Part of Mid-Atlantic Ridge: Geological Society of America Bulletin, v. 74, p. 1–8.

Renard, A.
1882 : On the petrology of St. Paul's Rocks, Appendix B: Narrative of the Challenger Report, v. 2, 29 p.

Richard, P., Shimizu, N., and Allegre, C. J.
1976 : $^{143}Nd/^{144}Nd$, a natural tracer: An application to oceanic basalts: Earth and Planetary Science Letters, v. 31, p. 269–278.

Richter, F. M., and Ribe, N. M.
1979 : On the importance of advection in determining the local isotopic composition of the mantle: Earth and Planetary Science Letters, v. 43, p. 212–222.

Rideout, M., and Schilling, J-G.
1983 : IPOD Leg 82: REE, $^{143}Nd/^{144}Nd$ and $^{87}Sr/^{86}Sr$: Mantle heterogeneities in time and space: Transactions of the American Geophysical Union, v. 64, p. 345.

Rothé, J. P.
1954 : La zone séismique médiane Indo-Atlantique: Proceedings, Royal Society of London, v. 221, p. 387–397.

Rowe, E. C., and Schilling, J-G.
1979 : Fluorine in Iceland and Reykjanes Ridge basalts: Nature, v. 279, p. 33–37.

Schilling, J-G.
196 : Rare Earth Fractionation in Hawaiian Volcanic Rocks: [Ph.D. Diss.], Massachusetts Institute of Technology, Cambridge, Massachusetts, 390 p.
1971 : Sea-floor evolution: rare earth evidence: Philosophical Transactions of the Royal Society of London, v. 268, p. 663–706.
1972 : Rare earths in basalts; in Encyclopedia of Earth Sciences, Geochemistry and Environmental Sciences, Volume 4A, ed. R. W. Fairbridge; New York, Reinhold Van Nostrand, v. 4A, p. 1029–1039.
1973a: Iceland mantle plume, geochemical evidence along Reykjanes Ridge: Nature, v. 242, p. 565–571.
1973b: Iceland mantle plume: Nature, v. 246, p. 141–143.
1985 : Upper-Mantle Heterogeneities and Dynamics: Nature, v. 314, p. 62–67.

Schilling, J-G., and Noe-Nygaard, A.
1974 : Faeroe-Iceland plume: Rare earth evidence: Earth and Planetary Science Letters, v. 24, p. 1–14.

Schilling, J-G., Bergeron, M. B., and Evans, R.
1980 : Halogens in the mantle beneath the North Atlantic: Philosophical Transections of the Royal Society of London, v. A297, p. 147–178.

Schilling, J-G., Meyer, P. S., and Kingsley, R. H.
1982 : Evolution of the Iceland hotspot: Nature, v. 296, p. 313–320.

Schilling, J-G., Zajac, M., Evans, R., Johnston, T., White, W., Devine, J. D., and Kingsley, R.
1983 : Petrologic and geochemical variations along the Mid-Atlantic Ridge from 29°N to 73°N: American Journal of Science, v. 283, p. 510–586.

Schilling, J-G., Thompson, G., Kingsley, R., and Humphris, S.
1985 : Hotspot—Migrating Ridge Interaction in the South Atlantic: Geochemical Evidence: Nature, v. 313, p. 187–191.

Schilling, J-JG., and Winchester, J. W.
1967 : Rare earth fractionation and magmatic processes; in Mantles of the Earth Terrestrial Planets, ed. S. K. Runcorn; London, Wiley, p. 267–283.

Shand, S. J.
1949 : Rocks of the Mid-Atlantic Ridge: Journal of Geology, v. 57, p. 89–91.

Shibata, T., Thompson, G., and Frey, F. A.
1979 : Tholeiitic and Alkali Basalts from the Mid-Atlantic Ridge at 43°N: Contributions to Mineralogy and Petrology, v. 70, p. 127–141.
Sigurdsson, H.
1981 : First-order major element variation in basalt glasses from the Mid-Atlantic Ridge: 29°N to 73°N: Journal of Geophysical Research, v. 86, p. 9483–9502.
Sigvaldason, G. E.
1974 : Basalts from the centre of the assumed Icelandic Mantle Plume: Journal of Petrology, v. 15, p. 497–524.
Sigvaldason, G. E., Steinthorsson, S., Oskarsson, N., and Imsland, P.
1974 : Compositional variation in recent Icelandic tholeiites and the Kverkfjoll hot spot: Nature, v. 251, p. 579–582.
Sleep, N. H.
1984 : Tapping of magmas from ubiquitous Mantle Heterogeneities, an alternative to mantle plumes: Journal of Geophysical Research, v. 89, p. 10,029–10,041.
Stakes, D., Shervais, J. W., and Hopson, C. A.
1984 : The volcano-tectonic cycle of the FAMOUS and AMAR valleys, Mid-Atlantic Ridge (36°47′N): Evidence from basalt glass and phenocryst compositional variations for a steady-state magma chamber beneath the valley midsections: Journal of Geophysical Research, v. 89, p. 6995–7028.
Sun, S.-S.
1980 : Lead isotope study of young volcanic rocks from mid-ocean ridges, ocean islands and island arcs: Philosophical Transactions of the Royal Society of London, v. A 297, p. 409–445.
Sun, S.-S., and Jahn, B.
1975 : Lead and strontium isotopes in post-glacial basalts from Iceland: Nature, v. 255, p. 527–530.
Sun, S.-S., Nesbitt, R. W., and Sharaskin, A. Y.
1979 : Geochemical characteristics of mid-ocean ridge basalts: Earth and Planetary Science Letters, v. 44, p. 119–138.
Sun, S.-S., Tatsumoto, M., and Schilling, J-G.
1975 : Mantle plume mixing along the Reykjanes Ridge axis: lead isotopic evidence: Science, v. 190, p. 143–147.
Talwani, M., and Eldholm, D.
1977 : Evolution of the Norwegian-Greenland Sea: Geological Society of America Bulletin, v. 88, p. 969–999.
Tatsumoto, M.
1978 : Isotopic composition of lead in oceanic basalt and its implication to mantle evolution: Earth and Planetary Science Letters, v. 38, p. 63–87.
Tatsumoto, M., Hedge, C. E., and Engel, A.E.J.
1965 : Potassium, Rubidium, Strontium, Thorium, Uranium, and the ratio of Strontium-87 to Strontium-86 in oceanic tholeiitic basalts: Science, v. 150, p. 886–888.
Termier, P.
1899 : Sur une tachylyte du fond de l'Atlantique Nord: Comptes Rendus de l'Académie des Sciences Francaises, v. 128, p. 849–851, p. 1256–1258.
Unni, C. K., and Schilling, J-G.
1978 : Cl and Br degassing by volcanism along the Reykjanes Ridge and Iceland: Nature, v. 272, p. 19–23.
van Andel, Tj. H., and Bowin, C. O.
1968 : Mid-Atlantic Ridge between 22° and 23° North Latitude and the tectonics of mid-ocean rises: Journal of Geophysical Research, v. 73, p. 1279–1298.
van Andel, Tj. H., Bowen, H., Sachs, P. L., and Siever, R.
1965 : Morphology and sediments of a portion of the Mid-Atlantic Ridge: Science, v. 148, p. 1214–1216.
Vine, F. J.
1966 : Spreading of the Ocean Floor: New Evidence: Science, v. 154, p. 1405–1415.
Vine, F. J. and Matthews, D. H.
1963 : Magnetic anomalies over oceanic ridges: Nature, v. 199, p. 947–949.
Vogt, P. R.
1971 : Asthenosphere motion recorded by the ocean floor south of Iceland: Earth and Planetary Science Letters, v. 13, p. 153–160.
Vogt, P. R., and Johnson, G. L.
1975 : Transform faults and longitudinal flow below the Mid-oceanic Ridge: Journal of Geophysical Research, v. 80, p. 1399–1428.
Walker, D., Shibata, T., and Long, S. E.
1979 : Abyssal tholeiites from the Oceanographer Fracture Zone II. Phase Equilibria and Mixing: Contributions to Mineralogy and Petrology, v. 70, p. 111–135.
Wandel, C. F.
1899 : Report of the Voyage in the Danish Ingolf expedition, Volume 1, Bianco Luno (F. Dreyer): Copenhagen, Printer to the Court, pt. 1, p. 1–21.
Wendlandt, R. F.
1982 : Sulfide saturation of basalt and andesite melts at high pressures and temperatures: American Mineralogist, v. 67, p. 877–885.
White, W. M.
1979 : Geochemistry of basalts from the FAMOUS area: a re-examination; in Department of Terrestrial Magnetism, Carnegie Institute Yearbook, Washington, D.C., p. 325–331.
White, W. M., and Bryan, W. B.
1977 : Sr-isotope, K, Rb, Cs, Sr, Ba, and rare-earth geochemistry of basalts from the FAMOUS area: Geological Society of America Bulletin, v. 88, p. 571–576.
White, W. M. and Hofmann, A. W.
1982 : Sr and Nd isotope geochemistry of oceanic basalts and mantle evolution: Nature, v. 296, p. 821–825.
White, W. M., and Schilling, J-G.
1978 : The nature and origin of geo-chemical variation in Mid-Atlantic Ridge basalts from the central North Atlantic: Geochimica et. Cosmochimica Acta, v. 42, p. 1501–1516.
Wilson, J. T.
1940 : The Love waves of the South Atlantic earthquake of August 28, 1933: Seismological Society of America Bulletin, v. 30, p. 273–301.
Wiseman, J.D.H.
1937 : Basalts from the Carlsberg Ridge, Indian Ocean; in the John Murray Expedition 1933–1934; British Museum of Natural History Scientific Reports, v. 3, p. 1–28.
Wood, D. A.
1979 : A variably veined suboceanic upper mantle—Genetic significance for mid-ocean ridge basalts from geochemical evidence: Geology, v. 7, p. 499–503.
Wood, D. A., Joron, J-L., Treuil, M., Norry, M., and Tarney, J.
1979 : Elemental and Sr isotope variations in basic lavas from Iceland and the surrounding ocean floor: Contributions to Mineralogy and Petrology, v. 70, p. 319–339.
Zindler, A., Hart, S. R., Frey, F. A., and Jakobsson, S. P.
1979 : Nd and Sr isotope ratios and rare earth element abundances in Reykjanes Peninsula basalts: Evidence for mantle heterogeneities beneath Iceland: Earth and Planetary Science Letters, v. 45, p. 249–262.

Manuscript Accepted by the Society November 12, 1984

ACKNOWLEDGMENT

I thank F. Frey, W. Bryan, G. Waggoner, R. Scott and W. Melson for reviews or unpublished MAR data, D. Stakes for unpublished information from the AMAR area and Renate Durig for an English translation of Correns' 1930 paper. I am also grateful to F. DiMeglio and his staff for neutron activation and facilities at the Rhode Island Nuclear Science Center. This work would not have been possible without the invaluable assistance of Richard Kingsley, Brian McCully, Holly Turton, and Grace Bode. Part of this work was supported by NSF under grants OCE8200137 and OCE8208014.

Printed in U.S.A.

The Geology of North America
Vol. M, The Western North Atlantic Region
The Geological Society of America, 1986

Chapter 10

The geology of North Atlantic transform plate boundaries and their aseismic extensions

Paul J. Fox
David G. Gallo
Graduate School of Oceanography, University of Rhode Island, Narragansett, Rhode Island 02882

INTRODUCTION

By the late 1960s the morphotectonic character (Menard and Dietz, 1952; Menard, 1955, Heezen and others, 1964a,b) and kinematic significance (Wilson, 1965; Sykes, 1963, 1965, 1967; Stover, 1966; McKenzie and Parker, 1967; Morgan, 1968; LePichon, 1968) of fracture zones had been established and it was generally accepted that fracture zones are comprised of two principal tectonic elements; a seismically active zone of strike-slip tectonism linking two offset ridge axes called the transform fault or ridge-transform-ridge (RTR) plate boundary, and two aseismic limbs, continuous with each end of the transform fault, that represent the fossil trace of the transform. With this basic tectonic framework developed, investigators turned their attention to elucidating the details about how RTR plate boundaries behave in time and space and how this fundamental plate boundary conditions the accretion of oceanic lithosphere. There are important kinematic and geometric variables that are likely to condition and govern the structural and petrologic processes operating along a RTR plate boundary. These variables will be reviewed before we summarize the salient properties of the slowly-slipping RTR plate boundaries found in the North Atlantic.

Kinematic considerations of the plate tectonic hypothesis mandate that strike-slip motion must occur along an oceanic transform fault at a rate equal to the rate of plate separation of the associated ridge segments. The total opening rates of the ridge segments comprising the mid-oceanic ridge system vary from less than 1.7 cm/yr (Arctic Ridge) to approximately 18 cm/yr (southern branch of the East Pacific Rise). Given the range in slip-rate, it is likely that the style, timing, location, and evolution of deformation and metamorphism developed within and along a transform will change in a progressive and systematic way as rates of slip range from low to high values (Fox and Gallo, 1984). Independent of slip-rate, the strike-slip environment of a RTR plate boundary will become more complex with temporal changes in tectonic geometry. For example, when the strike of a transform changes in response to changes in relative plate motion (Menard and Atwater, 1968, 1969; Dewey, 1975) a component of extension or compression will characterize the transform as its orientation changes to accommodate the new slip direction.

The geometry of RTR plate boundaries implies that an edge of oceanic lithosphere is juxtaposed against the truncated axis of accretion at the ridge-transform (RT) intersection. Given the monotonic thickening of young oceanic lithosphere with age, which can be modeled for young lithosphere (80 m.y.) equally well by a cooling plate model (Langseth and others, 1966; McKenzie, 1967; Sclater and others, 1971) or as a growing boundary layer (Turcotte and Oxburgh, 1967; Parker and Oldenburg, 1973), the thickness of a truncating edge of lithosphere will depend on its age, which varies as a function of transform slip-rate and the length of the transform offset. The lithosphere is comprised of an upper, brittle/elastic portion (mechanical lithosphere) characterized by temperatures less than approximately 975°C and an underlying portion that deforms in a ductile way (thermal lithosphere) characterized by temperatures greater than 975°C and less than 1300°C (Parsons and McKenzie, 1978). Actual brittle failure within the mechanical lithosphere will occur at much lower temperatures in the range of a few to several hundred degrees centigrade (Burr and Solomon, 1978). The thermal structure of the lithosphere in the vicinity of a RTR boundary is exceedingly variable because strong lateral and vertical temperature gradients across the plate boundary are caused by a dynamically maintained strike-slip fault system penetrating to the base of the mechanical lithosphere, which juxtaposes rock bodies of contrasting compositions and histories and is likely to act as a chimney for convective heat loss through the circulation of water. These conditions are temporally and spatially variable and lead to complex rheological behaviours. Although it is undoubtedly an oversimplification of the structure of the lithosphere at a RTR plate boundary, we take as a first approximation the total thickness of a truncating edge of lithosphere to be the thermal thickness as dictated by plate models (e.g. Parker and Oldenburg, 1973). Ridge axis segments with low total opening rates (<2 cm/yr–3 cm/yr) that abut against large-offset transforms (80

Fox, P. J., and Gallo, D. G., 1986, The geology of North Atlantic transform plate boundaries and their aseismic extensions; *in* Vogt, P. R., and Tucholke, B. E., eds., The Geology of North America, Volume M, The Western North Atlantic Region: Geological Society of America.

km–300 km) are juxtaposed against evolved lithosphere that is several tens of kilometers thick, but the contrast in the thickness of lithosphere across a ridge-transform intersection decreases markedly as the rate of accretion increases (Vogt and Johnson, 1975; Fox and Gallo, 1984).

With these first-order variables in mind, it seems intuitively clear that the oceanic lithosphere proximal to a RTR plate boundary is conditioned by the complex interplay between strain-rate, stability of the plate boundary geometry, and thermal ramifications of juxtaposing a cold edge of lithosphere against an accreting plate boundary. The primary focus of the following discussion is to review our understanding of the properties of the transforms along the Mid-Atlantic Ridge (MAR). These RTR plate boundaries are representative of that class of transform slipping at a slow rate (<3 cm/yr). Offset-lengths of the transforms are variable (20 to >100 km) and, therefore, thicknesses of the truncating edge of lithosphere will vary markedly.

CHARACTERISTICS OF SLOWLY-SLIPPING TRANSFORMS IN THE NORTH ATLANTIC

Distribution

On the basis of a compilation of widely-spaced sounding profiles in the North Atlantic, Heezen and Tharp (1968) predicted that the MAR is frequently offset by transform faults of variable length. Furthermore, their physiographic diagram suggested that the distinctive morphology of these transforms could be traced as a continuous belt across the flanks of the MAR. Subsequent surveys designed to investigate the distribution of transforms along the strike of the ridge (Fox and others, 1969a; Johnson and Vogt, 1973; Collette and others, 1974; Feden and others, 1975; Phillips and Fleming, 1978) and the continuity and morphotectonic character of fracture zones with distance away from the ridge axis (Heezen and Tharp, 1965; van Andel and others, 1969; Fox and others, 1969b; Fox, 1972; Collette and others, 1974; Vogt and Avery, 1974; Olivet and others, 1974; Rabinowitz and Purdy, 1976; Schroeder, 1977; Purdy and others, 1979) confirmed the importance of RTR plate boundaries as integral parts of the plate boundary geometry in the North Atlantic since the early Tertiary (GEBCO, 1982; Plate 2 this volume).

In the equatorial Atlantic the relief of fracture zone terrain is so great that it is relatively easy to trace fracture zone trends across the flanking basins to offsets in the outline of the conjugate continental margins (Heezen and Tharp, 1961). In general, however, at the distal edges of the flanks of the MAR in the North Atlantic it is difficult to trace the extensions of fracture zones across old oceanic basement because the igneous terrain is buried by a thickening wedge of sediments (Ewing and others, 1964). Schouten and Klitgord (1982), utilizing densely spaced aeromagnetic data for the western Central Atlantic, have defined a sequence of Mesozoic seafloor spreading lineations that are frequently offset, suggesting that the typical transform-segmented geometry of today's plate boundary characterized the Mesozoic as well. Schouten and Klitgord suggest that the segmented geometry of the ridge axis and the spacing of transforms along the MAR has remained remarkably uniform for long periods of time, surviving changes in poles of relative motion and periods of asymmetric spreading. The work of Vogt and Avery (1974), however, challenges this notion because their bathymetric and magnetic data from the flanks of the Reykjanes Ridge indicate that small offset transforms in that area form for a short period of time and then disappear never to reappear.

Topography and Structural Fabric

The hallmark of a fracture zone trace is a 10 to 30 km wide band of troughs and ridges that strike at a high angle to the ridge axis fabric and that, independent of age, stand at anomalous levels compared to surrounding sea floor. The fracture zone is defined by a central trough that strikes at a high angle to the ridge axis (Fig. 1a; e.g. Heezen and others, 1964b; van Andel and others, 1967; Fox and others, 1969a; Collette and others, 1974; Feden and others, 1975; Searle and Laughton, 1977; Phillips and Fleming, 1978). This broad elongate valley has regional inward-facing slopes that intersect to create a central axis of maximum depth. Subsidiary basins and ridges are often found along the valley and interestingly, the axis of maximum depth resides at depths several hundred (along small-offset transforms, <30 km) to a few thousand (along large-offset transforms, >100 km) meters below adjacent lithosphere of equivalent age. In some cases, one or both sides of the valley are defined by anomalously shallow, fracture-zone parallel ridges, called transverse ridges, that stand hundreds to thousands of meters above the sea floor to either side (e.g. south wall of Vema Transform—Heezen and others, 1964; van Andel and others, 1967) and that are several tens to several hundred kilometers in length. It is more typical, however, that as the fracture zone is approached along an isochron the seafloor is observed to deepen continuously over distances of 10 km to 20 km (Fig. 1a, e.g., northern flanks of the Oceanographer and Kane Transforms, Schroeder, 1977; Purdy and others, 1978). Furthermore, over this interval, high-resolution data show that ridge-axis parallel structures (ridges and troughs) can be traced without disruption or offset down toward the maximum depths observed along the fracture zone (Fig. 1a; Detrick and others, 1973; Renard and others, 1975; Laughton and Rusby, 1975; Searle, 1979; 1981; H. D. Needham, unpublished Sea Beam data for the Oceanographer, Kane, and Vema F. Z.s). Therefore, although the ridge-axis parallel structures remain undisrupted and continuous, indicating that the terrain has not been disrupted by strike-slip tectonism, the regional deepening of the sea floor toward the axis of the fracture zone clearly suggests that the properties of oceanic lithosphere are changing in a systematic way. This broad morphotectonic zone, which includes all the fracture-zone parallel topographic elements, as well as the ridge axis parallel elements that slope down towards the axis of the fracture zone, is called the transform domain (Fig. 1). The fundamental characteristics defining the transform domain indicate clearly that the

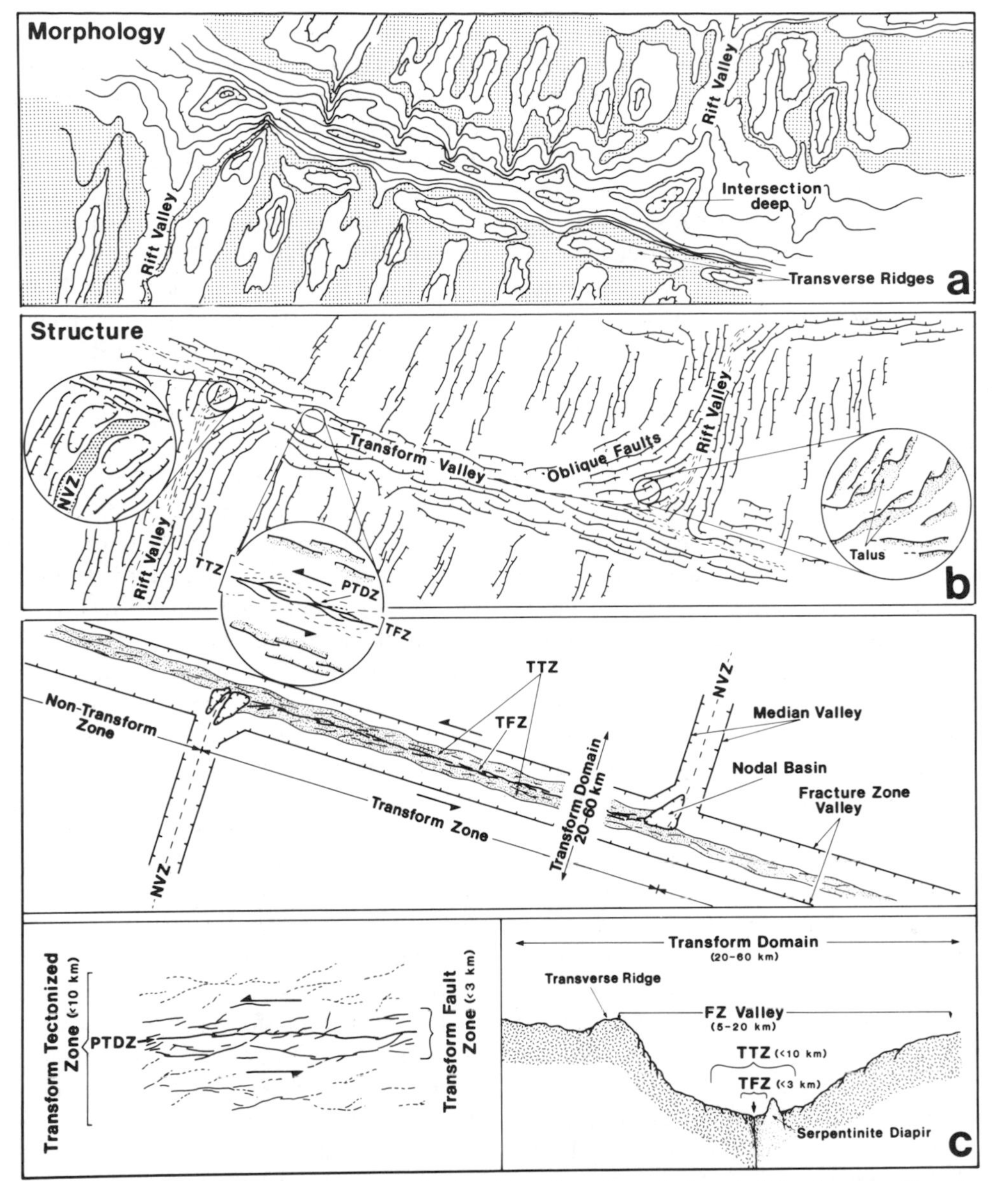

Figure 1. The Oceanic Lithosphere Proximal to Transform Boundaries. a). **Morphology.** The rift valleys in this example are taken to be offset by approximately 100km. The depth of the rift valley floor increases systematically toward the transform over distances of 30–40km (See 1b for location of transform valley). Ridge-transform intersections are characterized by deep (3000–6000m) closed contour depressions, rift valley walls truncated by a transform become progressively more oblique with proximity to the transform boundary. The transform valley strikes at high angles to the ridge axes (rift valleys), resides at depths of 1000m to several thousand meters below adjacent lithosphere and truncates ridge axis parallel topography. See text for discussion. b). **Structure.** Large scale terrain elements are the product of a large number of small-throw dip-slip faults. Note that on a finer scale (insert, right) the oblique rift valley walls that abut the transform consist of integrated ridge-parallel, transform-parallel, and oblique structures. The rift valley floor is occupied by the neo-volcanic zone (NVZ; insert left) consisting of most-recent volcanics and dissected by numerous fissures and small-throw faults whose trends generally follow the strike of the ridge. The transform valley contains the transform fault zone (TFZ); a narrow system of braided faults that accommodate strike-slip displacement (middle insert). At any given time, most of the slip will occur along a single, through-going fault strand (PTDZ). See text for discussion. c. **Tectonic elements of the transform domain.** The transform domain ranging in width from 20 to 60km, contains the fracture zone valley, the transform tectonized zone (TTZ, shaded), the transform fault zone (TFZ), and ultimately the most recent locus of strike-slip deformation the principal transform displacement zone (PTDZ). Insert (lower left) shows detail of TTZ. Insert (lower right) shows schematic cross-section of a transform domain with vertical exaggeration approximately 4:1. See text for discussion.

tectonic processes operating along RTR plate boundaries lead to the creation of distinctive lithosphere that contrasts markedly with lithosphere accreted at some distance from the RTR plate boundary (Fox and Gallo, 1984).

The oceanic lithosphere that comprises the transform domain is created at or proximal to one of the two RT intersections that define the ends of a RTR plate boundary. A slowly-slipping RT intersection is characterized by three morphotectonic provinces: the nodal basin that marks the center of the intersection where the rift valley meets the transform valley; the rift valley wall that faces the aseismic limb of the fracture zone; and the rift valley wall that faces the transform valley (Fig. 1a).

Well constrained mapping of a few RT intersections (FAMOUS area—Needham and Francheteau, 1974; Macdonald and Luyendyk, 1977; Ramberg and van Andel, 1977; Phillips and Fleming, 1978; Oceanographer Transform—Fox and others, 1969a; Schroeder 1977; Fox and others, in preparation; Kane Transform—Purdy and others, 1978; Purdy and others, 1979; Vema Transform—Macdonald and others, in press) document that these three morphotectonic provinces exhibit a distinctive topographic fabric that varies in a systematic way. As the RT intersection is approached over distances of 20 to 40 km the floor of the rift valley is observed to deepen continuously by magnitudes of several hundred to greater than 1000 m (Fig. 1a). The larger the age-offset of the transform truncating a ridge segment at a RT intersection, the greater the change in depth, the longer the interval over which the rift floor slopes down toward the intersection, and the steeper the longitudinal slope of the rift floor (Fox and Gallo, 1984; Parmentier and Forsyth, 1985). Furthermore, within 10 km of the intersection the slope down toward the intersection increases and the rift valley floor widens appreciably. High resolution investigations of the structures developed along the floor of the rift (Transform A—Detrick and others, 1973; Renard and others, 1975; Macdonald and Luyendyk, 1977; ARCYANA, 1975; Whitmarsh and Laughton, 1975, 1976; Oceanographer Transform western intersection—OTTER, 1984; Kane Transform—Karson and Dick, 1983; Vema Transform—Macdonald and others, in press) document that although volcanic constructional terrain can be traced continuously along the rift floor down toward the intersection, the degree of disruption of the extrusive carapace increases as the transform boundary is approached and the orientations of faults and fissures become more variable (Fig. 1b). At distances greater than 5 to 10 km away from the transform intersection, extensional structures (faults, fissures, horsts and graben) exhibit trends that are generally parallel with the regional strike of the ridge axis. With increasing proximity to the transform, however, structural orientations are observed that depart markedly from ridge-parallel trends. Small-throw (<5 m), dip-slip faults that strike perpendicular to the ridge axis and face the intersection are observed across the rift valley floor stepping the sea floor down into the intersection deep (OTTER, 1984; Macdonald and others, in press). Scarps and fissures that are oblique to the ridge axis and become more oblique as the transform intersection is approached are found to occur along that side of the rift valley bounded by the transform fault (Karson and Dick, 1983; OTTER, 1984; Macdonald and others, in press).

The intersection of the rift valley and the transform valley is punctuated by a closed-contour depression or nodal basin (Fig. 1a). In general, the nodal basin approximates the shape of a right-angle triangle and is oriented so that the hypotenuse falls along the oblique rift valley wall (Renard, and others, 1975; Phillips and Fleming, 1978; Karson and Dick, 1983; OTTER, 1984). There seems to be a well-defined relationship between the depth of the nodal basin and the age of the truncating edge of lithosphere juxtaposed against the rift axis at a RT intersection. The greater the age or thickness of the truncating edge, the greater the depth and size of the intersection depression (Fox and Gallo, 1984). High resolution investigations of a few intersections (Detrick and others, 1973; ARCYANA, 1975; Karson and Dick, 1983; OTTER, 1984; Macdonald and others, in press) indicate that recent extrusive products can be traced down the rift valley and into the nodal basin. In some cases, the fresh volcanic terrain at the rift valley floor is lost below a relatively thick mantle (up to 100 m at the eastern end of the Oceanographer Transform) (Schroeder, 1977; Fox and others, in preparation) of rubble and pelagic sediment that is concentrated in this depositional sink (OTTER, 1984) but, in most cases volcanic ridges can be traced across the nodal basin where they terminate abruptly against the steep, older transform-parallel wall of the nodal basin (Karson and Dick, 1983; Macdonald and others, in press). These steep-sided volcanic constructional edifices, draped with extrusive lavas, are flanked by the relatively flat sedimented floors of the nodal basin. The ridge axis parallel and oblique structures that disrupt the volcanic terrain can be traced as discontinuous strands down into and across the nodal basin where they are observed to truncate against transform-parallel structures (Fig. 1b). Although not well constrained by numerous observations, submersible data document that oblique structures along the rift floor accommodate dip-slip motion and there is no evidence for significant amounts of strike-slip displacements (ARCYANA, 1975; Karson and Dick, 1983; OTTER, 1984).

There is a marked structural polarity observed in and around the nodal basin. The side of the nodal basin flanking the transform valley is disrupted by numerous faults with variable orientations. Its steep slopes are scarred by evidence of mass wasting and slope decay. In contrast, the side of the nodal basin that flanks the aseismic limb of the fracture zone is less disrupted by faulting, slopes are more smoothly sedimented, and the orientation of ridges and faults is predominantly ridge parallel.

The rift valley wall truncating against the aseismic limb of the fracture zone is relatively simple in that ridge axis parallel trends can be traced into the fracture zone valley where they end abruptly, forming an almost right-angle corner against the fracture zone valley (FAMOUS area—Laughton and Rusby, 1975; Whitmarsh and Laughton, 1975, 1976; Renard and others, 1975; Phillips and Fleming, 1978; Oceanographer Transform—Schroeder, 1977; Williams and others, 1983; Fox and others, in preparation; Vema Transform eastern intersection—Macdonald and others, in press; Kurchatov Transform—Searle and Laughton, 1977; Charlie-Gibbs Transform—Searle, 1979, 1981; Lonsdale and Shor, 1979). Deep-towed camera investigations of intersections at the Oceanographer (OTTER, 1984) and Kane (Karson and Dick, 1983) fracture zones, as well as deep-towed side-scan sonar studies of the western intersection of Transform A (Detrick and others, 1973; Macdonald and Luyendyk, 1977) and the eastern intersection of the Vema Transform (Macdonald and others, in press) reveal that a large number of ridge-axis parallel, small throw, dip-slip faults create the relief of the rift wall. The strike of these predominantly rift-axis facing faults does not change significantly as the fracture zone is approached.

In marked contrast, the rift valley wall truncated by the transform exhibits a morphotectonic fabric oblique to the regional strike of the rise axis (Fig. 1b). At distances of 10 km or more away from the RT intersection the rift wall is composed of terrain elements that parallel the ridge axis but, with increasing proximity to the transform, oblique trends become better defined until the valley wall merges with and swings into the transform valley. When viewed in detail, the overall oblique character of this morphotectonic province is really the spatial integration of three structural elements: scarps that are 20° to 50° oblique to the strike of the ridge axis; scarps that are parallel to the ridge axis; and scarps that parallel the strike of transform. In situ investigations of this terrain with the submersible ALVIN at the eastern intersection of the Kane Transform (Karson and Dick, 1983) and the eastern intersection of the Oceanographer Transform (OTTER, 1984) document that irrespective of orientation there is no evidence to suggest any significant oblique-slip or strike-slip motions. Serrated surface traces comprising short (<1 m) linear segments, vertical slickensides, striae, and grooves on exposed material all document that the most recent motion was dip-slip. Although it is possible that only the results of some comparatively recent dip-slip motion on faults with complex displacement histories was observed in these two widely separated field areas, it is significant that there was no evidence, even on those faults with transform parallel strike, to suggest anything but dip-slip faulting.

The transform valley portion of a fracture zone is the seismically active segment linking the two ridge tips truncated at each RT intersection. In kinematic terms, a transform fault is relatively simple with strike-slip motion located along the faces of the opposing plate edges and with relative motion across the plate edges opposite to the sense of offset defined by the displaced ridge tips (Wilson, 1965; Sykes, 1967). Kinematic considerations aside, however, the considerable width of the transform valley, the morphologic complexity of the terrain elements found along a transform, and the great relief developed by the transform valley are striking and suggest that there are processes unique to this type of plate boundary setting, beyond simple strike-slip tectonism, that integrate in time and space to create the distinctive topography.

The first real constraints on the structural fabric of the transform valley were established by the submersible *Cyana* (ARCYANA, 1975; Choukroune and others, 1978) and, to a lesser extent, by Deep Tow (Detrick and others, 1973) during investigations of Transform A (20 km offset). More recently, these data have been supplemented by a submersible (*Alvin*) and deep-towed camera (ANGUS) study of the central portion of the Oceanographer Transform (OTTER, 1985) and by a Deep-Tow experiment that surveyed a 60 km long segment of the Vema Transform (Macdonald and others, in press; Kastens and others in press).

These data indicate that the zone of recent tectonism, called here the transform fault zone (TFZ), is generally centered around the axis of maximum depth and is characterized by a narrow (500 m to 2 km) belt of disrupted terrain broken by a series of short fault strands that can be traced the length of the transform (Fig. 1b, c). Individual faults, exposing a variety of rock types (consolidated pelagic sediment and talus, extrusive basalt, gabbroic, and ultramafic rocks), exhibit at least a dip-slip component of motion, creating steps in the sea floor several centimeters to several tens of meters in height that can integrate spatially to create very steep slopes with relief of hundreds of meters. Based on outcrop relationships, definitive evidence for strike-slip relative motion has been hard to document, but microtectonic indicators (e.g. sigmoidal extension gashes) associated with transform-parallel faults in Transform A indicate left-lateral displacements consistent with transform kinematics. Although not well documented, it appears from in situ investigations that fault strands anastomose along strike outlining lozenge shaped blocks and, using the structural relationships developed within continental shear zones as a viable analogue (e.g. Tchalenko, 1970; Tchalenko and Ambraseys, 1970), it is likely that at any given time a number of fault strands within the TFZ will link up to create a through-going strand. This single fault trace, called the principal transform displacement zone (PTDZ) will define the locus for most of the relative motion across the transform (Figures 1b, c). The faults developed within the TFZ exhibit a complex and variable pattern in terms of distribution, orientation, and density. Although the kinematics of these fault strands within the TFZ have yet to be adequately investigated at a regional scale, the characteristic orientations of fault strands measured at localities within a few North Atlantic transforms (Transform A—Choukroune and others, 1978; Transform B—Goud and Karson, 1984; Oceanographer—OTTER, 1985; Vema—Macdonald and others, in press) suggest that the TFZ is largely comprised of Reidel and primary shears developing a structural geometry similar to continental shear zones (e.g. Tchalenko, 1970; Tchalenko and Ambraseys, 1970). Compressional and extensional structures may form at oblique angles to the transform trend relaying strike-slip motion from one locality along the fault to another. When viewed along strike, the relatively fast accumulation of talus at the feet of fault scarps testifies to the rapid rates of slope degradation along the transform. However, the unequal distribution of talus wedges at the base of fault scarps within the TFZ documents that the timing of faulting and the relative age of structures is variable, signifying a continuously evolving fault geometry (Choukroune and others, 1978).

The small-scale structural complexities within the TFZ notwithstanding, it appears, based on only a few observations in various localities along the Vema Transform (Eittreim and Ewing, 1975; Macdonald and others, in press; Kastens and others, in press; H. D. Needham, unpublished data), the Oceanographer Transform (OTTER, 1985), and Transform A (Detrick and others, 1973; ARCYANA, 1975; Choukroune and others, 1978) that the 500 m to 2 km wide band of recent faulting (TFZ) can be traced along the transform valley from the transform-parallel side of one ridge-transform nodal basin to the other (Fig. 1c). Although none have been documented, marked departures (i.e. the development of large bends) from this relatively simple geometry would result in asperities along the TFZ that would lead to the development of basins (extensional strain) or ridges (compressive strain) analogous to the structures documented to form along the continental strike-slip faults (Wilcox and others, 1973; Crowell, 1974a, b). Such asperities could be caused by minor changes in the transform slip vector or by contrasts in the rheology of the two lithospheric plates creating the transform interface (Choukroune and others, 1978).

The relatively simple (regional perspective) TFZ geometry outlined above can only exist as long as the RTR geometry is stable with the pole of relative motion fixed for a long period of time (millions of years). The geometry will have to evolve, however, following a change in the pole of relative motion and, depending on the way that circles of rotation around the new pole cross the original transform orientation, the transform will experience a component of extension or compression (Menard and Atwater, 1968, 1969; van Andel and others, 1969; Dewey, 1975; Fox and others, 1976). Another possible variable that could initiate changes in the geometry of the TFZ would be the migration in time of the TFZ over some finite width producing a wide ensemble of tectonized terrain, the transform tectonized zone (TTZ). However, for a number of transforms (i.e. Transform A, Oceanographer, Kane, and Vema) undisrupted ridge axis parallel structures can be traced to within several kilometers of the TFZ, thereby documenting that the location of the TFZ does not migrate great distances in time creating a broad TTZ. For some small offset transforms, however, where the lithosphere is sufficiently thin to allow a truncated ridge axis to propagate into old lithosphere across a RT intersection, a relatively wide shear zone would be created. Such a geometry has been proposed for the 20 km offset Transform B in the FAMOUS area (Ramberg and others, 1977; Schouten and others, 1980). Goud and Karson (1984) have recently reviewed deep-towed camera (ANGUS) traverses across Transform B and document that strike-slip faulting has been distributed across a relatively wide (6 km) band. Recent fault scarps disrupt and expose semiconsolidated chalks defining asymmetric and cross-cutting shear fracture patterns with the development of what are interpreted to be Riedel and primary shears. This wide TFZ is compatible with the interpretation, based on regional morphologic relationships, that this plate boundary is presently unstable and strike-slip related strains are distributed over a relatively wide area.

The TTZ, which represents the time-averaged locations of strike-slip displacement, is flanked by inward facing slopes of the transform valley walls that rise up above the transform creating thousands of meters of relief (Fig. 1e). The true morphologic character of these walls was first shown by high resolution mapping (Renard and others, 1975; Philips and Fleming, 1978; H. D. Needham, unpublished Sea Beam data) and side-scan sonar data (GLORIA—Laughton and Rusby, 1975; Whitmarsh and Laughton, 1976; Searle and Laughton, 1977; Searle, 1979, 1981) to have ridge axis parallel lineaments protruding down into the transform valley walls creating a series of promontories and reentrants along the valley walls (Fig. 1a).

The individual structural elements that integrate spatially to create the overall morphotectonic fabric of the transform valley have only been defined at localities within Transform A (ARCYANA, 1975; Choukroune and others, 1978) and the Oceanographer Transform (OTTER, 1985). The structural fabric of the transform valley walls along these two transforms is controlled by small-throw, dip-slip faults whose orientation and distribution is variable as one traverses up and across the valley wall. Probably the most abundant structures recognized on the transform slopes are inward-facing transform-parallel fault traces that step the sea floor down towards the axis of the transform. It is the cumulative throw on these faults that creates the great relief of the fracture zone. Oblique trending scarps, which mimic the orientation of scarps observed at the ridge-transform intersection, as well as ridge-axis parallel scarps, are frequently observed creating steep walled buttresses and reentrants. The scarps observed on the transform valley walls, independent of orientation, can be traced for distances ranging from ten to several hundred meters and have a vertical component of displacement ranging from less than one to several tens of meters. There is no evidence to suggest that these faults have accommodated anything but dip-slip motion. Most of the relief generated by these faults must occur when the lithosphere is relatively young (a few millions of years) because the scarps appear to degrade relatively rapidly and are buried by the products of the mass wasting process—talus wedges and debris slides. In Transform A, where relatively young terrain (<1 m.y. old) on the valley walls was investigated, scarp faces exposing basement were numerous and the intervening steps between fault scarps are strewn with only a veneer of fresh rubble (ARCYANA, 1975; Choukroune and others, 1978). In the Oceanographer Transform, the valley walls investigated are approximately 5 m.y. old and the importance of mass wasting is profound. The lower flanks of the transform valley are choked with semiconsolidated rubble and pelagic carbonate; no basement outcrops were observed. Badly degraded outcrops were recognized high on the upper flanks of the transform valley, at elevations greater than 2000 m above the axis. There is no suggestion of recent faulting (i.e. disrupted talus; uneven coatings of Mn), and the rate of slope decay seems to be waning, because evidence for the generation of fresh talus is rare.

The aseismic extensions or non-transform extensions of the North Atlantic RTR plate boundaries have not been surveyed in great detail; survey lines across any given aseismic limb are widely spaced and topographic trends are typically extrapolated over tens of kilometers. Nevertheless, a number of surveys (Fox and others, 1969b; Fox, 1972; Johnson and Vogt, 1973; Collette and others, 1974; Olivet and others, 1974; Rabinowitz and Purdy, 1976; Schroeder, 1977; Purdy and others, 1979; Fox and others, in preparation) indicate that, although the basement relief and tectonic grain of the transform domain are muted beneath a thickening blanket of sediment (thicknesses in excess of 1 km), terrain similar to the physiography of the transform domain can be traced continuously across the flanks of the Mid-Atlantic Ridge (see Plate 2 of this volume). The morphology of the aseismic limbs of fracture zones is heterogeneous with the relief and shape of the flanking terrain and the apparent width of the fracture valley changing markedly along strike. In some instances the walls of the fracture zone valley are defined by high ridges that stand several hundred to over one thousand meters above the axis of maximum depth. These ridges can be traced along strike for several tens of kilometers (analogous to the transverse ridges of the transform domain). In other cases, the sea floor, over distances of 20 km to 40 km, becomes increasingly deeper by a thousand meters or more as the fracture zone axis is approached. The trace of a given fracture zone limb can change strike abruptly, recording a significant change in relative motion along the paleo-transform and these locations are often associated with marked changes in fracture zone morphology with the development, for example, of multiple troughs, seamount complexes, and/or elongate ridges. Independent of the along-strike morphotectonic variations, however, the salient and distinctive feature of the fracture zone persists independent of plate age: a deep linear trough striking at a high angle to the tectonic grain of the ridge flank resides at depths that are abnormally great for lithosphere of equivalent age and sediment burial.

A thick blanket of pelagic and continental detritus almost completely obscures the morphotectonic fabric of the basement near the continents but distinctive offsets in the outline of the continental margin (Heezen and Tharp, 1965; LePichon and Fox, 1971) and magnetic anomaly pattern indicate that some of fracture zones can be traced to the edges of the continents. Scattered seismic reflection profiles show that anomalously deep V-shaped basement depressions are located where magnetic data indicate an offset in the anomalies (Schouten and Klitgord, 1982). These data confirm, as the reconnaissance mapping of Heezen and Tharp first suggested (Heezen and Tharp, 1961; 1965; 1968), that fracture zones have been an essential component in the evolution of the North Atlantic from the earliest days of rifting to the present.

Distribution of Rocks

Francheteau and others (1976) summarized the dredging results for North Atlantic fracture zones and showed that gabbroic rocks and serpentinized peridotites are recovered from all levels of the transform valley making it difficult to support the proposition that basalts, gabbros, and ultramafic rocks exhibit a simple stratigraphy arranged sequentially down the face of a single, large-throw fault. Furthermore, based on observations made from submersible CYANA during traverses across Transform A (ARCYANA, 1975), these authors document that the relief of this transform valley is distributed on a large number of faults with a complex geometry and conclude that these structural relationships, if they are representative of slowly-slipping transforms in general, will not create structural windows deep into the oceanic crust. In order to explain the occurrence of deep-seated rocks, they suggest that the shallow intrusive and extrusive carapace must be very thin and/or that structural processes, not yet identified, are at work to tectonically thin the crust.

Recently, submersible investigations proximal to ridge-transform intersections (CAYTROUGH, 1979; Stroup and Fox, 1981; Karson and Dick, 1983; OTTER, 1984), within a large-offset transform (OTTER, 1985) and along a fracture zone extension (Auzende and others, 1978) confirm and expand upon the suggestion proposed by Francheteau and others (1976). These studies document that the high walls of the transform valley and the adjoining ridge-transform intersection are the result of many small-throw faults. These escarpments yield a diverse assemblage of rock similar to those dredged over the years at many localities (e.g. Shand, 1949; Quon and Ehlers, 1959; Melson and others, 1967; Bogdanov and Ploshko, 1967; Bonatti and others, 1971; Miyashiro and others, 1969; Fox and others, 1973; Thompson and Melson, 1972). The exposure of gabbroic and ultramafic rocks on small throw faults argues that the total crustal section created close to slowly-slipping transforms is anomalously thin (CAYTROUGH, 1979; Stroup and Fox, 1981; Karson and Dick, 1983; OTTER, 1984).

The most abundant ultramafic samples recovered from the walls of North Atlantic transforms are harzburgite and lherzolite although lesser amounts of dunite, clinopyroxenite and wehrlite have been reported (see Bonatti and Hamlyn, 1981). The majority of the ultramafic samples have a strong tectonite fabric similar to the microstructures associated with the basal ultramafic assemblage found in ophiolites (e.g. Nicolas and others, 1971) but some ultramafic samples exhibit cumulate textures suggestive of crystal fractionation in magmatic reservoirs. Fracture zone peridotites are typically medium to coarse grained and, although almost always overprinted by serpentinization, porphyroclastic, and mylonitic textures are generally observed. Geothermometric investigations of coexisting primary and recrystallized pyroxene pairs found in some ultramafic samples indicate that crystallization of some phases took place at depths in excess of 25 km before emplacement at shallow levels (Bonatti and Hamlyn, 1978). These textural and mineral chemistry data argue persuasively that the majority of ultramafic samples recovered from transform faults are tectonites of residual upper mantle derivation.

The most abundant plutonic rocks recovered from the walls of fracture zones are the metamorphosed and unmetamorphosed members of the gabbro clan (see Fox and Stroup, 1981). Olivine gabbro and two-pyroxene gabbro are the most commonly recognized, but "normal" gabbro, troctolite, Ti-ferrogabbro, norite, and anorthosite are recovered as well. These gabbroic rocks have a characteristic mineralogy and chemistry that distinguishes them as products of differentiation of a tholeitic magma. Coarsely crystalline late-stage differentiates of this series (i.e. quartz diorites, aplites, and trondhjemites) are sometimes recovered in minor amounts. In terms of primary igneous textures these gabbroic rocks are characterized by substantial heterogeneity of textures and mineral associations. This primary heterogenity is compounded in many samples that have experienced metamorphism and/or deformation. The effects of metamorphism are variable, producing a range from almost pristine rocks, in which only mafic phases have experienced incipient replacement, to extremely altered samples in which primary phases are completely replaced by hydrothermal minerals (e.g. zeolite, chlorite, actinolite, serpentine, quartz, epidote, and hornblende). The expressions of deformation vary in a continuum from high-temperature recrystallization and ductile deformation to low-temperature brittle deformation and the formation of low-temperature minerals in cracks and veins. A close interdependence between deformation and metamorphism is indicated by the association of hydrous alteration phases with zones of mylonitization and brittle fractures. All the investigations of gabbroic rocks and their distribution within fracture zones point toward a heterogenous assemblage, both in terms of primary and secondary characteristics, that is complexly distributed across and along the fracture zone walls.

In first order terms, the textural, mineralogic, and geochemical characteristics of basalts recovered in and along slowly-slipping transforms are indistinguishable from basalts sampled from outcrops distant from the fracture zone. However, recent detailed geochemical investigations of basalts collected proximal to fracture zones show that there are consistent and systematic changes in the major and trace element chemistry of basalts (Bryan and others, 1981; Sigurdsson, 1981; Schilling and others, 1983; Schilling, this volume; Melson and O'Hearn, this volume; Bryan and Frey, this volume). Basalt samples proximal to the transform generally have higher abundances of incompatible elements, as well as higher amounts of TiO_2 and FeO for a given MgO concentration. These data have been interpreted as reflecting lesser extents of partial melting as the fracture zone is approached (Bender and others, 1984; Langmuir and Bender, 1984). Finally, at a few localities (St. Paul F.Z.—Melson and others, 1967; Frey, 1970; Romanche F.Z.—Bonatti and others, 1970) basalts with alkali affinities and distinctive geochemistries have been recovered.

A wide range of consolidated and semi-consolidated sedimentary rocks are often recovered along with igneous and metamorphic rocks from fracture zones. Sedimentary conglomerates and fault-breccias containing clasts of variable composition mimic the heterogeneous distribution of exposed igneous basement and attest to the importance of tectonism and mass wasting (Fox, 1972; Bonatti and others, 1973, 1974). The matrix material of some of these rocks is generally recrystallized carbonate, indicating crystallization at elevated temperatures from circulating sea water. Of particular interest in terms of the tectonics of the transform domain is the occurrence of shallow water limestones on the crests of the transverse ridges that border the southern flank of the Vema Transform and the north flank of the Romanche Transform (Bonatti and others, 1977; Bonatti and others, 1979; Bonatti and Chermak, 1981; Bonatti and others, 1983). These samples clearly document the importance of vertical tectonism, both uplift and subsidence, in the evolution of the transverse ridges (Bonatti, 1978).

Regional Crustal Structure: Geophysical Data

Well constrained seismic refraction experiments along and proximal to two large North Atlantic fracture zones (Kane—Detrick and Purdy, 1980; Cormier and others, 1984; Vema—Detrick and others, 1982; Oceanographer—Sinha and Louden, 1983) and an apparent small-offset fracture zone (White and Matthews, 1980; White, 1984) have documented unequivocally that when compared to normal oceanic crust the velocity structure of these fracture zones is anomalous and varies markedly both along and across the strike (Purdy and Ewing, this volume). Analyses of seismic data from refraction lines oriented down the fracture zone reveal that velocity gradients in the crust are very steep, the crustal velocities are anomalously low, the crust-mantle boundary is seismically abrupt (transition width of 500 to 1000 m), and the crust is anomalously thin (2–3 km). Significant along-strike heterogeneity in these properties is indicated because travel time data are hard to model using the standard slope intercept methods. Indeed, delay time analyses of mantle arrivals show that crustal thickness and/or velocity structure must vary considerably over a distance of 10 to 20 km to explain large delay time differences. Assuming that the velocity remains constant laterally, solutions to these data indicate changes in crustal thickness ranging from 1 to 4 km. Results from refraction lines oriented perpendicular to the strike of the fracture zone reveal that over distances of 20 to 30 km the oceanic crust appears to thin gradually by 1–2 km as the fracture zone is approached; thinning and crustal heterogeneity become dramatic only within 10 km of the fracture zone axis with the very thin crust confined to a narrow ribbon several kilometers in width that flanks the axis of maximum depth (Detrick and Purdy, 1980; Sinha and Louden, 1983; Cormier and others, 1984).

Multichannel reflection data across a small western North Atlantic fracture zone define a strong subbottom reflector interpreted to represent the crust/mantle transition at a depth of only 2 to 2.4 km (Mutter and others, 1984). This fracture zone truncates Mesozoic magnetic anomalies that indicate a half spreading rate of approximately 3 cm/yr, and, based on the reconstructions of Schouten and Klitgord (1982), the offset along the fracture zone is only 20 km. Given this spreading rate and transform-offset length, the age contrast across this paleo-plate boundary at the time of crustal formation would be no more than one or two million years.

These seismic data establish without question that upper mantle velocities (7.7 km/sec–8.1 km/sec) are observed at generally shallow levels proximal to fracture zones and that the overlying crustal component is composed of rock types with low compressional wave velocities and steep gradients. It is inherent in the seismic reflection data that the resolution of crustal properties is on the scale of a few thousand meters (even more for refraction data), and therefore even the best constrained experiment provides insight into the crustal variations that occur only at wavelengths greater than a few thousand meters. Furthermore, it is difficult to uniquely interpret the velocity data because causes for these velocity variations cannot be isolated from these data alone. Crustal fracturing, metamorphism, and the distribution of rock types with contrasting velocities are three obvious variables that combine in time and space along a RTR plate boundary in a capricious way to create a complex velocity structure on all scales (Purdy and Ewing, this volume).

The analysis of gravity and geoid data collected on long traverses across fracture zones provides some insight about the long wavelength mass anomalies associated with fracture zones (Rabinowitz and Jung, this volume; Vogt, this volume, Ch. 14). Analyses show that large mass anomalies are generally associated with fracture zones (Cochran, 1973; Robb and Kane, 1975; Sibuet and Veyrat-Peinet, 1980; Louden and Forsyth, 1982; Detrick and others, 1982) but the interpretation of these data is equivocal because any number of mass distribution solutions can satisfy the data. Cochran (1973) and Robb and Kane (1975) propose, based on gravity data from equatorial fracture zones, that positive mass anomalies associated with fracture zone troughs can best be modeled by the emplacement of dense upper mantle rocks at shallow levels. Sibuet and Veyrat-Peinet (1980), however, challenge this interpretation and contend that mass anomalies are not required if mantle density is allowed to change with age across the fracture zone. A series of six long gravity profiles across the Kane fracture zone define large mass anomalies in association with this fracture zone. Mass anomaly patterns cannot be correlated from one widely spaced profile to another and if one assumes that the density of the crust and upper mantle remain the same as the fracture zone is approached no evidence for systematic crustal thinning is implied (Louden and Forsyth, 1982). Modeling of gravity and seismic refraction results from the Vema Transform suggest that the floor of the fracture zone is underlain by a thick assemblage of serpentinized ultramafic rocks (Detrick and others, 1982).

There are only a few investigations of the magnetic anomaly character of fracture zones in the North Atlantic. These results show that the fracture zones are associated with either a belt of reduced magnetization (Cochran, 1973; Collette and others, 1974; Twigt and others, 1979) or, in some cases, a zone of positive magnetization (Twigt and others, 1983). Marked changes in the magnetic properties (natural remnance and susceptibility) of oceanic rocks are caused by metamorphism and alteration of oceanic rocks (Luyendyk and Melson, 1967; Fox and Opdyke, 1973). Given the documented complex vertical and lateral distribution of rock bodies within the fracture zone, it is not surprising that the magnetic anomaly character of fracture zones is complex.

THE OPHIOLITE PERSPECTIVE

Marine geological and geophysical data for slowly-slipping RTR plate boundaries make a strong case for a complex crustal structure. This marine perspective is limited, however, because it is currently impossible to directly investigate the complexities developed at depth. Fortunately, segments of fracture zones are preserved in some ophiolite assemblages (Moores and Vine, 1971; de Wit and others, 1977; Saleeby, 1977; Karson and Dewey, 1978; Simonian and Gass, 1978; Smewing, 1980; Prinzhofer and Nicolas, 1980) and the relationships developed in these rock bodies provide important constraints on processes that shape this structural environment.

Probably the best preserved and documented RTR ophiolite assemblage is the Coastal Complex of the Bay of Islands Ophiolite found in Western Newfoundland (Karson and Dewey, 1978; Karson, 1982, 1984; Karson and others, 1983) and some of the most salient results will be briefly summarized here. Over a distance of approximately 10 km the structural geometry, metamorphic zonation and magmatic history of the oceanic rocks change dramatically as the paleo-transform is approached. The relatively stratiform crustal stratigraphy of the Bay of Islands ophiolite changes in close proximity to the Coastal Complex: the magmatic component thins markedly and cumulate ultramafic rocks, including dunite and wehrlite, thicken. These rocks lap up against a several kilometer wide belt of highly-strained metabasites and serpentinites that are heterogeneously arranged with lithologic contacts and metamorphic zones bounded by steeply dipping contacts that trend at a high angle to the paleo-spreading direction of the adjacent Bay of Islands ophiolite. At depth within the Coastal Complex, ultramafic rocks exhibit a history of strike-slip deformation with the development of a several kilometer wide high-strain zone characterized by lineations and subordinate foliations that trend parallel with the strike of the paleo-transform. High-temperature metamorphic rocks (amphibolites), suggesting complex partial melting relationships, and intrusive peridotite bodies are frequently observed at deep structural levels. At shallow structural levels, metamorphosed basalts, gabbros, and serpentinites are cut and disrupted by relatively narrow, anastomosing shear zones and faults. Felsic intrusive bodies occur locally and serpentinite diapirism is widespread. Diabase dikes cut the entire assemblage and feed basaltic lava flows that blanket the faulted terrain.

This brief summary of the internal constitution of the Coastal Complex is representative of the range of igneous, metamorphic, and structural relationships developed within fracture zone terrain preserved within ophiolites in general (see preceeding references) and illustrates the exceedingly complex and variable properties of oceanic lithosphere created within and proximal to a RTR plate boundary. Although not well constrained, the internal relationships within the adjacent Bay of Islands ophiolite complex suggest that these rocks were emplaced along a relatively fast accreting (>6 cm/yr whole rate) plate boundary characterized by a steady-state magma chamber (Casey and others, 1981; Casey and Karson, 1981). If this interpretation of paleo-plate boundary kinematics is correct, then the Coastal Complex records relationships typical of only fast-slipping RTR plate boundaries and these relationships are not necessarily directly applicable to the slowly-slipping end of the RTR plate boundary spectrum. This caveat notwithstanding, the potential for vertical and lateral complexities, at a range of scales, is clearly established and likely to be typical of the transform domain.

A GEOLOGIC AND TECTONIC MODEL FOR SLOWLY-SLIPPING RTR PLATE BOUNDARIES

In first-order terms, the tectonic setting of RTR plate boundaries must be considered in the context of the accretionary environment that characterizes the ridge axes adjoining the transforms. The theoretical thermal models of the conditions governing the accretion and evolution of oceanic lithosphere indicate that slowly accreting plate boundaries (1–3 cm/yr full rate) are approaching a lower limit below which emplacement of basaltic melts at shallow crustal levels can no longer take place (Sleep, 1975; Kuznir and Bott, 1976). In this environment, slower advection beneath the axis of slowly spreading ridges results in lower temperatures and, as a result, less partial melt migrates to the surface to form oceanic crust (Bottinga and Allegre, 1978). This model is consistent with some seismic refraction results that suggest a thinning of oceanic crust with a decrease in spreading rate (Reid and Jackson, 1981; Jackson and others, 1982). Petrologic and seismic refraction data are consistent with a model suggesting that basaltic melts are episodically emplaced in small (<2 km in width), isolated, and possibly ephemeral reservoirs along the axes of slowly accreting boundaries (Nisbet and Fowler, 1978). Moreover, significant small-scale crustal heterogeneity for slowly accreting environments is indicated by the recovery of gabbroic and ultramafic rocks at shallow crustal levels (<200 m; Aumento and others, 1977; Melson and others, 1978; Leg 82 Scientific Party, 1982) at six localities within the North Atlantic. All these sites were chosen on the basis of marine geological and geophysical data to be representative of "normal" oceanic crust. These data are all supportive of the notion that the thermal environment beneath and along slowly accreting plate boundaries is not robust and is hostile to the processes controlling the timing and volume of basaltic melts emplaced at shallow levels.

Given the subdued accretionary processes along slowly spreading plate boundaries, it seems reasonable to infer that the rising asthenospheric wedge beneath such boundaries will become colder and the properties within the wedge (viscosity and flow characteristics) will be perturbed as a transform boundary is approached. Sleep and Biehler (1970) propose that viscous interaction of the upwelling mantle beneath the ridge at the RT intersection will result in the emplacement of crust below its level of isostatic equilibrium. Parmentier and Forsyth (1985) have shown that the alterations of the flow pattern by this viscous interaction can produce significant effects over distances of tens of kilometers. A simple three dimensional thermal model for a RTR plate boundary that only considers conductive heat loss and simple flow characteristics indicates that conductive cooling modifies the temperature of upwelling material to a distance of 10 km from the fault (Forsyth and Wilson, 1984). More recent calculations that take into account the effect of terminating flow at the RT intersection suggest that the combined effects of conductive heat loss and perturbation of flow may extend out to distances of several tens of kilometers from the large offset transform boundaries (D. Forsyth, personal communication). The juxtaposition of a variably thick truncating edge of lithosphere against an axis of accretion will result in emplacement of smaller volumes of basaltic melt at shallow levels per unit time. The migration of smaller volumes of partial melt to shallow levels is a consequence of two complementary factors. First, smaller volumes of melt will be produced from the rising wedge of asthenosphere because of the edge-effect induced perturbations mentioned above (Bender and others, 1984; Langmuir and Bender, 1984). Second, some of the melt segregated in the upper mantle at depth will never reach shallow-level reservoirs but will be trapped as veins and pods in the relatively cool and viscous upper mantle (Fig. 2). Consequently, as slowly-slipping RTR plate boundaries are approached, the oceanic crust (i.e. magmatic component) will thin continuously, the underlying upper mantle will become increasingly more heterogeneous, and the depths to the base of the lithosphere will increase. In addition, the melt that does rise to shallow reservoirs along an accreting plate boundary proximal to a transform boundary will experience a crystallization history (i.e. cold environment; low magma replenishment rates) markedly different from melt emplaced distal to this interface. It is the profound changes in the properties (composition, distribution of mass, thickness and shape of lithologic units) that result in a systematic change in isostasy for the lithosphere as the fracture zone is approached leading to the creation of the distinctive morphologic signature of the fracture zone, a broad anomalously deep trough (Fox and Gallo, 1984).

The manifestations of this transform edge effect become more dramatic with the increasing thickness of the truncating transform edge. At the slow rates of accretion that characterize the Mid-Atlantic Ridge (⩽3 cm/yr, full rate), large offset transforms like Kane, Vema, or Oceanographer juxtapose thick (30 to 50 km) edges of lithosphere against the accreting plate boundary all but nullifying the processes that lead to the creation of normal oceanic lithosphere. In close proximity (several kilometers) to these boundaries, the emplacement of partial melts is discontinuous at best and the shallow structure of oceanic lithosphere is characterized by pods of basaltic rocks enclosed in a matrix of lherzolite and harzburgite (Fig. 2). With increasing distance from these thick edges, the volume of partial melt produced increases and the emplacement of these melts at shallow levels is more continuous leading to a thicker crust and a relatively more homogeneous crust and upper mantle. At relatively small offset (<20 km) North Atlantic RTR plate boundaries, the thickness of the truncating transform edge is not great (10 to 15 km) and, therefore, the transform edge effect should not be as profound (Fig. 2). The reduction in the volume of melt production will

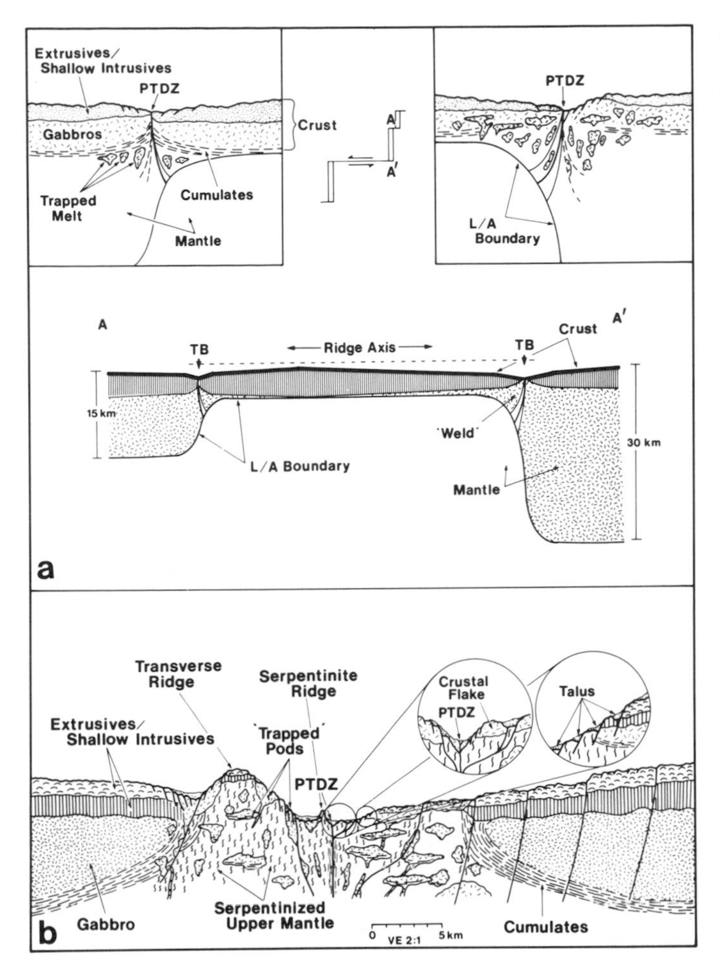

Figure 2. The influence of transforms on the structure of the Oceanic Lithosphere. a). Schematic cross-section across a ridge-transform-ridge system (location A-A shown above). L/A=Lithosphere/Asthenosphere Boundary. Note that as transform boundaries (TB) are approached, lithosphere thickens as crustal component thins. Magnitude of the influence of transforms on accretionary processes is related to thickness of the lithosphere at ridge-transform intersections (a function of spreading rate and transform offset) and is reflected by increase in depth to the sea floor as transform boundaries are approached. Inserts above show detail of ridge-transform intersections and progressive perturbation of ridge-axis processes with proximity to the transform boundary. Note that the influence is greater (crust thins, lithosphere thickens) for the longer offset transform (upper right). b). Cross-section. A schematic geologic cross-section across an oceanic transform fault. Vertical exaggeration is approximately 2:1. The effect of transform boundaries is to influence the processes responsible for the generation and evolution of oceanic lithosphere. As such, the ridge-transform-ridge systems generate swaths of anomalous lithosphere. For instance, within about 10km of the transform boundary the continuity of the petrologic units generated by magma chamber processes becomes disrupted and replaced by small blebs and pockets of magma, trapped in an upper mantle matrix that is subsequently serpentinized. Inserts show detail of PTDZ and transform walls. See text for discussion.

not be as pronounced and melt will be produced in a relatively continuous fashion. The crust will not thin as dramatically and the upper mantle will not be as heterogeneous. As a consequence, the contrast in the distribution of mass will not be as great and, as a small offset fracture zone is approached, the relief and width of the transform domain will be diminished. The total thickness of the crust produced near a small offset RTR plate boundary will not contrast markedly with normal crustal thickness, but the crustal structure (distribution of rock types) will be distinctive because thicker sequences of ultramafic cumulates are expected to dominate the basal portion of the crust near a RT intersection. The relatively cold reservoir environment at shallow levels caused by proximity to the cold transform edge (Karson, 1982, 1984) as well as enhanced crustal fracturing and consequent circulation of water near the transform boundary, lead to the production of large volumes of these rocks. In this environment, the crust will appear thin in seismic terms because high velocities (7.6 km/s) characteristic of the upper mantle will be recorded at shallow levels and reflect the development of a thick sequence of cumulate ultramafics. However, petrologic moho, the boundary between magmatic rocks and the underlying residual ultramafics of the upper mantle, will reside at greater depths and the thickness of the crust in magmatic terms will not be markedly different from magmatic volumes associated with normal accretionary environments.

To accommodate the changes in mass distribution as the fracture zone is approached, the oceanic lithosphere is broken by a large number of small throw dip-slip faults that step the sea floor down toward the axis of the fracture zone (Fig. 2). The faults disrupt the crust and expose a diverse suite of rock types that reflect the complex and variable igneous and metamorphic processes that operate at and along the fracture zone interface in time and space. Given the relatively thin crust that characterizes this environment, sea water migrates along permeable fault-produced pathways to hydrate ultramafic rocks at relatively shallow levels. Hydration leads to serpentinization and the upward migration of serpentinite screens and diapirs creating further complications in terms of crustal structure. Along the very large offset fracture zones where the crust is defined by discontinuous pods of basaltic rocks set in an ultramafic matrix serpentinization can occur on a grand scale, producing large serpentinite diapirs that rise vertically to create the elongate serpentinite ridges that are often the hallmark of slowly-slipping, large-offset fracture zones (Bonatti, 1976, 1978; Bonatti and Honnorez, 1976).

Although the transform domain is relatively broad and reflects the perturbation of accretionay processes by the transform edge and the complicating effects of vertical tectonism, strike-slip displacement is confined to a relatively narrow belt (<10 km), the transform tectonized zone (TTZ), that is centered around the axis of the transform valley (Fig. 1c). At any given time within the TTZ, displacement is concentrated along a swath (<3 km wide) of ongoing tectonism, the transform fault zone (TFZ), comprised of a braided network of fault strands. This belt of strike-slip tectonism can be traced down the length of the transform linking the two ridge tips. The structural geometry within the TFZ is complex and continually evolving in response to the interplay between rheology, transform-slip vector, and strain rate, and it is the wandering of the TFZ in time and space that creates the TTZ.

Based on the study of continental shear zones (Tchalenko, 1970; Tchalenko and Ambraseys, 1970), it is likely that at any given time within the TFZ, most of the strike-slip motion is accommodated along a single through-going strand called the principal transform displacement zone. The relatively narrow TTZ is a function of the thick edges of lithosphere that face each other across a slowly slipping transform and tend to confine the deformation associated with relative motion to a narrow, highly-strained interval. Along the transform an isotherm controls the lower limit of the area over which strained rock bodies will deform by brittle failure; below this isotherm strained rocks will deform in a ductile fashion. Burr and Solomon (1978) have presented a strong case that the bounds on this limiting isotherm are probably 50° and 300°C. The circulation of water along the TTZ should be greatly enhanced because strike-slip relative motion along the transform interface will be continually regenerating a network of permeable pathways magnifying the cold signature of RTR plate boundary environment making it difficult to construct simple thermal models. Focal depths as deep as six km for earthquakes along a few North Atlantic transforms indicate that brittle failure occurs at least down to these depths (Weidner and Aki, 1973; Tsai, 1969) and it is likely that even deeper depths may be recorded.

At RT intersections a thick wedge of lithosphere forms beneath the axis of accretion proximal to the transform boundary creating a weld with the cold, thick lithosphere across the transform interface (Fig. 2; Fox and Gallo, 1984). Part of the weld links newly formed lithosphere across the aseismic limb of the fracture zone and has little effect on intersection tectonics because the sense and magnitude of relative motion across the boundary are comparable. The weld that bonds the newly formed lithosphere with the cold lithosphere across the transform portion of the fracture zone creates a shear couple in the underlying lithosphere, resulting in the progressive reorientation of the maximum tensile stress from normal to the ridge axis at some distance from the ridge-transform intersection to an oblique angle near the boundary. The brittle carapace of oceanic crust overlying the mantle weld deforms accordingly with the development of oblique-trending, dip-slip faults. The distance over which the transform-generated shear couple affects the tectonics of the accreting plate boundary depends on the cross sectional area of the mantle weld, which in turn varies as a function of spreading rate and the thickness of the truncating edge of lithosphere at the transform boundary.

Slowly accreting plate boundaries (⩽3 cm/yr total rate) are typically characterized by the frequent occurrence of variably-sized transform faults that offset the ridge axis every 30 to 80 km (see Plate 2 of this volume). Although the relative length of the transform may change with time due to phases of asymmetric spreading, the ridge-transform geometry is remarkably stable through time, surviving numerous changes of the pole of relative motion (Schouten and Klitgord, 1982). Given the validity of our model and the temporal stability of slowly-slipping RTR plate boundaries, ribbons of anomalous lithosphere can be traced from the ridge axis across the ocean basin to the margins (Fig. 2). Each swath of anomalous lithosphere is composed, in first-order terms, of two parts. There is a relatively narrow band (<10 km) of highly-strained terrain, the paleo-TTZ, made up of pods of thin oceanic crust and partially serpentinized ultramafic rock representing the integrated wandering of the paleo-TFZ. (See Karson and Dewey, 1978; Prinzhofer and Nicolas, 1980; Karson, 1982, 1984 for a discussion of the complexities that characterize this interval.) This high-strain zone is generally narrow (except for times of changes in the pole of relative motion) because the strike-slip tectonism of the transform is constrained to a relatively narrow crush zone by the thick edges of opposing lithosphere facing each other across the transform fault. It is interesting to note that for small offset (<30 km) slowly-slipping RTR plate boundaries, the thickness of the lithosphere may be thin enough (<15 km) to permit the development of complex plate boundary geometries. The maps of the FAMOUS area (Phillips and Fleming, 1978) reveal a confused ridge axis-transform geometry showing overlapping ridge axes and discontinuous transform trends that have been interpreted to reflect an unstable plate boundary geometry involving the migration of transform fault zones and the change in length of en echelon ridge segments (Ramberg and others, 1977; Schouten and others, 1980). Analysis of bottom photographs from the short offset (<20 km) Transform B in the FAMOUS area documents a zone of recent deformation within the transform that is very wide (>6 km). Furthermore, times of major changes in the pole of relative motion would induce reorganization of the RTR geometry and, depending on the tectonic adjustments, the zone of strike-slip tectonism could widen considerably (Menard and Atwater, 1968, 1969; Dewey, 1975).

Flanking the transform tectonized zone will be a wide (10 to 30 km) band of anomalous oceanic lithosphere that becomes increasingly more heterogeneous with proximity to the fracture zone (Fig. 2). The width of these ribbons of anomalous lithosphere depends on the properties of the accretionary environment at the time of lithosphere formation which in turn is primarily a function of the thickness of the truncating edge of lithosphere at the RT intersection. The frequent occurrence of RT intersections along slowly accreting plate boundaries (3 cm/yr full rate), the temporal stability of this plate tectonic geometry, and the predicted manifestations of the cold-edge effect at RT intersections (Fig. 1c, 2) means that the time-integrated product of the accretion will lead to a very heterogeneous oceanic lithosphere.

REFERENCES CITED

ARCYANA
1975 : Transform fault and rift valley from bathyscaphe and diving saucer: Science, v. 190, p. 108–116.

Aumento, F., and Melson, W. G., eds.
1977 : Initial Reports of the Deep Sea Drilling Project, v. 37: Washington, D.C., U.S. Government Printing Office, v. 37, 998 p.

Auzende, J. M., Olivet, J. L., Charuet, J., LeLann, A., LePichon, X., Monteiro, H. J., Nicolas, A., and Ribeiro, A.

1978 : Sampling and observations of oceanic mantle and crust on Gorringe Bank: Nature, v. 273, p. 45–49.

Bender, J. F., Langmuir, C. H., and Hanson, G. N.
1984 : Petrogenesis of basalts from the Tamayo region, East Pacific Rise: Correlation of glass chemistry with distance from a transform fault: Journal of Petrology, v. 25, pp. 213–254.

Bogdanov, Yu.A., and Ploshko V. V.
1967 : Igneous and metamorphic rocks from the abyssal Romanche Depression: Doklady Akademii Nauk. U.S.S.R., v. 177, p. 173–176.

Bonatti, E.
1976 : Serpentinite protrusions in the oceanic crust, Earth and Planetary Science Letters, v. 32, p. 107–113.
1978 : Vertical tectonism in oceanic fracture zones: Earth and Planetary Science Letters, v. 37, p. 369–379.

Bonatti, E., Emiliani, C., Ferrara, G., Honnorez, J., and Rydell, H.
1974 : Ultramafic-carbonate breccias from the equatorial Mid-Atlantic Ridge: Marine Geology, v. 16, p. 83–102.

Bonatti, E., and Honnorez, J.
1976 : Sections of the earth's crust in the equatorial Atlantic: Journal of Geophysical Research, v. 81, p. 4104–4116.

Bonatti, E., Honnorez, J., and Ferrara, G.
1970 : Equatorial Mid-Atlantic Ridge: Petrologic and Sr isotopic evidence for alpine-type rock assemblage: Earth and Planetary Science Letters, v. 9, p. 247–256.
1971 : Peridotite-gabbro-basalt complex from the equatorial Mid-Atlantic Ridge: Philosophical Transactions of the Royal Society of London, v. 268, p. 385–402.

Bonatti, E., Honnorez, J., and Gartner, S.
1973 : Sedimentary serpentinites from the Mid-Atlantic Ridge: Journal of Sedimentary Petrology, v. 43, p. 728–735.

Bonatti, E., and Hamlyn, P. R.
1978 : Mantle uplifted block in the western Indian Ocean: Science, v. 201, p. 249–251.
1981 : Oceanic Ultramafic rocks; in The Sea, v. 7, The Oceanic Lithosphere; ed. C. Emiliani: New York, John Wiley and Sons, p. 241–283.

Bonatti, E., and Chermak, A.
1981 : Formerly emerging crustal blocks in the equatorial Atlantic; Tectonophysics, v. 72, p. 165–180.

Bonatti, E., Chermak, A., and Honnorez, J.
1979 : Tectonic and igneous emplacement of crust in oceanic transform zones: In Deep Drilling Results in the Atlantic: Ocean Crust, ed. M. Talwani, C. G. Harrison and D. E. Hayes, Maurice Ewing Series: American Geophysical Union, v. 2, p. 239–248.

Bonatti, E., Sartori, R., and Boersma, A.
1983 : Vertical crustal movements at the Vema fracture zone in the Atlantic: Evidence from dredged limestones: Teconophysics, v. 91, p. 213–232.

Bonatti, E., Sarntheim, M., Boersma, A. Gorini, M., and Honnorez, J.
1977 : Neogene crustal emersion and subsidence at the Romanche fracture zone, equatorial Atlantic: Earth and Planetary Science Letters, v. 35, p. 369–383.

Bottinga, Y., and Allegre, C. J.
1978 : Partial melting under spreading ridges: Philosophical Transactions of the Royal Society of London A., v. 288, p. 501–525.

Bryan, W. B., Thompson, G., and Ludden, J. N.
1981 : Compositional variation in normal MORB from 22°–25° N: Mid-Atlantic Ridge and Kane Fracture Zone: Journal of Geophysical Research, v. 86, p. 11, 815–11,836.

Burr, N. C., and Solomon, S. C.
1978 : The relationship of source parameters of oceanic transform earthquakes to plate velocity and transform length: Journal of Geophysical Research, no. 83, p. 1193–1205.

Casey, J. F., Dewey, J. F., Fox, P. J., Karson, J. A., and Rosenbrantz, E.
1981 : Heterogeneous nature of oceanic crust and upper mantle: a perspective from the Bay of Island Ophiolite complex, in The Sea, v. 7, The Oceanic Lithosphere, ed. C. Emiliani: New York, John Wiley and Sons, p. 305–338.

Casey, J. F., and Karson, J. A.
1981 : Magma chamber profiles from the Bay of Island ophiolite complex: Nature, v. 292, p. 295–301.

CAYTROUGH
1979 : Geological and geophysical investigation of the mid-Cayman Rise spreading center: Initial results and observations, in Deep drilling results in the Atlantic Ocean: Ocean crust, ed. M. Talwani, C. G. Harrison and D. E. Hayes: American Geophysical Union, Maurice Ewing Series 2, p. 66–93.

Choukroune, P., Francheteau, J., and LePichon, X.
1978 : In situ structural observations along Transform Fault A in the FAMOUS area, Mid-Atlantic Ridge: Geological Society of America Bulletin, v. 89, p. 1013–1029.

Cochran, J. R.
1973 : Gravity and magnetic investigations of the Guiana Basin, Western equatorial Atlantic: Geological Society of America Bulletin, v. 84, p. 3249–3268.

Collette, B. J., Schouten, J. M., Rutten K., and Slootweg, A. P.
1974 : Structure of the Mid-Atlantic Ridge Province between 12° and 18°N: Marine Geophysical Research, v. 2, p. 143–179.

Cormier, M. H., Detrick, R. S., and Purdy, G. M.
1984 : Seismic refraction studies of the Kane Fracture Zone: Journal of Geophysical Research, v. 89, p. 10,249–10,266.

Crowell, J. C.
1974a: Origin of Late Cenozoic Basins in Southern California; in Tectonics and Sedimentation, ed. W. R. Dickinson: Society of Economic Paleontologists and Mineralogists Special Publication 22, p. 190–204.
1974b: Sedimentation along the San Andreas Fault, California; in Modern and Ancient Geosynclinal Sedimentation, ed. R. M. Dott, Jr. and R. H. Shaver: Society of Economic Paleontologists and Mineralogists, Special Publication 19, p. 292–303.

Detrick, R. S., Cormier, M. M., Prince, R., and Forsyth, D. W.
1982 : Seismic constraints on the crustal structure within the Vema Fracture zone: Journal of Geophysical Research, v. 87, 10, 599–10, 612.

Detrick, R. S., Mudie, J. D., Luyendyk B. P., and Macdonald, K. C.
1973 : Near bottom observations of an active transform fault (Mid-Atlantic Ridge at 37°N): Nature Physical Sciences, v. 246, p. 59–61.

Detrick, R. S., and Purdy, G. M.
1980 : The crustal structure of the Kane fracture zone from seismic refraction studies: Journal of Geophysical Research, v. 85, p. 3759–3777.

Dewey, J. F.
1975 : Finite plate evolution: Implications for the evolution of rock masses at plate margins: American Journal of Science, v. 275-A, p. 260–284.

deWit, M. J. Dutch, S., Kligfield, R., Allen, R., and Stern C.
1977 : Deformation, serpentinization and emplacement of a dunite complex, Gibbs Island, South Shetland Islands: possible fracture zone tectonics: Journal of Geology, v. 85, p. 745–762.

Eittreim, S., and Ewing, J.
1975 : Vema Fracture Zone Transform Fault: Geology, v. 3, p. 555–559.

Ewing, M., Ewing, J., and Talwani, M.
1964 : Sediment Distribution in the Oceans: The Mid-Atlantic Ridge: Geological Society of America Bulletin, v. 75, p. 17–36.

Feden, R. M., Fleming, M. S., and Perry, R. K.
1975 : The Mid-Atlantic Ridge at 33 N: The Hayes Fracture Zone: Earth and Planetary Science Letters, v. 26, p. 292–298.

Forsyth, D. W., and Wilson, B.
1984 : Three-dimensional temperature structure of a ridge-transform-ridge system; Earth and Planetary Science Letters, no. 70, p. 355–362.

Fox, P. J.
1972 : The geology of some Atlantic fracture zones, Caribbean escarpments and the nature of the oceanic basement and crust: [Ph.D. Thesis], Columbia University, New York, 357 p.

Fox, P. J., and Gallo, D. G.
1984 : A tectonic model for Ridge-Transform-Ridge Plate Boundaries: Im-

plications for the Structure of Oceanic Lithosphere: Tectonophysics, v. 104, p. 205–242.
Fox, P. J., Lowrie, A., and Heezen, B. C.
1969a: Oceanographer fracture zone: Deep-Sea Research, v. 16, p. 55–66.
Fox, P. J., Pitman, W. C., III, and Shepard, F.
1969b: Crustal Plates in the Central Atlantic: Evidence for at least Two Poles of Rotation: Science, v. 165, p. 487–489.
Fox, P. J., and Opdyke, N. D.
1973 : Geology of the oceanic crust: magnetic properties of oceanic rocks: Journal of Geophysical Research, v. 78, p. 5139–5154.
Fox, P. J., Schreiber, E., and Peterson, J. J.
1973 : The geology of the oceanic crust: compressional wave velocities of oceanic rocks: Journal of Geophysical Research, v. 78, p. 5155–5172.
Fox, P. J., Schreiber, E., Rowlett, H., and McCamy, K.
1976 : The geology of the Oceanographer Fracture Zone: A model for fracture zones: Journal of Geophysical Research, v. 81, p. 4117–4128.
Fox, P. J., Schroeder, F. , Moody, R., and Pitman, W. C., III.
in preparation : The morphotectonic character of the Oceanographer Fracture Zone.
Fox, P. J., and Stroup, J. B.
1981 : The plutonic foundation of the oceanic crust; in the Sea, v. 7, the Oceanic Lithosphere; ed. C. Emiliani: New York, John Wiley and Sons, p. 119–218.
Francheteau, J. Choukroune, P., Hekinian, R., LePichon, X., and Needham, D.
1976 : Oceanic fracture zones do not provide deep sections into the crust: Canadian Journal of Earth Science, v. 13, p. 1223–1235.
Frey, F. A.
1970 : Rare earth and potassium abundances in St. Paul Rocks: Earth and Planetary Science Letters, v. 7, p. 351–360.
General Bathymetric Chart of the Oceans (GEBCO), 1982 North Atlantic Chart: Canadian Hydrographic Service, Ottawa, Canada.
Goud, M. R., and Karson, J. A.
1984 : Tectonic Significance of Sheared Chalks from FAMOUS Fracture Zone B: EOS, Transactions of the American Geophysical Union, v. 65, p. 275.
Heezen, B. C., Bunce, E. T., Hersey, J. B., and Tharp, M.
1964a: Chain and Romanche Fracture Zones: Deep Sea Research, v. 11, p. 11–33.
Heezen, B. C., Gerard, R. D., and Tharp, M.
1964b: The Vema Fracture Zone in the Equatorial Atlantic: Journal of Geophysical Research, v. 69, p. 733–739.
Heezen, B. C., and Tharp, M.
1961 : Physiographic Diagram of the South Atlantic, The Caribbean, the Scotia Sea, and the eastern margin of the South Pacific Ocean: Geological Society of America, New York.
1965 : Tectonic fabric of the Atlantic and Indian Oceans and continental drift; Philosophical Transactions, Geological Society of London, v. 258, p. 90–106.
1968 : Physiographic Diagram of the North Atlantic (new edition): Geological Society of America, New York.
Jackson, N. R., Reid, I., and Falconer, R.K.N.
1982 : Crustal structure near the Arctic Mid-Ocean Ridge: Journal of Geophysical Research, v. 87, p. 1773.
Johnson, G. L., and Vogt, P. R.
1973 : Mid-Atlantic Ridge from 47 to 51 N: Geological Society of America Bulletin, v. 84, p. 3443–3462.
Karson, J. A.
1982 : Reconstructed seismic velocity structure of the Lewis Hills massif and implications for oceanic fracture zones: Journal of Geophysical Research, v. 87, p. 961–978.
1984 : Variations in structure and petrology in the coastal complex, Newfoundland: The anatomy of an oceanic fracture zone; in Ophioliter and Oceanic Lithosphere, ed. I. G. Gass, S. J. Lippard, A. W. Sheldon, Oxford, Blackwell Scientific Publications, v. 13, p. 131–144.
Karson, J. A., and Dewey, J. F.
1978 : Coastal complex, western Newfoundland: An early Ordovician oceanic fracture zone: Geological Society of America Bulletin, v. 89, p. 1037–1049.
Karson, J. A., and Dick, H.
1983 : Tectonics of ridge-transform intersections at the Kane fracture zone, Marine Geophysical Research, v. 6, p. 51–98.
Karson, J. A., Elthon, D. L., and DeLong, S. E.
1983 : Ultramafic intrusions in the Lewis Hills Massif, Bay of Islands Ophiolite complex, Newfoundland: Implications for igneous processes at oceanic fracture zones: Geological Society of America Bulletin, v. 94, no. 1, p. 15–29.
Kastens, K. A., Macdonald, K. C., and Fox, P. J.
in press: Deep Tow survey along the principal transform displacement zone of the Vema Fracture Zone: Journal of Geophysical Research, in press.
Kuznir, N. J., and Bott, M.H.P.
1976 : A thermal study of the formation of oceanic crust: Geophysical Journal of the Royal Astronomical Society, v. 47, p. 83–95.
Langmuir, C. H., and Bender, J. F.
1984 : Chemical variations of ORB in the vicinity of transform faults; observations and implications: Earth and Planetary Science Letters, v. 69, p. 107–127.
Langseth, M. S., LePichon, X., and Ewing, M.
1966 : Crustal structure of mid-ocean ridges, 5. Heat flow through the Atlantic Ocean and convection currents: Journal of Geophysical Research, v. 71, p. 5321–5355.
Laughton, A. S., and Rusby, J.S.M.
1975 : Long-Range Sonar and Photographic Studies of the Median Valley in the FAMOUS area of the Mid-Atlantic Ridge near 37°N: Deep Sea Research, v. 22, p. 279–298.
Leg 82 Scientific Party
1982 : Elements traced in Atlantic: Geotimes, v. 27, p. 21–23.
LePichon, X.
1968 : Sea floor spreading and continental drift: Journal of Geophysical Research, v. 73, p. 3661–3697.
LePichon, X., and Fox, P. J.
1971 : Marginal Offsets, Fracture Zones, and the Early Opening of the North Atlantic: Journal of Geophysical Research, v. 76, p. 6294–6308.
Lonsdale, P., and Shor, A.
1979 : The Oblique Intersection of the Mid-Atlantic Ridge with Charlie Gibbs Transform Fault: Tectonophysics, v. 54, p. 195–209.
Louden, K. E., and Forsyth, D. W.
1982 : Crustal structure and isostatic compensation near the Kane fracture zone from topography and gravity measurements - I. spectral analysis approach: Geophysical Journal of the Royal Astronomical Society, v. 68, p. 725–750.
Luyendyk, B. P., and Melson, G. P.
1967 : Magnetic properties of rocks near the crest of the Mid-Atlantic Ridge: Nature, v. 215, p. 147–149.
Macdonald, K. C., Castillo, D. A., Fox, P. J., Kastens, K., and Miller, S. P.
in press : A Deep-Tow Survey of the Vema Transform and Eastern Intersection: Journal of Geophysical Research.
Macdonald, K. C., and Luyendyk, B. P.
1977 : Deep-tow studies of the structure of the Mid-Atlantic Ridge crest near 37°N (FAMOUS): Geological Society of America Bulletin, v. 88, p. 621–636.
McKenzie, D.
1967 : Some remarks on heat flow and gravity anomalies; Journal of Geophysical Research, v. 72, p. 6261–6273.
McKenzie, D. P., and Parker, R. L.
1967 : The North Pacific: an example of tectonics on a sphere: Nature, v. 216, p. 1276–1280.
Melson, W. G., Jorosewitch, E., Bowen, V. T., and Thompson, G.
1967 : St. Peter and St. Paul Rocks: a high temperature, mantle-derived intrusion: Science, v. 155, p. 1532–1535.
Melson, W. G., and Rabinowitz, P. D., eds.

1978 : Initial Reports of the Deep Sea Drilling Project, v. 45: Washington, D.C., U.S. Government Printing Office, v. 45, 717 p.

Menard, H. W.
1955 : Deformation of the Northeastern Pacific Basin and the West Coast of North America: Geological Society of America Bulletin, v. 69, p. 1149–1198.

Menard, H. W., and Atwater, T.
1968 : Changes in direction of sea floor spreading: Nature, v. 219, p. 463–467.
1969 : Origin of fracture zone topography: Nature, v. 222, p. 1037–1040.

Menard, H. W., and Dietz, R. S.
1952 : Mendocino submarine escarpments: Journal of Geology, v. 60, p. 266–278.

Miyashiro, A., Shido, F., and Ewing, M.
1969 : Composition and origin of serpentinites from the Mid-Atlantic Ridge, 24° and 30° N: Contributions to Mineralogy and Petrology, v. 32, p. 38–52.

Moores, E. M., and Vine, F. J.
1971 : The Troodos Massif, Cyprus and other ophiolites as oceanic crust: evaluation and implications: Philosophical Transactions of the Royal Society of London, v. 268A, p. 443–465.

Morgan, W. J.
1968 : Rises, trenches, great faults and crustal blocks: Journal of Geophysical Research, v. 73, p. 1959–1982.

Mutter, J. C. and Detrick, R. S., and North Atlantic Transect Study Group.
1984 : Multichannel seismic evidence for anomalously thin crust at Blake Spur fracture zone: Geology, v. 12, p. 534–537.

Needham, H. D., and Francheteau, J.
1974 : Some characteristics of the rift valley in the Atlantic Ocean near 36° 48°N: Earth and Planetary Science Letters, v. 22, p. 29–43.

Nicolas, A., Bouchez, J. L., Boudier, F., and Mercier, J. C.
1971 : Textures, structures and fabrics due to solid state flow in some European lherzolites: Tectonophysics, v. 12, p. 55–86.

Nisbet, E. G., and Fowler, C.M.R.
1978 : The Mid-Atlantic Ridge at 37 and 45 N: some geophysical and petrological constraints: Geophysical Journal of the Royal Astronomical Society, v. 54, p. 631–680.

Olivet, J-L., LePichon, X., Monti, S., and Sichler, B.
1974 : Charlie-Gibbs fracture zone: Journal of Geophysical Research, v. 79, p. 2059–2072.

OTTER Scientific Team
1984 : The geology of the Oceanographer transform: The ridge-transform intersection, Marine Geophysical Research, v. 6, p. 109–141.
1985 : The geology of the oceanographer transform: the transform domain, Marine Geophysical Research, in press.

Parker, R. L., and Oldenberg, D. W.
1973b: Thermal model of ocean ridges, Nature Physical Sciences, v. 242, p. 137–139.

Parmentier, E. M., and Forsyth, D. W.
1985 : Three-dimensional flow beneath a slow spreading ridge axis: A dynamic contribution to deepening of the median valley toward fracture zones; Journal of Geophysical Research, no. 90, p. 678–684.

Parsons, B., and McKenzie, D. P.
1978 : Mantle convection and the thermal structure of the plates; Journal of Geophysical Research, no. 83, p. 4485–4496.

Phillips, J. D., and Fleming, H. S.
1978 : Multi-beam sonar study of the Mid-Atlantic rift valley 36–37°N: FAMOUS: Geological Society of America, MC-19.

Prinz, M., Keil, K., Green, J. A., Reid, A. M., Bonatti, E., and Honnorez, J.
1976 : Ultramafic and mafic dredge samples from the equatorial Mid-Atlantic Ridge and fracture zones: Journal of Geophysical Research, v. 81, p. 4087–4103.

Prinzhofer, A., and Nicolas, A.
1980 : The Bogata peninsula, New Caledonia: A possible oceanic transform fault: Journal of Geology, v. 88, p. 387–398.

Purdy, G. M., Rabinowitz, P. D., and Schouten, H.
1978 : The Mid-Atlantic Ridge at 23°N: Bathymetry and magnetics; in Initial Reports of the Deep Sea Drilling Project, vol. 45, eds. W. G. Melson and P. D. Rabinowitz, Washington, D.C., U.S. Government Printing Office, p. 119–128.

Purdy, G. M., Rabinowitz, P. D., and Velterop, J.J.A.
1979 : The Kane fracture zone in the central Atlantic ocean: Earth and Planetary Science Letters, v. 45, p. 429–434.

Quon, S. H., and Ehlers, E. G.
1959 : Rocks of the northern part of the Mid-Atlantic Ridge, v. 74, p. 1–7.

Rabinowitz, P. D., and Purdy, G. M.
1976 : The Kane Fracture Zone in the Western Central Atlantic Ocean: Earth and Planetary Science Letters, v. 33, p. 21–26.

Ramberg, I. B., and van Andel, T. J.
1977 : Morphology and tectonic evolution of the rift valley at latitude 36 30°N, Mid-Atlantic Ridge: Geological Society of America Bulletin, v. 88, p. 577–586.

Ramberg, I. B., Gray, D. F., and Reynolds, R.G.M.
1977 : Tectonic evolution of the FAMOUS area of the Mid-Atlantic Ridge, latitude 35 50° to 37 20°N: Geological Society of America Bulletin, v. 88, p. 609–620.

Reid, I. D., and Jackson, H. R.
1981 : Oceanic spreading rate and crustal thickness: Marine Geophysical Research, v. 5, p. 165–172.

Renard, V., Schrumpf, B., Sibuet, J. C., and Carre, D.
1975 : Bathymetric Detaillee Dune partie de vallee du rift et de faille transformante, pres de 36 50°N dans l'ocean Atlantique: Paris, Centre National pour l'Exploration des Oceans.

Robb, J. M., and Kane, M. F.
1975 : Structure of the VEMA fracture zone from gravity and magnetic intensity profiles: Journal of Geophysical Research, v. 80, p. 4441–4445.

Saleeby, J. B.
1977 : Fracture zone tectonics, continental margin fragmentation and emplacement of the Kings-Kaweah Ophiolite Belt, Southwest Sierra Nevada, California; in North American Ophiolites, ed., by R. G. Coleman and W. P. Irwin: Oregon Department of Geology and Mineralogy Industries Bulletin, v. 95, p. 141–160.

Schilling, J-G., Zajac, M., Evans, R. Johnston, T., White, W., Devine, J. D., and Kingsley, R.
1983 : Petrologic and geochemical variations along the Mid-Atlantic Ridge from 29°N to 73°N: American Journal of Science, v. 283, p. 510–586.

Schouten, H., Karson, J., and Dick, H.
1980 : Geometry of transform zones: Nature, v. 288, p. 470–473.

Schouten, H., and Klitgord, K. D.
1982 : The memory of the accreting plate boundary and the continuity of fracture zones: Earth and Planetary Science Letters, v. 59, p. 255–266.

Schroeder, F. W.
1977 : A geophysical investigation of the Oceanographer fracture zone and the Mid-Atlantic ridge in the vicinity of 35°N: [Ph.D. Thesis]: Columbia University, 458 p.

Sclater, J. G., Anderson, R. N., and Bell, M. L.
1971 : Elevation of ridges and evolution of the central eastern Pacific; Journal of Geophysical Research, no. 76, p. 7888–7915.

Searle, R. C.
1979 : Side-scan sonar studies of North Atlantic fracture zones: Journal of the Geological Society of London, v. 136, p. 283–293.
1981 : The active part of the Charlie-Gibbs fracture zone: A study using sonar and other geophysical techniques: Journal of Geophysical Research, v. 86, p. 243–262.

Searle, R. C., and Laughton, A. S.
1977 : Sonar studies of the Mid-Atlantic ridge and Kurchatov Fracture zone: Journal of Geophysical Research, v. 82, p. 5313–5328.

Shand, S. J.
1949 : Rocks of the Mid-Atlantic Ridge; Journal of Geology, v. 57, p. 89–91.

Sibuet, J.-C., and Veyrat-Peiney, B.
1980 : Gravimetric model of the Atlantic Equatorial fracture zones: Journal of Geophysical Research, v. 85, p. 943–954.
Sigurdsson, H.
1981 : First-order major element variations in basalt glasses from the Mid-Atlantic Ridge: 29°N to 73°N: Journal of Geophysical Research, v. 86, p. 9483–9502.
Simonian, K. O., and Gass, I. G.
1978 : Arakapas fault belt, Cyprus: a fossil transform belt: Geological Society of America, v. 89, p. 1220–1230.
Sinha, M. C., and Louden, K. E.
1983 : The Oceanographer fracture zone 1. Crustal structure from seismic refraction studies; Journal of Royal Astronomical Society of London, v. 75, p. 713–736.
Sleep, N. H.
1975 : Formation of ocean crust: Some thermal constraints: Journal of Geophysical Research, v. 80, p. 4037–4042.
Sleep, N. H., and Biehler, S.
1970 : Topography and tectonics at the intersections of fracture zones and central rifts: Journal of Geophysical Research, v. 75, p. 2748–2752.
Smewing, J. D.
1980 : An upper Cretaceous ridge-transform intersection in the Oman Ophiolite; in Proc. Int. Ophiolite Symp., ed. A. Panayiotou; Cyprus Geological Survey Department, Nicosine, p. 407–413.
Stover, C. W.
1966 : Seismicity of the Indian Ocean: Journal of Geophysical Research, v. 71, p. 2575–2581.
Stroup, J. B., and Fox, P. J.
1981 : Geologic investigations in the Cayman Trough: Evidence for thin Oceanic Crust along the Mid-Cayman Rise; Journal of Geology, v. 89, p. 395–420.
Sykes, L. R.
1963 : Seismicity of the Indian Ocean: Journal of Geophysical Research, v. 71, p. 2575–2581.
1965 : The Seismicity of the Arctic: Bulletin of the Seismological Society of America, v. 55, p. 501–518.
1967 : Mechanism of earthquakes and nature of faulting on the Mid-Oceanic Ridges: Journal of Geophysical Research, v. 72, p. 2131–2153.
Tchalenko, J. S.
1970 : Similarities between shear zones of different magnitudes: Geological Society of America Bulletin, v. 81, p. 1625–1640.
Tchalenko, J. S., and Ambraseys, N. N.
1970 : Structural analysis of the Dasht-E-Bayaz (Iran) earthquake fractures: Geological Society of America Bulletin, v. 81, p. 41–60.
Thompson, G., and Melson, W. G.
1972 : The petrology of oceanic crust across fracture zones in the Atlantic Ocean: evidence of a new kind of sea-floor spreading: Journal of Geology, v. 80, p. 526–538.
Tsai, Y. B.
1969 : Determination of focal depth of earthquakes in mid-oceanic ridges from amplitude spectra of surface waves: [Ph.D. thesis]: Cambridge, Massachusetts Institute of Technology, 144 pp.
Turcotte, D. L., and Oxburgh, E. R.
1967 : Finite amplitude convective cells and continental drift; Journal of Fluid Mechanics, no. 28, p. 29–42.
Twigt, W., Slootweg, A. P., and Collette, B. J.
1979 : Topography and a magnetic analysis of an area south-east of Azores (36°N, 23°W): Marine Geophysical Research, v. 4, p. 91–104.
Twigt, W., Verhoef, J., Rohr, K., Mulder, Th.F.A., and Collette, B. J.
1983 : Topography, magnetics and gravity over the Kane fracture zone in the Cretaceous magnetic quiet zone (African Plate). Proceedings of the Koninklijke Nederlandse Akademie van Wetenschappen, Series B: Palaeontology, Geology, Physics and Chemistry, v. 86, p. 181–210.
van Andel, T. H., Corliss, J. B., and Bowen, V. T.
1967 : The intersection between the Mid-Atlantic Ridge and the Vema Fracture Zone in the North Atlantic: Journal of Marine Research, v. 25, p. 343–351.
van Andel, T. H., Phillips, J. D., and von Herzen, R. P.
1969 : Rifting origin for the Vema Fracture in the North Atlantic, v. 5, p. 296–300.
Vogt, P. R., and Avery, O. E.
1974 : Detailed magnetic surveys in the northeast Atlantic and Labrador Sea: Journal of Geophysical Research, v. 79, p. 363–389.
Vogt, P. R., and Johnson, G. L.
1975 : Transform faults and longitudinal flow below the Mid-Oceanic Ridge: Journal of Geophysical Research, v. 80, p. 1399–1428.
Weidner, D. J., and Aki, K.
1973 : Focal depth and mechanism of mid-ocean ridge earthquakes; Journal of Geophysical Research, no. 78, p. 1818–1831.
White, R. S.
1984 : Atlantic oceanic crust: Seismic structure of a slow spreading ridge; in Ophiolites and Oceanic Lithosphere: Ed. I. G. Gass, S. J. Lippard, A. W. Shelton, Geological Society of London Special Publication, p. 101–111.
White, R. S., and Matthews, D. H.
1980 : Variations in oceanic crustal structure in a small area of the northeastern Atlantic, Geophysical Journal of the Royal Astronomical Society, v. 61, p. 401–436.
Whitmarsh, R. B., and Laughton, A. S.
1975 : The fault pattern of a slow-spreading ridge near a fracture zone: Nature, v. 258, p. 509–510.
1976 : A long-range sonar study of the Mid-Atlantic ridge crest near 37°N (FAMOUS area) and its tectonic implications: Deep-Sea Res., v. 23, p. 1005–1023.
Wilcox, R., Harding, T., and Sealy, D. R.
1973 : Basic wrench tectonics: American Association of Petroleum Geology Bulletin, v. 57, p. 74–96.
Williams, C. A., Louden, K. E., and Tanner, S. J.
1983 : The Western Intersection of Oceanographer Fracture Zone with the Mid-Atlantic Ridge: Marine Geophysical Research, in press.
Wilson, J. T.
1965 : A new class of faults and their bearing on continental drift: Nature, v. 207, p. 343–347.

Manuscript Accepted by the Society March 11, 1985

ACKNOWLEDGMENTS

Many of the results summarized in this paper are the outgrowth of field experiments located at the Mid-Cayman Rise, and the Oceanographer, Vema, and Kane Transforms. We thank the many investigators who worked with us on these projects for their help as well as the National Science Foundation and the Office of Naval Research for their support. In particular, a number of colleagues have critiqued early drafts of this text and we would like to thank R. Detrick, J. Karson, K. Macdonald, and R. Searle. B. Tucholke and P. Vogt, the editors of this volume, and A. R. Palmer of the Geological Society of America have been very helpful throughout the preparation of this report and we are thankful for their constructive comments. We are particularly grateful to ONR for support that allowed us to review the literature and to synthesize and integrate our data from various field programs into a large framework (ONR N00014-81-0062).

Printed in U.S.A.

The Geology of North America
Vol. M, The Western North Atlantic Region
The Geological Society of America, 1986

Chapter 11

Hydrothermal activity in the North Atlantic

J. M. Edmond
Department of Earth, Atmospheric, and Planetary Sciences, E34-201, Massachusetts Institute of Technology, Cambridge, Massachusetts 02139

INTRODUCTION

One of the major developments in the earth sciences in the last ten years has been the recognition of the importance of hydrothermal circulation in the thermal and chemical evolution of the oceanic crust. Submarine hot springs with exit temperatures as high as 350°C have been located and sampled at several locations on the ridge systems of the Eastern Pacific (Edmond and others, 1982). Hydrothermal deposits of sulphide ores and pavements of pure MnO_2 have been recovered by dredging at numerous other sites. Exploration has been facilitated at these medium and fast spreading ridges by the relatively subdued topography. Success has a high probability, given the frequency of intrusive and eruptive events and the permanent presence of shallow magma chambers (Sleep and Rosenthal, 1979). On the Mid-Atlantic Ridge (MAR), none of these circumstances obtain and the results have been correspondingly meagre, with the obvious exception of Iceland. The strongest evidence for current hydrothermal activity on the MAR comes from the TAG area (Trans-Atlantic Geotraverse) at 26°N. While small temperature and ^{3}He anomalies have been observed in the water column in the axial valley, active venting was not observed during subsequent *Alvin* dives. However, there are abundant indications that hot springs have played an important role in the development of the MAR over the time-span of its existence.

Much of the evidence is for low-temperature activity of the kind observed on the Galapagos Spreading Center (GSC) in the Eastern Pacific (Edmond and others, 1979 a,b). It is known, in the latter instance, that the temperatures in the surface vents are artifacts of the extent of sub-surface mixing of a high temperature end-member (~350°C) with cold "groundwater" having properties indistinguishable from those of the overlying ambient seawater. It is reasonable to suggest that, given the great depth of the magma chambers and the extensive tectonism on the MAR, surface manifestations of hydrothermal activity generally will be of the Galapagos type. It is also possible that shallow convection cells of low-temperature conductively heated groundwater may exist analogous to those observed at the Galapagos Mounds on 500,000 year old crust off-axis on the GSC (Corliss and others, 1978) and as common features on Iceland. In what follows, the evidence for recent hydrothermal activity on the MAR will be summarized. The active Icelandic systems will then be discussed in detail.

SUBMARINE HYDROTHERMAL ACTIVITY ON THE MAR

Five known manifestations of hydrothermal activity are preserved in the oceanic record.

1. The most ubiquitous are the alteration products of the seawater-basalt reactions themselves (Corliss, 1971). Their mineralogy and composition are indicative of a wide range of temperatures of formation (Honnorez, 1981).

2. On the seafloor, sulphide deposits record the transport of material by solutions at temperatures greater than about 250°C. The Cyprus-type massive sulphide ore bodies characteristic of ophiolite terrains are deposited on the new basalts on sediment starved ridge axes. The compositionally more diverse sediment hosted bodies form above the buried spreading centers often observed in early opening environments, for example the Red Sea, the Gulf of California, or in back arc spreading systems. It is quite likely, although unproven, that such deposits are common at depth below the continental rise close to the passive margins of the North Atlantic.

3. Pure MnO_2 crusts or pavements are evidence of extreme sub-surface dilution of the primary, high temperature, ore transporting solutions by groundwater similar in composition to the ambient seawater. In these circumstances the sulphide forming elements are precipitated as vein deposits and, of the economically important elements, only Mn remains in solution. These deposits are probably precipitated from vents with exit temperatures of less than about 5°C by a combination of inorganic and bacterial oxidation reactions.

4. There appear to be no localized deposits uniquely characteristic of the intermediate temperature range, that is, deposits with only partial dilution of the high temperature fluids. The buoyant plumes of the "black smokers" and the lower energy vents rise to several hundred meters above the ridge axis. Con-

Edmond, J. M., 1986, Hydrothermal activity in the North Atlantic; *in* Vogt, P. R., and Tucholke, B. E., eds., The Geology of North America, Volume M, The Western North Atlantic Region: Geological Society of America.

comitantly, precipitation of their complement of ore forming elements as fine grained sulphides and oxides is induced by the vigorous turbulent entrainment of ambient seawater. These particulates are dissipated in the regional mid-depth circulation and precipitate out as the Fe, Mn, and trace element enriched metalliferous sediments commonly found as basal and flank deposits around spreading centers (Edmond and others, 1982).

5. Diffuse emission of conductively heated groundwaters has been observed off-axis at the GSC (Corliss and others, 1978). Here the waters have not experienced a high temperature environment and, although reducing and containing appreciable amounts of Fe and Mn, they appear to be very low in H_2S. In the Galapagos Mounds area the waters issue through a sediment column several tens of meters thick and build accumulations of iron nontronite capped with a MnO_2 crust. While such systems have not been identified in the axial zone of the GSC or EPR, they may be more prevalent on slow spreading ridges. Their manifestation as localized deposits containing both iron and manganese should be diagnostic.

Turekian and Imbrie (1966) first noticed that MAR flank sediments had higher concentrations of Cu, Co, Ni, and Mn than those in the deep basins. Bostrom and others (1969) presented a preliminary distribution map of Fe-Mn rich sediments for the North Atlantic and noted that these were neither as common nor as enriched in metals as those associated with the faster spreading ridges in the Indian and Eastern Pacific Oceans. Horowitz (1970) demonstrated high trace metal concentrations in crestal sediments between 24 and 29°N. Cronan (1972) reported relatively enriched metalliferous deposits from the median valley near 45°N. Information on the general distribution of these sediments was supplemented by Bostrom (1973). Betzer and others (1974) showed that the concentrations of particulate iron and manganese were higher in the mid-depth water column (2000–1500 meters depth) over the MAR axis than at locations off on the flanks, implying active injection of the elements at the ridge crest.

The earliest DSDP results demonstrated that basal metalliferous sediments were widespread in the North Atlantic (Horowitz and Cronan, 1976). The latter authors, in a comprehensive study, concluded that the flux of metals to these deposits from the ridge crest had been quite constant since the Cretaceous. Deposits characteristic of local hydrothermal activity were first reported from the TAG area of the ridge crest at 26°N (Scott and others, 1974). The dredged samples were almost pure MnO_2 crusts, extremely depleted in iron and the other transition metals. The distributions of the natural U and Th series radioactive isotopes suggested very rapid accumulation. The strong fractionation of iron from manganese depends on the fact that iron forms insoluble sulphides at intermediate temperature and pH. It is therefore removed very efficiently from the hydrothermal fluids during subsurface mixing with the ambient "groundwater." A detailed geological description of the area (Rona, 1976) indicates that the MnO_2 crust formed on a talus slope, an ideal setting for the entrainment of unreacted seawater. Subsequently, small anomalies in temperature and ^3He were observed in the water column above the site. However, *Alvin* dives in the area in 1982 failed to find any active venting of hydrothermal waters. No detectible chemical or isotopic anomalies were found in water samples collected by the submersible immediately above the bottom (Jenkins, personal communication, 1983; Edmond, unpublished data). South of the TAG area, Rona and others (1980) dredged samples of hydrothermal vein quartz. The oxygen isotope composition of this required deposition from solutions at greater than 200°C.

Thompson and others (1975) dredged large boulders of hydrothermal MnO_2 from the ridge crest at 23°N. These samples were closely similar to those from TAG. During the FAMOUS program, extinct hydrothermal deposits were observed and sampled by submersible (ARCYANA, 1975). In these mounds, the Fe:Mn ratio decreased rapidly with distance from the orifice. This suggests somewhat elevated solution temperatures but, given the minor dispersion, very low flow rates. These features are possible candidates for a shallow, conductively driven convection system (Galapagos Mound type).

No surficial sulphide deposits have yet been recovered from the MAR. However, disseminated mineralization is a common occurrence in the upper part of the crust sampled by dredging and drilling. Spectacular examples have been described by Bonatti and others (1976, a,b) and Delaney (in press) who, in dredged and submersible-collected samples from the MAR, found mineralization characteristic of the stockworks that occur below the massive sulphide deposits in ophiolites. Humphris and Thompson (1978) presented results of a systematic examination of the chemistry and mineralogy of hydrothermal alteration products from a large suite of dredged rocks. Bohlke and others (1980) have made similar studies on DSDP cores from relatively young crust (~9 m.y.) on the east flank of the MAR. Both papers reported evidence for pervasive reactions between the basalts and seawater over a broad range in temperatures.

In a pioneering series of papers, Hart and co-workers have provided valuable constraints on the timing and reaction chemistry of the various types of alteration that occur in the upper layers of the young crust of the MAR (Richardson and others, 1980, Staudigel and others, 1981 a,b). Using the isotope systematics of strontium and oxygen, they distinguished three hydrothermal stages. The first involves solutions that have already undergone extensive reactions with basalts, presumably at depth. It results in the formation of palagonite and is essentially contemporaneous with crustal formation. Stage II persists for not more than 3 m.y., involves solutions of similar compositions but at lower temperatures (15–80°C), and is characterized by extensive smectite formation. Stage III extends to perhaps 10 m.y. and results in vein-filling by calcite. Presumably, the calcium is derived from continuing low-temperature alteration of the basalt. Thereafter, the system "closes down," leaving compaction and dehydration of the most hydrous minerals as the only remaining significant processes.

In summary, it can be concluded that hydrothermal activity is a pervasive concomitant to crustal production at the slow-spreading MAR in the North Atlantic. Although active submarine hot-springs have not been discovered there, geochemical signals of such activity have been observed at numerous locations. While most active springs may be of the diffuse, low-temperature Galapagos types, it should be emphasized that little systematic exploration has been done using the techniques that have worked so well in the Pacific. Such operations will certainly be complicated by the great topographic roughness of this slow spreading ridge.

HYDROTHERMAL ACTIVITY ON THE AZORES

The history of eruptive and associated hydrothermal activity on the volcanic islands of the eastern Atlantic is summarized by Machado (1965). He reports fumarole activity on Lanzarote in the Canaries, and solfataras, hot springs and fumaroles on São Miguel, Faial, Pico, and Terceira in the Azores. No information appears to be available on the chemistry of these waters. A single 981 meter drill hole has been made on the flank of the historically active (A.D. 1563) volcano, Agua de Pau, on São Miguel. Temperatures of between 200 and 210°C were encountered at 550 meters (Muecke and others, 1974). Isotopic studies of calcite veins and vugs in the recovered basalts (Lawrence and Maxwell, 1978) indicate that equilibrium between the oxygen isotopes in the carbonates and the water at the reported down-hole temperatures is only achieved at depths of 600 to 700 meters. This is interpreted as being the location of the primary aquifer feeding the hole. No follow-up work on this system has been reported.

HYDROTHERMAL ACTIVITY IN ICELAND

Iceland, lying as it does athwart the actively spreading MAR, is the site of very extensive hydrothermal activity. The island itself can be regarded very crudely as a ridge crest segment bounded to the north by a fracture zone joining it to the Kolbeinsey Ridge and connected more or less directly to the Reykjanes –MAR system to the south (Ward, 1971; Fridleifsson, 1982). The spreading axis, referred to also as the neo-volcanic zone, runs as an arc, concave to the west, through the eastern half of the island (Fig. 1). It bifurcates about half way along its length with the

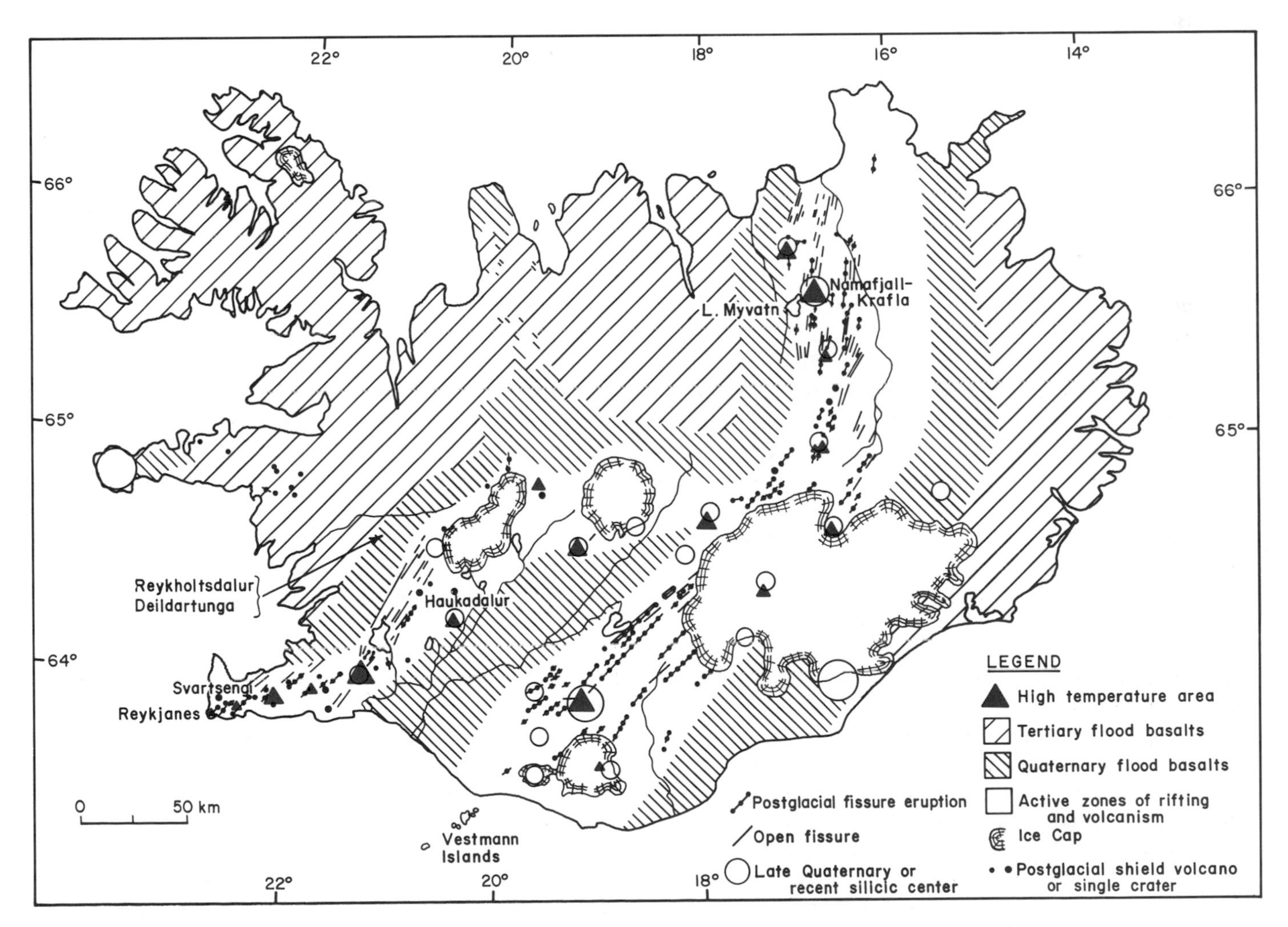

Figure 1. Schematic geologic map of Iceland. The occurrences of high-temperature hot springs are indicated by the filled red triangles.

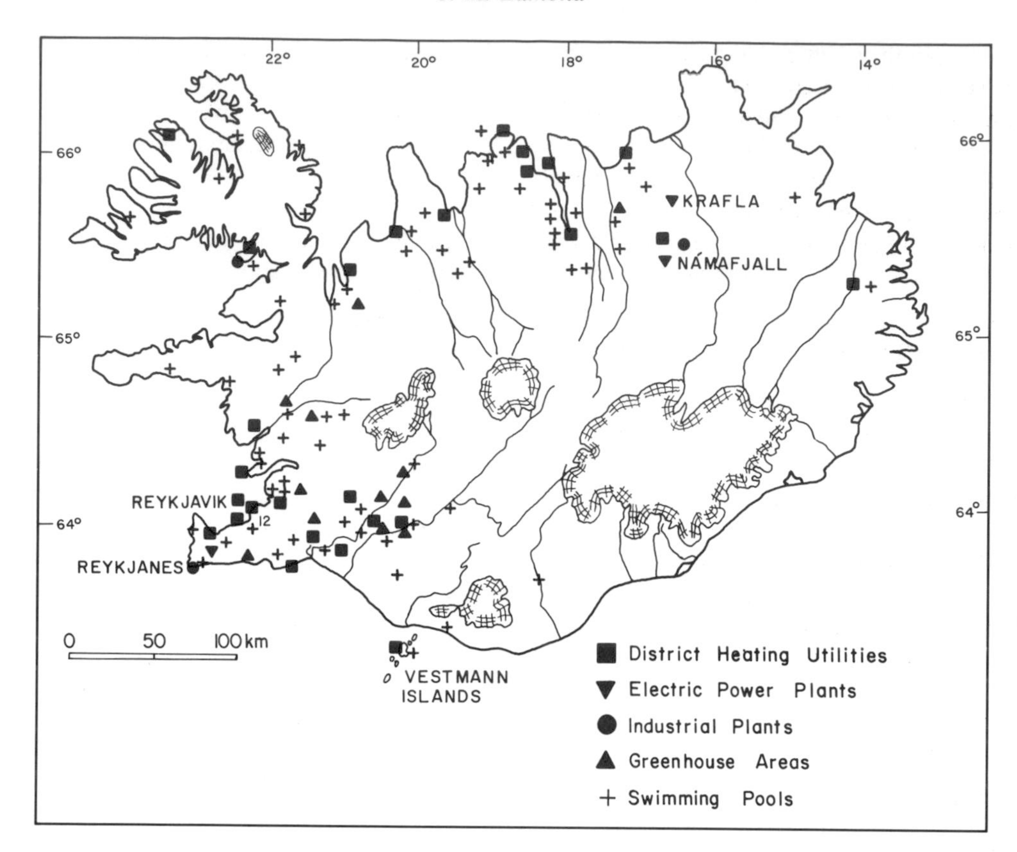

Figure 2. Location and type of utilisation of the high and low-temperature geothermal areas presently being exploited in Iceland. (From Gundmundsson, 1982).

eastern limb dying out near the south coast. The western limb runs north of west from this arcuate segment and then turns sharply southwest to join the oceanic ridge. This axial zone is bounded by Plio-Pleistocene basalts, which also occupy the region between the arms of the southern bifurcation. The extreme east and most of the northwestern part of the island are composed of Tertiary plateau basalts. Superimposed on this relatively regular age progression in basaltic volcanicity are a number of discrete silicic volcanic centers. These occur in all three age zones but the currently active ones are concentrated along the spreading axis.

Hydrothermal activity occurs in rocks of all ages on the island (Pálmason, 1974; Pálmason and others, 1979; Fridleifsson, 1979; Arnorsson, 1974). Previous workers have characterized two types of terrains (Figs. 1, 2); those where temperatures in excess of 200°C are encountered, by drilling, in the upper 1000 meters; and those where temperatures do not exceed 150°C in this zone. The high temperature areas, of which over 20 have been identified, are found only in the Late Quaternary and postglacial volcanic terrains, that is in the neo-volcanic zone and in the active silicic centers. Low temperature fields are found mainly in the Tertiary plateau basalts in the northwestern part of the island and in the southwestern part of the axial region. Such activity is minor in the Tertiary of the east coast of the island. While there is no correlation between the age of the terrains and the maximum observed temperatures in the low temperature regions, there is a marked decrease in the intensity of activity, that is in the flux of transported heat, with increasing crustal age. The inverse correlation between crustal age and heat flow is consistent with the plate tectonic model and the geology (Pálmason and Saemundsson, 1979).

Systematic studies of the isotope hydrology (D/H ratios only) of the island (Arnason and Sigurgeirsson, 1967; Arnason, 1976, 1977a) provide the basis for a delineation of the flow field of the groundwaters (Fig. 3). Two processes affect the D/H ratios in precipitation; the Rayleigh fractionation associated with progressive vapor loss from weather systems: the temperature/altitude effect caused by ground topography. In southern Iceland the Rayleigh effect amounts to a decrease in δD of about $0.2‰$ per kilometer from the coast (Arnason, 1977a). With increasing altitude, as the highlands and icecaps are approached, this gradient rises to as much as $1‰$ per lateral kilometer. Since δD has been measured to $\pm 0.7‰$, the range in the mean δD of precipitation from $-48‰$ on the Reykjanes Peninsula to below $-100‰$ on the interior icecaps provides a very sensitive tag for the origin of the hydrothermal waters. This signal is preserved even during the high-temperature water-rock interactions because of the extremely low abundance of "juvenile" hydrogen in the basalts and the relatively minor importance of fractionation associated with

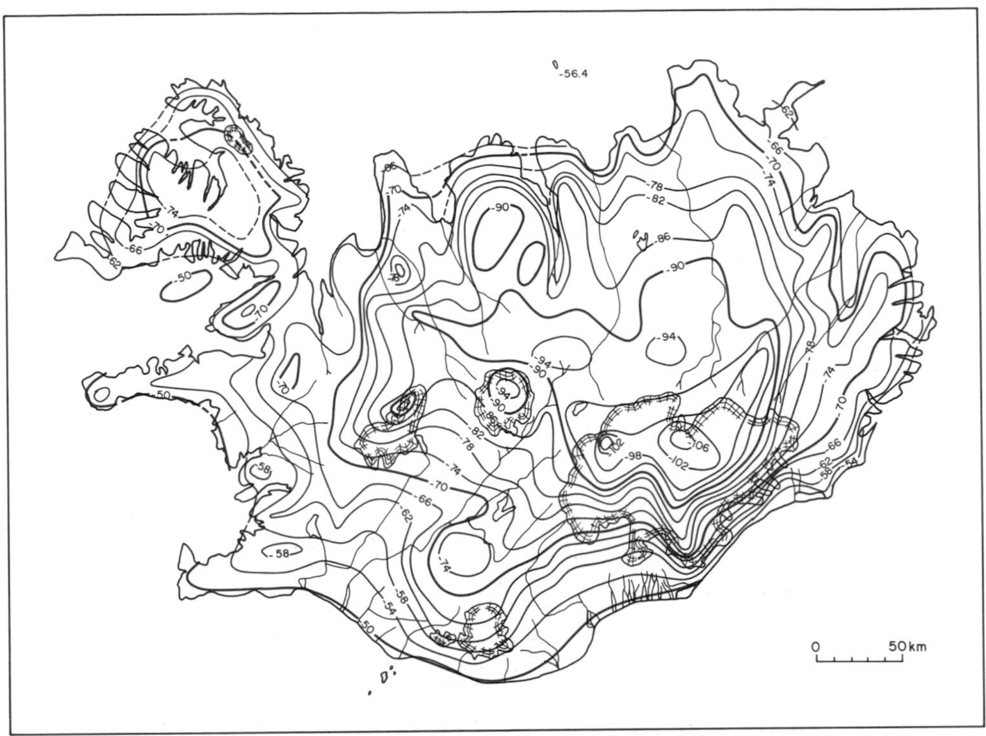

Figure 3. Distribution of Deuterium in precipitation in Iceland. (From Arnason, 1977).

the formation of hydrous minerals. Thus the low-temperature springs of the western Tertiary province derive their waters from the mountainous areas of the northwestern peninsulas and from the Central Highlands and icecaps (Fig. 4). The latter source also contributes to the southern part of the neo-volcanic zone. The rest of this segment is supplied both locally and from the southern icecap. Flow lines of over 50 km have been demonstrated (Arnason, 1977a) as has the importance, in some cases, of "paleo-waters" from the last glacial period. There is no detectable contribution of juvenile water to the Icelandic hot spring systems.

The geothermal gradient in the Tertiary terrains, as measured in drill holes, averages about 75°/km. P. Einarsson (1942) first proposed that the low-temperature springs originate from the regional geothermal heating of groundwaters as they percolate down from the recharge areas. Fridleifsson (1979) has pointed out that there is good correlation between the groundwater flow paths inferred by Arnason (1977a) and the geologic strike. The rarity of hot springs in the eastern Tertiary province can then be explained by the fact that the dominant (glacial) erosion directions are normal to the strike, truncating and localizing the groundwater circulation. Thus the water temperatures are generally quite low because of their short sub-surface residence times. In the western province, by contrast, erosion parallels strike leading to long, uninterrupted flow paths in aquifers predominantly located in the scoriaceous contacts between basalt flows and detrital debris. High temperatures can then result.

The total natural discharge from the Icelandic hot springs is estimated at 1825 liters per second with a weighted average temperature of about 67°C (Gudmundsson, 1982). The great majority of the individual springs discharge less than 5 1/sec. The main up-flow zones are generally controlled by dykes and faults; that is, through the enhancement of the local vertical permeability (Fridleifsson, 1979). The largest natural hot spring is at Deildartunga in Reykholtsdalur in western Iceland. With a flow of 180 1/sec of boiling water, it is the dominant feature in a 1.4 km linear trend of springs with a total production of 253 1/sec.

Along the spreading axis where rifting and associated intrusive activity are intense, the high temperature fields are associated with the central volcanic complexes and probably derive their heat from shallow intrusions (Pálmason and others, 1979). Magnetotelluric surveys have proved valuable in delineating, on a regional scale, the distributing of hydrothermal fluids in the upper crust and the configuration of the underlying zone of partial melting (Beblo and Björnsson, 1978, 1980; Thayer and others, 1981; Hersir and others, 1984). Relatively conducting crust at shallow depth (less than 5 km) is interpreted as reflecting high electrical conductivity in hydrothermal pore waters. A deep conducting layer, as shallow as 8 km below the axis of the neo-volcanic zone and plunging in the direction of spreading is modeled as a zone of partial melting. The volume fraction of the melt phase is estimated to be in the range of 10 to 20 percent and presumably represents the accumulation of molten magma that feeds the spreading axis.

Temperatures at the bottom of deep drill holes in the neo-volcanic zone range up to 350°. The fields vary greatly in area. Most are small, 1 to 20 km^2, but there are three that extend over at least 100 km^2. The total natural heat discharge from these high temperature fields is estimated at 4000 MW (Bodvarsson, 1961)

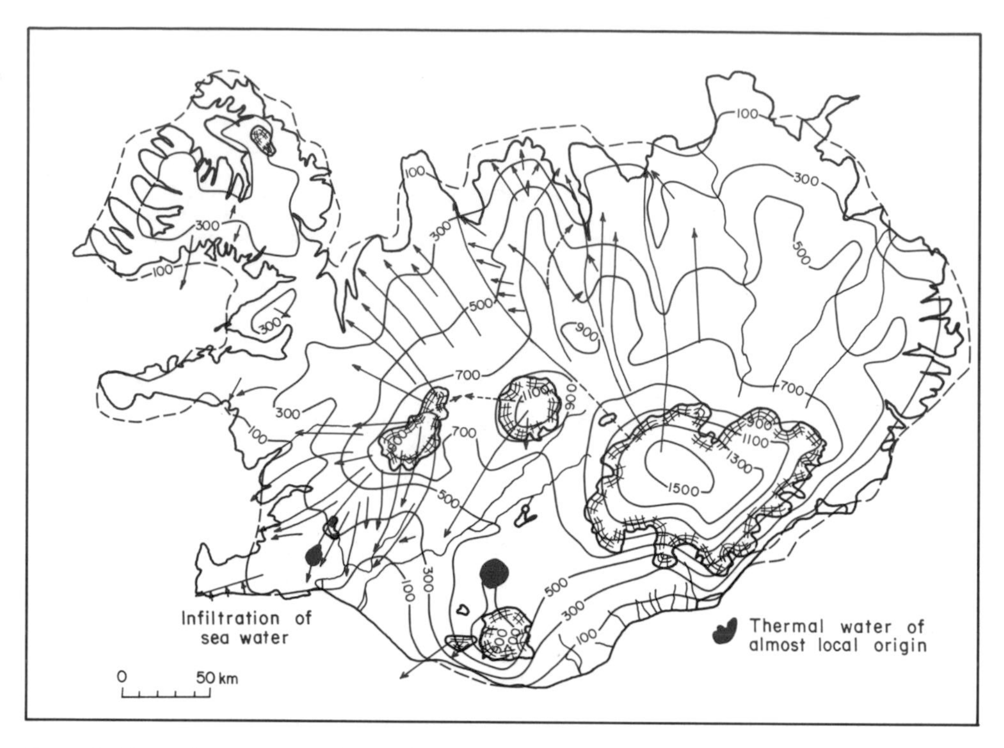

Figure 4. The recharge areas and general flow patterns for thermal groundwater systems in Iceland based on deuterium measurements. (From Arnason, 1977).

from an aggregate area of about 500 km^2 (Fridleifsson, 1979).

The response of the high-temperature discharge to the recent episode of rifting at the Krafla volcanic system in northern Iceland demonstrates the intimate association between shallow magmatic activity and hydrothermalism (Kristmannsdottir, 1983; Stefansson, 1981, Larsen and others, 1979). Hot spring flows inside the caldera have increased dramatically. Steam erosion has excavated craters up to 15 meters deep and 50 meters in diameter at the most active centers. In the drill holes of the Krafla geothermal plant, increases of the water levels of up to 80 meters were triggered by the deflation of the caldera floor as magma injection into the fissure swarm commenced. Pronounced changes in the gas content of the hydrothermal fluids have also been observed during the eruptions. Along the strike of the fissures dramatic increases in hydrothermal activity have also been recorded (Björnsson and others, 1979). There has even been a brief eruption of about 3 tonnes of basaltic scoria through a 1138 meter drill hole in the Namafjall steam field 9 km south of the caldera (Larsen and others, 1979).

As mentioned above, the isotopic studies indicate that recharge in the high temperature areas is much more local than in fields on older crust. Flashing of the ascending hot waters to steam can occur at depths of 1 km or more. This results in the stripping of the CO_2 and the H_2S into the gas phase. When this mixes with the local groundwater at shallow depths, the H_2S is oxidized to sulphur and sulphuric acid. Because of this, the surface rocks in the high temperature areas are often intensely altered (Fridleifsson, 1979).

The only occurrences of a major component of seawater in the high temperature fields are on the Reykjanes Peninsula at Reykjanes and Svartsengi. While compositionally these systems appear to be directly the result of seawater-basalt reactions at elevated temperatures, about 300°C, (Kristmannsdottir, 1983) their isotopic systematics suggest a more complex origin. The original data for $\delta^{18}O$ and $\delta^{19}D$ were reported by Olafsson and Riley (1978). The $\delta^{18}O$ data from drill holes range from –1.08 to +0.45$^0/_{00}$, that is, around the seawater value. However, the deuterium shows a strong relative fractionation at –23.0 to –16.2$^0/_{00}$. Olafsson and Riley (1978) interpreted these results as ruling out a direct seawater origin for the fluids and postulated instead a combination of vapour loss by boiling of waters of seawater origin plus addition of meteoric waters. In this scenario, the present chlorinity of the Reykjanes fluids is only coincidentally close to the seawater value. In addition, it is implied that there is very little net flux of water through the system given the delicacy of the isotopic and concentration balances. This is consistent with the general geochemical features of the system (see below). Kristmannsdottir (1983) states that a complex origin is not compatible with the geochemistry of the fluids but gives no compelling arguments. Since all that is involved is the balanced loss and gain of fresh water, it is difficult to see what diagnostic chemical features (apart from the isotopic systematics) could be produced by such a process. Gundmundsson and others (1981) propose that the deutrium fractionation is caused by formation of secondary minerals; however, a parallel effect is not seen in the $\delta^{18}O$ data. The question of the origin of the Reykjanes Brines has taken on a general importance since it has been used by many investigators as a direct analogue of the fluids produced by submarine, high temperature, seawater-basalt reactions in open convection systems. While there are some similarities, it is apparent from both compositional (see below) and isotopic data that such comparisons are far from exact.

Geysers are presently active in two of the low temperature fields, Reykholtsdalur and Reykjadalur in western Iceland (Fridleifsson, 1979). The eruptions are only to heights of a few meters. In the high temperature areas, Geysir, the type example of the phenomenon at Haukadalur in south Iceland, has been dormant for many decades. Flashing occurred at about 10 meters depth (120°C) in the 20 meter deep vertical conduit. A small geyser in the same field was activated by drilling in 1963 and erupts every 10 to 15 minutes. In the Reykjanes field a geyser was reactivated by an earthquake in 1967. It erupts every few minutes to a height of about 5 meters.

A number of the high-temperature areas have been investigated by deep drilling as part of the on-going exploration and development program outlined by Björnsson (1970). Geothermal power was first generated in the Namafjall field in northern Iceland adjacent to Lake Myvatn (Ragnars and others, 1970). Initial drilling in the area from 1945 to 1953 was limited to shallow wells. Small deposits of native sulphur are common in the area and the intention was to produce steam from which sulphur (as H_2S) could be extracted directly. This did not prove economical although it was feasible technically. Subsequently rich deposits of high-grade diatomite were discovered on the bottom of Lake Myvatn. Deep drilling was commenced to produce steam to dry the diatomite. The first producing well was completed in 1966 and by 1983 twelve holes to 200–300m depth had been drilled. In 1968 construction began of a 2.5 MW power station completed in 1969.

The Namafjall field lies on the Krafla fault swarm about 9 km south of the Krafla caldera and in the middle of the active spreading center. It is characterized by numerous mud-pools and fumaroles. The last eruption at the site occurred in 1728.

Exploration of the Krafla field itself began in 1970 and was greatly accelerated by the 1975 decision to construct a 60 MW power plant (Stefansson, 1981). The hydrology of the field is complex since there co-exist two separate geothermal zones. The upper one, extending to 1100 meters, is water dominated with a mean temperature of 205°C. Below this, to a depth of at least 2200 meters, is a boiling zone with temperatures ranging from 300°C up to 360°C. Oxygen isotopic fractionation between the co-existing secondary minerals, quartz and epidote, give estimates of the equilibrium alteration temperatures in the range 350° to 400°C (Hattori and Muehlenbacks, 1982). The two zones are connected by an up-flow channel located under a gully (actually a series of explosion craters) in the crater floor, which geophysical and chemical studies had originally identified as being the best prospect for drilling. Seismic monitoring of earthquakes indicates the presence of a magma chamber at a depth of between 3 and 7 km (Einarsson, 1978).

The upper zone is a typical water-dominated reservoir. Since the waters are saturated with calcite, flashing produces severe scaling. This problem, coupled with the relatively low temperatures, has led to attempts to avoid this aquifer by casing the hole. In the lower zone the initial flow is of a water-steam mixture. As extraction continues the higher permeation rate of steam has led to a decrease in the water-steam ratio until eventually superheated steam is produced. During the local eruptions in the caldera, pressure transients are observed in the upper aquifer. At the same time, magmatic gases are injected, particularly into the lower zone, leading to sharp decreases in pH (Armannsson and others, 1982). These events have been accompanied by extensive corrosion of drill hole casings and the discharge of "black water" laden with iron sulphide precipitates. Kristmannsdottir (1983) has summarized an extensive study of the alteration mineralogy and chemistry exhibited in cuttings and cores from the Krafla drill holes.

Originally five producing wells were drilled. These intersected aquifers between 600 and 1100 meters with the main steam producers at 638 to 680 meters. Over the course of their exploitation, the chemistry of the discharging fluids changed significantly (Arnorsson, 1977). This was attributed to the progressive extension of the zone of flashing into the aquifer as a result of the pressure drop associated with production. Silica concentrations dropped by as much as 40 percent as did those of the non-condensible gases. The hydrology was not significantly affected, however, by this silica precipitation.

The evolution of the field took a dramatic turn with the initiation of the major rifting event on the Krafla fault swarm in 1975 (Björnsson and others, 1977, 1979). New steam fields formed and the activity at the pre-existing fumaroles increased. This step-up in activity culminated in a magma injection event and the bore hole eruption mentioned previously. The wells were affected to a varying degree by the movements on the fissure zone. Two new deep wells were drilled in 1979-80 and these now supply the steam for the power plant (Pálmason, personal communication, 1983).

While the two fields in the Krafla area are dilute solution systems, the other two high-temperature areas that have been exploited, Reykjanes and Svartsengi on the Reykjanes Peninsula, have high ionic strength fluids, probably derived from connate sea water trapped in buried submarine haloclastites (Olafsson and Riley, 1978). The Reykjanes thermal area is very small (Björnsson and others, 1972). Visible activity on the surface—mud pools, steam vents, solfataras etc.—covers only about 1 km^2. There are three natural springs emitting hot saline waters. Drilling in the area has been concentrated on the axis of the Ridge as defined by seismic activity and recent volcanism. The main aquifers in the deepest hole are in altered basalt flows at between 1100 and 1700 meters depth. In the region of surface activity the brine, at temperatures of 250 to 290°C, ascends from the basaltic horizons through a thick layer of haloclastites. Here it cools and presumably sinks again. Only a small proportion of the circulating fluid actually discharges at the surface. Around the periphery of the

TABLE 1. THE CHEMICAL COMPOSITION OF WATER FROM DEEP DRILLHOLES IN ICELAND
(Concentrations in ppm. Location numbers refer to Figure 1.)

Location	Sample no.[a]	Temp. °C	Downhole sample m	pH/°C	SiO_2	Na	K	Ca	Mg	CO_2[e]	SO_4	H_2S[f]	Cl	f	Diss. Solids
Vik i Myrdal, 1	77-0153	21[b]		9.69/23	52	473	9.4	24.4	0.59	28.4	211	<0.1	648	4.01	1502
Vestmannaeyjar	73-0011	35[b]		6.55/20	148	6000	134	660	575	645	923	<0.1	9840	0.65	18994
Storolishvoll, 1	77-0003	43[b]	500	7.35/20	48	1082	27.8	406	2.61	21.1	293	<0.1	2161	1.19	4545
Laugaland in Holt, 3	77-0155	69[b]	212	9.78/21	80	106	1.64	6.3	0.07	21.1	102	<0.1	76.3	0.93	393
Reykjabol, 1	74-0028	152[c]		7.23/169[d]	233	87.4	6.6	1.0	0.02	128	72.6	2.9	29.2	1.80	518
Reykholt, 1	74-0031	132[c]		7.67/140[d]	167	105	4.8	2.2	0.06	51.4	74.4	1.7	73.4	2.49	480
Selfoss, 8	75-0082	82[b]		8.44/20	74	172	5.1	31.4	0.10	15.0	56.4	0.2	268	0.80	667
Arbaer, 1	75-0083	96[b]		9.67/20	95	68.2	1.4	1.9	0.05	29.7	42.9	0.7	35.2	0.80	311
Eyarbakki, 1	72-0169	70[c]	750	7.95/20	38	1360	11.0	1240	0.90	15.0	398	<0.1	2880	0.15	7617
Nesjavellir, 5	72-0193	268[d]		7.43/255[d]	555	133	24.3	1.1	0.04	278	31.8	163	13.4	1.74	950
Veragerdi, 4	73-0082	184[c]		6.99/199[d]	270	151	11.9	1.5	0.16	169	62.3	26.4	112	1.80	681
Thorlakshofn (Bakki),1	77-0023	126[c]		6.84/128[d]	131	391	21.1	74.9	0.03	13.7	124	0.4	596	0.39	1515
Reykjanes, 8	77-0147	282[d]		5.58/246[d]	501	9050	1271	1736	3.36	1234	68.5	20.6	19966	0.16	31572
Svartsengf, 4	74-0063	242[c]		5.47/237[d]	446	6444	974	987	5.16	436	34.0	1.5	12647	0.11	18777
Krisuvik, 5	71-0132	151[c]	800	8.85/20	164	233	16.7	16.5	0.51	63.0	325	1.3	52.0	0.60	896
Trolladyngja, 6	71-0106	218[c]	800	7.30/20	304	596	64.0	40.0	0.44	59.5	40.1	1.7	914	0.30	2020
Seltjarnarnes, 3	77-2052	102[b]		8.44/28	116	368	10.8	144	0.17	5.0	205	0.1	685.0	0.73	1631
Reykjavik, G-17	77-2008	132[b]		9.64/18	150	58.5	2.9	2.5	0.02	20.7	25.2	0.1	37.0	0.80	323
Reykjavik, G-23	77-2011	93[b]		9.70/20	94	41.8	1.3	2.4	0.01	17.2	19.2	0.1	17.4	0.42	219
Mosfellssveit, MG-16	77-2044	101[b]		9.65/23	110	50.8	1.5	2.3	0.01	20.5	28.6	1.4	24.2	0.95	230
Akranes, 4	67-	110[c]	810	7.20/20	100	1120	17.0	540	0.18	8.3	53.0	<0.1	2485	0.10	5600
Leira, 4	76-0017	128[b]		6.29/20	219	213	20.6	30.9	0.37	279	56.5	0.3	240	2.34	894
Hvalstod, 4	77-0045	12[b]		9.20/26	25	76.8	1.00	8.5	0.48	17.8	122	<0.1	50.4	3.20	290
Laugarholt, 1	77-0133	91[b]		9.16/22	116	108	3.25	14.5	0.02	12.9	74.7	0.5	125	1.99	438
Kolvidarnes, 4	77-0091	60[b]		7.97/28	77	129	2.5	24.9	0.25	40.2	52.0	<0.1	187	0.66	481
Reykholar, 5	75-0073	108[b]		9.61/20	140	58.3	1.9	2.7	0.03	15.3	43.4	0.7	33.7	0.50	302
Isafjordur, 2	76-0023	27[b]		9.90/19	59	90.3	0.6	4.2	0.01	9.6	53.4	<0.1	77.4	1.92	327
Sugandfjordur, 2	76-0111	64[b]		9.80/17	59	72.1	1.0	6.1	0.02	9.5	71.2	0.1	63.4	0.30	296
Reykir in Midfjordur,4	69-0151	96[b]		9.20/23	97	141	2.3	22.7	0.03	14.0	143	0.2	140	2.00	613
Reykir at Reykjar, 5	76-0191	69[b]		9.64/21	116	54.8	1.8	2.9	0.01	24.0	59.0	1.2	13.3	5.84	306
Saudarkrokur, 10	69-0145	70[b]		9.92/23	74	51.0	1.1	3.0	0.03	16.2	37.7	0.5	19.7	1.40	232
Siglufjordur, 8	77-0125	68[b]		10.08/25	99	38.3	1.0	2.1	0.01	15.4	14.5	<0.1	9.5	0.38	226
Olafsfjordur, 8	69-0125	56[b]		10.20/24	58	39.0	0.8	1.8	0.06	11.5	5.3	<0.1	2.9	0.10	150
Dalvik, 10	77-0148	64[b]		10.29/22	95	46.7	0.5	2.2	0.01	13.5	15.5	<0.1	10.1	0.50	220
Laugaland in Horg., 2	69-0121	88[b]		9.78/24	130	55.5	1.6	2.0	0.02	24.0	31.3	0.2	13.7	0.60	288
Laugaland in Eyjafj.,8	77-0173	92[b]	2300	9.99/23	116	72.1	2.6	3.9	0.09	27.7	63.9	<0.1	13.8	0.53	304
Storutjarnir, 2	69-0119	65[b]		9.69/24	117	51.5	1.5	2.8	0.06	20.0	30.9	0.9	18.2	0.80	253
Laugar, 2	75-0091	62[b]		10.58/20	90	45.5	0.5	2.0	<0.01	7.5	25.9	<0.1	4.7	0.76	198
Arnes, 1	75-0162	97[c]	1160	9.65/21	122	51.2	1.1	2.8	0.10	11.3	37.4	<0.1	11.9	0.69	282
Husavik, 1	66-	94[b]	500	7.40/20	80	840	18.7	17.6	10.6	5.0	82.0	-	1633	0.20	3600
Krafla, 9	77-1178	300[c]		7.12/303[d]	681	143	20.2	1.1	0.02	10013	73.1	92.4	11.8	0.46	1014
Namafjall, 10	76-0156	300[c]		7.31/272[d]	607	118	22.8	0.5	0.02	231.5	26.6	232.0	82.3	0.75	1036
Urridavatn, 3	76-0025	43[b]		9.95/20	56	63.4	0.9	5.4	0.03	8.5	47.6	<0.1	41.9	0.60	252

[a]year of sampling-sample no. [b]measured temperature of well discharge. [c]measured temperature at aquifer inflow. [d]computed pH of the reservoir water. [e]total carbonate ($H_2CO_3 + HCO_3^- + CO_3^{-2}$). [f]total suphide ($H_2S + HS^- + s^{-2}$).

active area the groundwater in the upper few hundred meters is cold sea water. This is isolated from the convecting system by an anhydrite seal produced by the precipitation of calcium sulphate from the sea water upon heating. Tomasson and Kristmannsdottir (1972) have shown that this cap is subject to fracturing during local earthquake activity. Some invasion of the cold seawater then occurs, accompanied by an increase in the activity of the hot springs and fumaroles. They suggest that the presence of high temperature alteration minerals such as epidote in presently low temperature environments and the occurrence of retrograde reactions reflect hydrologic changes brought about by these additions of cold water. At the present time, a pilot plant is in operation separating NaCl and $CaCl_2$ from the brines.

The Svartsengi field has similar characteristics to Reykjanes (Arnorsson, 1978). However, the temperatures are lower (about 235°C) and the waters less concentrated (29‰) indicating a similar evolution as at Reykjanes but with a greater relative admixture of meteoric waters. At present, the field is exploited to warm fresh water for district heating and domestic use. In addition, 8 MW of elective power are installed. Eleven holes up to 2000 meters deep have been drilled in the area.

Low temperature waters are used very extensively for space heating in Iceland (Fig. 2; Gudmundsson, 1982). This represents more than 30 percent (500 MWt) of the total energy consumption of the country. The oldest field to be extensively exploited is Reykir near Reykjavik. Production began in 1943 and has been progressively extended. To date over 400 million cubic meters of 86°C water has been extracted with no decrease in temperature or production rate.

CHEMISTRY OF THE HYDROTHERMAL FLUIDS

The major element chemistry of the hot springs in Iceland (mainly from drill hole samples) has been the subject of extensive study both for scientific reasons and with a view to predicting scaling and pollution problems associated with exploitation (Arnorsson, 1981). There are three controlling variables; the abundance and source of chloride, the water temperature, and the rock type. The latter is of minor importance save for incompatible elements such as fluoride since the entire island is composed of tholeiitic basalts and their differentiation products (Table 1; Arnorsson and others, 1982).

Chloride in Icelandic geothermal waters occurs in widely varying concentrations from below 3 to almost 20,000 ppm (Table 1). There appears to be no reported systematic study of the chloride concentration in precipitation over the island. Arnason and Tomasson (1970) state that the values in rain range up to 40 ppm. Junge's compilation for continental Europe (Junge, 1963) shows mean values of less than 10 ppm along the Scandanavian coast at similar latitudes. The chloride content of Icelandic basalts have been investigated in some detail by Sigvaldason and Oskarsson (1976). They find that it behaves as an incompatible element increasing along the fractionation trend in parallel with potassium and phosphorous. A histogram of the data (126 samples) shows a peak in the frequency of occurrence centered at about 175 ppm. Most values fall between 100 and 300 ppm with a tail up to about 800 ppm. A compilation of chemical data for water from deep drill holes shows that most values fall below 100 ppm and many below 50 ppm. There are only a few examples in the mid-range of 100 to 1000 ppm. There are several fields with concentrations greater than 1000 ppm. The dilute waters (Cl < 15 ppm) can be explained as heated meteoric waters. The high concentrations (Cl >1000 ppm) can only be derived from an admixture of sea water. The intermediate concentrations are probably due to leaching of chlorine from the basalts. However, there is only a very rough correlation between chloride concentration and temperature. As an extreme example the Krafla-Namafjall field, with temperatures of over 350°C the hottest yet observed in Iceland, has chloride levels as low as 12 ppm (Pálmason and others, 1979). Various authors have suggested that the intermediate chloride levels are caused by incorporation of relict sea water trapped in deposits either during submarine deposition or post-depositional invasion caused by thermal or tectonic subsidence or sea level oscillations. This phenomenon should be reflected in the isotopic composition of the waters. In one example where the deuterium and chloride have been measured systematically in the deep well in the Vestmann Islands (Tomasson, 1967) the concentration data require the mixing of sea water with a fresh water end-member with a δD of about $-120‰$. Although this value is about $20‰$ lower than any observed in contemporary precipitation it is in the range observed for paleo-waters at many locations on Iceland (Arnason, 1977b). Appeals to membrane filtration to explain the chloride enrichments (Tomasson, 1967) are invalid since this process necessitates extremely low permeabilities sustained over very long time periods (Graf and others, 1965), a situation that has never been documented in the region. In general, however, it seems likely that groundwater mixing is sufficient to explain the elevated chloride levels. Whether chloride is actually absorbed by the rocks during penetration by sea water at low temperatures remains to be demonstrated (Pálmason, and others, 1979). Active outgasing of chloride from cooling flows has been observed (Oskarsson, 1981). While this direct injection of chloride may be important in some local instances it cannot be of general significance given the very low chloride levels at Krafla-Namafjall.

The concentrations of the other anionic species—carbonate, bicarbonate, sulphate, sulphide, and fluoride—are subject to considerable variations (Table 1). Presentation of the data of Pálmason and others (1979) on a ternary diagram (Fig. 5) allows general evolutionary trends to be observed. When heated, sea water precipitates anhydrite, leading to a drastic decrease in the sulphate levels and to a chloride-dominated brine (e.g. Reykjanes, Svartsengi). At lower ionic strength, sulphate and the CO_2 species

Figure 5. Ternary diagram calculated from the data given in Table 1.

become increasingly significant components of the anion balance with some tendency to CO_2 dominated systems. All the springs contain significant quantities of sulphate co-existing with variable amounts of H_2S. Moore and Schilling (1973) and Gunnlaugsson (1977) report the sulphur concentrations in Icelandic volcanic rocks as showing a wide range, 50 to 900 ppm. This presumably reflects varying degrees of degassing. In the lower temperature springs inorganic reduction of sulphate by ferrous iron should be too slow to be of any significance and bacterial reduction should dominate. At high temperature inorganic processes should dominate. Neither mechanism appears to occur to any large extent. In fact, sulphate must be being produced in the hydrothermal aquifers. Concentrations are generally higher than can be accommodated by in situ oxidation of basaltic sulphides (about 15 ppm) given the initial saturation concentration of oxygen. While some sulphate may result from mixing with relict sea water as for chloride there is no strong correlation between the levels of the two species. In addition, the persistence of sulphate in the presence of appreciable concentrations of methane and hydrogen is puzzling (Sigvaldason, 1966; Arnorsson, 1974; Stefansson, 1981). While the hydrogen levels may be to some degree an artifact of electrolytic reactions with the bore hole casing and the well head installation (Arnason, 1977b), methane can be taken as an unambiguous indicator of strongly reducing conditions. Sakai and others, (1980) reported isotopic data for H_2S, SO_4 and pyrite in low and high temperature drill holes. The sulphide values in the dilute fluids were generally close to the number for igneous pyrite $\sim 0‰$. In the seawater derived systems, the values were heavier, up to $+8‰$, indicating partial reduction of the seawater sulphate ($+20‰$). When data were available, sulphide was always at isotopic disequilibrium with co-existing sulphate. Clearly more extensive work needs to be done on this important problem.

Similar problems exist for CO_2. The values are high and

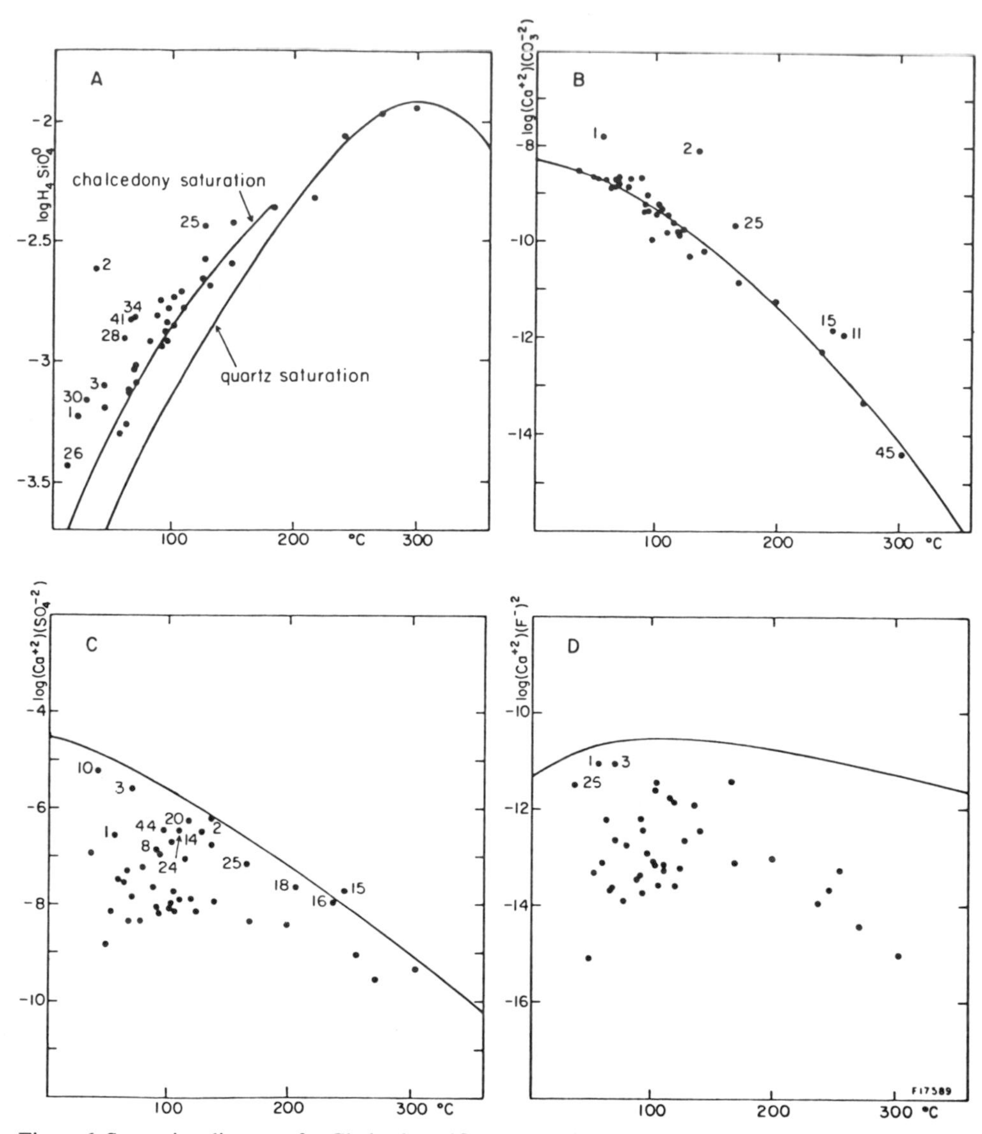

Figure 6. Saturation diagrams for Chalcedony/Quartz (A), Calcite (B), Anhydrite (C) and Fluorite (D) based on the data in Table 1 where the identification numbers are also listed.

variable and must be caused by the degassing of the basalt. It is remarkable that the extremely high concentrations observed in the lower aquifer at Krafla (1,357 to 19,048 ppm; Stefansson, 1981) are unaccompanied by substantial methane enrichments. Again systematic isotopic analyses would be valuable in the study of the carbon system.

Bodvarsson (1960) and Bodvarsson and Pálmason (1961) pioneered the attempt to estimate the underground temperature of thermal springs from their silica concentrations, a technique now in general use world-wide. Subsequently, Arnorsson (1970b) demonstrated a good correlation between the measured bore hole temperatures and the value computed from the silica data assuming equilibrium with quartz above 110°C and with chalcedony below this. In subsequent work with much more data, Arnorsson (1975) revised these estimates, and showed that in the interval between 110 and 180°C the waters are at equilibrium with neither phase (Fig. 6). This is often due to conductive cooling as waters ascend from the primary aquifer and also to the effects of mixing with cold groundwaters (Fig. 6). Where carefully interpreted, however, silica geothermometry is a valuable tool in geothermal prospecting (Arnorsson, 1975).

The distributions of the major cationic species—Na, K, Mg, Ca, H^+—have been shown to be controlled by mineral equilibria (Arnorsson and others, 1978; Pálmason and others, 1979). Well waters are generally at calcite saturation (Fig. 6). The calcium rich, saline waters and the high temperature, dilute fluids (e.g. Krafla) are at equilibrium with anhydrite (Fig. 6). Thermal waters in differentiated, acid volcanics enriched in fluoride may be sat-

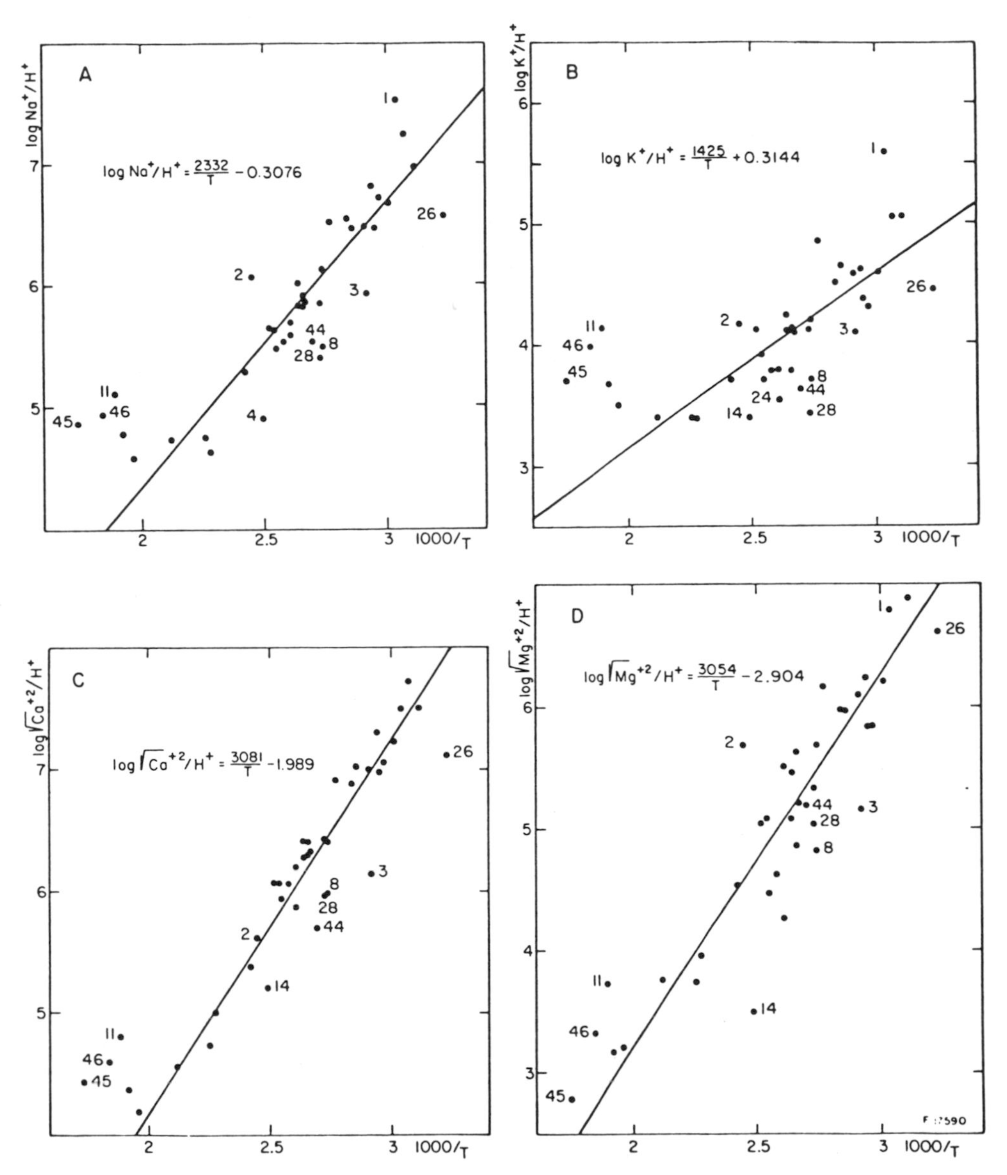

Figure 7. Van't Hoff plots for the major cations in Icelandic geothermal waters: A, sodium; B, potassium; C, calcium; D, magnesium. The formulae indicate the best fit lines through the data.

urated with fluorite but this is not commonly observed in well samples (Fig. 6). Overall control of the cation distributions is by ion exchange reactions of the type (Arnorsson and others, 1978):

$$(v\tfrac{1}{2}+x+y+z) * \tfrac{1}{2} * Mg^{2+} + \text{solid} = vNa^{+}\ xH^{+} + yK^{+} + z * \tfrac{1}{2} * Ca^{2+} + \text{solid}$$

The equilibrium constant can be expressed as

$$\log K = v\log\frac{Na^{+}}{H^{+}} + y\log\frac{K^{+}}{H^{+}} + z\log\frac{Ca^{2+}}{H^{+}} - (v+x+y+z)\log\frac{Mg^{2+}}{H^{+}}$$

where the stoichiometry depends on the mineral substrates involved. The chemical data from a wide range of fields follow the van't Hoff relation consistent with this mechanism, that is the cation/proton ratios show a log-linear relationship with the reciprocal of the absolute temperature (Fig. 7). Calculation of specific mineral equilibria (Helgeson, 1969) shows that alkali-feldspar, K-mica, and wollastonite are the important controlling phases of Na, K, and Ca respectively (Fig. 8; Pálmason and others, 1979). The Ca/Mg ratio is consistent with exchange control by montmorillonite. Arnorsson and others (1978) discuss the problems associated with these computations in some detail. Given the difficulties involved in the back-calculations of the original aquifer compositions from data on flashed well waters and with the estimation of the extent of non-equilibrium, conductive cooling and groundwater mixing the consistency and generality revealed by these calculations is remarkable.

Arnorsson and others (1983, a, b) have continued and extended this investigation by calculating the speciation of the dis-

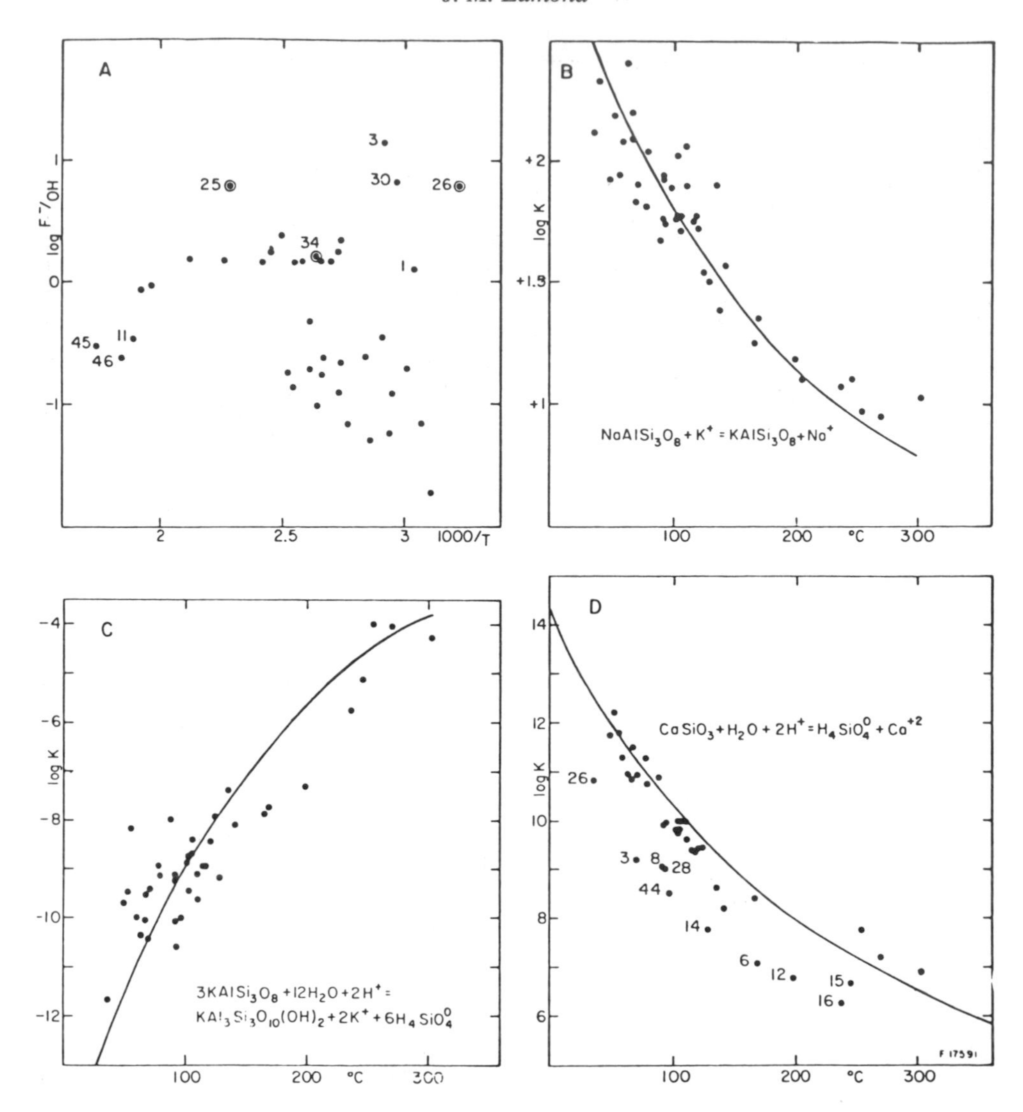

Figure 8. F/OH scatter plot (A) and equilibrium diagrams for alkali-feldspar (B), K-mica (C), and wollastonite (D).

solved constituents for temperatures ranging to 370°C. They have used this modeling capability to refine the calculations of solution/mineral equilibria and associated geothermometers (Arnorsson and others, 1982c) and have also applied it to dynamic situations involving boiling, degassing, and conductive cooling. These explicit models are very useful in identifying inadequacies in the primary data base. Particular problems appear to exist with that perennial bug-bear of the aqueous geochemist, the CO_2 system.

Gunnlaugsson and Arnorsson (1982) have applied a related model to the study of the chemistry of iron in the hydrothermal waters. They found that the dominant species in high temperature solutions is $Fe(OH)_4^-$. This is an artifact of their conventional choice of the H_2S/SO_4 ratio as a redox indicator. It is highly unlikely that the sulphur system is, in fact, at equilibrium (Sakai and others, 1980). The computations indicate a general equilibrium of the well waters with pyrrhotite and marcasite below 180°C and with pyrite and anhydrite at higher temperatures. At low temperatures the waters were found to be at equilibrium with amorphous iron oxide and, as the temperatures increase, progressively with lepidocrocite, maghemite, goethite, and hematite.

Trace element data from the Icelandic thermal waters are sparse. Problems of contamination by casing pipe and well head installations can be overcome only by in situ sampling. Apparently, this has not been attempted in the trace element investigations. Arnorsson (1970a) investigated the distributions of a number of elements in 135 springs chosen to cover the range of temperatures and chemical types observed. He found that gallium, germanium, molybdenum, and vanadium were generally present in microgram quantities. Chromium, cobalt, nickel, and zinc were detected only in acid waters. He was not able to detect bismuth, cadmium, or, surprisingly, copper. Olafsson and Riley

(1978) reported the concentrations of a wide variety of trace metals in the Reykjanes brines. They found substantial enrichments of rubidium, mercury, and arsenic consistent with their mobilization from the basalts.

HYDROTHERMAL ALTERATION OF THE BASALTS

Walker (1960) first established the now familiar zeolite facies metamorphism of Icelandic basalts based on sections through the exposed cores of Tertiary volcanic centers. Work on contemporary alteration processes in the thermal regions has been hampered by the absence of core samples. Until the completion of the IRDP hole in eastern Iceland in 1978 (Fridleifsson and others, 1982) only cuttings were available for study. Nevertheless, in a series of papers, Kristmannsdottir and colleagues (Kristmannsdottir, 1976, 1979, 1981, 1982; Pálmason and others, 1979), using this material and comparison with exposed paleohydrothermal zones, were able to erect an alteration model that compares very well with the much more detailed IRDP results (Viereck and others, 1982; Mehegan and others, 1982) when care is taken to account for late stage low temperature effects on the IRDP core material.

Drilling in the active systems established that in the upper levels, at temperatures below 100°C, the assemblage chabazite-scolecite-stilbite is characteristic. In hotter waters, these low-temperature zeolites transform to laumontite. At about 200°C laumontite itself is converted to wairakite. The occurrence of epidote is first noted at about 230°C and actinolite formation and albitization of the plagioclase in the basalts commences at about 280°C. In the zeolite facies rocks smectite is the characteristic clay mineral. This becomes progressively more inter-layered with increasing temperature converting to chlorite at around 230°C. The greenschist facies minerals epidote and actinolite persist metastably as thermal activity wanes. Their occurrence can therefore be used to estimate the maximum geothermal gradient experienced in a particular area (Kristmannsdottir, 1982).

CONCLUSIONS

Knowledge of the distribution and scale of hydrothermal activity in the North Atlantic is quite fragmentary. As the result of a sustained and intensive effort over many years, the geothermal fields on Iceland are now perhaps the best described and understood in the world. Given the important and increasing role that geothermal resources play in the national energy budget, the Icelandic program is an exemplary case of fundamental research providing the base for a large and successful industry. Based on the limited information available, similar developments could be expected for the volcanically active islands of the eastern Atlantic. However, in this case, systematic investigations have yet to begin.

The exciting discoveries in the eastern Pacific have resulted, for the time being, in the eclipse of the Mid-Atlantic Ridge as the focus of exploration efforts at accreting plate boundaries. Hence, there is still no direct evidence of active hydrothermal convection on slow-spreading submarine ridges, that is, observations of contemporary hot springs. Discovery of such systems awaits a substantial increase in programs of detailed exploration.

It has been common among marine geologists to regard the Reykjanes brines as an analogue to submarine hydrothermal fluids. That this is not valid can be appreciated by comparing the solution compositions themselves (Edmond and others, 1982, Table 2; Arnorsson, 1978). The Icelandic systems are of near neutral pH and contain substantial amounts of sulphate and low concentrations of H_2S. The submarine systems are acid, pH = 3.5, are completely depleted in sulphate and have H_2S concentrations ranging up to over 300 ppm. Unlike the Icelandic fields, they are associated with extensive ore mineralization. It would appear from the isotopic and chemical data, that fluid throughput at Reykjanes and Svartsengi is minimal, the convecting cells acting as closed systems. The long residence time of the waters in this situation is largely responsible for the chemical differences. The lack of transport precludes the formation of ore deposits.

REFERENCES

ARCYANA
1975 : Transform fault and rift valley from bathyscape and diving saucer: Science, v. 190, p. 108–116.

Armannsson, H., Geslason, G., and Hankson, T.
1982 : Magmatic Gases in well fluids and the mapping of the flow pattern in a geothermal system: Geochimica et Cosmochimica Acta, v. 46, p. 167–177.

Arnason, B., and Tomasson, J.
1970 : Deuterium and chloride in geothermal studies in Iceland: Geothermics, Special Issue no. 2, p. 1405–1415.
1976 : Groundwater systems in Iceland traced by deuterium: Publication of the Scientific Society of Iceland, v. 42, 276 pp.
1977a: The hydrogen-water isotope thermometer applied to geothermal areas in Iceland: Geothermics, v. 5, p. 75–80.
1977b: Hydrothermal systems in Iceland traced by deuterium: Geothermics, v. 5, p. 125–151.

Arnason, B., and Sigurgeirsson, Th.
1967 : Hydrogen isotopes in hydrological studies in Iceland: Proceedings of the Symposium, Isotopes-Hydrology, (IAEA, Vienna), p. 35–47.

Anorsson, S
1970a: The distribution of some trace elements in thermal waters in Iceland: Geothermics, Special Issue no. 2, p. 542–546.
1970b: Underground temperatures in hydrothermal areas in Iceland as deduced from the silica content of the thermal water, Geothermics, Special Issue no. 2, p. 536–541.
1974 : The composition of thermal fluids in Iceland and geological features related to the thermal activity; in Geodynamics of Iceland and the North Atlantic Area, ed. Kristjansson: Dordrecht, D. Reidel, p. 307–323.
1975 : Application of the silica geothermometer in low temperature areas in Iceland: American Journal of Science, v. 275, p. 763–784.
1977 : Changes in the chemistry of water and stream discharged from wells in the Namafjall geothermal field, Iceland, during the period 1970–76: Jokull, v. 27, p. 47–58.
1978 : Major element chemistry of the geothermal sea water at Reykjanes

and Svertsengi, Iceland: Mineralogical Magazine, v. 25, p. 209–220.
1981 : Mineral deposition from Icelandic geothermal waters: environmental and utilization problems: Journal of Petroleum Technology, p. 181–187.
Arnorsson, S., Gronvold, K., and Sigurdsson, L.
1978 : Aquifer chemistry of four high-temperature geothermal systems in Iceland, Geochemica et Cosmochimica Acta, v. 42, p. 523–536.
Arnorsson, S., Gunnlaugsson, E., and Svavarsson, H.
1983a: The chemistry of geothermal water in Iceland, II. Mineral equilibria and independent variables controlling water composition: Geochimica et Cosmochimica Acta, v. 47, p. 547–566.
1983b: The chemistry of geothermal water in Iceland, III: Chemical geothermometry in geothermal investigations: Geochimica et Cosmochimica Acta, v. 47, p. 567–577.
Arnorsson, S., Sigurdsson, S., and Svavarsson, H.
1982 : The chemistry of geothermal waters in Iceland, I. Calculation of aqueous speciation from 0° to 370°C: Geochimica et Cosmochimica Acta, v. 46, p. 1513–1532.
Beblo, M., and Bjornsson, A.
1978 : Magnetotelluric investigation of the lower crust and upper mantle beneath Iceland: Journal of Geophysics, v. 45, 1–16.
1980 : A model of electrical resistivity beneath NE-Iceland, correlation with temperature: Journal of Geophysics, v. 47, 184–190.
Betzer, P. R., Bolger, G. W., McGregor, B. A., and Rona, P. A.
1974 : The Mid-Atlantic Ridge and its effect on the composition of particulate matter in the deep ocean: American Geophysical Union Transactions, v. 55, p. 293.
Bjornsson, S.
1970 : A program for the exploration of high temperature areas in Iceland: Geothermics, Special Issue, no. 2, p. 1050–1054.
Bjornsson, S., Arnorsson, S., and Tomasson, J.
1970 : Exploration of the Reykianes thermal brine area: Geothermics, Special Issue, no. 2, p. 1640–1650.
1972 : Economic evaluation of Reykjanes thermal brine area, Iceland: American Association of Petroleum Geologists Bulletin, v. 56, p. 2380–2391.
Bjornsson, A., Saemundsson, Einarsson, P., Tryggvason, E., and Gronvold, K.
1977 : Current rifting episode in North Iceland: Nature, v. 266, 318–323.
Bjornsson, A., Johnsen, G., Siguvokson, S., Thorbergsson, G., and Tryggvason
1979 : Rifting of the plate boundary in North Iceland, 1975–1978: Journal of Geophysical Research, v. 84, p. 3029–3038.
Bodvarsson, G.
1960 : Exploration and exploitation of natural heat in Iceland: Bulletin Volcanologique, v. 23, p. 241.
1961 : Physical characteristics of natural heat resources in Iceland: Jokull, II, p. 29–38.
Bodvarsson, G., and Pálmason, G.
1961 : Exploration and subsurface temperatures in Iceland: U.N. Conference on New Sources of Energy, Rome, p. 39–48.
Bohlke, J. K., Honnorez, J., and Honnorez-Guerstein, B. M.
1980 : Alteration of basalts from Site 396B, DSDP, Petrographic and mineralogic studies: Contributions to Mineralogy and Petrology, v. 73, p. 341–364.
Bonatti, E., Guerstein-Honnorez, B. M., and Honnorez, J.
1976a: Hydrothermal pyrite concretions from the Romanche Trench (equatorial Atlantic): Metallogenesis in oceanic fracture zones: Earth and Planetary Science Letters, v. 32, p. 1–10.
1976b: Copper-iron sulphide mineralizations from the equatorial Mid-Atlantic Ridge: Economic Geology, v. 71, p. 1515 1525.
Bostrom, K., Peterson, M.N.A., Joensuu, O., and Fisher, D. E.
1969 : Aluminum-poor ferro-manganoan sediments on active oceanic ridges: Journal of Geophysical Research, v. 74, p. 3261–3270.
Bostrom, K.
1973 : The origin and fate of ferromanganoan active ridge sediments: Stockholm Contributions to Geology, v. 27, p. 149–243.
Corliss, J. B.
1971 : The origin of metal-bearing submarine hydrothermal solutions: Journal of Geophysical Research, v. 76, p. 8128–8136.
Corliss, J. B., Lyle, M., Bymond, J., and Crane, K.
1978 : The chemistry of hydrothermal mounds near the Galapagos Rift: Earth and Planetary Science Letters, v. 40, p. 12–24.
Cronan, D. S.
1972 : The Mid-Atlantic Ridge near 45°N, XVII: Al, As, Hg, and Mn in Ferrugininous Sediments from the Median Valley: Canadian Journal of Earth Sciences, v. 9, p. 319–323.
Delaney, J. R., Mogk, D. W., and Motte, M. J.
in press: Quartz-cemented, sulphide-bearing greenstone breccias from the Mid-Atlantic Ridge: samples of a high-temperature hydrothermal upflow zone: Science.
Edmon, J. M., Measures, C. I., McDuff, R. E., Chan, L. H., Collier, R., Grant, B., L. I., and Corliss, J. B.
1979a: Ridge crest hydrothermal activity and the balances of the major and minor elements in the ocean; the Galapagos data: Earth and Planetary Science Letters, v. 46, p. 1–18.
Edmond, J. M., Measures, C. I., Mangun, B., Grant, B., Sclater, R. R., Collier, R., Hudson, A., and Corliss, J. B.
1979b: On the formation of metal-rich deposits on ridge crests: Earth and Planetary Science Letters, v. 46, p. 19–30.
Edmond, J. M., Von Damm, K. L., McDuff, R. E., and Measures, C. I.
1982 : Chemistry of hot springs on the East Pacific Rise and their effluent dispersal: Nature, v. 297, p. 187–191.
Einarsson, T.
1942 : Uber das Wesen der heissen Quellen Islands: Publication of the Scientific Society of Iceland, v. 26.
Einarsson, P.
1978 : S-wave shadows in the Krafla caldera in NE-Iceland evidence for a magma chamber in the crust: Bulletin Volcanologique, v. 41, p. 1–9.
Fridleifsson, I. B.
1979 : Geothermal Activity in Iceland: Jokull, v. 29, p. 47–56.
1982 : The Iceland Research Drilling Project in relation to the geology of Iceland: Journal of Geophysical Research, v. 87, p. 6363–6370.
Fridleifsson, I. B., Gibson, I. L., Hall, J. M., Johnson, H. P., Christensen, N. I., Schmincke, H. U., and Schonharting, G.
1982 : The Iceland Research Drilling Project: Journal of Geophysical Research, v. 87, p. 6359–6362.
Graf, D. L., Friedman, I., and Meents, W. F.
1965 : The origin of saline formation waters 2: Isotopic fractionation by shale micropore systems: Illinois State Geological Survey Circular no. 393, 32 p.
Gundmundsson, J. S.
1982 : Low-temperature geothermal energy use in Iceland: Geothermics, v. 11, p. 59–68.
Gundmundsson, J. S., Hanksson, T., and Tomasson, J.
1981 : The Reykjaves geothermal field in Iceland: subsurface exploration and well discharge characteristics: Proceedings 7th Workshop on Geothermal Reservoir Engineering, Stanford, 61–69.
Gunnlaugsson, E.
1977 : The origin and distribution of sulphur in fresh and geothermally altered rocks in Iceland. [Ph.D. thesis]: Leeds University, 192 pp.
Gunnlaugsson, E., and Arnorsson, S.
1982 : The chemistry of iron in geothermal systems in Iceland: Journal of Volcanology and Geothermal Research, v. 14, p. 281–299.
Hattori, K., and Muehlenbachs, K.
1982 : Oxygen isotope ratios of the Icelandic crust: Journal of Geophysical Research, v. 87, p. 6559–6565.
Helgeson, H. C.
1969 : Thermodynamics of hydrothermal systems at elevated temperatures and pressures: American Journal of Science, v. 267, p. 729–804.
Hersir, G. P., Bjornsson, A., and Pedersen, L. B.
1984 : Magnetotelluric survey across the active spreading zone in Southwest Iceland: Journal of Volcanology and Geothermal Research, v. 20,

p. 253–265.
Honnorez, J.
1981 : The aging of the oceanic crust at low temperature: in The Sea, v. 7, The Oceanic Lithosphere, ed. C. Emiliani: New York, John Wiley & Sons, Inc., p. 525–587.
Horowitz, A.
1970 : The distribution of Pb, Ag, Sn, Tl and Zn in sediments on active oceanic ridges: Marine Geology, v. 9, p. 241–259.
Horowitz, A., and Cronan, D. S.
1976 : The geochemistry of basal sediments from the North Atlantic Ocean: Marine Geology, v. 20, p. 205–228.
Humphris, S. E., and Thompson, G.
1978 : Hydrothermal alteration of oceanic basalts by seawater: Geochimica et Cosmochimica Acts, v. 42, p. 107–125.
Jenkins, W. J., Rona, P. A., and Edmond, J. M.
1980 : Excess ^{3}He in the deep water over the mid-Atlantic ridge at 26°N: evidence of hydrothermal activity: Earth and Planetary Science Letters, v. 49, p. 39–44.
Junge, C. E.
1963 : Air Chemistry and Radioactivity: New York, Academic Press, 282 p.
Kristmannsdottir, H.
1976 : Types of clay minerals in hydrothermally altered basaltic rocks, Reykjanes, Iceland: Jokull, v. 26, p. 30–39.
1979 : Alteration of basaltic rocks by hydrothermal activity at 100–300°C; in International Clay Conference 1978, ed. M. M. Mostland and V. C. Farmer; New York, Elsevier, p. 359–367.
1981 : Wollastonite from hydrothermally altered basaltic rocks in Iceland: Mineralogical Magazine, v. 44, p. 95–97.
1982 : Alteration in the IRDP drill hole compared with other drill holes in Iceland: Journal of Geophysical Research, v. 87, p. 6525–6531.
1983 : Chemical evidence from Icelandic geothermal systems as compared to submerged geothermal systems, in Hydrothermal processes at Seafloor Spreading Centers, ed. P. A. Rona, K. Bostrom, L. Laubier, and K. L. Smith, Jr.: New York, Plenum Press, p. 291–320.
Larsen, G., Gronvold, K., and Thorarinsson, S.
1979 : Volcanic eruption through a geothermal borehole at Mamfjall, Iceland: Nature, v. 278, p. 707–710.
Lawrence, J. R., and Maxwell, S.
1978 : Geothermal exploration in the Azores: $^{18}O/^{16}O$ in calcites from volcanic rocks: Journal of Volcanology and Geothermal Research, v. 4, p. 219–223.
Machado, F.
1965 : Vulcanismo das Ilhas de Cabo Verde e das outras Ilhas Atlantidas, Junta de Investigacoes do Ultramar: Estudos Ensaios e Documentos, v. 117, 83 pp.
Mehegan, J. M., Robinson, P. T., and Delaney, J. R.
1982 : Secondary mineralisation and hydrothermal alteration in the Reydarfjordur drill core, eastern Iceland: Journal of Geophysical Research, v. 87, p. 6511–6524.
Moore, J. G. and Schilling, J-G.
1973 : Vesicles, Water and Sulfur in Reykjanes Ridge Basalts: Beitrage Mineralogie und Petrologie, v. 41, p. 105–118.
Muecke, G. K., Ade-Hall, J. M., Aumento, F., MacDonald, A., Reynolds, P. H., Hyndman, R. D., Quintino, J., Opdyke, N., and Lowrie, W.
1974 : Deep drilling in an active geothermal area in the Azores: Nature, v. 252, p. 281–285.
Olafsson, J., and Riley, J. P.
1978 : Geochemical studies on the thermal brine from Reykjanes (Iceland): Chemical Geology, v. 21, p. 219–237.
Oskarsson, N.
1981 : The chemistry of Icelandic lava incrustations and the latest stages of degasing: Journal of Volcanology and Geothermal Research, v. 10, p. 93–111.
Pálmason, G.
1974 : Heat flow and hydrothermal activity; in Iceland, Geodynamics of Iceland and the North Atlantic area, Kristjansson, L., Dordrecht, D. Reidel, p. 297–306.
Pálmason, G., and Saemundsson, K.
1979 : Summing of conductive heat flow in Iceland, in Terrestrial Heat Flow in Europe, V. Cermak and L. Ryback, Berlin, Springer Verlag, p. 218–220.
Pálmason, G., Arnorsson, S., Fridleifsson, I. B., Dristmannsdottir, H., Saemundsson, K., Stefansson, V., Steingrimsson, B., Tomasson, J., and Kristjansson, L.
1979 : The Iceland Crust: evidence from drill hole data on structure and processes, Ewing Symposium: American Geophysical Union, p. 43–65.
Ragnars, K., Saemundsson, K., Benediktsson, S., and Einarsson, S. S.
1970 : Development of the Namafjall area—northern Iceland: Geothermics, Special Issue no. 2, p. 925–935.
Richardson, S. H., Hart, S. R., and Staudigel, H.
1980 : Vein mineral ages of old oceanic crust: Journal of Geophysical Research, v. 85, p. 7195–7200.
Rona, P. A.
1976 : Pattern of hydrothermal mineral deposition: Mid-Atlantic Ridge crest at latitude 26°N: Marine Geology, v. 21, p. M59–M66.
Rona, P. A., Bostrom, K., and Epstein, S.
1980 : Hydrothermal quartz vug from the Mid-Atlantic Ridge: Geology, v. 8, p. 569–572.
Sakai, H., Gunnlaugsson, E., Tommason, J., and Rouse, J. E.
1980 : Sulphur isotope systematics in Icelandic geothermal systems and influence of seawater circulation at Reykjanes: Geochimica et Cosmochimica Acta, v. 44, p. 1223–1231.
Scott, M. R., Scott, R. B., Rona, P. A., Butler, L. W., and Nalwalk, A. J.
1974 : Rapidly accumulating manganese deposit from the median valley of the Mid-Atlantic Ridge: Geophysical Research Letters, v. 1, p. 355–358.
Sigvaldason, G. E.
1966 : Chemistry of thermal waters and gases in Iceland: Bulletin Volcanologique, v. 29, p. 589–604.
Sigvaldason, G. E., and Oskarsson, N.
1976 : Chlorine in basalts from Iceland: Geochimica et Cosmochimica Acta, v. 40, p. 777–789.
Sleep, N. H., and Rosendahl, B. R.
1979 : Topography and tectonics of Mid-oceanic ridge axes: Journal of Geophysical Research, v. 84, p. 6831–6839.
Staudigel, H., Hart, S. R., and Richardson, S. H.
1981a: Alteration of oceanic crust: processes and timing: Earth and Planetary Science Letters, v. 52, p. 311–327.
Staudigel, H., Muchlenbachs, K., Richardson, S. H., and Hart, S. R.
1981b: Agents of low temperature ocean crust alterations: Contributions to Mineralogy and Petrology, v. 77, p. 150–157.
Stefansson, V.
1981 : The Krafla Geothermal field, northeast Iceland; in Geothermal Systems: Principles and Case Histories, ed. L. Rybach and L.J.P. Muffler, New York, John Wiley & Sons, p. 273–294.
Thayer, R. E., Bjornsson, A., Alvarez, L., and Hermance, J. F.
1981 : Magma genesis and crustal spreading in the northern neovolcanic zone of Iceland: telluric, magnetotelluric constraints: Geophysical Journal of the Royal Astronomical Society, v. 65, p. 423–442.
Thompson, G., Woo, C. C., and Sung, W.
1975 : Metalliferous deposits on the Mid-Atlantic Ridge: Geological Society of America, Abstracts with Programs, v. 7, p. 1297–1298.
Tomasson, J.
1967 : On the origin of sedimentary water beneath Vestmann Islands: Jokull, v. 17, p. 300–310.
Tomasson, J., and Kristmannsdottir, H.
1972 : High temperature alteration minerals and thermal brines, Reykjanes, Iceland: Contributions to Mineral. Petrol., v. 36, p. 123–134.
Turekian, K. K., and Imbrie, J.
1966 : The distribution of trace elements in deep-sea sediments of the Atlantic Ocean: Earth and Planetary Science Letters, v. 1, p. 161 168.
Viereck, L. G., Griffin, B. J., Schmincke, H. U., and Pritchard, R. G.

1982 : Volcaniclastic rocks of the Reydarfjordur drill hole, eastern Iceland 2, Alteration: Journal of Geophysical Research, v. 87, p. 6459–6476.
Walker, G.P.L.
1960 : Zeolite zones and dyke distribution in relation to the structure of the basalts in eastern Iceland: Journal of Geology, v. 68, p. 575–528.
Ward, P. L.
1971 : New interpretation of the geology of Iceland: Geological Society of America Bulletin, v. 82, p. 2991–3012.

MANUSCRIPT ACCEPTED BY THE SOCIETY FEBRUARY 13, 1985

Note in Proof

In the summer of 1985 the long-term exploration of the TAG area on the MAR bore fruit with the discovery of an active "black smoker" field (Rona, 1985). In addition to video pictures of the vents, sulphides were recovered and large plumes of manganese were observed in the overlying water column. The vents are located at a depth of 3700 meters on the wall of the Rift. In ALVIN work planned for the summer of 1986 it is hoped to determine whether the heat source is from an off-axis intrusion emplaced at relatively shallow depth below the valley wall or whether it is the deep-seated magma chamber that drives the spreading process. If the latter, then the hydrostatic pressure in the hydrothermal reaction zone could approach one kilobar, about twice that estimated for the active systems already sampled in the Pacific.

In addition, shallow water activity (90 meters) was found on the crest of the Kolbeinsey Ridge about 200 km north of Iceland near the island of Kolbeinsey (67°06′N, 18°42′W) (Stefansson, 1983).

Rona, P. A.
1985 : Black smokers and massive sulfides at the TAG hydrothermal field, mid-Atlantic Ridge 26°N: EOS Transactions of the American Geophysical Union, v. 66, p. 936.
Stefansson, V.
1983 : Environment of hydrothermal systems in Iceland: in Hydrothermal processes at seafloor spreading centers, eds. P. A. Rona, K. Bostroem, L. Laubier and K. L. Smith, Jr., Plenum, NY, p. 321–360.

Printed in U.S.A.

The Geology of North America
Vol. M, The Western North Atlantic Region
The Geological Society of America, 1986

Chapter 12

The present plate boundary configuration

Peter R. Vogt
Naval Research Laboratory, Washington, D.C. 20375-5000

INTRODUCTION

The discovery of an elongated geological feature is soon followed by studies of how the feature varies along its length. This is true of the Mid-Oceanic Ridge (MOR), the longest feature on our planet, although there has been far more work done on transverse than on longitudinal variations. Since the MOR accreting plate boundary (the zero-age isochron) and older isochrons are offset by many transform faults, a precise *longitudinal* profile must be assembled from numerous detached segments. In practice such profiles are generally constructed from transverse crossings. By contrast, long continuous *transverse* profiles can be obtained (e.g., Klitgord and Schouten, this volume) by following flowlines of plate motion, keeping a constant distance from adjacent fracture zones (FZ). Early ignorance of the existence of fracture zones led to misinterpretation of transverse bathymetric profiles across the Mid-Atlantic Ridge (MAR), e.g., in terms of spurious elongated terraces (the "Upper," "Middle," and "Lower Step" of Heezen and others, 1959). Even the later Trans-Atlantic Geotraverse (TAG) profiles (Rona, 1980) were collected without knowledge of the late Cretaceous/early Tertiary FZ bend, and consequently cross fracture zones at low angles and jump from one flowline to another.

The first longitudinal profile of the MAR (both North and South Atlantic) showed regional variations in minimum crestal depth (Stocks and Wüst, 1935). Later, Heezen and others (1959) plotted rift valley depth, west and east rift mountain heights, and other bathymetric parameters from 14°N to 55°N. Heirtzler and LePichon (1965) examined the along-strike amplitude variation of the central magnetic anomaly. Other parameters followed with the onset of the plate tectonic era, as summarized below (see also Plates 8A, B). The point of such studies is no longer merely to describe, but to understand the physics and chemistry of the processes involved. The first step is to demonstrate correlations with "independent" variables such as total opening rate (as proposed by Reid and Jackson, 1981, for crustal thickness, and Jackson and Reid, 1983, for crustal magnetization), motion of the accreting plate boundary with respect to the deeper mantle, distance from the nearest hotspot, or, on more local scales, distance from the nearest fracture zone. However, physical models to explain the variability are still in their infancy.

PLATE BOUNDARY DESCRIPTORS

A large number of variables could be used to characterize an accreting plate boundary as a function of distance along its strike. The MAR is targeted here, but the approach would be similar for any other part of the MOR. Plate 8B shows various isotope and trace elements as a function of geographic latitude, and similar plots for major elements are shown by Melson and O'Hearn (this volume). Plate kinematics provides a more natural coordinate system (Plate 8A), in which horizontal distance is measured from the pole(s) of plate rotation, ideal transform faults being compressed into points on the x-axis. The Best Fitting Poles (BFP) for each of the three plate boundaries used in Plate 8A are those of Minster and Jordan (1978). Use of newer rotation poles (DeMets and others, 1986) would make no significant changes in the appearance of the graphs.

Besides the various geochemical and petrologic parameters, the present plate boundary can be characterized *morphologically* (Heezen and others, 1959; LeDouaran and Francheteau, 1981; Vogt, 1983; Vogt and others, 1982) in terms of average crestal depth, presence or absence of rift valley, rift valley width and depth, rift mountain depth and relief above valley floor, and rift mountain depth asymmetry. The *configuration* of the boundary can be described in terms of density and offset of fracture zones, and obliqueness of accretion axes. (A number of parameters can be used to define this configuration; see the next section).

The boundary could further be characterized *plate kinematically* by total opening rate, spreading asymmetry, and absolute motion. It could be characterized *seismically,* in terms of the frequency of epicenters per unit distance along the axis (Plate 8A) and other seismic parameters; it could also be characterized *magnetically* in terms of magnetization intensity or layer thickness. There might be along-strike variations in flexural rigidity, crustal structure, density of volcanic and hydrothermal systems, and development of magma chambers. This list is not complete and includes some criteria that are merely candidates for future work. Previous descriptions of the plate boundary have been limited to parts of the MAR; for many of the above parameters, the data are insufficient for meaningful synthesis. The parameters differ in their degree of time averaging, ranging from a few years for seismicity, to pehaps 10^4–10^5 years for "zero-age" basalts, to

Vogt, P. R., 1986, The present plate boundary configuration; *in* Vogt, P. R., and Tucholke, B. E., eds., The Geology of North America, Volume M, The Western North Atlantic Region: Geological Society of America.

10^5–10^6 years for rift valley/rift mountain morphology and average zero-age depth.

Except for parameters relating to rift valley/rift mountain morphology and earthquakes, the same parameters could in principle be examined along any definable isochron. Comparisons (Vogt and others, 1982) between conjugate parcels of crust (connected by flowlines) may reveal asymmetries either inherited from the accretion zone or subsequently imprinted, perhaps as a result of passage over a hotspot, or of different sedimentation/low temperature hydrothermal histories.

Space and data limitations preclude detailed examination of most parameters mentioned above. What is emphasized herein are the configuration (plan view shape) of the MAR plate boundary and the processes that may have influenced its evolution to the present state. Plates 8A, B show other parameters—geochemical, seismic, morphologic, and plate kinematic.

DEFINING THE PLATE BOUNDARY

A lithospheric plate has upper and lower boundaries as well as lateral ones that are subdivided into collisional (e.g., subduction), strike-slip (transform), and accreting (spreading). Many papers in this volume deal with the topography, sedimentation/erosion, and crustal structure of the top of the oceanic part of the North America plate, and a few deal with its hot and mushy underbelly (e.g., Sclater and Wixon, this volume). Of the lateral boundaries, Westbrook and McCann (this volume) describe subduction of Atlantic lithosphere under the Caribbean plate, Fox and Gallo (this volume) discuss transform boundaries, and Macdonald (this volume) and Pálmason (this volume) analyze the accretion process.

The trailing plate margin along the axis of the MAR represents a narrow, dynamic, and jagged, but somehow still orderly zone, where the upper and lower surfaces of the plate converge with the accreting/transform boundaries. Although such boundaries are often depicted as sharp continuous lines (e.g., Fig. 1a) separating perfectly rigid plates, in reality they are zones of finite width within which nonrigid behavior may be important. Although the lateral plate boundaries are vertical, or at least dipping, they are mostly defined on the basis of the plate upper surfaces (e.g., rift morphology) or of characteristics of the upper few kilometers of axial crust (magnetics and seismicity). (Properly, a line describing a plate boundary on a map [Fig. 1a] should be called a "trace," following usage on land). The lithosphere may be less than 10 km thick at the MAR axis, but processes such as partial melting and flow to 10 times that depth may still be an integral part of the plate boundary zone.

Definition of the "present" trace of the MAR plate boundary is a function of data density and accuracy. It is also a function of the type and resolving power of the data, and of what is meant by the "present." For example, the locus of historical seismicity may reveal that the west wall of a rift valley has been rupturing in recent years, while the most recent volcanics may have erupted 10^4 years ago on the east side of the valley. Yet magnetic anomalies and gross morphology may place the geologically "present" (1 Ma average) boundary in the middle of the valley. In terms of spatial resolution, a section of plate boundary appearing linear and oblique on the basis of teleseisms and magnetic profiles may be found upon detailed multi-beam surveying to consist of a series of en echelon overlapping ridges built from fissure eruptions.

Because of these and other differences among boundary descriptors, it is wise to distinguish, for example, the magnetic plate boundary from the morphologic and seismic. In addition to these three, the plate boundary can be defined at lower resolution by gravity, geoid, and heat flow highs. Provided enough data are available, the boundary might also be defined by age dating the igneous basement rock by radiometric or fission track methods. Another class of techniques is to date the basement *surface* from the age of the oldest sediment in contact with basement rocks, or from the *thickness* of such age-dependent accumulations as sediment, palagonitized basalt glass, or low-temperature chemical precipitates such as "manganese crust." However, all methods that merely date the basement surface are only dating the latest igneous body to be emplaced at the surface and may seriously underestimate (in Iceland by several million years) the average crustal age at a site (Pálmason, this volume). If the accretion process is symmetrical, this will not cause errors in the average location of the plate boundary. However, the discrete, locally random character of the eruptions would still cause mislocation of the local axis; for example, if the latest basalt flow "happened" to erupt preferentially on one side of the average axis. In principle, all such problems could be sorted out, but only with numerous deep drillholes! Due to space limitation, only the magnetic and morphologic MAR plate boundaries actually used are further elaborated (Fig. 1a, b).

With a few exceptions such as Iceland or parts of the Knipovich and Nansen Ridges (Vogt, this volume, Ch. 15) and close to the Equator, the MAR axis is demarcated by a prominent central magnetic anomaly, anomaly 1 of the sea-floor spreading sequence. To the extent that spreading has been symmetrical since the beginning of the Brunhes (chron 1, 0–0.73 Ma; Kent and Gradstein, this volume) the median line of anomaly 1—more correctly, the median line of the belt of normal magnetization obtained by inverting the data in the presence of topography—defines the **magnetic plate boundary** (MPB). At MAR opening rates from about 1 to 4 cm/yr, the central anomaly is 10 to 40 km wide. Even if one flank has been spreading 50 percent faster than the other (an extreme amount of asymmetry) since the last reversal, the plate boundary will only be mislocated by 2 to 7 km if symmetry is assumed. The MPB is the most accurate way of defining the accreting plate boundary by remote sensing, using an airborne magnetometer. Transform faults (FZ) are recognized as offsets of this boundary. The vertical measure of the MPB depends on the thickness of the magnetized layer *(sensu lato)* (Vogt, this volume, Ch. 15) but in no case can exceed the depth to the average Curie isotherm in Brunhes-aged crust.

One or several "excursions" or short intervals of reversed

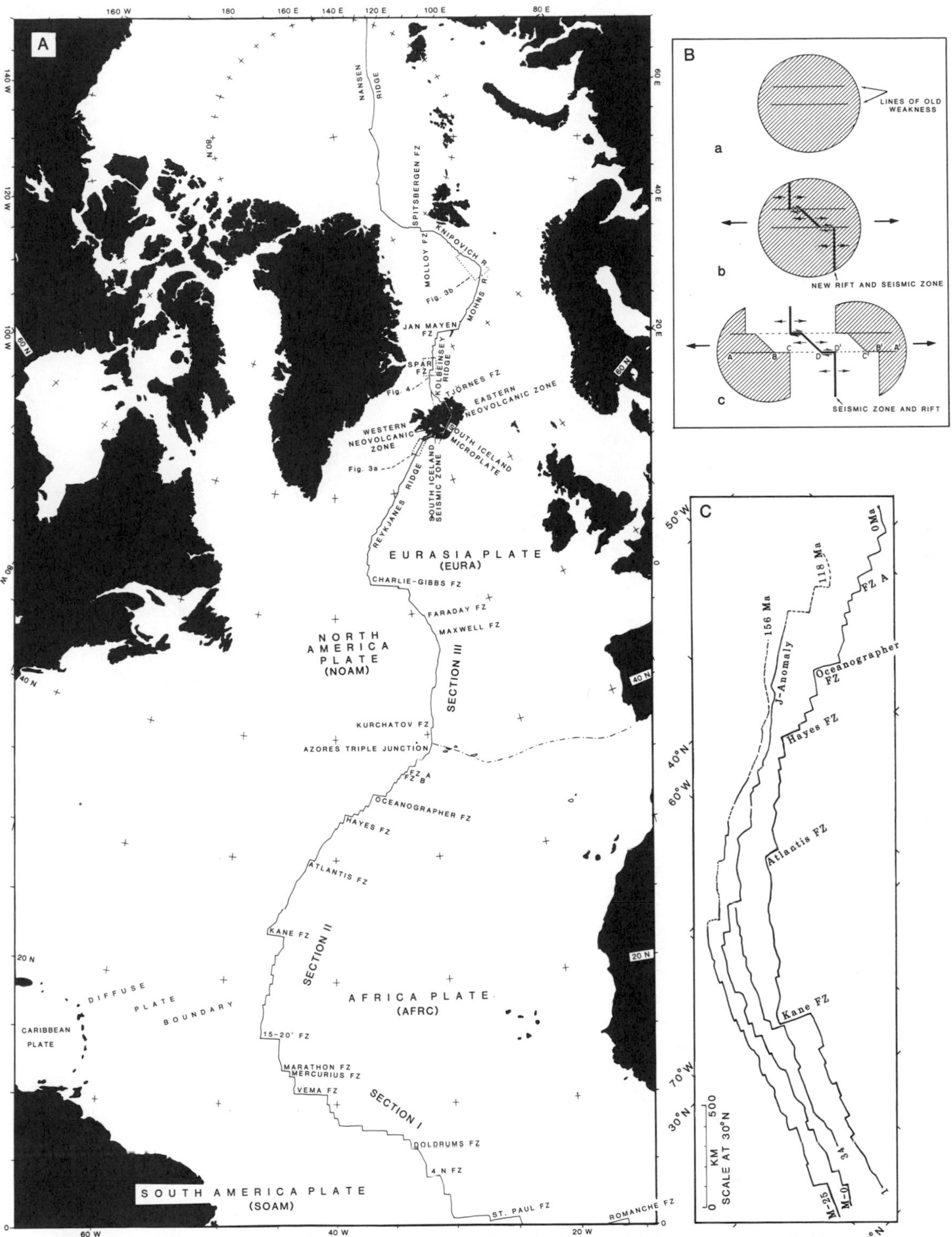

Figure 1a. A: Present plate boundary and coastlines with principal named fracture zones and ridges discussed in text. Section III is the NOAM/EURA boundary; II, AFRC/NOAM; I, SOAM/AFRC. The SOAM-AFRC-NOAM triple junction is nominally put at the 15°-20′ FZ. Boxes show locations of Figures 3A, 3B, and 4. B: Origin of transform faults according to Wilson (1965). The present plate boundary is a replica of the original break between two rifted continents (diagonally ruled). (There is no requirement for the old structural lineations AB–B′A′ to parallel the direction of plate separation, however). C: Comparison of present plate boundary shape (as in A, but rotated westward) with plate boundary shape at times of anomaly 34 (84 Ma), M-O (118 Ma), and M-25 (156 Ma), with M-O in present location on standard Mercator projection. Magnetic isochrons from Klitgord and Schouten (this volume, Ch. 22) and Vogt and Bracey (unpublished manuscript). See also Vogt (this volume, Ch. 24).

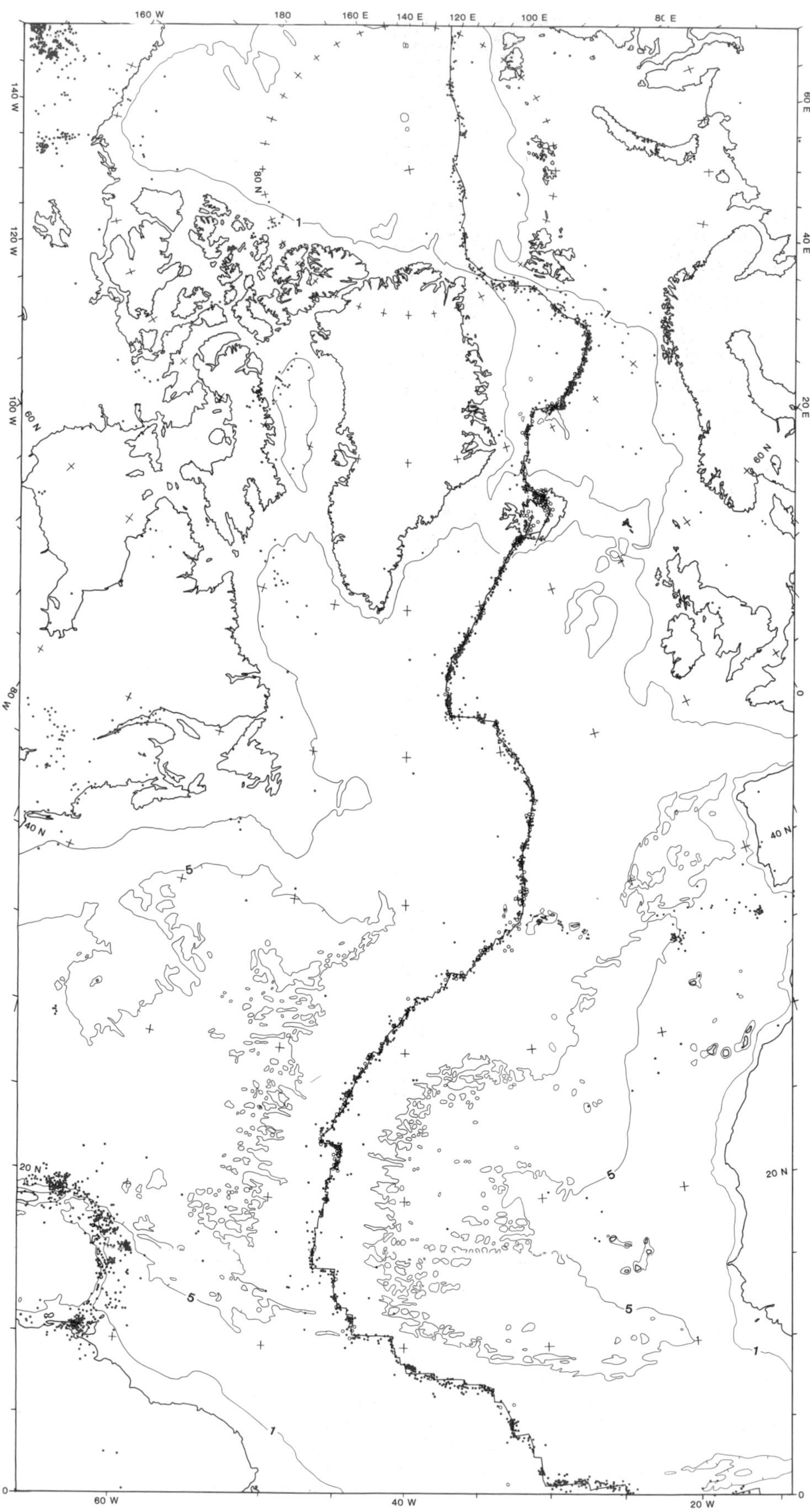

Figure 1b. Present Mid-Atlantic Ridge plate boundary (solid line from Fig. 1a) from bathymetric and magnetic data; for comparison earthquakes detected by the World Wide Seismograph Station Network (WWSSN) since 1961-1982 (red dots, from Nishenko and others, this volume, Plate 11; see also Einarsson, this volume Ch. 7) and location of fresh igneous rock samples (circles; see also Schilling (this volume, Ch. 9), Melson and O'Hearn (this volume, Ch. 8), and plates 8A, B). The 1- and 5-km isobaths are from Plate 2.

polarity (Harland and others, 1982) have occurred subsequent to 0.73 Ma, and their recognition as magnetic anomalies would sharpen resolution of the "present" MPB. Although such anomalies have been reported for the faster spreading Galapagos axis (e.g., the Emperor event at 0.47 Ma; Wilson and Hey, 1981), the distribution of basalts with such reverse polarity is probably too irregular (e.g., Schouten and Denham, 1979) and limited to be of much use along the MAR. More promising is the "central magnetization high" (Harrison, 1981) caused by high initial magnetization intensity, which decays with a time constant of about 0.5 m.y. as a result of maghematization. This central high may be conspicuous on deep-tow profiles (Macdonald, 1977) but along the MAR is not generally apparent at the sea surface (an exception being McGregor and others, 1977). In situations where the central anomaly itself is not apparent, the MPB can be estimated as the midpoint of the youngest pair of identifiable anomalies (positive or negative). The MPB of an extinct accretion axis can be found in the same manner. In general no "central magnetization high" will remain in the extinct axis; any central anomaly will be lower in amplitude due to posteruptive maghematization and greater water depth, and the axial anomalies may be unrecognizable if plate motion decelerated gradually.

The **morphologic plate boundary** along most of the MAR consists of: (1) a narrow braided network of recent faults (the Transform Fault Zone, or TFZ, of Fox and Gallo, 1984, and this volume); separated by (2) rift valleys, which generally become deeper on approach to FZ intersections. The bottom or center line of the rift valley is the obvious morphologic plate boundary on the normal MAR. Narrow ridges on the floor of the rift valley may define the locus of most recent volcanism (e.g., Mts. Pluto and Venus in the FAMOUS area; Macdonald, 1977). On the northern Reykjanes and southern Kolbeinsey ridges and on the MAR just south of the Azores triple junction, the rift valley is absent or poorly developed (Vogt, 1979, 1983), and in some areas a central ridge ("axial horst") may be present, similar to fast-spreading plate boundaries. Where neither ridge nor valley is prominent, the morphologic boundary may not be obvious and the magnetic boundary is more reliable.

PRESENT MAR PLATE BOUNDARY CONFIGURATION

The MAR plate boundary configuration (Fig. 1a, b) was estimated from bathymetric charts (e.g., Perry and others, 1981) south of about 60°N, and a combination of bathymetric and aeromagnetic data to the north. It is thus a morphologic plate boundary in the south, and a combined morphologic/magnetic plate boundary in the north. Since the NOAM/SOAM boundary appears diffuse, the NOAM/SOAM/AFRC triple junction is somewhat arbitrarily placed at the 15°20′N FZ. The boundary model (probably accurate in most areas to better than ±20 km) consists of 61 linear segments from 3.0°S, 12.0°W to 15.25°N, 44.9°W (the northern SOAM/AFRC boundary; hereafter, I), 74 segments (II) from the latter point to the Azores triple junction at 39.38°N, 29.72°W, and 200 segments (III) from there to the entry of the Nansen axis into the Siberian continental shelf (77.5°N, 128.3°E). This includes five extra segments required by the presence of two parallel rift zones on Iceland. The longest fracture zones were arbitrarily subdivided into linear segments about 100–200 km long, but otherwise the segments extend from one bending point to the next. Typical segment lengths are 50–125 km for section I, 25–75 km for II, and 15–100 km for III (Fig. 2). The closer spacing along sections II and III reflects a combination of greater data density and greater boundary complexity (closer FZ spacing) (Figs. 3, 4). Plate boundaries close to the one defined here were independently derived (although from mostly the same data) by Abbott (1986) and Sandwell (1986).

Seismicity was not used in defining the plate boundary and thus serves as an independent check. The seismicity (Fig. 1b) agrees closely with the boundary except near several FZ's south of 10°N, where errors in the boundary are suspected; systematic epicenter location errors are also possible (e.g., Sørnes and Fjeldskaar, 1980). The basalt samples from the MAR axis (Schilling, this volume) are tightly clustered along the plate boundary (Fig. 1b). Those located well away from the axis were deliberately collected there.

A synthesis of lengths and orientations of MAR Segments (Fig. 2) shows a northward decrease in the spacing between "minor fracture zones" (including numerous features such as the Kurchatov "Fracture Zone," which lacks a narrow transform fault zone and can equally be called a very oblique spreading axis). The characteristic FZ spacing in the equatorial area is 200 km (perhaps because small FZ's located between major ones remain undetected), whereas north of 15°N the most common spacings are 27 to 75 km, which is in agreement with estimates by Macdonald (this volume; 55 km between 10°N and 53°N), Schouten and others (1985; 50 ± 10 km between 23° and 39°N), Abbott (1986; 12 to 74 km, with a mean of 30 km between 15°N and 58°N), and Sandwell (1986). All estimates are probably upper limits because FZ's of small or even zero offset (e.g. Schouten and White, 1980) may have been missed.

A positive correlation is also expected between FZ spacing and the angle between the regional MAR trend and the spreading direction. Sandwell (1985) found this correlation to be weak. If the correlation were strong, section I would have the closest FZ spacing, not the widest, as observed. In the Arctic, the oblique Knipovich Ridge and the normal Nansen Ridge are very similar in their scale of segmentation. However, the short lengths of volcanic ridges in Figure 3A and Mohns Ridge rift valley segments in Figure 3B may reflect strongly oblique spreading.

The northward decrease in FZ spacing (Fig. 2) may be a function of the decreasing spreading rate. Schouten and others (1985) have proposed that FZ spacing (km) is related to total opening rate V (cm/yr) according to 31 $V^{1/3}$, whereas Sandwell (1986) has suggested a linear relationship, 12.56 V. The former model (A) supposes that MOR segmentation reflects the wavelength of gravitational instabilities in the partially molten subaxial mantle. Sandwell's model (B), following Collette (1974) and

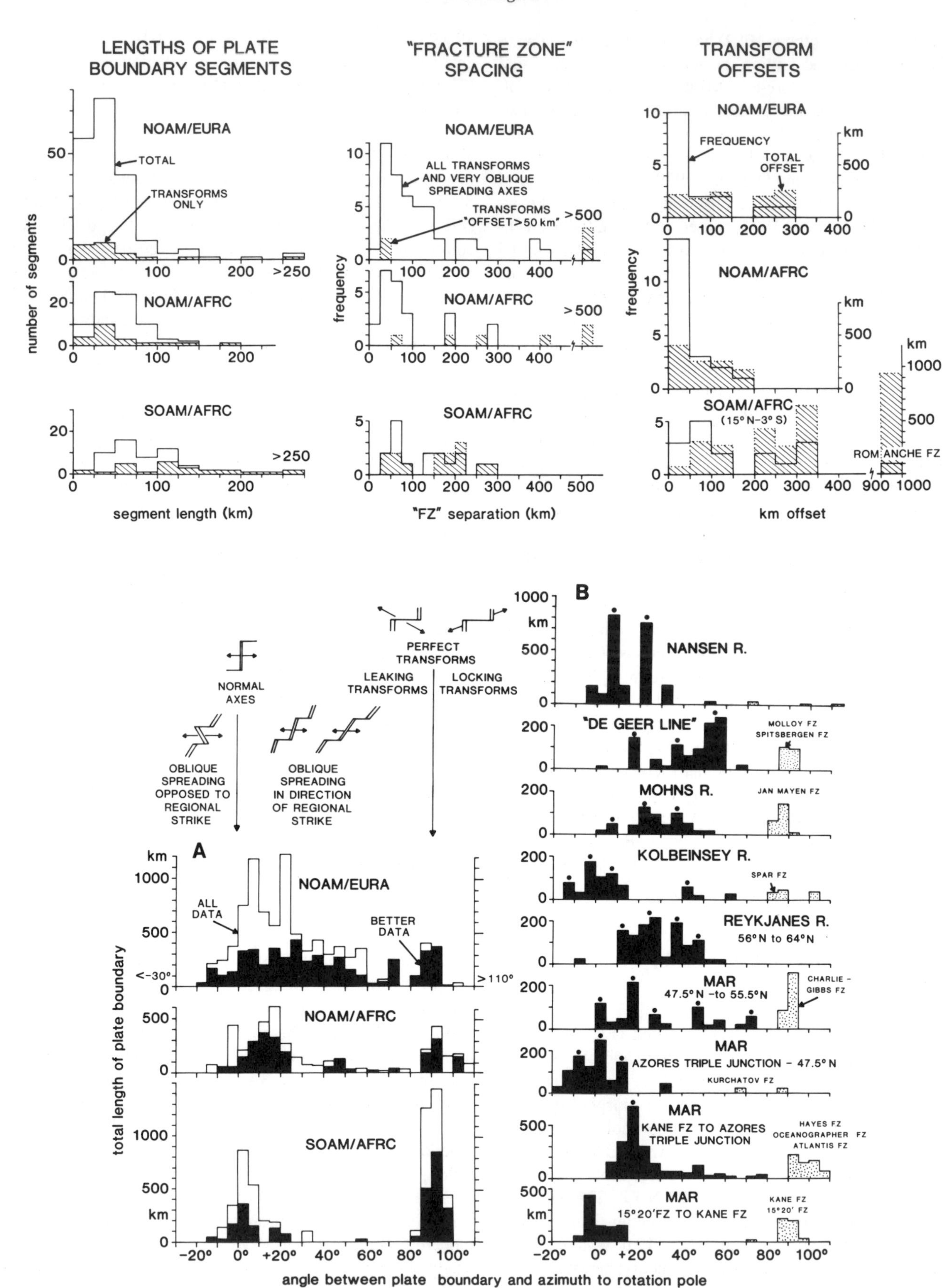

LENGTHS OF PLATE BOUNDARY SEGMENTS
"FRACTURE ZONE" SPACING
TRANSFORM OFFSETS
NOAM/EURA
TOTAL
TRANSFORMS ONLY
ALL TRANSFORMS AND VERY OBLIQUE SPREADING AXES
TRANSFORMS "OFFSET>50 km"
FREQUENCY
TOTAL OFFSET
NOAM/AFRC
SOAM/AFRC
(15° N-3° S)
ROMANCHE FZ
number of segments
frequency
segment length (km)
"FZ" separation (km)
km offset
PERFECT TRANSFORMS
LEAKING TRANSFORMS
LOCKING TRANSFORMS
NORMAL AXES
OBLIQUE SPREADING OPPOSED TO REGIONAL STRIKE
OBLIQUE SPREADING IN DIRECTION OF REGIONAL STRIKE
A
ALL DATA
BETTER DATA
total length of plate boundary
angle between plate boundary and azimuth to rotation pole
B
NANSEN R.
"DE GEER LINE"
MOLLOY FZ
SPITSBERGEN FZ
MOHNS R.
JAN MAYEN FZ
KOLBEINSEY R.
SPAR FZ
REYKJANES R. 56°N to 64°N
MAR 47.5°N -to 55.5°N
CHARLIE - GIBBS FZ
MAR AZORES TRIPLE JUNCTION - 47.5°N
KURCHATOV FZ
MAR KANE FZ TO AZORES TRIPLE JUNCTION
HAYES FZ
OCEANOGRAPHER FZ
ATLANTIS FZ
MAR 15°20'FZ TO KANE FZ
KANE FZ
15°20' FZ

Turcotte (1974), attributes FZ spacing to longitudinal thermal contraction of the axial lithosphere. The empirical relations A (and B) predict FZ spacings of 38 to 46 km (45 to 48 km) on section I, 26 to 36 km (40 to 44 km) on section II, and 9 to 29 km (27 to 41 km) on section III. (Values are calculated from present opening rate ranges of 3.0–3.7, 2.1–2.8 and 0.7–2.3 cm/yr along the three sections.) The predicted spacings are somewhat lower but overlap the observed ranges (Fig. 2).

However, neither model explains the spacings in the slow-spreading Arctic. For the southwestern Nansen Ridge (1.0–1.2 cm/yr opening rate) model A predicts 13–15-km spacings, and model B, 31–33 km, but available magnetic and bathymetric data (Feden and others, 1979) suggest 50–100 km (similarly on Knipovich Ridge; Fig. 3B), and the compilations by Sandwell (1986) and Abbott (1986) show little or no dependence on opening rate below about 6 cm/yr. In fact the 50–100 km spacing on the Knipovich and Nansen Ridges is close to the 58–95 km value observed by Abbott (1986) on the ultrafast (9–12 cm/yr) East Pacific Rise! At the same time, volcanic centers are spaced only 25 to 50 km on Iceland, which should be most like the East Pacific Rise.

In section III, two of the three large fracture zones are actually double, with a short spreading axis (45 km for the Charlie-Gibbs FZ, and 60 km for the Spitsbergen-Molloy FZ) between the transforms. The Ascension FZ at 7°S is also a pair, separated by 27 km (J. Brozena, personal communication, 1985). Both FZ members of these double FZ's are sharply lineated in the direction of plate motion, and thus are distinct from the complex, poorly lineated "minor FZ's" (oblique spreading axes?) such as the Kurchatov FZ. The MAR plate boundary comprises widely spaced single or double transform FZ's separated by long stretches with alternations of more normal and more oblique axes; whether the latter are (as here) included under "fracture zones" is a matter of definition.

Also shown for each MAR section (Fig. 2) is total offset for FZ's as a function of offset range. In the slow-spreading sections II and III, small offsets (<50 km) are far more common than larger ones, but the summed effect of many short ones is about the same as that of fewer large-offset FZ's. This is not true of section I (SOAM/ AFRC), where the less frequent larger offsets contribute most to the total offset.

The orientations of MAR boundary segments (Fig. 2) are defined in reference to the BFP poles of present plate motion (Minster and Jordan, 1978); use of other poles (e.g., DeMets and others, 1986) would make little difference. Section I is essentially a classic orthogonal stepped (staircase or zed) distribution, with FZ trends all within ±10° of predicted flowlines (mostly ±5°), and accretion axes also mostly within ±10° of pointing to the SOAM/AFRC rotation pole. (Note that FZ trend changes of only a few degrees occurring over the last few m.y. (Tucholke and Schouten, 1985) are not resolved in Figure 2). There is a slight systematic obliqueness (Θ) of about 5° in the direction of the regional MAR trend, and a few accretion axes with Θ exceeding 20°. Sections II and III have few accretion axis segments normal to the flowlines. In section II, Θ is most commonly 10°–20°, and in section III, values up to 50°–60° are still common.

The average strikes of major FZ's in section II, known to ±1°–2° (Vogt and Perry, 1981), depart systematically from plate kinemetic models, being turned 3°–4° clockwise in the north (Oceanographer FZ) to 3°–4° counterclockwise in the south (15°20′ FZ). This FZ "fanning has been attributed to thermal contraction and "ridge-push" asymmetry about the broad MAR bend (Roest and others, 1984) but may also relate to diffuse NOAM/SOAM motion.

A subdivision of sections II and III shows characteristic strike distributions for each portion of the central and northern MAR (Fig. 2B). Note that slow-spreading ridges are not necessarily oblique: the Nansen Ridge is no more oblique than the equatorial MAR. However, oblique spreading is associated with slow spreading where it does occur (Atwater and Macdonald, 1977). From the ratio of power dissipated at accretion axes to that on transform faults, Stein (1978) concluded that sin Θ should be roughly proportional to V^{-1}. If only the well-defined oblique parts of the MAR are considered (Θ = 5° in section I, 20° in section II, 35° on Reykjanes Ridge, and 45° on Knipovich Ridge), a V^{-2} or even V^{-3} dependence may better fit the data. (The generally lower obliqueness of individual fissure swarms—compared to the oblique axis as a whole, e.g., Figure 3A—probably reflects the stress field in the axial lithosphere).

The distributions of Θ along different parts of the NOAM/EURA boundary suggest several "modes" at intervals of 10°–20° (dots in Fig. 2B). This pattern hints that obliqueness may be "quantized" (in analogy with strikes and dips of faults and joints),

Figure 2 (previous page) Top—Frequency distributions: Lengths of individual linear MAR plate boundary segments used to construct Figure 1a; distance between neighboring transform faults; and offsets across transform faults. Separate distributions are shown for North America (NOAM)–Eurasia (EURA), North America–Africa (AFRC), and South America (SOAM)–Africa plate boundary (15° to 3°S only). The "plate boundary" histograms are subdivided into total and transform fault (hatched) categories. The "fracture zone spacing" histograms are shown both for all transforms and highly oblique spreading axes, and separately for transforms exceeding 50 km of offset (hatched). The "transform offset" histograms show both frequency and total offset for each offset range (hatched). Bottom—total length of plate boundary as distributed according to angle between plate boundary and azimuth (great circle) to BFP plate rotation pole of Minster and Jordan (1978). See also Vogt (this volume, Ch. 24). 0° is a spreading axis pointing to the rotation pole, and 90° is an ideal transform fault. The histograms at left (A) show subdivision according to the three MAR plate boundaries. Data believed most reliable are shown solid. Stack of histograms at right (B) is a subdivision of NOAM/EURA and NOAM/AFRC boundaries (text sections III and II and Fig. 1a) into separate provinces. In these histograms the "spreading" boundaries are solid and the "transform" boundaries are stippled. Dots show modes—possibly preferred orientations of oblique spreading axes. Schematic diagrams above "A" histograms show significance of different angular categories. Negative angles denote oblique spreading axes oriented in a sense opposed to the regional strike of the plate boundary. Angles slightly less than 90° denote "leaking," and angles above 90° are "locking" transforms.

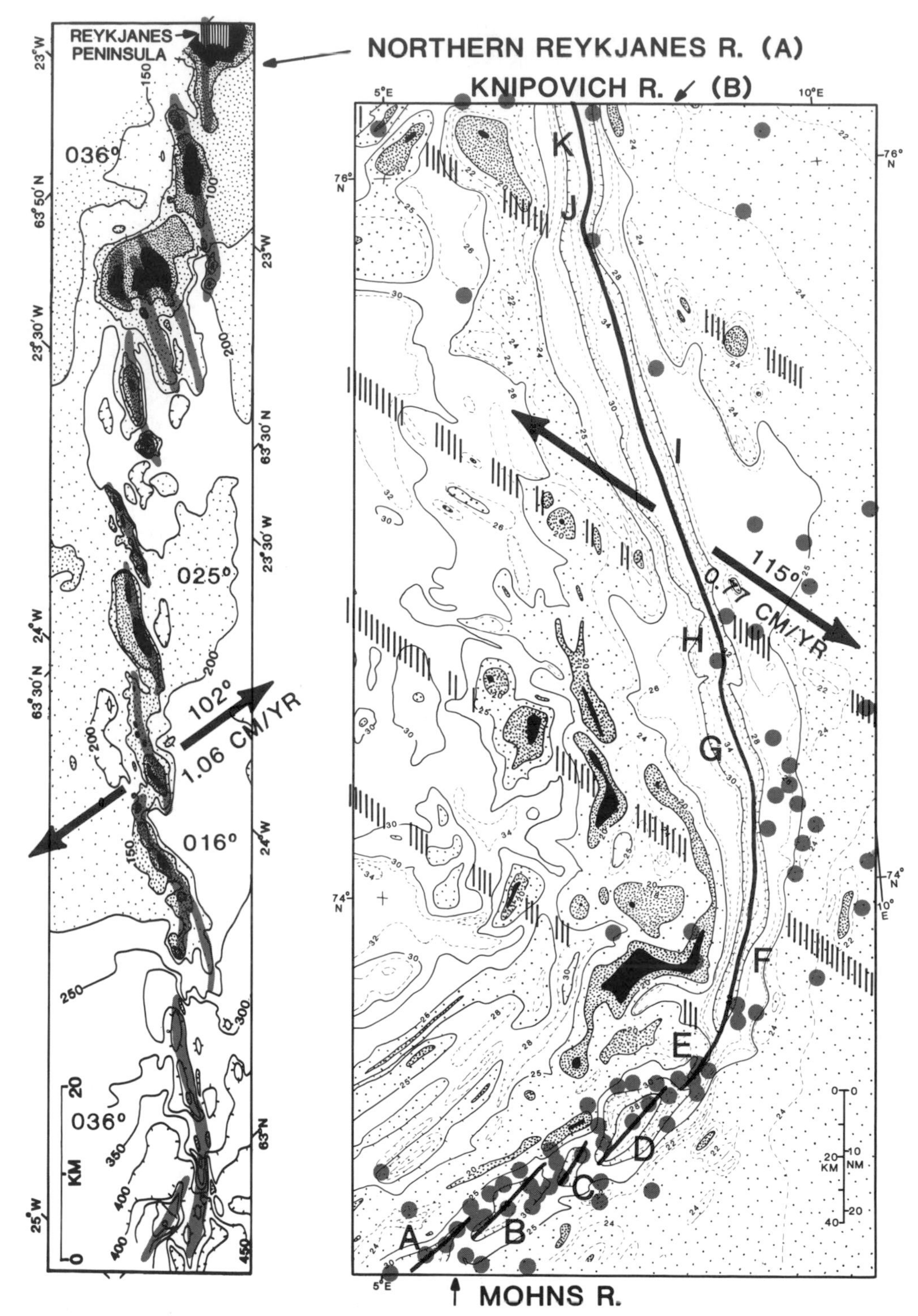

Figure 3. A: Detailed bathymetry (50-m contour interval) showing accreting plate boundary crossing the southwest Iceland shelf and average strikes of volcanic ridges. Depths less than 50 m are solid, and at 50 to 100 m are densely stippled. Adapted from Johnson and Jakobsson (1985). B: Bathymetry (in hundreds of meters, after Perry and others, 1980) and earthquake epicenters (1961-1982, after Plate 11 in Nishenko and others, this volume), in the area of Mohns-Knipovich bend. Thick line shows present plate boundary, with different segments lettered. Note arrangement of topographic highs in flowline-parallel bands 40-70 km apart (ruled lines connecting highs); difference in seismicity between Mohns Ridge (bottom) and Knipovich Ridge (center and top); E-W asymmetry in rift mountain depths (higher to west); and smooth nature of Mohns-Knipovich bend. Arrows show direction of relative plate motion predicted by "best fit" model of DeMets and others (1986).

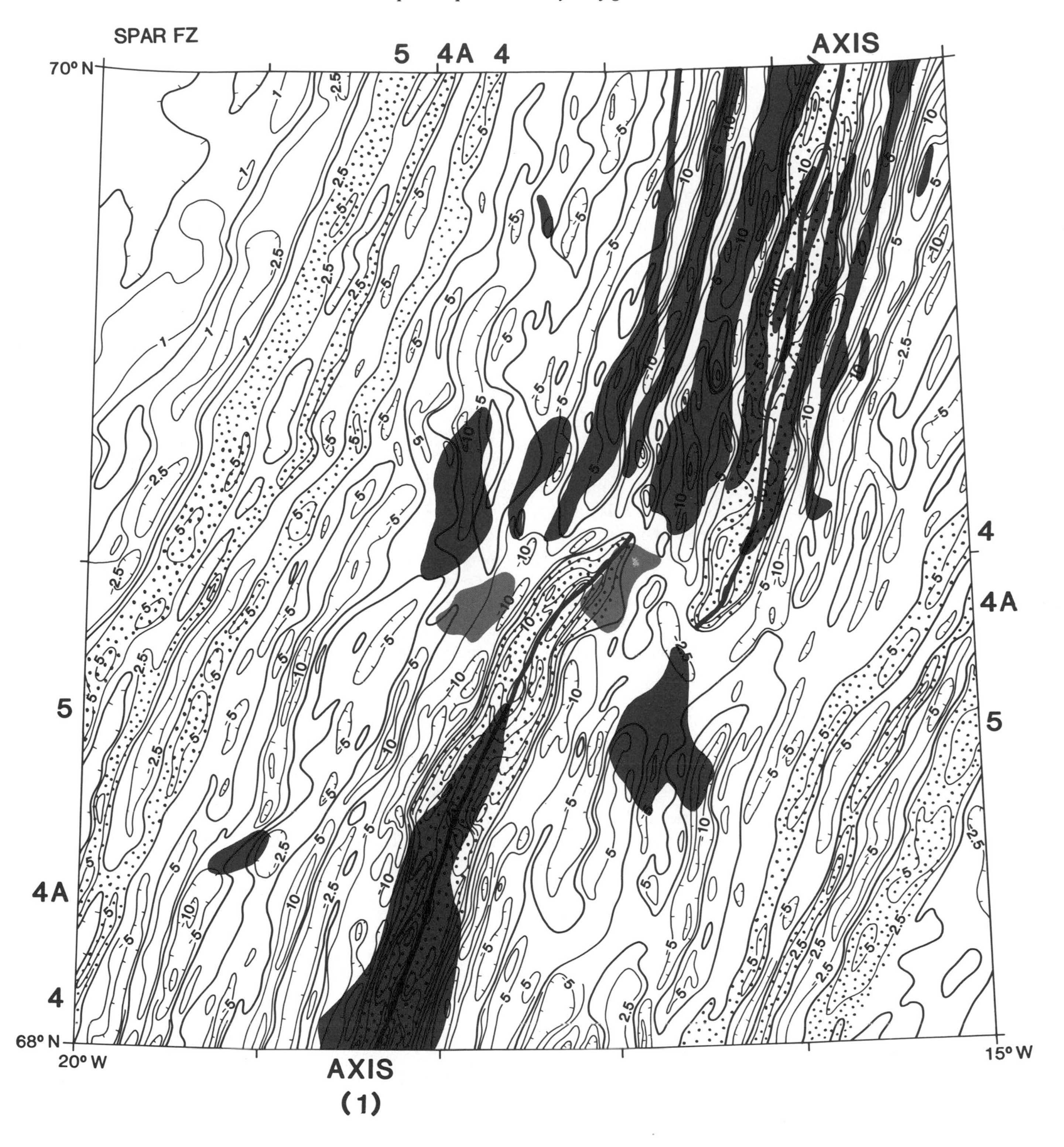

Figure 4. Magnetic anomaly contours (in hundreds of nT, from data reported by Vogt and others, 1980) with anomalies 1, 4, 4A, and 5 stippled; bathymetry (less than 1000 m in red, greater than 1600 m in gray) from Perry and others, 1980); and present plate boundary, based on median line of central magnetic anomaly, in area of Spar FZ on Kolbeinsey Ridge north of Iceland (see also Vogt, this volume, Ch. 15).

with more than two values of Θ preferred (besides just Θ = 0° and 90° as on ideal fast-spreading ridges). If shown to be real by additional work, the MAR Θ modes may constrain models of lithospheric stress in the axial domain.

THREE UNUSUAL CONFIGURATIONS

The MAR plate boundary is varied in its morphology and geometry, particularly section III. Figures 3A, B and 4 show relatively well-surveyed examples illustrating the scales of morphologic variability and the difficulties in specifying the plate boundary. In this volume, detailed examinations of parts of the MAR plate boundary are also found in Chapters 2, 4, 5, 15, and 24.

The MAR crosses the south Iceland shelf (Fig. 3A) in a narrow (3–8 km wide) band of 10–25 km long (average, 18 km) ridges. Each ridge probably represents fissure eruptions derived from a single volcanic system in the 10^4 years of postglacial time. Thus Figure 3A shows a geologically "instantaneous" view of axial volcanism. The accreting plate boundary is so shallow that geophysical data from surface ships (Johnson and Jakobsson, 1985) offer resolution comparable to deep-tow measurements. The axis appears as narrow on the Iceland shelf as on more typical parts of the MOR. (On nearby Iceland the neovolcanic zones are 50 km wide, in part a reflection of the multiply overlapping fissure swarms 75–150 km in length). In Figure 3A, several equidimensional highs are probably central volcanoes. Overall the axis executes a smooth regional bend (150-km radius of curvature).

The strike of individual ridges is variable (022° to 038°), and, although not normal to the direction of relative plate motion (102°), it tends to be less oblique than the regional trend of 035° to 038°. A similar pattern was found on other oblique ridges such as the Reykjanes Ridge farther south (Searle, 1978; Shih and others, 1978). Although not so treated in the plate boundary description (Figs. 1, 2), the individual ridges in Figure 3A might be considered "end members"—ultrashort spreading axes. In their slightly sigmoid trends and 3- to 6-km offset, the ridges resemble conjugate (overlapping) rifts (2–20-km spacing) on the East Pacific Rise (Macdonald and Fox, 1983). The EPR rifts are four to five times longer, in proportion to the higher opening rates. No transverse structures suggestive of transform faults are apparent between the individual ridges in Figure 3A.

The Spar (Fig. 4) and Tjörnes FZ north of Iceland exemplify conjugate rifts curving toward each other, with a bathymetric deep lying between the Spar FZ rift tips, as in EPR examples. The overlap and lack of transverse FZ trough make precise plate boundary definition somewhat arbitrary. The Spar FZ may be the first conjugate rift identified from magnetic anomalies. The Spar and Tjörnes rift tips are offset 30 and 100 km, more than typical on the EPR. The morphologic plate boundary changes from an axial high south of the Spar FZ to a shallow rift in the north (Perry and others, 1980). The anomaly pattern shows how the Spar FZ evolved subsequent to anomaly 4 time (7 Ma). Small ephemeral fracture zones also disturb the lineation pattern; rift propagation may be involved (Vogt and others, 1980), but not at the 3–5-cm/yr speeds observed on faster spreading ridges (Hey and others, 1980).

The Mohns-Knipovich bend (Fig. 3B; Vogt, 1986b) has not been surveyed by multi-beam bathymetry, but track spacing (ca. 10 km) is adequate to depict major features of the plate boundary: The northeastern Mohns boundary comprises several en echelon rift valley segments (A–E) each about 15–30 km long, like the ridges in Figure 3A, nearly perpendicular to the plate-separation flowlines. These segments were incorporated in the plate boundary description (Figs. 1, 2). A volcanic high in segment D has "split" the rift valley in half. At the available resolution, the Mohns-Knipovich bend appears continuous with a 100-km curvature radius. The Knipovich boundary (F–K) shows an inner rift valley 5–10 km wide within a broad outer valley 30 km wide. More oblique segments, G, I, and K, are associated with lower flank topography, compared to segments F, H, and J, which are evidently associated with bands of higher topography. These bands, 50–100 km apart, are oriented 124° to 135°, parallel to the 125° flowlines predicted by the pole of DeMets and others (1986). (Similar zigzag patterns on the central MAR have migrated along the axis at a few mm/yr; Rona and Gray, 1980). The west flanks of both the Knipovich and Mohns ridges are higher than the east flanks, particularly in the bend area. The east flank has been smoothed by thick terrigenous sediments from the Barents margin (Vogt, this volume, Ch. 15). Such sediments have poured into the Knipovich valley and are present locally **west** of the present axis, suggesting a recent switch of the axis or a period when no rift valley was present (Vogt, 1986a). The eastward displacement of the Knipovich epicenters may reflect network location errors, but the systematically lower and more scattered level of seismicity on Knipovich Ridge is real.

PLATE BOUNDARY EVOLUTION

How and why did the MAR plate boundary acquire its present shape? As originally postulated by Wilson (1965), an idealized MAR-type plate boundary between two separating continents consists of some configuration of normal and oblique spreading axes and transform faults (Fig. 1a-B). This plate boundary is inherited from, and in the classic model continuously replicates, the shape of the initial break. (A less strict model would require present fracture zones to correspond only in **location,** not in offset length, to offsets in the initial breakup.) To the extent that the classic model applies to the MAR, the transform faults and other irregularities are inherited from pre-breakup structures, such as older intracontinental faults or rift zones. (Such older faults—e.g., ABC–C′B′A′ in Fig. 1a-B—do not, however, necessarily parallel the direction of plate motion, as implied in Wilson's original diagram.)

How well does the classic model fit the present MAR plate boundary shown in Figure 1? This question could be addressed if the continent-oceanic crustal boundary were indeed a sharp boundary and its location precisely known. In general, this is not the case. In lieu of such a test, the present plate boundary can be compared to previous boundaries (isochrons) as defined by magnetic lineations or other parameters. Although relatively plentiful in the Atlantic, magnetic data (Plate 3) are mostly inadequate (or the pattern too confused) for unambiguous resolution of fine-scale features of crustal isochrons. Of the three MAR sections discussed here (Fig. 1a-A), the southern portion (I), is dominated by major fracture zones, some of which have been traced across the ocean to conjugate points on the African and South American margins (Fox, 1972; Sibuet and Mascle, 1978; Gorini, 1981). Continental breakup occurred sometime between M-4 and M-0 time (126–118 Ma) (Klitgord and Schouten, this volume). Post-breakup evolution of the plate boundary is not well known, owing to the low magnetic latitudes, frequent major fracture zones, and low data density. (However, Westbrook and McCann [this volume] have mapped anomalies 33/34 between the MAR and the West Indies.) The equatorial MAR probably comes closest to the Wilson model (Fig. 1a-B), although much more work needs to be done.

The northern MAR section (III) is the youngest of the three, the portion from about 57°N to Siberia having evolved only since breakup about 58 Ma (Srivastava and Tapscott, this volume, Ch. 23). The Nansen Ridge axis is offset by some minor fracture zones, the largest of which corresponds to a bend in the Eurasia margin and in the Lomonosov Ridge, a presumed microcontinent (Vogt and others, 1979). The plate boundary Mohns Ridge–Knipovich Ridge–Molloy Fracture zone–Molloy Ridge–Spitsbergen Fracture Zone–Lena Trough crudely resembles the line of breakup betweek Eurasia and Greenland, although the detailed shape (including the Molloy and Spitsbergen FZ's) probably developed as a result of the change in plate motion between Greenland and Eurasia (Crane and others, 1982). Kolbeinsey Ridge is the only MAR segment much closer to one margin than the other. It is, however, roughly midway between the Jan Mayen Ridge, a supposed microcontinent, and the Greenland margin, although some reconfiguration occurred starting 6–7 Ma (Fig. 4). The Iceland neovolcanic zones are somewhat east of the median line, but debate about the continent-ocean crustal boundary in the Faeroes area has not been settled (Vogt, 1983). The northern Reykjanes Ridge has changed its configuration twice (Vogt and Avery, 1974), but at present once more resembles the original linear break between Rockall–Hatton Bank and southeastern Greenland. The present axis bend at 57°N is a "memory" of the Greenland–Eurasia–North America triple junction. The Charlie Gibbs FZ (CGFZ) is easily traced back to anomaly 24 (Vogt and Avery, 1974), and probably to anomaly 34 (Srivastava and Tapscott, this volume), and its intersection with continental crust appears to correlate with pre-breakup structures (e.g., Wade and others, 1977). From the CGFZ to the Azores, the present boundary is crudely the same as it was at anomaly 34 time, although detailed resemblance to older isochrons and to the breakup line is questionable (e.g., Fig. 1 of Klitgord and Schouten, this volume). The gentle bend at 47°N may be a memory of the Biscay-MAR triple junction, or also a hotspot "anchoring" effect (see below). The present MAR bend at the Azores triple junction may also reflect hotspot influence, although the shape at anomaly 34 time was crudely similar to the present boundary.

The central MAR (II) is the oldest section, central to the theme of this volume, and therefore singled out for closer examination. Perhaps the best-mapped magnetic lineations are M-0 (118 Ma) and M-25 (156 Ma) in the western North Atlantic (Sundvik, 1986; Klitgord and Schouten, this volume). In Figure 1a-C the present plate boundary is placed adjacent to anomalies M-25, M-0, and 34 to allow comparison of shapes. Although the plate boundary to this day has retained the broad arcuate shape of the initial break between Africa and North America, the detailed shape of the boundary evidently changed from one isochron to the next, and it is questionable whether any of the present transform faults hark back to the configuration at initial breakup as in the Wilson model. The present boundary does appear more densely fractured (stepped) than earlier isochrons, particularly the section from the Hayes FZ to the Azores triple junction. However, the plate boundary at anomaly 33 time, rotated to the present boundary between 25°N and 37°N, seems as fractured as it is at present (Collette and others, 1984). Thus, the change in shape must have occurred largely between M-0 and anomaly 33 time. However, FZ's of small or even zero offset (e.g. Schouten and White, 1980) may exist in the apparently less densely fractured isochrons in Figure 1a–c.

How does an accreting plate boundary change its shape? Consider first the kinematic aspects of this question, specifically the FZ offset, which can change if (1) spreading half-rates are not the same on both sides of the FZ, i.e., there is asymmetric spreading on at least one side (asymmetric spreading proportionally constant along the plate boundary would not change its shape); (2) propagating rift(s) intercept the FZ from one or both sides; or (3) the accretion axis jumps on at least one side of the FZ. Mechanisms 2 and 3 change offsets abruptly, whereas asymmetric spreading, where examined, appears to be a continuous process at least down to the ca. 10-km resolution permitted by magnetic anomalies. Even with detailed data (lacking over much of the Atlantic), it is generally difficult to separate these three effects unless they are large.

According to Twigt and others (1983), the Kane FZ can possibly be traced back into the Keathley (M) anomalies, where anomaly M-4 (126 Ma) is offset 15 km and M-O, 38 km. During the 34 m.y. from anomalies M-O to 34, the offset increased further to 75 km, and its present offset 84 m.y. later is 160 km. These observations are consistent with, but do not prove, a gradual rate of FZ offset growth on the order 1 mm/yr. For comparison, the SEIR (Southeast Indian Ridge) between Australia and Antarctica was examined with detailed aeromagnetics back to 25 Ma (Vogt and others, 1983). Resolution is higher there compared to the MAR on account of higher opening rates (3 cm/yr). On the SEIR, as along the MAR, FZ offsets have tended to increase with time both by continuous differential asymmetric spreading (at about 3 mm/yr) and by propagating rifts. However, no measurable offset change was noted for several of the SEIR FZ's over 10 m.y. Similarly, the present total offset of the Charlie Gibbs FZ is 340–350 km, not significantly greater than the 330-km offset at 55–60 Ma (Vogt and Avery, 1974).

The MAR also provides examples of more rapidly changing FZ offsets. Where rates can be measured, the small fracture zones formed on the ancestral Reykjanes Ridge about 40–35 Ma increased their offsets at about 3–6 mm/yr, with spreading axes reorienting themselves at 1°–4°m.y. (Vogt and others, 1969; Vogt and Avery, 1974). Between 30 and 10 Ma, these fractures disappeared by reorientation of the accretion axes into the present continuous oblique axis. This reorientation occurred earlier closer to Iceland, later to the southwest. From 36° to 40°N, the present MAR is offset by numerous FZ's, e.g., FZ A and B (Fig. 1a-A), which began to form between 3.7 and 5.2 Ma (Macdonald, 1977). The present configuration evolved from a continuous 50° striking oblique axis that broke up largely between 5 and 3 Ma, with present axial segments striking 23° (Ramberg and others, 1977). This implies reorientation rates of the order 10°/m.y. In the last few m.y., differential asymmetric spreading has tended to reduce some FZ offsets and straighten the axis. For example, the offset across FZ B decreased from 33 to 23 km since 3.5 Ma, a rate of 3 mm/yr.

The MAR at 24°–27°N, just north of the Kane FZ (Rona and Gray, 1980) underwent an evolution opposite in sense from that of the FAMOUS area. In the last 6 m.y. (mainly about 4 Ma), axis rotations have replaced an originally stepped pattern with a zigzag pattern of oblique axes. Axis rotation rates up to 10°–15°/m.y. decreased FZ offsets at 3–15 mm/yr. The segment north of the Kane FZ rotated counterclockwise, away from the regional trend of the MAR. Far in the north (Fig. 3B) the Knipovich Ridge (74°–76°N) also shifted from a stepped to oblique geometry about 4 Ma or later, but whether it did so by ridge jump or continuous rotation is unclear (Vogt, 1986b). The Spar FZ (69°), a conjugate rift offsetting the present axis by 30 km, only began to form about 6–7 Ma (Fig. 4).

The above examples illustrate that the MAR plate boundary has changed its configuration in many different areas and at different times. The axis configuration may change without changes in FZ location or trend (Schouten and Klitgord, 1982). Fracture zone offsets may change at rates of 1–10 mm/yr, and accretion axis segments may rotate at 1°–10°/m.y. Zigzag (oblique) configurations evolve into stepped or crenulate configurations, or vice versa. Small discrete axis jumps have also been identified near the axis (Ramberg and others, 1977; Rona and Gray, 1980) and in the M-series (Sundvik, 1986), but the process is generally continuous at least down to the 5-km resolution of sea-surface data.

What can cause the plate boundary to change its configuration? Proposed processes include changes of opening pole, opening rate, hotspot (plume) influence, motion of spreading axis over the mantle, and other sources of stress acting on the axial lithosphere. Fracture zones must form (or change) in response to changes in plate rotation pole, particularly if an orthogonal (stepped) pattern is to be maintained. Such configuration changes should affect entire plate boundaries more or less coevally, and should be predictable from plate kinematic models if the pole change is large enough to be resolved by magnetic or FZ-trace data. Examples of FZ formation caused by pole changes are the transforms formed on the ancestral Reykjanes Ridge about 40–30 Ma as a result of annexation of the Greenland plate to North America (Vogt and Avery, 1974) and probably also the development of the Molloy and Spitsbergen FZ's.

As another consequence of this change in motion between Greenland and Eurasia, Nunns (1982) suggested that a northward-propagating rift, ancestral to the present Kolbeinsey Ridge, split the Jan Mayen microcontinent away from Greenland and created the present Jan Mayen FZ. Ridge reorientations by rift propagation, in response to change in spreading direction (such as observed south of Australia (Vogt and others, 1983)) have not yet been demonstrated in the Atlantic. Rift propagation may be involved in the reconfiguration in the Spar FZ area (Fig. 4), although perhaps without a change in plate rotation pole.

The Kane FZ trace exhibits numerous changes in orientation—typically a few degrees—and accompanying "blockages" by basement highs (Tucholke and Schouten, 1985) which are attributed to small NOAM/AFRC rotation pole changes, the most recent three occurring shortly after anomaly 3A (~5 Ma) near anomaly 2A (2.5 Ma), and subsequent to 0.5 Ma. The FZ trace complexities, if they indeed reflect plate rotation pole changes, will probably be found to relate to some of the changes in MAR configuration noted above.

As discussed earlier, if opening rates influence plate boundary configurations, this should be manifested by regional variations in present boundary shape; past configurations should reflect past opening rates (among other factors). Although the present MAR plate boundary appears more densely fractured, and oblique spreading more common in the slower spreading regimes of the North Atlantic and Arctic compared to the equatorial MAR (Figs. 1, 2), this may be partly an artifact of better data coverage in middle-higher latitudes. Identification of spreading rate dependence in MAR paleoplate boundaries is even more problematic. Opening rates may have been somewhat higher along the MAR around 3–7 Ma and 0–1 Ma (Fig. 5 of Vogt; this volume, Ch. 24), and perhaps these changes influenced the complex axis reorganizations observed, many of which were initiated during the period 3–8 Ma. However, no regionally consistent pattern of changes is yet apparent (e.g., toward wider fracture spacing and less oblique spreading as a result of faster spreading around 3–8 Ma). Furthermore, there is yet no evidence that the more rapid Atlantic spreading of the mid-Cretaceous (ca. 80–110 Ma) (Klitgord and Schouten, this volume) was associated with greater FZ separation, although this can be tested. Sandwell's (1985) model predicts FZ spacings of around 50–75 km on the present MAR should almost double in mid-Cretaceous crust. The diapiric model (Schouten and others, 1985) predicts fracture spacing only about 20 percent greater, which would be difficult to detect. Conjugate (overlapping) rifts constitute another feature that depends on opening rate, being common on the fast-spreading East Pacific Rise but not on the MAR (Fig. 4 shows an exception).

Mantle plumes (hotspots) may influence the plate boundary configuration up to more than 1000 km from hotspot centers (Vogt, 1974, 1976, 1983; Schilling, 1985). In some respects the plume influence may make a slowly opening plate boundary "look" like a faster one. This might explain the absence of a rift valley and the existence of conjugate rifts on the Reykjanes and Kolbeinsey Ridges near Iceland (Figs. 3A, 4), but does not explain the short length of the volcanic systems in Figure 3A. Since accreting plate boundaries are not generally fixed to mantle hotspots (Burke and Wilson, 1972), the latter would exert their greatest influence while the boundary is located over a hotspot. For hotspot diameters of 100 to 200 km (Morgan, 1981), and for proposed absolute motion models (Duncan, 1984), a spreading axis might take from a few to a few tens of m.y. to cross a hotspot. These times can be increased if there is channelized flow from the hotspot toward the axis (Vink, 1984; Schilling and others, 1985).

Hotspots have also been suggested capable of preferentially weakening the plate such as to "anchor" the accretion axis, even after the axis has migrated across the hotspot. This could be accomplished by asymmetric spreading or by one or more axis jumps back toward the hotspot, in either case changing the plate boundary. A possible example is the Iceland area (Plate 3), where the early Tertiary magnetic anomalies bulge westward around the Faeroes, whereas the late Tertiary lineations (and the present axis) bulge eastward into eastern Iceland, the spreading axis having jumped there from the west around 6–7 Ma. This "hourglass" magnetic pattern could reflect the Iceland hotspot having been located toward the west in the early Tertiary (Vink, 1984) and, as a result of westward motion of the MAR axis away from a relatively stationary Eurasia plate, having become stranded under this plate in later Tertiary time. The time of axis jump (6–7 Ma) corresponds to the first development of several hotspot-related features near Iceland (Vogt, 1983) and to a period of plate boundary reorganization along widely separated parts of the MAR, including the Kolbeinsey Ridge north of Iceland (Fig. 4), and to a pulse of relatively faster spreading (Vogt, this volume, Ch. 24).

The Reykjanes Ridge, however, maintained its oblique configuration from an earlier time, ca. 30 Ma near Iceland and later farther southwest, when an earlier stepped boundary disappeared. Vogt and Avery (1974) attributed this reconfiguration to an increase in subaxial asthenosphere flow that altered the stress field or weakened the lithosphere in the flow direction. Strong flow might also explain the linear M-0 south of the Grand Banks (Fig. 1a-C) in the area of the J-Anomaly Ridge (Tucholke and Ludwig, 1982). Ramberg and others (1977) suggested that the switch from continuous oblique spreading to the present stepped pattern (36°–37°N) starting about 6 Ma resulted from a weakening of subaxial flow from the Azores plume. Oldenburg and Brune (1972) suggested that oblique spreading could result either from low shear strength in the plate, on increased stress on transform faults. Perhaps the shear strength of the axial lithosphere is reduced in areas of hot, melt-rich mantle such as under the Reykjanes Ridge (Ramberg and others, 1977). Elsewhere, regional stress changes on the plates may be responsible for the observed reconfigurations from stepped to zigzag or vice versa. Two different ways to produce oblique spreading are seemingly needed to reconcile the conspicuously oblique Reykjanes Ridge with the generally orthogonal pattern on fast-spreading ridges, inasmuch as the Reykjanes Ridge, although spreading slowly, resembles faster spreading ridges in morphology.

If overlapping spreading centers are associated with hotspots or fast spreading (with plentiful along-axis magma transport in fissure swarms), the typical MAR "zigzag" configuration of short normal and oblique axes (Fig. 3B) may reflect **underlapping**, i.e., magma supplies adequate to maintain only short fissure systems normal to the least principal stress. Complex oblique spreading (e.g., Kurchatov FZ) would then connect the ends of the "underlappers."

The influence of other Atlantic hotspots (Morgan, 1983) on plate boundary evolution is poorly known. The eastward embayments of the plate boundary around 40°N (Azores) and 47°N both correspond to proposed hotspots, the latter principally from its geochemical signature (Plate 8B). These embayments, although perhaps related to active or extinct triple junctions, may also reflect the present or past "anchoring" influence of hotspots now stranded east of the plate boundary. Passage of the central MAR over the Verde hotspot about M-24 time may account for small eastward jumps recorded about 36°N, 65°W (Sundvik, 1986).

As mentioned earlier, differential asymmetric spreading from one side of a FZ to another results in fracture zone offset growth or contraction. There is another type of spreading asymmetry that, because of its regional and temporal extent, may have a different origin. Unless it uniformly affects an entire plate boundary, even regionally asymmetric spreading gradually changes the shape of the boundary. For example, spreading half-rates were about 5–15% lower on the NOAM flanks versus the EURA flanks of Mohns Ridge for the period 57–35 Ma, and similarly for the Nansen Ridge (0–57 Ma) (Vogt and others, 1982), whereas no regional asymmetry has been reported south of Iceland. With some exceptions the fast-spreading flanks are also a few hundred meters deeper (after correction for sediment loading) and smoother. The Arctic spreading rate asymmetry qualitatively conforms to a fluid-dynamic model (Stein and others, 1977), which predicts less lithosphere accreted to the more rapidly moving plate (over the mantle), in this case the North America and Greenland plates. However, spreading asymmetry south of Australia is opposite to model predictions, and even in the Arctic the model does not explain why the stationary flank is deeper. Possibly shear heating makes the moving plate slightly hotter and shallower (Vogt and others, 1982).

Comparably detailed searches for regional asymmetry have not been performed on the central and equatorial MAR, although local, temporary spreading asymmetry occurred at 36°–37°N

(Macdonald, 1977) and 23°–27°N (Rona and Gray, 1980). Analysis of a flowline profile across the central Atlantic (Klitgord and Schouten, this volume) shows faster AFRC spreading approximately 15–25 Ma, 40–45 Ma, and 55–70 Ma, nearly symmetrical spreading 0–15 Ma, and 45–55 Ma and faster NOAM spreading only 25–40 Ma. There is a tendency for greater asymmetry to correlate with relatively faster spreading, and for the more stationary flank to spread faster, as is the case also in the Arctic (Vogt and others, 1982). Differences in spreading rates on the two flanks ranged from 0 percent to 15–25 percent along this flowline.

CONCLUSION

This chapter has emphasized the plate boundary *configuration* versus other possible plate boundary descriptors, some of which are shown and discussed in other chapters and on Plates 8A, B. Although some theoretical models predict obliqueness, spreading asymmetry, and segment length (FZ spacing) as a function of "independent variables" such as opening rate or absolute motion, present understanding of the physics behind the shape of the MAR plate boundary is still primitive and a model that explains the host of along-strike variations (Plates 8A, B) in a unified manner is yet to be developed.

REFERENCES

Abbott, D.
1986 : The statistics of ridge crest offsets: Geophysical Research Letters (in press).

Atwater, T. M., and Macdonald, K. C.
1977 : Slowly spreading ridge crests; Are they perpendicular to their transform faults?: Nature, v. 270, p. 715–719.

Burke, K., and Wilson, J. T.
1972 : Is the African plate stationary?: Nature, v. 239, p. 387–389.

Collette, B. J.
1974 : Thermal contraction joints in a spreading seafloor as origin of fracture zones: Nature, v. 251, p. 299–300.

Collette, B. J., Slootweg, A. P., Verhoef, J., and Roest, W. R.
1984 : Geophysical investigations of the floor of the Atlantic ocean between 10° and 38°N (Kroonvlag - project): Proceedings of the Koninklijke Nederlandse Akademie van Wetenschappen, Series C, v. 87, p. 1–76.

Crane, K., Eldholm, O., Myhre, A. M., and Sundvor, E.
1982 : Thermal implications for the evolution of the Spitsbergen transform fault: Tectonophysics, v. 89, p. 1–32.

DeMets, C., Gordon, R., Stein, S., Argus, D., Engeln, J., Lundren, P., Quible, D., Stein, C., Wiens, D., Weinstein, S., and Woods, D.
1986 : Current plate motions: Journal of Geophysical Research (in press).

Duncan, R. A.
1984 : Age progressive volcanism in the New England seamounts and the opening of the Central Atlantic Ocean: Journal of Geophysical Research, v. 89, p. 9980–9990.

Feden, R. H., Vogt, P. R., and Fleming, H. S.
1979 : Magnetic and bathymetric evidence for the "Yermak Hot Spot" northwest of Svalbard in the Arctic Basin: Earth and Planetary Science Letters, v. 44, p. 18–38.

Fox, P. J.
1972 : The geology of some Atlantic fracture zones, Caribbean escarpments and the nature of the oceanic basement and crust [Ph.D. thesis]: New York, Columbia University, 357 p.

Fox, P. J., and Gallo, D. G.
1984 : A tectonic model for ridge-transform–ridge plate boundaries; Implications for the structure of the oceanic lithosphere: Tectonophysics, v. 104, p. 205–242.

Gorini, M. A.
1981 : The tectonic fabric of the equatorial Atlantic and adjoining continental margins; Gulf of Guinea to northeastern Brazil: Série Projet D Remac, no. 9; University of Rio de Janeiro, p. 11–116.

Harland, W. B., Cox, A. V., Llewellyn, P. G., Pickton, C.A.G., Smith, A. G., and Walters, R.
1982 : A geologic time scale: Cambridge Earth Science Series, Cambridge University Press, 131 p.

Harrison, C.G.A.
1981 : Magnetism of the oceanic crust, 5; in Emiliani, C., ed., The Sea, v. 7—The Oceanic Lithosphere: New York, John Wiley and Sons, p. 219–239.

Heezen, B. C., Tharp, M., and Ewing, M.
1959 : The Floors of the Oceans, I, The North Atlantic Ocean: Geological Society of America Special Paper 65, 122 p.

Heirtzler, J. R., and LePichon, X.
1965 : Crustal structure of the mid-ocean ridges, 3, magnetic anomalies over the Mid-Atlantic Ridge: Journal of Geophysical Research, v. 70, p. 4013–4033.

Hey, R., Duennebier, F. K., and Morgan, W. J.
1980 : Propagating rifts on mid-ocean ridges: Journal of Geophysical Research, v. 85, p. 3647–3658.

Jackson, H. R., and Reid, I.
1983 : Oceanic magnetic anomaly amplitudes; Variation with sea-floor spreading rate, and possible implications: Earth and Planetary Science Letters, v. 63, p. 368–378.

Johnson, G. L., and Jakobsson, S. P.
1985 : Structure and petrology of the Reykjanes Ridge between 62°00′N and 63°48′N: Journal of Geophysical Research, v. 90, p. 10073–10083.

LeDouaran, S., and Francheteau, J.
1981 : Intermediate to long-wavelength axial depth anomalies from 10° to 50°N along the Mid-Atlantic Ridge; Correlation with other mantle properties: Earth and Planetary Science Letters, v. 54, p. 29–47.

Macdonald, K. C.
1977 : Near-bottom magnetic anomalies, asymmetric spreading, oblique spreading, and tectonics of the Mid-Atlantic Ridge near 37°N: Geological Society of America Bulletin, v. 88, p. 541–555.

Macdonald, K. C., and Fox, P. J.
1983 : Overlapping spreading centres; New accretion geometry on the East Pacific Rise: Nature, v. 302, p. 55–58.

McGregor, B. A., Harrison, C.G.A., Lavelle, J. W. and Rona, P. A.
1977 : Magnetic anomaly patterns on Mid-Atlantic Ridge crest at 26°N: Journal of Geophysical Research, v. 82, p. 231–238.

Minster, J. G., and Jordan, T. H.
1978 : Present-day plate motions: Journal of Geophysical Research, v. 83, p. 5331–5354.

Morgan, W. J.
1981 : Hotspot tracks and the opening of the Atlantic and Indian Oceans; in Emiliani, C., ed., The Sea, v. 7—The Oceanic Lithosphere: New York, John Wiley and Sons, p. 443–488.

Morgan, W. J.
1983 : Hotspot tracks and the early rifting of the Atlantic: Tectonophysics, v. 94, p. 123–139.

Nunns, A.
1982 : Structure and evolution of the Jan Mayen Ridge and surrounding areas: Tulsa, American Association of Petroleum Geologists Memoir 34, p. 193–208.

Oldenburg, D. W., and Brune, J. N.
1972 : Ridge transform fault spreading pattern in freezing wax: Science, v. 178, p. 301–304.
Perry, R. K., Fleming, H. S., Cherkis, N. Z., Feden, R. H., and Vogt, P. R.
1980 : Bathymetry of the Norwegian-Greenland and Western Barents Sea (map): Geological Society of America, MC-21.
Perry, R. K., Fleming, H. S., Vogt, P. R., Cherkis, N. Z., Feden, R. H., Thiede, J., Strand, J. E., and Collette, B. J.
1981 : North Atlantic Ocean; Bathymetry and plate tectonic evolution: Geological Society of America, MC-35.
Ramberg, I. B., Gray, D. F., and Raynolds, R.G.H.
1977 : Tectonic evolution of the FAMOUS area of the Mid-Atlantic Ridge, lat. 35°50′ to 37°20′N: Geological Society of America Bulletin, v. 88, p. 609–620.
Reid, I., and Jackson, H. R.
1981 : Oceanic spreading rate and crustal thickness: Marine Geophysical Research, v. 5, p. 165–172.
Roest, W. R., Searle, R. C., and Collette, B. J.
1984 : Fanning of fracture zones and a three-dimensional model of the Mid-Atlantic Ridge: Nature, v. 308, p. 527–531.
Rona, P. A.
1980 : The central North Atlantic Ocean Basin and continental margin; Geology, geophysics, geochemistry, and resources, including the Trans-Atlantic Geotraverse (TAG): Miami, NOAA Atlas 3 Environmental Research Laboratories, National Oceanic and Atmospheric Administration, 99 p.
Rona, P. A., and Gray, D. F.
1980 : Structural behavior of fracture zones symmetric and asymmetric about a spreading axis; Mid-Atlantic Ridge (latitude 23°N to 27°N): Geological Society of America Bulletin, v. 91, p. 485–494.
Sandwell, D. T.
1986 : Thermal stress and the spacings of transform faults: Journal of Geophysical Research (in press).
Schilling, J.-G.
1985 : Upper mantle heterogeneities and dynamics: Nature, v. 314, p. 62–67.
Schilling, J.-G., Thompson, G., Kingsley, R., and Humphris, S.
1985 : Hotspot-migrating ridge interaction in the South Atlantic: Nature, v. 313, p. 187–191.
Schouten, H., and Denham, C. R.
1979 : Modeling the oceanic magnetic source layer; in Talwani, M., Harrison, C. G., and Hayes, D. E., eds., Deep Drilling Results in the Atlantic Ocean: Ocean Crust: American Geophysical Union, Maurice Ewing Series 2, p. 151–159.
Schouten, H., and Klitgord, K. D.
1982 : The memory of the accreting plate boundary and the continuity of fracture zones: Earth and Planetary Science Letters, v. 59, p. 255–266.
Schouten, H., and White, R. S.
1980 : Zero-offset fracture zones: Geology, v. 8, p. 175–179.
Schouten, H., Klitgord, K. D., and Whitehead, J. A.
1985 : Segmentation of mid-ocean ridges: Nature, v. 317, p. 225–229.
Searle, R. C.
1978 : Oblique spreading and fracture zones: Nature, v. 274, p. 187–188.
Shih, J.S.F., Atwater, T., and McNutt, M.
1978 : A near-bottom geophysical traverse of the Reykjanes Ridge: Earth and Planetary Science Letters, v. 39, p. 75–83.
Sibuet, J. C., and Mascle, J.
1978 : Plate kinematic implications of Atlantic equatorial fracture zone trends: Journal of Geophysical Research, v. 83, p. 3401–3421.
Sørnes, A., and Fjeldskaar, W.
1980 : The local seismicity in the Jan Mayen area: Norsk Polarinstitutt, Skrifter Nr. 172, p. 21–32.
Stein, S.
1978 : A model for the relation between spreading rate and oblique spreading: Earth and Planetary Science Letters, v. 39, p. 313–318.
Stein, S., Melosh, H. J., and Minster, J. B.
1977 : Ridge migration and asymmetric sea-floor spreading: Earth and Planetary Science Letters, v. 36, p. 51–62.
Stocks, T., and Wüst, G.
1935 : Die Tiefenverhältnisse des offenen Atlantischen Ozeans; Deutsche Atlantische Expedition *Meteor,* 1925–27: Wissenschaftliche Ergebnisse, v. 3, Teil 1, 1. Lieferung, 31 p.
Sundvik, M. T.
1986 : Plate tectonic framework of western North Atlantic derived from Keathley sequence magnetic anomaly pattern: Journal of Geophysical Research, in press.
Tucholke, B. E., and Ludwig, W. J.
1982 : Structure and origin of the J Anomaly Ridge, western North Atlantic Ocean: Journal of Geophysical Research, v. 87, p. 9389–9407.
Tucholke, B. E. and Schouten, H.
1985 : Global plate motion changes recorded in the Kane Fracture Zone: Abstracts with Programs, Geological Society of America, v. 17, no. 7, p. 737.
Turcotte, D. L.
1974 : Are transform faults thermal contraction cracks?: Journal of Geophysical Research, v. 79, p. 2573–2577.
Twigt, W., Verhoef, J., Rohr, K., Mulder, Th.F.A., and Collette, B. J.
1983 : Topography, magnetics and gravity over the Kane Fracture Zone in the Cretaceous Magnetic Quiet Zone (African Plate): Proceedings of the Koninklijke Nederlandse Akademie van Wetenschappen, Series B, v. 86, p. 181–210.
Vink, G. E.
1984 : A hotspot model for Iceland and the Vøring plateau: Journal of Geophysical Research, v. 89, p. 9949–9959.
Vogt, P. R.
1974 : The Iceland phenomenon; Imprints of a hotspot on the ocean crust, and implications for flow below the plates; in Kristjansson, L., ed., Geodynamics of Iceland and the North Atlantic Area: NATO Advanced Study Institute Series, Reidel, Dordrecht, p. 49–62.
1976 : Plumes, sub-axial pipe flow, and topography along the mid-oceanic ridge: Earth and Planetary Science Letters, v. 29, p. 309–325.
1979 : Global magmatic episodes; New evidence and implications for the steady-state mid-oceanic ridge: Geology, v. 7, p. 93–98.
1983 : The Iceland mantle plume; Status of the hypothesis after a decade of new work; in Bott, M.P.H., Saxov, B., Talwani, M., and Thiede, J., eds., Structure and Development of the Greenland-Scotland Ridge; Plenum Publishing Co., p. 191–213.
1986a: Seafloor Topography, Sediments, and Paleoenvironments, Ch. 10; in Hurdle, B. G., ed., The Nordic Seas: New York, Springer-Verlag, p. 237–410.
1986b: Geophysical and geochemical signatures and plate tectonics; Ch. 11, in Hurdle, B. G., ed., The Nordic Seas; New York, Springer-Verlag, p. 413–662.
Vogt, P. R., and Avery, O.
1974 : Detailed magnetic surveys in the northeast Atlantic and Labrador Sea: Journal of Geophysical Research, v. 79, p. 363–389.
Vogt, P. R., and Perry, R. K.
1981 : North Atlantic Ocean: Bathymetry and plate tectonic evolution: Geological Society of America, Text to accompany map MC-35, 21 p.
Vogt, P. R., Avery, O. E., Anderson, C. N., Bracey, D. R., and Schneider, E. D.
1969 : Discontinuities in sea-floor spreading: Tectonophysics, v. 8, p. 285–317.
Vogt, P. R., Kovacs, L. C., Johnson, G. L., and Feden, R. H.
1979 : The Eurasia Basin; in Norwegian Sea Symposium: Oslo, Norwegian Petroleum Society, paper NSS/3, 29 p.
Vogt, P. R., Johnson, G. L., and Kristjansson, L.
1980 : Morphology and magnetic anomalies north of Iceland: Journal of Geophysics, v. 47, p. 67–80.
Vogt, P. R., Kovacs, L. C., Bernero, C., and Srivastava, S. P.
1982 : Asymmetric geophysical signatures in the Greenland-Norwegian and

south Labrador seas and the Eurasian basin: Tectonophysics, v. 89, p. 95–160.

Vogt, P. R., Cherkis, N. Z., and Morgan, G. A.
1983 : Project Investigator - I; Evolution of the Australia-Antarctic Discordance deduced from a detailed aeromagnetic study; in Oliver, R. L., James, P. R., and Jago, J. B., Antarctic Earth Science: Canberra, Australian Academy of Science, p. 608–613.

Wade, J. A., Grant, A. C., Sanford, B. V., and Barss, M. S.
1977 : Basement Structure—Eastern Canada and adjacent areas (4 sheets), 1400A: Ottawa, Geological Survey of Canada.

Wilson, J. T.
1965 : A new class of faults and their bearing on continental drift: Nature, v. 207, p. 343–347.

Wilson, D. S., and Hey, R. N.
1981 : The Galapagos axial magnetic anomaly; Evidence for the Emperor event within the Brunhes and for a two-layer magnetic source: Geophysical Research Letters, v. 8, p. 1051–1054.

MANUSCRIPT ACCEPTED BY THE SOCIETY FEBRUARY 28, 1986

ACKNOWLEDGMENTS

I thank M. Peters (programming), I. Jewett (illustrations), and M. Whitney (word processing). Reviewed by P. J. Fox and J. G. Schilling.

Printed in U.S.A.

The Geology of North America
Vol. M, The Western North Atlantic Region
The Geological Society of America, 1986

Chapter 13

Gravity anomalies in the western North Atlantic Ocean

Philip D. Rabinowitz
Department of Oceanography, Texas A&M University, College Station, Texas 77843, and Lamont-Doherty Geological Observatory of Columbia University, Palisades, New York 10964
Woo-Yeol Jung
Department of Oceanography, Texas A&M University, College Station, Texas 77843

INTRODUCTION

The first accurate gravity measurements in the deep ocean were made in 1923 by Vening Meinesz with pendulum apparatus onboard submarines (Vening Meinesz, 1932). Since that time, and up until the late 1950s, many pendulum stations were occupied in order to obtain information about the earth's gravitational field over the vast oceanic regions. Though these measurements were accurate (generally 3 or 4 mgal; Ewing and others, 1957) and yielded valuable information concerning the earth's crust, upper mantle, and isostatic processes, the measurements were discrete and separated by large distances. For this reason, the construction of detailed gravity contour maps of large oceanic areas was virtually impossible. In 1958, continuous gravity measurements at sea aboard surface ships were initiated onboard, U.S.S. *Compass Island* (Worzel, 1959). Subsequently, gravity measurements have been routinely collected by several of the academic oceanographic institutions and industrial and government agencies.

During the 1960s, major technological advances enhanced the reliability of shipboard gravity measurements. These advances included the development of the satellite navigation system (Guier, 1966; Talwani and others, 1966a) to more reliably compute the Eötvös correction and theoretical gravity values, the development of cross-coupling and off-leveling computers to measure errors resulting from shipboard accelerations (Talwani and others, 1966b), the design of reliable temperature control systems for vibrating string instruments (Bowin and others, 1972), and the modification of the LaCoste-Romberg gravity meter for mounting on a stabilized platform (LaCoste, 1967). Accompanying these technological advances, computer systems were developed for rapid reduction, display, and interpretation of the gravity data (Bernstein and Bowin, 1963; Talwani, 1969). Gravity measurements have also been taken aboard aircraft (Thompson and others, 1960) but currently are virtually nonexistent for the oceanic realm. These measurements have promise for the future, especially for the more remote oceanic areas when the technological problems resulting from the aircraft motions are resolved (Brozena, 1984).

As a result of the technological advances, sufficient ocean gravity measurements have been accumulated to contour gravity maps over major ocean basins (e.g., Bowin and others, 1982). The surface ship gravity maps provide a means for determining the short-period undulations of the geoid over oceanic regions (Talwani and others, 1972) and are invaluable, especially when combined with other geophysical methods (e.g., magnetics, seismic reflection, and refraction), in providing information relating to the earth's structure and geologic history.

The recent developments in radar altimetry from satellites have allowed measurements of the sea surface elevation to within approximately 10 cm (Tapley and others, 1982). Since the main cause of the sea surface undulations is variations in the gravitational attraction due to the distribution of mass beneath the sea floor, these measurements will prove to be powerful for the construction of gravity maps at sea (Haxby, 1982; Vogt, this volume, Ch. 14; frontispiece).

REGIONAL GRAVITY ANOMALIES IN THE NORTH ATLANTIC OCEAN

Talwani and LePichon (1969) constructed regional free-air gravity anomaly maps of the North Atlantic Ocean by averaging surface ship gravity data over 20° and 5° squares, and they compared these maps with those from orbital motions of satellites (Kaula, 1966) and with sea-floor topography. They noted that a general correlation exists between topography and regional free-air gravity anomalies. East-west asymmetries are present in the gravity field, which they concluded arise from nonisostatic lateral inhomogeneities in the upper mantle.

Rabinowitz, P. D., Jung, W-Y., 1986, Gravity anomalies in the western North Atlantic Ocean; *in* Vogt, P. R., and Tucholke, B. E., eds., The Geology of North America, Volume M, The Western North Atlantic Region: Geological Society of America.

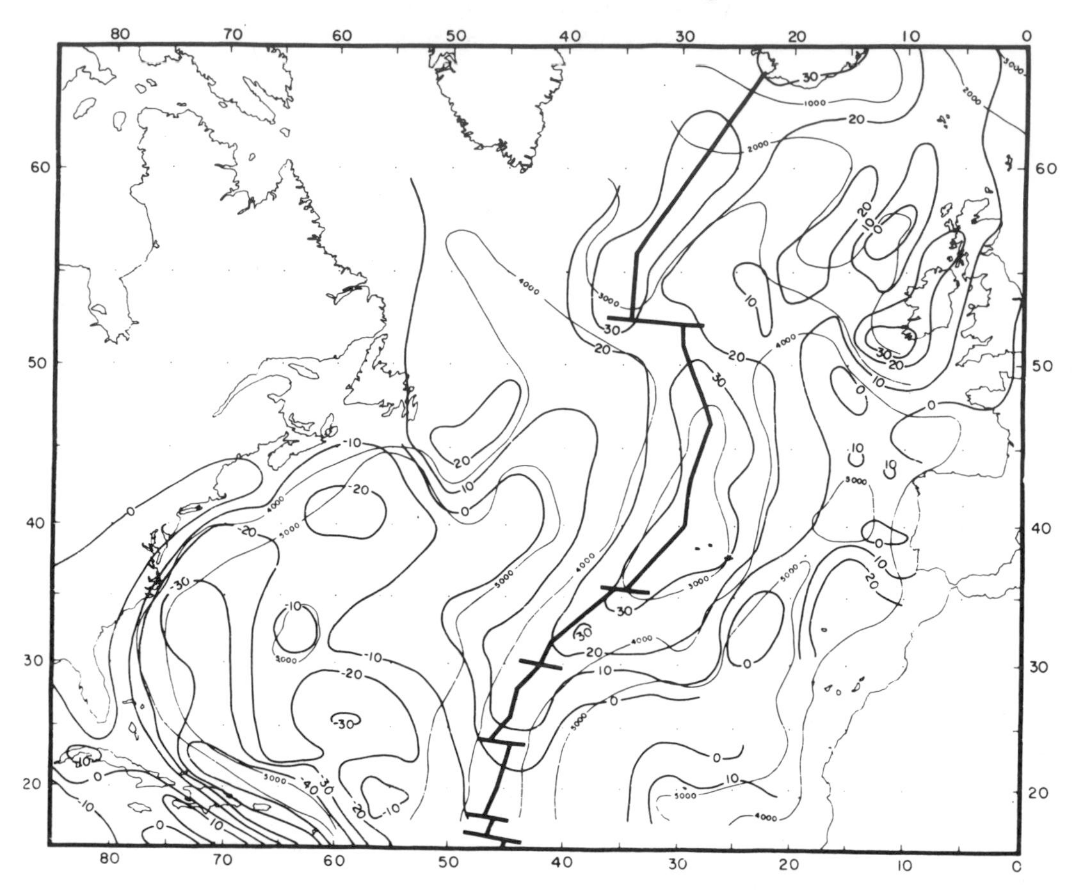

Figure 1. Map of 5° × 5° free-air gravity anomalies (10-mgal contour interval) and depths (1000-m contour interval) in the North Atlantic Ocean (after Cochran and Talwani, 1977).

Sclater and others (1975) suggested that the residual elevation anomalies (predicted depth for that age crust minus actual depth corrected for sediment loading) in the North Atlantic Ocean correlate well with variations in the long-wavelength (5° × 5°) free-air gravity field. Though some age versus gravity correlation exists, Cochran and Talwani (1977) utilizing a larger data base demonstrated that this correlation is not a general one and does not hold for the entire ocean. They noted that there is a 3000-m difference in ridge-crest depth between the Azores and Iceland, with only little change in the gravity field. They further noted that the large regional negative (more negative than –30 mgal) gravity anomaly in the western North Atlantic basin has no reflection in the residual depth anomaly. Cochran and Talwani (1977) concluded that the only places where a correlation exists between the regional and free-air gravity anomalies and residual depth anomalies are areas where extensive volcanism has occurred away from the Mid-Atlantic Ridge system.

In Figure 1 we show a 5° × 5° average map of free-air gravity anomalies and depths for the North Atlantic Ocean (after Cochran and Talwani, 1978). The most prominent features of this map are:

1. The western basin is characterized by regional negative free-air gravity anomalies with values, in places, more negative than –30 mgal. This contrasts with the eastern basin where free-air gravity values are close to 0 mgal. The negative anomalies do not appear to be related to surface features nor to residual depth anomalies (except near Bermuda).

2. Near 30°N, a rather distinct change in the free-air gravity field is observed in the vicinity of the Mid-Atlantic Ridge. North of 30°N the free-air gravity values are more positive than to the south and are associated with shallower water depths. Cochran and Talwani (1978) concluded that in order to account for the gravity anomalies in the shallow ridge area to the north of 30°N, a portion of the compensation must be distributed to depths of several hundred kilometers (within the asthenosphere) and that this mass deficiency is relatively uniform throughout the area. However, within this region, the more local features are compensated at much shallower depths within the lithosphere. South of ~30°N the Mid-Atlantic Ridge is typical of other mid-ocean ridges and is compensated at depths less than ~100 km.

3. Large negative free-air gravity anomalies (more negative than –40 mgal) are associated with the Puerto Rico Trench area. The locations of extensive volcanic activity away from the Mid-Atlantic Ridge, such as the Bermuda Rise, the Cape Verde Is-

lands, the Canary Islands, and Madeira, are all associated with relative regional (1000 to 2000-km wavelengths) free-air gravity highs.

FREE-AIR GRAVITY ANOMALY MAP OF THE WESTERN NORTH ATLANTIC OCEAN

We show in Figure 2 the locations of all gravity measurements used in compiling the 20-mgal gravity map given in Plate 4 (in pocket inside back cover). A source of error in constructing regional gravity maps concerns interpolating data in regions where relatively few gravity measurements have been made. Investigators have developed methods for interpolating gravity data in poorly surveyed regions utilizing the knowledge that small wavelength free-air anomalies arise principally from variations in water depths (Talwani and others, 1972; Fig. 3). McKenzie and Bowin (1976) applied methods of time series analysis and filtering to gravity and bathymetric observations and demonstrated that isostatic compensation generally becomes significant only when wavelength exceeds 100 km and becomes more pronounced with increasing wavelength. Thus, the shorter wavelengths should correlate well with topography. The bathymetry of the northwest Atlantic Ocean is reasonably well known. The Naval Research Laboratory charts (Perry and others, 1981; Vogt and Perry, 1981) have been utilized here as an aid in interpolating the gravity data. Excellent reviews of methods in errors in marine gravity measurements are given in Dehlinger and Chiburis (1972) and Dehlinger (1978).

The major physiographic provinces in the northwest Atlantic Ocean are discussed by Tucholke and others (this volume). Their corresponding gravity fields (Pl. 4) are:

The Mid-Atlantic Ridge and Adjacent Western Basins

The western North Atlantic basins have lows with minima more negative than –40 mgal. As we proceed east from the basins, the gravity values become less negative and reach positive values over the Mid-Atlantic Ridge. Relative negative values are present over the median rift valley. Further, a large north-south variation in the absolute values of the free-air gravity anomalies exists in the Mid-Atlantic Ridge area. As noted earlier, north of 30°N the regional gravity values (>100 km wavelengths) over the Mid-Atlantic Ridge are much more positive than to the south. Local, shorter wavelength gravity anomalies are observed within these regional fields. In the basins, the local anomalies are present with peak to trough amplitudes of 20 mgal; closer to the ridge the peak to trough local anomalies are generally greater (~40 to 80 mgal). These local gravity anomalies show a dependence on bottom topography, as noted earlier, and on subbottom topography as observed on seismic reflection profiler records (Rabinowitz, 1973). The smaller amplitudes of the local anomalies over the basins as compared to the ridge areas result from deeper basement topography in the basins as well as from the effects of density contrasts being less between sediments and basement rocks in the basins compared to water and basement rocks in the ridge areas. Further, the gravity field can show trends in basement morphology that are not necessarily manifested in the bathymetry, such as the oblique basement ridges observed south of Iceland (Vogt, 1983).

The bathymetry (Tucholke and others, this volume) shows the segmented nature of the Mid-Atlantic Ridge crest. It can be interpreted as a succession of relatively short spreading centers, separated and offset at 50-km intervals by mostly small offset transform zones. The spacing of available bathymetric data is, in places, too wide to accurately register fracture zone morphology that occurs at spacings less than ~50 km. In addition, the faulted "normal" sea floor in the slow spreading North Atlantic Ocean makes it, in most instances, difficult to identify fracture zones bathymetrically. We are thus confronted with a high noise level and pronounced spatial aliasing in identifying small offset fracture zones from available bathymetric data. The gravity presents a smoothed (low pass filter) version of the bathymetry for the mid-ocean ridge area and thus rather nicely displays some of the small fracture zone traces. In some cases, the gravity displays the continuation of fracture zones where the bathymetry does not. One example is the eastern extension of Fracture Zone B at 36.5°N, where a relative gravity anomaly of ~40 mgal is present without a major topographic expression. The major fracture zone traces such as Kane, Atlantis, Oceanographer, and Charlie Gibbs Fracture Zones are manifested as pronounced long continuous belts of predominantly negative gravity anomalies with peak to trough amplitudes greater than 100 mgal in places.

Within the Labrador Sea, seismic reflection profiles by Hinz and others (1979) show that relatively thick sediment buries an inferred extinct median rift valley. The free-air gravity anomalies here exhibit a relative low about –20 mgal which trends northwesterly around the center of the Labrador Sea. This negative trend conforms well with the extinct rift valley (Drake and others, 1963; Srivastava, 1978, 1979).

Only a few good crustal models deduced from gravity measurements exist over oceanic fracture zones and rift valleys. Cochran (1973) and Robb and Kane (1975) suggested that only about half of the gravity anomaly over the Romanche and Vema Fracture Zones can be assigned to topographic relief. They imply the presence of dense ultrabasic rocks at shallow depths under the fracture zones. Sibuet and others (1974) and Sibuet and Mascle (1978) have very nicely demonstrated that the variations in fracture zone gravity anomalies are associated with the variations in the density structure predicted by the thermal evolution of lithospheric plates. Collette and others (1980) showed that the negative free-air anomaly under the median valley arises in part from topography and in part from a mass deficiency beneath the floor of the median valley. The relative positive anomalies on the walls of the median valley are explained for the most part by a direct topographic effect. They suggested that the median valley is below its isostatic equilibrium position and that its walls are upheaved.

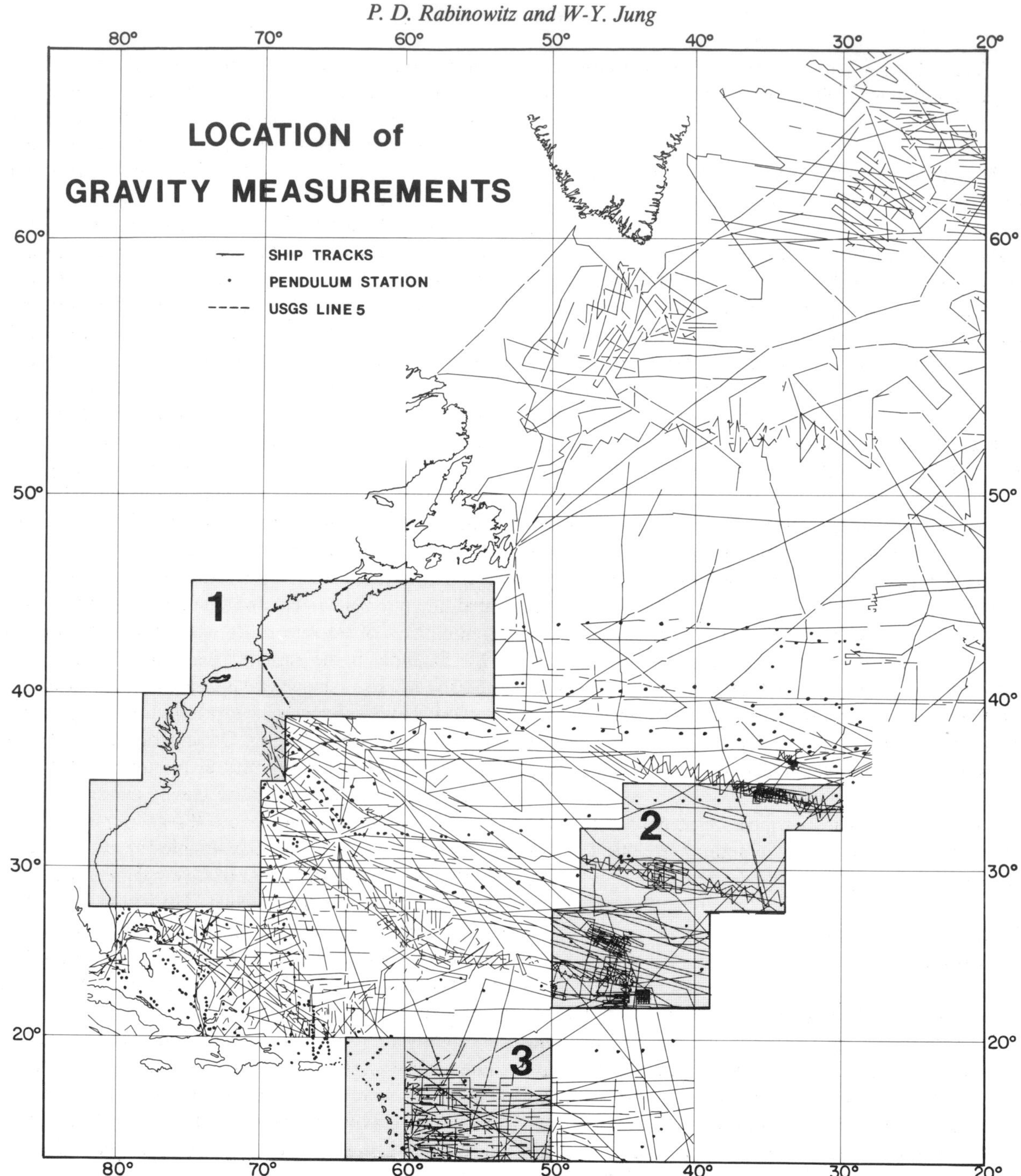

Figure 2. Locations of gravity measurements used to construct gravity map in Plate 4. The source of these measurements includes Lamont-Doherty Geological Observatory, Woods Hole Oceanographic Institution, National Oceanographic and Atmospheric Administration, National Geophysical and Solar-Terrestrial Data Center, Hawaii Institute of Geophysics, U.S. Naval Oceanographic Office, Vening Meinesz Laboratory (Netherlands), and P.O. Shrishow Institute of Oceanography (USSR). Other data were obtained from tabulations and existing free-air gravity maps, including Bowin, 1976; Bowin and others, 1969; Bowin and others, 1982; Dorman and others, 1973; Emery and others, 1970; Ewing (in press); Goodacre, 1964; Grow and others, 1976a; Grow and others, 1979a; Hinz and others, 1979; Hussong and others, 1978; Jacobi and Kristofferson, 1976; Keen and others, 1971; LeBis, 1975; Olivet and others, 1974; Orlin and others, 1965; Purdy and others, 1978; Rabinowitz, 1973, 1974, 1981; Rabinowitz and Ludwig, 1975, 1980; Rogan and Rabinowitz (in press); Rona, 1980; Sobczak, 1975; Srivastava, 1979; Talwani, 1959; Taylor and Greenewalt, 1976; van der Linden and Srivastava, 1974; Vening Meinesz, 1948; Westbrook (in press); Woodside, 1972; and Worzel, 1965. Shaded areas are regions where previous gravity maps of large areas were incorporated [1 after V. Ewing (in press); 2 after Rogan and Rabinowitz (in press); and 3 after Westbrook (in press)].

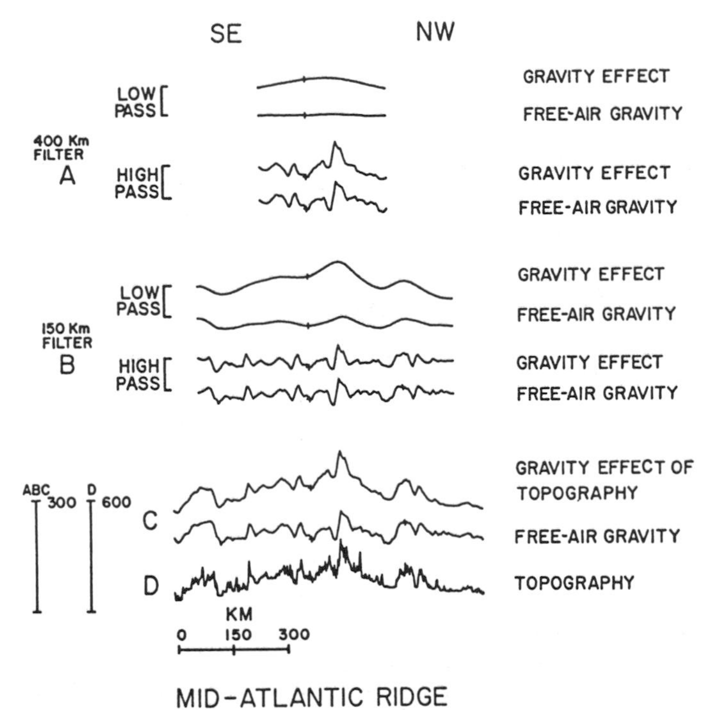

Figure 3. The bottom curve gives the topography over a section of the north Mid-Atlantic Ridge (Scale D is in meters). The measured free-air gravity curve is immediately above the topography curve (Scales A, B, and C are in milligals). The gravity effect of topography is computed using a density contrast of 2.67-1.03 g/cm^3. Both the free-air gravity and the gravity effect of topography are filtered, using two different Gaussian filters of widths 150 and 400 km. The filters are designed in such a way that the wavelength of unit amplitude after being filtered through the low pass Gaussian filter of width w has an amplitude exp $[-0.55\,(w^2/\lambda^2)]$. For example, for $\lambda = 2w$ and $w/2$ the respective filtered amplitudes are 0.96 and 0.11. We notice that for wavelengths of the order 400 km, the free-air gravity can be predicted well from topography in the region of the Mid-Atlantic Ridge (after Talwani and others, 1972).

Blake-Bahama Outer Ridge

The Blake-Bahama outer ridge, a sedimentary depositional feature (Ewing and Ewing, 1964), is a major topographic feature extending for over 1100 km to the southeast of the northern part of the Blake Plateau and has a maximum relief greater than 2000 m. A free-air gravity high is centered over the axis of maximum sediment deposition and attains a magnitude of 30 mgal. This gravity high is about 80 mgal greater than the regional value observed in the surrounding basin (i.e., the Hatteras abyssal plain and Blake-Bahama basin).

Bahamas

The physiography of the Bahamas is characterized by shallow banks broken by relatively narrow deep tongues or channels (Heezen and others, 1959; Uchupi and others, 1971). Positive free-air anomalies are generally associated with the Bahama Banks. These values increase from west to east and in places exceed 100 mgal. The channels are characterized by large negative anomalies with very steep gravity gradients between the banks and the channels. In particular, the Tongue of the Ocean and the NE Providence Channel have free-air anomalies more negative than −100 mgal. Gravity computations by Talwani (1959) indicate that the upper part of the crust forming the channels is composed of low-density material and that the channels are uncompensated locally.

Bermuda Rise

The Bermuda Rise is a broad, roughly elliptical swell with its long axis oriented northeast-southwest. Free-air gravity values more positive than the surrounding basins are observed over the Bermuda Rise. Many closures of the free-air contours are observed over the rise, which may be associated with variations in basement topography and sediment thickness. Isostatic geoid computations utilizing a Pratt model, over the Bermuda Rise suggest that it is compensated in the upper 100 km (Haxby and Turcotte, 1978).

The island of Bermuda, an atoll that is built on a volcanic peak, lies on the northeast by southwest flat-topped Bermuda Pedestal, roughly 90 to 150 km wide at its base and characterized by high positive free-air gravity anomalies. Free-air anomalies more positive than 350 mgal are associated with Bermuda; it is the largest positive anomaly observed in the Atlantic Ocean.

New England Seamounts

The New England Seamount chain, a row of extinct volcanoes, forms a belt 1200 km long from the upper continental rise off Cape Cod to beyond the northern edge of the Bermuda Rise. Included are at least 32 seamounts ranging in height from 400 to more than 4000 m (Uchupi and others, 1970). Numerous hypotheses have been given in the literature concerning the origin of this seamount chain. These include primarily fracture-zone traces (Drake and Woodward, 1963; Uchupi and others, 1970; Le Pichon and Fox, 1971; Pitman and Talwani, 1972) or hot-spot traces (Duncan, 1982; Morgan, 1981).

These seamounts are well defined gravimetrically. Free-air anomalies more positive than 100 mgal are common. The New England Seamount chain does not show a major expression in the 5° × 5° regional gravity map (Fig. 1). A small relative high is observed in the regional 1° × 1° gravity map (Rabinowitz, 1973; Cochran and Talwani, 1978). The local gravity highs most probably represent primarily the effect of local topographic relief. The negative gravity "moats" surrounding the seamounts may imply that the lithosphere beneath the seamounts bends due to loading, similar to observations in the Hawaiian-Emperor Seamount chain (Watts and Cochran, 1974).

Greater and Lesser Antilles

On the northern side of this island arc the free-air gravity anomalies over the Puerto Rico trench area exhibit the wide range in amplitudes common to other trenches (Westbrook, 1982, and in press). To the north of Puerto Rico, values more negative than –350 mgal are observed; these are the most negative free-air gravity values in the North Atlantic Ocean. On the adjacent Greater Antilles, free-air gravity anomalies more positive than 200 mgal are observed. Over the eastern part of the island arc the gravity anomalies also exhibit a wide range (150 to –250 mgal). Bowin (1976) concluded that the east-trending negative gravity anomaly belts along the Puerto Rico trench are a result of continuing compression across a zone of transform faulting, and that the negative belt east of the Lesser Antilles island arc is a result of underthrusting of the Atlantic plate beneath the island arc.

Continental Margin of Eastern North America

A prominent feature of the free-air gravity map is the gravity high which roughly parallels the east coast of North America at the seaward edge of the continental shelf. It is a continuous high, ranging in magnitude from values near 0 mgal to values more positive than 75 mgal. Seaward of the shelf-edge high, a trend of negative anomalies is observed on the continental slope and rise, with magnitudes ranging from about –25 mgal to values more negative than –125 mgal in the region of the precipitous Blake escarpment in the south. A rather broad wavelength (100 to 150 km) and small amplitude (5 to 20 mgal) relative gravity high is observed seaward of the continental slope/rise gravity low. Isostatic computations (Rabinowitz, 1973, 1974) demonstrate that the free-air shelf-edge high is not totally a simple "edge effect" resulting from the topography and its compensation. An isostatic gravity anomaly high is present close to the shelf edge.

Gravity maps by Grow and others (1976, 1979a) and Ewing (in press), published at smaller contour intervals for large parts of the eastern margin of the United States, allow inferences to be made with respect to fracture zone intersections with the coastlines. In particular, the locations where the shelf-edge gravity high changes rather abruptly in magnitude have been shown by Grow and others (1979a) to correspond closely with the boundaries between platform and basin areas on the continental shelf.

Numerous models have been given in the literature to explain the gravity anomalies bordering passive continental margins. The early gravity models suggested that the continental margins were in near isostatic equilibrium and that the "transition zone" between continental and oceanic thicknesses has a width between 50 to 300 km (e.g., Worzel, 1968). The isostatic anomalies were accounted for by changes in the crust-mantle interface.

Keen and Loncarevic (1966) and Scrutton (1979), among others, have presented crustal models in which vertical and/or lateral density changes have been assumed in the mantle to account for the gravity anomalies. Emery and others (1975) have modeled the gravity data with lateral changes in the crustal densities. Rabinowitz (1974) and Talwani and Eldholm (1972) have suggested intrabasement crustal density highs to account for parts of the observed gravity anomalies. Gravity models by Grow and

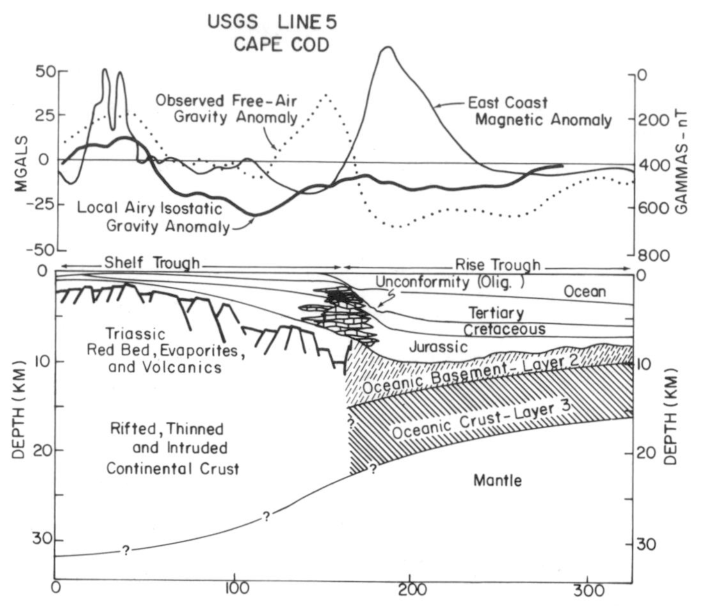

Figure 4. Schematic cross-section summarizing seismic measurements together with free-air and isostatic gravity and magnetic anomalies (modified after Grow and others, 1979b). Locations in Figure 2. This line crosses the eastern end of the Long Island Platform and is described by Grow and others, (1979b) as typical of a narrow transition zone between continental and oceanic crust.

others (1979a, 1979b), utilizing extensive multichannel seismic measurements (Schlee and others, 1976; Grow and Schlee, 1976) and refraction measurements (Sheridan and others, 1979), have suggested that the oceanic crust is ~11 to 18 km in thickness near the ocean/continent boundary as defined by the prominent east coast magnetic anomaly. A composite schematic model, showing crustal structure as well as magnetic and gravity anomalies, is shown in Figure 4 (after Grow and others, 1979b). Further, Grow and others (1979a) have noted that in the thickly sedimented margins, such as the east coast of the United States, the lateral density variations within sedimentary rocks may be an important effect.

Walcott (1972) modeled the gravity anomalies, flexural stress, and vertical displacement that can be expected at a continental edge due to loading by large accumulations of sediments. This flexural model assumes homogeneous elastic constants for both oceanic and continental plates and explains the shape and the thickness (up to 18 km) of sediments on wide delta-type passive margins. Turcotte and others (1977) extended the flexural model by allowing for sediment loading of decoupled oceanic and continental crust. They concluded that the observed free-air gravity anomalies were most satisfactorily explained by a continental margin fault system that remains active during much of the evolution of the continental margin. The continental margin isostatic anomaly, when associated with a magnetic-edge effect anomaly, has been modeled by Rabinowitz and LaBrecque (1977, 1979) as resulting from elevated thickened oceanic crust adjacent to the continent. These highs have been attributable to relicts of a transient phenomena of a higher ridge axis elevation during the early separation of the continents. Karner and Watts (1982) interpreted the same continental margin isostatic anomaly in terms of a flexure model of isostacy in which the flexural strength of the lithosphere increases with age. Clearly, more knowledge about the upper crustal and deeper structure is necessary to unequivocally determine the origin of the gravity anomalies here and learn more with respect to the zone between oceanic and continental basement.

CONCLUSIONS

The accuracy of gravity measurements at sea has been increased substantially since their inception. We now can produce contour maps of large oceanic areas with reliabilities better than ±5 mgal. A free-air gravity map has been compiled and described in detail for the western North Atlantic at a 20-mgal contour interval. The major observations are:

1. Negative regional free-air gravity anomalies (more negative than −30 mgal) that do not appear related to surface features or residual depth anomalies are observed over the basins in the western North Atlantic Ocean.

2. Local short-wavelength anomalies that are closely associated with bathymetry of subbottom basement features are present within the basins and the Mid-Atlantic Ridge area.

3. A marked change in the gravity field over the Mid-Atlantic Ridge with more positive gravity values observed north of 30°N is attributed to a portion of the compensation distributed to depths of several hundred kilometers.

4. Areas of extensive volcanic activity such as the Bermuda Rise are regions of relative regional free-air gravity highs.

5. The Mid-Atlantic Ridge rift valley is characterized by relative negative gravity anomalies. These negative anomalies are also observed over the buried extinct rift valley in the Labrador Sea.

6. Fracture zone traces are readily defined by elongate belts of gravity anomalies and/or disturbances in the linear pattern of the ridge crest.

7. The Blake-Bahama Outer Ridge, a major sedimentary ridge, is characterized by positive gravity anomalies.

8. The New England Seamounts are characterized by free-air gravity highs surrounded by negative "moats." These anomalies probably result from the flexure of the lithosphere beneath the seamounts due to its loading.

9. Elongate, very high amplitude free-air gravity anomaly belts are associated with the Greater and Lesser Antilles island-arc systems. These gravity anomalies have the largest peak to trough amplitudes (up to 550 mgal) in the northwest Atlantic Ocean.

10. A free-air gravity high is present at the seaward edge of the continental shelf. Isostatic calculations indicate that this high cannot be totally explained by isostatic edge effects.

REFERENCES

Bernstein, R., and Bowin, C. O.
1963: Real-time digital computer acquisition and computation of gravity data at sea; Institute of Electrical and Electronic Engineers Transactions on Geoscience Electronics, v. GE-1, no. 1, p. 2–10.

Bowin, C. O.
1976: Caribbean gravity field and plate tectonics: Geological Society of America Special Paper 169, p. 1–79.

Bowin, C. O., Aldrich, T. C., and Wertheimer, A.
1969: Gravity data obtained during Chain Cruise 75: Technical Report Reference No. 69-16.

Bowin, C. O., Aldrich, C., and Folinsbee, R. A.
1972: VSA gravity meter system: Tests and recent developments: Journal of Geophysical Research, v. 77, p. 2018–2033.

Bowin, C. O., Warsi, W., and Milligan, J.
1982: Free-air gravity anomaly atlas of the world: Geological Society of America Map and Chart Series MC-46.

Brozena, J. M.
1984: A preliminary analysis of the NRL airborne gravimetry system: Geophysics, v. 49, p. 1060–1069.

Cochran, J. R.
1973: Gravity and magnetic investigations in the Guiann Basin, western equatorial Atlantic: Geological Society of America Bulletin, v. 84, p. 3249–3268.

Cochran, J. R., and Talwani, M.
1977: Free-air gravity anomalies in the world's oceans and their relationship to residual elevation: Geophysical Journal of the Royal Astronomical Society, v. 50, p. 495–552.
1978: Gravity anomalies, regional elevation, and the deep structure of the North Atlantic: Journal of Geophysical Research, v. 83, p. 4907–4924.

Collette, G. J., Verhoef, J., and de Mulder, A.F.J.
1980: Gravity and a model of the median valley: Journal of Geophysics, v. 47, p. 91–98.

Dehlinger, P.
1978: Marine Gravity: New York, Elsevier North Holland, Inc., p. 1–322.

Dehlinger, P., and Chiburis, E. F.
1972: Gravity at sea: Marine geology, v. 12, p. 1–41.

Dorman, L. M., Bassinger, B. G., Bernard, E., Bush, S. A., Dewald, O. E., Lapine, L. A., Lattimore, R. K., and Peter, G.
1973: Caribbean Atlantic Geotraverse: NOIAA-IDOE 1971, Report No. 3, Gravity: NOAA Technical Report, ERL 277-AOML 11, p. 1–35.

Drake, C. L., and Woodward, H. P.
1963: Appalachian curvature, wrench faulting, and offshore structures: Transactions, New York Academy of Science, v. 26, p. 48–63.

Drake, C. L., Campbell, N. J., Sander, G., and Nafe, J. E.
1963: A Mid-Labrador Sea Ridge: Nature, v. 200, no. 4911, p. 1085–1086.

Duncan, R. A.
1982: The New England Seamounts and the absolute motion of North America since mid-Cretaceous time: EOS, American Geophysical Union, v. 63, p. 1103–1104.

Emery, K. O., Uchupi, E., Phillips, J. D., Bowin, C., Bunce, E., and Knott, S.
1970: Continental rise off eastern North America: American Association of Petroleum Geologists Bulletin, v. 54, no. 1, p. 44–108.

Emery, K. O., Uchupi, E., Bowin, C., Phillips, J., and Simpson, E.S.W.
1975: Continental margin off western Africa: Cape St. Francis (South Africa) to Walvis Ridge (South-West Africa): American Association of Petroleum Geologists Bulletin, v. 59, no. 1, p. 3–59.

Ewing, M., and Ewing, J.
1964: Distribution of oceanic sediments: *in* Studies of oceanography: Geophysical Institute, University of Tokyo, Japan, p. 525–537.
Ewing, M., Worzel, J. L., and Shurbet, G. L.
1957: Gravity observations at sea in U.S. submarines Barracuda, Tusk, Conger, Argonaut, and Medregal: Verhandel, Ned. Geol. Mijnbouwk, Genoot.; Geol. Series, v. 18, p. 49–96.
Ewing, V.
In press: Gravity anomalies in the continental margins of eastern North America: in OMD Atlas Areas I, III, and IV; Joint Oceanographic Institutions, Inc.
Goodacre, A. K.
1964: A shipborne gravimeter testing range near Halifax, Nova Scotia: Journal of Geophysical Research, v. 69, p. 5373–5381.
Grow, J. A., and Schlee, J. S.
1976: Interpretation and velocity analysis of U.S. Geological Survey multichannel reflection profile 4, 5, and 6, Atlantic continental margin: U.S. Geological Survey Miscellaneous Field Series, Map MF-808.
Grow, J. A., Bowin, C., Hutchinson, D. R., and Kent, K. M.
1976: Preliminary free-air gravity anomaly map along the Atlantic continental margin between Virginia and Georges Bank: U.S. Geological Survey Miscellaneous Series, Map MF-795.
Grow, J. A., Bowin, C., and Hutchinson, D. R.
1979a: The gravity field of the U.S. continental margin: Tectonophysics, v. 59, p. 27–52.
Grow, J. A., Mattrick, R. E., and Schlee, J. R.
1979b: Multichannel seismic depth sections and interval velocities over outer continental shelf and upper continental slope between cape Hatteras and Cape Cod: *in* Geological and Geophysical Investigations of Continental Margins, eds., J. S. Watkins, L. Montadert, and P. Dickerson; American Association of Petroleum Geologists Memoir 29, p. 65–83.
Guier, W. H.
1966: Satellite navigation using integral Doppler data, the AN/SRN-9 equipment: Journal of Geophysical Research, v. 71, p. 5903–5910.
Haxby, W., and LaBrecque, J.
1982: Geotectonic imagery: The application of Seasat altimetry to the tectonic evolution of the Indo-Atlantic Basin: EOS, American Geophysical Union, v. 63, p. 908.
Haxby, W., and Turcotte, D. L.
1978: On isostatic geoid anomalies: Journal of Geophysical Research, v. 83, p. 5473–5478.
Heezen, B. C., Tharp, M., and Ewing, M.
1959: The floors of the ocean: I. The North Atlantic: Geological Society of America Special Paper 65, p. 1–122.
Hinz, K., Schlüter, H. U., Grant, A. C., Srivastava, S. P., Umpleby, D., and Woodside, J.
1979: Geophysical transects of the Labrador Sea: Labrador to southwest Greenland: Tectonophysics, v. 59, p. 151–183.
Hussong, D. M., Fryer, P. B., Tuthill, J. D., and Wipperman, L. K.
1978: The geological and geophysical setting near DSDP site 395, North Atlantic Ocean: *in* Initial Reports of the Deep Sea Drilling Project, v. 45, eds., W. G. Melson, P. D. Rabinowitz; U.S. Government Printing Office, Washington, D.C., p. 23–28.
Jacobi, R., and Kristoffersen, Y.
1976: Geophysical and geological trends on the continental shelf off northeastern Newfoundland: Canadian Journal of Earth Sciences, v. 13, p. 1039–1051.
Karner, G. D., and Watts, A. B.
1982: On isostacy at Atlantic-type continental margins: Journal of Geophysical Research, v. 87, p. 2923–2948.
Kaula, W. M.
1966: Tests and combination of satellite determinations of the gravity field with gravimetry: Journal of Geophysical Research, v. 71, p. 5303–5314.
Keen, C., and Loncarevic, B. D.
1966: Crustal structure on the eastern seaboard of Canada: Studies on the continental margin: Canadian Journal of Earth Sciences, v. 3, p. 65–76.
Keen, J. J., Loncarevic, B. D., and Ewing, G. N.
1971: Continental margin of eastern Canada: Georges Bank to Kane Basin: *in* The Sea, Volume 4, ed., A. E. Maxwell; p. 251–291.
LaCoste, L.J.B.
1967: Measurement of gravity at sea and in the air; Reviews of Geophysics, v. 5, p. 477–526.
LeBis, A. P.
1975: Investigation of a gravity high—offshore Newfoundland: *in* Canada's Continental Margins and Offshore Petroleum Exploration, eds. C. J. Yorath, E. R. Parker, and D. J. Glass; Canadian Society of Petroleum Geologists Memoir 4, p. 169–180.
LePichon, X., and Fox, P. J.
1971: Marginal offsets, fracture zones and the early opening of the North Atlantic: Journal of Geophysical Research, v. 76, p. 6294–6308.
McKenzie, D. P., and Bowin, C. O.
1976: The relationship between bathymetry and gravity in the Atlantic Ocean: Journal of Geophysical Research, v. 81, p. 1903–1915.
Morgan, W. J.
1981: Hotspot tracks and the opening of the Atlantic and Indian Oceans, Ch. 13: *in* The Sea, Volume 7, ed. C. Emiliani; John Wiley and sons, New York.
Olivet, J. L., LePichon, X., Monti, S., and Sichler, B.
1974: Charlie Gibbs fracture zone: Journal of Geophysical Research, v. 79, p. 2059–2072.
Orlin, H., Bassinger, B. G., and Gray, C. H.
1965: Cape Charles-Wallops Island, Virginia, offshore gravity range: Journal of Geophysical Research, v. 70, p. 6525–6267.
Perry, R. H., Fleming, H. S., Vogt, P. R., Cherkis, N. Z., Feden, R. H., Thiede, J., Strand, J. E., and Collette, B. J.
1981: North Atlantic Ocean: Bathymetry and plate tectonics evolution: Geological Society of America Map and Chart Series MC-35.
Pitman, W. C., III, and Talwani, M.
1972: Sea-floor spreading in the North Atlantic: Geological Society of America Bulletin, v. 83, p. 619–646.
Purdy, G. M., Schouten, H., Crowe, J., Barrett, D. L., Falconer, R.K.H.,
Udintsev, G. B., Marova, N. A., Litvin, V. M., Valyashko, G. M.
1978: AT-6, a site survey; *in* Initial Reports of the Deep Sea Drilling Project, v. 45, eds., W. G. Melson, P. D. Rabinowitz: U.S. Government Printing Office, Washington, D.C., p. 119–120.
Rabinowitz, P. D.
1973: The continental margin of the Northwest Atlantic Ocean: A geophysical study [Ph.D. dissertation]: Columbia University, New York, 1981 p.
1974: The boundary between oceanic and continental crust in the western North Atlantic: *in* The Geology of Continental Margins, eds., C. A. Burk and C. L. Drake; Springer-Verlag, New York, p. 67–84.
1981: Gravity measurements bordering passive continental margins: *in* Dynamics of Passive Margins: ed., R. A. Scrutton; American Geophysical Union, Geodynamics Series, v. 6, p. 91–115.
Rabinowitz, P. D., and LaBrecque, J.
1977: The isostatic gravity anomaly: A key to the evolution of the ocean-continent boundary: Earth and Planetary Science Letters, v. 35, p. 145–150.
1979: The Mesozoic South Atlantic Ocean and evolution of its continental margins: Journal of Geophysical Research, v. 84, p. 5973–6002.
Rabinowitz, P. D., and Ludwig, W. J.
1975: Drill sites proposed for International Phase: Geotimes, v. 20, no. 10, p. 21–23.
1980: Geophysical measurements at candidate drill sites along an east-west flowline in the central Atlantic Ocean: Marine Geology, v. 35, p. 243–275.
Robb, J. M., and Kane, M. F.
1975: Structure of the Vema fracture zone from gravity and magnetic intensity profiles: Journal of Geophysical Research, v. 80, p. 441–445.

Rogan, A. M., and Rabinowitz, P. D.
In press: Gravity anomalies over the Mid-Atlantic Ridge, in OMD Atlas Area VII—Mid-Atlantic Ridge: Joint Oceanographic Institutions, Inc.
Rona, P. A.
1980: The central North Atlantic Ocean basin and continental margins: Geology, geophysics, geochemistry, and resources, including Trans-Atlantic Geotraverse (TAG): NOAA Atlas 3; U.S. Department of Commerce, NOAA, Environmental Research Laboratories, p. 7.
Schlee, J., Behrendt, J. C., Grow, J. A., Robb, J. M., Mattick, R. E., Taylor, P. T. and Lawson, B. J.
1976: Regional geologic framework off northeastern United States: American Association of Petroleum Geologists Bulletin, v. 60, no. 6, p. 926–951.
Sclater, J. G., Lawver, L. A., and Parsons, B.
1975: Comparison of long-wavelength residual elevation and free-air gravity anomalies in the North Atlantic and possible implications for the thickness of the lithospheric plate: Journal of Geophysical Research, v. 80, p. 1031–1052.
Scrutton, R. A.
1979: Structure of the crust and upper mantle at Goban Spur southwest of the British Isles—some implications for margin studies: Tectonophysics, v. 59, p. 201–215.
Sheridan, R. E., Grow, J. A., Behrendt, J. C., and Bayer, K. C.
1979: Seismic refraction study of the continental edge off the eastern United States: *in* Crustal Properties across Passive Margins, ed., C. E. Keen; Tectonophysics, v. 59, p. 1–26.
Sibuet, J. C., and Mascle, J.
1978: Plate kinematic implications of Atlantic equatorial fracture zone trends: Journal of Geophysical Research, v. 83, p. 3401–3421.
Sibuet, J. C., LePichon, X., and Goslin, J.
1974: Thickness of lithosphere deduced from gravity edge effects across the Mendocino Fault: Nature, v. 252, p. 676–3421.
Sobczak, L. W.
1975: Gravity anomalies and passive continental margins, Canada and Norway: *in* Canada's Continental Margins and Offshore Petroleum Exploration, eds., C. J. Yorath, E. R. Parker, and D. J. Glass; Canadian Society of Petroleum Geologists Memoir 4, p. 743–761.
Srivastava, S. P.
1978: Evolution of the Labrador Sea and its bearing on the early evolution of the North Atlantic: Geophysical Journal of the Royal Astronomical Society, v. 52, p. 313–357.
1979: Marine gravity and magnetic anomalies maps of the Labrador Sea: Geological Survey of Canada, Open-File Report No. 627.
Talwani, M.
1959: Gravity anomalies in the Bahamas and their interpretation [Ph.D. thesis]: Columbia University, New York, p. 1–89.
1969: A computer system for the reduction, storage and display of underway data acquired at sea: Technical Report #1, CU-1-69, N00014-67-A-0108-004; Lamont-Doherty Geological Observatory, Palisades, New York.
Talwani, M., and Eldholm, O.
1972: Continental margin off Norway: A geophysical study: Geological Society of America Bulletin, v. 83, p. 3575–3606.
Talwani, M., and LePichon, X.
1969: Gravity field over the Atlantic Ocean: *in* The Earth's Crust and Upper Mantle, ed. P. J. Hart; American Geophysical Union, Washington, D.C., p. 341–351.
Talwani, M., Dorman, J., Worzel, J. L., and Bryan, G. M.
1966a: Navigation at sea by satellite: Journal of Geophysical Research, v. 71, p. 5891–5902.
Talwani, M., Early, W. P., and Hayes, D. E.
1966b: Continuous analog computation and reading of cross-coupling and off-leveling errors: Journal of Geophysical Research, v. 71, p. 2079–2090.
Talwani, M., Poppe, H. R., and Rabinowitz, P. D.
1972: Gravimetrically determined geoid in the western North Atlantic: *in* Sea Surface Topography, Volume 2; NOAA Technical Report ERL-228-AOML7-2, p. 1–34.
Tapley, B. D., Born, G. H., and Parke, M. E.
1982: The SEASAT altimeter data and its accuracy assessment: Journal of Geophysical Research, v. 87, p. 3179–3188.
Taylor, P. T., and Greenewalt, D.
1976: Geophysical transitions across the northwest Atlantic magnetic quiet zone border: Earth and Planetary Science Letters, v. 29, p. 435–446.
Thompson, L.G.D., and LaCoste, L.J.B.
1960: Aerial gravity measurements: Journal of Geophysical Research, v. 65, p. 305–322.
Turcotte, D. L., Ahern, J. L., and Bird, J. M.
1977: The state of stress at continental margins: Tectonophysics, v. 42, p. 1–28.
Uchupi, E., Phillips, J. D., and Prada, K. E.
1970: Origin and structure of the New England Seamounts chain: Deep-Sea Research, v. 17, p. 483–494.
Uchupi, E., Milliman, J. D., Luyendyk, B. P., Bowin, C. O., and Emery, K. O.
1971: Structure and origin of Southeastern Bahamas: American Association of Petroleum Geologists Bulletin, v. 55, p. 687–704.
van der Linden, W.J.M., and Srivastava, S. P.
1974: The crustal structure of the continental margin off central Labrador: Geological Survey of Canada, Paper 7430, p. 233–245.
Vening Meinesz, F. A.
1932: Gravity expeditions at sea, 1923–1930, Volume I: Publication of Netherlands Geodetic Commission, Delft, p. 1–110.
1984: Gravity expeditions at sea, 1932–1938, Volume IV: Publication of Netherlands Geodetic Commission, Delft, p. 1–233.
Vogt, P. R.
1983: The Iceland mantle plume: Status of the hypothesis after a decade: *in* Structure and Development of the Greenland-Scotland Ridge, eds. Bott, Saxov, Talwani, and Thiede; Plenum Publishing Corporation, New York.
Vogt, P. R., and Perry, R. K.
1981: North Atlantic Ocean: Bathymetry and plate tectonics evolution: text to accompany Map and Chart Series MC-35; Geological Society of America.
Walcott, R. I.
1972: Gravity, flexure and growth of sedimentary basins at a continental edge: Geological Society of America Bulletin, v. 83, p. 1845–1948.
Watts, A. B., and Cochran, J. R.
1974: Gravity anomalies and flexure of the lithosphere along the Hawaiian-Emperor Seamount Chain: Geophysical Journal of the Royal Astronomy Society, v. 38, p. 119–141.
Westbrook, G.
In press: Gravity anomalies offshore Barbados, in OMD Atlas Area; Joint Oceanographic Institutions, Inc.
1982: The Barbados Ridge Complex: Tectonics of a mature forearc system: *in* Trench and Forearc Geology: Sedimentation and tectonics in ancient and modern plate margins, ed. J. K. Leggett; Special Publication of the Geological Society of London, no. 10, p. 261–276.
Woodside, J. M.
1972: The Mid-Atlantic Ridge near 45°N: The gravity field: Canadian Journal of Earth Science, v. 9, p. 942–961.
Worzel, J. L.
1959: Continuous gravity measurements on a surface ship with the Graf sea gravimeter: Journal of Geophysical Research, v. 64, p. 1299–1315.
1965: Pendulum gravity measurements at sea: 1936–1959: John Wiley and Sons, New York, p. 1–422.
1968: Advances in marine geophysical research of continental margins: Canadian Journal of Earth Science, v. 5, p. 963–983.

Manuscript Accepted by the Society January 18, 1984

ACKNOWLEDGMENTS

The gravity measurements used in the compilation of the gravity map were collected on numerous cruises aboard research vessels of various institutions. We thank the officers, crew, and scientists aboard the research vessels for their assistance in gathering the data. We are grateful to Carl Brenner, John Bryant, and Fabiola Byrne who assisted in the technical aspect of this manuscript. Financial support for this study was provided through the years by numerous grants from the National Science Foundation and contracts from the Office of Naval Research. This is Lamont-Doherty Geological Observatory Contribution #3615.

Printed in U.S.A.

The Geology of North America
Vol. M, The Western North Atlantic Region
The Geological Society of America, 1986

Chapter 14

Geoid undulations mapped by spaceborne radar altimetry

P. R. Vogt
Naval Research Laboratory, Washington, D.C. 20375

INTRODUCTION

Were the ocean and atmosphere at rest, the sea surface would describe a surface of constant gravitational potential. This equipotential surface is called the *geoid*. Removal of a simple reference surface (spheroid or ellipsoid) leaves geoid anomalies, or *undulations* of various wavelengths. Geoid undulations range in amplitude from many tens of meters for the broad high in the Azores-Iceland area (Fig. 1) down to a few centimeters for the local effect of a small seamount or minor fracture valley. The positive density anomaly (mass excess) represented by a seamount acts as a small "moon" on the ocean floor by pulling the sea towards itself, producing a bump (undulation) on the ocean surface. Conversely a sea-floor valley produces a dimple on the ocean surface.

The gravity field is the normal component of the gradient of the potential. Thus, gravity anomalies (Rabinowitz and Jung, this volume) are a measure of the spacing between the geoid and neighboring potential surfaces. The slope of the geoid is described by the Vertical Deflection (V.D.), a small angle (rarely more than 0.5′ arc) between the normal to the reference ellipsoid (a function of latitude only) and the normal to the geoid (the local astronomic vertical). Both geoid and V.D. can be calculated from gravity data (Rice, 1952; Brammer, 1979). Using a gyro-stabilized telescope, von Arx (1966) measured a V.D. of 60 to 90 (± 12) arc sec over the Puerto Rico Trench. This pioneering work in direct measurement of V.D. was supplanted by radar altimetry.

Satellite radar altimetry, an electromagnetic analogue of shipborne echo sounding, measures the round trip time of radar pulses reflected off the ocean surface. If the orbit of the satellite (ephemeris) is precisely known, this surface can be mapped after the travel times are corrected for time/space variations in the speed of the microwave pulses. Such variations are caused by the mass of air, its water vapor content, clouds and rain, and ionospheric electrons between the altimeter and the sea below (Tapley and others, 1982). Additional errors are introduced by altimeter noise, sea state, timing, and several kinds of orbital errors. The latter are generally of such long wavelength (10^4 km) that the shapes of undulations due to the Mid-Oceanic Ridge (MOR), a trench, or a seamount are not affected. Obtaining the geoid from measured sea surface topography (i.e. actual sea level) requires further corrections for dynamic effects like tides, storm surges, ocean currents, and eddies. The difference between the geoid and actual sea level is sometimes called "ocean topography" (Stewart, 1985).

First proposed in the 1960's, satellite radar altimetry has evolved through Skylab (McGoogan and others, 1974) and GEOS 3 (Stanley, 1979) to the 1978 SEASAT mission (Born and others, 1979; Townsend, 1980; Lame and Born, 1982; Tapley and others, 1982), which is the basis for much of the synthesis presented here (see also frontispiece). From 7 July through 17 August, 1978 an altimetry data set with an equatorial track spacing of 165 km was collected from a circular orbit 800 km above the earth's surface and covering the globe from 72°N to 72°S. Then the orbit was changed to repeat each ground track once every three days, producing bands of repeat tracks spaced 900 km at the equator. Track separations decrease at higher latitudes, e.g. the North Atlantic.

The radar altimeter, one of several oceanographic instruments carried, emitted 3-ns-long microwave (13.5 GHz) pulses. The field-of-view of the altimeter was 2.4 to 12 km depending on sea state. The precision of the height measurement was 8 cm for significant wave heights (SWH) less than 5 m (Lame and Born, 1982). The noise level ranged from 4 cm at very low SWH to 14 cm at 10-m SWH (Smith and Hicks, 1981). Random noise was reduced by averaging the pulse rate of 10/sec to obtain data points at 1 or 2/sec, representing one value per 6.6 or 3.3 km of track.

In addition to its primary mission of measuring geoid undulation, the SEASAT altimeter provided data valuable for marine meteorology (significant wave height and wind speed), oceanography (storm surges, tides, ocean currents and eddies; Apel, 1980, Cheney, 1982) and even glaciology/climatology (ice surface topography on Greenland; Zwally and others, 1983). Since the altimeter measures actual ocean surface topography, which is the sum of geoid and dynamic (oceanographic) effects, an accurate *independent* knowledge of the geoid, computed from gravity data, (see Chapman and Talwani, 1979) is necessary to remove dynamic effects. Since this computation is based on a surface integral over the entire earth, gravity data over land may be important for oceanographic applications of satellite altimetry. Conversely, ground-truth oceanographic data (e.g., current speeds, temperature and salinity data) are important for geodetic

Vogt, P. R., 1986, Geoid undulations mapped by spaceborne radar altimetry; *in* Vogt, P. R., and Tucholke, B. E., eds., The Geology of North America, Volume M, The Western North Atlantic Region: Geological Society of America.

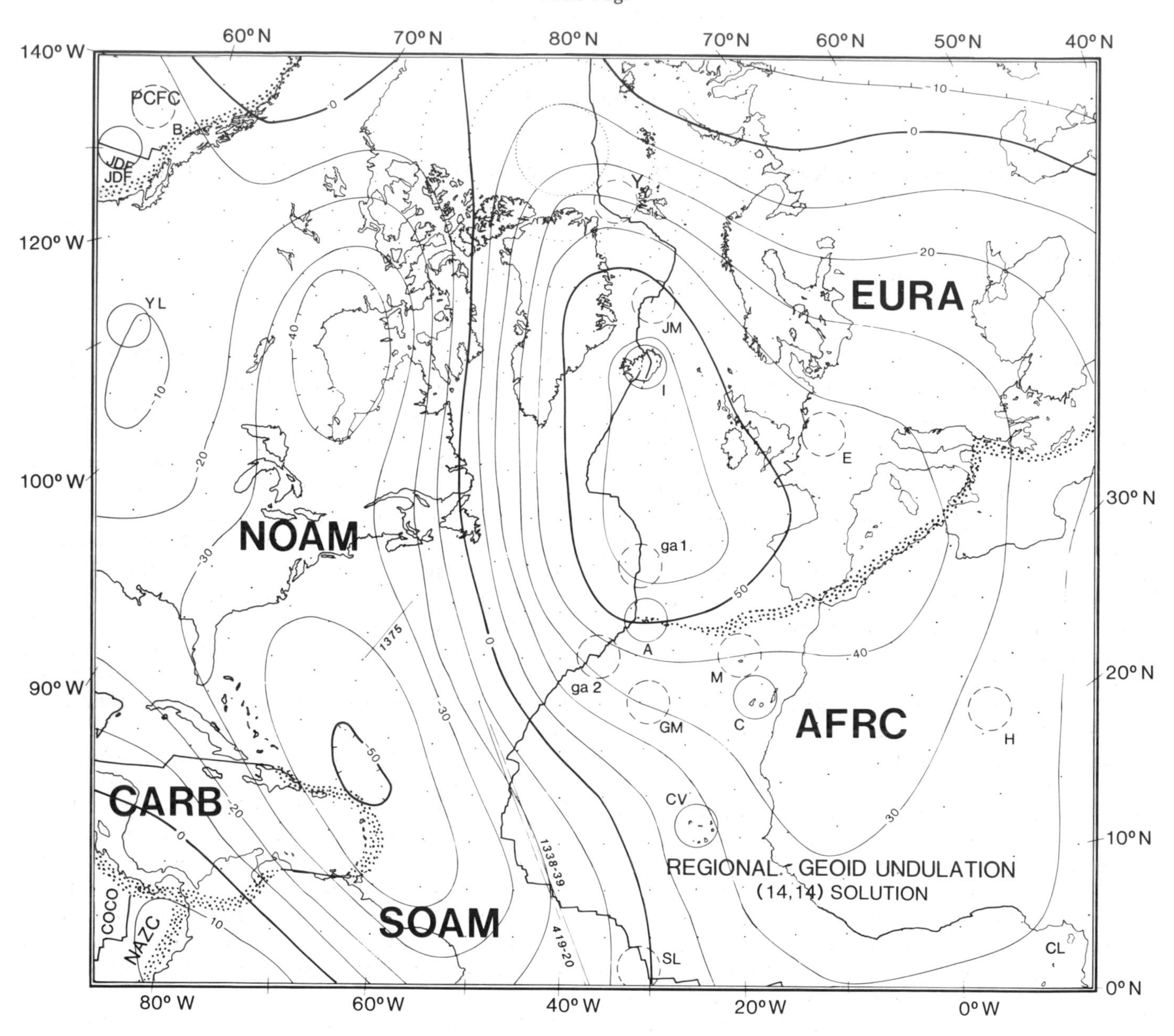

Figure 1. Long-wavelength geoid (to degree and order 14) at 10-m-contour interval, replotted from Hadgigeorge and others (1981) on Transverse Mercator Projection (40°W center meridian) showing plate boundaries, plate name abbreviations, and hotspots. Dashed circles designate minor hotspots and those (gal, ga2) suggested only by their geochemical anomalies (see Schilling, this volume). From top to bottom: B, Bowie Smt; JDF, Juan de Fuca; Y, Yermak; JM, Jan Mayen; YL, Yellowstone; I, Iceland, E, Eiffel; A, Azores; M, Madeira; C, Canary; H, Hoggar; GM, Great Meteor; CV, Cape Verde; SL, Sierra Leone; and FP, Fernando Poo (hotspot location uncertain). Portions of SEASAT revs followed by ship shown by thin lines labelled by revolution number (Figs. 2-4).

applications of altimetry. For example, the geoid undulation caused by a seamount under the Gulf Stream can only be accurately defined once the Gulf Stream's contribution to the ocean surface topography is precisely known and subtracted from the altimetry. The altimetric anomaly over a seamount may include the effect of eddies produced by flow across the seamount (Huppert and Bryan, 1976; Royer, 1978). The western North Atlantic is a challenging area for radar altimetry because of the presence of the Gulf Stream and its shifting patterns of meanders and cyclonic and anticyclonic eddies. Furthermore, the short-wavelength geoid undulations, other than over prominent seamounts, are generally of low amplitude (comparable to dynamic effects), as described below. This low amplitude partly reflects the age and consequent depth and burial of the basement topography by sediment.

Before examining these short-wavelength geoid undulations (40 - 300 km), we discuss the long (>3000 km) and intermediate

(300 - 3000 km) wavelength undulations which were reasonably well known prior to the SEASAT and GEOSAT missions. Dividing the geoid into these three spectral slices is not entirely arbitrary, since the long-wavelength undulations largely originate in the middle and deep mantle (Bowin, 1983), and the intermediate-wavelength ones in the lower lithosphere and asthenosphere; the shorter wavelengths are largely due to sea-floor and basement topography. Although a portion of the long-wavelength field may originate above the middle mantle, the sources of short-wavelength components cannot have a deep origin. There is overlap between neighboring spectral categories, and as will be seen there are also physical connections among them.

LONG-WAVELENGTH GEOID UNDULATIONS AND MANTLE CONVECTION

The long-wavelength structure of the gravity field and the geoid are commonly described for the earth as a whole by a set of spherical-harmonic coefficients. The set of coefficients is usually truncated at values of degree and order, somewhere between 5 and 15. In general, the lower the order the higher the ratio of geoid to gravity, and the greater the depth to the corresponding density anomaly (but see Bowin, 1983). The lowest-order feature of the geoid is the "reference ellipsoid" (West, 1982) which is subtracted from the observed geoid in order to emphasize geoid anomalies (undulations). Numerous spherical harmonic expansions for the geoid are in the literature. In Figure 1 we show the model of Hadgigeorge and others (1981) which incorporates SEASAT altimetry, satellite orbit information, and surface gravity measurements. Truncation at degree and order 14 [expressed as (14, 14)] describes the long-wavelength characteristics of the oceanic geoid without being unduly "contaminated" by intermediate-wavelength features like those associated with the Bermuda Rise. (A small contribution by the Mid-Atlantic Ridge and Puerto Rico Trench no doubt remains). Any recent solution in the range 10 to 14 would have produced similar results (Bowin, 1983); Bowin and others (1984), Jung and Rabinowitz (1986) and Watts and others (1985) removed geoids of degree 10, 10 and 12, respectively.

Two vast geoid anomalies dominate the Central and North Atlantic. A high, centered in the northeast Atlantic, including Iceland, extends from northern Greenland into the eastern equatorial Atlantic. The highest values are more than 60 m above the reference ellipsoid. In the western Atlantic there is a geoid low extending from eastern South America north across the eastern U.S. and into the Canadian Arctic. One minimum below −50 m occurs northeast of Puerto Rico, and a second, below −40 m, is located over Hudson's Bay. Both the northeast Atlantic high and the west Atlantic low are parts of still more extensive anomalies; the high extends southeast, by way of a saddle in the eastern South Atlantic, to another high in the southwest Indian Ocean and adjacent parts of Antarctica. The west Atlantic low extends across the western South Atlantic to the coast of Antarctica.

What is the origin of these long-wavelength undulations (and gravity anomalies)? Many authors have pointed out that the causative mass excesses and deficits cannot be supported by the lithosphere. The undulations also cannot arise from rigidly supported density anomalies within the deeper mantle. They must largely reflect dynamic effects, i.e. density differences between rising and sinking mantle, and associated deflections of density discontinuities such as the core/mantle boundary, the 670-km discontinuity, and the earth's surface (Pekeris, 1935; Richards and Hager, 1984). The long-wavelength geoid does not correlate systematically and globally with surface-geologic or topographic features such as continent-ocean crustal boundaries, the MOR (Mid-Oceanic Ridge), and attempts to correlate e.g. gravity anomalies with intraplate seismicity (Vogt, 1978) remain unconvincing. The long-wavelength geoid therefore seems inconsistent with whole-mantle convection closely coupled to plate motions.

Three important exceptions to this lack of correlation are the geoid low over Hudson's Bay—partly due to isostatic disequilibrium from the Wisconsin-age ice sheet (Cathles, 1975; Walcott, 1972), the geoid highs over subduction zones (e.g., the high in the southwest corner of Fig. 1), and the geoid highs over hotspots or groups of hotspots (Fig. 1). Chase (1979) modeled the effect of subducted slabs and removed this effect from the observed geoid. The remaining effects were modelled with point masses whose depths represent the maximum possible depths of the sources. Chase concluded that the northeast Atlantic *high* may be due to sources as deep as the core-mantle boundary (2900 km), whereas the west Atlantic *low* cannot have a source deeper than 1000 km. But this is not a unique result; both Chase (1979) and Bowin (1983) suggest that some of the very low-order geoid may be associated with low (of order 1 km) swells on the core-mantle boundary. Hager (1984) found that geoid degrees 4 to 9 are highly correlated with subducted slabs, particularly if the slabs extend into the lower mantle or "pile up" at the 670-km base of the upper mantle. Dziewonski (1984) showed a correlation between seismic P-wave structure in the lower mantle and the source of the very low-order (2-3) geoid. Woodhouse and Dziewonski (1984) showed that upper-mantle-velocity structure also correlates directly with the very low-order geoid, and inversely (high velocity with low density) for degrees 4-6. It remains to be seen how well these global results apply specifically to the mantle below the North Atlantic.

The central part of the northeast Atlantic high, from south of the Azores to north of Iceland, corresponds to a large area of anomalously shallow ocean crust associated with the major Azores and Iceland hotspots and several lesser ones (Vogt and others, 1981; Cochran and Talwani, 1978). There is in fact a global tendency for hotspots to be non-randomly clustered (Stefanick and Jurdy, 1984) and for these clusters to lie within major geoid highs or their flanks, rather than in the geoid lows (Chase, 1979; Vogt, 1981). (This association presumes hotspots can be identified by volcanism, and may not be valid if mid-plate epeirogeny (Crough, 1979b) *unaccompanied* by volcanism is also attributed to hotspots).

Hotspots (defined volcanically) are notably absent in the

west Atlantic-east North American geoid low (Fig. 1). The only possible exception might be Bermuda, which has sometimes been treated as an active hotspot (Crough, 1981a; Morgan, 1981), but the timing of the main phase of volcanism and the uplift of the Bermuda Rise (some time between 50 and 30 mybp; Vogt and Tucholke [1979], Tucholke and Vogt [1979]) leaves this in doubt. (If the long-wavelength geoid undulations and the Africa plate have been roughly stationary since then, Morgan's plate-motion model [1981] shows that Bermuda originated on the flanks of the geoid high, at least 500 km east of its present position within the geoid low.) At any rate, the overall correlation between low-order geoid highs and hotspot density suggests a physical interaction between the deep mantle and the earth's surface. This interaction may involve both broad convection and narrow plumes. Such interpretations, although plausible, will remain non-rigorous until the inherent non-uniqueness of gravity field inversion is reduced by other types of constraints about deep-mantle struacture, e.g. from seismic tomography (e.g., Dziewonski, 1984; Woodhouse and Dziewonski, 1984) and by forcing density structures to satisfy equations of fluid motion (Hager, 1984; Richards and Hager, 1984). It remains unclear to what extent the greater density of hotspots in the eastern Atlantic and Africa (vs. the western Atlantic and eastern North America) reflects processes in the deeper mantle and to what extent it is merely an expression of differences in plate thickness and speed (Gass and others, 1978).

INTERMEDIATE-WAVELENGTH GEOID UNDULATIONS: THERMAL ANOMALIES IN THE LITHOSPHERE/ASTHENOSPHERE

Anomalies in this part of the spectrum generally correlate with geological/topographic structures, for example subduction zones, mid-plate swells like the Bermuda and Cape Verde rises, massive aseismic ridges, and the MOR. In the area of emphasis in this volume, the two most prominent intermediate-wavelength anomalies are those over the Mid-Atlantic Ridge (Fig. 2) and the Bermuda Rise (Anderle, this volume; frontispiece). Subtraction of a low-order model (Fig. 1) from the raw altimetric geoid leaves behind, with little attenuation, the 1000-km wide, ca. 3 to 6 m amplitude geoid high (Haxby and Turcotte, 1978; Rapp, 1983) associated with the Bermuda Rise, and to some extent the high over the Mid-Atlantic Ridge.

Surface-ship gravity data (Fig. 4) show a regional high of amplitude +20 to +30 mgals centered over the MOR spreading axis and decaying into back-ground noise on about 40 Ma crust (Cochran and Talwani, 1977; Rabinowitz and Jung, this volume). The equivalent geoid anomaly, relatively less sensitive to "crustal clutter," can be traced to crustal ages of the order of 100 Ma (Sandwell and Schubert, 1980, 1982). The MOR geoid/gravity high, the topographic high (i.e. the MOR itself), and the total heat flow high (Parsons and Sclater, 1977) are all expressions of the same process—the decay, with increasing crustal age, of a thermal expansion effect continuously created in the mantle below the MOR axis. Although the MOR geoid high has most of its power at intermediate wavelengths, segmentation of upper-mantle thermal structure by close-spaced transform faults also creates geoid anomalies which have short wavelengths when traversed at a steep angle to the fracture zones (Fig. 2).

Theoretical relationships between geoid height and crustal age have been derived by Haxby and Turcotte (1978) for the cooling half-space (CH-S) model, in which plate thickness increases as the square root of crustal age, and for the cooling-plate CP model (Sandwell and Schubert, 1980, 1982) for which plate thickness is fixed. Parsons and Richter (1981) predict geoid slopes from upper-mantle convection models.

An approximate solution for the CH-S model (Haxby and Turcotte, 1978) is that the geoid slope is

$$\frac{\Delta N}{\Delta t} \cong \frac{2\pi G \rho_m \alpha (T_m - T_o) \kappa}{g} t$$

where G is the gravitational constant (6.67×10^{-8} cm^3/sec^2), g is the earth's surface acceleration (982 cm/sec^2), ρ_m is the density of the mantle (3.33 gm/cm^3), α is the volume coefficient of thermal expansion (3.2×10^{-5} °C^{-1}), ($T_m - T_o$) is the temperature difference across the thermal boundary layer (1350°C), κ is the thermal diffusivity (0.8×10^{-2} cm^2/sec), and t is age of the lithosphere. Using the above values for the constants, Haxby and Turcotte derived a value of 16 cm/m.y. for the decrease of geoid height with increasing crustal age. Sandwell and Schubert (1980) used GEOS-3 altimeter data to derive slopes of 9.4 ± 2.5, 13.1 ± 4.1, and 14.9 ± 2.8 cm/m.y. for the geoid/age relationship in the South Atlantic (10°S–50°S), Southeast Indian, and North Atlantic ocean regions (20°–50°N), respectively. These values apply to crust 0 to 80 Ma age. Bowin and others (1984) determined slopes from 8 to 22 cm/m.y. in the North Atlantic, with higher slopes correlating with positive geoid and depth anomalies (e.g., near the Azores). Bowin and others suggest the higher slopes reflect plume-related crustal thickness and density variations. Moreover, the Azores segment of the MAR may only have become hot and elevated in mid-Tertiary time, which would result in a non-steady-state profile and a steep slope.

For ages greater than 80 Ma, the geoid-age relation for the North Atlantic is nearly flat, indicating a reduction in the rate of boundary-layer thickening with age, plus or minus interference with the long-wavelength field of deeper origin discussed above. Geoid height vs. crustal-age curves based on the CP model could predict some of the observed flattening out at ages greater than 60-80 Ma (Sandwell and Schubert, 1980). GEOS altimetry across the Mendocino F.Z. (Crough, 1979a) was found compatible with the CH-S model if the lithosphere thickness is about 9.1 $\sqrt{t}$-km thick (t is crustal age in m.y.). However, Sandwell and Schubert (1982) concluded that the thermal structure of the lithosphere already begins to deviate from a CH-S model at 20-40 Ma crustal ages. If interpreted by the CP model, the geoid anomaly over the mid-oceanic ridge is more sensitive to the thermal thickness of the lithosphere than is either topography or heat flow and it may thus be possible to determine from altimetry whether parts of the Mid-Atlantic Ridge or adjacent basins are hotter or colder

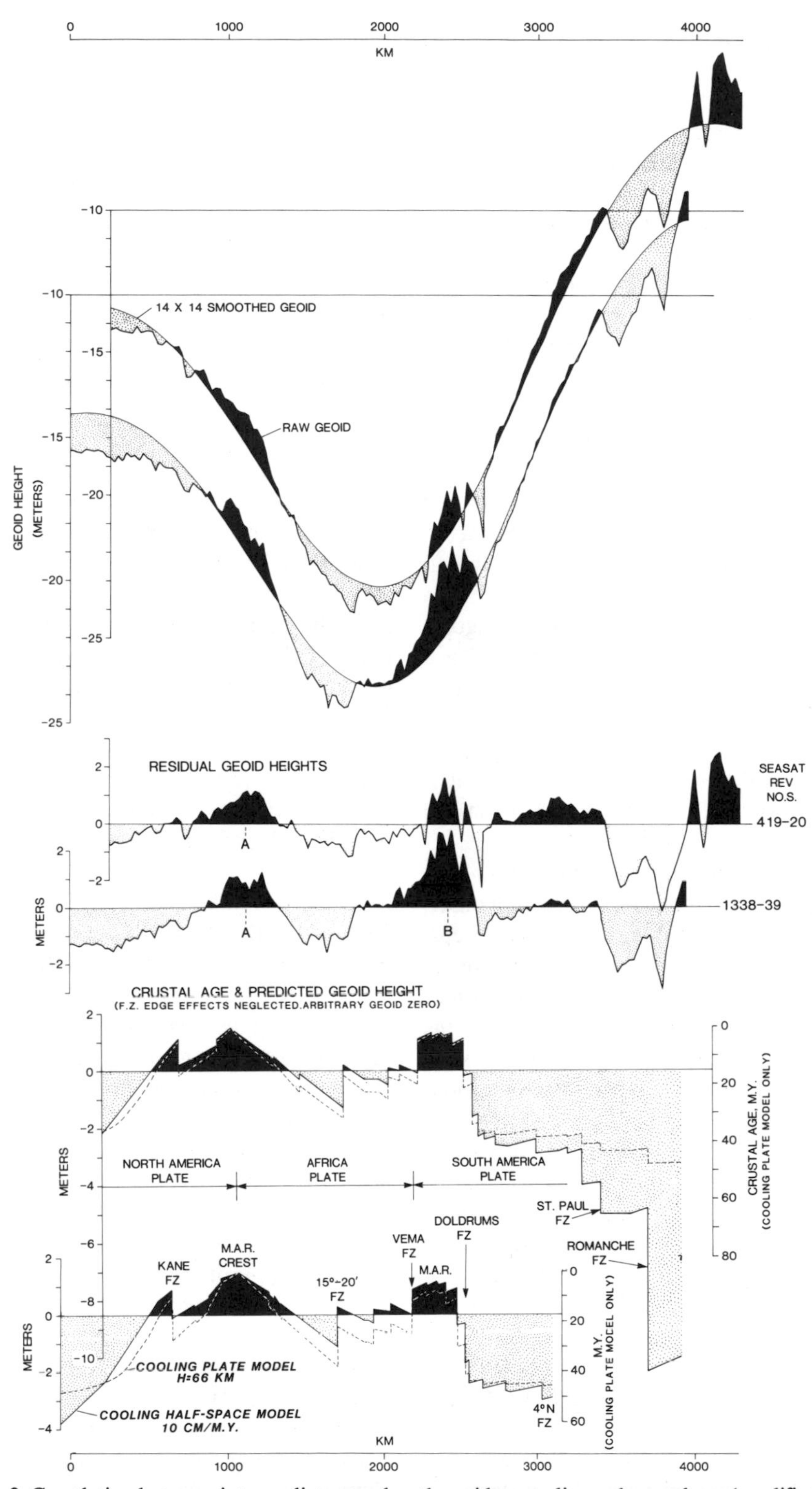

Figure 2. Correlation between intermediate wavelength geoid anomalies and crustal age (modified from Vogt and others, 1984). Raw, smoothed (Fig. 1), and residual altimetry (geoid heights) along portions of rev 419-20 and 1338-39; predicted crustal age along tracks derived from Perry and others (1981) and other sources. Geoid curves predicted from crustal age for two possible lithosphere models: the cooling half-space model with a constant geoid slope of 10 cm/m.y. (Eq. 1) (solid line), and the cooling-plate model with initial slopes of 15 cm/m.y., declining with age according to a 100-km-plate thickness (dashed line) constructed using Figure 9 of Sandwell and Schubert (1982). Crustal age scale applies only to *solid line.*

compared to other areas of comparable crustal age. Published estimates of thermal-plate thickness in different oceans range from about 70 to 125 km (Watts and others, 1985; Cazenave and others, 1983; Sandwell and Schubert, 1982). The apparent better agreement of geoid data with the CP model (Jung and Rabinowitz, 1986; also Fig. 2) does not invalidate the CH-S model, because the flattening out of the age/depth curve, and hence probably also the geoid/age curve, may simply result from the thermal anomalies old lithosphere is likely to have traversed, causing mid-plate swells of which only the more prominent may be apparent, e.g. the Bermuda Rise (Heestand and Crough, 1981).

Removal of a low-order geoid (Fig. 1) from "raw" SEASAT profiles (Fig. 2) allows analysis of the geoid/age relation to the extent that transform faults and crustal ages have been correctly mapped along the sub-satellite track. However, if the geoid continues to deepen at progressively lower rates out to crustal ages of 100 Ma or more (Fig. 2; also Sandwell and Schubert, 1982) the thermal expansion–induced geoid anomaly is so broad that the prior removal of the (14, 14) field will have also removed some of the "signal." The analysis in Figure 2 assumes the geoid changes abruptly across transform faults; actually this "step" is a smooth ramp function (Sandwell and Schubert, 1982). If a CH-S model is assumed, the best-fitting geoid-age slope is about 10 cm/m.y. However, this slope does not fit the profiles (Fig. 2) for crustal ages greater than about 40 Ma. A better overall fit is obtained by assuming a cooling plate about 100 km thick (dashed line in Fig. 2).

All the major fracture zones (or close-spaced groups of fracture zones) with offsets 10 m.y. and more, and some with offsets of order 5 m.y., clearly stand out as ramps (smoothed geoid steps) in the profile illustrated (Fig. 2). Because crustal-age maps of the North Atlantic (Perry and others, 1981; Vogt and Perry, 1982) are still far from perfect, some of the discrepancies between the magnitudes of observed and "predicted" geoid heights and steps undoubtedly reflect errors in supposed fracture-zone locations and in the age jumps across them.

We now turn briefly to the Bermuda Rise, a 30 to 50-m.y.-old northeast-trending mid-plate swell, about 1000 km in diameter, capped by a northeast-trending cluster of volcanoes dominated by the Bermuda edifice. The geoid anomaly over the Bermuda Rise—distant from the local effect of the Bermuda Pedestal—is about +3 to +6 m (Marsh and Chang, 1978; Anderle, this volume), depending on what is taken as the undisturbed norm. Correlation with the depth anomaly gives slopes from 3.5 to 10 m/km (Bowin and others, 1984) or 15 mgals/km Sclater and Wixon (this volume).

Several models have been advanced to explain mid-plate swells and their associated geophysical anomalies in terms of mantle convection. Sclater and others (1975) argue that the mid-plate swells in the North Atlantic are dynamically supported by mantle convection. Richter and Parsons (1975) postulate that the existence below the plates of second-order convection "rolls" orient parallel to the spreading direction. Sclater and Wixon (this volume) prefer a generalized convection model, with both upwelling and downwelling below a cooling plate, to a CH-S model perturbed by localized hotspots. In the former model convection is allowed to penetrate a lower, thermal boundary layer to the base of a 75 km-thick mechanical boundary layer (Parsons and Daly, 1983). According to this model, some of the intermediate-wavelength geoid lows (Anderle, this volume) may be sites of localized downwelling.

Swells could also be formed quite differently, for example, by underplating of depleted, and therefore actually less dense, mantle (Jordan, 1979). Crough (1981a) noted that convection models (e.g., Sclater and others, 1975) and Jordan's model imply that compensation depths should be substantially greater for continental, particularly cratonic, swells, which are developed in lithosphere at least 100 km (Crough and Thompson, 1976), and perhaps as much as 400 km thick (Jordan, 1979), than for oceanic swells like the Bermuda Rise. However, Crough (1981a) uses gravity data from west Africa to show that the swell associated with the Hoggar hotspot ("H" in Fig. 1) is compensated at a depth of only about 60 km, which is similar to depths calculated from the geoid for oceanic swells such as the Bermuda Rise (Crough, 1981a, b). The compensation depth of the latter was estimated to be 50 km by Crough (1978), 100 km by Haxby and Turcotte (1978), and 230 km by Bowin and others (1984), the latter two estimates presuming a Pratt-type compensation. Other kinds of data (e.g., teleseismic waveforms and travel times) are clearly needed to constrain the density structure and physical processes which form mid-plate swells such as the Bermuda Rise.

SHORT-WAVELENGTH GEOID UNDULATIONS: OCEANIC BASEMENT TOPOGRAPHY AND OTHER SHALLOW DENSITY DISTRIBUTIONS

The best correlations between sea-floor topography and geoid undulation (or gravity anomalies) are found in the wavelength domain from 300-400 km down to the resolution limit of the data which is about 30-50 km for altimetry. Brammer and Sailor (1980) and Watts and others (1985) derived limits of 33 and 45 km respectively. The resolution is better for high-quality surface-ship gravity since the signal-to-noise ratio is higher at short wavelengths (Chapman and Talwani, 1979). These limits vary geographically and temporally, as a function of sea state. On young oceanic crust some of the short-wavelength geoid structure as well as topography reflects the age-dependence of thermal structure, but most of the local geoid structure is permanently frozen into the plate.

Recent papers describing the short wavelength geoid–topography relationship (which, despite the newness of spacecraft altimetry, has long been known), include McKenzie and Bowin (1976), Watts (1979), Groeger (1981), Ribe (1982), Parke and Dixon (1982), White and others (1983), Dixon and others (1983), Souriau (1984), Watts and Ribe (1984), Vogt and others (1984), and Watts and others (1985). The formulas needed to compute geoid undulations from bodies of arbitrary shape are found in Chapman (1979) and Parker (1972).

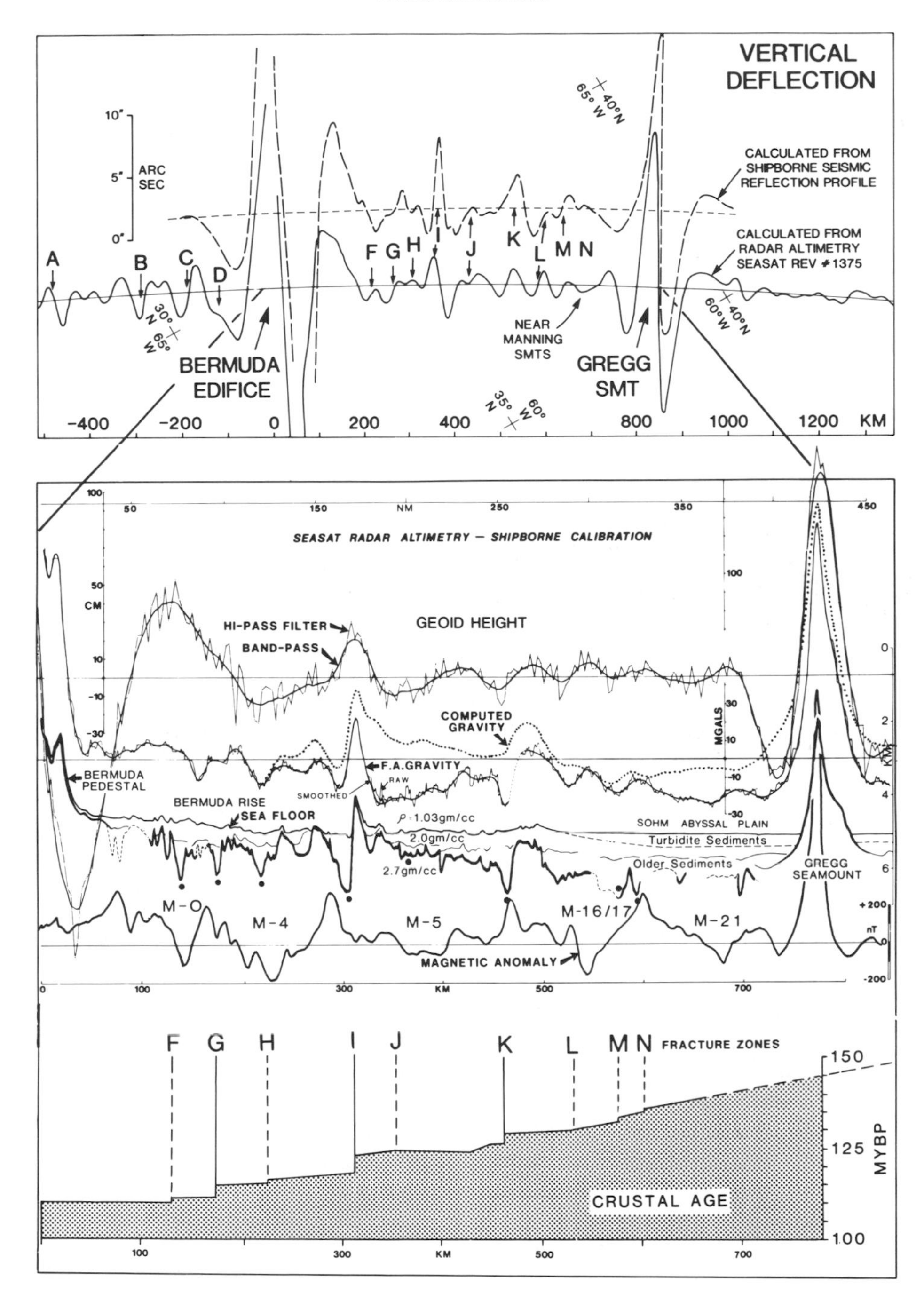

Figure 3 (top). Filtered (200-km cutoff) vertical deflection (V.D.) along rev 1375, derived from band-passed SEASAT altimetry (solid line) and independently (dashed line) from two-dimensional isostatically uncompensated model based on seismic-reflection data. Letters A thru N show locations of minor fracture zones identified from detailed magnetic anomaly contours (see Vogt, this volume). (bottom) Filtered (200-km cutoff) geoid height (top), gravity anomaly computed (dotted) from seismic data, measured free-air gravity, seismic-reflection profile, magnetic anomaly, and crustal age. From Vogt and others (1984).

There are several ways to emphasize the shorter-wavelength geoid contributions: (a) calculate the gravity anomaly from the geoid (Haxby and others, 1983; frontispiece); (b) calculate the slope of the geoid (vertical deflection) (Vogt and others, 1984); or (c) high-pass filter the geoid (Zondek, 1982; Vogt and others, 1984). The vertical deflection can only be calculated from the geoid (Fig. 3) if the high-frequency noise, partly due to ocean waves (Smith and Hicks, 1981), is first smoothed out of the data. Along-track high-pass filtering, with cutoffs of about 200 km (Fig. 3) or 400 km (Figs. 4, 5), effectively removes the long and

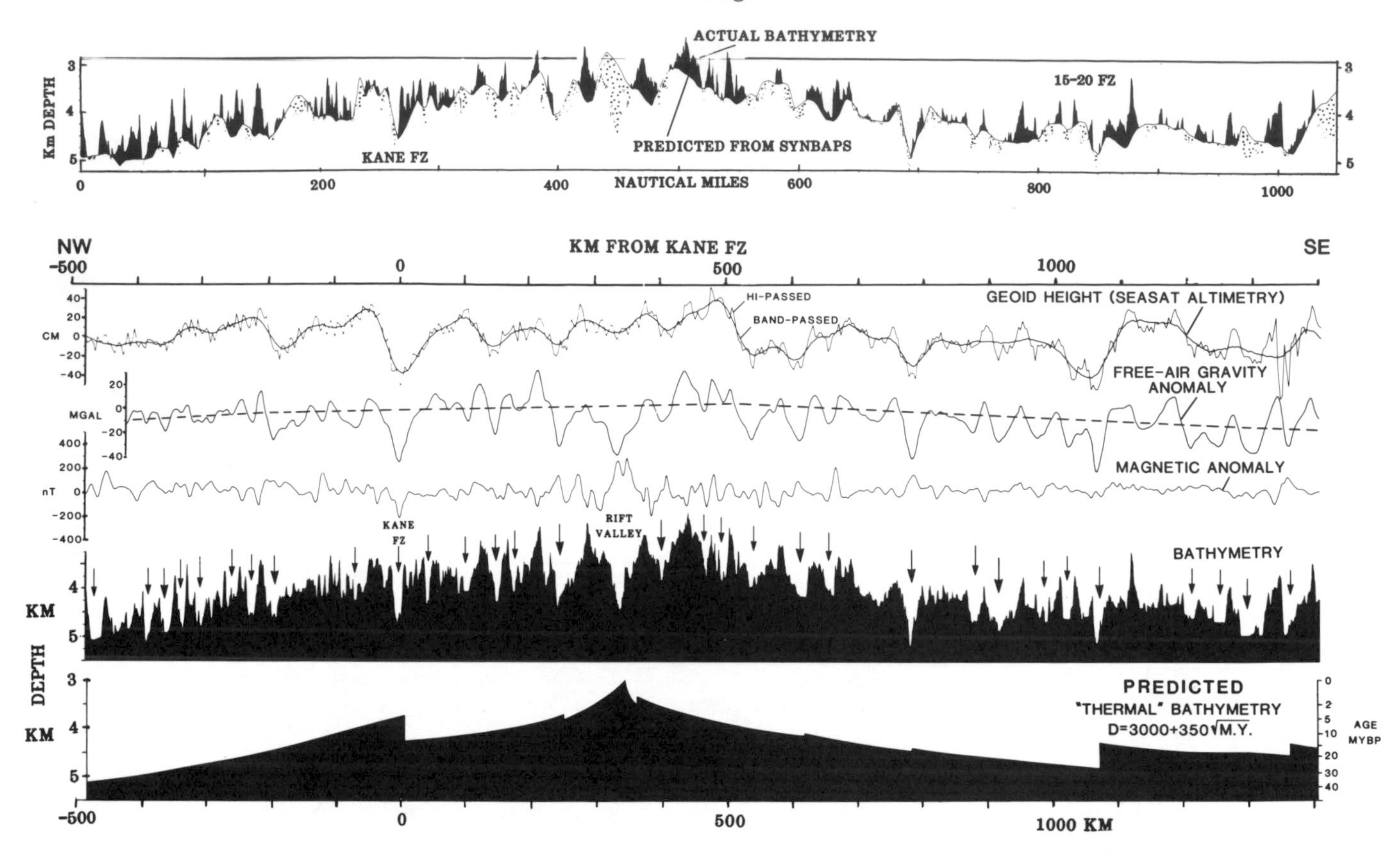

Figure 4. Filtered (400-km cutoff) SEASAT altimetry (rev 419), free-air gravity anomaly, magnetic anomaly, measured bathymetry (arrows indicate probable fracture zones), and "thermal bathymetry" (predicted from crustal age in m.y. according to the relation depth = 3000 + 350 $\sqrt{\text{m.y.}}$). Note broad gravity high over MAR (dashed). Crustal age (right scale) was estimated a priori from Perry and others (1981) and other sources. Distances (km) measured from Kane FZ (from Vogt and others, 1984).

intermediate-wavelength undulations (Figs. 1, 2) and strongly reduces dynamic effects such as Gulf Stream eddies whose power is concentrated at 400–700 km (Menard, 1983). Typical anomalies of 50–100 km wavelength due to basement topography are passed with negligible distortion, but large seamounts such as Bermuda or Gregg (Fig. 3) suffer some amplitude reduction and introduction of flanking negatives. High-pass filtering suppresses short-wavelength "ridge crest"—parallel ridges or valleys which look like longer wavelengths when crossed at a small angle.

The geoid anomaly amplitude associated with a seamount of fixed size depends on the age of the crust (i.e., the elastic plate thickness T_e) at the time of seamount volcanism and can be 2 to 3 times higher for a seamount emplaced on 100 to 150 Ma crust than for one erupted onto crust only 5–10 Ma (Dixon and others, 1983). Atlantic seamounts (Fig. 3) have not yet been analyzed, but results from Pacific seamounts (Watts and Ribe, 1984) would predict a T_e of the order 30 km for Bermuda and 10–20 km for the New England Seamounts. A number of studies using gravity or geoid undulations (Bowin and Milligan, 1985; Watts and others, 1985) suggest that basement topography created at the axis of the MOR reflects an axial T_e of about 5–13 km.

Bathymetric and other geophysical data coverage is so sparse over most oceanic areas that profiles constructed from contour charts along SEASAT-quality subsatellite tracks are generally not accurate enough to provide satisfactory ground-truth data for comparing with the short-wavelength altimetric geoid. To remedy this, Vogt and others (1984) followed, by ship, North Atlantic segments of three SEASAT revs (revolutions), totaling 9300 km. Comparisons between the sub-satellite marine geophysics and filtered altimetry along portions of two revs are shown in Figures 3 and 4. Most SEASAT revs in the Atlantic are oriented at a steep angle to transform fracture zones. Since the satellite was in a near-polar orbit and Atlantic fracture zones trend more EW than NS, revs paralleling flow lines and crustal ribbons do not exist in this ocean. Away from seamounts and continental margins the geophysical signatures along SEASAT tracks are therefore dominated by fracture zones and the crustal "ribbons" between them.

The portion of rev 1375 (Fig. 3) followed by ship could be analyzed exhaustively because it lies within a region where sediment thickness and basement depth (Tucholke and others, 1982), as well as M-Series magnetic lineations (Vogt and Einwich, 1979; Schouten and Klitgord, 1977) have been mapped. The magnetic data are dense enough to reveal even minor offsets or other

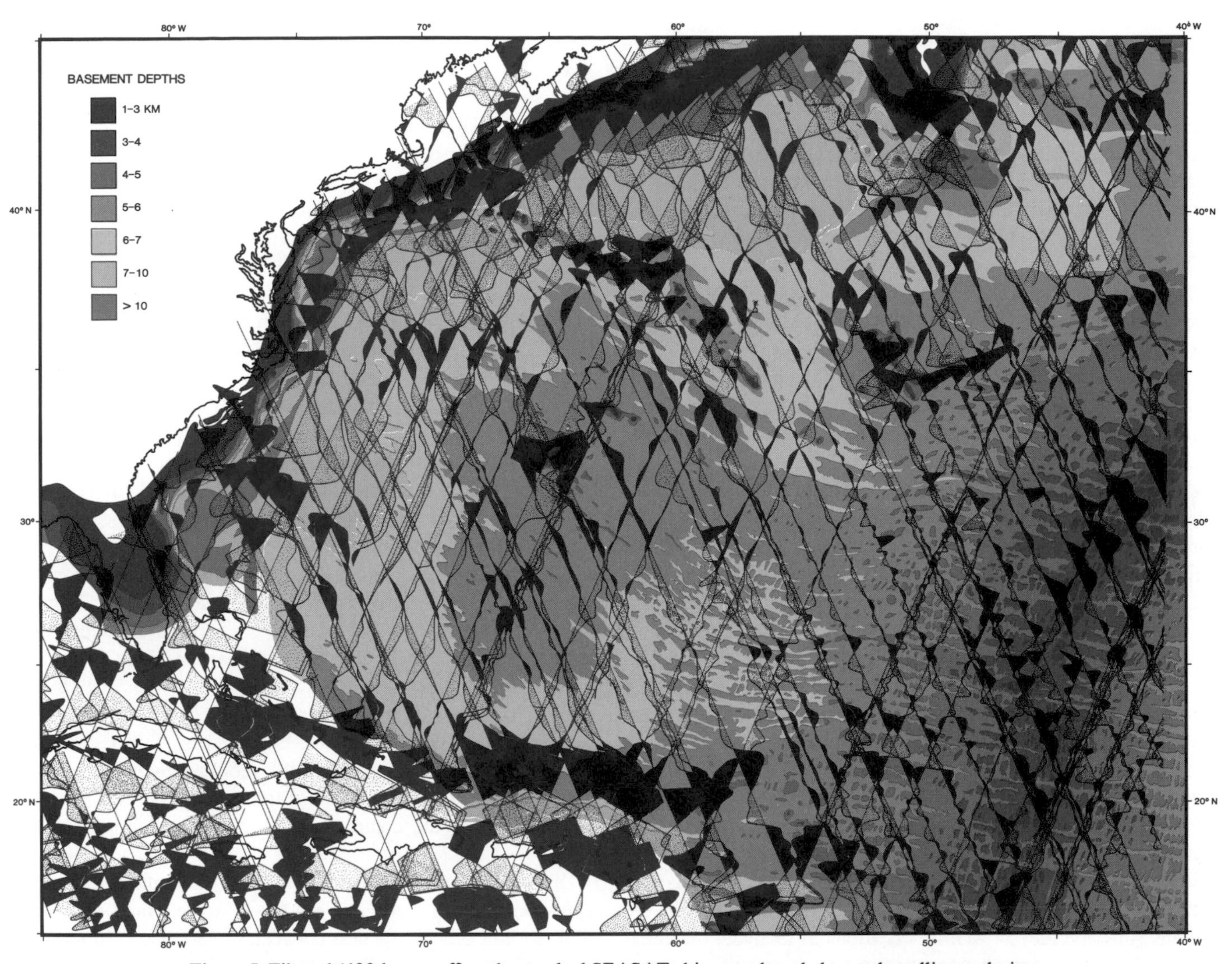

Figure 5. Filtered (400-km cutoff) and smoothed SEASAT altimetry plotted along subsatellite tracks in western North Atlantic. Positive anomalies black, negative stippled. Residuals are clipped at ± 100 cm, which gives the scale. Basement depths (Tucholke and others, 1982) shown in shades of red and gray.

irregularities indicative of small fracture zones. The line from Bermuda to Gregg Seamount is the first marine profile in the oceans along which every continuously measurable type of geophysical parameter (geoid, gravity, bathymetry, seismic reflection, and magnetics) has been recorded. Comparison of the various parameters (Fig. 3) shows that density contrasts between FZ topography and seawater are the primary source for the small (±5 to 10 cm) altimetric anomalies over the north flank of the Bermuda Rise, and by inference over much of the North Atlantic except near seamounts. Basement valleys (and therefore geoid lows) demarcate most of the fractures, but the prominent one (FZ "I" with a 50-km, 5-m.y. offset) also has a 1-km-high ridge with a 20-cm geoid high on its NE edge. SEASAT quality altimetry is even able to map some minor fracture zones where they are completely obscured by sediment.

Computing the slope (V.D.) of the band passed geoid undulation in the direction of the track emphasizes the shorter-wavelength, topographic signals. The V.D. anomalies shown in Figure 3 range from ±1 to 5 arc sec over the fracture zones to ±15 arc sec over Gregg Seamount and ±20 to 30 arc sec over the Bermuda pedestal. Similar values were derived from SEASAT profiles across the Pacific by Watts et al. (1984), calculated for Mid-Atlantic-Ridge topography by Fischer (1979), and measured and calculated for oceanic islands (Fischer and Wyatt, 1974). Figure 3 shows the gravity anomaly, geoid undulation and V.D., computed (in that order), assuming uncompensated two-dimensional structures. The correlation between calculated and "observed" V.D. is good. Differences between the two curves may be due to the lack of two-dimensionality and the presence of subbasement (e.g., FZ valley floor) density contrasts, including those due to flexural compensation.

The SEASAT altimetry over the present MAR (Fig. 4) cannot be analyzed as thoroughly as rev 1375 because detailed magnetic data are lacking. The major FZ's (e.g., Kane, 15°–20′, and an unnamed FZ at km 1070 in Fig. 4) and the rift valley can be reliably identified, as well as a few other FZ's suspected from previous data. Many other fracture zones, particularly minor ones, were crossed. Major unnamed FZ's suggested by observed basement valleys are located at kms –200, +140, 260, 400, 620, 770, 930, and 1360. Of those features the ones at kms 260, 400, 620, and 1360 were previously suspected, but the age offsets across the first three of these are probably underestimated in Figure 4. All major FZ's were also detected as geoid lows in the filtered altimetry.

Inspection of both Figures 3 and 4 shows good visual correlation between surface-ship free-air gravity and filtered geoid undulation. Since the gravity anomaly decays more rapidly than the geoid as a function of distance from the source, the long-wavelength and intermediate-wavelength gravity anomalies of deeper origin are not as pronounced as the corresponding geoid anomalies (Figs. 1, 2); therefore the gravity was left unfiltered (Figs. 3, 4). The relation between gravity and geoid anomalies (Chapman and Talwani, 1979) explains why the smoothed gravity profile is somewhat richer in shorter, "topographic" wavelengths compared to the smoothed geoid (Figs. 3, 4). Whereas both the geoid and the gravity field dip sharply over major FZ valleys, the filtered geoid low over the median rift valley is modest compared to the gravity low. The dominant wavelengths of both geoid and gravity are about 50–100 km, a reflection of FZ spacing.

The primary source for both gravity and filtered geoid anomalies is sea-floor topography (McKenzie and Bowin, 1976). However, crustal density and thickness contrasts also may be important, e.g. between FZ crust and "normal" crust (Purdy and Ewing, this volume) as well as variations in sediment thickness. Over relatively young crust some short-wavelength contributions arise from the segmentation of MOR thermal structure by FZ's (Fig. 4).

Having indicated the geologic significance of short-wavelength altimetry, we show (Fig. 5) the filtered altimetry plotted along subsatellite tracks over the entire western Atlantic, and an interpretation (Fig. 6). Major anomalies (e.g. the Puerto Rico Trench area and Bermuda) were known earlier from GEOS data. The remaining geoid features can be divided into the following categories: seamounts, sediment accumulations, continental margin anomalies, possible oceanographic effects, and oceanic basement topography.

The anomalies associated with known seamounts range up to over +100 cm after filtering, e.g. Gregg Seamount discussed earlier. Many seamounts show highs of only 25–50 cm and most of those represent flank crossings. A number of seamounts were missed by SEASAT; no new ones of large size (>3 km relief) were found in the study area.

Geoid highs due to sediment masses include the Laurentian Cone (+ 25 cm) and the Blake Ridge (50 to over 100 cm). The continental margin high exceeds +100 cm mainly north of the New England Seamounts and over the SE edge of the Grand Banks.

Some dynamic effects probably remain in the filtered altimetry, particularly the landward edge of the Gulf Stream, cold-core eddies southeast of the Stream (altimetric lows), and warm-core eddies to the northwest (altimetric highs) (Cheney, 1982). The Gulf Stream itself, associated with a slope on the ocean surface, seems to appear in the filtered profiles as a prominent low (Figs. 5, 6), but separation of geologic and dynamic effects is complicated by the unfortunate parallelism between the Gulf Stream and such geologic features as the continental margin, the western New England Seamounts, and the southeast Newfoundland Ridge.

Most of the observed anomalies (Figs. 5, 6) are due to oceanic basement topography. The rift mountains produce a characteristic double-peaked anomaly of about +50 cm, except along revs nearly parallel to the MAR crest, for which the apparent wavelength is long enough to be filtered out. The rift-valley low is subdued, partly because of filtering.

Numerous major and minor transform fracture zones can be identified and traced, particularly on crust younger than anomaly 34 and from the 15°–20′ FZ to the Kane FZ The now well-known WSW trend of fractures between anomaly 21 and 34 time is unmistakably revealed by the SEASAT altimetry. In several places, but particularly in the area 20°–26°N, 50°–58°W, the bathymetric trends shown on GEBCO sheet 5.08 (Searle and others, 1982), do not agree with the geoid anomaly trends (Vogt and others, 1984). In those areas the geoid lineations are more consistent with the contours of Tucholke and others (1982) (Fig. 5).

The geoid is very smooth over crust of about 100 to 80 Ma age, and fracture zones are not easily traced through it. The difference in geoid roughness between the 80–0 and 100–80 Ma belts can also be seen in the global study by Brown and others (1983). The smooth geoid correlates with smooth basement topography (the "Nares Smooth Belt" of Vogt and others, 1971) and fast spreading. Most Cenozoic Atlantic crust north of the equator was formed at 1.5 to 3.5 cm/yr total opening rates and exhibits a short-wavelength roughness of 30 to 50 cm. By contrast most South Atlantic crust, and North Atlantic crust 80–100 Ma old, was formed at 3.5 to 6 cm/yr and, according to Brown and others (1983), has a geoid roughness of only 10–30 cm. Thus the satellite altimeter can be used as an indirect measure of spreading rate.

Some nearly EW geoid trends in the 80–100 Ma zone southeast of Bermuda are not in harmony with bathymetric and magnetic trends, but the significance of this is uncertain. Farther west, over crust 100 to 120 Ma in age, the geoid again becomes rougher, in conformance with known rougher basement, and slow spreading (Sundvik and others, 1984). The Figure 4 profile lies within this province.

An interesting feature of the geoid pattern is the suggestion of "crustal ribbons" of alternately higher and lower geoid value (Figs. 5, 6). However, although in some cases associated with

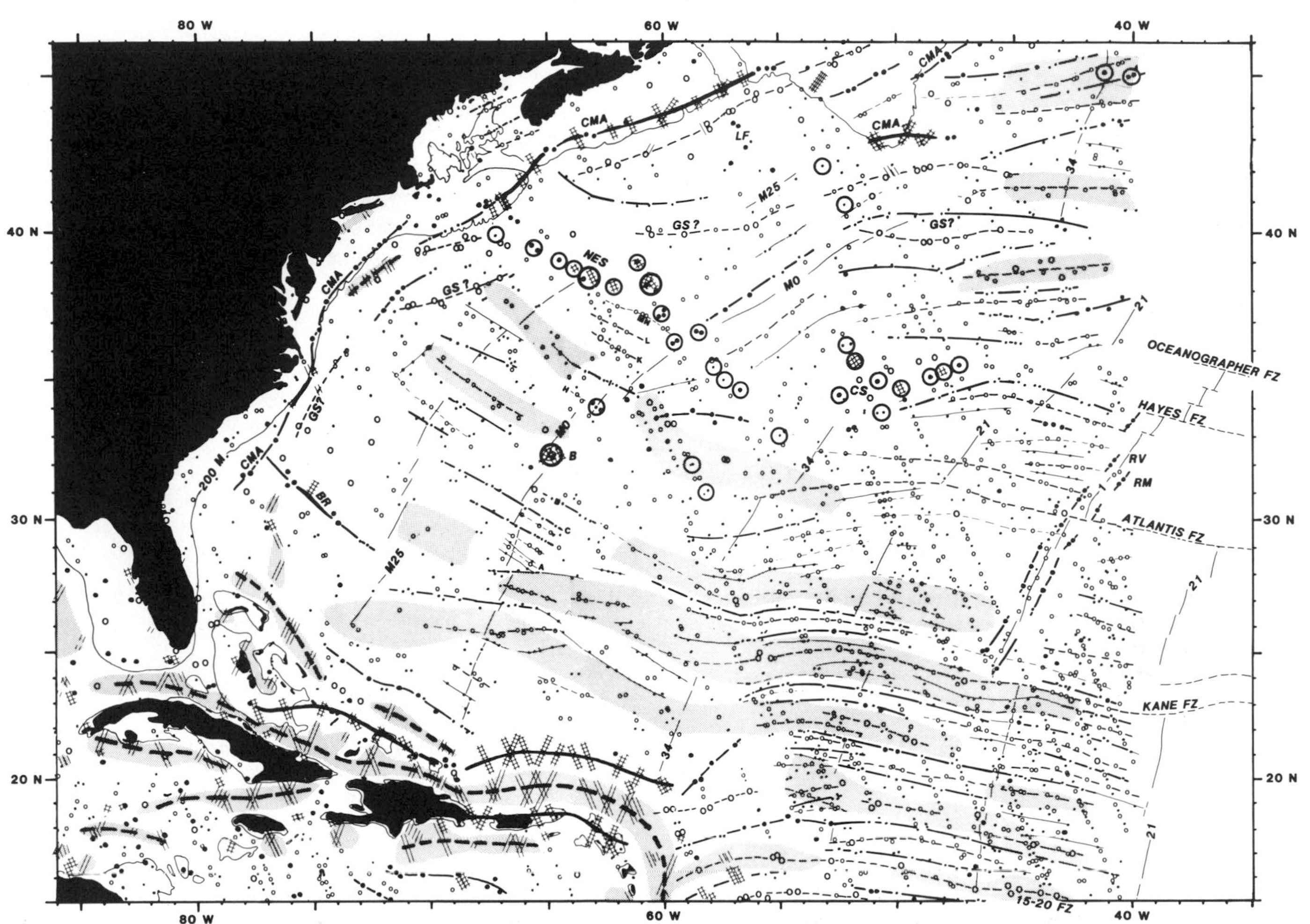

Figure 6. Locations of filtered geoid maxima (filled symbols) and minima (open symbols), connected by lines wherever apparently lineated, in relation to present plate boundary, major FZ's, and anomalies 21, 34, M-O, and M-25 from Perry and others (1981). Parallel bars show filtered values exceeding ± 100 cm; larger circles indicate > 100-cm extremes, and small circles less than 50 cm. Large open circles surround anomalies associated with known seamounts. Stippled bands delineate broader lows in western North Atlantic as well as deep lows (< 100 cm) in Caribbean area. Anomalies identified by letter include: CMA, Continental Margin Anomaly; CS, Corner Seamounts; GS, Gulf Stream; BR, Blake Ridge; NES, New England Seamounts; RV, Rift Valley; RM, Rift Mountains; B, Bermuda. Fracture Zones A-N identified using detailed magnetic data and Fig. 3 (from Vogt and others, 1984).

fracture zones and in general parallel to them, the ribbons are too wide to be single fracture zones. They could represent spreading axis segments which maintained some characteristic (e.g., obliqueness, magma chamber size, or melt production) over long periods of time. Alternatively, the ribbons reflect varying densities of minor fracture zones.

Several geoid anomalies or anomaly clusters trend oblique to both transform fracture zone and isochron trends (Figs. 5, 6). Examples of possible diachronous geoid lineations include: (1) broad 50 to 100-cm–high trends northeast from the eastern end of the Puerto Rico Trench. The anomaly is associated with only a minor sea-floor elevation. This geoid high ("C" in Figure 6) may indicate compression and crustal thickening within the Atlantic lithosphere as it approaches the plate boundary. Some of the large geoid anomalies south and southeast of this high may have a similar origin; (2) a northeast trending high in line with the Corner Seamounts might reflect a propagating rift (Hey and others, 1980) created when the New England/Corner Seamounts hotspot was near the ridge crest; and (3) a major EW high lies over the continental rise off Cape Hatteras; similar features trend NE to E in the area north of the Corner and New England Seamounts. Although probably in part a reflection of the Gulf Stream and its eddies, the anomalies may also indicate regions of distinctive crustal/upper-mantle density or crustal thickness.

So far most geoid and gravity studies have been necessarily one-dimensional (profile) in form. Since the geoid undulation differs for 1- vs 2-dimensional features (ridges vs seamounts) (Watts and Ribe, 1984), the complete analysis of an altimeter profile requires a data swath along the rev, not merely a single track as in Figures 3 and 4. In the near future (see Anderle, this volume) additional altimetry of SEASAT quality (GEOSAT) or better (TOPEX) (Stewart, 1985), as well as detailed aerogravity surveys (Brozena, 1984) will allow two-dimensional inversion-to-bottom (or basement) topography, and derivation of two-dimensional response functions and plate-loading studies. To be effective, such refined studies require accurate bathymetric and sediment-thickness maps and tectonic constraints, particularly the age of the crust at the time of loading. Notably in the western Atlantic, the altimetry will have to be averaged over many passes to remove transient dynamic effects. Contributions to the shape of the ocean and surface by the stationary invariant part of the circulation can only be identified by direct oceanographic observation or by subtracting the time-averaged *altimetry* from the *gravimetric* geoid, the latter computed from surface gravity measurements or satellite-to-satellite tracking.

REFERENCES

Apel, J. R.
1980 : Satellite sensing of ocean surface dynamics: Annual Review Earth and Planetary Science Letters, v. 8, p. 303–342.

Born, G. H., Dunne, J. A., and Lame, D. B.
1979 : SEASAT mission overview: Science, v. 204, p. 1405–1406.

Bowin, C.
1983 : Depth of principal mass anomalies contributing to the earth's geoidal undulations and gravity anomalies: Marine Geodesy, v. 7, p. 61–100.

Bowin, C., Thompson, G., and Schilling, J. G.
1984 : Residual geoid anomalies in Atlantic Ocean Basin: Relationship to mantle plumes: Journal of Geophysical Research, v. 89, p. 9905–9918.

Bowin, C., and Milligan, J.
1985 : Negative gravity anomaly over spreading rift valleys; Mid-Atlantic Ridge at 26°N: Journal of Geophysical Research, in press.

Brammer, R. F.
1979 : Estimation of the ocean geoid near the Blake escarpment using GEOS-3 altimetry data: Journal of Geophysical Research, v. 84, p. 3843–3852.

Brammer, R. F., and Sailor, R. V.
1980 : Preliminary estimates of the resolution capability of the SEASAT radar altimeter: Geophysical Research Letters, v. 7, p. 193–196.

Brown, R. D., Kahn, W. D., McAdoo, D. C., and Himwich, W. E.
1983 : Roughness of the marine geoid from SEASAT altimetry: Journal of Geophysical Research, v. 88, p. 1531–1540.

Brozena, J. M.
1984 : A preliminary analysis of the NRL airborne gravimetry system: Geophysics, v. 49, p. 1060–1069.

Cathles, L. M. III.
1975 : The Viscosity of the Earth's Mantle: Princeton Univ. Press, Princeton, NJ 386 p.

Cazenave, A., Lago, B., and Dominh, K.
1983 : Thermal parameters of the oceanic lithosphere estimated from geoid height data: Journal of Geophysical Research, v. 88, p. 1105–1118.

Chapman, M.E.D.
1979 : Techniques for interpretation of geoid anomalies: Journal of Geophysical Research: v. 84, p. 3793–3801.

Chapman, M.E.D., and Talwani M.
1979 : Comparison of gravimetric geoids with Geos-3 altimetric geoid: Journal of Geophysical Research: v. 84, p. 3803–3816.

Chase, C. G.
1979 : Subduction, the geoid, and lower mantle convection: Nature: v. 282, p. 464–468.

Cheney, R. E.
1982 : Comparison data for SEASAT altimetry in the western North Atlantic: Journal of Geophysical Research, v. 87, p. 3247–3253.

Cochran, J. R. and Talwani, M.
1977 : Free-air gravity anomalies in the world's oceans and their relationship to residual elevation: Geophysical Journal of the Royal Astronomical Society, v. 50, p. 495–552.

Cochran, J. R. and Talwani, M.
1978 : Gravity anomalies, regional elevation, and the deep structure of the North Atlantic: Journal of Geophysical Research, v. 83, p. 4907–4924.

Crough, S. T.
1978 : Thermal original of mid-plate hot-spot swells: Geophysical Journal of the Royal Astronomical Society, v. 55, p. 451–469.

Crough, S. T.
1979a: Geoid anomalies across fracture zones and the thickness of the lithosphere: Earth and Planetary Science Letters, v. 44, p. 224–230.

Crough, S. T.
1979b: Hotspot epeirogeny: Tectonophysics, v. 61, p. 321–333.

Crough, S. T.
1981a: Free-air gravity over the Hoggar Massif, Northwest Africa: Evidence for alteration of the lithosphere: Tectonophysics, v. 77, p. 189–202.

Crough, S. T.
1981b: The Darfur swell, Africa: Gravity constraints on its isostatic compensation: Geophysics Research Letters, v. 8, p. 877–879.

Crough, S. T., and Thompson, G. A.
1976 : Thermal model of continental lithosphere: Journal of Geophysical Research, v. 81, p. 4857–4862.

Dixon, T. H., Naraghi, M., McNutt, M. K., and Smith, S. M.
1983 : Bathymetric prediction from SEASAT altimeter data: Journal of Geophysical Research, v. 88, p. 1563–1571.

Dziewonski, A. M.
1984 : Mapping the lower mantle: determination of lateral heterogeneity in P velocity up to degree and order 6: Journal of Geophysical Research, v. 89, p. 5929–5952.

Fischer, I., and Wyatt, P., III.
1974 : Deflections of the vertical from bathymetric data: Defense Mapping Agency, Topographic Center, Washington, D.C., International Symposium on Applications of Marine Geodesy, p. 1–13.

Fischer, I.
1979 : The effect of the Mid-Atlantic Ridge in terms of gravity anomalies, geoidal undulations and deflections of the vertical: Marine Geodesy, v. 2, p. 215–217.

Gass, I. G., Chapman, D. S., Pollack, H. N., and Thorpe, R. S.
1978 : Geological and geophysical parameters of mid-plate volcanism: Philosophical Transactions of the Royal Society of London A, v. 288, p. 581–597.

Groeger, W. J.
1981 : An experimental computer algorithm for seamount model parameter estimation based on SEASAT-A satellite radar altimetry: NSWC TR 81-2000, Dahlgren, VA, 50 p.

Hadgigeorge, C., Blaha, G., and Rooney, T. P.
1981 : SEASAT altimeter reductions for detailed determinations of the oceanic geoid: Annual Geophysics, v. 37, p. 123–132.

Hager, B. H.
1984 : Subducted slabs and the geoid: constraints on mantle rheology and flow: Journal of Geophysical Research, v. 89, p. 6003–6015.

Haxby, W. F. and Turcotte, D. L.
1978 : On isostatic geoid anomalies: Journal of Geophysical Research, v. 83, p. 5473–5478.

Haxby, W. F., Karner, G. D., LaBrecque, J. L., and Weissel, J. K.
1983 : Digital images of combined oceanic and continental data sets and their use in tectonic studies: EOS-Transactions of the American Geo-

physical Union, v. 64, p. 995–1004.
Heestand, R. L. and Crough, S. T.
1981 : The effect of hotspots on the oceanic age-depth relation: Journal of Geophysical Research, v. 86, p. 6107–6114.
Hey, R., Duennebier, F. K., and Morgan, W. J.
1980 : Propagating rifts on mid-ocean ridges: Journal of Geophysical Research, v. 85, p. 3647–3658.
Huppert, H. E., and Bryan, K.
1976 : Topographically generated eddies: Deep Sea Research, v. 23, p. 655–679.
Jordan, T. H.
1979 : Mineralogies, densities and seismic velocities of garnet lherzolites and their geophysical implications, eds. F. Boyd and H. Meyer; Proceedings of the Second International Kimberlite Conference, v. II, American Geophysical Union, Washington, DC, p. 1–14.
Jung, W.-Y., and Rabinowitz, P.D.
1986 : Residual geoid anomalies of the North Atlantic Ocean and their tectonic implications: Journal of Geophysical Research, in press.
Lame, D. B. and Born, G. H.
1982 : SEASAT measurement system evaluation: Achievements and limitations: Journal of Geophysical Research, v. 87, p. 3175–3178.
Marsh, J. G. and Chang, E. S.
1978 : 5′ detailed gravimetric geoid in the northwestern Atlantic Ocean: Marine Geodesy, v. 1, 253–261.
McGoogan, J. T., Miller, L. S., Brown, G. S. and Hayne, G. S.
1974 : The S-193 radar altimeter experiment: IEEE Proceedings, v. 62(6), p. 793–804.
McKenzie, D. and Bowin, C.
1976 : The relationship between bathymetry and gravity in the Atlantic Ocean: Journal of Geophysical Research, v. 81, p. 1903–1915.
Menard, Y.
1983 : Observations of eddy fields in the northwest Atlantic and northwest Pacific by SEASAT altimeter data: Journal of Geophysical Research, v. 88, p. 1853–1866.
Morgan, W. J.
1981 : Hotspot tracks and the opening of the Atlantic and Indian Oceans; Ch. 13: The Sea, V. 7, ed. C. Emiliani, Wiley and Sons, NY, p. 443–487.
Parke, M. E. and Dixon, T. H.
1982 : Topographic relief from Seasat altimeter mean sea-surface, July 7 - December 10, 1978: Nature, v. 300, p. 317.
Parker, R. L.
1972 : The rapid calculation of potential anomalies: Geophysical Journal of the Royal Astronomical Society, v. 31, p. 447–455.
Parsons, B. and Sclater, J. G.
1977 : An analysis of the variation of ocean floor bathymetry and heat flow with age: Journal Geophysical Research, v. 82, p. 803–827.
Parsons, B. and Richter, F. M.
1981 : Mantle convection and the oceanic lithosphere: Sea: v. 7, p. 73–117.
Parsons, B. and Daly, S.
1983 : The relationship between surface topography, gravity anomalies and temperature structure of convection: Journal of Geophysical Research, v. 88, p. 1129–1144.
Pekeris, C. L.
1935 : Thermal convection in the interior of the earth: Monthly Notices of the Royal Astronomical Society, Geophysical Supplement, v. 3, p. 343–367.
Perry, R. K., Fleming, H. S., Vogt, P. R., Cherkis, N. Z., Feden, R., Thiede, J., Strand, J. E., and Collette, B. J.
1981 : North Atlantic Ocean: Bathymetry and plate tectonic evolution; Geological Society of America, Map and Chart Series MC-35.
Rapp, R. H.
1983 : The determination of geoid undulations and gravity anomalies from SEASAT altimeter data: Journal of Geophysical Research, v. 88, p. 1552–1562.
Ribe, N. M.
1982 : On the interpretation of frequency response functions for oceanic gravity and bathymetry: Geophysical Journal of the Royal Astronomical Society, v. 70, p. 273–294.
Rice, D. A.
1952 : Deflections of the vertical from gravity anomalies: Bulletin Geodesie, v. 25, p. 285–312.
Richards, M. A., and Hager, B. H.
1984 : Geoid anomalies in a dynamic earth: Journal of Geophysical Research, v. 89, p. 5987–6002.
Richter, F. M. and Parsons, B.
1975 : On the interaction of two scales of convection in the mantle: Journal of Geophysical Research, v. 80, p. 2529–2541.
Royer, T. C.
1978 : Ocean eddies generated by seamounts in the North Pacific: Science, v. 199, p. 1063–1064.
Sandwell, D. and Schubert, G.
1980 : Geoid height vs age for symmetric spreading ridges: Journal of Geophysical Research, v. 85, p. 7235–7241.
Sandwell, D. T. and Schubert, G.
1982 : Geoid height-age relation from SEASAT altimeter profiles across the Mendocino Fracture Zone: Journal of Geophysical Research, v. 87, p. 3949–3958.
Schouten, H. and Klitgord, K. D.
1977 : Map showing Mesozoic magnetic anomalies: Western North Atlantic, U.S. Geological Survey Miscellaneous Field Studies, Map MF 915, Reston, VA.
Sclater, J. G., Lawver, L. A., and Parsons, B.
1975 : Comparison of long wavelength residual elevation and free air gravity anomalies in the North Atlantic and possible implications for the thickness of the lithospheric plate: Journal of Geophysical Research, v. 80, p. 1031–1052.
Searle, R., Johnson, G. L., and Monahan, D.
1982 : General Bathymetric Chart of the Oceans (GEBCO) 5th Edition, chart 5.08 (North Atlantic): Canadian Hydrographic Survey, Ottawa, Canada (1 sheet).
Smith, S. L. III, and Hicks, T. I.
1981 : Evaluation of geodetic products produced by the NSWC reduction of SEASAT radar altimeter data, NSWC TR 81-260, Dahlgren, VA, 27 p.
Souriau, A.
1984 : Geoid anomalies over Gorringe Ridge, North Atlantic Ocean: Earth and Planetary Science Letters, v. 68, p. 101–114.
Stanley, H. R.
1979 : The GEOS-3 Project: Journal of Geophysical Research, v. 84(B8), p. 3779–3783.
Stefanick, M., and Jurdy, D. M.
1984 : The distribution of hot spots: Journal of Geophysical Research, v. 89, p. 9919–9925.
Stewart, R. H.
1985 : The NASA NSCAT AND TOPEX/POSEIDON programs: Ocean Engineering and the Environment, MTS/IEEE Conference Record, p. 256–263.
Tapley, B. D., Born, G. H., and Parke, M. E.
1982 : The SEASAT altimeter data and its accuracy assessment: Journal of Geophysical Research, v. 87, p. 3179–3188.
Townsend, W. F.
1980 : An initial assessment of the performance achieved by the SEASAT radar altimeter, IEEE Journal of Oceanic Engineering, v. OE-5(2), p. 80–92.
Tucholke, B. E. and Vogt, P. R.
1979 : Western North Atlantic: Sedimentary evolution and aspects of tectonic history, in Initial Reports of the Deep Sea Drilling Project, v. 43: eds. B. E. Tucholke, and P. R. Vogt, U.S. Government Printing Office, Washington, D.C., p. 791–826.
Tucholke, B. E., Houtz, R. E., and Ludwig, W. J.
1982 : Maps of sediment thickness and depth to basement in the western

North Atlantic Ocean Basin (two charts plus 15 p. text): American Association Petroleum Society, Tulsa, 1982.

Vogt, P. R., Johnson, G. L., Holcombe, T. L., Gilg, J. G., and Avery, D. E.
1971 : Episodes of sea-floor spreading recorded by the North Atlantic basement: Tectonophysics, v. 12, p. 211–234.

Vogt, P. R.
1978 : Long-wavelength gravity anomalies and intraplate seismicity: Earth and Planetary Science Letters, v. 37, p. 465–475.

Vogt, P. R. and Tucholke, B. E.
1979 : The New England Seamounts: Testing origins, in Initial reports of the Deep Sea Drilling Project, v. 43: eds. B. E. Tucholke and P. R. Vogt, U.S. Government Printing Office, Washington, D.C., p. 847–856.

Vogt, P. R. and Einwich, A. M.
1979 : Magnetic anomalies and sea-floor spreading in the western North Atlantic, and a revised calibration of the Keathley(M) geomagnetic reversal chronology, in Initial Reports of the Deep Sea Drilling Project, v. 43: eds. B. E. Tucholke, P. R. Vogt, et al., Initial reports of the Deep Sea Drilling Project, U.S. Government Printing Office, Washington, D.C., p. 857–876.

Vogt, P. R.
1981 : On the applicability of thermal conduction models to mid-plate volcanism: comments on a paper by Gass et al.: Journal of Geophysical Research, v. 86, p. 950–960.

Vogt, P. R., Perry, R. K., Feden, R. H., Fleming, H. S., and Cherkis, N. Z.
1981 : The Greenland-Norwegian Sea and Iceland environment: Geology and geophysics: Ch. 11: The Ocean Basins and Margins, v. 5, eds. A.E.M. Nairn, M. C. Churkin, Jr., and F. G. Stehli, Plenum NY, p. 493–598.

Vogt, P. R. and Perry, R. K.
1982 : North Atlantic Ocean: Bathymetry and plate tectonic evolution, text to accompany chart MC-35: Geological Society of America, 21 p.

Vogt, P. R., Zondek, B., Fell, P. W., Cherkis, N. Z., and Perry, R. K.
1984 : SEASAT altimetry, the North Atlantic geoid, and evaluation by shipborne subsatellite profiles: Journal of Geophysical Research, v. 89, p. 9885–9903.

von Arx, W. S.
1966 : Level-surface profiles across the Puerto Rico Trench: Science, v. 154, p. 1651–1653.

Walcott, R. J.
1972 : Late Quaternary vertical movements in eastern North America: Quantitative evidence of glacio-isostatic rebound: Reviews of Geophysics, v. 10, 849–884.

Watts, A. B.
1979 : On geoid heights derived from the GEOS 3 altimeter data along the Hawaiian - Emperor seamount chain: Journal of Geophysical Research, v. 84, p. 3817–3826.

Watts, A. B., and Ribe, N. M.
1984 : On geoid heights and flexure of the lithosphere at seamounts: Journal of Geophysical Research, v. 89, p. 11152–11170.

Watts, A. B., Horai, K., and Ribe, N. M.
1984 : On the determination of the deflection of the vertical by satellite altimetry: Marine Geodesy, v. 8, p. 85–127.

Watts, A. B., Cochran, J. R., Patriat, P., and Doucoure, M.
1985 : A bathymetry and altimetry profile across the southwest Indian Ridge crest at 31°S latitude: Earth and Planetary Science Letters, v. 73, p. 129–139.

West, G.
1982 : Mean earth ellipsoid determined from SEASAT 1 altimetric observations: Journal of Geophysical Research, v. 87, p. 5538–5540.

White, J. V., Sailor, R. V., Lazarewicz, A. R., LeSchack, A. R.
1983 : Detection of seamount signatures in SEASAT altimeter data using matched filters: Journal of Geophysical Research, v. 88, p. 1541–1551.

Woodhouse, J. H., and Dziewonski, A. M.
1984 : Mapping the upper mantle: Three-dimensional modeling of earth structure by inversion of seismic waveforms: Journal of Geophysical Research, v. 89, p. 5953–5986.

Zondek, B.
1982 : Highpass filtering of satellite altimeter data, NSWC/TR 82-427, Dahlgren, VA, 21 p.

Zwally, H. J., Bindschadler, R. A., Brenner, A. C., Martin, T. W., and Thomas, R. H.
1983 : Surface elevation contours of Greenland and Antarctic ice sheets: Journal of Geophysical Research, v. 88, p. 1589–1596.

MANUSCRIPT ACCEPTED BY THE SOCIETY OCTOBER 7, 1985

ACKNOWLEDGMENTS

For the sharing of ideas, effort, and data, I thank R. Anderle, P. Fell, W. Groeger, S. Smith, G. West and B. Zondek of the Naval Surface Weapons Center; W. Hadgigeorge and T. Rooney of the Air Force Geophysical Lab, R. Gebhardt of the Naval Oceanographic Office, and J. Brozena of the Naval Research Laboratory. Assistance at the computer, drafting table, and word processor was provided respectively by L. La Lumiere, I. Jewett, and M. Whitney. I thank our NRL Navigators J. Ostrander and A. Zuccaro and the officers and crew of USNS *Hayes*. Insightful reviews of this paper were provided by C. Bowin and W. Haxby. This study was in part supported by the Office of Naval Research.

Printed in U.S.A.

The Geology of North America
Vol. M, The Western North Atlantic Region
The Geological Society of America, 1986

Chapter 15

Magnetic anomalies and crustal magnetization

Peter R. Vogt
Naval Research Laboratory, Washington, D.C. 20375-5000

INTRODUCTION AND HISTORY

The study of geomagnetism is, after geodesy, the oldest of the geophysical sciences and, with the advent of plate tectonics, perhaps the most important. European mariners were using compasses, probably of the "floating lodestone" type, by no later than the treatise by Alexander Neckam in the year 1187. The GSA centennial thus also marks *eight* centuries of geomagnetic measurement. When Columbus measured magnetic declination (variation) at several points on his 1492 voyage, he made the first "geophysical transect" across the Atlantic Ocean, some 3½ centuries before the first bathymetric profile was drawn. Of course, Columbus was actually prospecting the earth's core, not the oceanic crust. Geomagnetic field strength was first measured by Gauss in 1832, but continuous, accurate measurement suitable for mapping magnetic anomalies of geologic origin was only achieved a century later. However, the first magnetic chart detailed enough to resolve magnetic anomalies of geologic origin in the western Atlantic (Bermuda; Cole, 1908) depicted anomalies of declination, *not* intensity. Cole referred to the anomalies merely as "sources of considerable local disturbance." Declination anomalies measured already in 1786 in Icelandic harbors by P. Lövenörn (published in 1799) were correctly attributed to magnetic effects of iron-rich minerals in nearby basalts (Kristjánsson, 1982).

Continuous magnetic field strength measurements began after the end of World War II, which had stimulated the development of airborne magnetometers to detect submarines. In 1948 a U.S. Navy self-orienting fluxgate magnetometer was converted to a "towed fish" at Lamont Observatory (Heezen and others, 1953). That year the first aeromagnetic flights profiled what are now called the East Coast Magnetic Anomaly (ECMA), the Jurassic Magnetic Quiet Zone (JMQZ), the M (Keathley) sequence, and the Bermuda edifice (Keller and others, 1954). The first shipborne magnetic anomaly profile, across the Atlantic from Dakar to Barbados, was collected in October–November 1948, on R/V *Atlantis*. This profile revealed the "conjugate" magnetic smooth zone—extending ca. 300 km west from Dakar—while the rest of the profile showed anomalies of 50–200nT amplitude and 10–75 km wavelength (Heezen and others, 1953). Although attributing the anomalies to a "heterogeneity of material," the authors noticed the lack of correlation with topography (later verified by statistical cross-correlation; Heirtzler and Le Pichon, 1965) and they inferred correctly that the top of the magnetized layer was not deep below the sea-floor. On October 23, 1948, *Atlantis* crossed what is now recognized as the Mid-Atlantic Ridge (MAR) rift valley, and recorded a 900nT (gamma) positive anomaly now known as the "central anomaly," #1 of the sea-floor spreading (SFS) sequence. The MAR rift valley was discovered by Marie Tharp in 1953 and the magnetic anomaly reported by Ewing and others in 1957. According to Heezen and others (1959, p. 100) "The Rift Valley is characterized by a large positive anomaly, while the adjoining Rift Mountains show negative anomalies of 300 to 500 gammas." These lows are now understood to mark crust formed during the Matuyama reversed chron (0.73–2.48 Ma). While the significance of marine magnetic anomalies went unnoticed in the 1950's, the new science of paleomagnetism was beginning to produce support for continental drift (Runcorn, 1956).

Notwithstanding Hospers' (1953) demonstration of both normally and reversely magnetized basalt units on Iceland, the possibility of reversely magnetized rocks in the ocean basins was not proposed until the work of Girdler and Peter (1960). Workers at the U.S. Naval Oceanographic Office (N.O.O.) noted that magnetic anomalies over the Pacific-Antarctic Ridge are lineated parallel to the rise crest, and they suggested remanent magnetization, locally reversed, among possible interpretations (Marine Surveys Division, 1962). However, even after the subsequent publication of the Vine-Matthews hypothesis (1963), independently proposed by Morley and LaRochelle (1964), this idea was rejected for a few more years. For example, Heirtzler and Le Pichon (1965) attributed the MAR central anomaly to induced magnetization. Heirtzler and others (1966) analyzed the detailed aeromagnetic survey flown by N.O.O. Project MAGNET across the Reykjanes Ridge south of Iceland. The anomaly pattern, soon to play a key role in the conversion of geologists to the sea-floor spreading hypothesis, was initially attributed to topographic effects, and, if so, "would indicate that the magnetic pattern is controlled primarily by the tectonic history of the ridge and not be reversals of the earth's magnetic field" (Heirtzler and others,

Vogt, P. R., 1986, Magnetic anomalies and crustal magnetization, *in* Vogt, P. R., and Tucholke, B. E., eds., The Geology of North America, Volume M, The Western North Atlantic Region: Geological Society of America.

1966, p. 434). Examining the rift valley and crestal mountains at 42.5–46°N, Vogt and Ostenso (1966) showed that many of the short-wavelength positive magnetic anomalies could result from topographic features of high remanence (ca. 5 A/m) similar to values they measured on dredge samples. (Such high magnetization of submarine basalts was already reported by Matthews [1961]). But, since "none of the [short-wavelength] anomalies required reversely polarized sources," Vogt and Ostenso cautioned that "there is yet no way to establish that the longer-wavelength anomalies [of Vine and Matthews, 1963] are not also the result of alternating intensities of magnetization." However, the symmetry of the magnetic anomaly pattern (e.g. on the Reykjanes Ridge) and its correlation to the fledgling geomagnetic polarity history of Cox and others (1963) soon provided an almost airtight case for the Vine-Matthews hypothesis (Vine, 1966). Heirtzler and his students at Lamont became converts to the hypothesis in 1966, and by 1967 massive reanalyses of available magnetic data in the major ocean basins convinced most marine geologists of the validity of the hypothesis. Marine magnetic data were used to construct a geomagnetic reversal chronology that extended back to anomaly 32 in the late Cretaceous (Heirtzler and others, 1968). Synthesis of these and other results, in terms of plate tectonics, followed immediately (Morgan, 1968; Le Pichon, 1968).

The first airborne and shipborne magnetometers (Keller and others, 1954; Heezen and others, 1953) were fluxgate instruments; refined three-component versions are still used on the Project MAGNET aircraft, as well as on MAGSAT as three-component (vector) magnetometers. Use of the proton precession magnetometer (Hurwitz and Nelson, 1960), first used as a towed marine instrument by the N.O.O. in late 1958, spread through the marine geophysics community by the early 1960s and today remains a standard shipborne total field magnetometer. The optically pumped helium-vapor magnetometer, initially developed for petroleum exploration, was first used at the N.O.O. in 1963; the first extensive detailed aeromagnetic survey extending far out over oceanic crust (Jurassic Magnetic Quiet Zone; Fig. 1) was performed from 1964–1966, partly with a helium magnetometer. Satellite total field instruments progressed from the proton precession types on the Vanguard (1959) and early Cosmos satellites (1964) through Rubidium-vapor types on OGO (1964–1971) to cesium vapor instruments on Cosmos 321 (1970) and MAGSAT (1979–80) where the accuracy had been improved to ±2nT (Taylor and others, 1983).

In this volume, the application of SFS (Sea Floor Spreading) magnetic anomalies to plate kinematics is discussed in Chapters 22 and 23. Here, after a summary of rock magnetism, the major SFS provinces (Anomalies 1-34, Cretaceous Magnetic Quiet, M-Series, and Jurassic Quiet [or smooth] Zone) are sumarized. The main part of this chapter discusses the features of Atlantic magnetic anomalies (SFS and otherwise) and implications for the structure, composition, and evolution of oceanic crust in the greater North Atlantic. The supporting figures, including Plate 3, are a selection of high-quality data sets best illustrating the phenomena discussed. The Iceland/northeast Atlantic area is over-represented (Fig. 1) by virtue of a wealth of data as well as phenomena discovered there.

MAGNETIC PROPERTIES OF THE OCEANIC CRUST

The magnetic properties of the oceanic crust — and, at some distance from the accretion axis also the upper mantle down to the Curie isotherm for magnetite (580°C) — have been deduced from magnetic field measurements (discussed later) and from measurements on rock samples, either from the present oceanic crust or from ophiolites, some of which may be obducted ancient oceanic crust (e.g., Banerjee, 1980). The "present crust" samples come from Iceland's surface, in the oceans from dredging outcrops and talus, and in both areas from boreholes. Some surface samples originated in the deeper crust: Along the MAR (Fox and Opdyke, 1973), particularly in fracture zones, faulting or diapirism have brought mafic or ultramafic intrusives to the sea floor. On Iceland, erosion has exposed basalts up to 1,500 m or more below the original surface. Drillholes up to 3,085 m deep (and to 4,300 m below the original surface) have returned chips whose magnetic properties could be measured (Pálmason and others, 1979). The Iceland Research Drilling Project (IRDP) drillsite was cored to its base 1,919 m below the ground, about 3,600 m below the original surface. Although the original top 600 m was removed by erosion, the lava section from that level to the top of the drillhole was sampled in updip outcrops. The magnetic properties of the IRDP samples are discussed by Bleil and others (1982), Hall (1985), and others.

Of the eight deepest boreholes into oceanic crust, all but one (1,076 m basement penetration, Site 504B, Anderson and others, 1982) are in the central North Atlantic. Recent syntheses of oceanic crustal magnetization include Harrison (1976, 1981) and Lowrie (1977). The magnetism of deeper crust and upper mantle rocks was studied by Kent and others (1978) (dredge samples) and Dunlop and Prévot (1982) (DSDP drill cores). In the latter study, 35 of 50 samples from two (334 and 395A) of the three drillsites are Atlantic. Figure 2 compares the magnetization (NRM) structure of the oceanic crust inferred in these two papers, as related to the average petrologic and seismic structure in terms of a layered crust. (As discussed in Chapter 19, however, there are probably no sharp, laterally continuous seismic discontinuities between the "layers" except at the 3B/4 boundary, the "geophysical Moho"). Based on dredge sampling at the MAR axis (Irving and others, 1970) and magnetic/topographic profiles along the Reykjanes Ridge (Talwani and others, 1971) it was long believed that SFS lineations were entirely due to a highly magnetized (10-20 A/m) pillow basalt layer only about 500 m thick. This result was challenged when DSDP drillholes demonstrated a rather low site mean (2.4 A/m for 50 sites reduced to the Equator, Harrison [1976]; 3.7 A/m for 50 sites unreduced, Lowrie [1977]) This is 4 to 5 times too weak to explain SFS lineation amplitudes if the magnetized layer is only 500 m thick. The presence of basalts of both polarities in a vertical section (e.g.,

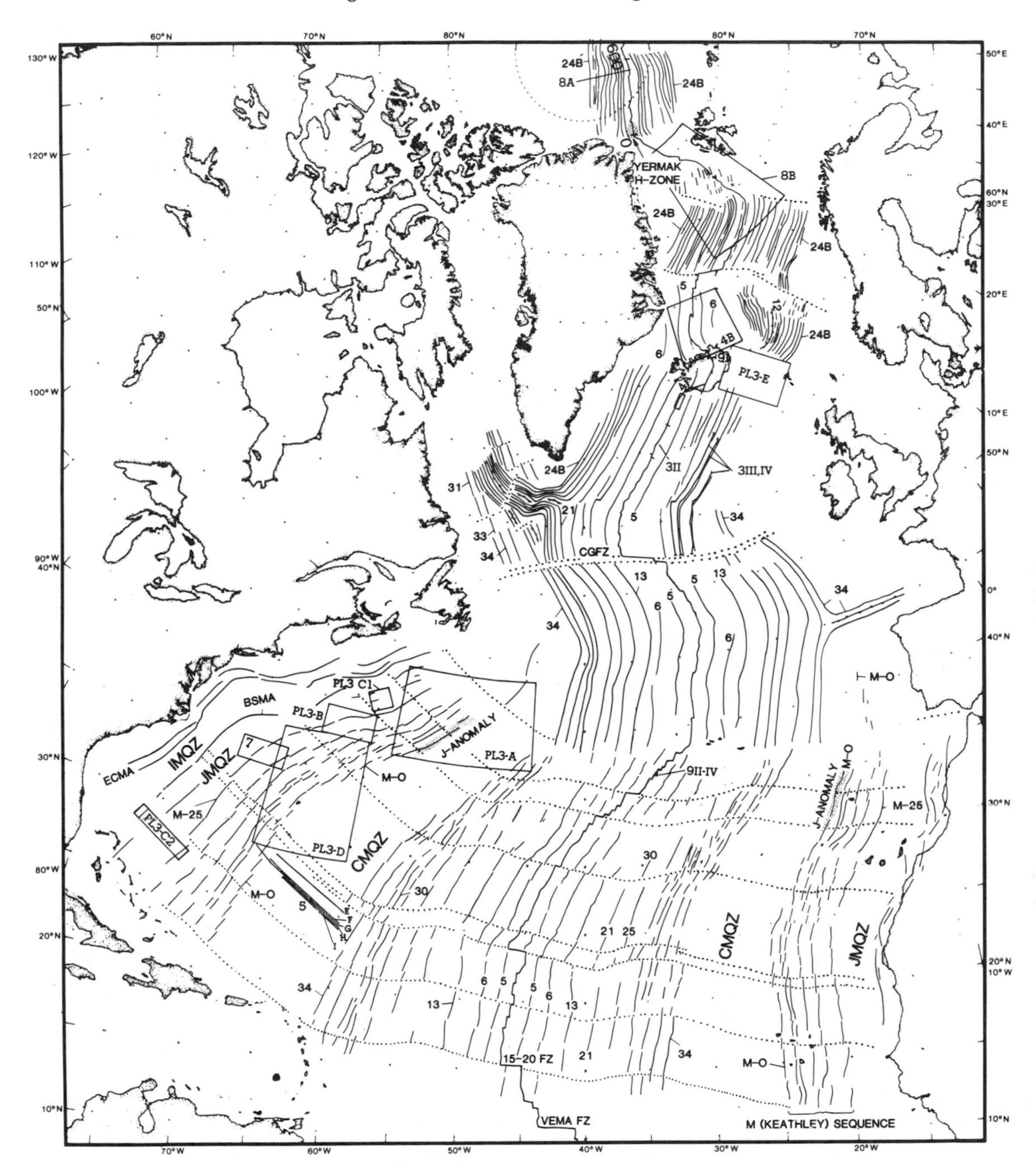

Figure 1. Magnetic lineations and transform FZs (dotted) in North Atlantic-Arctic area, simplified after Klitgord and Schouten (chapter 22) and other sources, and locations of data illustrated in subsequent figures and Plate 3. Transverse Mercator projection with 40°W center meridian.

DSDP sites 332A, 395A, 396B, and possibly 396) further complicates the problem (Harrison, 1981). Only locally (e.g., FAMOUS rift valley, Prévot and others, 1979) can the anomalies be explained by a 500 m thick basalt layer. As a group, the eight deepest DSDP boreholes show no downward trend in TRM intensity, at least in the top 600 m of crust sampled (Hall, 1985). A significant contribution by deeper intrusive rocks thus seems indicated, although not all agree (e.g., Bleil and Petersen, 1983). Greenschist-facies metabasalts ("greenstones") are almost devoid of magnetic minerals (Fox and Opdyke, 1973), but metadiabases preserve a significant fraction of their NRM (TRM) because the larger grains are more resistant to chloritization (Dunlop and Prévot, 1982). Sills and dikes in the upper part of layer 2B are still in the zeolite facies and, although much of the NRM is soft (VRM), Dunlop and Prévot (1982) — contrary to Kent and others (1978) — consider that the sills and sheeted dikes contrib-

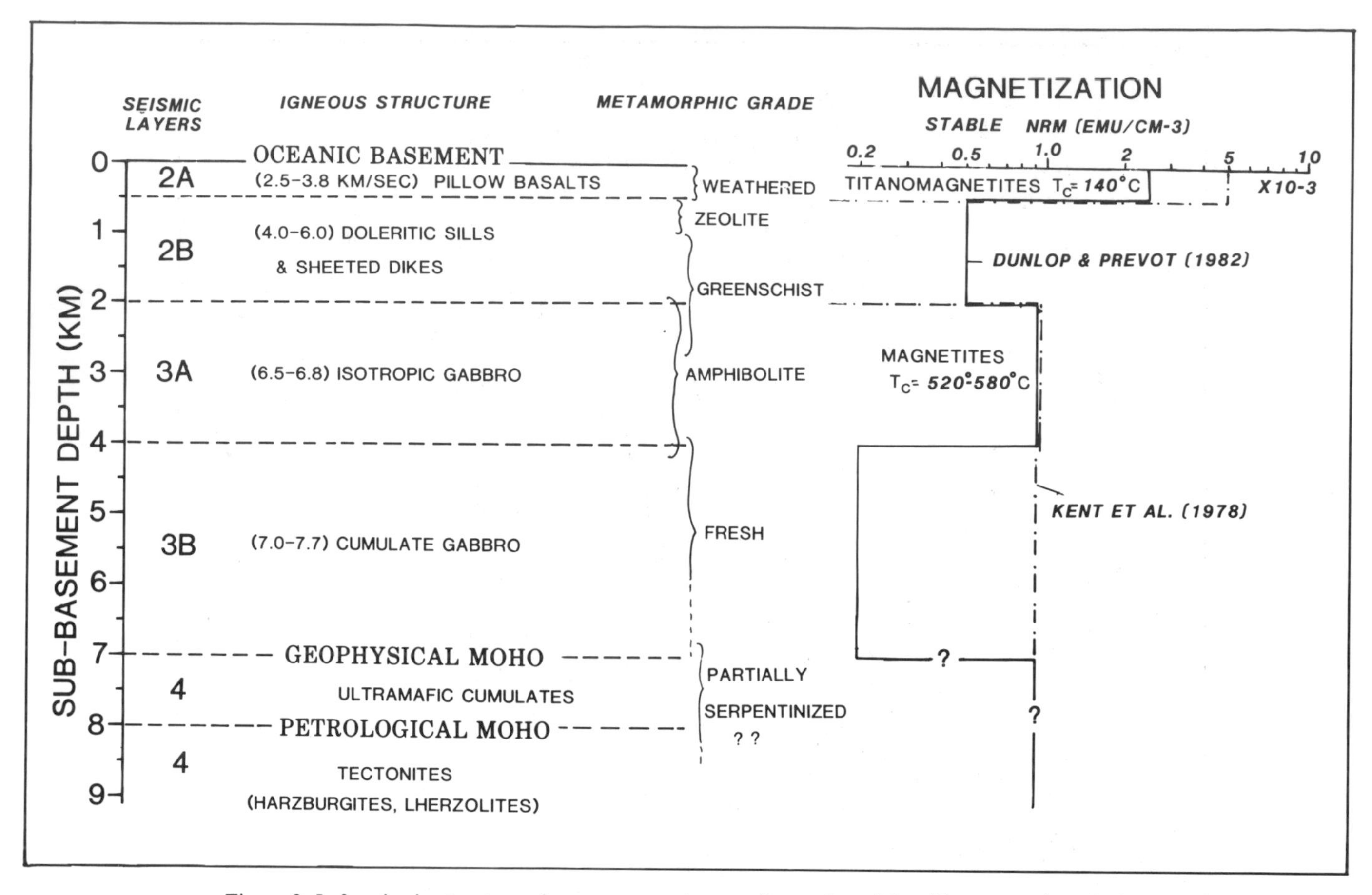

Figure 2. Left, seismic structure of average oceanic crust (layered model) with presumed equivalent igneous structure and metamorphic grade. Right, corresponding magnetization models according to Kent and others (1978) (dash-dot) and Dunlop and Prévot (1982) (solid line). T_c is Curie temperature.

ute significantly to the TRM (ca. 0.5 A/m), as is true in the Troodos complex. Cores from Bermuda suggest that hydrothermal remagnetization (2 A/m) is important in lower 2A and upper 2B (Rice and others, 1980).

Significant stable TRM (1 A/m) may reside in the greenschist-facies and in the amphibolite-facies "isotropic" gabbros. Based on dredged gabbros from Atlantic fracture zones, Kent and others (1978) assigned 1 A/m stable TRM to the "cumulate gabbro," but 10 fresh drillcore gabbros from site 334 yielded only ca. 0.2 A/m (Dunlop and Prévot, 1982). However, the heterogeneity of oceanic gabbros gives little confidence in such averages. Hydrothermal alteration of gabbros or peridotites at temperatures above 250–300°C may cause serpentinization of olivines and precipitation of secondary magnetite. The resultant stable CRM measured on site 334 serpentinized cumulate gabbros (>0.5 A/m) and peridotites (ca. 5 A/m) (Dunlop and Prévot, 1982) makes such rocks potentially significant contributors to SFS anomalies, provided (1) they are a significant (say 20%) constituent of the lower crust (see Harrison [1976] for pros and cons; Kent and others [1978] consider it unlikely) and (2) the CRM was acquired soon (<0.5 m.y.) after the overlying basalts and dikes acquired their TRM; otherwise the CRM would distort the SFS lineation pattern. Diapirically or tectonically intruded serpentinite (e.g., sites 334, 395A) would, in general, contribute to the magnetic anomaly "noise," which tends to be high in the Atlantic.

While NRM (generally TRM) is of greatest interest, induced and viscous remanent (VRM) magnetization may be locally important to explain magnetic anomalies other than SFS lineations. VRM may approach TRM in magnitude (Harrison, 1981) and might explain some magnetic smooth zones (Lowrie, 1973) and the global preponderance of normally magnetized seamounts (Williams and others, 1983). Massive flows and intrusive basalts, by their lower median destructive fields, probably carry more VRM than do pillow basalts. Induced magnetization is generally an order of magnitude weaker than NRM (the "Q"-ratio averages 7.8 for DSDP basalts; Lowrie, 1974) but may be of comparable magnitude in intrusives. Neither VRM nor induced magnetization can contribute to SFS lineations, but spatial variations in their magnitude add "noise."

The magnetic mineral of average submarine extrusives (flows) is titanomagnetite with a composition $xFe_2TiO_4 \times$

(1-x)Fe_3O_4 where x is typically 0.6. This mineral is finely disseminated, single-domain, and comprises typically 1 wt %. The Curie temperature averages 160 to 210°. Where Curie temperatures exceed 500°C in extrusives (e.g. Sites 384, 336, 338, and 342), a former subaerial environment is indicated, the magnetite being formed by subsolidus exsolution of ilmenite following high temperature oxidation during initial cooling. The principal carrier of NRM in most intrusive rocks is magnetite ($\theta_c \simeq$515–575°C) a product of high-temperature deuteric oxidation of titanomagnetite or pyroxene and plagioclase. In serpentinized rocks the NRM is a CRM, the carrier being magnetite crystallized between 250 and 500°C.

Low-temperature oxidation of seafloor basalts changes the titanomagnetite to titanomaghemites that are only weakly magnetized. As a result, the NRM of average pillow basalt is reduced over time from 24 to 0.86 A/m and that of massive flows from 6.5 to 1.2 A/m (Johnson and Hall, 1978). For crust older than 10–30 Ma, NRM increases back to 120 Ma (Bleil and Petersen, 1983). The corresponding average susceptibilities initially decrease from 35 to 0.86 S.I. units in basalt and from 8.4 to 1.3 in massive flows, while Curie temperatures rise from 145 to 340°C and from 147 to 290°C. The relatively weak initial NRM of massive flows reflects slower cooling and larger grain sizes, but this is compensated on older crust by the inaccessibility of massive flow interiors to sea water. Certain Fe-Ti–rich basalts (including "ferrobasalts"), reflecting intense shallow fractionation, exhibit high NRM (up to 50–100 A/m, corresponding to the "H" basalts of Marshall and Cox, 1971) (Vogt, 1979), with corresponding high weight percent titanomagnetite (up to 3%) and rather low Curie temperatures (Anderson and others, 1980). In the FAMOUS area and at Site 332, Prévot and others (1979) showed that olivine basalts, although low in titanomagnetite, are nevertheless, because of a smaller grain size, about twice as magnetic as surrounding plagioclase or pyroxene-rich basalts.

SEA-FLOOR SPREADING (SFS) ANOMALY PROVINCES IN THE NORTH ATLANTIC

A. Anomalies 1-34. Iceland provided early evidence for magnetic reversals (Hospers, 1953), and the symmetric magnetic anomaly pattern over the Reykjanes Ridge (Vine, 1966) was instrumental in proving Vine and Matthews (1963) correct. Nevertheless, the MAR as a whole has been difficult to interpret accurately in terms of the late Cretaceous and Cenozoic SFS lineations (1-34), which cover more than half of the North Atlantic–Arctic oceanic crust (Fig. 1). These difficulties led to early anomaly misidentifications. Except for #6 and #13, the anomalies between 5 and 20 are least clear. The main reason is the slow spreading, which, in combination with close-spaced fracture zones, has produced irregular, poorly resolved sublinear anomalies not easy to identify even where surveyed in detail. The clearest recording of SFS anomalies 1–24B is found in the area within about 500 to 1,500 km of Iceland, a regional influence by the Iceland hotspot discussed later. The along-strike stability of anomaly separation (Fig. 3IV) and amplitude (Fig. 3III), as well as the resolution (Fig. 4) and data density probably represent an optimum for slow spreading.

Examples of significant revisions in anomaly identifications are by Nunns and Peacock (1983) (Norway Basin, 7–24B revised from Talwani and Eldholm, 1977) and Vogt and others (1980) (Iceland Plateau, 5–7 revised from Talwani and Eldholm, 1977). Cande and Kristoffersen (1977), Kristoffersen (1978), and Srivastava and Tapscott (this volume) reidentified as 31, 33, and 34 (119–84 Ma) the anomalies first called 28, 31, and 32 by Pitman and Talwani (1972) and Srivastava (1978) in the North Atlantic and southern Labrador Sea. This revision resulted because anomalies 33 and 34 (the low between them first called "R" by Vogt and others, 1971) were not present in the original reversal chronology of Heirtzler and others (1968). Other features of SFS lineations 1–34 are discussed elsewhere in this chapter and in chapters 22 and 23.

B. The Cretaceous Magnetic "Quiet Zone" (CMQZ). Between SFS lineations M-0 and 34, the central North Atlantic is associated with a 500 to 1,000-km wide province of irregular or weakly lineated magnetic anomalies typically ±50–150nT in amplitude and 30–50 km in wavelength (Figs. 1, 5, Plate 3). These amplitudes are comparable to those in the adjacent SFS anomaly provinces, whereas the Jurassic Magnetic Quiet Zone is indeed "quiet" (±10–50nT amplitudes; Plate 3). Although Cretaceous Magnetic "Quiet Zone" (CMQZ) seems a misnomer, the term is used here for historical consistency, CMQZ magnetic anomalies are not totally chaotic, but somewhat lineated in association with fracture zones (Vogt and others, 1971; Vogt and Einwich, 1979, also the early survey by Jones and others, 1966). A detailed analysis of the CMQZ portion of the Kane FZ in the eastern Atlantic (Twigt and others, 1983) suggests that 10 to 20-km wide strips of more magnetized (or greater layer thickness) crust lie along the young edges of fracture zones (see later). The magnetized layer may also be thinned in narrow zones along FZ axes.

Magnetic profiles along flow lines (Fig. 5) through the CMQZ show weakly lineated anomalies of enigmatic origin. Vogt and Johnson (1971) suggested that some of them could represent reversed polarity intervals or other geomagnetic behavior. However, magnetostratigraphic studies on terrestrial and subseafloor sediments (Lowrie and others, 1980) indicate that geomagnetic polarity "was normal except possibly for a few short reversed polarity intervals, the durations of which were probably less than 0.03 m.y. and the ages of which are somewhat uncertain" (Harland and others, 1982). It seems unlikely that reversed events of duration 0.03 m.y. can be resolved within the central Atlantic CMQZ, even though spreading half-rates were relatively high (2.4 cm/yr; Ch. 22). If the CMQZ crust is magnetized with normal polarity, observed magnetic anomalies should correlate well with those calculated from basement topography (Vogt and Johnson, 1971; Vogt and Einwich, 1979; Fig. 5). The axial MAR profile (Fig. 3II) might be used as a standard, although the higher frequencies present would not be observed at sea level in the

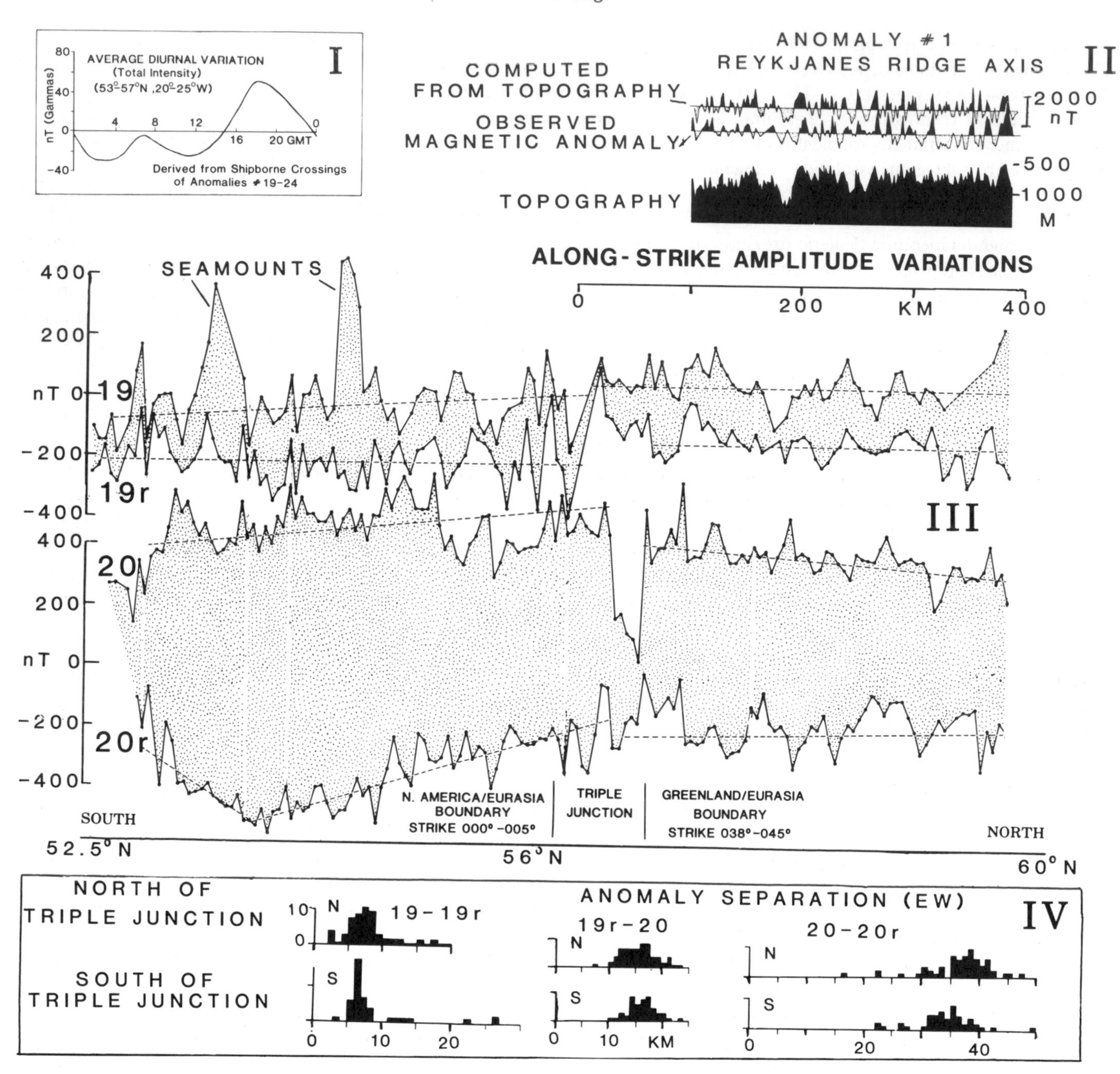

Figure 3. I: Average diurnal variation for the area 52–60°N, 20–30°W derived from III (see text); II: Along-strike magnetic, topographic, and computed magnetic profile (30 A/m) along Reykjanes Ridge axis (after Talwani and others, 1971); III: Along-strike profiles along crests of anomalies 19 and 20, and troughs east of 19 and 20, constructed from east-west crossings, with dashed lines showing average trends and stippling indicating magnetic anomaly relief (crest minus trough); IV: Distribution of magnetic anomaly crest-trough separations measured on east-west tracks (see Fig. 1 for location). Data in III and IV discussed in Vogt and Avery (1974).

CMQZ owing to the greater water depth. The correlation of observed and calculated anomalies is rather poor. For example, profile F shows some anomalies that correlate with basement features but many others that do not. Some of the lack of correlation may result from the topography and observed magnetic anomalies not being linear, but significant magnetization contrast within the basement is indicated. This contrast may have originated when the crust first formed, or it may reflect differential low temperature alteration (e.g., Whitmarsh, 1982), which could have increased anomaly amplitudes from what was originally

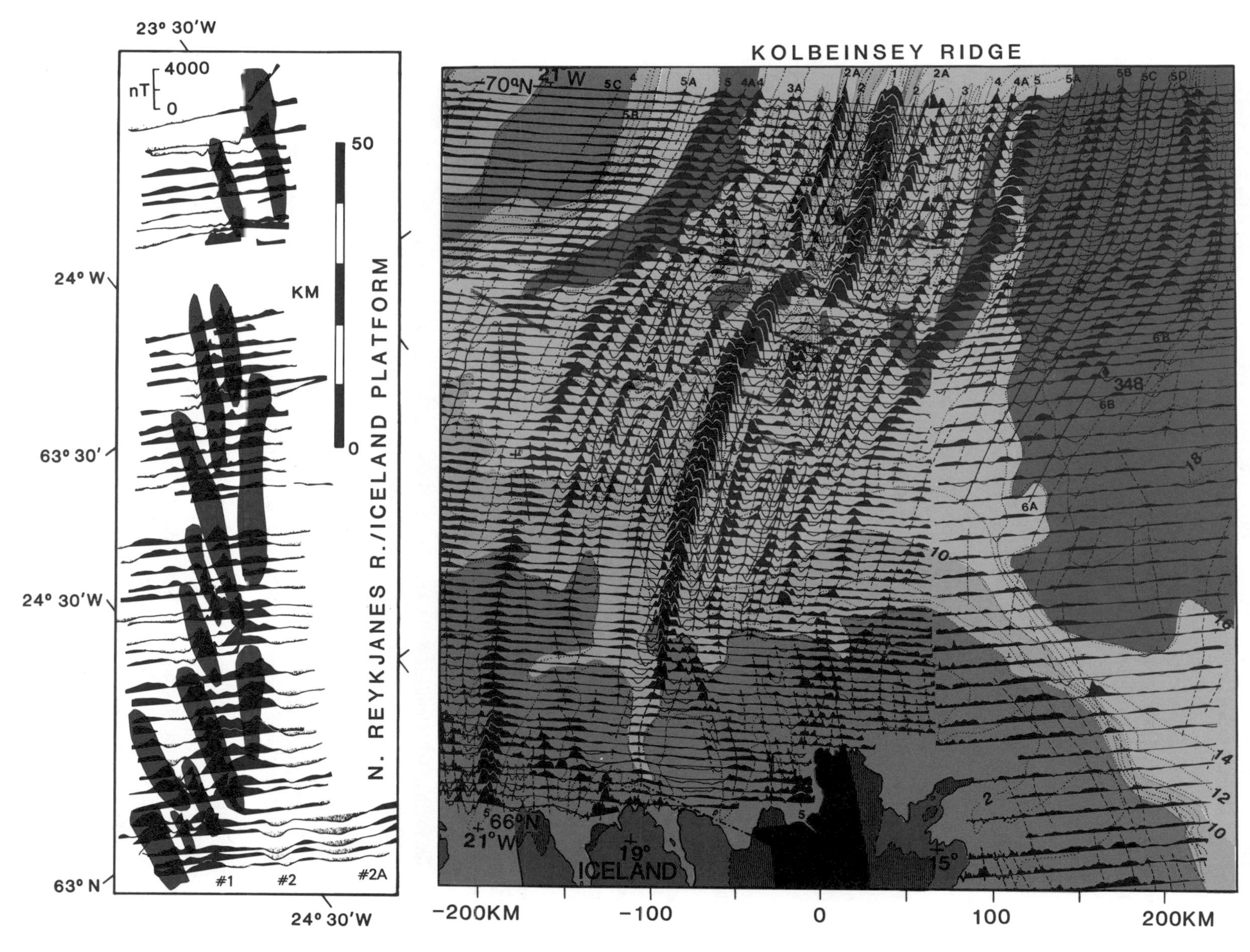

Figure 4. LEFT ("4A" in Fig. 1): Detailed shipborne magnetic anomaly profiles across central anomaly (#1) where it crosses shelf south of Iceland (after Johnson and Jakobsson, 1985). Black indicates positive anomaly: red indicates high-amplitude zones associated with fissure-swarm ridges. RIGHT ("4B" in Fig. 1): Magnetic anomaly profiles across Kolbeinsey Ridge north of Iceland, with anomaly correlations (solid and dashed) and contours (dotted) in hundreds of meters. Shades of red and pink demarcate Iceland and 400, 1,000, and 1,600 m isobaths (dotted). DSDP site 348 indicated by diamond.

present. However, where observed and calculated anomalies agree, the implied magnetization is of the order 10 A/m (Fig. 5), inconsistent with the idea that the upper crust, particularly basement highs, has lost its magnetization in the manner suggested by Blakely (1976, 1983) and Whitmarsh (1982). The high magnetization of some CMQZ basement topography is consistent with the CMQZ anomaly at satellite heights (LaBrecque and Raymond, 1985), with magnetizations of drillcore samples (Bleil and Petersen, 1983), and with the 10 A/m magnetization of the supposedly coeval Lower Pillow Lavas of the Troodos ophiolite (Vine and Moores, 1972). A high paleo-dipole intensity could explain the high CMQZ magnetization, but independent evidence is lacking (Merrill and McElhinny, 1983). More promising hypotheses include long-term low-temperature oxidation (Bleil and Petersen, 1983) and secondary CRM (Raymond and LaBrecque, 1986). CMQZ basement topography and magnetic anomalies exhibit higher relief in the younger half of the zone (Fig. 5). If basement roughness relates to spreading rate, the fastest spreading occurred later in the CMQZ period, about 100–84 Ma, and at rates exceeding the CMQZ average of 2.4 cm/yr.

C. The Keathley Sequence (M - Series). A distinctive 300 to 600 km wide band of SFS lineations called the M (or Keathley) sequence is associated with late Jurassic to early Cretaceous crust in the central North Atlantic (Fig. 1; Plate 3). The Keathley anomalies lie between the Jurassic and Cretaceous magnetic quiet zones (JMQZ and CMQZ). Two 1948 aeromagnetic lines from Cape May to Bermuda, and back to Long Island, first profiled what are now called the East Coast Magnetic Anomaly (ECMA, Figs. 1, 13), the JMQZ, and the M-series. Reporting on these data six years later, Keller and others (1954) concluded that the "shelf anomaly" (ECMA) could not be explained as a continent/ocean edge effect but was probably an intrusive mass 48 km wide, with a top 3.6 to 4.9 km deep. They attributed the eastward disappearance of the smooth field to a thinning of the (less magnetized) "sial." Noting the similarity of anomalies in the disturbed zone to anomalies in the Pacific south of the Aleutians, Keller and others suggested (correctly) that sial may be absent from both areas. The first detailed survey of what is now known as the middle of the M-Sequence northeast of the Bahamas (Bracey and Avery, 1963) was followed in the late 1960s by a systematic magnetic survey of the western North Atlantic (east-west lines 36 km apart). The magnetic profiles (Vogt and Tucholke, 1982) clearly showed the M-lineations (named the "Keathley Sequence" after the survey ship; Vogt and others, 1971) and bordering CMQZ and JMQZ. The more significant fracture zones, 100 to 300 km apart, were easily resolved by the Keathley data, first reported by Anderson and others (1969). More detailed aeromagnetic studies of the Keathley sequence followed (Schouten and Klitgord, 1977; Vogt and Einwich, 1979; Kovacs and others, 1980; Plate 3); the most recent tectonic analyses are found in Sundvik (1986) and in chapter 22 of this volume. Due to slow spreading, numerous minor fracture zones, and some ridge jumps, the Keathley (M) anomalies are not easily correlated on random, wide-spaced tracks. Some early attempts to correlate anomalies from track to track turned out to be incorrect (e.g. Emery and others, 1970; Laughton and Whitmarsh, 1974; Barrett and Keen, 1976) and even detailed data may allow different interpretations (e.g. anomaly M-9 of Sundvik, 1986; reinterpreted from Schouten and Klitgord (1977) and Vogt and Einwich (1979)). The first reversal time scale for the M-series (Keathley Sequence) (Vogt and others, 1971; Fig. 6) was determined from the average positions of correlatable magnetic peaks and troughs on the Keathley profiles. The beginning of the original reversal sequence (Vogt and others, 1971) was then dated at 150 Ma based on basement ages at DSDP sites 100 and 105 (Fig. 7). The end of the sequence was interpolated to be near the Jurassic/Cretaceous boundary, then 135 Ma, thus placing the entire sequence in the Jurassic (hence the "J" prefix of Vogt and others, 1971). The original Keathley sequence (now M-0 to M-23) extends from 119 Ma to 152 Ma on the DNAG time scale.

The present M-series reversal boundaries are based on magnetic anomaly data in the western Pacific, where polarity intervals of much shorter duration (down to about 0.03 m.y.) are resolved due to faster spreading. Even if polarity boundaries are sharp and vertical, resolution would improve with increasing spreading rate. However, resolution at slow spreading is further degraded by the rougher basement topography and greater volcanic/tectonic structural complexity of the crust (Schouten and Denham, 1979). The first reversal time scale using Pacific anomaly data (and both Atlantic and Pacific age data from DSDP sites) was by Larson and Pitman (1972). Their block sequence was added to (e.g. anomaly M-O) and otherwise modified by Larson and Hilde (1975). Whereas Vogt and others (1971) had correctly equated the old western Pacific anomalies of Hayes and Pitman (1970) with the Keathley lineations, they erred in the specific correlations, which assumed an origin of the Pacific crust north of the Equator. Larson and Pitman (1972) showed that the Pacific crust in question had been formed in the southern hemisphere. Thus magnetic highs in the western Pacific correspond to reverse polarity and therefore correlate with magnetic lows in the North Atlantic. Due to the traditional—but arbitrary—numbering of positive rather than negative features, the M-numbers (except for M-2 and M-4) assigned by Larson and Pitman (1972) correspond to reversed polarity and thus negative anomalies in the Atlantic. The difference in magnetic anomaly resolution between Atlantic (slow) and Pacific (fast) spreading rates is shown in Figure 6, where the original Keathley reversal sequence (B, from Vogt and others, 1971) is placed beside the Pacific based time scale (C; Vogt and Einwich, 1979; and A, Larson and Hilde, 1975, as modified by Sundvik, 1986). For times of frequent reversals (M-5 to M-15, and M-24 to M-29) the Atlantic-based block model (B) is poor. The two cross-hatched zones, interpreted as reversed intervals by Vogt and others (1971), do not correspond to reversals in the Pacific-based scale, although Sundvik (1986) agrees that reversals may be present (Fig. 6).

Errors associated with radiometric dating of oceanic basalt have required magnetic reversals to be dated from the age of the oldest sediment directly overlying basement. Deep drilling is

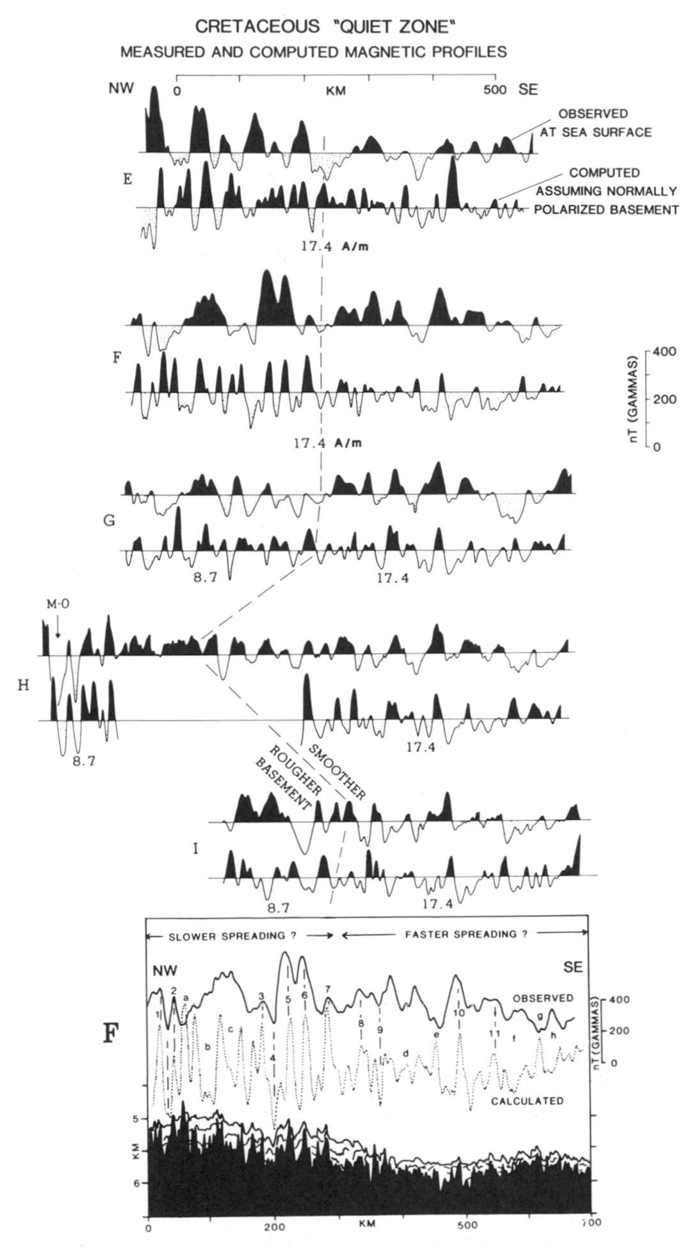

Figure 5. Magnetic profiles (Vogt and Johnson, 1971; Vogt and Einwich, 1979) across Cretaceous "Quiet Zone" (CMQZ) southeast of Bermuda (Fig. 1) compared to profiles computed from basement topography, assuming normal magnetization. Profile F repeated at bottom with seismic reflection profile.

practically the only way to recover this oldest sediment, or the basalt below it. Construction of magnetic reversal time scales is discussed elsewhere (Harland and others, 1982; Lowrie and Ogg, 1985; Kent and Gradstein, this volume). The latter three time scales are compared with the Vogt-Einwich time scale in Figure 6. The variations in isochron spacing along scales II-IV reflect variations in M-series spreading rate (e.g., A) implied for the Atlantic by these scales.

D. The Jurassic and Inner Magnetic Quiet Zones (JMQZ and IMQZ) and Blake Spur Magnetic Anomaly (BSMA). From the Bahamas to the Grand Banks, and in a conjugate position on the African side of the MAR, there are 250 to 350-km-wide bands of low-amplitude (±10-50 nT) anomalies called the "Jurassic Magnetic Quiet Zone" (JMQZ) (Figs. 1, 7, and Plate 3) (Vogt and Einwich, 1979; Klitgord and Schouten, this volume). The African JMQZ abuts the African continental margin, but on the North American side there is an additional, ca. 100-km-wide, "Inner Magnetic Quiet Zone" (IMQZ). The two smooth zones are separated by the 100nT Blake Spur Magnetic Anomaly (BSMA), which can be traced approximately from 28 to 36°N. South of 30°N the Blake Plateau and Bahamas Platform lies west of the BSMA. North of 30°N the IMQZ extends west to the East Coast Magnetic Anomaly (ECMA) (Fig. 1) whose source lies on or near the ocean-continent transition (Vogt and Einwich, 1979) but which may demarcate a Paleozoic suture zone (Nelson and others, 1985). Anomalies similar to the ECMA are locally present along the African margin (Klitgord and Schouten, this volume).

The JMQZ and IMQZ were once thought to demarcate Paleozoic crust of an ancient Atlantic (Drake and others, 1963). Heirtzler and Hayes (1967) and Emery and others (1970) suggested a Permian age (270–220 Ma), by which the apparent absence of SFS anomalies was attributed to the Permo-Carboniferous "superchron" of constant reversed polarity (now 320–250 Ma; Harland and others, 1982). Vogt and others (1970) suggested that reversals were present in the smooth zone crust, but that amplitudes were attenuated due to low paleomagnetic latitudes and/or to sedimentation near the accretion axis (as discussed later for the Knipovich Ridge, Fig. 8B). Other suggestions included metamorphism (Taylor and others, 1968) and viscous remagnetization (Lowrie, 1973). JMQZ basalt samples from DSDP site 105 exhibit no unusual magnetic properties (Taylor and others, 1973), although a global compilation suggests reduced drillcore NRM in the few Jurassic samples (Bleil and Petersen, 1983). Magnetic smooth zones (Poehls and others, 1973; Roots, 1976) can also be produced if the ratio of spreading half-rate to reversal frequency (i.e., the average block width of constant polarity) is sufficiently low (Le Pichon and others, 1971) (of the order 1 to 3 km depending on basement depth) or if there is oblique spreading and fragmentation of the accretion axis by close-spaced FZ's (Roots and Srivastava, 1984).

Vogt and others (1970) estimated the age of the transitional seaward edge of the JMQZ at about 190 Ma, but subsequent drilling at DSDP Site 105 (Fig. 7) reduced this to 150 Ma (155 Ma on the DNAG scale). The original Keathley reversal time scale (Vogt and others, 1971) extended to about M-23 within the transition zone. With the more precise Pacific-based M-reversal scale (Larson and Pitman, 1972; Larson and Hilde, 1975), anomalies out to M-25 could be identified in the Atlantic. Those workers considered the JMQZ older than M-25 to be of constant normal polarity, whereas Taylor and Greenewalt (1972) believed that the JMQZ magnetization is uniform and reversed. Cande

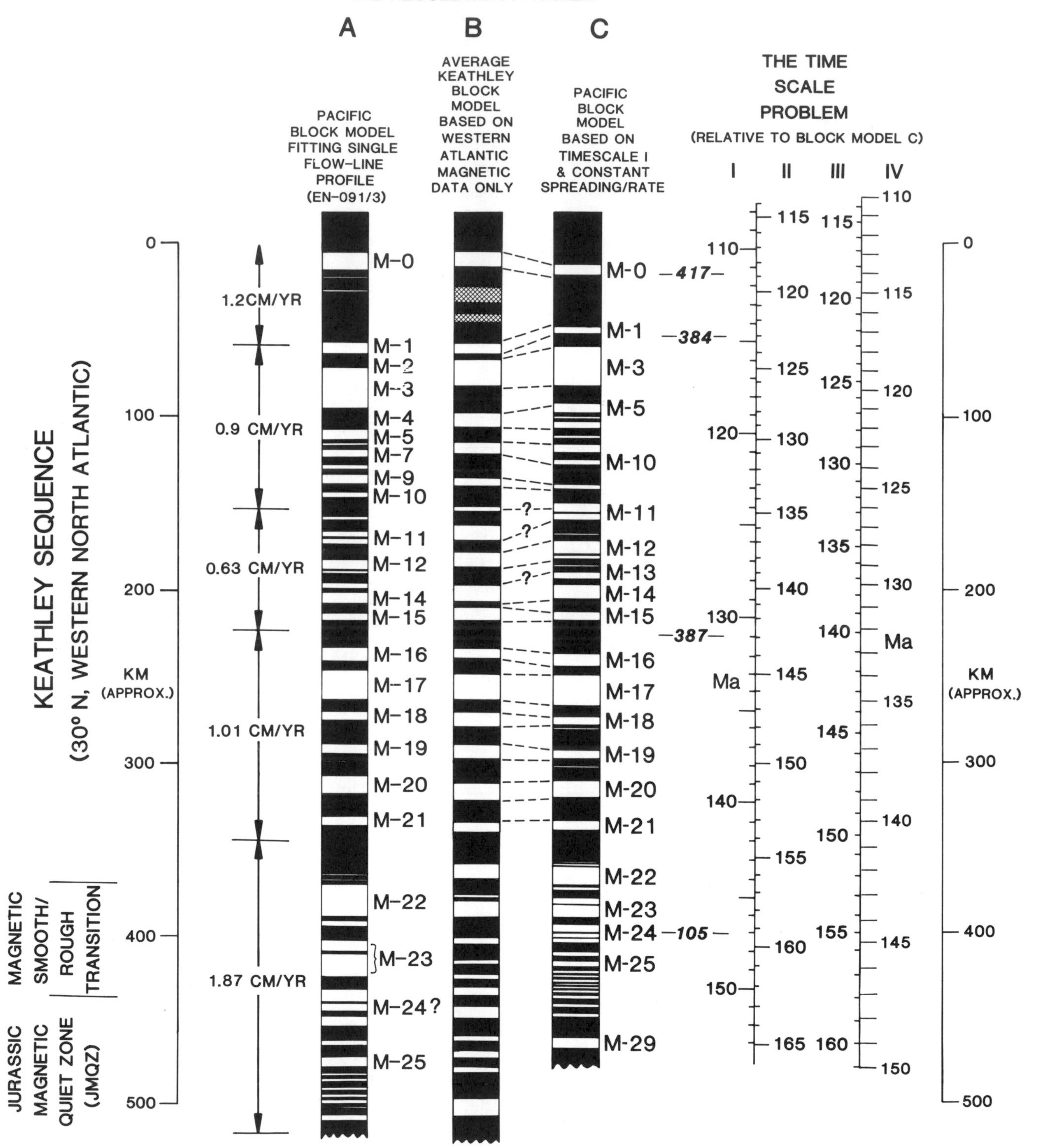

Figure 6. The Keathley (M) geomagnetic reversal sequence in the western North Atlantic. Block models: A (Sundvik, 1986), based on a single flow-line parallel profile (EN-91/3, passing through DSDP site 105 [Fig. 7], variable spreading rates, and time scale of Larson and Hilde (1975); B (Vogt and others, 1971, modified below M-21) based on average Atlantic profiles; and C (Vogt and Einwich, 1979) assuming constant spreading and time scale I. Km scale refers to B and C. Model A is fit to B at M-0 and M-21. The four time scales are, I; Vogt and Einwich, (1979); II; Harland and others (1982); III; Kent and Gradstein (this volume, Ch. 4); and IV; Lowrie and Ogg (1985).

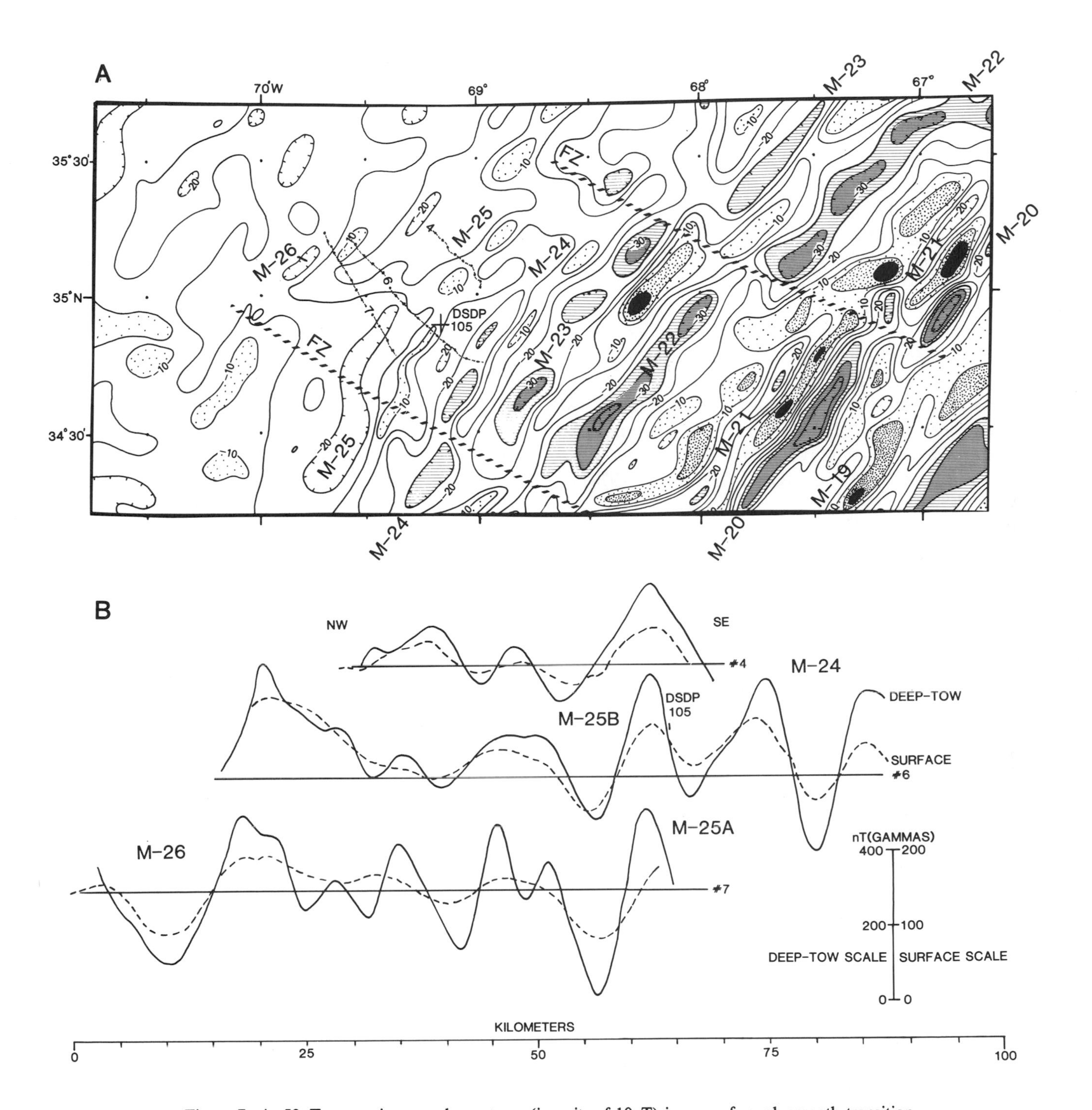

Figure 7. A: 50nT magnetic anomaly contours (in units of 10nT) in area of rough-smooth transition between Keathley (M) series and Jurassic Magnetic Quiet Zone (JMQZ). Zero-level is about ±150nT. Note locations of DSDP site 105 and deep-tow magnetic profiles no. 4, 6, and 7 of Taylor and Greenwalt (1976). B: Comparison of surface and deep-tow magnetic profiles; note difference in vertical scales.

and others (1978) extended the Pacific-based time scale back to M-29, whose age on the DNAG scale is 160 Ma (165 Ma on the Harland and others, 1982 scale). Although this extension provides a basis for interpreting the subdued anomalies in the JMQZ, the anomalies are too irregular for unambiguous correlation, and M-25 is the oldest SFS lineation consistently present in the Atlantic. Detailed contour maps, combined with deep-tow profiles like those of Taylor and Greenewalt (1976), could map the isochrons further into the JMQZ; near DSDP Site 105 (Fig. 7) the lineations can be traced out to M-26 and suggest a spreading half-rate of 1.5–2 cm/yr over the interval M-24 to M-26. Cande and others (1978) tentatively identified M-29 on one Atlantic profile. A detailed magnetic/seismic survey in the JMQZ (Plate 3, C1) showed several bands of alternating magnetization polarity (Barrett and Keen, 1976), and a similar conclusion was reached from a detailed survey across the southern JMQZ (Plate 3, C2) (Bryan and others, 1980). Both papers labelled anomalies "M-26" through "M-29" but this nomenclature should be avoided pending a clear correlation to the different Pacific anomalies M-26 to M-29 (Cande and others, 1978) or possibly to M-34 (D. Handschumacher, personal communication, 1985).

Magnetostratigraphic studies of Jurassic sediments (Ogg and Steiner, 1985) do not support constant polarity explanations for the JMQZ or IMQZ. However, if the older JMQZ and IMQZ correspond to crustal ages between early Callovian and early Bajocian, the low amplitudes may be explained by rapid reversals (ca. 4/m.y.; Ogg and Steiner, 1985). The younger JMQZ corresponds to a period of less frequent (ca. 2/m.y.) reversals (early Oxfordian–early Callovian, Ogg and Steiner, 1985). For spreading half-rates of about 1.5-2 cm/yr, this reversal frequency should produce resolvable magnetic lineations of moderate amplitude. Thus the low magnetic amplitudes, as well as the existence of an envelope of gradually decaying amplitudes in the older M-series, remain unexplained. Rapid polar wander or axial sedimentation (Vogt and others, 1970) would not explain the existence of a JMQZ (including amplitude envelope) in the western Pacific, although rapid eastward movement of the pole in the Oxfordian-Kimmeridgian (Steiner, 1983) may still have been a factor in the Atlantic. Two other geomagnetic explanations (Cande and others, 1978) involve (1) numerous short reversed intervals (say 0.05 m.y. duration) within longer chrons (say, 0.5 m.y.) with polarity bias, or (2) a lower JMQZ geomagnetic intensity, gradually increasing during mid and late Oxfordian to explain the amplitude envelope. Although lacking independent confirmation (e.g., from paleo-intensity measurements), the second hypothesis is attractive because some geomagnetic field models predict an inverse correlation between dipole strength and reversal frequency; this is consistent with the decreasing reversal frequency over the interval M-25 to M-22 which contains the amplitude envelope.

The Blake Spur Magnetic Anomaly (BSMA) resembles the more prominent M-anomalies in amplitude and linearity but is probably not simply a SFS lineation. It is associated with a west-dipping basement escarpment and probably marks the site of an eastward jump of the spreading axis that occurred some time during the interval 170–180 Ma (Vogt, 1973; Klitgord and Schouten, this volume).

REGIONAL VARIABILITY AND LOCAL MODULATION OF SEA-FLOOR SPREADING (SFS) MAGNETIC ANOMALIES

A. Regional Effects. The shape, amplitude, and local variability of SFS lineations depends on a number of factors, including the depth, thickness, and homogeneity of the source layer, the vectors of both present and paleomagnetic fields, and the bulk intensity of magnetization. Magnetic anomalies represent the integrated effect of an area of crust of magnitude H^2, where H is distance from observation level to basement. At typical oceanic depths, sea-surface magnetic anomalies average the magnetic characteristics of several tens of km^2 crustal area, roughly 10 km^3 crust. An instrument mounted on a submersible responds to the magnetization of a single boulder (Fig. 9 IV), while an earth satellite may average over 10^5 km^2 (e.g., Fig. 10). The averaging extent should be kept in mind when bulk (net) magnetization computed from anomalies is compared with local rock magnetization measured on, for example, DSDP core samples. We first consider regional (>10 km^3 scale) influences, and then expand on more local effects of greater geological interest.

The spatial variation of the past or present dipole fields does not significantly change anomaly amplitudes and shapes over tens or hundreds of kilometers (except at low magnetic latitudes, and then only where anomalies abruptly change strike). In general, geomagnetic influences on SFS anomalies are thus of regional character. Geological influences may be either local or regional. The "texture" of the SFS anomaly pattern (including "noisiness") and perhaps also magnetized layer thickness (Jackson and Reid, 1983) depend on spreading rate, which changes gradually, at least along strike, thus producing regional variations. Based on core sample analyses, Pecherskiy and Tikhonov (1983) postulate global time variations in basalt chemistry and magnetization. Hotspots also affect anomaly texture, for example, confused anomalies on Iceland itself but clear patterns at 500 to 1,500 km ranges. Another regional effect is caused by regionally varying source depth. Depth (source distance) is the "earth filter" (Schouten and McCamy, 1972) that smoothes and attenuates the SFS anomalies, any given spectral component of wavelength λ decaying as $e^{-H/\lambda}$. Since the top of the magnetized portion of the oceanic crust is generally the top of the basaltic basement, the gradual increase of basement depth with age and sediment burial causes regional changes in the magnetic anomalies. Progressive low-temperature alteration and maghematization of the upper part of the magnetic source layer with increasing crustal age is another regional influence (Blakely, 1976, 1983).

A global compilation of drillcore NRM reduced to the Equator (Bleil and Petersen, 1983) shows a minimum (1 A/m) for 10–30 Ma crust; NRM then increases with increasing age to 3–5 A/m at 80–120 Ma. This long term $\surd$-shaped NRM variation, also suggested by magnetic anomaly amplitudes, may reflect

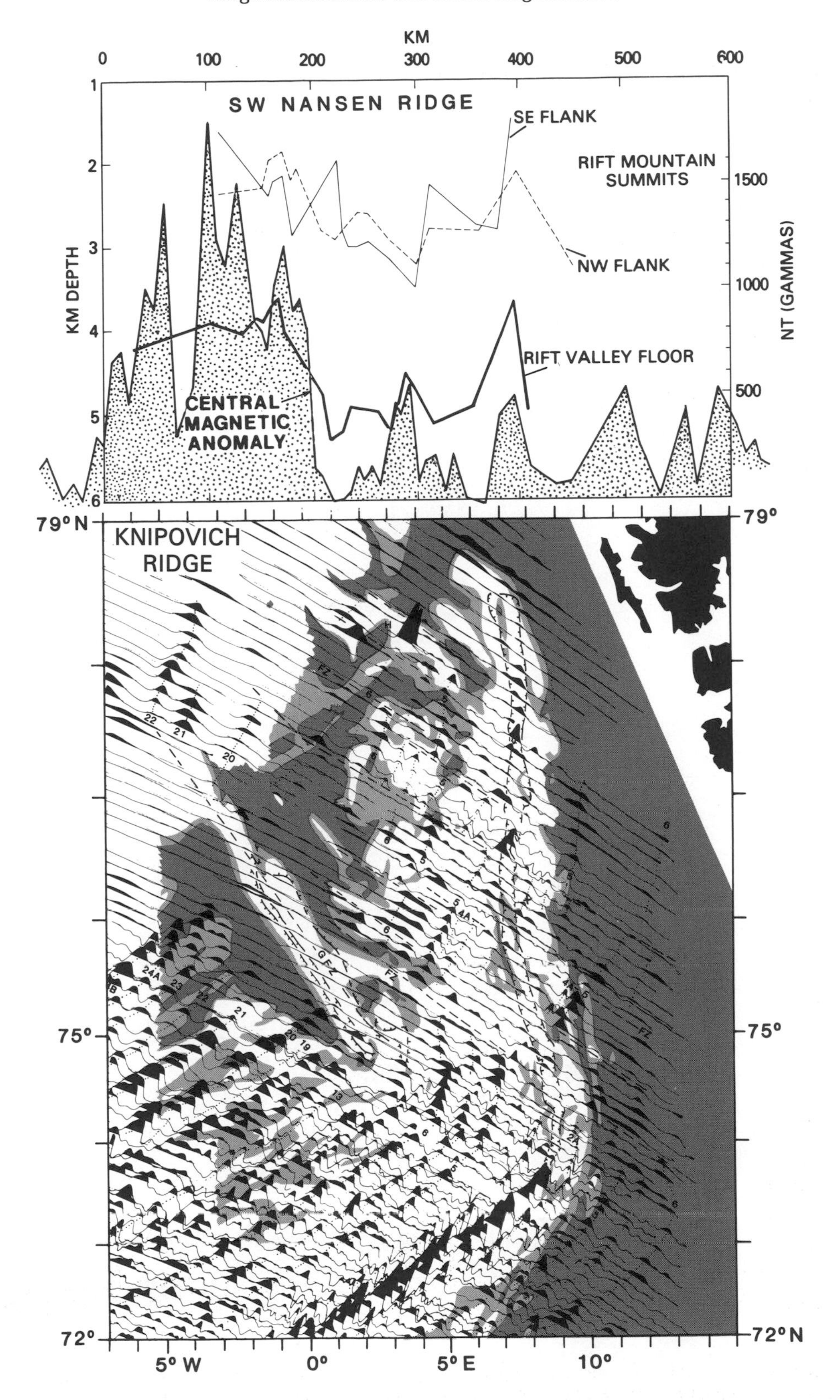

Figure 8. TOP: Longitudinal profile of central magnetic anomaly amplitude, rift valley depth, and minimum rift mountain depths along Nansen Ridge (after Feden and others, 1979; shown as 8A in Fig. 1). BOTTOM (shown as 8B in Fig. 1): Magnetic anomaly profiles and bathymetry in areas of Knipovich Ridge (based on Vogt and others, 1982, and Kovacs and Vogt, 1982). Solid red indicates sediments ≥1 km thick; pink, 0.6-1 km (Vogt, 1986a,b). DSDP Site 344 is marked by "x."

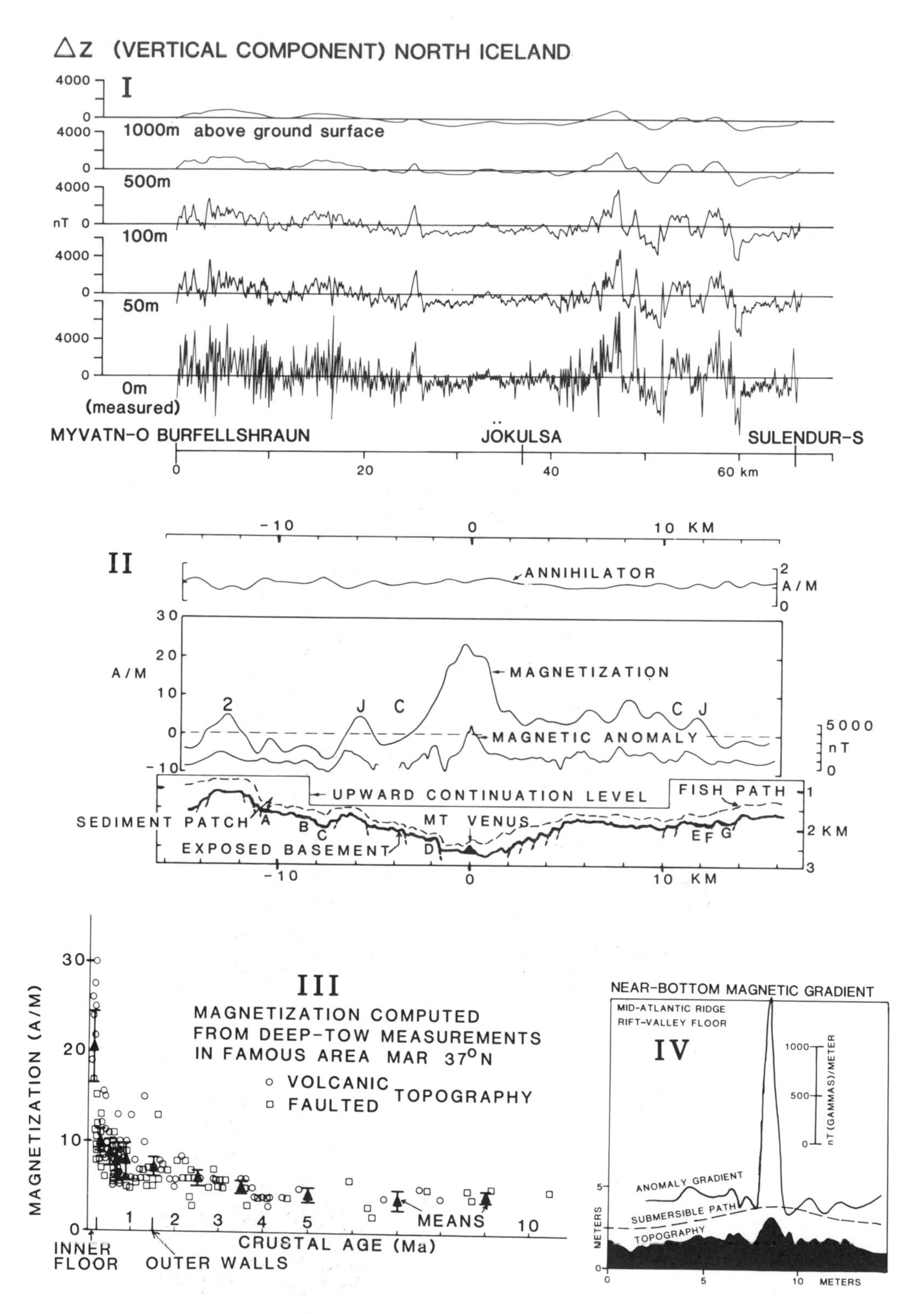

Figure 9. I: Ground-level, vertical-component magnetic anomaly profile in northern Iceland, with upward-continued profiles (Becker, 1980); II: Near-bottom magnetic anomalies and their inversion to give magnetization of topography on young oceanic crust in FAMOUS area (after Macdonald, 1977); III: Magnetization values from inversion of deep-tow data, plotted against crustal age (after Macdonald, 1977); IV: Magnetic gradient measured very near basalt outcrop near FAMOUS area on board submersible *Alvin* (T. Atwater, personal communication, 1985).

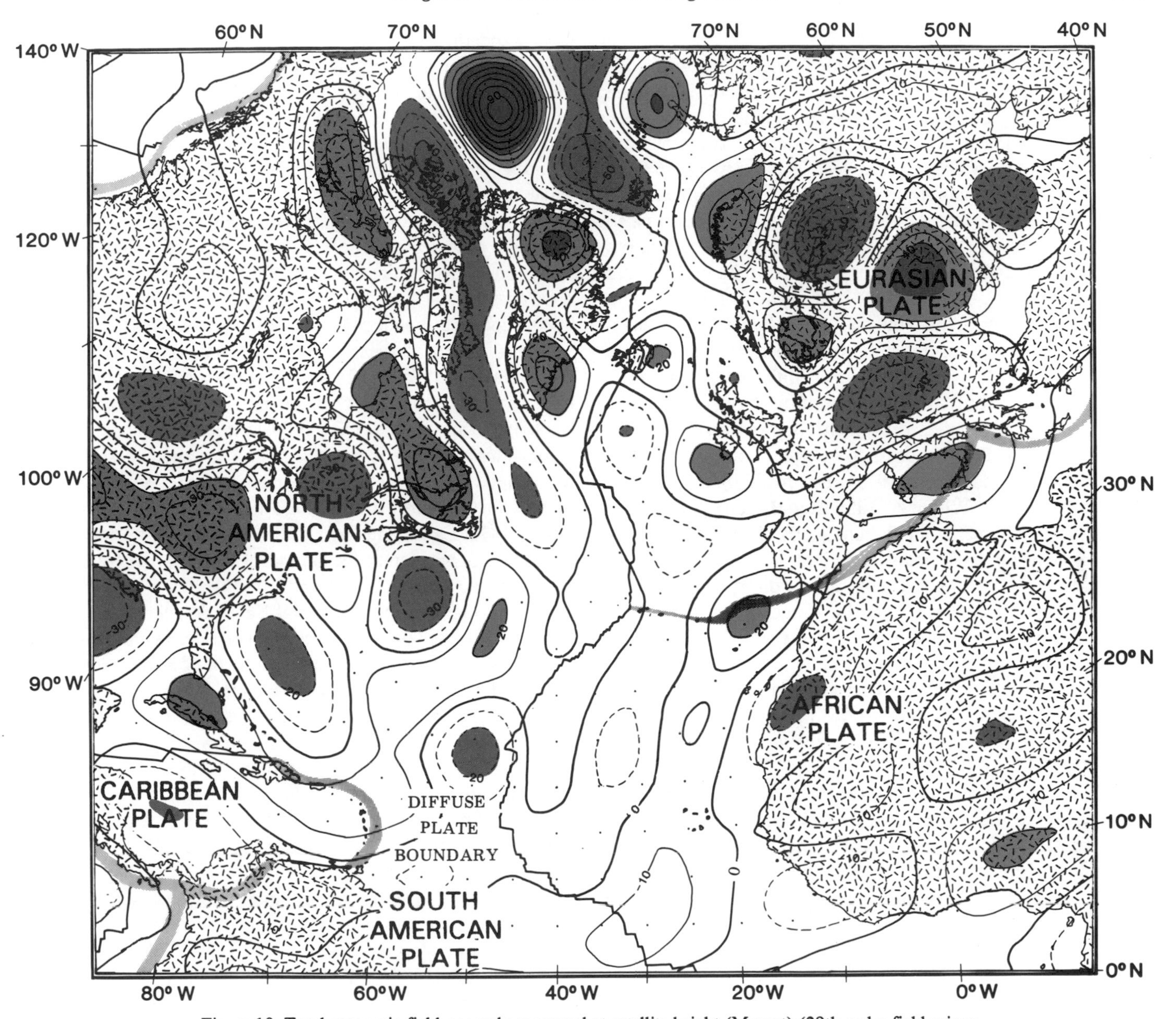

Figure 10. Total magnetic field anomaly measured at satellite height (Magsat) (29th order field minus 13th order field) from Cain and others (1984) (courtesy J. Cain). Contour interval 10nT, with negative contours dashed. Mid-Atlantic Ridge and other plate boundaries are also indicated on this Transverse Mercator projection. Red, >40 nT; pink, 20-40 nT, light gray, –20 to –40 nT; dark gray, <–40 nT.

progressive low-temperature titanomagnetite oxidation (Bleil and Petersen, 1983) or acquisition of secondary CRM over several m.y. in the ambient field direction (Raymond and LaBrecque, 1986). In the latter case, the $\surd$ shape of the NRM curve is determined by geomagnetic polarity history and the time scale of secondary CRM acquisition. Raymond and LaBrecque (1986) propose that CRM acquired within the first 20 m.y accounts for 80% of the NRM in older basalts. Basement depths in the North Atlantic vary as much along the strike of the MAR (on account of hotspots like the Azores and Iceland) as they do as a function of age. Axial depths range from above sea level on Iceland to over 5 km on the floor of the Nansen Ridge rift valley and near major fracture zones. Since magnetic measurements are made at fixed elevations, such large basement depth variations produce large differences in magnetic anomaly shape and amplitude. This was shown by Heirtzler and Le Pichon (1965), who examined the latitude variation of the central magnetic anomaly (now recognized as SFS #1 rather than due to induced magnetization). The peak-to-trough amplitude of the MAR central anomaly at sea level varies from more than 4000nT on the Reykjanes and Kolbeinsey ridges near Iceland to 1500nT on the southern Reykjanes Ridge, 600nT at 47°N, up to 2000nT near the Azores triple junction, and down to a few hundred nT in equatorial latitudes (Vogt, 1979). Most of this regional variation reflects differences in water depth, and, to a lesser degree, geomagnetic parameters such as dipole intensity, which increases by a factor of two from Equator to pole, and anomaly strike, which causes zero amplitudes for a NS strike at the Equator. In an early attempt to extract paleomagnetic vector information from SFS anomalies, Vogt and others (1971) concluded that the northward amplitude increase of anomalies J-17 to J-20 (chrons M-19n to M-22n) from 25°N to 34°N fits a paleomagnetic pole about 75°N, 90°E, near present estimates for the mid-Jurassic pole (Steiner, 1983). Other regional variations (e.g., due to regional differences in magnetic layer thickness or bulk rock magnetization) have not been isolated, although the corrected magnetization of basement topography (Harrison, 1981) appears to be higher on the Reykjanes Ridge (Talwani and others, 1971) than in the FAMOUS area (Macdonald, 1977). Within ca. 1 km from the top of the magnetized layer, the anomalies are dominated by effects of higher spatial frequency discussed later (e.g., Figs. 3, 4, 9). These local effects, due to features such as fault scarps, dikes, and lava flows, decay rapidly with distance. At greater ranges the amplitude of an isolated linear SFS anomaly decays as $1/H^2$. In the western Atlantic (5 to 7 km basement depth) a typical M-series lineation decays about 3% per 100 m (Vogt and others, 1971). The low M-anomaly amplitudes (±100–200nT) north of the New England Seamounts at least partly reflect the great basement depths there (7 to 8.5 km).

B. Magnetic Anomaly Skewness, Anomalous Skewness, Drillcore Magnetization, and Paleomagnetic/Tectonic Implications. Paleomagnetic and tectonic information can also be extracted from the shape of SFS anomalies and from drillcores. The shape of a linear anomaly is related to the phase (skewness) angle $\theta = I_0' + I_r' - 180°$, where I'_0 and I'_r are the effective present and remanent inclinations (the effective inclination is the inclination in a vertical plane perpendicular to the anomaly trend). SFS anomaly shape may also contain *tectonic* information: If newly magnetized crust in the MOR axial zone is tectonically rotated (e.g., by listric faulting) about an axis parallel to the anomaly trend, then I_r' is changed, as is θ. A change in θ is equivalent to phase-shifting the anomaly pattern. "Anomalous skewness" ($\Delta\theta$; Cande, 1976) is the difference between the measured skewness and that expected from the standard thin-layer block model, given the correct paleomagnetic pole.

Three methods to test for anomalous skewness (Cande and Kent, 1985) can, in principle, be applied in the Atlantic. (A) The skewness of the observed profile can be compared to that predicted from an independently determined paleomagnetic pole. This is done by reducing the observed anomaly to the way it would appear at the magnetic pole (Blakely and Cox, 1972), and comparing the reduced profile to a model profile calculated for $\theta = 0°$ (Schouten and Cande, 1976). For any parcel of oceanic crust in the western North Atlantic, the North American apparent polar wandering curve (e.g., Irving, 1979) describes the expected paleomagnetic pole locations back to the time the parcel became rigidly attached to the plate. (B) A method independent of a priori paleomagnetic knowledge is to compare "conjugate" anomalies formed at the same time and place on opposite sides of the accreting plate boundary. Without anomalous skewness, the θ values should differ only by I_o'. (C) For a given anomaly, the locus of possible paleomagnetic poles, all of which would describe the observed skewness, define a "lune of confidence." If several lunes can be determined using geographically well separated observations from the same anomaly on the same plate, the intersection area should coincide with the paleomagnetic pole.

It is difficult to determine the skewness of anomalies narrower than about 25 to 50 km, and North Atlantic studies have been limited to anomalies 33/34 and M-O/M-4 (Cande and Kent, 1985) and the CMQZ (LaBrecque and Raymond, 1985). Anomalous skewness is essentially zero for anomalies M-O to M-10 in the Pacific (method C; Cande and Kent, 1985) and M-0 to M-4 in the North and South Atlantic (methods A and B; Cande, 1978; Cande and Kent, 1985). However, 14° of anomalous skewness was found for anomalies 27-32 in the North and South Pacific (method C; Cande, 1976) and ± 30°–40° for anomalies 33-34 in the North and South Atlantic. Anomalous skewness is of opposite sign on opposite sides of the MAR. A CMQZ skewness of ± 35° improves the fit of model profiles to MAGSAT data (LaBrecque and Raymond, 1985).

Cande (1978) considered four models to explain anomalous skewness. Two models (dipole intensity decreasing gradually between reversals, and frequency and/or duration of undetected short polarity events increasing toward the end of long periods [>1 m.y.] of predominately one polarity) imply peculiar behavior of the geomagnetic field. Tectonic tilting (model 3) by 30–40° is much larger than the 3–8° observed at present spreading axes. In the fourth model, a slowly cooled, moderately magnetized lower crustal layer has broad, outward sloping transition zones reflecting the outward slope of isotherms (Blakely, 1976; Kidd, 1977; Cande and Kent, 1976). This model can reproduce the overall anomaly shape and skewness but fails to match the narrow positive anomaly over the seaward edge of the long Cretaceous normal polarity zone. A fifth model (Raymond and LaBrecque, 1986) related anomalous skewness to acquisition of secondary or CRM with a time constant of about 5 m.y. after crustal formation. Favoring models 1 and 2 is the (so far) consistent absence of skewness over crust of M-0 to M-10 age and its presence in anomalies 33-34 (Cande and Kent, 1985).

Drillcore magnetization can, in principle, constrain absolute plate motion and help test the different skewness models (e.g., Verosub and Moores, 1981). However, so far only the upper part of the magnetic source layer has been cored. Global compilations of TRM in DSDP boreholes (Pierce, 1976; Harrison, 1981) are generally consistent with other paleomagnetic information, although individual measurements show large scatter, even in the same borehole. The magnetization of sediments of varying ages above the basement (e.g. Tauxe and others, 1983) can be used to track the changing paleolatitude, that is the northward component of absolute plate motion, at any site. This was first tried by Sclater and Cox (1970) on cores from DSDP site 10; it showed an increase of paleolatitude from about 20 to 35° since the Late Cretaceous, presumably a reflection of the northward motion of the North America plate. All Virtual Geomagnetic Pole (VGP) data for a plate can be reduced to dip and inclination histories for any site by the method of Cordell (1974).

Some drillholes in North Atlantic crust (DSDP 332A & B, 410, 417D) show large departures in magnetization dip from that expected; Verosub and Moores (1981) explain these anomalous dips in terms of tectonic rotations (tilts) in the accretion zone. At sites 332A and 332B at 37°N, the 3.5 Ma upper crust has apparently been tilted by 55 and 70°, respectively. At site 410 on the MAR at 45°N, the implied tilts increase downward from 50° at the top to 84° at the bottom of the recovered section. A similarly large rotation (58°) is implied for 118 Ma (M-0) crust at 417D. While the sense and magnitude of such rotations is consistent with observations from ophiolites and present extensional regimes such as the Basin and Range province (Verosub and Moores, 1981), the inferred rotations near the MAR are much larger than the 3° to 8° determined for the FAMOUS area from the morphology (Macdonald and Atwater, 1978). (However, listric faulting implies increased rotation with depth, so large rotations are not expected at the sea floor.) The large rotation inferred at site 417D must be a local or shallow effect since anomalies M-O to M-4 at this site exhibit zero skewness (Cande and Kent, 1985). Effects other than tectonic rotation or paleomagnetic behavior may be important. For example, tabular bodies such as lava flows may exhibit self-demagnetization (reducing the dip) (Vogt, 1969) and the intense near-bottom anomalous fields due to young lavas with magnetization up to 100 A/m could deflect the total field vector seen by newly erupted lavas by as much as 10° to 20° for anomaly magnitudes of $1\text{-}2 \times 10^4$ nT (e.g., Kontis and Young, 1964).

C. Local Variability. A variety of effects, many not well understood, modulate the shape and amplitude of SFS anomalies at scales less than 10^3 km. The simplest effect is basement topography. Along the central anomaly of the Reykjanes Ridge (Fig. 3III), anomalies of ±1000nT amplitude and 10 km wavelength correlate directly with basement topography and even better with a model profile calculated from the topography with an assumed high (30 A/m) magnetization (Talwani and others, 1971). This axial magnetization high is discussed later. Similar profiles on older crust (reversed polarity interval between anomalies 4 and 4A) indicated weaker magnetization (12 A/m).

Detailed magnetic surveys (E-W tracks) conducted over still older crust (anomalies 19 to 24) east of the Reykjanes Ridge (Vogt and Avery, 1974) could also be analyzed for along-strike amplitude variations because adjacent tracks were close together. This is an area of unusually (for the Atlantic) well-developed, linear anomalies formed under the regional influence of the Iceland hotspot. Figure 3III shows anomalies 19 to 20. Two small, normally magnetized seamounts of unknown age produce 400nT spikes on anomaly 19; longer-wavelength variations extend over hundreds of kilometers (dashed lines). The amplitude difference between 20 and 20r reaches a maximum of 800–900nT about 53°–54°N, perhaps indicating the location of an axial magma center and thicker crust in this region. South of 53°–54°N the anomaly amplitudes decrease, perhaps reflecting a gradual thinning of the magnetized layer as a thermal response caused by nearness to the Charlie-Gibbs FZ at 52.5°N. The along-strike profiles also exhibit short-wavelength (<50 km) irregularities, largely artifacts of diurnal variation since the lineations were crossed at different times of the day. When the differences between observed values and the linear fits (dashed lines in Fig. 3III) are plotted against the time of day the magnetic crest or trough was crossed, the resulting data for lineations 19 through 24 yield an average diurnal curve (Fig. 3I) having a range of about 75nT, with a small peak near local sunrise and a major peak at local sunset. This variation differs from that observed at nearby land-based observatories (e.g., Valentia, Ireland) and demonstrates that temporal variations need to be measured in marine areas if accurate anomaly charts are to be constructed, particularly in magnetic smooth zones such as the JMQZ (Fig. 7; Plate 3). However, in the area of Fig. 3III the anomaly amplitudes are much larger than the temporal variation so the latter can be ignored for most purposes.

As the Reykjanes and Kolbeinsey (Fig. 4B) ridges approach Iceland, the SFS magnetic lineations become smoothed and attenuated (Talwani and others, 1971; Vogt and others, 1980) in areas where very high amplitudes would be expected on account of the shallow depths. Vogt and others (1980) suggested several mechanisms for the smoothing and/or amplitude reduction: (1) Partial removal of the magnetized layer by glacial, wave, and/or fluvial erosion; (2) degassing at shallow confining pressures, resulting in (a) higher oxidation state and thus reduced magnetization, and (b) high vesicularity and explosive brecciation of rocks, both facilitating hydrothermal and low-temperature alteration and magnetization loss (Even if rock magnetization were high, brecciation would randomly reorient rock fragments, reducing the bulk magnetization); (3) anomalously great subsidence; (4) broadening of the extrusion zone, as observed on Iceland itself; (5) insulating of the magnetic layer by thick sediments, thereby causing temperatures to rise and magnetization at depth in the crust to be lost; and (6) complex minor shifts in the spreading axis. Vogt and others (1980) considered processes 1, 2, 4, and 6 to be most important close to Iceland. At greater distances, process 2 may be most important at the axis and process 5 on crust older than 1 Ma. Whatever the processes, it is clear that magnetic smooth zones can form at or close to the MOR axis even when other circumstances (spreading rate, reversal frequency, water depth) are favorable for high-amplitude anomalies. This may bear on smooth zones associated with early-formed oceanic crust adjacent to continental margins (e.g. off the eastern U.S.), where shallow initial water depths and rapid sediment input may have created conditions similar to those presently observed near Iceland.

Amplitude variations observed in the vicinity of the Knipovich and Nansen ridges (Fig. 8) show that magnetic smooth zones can be formed along the MOR axis even at typical MOR water depths. To a certain extent, the amplitude variations in both areas parallel basement depth variations, but only a small part of the amplitude variation can be directly due to depth differences. In the case of the Knipovich Ridge (Fig. 8B), the magnetic smooth zone near the plate boundary is associated with very oblique spreading and thick (up to 1 km and more) sediments from the adjacent continental margin. Rapid sedimentation in the axial zone could reduce magnetization in two says. (1) Pillow lavas or deeper intrusives could lose their initial magnetization on account of reheating by thermal conduction brought on by the sediment lid, which shuts off seawater circulation while simultaneously allowing the temperature at all levels in the crust to rise by conduction. This would be most effective at the immediate axis where heat flow is high. Away from the axis, the main effect of a temperature rise would be to speed up maghematization, which takes 1 m.y. to complete at 0°C, but only 10^3y at 50° and a few years at 100°C (Ozima, 1971; Prévot and others, 1979). (2) Newly produced basalt melts rising from axial magma chambers would, on account of their high density compared to sediment, tend to flow out as sills rather than rising to the sea-floor. As intrusives the basalts would cool slowly, resulting in magnetization weaker than that for rapidly quenched pillow lava. An early explanation for the JMQZ (Vogt and others, 1970), this may still be valid for the landward parts of this smooth zone. One supporting datum is the weak basement magnetization (0.145 A/m)—probably from a 2–3 Ma sill—at DSDP Site 344 on crust of similar age, below 378 m sediment, on the east flank of Knipovich Ridge (Fig. 8B; Kent and Opdyke, 1978).

A different explanation for the Knipovich magnetic smooth zone is based on the oblique spreading, which results in a very low opening rate normal to the plate boundary. At such low opening rates the thermal structure of a spreading axis would preclude large quantities of basalt melt. Since oblique spreading axes can be considered "leaky fracture zones," seismic results from (or near) fracture zones are applicable. The basaltic crust in fracture zones seems to be thin or even non-existent (Purdy and Ewing; this volume). Isostasy demands that areas of thin crust are deeper, and indeed the only parts of Knipovich Ridge that are normal in magnetic anomaly amplitude are not only less oblique but also more elevated. Unfortunately, the oblique, deep areas have naturally collected more sediment, making it hard to separate the effects of oblique spreading and sediment cover.

The aplitude variation of the central anomaly along the Nansen Ridge axis (Fig. 8A) is as dramatic as along Knipovich Ridge. However, the Nansen axis is not oblique and there is no evidence for thick sediments in the rift valley. The variation of anomaly amplitude does parallel variations in depth of the rift valley and adjacent rift mountains (Fig. 8A), as in the case of Knipovich Ridge. Perhaps at very slow spreading rates—and in the absence of a nearby hotspot—basalt production is reduced and magma chambers occur only at intervals along the plate boundary (Feden and others, 1979). Near these chambers, normal thicknesses of basaltic crust are produced, while between them only small amounts of lava are erupted, perhaps traveling subhorizontally in dike-conduits from the magma chamber, as on Iceland. Thus, the average magnetized layer would be thin and anomaly amplitudes low. This may also explain the correlation with obliqueness observed on Knipovich Ridge, since oblique spreading is thermally equivalent to slow spreading. In support of the model, a compilation of MOR source-layer magnetization suggests that layer thickness (and/or bulk magnetization) decreases with spreading rate below about 1 to 2 cm/yr (Jackson and Reid, 1983). Crustal thickness may also decrease with decreasing spreading rate below about 2 cm/yr (Reid and Jackson, 1981), although this result is questionable (Purdy and Ewing, this volume).

A zone of high anomaly amplitudes at the southwestern end of the Nansen Ridge ("Yermak H-zone," Figs. 1, 8A) may reflect yet another process. Vogt and Johnson (1973), Anderson and others (1975), and Vogt (1979) attributed similar zones of anomalously high magnetic anomaly amplitudes ("H-zones") on the Juan de Fuca, central Galapagos, and Southeast Indian spreading axes to a greater abundance of Fe-rich basalts with higher titanomagnetite content and generally high TRM. The Fe-enrichment in the H-zone basalts appears mainly to reflect high degrees of crystal fractionation. Although a simple enrichment of iron in a basalt melt does not necessarily result in high TRM (Watkins, 1974), a few basalts dredged in the H-zone south of Tasmania do exhibit higher TRM, higher Fe-content, and a greater abundance of titanomagnetite (Anderson and others, 1980). A greater proportion of smaller-grain size titanomagnetite could also contribute to higher TRM (Anderson and others, 1975). Although not yet sampled, the Yermak H-zone resembles others in terms of the length of plate boundary (280 km), and area of crust (2.4×10^4 km^2) affected, and in its time-transgressive (diachronous) boundary with adjacent normal crust. The Yermak H-zone has expanded northeastward along the Nansen Ridge at about 1 cm/yr since 10 Ma (Feden and others, 1979), perhaps by rift propagation. The central anomaly amplitudes (above the flanking magnetic lows) are about 500–1000 nT in the Yermak H-zone vs. a median 250nT over "normal" crust to the northeast (Fig. 8A). The H-zone amplitudes exceed the normal by a factor of about 2-4, similar to the Juan de Fuca and Galapagos cases (Vogt, 1979). So far the Yermak H-zone is the only clear example of an "H-zone" along the MAR axis, although similar crust may be found in 10–30 km-wide strips along fracture zones (See later). If this phenomenon occurs at smaller scales (applitude enhancement and/or horizontal dimensions), detection would be difficult in the background of other amplitude-modulating processes.

Are there any H-zones on older Atlantic crust? An example may be the "J"-magnetic anomaly (Plate 3A), a prominent linear zone of high-amplitude (up to 1000nT) anomalies associated with ca. 120 Ma oceanic crust formed between magnetic anomalies M-O and M-1, along a segment of the paleo-Mid-Atlantic Ridge south of the Grant Banks (Rabinowitz and others, 1979; Vogt and Einwich, 1979; Tucholke and Vogt, 1979; Kovacs and others, 1980; Tucholke and Ludwig, 1982). Including the conjugate J-anomaly zone in the eastern Atlantic, a total crustal area of about 5×10^4 km^2 is involved (Vogt, 1979). Rabinowitz and others (1979) modelled the high-amplitudes by a saw-toothed magnetization high with a maximum, on crust just older than M-O, six times normal (60 vs 10 A/m for a 500-m-thick layer). The high amplitudes could also reflect an anomalously thick magnetized layer (Sullivan and Keen, 1978). Supporting this is the thick lower-velocity crustal section under the J-Anomaly Ridge (Tucholke and Ludwig, 1982) associated (but not directly correlated) with the magnetic anomaly.

The basaltic basement on the crest of the northern J-Anomaly Ridge (JAR) was cored at DSDP site 384 (Tucholke and Vogt, 1979). However, the crust there is of M-2 age, older than the magnetization high (Rabinowitz and others, 1979), so the undistinguished rock magnetization and chemistry measured on the cores was, in retrospect, to be expected. Although the basement ridge (becoming a west-facing escarpment to the southwest) is centered on somewhat older crust than the magnetization high, the two features are evidently related. Both basement ridge and magnetization high diminish towards the southwest (Plate 3A). The basement structure was first formed in the northeast about M-4 time (125 Ma) and progressively later towards the southwest (Tucholke and Vogt, 1979; Tucholke and Ludwig, 1982). The propagation rate of ca. 5 cm/yr is similar to the rates (5–20 cm/yr) measured on several morphologically similar basement features on the Reykjanes Ridge southwest of Iceland (Vogt, 1974). Kovacs and others (1980) used detailed aeromagnetic data (Plate 3A) to demonstrate that the magnetization (or layer thickness) high is also diachronous. One explanation for the J-Anomaly and associated basement structures is that an Iceland-like hotspot affected the area of the Grand Banks and Iberia about the time of breakup (Tucholke and Vogt, 1979; Tucholke and Ludwig, 1982). Thus the J-Anomaly and basement structures could have been formed by excess mantle plume flow under and along the accreting plate boundary, as proposed for the Reykjanes Ridge (Vogt, 1974). Although some hotspots near the MOR are associated with magnetic H-zones (Vogt, 1979) and the Iceland hotspot is associated with basement structures similar to the JAR, only in the latter area are high amplitude SFS magnetic lineations associated with such basement structures. The J-Anomaly zone is also narrower than other H-zones (and thus represents a short-lived perturbation) and its maximum magnetization enhancement (up to 6 times normal) exceeds that of more typical H-zones (1.5 to 3 times; Vogt, 1979). The high amplitudes are especially remarkable in view of seismic and other evidence that the northeastern part of the JAR was formed by subaerial volcanism like that on Iceland (Tucholke and Ludwig, 1982) where magnetic anomalies are relatively weak and/or complex. The subaerial lava flows implied by the seismic reflection data should be rather weakly magnetized and by flowing out would further reduce the sharpness of SFS magnetic lineation M-1, yet this anomaly appears sharp. Perhaps the demonstrated absence of high magnetization in the crest of the basement ridge results from its subaerial setting. However, the nature of the source magnetization responsible for the J-Anomaly remains unknown and intrusive sources cannot be ruled out. At its northern end the western J-Anomaly belt merges into a magnetically complex region (Plate 3A) with relatively shallow basement and scattered volcanic centers. A resemblance to Iceland or the Iceland-Faeroe Ridge (Plate 3E) in horizontal scale and magnetic character supports a hotspot origin.

North of Iceland, detailed magnetic data are available from both flanks of the MOR, and bulk magnetization can be examined for symmetry. In general, anomaly amplitudes for conjugate parcels of crust are lower east of Mohns and Nansen ridges than west (Vogt, 1986b). This systematic asymmetry, not entirely explained by different basement depths, may relate to the greater sediment thickness east of the ridge (reheating effects, previously discussed) or to the greater zero-age (backtracked) depth, which may indicate a thinner magnetized layer (Vogt and others, 1982). (The zero-age depth asymmetry remains unexplained.) On Kolbeinsey Ridge (Fig. 4B), anomalies are lower on the western flank (Vogt and others, 1980), an asymmetry, which near Iceland is developed on very young crust and to the northwest, near Greenland, may be a thermal blanketing effect of thick terrigenous sediment.

D. Small Scale Modulation. The previous examples revealed processes modulating magnetic anomaly amplitudes over scales of 50 to 1,000 km. Some of this variability at the shorter scales correlates with basement topography (e.g. Figs. 3II, 4A), principally because topographic highs represent local excesses of magnetized material (to a much lesser extent also because the highs are closer to the magnetometer). On older crust such as the CMQZ, the correlation is only fair between observed anomalies and those calculated from basement topography (Fig. 5). The effect of basement relief, even in excess of 500 m, was found subordinate to magnetization contrasts within Cretaceous crust near 36°N, 21°W (Twigt and others, 1979). The small scale "random" amplitude variability of SFS anomalies was modeled by Schouten and Denham (1979) using a simple two-parameter statistical process: 5 to 10 major extrusive units per km spreading emplaced with a standard deviation <2.5 km about the axis. The numerical simulations, which also explain deep-tow and DSDP borehole data, predict up to four-fold variations in source layer thickness, and account for the coring of mixed polarities (e.g., at DSDP sites 332A, 334, 395, 396B, 407, 410, and 418A in the Atlantic) far from the nearest polarity boundaries inferred from magnetic anomalies (Denham and Schouten, 1979).

Hydrothermal effects on magnetic amplitudes have been proposed by Rona (1978) and Whitmarsh (1982). Rona noted that a local (10 km diameter) 200nT low is associated with the TAG Hydrothermal Field (26.1°N, 44.7°W on the MAR), and a similar low overlies the Reykjanes hydrothermal field in Iceland. Rona proposed that such magnetic lows reflect high-temperature hydrothermal alteration of basalt to greenstone, which is only weakly magnetic (remanence 10^{-2} A/m) compared to basalt (1 to 10 A/m) (Fig. 2). However, magnetic lows measured at ground level over the 50–400 m diameter sulfide ore bodies in basalts of the Troodos ophiolite on Cyprus (Johnson and others, 1982) are too short in wavelength to be detected at ranges of several km. Whitmarsh (1982) noted reduced magnetic amplitudes associated with basement highs outcropping (or thinly sediment covered) on 45 Ma crust near 45°N, 21°W and also 40°N, 24°W. He suggested that the basement highs have (or had) acted as hydrothermal chimneys for many m.y., promoting more pervasive low temperature alteration—hence maghematization of the magnetic oxides, and reduction of crustal magnetization—compared to nearby areas sealed by a thicker sediment cover. This process may also explain the poor direct (and locally reverse) correlation of topography with magnetic/anomalies in the CMQZ (Fig. 5).

A special magnetization anomaly is associated with very young crust (<1 Ma) along the crest of the MOR (Fig. 9III). It was recognized early that the central, normally magnetized block (corresponding to the Brunhes, <0.73 Ma) had to be assigned about twice the magnetization as older crust to achieve a match between model and observed profiles (Vine, 1966, 1968). Dredging of basalts at the MAR axis at 45°N (Irving, 1970) and later at 37°N (Johnson and Atwater, 1977) revealed that very young basalts are most highly magnetized, as did a magnetic profile along the Reykjanes Ridge (Fig. 3II, Talwani and others, 1971), and, later, inversion of magnetic/topographic profiles in the FAMOUS (37°N) (Fig. 9II, Macdonald, 1977) and TAG (26°N) areas (McGregor and others, 1977). Similar results were reported from the MOR in the Pacific. Harrison (1981) corrected all measurements to the Equator to remove the effect of geographically varying dipole intensity (The equatorial intensity is $I_0 = I_\theta (1 + 3 \sin^2 \theta)^{-1/2}$ where θ is latitude). The combined dredge and inversion data for all areas suggest that I_0 averages 12 to 16 A/m at the axis, decays to about 7 A/m in the first 0.5 m.y. (1/3 of the TRM is lost in the first 0.2 m.y. in the FAMOUS area; Johnson and Atwater, 1977), then more slowly to values mostly less than 5 A/m, and approaching a DSDP site mean of 2.5-3 A/m (Harrison, 1976, 1981). Progressive low-temperature maghematization is held responsible for the decay (Irving, 1970), although in the FAMOUS area the magnetization high correlates with high TRM, fine grained olivine basalts (Prévot and others, 1979). The high magnetization at the immediate axis gives rise to a local narrow magnetic anomaly within the Brunhes. This axial high is commonly visible at the sea-surface for fast-spreading ridges; along the MAR it is generally seen only on deep-tow profiles (e.g., Fig. 9II).

FRACTURE ZONES AND SEAMOUNTS

Fracture Zone Magnetic Effects. The sea-floor trace of an idealized oceanic transform-type fracture zone is a line separating crust of different ages. The magnetic effect of such a boundary would be to offset SFS lineation patterns and, wherever the polarity (or more generally the magnetization function) is not identical across the FZ, to generate edge effect–type anomalies. However, FZ anomalies are not just edge effects (Rea, 1972; Detrick and Lynn, 1975). Prominent anomalies occur along western Atlantic CMQZ fracture zones where the crust on both sides has the same polarity (Vogt and others, 1971; Vogt and Einwich, 1979); a conspicuous linear FZ positive anomaly occurs just north of the Bahamas (Bracey, 1968). Analysis of magnetic profiles across the Kane and other fracture zones in the eastern Atlantic CMQZ (Twigt and others, 1983) revealed FZ-parallel strips up to 34 km wide and (nominally) 1 km thick with an extra magnetization (TRM) of 8 to 17 A/m concentrated on the young side of the FZ. Assuming a magnetization of ± 5 A/m for a 1 km thick layer in M-0 to M-4 aged crust, Twigt and others (1983) deduced magnetization intensities of 13 to 22 A/m (1 km thick layer) in the highly magnetized strips, also found in the western Atlantic CMQZ. Collette and others (1974) had originally inferred a narrow (up to 5 km) zone of zero magnetization along FZs on Neogene crust on the MAR 12° to 18°N (similarly Twigt and others [1979] or anomaly 33-34 crust south of the Azores). As a result of the Twigt and others (1983) study, the anomalies were reanalyzed (Collette and others, 1984) and found consistent with 10 to 20-km-wide zones (±4.5 A/m, vs. normally ± A/m) along the young-crust FZ edges. The strip widths (10 to 30 km) and typical Atlantic FZ spacings (50 to 100 km) imply that about 20% of the oceanic crust is anomalously magnetized!

The above findings call for reexamination of earlier results (e.g., Rea, 1972 and Cochran, 1973). However, ultrabasic and altered ultrabasic rocks crop out along major MAR fracture zones (Fox and Opdyke, 1973), supporting Cochran's (1973) model of a wide zone of reduced magnetization along the Romanche FZ with intruded bodies and induced magnetization generating local anomalies. Similarly, seismic results (Purdy and Ewing, this volume) suggest thinning or even absence of basaltic upper crust along Atlantic fracture zones. Serpentinite intrusives with considerable secondary magnetite would be expected to contribute to the FZ anomaly field. SFS anomaly amplitudes decline adjacent to some Atlantic FZs, e.g. the Charlie Gibbs (Fig. 3III) and Greenland FZs (Fig. 8B), perhaps reflecting reduced basalt production. Some large FZ ridges—for example, the Mendocino Ridge in the Pacific (Vogt, 1979) and the Greenland FZ ridge (Fig. 8B)—have a weak bulk magnetization, whereas the Hovgaard FZ ridge exhibits pronounced anomalies (Fig. 8B).

Twigt and others (1983) note that since the enhanced magnetization occurs on the young side of a FZ axis, the magnetically anomalous strips did not pass through the active transform domain. They suggest "flood basalts" as a source of high magnetization. Whatever the source rocks are, their emplacement about the tip of the spreading axis could have been originally symmetrical. The crust passing through the active zone could have lost its magnetization by active oblique/normal faulting in the inside corner of the FZ/ridge intersection, and by associated hydrothermal-generated reheating and maghematization. The magnetized strips have a similar width, magnetization excess, and preference for the young side of the FZ as the "Fracture High Amplitude Zones" (FHAZ) earlier described in the Pacific (Vogt and Byerly, 1976; Vogt, 1979) and therefore may have a common origin, perhaps a higher titanomagnetite concentration due to more extensive fractional crystallization. (On some fracture zones [Vogt, 1979] high magnetization also occurs on the side that passed through the active transform zone.) The high magnetization and FeTi basalts at the tips of Pacific propagating rifts (Vogt and de Boer, 1976; Sinton and others, 1983; Miller and Hey, 1986) may be another expression of the same phenomenon.

Seamount Magnetic Anomalies. Seamounts are the largest units of oceanic crustal material for which bulk magnetization (magnitude and direction) can be calculated from magnetic anomalies. This is not possible for SFS lineations. The Bermuda pedestal, the largest volcanic edifice in the western North Atlantic, and its satellite banks were early targets of aeromagnetic study. At the 1947 AGU, Press and Ewing (1952) first presented model profiles calculated over several oceanic structures, including Bermuda. As part of ONR's "Project Volcano," six low level (1,000 ft) aeromagnetic profiles were flown across the Bermuda area in 1948 (Keller and others, 1954). The 4000nT positive over Plantagenet Bank remains the highest known sea-level magnetic anomaly in the North Atlantic south of Iceland; a subsequent detailed vector survey (Young and Kontis, 1964) revealed inclination and declination anomalies up to 2° to 6°, respectively. The anomaly pattern over Bermuda is more complex than the simple dipole pattern of smaller seamounts, perhaps because rocks of both polarities exist in the edifice, which formed during 50–30 Ma, a time of frequent reversals (Vogt and Tucholke, 1979). A borehole with 859 m basement penetration recovered "Cretaceous" pillow basalts with original reverse polarity hydrothermally remagnetized (ca. 2A/m) by Tertiary sheets (2-5A/m) of normal polarity (Rice and others, 1980).

Between Bermuda and the New England Seamounts, a northwest-trending line of widely spaced, normally polarized, probably Cretaceous seamounts and knolls extends via Muir and Caryn seamounts to Knauss Knoll at 37.4°N, 70.8°W (Vogt and Perry, 1982). Caryn Seamount, surveyed by R/V *Vema* with a towed fluxgate magnetometer in 1954, was the first seamount subjected to detailed magnetic study (Miller and Ewing, 1956; Talwani, 1965). Anomaly highs range from 250nT (Knauss Knoll) and 350nT (Caryn Smt.) to 900nT (Muir Smt.), with lows of –200 to –300nT for the larger features. The conclusions of Miller and Ewing (1954) about Caryn Seamount (a rather high Cretaceous TRM, roughly parallel to the present geomagnetic field) have not been significantly refined in the subsequent three decades, and this holds for most western Atlantic seamounts. The only new development concerns the possible importance of VRM or CRM (Williams and others, 1983; Raymond and LaBrecque, 1986).

The New England (or Kelvin) Seamounts and their possible extension to the Corner Seamounts represent the longest intraplate volcanic chain in the North Atlantic (Uchupi and others, 1970; Vogt and Tucholke, 1979; Duncan, 1984). Magnetic anomalies over the New England seamounts correlate well with basement topography and are consistent either with normal-polarity TRM or VRM. In 1957 a vector aeromagnetic survey covered a swath from Kelvin to Bear (Kontis and Young, 1965); this is one of the few detailed vector data sets in the oceans. A shipborne total field survey at 9 km track separation (Walczak, 1963; Walczak and Carter, 1964) extended the coverage to 56°W. Gilliss Seamount is the most thoroughly studied member of the chain (Taylor and others, 1975); data include multi-beam bathymetry, seismic reflection, gravity, magnetics, bottom photographs, and dredge samples whose magnetic properties were measured. Cordell and Taylor (1971) used Poisson's Theorem to compare gravity and magnetic anomalies over Gilliss Seamount. Many of the "seamounts" actually comprise two or more distinct peaks (Pl. 3B). DSDP Sites 282 (Nashville Smt.) and 385 (Vogel Smt.) are the only deep submarine drill sites on western North Atlantic seamounts. Only 135 m and 55 m of volcaniclastics were penetrated on the lower flanks at these sites, inadequate for seamount magnetism studies. Anomaly amplitudes over the New England Seamounts range from about + 200 to + 2200nT (Atlantis II) for the positives, down to as low as –700nT for the associated negative anomalies (Plate 3B). The larger seamounts decrease the geomagnetic dip about 0.3° on the south and increase it by 1° on the north (Kontis and Young, 1965). All seamounts of the chain appear normally magnetized, consistent with formation during normal geomagnetic polarity, 118 to 84 Ma (Duncan, 1984). A third group of seamounts, the Fogo seamounts and associated basement highs, occurs just south of the Grand Banks (Plate 3A). Most of the peaks rise only to depths of 3 to 4 km and exhibit positive anomalies of a few hundred nT at most (Plate 3A). Association of these seamounts with the J-Anomaly Ridge and hotspot (Tucholke and Vogt, 1979) suggests a Barremian to Albian age. The topographic (or basement) shape of a seamount plus its observed magnetic anomaly can be used to estimate the bulk magnetization (magnitude and direction). Using two-dimensional models, Kontis and Young (1965) derived a high (and therefore remanent) bulk magnetization (8.6A/m) for Kelvin Seamount. Similar high magnetizations (7 to 10 A/m) were found for Bear, Physalia, Retriever, Picket, and Balanus seamounts ("Zone 1"). However, a second line of smaller seamounts (the Mytilus-Asterias-Panulirus line, or "Zone 2") exhibits lower amplitudes, not entirely accounted for by their greater depth and smaller size. Evidently the Zone 2 volcanoes have a bulk magnetization several times less than that of the main line. If verified by newer data, this would be an unusual example of a systematic difference in magnetic properties of two nearby (and evidently related) seamount groups.

A more refined approach is to approximate a seamount by a finite sum of rectangular prisms (Vacquier, 1962) or polygonal laminae (Talwani, 1965) and determine the best-fitting magnetization vector. Generally magnetization is assumed constant, but non-magnetic tops and lower flanks have been demonstrated for

some Pacific seamounts (e.g. Saeger and Keating, 1984).
Richards and others (1967) applied Vacquier's method to a number of seamounts, seven in the New England chain (the Kelvin, Atlantis II, and Gosnold clusters). Computed magnetizations range from 7.4 to 13.6 A/m, and the average VGP (Virtual Geomagnetic Pole) is at 76°N, 108°E. A 95% confidence radius of 21° makes this VGP consistent with the Cretaceous pole for North America, but is large enough to include the present Geographic Pole and therefore unable to prove that VRM is unimportant. Using the same data as Walczak (1963), Mason and others (1966) examined twelve members of the chain plus Caryn Seamount, and inferred a systematically decreasing paleolatitude, from 31° for Rehoboth to 12° for Bear. Talwani (1965) derived a magnetization of about 3 A/m for Caryn Seamount, essentially parallel to the present field.

In the eastern Atlantic, Verhoef (1984) derived a magnetization of about 5 A/m for Hyeres, Irving, Cruiser, Plato, Tyro, and Atlantis seamounts; this value also typifies Great Meteor Seamount and various Pacific seamounts (e.g., Sager and Keating, 1984). Verhoef could not discriminate between the small magnetization direction differences between the present and early or mid-Tertiary fields in the eastern Atlantic. However, the lack of reversely magnetized seamounts suggests Viscous Remanent Magnetization (VRM), not TRM as the principal magnetization (Williams and others, 1983; Verhoef, 1984).

THE NEAR FIELD AND THE FAR FIELD

The SFS lineations and seamount anomalies that dominate the anomalous magnetic field near sea level (1 to 10 km basement depths) are no longer apparent either at satellite heights (a few hundred km; Fig. 10), where they are too attenuated to resolve, or at ranges closer than about 1 km (Fig. 9), where they are swamped by high-frequency effects. The "near field" can be examined by magnetometers (1) lowered into boreholes, (2) towed at depth (Fig. 9II), (3) mounted on submersibles (Fig. 9IV), (4) read directly on the ground in areas of subaerial ocean crust like Iceland (Fig. 9I), and (5) towed behind surface ships (Fig. 4A) or (6) installed on low flying aircraft (Fig. 4B) in areas of very shallow water. On older crust where sediments are thick, deep-towed magnetometers cannot be towed close enough to the basement to sense the "near field," but such data still enhance the resolution of SFS anomalies (Fig. 7). Surveying rates are limited by platform speed, ranging from ca. 1 km/hr for deep-tow packages and submersibles to 15 to 30 km/hr for ships towing surface magnetometers, and 400 km/hr for fixed-wing aircraft.

Along the MAR accretion axis and on Iceland, the near field is dominated by anomalies of typically 0.1 to 5 km wavelength and $\pm$ 1 to 2×10^3nT amplitude. The shorter wavelengths (a few of which were caused by lightning strikes) attenuate very rapidly upward, decaying by over 50% in the first 50 m (Fig. 9I). Within a few meters of the basement in the MAR rift valley the gradients may reach 10^2 to 10^3 nT/m (Fig. 9IV). Anomaly variations of $\pm 10^4$nT in a few meters in basalts, but only $\pm 10^2$nT in thick tuffs were measured in a drillhole in Reykjavik (Jónsson, 1976). Similarly, variations of $\pm$ 2–5 $\times 10^3$nT with 2 m wavelength, and a 2×10^4nT change in 50 m across a polarity reversal, were logged in hole 418A (110 Ma) (Leg 102 Scientific Party, 1985). An exceptional anomaly of 1.8×10^4nT on the ground, a few tens of m above tholeiites of unusually high (10–20 × normal) magnetization occurs at Stardalur farm 24 km ENE of Reykjavik (Kristjansson, 1972). Near-field anomalies of similar amplitude undoubtedly occur at intervals along the MAR axis, particularly in association with pillowed ferrobasalts. Although topography causes some near-field anomalies (Macdonald, 1977), others must reflect spatial variations in magnetization intensity related to variable petrology or structure.

The "far field" observed at the 325 to 550 km elevation of MAGSAT (Fig. 10) exhibits minimum consistent wavelengths of 700 km, with a theoretical resolution limit of 250 km (Sailor and others, 1982). Satellites in lower orbits, for example, the 160 km for the proposed Geophysical Research Mission (Taylor and others, 1983) may drive the resolution below 100 km. Described as a spectrum of spherical harmonic coefficients, the earth's magnetic field changes slope at degree and order ca. 13 to 15 (Langel and Estes, 1982). Lower degrees presumably originate in the core (although crustal structures may contribute; Meyer and others, 1985). Higher degrees originate in the crust, but with possible contamination by core effects (Harrison and Carle, 1981; Carle and Harrison, 1982) and by persistent ionospheric currents above regions of high conductivity gradients (Hermance, 1982), for example, ocean-continent boundaries or the MAR. The part of the MAGSAT total-field anomaly of presumed crustal origin (Fig. 10) is dominated by ca. 2×10^3 km wavelength, $\pm$ 10–40nT anomalies, most of which do not clearly correspond to crustal geological features. The lack of contrast between ocean basins and continents may be due to the removal of longer-wavelength crustal wavelengths that spectrally overlap core effects (Meyer and others, 1985). As expected (LaBrecque and Raymond, 1985), the MAGSAT field contains no hint of SFS lineations; not even the J-anomaly (Pl. 3A) can be seen at this height. However, the two prominent northeast trending highs in the eastern and western central Atlantic are attributed by LaBrecque and Raymond (1985) to the Cretaceous Magnetic Quiet Zone (CMQZ; Fig. 1). In their model, the CMQZ crust is relatively strongly magnetized (15 A/m for a 500-m thick layer), with $\pm$ 35° anomalous skewness, both features a reflection of secondary CRM (Raymond and LaBrecque, 1986). The 90nT high near the North Pole is associated with the Alpha Ridge, probably an oceanic plateau with thick, normally magnetized (TRM, CRM, and VRM) crust, also of CMQZ age. A lesser magnetic positive over Iceland is still twice that accounted for by drillcore-based magnetization models (3 A/m to 4 km depth) (Coles, 1985). Along the central MAR there tends to be a high just east of the axis, and a low to the west (Fig. 10), possibly due to the decay of susceptibility with crustal age (from 1 A/m at the axis for a 2 km thick layer; LaBrecque and Raymond, 1985); no such effect is apparent north of the Azores. Since the anomaly field lacks an absolute zero level, the ocean basin lows (Fig. 10) can be considered zero, which is what magnetic model calculations predict for areas of SFS lineations if VRM and induced magnetizations are neglected.

CONCLUSION

While the magnetic anomaly data base (e.g., Plate 3) still falls far short of what has been achieved on the North American continent, reliable SFS anomaly identifications are already so plentiful (e.g., chapters 22 and 23) that additional data will serve merely to "fine tune" plate kinematic models. By contrast, anomaly surveys comparable to Plate 3 A-E in detail, but using GPS navigation, and perhaps also magnetic gradiometry, are needed elsewhere to map finer-scale volcano-tectonic structures (magma center spacing, minor FZs and axis jumps, hydrothermal effects, etc.). The magnetization structure of the crust cannot be uniquely determined from magnetic anomalies alone. However, simultaneous mapping of the magnetic field and basement topography—particularly by deep-tow surveys—leads to plausible solutions of bulk basement magnetization both in one (Parker and Huestis, 1974) and two dimensions (Miller and Hey, 1986). Some remaining ambiguities may be resolved by other geophysical experiments: For example, seismic refraction could test whether high magnetic amplitudes not accounted for by basement topography could reflect (1) a high percent of massive basalts or (2) a thick magnetized layer. The fine scales of crustal magnetization can only be resolved by sample analysis or borehole logging. Such data are still too sparse to estimate the scales of vertical and particularly horizontal variability of, for example, pillow versus massive basalts, ferrobasalts versus olivine basalts, or numbers of reversals per unit vertical section. If magnetic anomaly data are used to locate deep drill sites to test hypotheses of magnetization structure, the vast sea-surface magnetic anomaly data base (Plate 3) can be more meaningfully interpreted in terms of variations in crustal properties on scales resolved by the anomalies.

Specific unsolved problems raised by previous magnetic anômaly analyses in the Atlantic can be addressed by deep drilling and magnetic borehole logging, for example, into (a) anomaly 33/34-aged crust to understand the 30° to 40° anomalous skewness; (b) the J-anomaly, Plantagenet Bank, Yermak H-zone, and fracture-zone-parallel magnetization highs to determine whether layer thickness or rock magnetization (or both) are anomalous, and why (Fe-rich basalts due to intense fractionation?); and (c) magnetic smooth zones such as those near Iceland, the Knipovich Ridge and off the U.S. east coast, perhaps all of different origins. In addition, (d) a series of holes along a flow line through the Cretaceous Magnetic "Quiet Zone" could examine (1) possible spreading rate variations between anomalies M-0 and 34 and (2) the nature of CMQZ magnetization, including possible geomagnetic imprints such as paleointensity fluctuations or short reversals. Before the GSA's bicentennial, a set of standard "moholes" may have been drilled on representative crust of different ages, spreading rate, and hotspot influence. The story of oceanic crustal magnetization will not be complete until then.

REFERENCES

Anderson, C. N., Vogt, P. R., and Bracey, D. R.
1969 : Magnetic anomaly trends between Bermuda and the Bahama-Antilles Arc [abs.]: EOS Transactions of the American Geophysical Union; v. 50, p. 189.

Anderson, R. N., Clague, D. A., Klitgord, K. D., Marshall, M., and Nishimori, R. K.
1975 : Magnetic and petrologic variations along the Galapagos spreading center and their relation to the Galapagos melting anomaly: Geological Society of America Bulletin, v. 86, p. 683–694.

Anderson, R. N., Spariosu, D. J., Weissel, J. K., and Hayes, D. E.
1980 : The interrelation between variations in magnetic anomaly amplitudes and basalt magnetization and chemistry along the Southeast Indian Ridge: Journal of Geophysical Research, v. 85, p. 3883–3898.

Anderson, R. N., Honnorez, J., Becker, K., Adamson, A. C., Laverne, C., Mottl, M. J., and Newmark, R. L.
1982 : DSDP hole 504B, the first reference section over 1 km through layer 2 of the oceanic crust: Nature, v. 300, p. 589–594.

Banerjee, S. K.
1980 : Magnetism of the oceanic crust; Evidence from ophiolite complexes: Journal of Geophysical Research, v. 85, p. 3557–3566.

Barrett, D. L., and Keen, C. E.
1976 : Mesozoic magnetic lineations, the magnetic quiet zone, and sea-floor topography: Deep-Sea Research, v. 22, p. 883–892.

Becker, H.
1980 : Magnetic anomalies (ΔZ) in NE-Iceland and their interpretation based on rock-magnetic investigations: Journal of Geophysics, v. 47, p. 43–56.

Blakely, R. J.
1976 : An age-dependent, two-layer model for marine magnetic anomalies, *in* Sutton, G. H., Manghnani, M. H., Moberly, R., and McAffee, E. U., eds., The Geophysics of the Pacific Ocean Basin and its Margin: Geophysical Monograph 19, American Geophysical Union, Washington, D.C., p. 227–234.
1983 : Statistical averaging of marine magnetic anomalies and the aging of oceanic crust: Journal of Geophysical Research: v. 88, p. 2289–2296.

Blakely, R. J., and Cox, A.
1972 : Evidence for short geomagnetic polarity intervals in the early Cenozoic: Journal of Geophysical Research, v. 77, p. 7065–7072.

Bleil, U., and Petersen, N.
1983 : Variations in magnetization intensity and low-temperature titanomagnetite oxidation of ocean floor basalts: Nature; v. 301, p. 384–388.

Bleil, U., Hall, J. M., Johnson, H. P., Levi, S., and Schönharting, G.
1982 : The natural magnetization of a 3 km section of Icelandic crust: Journal of Geophysical Research, v. 87, p. 6569–6589.

Bracey, D. R.
1968 : Structural implications of magnetic anomalies north of the Bahamas-Antilles islands: Geophysics, v. 33, p. 950–961.

Bracey, D. R., and Avery, O. E.
1963 : Marine magnetic survey off the southern Bahamas: Project M-15, U.S. Naval Oceanographic Office Technical Report TR-160.

Bryan, G. M., Markl, R. G., and Sheridan, R. E.
1980 : IPOD site survey in the Blake-Bahamas Basin: Marine Geology, v. 35, p. 43–63.

Cain, J. C., Schmitz, D. R., and Muth, L.
1984 : Small-scale features in the earth's magnetic field observed by Magsat: Journal of Geophysical Research, v. 89, p. 1070–1076.

Cande, S. C.
1976 : A paleomagnetic pole from late Cretaceous marine magnetic anomalies in the Pacific: Geophysical Journal of the Royal Astronomical Society, v. 44, p. 547–566.
1978 : Anomalous behavior of the paleomagnetic field inferred from the skewness of anomalies 33 and 34. Earth and Planetary Science Letters,

v. 40, p. 275–286.

Cande, S. C., and Kent, D. V.
1976 : Constraints imposed by the shape of marine magnetic anomalies on the magnetic source: Journal of Geophysical Research, v. 81, p. 4157–4162.
1985 : Comment on "Tectonic Rotations in Extensional Regimes and Their Paleomagnetic Consequences for Ocean Basalts" by K. L. Verosub and E. M. Moores: Journal of Geophysical Research, v. 90, p. 4647–4651.

Cande, S. C. and Kristoffersen, Y.
1977 : Late Cretaceous magnetic anomalies in the North Atlantic: Earth and Planetary Science Letters, v. 35, p. 215–224.

Cande, S. C., Larson, R. L., and LaBrecque, J. L.
1978 : Magnetic lineations in the Jurassic Quiet Zone: Earth and Planetary Science Letters, v. 41, p. 434–440.

Carle, H. M., and Harrison, C.G.A.
1982 : A problem representing the core magnetic field of the earth using spherical harmonics: Geophysical Research Letters, v. 9, p. 265–268.

Cochran, J. R.
1973 : Gravity and magnetic investigations in the Guiana Basin, western Equatorial Atlantic: Geological Society of America Bulletin, v. 84, p. 3249–3268.

Cole, J. F.
1908 : Magnetic declination and latitude observations in the Bermudas: Terrestrial Magnetism and Atmospheric Electricity, v. 13, p. 49–56.

Coles, R. L.
1985 : Magsat scalar magnetic anomalies at northern high latitudes: Journal of Geophysical Research, v. 90, p. 2576–2582.

Collette, B. J., Schouten, J. A., Rutten, K., and Slootweg, A. P.
1974 : Structure of the Mid-Atlantic Ridge Province between 12° and 18°N: Marine Geophysical Researches, v. 2, p. 143–179.

Collette, B. J., Slootweg, A. P., Verhoef, J., and Roest, W. R.
1984 : Geophysical investigations of the floor of the Atlantic Ocean between 10° and 38°N (Kronvlag project); Proceedings of the Koninklijke Nederlandse Akademie van Wetenschappen, Series C, v. 87, p. 1–76.

Cordell, L.
1974 : Reconsideration of paleomagnetic data in terms of wandering continents and a stable magnetic pole: Geology, v. 2, p. 363–366.

Cordell, L., and Taylor, P. T.
1971 : Investigation of magnetization and density of a North Atlantic seamount using Poisson's Theorem: Geophysics, v. 36, p. 919–937.

Cox, A., Doell, R. R., and Dalrymple, G. B.
1963 : Geomagnetic polarity epochs and Pleistocene geochronometry: Nature, v. 198, p. 1049–1051.

Denham, C. R., and Schouten, H.
1979 : On the likelihood of mixed polarity in oceanic basement drillcores; in (Talwani, M., Harrison, C.G.A., and Hayes, D. E., eds.); Deep Drilling Results in the Atlantic Ocean: Ocean Crust, Maurice Ewing Series, v. 2, American Geophysical Union, p. 160–165.

Detrick, R. S., and Lynn, W. S.
1975 : The origin of high amplitude magnetic anomalies at the intersection of the Juan de Fuca Ridge and Blanco Fracture Zone: Earth and Planetary Science Letters, v. 26, p. 105–113.

Drake, C. L., Heirtzler, J., and Hirshman, J.
1963 : Magnetic anomalies off eastern North America: Journal of Geophysical Research, v. 68, p. 5259–5275.

Duncan, R. A.
1984 : Age progressive volcanism in the New England Seamounts and the opening of the Central Atlantic Ocean: Journal of Geophysical Research, v. 89, p. 9980–9990.

Dunlop, D. J., and Prévot, M.
1982 : Magnetic properties and opaque mineralogy of drilled submarine intrusive rocks: Geophysical Journal of the Royal Astronomical Society, v. 69, p. 763–802.

Emery, K. O., Uchupi, E., Phillips, J. D., Bowin, C. O., Bunce, E. T., and Knott, S. T.
1970 : Continental rise off eastern North America: American Association of Petroleum Geologists Bulletin, v. 54, p. 44–108.

Feden, R. H., Vogt, P. R., and Fleming, H. S.
1979 : Magnetic and bathymetric evidence for the "Yermak Hot Spot" northwest of Svalbard in the Arctic Basin: Earth and Planetary Science Letters, v. 44, p. 18–38.

Fox, P. J., and Opdyke, N. D.
1973 : Geology of the oceanic crust; Magnetic properties of oceanic rocks: Journal of Geophysical Research, v. 78, p. 5139–5154.

Girdler, R. W., and Peter, G.
1960 : An example of the importance of natural remanent magnetization in the interpretation of magnetic anomalies: Geophysical Prospecting, v. 8, p. 474–483.

Greenewalt, D., and Taylor, P. T.
1978 : Near-bottom magnetic measurements between the FAMOUS area and DSDP sites 332 and 333: Geological Society of America Bulletin, v. 89, p. 571–576.

Hall, J. M.
1985 : The Iceland Research Drilling Project crustal section; Variation of magnetic properties with depth in Icelandic-type oceanic crust: Canadian Journal of Earth Science, v. 22, p. 85–101.

Harland, W. B., Cox, A. V., Llewellyn, P. G., Pickton, C.A.G., Smith, A. G., and Walters, R.
1982 : A Geologic time scale: Cambridge Earth Science Series, Cambridge University Press, 131 p.

Harrison, C.G.A.
1976 : Magnetization of the oceanic crust: Geophysical Journal of the Royal Astronomical Society, v. 47, p. 257–283.
1981 : Magnetization of the oceanic crust, in Emiliani, C., ed., Ideas and Observations on Progress in the study of the Seas, v. 7, The Oceanic Lithosphere: New York, John Wiley and Sons, p. 219–338.

Harrison, C.G.A., and Carle, H. M.
1981 : Intermediate wavelength magnetic anomalies over ocean basins: Journal of Geophysical Research, v. 86, p. 11585–11599.

Harrison, C.G.A., McDougall, I., and Watkins, N. D.
1979 : A geomagnetic field reversal time scale back to 13.0 million years before present: Earth and Planetary Science Letters, v. 42, p. 143–152.

Hayes, D. E., and Pitman, W. C.
1970 : Magnetic lineations in the North Pacific, in Hays, J. D., ed., Geological Investigations of the North Pacific: Geological Society of America Memoir 126, p. 291–314.

Hayes, D. E., and Rabinowitz, P. D.
1975 : Mesozoic magnetic lineations and the magnetic quiet zone off northwest Africa: Earth and Planetary Science Letters, v. 28, p. 105–115.

Heezen, B. C., Ewing, M., and Miller, E. T.
1953 : Trans-Atlantic profile of total magnetic intensity and topography, Dakar to Barbados: Deep Sea Research, v. 1, p. 25–33.

Heezen, B. C., Tharp, M., and Ewing, M.
1959 : The floors of the oceans, I. the North Atlantic: Geological Society of America Special Paper 65, 122 p.

Heirtzler, J. R., and Hayes, D. E.
1967 : Magnetic boundaries in the North Atlantic Ocean: Science, v. 157, p. 185–187.

Heirtzler, J. R., and LePichon, X.
1965 : The structure of mid-ocean ridges and magnetic anomalies over the Mid-Atlantic Ridge: Journal of Geophysical Research, v. 70, p. 4013–4033.

Heirtzler, J. R., LePichon, X., and Baron, J. G.
1966 : Magnetic anomalies over the Reykjanes Ridge: Deep Sea Research, v. 13, p. 427–443.

Heirtzler, J. R., Dickson, G. O., Herron, E. M., Pitman, W. C. III, and LePichon, X.

1968 : Marine magnetic anomalies, geomagnetic field reversals, and the motions of the ocean floor and continents: Journal of Geophysical Research, v. 73, p. 2119–2136.

Hermance, J. F.
1982 : Model simulations of possible electromagnetic induction effects on Magsat activities: Journal of Geophysical Research, v. 9, p. 373–376.

Hospers, J.
1953 : Reversals of the main geomagnetic field, I-II: Proceedings of the Koninklijke Nederlandse Akademie van Wetenschappen, v. B56, p. 467–491.

Huestis, S. P., Parker, R. L.
1977 : Bounding the thickness of the oceanic magnetized layer: Journal of Geophysical Research, v. 82, p. 5293–5303.

Hurwitz, L., and Nelson, J. H.
1960 : Proton vector magnetometer: Journal of Geophysical Research, v. 65, p. 1759–1765.

Irving, E.
1970 : The Mid-Atlantic Ridge at 45°N. XIV; Oxidation and magnetic properties of basalt, review and discussion: Canadian Journal of Earth Science, v. 7, p. 1528–1538.
1979 : Pole positions and continental drift since the Devonian; in McElhinny, M. W., ed., The Earth: Its Origin, Structure and Evolution: London and New York, Academic Press, 567 p.

Irving, E., Robertson, W. A., and Aumento, F.
1970 : The mid-Atlantic ridge near 45°N, VI; Remanent intensity, susceptibility, and iron content of dredged samples: Canadian Journal of Earth Science, v. 7, p. 226–238.

Jackson, H. R., and Reid, I.
1983 : Oceanic magnetic anomaly amplitudes; Variation with sea-floor spreading rate, and possible implications: Earth and Planetary Science Letters, v. 63, p. 368–378.

Johnson, G. L., and Jakobsson, S. P.
1985 : Structure and petrology of the Reykjanes Ridge between 62°00′N and 63°48′N: Journal of Geophysical Research, v. 90, p. 10073–10083.

Johnson, H. P.
1979 : Magnetization of the oceanic crust: Reviews of Geophysics and Space Physics, v. 17, p. 215–226.

Johnson, H. P., and Atwater, T.
1977 : Magnetic study of basalts from the Mid-Atlantic Ridge, lat. 37°N: Geological Society of America Bulletin, v. 88, p. 637–647.

Johnson, H. P., and Hall, J. M.
1978 : A detailed rock magnetic and opaque mineralogy study of the basalts from the Nazca plate: Geophysical Journal of the Royal Astronomical Society, v. 52, p. 45–64.

Johnson, H. P., Karsten, J. L., Vine, F. J., Smith, G., and Schönharting, G.
1982 : A low-level magnetic survey over a massive sulfide ore body in the Troodos ophiolite complex, Cyprus: Marine Technology Society, v. 16, p. 76–80.

Jones, E.J.W., Laughton, A. S., Hill, M. N., and Davies, D.
1966 : A geophysical study of part of the western boundary of the Madeira–Cape Verde Abyssal Plain: Deep-Sea Research, v. 13, p. 889–907.

Jönsson, B. B.
1976 : A proton precession magnetometer for drillhole measurements: Reykjavik, University of Iceland, Science Institute, Report No. 17.

Keller, F. J., Meuschke, J. L., and Alldredge, L. R.
1954 : Aeromagnetic surveys in the Aleutian, Marshall, and Bermuda islands: EOS American Geophysical Union Transactions, v. 35, 558–572.

Kent, D. V., and Opdyke, N. D.
1978 : Paleomagnetism and magnetic properties of igneous rock samples; Leg 38, in White, S. M., and others, eds., Initial Reports of the Deep Sea Drilling Project: Washington, D.C., U.S. Government Printing Office, supplement to v. 38, 39, 40, and 41, p. 3–8.

Kent, D. V., Honnerez, B. M., Opdyke, N. D., and Fox, P. J.
1978 : Magnetic properties of dredged oceanic gabbros and the source marine magnetic anomalies: Geophysical Journal of the Royal Astronomical Society, v. 55, p. 513–537.

Kidd, R.G.W.
1977 : The nature and shape of the source of marine magnetic anomalies: Earth and Planetary Science Letters, v. 33, p. 310–320.

Klitgord, K. D., Huestis, S. P., Mudie, J. D., and Parker, R. L.
1975 : An analysis of near-bottom magnetic anomalies; Sea-floor spreading and the magnetized layer: Geophysical Journal of the Royal Astronomical Society, v. 43, p. 387–424.

Kontis, A. L., and Young, G. A.
1964 : Approximation of residual total magnetic intensity anomalies: Geophysics, v. 29, p. 623–627.
1965 : A study of aeromagnetic data–New England seamount area: U.S. Naval Oceanographic Office Technical Report TR-166, 18 p.

Kovacs, L. C., and Vogt, P. R.
1982 : Depth-to-magnetic source analysis of the Arctic Ocean region: Tectonophysics, v. 89, p. 255–294.

Kovacs, L. C., Cherkis, N. Z., and Vogt, P. R.
1980 : The western J-anomaly: Geological Society of America Abstracts with Programs, v. 12, p. 67.

Kristjansson, L.
1972 : On the thickness of the magnetic crustal layer in southwestern Iceland: Earth and Planetary Science Letters, v. 16, p. 237–244.
1982 : Paleomagnetic research on Icelandic rocks: A bibliographic review 1951-1981: Jökull, v. 32, p. 91–106.

Kristoffersen, Y.
1978 : Sea-floor spreading and the early opening of the North Atlantic: Earth and Planetary Science Letters, v. 38, p. 273–290.

LaBrecque, J. L., and Raymond, C. A.
1985 : Seafloor spreading anomalies in the MAGSAT field of the North Atlantic: Journal of Geophysical Research, v. 90, p. 2565–2575.

Langel, R. A., and Estes, R. H.
1982 : A geomagnetic field spectrum: Geophysical Research Letters, v. 9, p. 250–253.

Larson, R. L., and Hilde, T.W.C.
1975 : A revised time scale of magnetic reversals for the Early Cretaceous and Late Jurassic: Journal of Geophysical Research, v. 80, p. 2585–2594.

Larson, R. L., and Pitman, W. C.
1972 : World-wide correlation of Mesozoic magnetic anomalies, and its implications: Geological Society of America Bulletin, v. 83, p. 3645–3662.

Laughton, A. S., and Whitmarsh, R. B.
1974 : The Azores-Gibraltar plate boundary, in Kristjansson, L., ed., Geodynamics of Iceland and the North Atlantic Area: Dordrecht, Holland, Reidel, p. 63–81.

Leg 102 Scientific Party
1985 : Old hole yields new information: Geology, v. 30, p. 13–15.

Le Pichon, X. L.
1968 : Sea-floor spreading and continental drift: Journal of Geophysical Research, v. 73, p. 3661–3697.

LePichon, X. L., Hyndman, R. D., and Pautot, G.
1971 : Geophysical study of the opening of the Labrador Sea: Journal of Geophysical Research, v. 76, p. 4724–4743.

Lowrie, W.
1973 : Viscous remanent magnetization in oceanic basalts: Nature, v. 243, p. 27–29.
1974 : Oceanic basalt magnetic properties and the Vine and Matthews hypothesis: Journal of Geophysics, v. 40, p. 513–536.
1977 : Intensity and direction of magnetization in oceanic basalts: Journal of the Geological Society of London, v. 133, p. 61–82.

Lowrie, W., and Ogg, J. G.
1985 : A magnetic polarity time scale for the Early Cretaceous and Late Jurassic: Earth and Planetary Science Letters, v. 76, p. 341–349.

Lowrie, W., Channell, J.E.T., and Alvarez, W.

1980 : A review of magnetic stratigraphy investigations in Cretaceous pelagic carbonate rocks: Journal of Geophysical Research, v. 85, p. 3597–3605.
Macdonald, K. C.
1977 : Near bottom magnetic anomalies, asymmetric spreading oblique spreading and the tectonics of the Mid-Atlantic Ridge near lat. 37°N: Geological Society of America Bulletin, v. 88, p. 541–555.
Macdonald, K. C., and Atwater, T.
1978 : Evolution of rifted ocean ridges: Earth and Planetary Science Letters, v. 39, p. 319–327.
Marine Surveys Division
1962 : Operations Deep Freeze 61 (1960–1961): Marine Geophysical Investigations, TR-105, U.S. Navy Hydrographic Office, 111 p.
Marshall, M., and Cox, A. V.
1971 : Magnetism of pillow basalts and their petrology: Geological Society of America Bulletin, v. 82, p. 537–552.
1972 : Magnetic changes in pillow basalt due to sea floor weathering: Journal of Geophysical Research, v. 77, p. 6459–6469.
Mason, R. G., Richardson, A., and Watkins, N. D.
1966 : Paleomagnetism of the New England Seamounts and the Canary Islands as a test for oceanic crustal spreading: Transactions of the American Geophysical Union, v. 47, p. 79–80.
Matthews, D. H.
1961 : Lavas from an abyssal hill on the floor of the Atlantic Ocean: Nature, v. 190, p. 158–159.
McGregor, B. A., and Harrison, C.G.A.
1977 : Magnetic anomaly patterns on Mid-Atlantic Ridge crest at 26°N: Journal of Geophysical Research, v. 82, p. 231–238.
McGregor, B. A., Harrison, C.G.A., Lavelle, J. W., and Rona, P. A.
1977 : Magnetic anomaly pattern on Mid-Atlantic Ridge crest at 26°N: Journal of Geophysical Research, v. 82, p. 231–238.
Merrill, R. T., and McElhinney, M. W.
1983 : The Earth's Magnetic Field: London, Academic Press, 401 p.
Meyer, J., Hufen, J-H., Siebert, M., and Hahn, A.
1985 : On the identification of Magsat anomaly charts as crustal part of the internal field: Journal of Geophysical Research, v. 90, p. 2537–2541.
Miller, E. T., and Ewing, M.
1956 : Geomagnetic measurements in the Gulf of Mexico and in the vicinity of Caryn Peak: Geophysics, v. 21, p. 406–432.
Miller, S. P., and Hey, R. N.
1986 : Three-dimensional magnetic modeling of a propagating rift, Galapagos 95°30′W: Journal of Geophysical Research, v. 91, p. 3395–3406.
Morgan, W. J.
1986 : Rises, trenches, great faults and crustal blocks: Journal of Geophysical Research, v. 73, p. 1959–1982.
Morley, L. W., and Larochelle, A.
1964 : Paleomagnetism as a means of dating geological events; Royal Society of Canada, Special Publication 8, p. 512–521.
Moore, J. G., and Schilling, J. G.
1973 : Vesicles, water, and sulfur in Reykjanes Ridge Basalts: Contributions to Mineralogy and Petrology—Beiträge zur Mineralogie and Petrologie, v. 41, p. 105–118.
Nelson, J. H., Hurwitz, L., and Knapp, D. G.
1962 : Magnetism of the earth: Publication 40-1, U.S. Dept. Commerce, Coast and Geodetic Survey, Washington, D.C., U.S. Government Printing Office, 79 p.
Nelson, K. D., McBride, J. H., Arnow, J. A., Oliver, J. E., Brown, L. D., and Kaufman, S.
1985 : New COCORP profiling in the southeastern United States. Part II; Brunswick and east coast magnetic anomalies, opening of the north-central Atlantic Ocean: Geology, v. 13, p. 718–721.
Nunns, A. G., and Peacock, J. H.
1983 : Correlation, identification and inversion of magnetic anomalies in the Norway Basin-Earth Evolution Sciences, v. 2, p. 130–138.
Ogg, J. G., and Steiner, M. B.
1985 : Jurassic magnetic polarity time scale: current status and compilation; Special Paper IGCP Project 171: Circum-Pacific Jurassic, v. 8, La Jolla, California, Scripps Institution of Oceanography, 17 p.
Ozima, M.
1971 : Magnetic processes in oceanic ridges: Earth and Planetary Science Letters, v. 13, p. 1–5.
Pálmason, G., Arnorsson, S., Friedleifsson, I. B., Kristmannsdottir, H., Saemundsson, K., Stefansson, V., Steingrimsson, B., Tomasson, J., and Kristjansson, L.
1979 : The Iceland crust; Evidence from drillhole data on structure and process, in Talwani, M., Harrison, C.G.A., and Haye, D. E., eds., Deep Drilling Results in the Atlantic Ocean: Ocean Crust: Washington, D.C., American Geophysical Union Maurice Ewing Series, v. 2, p. 43–65.
Parker, R. L., and Huestis, S. P.
1974 : Inversion of magnetic anomalies in the presence of topography: Journal of Geophysical Research, v. 79, p. 1587–1593.
Pecherskiy, D. M., and Tikhonov, L. V.
1983 : Rock magnetic features of basalts of the Atlantic and Pacific oceans: Izvestiya, Earth Physics, v. 19, p. 305–311.
Pierce, J. W.
1976 : Assessing the reliability of DSDP paleolatitudes: Journal of Geophysical Research, v. 81, p. 4173–4187.
Pitman, W. C., and Talwani, M.
1972 : Sea-floor spreading in the North Atlantic: Geological Society of America Bulletin, v. 83, p. 619–645.
Poehls, K. A., Luyendyk, B. P., Heirtzler, J. R.
1973 : Magnetic smooth zones in the world's oceans: Journal of Geophysical Research, v. 78, p. 6985–6997.
Press, F., and Ewing, M.
1952 : Magnetic anomalies over oceanic structures: Transactions of the American Geophysical Union, v. 33, p. 349–355.
Prévot, M., and Lecaille, A., and Hekinian, R.
1979 : Magnetism of the Mid-Atlantic Ridge crest near 37°N from FAMOUS and DSDP results: A review: in Talwani, M., Harrison, C.G.A., and Hayes, D. E., eds., Deep Drilling Results in the Atlantic Ocean: Ocean Crust: American Geophysical Union Maurice Ewing Series, v. 2, p. 210–229.
Rabinowitz, P. D., Cande, S., and Hayes, D. E.
1979 : The J-Anomaly in the central North Atlantic Ocean, in Tucholke, B. E., and Vogt, P. R., eds., Initial Reports of the Deep Sea Drilling Project: U.S. Government Printing Office, Washington, D.C., v. 43, p. 879–885.
Raymond, C. A., and LaBrecque, J. L.
1986 : An alteration model to describe skewness and amplitude variation in the oceanic magnetic anomaly pattern: Journal of Geophysical Research (in press).
Rea, D. K.
1972 : Magnetic anomalies along fracture zones: Nature Physical Sciences, v. 236, p. 58–59.
Reid, I., and Jackson, H. R.
1981 : Oceanic spreading rate and crustal thickness: Marine Geophysical Research, v. 5, p. 165–172.
Rice, P. D., Hall, J. M., and Opdyke, N. D.
1980 : Deep Drill 1972; A paleomagnetic study of the Bermuda Seamount: Canadian Journal of Earth Science, v. 17, p. 232–243.
Richards, M. L., Vacquier, V., and Van Voorhis, G. D.
1967 : Calculation of the magnetization of uplifts from combining topographic and magnetic surveys: Geophysics, v. 32, p. 678–707.
Rohr, K., and Twigt, W.
1979 : Mesozoic complementary crust in the North Atlantic: Nature, v. 283, p. 758–760.
Rona, P. A.
1978 : Magnetic signatures of hydrothermal alteration and volcanogenic mineral deposits in oceanic crust; Journal of Volcanology and Geothermal Research, v. 3, p. 219–225.

Roots, W. D.
1976 : Magnetic smooth zones and slope anomalies; A mechanism to explain both: Earth and Planetary Science Letters, v. 31, p. 113–118.
Roots, W. D., and Srivastava, S. P.
1984 : Origin of the marine magnetic quiet zones in the Labrador and Greenland seas: Marine Geophysical Researches, v. 6, p. 395–408.
Runcorn, S. K.
1956 : Palaeomagnetism, polar wandering and continental drift; Geologie en Mijnbouw, v. 18, p. 253–256.
Ryall, P.J.C., and Ade Hall, J. M.
1975 : Radial variation of magnetic properties in submarine pillow basalt: Canadian Journal of Earth Science, v. 12, p. 1959–1969.
Saeger, W. W., and Keating, B. H.
1984 : Paleomagnetism of the Line Islands Seamounts; Evidence for Late Cretaceous and Early Tertiary volcanism: Journal of Geophysical Research, v. 13, p. 11135–11151.
Sailor, R. V., Lazarewicz, A. R., and Brammer, R. F.
1982 : Spatial resolution and repeatability of Magsat crustal anomaly data over the Indian Ocean: Geophysical Research Letters, v. 9, p. 289–292.
Schouten, H.
1971 : A fundamental analysis of magnetic anomalies over oceanic ridges: Marine Geophysical Research, v. 1, p. 111–144.
Schouten, H., and Cande, S. C.
1976 : Paleomagnetic poles from marine magnetic anomalies: Geophysical Journal of the Astronomical Society, v. 44, p. 567–575.
Schouten, H., and Denham, C. R.
1979 : Modelling the oceanic magnetic source layer, in Talwani, M., Harrison, C.G.A., and Hayes, D. E., eds. Deep Drilling Results in the Atlantic Ocean: Ocean Crust: American Geophysical Union Maurice Ewing Series, v. 2, p. 151–159.
Schouten, H., and Klitgord, K. D.
1977 : Map showing Mesozoic magnetic anomalies; Western North Atlantic: Reston, Virginia, U.S. Geological Survey Miscellaneous Field Studies, map MF915.
Schouten, H., and McCamy, K.
1972 : Filtering marine magnetic anomalies: Journal of Geophysical Research, v. 77, p. 7089–7099.
Schouten, H., and White, R. S.
1980 : Zero-offset fracture zones: Geology, v. 8, p. 175–179.
Sclater, J. G., and Cox, A.
1970 : Paleolatitudes from JOIDES deep sea sediment cores: Nature, v. 226, p. 9343–935.
Sinton, J. M., Wilson, D. S., Christie, D. M., Hey, R. N., and Delaney, J. R.
1983 : Petrologic consequences of rift propagation on oceanic spreading ridges: Earth and Planetary Science Letters, v. 62, p. 193–207.
Srivastava, S. P.
1978 : Evolution of the Labrador Sea and its bearing on the early evolution of the North Atlantic: Geophysical Journal of the Royal Astronomical Society, v. 52, p. 313–357.
Steiner, M. B.
1983 : Mesozoic apparent polar wander and plate motions of North America, in Reynolds, M. W., and Dolly, E. D., eds., Mesozoic paleogeography of west-central United States: Denver, Colorado, Rocky Mountain Section, Society of Economic Paleontologists and Mineralogists, p. 1–11.
Sullivan, K. D., and Keen, C. E.
1978 : On the nature of the crust in the vicinity of the Southeast Newfoundland Ridge: Canadian Journal of Earth Science, v. 15, p. 1462–1471.
Sundvik, M. T.
1986 : Plate tectonic framework of North Atlantic derived from Keathley sequence magnetic anomaly lineation pattern: Journal of Geophysical Research, in press.
Talwani, M.
1965 : Computation with the help of a digital computer of magnetic anomalies caused by bodies of arbitrary shape: Geophysics, v. 30, p. 797–817.
Talwani, M., and Eldholm, O.
1977 : Evolution of the Norwegian-Greenland Sea: Geological Society of America Bulletin, v. 88, p. 969–999.
Talwani, M., Windisch, C. C., and Langseth, M. G., Jr.
1971 : Reykjanes Ridge Crest; A detailed geophysical study: Journal of Geophysical Research, v. 76, p. 473–517.
Tauxe, L., Tucker, P., Peterson, N. P., and LaBrecque, L. L.
1983 : The magnetostratigraphy of leg 73 sedimentsF: Palaeogeography, Palaeoclimatology, and Palaeoecology, v. 42, p. 65–90.
Taylor, P. T., and Greenewalt, D.
1972 : Deep magnetic measurements in the West Atlantic quiet zone: Nature Physical Science, v. 237, p. 51–53.
1976 : Geophysical transitions across the northwest Atlantic magnetic quiet-zone border: Earth and Planetary Science Letters, v. 29, p. 436–446.
Taylor, P. T., Zietz, I., and Denis, L. S.
1968 : Geologic implications of aeromagnetic data from the eastern continental margin of the United States: Geophysics, v. 33, p. 755–780.
Taylor, P. T., Watkins, N. D., and Greenewalt, D.
1973 : Magnetic property analysis of basalt beneath the quiet zone: EOS Transactions of the American Geophysical Union, v. 54, p. 1030–1032.
Taylor, P. T., Stanley, D. J., Simkin, T., and Jahn, W.
1975 : Gillis Seamount; Detailed bathymetry and modification by bottom currents: Marine Geology, v. 19, p. 139–157.
Taylor, P. T., Keating, T., Kahn, W. D., Langel, R. A., Smith, D. E., and Schnetzler, C. C.
1983 : GRM; Observing the terrestrial gravity and magnetic fields in the 1990s: EOS Transactions of the American Geophysical Union, v. 43, p. 609–611.
Tucholke, B. E., and Ludwig, W. J.
1982 : Structure and origin of the J-anomaly ridge, western North Atlantic: Journal of Geophysical Research, v. 87, p. 9389–9407.
Tucholke, B. E., and Vogt, P. R.
1979 : Western North Atlantic; Sedimentary evolution and aspects of tectonic history, in Tucholke, B. E., Vogt, P. R., eds., Initial Reports of the Deep Sea Drilling Project: Washington, D.C., U.S. Government Printing Office, v. 43, p. 791–825.
Twigt, W., Slootweg, A. P., and Collette, B. J.
1979 : Topography and a magnetic analysis of an area southeast of the Azores (36°N, 23°W): Marine Geophysical Researches, v. 4, p. 91–104.
Twigt, W., Verhoef, J., Rohr, K., Mulder, Th.F.A., and Collette, B. J.
1983 : Topography, magnetics, and gravity over the Kane Fracture Zone in the Cretaceous Magnetic Quiet Zone (African Plate): Proceedings of the Koninklijke Nederlandse Akademie van Wetenschappen, Series B, v. 86, p. 181–210.
Uchupi, E., Phillips, J. D., and Prada, K. E.
1970 : Origin and structure of the New England Seamount Chain: Deep-Sea Research, v. 17, p. 483–494.
Vacquier, V.
1962 : A machine method for computing the magnitude and direction of magnetization of a uniformly magnetized body from its shape and a magnetic survey: Proceedings Benedum Earth Magnetism Symposium, p. 123–137.
1972 : Geomagnetism in Marine Geology: Elsevier, Amsterdam, 185 p.
Van der Voo, R., and French, R. B.
1974 : Apparent polar wandering for the Atlantic-bordering continents; Late Carboniferous to Eocene: Earth Science Reviews, v. 10, p. 99–119.
Verhoef, J.
1984 : A geophysical study of the Atlantis-Meteor seamount complex: Geologica Ultraiectina Univ. Utrecht, no. 38, 152 p.
Verosub, K. L., and Moores, E. L.
1981 : Tectonic rotations in extensional regimes and their paleomagnetic consequences for oceanic basalts: Journal of Geophysical Research, v. 86, p. 6335–6349.

Vine, F. J.
1966 : Spreading of the ocean floor; New evidence: Science, v. 154, p. 1405–1415.
1968 : Magnetic anomalies associated with mid-ocean ridges, in Phinney, R. A., ed., The History of the Earth's Crust: Princeton University Press, p. 73–89.
Vine, F. J., and Matthews, D. H.
1963 : Magnetic anomalies over oceanic ridges: Nature, v. 199, p. 947–949.
Vine, F. J., and Moores, E. M.
1972 : A model for the gross structure, petrology, and magnetic properties of oceanic crust, in Shagam, R., ed., Studies in earth and space sciences: Geological Society of America Memoir 132, p. 195–205.
Vogt, P. R.
1969 : Can demagnetization explain seamount drift?: Nature, v. 224, p. 574–576.
1973 : Early events in the opening of the North Atlantic, in Tarling, D. H., and Runcorn, S. K., eds., Implications of Continental Drift to the Earth Sciences: London, Academic Press, v. 2, p. 693–712.
1974 : The Iceland phenomenon; Imprints of a hotspot on the ocean crust, and implications for flow below the plates, in Kristjansson, L., ed., Geodynamics of Iceland and the North Atlantic Area: NATO Advanced Study Institute Series, Dordrecht, Holland, Reidel, p. 49–62.
1975 : Changes in geomagnetic reversal frequency at times of the lithosphere: Earth and Planetary Science Letters, v. 21, p. 235–252.
1979 : Amplitudes of oceanic magnetic anomalies and the chemistry of oceanic crust; Synthesis and review of "magnetic telechemistry": Canadian Journal of Earth Science, v. 16, p. 2236–2262.
1986a: Seafloor topography, sediments, and paleoenvironments, in Hurdle, B. G., ed., The Nordic Seas, New York, Springer-Verlag, p. 237–410.
1986b: Geophysical and geochemical signatures and plate tectonics, in Hurdle, B. G., ed., The Nordic Seas, New York, Springer-Verlag, p. 413–662.
Vogt, P. R., and Avery, O. E.
1974 : Detailed magnetic surveys in the northeast Atlantic and Labrador Sea; Journal of Geophysical Research, v. 79, p. 363–388.
Vogt, P. R., and Byerly, G. R.
1976 : Magnetic anomalies and basalt composition in the Juan de Fuca–Gorda Ridge area: Earth and Planetary Science Letters, v. 33, p. 145–163.
Vogt, P. R., and de Boer, J.
1976 : Morphology, magnetic anomalies, and basalt magnetization at the ends of the Galapagos high-amplitude zone: Earth and Planetary Science Letters, v. 33, p. 145–163.
Vogt, P. R., and Einwich, A. M.
1979 : Magnetic anomalies and sea-floor spreading in the western North Atlantic, and a revised calibration of the Keathley (M) geomagnetic reversal chronology, in Tucholke, B., and Vogt, P. R., eds., Initial Reports of the Deep Sea Drilling Project: Washington, D.C., U.S., U.S. Government Printing Office, v. 43, p. 857–876.
Vogt, P. R., and Johnson, G. L.
1971 : Cretaceous sea-floor spreading in the western North Atlantic: Nature, v. 234, p. 22–25.
1973 : Magnetic telechemistry of oceanic crust?: Nature, v. 245, p. 373–375.
Vogt, P. R., and Ostenso, N. A.
1966 : Magnetic survey over the Mid-Atlantic Ridge between 42°N and 46°N: Journal of Geophysical Research, v. 71, p. 4389–4411.
Vogt, P. R. and Perry, R. K.
1982 : North Atlantic Ocean: Bathymetry and plate tectonic evolution, text to accompany chart MC-35; Geological Society of America, Boulder, 21 p.
Vogt, P. R., and Tucholke, B. E.
1982 : The Western North Atlantic; in Palmer, A. R., ed., Perspectives in Regional Geological Synthesis, DNAG Special Publication 1, Geological Society of America, Boulder, p. 117–132.
Vogt, P. R., and Tucholke, B. E.
1979 : The New England Seamounts; Testing origins, in Tucholke, B. E., and Vogt, P. R., eds., Initial Reports of the Deep Sea Drilling Project, Washington, D.C., U.S. Government Printing Office, v. 43, p. 847–856.
Vogt, P. R., Anderson, C. N., Bracey, D. R., and Schneider, E. D.
1970 : North Atlantic magnetic smooth zones: Journal of Geophysical Research, v. 75, p. 2955–2968.
Vogt, P. R., Anderson, C. N., and Bracey, D. R.
1971 : Mesozoic magnetic anomalies, sea-floor spreading, and geomagnetic reversals in the southwestern North Atlantic: Journal of Geophysical Research, v. 76, p. 4796–4823.
Vogt, P. R., Johnson, G. L., and Kristjansson, L.
1980 : Morphology and magnetic anomalies north of Iceland: Journal of Geophysics, v. 47, p. 67–80.
Vogt, P. R., Kovacs, L. C., Bernero, C., and Srivastava, S. P.
1982 : Asymmetric geophysical signatures in the Greenland–Norwegian and southern Labrador seas and the Eurasia Basin: Tectonophysics, v. 89, p. 95–160.
Vogt, P. R., Cherkis, N. Z., and Morgan, G. A.
1983 : Project Investigator-I; Evolution of the Australia–Antarctic Discordance deduced from a detailed aeromagnetic study, in Oliver, R. L., James, P. R., and Jago, J. B., eds., Antarctic Earth Science: Canberra, Australian Academy of Science, p. 608–613.
Vogt, P. R., Bracey, D. R., and Kovacs, L. C.
1986 : Magnetic anomalies in the western North Atlantic Ocean, chart and manuscript in preparation.
Walczak, J. E.
1963 : A marine magnetic survey of the New England seamount chain: U.S. Naval Oceanographic Office, Technical Report 159, 37 p.
Walczak, J. E., and Carter, T.
1964 : A bathymetric and geomagnetic survey of the New England Seamount Chain: International Hydrographic Review, v. 16, p. 59–67.
Watkins, N. D.
1974 : Magnetic telechemistry is elegant but nature is complex: Nature, v. 251, p. 497–498.
Whitmarsh, R. B.
1982 : Along-strike amplitude variations of oceanic magnetic stripes; Are they related to low-temperature hydrothermal circulation?: Geology, v. 10, p. 461–465.
Williams, C. A., Verhoef, J., and Collette, B. J.
1983 : Magnetic analysis of some large seamounts in the North Atlantic: Earth and Planetary Science Letters, v. 63, p. 399–407.
Young, G. A., and Kontis, A. L.
1963 : A study of aeromagnetic component data-Plantagenet Bank: U.S. Naval Oceanographic Office Technical Report 144.

MANUSCRIPT ACCEPTED BY THE SOCIETY SEPTEMBER 19, 1985

ACKNOWLEDGMENTS

I thank Jim Hall, Steve Cande and Bob Higgs for reviews and the Office of Naval Research for support. Assistance was provided by Irene Jewett (illustrations) and Myra Whitney and Julie Perrotta (manuscript). Joe Cain kindly provided the contours in Figure 10.

The Geology of North America
Vol. M, The Western North Atlantic Region
The Geological Society of America, 1986

Chapter 16

The relationship between depth and age and heat flow and age in the western North Atlantic

John G. Sclater*
Lewellyn Wixon*
Department of Earth and Planetary Sciences, Massachusetts Institute of Technology, Cambridge, Massachusetts 02139

INTRODUCTION

There are two simple models, compatible with the theory of plate tectonics, that can account for the observed relations between depth and age and between heat flow and age for oceanic crust. The first is the cooling plate model (Langseth and others, 1966; McKenzie, 1967; Sclater and others, 1971), which has been best matched to the depth and heat flow observations by Parsons and Sclater (1977). The second is that of a growing boundary layer (Turcotte and Oxburgh, 1967; Parker and Oldenburg, 1973) that has been matched to the depth and heat flow data by Davis and Lister (1974) and Lister (1977). Both models give good agreement with observations in the range 0–80 Ma. At greater ages the observations of average depth and mean heat flow are best matched by the plate model.

The models are similar in that thermal cooling is the dominant mechanism. The difference between them results from concentrating on two different aspects of the observations. In the plate model, emphasis is placed on matching the exponential decay of heat flow and depth at older ages, whereas the boundary layer model was developed to explain the relation between subsidence and the square root of time on young ocean floor. Though the predictions of the plate model best match the average observations there is a problem with this model in that the constant temperature bottom boundary condition is physically unrealistic. Parsons and McKenzie (1978) have reconciled the two models and also developed a more realistic physical basis to the plate model by introducing the concept of a thermal as well as a mechanical boundary layer. The rheology of the plate is represented in a simple manner. Below a given temperature the material is assumed to move rigidly defining an upper mechanical boundary layer through which heat is lost by conduction. Beneath this layer is a viscous region where a thermal boundary layer is formed. As the lithosphere cools, both the mechanical and thermal boundary layers increase in thickness as the square root of the age. At a certain point the thermal boundary layer becomes unstable, giving rise to convective heat transport that maintains the rigid layer at a roughly constant thickness. The thickness of the combined mechanical and thermal boundary layer is greater than that given by the plate model, that is, on the order of 150-200 km, but matches the observations equally well.

As a result of increased information on depth, the gravity field, and the age of the ocean floor, coupled with measurements of the geoid by satellite altimetry, it is becoming obvious that these two simple models do not adequately account for the data when considered in total. Large areas of anomalous depth have been recognized in all three major ocean basins, which in the Pacific (McKenzie and others, 1980) and Indian Ocean (Roufosse and others, 1980) appear to correlate with geoid anomalies. In the Atlantic the correlation with the gravity field is not so obvious (Cochran and Talwani, 1977).

Anderson and others (1973) along the mid-ocean ridges and Sclater and others (1975) and Watts (1976) for the North Atlantic and Central Pacific considered the relationship between long-wavelength gravity and depth anomalies. They found a reasonable correlation between the two sets of anomalies. Further, they noted that the correlation was similar to that predicted by McKenzie and others (1974) based on numerical models of convection in the upper mantle.

In contrast, Haxby and Turcotte (1978) and Crough (1978) examined the relation between geoid anomaly from the GEOS-3 altimeter and depth over various features such as Hawaii and Bermuda associated with recent volcanism. They concluded that these features were supported by a mass deficiency at 40–70 km, which is shallower than the thermal thickness of the plate inferred from the age of the adjacent sea floor. Crough (1978) used these relations and observations of subsidence along the mid-ocean swells in the Pacific (Detrick and Crough, 1978; Crough and

*Present addresses: Sclater—The Institute for Geophysics, The University of Texas at Austin, 4920 North IH 35, Austin, Texas 78751; Wixon—The Department of Geological Sciences, The University of Chicago, Chicago, Illinois 60637

Sclater, J. G., and Wixon, L., 1986, The relationship between depth and age and heat flow and age in the western North Atlantic; *in* Vogt, P. R., and Tucholke, B. E., eds., The Geology of North America, Volume M, The Western North Atlantic Region: Geological Society of America.

Jarrard, 1981) to suggest that these features had been reheated by a hot spot in such a way that the effective thermal age had been reset to that of 25 million years old crust. The problem with this suggestion is that it would take an unrealistically large thermal anomaly to thin the crust by conductive processes (Detrick and Crough, 1978). However, the subsidence data, especially for Hawaii, is strong evidence that the anomaly has a thermal origin.

An answer to this dilemma comes from numerical studies on constant viscosity fluids by Parsons and Daly (1983). They have shown that for cellular convection, surface topography and gravity anomalies largely reflect temperature variations in the vicinity of the upper thermal boundary layer. If the lithospheric plates are comprised of an upper mechanical boundary layer and a lower thermal boundary layer, it is unlikely that the two layers are rigidly attached. Hence, the thermal thickness estimated from the subsidence of oceanic crust will include mantle material cooled by conduction from the surface but not moving with the plate. Parsons and Daly (1983) reason that the upwelling portion of a convection cell will penetrate into the thermal boundary layer and significantly raise the temperature in this region. Such convection can explain the geoid anomaly. The subsequent cooling of this reheated, thermally defined lithospheric plate could produce the observed subsidence.

In order to construct a relation between the depth of the ocean floor and age, it is necessary to measure the depth, remove the sediment load, and determine the age of the ocean floor. In the past such studies have concentrated on averaging the depth, corrected for sediment load, from individual profiles in areas undisturbed either by excess topographic relief near the ridge axis or by swells in the deep ocean (Sclater and others, 1971; Parsons and Sclater, 1977). With the advent of digitized depth information for the oceans, much improved sediment thickness charts, and good estimates of the age of the ocean floor (Sclater and others, 1980), such a simple approach is no longer adequate. It is necessary to examine the entire deep sea topographic data set and to compare the compilation of this set with the models of lithospheric creation.

The two models for the cooling of the oceanic lithosphere are reasonably similar. The only matter of dispute is whether or not small scale convection is required to flatten the plate. Detrick (1981) has argued from geoid information over a fracture zone in the Pacific that he can observe this flattening in an area unaffected by a residual depth anomaly. On the other hand, Heestand and Crough (1981) have suggested that in the North Atlantic the simple relation between depth and the square root of age holds for all ages and that the flattening is only apparent and results from averaging the elevated topography associated with a large number of hot spots on old ocean floor. It is important to examine currently available depth to determine if it is possible to resolve between these two explanations of the exponential decay of the subsidence and heat flow on old ocean floor.

Recently digitized versions of the bathymetry of the Atlantic at five minute intervals have become available (SYNBAPS; Van Wyckhouse, 1973). In addition, Tucholke and others (1982) have compiled and published sediment contour charts for the Western North Atlantic. In this paper, we use these two data sets to construct half-by-half degree averages of the basement depth corrected for sediment load in the Western North Atlantic between 45 and 15°N and 80 and 40°W. We examine the relation between corrected depth and age and compare the results with theoretical models. In addition, we plot the observed heat flow data against age for the Western North Atlantic. We construct a residual depth anomaly chart and compare it with the residual free air gravity field. We consider the Bermuda Rise in detail and discuss the implication of our observations for current models of residual depth anomalies and the flattening with age of the subsidence and heat flow curves.

THE RELATION BETWEEN DEPTH AND AGE

The area of the Western North Atlantic considered in this study is bounded to the west and north by the continental shelf of North America, to the south by the Bahama Platform and the Puerto Rico Trench, and to the east by the Mid-Atlantic Ridge. It encompasses the Hatteras, Nares, and Sohm abyssal plains. In the center lies the broad swell of the Bermuda Rise and to the north of this rise the New England Seamounts trend northwest by west. To the east of the seamounts, just on the edge of the rough topography associated with the Mid-Atlantic Ridge, lie the Corner Seamounts (Plate 2).

Basement Corrected Depth, from the SYNBAPS Compilation and Sediment Thickness Data, and Age.

We averaged the 5 interval depths from the Synbaps compilation into half-by-half degree elements and contoured them at 1000 m intervals (Fig. 1). The Mid-Atlantic Ridge, the Corner Seamounts, the Bermuda Rise, and the deep between the rise and the ridge are all visible on this chart.

The sediment thickness in the area is highly variable. Between the Mid-Atlantic Ridge and the Bermuda Rise, there is generally less than 500 m of cover. To the south and north of this rise there are large accumulations in the Sohm and Nares abyssal plains and on the Blake-Bahama Outer Ridge (Plate 6). We digitized the sediment thickness at half by half degree elements from the charts of Tucholke and others (1982). The relation between density and sediment thickness of LeDouran and Parsons (1982) was used to calculate the sediment load. Finally, the corrected depth of the basement that would be observed in the absence of a point load was computed.

The age of the ocean floor is best determined by constructing a chart of the isochrons from a compilation of the observed magnetic lineations and using a time scale that relates these lineations to age. For the Mesozoic lineations (magnetic anomalies M25-M0) we took the compilation of Schouten and Klitgord (1977) and for the Late Cretaceous to Present (magnetic anomalies 34-1) a more recent compilation by the same authors (Klitgord and Schouten, this volume). To convert the younger lineations to age we used the time scale of Harland and others (1983). For the older lineations we chose the time scale of Kent

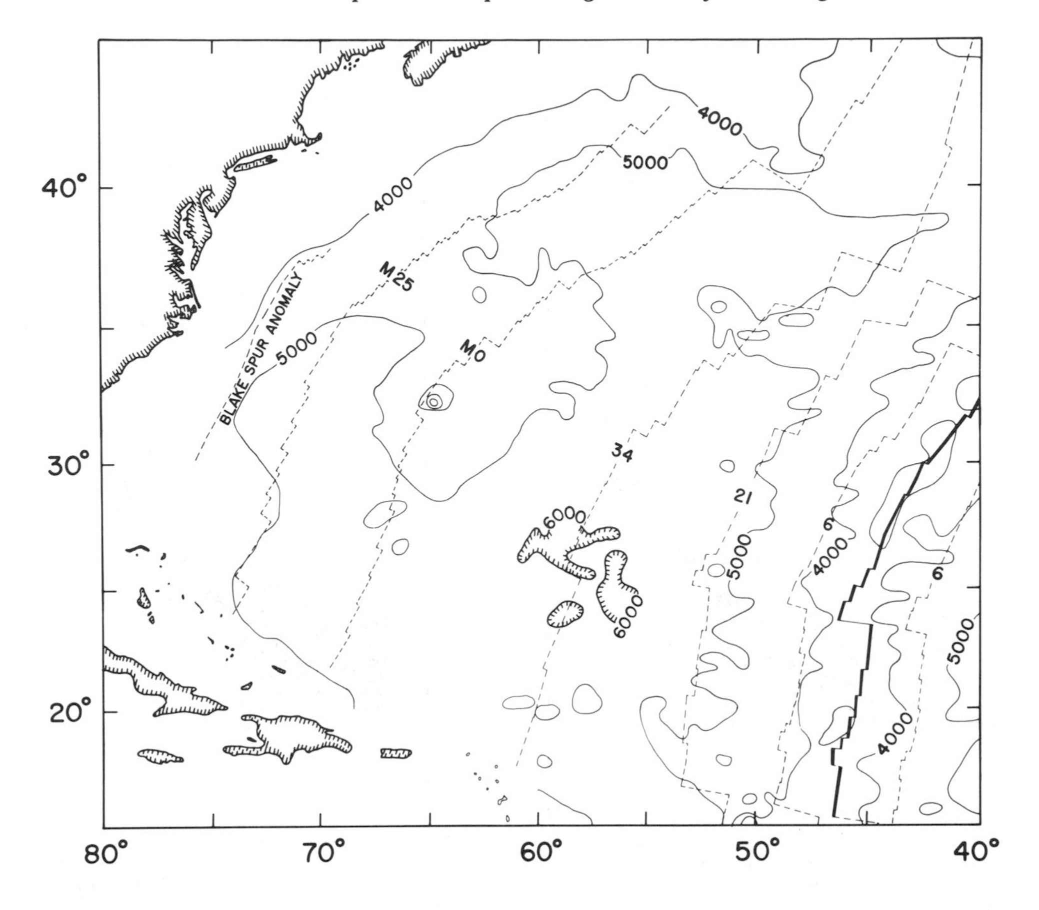

Figure 1. Bathymetric contours at 1000 m intervals, continuous lines, of the Western North Atlantic from half-by-half degree averages of the SYNBAPS data set (Van Wyckhouse, 1973). The heavy black lines mark the axis of the Mid-Atlantic Ridge and the dashed lines the position of specific magnetic anomaly lineations.

and Gradstein (this volume). This time scale is based on a constant rate of spreading in the Pacific. It has the advantage that it gives a sensible age progression between DSDP site 105 on magnetic anomaly M25 and the age of the oldest sediments at DSDP site 534 just east of the Blake Spur magnetic anomaly. Specific isochrons were superimposed on the bathymetric chart (Fig. 1) and an age was assigned to each half by half degree element within the area studied.

We compiled the corrected basement depth for each half-by-half degree element from the observed depth and the sediment load. These values were plotted against age (Figs. 2a and 2b). They were averaged for each million years and the mean and the standard deviation determined. All values lying more than three standard deviations from the mean were rejected. For ocean crust less than 100 Ma the subsidence with age is uniform and has little scatter. One standard deviation lies within ±150 m and two within ±300 m throughout most of the range in age. The only significant departure from this close grouping occurs between 70 and 90 Ma and is associated with the Corner Seamounts. The agreement between the mean depth and the square root of age out to 90 Ma is particularly striking. Beyond 100 Ma the mean rises, falls again around 110 Ma and rises again near the continental shelf on crust older than 160 Ma. The first uplift is the Bermuda Rise. The reason for the second is not understood. It could be real although it may also result from assigning incorrect sediment densities to sediments on the continental shelf and the Blake-Bahama Outer Ridge. For crust older than 100 Ma, the scatter is higher than that for the younger crust. It is ±250 m for one standard deviation and ±500 m for two standard deviations above the mean.

Basement Corrected Depth from DSDP Sites and Age

Since the start of the Deep Sea Drilling Project (DSDP) in 1968, 43 holes drilled in the North Atlantic have reached and dated basement material. The location of each site can be found on Plate 2 and the principal characteristics in Table 1. The basement depth at each site was corrected for sediment load using the density versus depth relation of LeDouran and Parsons (1982) and plotted versus the age of the ocean floor (Fig. 3).

All of the sites except two, 395 and 396 which are near fracture zones (Plate 2) lie above the depth versus age relation given by the plate model. However, roughly 30 of the sites lie within 500 m of the theoretical relation and the other thirteen are

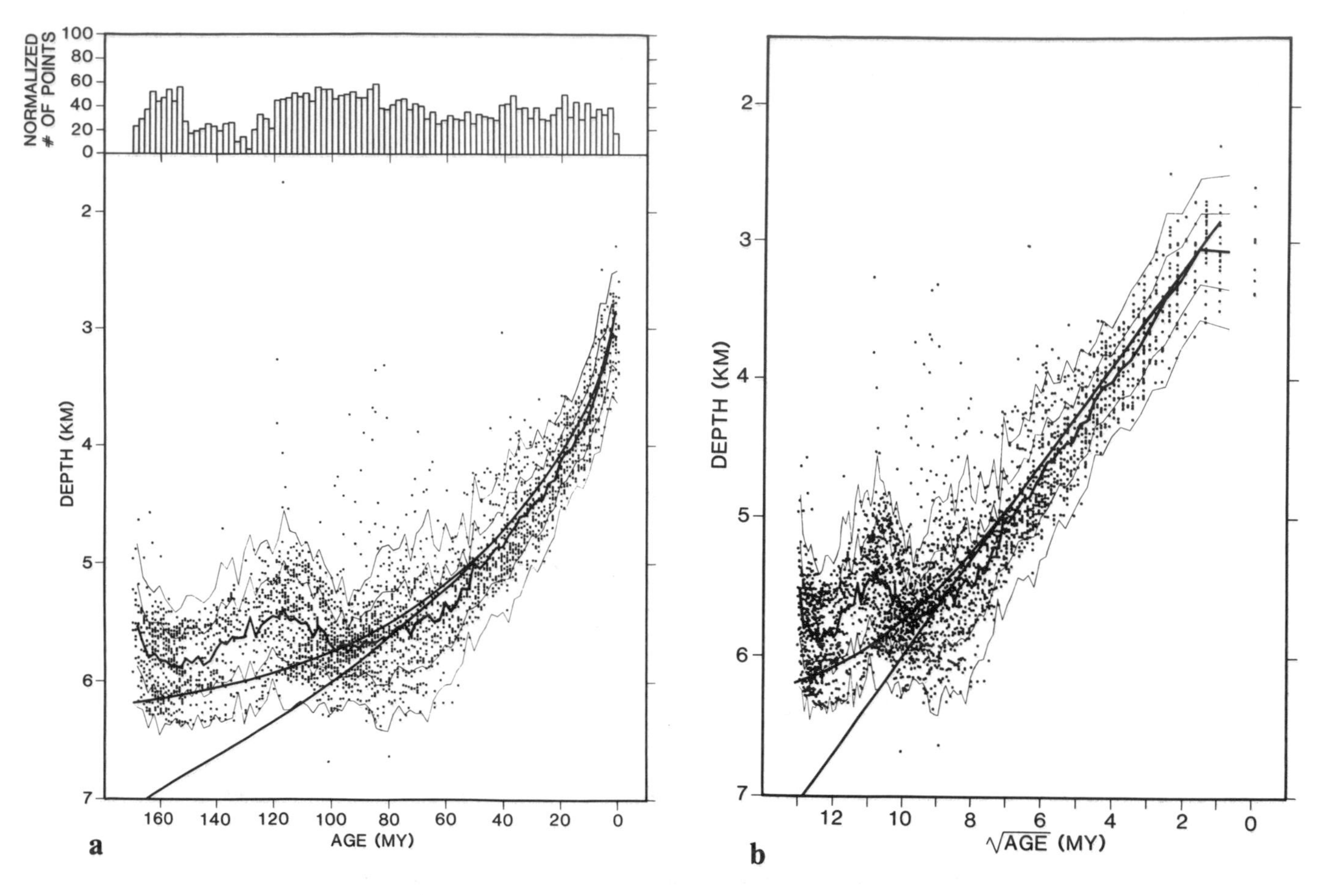

Figure 2. a) A plot of the half by half degree averages of the basement depth corrected for sediment load against age for the Western North Atlantic. The heavy jagged line is the mean of the data and the lighter lines are the one and two standard deviations about the mean. These lines were computed taking into account the effect of latitude on the half-by-half degree averages. The heavy continuous lines represent the theoretical for the plate (upper) and boundary layer models from Parsons and Sclater (1977). The number of values in each million years interval are shown above the data. b) A plot of the half by half degree averages of the basement depth corrected for sediment load against the square root of age. The individual lines are the same as those for Figure 2a.

on clearly anomalous older crust such as the New England Seamount Chain (395), the J-anomaly Ridge (384), the Cape Verde Plateau (136) and the continental shelf of western Europe (550B). All of the anomalous younger sites are associated either with the Mid-Atlantic Ridge just south of Iceland (112, 114, 407, 408, 409) or the Azores (332, 333). Given the sensible preference to drill on the shallowest place available it is not unexpected that the Deep Sea Drilling sites should in general be shallower than the average depth in the surrounding ocean.

Two results from the drilling are of interest to this present study. First, the age of the pronounced volcanic debris-related sediment layer around Bermuda, which is thought to date the volcanism at Bermuda and therefore the uplift of the swell, is mid-Oligocene (Tucholke and Vogt, 1979). Second, dates for the New England Seamounts from drilling and from dredging have been reported by Vogt and Tucholke (1979) and Duncan (1984). These authors have shown that the chain is 10 to 60 million years younger than the ocean floor on which it resides and that the age differential decreases to the southeast.

THE RELATION BETWEEN HEAT FLOW AND AGE

Sclater and others (1980) examined the relation between heat flow and age for the Pacific, Atlantic, and Indian Oceans using data compiled by Jessop and others (1976). Sclater and others (1980) showed that the heat flow was highly scattered in the younger regions of the ocean floor. The average value on young crust was less than that predicted by thermal models that matched the simple relation between depth and age. Both the scatter and the lower than expected heat flow are attributed to the effects of hydrothermal circulation in the porous oceanic crust and the loss of heat by advection at the basement/water interface (Lister, 1972; Williams and others, 1974). In the older regions the scatter decreases and the values lie close to but slightly above those predicted by the simple thermal models.

TABLE 1. DEPTH CORRECTED FOR SEDIMENT LOAD AT DEEP SEA DRILLING SITES IN THE NORTH ATLANTIC AND THE MAGNETIC ANOMALY OR BIOSTRATIGRAPHIC AGES

Site	Lat.	Long.	Depth	Penetration	Sediment Correction	Correct. Depth	Biostrat. Zone	Mag. Anom.	Age	Com.
	N	N	m	m	m	m			Ma	
9	32°46'	59°12'	4965	835	545	5511	-	-	102	-
10	32°51'	52°13'	4697	457	320	5017	-	33	77	-
11	29°57'	44°45'	3656	282	203	3760	-	5B	15	-
100	24°41'	73°48'	5325	317	230	5555	-	M23	154	-
105	24°54'	69°10'	5251	623	423	5675	-	M25	154	-
112	54°01'	46°36'	3657	662	447	4103	-	26	60	-
114	59°56'	26°48'	1927	623	423	2251	-	5	12	-
118	45°03'	09°01'	4901	757	502	5403	-	34	84	-
136	34°10'	16°18'	4169	308	221	4390	-	M16	142	-
137	25°56'	27°04'	5361	397	280	5642	-	-	106	-
138	25°55'	25°34'	5288	437	307	5595	-	-	115	-
332	36°53'	33°39'	1806	142	104	1911	-	2'-3	4	-
333	36°51'	33°40'	1666	219	160	1826	-	2'-3	4	-
334	37°02'	34°25'	2632	259	187	2820	-	5	9	-
335	37°18'	35°12'	3198	454	318	3516	-	>5	17	-
367	12°29'	20°03'	4748	1140	703	5460	a	-	150-163	-
382	34°25'	26°32'	5526	385	273	5799	b	-	103	1
384	40°22'	51°40'	3909	325	233	4142	-	M2	122	2
385	37°22'	60°09'	4936	338	242	5178	-	M11	134	1
386	31°11'	64°15'	4782	964	615	5397	c	-	105-113	-
387	32°19'	67°40'	5117	792	522	5639	-	M16	141	-
395	22°45'	46°05'	4484	92	69	4552	-	4	8	-
396	22°59'	43°31'	4450	96	71	4521	-	5	10	-
396B	22°59'	43°31'	4459	151	111	4570	-	5	10	-
407	63°56'	30°35'	2472	304	218	2691	-	13	35	-
408	63°23'	25°55'	1624	322	230	1855	-	6	20	-
409	62°37'	25°57'	832	80	60	892	-	2'	3	-
410	45°31'	29°29'	2975	340	243	3218	-	5	10	-
411	36°46'	33°23'	1935	74	55	1990	-	1	1	-
412	36°34'	33°10'	2609	160	118	2727	-	2	1.6	-
413	36°33'	33°10'	2598	110	81	2680	-	2	1.6	-
417A	25°07'	68°02'	5468	208	152	5620	-	MO	120	-
417D	25°07'	68°03'	5492	343	245	5737	-	MO	120	-
418A	25°02'	68°03'	5511	324	232	5743	-	MO	120	-
418B	25°02'	68°03'	5514	320	229	5744	-	MO	120	-
534A	28°21'	75°23'	4971	1635	908	5879	d	MO	166-169	-
550B	48°31'	13°26'	4432	686	461	4893	e	-	105-113	3
556	38°56'	34°41'	3672	462	323	3995	-	12	33	-
557	38°50'	32°34'	2143	460	321	2465	-	5D	18	-
558	37°46'	40°55'	3754	406	287	4040	-	12-13	34	-
559	35°08'	40°55'	3754	238	173	3927	-	12-13	34	-
560	34°43'	38°51'	3443	375	266	3709	-	5D	18	-
561	34°47'	39°02'	3459	412	290	3750	-	5E	19	-
562	33°09'	41°41'	3172	240	174	3346	-	5D	18	-
563	33°29'	43°46'	3746	364	259	4005	-	13	36	-
564	33°44'	43°46'	3820	284	204	4024	-	13	36	-

a = Oxfordian-Kimmeridgian; b = >Lower Campanian or older; c = Early Aptian; d = Basal Callovian; e = Lower to Upper Albian; 1 = New England Seamounts; 2 = J. Anomaly Ridge; 3 = Golan Spur.

Since the compilation of Jessop and others (1976) and the study of Sclater and others (1980), Galson and Von Herzen (1981) and Davis and others (1984) have carried out detailed heat flow surveys at four sites on older, probably undisturbed, ocean floor in the Nares Abyssal Plain and over the Blake-Bahama Outer Ridge. A chart showing the original data and the position of the new surveys in the form of symbols is presented as Fig. 4. The values show a concentration near the ridge axis where many are high ($>$80 mW/m^2). There is another concentration around Bermuda where the values are normal to low ($<$60 mW/m^2). Each of the values was assigned an age from the digitized elements and then plotted against age (Fig. 5a). The values near the ridge axis almost all lie below the theoretical and are very scattered. As has been mentioned earlier this is attributed to hydrothermal circulation. On crust 100 Ma and older the scatter decreases substantially and the values lie close to or above the theoretical for the plate model. When the data are averaged in groups of twenty million years (Table 2) the increase over the predicted values at the older ages is more obvious (Fig. 5b).

DISCUSSION OF THE RELATIONS BETWEEN DEPTH AND AGE AND BETWEEN HEAT FLOW AND AGE

The observed relation between sediment corrected basement depth and age is fit well by both the plate and boundary layer models for crust younger than 90 Ma. The match to the square root of time predicted by both models and the low scatter is striking. Beyond 90 Ma the observations do not increase substantially in depth, and the misfit between them and the simple square root of time predicted by the boundary layer model becomes increasingly larger (Fig. 2a). The plate model, which predicts an exponential flattening to a depth at infinite time of 6400 m, provides a better match to the observations. However, this model,

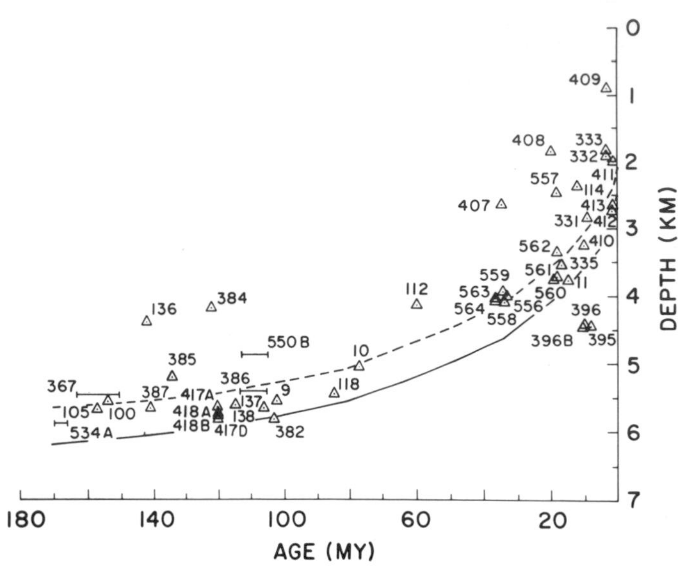

Figure 3. Basement depth corrected for sediment load plotted against magnetic age (triangles) or biostratigraphic age (horizontal bars) for DSDP sites in the North Atlantic that reached basement. The heavy continuous line is the theoretical depth versus age relation from the plate model (Parsons and Sclater, 1977) and the dashed line represents 500 m added to this relation.

constructed from carefully selected tracks in the Pacific and Atlantic by Parsons and Sclater (1977), does not account for the rise in the depth associated with the Bermuda Rise and is still 200 to 300 m deeper than the depth observed at the oldest ages.

The sediment-corrected basement depths at the DSDP sites tend to support the analysis of the SYNBAPS data. The scatter near the ridge axis is larger because of the selection of sites on the relatively elevated topography around and north of the Azores. On older ocean crust the observed basement depth is also significantly shallower than that predicted by the plate model (Fig. 3).

As has been pointed out by others, the heat flow is highly scattered and the observations have a mean that is much lower than expected (Fig. 5a, b) on crust younger than 60 Ma. For crust older than this, the heat flow remains remarkably constant, has a low scatter and actually becomes greater than that predicted either by simple cooling or the plate model (Williams and Poehls, 1975; Sclater and others, 1980).

The relatively abrupt flattening of the heat flow and the shallower than predicted depths are evidence that even more heat enters the old lithosphere than that which is predicted by the plate model. Two simple explanations of these observations are suggested by the data. One possibility is that the square root of time relation for the subsidence predicted by the simple boundary layer model is perturbed by a massive upwelling and reheating under Bermuda. An alternative hypothesis, based on the plate model, is that the plate thickness chosen by Parsons and Sclater (1977) is too great and that a thinner plate would give a better match to the heat flow and the average depths. In such a model the depth of crust between 80 and 100 Ma would lie below that predicted, and that between 110 and 130 Ma would lie above it. These anomalies would be related to the downwelling and upwelling of convection, all in the upper mantle, perturbing the average relations predicted by the plate model.

TABLE 2. HEAT FLOW STATISTICS FOR THE WESTERN NORTH ATLANTIC

Age Span Ma	Number of Stations	Mean	Heat Flow (mW/M^2) Std. dev.	Std. error
0-20	43	95	89	14
20-40	17	47	54	13
40-60	7	66	29	11
60-80	22	41	16	3
80-100	15	48	12	3
100-120	25	52	10	2
120-140	25	53	15	3
140-160	19	49	10	2
>160	25	46	7	1.4

RESIDUAL DEPTH ANOMALIES

The residual depth is defined as the difference between the observed depth, corrected for sediment load and the expected depth from some relation between depth and age. Clearly, to remove any preconceived bias, the depth/age relation would be empirically derived from the sum of all the actual data. However, although underway, such an empirical curve has not been developed and it is still necessary to use theoretical models that have been fit to a limited number of topographic profiles. The major argument against using the plate model has been a lack of physical concept on which to base this model. Now that Parsons and McKenzie (1978) have provided such a basis we have chosen to use the plate model of Parsons and Sclater (1977) as their predicted curves best match the mean of the observations in the Pacific and Western North Atlantic.

The residual depth anomaly for each half-by-half degree was computed by removing the theoretical depth for the age of element from the corrected basement depth. These depths were contoured in 200 m intervals (Fig. 6). To examine the long wavelength features of the residual depth field, two-and-a-half by two-and-a-half degree averages were constructed (Fig. 7). The striking positive features of the long wavelength chart are the Bermuda Rise, the Mid-Atlantic Ridge south of the Azores and Corner Seamount High. The major negative feature is the deep area to the east of Bermuda, which almost disappears by the axis of the Mid-Atlantic ridge. Two other, less pronounced features are of interest, a localized high area over the Blake-Bahama Outer Ridge that is clearly separated from the Bermuda Rise by a saddle (Fig. 6) and the slight negative depth anomaly in the Sohm abyssal plain north of the Bermuda Rise. One feature, which is prominent on the topographic chart but does not show up on the residual depth contours is the New England Seamount Chain.

On the older sea floor the area is dominated by the Bermuda Rise. This feature trends northeast by southwest and is about 1000 km long and about 400 km wide. There is an indication of a trend

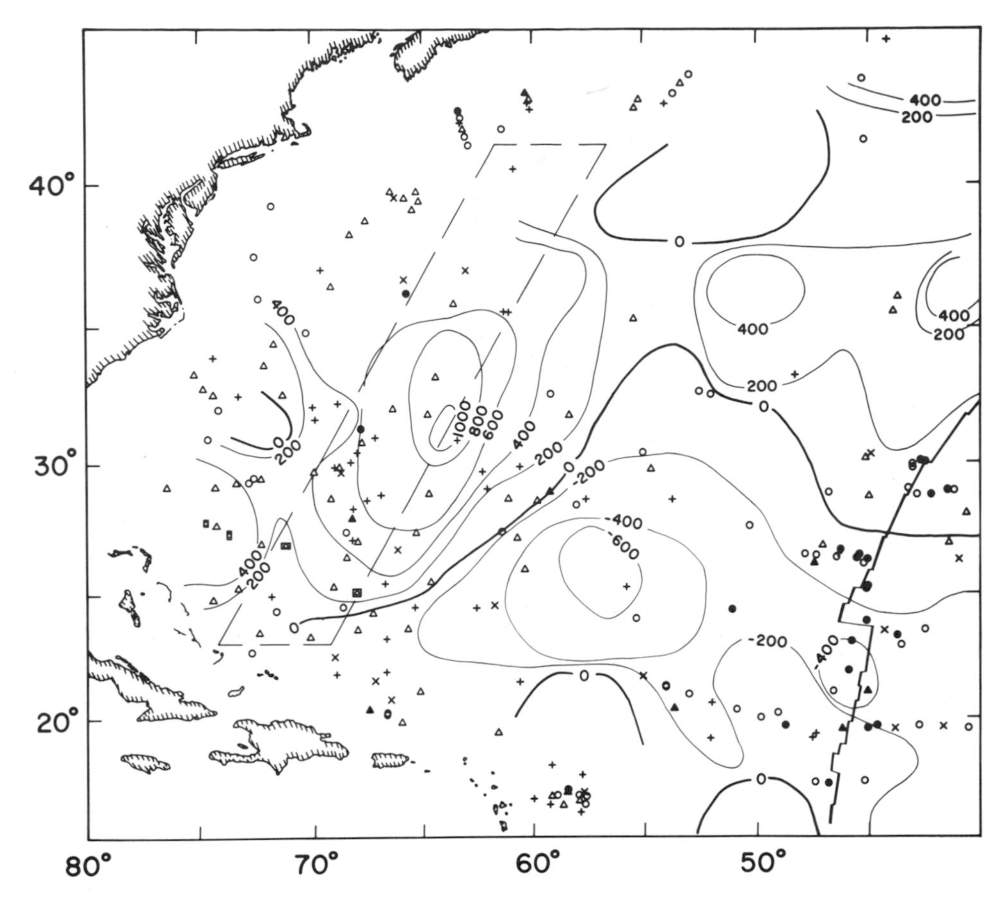

Figure 4. Heat flow data from Jessop and others (1976) and later published data in the Western North Atlantic. The data are in symbol form. The heat flow surveys discussed in Galson and Von Herzen (1981) and Davis and others (1984) are presented as shaded rectangles marking the survey area. The mean value is presented as a symbol within the rectangle. The symbols o, Δ, +, ×, ▲, and ● represent respectively heat flow values of <40, 40–49, 50–59, 60–69, 70–79, >80 mW/m^2. The continuous lines are contours of the two and a half degree averages of the residual depth. The dashed parallelogram, five degrees wide, represents the area covered by the heat flow data plotted against latitude in Figure 10.

away from Bermuda in a northwest direction, but it is slight and in fact the broadest region of elevated topography outlined by the 600 m contour extends at ninety degrees to this trend.

Another feature of the charts (Figs. 6 and 7) is the +200 m contour on crust older than 100 Ma. If we were to have removed the average depth for the region rather than the theoretical this contour would represent the line of zero anomaly. The effect would be to make the Bermuda Rise high less prominent but to emphasize the low to the east of Rise. It would demonstrate even more clearly that, for ocean floor older than 80 Ma, the dominant feature is the association of a residual depth high over Bermuda with a pronounced counter-balancing low to the southeast.

COMPARISON OF RESIDUAL DEPTH AND RESIDUAL GRAVITY ANOMALIES

It has been suggested that 500–1000 km wavelength regional depth anomalies in the oceans are associated with convection in the upper mantle. If this is the case, then it is to be expected that these convection cells will perturb both the geoid and free air gravity fields and that there should be a correlation between these features and depth. A medium-wavelength geoid anomaly field for the North Atlantic is under construction (Parsons and others, in preparation). In the absence of this data we decided to compare our two-and-a-half by two-and-a-half degree depth anomalies with the residual free air gravity field of Cochran and Talwani (1977) (Fig. 8). This field was constructed by averaging shipboard gravity values (Rabinowitz and Jung, this volume Fig. 2) in one by one degree elements, removing the effect on the gravity field of the cooling plate and then averaging these elements at five by five degree intervals. Caution has to be observed when considering the effect of the cooling plate on the gravity field as the field is dependent on the spreading rate. However, more than 500 km from the ridge axis, the effects are small (Cochran and Talwani, 1977; Fig. 5).

The free air gravity field in the North Atlantic is dominated by a large low in the vicinity of the Puerto Rico Trench (Fig. 1;

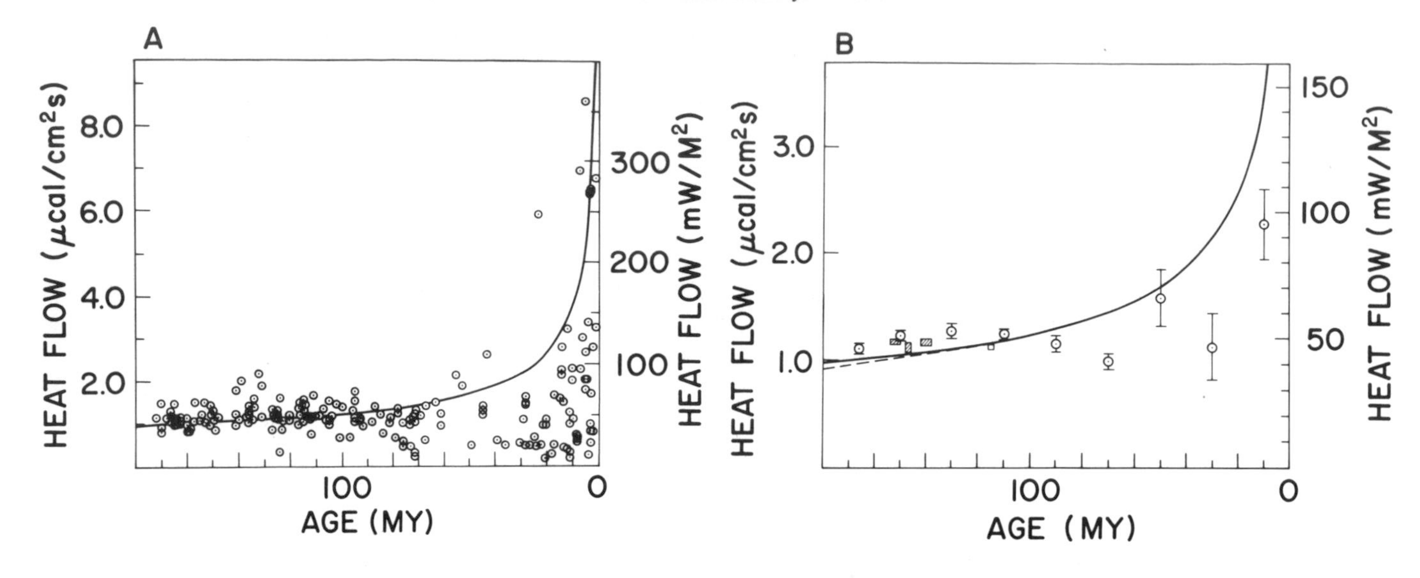

Figure 5. a) Heat flow versus age for the Western North Atlantic. The solid line is the theoretical heat flow from the plate model (Parsons and Sclater, 1977). b) Average heat flow and standard error in 20 million year intervals plotted against age. The filled in rectangles are the average and the standard error of the surveys discussed by Davis and others, 1984). The continuous line is the theoretical for the plate model and the dashed line that for the boundary layer model (Parsons and Sclater, 1977).

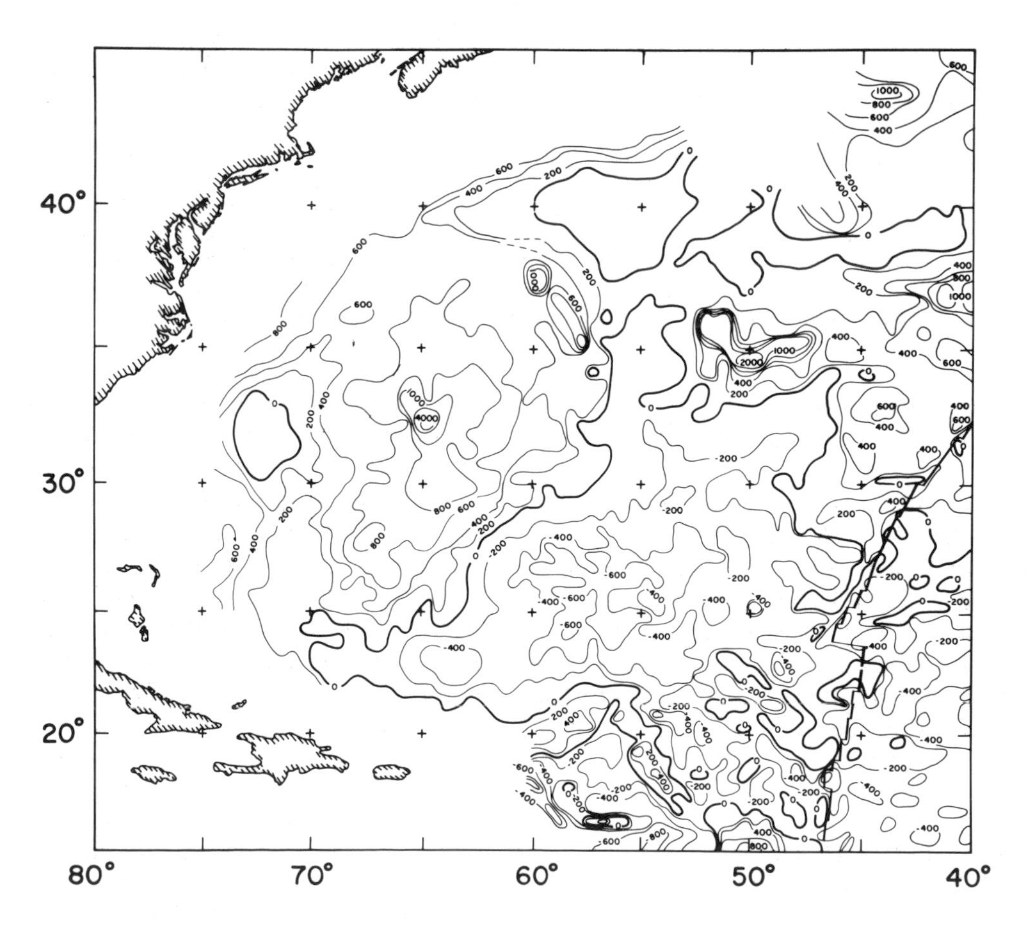

Figure 6. Half-by-half degree residual depth anomalies in the Western North Atlantic contoured at 200 m intervals. The heavy continuous lines mark the ridge axis.

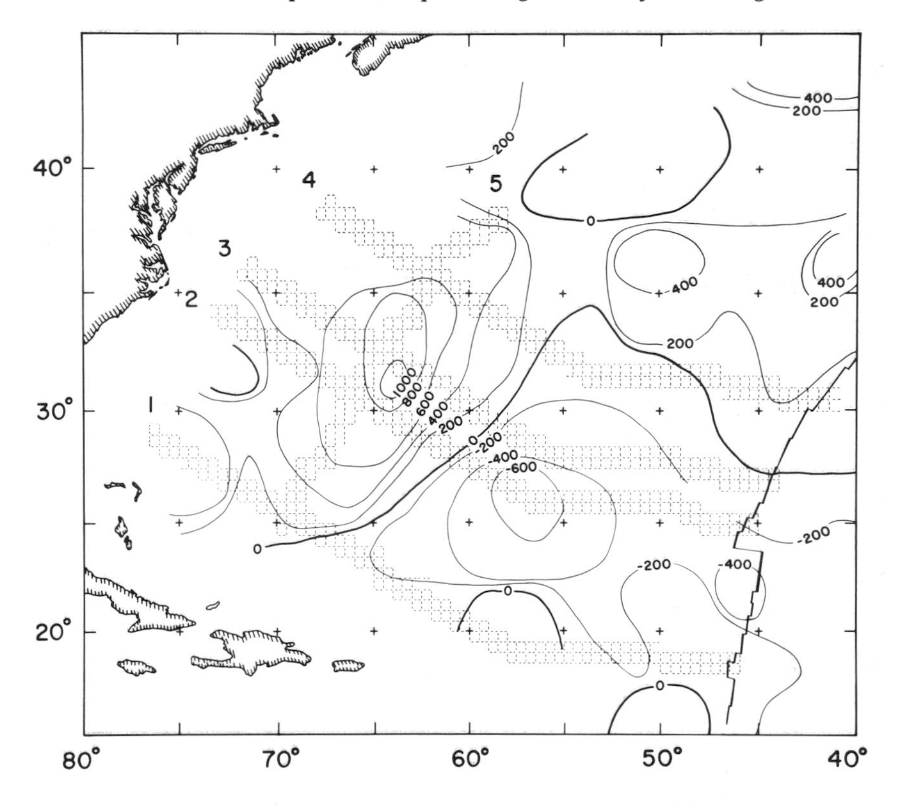

Figure 7. The two-and-a-half degree residual depth anomalies contoured at 200 m intervals. The dashed squares represent the area covered by the five profiles presented in Figures 9 (1–4) and 10 (5).

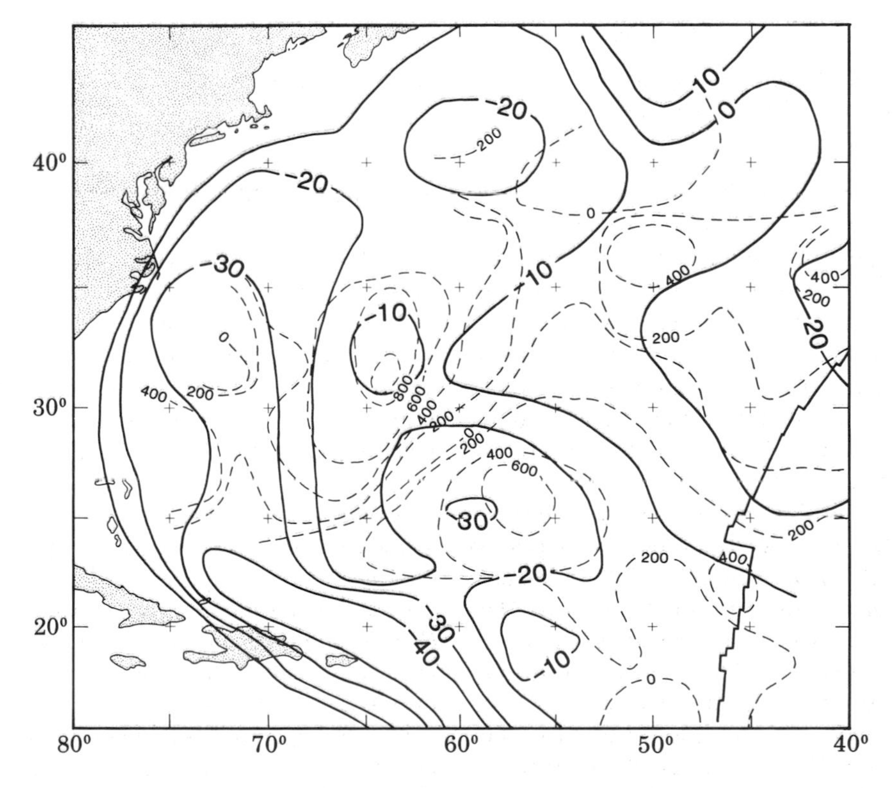

Figure 8. The five degree residual free air gravity field contoured at 10 mgal intervals from Cochran and Talwani (1977) compared with the two and a half degree residual depth anomalies contoured at 200 m.

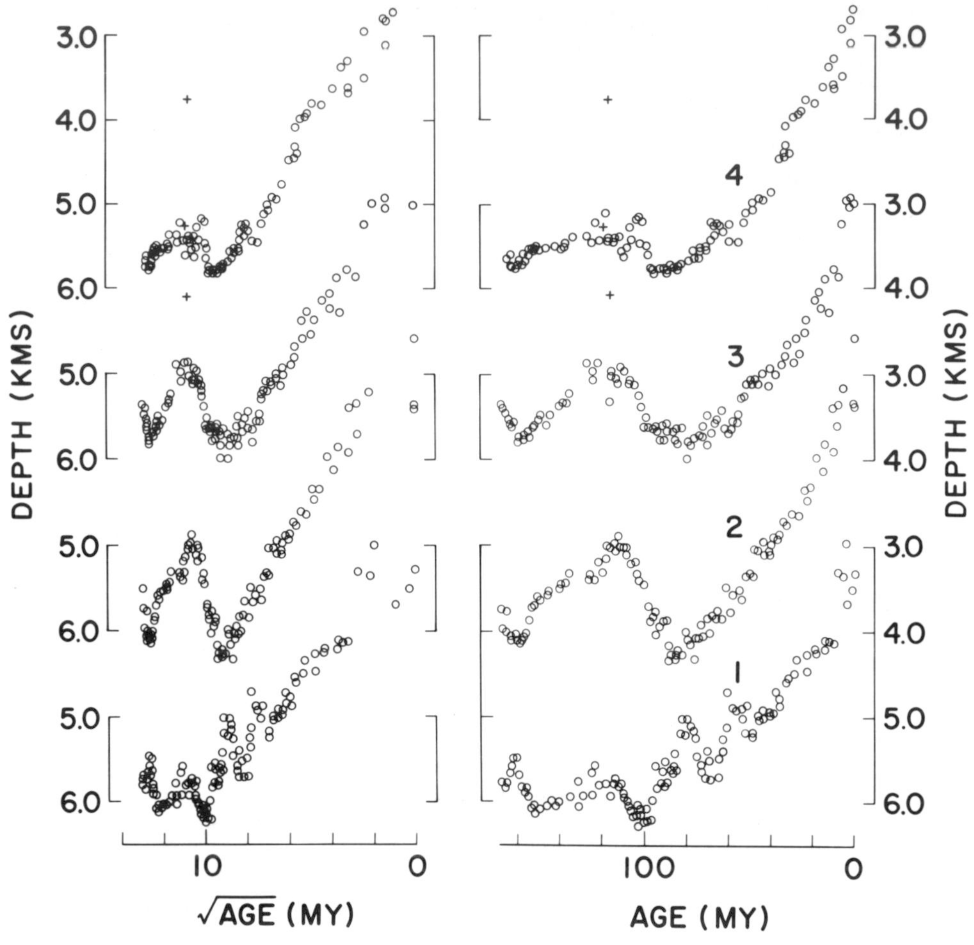

Figure 9. Depth versus age and the square root of age for the four profiles across the Bermuda Rise outlined as dashed lines on Figure 7.

Rabinowitz and Jung, this volume) which also shows up in the long wavelength geoid field (Fig. 1; Vogt, this volume) and in the intermediate-wavelength geoid (Anderle, this volume; Bowin and others, 1984 and Vogt and others, 1984). The regional gravity field decreases to the southwest from values of 0 mgals south of the Azores to values of greater than −40 mgals over the Puerto Rico Trench. If this trend is removed there is a striking visual correlation between the residual free air gravity field and the residual depths. For example there is a relative high of −10 mgals over the Bermuda Rise with a relative low of −30 mgals over the deep to the southeast (Fig. 8). The correlation between negative gravity anomaly and deeper depths does not appear on the original charts of Cochran and Talwani (1977). In addition, the Corner Seamounts and the southern tip of the Azores High are also associated with a positive residual gravity anomaly.

The correlation between the residual gravity and residual depth anomalies has a peak to trough wavelength of 1000 km and an amplitude of 15 mgal/km. This is lower than the figure given by Sclater and others (1975) from data collected elsewhere in the Atlantic but is similar to those reported by Watts (1976) for the Pacific.

THE BERMUDA RISE

The Bermuda Rise is the major topographic feature in the Western North Atlantic lying between the Mid-Atlantic Ridge and the coast of North America. To examine the rise and the low to the east, we constructed four profiles from the half-by-half degree elements. These profiles run from the ridge to the continental shelf and are parallel to the direction of motion of the plate boundary (Fig. 7). The half-by-half degree values of the corrected basement depth are plotted against both age and the square root of age (Fig. 9). As is to be expected the Bermuda Rise appears as a strikingly obvious topographic anomaly, especially on the two central profiles (Fig. 9, profiles 2 and 3). It is considerably less prominent on the southern and northern profiles (Fig. 9, profiles 1 and 4). In the age range 0 to 100 Ma the match of the subsidence to the square root of age on the southern three profiles is excellent. On the northern profile the fit to this relation is not as good because the profile starts on the beginning of the Azores High and runs close to the Corner Seamounts (Fig. 7). Between 80 and 100 Ma on all four profiles the topography starts to shallow. In the case of the second profile the

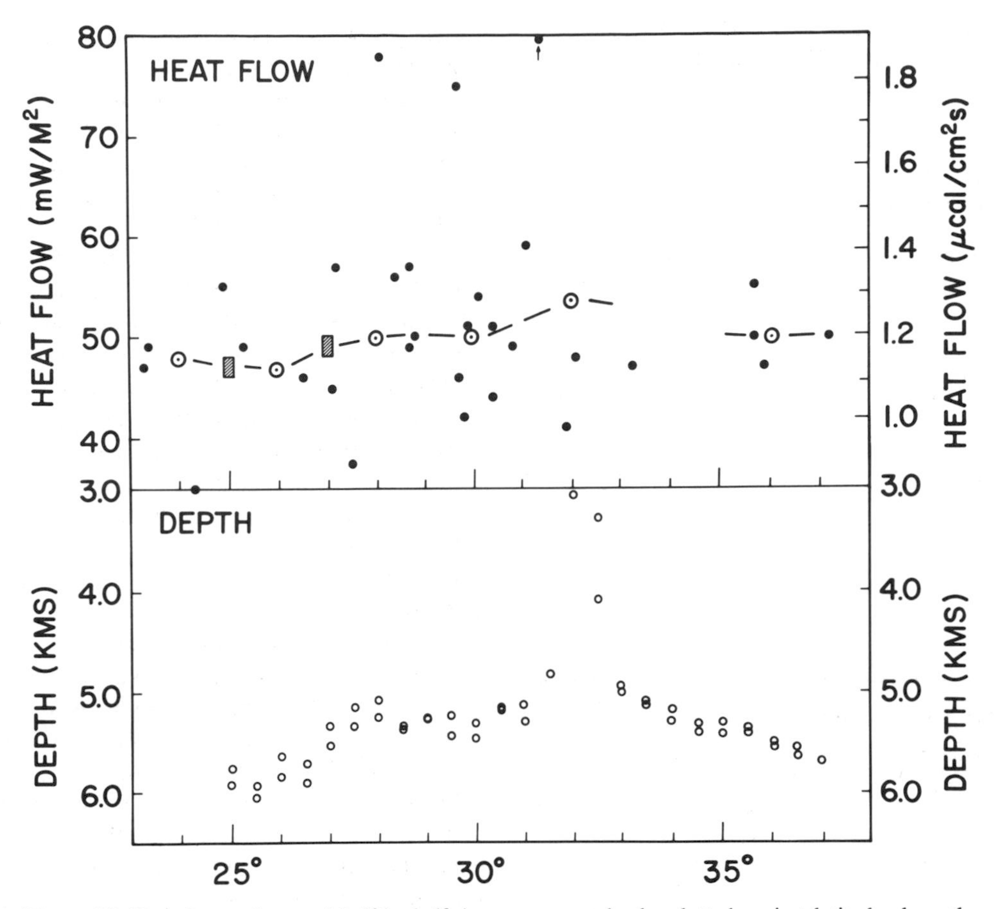

Figure 10. Heat flow values and half-by-half degree average depths plotted against latitude along the crest of the Bermuda Rise. The position of the depth profile is shown in Figure 7 and the window for the heat flow values in Figure 4. The dashed line through the circled points connects the median heat flow values calculated at two degree intervals.

effect is quite spectacular. The residual depth anomaly over the crest of the rise is on the order of 1 km. A second rise associated with the Blake-Bahama Outer Ridge is observed on profile 1. This feature is separated from the Bermuda Rise by a saddle in the residual depth field and is probably not related to the rise. As mentioned in a previous section its origin is a matter of debate.

We constructed a fifth residual depth profile along the crest to examine the southwest-northeast extent of the Bermuda Rise (profile 5; Figs. 7 and 10). In addition we compiled all the heat flow values within a five degree window about this profile (Figs. 4 and 10). The depth shallows abruptly at 27°N and tapers off slowly around 37°N. The heat flow, apart from three high values of questionable validity, shows no dramatic evidence for an increase over the rise. In fact, the median of the values, grouped by two degree elements, shows no detectable change along the profile.

DISCUSSION OF BOUNDARY LAYER AND PLATE MODELS

As mentioned earlier, two models have been advanced to explain residual depth anomalies such as Bermuda and the flattening of the depth/age relation on old ocean floor. The first involves simple boundary layer cooling and a fixed localized hot spot that resets the thermal structure of the lithosphere. The second views the lithosphere as a thermo-mechanical plate perturbed by the up and downwelling limbs of convection in the upper mantle. Both models are convective in concept but stress different aspects of the motion.

The major fact in favor of the first model (boundary layer thickening) is the excellent initial fit of the observed subsidence to the square root of age and the abrupt shallowing of the depth on crust older than 100 Ma. However, it is not supported by the heat flow data. If the plate thickness had been instantaneously reset to that of 25 Ma ocean floor 30 to 50 million years ago, then from the simple cooling model the heat flow should be between 65 and 55 mW/m^2 (1.5–1.3 μcal/cm^{2}s). The median values along the Bermuda Rise are uniformly lower than this. The topographic profiles either side of and close to the continental shelf present a second problem. They do not drop off sharply to the depth predicted by simple cooling; rather they appear to flatten exponentially with age. A final problem is that the rise is lineated in a

EFFECT ON DEPTH AND HEAT FLOW
OF CONVECTION CELL

RESULTS

1.	Δe=~+600 M ΔH=~+0.15 HFU	Δe=~-400 M ΔH=~-0.15 HFU
2.	Δe=~+400 M ΔH= 0.0 HFU	Δe=~-400 M ΔH= 0.0 HFU

MODEL

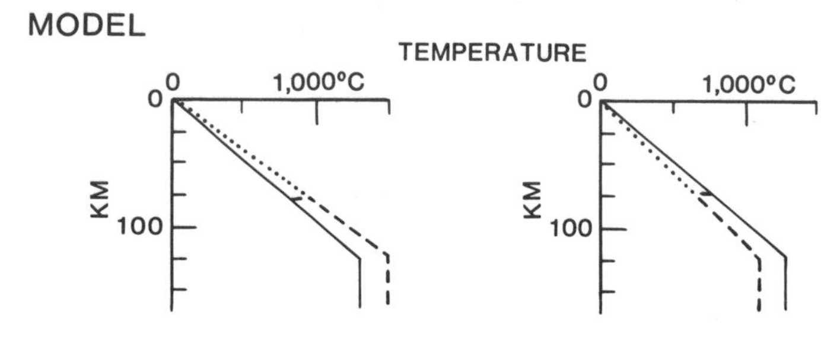

CONCEPT

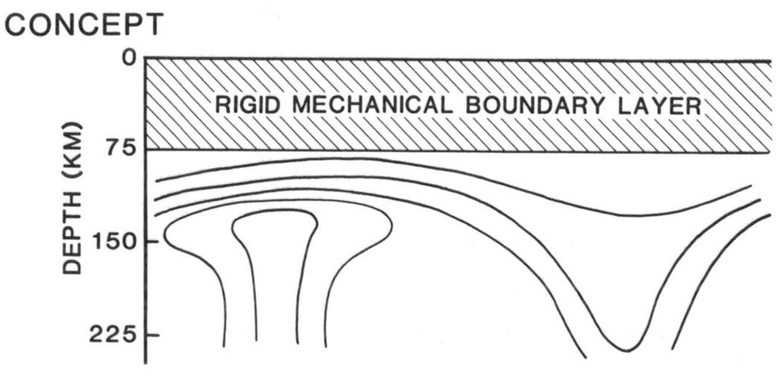

Figure 11. A schematic diagram showing the change in depth (Δe) and heat flow (ΔH, 1HFU = 1 $\mu cal/cm^2 s$) associated with an upper mantle convection cell beneath the lithosphere modeled as a thermomechanical plate. The concept is shown at the bottom of the diagram (after Parsons and Daly, 1983). Above is presented a model of the temperature structure in an old plate above the upwelling and downwelling limbs of a convection cell. These cells interact with the thermal boundary layer portion of the lithosphere and for simplicity are modeled as a ±200°C temperature difference at the base of a 125 km thick plate. The dotted line is the temperature structure for the cell and plate moving together and gives Result 1. The dashed line is for the plate moving fast with respect to the cell and gives Result 2. See text for further explanations.

northeast by southwest direction and lies at ninety degrees to the trend given by assuming the Pacific hot spot frame (Morgan, 1981). Clearly, if the hot spot concept applies to the Bermuda Rise, the rise must be moving with the North American plate and hence cannot be considered fixed. This final point is not really an argument against the hot spot model but rather against a fixed convective system in the mantle below Bermuda.

Two principle facts favor the second (plate) model. These are the flattening of the subsidence curves on either side and beyond the Bermuda Rise, and the correlation between residual depth and residual free air gravity anomalies. The wavelength and amplitude of the correlation over Bermuda and the deep to the southeast are similar to values obtained by McKenzie and others (1974) from numerical experiments on convection in the upper mantle. This similarity and the correlation between gravity and depth are evidence that the two features are caused by the upwelling and downwelling limbs of a convection cell in the upper mantle. The flattening of the subsidence curves either side of the rise, away from the regions of the gravity anomalies, is evidence that it is a plate that is being perturbed, not a cooling boundary layer. The major argument against this model is that the best match to the observations requires a thinner effective plate thickness (about 100 km) than that proposed by Parsons and Sclater (1977). Further work is needed, especially in the Pacific, to determine whether or not a thinner plate is justified.

An additional feature that favors the second model is that it is easy to demonstrate how a depth anomaly can be created that is not correlated with an observable heat flow anomaly. Consider the thermomechanical plate model of Parsons and McKenzie (1978) on top of a convection cell and that the mechanical portion of the plate is on the order of 75 km thick and that the effective plate thickness is 125 kms (concept, Fig. 11). If the convection cell moves with the plate the effect of this cell can be modelled as a +200° temperature anomaly above the upwelling limb and a −200° temperature anomaly above the downwelling limb (dotted lines model, Fig. 11). Such a perturbation would cause a variation in depth of ±600 m, and in heat flow of 6 mW/M^2 (.15 $\mu cal/cm^2$) (Result 1, Fig. 11). If the plate is now allowed to move with respect to the convection cell, then the effect of the upwelling and downwelling limbs on the mechanical portion of the plate is much reduced. However, it is still just as large in the thermal boundary layer (dashed line, model Fig. 11). This would significantly reduce the heat flow anomaly but still leave a depth anomaly of approximately ±400 m (Result 2, Fig. 11). These calculations are simplistic but they serve to illustrate that if the lithosphere is modelled as a thermomechanical plate then it is not necessary to expect heat flow anomalies over regions of anomalous depth.

Whether or not a moving plate model can be applied directly to the Bermuda Rise is questionable. The island of Bermuda which was created in the mid-Oligocene (Tucholke and Vogt, 1979) lies on the crest of the Rise. If it is assumed that the volcanism, which created the island and the Rise, have the same mantle source then the position of Bermuda implies little or no movement of the plate with respect to the source. However, the heat flow anomaly would still be small and any relative movement, however small, would reduce it.

CONCLUSIONS

The following conclusions can be drawn from the observations considered in this study of the Western North Atlantic.

1. The depth increases as the square root of age for ocean crust younger than 80 Ma.

2. On crust older than 100 Ma the depth decreases significantly. Much of this shallowing is associated with the Bermuda Rise.

3. Neither the boundary layer nor the thermo-mechanical plate models can account for this shallowing. However, the plate model gives a better overall match to the average depths on older ocean floor.

4. The heat flow data are highly scattered near the ridge axis but show less scatter on crust older than 100 Ma. The averages on old ocean floor are slightly higher than those expected from either model. There is no obvious heat flow anomaly associated with the Bermuda Rise.

5. The residual depth field shows a high over Bermuda and a low to the southeast. This high and low are correlated with a relative high and low in the residual free air gravity field.

The following inferences are drawn from the above observations:

1. The subsidence of the Mid-Atlantic Ridge and the residual depth anomalies can be accounted for either by a cooling boundary layer of which the thermal structure is reset by a localized convective jet, or by a two layer thermo-mechanical plate perturbed by upper mantle convection.

2. On balance the plate model is preferred because (a) it accounts for residual depth and residual gravity correlation, (b) no evidence is found on older ocean floor away from areas of relative free air gravity anomaly for a relation between depth and the square root of age, and (c) it can explain the absence of a heat flow anomaly over the Bermuda Rise.

Further work is needed in other areas of the ocean floor to investigate if these inferences are general in nature or just restricted to the Western North Atlantic.

REFERENCES

Anderson, R. N., McKenzie, D., and Sclater, J. G.
1973 : Gravity, bathymetry and convection in the earth: Earth and Planetary Science Letters, v. 18, p. 391–407.

Bowin, C., Thompson, G., and Schilling, J. G.
1984 : Residual geoid anomalies in Atlantic Ocean basin: relationship to mantle plumes: Journal of Geophysical Research, v. 89, B12, p. 9905–9918.

Cochran, J. R., and Talwani, M.
1977 : Free-air gravity anomalies in the world's oceans and their relationship to residual elevation: Geophysical Journal of the Royal Astronomical Society, v. 50, p. 495–552.

Crough, S. T.
1978 : Thermal origin of mid-plate hot-spot swells: Geophysical Journal of the Royal Astronomical Society, v. 55, p. 451–470.

Crough, S. T., and Jarrard, R. D.
1981 : The Marquesas-Line swell: Journal of Geophysical Research, v. 86, p. 11763–11771.

Davis, E. E., and Lister, C.R.B.
1974 : Fundamentals of ridge crest topography: Earth and Planetary Science Letters, v. 21, p. 405–413.

Davis, E. E., Lister, C.R.B., and Sclater, J. G.
1984 : Toward determining the thermal state of old ocean floor: Geophysical Journal of the Royal Astronomical Society, v. 78, p. 507–546.

Detrick, R. S.
1981 : An analysis of geoid anomalies across the Mendocino fracture zone: Implications for thermal models of the lithosphere: Journal of Geophysical Research, v. 86, p. 11751–11762.

Detrick, R. S., and Crough, S. T.
1978 : Island subsidence, hot spots and lithospheric thinning: Journal of Geophysical Research, v. 83, p. 1236–1244.

Duncan, R. A.
1984 : Age progressive vulcanism in the New England Seamounts and the opening of the Central Atlantic Ocean: Journal of Geophysical Research, v. 89, B12, p. 9980–9990.

Galson, D., and Von Herzen, R. P.
1981 : A heat flow survey on anomaly M-O south of the Bermuda Rise; Earth and Planetary Science Letters, v. 53, p. 296–306.

Harland, W. B., Cox, A. V., Llewellyn, P. G., Pickton, C.A.G., Smith, A. G., and Walters, R.
1983 : A geologic time scale: Cambridge, England, Cambridge University Press, 129 p..

Haxby, W. F. and Turcotte, D. L.
1978 : On isostatic geoid anomalies: Journal of Geophysical Research, v. 83, p. 5473–5478.

Heestand, R. L., and Crough, S. T.
1981 : The effect of hot spots on the ocean age-depth relation: Journal of Geophysical Research, v. 86, p. 6107–6114.

Jessop, A., Hobart, M., and Sclater, J. G.
1976 : World-wide compilation of heat flow data; Geothermal Series No. 5, Ottawa, Canada.

Langseth, M. S., LePichon, X., and Ewing, M.
1966 : Crustal structure of mid-ocean ridges, 5. Heat flow through the Atlantic Ocean and convection currents: Journal of Geophysical Research, v. 71, p. 5321–5355.

LeDouran, S., and Parsons, B.
1982 : A note on the correction of ocean floor depths for sediment loading: Journal of Geophysical Research, v. 87, p. 4715–4722.

Lister, C.R.B.
1972 : On the thermal balance of a mid-ocean ridge: Geophysical Journal of the Royal Astronomical Society, v. 26, p. 515–535.
1977 : Estimators for heat flow and deep rock properties based on boundary layer theory: in Heat Flow and Geodynamics, Tectonophysics, ed. A. M. Jessop, v. 41, p. 157–171.

McKenzie, D.
1967 : Some remarks on heat flow and gravity anomalies: Journal of Geophysical Research, v. 72, p. 6261–6273.

McKenzie, D. P., Roberts, J. M., and Weiss, N. O.
1974 : Towards a numerical simulation: Journal of Fluid Mechanics, v. 62, p. 465–538.

McKenzie, D. P., Watts, A., Parsons, B., and Roufosse, M.
1980 : Planform of mantle convection beneath the Pacific Ocean: Nature, v. 288, p. 442–446.

Morgan, W. J.
1981 : Hotspot tracks and the opening of the Atlantic and Indian Oceans; *in* The Sea, Volume 7, chap. 13 ed. C. Emiliani: New York, John Wiley and Sons, p. 443–487.

Parker, R. L., and Oldenburg, D. W.
1973 : Thermal model of ocean ridges: Nature, v. 242, p. 137–139.

Parsons, B., and Daly, S.
1983 : The relationship between surface topography, gravity anomalies and temperature structure of convection: Journal of Geophysical Research, v. 88, p. 1129–1144.

Parsons, B., and McKenzie, D. P.
1978 : Mantle convection and the thermal structure of the plates: Journal of Geophysical Research, no. 83, p. 4485–4496.

Parsons, B., and Sclater, J. G.
1977 : An analysis of the variation of ocean floor bathymetry and heat flow with age: Journal of Geophysical Research, v. 82, no. 5, p. 803–827.

Perry, R. H., Fleming, H. S., Vogt, P. R., Cherkis, N. Z., Feden, R. H., Thiede, J., and Strand, J. E.
1980 : North Atlantic Ocean: bathymetry and plate tectonic evolution; Naval Research Laboratory, Acoustics Division: Washington, D.C., Williams and Heintz.
Roufosse, M., Parsons, B., McKenzie, D., and Watts, A. B.
1981 : Geoid and depth anomalies in the Indian Ocean (abstract); EOS: Transactions of the American Geophysical Union, v. 62, p. 389.
Schouten, H., and Klitgord, K.
1977 : Mesozoic Magnetic Anomalies, Western North Atlantic; Miscellaneous Field Studies Map MF-915: United States Geological Survey, Reston, Virginia.
Sclater, J. G., Anderson, R. N., and Bell, M. L.
1971 : The elevation of ridges and the evolution of the central eastern Pacific: Journal of Geophysical Research, v. 76, p. 7888–7915.
Sclater, J. G., Jaupart, C., and Galson, D.
1980 : The heat flow through oceanic and continental crust and the heat loss of the earth: Reviews of Geophysics and Space Physics, v. 18, p. 269–311.
Sclater, J. G., Lawver, L. A., and Parsons, B.
1975 : Comparison of long-wavelength residual elevation and free air gravity anomalies in the North Atlantic and possible implications for the thickness of the lithospheric plate: Journal of Geophysical Research, v. 80, p. 1031–1052.
Tucholke, B. E., Houtz, R. E., and Ludwig, W. J.
1982 : Sediment thickness and depth to basement in western North Atlantic ocean basin: Bulletin of the American Association of Petroleum Geologists, v. 66, p. 1384–1395.
Tucholke, B. E., and Vogt, P. R.
1979 : Western North Atlantic: sedimentary evolution and aspects of tectonic history: *in* Initial Reports of the Deep Sea Drilling Project, v. 43, eds. B. E. Tucholke and P. R. Vogt; U.S. Government Printing Office, Washington, D.C., p. 791–825.
Turcotte, D. L., and Oxburgh, E. R.
1967 : Finite amplitude convection cells and continental drift: Journal of Fluid Mechanics, v. 28, p. 29–42.
Van Wyckhouse, R.
1973 : SYNBAPS (Synthetic Mathematic Profiling Systems); Technical Report TR-233: Washington, D.C., Naval Oceanographic Office.
Vogt, P. R., and Tucholke, B. E.
1979 : The New England Seamounts: Testing Origins, *in* Initial Reports of the Deep Sea Drilling Project, v. 43: eds. B. E. Tucholke and P. R. Vogt; U.S. Government Printing Office, Washington, D.C., p. 847–856.
Vogt, P. R., Zondek, B., Fell, P. W., Cherkis, N. Z., and Perry, R. H.
1984 : Seasat Altimetry, the North Atlantic Geoid and evaluation by shipborne subsatellite profiles: Journal of Geophysical Research, v. 89, B12, p. 9885–9903.
Von Herzen, R. P., Detrick, R. S., Crough, T., Epp, D., and Fehn, U.
1982 : Thermal origin of the Hawaiian swell: heat flow evidence and thermal models: Journal of Geophysical Research, v. 87, p. 6711–6723.
Watts, A. B.
1976 : Gravity and bathymetry in the central Pacific Ocean: Journal of Geophysical Research, v. 81, p. 1533–1553.
Williams, D. L., and Poehls, K. A.
1975 : On the thermal evolution of the oceanic lithosphere: Geophysical Research Letters, v. 2, p. 321–325.
Williams, D. L., Von Herzen, R. P., Sclater, J. G., and Anderson, R. N.
1974 : The Galapagos spreading center: Lithospheric cooling and hydrothermal circulation: Geophysical Journal of the Royal Astronomical Society, v. 38, p. 587–608.

MANUSCRIPT ACCEPTED BY THE SOCIETY FEBRUARY 25, 1985

ACKNOWLEDGMENTS

We would like to thank K. Klitgord and H. Schouten for allowing us to use charts of magnetic anomalies in the Western North Atlantic prior to publication and for help in selecting a least-erroneous Mesozoic Magnetic time scale. R. P. Von Herzen and B. Parsons made helpful suggestions when reviewing the original manuscript.

This research was supported by National Science Foundation Contract OCE80-24287 to M.I.T. and National Science Foundation Contract OCE83-20091 to U. T. Austin and a Shell Foundation Professorship to the senior author.

Printed in U.S.A.

The Geology of North America
Vol. M, The Western North Atlantic Region
The Geological Society of America, 1986

Chapter 17

Petrologic and geochemical evolution of pre–1 Ma western North Atlantic lithosphere

W. B. Bryan
Department of Geology and Geophysics, Woods Hole Oceanographic Institution, Woods Hole, Massachusetts 02543
F. A. Frey
Department of Earth, Atmospheric, and Planetary Sciences, Massachusetts Institute of Technology, Cambridge, Massachusetts 02139

INTRODUCTION

The Mid-Atlantic Ridge in the central North Atlantic was the source of the first deep-sea basement rocks to be studied in detail (Nicholls and others, 1964; Muir and Tilley, 1966; Miyashiro and others, 1969). The historical background and compositional details of basalts recovered from the modern Mid-Atlantic Ridge are discussed by Schilling and by Melson and O'Hearn in other chapters in this volume. Some of the first recoveries of old oceanic basement rocks were obtained by drilling in the western North Atlantic (Ayuso and others, 1976; Bryan and others, 1977). In this paper we summarize (1) the characteristics of pre–1 Ma lithosphere in the western North Atlantic and (2) inferences about its geologic evolution during the past 200 m.y. Such a review is subject to some serious limitations. Very large areas of the North Atlantic remain unsampled by drilling, and many crustal rocks that have been recovered may be too altered to yield definitive information about their magmatic geochemistry. Also, the deepest crustal penetrations in the North Atlantic have failed to reach the base of layer 2, which is composed almost entirely of extrusive basalts. The nature of the deeper levels of layer 2 and of the (presumably) gabbroic layer 3 must be inferred from samples recovered from a few holes apparently drilled in anomalously thin layer 2, from dredged samples recovered in fracture zones, or by extrapolation of observations at site 504 which was drilled into the deeper levels of layer 2 near the Costa Rica rift in the eastern Pacific (Anderson and others, 1982). Data on the presumed ultramafic layer 4 are even more indirect because sampling is limited to a few shallow recoveries in drill holes, and to a few dredges from fracture zones.

Our task is further complicated by the compositional diversity along the present-day Mid-Atlantic Ridge, as indicated by the rapidly growing data base for modern ocean ridge basalts (see discussions by Melson and O'Hearn and by J-G. Schilling, this volume). If the pattern of compositional variability now known to exist along the Mid-Atlantic Ridge axis had persisted in exactly the same way throughout the whole period of opening, then the pattern of variation of *initial* rock geochemistry in older seafloor should consist of a series of arcuate strips parallel to fracture zones and perpendicular to magnetic anomalies; boundaries between these strips would be either sharp or gradational, as on the modern ridge. However, data from intensely sampled modern ridge segments and from some drill holes in older crust indicate that even this fairly complex pattern is too simple. There appear to have been substantial local variations in some critical compositional parameters on a time scale of less than a few million years, and it is not clear how this variability can be extrapolated into older crust where no data exist. Also, it is known that weathering and metamorphism of oceanic crust, once thought to be a simple function of age, is very irregularly distributed and may differ greatly between two closely spaced drill holes. Given all of these variables, and the limited sampling of oceanic basement rocks in the western North Atlantic, it is certainly not possible to provide a three-dimensional compositional "map" of the oceanic crust for either primary or secondary variations in rock compositions. However, the data are sufficient to at least indicate the minimum degree of complexity that we could expect in such a map.

Within these serious limitations we attempt to extrapolate magmatism on the present-day Mid-Atlantic Ridge axis back in time, using data for older rocks to show the kinds of variability that can be expected, and the scale on which it may occur in space and time. We begin by discussing the stratigraphy, mineralogy, and compositions of basalts sampled by drilling in crust from a few m.y. to over 100 m.y. in age and we summarize the petrogenetic constraints that can be inferred from such data. We then review middle to late Mesozoic magmatism in eastern North America, the western North Atlantic, and western North Africa, and briefly consider Tertiary magmatic activity associated with rifting in the far North Atlantic.

For brevity in subsequent discussions we use some conventional abbreviations and simplifications which also are used and discussed in more detail in many of the cited references. These

Bryan, W. B., and Frey, F. A., 1986, Petrologic and geochemical evolution of pre-1 Ma western North Atlantic lithosphere; *in* Vogt, P. R., and Tucholke, B. E., eds., The Geology of North America, Volume M, The Western North Atlantic Region: Geological Society of America.

TABLE 1. WESTERN NORTH ATLANTIC DSDP AND IPOD DRILL SITES WITH SIGNIFICANT BASEMENT RECOVERY

Site	Location (Lat.,Long.)	Age (M.Y.)	Basement Penetration (m)	Comments
10	32°51.7'N 52°12.9'W	15.9(?)	2.5	Alkaline basalt sill.*
11A	29°56.6'N 44°44.8'W	15.0(?)	1.0(?)	Pyroxene-phyric normal MORB.
100	24°41.3'N 73°48.0'W	158	14	Plagioclase and pyroxene phyric pillow basalt and altered massive flows.
105	34°53.7'N 69°10.4'W	156	9.3	Moderately altered plagioclase-phyric pillow basalt.
112	54°01' N 46°36.2'W	58	1.0(?)	Highly altered MORB.
332A	36°52.7'N 33°38.5'W	3.5±.1	333	Aphyric to highly phyric basalt and interlayered sediment.
332B	36°52.7'N 33°38.5'W	3.5+.1	583	Aphyric to highly plagioclase-phyric or picritic basalt, fresh to moderately altered.
333A	36°50.5'N 33°40.1'W	3.5±.1	311	Brecciated olivine-phyric basalt.
334	37°02.1'N 34°24.9'W	9.5	124	Aphyric basalt overlying gabbro and serpentinized peridotite.
335	37°17.7'N 35°11.9'W	13	108	Aphyric to sparsely phyric pillow basalt.
382	35°25.0'N 56°32.3'W	85(?)	---	Basalt clasts and breccia interbedded with sediment.+
384	40°21.7'N 51°39.8'W	120(?)	5.5	Altered phyric basalt.+
385	37°22.2'N 60°09.5'W	21.0±3 (sill) 70–90 (Breccia)	55.1	Volcaniclastic sandstone and breccia;+ basalt sill.
386	31°11.2'N 64°14.9'W	125±25(?)	9.6	Moderately altered phyric basalt.**,+
387	32°19.2'N 67°40.0'W	126±20(?)	2.9	Altered phyric basalt sill.±
395	22°45.4'N 46°04.9'W	15	93	Aphric basalt, gabbro, and serpentinized peridotite.
395A	22°45.4'N 46°04.9'W	15	571	Aphyric and phyric pillow basalt and massive basalt; basalt breccia and cobbles.

Notes: Letters (A, B, etc.) adjacent to site numbers refer to multiple holes drilled at the same location. Ages are based on oldest overlying sediment age or basement magnetic anomaly age as indicated in relevant DSDP Initial Report volumes, adjusted as necessary to agree with standard DNAG timescale. Age uncertainty indicated by question mark reflects possible errors due to basalt-sediment interbedding, hiatus in core recovery, or possible complications from sill or dike intrusion; see footnotes for other data and sources.

* Fission track age (Frey and others, 1974).

** $^{40}Ar/^{39}Ar$ age reported as 126.1 ± 27.2 (Houghton and others, 1979).

+ Also see discussion of dredged rock ages from New England Seamounts (Duncan 1984).

include MORB (Mid-Ocean Ridge Basalt), LREE and HREE (Light- and Heavy-Rare Earth Elements), MAR (Mid-Atlantic Ridge) and DSDP (Deep Sea Drilling Project). Rare earth elements are discussed in terms of their chondrite-normalized enrichment values, rather than absolute values; for some purposes we also discuss ratios between chondrite-normalized rare earths or other trace elements. Such normalized ratios are shown with a subscript n (e.g., $[La/Sm]_n$). We use the term "incompatible element" for any trace element which is strongly partitioned into silicate liquid during crystallization or melting of common rock-forming minerals. Abundance ratios between incompatible elements behave much like ratios between isotopes in that they are relatively unaffected by shallow, low-pressure crystal-liquid fractionation and so are good indicators of mantle-derived magmatic character. Ocean ridge basalts originally attracted considerable interest because of their pronounced depletion in incompatible elements relative to continental or ocean island basalts; MORB showing these depleted characteristics are still considered to be "normal" or "typical." However, MORB associated with island platforms, such as the Azores or other "anomalous" ridge seg-

TABLE 1. (CONTINUED).

Site	Location (Lat.,Long.)	Age (M.Y.)	Basement Penetration (m)	Comments
407	63°56.3'N 30°34.6'W	38	161	Aphyric to sparsely olivine-phyric pillow basalt.
408	63°22.6'N 28°54.7'W	20	37.3	Aphyric and sparsely olivine-plagioclase phyric basalt and hyaloclastite.
409	62°37' N 25°57.2'W	2.3	239	Variable aphyric and phyric pillow basalt.
410A	45°30.5'N 29°28.6'W	8.5	47.5	Aphyric to sparsely ol.-plag. phyric.
410	45°30.5'N 29°28.6'W	10	49.0	Basalt breccia and hyaloclastite.
411	36°46' N 33°23.3'W	1	45.5	Olivine- and plagioclase-olivine phyric basalt.
417A	25°06.6'N 68°02.5'W	118	209	Highly altered pillowed and basalt and breccia.
417D	25°0607'N 68°02.8'W	118	366	Moderately fresh to very fresh pillowed and massive basalt.
418A	25°02.1'N 68°03.4'W	118	544	Aphyric to moderately phyric massive and pillowed basalt.
418B	25°02.1'N 68°03.5'W	118	10	Sparsely phyric pillow basalt.
534A	28°20.6'N 75°22.9'W	165	31	Altered aphyric basalt.
556	38°56.4'N 34°41.1'W	32	178	Basalt breccia, flows, gabbro and serpentinite.
557	38°50.0'N 32°33.6'W	18	3	Coarse aphyric basalt.
558	37°46.2'N 37°20.6'W	35	150	Aphyric basalt, basalt breccia, gabbro and serpentinite.
559	35°07.5'N 40°55.0'W	35	63	Aphyric pillow basalt.
560	34°43.3'N 38°50.6'W	18	47	Altered gabbro and serpentinite.
561	34°47.1'N 39°01.7'W	19	15	Aphyric pillow basalt.
562	33°08.5'N 41°40.8'W	18	90	Sparsely plag.-phyric pillow basalt.
563	33°38.5'N 43°46.0'W	35	18.5	Sparsely plag.-phyric pillow basalt.
564	33°44.4'N 43°46.0'W	35	81	Aphyric pillow basalt and massive flows.

ments, have been shown to be substantially enriched in incompatible elements, and like the subaerial island lavas they have radiogenic isotope ratios closer to bulk earth values. Although there is a complete gradation from the "most depleted" to the "most enriched" MORB, for simplicity we discuss these rocks as either 'depleted' MORB or as 'enriched' MORB. A convenient (but arbitrary) chemical discriminant is provided by $(La/Sm)_n$, with depleted MORB having $(La/Sm)_n < 1$, and enriched MORB having $(La/Sm)_n > 1$. Basalts with essentially "flat" normalized REE patterns ($[La/Sm]_n \cong 1$) may be called "transitional." Unaltered depleted MORB typically have initial $^{87}Sr/^{86}Sr < 0.7030$ and $^{143}Nd/^{144}Nd > 0.5130$; usually these ratios increase and decrease respectively with increasing incompatible element enrichment. In subsequent discussion we describe other chemical and mineralogical characteristics of these groups, and their variation relative to age, location, and degree of secondary alteration in the western North Atlantic. For locations of drill sites and other geographic features not otherwise documented, the reader should refer to Table 1 and Plate 2.

STRATIGRAPHY OF WESTERN NORTH ATLANTIC LITHOSPHERE: INFERENCES FROM DEEP DRILLING, FRACTURE ZONES, AND OPHIOLITES

The transition from conformable marine sediments to layer 2 basalts is well documented in many of the deep drill holes in the western North Atlantic. Usually the transition is anticipated by the appearance of intermixed marine sediment and basaltic rock fragments a few meters above the contact, followed by an abrupt transition to basalt, with scattered inclusions of indurated calcareous sedimentary fragments in the upper few meters of basalt. At site 332B, additional thin sedimentary intervals were penetrated throughout the upper 500 m of the 583 m penetrated, but this interbedding of basaltic and sedimentary units is not common at other sites where extrusive basement has been encountered.

The sheeted dikes anticipated in the deeper levels of layer 2 by analogy with ophiolite complexes have rarely been observed in rift valley walls, fracture zone scarps, or in deep drill penetrations. This may partly reflect the practical difficulty of recognizing

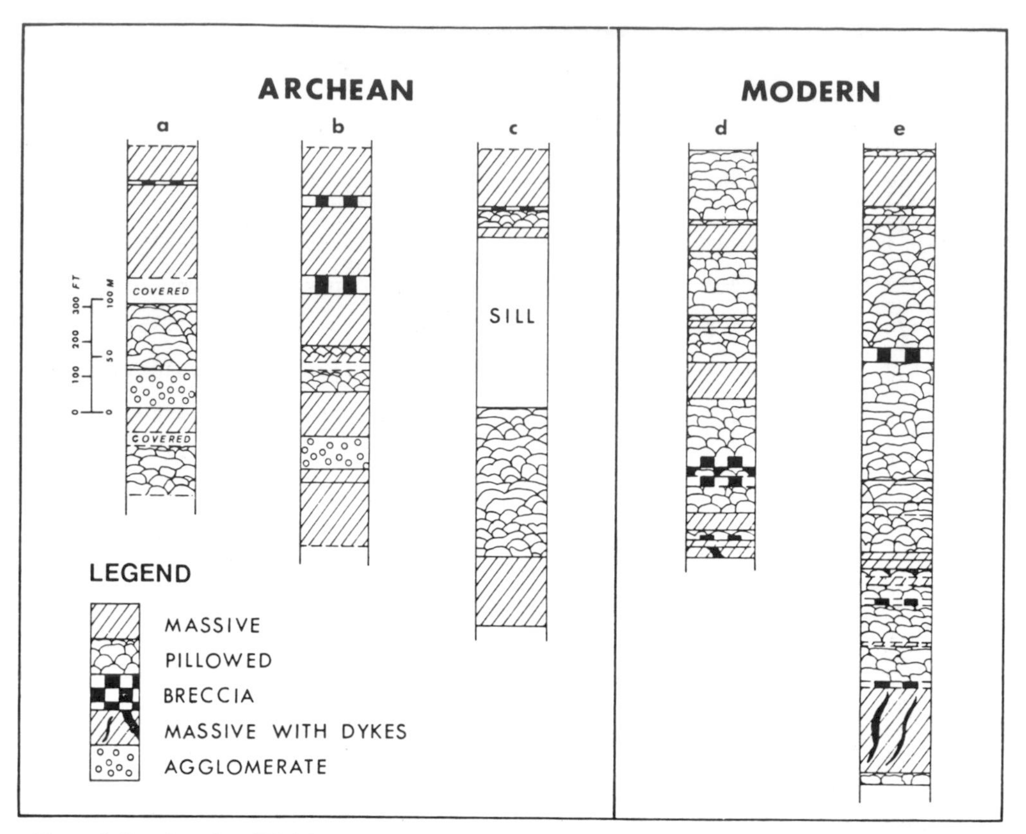

Figure 1. Stratigraphy of IPOD holes 417d (*d*) and 418a (*e*), compared to Archean lava sequence from the Abitibi district (*a, b, c*), after Wells and others, 1979. Note inter-bedding of pillowed and massive flows in both sequences. Boundaries between pillowed units are marked by thin sedimentary horizons or more commonly, by breccia and/or weathered zones.

vertical dike contacts in vertical drill holes, especially when core recovery is low and fragments are broken and rotated. In the deepest basement penetration yet achieved (hole 504B, Costa Rica Rift), dikes appear at about 600 m below top of basement, but are dominant only below 800 m (Anderson and others, 1982). The excellent core recovery at holes 417D and 418A revealed a few probable high-angle dike contacts in the lower part of each hole at 400–500 m below top of basement (Fig. 1). Other deep holes, as at sites 332 and 395, have not yielded cores of sufficient quality to evaluate diking, although the prevalence of quench textures suggests most of the material is pillow lava. At site 332B, possible intrusive rock was encountered at 180 and 350 m within layer 2 (Aumento and others, 1977). Submarine lavas are not invariably pillowed; at sites 417 and 418 massive flows form a significant proportion of layer 2. Wells and others (1979) showed that a similar morphology and sequence of pillowed and massive flows occurs in the Archean lavas of the Abitibi district in northern Quebec (Fig. 1).

The deeper layers of the western North Atlantic lithosphere may be represented by samples recovered from deep exposures along the walls of major fracture zones (Miyashiro and others, 1970; Dick and others, 1984). It is curious that gabbros presumably representative of layer 3, and peridotite and serpentinite possibly representing upper mantle, have not been penetrated at the lower levels of any of the deepest drill holes in the western North Atlantic, but have been encountered at a relatively shallow level (50–100 m below top of basement) at sites 395, 334, 556, 558, and 560. At site 395, serpentinized peridotite and gabbro encountered at 57 m below the top of layer 2 in hole 395 were followed by basalt to a depth of 95 m, where the hole terminated. At hole 395A, serpentinite was recovered in a thin interval at about 60 m within layer 2. Subsequent evaluation of bathymetry along with tectonic, stratigraphic, and geophysical data at most of these sites indicate "abnormal" stratigraphy due to normal faulting, talus slides from a nearby scarp, or proximity to a fracture zone. Very likely, the unusual abundances of gabbro and serpentinized peridotite exposed in fracture zones also reflect tectonic emplacement in part, and neither drill holes nor fracture zone surveys to date have documented unequivocally a continuous stratigraphic succession from layer 2 through layer 3 into upper mantle.

Plutonic rocks from ophiolites have been studied in more detail than those from the ocean floor, and they have been used to infer the structure and evolution of the deeper levels of the oceanic crust. However, direct application of ophiolite stratigraphy to older seafloor must take into account that (1) the ophiolite

Figure 2. Ordovician pillow lavas, Bay of Islands complex, Newfoundland. This coastal exposure at Green Gardens is separated from the main ophiolite complex by a transform fault. The interbedded lavas, sills, and tuffs dip very steeply east and strike approximately north-south. This view looking west is approximately perpendicular to the exposed upper surface of one of the pillowed units and shows tubular lava pillows branching from a central feeder dike. Figure is standing on contact with overlying bedded tuffs.

sequence of rocks can form in different tectonic settings; in particular, basalts associated with some ophiolites appear to have affinities with island arc basalts (Cameron and others, 1980; Dick and Bullen, 1984; Schmincke and others, 1983); (2) no ophiolite has been studied in enough detail to develop a thorough understanding of the petrogenetic relationships between the lavas and plutonic rocks; and (3) ophiolite sequences are generally thinner than the ocean crust inferred from seismic data (Anderson and others, 1982). The well-exposed Paleozoic Newfoundland ophiolites provide good examples of the difficulties encountered in using ophiolites to understand the evolution of oceanic lithosphere. The Betts Cove ophiolite contains some lavas compositionally distinct from MORB (Coish and Church, 1979; Coish and others, 1982; Dick and Bullen, 1984). In contrast, the Bay of Islands ophiolite contains pillow lavas similar to MORB (Fig. 2; Malpas, 1978; Suen and others, 1979), but the ultramafic cumulates that form the base of the plutonic section are not consistent with formation from MORB at low pressure (Elthon and others, 1982). Similar difficulties in relating the plutonic section of ophiolites to the overlying lavas were discussed by Dungan and Green (1980). Given all these compositional, stratigraphic, and petrogenetic uncertainties associated with the ophiolite model for ocean crust, it must be applied with caution to the western North Atlantic.

PETROGRAPHIC AND CHEMICAL VARIATIONS

Petrography of Layer 2 Basalts

The textures and phenocryst assemblages in older seafloor rock samples are important in deducing the original chemical characteristics of rocks that have been substantially altered in composition by hydrothermal activity or seafloor weathering. Certain mineral phases with characteristic skeletal quench morphology can be recognized by their unique forms even when pseudomorphed by clays and iron oxides. Larger phenocrysts and megacrysts may resist weathering and can be identified and analyzed by electron microprobe even when the surrounding matrix is totally altered. Nevertheless, many papers do not contain detailed petrographic descriptions of chemically analyzed rocks, and reliable modal analyses of mineral proportions are even more rarely supplied. Consequently, our discussion reflects the qualitative nature of the literature.

Bryan and others (1976) and Bryan and others (1979) summarized the textures, mineral growth forms, and mineral compositions of the first major penetrations of layer 2 in the

TABLE 2. PETROGRAPHIC CHARACTERISTICS OF UNALTERED SUBMARINE BASALT AND BASALT GLASS (after Bryan and others, 1976).

	Incompatible-Element Depleted		Incompatible-Element Enriched	
	Unfractionated	Fractionated	Unfractionated	Fractionated
Plagioclase	An_{90-80}; molecular Mg > Fe: K_2O < 0.02; zoning conspicuous.	An_{80-60}; molecular Mg = Fe; K_2O < 0.04 minor zoning.	An_{85-60}; molecular Mg < Fe; K_2O > 0.05; Marginal zoning conspicuous.	An_{60}; alkali feldspar molecular Mg < Fe; K2O > 0.10; prominent marginal zoning or alkali feldspar overgrowths.
Olivine	Fo_{90-85}; present as phenocrysts and groundmass phase and as skeletal crystals in glass.	Fo_{85-90}; present as phenocrysts, absent in groundmass or as quenched crystals in glass.	Fo_{85-90}; present as phenocrysts, rare in groundmass or as quenched skeletal crystals in glass.	Absent.
Pyroxene	Diopsidic augite(?), detectable only in groundmass, usually as skeletal crystals.	Aluminous diopsidic augite as phenocrysts and microphenocrysts, usually optically strained and sector-zoned.	Titaniferous augite, rarely as phenocrysts, and prominent in groundmass.	Titaniferous augite or ferro-augite as phenocrysts; distinct Ti and Al enrichment.
Spinel	Aluminous, magnesian spinel; microphenocrysts associated with olivine.	Absent or with distinct marginal reaction rims of magnetite overgrowths.	Rare or absent.	Absent.
Magnetite	Absent in glass; moderately abundant as skeletal groundmass crystals.	Absent in glass except as overgrowths on spinel; skeletal groundmass crystals may be abundant.	Rare as phenocrysts common to abundant in groundmass.	May form euhedral phenocrysts, prominent in groundmass.
Ilmenite	Absent	Absent	May be associated with magnetite in groundmass.	May be associated with magnetite in groundmass; rare as phenocrysts.

western North Atlantic. Data from more recent crustal penetrations by IPOD (International Phase of Ocean Drilling) legs 49 and 82 (Wood and others, 1979; Scientific Party, 1982) support earlier conclusions that most of the older basalt samples resemble those from the modern ocean ridge system. The similarities include textural and morphological growth features, such as sector zoning in plagioclase, ornate lattice and lantern-and-chain growth forms in olivine, and similarly ornate pinwheel or lattice growth forms characteristic of pyroxene (Bryan, 1972, 1974). Such features are almost unknown in air-quenched basaltic lavas and evidently are a consequence of higher cooling rates in water-quenched basalts (Fig. 3).

The mineralogy of "typical" MORB depends on its major element composition. Table 2 summarizes the mineralogy and petrography typical of fractionated and unfractionated depleted and enriched MORB. We emphasize that many samples are intermediate both in chemical characteristics and in degree of fractionation, with corresponding intermediate petrographic characteristics. Many of these transitional basalts are included in IPOD leg 37, leg 49, and leg 82 drill sites south and west of the Azores platform. Clinopyroxene composition and occurrence is especially sensitive both to degree of fractionation and to chemical type. Clinopyroxene appears as phenocrysts and microphenocrysts only in the relatively more fractionated basalts and becomes distinctly more enriched in TiO_2 and Al_2O_3 as the concentration of incompatible elements increases. Ca-poor pyroxene (pigeonite) reported in some highly fractionated ferrobasalts near Iceland (Sigurdsson, 1981) has yet to be found in older samples in the western North Atlantic. Clinopyroxene seems to occur at earlier stages of fractionation, and in greater abundance, in basalts enriched in incompatible elements (Schilling and others, 1982; Bryan, 1983). Olivine enrichment, leading to a picritic bulk composition, also is more common in these basalts (Bryan, 1983).

Petrography of Plutonic Rocks

The mineralogy and mineral compositions in gabbro and peridotite from the deep sea floor are similar to some, but not all, plutonic rocks from ophiolite complexes (Melson and Thompson, 1971; Prinz and others, 1976; Sinton, 1978; Dick and Bullen, 1984; Dick and others, 1984). The layered texture of some gabbros suggests that they may be residual cumulates associated with fractionation of basalt in deep chambers (Prinz and others, 1976; Elthon and others, 1982; Hodges and Papike, 1976; Dostal and Muecke, 1978), but some gabbros do not have a simple genetic relation to associated basalts (Dostal and Muecke, 1978; Tiezzi and Scott, 1980). Mineral assemblages in the gabbros usually differ from phenocryst assemblages in MORB in the following ways: pyroxene usually consists of both clino- and orthopyroxene; olivine and pyroxene are often more iron-rich than corresponding phenocrysts in MORB; and iron-titanium oxides form distinct, blocky crystals. Other phases, such as hornblende and apatite, may form discrete crystals (Hodges and Papike, 1976; Tiezzi and Scott, 1980). These characteristics partly reflect slower cooling at depth, leading to more complete crystallization

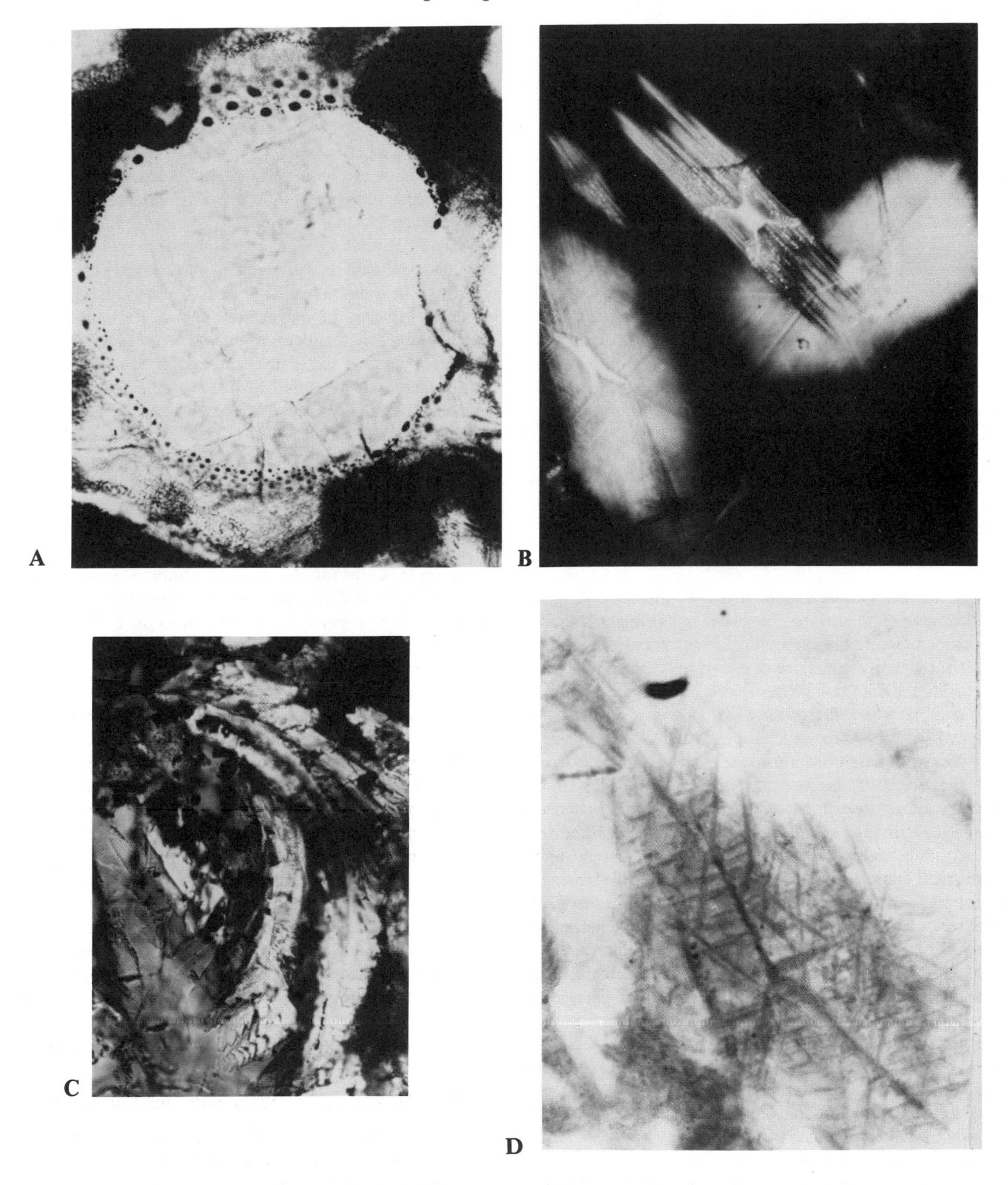

Figure 3. Morphology of quench mineral phases in ancient ocean ridge basalt from DSDP site 105. *A*, vesicle lined with oxidized sulfide globules and filled by calcite; *B*, quench olivine in fresh glass; *C*, pyroxene 'feathers' in crypto-crystalline groundmass; *D*, relict quench olivine in weathered glass (palagonite).

and equilibration compared to the early stages of crystallization quenched in as phenocrysts in extrusive basalts. Their relatively iron-enriched compositions and cumulate textures imply that most gabbros are cumulates related to fractionated MORB (Table 2).

Abyssal peridotites are mineralogically similar in many respects to alpine peridotites; the greater mineralogical diversity of the latter suggests they represent sub-continental and island arc environments as well as some abyssal settings (Dick and Fisher, 1984). Mineralogical evidence, and apparent correlations between abyssal peridotite mineralogy and the composition of associated MORB, suggests that at least some of the peridotites are true mantle samples which are residual products of partial melting (Sinton, 1978; Dick and Bullen, 1984; Dick and others, 1984). Dick and Fisher (1984) also noted that North Atlantic peridotites are more depleted (in the sense of being pyroxene and alumina-poor) than average abyssal peridotite. Dick and others (1984) show that this may be a consequence of the proximity of many of the North Atlantic peridotite localities to mantle plumes, or "hot spots," such as the Azores, that may have resulted in relatively high degrees of melting and melt extraction at these localities.

Special Mineralogical and Textural Features

Certain petrographic features in MORB that have been cited as indicators of tectonic setting or spreading rate might be used to constrain inferences about spreading rate or tectonic settings during earlier stages of the opening of the North Atlantic. It has been suggested that plagioclase phenocrysts are much more abundant in Atlantic MORB and that these phenocrysts are more calcic and may have crystallized at relatively greater depths, compared to plagioclase in basalts from the East Pacific Rise (Scheidegger, 1973; Flower, 1980). However, North Atlantic MORB are not invariably plagioclase phyric (ARCYANA, 1977; Bryan and Moore, 1977; Bryan, 1983), and basalts recovered in the very slow-spreading center in the Cayman Trough are sparsely phyric to aphyric (Thompson and others, 1978). Also, plagioclase-phyric basalts are interstratified with aphyric basalts at sites 332B, 395, 418A, and 417D (Blanchard and others, 1976; Dungan and others, 1978; Flower and Robinson, 1981). Therefore, although basalts enriched in plagioclase do seem more common in the Atlantic, plagioclase abundance probably is not an efficient discriminant for spreading rates. Staudigel and Bryan (1981) discussed crystal-liquid sorting in pillow-sized cooling units from sites 417D and 418A, and concluded that late-stage gravity settling, combined with dynamic sorting in flowing lava tubes, produce selective crystal accumulation. They caution that bulk rock petrography and compositional variations may be a poor indicator of *deep* crustal processes.

The magnetic signature in layer 2 (See also Vogt, this volume, Ch. 15) is almost entirely attributed to very finely divided titaniferous magnetite or its oxidation products. In spite of its importance in creating the magnetic signature, magnetite is almost never present as phenocrysts and the tiny groundmass granules are rarely analyzed. Data from western North Atlantic drill sites (Petersen and others, 1979) indicate that unaltered magnetite varies little in composition, being close to the formula:

$$(Fe^{2+})_{1.5}\,(Fe^{3+})_{0.77}\,Ti_{0.58}\,Al_{0.07}Mg_{0.06}\,Mn_{0.02}\,O_4.$$

Oxidation in older North Atlantic crust extends at least to the greatest depths penetrated (600 m), with the typical oxidation product being a titanomaghemite. This mineral is relatively enriched in Ti compared to the original magnetite, due to progressive iron loss during oxidation. Ilmenite is sparingly present as a primary oxide in some seafloor basalts, and may also appear as exsolution lamellae in magnetite oxidized at relatively high temperature. The intensity of magnetization of layer 2 decreases with increasing grain size of magnetite and with increasing Ti/Fe in the magnetite. In a detailed discussion of zero-age basalt at the FAMOUS (French-American Midocean Undersea Study) ridge axis and 3 my-old basalt at IPOD site 332, Prèvot and others (1979) showed that magnetic intensity in the FAMOUS area is reduced in the older basalts from 1/3 to 1/4 the intensity in the fresh basalts at the central volcanic axis. The difference is shown to result from both more intensive oxidation and higher Ti/Fe in the flank lavas. The very fine grain size of the disseminated magnetites in basalts from the central volcanic axis also contributes to this difference. Although as a group the oxidized site 332 basalts have lower magnetic intensities than the FAMOUS basalts, differences in Ti/Fe ratios and in grain size of oxide phases leads to the same relative differences in magnetic intensities between petrographic types that were found in the FAMOUS area. Prèvot and others (1979) also noted that, given the high magnetic intensity of the zero-age basalts, a layer 2 thickness of only about 500 m is needed to account for the magnetic amplitudes measured at the sea surface, while the much lower magnetic intensities at site 332 would require about 2 km of layer 2 to give the magnetic amplitude observed there. Petersen and others (1979), however, attributed most of the variation in magnetic amplitude to progressive oxidation of magnetite. They found that with increasing oxidation, saturation magnetization first decreases to a minimum value, then rises again, apparently at the point where Fe begins to be lost from tetrahedral instead of octahedral sites. They make the interesting suggestion that in the western North Atlantic this oxidation process may explain the high amplitudes of Cretaceous magnetic anomalies, while the Jurassic quiet zone may reflect the inversion of totally oxidized titanomagnetites to non-magnetic oxides.

Off-Ridge Igneous Events, Seamounts, and Fracture Zones

Very few recoveries of older seafloor in the western North Atlantic include material that is demonstrably an off-ridge intrusive or extrusive unit. The best examples include Leg 2 DSDP site 10, an alkaline basalt (or dolerite) substantially younger (as dated by overlying sediments) than the spreading age inferred by distance from the Mid-Atlantic Ridge and the local magnetic signature (Frey and others 1974); basalts and mafic intrusive rock recovered from rift mountains adjacent to the median valley (Aumento, 1971); samples recovered by drilling on Bermuda (Reynolds and Aumento, 1974); and samples collected by submersible and by drilling on and near the New England Seamounts (Houghton, 1979; Duncan, 1984).

Many igneous rocks recovered from off-ridge intrusive and extrusive settings resemble the most enriched MORB in composition and mineralogy (Frey and others, 1974; Shibata and Fox, 1975; Hekinian and Thompson, 1976). For example, the characteristic pyroxene is an aluminous, titaniferous augite, and plagioclase tends to be relatively sodic and contains significantly more K_2O than is usual for depleted MORB (Table 2). These alkaline basalts and their intrusive equivalents tend to have higher total FeO, a higher Fe_2O_3/FeO ratio, and higher TiO, than depleted MORB. These characteristics may be reflected in more iron-rich olivine and pyroxene, and in the appearance of magnetite and/or ilmenite as a phenocryst in the more fractionated varieties. Alkaline basalts also tend to have higher primary water contents; mineralogically this may be reflected in the presence of hornblende or biotite in addition to olivine and pyroxene. Also, the higher TiO_2, Fe_2O_3, and alkali contents result in significant compositional variety in these minerals; thus, amphiboles may include kaersutite, arfvedsonite, or riebeckite components; pyroxenes may include acmite.

It has been suggested that alkaline volcanism is characteristic of basalts erupted in fracture zones (Melson and Thompson, 1973). Relatively alkaline basalts are associated with the Oceanographer Fracture Zone and with small fracture zones in the FAMOUS area and at 43°N (Shibata and Fox, 1975; Bryan, 1979; Shibata, DeLong, and Walker, 1979; Shibata, Thompson, and Frey, 1979). However, basalts recovered along the transform sections of the Gibbs and Kane Fracture Zones are typical tholeiitic basalts (Hekinian and Aumento, 1973; Bryan and others, 1981). It is usually difficult to establish whether such basalts have been tectonically transported out along the fracture zone from the adjacent spreading centers, or whether they represent magmatic events located within the transform zone.

Compositional Variations in Layer 2 Basalts

Basalts formed along the modern MAR spreading center show substantial along-ridge compositional variation and this variability is especially well documented in the North Atlantic (Schilling and others, 1983; Melson and O'Hearn, this volume; Schilling, this volume). In general, depleted MORB characterize the MAR from 10°–34°N, 40°–43°N, 50°–60°N, and about 67°–71°N, whereas enriched MORB are found over the Reykjanes Ridge (60°–65°N), and near Jan Mayen (70°–73°N). The differences in abundance ratios of incompatible elements (e.g., La/Th, La/Sm, and Zr/Nb), and isotopic ratios (Sr, Nd, and Pb) along the MAR axis are believed to reflect compositional differences in the mantle sources. It is obviously important to determine whether these present-day along-ridge variations can be projected back in time along spreading "flow lines," or whether the geochemical differences vary rapidly in both time and space.

There are three regions in the western North Atlantic where there has been sufficient recovery of basalt to evaluate compositional variations as a function of age. (Note: Ages are inferred from magnetic anomalies, and the age of overlying sediment at each drilling site. Because of difficulties in obtaining reliable isotopic ages for submarine basalts, we assume that this method gives a reasonable estimate of the extrusion age.) From north to south the regions are: (1) 63°N—DSDP leg 49, site 407 (≅35 my), 408 (≅20 my), and 409 (≅2.3 my), which form a flow line emanating from the Reykjanes Ridge near the mid-point of the geochemical gradient defined by axial dredged basalts; (2) 33°–39°N, latitudes similar to and south of the Azores: DSDP leg 37, sites 332–335; DSDP leg 49, sites 411, 412 and 413; and DSDP leg 82, sites 556–564, ranging in age from ≅1 my to 35 my; (3) 22°–25°N: DSDP legs 45/46, sites 395 and 396 (7–9 my); DSDP leg 51/52/53, sites 417 and 418 (≅110 my); DSDP leg 43, site 386 (≅105 my); and DSDP leg 11, sites 100 and 105 (≅150 my) (see Table 1 and Plate III for site locations). Even older basalt (154 my) with major element chemical and petrographic characteristics of typical MORB, has been recovered at site 534 (Sheridan and others, 1982); it has not been included in our summary because detailed trace element data are not available.

Reykjanes Ridge Transect at ~63° N. Drill cores from this transect perpendicular to the Reykjanes Ridge (Plate III) show significant variation both between sites and among basalts recovered from a single site (Fig. 4; Wood and others, 1979; Tarney and others, 1979; Tarney and others, 1980). At site 407, the most distant from Iceland, all basalts have $(La/Sm)_n > 1$, and $^{87}Sr/^{86}Sr$ of acid-leached powders exceeds 0.7033. These characteristics of site 407 basalts, which are inferred to be older than 35 my, are similar to the majority of recent Icelandic basalts (O'Nions and others, 1976; Wood, 1978; Schilling and others, 1978; Zindler and others, 1979). However, at site 409, closest to Iceland, the basalts have the geochemical characteristics of depleted MORB (i.e., $(La/Sm)_n < 1$ and $^{87}Sr/^{86}Sr < 0.7030$). Consequently, it is apparent that the geochemical characteristics of uppermost Layer 2 basalts along this transect do not correlate simply with age (Fig. 4).

Moreover, some within-site geochemical variations are nearly as large as differences between sites along this transect. For example, at the youngest site, 409 (62°37′N, 25°57′W), approximately 58 flows in the 240 m basalt core can be divided into four stratigraphic units with differences in LREE/HREE abundance ratios and $^{87}Sr/^{86}Sr$ which probably reflect mantle source differences. The range in La/Sm within this site is greater than that along the Reykjanes Ridge axis from 62°30′N to 63°30′N (Fig. 4 and Schilling, this volume). Similar local heterogeneities occur in subaerial basalts on the Reykjanes Peninsula, Iceland, where nearly adjacent and coeval basalts range widely in incompatible element ratios and in some isotopic ratios (Fig. 4 and Wood, 1978; Schilling and others, 1978; Zindler and others, 1979). In summary, along this 63°N transect there is no simple geochemical

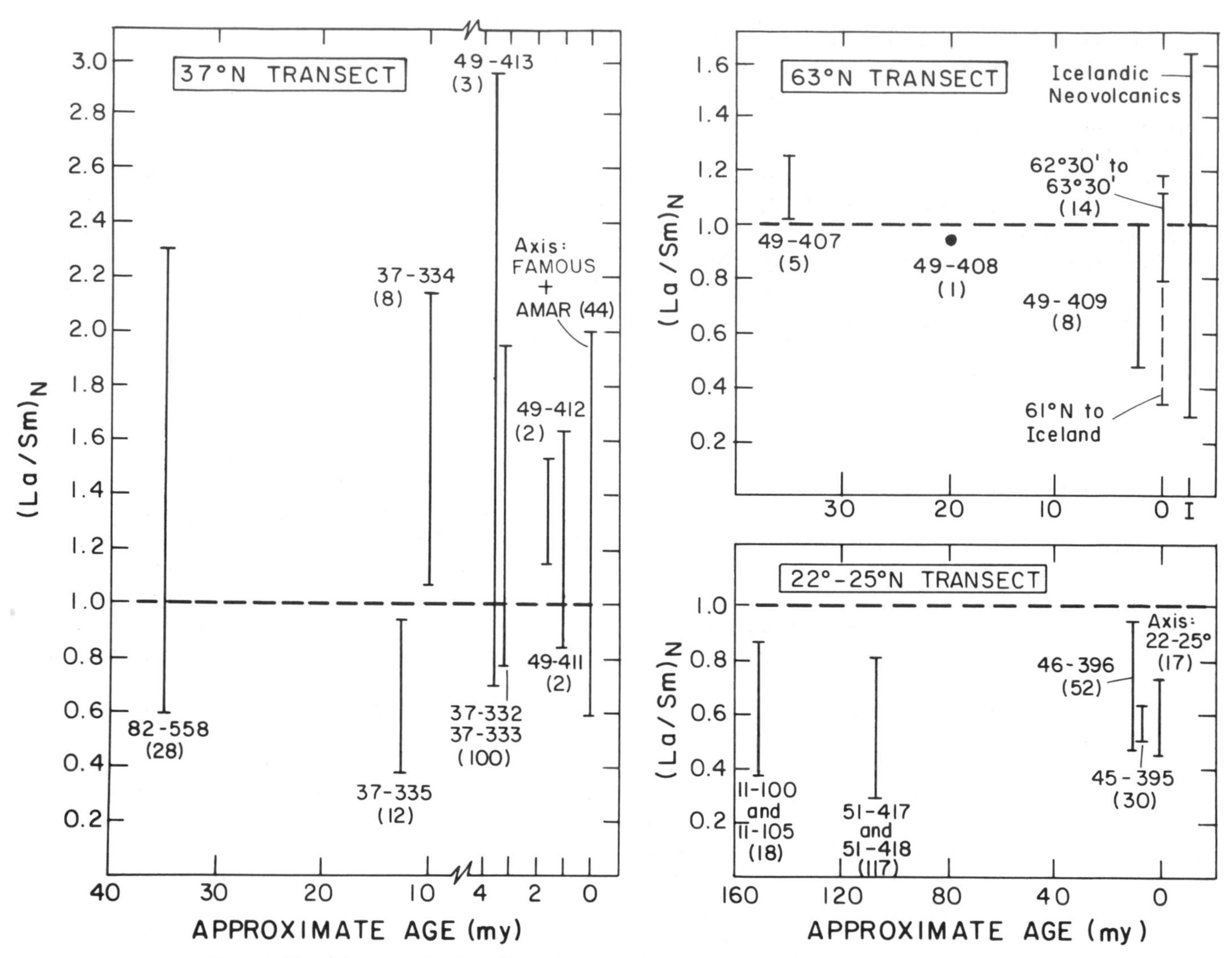

Figure 4. Chondrite normalized La/Sm ratios along three age transects in the western North Atlantic. Enriched MORB have $(La/Sm)_n > 1$ whereas depleted MORB have $(La/Sm)_n < 1$. Note that only depleted MORB occur along the 22°–25°N transect whereas both MORB types occur on the other transects. In particular, basalts from along the 37°N transect, south of the Azores, have a wide variety of $(La/Sm)_n$ ratios.

Data from pertinent Initial Reports of the Deep Sea Drilling Project plus axial FAMOUS and AMAR data from Langmuir and others (1977) and Frey (unpublished); axial 22°–25°N data from Bryan and others (1977); axial basalt data for Reykjanes Ridge from Schilling (1973) and data for neovolcanics from Iceland (designated by *I* along age axis) from Schilling and others (1978) and Zindler and others (1979). Length of bar indicates range of variation at each site; number of samples used to define range is indicated within parentheses.

evolution with time either on the time scale of a single hole (0.5 to 1 my) or for the 35 m.y. spanned by the transect.

MAR Transects Near the Azores. As of 1984, the most sophisticated sampling to determine compositional variations of Layer 2 basalts as a function of age and latitude was carried out by DSDP Leg 82 (Sept.–Nov. 1981). (Note: As of June 1984, the Leg 82 preliminary report has not been published. Our review of results from this leg is based on preprints for the initial report that were available in May, 1984.) This leg was designed to study the spatial and temporal extent of compositional heterogeneities in the central North Atlantic near the Azores platform (Scientific Party, 1982). Specifically, ten holes drilled west of the ridge axis combined with data from DSDP Legs 37 and 49 plus dredge samples from the MAR axis, form a coarse grid ranging in age from 0 to ~35 my along four flow lines emanating from the Azores platform, the FAMOUS region, ~35°N, and south of the Hayes fracture zone at ~33°N (Plate III). Although limited to <35 my, the most detailed transect is in the vicinity of the FAMOUS axial region south of the Azores (8 DSDP holes from 36°–38°N). In this area, the axial valleys have been intensively

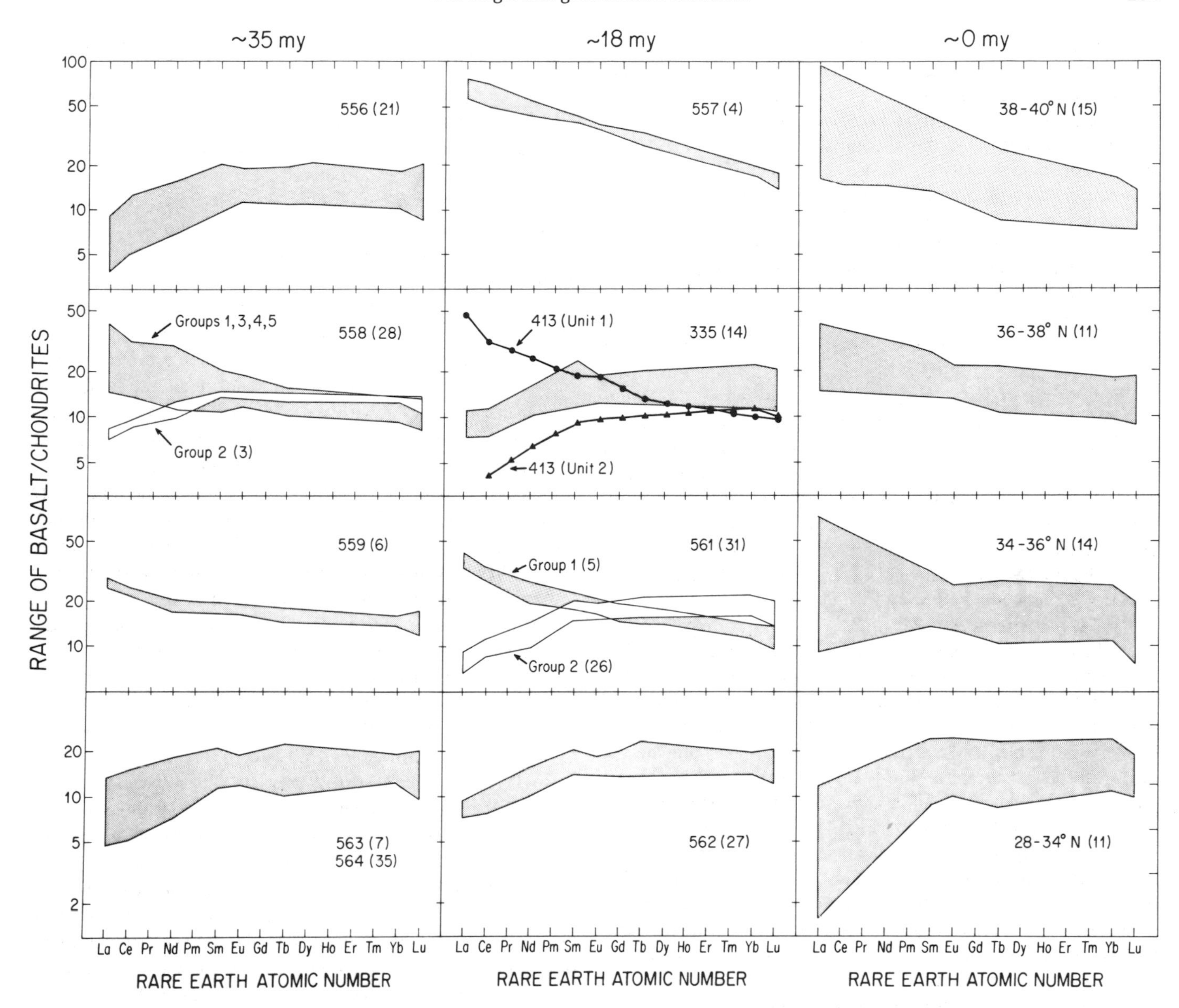

Figure 5. Range of chondrite-normalized REE patterns (shaded, or shaded and clear) for sites near and south of the Azores. Number of samples to define range is indicated within parentheses. Variations along a horizontal row represent temporal variations along a flow line from the spreading ridge axis (latitude indicated in right column for axial basalts). Intersite variations along a vertical column represent spatial variations along the ridge axis at a particular time (35 my, 18 my, and 0 my). Note that enriched and depleted ($[La/Sm]_n > 1$ and <1, respectively) basalt occurs along the 36°–38° transect at 3.5 my (leg 49, site 413 plotted in 18 m.y. column), 35 m.y. (leg 82, site 558), and along the 34°–36° transect at 19 m.y. (leg 82, site 561). Data sources: axial basalts, Schilling and others, 1983; DSDP sites, DSDP Initial Reports for Legs 37, 49, and 82.

studied and a geochemical gradient similar to that existing along the Reykjanes Ridge is well-defined (Schilling and Melson, this volume). The transect formed by DSDP sites 412 and 413 east of the median valley and sites 332–335, 411, and 558 west of the median valley, intersects the ridge axis in a region with geochemical characteristics transitional between depleted and enriched MORB.

The basalts along this transect illustrate the complexity that must be explained by petrogenetic models for MORB. For example, some basalts from each site, except site 335, have enriched geochemical characteristics, such as $(La/Sm)_n > 1$, akin to the enriched axial basalts (Figs. 4 and 5). However, as in the Icelandic transect, the variations in key geochemical parameters are not a simple function of age; for example, basalts from site 335 ($\cong$13 m.y.) are depleted MORB, (i.e., $[La/Sm]_n < 1$), whereas the older ooroo from oito 558 (~35 my) contain both depleted and enriched

MORB (Figs. 4 and 5). The relatively short core (39.5 m) at site 413 exemplifies the geochemical variability that is found at some DSDP sites; units 413-1 and 413-2 have similar and relatively high MaO (≅10.5%), Ni, and Cr contents, but differ by a factor of five in abundances of highly incompatible elements (Fig. 5 and Wood and others, 1979). At leg 37, site 332, there is no correlation of lava composition with stratigraphic position (Byerly and Wright, 1977; Flower and Robinson, 1979). Detailed petrologic and geochemical studies of 70 samples (Blanchard and others, 1976) show that generation of multiple magmas from heterogeneous mantle source regions is required to account for the range of basalt compositions within hole 332B.

Slightly north of the FAMOUS transect at ~38°–40°N, Leg 82 sites 556 and 557 combined with dredged samples from the ridge axis form another transect. Basalts from site 557 (~18 my) are similar to the enriched basalts found at the ridge axis, but basalts from site 556 are depleted MORB (Fig. 5). In contrast, all three localities forming the transect at 34°–36°N contain enriched MORB (Fig. 5), although at site 561, only the upper 3 meters of the 15 m core contain enriched MORB; most of the site 561 cores are depleted. Finally, south of the Hayes fracture zone (28°–34°N), basalts ranging in age from very recent at the axis to ~35 my (Leg 82 sites 563 and 564) are all depleted MORB (Fig. 5).

MAR Transect at ≅22°–25°N. The largest range in age is formed by an approximate transect consisting of near axis DSDP sites 395 (≅7 my, west of the axis) and 396 (≅10 my, east of the axis): and the distant sites 417 and 418 (≅110 my); DSDP sites 100, 105 (≅150 my); and site 534 (≅154 my), which are all west of the axis (Table 1; Plate III). Sites 100 and 105 provide the oldest samples of western North Atlantic seafloor for which detailed data are available. All basaltic samples obtained along this transect are depleted MORB (Fig. 4). This demonstrates that magmas with depleted geochemical characteristics erupt early in the construction of an oceanic plate (Ayuso and others, 1976; Bryan and others, 1977; Bryan and others, 1979). There are no indications that spreading-center basalts formed during the early history of the Atlantic are necessarily enriched in incompatible elements relative to basalts forming more than 150 my later along the same ridge axis. Therefore, a renewable magma source is required at the spreading ridge axis.

The extensive basement penetration at sites 395 and 396 (255 m and 355 m) and at sites 417 and 418 (206 m, 365.5 m, and 544 m) permit more detailed evaluation of compositional changes in depleted MORB as a function of extrusive age. As at site 332b, there are no systematic compositional trends with depth but crystal segregation and mixing appear to have modified several distinct parental liquids (Dick and Bryan, 1978; Staudigel and Bryan, 1981; Flower and Robinson, 1981). Detailed studies of basalts from sites 395 and 396 provide convincing evidence that magma mixing was a major petrogenetic process (Dungan and Rhodes, 1978; Rhodes and others, 1979; Kuo and Kirkpatrick, 1982).

Chemical and Mineralogical Alteration of Layer 2

Early studies of dredged basalts from the western North Atlantic suggested a simple, progressive "alteration aging" with time, and led to the view that "old" ocean crust must inevitably be heavily altered (e.g., Hekinian, 1971). This simple concept has been drastically complicated by studies of basalts recovered in deep drill holes. Honnorez (1981) concluded that there is no single, simple chemical or mineralogical index of the "degree" or intensity of alteration, since the nature of the changes depends on complex and variable interactions between temperature, rock/water ratios, and length of time the rock is exposed to seawater. In particular, rocks in older crust accessible to dredging must necessarily be atypical, superficial exposures that have had much longer continuous exposure to seawater than the deeper crustal layers covered by marine sediments. Drill core studies have shown that the amount of low temperature alteration does not vary in any simple way with age or depth in the crust (Honnorez, 1981; Thompson, 1983). This is illustrated most dramatically at site 417, in 110 Ma crust south of Bermuda. Hole 417A was drilled 206 m into a highly altered basement while hole 417D, only 450 m from 417A, recovered fresh to only moderately altered basalt to a depth of 365 m. The highly altered core at site 417A reflects localized intense low temperature alteration, which required interaction with large amounts of water, perhaps as a result of passive, upward hydrothermal circulation on the ridge flank (Donnelly and others, 1979; Thompson, 1983).

In general, there are only minor differences in type and degree of low temperature alteration in basalts from the relatively young DSDP sites 332 (3.5 my) and 335 (13 my) and the older (110 my) basalts from holes 417D and 418A. The deep core at hole 418A has been used extensively to evaluate the effects of low temperature basalt-seawater interactions. For example: (1) The abundant, unaltered glass in this core shows that glass may survive for >100 my at depths of tens to hundreds of meters (Byerly and Sinton, 1979). This allows the unaltered major element compositions of basaltic liquids emplaced more than 100 my ago to be determined. (2) The abundance of glass and its complex alteration product "palagonite" in the core enabled detailed evaluation of elemental fluxes created by alteration of glass (Staudigel and Hart, 1983). (3) The effects and duration of low temperature alteration beneath the sediment-basalt interface have been estimated (Richardson and others, 1980; Hart and Staudigel, 1982; Thompson, 1983).

Extensive metamorphism of older oceanic crust was anticipated by analogy with ophiolites, and from greenstones dredged from the walls of the Mid-Atlantic Ridge rift valley and rift mountains, especially at 22°N and 45°N (Melson and others, 1968; Aumento and others, 1971). However, surprisingly few metamorphosed basalts have been recovered by drilling (Cann, 1979; Thompson, 1983) nor do the deepest holes at sites 332, 395, and 418 show any consistent increase in "grade" of metamorphism with increasing depth. To date in the northwest Atlantic, rocks metamorphosed within or above the greenschist facies have been recovered locally only in and near the median valley, and along scarps in major fracture zones (e.g., Thompson, 1983). The absence of metamorphic rocks in the northwest Atlantic drill

TABLE 3. METAMORPHIC FACIES AND REPORTED MINERALS OBSERVED IN EACH FACIES (AFTER THOMPSON 1983).

Facies	Mineralogy
Halmyrolysis	Celadonite, phillipsite, palagonite, saponite, montmorillonite, nontronite, Fe-Mn hydroxide, orthoclase.
Zeolite	Analcite, stilbite, heulandite, natrolite-mesolite-scolectite, chlorite-smectite, saponite.
Prehnite-pumpellyite	Prehnite, chlorite, calcite, laumontite, epidote.
Greenschist	Albite, actinolite, chlorite, epidote, quartz, sphene, hornblende, tremolite, talc, magnetite, nontronite.
Amphibolite	Hornblende, plagioclase, actinolite, leucoxene, quartz, chlorite, apatite, biotite, epidote, magnetite, sphene.

holes is almost certainly a result of their shallow basement penetration (550 m). The deepest DSDP hole, 504B on the Costa Rica rift penetrated 1075 m into basement rock and encountered greenschist facies basalt at depths >889 m below the seafloor (Anderson and others, 1982, Fig. 4).

Mineralogical Changes Resulting From Alteration and Metamorphism. The interaction of crystalline basalt and seawater at low temperatures leads to the following mineralogical changes (Honnorez, 1981): (1) Oxidation of magnetite, with formation of maghemite; (2) Precipitation of celadonite and nontronite in vesicles and along walls of fractures; (3) Oxidation of olivine to "iddingsite," a complex mixture of clays and iron oxides; (4) Continued oxidation with formation of smectites, and deposition of carbonates and zeolites in vesicles and fissures; (5) As fissures and voids become sealed, a more pervasive, mildly oxidative alteration continues, and may involve growth of secondary sulfides, sulfates, and magnesium-rich carbonates.

Low temperature alteration of basaltic glass differs from alteration of crystalline basalt. Alteration of glass produces "palagonite," a mixture of zeolite, phillipsite, smectites, and Fe-Mn oxides (Honnorez, 1981). Although glass in some dredged pillow basalts with ages of only a few million years has been extensively palagonitized (Thompson, 1983), unaltered glass occurs even in some of the oldest Atlantic DSDP cores (Leg 11 site 105, 150 my; and sites 417 and 418 of Legs 51–53, 110 my).

Hydrothermally altered rocks have not been recovered from older parts of the western North Atlantic, although they are presumed to exist at depth. Cann (1979) notes that most altered seafloor rocks belong to what he calls the "brownstone" facies, named for the brownish or yellowish tint imposed by oxidative low temperature alteration. With increasing temperature the "brownstone" facies passes into the zeolite facies, characterized by saponite-chlorite replacement of olivine, and replacement of plagioclase both by zeolite and by saponite-chlorite assemblages (Table 3). The greenschist facies and the development of true greenstones is marked by the appearance of albite and chlorite to the exclusion of zeolites and saponite; plagioclase may be partly replaced by epidote, and augite by actinolite (Table 3); in gabbros or peridotites, which contain large accumulations of olivine, the olivine may be largely or completely replaced by talc or serpentine. In contrast to the common development of true schistose texture in metamorphosed continental volcanic rocks, deep-sea basalts and plutonic rocks usually preserve relict igneous textures (Cann, 1979). Metamorphism extending to the amphibolite facies has been found only in dredged gabbros, probably reflecting their greater depth of formation and residence within layer 3. In these samples, pyroxene is replaced in whole or in part by green hornblende, while plagioclase retains its original igneous character.

Chemical Changes Resulting From Alteration and Metamorphism. Studies of western North Atlantic DSDP drill cores have shown that interactions between basalt and seawater are highly variable in space and time, and that the chemical composition of ocean floor basalt also changes significantly during alteration and metamorphism. The interactions of basalt with seawater and hydrothermal fluids also may be significant in the geochemical cycles of several elements. In order to evaluate these reactions as a net sink or source for various elements in seawater, elemental fluxes must be determined for a range of temperatures and for different water/rock ratios. It is difficult to determine elemental fluxes for the basalt-seawater system (net transport of elements/unit time between rocks and seawater) because the volume of ocean floor affected by each reaction in a given time interval must be known. At present this can only be evaluated by gross approximations (Thompson, 1983; Thompson, 1984). Table 4 illustrates two attempts at making such estimates.

In general, low temperature alteration of both crystalline basalts and their quenched glasses leads to loss of SiO_2, CaO, and MgO and enrichment in alkalies and various incompatible trace elements such as Rb, Cs, B, and Ba. H_2O, CO_2, Fe_2O_3/FeO, and radiogenic isotopes such as ^{87}Sr and ^{18}O are added to the rock

TABLE 4a. ESTIMATES OF HYDROTHERMAL FLUXES BETWEEN OCEANIC BASEMENT AND SEAWATER (AFTER THOMPSON, 1983).

	x 10^{14} g/yr				x 10^{10} g/yr			
	Si	Ca	Mg	K	B	Li	Rb	Ba
Case A								
Surface	-0.006	-0.045	-0.03	+0.013	+0.45	+0.44	+0.14	+0.43
Basins	-0.52	-0.082	-0.26	+0.09	+2.69	+2.42	+1.37	+2.73
Flanks	-0.2	-0.47	-0.11	+0.22	+5.12	+3.7	+4.23	+1.1
Axis	-0.87	-1.3	-1.87	-0.49	-?	-111	-20.5	-46
Total	-1.60	-1.90	-1.47	-0.17	+8.26	-104.54	-14.6	-41.74
River	-1.99	-4.88	-1.33	-0.74	-47.0	-9.4	-3.2	-137.3
Basement Flux as % of River Flux	80.4	38.9	110.5	23.0	17.6	1112	461	30.4
Case B								
Surface	-0.006	-0.045	-0.03	+0.013	+0.45	+0.44	+0.14	+0.43
Basins	-0.52	-0.082	-0.26	+0.09	+2.69	+2.42	+1.37	+2.73
Flanks	-0.2	-0.47	-0.11	+0.22	+5.12	+3.7	+4.23	+1.1
Axis	-0.087	-0.13	+1.0	-0.049	-?	-11.1	-2.05	-4.6
Total	-0.82	-0.73	+0.6	+0.27	+8.26	-4.54	+3.69	-0.34
River	-1.99	-4.88	-1.33	-0.74	-47.0	-9.4	-3.2	-137.3
Basement Flux as % of River Flux	41.2	14.9	45.1	36.5	17.6	48.3	115.3	0.2

Note: Negative sign indicates removal of element from the crust; positive sign indicates addition of element to crust. Total flux is computed as the sum of fluxes in four environments: Surface represents flux associated with low-temperature surface alteration of exposed basement rocks (uppermost 10 meters); Basin represents low-temperature alteration of buried basement (upper 500 meters); Flank represents intermediate-temperature alteration at ridge flanks and basement highs affected by off-ridge hydrothermal activity; and Axis represents high-temperature hydrothermal alteration at ridge axis. Case A is identical to Case B, except that Axis flux in Case B is an order of magnitude less than in Case A, for elements removed from crust, as suggested by Hart ahd Staudigel (1982) and others. Note that this substantially changes total flux for most elements, illustrating the need to clarify parameters on which such calculations are based.

(Frey and others, 1974; Thompson, 1983; Staudigel and Hart, 1983). The overall effect is to give the sample a more "alkaline" composition, and a mildly altered depleted MORB might be confused with an enriched MORB unless the data are carefully evaluated. Under mild alteration rare earth elements are very little affected, but more intense alteration may affect both their absolute concentrations and relative abundances (Ludden and Thompson, 1979; Staudigel and Hart, 1983). Other elements such as Al, Ti, Y, and Zr are relatively immobile during low temperature alteration (Frey and others, 1974; Thompson, 1983).

However, in detail the effects of alteration depend on the temperature of alteration, rock/water ratios, and whether the original material is glass or crystalline (Honnorez, 1981; Mottl, 1983; Staudigel and Hart, 1983; Thompson, 1983). At higher temperatures, MgO may be gained by the rock, while K_2O and elements such as Li and B may be lost (Humphris and Thompson, 1978). Water/rock ratios can modify or even reverse the effects of temperature. Mottl (1983) showed that Na_2O is added by hydrothermal fluids at low water/rock ratios, but it is leached from the rock at high water/rock ratios. He also showed that the greenschist facies quartz-epidote-albite-chlorite mineral assemblages simplify to chlorite-quartz as water/rock ratios increase (Fig. 6). Further complications are introduced by the possibility that some rock masses may be subjected to several different alteration environments as the seafloor evolves. For example, an initial intense high temperature hydrothermal alteration in the median valley could be followed by exposure to cold seawater due to faulting and uplift in the rift mountains, and then by burial under accumulating sediment; leading to further alteration at moderate temperatures and low water/rock ratios. Given the demonstrable variability of "brownstone" alteration (or metamorphism) within single drill holes and between closely associated drill holes, the nature and extent of greenschist and amphibole facies metamorphism within layers 2 and 3 can not be inferred.

Applying the principles discussed above, however, some general conclusions are possible about the nature of alteration in the western North Atlantic (Table 4). Compared to fresh ocean

TABLE 4b. NET EXCHANGE BETWEEN BASALT AND SEAWATER CONSIDERING THE FULL RANGE OF TEMPERATURE REACTIONS.

Basalt as Source	Basalt as Sink
Si	Mg
Ca	K
Ba	B
Li	Rb
-----x	-----x
Fe	H2O
Mn	Cs
Cu	U
Ni	
Zn	

Note: Element fluxes above line "x" are calculated; fluxes below the line are estimated (after Thompson, 1983).

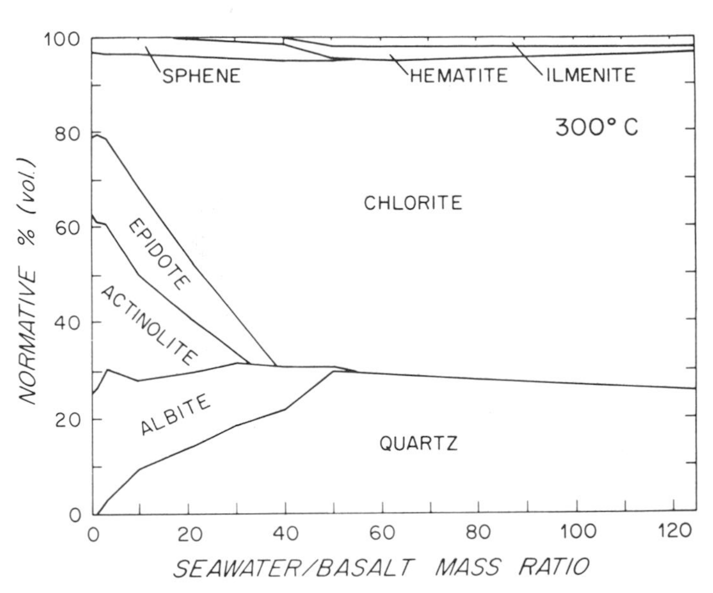

Figure 6. Mineralogical changes in greenschist-facies basalt or gabbro as a function of rock-water ratio, after Mottl, 1983.

ridge basalt, the older parts of layer 2 will have generally higher alkalies, H_2O, and Fe_2O_3; while SiO_2, CaO, and MgO may be lower; and Sr isotopic ratios will be more radiogenic. Hydration and oxidation produce mineralogical changes that will lower rock densities, but filling of cracks by carbonate and smectite will compensate this to some extent. Although most of the major alteration effects will have been completed in the first 10 my of crustal evolution, the changes will not be uniformly pervasive, and relatively fresh basement rocks may persist at shallow levels within layer 2, even in Cretaceous and Jurassic seafloor adjacent to the continental slope. Alteration effects associated with seawater circulation extend to at least the greatest depths penetrated into layer 2 (≅584 m), but show no progressive, systematic changes with depth, except for a tendency to a more oxidized state in the uppermost few meters.

PETROGENETIC INTERPRETATION OF BASALT GEOCHEMISTRY IN THE WESTERN NORTH ATLANTIC

Differences in geochemical parameters, such as isotopic ratios and abundance ratios of highly incompatible elements, which are not readily affected by crustal processes, provide compelling evidence for substantial compositional variations in the upper mantle beneath the western North Atlantic. Below, we summarize: (1) the geochemical variations in specific drill sites and along western North Atlantic transects; (2) interpretations for the cause of these geochemical variations; and (3) implications for the oceanic mantle beneath the western North Atlantic.

Geochemical Variations Within a Basaltic Drill Core

MORB with diverse trace element and isotopic characteristics occur in several DSDP sites, especially, 409, 413, 558 and 561 (Figs. 4, 5, 7, 8 and 9). There are three general ways to explain compositional diversity in basalts from a single drill core or dredge haul: (1) the mantle source of the basaltic magmas may have been heterogeneous; (2) the melting process may have been complex so that a wide variety of compositions was derived from a compositionally homogeneous source; (3) the magmas may have been affected by a variety of crustal processes such as segregation of solids from melts, magma mixing, and assimilation of wall rock within a relatively shallow magma chamber. These mechanisms are not mutually exclusive.

O'Hara and Mathews (1981) presented a theoretical evaluation of the effects of complex magma chamber processes on geochemical characteristics of erupted magmas. There is no doubt that such processes have affected the compositions of some ocean floor basalts; however, it is unlikely that crustal processes created the observed variability in Sr, Nd, and Pb isotopic ratios and in abundance ratios of highly incompatible elements. This is particularly true in regions where the isotopic and incompatible element differences occur in relatively primitive basalts (i.e., Mg/(Mg+Fe) = 0.65–0.70, Ni > 200 ppm) as at DSDP sites 332, 413, 561 and the FAMOUS axial region (Blanchard and others, 1976; Langmuir and others, 1977; Wood and others, 1979; Bougault and others, 1985).

Usually, heterogeneity in abundance ratios of incompatible trace elements is accompanied by isotopic heterogeneity; for example, basalts from the western North Atlantic have a general positive correlation between $^{143}Nd/^{144}Nd$ and $(Sm/Nd)_n$ (Fig. 7). However, there are important deviations from this positive trend; (1) the approximately 108 my old basalts from site 417D and 418A have relatively high but variable ϵ_t^{CHUR} (6.9 to 9.5; see Fig. 7 caption for "ϵ_t" definition) despite a nearly uniform $(Sm/Nd)_n$ ratio of 0.36 to 0.39; (2) in contrast, four site 558 samples with $\epsilon_t^{CHUR} \simeq 7$ have widely varying $(Sm/Nd)_n$ (0.26 to 0.42). Also, the basaltic suite from Leg 82 shows poor coherence between Pb isotopic ratios and incompatible element abundances (Bougault and others, 1985). In addition, in the vicinity of the

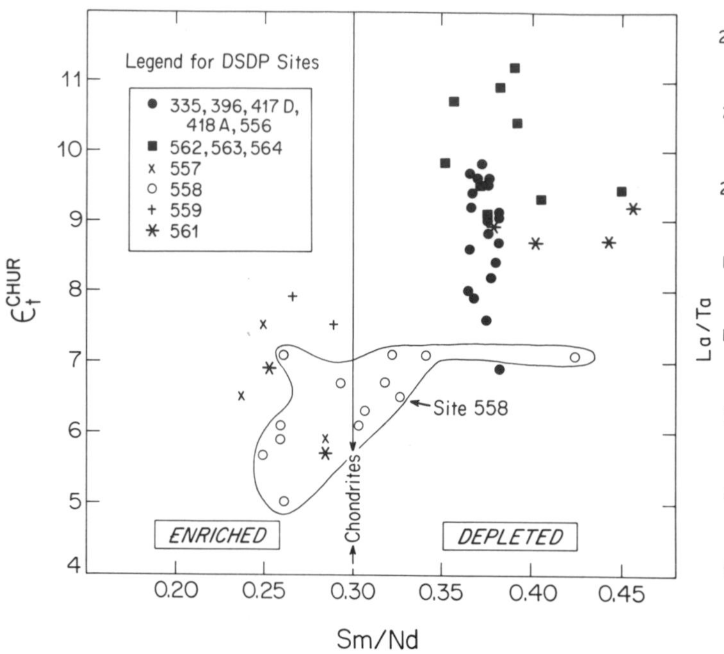

Figure 7. ϵ_t^{CHUR} versus Sm/Nd showing the general correlation of Sm/Nd with $^{143}Nd/^{144}Nd$ in basalts recovered by deep-sea drilling in the western North Atlantic. Note that the positive ϵ_t^{CHUR} reflects a time integrated $(Sm/Nd)_n > 1$ for all these MORB; that is, even the enriched MORB with Sm/Nd <0.3 ($[Sm/Nd]_n < 1$) evolved in sources with $(Sm/Nd)_n > 1$. The notation is the initial isotopic ratio of a sample formed "t" years ago as fractional deviations from the isotopic ratio of a uniform reservoir at the same time. "CHUR" means "chondritic meteorites uniform reservoir"; there is good evidence that the bulk earth has chondritic Sm/Nd and $^{143}/^{144}Nd$ ratios.

$$\epsilon_t^{CHUR} = \frac{(^{143}Nd/^{144}Nd)^{MORB} - (^{143}Nd/^{144}Nd)^{CHUR}}{(^{143}Nd/^{144}Nd)_t^{CHUR}} \times 10^4$$

Data from Jahn and others, 1980; Cohen and O'Nions, 1982; and Bougault and others, 1985.

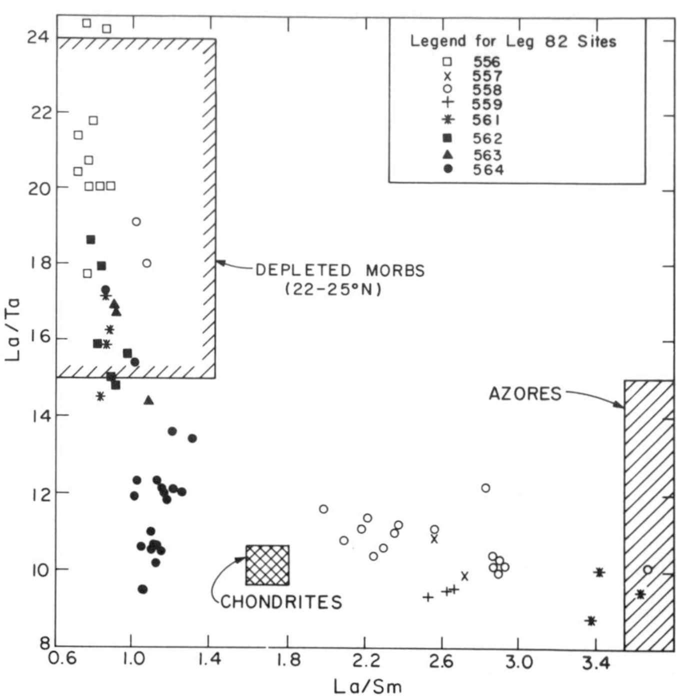

Figure 8. La/Ta versus La/Sm for leg 82 basalts illustrating overall correlation between the ratio of two highly incompatible elements, La/Ta, and a ratio, La/Sm, involving a moderately incompatible and a highly incompatible element. Typical depleted MORB has La/Sm <1.6 ($[La/Sm]_n < 1$) and $(La/Ta)_n > 15$. Shown for comparison are: (1) range for basalts from the Azores (43 samples plus 1 sample, not plotted, from Terceira that has La/Ta = 20.5; Flower and others, 1976; White and others, 1979); (2) range for most MORB in the 22°–25°N transect, data from figures 4 and 9; (3) estimated range for chondrites (Bougault and others, 1985). Data for leg 82 from DSDP Initial Report.

Azores large differences in ratios of incompatible element abundances occur among basalts which are nearly isotopically homogeneous. This is particularly well established in the FAMOUS region where the basalts are unaltered (White, 1979; Dupré and others, 1981). In older, somewhat altered basalts, $^{87}Sr/^{86}Sr$ is not a reliable indicator of magmatic characteristics because it is affected by rock-seawater reactions (e.g., O'Nions and Pankhurst, 1976; Jahn and others, 1980; Cohen and O'Nions, 1982); however, studies of Leg 37 basalts suggest that large variations in LREE/HREE are not always accompanied by variations in $^{87}Sr/^{86}Sr$ (O'Nions and Pankhurst, 1976). Thus, there is considerable complexity within the general correlation between radiogenic isotopic ratios and incompatible element abundance ratios (e.g., Fig. 7).

These results have led to models for complex melting processes such as dynamic melting (Langmuir and others, 1977) whereby melts are continuously generated and segregated from an ascending mantle diapir that was initially homogeneous in composition. This continuous segregation of melts, combined with incomplete removal of melt, creates compositional heterogeneity in the diapir so that later magmas derived from the diapir reflect compositionally distinct sources. Because the complex melting process is rapid relative to the half-lives of long-lived, naturally occurring radiogenic Rb, Sm, Th, and U isotopes, this process can lead to basalts isotopically homogeneous in Sr, Nd, and Pb, with incompatible element abundance ratios spanning the range from depleted to enriched MORB. Because dynamic melting is capable of producing a wide range of enrichment, it has been invoked to explain complex geochemical variations in several basalt suites (e.g. Wood, 1979; Wood, 1981). Although it is probably more realistic than simple batch or fractional melting models, the necessity of dynamic melting to explain geochemical variations in basalts at FAMOUS and site 558 has been questioned (Flower, 1981; Le Roex and others, 1981; Bougault and others, 1985).

At some DSDP sites, the basalts are isotopically heterogeneous. For example, site 561 contains enriched and depleted MORB that range from 0.7029 to 0.7040 in $^{87}Sr/^{86}Sr$; from 0.51312 to 0.51294 in $^{143}Nd/^{144}Nd$; from 38.31 to 39.28 in $^{208}Pb/^{204}Pb$; and from 18.76 to 19.53 in $^{206}Pb/^{204}Pb$ (Fig. 7 and

Bougault and others, 1985). Such localized heterogeneities have been interpreted as reflecting distinct mantle sources either as (1) incompletely mixed large-scale mantle components, such as depleted asthenosphere and enriched plume material (Bougault and others, 1985; Schilling, this volume); or (2) a mantle source that has been veined by introduction and crystallization of a fluid phase (Hanson, 1977; Wood, 1979; Tarney and others, 1980). Veining is a plausible process for creating long-lived, localized compositional and isotopic heterogeneities in the mantle. During melting of the veined mantle, the vein/host proportions may differ in various magma batches, thereby creating compositionally and isotopically heterogeneous magmas. If these distinct magmas do not subsequently mix, these heterogeneities will be preserved and localized heterogeneities will result, such as in the site 561 drill core.

Although a veined source commonly has been invoked to explain the occurrence of enriched MORB, depleted MORB also have considerable compositional variations. For example, highly depleted MORB are characterized by $(La/Ta)_n$ of 15 to 24 and $(La/Sm)_n$ <1 while enriched MORB have $(La/Ta)_n$ ~9 with $(La/Sm)_n$ >1 (Fig. 8 and Bougault and others, 1979; Bougault and others, 1985). However, MORB with $(La/Sm)_n$ <1 at Leg 82, site 564 have $(La/Ta)_n$ = 9 to 14 that overlaps with the ratios in enriched MORB (Fig. 8). Also, Jahn and others, (1980) found relatively small but significant variations in $^{143}Nd/^{144}Nd$ and $(La/Sm)_n$ among depleted site 417–418 basalts, which they attributed to mantle heterogeneity. A veined mantle can conceptually explain compositional variations within both depleted and enriched MORB because the amount and composition of the veins may vary on a local scale. Unfortunately, like dynamic melting, the wide variety of geochemical characteristics conceptually permitted by a veined mantle make this model difficult to test. However, mantle samples exposed in the crust are commonly veined (Wilshire and Pike, 1975) so the model is realistic.

Geochemical Variation as a Function of Latitude

It is apparent from the previous discussion that evaluation of geochemical variations as a function of latitude (see also Plates 8A, 8B) is complicated by the extreme geochemical variability occurring within some individual drill sites. For example, compositional variations with latitude do not greatly exceed those occurring along a transect or even variations within a single drill core such as at site 561 (Figs. 4, 5, 7, 8 and 9). Extensive sampling in three dimensions will be required to fully evaluate geochemical variations as a function of location and age. Nevertheless, it is already apparent that geochemical variations with latitude are present in older MAR basalts.

In order to distinguish the effects of fractionation and melting processes from source heterogeneities, abundance ratios of highly incompatible elements and isotopic ratios involving radiogenic isotopes must be utilized. Evidence that the $(La/Sm)_n$ and LREE/HREE variations in MORB from the western North Atlantic (Figs. 4 and 5) reflect mantle heterogeneities is provided by the correlation between $(Sm/Nd)_n$ and $^{143}Nd/^{144}Nd$ (Fig. 7). However, for MORB older than 1 Ma, the data base for trace element abundances is significantly larger than that for isotopic ratios; consequently, incompatible trace element abundances have been extensively used to evaluate mantle heterogeneity. During the petrogenesis of MORB, the following order of decreasing compatibility has been proposed:

$$D^{Y} \cong D^{Tb} > D^{Ti} > D^{Sm} \cong D^{Zr} \cong D^{Hf} > D^{La} = D^{Ta} \cong D^{Nb} > D^{Th}$$

(i.e., the bulk solid/melt partition coefficient (D) for Th is smaller than all others; Bougault and others, 1979; Sun and others, 1979; Wood, 1979). The large amount of data for incompatible element abundance ratios in older MORB define geochemical trends as a function of latitude. Ratios involving elements of very similar incompatibility, such as Nb/Ta, Zr/Hf, and Y/Tb are not correlated with latitude in the western North Atlantic (Fig. 9a); however, ratios involving elements of moderately different incompatibility do vary with latitude (e.g., Hf/Ta and Zr/Nb in Fig. 9b) and even ratios among incompatible elements with only slightly different degrees of incompatibility vary as a function of latitude. Specifically, the average La/Th and La/Ta ratios increase from 45°N, site 410, to the Azores transect (37°N), to the Icelandic transect (63°N), with the highest ratios in the 22°–25°N transect (Fig. 9b).

There is no evidence that these latitudinal variations in abundance ratios among highly incompatible elements such as Th, La, Ta, and Nb are caused by oceanic crustal processes, and in theory they can be caused by partial melting processes only if the bulk-solid/melt partition coefficient (D) is similar to or greater than the degree of melting. Because there is good evidence that D's for these elements are <<0.1, fractionation of these element ratios can only be caused by very small degrees of melting (<1%). Moreover, these ratio differences persist despite large variations in trace element abundances (Fig. 9b), which is not in accord with melting a homogeneous source to different degrees. Therefore, the consensus opinion is that the variations in La/Ta and La/Th in western North Atlantic MORB reflect large-scale geochemical variations in the mantle source of these MORB.

Schilling (this volume) summarized the geochemical characteristics of recent axial basalts in the North Atlantic and emphasized the systematic long wavelength variations in geochemical parameters, such as $(La/Sm)_n$ and $^{87}Sr/^{86}Sr$, as a function of latitude. Because large-scale gradients persisting over hundreds of kilometers are not readily explained by models that invoke complex melting, crustal processes, or veined sources, the generally accepted model for such gradients is mixing between plume-related enriched mantle and depleted asthenosphere. However, in most oceanic regions more than two isotopically distinct components are required to account for all the isotopic variations observed; perhaps each of the major components is locally heterogeneous (e.g., Langmuir and others, 1978; Zindler and others, 1979; Bougault and others, 1985). The geochemical trends for older MORB shown in Figure 9 are generally consistent with this model; that is, (1) the relatively enriched basalts at 63°N and 37°N can be associated with the Icelandic and Azores hot spots,

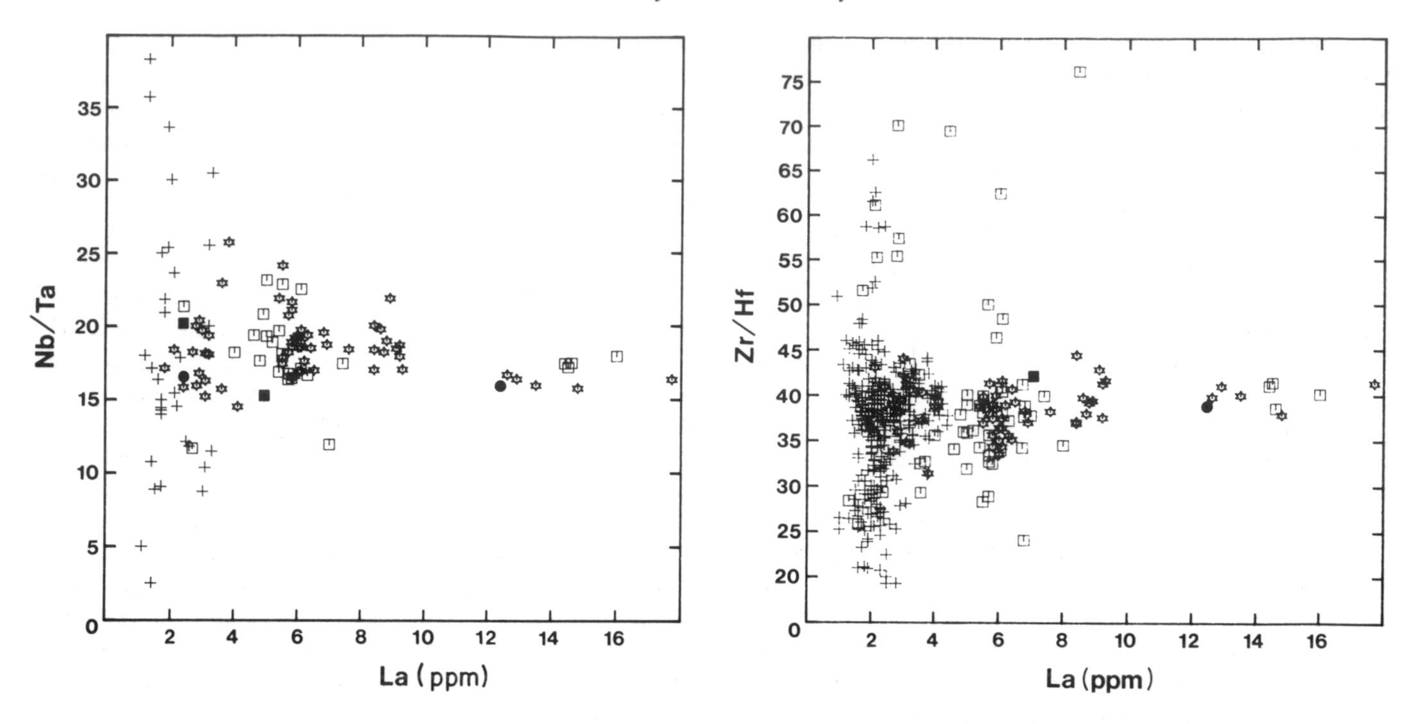

Figure 9a. Abundance ratios, Nb/Ta and Zr/Hf, of highly incompatible elements as a function of La content for the DSDP sites plotted in figure 4, except that data for individual rocks from site 558 are not plotted. *Plus,* 22°–25°N; *star,* 63°N; *open square,* 37°N. *Solid circles* and *solid squares* are averages of diverse geochemical groups from Leg 82 sites 561 and 558, respectively. Note that there are no systematic variations of Nb/Ta and Zr/Hf with latitude, and that depleted and enriched MORB have similar ratios. The dispersion of ratios at low La content reflects analytical errors associated with the correspondingly low abundances of Nb, Ta, Zr, and Hf.

respectively, and as noted by Schilling, the enriched components are geochemically different in these two areas; and (2) the relatively depleted MORB in the 22°–25°N region are consistent with formation on a ridge axis free from the influence of hot spot volcanism.

Geochemical Variations Along a Transect

As discussed earlier, transects perpendicular to the ridge crest in the western North Atlantic establish that $(La/Sm)_n$ and other geochemical parameters vary with age in a non-systematic manner (Figs. 4 and 5). In some regions, such as the approximately 18 Ma and 35 Ma basalts west and southwest of the Azores, the geochemical variations of axial basalts are not matched by the older basalts (Fig. 5). In other regions, there is a close correspondence; for example, from 22°N to 34°N only depleted MORB have been recovered, both in older DSDP sites and along the ridge axis (Figs. 4 and 5 and Schilling, this volume). Typically, the widest range of geochemical characteristics is found in MORB near oceanic islands. This diversity occurs either as a gradual transition from enriched to depleted MORB, such as along the ridge axis south of Iceland and the Azores (Schilling, this volume), or as extensive heterogeneity at a specific site such as at DSDP sites 558 and 561 south of the Azores (Figs. 4, 5, 7, 8, 9); the Reykjanes Peninsula, Iceland (Zindler and others, 1979); and dredges near Bouvet Island in the south Atlantic (Le Roex and others, 1983). These geochemical variations could arise from the same three general processes that were evaluated as possible causes of geochemical variations within individual drill cores.

Data for these transects are especially useful for evaluating the model discussed by Schilling (this volume) that much of the compositional variation in MORB results from the mixing of two large-scale components, that is, depleted asthenosphere and enriched mantle plumes. Recent developments of this model have incorporated complexities, such as intermittent mixing of depleted mantle with enriched ascending plumes formed of several discrete blobs; incomplete mixing which enables both end-members to survive in some environments; and damming of mantle flow by fracture zones which impedes mixing. These complexities give the model great flexibility, thereby making it difficult to formulate rigorous tests. However, this model always invokes a plume component to explain the occurrence of enriched basalts (see Schilling, this volume). Therefore, there must be some manifestation of a plume wherever enriched basalts occur and the transects in the western North Atlantic provide a basis for testing this model.

The least geochemical variability occurs in the ~33°N transect (south of the Hayes fracture zone) and along the 22°–25°N transect. These transects are distant from postulated mantle plumes and include only depleted MORB. However, along the 33°N transect, basalts have the isotopic and REE characteristics

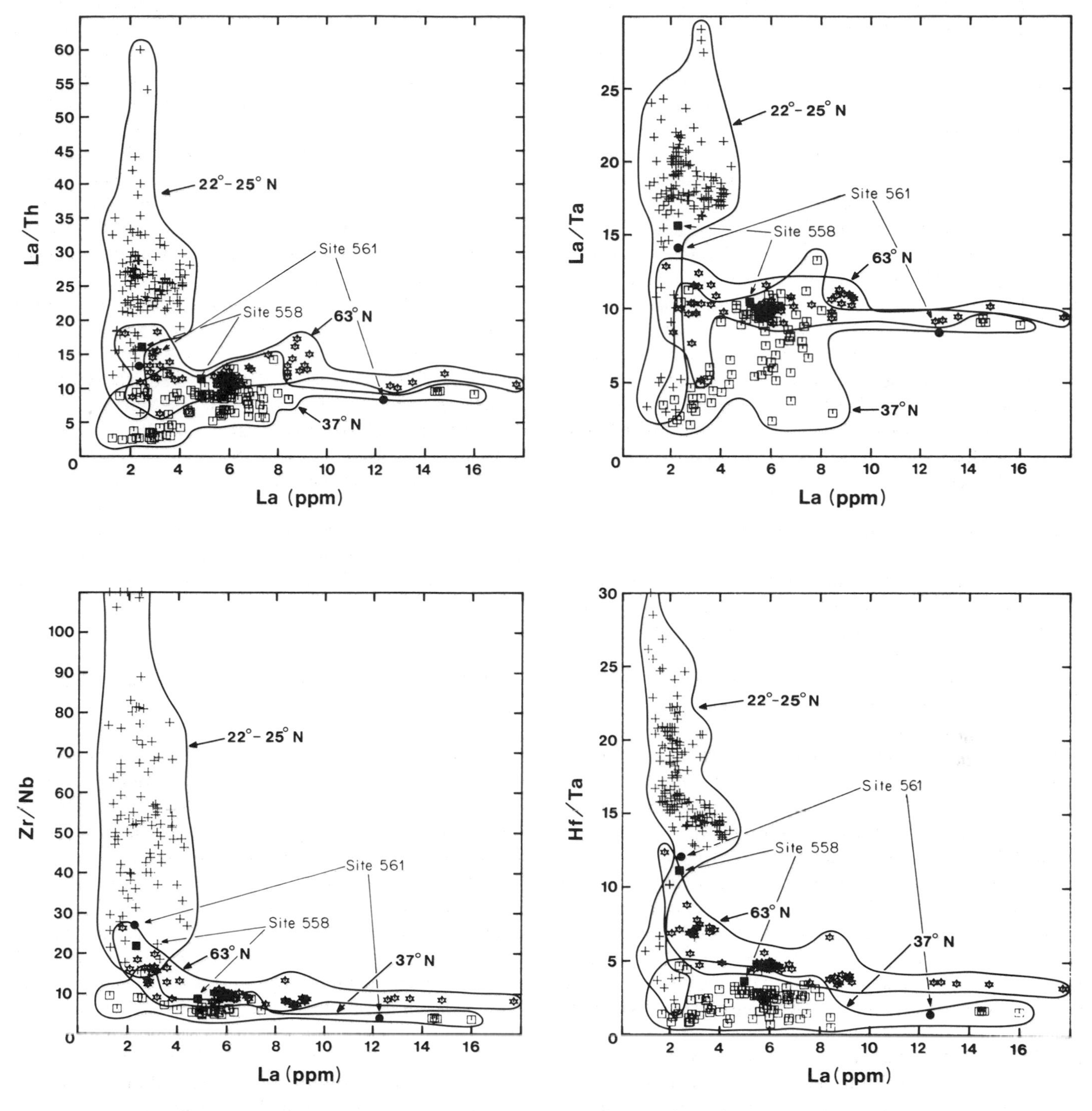

Figure 9b. Abundance ratios of La/Th, La/Ta, Zr/Nb, and Hf/Ta as a function of La content for the DSDP sites plotted in figure 9a. The mean values for La/Ta, Zr/Nb, La/Th, and Hf/Ta (ratios involving elements with different degrees of incompatibility) vary with latitude, increasing from 37°N to 63°N to 22°–25°N. The vertical trends for the 22°–25°N transect in all plots probably reflect large analytical errors at the low levels of Th, Nb, Ta, and Hf in these basalts. Additional complexities are the large within-site variations of La/Ta, Zr/Nb, La/Th, and Hf/Ta at leg 82 sites 558 and 561. Leg 11 data from Ayuso and others, 1976 and Bryan and others, 1977; all other data from relevant Initial Reports of the Deep Sea Drilling Project.

of depleted MORB (Figs. 5 and 7), but La/Ta ratios encompass the range from depleted to enriched MORB (Fig. 8). Moreover, there are important differences in major element compositions and in some trace element ratios between sites 417/418 and younger basalts from sites 395/396 and from the ridge axis (Byerly and Sinton, 1979; Flower and Bryan 1979; Joron and others, 1980; Bryan and Dick, 1982). Overall, the geochemical characteristics of young and old MORB from 22°–33°N in the western North Atlantic reflect derivation from compositionally heterogeneous but generally depleted sources.

The occurrence of enriched basalts along transects emanating from axial regions containing enriched basalts is consistent with a plume model, but the juxtaposition of enriched and depleted basalts at some sites requires inefficient mixing between enriched and depleted sources. In addition, reconstructed hot spot traces indicate that many occurrences of enriched MORB may be related to hot spot volcanism, but some locations of enriched MORB, as at IPOD site 558 (Plate III) are not readily related to hot spots (Duncan, 1984); apparently, enriched MORB is not confined to regions of hot spot volcanism (also see Zindler and others, 1984).

Mantle Heterogeneity in the Western North Atlantic

The concepts discussed above provide the basis for speculation about the composition of mantle sources in the western North Atlantic. The heterogeneity inferred for sources of younger basalts probably occurs to the same degree and on the same scale in space and time in the oldest parts of the seafloor. The major unresolved problems are: how were mantle heterogeneities created and how are they distributed in the mantle? The large-scale gradients occurring over hundreds of kilometers along the MAR axis have led to the development of models involving interactions and mixing between ascending enriched mantle plumes and depleted, convecting asthenosphere (Schilling, this volume). In detail, both end-members of this model are likely to be heterogeneous, but a major aspect of this model is large-scale mantle heterogeneity. Difficulties with this model as a general explanation for geochemical variations in MORB are twofold: (1) not all enriched MORB can be readily correlated with the presence of a mantle plume (e.g., site 558); and (2) a wide range of geochemical characteristics, from depleted to enriched MORB can occur in a single location; such as at DSDP site 561, at 53°–55°S near Bouvet Island in the south Atlantic (Le Roex and others, 1983), the Reykjanes Peninsula, Iceland (Zindler and others, 1979), and at Pacific ocean seamounts (Zindler and others, 1984). A significant consequence of these localized geochemical heterogeneities is that surficial magmas retain source heterogeneities; therefore, large, long-lived, well-mixed, magma chambers are unlikely to have been important in regions such as FAMOUS or at DSDP sites such as 558 and 561.

These localized heterogeneities in MORB can be explained only by the enriched-plume/depleted-asthenosphere mixing model if the efficiency of mixing varies, from high along the current northern MAR axis where the geochemical variations change gradually with latitude, to poor at individual DSDP sites such as 561 which contain both enriched and depleted MORB. An even more plausible alternative involves small, localized heterogeneities in the mantle that are tapped by small-scale partial melting. This model is frequently referred to as the veined mantle model. Because the proportion, composition, and formation time of the veins may vary, this model can explain the localized heterogeneties found in MAR basalts. However, a veined mantle model does not readily explain large, hundreds of kilometer–scale gradients unless the density of veins varies substantially with location; in this case, this model grades into the enriched-plume/depleted-asthenosphere model.

At present, the density of sampling, even in the western North Atlantic which is the most extensively sampled oceanic floor, is not sufficient to choose between these alternative models, which are not mutually exclusive. It is very likely that the mantle has both small and large-scale compositional heterogeneities. The current geochemical base for MORB is a tantalizing mix of geochemical systematics with superimposed complexities. Some relatively simple and potentially significant models can account for much of the data; however, in order to explain all the data, important but poorly constrained complexities must be incorporated into the models. Considerably more data from diverse locations of varying age will be required to constrain and evaluate such models.

INITIAL ATLANTIC RIFTING AND VOLCANISM

Typical pre-drift reconstructions of North America and Africa place the northwest coast of Africa (Morocco) against Nova Scotia (i.e., Jansa and Wade, 1975; Jansa and Wiedmann, 1982; Tucholke and Ludwig, 1982). Starting with this assumption, and also assuming fracture zone traces mark seafloor "flow lines" (Schouten and Klitgord, 1977), it is possible to relate geologic features on the African and North American continental margins to one another and to specific features within the Atlantic seafloor (Plate 5). On this basis, the present Straits of Gibralter and the Azores Platform at 39°N may be "compressed" to a position directly south of the Grand Banks and Newfoundland. Similarly, the Canary Islands, Great Meteor Seamount, the Corner Rise, and New England (Kelvin) Seamounts all project to a point southeast of Nantucket and would conform to a position of about 30°N on the present Mid-Atlantic Ridge. Similarly, Bermuda would project to about 26°N, and the Cape Verde Islands would correspond to about 17°N on the modern ridge. Neither Bermuda or the Cape Verde Islands have obvious counterparts symmetrically disposed on the opposite side of the Atlantic. Of most importance to the following discussion is that Mesozoic volcanism in Morocco was physically adjacent to New England and Nova Scotia, and both of these margins now correspond to the segment of Mid-Atlantic Ridge approximately between the Atlantis Fracture Zone and the Azores Islands (Plate 5).

Rifting Between North America and Africa

Several studies have suggested that "plume"-type volcanism, similar to that presumed to be associated with the Azores platform, or New England Seamounts, may have been active during the Mesozoic and could even reflect deep mantle convection instrumental in triggering, and perhaps maintaining, the rifting and subsequent spreading of the North Atlantic (Morgan, 1973). Comparison of age relations and major element compositions of Mesozoic volcanism in Africa and eastern North America indicate that 196 Ma basalts of the Moroccan High Atlas province are stratigraphic equivalents of the First and Second Watchung basalts, the York Haven basalt, and Quarryville basalt in New Jersey and Pennsylvania; the Holyoke basalts of New England, and the North Mountain basalts of Nova Scotia (Manspeizer and others, 1978). The 186 Ma basalts of the Moroccan Meseta are correlated by age and composition to the Rossville basalt of Pennsylvania, and by age with the Third Watchung basalt of New Jersey. Recent geochemical studies of early Mesozoic basalt from Morocco (Bertrand and others, 1982) further confirm their qualitative similarities in chemical composition to similar age volcanics in eastern North America. However, there have been no attempts to correlate isotopic (Sr, Nd, Pb) ratios in Mesozoic basalts between eastern North America and Morocco; without such data it cannot be determined if these stratigraphic correlations also imply a common mantle source for basalts now on separate continents.

There is evidence both in eastern North America (Smith and others, 1975; Puffer and Lechler, 1980) and in Morocco (Bertrand and others, 1982) for chemical evolution from relatively incompatible-element enriched to relatively depleted compositions with decreasing age of eruption. In addition, Bryan and others, (1977) noted that trace element data for Triassic and Jurassic diabases from the eastern North American continental margin show progressive enrichments in incompatible elements from south to north that are broadly similar to those observed along the corresponding section of the modern Mid-Atlantic Ridge. However, there is an apparent "overprint" of the incompatible-element enrichment characteristic of continental flood basalts. McHone and Butler (1984) suggested that these continental lavas also change from quartz tholeiitic to alkalic basalt with increasing distance from the line of continental breakup. Unfortunately, isotopic data (Sr, Nd, Pb) critical to evaluating mantle source characteristics have been very limited for these continental basalts. In a recent Sr, Nd, and Pb isotopic and geochemical study of Mesozoic Appalachian tholeiites, Pegram (1983) concluded that these continental tholeiites experienced relatively little contamination by continental crust. Instead, he attributes their distinctive geochemistry to a subcontinental mantle source isotopically distinct from present-day MORB sources.

Magmatic evolution on the continental shelves is recorded by wells drilled for petroleum exploration (Jansa and Weidmann, 1982; Jansa and Pepiper, 1985). Middle Jurassic volcanism, possibly related to early activity of the New England Seamounts or White Mountain Magma Series (Foland and Faul, 1977) is recorded in breccia and ash layers on Georges Bank. Volcanic and intrusive igneous activity extending into the Cretaceous has been recorded from the Baltimore Canyon Trough and Scotia Basin; these magmas tend to be alkalic or andesitic in character. Also, alkalic pillow lavas of late Jurassic to Eocene age have been recovered adjacent to the African coast. It is interesting that alkalic volcanism as represented by the Canary and Cape Verde islands, has continued through the Tertiary to the present time off Africa; while volcanism along the western North Atlantic margins appears to have ceased in the Cretaceous or early Tertiary (Vogt and Perry, 1982) and certainly is not active today. Possible late Tertiary volcanism may remain to be discovered among the Newfoundland or New England Seamounts. Jansa and Pepiper (1985) suggested that volcanic episodes re-occur at 50–70 my intervals, reflecting concurrent mantle pulses that drive the plates.

By late Jurassic (Oxfordian) time, rifting of the Atlantic was evidently well established, and formation of typical "layer 2" volcanic basement is recorded at DSDP sites 100 and 105 (Ayuso and others, 1976; Bryan and others, 1977) and 534 (Sheridan and others, 1982). Mineralogy, textures, and major and trace element compositions define these basalts as typical water-quenched depleted oceanic basalts; as noted in earlier discussions, spreading "flow lines" in the western North Atlantic also project these sites back to geochemically depleted ridge segments.

As spreading continued, however, subsequent continental volcanism reverted to an alkalic or calc-alkalic chemistry more typical of ocean islands or island arcs. A linear, transverse zone of alkalic volcanism along the White Mountain–New England Seamount lineament became active at about 100 Ma. Taras and Hart (1983) found that the alkali basalts forming the New England Seamounts have Sr, Nd, and Pb isotopic ratios similar to alkalic basalts from the Azores and Canary Islands. The age of the New England Seamounts increases from southeast (82 my) to northwest (103 my); therefore, the New England Seamounts along with the younger (70–75 my) Corner Seamounts to the east and the youngest phase of the White Mountain igneous province (100–124 my) to the northwest can be interpreted as reflecting northwest motion of the North American plate over the New England 'hot spot' now located southwest of Great Meteor Seamount (Duncan, 1984).

At about 150 Ma, late-stage dolerite dikes in eastern North America were emplaced in tensional fractures oriented perpendicular to the line of rifting, in contrast to earlier rift-parallel orientations (DeBoer and Snider, 1979; McHone and Butler, 1984). Alkaline volcanism along the White Mountain–New England Seamount lineament apparently ended in late Cretaceous to early Tertiary time, but similar activity continues to the present time in the Azores Rift and in the Canary Islands. Possibly, a clockwise rotational adjustment in spreading direction south of the Pico Fracture Zone tended to close fissures in the western

Atlantic, but opened or at least maintained them in the eastern Atlantic. Alternatively, the eastern North Atlantic may have become stationary over a zone of active "hot spots" (Vogt and Perry, 1982).

Other obvious complications in the subsequent magmatic history of the western North Atlantic include the possible voluminous alkalic volcanism along the J-anomaly Ridge and Madeira-Tore Rise (Tucholke and Ludwig, 1982), and the mechanism by which original tholeiitic volcanism on the Bermuda Seamount was renewed by intrusion of unusual alkalic limburgite sills after a hiatus of some 60 my (Reynolds and Aumento, 1974). The J-anomaly event could be explained by injection of an active upwelling deep-mantle-derived "blob" along the spreading center in J-anomaly time, analogous to the enrichment events along the modern Reykjanes Ridge or Azores Platform. The renewed volcanism in Bermuda suggests a fortuitous encounter with a latent "hot spot." Proper evaluation of such speculative hypotheses clearly will require much more detailed sampling and comprehensive laboratory analysis.

In the far North Atlantic, rifting was initiated at about 60 Ma, much later than in the central North Atlantic. In contrast to continental basalts from eastern North America and Morocco, trace element and isotopic studies of basalts and dolerite dikes associated with opening of the far North Atlantic show that some subaerial basalts from these areas originated from mantle sources compositionally similar to the present sub-oceanic mantle (O'Nions and Clarke, 1972; Schilling and Noe-Nygaard, 1974; Nielsen, 1978; Carter and others, 1979; Thompson, 1982). Other basalts in these areas have geochemical characteristics reflecting crustal contamination (Carter and others, 1978; Thompson, 1982). Although many of the continental tholeiitic basalts associated with initial rifting in the far North Atlantic are compositionally similar to MORB, as rifting proceeded there was a reversion to alkaline volcanism in East Greenland (Nielsen, 1978) similar to that on the continental margins of eastern North America and Morocco.

Clearly, magmatic activity associated with initial continental rifting deserves much more detailed study, and the limited data available suggest that properly integrated major element, trace element, and isotopic analyses may provide much insight into mantle sources, the role of crustal contamination, and the timing of changes in crustal setting and tectonic regime. Comparisons of magmatic sequences between opposing continental blocks, and of these sequences with MORB sampled from intervening seafloor, are necessary. Unfortunately, rigorous comparison and interpretation of even the limited data available are compromised by substantial differences in the data sets presented by different investigators. Also, an important caveat emerges from the recent intensive sampling and analysis of deep-sea drill cores which have demonstrated significant geochemical contrasts in closely associated "magma batches" within the same drill hole, or between nearby drill holes. Similar diversity has emerged from detailed sampling and analyses of closely-spaced dredge or submersible samples (e.g., Langmuir and others, 1977; Le Roex and others, 1983). Given this diversity over small intervals of space and time, it would appear that extreme caution must be exercised in accepting apparent geochemical correlations across ocean basins, or between ancient continental rocks and those of the modern ocean ridges.

REFERENCES

Anderson, R. N., Honnorez, J., Becker, K., Adamson, A.C., Alt, J. C., Emmerman, R., Kempton, P. D., Kinoshita, H., Laverne, C., Mottl, M. J., and Newmark, R. L

1982 : DSDP hole 504B, the first reference section over 1 km through Layer 2 of the oceanic crust: Nature, v. 300, p. 589–594.

Arcyana

1977 : Rocks collected by bathyscaph and diving saucer in the FAMOUS area of the Mid-Atlantic Rift Valley: Petrological diversity and structural setting: Deep-Sea Research, v. 24, p. 565–589.

Aumento, F., Loncarevic, B., and Ross, D. I.

1971 : Hudson geotraverse: Geology of the Mid-Atlantic Ridge at 45°N: Philosophical Transactions of the Royal Society of London, v. A268, p. 623–650.

Aumento, F., and Melson, W. G., eds.

1977 : Initial reports of the Deep Sea Drilling Project, Vol. 37: U.S. Government Printing Office, Washington, D.C., 998 p.

Ayuso, R. A., Bence, A. E., and Taylor, S. R.

1976 : Upper Jurassic tholeiitic basalts from DSDP leg 11: Journal of Geophysical Research, v. 81, p. 4305–4325.

Bertrand, H., Dostal, J., and Dupuy, C.

1982 : Geochemistry of Early Mesozoic tholeiites from Morocco: Earth and Planetary Science Letters, v. 58, p. 225–239.

Blanchard, D. P., Rhodes, J. M., Dungan, M. A., Rodgers, K. V., Donaldson, C. H., Brannon, J. C., Jacobs, J. W., and Gibson, E. W.

1976 : The chemistry and petrology of basalts from leg 37 of the Deep Sea Drilling Project: Journal of Geophysical Research, v. 81, p. 4231–4246.

Bougault, H., Joron, J-L., and Treuil, M.

1979 : Alteration, fractional crystallization, partial melting, and mantle properties from trace elements in basalts recovered in the North Atlantic; in Deep drilling results in the Atlantic Ocean: Ocean crust, eds. M. Talwani, C. G. Harrison and D. E. Hayes; Maurice Ewing Series, v. 2, American Geophysical Union, Washington, D.C., p. 352–368.

Bougault, H., and Cande, S. C., eds.

1985 : Initial Reports of the Deep Sea Drilling Project, Volume 82: in press.

Bryan, W. B.

1972 : Morphology of quench crystals in submarine basalts: Journal of Geophysical Research, v. 77, p. 5812–5819.

1974 : Fe-Mg relationships in sector-zoned submarine basalt plagioclase: Earth and Planetary Science Letters, v. 24, p. 157–165.

1979 : Regional variation and petrogenesis of basalt glasses from the FAMOUS area, Mid-Atlantic Ridge: Journal of Petrology, v. 20, p. 293–325.

1983 : Systematics of modal phenocryst assemblages in submarine basalts: Petrologic implications: Contributions to Mineralogy and Petrology, v. 83, p. 62–74.

Bryan, W. B., and Dick, H.J.B.

1982 : Contrasted abyssal basalt liquidus trends: Evidence for mantle major element heterogeneity: Earth and Planetary Science Letters, v. 58,

p. 15–26.
Bryan, W. B., Frey, F. A., Thompson, G.
1977 : Oldest Atlantic seafloor: Mesozoic basalts from western North Atlantic margin and eastern North America: Contributions to Mineralogy and Petrology, v. 64, p. 223–242.
Bryan, W. B., and Moore, J. G.
1977 : Compositional variations of young basalts in the Mid-Atlantic Ridge rift valley near 36°49′N: Geological society of America Bulletin, v. 88, p. 556–570.
Bryan, W. B., Thompson, G., and Frey, F. A.
1979 : Petrologic character of the Atlantic crust from DSDP and IPOD drill sites; in Deep drilling results in the Atlantic Ocean: Ocean crust, eds. M. Talwani, C. G. Harrison and D. E. Hayes; Maurice Ewing Series, v. 2, American Geophysical Union, Washington, D.C., p. 273–284.
Bryan, W. B., Thompson, G., Frey, F. A., and Dickey, J. S.
1976 : Inferred geologic settings and differentiation in basalts from the Deep Sea Drilling Project: Journal of Geophysical Research, v. 81, p. 4285–4304.
Bryan, W. B., Thompson, G., and Ludden, J. N.
1981 : Compositional variation in normal MORB from 22-25°N: Mid-Atlantic Ridge and Kane Fracture Zone: Journal of Geophysical Research, v. 86, p. 11815–11836.
Byerly, G. R., and Sinton, J. M.
1979 : Compositional trends in natural basalt glasses from DSDP holes 417D and 418A; in Initial Reports of the Deep Sea Drilling Project, eds. T. W. Donnelly and J. Francheteau; Washington, D.C., U.S. Government Printing Office, v. 51, 52, 53, p. 957–971.
Byerly, G. R., and Wright, T. L.
1977 : Origin of major element chemical trends in DSDP leg 37 basalts, Mid-Atlantic Ridge: Journal of Volcanism and Geothermal Research, v. 3, p. 229–279.
Cameron, W. E., Nisbet, E. G., and Dietrich, V. J.
1980 : Petrographic dissimilarities between ophiolitic and ocean floor basalts; in Ophiolites, ed. A. Panayitou; Proceedings of the International Ophiolite Symposium, Geological Survey Department, Nicosia, Cyprus, p. 192.
Cann, J. R.
1979 : Metamorphism in the ocean crust; in Deep drilling results in the Atlantic Ocean: Ocean crust, eds. M. Talwani, C. G. Harrison and D. E. Hayes; Maurice Ewing Series, v. 2, American Geophysical Union, Washington, D.C., p. 230–238.
Carter, S. R., Evensen, N. M., Hamilton, P. J., and D'Nions, R. K.
1978 : Neodymium and strontium isotope evidence for crustal contamination of continental volcanics: Science, v. 202, p. 743.
1979 : Basalt magma sources during the opening of the North Atlantic: Nature, v. 281, p. 28–30.
Cohen, R. S., and O'Nions, R. K.
1982 : the lead, neodymium and strontium isotopic structure of ocean ridge basalts: Journal of Petrology, v. 23, p. 299–324.
Coish, R. A., and Church, W. R.
1979 : Igneous geochemistry of mafic rocks from the Betts Cove ophiolite, Newfoundland: Contributions to Mineralology and Petrology, v. 70, p. 29–39.
Coish, R. A., Hickey, R., and Frey, F. A.
1982 : Rare earth element geochemistry of the Betts Cove ophiolite, Newfoundland: Complexities in ophiolite formation: Geochim Cosmochim Acta, v. 46, p. 2117–2134.
De Boer, J., and Snider, F. G.
1979 : Magnetic and chemical variations of Mesozoic diabase dikes from eastern North America: Evidence for a hotspot in the Carolinas?: Geological Society of America Bulletin, v. 90, p. 185–198.
Dick, H.J.B., and Bryan, W. B.
1978 : Variation of basalt phenocryst mineralogy and rock compositions in DSDP hole 396B; in Initial reports of the Deep Sea Drilling Project, eds. L. Dmitriev and J. Heirtzler; Washington, D.C., U.S. Government Printing Office, v. 46, p. 215–225.
Dick, H.J.B., and Bullen, T.
1984 : Chromian spinel as a petrogenetic indicator in abyssal and alpine-type peridotites and spatially associated lavas: Contributions to Mineralogy and Petrology, v. 86, p. 54–76.
Dick, H.J.B., and Fisher, R. L.
1984 : Mineralogical studies of the residues of mantle melting: Abyssal and alpine-type peridotites; in Kimberlites. II. The mantle and crust-mantle relationships, ed. J. Kornprobst; Elsevier, Amsterdam, p. 295–308.
Dick, H.J.B., Fisher, R. L., and Bryan, W. B.
1984 : Mineralogical variability of the uppermost mantle along mid-ocean ridges: Earth and Planetary Science Letters, v. 69, p. 88–106.
Donnelly, T. W., Thompson, G., and Salisbury, M. H.
1979 : The chemistry of altered basalts at site 417, Deep Sea Drilling Project leg 51; in Initial Reports of the Deep Sea Drilling Project, , eds. T. W. Donnelly, and J. Francheteau; Washington, D.C., U.S. Government Printing Office, v. 51, 52, 53, p. 1319–1330.
Dostal, J., and Muecke, G. K.
1978 : Trace element chemistry of the peridotite-gabbro-basalt suite from DSDP leg 37: Earth and Planetary Science Letters, v. 40, p. 415–422.
Duncan, R. A.
1984 : Age progressive volcanism in the New England Seamounts and the opening of the central Atlantic Ocean: Journal of Geophysical Research, v. 89, p. 9980–9990.
Dungan, M. A., and Green, D. H.
1980 : The role of multi-stage melting in the formation of oceanic crust: Geology, v. 8, p. 22–28.
Dungan, M. A., Long, P. E., and Rhodes, J. M.
1978 : The petrography, mineral chemistry, and one-atmosphere phase relations of basalts from site 395; in Initial reports of the Deep Sea Drilling Project, Volume 45, eds. W. G. Melson, P. D. Rabinowitz, Washington, D.C., U.S. Government Printing Office, v. 45, p. 461–477.
Dungan, M. A., and Rhodes, J. M.
1978 : Residual glasses and melt from DSDP legs 45 and 46: Evidence for magma mixing: Contributions to Mineralogy and Petrology, v. 67, p. 417–431.
Dupré, B., Lambret, B., Rousseau, D., and Allègre, C. J.
1981 : Limitations on the scale of mantle heterogeneities under ocean ridges: Nature, v. 294, p. 552–554.
Elthon, D., Casey, J. F., and Komor, S.
1982 : Mineral chemistry of ultramafic cumulates from the North Arm Mountain massif of the Bay of Islands ophiolite: Evidence for high pressure crystallization of oceanic basalts: Journal of Geophysical Research, v. 87, p. 8717–8734.
Flower, M.F.J.
1980 : Accumulation of calcic plagioclase in ocean-ridge tholeiite: an indication of spreading rate?: Nature, v. 287, p. 530–532.
1981 : Thermal and kinematic control on ocean-ridge magma fractionation: Contrasts between Atlantic and Pacific spreading axes: Journal of the Geological Society of London, v. 138, p. 695–712.
Flower, M.F.J., and Bryan, W. B.
1979 : DSDP sites 417 and 418: A petrogenetic synthesis; in Initial reports of the Deep Sea Drilling Project, eds. T. Donnelly and J. Francheteau; Washington, D.C., U.S. Government Printing Office, v. 51, 52, 53, p. 1557–1562.
Flower, M.F.J., and Robinson, P. T.
1979 : Evolution of the 'FAMOUS' ocean ridge segment: Evidence from submarine and deep sea drilling investigations; in Deep drilling results in the Atlantic Ocean: Ocean crust, eds. M. Talwani, C. G. Harrison and D. E. Hayes; Maurice Ewing Series, v. 2, American Geophysical Union, Washington, D.C., p. 314–330.
1981 : Basement drilling in the western Atlantic Ocean. 1: Magma fractionation and its relation to eruptive chronology: Journal of Geophysical

Research, v. 86, p. 6273–6298.

Flower, M.F.J., Schmincke, H.-U., and Bowman, H.
1976 : Rare earth and other trace elements in historic Azorean lavas: Journal of Volcanology and Geothermal Research, v. 1, p. 127–147.

Foland, K. A., and Faul, H.
1977 : Ages of the White Mountain intrusives—New Hampshire, Vermont, and Maine, USA: American Journal of Science, v. 277, p. 88–904.

Frey, F. A., Bryan, W. B., and Thompson, G.
1974 : Atlantic Ocean floor: Geochemistry and petrology of basalts from Legs 2 and 3 of the Deep Sea Drilling Project: Journal of Geophysical Research, v. 79, p. 5507–5527.

Hanson, G. N.
1977 : Geochemical evolution of the sub-oceanic mantle: Journal of the Geological Society of London, v. 134, p. 235–253.

Hart, S. R., and Staudigel, H.
1982 : The control of alkalies and uranium in sea water by ocean crust alteration: Earth and Planetary Science Letters, v. 58, p. 202–212.

Hekinian, R.
1971 : Chemical and mineralogical differences between abyssal hill basalts and ridge tholeiites in the eastern Pacific Ocean: Marine Geology, v. 11, p. 77–91.

Hekinian, R., and Aumento, F.
1973 : Rocks from the Gibbs Fracture Zone and the Minia seamount near 53°N in the Atlantic Ocean: Marine Geology, v. 14, p. 47–72.

Hekinian, R., and Thompson, G.
1976 : Comparative geochemistry of volcanics from rift valleys, transform faults and aseismic ridges: Contributions to Mineralogy and Petrology, v. 57, p 145–162.

Hodges, F. N., and Papike, J. J.
1976 : DSDP site 334: magmatic cumulates from oceanic layer 3: Journal of Geophysical Research, v. 81, p. 4135–4151.

Honnorez, J.
1981 : The aging of the oceanic crust at low temperature; in The sea: The oceanic lithosphere, ed. C. Emiliani; John Wiley, New York, p. 525–587.

Houghton, R. L.
1979 : Petrology and geochemistry of basaltic rocks recovered in leg 43 of the Deep Sea Drilling Project; in Initial reports of the Deep Sea Drilling Project, eds. B. E. Tucholke and P. R. Vogt, Washington, D.C., U.S. Government Printing Office, v. 43, p. 721–738.

Houghton, R. L., Thomas, J. E. Jr., Diecchio, R. J., and others.
1979 : Radiometric ages of basalts from DSDP Leg 43; sites 382 and 385 (New England Seamounts), 384 (J. Anomaly), 386 and 387 (central and western Bermuda Rise): in Initial Reports of the Deep Sea Drilling Project, eds. B. E. Tucholke, P. R. Vogt; Washington, D.C., U.S. Government Printing Office, v. 43, p. 739–753.

Humphris, S. E., and Thompson, G.
1978 : Trace element mobility during hydrothermal alteration of oceanic basalts: Geochim et Cosmochim Acta, v. 42, p. 127–136.

Jahn, B., Bernard-Griffiths, J., Charlot, R., Cornichet, J., and Vidal, F.
1980 : Nd and Sr isotopic compositions and REE abundances of Cretaceous MORB (holes 417D and 418A, legs 51, 52 and 53): Earth and Planetary Science Letters, v. 48, p. 171–184.

Jansa, L. F., and Pe-piper, G.
1985 : Early Cretaceous volcanism on the northeastern American margin and implications for plate tectonics: Geological Society of America Bulletin, v. 96, p. 83–91.

Jansa, L. F., and Wade, J. A.
1975 : Geology of the continental margin off Nova Scotia and Newfoundland: Geological Survey of Canada Paper 74-30, v. 2, p. 51–106.

Jansa, L. F., and Wiedmann, J.
1982 : Mesozoic-Cenozoic development of the eastern North American and northwest African continental margins: A comparison; in Geology of the northwest African continental margin, eds. U. von Rad, K. Hina, and others, Springer, New York, p. 215–269.

Joron, J. L., Bollinger, C., Qusefit, J. P., Bougault, H., and Treuil, M.
1980 : Trace elements in basalts at 25°N, old crust, in the Atlantic Ocean: Alteration, mantle, and magmatic processes; in Initial Reports of the Deep Sea Drilling Project, eds. T. W. Donnelly and J. Francheteau, Washington, D.C., U.S. Government Printing Office, v. 51, 52, 53, p. 1087–1098.

Kuo, L-C., and Kirkpatrick, R. J.
1982 : Pre-eruption history of phyric basalts from DSDP legs 45 and 46: Evidence from morphology and zoning patterns in plagioclase: Contributions to Mineralogy and Petrology, v. 79, p. 13–27.

Langmuir, C. H., Bender, J. F., Bence, A. E., and Hanson, G. N.
1977 : Petrogenesis of basalts from the FAMOUS area, Mid-Atlantic Ridge: Earth and Planetary Science Letters, v. 36, p. 133–156.

Langmuir, C. H., Vocke, R. D., Hanson, G. N., and Hart, S. R.
1978 : A general mixing equation with applications to Icelandic basalts: Earth and Planetary Science Letters, v. 37, p. 380–392.

Le Roex, A. P., Dick, H.J.B., Erlank, A., Reid, A. M., Frey, F. A., and Hart, S. R.
1983 : Geochemistry, mineralogy, and petrogenesis of lavas erupted along the Southwest Indian Ridge between the Bouvet triple junction and 11°East: Journal of Petrology, v. 24, p. 267–318.

Le Roex, A. P., Erlank, A. J., Reid, A. M., and Needham, H. D.
1981 : Geochemical and mineralogical evidence for the occurrence of at least three distinct magma types in the 'FAMOUS' region: Contributions to Mineralogy and Petrology, v. 77, p. 24–37.

Ludden, J. N., and Thompson, G.
1979 : An evaluation of the behavior of the rare earth elements during the weathering of the seafloor basalt: Earth and Planetary Science Letters, v. 43, p. 85–92.

Malpas, J.
1978 : Magma generation in the upper mantle: Field evidence from ophiolite suites, and application to the generation of oceanic lithosphere: Philosophical Transactions of the Royal Society of London, v. A288, p. 527–546.

Manspeizer, R. W., Puffer, J. H., and Cousminer, H. L.
1978 : Separation of Morocco and eastern North America: A Triassic-Liassic stratigraphic record: Geological Society of America Bulletin, v. 89, p. 901–920.

McHone, J. G., and Butler, J. R.
1984 : Mesozoic igneous provinces of New England and the opening of the North Atlantic: Geological Society of America Bulletin, v. 95, p. 757–765.

Melson, W. G., and Thompson, G.
1971 : Petrology of a transform fault and adjacent ridge segments: Philosophical Transactions of the Royal Society of London, v. A268, p. 423–441.
1973 : Glassy abyssal basalts, Atlantic seafloor near St. Paul's Rocks: Petrography and composition of secondary clay minerals: Geological Society of America Bulletin, v. 84, p. 703–716.

Melson, W. G., Thompson, G., and VanAndel, Tj.H.
1968 : Volcanism and metamorphism in the Mid-Atlantic Ridge, 22°N latitude: Journal of Geophysical Research, v. 73, p. 5925–5941.

Miyashiro, A., Shido, F., and Ewing, M.
1969 : Diversity and origin of abyssal tholeiite from the Mid-Atlantic Ridge near 24° and 30°N latitude: Contributions to Mineralogy and Petrology, v. 23, p. 38–52.
1970 : Crystallization and differentiation in abyssal tholeiites and gabbros from mid-ocean ridges: Earth and Planetary Science Letters, v. 7, p. 361–365.

Morgan, W. J.
1973 : Plate motions and deep mantle convection; in Studies in earth and space sciences, ed. R. Shagam; Hess volume, Geological Society of America Memoir 132, p. 7–22.

Mottle, M. J.

1983 : Metabasalts, axial hot springs, and the structure of hydrothermal systems at mid-ocean ridges: Geological Society of America Bulletin, v. 94, p. 161–180.

Muir, I. D., and Tilley, C. E.
1966 : Basalts from the northern part of the Mid-Atlantic Ridge: Journal of Petrology, v. 7, p. 193–201.

Nicholis, G. D., Nawalk, A. J., and Hays, E. E.
1964 : The nature and composition of rock samples dredged from the Mid-Atlantic Ridge between 22°N and 52°N: Marine Geology, v. 1, p. 333–343.

Nielsen, T.F.D.
1978 : The Tertiary dike swarms of the Kangerdlugssuaq area, East Greenland: Contributions to Mineralogy and Petrology, v. 67, p. 63–78.

O'Hara, M. J., and Mathews, R. E.
1981 : Geochemical evolution in an advancing, periodically replenished, periodically tapped, continuously fractionated magma chamber: Journal of the Geological Society of London, v. 138, p. 237–277.

O'Nions, R. K., and Clarke, D. B.
1972 : Comparative trace element geochemistry of Tertiary basalts from Baffin Bay: Earth and Planetary Science Letters, v. 15, p. 436–444.

O'Nions, R. K., and Pankhurst, R. J.
1974 : Petrogenetic significance of isotope and trace element variations in volcanic rocks from the Mid-Atlantic: Journal of Petrology, v. 15, p. 603–634.
1976 : Sr isotope and rare earth element geochemistry of DSDP Leg 37 basalts: Earth and Planetary Science Letters, v. 31, p. 255–261.

O'Nions, R. K., Pankhurst, R. J., and Gronvold, K.
1976 : Nature and development of basalt magma sources beneath Iceland and the Reykjanes Ridge: Journal of Petrology, v. 17, p. 315–338.

Pegram, W.
1983 : Isotope characteristics of the Mesozoic Appalachian tholeiites: Geological Society of America Abstracts with Programs, v. 15, p. 660.

Petersen, N., Eisenach, P., and Bleil, U.
1979 : Low temperature alteration of the magnetic minerals in ocean floor basalts; in Deep drilling results in the Atlantic Ocean: Ocean crust, eds. M. Talwani, C. G. Harrison and D. E. Hayes; Maurice Ewing Series, v. 2, American Geophysical Union, Washington, D.C., p. 169–209.

Prèvot, M., Lecaille, A., and Hekinian, R.
1979 : Magnetism of the Mid-Atlantic Ridge crest near 37°N from FAMOUS and DSDP results: A review; in Deep drilling results in the Atlantic Ocean: Ocean crust, eds. M. Talwani, C. G. Harrison and D. E. Hayes; Maurice Ewing Series, v. 2, American Geophysical Union, Washington, D.C., p. 210–229.

Prinz, M., Keil, K., Green, J. A., Reid, A. M., Bonnatti, E., and Honnorez, J.
1976 : Ultramafic and mafic dredge samples from the equatorial Mid-Atlantic Ridge and fracture zones: Journal of Geophysical Research, v. 81, p. 4087–4103.

Puffer, J. H., and Lechler, P.
1980 : Geochemical cross sections through the Watchung basalt of New Jersey: Geological Society America Bulletin, v. 91, p. 156–191.

Reynolds, P. R., and Aumento, F. A
1974 : Deep Drill 1972: Potassium-argon dating of the Bermuda drill core: Canadian Journal of Earth Science, v. 11, p. 1269–1273.

Rhodes, J. M., Dungan, M. A., Blanchard, D. P., and Long, P. E.
1979 : Magma mixing at mid-ocean ridges: Evidence from basalts drilled near 22°N on the Mid-Atlantic Ridge: Tectonophysics, v. 55, p. 35–61.

Richardson, S. H., Hart, S. R., and Staudigel, H.
1980 : Vein mineral ages of old oceanic crust: Journal of Geophysical Research, v. 85, p. 7195–7200.

Scheidegger, K. F.
1973 : Temperatures and compositions of magmas ascending along mid-ocean ridges: Journal of Geophysical Research, v. 78, p. 3340–3355.

Schilling, J.-G.
1973 : Iceland mantle plume, geochemical evidence along Reykjanes Ridge: Nature, v. 242, p. 565–571.

Schilling, J.-G., Kingsley, R. H., Devine, J. D.
1982 : Galapagos hot spot—spreading center system: I. Spatial petrological and geochemical variations (83°W–101°W): Journal of Geophysical Research, v. 87, p. 5593–5610.

Schilling, J.-G., and Noe-Nygaard, A.
1974 : Faeroe-Iceland plume: Rare earth evidence: Earth and Planetary Science Letters, v. 24, p. 1–14.

Schilling, J.-G., Sigurdsson, H., and Kingsley, R. H.
1978 : Skaergaard western neovolcanic zones in Iceland. 2. Geochemical variations: Journal of Geophysical Research, v. 83, p. 3983–4002.

Schilling, J.-G., Zajac, M., Evans, R., Johnston, T., White, W., Devine, J. D., and Kingsley, R.
1983 : Petrologic and geochemical variations along the Mid-Atlantic Ridge from 29°N to 73°N: American Journal of Science, v. 283, p. 510–586.

Schmincke, H.-U., Rautenschlein, M., Robinson, P. T., and Mehegan, J. M.
1983 : Troodos extrusive series on Cyprus: A comparison with oceanic crust: Geology, v. 11, p. 405–409.

Schouten, H., and Klitgord, K. D.
1977 : Map showing Mesozoic magnetic anomalies: Western North Atlantic: U.S. Geological Survey Miscellaneous Field Studies, Map MF-915.

Scientific Party
1982 : Elements traced in Atlantic: Geotimes, v. 27, p. 21–23.

Sheridan, R. E., Gradstein, F. M., et al.
1982 : Early history of the Atlantic Ocean and gas hydrates on the Blake Outer ridge: Results of the Deep Sea Drilling Project leg 76: Geological society of America Bulletin, v. 93, p. 876–885.

Shibata, T., DeLong, S. E., and Walker, D.
1979 : Abyssal tholeiites from the Oceanographer Fracture Zone: I. Petrology and fractionation: Contributions to Mineralogy and Petrology, v. 70, p. 89–102.

Shibata, T., and Fox, P. J.
1975 : Fractionation of abyssal tholeiites: Samples from the Oceanographer Fracture Zone (35°N, 35°W): Earth and Planetary Science Letters, v. 27, p. 62–72.

Shibata, T., Thompson, G., and Frey, F. A.
1979 : Tholeiitic and alkali basalts from the Mid-Atlantic Ridge at 43°N: Contributions to Mineralogy and Petrology, v. 70, p. 127–141.

Sigurdsson, H.
1981 : First-order major element variation in basalt glasses from the Mid-Atlantic Ridge: 29°N to 73°N: Journal of Geophysical Research, v. 86, p. 9483–9502.

Sinton, J. M.
1978 : Petrology of (alpine-type) peridotites from site 395, DSDP leg 45; in Initial reports of the Deep Sea Drilling Project, eds. W. G. Melson and P. Rabinowitz; Washington, D.C., U.S. Government Printing Office, v. 45, p. 595–604.

Smith, R. C., Rose, A. W., and Lanning, R. M.
1975 : Geology and geochemistry of Triassic diabase in Pennsylvania: Geological society of America bulletin, v. 86, p. 943–955.

Staudigel, H., and Bryan, W. B.
1981 : Contrasted glass—whole rock compositions and phenocryst redistribution, IPOD sites 417 and 418: Contributions to Mineralogy and Petrology, v. 78, p. 255–262.

Staudigel, H., and Hart, S. R.
1983 : Alteration of basaltic glass: Mechanisms and significance for the oceanic crust-seawater budget: Geochim Cosmochim Acta, v. 47, p. 337–350.

Suen, C. J., Frey, F. A., and Malpas, J.
1979 : Bay of Islands ophiolite suite, Newfoundland: Petrologic and geochemical characteristics with emphasis on rare earth element geochemistry: Earth and Planetary science Letters, v. 45, p. 337–348.

Sun, S. S., Nesbitt, R. W., and Sharaskin, A. Y.
1979 : Geochemical characteristics of midocean ridge basalts: Earth and

Planetary Science Letters, v. 44, p. 119–138.
Taras, B., and Hart, S. R.
1983 : Sr, Nd and Pb isotopic compositions of the New England Seamount Chain: EOS, v. 64, p. 907.
Tarney, J., Wood, D. A., Saunders, A. D., Cann, J. R., and Varet, J.
1980 : Nature of mantle heterogeneity in the North Atlantic: Evidence from deep sea drilling: Philosophical Transactions of the Royal Society of London, v. A297, p. 179–202.
Tarney, J., Wood, D. A., Varet, J., Saunders, A. D., and Cann, J. R.
1979 : Nature of mantle heterogeneity in the North Atlantic: Evidence from leg 49 basalts; in Deep drilling results in the Atlantic Ocean crust, eds. M. Talwani, C. G. Harrison and D. E. Hayes; Maurice Ewing Series, v. 2, American Geophysical Union, Washington, D.C., p. 285–301.
Thompson, G.
1983 : Hydrothermal fluxes in the ocean; in Chemical Oceanography, J. P. Riley, ed. R. Chester; Academic Press, London, v. 8, p. 272–337.
1984 : Basalt-seawater interaction; in Hydrothermal processes at seafloor spreading centers, eds. P. A. Rona, K. Boström, L. Laubier and K. L. Smith, Jr.; Plenum, New York, p. 225–278.
Thompson, G., Bryan, W. B., Frey, F. A., and Dickey, J. S.
1978 : Basalts and related rocks from Deep Sea Drilling sites in the central and eastern Indian Ocean: Marine Geology, v. 26, p. 119–138.
Thompson, R. N.
1982 : Magmatism of the British Tertiary volcanic province: Scottish Journal of Geology, v. 18, p. 49–107.
Tiezzi, L. J., and Scott, R. B.
1980 : Crystal fractionation in a cumulate gabbro, MAR 26°N: Journal of Geophysical Research, v. 85, p. 5438–5454.
Tucholke, B. E., and Ludwig, W. J.
1982 : Structure and origin of the J Anomaly Ridge, western North Atlantic Ocean: Journal of Geophysical Research, v. 87, p. 9389–9407.
Vogt, P. R., and Perry, R. K.
1982 : North Atlantic Ocean: Bathymetry and plate tectonic evolution: Geological Society of America Map and Chart series MC35.
Wells, G., Bryan, W. B., and Pearce, T. H.
1979 : Comparative morphology of ancient and modern pillow lavas: Journal of Geology, v. 87, p. 427–440.
White, W. M.
1979 : PB isotope geochemistry of the Galapagos Islands: Carnegie Institution of Washington Yearbook, v. 78, p. 331–335.
White, W. M., Tapia, M.D.M., and Schilling, J.-G.
1979 : The petrology and geochemistry of the Azores Islands: Contributions to Mineralogy and Petrology, v. 69, p. 201–213.
Wilshire, H. G., and Pike, J. E.
1975 : Upper mantle diapirism: Evidence from analogous features in alpine peridotite and ultramafic inclusions in basalt: Geology, v. 3, p. 467–470.
Wood, D. A.
1978 : Major and trace element variations in the Tertiary lavas of eastern Iceland and their significance with respect to the Iceland geochemical anomaly: Journal of Petrology, v. 19, p. 393–436.
1979 : A variably veined sub-oceanic upper mantle: Genetic significance for mid-ocean ridge basalts from geochemical evidence: Geology, v. 7, p. 499–503.
1981 : Partial melting models for the petrogenesis of Reykjanes Peninsula basalts, Iceland: Implications for the use of trace elements and strontium and neodymium isotope ratios to record inhomogeneities in the upper mantle: Earth and Planetary Science Letters, v. 52, p. 183–190.
Wood, D. A., Tarney, J., Varet, J., Saunders, A. D., Bougault, H., Joron, J. L., Treuil, M., and Cann, J. R.
1979 : Geochemistry of basalts drilled in the North Atlantic by IPOD leg 49: Implications for mantle heterogeneity: Earth and Planetary Science Letters, v. 43, p. 77–97.
Zindler, A., Hart, S. R., Frey, F. A., and Jakobsson, S. P.
1979 : Nd and Sr isotope ratios and rare earth element abundances in Reykjanes Peninsula basalts: Evidence for mantle heterogeneity beneath Iceland: Earth and Planetary Science Letters, v. 45, p. 249–262.
Zindler, A., Staudigel, H., and Batiza, R.
1984 : Isotope and trace element geochemistry of young Pacific seamounts: Implications for the scale of upper mantle heterogeneity: Earth and Planetary Science Letters, v. 70, p. 175–195.

MANUSCRIPT ACCEPTED BY THE SOCIETY JULY 15, 1985

Acknowledgments

This manuscript has benefitted from discussions with other authors in this volume, especially W. G. Melson, J.-G. Schilling, L. Jansa, and B. Tucholke. We are especially grateful to J.-G. Schilling and H. Bougault for preprints of leg 82 papers. We thank D. Gerlach and R. Hickey for assistance in preparing the figures. We also thank S. Humphris, C. H. Langmuir, and an anonymous reviewer for detailed comments on an earlier version of the manuscript.

Printed in U.S.A.

The Geology of North America
Vol. M, The Western North Atlantic Region
The Geological Society of America, 1986

Chapter 18

Mid-plate stress, deformation, and seismicity

Mary Lou Zoback
U.S. Geological Survey, MS 977, 345 Middlefield Road, Menlo Park, California 94025
Stuart P. Nishenko
National Earthquake Information Center, U.S. Geological Survey, MS 967, Denver Federal Center, Denver, Colorado 80225
Randall M. Richardson
Department of Geosciences, University of Arizona, Tucson, Arizona, 85721
Henry S. Hasegawa
Earth Physics Branch, Energy, Mines, and Resources Canada, Ottawa, Ontario K1A 0Y3, Canada
Mark D. Zoback
Department of Geophysics, Stanford University, Stanford, California 94305

INTRODUCTION

The state of stress in the lithosphere is the product of a variety of forces that act on the plates. The interior region of a plate may be stressed by large-scale plate-driving forces, local density contrasts in the plate, regional thermal anomalies, the residual effects of past tectonic events, or load redistribution caused by glaciation, volcanism, or erosion/sedimentation. Investigation of the tectonic in-situ stress field can enhance our understanding of mid-plate deformation and may also elucidate the relative importance of the various forces acting on the plates.

To this end we have assembled deformation, seismicity, and stress data for the mid-plate region of the North American plate. Our area of study covers continental North America and the oceanic portions of the North American plate, excluding zones of deformation along the plate boundaries. Excluded are the Mid-Atlantic Ridge and the shear and subduction boundaries associated with the Caribbean plate (covered by other chapters in this volume) and the entire western Cordillera of Canada and the United States (from the eastern Rocky Mountains front to the Pacific coast). The state of stress in the broad, intensely deformed western part of the North American plate is considered to be influenced primarily by local processes related to Pacific–North American plate interaction.

Numerous techniques and stress indicators provide information on the principal tectonic stresses in the crust and, for the oceans, the mantle lithosphere. Compilations of available data show that stress orientations inferred from different methods or techniques quite often agree within the accuracy of the individual determinations (Zoback and Zoback, 1980). Good correlations are particularly significant in view of the vastly different depths sampled by the various methods (e.g., in-situ stress measurements are made in the upper few kilometers, whereas earthquake focal mechanisms may indicate the state of stress at depths of several tens of kilometers). The implication is that over fairly broad regions, a relatively uniform stress field may exist throughout the upper crust. Even at relatively shallow depths in the crust, typically less than a few kilometers and in some cases even significantly closer to the surface, the stress field may be controlled to a greater degree by regional stresses than by strain energy in the rock from remanent or local phenomena (e.g., Zoback and Zoback, 1980; Sbar and others, 1984).

To determine the extent to which the stress pattern within a plate is correlated with plate driving forces, the stresses derived from these forces can be modeled using the configuration of plate boundaries together with physically plausible models of the driving mechanism. The predicted principal stress pattern can then be compared with the observed data to test the respective contributions of plate tectonic and local forces to the mid-plate stress field.

Our goal is to examine the major sources of intraplate stress, which for the North American plate may be largely linked to its spreading boundary, and also to compare mid-plate deformation in oceanic areas to deformation in the continental parts of the same plate. Thus, it is useful to include a discussion of continental stresses and seismicity in this oceanic volume.

MID-PLATE DEFORMATION

Along plate boundaries, both the deformation and state of stress are dominated by plate interaction. This plate boundary deformation generally occurs on one or more well-defined structures (e.g., rift valley normal faults, transform faults, or Benioff

Zoback, M. L., Nishenko, S. P., Richardson, R. M., Hasegawa, H. S., Zoback, M. D., 1986, Mid-plate stress, deformation, and seismicity; *in* Vogt, P. R., and Tucholke, B. E., eds., The Geology of North America, Volume M, The Western North Atlantic Region: Geological Society of America.

zones) that, over geologic time scales, may have large cumulative displacements. In contrast, deformation in the interior regions of a plate is typically diffuse and occurs on structures with little evidence for repeated long-term neotectonic movement. Recognizable fault scarps in mid-plate areas are rare, and deformation rates are undoubtedly much lower than those observed along plate boundaries.

The diffuse and sporadic nature of mid-plate deformation makes it difficult to study. The best indicator of active intraplate deformation is seismicity. Plate 11 (in pocket) and Figure 1 show the pattern of seismicity for the North American plate. These data represent a complete compilation of epicenters on a plate-wide basis for the time period 1900–1982. To provide a more internally consistent data set, we selected a subset of events with threshold magnitudes that vary with time (see figure caption). Without appropriate cut-offs, the pattern of seismicity would be strongly biased by the temporal variations in the detection threshold of smaller events and the contribution from detailed local micro-earthquake networks on land. Habermann (1982) reviewed the consistency of worldwide teleseismic data since 1963 and determined that, on the average, events with m_b of 4.5 to 5.0 are near the limit of uniform teleseismic detection for that time period. Even so, it is likely that some oceanic events of m_b 4.5 or greater went undetected in 1960–1982, but it is unlikely that any land event of this magnitude was missed.

An additional problem with the seismicity map is that it is dominated by small (i.e., $m_b < 6.0$) instrumentally recorded continental events from the time period 1960–1982. Because of the low rate of deformation in this region, such a short period of recording probably does not provide an accurate picture of either the long-term distribution of seismicity or the return period of larger earthquakes. A somewhat longer time window of activity can be obtained for the continental areas of eastern North America, where written records exist for the last 300–350 years (see York and Oliver, 1976, Fig. 1). The historical earthquake record and the instrumentally recorded events show similar patterns of seismicity; however, as discussed below, the historical record provides a better estimate of the frequency of major events.

Western North Atlantic

The occurrence of earthquakes clearly indicates mid-plate deformation in the western North Atlantic, which we define herein as extending from the continental slope of eastern North America to the Mid-Atlantic Ridge province. During the brief time period of recording, much of the western North Atlantic was aseismic (above the probable detection threshold of m_b 4.5-5.0). The earthquakes that did occur were commonly clustered around major structural and/or topographic features.

The discussion in this chapter excludes those earthquakes associated with the major plate boundaries in the western North Atlantic region (i.e., the Mid-Atlantic Ridge and the shear and subduction boundaries of the Caribbean plate), and concentrates instead on the intraplate seismicity of this region. For the purpose of clarity, we have subdivided the family of intraplate earthquakes into two subgroups: (1) events occurring in the vicinity of (and possibly influenced by processes occurring at) the above named plate boundaries, and (2) events located at some distance from these plate boundaries that appear to occur in a "true" mid-plate tectonic environment. The earthquakes in the western North Atlantic that fall into the second category are rare and include primarily events located in the vicinity of the Bermuda Rise. These and other oceanic and continental margin events on the North American plate are discussed in later sections.

Examples of intraplate earthquakes in the western North Atlantic in the first category include near-ridge earthquakes (events that occur off the axis of the Mid-Atlantic Ridge in lithosphere less than 35 m.y. old), subduction zone-influenced earthquakes near the Greater and Lesser Antilles, and earthquakes related to an incipient or diffuse plate boundary between the North and South American plates in the region between the Antilles and the Mid-Atlantic Ridge.

Globally, the majority of seismic moment release associated with near-ridge earthquakes occurs in lithosphere of 0 to 15 m.y. in age, at depths of 5 to 20 km (i.e., beneath the Moho to depths associated with the 750°–800°C isotherm in standard cooling models—Bergman and Solomon, 1984; Wiens and Stein, 1984). The wide variety of faulting mechanisms occurring at all ages in the 0 to 35 m.y. age band (from shallow, <5 km, crustal thrust faults to intermediate and deeper, 5–20 km, strike-slip and normal faults) appear to discount the observation of Sykes and Sbar (1974) of an orderly age-dependent transition from a tensional to compressional stress regime at ~20 m.y. Thermoelastic stresses associated with plate cooling and the physical evolution of young lithosphere (<35 m.y.) are probably the predominant forces resulting in strain release in the near-ridge environment (Bergman and Solomon, 1984; Wiens and Stein, 1984). The relevant source parameter data for the earthquakes that have been studied in this environment are presented in Table 1; see Figure 1 and Plate 11 (pocket) for locations of events discussed.

Examples of other near-plate boundary earthquakes in the western North Atlantic are the December 26, 1969–January 7, 1970 sequence in the Demerara Abyssal Plain (seaward of the Lesser Antilles arc, location shown by hachured region on Fig. 2) and the events located in the Barracuda Abyssal Plain near the Barracuda Ridge (see Table 1). The largest event (M_S 7.5) in the 1969–1970 Demerara Abyssal Plain sequence is a normal faulting earthquake that appears to have occurred in response to lithospheric flexure associated with the subduction of the western Atlantic beneath the Lesser Antilles island arc (Stein and others, 1982). Both of the Barracuda Abyssal Plain events yield thrust fault focal mechanisms with a component of strike-slip motion and appear to be rather deep for intraplate shocks (20 to 40 km). The sense of motion associated with these events may reflect deformation along a broad, diffuse boundary between the North and South American plates (Stein and others, 1982). It is of interest to note that the Nares Abyssal Plain event of September 3, 1968 (M_S 5.9, see Table 1), which occurred ~150 km north of

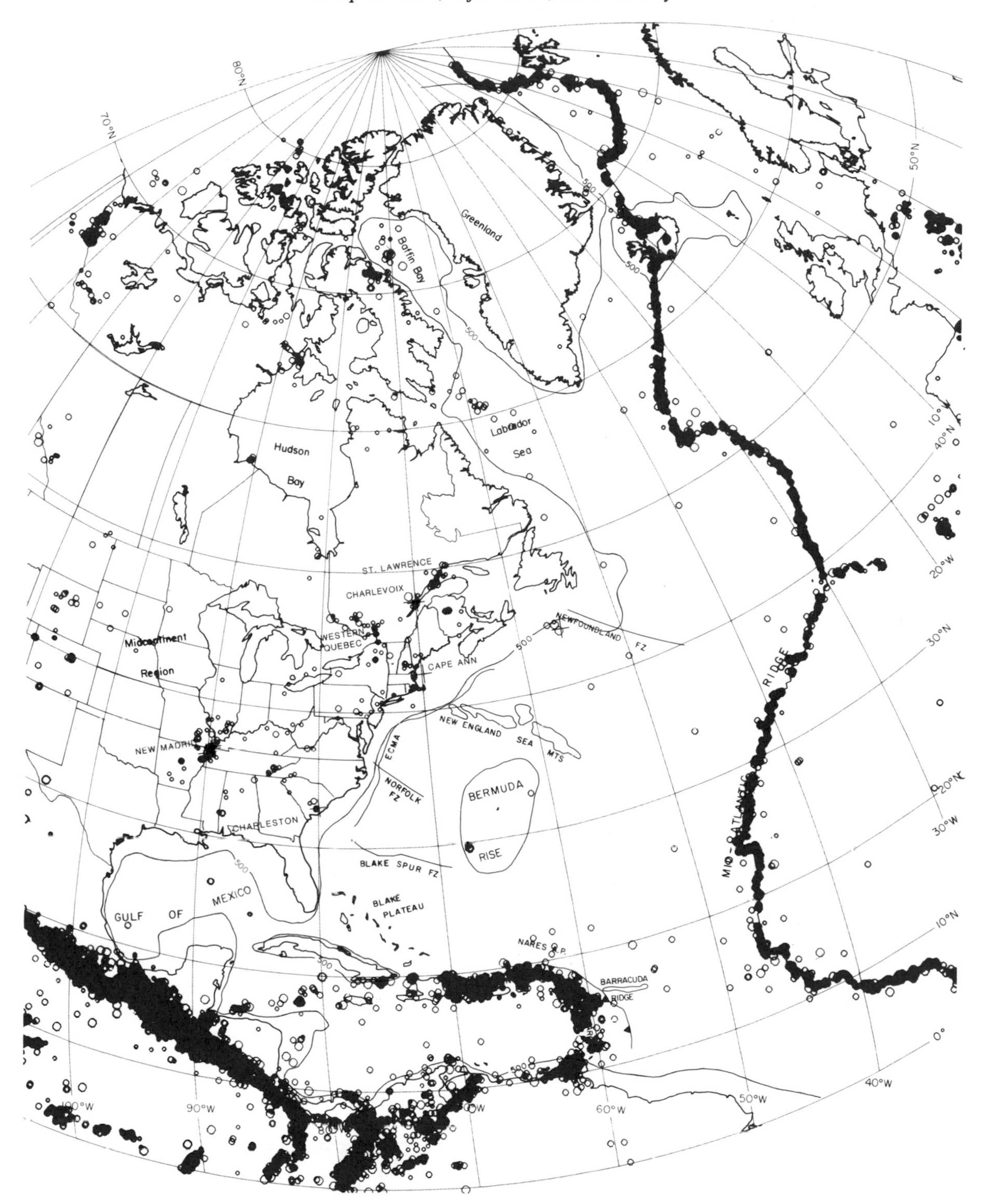

Figure 1. Seismicity in eastern North America and the western Atlantic region for the period 1900–1982. Symbol size indicates magnitude range: Smallest symbols are magnitudes between 3.0–4.5, intermediate size between 4.5–6.0, and large symbols for magnitudes 6.0 and greater. Areas of high density of seismicity appear black. *Red crosses* indicate locations of significant historical earthquakes mentioned in text. Cutoff magnitudes vary by time interval and coverage: 1900–1959, M $\geqslant$6.0; 1960–1969, $m_b \geqslant 4.5$ and greater than ten stations reporting; and 1970–1982, $m_b \geqslant 3.0$ and more than ten stations reporting. Data from the International Seismological Center, the U.S. Geological Survey, the National Ocean Survey, and the Department of Energy, Mines, and Resources, Canada. Red 500 m isobath marks edge of the continent. ECMA—East Coast Magnetic anomaly, NARES A.P.—Nares abyssal plain.

TABLE 1. FOCAL MECHANISM DATA FOR OCEANIC EVENTS

Event Number/ Location	P-axis Azimuth/ Plunge	Comments	References
		NEAR MID-ATLANTIC RIDGE	
WA-3.1 32.36°N, 41.03°W	275°/59°	6 Aug. 1962. Normal fault with some strike-slip component, constrained by body wave modeling. Depth: 10 km below sea floor.	Bergman and Solomon (1984)
WA-3.2 44.58°N, 31.34°W	284°/14°[1] 301°/18°[2]	17 Sept. 1964 m_b 5.5. Reverse fault, partially constrained by first motion data and body and surface wave modeling. Depth 9 km below sea floor[2].	Sykes and Sbar (1974)[1] Bergman and Solomon (1984) Wiens and Stein (1984)[2]
WA-3.3 72.66°N, 8.14°E	146°/07°[1]	21 Nov. 1967 m_b 5.4 Reverse fault with strike-slip component constrained by body and surface wave modeling. Depth 6 km below sea floor[1].	Wiens and Stein (1984)[1]
WA-3.4 24.87°N, 46.30°W	131°/77°[1]	13 Sept. 1981 m_b 5.8, M_s 5.5 Normal fault constrained by body wave modeling. Depth: 11 km below sea floor.	Dziewonski and Woodhouse (1983) Bergman and Solomon (1984)[1]
		DEMERARA ABYSSAL PLAIN	
WA-7 15.79°N, 59.64°W	058°/58°	25(2), 26, 29 Dec. 1969, and 7 Jan. 1970, m_b 6.4, 5.8, 5.3, 5.5. Combination of normal faulting and strike-slip events, constrained by body wave first motions and surface waves. Depths: 37 to 42 km.	Stein and others (1982)
WA-8 16.08°N, 59.79°W	155°/03°		
WA-9 15.79°N, 59.56°W	162°/17°		
WA-10 16.18°N, 59.74°W	090°/71°		
WA-11 15.86°N, 59.78°W	070°/24°		
		BARRACUDA ABYSSAL PLAIN	
WA-4 19.8°N, 56.1°W	165°/03°[1]	23 Oct. 1964. m_b 6.2, M_s 6.8. Reverse faulting with large strike-slip component, constrained by surface wave and body wave modeling. Depth: 23 km[1], 35 km[2].	Molnar and Sykes (1969) Sykes and Sbar (1974) Liu and Kanamori (1980)[1] Stein and others (1982)[2]
WA-5 17.53°N, 54.91°W	359°/12°[1]	13 Dec. 1977, m_b 5.7, M_s 6.9. Reverse faulting with strike-slip component, constrained by surface wave and body wave modeling. Depth: 25 km below sea floor.	Bergman and Solomon (1980)[1] Stein and others (1982)[2]
WA-6 17.44°N, 54.83°W ISC	354°/12°	6 Dec. 1978, m_b 5.4. Reverse faulting with strike-slip component, constrained by short period first motion data. Depth: 10 km.	Stein and others (1982)
		NARES ABYSSAL PLAIN	
WA-12 20.58°N, 62.30°W	299°/10°[1]	3 Sept. 1968, m_b 5.6, M_s 5.9. Strike-slip event constrained by first motions and body wave modeling. Depth: 27 km below sea floor.[1]	Sykes and Sbar (1974)[1] Bergman and Solomon (1980)
		BERMUDA RISE	
WA-1 33.01°N, 61.66°W	268°/05°	24 Nov. 1976, m_b 5.1. Reverse faulting with strike-slip component, one nodal plane constrained by surface wave data. Depth: 10 km below sea floor, 15 km below sea level.	Nishenko and Kafka (1982)
WA-2 29.8°N, 67.4°W	250°/10°[1]	24 Mar. 1978, m_b 6.0, M_s 6.0. Reverse faulting event constrained by surface wave and body wave modeling. Depth: 6 km below sea floor.[1]	Stewart and Helmberger (1981) Nishenko and Kafka (1982)[1] Nishenko and others (1982)
		GULF OF MEXICO	
WA-13 26.49°N, 88.79°W	157°/00°	25 July 1978, m_b 5.0. Reverse faulting with strike-slip component, partially constrained by short period first motion data. Depth: 15 km below sea level.	Frohlich (1982)
		BAFFIN BAY/ LABRADOR SEA	
WA-14 73.30°N, 70.7°W	170°/18°	20 Nov. 1933, M_s 7.3. Reverse faulting event, poorly constrained by surface waves and body wave first motions. Depth originally reported as 65 km, may be only 25–30 km.	Stein and others (1979) N.H. Sleep, oral comm. (1983)
WA-15 60.5°N, 58.7°W	137°/25°	24 Nov. 1969, m_b 5.0. Reverse faulting constrained only by body wave first motions. Depth: 33 km.	Sykes and Sbar (1974)
WA-16 55.0°N, 54.3°W	245°/03°	7 Dec. 1971, m_b 5.4, M_s 5.3. Reverse faulting events, constrained by surface waves and body wave first motions. Depth: 16 km.	Hashizume (1977)
WA-17 72.4°N, 70.2°W	46°/05°	12 Nov. 1976, m_b 5.4, M_s 5.1. Reverse faulting event, constrained by surface waves and body wave first motions. Focal depth fixed at 33 km.	Stein and others (1979)

the Puerto Rico trench, may represent deformation along a western extension of this broad diffuse plate boundary. Both the focal mechanism and location of this event differ from that expected for a near-subduction zone interaction (compare to the 1969–1970 Lesser Antilles sequence) and the focal mechanism is more similar to those events in the Barracuda Abyssal Plain.

Bermuda Rise

Of all the Atlantic intraplate shocks, the 24 March 1978 M_s 6.0 Bermuda Rise event has been studied in the greatest detail, both in terms of mainshock parameters and aftershock locations (Stewart and Helmberger, 1981; Nishenko and Kafka, 1982; Ni-

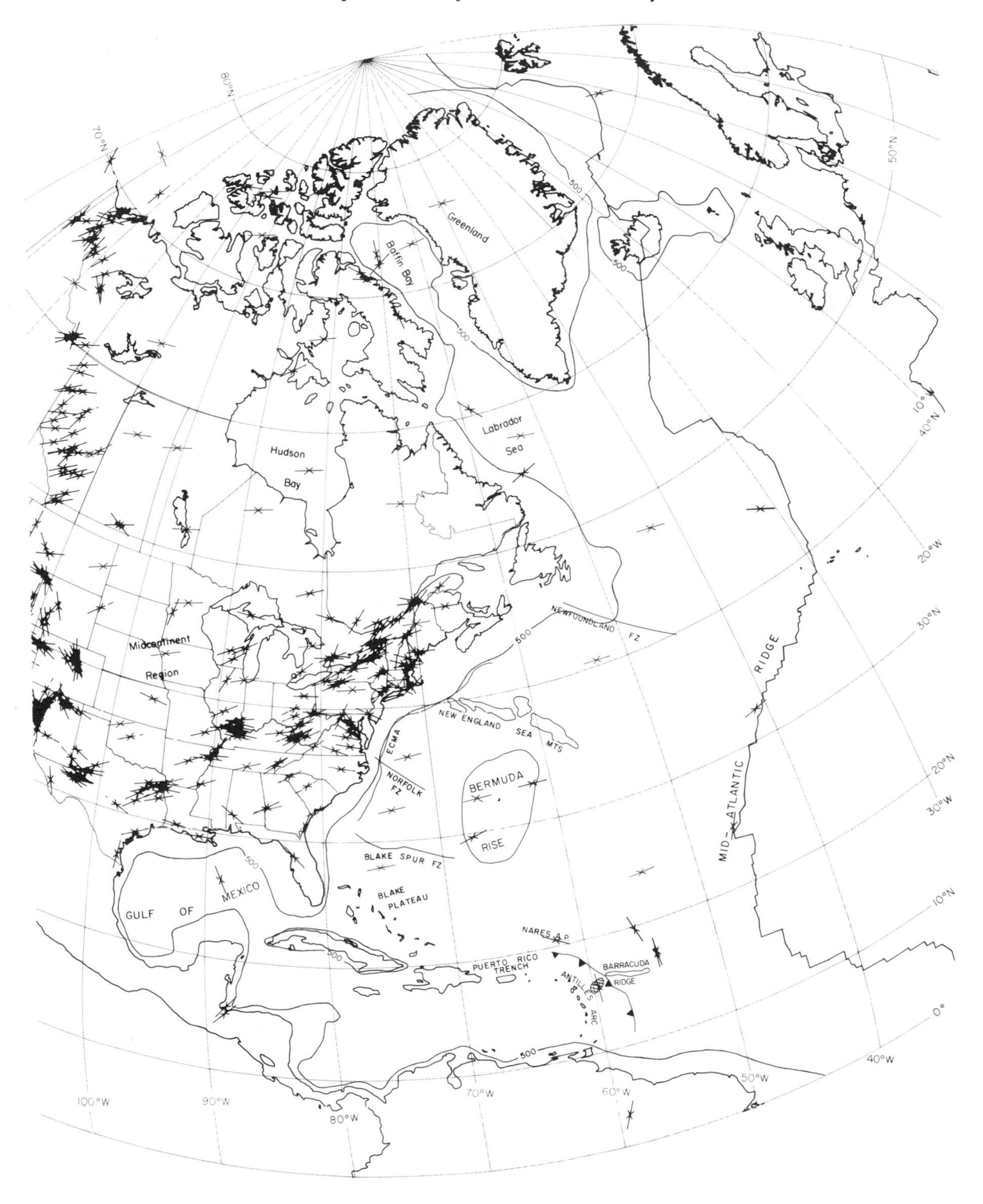

Figure 2. Stress map of the North American plate. Maximum horizontal principal stress orientations are plotted in black. Stress data in the western Cordillera are not considered here. *Red arrows* are predicted maximum horizontal stresses for an elastic single-plate finite-element model of the North American plate with ridge push forces. Oceanic stress data are summarized on Table 1. Primary sources for the stress data are Zoback and Zoback (1980), Hasegawa and Adams (1981), Bell and Gough (1981), Gough and others (1983), and Zoback and others (1984). See Figure 1 for further explanation of labelled localities.

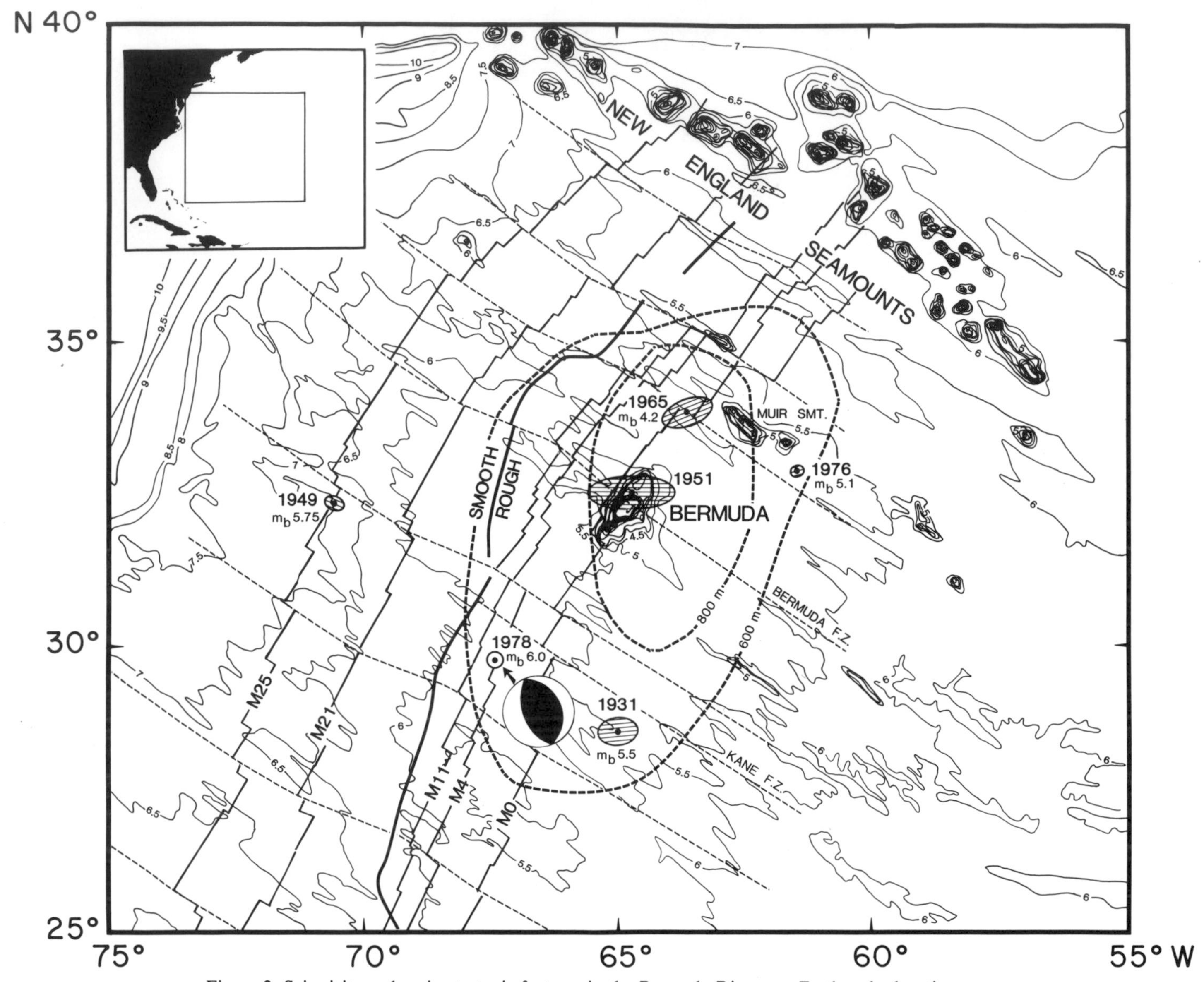

Figure 3. Seismicity and major tectonic features in the Bermuda Rise area. Earthquake locations are based on the method of joint hypocenter determination, using the 1978 earthquake as a master event. *Hachured ellipses* show the 95 confidence limits for these locations. Focal mechanism for the 1978 event (lower hemisphere projection, compressional quadrant shaded) is from Nishenko and Kafka (1982). 600 and 800 m contours (*thick dashed lines*) of residual seafloor depth anomalies (from Sclater and Wixon, this volume) show the long-wavelength topographic expression of the Bermuda rise. The rough/smooth boundary (*solid thick lines*) from Sundvik and others (1984) marks the change in roughness of the oceanic crust in this area, reflecting variations in rates of crustal accretion and directions of plate motion. Depth to basement (contour interval 500 m) from Tucholke and others (1982). Magnetic anomalies and fracture zone trends after Schouten and Klitgord (1982). Inset shows location of map area with respect to eastern North America.

shenko and others, 1982). As shown in Figure 3, the 1978 shock occurred approximately 380 km southwest of Bermuda, near magnetic anomaly 44 (126 Ma, based on the timescale of Kent and Gradstein, this volume). The focal mechanism of this event indicates pure reverse faulting along NNW-trending fault planes (340° ± 10°). In the vicinity of the mainshock, however, the major fracture zone trends are 300° ± 5°. This observation and the location of the mainshock and subsequent aftershocks 10 to 20 km from the nearest fracture zone lead Nishenko and Kafka (1982) and Nishenko and others (1982) to suggest that faulting occurred along a subsidiary fault rather than along a major fracture zone.

Earthquakes that occurred in the Bermuda rise and vicinity prior to 1978 were relocated using the method of joint hypocenter determination (Dewey, 1972) with the 1978 event as a reference or master event. These new locations and associated 95 confidence ellipses are shown in Figure 3. Many of these events appear to be located near fracture zones and suggest a simple causal relationship. The degree of correlation with specific tectonic features, however, is limited by the absence of focal mechanism data and lack of precision in the earthquake locations. Fracture zone spacing in this area is on the order of 50 to 100 km, well within the range of location errors for many of the events shown in Figure 3. Hence, it is probably more instructive to focus on the regional correlation of earthquake occurrence with the Bermuda rise, rather than on detailed local structural and bathymetric features.

The Bermuda rise is an anomalous topographic feature in the western North Atlantic believed formed by mid-plate thermal uplift accompanied by magmatism. At the Deep Sea Drilling Project (DSDP) site 386, volcaniclastic sediments indicate that the main shield-building stage occurred at 50–40 Ma. Boreholes on the island of Bermuda have recovered lamprophyres with K-Ar ages of 35–30 Ma (Reynolds and Aumento, 1974). The abrupt termination of turbidite deposition on the flanks of the rise indicates as much as 700 m of uplift during the middle Eocene (Ewing and others, 1969; Tucholke and Mountain, 1979). Satellite altimetry data (Haxby and others, 1969; Tucholke and Mountain, 1979). Satellite altimetry data (Haxby and Turcotte, 1978) have been interpreted to indicate that the rise is isostatically compensated in a Pratt manner at a depth of about 50 to 100 km. At present, two major hypotheses have been presented to account for the Bermuda rise. One suggests that the rise marks the track of a faint hotspot or mantle plume (Crough, 1978; Morgan and Crough, 1979; Morgan, 1981), while the other hypothesis suggests that the rise represents the surface manifestation of an upwelling convection cell in the western North Atlantic (Parsons and Daly, 1983; Sclater and Wixon, this volume).

The topographic expression of the Bermuda rise, as defined by residual depth anomalies (see Sclater and Wixon, this volume) is shown in Figure 3 by the +600 and +800 m contours. The eastern boundary of the rise is well defined by the sharp deviation of seafloor depths from the square root of age relationship at 100 Ma. The western boundary of the rise is coincident, in part, with the location of the rough/smooth basement boundary in the western North Atlantic (Sundvick and others, 1984). The rough/smooth boundary represents a distinct change in the average relief of acoustic basement (from 150–450 m in the smooth zone to 450–1200 m in the rough zone) and is considered to have a primary seafloor spreading origin in response to changes in spreading rates and directions of plate motion between anomalies M21 and M11 (150–133 Ma; Vogt and others, 1971; Sundvick and others, 1984).

The observation that the majority of events in Figure 3 occur within the topographic expression of the rise suggests that the overall pattern of strain release may be influenced by the rise itself. Historically, the islands of Bermuda have experienced earthquakes since at least the end of the nineteenth century (Sieberg, 1932). One event in 1883 was felt at the MM VI–VII level throughout the islands (M. Brewer, Bermuda College, pers. com., 1982). Hence, it appears that the instrumental record for the last 50+ years is consistent with the historic record for the last 100+ years.

The spatial association of mid-plate seismicity in the western North Atlantic with anomalous topography and young tectonism is an appealing correlation at present. However, as discussed later and shown in Figure 1, the axes of maximum principal stress determined from the earthquake focal mechanisms (see Table 1), are parallel to the absolute plate motion of North America as determined by Minster and Jordan (1978) as well as to the predicted stress orientation for finite-element models of the North American plate with only ridge push forces.

Other Oceanic and Continental Margin Regions of Seismicity

Gulf of Mexico. The majority of events occurring in the Gulf of Mexico are so small that reliable hypocenters are usually not obtainable. An event on 24 July 1978 (m_b 5.0), however, was well recorded. The depth and focal mechanism for this event, as determined by short-period body waves, indicates reverse faulting with a possible strike-slip component at a depth of 15 km, near the local depth to Moho. The P-axis, while not well constrained, indicates N-S compression (Frohlich, 1982).

The basement of the Gulf of Mexico was extensively rifted during Triassic and Jurassic times (Martin and Case, 1975). In addition to possible ancient structural controls, flexure of the basement due to sediment loading by the Mississippi Fan could produce N-S compressive stresses in the shallow seaward portion of the Gulf crust (Frohlich, 1982).

Grand Banks/Laurentian Slope. Much of the earthquake activity along the Nova Scotia–Newfoundland continental margin appears to be confined to the mouth of the Laurentian Channel, where a NE-trending rifted continental margin intersects a NW-trending transform margin. Both margins were formed by the African–North American plate separation (see Keen, 1982 for summary). This seismogenic area is located seaward of a major change in strike of Appalachian tectonic trends that may have also controlled the Jurassic breakup geometry (Sykes, 1978). During the past 30 years, four earthquakes with magnitude greater than 5 have occurred in this region. The most significant historical event in this area is the 1929 M 7.2 "Grand Banks earthquake" (Doxsee, 1948), which was felt over a wide area and was associated with a submarine landslide that generated a destructive tsunami; which also became a turbidity current that broke numerous trans-Atlantic telegraph cables (Heezen and Ewing, 1952).

Possible local sources of stress in this region are loading due to sedimentation in the Laurentian fan and density-driven forces due to the contrast between oceanic and continental crustal structure. However, accurate fault-plate solutions and focal depths of earthquakes occurring along the Laurentian slope are required to determine the nature of the stress field and the zones of weakness that are being reactivated.

Labrador Sea/Davis Strait. Seismic reflection data and magnetic lineations in the Labrador Sea have been used to identify the extinct Mid–Labrador Sea Ridge and associated fracture zones (Drake and others, 1963; Roots and Srivastava, 1984). Seafloor spreading commenced between 70–65 m.y. ago [Srivastava, 1978) and ceased about 40 m.y. ago (Kristoffersen and Talwani, 1977). Davis Strait may have been the site of either a hot spot or a leaky transform fault in early Tertiary time (Srivastava, 1978; Keen and others, 1974; Hyndman, 1975; Le Pichon and others, 1971; Menzies, 1982) or may be underlain by continental crust (see van der Linden, 1975; Grant, 1975; Keen and Hyndman, 1979).

Seismicity in the Labrador Sea manifests two trends, one along the ocean-continent boundary and another near the Labrador Sea Ridge (Basham and Adams, 1982). The region between the central ridge and Greenland appears aseismic (Basham and others, 1977; Gregersen, 1982); Davis Strait also appears to be largely aseismic.

The P-axes of the two Labrador Sea earthquake fault-plane solutions (WA-15 and WA-16) are perpendicular to each other. Near the northwest boundary between the Labrador Sea and Davis Strait, WA-15 indicates NW-SE compression approximately parallel to the coastline (Sykes and Sbar, 1974), and further south, WA-16 indicates NE-SW compression perpendicular to the coastline (Hashizume, 1977). Both events occurred seaward of the 1000 m bathymetric contour.

Baffin Bay. Baffin Bay is characterized by a moderate level of seismicity. Although there are different views as to the nature and evolution of Baffin Bay (for a complete review see Srivastava and Falconer, 1982), plate tectonic reconstructions suggest that both Baffin Bay and the Labrador Sea formed during early Tertiary time, with seafloor spreading terminating about 40 mybp. Geophysical data in Baffin Bay indicate that the central region is oceanic crust (e.g., see Menzies, 1982).

Seismicity in Baffin Bay occurs landward of the 2000 m bathymetry contour. The 1933 M 7.3 earthquake in northern Baffin Bay is the largest from northeastern Canada. Although Stein and others (1979) calculated a focal depth of 65 km for this event, a revised depth based on additional data is only 25–30 km (N. H. Sleep, oral com., 1983). Revised fault-plane solutions for the 1933 earthquake are essentially the same as the original solutions by Stein and others (1979; S. Stein, written com., 1983) and indicate reverse faulting resulting from approximately N-S compression.

In sharp contrast to the compressional nature of Baffin Bay seismicity, earthquakes in the continental crust beneath adjacent Baffin Island have fault-plane solutions that indicate normal faulting (Sykes, 1970; Hashizume, 1973; Stein and others, 1979; Liu and Kanamori, 1980). Stein and others (1979) suggested that flexure due to glacial unloading can account for both the normal-fault seismotectonic regime along northeastern Baffin Island and the thrust-fault regime under Baffin Bay. However, this interpretation may not be unique since other density-derived edge effects at continent-ocean boundaries could produce a similar pattern of deformation (Bott and Dean, 1972; Artyushkov, 1973). Moreover, finite element calculations by Quinlan (1981) suggest that crustal stresses induced by glacial unloading in northeastern Canada are rarely large enough to dictate the type of earthquake mechanism.

Continental North America

Instrumental and historical seismicity data in the eastern half of continental North America (i.e., east of the Rocky Mountain front) delineate several large clusters and a few diffuse trends (Fig. 1; Plate 11—in pocket). The more prominent seismic zones include New Madrid, western Quebec, Cape Ann, Charlevoix, and lower St. Lawrence. Smaller pockets of seismicity occur in central New Brunswick and Charleston, South Carolina. A diffuse trend is apparent along the Appalachians and a curvilinear trend along the Boothia Uplift–Bell Arch region in northern Canada.

Major historical earthquakes in the continental mid-plate regions of North America and their estimated magnitudes are shown on Figure 1 and include: the 1811–1812 New Madrid, Missouri earthquakes (3 events with $M \geqslant 7.0$); the 1886 $M \geqslant 7.0$ Charleston, South Carolina earthquake; the 1638 and 1755 earthquakes off-shore from Cape Ann, Massachusetts, (M uncertain); and the 1663 and 1925 St. Lawrence Valley earthquakes (both $M \sim 7.0$). The above data indicate how important historical seismic data are in providing information about major earthquakes in eastern North America. Two of these areas, Charleston and the Cape Ann region, currently have low levels of seismic activity and would probably be overlooked as seismically hazardous regions without the historical evidence of earthquakes.

The above continental seismic zones and the major historic earthquakes have been discussed at length by various authors (Basham and others, 1977; Basham and others, 1979; Rankin, 1977; Sykes, 1978; McKeown and Pakiser, 1982; Gohn, 1983). As the focus of this chapter is on the western North Atlantic, only a few general characteristics of these seismic zones are described here. It is perhaps noteworthy, however, that three of the aforementioned sites of major earthquakes lie near the eastern continental margin.

A significant characteristic of continental mid-plate deformation is the apparent long return time between major events, which implies lower rates of strain and stress build-up in mid-plate regions as compared with plate boundary regimes. Several lines of evidence suggest that continental seismic zones may coincide with zones that have experienced repeated episodes of deformation throughout geologic time (Sykes, 1978; Zoback and Zoback, 1981). Despite apparent high historic deformational rates suggested by the frequency and estimates of average dislocation (e.g., see Nuttli, 1983) of major events, it is unlikely that such high rates persist at any one site for time periods on the order of a million years or more. If this were the case, geologic evidence of on-going deformation, such as fault scarps, would be conspicuous. Seismic reflection data from sites of major historic earthquakes (primarily Charleston and New Madrid) indicate very low long-term deformation rates, orders of magnitude lower than might be inferred from current and historic seismicity (Zoback and others, 1980; Hamilton and others, 1983). In sharp contrast, trenching studies in the New Madrid area (Russ, 1979), suggest that the high historic rate of deformation may have persisted throughout Holocene time.

During the last two decades, several hypotheses have been proposed that suggest a causal relation between certain continental seismic zones, oceanic fracture zones, and faulting that is related to proto-Atlantic rifting and ocean basin formation (e.g., Sykes, 1978). The Grand Banks earthquake of 1929, discussed earlier, occurred near the Newfoundland fracture zone, which is a transform boundary between continental and oceanic crust (Fletcher and others, 1978). Farther south, the apparent correlation among the trends of the New England Seamounts, the Boston-Ottawa seismic zone (a NW-SE trending zone of earthquakes), Mesozoic intrusions in New England, and the Ottawa-Bonnechere graben was pointed out by Diment and others (1972), Sbar and Sykes (1973) and Fletcher and others (1978). However, this correlation has been questioned by Yang and Aggarwal (1981) and Forsyth (1981). The landward extension of the Norfolk fracture zone has been correlated with the central Virginia seismic zone, Eocene intrusives, and Cenozoic folding and faulting along the James River (Sykes, 1978). The epicenter of the 1886 Charleston earthquake lies close to a postulated extension of the Blake Spur fracture zone and the northern edge of a major Triassic basin beneath the southeastern coastal plain (Fletcher and others, 1978; Nishenko and Sykes, 1979).

Beneath the continental shelf, Klitgord and Behrendt (1979) demonstrate that a number of fracture zones, as defined by major offsets in the East Coast Magnetic Anomaly and the Blake Spur Anomaly, coincide with the boundaries of major sedimentary basins. These crustal boundaries most probably mark the location of initial offsets in the separation of Africa from North America that were propagated as transform faults into the adjacent ocean basin. Tracing these initial crustal offsets onland, however, has been difficult and questionable. In perhaps the best-studied example, on the continental margin off Charleston, South Carolina, seismic reflection data have been used to delineate a shallow ridge that extends landward along the Blake Spur fracture zone (Dillon and others, 1983, p. N5). However, a landward extension of this ridge would intersect the coast near Savannah, Georgia. Thus, detailed geophysical data from the continental shelf do not support a link between the Blake Spur fracture zone and the Charleston seismic zone.

The relatively high levels of seismicity in the vicinity of the Ramapo fault zone in southeastern New York and New Jersey and in the Charlevoix region are thought by some to be examples of reactivation of proto-Atlantic rift faults. However, a detailed correlation of regional fault patterns and epicentral data in the Ramapo area does not support such a simple interpretation; the modern seismicity and also possibly the Triassic Ramapo fault may be localized by major zones of weakness ("crustal shear zones") of early Paleozoic and Precambrian age (Ratcliffe, 1980, 1981). Furthermore, recent analysis of focal mechanisms in the vicinity of the Ramapo suggest that much of the current seismic activity is occurring on faults that strike obliquely to the Ramapo fault zone (Seborowski and others, 1982; Statton and others, 1982).

Comparison of Mid-plate Continental and Oceanic Deformation

Perhaps the most striking similarity between the continental and oceanic seismicity data is the pattern of seismicity, consisting of concentrated zones of activity in which major events appear in clusters. The instrumentally recorded data are also beginning to verify the historic continental pattern of concentrated zones of deformation in which several large events have occurred in close proximity. It is significant, however, that no surface fault scarps or evidence of major repeated long-term deformation have been positively identified in any of the continental seismic zones. This suggests that the zones of clustered activity move around (possibly on time scales less than about a million years). The available instrumental data also indicate approximately the same number of moderate to large events ($M > 4.5$) for the comparably sized continental and oceanic intraplate regions of the North American plate. Only a few major events ($M \geqslant 7.0$) have occurred in either area during the period of instrumental recording, although historic data indicate about 6 or 7 major events on land in the last 200 years. Obviously the oceanic record is incomplete for this time period.

Another similarity between oceanic and continental seismicity is a possible correlation of seismicity with major intraplate structural and topographic features (Sykes, 1978; Stein, 1979; Bergman and Soloman, 1980). This correlation can be demonstrated more convincingly in the oceanic regions (e.g., Bermuda Rise); the greatly deformed continental crust generally affords a multitude of possible pre-existing structures that may be susceptible to reactivation.

An important difference between the continental and oceanic seismicity is the lithospheric level of the respective events. Depth ranges for oceanic events are commonly 10–40 km, which places many of the earthquakes in the upper mantle, whereas continental events (generally restricted to depths less than 25 km—see Herrmann, 1979, and Basham and others, 1979) occur within the highly deformed crust. In independent analyses of the depth distribution of intraplate earthquakes, Chen and Molnar (1983) and Wiens and Stein (1983) note that the maximum focal depths, particularly those for oceanic events, increase with increasing age of the lithosphere. They suggest that the depth dependence of crust and mantle rheology, as influenced by the regional thermal gradient, is probably the critical factor in determining whether seismic (brittle) deformation occurs or not.

Information on long term deformation must come from geologic data. As noted above, there are no positively identified surface fault scarps or other geomorphic manifestations in any of the currently active seismic areas in the continental mid-plate region. Perhaps this is partly because young sediments are rare and generally quite restricted laterally. It is thus generally impossible to date bedrock offsets in this area, although minor fault offsets of glacially striated bedrock surfaces have been noted. These occurrences of faulted glacial striations are generally restricted to highly foliated metamorphic rocks and probably reflect shallow adjustment to glacial drag and regional adjustment to removal of the ice load.

Numerous reverse fault offsets of Tertiary sediments occur along the Atlantic Coastal Plain (Prowell, 1983). Also, seismic reflection data have revealed numerous small faults with evidence of repeated movement in several of the currently active seismic zones (Zoback, 1979; Behrendt and others, 1983; Hamilton and others, 1983). Where datable, the vertical offsets on these faults indicate extremely low long-term deformation rates (e.g., Zoback and others, 1980); although, such an inference may not be valid if much of the deformation is strike-slip in nature. Earthquake-induced sand blows in Quaternary sediments near Charleston, South Carolina suggest recurrent moderate to large earthquakes in this area during the last 5000 + years (Obermeier and others, 1984).

Offshore reflection data on the continental shelf have also revealed the presence of faulting. Behrendt and others (1983) discuss evidence for limited amounts of NE-trending Cenozoic reverse faulting in the area immediately off Charleston, South Carolina. Sheridan and Knebel (1976) found post-Pleistocene faulting offshore from New Jersey, along a fault zone whose position and trend coincided with the East Coast Magnetic Anomaly. The offset was 1.5 m at shallow levels and about 90 m at deeper levels. Hutchinson and others (1983) reported a fault zone that offsets a Quaternary section about 10 m in the New York Bight near Long Island. This N- to NE-trending fault offsets the basement about 85 m.

Thus, both the offshore (shelf) and onshore data provide evidence for repeated offsets on faults. No seismic evidence for mid-plate sediment deformation has been found in the deep-water western North Atlantic, but small offsets would be hard to resolve. Despite seemingly large short-term deformation rates (up to 10 m Quaternary offset in the New York Bight), the long-term rates (85 m post-Cretaceous basement offset at the same site) are small. Thus the nature of faulting may be an indication that the locus of deformation may migrate sporadically and return to a specific fault or fault zone only after very long time intervals.

MID-PLATE STATE OF STRESS

Maximum horizontal principal stress orientations can be inferred from various types of data. Stress field indicators, and their associated uncertainties, have been discussed at length by numerous authors (for detail, refer to: Sbar and Sykes, 1973; Zoback and Zoback, 1980; Hasegawa and others, 1985). Stress orientations for the current study were inferred from four main indicators: earthquake focal mechanisms, geologic evidence of recent deformation, hydraulic fracturing in-situ stress measurements, and measurement of stress-induced borehole elongation ("breakouts"). If all the problems with the application and interpretation of all these techniques are taken into consideration, then the accuracy of any given orientation is probably on the order of 10°–20° (Zoback and Zoback, 1980). Near-surface overcoring stress measurements often show considerable scatter (e.g., see Hasegawa and Adams, 1981, for a compilation of stress measurements in eastern Canada, particularly for the Canadian Shield) and are not included in the current compilation.

Patterns of Mid-plate Stress and Stress Regime

The stress data for the mid-plate region of the North American plate is given on Plate 11 and Figure 2. Details for the oceanic stress data points are given in Table 1. A detailed tabulation of all of the continental stress data is beyond the scope of this article; the data are principally taken from Zoback and Zoback (1980), Hasegawa and Adams (1981), Bell and Gough (1979), Gough and others (1983), and Zoback and others (1984).

Geological, seismological, and available in-situ stress data all suggest a NE to ENE compressive stress regime (characterized by strike-slip or reverse faulting) throughout the mid-plate region. While stress orientations from focal mechanisms in the oceanic regions are sparse, the available data suggest that the oceanic part of the plate is also characterized by the same NE to ENE orientation of the maximum horizontal compressive stress. The position of the transition between extensional events in young oceanic crust and compressional events at older ages is currently controversial. The compressional event closest to the Mid-Atlantic Ridge (event WA-3.2, Table 1) suggests that the boundary between the mid-plate and inter-plate deformation in the North American plate lies approximately 230 km east of the Mid-Atlantic Ridge in the western North Atlantic.

Exceptions to this general pattern of NE to ENE compressive stress are found in the vicinity of the Antilles arc and along the margins of continental North America. The state of stress in the southwestern region of the mid-continent of the United States also appears to represent an exception to the regional pattern. The state of stress in this area is probably profoundly influenced by its proximity to the actively extending Rio Grande rift, which is, in turn, related to active deformation throughout the western Cordillera and, consequently, is outside the scope of this paper (refer to Zoback and Zoback, 1980 and McGarr, 1982 for a discussion of the state of stress in the southwestern mid-continent region).

Along the eastern seaboard of the northeastern United States (between about 40° and 45°N latitude), focal mechanism studies by several workers (Yang and Aggarwal, 1981; Graham and Chiburis, 1980) suggest an approximately 100 km wide zone of NW compressive stress along the coast. A similar zone of NW compression was proposed by Zoback and Zoback (1980) and Wentworth and Mergner-Keefer (1983) to extend southward into the southeastern United States based on a compilation of fault offsets of Tertiary sediments along the Atlantic coastal plain by Prowell (1983). These data indicate reverse offsets along generally NE-striking, steeply-dipping faults or fault zones. In areas where fault planes were exposed, the sense of offset was nearly pure reverse dip-slip (D. C. Prowell, oral com., 1984).

The evidence for NW compression in the southeastern United States, however, is restricted to the fault data. A few focal mechanisms are now available for the southeast, primarily in South Carolina (see Talwani, 1982; Talwani and others, 1980) and all indicate a NE to N 60° E maximum compressive stress orientation. These results are consistent with orientations inferred from borehole elongations (breakouts) and in-situ stress measurements in two wells in active seismic areas in South Carolina (Hickman and Zoback, written com., 1983).

Recent focal mechanism investigations by Statton and others (1982) and C. T. Statton (written com., 1984) in southeastern New York and northern New Jersey failed to reproduce the evidence of NW compressive stress in this region reported by Yang and Aggarwal (1981). Upon reanalysis of Yang and Aggarwal's data, Statton found many of the focal mechanisms to be possible but not unique solutions for the reported events. In his reanalysis Statton utilized a grid-search methodology to determine all valid fault plane solutions for a given set of first motion data and gave careful consideration to the effects of uncertainty in focal depth as well as the validity and/or reliability of individual first motion picks. These results combined with new, well-constrained focal mechanism solutions suggest a compressive stress regime with a NE to ENE maximum compressive stress direction for the northern New Jersey and southeastern New York area (Seborowski and others, 1982; Houlday and others, 1984; Quittmeyer and others, 1985). Only stress orientations inferred from P-axes of these newer and better constrained mechanisms are included on the map in Figure 2.

Focal mechanism solutions for the coastal New England area were done by Yang and Aggarwal (1981), Graham and Chiburis (1980), and Pulli and Toksoz (1981). Most of these events are poorly constrained because of poor station coverage. In fact, comparison of one event studied by both Yang and Aggarwal and Graham and Chiburis indicated solutions whose P-axes differed by about 90°. Only the most reliable events are plotted on Figure 2 and, yet, the P-axes have azimuths over nearly a 360° range.

A large group of new stress data from analysis of borehole elongation in wells in the northeastern United States and southeastern Canada also indicate a relatively uniform NE to ENE compressive stress field throughout this region (Cox, 1983; Plumb and Cox, 1985). Data from southeasternmost Canada and the Scotian shelf yield a consistent compressive stress orientation of N 50°–55° E. Data from the Appalachian basin from Kentucky to New York indicate a mean horizontal compressive stress direction of about N 55°–60° E. A similar orientation (about N 50° E) was also obtained by Plumb and Cox for 6 wells within the Appalachian fold belt proper in West Virginia, Virginia, and northernmost Tennessee. These new borehole elongation data are not shown on the map but are included in a manuscript in preparation by Plumb and Cox (see abstract, Plumb and Cox, 1985).

These new data appear to cast serious doubt on the existence of a previously proposed, regionally extensive Atlantic Coast stress province characterized by NW compression. Focal mechanism data in the New York–New Jersey–New England area still appear equivocal but are suggestive of uniform NE to ENE compression. The bulk of the evidence for NW compression lies with the small reverse fault offsets of Tertiary age, primarily in the southeastern U.S. In detail, this fault data does not agree with the best available nearby focal mechanism and borehole elongation data. The significance of the young fault offsets as reliable indicators of the modern stress field must be questioned. No satisfactory explanation of all the data, however, has been proposed.

Thus, with few exceptions, the state of stress in the interior portions of the continental North American plate indicates a compressional stress regime with a maximum horizontal stress oriented NE to ENE. Sparse data from the western North Atlantic suggest that within the accuracy of stress orientations inferred from focal mechanisms a similar stress field may exist throughout the entire mid-plate region; the one exception in the western Atlantic basin is the seismicity in the Caribbean region near the plate boundaries of the North American, South American, and Caribbean plates. This deformation may be best categorized as a zone of diffuse plate-boundary deformation.

SOURCES OF MID-PLATE STRESS

Plate Driving Forces

The relative magnitudes of the forces that drive and resist plate motions remain difficult to constrain. For North America, essentially without any attached subducting slab, the most likely plate driving forces are ridge push and basal drag due to relative motion between the lithosphere and asthenosphere in a direction essentially parallel to absolute plate motion.

The ridge-push force due to the thermal elevation of mid-ocean ridges and the associated thickening of oceanic lithosphere is easily modeled in terms of reasonably well-known values for the density, bathymetry, and thickness of young oceanic lithosphere. The magnitude of the ridge force is on the order of 3×10^{12} Nm^{-1}. If this force is supported across the full 100 km thickness of the plate it is equivalent to a compressive stress of 30 MPa (300 bars). If, as is more likely, stress is concentrated in the stronger portions of the plate near the surface, then compressive stresses in the upper few tens of km due to ridge push could be on the order of 600 bars.

Shear forces on transform boundaries may resist plate motion, stress the plates, and hence play a role in the driving mechanism. The eastern U.S. stress field is probably not influenced by the San Andreas fault because the Basin and Range Province, probably the weakest part of the North American plate, would deform and relax any shear stress transmitted from the San Andreas fault.

The role of basal shear stresses on the plate is perhaps the least well constrained of any of the potential forces in the driving mechanism. Small basal shears, on the order of a few bars, can produce significant forces when typical horizontal plate dimensions of a few thousand km are considered.

Several possible generalized models of mantle flow patterns have been proposed. In one model mantle material flows essentially in a SW direction beneath the North American plate, or parallel to the absolute plate velocity. In another proposed model the mantle flow pattern is determined by the flow of material from subduction zones to mid-ocean ridges globally and results in essentially a SE-directed flow beneath North America. For both models, velocities of the flow are relatively small beneath the North American plate as compared to other plates, especially when compared to largely oceanic plates. In yet other models one may assume that the mantle is passively providing resistance and a balancing torque to plate motion.

Localized Sources of Stress

Lateral variations in lithospheric density structure, lithospheric flexure resulting from sediment loading or unloading by erosion, glaciation-deglaciation, and igneous intrusions can all produce localized crustal-scale stress fields. However, both lateral density variations at the continent-ocean boundary and lithospheric flexure models for sediment loading or erosional unloading predict deviatoric extensional stresses in the continental regions, not deviatoric compression as observed (Zoback and Zoback, 1980; Zoback and Zoback, 1981). Glacial loading can create deviatoric horizontal compression under the ice load and deviatoric horizontal extension ahead of the glacier (Walcott, 1970). Conversely, glacial unloading can induce stresses in the crust that are opposite to those induced during loading (see Stein and others, 1979). There is evidence to support the view that induced stresses related to glaciation-deglaciation may contribute to the seismicity in much of the eastern Canada (Basham and others, 1977; Wetmiller and Forsyth, 1982; Price, 1979). However, Sykes (1978) could determine no apparent correlation in the northeastern United States between glaciation effects and seismicity. Documentation of faults with postglacial offsets in some of the exposed areas north of the Great Lakes and in the Canadian Maritime Provinces indicates that faults tend to be parallel to the retreating ice fronts (Adams, 1981).

Most mid-plate regions with stress orientations inconsistent with regional NE to ENE compression occur along the continental margin of eastern North America. These regions include the Baffin Bay area of northern Canada (possible glacial effects) and the Gulf coastal plain (sediment loading), both discussed previously. As discussed above, the available data along the Atlantic Seaboard are still somewhat equivocal, yet now appear to indicate NE compression and contradict a previously proposed, regionally extensive NW compressive stress province. Difficulties with localized sources for producing a NW compressive stress regime along the Atlantic Coast area have been addressed in detail by Zoback and Zoback (1980, 1981).

Finite-Element Modeling of the Intraplate Stress Field

To understand the deformation and state of stress within plates, it is necessary to quantitatively model the relationship between plate geometry, rheology, and the forces acting on the plates. With finite-element modeling it is possible to include complex geometry and physical properties that may vary spatially. Richardson and others (1979) used an elastic finite-element analysis to study the driving mechanism for plate tectonics; their results were tested against a global data set of intraplate stress data. Richardson (1983), applying the correspondence principle, extended the finite-element analysis using a viscous formulation. The major advantage of the viscous formulation is that it permits the utilization of both the intraplate stress field and the relative motions between the plates. Based on both the elastic and viscous global finite-element modeling, ridge push forces are the best predictor of both observed relative plate velocities and mid-plate stress orientations.

State of stress in the North American plate has been investigated using a single plate, elastic, finite-element analysis of plate-driving forces acting on the plate (Reding, 1984). Two classes of models were considered, ridge push models (in which the ridge push forces are distributed over the oceanic lithosphere) and driving basal drag models. To balance the net torque on the plate due to ridge forces, resistance was provided by either pinning the western plate margin or with resistive basal drag forces. Predicted stress orientations are nearly identical in either case and are plotted in red on Figure 2. Stress magnitudes are fairly constant in the former case, and decrease from east to west in the latter case. In either case, ridge push forces account for the observed stress orientations. The second class of models, driving drag models, also require an additional force to balance the torque, in this case modeled by pinning the western plate boundary. Such models also predict a general NE to ENE orientation of maximum horizontal compressive stress.

The major difference between the ridge push and the driving drag models is not the orientation of the predicted stresses, but rather that ridge-produced stresses show either very little change in magnitude with distance or a decrease from east to west, while the drag-induced stresses are at a minimum at the ridge and increase essentially linearly across the plate from east to west. Thus, one way to distinguish between the models is with accurate stress magnitude data. Unfortunately, stress magnitude data are quite limited both in number and in depth (compared to crustal or plate thickness). Even if a more complete stress magnitude data set were available, it is unlikely that these data could conclusively distinguish between the ridge push and driving drag models. These simple models do not take into account any local variations within the plate itself, either due to local geologic setting or to additional localized sources of stress. Simple plate driving drag is not likely to be a dominant plate driving force because, for very large plates such as the Pacific, it predicts almost a factor of one hundred in the stress amplification across the plate that is not reflected in the intraplate stress field or tectonics. Further, driving drag forces should be proportional to plate area, but plates of very different areas such as the Pacific and Cocos plates have very similar velocities. Thus, based on global considerations, the ridge push models are preferred for the North American plate.

CONCLUSIONS

The state of stress in much of the mid-plate region of the North American plate is compressive, with a NE to ENE oriented maximum horizontal principal stress. This mid-plate compressive stress field appears to extend from the Rocky Mountain front eastward across the continental margin to within 230 km of the Mid-Atlantic Ridge as indicated by the closest studied event in the vicinity of the ridge. While localized stresses may contribute to failure, the overall uniformity in the mid-plate stress pattern suggests a far field source. The currently available stress data do not support the existence of a previously proposed distinct stress province along the Atlantic seaboard characterized by NW compression (i.e., perpendicular to the continental margin). A likely source for the plate-wide NE to ENE compressive stress field is a distributed ridge push force that originates along the Mid-Atlantic Ridge plate boundary. This ridge push force may be balanced by normal forces across boundaries between the North American and surrounding plates and by a small, resistive basal drag force.

The level and pattern of mid-plate deformation as inferred from seismicity appears comparable in the oceanic and continental regions. Major events are often clustered and are occurring in widely spaced seismic zones. Available data on longer-term deformation rates (from wells and seismic reflection profiling) indicate much lower average long-term rates than might be inferred from Holocene or Quaternary deformation in a given area.

A correlation of seismicity with pre-existing structural or topographic features can be inferred in both continental and oceanic regions. This correlation is stronger for the oceanic events; the highly deformed continental crust affords a multitude of pre-existing structures that could be reactivated.

Evidence for linking current continental deformation to an onshore projection of oceanic fracture zones is weak. Strictly speaking, of course, the fracture zones cannot exist beyond the boundaries of the oceanic and transitional crust in which they formed. There is, however, good evidence, albeit much of it is still circumstantial, that major zones of weakness in the continental crust influenced the breakup geometry of the North Atlantic and the initial development of fracture zones. Whether these same zones are responsible for the present day patterns of intraplate seismicity as originally envisioned, however, is still not clear. In perhaps the best studied example, detailed geophysical investigations on the shelf offshore from Charleston, South Carolina fail to support the link between the oceanic fracture zone and the onshore seismicity.

The range of focal depths of oceanic and continental intraplate earthquakes have considerable overlap (0–25 km for continental events, 10–40 km for oceanic events); however, they correspond to different lithospheric levels. The oceanic events occur largely in a seemingly rather homogeneous upper mantle (based on velocity structure) whereas the continental events occur within the upper two-thirds of the highly deformed crust. While the occurrence of brittle deformation is strongly influenced by thermal regime (affecting rheology; e.g., Chen and Molnar, 1983; Wiens and Stein, 1983), the similarities between oceanic and continental seismicity suggests that the oceanic upper mantle may be much more heterogeneous than is commonly believed.

REFERENCES

Abe, Katsuyaki
1981 : Magnitudes of large shallow earthquakes from 1904 to 1980: Physics of Earth and Planetary Interiors, v. 27, p. 72–92.

Abe, Katsuyaki and Noguchi, Shin'ichi
1983 : Revision of magnitudes of large shallow earthquakes, 1897–1912: Physics of Earth and Planetary Interiors, v. 33, p. 1–11.

Adams, J.
1981 : Postglacial faulting: A literature survey of occurrences in eastern Canada and comparable glaciated areas: Atomic Energy of Canada, Technical Record no. 142, 63 p.

Artyushkov, E. V.
1973 : Stresses in the lithosphere caused by crustal thickness inhomogeneities: Journal of Geophysical Research, v. 78, p. 7675–7708.

Basham, P. W., and Adams, J.
1982 : Earthquake hazards to offshore development on the eastern Canadian continental shelves: Proceedings Second Canadian Conference on Marine Geotechnical Engineering, Dartsmouth, Nova Scotia, 6 p.

Basham, P. W., Forsyth, D. A., and Wetmiller, R. J.
1977 : The seismicity of northern Canada: Canadian Journal of Earth Sciences, v. 14, p. 1646–1667.

Basham, P. W., Weichert, D. H., and Berry, M. J.
1979 : Regional assessment of seismic risk in eastern Canada: Bulletin of the Seismological Society of America, v. 69, p. 1567–1602.

Behrendt, J. C., Hamilton, R. M., Ackerman, H. D., and Henry, V. J.
1981 : Cenozoic faulting in the vicinity of the Charleston, South Carolina, 1886 earthquake: Geology, v. 9, p. 117–122.

Behrendt, J. C., Hamilton, R. M., Ackermann, H. D., Henry, V. J., and Bayer, K. C.
1983 : Marine multi-channel seismic-reflection evidence for Cenozoic faulting and deep crustal structure near Charleston, South Carolina: U.S. Geological Survey Professional Paper 1313, p. J1–J29.

Bell, J. S., and Gough, D. I.
1979 : Northeast-southwest compressive stress in Alberta: evidence from oil wells: Earth and Planetary Science Letters, v. 45, p. 475–482.
1981 : Intraplate stress orientations from Alberta oil wells; in Evolution of the Earth, Geodynamics Series, v. 5; American Geophysical Union, p. 96–104.

Bergman, E. A., and Solomon, S. C.
1980 : Oceanic intraplate earthquakes: Implications for local and re-

gional intraplate stresses: Journal of Geophysical Research, v. 85, p. 5389–5410.
1984 : Source mechanisms of earthquakes near mid-ocean ridges from body waveform inversion: implications for the early evolution of oceanic lithosphere: Journal of Geophysical Research, v. 89, p. 11,415–11,441.
Bott, M.H.P., and Dean, D. S.
1972 : Stress systems at young continental margins: Nature, v. 235, p. 23–25.
Chen, W.-P., and Molnar, P.
1983 : Focal depth of intracontinental and intraplate earthquakes and their implications for the thermal and mechanical properties of the lithosphere: Journal of Geophysical Research, v. 88, p. 4183–4214.
Cox, J. W.
1983 : Long axis orientation in elongated boreholes and its correlation with rock stress data: Society of Professional Well Logging Analysts Twenty-fourth Annual Logging Symposium Transactions, v. 1, p. 1–17.
Crough, S. T.
1978 : Thermal origin of mid-plate hot-spot swells: Geophysical Journal of the Royal Astronomical Society, v. 55, p. 451–469.
Dewey, J. W.
1972 : Seismicity and tectonics of western Venezuela: Seismological Society of America Bulletin, v. 62, p. 1711–1751.
Dillon, W. P., Klitgord, K. D., and Paull, C. K.
1983 : Mesozoic development and structure of the continental margin of South Carolina; in Studies related to the Charleston, South Carolina, earthquake of 1886—tectonics and seismicity, ed. G. S. Gohn; U.S. Geological Survey Professional Paper 1313, p. N1–N16.
Diment, W. H., Urban, T. C., and Revetta, F. A.
1972 : Some geophysical anomalies in the eastern United States; in The Nature of the Solid Earth, ed. E. Robertson; McGraw-Hill, New York, p. 544–574.
Doxsee, W. W.
1948 : The Grand Banks earthquake of November 18, 1929: Publications of the Dominion Observatory Ottawa, v. 7, p. 323–335.
Drake, C. L., Campbell, N. J., Sander, G., and Nafe, J. E.
1963 : A mid-Labrador Sea Ridge: Nature, v. 200, p. 1085.
Dziewonski, A. M., and Woodhouse, J. H.
1983 : An experiment in the systematic study of global seismicity: Centroid-moment tensor solutions for 201 moderate and large earthquakes of 1981, Journal of Geophysical Research, v. 88, p. 3247–3272.
Ewing, M., Worzel, J. L., Beall, A. O., Berggren, W. A., Burky, D., Burke, C. A., Fisher, A. G., and Pessagno, E. A., eds.
1969 : Initial Reports of the Deep-Sea Drilling Project, Volume 1, U.S. Government Printing Office, Washington, D.C., 672 p.
Fletcher, J. B., Sbar, M. L., and Sykes, L. R.
1978 : Seismic trends and travel time residuals in eastern North America and their tectonic implications: Geological Society of American Bulletin, v. 89, p. 1656–1676.
Forsyth, D. A.
1981 : Characteristics of the western Quebec seismic zone: Canadian Journal of Earth Science, v. 18, p. 103–119.
Frohlich, C.
1982 : Seismicity of the central Gulf of Mexico: Geology, v. 10, p. 103–106.
Gohn, G. S.
1983 : Studies related to the Charleston, South Carolina, earthquake of 1886—tectonics and seismicity: U.S. Geological Survey Professional Paper 1313, 375 p.
Gough, D. I., Fordjor, C. K., and Bell, J. S.
1983 : A stress province boundary and tractions on the North American plate: Nature, v. 305, p. 619–621.
Graham, T., and Chiburis, E. F.
1980 : Fault plane solutions and the state of stress in New England: Earthquake Notes, v. 51, p. 3–12.
Grant, A. C.
1975 : Structural modes of the western margin of the Labrador Sea; in Offshore geology of eastern Canada, Volume 2, eds. W.J.M. Van der Linden, and J. A. Wade; Geological Survey of Canada Paper 74-30, p. 217–231.
Gregersen, S.
1982 : Earthquakes in Greenland: Bulletin of Geological SOciety of Denmark, v. 31: Copenhagen, p. 11–27.
Habermann, R. E.
1982 : Consistency of teleseismic reporting since 1963: Bulletin of the Seismological Society of America, v. 72, p. 93–111.
Hamilton, R. M., Behrendt, J. C., and Ackerman, H. D.
1983 : Land multi-channel seismic-reflection for tectonic features near Charleston, South Carolina; in Studies related to the Charleston, South Carolina, earthquake of 1886—tectonics and seismicity, ed. G. S. Gohn; U.S. Geological Survey Professional Paper 1313, p. I1–I18.
Hasegawa, H. S., and Adams, J.
1981 : Crustal stresses and seismotectonics in eastern Canada: Energy, Mines and Resources Canada, Earth Physics Branch Open File Report 81-12, Ottawa, Canada, 62 p.
Hasegawa, H. S., Adams, J., and Yamazaki, K.
1985 : Upper crustal stresses and vertical stress migration in eastern Canada: Journal of Geophysical Research, v. 90, p. 3637–3648.
Hashizume, M.
1973 : Two earthquakes on Baffin Island and their tectonic implications: Journal of Geophysical Research, v. 78, p. 6069–6081.
1977 : Surface-wave study of the Labrador Sea earthquake 1971 December: Geophysical Journal of the Royal Astronomical Society, v. 51, p. 149–168.
Haxby, W. F., and Turcotte, D. L.
1978 : On isostatic ground anomalies: Journal of Geophysical Research, v. 83, p. 5473–5478.
Heezen, B. C., and Ewing, M.
1952 : Turbidity currents and submarine slumps, and the 1929 Grand Banks earthquake: American Journal of Science, v. 250, p. 849–873.
Herrmann, R. B.
1979 : Surface wave focal mechanisms for eastern North American earthquakes with tectonic implications: Journal of Geophysical Research, v. 84, p. 3543–3552.
Houlday, M., Quittmeyer, R., Mrotek, K., and Statton, C. T.
1984 : Recent seismicity in north- and east-central New York State: Earthquake Notes, v. 55, no. 2, p. 16–20.
Hutchinson, D. R., Grow, J. A., and Hulit, M. B.
1983 : New York Bight Fault: Geologic Society of America, Abstracts with Programs, 1983, North East Section, p. 138.
Hyndman, R. D.
1975 : Marginal basins of the Labrador Sea and the Davis Strait hot spot: Canadian Journal of Earth Sciences, v. 12, p. 1041–1045.
Keen, C. E.
1982 : The continental margins of eastern Canada: a review; in Dynamics of Passive Margins, ed. R. A. Scruton; Geodynamics Series, v. 6, p. 45–58.
Keen, C. E., and Hyndman, R. D.
1979 : Geophysical review of the continental margins of eastern and western Canada: Canadian Journal of Earth Sciences, v. 16, p. 712–747.
Keen, C. E., and Keen, M. J., Ross, D. I., and Lack, M.
1974 : Baffin Bay: Small ocean basin formed by sea-floor spreading: American Association of Petroleum Geologists Bulletin, v. 58, p. 1089–1108.
Klitgord, K. D., and Behrendt, J. C.
1979 : Basin structure of the U.S. Atlantic Margin: American Association of Petroleum Geologists Memoir 29, p. 85–112.
Kristoffersen, Y., and Talwani, M.
1977 : Extinct triple junction south of Greenland and the Tertiary motion of Greenland relative to North America: Geological Society of America Bulletin, v. 88, p. 1037–1049.
LePichon, X., Hyndman, R. D., and Pautot, G.
1971 : Geophysical study of the opening of the Labrador Sea: Journal of

Geophysical Research, v. 76, p. 4724–4743.
Liu, H. L., and Kanamori, H.
1980 : Determination of source parameters of mid-plate earthquakes from the waveforms of body waves: Bulletin of the Seismological Society of America, v. 70, p. 1989–2004.
Martin, R. G., and Case, J. G.
1975 : Geophysical studies in the Gulf of Mexico; in The Ocean Basins and Margins, Volume 3, eds. A.E.M. Nairn, and F. G. Stehli; Plenum, New York, p. 66–69.
McGarr, A.
1982 : Analysis of state of stress between provinces of constant stress: Journal of Geophysical Research, v. 87, p. 9278–9288.
McKeown, F. A., and Pakiser, L. C.
1982 : Investigations of the New Madrid, Missouri, earthquake region: U.S. Geological Survey Professional Paper 1236, 201 p.
Menzies, A. W.
1982 : Crustal history and basin development of Baffin Bay; in Nares Strait and the drift of Greenland: a conflict in plate tectonics: eds. P. R. Dawes, and J. W. Kerr; Meddelelser om Grønland, Geoscience, v. 8, p. 295–312.
Minster, J. B., and Jordan, T. H.
1978 : Present day plate motions: Journal of Geophysical Research, v. 83, p. 5331–5354.
Morgan, W. J.
1981 : Hotspot tracks and the opening of the Atlantic and Indian Oceans, in The Sea, Volume 7, ed. C. Emiliani; J. Wiley and Sons, New York, p. 443–487.
Morgan, W. J., and Crough, S. T.
1979 : Bermuda hotspot and the Cape Fear arch: EOS, (Transactions American Geophysical Union), v. 60, p. 392–393.
Molnar, P., and Sykes, L. R.
1969 : Tectonics of the Caribbean and Middle America regions from focal mechanisms and seismicity: Geological Society of America Bulletin, v. 80, p. 1639–1684.
Nishenko, S. P., and Kafka, A. L.
1982 : Earthquake focal mechanisms and the intraplate setting of the Bermuda Rise: Journal of Geophysical Research, v. 87, p. 3929–3941.
Nishenko, S. P. and Sykes, L. R.
1979 : Fracture zones, Mesozoic rifts and the tectonic setting of the Charleston, South Carolina earthquake of 1886: EOS, (Transactions American Geophysical Union), v. 60, p. 887.
Nishenko, S. P., Purdy, G. M., and Ewing, J. I.
1982 : Microaftershock survey of the 1978 Bermuda Rise earthquake: Journal of Geophysical Research, v. 87, p. 10624–10636.
Nuttli, O. W.
1983 : Average seismic source-parameter relations for mid-plate earthquakes: Bulletin of the Seismological Society of America, v. 73, p. 519–535.
Obermeier, S. F., Goh, G. S., Weems, R. E., Gelinas, R. L., and Rubin, M.
1984 : Geologic evidence for recurrent moderate to large earthquakes near Charleston, South Carolina: Science, v. 227, p. 408–411.
Parsons, B., and Daly, S.
1983 : The relationship between surface topography, gravity anomalies and the temperature structure of convection: Journal of Geophysical Research, v. 88, p. 1129–1144.
Plumb, R. A., and Cox, J. W.
1985 : Deep crustal stress directions in eastern North America determined from borehole elongation: EOS (Transactions American Geophysical Union), v. 65, p. 1081–1082.
Price, N. J.
1979 : Fracture patterns and stresses in granites: Geoscience Canada, v. 6, p. 209–212.
Prowell, D. G.
1983 : Index of Cretaceous and Cenozoic faults in the eastern United States: U.S. Geological Survey Miscellaneous Field Studies Map, MF-1269.
Pulli, J. J., and Toksoz, M. N.
1981 : Fault plane solutions for northeastern United States earthquakes: Bulletin of the Seismological Society of America, v. 71, p. 1875–1882.
Quinlan, G. M.
1981 : Numerical models of postglacial relative sea level change in Atlantic Canada and the eastern Canada Artic: [unpublished Ph.D. thesis], Dalhousie University, 499 p.
Quittmeyer, R. C., Statton, C. T., Mrotek, K. A., and Houlday, M.
1985 : Possible implications of recent microearthquakes in southeastern New York state: Earthquake Notes, in press.
Rankin, D. W.
1977 : Studies related to the Charleston, South Carolina, earthquake of 1886—A preliminary report: U.S. Geological Survey Professional Paper 1028, 204 p.
Ratcliffe, N. M.
1980 : Brittle faults (Ramapo fault) and phyllomitic ductile shear zones in the basement rocks of the Ramapo seismic zones, New York and New Jersey, and their relationship to current seismicity; in Field studies of New Jersey geology and guide to field trips, ed. W. Manspeizer; 52nd annual meeting of the New York State Geological Association, p. 278–312.
1981 : Reassessment of the Ramapo fault system as control for current seismicity in the Ramapo seismic zone and the New York recess: Geological Society of America Abstracts with Programs, v. 13, p. 171.
Reding, L. M.
1984 : North American plate stress modeling: A finite element analysis: [unpublished M.S. thesis], University of Arizona, Tucson, AZ, 111 p.
Reynolds, P. H., and Aumento, F.
1974 : Deep Drill 1972: Potassium-Argon dating of the Bermuda drill core: Canadian Journal of Earth Sciences, v. 11, p. 1269–1273.
Richardson, R. M.
1983 : Inversion for the driving forces of plate tectonics: Proceedings IEEE International Geoscience and Remote Sensing Symposium, II, p. FA 2.3.6.
Richardson, R. M., Solomon, S. C., and Sleep, N. H.
1979 : Tectonic stress in the plates: Reviews of Geophysics and Space Physics, v. 17, p. 981–1019.
Roots, W. D., and Srivastava, S. P.
1984 : Origin of the marine magnetic quiet zones in the Labrador and Greenland Seas: Marine Geophysical Researches, v. 6, p. 395–408.
Russ, D. P.
1979 : Late Holocene faulting and earthquake recurrence in the Reelfoot Lake area, northwestern Tennessee: Geological Society of America Bulletin, v. 90, p. 1013–1018.
Sbar, M. L., Richardson, R. M., Flaccus, C. E., and Engelder, T.
1984 : Near-surface in situ stress, 1: strain relaxation measurements along the San Andreas fault in southern California: Journal of Geophysical Research, v. 89, p. 9323–9332.
Sbar, M. L., and Sykes, L. R.
1973 : Contemporary compressive stress and seismicity in eastern North America: an example of intra-plate tectonics: Geological Society of America Bulletin, v. 84, p. 1861–1882.
Schouten, H. and Klitgord, K. D.
1982 : The memory of the accreting plate boundary and the continuity of fracture zones: Earth and Planetary Sciences Letters, v. 59, p. 255–266.
Seborowski, D. D., Williams, G., Kelleher, J. A., and Statton, C. T.
1982 : Tectonic implications of recent earthquakes near Annsville, New York: Bulletin Seismological Society of America, v. 72, p. 1606–1609.
Sheridan, R. E., and Knebel, H. J.
1976 : Evidence of post-Pleistocene faults on New Jersey outer continental shelf: American Association of Petroleum Geologists Bulletin, v. 60, p. 1112–1117.
Sieberg, A.
1932 : Earthquake geography; in Handbuch der Geophysik, v. 4, ed. B.

Gutenberg; Berlin, p. 687–1005.
Srivastava, S. P.
1978 : Evolution of the Labrador Sea and its bearing on the early evolution of the North Atlantic: Geophysical Journal of the Royal Astronomical Society, v. 52, p. 313–357.
Srivastava, S. P., and Falconer, R.K.H.
1982 : Nares Strait: a conflict between plate tectonic predictions and geological interpretations; in eds. P. R. Dawes, and J. W. Kerr; Meddelelser om Grønland, Geoscience, v. 8, p. 339–352.
Statton, C. T., Quittmeyer, R. C., and Houlday, M.
1982 : Contemporary stress as inferred from recent seismicity in New York and New Jersey: Earthquake Notes, v. 53, p. 36.
Stein, S.
1979: Intraplate seismicity on bathymetric features: the 1968 Emperor Trough earthquake: Journal of Geophysical Research, v. 84, p. 4763–4768.
Stein, S., Sleep, N. H., Geller, R. J., Wang, S. C., and Kroeger, G. C.
1979 : Earthquakes along the passive margin of eastern Canada: Geophysical Research Letters, v. 6, p. 537–540.
Stein, S., Engeln, J. F., Weins, D. A., Speed, R. C., and Fujita, K.
1982 : Subduction seismicity and tectonics in the Lesser Antilles arc: Journal of Geophysical Research, v. 87, p. 8642–8664.
Stewart, G. S., and Helmberger, D. V.
1981 : The Bermuda earthquake of March 24, 1978: a significant oceanic intraplate event: Journal of Geophysical Research, v. 86, p. 7027–7036.
Sundvik, M., Larson, R. L., and Detrick, R. S.
1984 : Rough-smooth basement boundary in the western North Atlantic basin: evidence for seafloor-spreading origin: Geology, v. 12, p. 31–34.
Sykes, L. R.
1970: Focal mechanism solutions for earthquakes along the world rift systems: Bulletin of the Seismological Society of America, v. 60, p. 1749–1752.
1978 : Intraplate seismicity, reactivation of preexisting zones of weakness, alkaline magmatism, and other tectonism postdating continental fragmentation: Reviews of Geophysical and Space Physics, v. 16, p. 621–688.
Sykes, L. R., and Sbar, M. L.
1974 : Focal mechanism solutions of intraplate earthquakes and stresses in the lithosphere; in Geodynamics of Iceland and the North Atlantic area, ed. L. Kristjansson; D. Reidel Publishing Company, Boston, Massachusetts, p. 207–224.
Talwani, P. D.
1982 : Internally consistent pattern of seismicity near Charleston, South Carolina: Geology, v. 10, p. 654–658.
Talwani, P. D. Rastogi, B. K., and Stevenson, D.
1980 : Induced seismicity and earthquake prediction studies in South Carolina, Tenth Technical Report, Contract 14-08-0001-16809: U.S. Geological Survey, Reston, Virginia.
Tucholke, B. E., and Mountain, G. S.
1979 : Seismic stratigraphy, lithostratigraphy and paleosedimentation patterns in the North Atlantic basin; in Deep Drilling Results in the Atlantic Ocean, Continental Margins and Paleoenvironments, eds. M. Talwani, W. Hay, and W.B.F. Ryan; Maurice Ewing Series, vol. 3: American Geophysical Union, Washington, D.C., p. 58–86.
Tucholke, B. E., Hougtz, R. E., and Ludwig, W. J.
1982 : Sediment thickness and depth to basement in western North Atlantic basins: American Association of Petroleum Geologists Bulletin, v. 66, p. 1384–1395.
van der Linden, W.J.M.
1975 : Crustal attenuation and sea-floor spreading in the Labrador Sea: Earth and Planetary Science Letters, v. 27 p. 409–423.
Vogt, P. R., Johnson, G. L., Holcombe, T. L., Gilg, J. G., and Avery, D. E.
: 1971: Episodes of seafloor spreading recorded by North Atlantic basement: Tectonophysics, v. 12, p. 211–234.
Walcott, R. I.
1970 : Isostatic response to loading of the crust, in Canada: Canadian Journal of Earth Science, v. 7, p. 716–727.
Wiens, D. A., and Stein, S.
1983 : Age dependence of oceanic intraplate seismicity and implications for lithospheric evolution, Journal of Geophysical Research, 88, p. 6455–6468.
1984 : Intraplate seismicity and stresses in young oceanic lithosphere, Journal of Geophysical Research, v. 89, p. 11,422–11,464.
Wentworth, C. M., and Mergner-Keefer, M.
1983 : Regenerate faults of small Cenozoic offset: probable earthquake sources in the southeastern United States; in Studies related to the Charleston, South Carolina, earthquake of 1886–tectonics and seismicity, ed. G. S. Gohn; U.S. Geological Survey Professional Paper 1313, p. S1–S20.
Wetmiller, R. J., and Forsyth, D. A.
1982 : Review of seismicity and other geophysical data near Nares Strait, in Nares Strait and the drift of Greenland: a conflict in plate tectonics: eds., P. R. Dawes, and J. W. Kerr; Meddelelser om Grønland, Geoscience, v. 8, p. 261–274.
Yang, J. P., and Aggarwal, Y. P.
1981 : Seismotectonics of northeastern United States and adjacent Canada: Journal of Geophysical Research, v. 86, p. 4981–4998.
York, J. E., and Oliver, J. E.
1976 : Cretaceous and Cenozoic faulting in eastern North America: Geological Society of America Bulletin, v. 87, p. 1105–1114.
Zoback, M. D.
1979 : Recurrent faulting in the vicinity of Reelfoot Lake, northwestern Tennessee: Geological Society of America Bulletin, v. 90, p. 1019–1024.
Zoback, M. D., Hamilton, R. M., Crone, A. J., Russ, D. P., McKeown, F. A., and Brockman, S. R.
1980 : Recurrent intraplate tectonism in the New Madrid seismic zone: Science, v. 209, p. 971–976.
Zoback, M. D., and Zoback, M. L.
1981 : State of stress and intraplate earthquakes in the United States: Science, v. 213, p. 96–104.
Zoback, M. L., and Zoback, M. D.
1980 : State of stress in the conterminous United States: Journal of Geophysical Research, v. 85, p. 6113–6156.
Zoback, M. L., Zoback, M. D., and Schiltz, M. E.
1984 : Index of stress data for the North American and parts of the Pacific plate: U.S. Geological Survey Open-File Report, 84-157, 62 p.

Manuscript Accepted by the Society July 1, 1985
Energy, Mines, and Resources Canada Contribution from the Earth Physics Branch No. 1157

ACKNOWLEDGMENTS

We thank J. Adams, P. W. Basham, M. J. Berry, J. W. Dewey, A. G. Green, and W. R. Thatcher for helpful comments on the manuscript. Mara E. Schiltz generated the computer-plotted versions of the stress and seismicity maps and W. E. Shannon generated a magnetic tape of eastern Canada seismicity. Paloma Nishenko drafted Plate 11.

Printed in U.S.A.

The Geology of North America
Vol. M, The Western North Atlantic Region
The Geological Society of America, 1986

Chapter 19

Seismic structure of the ocean crust

G. M. Purdy
John Ewing
Woods Hole Oceanographic Institution, Woods Hole, Massachusetts 02543

INTRODUCTION

Several notable papers have been published recently that review our knowledge of the seismic structure of oceanic crust (e.g. Christensen and Salisbury, 1975; Tréhu and others, 1976; Kennett, 1977; Lewis, 1978; Ewing and Houtz, 1979; Spudich and Orcutt, 1980a; Houtz, 1980; White, 1984). To reduce duplication of those previous efforts, this paper will emphasize evaluation of the resolving power of existing seismic data in the Western Atlantic Ocean and will compare this evaluation with the scale of local and regional variations in crustal structure predicted by reasonable geological models. We shall not focus attention on the relationship between the seismic and petrological structure of the crust and will use the terms Layer 2, Layer 3, and Moho only in their original seismic sense. Layer 2 is a region of high velocity gradients (perhaps as high as $3s^{-1}$ at its upper surface), with velocities in the range 3 to 6.5 km/s that extends from the seafloor to depths of 2–3 km. (We neglect sediments in this chapter, restricting ourselves to the igneous crustal column.) Layer 3 is typically about 4 km in thickness, has velocities of 6.7 to 7.2 km/s, and is characterized by low velocity gradients, generally less than 0.1 s^{-1}. The Mohorovicic discontinuity (Moho) is actually a transition zone varying in thickness from a few hundred meters (or less) to 1–2 km and is almost always the most prominent seismic characteristic of the ocean crust. Velocities increase by about 1 km/s across this transition to the typical values of about 8 km/s found in the upper mantle. The uppermost 5–10 kilometers of oceanic upper mantle can be defined as a homogeneous (but sometimes recognizably anisotropic) region with velocities within a few tenths of a kilometer per second of 8 km/s, and low velocity gradients that are certainly less than 0.1 s^{-1} and are perhaps negative in some locations.

The primary source of data on the seismic structure of the Western Atlantic ocean crust is provided by almost two hundred seismic refraction and wide angle reflection experiments carried out during the past thirty years. (Ewing, 1963). Many techniques were used to collect these data: the classic two ship method (Shor, 1963); ratio telemetering or internally recording sonobuoys (Hill, 1963; LePichon and others, 1968); ocean bottom seismometers or hydrophones (Ewing and Ewing, 1961; Purdy, 1983) and, most recently, expanding spread profiles using a multichannel hydrophone array (Stoffa and Buhl, 1979) and oblique seismic experiments using borehole seismometers (Stephen and others, 1980; Stephen, 1981). Sound sources for these experiments vary from 40 cu. in. (0.66 l) airguns fired at 50m intervals to 500 kg explosive charges detonated at spacings of as much as 6 km. Many of these experiments were carried out before the advent of satellite navigation, before the widespread use of airgun reflection profiling to routinely map sub-seafloor basement topography, before the seafloor spreading process was fully understood and certainly before the tectonic framework of the ocean basin was known well enough to correctly influence the choice of experiment location.

The interpretation methods applied to these various data sets have varied from the 'slope-intercept' technique that provides isovelocity layered solutions (e.g. Ewing and Ewing, 1959), through the delay time function approach to study upper mantle anisotropy and the two dimensional form of the Moho (e.g. Keen and Tramontini, 1970), to forward modelling of the phase and amplitude of individual waveforms by calculating synthetic seismograms (e.g. Fowler, 1976). Most recently, particle motion studies of three component borehole seismometer data have been used to define anisotropy in the shallow crust (Stephen, 1981). This wide range of experiment type, quality, and interpretation approach is perhaps one of the primary reasons that no simple model has been accepted for the seismic structure of Western Atlantic ocean crust.

A seismic refraction experiment, despite its complexity and expense, typically results in a single compressional-wave velocity versus depth function from which inferences concerning the crustal structure may be drawn. How much, or how little, information is contained in this single measurement of seismic structure? What are the primary parameters that determine crustal seismic velocity structure and what can these parameters tell us about the geology of the ocean crust? A simplistic answer to this question is given in Figure 1.

The compressional wave velocity at a particular location

Purdy, G. M., and Ewing, J., 1986, Seismic structure of the ocean crust; *in* Vogt, P. R., and Tucholke, B. E., eds., The Geology of North America, Volume M, The Western North Atlantic Region: Geological Society of America

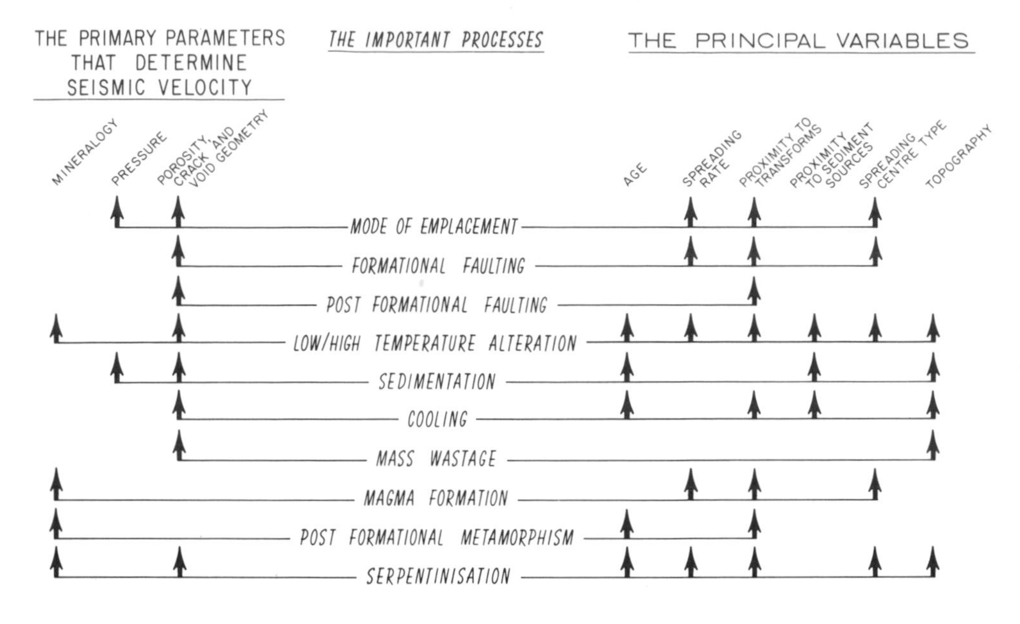

Figure 1. Factors that determine the seismic velocity structure of oceanic crust. See text for discussion.

within the oceanic crust is principally dependent upon the types of minerals that make up the rock matrix (e.g. Christensen and Salisbury, 1975), the ambient external pressure and the pore fluid pressure (e.g. Spudich and Orcutt, 1980a), and the porosity and crack and void geometry (Hyndman and Drury, 1976; Schreiber and Fox, 1977; Spudich and Orcutt, 1980b). One can conceive of many processes that determine these parameters: for example, the emplacement process controls whether the rock is formed as an extrusive pillow lava (with high porosity and random void geometry) or as dikes (with the possibility of more orderly large pore or crack geometry and lower porosity). At a slow spreading ridge the crust, immediately after its emplacement, is subjected to extensive normal faulting to raise it from the median valley floor into the crestal mountains. This faulting could play an important role in modifying the porosity and crack geometries. The same applies, of course, to faulting associated with other tectonic events, for example, transform faulting or seamount formation. The complex processes of geochemical alteration of oceanic crust are not fully understood but it is clear that they influence seismic velocity, not only through the sealing of cracks and pores, but also by mineralogical changes to the rock matrix (e.g. Christensen and Salisbury, 1972, 1973; Humphris and Thompson, 1978; Staudigel and others, 1981). Sediment cover will not only directly modify the porosity and, by its load, the external pressure within the crust, but can also seal the seafloor and thus influence the convective hydrothermal circulation that is a major control on the geochemical alteration process. In the case of a young, narrow ocean formed at the initiation of rifting, the insulating effect of thick sediment cover may be the dominant factor influencing the crustal formation. As the newly formed lithosphere cools and subsides, its velocity structure is not only modified by the direct relationship to temperature but also by the contribution to the total porosity provided by cooling-related fissuring. Submersible observations over young crust show that in regions of steep topography, debris flows are common and are an important process, moving loose high porosity extrusive materials into topographic lows. The mineralogy of the magma source is different for a back arc basin spreading center compared with, for example, the Mid-Atlantic Ridge. Metamorphic alteration of the crust after its formation may be associated with ridge jumps, seamount formation, or transform fault related processes and will modify the petrological makeup of the crust. Penetration of water into the crust will inevitably result in serpentinization to degrees varying from bands centered on fault traces (Francis, 1981) to the formation of diapiric structures. The principal controls on these various processes are the crustal age, the spreading rate and type of spreading center at which the crust was formed, its proximity to transforms and sediment sources and lastly, simply, the topography around the site in question.

The function of this discussion and Figure 1 is to emphasize the breadth and complexity of the problem we are tackling. An important component of this complexity is the many widely differing scales on which the principal variables determining the seismic velocity structure of oceanic crust operate. The scales range from thousands of kilometers (e.g. spreading center type), through hundreds of kilometers (e.g. age or proximity to sediment source) to tens of kilometers (e.g. proximity to transforms), and finally down to kilometers or less (e.g. topography). One of the primary difficulties to be faced in reviewing the seismic data in the Western Atlantic is simply the fact that it has all been collected on a scale of a few tens of kilometers, with consideration given only rarely to tailoring the scale of the experiment to the scale of the process that it is intended to elucidate.

THE OCEAN CRUST

The locations of the seismic refraction stations in the

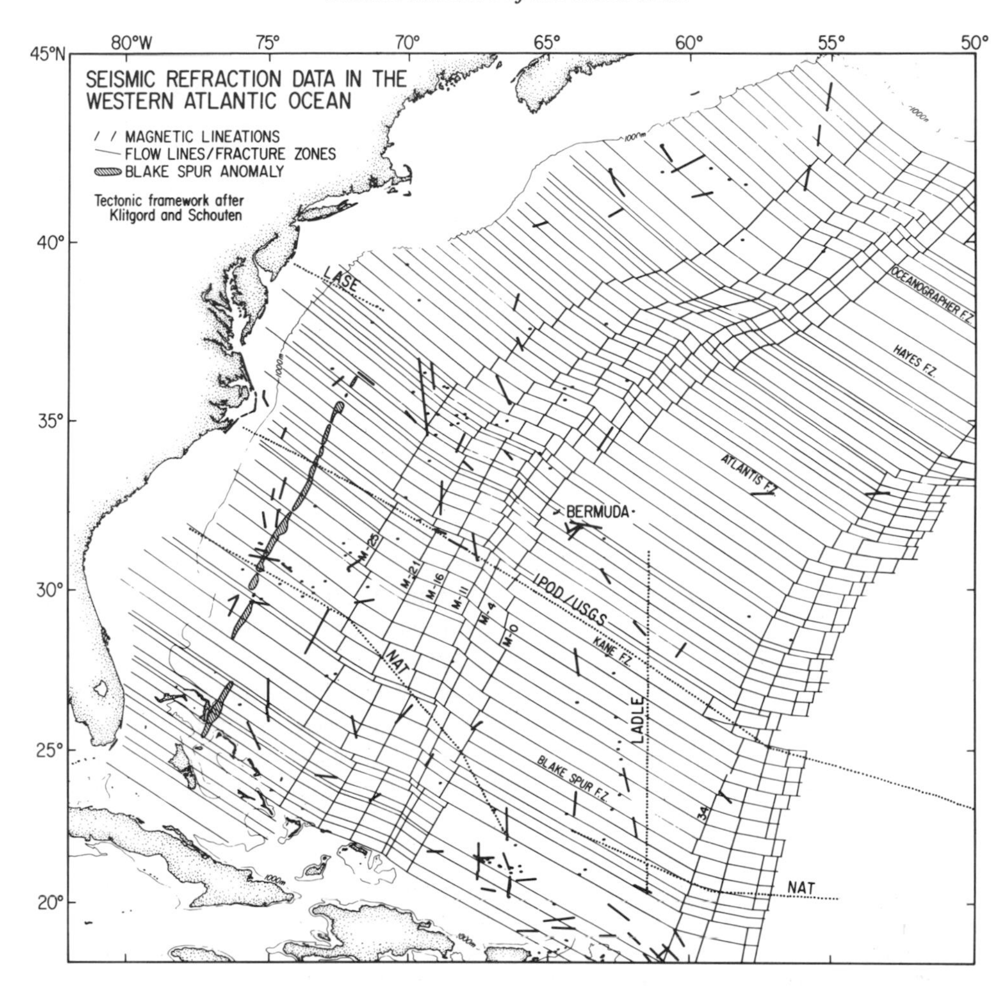

Figure 2. The locations of the seismic refraction experiments, the results from which were used in the compilation of the histogram shown in Figure 3. The tectonic framework is that of Klitgord and Schouten (this volume) and shows the principal magnetic lineations and the fracture zone locations as inferred from flow line calculations. Heavy lines denote reversed refraction profiles. The short lines denote the receiver location for unreversed lines. Also shown are the locations of the IPOD/USGS multichannel reflection line (Grow and Markl, 1977), the North Atlantic Transect (NAT: Buhl and others, 1983), Large Aperture Seismic Experiment (LASE: LASE Study Group, n.d. a and b) and the Lesser Antilles Deep Lithosphere Experiment (LADLE: LADLE Study Group, 1983).

Western Atlantic Ocean that provide the base data for this study of its structure are shown in Figure 2 and a compilation of the results in the form of a histogram is shown in Figure 3. The most obvious features of the experiment location chart are the bias of the data sampling toward the older crust and the large proportion of experiments that cross or are close to fracture zones. The prominent feature of the data histogram is the broad scatter of velocities from 3 to 8 km/s, with the only clear peaks being those at the Layer 3 velocity (6.8 km/s) and at the upper mantle velocity of about 8 km/s. No experiments specifically designed to study the Mid-Atlantic Ridge, fracture zone, or continental margin structures have been included in this compilation; they will be treated separately in the following sections.

The historical dataset is adequate to define the mean crustal structure, but it has not yet revealed any substantial or well determined systematic changes in the complete crustal column. It is the mapping of systematic changes in structure that can provide useful data upon which new models for crustal processes could be based, and thus the search for such trends should be a primary objective of crustal structure studies. Two efforts, notable because of their completeness and rigor, that searched specifically for trends related to age but without significant success, are those of Tréhu and others (1976) and Houtz (1980). The variability revealed by these studies is random, with the exception of the increase in upper Layer 2 velocities as the crust ages (Houtz and Ewing 1976) and the apparently greater thickness of the late Jurassic crust. This stimulates several questions: Are the primary structural changes random in their nature and distribution, or do systematic patterns exist that are beyond the resolution of the currently available data? What is the resolving power of the available dataset? Does the historical dataset provide a true representation of the variability in normal oceanic crustal structure? To investi-

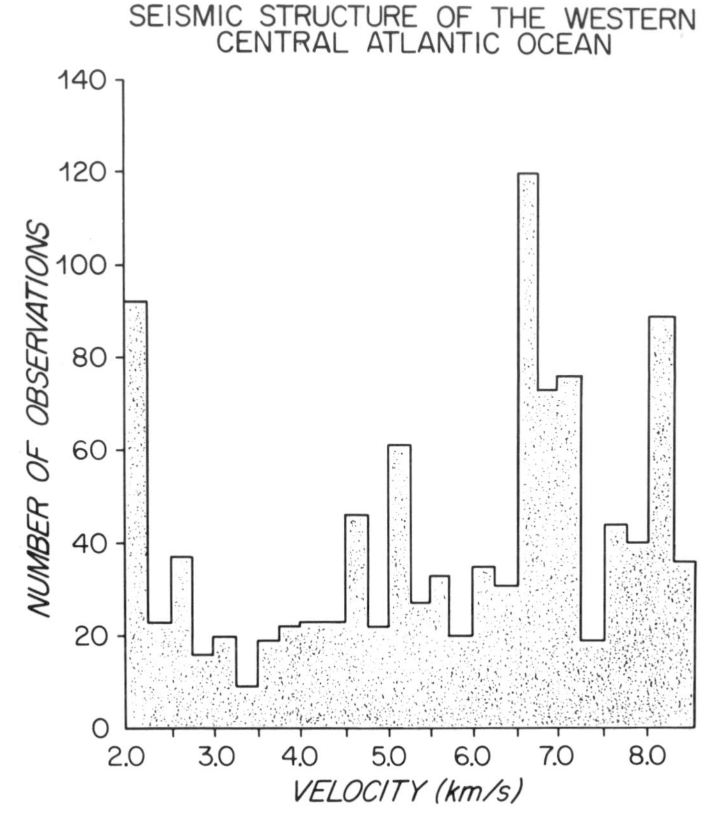

Figure 3. Histogram of recorded seismic velocities in the Western Atlantic Ocean. Locations of the experiments are shown in Figure 2. Data used from: Bentley and Worzel, 1956; Bunce and Fahlquist, 1962; Bunce and others, 1969; Detrick and Purdy, 1980; Ewing and Ewing, 1959; Ewing and Houtz, 1979; Ewing and others, 1952; Ewing and others, 1950; Gaskell and others, 1958; Gaskell and Swallow, 1951; Hersey and others, 1959; Houtz, 1980; Houtz and Ewing, 1963 and 1964; Katz and Ewing, 1956; McConnell and others, 1966; Northrop and Ransone, 1962; Officer and Ewing, 1954; Officer and others, 1952; Officer and others, 1959; Sheridan and others, 1966; Sheridan and others, 1979; Tolstoy and others, 1953; Trehu and others, 1976.

gate these questions we set up a simplistic model of oceanic crust, the primary characteristic of which is that the isovelocity contours are conformal with the basement surface.

These model calculations were carried out as described by Purdy (1982), with modifications of adding a flat-topped isovelocity sediment layer above the model and introducing a second analytical function at the base of the crust to represent the Moho and upper mantle (Fig. 4). Because the origin for the function shown in this figure is always maintained at the basement surface, the isovelocity contours within the igneous crust are always conformal with the basement surface, and accurate ray tracing is possible through a velocity field that can be completely described analytically. The spline smoothing function describing the basement surface was computed from some randomly chosen young Western Atlantic ocean bathymetry data (Fig. 4).

In this way we can calculate error-free travel time datasets over a reasonable model that may be representative of true oceanic crustal structure. We can carry out seismic refraction experiments in the computer and vary the shot spacing, the experiment location, or the type of experiment, and study how our results vary. In this paper we shall present the results of four such experiments; two used a 1 km shot spacing and two used a 3 km shot-spacing. For each shot spacing two sea surface locations were used for the receiver; one at a horizontal coordinate 3 km and one a horizontal coordinate 20 km (Fig. 4b). But at no time was the velocity model changed in any way. We will simply be interpreting four experiments, two of which are located less than 20 km from the other two and have shot spacings that differ from 1 to 3 km.

In our first experiments we take the two sets of data (i.e. one each for the 3 km and 20 km receiver coordinate) and pick the first arrival in each 3 km range window to simulate a typical shooting pattern of some of the early experiments. These data were then corrected for changes in sediment thickness using a horizontal datum 0.85 km beneath the seafloor and assuming a sediment velocity of 2 km/s and an upper basement velocity of 4 km/s. We allowed ourselves a correct knowledge of the basement morphology but not of the ray entry points. Straight line segments were then placed through the data (Fig. 5a) and solutions calculated using the straightforward slope-intercept method. The travel time data and slope intercept interpretation for the first experiment is shown in Figure 5a. The velocity depth functions are shown in Figure 4. Before we discuss the differences between these two results, we will describe the third and fourth experiment. In this case we allowed ourselves a 1 km shot spacing, the sediment thickness corrections were computed as before, but the increased data density allowed the fitting of smooth curves through the data (Fig. 5b). The two resulting velocity-depth functions are shown in Figure 4. These four experiments, that strictly contain no errors, yield very different results for what is geometrically and geologically the same structure, that is, a faulted basement surface beneath which exists a constant velocity depth function. This is despite the fact that we have allowed no errors in travel time or range (inevitable in reality) and have allowed ourselves a perfect knowledge of basement morphology. The variability shown in Figure 4a is thus a conservative minimum estimate of what we would expect to see in practice (given the validity of our crustal model). Total crustal thicknesses are seen to vary by about 2 km, Layer 3 velocities seem to be reasonably constant but the uppermost two kilometers of the crust are extremely variable. In the slope-intercept solutions it was necessary to assume the existence of an uppermost basement layer, and the velocity used for this was 4 km/s. In the travel time inversions the linear gradient assumption of Ewing and Purdy (1982) was used to model the shallowest section of Layer 2 from which no arrivals are observed. There are three primary causes of this variability that we arrange qualitatively in order of decreasing importance:

(i) Inaccurate corrections for sediment thickness variations due to lack of knowledge of phase velocity through the water column and ray entry point into the basement.

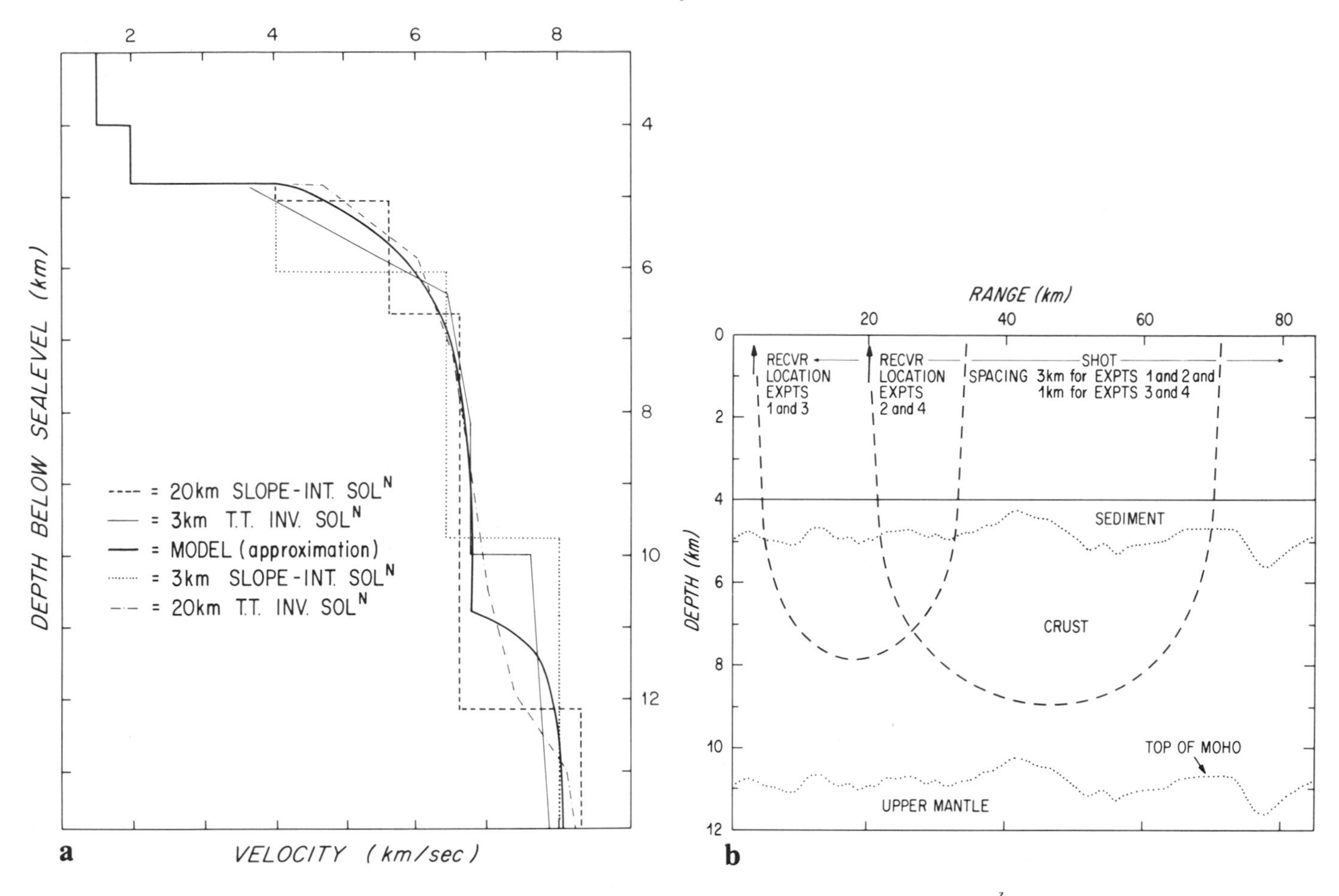

Figure 4. a. The solid line is the model velocity vs. depth function defined by $V = V_0 + G_0\int_0^z \exp\text{-}\alpha z dz$ for $0 < z < 6$ km, $V_0 = 4.0$ km/s, $G_0 = 2.8\text{s}^{-1}$, and $\alpha = 1.0\ \text{km}^{-1}$ for $z > 6$ km, $G_0 = 2.5\text{s}^{-1}$ and $\alpha = 2.0\text{km}^{-1}$. Z is measured downward from the basement surface as defined by the spline function illustrated in Figure 4b. V_0 and G_0 are the velocity and gradient at the seafloor, and α is an arbitrary constant that determines the rate of decrease of gradient with depth. The four solutions result from model calculations with the receiving location at 3 km and 20 km range coordinate in Figure 4b using 1 km shot spacing to give the gradient solutions and 3 km shot spacing to give the slope-intercept solutions. b. The spline fit to a section of typical Atlantic ocean basement topography used in the model ray tracing calculations; sea floor is at 4 km depth and is flat and horizontal. The receiver locations at coordinates 3 and 20 km are marked by the arrows.

(ii) Misidentification of phases during interpretation due to large shot spacing (a classic example of spatial alisasing.)
(iii) Lack of direct observations on the shallowmost crust due to the geometry of the experiment (Ewing and Purdy, 1982) thus requiring some assumption.

We contend these simple results are sufficient cause to question whether the histogram shown in Figure 3 is truly representative of Western Atlantic structure. Given the type of experiments that constitute the bulk of the historical data set, the variability in structure may largely be an artifact and the real structure could be (though we doubt that it is) as simple as that used in our model calculations. The model calculations described here misrepresent the resolving power of data that have been interpreted using techniques that take account of amplitudes. Bratt and Purdy (1984), using a dataset of 500 seismograms from an experiment on the flanks of the East Pacific Rise show that systematic patterns in the amplitude distribution define lateral changes in structure that are not observable in the travel times.

Another important cause of the scatter in velocity structure observed in the Western Atlantic Ocean may be the almost random location of the experiments relative to the inferred fracture zone traces (Fig. 2). As will be discussed later in this chapter, fracture zones are regions of significant heterogeneity in seismic structure, with crustal thicknesses reduced to as little as 2–3 km. Because of the many closely spaced fracture zones in the Western Atlantic Ocean any attempt to characterize the variability of normal crustal structure must only use those experiments that have been carefully located within the tectonic framework (Purdy 1983). We contend that too few good experiments exist for this to be a meaningful process. We seem to be only too ready to accept

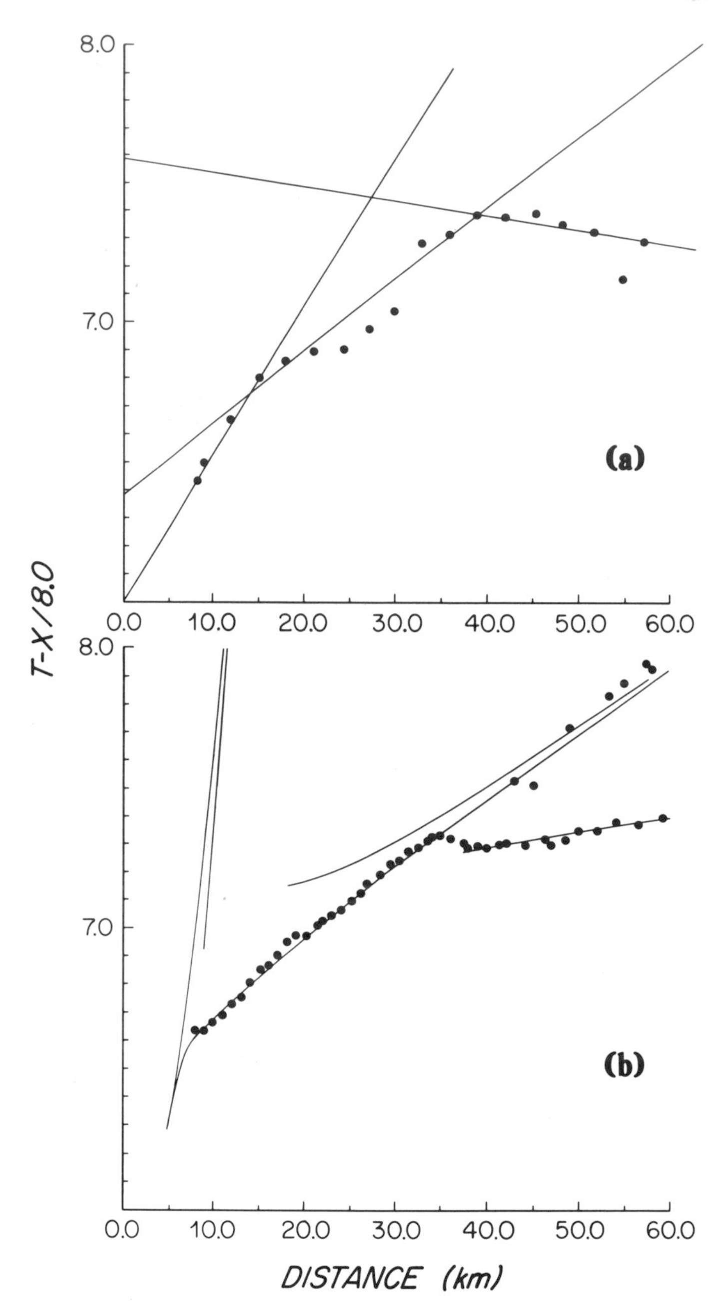

Figure 5. a. Computed travel time data for receiver at 3 km coordinate and 3 km shot spacing (dots) with best fit slope-intercept segments shown. Solution is shown in Figure 4. b. Computed travel time data for receiver at 20 km coordinate and 1 km shot spacing (dots) with best fit model travel time curves. Solution is shown in Figure 4.

the notion, supported by the histogram in Figure 3, that the seismic structure of oceanic crust is randomly variable with only the Layer 3 and upper mantle velocities exhibiting any uniformity. We are encouraged to this conclusion by the geological evidence concerning the structure of oceanic crust. Evidence for considerable structural heterogeneity exists in data collected by ocean floor photography and manned submersibles (e.g. Ballard and others, 1975), deep ocean drilling (e.g. Aumento and others, 1977) and from inferences drawn from studies of ophiolite complexes (e.g. Casey and others, 1981). But the relationship between these results and those illustrated in Figure 3 is not straightforward primarily because of the very different scales for which the various datasets have validity. Although the wavelength of the seismic energy used to make the determinations under discussion is 0.5–1.0 km, a typical refraction experiment will not resolve structural changes on this scale. Bratt and Purdy (1984) present a qualitative discussion of this issue and conclude that heterogeneities must reach several tenths of kilometers per second and exist over lateral distances of many kilometers before they can be truly resolved and defined by a typical seismic refraction data set.

One well controlled and precisely located experiment on 140 m.y. old crust southwest of Bermuda deserves special mention because it provides a data set that is large enough to average out errors and heterogeneities too small to be defined. It is supported by sufficient auxiliary data to allow adequate corrections for basement topography to be made and it is positioned within a tectonic framework well defined by aeromagnetic data. This experiment also recorded converted shear waves that were consistent with an average Poisson's ratio throughout the igneous crust of 0.28. This was not a well constrained determination but is worthy of mention because few others exist for the Western Atlantic Ocean (cf. Detrick and Purdy, 1980; Hyndman, 1979). The compressional velocity depth function resulting from this experiment is shown in Figure 6 (from Purdy, 1983) and represents our best estimate of the typical structure of old Atlantic crust. The principal features of the section are described in the figure caption, and a comparison with the structure of young Atlantic crust is given in the discussion at the end of this chapter.

An exceptional data set available for the study of the uppermost crustal section is that collected using airguns and sonobuoys. These are valuable data not only because of the relatively large number of experiments that have been carried out, but also because the 50m shot spacing allows more confident determinations of phase velocities to be made without so much concern with spatial aliasing problems. The experiments sample the crust over lateral distances of often as little as 5km. It is notable that these are the only data that have provided evidence of order in the seismic structure of oceanic crust. These are the data used by Houtz and Ewing (1976) to show that the seismic velocity of the uppermost kilometer increases by 1–2 km/s during the first few tens of millions of years due, perhaps, to the sealing of cracks and pores in the extrusive lavas. This is the best documented systematic process determining one particular characteristic of the seismic structure. If more seismic experiments were carried out on a scale appropriate to the scale of the process they were intended to elucidate, then a simpler and more systematic view of the seismic structure of oceanic crust might be revealed.

A second unique data set is that provided by multichannel seismic reflection profiling systems. The single largest published source of data of this type in the Atlantic is the IPOD/USGS line that extends from Cape Hatteras to the Mid-Atlantic Ridge at 22°N (Grow and Markl, 1977). The location of this line is shown in Figure 2. A reflector associated with the Moho can be identified along most of the line that is over oceanic crust older than approximately 80 m.y. No reflectors within the crust can be traced over lateral distances greater than a few kilometers. This is in contrast with the recently completed North Atlantic Transect Experiment (Hinz and others, 1982) that, with the application of a large airgun array and synthetic (10 km) aperture, common depth point reflection profiling techniques, succeeded in mapping Moho reflections essentially continuously over crust older than 60 my, and identified important reflections within the crust at ages greater than about 100–110 my (Mutter, Detrick and others, 1984). These new reflection techniques will have a major impact on our knowledge of crustal structure because of their innate capability (when reflectors are present) of resolving lateral structural changes on a much smaller scale and with far superior accuracy than the highest quality refraction experiment.

The emphasis of this chapter has been on the variability in structure of the crust, with little discussion of the upper mantle. This is dictated by the type of data available and not by any scientific priority. An understanding of the structure of the subcrustal lithosphere is the key to advancing our knowledge of plate formation and evolution, and yet only four experiments in the Atlantic Ocean have addressed this goal: two attempts to define upper mantle velocity anisotropy have been made in the Atlantic by Keen and Tramontini (1970) and Whitmarsh (1971). The former presents a poorly constrained determination of ±0.25 km/s velocity deviation and the latter finds no evidence to support the existence of velocity anisotropy. Two experiments have been carried out to determine deep upper mantle velocity structure. Steinmetz and others (1977) presented a complex model of lithosphere evolution based on their data from near the Azores Islands. The best constrained measurement is that of the LADLE Study Group (1983) based on data from a 1000 km long array of eighteen ocean bottom seismometers along a 80–95 Ma isochron north of the Lesser Antilles (Fig. 2). Their principal finding was a sharp boundary at approximately 50 km depth that separated 8.2 km/s from high velocity 8.5–8.6 km/s material beneath it. This is obviously an area within which we lack even the crudest constraints on structural models. There is a clear need for more data collection.

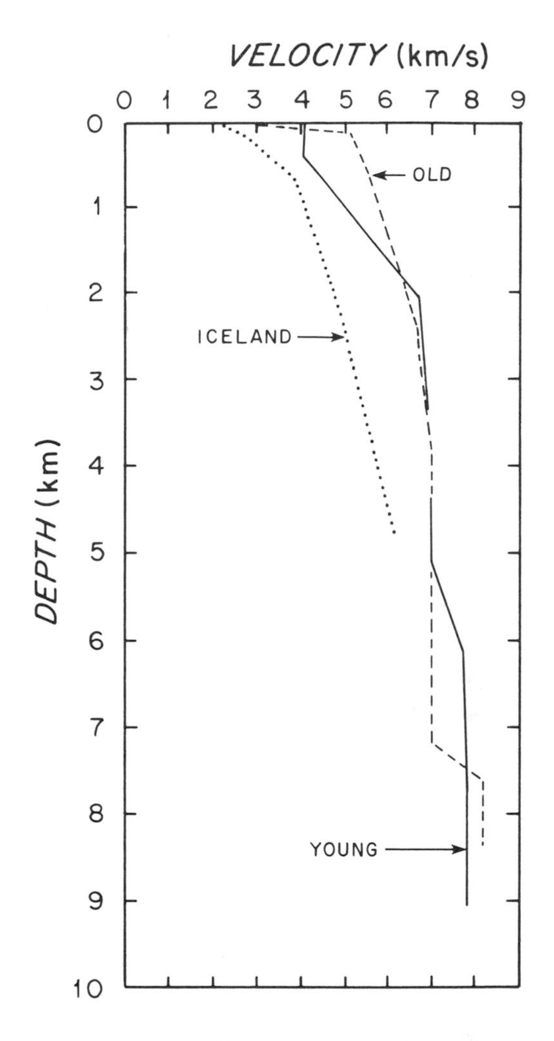

Figure 6. Comparison of well-determined velocity versus depth functions from crust not disrupted by fracture zones, on 140 my old crust (Purdy, 1983) shown by the broken line and 10 my old crust (Detrick and Purdy, 1980) in the central Atlantic Ocean. The significant differences exist in the total crustal thickness (7.5 km for the old crust versus 6 km for the young) and in the velocity structure of the uppermost 2 km of Layer 2, the old 'well sealed' crust exhibiting the higher velocities and smaller gradients. Also shown by the dotted line is the mean of the solutions by Flóvenz (1980) for the shallow crust of Iceland.

THE MID-ATLANTIC RIDGE

The single most prominent tectonic feature of the Atlantic Ocean is obviously the Mid-Atlantic Ridge (MAR). Knowledge of its seismic structure would constrain models of thermal, mechanical, and volcanic processes. However, we cannot discuss MAR seismic structure only by compiling and reviewing the results of the few pertinent refraction experiments. The nine major experiments (Table 1) must first be reviewed because their design, extent, location, and interpretation method strongly bias the structural solutions. Existing knowledge of ridge structure is poor and limited primarily to three small areas at latitudes 37°N, 45°N, and 60°N. Further research combining data collection by well positioned and precisely controlled experiments with improved three dimensional interpretation methods is of high priority.

The LePichon and others (1965) and Talwani and others (1971) papers are the only two that consist of many discrete and individually interpreted profiles. The LePichon and others (1965) work can be criticized for the major spatial aliasing problems resulting from the wide shot separations as well as the lack of control on experiment location relative to the ridge and fracture zones; knowledge of this tectonic fabric was limited in the early 1960s. However, the signal-to-noise ratio on some of these early two ship experiments was often high and most of the LePichon experiments are reversed with good reversed point agreement.

TABLE 1. MID-ATLANTIC RIDGE SEISMIC EXPERIMENTS

Reference	Area	Comments
LePichon, Houtz, Drake, and Nafe, 1965	Randomly distributed, widely spaced stations between Latitudes 30°N and 60°N	Classical two ship explosive refraction data. Shot spacing often greater than 5 km, but some good data giving consistent reversed slope-intercept solutions.
Keen and Tramontini, 1970	Within the Canadian detailed survey area at 45°N (Aumento and others, 1971)	Probably the single largest travel time dateset (explosive sources with moored sonobouys and a ship as receivers) in existence from the MAR. A two dimensional experiment interpreted using the delay time function method: however, little or no data from the median valley itself.
Talwani, Windisch, and Langseth, 1971	On the Reykjanes Ridge between 60°N and 63°N	Fourteen individual profiles collected with expendable sonobouys and airgun or small (<4kg) explosive charges. No corrections for seafloor topography were applied and solutions produced using unreversed slope-intercept method.
Whitmarsh, 1975	Median valley and eastern crestal mountains. FAMOUS area near 37°N	An experiment specifically designed to investigate median valley structure using large airgun and explosive sources with ocean bottom hydrophones. Solutions from detailed analysis of travel times.
Fowler, 1976	Median valley and western and eastern crestal mountains, FAMOUS area near 37°N	An experiment designed to study median valley and flank structure; OBS receivers with explosive shots. The first application of synthetic seismogram methods to this problem.
Steinmetz, Whitmarsh, and Moreira, 1977	Ridge flanks north of the Azores Islands	Long (100-400km) lines of large explosives recorded by ocean bottom hydrophones and island seismometers. Little data on the crust: emphasis on synthetic seismogram and ray trace modelling of upper mantle structure.
Fowler, 1978	Within the Canadian detailed survey area at 45°N	OBS microearthquake network used to monitor explosive shots along median valley and across eastern crestal mountains. Careful analysis using synthetic seismogram modelling.
Fowler and Keen, 1979	Within the Canadian detailed survey area at 45°N	Reinterpretation of some of the Keen and Tramontini (1970) data using synthetic seismogram modelling.
Bunch and Kennett, 1980	Reykjanes Ridge at 59°30'N	Three lines over 0, 3, and 9 m.y. old crust: explosives and recording sonobouys. Travel time inversion and synthetic seismogram analysis.

The major conclusion was the definition of an approximately 500 km wide zone centered on the ridge that was underlain by ~7.3 km/s material although two (of the total of 13) stations on the youngest crust did determine normal 8.3 km/s upper mantle velocities. The Talwani and others (1971) data on the Reykjanes Ridge again defined a presumed low upper mantle velocity of 7.4 km/s but its extent could not be well defined.

Although carried out fifteen years ago, the Keen and Tramontini (1970) experiment remains one of the largest and most rigorously interpreted (travel time) datasets. This was the first

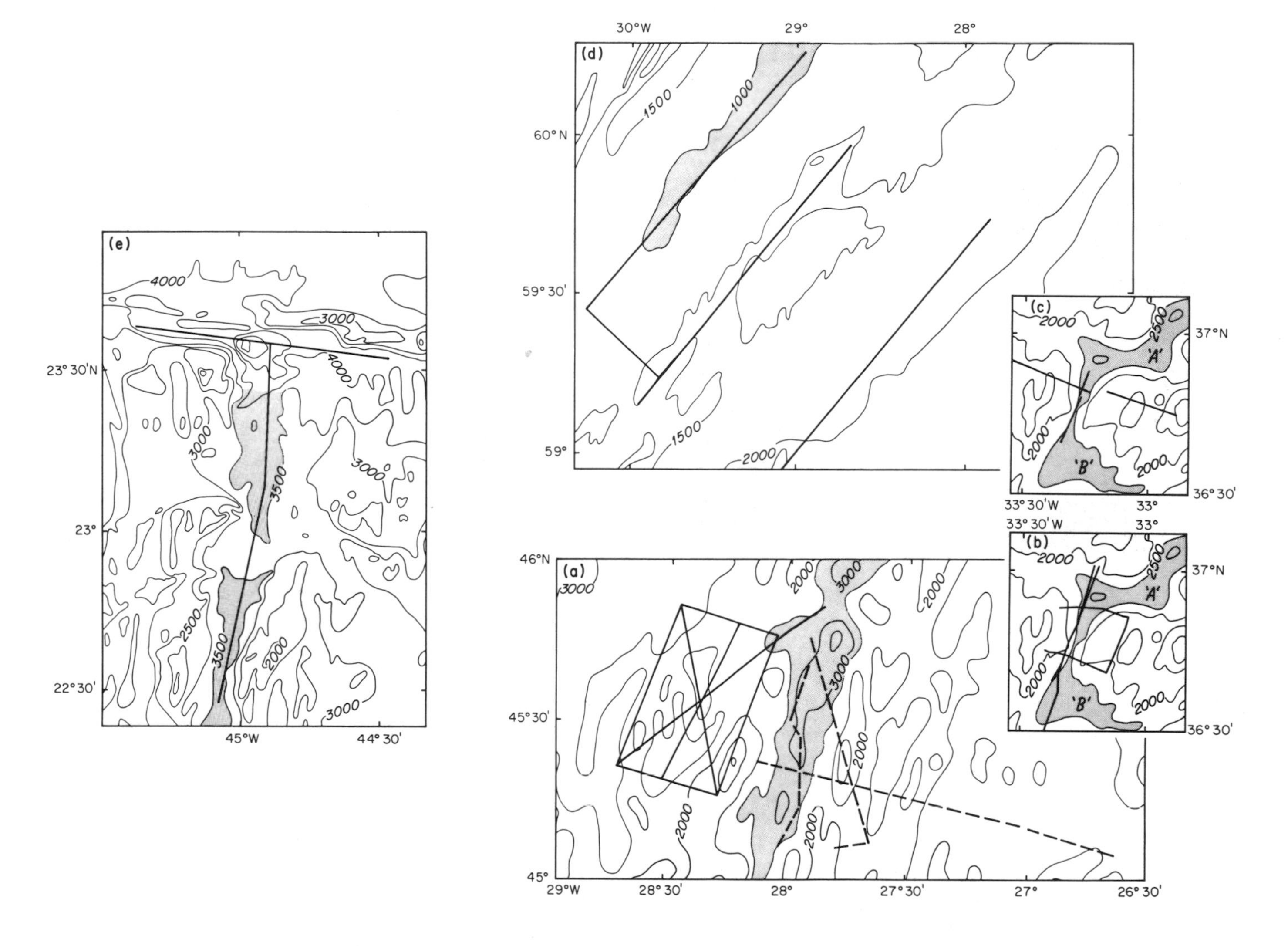

Figure 7. A comparison of six of the major refraction experiments carried out on the mid-Atlantic Ridge and referred to in Table 1. The contour interval is 500 m everywhere and the scale is approximately constant. (a) The Mid-Atlantic at 45°N: the median valley is delineated approximately by the shading within the 2500m contour. The solid lines are the shooting tracks of Keen and Tramontinin (1970); the dashed lines are Fowler (1978). (b) The FAMOUS area: depths greater than 2500m are shaded and approximately delineate the median valley. 'A' and 'B' denote fracture zones A and B. Solid lines are the shooting tracks of Whitmarsh (1975). (c) The FAMOUS area: the shooting tracks are of Fowler (1976). (d) The Reykjanes Ridge: the ridge crest is located approximately by the depths less than 1000m that are shaded. The three lines shot by Bunch and Kennett (1980) are shown. The velocity vs. depth function resulting from their line 'X', located along the ridge crest, is shown in Figure 10. (e) The Mid-Atlantic Ridge south of the Kane Fracture Zone: from Cormier and others (1984). The fracture zone is at latitude 23°35′N and the median valley is approximately defined by the shading within the 3500m contour. The shooting line, along which five receivers were deployed, runs down the center of the median valley.

work executed within a detailed survey area and in two dimensions (Figs. 7 and 8). However, the experiment was centered 30–40 km west of the median valley and the mean crustal structure was found to be essentially normal. Tenuous evidence for mantle anisotropy was found, the well determined mean upper mantle velocity was 7.9 km/s, and the mean crustal thickness was 5 km. Only one line, paralleling and 15–20 km west of the median valley, yielded an anomalously low 7.5 km/s upper mantle velocity. The crust was interpreted as a simple two layer structure with mean velocities of 4.5 ± 0.7 km/s and 6.6 ± 0.3 km/s. There was no evidence from the travel times alone for a higher velocity Layer 3B at the base of the crust. This careful work cast doubt on earlier conclusions and suggested that any zone of anomalously low (7.3–7.5 km/s) upper mantle velocities could be no wider than a few tens of kilometers.

The first work with ocean bottom receiving instruments on

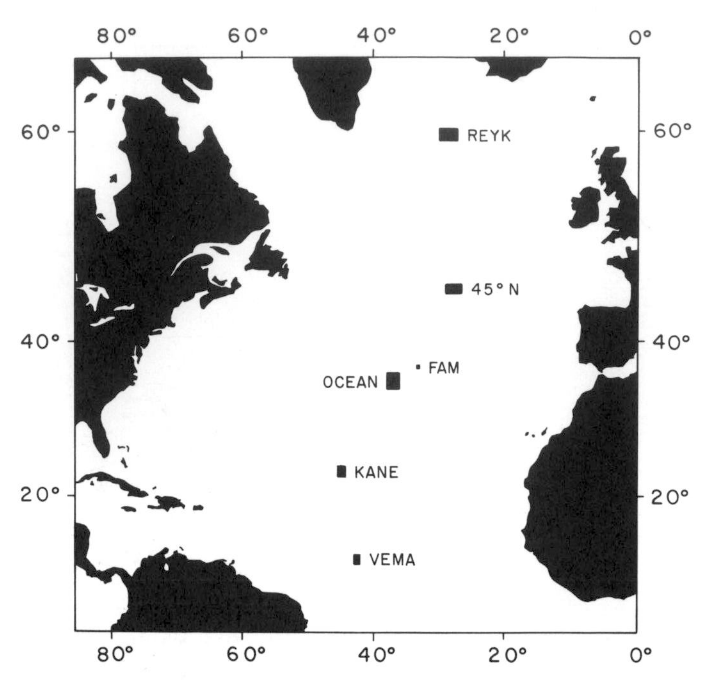

Figure 8. Shaded boxes denote location of major experiments referred to in Tables 1 and 2.

the MAR was reported by Whitmarsh (1973 and 1975). This was an experiment of great significance because it combined fixed receivers well positioned in the center of the median valley within an adequately known tectonic framework (FAMOUS area) with both the benefit of the small 250 m shot separation of a 1000 cubic inch airgun and careful and apparently objective topographic corrections (Figs. 7 and 8). The principal findings are a widespread anomalously low Layer 3 velocity of 6.2 km/s and a 2–3 km wide axial zone with low Layer 2 velocities of 3.2 km/s outside of which there was a poorly constrained upper mantle velocity determination of 8.1 ± 0.4 km/s. If the lower Layer 3 velocity of 6.2 km/s is in fact due simply to undetected refractor dip, as suggested by the author, and the upper mantle velocity determination is discarded as unreliable, then no significant difference exists between these results and those of Keen and Tramontini (1970). The new observation provided by this work is the presence of a narrow axial strip (presumed to be the zone of intrusion) characterized by low (3.2 km/s) Layer 2 velocities.

A landmark paper in the study of MAR structure using refraction methods is Fowler (1976). This work was carried out in the same location as that of Whitmarsh (see Figs. 7 and 8) using ocean bottom seismometer and recording sonobuoy receivers with explosive sources, and is outstanding because it constitutes the first particularly detailed application of synthetic seismogram modelling to this problem.

The principal results of Fowler's paper are compared with those of Whitmarsh in Figure 9. One important difference is the addition by Fowler of a 2–3 km thick 7.2 km/s layer at the base of the crust required to obtain reasonable agreement between computed and observed waveforms. Fowler (1976) states that there can be no 8 km/s material beneath the median valley because of its effect on the amplitudes, but the axial upper mantle velocity of 7.6 km/s shown in Figure 9 is very poorly constrained by the modelling. Although exceptionally precise agreement between computed and observed waveforms result from Fowler's model in some cases (see Figs. 13 and 15 in Fowler, 1976), the dataset seriously lacks the long range (>30 km) data crucial to constraining lower crustal and upper mantle structure. This is because the experiment was sensibly restricted to the short ridge segment bounded by the FAMOUS area's fracture zones A and B (see Fig. 7). Fowler's later work with OBS data at 45°N (Fowler, 1978) and the reinterpretation of the Keen and Tramontini (1970) data using waveform modelling (Fowler and Keen, 1979) are both consistent with these results from the FAMOUS area (Fowler, 1976). The principal conclusions of these three papers may be summarized as follows:

(1) Because shear waves are observed to propagate across the axis, no substantial magma chamber can be present at shallow depth. But it cannot be proven that pockets of melt 1–2 km in diameter or smaller do not exist.

(2) The formation of a normal 6–7 km thick crustal section with a 7.2 km/s basal layer and an 8.1 km/s upper mantle occurs within 10 km of the axis.

The dataset used by Steinmetz and others (1977) consists of large explosives at ranges up to 400 km recorded by stations on the Azores Islands and by ocean bottom hydrophones. Their conclusions are restricted to (1) no evidence for an axial crustal magma chamber is found and (2) although energy propagates across the ridge axis within the crust, the axial region marks a distinct barrier to propagation within the mantle.

The most recently completed study is that of Bunch and Kennett (1980) on the Reykjanes Ridge (Figs. 7 and 8). Only one of their three experiments was located over the ridge crest, but the work is worthy of particular mention because of the care and objectivity applied to its interpretation. The caption for Figure 10 provides details of the velocity depth structure at the ridge crest, the outstanding features being the small (6.8 to 6.6 km/s) velocity inversion at 4 km beneath the seafloor, and the 7.1 km/s upper mantle velocity. This low upper mantle velocity has been a property common to several of the experiments described here but this is the only documented case of a velocity inversion within the crust.

The full interpretation of the final experiment to be discussed here (Fig. 7, Purdy and others, 1983) is as yet unavailable, but because of the size of the effort, a brief description of the experiment's principal features is justified here. As can be seen in Figure 7 the shooting line extended for over 120 km along the axis of the median valley south of the intersection of the MAR with the Kane Fracture Zone. Because ocean bottom hydrophone receivers were positioned at 30 km spacing on this line, along-axis changes in structure could be recognized and defined. Qualitatively, two important changes are revealed: a thinning of the crust from south to north as the fracture zone is approached and a large and sudden change in crustal structure occurring over a lateral

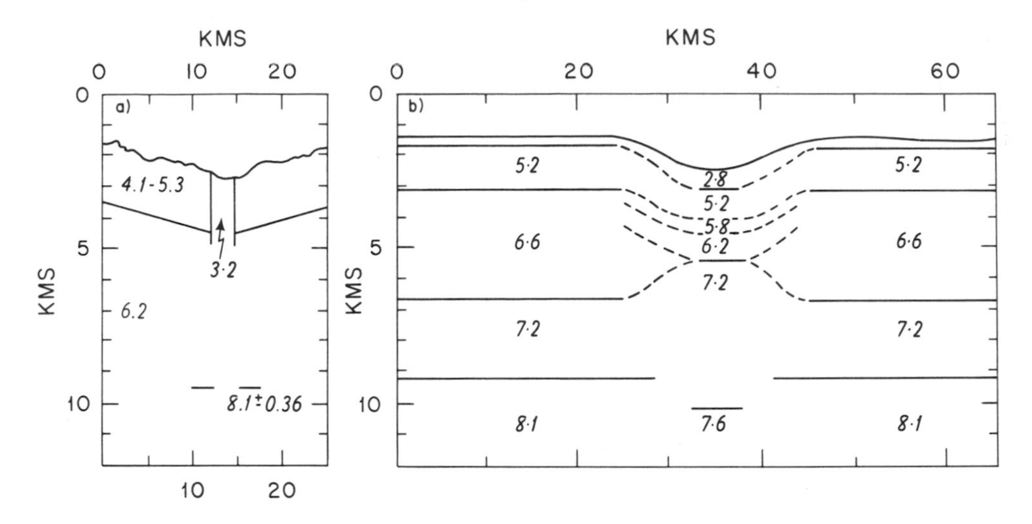

Figure 9. Cross-sections (on the same scale) of the Mid-Atlantic Ridge at the FAMOUS area, based on the experiments of Whitmarsh (1976) and Fowler (1976). Location shown in Figures 7b, 7c, and 8.

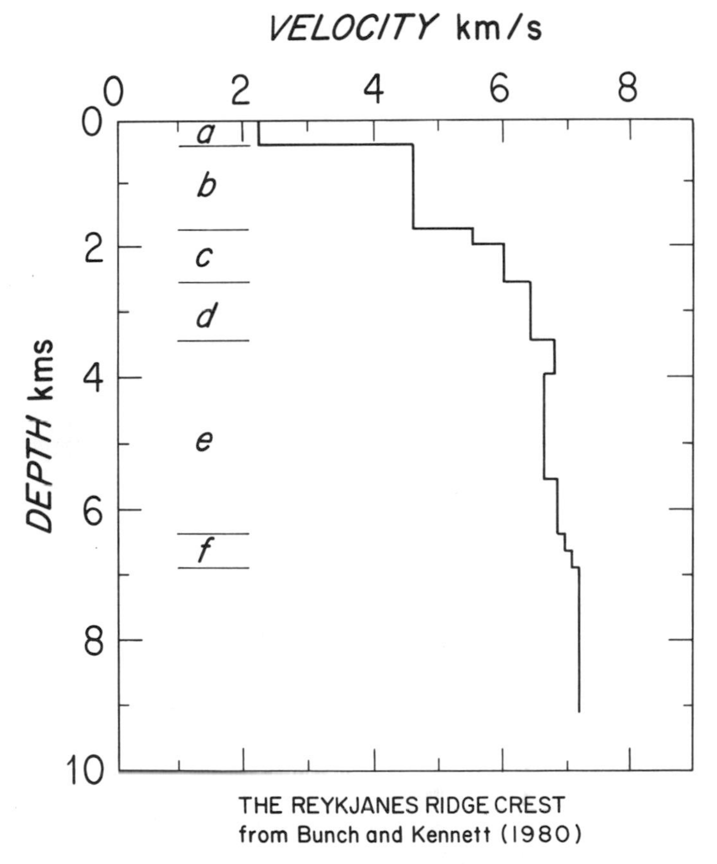

Figure 10. The velocity-depth function resulting from the refraction experiment along the crest of the Reykjanes Ridge shown in Figure 7d (Bunch and Kennett, 1980). The layers a through c are isovelocity representations of the steep velocity gradients within Layer 2 that extend from the 2.2 km/s seafloor velocity to 6.2 km/s at the base of layer c at 2.5 km depth. Layer d is an 800m thick 6.4 to 6.6 km/s unit that overlies the 500m thick 6.8 km/s lid of the 6.6 km/s low velocity zone that constitutes most of layer e. The small transition zone to the 7.1 km/s upper mantle velocity is denoted by f.

distance of less than 2 km, 50 km south of the fracture zone. The quantitative definition of these changes awaits further work, but the important and unsurprising result is that major along-axis changes in structure do occur.

Thus any view of MAR structure culled from the small number of widely separated spot measurements presented in this review is most likely misleading.

The single greatest along-axis change in the MAR occurs at Iceland, the seismic structure of which has been studied in detail. It is well established that the whole island is underlain by anomalously low velocity mantle material. Pálmason (1971) and Gebrande and others (1980) calculated velocities of 7.2 to 7.4 km/s and observations of teleseismic travel time residuals (Tryggvason, 1964; Long and Mitchell, 1970) suggest that anomalously low velocities extend down to at least 250 km depth. Iceland's crust is significantly thicker than that found in the ocean and is characterized by lower velocities in Layer 3 (6.5 km/s). Flóvenz (1980) carried out vigorous amplitude and travel time inversions of selected refraction profiles from Pálmason (1971) and showed that depths to the essentially isovelocity 6.5 km/s layer for an uneroded basalt pile were typically 5–6 km, above which steep continuous velocity gradients exist with surficial velocities as low as 2.0 km/s (Fig. 6). There exists some evidence for crustal thickening from 8–10 km near the axial rift in the southwest to approximately 10–15 km outside the rift zone in north- and southeast Iceland (Pálmason, 1971). This observation has been explained by Hermance (1981) in terms of a process of crustal thickening with age involving the accretion of new material to the base of the crust. Recent reflection and refraction surveys by Zverev and others (1980) have revealed important evidence for reflectors within the crust dipping toward the spreading axis at angles of up to 30° at depths of more than 3 km.

A commonly quoted feature of the MAR structure is the presence of low velocity upper mantle material (~7.2 km/s) beneath the median valley. If the dataset is restricted to that which is well constrained by careful waveform modelling then only two determinations exist; the Fowler (1976) work in the FAMOUS area (that can be criticized for its lack of data beyond 30 km range) and the Bunch and Kennett (1980) experiment on the Reykjanes Ridge (that seems a tightly controlled result). The conclusion must be that the universal presence of low velocity upper mantle material beneath the MAR is not well determined.

The pioneering work of Orcutt and others (1976) and Rosendahl and others (1976) provided the first observations of a low velocity zone beneath the East Pacific Rise and stimulated the search for a comparable feature beneath the MAR median valley. None of the experiments carried out to date support the existence of a magma chamber beneath the MAR. Indeed, observations of shear wave propagation across the ridge crest specifically disallow the presence of a substantial magma body (Fowler, 1976, 1978) although because of the limited resolving power of the conventional refraction experiment "pockets of melt up to perhaps 1–2 km in diameter cannot be shown to be absent" (Fowler, 1978).

Much of the data described here were gathered over young crust on the ridge flanks and provided little information on the structure of the accretion zone. Only in recent years have geologists appreciated the importance of hydrothermal circulation as a mechanism for very rapid cooling of newly formed crust. Structures associated with the process of new crust creation (as opposed to the structure of a young but fully formed crustal section) will be restricted to zones only a few kilometers in width. Only four studies (Whitmarsh, 1975; Fowler, 1976, 1978; Bunch and Kennett, 1980) included data that were suitably located within a sufficiently well determined tectonic framework to provide some coverage of the narrow accretion zone. It is clear that the data base available for the study of the accretion zone of the MAR is extremely small. The extent of our knowledge is further lessened if it is accepted that the creation of crust at a slow spreading ridge is not a steady state process. Thus it is reasonable to expect major changes in structure along the ridge axis as the boundaries are crossed between ridge segments that are at different stages in their accretionary cycle. This question can only be addressed by obtaining more coverage along axis with well located experiments supported by a clear and detailed knowledge of the tectonic framework. Without knowledge of the location of an experiment relative to the principal local and regional tectonic features, the resulting structural information is uninterpretable in terms of physical models of accretion, and thus is of limited value.

It is striking that many of the experiments over crust a few million years in age on the flanks of the MAR yield total crustal thicknesses of only 5–6 km and upper mantle velocities of ~7.7 km/s compared with the more commonly determined ~7 km thicknesses and ~8.1 km/s velocities found in the Mesozoic and older crust. No statistical significance can be attached to this subjective observation, but we believe it deserves further investigation.

Practical considerations limit the seismologist to using two-dimensional techniques to attack clearly three dimensional problems. Although in most cases the magnitude of the vertical velocity gradients in the crust are so much larger than any lateral gradients that this is a reasonable simplification, it remains a profound approximation. Progress to full three-dimensional imaging methods awaits the capability to sample the seismic wavefield at the sea surface in two dimensions with sufficient precision and spatial density.

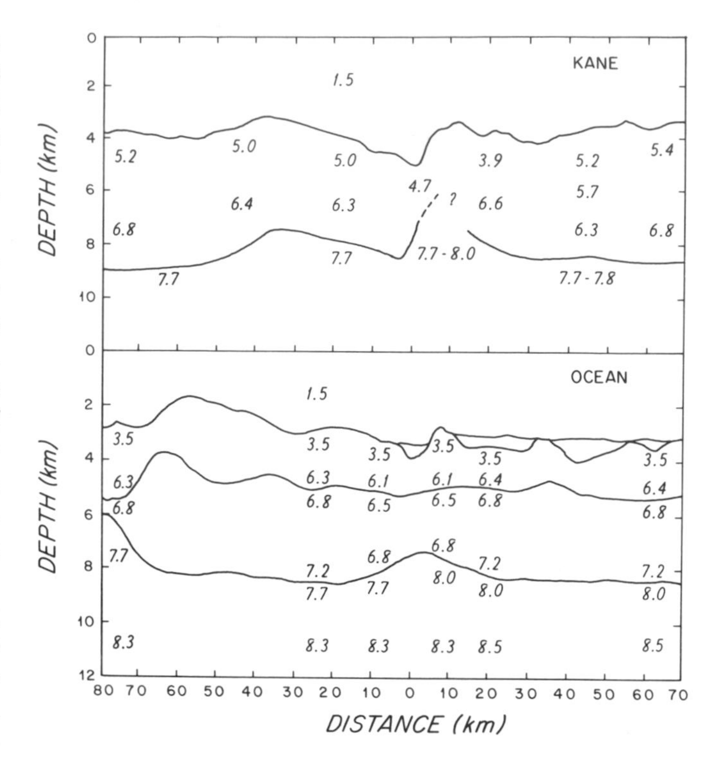

Figure 11. Cross-sections of the Oceanographer and Kane Fracture Zones redrawn on the same scale from Sinha and Louden (1983) and Detrick and Purdy (1980).

FRACTURE ZONES

Few experiments have been carried out to study the structure of fracture zones, and most have been carried out during the past three to four years (Table 2). Not until it was realized that the hundreds of closely spaced fracture zones existing in the Atlantic Ocean dominate the tectonic framework (see Klitgord and Schouten, this volume) was sufficient motivation generated to tackle the difficult problem of determining their structure. Just as with the Mid-Atlantic Ridge studies, where the focus of the efforts was to detect and define an axial magma chamber, then the focus of the fracture zone experiments has been the confirmation and definition of the thin crust that was predicted to underlie the axial trough (Fox, 1978). Of the seven principal papers, three report substantial experiments designed specifically to tackle this problem: Detrick and Purdy, 1980; Sinha and Louden, 1983; and Cormier and others, 1982, 1984.

The Detrick and Purdy study (1980) constitutes the first convincing measurement of thin (2–3 km) crust beneath a fracture zone trough (Fig. 11). This work was expanded by

TABLE 2. ATLANTIC FRACTURE ZONE SEISMIC EXPERIMENTS

Reference	Fracture Zone	Comments
Fox, Schreiber, Rowlett, and McCamy, 1976	Oceanographer	One 35 km longline: poorly constrained two layer slope-intercept solution.
Detrick and Purdy, 1980	Kane	Well constrained determination by two-dimensional experiment of 2-3 km thick crust beneath 50 km length of fracture zone trough.
White and Matthews, 1980	Small offset at 45.5'N, 21oW	Delay time experiment using OBH and sonobuoys show 0.2s decrease in Layer 2 delay time over trough.
Ludwig and Rabinowitz, 1980	Vema	Airgun and explosive with sonobuoys: unreversed lines showed significantly thinned Layer 3 beneath the trough but total crustal thickness of 5 km.
Detrick, Cormier, Prince, Forsyth, and Ambos, 1982	Vema	Two OBS with one explosive line along trough give 5-6 km thick crust of almost constant vertical velocity gradient.
Sinha and Louden, 1983	Oceanographer	Four 100 km long explosive with sonobuoy lines interpreted by two dimensional ray tracing show crustal thinning extending >30 km either side of fracture zone trough.
Cormier, Detrick, and Purdy, 1984	Kane	Extensive OBH - explosive and airgun experiments along 250 km length of fracture zone axis show thin crust of variable thickness.

Cormier and others (1982, 1984) to yield coverage over a 250 km length of the Kane FZ (KFZ). The conclusions of both these papers are based on travel-time interpretations from shots and ocean bottom hydrophones along the trough, the recognition of a sudden decrease in mantle delay times at the trough, and the confirmation of the presence of normal crustal structures outside the fracture zone region both by travel time and synthetic seismogram interpretations. Furthermore, Cormier and others (1984) recognize and model the waveforms of the Moho triplication point at 10 km range (clearly recognizable from airgun shots along the trough axis) using crustal thicknesses of approximately 3 km. There seems no doubt that thin crust exists along a substantial length of the KFZ and, because of the supportive determinations from other areas, it seems reasonable that this constitutes a common feature of all Atlantic transforms. Confirmation of this comes from the work of Sinha and Louden (1983) whose careful two-dimensional ray tracing analysis, although admittedly non-unique, shows comparably thin crust beneath both the Oceanographer Fracture Zone and an adjacent, small offset transform. The crustal sections resulting from the Kane and Oceanographer studies are compared in Figure 11. Sinha and Louden (1983) appropriately emphasize, as shown in Figure 12, that two types of thin crust have been found. The first (Type A) is attenuated (4–5 km) oceanic crust (i.e. crust with a similar velocity structure but with all its components thinned) formed as such because of the decrease in crustal production by the independent ridge segments near fracture zones. This is suggested by Schouten and White (1980) to be a fundamental property of the ridge segments in the Central Atlantic, perhaps a consequence of lateral along-axis transport of magma from some restricted zone centrally placed in the ridge segment. It may not only be a consequence of the anomalous thermal regime resulting from the offset at the transform because the observations of Sinha and Louden (1983) and Mutter, Detrick and others (1984) support the existence of comparably thinned crust beneath the smallest offset fracture zones. This latter observation is of particular significance because it is based on synthetic aperture multichannel reflection profiling data that show a strong Moho reflector rising to within 0.8s of the basement reflector beneath the Blake Spur fracture zone at 70°W. This corresponds to a crustal thickness of perhaps as little as 2 to 2.4 km beneath a fracture zone of only 20 km offset.

The second type of thin crust (Type B) has two primary characteristics: (1) it can be as thin as 2–3 km and (2) its velocity structure is wholly different from that of normal oceanic crust (see Fig. 12). We have no substantial clues as to its composition or its process of accretion. Sinha and Louden (1983) suggest an explanation for the existence of two types of thin crust (sometimes within the same fracture zone) in terms of some critical minimum half spreading rate required to support normal crustal production at a spreading center. When the rate falls below this critical value (as it may in the Central Atlantic at ridge-fracture zone intersections) then crustal magmatic processes would cease. The second type of thin crust (Type 'B') would then be created, which probably consists of a thin lid of fractured and hydrothermally altered ultramafics, with possibly minor amounts of intrusive and extrusive mafic rocks, directly overlying unaltered upper mantle (Sinha and Louden, 1983).

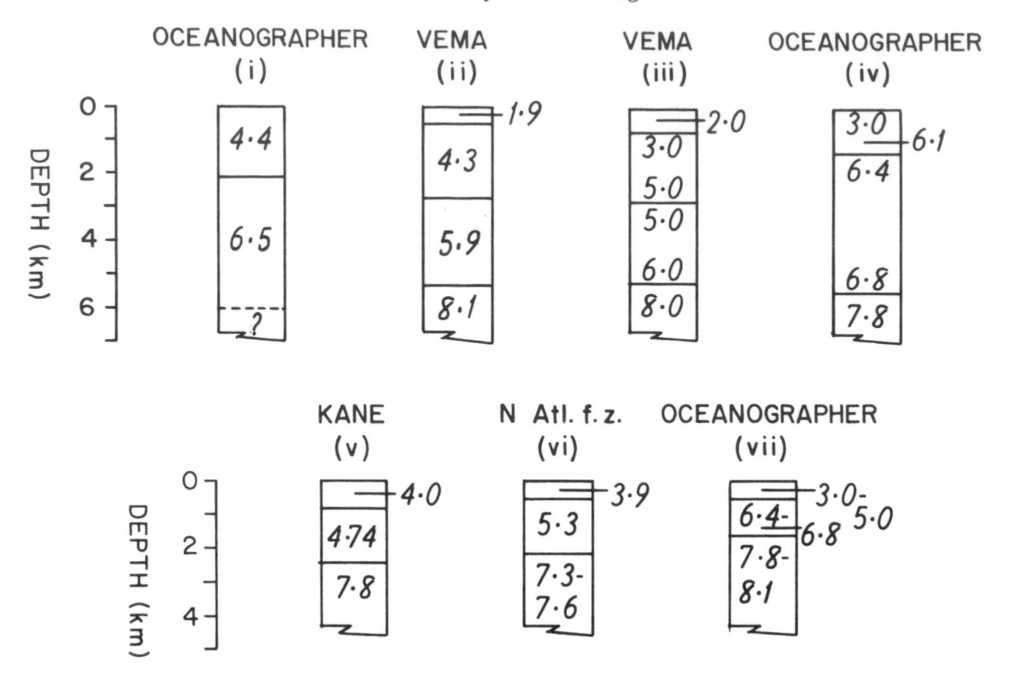

Figure 12. Sections of thin crust determined by refraction experiments over various fracture zones. Sections (i) through (iv) typify attenuated normal crust (Type A) and (v) through (vii) anomalous crust (Type B). Redrawn from Sinha and Louden (1983). The sources for the structure sections are as follows: (i) Fox and others, 1976; (ii) Ludwig and Rabinowitz, 1980; (iii) Detrick and others, 1982; (iv) Sinha and Louden, 1983; (v) Detrick and Purdy, 1980; (vi) White and Matthews, 1980; (vii) Sinha and Louden, 1983.

While agreeing in general with these conclusions, we suggest that insufficient evidence exists to preclude existence of Type 'B' thin crust everywhere on all fracture zones. The reason it is not always observed is simply because the zone of thin crust is extremely narrow (perhaps only a few kilometers) and a combination of inadequate spatial sampling and imprecise location of the refraction experiments relative to the true fossil trace of the transform results in the few observations so far made. Other important characteristics are that the transition from normal to Type 'A' thin crust is gradual, whereas the transition to Type 'B' crust is sudden, as is shown by the sudden decrease in mantle delay times of several tenths of a second over a distance of only five kilometers (Detrick and Purdy, 1980; Cormier and others, 1984).

The observed along-axis variability in structure of the thin crust beneath the trough (Sinha and Louden, 1983; Cormier and others, 1984) would then be a straightforward consequence of either the sinuous nature of the boundary between Type A and B crust (perhaps as predicted by Schouten and others, 1980) or simply the unavoidable deviations of the shooting track from the precise trend of the fracture zone axis. There is a need for small-scale high-resolution measurements within a few kilometers of the fracture zone axis that could confirm or deny the omnipresence of Type 'B' crust and define the character of its boundaries with Type 'A' or normal crust on either side. The expected asymmetry of fracture zone structure crucial to the understanding of the processes occurring within transform zones is not well defined by existing data. The crust on only one side of the fracture zone has passed through the transform zone and there exists crust on opposite sides of all fracture zones that has undergone grossly different thermal and mechanical histories. One speculation may be that although crustal thickness (or rather thinness) is not dependent upon transform offset, the degree of asymmetry may well be.

In summary, the existence of thin crust beneath several Atlantic fracture zones is well established. The exact geometry of the zone of thin crust, its relationship with other major tectonic features observed along fracture zones (e.g. transverse ridges, depression at ridge fracture intersection) or the along-axis structural variability is not well understood. The importance of further work in this area was made clear by Detrick and Purdy (1980) when they stated that if thin crust were a feature common to all Atlantic fracture zones "then our view of the gross structure of the ocean basins must change. These narrow zones of anomalously thin crust associated with fracture zones would be a major feature of the seismic structure of the ocean basins. Ribbons of relatively homogeneous, so called normal crust 50–100 km wide will extend away from spreading centers, each ribbon formed at a discrete ridge crest segment and each separated by the thin crust underlying the fracture zone troughs."

THE BOUNDARY WITH THE CONTINENT

Without doubt, the single largest lateral change in seismic structure in the Western Atlantic occurs at the boundary between oceanic and continental crust where, over a lateral distance several tens of kilometers, the crustal thickness increases from ~8 to >30 km. The precise location and width of this boundary is unknown. Some of the earliest refraction experiments were directed toward investigating this problem (e.g. Ewing and Ewing, 1959) and it has been a focus for continuing efforts (e.g. Sheridan and others, 1979) culminating with the recently completed Large Aperture Seismic Experiment (LASE) during which five Expanding Spread Profiles (ESP) (Stoffa and Buhl, 1979) provided the most detailed determinations of deep structure available to date. While our knowledge of the sedimentary structure of the margin has vastly improved during the past ten years because of the thousands of kilometers of multichannel reflection data collected by the United States Geological Survey and by industry, only modest progress has been made in furthering our knowledge of the deep sub-basement structures. The current state of knowledge is best summarized by the two recent papers by the LASE Study Group (n.d.a and b).

DISCUSSION

We are unable to quantify some of the most fundamental characteristics of ocean crustal structure. We cannot claim knowledge of whether or not the crust thickens with age. Nor can we be sure whether or not crustal thickness is dependent upon spreading rate (despite the tenuous correlations of Reid and Jackson, 1981). Numerous statistical compilations have been made to tackle these questions (e.g. Tréhu and others, 1976; Houtz, 1980) but the results are unconvincing. If one simple-mindedly compares two well-determined velocity depth functions from young and old crust, as in Figure 6, then one may conclude that thickening with age does exist. That this could be invalid is obvious from Figure 1, which shows the wide range of processes and variables that could explain these differences in structure. The earth acts as a broad band source issuing signals in the form of structural changes (that are clues to its evolutionary processes) at a wide range of temporal and spatial scales. We are effectively trying to define the characteristics of this source (the processes) by monitoring it with a very narrow band receiver (our methods), and not surprisingly, we are having difficulty.

To further our understanding of the evolution of the shallow crust, where most of the significant age related changes are known to take place, we need to carry out seismic experiments on two different scales, both of which differ from that of the historical data set. For example, small scale experiments are required that are capable of sampling individual fault blocks to allow correlations of structure with hydrothermal circulation cells. Conversely, large scale continuous mapping of velocity structure is needed to reveal long-term trends related to aging or spreading rate while averaging out effects of topography and other small scale heterogeneities. For studies of the complete crustal column a number of type sections need to be established by well-controlled and carefully located experiments. Variations can then be quantified and understood relative to some standard. Oceanic crust is a stack of gradient zones of varying thicknesses and velocities. It is considerably more representative to characterize crustal structure in terms of the magnitude of the gradients within these zones and their thicknesses than by artificially averaged isovelocity layers. The changing thermal gradients within the crust will modify the velocity gradients, as will the changes in porosity caused by progressive crack infilling. It follows that the exact definition of the magnitude of the gradients is a primary aim of any seismic experiment. For this purpose the resolving power of travel time analyses alone is notoriously poor, but fortunately the amplitude versus range distribution is intimately related to the gradients and it is the forward modelling of this relationship, using synthetic seismograms, that provides the strongest constraints. Unfortunately, no quantitative estimate of uncertainty can be made of the result. Existing economically practical modelling methods depend upon the assumption of laterally homogeneous horizontal layering; thus the major source of error is in the contradiction of this assumption. The magnitude of the error produced by a heterogeneity of a given size and extent is unknown. Obviously, the topography on the upper surface of Layer 2 is one of the largest heterogeneities, and qualitative assessments of its effect, using ray tracing suggests that its effect is substantial (e.g. Purdy, 1982; White and Purdy, 1983). However, confirmation of this, along with quantitative determinations of the effects, must await more rigorous calculations (e.g. Stephen, 1983).

The way forward in studies of the oceanic crust now lies with substantial experiments or sets of experiments of widely differing scales designed specifically to test and refine models of formational and evolutionary processes. The days of exploration essentially have passed and sufficient knowledge and understanding have been gained so that reasonably detailed hypotheses can be posed and tested. Such is not the case for the sub-crustal lithosphere. Our knowledge of the structure of the upper several tens of kilometers of the earth's mantle is extremely poor. Ignorance of upper mantle structure is especially serious because it is the interaction between lithospheric plates and not only the thin crustal skin that shapes the major features of the earth's surface. A critical mass of data needs to be collected and structural interpretations determined before substantial progress can be made.

Studies of the crust and upper mantle cannot be separated from those of mid-ocean ridges, fracture zones, and continental margins. They constitute a single system with processes and controlling variables that are intimately interwoven. There can be no complete understanding of one part without understanding the whole. Research in the future should be balanced in its progress to optimize the cross flow of ideas and understanding between studies of ridges, fracture zones, margins, and the oceanic lithosphere.

REFERENCES

Aumento, F., Loncarevic, B. D., and Ross, D. I.
1971 : Hudson Geotraverse: Geology of the Mid-Atlantic Ridge at 45°N: Philosophical Transactions of the Royal Society of London, v. 268, p. 623–650.

Aumento, F., and Melson, W. G., eds.
1977 : Deep Sea Drilling Project, v. 37: U.S. Government Printing Office, Washington, D.C., 998 p.

Ballard, R. D., Bryan, W. B., Heirtzler, J. R., Keller, G., Moore, J. G., and van Andel, T. H.
1975 : Manned submersible operations in the FAMOUS area, Mid-Atlantic Ridge: Science, v. 90, p. 103–108.

Bentley, C. R., and Worzel, J. L.
1956 : Geophysical investigations in emerged and submerged Atlantic coastal plain: Geological Society of America Bulletin, v. 67, p. 1–18.

Bratt, S. R., and Purdy, G. M.
1984 : Structure and variability of oceanic lithosphere on the flanks of the East Pacific Rise between 11° and 13°N: Journal of Geophysical Research, v. 89, p. 6111–6125.

Buhl, P., Mutter, J. C., Alsop, J. M., Stoffa, P. L., Diebold, J. B., Hinz K., Phillips, J. D., and Detrick, R.
1983 : North Atlantic transect: Wide aperture CDP data (abs.): EOS, Transactions of the American Geophysical Union, v. 63, p. 1031.

Bunce, E. T., and Fahlquist, P. A.
1962 : Geophysical investigation of the Puerto Rico trench and outer ridge: Journal of Geophysical Research, v. 67, p. 3955.

Bunce, E. T., Fahlquist, D. A., and Clough, J. W.
1969 : Seismic refraction and reflection measurements—Puerto Rico outer ridge: Journal of Geophysical Research, v. 74, p. 3082–3094.

Bunch, A., and Kennett, B.L.N.
1980 : The crustal structure of the Reykjanes Ridge at 59°30′N: Geophysical Journal of the Royal Astronomical Society, v. 61, p. 141–166.

Casey, J. F., Dewey, J. F., Fox, P. J., Karson, J. A., and Rosencrantz, E.
1981 : Heterogeneous nature of oceanic crust and upper mantle: A perspective from the Bay of Islands ophiolite complex: in The Sea, Volume 7, ed., C. Emiliani, Wiley-Interscience, New York, p. 305–338.

Christensen, N. I., and Salisbury, M. H.
1972 : Sea floor spreading, progressive alteration of Layer 2 basalts, and associated changes in seismic velocities: Earth and Planetary Science Letters, v. 15, p. 367–375.
1973 : Velocities, elastic modulii and weathering age relations for Pacific layer 2 basalts: Earth and Planetary Science Letters, v. 19, p. 461–470.
1975 : Structure and constitution of the lower oceanic crust: Reviews of Geophysics and Space Physics, v. 13, p. 57–86.

Cormier, M. H., Detrick, R. S., and Purdy, G. M.
1982 : Seismic constraints on crustal thickness variations along the Kane Fracture Zone, (abs): EOS, v. 63, p. 1100.
1984 : Anomalously thin crust in oceanic fracture zones: New seismic constraints from the Kane Fracture Zone: Journal of Geophysical Research, v. 89, p. 10249–10266.

Detrick, R. S., Cormier, M. H., Prince, R. A., Forsyth, D. W., and Ambos, E. L.
1982 : Seismic constraints on the crustal structure of the Vema fracture zone: Journal of Geophysical Research, v. 87, p. 10599–10612.

Detrick, R. S. and Purdy, G. M.
1980 : The crustal structure of the Kane Fracture Zone from seismic refraction studies: Journal of Geophysical Research, v. 85, p. 3759–3777.

Ewing, J. I.
1963 : Elementary theory of seismic refraction and reflection measurements: in The Sea, Volume 3, ed., M. N. Hill: Wiley-Interscience, New York.

Ewing, J. I., and Ewing, M.
1959 : Seismic refraction measurements in the Atlantic Ocean basins, in the Mediterranean Sea, on the Mid-Atlantic Ridge and in the Norwegian Sea: Geological Society of America Bulletin, v. 70, p. 291–318.
1961 : A telemetering ocean bottom seismograph: Journal of Geophysical Research, v. 66, p. 3863–3878.

Ewing, J. I., and Houtz, R.
1979 : Acoustic stratigraphy and structure of the oceanic crust: in Deep Drilling Results in the Atlantic Ocean: in Ocean Crust, eds., Talwani, M., Harrison, C. G., and Hayes, D. E.: American Geophysical Union, Washington, D.C.

Ewing, J. I., and Purdy, G. M.
1982 : Upper crustal velocity structure in the ROSE area of the East Pacific Rise: Journal of Geophysical Research, v. 87, p. 8397–8402.

Ewing, M., Sutton, G. H., and Officer, C. B.
1952 : Seismic refraction measurements in the Atlantic Ocean, Part VI: Typical deep stations, North America basin: Seismological Society of America Bulletin, v. 44, p. 21–38.

Ewing, M., Worzel, J. L., Hersey, J. B., Press, F., and Hamilton, G. R.
1950 : Seismic refraction measurements in the Atlantic Ocean basin (Part One): Seismological Society of America Bulletin, v. 40, p. 233–242.

Flóvenz, O. G.
1980 : Seismic Structure of the Icelandic Crust above Layer 3 and the relation between body wave velocity and the alteration of the basaltic crust: Journal of Geophysics, v. 47, p. 211–220.

Fowler, C.M.R.
1976 : Crustal structure of the Mid-Atlantic ridge crest at 37°N: Geophysical Journal of the Royal Astronomical Society, v. 47, p. 459–491.
1978 : The Mid-Atlantic Ridge: Structure at 45°N: Geophysical Journal of the Royal Astronomical Society, v. 54, p. 167–183.

Fowler, C.M.R., and Keen, C. E.
1979 : Oceanic crustal structure Mid-Atlantic Ridge at 45°N: Geophysical Journal of the Royal Astronomical Society, v. 56, p. 219–226.

Fox, P. J.
1978 : The effect of transform faults in the character of the oceanic crust: Geological Society of America Abstracts with Programs, v. 7, p. 403.

Fox, P. J., Schreiber, E., Rowlett, H., and McCamy, K.
1976 : The Geology and the Oceanographer Fracture Zone: A model for fracture zones: Journal of Geophysical Research, v. 81, p. 4117–4128.

Francis, T.J.G.
1981 : Serpentinization faults and their role in the tectonics of slow spreading ridges: Journal of Geophysical Research, v. 86, p. 11616–11622.

Gaskell, T. F., Hill, M. N., and Swallow, J. C.
1958 : Seismic measurements made by the HMS Challenger in the Atlantic Pacific and Indian Oceans and in the Mediterranean Sea, 1950–1953: Philosophical Transactions of the Royal Society of London, v. 251, p. 23–83.

Gaskell, T. F., and Swallow, J. C.
1951 : Seismic refraction experiments in the North Atlantic: Nature, v. 167, p. 723–724.

Gebrande, H., Miller, H., and Einarsson P.
1980 : Seismic structure of Iceland along RRISP Profile 1: Journal of Geophysics, v. 47, p. 239–249.

Grow, J. A., and Markl, R.
1977 : IPOD-USGS multichannel seismic reflection profile from Cape Hatteras to Mid-Atlantic Ridge: Geology, v. 5, p. 625–630.

Hermance, J. J.
1981 : Crustal genesis in Iceland: Geophysical constraints on crustal thickening with age: Geophysical Research Letters, v. 8, p. 203–206.

Hersey, J. B., Bunce, E. T., Wyrick, R. F., and Dietz, F. T.
1959 : Geophysical investigation of the continental margin between Cape Henry, Virginia, and Jacksonville, Florida: Geological Society of America Bulletin, v. 70, p. 437–466.

Hill, M. N.
1963 : Single ship refraction shooting: in The Sea, Volume 3, ed., M. N. Hill: Wiley-Interscience, New York, p. 39–46.

Hinz, K., Meyer, H., Krause, W., Popovici, A., Austin, J. A., Phillips, J. D., Buhl, P., Mutter, J., Mithal, R., Yang, J., Detrick, R., Diebold, J., Houtz, R., and Stoffa, P.
1982 : A wide aperture CDP Transect across the Western North Atlantic: EOS, Transactions of the American Geophysical Union, v. 63, p. 427 (abs.).
Houtz, R. E.
1980 : Crustal structure of the North Atlantic on the basis of large airgun-sonobuoy data: Geological Society of America Bulletin, v. 91, p. 406–413.
Houtz, R., and Ewing, J. I.
1963 : Detailed sedimentary velocities from seismic refraction profiles in the Western North Atlantic: Journal of Geophysical Research, v. 68, p. 5233–5258.
1964 : Sedimentary velocities of the Western North Atlantic margin: Seismological Society of America Bulletin, v. 54, p. 867–895.
1976 : Upper crustal structure as a function of plate age: Journal of Geophysical Research, v. 81, p. 2490–2498.
Humprhis, S., and Thompson, G.
1978 : Hydrothermal alteration of oceanic basalts by seawater, Geochimica et Cosmochimica Acta, v. 42, p. 107–125.
Hyndman, R. D.
1979 : Poisson's ratio in the oceanic crust - a review: Tectonophysics, v. 59, p. 321–333.
Hyndman, R. D., and Drury, M. J.
1976 : The physical properties of oceanic basement rocks from deep drilling on the Mid-Atlantic Ridge: Journal of Geophysical Research, v. 81, p. 4042–4060.
Katz, S., and Ewing, M.
1956 : Seismic refraction measurements in the Atlantic Ocean, Part VII: Atlantic Ocean basin, west of Bermuda: Geological Society of America Bulletin, v. 67, p. 475–510.
Keen, C., and Tramontini, C.
1970 : A seismic refraction survey on the Mid-Atlantic Ridge: Geophysical Journal of the Royal Astronomical Society, v. 20, p. 473–491.
Kennett, B.L.N.
1977 : Towards a more detailed seismic picture of the oceanic crust and mantle: Marine Geophysical Researches, v. 3, p. 7–42.
LADLE Study Group
1983 : A lithospheric seismic refraction profile in the Western North Atlantic Ocean: Geophysical Journal of the Royal Astronomical Society, v. 75, p. 23–70.
LASE Study Group
n.d.a. : The large aperture seismic experiment Part I: Data acquisition and analysis: submitted to Geophysics.
n.d.b. : Deep structure and evolution of the continental margin off New Jersey: Seismic results from LASE: in preparation.
LePichon, X., Ewing, J., and Houtz, R. E.
1968 : Deep sea sediment velocity determination made while reflection profiling: Journal of Geophysical Research, v. 73, p. 2597–2614.
Le Pichon, X., Houtz, R. E., Drake, C. L., and Nafe, J. E.
1965 : Crustal structure of the mid-ocean ridges: Seismic refraction measurements: Journal of Geophysical Research, v. 70, p. 319–338.
Lewis, B.T.R.
1978 : Evolution of ocean crust seismic velocities: Annual Review of Earth and Planetary Science, v. 6, p. 377–404.
Long, R. E., and Mitchell, M. G.
1970 : Teleseismic P-wave delay time in Iceland: Geophysical Journal of the Royal Astronomical Society, v. 20, p. 41–48.
Ludwig, W. J., and Rabinowitz, P. D.
1980 : Structure of the Vema Fracture Zone: Marine Geology, v. 35, p. 99–110.
McConnell, R. K., Gupta, R. N., and Wilson, J. T.
1966 : Compilation of deep crustal seismic refraction profiles: Reviews of Geophysics and Space Physics, v. 4, p. 4–100.
Mutter, J., Detrick, R. S., and NAT Study Group
1984 : Multichannel seismic evidence for anomalously thin crust at Blake Spur Fracture Zone: Geology, v. 12, p. 534–537.
Mutter, J. C., and North Atlantic Transect Study Group
n.d. : Intra-crustal reflections in the Mesozoic crust of the Western North Atlantic; in preparation.
Northrop, J., and Ransone, M.
1962 : Some seismic profiles near the western end of the Puerto Rico trench: Journal of General Physics, v. 45, p. 243–251.
Officer, C. B., and Ewing, M.
1954 : Geophysical investigations in emerged and submerged Atlantic coastal plain: Part VII, Continental shelf, continental slope, and continental rise south of Nova Scotia: Geological Society of America Bulletin, v. 65, p. 653–670.
Officer, C. B., Ewing, J. I., Hennion, J. F., Harkrider, D. G., and Miller, D. E.
1959 : Geophysical investigations of the eastern Caribbean: Summary of 1955 and 1956, cruises: in Physics and Chemistry of the Earth, v. 3, eds., Ahrens, L. H., and others; London, Pergamon Press, p. 17–109.
Officer, C. B., Ewing, M., and Wuenschel, P. C.
1952 : Seismic refraction measurements in the Atlantic Ocean Part IV: Bermuda, Bermuda Rise, and Nares Basin: Geological Society of America Bulletin, v. 63, p. 777–808.
Orcutt, J. A., Kennett, B.L.N., and Dorman, L. M.
1976 : Structure of the East Pacific Rise from an ocean bottom seismometer survey: Geophysical Journal of the Royal Astronomical Society, v. 45, p. 305–320.
Pálmason, G.
1971 : Crustal structure of Iceland from explosion seismology: Societas Scientarium Islandica Rit, v. 40, p. 187.
Purdy, G. M.
1982 : The variability in seismic structure of Layer 2 near the East Pacific Rise at 12°N: Journal of Geophysical Research, v. 87, p. 8403–8416.
1983 : The seismic structure of 140 my old crust in the Western Central Atlantic Ocean: Geophysical Journal of the Royal Astronomical Society, v. 72, p. 115–137.
Purdy, G. M., Detrick, R. S., and Cormier, M. H.
1983 : Seismic structure of the crust beneath the Mid-Atlantic Ridge median valley and its intersection with the Kane Fracture Zone: International Union of Geodesy and Geophysics, XVIII General Assembly, Hamburg FRG, Programme and Abstracts, v. 2, p. 547.
Reid, I., and Jackson, H. R.
1981 : Oceanic spreading rate and crustal thickness: Marine Geophysical Researches, v. 5, p. 165–172.
Rosendahl, B. R., Raitt, R. W., Dorman, L. M., Bibee, L. D., Hussong, D. M., and Sutton, G. H.
1976 : Evolution of oceanic crust. 1. A physical model of the East Pacific Rise crest derived from seismic refraction data: Journal of Geophysical Research, v. 81, p. 5294–5304.
Schouten, H., Karson, J., and Dick, H.
1980 : Geometry of transform zones: Nature, v. 288, p. 470–473.
Schouten, H., and White, R. S.
1980 : Zero-offset fracture zones: Geology, v. 8, p. 175–179.
Schreiber, E., and Fox, P. J.
1977 : Density and P-wave velocity of rocks from the FAMOUS region and their implication to the structure of the oceanic crust: Geological Society of America Bulletin, v. 88, p. 600–608.
Sheridan, R. E., Drake, C. L., Nafe, J. E., and Hennion, J.
1966 : Seismic refraction study of continental margin and east of Florida: Bulletin of the American Association of Petrology Geologists, v. 50, p. 1972–1991.
Sheridan, R. E., Grow, J. A., Behrendt, J. C., and Bayer, K. C.
1979 : Seismic refraction study of the continental edge off the eastern United States: Tectonophysics, v. 59, p. 1–26.

Shor, G. G.
1963 : Refraction and reflection techniques and procedure: in The Sea, Volume 3, ed., M. N. Hill: Wiley-Interscience, New York, p. 20–38.
Sinha, M. C., and Louden, K. E.
1983 : The Oceanographer Fracture Zone 1. Crustal structure from seismic refraction studies: Geophysical Journal of the Royal Astronomical Society, v. 75, p. 713–736.
Spudich, P., and Orcutt, J.
1980a: A new look at the seismic velocity structure of the oceanic crust: Reviews of Geophysics and Space Physics, v. 18, p. 627–645.
1980b: Petrology and porosity of an oceanic crustal site: Results from wave form modelling of seismic refraction data: Journal of Geophysical Research, v. 85, p. 1409–1433.
Staudigel, H., Hart, S. R., and Richardson, S. H.
1981 : Alteration of the oceanic crust: Processes and timing: Earth and Planetary Science Letters, v. 52, p. 311–327.
Steinmetz, L., R. B. Whitmarsh, and Moreira, V. S.
1977 : Upper mantle structure beneath the Mid-Atlantic Ridge north of the Azores based on observations of compressional waves: Geophysical Journal of the Royal Astronomical Society, v. 50, p. 353–380.
Stephen, R. A.
1981 : Seismic anisotropy observed in upper oceanic crust: Geophysical Research Letters, v. 8, p. 865–868.
1983 : A comparison of finite difference and reflectivity seismograms for marine models: Geophysical Journal of the Royal Astronomical Society, v. 72, p. 39–58.
Stephen, R. A., Louden, K. E., and Matthews, D. H.
1980 : The oblique seismic experiment on DSDP Leg 52: Geophysical Journal of the Royal Astronomical Society, v. 60, p. 289–300.
Stoffa, P. L., and Buhl, P.
1979 : Two ship multichannel seismic experiments for deep crustal studies: Expanded spread and constant offset profiles: Journal of Geophysical Research, v. 84, p. 7645–7660.
Talwani, N., Windisch, C. C., and Langseth, M. G.
1971 : Reykjanes Ridge Crest: A detailed geophysical study: Journal of Geophysical Research, v. 76, p. 473–517.
Tolstoy, I., Edwards, R. S., and Ewing, M.
1953 : Seismic refraction measurements in the Atlantic Ocean (Part 3): Seismological Society of America Bulletin, v. 43, p. 35–48.
Tréhu, A. M., Sclater, J. G., and Nabalek, J.
1976 : The depth and thickness of the ocean crust and its dependence upon age: Bulletin Societe Geologique de France, v. 7, p. 917–930.
Tryggvason E.
1964 : Arrival times of P-waves and upper mantle structure: Seismological Society of America Bulletin, v. 54, p. 727–736.
White, R. S.
1984 : Atlantic oceanic crust: Seismic structure of a slow spreading ridge: in Ophiolites and Oceanic Lithosphere: Geological Society of London, special publication.
White, R. S., and Matthews, D. H.
1980 : Variations in oceanic upper crustal structure in a small area of the northeastern Atlantic: Geophysical Journal of the Royal Astronomical Society, v. 61, p. 401–431.
White, R. S., and Purdy, G. M.
1983 : Crustal velocity structure on the flanks of the Mid-Atlantic Ridge at 24°N: Geophysical Journal of the Royal Astronomical Society, v. 75, p. 361–386.
Whitmarsh, R. B.
1971 : Seismic anisotropy of the uppermost mantle absent beneath the east flank of the Reykjanes Ridge: Seismological Society of America Bulletin, v. 61, p. 1351–1368.
1973 : Median valley refraction line: Mid-Atlantic Ridge at 37°N: Nature, v. 246, p. 297–299.
1975 : Axial intrusion zone beneath the median valley of the Mid-Atlantic Ridge at 37°N detected by explosion seismology: Geophysical Journal of the Royal Astronomical Society, v. 42, p. 189–215.
Zverev, S. M., Litvinenko, I. V., Palmason, G., Yaroshevskaya, G. A., and Osoki, N. N.
1980 : A seismic crustal study of the axial rift zone in SW Iceland: Journal of Geophysical Research, v. 47, p. 202–210.

Manuscript Accepted by the Society September 1,1984
Woods Hole Oceanographic Institution Contribution No. 2730

Acknowledgments

The authors acknowledge the continued support of the National Science Foundation and the Office of Naval Research.

The Geology of North America
Vol. M, The Western North Atlantic Region
The Geological Society of America, 1986

Chapter 20

Structure of basement and distribution of sediments in the western North Atlantic Ocean

Brian E. Tucholke
Woods Hole Oceanographic Institution, Woods Hole, Massachusetts 02543

INTRODUCTION

During the past 20 years the morphology of the seafloor has become relatively well known along the crest and shallow flanks of the Mid-Atlantic Ridge through echo sounding and seismic reflection profiling (see Plate 2). In this area sediment cover is so thin, rarely more than 50-100 m, that high frequency echo sounding (12 kHz) essentially maps the surface morphology of oceanic crust. Although our knowledge of this crustal morphology still is far from being detailed or complete, the observed patterns and spacing of fracture zones and crustal blocks provide a general model that can be used to understand the basement morphology in more thickly sedimented regions away from the ridge crest. Structure contours on basment for most of the western North Atlantic recently have been compiled by Tucholke and others (1982) and Tucholke and Fry (1985). Their mapping used the crustal "template" of the mid-ocean ridge together with data from all available seismic reflection profiles in the western Atlantic basins. With some additional mapping, I have extended these structure contours to cover the area from the Mid-Atlantic Ridge crest west to the continental margin between about 8°N and 70°N (Plate 5).

Until the basement morphology is completely masked by sediments, it exerts strong control on how sediments are distributed in the ocean basin. Gravity-controlled mass movements such as debris flows and turbidity currents deposit their sediment load in the deepest accessible basins; consequently, thick ponded sediments often are observed in fracture-zone troughs (Plate 5, Profile F; Plate 6, Profile J. (Note: All subsequent citations of profiles refer by letter to seismic profiles A-P on Plates 5 and 6). Basement peaks and ridges also can influence the path and intensity of deep currents and thereby cause differential deposition over distances of a few kilometers to hundreds of kilometers (McCave and Tucholke, this volume; see Profiles J, L). Thus, once the morphology of the basement is understood, it becomes possible to map sediment thickness in a meaningful way (Plate 6) and to study more effectively the geologic processes that control sediment distribution.

The data base of seismic reflection profiles in the western North Atlantic is highly variable in both line spacing and quality. Most profiles in the deep basin are single-channel data recorded along randomly spaced and variably oriented track lines several tens to many hundreds of kilometers apart. Exceptions include a score of detailed surveys designed to study special features such as segments of fracture zones or to delineate the structure and stratigraphy around deep-sea drillsites. Line spacing in these surveys generally is 5-50 kilometers and the surveys cover one to several square degrees. Along the continental margins numerous multichannel profiles have been acquired, principally by industry and by government agencies. Enough of these data are open-filed that profiles spaced less than ten to a hundred kilometers apart are available, and the general structural and stratigraphic framework of the margins can be outlined in fair detail. The overall data density in the western Atlantic is adequate to study crustal and sedimentary structure at scales of about 1:4 to 1:5 million. The simplified maps at the back of this volume (Plates 5 and 6) are at more than twice this scale (~1:10 million), but they show the principal features that are important in a general synthesis. The ensuing summary deals with the geologic significance of these larger-scale patterns of basement morphology and sediment thickness in the western North Atlantic.

SEDIMENT VELOCITY STRUCTURE

Sound velocity in sediments increases rapidly with depth below the seafloor. Within the upper 10-20 m of the seabed, velocities typically are 1.48 to 1.52 km/sec, up to about 0.08 km/sec slower than sound velocities in the overlying bottom water (Fig. 1, A). Increases to 1.7 to 1.8 km/sec usually occur within the upper 100 m of the seabed, and sound velocity then increases more slowly with increasing depth beyond these limits (Fig. 1, B and C). Because seismic reflection profiles record subbottom reflecting interfaces, including basement, in time rather than in depth, and because of the rapid increase in sound

Tucholke, B. E., 1986, Structure of basement and distribution of sediments in the western North Atlantic Ocean; *in* Vogt, P. R., and Tucholke, B. E. eds., The Geology of North America, Volume M, The Western North Atlantic Region; Geological Society of America.

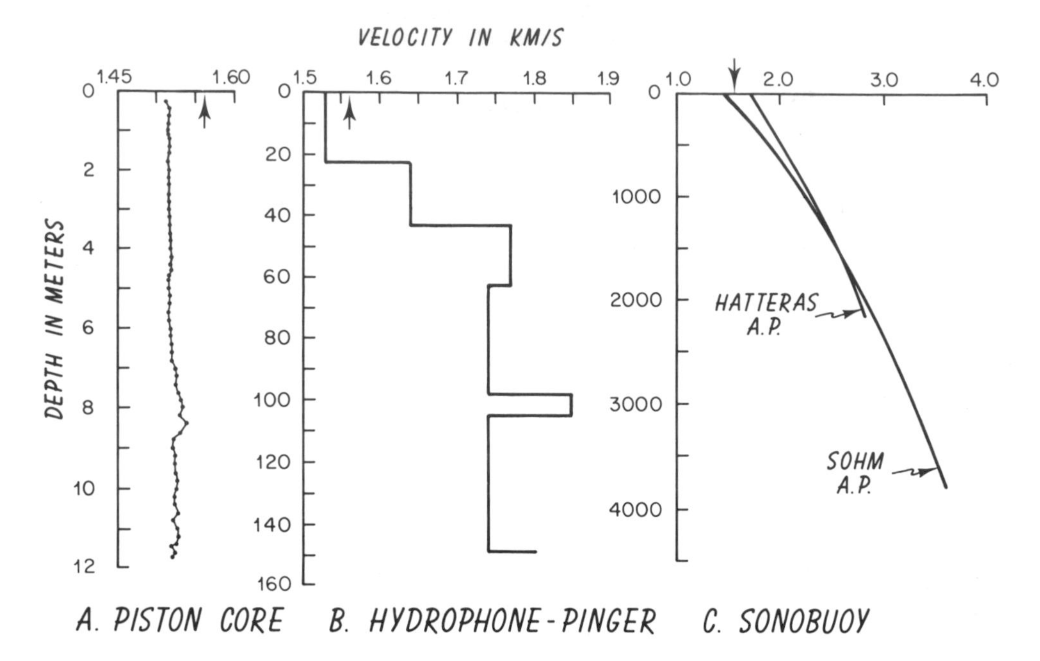

Figure 1. Sound velocity in seafloor sediments determined at three different scales. A - Velocities (dots) measured on a 12-m piston core from the northeastern Nares Abyssal Plain; values are corrected to *in situ* pressure and temperature (Tucholke, 1980). B - Velocities determined for the upper 150 m of the seabed in the northeastern Nares Abyssal Plain by means of a near-bottom, wide-angle reflection experiment (Bryan, 1980). C - Velocity versus depth for the Sohm and Hatteras Abyssal Plains, determined by best-fit regression lines for numerous sonobuoy-interval-velocity solutions from each area (from Houtz, 1980). The seafloor-intercept velocities are an artifact of the solution and have no significance. Arrows show bottom-water velocity.

velocity with subbottom depth, reflection records can give a distorted picture of structural and stratigraphic relations, particularly in areas of thick sediments or buried, strong basement relief. Consequently, the maps in Plates 5 and 6 are contoured in true depth and thickness, rather than in reflection time. Conversion from reflection time to thickness was accomplished by applying appropriate velocity functions throughout the basins.

The velocity functions are based on interval velocities derived from airgun-sonobuoy measurements (wide-angle reflection/refraction) taken in numerous locations across the western North Atlantic. The interval velocities were calculated using a modified X^2-T^2 method (LePichon and others, 1968; Houtz, 1980). Solutions were grouped into geologic regions, each with a generally uniform morphology and seismic-reflection character (Plate 6, Back). A velocity function for each region was derived by least-squares fit of a first-order polynomial to the interval velocities at the mid-points of the layers resolved. The region boundaries were then adjusted to include or exclude certain sonobuoy-velocity solutions and thus optimize the interval-velocity groupings. The velocity regression equations take the form

$$V = V_0 + Kt \pm \Delta V.$$

V_0 is the intercept velocity at t = O, K is a constant acceleration, t is *one-half* the reflection (two-way-travel) time, and ΔV is the standard error of estimate. Thickness (H) can be computed when this expression is integrated to

$$H = V_0 t + 0.5\,(Kt^2).$$

Table 1 lists the values of V_0 and K for the velocity provinces shown on the back of Plate 6.

BASEMENT STRUCTURE

Ocean Crust

The Mid-Atlantic Ridge forms the most prominent large-scale basement feature of the North Atlantic (Plate 5). Average ridge-crest elevation is on the order of 2.7 km although this is strongly perturbed in the regions of the Azores and Iceland where upwelling mantle plumes or hot spots are thought to exist (Morgan, 1981). Average depth of the Azores plateau is slightly less than 2 km, and the average depth of the Iceland plateau is near sea level. Ridge-crest elevation decreases away from these areas to normal elevations at an average gradient of about 1:700 (see Plate 8A). The axial rift valley has similar trends in depth, although greater absolute depths, and it locally reaches to more than 4 km, as observed for example north of the Kane Fracture Zone (Plate 5). On the Reykjanes Ridge north of about 59°N, the axial rift valley

TABLE 1. VELOCITY DATA FOR SEDIMENTARY PROVINCES IN THE WESTERN NORTH ATLANTIC OCEAN*

Area	V (km/s)	Standard Deviation S_V (km/s)	V_o** (km/s)	K (km/s^2)	Std. Error of Estimate ΔV (km/s)	Correlation Coefficient r	Number of Determinations n	Reflection Time† T (sec)
A	2.14	0.21					8	<1.0
B			1.53	1.30	0.18	0.83	39	2.6
C			1.60	1.00	0.16	0.89	69	4.0
D			1.73	1.16	0.15	0.69	88	1.8
E			1.69	0.98	0.12	0.94	51	3.0
F			1.68	1.16	0.24	0.74	50	2.5
G			1.62	1.19	0.19	0.88	55	3.0
H			1.49	1.43	0.23	0.93	62	3.0
I	2.03	0.23					8	<1.0
J	1.88	0.20					7	<1.0
K			1.69	0.91	0.17	0.65	16	1.8
L	2.26	0.16					8	<1.0
M	2.00	0.20					27	<1.0
N	2.00[§]							<1.0
O			1.67	1.46	0.25	0.86	32	2.2
P			1.66	1.58	0.22	0.87	52	1.4
Q			1.78	0.98	0.23	0.78	42	2.8
R	2.19	0.28					24	<1.0
S	2.14	0.28					15	<1.0

*From Houtz (1980), Houtz (pers. comm., 1980) and Tucholke and others (1982). Bounds of areas are shown on the back of Plate 6.
**V_o is the intercept velocity at t=0, where t is one-way vertical travel time.
†Maximum depth, in seconds two-way travel time, to which the functions can be applied confidently.
§Assumed velocity; data are insufficient or poorly sampled.

is replaced by a central horst (Profile A). This segment of the Mid-Atlantic Ridge also is unusual in that no significant fracture zones are developed. Both of these conditions probably are related to asthenospheric flow from the adjacent Iceland hot spot toward the southwest beneath the axis of Reykjanes Ridge (e.g. Vogt and Avery, 1974).

Away from the axis of the Mid-Atlantic Ridge, oceanic crust deepens with increasing age (Profiles C, H). Depth generally follows a trend proportional to the square root of crustal age ($t^{1/2}$) out to about 80 m.y. old crust, and the depth increase decays exponentially to a maximum value at greater ages (see Sclater and Wixon; Tucholke and McCoy, both in this volume). This general age-depth relation is observed in the distribution of basement depth in the several different basins of the western North Atlantic, each of which has a different maximum age, despite the fact that the basement depths are significantly modified by sediment loading. Of course, older crust generally has accumulated a thicker sediment load, so that crustal age, depth, and sediment thickness all show similar trends. The main North American basin south of the Grand Banks has ocean crust dating to about 175-180 Ma at its western margin, and it also has large areas with crustal depths greater than 6 km. The Flemish Basin east of the Grand Banks opened about 118 Ma (anomaly MO; Aptian), and it has a much smaller area of basement depths below 6 km. In the Labrador Sea, ocean crust formed between about 82 and 37 Ma (anomalies 34 to 13). Only very small areas of basement reach 6 km there, and those depths are primarily a result of loading by thick sediments. The crust along the former spreading ridge in the Labrador Sea (Mid-Labrador Sea Ridge) is now at an average depth of nearly 5 km. The very deep basins along the Labrador margin probably are formed in extended continental crust (e.g. Srivastava, 1978). The most youthful basin west of the Mid-Atlantic Ridge is the Irminger Basin between Greenland and the Reykjanes Ridge. The oldest ocean crust there dates to about 56 Ma (anomaly 24) and is 4 to 4.5 km deep. These depths are more than a kilometer shallower than predicted from age-depth relations when sediment loading is taken into account. Thus the crust in Irminger Basin probably was influenced by the Iceland mantle plume starting from the time that the basin first began to open.

The general age-depth relation in oceanic crust is perturbed at intermediate scales (500-1500 km) in several places. The largest positive depth anomaly is the Bermuda Rise, which is elevated up to 0.7-0.8 km above the normal depth for North Atlantic crust of this age (Plate 5; Profile E). Studies of seismic stratigraphy and its correlation to deep-sea drilling results (Tucholke, 1979) indicate that the uplift that initially formed the rise occurred in the late Eocene. Sclater and Wixon (this volume) suggest that the elevation has since been maintained by convective upwelling in the mantle, with correlative downwelling beneath an area of anomalously great depths (>6 km) immediately to the southeast (Plate 5).

Smaller positive depth anomalies that probably are associated with hot-spot volcanism occur in four places. The New England Seamount Chain (e.g. Profile D) and Corner Seamounts are thought to have been formed as the western Atlantic crust of the North American plate drifted northwest over a "fixed" mantle plume between about 100 and 75 Ma; the available radiometric ages suggest that this relative motion was at a rate of about 4.7

cm/yr (Duncan, 1984). Immediately to the south and parallel to the New England Seamounts is another, less well defined "chain" of seamounts. This chain begins with a large, buried peak beneath the U.S. continental rise north of Knauss Knoll, and it extends as a series of widely spaced seamounts southeast across the northern Bermuda Rise (e.g. Profile E). It continues as a positive depth anomaly (less than 6 km) that links the northern Bermuda Rise to the flank of the Mid-Atlantic Ridge (Plate 5). It is not known whether this series of seamounts and the associated depth anomaly is related to hot-spot volcanism. If it is, its parallelism with the New England-Corner Seamount group would suggest that it formed at about the same time as the New England chain. Farther north, along the southeastern margin of the Grand Banks, the J-Anomaly Ridge is a large, well developed basement feature (Profile L). It probably was formed by hotspot volcanism at the axis of the Mid-Atlantic Ridge as initial drift began between the Grand Banks and Iberia in early Aptian time (anomaly MO; Tucholke and Ludwig, 1982). The J-Anomaly Ridge has a conjugate basement ridge in the eastern North Atlantic (Madeira-Tore Rise), and in the Aptian this composite ridge system was at or above sea level (see Plate 9, E). Still farther north is a shallow basement trend extending east from the Newfoundland Seamounts through the Milne Seamounts and Altair Seamount. This trend forms a fourth depth anomaly that probably is related to hot-spot volcanism. Except for very limited radiometric and biostratigraphic age control (middle Cretaceous) in the Newfoundland group (Sullivan and Keen, 1977), the ages of these seamounts are unknown. A hot-spot model by Duncan (1984) suggests that the seamounts may have formed as the North America plate passed over the Madeira hot spot between about 120 and 80 Ma.

Unusually deep (8-10 km) oceanic crust occurs beneath the western Sohm Abyssal Plain north of the New England Seamount Chain (Plate 5). When corrected for sediment loading, this 125-155 m.y. old crust falls 300 to 700 meters below the theoretical elevation predicted by a cooling plate model (Parsons and Sclater, 1977). However, it does match a linear relationship of depth vs. $t^{1/2}$. Heestand and Crough (1981) argued that most North Atlantic crust has been affected by hot spots and that unaffected crust should follow such a linear depth vs. $t^{1/2}$ relation. The crust in the Sohm Basin may therefore be an example of old North Atlantic seafloor that has not experienced any significant elevation change because of passage over a hotspot.

A positive depth anomaly not related to mantle plumes occurs in the oceanic crust along the northern edge of the Bahama Banks and extends eastward just north of the Puerto Rico Trench (Plate 5). The crustal swell along the Bahamas most likely is a peripheral bulge caused by the crustal load of the adjacent carbonate banks; the best known analog is a crustal bulge that is developed along the flanks of the Hawaiian-Emperor Seamount chain and is caused by the load of the volcanoes (Watts and Cochran, 1974). The crustal bulge north of the Puerto Rico Trench (Profile G) is an "outer high" similar to those formed seaward of Pacific trenches (Watts and Talwani, 1974). Although the movement between the Caribbean and North American plates is largely strike-slip in the Puerto Rico Trench, there may be a small component of underthrusting in the western part of the trench and beneath Hispaniola (Molnar and Sykes, 1969); this compression probably maintains the depth anomalies of both the Puerto Rico Trench and its outer high.

Along the Lesser Antilles, thrusting of the North American plate westward beneath the Caribbean carries Atlantic oceanic crust more than 20 km below sealevel before it is overridden by "basement" of the island arc (Emery and Uchupi, 1984; Plate 5). The crust is loaded by a very thick sedimentary section that has been contributed mostly by the Orinoco and Amazon Rivers and that is tectonically thickened in the thrust zone.

Of the finer-scale features in the western Atlantic ocean crust, fracture zones are by far the most pervasive, regularly spaced, and coherently developed over large distances (Plate 5). Individual fracture zones are on the order of 10-30 km wide, and there is a nearly uniform spacing between the fracture zones on the order of 50 km. The basement-structure map suggests that this spacing is consistent from the Mid-Atlantic Ridge to the continental margin and that many individual fracture zones can be traced more or less continuously over this distance. The consistency of this pattern and of a similar pattern in detailed magnetic anomaly maps led Schouten and Klitgord (1982) to suggest that the accreting plate boundary contains a series of spreading cells (average spacing 50 km), and that a "memory" of these cells is locked into the plate boundary because of a different timing of cycles of plate accretion in each cell. Consequently, the fracture zones follow synthetic flow lines of relative plate motion quite closely (see Plate 5, Back). In more recent studies, Schouten and others (1985) have suggested that the spacing of fracture zones is controlled by the dynamics of melt generation beneath the ridge axis and that the spacing is spreading-rate dependent. In this model, predicted spacing for North Atlantic fracture zones varies from 40 to 50 km, given the spreading-rate history shown in Plate 1.

Not all the North Atlantic fracture zones are equally well developed. About three-quarters of the fracture zones only intermittently have fracture valleys that are well defined along their entire length; this is readily observed, for example, in the distribution of the 4 km basement contour along the Mid-Atlantic Ridge (Plate 5). The remaining fracture zones have much better developed fracture valleys at the ridge crest, and these are mostly fracture zones that offset the axis of the Mid-Atlantic Ridge. They tend to be spaced several hundred kilometers apart, although closer spacing exists for the large-offset transforms in the equatorial Atlantic. These fracture valleys can be traced more or less continuously to the continental margins, and they therefore probably incorporated significant ridge-crest offsets throughout their history. These fracture zones tend to separate the ocean crust into wide (400-500 km) crustal bands. Crustal depth is relatively consistent in an isochron-parallel direction across each of these bands, but the depth in one band can vary significantly from that in adjacent bands (Fig. 2). The depth offsets between bands are not related simply to crustal-age offsets, although this effect is

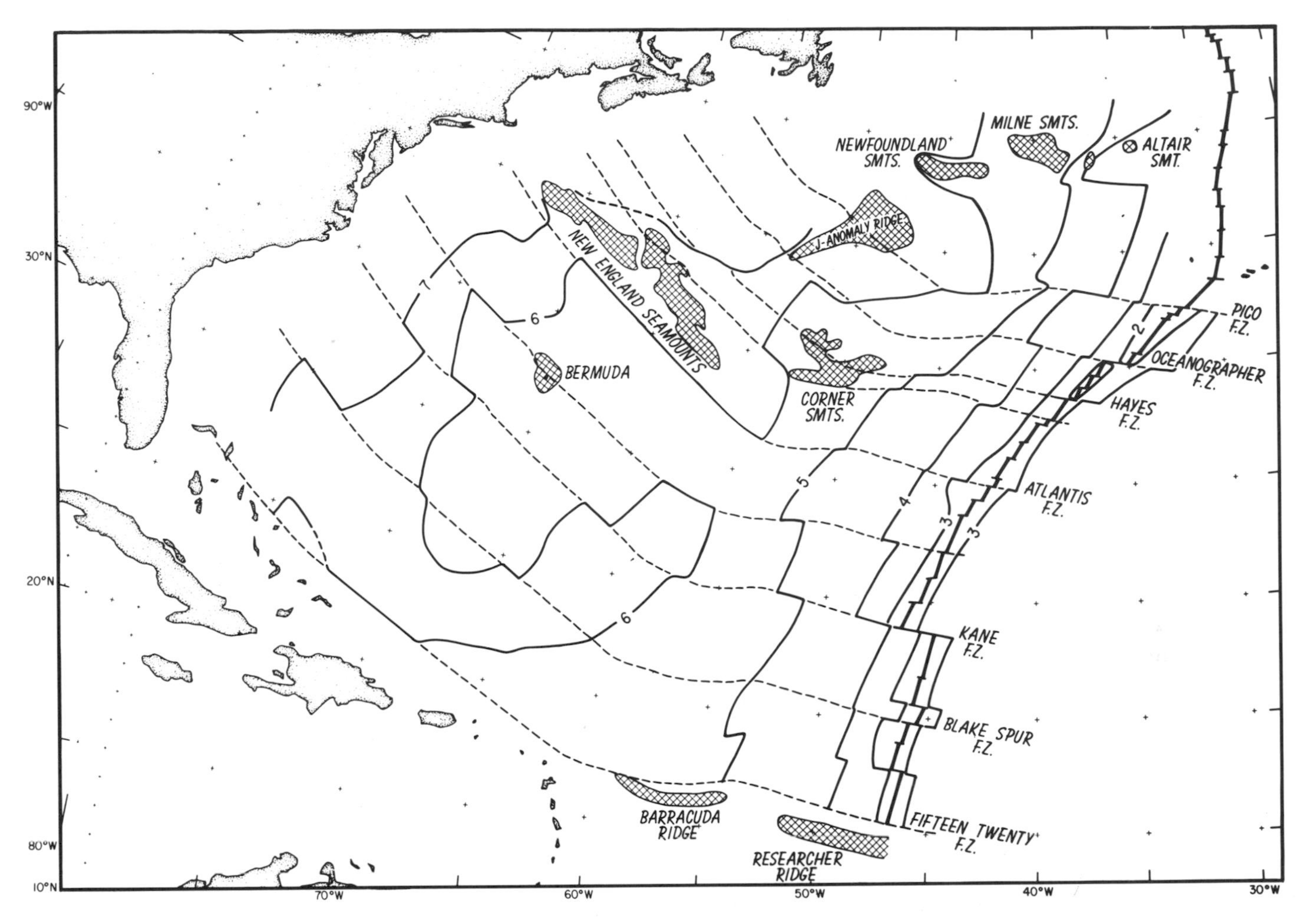

Figure 2. Highly simplified contours of basement depth, in kilometers below sea level, for the western North Atlantic. Dashed lines are synthetic flow lines of relative plate motion, generally equivalent to fracture-zone traces, from Schouten and Klitgord (this volume). Note that any crustal band between a pair of these flow lines has depth variations that are relatively independent of depths in adjacent bands. Transverse Mercator projection, prime meridian 40°W.

locally apparent, for example on the Mid-Atlantic Ridge at the Kane Fracture Zone. Thus, just as there appear to be some fundamental processes that control primary spacing of fracture zones, there may also be some additional larger-scale mantle processes that affect crustal formation at scales of several hundred kilometers along the axis of the mid-ocean ridge.

Continental Margins

Near the continental margins oceanic crust deepens markedly because of the thick sediment load. The position of the ocean-continent boundary along the margins is generally not well known, but reasonable inferences have been made from a variety of geophysical and geological data. The basis for the inferred position of the boundary, shown as a dotted line in Plate 5, is summarized in Table 2. Landward of this boundary the "basement" mapped generally is equivalent to the top of pre-rift rocks. However, there is substantial uncertainty in many places about the stratigraphic assignment of these deeply buried (and mostly unsampled) rocks and the age of structural disturbances observed in seismic reflection profiles across the margin rift basins. Furthermore, the pre-rift rocks in some places are covered with syn-rift or post-rift volcanics, and these form an effective acoustic basement in many reflection profiles. Consequently, the structural contours along the continental margins are best used as guides to structural trends rather than for absolute depths to specific, age-diagnostic horizons. A summary of the presumed nature of the mapped basement along various segments of the continental margins is given in Table 2.

The principal basins in continental crust and their minimum depths along the U.S. Atlantic margin are the Blake Plateau Basin (14 km), Carolina Trough (11 km), Baltimore Canyon Trough (15 km) and Georges Bank Basin (16 km). The basement contours shown in Plate 5 seaward of these basins suggest that a

TABLE 2. SUMMARY OF STRUCTURE AND STRATIGRAPHY OF WESTERN NORTH ATLANTIC CONTINENTAL MARGINS*

Area	Presumed Continent-Ocean Boundary Position	Basement Age of Adjacent Continental Crust	Age of Oldest Overlying Sediments
Blake Plateau	East: Inferred, beneath Blake Escarpment North: Blake Spur F.Z.	Paleozoic	Inferred ?Triassic-Jurassic
Atlantic Margin North to Newfoundland	East Coast Magnetic Anomaly	Paleozoic	Triassic and Jurassic
Grand Banks	South: Transform fault trace Flemish Basin: J Anomaly (?MO) Flemish Cap/Orphan Knoll: Break in basement slope, with associated positive magnetic anomaly; Charlie-Gibbs transform trace on north	Paleozoic, locally Precambrian	Triassic and Jurassic
Labrador Margin	Inferred from gravity/magnetic anomalies and basement structure	Precambrian nearshore, Paleozoic offshore, locally basement is Lower Cretaceous basalt	Lower Cretaceous sampled in wells, older sediments probable
Baffin Margin, Davis Strait	East edge of Davis Strait High; probable mixed oceanic-continent slivers to west	Paleozoic; locally Cretaceous (and Paleocene?) basalt	Middle Cretaceous sampled; older sediments probable
West Greenland Margin	Inferred at break in basement slope	Precambrian (locally Paleozoic?)	Inferred Lower Cretaceous and older
East Greenland Margin	Inferred from magnetic/seismic data (Larsen, 1984)	Precambrian (locally Paleozoic?); Paleogene basalt near Denmark Strait	Eocene and older

*Based on Tucholke and others (1982), Tucholke and Fry (1985), and references therein.

pronounced ridge separates the basins from the adjacent oceanic crust. However, it should be noted that this ridge is based on aeromagnetic depth-to-basement estimates of Klitgord and Behrendt (1979), and it may be an artifact of the assumptions about basement magnetization. Existing seismic reflection profiles do not resolve basement in this zone because it lies beneath a Jurassic-Cretaceous carbonate reef-bank edge; however, fairly flat basement very close to the possible ridge location suggests that either the ridge may not exist or it is poorly developed (see Grow and others, 1983; Mountain and Tucholke, 1985).

The numerous rift basins along the Canadian margin mostly follow the trend of the continental edge. In contrast, two large basins, Sydney and St. Anthony, trend east-west to the south and north of Newfoundland; however, they probably are Paleozoic features (Wade and others, 1977). Highly extended continental crust forms a series of basins and highs west of Orphan Knoll (O. K., Plate 5) and Flemish Cap. Keen and Barrett (1981) conducted seismic refraction work in this area and determined that the crust has been thinned by about 50%. To the northwest, attenuated continental crust presumably also underlies the deep (11-12 km) Hopedale and Saglek basins, but Upper Cretaceous magnetic anomalies identified over Hawke Basin indicate that this basin is floored by oceanic crust (Srivastava and Tapscott, this volume).

Basement in the Davis Strait between Baffin Island and Greenland has a very complex structure. During the early spreading history of Labrador Sea-Baffin Bay from about anomaly 34 time (Santonian), Davis Strait was a locus of extension; later (anomaly 25 to about anomaly 13) this region became a zone of shear, commonly referred to as the Ungava transform complex (see Srivastava and Tapscott, this volume). Basement in Davis Strait therefore may include slices of both oceanic crust and Precambrian and Paleozoic rocks that locally are covered by Cretaceous and Paleocene basalts (Tucholke and Fry, 1985).

Basement beneath the southwest and southeast Greenland continental shelves is probably Precambrian metamorphic rock that is relatively unfaulted (Henderson and others, 1981; Larsen, 1983). At the shelf edge this basement drops abruptly to oceanic depths, and along the southwest Greenland margin it is thought to abut oceanic crust near the base of the continental slope. Thus there is a striking asymmetry with the conjugate Labrador margin where wide rift basins possibly containing stretched and thinned continental crust are present. When seafloor spreading began in the Labrador Sea, the spreading axis appears to have developed along the northern edge of the continental rift zone, thus stranding the rift basins on the Labrador margin (Hinz and others, 1979).

On the southeast Greenland margin south of 65°N, Featherstone and others (1977) mapped a 20-60 km wide zone of sub-basement dipping reflectors, centered approximately on the 3 km basement contour, that they interpreted as Mesozoic sediments on subsided continental crust. The oldest identified magnetic anomaly adjacent to this zone south of 63°30′N is anomaly 24, but to the north the zone is bordered by anomaly 22. This could indicate that the ocean-continent boundary is diachronous in this area. However, Larsen (1984) suggested that anomaly 24 can be traced farther north and he proposed that the ocean-continent boundary is much farther landward (Plate 5). This position of the boundary would indicate that there are large areas of unusually shallow oceanic crust off eastern Greenland. However, it also allows the zone of sub-basement dipping reflectors to be explained as seaward-offlapping basalt flows emplaced subaerially during the initial opening of the basin, in much the same manner as has been proposed for the Voring Plateau (Mutter and others, 1982).

SEDIMENT DISTRIBUTION

At scales of thousands of kilometers, the most obvious variations in sediment thickness are related to the age of underlying oceanic crust in the North Atlantic (Plate 6). Crust younger than about 5 Ma along the crest of the Mid-Atlantic Ridge uniformly has less than 100 m of sediment cover, except for small sediment ponds (Profiles A, C, H). With rare exceptions the sediment cover increases steadily on progressively older crust toward the continental margins. Superimposed on this age-thickness relationship is a large area of the central Atlantic, centered near 25°N, 50°W, with very thin sediments. This seafloor region underlies the unproductive surface waters of the Sargasso Sea, which contains the central waters of the main circulation gyre of the North Atlantic (see Plate 10, L). This gyre probably was established by Late Jurassic time (Tucholke and McCoy, this volume); consequently, the effects of this region of minimal input of pelagic sediments can be seen extending well to the west, across the southeastern Bermuda Rise. As an example of the very limited sediment accumulation in this region, we can look at the area of <100 m sediment thickness immediately southeast of Bermuda. The crust there is 96 m.y. old, yielding an average sediment accumulation rate of about 1 m/my (1 mm/ky). This rate is comparable to accumulation rates of fine-grained pelagic clays in the central gyre of the North Pacific Ocean, and it is among the lowest known rates of continuous sediment accumulation in the world's oceans.

Nearer the continental margins, sediment thickness increases rapidly both because the underlying crust is older and because of proximity to the major sources of detrital materials. The deposition of detrital sediment in the basin near the continental margins is effected principally by two mechanisms, turbidity currents and contour-following bottom currents. Because turbidity currents tend to form flat, ponded deposits, the control of basement structure on sediment thickness patterns is especially evident beneath and along the perimeters of the modern and former abyssal plains (Profiles D, E, K, M). The seaward limit to which turbidity currents have significantly affected sediment accumulation is roughly equivalent to the 0.5 km sediment thickness contour (Plate 6). Comparison with the bathymetry in Plate 2 shows that this limit is near the boundary of the present abyssal plains, but there are some exceptions. On the western Bermuda Rise, for example, turbidites were deposited during the Paleocene and early Eocene, but after the rise was uplifted in late Eocene time it accumulated only pelagic and hemipelagic sediments (Tucholke, 1979). On the northeastern Bermuda Rise sediments thicker than 0.5 km are all pelagic and current-controlled hemipelagic deposits, and this area has never been affected by turbidity currents (Profile E).

Most significant current-controlled deposits (sediment drifts) have sediment thicknesses of one kilometer and greater (compare Plates 2 and 6), and most occur near the continental margins. However, some sediment drifts are totally isolated from the continental margins. The most important of these are the Bermuda Rise Drift on the northeastern Bermuda Rise (e.g. Profile E; sediment thicknesses locally >1.5 km) and the Gloria Drift in the east central Labrador Sea (Profile J; thicknesses >1 km). A smaller patch of current-deposited sediments thicker than 1 km and exhibiting well defined sediment waves occurs just northeast of the eastern end of the New England Seamounts (Plate 6). The abyssal current patterns responsible for deposition of these drifts are very poorly known, but it is likely that basement structure has had a direct and long-term influence in stabilizing the flow so that coherent patterns of deposition could develop. For example, the New England Seamount Chain probably has been important in controlling the position of deep circulation gyres that affect the northern Bermuda Rise and southern Sohm Abyssal Plain; similarly, farther north, deep flow exiting west from the basement trough of Charlie-Gibbs Fracture Zone probably is responsible for the construction of Gloria Drift (see McCave and Tucholke, this volume).

Along the continental margins, the most extensively developed sediment drifts are the Blake-Bahama outer ridge system (Profile N), the Eirik Ridge (Profile I), and the southeast Newfoundland Ridge (Plates 2, 6). All have sediment thicknesses in excess of 2 kilometers. The Blake Outer Ridge is thought to have developed initially by deposition beneath the crossover of the Gulf Stream and the southerly flowing deep Western Boundary Undercurrent along the continental rise (Bryan, 1970). As it grew, its sister drift, the Bahama Outer Ridge, developed in response to interaction of currents flowing around the contours of the Blake ridge. Basement is relatively flat beneath both of these drifts and had no effect in controlling the current patterns that led to their construction. The Eirik Ridge also formed above relatively flat-lying reflectors, apparently as a depositional response to deceleration of bottom currents as they rounded the southern tip of the Greenland continental margin. The Southeast Newfoundland Ridge, on the other hand, is mantled by a drift constructed above a basement ridge. The deep Western Boundary Undercurrent flowing south along the Grand Banks margin is diverted eastward along the north flank of this ridge and then flows west along its south flank. The interaction of these opposing flows over the crest of the ridge probably is responsible for deposition of the thick sediments that form the drift. A similar interaction of opposing flows probably occurs around the J-Anomaly Ridge (Profile L). In addition, the Gulf Stream flows northeast across this region, and it is possible that interaction of the Western Boundary Undercurrent with the base of the Gulf Stream flow has accentuated deposition on the Southeast Newfoundland Ridge and J-Anomaly Ridge. Although other sediment drifts are morphologically well defined along the western Atlantic continental margins (Profiles M, O, P; see also Plate 2), they generally are not large enough to greatly perturb the sediment thickness patterns created by downslope introduction of sediment derived from the adjacent landmasses.

The thickest sediments in the western North Atlantic, of course, are located in the shallow-water regions of the continental margins where largely terrigenous deposits have filled and covered rift basins that were formed during the early opening of the Atlantic. Most major sedimentary basins along the margins contain more than 10 km of fill in their deepest parts, but the relative proportions of syn-rift and post-rift sediments are not yet well known. The Atlantic margins probably experienced several cycles of rifting during the Late Paleozoic and Early Mesozoic before continental drift finally occurred (e.g. McWhae, 1981), and at least the deeper margin basins probably contain relatively great thickness of syn-rift sediments that accumulated during these cycles.

The one active, underthrust margin present in the North Atlantic, the Lesser Antilles arc, is also the site of very thick accumulations of sediments. As noted earlier, both the Orinoco and Amazon rivers have contributed large volumes of sediment to the continental margin of northeastern South America, and the tectonic thickening of these deposits in successive thrust sheets has built a sediment wedge beneath Barbados that probably is nearly 20 km thick (Emery and Uchupi, 1984).

CONCLUSIONS

The syntheses of basement structure and sediment distribution in Plates 5 and 6 have enough detail that they show many geologic patterns that previously were poorly resolved or unrecognized. The observed basement structure also provides tests for the adequacy of plate-kinematic reconstructions of the North Atlantic by allowing comparison of fracture zone traces with modelled plate flow lines and by providing structural control on how the Atlantic plate boundaries must have evolved through time. At the present level of study, broad conceptualizations such as the currently available plate-kinematic models of the North Atlantic are in reasonably good agreement with the observed structure of basement, and the general distribution of sediments accords with our understanding of how sedimentary processes control sediment accumulation in space and time. However, as with any scientific endeavor, numerous new questions are raised and must be dealt with by study in more detail and at different scales.

For example, while we know that many of the well defined fracture zones appear to be nearly continuous structural features from the U.S. Atlantic continental margin to the Mid-Atlantic Ridge, we have little understanding of their detailed morphology. Are there structural discontinuities along fracture zones in the western Atlantic that are mirrored in their eastern Atlantic conjugates, and do these record important plate-tectonic events? Do the fracture-zone traces follow smooth trends or are they, for example, a series of straighter, tandem lineaments that can give us more detailed information on relative plate motions? Are fracture zones traces and "ridge-parallel" structural fabric essentially orthogonal features, and if not, under what conditions does the concept of orthogonality break down? How do the variously developed fracture zones relate to original structural offsets at and near the ocean-continent boundary?

At a larger scale, we will be able to learn more about formation and modification of the lithosphere by study of age-depth relations in the oceanic crust. Sclater and Wixon (this volume) have proposed that the Bermuda Rise and the adjacent negative depth anomaly to the southeast are a result of mantle upwelling and downwelling, respectively. By mapping the basement structure in the conjugate part of the eastern North Atlantic, we will be able to determine whether this lithosphere, when formed at the Mid-Atlantic Ridge, assumed a structure that predisposed it to the observed depth modifications. We also will be able to determine whether the 400-500 km-wide crustal bands that seem to have internally consistent age-depth relations are a persistent and important feature of the North Atlantic crust.

Among the outstanding problems in explaining sediment distribution are the age and early depositional history of sediment drifts and of marginal basins. Most of the available seismic data in these areas is limited by depth of signal penetration and/or resolution. Although detailed study of existing data will greatly clarify the geologic record, more technologically advanced seismic acquisition and processing techniques will pay big dividends in the future.

REFERENCES

Bryan, G. M.
1970 : Hydrodynamic model of the Blake Outer Ridge: Journal of Geophysical Research, v. 75, p. 4530–4537.
1980 : The hydrophone-pinger experiment: Journal of the Acoustical Society of America, v. 68, p. 1403–1408.

Duncan, R. A.
1984 : Age progressive volcanism in the New England Seamounts and the opening of the central Atlantic Ocean: Journal of Geophysical Research, v. 89, p. 9980–9990.

Emery, K. O. and Uchupi, E.
1984 : The Geology of the Atlantic Ocean: Springer Verlag, New York, 1050 p.

Featherstone, P. S., Bott, M.H.P. and Peacock, J. H.
1977 : Structure of the continental margin of south-eastern Greenland: Geophysical Journal of the Royal Astronomical Society, v. 48, p. 15–27.

Grow, J. A., Hutchinson, D. R., Klitgord, K. D., Dillon, W. P. and Schlee, J. S.
1983 : Representative multichannel reflection profiles over the U.S. Atlantic continental margin; in Seismic Expression of Structural Styles, v. 2, ed. A. W. Bally; American Association of Petroleum Geologists Studies in Geology Series 15, p. 2.2.3-1 through 2.2.3-19.

Heestand, R. L. and Crough, S. T.
1981 : The effect of hot spots on the oceanic age-depth relation: Journal of Geophysical Research, v. 86, p. 6107–6114.

Henderson, G.
1976 : Petroleum geology; in Geology of Greenland, eds. A. Escher and W. S. Watt; Geological Survey of Greenland, Denmark, p. 489–505.

Henderson, G., Schiener, E. J., Risum, J. B., Croxton, C. A. and Andersen, B. B.
1981 : The West Greenland Basin; in Geology of the North Atlantic Borderlands, eds. J. W. Kerr and A. J. Fergusson; Canadian Society of Petroleum Geologists Memoir 7, p. 399–428.

Hinz, K., Schlüter, H. U., Grant, A. C., Srivastava, S. P., Umpleby, D. and Woodside, J.
1979 : Geophysical transects of the Labrador Sea: Labrador to southwest Greenland: Tectonophysics, v. 59, p. 151–183.

Houtz, R. E.
1980 : Comparison of velocity-depth characteristics in western North Atlantic and Norwegian Sea sediments: Journal of the Acoustical Society of America, v. 68, p. 1409–1414.

Keen, C. E. and Barrett, D. L.
1981 : Thinned and subsided continental crust on the rifted margin of eastern Canada: crustal structure, thermal evolution and subsidence history: Geophysical Journal of the Royal Astronomical Society, v. 65, p. 443–465.

Klitgord, K. D. and Behrendt, J. C.
1979 : Basin structure of the U.S. Atlantic margin; in Geological and Geophysical Investigations of Continental Margins, eds. J. S. Watkins, L. Montadert and P. W. Dickerson; American Association of Petroleum Geologists Memoir 29, p. 85–112.

Larsen, B.
1983 : Geology of the Greenland-Iceland Ridge in the Denmark Strait; in Structure and Development of the Greenland-Scotland Ridge, eds. M.H.P. Bott, S. Saxov, M. Talwani, and J. Thiede; Plenum Press, New York, p. 425–444.

Larsen, H. C.
1984 : Geology of the east Greenland shelf; in Petroleum Geology of the North European Margin, eds., A. M. Spencer, E. Holter, S. O. Johnsen, A. Mørk, E. Nysaether, P. Songstad, and Å Spinnangr; Norwegian Petroleum Society, Graham and Trotman Ltd., London, p. 329–339.

Le Pichon, X., Ewing, J. I. and Houtz, R.
1968 : Deep-sea sediment velocity determination while reflection profiling: Journal of Geophysical Research, v. 73, p. 2597–2614.

McWhae, J.R.H.
1981 : Structure and spreading history of the northwestern Atlantic region from the Scotian shelf to Baffin Bay; in Geology of the North Atlantic Borderlands, eds. J. W. Kerr and A. J. Fergusson; Canadian Society of Petroleum Geologists Memoir 7, p. 299–332.

Molnar, P. and Sykes, L. R.
1969 : Tectonics of the Caribbean and Middle America regions from focal mechanisms and seismicity: Geological Society of America Bulletin, v. 80, p. 1639–1684.

Morgan, W. J.
1981 : Hotspot tracks and the opening of the Atlantic and Indian oceans; in The Sea, v. 7, The Oceanic Lithosphere, ed. C. Emiliani; John Wiley and Sons, New York, p. 443–488.

Mountain, G. S. and Tucholke, B. E.
1985 : Mesozoic and Cenozoic geology of the U.S. Atlantic continental slope and rise; in Geologic Evolution of the United States Atlantic Margin, ed. C. W. Poag; Van Nostrand Reinhold, New York, p. 293–341.

Mutter, J. C., Talwani, M. and Stoffa, P. L.
1982 : Origin of seaward-dipping reflectors in oceanic crust off the Norwegian margin by "sub-aerial seafloor spreading": Geology, v. 10, p. 353–357.

Parsons, B. and Sclater, J. G.
1977 : An analysis of the variation of ocean floor bathymetry and heat flow with age: Journal of Geophysical Research, v. 82, p. 803–827.

Schouten, H. and Klitgord, K. D.
1982 : The memory of the accreting plate boundary and the continuity of fracture zones: Earth and Planetary Science Letters, v. 59, p. 255–266.

Schouten, H., Klitgord, K. D. and Whitehead, J. A.
1985 : Segmentation of mid-ocean ridges: Nature, v. 317, p. 225–229.

Srivastava, S. P.
1978 : Evolution of the Labrador Sea and its bearing on the early evolution of the North Atlantic: Geophysical Journal of the Royal Astronomical Society, v. 52, p. 313–357.

Sullivan, K. D. and Keen, C. E.
1977 : Newfoundland Seamounts; in Volcanic Regimes of Canada, eds. W.R.A. Baragar, L. C. Coleman and J. M. Hall; Geological Association of Canada Special Paper 16, p. 461–476.

Tucholke, B. E.
1979 : Relationships between acoustic stratigraphy and lithostratigraphy in the western North Atlantic basin; in Initial Reports of the Deep Sea Drilling Project, v. 43, eds. B. E. Tucholke and P. R. Vogt; Washington, D.C., U.S. Government Printing Office, p. 827–846.
1980 : Acoustic environment of the Hatteras and Nares abyssal plains, western North Atlantic Ocean, determined from velocities and physical properties of sediment cores: Journal of the Acoustical Society of America, v. 68, p. 1376–1390.

Tucholke, B. E. and Fry, V. A.
1985 : Basement structure and sediment distribution in northwest Atlantic Ocean: American Association of Petroleum Geologists Bulletin, v. 69, p. 2077–2097.

Tucholke, B. E., Houtz, R. E. and Ludwig, W. J.
1982 : Sediment thickness and depth to basement in western North Atlantic Ocean basin: American Association of Petroleum Geologists Bulletin, v. 66, p. 1384–1395.

Tucholke, B. E. and Ludwig, W. J.
1982 : Structure and origin of the J Anomaly Ridge, western North Atlantic Ocean: Journal of Geophysical Research, v. 87, p. 9389–9407.

Vogt, P. R. and Avery, O. E.
1974 : Detailed magnetic surveys in the northeast Atlantic and Labrador Sea: Journal of Geophysical Research, v. 79, p. 363.

Wade, J., Grant, A. C., Sanford, B. V. and Barss, M. S.
1977 : Basement structure eastern Canada and adjacent areas: Geological Survey of Canada Map 1400A (4 sheets), scale 1:2 million, Lambert

Conformal Conic projection, Ottawa.
Watts, A. and Cochran, J.
1974 : Gravity anomalies and flexure of the lithosphere along the Hawaiian-Emperor Seamount Chain: Geophysical Journal of the Royal Astronomical Society, v. 38, p. 119–141.
Watts, A. and Talwani, M.
1974 : Gravity anomalies seaward of deep-sea trenches and their tectonic implications: Geophysical Journal of the Royal Astronomical Society, v. 36, p. 57–90.

MANUSCRIPT ACCEPTED BY THE SOCIETY JUNE 25, 1985

ACKNOWLEDGMENTS

My studies of the basement structure and sedimentation patterns in the western North Atlantic were supported by Office of Naval Research Contracts N00014-79-C-0071 and N00014-82-C-0019, and by Sandia Laboratories Contract 15/9944.00 to Woods Hole Oceanographic Institution. Preparation of this paper was accomplished with the support of a Mellon Senior Study Award at Woods Hole Oceanographic Institution. I thank J. I. Ewing, D. S. Sawyer and E. Uchupi for reviewing the manuscript and making helpful comments. Contribution No. 5972 of Woods Hole Oceanographic Institution.

Printed in U.S.A.

The Geology of North America
Vol. M, The Western North Atlantic Region
The Geological Society of America, 1986

Chapter 21

Subduction of Atlantic lithosphere beneath the Caribbean

G. K. Westbrook*
Department of Geological Sciences, University of Durham, South Road, Durham DH1 3LE, England
W. R. McCann
Lamont-Doherty Geological Observatory of Columbia University, Palisades, New York 10964

INTRODUCTION

The Caribbean Plate lies between the North American and South American plates and moves relatively eastward with respect to them. The interior of the plate is predominantly oceanic in origin, but along its margins there are island arcs and continental fragments. The plate's boundaries are often diffuse and ill-defined, especially that with South America.

The boundary between the Caribbean and North American plates exhibits great variety. In the Cayman Trough at the western end of the boundary, a small spreading center (Holcombe and others, 1973) links two major systems of transform faults that extend westward along the southern margin of the trough and eastward along its northern margin and also, less clearly, on its southern margin. The eastward transform faults develop into a complex zone of strike-slip faults in the region of Hispaniola which may also have a component of subduction/underthrusting (Bracey and Vogt, 1970), before passing into a zone of oblique subduction at the Puerto Rico Trench (Fig. 1). Some proportion of the motion across the boundary zone is distributed along the southern margin of Hispaniola and Puerto Rico, with associated deformation of sediments in the Muertos Trough (Ladd and Watkins, 1978). As the plate boundary curves round east of the Lesser Antilles island arc, normal subduction becomes predominant, with active volcanism in the Lesser Antilles. Somewhere along the Lesser Antilles segment of the Caribbean plate boundary, probably at about 14N, lies the triple junction between the Caribbean and the North and South American plates. Its position is poorly defined because of the currently cryptic nature of the boundary between the North and South American plates.

The present rate of movement along the boundary between the Caribbean and North American plates was estimated to be about 20 km/m.y. from the spreading rate of the ridge in the Cayman Trough since 2.4 Ma (Macdonald and Holcombe, 1978; Minster and Jordan, 1978) and with predominantly strike-slip motion except in front of the Lesser Antilles. This estimate has been revised by Sykes and others (1982), who have pointed out that seismicity on the southern boundary of the Cayman Trough east of the spreading center and strike-slip faulting in Jamaica (Burke and others, 1980) indicate motion between the plates that is not accounted for by spreading. From a consideration of the shape and depth of the Benioff zone and the rate of spreading in the Cayman Trough before 2.4 Ma, they derived a rate of motion of 37 km/m.y. with a significant component of north-south convergence between the plates.

The geology of the Greater and Lesser Antilles and the Aves Ridge reflects a history of subduction beneath the northern and eastern Caribbean since the Cretaceous. Reconstructions of the motion along the Cayman Trough suggest that the Caribbean has migrated 1100 km into the Atlantic between North and South America since Eocene times (e.g., Pindell and Dewey, 1982). Before then, the position of the Caribbean Plate is uncertain except that it was in the Pacific region somewhere.

The geology and tectonics of the Caribbean region are graphically summarised in the map of Case and Holcombe (1980), and will be treated in detail in the volume in this series on the Caribbean region (Dengo and Case, eds., in preparation). In this short chapter, we confine ourselves to the consideration of the subduction of Atlantic oceanic lithosphere beneath the Caribbean Plate between Hispaniola in the northwest and Trinidad in the southeast.

SEISMICITY AND DYNAMICS OF THE SUBDUCTION ZONE

The seismicity along the plate boundary has been considered

*Present address: Department of Geological Sciences, The University of Birmingham, P.O. Box 363, Birmingham B15 2TT, England

Westbrook, G. K., and McCann, W. R., 1986, Subduction of Atlantic lithosphere beneath the Caribbean, *in* Vogt, P. R., and Tucholke, B. E., eds., The Geology of North America, Volume M, The Western North Atlantic Region: Geological Society of America.

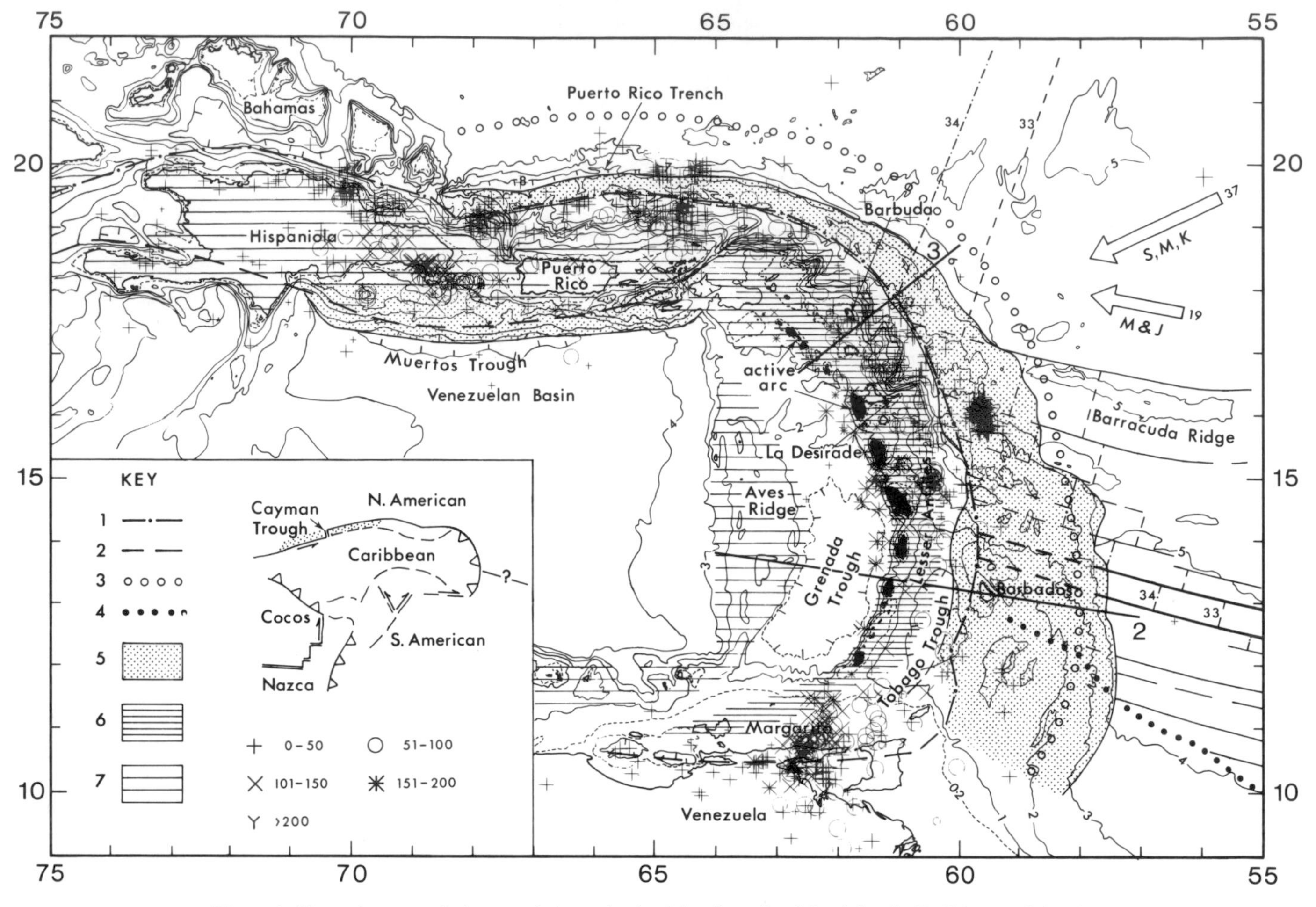

Figure 1. Tectonics, crustal characteristics and seismicity (in red) of the Atlantic-Caribbean subduction zone.

1. Trace of the contact between crystalline crust of the Caribbean and the Atlantic lithosphere.
2. Other zones of significant displacement (plate boundaries?) south of Greater Antilles and on southern margin of Caribbean.
3. Trench outer rise, shown by positive gravity anomaly.
4. Boundary between Jurassic-Early Cretaceous ocean crust (to SW) formed during opening of central Atlantic, and mid-Late Cretaceous crust (to NE) formed since opening of South Atlantic.
5. Accretionary complexes.
6. Thickened crust beneath the Lesser Antilles arc. Active volcanic islands are black.
7. Thickened crust beneath Aves Ridge and Greater Antilles (old island arc terranes).

Also shown are: the vectors of convergence (rate in km/m.y. between the Caribbean and the North American plates from Sykes, McCann, and Kafka (1982) and Minster and Jordan (1978); positions of anomalies 33 and 34; fracture zones, of which the two with extra emphasis are considered to be close to the North American/South American plate boundary; line of sections of Figures 2 and 3; bathymetric contours in km. Depth ranges of earthquake hypocentres in km are shown by different symbols. Magnitude is proportional to size.

by several authors (Sykes and Ewing, 1965; Molnar and Sykes, 1969; Tomblin, 1975; Schell and Tarr, 1978; Dorel, 1981; Stein and others, 1982; Sykes and others, 1982). The Benioff zone beneath the Lesser Antilles, which has a maximum depth extent of around 200 km, continues westward beneath Puerto Rico to the eastern part of Hispaniola where there is locally increased activity associated with a change in trend of the plate boundary and possibly subduction (Bracey and Vogt, 1970).

The dip of the Benioff zone below 60 km is typically 45° beneath the Lesser Antilles, but it is locally steeper (50° to 60°) beneath the center of the arc and shallower (35°) beneath the northeastern quadrant. The zone has its maximum curvature at about 40 km depth. Beneath Puerto Rico and Hispaniola, the zone is not clearly defined but appears to be steep (60°).

The age of the lithosphere currently entering the subduction zone ranges from 120 Ma near Hispaniola to 80 Ma off Barbuda in the Lesser Antilles, so that variations in the initial temperature, and consequently the buoyancy, of the subducted lithosphere are unlikely to be the cause of major variations in the dip of the Benioff zone.

The level of seismicity is low compared with many subduction zones; and instrumental data for the past 30 years show many gaps, some of which are known to have been the sites of historical earthquakes (McCann and others, 1979). Both instrumental data and historical records indicate that seismic activity south of about 14°N is much less than to the north. The reason for this is not established; but in the south the complex of accreted sediment in front of the arc is very thick, and it is probable that South American rather than North American lithosphere is being subducted. The plate motions of Minster and Jordan (1978) predict that the rate of subduction of the South American plate should be 3 km/m.y. faster than the North American plate. Slip rates calculated from seismicity of the whole arc give values of around 5 km/m.y., much less than the 19 km/m.y. and 37 km/m.y. rates from the models of Minster and Jordan (1978) and Sykes and others (1982). Consequently, although slip rates calculated from cumulative seismic moments are often unreliable when rates are low and the speeds predicted by plate motion models may be in error by 50 to 100 percent, it appears that much of the slip takes place aseismically.

Fault plane solutions indicate down-dip tension in the subducted lithosphere. Few thrust mechanisms have been determined, although it is possible that they were important in large historical earthquakes. In the upper and outer part of the Benioff zone, normal and strike-slip fault plane solutions indicate possible reactivation of faults, including transform faults in the oceanic lithosphere as it flexes into the subduction zone (Schell and Tarr, 1978; Stein and others, 1982). A strike-slip solution at intermediate depth (140 km) may indicate continued reactivation of major faults in the subducted lithosphere (Stein and others, 1982). Along the Greater Antilles, there is a predominance of strike-slip and oblique-slip motions.

The region above the Benioff zone, beneath the Lesser Antilles, has been shown by Rial (1976) to be a zone of high attenuation of seismic waves, in common with many other arcs.

The dynamics of subduction of Atlantic lithosphere are reflected not only in the seismicity, but also in the bathymetry and gravity anomalies. Associated with the Puerto Rico Trench and its outer trench high, a flexural bulge in the oceanic lithosphere (Bunce and others, 1974; Watts and Talwani, 1974; Parsons and Molnar, 1976), are a very strong negative gravity anomaly (free air, Bouguer, and isostatic) and a smaller, but pronounced, positive anomaly. The same features are seen in the geoid as mapped by radar altimetry (Vogt, this volume). The negative anomaly (Bowin, 1976; Speed and others, 1984) does not follow the trench as it continues southward; but it remains roughly parallel to the arc as far as Tobago (Fig. 1), from where, after some local disturbance and diminution, it continues into eastern Venezuela. The axes of the negative Bouguer and isostatic anomalies lie close to the suture along which Atlantic lithosphere passes beneath the leading edge of the crystalline crust of the Caribbean. Along this line of suture, seismic refraction and reflection data show that the surfaces of the igneous basements of the Atlantic and Caribbean plates are depressed to their maximum depth, the level of seismicity shows a significant increase, and magnetic anomalies associated with oceanic crust are truncated (Speed and others, 1984). The outer trench high is not easy to follow far south of the Barracuda Ridge because the feature is covered by the wide accretionary prism, but calculation of mass anomalies shows its continued presence at the same distance from the negative anomaly (Westbrook, 1975). Molnar (1977) has shown that the anomalies and the flexural bulge in the lithosphere off Puerto Rico can be accounted for by the effect of dense subducted lithosphere extending to a depth of 150 km. Isostatic anomalies (Bush and Bush, 1969; Kearey, 1974) and residual geoid anomalies (Bowin, 1982) show that the island arc is positively out of equilibrium, particularly strongly so in the north. This is a consequence of uplift produced by convergence of the two plates at the subduction zone, and conversely, the arc can be considered to contribute to the load which bends down the Atlantic lithosphere in the subduction zone and produces the flexural bulge of the outer trench high.

STRUCTURE AND GEOLOGY OF THE EDGE OF THE CARIBBEAN PLATE

The development of the crust along the margin of the Caribbean Plate has been dependent upon the constructive processes of plutonism, volcanism, sedimentation and tectonic accretion at the subduction zone, counterbalanced to some extent by tectonic erosion, and subjected to the modifying processes of tectonic displacement and deformation. The crust of the island arc terrane bordering the Caribbean Plate is much thicker than the crusts of the Venezuelan Basin, Grenada Trough, and the external fringe of forearc (Figs. 2 and 3). The crust beneath the Lesser Antilles has an average maximum thickness of about 30 km, being slightly thinner in the north (Westbrook, 1975; Boynton and others, 1979; Bowin, 1976). Seismically, it is divided into upper and lower crustal layers ($V_p = 6.2$ and $6.9 \text{ km} \cdot \text{s}^{-1}$, respectively), but the relative thickness of these layers varies considerably along the arc, as does the velocity of the upper crustal layer (Boynton and others, 1979). The upper crustal layer of the arc appears to be quite heterogeneous, composed predominantly of intrusive rocks of a variety of compositions from mafic to silicic. The lower crustal layer is probably formed partly of the original, presumably oceanic, crust of the Caribbean, and partly of mafic intrusives, in which igneous cumulates occur (Lewis, 1973; Powell, 1978).

The thickness of the crust beneath the Aves Ridge, the com-

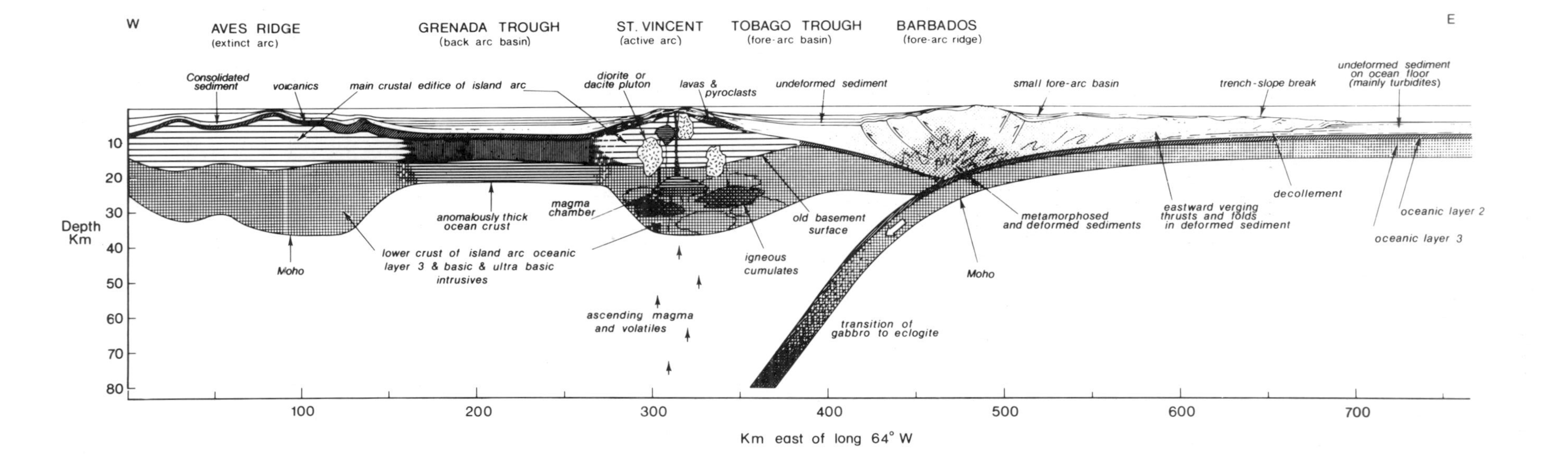

Figure 2. Cross section of the Lesser Antilles arc at latitude 13°30′N (modified from Westbrook, 1982), based upon gravity modelling, seismic refraction and reflection data.

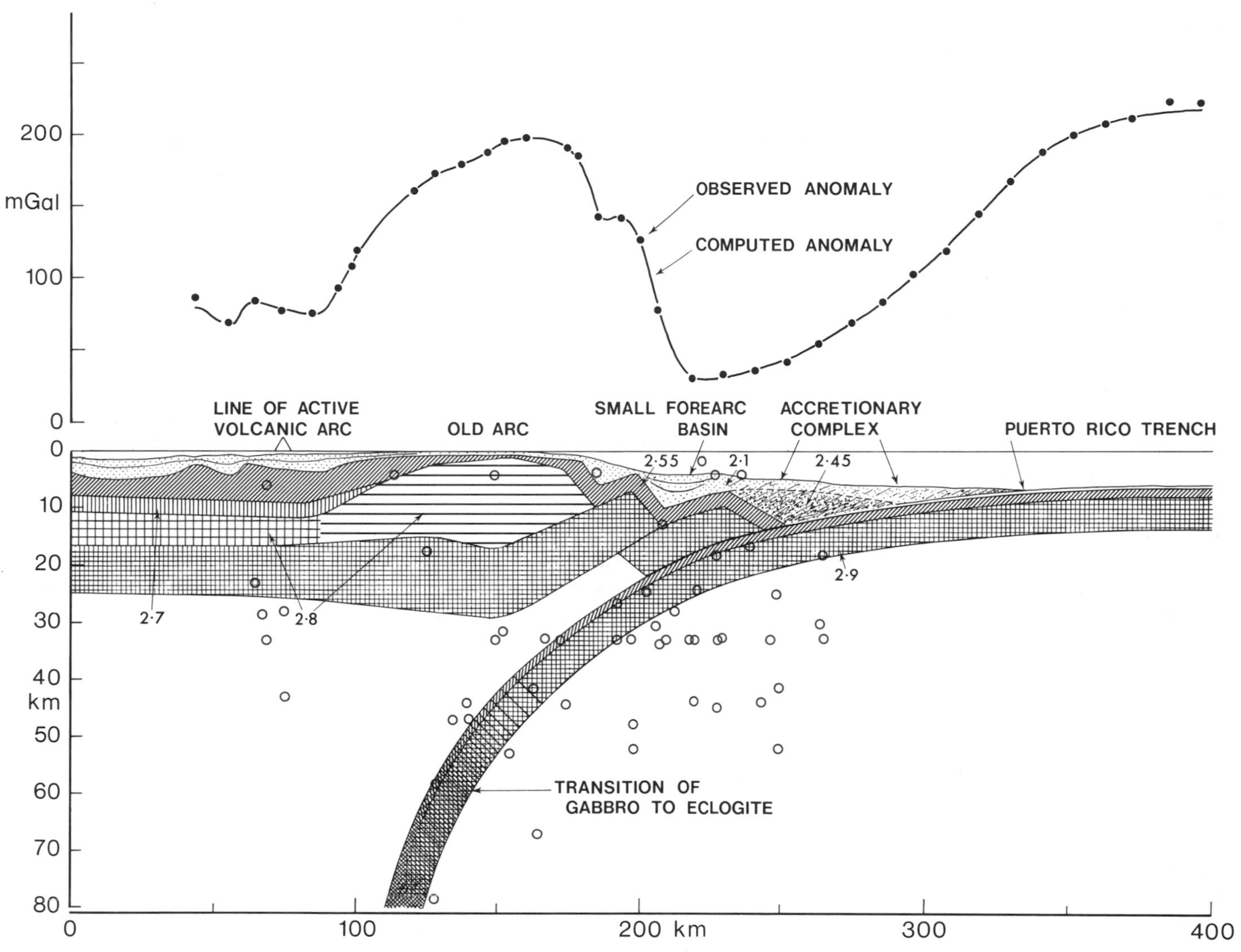

Figure 3. Cross section of the northeastern Lesser Antilles and Puerto Rico Trench. The currently active arc sits on the southwestern flank of the main crustal edifice associated with Late Cretaceous-Early Miocene volcanism. The forearc is significantly shortened by comparison with the arc further south (see Fig. 2), and this may have been caused by the displacement of crust from the forearc by subducted transform ridges. Cross section is derived from a gravity model constrained by seismic refraction and reflection data. The densities of the layers in gm/cm^3 are indicated and the fit of the computed gravity anomaly (line) to the observed (dots) is shown above the section (Bouguer anomaly with 2-D correction for bathymetry of 2.1 gm/cm^3). Positions of earthquake focii are shown in red.

positions and ages of rocks dredged from it (Fox and others, 1971; Nagle, 1972), and the similarity between the curvature of its eastern margin and that of the Lesser Antilles island arc suggest that it is an old island arc active in the Late Cretaceous and earliest Tertiary time. The intervening Grenada Trough, with a crustal structure similar to that of the Venezuelan Basin, may be a backarc (interarc) basin formed by the splitting away of the nascent Lesser Antilles. However, Bouysse and Martin (1979) have suggested that it is an old forearc to the "Aves arc" which was abandoned by a seaward "jump" at the subduction zone to form the Lesser Antilles.

The crystalline crust of the forearc to the Lesser Antilles is thinner than that of the arc, but it is thicker than normal oceanic crust. It has probably undergone modification by magmatism associated with the island arc or by deformational processes associated with the subduction zone. Close to the arc, the forearc crust thickens arcward, its surface dips gently oceanward, and it could be considered part of the arc. Further from the arc, the crust is much thinner and dips more steeply to where it meets the Atlantic lithosphere (Fig. 2). In the north, the surface of inner forearc crust (outer flank of the arc) is very shallow and incorporates the islands of Barbuda and La Desirade. Towards the south, it becomes progressively deeper, being buried to a depth of 7 km beneath sediment in the Tobago Trough.

If a seaward jump of the subduction zone did take place in early Tertiary times to shift the line of the volcanic arc from the Aves Ridge to the Lesser Antilles, then the forearc crust is likely to be trapped Atlantic crust which had been formed by sea floor spreading between North America and South America during the Jurassic and Cretaceous. However, the southward continuation of the basement of the Lesser Antilles to the island of Margarita, which has Cretaceous igneous rocks, suggests that a Cretaceous arc existed along the present line of the Lesser Antilles. Consequently, backarc spreading is the most probable origin of the Grenada Trough, and the forearc is original Caribbean crust.

The crust of the northern Lesser Antilles and that of the Virgin Islands and Puerto Rico are similar, despite the intervening dislocation of the Anegada Passage (a fault zone with probable left-lateral displacement; Hess, 1966; Murphy and McCann, 1979), in terms of their thickness and seismic layering (Officer and others, 1959; Talwani and others, 1959; Molnar, 1977).

The small island of La Desirade has been the subject of extensive study, because of the late Jurassic age obtained for a trondhjemite of the basement igneous complex (Mattison and others, 1973). This age has been supported by the Lower Cretaceous age obtained from biostratigraphic dating of cherts in the formation overlying the trondhjemite (Bouysse and others, 1983). The presence of pillow lavas has been taken as suggestive of an ophiolitic origin for the basement complex of La Desirade, but the age relations of the rocks (Bouysse and others, 1983) and their geochemical similarity to the primitive island arc series of Puerto Rico (Donnelly and Rogers, 1980) imply that La Desirade is part of the earliest island arc formed at the advancing edge of the Caribbean Plate.

Since the Eocene, only the Lesser Antilles have been volcanically active, and until the Lower Miocene, the arc ran through eastern Guadeloupe, Antigua, St. Martin and Anguilla in the north (Martin-Kaye, 1969; Briden and others, 1979). The Pliocene-Recent volcanic arc lies west of the old arc and runs from western Guadeloupe through St. Kitts to Saba. From Martinique southward, the younger volcanics are superimposed upon the older.

In the north, geophysical data show that the most recent phase of magmatism has contributed relatively little to the crust of the arc as a whole (Fig. 3). The young volcanic islands sit on the comparatively thin flank of the arc, whereas the main crustal mass underlies the older islands, reflecting a much longer history of magmatism.

The igneous rocks of the arc are predominantly of the calc-alkaline suite, although there does appear to be local separation into predominantly basaltic or andesitic centres (Smith and others, 1980). The southernmost part of the arc (Grenadines and Grenada), however, is characterized by an undersaturated alkali basaltic suite (Arculus, 1976; Brown and others, 1977).

Ash from volcanic eruptions in the arc has been distributed widely in the region east of the arc by winds in the troposphere to distances of 900 km from the arc (Sigurdsson and Carey, 1981). The oldest ash found in the holes of DSDP Leg 78A, 275 km east of the arc, was of Oligocene age (Moore and others, 1982). If the distribution of ash relative to the arc in the Oligocene was the same as in the Quaternary, then the Lesser Antilles were 600 km further west and have subsequently moved eastward at about 20 km/m.y.

Accretionary Complex

East of the Lesser Antilles lies an accretionary complex which increases in its width (40 to 300 km) and maximum thickness (7 to 20 km) from north to south, and most of which overlies Atlantic ocean crust (Westbrook, 1982).

The shape, size, and structural style of the accretionary complex shows a strong correlation with the thickness of sediment on the ocean floor. This increases from about 300 m in the north to 7 km or more in the south, where the greater part of the sedimentary sequence is composed of Neogene turbidites from the Orinoco River. However, it is the level at which a decollement develops between the horizons accreted at the leading edge of the complex and those which pass undeformed beneath it that provides the direct control on the wavelength of the structures.

Off the northern part of the Lesser Antilles and Puerto Rico, the complex is narrow and sufficiently internally deformed to render it incoherent and somewhat opaque on a seismic reflection section, although the oceanic basement and sedimentary horizons can be traced beneath it to varying extents, by as much as 20 km from the trench (Marlow and others, 1974). Fine-scale imbrication of accreted horizons by thrusting is characteristic of the deformation of the thin layer of accreted sediment. Further south, with increasing thickness of the accreted layer, the thrust spacing increases and broad folding is prominent. This variation in scale is also expressed in the topography of the sea bed which develops ridges and troughs that have a wavelength equal to the spacing of the thrusts, and are clearly shown by long-range, side-scan sonar (Stride and others, 1982) and SEABEAM mapping (Biju-Duval and others, 1982).

Most of the sediments on the ocean floor initially pass under the front of the complex without deformation. The extent to which these undeformed sediments can be traced westward beneath the accretionary wedge exceeds 82 km (Westbrook and Smith, 1983). The decollement does not always maintain a constant stratigraphic level and in some sections can be seen to change level along ramps (Fig. 2). The eastward migration of these ramps incorporates into the wedge undeformed sediment from beneath. It remains a possibility that some of the sediment is subducted beneath the Caribbean Plate.

In its wider parts, the complex can be divided into three broad zones: (1) a zone of accretion beneath the most steeply sloping outer part of the complex (30 to 50 km wide) where the strain rates are high; (2) a zone of stabilization where the slope of the surface of the complex is gentle and may locally dip westward. It is covered by an apron of sediment that in areas in the south can be up to 3 km thick. Deformation is localised and increases downward through the apron sediments; (3) The Barbados Ridge uplift zone, including the island of Barbados in which the accretionary complex is raised with only a thin covering of sediment which is absent in places (Westbrook and others, 1984). The western margin of the accretionary complex is slowly overriding the forearc basins between it and the island arc, developing westward vergent structures.

The forearc basins vary greatly in character around the arc. That underlying the Tobago Trough in the south contains more than 7 km of sediment, and its upper horizons overstep and are uplifted by the Barbados Ridge. North of Guadeloupe, small basins are locally developed in V-shaped re-entrants into the island arc massif. Off Puerto Rico, the basins become more prominent, although they still show lateral division by underlying basement features (Murphy and McCann, 1979).

Several aspects of the deformation of the accretionary complex indicate the presence of high pore fluid pressures. The small angle of slope of its surface indicates that shear stresses on its base are very low, which could be brought about only if pore fluid pressure were very high (90 to 95 percent of the lithostatic load) (Westbrook and others, 1982; Davis and others, 1983). The southern part of the complex is covered by a great many mud diapirs (Biju-Duval and others, 1982).

INFLUENCE OF BASEMENT MORPHOLOGY OF THE ATLANTIC LITHOSPHERE

The morphology of the surface of the Atlantic igneous oceanic crust is dominated by WNW and NW trending ridges and troughs associated with transform-fault traces. Some of the ridges have a bathymetric expression, such as the Barracuda Ridge and Tiburon Rise (Fig. 4), but most are buried features. The gravity and magnetic anomalies associated with them show that they extend beneath the accretionary complex to the edge of the crystalline crust of the Caribbean Plate. The ridges have a two-fold influence upon the accretionary complex, which suffers major changes in elevation across the axes of the ridges. By controlling the thickness of sediment upon the ocean floor, they control the thickness and forward growth of the complex, and because their trend is oblique to the direction of subduction, they have a snowplow effect on the complex, producing compressional features on their southern flanks (Westbrook, 1982; Westbrook and others, 1984). Off the northernmost Lesser Antilles and Puerto Rico, the low angle between the trend of the ridges and the front of the accretionary complex, coupled with the oblique-slip motion, is probably responsible for the particularly intense deformation in the narrow accretionary complex there.

The oceanic basement ridges have a dynamic effect on the outer part of the Caribbean Plate (Vogt and others, 1976). In those regions where they pass beneath the plate, inter- and intraplate seismicity is increased (Dorel, 1981; McCann and Sykes, 1984), and they locally cause uplift of the outer part of the arc. The Barracuda Ridge and Tiburon Rise pass beneath Barbuda and La Desirade, respectively. Both have positive free-air gravity anomalies of nearly 250 mGal, far higher than the rest of the arc, which is generally about 150 mGal.

The possible influence of the subduction of the basement ridges upon the development of the plate boundary since mid-Miocene times is shown by reconstructions from 10.5 Ma to the present (Fig. 4) based on the direction and rates of plate motion determined by Sykes and others (1982). The two major oceanic features considered are the NW extensions along sea-floor spreading flow lines of the Barracuda Ridge and the Tiburon Rise. The latter aligns with the edge of the Bahama Platform.

The eastern tip of the Bahama Platform lies to the north of Puerto Rico in the 10.5 Ma reconstruction (Fig. 4a). The anomalously shallow block that is now situated on the inner wall of the Puerto Rico Trench near Mona Passage has been moved along with the North American plate in this reconstruction. It is interpreted as a segment of the Bahama platform which interfered with the subduction process in western Puerto Rico and sutured on to the Caribbean Plate, thereby causing the plate boundary to jump about 60 km to the northeast at about 3 Ma (McCann and Sykes, 1984). It is possible that other, smaller blocks extended the Bahama Platform southeastwards along this trend, along strike with the Tiburon Rise, and were incorporated onto the Caribbean Plate as anomalous blocks in the northern Lesser Antilles, the Virgin Islands, and Puerto Rico, beginning about 7 Ma (Fig. 4b) when this feature first reached the subduction zone at the northeastern corner of the Caribbean Plate.

In pre-Miocene time, volcanism occurred along the outer arc as shown in Figure 4a, b. In Late Miocene time, the locus of volcanism shifted some 50 km to the west. This event most strongly affected the northern Lesser Antilles. The amount of westward shift decreased southward. McCann and Sykes (1984) proposed that the westerly shift in the volcanic front may have occurred in response to the arrival in the subduction zone of a relatively buoyant piece of crust in the Bahama-Tiburon feature, which, by buoying up the subducted lithosphere, caused it to descend at a significantly shallower angle (35° instead of 55°). The present downgoing seismic zone lies at only 60 km depth beneath the pre-Miocene volcanic arc and underlies the present one at a depth of 110 km.

Alternatively, the shift could have been caused by tectonic erosion of the leading edge of the Caribbean Plate by the basement ridges, producing arcward movement of the locus of subduction and its related features, and shortening the forearc of the northern Lesser Antilles (compare Figs. 2 and 3).

About 3.5 Ma the extension of the Barracuda Ridge entered the subduction zone, producing the uplift of the forearc of the northern Lesser Antilles.

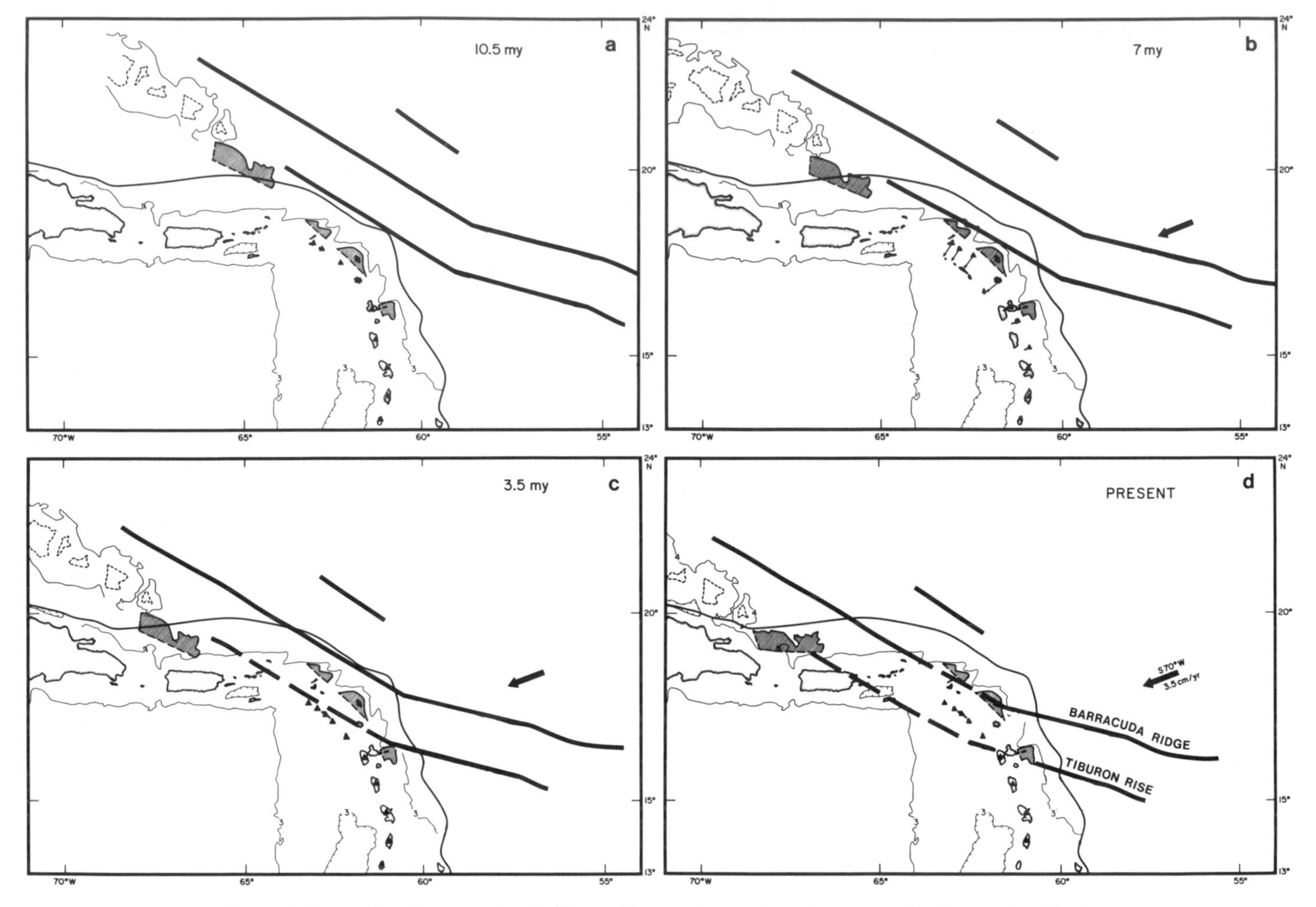

Figure 4. Interactions between the Caribbean Plate and anomalous features on the downgoing North American Plate (solid lines). Shading denotes regions of anomalous high topography along outer edge of present Caribbean Plate.

(a) In Middle Miocene (10.5 Ma) time, a major fracture zone ridge along the southeastern extension of the Bahama Bank entered the subduction zone in the northern Lesser Antilles.

(b) 7 Ma; the volcanic chain along the northern end of the Lesser Antilles shifted (arrows) to the west some 50 km.

(c) 3 Ma; an extension of the Bahama Bank (westernmost shaded region) arrived at the Puerto Rico Trench off northwestern Puerto Rico and was accreted onto the Caribbean Plate causing a small northeastward jump in the subduction zone. The extension of the Barracuda Ridge entered the northern Lesser Antillean trench.

(d) Present; the portion of the Bahama Bank is now attached to the Caribbean Plate. The Barracuda Ridge's extension lies beneath the forearc of the northeastern Lesser Antilles.

CONCLUSION

Atlantic lithosphere has been subducted beneath the eastern margin of the Caribbean Plate during most of the Tertiary period. The history of subduction has been episodic, in that it has brought about changes in the crust of the leading edge of the Caribbean Plate such as the shift of the axis of volcanism in the Lesser Antilles at the beginning of the Pliocene. In addition to the gross effects of the subduction, local variations in the structure of the oceanic lithosphere, such as fracture zones, have had profound effects on the island arc and the accretionary complex.

REFERENCES

Arculus, R. J.
1976: Geology and Geochemistry of the alkali basalt-andesite association of Grenada, Lesser Antilles island arc: Bulletin of the Geological Society of America, v. 87, p. 612–624.

Biju-Duval, B., Le Quellec, P., Mascle, A., Renard, V., and Valery, P.
1982: Multibeam bathymetric survey and high resolution seismic investigations on the Barbados Ridge Complex (eastern Caribbean): a key to the knowledge and interpretation of an accretionary wedge: Tectonophysics, v. 86, p. 275–304.

Bouysse, P., and Martin, P.
1979: Caracteres morphostructuraux et evolution geodynamique de l'arc in-

sulaire des Petites Antilles: Bulletin de B.R.G.M. (deuxieme serie), Section IV, no. 3/4, p. 185–210.
Bouysse, P., Schmidt-Effing, R., and Westercamp, D.
1983: La Desirade Island (Lesser Antilles) revisited: Lower Cretaceous radiolarian cherts and arguments against an ophiolitic origin for the basal complex: Geology, v. 11, p. 244–247.
Bowin, C.
1976: Caribbean gravity field and plate tectonics: Special Paper of the Geological Society of America, no. 169, 79 p.
1982: Gravity and geoid anomalies of the Caribbean. Transactions of the 9th Caribbean Geological Conference, Santo Domingo, Dominican Republic 1980, p. 527–538.
Boynton, C. H., Westbrook, G. K., Bott, M.H.P., and Long, R. E.
1979: A seismic refraction investigation of crustal structure beneath the Lesser Antilles island arc: Geophysical Journal of the Royal Astronomical Society, v. 58, p. 371–393.
Bracey, D. R., and Vogt, P. R.
1970: Plate tectonics in the Hispaniola area: Bulletin of the Geological Society of America, v. 81, p. 2855–2870.
Briden, J. C., Rex, D. C., Faller, A. M., and Tomblin, J. F.
1979: K-Ar geochronology and palaeomagnetism of volcanic rocks in the Lesser Antilles island arc: Philosophical Transactions of the Royal Society of London Series A, v. 291, no. 1383, p. 485–528.
Brown, G. M., Holland, J. G., Sigurdsson, H., Tomblin, J. F., and Arculus, R. J.
1977: Geochemistry of the Lesser Antilles volcanic island arc: Geochimica and Cosmochimica Acta, v. 41, p. 785–801.
Bunce, E. T., Phillips, J. D., and Chase, R. L.
1974: Geophysical study of Antilles Outer Ridge and northeast margin of the Caribbean Sea: Bulletin of the American Association of Petroleum Geologists, v. 58, no. 1, p. 106–123.
Burke, K., Grippi, J., and Sengor, A. M.
1980: Neogene structures in Jamaica and the tectonic style of the northern Caribbean plate boundary zone: Journal of Geology, v. 88, p. 375–386.
Bush, S. A., and Bush, P. A.
1969: Isostatic gravity map of the eastern Caribbean: Transactions of the Gulf Coast Association of Geological Societies, v. 19, p. 281–285.
Case, J. E., and Holcombe, T. L.
1980: Geologic-tectonic map of the Caribbean region: United States Geological Survey Miscellaneous Investigations Series Map, I-1100.
Chase, R. L., and Bunce, E. T.
1969: Underthrusting of the eastern margin of the Antilles by the floor of the western North Atlantic Ocean and origin of the Barbados Ridge: Journal of Geophysical Research, v. 74, p. 1413–1420.
Davies, D., Suppe, J., and Dahlen, F. A.
1983: Mechanics of fold-and-thrust belts and accretionary wedges: Journal of Geophysical Research, v. 88, p. 1153–1172.
Dorel, J.
1981: Seismicity and seismic gap in the Lesser Antilles arc and earthquake hazard in Guadeloupe. Geophysical Journal of the Royal Astronomical Society, v. 67, p. 679–695.
Fox, P. J., Schreiber, E., and Heezen, B. C.
1971: The geology of the Caribbean crust: Tertiary sediments, granitic and basic rocks from the Aves Ridge: Tectonophysics, v. 12, p. 88–109.
Hess, H. H.
1966: Caribbean research project 1965, and bathymetric chart, in Caribbean Geological Investigations, ed. H. H. Hess; Memoir of the Geological Society of America, no. 98, no. 1–10.
Holcombe, T. L., Vogt, P. R., Matthews, J. E., and Murchison, R. R.
1973: Evidence for sea-floor spreading in the Cayman Trough: Earth and Planetary Science Letters, v. 20, p. 357–371.
Kearey, P.
1974: Gravity and seismic reflection investigations into the crustal structure of the Aves Ridge, eastern Caribbean: Geophysical Journal of the Royal Astronomical Society, v. 38, p. 435–448.
Ladd, J. W., and Watkins, J. S.
1978: Active margin structures within the north slope of the Muertos trench: Geologie en Mijnbouw, v. 57, p. 255–260.
Lewis, J. F.
1973: Petrology of ejected plutonic blocks of the Soufrience volcano, St. Vincent, West Indies: Journal of Petrology, v. 14, p. 81–112.
Macdonald, K. C., and Holcombe, T. L.
1978: Inversion of magnetic anomalies and seafloor spreading in the Cayman Trough; Earth and Planetary Science Letters, v. 40, p. 407–414.
Marlow, M. S., Garrison, L. E., Martin, R. G., Trumbull, J.V.A., and Cooper, A. K.
1974: Tectonic transition zone in the northeastern Caribbean: Journal of Research of the U.S. Geological Survey, v. 2, p. 289–302.
Martin-Kaye, P.H.A.
1969: A summary of the geology of the Lesser Antilles: Overseas Geology and Mineral Resources, v. 10, no. 2, p. 172–206.
Mattinson, J. M., Fink, L. K., and Hopson, C. A.
1973: Age and origin of ophiolitic rocks on La Desirade Island, Lesser Antilles arc: Annual Report of Carnegie Institution, Washington 1972-73, p. 616–623.
McCann, W. R., Nishenko, S., Sykes, L. R., and Krause, J.
1979: Seismic gaps and plate tectonics: Seismic potential for major plate boundaries: Pure and Applied Geophysics, v. 117, p. 1083–1104.
McCann, W. R., and Sykes, L. R.
1984: Subduction of aseismic ridges beneath the Caribbean Plate: implications for the tectonics and seismic potential of the northwestern Caribbean: Journal of Geophysical Research, v. 89, p. 4493–4519
Minster, J. B., and Jordan, T. H.
1978: Present-day plate motions: Journal of Geophysical Research, v. 83, p. 5331–5354.
Molnar, P.
1977: Gravity anomalies and the origin of the Puerto Rico trench: Geophysical Journal of the Royal Astronomical Society, v. 51, p. 701–708.
Molnar, P., and Sykes, L. R.
1969: Tectonics of the Caribbean and Middle American regions from focal mechanisms and seismicity: Bulletin of the Geological Society of America, v. 80, p. 1639–1684.
Moore, J. C., Biju-Duval, B., Bergen, J. A., Blackington, G., Claypool, G. E., Cowan, D. S., Duennebier, F., Guerra, R. T., Hemleben, C.H.J., Hussong, D., Marlow, M. S., Natland, J. H., Pudsey, C. J., Renz, G. W., Tardy, M., Willis, M. E., Wilson, D., and Wright, A. A.
1982: Offscraping and underthrusting of sediment at the deformation front of the Barbados Ridge: Deep Sea Drilling Project Leg 78A: Bulletin of the Geological Society of America, v. 93, p. 1065–1077.
Murphy, A., and McCann, W. R.
1979: Preliminary results from a new seismic network in the north east Caribbean: Bulletin of the Seismological Society of America, v. 69, p. 1497–1513.
Nagle, F.
1972: Rocks from the seamounts of escarpments on the Aves Ridge: in Translations of the 6th Caribbean Geological Conference: Impreso por Cromotip, Caracas, Venezuela, p. 409–413.
Officer, C. B., Ewing, J. I., Hennion, J. F., Hardrider, G. D., and Miller, D. E.
1959: Geophysical investigations in the eastern Caribbean: summary of 1955 and 1956 cruises: in Physics and Chemistry of the Earth, v. 3, eds. C. H. Ahrens, F. Press, K. Rankama, and S. K. Runcorn; Pergamon Press, London, p. 17–109.
Parsons, B., and Molnar, P.
1976: The origin of outer topographic rises associated with trenches: Geophysical Journal of the Royal Astronomical Society, v. 45, p. 707–712.
Pindell, J., and Dewey, J. F.
1982: Permo-Triassic reconstruction of Western Pangea and the evolution of the Gulf of Mexico/Caribbean region: Tectonics, v. 1, p. 179–212.
Powell, M.

1978: Crystallisation conditions of low-pressure cumulate nodules from the Lesser Antilles island arc: Earth and Planetary Science Letters, v. 39, p. 162–172.

Rial, J.
1976: Seismic wave attenuation across the Caribbean plate: high concentration on the concave side of the Lesser Antilles arc: Bulletin of the Seismological Society of America, v. 66, p. 1905–1920.

Schell, B. A., and Tarr, A. C.
1978: Plate tectonics of the northeastern Caribbean Sea region: Geologie en Mijnbouw, v. 57, p. 319–324.

Sigurdsson, H., and Carey, S.
1981: Marine tephrochronology and Quaternary explosive volcanism in the Lesser Antilles arc; in Tephra Studies, eds. Self and R.S.J. Sparks; D. Reidel Pub. Co., Dordrecht, Holland, p. 255–280.

Smith, A. L., Roobol, M. J., and Gunn, B. M.
1980: The Lesser Antilles—a discussion of the island arc magmatism: Bulletin of Volcanology, v. 43-2, p. 287–302.

Speed, R. C., Westbrook, G. K., Biju-Duval, B., Ladd, J. W., Mascle, A., Moore, J. C., Saunders, J. B., Schoonmaker, J. E., and Stein, S.
1984 : Lesser Antilles arc and adjacent terranes. Atlas 10, Ocean Margin Drilling Program, Regional Atlas Series: Marine Science International, Woods Hole, 27 sheets.

Stein, S., Engeln, J. F., Wiens, D. A., Fujita, K., and Speed, R. S.
1982: Subduction seismicity and tectonics in the Lesser Antilles arc: Journal of Geophysical Research, v. 87, p. 8642–8664.

Stride, A. H., Belderson, R. H., and Kenyon, N. H.
1982: Structural grain, mud volcanoes and other features on the Barbados Ridge Complex revealed by GLORIA long range side-scan sonar: Marine Geology, v. 49, p. 187–196.

Sykes, C. R., McCann, W. R., and Kafka, A. L.
1982: Motion of the Caribbean Plate during the last 7 million years and implications for earlier Cenozoic movements: Journal of Geophysical Research, v. 87, p. 10656–10676.

Sykes, L. R., and Ewing, M.
1965: The seismicity of the Caribbean region: Journal of Geophysical Research, v. 70, p. 5065–5074.

Talwani, M., Sutton, G. H., and Worzel, J. L.
1959: A crustal section across the Puerto Rico trench: Journal of Geophysical Research, v. 64, p. 1545–1555.

Tomblin, J. F.
1975: The Lesser Antilles and Aves Ridge; in the Gulf of Mexico and the Caribbean: The Ocean Basins and Margins, eds. A.E.M. Nairn and F. G. Stehli; v. 3, Plenum Press, New York, p. 476–500.

Vogt, P. R., Lowrie, A., Bracey, D. R., and Hey, R. N.
1976: Subduction of aseismic oceanic ridges: effects on shape, seismicity and other characteristics of consuming plate boundaries: Special Paper of the Geological Society of America, no. 172, 59 p.

Watts, A. B., and Talwani, M.
1974: Gravity anomalies seaward of deep-sea trenches and their implications: Geophysical Journal of the Royal Astronomical Society, v. 36, p. 57–90.

Westbrook, G. K.
1975: The structure of the crust and upper mantle in the region of Barbados in the Lesser Antilles: Geophysical Journal of the Royal Astronomical Society, v. 43, p. 201–242.
1982: The Barbados Ridge Complex: tectonics of a mature forearc system: in Trench and Forearc Geology: sedimentation and tectonics in ancient and modern plate margins, ed. J. K. Leggett; Special Publication of the Geological Society of London, no. 10, p. 261–276.

Westbrook, G. K., Mascle, A., and Biju-Duval, B.
1984 : Geophysics and the Structure of the Lesser Antilles Forearc: in Initial Reports of the Deep Sea Drilling Project, v. 78A, eds., B. Biju-Duval and J. C. Moore; U.S. Government Printing Office, Washington, D.C., p. 23–38.

Westbrook, G. K., and Smith, M. J.
1983: Long decollement and mud volcanoes; evidence from the Barbados Ridge Complex for the role of high pore water pressures in the development of an accretionary complex: Geology, v. 11, p. 279–283.

Westbrook, G. K., Smith, M. J., Peacock, J. H., and Poulter, M. J.
1982: Extensive underthrusting of undeformed sediment beneath the accretionary complex of the Lesser Antilles subduction zone: Nature, v. 300, no. 5893, p. 625–628.

Manuscript Accepted by the Society May 3, 1984

Printed in U.S.A.

The Geology of North America
Vol. M, The Western North Atlantic Region
The Geological Society of America, 1986

Chapter 22

Plate kinematics of the central Atlantic

Kim D. Klitgord
U.S. Geological Survey, Woods Hole, Massachusetts 02543
Hans Schouten
Woods Hole Oceanographic Institution, Woods Hole, Massachusetts 02543

INTRODUCTION

Opening of the central Atlantic Ocean basin during the past 200 Ma separated North America from Africa and created a classic example of plate tectonic divergent motion and associated geologic features (LePichon, 1968; Morgan, 1968). The entire history of relative motion of these two plates is preserved in the fabric of sea-floor spreading (SFS) recorded by magnetic lineation and fracture zone (FZ) patterns (Fig. 1) (Vine and Matthews, 1963; Heezen and Tharp, 1965) on both flanks of the Mid-Atlantic Ridge. Age calibration of the SFS magnetic anomaly pattern (Cox, 1973; Harland and others, 1982; Kent and Gradstein, this volume) enables us to treat SFS lineations as isochrons of sea-floor crustal ages. FZs mark the path of spreading center offsets (transform faults) through time, providing an approximate flowline trace of the motions that separated the North American and African plates.

Reconstruction poles of rotation and stage poles of motion can be determined from the SFS lineation and FZ data sets (Bullard and others, 1965; McKenzie and Parker, 1967; McKenzie and Sclater, 1971; Harrison, 1972). The kinematic history described by these poles provides a framework for examining major tectonic events, anomalous plate behavior, geologic phenomena, paleooceanographic events, etc. (e.g., Vogt and others, 1969; Tarling and Runcorn, 1973; Dewey and others, 1973; Vail and others, 1977; Sclater and others, 1977; Pitman, 1978; Rona and Richardson, 1978; Schwan, 1980; Kerr and Fergusson, 1981; Harland and others, 1982).

Reconstruction of past positions of the continents (Wegener, 1924; du Toit, 1937; Carey, 1958) is one of the primary products derived from plate kinematic history. The works of Bullard and others (1965), McKenzie and Parker (1967), Morgan (1968), Funnell and Smith (1968), and Pitman and Talwani (1972) established the technique of using poles of rotation to reconstruct the past positions and motions of the continents. Since the middle 1960s, numerous closure reconstructions of the Atlantic have been published, but it is difficult to evaluate the compatibility of many reconstructions with diverse sets of geologic information because pole-of-rotation information commonly is not given. In most examples, plate positions are based on previously published reconstructions with a graphic shift added to fit some new information or idea. We discuss here only those few reconstructions that have served as the basis for most other Atlantic closure models and for which pole information is available.

There have been only a few studies of the post-Triassic opening history of the Atlantic. Since the initial studies of Heirtzler and others (1968), LePichon and Fox (1971), Pitman and Talwani (1972), and Francheteau (1973), the only comprehensive studies have been those of Sclater and others (1977) and Olivet and others (1984). The series of reconstructions presented by Sclater and others (1977) is based on the studies of LePichon and Fox (1971) and Francheteau (1973). This series of paleo-Atlantic configurations is the most commonly used set at the present time (e.g., Pindell and Dewey, 1982).

Reconstructions for the north and central Atlantic published by Olivet and others (1984) are based on a detailed study of northeast Atlantic data (Olivet, 1978), a re-examination of some central and north Atlantic SFS magnetic lineations, and a re-evaluation of previous published works to remove inconsistencies in north and central Atlantic reconstructions. Our study of central Atlantic SFS data (Fig. 2) involves a comprehensive re-examination of magnetic lineations and bathymetric patterns and determination of a set of finite-difference poles to describe the kinematic history (Tables 1 and 2). The SFS data set used here is more extensive than that of Olivet and others (1984), and provides better control for identifying the locations of corresponding FZs on opposite flanks. Our reconstruction poles are slightly

Figure 1 (following pages). Sea-floor-spreading magnetic lineation and fracture zone pattern in the central Atlantic and important plate boundaries. Only selected lineations and fracture zones (FZ) are shown: East Coast Magnetic Anomaly (ECMA), Blake-Spur Magnetic Anomaly (BSMA), West African Coast Magnetic Anomaly (WACMA), SFS lineations (M-25, M-21, M-16, M-10N, M-4, M-0, 34, 33_o, 33_y, 32, 30, 25, 21, 13, 6, 5), and Mid-Atlantic Ridge. Other tectonic features discussed in text are also labeled. The 200-m, 2000-m and 4000-m bathymetric contours are indicated.

Klitgord, K. D., and Schouten, H., 1986, Plate kinematics of the central Atlantic; *in* Vogt, P. R., and Tucholke, B. E. eds., The Geology of North America, Volume M, The Western North Atlantic Region: Geological Society of America.

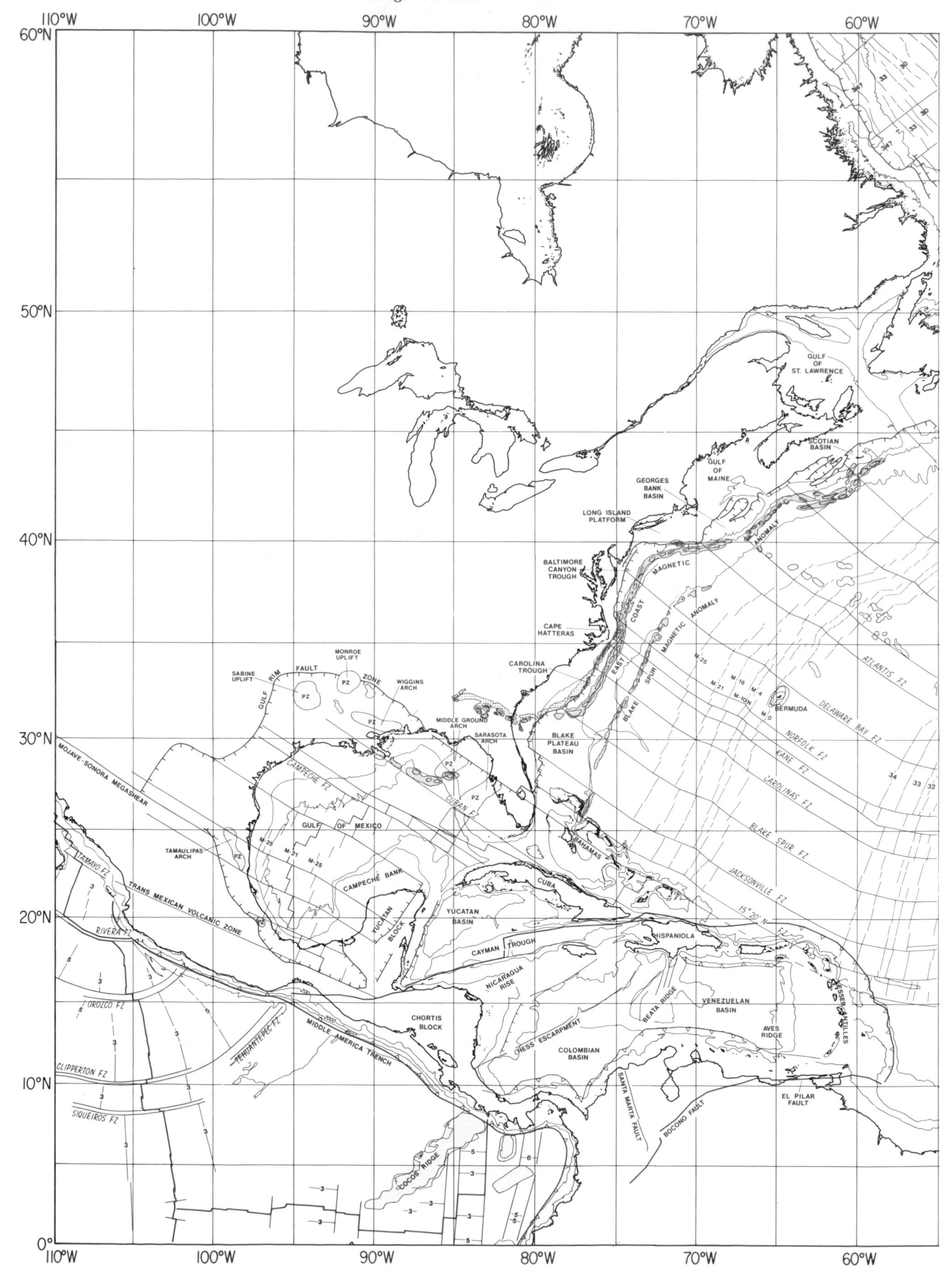

110°W
100°W
90°W
80°W
70°W
60°W
60°N
50°N
40°N
30°N
20°N
10°N
0°
GULF OF ST. LAWRENCE
SCOTIAN BASIN
GULF OF MAINE
GEORGES BANK BASIN
LONG ISLAND PLATFORM
BALTIMORE CANYON TROUGH
CAPE HATTERAS
CAROLINA TROUGH
EAST COAST MAGNETIC ANOMALY
BLAKE SPUR MAGNETIC ANOMALY
BLAKE PLATEAU BASIN
BERMUDA
ATLANTIS FZ
DELAWARE BAY FZ
NORFOLK FZ
KANE FZ
CAROLINAS FZ
BLAKE SPUR FZ
JACKSONVILLE FZ
15°20' N FZ
MONROE UPLIFT
SABINE UPLIFT
GULF RIM FAULT ZONE
WIGGINS ARCH
MIDDLE GROUND ARCH
SARASOTA ARCH
PZ
MOJAVE-SONORA MEGASHEAR
CAMPECHE FZ
CUBAN FZ
GULF OF MEXICO
TAMAULIPAS ARCH
CAMPECHE BANK
YUCATAN BLOCK
YUCATAN BASIN
CUBA
BAHAMAS
HISPANIOLA
CAYMAN TROUGH
NICARAGUA RISE
CHORTIS BLOCK
BEATA RIDGE
VENEZUELAN BASIN
AVES RIDGE
LESSER ANTILLES
HESS ESCARPMENT
COLOMBIAN BASIN
SANTA MARTA FAULT
BOCONO FAULT
EL PILAR FAULT
TAMAYO FZ
TRANS MEXICAN VOLCANIC ZONE
RIVERA FZ
OROZCO FZ
TEHUANTEPEC FZ
MIDDLE AMERICA TRENCH
CLIPPERTON FZ
SIQUEIROS FZ
COCOS RIDGE
M-25
M-21
M-16
M-4
M-10N
M-0
34
33
32

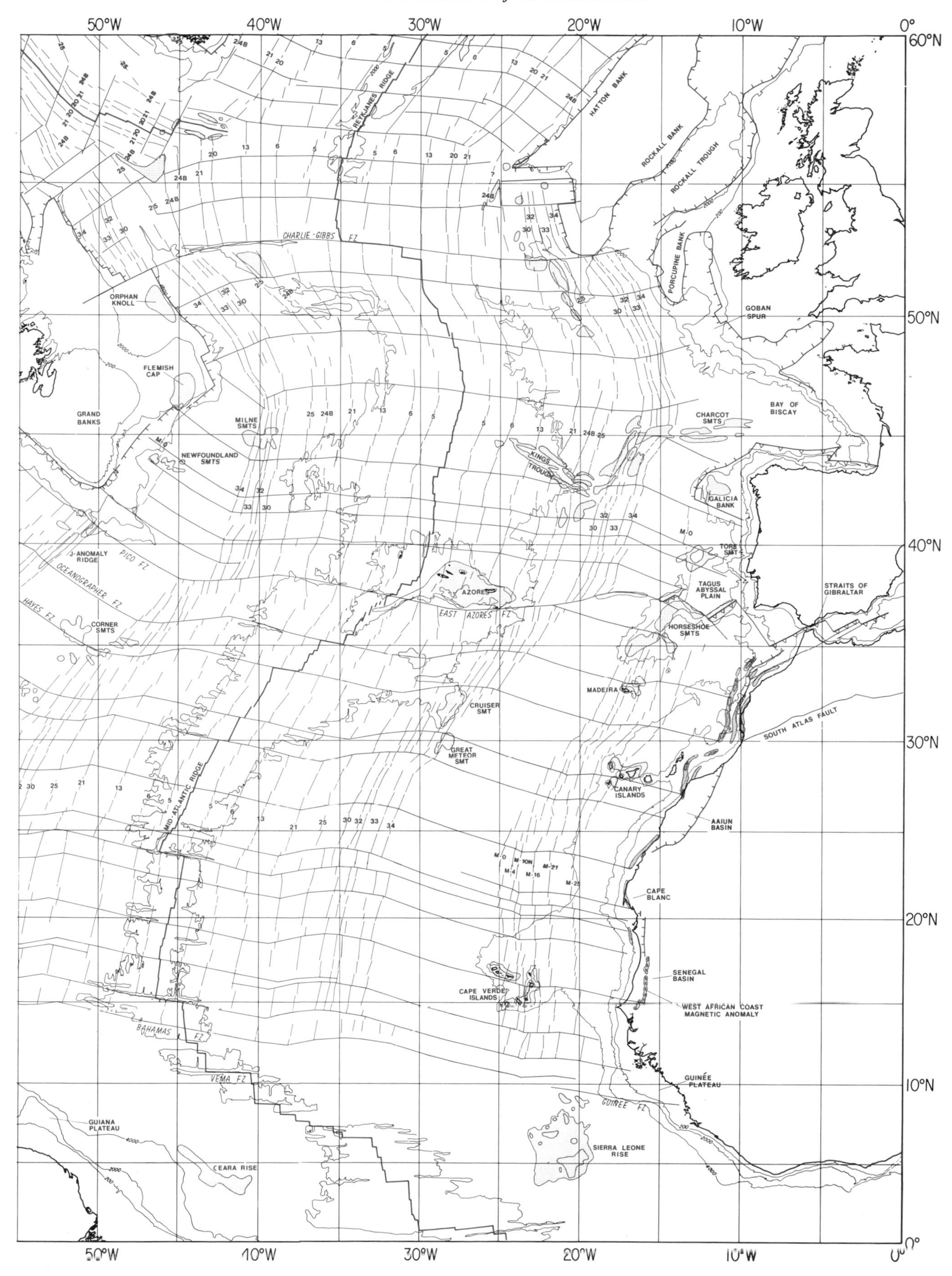
50°W
40°W
30°W
20°W
10°W
0°
60°N
50°N
40°N
30°N
20°N
10°N
REYKJANES RIDGE
HATTON BANK
ROCKALL BANK
ROCKALL TROUGH
CHARLIE-GIBBS FZ
PORCUPINE BANK
GOBAN SPUR
ORPHAN KNOLL
FLEMISH CAP
GRAND BANKS
MILNE SMTS
NEWFOUNDLAND SMTS
KINGS TROUGH
CHARCOT SMTS
BAY OF BISCAY
GALICIA BANK
TORE SMT
J-ANOMALY RIDGE
PICO FZ
OCEANOGRAPHER FZ
HAYES FZ
CORNER SMTS
AZORES
EAST AZORES FZ
TAGUS ABYSSAL PLAIN
HORSESHOE SMTS
STRAITS OF GIBRALTAR
MADEIRA
CRUISER SMT
SOUTH ATLAS FAULT
GREAT METEOR SMT
CANARY ISLANDS
MID-ATLANTIC RIDGE
AAIUN BASIN
CAPE BLANC
SENEGAL BASIN
CAPE VERDE ISLANDS
WEST AFRICAN COAST MAGNETIC ANOMALY
BAHAMAS FZ
VEMA FZ
GUINÉE PLATEAU
GUINÉE FZ
GUIANA PLATEAU
CEARA RISE
SIERRA LEONE RISE

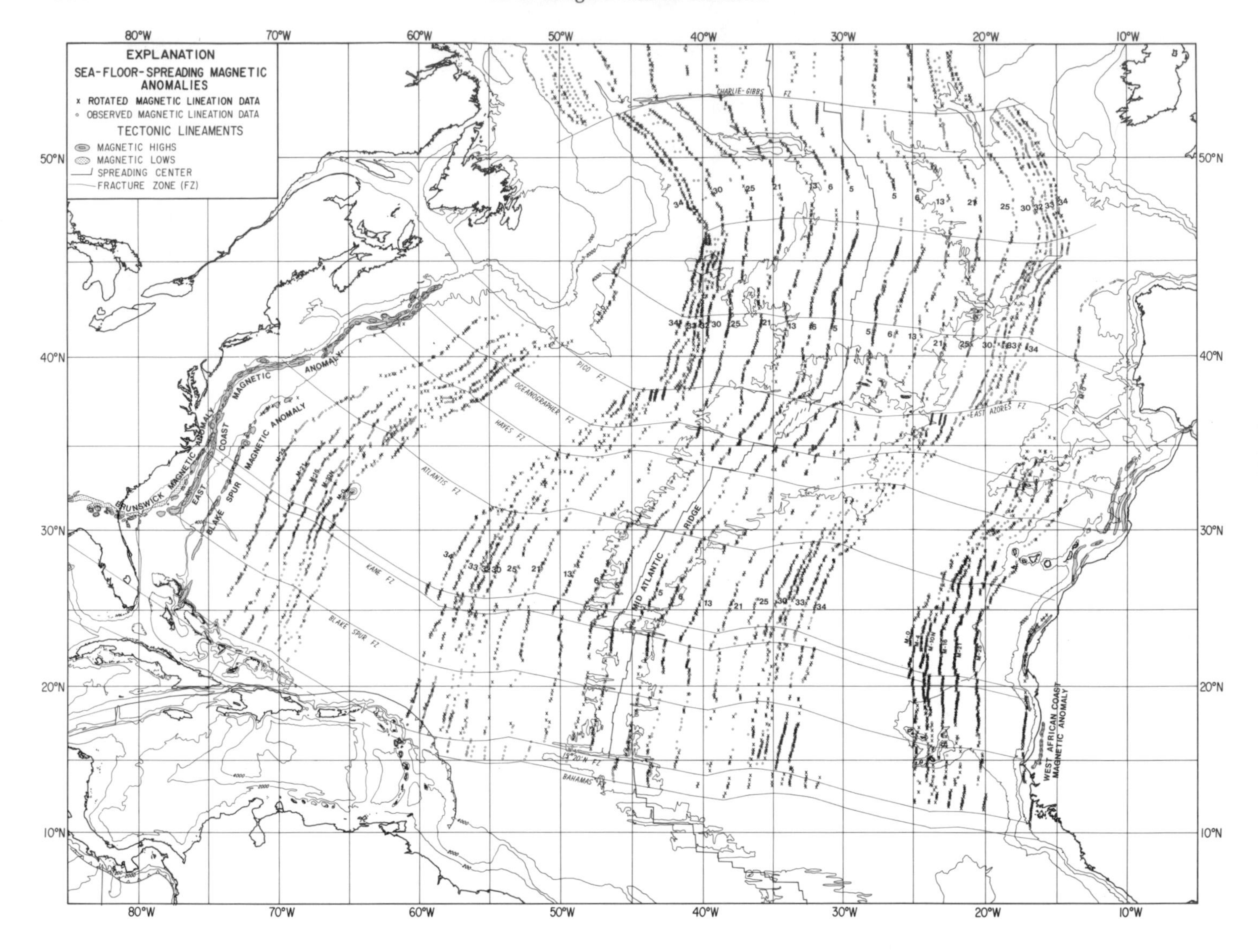

Figure 2. Distribution of sea-floor-spreading magnetic-lineation data points used in this study. Digitized anomaly points are for the SFS lineation set listed in Figure 1. Each of the red dots correspond to a track crossing of a particular magnetic anomaly (observed SFS data). Each of the black x's corresponds to an SFS data point rotated from the opposite flank using the reconstruction poles in Table 1 (Rotated SFS data). The Mid-Atlantic Ridge, selected fracture zones and 200-m, 2000-m and 4000-m bathymetric contours are also indicated.

different from those of Olivet and others because our better constraint on fracture zone locations produced a different latitudinal match in the SFS lineation patterns.

Kinematic information of the north and south Atlantic systems is derived from other studies. Evolution of the south Atlantic is based on the works of Ladd (1974) and Rabinowitz and LaBrecque (1979), modified by Sibuet and Mascle (1978), and Martin and others (1982). South Atlantic kinematics are poorly constrained because the data base is sparse. Only a few SFS magnetic lineations and fracture zones have been identified and mapped, and most of the rotations are interpolated (Table 3). Caribbean plate kinematics are constrained primarily by geologic structures on plate margins (Bonini and others, 1984) and relative motions of South America to North America determined from central and south Atlantic kinematic studies (e.g., Ladd, 1976; Burke and others, 1984). Plate kinematics for the north Atlantic are discussed by Srivastava and Tapscott in Chapter 23 of this volume. Studies of the magnetic lineations just north of the Azores have produced a variety of poles describing the Iberia-North American plate positions; see the summaries in Srivastava (1978), Kerr and Fergusson (1981), Srivastava and Tapscott (this volume), and Vogt (this volume, Ch. 15). Our recent reexamination of the Iberia–North America magnetic lineation data, considered in the framework of the entire central Atlantic

TABLE 1. CENTRAL ATLANTIC
AFRICAN PLATE TO NORTH AMERICAN PLATE
FINITE-DIFFERENCE POLES OF ROTATION: RECONSTRUCTION POLES

Lineation	Age (Ma)*	Latitude	Longitude	Angle	Source**
Anomaly 5	10.0	79.08°N	77.95°E	-2.41°	1
Anomaly 6	20.0	79.57°N	37.84°E	-5.29°	1
Anomaly 13	35.5	76.41°N	7.12°E	-9.81°	1
Anomaly 21	49.5	74.51°N	4.83°W	-15.32°	1
Anomaly 25	59.0	80.60°N	0.50°W	-18.07°	1
Anomaly 30	67.5	82.51°N	0.63°W	-20.96°	1
Anomaly 32	72.5	81.35°N	9.15°W	-22.87°	1
Anomaly 33y	74.3	80.76°N	11.76°W	-23.91°	1
Anomaly 33o	80.2	78.30°N	18.35°W	-27.06°	1
Anomaly 34y	84.0	76.55°N	20.73°W	-29.60°	1
Cenomanian/ Turonian	91.0	73.82°N	19.48°W	-34.28°	1
Anomaly M-0	118.0	66.30°N	19.90°W	-54.25°	1
Anomaly M-4	126.0	66.13°N	19.00°W	-56.39°	1
Anomaly M-10N	131.5	65.95°N	18.50°W	-57.40°	1
Anomaly M-16	141.5	66.10°N	18.40°W	-59.79°	1
Anomaly M-21	149.5	66.50°N	18.10°W	-61.92°	1
Anomaly M-25	156.5	67.15°N	16.00°W	-64.70°	1
BSMA	170.0	67.02°N	13.17°W	-72.10°	1
Closure (min)	175.0	66.97°N	12.34°W	-74.57°	1
Closure (max)	175.0	66.95°N	12.02°W	-75.55°	1
Closure		67.6°N	14.0°W	-74.8°	2
Closure		66.0°N	12.0°W	-74.8°	3
Closure		66.2°N	12.4°W	-71.8°	4
Closure		65.95°N	13.35°W	-76.74°	5
Closure		64.10°N	15.74°W	-78.40°	6

*Kent and Gradstein (this volume).

**(1) Klitgord and Schouten (this chapter), (2) Bullard and others (1965), (3) LePichon and Fox (1971), (4) LePichon and others (1977), (5) Lefort and Van der Voo (1981), (6) Wissmann and Roeser (1982).

TABLE 2. CENTRAL ATLANTIC FINITE-DIFFERENCE POLES OF ROTATION:
STAGE POLES OF MOTION

	North American Plate			African Plate		
Stage	Latitude	Longitude	Angle	Latitude	Longitude	Angle
Present*	82.68°N	17.68°W	0.275°/My	82.68°N	17.68°W	-0.275°/My
Present**	80.43°N	56.36°E	0.258°/My	8.43°N	56.36°E	-0.258°/My
Anom 5-0	79.08°N	77.95°E	1.205°	79.08°N	77.95°E	-1.205°
Anom 6-Anom 5	76.81°N	11.26°E	1.458°	76.39°N	12.84°E	-1.458°
Anom 13-Anom 6	70.57°N	12.86°W	2.303°	69.85°N	9.41°W	-2.303°
Anom 21-Anom 13	70.60°N	21.26°W	2.78°	69.58°N	17.42°W	-2.78°
Anom 25-Anom 21	68.62°N	151.71°E	1.64°	69.39°N	176.94°E	-1.64°
Anom 30-Anom 25	85.51°N	153.13°E	1.481°	85.43°N	151.27°W	-1.481°
Anom 32-Anom 30	66.31°N	45.94°W	1.006°	64.66°N	29.90°W	-1.006°
Anom 33y-Anom 32	67.30°N	40.67°W	0.542°	65.92°N	25.29°W	-0.54°
Anom 33o-Anom 33_y	60.18°N	41.88°W	1.685°	53.79°N	24.30°W	-1.685°
Anom 34_y-Anom 33o	59.16°N	38.82°W	1.345°	58.06°N	21.05°W	-1.345°
Anom M0-Anom 34_y	55.30°N	29.00°W	12.815°	55.45°N	10.06°W	-12.815°
Anom M4-Anom M0	57.84°N	8.22°W	1.086°	64.77°N	4.23°E	-1.084°
Anom M10N-Anom M4	52.25°N	10.79°W	0.522°	59.59°N	9.75°E	-0.522°
Anom M16-Nom M10N	68.48°N	11.49°W	1.198°	69.54°N	20.92°W	-1.198°
Anom M21-Anom M16	72.05°N	9.39°E	1.086°	76.96°N	30.01°W	-1.086°
Anom M25-Anom M21	60.58°N	31.20°E	1.496°	82.20°N	52.51°E	-1.496°
BSMA-Anom M25	60.00°N	0.00°W	3.752°	67.73°N	10.62°E	-3.752°
Closure-BSMA	60.00°N	0.00°W	2.50°	69.69°N	10.85°E	-2.50°

*NOAM-AFRC best fit model of Minster and Jordan (1978, Table 3).
**RM2 model of Minster and Jordan (1978, Table 2).

TABLE 3. SOUTH ATLANTIC SOUTH AMERICAN PLATE TO AFRICAN PLATE FINITE-DIFFERENCE POLES OF ROTATION: RECONSTRUCTION POLES

Lineation	Age (Ma)*	Latitude	Longitude	Angle	Source**
Anomaly 5	10.0	57.40°N	37.50°W	3.78°	interpolated
Anomaly 6	20.0	57.40°N	37.50°W	7.55°	interpolated
Anomaly 13	35.5	57.40°N	37.50°W	13.40°	1
Anomaly 21	49.5	61.16°N	37.73°W	19.31°	interpolated
Anomaly 25	59.0	62.59°N	37.83°W	23.19°	interpolated
Anomaly 30	67.5	63.56°N	37.90°W	26.89°	interpolated
Anomaly 32	72.5	64.00°N	37.94°W	28.96°	interpolated
Anomaly 33_y	74.3	64.12°N	37.95°W	29.57°	interpolated
Anomaly 33_o	80.2	64.59°N	37.99°W	32.24°	interpolated
Anomaly 34_y	84.0	64.84°N	38.01°W	33.90°	2
West African Closure	116.0	55.10°N	35.70°W	50.90°	3
Anomaly M-0	118.0	51.78°N	34.74°W	52.51°	2
Anomaly M-4	126.0	49.33°N	33.67°W	54.30°	2
Anomaly M-10	130.0	47.53°N	32.94°W	54.74°	2
Closure M-10N	131.5	44.1°N	30.3°W	56.1°	1
Closure M-10N	131.5	46.75°N	32.65°W	56.40°	2
Closure M-10N	131.5	45.50°N	32.20°W	57.50°	3
Closure M-10N	131.5	44.0°N	30.6°W	57.0°	4

*Kent and Gradstein (this volume).

**(1) Sibuet and Mascle (1978), (2) Martin and others (1982), (3) Rabinowitz and LaBrecque (1979), (4) Bullard and others (1965).

SFS data set (Schouten and others, in press), has defined reconstruction poles that simplify the Iberian tectonic history.

PLATE KINEMATICS

Plate kinematics must be tied into a fixed framework; in most of our discussion this will be the North American plate. Absolute plate motions relative to the mantle or hot spot reference frame are discussed by Vogt (this volume, Ch. 24) and Morgan (1983). The discussion that follows on plate kinematics is meant to cover the terminology, rationale, and importance of various components of the topic. Explanations of the mathematical techniques can be found elsewhere (e.g., McKenzie and Parker, 1967; McKenzie and Sclater, 1971; Harrison, 1972; LePichon and others, 1973, Ch. 4; Hellinger, 1981; Stock and Molnar, 1983).

Poles of Rotation

A *pole of rotation* is the position on the earth's surface (latitude and longitude) of a rotation vector (Fig. 3a) describing the transfer of a plate from one position to another (Harrison, 1972). *Reconstruction poles* (finite-difference poles of rotation), including an angle of rotation, describe the *rotation necessary to reconstruct* past relative plate positions, regardless of how the plates actually moved in detail (Fig. 3b) (Bullard and others, 1965). A reconstruction pole describes the total rotation necessary to superimpose the magnetic lineations and fracture zones of the same age on opposite sides of a spreading center. *Poles of motion* (instantaneous poles of rotation), including an angular rate, describe the relative *motion* between two plates at a given time. *Stage poles* (finite-difference poles of rotation), including an angle of rotation, are an approximation to poles of motion in which we assume that the pole remained fixed, with respect to the two plates, for a finite period of time (Pitman and Talwani, 1972). Plate kinematics are described ideally by an infinite number of poles of motion, but in actuality only a small, finite number of stage poles are ever determined. Because we assume a certain amount of pole stability when representing plate motions by a finite series of stage poles of motion, the accuracy of these estimates is dependent on our ability to identify periods of stable (constant) spreading.

Stability of a pole representing plate motion refers to both the pole position and angular rate used to describe such a motion. We assume here that any significant plate motion change results in both pole position and angular rate changes. A major spreading direction (pole position) change will always cause a change in linear spreading rate along a ridge axis, even if the angular rate remains constant. This is because the linear distance from the new pole to any point on the spreading axis will be different from that of the old pole. An angular rate change without a spreading direction (pole position) change would imply that plate driving forces and impulse changes in these forces act in exactly the same direction. The implications of our assumption concerning pole stability need to be carefully considered when discussing plate driving forces, but in our present study of plate kinematics this assumption is useful for estimating at what ages to calculate reconstruction poles. Periods of constant linear spreading rates (relative or absolute) can be identified from single flowline magnetic profiles, with rate changes implying pole changes. Comparison of anomaly spacings and fracture zone trends along the length of the spreading center with those predicted by the stage poles readily

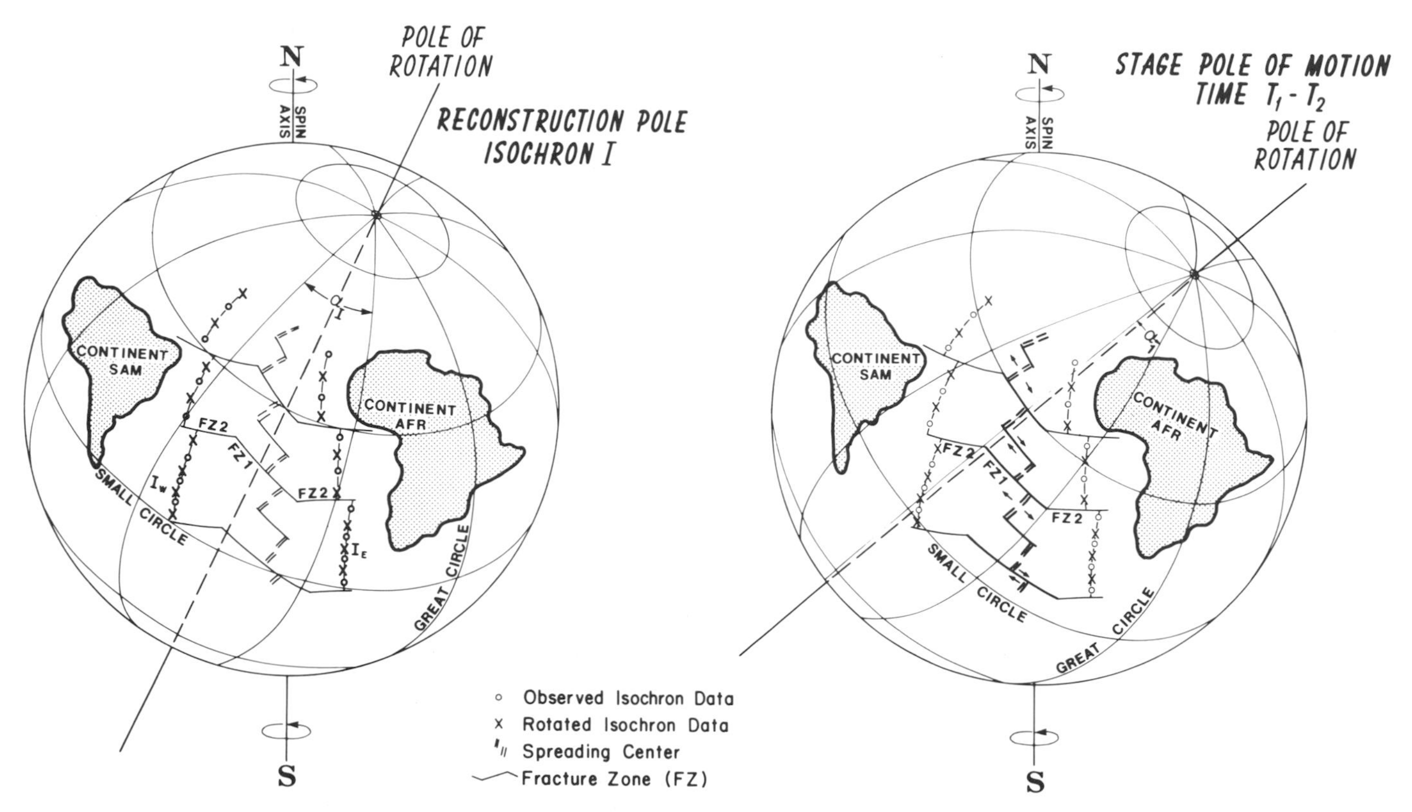

Figure 3. Schematic diagrams showing poles of rotation associated with reconstruction poles (A) and stage poles of motion (B). The reconstruction pole and angle α_I in (A) rotate observed isochron data on east flank (I_E) onto isochron data on west flank (Iw) of a midocean ridge system and can be used to reconstruct the position of continent AFR relative to continent SAM at time I. The stage pole of motion and angle α, in (B) represent the actual plate motion between times T_1 and T_2 of plate AFR relative to plate SAM. The fracture zone (flowline) FZ, is along a small circle about the stage pole of motion. A different stage pole of motion describes the plate motions between times T_2 and T_I and spreading generated fracture zones FZ_2 along a different set of small circles. Note that the small circle trends for the reconstruction pole (A) do not predict fracture zone trends.

identify other points where reconstruction poles need to be calculated. Data quality, density, and along-axis distribution usually determine our ability to identify stable poles in a given region.

Stage Poles, Flow Lines, and Fracture Zones

Stage poles are calculated on both ridge flanks using the reconstruction poles determined from SFS data at isochrons when major plate motion changes occurred. Except for the present spreading stage pole, these stage poles are different on opposite flanks for the same time period. This is because a stage pole describes the motion that created the crust on a particular plate, and therefore the frame of reference for the stage pole is that plate. For example, the stage pole in the central Atlantic between anomalies 25 and 21 (Table 2) is different for the North American and African plates. The North American plate stage pole (68.62°N, 151.71°E) describes the motion in the frame of reference of the North American plate. The African plate stage pole (69.39°N, 176.94°E) is in the African plate frame of reference and is used to predict flow lines in crust on the African plate. A stage pole moves with the particular plate and can be reconstructed to its past position relative to some other frame of reference, just like a continental coastline on that plate. The anomaly 25-21 stage pole for the African plate can be recalculated in the North American plate frame of reference at anomaly 21 or 25 time (using the anomaly 21 or 25 reconstruction poles in Table 1), and it will be identical to the North American plate anomaly 25-21 stage pole.

A *flow line* is the trace on each plate of plate motion through time with respect to the spreading center (Pitman and Talwani, 1972). Ideally this should correspond to the fracture zone trace of a transform fault, but large changes in plate motions, ridge jumps, finite-width transform zones, propagating spreading centers, etc., make the fracture zone a valid trace of a flow line for only short distances (a few hundred kilometers). A flow line between two isochrons, for which reconstruction poles have been determined, is approximated by a small circle about the stage pole calculated from the reconstruction poles.

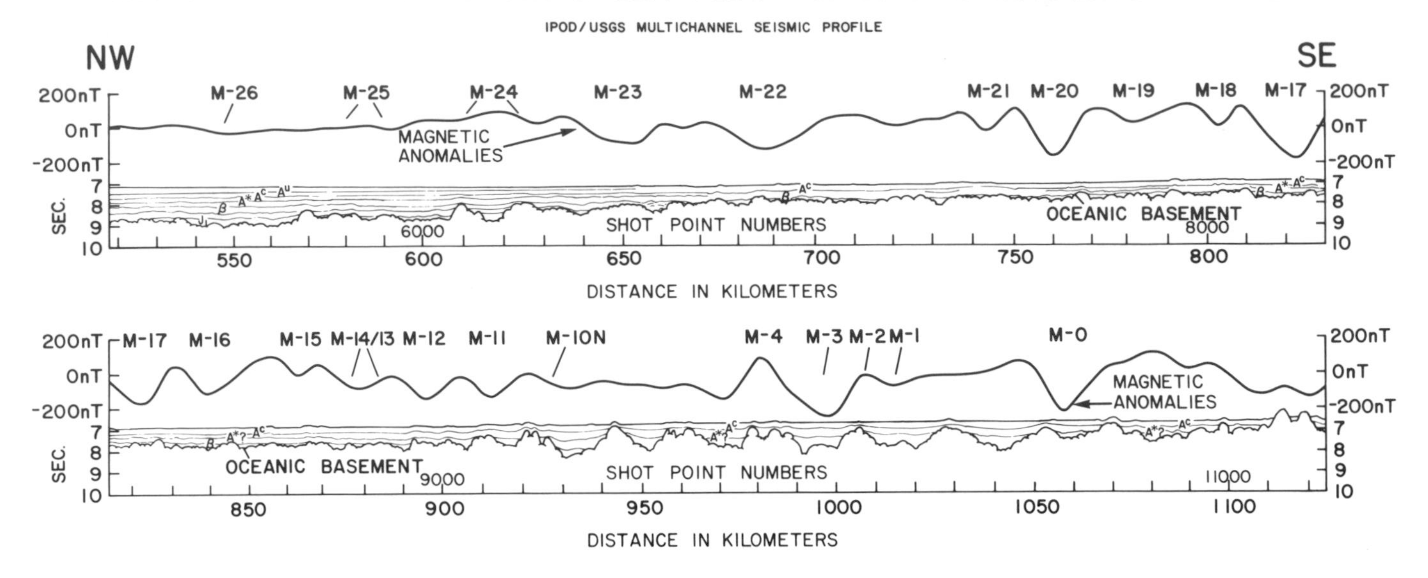

Figure 4a. Magnetic anomaly and seismic reflection profile from IPOD-USGS line across the Mesozoic sea-floor–spreading magnetic anomaly sequence (Larson and Hilde, 1975) in the western Atlantic.

Flow line traces of plate motions between North America and Africa can be estimated from fracture zone trends or calculated with the stage poles (Table 2). Transform-fault offsets of spreading centers leave in oceanic crust a distinctive pattern of offset magnetic lineations and of bathymetric relief along fracture zones (Heezen and Tharp, 1965). If we assume that these offsets are fixed and do not migrate along the spreading center, then the FZs represent flow lines of the plate away from the spreading center (van Andel, 1971; Schouten and Klitgord, 1982; Vogt and Perry, 1982). Small-offset transform faults usually produce coherent bathymetric and magnetic traces in the sea floor that are a reliable representation of flow lines. The small offset is an important characteristic because the relief is produced during only a short period of time in the transform zone. A small-offset transform zone can also adjust to plate motion variations without significant changes in position relative to adjacent ridge segments. Large-offset transform-fault zones may have greater topographic expressions (Vogt and Perry, 1982), but their structure is usually a composite of structural trends generated at both spreading center intersections and in the active transform zones over a long period of time. Deformation of older crust or renewed magmatic activity of leaky transform zones can occur when plate motions change (Menard and Atwater, 1968), complicating the structure along large-offset transform faults and generating composite bathymetric trends that do not accurately describe the plate motion.

Mapping of fracture zones is possible in regions of dense magnetic and poor seismic coverage by locating discontinuities in the magnetic-lineation patterns (Schouten and Klitgord, 1982). Small discontinuities in the lineation pattern are usually hard to identify on a single profile (e.g., Fig. 4a). When data density is sufficient to produce contour maps, disruptions in these lineations are easily identified (Fig. 4b), and many of them have been found to correspond with fracture zone locations mapped with seismic reflection data, offsets in SFS magnetic lineations, and with flow lines predicted from poles of motion. This association between fracture zones and disruptions in the magnetic anomalies allows FZs to be mapped in areas where SFS magnetic lineations do not exist (Klitgord and others, 1984).

Small-circle flowline information estimated from fracture zone trends can be used to improve stage pole determination calculated from reconstruction poles or to calculate stage poles when SFS data is available on only one ridge flank. In the latter case, stage poles are determined from the variation along axis in isochron spacing and small-circle trends between these isochrons (McKenzie and Sclater, 1971). Assuming symmetric spreading (or a percentage of asymmetric spreading), a series of stage poles can be added to estimate reconstruction poles or added to previously determined reconstruction poles to estimate additional reconstruction poles.

Spreading Rates

Spreading rates can be described as either relative or absolute rates. Absolute spreading rates along a flowline profile are determined by comparing the SFS magnetic anomaly spacing with a geomagnetic time scale (Figs. 5 and 6) (e.g., Heirtzler and others, 1968; Larson and Pitman, 1972; Larson and Hilde, 1975; LaBrecque and others, 1977, 1983; Cohee and others, 1978; Ness and others, 1980; Harland and others, 1982; Kent and Gradstein, this volume). The list of geomagnetic time scales is long, but most time scales are modifications of earlier ones with the addition of

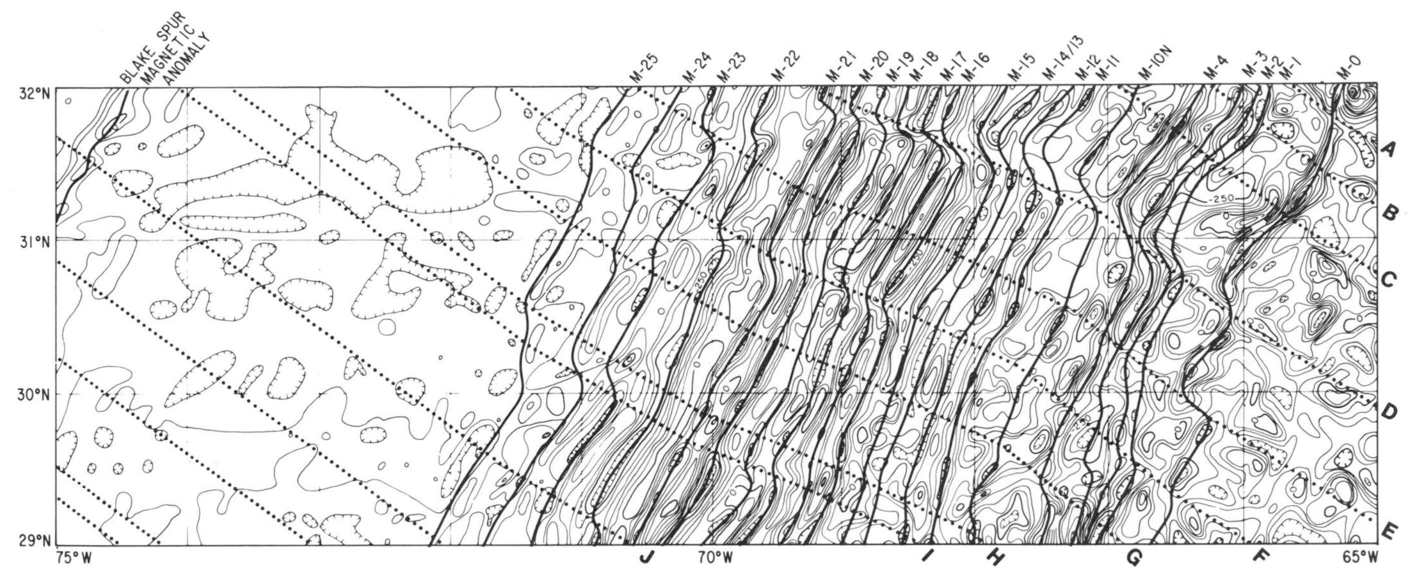

Figure 4b. Magnetic anomaly contour map from Project Magnet Survey of Mesozoic sea-floor–spreading magnetic anomalies southwest of Bermuda; contour interval is 50 nT. Superimposed are Mesozoic magnetic anomaly isochrons (solid lines) and flowlines (dotted lines) of Mesozoic relative plate motion calcaluted from central North Atlantic stage poles of rotation, as discussed in text. From Schouten and Klitgord (1982, Fig. 4).

new paleomagnetic, biostratigraphic-age, or radiometric-age information. A discussion of the concept and calibration of the geomagnetic time scale by Kent and Gradstein can be found in Chapter 3 of this volume; we confine our discussion here to the correlation of their time scale with Atlantic SFS data. We use the term *anomaly* when referring to the position of SFS magnetic anomaly lineations and the term *chron* (Harland and others, 1982) when referring to the age of that SFS lineation. For example, we refer to anomaly M-25 or chron M-25 when discussing the position or age of SFS magnetic lineation M-25. When referring to the older or younger age boundaries of a chron, we have used the subscripts o and y.

Absolute spreading rates are one of the desired kinematic parameters, but dating errors in constructing geomagnetic time scales can generate spurious spreading rates and rate variations. We have therefore relied only on relative-rate determinations (independent of any time scale) to estimate periods of constant spreading in the central Atlantic (Figs. 7 and 8). Reconstruction poles (Table 1) were calculated for the times when spreading rates changed, assuming that these times correspond to when poles of motion changed.

Relative rates between two or more ridge flanks are determined by comparing the anomaly spacing found on different ridges with each other or with standard profiles (Figs. 7 and 8) (Heirtzler and others, 1968; Klitgord and others, 1975; LaBrecque and others, 1977; Schouten and Klitgord, 1982). Comparison of anomaly spacings on opposite flanks of a mid-ocean ridge is useful for estimating the amount of asymmetric spreading on any ridge. In this comparison technique for estimating relative spreading rates (Klitgord and others, 1975), the magnetic anomaly patterns have to be phase shifted (Schouten and McCamy, 1972) to ensure that equivalent points of crust on ridge flanks are chosen. The graph of anomaly spacing for two ridge flanks with different but constant spreading rates will be a straight line whose slope is the ratio of their spreading rates. Any change in spreading rate will cause a change in slope of the graph (e.g., at point 24/25 in Fig. 8), unless both ridge flanks underwent a rate change of exactly the same percentage and at the same time. A comparison between anomaly spacings on three or more ridges is necessary to establish on which ridge a rate change occurred. This technique is most useful in defining times of pole of motion changes and periods of constant spreading on a particular spreading center. This usually results in more stage poles being calculated for a given system than is necessary (the rate change may not have been on the spreading system of interest).

ATLANTIC SEA-FLOOR–SPREADING DATA

Atlantic Sea-Floor–Spreading Magnetic Lineations

The primary data for determining reconstruction poles are sea-floor–spreading magnetic lineations (Heirtzler and others, 1968; Vogt, this volume, Ch. 15) from both flanks of the Mid-Atlantic Ridge. We have constructed a comprehensive set of SFS magnetic lineations for the entire central Atlantic (Fig. 2), derived from studies of the Mesozoic anomaly sequence (Figs. 4a, 5) and the Cenozoic magnetic anomaly sequence (Fig. 6) in the western and eastern Atlantic. We have re-examined all of the magnetic anomalies on profiles available to us and identified the SFS lineations in the framework of a single data set between the Bahamas

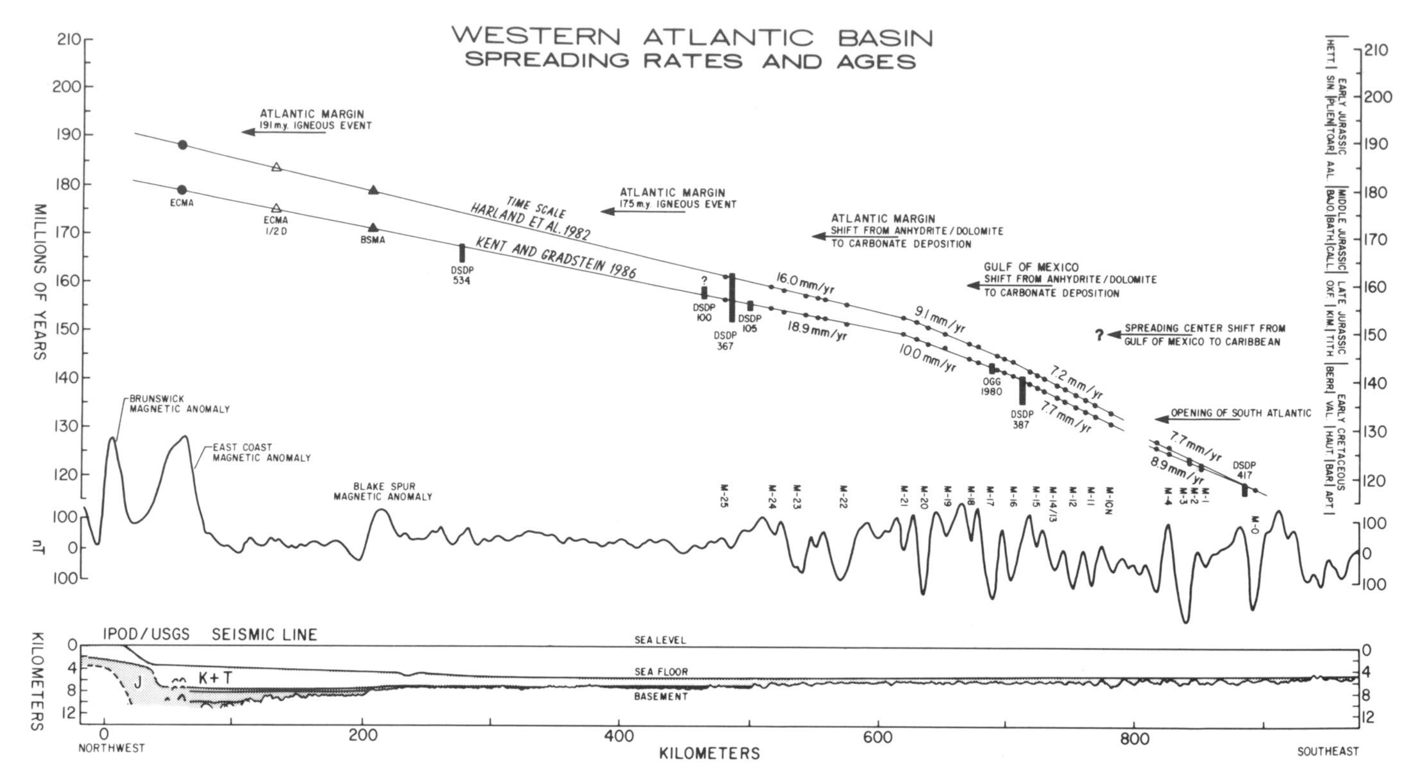

Figure 5. Comparison of distance along track of Mesozoic sea floor spreading magnetic lineations on IPOD-USGS line (Fig. 4) and the geomagnetic time scales of Harland and others (1982) and Kent and Gradstein (this volume). Geologic ages given on the right are from the DNAG geologic time scale (Palmer, 1983) and critical isochron-age control points are shown. Various tectonic events are indicated for reference. The position ½ ECMA is one-half the distance between the BSMA and ECMA. If one assumes that a ridge jump at BSMA time (Vogt, 1973) isolated between the ECMA and BSMA, crust generated on both flanks during initial opening, then the ½ ECMA distance is the appropriate distance to use in determining spreading half-rates from one flank of a spreading center.

FZ and the Charlie-Gibbs FZ. Our identification of various SFS magnetic anomalies is generally the same as that of the more recent studies, as discussed below. The shape of individual anomalies and series of anomalies is the most important characteristic of recorded magnetic anomaly profiles for determining correct correlations with the worldwide anomaly pattern and geomagnetic time scale. Anomaly shapes can be altered significantly by changes in ridge orientation and in present and past magnetic field parameters. This effect, called skewness (Schouten, 1971; Schouten and McCamy, 1972), has been removed from the original data on selected profiles to facilitate the identification of magnetic lineations.

We shall discuss the Atlantic SFS magnetic anomalies using two profiles, IPOD-USGS line and ATLANTIS II-93.21 (Figs. 4a, 5, and 6), that lie along flowlines of plate separation and do not cross major fracture zones. The IPOD-USGS profle is located just north of the Kane FZ, and the ATLANTIS II-93 profile is midway between the Kane and Atlantis FZs (Fig. 2). Magnetic anomalies on the IPOD-USGS line (Figs. 4a, 5) are representative of the Atlantic Mesozoic pattern (Schouten and Klitgord, 1982), and anomalies on the ATLANTIS II-93 line (Fig. 6) are representative of the Cenozoic pattern.

Mesozoic magnetic anomalies (M-0 to M-25) are easily identified in the western Atlantic south of the New England Seamounts (Fig. 4a), but lower anomaly amplitudes and anomaly skewness have caused some misidentifications north of the seamounts and in the eastern Atlantic. We have used the western Atlantic pattern south of the seamounts as our primary control for anomaly identifications and anomaly spacings. The M-anomalies used here are nearly identical with those of Schouten and Klitgord (1977), Vogt and Einwich (1979), Perry and others (1981), and Olivet and others (1984), with a minor shift to the west of anomalies M-11 to M-13 (Sundvik and others, 1984). A slow-down in spreading rate and a ridge reorganization make it difficult to identify individual anomalies between M-10N and M-4 (Schouten and Klitgord, 1977, 1982). North of the New England Seamounts, most of the short wavelength components of the anomaly signature have been removed, resulting in some previous misidentifications. Our M-0 to M-25 sequence corresponds to the M-17 to M-25 sequence of Barrett and Keen (1976) and is in complete agreement with the J-anomaly drilling results and adjacent SFS magnetic lineations (M-0 to M-4) (Rabinowitz and others, 1979; Tucholke and Ludwig, 1982). Our older sequence, M-10N to M-23, is similar with that of Perry and others

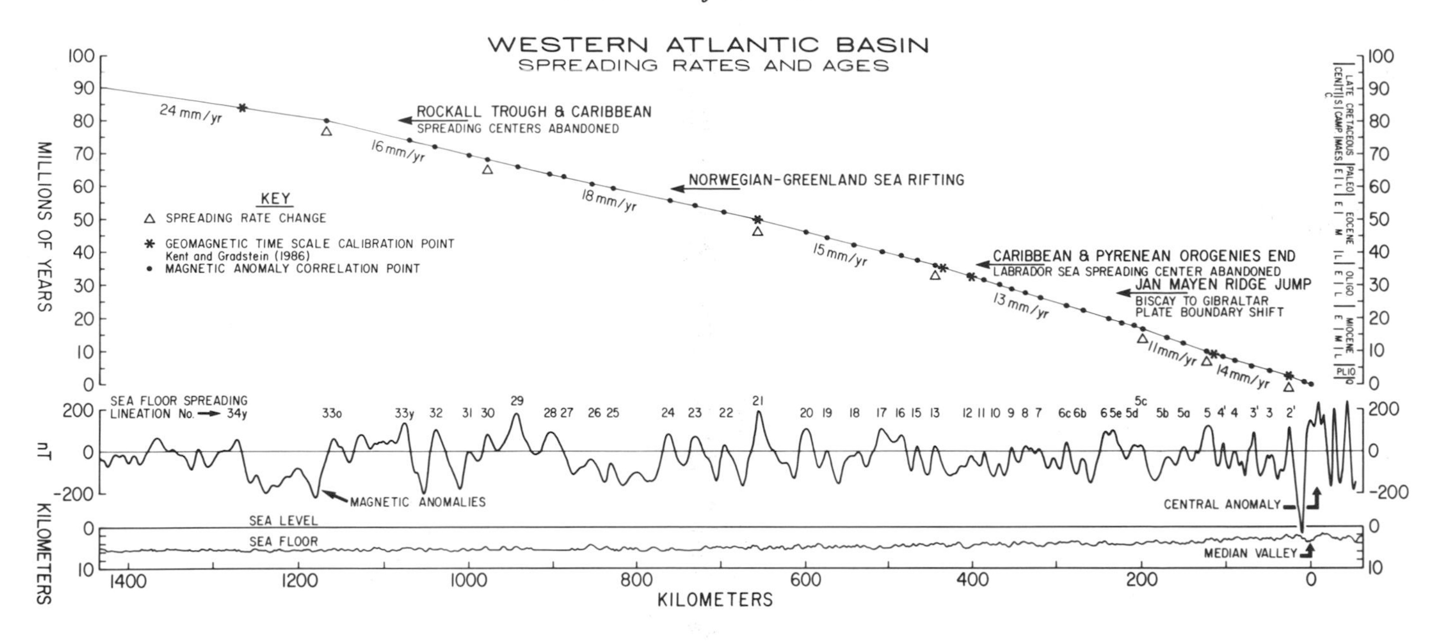

Figure 6. Comparison of distance along track of Late Cretaceous and Cenozoic sea-floor–spreading magnetic lineations on *Atlantis* II-93 profile and the geomagnetic time scale of Kent and Gradstein (this volume).

(1981) and Olivet and others (1984), but our M-25 is nearly 100 km west of the M-25 of Perry and others. Anomaly amplitudes are very low for M-24 and M-25 (Vogt and others, 1971) on both flanks, making them difficult to identify. Our M-25 choice was based on an attempt to remain consistent in anomaly patterns north and south of the seamounts and on both flanks. The M-25 pattern of Perry and others requires different reconstruction poles north and south of the seamounts.

In the eastern Atlantic, the SFS M-anomaly pattern is not as clear as in the western Atlantic, partially because of the skewed anomaly shapes and partially as a result of a lower data density. Our M-11 to M-23 pattern is nearly identical with that of Hayes and Rabinowitz (1975), Perry and others (1981), and Olivet and others (1984). The disruption in spreading between M-10N and M-4 was not identified by Hayes and Rabinowitz (1975) in the eastern Atlantic, and their M-0 to M-4 anomalies—used by Perry and others (1981)—are too far to the west. Our M-0 corresponds to their M-3, and our M-0 to M-4 sequence is identical with the most recent identifications by Rabinowitz and others (1979) for the J-anomaly sequence and with the sequence used by Olivet and others (1984). Low anomaly amplitudes make it difficult to identify anomalies M-24 and M-25 (Hayes and Rabinowitz, 1975). Our M-25 is a little east of that of Hayes and RAbinowitz because we have chosen M-25 as the magnetic low rather than the peak, as they did, so that our M-25 would be consistent with the phase-shifted anomalies from the western Atlantic. Our anomaly M-0 north of the Azores-Gibraltar transform zone is the same as that of Tucholke and Ludwig (1982) and only slightly modified from that of Sibuet and others (1980).

The Late Cretaceous SFS anomaly sequence, anomalies 34 to 30, is clearly recorded on both flanks of the Mid-Atlantic Ridge, and we have used the same identifications as Cande and Kristoffersen (1977) and Olivet and others (1984). The skewness of the anomalies in the eastern Atlantic is very large (Cande, 1978), and anomalies are nearly inverted, with anomaly peaks in the west corresponding to anomaly troughs in the east. Previous studies of this part of the anomaly sequence had identified the 30-34 sequence as anomalies 30-32 (Pitman and Talwani, 1972; Srivastava, 1978), because they were comparing the Atlantic pattern with the Pacific pattern (e.g., Fig. 8) where anomaly 34 usually is not found because of a major spreading system reorganization (Molnar and others, 1975).

Within the Cenozoic SFS anomaly sequence, anomalies are difficult to identify because of the slow spreading rate in the Atlantic. Anomalies 20 to 24 are broad enough to be clearly recognized, and our identifications are the same as Pitman and Talwani (1972), Perry and others (1981), and Olivet and others (1984). Anomalies 25 and 26 are easily confused with fracture zone anomalies, and the spacings of anomalies 27 to 29 are too close to resolve on most profiles. Our anomaly 25 coincides with that of Perry and others (1981), where the data are voluminous and our two sets of anomaly 25s are scattered near each other in the regions of sparce track lines. Although we agree with both Pitman and Talwani (1972) and Perry and others (1981) on what should be called anomalies 5, 6, and 13, it is not easy to identify them because of the slow spreading rate. We have reidentified the entire anomaly 5-6-13 sequence (Schouten and others, 1984) using rotated data sets from both flanks to sort out fracture zone anomalies. In general, our anomaly 13 coincides with that of Pitman and Talwani (1972), Perry and others (1981), and Olivet

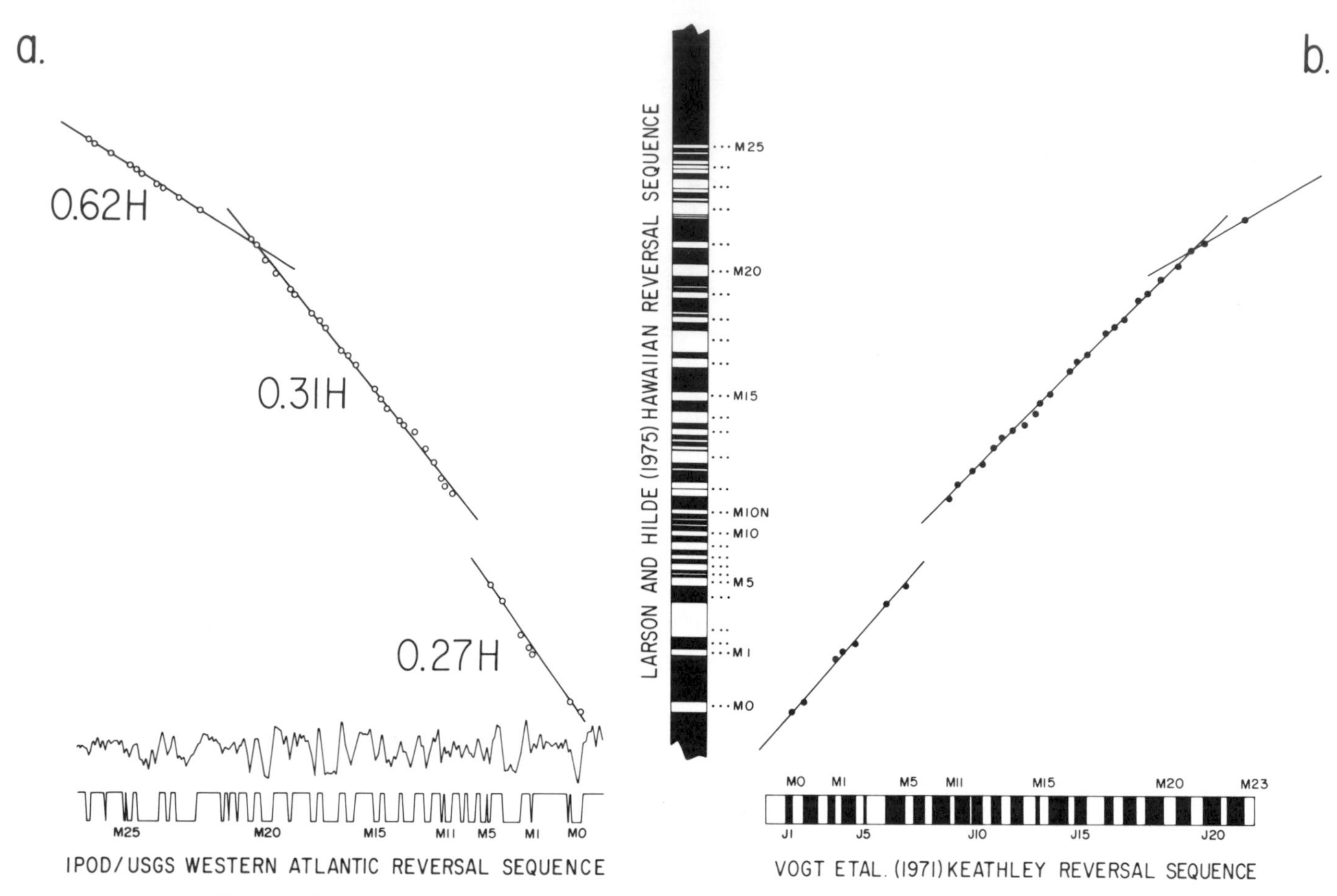

Figure 7. The Hawaiian M-anomaly model (Larson and Hilde, 1975) plotted: a: against the inversion solution and polarity transitions derived from the IPOD-USGS Mesozoic magnetic anomaly profile (Fig. 4a), and b: against the western Atlantic Keathley anomaly model of Vogt and others (1971). A constant spreading rate was assumed by Larson and Hilde (1975) for sea-floor–spreading on the Hawaiian limb of the Pacific spreading system; thus this model represents the Hawaiian magnetic anomaly pattern independent of a geomagnetic time scale. From Schouten and Klitgord (1982, Fig. 3).

and others (1984), with most differences occurring south of the Kane FZ where the anomaly amplitudes are much lower. There are major differences of as much as 50 km between our anomaly 5 positions and those of Perry and others (1981). For some spreading segments our identifications are identical on one, and sometimes both, flanks (near 15°-17°N, 22°N, 26°N, and 35°-37°N). These segments of agreement are usually where there is the greatest data density, whereas the segments with poor agreement of anomaly 5 locations are where Perry and others (1981) had a smaller data set to use than we did.

Other identification changes involve only single profiles where the comprehensive framework enables us to identify anomaly patterns that are partially disturbed by fracture zone anomalies. This procedure has resulted in a small change in the isochron position for many SFS magnetic anomalies and thus in our calculated reconstruction poles. Tectonic maps appearing in the Ocean Margin Drilling program's Eastern North American Continental Margin Atlas tectonic maps (e.g., Uchupi and others, 1983) are based on our preliminary interpretation of this data set, and the summary study by Colette and others (1984) incorporates many of our revised identifications.

Selected magnetic lineations have been digitized at track crossings to construct a digital data base suitable for evaluating poles of rotation. The distribution of SFS magnetic anomaly lineations digitized for the central Atlantic is shown as red dots in Figure 2. These data represent the points where ship or plane tracks crossed selected lineations on either flank and the black x's are the same points rotated to the opposite flank using the reconstruction poles in Table 1. The numerous offsets in the lineations provide a distinctive pattern by which complementary isochrons on opposite flanks can be matched for calculating reconstruction poles.

Atlantic Fracture Zones

Relief associated with FZs can be readily seen in both bathy-

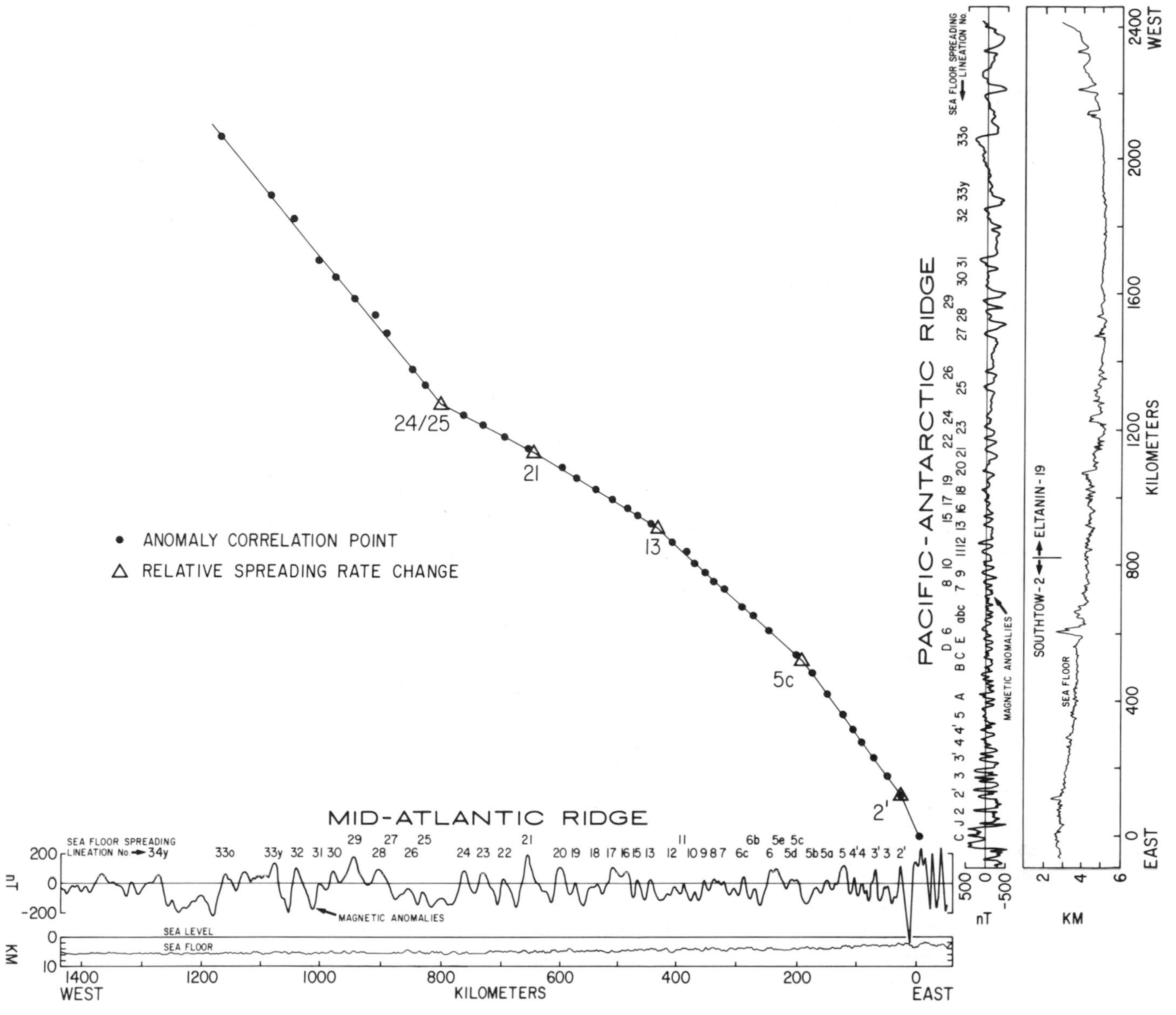

Figure 8. Magnetic anomalies on *Atlantis* II-93 cruise crossing the western flank of the Mid-Atlantic Ridge plotted against magnetic anomalies from profiles *Eltanin*-19 and *Southtow*-2 on the western flank of the Pacific-Antarctic Ridge in the southeastern Pacific (Molnar and others, 1975). The graph indicates distance along track to the same magnetic lineation on both mid ocean ridges. The slope of the lines connecting these points represents the relative spreading rates between the two spreading centers.

metric and basement surface contour maps for the Atlantic (Heezen and Tharp, 1965; Johnson and Vogt, 1973; Phillips and Fleming, 1978; Vogt and Perry, 1982; Tucholke and others, 1982; Vogt and Tucholke, this volume). The large relief generated by a slow-spreading rate for the Cenozoic oceanic crust on the flank of the Mid-Atlantic Ridge makes it difficult to physiographically identify FZ traces on individual bathymetric profiles. Bathymetric contour charts are more useful for mapping the prominent bathymetric troughs that characterize FZs on Cenozoic crust of the Mid-Atlantic Ridge. On Mesozoic-age crust, sediment fill has masked most troughs but the rugged relief of a FZ contrasts with the smooth relief of oceanic crust sufficiently for FZs to be mapped with seismic reflection data (Klitgord and Grow, 1980; Tucholke and others, 1982; Tucholke, this volume).

Within the area landward of anomaly M-25, we have relied on magnetic data to trace fracture zones from the SFS lineation offsets to the edge of oceanic crust (near the ECMA) at the continental margin. In the western Atlantic, the contoured

PROJECT MAGNET aeromagnetic data and marine magnetic data (Vogt and others, 1971; Vogt and Einwich, 1979) display numerous disruptions in the magnetic lineation pattern that can be correlated with fracture zones within and just west of the M-anomaly sequence (Schouten and Klitgord, 1982). These same fracture zones can be traced across the Jurassic magnetic quiet zone from the BSMA to the ECMA using contoured data from a U.S. Geological Survey aeromagnetic survey of the U.S. Atlantic margin (Klitgord and Behrendt, 1977, 1979; Klitgord and Schouten, 1986a, b). Magnetic data have also been used in the eastern Atlantic to trace fracture zones between anomaly M-25 and the margin off Morocco (Roeser, 1982) and off the Guinée coast of West Africa [Liger, 1980; Jones and Mgbatogu, 1982].

Atlantic Spreading Rates

The kinematic history of the central Atlantic is one of long periods of constant spreading and fairly stable poles of motion, interrupted by short periods of plate motion changes (Figs. 5–8, Table 4) (Vogt and others, 1969; Schouten and Klitgord, 1982). These long periods of constant spreading are the cause of the straight-line segments in the graphs comparing Atlantic anomaly spacings with those in the Pacific (Figs. 7, 8). In the Atlantic, the most important of these changes (Table 4) are associated with the propagation of spreading into the south and north Atlantic, or with spreading center jumps that eliminated long transform-fault offsets.

Absolute spreading rates (Figs. 5 and 6) were determined using the geomagnetic time scales of Harland and others (1982) and Kent and Gradstein (this volume). The discrepancies between these scales reflect corrections to the paleomagnetic data base applied by Kent and Gradstein; both geomagnetic calibrations used virtually the same geologic time scale. The crucial difference in the two time scales is the addition by Kent and Gradstein of Jurassic–Early Cretaceous ages determined recently in the Italian Alps (Channell and others, 1982; Ogg, 1983; Lowrie and Channell, 1984) and from DSDP site 534 in the western Atlantic (Sheridan and others, 1983). Slight differences in the Cenozoic time scale are caused by their choices of age-calibration tie points. Spreading rates discussed below are half-spreading rates determined on the IPOD/USGS and ATLANTIS II track lines (between the Kane and Atlantis fracture zones), using the Kent and Gradstein geomagnetic time scale. Spreading rates on the central Atlantic spreading system are slightly lower to the north toward the rotation pole and slightly higher to the south of this region. Small variations in half-spreading rates occur along the entire ridge axis as a result of local asymmetric spreading.

Jurassic spreading in the central Atlantic generated crust older than the M-anomaly SFS reversal sequence, so ages and spreading rates cannot be determined from SFS lineations. This older crust on both flanks is referred to as the Jurassic magnetic quiet zone, but it does contain two prominent magnetic lineations, the Blake Spur Magnetic anomaly (BSMA) and the East Coast Magnetic Anomaly (ECMA) (Vogt, 1973). The ECMA is located very near the landward edge of oceanic crust in the western Atlantic (Klitgord and Behrendt, 1979), whereas the BSMA is about a third of the way between the ECMA and anomaly M-25. Extrapolation of spreading rates and ages back to the BSMA and ECMA, assuming a ridge jump at chron BSMA leaving crust from both ridge flanks between the BSMA and ECMA (Vogt, 1973), gives ages of 170 Ma and 175 Ma respectively (Fig. 5).

Recent age assignments to the start of this early opening phase by Vogt and Einwich (1979) (170-190 Ma), Klitgord and Grow (1980) (180 Ma) and Sheridan (1983) (155-185 Ma) depended on interpolation and extrapolation techniques similar to the ones used in Figure 5. The different ages estimated for the BSMA and ECMA by these authors are a consequence of the use of different geologic or geomagnetic time scales and the age selected for the Late Jurassic spreading rate change in the Atlantic Mesozoic SFS sequence. Our analysis of relative spreading rates indicates that the change in spreading occurred at chron M-21 (Fig. 5), whereas Vogt and Einwich (1979), Bryan and others (1980), and Sheridan (1983) used M-22, and Vogt and Einwich also used M-0 as an alternative. Comparison of DSDP age data with the distance to anomalies older than M-21 (Fig. 5) indicates a constant half rate of 1.9 cm/yr between M-21 and the BSMA.

Although we assume that a ridge jump at chron BSMA left oceanic crust from both flanks of the earliest spreading center between the BSMA and ECMA (Vogt, 1973), there are no data except the Early Jurassic dike ages onshore for estimating spreading rates prior to chron BSMA. Therefore, we have assumed the same constant spreading rate prevailed prior to chron BSMA to provide an age estimate for initiation of sea-floor spreading based only on SFS data. Our initial opening age of 175 Ma coincides with the last of two major pulses of Jurassic igneous activity (190 and 175 Ma) recorded along the Atlantic margin (Sutter and Smith, 1979), although recent re-evaluation of these data suggest that only the 190-Ma age is reliable (Sutter, 1985). Much slower spreading prior to chron BSMA would imply that the initiation of sea-floor spreading might coincide with the earlier (190 Ma) igneous event onshore (Vogt and Einwich, 1979; Sheridan, 1983). The BSMA ridge jump in the latest Bathonian or earliest Callovian (170 Ma) and the spreading center shift from the Gulf of Mexico to the proto-Caribbean at chron M-21 (Tithonian—150 Ma) (Klitgord and others, 1984) bracket a long period of constant, medium-rate spreading at 1.9 cm/yr half-rate.

The Cretaceous spreading rate in the Atlantic was initially much slower than the rate in the Jurassic. Between chrons M-21 and M-16 (150 Ma to 141 Ma), the half-rate was only 1.0 cm/yr. The half-rate from chrons M-16 to M-10N (141 Ma to 132 Ma) slowed to about 0.7 cm/yr (Sundvik and others, 1984). A significant ridge reorganization caused variable asymmetric spreading between chrons M-10N and M-4 (132 Ma to 126 Ma) as rifting and sea-floor spreading commenced in the south Atlantic and between Grand Banks and Iberia. This reorganization makes spreading rate determination inaccurate for this time period. The central Atlantic half-rate was 0.9 cm/yr between chrons M-4 and

TABLE 4. ATLANTIC PLATE MOTION CHANGES

Anomaly	Age (Ma)	Atlantic Tectonic Events
2'	2.5	Reorganization of spreading direction and rate along the entire Atlantic sea-floor-spreading (SFS) system.
5	10	No major change identified.
5C	17	Jan Mayen ridge jump; Gulf of Lions SFS abandoned; central Atlantic slow-down in spreading rate.
9	28	African plate boundary jump (King's Trough-Pyrenees to Gibraltar); Gulf of Lions SFS starts.
13	36	Caribbean plate eastward movement with Lesser Antilles subduction; Caribbean Orogeny (north coast South America) and Pyrenean Orogeny end; Labrador Sea SFS abandoned.
21	50	Major change in central Atlantic spreading direction; King's Trough rifting starts.
24/25	59	South Atlantic-Central Atlantic-North Atlantic (Falkland FZ to Siberia) SFS; North Atlantic (Reykjanes-Norwegian-Greenland Sea-Arctic) SFS starts; Labrador Sea SFS reorganization.
30	67	South Atlantic-Central Atlantic-North Atlantic (Falkland FZ to Labrador Sea) SFS; central Atlantic SFS direction change.
33_o	80	South Atlantic-Central Atlantic-North Atlantic (Falkland FZ to Labrador Sea) SFS; Rockall Trough rifting ends; Labrador Sea SFS starts; Biscay and Caribbean SFS abandoned; northern Iberian-African plate boundary subduction starts.
M-0	119	South Atlantic-Central Atlantic-North Atlantic (Falkland FZ to Biscay) and Caribbean SFS; North Atlantic (Biscay-Rockall-Norwegian Sea) rifting.
M-10N	132	Caribbean-Central Atlantic-Ligurian Tethys (Pacific to Cuba to Gibraltar to Italy) SFS; North Atlantic (Gibraltar to Rockall) rifting and South Atlantic (Benué to Falkland FZ) SFS starts.
M-21	150	Caribbean-Central Atlantic-Ligurian Tethys SFS (Pacific to Cuba to Gibraltar to Italy) SFS; Gulf of Mexico to Caribbean ridge jump.
BSMA	170	Gulf of Mexico-Central Atlantic-Ligurian Tethys (Mojave-Sonora megashear to Florida to Gibraltar to Italy) SFS; Central Atlantic ridge jump to east.
	175	Gulf of Mexico-Central Atlantic-Ligurian Tethys SFS starts.

M-0 (126 Ma to 118 Ma) as sea-floor spreading offshore of Iberia (Tucholke and Ludwig, 1982) propagated northward into the Bay of Biscay, and rifting began to join the south Atlantic and central Atlantic spreading systems. The rate increased again to about 2.4 cm/yr (Vogt and Einwich, 1979) between chrons M-0 and 33_o (118 Ma to 80 Ma) as a spreading system evolved from the southern tip of Africa to Rockall Trough. A major slowing of central Atlantic spreading occurred just after chron 33_o, as Caribbean spreading was abandoned and spreading shifted from Rockall Trough to the Labrador Sea. The remainder of the Late Cretaceous time was a period of slower spreading, 1.6 cm/yr half-rate between anomalies 33_o and 30 (80 Ma to 67 Ma), as rifting propagated northward into the Labrador Sea and Baffin Bay.

Because Atlantic spreading rates during the Cenozoic were slow and geomagnetic reversal rates were higher than in the Mesozoic (Harland and others, 1982), anomaly identifications between anomalies 5 and 6 and between 6 and 13 remain tentative. Spreading rates between chrons 5 and 13 must therefore be treated with caution. Anomalies 13 to 34_y form a clear, identifiable lineation pattern on both flanks of the ridge. Absolute spreading half-rates between chrons 33_o and 13 (Fig. 6) decrease from 2.4 to 1.3 cm/yr, but data scatter in Figure 6 precludes determining distinctly when the rate changes occur. Comparison of Atlantic with Pacific SFS lineations (Fig. 8) demonstrates that the most significant rate changes occurred between anomalies 25 and 24 and at anomalies 13, 5C and 2′, with a small change at anomaly 21. There is little indication of a spreading rate change between chrons 25 and 33_o, but a distinct change in fracture zone trend (pole position) occurs during this time (Rabinowitz and Purdy, 1976).

Atlantic spreading half-rates in Cenozoic time slowed from 1.8 to 1.1 cm/yr. A half-rate of 1.8 cm/yr prevailed from chron 30 until chron 25 (29 Ma) when spreading began between Greenland and Eurasia. This rate continued until chron 21 (~50 Ma), when it slowed to 1.5 cm/yr as compression developed along the African northern plate boundary—the King's Trough, Azores-Biscay Rise, North Spanish margin, and Pyrenees. At chron 13 (35 Ma), the rate slowed again to 1.3 cm/yr as the Pyrenean and Caribbean Orogenies peaked and waned, and spreading in the Labrador Sea spreading was abandoned. The northern plate boundary of Africa shifted southward to the Straits of Gibraltar (Purdy, 1975; Schouten and others, in press) after chron 10 (30 Ma), with no apparent spreading rate change. A period of very slow spreading (1.1 cm/yr half-rate) between chrons 6 and 5 (20 to 10 Ma) makes anomaly identifications difficult. Between chrons 5 and 2′ (10 to 2.5 Ma), the rate increased to 1.4 cm/yr; the present spreading half-rate is about 1.2 to 1.3 cm/yr (Minster and Jordan, 1978).

Atlantic Poles of Rotation

Reconstruction poles of rotation for the central Atlantic (Table 1) have been determined for each of the SFS lineations that correspond to times of major plate reorganizations (closure, BSMA, M-21, M-10N, M-4, M-0, 33_o, 33_y/32/30, 25, 21, and 13). Reconstruction poles were not calculated for chrons 9, 5C, and 2′ because these lineations have not been reliably mapped on a regional basis. Additional reconstruction poles were calculated at chrons M-25 (oldest SFS lineation identified on both flanks), M-16, Cenomanian/Turonian boundary, 34_y, 6, and 5 (youngest SFS lineation identified and mapped on both flanks). Stage poles of motion (Table 2) were calculated from these reconstruction poles, assuming stable instantaneous poles of motion existed between each reorganization. The reconstruction poles are constrained by the SFS lineation and fracture zone patterns and are independent of the geomagnetic time-scale uncertainties. The accuracy of our reconstruction poles can be evaluated qualitatively by examination of the fit in Figure 2 of rotated to observed SFS lineation data points.

Poles of rotation between North and South America and between Africa and Eurasia cannot be calculated directly because there were no long spreading centers between these plates generating SFS magnetic lineations. The reconstruction and stage poles are determined from the vector differences between the North American-African (Table 1), South American–African (Table 3), and North American–Eurasian poles (Srivastava and Tapscott, this volume).

Kinematics of the Caribbean and Mediterranean areas are the result of small differential motions between the north, central, and south Atlantic spreading systems, and pole of motion determinations for these two areas are sensitive to errors in estimating the kinematics of any of three Atlantic systems. Our study of the central Atlantic has improved estimates of its kinematic history so that more realistic determinations are possible for its contribution to Caribbean and Mediterranean tectonic activity. Unfortunately, a similar level of plate kinematic information does not exist for the north or south Atlantic spreading systems. Knowledge about the north Atlantic, with its complex pattern of spreading centers, would benefit from a more rigorous evaluation of existing data. A comprehensive evaluation of existing south Atlantic data has not been undertaken in over 10 years; such studies are now in progress at several research institutions. More magnetic lineation data are required to achieve a comparable level of understanding as in the central Atlantic, but this is being partially balanced by the use of SEASAT data to constrain flow line trends (e.g., Vogt and others, 1984).

Locating the past positions of the continents is improved by the new reconstruction poles for the central Atlantic (Tables 1 and 3), but the major plate motions still are not adequately determined to predict a regional kinematic framework for the Caribbean or Mediterranen regions with any confidence. The observation of Ladd (1976) that "major plate rotations do not fully explain Caribbean tectonics" is still valid in both the Caribbean and Mediterranean. Rather than predict plate motions between Iberia and Eurasia and between North and South America using the inadequately constrained North and South Atlantic poles, we have indicated what motions are most consistent with geologic and major plate motion information.

PLATE RECONSTRUCTIONS, BOUNDARIES, AND MOTIONS

Evolution of the central Atlantic is summarized by the series of reconstructions in Figures 10, 11, and 12, and in Plate 10 (I, J, K). Abandoned, active, and future plate boundaries are shown, along with relative plate motions at the active boundaries and absolute motions of the major plates relative to a fixed North America. Atlantic SFS data sets provide primary control for the North American, South American, African, Iberian, Eurasian, and Greenland plates. Although the positions of the Central American microplates are speculative, they are consistent with kinematics of the major plates. Positions of the Iberian plate are based on SFS data, and its motion history has important implications on the interpretation of Central American microplate history.

Plate reconstructions in Figures 9, 10, 11, and 12 and Plate 10 (I, J, K) are shown with present-day latitude-longitude grids and coastline configurations. This procedure does not imply that modern coastal outlines existed in the past, but rather, it provides markers useful in evaluating these reconstructions with other geologic or geophysical data. Plate boundaries in Central America

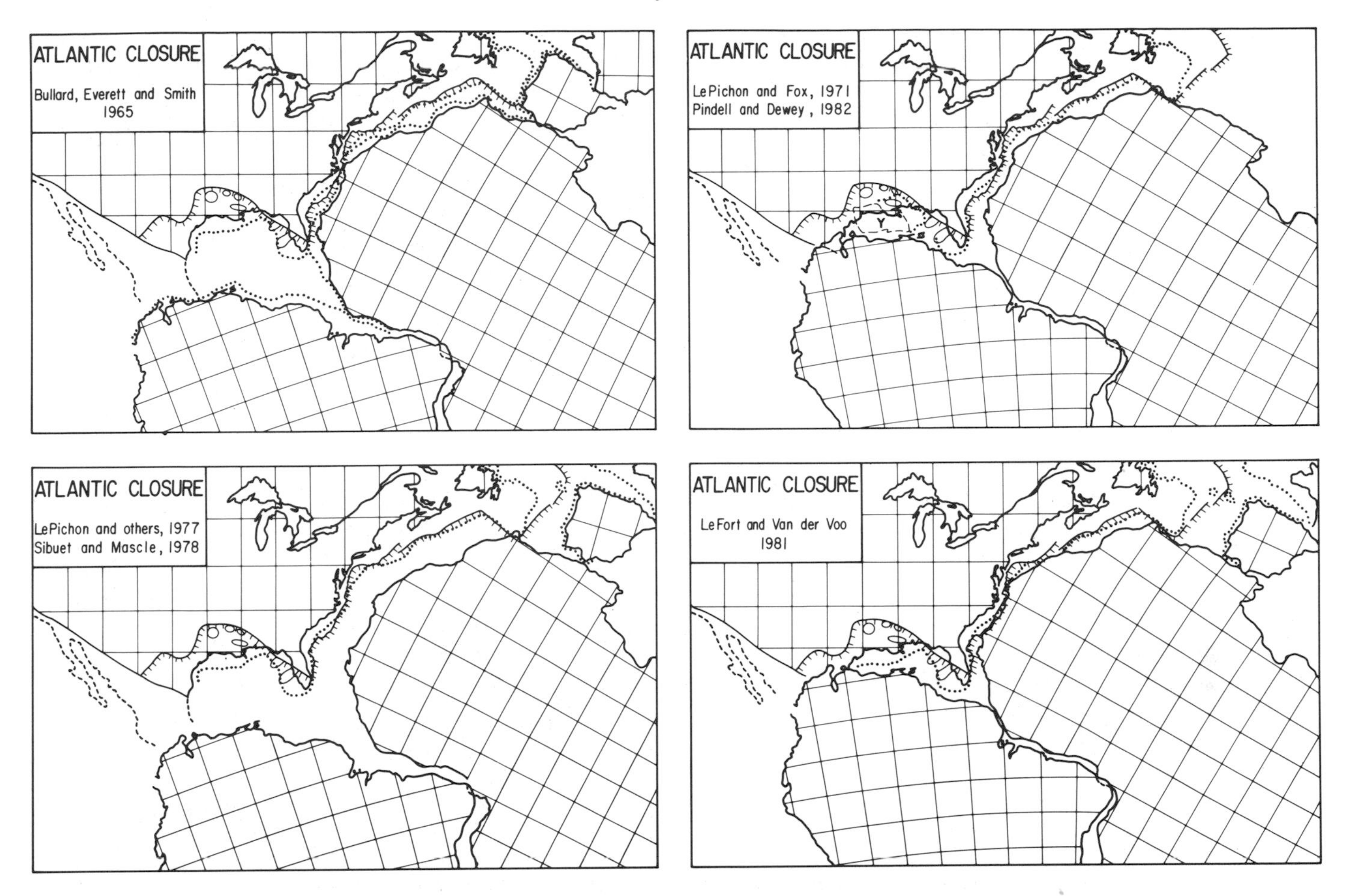

Figure 9. Atlantic closure reconstructions of continental positions: a: Bullard and others (1965); b: LePichon and Fox (1971) for the Central Atlantic, and Pindell and Dewey (1982) for the South Atlantic using the 107 Ma isochron [116 Ma with the Kent and Gradstein (this volume) timescale] of Rabinowitz and LaBrecque (1979) and with Pindell and Dewey's Yucatan microplate (Y) added; c: LePichon and others (1977); and d: Lefort and van der Voo (1981). The North American continental edge has been added to all four reconstructions to facilitate comparisons.

and northwestern South America represent only schematically the types of plate boundaries expected. For example, the rifted margin on either side of the proto-Caribbean must have existed, and there is evidence for one in the Eastern Cordillera and Sierra de Perija on the northwest edge of South America (Mooney, 1980; Maze, 1984). Paleozoic blocks in the Central Cordillera and Sierra Nevada de Santa Marta are probably similar to Flemish Cap, Galicia Bank, or the stranded blocks in the northern Gulf of Mexico. Proper reconstruction of the continental edge of northwestern South America depends on determining the kinematic histories of these Paleozoic blocks which are still uncertain.

Microplate or Terrane Boundaries

Plate boundaries shifted throughout Mesozoic and Cenozoic time, and the motions of microplates in the Mediterranean and Caribbean regions are hard to verify. The Iberian plate is an exception because a good pattern of SFS magnetic anomalies between the Azores and Charlie-Gibbs FZs recorded its motion relative to North America (Roberts and Jones, 1980). A re-examination of SFS data south of Kings Trough indicates that for its entire post-rifting history, Iberia was attached to either the African or to the Eurasian plate (Schouten and others, 1986). These attachments imply that Iberia acted as an accreted terrane on more than one plate. Our closure reconstruction for Iberia is its position at chron M-4 (at the shift from rifting to sea-floor spreading). We have not attempted to estimate the amount of Iberian plate movement during the long rifting phase prior to the separation of Iberia and Grand Banks. Iberia was attached to Africa from postrifting (~chron M-4) to chron 10 (Figs. 10a-d, 11a, b); the northern plate boundary extended along Kings Trough, Azores–Biscay Rise, into the Bay of Biscay, and along the Pyrenees (Whitmarsh and others, 1982; Grimaud and others, 1982). After chron 10, Iberia was part of Eurasia; the northern plate boundary of Africa shifted to the Azores transform zone,

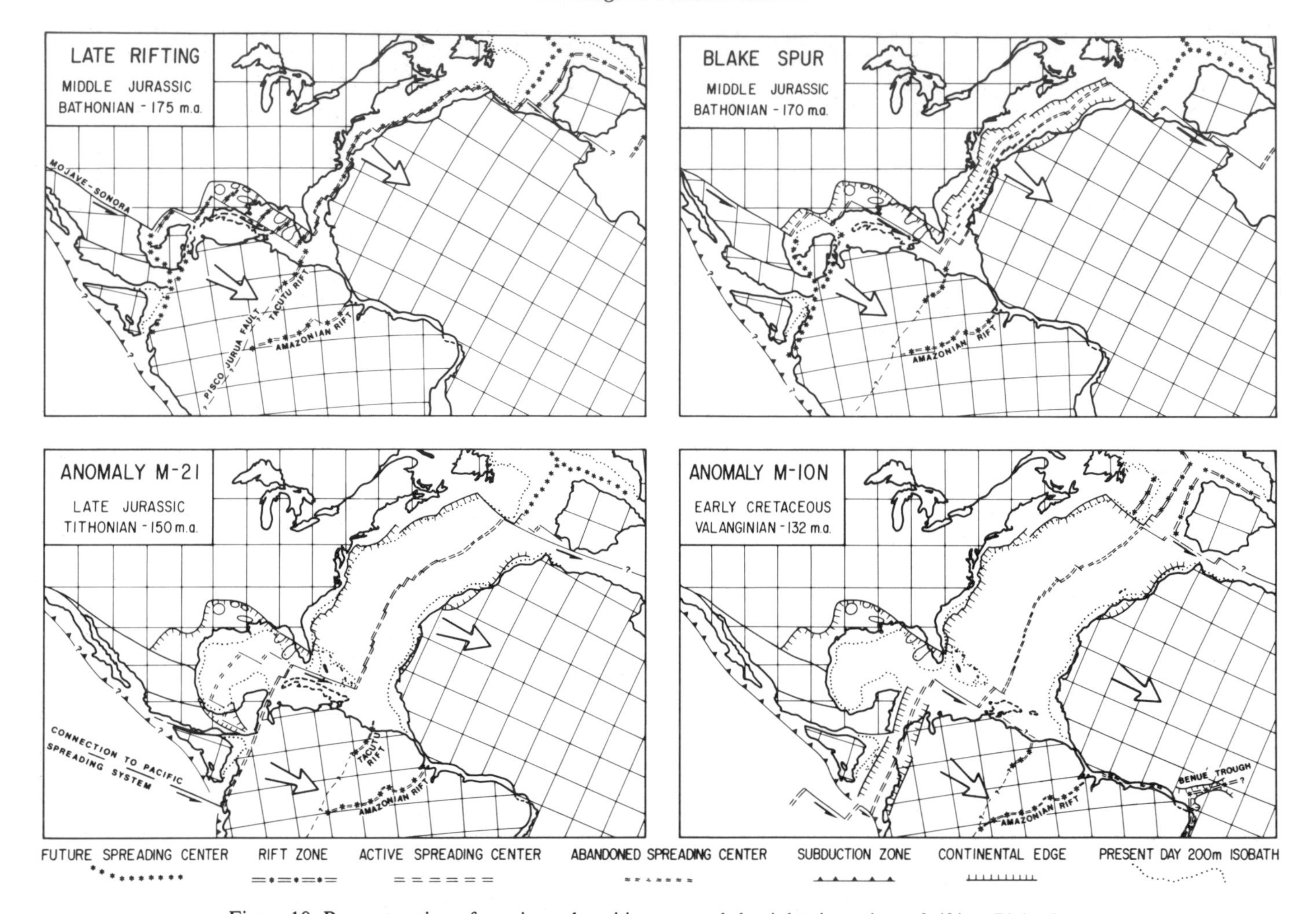

Figure 10. Reconstruction of continental positions around the Atlantic at time of rifting, Blake Spur Magnetic Anomaly, anomaly M-21, and anomaly M-10N. Abandoned, present, and future spreading centers, major transform zones, and subduction zones are indicated. The small arrows indicate relative plate motions across plate boundaries and large arrows indicate plate motions relative to North America. African plate rotated using poles in Table 1, Eurasian-Greenland plate using poles of Srivastava and Tapscott (this volume), Iberian plate rotated using poles from Schouten and others (1986), and South American plate rotated using poles in Table 3. South American rift zones are from Szatmari (1983).

passing through the Straits of Gilbraltar (Purdy, 1975; Schouten and others, 1986).

In all of our reconstructions we have assumed the absence of independent motion for the microplates that form Central America (Anderson and Schmidt, 1983). The positions of these microplates are important elements in understanding the tectonic history of the Gulf of Mexico–Caribbean (e.g., Burke and others, 1984). Recent studies of the region by Case and Holcombe (1980), Anderson and Schmidt (1983), and Bonini and others (1984) provide a framework for examining plate boundary tectonic activity and microplate motions of the Central American region.

The simplest model for the Jurassic kinematics of the circum-Atlantic plates is the assumption that Central America, South America, and Africa moved as a single plate away from the North American–Eurasian plate. Anderson and Schmidt (1983) made this assumption for most of the Central American microplates with the exception of the Yucatan block. Pindell and Dewey (1982) have presented a more complicated model and indicated very schematically a system of spreading centers rotating the Yucatan block out of the northern Gulf of Mexico. The system of spreading centers postulated by Pinedell and Dewey (1982, Figs. 13, 14) in the Gulf of Mexico at 165 Ma and 150 Ma implies plate driving forces that require tensional forces transmitted across a parallel set of spreading centers. Most parallel sets of spreading centers are associated with propagating rifts (e.g., Hey and others, 1980; Klitgord and Mammerickx, 1982), not stable systems. Pindell and Dewey (1982) also moved the Chortis block a little farther west than the position we have chosen, but there is little geologic control on the motions or original positions of these microplates. Movement of these microplates that is significantly greater than the motions assumed by us or by Pindell and Dewey

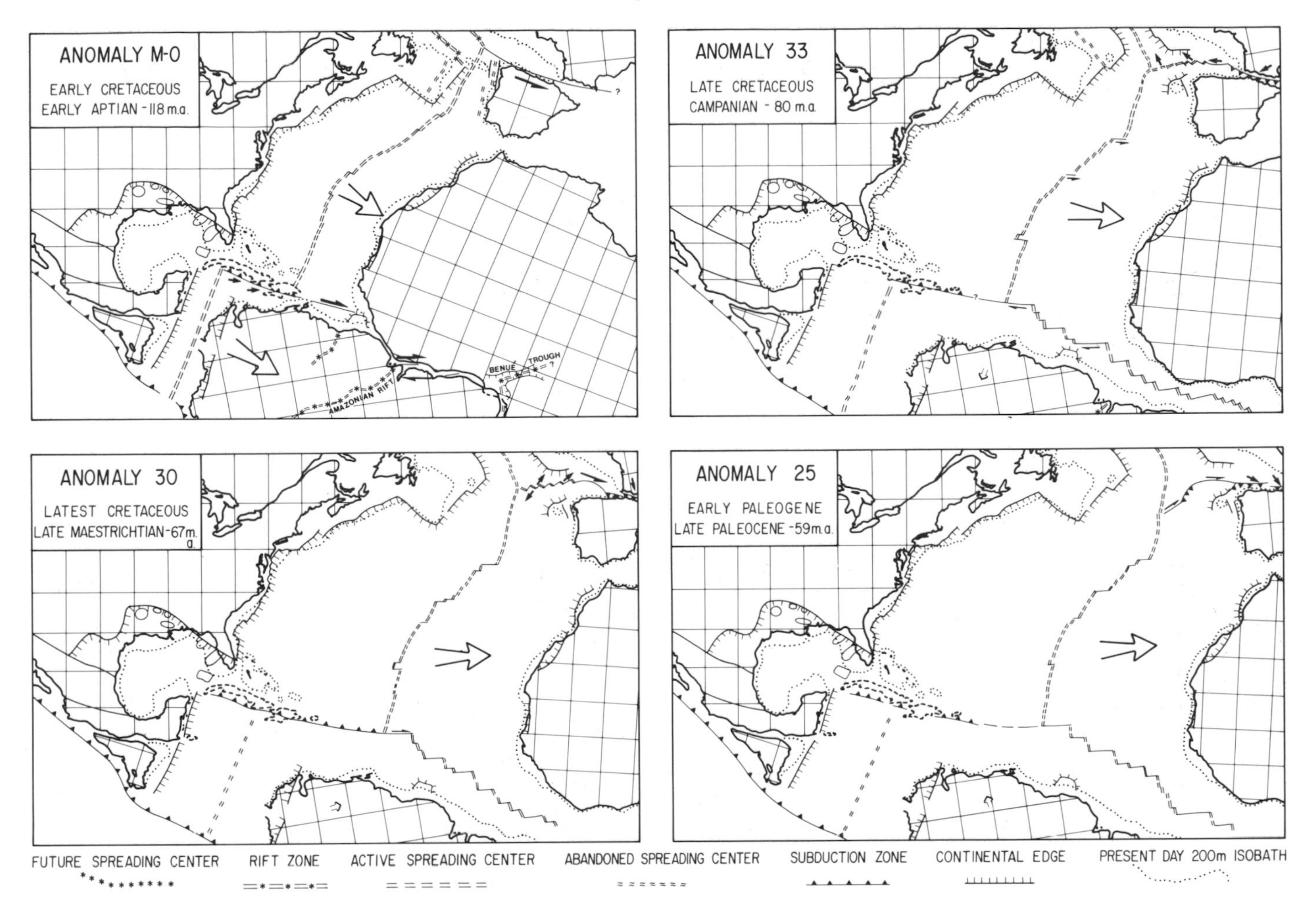

Figure 11. Reconstruction of continental positions around the Atlantic at time of anomaly M-0, anomaly 33_o, anomaly 30, and anomaly 25. See Figure 10 for explanation.

has been inferred from paleomagnetic data (Gose and others, 1982). Because the plausibility of some tectonic models for the Caribbean is dependent on the correct determination of the original location of these microplates, the accuracy of these models must still be treated cautiously.

As noted, we assume that Central American microplates never moved independently but moved attached to either the North American or South American plates. The motion of any of the microplates relative to North America is the same as that of South America, with each microplate being left accreted to North America as it reaches its present-day position. This model correctly predicts the motions recorded along known shear zones (Anderson and Schmidt, 1983), the Callovian rifting of Yucatan away from Mexico with development of an Upper Jurassic basin in the Reforma–Campeche Shelf area (Meyerhoff, 1980), and the evolution of a Tithonian-Kimmeridgian rift zone between the Yucatan-Chortis blocks and the northwest corner of South America (Mooney, 1980; Maze, 1984). All of these microplates are then left attached to North America until 36 Ma (chron 13). The reconstructed position of the Chortis block at chron 5 (10 Ma) is based on the work of Macdonald and Holcombe (1978). The Chortis block has been moved back along the Bartlett transform zone for the anomaly 13 reconstruction, similar to the motion used by Pindell and Dewey (1982). Further evaluation is needed to understand the complex history of Central America and to integrate paleomagnetic data from the several Central American blocks.

Closure and Initial Opening

Determinations of closure poles for the continents around the Atlantic have been based on: (1) geometric fits of coastlines, isobaths, or other geophysical-geologic lineaments, e.g., Figure 9a (Bullard and others, 1965; Dietz and Holden, 1970; Walper and Rowett, 1972; Wissmann and Roeser, 1982); (2) geologic fits using pre-breakup data such as paleomagnetic records and lithotectonic assemblages, e.g., Figure 9d (McElhinny, 1973; Van der Voo and French, 1974; Pilger, 1978; Morel and Irving, 1981;

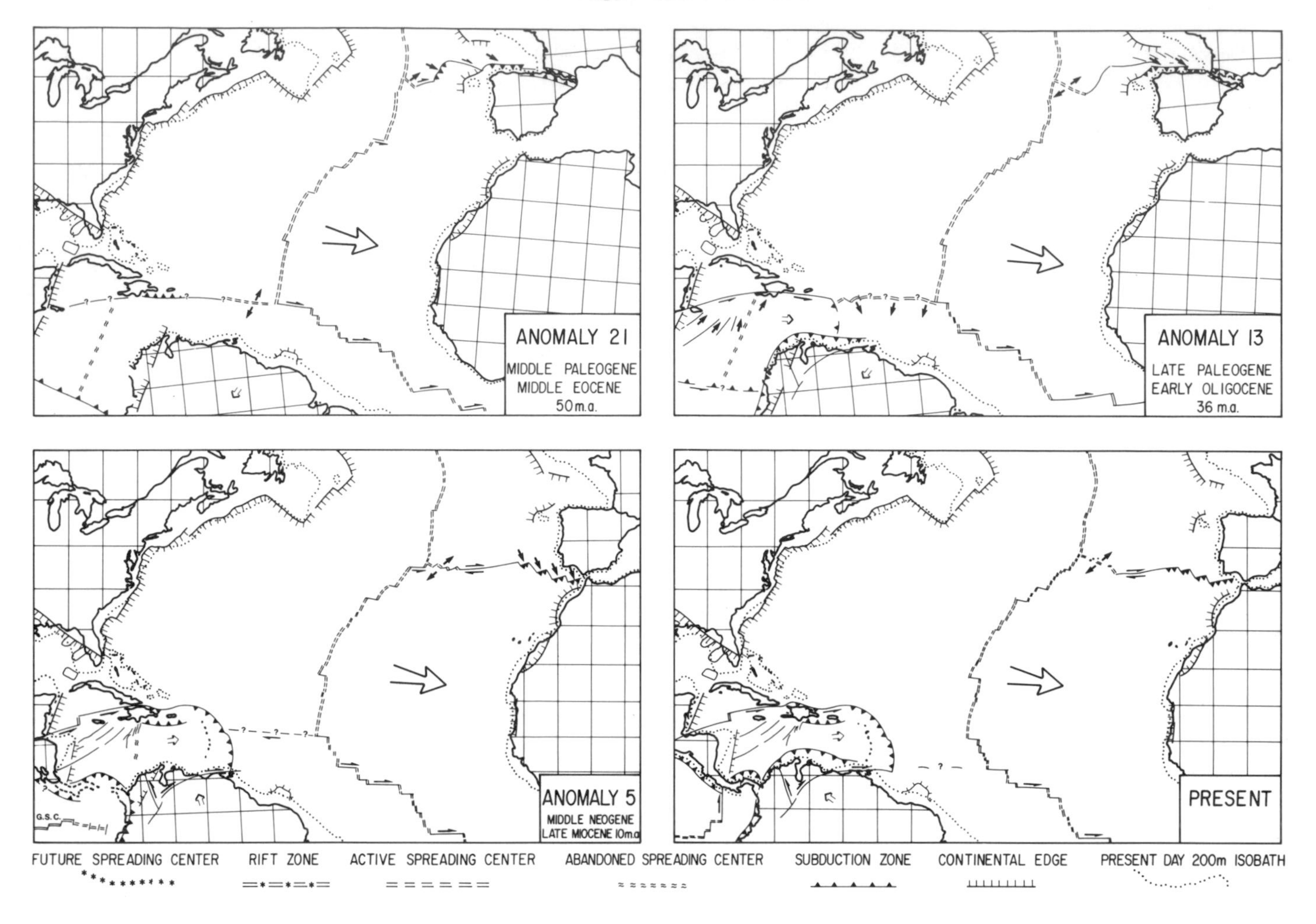

Figure 12. Reconstruction of continental positions around the Atlantic at time of anomaly 21, anomaly 13, anomaly 5, and present day. See Figure 10 for explanation.

Lefort and Van der Voo, 1981; Van der Voo, 1983); (3) post-breakup SFS data (LePichon and Fox, 1971; Pitman and Talwani, 1972; Vogt, 1973; Francheteau, 1973; LePichon and others, 1977; Sibuet and Mascle, 1978; Klitgord and Schouten, 1982; Pindell and Dewey, 1982; or (4) integration of post-breakup SFS data and paleomagnetic data (Smith and others, 1981; Briden and others, 1981). Reconstructions using pre-breakup data, e.g., Figure 9d, do not automatically correlate with the Atlantic breakup geometry. This is because the sparse Permo-Triassic data obscure the final tectonic history of Pangea, including rifting, prior to its early Mesozoic breakup (e.g., Morel and Irving, 1981; Van der Voo, 1983). We consider below those reconstructions based on post-breakup SFS data. Studies of Atlantic continental margins using geophysical data (e.g., Haworth and Keen, 1979; Klitgord and Behrendt, 1979) have eliminated the need for geometric fits that use isobaths for closure control, but we include a brief discussion of the reconstruction of Bullard and others (1965) because it has been so widely used.

Closure modeling for the central Atlantic requires assumptions about the landward limit of oceanic crust and crustal extension during rifting. Recent seismic studies of the Atlantic margin have shown that typical oceanic crust extends west of the BSMA nearly to the ECMA (Sheridan and others, 1979; Klitgord and Grow, 1980; Klitgord and Schouten , 1986a, b). Geophysical studies on the West African margin (Hinz and others, 1982; Wissmann and Roeser, 1982; Roussel and Liger, 1983) have mapped the basement hinge zone and the approximate landward edge of oceanic crust. This edge of oceanic crust is marked by a chain of salt diapirs and the hinge zone is marked by the West African coast magnetic anomaly (WACMA). Our minimum closure pole eliminates oceanic crust seaward of the ECMA and WACMA, and our maximum closure pole eliminates part of the space occupied by rift-stage crust. The Bullard fit (Fig. 9a) matches 200-m isobaths and exhibits a large gap in continental crust at the North American–South American–African junction. There also are numerous small overlaps of continental crust and gaps where ocean crust is known to exist. This reconstruction uses the single African plate assumption and creates a large gap in the

Gulf of Mexico–Caribbean region to which many of the Central American microplates can be located. LePichon and Fox (1971) determined a closure (Fig. 9b) controlled primarily by four major fracture zones near the continental margin and the 200-m isobath. The resulting pole is close to our minimum closure pole (Table 1), demonstrating the importance of fracture zone control and the close correspondence between the 200-m isobath and the landward edge of oceanic crust. A revised estimate of the extent of oceanic crust led LePichon and others (1977) to propose a fit (Fig. 9c) that is close to our BSMA reconstruction.

The closure fit presented here, in Figure 10a, (Klitgord and Schouten, 1980, 1982) assumes that: (1) the Jurassic marginal basins on the North American and African margins (Klitgord and Behrendt, 1979; Hinz and others, 1982) overlie transitional or rift-stage crust; and (2) Africa did not act as a single plate during Early Cretaceous time (Burke and Dewey, 1974). We have set limits that the maximum possible closure must not overlap the basement hinge zone along the landward edge of the marginal basins, and that a minimum possible closure must still eliminate all oceanic crust. Our minimum closure places the East Coast Magnetic Anomaly (ECMA) against the seaward edge of the west African salt diapir province (Jansa and Wiedmann, 1982). Both the maximum and minimum poles are given in Table 1, and the paleogeographic reconstruction based on the minimum closure pole is shown in Figure 10a.

Two Atlantic closure configurations recently published are similar to the one presented in Figure 10a. Lefort and van der Voo (1981) made a slight shift in our maximum closure fit (Klitgord and Schouten, 1980, 1982) when they determined a closure that would fit the paleomagnetic data and the Paleozoic structure lineaments split by Mesozoic rifting (Fig. 9d). In their recent study of the Gulf of Mexico and Caribbean, Pindell and Dewey (1982) used the central Atlantic closure of LePichon and Fox (1971) and assumed the two-part African plate to determine the closure shown in Figure 9b. Differences between the closure reconstructions displayed on Figures 9b, d, and 10a are slight and important only for problems in which shifts of a few tens of kilometers would lead to different conclusions or when calculating stage poles.

The assumption of a two-part African plate during the mid-Cretaceous (Burke and Dewey, 1974) does have a significant influence on our Jurassic and Early Cretaceous reconstructions of the Gulf of Mexico and Caribbean region. A single, rigid African plate (Bullard and others, 1965; Rabinowitz and LaBrecque, 1979) will usually result in closure reconstructions with a larger gap between the three major plates surrounding the region (Figs. 9a, c) than the gap when using a two-part plate (Figs. 9b, d, 10a). Ages of rifting and of sea-floor spreading in the northernmost South Atlantic and Benué trough are not well constrained, with most dates post-chron M-0. The oldest dated postrift sedimentary units are Albian-age marine shales that overlie a nonfossiliferous, arkosic sandstone unit in the Benué trough (Short and Stauble, 1967) and Neocomian continental clastics and Aptian (pre-Aptian?) salt on the African and Brazilian margins (Pautot and others, 1973; Leyden and others, 1976). Pre-chron M-0 rifting (Hauterivian to Barremian, M-10N to M-0) could have extended either northeast through the Benué trough (Burke and Dewey, 1974) or northwest between West Africa and northeastern South America (Rabinowitz and LaBrecque, 1979). Our post-chron M-0 reconstruction for the South Atlantic is that of Rabinowitz and LaBrecque (1979), and we assume a single, rigid African plate from this time onward.

The oldest oceanic crust on both flanks of the Mid-Atlantic Ridge was generated within the Jurassic magnetic quiet zone (Vogt, 1973), and there are no SFS magnetic lineation isochrons identified in this region for calculating the initial opening reconstruction pole. The oldest isochron on both flanks is chron M-25 (156 Ma). Older SFS lineations have been mapped in the western Atlantic (Bryan and others, 1980), but the conjugate set in the eastern Atlantic has not been identified. Fracture zone trends are used to calculate the poles of motion between the ECMA/WACMA and anomaly M-25. Lateral discontinuities in magnetic anomaly contours (Fig. 4b) (Vogt and Einwich, 1979, Klitgord and Schouten, 1986a, b) and basement relief (Tucholke and others, 1982) are used to map the fracture zone trends.

Assuming that these trends represent flowlines, we have used them to calculate a stage pole of motion for initial opening (60.0°N., 0.0°W.) and a reconstruction pole (Klitgord and Schouten, 1982). A series of poles, lying along a great circle perpendicular to the fracture zone traces, is satisfied by this flowline data set; we have chosen the pole that is most consistent with the data in the zone between the Blake Spur and Long Island FZs. This stage pole for initial opening predicts the early movements along the Mojave-Sonora megashear (Fig. 10a) (Anderson and Schmidt, 1983) and accounts for the pattern of lineations and discontinuities in the magnetic and gravity data of the Florida region (Klitgord and others, 1984). Choice of another pole of motion along the great circle determined from the fracture zone trends would result in closure reconstruction poles that produce continental fits with either more or less continental separation near the Grand Banks. All poles that we considered predicted about the same amount of continental separation between the Carolinas and West Africa.

Rifting produces a series of rift basins (later filled with salt) from the Gulf of Mexico to the Bay of Biscay and into the Rockall–North Sea and western Mediterranean regions. Rifting occurred in the Atlantic and Ligurian Tethys during at least Late Triassic and Early Jurassic time (Manspeizer and others, 1978; Lemoine, 1983, 1984). True sea-floor spreading and plate separation in the Atlantic during the Middle Jurassic time was confined between the Mojave-Sonora megashear and the Gibraltar transform zone. Studies of the Iberian margin (Boillot and others, 1979) and Western Alps (Lemoine, 1983, 1984) indicate that rifting extended northward between Galicia Bank and Iberia, linking the central Atlantic rift system with the western European margin rift system and the Tethys.

Early to Middle Jurassic rifting may also have extended southward across the Guyanan Shield of northern Brazil along the Pisco-Jurua fault (Szatmari, 1983). This rift branch, like the onshore Late Triassic–Early Jurassic rift basins of eastern North America, became inactive when sea-floor spreading began in the Gulf of Mexico and Central Atlantic. Spreading center (or rift zone) shifts stranded blocks of Paleozoic crust adjacent to salt basins between the Bahamas and Cuban FZ's in the northern Gulf basin (Klitgord and others, 1984) and off Iberia (Montadert and others, 1974; Boillot and others, 1979). At the northern end of the Atlantic spreading center, the initial opening pole predicts Middle Jurassic plate motions that require a small spreading segment adjacent to the southern Grand Banks (Fig. 10a, b). This northern spreading segment may have generated the sea floor in the Tagus Abyssal Plain off southwestern Iberia before a plate motion shift in Late Jurassic time (Fig. 10c) caused it to be abandoned. At the time of early Atlantic opening, the Gibraltar transform zone connected the central Atlantic spreading system with the Ligurian Tethys spreading system.

Mesozoic Motions

Middle Jurassic reconstructions (Figs. 10a, b) indicate a narrow Atlantic Basin opening into the Gulf of Mexico via a strait across southern Florida (Klitgord and others, 1984) and into the Ligurian Tethys via the Straits of Gibraltar (Bullard and others, 1965; Pitman and Talwani, 1972; Olivet and others, 1984; Lemoine, 1983, 1984; Schouten and others, 1986). The formation of salt occurred along the entire Atlantic-Gulf basin during Middle Jurassic; Bathonian or older salt has been identified in the northern Atlantic (Jansa and Wade, 1975; Barss and others, 1979; Jansa and others, 1980; Jansa and Wiedmann, 1982), and Callovian salt has been found in the Gulf of Mexico (Imlay, 1980).

Reorganization of spreading in latest Bathonian time (Fig. 10b) (Vogt, 1973; Klitgord and Grow, 1980) resulted in the formation at the BSMA of a new spreading system characterized by smaller ($<$100 km) transform offsets. This change marked the end of restricted water circulation within the Atlantic Basin and caused a shift in margin sedimentation from anhydrite-dolomite deposits (Iroquois Formation) to limestone (e.g., Scatarie Limestone) (Jansa and Wade, 1975; Jansa, this volume, Ch. 35). As the plates moved farther apart during the Late Jurassic, an ocean basin surrounded by carbonate platforms evolved (Jansa, 1981).

The Late Jurassic was a time of significant plate reorganizations. By Oxfordian time, the Mojave-Sonora megashear became inactive and the Yucatan block moved away from Mexico, this latter separation forming the petroliferous Reforma-Campeche basin (Fig. 9c). Rifting began between the Yucatan-Chortis blocks and South America, forming Upper Jurassic basins that were later deformed in the Eastern Cordillera of Colombia and Sierra de Perija of Venezuela (Mooney, 1980; Maze, 1984). The small spreading segment in the Tagus Abyssal Plain off southwestern Iberia was abandoned at this time. Shift of the spreading center from the Gulf of Mexico into the proto-Caribbean in Tithonian time (chron M-21; Fig. 10c) (Klitgord and others, 1984) rifted Central America from South America and reduced the length of the Cuban transform zone offset to 1000 km. This same shift also initiated subduction along the Cuban transform zone.

The Early Cretaceous began with a long period of relatively stable plate motions, but it ended with a major plate reorganization tha split South America from Africa and Iberia from North America–Eurasia. The Atlantic–proto Caribbean Basin gradually widened during Neocomian time (Fig. 10d), and carbonate banks continued to form along the basin margins. A continuing shift in plate motions maintained a component of compression along the Cuban transform-subduction zone (Ladd, 1976). A gradual eastward migration of the spreading centers at both ends of the Cuban transform zone produced a corresponding shift in subduction activity to the east.

At the end of the Neocomian (chron M-10N), a major tectonic event shifted plate motions around the world. Africa began rifting from South America at about chron M-10N (Burke and Dewey, 1974; Rabinowitz and LaBrecque, 1979). As mentioned previously, we assume that the initial rifting and sea-floor spreading included the zone between South America and southern Africa, and extended into central Africa along the Benué Trough (Klitgord and others, 1984). The break between West Africa and South America, linking the south and central Atlantic spreading systems, occurred about chron M-0 and produced the fracture zone trends mapped by Sibuet and Mascle (1978).

As the central Atlantic spreading system propagated northward at chron M-10N (Fig. 10d), rifting formed the petroliferous Jeanne d'Arc basin in the Hibernia area of the Grand Banks (Arthur and others, 1982) and may have partially separated Galicia Bank from Iberia and Iberia from Eurasia (Boillot and others, 1979; Grimaud and others, 1982). A sharp decrease in the rate of Atlantic opening occurred between the formation of anomalies M-10N and M-4, from 132 to 126 Ma (Schouten and Klitgord, 1977, 1982), just prior to spreading extending from north of Iberia to the southernmost South Atlantic (Talwani and Eldholm, 1977; Kristoffersen and Talwani, 1977; Kristoffersen, 1978; Srivastava, 1978; Rabinowitz and LaBrecque, 1979). Iberia shifted from the Eurasian plate to the African plate between chron M-10N and M-4 (Fig. 11a) (Schouten and others, 1986). Chron M-4 to M-2 (126 to 123 Ma) was a time of transition from rifting to sea-floor spreading between Iberia and Grand Banks (Tucholke and Ludwig, 1982) and between Iberia and Eurasia. This reorganization involved the abandonment of the Ligurian Tethys spreading center (Lemoine, 1983, 1984).

Rifting continued to propagate northward during the Late Cretaceous, first into Rockall Trough and then into the Laborador Sea (Srivastava, 1978; Roberts and others, 1979; Srivastava and Tapscott, this volume). A ridge-ridge-ridge triple junction linked the central Atlantic, north Atlantic, and Bay of Biscay spreading centers from chron M-4 to chron 30. The northern end of the North American–African plate boundary remained in the Bay of Biscay until the end of the Cretaceous (Fig. 11b, c). At the southern end of the central Atlantic spreading system, the Caribbean spreading center was abandoned at chron 33_0 ($\sim$80 Ma).

The southern triple junction became more of a ridge-ridge-shear zone triple junction (central Atlantic spreading system–south Atlantic spreading system–Cuban shear zone) involving only minor motions between the North and South American plates. A major direction change in the central Atlantic between chrons 33_o and 30 is recorded in the fracture zone pattern (Fig. 1 and Plate A) and corresponds to the initial opening phase of the Labrador Sea (~chron 34_y-30 time).

Cenozoic Motions

Cenozoic reconstructions at chrons 25, 21, 13, and 5 (Figs. 11, 12) involve (1) a period of reorganization between the North American, Eurasian and African plates; (2) abandonment of the Labrador Ridge; and (3) initiation of opening of the Norwegian and Arctic Seas (chron 24-25) (see Srivastava and Tapscott, this volume). Major changes in deep water circulation occurred as a result of these changes which permitted cold polar waters to move southward.

During the early Tertiary, the northern African plate (plus Iberia) boundary slowly shifted from shear and extension in the Bay of Biscay region (chrons M-0 to 25) to a compressional boundary during the Eocene (chrons 25 to 13) along the north coast of Iberia (see Figs. 11c, d, 12a, b) (Grimaud and others, 1982). King's Trough formed after chron 21 as an extensional boundary (Kidd and others, 1982), while compression continued along the north coast of Iberia (Figs. 12a, b). Comparisons between Alpine orogenic events and Atlantic kinematics have been attempted (e.g., Dewey and others, 1973; Schwan, 1980; Grimaud and others, 1982; Olivet and others, 1984; Livermore and Smith, in press). Uncertainties, however, in previously used north and central Atlantic poles of motion have made all estimated Mediteranean motions tenuous, because the involved movements of Africa and Eurasia are very sensitive to the much larger north and central Atlantic motions.

SUMMARY

Plate kinematics between North America and Africa during the past 175 m.y. have been described by a series of reconstruction poles of rotation and stage poles of motion determined from a re-examination of sea-floor–spreading magnetic anomaly and fracture zone patterns on both flanks of the Mid-Atlantic Ridge. Comparison of these data with Pacific SFS magnetic anomalies indicates that long periods of constant spreading in the central Atlantic were interrupted by short periods of plate motion adjustment. Atlantic SFS magnetic anomalies have been compared with two recently published geomagnetic time scales (Harland and others, 1982; Kent and Gradstein, this volume) to determine absolute spreading rates for the central Atlantic.

Spreading rates, determined using the Kent and Gradstein time scale, fluctuate in response to shifts in plate boundaries as spreading centers are abandoned, transform offsets shortened, and new spreading centers propagated northward and southward from the central Atlantic. The combination of these poles of rotation with those for the north Atlantic and south Atlantic yields a revised set of Atlantic region plate reconstructions, (Figs. 10, 11, 12; Plate 10 (I, J, K) illustrating plate motions and evolving crustal configurations during the Mesozoic and Cenozoic. Throughout this history, it is the tectonic activity at either end of the Central Atlantic spreading center, the Iberian–Grand Banks segment, or the Gulf of Mexico–Caribbean segment, that has the most important geologic ramifications.

REFERENCES

Anderson, T. H., and Schmidt, V. A.
1983 : The evolution of Middle America and the Gulf of Mexico–Caribbean Sea region during Mesozoic time: Geological Society of America Bulletin, v. 94, p. 941–966.

Arthur, K. R., Cole, D. R., Henderson, G.G.L., and Kushnir, D. W.
1982 : Geology of the Hibernia discovery; in The Deliberate Search for the Suble Trap, ed. M. T. Halbouty; American Association of Petroleum Geologists, Memoir 32, p. 181–195.

Barrett, D. L., and Keen, C. E.
1976 : Mesozoic magnetic lineations, the magnetic quiet zone, and sea floor spreading in the northwest Atlantic: Journal of Geophysical Research, v. 81, no. 26, p. 4875–4884.

Barss, M. S., Bujak, J. P., and Williams, G. L.
1979 : Palynological zonations and correlations of sixty-seven wells, eastern Canada: Geological Survey of Canada Paper 78-24, 118 p.

Boillot, G., Auxietre, J. L., and Dunand, J. P.
1979 : The northwestern Iberian Margin: A Cretaceous passive margin deformed during Eocene; in Deep Drilling Results in the Atlantic Ocean: Continental Margins and Paleoenvironment, eds. M. Talwani, W. Hay and W.B.F. Ryan; American Geophysical Union, Maurice Ewing Series, v. 3, p. 138–153.

Bonini, W. E., Hargraves, R. B., and Shagam, R., eds.
1984 : The Caribbean–South American Plate Boundary and Regional Tectonics: Geological Society of America Memoir 162, 421 p.

Briden, J. C., Hurley, A. M., and Smith, A. G.
1981 : Paleomagnetism and Mesozoic-Cenozoic paleocontinental maps: Journal of Geophysical Research, v. 86, no. B12, p. 11631–11656.

Bryan, G. M., Markl, R. G., and Sheridan, R. E.
1980 : IPOD site survey in the Blake-Bahamas Basin: Marine Geology, v. 35, p. 43–63.

Bullard, E. C., Everett, J. E., and Smith, A. G.
1965 : Fit of the continents around the Atlantic; in A Symposium on Continental Drift, eds., P.M.S. Blackett, E. C. Bullard and K. S. Runcorn; Philosophical Transactions of the Royal Society of London, series A, v. 258, p. 41–51.

Burke, K., and Dewey, J. F.
1974 : Two plates in Africa during the Cretaceous?: Nature, v. 249, p. 313–316.

Burke, K., Cooper, C., Dewey, J. F., Mann, P., and Pindell, J. L.
1984 : Caribbean Tectonics and relative plate motions; in The Caribbean-South American Plate Boundary and Regional Tectonics, eds. W. E. Bonini, R. B. Hargraves and R. Shagam; Geological Society of America, Memoir 162, p. 31–63.

Cande, S. C.

1978 : Anomalous behavior of the paleomagnetic field inferred from the skewness of anomalies 33 and 34: Earth and Planetary Science Letters, v. 40, p. 275–286.

Cande, S. C., and Kristoffersen, Y
1977 : Late Cretaceous magnetic anomalies in the North Atlantic; Earth and Planetary Science Letters, v. 35, p. 215–224.

Carey, S. W.
1958 : A tectonic approach to continental drift; in Continental Drift, A Symposium, ed. S. W. Carey; University of Tasmania Geology Department Symposium 5, Hobart, p. 177–355.

Case, J. E., and Holcombe, T. L.
1980 : Geologic-tectonic map of the Caribbean region; U.S. Geological Survey Miscellaneous Investigation Series, Map I-1100, scale 1:2,500,000.

Channell, J.E.T., Ogg, J. G., and Lowrie, W.
1982 : Geomagnetic polarity in the Early Cretaceous and Jurassic; Philosophical Transactions of the Royal Society of London, series A, v. 306, p. 137–146.

Cohee, G. V., Glaessner, M. F., and Hedberg, H. D., eds.
1978 : Contributions to the geologic time scale; Studies in Geology, no. 6; Tulsa, American Association of Petroleum Geologists, 388 p.

Collette, B. J., Slootweg, A. P., Verhoef, J., and Roest, W. R.
1984 : Geophysical investigations of the floor of the Atlantic Ocean between 10° and 38°N (Kroonvlag project); Proceedings of the Koninklijke Nederlandse Akademie van Wetenschappen, series C, v. 87, no. 1, p. 1–76.

Cox, A., ed.
1973 : *Plate Tectonics and Geomagnetic Reversals;* San Francisco, Freeman and Company, 702 p.

Dewey, J. F., Pitman, W. C., III, Ryan, W.B.F., and Bonnin, J.
1973 : Plate tectonics and the evolution of the Alpine system; Geological Society of America Bulletin, v. 84, p. 3137–3180.

Dietz, R. S., and Holden, J. C.
1970 : Reconstruction of Pangea: Breakup and dispersion of continents, Permian to present; Journal of Geophysical Research, v. 75, p. 4939–4956.

du Toit, A. A.
1937 : *Our Wandering Continents;* Oliver & Boyd, Edinburgh, 366 p.

Francheteau, J.
1973 : Plate tectonics model of the opening of the Atlantic Ocean south of the Azores; in Implications of Continental Drift to the Earth Sciences, eds. D. H. Tarling and S. K. Runcorn; Academic Press, New York, p. 197–202.

Funnell, B. M., and Smith, A. G.
1968 : Opening of the Atlantic Ocean; Nature, v. 219, p. 1328–1333.

Gose, W. A., Belcher, R. C., and Scott, G. A.
1982 : Paleomagnetic results from northeastern Mexico: Evidence for large Mesozoic rotations; Geology, v. 10, p. 50–54.

Grimaud, S., Boillot, G., Collette, B. J., Mauffret, A., Miles, P. R., and Roberts, D. B.
1982 : Western extension of the Iberian-European plate boundary during early Cenozoic (Pyrenean) convergence: A new model; Marine Geology, v. 45, p. 63–77.

Harland, W. B., Cox, A. V., Llewellyn, P. G., Pickton, C.A.G., Smith, A. G., and Walters, R.
1982 : A Geologic Time Scale; Cambridge Earth Science Series; Cambridge, Cambridge University Press, 131 p.

Harrison, C.G.A.
1972 : Poles of rotation; Earth and Planetary Science Letters, v. 14, p. 31–38.

Haworth, R. T., and Keen, C. E.
1979 : The Canadian Atlantic Margin: A passive continental margin encompassing an active past; Tectonophysics, v. 59, p. 83–126.

Hayes, D. E., and Rabinowitz, P. D.
1975 : Mesozoic magnetic lineations and the magnetic quiet zone off northwest Africa; Earth and Planetary Science Letters, v. 28, p. 105–115.

Heezen, B. C., and Tharp, M.
1965 : Tectonic fabric of the Atlantic and Indian Oceans and continental drift; in A Symposium on Continental Drift, eds. P.M.S. Blackett, E. C. Bullard and K. S. Runcorn; Philosophical Transactions of the Royal Society of London, series A, v. 258, p. 90–106.

Heirtzler, J. R., Dickson, G. O., Herron, E. M., Pitman, W. C., III, and LePichon, X.
1968 : Marine magnetic anomalies, geomagnetic field reversals, and motions of the ocean floor and continents; Journal of Geophysical Research, v. 73, no. 6, p. 2119–2136.

Hellinger, S. J.
1981 : The uncertainties of finite rotations in plate tectonics; Journal of Geophysical Research, v. 86, p. 9312–9318.

Hey, R. N., Duennebier, F. K., and Morgan, W. J.
1980 : Propagating rifts on mid-ocean ridges: Journal of Geophysical Research, v. 85, p. 3647–3658.

Hinz, Karl, Dostmann, Hans, and Fritsch, Jurgen
1982 : The continental margin of Morocco. Seismic sequences, structural elements and geological development; in Geology of the Northwest African Continental Margin, eds. V. von Rad, K. Hinz, M. Sarnthein and E. Siebold; Springer-Verlag, New York, p. 34–60.

Imlay, R. W.
1980 : Jurassic paleobiogeography of the conterminous United States in its continental setting; U.S. Geological Survey Professional Paper 1062, 125 p.

Jansa, L. F.
1981 : Mesozoic carbonate platforms and banks of the eastern North American margin; Marine Geology, v. 44, p. 97–117.

Jansa, L. F., Bujak, J. P., and Williams, G. L.
1980 : Upper Triassic salt deposits of the western North Atlantic; Canadian Journal of Earth Sciences, v. 17, p. 547–559.

Jansa, L. F., and Wade, J. A.
1975 : Paleogeography and sedimentation in the Mesozoic and Cenozoic, southeastern Canada; in Canada's Continental Margins and Offshore Petroleum Exploration, eds. C. J. Yorath, E. R. Parker, and D. J. Glass; Calgary Canadian Society of Petroleum Geologists, Memoir 4, p. 79–102.

Jansa, L. F., and Wiedmann, J.
1982 : Mesozoic-Cenozoic development of the eastern North American and Northwest African continental margins: A comparison; in Geology of the Northwest African Continental Margin, eds. U. von Rad, K. Hinz, M. Sarnthein and E. Siebold; Springer-Verlag, New York, p. 215–269.

Johnson, G. L., and Vogt, P. R.
1973 : Mid-Atlantic Ridge from 47° to 51° N; Geological Society of America Bulletin, v. 84, p. 3443–3462.

Jones, E.J.W., and Mgbatogu, C.C.S.
1982 : The structure and development of the West African continental margin off Guinee Bissau, Guinee and Sierra Leone; in The Ocean Floor, eds. R. A. Scrutton and M. Talwani; John Wiley and Sons, New York, p. 165–202.

Kerr, J. W., and Fergusson, A. J., eds.
1981 : Geology of the North Atlantic borderlands; Canadian Society of Petroleum Geologists, Memoir 7, 743 p.

Kidd, R. B., Searle, R. C., Ramsay, A.T.S., Prichard, H., and Mitchell, J.
1982 : The geology and formation of King's Trough, northeast Atlantic Ocean; Marine Geology, v. 48, p. 1–30.

Klitgord, K. D., and Behrendt, J. C.
1977 : Aeromagnetic anomaly map—U.S. Atlantic continental margin: U.S. Geological Survey Miscellaneous Field Studies Map MF-913, 2 sheets, scale 1:1,000,000.
1979 : Basin structure of the U.S. Atlantic margin; in Geological and Geophysical Investigations of Continental Margins, eds. J. S. Watkins, L. Montadert, and P. W. Dickerson; American Association of Petroleum

Geologists, Memoir 29, p. 85–112.
Klitgord, K. D., and Grow, J. A.
1980 : Jurassic seismic stratigraphy and basement structure of western North Atlantic magnetic quiet zone; American Association of Petroleum Geologists Bulletin, v. 64, no. 10, p. 1658–1680.
Klitgord, K. D., and Mammerickx, J.
1982 : Northern East Pacific Rise: Magnetic anomaly and bathymetric framework; Journal of Geophysical Research, v. 87, no. B8, p. 6725–6750.
Klitgord, K. D., and Schouten, H.
1980 : Mesozoic evolution of the Atlantic, Caribbean and Gulf of Mexico (abs.); in Proceedings of a Symposium, The Origin of the Gulf of Mexico and the Early Opening History of the Central North Atlantic, ed. R. H. Pilger; Louisiana State University, Baton Rouge, p. 100–101.
1982 : Early Mesozoic Atlantic reconstructions from seafloor spreading data (abs.); EOS, Transactions, American Geophysical Union, v. 83, no. 18, p. 307.
1986a: Tectonic and magnetic structure: Baltimore Canyon trough and adjacent magnetic quiet zone; U.S. Geological Survey Miscellaneous Field Studies Map MF-XX, scale 1:1,000,000, in press.
1986b: Tectonic and magnetic structure: Carolina trough and adjacent magnetic quiet zone; U.S. Geological Survey Miscellaneous Field Studies Map MF-XX, scale 1:1,000,000, in press.
Klitgord, K. D., Heustis, S. P., Mudie, J. D., and Parker, R. L.
1975 : An analysis of near-bottom magnetic anomalies: Sea-floor spreading and the magnetized layer; Geophysical Journal of the Royal Astronomical Society, v. 83, p. 387–424.
Klitgord, K. D., Popenoe, P., and Schouten, H.
1984 : Florida: A Jurassic transform plate boundary; Journal of Geophysical Research, v. 89, no. B9, p. 7753–7772.
Kristoffersen, Y., and Talwani, M.
1977 : Extinct triple junction south of Greenland and the Tertiary motion of Greenland relative to North America; Geological Society of America Bulletin, v. 88, p. 1037–1049.
Kristoffersen, Y.
1978 : Seafloor spreading and the early opening of the North Atlantic; Earth and Planetary Science Letters, v. 38, p. 273–290.
LaBrecque, J. L., Kent, D. V., and Cande, S. C.
1977 : Revised magnetic polarity time scale for Late Cretaceous and Cenozoic time; Geology, v. 5, p. 330–335.
LaBrecque, J. L., Hsu, K. J., and 11 others
1983 : DSDP Leg 73: Contributions to Paleogene stratigraphy in nomenclature, chronology and sedimentation rates; Paleogeography, Paleoclimatology, Paleoecology, v. 42, p. 91–125.
Ladd, J. W.
1974 : South Atlantic seafloor spreading and Caribbean tectonics [Ph.D. Thesis]: New York, Columbia University, 251 p.
1976 : Relative motion of South America with respect to North America and Caribbean tectonics; Geological Society of America Bulletin, v. 87, p. 969–976.
Larson, R. L., and Hilde, T.W.C.
1975 : A revised time scale of magnetic reversals for the Early Cretaceous and Late Jurassic; Journal of Geophysical Research, v. 80, p. 2585–2594.
Larson, R. L., and Pitman, W. C., III
1972 : World-wide correlation of Mesozoic magnetic anomalies, and its implications; Geological Society of America Bulletin, v. 83, p. 3645–3662.
Lefort, J. P., and Van der Voo, R.
1981 : A kinematic model for the collision and complete suturing between Gondwanaland and Laurussia in the Carboniferous; Journal of Geology, v. 89, no. 5, p. 537–550.
Lemoine, M.
1983 : Rifting and early drifting: Mesozoic central Atlantic and Ligurian Tethys; in Initial Reports of the Deep Sea Drilling Project, v. 76, eds. R. E. Sheridan and F. M. Gradstein; U.S. Government Printing Office, Washington, D.C., p. 885–895.
1984 : Mesozoic evolution of the Western Alps; Annales Geophysicae, v. 2, p. 171–172.
LePichon, X.
1968 : Seafloor spreading and continental drift; Journal of Geophysical Research, v. 73, p. 3661–3697.
LePichon, X., and Fox, J. P.
1971 : Marginal offsets, fracture zones, and the early opening of the North Atlantic; Journal of Geophysical Research, v. 76, p. 6294–6308.
LePichon, X., Francheteau, J., and Bonnin, J.
1973 : Plate Tectonics; Development in Geotectonics, v. 6, New York, Elsevier, 300 pp.
LePichon, X., Sibuet, J. C., and Francheteau, J.
1977 : The fit of the continents around the North Atlantic Ocean: Tectonophysics, v. 38, p. 169–209.
Leyden, R., Asmus, H., Zembruscki, S., and Bryan, G.
1976 : South Atlantic diapiric structures; American Association of Petroleum Geologists Bulletin, v. 60, no. 2, p. 196–212.
Liger, J. L.
1980 : Essai de comparison des marges continentales-est-americaine et ouest-africaine; Marseille, Travaux des laboratoires des Sciences de la terre, Saint-Jerome, series X, no. 36, 37 p.
Livermore, R. A., and Smith, A. G.
1985 : Some boundary conditions for the evolution of the Mediterranean region; NATO Advanced Research Institute, in press.
Lowrie, W., and Channel, J.E.T.
1984 : Magnetostratigraphy of the Jurassic-Cretaceous boundary in the Maiolica Limestone (Umbria, Italy); Geology, v. 12, p. 44–47.
McElhinny, M. W.
1973 : Paleomagnetism and Plate Tectonics; Cambridge University Press, New York, 386 p.
McKenzie, D. P., and Parker, R. L.
1967 : The North Pacific: An example of tectonics on a sphere; Nature, v. 216, no. 5122, p. 1276–1280.
McKenzie, D. P., and Sclater, J. G.
1971 : The evolution of the Indian Ocean since the Late Cretaceous; Geophysical Journal of the Royal Astronomical Society, v. 25, p. 437–528.
Macdonald, K. C., and Holcombe, T. L.
1978 : Inversion of magnetic anomalies and seafloor spreading in the Cayman Trough; Earth and Planetary Science Letters, v. 40, p. 407–414.
Manspeizer, W., Puffer, J. H., and Cousminer, H. L.
1978 : Separation of Morocco and eastern North America: A Triassic-Liassic stratigraphic record; Geological Society of America Bulletin, v. 89, p. 901–920.
Martin, A. K., Goodlad, S. W., Hartnady, C.J.H., and du Plessis, A.
1982 : Cretaceous paleopositions of the Falkland Plateau relative to southern Africa using Mesozoic seafloor spreading anomalies; Geophysical Journal of the Royal Astronomical Society, v. 71, p. 567–579.
Maze, W. B.
1984 : Jurassic LaQuinta Formation in the Sierra de Perija, northwestern Venezuela: Geology and tectonics environment of the red beds and volcanic rocks; in The Caribbean–South American Plate Boundary and Regional Tectonics, eds. W. E. Bonini, R. B. Hargraves and R. Shagam; Geological Society of America, Memoir 162, p. 263–282.
Menard, H. W., and Atwater, T.
1968 : Changes in direction of seafloor spreading; Nature, v. 219, p. 463–467.
Meyerhoff, A. A.
1980 : Geology of Reforma–Campeche Shelf; Oil and Gas Journal, v. 78, no. 16, p. 121–124.
Minster, J. B., and Jordan, T. H.
1978 : Present-day plate motions; Journal of Geophysical Research, v. 83, no. B11, p. 5331–5354.

Molnar, Peter, Atwater, Tanya, Mammerickx, Jacqueline, and Smith, S. M.
1975 : Magnetic anomalies, bathymetry and the tectonic evolution of the South Pacific since the Late Cretaceous; Geophysical Journal of the Royal Astronomical Society, v. 40, p. 383–420.
Montadert, L., Winnock, E., Deltiel, J. R., and Grau, G.
1974 : Continental margins of Galicia-Portugal and Bay of Biscay; in The Geology of Continental Margins, eds. C. A. Burk and C. L. Drake; Springer-Verlag, New York, p. 323–342.
Mooney, W. D.
1980 : An East Pacific-Caribbean ridge during the Jurassic and Cretaceous and the evolution of western Colombia; in Proceedings of a Symposium, The origin of the Gulf of Mexico and the Early Opening History of the Central North Atlantic Ocean, ed. R. H. Pilger; Louisiana State University, Baton Rouge, p. 55–73.
Morgan, W. J.
1968 : Rises, trenches, great faults, and crustal blocks; Journal of Geophysical Research, v. 73, no. 6, p. 1959–1982.
1983 : Hot spot traces and early rifting of the Atlantic; Tectonophysics, v. 94, p. 123–139.
Morel, P., and Irving, E.
1981 : Paleomagnetism and the evolution of Pangea; Journal of Geophysical Research, v. 86, p. 1858–1872.
Ness, G., Levi, S., and Couch, R.
1980 : Marine magnetic anomaly time scales for the Cenozoic and Late Cretaceous: a precis, critique, and synthesis; Reviews of Geophysics and Space Physics, v. 18, no. 4, p. 753–770.
Ogg, J. G.
1983 : Magnetostratigraphy of Upper Jurassic and Lower Cretaceous sediments, DSDP site 534, western North Atlantic; in Initial Reports of the Deep Sea Drilling Project, eds. R. E. Sheridan and F. M. Gradstein; v. 76, U.S. Government Printing Office, Washington, D.C., p. 685–699.
Olivet, J. L.
1978 : Nouveau modle d'evolution de l'Atlantique nord et central [Ph.D. Thesis], Université de Pierre et Marie Curie, Paris, 150 p.
Olivet, J. L., Bonnin, J., Beuzart, P., and Auzende, J. M.
1984 : Cinematique de l'Atlantique nord et central; Centre National pour l'exploration des oceans, Rapports scientifiques et techniques No. 54, 108 p.
Palmer, A. R.
1983 : The decade of North American geology 1983 Geologic Time Scale; Geology, v. 11, no. 9, p. 503–504.
Pautot, G., Renard, V., Daniel, J., and Dupont, J.
1973 : Morphology, limits, origin, and age of salt layer along South Atlantic African margin; American Association of Petroleum Geologists Bulletin, v. 57, no. 9, p. 1658–1671.
Perry, R. K., Fleming, H. S., Vogt, P. R., Cherkis, N. Z., Feden, R. H., Thiede, J., Strand, J. E., and Collette, B. J.
1981 : North Atlantic Ocean; Bathymetry and plate tectonic evolution: Geological Society of America, MC-35.
Phillips, J. D., and Fleming, H. S.
1978 : Multi-beam sonar study of the Mid-Atlantic Ridge rift valley, 36°-37°N: Geological Society of America, MC-19.
Pilger, R. H., Jr.
1978 : A closed Gulf of Mexico, pre-Atlantic Ocean plate reconstruction and the early rift history of the Gulf and North Atlantic; Gulf Coast Association of Geological Societies Transactions, v. 28, p. 385–393.
Pindell, J. L., and Dewey, J. F.
1982 : Permo-Triassic reconstruction of western Pangea and the evolution of the Gulf of Mexico/Caribbean region; Tectonics, v. 1, p. 179–211.
Pitman, W. C., III
1978 : Relationship between eustacy and stratigraphic sequences of passive margins; Geological Society of America Bulletin, v. 89, no. 9, p. 1389–1403.
Pitman, W. C., III, and Talwani, M.
1972 : Seafloor spreading in the North Atlantic; Geological Society of America Bulletin, v. 83, p. 619–646.
Purdy, G. M.
1975 : The eastern end of the Azores-Gibraltar plate boundary; Geophysical Journal of the Royal Astronomical Society, v. 43, p. 973–1000.
Rabinowitz, P. D., and LaBrecque, J.
1979 : The Mesozoic South Atlantic Ocean and evolution of its continental margin; Journal of Geophysical Research, v. 84, p. 5973–6002.
Rabinowitz, P. D., and Purdy, G. M.
1976 : The Kane Fracture zone in the western Central Atlantic; Earth and Planetary Science Letters, v. 33, p. 21–26.
Rabinowitz, P. D., Cande, S. C., and Hayes, D. E.
1979 : The J-anomaly in the central north Atlantic Ocean; in Initial Reports of the Deep Sea Drilling Project, eds. B. E. Tucholke and P. R. Vogt; v. 43, U.S. Government Printing Office, Washington, D.C., p. 879–885.
Roberts, D. G., and Jones, M. T.
1980 : Magnetic anomalies in the northeast Atlantic: Mid-Atlantic Ridge to southwest Europe; Wormley, Surrey, Institute of Oceanographic Sciences, 2 sheets.
Roberts, D. G., Montadert, L., and Searle, R. C.
1979 : The western Rockall Plateau: Stratigraphy and structural evolution; in Initial Reports of the Deep Sea Drilling Project, eds. L. Montadert and D. G. Roberts; v. 48, U.S. Government Printing Office, Washington, D.C., p. 1061–1088.
Roeser, H.
1982 : Magnetic anomalies in the magnetic quiet zone off Morocco: in Geology of the Northwest African Continental Margin, eds., U. von Rad, K. Hinz, M. Sarnthein and E. Siebold; Springer Verlag, New York, p. 60–68.
Rona, P. A. and Richardson, E. S.
1978 : Early Cenozoic global plate reorganizations; Earth and Planetary Science Letters, v. 40, p. 1–11.
Roussel, J., and Liger, J. L.
1983 : A review of deep structure and ocean-continent transition in the Senegal basin (West Africa); Tectonophysics, v. 91, p. 183–211.
Schouten, J. A.
1971 : A fundamental analysis of magnetic anomalies over oceanic ridges; Marine Geophysical Researches, v. 1, p. 111–144.
Schouten, H., and Klitgord, K. D.
1977 : Map showing Mesozoic magnetic anomalies, western North Atlantic; U.S. Geological Survey Miscellaneous Field Studies Map MF-915, scale 1:2,000,000.
1982 : The memory of the accreting plate boundary and the continuity of fracture zones; Earth and Planetary Science Letters, v. 59, p. 255–266.
Schouten, H., and McCamy, K.
1972 : Filtering marine magnetic anomalies; Journal of Geophysical Research, v. 77, no. 35, p. 7089–7099.
Schouten, H., Cande, S. C., and Klitgord, K. D.
1984 : Magnetic anomaly profile and generalized tectonic framework; in Mid-Atlantic Ridge Between 22° and 38°N, eds. P. D. Rabinowitz and H. Schouten; Ocean Margin Drilling Program Atlas Series, v. 11.
Schouten, H., Klitgord, K. D., and Srivastava, S. P.
1986 : Iberian plate kinematics: The African connection; Nature, in review.
Schwan, W.
1980 : Geodynamic peaks in Alpinotype orogenies and changes in ocean-floor spreading during Late Jurassic–Late Tertiary time; American Association of Petroleum Geologists Bulletin, v. 64, no. 3, p. 350–373.
Sclater, J. G., Hellinger, S., and Tapscott, C.
1977 : The paleobathymetry of the Atlantic Ocean from Jurassic to present; Journal of Geology, v. 85, p. 509–552.
Sheridan, R. E.
1983 : Phenomena of pulsation tectonics and early rifting of the North American continental margin; in Initial Reports of the Deep Sea Drilling

Project, eds. R. E. Sheridan and F. M. Gradstein; v. 76, U.S. Government Printing Office, Washington, D.C., p. 879–909.
Sheridan, R. E., Grow, J. A., Behrendt, J. C., and Bayer, K. C.
1979 : Seismic refraction study of the continental edge off the eastern United States; Tectonophysics, v. 59, p. 1–26.
Sheridan, R. E., Gradstein, F. M., eds.
1983 : Initial Reports of the Deep Sea Drilling Project, v. 76; Washington, D.C., U.S. Government Printing Office, 949 p.
Short, K. C., and Stauble, A. J.
1967 : Outline of geology of Niger Delta; American Association of Petroleum Geologists Bulletin, v. 51, no. 5, p. 761–779.
Sibuet, J. C., and Mascle, Jean
1978 : Plate kinematic implications of Atlantic equatorial fracture zone trends; Journal of Geophysical Research, v. 83, no. B7, p. 3401–3421.
Sibuet, J. C., Ryan, W.B.F., and 17 others
1980 : Deep drilling results of Leg 47b (Galicia Bank area) in the framework of the early evolution of the North Atlantic Ocean; Philosophical Transactions of the Royal Society of London, v. A294, p. 51–61.
Smith, A. G., Hurley, A. M., and Briden, J. C.
1981 : Phanerozoic Paleocontinental World Maps; New York, Cambridge University Press, 102 p.
Srivastava, S. P.
1978 : Evolution of the Labrador Sea and its bearing on the early evolution of the North Atlantic; Geophysical Journal of the Royal Astronomical Society, v. 52, p. 313–357.
Stock, J. M., and Molnar, P.
1983 : Some geometrical aspects of uncertainties in combined plate reconstructions; Geology, v. 11, no. 12, p. 697–701.
Sundvik, M., Larson, R. L., and Detrick, R. S.
1984 : Rough-smooth basement boundary in the western North Atlantic basin: Evidence for a seafloor-spreading origin; Geology, v. 12, no. 1, p. 31–34.
Sutter, J. F., and Smith, T. E.
1979 : $^{40}Ar/^{39}Ar$ ages of diabase intrusions from Newark trend basins in Connecticut and Maryland: initiation of central Atlantic rifting; American Journal of Science, v. 279, p. 808–831.
Sutter, J. F.
1985 : Progress on geochronology of Mesozoic diabases and basalts; U.S. Geological Survey Circular 946, ch. 21, p. 110–114.
Szatmari, P.
1983 : Amazon rift and Pisco-Jurua fault: Their relation to the separation of North America from Gondwana; Geology, v. 11, no. 5, p. 300–304.
Talwani, M., and Eldholm, O.
1977 : Evolution of the Norwegian-Greenland Sea; Geological Society of America Bulletin, v. 88, p. 969–999.
Tarling, D. H., and Runcorn, S. K., eds.
1973 : Implications of Continental Drift to the Earth Sciences; Academic Press, New York, 2 vols., 1184 p.
Tucholke, B. E., and Ludwig, W. J.
1982 : Structure and origin of the J Anomaly Ridge, western North Atlantic Ocean: Journal of Geophysical Research, v. 87, no. B11, p. 9389–9407.
Tucholke, B. E., Houtz, R. E., and Ludwig, W. J.
1982 : Sediment thickness and depth to basement in western North Atlantic Ocean Basin; American Association of Petroleum Geologists Bulletin, v. 66, no. 6, p. 1384–1395.
Uchupi, E., Bolmer, S. T., and 5 others
1983 : Tectonic Features; in Eastern North America Continental Margin and Adjacent Ocean Floor, 34° to 41°N and 68° to 78°W, eds. J. I. Ewing and P. D. Rabinowitz; Ocean Margin Drilling Program, Regional Atlas Series 4.
Vail, P. R., Mitchum, R. M., Jr., and 6 others
1977 : Seismic stratigraphy and global changes of sea level; in Seismic Stratigraphy—Applications to Hydrocarbon Exploration, ed. P. E. Payton; American Association of Petroleum Geologists, Memoir 26, p. 49–212.
Van Andel, Tj. H.
1971 : Fracture zones; Comments on Earth Sciences: Geophysics, v. 1, p. 159–166.
Van der Voo, R.
1983 : Paleomagnetic constraints on the assembly of the Old Red Continent; Tectonophysics, v. 91, p. 271–283.
Van der Voo, R., and French, R. B.
1974 : Apparent polar wander for the Atlantic-bordering continents: Late Carboniferous to Eocene; Earth Science Reviews, v. 10, p. 99–119.
Vine, F. J., and Matthews, D. H.
1963 : Magnetic anomalies over oceanic ridges; Nature, v. 199, p. 947–949.
Vogt, P. R.
1973 : Early events in the opening of the North Atlantic; in Implications of Continental Drift to the Earth Sciences, v. 2, eds. D. H. Tarling and S. K. Runcorn; Academic Press, London, p. 693–712.
Vogt, P. R., and Einwich, A. M.
1979 : Magnetic anomalies and seafloor spreading in the western North Atlantic, and a revised calibration of the Keathley (M) geomagnetic reversal chronology; in Initial Reports of the Deep Sea Drilling Project, v. 43, eds. B. E. Tucholke and P. R. Vogt, U.S. Government Printing Office, Washington, D.C., p. 857–876.
Vogt, P. R., and Perry, R. K.
1982 : North Atlantic Ocean: Bathymetry and plate tectonic evolution; Text to accompany Geological Society of America Map and Chart Series MC-35, 21 p.
Vogt, P. R., Anderson, C. N., and Bracey, D. R.
1971 : Mesozoic magnetic anomalies, seafloor spreading, and geomagnetic reversals in the southwestern North Atlantic; Journal of Geophysical Research, v. 76, p. 4796–4823.
Vogt, P. R., Avery, O. E., Anderson, C. N., Bracey, D. R., and Schneider, E. D.
1969 : Discontinuities in seafloor spreading; Tectonophysics, v. 8, p. 285–317.
Vogt, P. R., Zondek, B., Fell, P. W., Cherkis, N. Z., and Perry, R. K.
1984 : SEASAT altimetry, the North Atlantic geoid, and evaluation of shipborne subsatellite profiles; Journal of Geophysical Research, v. 89, no. B12, p. 9885–9903.
Walper, J. L., and Rowett, C. L.
1972 : Plate tectonics and the origin of the Caribbean and the Gulf of Mexico; Gulf Coast Association of Geological Societies Transactions, v. 22, p. 105–116.
Wegener, A.
1924 : Origin of Continents and Oceans; London; Methuen [original German edition Die Entstehung der Kontinente und Ozeane, 1915], Dutton and Co., 212 p.
Whitmarsh, R. B., Ginsburg, A., and Searle, R. C.
1982 : The structure and origin of the Azores-Biscay Rise, Northeast Atlantic Ocean; Geophysical Journal of the Royal Astronomical Society, v. 70, p. 79–107.
Wissmann, G., and Roeser, H. A.
1982 : A magnetic and halokinetic structural Pangaea fit of Northwest Africa and North America; Geologisches Jahrbuch, Reihe E, Geophysik, v. 23, p. 43–61.

MANUSCRIPT ACCEPTED BY THE SOCIETY FEBRUARY 28, 1986

Acknowledgments

Our synthesis of central Atlantic data benefited greatly from access to unpublished data provided by B. J. Collette, P. Vogt, S. Srivastava, P. Rabinowitz, S. Cande, and B. Tucholke. T. Atwater and P. Molnar provided magnetic and bathymetric data from the Pacific. Discussions with J. Morgan, S. Srivastava, B. J. Collette, B. Tucholke, and S. Cande and comments from M. Talwani and J.

LaBrecque were most appreciated. Extensive critical reviews by D. Scholl and D. Hutchinson led to major improvements in the manuscript. We also acknowledge a revised identification of the M-15 to M-4 anomaly sequence suggested by M. Sundvik. Computer programming support for our rotation mapping system was provided by E. Coward and L. Gilbert, graphic support by P. Forrestel and J. Zwinakis, and typing by P. Mons-Wengler and K. DeMello. Research support for H.S. was from National Science Foundation grants 20/22523 and 20/9027.00.

Printed in U.S.A.

The Geology of North America
Vol. M, The Western North Atlantic Region
The Geological Society of America, 1986

Chapter 23

Plate kinematics of the North Atlantic

S. P. Srivastava
Atlantic Geoscience Centre, Geological Survey of Canada, Bedford Institute of Oceanography, Dartmouth, Nova Scotia B2Y 4A2, Canada
C. R. Tapscott
Exxon Production Research Company, P.O. Box 2189, Houston, Texas 77001

INTRODUCTION

The problem of determining the past relative positions of continents now situated around the North Atlantic has been a challenge to geologists and geophysicists for many years. Yet a solution acceptable to all does not exist. Since the acceptance of the idea of continental drift, a number of reconstructions has been published. The earlier reconstructions, prior to the birth of the plate tectonic theory, were done qualitatively in order to match morphological features of the continental margins (Choubert, 1935) or geological boundaries on land. The first quantitative fit of the North Atlantic was published by Bullard and others (1965) who described the motion between continents or plates as angular rotation about a set of poles. Since then a number of papers have been published showing reassembly of continents during various stages of evolution of the North Atlantic using magnetic anomalies in the oceanic regions as isochrons or plate boundaries and fracture zones as the direction of motion between plates. The most notable of these are by Pitman and Talwani (1972), Laughton (1971, 1972), Williams (1975), LePichon and others (1977), Sclater and others (1977), Kristoffersen and Talwani (1977), Talwani and Eldholm (1977), Kristoffersen (1978), Srivastava (1978, 1985), Olivet and others (1981), Unternehr (1982), Nunns (1983), and Vink (1982, 1984).

Most of these reconstructions are based on a limited amount of geophysical information from one or more regions of the North Atlantic available at the time of publication. Thus different criteria and assumptions had to be made to derive satisfactory configurations between plates for the entire North Atlantic. Consequently these reconstructions differ both in data base and working assumptions and some deal only with a small part of the North Atlantic. The attempt to find a solution for the entire North Atlantic by combining solutions from different regions is not possible without making further assumptions.

The Problem

It is well known that the spreading in the North Atlantic is now taking place across the Mid-Atlantic Ridge which, for the region discussed in this chapter, extends from the Azores-Gibraltar Fracture Zone through the Greenland Sea to the Eurasia Basin in the Arctic (Fig. 1). Thus its opening is confined between the North American plate, which includes the whole of North America, the western Arctic Ocean, including the Lomonosov Ridge and Greenland, and the Eurasian Plate, which includes Europe, Asia and Iberia. It is also known, from detailed geophysical measurements in the Labrador sea (Kristoffersen and Talwani, 1977; Srivastava, 1978), Norwegian-Greenland seas (Talwani and Eldholm, 1977), the Eurasia Basin (Vogt and others, 1979), Rockall Trough (Roberts, 1975; Roberts and others, 1981) and the Bay of Biscay (Williams, 1975; Kristoffersen, 1978) that Greenland, Jan Mayen, the Lomonosov Ridge, Rockall and Iberia have had motion independent of the plates to which they are now attached. The questions which need to be answered are when these motions took place and how. Previous reconstructions of the North Atlantic also addressed this problem but they differ significantly in their treatments of the relative motions between these blocks. For example, the Lomonosov Ridge has been considered as part of the Greenland plate (LePichon and others, 1977), as an independent plate (Sclater and others, 1977; Phillips and Tapscot, 1981) and as part of the North American plate (Pitman and Talwani, 1972; Vink, 1982, 1984; Srivastava, 1985). Similarly Rockall has been treated as part of the Greenland plate to explain the formation of Rockall Trough during the early stages of opening of the North Atlantic (Kristoffersen, 1978; Srivastava, 1978) and as an independent plate from anomalies 20 to 24 times (Phillips and Tapscot, 1981). Such assumptions were made in order to match plate boundaries defined by the magnetic anomalies or the bathymetry contours and in most cases these could not be supported independently due to the lack of detailed geological or geophysical observations. Until independent evidence can be found to negate these assumptions, or better explanations can be suggested, any further refinement of these models is not going to produce a unique solution to the plate kinematics of the North Atlantic.

Srivastava, S. P., and Tapscott, C. R., 1986, Plate kinematics of the North Atlantic; *in* Vogt, P. R., and Tucholke, B. E. eds., The Geology of North America, Volume M, The Western North Atlantic Region: Geological Society of America.

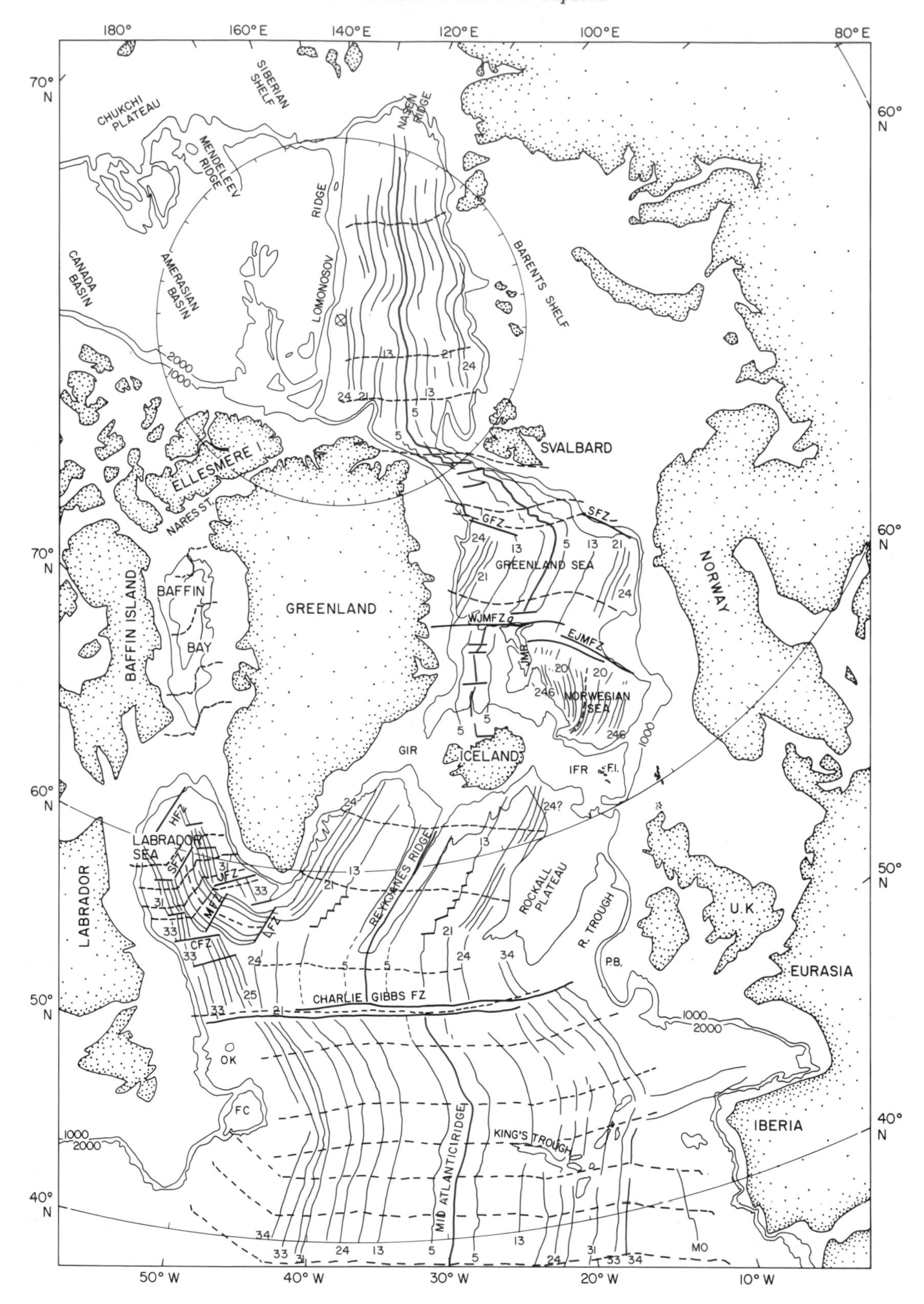

Figure 1. Magnetic lineations and fracture zones in the North Atlantic Ocean, Labrador Sea, Norwegian-Greenland Sea and Eurasian Basin. Also shown are 1000 m and 2000 m isobaths and flow lines (dotted lines). Transverse mercator projection.

It has generally been assumed that the plates surrounding the North Atlantic have behaved rigidly during most of the period of its evolution, though geological observations on some of these plates or along their boundaries do suggest that a portion of them have undergone deformation at some time. Typical examples of such interactions are the formation of the Pyrenees between Iberia and Eurasia during the Eocene and the folding and faulting of some formations in the Canadian Arctic Islands during the Eurekan Orogeny in Eocene-Oligocene times. However, it is not yet established if these deformations have taken place within the plates themselves or at their boundaries. To obtain a satisfactory solution of the entire North Atlantic such interactions between plates and the resulting deformations must be taken into account.

The Approach Followed Here

The idea that the region under study evolved due to the separation of Greenland and Eurasia from North America, since Cretaceous time, by a sea-floor–spreading mechanism is now well accepted. It has further been established that there was simultaneous spreading in the Eurasia Basin, the Norwegian-Greenland Sea, the Labrador Sea, and in the region south of Greenland during anomalies 24 to 20 time (Kristoffersen and Talwani, 1977; Talwani and Eldholm, 1977; Srivastava, 1978; and Vogt and others, 1979). Thus there was spreading all around Greenland during this period and this provides us with the biggest constraint in deriving a satisfactory solution for this period. Prior to this period there was spreading only between Greenland and North America, mainly in the Labrador Sea (Srivastava, 1978), though some spreading may have taken place in the Amerasia Basin, too (Vogt and others, 1982). Since anomaly 13 time, spreading has been confined between the North American and Eurasian plate, Greenland moving with the North American plate. The part which is not definite is the relation of the Lomonosov Ridge to the North American and Greenland plates during these times. As mentioned earlier several solutions have been given, but the one which is followed here treats the Lomonosov Ridge as part of the North American plate for reasons given later in describing the poles of rotations. Further, we have regarded the region of deformation within the Canadian Arctic Islands, including Nares Strait, as a diffuse boundary between the North American and Greenland plates, and the north Spanish Trough and Pyrenees as the boundary between Eurasian and Iberian plates. We have, thus, treated Iberia as a separate plate and not as part of Eurasia, until Mid-Oligocene time. This was also required in order to obtain a reasonable match between the rotated and unrotated magnetic anomalies from one side of the ridge axis to the other, as well as to explain the formation of some of the prominent bathymetric features observed in the North Atlantic south of the Charlie-Gibbs Fracture Zone. Such a treatment of Iberia was further strengthened when we considered it to be a part of Africa during this time (Schouten and others, 1986).

To determine the initial position of continents located on either side of the region formed by sea floor spreading, it is necessary to delineate their lines of initial opening. However, because these boundaries are complex (Keen and Hyndman, 1979) and cannot be delineated precisely everywhere, different reconstructions have used different criteria in deriving the initial position of continents. For example, Bullard and others (1965) used 500 fm isobath as the boundary in their reconstruction, while LePichon and others (1977) recognized that one particular isobath did not necessarily mark the true boundary everywhere and used different isobaths in different regions in their reconstructions. An important assumption which is intrinsic in all these reconstructions is that plates separate instantaneously and hence one should be able to match their boundaries with one pole of rotation. Vink (1982) and Courtillot (1982), on the other hand, have demonstrated that plates do not rift instantaneously, but rather they open by propagating rifts. Thus, even if these boundaries could be located precisely everywhere, their superposition cannot be obtained with one pole of rotation. We concur with Vink's and Courtillot's interpretation and have done our reconstructions to the time of initial openings by fitting such boundaries at the southernmost region of the plates wherever possible, and we have merely shown 2000- or 3000-m isobaths along the plates to show their overlaps.

MAGNETIC ANOMALIES AND FRACTURE ZONES

Sea-Floor–Spreading Magnetic Anomalies

Sea-floor–spreading (SFS) magnetic anomalies and fracture zones are the key parameters needed in determining the past plate motions. The accuracy with which the past plate configuration can be determined depends heavily on the accuracy of these two parameters. Magnetic anomalies in the North Atlantic have been identified and correlated between tracks by several workers. A detailed description of these is given by Vogt (this volume, Ch. 15; Plate 3) however, a brief description of the regions where we have re-identified anomalies has been included here. These are the Northeast Atlantic, Northwest Atlantic, and the Labrador Sea.

Northeast Atlantic. Considerable shipborne and airborne magnetic data now exist for the Northeast Atlantic. Figure 2 shows the contoured map of the Northeast Atlantic between the Charlie-Gibbs Fracture Zone and the Azores-Gibraltar Fracture Zone by Roberts and Jones (1985). Shown in the figure are the positive bands of the anomalies and a few of the E-W tracks along which magnetic anomalies have been plotted to show their characteristic shapes. The correlations of the anomalies which follow the positive bands are shown by dotted lines. Also shown is a synthetic profile at 40°N. This profile was generated using Kent and Gradstein's (this volume) time scale which has been followed throughout this chapter.

We have used Cande and Kristoffersen (1977) rates of spreading primarily because of their re-identifications of anomalies 27 to 34 in this region. The rates of spreading for younger anomalies are similar to those given by Pitman and Talwani (1972) and Williams and McKenzie (1971).

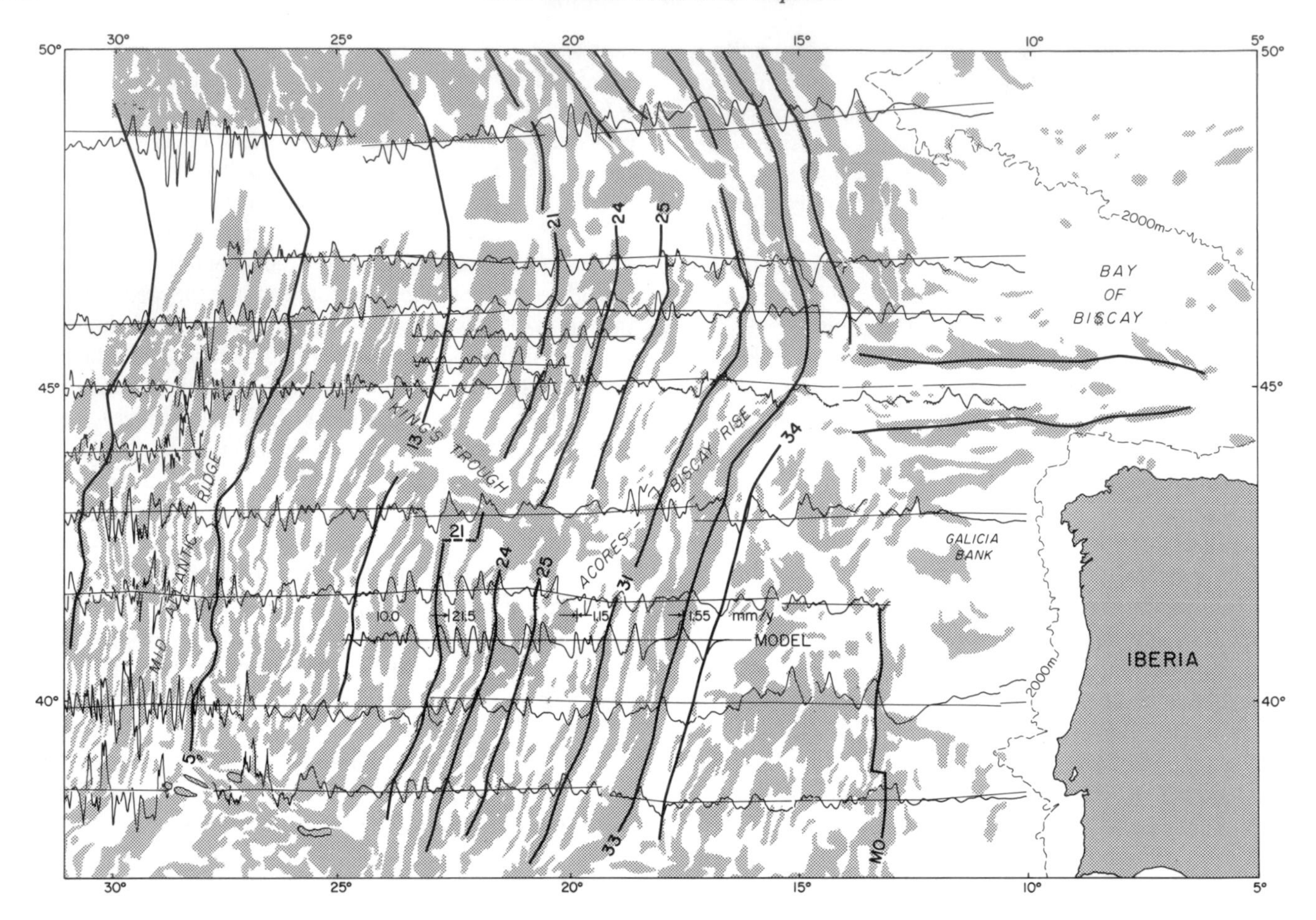

Figure 2. Magnetic lineations in the NE Atlantic. Also shown are magnetic anomalies plotted along selected tracks and a synthetic profile. The positive bands of anomalies from Roberts and Jones' (1985) map are shown in red. Mercator projection.

Because the correlation of anomalies shown in Figure 2 is based on the contoured map of Roberts and Jones (1985), the lineations are no longer continuous from north to south, as was implied by profile data (Pitman and Talwani, 1972, Williams and McKenzie, 1971), but show several small offsets in addition to the large offset at the Charlie-Gibbs FZ. Small offsets in anomalies 21 and 24 can be seen near King's Trough (43°N, 23°W) and also at 45°N and 22°W. The latter was observed by White and Matthews (1980) from their detailed survey. Johnson and Vogt (1973) used detailed bathymetry to suggest several offsets in anomalies 21 an 24 immediately south of the Charlie-Gibbs FZ. These are not prominent on the contoured map, Figure 2, and are perhaps only minor.

In our identification of anomalies 30-34 in Figure 2 we have used Cande and Kristoffersen (1977) and Kristoffersen (1978) identifications of these anomalies. We have done this because of the presence of two major positive bands of anomalies off Iberia between old anomaly 32 (called here anomaly 34) and anomaly 26, whose shapes seem to match with anomalies 31 and 32 recognised by Cande and Kristoffersen (1977) to the south. Presence of anomaly 34 in the south demands its continuation to the north. Moreover, Miles and Parson (1982) and more recently Masson and Miles (1984) have also shown presence of these anomalies off Iberia and off the Grand Banks of Newfoundland. With this identification, old anomalies 28, 31 and 32 become anomalies 31, 33 and 34, respectively. This implies a much slower rate of spreading (<2 cm/yr) since anomaly 27 time than had been used in the past.

Northwest Atlantic. This comprises the area on the western flank of the Mid-Atlantic Ridge between the Newfoundland and Charlie-Gibbs Fracture Zones. The region is not as extensively covered by shipborne measurements as the area to the east of the Mid-Atlantic Ridge (Fig. 2), but enough data exist to allow correlation of anomalies between tracks (Pitman and Talwani, 1972; Cande and Kristoffersen, 1977; Kristoffersen, 1978; and Srivastava, 1978). Figure 3 shows correlation and identification of anomalies based on the published (Srivastava, 1978) and unpublished (Miles and Parson, 1982) data from this region. Variability in the signatures of the anomalies from track to track does not prevent correlation of anomalies older than 13 (Fig. 3), but the variability of younger anomalies (between 5 and 13) is so pronounced that it becomes difficult to correlate them between widely spaced tracks. Identical anomalies have been joined by smooth lines to show their trends, though it is very likely that they are offset by a number of fracture zones similar to the ones observed in the Northeast Atlantic (Fig. 2). The paucity of data does not allow the definition of such fracture zones in Figure 3.

Calculated anomalies based on the two different rates of spreading are shown in Figure 3. Model A uses rates of spreading determined by Pitman and Talwani (1972) and used by Srivastava (1978), with slight modification in rate for anomalies older than 26. Model B, which is used here, is based on the rates of spreading proposed by Kristoffersen (1978) and Cande and Kristoffersen (1977) where they identified anomaly 31 as 33. Anomalies 24 to 34 can be correlated easily south of 45 N while to the north their correlation becomes weaker. Similar correlation of these anomalies has also been published recently by Masson and Miles (1984).

South of Flemish Cap (Fig. 3), a large number of poorly lineated anomalies can be seen west of anomaly 34. These anomalies seem different from those to the north. They are larger in amplitude and at places show similarity to the "J" anomaly observed farther to the south (Tucholke and Ludwig, 1982; Sullivan, 1983). Location of this anomaly is shown in Figure 3.

Labrador Sea. A large quantity of shipborne and airborne data collected in the Labrador Sea has been described in detail by Srivastava (1978). Figure 4 shows a simplified map of the Labrador Sea with magnetic anomalies plotted along ships' tracks. Identification of anomalies in this region was done by Srivastava by correlating the observed anomalies with synthetic anomalies. Two models with different rates of spreading were considered, one using the rate of spreading comparable to those used for the North Atlantic south of the Charlie-Gibbs FZ for anomalies 25 to 31 by Pitman and Talwani (1972), and the other using a lower rate of spreading for anomalies 27 to 32 similar to those of Cande and Kristoffersen (1977) for the Northwest Atlantic. For the reasons mentioned in the previous section, enough evidence now exists to substantiate Cande and Kristoffersen's re-identification of anomaly 31 as 33 in the Northwest Atlantic. It follows that a similar re-identification can be carried out for the Labrador Sea. Admittedly, anomalies older than 27 are not well developed in the Labrador Sea and can be interpreted as anomalies 31 and 33 equally well. On the other hand, if anomaly 33 was formed immediately north of the Charlie-Gibbs FZ as shown previously, then it would be expected in the Labrador Sea as well. Complete absence of anomalies 31 or 33 in the Labrador Sea would imply a jump or jumps of the ridge axis between the Labrador Sea and the Rockall-Hutton Bank. We find no evidence that this is the case. Thus, having recognised anomalies 33 and 34 in the Northwest Atlantic south of the Charlie-Gibbs FZ we have re-identified anomalies 28, 31 and 32 as anomalies 31, 33 and 34, respectively, in the Labrador Sea. Figure 4 shows the magnetic lineations in the Labrador Sea according to this new numbering scheme.

Fracture Zone

A number of prominent fracture zones exists in the region between the Azores-Gibraltar Ridge and the Arctic Basin. A majority of these fracture zones has been delineated from detailed bathymetric, seismic, and other geophysical measurements in different regions of the North Atlantic. A detailed description of these fracture zones is beyond the scope of the present chapter; those interested in finding more about them should refer to the publications describing the evolution of the region concerned. The following fracture zones were used in determining the poles of rotation for different regions; for the North Atlantic: Azores-Gibraltar and Charlie-Gibbs Fracture Zones; for the Labrador Sea: Leif, Minna, Snorri, Hudson, Cartwright, and Julianehaab Fracture Zones; for the Norwegian-Greenland Sea: Jan Mayen, Greenland, Senja, and Spitsbergen Fracture Zones.

POLES OF ROTATION

Poles of rotation between plates, including their angle of rotation, merely describe the rotation necessary to transform a given magnetic anomaly or the plate boundary from one side of the ridge axis into coincidence with the corresponding anomaly or boundary on the other side of the ridge. Thus, past plate positions can be adequately described by a series of poles of rotation. The differences between these poles, known as finite-difference poles, describe the direction and rate of motion between plates over a given period of time during which they have remained constant. However, the directions and rates do change from time to time and the period over which they remain constant needs to be carefully established before their finite-difference poles of rotation can be determined. In the North Atlantic, with its complex system of spreading, the problem becomes even more difficult because of the simultaneous spreading in many of its parts, as well as growth and decay of micro-plates at various times, all of which are not fully known. In determining the poles of rotation for each branch of the spreading center, we have tried to keep the number of plates to a minimum while meeting the necessary boundary conditions.

The SFS data, as summarised in the preceeding sections and shown in Figure 1, show that at anomalies 5, 6, 13, 21, 24, 25, 30-31, 34, M0, M4, and possibly M11 times there were major changes either in the directions or rates of spreading, or both, in the North Atlantic. Thus we have determined the poles of rotation for these anomalies as described below.

Finite poles of rotation have been published for the Labrador Sea (LePichon and others, 1971; Kristoffersen and Talwani, 1977; Srivastava, 1978; and Vink, 1982), the Norwegian-Greenland Sea (Talwani and Eldholm, 1977; LePichon and others, 1977; Nunns, 1983; Unternehr, 1982; Olivet and others, 1981; Phillips and Tapscott, 1981; and Vink, 1982, 1984), the Eurasia Basin (LePichon and others, 1977; Phillips and Tapscott, 1981; Vink, 1982, 1984; Srivastava, 1985) and for the region south of Greenland (Pitman and Talwani, 1972; LePichon and others, 1977; Kristoffersen, 1978; Srivastava, 1978; Phillips and Tapscott, 1981; Olivet and others, 1981). Most of these poles fit the SFS data well in individual regions, but when used collectively they do not fit equally well the entire North Atlantic SFS

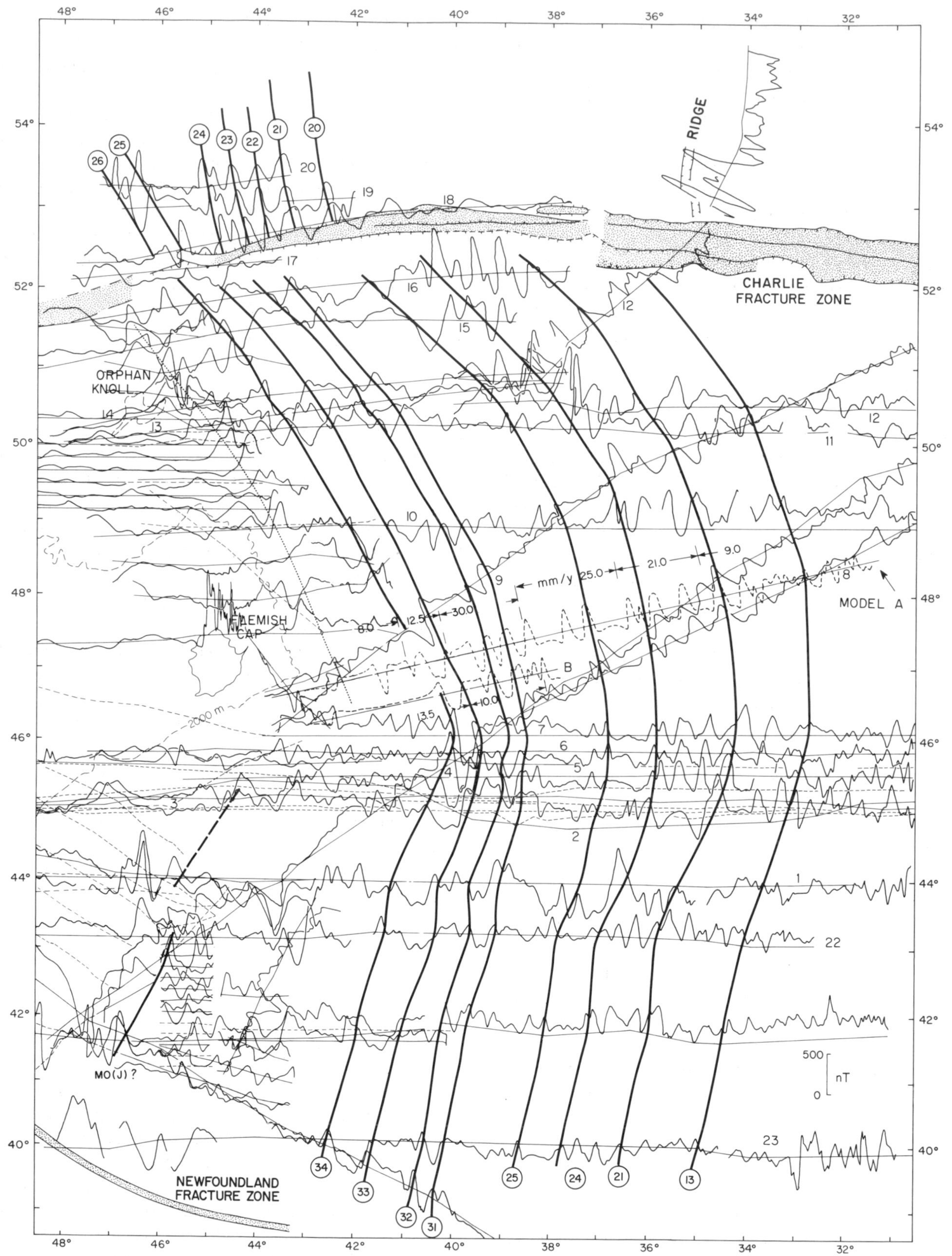

Figure 3. Magnetic lineations in the NW Atlantic. Also shown are magnetic anomalies plotted along selected tracks. The synthetic profiles based on different rates of sea-floor spreading (Models A and B) are shown as dashed profiles. Mercator projection.

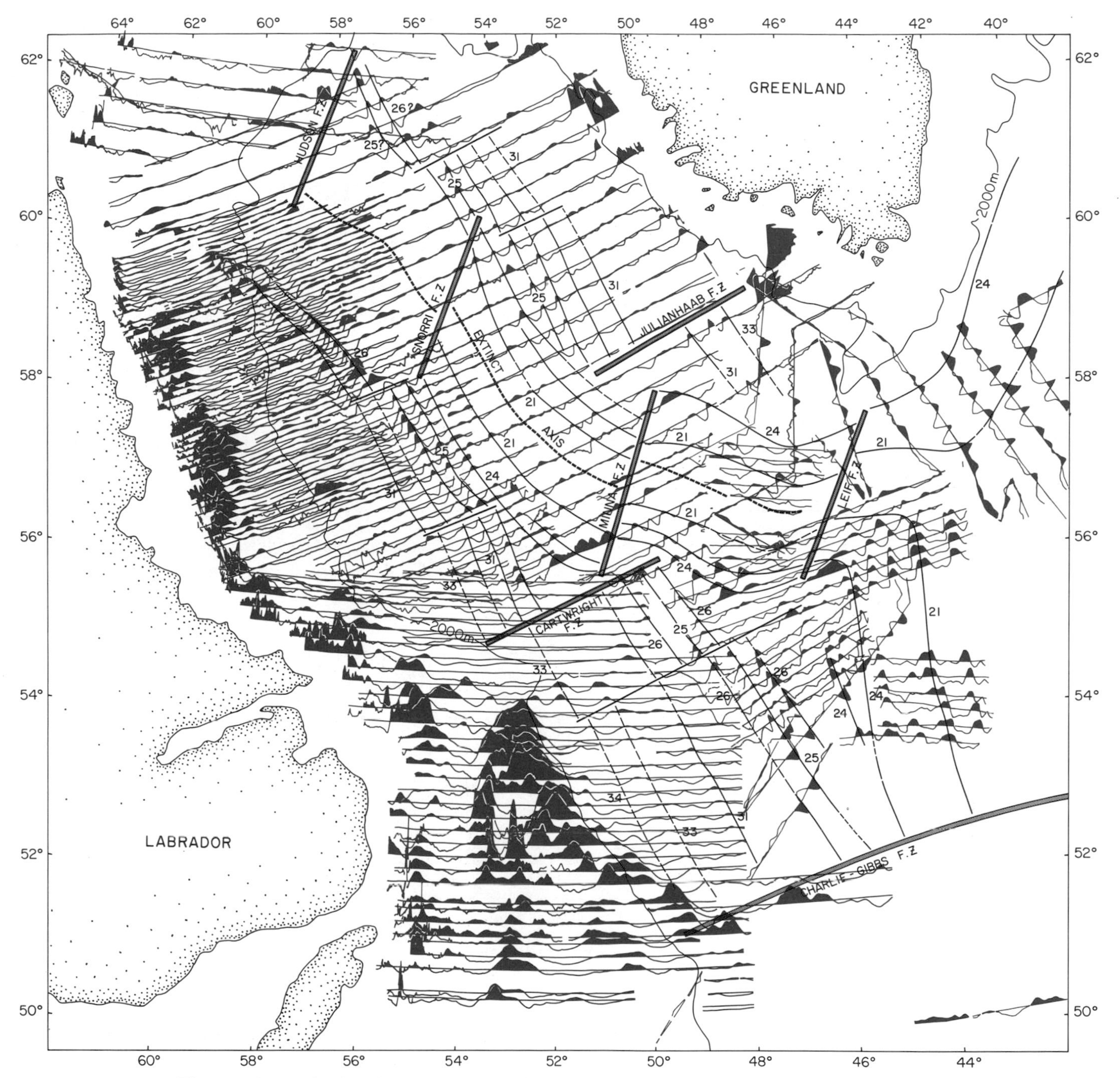

Figure 4. Magnetic lineations in the Labrador Sea as obtained from correlations of the magnetic anomalies. Also shown are the fracture zones and the extinct ridge axis. Lambert conformal projection.

data from the Eurasia Basin to the Azores-Gibraltar Fracture Zone. This is because of the differences in the data sets as well as in the plate configurations used in deriving various pole positions. The presence of an extinct axis and of a continental(?) fragment in the Norwegian-Greenland Sea (Fig. 1) clearly shows the complexity of its spreading history in comparison to the two-plate situation which presently exists there. Thus, in determining the poles of rotation for various regions in the North Atlantic, we went backward from the present simple two-plate configuration to a more complicated multiplate system, as spreading shifted from place to place. New poles of rotation were calculated and old ones modified wherever possible, using our SFS data for the entire North Atlantic, and these are given in Tables 1 and 2. In spite of the vast amount of SFS data which is used in our determinations of these pole positions, certain assumptions had to be made in order to

TABLE 1A. POLES OF TOTAL OPENING WITH RESPECT TO NORTH AMERICA

Anomaly	Age (m.y.)	Greenland			Eurasia			Porcupine			Iberia		
		(a)	(b)	(c)	(a)	(b)	(c)	(a)	(b)	(c)	(a)	(b)	(c)
5	10		-		68.00	137.00	-2.50	68.00	137.00	-2.50	68.00	137.00	-2.50(1)
6	20		-		68.00	138.20	-4.75	68.00	138.20	-4.75	68.00	138.20	-4.75(2)
13	36		-		68.00	129.90	-7.78	68.00	129.90	-7.78(3)	48.06	143.50	-7.20*
21	50	62.80	-84.00	-3.00(4)	67.12	137.28	-10.94	62.55	142.01	-10.25	73.38	129.05	-11.04*
24	56	58.10	-104.60	-4.09	62.28	140.37	-12.68	55.51	145.54	-11.76	72.94	133.66	-13.15*
25	59	61.67	-122.79	-3.93	63.25	143.89	-14.15	57.15	148.19	-13.22	72.72	135.50	-14.29*
30	67.6	68.89	-115.10	-7.21(5)	69.82	145.61	-17.10(5)	65.12	149.84	-16.01(5)	77.79	134.80	-17.66*
31	69	69.46	-114.25	-7.74	70.66	145.91	-17.59	66.14	150.12	-16.48	78.62	133.88	-18.01*
33	80	71.17	-111.35	-10.02	74.52	147.69	-20.30	70.78	151.77	-19.11	85.49	110.28	-22.41*
34	84	72.66	-109.32	-11.64	76.23	148.80	-21.83	72.82	152.76	-20.60	87.18	57.43	-24.67*
CLBS	95(?)	73.97	-107.20	-13.62	78.00	150.27	-23.70	74.92	154.07	-22.44	83.40	6.90	-28.94*
CRT	105		-		79.50	151.92	-25.59	76.70	155.50	-24.30	74.92	-10.42	-39.80*
M-O	118		-			-			-		71.17	-12.60	-48.00*
M-4	126		-			-			-		70.90	-11.98	-50.07*
M-11	133		-			-			-		70.33	-11.08	-51.09*

(1) Pitman and Talwani (1972)
(2) Olivet and others (1981) modified
(3) Talwani and Eldholm (1977)
(4) Vink (1982) modified
(5) Interpolated

* Calculated from Africa/North America poles of Klitgord and Schouten (1985)
(a) Lat. - N. positive, (b) Long.- E. positive, (c) Rotation to east positive.

CLBS-Close Labrador Sea to Initial Opening
CRT-Close Rockall Trough to Initial Opening

TABLE 1B. POLES OF TOTAL OPENING FOR NORWEGIAN-GREENLAND SEA BETWEEN EURASIA AND GREENLAND RELATIVE TO GREENLAND

Anomaly	Age (m.y.)	Eurasia			Reference
		(a)	(b)	(c)	
5	10	68.00	137.00	-2.50	Pitman and Talwani (1972)
6	20	68.00	138.20	-4.75	Olivet et al. (1981) modified
13	36	68.00	129.90	-7.78	Talwani and Eldholm (1977)
21	50	53.78	129.71	-9.14	Nunns (1983) modified
24	56	46.78	126.85	-10.50	Srivastava (1985)
25	59	52.87	130.03	-11.38	Calculated from Table 1A
30	67.6	52.79	127.82	-11.55	Calculated from Table 1A
31	69	52.77	127.49	-11.57	Calculated from Table 1A
33	80	54.16	125.85	-12.04	Calculated from Table 1A
34	84	54.79	126.15	-12.17	Calculated from Table 1A
CLBS	95	54.77	126.13	-12.16	Calculated from Table 1A
CRT	105	60.44	126.35	-13.71	Calculated from Table 1A

(a) Lat. - N. positive
(b) Long. - E. positive
(c) rotation positive to east

CLBS-Close Labrador Sea to Initial Opening
CRT-Close Rockall Trough to Initial Opening

obtain a satisfactory match between anomalies. These are described below.

Anomalies 5 to 13

Published poles of rotation for anomalies younger than 7 were found to be satisfactory to transpose anomalies from one side of the ridge axis to the other, indicating that the present two-plate situation has persisted at least since anomaly 7 time. However, a different situation existed for older anomalies, as evidenced by the fanning of these anomalies in the Norway Basin and their interruption by King's Trough in the North Atlantic (Fig. 1). In determining the poles of rotation for anomalies 13 and older, we assumed that (a) the fanning of magnetic anomalies in the Norway Basin can be explained adequately by the rotation and separation of Jan Mayen block from Greenland between anomalies 7 and 20 times (Nunns, 1983), and (b) the interruption of anomalies across King's Trough is caused by the presence of a plate boundary through this region during this time (Grimaud and others, 1982).

It has long been regarded that, since the termination of spreading in the Labrador Sea (prior to anomaly 13 time, Kristoffersen and Talwani, 1977), the spreading in the rest of the North Atlantic has mainly been a two-plate motion, North American and Eurasian plates. Two sources of information suggest that this may not have been the case; one, the published pole for anomaly 13 (Talwani and Eldholm, 1977) fits this anomaly well everywhere except in the region of King's Trough, and two, anomaly 6, not 13, is the first recognisable, uninterrupted anom-

TABLE 2A. (NORTH ATLANTIC) FINITE DIFFERENCE POLES BETWEEN THE NORTH AMERICAN, EURASIAN AND PORCUPINE PLATES

Anomalies	Time Span (m.y.)	Eurasia/ North America*			North America/ Eurasia			North America/ Porcupine		
		(a)	(b)	(c)	(a)	(b)	(c)	(a)	(b)	(c)
0-5	10	68.00	137.00	-2.50	68.00	137.00	2.50	68.00	137.00	2.50
5-6	10	67.97	139.53	-2.25	68.01	139.53	2.25	68.01	139.53	2.25
6-13	16	67.50	117.15	-3.05	66.87	117.60	3.05	66.87	117.60	3.05
13-21	14	63.20	151.70	-3.20	64.32	153.46	3.20	44.45	162.45	2.73
21-24	6	34.22	144.66	-2.02	35.03	151.87	2.02	17.53	157.88	2.05
24-25	3	65.00	180.00	-1.54	68.78	-179.19	1.54	68.64	176.73	1.54
25-30	8.6	80.00	-80.00	-3.45	82.10	-28.22	3.45	82.63	-23.63	3.45
30-31	1.4	80.00	-80.00	-0.55	82.10	-28.22	0.55	82.63	-23.63	0.55
31-33	11	80.00	-80.00	-3.00	82.10	-28.22	3.00	82.63	-23.63	3.00
33-34	4	80.00	-80.00	-1.65	82.10	-28.22	1.65	82.63	-23.63	1.65
34-CLBS	11(?)	80.00	-80.00	-2.00	82.10	-28.22	2.00	82.63	-23.63	2.00
CLBS-CRT	10(?)	80.00	-80.00	-2.00	82.10	-28.22	2.00	82.63	-23.63	2.00

* Poles between Eurasia and North America relative to North America.

(a) Lat. - N. positive
(b) Long. - E. positive
(c) rotation to east positive

CLBS - Close Labrador Sea to Initial Opening
CRT - Close Rockall Trough to Initial Opening

TABLE 2B. (NORTH ATLANTIC) FINITE DIFFERENCE POLES BETWEEN THE NORTH AMERICAN AND IBERIAN PLATES

Anomalies	Time Span (m.y.)	Iberia/ North America*			North America/ Iberia		
		(a)	(b)	(c)	(a)	(b)	(c)
0-5	10	68.00	137.00	-2.50	68.00	137.00	2.50
5-6	10	67.97	139.53	-2.25	68.01	139.53	2.25
6-13	16	17.27	144.67	-3.19	17.53	148.52	3.19
13-21	14	70.66	-21.24	-5.56	68.70	-3.97	5.56
21-24	6	68.62	151.71	-2.13	69.94	154.72	2.13
24-25	3	68.62	151.71	-1.15	69.94	154.72	1.15
25-30	8.6	81.82	-61.09	-3.65	81.45	-18.41	3.65
30-31	1.4	64.75	-40.23	-0.44	63.64	-15.61	0.44
31-33	11	66.83	-38.90	-5.11	65.63	-13.66	5.11
33-34	4	59.19	-38.78	-2.69	58.00	-14.35	2.69
34-CLBS	11(?)	57.37	-25.30	-4.92	56.23	-1.27	4.92
CLB-CRT	10	54.57	-30.60	-12.06	53.17	-5.93	12.06
CRT-MO	13	55.30	-29.00	-8.68	54.14	-4.81	8.68
MO-M4	8	63.47	-8.65	-2.10	66.39	4.75	2.10
M4-M11	7	42.30	-10.00	-1.16	47.35	22.45	1.16

* Poles between Iberia and North America are relative to North America.

(a) Lat. - N. positive
(b) Long. - E. positive
(c) rotation to east positive

CLBS - Close Labrador Sea to Initial Opening
CRT - Close Rockall Trough to Initial Opening

aly which lies west of the King's Trough (Fig. 1). If the King's Trough feature was formed due to interactions between the Eurasian and Iberian plates (Grimaud and others, 1982), then it is very likely that at anomaly 13 time Iberia was still acting as a separate plate and not as part of the Eurasian plate. In determining the pole of rotation for anomaly 13, we regarded Iberia as part of the African plate (Schouten and others, 1986) and not of the Eurasian plate, for reasons discussed later.

Anomalies 21 to 24

Because the pole for anomaly 13 can satisfactorily match this anomaly everywhere from Eurasia Basin to the King's Trough region, it clearly shows that the Lomonosov Ridge was still moving as a part of the North American plate (Fig. 9). However, different assumptions have been made for its association with the North American plate for earlier times. The pres-

TABLE 2C. (LABRADOR SEA) FINITE DIFFERENCE POLES BETWEEN THE NORTH AMERICAN AND GREENLAND PLATES

Anomalies	Time Span (m.y.)	Greenland/ North America*			North America/ Greenland		
		(a)	(b)	(c)	(a)	(b)	(c)
13-21	14	62.80	-84.00	-3.00	62.80	-84.00	3.00
21-24	6	39.28	-134.37	-1.28	38.25	-132.42	1.28
24-25	3	00.00	-50.00	+0.70	2.00	-46.62	-0.70
25-30	8.6	76.00	-95.00	-3.36	76.81	-98.50	3.36
30-31	1.4	76.00	-95.00	-0.54	76.81	-98.50	0.54
31-33	11	76.00	-95.00	-2.30	76.81	-98.50	2.30
33-34	4	80.00	-80.00	-1.65	81.47	-88.30	1.65
34-CLBS	11(?)	80.00	-80.00	-2.00	81.47	-88.30	2.00

(a) Lat. - N. positive
(b) Long. - E. positive
(c) rotation to east positive

CLBS - Close Labrador Sea to Initial Opening

* Poles between Greenland and North America are relative to North America.

TABLE 2D. (NORWEGIAN-GREENLAND SEA) FINITE DIFFERENCE POLES BETWEEN THE GREENLAND AND EURASIAN PLATES.

Anomalies	Time Span (m.y.)	Eurasia/ Greenland*			Greenland/ Eurasia		
		(a)	(b)	(c)	(a)	(b)	(c)
0-5	10	68.00	137.00	-2.50	68.00	137.00	2.50
5-6	10	67.97	139.53	-2.25	68.01	139.53	2.25
6-13	16	67.50	117.15	-3.05	66.87	117.60	3.05
13-21	14	3.79	125.98	-2.49	3.76	133.00	2.49
21-24	6	9.27	115.04	-1.83	8.20	121.60	1.83
24-25	3	67.57	-111.58	-1.51	72.94	-90.44	1.51
25-30	8.6	26.63	58.35	-0.32	20.28	66.30	0.32
30-31	1.4	20.72	54.28	-0.04	14.19	62.71	0.04
31-33	11	72.74	48.19	-0.58	65.92	53.18	0.58
33-34	4	72.11	-114.31	-0.19	77.27	-86.68	0.19
34-CLBS	11	69.84	174.78	-0.01	74.92	169.31	0.01
CLBS-CRT	10	81.60	-89.04	-2.00	82.10	-28.22	2.00

(a) Lat. - N. positive
(b) Long. - E. positive
(c) rotation to east positive

CLBS - Close Labrador Sea to Initial Opening
CRT - Close Rockall Trough to Initial Opening

* Poles between Greenland and North America are relative to North America.

ence of anomalies 21 to 24 all around Greenland (Fig. 1) clearly shows that there was simultaneous spreading in these regions during this time. Pole positions for these anomalies have been given by many workers but none of them satisfactorily match these anomalies for the entire North Atlantic. Pole positions given by Srivastava (1978) and Kristoffersen and Talwani (1977) based on overall fits of these anomalies south of the Eurasia Basin, give large overlaps between these anomalies in the Eurasia Basin. These overlaps would then indicate that either the Lomonosov Ridge was not moving as part of the North American plate during these times (Phillips and Tapscott, 1981), and hence there may have been some spreading west of the Lomonosov Ridge, or the published poles of rotations for these regions are in error. Aeromagnetic data in the Arctic Basin do not support the presence of these anomalies anywhere except in the Eurasia Basin (Vogt and others, 1979); hence the Lomonosov Ridge did not move as a separate plate.

Recently Vink (1982, 1984) has shown that it is possible to get an "overall" satisfactory fit for anomalies 21 and 23 in all regions north of the Charlie-Gibbs Fracture Zone by treating the Lomonosov Ridge as part of the North American plate. For this he calculated a new set of poles for the Labrador Sea. Because anomalies 21-23 are not well developed and are difficult to recognize, especially in the northern Labrador Sea, some freedom exists in determining the poles of rotation for this region. Though Vink's poles fit these anomalies fairly well in the regions north of the Charlie-Gibbs FZ, they give rise to large overlaps between these anomalies to the south. Some overlaps are also observed between Svalbard and northern Greenland. The latter is explained (Vink, 1982) by the propagation of the rift in this region prior to active sea-floor spreading. The question to be answered is which overlap is more acceptable, the one obtained in the north in the Eurasia Basin (using the poles as given by Srivastava, 1978, and Kristoffersen and Talwani, 1977), or the one in the south in the North Atlantic (using Vink's poles). Enough geological and geophysical evidence now exists to support the idea that the Lomonosov Ridge may have remained attached to the North American plate since the start of sea-floor spreading in this region (Srivastava, 1985). Thus we favour Vink's interpretation and have interpreted the overlap for anomalies 21 and 23 in the south as resulting from the formation of a new plate, called here the Porcupine Plate, between the Charlie-Gibbs FZ and the Iberian plate, as explained later.

In our treatment we have, thus, regarded the Lomonosov Ridge as part of the North American plate and have determined the poles of rotation for anomalies 21 to 24 for different regions of the North Atlantic. For anomaly 21 we modified slightly Vink's (1982) pole for the Labrador Sea and Nunns' (1983) pole for the Norwegian Sea (Table 1) to get a better fit in the Labrador Sea and less overlap in the North Atlantic (Fig. 1). Pole positions for anomaly 24 were determined for the Labrador Sea and Norwegian-Greenland Sea, which gave satisfactory fits in these regions as well as in the Eurasian Basin. However, this also resulted in a large overlap south of the Charlie-Gibbs FZ (Fig. 1) which we interpret as due to the presence of another small plate in this region, as described below.

IBERIA as a part of AFRICAN Plate

The gradual increase in the overlaps for latitudes south of the Charlie-Gibbs FZ, as obtained for anomalies 21 (Fig. 5) and 24, could arise from (a) gross error in identification of these anomalies in this region, (b) wrong assumption that the Lomonosov Ridge was moving with the North American plate during these times and (c) the presence of microplates in this region. It is most unlikely that (a) is the cause because of the high density of data which have been used in the identification of these anomalies in this region (Fig. 2). Possibility (b) can not be ruled out, but consideration that the poles of rotation determined for the three plates (Greenland, North American, and Eurasian) for anomalies 21 and 24 not only give best overall fits for these anomalies near the triple junctions north and south of Greenland, but also follow the direction of the fracture zones in each region (Fig. 1), argues strongly that the Lomonosov Ridge must have remained a part of North America during these times. Thus, the only other alternative left is to assume (c), the presence of a microplate southeast of British Isles.

It has long been speculated that Iberia may have acted as a separate plate during the Cenozoic, thereby giving rise to the Pyrenees and the north Spanish Trough due to its interaction with the Eurasian plate (see LePichon and Sibuet, 1971). This implies that these features lie along the boundary between these two plates. The difficulty in such an assumption is in deciphering the offshore extension of this boundary. Two mutually exclusive interpretations for the western extention of this boundary have been proposed. LePichon and Sibuet (1971) proposed that the Spanish Trough, Azores-Biscay Rise, and the King's Trough (Fig. 1) form such a boundary. A second but similar westward continuation of this boundary has been proposed by Grimaud and others (1982). The two differ in their implied motions across the boundaries which are not in complete agreement with the detailed geophysical measurements carried out in this region (Searle and Whitmarsh, 1978; Whitmarsh and others, 1982; and Kidd and others, 1982). Even if such a boundary existed it could not account for the resulting overlap between the anomalies south of the Charlie-Gibbs FZ (Fig. 5) and north of this boundary.

We are faced with two possibilities: one, that this boundary lies further north, perhaps associated with one of the mid-Atlantic Ridge fracture-zone fault systems; two, that the King's Trough and Azores–Biscay Rise complex are associated with the Iberia-Eurasia plate boundary and the overlaps observed north of this boundary arise due to the presence of another microplate to the north. Such a plate would then be confined between this new boundary in the south and the Charlie-Gibbs FZ in the north (Fig. 5). Though the first possibility can not altogether be ruled out, the lack of any major fracture zone or fault system makes it difficult to locate the Iberia-Eurasia plate boundary in this region. We therefore propose the existence of a microplate in this region as a somewhat radical working hypothesis and name it Porcupine plate, as it lies immediately east of the Porcupine bank (Fig. 5). We further propose that the north Biscay margin, including the ocean-continent boundary farther to the north, forms a boundary between the Porcupine and the Eurasian plates. Alternately, one could regard it as a part of the Iberian plate. However, the idea that Iberia was moving as a part of the African plate (Schouten and others, 1986) right from the beginning to anomaly 10 time is more appealing because not only can it account for the observed overlap in the anomalies 21 and 24 in this region but equally well the formation of King's Trough and Azores-Biscay Rise.

Because anomaly 13 can be fitted well for the entire North Atlantic from Eurasia Basin to the King's Trough region, using a single-pole position, we interpret that the Porcupine plate only existed until anomaly 13 time. The geological and geophysical observations along the north Biscay margin and north Spanish Trough (Boillot and Capdevila, 1977; Montadert and others, 1979) show compression, strike-slip motion, and subduction, respectively. Most of these motions seem to have taken place during Eocene, which roughly corresponds to anomalies 20-25 times (Kent and Gradstein, this volume). It thus seems likely that the Porcupine plate came into existence at this time mainly due to the interactions between two major plates, the Iberian and Eurasian. A portion of the old Eurasian plate in the vicinity of the north Biscay margin and north Spanish Trough was deformed at this time, giving rise to a slightly different orientation to the crust (anomaly) which had already formed. Therefore, when these anomalies are rotated as part of the Eurasian plate, they result in the observed overlaps.

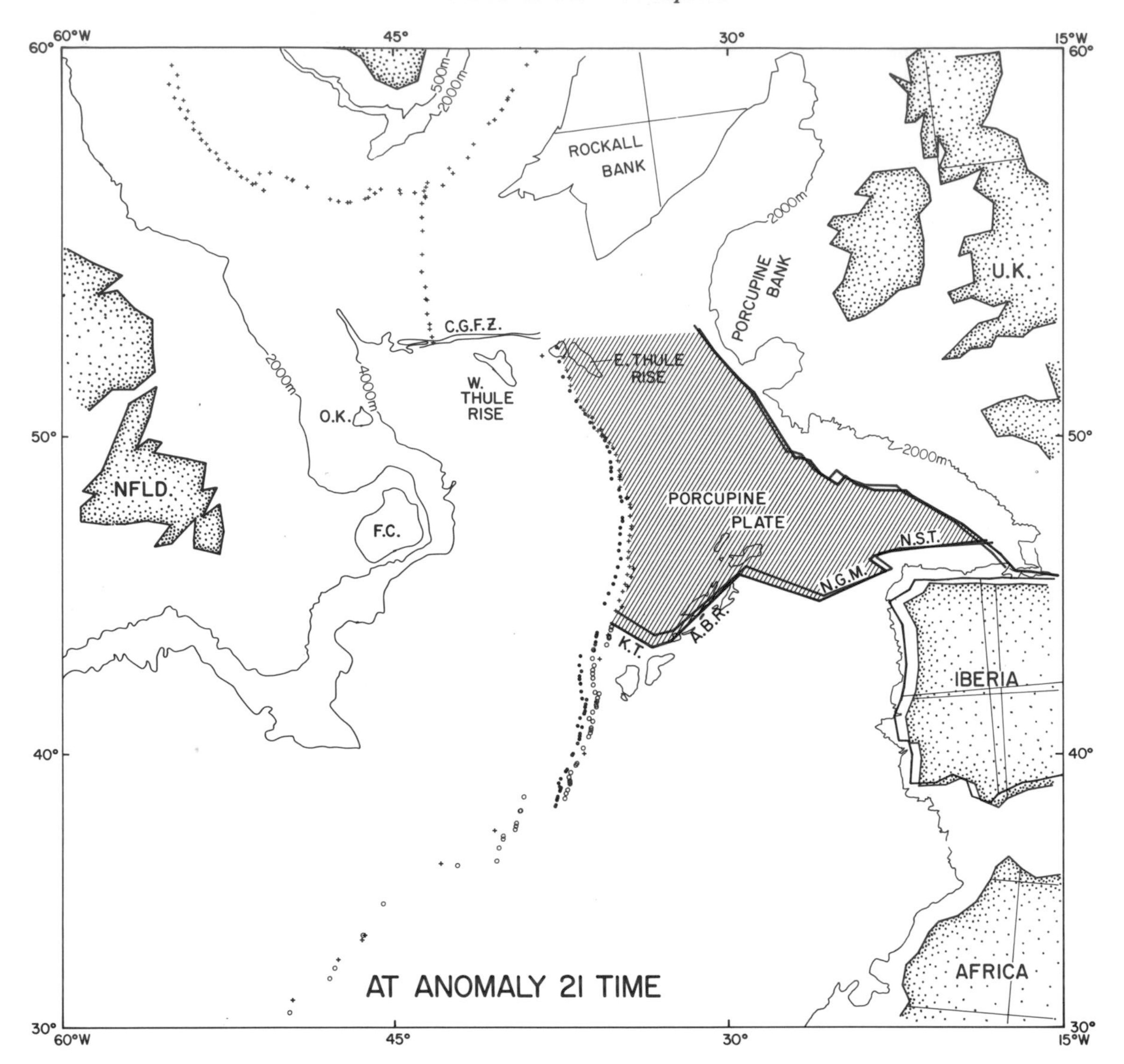

Figure 5. Reconstruction at anomaly 21 time relative to North America. The reconstruction is obtained by rotating the Greenland and Eurasian plates and their anomalies to the west using the poles of rotation from Table 1. Black crosses show the positions of anomaly 21 on the North American plate and of the rotated anomaly 21 on the east side of Greenland; red crosses show the positions of the rotated anomalies to the west. South of the Charlie-Gibbs Fracture Zone the black dots and the black outline of Iberia show the positions of anomaly 21 east and of Iberia when rotated to west as part of the Eurasian plate. The Porcupine plate is shown as a black stippled region and its boundaries with the Eurasian and Iberian-African plates are shown by black lines. Locations of these boundaries when rotated with the Eurasian and Iberian-African plates are shown in red. Red crosses and circles south of the Charlie-Gibbs FZ show the locations of rotated anomalies using the poles of rotation for the Porcupine and Iberia-African plates, respectively. KT—King's Trough, ABR—Azores Biscay Rise, NGM—North Galacia Margin, NST—North Spanish Trough. Mercator projection.

Support for the idea that Iberia moved with Africa from the start of spreading between Newfoundland and Iberia to anomaly 10 time comes from the observation that the poles of rotation between Africa and North America (Klitgord and Schouton, this volume) for these anomalies can be used successfully to rotate corresponding anomalies between Iberia and Newfoundland. However, a correction has to be first applied to these anomalies in order to bring them in the Africa–North America frame at anomaly 10 time (Schouten and others, 1985). Furthermore the correction results in positions of Iberia at various anomaly times very close to those it would have if it had been considered part of the Porcupine plate during these times. Thus, in determining the poles of rotation for various anomalies, we have treated Iberia as part of Africa until anomaly 10 time.

Anomalies 25 to 34

Further support for the idea that the Porcupine plate came into existence only after anomaly 25 time comes from the fact that a reasonably good fit between the rotated and unrotated anomalies 25 to 34 could be obtained by treating Porcupine as part of the Eurasian plate. This also resulted in obtaining minimum motion between Greenland and Eurasia during this period. Separate poles of rotation had to be determined for the Labrador Sea and North Atlantic for these anomalies (Tables 1 and 2), as no single pole could be found which would match the plate boundaries in these regions adequately as well as give the right direction of movement between the plates. The poles so determined for the Labrador Sea differ only slightly from previously published positions (Srivastava, 1978). Significant differences exist in the poles of rotation determined for the North Atlantic from the published poles because of the inclusion of the additional SFS data from the Eurasia Basin as well as the introduction of micro plates in the present analysis.

Initial Opening

To determine the poles of rotation for the time of initial opening we followed the propagating-rift model of Vink (1982) and Courtillot (1982). According to this model, continents do not rift instantaneously, rather they open by a propagating rift. Thus, there would be variable amounts of crustal extension along the continental-oceanic boundary prior to actual seafloor spreading. These boundaries, therefore, cannot be regarded as isochrons, but merely the lines along which plates ultimately separated. Thus, superposition of this boundary between continents where spreading first started would invariably result in overlaps in regions formed later. In our reconstruction, therefore, we have tried to match initial plate boundaries wherever possible from younger to older regions, i.e. from north to south. This was accomplished by making the following assumptions.

(a) The initial opening in the Labrador Sea took place prior to anomaly 34 time but after the opening of the Rockall Trough. The band of large magnetic anomalies west of anomaly 34 along the 2000-m isobath marks the approximate location of the initial plate boundary in this region (Fig. 4). A similar band also lies south of Rockall Bank.

(b) Rockall Trough opened at about the same time as the region to the south between the Porcupine Bank and Orphan Knoll, during the Cretaceous normal polarity. The position of initial plate boundaries in the Rockall trough are those given by Roberts and others (1981), and off Newfoundland and Porcupine Bank as those given by Kristoffersen (1978) and Srivastava (1978).

(c) The finite-difference pole of rotation between anomalies 33 and 34 remained the same as between anomaly 34 and the time of initial opening.

(d) Iberia separated from the Grand Banks of Newfoundland some time prior to anomaly M4 time and moved with Africa till anomaly 10 time.

Based on these assumptions, poles of rotation for different regions were then determined to their times of initial opening (Tables 1 and 2).

Flow Lines

Finite-difference poles of rotation between anomalies give the rate and direction of plate movement over a certain period of time. To demonstrate that the poles of differential rotation determined here for different regions depict accurately the direction of motion between plates, synthetic flow lines were generated for each region and these are shown in Figure 1 by dotted lines. Parallelism between the flow lines and the fracture zones clearly shows that the direction of movement between plates is represented accurately by the differential poles so determined. Similarly, to demonstrate how well the finite poles of rotation for different anomalies in different regions fit the observed data, we have plotted in Figures 6, 7, 8, and 9 the positions of the rotated and unrotated anomalies. The anomalies on the eastern side of the axis of symmetry are rotated to the west using the poles of rotation from Table 1. Because the rotated anomalies fall on their corresponding unrotated anomalies, this indicates clearly that the openings in different regions are represented accurately by the finite poles of rotation.

PLATE RECONSTRUCTIONS

Successive stages of the evolution of the North Atlantic are shown in Figures 6, 7, 8, and 9 and in Plates 9 and 10 (in pocket) where paleogeographic positions of Greenland, Europe, and Iberia are plotted relative to North America at various times. The evolution of different regions and their relationship to each other are summarized in Table 3, and are briefly described below. Detailed descriptions of the evolution of the Eurasia and Arctic Basins and Norwegian-Greenland Sea are given in Volume L, which is devoted to the geology and geophysics of the Arctic.

The reconstructions in Figures 6 to 9 and on Plates 9 and 10 show the present coastlines with five-degree grids as useful markers for those interested in evaluating these reconstructions with other geological and geophysical data. Also shown are the plate boundaries as defined by the fit of the rotated and unrotated anomalies at various times and the presence and absence of microplates.

Initial Opening

Based on the superposition of magnetic anomalies from one side of the ridge axis onto the other side, we have developed a scenario for the history of the opening of the North Atlantic north of the Azores-Gibraltar Fracture Zone. Klitgord and Schouten (this volume) have established a similar scenario for the region to the south. The SFS data off Grand Banks and off Iberia show that a shift from rifting to spreading began in this region at about anomaly M-4 time, though rifting may have started as early as

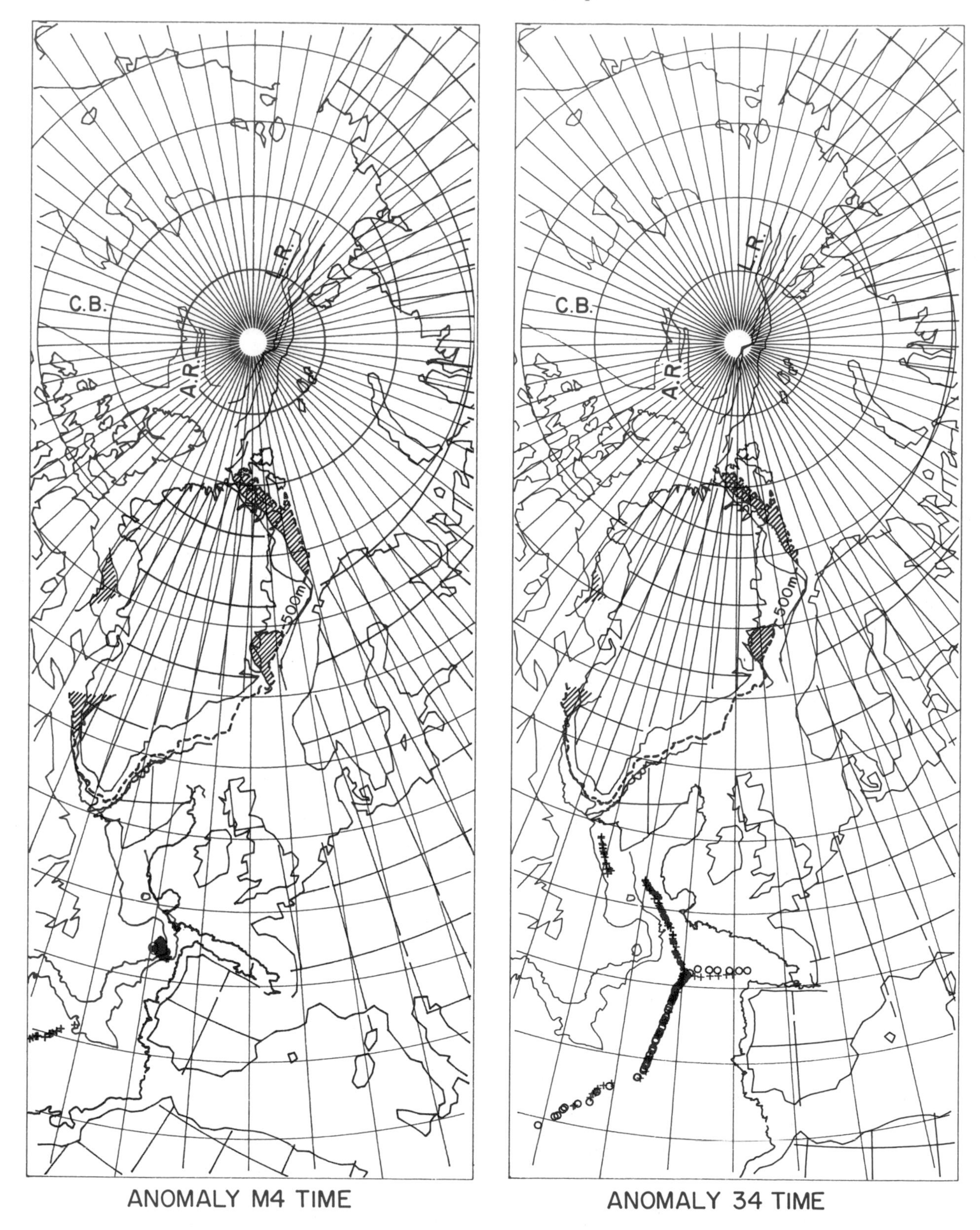

Figure 6. Reconstruction at anomalies M-4 and 34 time showing the positions of the rotated (red) and unrotated (black) anomalies and boundaries in each region of the North Atlantic, except for the black lines which mark the locations of the Porcupine plate boundaries. Also shown are 2000 m and 500 m isobaths in different regions, and the five-degree grids for each plate. The overlaps between the isobaths are shown by stippling. For a detailed explanation of the symbols see Fig. 5. L.R.—Lomonosov Ridge, A.R.—Alpha Ridge, C.B.—Canada Basin. Polar stereographic projection.

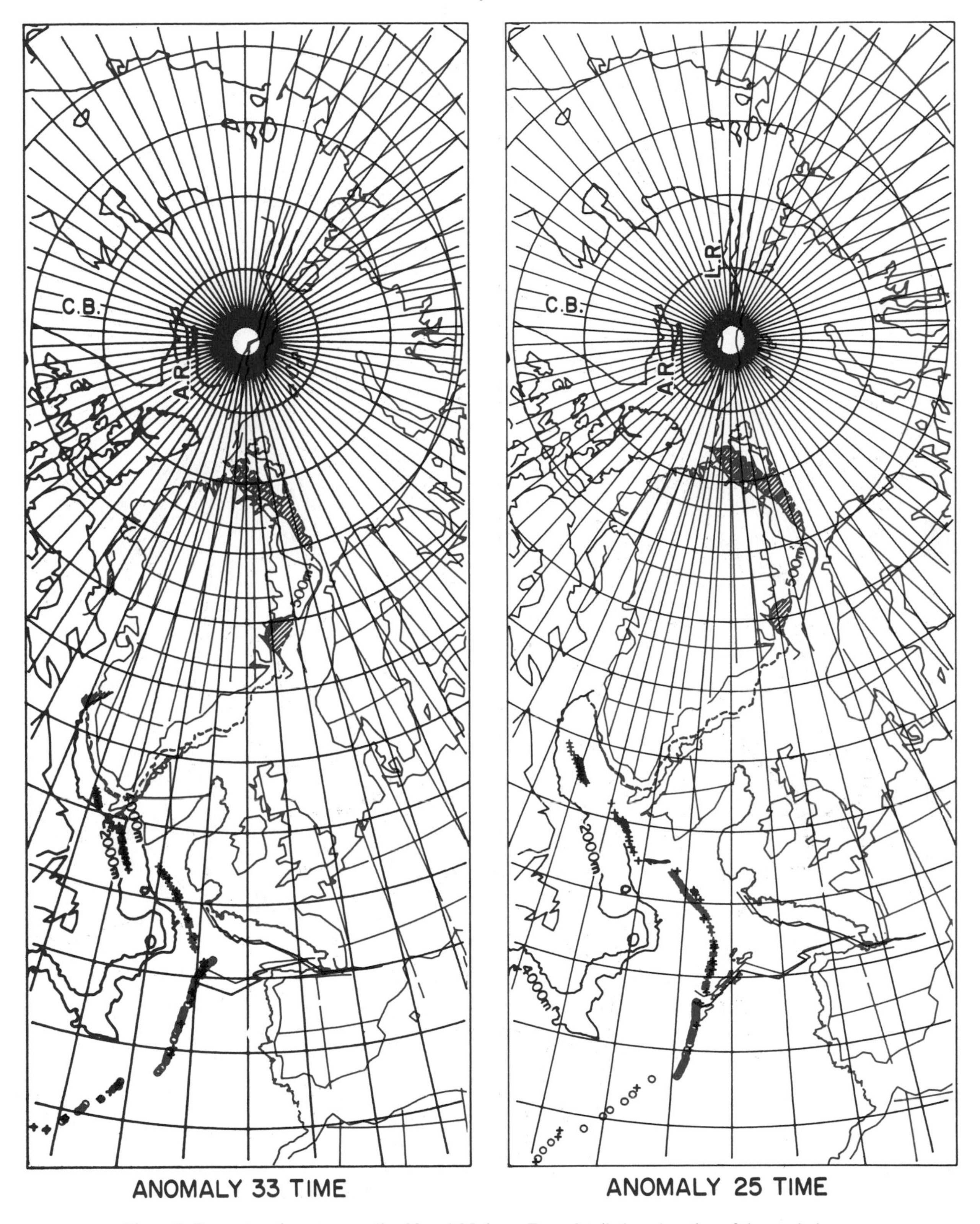

Figure 7. Reconstruction at anomalies 33 and 25 times. For a detailed explanation of the symbols see Figs. 5 and 6.

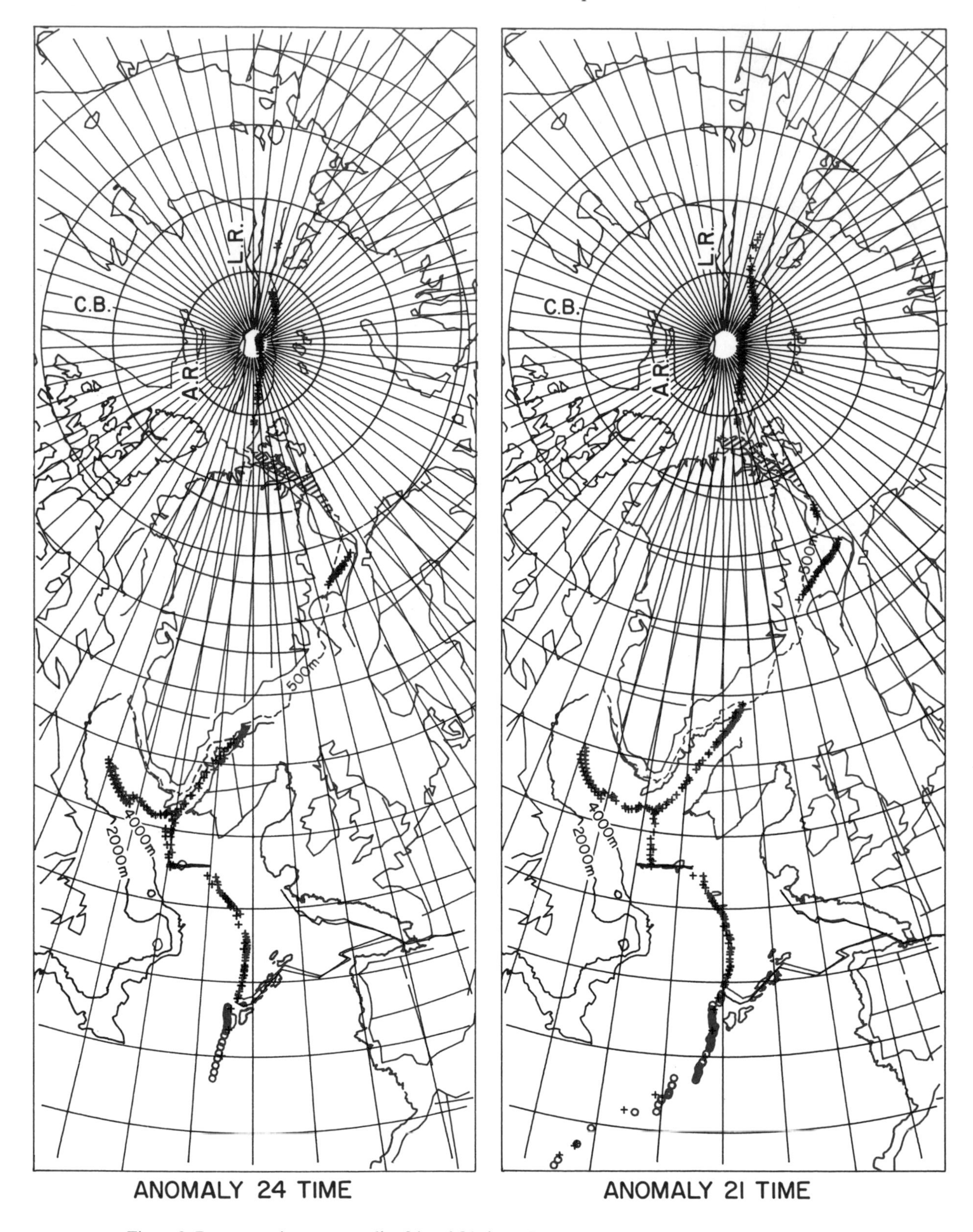

Figure 8. Reconstruction at anomalies 24 and 21 times. For a detailed explanation of the symbols see Figs. 5 and 6.

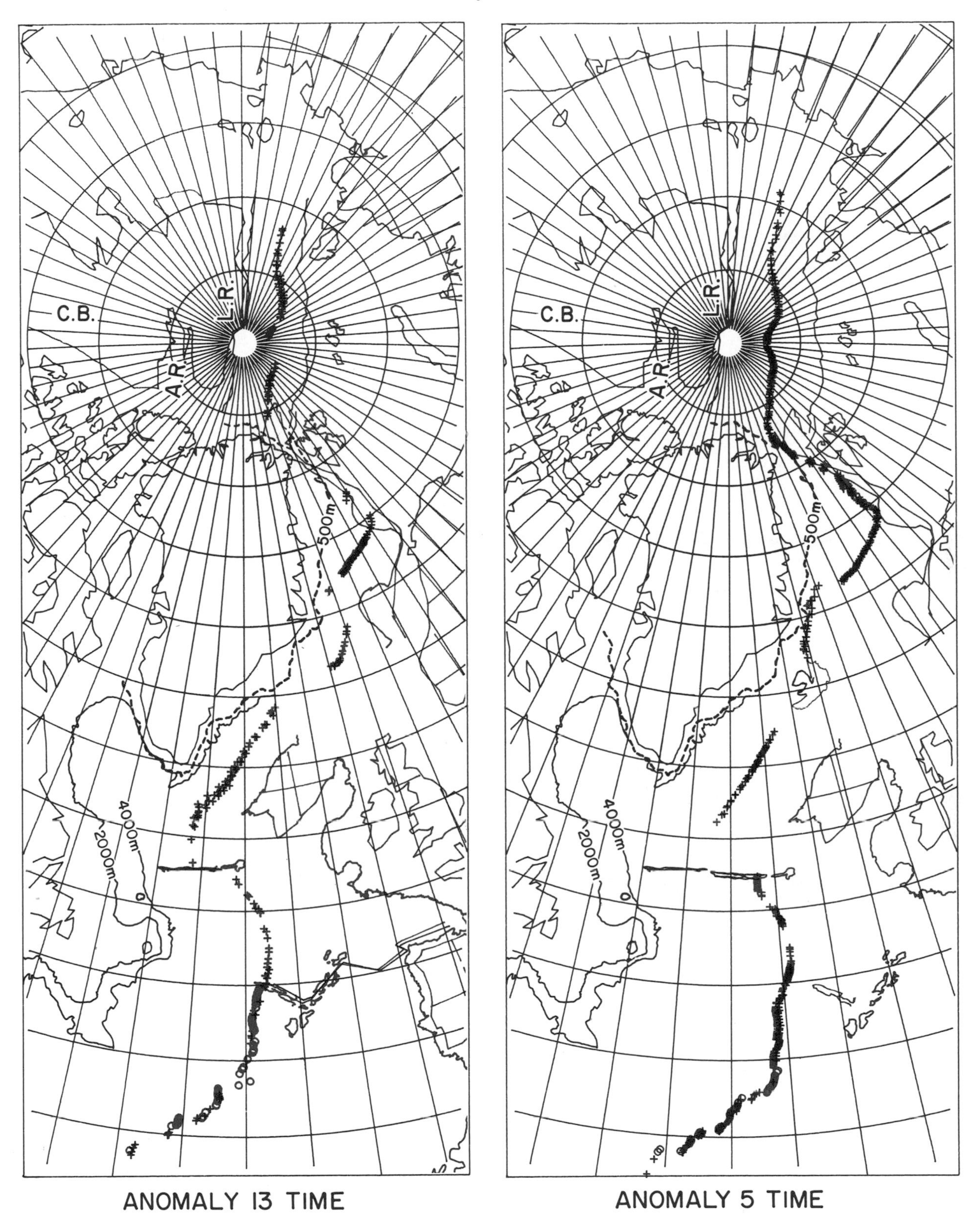

Figure 9. Reconstruction at anomalies 13 and 5 times. For a detailed explanation of the symbols see Figs. 5 and 6.

TABLE 3. SUMMARY OF THE TECTONIC EVENTS IN THE NORTH ATLANTIC, LABRADOR SEA, EURASIAN BASIN, ARCTIC BASIN, NORWEGIAN-GREENLAND SEA, AND BAY OF BISCAY

Anomaly	Age (Ma)	Bay of Biscay (Iberia)	North Atlantic	Arctic Basin Siberian Platform
M11	Valanginian (133)	Counterclockwise rotation of Iberia with Africa; initiation of opening in Bay of Biscay and in Newfoundland Basin.	---	Opening of the Canada Basin[1,2].
MO	Aptian (118)	Active sea-floor spreading in Newfoundland Basin and Bay of Biscay. Iberia moving with Africa.	Initiation of rifting between Porcupine Bank and Orphan Knoll and in Rockall Trough.	Near extinction of spreading in Canada Basin[1].
MO-34	Cenomanian (95)	Slowing down of spreading in Bay of Biscay.	Initiation of active sea-floor spreading in the North Atlantic and Rockall Trough (all regions south of the Charlie Gibbs FZ[3,4,5]).	Initiation of compression between Lomonosov Ridge and on Siberian Platform possibly along Verkhoyansk fold belt[9].
34	Campanian (84)	Stopping of spreading in the Bay of Biscay.	Active sea-floor spreading in all regions south of Greenland, jump of the ridge axis to the west and spreading stopped in Rockall Trough[3,4].	Some spreading in the Makarov Basin[2]. Compression along Verkhoyansk foldbelt[9].
31	Maastrichtian (68)	---	Active sea-floor spreading continuing[9].	Some spreading in the Makarov Basin.[2]
25	Thanetian (59)	Birth of Porcupine Plate and start of its north westward movement relative to Eurasia. Start of Pyrenian Orogeny.	Volcanism forming Thulean Rise[11]. Formation of Azores-Biscay Rise.	Shearing in northern Siberian platform due to separation of Lomonosov Ridge from Barents Shelf[4].

TABLE 3 (CONTINUED)

Anomaly	Age (Ma)	Labrador Sea	Eurasian Basin	Greenland-Norwegian Sea
M11	Valanginian (133)	---	---	---
MO	Aptian (118)	Volcanism on the southern Labrador Shelf and on land associated with the initial stages of sea-floor spreading in the Labrador Sea[6].	---	---
MO-34	Cenomanian (95)	Rifting between Greenland and Labrador, volcanism on the outer shelf[4,7].	---	---
34	Campanian (84)	Active sea-floor spreading in the south Labrador Sea and rifting in the north Labrador Sea and Baffin Bay. Counterclockwise rotation of Greenland relative to Ellesmere Island and start of compression in northern Sverdrup Basin[4].	Compression between Lomonosov Ridge and Alpha Ridge or subduction below Alpha Ridge[9].	Rifting in the Greenland Norwegian Sea[4].
31	Maastrichtian (68)	Active sea-floor spreading in the north Labrador Sea and rifting in the northern Baffin Bay. Compression in the Sverdrup Basin and start of Eurekan Orogeny[10].	-do-	-do-
25	Thanetian (59)	Volcanism in Davis Strait & surrounding regions, change in direction of motion between Greenland and North America; Greenland begins to move at an angle to Nares Strait[4].	Volcanism north of Greenland on Morris Jessop Rise and Yermark Plateau. Rifting between Lomonosov Ridge and Barents Shelf[12].	Volcanism in eastern Greenland near Scoresby Sund and on Voring Plateau, Faeroe Islands and initiation of Greenland-Scotland Ridge[4,13].

TABLE 3. (CONTINUED)

Anomaly	Age (Ma)	Bay of Biscay (Iberia)	North Atlantic	Arctic Basin Siberian Platform
24	Ypresian (56) (E. Eocene)	Compression and strike-slip movement along North Biscay margin, compression along Pyrenian fold belt.	Compression and strike-slip movement along Iberia-Africa and Eurasian boundary.	Dilation in the northern part of the platform and compression in the southern part, possibly along Verkhoyansk foldbelt.
21	Lutetian (50) (M. Eocene)	-do-	As above and start of formation of King's Trough.	-do-
13	Rupelian (36) (E. Oligocene)	Porcupine plate moving with Eurasia. Compression and subduction along north Spanish Trough.	Motion along King's Trough continuing.	Shearing motion on Siberian platform among Leona River.
7	Chattian (26) (L. Oligocene)	Iberia begins to move with Eurasia.	Motion along King's Trough stopped. Plate boundary between Eurasia and Africa shifts to the south along Azores-Gibraltor FZ.	-do-
0	Present	---	Spreading taking place across MAR.	Strike-slip motion along the intercontinental boundary between Eurasian and North America on Siberian platform.

TABLE 3. (CONTINUED)

Anomaly	Age (Ma)	Labrador Sea	Eurasian Basin	Greenland-Norwegian Sea
24	Ypresian (56) (E. Eocene)	Change in direction of motion between Greenland and North America. Oblique spreading in Baffin Bay and Davis Strait resulting in mainly shear motion[4,10].	Active sea-floor spreading between Lomonosov Ridge and Barents Shelf across Nansen Ridge. Strike-slip motion between Svalbard and Greenland[13,15,16].	Active sea-floor spreading in Norwegian and Greenland Seas[13].
21	Lutetian (50) (M. Eocene)	Change in direction of motion between Greenland and North America. Motion between Greenland and Ellesmere Island is mainly compressional[4].	Strike-slip motion between Svalbard and Greenland[13].	Formation of a large part of Iceland-Faeroe-Ridge. Possible jump of the ridge axis to the west[8,17].
13	Rupelian (36) (E. Oligocene)	Extinction of sea-floor spreading in Baffin Bay and Labrador Sea. Greenland started to move with North America[10].	Break of land bridge between Svalbard and Greenland and active seafloor spreading continues in Eurasian Basin[13].	Sea-floor spreading in Norwegian Sea and Greenland Seas continuing. Simultaneous spreading E and W of Jan Mayan Ridge[13].
7	Chattian (26) (L. Oligocene)	---	-do-	Jan Mayen Ridge separates from Greenland and spreading starts across Kolbeinsey Ridge. Spreading in the Norwegian Sea and on Iceland Faeroe Ridge terminates and starts in Iceland[13,11,17].
0	Present	---	Spreading taking place across Nansen Ridge.	Spreading taking place across Reykjanes, Kolbeinsey, Mohn and Knipovich Ridges and in Iceland.

[1] Vogt and others 1982
[2] Taylor and others 1981
[3] Kristofferson 1978
[4] Srivastava 1978
[5] Sclater and others 1977
[6] McWhae and Michel 1975
[7] Johnson and others 1982
[8] Voppel and others 1979
[9] Pitman and Talwani 1972
[10] Kristofferson and Talwani 1977
[11] Vogt and Avery 1974
[12] Vogt and others 1979
[13] Talwani and Eldholm 1977
[14] Nunns 1983
[15] Vink 1982
[16] Srivastava 1985
[17] Vink 1984

anomaly M-11 time. We have not tried to estimate here the amount of rifting which may have taken place during this time and have thus regarded the initial position for Iberia as its position at anomaly M-11 time relative to North America. By anomaly M-11 time the spreading between Africa and North America had progressed far enough north (details see Klitgord and Schouten, this volume) that rifting between Iberia and Grand Banks began and resulted in the formation of some of the basins like the Jeanne d'Arc Basin near the Hibernia area on the Grand Banks (Arthur and others, 1982) and may have also partly separated Galicia Bank from Iberia (Boillot and others, 1979). At anomaly M-4 time the rifting changed to spreading and Iberia started to move as part of Africa instead of as part of Eurasia (Schouten and others, 1986). This is also marked by a very prominent ridge in this region known as "J" anomaly ridge (M-2, Tucholke and Ludwig, 1982). Iberia continued to move with Africa until anomaly 10 time, with a plate boundary between Iberia-Africa and Eurasia extending along King's Trough, Azores-Biscay Rise, into the Bay of Biscay and the Pyrenees (Grimaud and others, 1982).

Rifting continued to move north during Late Cretaceous, first into the Rockall Trough (post M-0 and pre 34 anomalies) and then into the Labrador Sea (pre 34) (Fig. 6). A ridge-ridge-ridge triple junction existed west of the Bay of Biscay till anomaly 33 time (Fig. 7a), when the spreading stopped in the Bay of Biscay and shifted to the west and has been continuing ever since.

Rockall Trough

The origin of Rockall Trough has been debated for a number of years and little consensus exists about its age (Permian according to Russell, 1976; Russell and Smythe, 1978; Cretaceous according to Roberts, 1975; Roberts and others, 1981). We believe that the majority of the Rockall Trough was formed during the Cretaceous quiet period (Kristoffersen, 1978). Support for the interpretation comes from anomaly 34 south of the Charlie-Gibbs FZ (Map A in the pocket) which bypasses the Trough. Anomaly 34 is dislocated to the west across the Charlie-Gibbs FZ and lies south of the Rockall Plateau. This implies that spreading in the Rockall Trough had ceased prior to anomaly 34 time and started in the Labrador Sea. In the absence of any anomalies during this time it is difficult to decipher the time when such a change in the spreading system took place. In the reconstruction shown in Figure 6a we rotated Rockall and Greenland together with Eurasia after anomaly 34 time until a match between the large zone of magnetic anomalies off Labrador and Rockall Bank was obtained. This is interpreted as the time of initial opening for the Labrador sea (90–95 m.y.). Assuming that Rockall Trough opened at the time when Porcupine Bank separated from Orphan Knoll off Newfoundland, the closure position was obtained by rotating Eurasia further to the west until a match between the ocean-continent boundaries in this region and in the Rockall Trough (Roberts et al., 1981) were obtained. Based on the rate of spreading between anomalies 33 and 34, we estimate that spreading in Rockall Trough took place in Albian time (100-105 m.y.), which compares favourably with other estimates for the regions further to the north (Price and Rattey, 1984; Hanisch, 1984).

The above reconstruction has serious implications for the regions to the north, especially in the Arctic region. Enough geological evidence now exists to support the idea that rifting, which formed Rockall Trough, continued to the north, creating deep basins in the North Sea and off Norway during these times (Price and Rattey, 1984). If plates behave rigidly, then the question which arises is what happened north of Svalbard during this time. The answer lies in the location of plate boundary between Eurasia and North America in this region at that time. This is discussed later in this chapter.

The positions of Greenland, Europe, and Iberia relative to North America as obtained here at the time of initial opening, Figure 6, differ significantly from the previous reconstructions of Bullard and others (1965), Laughton (1971), Pitman and Talwani and others (1972), Kristoffersen (1978), Sclater and others (1977), and Olivet and others (1981). It is beyond the scope of the present chapter to describe these differences in detail; only major differences are covered here. The Bullard and others reconstruction places Greenland much farther east and south of Baffin Island than the reconstruction presented here, leaving a large portion of Baffin Bay open. In addition, it places Europe farther north, resulting in a wide separation between northern Greenland and Svalbard and a very close fit between the British Isles and southern Greenland, with Rockall Plateau overlying Greenland. In Laughton's reconstruction for the Late Jurassic, Europe and Iberia lie about 270 km south of the positions shown in our reconstruction. Pitman and Talwani place Europe about 100 km south of our position. Smaller differences exist between the relative positions of North America and Europe in Kristoffersen's reconstruction and our reconstruction, but the differences become significant for the positions of Greenland and northern Eurasia relative to North America. Kristoffersen's reconstruction gives a very tight fit between North America and Greenland, with a large overlap between the two. In the reconstruction of LePichon and others, Europe lies further to the east and north of our position. Sclater and others show a position of Eurasia which is in good agreement with ours. However we differ from their positions for Greenland and Iberia. In the Olivet and others reconstruction, Iberia lies south of its position obtained in our reconstruction. The position of British Isles relative to Newfoundland in their reconstruction is the same as we obtained but the position of Svalbard relative to Greenland is east of our position.

A considerable overlap is obtained in our reconstruction at anomaly M-11 time between Flemish Cap and Galicia Bank (Plate 9). This overlap can be explained if we consider that Galicia Bank and Flemish Cap did not occupy their present position prior to their separation but were displaced seaward during the initial stages of their separation. In a similar reconstruction of this region, based on the superposition of magnetic anomalies older than 34 between Newfoundland Basin and Iberia, Masson and Miles (1984) have obtained a pre-drift position between

Galicia Bank and Flemish Cap where very little overlap between the two is obtained. This is achieved by implying a motion along a transform fault located west of Flemish Cap and east of Galicia Bank. It is an interesting idea, but unless the motion of Iberia can be well constrained for anomalies older than 34, it raises an important question why Iberia should move independently of Africa. It is for this reason we favour motion of Iberia with Africa and propose that the overlap as we see between Flemish Cap and Galicia Bank is due to the crustal stretching during initial rifting between the two. Evidence for this stretching exists but not to the same extent as implied here (Keen and Barrett, 1981). We do not claim that the reconstructions presented here are entirely correct but, as they are constrained by the plate kinematics of the entire North Atlantic, they must be close. As additional data are collected in different parts of the ocean, allowing magnetic anomalies to be delinated more precisely, further refinements in the reconstructions are expected.

Anomalies 34-25 Reconstruction

With a shift in spreading to the west of Rockall Trough just prior to anomaly 34 time, spreading terminated in the Trough and active sea-floor spreading started in the Labrador Sea. A triple junction which had existed west of the Bay of Biscay terminated at anomaly 33 time and the spreading continued to the west of it as seen from the continuation of anomaly 33 west of the triple junction. The North Spanish Trough then became the boundary between the Iberian-African and Eurasian plates. Spreading continued in the Labrador Sea at a constant rate (Srivastava, 1978) until anomaly 25 time, when there was a major reorganization of the plates, with spreading starting in the Norwegian-Greenland Sea and in the Eurasia Basin and formation of a micro plate (Porcupine plate) in the North Atlantic. In the absence of any recognizable anomalies in the Baffin Bay region (though presence of anomalies 21 to 25 has been suggested by Srivastava, 1978; and Jackson and others, 1979), it is not certain how much spreading and how much rifting or stretching took place in this region, but the flow lines predict a maximum of 150 km of opening during this time. The reorganization of plates at anomaly 25 coincides with a major volcanic episode during which a large quantity of basalt was erupted across Davis Strait and north of the British Isles.

Motion of Greenland remained different from that of Eurasia relative of North America from anomaly 34 to 25 time, as poles of rotation (Table 1) suggest. Reconstructions for these times (Figs. 6 and 7) show hardly any motion between Eurasia and Greenland north of British Isles. This explains the absence of any sea-floor-generated anomalies older than anomaly 24 between the two. The motion which took place between the Rockall Plateau and Greenland during this period was not large enough to generate any recognizable anomalies either. Thus a triple junction which existed south of Greenland since anomaly 34 time became extinct only when the sea-floor spreading stopped in the Labrador Sea some time prior to anomaly 13 time. Active sea-floor spreading started in the Norwegian-Greenland Sea at anomaly 24 time. Between anomalies 25 and 34 time, motion along the Eurasian and Iberian plate boundary was mainly shear and extensional through the Bay of Biscay.

Anomalies 24 to 7 Time

This encompasses a period of major changes in the configuration of the plates in the North Atlantic. With a change in the direction of spreading in the Labrador Sea at anomaly 24 time, active sea-floor spreading started in the Norwegian-Greenland Sea and the Eurasian Basin. In the Bay of Biscay, motion along the Iberian Eurasian plate boundary changed to compressional from shear and extension, resulting in deformation of the Eurasian plate immediately to the north (Fig. 8). This correlates well with the Cenozoic deformation of the north Biscay margin, which shows the presence of numerous folds, faults and strike-slip movements (Montadert and others, 1979). Geological observations on land and in offshore regions immediately north of Iberia show that some subduction of the Biscay oceanic floor took place at least until the late Eocene-Oligocene time (Boillot and others, 1979). Further evidence of this deformation is seen on land in the formation of the Pyrenees. Also, when magnetic anomalies 21 and 24 are rotated from one side of the ridge axis onto the other side, they show large overlaps south of the Charlie-Gibbs FZ (Fig. 5). This we interpret to be indicative of the deformation of a small portion of the Eurasian plate within an area called here the Porcupine plate.

Further evidence of this interaction between plates can be seen in the dislocation of anomaly 24 immediately south of Charlie-Gibbs FZ (Johnson and Vogt, 1973) and to some extent in anomaly 21 at several places further to the south (Fig. 2). Several small fracture zones like the Faraday and Maxwell seem to have started at this time in the North Atlantic (GEBCO chart 5.04). Even the Charlie-Gibbs FZ shows a large breadth near this anomaly (Olivet and others, 1974).

The relative motion between the Eurasian and Iberian plates had decreased considerably by anomaly 13 time as evidenced by the decrease in the overlap when this anomaly is rotated from one side of the ridge to the other (Fig. 9). At this time Greenland had started to move with North America and the opening in the North Atlantic had been reduced to a two-plate situation. However, Iberia was still moving with Africa and kept moving so until anomaly 10 time, when it started to move with Eurasia instead, and the plate boundary between Eurasia and Iberia-Africa jumped to the south along the Azores-Gibraltar FZ.

The formation of King's Trough can thus be related directly to the interactions between the Eurasian and Iberian-African plates. Finite-difference poles of rotation between the Eurasian and Iberian-African plates (Table 2) show that the King's Trough region was formed as an extensional feature along the Iberia-Eurasia plate boundary between anomalies 24 and 10 time (Figs. 8 and 9). At anomaly 10 time, when Iberia stopped moving with Africa, the boundary shifted to the south along the Azores-Gibraltar FZ. Since then Iberia has been moving with Eurasia. A detailed description of the kinematics of this region are beyond the scope of this chapter and will be discussed elsewhere (Srivastava and others, in preparation).

With a change in direction of motion between Greenland and North America at anomaly 24 time, the spreading in the Labrador Sea became very oblique to the ridge axis, resulting in the formation of a quiet magnetic zone in the central part of the Labrador Sea (Roots and Srivastava, 1984). Further to the north in Baffin Bay, relative motion between Greenland and North America became mainly shear (Fig. 1) and little or no spreading occurred. The spreading in the Labrador Sea slowed down considerably after anomaly 21 time and stopped completely prior to anomaly 13 time.

Several changes took place between anomaly 24 and anomaly 7 times in the Norwegian-Greenland Sea and these have been described in detail by Talwani and Eldholm (1977), Nunns (1982, 1983), and Vink (1982, 1984). The poles of rotation given for these regions (Tables 1 and 2) agree with those obtained in earlier studies in general but differ in details, as mentioned earlier. Combining the present findings with the earlier studies, the following scenario emerged (for a detailed account of this region, the readers should consult the references cited above):

(1) Since the initiation of the spreading between Greenland and Eurasia, the region south of the Greenland-Scotland Ridge (GSR) has evolved symmetrically relative to the present axis of the Reykjanes Ridge. In the northern part of the region, closer to GSR, the double anomaly 24 observed along the Rockall margin (Voppel and others, 1979) suggests a small ridge jump toward the west between anomaly 24 and anomaly 23 times.

(2) In the Norway Basin the spreading took place in the normal manner until anomaly 20 time, when there was a slight shift in the direction of spreading resulting in a compression across the transform fault linking the Reykjanes and Aegir axes (Nunns, 1983). As a result, a rift propagated northward from the Reykjanes axis and split off the Jan Mayen block. There was simultaneous complementary fan-shaped spreading east and west of the Jan Mayen block until it had pivoted out from Greenland sufficiently enough to allow free spreading in the new direction about the Kolbeinsey Ridge. This happened about anomaly 7 time (Nunns, 1983). The spreading has been continuing since then across the Kolbeinsey, Reykjanes, Mohns, and Knipovich Ridges.

(3) Because of the difficulty in identifying magnetic anomalies over the GSR, it has not been possible to satisfactorily reconstruct its evolution, though several hypothesis have been proposed which imply jumps of spreading center across it (Bott, 1974; Bott and others, 1974; Fleischer and others, 1974; Talwani and Eldholm, 1977; Voppel and others, 1979; and Nunns, 1983). Furthermore, its elevated nature and its continuation across Iceland led many to suggest that it must have been formed by the presence of a hotspot in this region (Talwani and Eldholm, 1977; and Cochran and Talwani, 1978) which at present lies under Iceland. However, such explanations for the GSR which involve tracking of the lithosphere over a relatively fixed-mantle plume, are inconsistent with its age progression, orientation and location. Recently Vink (1984) proposed that this and another plateau, Voring, further to the north, were formed due to the channelling of material from a hot spot to the spreading axes under these regions, thereby producing more basalt and thicker crust at these locations. Furthermore, the geometric constraints of such a model predict unique orientation, location, and age-progression of the plateaus, and these fit remarkably well with the observations. By combining the poles of rotation for the hotspot for this region (Morgan, 1983) with a set of poles of rotation for the Norwegian-Greenland Sea, Vink was able to show satisfactorily the formation of these two features and at the same time their relationship with the surrounding regions. However Vink's model requires an extinct axis on the GSR and oceanic crust under the Faeroes, neither of which are proven.

In the Eurasia Basin the spreading has been confined mainly between the Lomonosov Ridge and the Barent Sea shelf since anomaly 24 time. Reconstructions until anomaly 13 time (Figs. 8 and 9) show a considerable overlap between northern Greenland and Svalbard. This is because the region between Svalbard and Greenland was undergoing strike-slip motion until anomaly 13 time, while the region east of Svalbard was subjected to considerable stretching and thinning (Vink, 1982; Srivastava, 1978) during this time. True sea-floor spreading started between the two at anomaly 13 time.

IMPLICATIONS FOR THE NARES STRAIT AND ARCTIC BASIN

The plate kinematics of the North Atlantic, as described in the previous sections, involves relative motion among four plates, the North American, Greenland, Eurasian and Iberian-African. The movements among these plates have profound effects on the surrounding regions, especially across plate boundaries. Two of such regions are the Nares Strait and the Arctic Basin. These are discussed briefly here.

Nares Strait

In our reconstructions (Figs. 6 to 9) we have regarded Nares Strait as the boundary between the North American and Greenland plates. Many geologists have argued against doing so (see Dawes and Kerr, 1982) because of the continuation of some of the geological boundaries from Greenland to Ellesmere Island. Previous and present plate reconstructions for the North Atlantic have implied movement along this boundary of 200 km or more. A serious disagreement thus exists among thc gcologists and plate tectonists concerning movement in this region (Dawes and Kerr).

In view of the changes in poles of rotation between Greenland and North America from those published earlier (Srivastava, 1978; Kristoffersen, 1978), the question was reexamined in some detail by Srivastava (1985) and a brief summary of it is included here for the sake of continuity.

The direction of movement of Greenland relative to North America at several places in the Arctic Archipelago is shown in

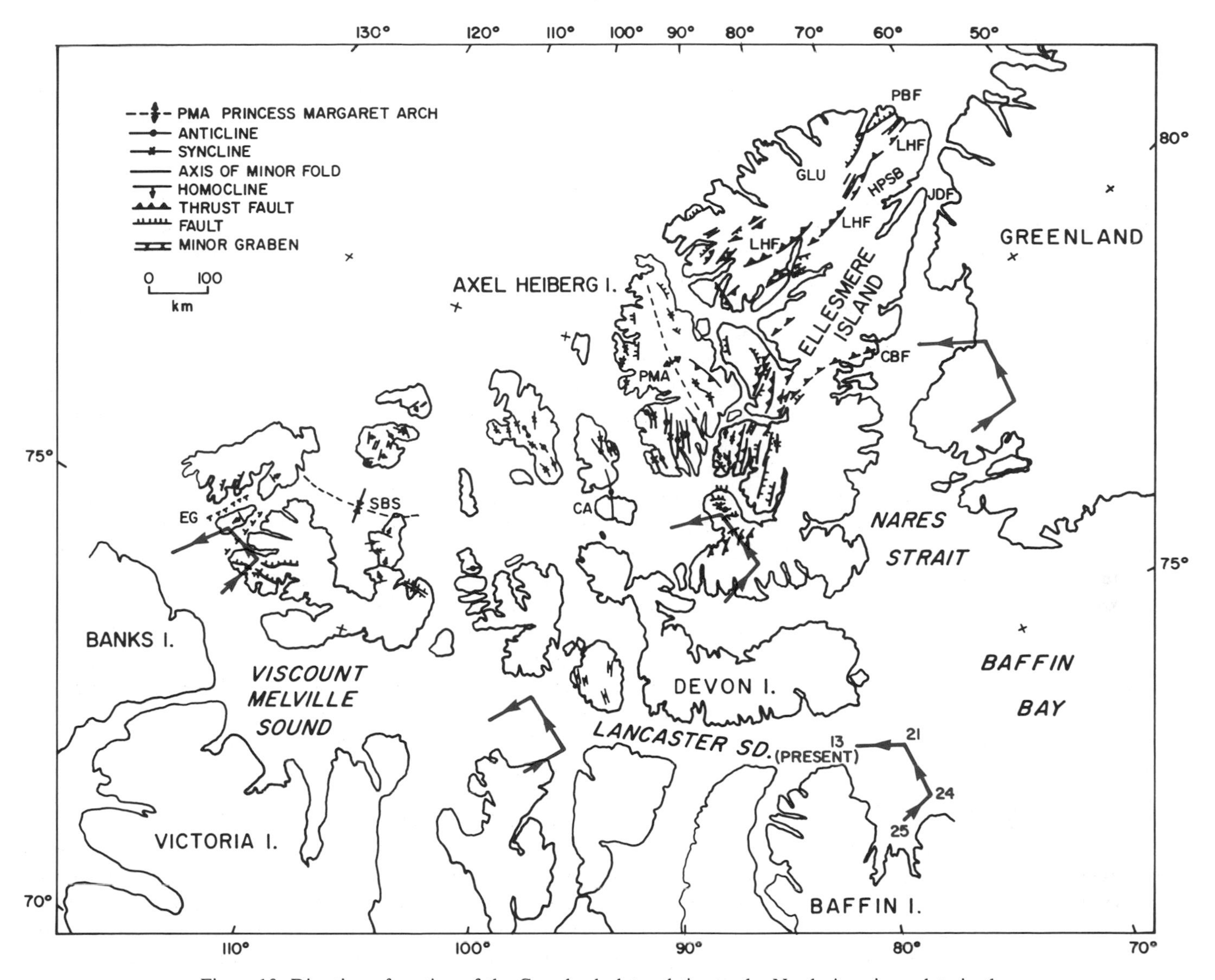

Figure 10. Direction of motion of the Greenland plate relative to the North American plate in the Canadian Arctic at various anomaly times. Polar stereographic projection.

Figure 10. It shows clearly that the relative movement between Greenland and the Arctic Islands was orthogonal to many of the fold belts after anomaly 25 time. Such compressive movement in these regions could thus be correlated with some of the deformation which seems to have taken place during the Eurekan Orogeny in this region (Balkwill, 1978). Furthermore, the plate kinematics suggest a net displacement of Greenland of about 150 km in the direction of Nares Strait relative to Ellesmere Island. It is possible that such a displacement between the two may not have been completely confined along the Nares Strait but distributed within the Ellesmere-Greenland fold belt, as suggested by Miall (1983, 1984).

Prior to anomaly 25 time, movement of Greenland relative to the Canadian Arctic Islands was rotational, with a pole of rotation located on Boothia Arch north of Cornwallis Island (76.0 N, 95.0 W). Such a movement would result mainly in compressional movement between northern Ellesmere Island and Greenland. Equally likely, some fragmentation within the Arctic Islands may have occurred if they remained part of Greenland and rotated in a counterclockwise direction relative to North America. Rotational motion of Ellesmere Island is also derived from paleomagnetic observations (Wynn and others, 1983).

Arctic Basin

Vogt and others (1982) have reviewed the evolution of the Amerasia Basin in light of the most recent aeromagnetic data. This is further discussed in Volume L, which deals with the geology and geophysics of the Arctic Basin. We will not review the presently held views on the origin of this basin but merely discuss the implications our reconstructions of the North Atlantic have on the development of this basin.

The fact which first needs to be ascertained is the position of the boundary between the Eurasian and North American plates on the Russian platform. On the basis of the seismicity, Chapman and Solomon (1976) have suggested that at present it extends from the Nansen Ridge through a broad zone of deformation in the Siberian platform to the Sea of Okhotsk and thence southward through Sakhalin and Hokkaido to a triple junction in the Kuril-Japan trench. This boundary follows more or less the Verkhoyansk foldbelt in northeast USSR. It is not certain where this boundary lay during the Mesozoic, though it has been associated with the Cherskiy foldbelt further to the east (Herron and others, 1974). The positions of this boundary during various times are shown in Figures 6 to 9. The motion has mainly been strike-slip since anomaly 13 time but was compressional prior to that. The reconstructions further show that the Arctic Basin was about 600 km wider near the Siberian platform at the time of initial opening (M-4 time) than it was at anomaly 25 time, assuming that the compressional zone was colinear with the extensional zone. This implies that there must have been a tremendous amount of compression between, presumably, the Alpha-Mendeleev Ridge and the Lomonosov Ridge. A similar reconstruction of this region was also obtained by Pitman and Talwani (1972). They suggested that the resulting compression was perhaps taken up in the form of subduction beneath the Alpha-Mendeleev Ridge system. Other interpretations of this region have been proposed (Vogt and others, 1982) and until we get more geophysical measurements from this region they all will remain speculative in nature.

SUMMARY AND CONCLUSIONS

Using the sea-floor spreading magnetic anomalies and the fracture zones in the North Atlantic, Labrador Sea, Norwegian-Greenland Sea and the Eurasia Basin, we have determined a set of poles of rotation to describe the plate kinematics of these regions. This has resulted in obtaining a comprehensive picture for the entire region north of the Azores-Gibraltar Fracture Zone to the Laptev Sea. This is not the first time a study of this nature has been carried out; previous work and the resulting differences are described in earlier sections. The present study differs from previous ones not only in the greater amount of data on which it is based but also in its consideration for the first time of the entire boundary between the Eurasian and North American plates in determining the poles of rotation for the whole North Atlantic.

The following sequence of events (Table 3) summarizes the evolutionary history of the North Atlantic:

(1) The opening of the North Atlantic (north of Azores) began in early Cretaceous (Valanginian), perhaps during anomaly M-11 time, with the separation of Iberia from Newfoundland. During this time, Iberia was moving with Africa, opening the Bay of Biscay at the same time.

(2) Iberia kept moving with Africa until mid-Oligocene time (anomaly 10 time), when it started to move with Eurasia.

(3) Prior to anomaly 34 time, in Late Albian (105 Ma), the spreading extended north of the Bay of Biscay into the Rockall Trough. Most of the Rockall Trough and part of the Bay of Biscay opened prior to the jump of the ridge axis to the west, when spreading commenced in the Labrador Sea in Late Santonian time (85-90 Ma).

(4) With a change in direction of motion between Greenland and North America and between Greenland and Eurasia, just after anomaly 25 time, active sea-floor spreading started in the Eurasia Basin and in the Norwegian-Greenland Sea. It was also at this time that the relative motion between Eurasia and Iberia had changed, resulting in some deformation of the Eurasian plate immediately north of Iberia. This motion seems to have propagated northward, giving rise to compression and strike-slip movement along the north Biscay margin and some subduction of old Biscay oceanic floor along the north Iberian margin. Such relative motion across the plate boundary seems to have continued until anomaly 20 time.

(5) King's Trough, Azores-Biscay Rise, and north Spanish Trough on the oceanic side, and the Pyrenees on land, were formed along the Eurasian and Iberian-African plate boundary between anomaly 25 and 10 time.

(6) Spreading in the Labrador Sea ceased prior to anomaly 13 time, which resulted in the change of direction of motion in the Norwegian-Greenland Sea. The Jan Mayen Ridge separated from Greenland between anomalies 20 and 7 times.

(7) A large quantity of basalt erupted on either side of Greenland, in Davis Strait, and in the Greenland-Scotland region at about anomaly 25 time (59 Ma, Paleocene). A similar event at this time may have begun the formation of the Morris Jesup Rise and Yermak Plateau in the Arctic.

(8) The Greenland-Scotland Ridge was formed by the channeling of asthenosphere material along the ridge axis since the start of sea-floor spreading in this region at anomaly 24 time. The center of spreading now lies under Iceland.

(9) Spreading in the Eurasia Basin has been along the Nansen Ridge since the beginning of spreading in this region (anomaly 25 time), with little or no offset of the ridge axis. The motion between Svalbard and Greenland was strike-slip until anomaly 13 time and then changed to spreading.

(10) The Makarov Basin was subjected to a large amount of compression prior to the opening of the Eurasia Basin. It is possible that part of this compression may have been taken up within the Lomonosov Ridge, but a more likely mechanism is by subduction under the Alpha—Mendeleev Ridge.

REFERENCES

Arthur, K. R., Cole, D. R., Henderson, G.G.L., and Kushnir, D. W.
1982 : Geology of the Hibernia discovery; in M. T. Halbouty, ed., 'The deliberate search for the subtle trap:' American Association of Petroleum Geologists Memoir 32, p. 181–196.

Balkwill, H. R.
1978 : Evolution of Sverdrup basin, Arctic Canada: American Association of Petroleum Geologists Bulletin, v. 62, p. 1004–1028.

Boillot, G. and Capdevila, R.

1977 : The Pyrenees: subduction and collision?: Earth and Planetary Science Letters, v. 35, p. 151–160.

Boillot, G., Auxietre, J. C. and Dunand, J. P.
1979 : The northwestern Iberian margin: A Cretaceous passive margin deformed during Eocene; in Deep Drilling Results in the Atlantic Ocean: continental margins and paleoenvironment, eds. M. Talwani, W. Hay, and W.B.F. Ryan, American Geophysical Union Monograph, Vol. 3, 437 p.

Bott, M.H.P., Sunderland, J., Smith, P. J., Casten, U. and Saxou, S.
1974 : Evidence for continental crust beneath the Faeroe Islands: Nature, v. 248, p. 202–204.

Bott, M.H.P.
1974 : Deep Structure, evolution, and origin of the Icelandic transverse ridge; in Geodynamics of Iceland and the north Atlantic area, ed. L. Kristjansson: D. Reidel Publishing Company, Boston, p. 33–47.

Bullard, J. C., Everett, J. E. and Smith, A. G.
1965 : The fit of the continents around the Atlantic: in A symposium on continental drift: Royal Society of London, Philosophical Transactions, Series A, v. 258, p. 41–51.

Cande, S. C. and Kristoffersen, Y.
1977 : Late Cretaceous magnetic anomalies in the north Atlantic: Earth and Planetary Science Letters, v. 35, p. 215–224.

Chapman, M. E. and Solomon, S. C.
1976 : North American–Eurasian plate boundary in northeast Asia: Journal of Geophysical Research, v. 81, p. 921–930.

Choubert, B.
1935 : Recherche sur la ginèse des chaînes paléozoiiques et antécambriennes: Review Geography et Physical Geologie Dynamique, v. 8, p. 5–50.

Cochran, J. R. and Talwani, M.
1978 : Gravity anomalies, regional elevation, and the deep structure of the North Atlantic: Journal of Geophysical Research, v. 83, p. 4907–4924.

Courtillot, V.
1982 : Propagating rifts and continental breakup: Tectonics, v. 1, p. 239–250.

Dawes, P. R. and Kerr, J. W.
1982 : Nares Strait and the drift of Greenland: a conflict in plate tectonics: Meddelelser om Gronland: Geoscience, v. 8, 392 p.

Fleischer, U., Holzkamm, F., Vollbrecht, K. and Voppel, D.
1974 : Die Strukturdes island-Faroer-Ruckens aus geophysikalischen Messungen: Deutsche Hydrographische Zeitschrift, v. 27, p. 97–113.

Grimaud, S., Boillot, G., Collette, B. J., Mauffret, A., Miles, P. R. and Roberts, D. G.
1982 : Western extension of the Iberian-European plate boundary during the early Cenozoic (Pyrenean) convergence: a new model: Marine Geology, v. 45, p. 63–77.

Hanisch, J.
1984 : The Cretaceous opening of the Northeast Atlantic: Tectonophysics, v. 101, p. 1–23.

Herron, E. M., Dewey, J. F. and Pitman III, W. C.
1974 : Plate tectonics model for the evolution of the Arctic: Geology, v. 2, p. 377–380.

Jackson, H. R., Keen, C. E., Falconer, R.K.H., and Appleton, K.
1979 : New geophysical evidence for sea-floor spreading in central Baffin Bay: Canadian Journal of Earth Sciences, v. 11, p. 2122–2135.

Johnson, G. L. and Vogt, P. R.
1973 : Mid-Atlantic Ridge from 47° to 51° North: Geological Society of American Bulletin, v. 84, p. 3443–3462.

Johnson, G. L., Vogt, P. R. and Schneider, E. D.
1971 : Morphology of the northeastern Atlantic and Labrador Sea: Deutsche Hydrographische Zeitschrift, v. 24, p. 49–73.

Keen, C. E. and Barrett, D. L.
1981 : Thinned and subsided continental crust on the rifted margin of eastern Canada: crustal structure, thermal evolution and subsidence history: Geophysical Journal of Royal Astronomical Society, v. 65, p. 443–465.

Keen, C. E. and Hyndman, R. D.
1979 : Geophysical review of the continental margins of eastern and western Canada: Canadian Journal of Earth Sciences, v. 16, p. 712–747.

Kidd, R. B., Searle, R. C., Ramsay, A.T.S., Prichard, H., and Mitchell, J.
1982 : The geology and formation of King's Trough, northeast Atlantic Ocean: Marine Geology, v. 48, p. 1–30.

Kristoffersen, Y.
1978 : Sea-floor spreading and the early opening of the North Atlantic: Earth and Planetary Science Letters, v. 38, p. 273–290.

Kristoffersen, Y. and Talwani, M.
1977 : Extinct triple junction south of Greenland and the Tertiary motion of Greenland relative to North America: Geological Society of America Bulletin, v. 88, p. 1037–1049.

Laughton, A. S.
1971 : South Labrador Sea and the evolution of the North Atlantic: Nature, v. 232, p. 612–617.

Laughton, A. S.
1972 : The southern Labrador Sea—Key to the mesozoic and early Tertiary evolution of the north Atlantic: in Initial Reports of the Deep Sea Drilling Project, U.S. Government Printing Office, Washington, D.C., v. 12, p. 1155–1179.

LePichon, X., Bounin, J., Francheteau, J. and Sibuet, J. C.
1971 : Une hypothese d'evolution tectonique du Golfe de Gascogne: in, l'Histoire structurale du Golfe de Gascogne, Editions Technips, "Coll. et seminaires" No. 22, p. 11–44.

LePichon, X. and Sibuet, J. C.
1971 : Western extension of boundary between European and Iberian plates during the Pyrenean Orogeny: Earth and Planetary Science Letters, v. 12, p. 83–88.

LePichon, X., Sibuet, J. C. and Francheteau, J.
1977 : The fit of the continents around the North Atlantic Ocean: Tectonophysics, v. 38, p. 169–209.

Masson, D. G. and Miles, P. R.
1984 : Mesozoic seafloor spreading between Iberia, Europe and North America; Marine Geology, v. 56, p. 279–289.

McWhae, J.R.H. and Michel, W.F.E.
1975 : Stratigraphy of Bjarni H-81 and Leif M-48, Labrador Shelf: Bulletin of Canadian Petroleum Geology, v. 23, p. 361–382.

Miall, A. D.
1983 : The Nares Strait problem: a re-evaluation of the geological evidence in terms of a diffuse, oblique-slip plate boundary between Greenland and the Canadian Arctic Islands, in M. Friedman and M. N. Toksöz eds., Continental Tectonics: Structure, Kinematics and Dynamics: Tectonophysics, v. 100, p. 229–239.

Maill, A. D.
1984 : Sedimentation and tectonics of a diffuse plate boundary: the Canadian Arctic Islands from 80 Ma B.P. to present: Tectonophysics, v. 107, p. 261–277.

Miles, P. R. and Parson, L. M.
1982 : Mesozoic magnetic anomaly identification in the northwest Atlantic: EOS, Transaction American Geophysical Union, v. 63, p. 1281, (Abs.).

Montadert, L., Robert, D. G., de Charpal, O. and Guennoc, P.
1979 : Rifting and subsidence of the northern continental margin of the Bay of Biscay: in Initial reports of the Deep Sea Drilling Project, v. 48, eds. L. Montadert and D. G. Roberts, U.S. Government Printing Office, Washington, D.C., p. 1025–1060.

Morgan, W. J.
1983 : Hot spot tracks and the early rifting of the Atlantic: Tectonophysics, v. 94, p. 123–139.

Nunns, A. G.
1982 : The structure and evolution of the Jan Mayen Ridge and surrounding areas: American Association of Petroleum Geologists Memoir, v. 34, p. 193–208.

Nunns, A. G.
1983 : Plate tectonic evolution of the Greenland-Scotland Ridge and surrounding regions: in Structure and development of the Greenland-Scotland Ridge, new methods and concepts, eds. M.P.H. Bott, S. Saxou, M. Talwani and J. Thiede: Plenum Publishing Corporation, New York, p. 11–30.
Olivet, J. L., LePichon, X., Monti, S. and Sichler, B.
1974 : Charlie-Gibbs Fracture Zone: Journal of Geophysical Research, v. 79, p. 2059–2072.
1981 : Cinematique de L'Atlantic nord et central: CNEXO, Cartes edities par le BEICIP.
Phillips, J. D. and Tapscott, C.
1981 : The evolution of the Atlantic Ocean north of the Azores or Seven easy pieces: Unpublished manuscript, Woods Hole Oceanographic Institute.
Pitman, W. C. and Talwani, M.
1972 : Sea-floor spreading in the North Atlantic: Geological Society of America Bulletin, v. 83, p. 619–649.
Price, I. and Rattey, R. P.
1984 : Cretaceous tectonics off mid-Norway: implications for the Rockall and Faeroe-Shetland troughs: Journal of Geological Society of London, v. 141, p. 985–992.
Roberts, D. G.
1975 : Marine geology of the Rockall Plateau and Trough: Philosophical Transaction of the Royal Society of London, Series A. v. 278, p. 447–509.
Roberts, D. G. and Jones, M. T.
1985 : Magnetic anomalies in the northeast Atlantic, Sheets 1 and 2 Report No. 207: Institute of Oceanographic Sciences, U.K.
Roberts, D. G., Masson, D. G. and Miles, P. R.
1981 : Age and structure of the southern Rockall Trough: New evidence: Earth and Planetary Science Letters, v. 52, p. 115–128.
Roots, W. D. and Srivastava, S. P.
1984 : Origin of the marine magnetic quiet zones in the Labrador and Greenland Seas: Marine Geophysical Researches, v. 6, p. 395–408.
Russell, M. J.
1976 : A possible Lower Permian age for the onset of ocean floor spreading in the northern North Atlantic: Scottish Journal of Geology, v. 12, p. 315–323.
Russell, M. J. and Smythe, D. K.
1978 : Evidence for an early Permian oceanic rift in the northern North Atlantic: in Petrology and geochemistry of continental rifts, eds. E. R. Neumann and I. B. Ramberg: Reidel Publishing Co., Dordrecht, p. 173–179.
Schouten, H., Srivastava, S. P. and Klitgord, K. D.
1986 : Iberian plate kinematics: the African connection: Nature, in press.
Sclater, J. G., Hellinger, S. and Tapscott, C.
1977 : The paleobathymetry of the Atlantic Ocean from the Jurassic to the present: Journal of Geology, v. 85, p. 509–552.
Searle, R. C. and Whitmarsh, R. B.
1978 : The structure of King's Trough, northeast Atlantic, from bathymetric, seismic and gravity studies: Geophysical Journal of Royal Astronomical Society, v. 53, p. 259–287.
Srivastava, S. P.
1978 : Evolution of the Labrador Sea and its bearing on the early evolution of the North Atlantic: Geophysical Journal of Royal Astronomical Society, v. 52, p. 313–357.
Srivastava, S. P.
1985 : Evolution of the Eurasian Basin and its implications to the motion of Greenland along Nares Strait: Tectonophysics, v. 114, p. 29–53.
Sullivan, K. D.
1983 : The Newfoundland Basin: ocean-continent boundary and Mesozoic sea-floor spreading history: Earth and Planetary Science Letters, v. 62, p. 321–339.
Talwani, M. and Eldholm, O.
1977 : Evolution of the Norwegian-Greenland Sea: Geological Society of America Bulletin, v. 88, p. 969–999.
Tucholke, B. E. and Ludwig, W. G.
1982 : Structure and origin of the J anomaly ridge, Western North Atlantic Ocean: Journal of Geophysical Research, v. 87, p. 9389–9407.
Unternehr, P.
1982 : Etude structurale et cinematique de la mer de Norvege et du Groenland—evolution du microcontinent de Jan Mayen; Diplome de Doctoral de 3 eme cycle: L'Universite de Bretagne Occidentale, Brest, France, 228 p.
Vink, G. E.
1982 : Continental rifting and the implications for plate tectonic reconstructions: Journal of Geophysical Research, v. 87, p. 10677–10688.
Vink, G. E.
1984 : A hotspot model for Iceland and the Voring Plateau: Journal of Geophysical Research, v. 89, p. 9949–9959.
Vogt, P. R. and Avery, O. E.
1974 : Detailed magnetic surveys in the Northeast Atlantic and Labrador Sea: Journal of Geophysical Research, v. 79, p. 363–389.
Vogt, P. R., Taylor, P. T., Kovacs, L., and Johnson, G. L.
1979 : Detailed aeromagnetic investigation of the Arctic Basin: Journal of Geophysical Research, v. 84, p. 1071–1089.
Vogt, P. R., Taylor, P. T., Kovacs, L. C. and Johnson, G. L.
1982 : The Canadian Basin: Aeromagnetic constraints on structure and evolution: Tectonophysics, v. 89, p. 295–336.
Voppel, D., Srivastava, S. P. and Fleischer, U.
1979 : Detailed magnetic measurements south of the Iceland-Faeroe Ridge: Deutsche Hydrographische Zeitschrift, v. 32, p. 154–172.
White, R. S. and Mathews, D. H.
1980 : Variations in oceanic upper crustal structure in a small area of the northeastern Atlantic: Geophysical Journal of Royal Astronomical Society, v. 61, p. 401–436.
Whitmarsh, R. B., Ginzburg, A. and Searle, R. C.
1982 : The structure and origin of the Azores-Biscay Rise, northeast Atlantic Ocean: Geophysical Journal of Royal Astronomical Society, v. 70, 79–107.
Williams, C. A.
1975 : Sea-floor spreading in the Bay of Biscay and its relationship to the North Atlantic: Earth and Planetary Science Letters, v. 24, p. 440–456.
Williams, C. A. and McKenzie, D.
1971 : The evolution of the Northeast Atlantic: Nature, v. 232, p. 168–173.
Wynn, P. J., Irving, E., and Ozadetz, K.
1983 : Paleomagnetic evidence for 30° anticlockwise rotation of northern Ellesmere Island, possible key to solution of Nares Strait problem: Geological Association of Canada. Annual Meeting Programme Abstract, v. 8, p. A75.

Manuscript Accepted by the Society October 22, 1985

ACKNOWLEDGMENTS

One of us (SPS) would like to extend his thanks to his wife for her assistance in the work reported here. We thank Allan Nunns for making available to us copies of his digital files of the magnetic lineations, fracture zones and bathymetry of the Norwegian-Greenland Sea; Peter Miles of IOS for supplying us magnetic data collected off Newfoundland; Ruth Jackson and John Woodside for critically reading this manuscript. We benefitted greatly from the observations and suggestions from the two referees of this paper, Allan Nunns and Gregory Vink, and would like to thank them both. Illustrations shown in this paper were prepared by the drafting, illustration, and photographic section of the Bedford Institute of Oceanography, to whom we extend our thanks. We also thank the typing staff at the Bedford Institute of Oceanography for typing this manuscript many times.

Printed in U.S.A.

The Geology of North America
Vol. M, The Western North Atlantic Region
The Geological Society of America, 1986

Chapter 24

Plate kinematics during the last 20 m.y. and the problem of "present" motions

P. R. Vogt
Naval Research Laboratory, Washington, D.C. 20375-5000

INTRODUCTION

While increasingly precise plate reconstructions are being made at intervals of the order 10 m.y. for times corresponding to well-defined magnetic lineations (Klitgord and Schouten, this volume; Srivastava and Tapscott, this volume), much less is known about possible irregularities in plate motions over shorter time scales. Obviously the most recent geologic past offers the best hope for detecting any "kinematic fine structure." This paper discusses the last 10–20 m.y., with special emphasis on the "present," which, of course, is not an instant but involves a finite averaging interval, whether several years for geodetic or earthquake measurements, or several million years for a measurement based on magnetic anomalies. The chapter begins with Wegener's attempts to measure continental drift, reviews the geodetic measurements on Iceland where the MAR (Mid-Atlantic Ridge) plate boundary is exposed, and discusses the latest attempts to measure plate motions over the "historical" present (several years time scale). The latter field is just being launched (see Anderle, this volume), and whatever is written today will soon be outdated. The main part of this paper is devoted to the question of relative and absolute plate motions over the "geological" present (m.y. time scale) resolved by magnetic lineations, transform faults, and hotspot traces. There is very little information about variability of plate motion over time scales intermediate between the "historical" and "geological" present. A case is made for significant geologically "short-period" (a few m.y.) global fluctuations in rates of plate motions over the last 10–20 m.y. The latest episode began only about 1 Ma and the mid-to-late Quaternary appears anomalous in several respects. This throws some doubt on the steady-state character of the MAR rift valley presently observed. Insofar as mid-plate stresses are related to plate motions (e.g., Zoback and others, this volume), it may also be hazardous to lump late Quaternary and recent tectonic, earthquake, and stress data together with early Quaternary or "late Tertiary" fault displacement.

WEGENER'S THINKING

Given that continents have drifted apart, did this drift begin and end long ago, or is it still in progress, waiting to be measured and thereby to be demonstrated? Alfred Wegener (1929, p. 23) reasoned, "If continental displacement was operative for so long a time, it is probable that the process is still continuing, and it is just a question of whether the rate of movement is enough to be revealed by our astronomical measurements in a reasonable period of time." Wegener was misled by his own errors—and those in the fledgling absolute geologic time scale—to believe that drift rates were high enough to be measurable. Curiously for a meteorologist and a student of the Greenland ice sheet, Wegener matched the southern limits of Quaternary glaciation on a reconstructed North Atlantic, and concluded that Greenland, North America, and Europe had moved apart *after* the glaciation, that is, from the dating available then, in the last 50,000–100,000 years. This implied drift rates of the order 10 to 40 m/year. Longitude differences between northeastern Greenland and Europe at various times from 1823 to 1927 appeared to support the sense and magnitude of the predicted rates, misleading Wegener into what, in hindsight, was a lot of wasted effort in a measurement that would still be difficult a half-century later. Wegener was only one order of magnitude too high in his estimate of present drift rates in other oceans (e.g., the central and South Atlantic and Indian oceans). For these oceans he had rather correctly estimated the *geologic* times of breakup, but radiometric dating had just begun and the absolute ages were underestimated. Wanach (1926) examined successive determinations of longitude between Europe and North America (1921–1925) with the help of radio time signals. From his measured westward drift rate of 0.6 ± 2.4 m/yr, Wanach correctly concluded that "any displacement of America with respect to Europe of appreciably more than 1 m/year is most improbable." Yet, only two years later, Littell and Hammond (1928) concluded that the longitude difference between Washington and Paris had increased between 1913/14 and 1927, indicating an average annual rate of 0.32 ± .08 m/yr, of the same magnitude of Wegener's prediction for the central Atlantic.

As the absolute geologic time scale evolved, it became ap-

Vogt, P. R., 1986, Plate kinematics during the last 20 m.y. and the problem of "present" motions, *in* Vogt, P. R., and Tucholke, B. E., eds., The Geology of North America, Volume M, The Western North Atlantic Region: Geological Society of America.

parent that drift rates could have been only on the order 1 to 10 cm/year. Thus the hope of geodetic measurement of continental drift faded. Yet, had the significance of Iceland been appreciated by the early advocates of drift, a rough geodetic measurement of drift might have been possible much earlier. It is an enduring mystery how Wegener, who traversed Iceland in 1912, believed that this seismically and volcanically active island, with its gaping fissures and normal faults, could be a mere continental fragment. (Ironically this same belief has been promoted in modern times by opponents of plate tectonics; e.g., Zverev and others, 1976.)

GEODETIC MEASUREMENTS ACROSS THE NOAM-EURA BOUNDARY IN ICELAND

The modern view of Iceland can be said to have begun with Nils Nielsen, who sponsored Wegenerian drift during a time when it was hotly debated (Drake and Girdler, 1982). Unlike Wegener, however, Nielsen realized the importance of the fissure eruptions and fault structures that were "the result of a pull from east to west which has simply split the land into innumerable fissures" (Nielsen, 1930). The first effort to measure the rate of these tensional movements directly, by geodetic means, was a 1938 German expedition to Iceland led by O. Niemczyk. A network of precisely measured benchmarks was established across the rift zone in North Iceland (Niemczyk and Emschermann, 1943). Accompanying the 1938 expedition was F. Bernauer, who noted that fissures splitting post-glacial lavas had been widening at an average rate of 3.56 m per km per 1,000 years (Bernauer, 1943). These measurements were made in the Krafla fissure swarm, which is some 3 to 5 km wide. As pointed out by Tryggvason (1982), Bernauer's result indicates that the Krafla swarm has widened by 10 to 15 m per 1,000 years in the 5,000 to 10,000 years since deglaciation. This works out to 1.0 to 1.5 cm/yr. Since there are several other fissure swarms as well as normal faults, in the Krafla region, Bernauer's result is not inconsistent with a plate separation rate of 2.0 cm/yr over the last few million years (Vogt and others, 1980). The aggregate widths of post-glacial fissures provides one of the few available estimates of plate motion on a time scale of 10^4 years, intermediate between the measurements using magnetic anomalies (7×10^5 years and longer), and geodetic measurements (<50 years). Bernauer's views on the significance of Iceland in terms of the continental drift hypothesis were similar to those of Holmes (1944, and earlier papers) and anticipated modern concepts of seafloor spreading and plate tectonics (Schwarzbach, 1980; Drake and Girdler, 1982).

Niemczyk's geodetic network across the northern Neo-Volcanic Zone was finally reoccupied and improved starting in 1964–1965 (Gerke, 1974; Wendt and others, 1985). Distance measurements began in south Iceland in the late 1960s: across the tip of Reykjanes Peninsula in 1968 and 1972 (Brander and others, 1976); across the Thingvellir graben (western Neo-Volcanic Zone) by separate British (1968, 1972), German (1967, 1971), and U.S. (1967, 1970, 1973) teams (Brander and others, 1976; Gerke, 1974; Decker and others, 1971, 1976); and across the southern part of the eastern Neo-Volcanic Zone, just north of the volcano Hekla (in 1967, 1970, and 1973; Decker and others, 1976).

On December 20, 1975, the first of numerous volcano/tectonic events, each consisting of slow inflation and rapid deflation of the Krafla caldera, affected the Krafla fissure swarm in North Iceland. The twentieth event (18–23 November 1981) was thought to be the last, but the occurrence of yet another (4–18 September 1984) indicates that further activity cannot be entirely ruled out. Only nine of the 21 events (six of the last eight) resulted in lava extrusion (Björnsson, 1985), but all events involved injection of magma north or south into fissures. (Whether by coincidence or not, the Afar area also experienced a fissure eruption about the same time—November 1978; Tryggvason, 1982.) The Krafla episode stimulated further geodetic work, most recently reviewed by Wendt and others (1985), Björnsson (1985), and Torge and Kanngieser (1985). As summarized by Tryggvason (1982), the horizontal distance measurements—emphasized in this chapter because of their relevance to "present-day plate motion"—were accompanied by precision leveling and measurements of tilt, gravity, fissure width, and changes of river courses and lake shores. Between volcano-tectonic episodes, the rift zone and its flanks are stretched and subside; during episodes, the flanks contract and rebound essentially to their previous state while the rift zone subsides by 1–2 m (Tryggvason, 1982; Björnsson, 1985). Notwithstanding the large amount of geodetic work on Iceland, the rate of present plate motion has not yet been reliably established by direct measurements there. However, the net result of the Krafla episode has been to widen the Krafla fissure zone by an additional approximate 7.5 m (Wendt and others, 1985) since the previous ("Myvatn fires") episode that occurred in the same area in 1724–1730 (Tryggvason, 1982). This is equivalent to about 3 cm/yr rate of plate separation, slightly higher than the 2 cm/yr for the last few million years measured from magnetic anomalies near Iceland (e.g., Vogt and others, 1980). However, such a comparison is only valid if there has been no net change since 1730 in the integrated strain (besides the 7.5 m fissure opening) on an east-west line crossing the entire Neo-Volcanic Zone in the Krafla area.

MODERN EFFORTS TO MEASURE PRESENT PLATE MOTION

Advances in radio-electronics, computers, and rocketry made it possible to contemplate the direct measurement of relative plate motion starting about the time the reality of plate tectonics was established in the 1960s. Movements between plates—as well as possible intraplate deformation—can now be measured at least in principle by laser-range or radio-frequency Doppler measurements on artificial satellites, laser range measurements to laser reflectors on the moon, short and medium base line electronic phase difference measurements using artificial satellites, and very long base line (VLBI) observations of quasars (Anderle, this volume; Committee on Geodesy, 1985). Although the lunar laser, VLBI, and GPS (Global Positioning Satellite) techniques are potentially the most accurate, the radio-frequency Doppler data are most abundant and extend back to 1963. Anderle and Malyevac (1983) analyzed the data from one of the TRANSIT navigation satellites for the decade 1973–1983. The calculated plate speeds generally agreed in magnitude and sign with those measured over the "geological present" (e.g., Minster and Jordan, 1978), suggesting that plate motion was actually measured—about a half century after Wegener first contemplated it—and also using radio waves, but in a way that could hardly have been imagined in 1930. The more accurate satellite laser ranging (SLR) measurements (1979–1982) from ground stations on the North American (NOAM), South American (SOAM), Pacific (PCFC), and Australian plates to the LAGEOS (Laser Geodynamics Satellite) showed overall agreement with the geologic rates (Tapley and others, 1985; Christodoulidis and others, 1985). In the North Atlantic, data from VLBI observatories at Westford (Massachusetts), Onsala (Sweden), and Wettzell (West Germany; Fig. 1) suggest that significant plate separation has occurred even during the short observation periods 1981–1985 (Onsala-Westford) and 1984–1985 (Wettzell-Westford; Carter and others, 1985). Different methods yield lengthening rates of 2.0 ± 0.2, 2.3 ± 0.3, and 3.2 ± 0.6 cm/yr for the 5,600 km long Onsala-Westford baseline and 1.2 ± 0.4 cm/yr for Wettzell-Westford. The true error is probably about 1 cm/yr, larger than the formal errors because of systematic errors introduced by unmodeled complexities in the atmosphere (chiefly H_2O content) and in the earth's rotation. At present the data suggest global plate motions are mostly within 1 or 2 cm/yr of the geological rates, but several more years will be required to test whether, for example, plate motion has tended to be 10% to 30% faster over the last million years (and hence perhaps at the present time) compared to rates averaged over the last four or more million years (See later section, "Evidence for Time Variations in Relative Plate Motions during the Last 10-20 m.y.") Apparently significant intraplate deformation—at rates comparable to inter-plate motions—were reported in the initial LAGEOS and VLBI studies (Christodoulidis and others, 1985; Carter and others, 1985). Crustal shortening of about 1 cm/yr between Westford and Texas (Carter and others, 1985; Musman and Schmidt, 1986) is consistent with stress measurements indicating much of the North American plate is under compression (Zoback and others, this volume). However, intraplate deformation at such high rates cannot have been sustained for geologically significant times (1 cm/yr is 100 km/10 m.y.) without being evident as orogenies or misfits in plate tectonic reconstructions. Therefore, if the intraplate rates are correct in magnitude, the "historical present" (and perhaps Quaternary) are atypical. This may also be true of vertical motions in some areas. For example, the Rhine graben is presently subsiding ten times as fast as the post-Miocene average (Illies and others, 1979). However, the influence of late Quaternary glaciation and deglaciation on vertical motions in plate interiors needs to be understood before geodynamic effects can be isolated.

RELATIVE PLATE MOTIONS DURING THE "GEOLOGICAL PRESENT"

Many papers have reported opening rates, fracture zone strikes, and earthquake slip vectors at specific localities along the Mid-Atlantic Ridge. There have been only a few attempts to integrate this information to find what pole location and angular opening rate best fits all the data (Minster and others, 1974; Minster and Jordan, 1978, 1980; Savostin and Karasik, 1981; DeMets and others, 1986). The rotation poles are plotted in Figure 2 and several are listed, together with rotation rates and related information, in Table 1. Savostin and Karasik (1981) examined only NOAM/EURA (North American plate/Eurasian plate) motion, but the other studies represent simultaneous inversions of global data sets and data sets for individual boundaries. The magnetic anomaly (spreading rate) and bathymetric (transform azimuth) data on "present" plate motion necessarily represent an average over the last few million years. By contrast, earthquake slip vectors represent relative plate motion during the "historical present." (The possibility that plate motions have not been constant even over the last 1–2 m.y. is examined later.)

The best-fitting global model of Minster and others (1974) was called RM1 ("RM" for relative motion); RM2 (Minster and Jordan, 1978, 1980) was a refined and updated model derived by the same methods but using 330 data points from all the major accreting and transform plate boundaries. Opening rates were based on anomalies 2 and 2A (1.8 and 3 Ma, see Kent and Gradstein, this volume) except for a few slow-spreading ridges where anomaly 3 was used. Thus RM2 is more "present" than RM1, which used anomalies 3 and 5. Both models are based on the reversal time scale of Talwani and others (1971). For the three plate boundaries represented by the MAR, RM2 used the following data: EURA/NOAM, 8 opening rates, 4 transform azimuths, and 7 fault-plane solutions (slip vectors); AFRC/-NOAM, 5 opening rates, 4 transform azimuths, and 2 slip vectors; and AFRC/SOAM, 10 opening rates, 15 transform azimuths, and 4 slip vectors. These data sets allowed a separate inversion for each plate boundary, giving a set of best fitting poles (BFP). A more recent set of models ("NUVEL-1", DeMets and others, 1986) incorporates newer, more accurate anomaly and transform-strike data but is based on similar time intervals (anomaly 5 for EURA/NOAM and 2A for all others) and the reversal time scale of Harland and others (1982). Plate rotation poles were used to define the x-axis in the zero-age synthesis chart (Plate 8A) and in the analysis of the present plate boundary shape (Vogt, this volume, Ch. 12). BFP-derived opening rates are compared to measured rates in Plate 8A. Both RM2 and BFP solutions and the newer NUVEL models generally fit the rate data to within a few mm/yr and the azimuth data to within a few degrees.

Differences between the DeMets and others (1986) and

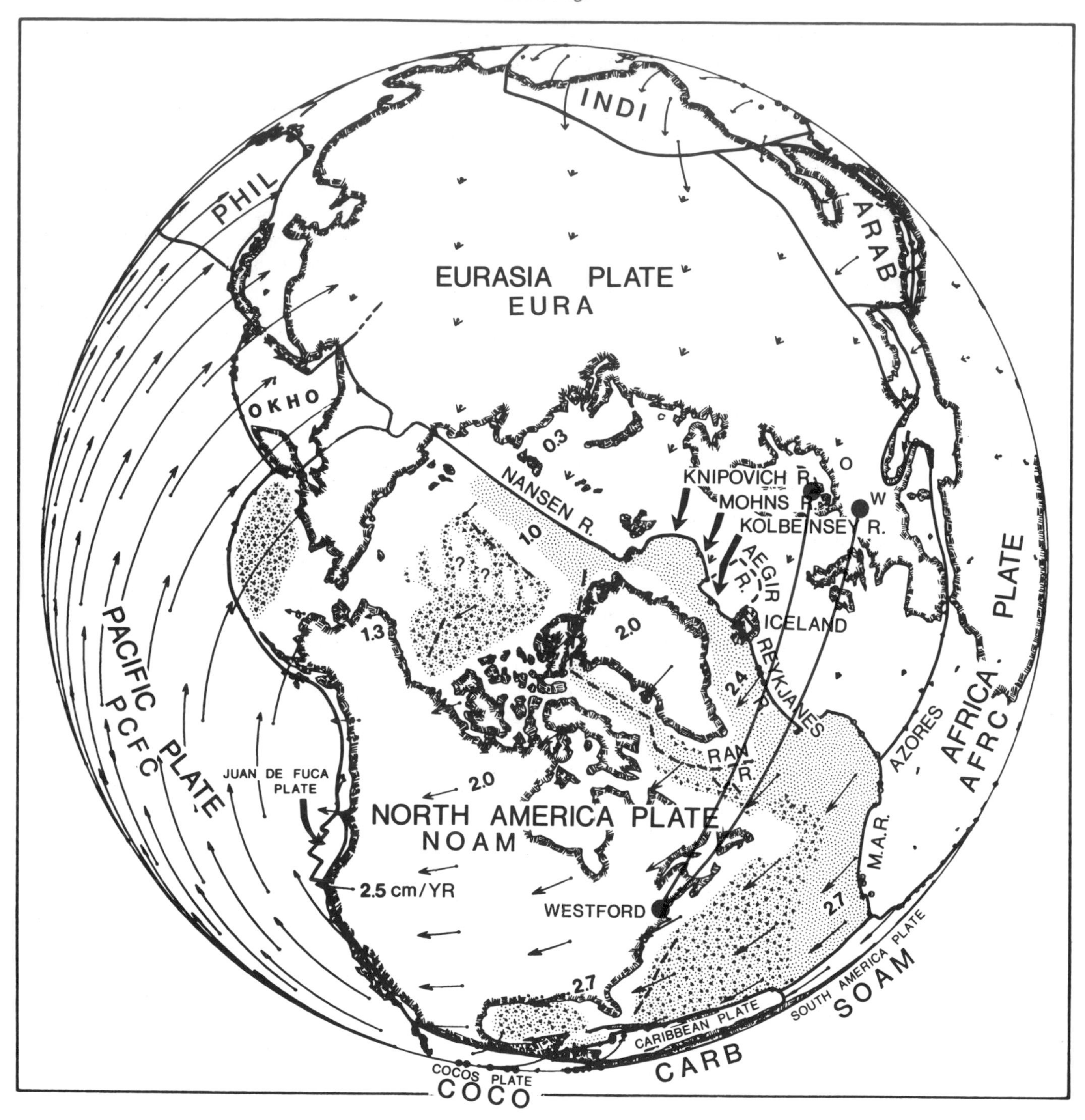

Figure 1. Polar view showing lithospheric plates, plate name abbreviations, and present absolute motion vectors in a fixed hotspot frame. No significance can be attached to absolute motion directions on the Eurasia plate. Oceanic lithosphere included in North America plate is stippled, with coarse stippling denoting Mesozoic oceanic crust. Great circle arcs (red) show baselines connecting VLBI observatories at Westford, Massachusetts (and nearby Haystack Observatory) with observatories at O, Onsala, Sweden; and W, Wettzell, West Germany; (Carter and others, 1985). Modified from Minster and Jordan (1980).

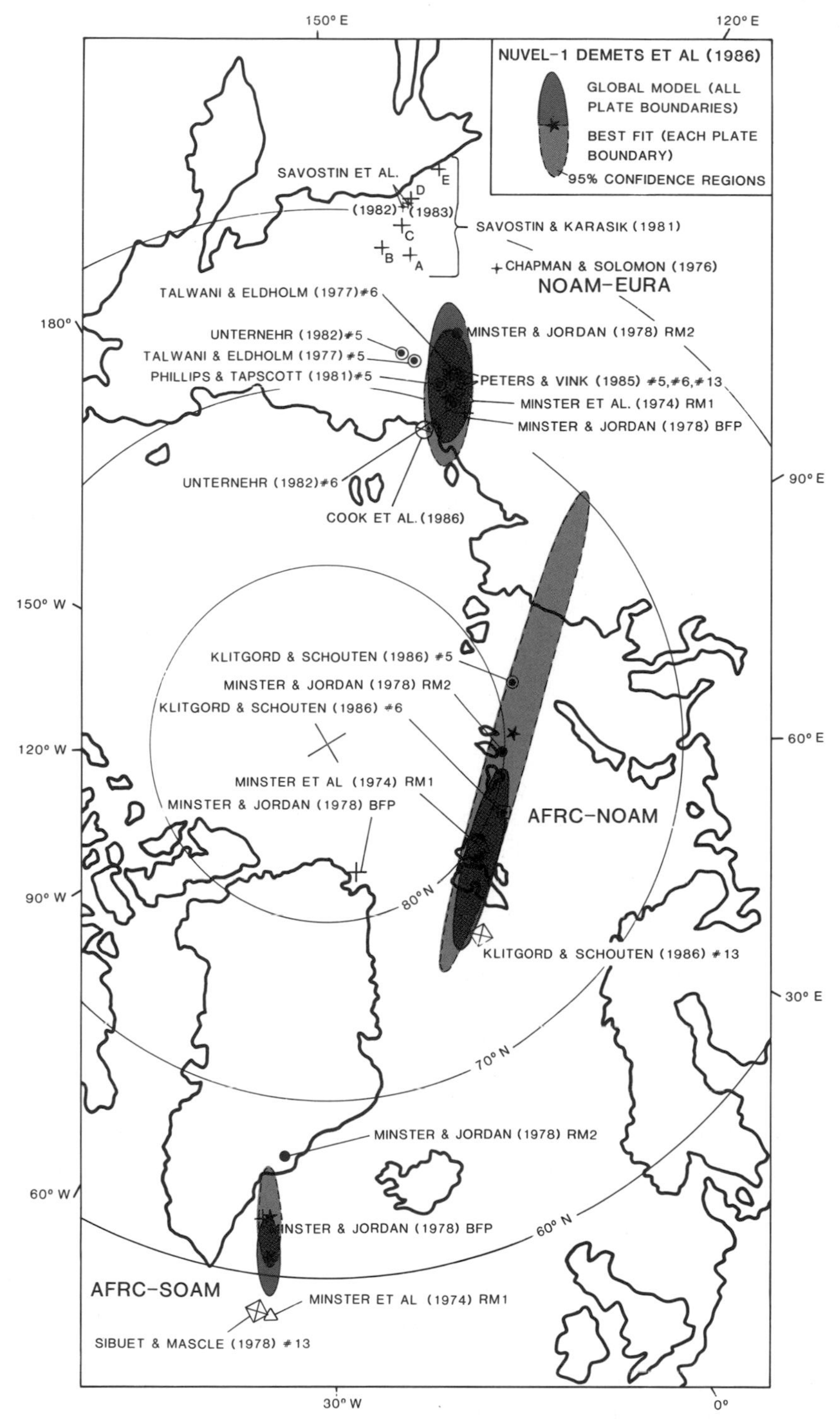

Figure 2. Poles of opening for the Africa–South America (AFRC/SOAM), Africa–North America (AFRC/NOAM), and Eurasia–North America plates (EURA/NOAM). "Present" poles are from Minster and others (1974; model RM 1), Minister and Jordan (1978; models RM2 and BFP), (DeMets and others (1986); global and best fitting poles and confidence regions), and Cook and others (1986; based on fault-plane solutions). Closure poles are indicated by anomaly number (e.g., #5). Poles labelled A through E (crosses) are estimates of present EURA/NOAM motion using the following combinations of data sets (Savostin and Karasik, 1981; see text for discussion): A, from the Knipovich Ridge south to the Reykjanes Ridge; B, from the strikes of supposed transform faults in the Cherskiy Ridge in Siberia; C, combined A and B data; D, A and B together with the eastern half of Nansen Ridge; E, all data, including the Spitsbergen-Molloy fracture zones and southwestern Nansen Ridge. "Klitgord and Schouten (1986)" refers to Klitgord and Schouten, this volume.

TABLE 1. ROTATION POLES AND ANGULAR RATES FOR "PRESENT" MOTIONS OF NORTH AMERICAN (NOAM), SOUTH AMERICAN (SOAM), AND AFRICAN (AFRC) PLATES

Global Inversion

Plate Pair	Rotation Pole (Euler Vector) RM-2	NUVEL-1	Rotation Rate (w) RM-2 (°/m.y.)	NUVEL-1 (°/m.y.)	Crustal area acccretion rate (km^2/yr)	% of total global accretion rate**	Crustal volume production rate*** (km^3/yr)	Mean opening rate* (cm/yr)	Range* (cm/yr)
EURA-NOAM	65.85°N,132.44°E	67.96°N,131.68°E	0.231	0.255±.032	0.1223	4.0	0.79	1.76	0.47-2.68
AFRC-NOAM	80.43°N, 56.36°E	79.06°N, 23.16°E	0.258	0.263±.025	0.0716	2.4	0.47	2.45	2.05-2.75
AFRC-SOAM	66.56°N, 37.29°W	61.10°N, 37.28°W	0.356	0.325±.010	0.2725	9.1	1.77	3.34	2.59-3.59

Best Fit to Plate Boundary

Plate Pair	Rotation Pole (Euler Vector) RM-2	NUVEL-1	Rotation Rate (w) RM-2 (°/m.y.)	NUVEL-1 (°/m.y.)	Crustal area acccretion rate (km^2/yr)	% of total global accretion rate**	Crustal volume production rate*** (km^3/yr)	Mean opening rate* (cm/yr)	Range* (cm/yr)
EURA-NOAM	69.70°N,127.37°E	69.30°N,130.56°E	0.252	0.266±.035	0.1239	4.1	0.81	1.78	0.42-2.77
AFRC-NOAM	82.68°N, 17.68°W	79.23°N, 62.93°E	0.275	0.243±.045	0.0711	2.4	0.46	2.42	2.12-2.63
AFRC-SOAM	62.98°N, 39.14°W	63.14°N, 37.53°W	0.357	0.322±.010	0.2707	9.0	1.76	3.32	2.65-3.56

Absolute Motions (AM1-2)

Plate	Pole	Rotation Rate (o/m.y.)	Error Ellipse Semi-major axis strike	Semi-major axis length	Semi-major axis length	Rotation rate error
NOAM	58.31°S, 40.67°W	0.247	S57°E	23.12°	12.14°	0.080
EURA	0.70°N, 23.19°W	0.038	S67°E	151.10°	118.90°	0.057
AFRC	18.67°N, 21.76°W	0.139	S73°E	40.40°	33.24°	0.055
SOAM	82.28°S, 75.67°E	0.285	N03°E	19.28°	11.38°	0.084

*Based on NUVEL-1 (DeMets and others, 1986) solutions; assumed: northern limit of EURA-NOAM accretion at 77.5°N, 128.3°E; EURA-NOAM-AFRC (Azores) triple junction at 39.38°N, 29.72°W; NOAM-SOAM-AFRC triple junction at 15.25°N, 44.9°W; and SOAM-AFRC-ANTA (Bouvet) triple junction at 55°S, 1°W. AFRC-SOAM rotation equator is near 21.5°S, 11.3°W (global model) and 24.9°S, 13.7°W (best fitting model).

**Based on total global accretion rate of 2.995 km^2/yr (Parsons, 1981) and NUVEL-1.

***Based on crustal thicknesss of 6.5 km.

Minster and Jordan (1978) pole locations are generally modest—within or close to each other's 95% confidence limits (Fig. 2). The only notable exception occurs in the FAMOUS area (Fig. 3) where RM2 and the NUVEL-1 poles predict transform trends of 102°–104°, but the observed bathymetry (Fig. 3; Phillips and Fleming, 1978) shows fracture zones trending about E–W, a 10°–15° discrepancy. Minster and Jordan considered a possible recent change in plate motion but rejected this because the earthquake slip vector data, which represent the "historical present," do not support such a change (However, as they note, a slip vector of 090° on the Oceanographer FZ (Udias and others, 1976) is also 12° from the RM2-predicted trend). The BFP pole naturally fits the "E–W" FZ trends in the FAMOUS area better, but the resulting pole is near Greenland (Fig. 2), not on a great circle connecting the EURA-NOAM and AFRC-EURA poles as required by the closure condition about the Azores triple junction. This pole is also some distance from NOAM/AFRC closure poles for anomalies 13, 6 and 5 (Fig. 2). As noted by Minster and Jordan, it is possible that FZ'S A and B are actually oblique spreading axes ("leaking" fracture zones) and do not record the correct azimuth of relative plate motion. Detailed bathymetric charts of other parts of the MAR (e.g., Fig. 4; also Rona and Gray, 1980) generally support this interpretation: slowly accreting plate boundaries seem to favor "underlapping" rather than overlapping spreading centers, the "underlappers" being connected by highly oblique spreading axes (see also Vogt, this volume, Ch. 12). (Macdonald, this volume, explains the oblique spreading axes in terms of coalescing oblique faults associated with ridge/transform intersections). A section of detailed bathymetry within the E-W-striking FZ A (Fig. 3A) reveals bathymetric trends with a nearly "correct" (RM2 and NUVEL-predicted) trend of about 103°–105°, which may represent the actual present transform motion. At the same time, independent evidence for significant changes in relative plate motion over the last few million years (discussed later) cannot be ignored, and the issue thus remains unresolved.

Using an almost entirely independent data set, Savostin and Karasik (1981) reanalyzed the "present" EURA/NOAM-motion. Their rotation poles (Fig. 2) are based on different combinations of the following data set: 23 focal mechanism solutions for earthquakes, 38 azimuths of slip vectors obtained by matching symmetric mountain pairs on both sides of Knipovich and Nansen ridge, and 14 azimuths of strike-slip faults (presumed transform faults) within the Cherskiy Ridge of Siberia. However, Cook and others (1986) found thrust faulting for events in the Cherskiy Ridge area and suggested that after the Arctic rift propagated through northeast Siberia to the Pacific—creating a separate Okhotsk plate—the NOAM/EURA pole moved back north to the Lena delta area sometime after 3 Ma. Savostin and Karasik (1981) suggested that the Spitsbergen corner of the Eurasia plate behaves as a separate microplate. Their conclusions depend on the bathymetric data used to define the "symmetric mountain pairs" and these data have not been made available. Independent published charts of the Knipovich Ridge confirm that rift mountain peaks line up along "flow lines" of relative plate motion (Vogt, this volume, Ch. 12, Fig. 3B,). However, in the North Atlantic (Johnson and Vogt, 1973; Fig. 4) and central Atlantic (Rona and Gray, 1980), bands of rift mountain peaks do not exactly parallel directions of relative plate motion. This implies an along-strike component of motion (1–2 mm/yr) between

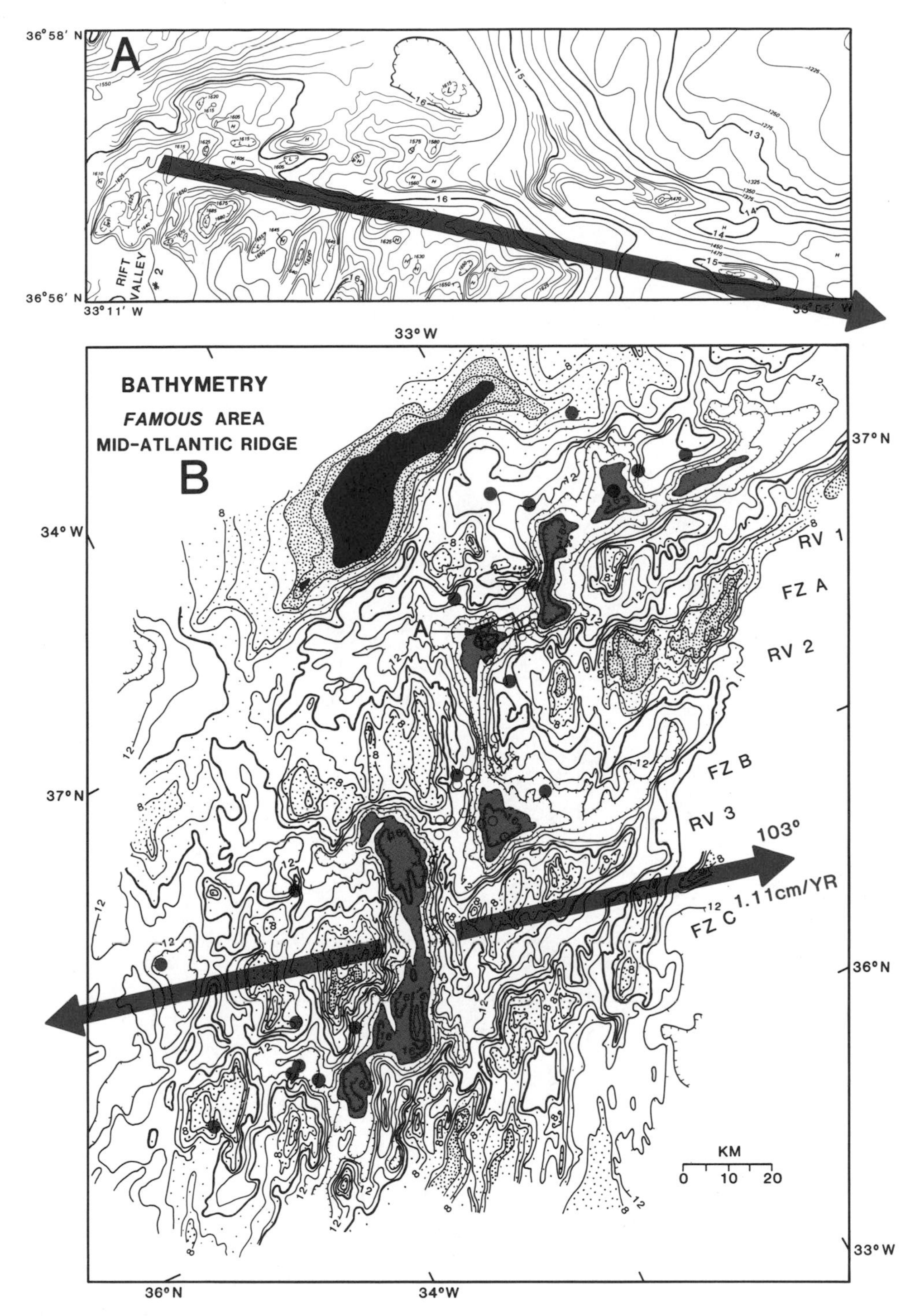

Figure 3. Bathymetry of FAMOUS area of Mid-Atlantic Ridge, based largely on multibeam surveys: A, detailed (5 fm = 9.2 m) contours within part of FZ A of FAMOUS area (from Phillips and Fleming, 1978). Red arrow shows direction of relative NOAM/AFRC plate motion predicted by best-fit model of DeMets and others (1986). B, FAMOUS area at 100 fm (183 m) contour interval, modified from Phillips and Fleming (1978) by Ramberg and van Andel (1977). Black, less than 300 fms deep; coarse stippling, 300–500 fms; moderate stippling, 500–800 fms; light stippling, 800–1,000 fms; unstippled, 1,000–1,500 fms; gray, >1,500 fms deep. Large red closed circles are teleseismic epicenters (Nishenko and others, this volume, Plate 11); small dots, crosses, and open circles are microearthquakes located using local seafloor networks by Francis and others (1977), Spindel and others (1974), and Reid and Macdonald (1973). Arrows and opening rate are predicted plate motion according to "best-fit" model of DeMets and others (1986). Box shows location of detailed map (A).

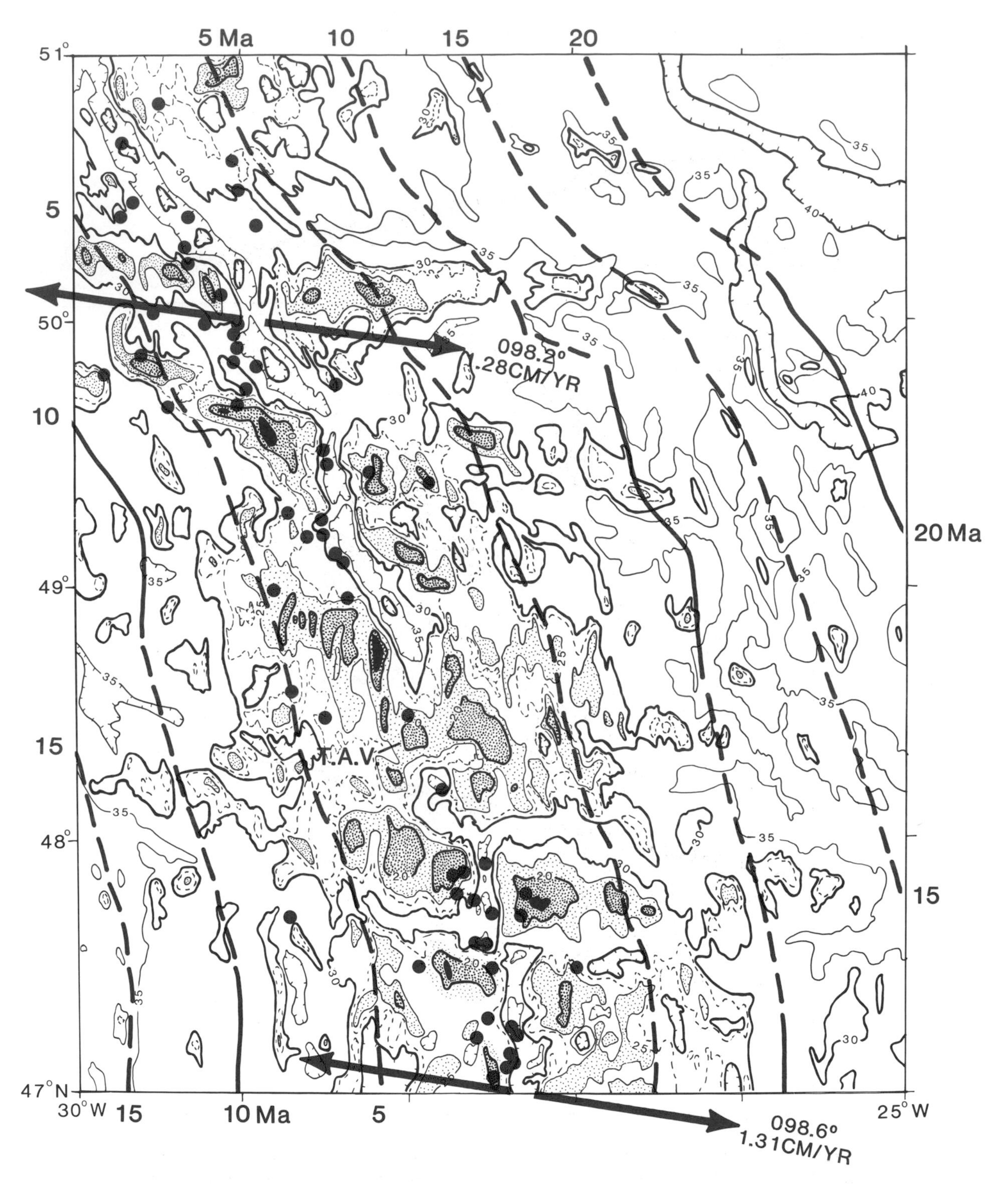

Figure 4. Mid-Atlantic Ridge from 47° to 51°N, showing bathymetric contours at 500 m interval (locally also 250 m, dashed lines; redrawn from Johnson and Vogt, 1973); teleseismic earthquake epicenters (Nishenko and others, this volume, Plate 11); and direction and rate of plate separation predicted by "best-fitting" model of DeMets and others (1986; See Table 1). Depths less than 1,500 m are solid, 1,500–2,000 m thickly stippled, 2,000–2,500 m lightly stippled. Note relatively aseismic zone 48°–49°N associated with "Telegraph Axial Volcano" (labeled TAV). Thin lines are crustal isochrons, dashed where extrapolated primarily from Klitgord and Schouten, and Srivastava and Tapscott, this volume). See Macdonald (this volume) for an explanation of the deep, oblique segments in terms of coalescing transform-ridge boundary effects.

MAR subaxial magma sources and the overlying axial lithosphere (Johnson and Vogt, 1973).

AREA AND VOLUME BUDGETS FOR THE MID-ATLANTIC RIDGE AND BOUNDING PLATES

Plate tectonic models allow ready calculation of areas and volumes of new oceanic crust created (or destroyed) per unit time along each plate boundary, once the relative rotation poles and angular rotation rates, and the approximate configuration of the boundary are known (e.g., Parsons, 1981). Table 1 shows the area generation rates for the three segments of the MAR, computed from the NUVEL model of DeMets and others (1986). Similar values were derived by Parsons (1981) based on the Minster and Jordan model (1978). Crustal volume generation rates (Table 1) are based on a nominal crustal thickness of 6.5 km; lithosphere production rates could be similarly computed using, say, 100 km for equilibrium plate thickness.

The NOAM/EURA, NOAM/AFRC, and SOAM/AFRC segments of the Mid-Atlantic Ridge have been generating 0.12, 0.07, and 0.27 km^2/year (or million km^2/m.y.) for the last few million years (global model, Table 1). The total for the entire MAR from the continental slope off the Laptev Sea in Siberia down the Bouvet triple junction is the sum of these, or 0.47 km^2/year. Of this, only 0.19 km^2/yr is formed by spreading along the eastern edge of the North American plate, while a larger part, 0.27 km^2/yr, is formed by spreading from 15°N south to the Bouvet triple junction. If crustal area is assumed to be accreted evenly to each plate (i.e., spreading is on the average symmetric), the accretion rates are 0.060 km^2/yr to the Eurasia plate, 0.097 km^2/yr to the North American plate, 0.136 km^2/yr to the South American plate, and 0.172 km^2/yr to the African plate. Mean opening rates range from 1.76 cm/yr (NOAM/EURA) to 3.34 cm/yr (SOAM/AFRC), with an all-Atlantic mean of 2.59 cm/yr being about half of the 5.0 cm/yr global mean of Parsons (1981). The entire Mid-Atlantic Ridge system contributes only 11.5% of the 3.0 km^2/yr (Parsons, 1981) produced and consumed globally.

If the "present" opening rates of the Atlantic Ocean were to be maintained until all the other ocean basins have somehow closed and a new single supercontinent has formed, the time of such a "maximum Atlantic" would lie about 470 m.y. in the future. Dividing the present area of a rifted ocean basin by the present opening rates gives a "present-rate age" that will be greater or lesser than the actual time of continental break-up, depending on whether the present opening rates are respectively lesser or greater than the long term average rates. For the entire Atlantic rifted basin (ca., 90×10^6 km^2 oceanic crust), the "present-rate age" is 190 Ma, which exceeds the ages of breakup. High opening rates during the mid- and late Cretaceous (e.g., Klitgord and Schouten, this volume) explain why the "present-rate age" of the Atlantic is greater than it should be. (Note that the formation of a separate North American plate was not completed until about 58 Ma, with the initiation of spreading in the Nordic and Eurasia basins; Vogt, 1986a.)

The North American plate (48×10^6 km^2) presently includes about 14×10^6 km^2 oceanic crust (6×10^6 km^3 Cenozoic-aged crust and 8×10^6 km^2 Mesozoic) and 34×10^6 km^2 continental crust, including about 3×10^6 km^2 of eastern Siberia and its Arctic shelf. Small amounts of NOAM are being subducted under the Caribbean plate, and small amounts are being gained by spreading at the Mid-Cayman Rise. However, spreading along the Mid-Atlantic Ridge is the main process changing (increasing) the area of the North America plate. The "present" accretion to the plate (0.10×10^6 km^2/m.y.) therefore represents a fractional growth rate of about 1.9×10^{-3}/m.y. or 1.9×10^{-9}/yr.

THE "TRANS-POLAR TREND"

Three pairs of plates (SOAM/AFRC, NOAM/AFRC, and NOAM/EURA) share the Mid-Atlantic Ridge as an accreting plate boundary. A comparison of published "present" (last 3 m.y.) and "recent" (last 36 m.y.) poles of relative motion for the three plate pairs (Fig. 2) suggests an interesting pattern, here called the "Trans-Polar Trend" (TPT). There are three elements to the TPT pattern: (A) The "present" and "recent" poles all lie along a "transpolar" trend aligned more or less along the extension of the Atlantic rifted ocean basin across the Arctic and into Siberia; (B) The more northerly the plate boundary, the farther along the TPT is the pole of relative opening (i.e., the NOAM/EURA pole is in Siberia whereas NOAM/AFRC is in the northern Barents Sea, and the SOAM/AFRC pole is in southeast Greenland); and (C) For any plate pair the total opening poles have been migrating along the TPT with time. When probable errors in pole positions are taken into account, observations A and B appear robust whereas C is rather tentative. (Note that since the confidence regions are also unfortunately elongated along the TPT, a set of randomly varying poles for any plate pair would generally lie along the TPT.) For the South Atlantic (SOAM/AFRC), separate poles for anomalies 6 and 5 are not yet published and the anomaly 13 pole (Sibuet and Mascle, 1978) is based on only a few data. Nevertheless the latter pole lies outside the confidence oval of the RM2 and NUVEL poles. The NOAM/AFRC total opening poles for anomalies 13, 6, and 5 (Klitgord and Schouten, this volume) show a clear TPT although only one lies outside the 95% confidence region for the NUVEL poles. For the NOAM/EURA boundary the results are conflicting. The closure poles for anomalies 5, 6, and 13 given by various investigators all cluster in a small area and are statistically indistinguishable from the poles of present opening of Minster and Jordan (1978) and DeMets and others (1986; Fig. 2). This would be consistent with the hypothesis that the NOAM/EURA pole has not moved in the last 36 million years (Peters and Vink, 1985). Contrastingly the "present" (last few million years) opening poles for NOAM/EURA derived by Savostin and Karasik (1981) under a variety of assumptions all lie well outside the 95% confidence regions of the RM2 and NUVEL models. The

Savostin-Karasik poles are consistent with a long-term migration of the NOAM/EURA pole along the TPT. Indeed, Savostin and Karasik (1981) and Savostin and others (1984) independently observed that the NOAM/EURA pole has tended to move progressively from the Eurasia Basin south-southeastward across Siberia since the early Tertiary, and has only recently reached the coast of the Sea of Okhotsk. The difference between the RM2/NUVEL and Savostin-Karasik models for "present-day" NOAM/EURA plate motion appears significant. One possible explanation for this difference is suggested by the difference in data sets. Most of the 19 data points describing NOAM/EURA motion in RM2 lie in the Atlantic, south of the majority of the 75 data points (in different combinations) used by Savostin and Karasik (1981). There may therefore be a diffuse plate boundary or intraplate deformation in the Arctic, the result of which is to make the Arctic portion of the NOAM/EURA boundary describe motion about a somewhat different pole. This could be analogous with the NOAM/AFRC and SOAM/AFRC poles that are statistically distinct (Fig. 2) even though no distinct NOAM/SOAM plate boundary has been identified (see also Westbrook and McCann, this volume).

No physical theory for the "Trans-Polar Trend" pattern (Fig. 2) has been offered and there is no geometrical requirement for its existence. The TPT does suggest a tendency for the motions of the four plates to be coupled. Such coupling is also indicated by spreading rate changes that tend to correlate north and south of the Azores triple junction (Pitman and Talwani, 1972; see also later). The movement of total opening poles at least since anomaly 13 time (Fig. 2) appears to have been such that the poles remained more or less fixed to each other as they migrated slowly in a transpolar direction towards Japan. (As noted earlier, the NOAM/EURA pole may have jumped back to its long-term average position subsequent to 3 Ma; Cook and others, 1986.) To a first-order approximation, the entire Atlantic rift has tended to open as if there were only two "megaplates" ("American" and "Eurafrican"), not four. The relation of the angular opening rates to rotation poles has been such as to attempt to maintain a relatively constant spreading rate along much of the MAR: although the rates decrease northward toward the rotation pole for each plate pair, they jump up again at each of the two (three according to Savostin and Karasik, 1981) triple junctions.

One possible physical reason for the TPT and the coupling between the plates is friction along the east-west boundaries between EURA and AFRC and between NOAM and SOAM. If the resistance to movement along these particular boundaries is high, the plates will adjust their motion in a way to minimize the movements along the "sticky" boundaries. If resistance to compression is much higher than to shear, the poles will tend to lie in a line along the rift or its extension, as observed (Fig. 2). Thus, if interplate boundary forces are important, a change in movement between one pair of separating plates (e.g., NOAM/EURA) would be transmitted to movements between an adjacent pair (NOAM/AFRC) as seems to be observed. A physical process that may be important in explaining the migration of the poles along the Trans-Polar Trend is the propagation of Arctic rifting into Siberia. Savostin and Karasik (1981) note that the initiation of NOAM/EURA rifting in Siberia migrated southeastward with time—from the end of the Eurasia Basin (55 Ma) to the Yana-Lena interfluve (Kendeiski graben; 30–25 Ma), to the Momo-Selenakhskaya Depression (5–2 Ma), to the small grabens of the Cherskiy Ridge (0 Ma, or to some time after 3 Ma when tension was replaced by compression; Cook and others, 1986). Thus at least until very recent geological times, rifting has been propagating southeast into and across Siberia at a long term average rate of the order 4–5 cm/yr, comparable to rift propagation rates measured elsewhere (e.g., Courtillot, 1982), even along the Mid-Oceanic Ridge (Hey and others, 1980). Perhaps Siberia's resistance to being split apart is a controlling factor in determining the migration rate of poles of opening for the three pairs of Atlantic plates.

ANGULAR RATES OF OPENING

The angular opening rates (Table 1) calculated by Minster and Jordan (1978) are based on the magnetic reversal time scale of Talwani and others (1971) and the NUVEL rates are based on the scale by Harland and others (1982). If the rates are recomputed to the DNAG reversal time scale (Kent and Gradstein, this volume) by using an age of 3.88 Ma for the younger edge of the youngest normal polarity interval in anomaly 3 (vs. 3.78 Ma used by Talwani and others [1971]) all Minster-Jordan rates are decreased by 2.7%. (This revision mainly reflects the revised radioactive isotope decay constants of Steiger and Jäger [1977].) Since anomalies 2A and 5 have the same ages on the DNAG and Harland and others (1982) scales, the NUVEL-based rates remain unchanged.

The NOAM/EURA and NOAM/AFRC rotation rates are virtually identical according to the Minster and Jordan (1978) and DeMets and others (1986) models. Using a different data set with magnetic data (opening rates) exclusively from the Eurasia Basin, Savostin and Karasik (1981) determined a rate of 0.189°/m.y. for NOAM/EURA motion. Subsequent estimates include 0.21°/m.y. (Savostin and others, 1982) and 0.219°/m.y. (Savostin and others, 1983). These low rates lie outside the error bars for other estimates for the present rotation rate of these plates (Table 1). The discrepancy may indicate non-rigid plate behavior within the North American and/or Eurasia plates, or the existence of "diffuse" plate boundaries. As noted earlier (Fig. 2), the Savostin-Karasik, poles (also Savostin and others, 1982, 1983) are significantly displaced southward from most other estimates. This is not surprising since pole positions and angular opening rates are highly correlated in any best fit to a plate boundary data set: the smaller angular opening rates produce about the same opening rates (spreading rates) on some part of the MAR axis as do higher opening rates around "closer" rotation poles (e.g., the NUVEL poles and rates). If correct, the NOAM/EURA motion models by the Soviet workers may thus be valid only until some rather recent time (post 3 Ma; Cook and others, 1986).

How do "present" angular opening rates compare with longer-term averages? The average rotation rates since anomaly 5 can be computed for the NOAM/EURA boundary from four estimates of rotation angle: 2.50° (Talwani and Eldholm, 1977; also Srivastava and Tapscott, this volume); 2.47° (Phillips and Tapscott, 1981, unpublished); 2.60° (Unternehr, 1982); and 2.53° (Peters and Vink, 1985). If an age of 9.67 Ma is chosen for the middle of the normal polarity responsible for anomaly 5 (Kent and Gradstein, this volume) the four rates are 0.259°, 0.255°, 0.269° and 0.262°/m.y. The corresponding rate for the NOAM/AFRC boundary (based on Klitgord and Schouten, this volume) is 0.249°/ m.y. Given the standard error in the present rotation rates (DeMets and others, 1986) and the spread in values for anomaly 5 closure, there are no significant differences between the corrected "geological present" rates (listed earlier) and the average rate since anomaly 5. (For EURA/NOAM, the NUVEL "geological present" is based on anomaly 5.)

In light of the above discussion it is tempting to accept the conservative "null" hypothesis that there has been no change in rotation pole or angular rotation rates for any of the three MAR plate pairs in the last 10 m.y. (Although there may have been no major pole changes since anomaly 13 time [35.6 Ma; Fig. 2; Peters and Vink, 1985] the rotation rate was definitely slower [2.18°/m.y.] from anomaly 5 to 6 [19.9 Ma] and still lower [1.95°/m.y.] between 6 and 13.) "No pole change in the last 10 m.y." is certainly allowed by the "present" motion models (Minster and Jordan, 1978; DeMets and others, 1986) together with the anomaly 5 reconstructions cited above. However, there is increasing evidence—elaborated in the following section—that plate motions—including those considered here—have actually *not* been constant over the last 10 million years. We are therefore faced with the possibility that the close agreement between the 10 Ma and "present" poles (Fig. 2) and angular rates is to some extent a fortuitous result of the averaging interval.

EVIDENCE FOR TIME VARIATIONS IN RELATIVE PLATE MOTION DURING THE LAST 10–20 M.Y.

It is remarkable that the central North Atlantic has been widening for more than 150 million years with "average" opening rates varying by only a factor of about two (Klitgord and Schouten, this volume). However, the age versus distance curves are actually based on a small number of key magnetic lineations whose age is presumed known (e.g., anomalies 2A, 5, 12, 13, 21, and 34 in the DNAG scale of Kent and Gradstein, this volume). Are opening rates really constant for long periods of time or is there in fact a plate kinematic "fine structure" that has not yet been resolved? Consider three successive marker anomalies n_i, n_{i+1}, n_{i+2} (for example, anomalies 1, 5, and 13). In general it is observed (in the Atlantic and elsewhere) that the average spreading rates for the $n_i - n_{i+1}$ interval are not the same as for the $n_{i+1} - n_{i+2}$ interval, but this cannot be taken as evidence for a spreading rate change at anomaly n_{i+1}-time: the marker anomalies happen to be easy to recognize and do not necessarily have any tectonic significance. An infinity of opening rate versus age relations could be found which when integrated would predict the correct locations and ages for lineations n_i, n_{i+1}, and n_{i+2}! Even if solutions with short term "pulses" of very slow or very fast spreading are rejected as physically unreasonable, the envelope of possible spreading rate histories remains wide.

The question of kinematic "fine structure" could be addressed by plotting distance to magnetic lineations on one accreting plate boundary against distance to the same lineations on another. The distances should be measured on *both* sides of the boundary to discriminate between changes in opening rates and changes in spreading asymmetry, which may be considerable, e.g., Vogt and others, 1983.) The accuracy of such "interplate" distance versus distance plots (e.g., Klitgord and Schouten, this volume) is limited only by the accuracy with which an anomaly can be identified on a map, not by the accuracy of its age. A bend in the curve at anomaly n_{i+1} time would show that relative motion of one plate pair changed with respect to the other. There are two problems in applying this method to the Central and North Atlantic. First, the magnetic anomalies are mostly rather "noisy" and only a few key lineations can be reliably identified. A more general problem, mostly ignored in the literature, concerns the possibility of regional or global synchronism in plate motions. To the extent that the rate of opening between two plates in each of two plate pairs (e.g., PCFC-NAZC [Pacific-Nazca] and SOAM-AFRC) changes by the same fractional amount, an interplate lineation plot would not reveal any bend, and no acceleration or deceleration would be suspected (or, at least, the magnitude of the change would be larger than suspected).

"Global synchronism" would also introduce errors in geomagnetic reversal chronologies for times prior to about 4 Ma. Because these time scales (e.g., Kent and Gradstein, this volume) are necessarily erected on the premise that seafloor spreading has been constant for long periods of time across certain parts of the Mid-Oceanic Ridge, any synchronized component in the "fine structure" of opening rates would represent an error signal in the reversal time scale. In fact, reversal chronologies are deliberately (i.e., conservatively) biased toward constant spreading: "[calibration] ages are used for age calibration in such a way as to minimize apparent acceleration in sea-floor spreading history' (Kent and Gradstein, this volume). Even if the central Atlantic Ocean contained a "perfect" recording of magnetic lineations, a precise calculation of spreading rates would be inaccurate to the extent that spreading rates in the central Atlantic fluctuated in the same way as those in the ocean basins that have been used to erect the reversal chronology.

The youngest (post-4 Ma) part of the reversal time scale, which is based on absolute ages of lava flows (Mankinen and Dalrymple, 1979) and not on assumptions about spreading rate, is obviously the most trustworthy basis for investigating possible kinematic fine structure. In this chapter "geologically recent" plate motions are compiled (Fig. 5) in some cases back to 20 Ma for the best studied accreting plate margins. The curves vary in data quality and quantity, resolving power (i.e., spreading rate), and averaging interval. Since the curves are first derivatives of the age versus distance plots, any errors in the latter are amplified and the detailed form of the calculated spreading rate variations (Fig. 5) is unreliable. Several different reversal time scales were used. For the East Pacific Rise and Southeast Indian ridges, spreading rates are computed according to two different time scales. The choice of one or another time scale influences the curves (Fig. 5) but the basic fluctuation patterns remain very similar. This is also true of longer-term plate speed fluctuations in the North Atlantic and North Pacific from the mid-Jurassic to present (Rich and others, 1986). Data from all but one of the independent accreting plate boundaries show that opening rates have been faster over the period 1–0 Ma than the period 1–2 Ma. For the better-quality data, the amount of increase in opening rate from the early Quaternary low to the late Quaternary high ranges from 0.2 to 1.7 cm/yr (mean, 0.8 cm/yr). For the three fast-spreading ridges the increases are 1.1 to 1.7 cm/yr (mean, 1.3 cm/yr), and for the Mid-Atlantic Ridge and its Arctic extension the increases range from 0.2 to 0.7 cm/yr (mean, 0.4 cm/yr). Expressed as percent increase the total range is +7% to +43% (mean, +22%), which can be subdivided into the three fast ridges (+7% to +43%, mean +22%), and the MAR (+9% to +35%, mean +22%). Magnetic data cannot resolve details of acceleration from the early to late Quaternary, and 1 Ma is only an estimate for the mid-point of the accelerating phase. However, the data from the three fast-spreading ridges where the Jaramillo event (0.73–0.91 Ma) is recorded suggest that much or all the acceleration had occurred prior to that time. Beyond that it cannot be determined how the present rates compare to the post-1 Ma (or post-0.73 Ma) average. If rates actually vary in sawtooth fashion (rather than as periods of constant spreading separated by abrupt acceleration[s]), then the actual range of rates from a sharp minimum about 1.5 Ma to the present would be double the values quoted above, and present rates of plate separation would be significantly higher than the post 1 Ma averages shown in Figure 5. The possibility of rather fast present plate motions will be verified or rejected in the next few years using the VLBI, SLR, and GPS techniques discussed earlier. Data available to date probably exclude present rates being twice or more the "geological present" average but smaller increases (e.g., 25%) cannot yet be resolved.

Figure 5 also suggests fluctuations in relative plate speeds prior to the minimum about 1–2 Ma. (Not all curves extend back before 5 Ma, however.) Episodes of more rapid plate separation occurred about 3–8 Ma, and with less certainty at 14–16 Ma. The kinematic "fine structure" (Fig. 5) suggests that the near equality of 10–0 Ma and 4–0 Ma averages—noted in the previous section—may be an artifact of the averaging interval rather than an indication of constant spreading. The 1–0 Ma averages are somewhat higher than the 4–0 Ma averages in the central North Atlantic and about the same north of the Azores (Fig. 5). In general the episodes of enhanced plate mobility (0–1, 3–8, and perhaps 14–16 Ma) correlate with increased ash output and magmatism at subducting margins and increased hotspot-type volcanism (Vogt, 1972, 1979; Kennett and Thunell, 1975, 1977;

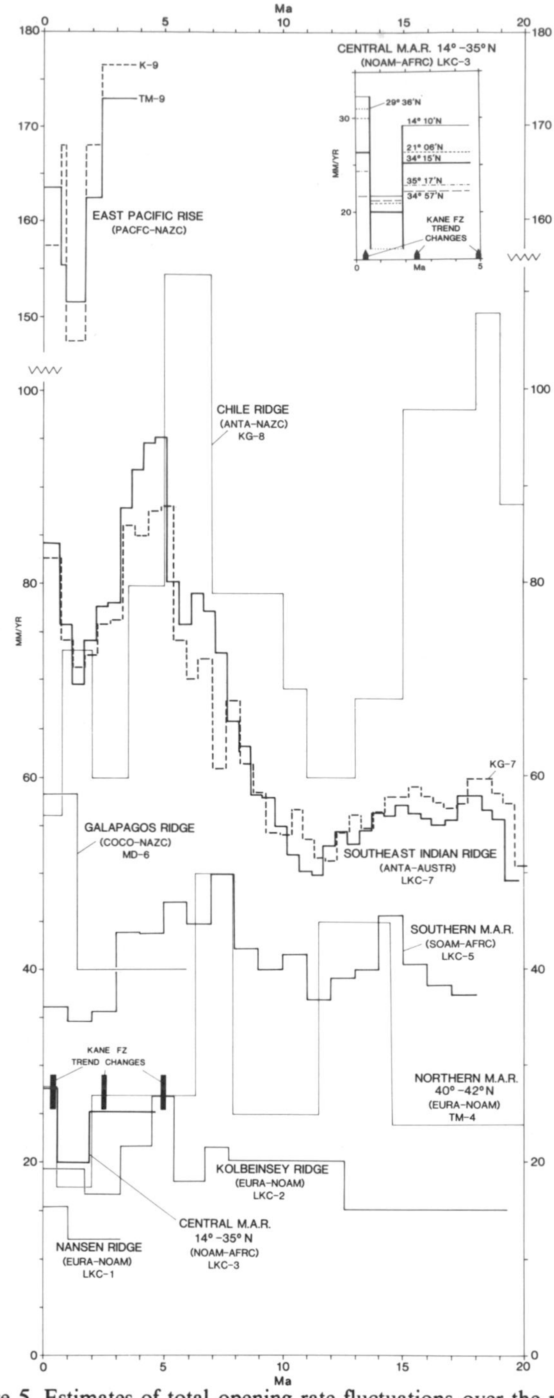

Figure 5. Estimates of total opening rate fluctuations over the past 20 m.y. across most major accreting plate boundaries. Thicker lines indicate more reliable data. Letters denote geomagnetic reversal time scales: LKC, LaBrecque and others (1977); K, Klitgord (1974); TM, Tarling and Mitchell (1976); MD, Mankinen and Dalrymple (1979); KG, Kent and Gradstein (this volume). Source publications, from bottom to top: 1, Vogt and others (1982), see also Coles and others (1978); 2, Vogt and others (1980); 3, Le Douaran and others (1982); 4, Searle (1977); 5, Brozena (1985); 6, Hey and others (1980); 7, Vogt and others (1983); 8, Cande (1984); 9, Rea and Scheidegger (1979). Analyses of individual profiles from Le Douaran and others (1982) shown in upper right. Solid bars mark NOAM/AFRC direction changes suggested by Kane FZ bathymetry (Tucholke and Schouten, 1985).

Kennett and others, 1977; Rea and Scheidegger, 1979). The effects at island arcs and hotspots are relatively more dramatic than the changes in plate speeds, however, suggesting strongly nonlinear dependence of such volcanism on rates of plate motion. However, since "present" island arc volcanism does not correlate strongly (geographically) with "present" subduction rate, the mechanisms behind the correlations over time are not obvious, and may actually represent transient responses (e.g., to accelerations rather than to the rate themselves). Vogt (1972, 1979) speculated that mantle plumes associated with hotspots like Iceland and Hawaii are the forcing function responsible for such global synchronism, but there are other possible "causality chains," and exploration of this topic is beyond the scope of this chapter.

In the Atlantic domain (Vogt, 1979), the three major episodes are "recorded" by several geological/geophysical features. The middle Miocene episode has been recorded by increased mid-plate volcanism at various sites in the North and South Atlantic, and by the first development of prominent basement ridges (See Plate 8A) at the Azores triple junction (heavy dashed lines in lower part of Figure 6A) and on the Reykjanes Ridge southwest of Iceland (Vogt, 1974, 1983). The escarpments are diachronous—becoming younger towards the southwest—and "middle Miocene" covers both the initiation and average age of the escarpments. The lava pile in northwest Iceland tentatively suggests a high extrusion rate 12–14 Ma (Saemundsson, this volume).

The 3–8 Ma episode (first recognized by Searle [1977] as a pulse of rapid spreading peaking ca. 7 Ma on the MAR at 40°–42°N) corresponds to the initiation of volcanism on the Azores Islands, to the development starting 7 Ma of a second pair of prominent diachronous basement escarpments on the Reykjanes Ridge (Vogt, 1974, 1983), and to the 6–7 Ma axis jumps on Iceland and apparent anomalously high extrusion rates 6–7 Ma in eastern Iceland (Saemundsson, this volume). In areas where highly detailed magnetic data exist, the configuration of the plate boundary changed during this period, for example, at 23°–27°N (Rona and Gray, 1980), at 37°N (Macdonald, 1977), at 68°–70°N (Vogt and others, 1980), and at 76°N (Vogt and others, 1982). In the 23–27°N area most of the reorganization took place during the period 6–4 Ma, with a change in spreading half-rates occurring about 3–4 Ma. A change in spreading direction at about 5 Ma was recorded by the Kane FZ (Tucholke and Schouten, 1985). In the 37°N area (FAMOUS) the fracture zones A and B (Fig. 3B) began to develop sometime during the period 3.7 to 5.2 Ma, with an E-W strike possibly indicating a change in plate motion (Macdonald, 1977). In the 68°–70°N area, fracture zones began to develop starting 7.7 Ma, with the present Spar FZ dating from 3 Ma. At 76°N the Knipovich Ridge developed its present oblique configuration after 4 Ma (Vogt and others, 1982); the outer limit of high "rift mountain" summits along the Mohns and Knipovich ridges occurs more or less abruptly at 3–4 Ma, suggesting that some tectonic event influenced the accretion process at that time. The amplitude of rift mountain topography increased about 6–8 Ma (Fig. 4) although here as elsewhere the ridge flank topography is irregular and "topochrons" (if any) are not obvious. Towards the end of the episode NOAM/AFRC separation slowed at about 2 Ma (Le Douaran and others, 1982) and perhaps changed direction (2.5 Ma; Tucholke and Schouten, 1985).

What about the late Quaternary (0–1 Ma) episode, most germane to the issue of "present" plate motion? in terms of ash production—largely from the Caribbean island-arc volcanoes—a Quaternary pulse is well recorded (Kennett and Thunell, 1975), and a few hotspot type edifices are mainly or entirely of Quaternary age (e.g., Bouvet, Madeira, and Jan Mayen). Simple correlation of island arc and hotspot volcanism with plate motion requires the volcanism to have increased specifically in the *second* half of the Quaternary, but this level of time-resolution is not yet available for the volcanism. On Iceland, the late Quaternary is marked by rapid southward propagation of the eastern Neovolcanic Zone, the development of the east-west South Iceland Seismic Zone (Einarsson, this volume), and a change towards light rare-earth–depleted or transitional basalts (Schilling and others, 1983). About 1 Ma the transform plate boundary off North Iceland jumped from the Husavik faults to the *en echelon* rifts throug the Grimsey Island area (Saemundsson, this volume). The Upper Pleistocene (post-0.7 Ma) volcanics transgress unconformably across the Plio-Pleistocene series, suggesting an increase or other change in volcanic activity about at that time. In the Vema FZ, present transform motions (094°) differ significantly from the overall 091° trend of the fracture valley (Kastens and others, 1986). A recent eastward motion of the SOAM/AFRC pole is one of several possible explanations. The Kane FZ topography suggests the most recent change in direction of NOAM/AFRC motion occurred after 0.4 Ma (Tucholke and Schouten, 1985). At the Asian end of the North America plate, the finger-like northeast Japan microplate may have been annexed by NOAM since 1–2 Ma; the northern tip of the Philippine plate was converging northwards towards this microplate until about 1 Ma, when the direction changed to northwestward (Nakamura and others, 1984). Alternatively the trans-Siberia rift may have reached the Pacific sometime after 3 Ma, creating a separate Okhotsk plate (Cooke and others, 1986). Plausibly this also occurred about 1 Ma.

Along the MAR, the acceleration of plate motion about 1 Ma is so recent that anything "recorded" in the topography or crustal structure is still within the active plate boundary zone. As we examine the accretion zone today, how do we separate the "record" from the "recording process"? Along most of the MAR a central rift valley is present, and the 1 Ma isochron lies within the area of the outer walls of the rift valley. This raises the possibility (Vogt and others, 1969; Vogt, 1979) that the rift valley as observed on most slow-spreading ridges today may not entirely be a steady-state feature but instead may in part be a record of the 1 Ma plate acceleration event indicated in Figure 5. Most authors have assumed—usually implicitly—that the rift valley is a steady-state feature and tailored their models accordingly (e.g., Deffeyes, 1970; Sleep and Rosendahl, 1979; Macdonald, this volume; see

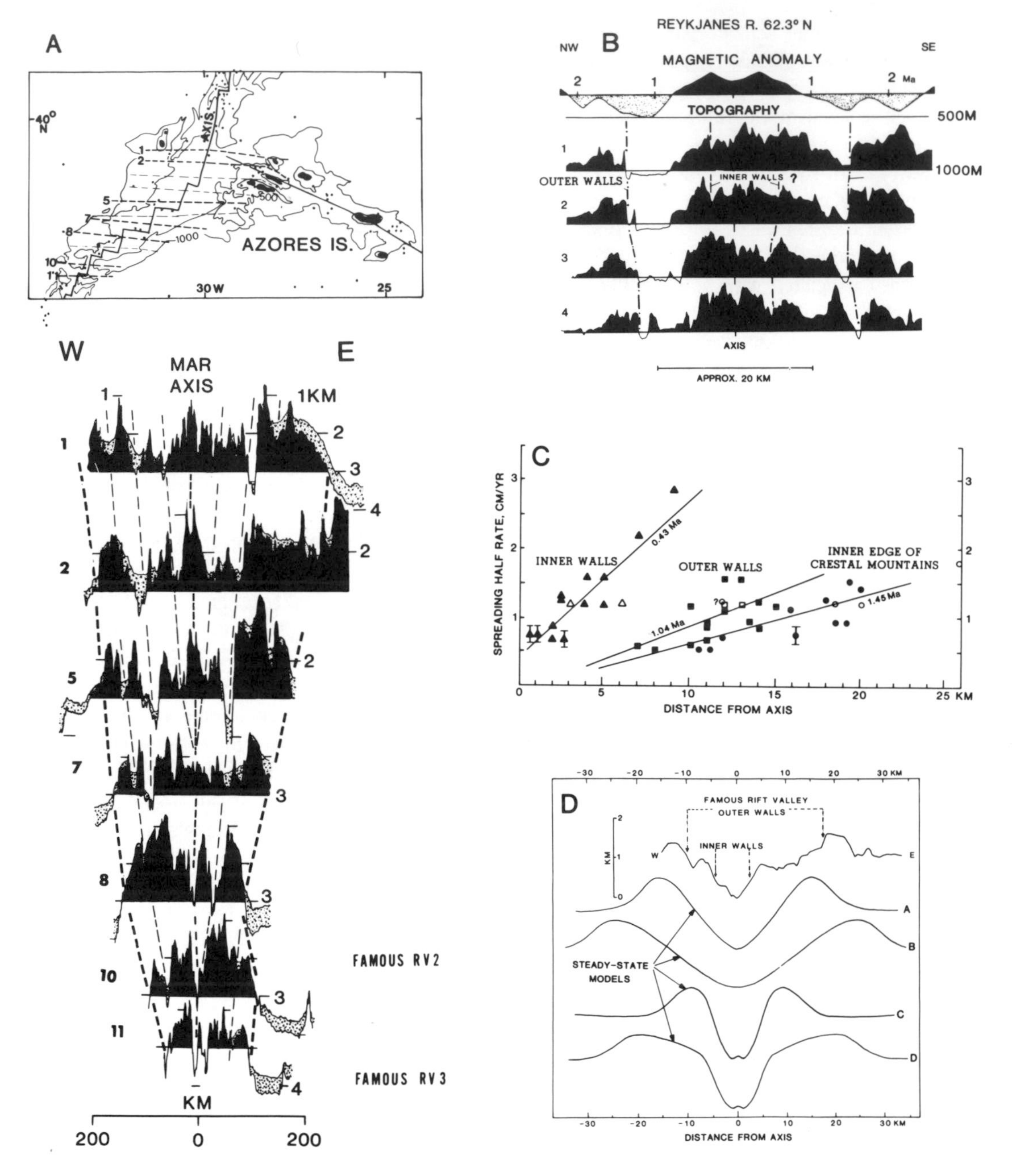

Figure 6. A, Bathymetric profiles across Mid-Atlantic Ridge from Azores triple junction southward to the FAMOUS area. Note progressive northward widening of distance between inner edge of rift mountains, and possible time-transgressive trends on younger crust. From Vogt (1979). B, Bathymetric profiles, 4.5 km apart, across Reykjanes Ridge axis about 62.3°N, 26.2°W (after Jacoby, 1980). Outer pair of dash-dot correlation lines show clear although subdued) main rift valley walls even in this area, which is influenced by the Iceland hotspot and lacks the normally well-developed MAR rift valley. The presence of "inner walls" is more conjectural. C, Distance from present accretion axis to walls of inner rift valley (triangles) and main rift valley (squares) against spreading half-rate. Solid symbols are from Mid-Atlantic Ridge from 8°S into the Arctic Basin. Open circles are from the Gorda (northeast Pacific) and Carlsberg (northwest Indian) ridges. Based on Ramberg and van Andel (1977) and Kristoffersen (1982), with Knipovich Ridge points from Vogt and others (1982). Isochrons are the least squares best fits to the data. D, Steady-state rift valley models A–D of Sleep and Rosendahl (1979) compared to one observed topographic profile (Needham and Francheteau, 1974) across FAMOUS rift valley. Models calculated for 1 cm/yr half-rate. A, low asthenosphere viscosity (1.6×10^{20}P) without axial magma chamber; B, high viscosity (8×10^{20}P) without magma chamber; C, low viscosity with magma chamber; D, high viscosity with magma chamber.

also Pálmason, this volume). It is true that quantitative steady-state models can be derived that fit the average shape of the rift valley (e.g., Fig. 6D), although there is no consensus about which models are correct. The presence of rift valleys on fast-spreading ridges close to transform faults (Macdonald, this volume) certainly supports a steady-state model insofar as the crust in a fast-spreading rift valley is much younger than 1 Ma. Even the slow-spreading rift valley cannot simply be the record of "something" that occurred 1 Ma, because faulting and seismicity still occur under the rift valley walls today. The presence of turbidites on steps along the Gorda Ridge rift valley walls has been adduced as evidence for a more or less steady state "escalator" from valley floor up into the rift mountains (Fowler and Kulm, 1970).

The possibility of a partly non-steady state MAR rift valley was first discussed by Vogt and others (1969) and revived later (Vogt, 1979, 1986b) when it had become more apparent that the Quaternary has been atypical in terms of ash production at subduction zones (Kennett and Thunell, 1975, 1977), hotspot basalt discharge (e.g., at Hawaii [Fig. 7A]), and climate (the most severe northern Hemisphere glaciations have occurred only since 0.9 Ma, to judge from the amplitudes of ^{18}O anomalies). These observations do not disprove steady-state rift valley models but they caution against the uncritical assumption of this class of models. A non-steady state MAR rift valley model was also proposed by Francis (1974), who used the 1968 caldera collapse event on Fernandina Island in the Galapagos as an analog.

As a result of the detailed FAMOUS investigation at 37°N it became apparent that the MAR magmatic/tectonic processes must be episodic on time scales of 10^4–10^5 years (e.g., Ramberg and van Andel, 1977) but the prevailing view is that each 30–100 km long segment of accretion axis (between fracture zones) is a more or less self-contained system. According to this argument the examination of many rift valleys is bound to sample some in each stage of development, and whatever features are shared by all would be the steady state "signal" on which local episodicity is the over-printed noise. However, the observed local episodicity is not inconsistent with occasional events that simultaneously affect the entire accreting plate boundary.

Does the morphology of the rift valley itself provide any evidence for non-steady state behavior? Several authors have noted that many rift valleys exhibit a narrow inner rift valley, separated by a terrace from the main outer rift valley walls, which can be considered to extend up to the inner edge of the rift (crestal) mountains. Ramberg and van Andel (1977) plotted the distance of the inner and outer walls (measured from the present axis) against spreading rate. This plot was augmented by Kristoffersen (1982) with the addition of points for "the inner edge of crestal mountains" and data from the slow-spreading Nansen Ridge. Figure 6C is a slightly revised version of Kristoffersen's plot. The distance of the inner walls from the axis correlates positively and linearly with spreading rate, as already noted by Ramberg and van Andel. Those authors interpreted the relation in terms of a "dependence on parameters such as lithospheric thickness and rheology." They were thinking of steady-state models. Vogt (1979) noted that the linear relationship could equally be interpreted in terms of a tectonic "event," which precipitated faulting close to the axis at a time given by the slope of the best-fitting line (recalculated to be 0.43 Ma in Fig. 6C). If so, the inner walls are "topochrons." The faster the spreading, the farther from the axis is the fault escarpment at present, since more crust would have been produced between the escarpments. (The subsequent continuation of faulting along the same escarpment does not affect the result—Fig. 6C shows only the *distances* to the walls, not their relief!) The inner wall event may correspond to the most recent (<0.5 Ma) change in NOAM/AFRC direction recorded by the Kane FZ (Tucholke and Schouten, 1985). Inclusion of additional data (e.g., Kristoffersen, 1982) shows that the outer walls and crestal escarpment edges also define isochrons (with considerable scatter), the best-fitting slopes being 1.04 Ma and 1.45 Ma. A non-steady state interpretation is that a tectonic "event" began to affect the accretion axis about 1.45 Ma and was in full swing by 1.04 Ma. This "event" is plausibly associated with the approximately 1 Ma acceleration discussed earlier (Fig. 5). As speculated by Vogt (1979), an abrupt increase in spreading rate would first be accommodated by faulting and rift valley formation, with volcanism taking time to catch up. The inner wall "event" at 0.43 Ma occurred subsequent to the last polarity reversal (0.73 Ma) and therefore magnetic anomalies are of no use in testing for a change in plate motion at that time. It is not easy to see how a steady-state model could explain the data in Figure 6C. Such a model would have to elevate the crust in two separate tectonic escalators. The theoretical models so far presented (e.g., Sleep and Rosendahl, 1979) do not predict a strong linear dependence of steady-state rift valley width on spreading rate. Furthermore, if a new rift valley forms, one should expect it to be wider at *slow* spreading rates, in response to the thicker axial lithosphere.

Some parts of the MAR lack a prominent median valley and look morphologically more like fast-spreading ridges. Since these anomalous accretion axes are located near the Iceland and Azores hotspots (e.g., profile 1 in Fig. 6A) it is natural to attribute the lack of a rift valley to the extra heat or/and basalt melt produced in the hotspot area (Vogt, 1974, 1983). Rarely the rift valley is replaced by an axial high in areas distant from known hotspots. One such place is "Telegraph Axial Volcano" at 48.5°N (Fig. 4). It is not clear whether this is simply a very weak transient version of a hotspot like Iceland, or whether it is a random, temporary condition at the accretion axis. In the latter case it may be an example of a MAR condition more common prior to about 1 Ma, according to the non-steady state hypothesis model of Vogt and others (1969).

Where the usual prominent rift valley is lacking, a test between steady-state and non-steady state models is possible, provided a subdued rift valley is present. The hotspot-influenced axis will have different rheological and thermal parameters (e.g., Sleep and Rosendahl, 1979) and therefore if a weakly developed steady-state rift valley is present at all it would be unlikely to be the same width as the normal rift valley. If the relations in Figure

6C are considered steady-state, one should empirically expect a hotter ridge to have a *wider* rift valley, appropriate to a faster spreading ridge. The non-steady state model predicts that a weakly developed rift valley along a hotspot-influenced axis has the *same width* as a "normal" axis with the same spreading rate, since the width would only be related to spreading rate, not thermal parameters. Jacoby (1980) published bathymetric profiles across the hotspot-influenced Reykjanes Ridge at 62.3°N, where a prominent rift valley is lacking (Vogt, 1974). These profiles (Fig. 6B) show that outer rift valley walls—and perhaps even an inner rift valley—are developed in subdued form at about the same distance from the axis as for the normal MAR with its prominent rift valley. This observation supports the non-steady state hypothesis. Perhaps the accretion axis is quite narrow on the Reykjanes Ridge and the "outer walls" (Fig. 6B) are "topochrons"—a "fossil" record, no longer being tectonically modified, of the plate acceleration event. Another example of non-steady state "fossil" rift valley walls is found south of the Azores triple junction (Fig. 6A) where the inner edge of the major diachronous basement ridges is an apparent faulted escarpment that, however, formed successively later with increasing distance northwest. (Note the similarity, except for scale, between profiles in Fig. 6B and profiles 1, 2, and 5 in Fig. 6A.) To the extent that the MAR median rift valley is a non-steady state feature, profiles 11 to 1 (Fig. 6A) give a rough idea of how the MAR topography might evolve from the present (profiles 10, 11) to 12–15 m.y. in the future (profile 1).

ABSOLUTE PLATE MOTIONS

The term "absolute" motion (as opposed to relative) means plate motion in a reference frame attached to the main body of the earth's mantle, which presumably has internal motions (other than vertical motions within possible narrow plumes) small compared to the motions of the plates. Once the absolute motion of one plate is measured or specified, the absolute motions of all the other plates can be calculated from a relative motion model such as RM2 or NUVEL by vector addition. Minster and Jordan (1978) considered four possible kinematic conditions, each of which specifies an absolute reference frame: AM0-2, the system of plates has no net rotation; AM1-2, the reference frame is defined to best fit five hotspot-trace age gradients and azimuths (Table 1); AM2-2, the Africa plate is fixed relative to the deeper mantle; and AM3-2, the Caribbean plate is fixed relative to the deeper mantle. Figure 1 shows absolute motions predicted for the "hotspot" frame based on the parameters in Table 1. (Improved absolute motion models will become available when new hotspot-trace age dates [Duncan and Clague, 1985] are combined with new relative motion models [DeMets and others, 1986].) Five of the hotspots used by Minster and Jordan (1978; Hawaii, Marquesas, Tahiti, MacDonald, and Pitcairn), including all that provided both azimuths and rates, are within the Pacific plate, and one (Juan de Fuca) is on its edge. The Galapagos hotspot has left traces on the Cocos and Nazca plates. Yellowstone (45°N, 110°W) is the only North American plate hotspot in the data set (azimuth S60°W ± 20°). Although there are numerous hotspots on slow-moving plates (e.g., the Africa plate; Burke and Wilson, 1972) the age and azimuth data are poor for resolving "present" plate motions. Therefore the absolute motion of the North American plate is primarily known from of (1) PCFC/NOAM relative motion, and (2) PCFC/hotspot motion (Fig. 7). Errors in either or both of these motions will combine to determine errors in NOAM/hotspot (absolute) motion. Even the Pacific hotspot traces need to be examined for not less than the last 10 m.y. to obtain reliable estimates of age gradient and azimuth along the chain; this exceeds by over a factor of 3 the mean averaging interval for the spreading rates. Thus, spreading rate estimates back to 10 Ma should more properly have been considered "present" for purposes of absolute motion models. Uncertainties arising from different averaging intervals will only be important if average plate motions since 10 Ma are significantly different from the average motions since 3 Ma; as discussed previously the relative motions NOAM/EURA and NOAM/AFRC appear to have been about the same over the two averaging intervals. However, if relative motions have changed over the last 3 m.y. (Fig. 5), it is reasonable to expect absolute motions to have changed as well. The topography along the best-studied hotspot trace (Hawaii; Fig. 7A) suggests a significant (clockwise) change in PCFC/hotspot direction and increase in basalt output since 1–2 Ma (Rea and Scheidegger, 1979; Vogt, 1979). Age dates along the Hawaii and other Pacific hotspot traces (McDougall and Duncan, 1980; Duncan and Clague, 1985) can be interpreted in a way consistent with the published episodes of faster motion approximately 0–1 and 6–8 Ma (Fig. 7B, C), implying changes in absolute motion also of the North American plate. However, the age-date data set by itself is still clearly inadequate to rule out a simpler model, for example, 9.49 ± 0.40 cm/yr constant motion over the Hawaii hotspot since 12 Ma (Duncan and Clague, 1985).

CONCLUSION

The kinematics of the larger plates was described already in the first few years of the plate tectonic revolution—but only on time scales of at least 10 m.y. Resolution of possible variability at 1–10 m.y. scales has improved only slowly, but a case can now be made for significant fluctuations in relative plate speed (Fig. 5) and direction (Tucholke and Schouten, 1985). Similar variability in "absolute" motion therefore appears probable, even though the speed changes cannot yet be verified by age gradients along hotspot traces, even those on fast moving plates (Fig. 7B, C). Direction changes (rotation pole movements) over 1–10 m.y. time scales may have been recorded by Pacific hotspot traces (Epp, 1984) but the resolution is marginal. It seems unlikely that hotspot traces on slow moving plates (like NOAM) could have resolved such short-lived speed or direction changes.

How can the evidence for plate-motion irregularities at 1–10 m.y. scales be reconciled with long-term stability of the average

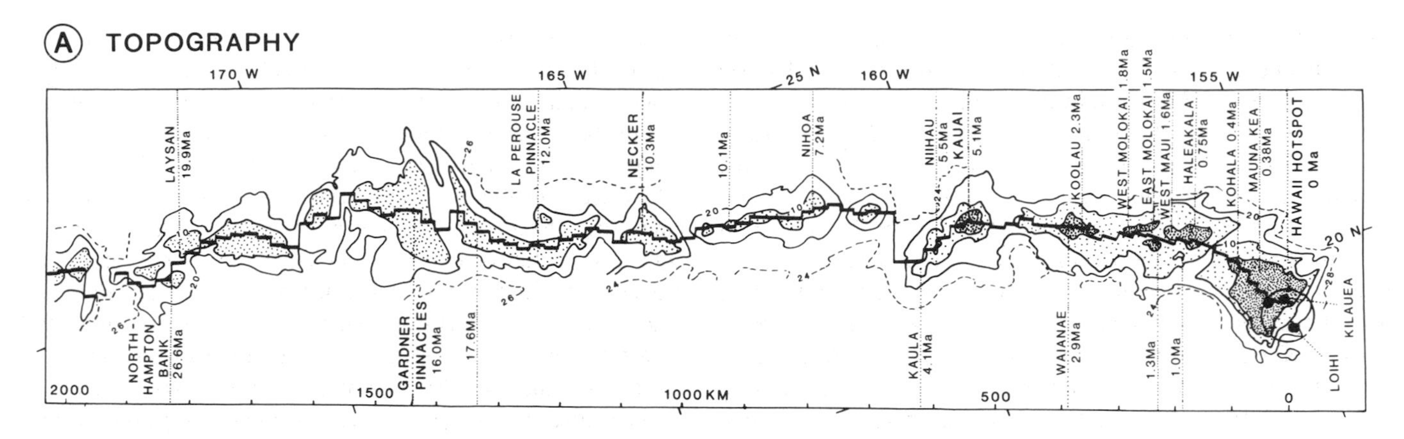

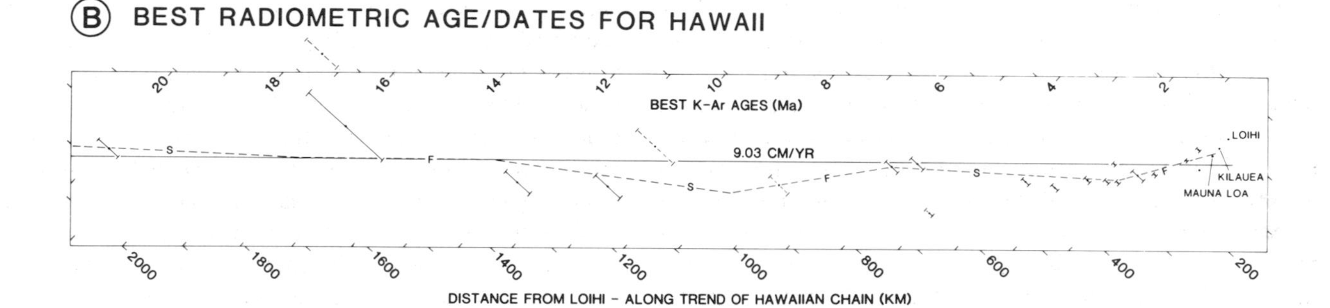

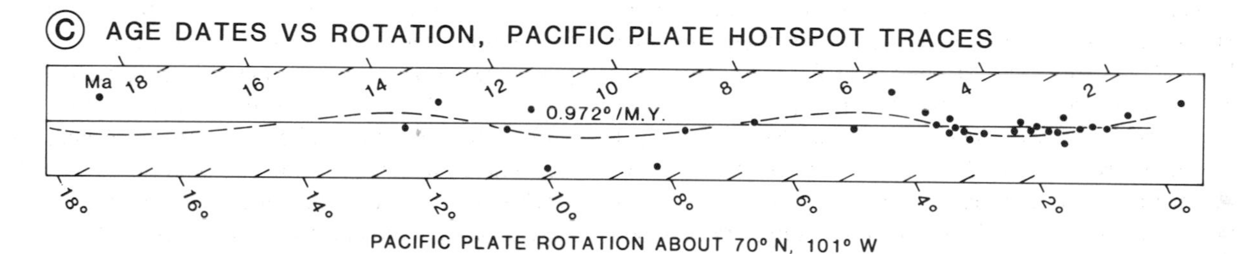

Figure 7. (A), Topography of southeastern Hawaii-Emperor chain in units of 100 fm (183 m) showing most reliable age dates (Duncan and Clague, 1985) and center of mass (crenulated line along axis); (B), age versus distance plot redrawn from Duncan and Clague (1985). Best-fit for the last 12 m.y. is 9.49 ± 0.4 cm/yr for all data (straight line), and 8.6 ± 0.2 cm/yr using only the most reliable data (Duncan and Clague, 1985). Segmented dashed line shows type of pattern expected from other data, that is, variable plate speeds (F, fast; S, slow). Variable plate speeds are not required by the presently available age dates. (C), Plot of average ages of volcanos from six Pacific seamount chains against distance in degrees of longitude (rotation) about a pole at 70°N, 101°W describing motion of Pacific plate over mantle. Simplified from Figure 6 of McDougall and Duncan (1980). Straight line has a slope of 0.998 ± 018°/m.y. This rate is reduced by about 2.6% if the revised isotope decay constants of Steiger and Jäger (1977) are used. Curved dash line is the type of pattern expected from other data (e.g., Fig. 5). Whether plate speeds change abruptly (dashed line in C) or gradually (dashed line in D) is unknown.

rotation poles, such as the Atlantic-area relative motion poles in the last 35–40 m.y. (Fig. 2) or the Pacific/hotspot motion as recorded from Hawaii to the Hawaii-Emperor bend? Presumably the minor changes in rotation pole (e.g., Tucholke and Schouten, 1985) represent perturbations from a stable state. Major transform boundaries may help anchor the rotation poles. Occasional major plate reorganizations (e.g., as recorded by the Hawaii-Emperor Bend) may then represent response to a threshold exceeded and a rapid switch to a new stable state. To the extent that both major and minor variability in plate kinematics has a global component, both kinds of signals may have been recorded in continental crust, offering the possibility of a record extending much farther back than 180 Ma. For example, the sealevel fluctuations of 1–10 m.y. scale responsible for the "Grand Cycles" of the Cambrian (Palmer and Halley, 1979) may ultimately reflect plate kinematic episodes (A. R. Palmer, personal communication,

1985). Might the correlation between marine taxonomic diversity and Atlantic and Pacific spreading rates since the mid-Jurassic (Rich and others, 1986) also extend farther back in time?

Unless the last 20 m.y. are atypical, observed correlations at 1–10 m.y. scales among hotspot and island arc volcanism, global climate, and plate motions (e.g., Fig. 5) suggest a single interrelated system (e.g., Kennett and Thunell, 1975, 1977; Vogt, 1979) but little progress has been made even on separating causes from effects. Since causes cannot be preceded by their effects, more accurate dating of seemingly coeval events would help pinpoint causalities. The northward motion of the Australian plate (Fig. 5) seems to have accelerated gradually from 10 to 5 Ma, implying that the 8–3 Ma global event was not caused abruptly and that anything occurring as late as 5–3 Ma was probably an effect. The volcanic discharge curve for the Hawaii hotspot (Vogt, 1979) resembles the history of island-arc volcanism and spreading rates (and therefore is somehow related) but does not lead these two indicators of relative plate motion. Thus, one cannot decide whether hotspot volcanic discharge episodicity is a direct measure of episodic mantle convection (and that such convection explains the observed plate speed fluctuations), or whether the discharge episodicity is merely an effect, for example, of changing mid-plate stresses. (Of course the plate speed fluctuations are themselves a measure of episodic convection insofar as the plates form the upper boundary layer of a mantle convection system.) It is not clear whether the diachronous escarpments near Iceland (Vogt, 1983) or the frequent shifts of the volcanic zones there (Helgason, 1984) could be a response to changes in EURA/NOAM motion, to changes in activity of an Iceland mantle plume, or to both occurring simultaneously. The situation in the Azores area is still more complicated since the V-shaped structures observed (Fig. 6A) might reflect propagation caused by instability of the triple junction (e.g., Patriat and Courtillot, 1984) rather than Azores mantle plume episodicity (Vogt, 1979).

If the measurables related to plate speed or direction are evaluated in terms of their resolving power at 1–10 m.y. scales, relations such as age versus depth and age versus gravity or geoid anomaly are low on the list. Basement roughness as a paleo-spreading rate index may be of value only near threshold rates (ca. 5–8 cm/yr opening rate) separating "slow" from "fast" spreading. Although hotspot traces (Fig. 7) have only modest resolving power for plate motion irregularities at 1–10 m.y. scales, they are generally superior to other measures, such as paleomagnetism. Multibeam bathymetric surveys of major transform faults (in some cases even single-beam derived charts; Vogt and Perry, 1981) reveal narrow, regular troughs whose strikes can be measured to an accuracy of ±1°–2°. If the set of FZ's along a plate boundary represents a significant angular spread about the rotation pole, the latter can be located accurately (e.g., the SOAM/AFRC pole, Fig. 2) even though the total kinematic data set is modest. Major FZ traces may also provide a high-resolution registration of small movements in rotation pole (e.g., Tucholke and Schouten, 1985). Whether kinematic information is also recorded by the inner and outer rift valley walls on the MAR (Fig. 6) and by inference by some of the ridge/valley topography on the MAR flanks (Fig. 4) is still speculative. Earthquake first motion (fault-plane) solutions along FZ's reveal only "present" motion but together with global geodetic data are valuable to compare the "present" (recent years) with the "geological present" (recent few m.y.). Earthquakes also provide other plate kinematic information of lower resolving power (e.g., normal-fault first motions—related to pole location—and maximum moment and fault width, which depend on slip rate; Burr and Solomon, 1978). The global geodetic measurement of inter- and intra-plate deformation is a new and rapidly evolving field (e.g., Carter and others, 1985). Present plate motion has indeed been demonstrated, but longer observation times and greater accuracies are needed just to bring the errors down to ±1 mm/yr as for "geologic" rates. If the geodetic rates are found to be close to the geologically present values (Fig. 5), the likelihood of fluctuations at scales much less than 1 m.y. will be much reduced. The geodetic observatory grid—eventually including laser and acoustic ranging networks, installed on the ocean floor but tied to satellite-based geodesy—will have to be dense enough to measure the more irregular ground motions near plate boundaries (Committee on Geodesy, 1985). Are motions in plate interiors even partially governed by short-time-scale physical processes at plate boundaries (and therefore perhaps also irregular) or are they smooth over human time scales, due to the dominance of plate driving forces?

Finally, magnetic lineations are still the primary data source for rotation rates (Table 1F) and a major data source for rotation poles (Fig. 2). The resolving power for pole motions depends on conjugate kinks or offsets in the anomaly pattern (Peters and Vink, 1985) and future improvements are possible if localized magnetic, bathymetric, or other markers can be found. If globally synchronized episodes of faster and slower plate motion (Fig. 5) are typical over geologic time, there may exist systematic errors, difficult to remove, in reversal time scales that assume constant spreading rates in at least one reference section of oceanic crust.

Whereas this paper has emphasized relative and absolute plate motion, the oceanic crust is imprinted with several other kinds of kinematic information that may help constrain models of plate/asthenosphere properties and perhaps asthenosphere flow. Propagating rifts have been identified in many oceanic areas (Hey and others, 1980; Vogt and others, 1983; see Vogt, this volume, Ch. 12, for possible Atlantic examples). Rift propagation is commonly in the regional down-slope direction at 2–10 cm/yr. By contrast minor MAR fracture zones appear to be propagating towards the Azores (in the upslope direction!) at 1–2 mm/yr (Johnson and Vogt, 1973; Rona and Gray, 1980). Small seamount chains east of the Reykjanes Ridge suggest asthenosphere motion at 1–2 cm/yr away from the Iceland hotspot (Plate 8A; Vogt, 1971). Diachronous (time-transgressive) basement ridges and outward-facing escarpments bracketing the Rykjanes Ridge (Vogt, 1971; Johansen and others, 1984) imply propagation of "something" along the plate boundary at rates of 5–20 cm/yr. Basement topography may thus be a record of asthenosphere flow speed away from hotspots, and simultaneously a record of hotspot (plume) episodicity as noted earlier. As plate motions come into ever sharper focus, ignorance of sublithospheric mantle motions has become the major barrier to geodynamic understanding.

REFERENCES

Anderle, R. J., and Malyevac, C. A.
1983 : Current plate motions based on Doppler satellite observations: Geophysical Research Letters, v. 10, p. 67–70.

Bernauer, F.
1943 : Junge Tektonik auf Island und ihre Uhrsachen, in Spalten auf Island, ed. O. Niemizyk; Verlag von Konrad Wittwer, p. 14–64.

Björnsson, A.
1985 : Dynamics of crustal rifting in NE Iceland: Journal of Geophysical Research, v. 90, p. 10, 151-10, 162.

Brander, J. L., Mason, R. G., and Calvert, R. W.
1976 : Precise distance measurements in Iceland: Tectonophysics, v. 31, p. 193–206.

Brozena, J.
1985 : Temporal and spatial variability of sea-floor spreading processes in the northern South Atlantic: Journal of Geophysical Research, v. 91, p. 497–510.

Burke, K., and Wilson, J. T.
1972 : Is the African plate stationary?: Nature, v. 239, p. 387–390.

Burr, N. C., and Solomon, S. C.
1978 : The relationship of source parameters of oceanic transform earthquakes to plate velocity and transform length: Journal of Geophysical Research, v. 83, p. 1193–1205.

Cande, S. C.
1984 : Nazca–South America plate interactions since 50 m.y. B.P.; in Peru-Chile Trench offshore Peru, Atlas 9; eds. D. M. Hussong, S. P. Dang, L. D. Kulm, R. W. Couch and T.W.C. Hilde; Ocean Margin Drilling Program Regional Atlas Series, 1:1 million, Marine Science International, Woods Hole, Massachusetts, sheet 14.

Carter, W. E., Robertson, D. S., and MacKay, J. R.
1985 : Geodetic radio interferometric surveying; Applications and results: Journal of Geophysical Research, v. 90, p. 4577–4587.

Chapman, M. E., and Solomon, S. C.
1976 : North American Eurasian plate boundary in northeast Asia: Journal of Geophysical Research, v. 81, p. 921–930.

Christodoulidis, D. C., Smith, D. E., Kolenkiewicz, R., Klosko, S. M., Torrence, M. H., and Dunn, P. J., 1985: Observing tectonic plate motions and deformations from satellite laser ranging: Journal of Geophysical Research, v. 90, p. 9249–9263.

Coles, R. L., Hannaford, W., and Haines, G. V.
1978 : Magnetic anomalies and evolution of the Arctic: Arctic Geophysical Review, v. 45, p. 51–66.

Committee on Geodesy,
1985 : Geodesy—A Look to the Future: National Academy Press, Washington, D.C., 179 p.

Cook, D. B., Fujita, K., and McMullen, C. A.
1986 : Present-day plate interactions in northeast Asia; North America, Eurasian, and Okhotsk plates: Journal of Geodynamics (in press).

Courtillot, V.
1982 : Propagating rifts and continental breakup: Tectonics, v. 1, p. 239–250.

Decker, R. W., Einarsson, P., and Mohr, P. A.
1971 : Rifting in Iceland: New geodetic data; Science, v. 173, p. 530–532.

Decker, R. W., Einarsson, P., and Plumb, R.
1976 : Rifting in Iceland: Measuring horizontal movements; Societas Scientarium Islandica, v. 5, p. 61–71.

Deffeyes, K. S.
1970 : The axial valley: A steady-state feature of the terrain; in Megatectonics of continents and oceans, eds. H. Johnson and B. L. Smith; Rutgers University Press, New Brunswick, p. 194–222.

DeMets, C., Gordon, R., Stein, S., Argus, D., Engeln, J., Lundgren, P., Quible, D., Stein, C., Weins, D., Weinstein, S., and Woods, D.
1986 : Current plate motions: Journal of Geophysical Research (in press).

Drake, C. L., and Girdler, R. W.
1982 : Continental and oceanic rifts; in Continental and oceanic rifts, ed. G. Pálmason; Geodynamics Series, v. 8, American Geophysical Union, Washington, p. 5–15.

Duncan, R. A., and Clague, D. A.
1985 : Pacific plate motion recorded by linear volcanic chains, in The Ocean Basins and Margins, v. 7A, eds. A.E.M. Nairn, F. G. Stehli, and S. Uyeda; Plenum, New York, p. 89–121.

Epp, D.
1984 : Possible perturbations to hotspot traces and implications for the origin and structure of the Line Islands: Journal of Geophysical Research, v. 89, p. 11273–11286.

Fowler, G. A., and Kulm, L. D.
1970 : Foraminiferal and sedimentological evidence for uplift of the deep-sea floor, Gorda Rise, northeastern Pacific: Journal of Marine Research, v. 28, p. 321–329.

Francis, T.J.G.
1974 : A new interpretation of the 1968 Ferdinandina caldera collapse and its implications for mid-ocean ridges: Geophysical Journal of the Royal Astronomical Society, v. 39, p. 301–318.

Francis, T.J.G., Porter, I. T., and McGrath, J. R.
1977 : Ocean-bottom seismograph observations on the Mid-Atlantic Ridge near lat 37°N: Geological Society of America Bulletin, v. 88, p. 664–677.

Gerke, K.
1974 : Crustal movements in the Myvatn- and in the Thingvallavatn-area, both horizontal and vertical; in Geodynamics of Iceland and the North Atlantic area, ed. L. Kristjansson: Reidel, Dordrecht, Holland, p. 263–275.

Harland, W. B., Cox, A. V., Llewellyn, P. G., Pickton, C.A.G., Smith, A. G., and Walters, R.
1982 : A geologic time scale: Cambridge University Press, Cambridge, 131 p.

Helgason, J.
1984 : Frequent shifts of the volcanic zone in Iceland: Geology, v. 12, p. 212–216.

Hey, R. N., Duennebier, F. K., and Morgan, W. J.
1980 : Propagating rifts on mid-ocean ridges: Journal of Geophysical Research, v. 85, p. 2647–2658.

Holmes, A.
1944 : Principles of physical geology: Nelson, London, 532 p.

Illies, J. H., Claus, P., Schminke, H.-U., and Semmel, A.
1979 : The Quaternary uplift of the Rhenish Shield in Germany: Tectonophysics, v. 61, p. 197–225.

Jacoby, W. R.
1980 : Morphology of the Reykjanes Ridge crest near 62°N: Journal of Geophysics, v. 47, p. 81–85.

Johansen, B., Vogt, P. R., and Eldholm, O.
1984 : Reykjanes Ridge; Further analysis of crustal subsidence and time-transgressive basement topography: Earth and Planetary Science Letters, v. 68, p. 249–258.

Johnson, G. L., and Vogt, P. R.
1973 : Mid-Atlantic Ridge from 47° to 51°N: Geological Society of America Bulletin, v. 84, p. 3443–3462.

Kastens, K. A., Macdonald, K. C., Miller, S. P., and Fox, P. J.
1986 : Deep-tow studies of the Vema Fracture Zone; 2. Evidence for tectonism and currents in the sediments of the transform valley floor: Journal of Geophysical Research, v. 91, p. 3355–3367.

Kennett, J. P., and Thunell, R. C.
1975 : Global increase in Quaternary explosive volcanism: Science, v. 197, p. 497–503.
1977 : On explosive Cenozoic volcanism and climatic implications: Science, v. 194, p. 889–906.
Kennett, J. P., McBirney, A. R., and Thunell, R. C.
1977 : Episodes of Cenozoic volcanism in the circum-Pacific region: Journal of Volcanological and Geothermal Research, v. 2, p. 145–163.
Klitgord, K. D.
1974 : Near-bottom geophysical surveys and their implications on the crustal generation process, sea-floor spreading history of the Pacific, and the geomagnetic time scale, 0 to 6 m.y. B.P. [Ph.D. thesis]: University of California, San Diego, 177 p.
Kristoffersen, Y.
1982 : The Nansen Ridge, Arctic Ocean; Some geophypsical observations of the rift valley at slow spreading rate: Tectonophysics, v. 89, p. 161–172.
LaBrecque, J. L., Kent, D. V., and Cande, S. C.
1977 : Revised magnetic polarity time scale for Late Cretaceous and Cenozoic time: Geology, v. 5, p. 330–335.
Le Douaran, S., Needham, H. D., and Francheteau, J.
1982 : Pattern of opening rates along the axis of the Mid-Atlantic Ridge: Nature, v. 300, p. 254–257.
Littell, F. B., and Hammond, J. C.
1928 : World longitude operation: Astronomical Journal, v. 38, p. 185.
Macdonald, K. C.
1977 : Near-bottom magnetic anomalies, asymmetric spreading, oblique spreading, and tectonics of the Mid-Atlantic Ridge near latitude 37°N: Geological Society of America Bulletin, v. 88, p. 541–555.
Mankinen, E. A., and Dalrymple, G. B.
1979 : Revised geomagnetic polarity time scale for the interval 0–5 m.y. B.P.: Journal of Geophysical Research, v. 84, p. 615–626.
McDougall, I., and Duncan, R. A.
1980 : Linear volcanic chains; Recording plate motions?: Tectonophysics, v. 63, p. 275–295.
Minster, J. B., and Jordan, T. H.
1978 : Present-day plate motions: Journal of Geophysical Research, v. 83, p. 5331–5354.
1980 : Present-day plate motions; Summary: Paris, Centre National de la Recherche Scientifique, p. 109–124.
Minster, J. B., Jordan, T. H., Molnar, P., and Haines, E.
1974 : Numerical modeling of instantaneous plate tectonics: Geophysical Journal of the Royal Astronomical Society, v. 36, p. 541–576.
Musman, S. A., and Schmidt, T.
1986 : The relationship of intraplate seismicity to continental scale strains: EOS Transactions of the American Geophysical Union, v. 67, p. 307.
Nakamura, K., Shimazaki, K., and Yonekura, N.
1984 : Subduction, bending, and education; Present and Quaternary tectonics of the northern border of the Philippine Sea plate: Bulletin Societie Geologique de France, v. 7, p. 221–243.
Needham, H. D., and Francheteau, J.
1974 : Some characteristics of the rift valley in the Atlantic Ocean near 36°48′ north: Earth and Planetary Science Letters, v. 22, p. 29–43.
Nielsen, N.
1930 : Tektonik und Vulkanismus Islands unter Berücksichtigung der Wegener-Hypothese: Geologische Rundschau, v. 21, p. 347–349.
Niemczyk, O., and Emschermann, E.
1943 : Sonderdreiecksmessung auf Island zur Feststellung feinster Erdkrustenbewegungen; in Spalten auf Island, ed. O. Niemczyk; Verlag von Konrad Wittwer, Stuttgart, p. 80–113.
Palmer, A. R., and Halley, R. B.
1979 : Physical stratigraphy and trilobite biostratigraphy of the Carrara Formation (Lower and Middle Cambrian) in the southern Great Basin: U.S. Geological Survey Professional Paper 1047, 162 p.
Parsons, B.
1981 : The rates of plate creation and consumption: Geophysical Journal of the Royal Astronomical Society, v. 67, p. 437–448.
Patriat, P., and Courtillot, V.
1984 : On the stability of triple junctions and its relation to episodicity in spreading: Tectonics, v 3, p. 317–332.
Peters, M. F., and Vink, G. E.
1985 : Plate reconstruction poles determined by a least squares algorithm, EOS, Transactions of the American Geophysical Union, v. 66, p. 1061.
Phillips, J. D., and Fleming, H. S.
1978 : Multi-beam sonar study of the Mid-Atlantic rift valley, 36°–37°N: Geological Society of America, Map and Chart Series MC-19, 5 p. plus charts at 1:36,457 scale.
Phillips, J. D., and Tapscott, C.
1981 : The evolution of the Atlantic Ocean north of the Azores: unpublished manuscript, Woods Hole Oceanographic Institution, Woods Hole, MA.
Pitman, W. C., and Talwani, M.
1972 : Seafloor spreading in the North Atlantic: Geological Society of America Bulletin, v. 83, p. 619–646.
Ramberg, I. B., and van Andel, T. H.
1977 : Morphology and tectonic evolution of the rift valley at latitude 36°31′N, Mid-Atlantic Ridge: Geological Society of America Bulletin, v. 88, p. 577–586.
Rea, D. K., and Scheidegger, K. F.
1979 : Eastern Pacific spreading rate fluctuation and its relation to Pacific area volcanic episodes: Journal of Volcanological and Geothermal Research, v. 5, p. 135–148.
Reid, I., and Macdonald, K. C.
1973 : Microearthquake study of the Mid-Atlantic Ridge near 37°N, using sonobuoys: Nature, v. 246, p. 88–90.
Rich, J. E., Johnson, G. L., Jones, J. E., Campie, J.
1986 : A significant correlation between fluctuations in sea-floor spreading rates and evolutionary pulsations: Paleoceanography, v. 1, p. 85–95.
Rona, P. A., and Gray, D. F.
1980 : Structural behavior of fracture zones, symmetric and asymmetric about a spreading axis; Mid-Atlantic Ridge (latitude 23°N to 27°N): Geological Society of America Bulletin, v. 91, p. 485–494.
Savostin, L. A., and Karasik, A. M.
1981 : Recent plate tectonics of the Arctic Basin and of northeastern Asia: Tectonophysics, v. 74, p. 111–145.
Savostin, L., Verzhbitskaya, A. I., and Baranov, B. V.
1982 : Holocene plate tectonics of the Sea of Okhotsk region: Doklady Academy of Science USSR, Earth Science Section, v. 226, p. 62–65.
Savostin, L., Zonenshain, L., and Baranov, B.
1983 : Geology and plate tectonics of the Sea of Okhotsk: American Geophysical Union, Geodynamics series, v. 11, p. 189–221.
Savostin, L. A., Karasik, A. M., and Zonenshain, L. P.
1984 : Istoriya raskrytiya Evraziiskogo basseina Arktiki (in Russian): Doklady Akademii Nauk SSSR, v. 275, p. 1156–1161.
Schilling, J-G., Meyer, P. S., and Kingsley, R. H.
1983 : Rare earth geochemistry of Iceland basalts; Spatial and temporal variations; in Structure and Development of the Greenland-Scotland Ridge, eds. M.H.P. Botts, S. Saxov, M. Talwani and J. Thiede; Plenum, New York, p. 319–342.
Schwarzbach, M.
1980 : Alfred Wegener und die Drift der Kontinente: Stuttgart, Wissenschaftliche Verlagsgesellschaft, 160 p.
Searle, R.
1977 : Geophysical studies of the Atlantic sea floor near 40°N, 24°W, and its relation to King's Trough and the Azores: Marine Geology, v. 25, p. 299–320.
Sibuet, J.-C., and Mascle, J.
1978 : Plate kinematic implications of Atlantic equatorial fracture zone trends: Journal of Geophysical Research, v. 83, p. 3401–3421.
Sleep, N. H., and Rosendahl, B.

1979 : Topography and tectonics of ridge areas: Journal of Geophysical Research, v. 84, p. 6831–6840.
Spindel, R. C., Davis, S. B., Macdonald, K. C., Porter, R. P., and Phillips, J. D.
1974 : Microearthquake survey of the median valley of the Mid-Atlantic Ridge at 36°30′N: Nature, v. 248, p. 577–579.
Steiger, R. H., and Jäger, E.
1977 : Subcommission on Geochronology; Convention on the use of decay constants in geo- and cosmochronology: Earth and Planetary Science Letters,v . 36, p. 359–362.
Talwani, M., Windisch, C. C., and Langseth, M. G.
1971 : Reykjanes Ridge crest: A detailed geophysical study: Journal of Geophysical Research, v. 76, p. 473–517.
Talwani, M., and Eldholm, O.
1977 : Evolution of the Norwegian-Greenland Sea: Geological Society of America Bulletin, v. 88, p. 969–999.
Tapley, B. D., Schutz, B. E., and Eanes, R. J.
1985 : Station coordinates, baselines, and earth rotation from LAGEOS laser ranging, 1976–1984: Journal of Geophysical Research, v. 90, p. 9235–9248.
Tarling, D. H., and Mitchell, J. G.
1976 : Revised Cenozoic polarity time scale: Geology, v. 4, p. 133–136.
Tryggvason, E.
1982 : Recent ground deformation in continental and oceanic rift zones; in Continental and Oceanic Rifts, ed. G. Pálmason; American Geophysical Union, Geodynamics Series, v. 8, p. 17–29.
Torge, W., and Kanngieser, E.
1985 : Regional and local vertical crustal movements in northern Iceland, 1965–1980: Journal of Geophysical Research, v. 90, p. 10173–10177.
Tucholke, B. E., and Schouten, H.
1985 : Global plate motion changes recorded in the Kane fracture zone: The Geological Society of America Abstract with Programs, v. 17, p. 737.
Unternehr, P.
1982 : Etude structurale et cinematique de la Mer de Norvege et du Groenland - Evolution du Microcontinent de Jan Mayen [Ph.D. thesis]: L'Universite de Bretagne Occidentale, Brest.
Vogt, P. R.
1971 : Asthenosphere motion recorded by the ocean floor south of Iceland: Earth and Planetary Science Letters, v. 13, p. 153–160.
1972 : Evidence for global synchronism in mantle plume convection and possible significance for geology: Nature, v. 240, p. 338–342.
1974 : The Iceland phenomenon; Imprints of a hot spot on the ocean crust, and implication for flow below the plates; in Geodynamics of Iceland and the North Atlantic area, ed. L. Kristjansson, D. Reidel, Dordrecht, Holland, p. 105–126.
1979 : Global magmatic episodes; New evidence and implications for the steady-state mid-oceanic ridge; Geology, v. 7, p.93–98.
1983 : The Iceland mantle plume; Status of the hypothesis after a decade of new work; in Structure and Development of the Greenland-Scotland Ridge, eds. M.H.P. Bott, S. Saxov, M. Talwani and J. Thiede; Plenum, New York, p. 191–216.
1986a: Geophysical and geochemical signatures and plate tectonics; in The Nordic Seas, ed. B. G. Hurdle; Springer, New York, p. 413–662.
1986b: Seafloor topography, sediments, and paleoenvironments; in The Nordic seas, ed. B. G. Hurdle; Springer, New York, p. 237–410.
Vogt, P. R., and Perry, R. K.
1981 : North Atlantic Ocean; Bathymetry and plate tectonic evolution, with chart: Geological Society of America MC-35, 21 p.
Vogt, P. R., Schneider, E. D., and Johnson, G. L.
1969 : The crust and upper mantle beneath the sea; in The Earth's crust and upper mantle, ed. P. Hart; American Geophysical Union Geophysical Monograph 13, p. 557–617.
Vogt, P. R., Johnson, G. L., and Kristjansson, L.
1980 : Morphology and magnetic anomalies north of Iceland: Journal of Geophysics, v. 47, p. 67–80.
Vogt, P. R., Kovacs, L. C., Bernero, C., and Srivastava, S. P.
1982 : Asymmetric geophysical signatures in the Greenland-Norwegian and southern Labrador seas and the Eurasia Basin; Tectonophysics, v. 89, p. 95–150.
Wanach, B.
1926 : Ein Beitrag zur Frage der Kontinentalverschiebung: Zeitschrift für Geophysik, v. 2, p. 161–163.
Wegener, A.
1929 : The Origin of Continents and Oceans, translated from the fourth revised 1929 German edition by Biram, J.: Dover, New York, 246 p.
Wendt, K., Möller, D., and Ritter, B.
1985 : Geodetic measurements of surface deformations during the present rifting episode in NE Iceland: Journal of Geophysical Research, v. 90, p. 10163–10172.
Zverev, S. M., Kosminskaya, I. P., Krasilschikova, G. A., and Mikhota, G. G.
1976 : The crustal structure of Iceland and of the Iceland-Faeroes-Scotland region: Societas Scientarium Islandica, v. 5, p. 72–95.

MANUSCRIPT ACCEPTED BY THE SOCIETY JUNE 23, 1986

ACKNOWLEDGMENTS

I thank J. B. Minster and D. Engebretson for thoughtful reviews. C. DeMets and D. Argus kindly provided results of the NUVEL present plate motion model prior to publication. I. Jewett was the illustrator and M. Whitney, J. Perrotta, and C. Williams prepared the manuscript.

Printed in U.S.A.

The Geology of North America
Vol. M, The Western North Atlantic Region
The Geological Society of America, 1986

Chapter 25

Surficial sedimentary processes revealed by echo-character mapping in the western North Atlantic Ocean

E. P. Laine*
Graduate School of Oceanography, University of Rhode Island, Narragansett Bay Campus, Narragansett, Rhode Island 02882-1197
J. E. Damuth*
Lamont-Doherty Geological Observatory of Columbia University, Palisades, New York 10964
Robert Jacobi
Department of Geological Sciences, State University of New York at Buffalo, Amherst, New York 14226

INTRODUCTION

The development of high-resolution precision depth recorders (PDR) in the 1950s (Luskin and others, 1954; Knott and Hersey, 1956) not only permitted continuous and accurate measurement of ocean depths, but also provided a new tool for the study of deep-sea sedimentation processes. Changes in acoustic reflectivity and microtopography of the seafloor as observed on these early 12-kHz PDR records were found in many cases to be caused by erosional/depositional processes (Heezen and others, 1959). One of the first studies helped to confirm the existence of the Western Boundary Undercurrent (WBUC), a thermohaline, contour-following bottom current which is important in shaping the continental rise off eastern North America (Heezen and others, 1966; see also McCave and Tucholke, this volume). As part of this study, Hollister (1967) constructed an "echo character" map of the continental rise off Nova Scotia by classifying and mapping the regional changes of acoustic reflectivity observed on 12-kHz echograms. This map suggested that certain types of echoes (e.g. hyperbolae and prolonged echoes) were apparently reflected from sand waves or other depositional bedforms created by the WBUC. Regionally, these echo types were thought to be aligned parallel to bathymetric contours beneath the WBUC axis (Heezen and others, 1966; Hollister, 1967; Hollister and Heezen, 1972). Although subsequent and more detailed studies using much better quality 3.5-kHz echograms have shown that the regional pattern of echo character along the Nova Scotian margin is primarily downslope rather than along slope, this pioneering work by Hollister and Heezen demonstrated the usefulness of echo-character mapping as a tool to infer sedimentary processes and environment. Several subsequent studies in the western North Atlantic included mapping and/or analysis of 12-kHz echograms in order to study the paths of flow and erosional/depositional effects of the WBUC and the Antarctic Bottom Water (AABW) on the seafloor (Clay and Rona, 1964; Schneider and others, 1967; Schneider and Heezen, 1966; Bryan and Markl, 1966; Rona and others, 1967; Fox and others, 1968; Rona, 1969).

In the late 1960s many oceanographic institutions began to collect continuous 3.5-kHz PDR echograms in addition to 12-kHz recordings. 12-kHz echograms generally show limited (<10 m) penetration of acoustic energy below the seafloor because of the preferred attenuation of high-frequency sound by both the water column and the seabed, but the lower-frequency 3.5-kHz records generally obtain 20 to ~100 m of sub-bottom penetration. Thus 3.5-kHz echograms reveal much more information about the nature and orientation of physical stratification (e.g. presence or absence of reflectors, truncation or migration of reflectors, etc.) in the upper part of the seabed (Damuth, 1975; Embley, 1975). Consequently, most studies of echo character since 1970 have used primarily 3.5-kHz echograms.

Such echo-character mapping studies are now widely used to evaluate regional seafloor sedimentation processes in many parts of the world's oceans, including portions of the western North Atlantic (see Damuth, 1980, for a review). The information obtained from 3.5-kHz and 12-kHz echograms generally can be used to define both the types and the regional influences of sedimentation processes such as contour-following bottom currents, turbidity currents, and mass-transport processes. These data are greatly enhanced when combined with other data such as sediment cores, bottom photographs, hydrographic and nephelometer measurements, and low-frequency (<100 Hz) seismic reflection profiles.

Present addresses: Laine, Environmental Studies Program and Department of Geology Bowdoin College, Brunswick, Maine 04011; Damuth, Mobile Research and Development Corporation, Dallas Research Laboratory, P.O. Box 819047, Dallas, Texas 75381.

Laine, E. P., Damuth, J. E., and Jacobi, R. D., 1986, Surficial sedimentary processes revealed by echo character mapping in the western North Atlantic; *in* Vogt, P. R., and Tucholke, B. E., eds., The Geology of North America, Volume M, The Western North Atlantic Region: Geological Society of America.

CLASSIFICATION AND DISTRIBUTION OF ECHO TYPES

Introduction

The principles and methods of echo-character mapping have been described in detail in a series of previous papers (see Damuth, 1980 and references therein), and they are not repeated here except for a brief discussion of geometrical effects. The classification of echo types used here is based on Damuth (1975; 1978; 1980) and Damuth and Hayes (1977). To produce a legible map for publication at the required scale (Plate 7), we simplified the more detailed original maps used for this compilation. In addition, the echo-character classification used here is, of necessity, a combined and simplified version of the original classification. Although the map consequently loses local detail, it allows us to illustrate the regional variation in echo character and sedimentary processes.

The same physical and geometrical principles which cause the distortion of conventional land-based seismic reflection profiles and thereby make their interpretation difficult (Hagedoorn, 1954; Hilterman, 1970; Smith, 1977) also influence the echograms on which this study is based (Flood, 1980). Because of the great depths in most instances to seafloor features, and the wide (30° to 60°) beam width of conventional surface-ship echo sounders, the echograms generated by such systems can appear significantly different from true seafloor morphology. This is because wide-beamwidth echo sounders insonify large areas of the seafloor that are not directly beneath the ship. Echoes returning from these areas can then confuse or distort the recorded echogram. In the simplest case, slopes measured on echograms taken over a sloping seafloor will be less than the true slope (Krause, 1962). More difficult to interpret than this simple distortion of shape are echograms taken over peaks or sharp changes in slope. These are generally recorded as hyperbolic echoes (Hoffman, 1957; Bryan and Markl, 1966; Flood, 1980). Echograms taken over sinusoidal or periodic bedforms may also be distorted (Flood, 1980).

The speed of the ship, its course (when passing over linear features), and water depth all play a role in determining the exact nature of the distortions appearing in any given echogram. For example, Flood (1980) has demonstrated that a trough between migrating sediment waves (Echo Type V, Plate 7) may appear very different in echograms having different water depths. In shallower water (<2.5 km) the troughs will only be slightly distorted, the main effects being altered slopes within the limbs and intensification of the echoes from beneath the trough. At significantly greater depths (~5.5 km) the same feature appears as a V-floored "valley" with strong and confused echo returns from beneath the valley.

The classification and distribution of echo types in the western North Atlantic Ocean are discussed below. For reference to geographic locations, the reader should consult Plate 2 at the back of the volume.

Distinct and Indistinct Echoes

Among the most commonly observed echo types within the western North Atlantic are those with distinct to indistinct bottom echoes. These echo types are commonly found within all of the well-mapped provinces from the continental shelf across the continental margin to the abyssal plains and basinal plateaus.

Echo Type IA. Distinct, sharp bottom echoes with several sharp, parallel sub-bottom reflectors which are continuous for tens of kilometers: seafloor is flat to undulating (Plate 7). This echo type is one of the most widespread and is recorded from many portions of the continental rises and abyssal plains, as well as from the plateaus of the Bermuda Rise Drift northeast of Bermuda. On the abyssal plains, these echoes are generally recorded from more distal portions, especially on the Demerara and Ceara abyssal plains in the Guiana Basin (Damuth, 1975; 1980). On the Bermuda Rise flanks, the seafloor is generally undulating to hilly where these echoes are recorded. In areas returning Type IA echoes, the maximum acoustic penetration observed in the western North Atlantic (often 100 m or more) is generally achieved.

Echo Type IB. Distinct echoes with no sub-bottom reflectors (Plate 7). This echo type is the most common return from the consolidated sediments of the continental shelves. Distinct, sharp echoes with no sub-bottoms are observed because most of the sound energy is reflected and acoustic penetration is limited to a few meters. Isolated examples are also found in the deep basin in areas where Quaternary cover has been removed, thus exposing reflective, semi-consolidated Neogene sediments (e.g. the northern Bermuda Rise).

Echo Type IIA. Indistinct, semi-prolonged bottom echoes with intermittent or discontinuous, indistinct (fuzzy), sub-bottom reflectors: seafloor is flat to undulating (Plate 7). This echo type is recorded from extensive regions of the continental rise and abyssal plains. It is apparently the result of sound-pulse reflections from laterally discontinuous sub-bottom sedimentary beds.

Echo Type IIB. Very prolonged bottom echoes with no sub-bottom reflectors: seafloor is flat to undulating (Plate 7). Prolonged echoes are recorded from many locations on the continental rises and abyssal plains, although they are commonly less extensive than Type IA and IIA echoes. Prolonged echoes also are recorded from the floors of submarine canyons and channels such as the Northwest Atlantic Mid-Ocean Channel (Plate 2). These echoes characterize the entire lower part of the Amazon deep-sea fan and adjacent Demerara Abyssal Plain (5°–11°N, 45°–50°W; Plate 7). Prolonged echoes similar to Type IIB are also recorded from debris-flow deposits; however, in this paper we have classified echoes from debris flows separately (Echo Type IV) in order to distinguish these important deposits (Plate 7).

In reality, there are wide variations in appearance of echo Types IA, IIA, and IIB, mainly in the depth of acoustic penetration below the seafloor, and the thickness, number, and clarity of sub-bottom reflectors. Thus distinct echoes with continuous sub-bottoms (IA) and very prolonged echoes with no sub-bottoms (IIB) form two end members of echo character. Numerous gradational forms (mostly variations of IIA) occur between these end members (see Damuth, 1980, and references therein for other examples). The relationship of these three echo types to the distribution of coarse terrigenous sediment is discussed later.

Hyperbolic Echoes

Hyperbolic echoes are the result of reflections from point or line reflection sources on or sometimes beneath the sea floor. Point reflection sources are most commonly rugged basement topography such as that found along the mostly unsedimented crest and flanks of the Mid-Atlantic Ridge. Line sources are commonly sedimentary features such as sea floor furrows or scarps associated with sediment mass movements.

Echo Type IIIA. Large, irregular, overlapping and single hyperbolae with widely varying vertex elevations: seafloor morphology is hilly to very rugged (Plate 7). These echoes are recorded from isolated seamounts, knolls, and basement outcrops; seamount chains such as the New England Seamounts; oceanic ridges and rises such as the Mid-Atlantic Ridge, Barracuda Ridge, and Ceara Rise; fracture-zone ridges; tectonized sediment wedges such as the Barbados Ridge; and steep, rugged areas of the continental slope. The morphology of such regions (excepting the continental slope and the Barbados Ridge) is controlled by the configuration of oceanic basement; the basement normally is rugged and either crops out or is covered by only a thin veneer of pelagic sediments. The rugged morphology of the basement rocks (numerous projections of unequal height above the seafloor) gives rise to the large irregular hyperbolae recorded on 3.5-kHz records. Although this echo type is generally not created by sedimentation in such areas, it indirectly suggests the nature of localized, small-scale sedimentation processes. Pelagic sedimentation, slumping, and debris flows all are important within steep-sloped regions such as these; however, these features are of such small scale that they are not resolved in our mapping (Plate 7), and often they are not even recorded in broad-beamwidth 3.5-kHz and 12-kHz echograms.

Along the western Atlantic continental slopes, the rugged topography creating echo Type IIIA is formed from semiconsolidated and consolidated sedimentary rocks. Down-cutting by turbidity currents, local and regional slope failure, and other forms of canyon and gully formation all have contributed to the dissection of these sedimentary rocks to create irregular topography. Beneath the Barbados Ridge, sediments and sedimentary rocks have been tectonized to form rugged topography. Thrusting, folding, and mud volcanism associated with underthrusting of the North Atlantic plate beneath the Caribbean plate all have contributed to the deformation of this sedimentary wedge and the creation of an irregular surface with steep slopes.

Echo Type IIIB. Regular overlapping hyperbolae with vertices approximately tangent to the seafloor (Plate 7). Amplitudes of these echoes are generally less than 50 m and wavelengths are short (100-500 m). The hyperbolae are recorded mainly from relatively small, scattered patches along the continental margin and in the basin interior where the seafloor has been affected by contour-following currents (Plate 7). The largest regions of these echoes are observed on the Blake-Bahama Outer Ridge system. Type IIIB hyperbolae are generally associated with regions of Type IIIC hyperbolae and with sediment waves or drift deposits (Echo Type V, see below). Some Type IIIB hyperbolae show changes in wavelength with profile azimuth, indicating that the bedforms from which they are reflected have a regular orientation (Damuth, 1975, 1980; Flood, 1980). In some areas such hyperbolae are observed not only at the seafloor but also along one or more discrete sub-bottom horizons.

Echo Type IIIC. Regular overlapping hyperbolae with varying vertex elevations above the seafloor; conformable sub-bottom reflectors may be present (Plate 7). Wave lengths are generally less than one kilometer although longer ones are occasionally observed. Amplitudes generally range from 10 to 100 m. These Type IIIC hyperbolae are recorded mainly from isolated locations on the continental rise and rarely in the basin interior (Plate 7). The most extensive occurrences of Type IIIC hyperbolae are observed on the Blake-Bahama Outer Ridge and adjacent portions of the continental rise. Type IIIB and IIIC hyperbolae are reflected from erosional/depositional bedforms that were created by thermohaline flows of bottom water (contour currents).

Type IIID. Regular overlapping hyperbolae with varying vertex elevations above the seafloor (Plate 7). These hyperbolae are very similar in appearance and size to IIIC hyperbolae. However, Type IIID hyperbolae are slightly less regularly spaced and vertically more variable than Type IIIC hyperbolae, and they are often found in association with debris-flow deposits (Echo Type IV) in isolated locations. They apparently represent displaced slump and slide material at the heads of slump and debris-flow deposits. Type IIID hyperbolae generally have no sub-bottom reflectors.

Miscellaneous Echo Types

Echo Type IV. Transparent lenses with prolonged bottom echoes (Plate 7). These echoes have a seafloor reflection similar to Type IIB echoes but are distinguished from them by the presence of a transparent lens. They are the deposits from submarine debris flows. Such debris-flow deposits generally appear as thin (<100 m), acoustically non-laminated lenses which extend for tens to hundreds of km down the continental rise (Embley, 1976, 1980; Embley and Jacobi, 1977; Jacobi, 1976; Jacobi and Hayes, 1982; Damuth and Embley, 1981). Well-stratified, undisturbed sediments occasionally are observed beneath the debris flows, but generally internal reflections within the lens are not observed because the sediments are homogenized during the transport process. Failure to observe these transparent lenses is not necessarily diagnostic of the *absence* of debris flows because the deposits can be so thick that underlying reflectors are not observed. In this case only a prolonged surface echo is recorded and no transparent "lens" is observed. Type IV echoes trend downslope at many locations on the continental rise off eastern North America (Plate 7). Two large zones of such echoes are also observed extending seaward of irregular, Type III D echoes on the Amazon Deep-Sea Fan (4°-6° N; Plate 7).

Echo Type V. Broad distinct, sediment waves with distinct conformable to unconformable sub-bottom reflectors: seafloor is undulating to hummocky. These echoes are recorded from large-scale sediment waves. They are found on drift deposits created by contour currents, or on levees deposited on the flanks of submarine canyons (Plate 7). The waves contain unconformable to conformable sub-bottom reflectors that often show erosional truncation and lateral migration. The sediment waves and associated drift deposits are widely distributed off North and South America, both along the continental margins and in the interior of the basins.

SEDIMENTATION PROCESSES REVEALED BY ECHO CHARACTER

Compilation of the Echo Character Map (Plate 7)

The echo character map of the western North Atlantic at the back of this volume (Plate 7) was compiled mainly from maps which were constructed for smaller, regional studies. Data sources include published maps (Damuth, 1975; Silva and others, 1976; Damuth and Hayes, 1977; Shipley, 1978; Mullins and others, 1979; Laine and others, 1983; Vasallo and others, 1984a, b; and Damuth and others, 1986), unpublished theses (Kristofferson, 1977; Flood, 1978; McCreery, 1983), and unpublished compilations (Damuth, Flood, and Laine). The eastern continental margins of both North and South America are well-studied, as are the adjacent abyssal plains. Other areas that are characterized by thick sediment accumulations and that are affected by contour currents (Blake-Bahama Outer Ridge, northern Bermuda Rise, and portions of Labrador Basin) have also been studied in detail.

Where possible, we have also examined echograms and mapped the echo character in regions of the western North Atlantic where no previous mapping had been done. In this way we attempted to achieve continuity between existing maps and to map as many uncharted regions as time and data permitted. In many cases, the data spacing in unmapped regions was too wide to permit meaningful determination of echo character distributions. As a result, the abyssal hills province on the flanks of the Mid-Atlantic Ridge, the ridge crest itself, and fracture zones were not systematically mapped in this study. Although these are regions of generally thin sediment cover, previous small-scale studies (van Andel and Komar, 1969; Shipley, 1978) have shown that pelagic sedimentation, gravity flows (turbidity currents and mass movements), thermohaline currents, and tectonic processes all shape the sedimentary deposits within these provinces. In these poorly mapped areas of the western Atlantic, we have shown generalized echo character, based on examination of scattered track lines and our knowledge of seafloor morphology. Province boundaries in these regions are based for the most part on the boundaries given by Emery and Uchupi (1984, Plate X). The resulting generalizations imply no rigorous interpretation of sedimentary processes in these areas, but they provide the reader an overview of the relative areal importance within the western North Atlantic of both the various echo types and their associated sedimentary processes.

The continental margins and adjacent abyssal plains and plateaus of the western North Atlantic are the areas of densest and most reliable echogram coverage. Three sedimentary processes are dominant there: deposition from turbidity currents (Pilkey and Cleary, this volume), deposition and erosion by bottom currents (McCave and Tucholke, this volume), and mass movements (Embley and Jacobi, this volume). Along the flanks, foothills and crest of the Mid-Atlantic Ridge where we have the lowest density of echograms, pelagic sedimentation and local sediment mass movements are dominant (van Andel and Komar, 1969; Shipley, 1978).

For each sedimentary process, several echo types are usually closely associated. Thus, it is the mapped assemblage of echo types rather than any one echo type which is most diagnostic of sedimentary processes within an area. For example, Echo Type V is created either by contour-following bottom currents, or by turbidity currents in a region of channel-bank overflow. When found in association with Echo Types IIIB and IIIC on sediment drift deposits, these echoes are usually interpreted as having a bottom-current origin. When found on the levees of submarine canyons in association with Echo Types IA, IIA, and IIB they are interpreted as having a turbidity-current origin.

Deposition from Turbidity Currents

Turbidity currents are important in depositing sediments along the continental rises, submarine canyon and fan systems, and abyssal plains of the western North Atlantic (Pilkey and Cleary, this volume). In almost all cases except the Blake-Bahama Abyssal Plain, turbidites containing terrigenous sediments of gravel to clay size have been deposited by these turbidity currents.

Previous echo-character studies in the Atlantic (Damuth, 1975, 1978, 1980; Damuth and Hayes, 1977) demonstrated that a qualitative relationship exists between the relative abundance of coarse (silt, sand, gravel), bedded sediment in the upper few meters of the seafloor, and the degree of development of Echo Types IA, IIA, and IIB. Regions returning distinct echoes with continuous sub-bottom reflectors (Type IA; Plate 7) generally contain little or no bedded silt/sand (normally <5%); regions of semi-prolonged echoes with intermittent sub-bottom reflectors (Type IIA) have low to moderate amounts of bedded silt/sand (normally 5–30%); and regions returning very prolonged echoes with no sub-bottom reflectors (Type IIB) exhibit large amounts of bedded silt/sand (up to 100%). The reasons for the observed relationship between echo type and quantity of bedded silt/sand are complex, but the relation probably is due to changing signal interference patterns in response to changes in thickness and frequency of silt/sand beds and to variable development of small-scale bedforms at and below the seafloor. These sediment-echo correlations have been discussed in detail in previous reports (Damuth, 1980, and references therein) but they still are not well understood in a quantitative sense. However, they do provide an empirical method of predicting concentration and distribution of coarse terrigenous sediment over large regions.

An example of the relationship between echo type and distribution of coarse terrigenous sediment is the region of the Amazon Deep-Sea Fan and adjacent abyssal plains off northeast Brazil (Plate 7; 3°–10°N; 43°–52°W). Relationships have been confirmed by extensive piston coring (Damuth, 1975, 1980; Damuth and Kumar, 1975). Most coarse sediment from the Amazon River bypasses the upper and middle fan via large distributary channels (<4.2 km depth) and it is deposited across the lower fan and on the adjacent Demerara and Ceara abyssal plains. The sediments of the upper and middle fan consequently return distinct echoes with continuous sub-bottom reflectors (Type IA). In contrast, the lower fan (4.2–4.8 km) and the adjacent portion of the abyssal plain to the north return very prolonged echoes (Type IIB). Around the perimeter of this depocenter the prolonged echoes give way to semi-prolonged echoes with intermittent sub-bottoms (Type IIA) which indicate moderate amounts of bedded silt/sand. Farther downslope, to the east and west on the distal parts of the Ceara and Demerara abyssal plains, distinct echoes with continuous sub-bottoms (Type IA) are again returned from seafloor sediments containing little coarse sediment. The echo-character pattern of the Amazon Fan region thus reflects the sedimentation patterns on the fan. Coarse sediments bypass the upper and middle fan, are deposited mostly across the lower fan and proximal portions of abyssal plains, and become progressively less abundant radially outward from the lower fan towards distal regions (Damuth 1975, 1980; Damuth and Kumar, 1975).

Farther north along the margin of the eastern United States are two large provinces of turbidity current deposition, one principally along the continental rise to depths of about 4.5 km and a second across the Hatteras Abyssal Plain. The major turbidite pathways across the continental slope and rise appear as narrow zones of prolonged bottom echoes with no sub-bottom reflectors (Echo Type IIB), suggesting deposition of coarse sediments along these pathways. The flanking levee and inter-canyon regions (where not disrupted by debris flows) appear as sharp, continuous bottom echoes with continuous sub-bottom reflectors (Echo Type IA) or as migratory sediment waves (Echo Type V). In many instances these echo types reflect spillover of fine material from turbidity currents moving down Hudson, Wilmington and Norfolk Canyons. In other areas they represent deposition from bottom currents, the sediment source being either the bottom nepheloid layer or material pirated from turbidity currents traversing the continental rise. It is important to note that the migrating sediment waves (Echo Type V) found on canyon levees are very similar in appearance to those formed by thermohaline flow in this region. Their occurrence in association with canyon pathways of turbidity currents demonstrates that overbank flow of turbidity currents can form migratory sediment waves similar to those formed by thermohaline flow (Embley and Langseth, 1977; Damuth, 1978, 1980).

Near the base of the Hatteras and Wilmington fan systems, which are two major entry points to the Hatteras Abyssal Plain (Pilkey and Cleary, this volume), semi-prolonged echoes (Type IIA) with intermittent sub-bottom reflectors are observed, suggesting deposition of moderate amounts of bedded silt/sand. Away from these entry points and southward to about 24°N, distinct echoes with continuous sub-bottom reflectors are observed beneath the Hatteras Abyssal Plain; this reflects a fining of sediment grain size away from the source areas. Farther southeast, the Nares Abyssal Plain receives distal turbidites from the Hatteras Abyssal Plain through Vema Gap (Plate 2). Very well-developed Type IA echoes are found there, indicating deposition of turbidite clays with occasional silt beds (Tucholke, 1980).

The Sohm Abyssal Plain is the primary depocenter for Laurentide-derived glacial-age terrigenous sediments in the western North Atlantic (Laine, 1980). The echo character of the western, well mapped part of the Sohm Abyssal Plain is consistent with a massive influx of coarse, terrigenous sediments from the Hudson Fan and Laurentian Fan; very prolonged bottom echoes with no sub-bottom reflectors are found beneath the northwest limb of the Sohm Abyssal Plain. Similar echo character probably extends farther east around the lower perimeter of the Laurentian Fan, although this remains to be mapped.

Further to the north, in the Labrador and Flemish Basins, the most conspicuous feature related to turbidity currents is the Northwest Atlantic Mid-Ocean Channel which funnels turbidity currents southward onto the Sohm Abyssal Plain (Cleary and Pilkey, this volume). Very prolonged echoes with no sub-bottom reflectors (Echo Type IIB) are found beneath the axis of this channel, and semi-prolonged echoes with intermittent sub-bottom reflectors (Echo Type IIA) flank the channel. This suggests deposition of coarse sands within the channel and spillover and deposition of smaller amounts of sand and silt in adjacent areas.

Bottom-Current Processes

Three current systems have an influence on sea-floor sedimentary processes in the western North Atlantic Ocean: the Western Boundary Undercurrent (WBUC), the deep Antarctic Bottom Water current, and the Gulf Stream system (McCave and Tucholke, this volume). Their influence is most readily seen in echograms across regions which are protected or distant from the masking influence of turbidity currents and sediment mass movements. These areas include portions of the continental rise deeper than 4.5 km off the eastern United States (e.g. Hatteras Outer Ridge); the continental rise off eastern Canada north of about 43°N; outer ridge (drift) deposits such as the Blake, Bahama, Greater Antilles, Eirik, and Gulf Stream ridges; and other sediment-drift deposits including the Bermuda Rise Drift and particularly the Gloria Drift (Plate 2). At least five separate echo types are observed in these provinces (Plate 7): distinct echoes (IA and IB), hyperbolic echoes (IIIB and IIIC), and sediment waves (V).

Both Type IIIB and IIIC hyperbolic echoes are reflected from bedforms below the resolving capability of standard shipboard, broad-beam echosounding systems. It is necessary either to collect near-bottom observations or to migrate surface-ship echograms in order to determine the true morphology of the bedforms that create hyperbolic echoes. Where such studies have been done they have shown that hyperbolic echo Types IIIB and IIIC are generally caused by a series of erosional/depositional bedforms termed "furrows" (Hollister and others, 1974a; Flood, 1978, 1980, 1983; Flood and others, 1979; Lonsdale, 1978; Tucholke, 1979; Embley and others, 1980; McCave and others, 1982). Furrows are found in the deep North Atlantic as fields of regularly-spaced, parallel grooves in cohesive sediments. They range from 1-100 m in width and 0.5-20 m in depth, and are spaced 20-350 m apart (Flood, 1978; 1983). While the origin of at least some furrows is erosional, developing through secondary circulation patterns in the turbulent oceanic bottom boundary layer (Flood, 1983), others are thought to be syndepositional (Tucholke, 1979). The balance between erosional and depositional episodes over time determines whether individual furrows enlarge through erosion, maintain their form during sedimentary upbuilding, or are smoothed over by sedimentation (Embley and others, 1980; Flood, 1983).

While hyperbolic echoes are widely distributed throughout the basin wherever bottom currents have an effect, they are most extensively developed and studied on the Blake-Bahama Outer Ridge. Type IIIC hyperbolae cover the northeast flank of the Blake Outer Ridge and the west flank of the Bahama Outer Ridge. The smaller Type IIIB hyperbolae are found in similar proportion on these ridges, with major fields developed in the central areas of both features. This distribution of Type IIIB and IIIC hyperbolic echoes suggests that erosion is greatest on the perimeter of these drifts, with less erosion or perhaps syndepositional development occurring within their central regions. This implies more intense bottom currents flowing along the perimeter with more tranquil flow above the central regions of the Blake and Bahama Outer Ridges.

Echo Type V represents another class of bedform, sediment waves, commonly found within provinces influenced by bottom currents. These bedforms are quasi-periodic undulations of the seafloor with characteristic amplitudes of 10-100 m and wavelengths of 2-11 km. Some sediment waves are regular and almost sinusoidal with conformable sub-bottom reflectors; others are systematically distorted, with differing slopes on each limb and internal thickness variations between reflectors that indicate migration of the bedform (Plate 7). When the true orientation of the wave crests is known, it is generally oblique to regional contours, oriented 35° - 45° upslope in a clockwise direction from the regional contours (Clay and Rona, 1966; Hollister and others, 1974b; Flood, 1978; Embley and Langseth, 1977). Where the migration direction can be determined the waves appear to migrate upslope and in most cases upcurrent (Embley and Langseth, 1977; Flood 1978).

In the North Atlantic a 1200-km zone of sediment waves extends discontinuously north from the southern tip of the Bahama Outer Ridge near 25°N to the vicinity of the Hudson Canyon near 36°N (Plate 7) between about 4.5 and 5.2 km water depths. Sediment waves also extend eastward about 1000 km onto the northern Bermuda Rise. North of the New England Seamounts, waves occur in the same depth range on the Nova Scotian continental rise between 40° and 42°N.

Farther north, several fields of sediment waves are found within the Flemish and Labrador basins. In the Labrador Basin these waves are associated with the counterclockwise flow of the Western Boundary Undercurrent as it moves southward from its Norwegian Sea source area. Extensive wave fields are developed on Gloria Drift. The smaller wave fields on the eastern margin of the Flemish Basin are poorly studied; however their position away from the continental margin suggests the influence of some kind of mid-basin circulation such as that above the Bermuda Rise Drift.

Sediment waves observed deeper than 4.5 km along the continental margin and Blake-Bahama Outer Ridge are associated with deep, thermohaline flow in the lower part of the Western Boundary Undercurrent. It has been suggested that at these depths the Western Boundary Undercurrent contains a high-velocity core (Bulfinch and others, 1982). At these depths along both the continental rise between 33° and 36°N and on the Blake-Bahama Outer Ridge, the wave deposits comprise a significant proportion of the sedimentary section (Markl and Bryan, 1983; Tucholke and Laine, 1983). The waves have actively accreted since the beginning of the last phase of contour-current deposition during the middle Miocene (Tucholke and Mountain, this volume).

All recent models for the sedimentary evolution of the northern Bermuda Rise (Laine and Hollister, 1981; Ayer and Laine, 1982; Tucholke and Laine, 1983) suggest that currents at the base of the Gulf Stream system were important in shaping the seafloor during the Neogene and Quaternary, although models differ as to the specific pattern of currents. The arcuate distribution of sediment waves across the northern rise and their apparent continuity with waves along the U.S. Atlantic margin suggest a possible eastward extension of bottom-sensing flow of the Gulf Stream during the Neogene (Tucholke and Laine 1983). However, the apparent sediment-wave distribution may be an artifact of preservation, because large portions of the northern Bermuda Rise have been strongly eroded during the Neogene and Quaternary (Laine and others, 1983).

Distinct Type IA echoes with sub-bottom reflectors are found in all provinces influenced by bottom currents. In these regions the echo character results from deposition of fine-grained sediments from the bottom nepheloid layer. However, this echo type is not diagnostic of deposition from currents. For example, in well studied regions of the Pacific, echoes of similar character are recorded from entirely pelagic sediments (Mayer, 1979; Damuth and others, 1983), and the same kinds of echoes occur in distal abyssal plain areas such as the eastern Nares Abyssal Plain (Tucholke, 1980). Thus the interpretation of current-controlled deposition is usually made with supporting evidence of high sedimentation rates, terrigenous lithology, direct current observations, and particularly association with other diagnostic echo types such as sediment waves (Laine and Hollister, 1981; McCave and others, 1982).

Type IA echoes commonly are found in association with hyperbolic echoes and sediment waves. In many instances the hyperbolae are developed on the surface of Type IA, well-stratified sediments, and the hyperbolae degrade the resolution and continuity of the sub-bottom reflectors. Where found adjacent to sediment waves, the sub-bottom reflectors in the Type IA echoes are often continuous with those in the sediment waves.

Mass Wasting Processes

Mass wasting in the western North Atlantic is an extremely important down-slope sediment transport mechanism, primarily along the continental margins and to a lesser extent within the center of the basin. The widespread development of mass wasting, however, only became clear with careful study of 3.5-kHz echograms and more advanced observational techniques (e.g. side-scan sonar and submersibles) (Walker and Massingill, 1970; Embley, 1976, 1980; Embley and Jacobi, 1977; Embley and Jacobi, this volume; McGregor and Bennett, 1979; Jacobi and Hayes, 1982; Popenoe and others, 1982; Ryan 1982; Farre and others, 1983).

These studies show that mass wasting is accomplished by slumps and/or sediment slides. Slumps are blocks that have undergone downslope translation along a slump fault, but the blocks have only minor internal deformation. In contrast, sediment slides have experienced sufficient liquefaction to promote internal flow and deformation, commonly resulting in pebbly mudstones (debris flows) and other highly deformed deposits (Embley and Jacobi, 1977). Large slump blocks (ca. 10-50 km width) are not as easily recognized in 3.5-kHz echograms as they are in low frequency (<100 H_z) seismic reflection profiles. However, smaller slump blocks (ca. 1 km width) can be identified readily in echograms, primarily on the basis of the slump fault and the surface morphology of the block (Jacobi, 1976; Embley, 1980; Popenoe and others, 1982).

On echograms the slump faults are observed as reflector dislocations, and/or as Type IIID hyperbolae (Plate 7) that are reflections from the fault surfaces; they also are sometimes observed as ductile faults (Jacobi and Hayes, 1982). Surface morphology over small slump blocks consists of Type IIID hyperbolae (Plate 7), isolated areas of elevated multiple reflectors (for smaller slump blocks), or variable wave forms on blocks containing ductile faults.

Perhaps the "best-known" slump block is the Grand Banks slump (Heezen and Ewing, 1952) located on the continental slope at the head of the Laurentian Fan. However, preliminary analysis (Piper and Normark, 1982a, b) suggests that the proposed slump block probably is a channel-levee complex. Similar problems of interpretation may exist for apparently large slump blocks along the central U.S. east coast margin (McGregor and Bennett, 1977). Nevertheless, smaller slump blocks do exist and they are well documented along the upper continental slope and on walls and floors of submarine canyons and channels (Malahoff and others, 1980; Twitchell and Roberts; 1982; Farre and others, 1983). They occur more rarely on the lower continental slope and rise and in mid-basin areas (Silva and others, 1976). No slumps are shown on Plate 7 because of the small map scale.

Sediment slides consist of a zone of sediment removal bounded by one or more slide scars, and a downslope zone of deposition. The main glide plane generally parallels a prominent reflector. The resulting scarp generally has about 50 m relief but rarely may reach 250 m. The slide scarps are recognized on echogram and low-frequency seismic reflection profiles primarily by the truncation of reflectors. The zone of deposition contains a number of echo types, including Type IIID hyperbolae and Type IV transparent lenses (Plate 7). Slide-generated turbidity flows can also occur but they are not considered part of the slide complex.

Type IIID hyperbolae with high relief (>40 m valley to vertex) are termed hummocky terrane (Jacobi, 1976). They generally result from (1) large slide blocks (olistoliths), (2) areas of relatively limited and discontinuous sediment removal, or (3) more rarely, large piles of highly deformed slide material. Both of the first two kinds of occurrences have been observed in the eastern North Atlantic (Jacobi, 1976; Flood and others, 1979), although most large Type IIID hyperbolae in the western North Atlantic appear to result from slide blocks (Embley, 1980; Ryan, 1982; Vasallo and others, 1984 a, b). Smaller Type IIID hyperbolae (with valley to vertex relief <40 m) are termed blocky terrane (Jacobi, 1976) and probably result from smaller slide blocks and piles of highly deformed slide material. Both hummocky and blocky terranes are combined in Type IIID hyperbolae on Plate 7.

Type IV transparent lenses (Plate 7) are generally the most prevalent of the slide-associated seismic facies and they consist of debris-flow material (Embley, 1976; Jacobi, 1976). Piston cores retrieved from these echo types confirm the disturbed nature of the sediment (Embley, 1976, 1980; Jacobi, 1976; Jacobi and Mrozowski, 1979).

Small slide complexes are common along submarine canyons and channels in the mapped area (Plate 7) and in these locations they are a response to slope oversteepening along canyon walls. Such slides may also be responsible for most canyon development (Farre and others, 1983). In contrast, large sediment-slide complexes often occur on the continental slope and rise independent of canyon position. These slide complexes are up to 700 km long, and some stretch from the upper slope to the abyssal plain (Plate 7). Three of the larger slide complexes on the U.S. east coast margin are the Blake and Hudson slides (Embley and Jacobi, this volume), and the Grand Banks slide (Jacobi, in prep.). The Blake slide complex (Plate 7) extends to the abyssal plain and exhibits a large zone of blocky terrane (Echo Type IIID). The debris-flow material (Echo Type IV) surrounding the hyperbolated zone may be more liquefied sediment mobilized during the same slide event, or it may be a younger slide event. In contrast, both the Hudson and Grand Banks slide complexes consist almost entirely of debris-flow material (Echo Type IV, Plate 7).

On the U.S. east coast margin, few large slides are observed south of 31°N, probably because of relatively low sediment accumulation rates. The low rates are due to two factors: low sediment supply in areas south of the margin of the Laurentide Ice Sheet, and the barrier formed by the Gulf Stream to direct seaward transport of sediment in this area (Emery and Uchupi, 1972).

SUMMARY AND CONCLUSIONS

In well mapped portions of the western North Atlantic Ocean, echo character shows that three sedimentary processes are of primary importance along the continental margins, abyssal plains, and basinal plateaus of the basin. These are deposition from turbidity currents, sediment mass movements, and deposition and erosion by bottom currents.

Turbidity currents have been the areally most important sedimentary process within the well mapped portions of the western North Atlantic Ocean. This dominance is reflected in the distribution of abyssal plains all along the perimeter of the continental margins from the Labrador Basin to the Guiana Basin. On the continental margins themselves, downslope sediment transport by turbidity currents is reflected by cross-contour patterns in the echo character. Canyon axes and levee deposits are the predominant sedimentary features.

Mass movements compete with turbidity currents as the principal mechanism of sediment transport along the continental margins. Debris flows and other mass movements also display cross-contour patterns in echo character.

The strong influence of both of these processes along the margins reflects the importance of steep slopes and an abundant source of sediments from adjacent landmasses in controlling the sedimentary regime along the continental margin. Significantly, off Newfoundland and Labrador where a wide shelf and a lack of major rivers have restricted sediment supply to the continental slope and rise, the cross-slope patterns of echo character are not observed.

Bottom-current effects dominate the seafloor echo character within those well mapped portions of the western North Atlantic that are not influenced by downslope processes. These regions include parts of the continental rise off eastern North America which are protected from the masking effects of turbidity currents and mass movements, outer-ridge deposits which trend obliquely away from the continental rise, and other drift deposits which lie in the central portions of the basin completely separated from the continental rise.

In the less well mapped areas of the western North Atlantic where we have presented only a generalized echo character map, there are echo character patterns that deserve future study. Our mapping and that of Emery and Uchupi (1984, Plate X) suggests a large, semi-continuous band of current-controlled deposition trending northeastward from the southern portions of the Bermuda Rise into the Labrador Basin. This band roughly parallels the flanks of the Mid-Atlantic Ridge. If this is indeed a distinct and continuous province, it would suggest an organized current system, perhaps Antarctic Bottom Water, flowing northward along the western flank of the Mid-Atlantic Ridge. However, if this pattern is not continuous but is comprised of individual, small-scale drift deposits, it may imply that small-scale deep circulation cells are active within the basin. Along the flanks and crest of the Mid-Atlantic Ridge, modern studies of echo character are non-existent. We have presumed on the basis of morphology and very limited sedimentary studies that pelagic sedimentation, sediment mass movements, and turbidity currents all play roles within this region; however, until the proper echo character studies are completed we will not be able to assess their relative importance.

REFERENCES

Ayer, E. A., and Laine, E. P.
1982 : Seismic stratigraphy of the northern Bermuda Rise: Marine Geology, v. 49, p. 169–186.

Bryan, G., and Markl, R.
1966 : Microtopography of the Blake-Bahama region: Lamont-Doherty Geological Observatory Technical Report, No. 8, CU-8-66-Nobsr 85077, 44 p.

Bulfinch, D. L., Ledbetter, M. T., Ellwood, B. B., and Balsam, W. L.
1982 : The high velocity core of the Western Boundary Undercurrent at the base of the U.S. continental rise: Science, v. 215, p. 970–973.

Clay, C. S., and Rona, P. A.
1964 : On the existence of bottom corrugations in the Blake-Bahama Basin: Journal of Geophysical Research, v. 69, p. 231–234.

Damuth, J. E.
1975 : Echo character of the western equatorial Atlantic and its relationship to the dispersal and distribution of terrigenous sediments: Marine Geology, v. 18, p. 17–45.

Damuth, J. E.
1978 : Echo character of the Norwegian-Greenland Sea; relationship to Quaternary sedimentation: Marine Geology, v. 28, p. 1–36.
1980 : Use of high-frequency (3.5-12 kHz) echograms in the study of near-bottom sedimentation processes in the deep sea; a review: Marine Geology, v. 38, p. 51–75.

Damuth, J. E., and Embley, R. W.
1981 : Mass-transport processes on Amazon Cone; western equatorial Atlantic: American Association of Petroleum Geologists Bulletin, v. 65, p. 629–643.

Damuth, J. E., and Hayes, D. E.
1977 : Echo character of the east Brazilian continental margin and its relationship to sedimentary processes: Marine Geology, v. 24, p. M73-M95.
Damuth, J. E., Jacobi, R. D., and Hayes, D. E.
1983 : Sedimentation processes in the northwest Pacific basin revealed by echo-character mapping studies: Geological Society of America Bulletin, v. 94, p. 381–395.
Damuth, J. E., and Kumar, N.
1975 : Amazon Cone; morphology, sediments, age and growth pattern: Geological Society of America Bulletin, v. 86, p. 863–878.
Damuth, J. E., Tucholke, B. E., and Shor, A. N.
1986 : Echo character of the Scotian Rise: Marine Geology (in prep).
Embley, R. W.
1975 : Studies of deep-sea sedimentation processes using high-frequency seismic data [Ph.D. thesis]: New York, Columbia University, 334 p.
1976 : New evidence for occurrence of debris-flow deposits in the deep sea: Geology, v. 4, p. 371–374.
1980 : The role of mass transport in the distribution and character of deep-ocean sediments with special reference to the North Atlantic: Marine Geology, v. 38, p. 23–50.
Embley, R. W., and Jacobi, R.
1977 : Distribution and morphology of large sediment slides and slumps on Atlantic continental margins: Marine Geotechnology, v. 2, p. 205–228.
Embley, R. W., and Langseth, M. G.
1977 : Sedimentation processes on the continental rise of northeastern South America: Marine Geology, v. 25, p 279–297.
Embley, R. W., Hoose, P. J., Lonsdale, P., Mayer, L., and Tucholke, B. E.
1980 : Furrowed mud waves on the western Bermuda Rise: Geological Society of America Bulletin, v. 91, p. 731–740.
Emery, K. O., and Uchupi, E.
1972 : Western North Atlantic Ocean; Topography, Rocks, Structure, Water, Life, and Sediments: American Association of Petroleum Geologists Memoir 17, 532 p.
1984 : The Geology of the Atlantic Ocean: New York, Springer-Verlag, 1050 p.
Farre, J. A., McGregor, B. A., Ryan, W.B.F., and Robb, J. M.
1983 : Breaching the shelf break: passage from youthful to mature phase in canyon evolution: in The Shelfbreak; Critical Interface on Continental Margins: eds. D. J. Stanley and G. T. Moore; Society of Economic Paleontologists and Mineralogists Special Publication No. 33, p. 25–39.
Flood, R. D.
1978 : Studies of deep-sea sedimentary microtopography in the North Atlantic Ocean [Ph.D. thesis]: Woods Hole, Massachusetts, Woods Hole Oceanographic Institution, 395 p.
1980 : Deep-sea sedimentary morphology; modeling and interpretation of echo sounding profiles: Marine Geology, v. 38, p. 77–92.
1983 : Classification of sedimentary furrows and a model for furrow initiation and evolution: Geological Society of America Bulletin, v. 94, p. 630–639.
Flood, R. D., Hollister, C. D., and Lonsdale, P.
1979 : Disruption of the Feni sediment drift by debris flows from Rockall Bank: Marine Geology, v. 32, p. 311–334.
Fox, P. J., Heezen, B. C., and Harian, A. M.
1968 : Abyssal anti-dunes: Nature, v. 220, p. 470–472.
Hagedoorn, J. G.
1954 : A process of seismic reflection interpretation: Geophysical Prospecting, v. 2, p. 85–127.
Heezen, B. C., and Ewing, M.
1952 : Turbidity currents and submarine slumps, and the 1929 Grand Banks earthquake: American Journal of Science, v. 250, p. 873.
Heezen, B. C., Tharp, M., and Ewing, M.
1959 : The Floors of the Oceans, 1. The North Atlantic: Geological Society of America Special Paper 65, 122 p.
Heezen, B. C., Hollister, C. D., and Ruddiman, W. F.
1966 : Shaping of the continental rise by deep geostropic contour currents: Science, v. 152, p. 502–508.
Hilterman, F. J.
1970 : Three-dimensional seismic modelling: Geophysics, v. 35, p. 1020–1037.
Hoffman, J.
1957 : Hyperbolic curves applied to echo sounding: International Hydrographic Review, v. 34, p. 45–55.
Hollister, C. D.
1967 : Sediment distribution and deep circulation in the western North Atlantic [Ph.D. thesis]: New York, Columbia University, 368 p.
Hollister, C. D., and Heezen, B. C.
1972 : Geologic effects of ocean bottom currents; western North Atlantic; in Studies in Physical Oceanography, ed. A. L. Gordon; Gordon and Breach, London, p. 37–66.
Hollister, C. D., Flood, R. D., Johnson, D. A., Lonsdale, P., and Southard, J.
1974aa: Abyssal furrows and hyperbolic echo traces on the Bahama Outer Ridge: Geology, v. 2, p. 395–400.
Hollister, C. D., Johnson, D. A., and Lonsdale, P. F.
1974b: Current-controlled abyssal sedimentation; Samoan Passage, equatorial West Pacific: Journal of Geology, v. 82, p. 275–300.
Jacobi, R. D.
1976 : Sediment slides on the northwestern continental margin of Africa: Marine Geology, v. 22, p. 157–173.
Jacobi, R. D., and Hayes, D. E.
1982 : Bathymetry, microphysiography, and reflectivity characteristics of the West African margin between Sierra Leone and Mauritania; in Geology of the Northwest African Continental Margin, eds. U. von Rad, K. Hinz, M. Sarnthein, and E. Seibold; Springer-Verlag, New York, p. 182–212.
Jacobi, R. D., and Mrozowski, C. L.
1979 : Sediment slides and sediment waves in the Bonin Trough, Western Pacific: Marine Geology, v. 29, p. M1–M9.
Knott, S. T., and Hersey, J. B.
1956 : Interpretation of high-resolution echo sounding techniques and their use in bathymetry, marine geophysics, and geology: Deep-Sea Research, v. 14, p. 36–44.
Krause, D. C.
1962 : Interpretation of echo sounding profiles: International Hydrographic Review, v. 39, p. 65–123.
Kristofferson, Y.
1977 : Labrador Sea; A geophysical study [Ph.D. thesis]: New York, Columbia University, 184 p.
Laine, E. P.
1980 : New evidence from beneath the western North Atlantic for the depth of glacial erosion in Greenland and North America: Quaternary Research, v. 17, p. 188–198.
Laine, E. P., and Hollister, C. D.
1981 : Geological effects of the Gulf Stream system in the North American Basin: Marine Geology, v. 39, p. 277–310.
Laine, E. P., Heath, G. R., Ayer, E., and Kominz, M.
1983 : Evaluation of the geological stability and predictability of northern Bermuda Rise sediments for the Subseabed Disposal Program: Marine Geotechnology, v. 5, p. 215–233.
Lonsdale, P.
1978 : Bedforms and the benthic boundary layer in the North Atlantic; a cruise report of INDOMED Leg II: Scripps Institution of Oceanography Reference, no. 78-30, 15 p.
Luskin, B., Heezen, B. C., Ewing, M., and Landisman, M.
1954 : Precision measurements of ocean depth: Deep-Sea Research, v. 1, p. 131–140.
Malahoff, A., Embley, R. W., Perry, R. B., and Fefe, C.
1980 : Mass-wasting on the continental slope and upper rise south of Baltimore Canyon: Earth and Planetary Science Letters, v. 49, p. 1–7.
Markl, R. G., and Bryan, G. M.

1983 : Stratigraphic evolution of the Blake Outer Ridge: American Association of Petroleum Geologists Bulletin, v. 67, p. 666–683.
Mayer, L.
1979 : The origin of fine-scale acoustic stratigraphy in deep-sea carbonates: Journal of Geophysical Research, v. 84, p. 6177–6184.
McCave, I. N., Hollister, C. D., Laine, E. P., Lonsdale, P. F., and Richardson, M. J.
1982 : Erosion and deposition on the eastern margin of the Bermuda Rise in the Late Quaternary: Deep-Sea Research, v. 29, p. 535–561.
McCreery, C. J.
1983 : Acoustic structure and echo character of surficial sediments of the northern Hatteras Abyssal Plain [M.S. thesis]: Narragansett, Rhode Island, Graduate School of Oceanography, University of Rhode Island, 177 p.
McGregor, B. A., and Bennett, R. H.
1977 : Continental slope sediment instability northwest of Wilmington Canyon: American Association Petroleum Geologists Bulletin, v. 61, p. 918–928.
1979 : Mass movements of sediment on the continental slope seaward of the Baltimore Canyon Trough: Marine Geology, v. 33, p. 163–174.
Mullins, H. T., Bordman, M. R., and Neumann, A. C.
1979 : Echo character of off-platform carbonates: Marine Geology, v. 32, p. 251–268.
Piper, D.J.W., and Normark, W. R.
1982a: Effects of the 1929 Grand Banks Earthquake on the continental slope off eastern Canada: in Current Research, Part B; Geological Survey of Canada Paper 82-1B, p. 147–151.
1982b: Acoustic interpretation of Quaternary sedimentation and erosion on the channeled upper Laurentian Fan, Atlantic margin of Canada: Canadian Journal of Earth Sciences, v. 19, p. 1974–1984.
Popenoe, P., Coward, E. L., and Cashman, K. V.
1982 : A regional assessment of potential environmental hazards to and limitations on petroleum development of the southeastern United States Atlantic continental shelf, slope, and rise, offshore North Carolina: U.S. Geological Survey Open-File Report, v. 82-136, 67 p.
Rona, P. A.
1969 : Linear "lower continental rise hills" off Cape Hatteras: Journal of Sedimentary Petrology, v. 39, p. 1132–1141.
Rona, P. A., Schneider, E. D., and Heezen, B. C.
1967 : Bathymetry of the continental rise off Cape Hatteras: Deep-Sea Research, v. 14, p. 625–633.
Ryan, W.B.F.
1982 : Imaging of submarine landslides with wide-swath sonar; in Marine Slides and Other Mass Movements; eds. S. Saxov and J. K. Nieuwenhuis; Plenum Publishing Corp., New York, p. 175–188.
Schneider, E. D., and Heezen, B. C.
1966 : Sediments of the Caicos Outer Ridge, the Bahamas: Geological Society of America Bulletin, v. 77, p. 1381–1389.
Schneider, E. D., Fox, P. J., Hollister, C. D., Needham, H. D., and Heezen, B. C.
1967 : Further evidence of contour currents in the western North Atlantic: Earth and Planetary Science Letters, v. 2, p. 351–359.
Shipley, T. H.
1978 : Sedimentation and echo characteristics in the abyssal hills of the west-central North Atlantic: Geological Society of America Bulletin, v. 89, p. 397–408.
Silva, A. J., Hollister, C. D., Laine, E. P., and Beverly, B.
1976 : Geotechnical properties of deep-sea sediments; Bermuda Rise: Marine Geotechnology, v. 1, p. 195–232.
Smith, S. G.
1977 : A reflection profile modelling system: Geophysical Journal of the Royal Astronomical Society, v. 49, p. 723–737.
Tucholke, B. E.
1979 : Furrows and focussed echoes on the Blake Outer Ridge: Marine Geology, v. 31, p. M13–M20.
1980 : Acoustic environment of the Hatteras and Nares Abyssal Plains, western North Atlantic, determined from velocities and physical properties of sediment cores: Journal of the Acoustical Society of America, v. 68, p. 1376–1390.
Tucholke, B. E., and Laine, E. P.
1983 : Neogene and Quaternary development of the lower continental rise off the central U.S. East Coast; in Studies in Continental Margin Geology, eds. J. S. Watkins and C. L. Drake; American Association of Petroleum Geologists Memoir 34, p. 295–305.
Twitchell, D. C., and Roberts, D. G.
1982 : Morphology, distribution, and development of submarine canyons on the United States continental slope between Hudson and Baltimore Canyons: Geology, v. 10, p. 408–411.
van Andel, T. H., and Komar, P. D.
1969 : Ponded sediments of the Mid-Atlantic Ridge between 22°-23° north latitude: Geological Society of America Bulletin, v. 80, p. 1163–1190.
Vassallo, K., Jacobi, R. D., and Shor, A. N.
1984a: Echo character, microphysiography, and geologic hazards; in Eastern North American Continental Margin and Adjacent Ocean Floor, 28° to 36°N and 70° to 82°W, eds. G. M. Bryan and J. R. Heirtzler; Ocean Margin Drilling Program, Regional Atlas 5, map 40.
1984b: Echo character, microphysiography, and geologic hazards; in North American Continental Margin and Adjacent Ocean Floor, 34° to 41°N and 68° to 78°W, eds. J. E. Ewing and P. D. Rabinowitz; Ocean Margin Drilling Program, Regional Atlas 4, map 31.
Walker, J. R., and Massingill, J. V.
1970 : Slump features on the Mississippi fan, northeastern Gulf of Mexico: Geological Society of America Bulletin, v. 81, p. 3101–3108.

MANUSCRIPT ACCEPTED BY THE SOCIETY FEBRUARY 11, 1986

ACKNOWLEDGMENTS

Support for echo-character studies and research has been provided by contracts from the Office of Naval Research to Lamont-Doherty Geological Observatory (N00014-75-C-0210) and to the University of Rhode Island (N00014-81-CO0062). Additional funds have been made available to the Graduate School of Oceanography and to Lamont-Doherty Geological Observatory by the Department of Energy through Sandia National Laboratory. Lamont-Doherty Geological Observatory Contribution No. 4003.

Printed in U.S.A.

The Geology of North America
Vol. M, The Western North Atlantic Region
The Geological Society of America, 1986

Chapter 26

Turbidite sedimentation in the northwestern Atlantic Ocean basin

Orrin H. Pilkey
Department of Geology, Duke University, Durham, North Carolina 27708
William J. Cleary
Department of Earth Sciences, University of North Carolina, Wilmington, North Carolina 28403

INTRODUCTION

Turbidity currents account for much of the sediment deposited on the western North Atlantic Ocean basin floor. The most obvious expression of the importance of this sedimentation mechanism is the large area of the sea floor occupied by the abyssal plains, which are large ponds of turbidity-current-derived sediment (Plate 2). In fact the Hatteras and Sohm Abyssal Plains are among the largest active depositional basins in the world's ocean basins. Other than the abyssal plains, the major feature of the western North Atlantic basin which is affected by turbidity current erosion and deposition is the North American continental rise. Turbidity currents are responsible for moving huge volumes of sediment off the continental shelf edge and continental slope. Some individual turbidity currents crossing the North American rise are known to exceed 100 km^3 in volume (e.g. Elmore and others, 1979). Both the amount of sediment transported and the frequency of turbidity-current events are assumed to be closely related to Pleistocene sea-level fluctuations; low sea levels produce more or at least larger turbidity currents because of direct contribution of sediment to the upper slope by rivers. High sea-level stands of geologically brief duration, such as the present one, result in trapping of most fluvial sediment in estuaries.

The genetic term turbidite is defined as the deposit of a short-lived single turbidity current. The concept of turbidity currents dates back to the observations by Forel (1887) on the cold, sediment-laden undercurrent formed by the Rhone River upon entering Lake Leman. Observations by engineers and geologists in Lake Mead in the southwest United States indicated that turbidity currents could deposit sediments on the extremely flat gradient of the lake floor for over 100 km (Grover and Howard, 1938). Daly (1936) was the first to suggest that turbidity currents were responsible for cutting submarine canyons in consolidated material off the east coast of the United States.

Experimental studies of turbidity currents began with Kuenen in the mid 1930's. These early experiments, which tested the ability of turbidity currents to cause erosion, met with little success. The focus of Kuenen's flume experiments had changed by 1950 and became centered on the capacity of turbidity currents to transport sand into the deep sea and to form graded beds similar to ancient greywackes (Kuenen, 1950; Kuenen and Migliorini, 1950). Results from these experiments were used to explain the graded-bed sequences in the Apennines.

Turbidity currents are thought to be surges that move downslope as channelized flows mostly through submarine canyons or fan valleys. However, on portions of the ocean floor, such as the distal fan regions or basin plains where poorly defined channels or no channels exist, the current is unconfined and spreads laterally. Turbidity currents flow as long as there is sediment contained in suspension. They move downslope as the result of gravity acting on the entrained suspended particles. Experimental studies have shown that an increase in density generally produces a concomitant increase in the velocity of the downslope flow.

In profile, a turbidity current consists of a head, body, and tail (Middleton, 1966). Experimental work by Keulegan (1957, 1958) and Middleton (1966) indicates that the head moves more slowly than the body of the flow and may be a region of erosion where the coarsest material is progressively concentrated. The head of a turbidity current consists of a series of lobes and re-entrants, and it has an "overhanging" front (Allen, 1971). Water trapped in the re-entrants may lead to dilution of the mixture (Middleton and Hampton, 1976).

The origins of turbidity currents are directly related to one or more triggering mechanisms. Turbidity currents often seem to be associated with peak river-discharge periods (Keller and Shepard, 1978). During high-discharge periods, a number of telegraph cables have been broken in front of major river deltas. The best documented examples of such breaks are off the Congo

Pilkey, O. H., and Cleary, W. J., 1986, Turbidite sedimentation in the northwestern Atlantic Ocean basin, *in* Vogt, P. R., and Tucholke, B. E., eds., The Geology of North America, Volume M, The Western North Atlantic Region: Geological Society of America.

(Ewing and others, 1970; Heezen and Hollister, 1971) and the Magdeline River (Heezen and Hollister, 1971). Earthquakes also are commonly assumed to be important "triggers," especially in areas where unstable strata and slopes occur—also off major river deltas. For example, the earthquakes that occurred in 1929 off the Grand Banks of Newfoundland triggered a flow that travelled 70 km onto the distal portion of the Sohm Abyssal Plain (Heezen and Ewing, 1952; Heezen and Hollister, 1971). Probably originating as a series of slumps, the turbidity current broke twelve submarine cables in approximately thirteen hours. Based on time-distance curves for these breaks, the velocity of the flow was estimated to be approximately 45 knots. A similar time-distance graph for the 1954 Orleansville turbidity current off Algeria indicated a probable velocity of 40 knots (Heezen and Ewing, 1955).

Bouma (1962) summarized much of the existing data on turbidity currents and developed a turbidite facies model. Harms and Fahnestock (1965) and Walker (1965) applied laboratory flow-regime studies to this model. Today the "Bouma sequence" is used as the norm for classical turbidite studies (Walker, 1979).

There are a variety of features that characterize turbidite sequences. These characteristics can be seen to varying degrees in both modern and ancient turbidites. Graded sandstone, current-rippled beds, and laminated beds alternating with pelagic units are typical features of turbidite sequences. Sole markings are common at the base of turbidite units and they result from the scouring action of the flow, particularly at the head. Tool marks are also common features.

Bouma (1962) showed that deposition from a single turbidity current results in an orderly sequence of sedimentary structures that record the waning of the current. This idealized turbidite unit consists of five intervals, each with diagnostic structures. The basal unit (Ta) is generally massive and graded. This interval contains the coarsest sediment, is usually the thickest, and is interpreted to have been deposited rapidly under upper-flow-regime conditions. The overlying, thinner Tb unit consists of parallel, laminated sediments. Texturally these are finer grained and represent deposition associated with the plane-bed mode of the upper flow regime. Division Tc is a relatively thick unit of rippled, cross-laminated, fine sand or mud deposited under conditions associated with the low flow regime. Interval Td is composed of laminated muds. The origin of this unit is not well understood and in many instances is difficult to distinguish from the overlying Te, fine-mud unit (Walker, 1979). This uppermost portion of the sequence represents deposition from the waning stages of the turbidity current and from subsequent pelagic sedimentation.

Individual turbidites do not always contain all units in the Bouma model. The model can be used as a norm, against which sections lacking the idealized sequence can be compared for purposes of hydrodynamic interpretations (Walker, 1979).

The distribution of turbidite deposits is closely related to physiography. Because turbidity currents transport material downslope under the influence of gravity, deposits are likely to occur along the distributary network of submarine canyons incised on the continental slope, in valleys on the continental rise and in submarine fans, on abyssal plains, and in the connecting mid-ocean channels and abyssal gaps. In addition to these sites of accumulation, turbidites have been identified in marginal/reentrant basins, in perched basins both along the continental margin and on the flanks of the Mid-Atlantic Ridge, and within the narrow fracture zones of the central Atlantic (see Plates 5, 6).

THE CONDUITS

Submarine canyons appear to be the principal cross-continental margin conduits that control basin-plain entry points for turbidity currents. Piston coring of the Hatteras Abyssal Plain near the base of the continental rise shows that turbidites, other than those introduced via canyons, are both infrequent and fine-grained (silty, clayey). The nature of turbidite sedimentation associated with canyons is discussed in more detail in the continental-rise section of this chapter.

Not all submarine canyons appear to originate by turbidity current erosion, nor do they act as conduits for turbidity currents. Numerous small canyons incised on the U.S. continental slope appear to be forming by headward erosion, probably slumping and sliding (Twichell and Roberts, 1982). These small canyons do not reach the shelf break, and they have no apparent relation to off-shelf sediment transport.

The larger canyons that cut the continental shelf edge and cross the entire continental slope and rise to debouch on the abyssal plains include the Laurentian Canyon (Piper, 1975; Stow, 1981), the Gully (Stanley, 1970), Hudson and Baltimore-Wilmington Canyons (Pratt, 1967; Stanley, 1970; Stanley and others, 1971), Norfolk-Washington Canyon (Stanley and others, 1971); Kelling and Stanley, 1971) and the Hatteras-Albemarle Canyon systems (Rona and others, 1967; Newton and Pilkey, 1969; Newton and others, 1971).

These canyon systems may be grouped into those formed as extensions of rivers and those formed largely by glacial processes. The major canyons of the mid-Atlantic states which are associated with land drainage include the Hudson, Baltimore-Wilmington, Norfolk-Washington and possibly the Hatteras Canyons (Pratt, 1967). Canyons north of the Hudson Canyon are to a great extent influenced by Pleistocene glacial erosion and glacial-outwash drainage on the shelf during ice-sheet maxima, and by associated mass movement on the adjacent continental slope.

Several cycles of excavation are reported for the deeper and larger east-coast canyons (Pratt, 1967; Ryan and others, 1978). Extensions of Hudson, Wilmington, and Washington Canyons have deep gorges incised into the central continental rise (3200-3400 m), testifying to the latest, Pleistocene phase of excavation (Pratt, 1967; Stanley and others, 1971).

The North American continental rise (Heezen and others, 1959) stretches from off Baffin Bay to the Cape Hatteras region. This province typically begins with a sharp change in slope at a depth of about 2000 m. The seaward boundary of the rise is generally in the vicinity of the 5,200 m isobath in the North American basin, but it is difficult to delineate because lower-fan systems gently merge with the Sohm and Hatteras abyssal plains.

The broad, gently sloping continental rise was originally thought to have been deposited largely from turbidity currents (Emery and others, 1970; Drake and others, 1968) as deep sea fans coalesced at the termination of submarine canyons. Embley (1980) suggested that large debris flows also have been important in shaping the rise. Such a mode of formation by downslope processes, primarily turbidity currents, is indeed the case for portions of the North American Pacific rise (e.g. Normark, 1978; Piper, 1978).

An additional mechanism of Atlantic rise development was proposed by Heezen and others (1966). They suggested that a major portion of the upper Quaternary sediment column off eastern North America is not of turbidity-current origin but instead consists of layered silts and homogeneous silty clays (contourites) deposited by southward-moving, contour-parallel bottom currents. The role of bottom currents in shaping the sedimentary record in the western North Atlantic is discussed by McCave and Tucholke (this volume).

Contour currents and turbidity currents may be interactive processes, because contour currents can derive their sediment load from the tails of canyon-confined turbidity currents. Distinction between layers of sediment deposited from bottom and turbidity currents is not always unequivocal, but Hollister and Heezen (1972) and Fritz and Pilkey (1975), among others, have developed a number of useful parameters for this purpose. These criteria are megascopic and rely on the abundances and thicknesses of silt/sand beds. Available data indicate that: (1) bedded silt/sand comprises less than 20% of the length of contourite-dominated cores, but turbidite-dominated cores are typically more than 20% bedded silt/sand; (2) contourite-dominated cores have numerous silt/sand beds (50 to 500 beds per 10 m of core) and the beds are always extremely thin (the thickest bed is 20 cm and the average bed thickness is 1 cm); in contrast, silt/sand beds in turbidite-dominated cores are thicker (usually 20 cm) and not as abundant (<50 beds per 10 m of core). (3) Turbidite sands also are often graded and contain appreciable amounts of calcareous material.

Based on the morphology of canyon systems and the areal distribution of fine-grained sediments, three "fans" can be described for the North American continental rise. These are the small Hatteras Fan off Cape Hatteras (Cleary and Conolly, 1974a), the Wilmington Fan off the mid-Atlantic, U.S. (Ayers and Cleary, 1980), and the Laurentian Fan off the Grand Banks (Stow, 1981). These fans contain some of the elements of fan models developed for the Pacific North American margin (Normark, 1974; Walker, 1978; Mutti and Ricci-Lucci, 1972). For example, they can be subdivided into upper, middle, and lower fans based on morphology and sediment texture. However, they differ from the Pacific model in regard to slope angles, canyon types, sediment loads, and particularly the relative importance of bottom currents. Thus, Atlantic fans can differ substantially from previously studied fan systems. Other "fans" probably are developed on the eastern North American continental rise (for example along the Hudson Canyon) but their coalescence with numerous adjacent fans of variable sizes, and strong modification by bottom currents, makes their definition and study difficult.

Hatteras Fan

The Hatteras Fan lies off Cape Hatteras and is an unusual and very small fan which has prograded onto the Hatteras Abyssal Plain from the mouth of the Hatteras Canyon at the base of the continental rise (5150 m). In contrast to the Wilmington and Laurentian Fans (Fig. 1) which contain distributary systems that disgorge sediment onto the abyssal plains, the tributaries of the Hatteras Canyon are forced to coalesce and are diverted southward by the 400 meter-high Hatteras Outer Ridge at 4900 m depth (Plate 2). Downslope, a solitary 4-5 km wide channel seaward of 5,200 m disperses sediment and produces a 100 km long fan (Cleary and Conolly, 1974a; Cleary and others, 1977). Seaward of the canyon mouth the channel narrows to 2 km and is 25 m deep; it forms an anastomosed system of braided channels and interchannel banks. On the distal fan margin these channels merge imperceptibly with the Hatteras Abyssal Plain. Turbidites within the braided channel system are characterized by thick (50 cm to 3 m), graded quartzose sands. Numerous thin (<25 cm) sand units interbedded with gray lutites typify the levees and interchannel banks. Toward the seaward margin of the fan within interchannel areas, thick (up to 2 m) olive gray lutites occur interbedded with cross-bedded and convolute-bedded silt and fine sands.

The fine to medium sands are generally subarkosic. The composition of the terrigenous suite suggests derivation from the Appalachian Piedmont crystalline rocks. The carbonate fraction (15%) contains shallow-water skeletal components, planktonic foraminifera, and occasional ooids derived from the continental shelf both north and south of Cape Lookout (Cleary and Conolly, 1974a).

Wilmington Fan

The Wilmington Canyon system occupies a broad area between the Hatteras and Hudson Canyons and consists of the Norfolk, Washington, Baltimore and Wilmington Canyons which coalesce into a single channel on the lower continental rise (Fig. 1). This portion of the rise is also the site of several large fans formed in association with each of the canyons. Sufficiently detailed seismic surveys are lacking to delineate the boundary of each fan. The lower continental rise here is an area of ponded turbidites that form a "perched" terrace. The perched basin began to fill during early Pleistocene, when the troughs and basins landward of the Hatteras Outer Ridge (Plate 2) were flooded by turbidity currents. The basin filled 0.5 m.y. ago, allowing flows to spill over the ridge crest (Tucholke and Laine, 1982).

WESTERN ATLANTIC FANS

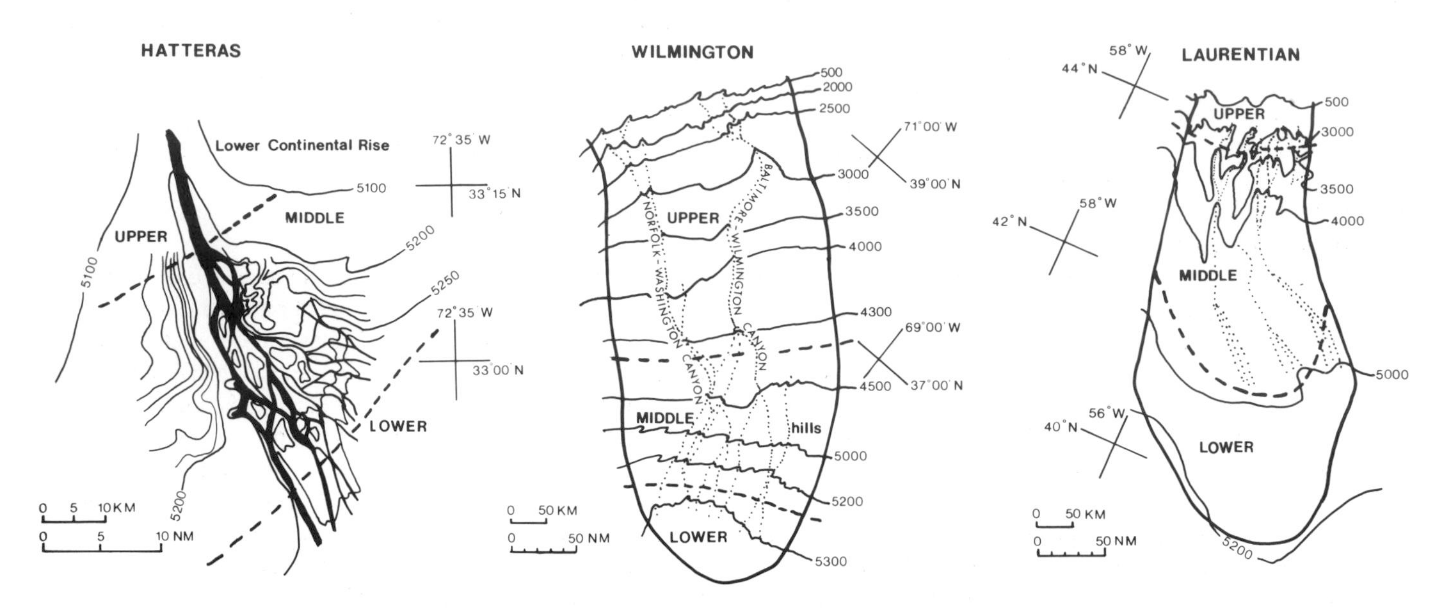

Figure 1. Comparison of subdivisions of the Hatteras Fan (Cleary and Conolly, 1974a), Wilmington Fan (Cleary and Pilkey, 1983), and Laurentian Fan (Stow, 1981).

Previously unpublished data from cores raised from this section of the rise, combined with previously published data on sand distribution of the lower rise at 4300 to 5300 m (Ayers and Cleary, 1980), reveal a distinct pattern of sand-layer distribution related to the effects of turbidity currents crossing the rise (Fig. 2). It is immediately apparent that sand layers are confined to areas near canyon axes and across the coalescing fans that form the lower continental rise terrace. Above 4300 m the sand beds are generally less than 20 cm thick. For the most part those units greater than 10 cm thick have the internal characteristics of turbidites. The highest frequency of turbidite units occurs within the small levees. Generally, sand thickness and frequency of sand layers increase with depth down the fan surface from 2000 to 4300 m. At 2000 m, sand layers were not encountered. At depths of 3500 m, sand units are generally less than 1 cm thick and occur within 7 km of canyons. At 4200 m sand-layer thickness increases to an average of 12 cm, and a number of sand layers >50 cm thick were penetrated. The downslope increase in frequency of sand layers ranges from a low of 2 per 10 m at 3000 m to an average of 9 per 10 m at 4200 m (Cleary and Pilkey, 1983).

Below 4300 m, where the canyons emerge onto the lower continental rise terrace, turbiditic sand distribution is quite complicated because of a complex distributary system. Farther seaward, these low-relief channels cross over the crest of the Hatteras Outer Ridge near 4500 m. At depths between 4500 and 4750 m, relief is relatively low and the channels are poorly confined to paths between sediment waves forming the "lower continental rise hills" on the ridge flank. Between 4750 and 5300 m, the channels are strongly confined within these troughs. Turbidites were locally deposited among the sediment waves to produce flat-floored sediment ponds, and some ponds eventually buried portions of the wave crests. Larger turbidity currents followed channel extensions completely through the sediment waves; they eventually disgorged their sand onto the Hatteras Abyssal Plain across comparatively narrow fronts, the width of each being dictated by the trough width.

Based on these data, the distribution of sand across the Mid-Atlantic continental rise appears to occur in long funnel-shaped patterns that widen seaward. The funnels widen most rapidly at approximately 4300 m where sediment dispersal occurs across a broad front. Comparing this modified "fan" system with the Pacific models noted earlier, and with the Laurentian Fan described below, upper, middle and lower fan subdivisions can be made as shown in Figure 1.

A relatively large upper-fan surface extends from the base of the continental slope to depths of 4300 m. The upper region of the fan is characterized by short slump-excavated canyons and the main leveed canyons (Baltimore-Wilmington, and Norfolk-Washington Canyons). The mid-fan region is complex and consists of a terrace-like area (4300-4500 m) that formed by ponding of turbidites behind the Hatteras Outer Ridge. Channels are wide and flat-bottomed, and they may have small levees developed on the major channel extensions. The lower mid-fan region is complicated by sediment waves of thc Hatteras Outer Ridge which surround flat-floored, partially filled troughs. This region extends to depths of approximately 5300 m. The exact boundary of the lower fan is difficult to delineate; in depths greater than 5300 m wide, low relief, sand-filled channels extend across the basin plain (Cleary and others, 1977; Ayers and Cleary, 1980).

Turbidite sands of the Wilmington system are arkoses or lithic arkoses. Lithic fragments (10%) are typical of a piedmont-metamorphic terrain. Carbonate is not a significant component of these sands, but planktonic foraminifera (2%) and shallow-water allochemical grains do occur in small amounts.

Laurentian Fan

Stow (1981) suggests that the Laurentian Fan (water depths of 2000-5000 m) off the Nova Scotian margin has been influenced greatly by the glacial history of the Maritime Provinces for the past 2 to 3 m.y. The fan extends 600 km south from the base of the continental slope and has a maximum width of 300 km (Fig. 1, Plate 2).

The rugged, slump-scarred, and small upper fan consists of a network of various-sized tributaries (Fig. 2). Major tributaries tend to be "V" shaped, reflecting the predominance of erosion on the upper fan. Distance between channels is often less than 10 km. The convex-upward mid fan, the largest of the divisions, extends from approximately 3000 to 5000 m water depth. The undulating surface is marked by three large meandering channels contained within asymmetric levees. Typically the levees are larger on the western edges of channels. Maximum channel relief is 840 m. Most channels are "U" shaped and have widths approaching 10 km. The central and largest channel is 26 km wide at its widest point and it contains several thalwegs. Interchannel distances are generally 40 km. A supra fan (depositional lobe) with small distributary channels has developed on the lowermost mid fan at the terminus of the main channel. Low-relief (50 m) channels extend between the depositional lobes (Stow, 1981). Normark and others (1983) discussed Quaternary development of the mid fan and noted a deflection of the major western fan valley due to a recent large slump or debris flow. The broad, concave-upward, lower fan has a low-relief, undulating surface cut by two main channels that extend onto the adjacent Sohm Abyssal Plain.

Thick, graded gravel and sand units, commonly 1-2 m thick and up to 7 m thick, characterize the channel systems and the depositional lobes on the lowermost mid fan. The gravels are predominantly sedimentary rocks and are derived mostly from the Gulf of St. Lawrence. Heavy minerals from the sand units reflect multiple sources, including the Scotian Shelf, Gulf of St. Lawrence and Laurentian Channel, and the Grand Banks Shelf (Piper, 1975; Stow, 1981).

A thick, red-brown mud interlaminated with thin silt beds is volumetrically the most important sediment type across the fan. The muds are thought to have been deposited from thick (1000 m), dilute, and low-velocity turbidity currents (Stow, 1981). The muds were originally transported to the shelf break by glaciers and proglacial streams and deposited on the continental slope. During the Wisconsin glacial epoch sedimentation rates were as high as 30 cm/1000 years on the western regions of the fan.

Silt laminae are thicker, coarser and more frequent on the upper fan, near levees, and in the depositional lobes on the lowermost mid fan. The laminae are more irregular and thinly bedded downslope and away from channel axes.

Due to the large volume of fine-grained material deposited, the fan has extended itself onto the abyssal plain. During this progradation the fan surface has built up more or less continually in interchannel areas, while channels episodically have filled and then deepened. A general increase in channel depth has enabled coarse sands to bypass much of the fan and ultimately accumulate both on the lowermost mid fan and on the adjacent abyssal plain (Stow, 1981).

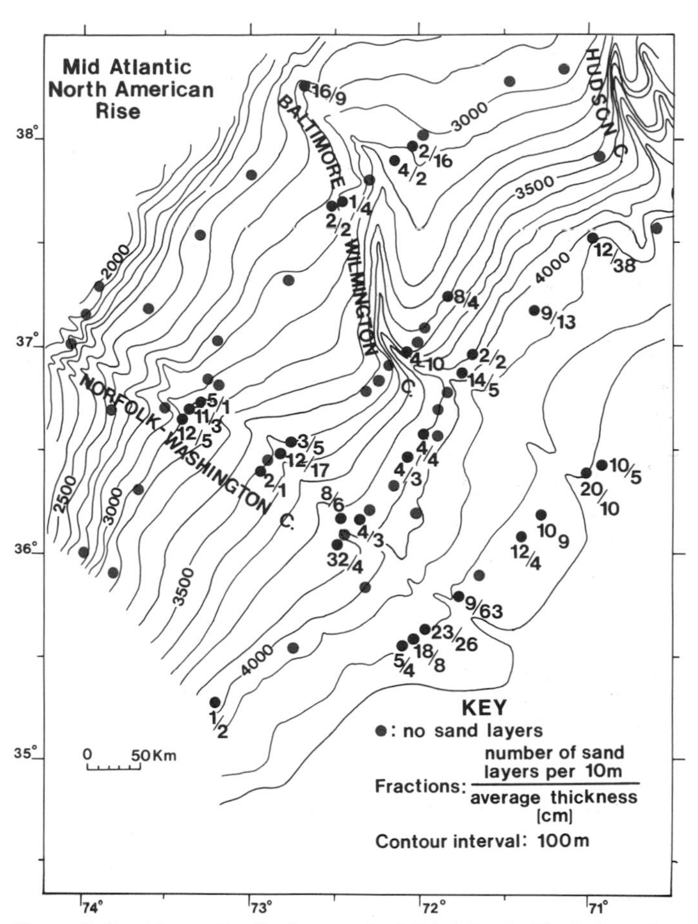

Figure 2. Sand-layer distribution on the Mid-Atlantic North American continental rise. Red circles designate locations of cores without sand layers. Blackened circles designate cores containing sand layers which are described by the fractions drawn beside each point.

ABYSSAL PLAINS

The deepest repositories of sediment in the western North Atlantic basin typically are the abyssal plains. A larger proportion of the ocean basin floor is occupied by abyssal plains in the western Atlantic than in any other ocean basin. Speaking broadly, the Pacific has few such abyssal plains because sediment is trapped in the fringing trenches. In the Indian Ocean, two of the world's major rivers, the Indus and Ganges, form huge deep sea fans rather than abyssal plains; the high proportion of silt and clay in the river-derived sediment promotes levee formation which channelizes turbidity currents rather than allowing them to spread out to form flat plains.

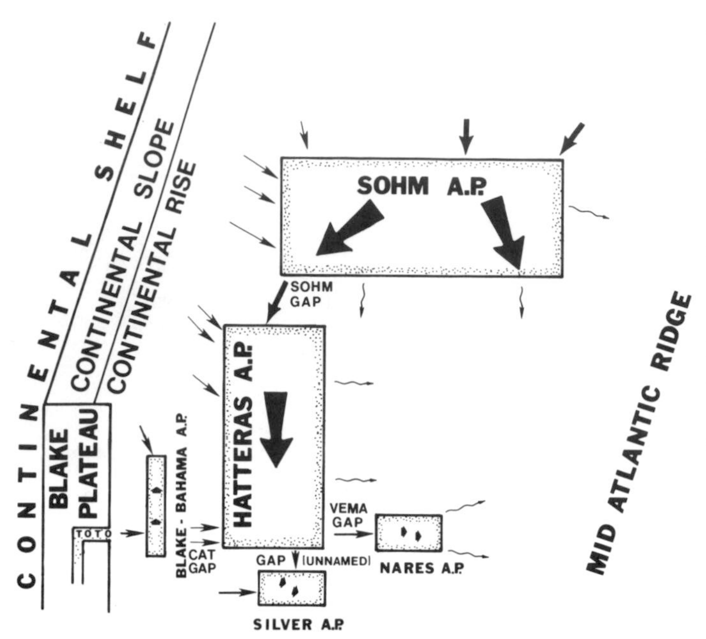

Figure 3. Diagrammatic illustration of the major depositional basin plains in the western North Atlantic Ocean basin. Arrows show dominant direction of sediment dispersal.

The abyssal-plain province is defined as that part of the deep ocean floor that is flat and has a gradient of less than 1:1000 (1 m/1 km). Abyssal plains are among the flattest features on the surface of the earth (Heezen and others, 1959). Cores raised from abyssal plains contain coarse sand and shallow-water fossil assemblages that are derived from the adjacent continent and its bordering continental shelves. Seismic reflection profiles show that abyssal plains are underlain by flat-lying reflectors that extend laterally for hundreds of kilometers.

Four major abyssal plains (Sohm, Hatteras, Nares, and Silver) are found in the northwestern Atlantic Ocean basin (Plate 2). Two additional, smaller abyssal plains are found in the Irminger and southern Labrador Basins. Figure 3 shows the relative size and location of each of the above. Also shown are major basin-entry points, abyssal gaps, and the Blake-Bahama Marginal Basin. Three of the major abyssal plains are interconnected. The Sohm and Vema Gaps connect the Sohm and Hatteras Abyssal Plains and the Hatteras and Nares Abyssal Plains, respectively. Some turbidity currents also flow through Cat Gap from the Blake-Bahama Basin to the Hatteras Abyssal Plain. A broad, gentle, and smooth-surfaced ridge with relief of about 100 meters separates the Silver and Hatteras Abyssal Plains.

Early descriptive investigations of abyssal plains in the western Atlantic Ocean include those of Ericson and others (1951). The problems of sand distribution, provenance, and general dispersal patterns were addressed by Ericson and others (1961), Hubert (1962), Hubert (1964), Hubert and Neal (1967), Fruth (1965), Hollister (1967), Horn and others (1971), Cleary and Conolly (1974b), Egloff and Johnson (1975), Johnson and others (1975), Cleary and others (1978), Pilkey and others (1980).

Unnamed Abyssal Plains (Irminger and Labrador Basins)

In the Irminger Sea and eastern Labrador Sea, an elongate, irregularly shaped abyssal plain (3200 m) adjoins the leveed Imarssuak Mid-Ocean Channel. The plain, which is over 800 km long, varies from less than 10 km to more than 175 km in width. In places, seamounts of various dimensions and exposed basement ridges interrupt the plains's surface (Egloff and Johnson, 1975).

Two major sources contribute turbiditic sediment to the plain: the continental margin of eastern Greenland and the volcanic Iceland plateau. Seismic reflection profiles indicate 200-300 m of highly reflective fill underlies the surface of the plain (Johnson and others, 1975). The Deep Sea Drilling Project Site 113 located on the western margin of the plain (56°48′N, 48°20′W; Plate 2) encountered approximately 550 m of mid-Pliocene to Pleistocene turbidites (Laughton and others, 1972). Sediments recovered include a mixture of clay, basaltic rock fragments, heavy minerals and volcanogenic silts and sands.

A second and more extensive unnamed abyssal plain occurs in the Labrador Sea. This elongate, northwest-southeast trending plain lies on both sides of the Northwest Atlantic Mid-Ocean Channel. The abyssal plain varies in its development and continuity along the entirety of the mid-ocean channel (Profiles I and J, Plate 6). The most detailed information about the plain, based on seismic studies, comes from the southern Labrador Sea (50°-57°N). Immediately south of the confluence of the Imarssuak and Northwest Atlantic Mid-Ocean Channels (56°N, 49°W), the plain is approximately 100-120 km wide. At 52°N the plain is offset by the Charlie-Gibbs Fracture Zone, and to the southeast the plain widens to 240 km (Egloff and Johnson, 1975).

Turbidite fill in the plain averages about 390 m (Laine, 1980). The major source for sandy sediment south of the Charlie-Gibbs Fracture Zone appears to be the mid-ocean channel, while north of 52°N, the Labrador Rise provides an additional source (Egloff and Johnson, 1975).

Sohm Abyssal Plain

The Sohm Abyssal Plain (5200-5400 m) has an areal extent of 550,300 km^2 and is the largest of the basin plains. The T-shaped plain lies south of the Grand Banks (Plate 2). Three major source areas feed dominantly terrigenous sediment onto the plain. These sources include the continental margin off the Gulf of Maine (west), the Laurentian Fan (north) and the Northwest Atlantic Mid-Ocean Channel (east) (Horn and others, 1971). A fourth source may include the Hudson Canyon system (west). It is not yet clear whether the Hudson Canyon dumps its load entirely onto the Sohm or the Hatteras Abyssal Plain or whether both basins receive sediment from this source.

Sediment facies patterns in the abyssal plain are complex and reflect (1) the large number and variability of sources, (2) the amount and type of sediment input and (3) the many knolls and seamounts that protrude above the plain floor and divert the flow paths of turbidity currents. Piston cores show that the greatest sand-layer thicknesses (Fig. 4) occur adjacent to the principal sources along the upper arm of the T-shaped plain and, generally speaking, the sand-layer thickness decreases away from the sources in a southerly direction. A major dispersal route extends southeast from the Laurentian Fan for approximately 1000 km. Abnormal sand thicknesses in the southeast may reflect ponding against the bordering abyssal hills. These beds are not graded, suggesting high-density, high-velocity currents. A 290 cm thick sand layer was recovered 1000 km from the nearest basin entry point (Horn and others, 1971). The flow responsible for deposition of this unit must have been comparable to the flow associated with the 1929 Grand Banks earthquake discussed previously.

Proximal/distal textural differences in the Sohm Abyssal Plain are not as great as in the other plains because relatively thick sand layers are widespread. Sand-layer thicknesses average about 122 cm in proximal areas and 93 cm in distal regions. The difference in frequency of sand layers is also slight: 0.4 per meter for proximal areas and 0.22 per meter for distal areas (Pilkey and others, 1980).

Petrographic studies of the sand fraction (Hubert, 1962) indicate the shelf-derived sands of the western half of the abyssal plain are feldspar-poor, glauconitic, and quartzose. In the eastern and southern abyssal plain, sands are highly feldspathic and contain a diagnostic suite of carbonate lithic fragments that are probably derived north of Newfoundland (Hubert, 1962). Long (60 to 80 km) isolated sediment fingers that represent the distal reaches of the abyssal plain extend well into the abyssal hills of the lower flank of the Mid-Atlantic Ridge. Subbottom profiles across these features at the southern end of the Sohm Abyssal Plain indicate the top 300 m consists of stratified sediments (McGregor, 1968).

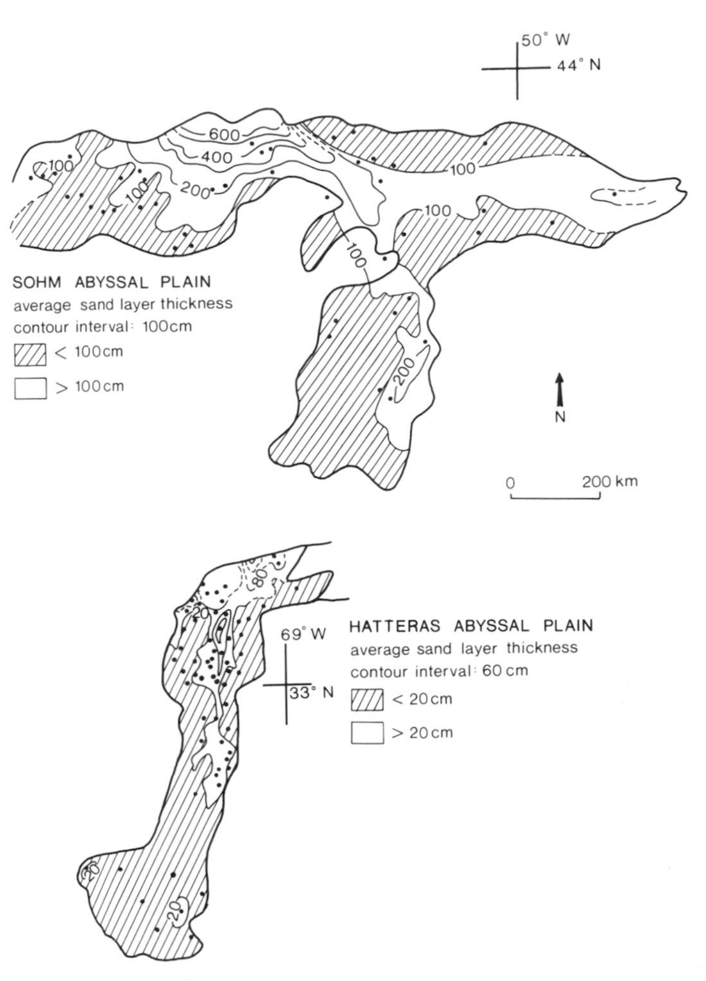

Figure 4. Average sand-layer thickness in the Sohm and Hatteras Abyssal plains based on piston-core studies.

Hatteras Abyssal Plain

The Hatteras Abyssal Plain is a north-south trending elongate basin that covers more than 180,000 km^2 (Plate 2). Gentle gradients (1:1000-1:10,000) characterize the basin floor over much of its 1000 km length (Profiles M and O, Plate 6). The plain is up to 260 km wide and ranges in depth from 5100 in the north to 5500 m in the south (Heezen and others, 1959; Horn and others, 1971). The plain floor, particularly in the south and southeast, is interrupted by low relief (30 m) abyssal hills and channels. The channels are particularly well developed on the southeastern margin near the Vema Gap which connects the Hatteras to Nares Abyssal Plain (Pratt, 1967; Cleary and others, 1978). The northwestern margin of the Hatteras Abyssal Plain is bordered by sediment waves on the lower continental rise (Hatteras Outer Ridge). The western and southern margins are formed by the Blake-Bahama Outer Ridge and Caicos Outer Ridge, respectively. The eastern margin is the Bermuda Rise, which exhibits undulating sedimentary topography and irregular abyssal hills of oceanic basement. Three major sources feed sediment onto the abyssal plain, all in the northern third. These include the Hatteras and Wilmington Canyon systems and the Sohm Gap (Cleary and others, 1977). An additional minor source, Cat Gap, funnels calcareous material from the Blake-Bahama Basin onto the southern extremities of the plain (Cleary and others, 1978). Patterns of turbidite sedimentation on the Hatteras Abyssal Plain are not as complex as those of the Sohm Abyssal Plain (Fig. 4). Previous studies (Horn and others, 1971; Cleary and Conolly, 1974b; Cleary and others, 1978; Pilkey and others, 1980) indicate a general north to south dispersal of sediment. Sands are medium grained at the basin entry points, and they grade to very fine sands away from the sources and primary flow paths.

Turbidite fill is thickest (350 m) in the region seaward of the Hatteras Fan (32°N) and south of the Wilmington Fan (Horn and others, 1971). Sand layers extend over approximately the northern half of the basin plain area, and the southern half of the plain is essentially sand-free (Cleary and others, 1978). In a broad sense, the distribution of sand layers in the basin is tongue-shaped. Maximum thickness of individual sand layers is along the axis of the basin (Fig. 4). Sand layers thin both laterally and southward. X-radiography (Sparks, 1979) shows that Bouma Ta units characterize the sands nearest the basin entries and along the basin axis. South of the Hatteras Fan and east of the basin axis, thin sand layers are frequent and characteristically have a basal Tb unit. In distal areas the sediment is typical silt or silty clay; turbidites grade upward from basal Tc or Td units into pelagic sediments.

Individual sand layers can be correlated across the northern part of the basin over north-south distances of 20 to 500 km and across widths up to 40 km. About 75% of the sand layers in the 1200-km-long basin cover only 30% of the basin floor. The maximum linear extent of any sampled sand layer in the basin is the 500-km-long "Black Shell Turbidite" which has a minimum volume of 100 km^3 (Elmore and others, 1979). The total volume of this turbidite could be significantly larger; the flow could be traced only by correlation of the basal sand layer. The diverse nature of the terrigenous sands indicates an ultimate derivation from piedmont streams north of Cape Hatteras. Allochems and minor amounts of ooids also are present and suggest an additional, minor source south of Cape Hatteras (Cleary and others, 1978).

Along the seaward margin of the Hatteras Abyssal Plain numerous flat-floored basins receive episodic influxes of turbidites. Transport and filling take place either through connected irregular channels among the abyssal hills or at times when the turbidity current thickness exceeds the local relief. The few cores recovered from these small basins show that the upper 10-12 m of sediment is clay or silty clay (Cleary and others, 1978).

Silver Abyssal Plain

The horseshoe shaped Silver Abyssal Plain (5500 m) is the smallest of the deep-sea plains, covering an area of only 19,050 km^2. The plain is bordered on the north and northeast by the Greater Antilles Outer Ridge and by a low relief, broad sill that separates the Silver and Hatteras abyssal plains (Plate 2; Profile P, Plate 6). The steep Bahama Escarpment forms the southern border, and the western margins are contained by the flanks of the Caicos Outer Ridge.

The principal source for the Quaternary basin fill (75 m thick) is clay and silty-clay spillover from turbidity currents crossing the Hatteras Abyssal Plain, (Horn and others, 1971; Van Tassel, 1981). The basin fill has a small percentage of sand layers but it has thick clay layers. One lobate turbidite unit (sand + mud) is 1.0-1.5 m thick and covered a majority of the basin floor (Van Tassel, 1981). This is the only sand layer within range ($\sim$10 m) of piston cores on the plain floor. Calcareous components indicate the sediment source was the Silver Bank area to the west (Van Tassel, 1981).

Nares Abyssal Plain

The Nares Abyssal Plain, with an areal extent of 93,600 km^2, is the deepest (5800-5900 m) abyssal plain in the northwest Atlantic (Plate 2). The eastward-sloping basin plain is a distal clay-turbidite pond (Horn and others, 1971). The surface of the plain is flooded by turbidity currents that cross the Hatteras Abyssal Plain and enter via Vema Gap. A small amount of bioclastic material is fed to the area from the Blake-Bahama Basin through Cat Gap and Vema Gap (Cleary and others, 1978; Tucholke, 1980).

Because the Nares Abyssal Plain received only distal flows, cores recovered from the plain contain only silt- and clay-sized material. At the western end of the plain, silt beds are thickest ($<$2 cm); they become thinner and more numerous toward the east. In the distal parts, silt laminae are seldom found (Cleary and others, 1978, Tucholke, 1980).

Marginal/Re-entrant Basins

Marginal basins, by definition, lie at the foot of the continental slope or at the base of a steep marginal escarpment. The floors of these basins typically are shallower than the adjacent ocean-basin floor. An outer ridge usually borders the seaward edge (Heezen and others, 1959). The term "re-entrant basin" is defined here as a semi-enclosed deep trough incised into a marginal platform. The "V" or "U" shaped basins are typically rimmed by gullied, steep walls.

Blake-Bahama Abyssal Plain. The Blake-Bahama Abyssal Plain (5000 m) is a north-south trending elongate feature that lies at the base of Blake Escarpment. It is bordered on the east by the Blake-Bahama Outer Ridge and to the southwest by the Bahama Banks. This marginal-basin abyssal plain has an areal extent of 34,750 km^2. The major source areas of the Bahama Banks and adjacent slopes feed calcareous sediments onto the plain floor. Two major canyons, Great Bahama and Great Abaco, are conduits for large volumes of turbidites that cross the abyssal plain. Smaller turbidity currents originate from the Blake and Bahama escarpments (Schaeffer, 1984).

Light olive gray to pale yellow brown calcareous turbidites comprise 10-50% of the upper 10 m of sedimentary fill. Major turbidity currents fill the basin longitudinally, entering the plain from the south and moving northward along its 400 km length. Individual sand layers range in maximum thickness from 200 cm in the southern region to 100 cm in some cores raised from the northern part. The ratio of sand to clay and the percentage of sand both decrease northward and away from the Blake Escarpment.

The fine- to coarse-sand fraction of the basal units in the calcareous turbidites is composed principally of the tests of planktonic organisms. Foraminifera, mostly globigerinids, account for more than 50% of the sand fraction. Fragmented tests of foraminifera and pteropods comprise less than 20%, and pteropod shells comprise an additional 15%. Shallow-water components such as *Halimeda* average less than 2% (Schaeffer, 1984).

Columbus Basin. The horseshoe-shaped Columbus Basin (2200-2500 m) is a deep re-entrant in the southern Great Bahama Bank and covers approximately 6900 km^2 (Plate 2). Steep slopes that abruptly terminate the broad, shallow Bahama Banks border the basin on all sides.

Turbidity currents reach the basin from three sides and are

uniformly small (Pilkey and others, 1980). The largest turbidite unit measured was 25 km long and a maximum of 10 km wide (Bornhold and Pilkey, 1971). Frequency of turbidity currents is on the order of one per 3000–6000 years. Two types of turbidites have been identified on the basis of sand constituents. The first type is characterized by grains indicating a shallow-water origin (upper slope/banks), dominated by *Halimeda* fragments and ooliths. The second type consists of exclusively planktonic foraminifera and pteropods and is presumed to originate on mid-slope areas.

Sand layers range from 5 to 250 cm in thickness and comprise a maximum of 40% of the section raised in piston cores. This maximum coincides with the axes of turbidity currents crossing the basin (Bornhold and Pilkey, 1971).

X-radiographs show that Bouma Ta, Tb, and Te sequences are common. Toward the center of the basin turbidites often begin with Tb units and have thicker sections of Tb, Tc, and Te units than those located near the basin slopes. There is a paucity of fine sediments in the basin. Presumably this is due to the short distance (4-5 km) the currents travel to reach the basin floor. This short distance inhibits significant erosion and incorporation of fine-grained material. The lack of fine material accounts for the relatively thin Te units observed in most of the piston cores retrieved (Bornhold and Pilkey, 1971).

Hispaniola-Caicos Basin. The Hispaniola–Caicos Basin (4160 m) is a relatively small basin with an areal extent of 9500 km^2. The basin is bordered by Hispaniola to the south, by the Caicos Islands and Banks to the northeast, and by Great Iguana Island to the northwest (Plate 2). Basin-margin slopes are steep and have high relief (Bennetts and Pilkey, 1976). Two deep-sea fans occur in the basin, the broad Caicos Fan (50 × 75 km) in the north and the smaller (25 × 40 km), steeper Hispaniola Fan in the southern extremity of the basin. Little published information exists on these features. The southern and central portion of the basin is a true abyssal plain with gradients of 1 m/1 km.

Turbidity currents play a dominant role in filling the basin as evidenced by abundant graded sand layers. Sands constitute 50% by volume of the upper 5 m of fill. Turbidites are a mixture of terrigenous and carbonate debris. Calcareous flows are derived from Great Iguana and the Caicos Islands. However, the most important source areas are on Hispaniola, which contributes larger and more frequent flows across a broad front. These flows contain dark-colored igneous and metamorphic lithic fragments, abundant heavy minerals, and substantial amounts of insular shelf and slope carbonates. Some flows may also be derived from Cuba via a broad valley that enters the basin at its extreme western end (Bennetts and Pilkey, 1976; Ditty and others, 1977). A number of turbidites have been correlated between cores; they cover 2,700 to 4,200 km^2 and range in volume from 8.94×10^8 m^3 to 30.8×10^8 m^3. Maximum distances traversed are 90-100 km (Bennetts and Pilkey, 1976). Turbidites generally thin away from their entry points but thicken locally by ponding in minor basin depressions. On the plain proper, the complete Bouma sequence (Ta-e) is common.

Because the basin is filling from all sides and the source area is relatively large, sand layers can be deposited across the entire plain floor. Continuity of sand layers is high, and proximal-distal differences are low. The proportion of sand to clay increases slightly in a distal direction, and the frequency of sand layers is lower (Pilkey and others, 1980). Rough estimates of the coarse-layer volume show that the larger flows contain 10^9 m^3 of sediment, while the smaller flows contain less than 10^6 m^3 (Ditty and others, 1977).

Puerto Rico Trench Abyssal Plain. The Puerto Rico Trench (8450 m) represents the deepest depression in the Atlantic Ocean floor (Profile G, Plate 5). The trench is filled with up to 1.7 km of sediments (Ewing and Ewing, 1962; Conolly and Ewing, 1967).

Twelve small rivers drain onto the northern insular shelf of Puerto Rico where a large number of submarine canyons are incised in water depths as shallow as 25 m. A number of these features lie less than 0.5 km off river mouths (Fuerst, 1979). The southern slope of the trench is topographically complex and contains a number of perched basins, some of which are filled. Presumably a number of the canyons terminate in these basins and do not extend downslope to the trench floor.

Two major abyssal plains are found on the trench floor. The main east-west trending plain is approximately 400 km long and ranges from 6 to 30 km in width. The axis of the plain floor lies approximately 110 km north of the Puerto Rico-Virgin Island shelf edge. A separate, elevated, smaller abyssal plain (21 × 85 km) is located to the southeast some 250 m above the main plain. An irregular abyssal gap connects the two abyssal plains. Due to the complex, poorly known topography it is difficult to determine the exact number of basin entry points. On the basis of piston-core data at least two major basin entries have been delineated for the elevated plain and six additional entries for the deeper, main trench floor (Conolly and Ewing, 1967; Doull, 1983). The exact source areas that feed the east and west extremities of the main plain are unknown.

Turbidites recovered in piston cores from both plains range from 6 cm to more than 600 cm in thickness. The most common Bouma sequence present is Ta, Tb, Te. Intercalated with these sequences are mud turbidites of variable thicknesses (Conolly and Ewing, 1967; Doull, 1983).

Turbidite sands in both plains are similar in color and composition. The calcareous components consist mainly of planktonic foraminifera and pteropods (20-50%). Shallow-water carbonate sands include lithoclasts, molluscs, echinoids and ooliths. Terrigenous-sand components (3-24%) consist primarily of quartz, plagioclase, serpentine, and volcanic lithic fragments (Conolly and Ewing, 1967; Doull, 1983).

In the upper 10 m of sediment in the elevated plain, sand units comprise 20-50% of the section. The highest frequency of sand layers (20/m) is recorded in the elevated abyssal plain. Generally, the sand layer frequency decreases to the west. Sand units in the deeper main abyssal plain comprise less than 20% of the section. A maximum of 10 sand layers/m are present at the extremities of the plain. In most regions of the main abyssal plain, however, there are 3 sand-units/m.

Based on piston-core data, four turbidites have been correlated in the elevated plain and three have been correlated in the main abyssal plain (Conolly and Ewing, 1967; Doull, 1983). Only one turbidite has been traced through both abyssal plains. Turbidite sand volume in these flows has been calculated to range from .015 to 2.0 km^3, with linear dimensions of 33-250 km.

ABYSSAL GAPS AND MAJOR CHANNELS

Gaps

An abyssal gap is a constricted passage connecting two abyssal plains that lie at topographically different levels. The ocean floor within the gap has a steeper gradient than either of the two plains. Within the western North Atlantic Ocean basin, four major abyssal gaps occur. These are the Newfoundland, Sohm, Vema and Cat Gaps (Plate 2). Two unnamed gaps also occur. One occurs at the southern end of the Imarssuak Mid-Ocean Channel between the Eirik Ridge and the Gloria Drift; it connects the Irminger Abyssal Plain to the Labrador Abyssal Plain south of Greenland. A second unnamed gap connects the Hatteras and Silver Abyssal Plains.

Newfoundland Gap. The Newfoundland Abyssal Gap connects the Northwest Atlantic Mid-Ocean Canyon with the northeast section of Sohm Abyssal Plain. Little published data exists on this feature (Heezen and others, 1959). Arkosic sands are transported through this passage to Sohm Abyssal Plain from the glaciated regions north of Newfoundland (Hubert, 1962).

Sohm Gap. Sohm Gap separates the Sohm Abyssal Plain to the north and the Hatteras Abyssal Plain to the south. It can generally be regarded as the area between 34°30′N and 36°30′N from longitude 66°30′W to 60°00′W. This wide area is an undulating surface marked with many depressions, intervening channels, and asymmetric hills that resemble levees associated with sediment spillovers (Pratt, 1965). An early interpretation (Pratt, 1965) suggests that the westernmost distributaries of the Hudson Fan have been responsible for building a depositional lobe eastward into the gap area. Published seismic profiles and records from R/V EASTWARD cruises reveal that channels and levee-like hills abound in the area. Many of the large channels appear to meander (Pratt, 1965); they typically are flat-floored, 1-2 km in width, and have relief of up to 40 m. Small channels with less than 10 m relief occur, along with much broader (5 km), very low-relief "channels." These channels presumably represent an anastomosed distributary system. Further south on the Hatteras Abyssal Plain, relief is extremely subdued so that delineation of very small channels and levees is impossible (Pratt, 1965; Cleary and others, 1977).

Abundant sand layers in cores indicate that Sohm Gap is a transport path for sandy sediment. A 30-70 cm thick layer of light to dark silty clay is found at the top of most cores. Sand is most often found within topographically low areas and on the eastern side of the gap. Thicknesses of individual sand layers range from 15 cm to as much as 3 m. Along ridges away from the main channels, thick sequences of clay and silty clay are found.

Large textural differences occur in the sands, and this is also reflected in their mineralogy. Fine sands usually contain abundant feldspars, and they are lithic arkoses to arkoses. Coarse sands at the base of massive units usually contain more lithic fragments and many large polycrystalline quartz grains. Minor calcareous components and detrital glauconite suggest reworking of shelf-derived material. Few carbonate grains are present (2% average).

Cat Gap. Cat Gap (5200-5300 m) acts as a conduit for calcareous sediment entering the Blake-Bahama Basin to reach the southern limits of the Hatteras Abyssal Plain. The gap is bordered by the Bahama Outer Ridge to the north and the Bahama Banks to the south, and it is a region of relatively high relief (Heezen and others, 1959). Narrow, complex channels and interchannel highs are present.

Limited piston cores raised from Cat Gap reveal thick (up to 180 cm), graded, manganiferous, calcareous sands and intercalated calcilutites. Relatively thick (80 cm) silt layers occur seaward of the gap on the Hatteras Abyssal Plain. Silt layers are frequent (4-8/m and comprise as much as 35% of the core lengths recovered (Cleary and others, 1978).

Vema Gap. Vema Gap occurs at the southeastern extremity of the Hatteras Abyssal Plain and connects this plain with the deeper Nares Abyssal Plain (Heezen and others, 1959). The gap is approximately 35-50 km wide and 115 km long, with its long axis oriented approximately east-west. The gradient within the gap is 1:300. Several channels up to 40 m deep and 21 km wide are incised into the Hatteras plain surface near the gap, and they converge on the gap (Heezen and others, 1959). Piston-core data are sparse in this topographically complex area. Existing data indicate that no sand layers are present. Sediment transferred through the gap is silt or clay (Horn and others, 1971; Cleary and others, 1978; Tucholke, 1980). The high proportion of clay mixed with silt suggests the turbidity currents entering the gap are sluggish and of low competency (Horn and others, 1971).

Mid-Ocean Channels

Mid-ocean channels (canyons) were originally defined as flat-floored, steep-walled, linear sedimentary depressions. Typically these features are 1.5 to 8 km wide with 2 to 300 m of relief (Heezen and others, 1959). Mid-ocean channels described in the North Atlantic are all associated with abyssal gaps. They include the Northwest Atlantic Mid-Ocean Channel, Imarssuak Mid-Ocean Channel, Mid-Ocean Channel No. 2, and an unnamed and poorly known channel on the western margin of the Hatteras Abyssal Plain (Plate 2). Only the Northwest Atlantic Mid-Ocean Channel, and to a lesser extent the Imarssuak Mid-Ocean Channel, are well documented.

Northwest Atlantic Mid-Ocean Channel. The Northwest Atlantic Mid-Ocean Channel (canyon) is an unusual feature which channels turbidity currents south for 4,000 km from the

Hudson Strait (61°N) through the Labrador Sea, eventually disgorging turbidites onto the eastern Sohm Abyssal Plain (Profile J, Plate 6; Ewing and others, 1953; Chough, 1976; Chough and Hesse, 1976). Several tributaries connect with the channel; the most important is the Imarssuak Mid-Ocean Channel between Greenland and the Reykjanes Ridge. Levees extend on both sides of the Northwest Atlantic Mid-Ocean Channel along its entire length, and the western side typically is higher by 18-87 m. Channel widths along the upper third of its length range from 1.5 to 7.5 km, and it is between 100 and 200 m deep (Chough and Hesse, 1976). The channel gradient here (54° N latitude) is 1:1500. Studies by Egloff and Johnson (1975) and Chough and Hesse (1976) described a meandering thalweg 0.4 to 1.2 km wide and 4-10 m deeper than the main channel. Meander lengths range from 50-160 km in the region studied.

Piston cores and gravity cores recovered along the channel's length show that the floor of the channel contains graded coarse gravels and sand capped by bioturbated calcareous ooze (Ewing and others, 1953; Chough and Hesse, 1976). The graded beds range in thickness from 2.5 to 9 m (Heezen and others, 1969; Chough and Hesse, 1976). Cores raised from the levees contain thinly laminated terrigenous silts and interbedded silt to coarse sand. Overbank spill from both the head and body of the channelized flows has been invoked to explain the levee sequences.

Imarssuak Mid-Ocean Channel. The Imarssuak Mid-Ocean Channel, a 1000 km long major tributary to the Northwest Atlantic Mid-Ocean Channel, is found in the Irminger Basin. This continuous channel winds westward for part of its course through an abyssal plain that overlies the rift valley of the relict Ran Ridge. In the vicinity of the Runa Seamount (57°N, 45°W) the channel's course is offset due to crustal faulting (Egloff and Johnson, 1975). West of this point the channel broadens into a system of braided channels. Seismic profiles reveal the presence of a minimum of four low-gradient, shallow (18-45 m) channels (Laughton and others, 1972). The Imarssuak Mid-Ocean Channel flows into the Northwest Atlantic Mid-Ocean Channel at approximately 56°N and 49°W (Egloff and Johnson, 1975).

Other Channels. Mid-Ocean Channel No. 2 is similar in shape and appearance to the Northwest Atlantic Mid-Ocean Channel. The channel can be traced across the northwestern Sohm Abyssal Plain, northeast of the New England Seamounts, for approximately 575 km (Heezen and others, 1959).

In the southwestern extremity of the Hatteras Abyssal Plain are several channels 40-50 m deep and 2 km wide. These flat-floored channels are contained within levee-like features. Seismic profiles show that the channel fill is stratified (Cleary and others, 1978).

OTHER PONDED TURBIDITES

Mid-Atlantic Ridge

Seismic reflection profiles obtained across the axis of the Mid-Atlantic Ridge indicate the sediment cover is very thin and rock outcrops are common (Van Andel and Komar, 1969; Ewing and others, 1973). The thin veneer of pelagic sediment extends out to 75 km from the ridge axis where a marked increase in thickness is observed (up to 40 m). At larger distances, sediment thickness gradually increases as the ridge flank merges into the abyssal plains. Regionally the sediment cover is generally less than 250 m (Ewing and others, 1973; Tucholke and others, 1982).

In areas of locally high relief on the ridge flanks, many sediment ponds occur (Profiles C and H, Plate 5). These flat-floored intermontaine basins contain stratified, remobilized pelagic deposits derived from adjacent highs (Van Andel and Komar, 1969; Ewing and others, 1970). Two such valleys on the flanks of the Mid-Atlantic Ridge at 22°N have been investigated to show the nature of the pond fill (Van Andel and Komar, 1969). These elongate valleys measure 5-13 km wide and are surrounded by sediment-capped hills that rise up to 1500 m above the valley floors. The ponds contain as much as 500 m of stratified calcareous fill, consisting of fine, tan oozes interbedded with thin foraminiferal sands. Intercalated coarse layers are composed of graded, well-sorted planktonic foraminifera. Sediments are mixtures of Quaternary and Upper Tertiary deposits (Van Andel and Komar, 1969). Nine meters of Upper Quaternary fill has been recovered. Petrology of the coarse fraction indicates that all sand units contain clay-size sediment, up to 14% in some layers.

Based on calculated flow densities of 1.02-1.22 g/cm^3, the flows originating from the bordering hills were probably as much as 50 m thick. Due to the confined nature of the basins, rebounding of the flows off valley walls was commonplace and it produced double-graded sands and are very well sorted textural types.

The volume of sediment in these valleys correlates well with the size of the bordering catchment areas. Van Andel and Komar (1969) calculated that an average of 15 m of pelagic sediment has been removed from the surrounding hills and redeposited in the adjacent "V" shaped valleys. Their data from cores recovered from one valley (northernmost) indicate a high frequency of recent flows.

Fracture Zones

Fracture zones are narrow, deep troughs cutting orthogonally across the flanks of the mid-ocean ridge, and they consequently are ideal loci for episodic turbidity current deposition. Flows originate locally on steep adjacent scarps where mass wasting of pelagic sediment occurs. Earthquakes are probably important triggering mechanisms for sediment failure within transform regions offsetting the ridge crest. Also, eventual oversteepening of slopes by pelagic accumulation probably triggers turbidity currents both in the transform region and within the inactive fracture troughs away from the ridge axis. Cores raised from flat floors of fracture zones in many oceanic areas contain turbidites intercalated with thick pelagic units that represent thousands of years of sedimentation (Ewing and others, 1970). Little detailed published information exists on the nature of turbidites within these narrow fracture valleys near the Mid-Atlantic Ridge crest.

Within the landward extensions of the fracture zones, turbidity currents originating more than 1000 km away on the continental margin may ultimately deposit fine-grained sediments and lead to rapid filling of the fracture valleys (Ewing and others, 1970). One such region which has received some attention in the northwest Atlantic is the abyssal hill province at the northeastern edge of Nares Abyssal Plain (Shipley, 1978; Tucholke, 1980). Two fracture zones lie next to the distal Nares Abyssal Plain. Ponded turbidites occur within both the Kane Fracture Zone (6000 m) and the Nares Deep Fracture Valley (6110 m) to the north. The flat floor of the Kane Fracture Zone is approximately 200 m below the floor of the Nares Abyssal Plain (Profile F, Plate 5). Of several main sedimentation units in this region, one unit interpreted to be a turbidite consists of a gray clay which contains an average of 28% silt. The unit occurs within the Nares Abyssal Plain, Kane Fracture Zone, Nares Deep Fracture Valley, and the interconnecting lows. The gray clays represent the distal deposits of turbidity currents that have flowed through Vema Gap, across the Nares Abyssal Plain, and ultimately into the fracture valleys. A 0.5 m thick pelagic brown clay caps the turbidite unit and indicates that turbidity currents have not entered the region for a minimum of 300,000 years (Shipley, 1978).

REFERENCES

Allen, J.R.L.
1971 : Mixing at turbidity current heads, and its geologic implications: Journal of Sedimentary Petrology, v. 41, p. 97–113.

Ayers, M. W. and Cleary, W. J.
1980 : Wilmington Fan: Mid-Atlantic Lower Rise development: Journal of Sedimentary Petrology, v. 50, p. 235–245.

Bennetts, K.R.W. and Pilkey, O. H.
1976 : Characteristics of three turbidites, Hispanola-Caicos Basin: Geological Society of America Bulletin, v. 87, p. 1291–1300.

Bornhold, B. D. and Pilkey, O. H.
1971 : Bioclastic turbidte sedimentation in Columbus Basin, Bahamas: Geological Society of America Bulletin, v. 82, p. 1341–1354.

Bouma, A. H.
1962 : Sedimentology of some flysch deposits: Elsevier, Amsterdam-New York, 168 p.

Chough, S. K.
1976 : Morphology, sedimentary facies and processes of the Northwest Atlantic Mid–Ocean Channel between 61° and 51° N, Labrador Sea: unpublished Ph.D. thesis, McGill University, 167 p.

Chough, S. K. and Hesse, R.
1976 : Submarine meandering thalweg and turbidity currents flowing for 4,000 km in the Northwest Atlantic Mid-Ocean Channel, Labrador Sea: Geology, v. 4, no. 9, p. 529–533.

Cleary, W. J. and Conolly, J. R.
1974a: The Hatteras Deep Sea Fan: Journal of Sedimentary Petrology, v. 44, p. 1140–1154.
1974 : Petrology and origin of the deep-sea sands: Hatteras Abyssal Plain: Marine Geology, v. 17, p. 263–279.

Cleary, W. J. and Pilkey, O. H.
1983 : Quaternary fan development: the mid-Atlantic continental rise: Southeastern Sectional meeting, Geological Society of America, Tallahassee, Florida, Abstract with Programs, p. 68.

Cleary, W. J., Pilkey, O. H., and Ayers, M. W.
1977 : Morphology and sediments of three ocean basin entry points, Hatteras Abyssal Plain: Journal of Sedimentary Petrology, v. 47, no. 3, p. 1157–1170.

Cleary, W. J., Pilkey, O. H., Curran, A. H. and Neal, W. J.
1978 : Patterns of turbidite sedimentation on a trailing plate margin: Hatteras Abyssal Plain, Western North Atlantic Ocean: Proceedings Tenth International Congress on Sedimentology, v. 1, p. 126–127.

Conolly, J. R. and Ewing, M.
1967 : Sedimentation in the Puerto Rico Trench: Journal of Sedimentary Petrology, v. 37, p. 44–59.

Ditty, P. S., Harmon, C. S., Pilkey, O. H., Ball, M. M., and Richardson, E. S.
1977 : Mixed Terrigenous-Carbonate Sedimentation in the Hispaniola-Caicos Turbidite Basin: Marine Geology, v. 24, p. 1–20.

Doull, M. E.
1983 : Turbidite sedimentation in the Puerto Rico Trench Abyssal Plain: unpublished Masters Thesis, Duke University, 124 p.

Drake, C. L., Ewing, J. I. and Stockard, H. P.
1968 : The Continental Margin of the Eastern United States: Canadian Journal of Earth Sciences, v. 5, p. 933

Egloff, J. and Johnson, G. L.
1975 : Morphology and stratigraphy of the Southern Labrador Sea: Canadian Journal of Earth Sciences, v. 12, p. 2111–2133.

Elmore, R. D., Pilkey, O. H., Cleary, W. J. and Curran, H. A.
1979 : Black Shell Turbidite, Hatteras Abyssal Plain, Western North Atlantic Ocean Basin: Geological Society of America Bulletin, v. 90, p. 1165–1176.

Embley, R. W.
1980 : The role of mass transport in the distribution and character of deep-ocean sediments with special reference to the North Atlantic: Marine Geology, v. 38, p. 23–50.

Emery, K. O., Uchupi, E., Phillips, J. D., Bowin, C. O., Bunce, E. T. and Knott, S. T.
1970 : Continental Rise of eastern North America: American Association of Petroleum Geologists Bulletin, v. 54, p. 44–108.

Ericson, D. B., Ewing, M. and Heezen, B. C.
1951 : Deep-sea sands and submarine canyons: Geological Society of America Bulletin, v. 62, p. 961–966.

Ericson, D. B., Ewing, M., Wollin, G., and Heezen, B. C.
1961 : Atlantic deep-sea sediment cores: Geological Society of America Bulletin, v. 72, p. 193–286.

Ewing, J. and Ewing, M.
1962 : Reflection profiling in and around the Puerto Rico Trench: Journal of Geophysical Research, v. 67, no. 12, p. 4729–4739.

Ewing, M., Carpenter, G., and Windisch, C.
1973 : Sediment distribution in the oceans: The Atlantic: Geological Society of America Bulletin, v. 84, p. 71–88.

Ewing, G. M., Ryan, W.B.F., Needham, H. D. and Schreiber, B. C.
1970 : Turbidity Currents on the ocean bottom: U.S. Naval Civil Engineering Laboratory Report, Port Hueneme, California, 151 p.

Ewing, G. M., Heezen, B. C., Ericson, D. B., Northrup, J. and Dorman, J.
1953 : Exploration of the Northern Atlantic Mid-Ocean Canyons: Geological Society of America Bulletin, v. 64, p. 865–868.

Forel, F. A.
1887 : Le Ravin Sous—Lacustre de Rhone dans le Lac LeMan: Bulletin de la Societe Vaudoise Science Naturelle, v. 23, p. 85–107.

Fritz, S. J. and Pilkey, O. H.

1975 : Distinguishing bottom and turbidity current coarse layers on the Continental rise: Journal of Sedimentary Petrology, v. 45, p. 57–62.
Fruth, L. S.
1965 : The 1929 Grand Banks Turbidite and the sediments of the Sohm Abyssal Plain: Masters Thesis, Columbia University, 257 p.
Fuerst, S. I.
1979 : Sediment transport on the insular slope of Puerto Rico off the La Plata River: unpublished thesis, Duke University, 101 p.
Grover, N. C. and Howard, C. S.
1937 : The passage of turbid water through Lake Mead (Arizona and Nevada): American Society of Civil Engineers, Proceedings, Vol. 63, part 1, p. 643–655.
Harms, J. C. and Fahnestock, R. K.
1965 : Stratification, bed forms and flow phenomena (with an example from the Rio Grande): in Primary Sedimentary Structures and their Hydrodynamic Interpretation, ed. G. V. Middleton; Society of Economic Paleontologists and Mineralogists Special Publication No. 12, p. 84–115.
Heezen, B. C. and Ewing, M.
1952 : Turbidity currents and submarine slumps and the 1929 Grand Banks earthquake: American Journal of Science, v. 250, p. 849–878.
Heezen, B. C. and Hollister, C. D.
1971 : The face of the deep: Oxford University Press, New York, 659 p.
Heezen, B. C., Tharp, M. and Ewing, W. M.
1959 : The floors of the oceans: Geological Society of America Special Paper 65, 122 p.
Heezen, B. C., Hollister, C. D. and Ruddiman, W. F.
1966 : Shaping of the Continental Rise by deep geostrophic contour currents: Science, v. 152, p. 502–508.
Heezen, B. C., Johnson, G. L. and Hollister, C. D.
1969 : The Northwest Atlantic Mid-Ocean Canyon: Canadian Journal of Earth Sciences, v. 6, p. 1441–1453.
Hollister, C. D.
1967 : Sediment distribution and deep circulation in the Western North Atlantic: [Ph.D. Dissertation]; Columbia University, 265 p.
Hollister, C. D. and Heezen, B. C.
1972 : Geologic effects of ocean bottom currents: western North Atlantic: in Studies in Physical Oceanography, Volume 2, ed. A. L. Gordon; Gordon and Breach, p. 37–66.
Horn, D. R., Ewing, M., Horn, B. M., and Delach, M. N.
1971 : Turbidites of the Hatteras and Sohm Abyssal Plains, western North Atlantic: Marine Geology, v. 11, p. 287–323.
Hubert, J. P.
1962 : Depositional pattern of Pleistocene sands on the North Atlantic Deep-Sea floor: Science, v. 136, p. 383–384.
Hubert, J. P.
1964 : Textural evidence for deposition of many Western North Atlantic deep-sea sands by ocean bottom currents rather than turbidity currents: Journal of Geology, v. 72, p. 757–785.
Hubert, J. P. and Neal, W. J.
1967 : Mineral composition and dispersal patterns of deep-sea sands in the Western North Atlantic Petrologic Province: Geological Society of America Bulletin, v. 78, p. 749–772.
Johnson, G. L., Sommerhoff, G., and Egloff, J.
1975 : Structure and morphology of the West Reykjanes Ridge Basin and the Southeast Greenland Continental Margin: Marine Geology, v. 18, p. 175–196.
Keller, G. H. and Shepard, F. P.
1978 : Currents and sedimentary processes in submarine canyons off the northeast United States: in Sedimentation on Submarine Canyons, Fans, and Trenches, eds. D. J. Stanley and G. Kelling, Dowden, Hutchinson and Ross, Stroudsburg, p. 15–32.
Kelling, G. and Stanley, D. J.
1970 : Morphology and structure of Wilmington and Baltimore Submarine Canyons, Eastern U.S.A.: Journal of Geology, v. 78, p. 637–660.
Keulegan, G. H.
1957 : Twelfth Progress Report on Model Laws for Density Currents; An experimental study of motion of saline water from locks into fresh water channels: U.S. National Bureau of Standards Report, Report no. 5168, 21 p.
1958 : Thirteenth Progress Report on Model Laws for Density Currents. The Motion of Saline Water Fronts in Still Water: U.S. National Bureau of Standards Report, Report no. 5168, 21 p.
Kuenen, Ph.H.
1950 : Turbidity Currents of High Density: Eighteenth International Geological Congress, London, Report, Part 8, pp. 45–52.
Kuenen, Ph.H. and Migliorini, C. I.
1950 : Turbidity Currents as a Cause of graded bedding: Journal of Geology, v. 58, p. 91–127.
Laughton, A. S., Berggren, W. A., Benson, R. N., Davies, T. A., Franz, U., Musich, L. F., Perch-Nielsen, K., Ruffman, A. S., van Hinte, J. E., and Whitmarsh, R. B.
1972 : The Southern Labrador Sea - a key to the Mesozoic and Tertiary Evolution of the North Atlantic: in Initial Reports of the Deep Sea Drilling Project, v. 12, U.S. Government Printing Office, Washington, D.C., p. 1155–1169.
McGregor, B. P.
1968 : Features at Sohm Abyssal Plain Terminus: Marine Geology, v. 6, p. 401–414.
Middleton, G. V.
1966 : Experiments on Density and Turbidity currents, III, Deposition of Sediment: Canadian Journal of Earth Sciences, v. 4, p. 475–505.
Middleton, G. V. and Hampton, M. A.
1976 : Subaqueous sediment transport and deposition by sediment gravity flows; in Marine sediment transport and environmental management, Chapter 11, eds. D. J. Stanley and D.J.P. Swift; John Wiley and Sons, New York, p. 197–216.
Mutti, E. and Ricci-Lucchi, F.
1972 : Le torbiditi dell' Appenino settentrionale: introduzione all' analisi di facies: Geolia Italiana Memoires, v. 11, p. 161–199.
Newton, J. G. and Pilkey, O. H.
1969 : Topography of the continental margin off the Carolinas: Southeastern Geology, v. 10, p. 87–92.
Newton, J. G., Pilkey, O. H. and Blanton, J. O.
1971 : An oceanographic atlas of the Carolina Margin: North Carolina Department of Conservation and Development, 57 p.
Normark, W. R.
1974 : Submarine canyons and fan valleys: Factors affecting growth patterns of deep-sea fans: in Modern and ancient geosynclinal sedimentation: Society of Economic Paleontologists and Mineralogists Special Publication No. 19, p. 56–68.
1978 : Fan valleys, channels and depositional lobes on modern submarine fans: characters for recognition of sandy turbidite environments: American Association of Petroleum Geologists Bulletin, v. 62, p. 912–931.
Normark, W. R., Piper, D.J.W. and Stow, D. A.
1983 : Quaternary development of channels, levees and lobes on Middle Laurentian Fan: American Association of Petroleum Geologists Bulletin, v. 67, p. 1400–1409.
Pilkey, O. H., Locker, S. and Cleary, W. J.
1980 : Comparison of sand layer geometry on flat floors of 10 modern depositional basins: American Association of Petroleum Geologists Bulletin, v. 64, p. 841–856.
Piper, D.J.W.
1975 : Late Quaternary deep water sedimentation off Nova Scotia and western Grand Banks: Canadian Society of Petroleum Geologists Memoir 4, p. 195–204.
1978 : Turbidite muds and silts on deep sea fans and abyssal plains: in Sedimentation in Submarine Canyons, Fans and Trenches, eds. D. J. Stanley and G. Kelling; Academic Press, p. 163–177.

Pratt, R. M.
1965 : Ocean-bottom topography: The divide between the Sohm and Hatteras Abyssal Plains: Science, v. 48, p. 1598–1599.
1967 : The seaward extension of submarine canyons off the Northeast Coast of the United States: Deep Sea Research, v. 14, p. 409–420.
Rona, P. A., Schneider, E. D. and Heezen, B. C.
1967 : Bathymetry of the continental rise off Cape Hatteras: Deep Sea Research, v. 14, p. 625–633.
Ryan, W.B.F., Cita, M.B., Miller, E. L., Hanselman, D., Nesteroff, W. D., Hecker, B. and Nibbelink, M.
1978 : Bedrock geology in New England submarine canyons: Oceanologica Acta, vol. no. 2, p. 233–254.
Schaeffer, C. L.
1984 : Sedimentation in the Blake-Bahama Abyssal Plain: unpublished Masters Thesis, Duke University, 84 p.
Shipley, T. H.
1978 : Sedimentation and echo characteristics in the Abyssal Hills of the West Central Atlantic: Geological Society of America Bulletin, v. 89, p. 397–408.
Sparks, T. H.
1979 : X-radiography of piston cores from the Hatteras Abyssal Plain: unpublished thesis, Duke University, 89 p.
Stanley, D. J.
1970 : Flyschoid sedimentation on the outer margin off northeast North America: in Flysch Sedimentology in North America, ed. J. Lajoie; Geological Association of Canada Special Paper, v. 7, p. 179–210.
Stanley, D. J., Sheng, H. and Pedaza, C. P.
1971 : Lower Continental Rise East of Middle Atlantic States: Predominant dispersal perpendicular to isobaths: Geological Society of America Bulletin, v. 82, p. 1831–1840.
Stow, D.A.V.
1981 : Laurentian Fan: morphology, sediments, processes and growth pattern: American Association of Petroleum Geologists Bulletin, v. 65, p. 375–393.
Tucholke, B. E.
1980 : Acoustic environment of the Hatteras and Nares Abyssal Plains, western North Atlantic Ocean, determined from velocities and physical properties of sediment cores: Journal of Acoustical Society of America, v. 68, no. 5, p. 1376–1390.
Tucholke, B. E. and Laine, E. P.
1982 : Neogene and Quaternary Development of the Lower Continental Rise off the Central U.S. East Coast; in Continental Margin Geology, eds. J. S. Watkins and C. L. Drake; American Association of Petroleum Geologists Memoir No. 34, p. 295–305.
Tucholke, B. E., Houtz, R. E. and Ludwig, W. J.
1982 : Sediment thickness and depth to basement in Western North Atlantic Ocean Basin: American Association of Petroleum Geologists Bulletin, v. 66, n. 99, p. 1384–1395.
Twichell, D. C. and Robert, D. G.
1982 : Morphology, distribution and development of submarine canyons on the United States Atlantic continental slope between Hudson and Baltimore Canyons: Geology, v. 10, p. 408–412.
Van Andel, T. H. and Komar, P. D.
1969 : Ponded sediments of the Mid-Atlantic Ridge between 22° and 23°N latitude: Geological Society of America Bulletin, v. 80, p. 1103–1190.
Van Tassell, J.
1981 : Silver Abyssal Plain carbonate turbidite: flow characteristics: Journal of Geology, v. 89, p. 317–333.
Walker, R. G.
1965 : The origin and significance of the internal sedimentary structures of turbidites: Yorkshire Geological Society Proceedings, v. 35, p. 1–32.
Walker, R. G.
1978 : Deep water sandstone facies and submarine fans: models for exploration for stratigraphic traps: American Association of Petroleum Geologists Bulletin, v. 62, p. 932–966.
Walker, R. G.
1979 : Turbidites and associated coarse clastic deposits: in Facies Models, ed. R. G. Walker; Geoscience Canada, Reprint Series 1, Geological Association of Canada, p. 91–103.

Manuscript Accepted by the Society June 26, 1985

ACKNOWLEDGMENTS

We wish to express our gratitude to a number of former students at Duke University whose masters theses contributed substantially to this paper. Most of the theses are already published; others will be published in the near future. These individuals include Claudia Hokanson, Lisa Doull, Pat Ditty, Nikki Schaeffer, and Thomas Sparks, among others. Our basin studies which are summarized in this paper were funded by the National Science Foundation.

Printed in U.S.A.

The Geology of North America
Vol. M, The Western North Atlantic Region
The Geological Society of America, 1986

Chapter 27

Deep current-controlled sedimentation in the western North Atlantic

I. N. McCave
Department of Earth Sciences, University of Cambridge, Downing Street, Cambridge CB2 3EQ, United Kingdom, and Woods Hole Oceanographic Institution
Brian E. Tucholke
Woods Hole Oceanographic Institution, Woods Hole, Massachusetts 02543

INTRODUCTION

With its multiple sources of sediment and bottom water, the North Atlantic experiences more active redistribution of sediments than most other ocean basins. North Atlantic bottom waters originate around Antarctica, in the Norwegian Sea, the Mediterranean, and the Labrador Sea. Sediment is supplied from continental land-masses, oceanic islands (notably Iceland) and from surface biological production. The water movements are controlled principally by differing densities of the water masses, and the currents tend to follow the contours of the sea floor. Where interfaces of steep density gradient intersect the seafloor, internal tides and high-frequency internal waves may also resuspend sediments. The Gulf Stream, and warm-core and cold-core mesoscale eddies with clockwise (anti-cyclonic) and anti-clockwise (cyclonic) circulation, apparently contribute to the variability of current velocity at great depths and help to erode and redistribute sediments.

The most important depositional products of this current activity are the great "sediment drifts" of the region (see Plate 2). These features probably contain a detailed but as yet poorly known record of the fluctuations in bottom-current activity of the North Atlantic. They have formed principally since the beginning of the Oligocene, when strong abyssal circulation began in the North Atlantic (Ewing and Hollister 1972; Tucholke and Mountain, 1979; Miller and Tucholke, 1983). The precise relationship of the drifts to the present abyssal current regime is not clear because details of that regime are poorly known. Few long-term, deep current-meter records have been taken and most of these are not adjacent to drifts. The drifts are mantled by smaller-scale bedforms including mud waves, furrows, longitudinal and transverse ripples, biogenic mudmounds with depositional tails, and smaller current-produced lineations. The drifts and their bedforms constitute the result of what we earlier termed "plastering and decorating" on the sides of the North Atlantic ocean basin (Hollister and others, 1978). We here give a summary account of this decorative distribution of material and its control by the deep circulation. Localities discussed in the text are shown in Plate 2 and Figure 1.

REGIONAL DEEP CIRCULATION

Hydrography—Water Masses

Worthington and Wright's (1970) atlas shows clearly that, with decreasing water depth, the floor of the Atlantic is successively invaded by warmer water masses that enter from different sides of the stage (Fig. 1). The deepest water mass, entering the western North Atlantic from the south between the Mid-Atlantic Ridge and northern South America, is Antarctic Bottom Water (AABW). By the time it arrives in the North Atlantic, it is rather warmer and saltier than at its source in the Weddell Sea. At its source near 65°S latitude, the water has a potential temperature (the temperature it would have if adiabatically raised to the surface) of $\Theta < -0.6°C$ and salinity $S \sim 34.6‰$, and when it enters the North Atlantic it has a potential temperature of 0.5°Θ and salinity of 34.75‰. At 4°N the flux of Antarctic Bottom Water at <1.9°Θ into the North Atlantic is 0.7 to 1.1×10^6 m^3/s (by current meter) or $2 \times 10^6 m^3/s$ (by geostrophic calculation) (Whitehead and Worthington, 1982). On entering the North Atlantic the water mass has a silicate content of about 100 m-mol/m^3, but this decreases northward to <50 m-mol/m^3 by mixing with overlying, silica-poor North Atlantic Deep Water (NADW) (Mantyla and Reid, 1983).

On the eastern side of the North Atlantic, Antactic Bottom Water flows north after entering the basin through Vema, Romanche, and similar fracture zones at and just north of the equator. This probably forms a weak, deep eastern boundary current that extends up to Rockall Trough (Fig. 1). At this latitude the original water mass is greatly diluted and warmed by mixing with overlying waters, but it still has a silicate content of ~50 m-mol/m^3. Worthington and Wright (1970) show that this water at 2.2° Θ reaches 53°N latitude.

McCave, I. N., and Tucholke, B. E., 1986, Deep current-controlled sedimentation in the western North Atlantic; *in* Vogt, P. R., and Tucholke, B. E., eds., The Geology of North America, Volume M, The Western North Atlantic Region: Geological Society of America.

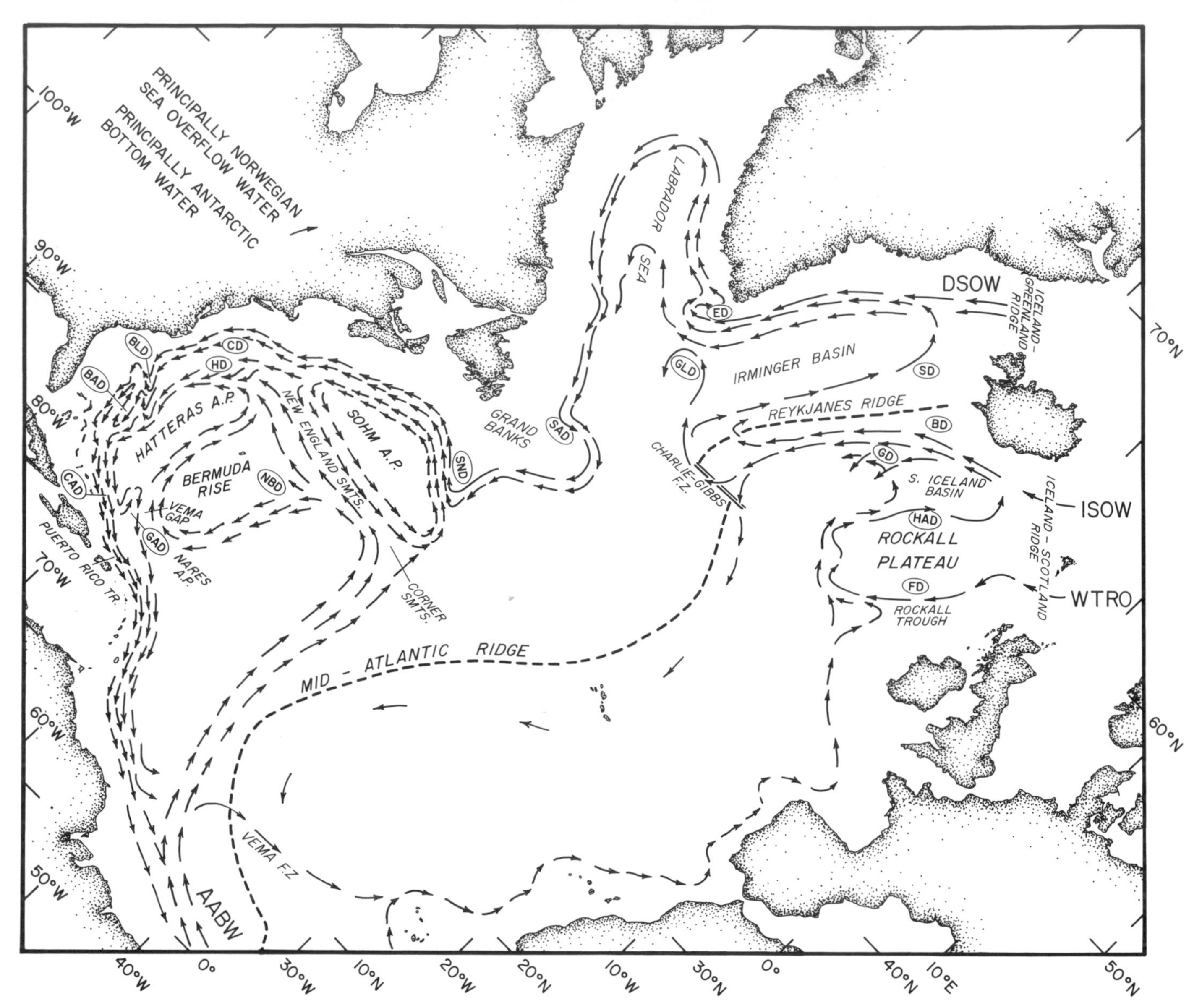

Figure 1. Generalized circulation of bottom water in the North Atlantic Ocean. DSOW - Denmark Strait overflow water, ISOW - Iceland-Scotland overflow water, WTRO - Wyville-Thomson Ridge overflow, AABW - Antarctic Bottom Water. Circled labels give locations of sediment drifts in Table 1 (see also Plate 2). Red arrows show currents with a significant component of Antarctic Bottom Water, and black arrows show the circulation of principally Norwegian Sea overflow water (North Atlantic Deep Water). Map is Mercator projection.

The Norwegian Sea is the source of most North Atlantic Deep Water by overflow between Iceland and Greenland through Denmark Strait and between Iceland and Scotland. The Denmark Strait water has the coldest potential temperatures with small amounts of water between 0° and 0.7°Θ, and then significant volumes of saline water (S ~34.90‰) of potential temperature 0.7°–1.8°Θ; the total flow is about 4×10^6 m^3/s. The Iceland-Scotland overflow water leaves the Norwegian Sea at much higher temperatures; the coldest water is 1.8°Θ, but the bulk is warmer than 2.2°Θ with high salinity, 34.99 to 35.00‰. The total volume of flow is 2×10^6 m^3/s. The majority of this water flows over the Iceland-Faeroe Ridge near Iceland and through Faeroe Bank Channel, a 600m-deep trough in the ridge, to enter the South Iceland Basin (Fig. 1). A small amount of water may also escape from the Norwegian Sea over the Wyville-Thomson Ridge (Fig. 1) and flow southwest into Rockall Trough (Ellett and Martin, 1973; Roberts, 1975). When entrainment of overlying water is taken into account, the two main flows over the Greenland-Scotland ridge each comprise underflows of 5×10^6 m^3/s, and they eventually give rise to a combined flow of about 10×10^6 m^3/s along the east Greenland continental rise (Fig. 1; Worthington, 1976).

Two other sources of deep water in the North Atlantic are less important to our discussion because they do not bathe significant areas of the seabed in the western North Atlantic. These sources are 1) Labrador Sea Water produced by sinking of water cooled in the Labrador Sea, and 2) a saline outflow of Mediterranean Water. The Labrador Sea Water (3.2°-3.6°Θ) spreads well above bottom in the deeper basin (1500-2000 m), as does the Mediterranean Water which is deepest (3300 m) off Portugal. The main tongue of Mediterranean Water is shallower than 3000 m, and in the west-central Atlantic it touches the seabed only around the crest of the Mid-Atlantic Ridge.

Deep Current Systems

Because there are two principal sources of bottom water and because the North Atlantic seafloor topography is complex, bottom current patterns also are complex and not very well known. In the last 9 years there have been at least three quite different circulation schemes proposed for the North American Basin. We here discuss the general patterns of deep circulation, starting with the flow paths of the various Norwegian Sea overflows in the northeast Atlantic.

The Basins North of 45° N (Fig. 1). In Rockall Trough, Norwegian Sea water leaks intermittently over the Iceland-Scotland Ridge and flows southwest over a stratum of dilute Antarctic Bottom Water which has come up from the south (Ellett and Martin, 1973; Ellett and Roberts, 1973; Roberts, 1975; Lonsdale and Hollister, 1979). The dilute Antarctic Bottom Water is found on the bottom below 2500 m and is thought to circulate cyclonically around the basin (Lonsdale and Hollister, 1979). The overlying Norwegian Sea water, comprising Ellett and Martin's (1973) Northeast Atlantic Deep Water, is presumed to flow southwest along the side of Rockall Plateau. The crest of Feni Drift (Fig. 1; Plate 2) is at about 2500 m so it might be supposed that its lower flanks underlie Antarctic Bottom Water while the upslope region is under Northeast Atlantic Deep Water, with both water masses flowing south; this picture agrees with a photographic traverse showing unequivocal scour crescents pointing south (Lonsdale and Hollister, 1979). However, Dickson and others (1985) recorded currents for more than a year in a transect across the southern Feni Drift, and they found northward flow along the south flank of Rockall Plateau and in the interior waters above 2500 m, with southward flow only along the lower flanks of the Feni Drift. Whether these measurements are representative of longer-term conditions presently is uncertain.

In the South Iceland Basin, bottom photographs and short-term current measurements suggest a northward flow along Hatton Drift (Fig. 1; McCave and others, 1980). Some of this flow is dilute Antarctic Bottom Water, but this is not the deepest water in the basin; the Antarctic water occurs at intermediate levels and overlies denser Iceland-Scotland Overflow Water (Mann and others, 1973; McCave and others, 1980).

The main Iceland-Scotland Overflow Water travels south of Iceland between about 1300 and 2200 m depth along the eastern flank of Reykjanes Ridge and over Gardar Drift (Fig. 1; Plate 2) (Steele and others, 1962; Worthington and Volkmann, 1965; Shor, 1978). This flow descends the slope, getting deeper to the south where it passes west through Charlie-Gibbs Fracture Zone with a sill depth of about 3600 m (Worthington and Volkmann, 1965). Some current meters have recorded easterly flow in the fracture zone trough, but there is no evidence of *net* eastward flux of water (Schmitz and Hogg, 1978; Dickson and others, 1980; Harvey, 1980; Shor and others, 1980). Most of the 5×10^6 m^3/s of overflow plus entrained water probably flows west through the trough; a small volume spills west over the crest of the Reykjanes Ridge (Vogt and Johnson, 1972), and some of the flow passes the entrance to the trough and continues south along the eastern flank of the Mid-Atlantic Ridge (Worthington and Volkmann, 1965).

In the Irminger and Labrador Seas the deep flow is comprised entirely of components of Norwegian Sea origin. The deepest flow exiting the Charlie-Gibbs Fracture Zone at 40°-45°W is presumed to flow to the northwest (Fig. 1), forming a cyclonic gyre over the crest of Gloria Drift (Egloff and Johnson, 1979). There are presently no hydrographic or current measurements to support this inference. At slightly shallower depths ($\lesssim$3200 m), about 5×10^6 m^3/s of water are presumed to flow north along the west flank of Reykjanes Ridge (Worthington, 1976), although Worthington and Volkmann (1965) show only 2×10^6 m^3/s flowing north through a hydrographic section there. The flow loops westward around the northern end of the Irminger Basin and overlies about 5×10^6 m^3/s of Denmark Strait overflow water that flows southwest along the east Greenland margin (Smith, 1975). This augmented, 10×10^6 m^3/s flow continues around Eirik Drift (which it constructs) at the southern tip of Greenland. No details are known of the flow around Eirik Drift, though Swallow and Worthington (1969) have a current measurement by neutrally buoyant float of 0.10 m/s, and Rabinowitz and Eittreim (1974) show several point speeds (measured by a photographic current meter) in excess of 0.20 m/s between 2200 m and 3300 m. This flow, which constitutes what has been called the Western Boundary Undercurrent, follows the bathymetric contours around the Labrador Basin (Fig. 1). Current measurements and bottom photographs on the lower continental slope off Labrador suggest that the core of the undercurrent is near 2800 m, with a current-influenced zone from 2300 m to below 3000 m (Swallow and Worthington, 1969; Rabinowitz and Eittreim, 1974; Carter and others, 1979; Carter and Schafer, 1983). This current is shown by Worthington (1976) to continue south around the eastern flanks of the Grand Banks and into the North American Basin.

The North American Basin South of 45°N (Fig. 1). The deepest water in this area is colder than 1.78°Θ and contains a significant proportion of Antarctic Bottom Water. South of Nova Scotia the water flux is poorly known, but Hogg (1983) has proposed a recirculation in a cyclonic gyre north of 36°N (Figure 1). Recirculating loops in this area have been prominent features of the models of Worthington (1976), Schmitz (1977), Wunsch and Grant (1982) and Hogg (1983). Worthington (1976) suggested a large *anticyclonic* recirculation of up to 62×10^6 m^3/s associated with the Gulf Stream between 32°N and 41°N. The schemes of Wunsch and Grant, based on inversion of data similar to that used by Worthington, and of Hogg, based on current meter data, both show a *cyclonic* deep gyre north of 36° to 37°N latitude. Further south, Wunsch and Grant's scheme is at variance with current meter results, and Hogg's has a very narrow anticyclonic gyre at 36° to 37°N sandwiched between a general westward flow over the northern Bermuda Rise and the main easterly flowing recirculation above 37°N.

The general circulation scheme suggested in Figure 1 contains elements of both Wunsch and Grant's, and Hogg's models, although we propose a broader west and southwest flow on the northeast Bermuda Rise. Data from a towed sensor show water colder than 1.78°Θ at depths as shallow as 4400 m depth on the northeast Bermuda Rise, suggesting that the rise is bathed in Antarctic Bottom Water (McCave and others, 1982). We suggest that this water is derived from a generally northward flow along the lower, western flank of the Mid-Atlantic Ridge; this flow is both deflected west near the Corner Rise seamounts at 36°N and feeds the deep gyre north of 37°N (Fig. 1). This circulation scheme agrees with available hydrographic and current-meter measurements; it is also based on the notion that the flow is generally forced to follow bathymetric trends, including the New England seamount chain which lies between the Sohm and Bermuda Rise circulation cells (Fig. 1).

South of 35°N a branch of Antarctic Bottom Water flows strongly southward along the eastern Bermuda Rise (McCave and others, 1982; Bird and others, 1982) and southwest around the southern end of Bermuda Rise (Fig. 1). A minimum westward transport of 1.6×10^6 m^3/s through Vema Gap into the area of the Hatteras Abyssal Plain has been documented by hydrographic measurements (Tucholke and others, 1973). Several workers have suggested a north to northeast flow of Antarctic Bottom Water on the western flank of the Bermuda Rise (Heezen and others, 1966; Schneider and others, 1967; Embley and others, 1980). At the northern end of Hatteras Abyssal Plain there is thus a convergence of deep flows coming from the south, from the Nova Scotian continental rise, and possibly across the northern Bermuda Rise. These are thought to form a combined flow that turns west and south and flows along the U.S. continental rise beneath the Norwegian Sea overflow water that forms the classic Western Boundary Undercurrent (black arrows, Fig. 1; Amos and others, 1971; Biscaye and Eittreim, 1974).

On the northeast flank of the Blake Outer Ridge, the deep western boundary current system includes Norwegian Sea water, as well as Antarctic Bottom Water at its base, and the current flows strongly southeastward between 3000 m and 4900 m (Heezen and others, 1966; Amos and others, 1971). Jenkins and Rhines (1980) demonstrated with tritium as a tracer that some of the water there had made the transit from the Norwegian Sea in less than about 15 years. They also measured mean currents of 22 cm/s at 3600 m (200 m above bottom). Freon measurements suggest that the core of the Western Boundary Undercurrent is less than 5 years old north of 35°N (Smethie and Trumbore, 1983). The water below about 5100 m on the northeast flank of the Blake Outer Ridge flows southeast and contains >20% Antarctic Bottom Water (Amos and others, 1971). This deep flow component continues south along the Bahama Outer Ridge and southeast along the face of the Bahama Banks and across the Caicos Outer Ridge. The shallower part of the flow, which contains principally Norwegian Sea overflow water, follows a circuitous path along the contours of the Blake and Bahama Outer Ridges (Drifts), thence along the flank of the Blake Escarpment until it joins the top of the Antarctic Bottom Water flow along the Bahama Banks.

At the eastern end of the Bahama Banks, the southward flow again splits. Deeper water, a mixture of Norwegian Sea and Antarctic components, flows clockwise around the Greater Antilles Outer Ridge (Tucholke and others, 1973) and onward to the east where part may recirculate with incoming, fresh Antarctic Bottom Water from the South Atlantic. A shallower branch of Norwegian Sea overflow water and Antarctic Bottom Water continues southeast across Navidad Sill at 5200 m and along the south flank of the Puerto Rico Trench (Plate 2; Tucholke and Eittreim, 1974). This flow presumably continues into the South Atlantic, joining the general southward flow of North Atlantic Deep Water and no longer forming a strong western boundary current.

The Gulf Stream

The Gulf Stream directly affects the seabed across the Blake Plateau (e.g. Pinet and others, 1981), but northeast of the plateau the flow generally is detached from the bottom (Fig. 2, Plate 2). At the point where the Gulf Stream crosses over the Western Boundary Undercurrent, just south of Cape Hatteras at about 35°N, Barrett (1965) and Richardson (1977) recorded southwestward flow at the bottom from seafloor depths of 500 m to more than 3000 m. However, there are occasional reversals of flow. For example, episodes of northerly flow within 200-400 m of the bottom have been recorded at 4000-4400 m on the continental rise at 37°N (Hogg, 1983). These may represent intermittent near-bottom influence of the Gulf Stream.

The Gulf Stream, including its warm- and cold-core rings, can exert an influence on the seabed either directly or by its interaction with the deeper flow and its generation of Rossby waves (Fig. 2). In terms of its effect on sedimentation dynamics, the variability of the bottom flow is as important as the steady current component. Long-term current meter records, like any record of a fluctuating quantity, can be characterized by a mean and its associated variance. For currents this variance is termed the eddy kinetic energy per unit mass (K_E). Regions such as the Gulf Stream that have fast mean surface currents also display high K_E (Wyrtki and others, 1976; Richardson, 1983a). For reasons which are not yet clear, the eddy kinetic energy propagates downward, so that regions of high variability in bottom currents underlie regions of highly variable surface currents (Dickson,

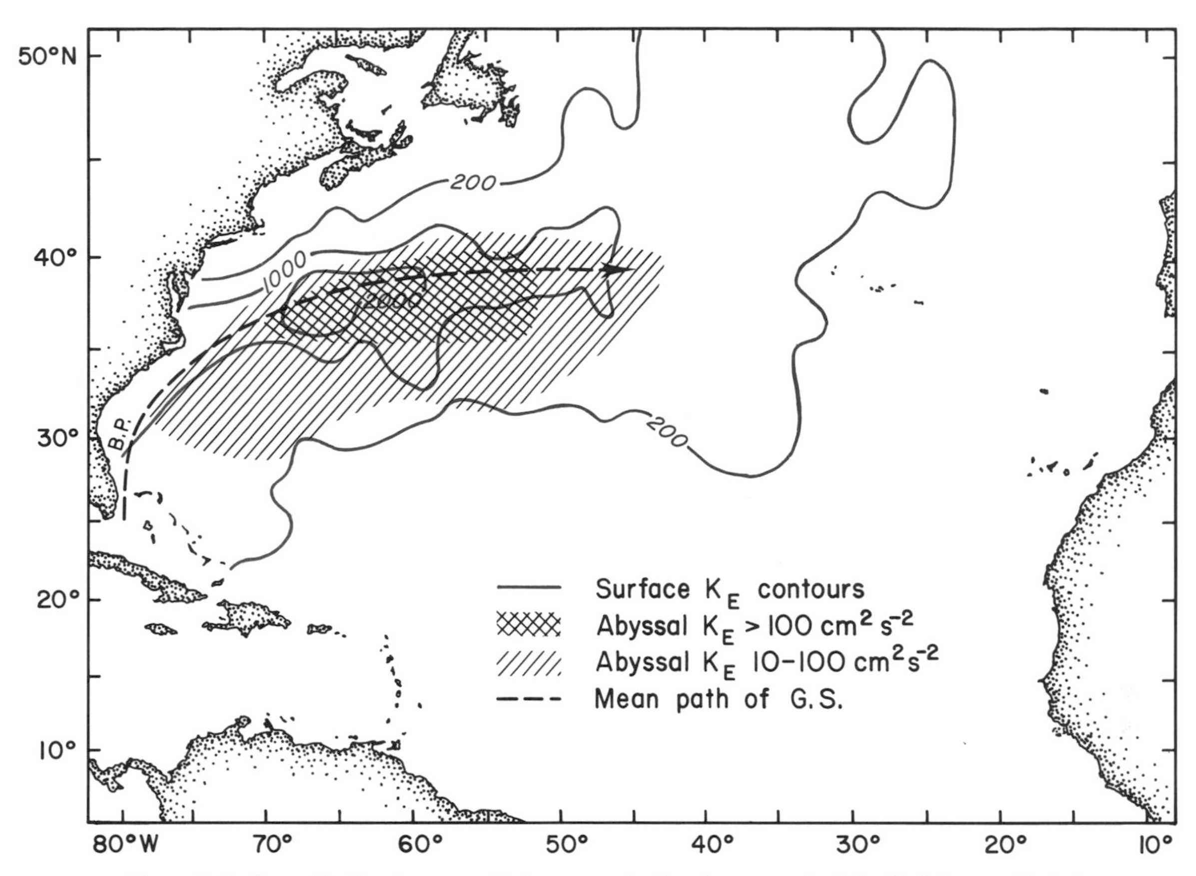

Figure 2. Surface eddy kinetic energy (K_E) contours (red) and mean path of the Gulf Stream (dashed arrow). Patterned areas show abyssal K_E values >100 cm^2/s^2 and 10-100 cm^2/s^2 centered under the Gulf Stream. After Richardson (1983a) and Schmitz (1984). B.P. is Blake Plateau.

1983; Richardson, 1983b; Schmitz, 1984). This abyssal variability, if added to an already substantial deep mean flow, such as in some areas of the deep western boundary current, can result in extremely strong currents alternating with very slow ones, or even periods of current reversal (Weatherly, 1984; Grant and others, 1985). Regions that are characterized by this combination of motions and that have an adequate supply of mud display extreme variability in rates of deposition or erosion, bedform development, and turbidity. These areas are termed "stormy" by Hollister and McCave (1984). In the North Atlantic, data from the Seasat altimeter and drifting buoys show that the highest surface K_E, and thus abyssal K_E, associated with the Gulf Stream extends as far north as the Nova Scotian continental rise, where it may intermittently augment high mean bottom flows directed to the southwest (Fig. 2). "Abyssal storms" are known to resuspend sediments and correspondingly cause very high turbidities in bottom water on the Nova Scotian rise (Hollister & McCave, 1984); records of high turbidity have also been taken elsewhere in this region of high abyssal K_E (Biscaye and Eittreim, 1977; Gardner and Sullivan, 1981).

Other Currents and Waves

Currents due to other causes are important locally and possibly regionally. Examples are unsteady currents found flowing up and down submarine canyons (Shepard and others, 1979), and internal waves breaking on the continental slope and in canyons (Wunsch and Hendry, 1972; Hotchkiss and Wunsch, 1982). These flows can be focussed by the morphology of submarine canyons, and they commonly are associated with sediment resuspension and generation of nepheloid layers (Johnson and Lonsdale, 1976; Dickson and McCave, 1982). There is some evidence that nepheloid layers detached from the seafloor are present along the canyon-incised continental slope of the eastern U.S. (Hunkins and others, 1983). These may represent sediment "pumped" from the canyons along isopycnal (density) surfaces, and they could be important in the supply of sediment to the basin.

Topographically intensified currents around seamounts and other seafloor irregularities also are important for local sediment distribution (Hogg, 1973; Gould and others, 1981). Suspended material is deposited in the low-velocity mixing zone downstream of seamounts. Local intensification of steady currents around seamounts also attenuates deposition, resulting in a "moat." The moats are asymmetrical, with reduced deposition on the left side (facing downstream in the northern hemisphere) and enhanced deposition on the right side (Roberts and others, 1974).

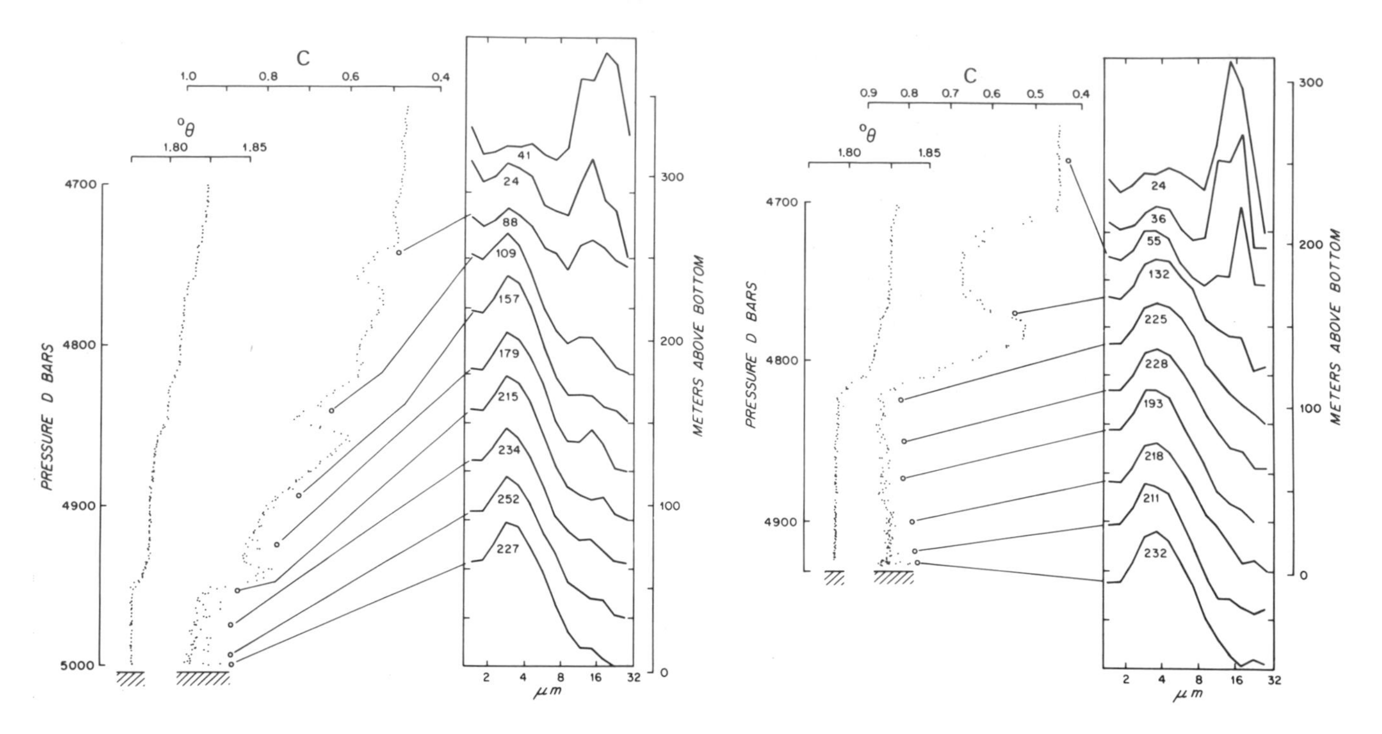

Figure 3. Two detailed nepheloid-layer profiles (Nova Scotian continental rise) of optical attenuation coefficient (C), with potential temperature (°Θ) and grain-size distribution of sampled suspended sediments (from McCave, 1983). Note that 1) inversions in the turbidity (C) profiles correspond to contacts between well mixed layers (uniform Θ) above the bottom, and 2) mean grain-size is coarser in the upper, weaker part of the nepheloid layer.

NEPHELOID LAYER

Origin and Behavior of Nepheloid Layers

The bottom nepheloid layer, defined as the zone of increased turbidity below a mid-water minimum, varies from a few hundred to about 2000 meters in thickness (e.g. Fig. 3). Early studies suggested that this distribution of suspended sediment represented a balance between vertical turbulent diffusion and particle settling, in the manner of a turbulent boundary layer. This analogy is not particularly good because the diffusivities required (20-800 cm^2/s; Eittreim and Ewing, 1972) are too high by one to two orders of magnitude. Even with a diffusivity of 10 cm^2/s, particle transfer up to 1000 m above the seafloor takes 30 years. Distinctly layered density structure of the lower water column in some regions having well developed nepheloid layers suggested to Armi (1978) that the best alternate to real cross-isopycnal diffusivity lay in vigorous boundary-layer mixing followed by detachment of mixed layers and their advection into the ocean interior. This explanation also seemed to explain fine-scale nepheloid-layer structure in records from high-resolution instruments (Fig. 3). These records often show "turbidity inversions" or upward increases in turbidity superimposed on a general decrease in turbidity. The inversions are seen to correspond to sections of the water column that are well mixed in temperature and salinity; consequently they may be "fossil" bottom mixed layers which have become detached and have been injected at an appropriate density level (Fig. 3). It has been proposed (McCave, 1983) that the whole deep nepheloid layer is dominated by this process, with layers of varying turbidity contributing their suspended load which slowly settles out and is also scavenged by larger sinking particles. Thick nepheloid layers were first explained this way by McCave and others (1980) who showed that a portion of the nepheloid layer in the South Iceland Basin originated as a bottom layer on Hatton Drift. Work by Armi and D'Asaro (1980) along Hatteras Abyssal Plain also showed detached and detaching turbid bottom layers contributing to the nepheloid layer. Studies of the Nova Scotian lower continental rise have reinforced this notion; turbidity steps and inversions within an "injection zone" above the bottom mixed layer are readily recognized there (Spinrad and Zaneveld, 1982; Weatherly and Kelley, 1981; McCave, 1983).

As detached mixed layers move away from their point of origin, they lose particles and decrease in turbidity. On average, points high in the water column receive layers that have come farther and have been detached from the bottom longer; thus they are less turbid than deeper layers. This causes the overall upward decrease in turbidity. Recent physical oceanographic calculations suggest that there is a small diffusivity of 1 to 4 cm^2/s across density surfaces (Whitehead and Worthington, 1982; Hogg and others, 1982). An apparent diffusivity can be calculated from the upward decrease in turbidity shown by the slope of light scatter-

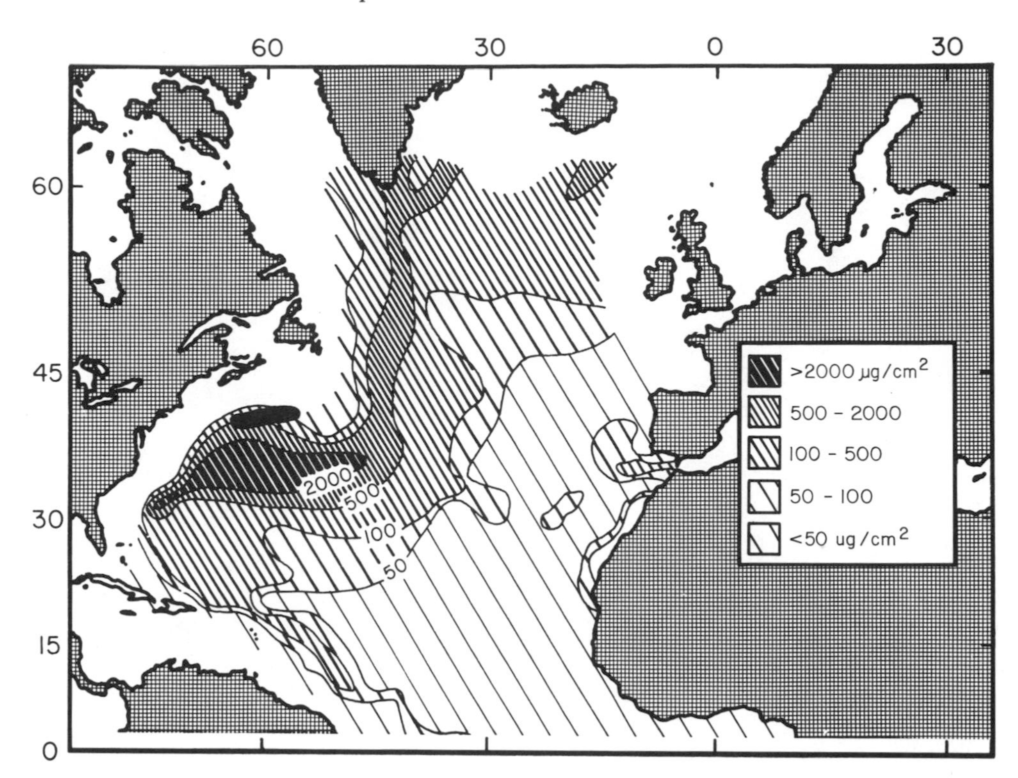

Figure 4. Net load (mass per unit area) of suspended sediment in the bottom nepheloid layer (adapted from Biscaye and Eittreim, 1977). The black area marks the extremely high loads (>25,000 $\mu g/cm^2$) found over the Nova Scotian continental rise (Amos and Gerard, 1979; Biscaye and others, 1980). Note the strong correlation between net load in the bottom nepheloid layer and the distribution of maximum abyssal K_E values in Figure 2.

ing profiles (Eittreim and Ewing, 1972). Most of the scattered-light signal comes from very small particles (<5 μm). If we make calculations using 2 μm particles, we obtain apparent diffusivities of ~10 cm^2/s. The difference between values of this order and the 1-4 cm^2/s cited above must be accounted for by lateral injection of turbid bottom layers.

Within active zones of sediment resuspension there is also rapid deposition. Chemical results show similarity of suspended and bed material in such areas (Bishop and Biscaye, 1982). In other areas where flows are more sluggish, the composition of the nepheloid layer and the bed may be dissimilar because the bed results largely from pelagic sediment flux, and the weak, 'old' nepheloid layer contributes little to the deposit. Initially, at high concentrations, there is rapid aggregation and deposition of particles; older suspensions at low concentration lose material slowly, and weak nepheloid layers may contain sediment that was resuspended years ago (McCave, 1983, 1984a). However, the *mean size* of particles in older, weaker nepheloid layer is coarser than in younger, more turbid layers. This is because coarse particles settling from the sea surface constitute a greater proportion of the total particle population (McCave, 1983). A weak nepheloid layer (e.g. 20 mg/m^3 concentration) contributes very little to bottom sediments (about 0.3 $kg/m^2/ky$ or 0.3 mm/ky, less than pelagic sedimentation rates). This contrasts with "stormy" regions where short-term sedimentation rates may be more than 10 mm per month, although *net* Holocene accumulation is on the order of 60 mm/ky (Hollister and McCave, 1984; McCave and others, 1984). The important conclusion is that one form of current-controlled sedimentation is rapid deposition from concentrated nepheloid layers, occurring where currents are strong. In contrast, very slow deposition occurs from widespread, weaker nepheloid layers which spread into quieter regions of the ocean; in those regions, pelagic sedimentation dominates (i.e. sediment settling from surface and mid-waters). Between these extremes, weak to moderate currents are guided by seafloor topography and control preferential deposition at intermediate rates from weakly to moderately developed nepheloid layers.

Overall Distribution

The distribution of suspended sediment in the bottom water of the North Atlantic shows a first-order correlation with the location of deep western boundary currents (Biscaye and Eittreim, 1977) (Fig. 4). The region of highest suspended-sediment content in the North American Basin occurs north of about 32°N. Recent studies of the Nova Scotian continental rise show that 1) this region of high sediment load extends northward to 41°N, and 2) the maximum (>500 g/m^2) occurs over the

TABLE 1. NORTH ATLANTIC SEDIMENT DRIFTS AND THEIR PRINCIPAL CHARACTERISTICS*

Drift	Symbol (Fig. 1)	Plastered	Detached	Separated	Presence Related to Local Sediment Supply	Presence Related to Two-Current Interaction
Feni (northern)	FD	X				
Feni (southern)	FD		X			?
Hatton	HAD	X				
Bjorn	BD	X			X	
Gardar	GD	X		?	X	
Gloria	GLD	X	?			?
Snorri	SD	X				
Eirik (eastern)	ED	X			X	
Eirik (western)	ED		X		X	
Sackville Spur	SAD		X		X	
Southeast Newfoundland Ridge	SND	X	X		X	?
Chesapeake	CD	X			X	?
Hatteras	HD			X	X	X
(Gulf Stream Outer Ridge)			X		?	?
Blake	BLD	X	X		X	X
Bahama	BAD		X	X	?	
Caicos	CAD	X	X	X		
Greater Antilles	GAD		X			
Northern Bermuda Rise	NBD	X				?

*Drifts are listed in order along the path of the abyssal boundary currents, from north to south. For exact drift locations, consult Plate 2 in slip case accompanying this volume.

Nova Scotian continental rise (Biscaye and others, 1980). The high-load zone has not only great areal extent but also large vertical extent, up to water depths of about 2000 m (Brewer and others, 1976). This high-load zone corresponds generally with the northern deep circulation cell over the Sohm Abyssal Plain and with the northern part of the cell over the Bermuda Rise (Fig. 1). Thus sediment suspended along the Nova Scotian margin may be a primary source for deposits on the northern Bermuda Rise. In addition, high abyssal K_E superimposed on strong, deep flows across the Bermuda Rise (Fig. 2) may intermittently cause strong sediment resuspension there.

LARGE-SCALE SEDIMENTARY TOPOGRAPHY

The North Atlantic contains many large sedimentary bodies having widths of tens of kilometers, lengths up to hundreds of kilometers and relief of 200 to about 2000 m (Johnson and Schneider, 1969). More than a dozen such "sediment drifts" have been named (Fig. 1, Plate 2, Table 1), and recognition and analysis of new ones continues to the present time (Laine and Hollister, 1981; Tucholke and Laine, 1983). The drifts are distributed mainly along the path of the Norwegian Sea overflow water (Fig. 1). Other features, such as the "Gulf Stream Drift" and the thick sediments on the northern Bermuda Rise, have an uncertain but probably mixed genesis related to interaction of the Gulf Stream with the deep circulation cells. Several smaller drifts around the margins of Labrador and the Grand Banks probably are related to the shallow (<2000 m) flow of the Labrador Current (Plate 2). In addition, it has recently been suggested that a deep north-flowing boundary current controls sedimentation at the foot of the Iberian continental margin in the eastern Atlantic (Roberts and Kidd, 1984). Thus a substantial part of the continental margins of the eastern and western North Atlantic is dominated by current-controlled redistribution of sediments.

The sedimentary facies associated with drifts and regions of current-controlled sedimentation along basin margins have been examined by McCave (1982; Hatton Drift) and Carter and Schafer (1983; Labrador rise) among others. They observed rippled sands under the axis of the currents and mud accumulation to the side. These zones correspond to sandy contourite and muddy contourite facies, respectively, recognized by Stow and Lovell (1979). "Contourite" was the term first applied by Hollister and Heezen (1972) to sediment deposited from contour-following currents in the deep ocean. Most contourites are dominated by fine silt and clay sizes and they are macroscopically homogeneous or burrow mottled. These "muddy contourites" contrast with thin sandy and coarse silt interbeds showing cross-lamination. The latter, "sandy contourites" were presented by Hollister and Heezen (1972) as typical contourites, but actually they appear to be minor components of sediment drifts, being most common under

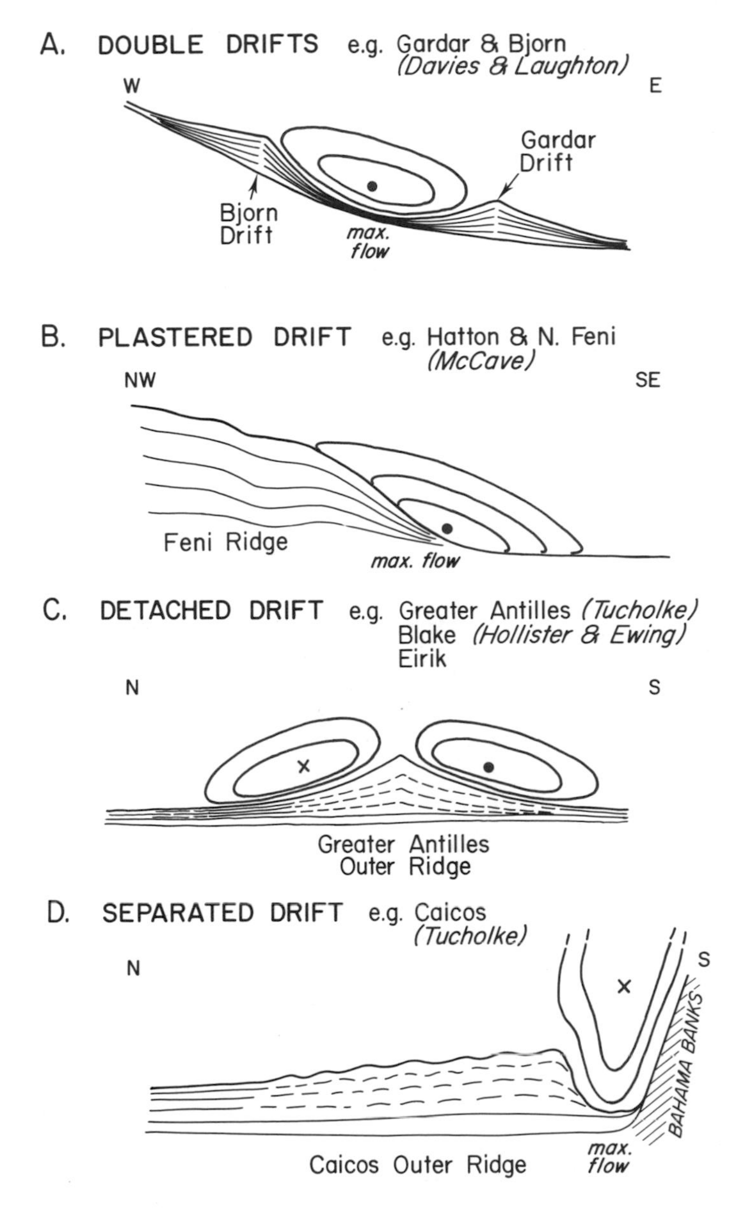

Figure 5. Summary of postulated relationships between current velocity and sediment-drift formation (Davies and Laughton, 1972; Tucholke and Ewing, 1974; McCave, 1982). Heavy red lines schematically show isotachs; dots represent current toward reader, x is current away from reader.

the axis of strongest flow and near submarine canyons, where coarser sediments are both deposited on levees by overflowing turbidity currents and are pirated from the turbidity currents by the contour-parallel flow. Muddy contourites generally contain 0-50% carbonate and 40-60% in clay (<4 μm) sizes. However, drifts far from sediment sources can be very fine grained. For example, the Greater Antilles Outer Ridge contains fine silt and mostly (70-90%) clay-size sediment, excluding coarser pelagic carbonate (Tucholke, 1973). These extremes in mean grain-size distribution of current-reworked and current-deposited sediments make recognition of contourites by grain-size criteria alone a nearly impossible task.

On the large scale, drift deposits are recognized by several features: 1) their surface and internal reflectors do not conform to deeper surfaces, 2) their thickness exceeds that of adjacent sedimentary cover, 3) bedding thickens at the drift axis and thins either at both drift margins or on one side of the drift where flow is fastest, 4) internal reflectors often are very weak, especially if the drift contains fine-grained sediments deposited far from a source area, and 5) drifts are commonly mantled by mud-waves and/or contain undulating reflectors that indicate earlier development of mud-waves (e.g., Jones and others, 1970; Le Pichon and others, 1971; Roberts and Kidd, 1979). The principal sediment drifts in the North Atlantic are summarized in Table 1 along with their general morphologic classification and factors affecting formation of the drifts.

Although we can recognize drifts with relative ease, it is much more difficult to explain why they have the particular morphologies and locations that we observe. The principal control on the *morphology* of the drift appears to be the form of the pre-existing seafloor, which controls and directs the various contour-following threads of the bottom current. The simplest case is that of a flow along the contours of a uniform slope. Generally, currents along steeper slopes will be topographically intensified and currents along gentler slopes will be weaker. Conventional wisdom is that sediment accumulates in comparatively tranquil zones on one or both sides of a higher-speed core that follows a "steep" slope; the depositional locus also is determined by the distribution of suspended matter in the flow (Davies and Laughton, 1972; Tucholke and Ewing, 1974; McCave, 1982) (Fig. 5, A and B). Rate of deposition is inversely related to flow speed, and the edges of deep boundary currents have both the low flow speeds and high suspended load necessary to maximize deposition rate. The core of the current has a stronger mean flow (thus slower deposition); it also has higher K_E (Schmitz, 1984) so that the seafloor there is intermittently affected by strong currents that cause erosion and thus reduce net accumulation. This axis/margin combination in a flow produces several forms of drifts. Along seafloor of relatively uniform slope, "plastered" drifts accrete at the current margin(s), or beneath the axis of main flow if the flow is weak enough and/or the slope gentle enough. The surface of these drifts blends smoothly and gently into the adjacent seafloor (Fig. 5B; see also Hatton Drift, Plate 5, Profile B), and double ridges may form (Fig. 5A). If the seafloor contains an abrupt slope change, the drift may be "separated" from the adjacent slope by a distinct zone of erosion or non-deposition (Caicos Outer Ridge, Fig. 5D; Plate 6, Profiles O, P).

If we depart from the simple-slope case and consider more complex seafloor morphology, then the interplay of the various threads of the current, which tend to flow along bathymetric contours, can significantly alter the growth pattern of the drift (Fig. 6). For example, deposition rate will increase significantly on the axis of a drift if the drift develops to a point where oppositely directed currents begin to flow along its two flanks

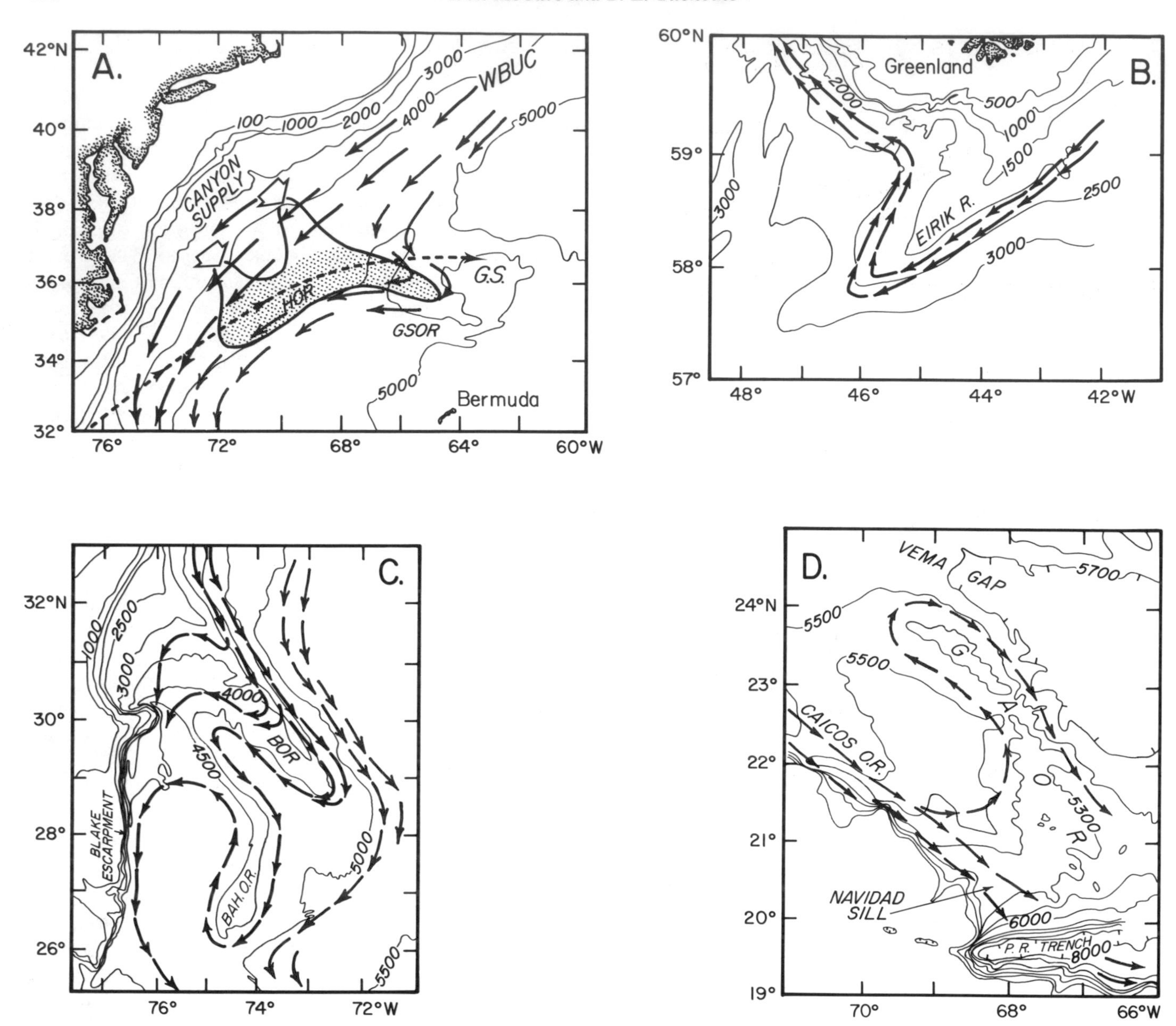

Figure 6. Patterns of bottom currents (and Gulf Stream - dashed) around North Atlantic sediment drifts. Red arrows show currents with a significant component of Antarctic Bottom Water. A) Hatteras (HOR) and Gulf Stream (GSOR) drifts (shaded) with inferred flow in the late Pliocene; the drift was subsequently cut by the deep western boundary current at 36°N, 67°W (see present 5000 m contour). The drift probably formed by interaction of the Gulf Stream and the bottom current; the eastern, "Gulf Stream drift" portion of the ridge was "detached." Modern contour-following bottom flow is shown around B) the "detached" Eirik drift, C) the Blake and Bahama Outer Ridges (both "detached"), and D) the Caicos Outer Ridge ("separated" segment shown) and Greater Antilles Outer Ridge ("detached"). All contours in corrected meters.

(Figs. 5, 6). The simplest example occurs where the current flows around a corner, building a "detached drift" that progrades in the direction of the initial flow. Eirik Ridge is representative of this situation (Fig. 6; Plate 6, Profile I).

A still more complex case occurs when two separate flows interact to form a shear zone where deposition can initiate the nucleation of a drift. The Blake Outer Ridge, south of Cape Hatteras, is the best example (Fig. 6; Plate 6, Profile N). This drift appears to have first formed at the locus where the Gulf Stream passed over the Western Boundary Undercurrent. Bryan (1970) modelled the interaction of the two flows and suggested that the over-riding Gulf Stream deflected the undercurrent and thus controlled the direction of the drift's axis. However, the supply of material swept off Blake Plateau to the region of the western end of the drift, especially during sea-level lowstands (Pinet and Popenoe, 1985), has also been an important factor in the drift's origin and development. Other drifts that possibly formed by the interaction of different current systems are the Hatteras-Gulf Stream Outer Ridge system (Fig. 6; Plate 6, Profile M) and possibly the drift on the northeastern Bermuda Rise.

Like *morphology* of drifts, *location* of drifts also is controlled to some degree by seafloor shape and incident currents. However, the location and quantity of sediment supply also are critical factors in drift location. The Bjorn and Gardar Drifts (Plate 5, Profiles A, B), for example, are immediately downstream from volcanic and glacial input from Iceland. If we follow the abyssal currents around to similar depths and seafloor slope/shape on the western side of the Reykjanes Ridge (Fig. 1), no ridges are observed there because of a lack of sediment supply. In the instance of the Blake Outer Ridge, this drift would not be nearly so well developed if sediment were not scoured from the adjacent Blake Plateau and carried to the head of the drift by the Gulf Stream.

The presence of significant sediment supply in itself does not guarantee formation of a drift, and in fact may have the opposite effect. For example, there is no drift on the Nova Scotian continental rise downstream from the Laurentian Fan, even though grain-size studies of the Holocene sediments there show a flow-axis/flow-margin distribution like that observed on developed drifts (Driscoll and others, 1985). While it is possible that the high-energy sediment redistribution on the Nova Scotian rise has precluded accumulation of a morphologically distinct body, it appears more likely that the continual sediment supply to the rise by downslope movement has prevented development of any drift morphology. The formation of drift morphology–a localized, elongate lens or ridge of sediment–requires millions of years at typical accumulation rates of tens of meters/m.y. Where this slow process of accumulation is disrupted by intermittent flux and deposition of material from upslope, the drift morphology does not develop; however, in the phases between downslope events, current reworking and deposition may establish the grain-size patterns characteristic of current-controlled deposition. The most recent phase on the U.S. Atlantic margin has been current-controlled reworking, but shallow seismic evidence shows that large seafloor areas are underlain by debris flows (Laine and others, this volume). This interplay of downslope supply and current-controlled redistribution, first discussed by Heezen and others (1966), is an important mechanism of continental rise construction.

If we rank the large-scale depositional products of abyssal currents in terms of the interaction between downslope and along-slope processes, then the continental rise off eastern North America can be considered an end member, wherein downslope processes play an important if not locally dominant role. The extreme opposite end member is the seismically non-laminated drift which contains terrigenous sediments of only very fine grain size and which is deposited exclusively from currents far from a sediment source (e.g., Greater Antilles Outer Ridge, Plate 6, Profile P).

SMALLER-SCALE SEDIMENTARY TOPOGRAPHY

From drifts, one descends one to two orders of magnitude in scale to arrive at mud waves, the largest of the superimposed bedforms. Progressively smaller forms are furrows, transverse and longitudinal ripples, mounds and crags with tails, and other small current-produced features observed in bottom photographs. We discuss them here in order of decreasing size.

Mud Waves

Mud waves (or sediment waves) are regular undulations of the sediment surface with wavelengths of about 0.5 to 3 km and heights of 10 to 100 m (see Laine and others, this volume; Plate 7). Most mud waves are very nearly symmetrical, but they have subsurface layering that indicates migration. Observed migration usually is upslope and upcurrent, but instances of downcurrent migration have been observed (Roberts and Kidd, 1979).

Under a simple flow, one expects maximum shear stress on the upstream face of a wavy bedform and lower shear stress, with greater deposition rate, on the downstream side. This would give downstream migration of the wave. Commonly observed upstream migration has suggested to some (e.g. Fox and others, 1968) that mud waves are analogous to fluvial antidunes developed under a flow with Froude number $\gtrsim 1$. In fact, for some deep-sea conditions over mud waves, the relevant densimetric Froude number [$Fr = (U^2 \rho/gh\Delta\rho)^{1/2}$] probably is $\gtrsim 1$, given reasonable values of flow speed $U = 0.05$ to 0.10 m/s, density $\rho = 1028$ kg/m^3, density difference between lower and upper layer $\Delta\rho = 10^{-2}$ to 10^{-3}, flow thickness $h = 100$ to 1000 m, and $g = 9.81$ m/s^2. In one particular case, Kolla and others (1980) estimated the Froude number on the interface of the benthic thermocline to be 0.91. In this case U was 0.08 to 0.1 m/s, which is comparable to the phase velocity of an internal wave on the interface; thus standing internal waves appear possible.

An alternative suggested by Flood (1978) is that the mud waves form under internal lee-waves initially triggered by an upstream topographic disturbance, without the necessity for a critical Froude number. The frequency of an internal wave in a stratified fluid is related to a vertical wave number $2\pi/M$, where M is four times the height at which a 90° phase shift occurs between the internal wave and the bed. Through analysis of temperature data over mud waves, Flood (1978) showed that such an upstream phase shift does occur, that the implicit internal-wave phase velocity is 0.05m/s, and that this is very similar to the measured flow velocity, as required for the internal wave to be stationary. He also showed that the thermal structure of the water column over mud waves closely matches numerical predictions (Huppert, 1968) of the stream-lines of lee waves that are formed by flow over a semicircular ridge. The flow pattern over the waves has widely spaced streamlines (small velocity

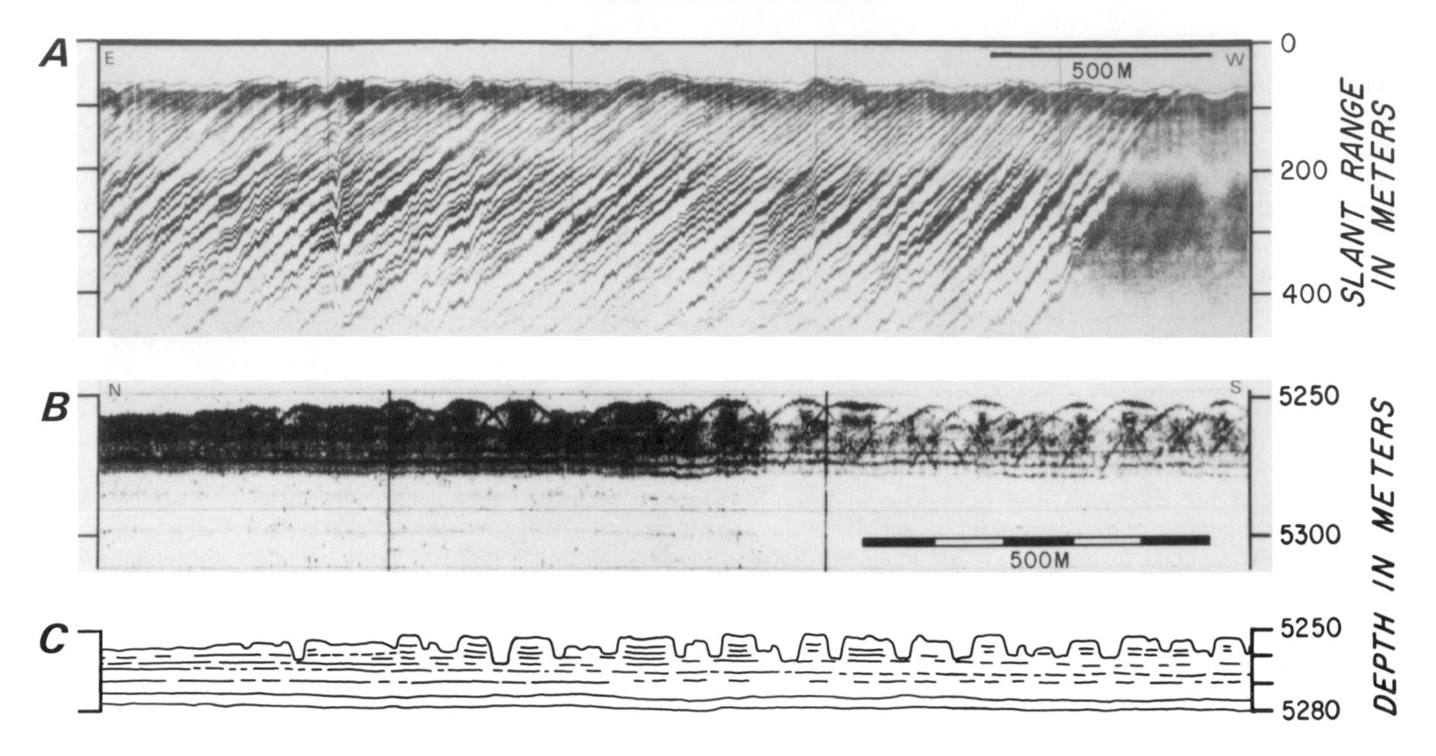

Figure 7. A) Deep-towed side-scan sonar record of furrows in muddy sediment on the southwestern Bermuda Rise (seafloor profile at top, far field at bottom) near 29°50′N, 68°52′W. Rolling and yawing of the towed instrument is responsible for the zig-zag pattern of the echoes in the far field. Current direction is parallel to furrows, from upper right to lower left (toward northeast). B) Near-bottom, 4-kHz echo profile across furrows at a location 8 km northeast of side-scan record in A. Note the characteristic hyperbolic reflections from the corners of the furrows. C) Line drawing of the seafloor profile in B, based on a near-bottom, narrow-beam 125-kHz echo-sounder record. Figure adapted from Embley and others (1980).

gradient and shear stress [= high deposition rate]) on the upstream slope, and the opposite on the downstream slope. This would give upstream mud-wave migration and is the most plausible and best documented explanation currently available for development of mud waves.

Furrows

Sedimentary furrows are long, narrow seafloor depressions that parallel the mean current direction (Fig. 7), and they are formed mainly in cohesive sediments. They were first discovered using side-scan sonar in shallow water (Dyer, 1970) and shortly after that they were observed in the deep Pacific and North Atlantic basins (Hollister and others, 1974). Furrows typically are in the range of 1 to 10 km long and they occur on seafloor slopes ranging from flat to 1:20. Detailed investigations of furrow morphology and associated sediments have been carried out by Flood (1981a) in shallow water and a general discussion is provided by Flood (1983).

There are two main types of furrows in the Atlantic. Relatively shallow furrows are 1–5 m deep and 1–20 m wide; they are spaced at 5 to 15 times furrow width and may be symmetrical or asymmetrical. Deep (5 to 30 m) furrows, spaced at about 2 times their width of 10 to 150 m, are known to occur on the Bahama Outer Ridge. Carbonates may also be furrowed, the forms being symmetrical but with indistinct troughs and spacing 5 to 15 times their width. At least some of the larger furrows are clearly erosional because they contain outcrops of the adjacent, horizontally bedded sediments in their walls (Hollister and others, 1974). However, smaller furrows often show evidence of continuing net accumulation both in core samples and acoustic profiles (Tucholke, 1979; Flood and Hollister, 1980; McCave and others, 1982). In plan view the furrows parallel one another (Fig. 7), but some join downstream in "tuning fork" junctions.

Flood (1983) explained furrow formation by appealing to the facts that slightly coarser sediments are found in available samples of furrow troughs and that helical patterns of flow tend to concentrate coarser material into strips, for example sand ribbons (Allen, 1968; McLean, 1981). Transport of concentrations of coarser material (e.g. foraminiferal sand and Sargassum weed in the deep sea) can abrade mud beds (Allen, 1971). Flood considers that furrows are initiated by corrasion under the localized coarse load; they may develop by continued erosion, thus giving wide deep furrows, or by net sedimentation giving narrow shallow furrows. Very rapid deposition would tend to fill them, so there must be a critical balance between deposition and relative scour for furrow maintenance. Furrows in Pacific calcareous sediments may also be initiated by corrasion, but it is possible that differential solution related to secondary circulation also plays a role (see Allen, 1971, p. 272). The principal difficulty with the corrasion hypothesis is that the initial stage of formation (ribbon-like concentrations of coarse material) has yet to be observed in the deep sea. Other mechanisms, including initial formation by differential *deposition,* are possible.

Longitudinal Ripples

Longitudinal ripples are analogous to furrows in that they are elongate features parallel to the depositing flow, and they probably have helical secondary circulation involved in their formation (Fig. 8). They are 5 to 15 cm high, 0.25 to 1 m wide and up to 10 m long, and have a generally symmetrical cross-section with sides slightly concave upwards. Two ripples sampled to date have dominantly muddy composition with up to 20% foraminiferal sand (Flood, 1981b; McCave and others, 1984). In many cases the ripples have a mound of biological origin at the upstream end. Surface markings on some ripples demonstrate the action of oblique flows with flow separation and a zone of helical reversed flow on the lee side (Tucholke, 1982).

These combined features were discussed by Tucholke (1982). He proposed that the ripples grow like oblique transverse ripples under the action of flows that varied in direction but had a mean direction coincident with the trend of the bedform. This mode of growth would be slow enough (months to years) that the ripples would indicate a long-term, mean flow direction. On the Nova Scotian rise this trend was found to be closely parallel to the bathymetric contours and the mean current vectors. Flood (1981b), however, mapped variation in ripple orientation in the Blake-Bahama Basin and considered it indicative of formation over a single short-lived high-velocity event. The longitudinal ripples would thus be parallel to "the latest, strongest current, not necessarily to the mean current direction" (Flood, 1981b).

Recently McCave and others (1984) sampled a longitudinal ripple obtained in a box core. Th-234 dating shows that the bedform was younger than about 3 months and was rapidly deposited at an average of ~1.5 cm per month. X-radiographs of internal sedimentary structures, including thin lenses of foraminiferal sand, indicate several episodes of growth. The bulk grain-size is principally fine silt (2-20 μm) with ~35% clay and <10% sand. McCave and others (1984) found no pellets although pellets are seen on the seafloor in photographs. This suggested to McCave and others that the longitudinal ripples formed by deposition from suspension and that this occurred in a few episodes of very rapid deposition following deep-sea storms. However, the possibility of pelletisation and the observation of sand lenses in the ripple means that bed-load transport is feasible, and the ripples could grow beneath variably directed currents by some combination of bed-load transport and deposition from suspension in the manner suggested by Tucholke (1982). The time scale also is allowable, especially if the box core sample was taken toward the downstream end of the ripple. Head erosion and tail growth of a longitudinal ripple (Tucholke, 1982) would create a significant age gradient along the length of the bedform. Longitudinal ripples should be sampled along their length to resolve these contrasting hypotheses of catastrophist versus gradualist views.

Smaller Features

Numerous smaller, current-controlled bedforms are revealed by deep-sea photography. Because we can show but a few of these here (Fig. 8), the reader is advised to spend a pleasant hour or two perusing *The Face of the Deep* (Heezen and Hollister, 1971), particularly Chapter 9. The photographed features of the sea bed can be arranged in a sequence indicative of increasing flow speed, as was first shown by Heezen and Hollister (1971, p. 357). A recent, more detailed appraisal of relative current speed indicated by photographed features has been made by Hollister and McCave (1984), and by Tucholke and others (1985), the latter shown here as Table 2. The progression is from tranquil seafloor (biological mounds, tracks, trails, and feces), through increasing overprinting by current effects, to features showing clear evidence of erosion. Biologic activity is almost ubiquitous so that a smoothed surface is indicative of an appreciable current, sufficient to remove the surface effects of biota. The table is essentially uncalibrated, but Hollister and McCave (1984) do observe that tranquil seafloor is associated with currents that are less than 5 cm/s, and that the other end of the scale correlates with speeds >40 cm/s. It is not presently possible to interpolate speeds for intervening classes with any confidence.

DISCUSSION

Resuspension of deep-sea sediments can be produced by currents of about 20 cm/s [about 10 cm/s near (1 m) bottom, (Grant and others, 1985)], but significant removal of material requires much faster currents. Deposition of components with a settling velocity $\geq$0.01 cm/s can be accomplished under currents $\leq$20 cm/s (McCave, 1984b). Most of the fine, current-deposited material in sediment drifts, however, probably has settling velocities less than 0.004 cm/s, requiring currents less than 15 cm/s for deposition. Measured currents in the vicinity of sediment drifts fall mainly between 5 and 15 cm/s but there are significant excursions above and below that range. Thus deposition is prevalent, but it is punctuated by brief episodes of erosion. Because the rate of deposition is controlled by the amount by which the shear stress is less than the deposition threshold (McCave and Swift, 1976), areas on the edges of current systems have the optimum combination of suspended-sediment concentration and low shear stress that yield local maxima of net accumulation. The additional necessity for a suspended-sediment load may also mean that deposition is maximised downstream from input points. Areas which do not have sediment drifts thus may *lack* at least one of: a deep current system, stable location of the current system, or a supply of sediment. Or, there may be a too vigorous

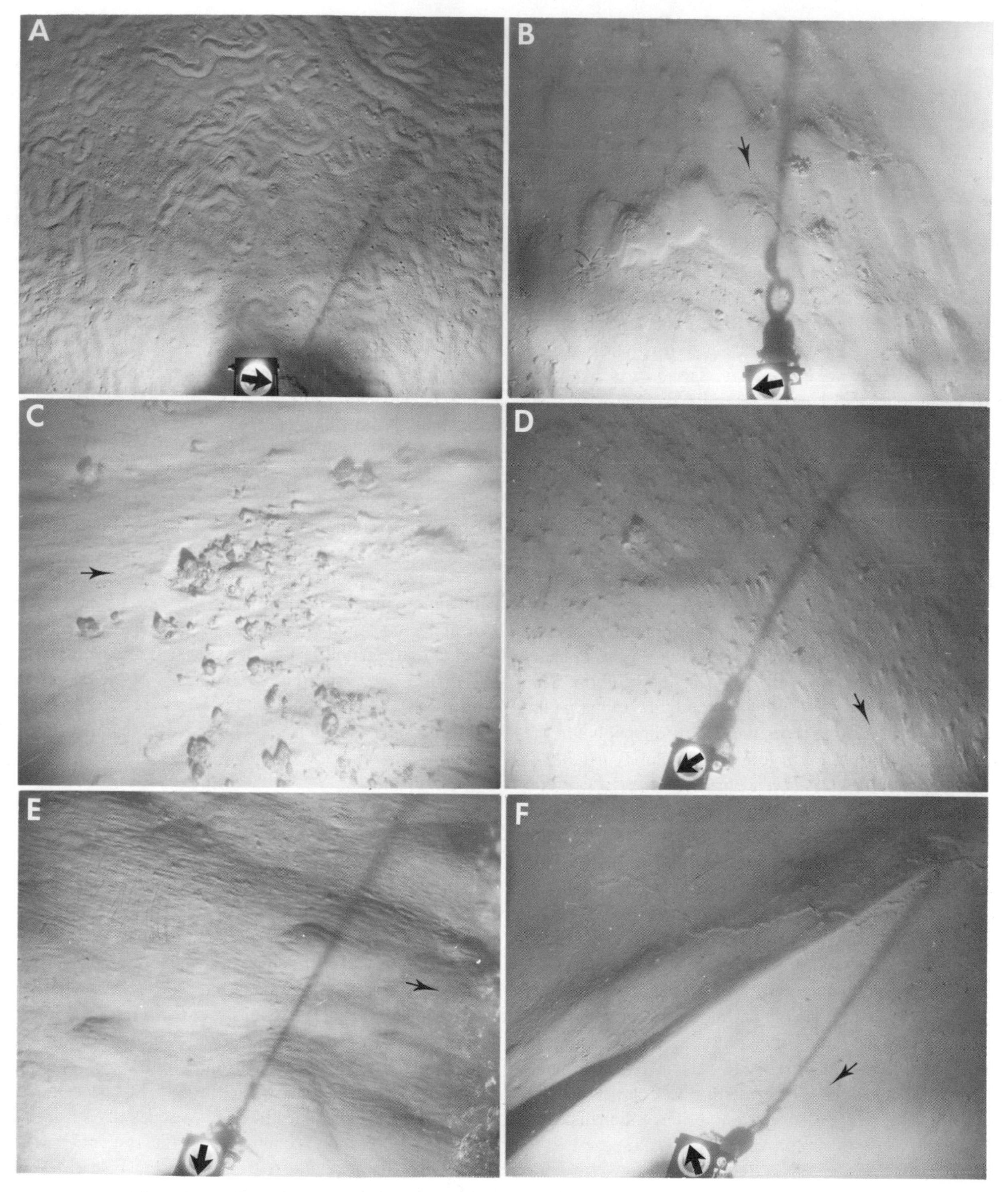

Figure 8. Sea-floor photographs of smaller bedforms. Small arrows show current direction. Bold arrows on compass give true north. A) Tranquil, biologically tracked seafloor (Nares Abyssal Plain, 5828 m). B) Biologic mud-mounds eroded by moderate current; note slip faces formed on downstream sides of mounds (Nova Scotian continental rise, 4244 m). C) Current scour around glacial-erratic dropstones; depositional tails are formed in lee of some pebbles (U.S. continental rise, 4620 m). D) Smooth, striated seafloor with crag-and-tail bedforms (Nova Scotian rise, 4485 m). E) Scoured, plucked, mound-and-tail bedforms with seafloor lineated by strong bottom currents (Nova Scotian rise, 4690 m). F) Longitudinal ripple and current-smoothed seafloor; note fresh (recent) animal tracks (Nova Scotian rise, 4615 m).

TABLE 2. INDICATORS OF MAXIMUM CURRENT STRENGTH FROM SEAFLOOR PHOTOGRAPHS

Currents	Degree of Bedform Development	Auxiliary Data
Tranquil	1. Flocs of organic/mineral debris; undisturbed animal tracks.	Clear bottom water
↓	2. Subtle smoothing; rare flocs.	↓
	3. Weak lineations; appearance of tool marks (poorly developed); flocs of debris in lee of obstacles.	
	4. Small crags with tails of mineral debris.	
	5. Appearance of barchan ripples of unconsolidated silt/sand; crags and tails; weak scour crescents in front of obstacles.	
	6. Mounds and tails; longitudinal ripples; common cornices; crags and tails widespread.	
	7. Well developed crags and tails very common; well developed scour crescents around obstacles; some erosional plucking of seabed and of existing bedforms.	
Strong	8. Strong and widespread development of erosional plucking, tool marks, and scour around obstacles; cohesive sediment exposed and unconsolidated silt/sand absent except in protected areas.	Very cloudy bottom water

Note: Scale not linear; the lowest speeds (scale #1) are <5 cm/s whereas the fastest (scale #8) are probably >40 cm/s.

deep circulation (frequent peak speeds over 20 cm/s) or a major, laterally injected sediment supply; this supply can be so large that it relegates current effects to production of thin reworked interbeds and prevents development of a distinctive drift morphology.

It is possible that the existence of larger bedforms like mud waves and furrows may not tell us much about present depositional conditions. Many furrows and mud waves could have originated long ago, perhaps millions of years in the case of mud waves. Although the forms persist, it is not certain that they presently have much active accretion. Sedimentological studies of carefully located samples across these features need to be done to answer the question. The smaller photographed features are ephemeral. Some "stormy" areas may fall at the tranquil end of the scale in Table 2 on a given occasion but at the other end several months later. With this sort of variability in local deposition rate it is difficult to evaluate the processes of *net* accumulation which leads over millions of years to the substantial features we call drifts. Ultimately the origin and history of drifts must be assessed through detailed sedimentological examination of deep-sea drill-cores. What we can provide from studies of modern sedimentation is a framework for that assessment. The historical work is well under way with drill holes on the Blake, Hatteras, Feni, Hatton, Bjorn and Gardar Drifts.

REFERENCES

Allen, J.R.L.
1968 : Current Ripples: their relation to patterns of water and sediment motion: North Holland Publishing Co., Amsterdam, 433 p.
1971 : Transverse erosional marks of mud and rock: their physical basis and geological significance: Sedimentary Geology, v. 5, p. 167–385.

Amos, A. F. and Gerard, R. D.
1979 : Anomalous bottom water south of the Grand Banks suggests turbidity current activity: Science, v. 203, p. 894–897.

Amos, A. F., Gordon, A. L. and Schneider, E. D.
1971 : Water masses and circulation patterns in the region of the Blake-Bahama Outer Ridge: Deep-Sea Research, v. 18, p. 145–165.

Armi, L.
1978 : Some evidence for boundary mixing in the deep ocean: Journal of Geophysical Research, v. 83, p. 1971–1979.

Armi, L. and D' Asaro, E.
1980 : Flow structures of the benthic ocean: Journal of Geophysical Research, v. 85, p. 469.

Barrett, J. R., Jr.
1965 : Subsurface currents off Cape Hatteras: Deep-Sea Research, v. 12, p. 173–184.

Bird, A. A., Weatherly, G. L. and Wimbush, M.
1982 : A study of the bottom boundary layer over the Eastward Scarp of the Bermuda Rise: Journal of Geophysical Research, v. 87, p. 7944–7954.

Biscaye, P. E. and Eittreim, S. L.
1974 : Variations in benthic boundary layer phenomena: nepheloid layer in the North American Basin; in Suspended Solids in Water, ed. R. J. Gibbs; Plenum Press, New York, p. 227–260.

Biscaye, P. E. and Eittreim, S. L.
1977 : Suspended particulate loads and transports in the nepheloid layer of the abyssal Atlantic Ocean: Marine Geology, v. 23, p. 155–172.

Biscaye, P. E., Gardner, W. D., Zaneveld, J.R.V., Pak, H. and Tucholke, B.
1980 : Nephels! Have we got Nephels!: EOS Transactions American Geophysical Union, v. 61, p. 1014.
Bishop, J.K.B. and Biscaye, P. E.
1982 : Chemical characterization of individual particles from the nepheloid layer in the Atlantic Ocean: Earth and Planetary Science Letters, v. 58, p. 265–275.
Brewer, P. G., Spencer, D. W., Biscaye, P. E., Hanley, A., Sachs, P. L., Smith, C. L., Kadar, S. and Fredericks, J.
1976 : The distribution of particulate matter in the Atlantic Ocean: Earth and Planetary Science Letters, v. 32, p. 393–402.
Bryan, G. M.
1970 : Hydrodynamic model of the Blake Outer Ridge: Journal of Geophysical Research, v. 75, p. 4530–4537.
Carter, L. and Schafer, C. T.
1983 : Interaction of the Western Boundary Undercurrent with the continental margin off Newfoundland: Sedimentology, v. 30, p. 751–768.
Carter, L., Schafer, C. T. and Rashid, M. A.
1979 : Observations on depositional environments and benthos of the continental slope and rise, east of Newfoundland: Canadian Journal of Earth Sciences, v. 16, p. 831–846.
Davies, T. A. and Laughton, A. S.
1972 : Sedimentary processes in the North Atlantic; in Initial Reports of the Deep Sea Drilling Project, v. 12, eds.; A. S. Laughton and W. A. Berggren; U.S. Government Printing Office, Washington, D.C., p. 905–934.
Dickson, R. R.
1983 : Global summaries and intercomparisons: flow statistics from long-term current meter moorings; in Eddies in Marine Science, ed. A. R. Robinson; Springer-Verlag, New York, p. 278–353.
Dickson, R. R., Gould, W. J., Müller, T. J. and Maillard, C.
1985 : Estimates of the mean circulation in the deep (>2000 m) layer of the eastern North Atlantic; in Progress in Oceanography, v. 14, eds. J. Crease et al.; Pergamon, New York, p. 103–127.
Dickson, R. R., Gurbutt, P. A. and Medler, K. J.
1980 : Long-term water movements in the southern trough of the Charlie-Gibbs Fracture Zone: Journal of Marine Research, v. 38, p. 571–583.
Dickson, R. R. and McCave, I. N.
1982 : Properties of nepheloid layers on the upper slope of west Porcupine Bank: International Council for the Exploration of the Sea, Committee Memorandum C. M. 1982/C:2 Hydrography Committee, 9 p.
Driscoll, M. L., Tucholke, B. E., and McCave, I. N.
1985 : Seafloor zonation in sediment texture on the Nova Scotian lower continental rise: Marine Geology, v. 66, p. 25–41.
Dyer, K. R.
1970 : Linear erosional furrows in Southampton water: Nature, v. 225, p. 56–58.
Egloff, J. and Johnson, G. L.
1979 : Erosional and depositional structures of the southwest Iceland insular margin; in Geological and Geophysical Investigations of Continental Margins, eds. J. S. Watkins, L. Montadert and J. W. Dickerson; American Association of Petroleum Geologists Memoir 29, p. 43–63.
Eittreim, S. L. and Ewing, M.
1972 : Suspended particulate matter in the deep waters of the North American Basin; in Studies in Physical Oceanography, ed. A. L. Gordon; Gordon and Breach, New York, p. 123–167.
Ellett, D. J. and Martin, J.H.A.
1973 : The physical and chemical oceanography of the Rockall Channel: Deep-Sea Research, v. 20, p. 585–625.
Ellett, D. J. and Roberts, D. G.
1973 : The overflow of Norwegian Sea deep water across the Wyville-Thomson Ridge: Deep-Sea Research, v. 20, p. 819–835.
Embley, R. W., Hoose, P. J., Lonsdale, P., Mayer, L. and Tucholke, B. E.
1980 : Furrowed mud waves on the western Bermuda Rise: Geological Society of America Bulletin, v. 91, p. 731–740.
Ewing, J. I. and Hollister, C. D.
1972 : Regional aspects of deep sea drilling in the western North Atlantic; in Initial Reports of the Deep Sea Drilling Project, v. 11, eds. C. D. Hollister and J. I. Ewing; U.S. Government Printing Office, Washington, D.C., p. 951–973.
Flood, R. D.
1978 : Studies of deep sea sedimentary microtopography in the North Atlantic Ocean: Ph.D. Thesis, Massachusetts Institute of Technology/Woods Hole Oceanographic Institution, Woods Hole Oceanographic Institution Report WHOI-78-64, 395 p.
1981a: Distribution, morphology, and origin of sedimentary furrows in cohesive sediments, Southampton Water: Sedimentology, v. 28, p. 511–529.
1981b: Longitudinal triangular ripples in the Blake-Bahama Basin: Marine Geology, v. 39, p. M13–M20.
1983 : Classification of sedimentary furrows and a model for furrow initiation and evolution: Geological Society of America Bulletin, v. 94, p. 630–639.
Flood, R. D. and Hollister, C. D.
1980 : Submersible studies of deep-sea furrows and transverse ripples in cohesive sediments: Marine Geology, v. 36, p. M1–M9.
Fox, P. J., Heezen, B. C. and Harian, A. M.
1968 : Abyssal antidunes: Nature, v. 220, p. 470–472.
Gardner, W. D. and Sullivan, L. G.
1981 : Benthic storms: temporal variability in a deep ocean nepheloid layer: Science, v. 213, p. 329–331.
Gould, W. J., Hendry, R. and Huppert, H. E.
1981 : An abyssal topographic experiment: Deep-Sea Research, v. 28A, p. 409–440.
Grant, W. D., Williams, A. J. and Gross, T. F.
1985 : A description of the bottom boundary layer at the HEBBLE site: Low frequency forcing, bottom stress and temperature structure: Marine Geology, v. 66, p. 219–241.
Harvey, J. G.
1980 : Deep and bottom water in the Charlie-Gibbs Fracture Zone: Journal of Marine Research, v. 38, p. 173–182.
Heezen, B. C. and Hollister, C. D.
1971 : The Face of the Deep: Oxford University Press, London, 659 p.
Heezen, B. C., Hollister, C. D. and Ruddiman, W. F.
1966 : Shaping of the continental rise by deep geostrophic contour currents: Science, v. 152, p. 502–508.
Hogg, N. G.
1973 : On the stratified Taylor column: Journal of Fluid Mechanics, v. 58, p. 517–537.
1983 : A note on the deep circulation of the western North Atlantic; its nature and causes: Deep-Sea Research, v. 30, p. 945–961.
Hogg, N., Biscaye, P., Gardner, W. and Schmitz, W. J.
1982 : On the transport and modification of Antarctic bottom water in the Vema Channel: Journal of Marine Research, v. 40 (supplement), p. 231–263.
Hollister, C. D., Flood, R. D., Johnson, D. A., Lonsdale, P. and Southard, J. B.
1974 : Abyssal furrows and hyperbolic echo traces on the Bahama Outer Ridge: Geology, v. 2, p. 395–400.
Hollister, C. D., Flood, R. and McCave, I. N.
1978 : Plastering and decorating in the North Atlantic: Oceanus, v. 21, no. 4, p. 5–13.
Hollister, C. D. and Heezen, B. C.
1972 : Geologic effects of ocean bottom currents: western North Atlantic; in Studies in Physical Oceanography, v. 2, ed. A. L. Gordon; Gordon and Breach, New York, p. 37–66.
Hollister, C. D. and McCave, I. N.
1984 : Sedimentation under deep-sea storms: Nature, v. 309, p. 220–225.
Hotchkiss, F. S. and Wunsch, C.
1982 : Internal waves in Hudson Canyon with possible geological implica-

tions: Deep-Sea Research, v. 29, p. 415–442.
Hunkins, K., Gardner, W. D., Stepien, J. C., Hecker, B., Logan, D. T. and Grandanillas, F. E.
1983 : Fourth interim report for the canyon and slope processes study: Bureau of Land Management Contract MMSAA851-CTO-59, 405 pp.
Huppert, H.
1968 : Appendix to J. W. Miles, Lee waves in a stratified flow. Part 2. Semi-circular obstacle: Journal of Fluid Mechanics, v. 33, p. 803–814.
Jenkins, W. J. and Rhines, P. B.
1980 : Tritium in the deep North Atlantic Ocean: Nature, v. 286, p. 877–880
Johnson, D. A. and Lonsdale, P. F.
1976 : Erosion and sedimentation around Mytilus Seamount, New England continental rise: Deep-Sea Research, v. 23, p. 429–440.
Johnson, G. L. and Schneider, E. D.
1969 : Depositional ridges in the North Atlantic: Earth and Planetary Science Letters, v. 6, p. 416–422.
Jones, E.J.W., Ewing, M., Ewing, J. and Eittreim, S. L.
1970 : Influence of Norwegian Sea overflow water on sedimentation in the northern North Atlantic and Labrador Sea: Journal of Geophysical Research, v. 75, p. 1655–1680.
Kolla, V., Eittreim, S., Sullivan, L., Kostecki, J. A. and Burckle, L. H.
1980 : Current-controlled abyssal microtopography and sedimentation in Mozambique Basin, southwest Indian Ocean: Marine Geology, v. 34, p. 171–206.
Laine, E. P. and Hollister, C. D.
1981 : Geological effects of the Gulf Stream system on the northern Bermuda Rise: Marine Geology, v. 39, p. 227–310.
Le Pichon, X., Eittreim, S. and Ewing, J.
1971 : A sedimentary channel along Gibbs Fracture Zone: Journal of Geophysical Research, v. 76, p. 2891–2896.
Lonsdale, P. and Hollister, C. D.
1979 : A near-bottom traverse of Rockall Trough: hydrographic and geologic inferences: Oceanologica Acta, v. 2, p. 91–105.
Mann, C. R., Coote, A. R. and Garner, D. M.
1973 : The meridional distribution of silicate in the western Atlantic Ocean: Deep-Sea Research, v. 20, p. 791–801.
Mantyla, A. W. and Reid, J. L.
1983 : Abyssal characteristics of the world ocean waters: Deep-Sea Research, v. 30, p. 805–833.
McCave, I. N.
1982 : Erosion and deposition by currents on submarine slopes: Institute Geologie Bassin d'Aquitaine Bulletin, v. 31, p. 47–55.
1983 : Particulate size spectra, behavior and origin of nepheloid layers over the Nova Scotian continental rise: Journal of Geophysical Research, v. 88, p. 7647–7666.
1984a: Size spectra and aggregation of suspended particles in the deep ocean: Deep-Sea Research, v. 31, p. 329–352.
1984b: Erosion, transport and deposition of fine-grained marine sediments; in Fine Grained Sediments: Deep Sea Processes and Facies; eds. D.A.V. Stow and D.J.W. Piper; Geological Society London Special Publication 15, p. 35–69.
McCave, I. N., Hollister, C. D., DeMaster, D. J., Nittouer, C. A., Silva, A. J. and Yingst, J. Y.
1984 : Analysis of a longitudinal ripple: Marine Geology, v. 58, p. 275–286.
McCave, I. N., Hollister, C. D., Laine, E. P., Lonsdale, P. F. and Richardson, M. J.
1982 : Erosion and deposition on the eastern margin of the Bermuda Rise in the Late Quaternary: Deep-Sea Research, v. 29, p. 535–561.
McCave, I. N., Lonsdale, P. F., Hollister, C. D. and Gardner, W. D.
1980 : Sediment transport over the Hatton and Gardar contourite drifts: Journal of Sedimentary Petrology, v. 50, p. 1049–1062.
McCave, I. N. and Swift, S. A.
1976 : A physical model for the rate of deposition of fine-grained sediments in the deep sea: Geological Society of America Bulletin, v. 87, p. 541–546.
McLean, S. L.
1981 : The role of non-uniform roughness in the formation of sand ribbons: Marine Geology, v. 42, p. 49–74.
Miller, K. G. and Tucholke, B. E.
1983 : Development of Cenozoic abyssal circulation south of the Greenland-Scotland Ridge; in Structure and Development of the Greenland-Scotland Ridge, eds. M.H.P. Bott, S. Saxov, M. Talwani and J. Thiede; Plenum Press, New York, p. 549–589.
Pinet, P. R. and Popenoe, P.
1985 : A scenario of Mesozoic-Cenozoic ocean circulation over the Blake Plateau and its environs: Geological Society of America Bulletin, v. 96, p. 618–626.
Pinet, P. R., Popenoe, P. and Nelligan, D. F.
1981 : Gulf Stream: reconstruction of Cenozoic flow patterns over the Blake Plateau: Geology, v. 9, p. 266–270.
Rabinowitz, P. D. and Eittreim, S. L.
1974 : Bottom current measurements in the Labrador Sea: Journal of Geophysical Research, v. 79, p. 4085–4090.
Richardson, P. L.
1977 : On the crossover between the Gulf Stream and the Western Boundary Undercurrent: Deep-Sea Research, v. 24, p. 139–159.
1983a: Eddy kinetic energy in the North Atlantic from surface drifters: Journal of Geophysical Research, v. 88, p. 4355–4367.
1983b: A vertical section of eddy kinetic energy through the Gulf Stream system: Journal of Geophysical Research, v. 88, p. 2705–2709.
Roberts, D. G.
1975 : Marine geology of the Rockall Plateau and Trough: Philosophical Transactions of the Royal Society of London, v. 278, p. 447–509.
Roberts, D. G., Hogg, N. G., Bishop, D. G. and Flewellyn, C. G.
1974 : Sediment distribution around moated seamounts in the Rockall Trough: Deep-Sea Research, v. 21, p. 175–184.
Roberts, D. G. and Kidd, R. B.
1979 : Abyssal sediment wave fields on Feni Ridge, Rockall Trough: long-range sonar studies: Marine Geology, v. 33, p. 175–191.
1984 : Sedimentary and structural patterns on the Iberian continental margin: an alternative view of continental margin sedimentation: Marine and Petroleum Geology, v. 1, p. 37–48.
Schmitz, W. J.
1977 : On the deep general circulation in the western North Atlantic: Journal of Marine Research, v. 35, p. 21–28.
1984 : Abyssal eddy kinetic energy in the North Atlantic: Journal of Marine Research, v. 42, p. 509–536.
Schmitz, W. J. and Hogg, N. G.
1978 : Observations of energetic low frequency current fluctuations in the Charlie-Gibbs Fracture Zone: Journal of Marine Research, v. 36, p. 725–734.
Schneider, E. D., Fox, P. J., Hollister, C. D., Needham, H. D. and Heezen, B. C.
1967 : Further evidence of contour currents in the western North Atlantic: Earth and Planetary Science Letters, v. 2, p. 351–359.
Shepard, F. P., Marshall, N. F., McLoughlin, P. A. and Sullivan, G. G.
1979 : Currents in Submarine Canyons and Other Seavalleys: American Association of Petroleum Geologists Studies in Geology, no. 8, Tulsa, 173 p.
Shor, A. S.
1978 : Bottom currents on East Katla Ridge, NW Iceland Basin: International Council for the Exploration of the Sea 1978/C:60 Hydrography Committee (unpubl. ms.).
Shor, A., Lonsdale, P., Hollister, C. D. and Spencer, D.
1980 : Charlie-Gibbs Fracture Zone: bottom-water transport and its geological effects: Deep-Sea Research, v. 27, p. 325–345.
Smethie, W. M., Jr. and Trumbore, S.
1983 : Chlorofluromethanes (F-11 and F-12) in the western North Atlantic

Ocean and the deep western boundary undercurrent: EOS Transactions American Geophysical Union, v. 64, p. 1089.
Smith, P. C.
1975 : A stream tube model for bottom boundary currents in the ocean: Deep-Sea Research, v. 22, p. 853–873.
Spinrad, R. W. and Zaneveld, J.R.V.
1982 : An analysis of the optical features of the near-bottom and bottom nepheloid layers in the area of the Scotian Rise: Journal of Geophysical Research, v. 87, p. 9553–9561.
Steele, J. H., Barrett, J. R. and Worthington, L. V.
1962 : Deep currents south of Iceland: Deep-Sea Research, v. 9, p. 465–474.
Stow, D.A.V. and Lovell, J.P.B.
1979 : Contourites: their recognition in modern and ancient sediments: Earth Science Reviews, v. 14, p. 251–291.
Swallow, J. C. and Worthington, L. V.
1969 : Deep currents in the Labrador Sea: Deep-Sea Research, v. 16, p. 77–84.
Tucholke, B. E.
1973 : The history of sedimentation and abyssal circulation on the Greater Antilles Outer Ridge; Ph.D. Thesis, Massachusetts Institution of Technology/Woods Hole Oceanographic Institution, WHOI-74-1, 314 p.
1979 : Furrows and focussed echoes on the Blake Outer Ridge: Marine Geology, v. 31, p. M13–M20.
1982 : Origin of longitudinal triangular ripples on the Nova Scotian continental rise: Nature, v. 296, p. 735–737.
Tucholke, B. E. and Eittreim, S.
1974 : The Western Boundary Undercurrent as a turbidity maximum over the Puerto Rico Trench: Journal of Geophysical Research, v. 79, p. 4115–4118.
Tucholke, B. E. and Ewing, J. I.
1974 : Bathymetry and sediment geometry of the Greater Antilles Outer Ridge and vicinity: Geological Society of America Bulletin, v. 85, p. 1789–1802.
Tucholke, B. E. and Laine, E. P.
1983 : Neogene and Quaternary development of the lower continental rise off the central U.S. east coast; in Studies in Continental Margin Geology, eds. J. S. Watkins and C. L. Drake; American Association of Petroleum Geologists Memoir 34, p. 295–305.
Tucholke, B. E. and Mountain, G. S.
1979 : Seismic stratigraphy, lithostratigraphy and paleosedimentation patterns in the North American Basin; in Deep Drilling results in the Atlantic Ocean, Continental Margins and Paleoenvironment, eds. M. Talwani, W. Hay and W.B.F. Ryan, Maurice Ewing Series 3; American Geophysical Union, p. 58–86.
Tucholke, B. E., Wright, W. R. and Hollister, C. D.
1973 : Abyssal circulation over the Greater Antilles Outer Ridge: Deep-Sea Research, v. 20, p. 973–995.
Tucholke, B. E., Hollister, C. D., Biscaye, P. E. and Gardner, W. D.
1985 : Abyssal current character determined from sediment bedforms on the Nova Scotian continental rise: Marine Geology, v. 66, p. 43–57.
Vogt, P. R. and Johnson, G. L.
1972 : Seismic reflection survey of an oblique aseismic basement trend on the Reykjanes Ridge: Earth and Planetary Science Letters, v. 15, p. 248–254.
Weatherly, G. and Kelley, E. A.
1981 : An analysis of hydrographic data from Knorr cruise 74 in HEBBLE area, September-October, 1979: Technical Report, Department of Oceanography, Florida State Univ., Tallahassee, 38 p.
Weatherly, G. L.
1984 : An estimate of bottom frictional dissipation by Gulf Stream fluctuations: Journal of Marine Research, v. 42, p. 289–301.
Whitehead, J. A. and Worthington, L. V.
1982 : The flux and mixing rates of Antarctic Bottom Water within the North Atlantic: Journal of Geophysical Research, v. 87, p. 7903–7924.
Worthington, L. V.
1976 : On the North Atlantic circulation: Johns Hopkins Series in Oceanography no. 6, Johns Hopkins University Press, Baltimore, 110 p.
Worthington, L. V. and Wright, W. R.
1970 : North Atlantic Ocean Atlas of Potential Temperature and Salinity in the Deep Water including Temperature, Salinity and Oxygen Profiles from the Erika Dan Cruise of 1962: Woods Hole Oceanographic Institution Atlas Series, v. 2; Woods Hole, Massachusetts, 24 p + 58 plates.
Worthington, L. V. and Volkmann, G. H.
1965 : The volume transport of the Norwegian Sea overflow water in the North Atlantic: Deep-Sea Research, v. 12, p. 667–676.
Wunsch, C. and Grant, B.
1982 : Towards the general circulation of the North Atlantic Ocean: Progress in Oceanography, v. 11, p. 1–59.
Wunsch, C. and Hendry, R.
1972 : Array measurements of the bottom boundary layer and the internal wave field on the continental slope: Geophysical Fluid Dynamics, v. 4, p. 101–145.
Wyrtki, K., Magaard, L. and Hager, J.
1976 : Eddy energy in the oceans: Journal of Geophysical Research, v. 81, p. 2641–2646.

Manuscript Accepted by the Society June 18, 1985

ACKNOWLEDGMENTS

We are pleased to acknowledge support by Office of Naval Research contract N00014-82-C-0019 to Woods Hole Oceanographic Institution during preparation of this paper. We thank S. L. Eittreim, C. D. Hollister, G. Jones and A. N. Shor for reviewing the manuscript. We also thank P. Barrows and P. Foster for typing the manuscript and L. A. Raymond for drafting the figures. Contribution No. 5970 of Woods Hole Oceanographic Institution.

Printed in U.S.A.

The Geology of North America
Vol. M, The Western North Atlantic Region
The Geological Society of America, 1986

Chapter 28

Oceanic particles and pelagic sedimentation in the western North Atlantic Ocean

Susumu Honjo
Woods Hole Oceanographic Institution, Woods Hole, Massachusetts 02543

INTRODUCTION

Deep-sea sediment is formed by the accumulation of oceanic particles originally generated elsewhere. A large part of oceanic sedimentation is due to biological activity in the oceanic surface layers. Some lithogenic particles, however, are transported long distances from interiors and edges of continents into the pelagic environment. Transportation pathways of oceanic particles usually are complex, involving biological, chemical, and physical processes. The time necessary to complete such pathways varies from instantaneous to hundreds of years. Not all oceanic particles settle as permanent members of the deep-sea sediment. Biogenic detritus in particular can be dissolved, decomposed, or changed in its properties. Refractory particles increase their relative proportions in pelagic sediments as a result of such alteration through time.

There are two major kinds of marine particles in oceans, lithogenic and biogenic. Lithogenic particles are mostly layered or framework aluminosilicate minerals such as clay, quartz, and plagioclase. Only fine particles—clay and silt-size—are transported to the pelagic ocean floor in quantity, except for some large ice-rafted boulders. Lithogenic particles are refractory and therefore become permanent constituents of the pelagic marine sediment without significant chemical alteration. They are transported from American, European, and African landmasses to the open North Atlantic Ocean by processes such as river outflow, coastline erosion, and glacial erosion. Significant amounts of fine refractory particles are transported from the continental shelves and slopes of both sides of the North Atlantic by advective processes to the interior of the ocean basins (e.g., McCave and Tucholke, this vol.). In boreal and Arctic latitudes, particularly in the Labrador Sea, Newfoundland Basin, and Greenland Sea, ice-transport contributes significantly to open-ocean sedimentation. Aeolian dust, transported as fine refractory particles directly from arid surfaces such as the Sahara Desert, contributes significantly to deep-ocean sedimentation in the tropical and subtropical North Atlantic.

Biogenic (non-refractory) particles are produced only in ocean surface waters. They are micro- or macroscopic skeletal remains of plankton such as planktonic foraminifera shells, pteropod shells, radiolarian tests, diatom frustules, and coccoliths. The shells and tests are made of crystalline calcium carbonate or amorphous opaline silicate (opal). Biogenic particles are by far the largest source of particles for deep-sea sediment; however, this may not be reflected in seafloor sediments because their final state of preservation is controlled by their chemical reaction to the bottom and pore waters. Biogenic particles are also susceptible to dissolution and decomposition in the water column. For example, most cells of organisms, or subcellular particles, are consumed in near-surface waters. Only a small percentage of organic matter settles to the seafloor, and there it is largely decomposed and consumed by the benthos. Thus, only a very small percentage of organic compounds produced by primary production in the surface, euphotic layer is preserved in seafloor sediment. There are other categories of ocean particles, but on a regional scale they are minor compared to lithogenic and biogenic particles (Table 1).

Large, macroscopic oceanic particles such as planktonic foraminifera tests settle vertically at rates of a few hundred to thousands of meters a day. Ice-rafted debris arrives at the deep-sea floor almost instantaneously and is little perturbed from its vertical trajectory by currents. Fine oceanic particles settle more slowly and may take complex advective or vertical pathways while they are being transported. Dissolution of susceptible particles is partly a function of residence time in the water column, which is derived from the sinking speed. There are theoretical approaches to estimating sinking speeds of oceanic particles, but the inherent assumptions (e.g., particles settle as quartz spheres) often lead to significant errors when applied to oceanic particles which deviate greatly from sphericity.

In reality, vertical settling of all fine oceanic particles occurs by co-aggregation with larger, faster-settling particles. Thus, set-

Honjo, S., 1986, Oceanic particles and pelagic sedimentation in the western North Atlantic; *in* Vogt, P. R., and Tucholke, B. E., eds., The Geology of North America, Volume M, The Western North Atlantic Region: Geological Society of America.

TABLE 1. OCEANIC PARTICLES IN THE PELAGIC REALM

Biogenic Particles	**Lithogenic Particles**
Individual Particles	Layered Aluminosilicates (clay)
Labile particles	Framework Silicates (rock-forming minerals)
-subcellular organels	-quartz particles
-pigmented particles	-plagioclase particles
-wax/fat particles	Volcanic Shards and Pumice
-zooplankton remains	Micrometeorites
-crustacean molt products	Glacial Till
-macrozooplankton	-glacial debris
	-sea-ice/airborne debris
Skeletal particles	
-shells and tests of micro-	**Authigenic Particles** (resuspended products)
zooplankton	Low temperature minerals
-frustules	-Smectite, palygorskite, zeolite, etc.
-coccoliths	Hydrothermal minerals
-land-derived phytoliths	
-and fresh water diatoms	**Anthropogenic Particles**
	Industrial particles
Aggregates	-fly ash
-large amorphous aggregates	-soot from ships
(marine snow)	-smelter's spheres
-fecal pellets	Pollutant particles
	-plastic debris
Large Parcels	-polychlorinated vinyl sheets
-large nekton	-polystyrene spheres
-marine animal corpses	-nylon fabric
-drifting logs	-refractory litter
-macrophytes	-crude oil and bunker oil slicks
Cosmogenic Particles	
-metallic spherules	
-stony spherules	
-microtektites	

tling of fine oceanic particles can be accelerated by many orders of magnitude. In fact, the only possible way for fine particles, less than 62 μm, to arrive at the seafloor is to be aggregated into larger particles.

In order to understand the settling of fine particles in the deep sea, it is necessary to recognize the different functions of *suspended* and *settling* particles in ocean systems. The role of settling particles appears to be far more important for marine sedimentation (McCave, 1975). A suspended particle is often a discrete material such as a clay flake or a coccolith. Stokesian vertical settling velocity of such particles is insignificant and their theoretical residence time is more than several hundred years. Settling particles, on the other hand, are aggregates of macroscopic size made up of numerous particles. The host or matrix of fine particles is usually of biogenic origin. Feces of microzooplankton—fecal pellets—and large amorphous aggregates (LAA; often called "marine snow") are common host particles in the pelagic ocean. They settle through the water column at high rates and arrive at the seafloor before being seriously altered by chemical or biochemical reactions.

To discuss pelagic sedimentation of oceanic particles, several terminologies must first be explained: *oceanic particles* are nonliving solids or semi-solids distributed between the sea surface and the seafloor regardless of their size. *Pelagic sediment* is formed on the seafloor by the accumulation and alteration processes of oceanic particles through time. *Oceanic mass flux* or *flux* is the rate of arrival of oceanic particles at certain depths (mass per unit area per unit time, i.e., mg $m^{-2}day^{-1}$ or $g/cm^2/1{,}000$ yr). Flux is a rate term but is often misused by calling it a mass; the mass of newly settled oceanic particles is called *new sediment* here. *Residence time* is the time an oceanic particle, independent or aggregated, takes to settle through a designated layer of water. *Accumulation rate* is the rate of burial of new sediment following dissolution and decomposition of susceptible particles.

The pelagic ocean contains layers of *ecological zones,* characterized by specific biological, geochemical, and physical properties. The *euphotic zone* is the topmost layer of the ocean where all primary production occurs. Its thickness varies according to the depth of solar penetration—from a few hundred meters in the nutrient-poor subtropical ocean, such as the Sargasso Sea, to merely a few meters in the Arctic summer. The *mesopelagic zone* underlies the euphotic layer. Light does not penetrate the mesopelagic zone enough to support photosynthesis but zooplankton are active. The bottom of the mesopelagic layer is usually not well defined, but often the oxygen-minimum zone delineates its lower boundary. The thickness of the mesopelagic zone, depending upon surface productivity and other factors, is usually several hundred to a thousand meters thick. Again, in the Arctic and boreal Atlantic the mesopelagic zone is as thin as several tens of meters. In the underlying *bathypelagic zone* zooplankton activity is subdued. Only a small amount of particulate food matter arrives in this zone after being "filtered" by the upper two layers. This zone contains the most stable, low-energy environment compared to the above two layers, but currents are found in most of the North Atlantic bathypelagic layers. In most areas of the North Atlantic a *nepheloid layer* persists several hundred meters or more above the ocean floor; it is characterized by a high density of resuspended and/or laterally transported refractory particles (McCave and Tucholke, this volume).

It should be emphasized that the existing basic data on oceanic particles is much too incomplete to attempt to synthesize the genesis, processes, and fate of oceanic particles and to discuss quantitative oceanic sedimentation in the North Atlantic. This article will be radically improved upon by the future acquisition of more quantitative information. This paper does not deal with a detailed geographic description and characterization of oceanic floor sediments. Excellent descriptive syntheses on biogenic and lithogenic contents of North Atlantic sediments are available in, for example, Lisitzin (1972) and Emery and Uchupi (1984).

In a general way, pelagic sedimentation in the North Atlantic Ocean can be categorized regionally: (1) Significant ice-rafting in the northwest North Atlantic. (2) Extensive biogenic sedimentation, particularly of carbonate, in the temperate and tropical North Atlantic Ocean. (3) Less development of biogenic silica sediment compared to other oceans, although production of opal is expected to be higher. (4) Vigorous redistribution of fine refractory particles by currents along the western Atlantic continental margin. (5) Sediment rich in terrigenous illite and far less contribution by authigenic smectite throughout the North Atlantic. (6) Significant influence of Saharan dust fallout over latitudes 5°N to 35°N across the North Atlantic Ocean.

BIOGENIC PARTICLES

Biogenic particles comprise over 90% of the total mass flux in the mesopelagic layer and 70% to 90% of the particle flux in the bathypelagic layer at temperate and subtropical North Atlantic locations (Honjo, Manganini, and Cole, 1982). Lithogenic matter makes up the balance of the particulate flux. Biogenic particles are divided into three basic material groups and they are all secreted by zooplankton and phytoplankton during their life cycles: (a) amorphous organic particles, (b) carbonate, and (c) opal. Labile organic matter settles through the pelagic water column as cells, cell organelles, and parts of plankton, and these are susceptible to ingestion and decomposition. Highly resistant organic material such as dinoflagellate remains and land pollen are often well preserved in marine sediment. Practically all carbonate particles are calcite (planktonic foraminifera tests, coccoliths) or aragonite (pteropods, heteropod shells, and planktonic gastropods) and all are built in microcrystalline form. The major sources of opaline particles are radiolaria shells, diatom frustules, and silicoflagellate skeletons. These skeletons are hydrated amorphous and have a complex architecture. Siliceous endoskeleton-bearing dinoflagellates, sponge spicules, and land-driven biogenic particles are found in trace quantities in marine sediments.

Planktonic Foraminifera Tests

The puzzlingly complicated life cycle of planktonic foraminifera and formation of their tests has been greatly clarified in recent years (Hemleben and Spindler, 1983). However, our knowledge is still limited to only a few species. Before the foraminifera is fully grown, it starts to descend into deeper layers. When an individual is fully grown, it undergoes gametogenesis (asexual splitting of cells), resulting in numerous gametes which swim away to the shallower layers. The empty shell, no longer supported by buoyancy, falls through the water column with a greater speed. Only a small fraction of the gametes successfully return to the surface and the calcification cycle thus continues. The turnover time or production cycle of a planktonic foraminifera test is not clearly known but it is assumed to be a few weeks.

The standing stock of living foraminifera is highly variable and it is difficult to generalize about its volume. Previous work cites up to about 10^4 individuals per 10^3 meters of euphotic layer throughout the temperate and tropical zones of the North Atlantic (Fairbanks and others, 1980).

Planktonic foraminifera tests contribute 50% to 80% of the carbonate flux to the abyssal depths in the temperate Atlantic (Honjo, 1980). Over the Demerara Abyssal Plain where surface productivity is relatively high, the flux of foraminifera at 5,068 m is about 7 to 13 mg m^{-2} day^{-1}. In a large ocean-gyre system or "marine desert," such as the Sargasso Sea, the flux of foraminiferal tests is lower, approximately 2 mg m^{-2} day^{-1} (Deuser and others, 1981).

The average weight of a foraminifera test in the tropical Atlantic is approximately 14 μg, while in the less nutrient-rich oceanic interior it is approximately 6 μg (Thunell and Honjo, 1981). The observed sinking speed of the adult test is from 150 m to 1,000 m per day; thus, the residence time of the tests in a deep water column (5,000 m) is a week to a month (Takahashi and Bé, 1984).

Adult tests are rarely incorporated into fecal pellets of common macrozooplankton such as salps and larger crustaceans. One explanation is that many planktonic foraminifera are equipped with long spines that extend their size up to several millimeters while living, and thus they may be too large to be consumed by common metazoans.

Pteropod Shells

Almost all of the aragonite particles in the open ocean are euthecothomatian pteropod shells which are increasingly recognized as an important component of oceanic carbonate systems (Byrne and others, 1984). Pteropods are ubiquitous in the North Atlantic Ocean, including the boreal Atlantic, and they live in the photic and mesopelagic zones. Large spatial and seasonal variability makes flux estimates difficult but it is at least 12% of the total carbonate flux in the subtropical North Atlantic (Berner and Honjo, 1981). A much larger contribution to the carbonate flux has been reported from the North Pacific (Betzer and others, 1984). Pteropod shells range from less than a millimeter up to 1 cm, and their sinking speeds range from a few hundred meters to several thousand meters per day. Although the dissolution rate of aragonite is faster than that of calcite, dissolution of large shells while setting through the Atlantic water column is insignificant; small shells, however, dissolve before reaching the bottom. It seems that shells arriving at the seafloor lack effective organic coating and the fine, laminar microcrystals are exposed during an early stage of dissolution. Thus, large pteropod shells on the deep-sea floor disintegrate and dissolve at an accelerated rate. Betzer and others (1984) believe that in the North Pacific where the carbonate saturation level is much shallower than in the Atlantic, shells of all sizes dissolve in the upper 2.2 km of the water column.

Coccoliths

The phytoplankton production in the pelagic North Atlantic Ocean is characterized more by abundant coccolithophorids than by diatoms. Depending upon the season, coccoliths contribute 25% to 60% of the carbonate flux to the deep North Atlantic Ocean. These calcite platelets cover unicellular algal coccolithophores like scales to make a coccosphere. The number of plates on a cell, generally several to a hundred, differs with growth stage, species, and nutrient/physiological state. The cell continuously manufactures new coccoliths and casts off old ones during the cell's mature life span. Species diversity is at a maximum in the Sargasso Sea (up to 100 species), less in the tropical Atlantic, and is almost monospecific, *Emiliania huxleyi,* in the Norwegian-Greenland Sea (Okada and McIntyre, 1979).

The sinking rate of individual coccoliths is merely a few micrometers per second, measured under laboratory conditions (Honjo, 1976). Thus, the estimated residence time in 5,000 m of water column is as long as 100 years. A species assemblage of coccoliths should not be accumulated on the seafloor because of horizontal diffusivity and advection while the coccoliths are slowly sinking. In reality, however, coccolith assemblages which represent latitudinal climatic zones within the productive surface water are duplicated in the bottom sediment lying underneath.

The accelerated vertical transport of coccoliths can be explained by their inclusion in large particles or aggregates which settle quickly, such as fecal pellets produced by zooplankton, and large amorphous aggregates. The sinking speeds of these media are similar to those for planktonic foraminifera tests, about a few hundred meters a day. Thus, surface production patterns and species combinations of coccoliths are "copied" with high resolution in deep-ocean sediments.

Radiolarian Skeletons

Radiolarian shells contribute most (80%–90%) particulate silica to the water column and bottom sediment in the tropical and temperate North Atlantic. There are probably more than 500 species of extant radiolarians in the North Atlantic.

The flux of radiolarian shells ranges from 16×10^3 to 24×10^3 shells m^{-2} day^{-1} in the subtropical/temporate pelagic North Atlantic Ocean (Takahashi and Honjo, 1983). Settling shells contribute 2 to 3 mg $m^{2-}day^{-1}$ of biogenic opal to the ocean floor over the Demerara Abyssal Plain, and 0.5 to 1 mg $m^{2-}day^{-1}$ over the Sohm Abyssal Plain; 95% of this opal is radiolarian shells.

The residence time of radiolarian shells ranges from 2 weeks to 14 months in a 5-km water column. Large skeletons probably spend only a few weeks in the water column and reach the abyssal floor essentially intact despite their soluble skeletons, while small phaeodarian (the most dissolution susceptible group of radiolaria) shells may be dissolved before reaching the seafloor.

There is a large discrepancy between the number of radiolarian species preserved in sediment and the number living in the water, suggesting that there is intense dissolution of shells at the seafloor. For example, 208 species of radiolarians were caught by a sediment trap over the Demerara Abyssal Plain but only 15 species were preserved in coretop sediment taken from beneath the sediment trap (Takahashi and Honjo, 1983). Although the radiolarian-skeleton content in surface sediments is relatively small in the North Atlantic compared to other oceans, there are some anomalies. Bjorklund (1977), for example, reported an area of abundance in the area between Faeroe Plateau and Reykjanes Ridge.

Acantharia, a group of protozoans related to radiolarians, are more abundant than radiolaria and foraminifera in the euphotic layers of the North Atlantic, particularly in the southern Sargasso Sea (Bottazi and others, 1971). The celestite ($SrSO_2$) shells of these plankton are extremely soluble after termination of the life cycle, and they do not reach to the bathypelagic layer.

Diatom Frustules and Silicoflagellates

Diatoms are unicellular algae and their opaline frustules range in size from a few micrometers to a few millimeters. Compared to their coastal counterparts, pelagic diatoms have weakly silicified frustules and less tendency to form resting spores. About 500 species are important members of Recent and fossil marine diatoms (Schrader and Schuette, 1981). Since diatoms are primary producers, production of frustules is closely related to the production of organic carbon. Lisitzin (1972) indicated that the ratio of diatomaceous opal is 2.3 greater than organic carbon. Primary production in the southern Labrador Sea, Norwegian Sea, and southern Barents Sea is particularly high, as much as 150 to 250 mg C $m^{-2}day^{-1}$. Lisitzin (1972) estimated that the production rate of opal in this area exceeds 0.5 g $m^{-2}day^{-1}$, assuming that the majority of opal is produced as diatom frustules.

In North Atlantic and Arctic bottom sediments, opaline content is far smaller than that in other major oceans (Heath, 1974). Opal content in dry sediment weight is less than 5% in the temperate to arctic North Atlantic and less than 10% in the tropical Atlantic (Lisitzin, 1972). A zone, approximately 12° wide and stretching from Nova Scotia to the northern end of Norway through the east coast of Iceland, is particularly high in primary production, with a high rate of dissolved silica uptake (Heath, 1974). It is also known that a receding ice edge, such as in the Greenland Sea, enhances production of diatoms (Smith and Nelson, 1985). However, the surface sediments through this area yield less than 5% of opaline content. In contrast, opal content of North Pacific sediments often exceeds 30%. In general, production of opal particles per unit area in the northern Atlantic is equal to or greater than production in the Indian, Pacific, or South Atlantic oceans. The basin-to-basin fractionation model of Berger (1970) suggests that the North Atlantic is the least favorable for the *preservation* of biogenic opal.

Amorphous Organic Particles and Vertical Supply of Nutrients

Since the *Challenger* expedition near the end of the last century, it has been known that many animals live in and on the abyssal seafloor. In order to feed this benthos, energy and nutrients have to be supplied from the surface euphotic layer in the form of nutrient particles. The benthos feed on labile organic particles such as cells and subcellular material, including wax/fat originally secreted by plankton.

At a tropical North Atlantic sediment trap station (the Demerara Abyssal Plain, Plate 2), the productivity of organic carbon in the euphotic layer was about 60 mg $m^{-2}day^{-1}$. The flux of particulate organic carbon from the euphotic layer was 9 mg $m^{-2}day^{-1}$ and rapidly decreased to 4 mg $m^{-2}day^{-1}$ at the bottom of the mesopelagic layer; about 2.5 mg $m^{-2}day^{-1}$ of organic carbon reaches the 5-km deep-sea floor. In the general relationship between organic carbon flux and depth in the North Atlantic, about 5% of production leaves the euphotic layer. This is reduced by feeding and decomposition during settling through the mesopelagic layer so that only 1% to 2% of the organic carbon arrives at abyssal depths (Honjo, Mangenini, and Cole, 1982).

TABLE 2. NON-BIOGENIC PARTICLES: PROCESSES OF PELAGIC SEDIMENTATION

Mechanism	Particle Origin	Particle Size	Horizontal Distance (in water)	Area in North Atlantic
Advective transport	continental margins, basin slopes	clay	3-4,000 km	ubiquitous
Aeolian transport	lithogenic; soil constituents	clay/silt	~10,000 km	30°N to 0°
Ice rafting	windblown, till, and beaches	all sizes	~1,000 km	north of ~40°N west of ~40°W
Organic rafting	nearshore	all sizes mainly >10 kg	500 km	20°N to 60°N
Cosmogenic source	cosmic, lunar-earth orbit	10μm to 0.5 mm	global	ubiquitous
Volcanogenic	volcanoes, submarine volcanism	all sizes	local to global	subaerial volcanic centers, spreading centers
Authigenic	precipitation		ubiquitous	ubiquitous

NON-BIOGENIC REFRACTORY PARTICLES

Many discrete mechanisms are involved in the supply and distribution of non-biogenic sediments in the deep ocean, including: (1) direct aeolian transport, (2) horizontal transport from continental margins by currents, (3) transport by organic debris, (4) volcanogenic contributions, (5) authigenic production at the ocean floor, (6) ice rafting, (7) cosmogenic contributions, and (8) mass wasting and turbidity currents (Embley and Jacobi, this volume; Pilkey and Cleary, this volume); these last, gravity-controlled processes are limited mostly to ocean margins. Processes of pelagic sedimentation are summarized in Table 2. There is no complete explanation for the accumulation of refractory sediment in the deep basins. The distribution, mineralogy, and geochemistry of oceanic clays are not treated here, as they are thoroughly discussed in other monographs (e.g., Biscaye, 1965, and Griffin and others, 1968).

Origin and Transportation of Aeolian Dust Over the North Atlantic Ocean

Sea-surface mineral aerosol flux in the North Atlantic, north of the area influenced by trade winds, has been estimated at 10^{-6} g $cm^{-2}yr^{-1}$ and the total annual fallout is estimated at 12×10^{12} g (Prospero, 1981). The flux is about equal to that of the total Pacific average. The area of prevailing trade winds over the North Atlantic Ocean is strongly characterized by dust fallout from the Sahara Desert.

North Africa, from approximately 20°N latitude north to the Mediterranean Sea is the largest arid land mass on earth and it significantly affects aeolian sedimentation in the North Atlantic. Dust outbreak is generated by interaction between the easterly mid-tropospheric jet stream and meridional meandering of the Inter-Tropical Convergence Zone. Airborne dust is transported throughout the North Atlantic by northeast trade winds blowing off North Africa between 5° and 35°N. For example, an infrared sensor on board the GEOS-East satellite recorded an outbreak of sand/dust from the western portion of the Sahara Desert on July 23, 1982. This dust storm was traced through its trans-Atlantic passage and was recorded reaching Florida on July 30, 1982 (D'Aguanno, 1983).

Estimates by Prospero (1981) indicate that 6 to 26×10^{13} g yr^{-1} of dust is entrained in winds over the Sahara. Compared to the estimated global production of mineral dust at 6 to 36×10^{13} g yr^{-1} (Judson, 1968), it is clear that the Sahara Desert contributes significantly to the global atmospheric mineral-dust load. Schutz and others (1981) estimates that 1.73×10^{13} yr^{-1} falls within 1,000 km of the coast of Mauritania, as far as Cape Verde. The flux of dust at the sea surface in this area would be on the order of 0.3 to 0.5 g $m^{-2}day^{-1}$. In an area between 1,000 km and 4,000 km from the west African coast line across the Atlantic, 8×10^{13} g yr^{-1} falls to the sea surface. Further to the west, near the Lesser Antilles, a distance of about 5,000 km from Africa, approximately 0.3×10^{13} yr^{-1} reaches the ocean surface between 5° and 35° north latitude.

The aerosol size distribution of Saharan dust is between 1.3 μm and 10 μm with a peak at 3.5 μm (Prospero, 1981), and it is essentially identical to that of soil aerosols measured in other regions. Johnson (1979) reported that a typical mineral assemblage of truly airborne particles over the central Atlantic included 46% total clay minerals, 12.5% quartz, 5% plagioclase, and 0.8% dolomite; an assemblage collected on board ship near the west African coast line across the Atlantic, 8% dolomite; an assemblage collected on board ship near the west African shore was richer in quartz and plagioclase at the expense of the clay constituent. More than 50% of the clay was micas and 6% to 7% was kaolinite (Prospero, 1981). Because of differential particle settling during trans-Atlantic transportation, dust at the western side of the Atlantic is compositionally much different. Dust reaching the snow fields of the Andes, for example, is strongly depleted in quartz and feldspar particles and enriched in fine clay particles (Windom, 1969).

Griffin and others (1968) estimated the accumulation rate of Saharan aeolian particles at up to 6 mm 10^{-3} yr within 1,000 km of the south Moroccan coast, 2 to 7 mm 10^{-3} yr just east of the Mid-Atlantic Ridge, and 0.3 to 0.4 mm 10^{-3} yr between the ridge and Barbados. Johnson (1979) also estimated the accumulation rate, based on a more detailed model, and he predicted rates about twice those of Griffin and others. The total mass flux measured by a sediment trap deployed at 389 m depth in the western Demerara Abyssal Plain (13°30′N, 54°00′W) was 69.4 mg $m^{-2}day^{-1}$. The flux of refractory silicate was 4.4 mg $m^{-2}day^{-1}$ (6.4% of the total flux). Supposing that all of the refractory particles caught in this 389 m deep trap were transported to the seafloor underneath, the vertical accumulation rate would be 0.8 mm 10^{-3} yr if we use a packing density of 2 g cm^{-2}; and 1.4 10^{-3} yr with a packing density of 1.15 g cm^{-2}. We assume a realistic value would be within this range. The lower range accumulation rate, 0.8 mm 10^{-3} yr, coincides with Johnson's (1979) estimated values of 0.7 to 1.4 mm 10^{-3} yr.

Transportation of aerosols for extremely long distances, such as from the Asian interior to the arctic Atlantic, has been observed. There is the possibility of a major pathway of aerosol transport in westerly winds from the American northeast industrial region to the Arctic through Bermuda, Newfoundland, and Iceland (Rahn, 1981). Particles which may be industrial in origin were collected in a sediment trap off Bermuda (Deuser and others, 1983). However, there is no arid desert in North America to supply significant dust to North Atlantic ocean sediment, and the aeolian component in sediments north of 35°N is extremely diluted by other detritus.

Vertical Transport of Fine Refractory Particles in the Pelagic Water Column

Recent sedimentological studies show that the total flux of refractory material in pelagic water increases with increasing depth (Honjo, Manganini, and Poppe, 1982). This suggests large-scale lateral transport of refractory fine particles by advection both locally and over long distances. This trend is most noticeable at stations relatively near to the continental margin where currents are most active. However, even in mid-oceanic areas, such as the central Sargasso Sea, refractory flux increases in deeper layers. Flux of particles larger than 62 μm (fecal pellets, planktonic foraminifera shells, and radiolarian tests) is usually constant; these are of surface-layer origin and sink quickly.

Clay-sized refractory particles have a long Stokesian residence time, on the order of several hundred years. Therefore, fine particles falling on the ocean surface or resuspended from the sea bottom could be transported by advection/diffusion and dispersed through a very wide area before arriving at the seafloor. However, as in the case of the "coccolith paradox," the mineral assemblage of the airborne fine refractory particles is not lost while settling through a long water column. For example, the general shape of the Saharan dust corridor is "rubber stamped" on the seafloor across the tropical Atlantic (Rex and Goldberg, 1958). Therefore, fine refractory particles must settle quickly to the seafloor without significant dispersion. This appears to be accomplished by large amorphous aggregates.

The biologically produced large amorphous aggregates settle through the deep water column and scavenge fine suspended refractory particles. The settling speed of aggregates is thus accelerated by a ballasting effect, and the fine particles reach the abyssal seafloor within several weeks; this is similar to the settling of aggregated coccoliths, but with large amorphous aggregates the scavenging happens throughout the water column. The aeolian pathway over the open North Atlantic Ocean generally provides a much smaller quantity of refractory sediments to the ocean floor than the deep advective pathway. For example, in the Demerara station, aeolian flux close to the seafloor is estimated to be about 12% of the total flux and the rest is advected mostly from the South American continental margin. Similarly, it appears that the aeolian input in the Sohm Abyssal Plain is less than 5% of the total flux, and the rest is probably advected from the North American continental margin.

Amorphous aggregates generated in surface waters are visible near the abyssal seafloor as "bathypelagic marine snow." The standing stock of large amorphous aggregates (>0.5 mm diameter) was about 0.2 per liter at several stations in the western Sargasso Sea (Honjo and others, 1984). These are one-time measurements, and significant variability is expected in time and space. Amorphous aggregates are made up of interim products of decomposing zooplankton and phytoplankton. They contain a large number of fibrous and elongated objects produced by plankton, which are bundled up and often loosely agglutinated together with viscous organic matter. Varieties of particles, including clays, lithogenic particles, and coccoliths, are trapped or scavenged by such aggregates while settling.

Organic Transport of Pebble and Cobble Sized Sediment

Large seaweeds, particularly three genera of brown algae, *Macrocystis, Pelagophycus,* and *Nereocystis* that reach from 30 m to 200 m in length, anchor in energy-rich, shallow coastal waters with holdfasts on seafloor rocks. When the balance between the buoyancy of the frond and the anchor weight is lost, especially after storms, some plants drift seaward, often up to 200 km; some have been sighted as far as 500 km offshore. Drifting plants stop growing in low-energy environments, partially decompose, and drop to the seafloor with their rocks. These kelp-rafted pebbles and cobbles are usually encrusted with shallow-water organisms and often are bored. Large kelps grow around all of the northern North Atlantic coasts, from New England to south Greenland, Iceland, and northern European shorelines. Quantitative information on the contribution of organic rafting to this area is not well known. A similar exotic transport of land-based rocks attached to the roots of drifting trees may also occur in the North Atlantic, but to a lesser degree than in other oceans because severe floods which erode wooded highlands are rare in eastern North America.

Volcanogenic and Authigenic Particles

Ash layers in deep-sea sediment provide isochronous, and often readily traceable, levels for inter-core correlations. In the North Atlantic, the number of subaerial volcanoes that potentially supply pyroclastics is relatively limited. The Atlantic volcanoes, such as the Azores, Canary Islands, and Jan Mayen Island, are well known. However, Iceland has contributed most to the supply of pyroclastic sediment in the North Atlantic Ocean. Lisitzin (1972) stated that, since 1500 AD, 19 km^3 or almost half of the world's mid-ocean pyroclastic particles were produced by eruptions on Iceland. Volcanic glass in the surface sediments of the North Atlantic is most concentrated in the area surrounding Iceland, Jan Mayen Island, and bordered by the Norwegian continental slope.

Volcanic glass is a major source of authigenic minerals in the ocean. Basaltic glass is altered to palagonite, and further hydration leads to formation of smectite, palygorskite, or zeolite. Small amounts of smectite also can be formed in association with siliceous tests and frustules (van Bennekom and van der Gaast, 1976). Like any other terrigenous matter, authigenic products are resuspended and redistributed in a wide horizontal range (Table 2). They are mixed and diluted with common land-based clays such as illite when they reach their ultimate site of deposition.

Precipitation of solids from pore-water solutions and from hydrothermal solutions play a role in formation of authigenic (formed *in-situ*) silica. It is now widely agreed, however, that biogenic silica is the most important precursor of authigenic silica, formed as chert and porcelanite through the diagenetic process from opal-A through the opal-CT stage to quartz (chalcedony). This normally occurs under relatively high temperatures due to burial (300 to 500 m deep) in the sediment column. Authigenic silica, preserved in deep-sea sediment, represents at most only 1% to 2% of the originally deposited siliceous skeletal remains (Wollast, 1974).

Other common authigenic particles in the North Atlantic Ocean are iron-manganese micronodules, nodules, and crusts, and phosphorite deposits. Of these, only the micronodules are widely distributed; they and larger nodules occur mostly in areas of slow sediment accumulation (Plate 2) where they are not buried faster than the rate at which they grow by precipitation from solution. Iron-manganese crusts are common on exposed basaltic ocean crust along the Mid-Atlantic Ridge and especially on the steep unsedimented flanks of seamounts. Both crusts and phosphorite deposits are common across the northern Blake Plateau. The genesis of these authigenic deposits and their economic potential is fully discussed by Dillon and others (this volume).

Ice-Rafted Sediment

Ice rafting transports a large volume of lithogenic particles over long distances (up to about 1,000 km) and it contributes significantly to pelagic sedimentation in the northwestern North Atlantic Ocean. Ice-rafted detritus is usually identified as fine to course sand (63 μm to 250 μm) in seafloor sediment, although it truly includes all sizes of particles.

Icebergs. The majority of icebergs in transit through the Davis Strait were calved from west Greenland coastal glaciers. The southeastern Greenland coast produces almost no full-sized icebergs. Annual production fluctuates but on average 20 to 30 × 10^3 icebergs are discharged from those areas every year (Murray, 1968). Svalbard and the Norwegian fjords provide an insignificant number of icebergs to the North Atlantic Ocean because of the northeastward flowing Norwegian coastal current. The average size of newly-produced icebergs along the western Greenland coast is estimated at 300 cubic meters, or at a specific gravity of 0.85, 23 × 10^6 tons. When these icebergs have drifted to near Labrador, the average volume decreases to 160 cubic meters or 3.5 × 10^6 tons (Fillon and others, 1981). During the winter, the majority of icebergs are entrapped in sea ice along the southeast Baffin and Labrador coasts, north of 51°N. 2,500 icebergs pass south of 60°N latitude, and 400 icebergs travel south of 48°N each year. Iceberg loss increases dramatically due to melting south of 58°N (Murray, 1968). However, the southernmost iceberg sighting recorded was at 28°N, 48°W; 350 km *southeast* of Bermuda. The zone of increased melting coincides closely with an area off northern Newfoundland where surface temperatures increase from 4° to 9°C (Fillon, 1977). This is the area where the major sedimentation of ice-rafted detritus to the deep-sea floor presently occurs. Three types of glacial sediments included in a typical iceberg are known: (a) bimodal tillite, including boulders, (b) unsorted diamicton, and (c) a relatively small amount of sorted sand which originates in water streams formed on the iceberg during the melting stage. In addition, icebergs potentially redistribute dust fallout, volcanic ash, cosmospheres, and exotic plant matter to southern locations.

The fluxes of ice-rafted detritus in the Labrador Sea and Baffin Bay were recently estimated at 40 to 150 mg $m^{-2}day^{-1}$ and 500 to 8,000 mg $m^{-2}day^{-1}$, respectively, from the concentration of supraglacial gravel and basal debris in icebergs (Fillon, pers. com., 1984). The average flux of iceberg ice-rafted detritus, estimated from studies of core material, has been 0.6 to 2 g $cm^{-2}yr^{-1}$ in the western North Atlantic over the last 10,000 years (Ruddiman, 1977; Fillon, 1977). However, the patchy nature of distribution, post depositional removal by current erosion and mass wasting, and masking by turbidites hinders a truly accurate estimation of flux.

Sea Ice. Multiyear sea ice develops along the east coast of Greenland and contributes significantly to marine sedimentation in the Greenland Sea and along the eastern margin of Greenland as far south as 63°N. First year ice, often developing as far south as south of Iceland, has no significant impact on marine sedimentation. The majority of the debris found in multiyear sea ice consists of fragmented rocks, minerals, and ash blown from exposed areas on coastal Greenland, Svalbard, and other Arctic islands. Relatively low precipitation rates and a lack of vegetation combined with strong winds in circum-Arctic land areas result in a dense covering of air-borne particles over the nearby ice fields. According to calculations by Shaw and others (1974), if fine sand is lifted to an initial height of 6,500 m, the mean North Atlantic wind velocities are capable of providing up to 600 km of horizontal transport; dust can cross the Fram Strait (a straight path between Svalbard and Greenland) and cover the entire multiyear ice belt along the eastern Greenland margin. Thus, multiyear sea ice in these areas generally has a distinct brown color. Plankton, especially diatoms, develop during the summer period on the bottom surface of sea ice and in ice cracks. Planktonic foraminifera tests, living in small openings between and underneath ice floes, also are often entrapped by ice as a result of ice turnover.

Most multiyear sea ice supplied to the Norwegian-Greenland Sea forms in the Arctic Ocean and then is pushed southward through the Fram Strait by transpolar current systems; discharge rate of the ice is 0.75 to 1.3 × 10^6 km^2y^{-1} (Vinje, 1976). The average thickness of multiyear sea ice is about 3 m and the average discharge volume at the outlet in Fram Strait is estimated as 0.3 × 10^6 m^3sec^{-1}. Rapid melting of multiyear sea ice occurs approximately north of 70°N and east of 5°W, but, depending upon winter weather conditions, multiyear sea ice may reach south of the Denmark Strait. The area of multiple year sea-ice melt is estimated as $0.5 \times 10^6 km^2$ in the Greenland Sea. Accumulation rates of sea-ice-rafted detritus in these areas have not been estimated.

Extraterrestrial Particles

Cosmic spheres are well preserved in the deep-sea sediment and are important in recording the earth's magnetic behavior and its relationship to cosmic events. The rate of arrival of cosmogenic debris at the sea surface is 3 to 4 orders of magnitude less than the production of windblown dust. Red clay sediment in the western Pacific, which accumulates at very low rates (~1 mm 10^{-3}yr), contains one to five cosmic spherules per gram (carbonate free basis; Pettersson and Fredirikson, 1958). However, such low accumulation rates are rare in the North Atlantic Ocean, and concentrations of cosmic particles are consequently much lower. Manganese nodules are known to concentrate cosmic spheres in quantities 3 to 4 orders of magnitude higher than surrounding red clay.

It has been known since the *Challenger* expedition that deep-sea sediment contains extraterrestrial spherules of macro- and microscopic size. The most readily extractable cosmic particles in sediments are 0.5 μm to 1 mm in diameter with a median mode of about 35 μm. "Cosmic spheres" in most cases appear to be shiny black solidified spherules of melted, partially oxidized, iron-nickel magnetite particles produced by meteoroids during their high-velocity entry into the earth's atmosphere (Brownlee, 1981). Other types of spherules consist of crystalline olivine and pyroxene, which are probably original cosmic particles or fragments of chondritic meteorites. The total mass of meteoroids accreted by the earth is estimated to be on an order of 10^4 tons yr^{-1} (Milliman, 1975); however, this estimate differs among various authors by a large margin. If one takes Milliman's estimate, the average flux over the earth's surface is on the order of 1 $\mu g\ m^{-2} day^{-1}$ or a few spherules $m^{-2} day^{-1}$. Microtectites (silica-rich macroscopic glass spheres) have been discovered in deep-sea sediment cores of specific location and age, although none have been reported in the North Atlantic. Isotopic studies indicate that, unlike iron or stony meteorites, microtektites cannot have originated outside the earth-moon system (O'Keefe, 1976).

TABLE 3. FLUX AND NET SEDIMENTATION RATES ESTIMATED AT TWO SEDIMENT TRAP STATIONS IN THE NORTH ATLANTIC

Sohm Station: Sediment trap, 3,694 m; bottom sediment samples, 5,581 m; 31°32'N, 55°55'W, Fall 1977

Particles	Mass flux $mg\ m^{-2}d^{-1}$	% in trapped sed.	% in bottom sed.	Net sed. rate $mg\ m^{-2}d^{-1}$	% remaining in bottom sediment
Refractory	2.26	12.29	78.57	2.26	100
Carbonate	12.43	67.59	15.56	0.45	3.62
Opal	1.70	9.24	0.82	0.024	1.41
Combustible	1.99	10.88	4.89	0.14	7.04
Organic C	0.88	4.76	0.52	0.015	1.70
Organic N	0.10	0.54	0.07	0.0020	2.00
Other	1.02	5.58	4.30	0.12	11.76
Total	18.39	100.00	99.84	2.87	

Demerara Station: Sediment trap, 3,755 m; bottom sediment sample, 5,005 m; 13°30'N, 54°00'W; Winter 1977-78.

Particles	Mass flux $mg\ m^{-2}d^{-1}$	% in trapped sed.	% in bottom sed.	Net sed. rate $mg\ m^{-2}d^{-1}$	% remaining in bottom sediment
Refractory	11.42	24.78	62.37	11.49	100
Carbonate	26.11	56.21	28.00	5.16	19.76
Opal	4.06	8.75	1.14	0.21	5.17
Combustible	4.76	10.26	8.48	1.56	32.77
Organic C	1.73	3.73	0.33	0.061	3.53
Organic N	0.19	0.41	0.03	0.0055	2.89
Other	2.84	6.12	8.12	1.50	52.82
Total	46.42	100.00	99.99	18.42	

DISSOLUTION OF BIOGENIC PARTICLES AND NET ACCUMULATION OF SEDIMENT IN THE NORTH ATLANTIC

Seawater in the pelagic North Atlantic realm is undersaturated with respect to silica, and the degree of undersaturation is far greater than that in the North Pacific. Considering the temperature profile of the North Atlantic, biogenic silica should dissolve at a high rate in shallow water and the majority of the silica should be recycled immediately into the euphotic zone. As explained before, however, a significant proportion of biogenic silica reaches the seafloor by complex vertical transport mechanisms, and it is dissolved there. Thus the major dissolution of opal occurs in the upper few hundred meters of the water column and at the sea bottom.

In many ways, the chemical environment of carbonate in the North Atlantic is the reverse of that of opal. Most of the North Atlantic water column is saturated with respect to calcite, and it is undersaturated only at depths greater than about 4,000 m (Takahashi, 1975). Carbonate dissolution occurs mostly beyond the chemical lysocline where the rate of dissolution accelerates. Biogenic carbonate particles are better preserved in the North Atlantic than in the North Pacific; however, this relation is reversed in the case of opal.

The solubility of biogenic particles depends on many chemical and sedimentological factors. Among them are: (1) the undersaturation factor determined by kinetic ionic dissociation, (2) the stoichiometric factor mainly caused by local, microbial activity which produces acids as a result of oxidation of organic matter, particularly labile carbon, (3) microstructure and extent of surface coatings which may protect minerals, and (4) speed of burial in the sea floor. The balance between carbonate fluxes and the rate of dissolution decides the "carbonate compensation depth" (CCD), where the carbonate content of seafloor sediments drops to less than about 10%. The solubility of aragonite is higher and pteropod shells disappear in shallower sediments. In North Atlantic sediment, surface-supplied biogenic aragonite is rarely preserved in layers deeper than 3,500 m. In the central North Atlantic the depth of the CCD is about 5,600 m compared to 4,000 m or less in the North Pacific (Berger and Winterer, 1974). Therefore, pelagic red clay does not develop in the North Atlantic except in relatively small, very deep, far offshore basins such as the northeasternmost Nares Abyssal Plain and southernmost Sohm Abyssal Plain.

Because the particle population reaching the deep-sea floor suffers significant alteration before it is buried, the flux does not always represent the net accumulation rate of sediment. Table 3 shows the difference between new sediment and core-top sediment collected at some locations in the North Atlantic. One can assess the abundance of individual sedimentary constituents against the mass of refractory particles in the preserved seafloor sediment; this allows an estimate of the net sedimentation rate and the percentage of new sediment that remains permanently in ocean floor sediments (assuming that post-sedimentation removal of refractory particles is negligible). Using available sediment-trap data, we have estimated that the total net sedimentation rate at the Demerara station and Sohm station is 18 and 3 mg $m^{-2}day^{-1}$, respectively; this is 16% and 40% of the new sediment near the bottom at these stations. According to these same estimates, at the Sohm station 96% of the carbonate, 99% of the opal, and 98% of organic carbon arriving at the bottom is dissolved and recycled into water. At the Demerara station 80% of the carbonate, 95% of the opal, and 96% of organic carbon flux is lost before permanent burial of the sedimentary constituents.

Clearly, the sedimentary record recovered in sediment cores from the North Atlantic Basin does not contain an accurate record of how marine biofacies evolved in space and time, and the thanatocoenose of fossil organisms must be considered with circumspection in reconstructing paleoceanographic patterns. Future studies of how pelagic sediment is delivered to the seafloor, and how it is altered or preserved, will provide a sound quantitative basis for interpreting the historical record locked in seafloor sediments of the North Atlantic Ocean.

REFERENCES

Berger, W. H.
1970 : Biogenous deep-sea sediments: Fractionation by deep-sea circulation: Geological Society of America Bulletin, v. 81, p. 1385–1402.

Berger, W. H., and Winterer, E. L.
1974 : Plate stratigraphy and the fluctuating carbonate line: in Pelagic Sediments on Land and Under the sea, eds. K. J. Hsü, and H. Jenkyns; Special Publication Institute Associated Sedimentologists 1, p. 11–48.

Berner, R. A.
1977 : Sedimentation and dissolution of pteropods in the ocean: in The Fate of Fossil Fuel CO_2 in the Oceans, eds. N. R. Anderson, and A. Malahoff; Plenum Press, New York, p. 243–260.

Berner, R. A., and Honjo, S.
1981 : Pelagic sedimentation of aragonite: Its geochemical significance: Science, v. 211, p. 940–942.

Betzer, P. R., Byrne, R. H., Acker, J. G., Lewis, C. S., Jolley, R. R., and Feeley, R. A.
1984 : The oceanic carbonate system: A reassessment of biogenic controls: Science, v. 226, p. 1074–1076.

Biscaye, P. E.
1965 : Mineralogy and sedimentation of recent deep-sea clay in the Atlantic Ocean and adjacent seas and oceans: Geological Society of America Bulletin, v. 76, p. 803–832.

Bjorkland, K. R.
1977 : *Actinomma haysi* n. sp., its Holocene distribution and size variation in Atlantic Ocean sediment: Micropaleontology, v. 23, p. 114–126.

Bottazzi, E. M., Schreiber, B., and Bowen, V. T.
1971 : Acantharia in the Atlantic Ocean, their abundance and preservation: Limnology and Oceanography, v. 16, p. 677–684.

Brownlee, D. E.
1981 : Extraterrestrial components: in The Oceanic Lithosphere, ed. E. Cesare, John Wiley and Sons, New York, p. 733–762.

Byrne, R. H., Acker, J. G., Betzer, P. R., Feely, R. A., and Cates, M. H.
1984 : Water column dissolution of aragonite in the Pacific Ocean: Nature, v. 312, p. 321–326.

D'Aguanno, J. A.
1983 : 1982 Saharan dust outbreak: Marine Weather Log, U.S. Navy, v. 27, p. 199–200.

Deuser, W. G., Ross, E. H., and Anderson, F. R.
1981 : Seasonality in supply of sediment to the deep Sargasso Sea and implications for the rapid transfer of matter to the deep ocean: Deep Sea Research, v. 28, p. 495–505.

Deuser, W. G., Emeis, K., Ittekot, V., and Degens, E. T.
1983 : Fly-ash particles intercepted in the deep Sargasso Sea: Nature, v. 305, p. 216–218.

Emery, K. O.
1963 : Organic transportation of marine sediments: in The Sea, Volume 3, ed. M. N. Hill; Wiley-Interscience, New York, p. 776–793.

Emery, K. O., and Uchupi, E.
1984 : The Geology of the Atlantic Ocean: Springer-Verlag, New York, 1,050 p. (plus maps).

Fairbanks, R. D., Wiebe, P. H., and Bé, A.W.H.
1980 : Vertical distribution and isotopic composition of living planktonic foraminifera in the western North Atlantic: Science, v. 207, p. 61–63.

Fillon, R. H., Miller, G. H. and Andrews, J. T.
1981 : Terrigenous sand in Labrador Sea hemipelagic sediments and paleoglacial events on Baffin Island over the last 100,000 years: Boreas, v. 10, p. 107–124.

Ganssen, G., and Lutze, G.
1982 : The aragonite compensation depth at the northeastern Atlantic continental margin: "Meteor" Forschungsergebnisse, Series C, 36, p. 57–59.

Griffin, J. J., Windom, H. L., and Goldberg, E. D.
1968 : The distribution of clay minerals in the world ocean: Deep Sea Research, v. 15, p. 433–459.

Heath, G. R.
1974 : Dissolved silica and deep-sea sediments; in Studies in Paleo-oceanography, ed. W. W. Hay; Society of Economic Paleontologists and Mineralogists Special Publication, v. 20, p. 77–93.

Heimdal, B. R., and Gaader, K. R.
1980 : Coccolithophorids from the northern part of the eastern central Atlantic, I. Holococcolithophorids. Meteor Forsch-Ergebnisse, Series D., v. 32, p. 1–14.
1981 : Coccolithophorids from the northern part of the eastern central Atlantic, II. Heterococcolithophorids: Meteor Forsch-Ergebnisse, Series D., v. 33, p. 37–69.

Hemleben, C. and Spindler, M.
1983 : Recent advances in research on living planktonic foraminifera; in Reconstruction of Marine Paleoenvironments, ed. J. E. Meulenkamp; Utrecht Micropaleontological Bulletins, v. 30, p. 141–170.

Honjo, S.
1976 : Coccoliths: production, transportation and sedimentation: Marine Micropaleontology, v. 1(1976), p. 65–79.

1980 : Material fluxes and modes of sedimentation in the mesopelagic and bathypelagic zones: Journal of Marine Research, v. 38, p. 53–97.
Honjo, S., Doherty, K., Agrawal, Y., and Asper, V.
1984 : Direct optical assessment of large amorphous aggregates in the deep ocean: Deep Sea Research, v. 31, p. 67–76.
Honjo, S., Manganini, S. J., and Cole, J. J.
1982 : Sedimentation of biogenic matter in the deep ocean: Deep Sea Research, v. 29, p. 609–625.
Honjo, S., Manganini, S. J., and Poppe, L. J.
1982 : Sedimentation of lithogenic particles in the deep ocean: Marine Geology, v. 50, p. 199–220.
Johnson, L. R.
1979 : Mineralogical dispersal patterns of North Atlantic deep-sea sediments with particular reference to eolian dust: Marine Geology, v. 29, p. 335–345.
Judson, S.
1968 : Erosion of the land, or what's happening to our continents?: American Scientist, v. 56, p. 356–374.
Kennett, J. P.
1981 : Marine tephrochronology: in The Sea, Volume 7, ed. E. Emiliani; Wiley-Interscience, New York, p. 1373–1436.
Lisitzin, A. P.
1972 : Sedimentation in the World oceans: Society of Economic Paleontologists and Mineralogists Special Publication, No. 17, 218 p.
McCave, I. N.
1975 : Vertical flux of particles in the ocean: Deep Sea Research, v. 22, p. 491–502.
Milliman, P. M.
1975 : Dust in the solar system: in The Dusty Universe, eds. G. B. Field and A.G.W. Ameron; Neal Watson, New York, p. 185–209.
Murray, J. E.
1968 : The drift, deterioration and distribution of icebergs in the North Atlantic Ocean: in Ice Seminar: A conference sponsored by the Petroleum Society of the Canadian Institute of Mining and Metallurgy (CIMM), Calgary, Alberta, Canada; Canadian Institute of Mining and Metallurgy Special Vol. 10, p. 3–18.
Okada, H. and McIntyre, A.
1979 : Seasonal distribution of modern coccolithophores in the western North Atlantic Ocean: Marine Biology, v. 54, p. 319–328.
O'Keefe, J. A.
1976 : Tektites and Their Origin: Elsevier, Amsterdam, p. 147–165.
Patterson, E. M., and Gillette, P. A.
1977 : Commonalities in measured size distribution for aerosols having a solid derived component: Journal of Geophysical Research, v. 82, p. 2074–2082.
Pettersson, H., and Fredirikson, K.
1958 : Magnetic spherules in deep-sea deposits: Pacific Science, v. 12, p. 71–81.
Prospero, J. M.
1981 : Eolian transport to the world ocean: in The Sea, Volume 7, ed. C. Emiliani; Wiley-Interscience, New York, p. 801–874.
Rahn, K. A.
1981 : Relative importance of North America and Eurasia as sources of Arctic Aerosol: Atmospheric Environment, v. 15, p. 1447–1455.
Rex, R. W., and Goldberg, E. D.
1958 : Quartz contents of pelagic sediment of the Pacific Ocean: Tellus, v. 10, p. 153–159.
Ruddiman, W. F.
1977 : Late Quaternary deposition of ice rafted sand in the subpolar North Atlantic (latitude 40° to 65°N): Geological Society of America Bulletin, v. 88, p. 1813–1827.
Schrader, H. J., and Schuette, G.
1981 : Marine diatoms: in The Sea, Volume 7, ed. C. Emiliani; Wiley-Interscience, New York, p. 1179–1232.
Schutz, L. Jaenicke, R., and Pietrek, H.
1981 : Saharan dust transport over the North Atlantic Ocean: in Desert Dust, Origin, Characteristics, and Effect on Man, ed. T. L. Pewe; Special Paper, Geological Society of America, v. 186, p. 87–100.
Shaw, D. M., Watkins, N. D., and Huang, F. C.
1974 : Atmospherically transported volcanic glass in deep sea sediments: theoretical considerations: Journal of Geophysical Research, v. 79, p. 3087–3094.
Smith, W. O., and Nelson, D. M.
1985 : Phytoplankton bloom produced by a receding ice edge in the Ross Sea: Spatial coherence with the density field: Science, v. 227, p. 163–166.
Takahashi, K., and Bé, A.W.H.
1984 : Planktonic foraminifera: factors controlling sinking speeds: Deep-Sea Research, v. 31, p. 1477–1500.
Takahashi, K., and Honjo, S.
1983 : Radiolarian skeletons; size weight, sinking speed, and residence time in tropical pelagic oceans: Deep Sea Research, v. 30, p. 543–568.
Takahashi, T.
1975 : Carbonate chemistry of sea water and the calcite compensation depth in oceans: in Dissolution of Deep-sea Carbonates, eds. W. Sliter, A. Be and W. Berger; Cushman Foundation for Foraminiferal Research, v. 13, p. 11–26.
Thunell, R. C., and Honjo, S.
1981 : Planktonic foraminiferal flux to the deep ocean: Results from the equatorial Atlantic and the central Pacific: Marine Geology, v. 40, p. 237–253.
van Bennekom, A. J., and van der Gaast, S. J.
1976 : Possible clay structures in frustules of living diatoms: Geochimica et Cosmochimica Acta, v. 40, p. 1149–1152.
Vinje, T. E.
1976 : The drift pattern of sea ice in the Arctic with particular reference to the Atlantic approach: in The Arctic Ocean, The hydrographic environment and the fate of pollutants, ed. L. Rey; McMillan, London, p. 83–95.
Windom, H. L.
1969 : Atmospheric dust records in permanent snowfields: implication to marine sedimentation: Geological Society of America Bulletin 80, p. 761–782.
Wollast, R.
1974 : The silica problem; in The Sea, Volume 5, ed. E. D. Goldberg; Wiley-Interscience, New York, p. 359–392.

Manuscript Accepted by the Society July 26, 1985

ACKNOWLEDGMENTS

I am indebted to Drs. B. E. Tucholke and K. O. Emery for discussions and constructive criticism. Drs. P. G. Brewer, W. B. Curry, J. J. Cole, R. Fillon, K. Takahashi, and S. J. Manganini, V. L. Asper and numerous other colleagues gave invaluable suggestions for which I am most grateful. E. Evans typed and edited the manuscript.

This research was partially supported by National Science Foundation Grant OCE-8208736. Woods Hole Oceanographic Institution Contribution No. 5664.

Printed in U.S.A.

The Geology of North America
Vol. M, The Western North Atlantic Region
The Geological Society of America, 1986

Chapter 29

Mass wasting in the western North Atlantic

Robert M. Embley
NOAA/MRRD Hatfield OSU Marine Science Center, Newport, Oregon 97365
Robert Jacobi
Department of Geological Sciences, State University of New York at Buffalo, Amherst, New York 14226

INTRODUCTION

The balance between processes of slope degradation and slope aggradation on the ocean floor is constantly shifting. Along "active" continental margins, tectonic effects can provide large amounts of sediments to submarine slopes and they also can trigger frequent large slope displacements. Along "passive" continental margins such as that of eastern North America, the aggradation and degradation processes probably operate at a slower rate; they are stimulated primarily by changes in sediment supply caused by climatic changes or by occasional intraplate earthquakes.

The North American continental margin has long been a "type area" for studies of submarine sedimentary features and the processes that formed them, including canyon incision (Veatch and Smith, 1939), turbidity currents (Heezen and Ewing, 1952), and contour currents (Heezen and others, 1966). Although slumping has always been considered an important process on the upper continental slope and in canyon heads, only within the past few years has it been recognized that slumps and slides are ubiquitous on the continental slope and upper rise and that mass-wasting products are an important component of the deep-water Pleistocene sedimentary sequence (Embley and Jacobi, 1977; Knebel and Carson, 1979; Embley, 1980). It also has only recently been appreciated how important such factors as jointing and biological erosion are to failure of consolidated sediments on steep slopes in the deep ocean. This chapter describes our present state of knowledge of mass-wasting processes in the provinces of the North American margin and basin.

PHYSICS OF MECHANISM AND DEFINITIONS

With increasing oil exploration in deeper water, the subject of submarine-slope failure has received increasing practical interest. The basic approach to the analysis of submarine-slope stability is usually based on the method of infinite-slope analysis, and has been summarized by a number of workers, including Moore (1961), Morgenstern (1967), and Prior and Suhayda (1979). The basic assumption of this analysis is that the slope is of large lateral extent compared to the thickness of failed sediments. Failure of a sediment layer occurs when the effective shear stress equals the effective shear resistance along the same surface. However, the effective shear strength in a sedimentary column at a given point in time and space depends on many variables, such as sedimentation rate, accelerations in addition to gravity, and presence or absence of gas. Furthermore, it is difficult to measure *in situ* shear strength, particularly in deep water and within the sediments. Thus, the actual analysis is rather complex. Use of Mohr circles and other techniques of analysis are discussed in detail by Moore (1961), Morgenstern (1967), Almagor and Wiseman (1977) and Karlsrud and Edgers (1982). Two types of slumping can be considered; drained and undrained (Morgenstern, 1967). In drained slumping, no excess pore pressure develops and sediment failure occurs as a function of the slope angle and internal cohesion of the sediments. In general, as long as the angle of internal friction is greater than the slope angle, sediment accumulation does not disturb the stability of the slope. As Moore (1961) originally pointed out, and as has been confirmed by numerous later measurements, the angles of internal friction in most ocean sediments under drained conditions are on the order of 15° to 40°, greatly in excess of the average value of the steepest submarine slopes (~6°). Therefore, most workers have concluded that drained slumping is a relatively unimportant process on the ocean floor. In contrast, undrained slumping results from a build-up of excess pore pressures by a variety of mechanisms (e.g., gas accumulation, rapid sedimentation, earthquakes) which reduce the shear strength at a particular depth so that the angle at which failure can occur is reduced. Undrained failure is generally assumed to be the most important process for mass-wasting on submarine slopes.

The basic theoretical models and analytical tools for predicting slope failure exist, but the physical data necessary to test these models are generally lacking, particularly in deep water. Such critical measurements as *in situ* pore pressure, and sea-floor acceleration caused by events like earthquakes, waves, tides and currents are not generally available.

Embley, R. W., and Jacobi, R., 1986, Mass wasting in the western North Atlantic; *in* Vogt, P. R., and Tucholke, B. E., eds., The Geology of North America, Volume M, The Western North Atlantic Region: Geological Society of America.

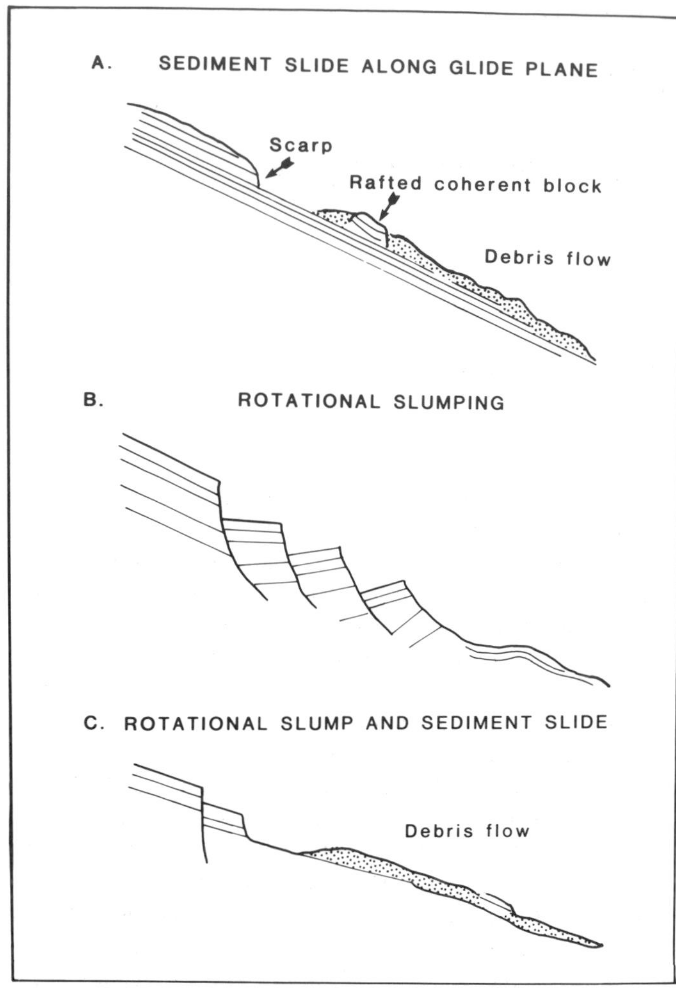

Figure 1. Examples of the geological effects of mass wasting on continental margins.

The physical manifestations of the failure of a submarine slope can take different forms depending on depth, slope angle, and areal extent of the failure. Some of the possible geometries of failure are illustrated in Figure 1. One can establish two end members: (1) a shallow failure which generates a highly mobile flow that translates over large distances into deep water (sediment slide) (Fig. 1A), and (2) a deep-seated rotational block that has experienced minimum translation and deformation (slump) (Fig. 1B). It is also possible to combine A and B, with the result that a large rotational slump generates a shallow slide near the surface (Fig. 1C).

The mechanisms of submarine mass transport and the classification of the depositional products have been summarized by Dott (1963), Carter (1975) and Middleton and Hampton (1973). Generally, increasing slope and amount of translation result in increasing mobility; this is manifested in a gradation from a simple deformation to plastico-viscous laminar flow (debris flow). This sequence can be recognized in the seafloor morphology discussed below. Hampton (1972) has suggested that submarine-debris flows may also generate turbidity currents and there is a significant amount of field evidence to support this hypothesis (e.g., in Embley, 1980).

RECOGNITION AND MAPPING OF SUBMARINE MASS WASTING

Prior to the development of modern instrumentation the occurrence and effects of various deep-sea sedimentary processes were inferred from bottom samples and seabed morphology studies based on echo-sounding records. The development of subbottom profiling and sidescan sonar systems during the 1960s has revolutionized these studies. Although early workers such as Arkhanguelsky (1930) and Shepard (1955) inferred the presence of mass wasting in a few areas of the deep ocean, the widespread nature of the phenomenon became clear with the increasing use of the new tools (Walker and Messingill, 1970; Embley, 1976, 1980, 1982; Jacobi, 1976; Jacobi and Hayes, 1982; Embley and Jacobi, 1977; McGregor and Bennett, 1977; Ryan, 1982, Farre and others, 1983). Vertical-incidence, subbottom profiling systems are capable of resolving the head scarps (Fig. 2) and the deposits (Fig. 3). Sidescan systems, particularly the new generation of digital systems, can resolve variations in the surface morphology of the slides over large areas, thereby providing valuable information on the mechanics of the slides, the relative ages of multiple events, and the relationships of the different facies. For example, Figure 4 shows a mid-range sidescan and 3.5-kHz record over a section of the Georges Bank continental slope; it illustrates the difficulty inherent in interpreting mass-wasting features with high resolution, vertical incidence seismic records alone. The record shows a few of the relationships, among the wide range possible, between back-tilted slump blocks, slide scars and debris from the slide scars. The prominent slide scar in the right center of the record may have been coupled to the back-tilted, rotational block just upslope. It is clear that the interpretation of the various features' relationships based on a seismic-reflection record alone would depend heavily on where the track crossed the region.

Sampling of mass-wasting deposits has usually been accomplished using the standard 2.5 in (6.35 cm) diameter piston corer. Because many of the easily recognizable primary sedimentary structures caused by mass wasting are not sampled in such a small diameter corer, workers studying continental-rise sediments have underestimated the importance of these facies (Embley, 1980). Large-diameter box cores are best for study of the facies (Fig. 5) but these corers have been infrequently used to date. Generally, two types of primary sedimentary structures are seen in cores: deformed contacts and a clast/matrix structure. The former is generally considered to be indicative of a slide and the latter of a debris flow.

The slide location map (Fig. 6) was constructed by examining all readily available 3.5 kHz records in the area and mapping the echo character (Damuth, 1975; Jacobi and others, 1983; Vassallo and others, 1983). Based on this mapping, Figure 6 shows all areas with echo character interpreted as being caused

36°10.5′ N
74°38′ W

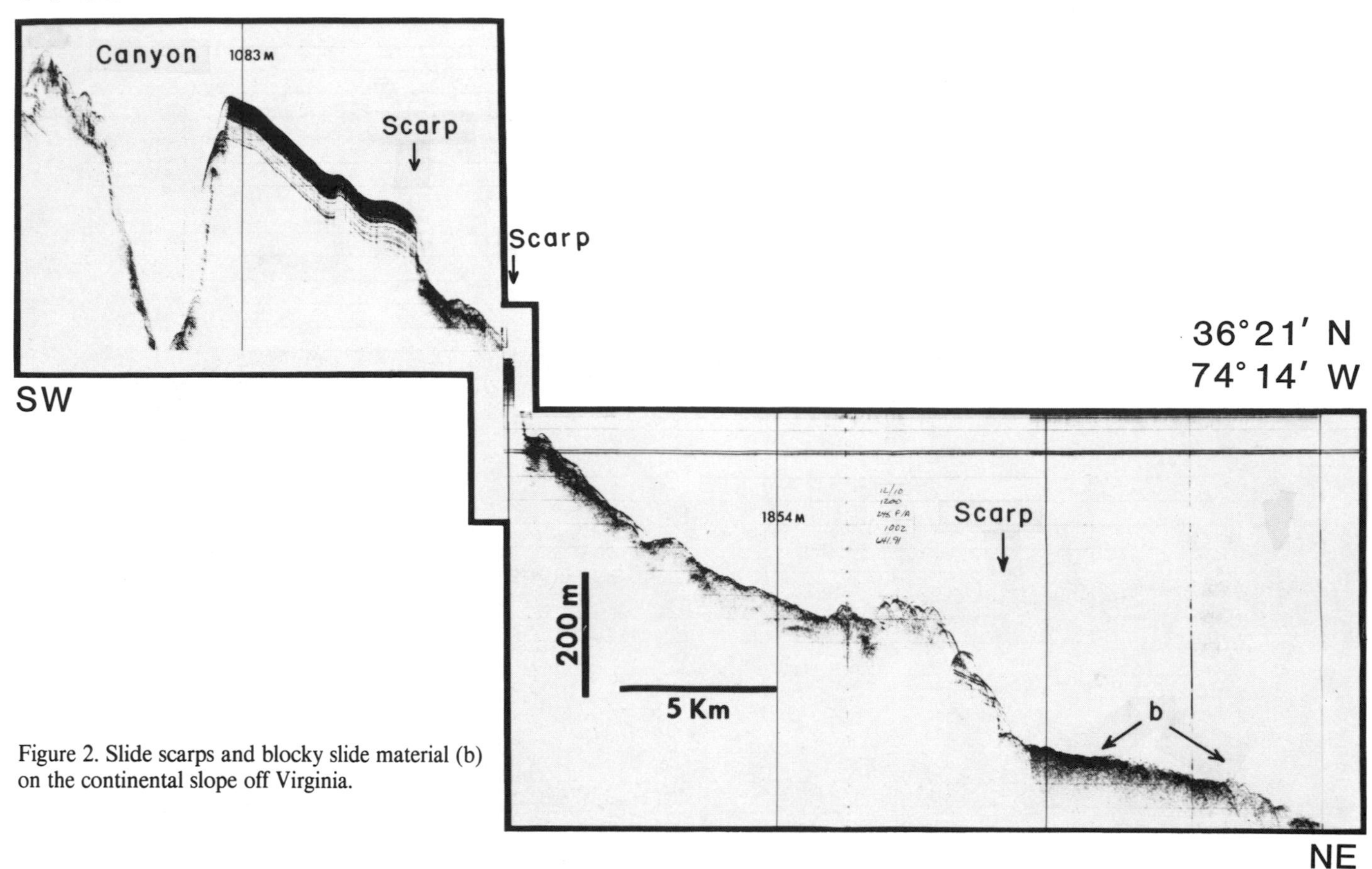

Figure 2. Slide scarps and blocky slide material (b) on the continental slope off Virginia.

by mass-wasting processes. The limitations in this data set are as explained above in reference to Figure 4.

DISTRIBUTION OF MASS-WASTING EVENTS IN THE WESTERN NORTH ATLANTIC

The object of this chapter is to summarize the important occurrences of mass wasting throughout the western North Atlantic basin. However, there are gross variations in data density and in research effort across the region. Thus, this discussion is concentrated on the North American margin from 32° to 40°, where the best data set and most detailed studies exist. Other regions in the basin are discussed where there are sufficient data and published results.

Blake Outer Ridge to Hatteras Canyon

Mappable sediment slides on the continental margin south of 32°N are rare and are of only local extent. One occurs at 32°N, 76.6°W on the Blake Escarpment; rotational slump blocks are observed upslope of sediment slides at 32°N, 76.6°W (Vassalo, 1983; Vassalo and others, 1983).

An extensive sedimentary slide complex covering more than 24,000 km occurs on the continental margin north of the Blake Outer Ridge (slide complex 1c, Fig. 6). This slide zone is particularly intriguing because (1) the deposits onlap the Hatteras Abyssal Plain to water depths as great as 5400 m, (2) the area contains variable acoustic character which can be mapped as distinct units, (3) the area is contiguous to South Carolina where the largest known east-coast earthquake occurred at Charleston in 1886 (Bollinger, 1977), and (4) there are a number of salt domes in a band along the lower slope and upper continental rise (Popenoe and others, 1982) which may have triggered some of the mass movements. Slide complex lc marks the southern extent of major slide complexes along the east-coast margin.

Surveys in this region (Embley, 1980; Popenoe and others, 1982) have delineated a series of rotational slumps on the slope and rise upslope from the slide complex (areas 1a and 1b). Several scarps have been traced for long distances along the slope in area 1a; for example, a scarp at 2400–2600 m depth is 50 km long; and another scarp at 1000 m water depth is 10 km long. These slumps are associated with very little surficial slide material except in the northern part of the area 1a. The source of most of the slide sediments there appears to be a slide scar which has a generally continuous scarp parallel to, and downslope from, the slump scarp.

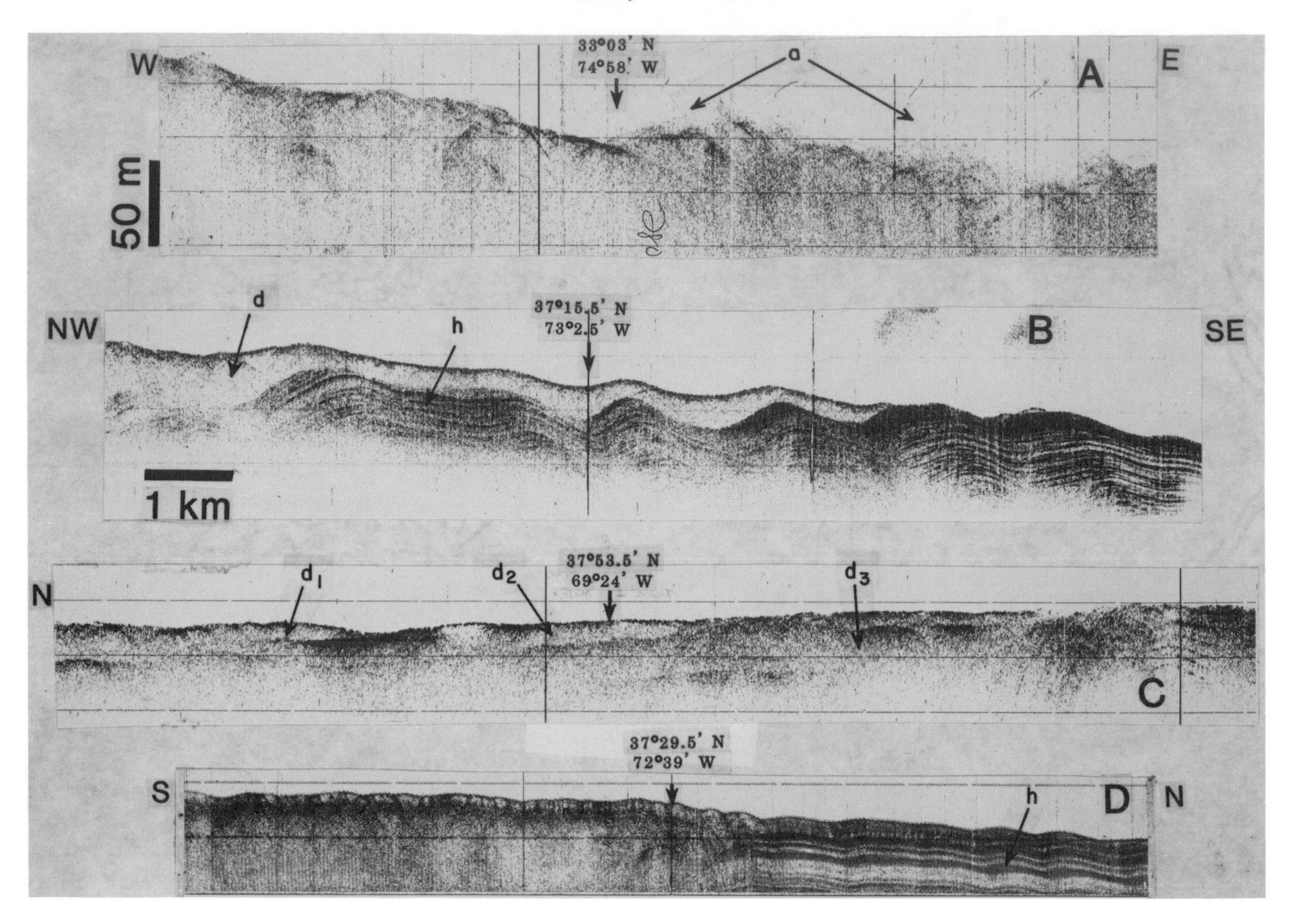

Figure 3. 3.5 -kHz records showing examples of mass-transport facies on the continental rise of eastern North America. Blocky slide material is shown as b, d is debris flows (note three successive debris flows in C), and h represents hemipelagic continental rise sediments. Note buried debris flow beneath 5–10 meter layer of hemipelagic sediments on left side of D.

The echo returns from the slide complex are highly variable, ranging from an irregular hyperbolated surface, to a smooth echo return, to an extremely diffuse return. The extreme variation in the echo return probably relates to differences in: 1) viscosity within any one depositing flow, 2) character of different flows, or 3) dynamic stages in the same flow. For example, a halo showing a distinct bottom echo (debris flow) occurs around an area characterized by a less coherent acoustic return in the deeper part of the Blake slide zone (1c, Fig. 6). The more distinct return could represent an earlier slide deposit overridden by a later slide, or both echoes could represent rheologically distinct units in a single slide event.

Cores taken within these slide zones contain disturbed layering such as inclined beds, small convolutions, and clasts. Silt turbidite beds directly overlie the mass flow deposits in two cores. Other cores containing mass flow deposits are capped by up to one meter of brown and olive-gray clay. Cores taken on the adjacent, normally stratified continental rise are generally homogeneous gray silty clays, with some thin silt laminae (Embley, 1980).

Slide complex 2 (Fig. 6) consists entirely of an acoustically transparent wedge with a weak, diffuse echo (i.e., "debris flow" material). Upslope from the wedge are numerous rotational slump blocks. The downslope extent of the debris flow is poorly determined with existing seismic control.

Slide complex 3a is confined to morphologic channels within the continental rise. The echo hyperbolae which delineate this slide zone could be interpreted as complex, channel-wall echoes, but profiles near 73°W suggest that at least part of the echo results from blocky slide material within the channel. Slide complex 3b consists of an acoustically very transparent wedge of sediment with uncertain origin. The diffuse echo could result from a thick, fine-grained turbidite, contourite, or debris flow.

Hatteras Canyon to Hudson Canyon

Sediment slide complex 4 consists of an acoustically trans-

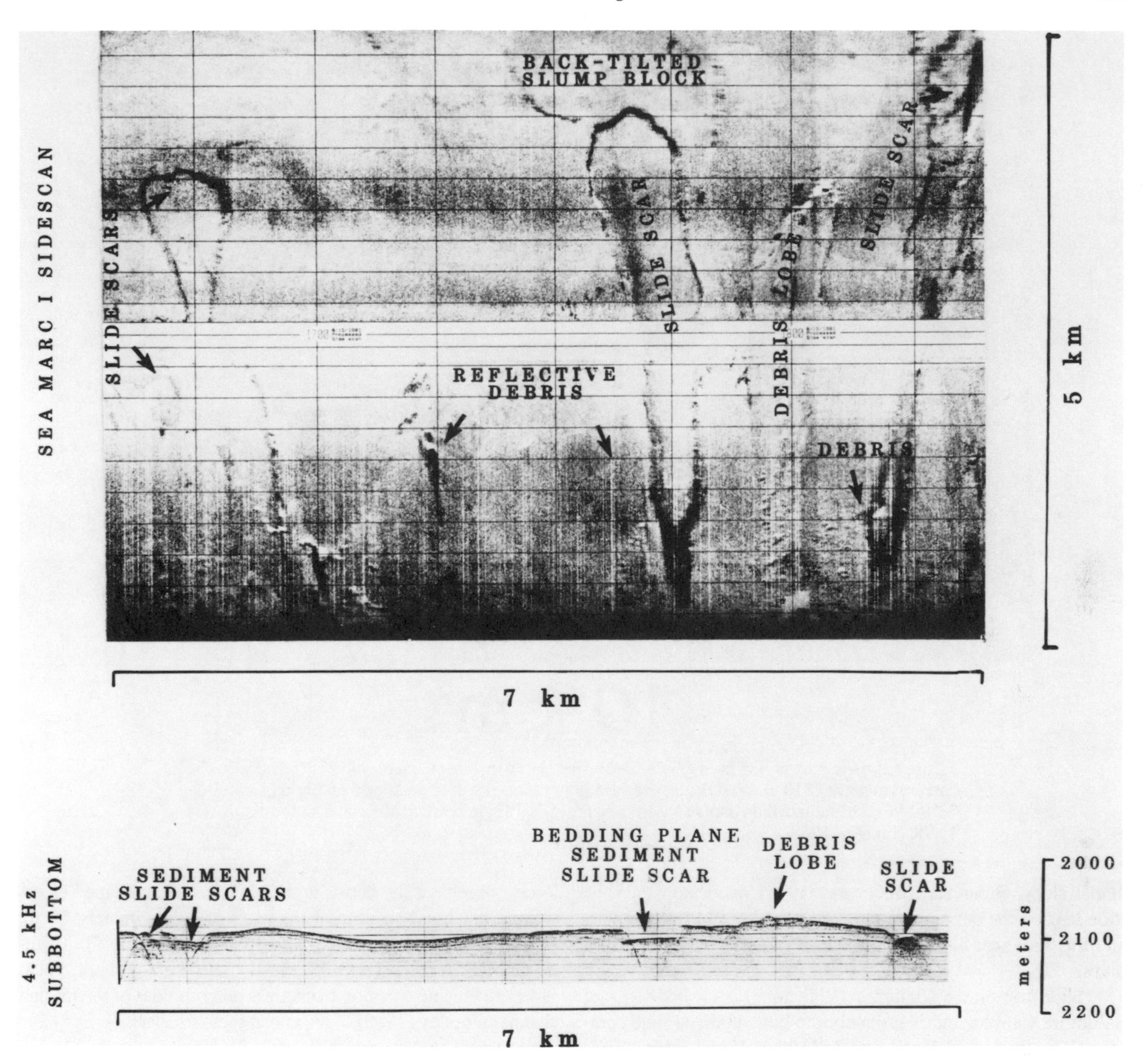

Figure 4. Top-Sea MARC I sidescan image of sediment slides on Georges Bank continental slope. Sidescan frequency is 30-kHz and the instrument was "flown" about 100 m above seafloor. Bottom-subbottom profile recorded from same vehicle along center line of sidescan record. The profile is approximately parallel to the regional contours. Data courtesy of John Farre (Lamont-Doherty Geological Observatory), collected jointly by Lamont-Doherty and the United States Geological Survey, Woods Hole, Mass.

parent wedge (debris-flow deposit) that has two narrow toes. Ductile faults and rotational slump blocks are observed upslope from the wedge.

Slide complex 5 (Norfolk slide zone) occurs on the continental margin south of Norfolk Canyon and seaward of Albemarle Sound. The mass-flow deposits extend down to at least 4300 m, and some of the flows may have entered the head of the Hatteras Tranverse Canyon. The total area covered by the deposits is approximately 11,000 km. Multiple slide deposits may occur in the lower portion of the complex; however, the echo-character data for the lower portion of the complex are limited and very difficult to interpret. The shallow part of this slide zone also is very poorly defined with our data set, but there are large (50 to 100 meter) scarps observed on the lower conti-

Figure 5. Lower part of box core RC19-49 showing clast structure characteristic of debris flows. The core was taken at 2810 m water depth near the northwestern edge of Slide Zone 6 on Fig. 6 (37°35.5′N, 73°00′W). The radiocarbon date of a clast from this core was greater than 40K whereas the matrix was 11.9K (Embley, 1980).

nental slope. Bunn and McGregor (1978) delineate a massive slide scar on the continental slope near the head of the complex, and this slide scar also shows up clearly in detailed bathymetric maps.

Slide complex 6 (Baltimore slide zone) is located south of Baltimore Canyon and is probably the best surveyed slide complex along the U.S. Atlantic margin (Malahoff and others, 1980; Embley, 1980; Malahoff and others, 1982; McGregor and others, 1979). The slides generally affect the uppermost 50 meters of sediment and the slip plane is observed as a well-defined acoustic reflector. Although there are several large slumps at the shelf edge (Malahoff and others, 1980), the dominant sedimentary process on the continental slope and upper rise appears to have been movement of small debris flows into and down the valleys.

Farther downslope, available 3.5-kHz profiles delineate an extensive area of the continental rise which has apparently been overridden by mass-flow deposits. These acoustically unstratified sediments lap onto the laminated continental-rise sediments to a depth of at least 3100 m. The deposits may be continuous down to a water depth of 4200 m, but the narrow neck in the middle of the complex is difficult to detect on some crossings. Where observed, the debris in the neck has a lens-shaped cross-section and has a maximum thickness of approximately 20 m. Near depths of 3600 m, the deposits fill or partially fill a submarine channel. Echograms clearly show that small, local slides feed the neck region from areas on both sides of the infilled channel. Embley (1980) suggests that these small slides may have been triggered by the arrival of mass flows from upslope.

A large number of piston and gravity cores from this region show ubiquitous slump masses and debris flows both in canyon floors on the continental slope and upper rise, and on the mid-continental rise (Embley, 1980, 1982; McGregor and others, 1979; Doyle and others, 1979; Embley, unpub. data). The existing data set of more than 40 corcs suggests that all of the mass-wasting events above 3000 m are of Wisconsin or early Holocene age. Many of the cores contain carbonate clasts of mid-Wisconsin age or older, intercalated with material of late Wisconsin age (Embley, 1980; Doyle and others, 1979; Fig. 5). Undated turbidite sand layers are present directly overlying the debris-flow deposits in several of the cores.

Slide complex 7 consists of a number of coalescing blocky slide and debris-flow deposits, some of which are buried. Analysis

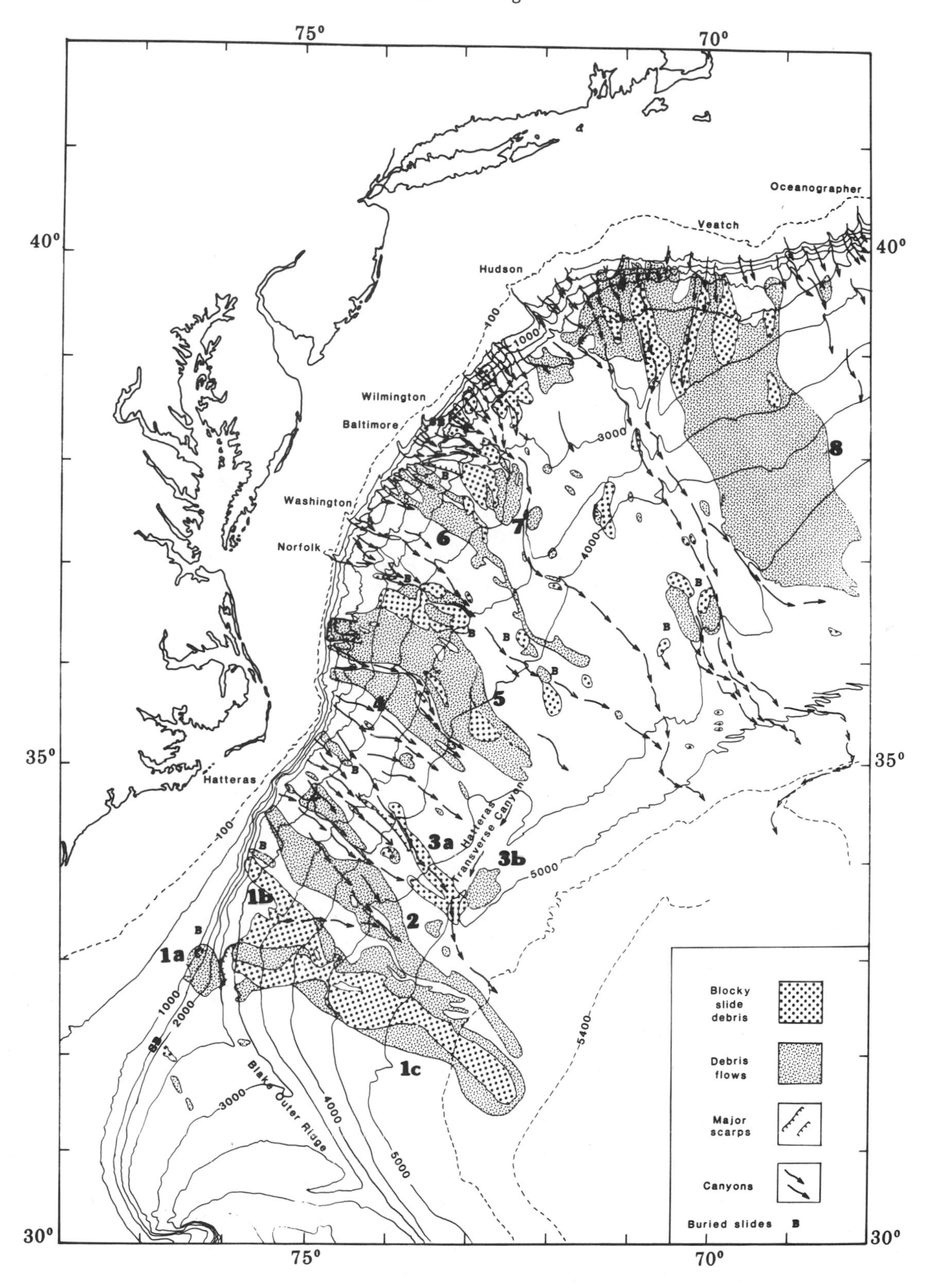

Figure 6. Summary map of mass-wasting areas on the U.S. Atlantic continental margin. Map is modified and simplified from Jacobi and others (1983) and Vassalo and others (1983). Blocky slide debris is characterized by uneven surface with numerous hyperbolic reflections with vertexes at irregular spacing (Fig. 3A). Debris flows are generally characterized by wedges of acoustically transparent sediment, in some places with prolonged fuzzy surface echo (Fig. 3, B and C).

of the complicated relations in this slide demands much more complete data than are presently available. The southwestern head of the slide complex is located along the Baltimore Canyon. The northwesternmost section of the slide complex is located along the Wilmington Canyon in the area of a proposed deep-seated slump block (McGregor and Bennett, 1977). This area may not be contiguous with the slide deposits to the east. The eastern extent of the slide complex is unknown, but debris flows may have entered the Baltimore/Wilmington Canyon system at 37.5°N.

Small patches of slide material are observed along the Baltimore/Wilmington Canyon system (e.g., 37°N-38°N; 72°W), but these deposits generally are isolated from slide complex 7. The lack of observed slide material beyond the immediate area of the canyon suggests that the deposits result from small, discrete slides on the flanks of the canyon.

In addition to the large slide zones already discussed, the upper continental rise and slope between 35°N and Hudson Canyon are characterized by numerous small slides which apparently either failed to generate large mass flows or occurred during an earlier cycle so that their deposits are buried. Sparker records across the upper rise and slope between Norfolk and Washington Canyons show numerous small scarps on the flanks of the canyons (Embley, 1980).

Also, detailed bathymetric, seismic, and mid-range (5 km) sidescan-sonar surveys of the continental slope in this region have revealed dramatic evidence for ubiquitous mass wasting (Ryan, 1982; Farre and others, 1983).

There are several large areas of the continental rise where 3.5-kHz records reveal small, buried slide deposits. These zones are primarily between slide complexes 5 and 6 and northeast of slide complex 7. The 3.5-kHz profiles clearly show a 5–10 meter layer of stratified, relatively undisturbed sediment overlying a mass of acoustically incoherent sediments. These flows probably were deposited during an older cycle of mass wasting. It is not clear whether or not the observed small zones of buried slide material comprise a single, much more extensive, buried slide sheet.

Hudson Canyon to Eastern Georges Bank

Uchupi (1967) and MacIlvaine (1973) discussed a large slump and series of slides on the southern margin of Georges Bank (slide complex 8, Fig. 6). Existing 3.5-kHz profiles through this region (Embley, 1980, Jacobi and others, 1983) demonstrate the existence of a very extensive slide complex which covers an area in excess of $4 \times 10^4 km^2$. The 3.5-kHz profiles are similar to those from other areas discussed above. Hyperbolated zones or transparent wedges are common on the continental slope and upper rise, while a transparent wedge typifies the lower portions of the slide complex. The track coverage in this area is relatively poor, with the result that the boundaries on most of the seismic facies are only approximate. Buried deposits can be identified within the slide complex and they attest to multiple movements of slide material.

Seismic profiles (MacIlvaine, 1973) and medium-range sidescan-sonar surveys over the northern portion of the slide (W.B.F. Ryan, pers. comm.) suggest that the geometries of the slides are similar to those south of Baltimore Canyon; a series of slides on the continental slope coalesced into the larger slide zone on the continental rise. Cores obtained from slide complex 8 commonly show disturbed bedding indicative of mass transport (Embley, 1980).

Canadian-Greenland Margins

The Nova Scotian continental rise was the classic area in Hollister's (1967) original study of contour-current sedimentation processes. Studies of this region for HEBBLE (High Energy Benthic Boundary Layer Experiment), however, show significant cross-slope movement of sediment (see Plate 7). The surveys (J. E. Damuth, pers. comm.) reveal that there are three large slide complexes at 40.5°N, 65°W; at 41.5°N, 64°W; and at 41°N, 63°W. The western and central complexes consist primarily of "blocky" slide material. The eastern complex is the largest of the three (approx. 1 × 10 km) and consists primarily of acoustically transparent wedges that probably are debris-flow deposits.

Slightly to the northeast, Hill and others (1982) documented the existence of debris flows around 42.7°N, 63.6°W; at least one of these occurred in late Wisconsin time (approximately 18,500 years B.P.). Stanley and Silverberg (1969) also documented late Wisconsin mass wasting on the continental slope south of Sable Island.

Although it is the only place along the North American margin where a documented historical mass-wasting and turbidity-current event occurred (in 1929; Heezen and Ewing, 1952), the Grand Banks sector of the margin has received relatively little attention in the literature. There have been several published seismic profiles (Heezen and Drake, 1964; Emery and others, 1970) which show some large slump scars, and a GLORIA survey (D. Roberts, pers. commun.) which shows large-scale mass wasting in this region, but there are no detailed coring/acoustic stratigraphic studies which could document the extent of historical events in this region.

However, a mid-range sidescan survey (Sea MARC I) in 1983 (Piper and others, 1985) covered the continental slope in the immediate vicinity of the 1929 earthquake. The survey revealed numerous thin debris-flow units generally downslope from irregularly shaped blocks that have apparently slipped short distances downslope. The implication is that large blocks rapidly disintegrate through grinding (shear) at their margins. Many of the debris-flow units appear to be very recent and they could be related to the 1929 earthquake. On the mid-Laurentian fan, massive debris-flow units that appear to be extremely young also are recognized.

Except for very broad reconnaissance studies, there is very little known about the Pleistocene to Recent sedimentary history of the northern Canadian and Greenland continental margins. In a regional summary of the area, Kristoffersen (1977) mapped two large slide zones (10 km^2 each) on the southern Greenland margin (centered about 58°55′N, 45°00′W and 58°00′N, 42°50′W). Hill and others (1982) also described Wisconsin-age debris flows in cores from northern Baffin Bay (72°–75°N).

Mass Wasting in Submarine Canyons

Numerous submarine canyons indent the continental margins and provide locally steepened slopes across which mass-wasting processes have occurred. Localized, small scale slumping along canyon walls has been documented by a number of submersible studies in canyons of eastern North America (e.g., Dillon and Zimmerman, 1970; Malahoff and others, 1982; Valentine and others, 1980). This localized slumping is an ongoing process due at least partly to the continuous undercutting and mining of the canyon walls by burrowing and digging benthic fauna. Some aspects of this process have been described by Warme and others (1978), Hecker (1982) and Malahoff and others (1982).

The rate and overall geologic importance of mass wasting as a secondary modifier of submarine canyons is unknown, but its existence as a process is well documented. More controversial is the possible role of mass wasting as a progenitor of submarine canyons. Recent work with long- and medium-range sidescan sonar along the U.S. Atlantic margin (Twitchell and Roberts, 1982; Popenoe and others, 1982; Ryan, 1982; Farre and others, 1983) has documented the ubiquity of second- and third-order channel systems in water depths too great to be explained by shelf-edge runoff during sea-level lowstands. It is suggested that these features may be bottleneck slides (Farre and others, in press) similar to those found on the fore-slope of the Mississippi delta (Prior and Coleman, 1977). Farre and others (1983) also suggest that large-scale mass wasting may have initiated formation of most or all submarine canyons along the U.S. eastern continental margin, while later channelized runoff (during sea-level lowstands) and turbidity flows modified and enlarged the canyons. One of the major lines of evidence cited in support of this hypothesis is the presence of numerous canyon-like incisions which extend only part way up the continent's slope between the major canyons; these features could not have been cut by turbidity currents originating near the shelf edge.

Florida and the Bahamas

The major escarpments along the Atlantic margins of the Bahamas platform and the Blake Plateau may be formed by erosional retreat through mass wasting. Up to 5 km of retreat has been proposed for the Bahama Escarpment (Freeman-Lynde, and others, 1981), and retreat of as much as 15 km may have occurred on the Blake Escarpment (Paull and Dillon, 1980). The major unanswered questions are 1) the relative importance of erosional mechanisms (spalling, slumping, biologic erosion), and 2) the fate of the thousands of cubic kilometers of sedimentary rock that have been removed.

Mullins and Neumann (1979) have documented the importance of mass-wasting processes along the carbonate margins of the northern Bahamas. Because of the large variations in slopes in this type of setting (1°–40°), all types of mass wasting are found, ranging from rockfalls and grainflows on the steepest slopes to debris flows on the gentler slopes.

Mid-Ocean Areas

Slides have also been documented in mid-oceanic areas of the western North Atlantic, in particular on the Bermuda Rise (Silva and others, 1976) and in the abyssal hills of the eastern Nares Abyssal Plain (Abbott, 1982). The ponded nature of the sediments and the occurrence of thick turbidite sequences in these ponds suggest that mass wasting of pelagic sediments is probably a primary mechanism of local redistribution of sediments in many areas of the Mid-Atlantic Ridge and abyssal hill provinces (Van Andel and Komar, 1969).

On the Mid-Atlantic Ridge, extensive talus piles of pillow basalts and other basic and ultrabasic rocks are generated by listric faulting and by earthquakes associated with the long-term shearing of the opposing plate boundaries. OTTER (1985) and Choukroune and others (1978), have documented the importance of these processes in the Oceanographer transform fault and in transform A (Mid-Atlantic Ridge axis, 37°N), respectively.

DISCUSSION

Moore (1961) and Morgenstern (1967) have shown that, as long as no excess pore pressures are present within the sedimentary column, slope failure of submarine sediments does not occur on slopes less than about 20°. However, slumping does occur on gentler slopes when excess pore pressures develop suddenly in the sediment column and the effective shear strength is reduced below the failure limit for the slope angle.

The classic Grand Banks study (Heezen and Ewing, 1952) established a definite relationship between the generation of a large turbidity current and an earthquake, but no slump or slide scar was ever found which could be definitely shown to have been contemporaneously formed with the turbidity current. Very large slump masses were thought to be present in the epicentral region, based on observations in seismic reflection profiles; however, these masses are now generally thought to be in-place sedimentary sections that were misinterpreted (Masson and others, 1985). It appears likely, particularly in view of Hampton's (1972) work that shows the transition between debris flows and turbidity currents, that large slide blocks (olistoliths) were triggered by the earthquake and that these mass movements subsequently generated a turbidity current. Relationships between several historical earthquakes and the generation of turbidity currents and/or slumps are well documented through the use of submarine cable-break records. The 1929 Grand Banks event, the 1886 Charleston earthquake (Bollinger, 1977), and a strong earthquake near Cape Ann, Massachusetts in 1755 (Coffman and Von Hake, 1973) demonstrate that significant seismic events do occur periodically even on passive continental margins. Because of the rare occurrence of such events, the underlying causes and geographic patterns of seismic events in intraplate areas is poorly understood, even though recent work has been increasing in this field (e.g., Sykes, 1978; Stein and others, 1979). There is some evidence that moderate-size earthquakes (approx. Richter magnitude 6) preferentially occur along old lines of weakness within a plate (Sykes, 1978). Examples of such lines of weakness include old sections of transform faults and the old boundary faults associated with original rifting of the Atlantic. Jacobi (1976) and Embley (1982) propose a relationship between the position of old fracture zones and the occurrence of sediment slides on the northwest African margin. Stein and others (1979) suggest that many of the large historical earthquakes along the northern Atlantic margins (e.g., the 1929 Grand Banks event and a 1933 Baffin Bay event) resulted from faulting on ancient rifts because of the stress drop associated with deglaciation. Stein also proposes (pers. com., 1983) that sediment loading outside previously glaciated areas could generate enough stress to trigger sediment movement. To date, all of these causal mechanisms for passive margin earthquakes are predicated on rather circumstantial evidence, primarily because the long recurrence time of intraplate earthquakes makes it difficult to see a "true" seismicity pattern. It is clear, however, that with the present state of understanding of intraplate seismicity, no area can be judged aseismic over a time span of thousands of years (see Zoback and others, this volume).

Although there have been several direct or indirect observations which correlate large-scale submarine slope failures to earthquakes (e.g. Heezen and Ewing, 1952), predicting the magnitude, distance from epicenter, and other factors that will cause a particular submarine slope to fail is a complex problem. Morgenstern (1967), Almagor and Wiseman (1977) and Marks (1980) have attempted to use actual observations to relate slope failure and earthquakes in a quantitative way. However, as Spudich and Orcutt (1982) point out, the magnitude of earthquake-induced seabed motion still is poorly understood. Marks (1980) worked on geotechnical data concentrated in an area of the continental slope north of Wilmington Canyon, and she suggests that earthquakes are not likely primary trigger mechanisms because 1) such regional shocks would tend to reduce all slopes in the area and 2) no such regional slope reduction is observed. However, this conclusion assumes lateral homogeneity in ground motion and sediment properties, and the latter are almost certainly not uniform (Silva and Booth, this volume).

Shifting in the level of the boundary between gas hydrate (clathrate) and the underlying free-gas boundary has also been suggested as a mechanism to induce slope instability. McIver (1977) and Summerhayes and others, (1979) proposed that pressure and temperature changes related to sea-level variations could change the sub-seafloor depth at which clathrate forms, thereby inducing instabilities. For example, a rise in the phase boundary could free gas from formerly hydrated sediment and cause overpressuring that leads to sediment failure.

Although associations of gas-saturated sediments and slope instabilities have been commonly noted in studies of the shallow regions of the Mississippi delta area (e.g., Roberts and others, 1976), it is unclear whether instability resulting from migration of the gas/hydrate phase boundary is a generally important triggering mechanism for submarine slides. Carpenter (1981) shows a direct correlation between the location of a clathrate horizon and a sediment slide. Thus there appears to be a relationship between clathrate occurrence and submarine slides at least in some areas. Furthermore, Dillon and others (this volume) mapped widespread occurrences of gas hydrate in sediments beneath the U.S. Atlantic margin, so hydrates could be a significant factor in mass movements over much wider areas. However, areas of seafloor above widespread hydrate layers lying beneath the crest of the Blake Outer Ridge show no apparent sediment instability, suggesting that mass movements associated with hydrated sediment are a multivariate phenomenon.

Undercutting at the base of a slope and subsequent failure in the form of progressive slumping is a common cause of subaerial-slope failure; this mechanism has also been suggested as a probable cause of failure of some submarine slopes (Berger and Johnson, 1976; Arthur and others, 1979). A similar mechanism is one possible way to explain the multiphase slumping south of Baltimore Canyon (Embley, 1982).

The majority of cores taken in areas affected by mass-wasting on the U.S. Atlantic continental slope and rise show that the debris flows and/or slumped deposits are covered with a pelagic or hemipelagic cap of sediments (Embley, 1980, 1982; Prior, and others, 1984). Radiocarbon dates of the base of the cap indicate that some mass-wasting events occurred within the Holocene (Embley, 1982; Stanley and others, 1984). We suggest that the continental slope is still coming to equilibrium after the "dumping" of sediments onto the margin during the last sea-level lowstand; this equilibration probably takes thousands of years and has occurred repeatedly on the U.S. Atlantic margin in response to sea-level cycles.

The overall importance of the mass-wasting process in the redistribution of sediment through geologic time still is unclear (Embley, 1980). The greatest insight into these problems probably will come with detailed spatial sampling by deep drilling and hydraulic piston coring, with long-term measurements of strong ground motion at the seafloor, and with expanded coverage by mid-range sidescan and multibeam profiling techniques.

REFERENCES

Abbott, D.
1982 : Processes Influencing Slope Stability in Marine Sediments: Rapid Deposition and Hydrothermal Circulation: Ph.D. Thesis, Columbia University, N.Y., 203 pp.

Almagor, G., and Wiseman, G.
1977 : Analysis of submarine slumping on the continental slope of the southern coast of Israel: Marine Geotechnology, v. 2, p. 349–388.

Arkhanguelsky, A. D.
1930 : Slides of sediments on the Black Sea bottom and the importance of this phenomenon for geology: Bull. Soc. Nat. Moscow, v. 38, p. 38–80.

Arthur, M., Von Rad, V., Cornford, C., McCoy, F., and Sarnthein, M.
1979 : Evolution and sedimentary history of the Cape Bajador continental margin, Northwestern Africa: in Initial Reports of the Deep Sea Drilling Project, v. 47, pt. 1, eds. U. Von Rad, and W.B.F. Ryan; U.S. Government Printing Office, Washington, D.C., p. 773–816.

Berger, W., and Johnson, T. C.
1976 : Deep-sea carbonates; dissolution and mass-wasting on the Ontong-Java Plateau: Science, v. 192, p. 785–787.

Bollinger, G. A.
1977 : Reinterpretation of the intensity data for the 1886 Charleston, South Carolina earthquake of 1886—a preliminary report: U.S. Geological Survey Professional Paper 1028, p. 17–32.
Bunn, A. R., and McGregor, B. A.
1978 : Morphology of the North Carolina continental slope, Western north Atlantic, shaped by deltaic sedimentation and slumping: Marine Geology, v. 37, p. 253–266.
Carpenter, G.
1981 : Coincident slump-clathrate complexes on the U.S. continental slope: Geo-Marine Letters, v. 1, p. 29–32.
Carter, R. M.
1975 : A discussion and classification of subaqueous mass-transport with particular application to grain-flow, slurry-flow and fluxoturbidites: Earth Science Review, v. 11, p. 145–177.
Choukroune, P., Francheteau, J., and LePichon, X.
1978 : In situ structural observations along transform fault A in the FAMOUS area, Mid-Atlantic Ridge: Geological Society of America Bulletin, v. 89, p. 1013–1029.
Coffman, J. L., and Von Hake, C. A.
1973 : Earthquake History of the United States: U.S. Dept. of Commerce Pub. 41-I (Revised Ed. Through 1970), 197 p.
Damuth, J. E.
1975 : Echo character of the Western Equatorial Atlantic Floor and its relationship to the dispersal and distribution of terrigenous sediments: Marine Geology, v. 18, p. 17–45.
Dillon, W. P., and Zimmerman, H. B.
1970 : Erosion by biological activity in two New England submarine canyons: Journal of Sedimentary Petrology, v. 40, p. 542–547.
Dott, R. H.
1963 : Dynamics of subaqueous gravity depositional processes: Bulletin American Association Petroleum Geologists, v. 47, p. 104–128.
Doyle, L. J., Pilkey, O. H., and Woo, C. C.
1979 : Sedimentation on the eastern United States continental slope: in Geology of Continental Slopes, ed. L. J. Doyle and O. H. Pilkey; Society Economic Paleontologists and Mineralogists Special Publication 27, p. 119–127.
Embley, R. W.
1976 : New evidence for the occurrence of debris-flow deposits in the deep sea: Geology, v. 4, p. 371–374.
Embley, R. W.
1980 : The role of mass transport in the distribution and character of deep-ocean sediments with special reference to the North Atlantic: Marine Geology, v. 38, p. 23–50.
Embley, R. W.
1982 : Anatomy of some Atlantic-margin sediment slides and some comments on ages and mechanisms: in Marine Slides and Other Mass Movements, eds. S. Saxov and J. K. Nieuwenhuis: Plenum, New York, p. 189–214.
Embley, R. W., and Jacobi, R.
1977 : Distribution and morphology of large submarine-sediment slides and slumps on Atlantic continental margins: Marine Geotechnology, v. 2, p. 205–228
Embley, R. W., and Morley, J.
1980 : Quaternary sedimentation and paleo-environmental studies off Namibia (Southwest Africa): Marine Geology, v. 36, p. 183–204.
Emery, K. O., Uchupi, E., Phillips, J. D., Bowin, C. O., Bunce, E. J., and Knott, S. T.
1970 : Continental rise of eastern North America: American Association Petroleum Geologists Bulletin, v. 54, p. 44–108.
Farre, J. A., McGregor, B. A., Ryan, W.B.F., and Robb, J.
1983 : Breaching the shelfbreak; passage from youthful to mature phase in submarine canyon evolution; in Shelf to Slope Break, eds. D. Stanley and D. Moore; Society Economic Paleontologists and Mineralogists Spec. Pub. No. 33, p. 25–39.
Freeman-Lynde, R. P., Cita, M. B., Jadoul, F., Miller, E. L., and Ryan, W.B.F.
1981 : Marine geology of the Bahama Escarpment: Marine Geology, v. 44, p. 119–156.
Hampton, M.
1972 : The role of subaqueous-debris flows in generating turbidity currents: Journal Sedimentary Petrology, v. 42, p. 775–793.
Hecker, B.
1982 : Possible benthic fauna and slope instability relationships; in Marine Slides and Other Mass Movements, eds. S. Saxov and J. K. Niewenhuis; Plenum, New York, p. 335–347.
Heezen, B. C., and Ewing, M.
1952 : Turbidity currents and submarine slumps and the 1929 Grand Banks earthquake: American Journal of Science, v. 250, p. 849–873.
Heezen, B. C., and Drake, C.
1964 : Grand Banks slump: American Association of Petroleum Geologists, v. 48, p. 221–225.
Heezen, B. C., Hollister, C. D., and Ruddiman, W. F.
1966 : Shaping of the continental rise by deep geostrophic contour currents: Science, v. 153, p. 502–508.
Hill, P. R., Aksu, A. E., and Piper, J. W.
1982 : The deposition of thin-bedded subaqueous debris flow deposits: in Marine Slides and Other Mass Movements, eds. S. Saxov and J. K. Nieuwenhuis; Plenum Publishing Co., New York, p. 273–287.
Hollister, C. D.
1967 : Sediment Distribution and Deep Circulation in the Western North Atlantic: Ph.D. Thesis, Columbia University, 368 pp.
Jacobi, R.
1976 : Sediment slides on the northwestern continental margin of Africa: Marine Geology, v. 22, p. 157–173.
Jacobi, R., and Hayes, D. E.
1982 : Bathymetry, microphysiography and reflectivity characteristics of the West African margin between Sierra Leone and Mauritania: in Geology of the Northwest African Continental Margin, eds. U. Von Rad, K. Hinz, M. Sarntheim and E. Seibold; Springer-Verlag, Berlin, p. 182–212.
Jacobi, R., Vassallo, K. L., and Shor, S.
1983 : Echo character, microphysiography and geologic hazards: in OMD regional data synthesis Series Atlas 4, eds. G. M. Bryan and J. R. Hiertzler, Marine Science International, Woods Hole, MA.
Karlsrud, K., and Edgers, L.
1982 : Some aspects of submarine slope stability: in Marine Slides and Other Mass Movements, eds. S. Saxov and J. K. Nieuwenhuis; Plenum, New York, p. 63–82.
Knebel, H., and Carson, B.
1979 : Small-scale slump deposits, middle Atlantic continental slope off eastern United States: Marine Geology, v. 29, p. 221–236.
Kristoffersen, Y.
1977 : Labrador Sea: A Geophysical Study: Ph.D. Thesis, Columbia University, New York, N.Y., 184 p.
MacIlvaine, J. C.
1973 : Sedimentary processes on the continental slope off New England: Ph.D. Thesis, MIT-WHOI, WHOI Technical Report, WHOI 73-58, 211 p.
Malahoff, A., Embley, R. W. Perry, R., and Fefe, C.
1980 : Submarine mass wasting of sediments on the continental slope and upper rise south of Baltimore Canyon: Earth Planetary Science Letters, v. 49, p. 1–7.
Malahoff, A., Embley, R. W., and Fornari, D.
1982 : Geomorphology of Norfolk and Washington Canyons and the surrounding continental slope and upper rise as observed from DSRV ALVIN: in The Ocean Floor, eds. R. A. Scrutton and M. Talwani; Wiley and Sons, New York, p. 97–112.
Marks, D. L.
1980 : Slope Stability Analysis of the Baltimore Canyon Region: Western Atlantic Ocean [M.S. thesis]: Cornell University, 99 p.
Masson, D. G., Gardner, J. V., Parson, L. M., and Field, M. E.

1985 : Morphology of upper Laurentian Fan using GLORIA long-range side-scan sonar: American Association Petroleum Geologists Bulletin, v. 69, p. 950–959.
McGregor, B. A., and Bennett, R. H.
1977 : Continental slope sediment instability northwest of Wilmington Canyon: American Association Petroleum Geologists Bulletin, v. 61, p. 918–928.
McGregor, B. A., Bennett, R. H., and Lambert, D. N.
1979 : Bottom processes, morphology and geotechnical properties of the Continental slope south of Baltimore Canyon: Applied Ocean Research, v. 1, p. 177–187.
McIver, R. D.
1977 : Hydrates of natural gas important agents in geologic processes: Geological Society of America Abstracts with Programs, v. 9, p. 1089.
Middleton, G. V., and Hampton, M. A.
1973 : Sediment gravity flows; Mechanics of flow and deposition: in Turbidites and Deep-water Sedimentation, eds. G. V. Middleton and A. H. Bouma; Society of Economic Paleontologists and Mineralogists Short Course Lecture Notes, p. 1–38.
Moore, D. G.
1961 : Submarine slumps: Journal of Sedimentary Petrology, v. 31, p. 343–357.
Morgenstern, N. R.
1967 : Submarine slumping and the initiation of turbidity currents: in Marine Geotechnique, ed. A. Richards; University of Illinois Press, Urbana, p. 189–220.
Mullins, H. T., and Newmann, A. C.
1979 : Deep carbonate margin structure and sedimentation in the Northern Bahamas; in Continental Slopes, eds. L. Doyle and O. H. Pilkey: Society of Economic Paleontologists and Mineralogists Special Publication 28, p. 165–192.
OTTER
1985 : Geology of the Oceanographer Transform: The Transform Domain: Marine Geophysical Researches, in press.
Paull, C. K., and Dillon, W. P.
1980 : Erosional origin of the Blake Escarpment: an alternative hypothesis: Geology, v. 8, p. 538–542.
Piper, D. J., Shor, A. N., Farre, J. A., O'Connel, S., and Jacobi, R.
1985 : Sediment Slides and Turbidity Currents on the Laurentian Fan: Sidescan Sonar Investigations Near the Epicenter of the 1929 Grand Banks Earthquake: Geology, v. 13, p. 538–541.
Popenoe, P., Coward, E. L., and Cashman, K. V.
1982 : A regional assessment of potential environmental hazards and limitations of petroleum development of the southeastern United States Continental shelf, slope and upper rise, offshore North Carolina: U.S. Geological Survey Open File Report, p. 82–136.
Prior, D. B., and Coleman, J. M.
1977 : Disintegrating retrogressive landslides on very-low-angle subaqueous slopes, Mississippi Delta: Marine Geotechnology, v. 14, p. 1–10.
Prior, D. B., and Suhayda, J. N.
1979 : Application of infinite slope analysis to subaqueous sediment instability, Mississippi Delta: Engineering Geology, v. 14, p. 1–10.
Prior, D. B., Coleman, J. M., and Doyle, E. H.
1984 : Antiquity of the continental slope along the Middle-Atlantic Margin of the United States: Science, v. 223, p. 926–298.
Roberts, H., Cratsley, D., Whelen, T., and Coleman, J.
1976 : Stability of Mississippi Delta sediments as evaluated by analysis of structural features in sediment borings: Eighth Annual Offshore Technology Conference Proceedings, v. 1, p. 9–28.
Ryan, W.B.F.
1982 : Imaging of submarine landslides with wide-swath sonar: in Marine Slides and Other Mass Movements, eds. S. Saxov and J. K. Nieuwenhuis; Plenum, New York, p. 175–188.
Shepard, F. P.
1955 : Delta front valleys bordering the Mississippi distributaries: Geological Society America Bulletin, v. 66, p. 1489–1498.
Silva, A. J., Hollister, C. D., Laine, E. P., and Beverly, B. E.
1976 : Geotechnical properties of deep-sea sediments: Bermuda Rise: Marine Geotechnology, v. 1, p. 195–232.
Spudich, P., and Orcutt, J.
1982 : Estimation of earthquake ground motions relevant to the triggering of marine mass movements: in Marine Slides and Other Mass Movements, eds. S. Saxov and J. K. Nieuwenhuis; Plenum, New York, p. 219–231.
Stanley, D. J., Nelson, T. A., and Stuckenrath, R.
1984 : Recent sedimentation on the New Jersey slope and rise: Science, v. 226, p. 125–133.
Stanley, D., and Silverberg, N.
1969 : Recent slumping on the continental slope off Sable Island Bank, Southeast Canada: Earth Planetary Science Letters, v. 6, p. 123–133.
Stein, S., Sleep, N. H., Galler, R. J., Wang, S., and Droeger, G. C.
1979 : Earthquakes along the passive margin off eastern Canada: Geophysical Research Letters, v. 6, p. 537–540.
Sykes, L. R.
1978 : Intra-plate seismicity, reactivation of pre-existing zones of weakness, alkaline magmatism and other tectonism postdating continental fragmentation: Reviews of Geophysics and Space Physics, v. 4, p. 621–688.
Summerhayes, C., Bornhold, B. D., and Embley, R. W.
1979 : Surficial slides and slumps on the continental slope and rise of southwest Africa: Marine Geology, v. 31, p. 265–277.
Twitchell, D. C., and Roberts, D. G.
1982 : Morphology, distribution and development of submarine canyons on the United States continental slope between Hudson and Baltimore Canyons: Geology, v. 10, p. 408–411.
Uchupi, E.
1967 : Slumping on the continental margin southeast of Long Island, New York: Deep-Sea Research, v. 14, p. 635–639.
Valentine, P. C., Uzmann, J. R., and Cooper, R. A.
1980 : Geologic and biologic observations in Oceanographer submarine canyon; descriptions of dives aboard the research submersibles Alvin (1967, 1978) and Nekton Gamma (1974): U.S. Geological Survey Open File Report 80-76, 40 p.
Van Andel, T., and Komar, P.
1969 : Ponded sediments of the mid-Atlantic Ridge: Geological Society of America Bulletin, v. 80, p. 1163–1190.
Vassallo, K. L.
1983 : Bottom reflectivity, microphysiography, and their implications for near-bottom sedimentation processes, western North Atlantic: Ph.D. Thesis, State University of New York at Albany, 201 pp.
Vassallo, K. L., Jacobi, R., and Shor, A.
1983 : Echo character, microphysiography and geological hazards: in OMD Regional Data Synthesis Series, Atlas 1, eds. J. I. Ewing, and P. D. Rabinowitz: Marine Science International, Woods Hole, MA, Map 40.
Veatch, A. C., and Smith, P. A.
1939 : Atlantic submarine valleys of the United States and the Congo Submarine Valley: Geological Society of America Special Paper 7, p. 100.
Walker, J. R., and Messingill, J. V.
1970 : Slump features on the Mississippi fan, Northeastern Gulf of Mexico: Geological Society of America Bulletin, v. 81, p. 3101–3108.
Warmer, J. E., Slater, R. A., and Cooper, R. A.
1978 : Bioerosion in submarine canyons: in Sedimentation in submarine Canyons, Fans and Trenches, eds. D. Stanley and G. Kellig; Hutchinson, Dowden and Ross, p. 65–70.

Manuscript Accepted by the Society October 15, 1985

ACKNOWLEDGMENTS

We thank Joyce Miller, Nancy Brown, and Susan Hanneman for their contributions to the preparation of this manuscript.

Printed in U.S.A.

The Geology of North America
Vol. M, The Western North Atlantic Region
The Geological Society of America, 1986

Chapter 30

Seabed geotechnical properties and seafloor utilization

Armand J. Silva
Departments of Ocean and Civil Engineering, University of Rhode Island, Narragansett, Rhode Island 02882
James S. Booth
U.S. Geological Survey, Woods Hole, Massachusetts 02543

INTRODUCTION

This chapter presents geotechnical physical property data on sediment in the western North Atlantic in the area of 20° to 45°N, and 50° to 85°W. Available data are synthesized to characterize major physiographic regimes in terms of pertinent engineering properties of the sediment, then these properties are interpreted in relation to utilization problems that may develop because of certain sedimentary processes or unfavorable sediment-deformational behavior. The data base for this synthesis, although still relatively meager in most areas, has been expanded considerably since the initial such synthesis by Keller and Bennett (1968) and the subsequent work by Horn and others (1974). In keeping with the purpose of the volume, this paper is a summary review and is not intended to exhaustively document all published and unpublished data.

Areas considered are those beyond the shelf break and include the continental slope, rise, and abyssal basin. The most abundant data are for the continental slope and rise, but studies are so concentrated locally that coverage is non-uniform. Available data for the deep-sea regime are sparse and most data are for only the upper two meters of sediment. Because of these limitations, this data base should not be used directly for particular site-specific applications. However, the preliminary generalizations resulting from this study are helpful in understanding processes and they provide a basis for planning more detailed studies.

In decreasing order of abundance, the following geotechnical data are available: water content and/or wet unit weight, grain size analysis, undrained shear strength (vane measurements), Atterberg limits (liquid and plastic), grain specific gravity, compressibility parameters, permeability, and strength parameters (from triaxial compression tests). The index properties (Table 1) are used in classifying sediments in an engineering context whereas the behavioral characterizations, such as the compressibility and strength parameters, are necessary for quantitative analysis of processes and for engineering calculations of seafloor applications.

Various combinations of the suite of geotechnical properties can be helpful in understanding geological processes such as mass wasting, erosion, furrowing, earthquake effects, intrusions, heat transfer, and diagenesis. The proper interpretation of bottom and subbottom acoustic characteristics commonly relies on knowledge of sediment physical properties, especially unit weight, velocity, grain-size distribution, and permeability. Utilization of the seafloor for civilian, scientific, and military purposes such as pipeline/cables, navigation aids, bottom-supported structures, mining operations, waste disposal, and instrumentation installations requires engineering calculations involving many of the properties mentioned above. For many applications it is imperative that an integrated approach be taken so that all the important environmental and imposed conditions are intercorrelated and incorporated into the analysis. For example, in the design of a structure or instrument that is supported on or in the sediment one must consider the sometimes very complex interactions amongst the ocean, the structure, the sediments (including earthquake effects) and also the atmosphere if the structure projects above the water. In addition to these interactions, the time element must also be considered since the properties of the sediments may change significantly over the life of the installation.

Of considerable importance to any engineering analysis and possible utilization of the seafloor is the stress history or state of consolidation throughout the sediment column. The usual measure is to compare the results of laboratory one-dimensional consolidation tests with the *in situ* stresses determined from physical property data, by determining the overconsolidation ratio, OCR (Table 1). Several investigators have reported very high OCRs for surficial deep-sea clays. The high OCR values are due to intrinsic interparticle bonding forces and/or cementation present in these very fine-grained sediments rather than true overconsolidation due to removal of overburden (Richards, 1967; Silva and Jordan, 1984). This zone of "apparent" overconsolidation can have important implications in the settlement calculations for bottom supported structures. In such situations, at least a part of the compression will be along a reload type of curve rather than along the steeper "virgin" compression portion of the curve; and the settlement will be much less (up to a factor of about 10) than

Silva, A. J., and Booth, J. S., 1986, Seabed geotechnical properties and seafloor utilization; *in* Vogt, P. R., and Tucholke, B. E., eds., The Geology of North America, Volume M, The Western North Atlantic Region: Geological Society of America.

TABLE 1. SYMBOLS USED FOR INDEX PROPERTIES AND BEHAVORIAL CHARACTERIZATION PARAMETERS

Symbol	Definition
	Index Properties
A	Activity = I_p/%<2μm.
G_s	Specific Gravity of mineral grains.
I_L	Liquidity Index = $(w-w_p)/I_p$, where w is the natural water content.
I_p	Plasticity Index = w_L-w_p.
e	Void ratio (e_o is the in situ value); ratio of void to solid volume.
w	Water content, expressed as percentage, based on dry weight (except as noted), usually corrected for 35 ppt salt content.
w_L	Liquid limit; minimum water content for "fluid" type behavior (remolded).
w_p	Plastic limit; minimum water content for "plastic" type behavior (remolded).
γ_b	bulk wet unit weight (gm/cm^3).
	Common Behavorial Parameters
C_c	Compression Index; slope of e log σ' curve in "normally consolidated" region.
C_s	Swell Index; the slope of the rebound-reload region of e log σ' curve.
OCR	Overconsolidation Ratio; ratio of preconsolidation stress, determined from test, to existing overburden stress.
S_t	Sensitivity; ratio of "undisturbed" to remolded strength.
S_u	Undrained shear strength (usually in kPa).
S_u/σ'_v	Ratio of undrained strength to effective overburden stress (useful in studies of stress history).
c'	Effective unit cohesion, (Mohr-Coulomb parameter, in kPa).
c_v	Coefficient of Consolidation; a measure of the rate of compression (L^2/T).
k	Coefficient of Permeability (or Hydraulic Conductivity); a measure of the resistance to water flow (usually in cm/s).
φ'	Effective angle of internal friction, Mohr-Coulomb parameter, (degrees).
σ'	Effective stress (usually in kPa).

if this condition were not taken into account (Silva and Jordan, 1984).

PROPERTIES, BEHAVIOR, AND PROCESSES

Overview

Locations of samples with geotechnical data are shown in Figure 1, but additional general information derived from publications is also discussed. There are 270 cores and borings represented in the data set. The geotechnical properties vary greatly and care must be exercised in the use of averages. There is much areal variability between and within provinces, and the vertical variability of some parameters at a given core location can be nearly as great as that between provinces.

Some uncertainties arise because of differences in coring techniques, sample handling, analytical techniques, and data-reporting methods among investigators. Shear strength is often of particular geotechnical interest, but unfortunately vane-shear determinations can vary significantly due to the equipment and procedures used; in some instances (e.g., coarser materials) vane measurements may be essentially meaningless.

Continental Slope

The continental slope is a relatively steep and narrow rampart between the Atlantic Ocean basin and the North American continent. It extends from the 75-200 m isobath at the edge of the continental shelf, to a depth of approximately 2000 m, where it is bounded to seaward by the continental rise. Regional slope angles may be as low as one or two degrees, but they commonly approach 10° (Keller and others, 1979) and may be greater than 15°, as on the Blake Escarpment (Emery and Uchupi, 1972). Locally, declivities frequently exceed 20° and some canyon walls are essentially vertical (see, for example, O'Leary and Twichell, 1981). Canyons, which may be hundreds of meters deep, often have rock outcrops and mass-wasting features which contribute to the rugged nature of the terrain. However, some intercanyon sections of the slope are typified by low relief.

The slope has been extensively modified through time (see Schlee and others, 1979), and during the Tertiary and Quaternary the prograded slope was subjected to large-scale net erosion as a result of a series of transgressive-regressive sea level cycles. Quaternary sediment is mostly terrigenous silt and clay although calcareous sediment is locally dominant off the southern United

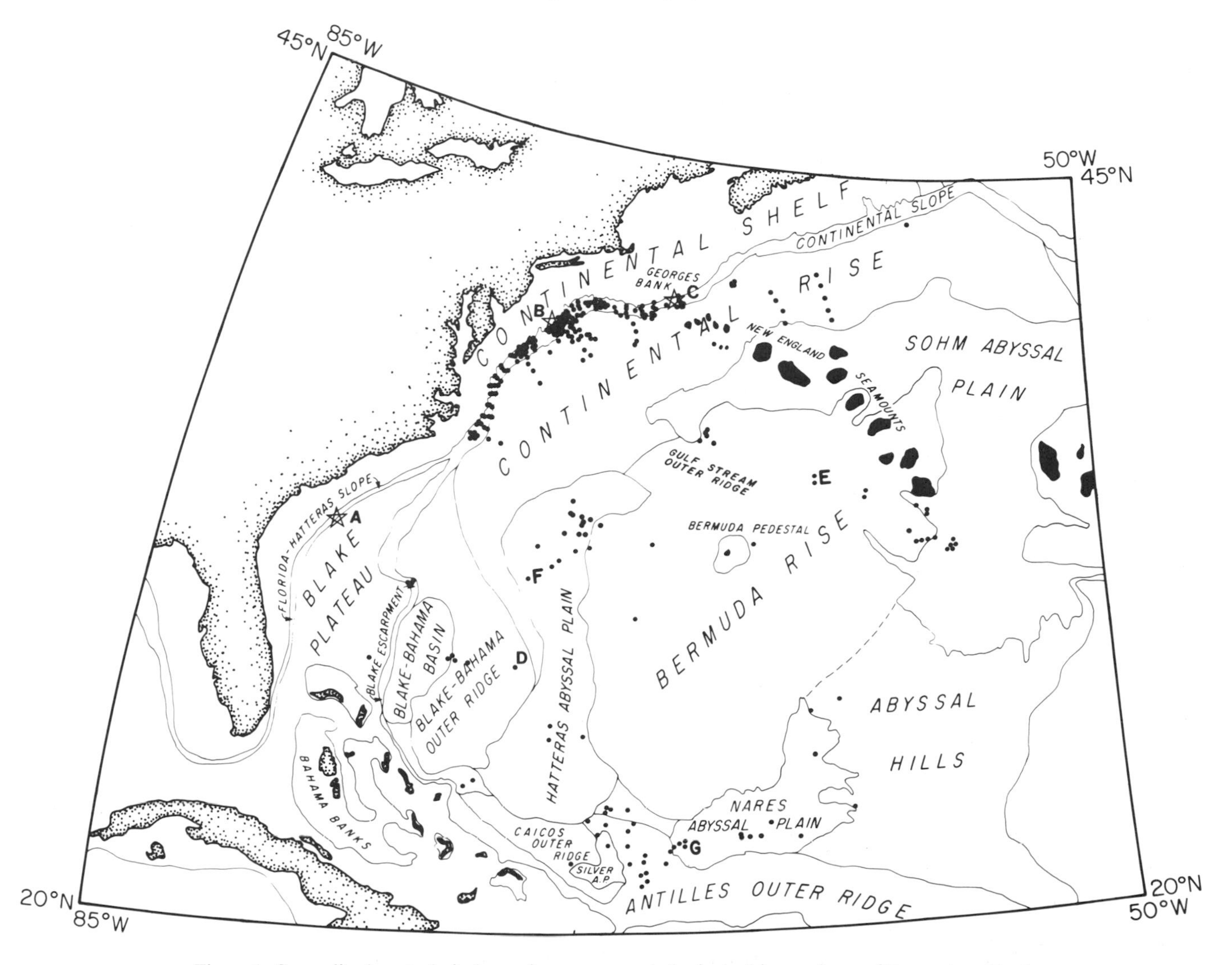

Figure 1. Generalized geotechnical sample coverage and physiographic provinces of the western North Atlantic. Sites *A, B,* and *C* refer to AMCOR sites 6004, 6021, and 6013, respectively (see Hathaway, 1979). Sites *D–G* locate cores in figures 5–7.

States. The Quaternary sediment varies in thickness from more than 400 m on parts of the upper slope to nil on parts of the lower slope and in some canyon axes. At present, the slope contains sites of both active deposition and degradation.

Slope North of Cape Hatteras. The slope north of Cape Hatteras has received by far the most geotechnical attention of any area in the western North Atlantic (Fig. 1). Properties of the surficial sediments (upper 2 m) in this subprovince are summarized in Table 2. Illite and chlorite are the dominant clay minerals in the region, but smectite, kaolinite, and other types are also present in small quantities (Hathaway, 1972). Quartz and feldspar are the most abundant of the non-clay minerals. Calcium carbonate, chiefly as planktonic foraminiferal tests, is quantitatively important (>10%) in some areas and organic matter may constitute as much as several percent of the sediment (Doyle and others, 1979). Texturally, silt and clay are the dominant size fractions, and they exist in about equal amounts in the sediment. Sand content, however, is not insignificant, and even larger grain sizes are locally present. Glacially derived gravels, ice-rafted coarser materials (off the northern U.S. and the Canadian Maritime provinces), and allogenic formational clasts have been identified. Note in Table 2 that any one of the three primary size classes can make up nearly all of a particular sample.

The wide ranges shown in the index properties (Table 2) are largely a reflection of the textural variability. This effect is most obvious in the water content (w) data, which span nearly an order of magnitude and show a strong negative correlation with mean grain size. Similarly, void ratio (e) and wet bulk unit weight (γ_b) show marked variability. All of these properties show a downslope trend, with w and e increasing and γ_b decreasing (Keller and others, 1979; Booth and others, 1985a, 1985b).

The other index properties also exhibit a wide range of values. The plasticity indices (I_P), which vary from a minimum of 3 to a maximum of 80, particularly exemplify the wide range in

TABLE 2. GEOTECHNICAL PROPERTIES OF SEDIMENTS ON THE U.S. CONTINENTAL MARGIN AND BLAKE-BAHAMA OUTER RIDGE

	Province Characteristics (upper 2 meters)														
	Continental Slope north of Cape Hatteras					Continental Rise					Blake-Bahama Outer Ridge				
Slope Declivities:	1°-11° (regional); 15°-30° (local)[§]					0.1°-2°					0.1°-3°				
Sediment type:	Clayey silt					Silty clay					Calcareous silty clay				
Dominant minerals:	Illite, quartz, chlorite, feldspar					Illite, quartz, chlorite					Illite, smectite, quartz				
Property	C*	n†	min	max	mean	C*	n†	min	max	mean	C*	n†	min	max	mean
Sand (%)	~145	~600	0	75	11	~35	~250	0	98	7	4	13	0	4	1
Silt (%)	~145	~600	17	88	46	~35	~250	2	80	42	4	13	23	41	32
Clay (<2μ) (%)	~145	~600	4	73	44	~35	~250	0	85	50	4	13	58	74	66
w (%)	~165	~1300	23	182	87	~35	~700	28	172	93	4	43	86	171	138
Gs	~145	~750	2.59	2.84	2.75	~35	~400	2.56	2.87	2.78					
γ_b (g/cm^3)	~145	~800	1.41	2.05	1.58	~35	~700	1.34	1.99	1.53					
e	~135	~700	.8	3.6	2.3	~35	~700	.7	4.33	2.3					
A	~20	~100	.4	10.2	.85	~10	22	.6	3.2	1.8					
w_P (%)	~145	~700	15	59	34	~35	~250	16	54	32	2	3	32	44	37
w_L (%)	~145	~700	24	135	76	~35	~250	26	109	73	2	3	82	136	104
I_P (%)	~145	~700	3	80	42	~35	~250	12	81	41	2	3	47	92	67
I_L	~145	~700	.2	4.8	1.3	~35	~250	.8	2.9	1.4	2	3	1.5	1.6	1.6
S_u (kPa)	~165	~700	.6	90.0	8.7	~35	~400	<.5	19.8	6.1	4	39	<.5	19.2	3.8
S_t	~165	~600	1	15	3	~35	~200	1	22	6					
c' (kPa)	14	14	0	20	7	4	4	0	3	2					
ϕ' (o)	14	14	14	33	27	4	4	20	31	26					
S_u/σ'_v	2	2	.30	1.00	.65	4	4	.20	1.00	.50					
OCR	16	16	.75	372	4.2**	1	1	---	---	3.0	2	2	1.23	7.01	4.1
C_c	17	17	.06	1.02	.41	1	1				2	2	1.31	2.22	1.77
C_s	12	12	.01	.09	.05	---	---	---	---	---					
c_v (cm^2/s)	3	3	.0007	.001	.0009	1	1	---	---	.003					
k x 10^{-6} (cm/s)	3	3	.07	.3	.1	1	1	---	---	.6					

*Number of cores used for evaluation.
† Number of measurements used to estimate mean.
§ Canyon walls may approach vertical.
**Mean value excludes data from obvious erosional surfaces.

Note: Data are salt corrected

Data sources: Booth and others (1981a, 1981b), Booth and others (1985a, 1985b); Keller and others (1979); Lambert and others (1981); McGregor and others (1979); Olsen and others (1981); Olsen and others (1982); Richards (1962); Silva and Hollister (1979); Silva and others, unpub. reports, (1981, 1982, 1983).

engineering characteristics of these sediments (the plasticity index is an indication of the range of water content in which the material behaves plastically, i.e., will deform easily). In general, however, these slope sediments are designated as inorganic clays of high plasticity with a Unified Soil Classification symbol of CH (see Wagner, 1957, for details on this engineering classification system). The slope sediments tend to have a low strength and have very low permeability. However, strengths may actually exceed 90 kPa on erosional surfaces, which occur in canyons and on large sections of the lower slope off the central United States. Samples recovered from these surfaces display exceedingly high OCRs as well, indicating that thick sections of overburden have been removed. The average overconsolidation (OCR=4.2) of the slope surficial sediments is presumably due to "apparent" overconsolidation, as noted earlier. The compression indices (C_c), hydraulic conductivities (k), angles of internal friction (ϕ'), and other data are typical of sediments of this texture and composition, namely a fine-grained material dominated by illite and less active minerals. According to the classification of Rosenqvist (1953), these sediments are characteristically of medium sensitivity (S_t) in terms of loss in strength due to reworking.

Two Atlantic Margin Coring Project (AMCOR) borings on the slope, 6021 and 6013, illustrate the vertical variability in properties. Hole 6021 (Fig. 2), located on the upper slope off New Jersey, shows little change in texture and consists of silty clay throughout. However, the shear-strength profile shows significant changes in strength occurring over very short vertical distances. There is a major reduction in strength centered near 70 m, where the value is about 10 kPa. Methane gas, which has been identified in several locales on the upper slope (e.g., Hall and Ensminger, 1979), and was analytically measured in this boring (Hathaway, 1979), may have escaped from and disturbed the sediment, accounting for this low value. The plasticity index varies by a factor of 2 in the upper 100 m, and the liquidity index is greater than 1

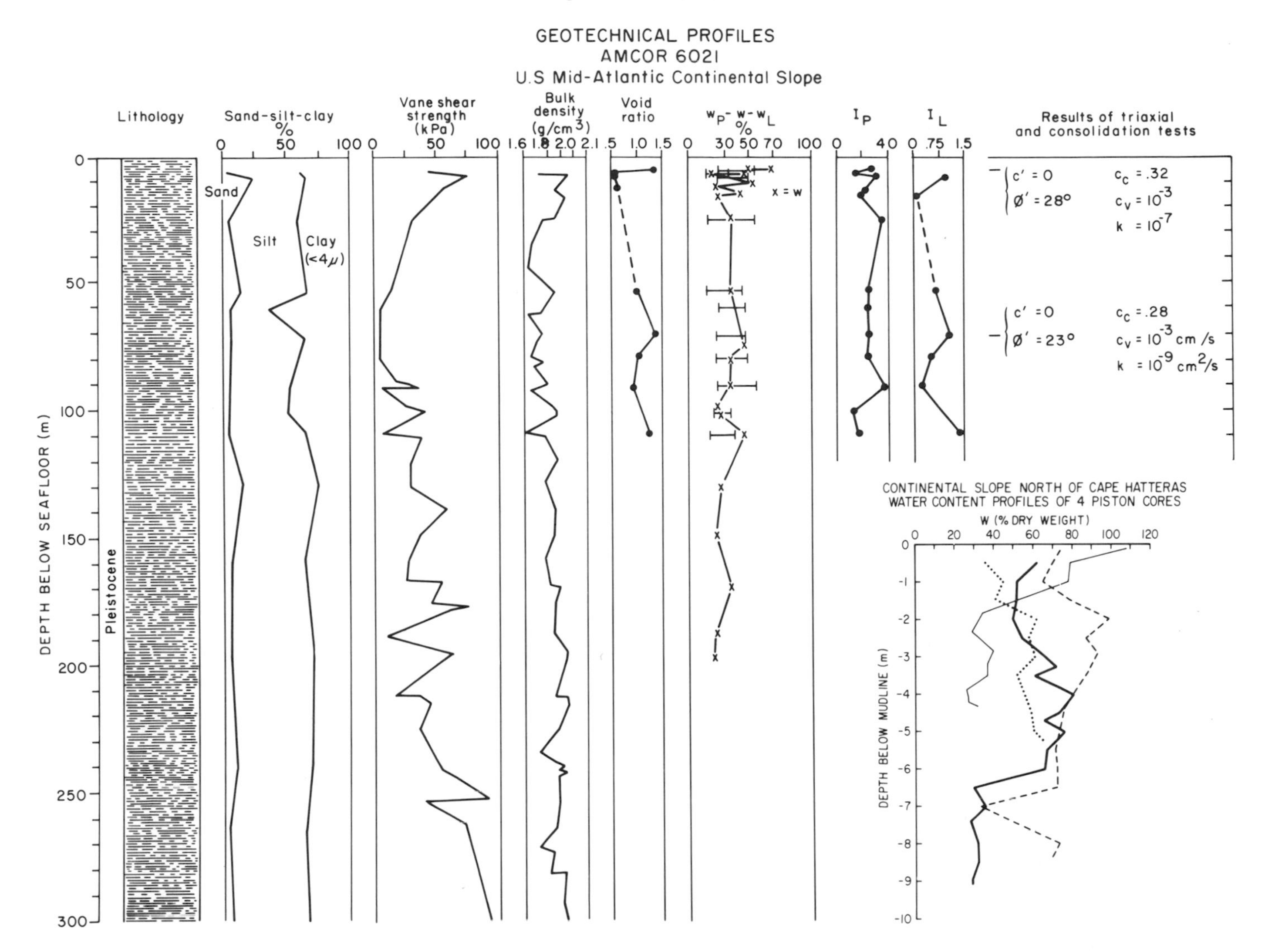

Figure 2. Geotechnical profiles and data from AMCOR borehole 6021 (Sources: Ferrebee [1981], Richards [1978], Swanson and Brown [1978]. Strength measurements were made with a hand-held vane. Location of AMCOR 6021 labelled *B* on figure 1. *Inset:* Water content profiles of 4 piston cores. Note that within a given core water content values may increase, decrease, fluctuate, and/or span a wide range over a given stratigraphic depth (data from Booth and others, 1985b).

at the 100 m depth. The plot of water content values, although relatively smooth, is not detailed enough to show the true variability. The inset in Figure 2 gives much greater resolution and thus a more valid indication of this variability. The triaxial and consolidation test results are similar to those reported for nearby surface (upper 2 m) sediments.

Data from AMCOR Hole 6013 further emphasizes the variability that exists on the continental slope. This boring, which is on the upper continental slope off Georges Bank, has very high sand content and three notable lithologic changes (Fig. 3). The wet bulk unit weight *decreases* down hole, due to a decrease in grain size of the sand fraction. The generally low w_P, w, and w_L values are in accord with the sandy nature of the sediment, as are the low compression indices (C_c) and the high coefficients of consolidation (c_v). However, the ϕ' values (28°) are somewhat low for this type of sediment.

From data on these two AMCOR borings, it is evident that any area of the slope can present a diversity of engineering conditions. The non-uniformity observed both laterally and vertically testifies that the continental slope north of Cape Hatteras has been a complex and dynamic geologic environment, with depositional processes varying in type and relative importance in recent time.

The canyons within this slope province, because they are so numerous and prominent, warrant special comment. Most canyon floors are essentially erosional surfaces but they can be covered by a veneer of hemipelagic sediment, sand, and/or

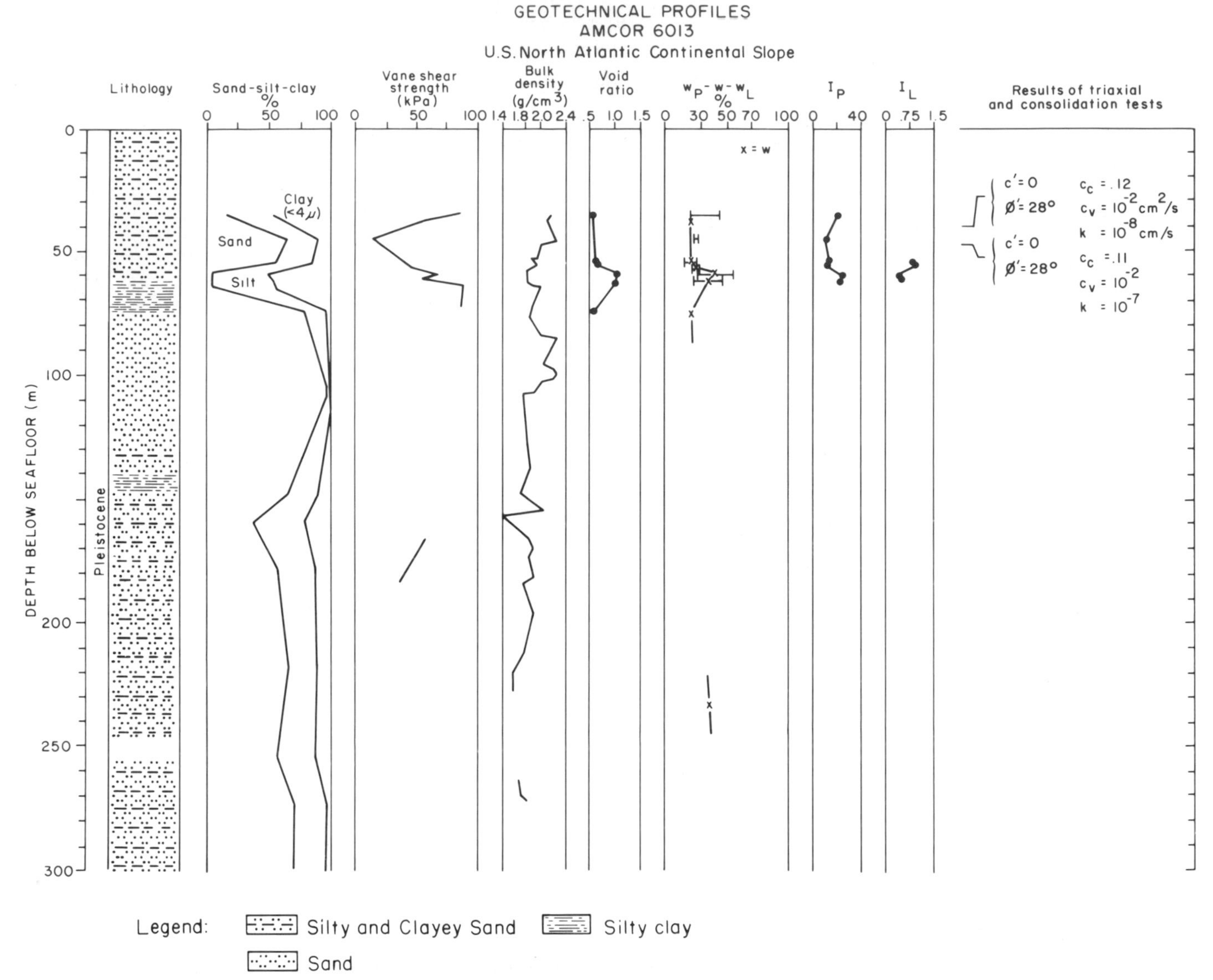

Figure 3. Geotechnical profiles and data from AMCOR borehole 6013 (Sources: Ferrebee [1981], Richards [1978], Swanson and Brown [1978]. Strength measurements were made with a hand-held vane. Location of borehole labelled *C* on figure 1.

gravel. Other mass wasting products, such as blocks and rubble, may also be present. The geotechnical properties of the erosional surfaces are poorly established to date because of sampling difficulties. The few available data indicate that shear strength can approach or exceed 100 kPa, and OCR values can exceed 100. The Quaternary, Tertiary, and possibly Cretaceous sediment that crops out in the canyons has significant compositional differences and corresponding major differences in engineering classifications. Thus there may be a great spatial contrast in geotechnical properties due to the presence of canyons alone, but this contrast is at least partially predictable on the basis of local relief.

Slope South of Cape Hatteras. Three distinct physiographic provinces (Florida-Hatteras Slope, Blake Plateau, and outer slope) are present and canyons are rare. The surficial sediments throughout this region generally share two characteristics: a high content of sand-size material and the presence of significant amounts of carbonate.

On the Florida-Hatteras slope (with a gradient of ~2–3°), Doyle and others (1979), Doyle and others (1981), and Ayers and Pilkey (1981) have shown that sand content is commonly over 90% and carbonate commonly exceeds 70%. The dominant clay minerals are kaolinite and smectite (Hathaway, 1972). The sediment becomes fine toward the base of this slope, where a "mini-rise" has been formed (Ayers and Pilkey, 1981). The geotechnical properties of the Florida-Hatteras Slope are represented by boring AMCOR 6004 (Fig. 4). The erratic shear strength profile, generally centered around 25 kPa, shows no general increase in strength until a depth of nearly 160 m, where a silty clay

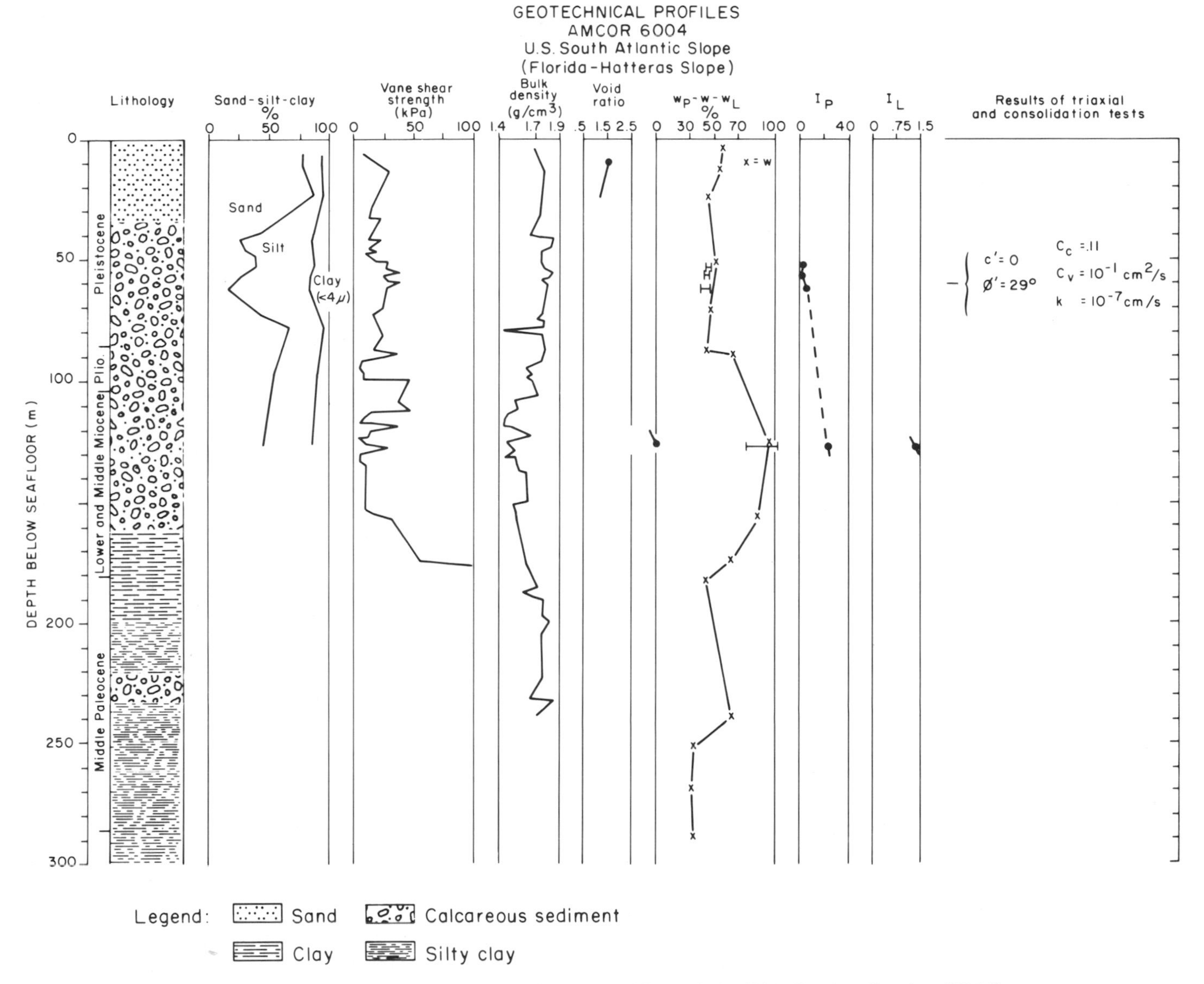

Figure 4. Geotechnical profiles and data from AMCOR borehole 6004 (Sources: Ferrebee [1981], Richards [1978], Swanson and Brown [1978]). Strength measurements were made with a hand-held vane. Location of borehole labelled *A* on figure 1.

is present. The very low I_P values in the upper part of the section underscore the effect of the sand component, as does the low compressibility (C_c=0.11). The effective stress friction angle, ϕ' = 29°, is low for a sediment that has such a high sand content.

The Blake Plateau (Fig. 1) also is covered by sands and carbonate debris, although a hard phosphorite-ferromanganese pavement covers a significant portion of the northern part. Elsewhere, foram sands occupy smooth areas, and coral debris is found in areas marked by irregular topography (Emery and Uchupi, 1972). Grain size tends to become finer to the south (Hollister, 1973). Results of geotechnical tests from a location in the southeast portion of the plateau (Lee, 1976) showed an average water content of 60% in the upper 2 m and a grain specific gravity of 2.71. Sensitivity is fairly high (5–10), as is shear strength (average ~20 kPa). Triaxial tests showed that ϕ' is about 32° and cohesion is nil.

The upper slope of the Blake Escarpment is also rich in carbonate. The surficial sediment is characteristically a clayey silt, although some sections have a relatively high sand content (Hollister, 1973). Sand averages more than 20% in cores from the vicinity of the Blake Spur (Richards, 1962; Richards and Hamilton, 1967), and this is reflected in the geotechnical data (w_L: 35%–51%; I_P: 2–11; A: 0.06–0.29). In general γ_b is abnormally low (~1.5 g/cm^3); and w (~90%) and e (~2.5) are high for a sand-dominated sediment. This unusual combination of properties probably reflects the fact that the sand is comprised of foram tests, which can have a relatively high volume of internal void space. Regarding stress-strain properties, S_u is reltively high in the upper 0.5 m (avg. 12 kPa) and the sediment exhibits "apparent overconsolidation." Compressibility is high ($C_c = 0.75$), perhaps due to crushing of hollow foram tests during testing; this may be a significant factor for engineering on this type of seafloor.

As the basic sediment type in this province is significantly different from that found north of Cape Hatteras, so are the dominant processes. In particular, the Gulf Stream and Antilles Current have a major influence on sediments in the area, sweeping sediment across and off much of the Blake Plateau; the Florida-Hatteras slope is modified to some degree by deposition of fines from a Gulf Stream countercurrent; and the Blake Escarpment is affected by the Western Boundary Undercurrent (e.g., see Hollister, 1973).

Continental Rise

The continental rise is a broad, gently sloping depositional apron extending east from the base of the continental slope (Fig. 1). It extends from a depth of about 2 km to more than 5 km, where it generally merges with the abyssal plain. Gradient generally ranges from 1° to 2° near the slope to less than 0.1°seaward.

Most surficial sediments of the rise are silty clay. Both terrigenous and biogenous sands are present in variable amounts. As shown in Table 2, the proportions of sand, silt, and clay vary to the extremes. Gravel and coarser material are present locally. The mineralogical suite is similar to that of the adjacent continental slope with illite being the most abundant clay mineral (Hathaway, 1972). Chlorite is dominant in some areas of the Grand Banks region. Smectite and kaolinite are also present as minor constituents. Biogenic carbonate constitutes up to 25% of the sediment (Emery and Uchupi, 1972).

Like the northern continental slope, the index-property data show marked variability, and we assume a correlation to texture. Also like the northern continental slope, the sediment is classified as "CH," but considering the range in I_P, many different types of engineering material are present. The most salient difference from the slope sediment is the activity, A, which is much higher (mean value of 1.8), indicating a more plastic type of stress-strain behavior. Except for S_u and S_t, the limited amount of data on stress-strain properties precludes generalizations. The sediments are very sensitive according to the classification of Rosenqvist (1953) and have a low shear strength.

The surface of the continental rise is continuously modified by the Western Boundary Undercurrent, which intermittently exceeds threshold velocities for grain movement (McCave and Tucholke, this volume). Thus, the current can scour the bottom, create bedforms, and redistribute sediment. Turbidity currents also modify the surface by eroding, depositing, and being primarily responsible for the presence of terrigenous sand in the province (Pilkey and Cleary, this volume). Also, as a result of slides generated on the continental slope (Plate 7), mass-wasting products are observed on the rise. Recent studies on the Nova Scotian rise show that brown mud with very low shear strength (~0.4 kPa) can overlay a much stiffer (~4 kPa) foraminiferal ooze over parts of the area (McCave and others, 1984). This very soft mud can be several centimeters thick, and it evidently has been eroded, transported, and redeposited as a result of intense, episodic bottom currents.

Outer Ridges and Bermuda Rise

Blake-Bahama Outer Ridge. The 1100 km-long Blake-Bahama Outer Ridge system is a seaward extension of the continental rise extending from the base of the continental slope near Cape Hatteras. The southern (Bahama Ridge) segment forms the seaward boundary of the Blake-Bahama Basin (Fig. 1). It ranges in depth from 2 to 5 km and has declivities from less than 0.1° to greater than 3°. The ridge is formed of sediments transported and deposited by deep boundary currents (McCave and Tucholke, this volume).

A geotechnical investigation of the surficial sediments of this province was conducted by Silva and Hollister (1979), whose results are summarized in Table 2. All four cores represented in the table are from the southern part of the area (Bahama Outer Ridge). In contrast to parts of the continental slope, the upper few meters of outer-ridge sediments are essentially devoid of sand and clay-sized material is dominant. Illite is the dominant clay mineral, but smectite, chlorite, and kaolinite are also present. Carbonate constitutes about 10% of the sediment.

The index properties reflect the texture and mineralogical suite. Average water content is high (mean of 138%), as is average w_P, w_L and I_P, in comparison with coarser, continental-slope sediment. Typically, I_L is greater than 1. High C_c values are the most salient aspect of the behavioral properties; at an average value of 1.77, they are greater than most terrestrial soils by a factor of three.

Geotechnical properties in profile are shown in Figure 5. There are consistent sand, silt, and clay percentages downcore; an overall, though erratic, decrease in water content; and a steady, classic increase in strength. Not shown in profile, but reported by Silva and Hollister (1979), is a tendency for the sediment in this region to become increasingly underconsolidated with depth below the seafloor (meaning that the sediment is not completely consolidated for the existing overburden). In general, a state of underconsolidation appears to be common in the region.

Bermuda Rise and Gulf Stream Outer Ridge. The sediment of the northern Bermuda Rise consists primarily of hemipelagic, current-deposited clay with thicknesses exceeding 1000 m, thereby effectively smoothing most basement topography (Laine and others, 1984). The upper 200–300 m in some areas are acoustically highly stratified and consist of illitic clays with variable calcium carbonate content up to 40%. Elsewhere, including outcrops along the edges of the region, there are acoustically transparent pelagic brown clays. Mass wasting has occurred along some of the steeper outcrop areas.

Almost all the cores from the stratified region show extreme

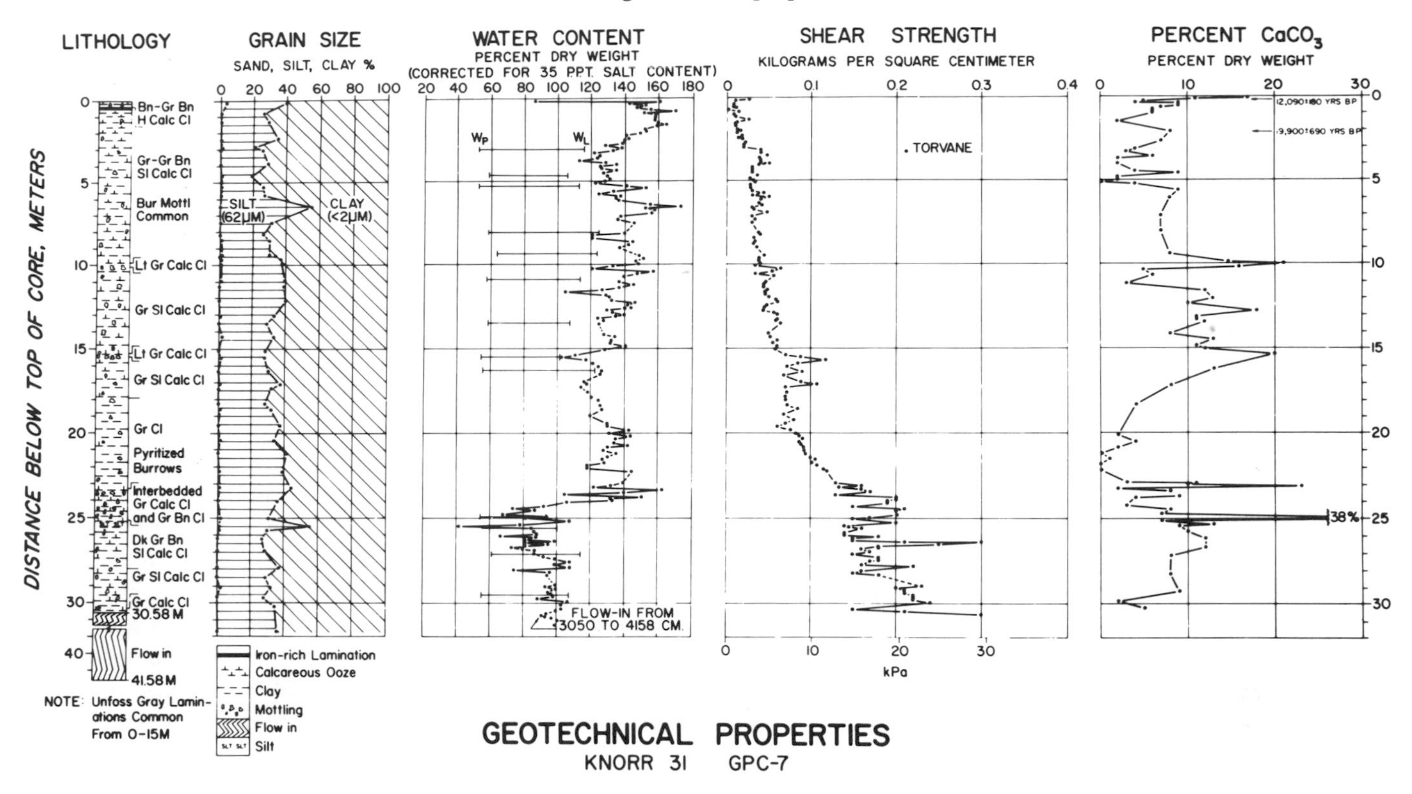

Figure 5. Geotechnical profiles of giant piston-core GPC-7 from the Bahama Outer Ridge (Silva and Hollister, 1979). Location of core sample labelled *D* on figure 1.

vertical variability in water content, with excursions from about 80% to over 200% sometimes occurring within a few centimeters (Fig. 6). Corresponding variations in wet bulk unit weight and sound velocity probably account for the presence of many, closely spaced subbottom acoustic reflectors. The variation usually is not caused by thin silt layers with lower water content (although these occur) but is rather due to fairly thick (several centimeters) sequences of clay with alternately high and low water content. A persistent zone of higher water content (>180%) and low unit weight begins at 1.0 to 3.0 m subbuttom and is 0.5 to 2.0 m thick in various cores; it seems to pervade the northern Bermuda Rise. Below this zone of high water content the value fluctuates by sometimes more than 100%, but there is a gradual decrease with depth. In the upper 2 m there appears to be a west to east increase in water content (from 90% to 110%).

The Atterberg limits (w_L and w_P, Table 1) follow the same general trends as water content. Liquidity index is substantially greater than one in the upper 15 m (more than 2.0 in the upper zone of high water content), and it appears to approach unity at about 25 m. The compression index varies from about 0.6 for the upper 2 m to about 1.8 in some deeper zones. Extensive consolidation data is available for one large-diameter core from the stratified zone (Silva and others, 1976). Even accounting for possible sample-disturbance effects, there is strong evidence that the sediment is underconsolidated (average OCR of ten tests below 2 m is 0.42). Additional research is warranted to check this result both experimentally and analytically.

There is evidence of episodic erosion with furrowing and mass wasting on the northeastern flanks of the Bermuda Rise (Laine and others, 1984; Silva and others, 1976). Slopes are locally quite steep (approximately 16°), and stiff, acoustically transparent sediment underlying the stratified sediment is just below the seafloor. Slump deposits are present at the toe of the slope bordering the Sohm Abyssal Plain. The geotechnical properties of the outcrops are quite different from the overlying laminated sediments. The water content profile shows a relatively constant value of about 100% between 5 and 15 m, the shear strength increases almost linearly with depth from about 2 kPa at the surface to over 20 kPa at 15 m, and the sediment is normally consolidated.

Elsewhere on the Bermuda Rise, the upper few meters of sediments show "apparent" overconsolidation with very high OCR values (>40) near the surface with rapidly decreasing values approaching unity at depth. On the other hand, there is considerable evidence showing that sediments over large areas of the Bermuda Rise are underconsolidated at depths below 8 m (Silva and Jordan, 1984; Silva and others, 1976). That is, the strata below the apparently overconsolidated surficial sediments are not fully consolidated under the existing overburden conditions; this is probably due to a combination of high sedimentation rates, low permeability, and the presence of the zone of "apparent" overconsolidation (which has the effect of lengthening the drainage path of underlying zones).

The Gulf Stream Outer Ridge occurs on the western side of

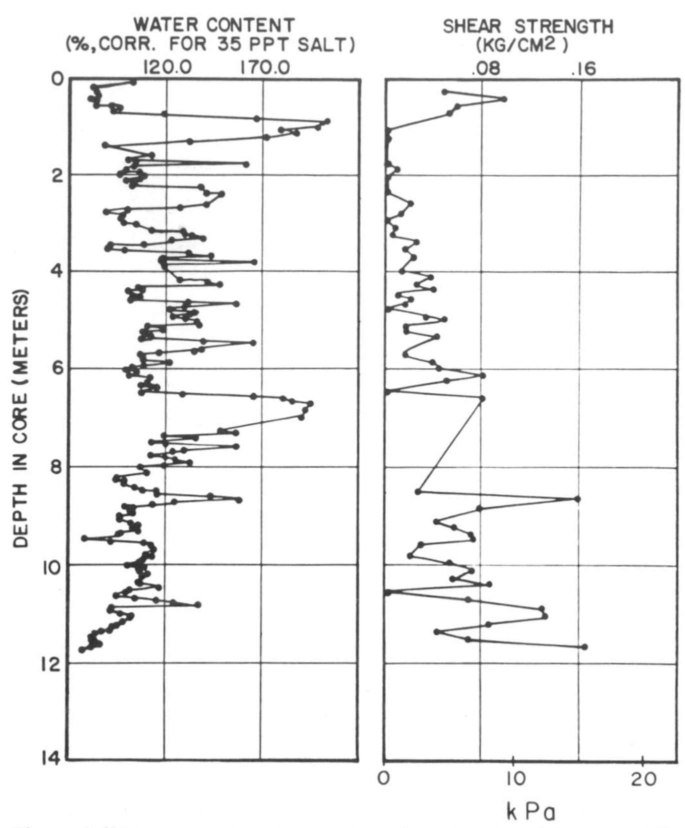

Figure 6. Water content and shear-strength profiles for piston core EN-023, LPC-01 (Silva and Calnan, 1980). Location of core sample labelled *E* on figure 1.

the Bermuda Rise but probably was once part of the Hatteras Outer Ridge along the lower continental rise (Tucholke and Laine, 1982). On the eroded western edge of the Gulf Stream Outer Ridge, "outcropping" reflectors are capped by recent sediments, which have properties similar to those of the surficial sediments on the northeast Bermuda Rise. Only one 4.2 m gravity core (EN-023, LGC-20 at 35°54.8′ N; 66°23.3′ W) apparently penetrated this veneer. Consolidation-test results indicate that at least 6–7 m of material has been eroded from that location (Silva and Jordan, 1984).

Greater Antilles Outer Ridge. The Greater Antilles Outer Ridge is a broad topographic swell that rises from the southeastern edge of the Hatteras Abyssal Plain, extends generally southeast to the Puerto Rico Trench, and continues east along the northern edge of the trench. It is composed of a large accumulation of acoustically transparent sediment up to 800 m thick (Tucholke and Ewing, 1974). Sediment from the Outer Ridge contains an average of less than 15% biogenous material.

The data base for this region is rather limited (Table 3) and therefore only a few comments are warranted. The upper few meters consist of silty clay with water content quite high at an average of 118%. The high liquidity index (~1.8) and fairly high shear strength (3.0 kPa) would indicate that a condition of apparent overconsolidation exists in the surficial sediment. The Atterberg limits are indicative of an illitic clay, and overall the geotechnical properties appear to be very similar to those of the Bermuda Rise. Below 2 m depth there is a decrease in water content and an increase in shear strength (table 3).

Abyssal Plains

The three major abyssal plains (A.P.) in the western North Atlantic are the Sohm, Nares (with which we here include the Silver A.P.), and the Hatteras. The sediments consist of turbidite sequences with alternating clays, silts, and sands of varying thickness. The clay sized material is occasionally calcareous (Pilkey and Cleary, this volume). Abyssal hills, either capped with pelagic clays or covered only with a thin veneer of turbidites, project through distal parts of the abyssal plains. Usually the physical properties directly reflect textural changes. With increasing distance from sediment source areas there is a general but not ubiquitous trend toward finer texture and a decrease in vertical variability (Tucholke, 1980). For example, average water content (w) of surficial sediments increases from 76% for the Hatteras A.P. to 99% for the Nares A.P. with a concurrent change in texture (sand decreases from 29% to a trace; Table 4). However some cores, even in distal parts of the abyssal plains, show a considerable downcore variability in water content and bulk density (Tucholke, 1980).

Surficial sediments in the distal reaches of the northeastern Nares A.P. are very fine clays with low shear strength (less than 2 kPa) and high water content (over 130%); however, available data are very sparse. In one piston core (Buckley and Cranston, 1982) the shear strength increases from about 5 kPa at 1 m to 15 kPa at 8 m depth, but there is essentially no data on stress history or stress-strain parameters. Nonetheless this strength profile suggests that the sediment column probably is normally consolidated.

Two profiles for the Hatteras and Nares abyssal plains are shown in Figure 7. The Hatteras profile exhibits more variation in all measured properties (Tucholke and Shirley, 1979). However, examination of other associated profiles reveals that some areas farther east in the Nares A.P. have almost as much variation as observed in the Hatteras A.P. It therefore appears that the abyssal plain regimes are also geotechnically complex and it would be misleading to attempt any generalizations.

Most of the data available for the Sohm A.P. are toward the southern regions and therefore they are not representative of the large areas of the abyssal plain that are proximal to the continental margin. The surficial sediment in the distal areas consist of fine clays and clayey silts with high water contents (over 150%) and very low strengths (Table 4). Layers of stiff, sand-sized material at 3 to 10 meters depth (sand content of over 60%) indicate that coarse turbidites have been episodically transported over great distances.

TABLE 3. GEOTECHNICAL PROPERTIES OF SEDIMENTS ON BERMUDA RISE GULF STREAM OUTER RIDGE AND GREATER ANTILLES OUTER RIDGE

	Bermuda Rise					Gulf Stream Outer Ridge					Greater Antilles Outer Ridge				
	Depth (m)	Min	Max	Ave	No. Cores	Depth (m)	Min	Max	Ave	No. Cores	Depth (m)	Min	Max	Ave	No. Cores
w (%)	0-2	81	122	100	10	0-2	71	124	102	4	0-2	94	143	118	16
	2-32	88	128	105	10	2-4	86	122	106	2	2-22	77	145	109	8
γ_b (g/cm^3)	0-2			1.52	1	0-2	2.65	2.74	2.69	4	0-2	1.38	1.50	1.44	5
	2-4	1.47	1.52	1.48	1	2-4	2.71	2.82	2.77	1	2-12	1.32	1.58	1.47	3
e_o	0-2	2.5	2.9	2.8	3	0-2	2.5	1.8	3.0	3					
	2-20	2.2	3.4	2.7	4	2-4	2.9	3.2	3.0	1					
G_s	0-2	2.78	2.87	2.80	2	0-2	2.64	2.79	2.73	3					
	2-20	2.72	2.80	2.75	3	2-4	2.83	2.83	2.83	1					
Sand (%)	0-2	0	3	1	6	0-2	2	8	5	2					
	2-32			*	7	2-4	-	-	*	1	0-2	*	4	1	2
Silt (%)	0-2	25	47	38	6	0-2	34	49	30	2	0-2	11	22	17	2
	2-32	25	39	31	7	2-4	-	-	44	1					
Clay (%)	0-2	50	72	60	6	0-2	43	64	54	2	0-2	73	89	82	2
	2-32	60	74	69	7	2-4	-	-	55	1					
w_L (%)	0-2	70	70	70	2	0-2	57	124	97	4	0-2	65	88	73	2
	2-32	84	121	102	6	2-4	63	100	80	1					
w_P (%)	0-2	25	30	27	2	0-2	27	50	37	3	0-2	31	39	34	2
	2-32	34	46	40	6	2-4	31	42	38	1					
I_P (%)	0-2	40	45	42	2	0-2	30	72	56	4	0-2			38	2
	2-32	47	77	62	6	2-4	3	60	45	1					
I_L	2-32	1.0	1.7	1.3	3	0-2	1.1	1.3	1.2	3	0-2	1.6	2.3	1.8	2
						2-4	-	-	1.8	1					
A	2-32	5.0	7.6	6.0	3	0-2	0.80	1.10	.95	4					
						2-4	-	-	.85	1					
S_u (kPa)	0-2	1.0	6.5	2.8	7	0-2	0	17.6	7.6	4	0-2	1.8	5.4	3.0	12
	2-20	2.0	35.0	9.2	8	2-4	8	22.0	12.0	3	2-22	2.9	8.0	5.9	5
S_t	2-32	17	57	32	3						0-2	2.3	8.0	3.9	2
C_c	0-2	0.63	1.12	0.78	10	0-2	0.9	1.5	1.2	4					
	2-26	0.47	1.28	2.86	6	2-4	0.9	1.1	1.0	2					
OCR	0-2	1.01	1.41	1.21	2	0-2	2.78	17.5	6.7	6					
	2-32	0.13	0.99	0.77	3	2-4	0.74	2.5	1.6	5					

* = Trace

Data Sources: Silva and others (1976); Silva and others (1981, 1982, 1983); Silva and Jordan (1984); Silva and Calnan (1980); Tucholke (1973).

SEAFLOOR UTILIZATION AND ENGINEERING CONSIDERATIONS

General Comments

Many geotechnical properties and data on engineering behavior of sediments can be used to gain insight into geologic processes. For example, it is now obvious that mass wasting on the continental slope is a major factor in shaping the slope and in providing sediment to deeper areas (Embley and Jacobi, this volume). Geotechnical analyses can be used to develop quantitative models of these processes. On the other hand, knowledge of geotechnical behavior in itself is necessary to calculations for engineering utilization of the seafloor and for assessment of natural hazards. It is not possible in this short treatment to discuss details of the interrelations, but since a greater use of the seafloor is almost inevitable, it is of interest to explore some geotechnically important problems and concepts in three of the physiographic regimes for which data have been presented (Table 5). Further detail on design and analysis is available in many excellent papers and monographs (e.g., Sangrey, 1977; Almagor and Bennett, 1984).

We emphasize that design for most engineering installations requires detailed, site-specific information; thus the data presented in this paper can be viewed only as a framework for planning detailed studies. Some of the factors listed in Table 5 can be

TABLE 4. GEOTECHNICAL PROPERTIES OF SEDIMENTS IN HATTERAS, NARES, AND SOHM ABYSSAL PLAINS

	Hatteras Abyssal Plain					Nares Abyssal Plain					Sohm Abyssal Plain				
	Depth (m)	Min	Max	Ave	No. Cores	Depth (m)	Min	Max	Ave	No. Cores	Depth (m)	Min	Max	Ave	No. Cores
w (%)	0-2	38	110	76	18	0-2	71	144	111	10	0-2	70	210	100	12
	2-11	50	90	71	15	2-12	41	86	66	9	2-13	60	120	86	9
γ_b (g/cm^3)	0-2	1.4	1.86	1.56	13	0-2	1.40	1.64	1.45	9	0-2	0.48	1.48	.97	4
	2-11	1.24	1.75	1.60	12	2-12	1.46	1.68	1.52	9	2-13	0.58	1.51	.87	4
e_o											0-2	2.5	4.6	3.4	6
											2-13	2.4	3.4	2.8	6
G_s						0-2	2.72	2.88	2.79	1	0-2	2.71	2.82	2.72	4
											2-13	2.77	2.79	2.78	3
Sand (%)	0-2	5	48	25	6				*	4	0-2			*	5
	2-8	3	53	28	6						2-13	0	16	7	4
Silt (%)	0-2	12	44	32	6	0-2	7	51	25	4	0-2	25	54	35	5
	2-8	15	34	27	6						2-13	20	42	34	4
Clay (%)	0-2	34	78	47	6	0-2	48	89	75	4	0-2	46	75	64	5
	2-8	21	76	46	6						2-13	0	80	59	4
w_L (%)	0-2	32	80	68	4	0-2	51	86	67	4	0-2	51	74	66	5
	2-4	28	79	62	3						2-13	46	75	62	4
w_P (%)	0-2	26	30	28	4	0-2	24	39	30	4	0-2	23	36	37	5
	2-4	13	29	23	3						2-13	18	32	28	4
I_p (%)	0-2	6	54	40	4	0-2	27	47	38	4	0-2	20	50	35	5
	2-4	15	51	39	3						2-13	14	43	34	4
I_L	0-2	0.7	6.8	2.4	4	0-2	1.4	2.8	2.1	4				3.2	1
A											0-2	0.5	0.7	0.6	3
											2-13	0.5	0.6	0.55	4
S_u (kPa)	0-2	1.3	8.7	3.9	10	0-2	1.9	8.4	4.0	7	0-2	0	3.0	1.8	8
	2-5	2.7	15.4	5.7	5	2-12	5.5	13.5	9.2	4	2-13	0	3.5	2.4	9
S_t	0-2	3.5	7	5.2	4	0-2	1.8	7.5	3.8	3					
C_c											0-2	1.11	2.20	1.27	5
											0-12	0.77	1.83	1.25	4
C_s											0-2	0.04	0.13	0.08	2
											2-12	0.11	0.22	0.16	4
OCR											0-2	2.1	9.1	4.05	4
											2-12	0.3	2.2	1.08	5
k @ σ'_o (cm/s) x 10^{-6}											0-2	0.16	2.16	1.44	3
											2-12	0.12	3.41	1.20	4

* = Trace

Data Sources: Tucholke (1980); Laine and others (1982); Levy and others (1983); Silva and Calnan (1980).

thought of as environmental hazards (i.e., mass wasting, creep, seismic effects) but in an engineering sense they are viewed as factors that must be considered in a proposed utilization.

We hesitate to preclude certain engineering utilizations based on what we know now and can reasonably predict in the future. Almost anything is feasible from an engineering point of view, provided there is the commitment to do it and the financial resources are available. What was considered to be almost impossible in offshore operations just a few decades ago is actually now being accomplished. Therefore we purposefully treat seafloor utilization in very general terms here and encourage continued, detailed studies oriented toward more specific possible applications.

Finally, a few comments seem appropriate on the nature of marine sediments from a geotechnical perspective. The geotechnical principles, which were developed primarily through work on terrestrially based soils, are valid for marine sediment. This has very important implications because it means we can use the basic geotechnical methodologies, such as the effective stress principle, in offshore studies and applications. To be sure, there are some important differences in the behavior of most marine

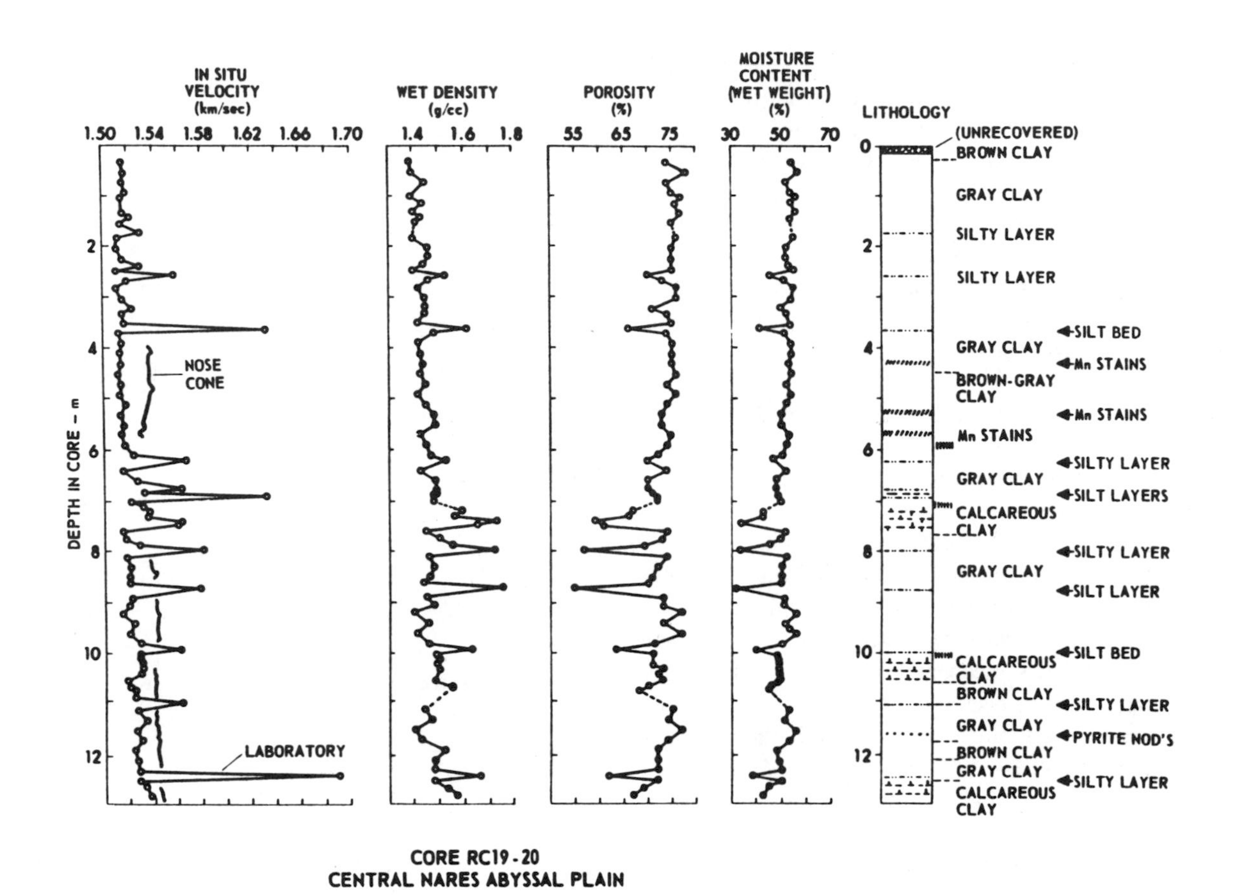

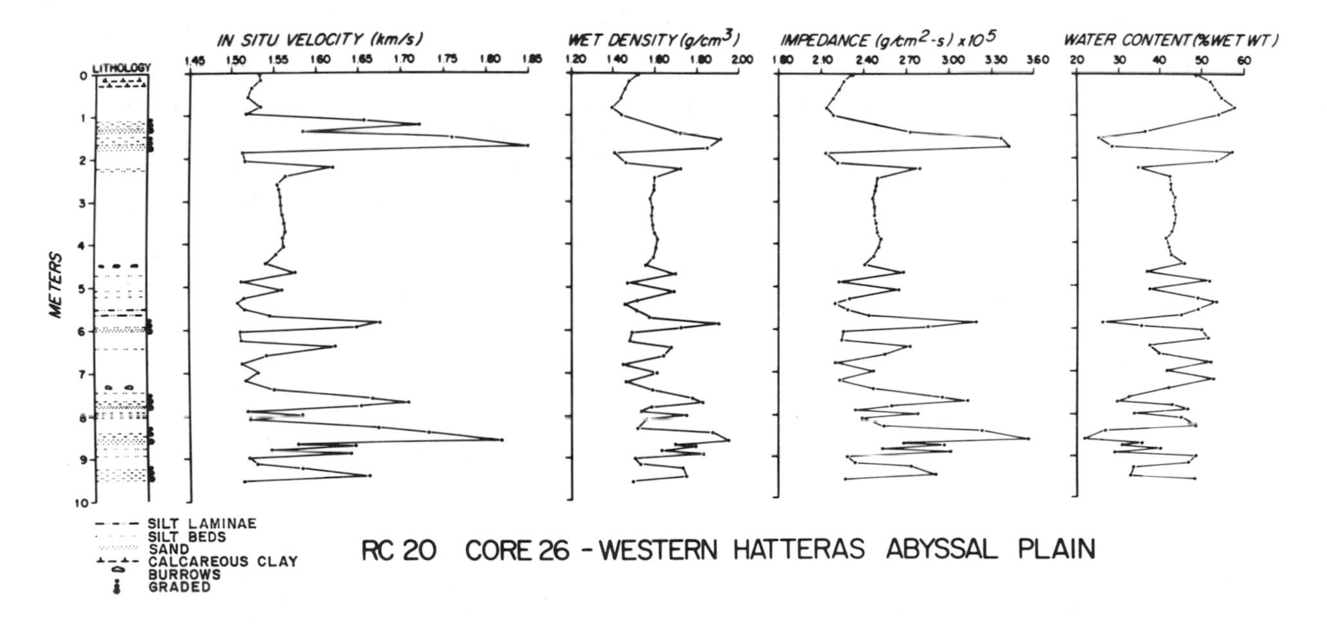

Figure 7. Geotechnical profiles from piston cores RC19-20 (Nares Abyssal Plain) and RC20-26 (northwestern Hatteras Abyssal Plain). Locations of piston cores labelled *G* and *F* on figure 1. From Tucholke and Shirley (1979) and Tucholke (1980). Note that water content is wet-weight basis.

sediments as compared to most terrestrial soils; for example, marine sediments tend to have higher void ratios and are usually more compressible. However, the differences are more in terms of degree rather than in basic phenomena, and the behavior can generally be predicted from known geotechnical relationships.

One of the problems of studying seafloor properties and processes is the difficulty in obtaining suitable relatively undisturbed samples and *in situ* measurements, both of which are necessary for many geotechnical studies.

Continental Slope and Rise

Most foreseeable activity will be concentrated in this very complex area which has perhaps the greatest variety of possible constraints and potential hazards to engineering use. Several investigators (e.g., Embley and Jacobi, 1977; McGregor, 1977; MacIlvaine and Ross, 1979; O'Leary and Twichell, 1981; Robb and others, 1981; and Popenoe and others, 1982) have identified numerous products of mass wasting throughout the U.S. continental slope and rise, and have presented evidence for various types of slides, debris flows, creep, and other degradational processes. Creep processes have been reported by Bennett and others (1977) in an area near Wilmington Canyon and many investigators (e.g., Lambert and others, 1981) have identified turbidites on the continental rise. The findings of all these authors are reflected in Table 5.

TABLE 5. SEAFLOOR UTILIZATION AND ENGINEERING CONSIDERATIONS

Factor	Upper Cont. Slope	Lower Cont. Slope and Rise	Ocean Basins
Uses			
Projecting Structure	Yes	No	No
Tension Leg	Yes	Possibly	No
Bottom Completion	Yes	Yes	No
Anchor/Moorings	Yes	Yes	Possibly
Pipelines	Yes	Yes	No
Cables	Yes	Yes	Yes
Instruments	Yes	Yes	Yes
Mining	Possibly	Yes	Possibly
Waste Disposal	No	Yes	Possibly
Engineering Considerations			
Wind forces	Yes	No	No
Surface Waves: Direct	Yes	No	No
Indirect	Yes	No	No
Internal Waves	Yes	Possibly	No
Gas	Yes	Possibly	No
Settlement	Yes	Yes	Yes
Scour	Yes	Yes	No
Erosion	Yes	Yes	Locally
Seismic Effects	Yes	Possibly	Possibly
Liquefaction	Possibly	Locally	Locally
Mass Wasting			
Creep	Yes	Yes	Locally
Slides, Slumps	Yes	Yes	Locally
Flows: Viscous	Yes	Yes	No
Flows: Fluid	Yes	Yes	No

Preliminary attempts to establish the mass-movement potential of the slope and rise (e.g., Booth and others, 1984; Almagor and Bennett, 1984) have shown that the area is inherently stable with respect to regional slope angles (<10°) and static analysis, but horizontal earthquake accelerations of ~0.05–0.10 g may cause certain areas to fail. Booth and others (1984) also show that canyon walls (headwalls and sidewalls) with declivites >15° may be near failure without calling on the effects of triggering mechanisms.

Creep may also be an active process in this environment, but it is difficult to assess. Initial results of laboratory creep tests of a "typical" slope-rise silty clay south of Cape Cod suggest that displacements over one year's time may be significant (e.g., several centimeters per year) for average slope gradients (Booth and others, 1984), but additional analyses are required before more definitive statements can be made. Other processes that may affect offshore hardward include liquefaction, debris-flows, and turbidity currents; these have not been extensively investigated.

Ocean Basins

Utilization of the ocean-basin regime seaward of the continental margins will probably be quite limited in the near future, but the sedimentary processes appear to be less complex than those in the continental margin regimes (Table 5). Except for localized areas that have experienced erosion (such as some areas of the Gulf Stream Outer Ridge), sediment shear strength of the upper two meters of the seabed tends to be quite low. This has important implications for the design of bottom-supported structures or instrument packages and for bottom-crawling vehicles such as might be considered in mining systems. On the other hand, the existence of a fairly thick zone (2–4 m) of material that exhibits apparent overconsolidation helps to reduce settlement due to consolidation because all or part of the imposed stresses would occur along a much lower slope of the compression curve.

The existence of an underconsolidated condition below 6–8 m depth in the acoustically laminated sediments on the Bermuda Rise and Blake Bahama Outer Ridge has important implications on the possible use of such areas for long-term disposal of toxic wastes within the seabed. For example, the natural convection due to dissipation of excess pore pressures would have to be factored into calculations to predict rates of pore fluid migration away from waste packages.

There are some localized areas of deep-sea regimes that experience erosion and slumping/sliding processes, such as the eastern scarp or slope of the Bermuda Rise. This area is evidently being heavily scoured and is experiencing episodic slumping along its steeper slopes. Therefore, any proposed deep-sea utilizations must consider all possible processes and assess their effects through detailed, site-specific studies.

REFERENCES

Almagor, G., and Bennett, R. H.
1984 : Analysis of slope stability, Wilmington to Lindenkohl Canyons, eastern U.S. Margin; in Proceedings International Union of Theoretical and Applied Mechanics, Symposium on Seabed Mechanics, ed. B. Denness; Newcastle Upon Tyne, England, 1983, p. 77–86.

Ayers, M. W., and Pilkey, O. H.
1981 : Piston core and surficial sediment investigations of the Florida-Hatteras slope in inner Blake Plateau, in Environmental geologic studies of the southeastern Atlantic Outer Continental Shelf. 1977–1978, ed. P. Popenoe; U.S. Geological Survey Open-File Report 81-582A, p. 5-1 to 5-89.

Bennett, R. H., Lambert, D. N., and Hulbert, M. H.
1977 : Geotechnical properties of a submarine slide area on the U.S. Continental Slope northeast of Wilmington Canyon: Marine Geotechnology, v. 2, p. 245–262.

Booth, J. S., Circe, R. C., and Dahl, A. G.
1985a: Geotechnical characterization and mass-movement potential of the United States North Atlantic Continental Slope and Rise: U.S. Geological Survey Open-File Report 85-123, 59 p.
1985b: Geotechnical characterization and mass-movement potential of the United States Mid-Atlantic Continental Slope and Rise: U.S. Geological Survey Open-File Report 85-351, 124 p.

Booth, J. S., Farrow, R. A., and Rice, T. L.
1981a: Geotechnical properties and slope stability analysis of surficial sediments on the Baltimore Canyon Continental Slope: U.S. Geological Survey Open-File Report 81-733, 255 p.
1981b: Geotechnical properties and slope stability analysis of surficial sediments on the Georges Bank Continental Slope: U.S. Geological Survey Open-File Report 81-566, 89 p.

Booth, J. S., Silva, A. J., and Jordan, S. A.
1984 : Slope stability analysis and creep susceptibility of Quaternary sediments on the northeast U.S. Continental Slope; in Proceedings International Union of Theoretical and Applied Mechanics Symposium on Seabed Mechanics, ed. B. Denness; Newcastle Upon Tyne, England, 1983, p. 65–75.

Buckley, D. E. and Cranston, R. E. eds.
1982 : Report of cruise No. 82-108, CSS Hudson: Internal Report of the Bedford Institute of Oceanography, Dartmouth, Nova Scotia, 37 p.

Doyle, L. J., Pilkey, O. H., and Woo, C. C.
1979 : Sedimentation on the eastern United States Continental Slope; in Geology of the continental slopes, eds. L. J. Doyle and O. H. Pilkey; Society of Economic Paleontologists and Mineralogists Special Publication No. 27, p. 119–129.

Doyle, L. J., Wall, F. M., and Schroeder, P.
1981 : Sediments and sedimentary processes as interpreted from piston cores and grab samples from the Continental Slope of the south eastern United States; in Environmental geologic studies of the southeastern Atlantic Outer Continental Shelf, 1977–1978, ed. P. Popenoe; U.S. Geological Survey Open-File Report 81-582-A, p. 4-1 to 4-45.

Embley, R. W., and Jacobi, R. D.
1977 : Distribution and morphology of large sediment slides and slumps on Atlantic Continental margins: Marine Geotechnology, v. 2, p. 205–228.

Emery, K. O., and Uchupi, E.
1972 : Western North Atlantic Ocean: topography, rocks, structure, water, life, and sediments: American Association of Petroleum Geologists Memoir 17, 532 p.

Ferrebee, W. M.
1981 : Grain size distribution and textural parameters; in Data file: Atlantic Margin Coring Project (AMCOR) of the U.S. Geological Survey, ed. L. J. Poppe; U.S. Geological Survey Open-File Report 81-239, p. 12–16.

Hall, R. W., and Ensminger, H. R., ed.
1979 : Potential geologic hazards and constraints for blocks in proposed Mid-Atlantic Oil and Gas Lease Sale 49: U.S. Geological Survey Open-File Report 79-264, 189 p.

Hathaway, J. C.
1972 : Regional clay mineral facies in estuaries and continental margin of the United States east coast; in Environmental Framework of the Coastal Plain Estuaries, ed. B. W. Nelson; Geological Society of America Memoir 133, p. 293–317.

Hathaway, J. C.
1979 : U.S. Geological Survey Core Drilling on the Atlantic Shelf: Science, v. 206, p. 515–527.

Hollister, C. D.
1973 : Atlantic Continental Shelf and Slope of the United States—texture of surface sediments from New Jersey to southern Florida: U.S. Geological Survey Professional Paper 529-M, 23 p.

Horn, D. H., Deloch, M. N., and Horn, B. M.
1974 : Physical properties of sedimentary provinces, North Pacific and North Atlantic Oceans; in Deep-Sea Sediments: Physical and Mechanical Properties, ed. A. L. Inderbitzen; Plenum Press, New York, p. 417–441.

Keller, G. H., and Bennett, R. H.
1968 : Mass physical properties of submarine sediments in the Atlantic and Pacific basins: XXIII International Geological Congress, v. 8, p. 33–50.
1970 : Variations in the mass physical properties of selected submarine sediments: Marine Geology, v. 9, p. 215–223.

Keller, G. H., Lambert, D. N., and Bennett, R. H.
1979 : Geotechnical properties of continental slope sediments—Cape Hatteras to Hydrographer Canyon; in Geology of the continental slopes, eds. L. J. Doyle and O. H. Pilkey; Society of Economic Paleontologists and Mineralogists Special Publication No. 27, p. 131–151.

Laine, E. P., Silva, A. J., McCreery, C., Lemmond, P. and Siciliano, R.
1982 : Low Level Waste Ocean Disposal Program, Report URI-7, Sandia National Laboratories, Contract No. 16-9963, 82 p.

Laine, E. P., Heath, G. R., Ayer, E., and Kominz, M.
1984 : Evaluation of the geological stability and predictability of sediment of the Northern Bermuda Rise by the Subseabed Disposal Program: Marine Geotechnology, v. 5, No. 3–4, p. 215–233.

Lambert, D. N., Bennett, R. H., Sawyer, W. B., and Keller, G. H.
1981 : Geotechnical properties of Continental upper rise sediments—Veatch Canyon to Cape Hatteras: Marine Geotechnology, v. 4, p. 281–306.

Lee, H. J.
1976 : DOSIST II—an investigation of the in-place strength behavior of marine sediments, Technical Note N-1438: sponsored by Naval Facilities Engineering Command, Civil Engineering Laboratory, Naval Construction Battalion Center, Port Hueneme, California 93043, 15 p.

Levy, W. P., Laine, E. P., Silva, A. J., Dickson, S. M., and Friedrich, N. E.
1983 : Low Level Waste Ocean Disposal Program, Report URI-16, Sandia National Laboratories Contract No. 16-9963, 61 p.

MacIlvaine, J. C., and Ross, D. A.
1979 : Sedimentary processes on the Continental Slope of New England: Journal of Sedimentary Petrology, v. 49, p. 563–574.

McCave, I. N., Hollister, C. D., DeMaster, D. J., Nittrouer, C. A., Silva, A. J., and Yingst, J. Y.
1984 : Analysis of a longitudinal ripple from the Nova Scotian Continental Rise: Marine Geology, v. 58, p. 275–286.

McGregor, B. A.
1977 : Geophysical assessment of submarine slide northeast of Wilmington Canyon: Marine Geotechnology, v. 2, p. 229–244.

McGregor, B. A., Bennett, R. H., and Lambert, D. N.
1979 : Bottom processes, morphology, and geotechnical properties of the continental slope south of Baltimore Canyon: Applied Ocean Research, v. 1, p. 177–187.

O'Leary, D. W., and Twichell, D. C.
1981 : Potential geologic hazards in the vicinity of Georges Bank Basin; in Summary report of the sediments, structural framework, petroleum potential and environmental conditions of the United States middle and northern continental margin in area of proposed Oil and Gas Lease Sale No. 76, ed. J. A. Grow; U.S. Geological Survey Open-File Report 81-765, p. 48–68.

Olsen, H. W., Booth, J. S., Robb, J. M., Gardner, W. S., Singh, R. D., Swanson, P. G., Mayne, P. W., and Hamodock, R. G.
1981 : Geotechnical test results on piston core samples taken from the mid-Atlantic upper continental slope by the U.S. Geological Survey during September, 1979: U.S. Geological Survey Open-File Report 81-366, 744 p.

Olsen, H. W., McGregor, B. A., Booth, J. S., Cardinell, A. P., and Rice, T. L.
1982 : Stabilities of near surface sediment on the mid-Atlantic upper continental slope; in Proceedings Fourteenth Annual Offshore Technology Conference, Houston, Texas; p. 21–35.

Popenoe, P., Coward, E. L., and Cashman, K. V.
1982 : A regional assessment of potential environmental hazards to and limitations on petroleum development of the southeastern United States Atlantic Continental Shelf, Slope, and Rise, Offshore North Carolina: U.S. Geological Survey Open-File Report 82-1365, 67 p.

Richards, A. F.
1962 : Investigation of deep-sea sediment cores, II. Mass physical properties: U.S. Navy Hydrographic Office Technical Report TR-106, 146 p.
1978 : Atlantic Margin Coring Project 1976: Preliminary report on shipboard and some laboratory geotechnical data: U.S. Geological Survey Open-File Report 78-123, 160 p.
1967 : Investigation of deep-sea sediment cores, III. Consolidation; in Marine Geotechnique, ed. A. F. Richards; University of Illinois Press, Urbana, p. 93–117.

Robb, J. M., Hampson, J. C., Kirby, J. R., and Twitchell, D. C.
1981 : Geology and potential hazards of the Continental Slope between Lindenkohl and South Toms Canyons, offshore Mid-Atlantic United States: U.S. Geological Survey Open-File Report 81-600, 33 p.

Rosenqvist, I. Th.
1953 : Considerations on Sensitivity of Norwegian quick clays: Geotechnique, v. 3, p. 195–200.

Sangrey, D. A.
1977 : Marine Geotechnology—State of the Art: Marine Geotechnology, v. 2, p. 45–80.

Schlee, J. S., Dillon, W. P., and Grow, J. A.
1979 : Structure of the Continental Slope off the eastern United States; in Geology of continental slopes, eds. L. J. Doyle and O. H. Pilkey, Society of Economic Paleontologists and Mineralogists, Special Publication No. 27, p. 95–118.

Silva, A. J.
1979 : Geotechnical properties of ocean sediments: Annual Report, Office of Naval Research Contract N00014-76-0226.

Silva, A. J., and Calnan, D. I.
1980 : Geotechnical aspects of subsurface seabed disposal of high level radioactive wastes: Annual Progress Report No. 6, Department of Energy/SLA Contract Nos. 13-2561 and 13-9927.

Silva, A. J., and Hollister, C. D.
1979 : Geotechnical properties of ocean sediments recovered with the giant piston core: Blake-Bahama Outer Ridge: Marine Geology, v. 29, p. 1–22.

Silva, A. J., Hollister, C. D., Laine, E. P., and Beverly, B. E.
1976 : Geotechnical properties of deep-sea sediments: Bermuda Rise: Marine Geotechnology, v. 1, p. 195–232.

Silva, A. J., and Jordan, S. A.
1984 : Consolidation properties and stress history of some deep sea sediment; in Proceedings of IUTAM Symposium, 1983, University of Newcastle Upon Tyne, ed. B. Denness, p. 25–39.

Silva, A. J., Jordan, S. A., and Levy, W. P.
1981 : Geotechnical aspects of subseabed disposal of high level radioactive wastes: Annual Report No. 7, Department of Energy/SLA Contract Nos. 13-9927 and 74-1098.
1982 : Geotechnical studies for subseabed disposal of high level radioactive wastes, Annual Report No. 9, Department of Energy/SLA Contract No. 16-3110.
1983 : Geotechnical properties of ocean sediments: Annual Reports 7, 8, 9, January, 1981–December, 1983, ONR Contract N00014-760-226.

Swanson, P. G., and Brown, R. E.
1978 : Triaxial and consolidation testing of cores from the 1976 Atlantic Margin Coring Project of the United States Geological Survey: U.S. Geological Survey Open-File Report 78-124, 144 p.

Tucholke, B. E.
1973 : The history of sedimentation and abyssal circulation on the Greater Antilles Outer Ridge: [Ph.D. Thesis], Massachusetts Institute of Technology—Woods Hole Oceanographic Institution, unpublished manuscript.
1980 : Acoustic environment of the Hatteras and Nares Abyssal Plains, western North Atlantic Ocean, determined from velocities and physical properties of sediment cores: Journal of Acoustical Society of America, v. 68, p. 1376–1390.

Tucholke, B. E., and Ewing, J. I.
1974 : Bathymetry and sediment geometry of the Greater Antilles Outer Ridge and vicinity: Geological Society America Bulletin, v. 85, p. 1789–1802.

Tucholke, B. F., and Laine, E. P.
1982 : Neogene and Quaternary development of the Lower Continental Rise off the central U.S. East Coast: American Association of Petroleum Geologists, Hedberg Research Conference on Continental Margin Processes, unpublished.

Tucholke, B. E., and Shirley, D. J.
1979 : Comparison of laboratory and in situ compressional wave velocity measurements in sediment cores from the western Nort Atlantic: Journal of Geophysical Research, v. 84, p. 687–695.

Wagner, A. A.
1957 : The use of the unified soil classification system by the Bureau of Reclamation: International Conference on Soil Mechanics and Foundation Engineering, 45th, London, v. 1, p. 125.

Manuscript Accepted by the Society July 26, 1985

Acknowledgments

B. Fagan, graduate student at URI, and A. Dahl at USGS/Woods Hole were of great assistance in the gathering and analysis of data. We appreciate the constructive criticisms of an early draft by M. Hampton and G. Keller as well as a very detailed review by B. Tucholke.

The Geology of North America
Vol. M, The Western North Atlantic Region
The Geological Society of America, 1986

Chapter 31

Northwestern Atlantic Mesozoic biostratigraphy

F. M. Gradstein
Atlantic Geoscience Centre, Bedford Institute of Oceanography, Dartmouth, Nova Scotia B2Y 4A2, Canada

INTRODUCTION

The marine Mesozoic sedimentary and paleoceanographic history of the western North Atlantic region is buried in the 0.5–2 km thick sediments of the ocean basin and in the 2–15 km thick sedimentary wedges forming the Atlantic continental margin and the Atlantic Coastal Plain. In the Atlantic Coastal Plain itself there are outcrops of shallow marine Upper Cretaceous strata for which Olsson (1978) has provided a good biostratigraphic summary. North of New Jersey, Paleozoic strata in the Appalachian orogenic belt are present along the shoreline. These Paleozoic strata form part or all of the basement beneath the shelf as far north as the Grenville Front between Newfoundland and Labrador (Wynne-Edwards, 1972). In a general sense, the present eastern coastline of Newfoundland, Nova Scotia, and the New England states matches that of the late Jurassic and late Cretaceous. This study surveys the types and distribution of Mesozoic microfossils to further our understanding of Atlantic Mesozoic stratigraphy.

Micropaleontology, the microscopic study of fossils from about 0.001 to several millimeters in longest dimension, plays an important part in the reconstruction of the geological history of ocean basins. These small fossils are recovered, unbroken, in large numbers in core samples. The types and distribution in time and space of these fossils help the micropaleontologists to determine the relative age and original environment of the rocks in which they are found, as well as to reconstruct paleoceanographic conditions.

Different types of microfossils found in the marine Mesozoic strata offshore are nannofossils (largely coccoliths), radiolarians, planktonic and benthic foraminifera, ostracods, spores and dinoflagellates, (Fig. 1). Rare macrofossils include juvenile ammonites, aptychi and, in carbonate-bank environments, rudists and molluscs.

Coccoliths are 1–15 μm, circular or elliptical calcareous shields that form a framework to envelope unicellular, autotro-

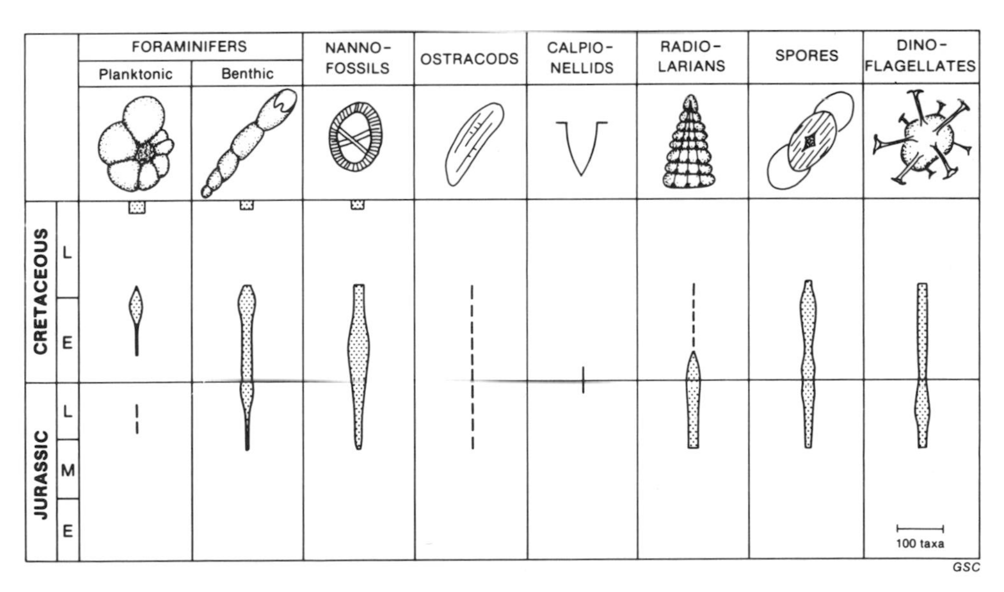

Figure 1. Schematic illustration of stratigraphic distribution and taxonomic diversification of the Mesozoic open-ocean microfossil record.

Gradstein, F. M., 1986, Northwestern Atlantic Mesozoic biostratigraphy; *in* Vogt, P. R., and Tucholke, B. E., eds., The Geology of North America, Volume M, The Western North Atlantic Region; Geological Society of America.

phic marine algae known as coccolithophores. These planktics live in the photic zone, down to about 150 m depth in shelf or open-ocean environments. Some nannofossil taxa are more resistant to dissolution than others, but in general nannofossils can be expected to dissolve below the calcite compensation depth (CCD) and never reach the seafloor in deeper ocean basins. However, many nannofossils probably were incorporated into faecal pellets of planktonic grazers; this sped their settling to the seafloor and helped protect them against dissolution. Nannofossils, which include other related but morphologically deviating groups of organisms, provide detailed Mesozoic zonations at or below the stage level. The zonation used in this paper for Late Cretaceous is standardized, which means it is applicable worldwide (in lower latitudes).

Radiolarians are exclusively open marine, pelagic organisms. They feed on other plankton, but algal symbionts may aid in metabolism. Their evolutionary history goes back into Paleozoic time, although we know from the modern record that only one major group, containing the radially symmetrical polycystinids and the helmet-shaped nasselarians, is preserved. Test size is 60–200 μm. Amorphous (opaline) silica is the skeletal building block. Post-mortem dissolution during sedimentation and diagenetic dissolution and mineral replacement after burial tend to make the stratigraphic and geographic record patchy. When well preserved Jurassic radiolarian fauna may be rich in both species and specimens. Jurassic and Early Cretaceous radiolarian zonations are important for Atlantic ocean geology. Stratigraphic resolution can reach the stage level.

The foraminifers are omnivorous, single-celled marine animals that secrete a calcareous test or build an arenaceous test. Mature tests are generally 100–300 μm in size. The agglutinating, arenaceous mode of test building, understandably, is confined to bottom dwellers. Construction varies from single tubes or sacs to highly complex multichambered labyrinthic structures. Many taxa have a test consisting of increasingly larger chambers, often wound in a spire, and with a lateral or radial symmetry.

The planktonic foraminifer record goes back to Middle Jurassic. The high resolution, standard planktonic zonation for Middle and Upper Cretaceous strata applies in the Atlantic. Benthic foraminifera already abounded in the Paleozoic, and many regionally recognized biostratigraphic events and zones aid in the study of Jurassic and Cretaceous oceanic and particularly continental margin stratigraphy. Particularly in shelf and slope marine sediments, paleoecologic interpretations rely on benthic foraminiferal distribution patterns.

Ostracods are primarily bottom dwellers that have a geologic range from Paleozoic through Recent. This microfossil group represents a higher order, multicellular phylum, in this case Crustaceans. The fossilized calcitic parts are the two shelly valves, often found separately. Provincialism is strong in the record; highest diversity is in shallow marine environments. The shallow-marine Mesozoic record is important paleoecologically and stratigraphically.

Spores, which exist today, are the progeny of primitive land plants and result from asexual reproduction. They first appeared in Silurian time. Pollen is the product of advanced land plants. Both organic-walled morphogroups are carried as sediment particles to the oceans where they are well preserved. Differentiation of these morphogroups in time is useful for broad stratigraphic interpretation in the Mesozoic of the western Atlantic margin and ocean sediments.

Dinoflagellates are one-celled, organic-walled organisms somewhat smaller than foraminifera. They resemble both animals (because they ingest food and are mobile), and plants (because they photosynthesize); they are grouped with the algae and occur in aqueous environments. Most forms are pelagic. The fossil part is an organic-walled (dino)cyst, a sessile resting stage in the life cycle. Cysts display an often complex morphology that evolved through geologic time. This morphogroup is known since the Permian. Provincialism strongly influences the record, but regionally recognized events and zones provide Middle Jurassic through Cretaceous correlation at approximately stage level.

Calpionellids are tiny, basket-shaped calcareous skeletons up to about 100 μm long. They typically occur in very fine grained, micritic limestones that are otherwise rich in nannofossils. The stratigraphic record of these incertae sedi (groups with unknown taxonomic affinity) is Tithonian through Valanginian, and a standard zonation exists. The record is of value for the western Atlantic regional stratigraphy.

Late Jurassic and Early Cretaceous correlations from Tethyan land sections into western Atlantic Ocean sites are supported by rare finds of aptychi and juvenile ammonites. Aptychi are concentrically grooved calcite or horny plates, 1–2 cm across, that probably were a kind of protective cover plate (operculum) of cephalopods. The ammonite shell itself is aragonitic and rarely is preserved in deep-sea sediments.

HISTORY OF INVESTIGATION

Published literature on offshore Mesozoic geology of the Atlantic has accumulated mostly during the last fifteen years. Before that time, in the pre-drilling era, our knowledge of Mesozoic (and Cenozoic) strata offshore was based on accounts of dredge samples and piston cores. For example, Cushman (1936) and Stephenson (1936) described Upper Cretaceous fossils of the Georges Bank canyons. The middle Cretaceous dark shale of the oceanic Hatteras Formation (Jansa and others, 1979) was first sampled by piston cores in the Cat Gap area (Plate 2) and dated by palynomorphs (Saito and others, 1966; Habib, 1968; 1970).

Piston cores and dredges along the Blake Escarpment, which separates the 1000 m level of the Blake Plateau from the 5000 m level of the Blake-Bahama Basin, sampled Neocomian to Aptian algal limestone and (?) Albian to Cenomanian pelagic clay (Sheridan and Osburn, 1975; Saito and others, 1966; Loeblich and Tappan, 1961). Other than *Orbitolina* and miliolids, no foraminifer were found in the limestones. In the pelagic clay, Loeblich and Tappan (1961) identified the following planktonic foraminifers: *Favusella washitensis* (Carsey), *Hedbergella planispira* (Tappan), *H. trocoidea* (Gandolfi), *Globigerinelloides bentonensis* (Morrow), *G. eaglefordensis* (Moreman), *Praeglobotruncana delrioensis* (Plummer) and *Rotalipora greenhornensis* (Morrow). The authors postulated a mid-Cenomanian age for the cored sediment. It more likely is late Albian to middle Cenomanian.

The Esso Hatteras Lighthouse No. 1 well, drilled in 1948 on Cape Hatteras (Plate 2) did much to elucidate regional stratigraphy. It penetrated a Lower Cretaceous sequence of shallow marine clastics and bottomed at approximately 3300 m in *Anchispirocyclina lusitanica*-bearing limestone of Tithonian age, overlying granitic basement (Gradstein n.d.).

Dredges on Flemish Cap, a 300–500 m shoal area beyond Grand Banks (Plate 2), recovered Paleozoic granite and middle Cretaceous shallow-water limestones rich in orbitolinids (see below) (Sen Gupta and Grant, 1971). The find dramatically illustrated that in Mesozoic time the northwestern Atlantic was part of the western Tethyan faunal province, which extended from the Mediterranean into the Caribbean. The find also supports the concept that the Cretaceous paleolatitude of Flemish Cap may only have been 35°N, rather than 45°N.

Ryan and others (1978), using the submersible *Alvin,* sampled some of the oldest outcropping sedimentary rocks in the Georges Bank canyons and found calpionellid limestones of Valanginian age. Later deep drilling on the shelf confirmed this regional stratigraphic record (Scholle and Wenkam, 1982).

On the Canadian continental margin, a variety of drilling platforms started industrial exploration in 1966 on behalf of numerous oil companies. Initial coring of Albian through Miocene marine sediments on the Grand Banks was by the D/V *Caldrill* on commission in 1965 to the Pan American Oil Company (Amoco) and Imperial Oil Enterprises.

The bulk of Mesozoic sample material from the Canadian Shelf comes from about two dozen (industry) deep wells off Labrador (Cretaceous), more than 50 wells on the Grand Banks, and about 60 wells on the Scotian shelf. Discovery of the Hibernia oil field in Lower Cretaceous and Upper Jurassic sands on the northern Grand Banks and of the Venture gas and condensate field, in Lower Cretaceous and Upper Jurassic sands near Sable Island, has given new impetus to drilling offshore.

Joides (Joint Oceanographic Institutions Deep Earth Sampling) drilling started in 1965 when D/V *Caldrill,* enroute to the Grand Banks, was chartered for Blake Plateau drilling. Paleocene bathyal strata were the oldest sediments cored (Joides, 1965). Along the United States eastern seaboard, commercial stratigraphic test holes intentionally drilled off-structure are: COST B-2 and COST B-3, off of New Jersey; COST GE-1, inside the Blake Plateau; and COST G-1 and COST G-2 on Georges Bank (Plate 2; Scholle and others, 1977; 1979; 1980; Scholle and Wenkam, 1982). More than 30 other commercial wells have been drilled to date, for which data are now publicly available.

Since 1969, D/V *Glomar Challenger* has completed seven Deep Sea Drilling Project (DSDP) legs in the western North Atlantic Ocean: Leg 2 (Peterson and others, 1970), Leg 11 (Hollister and others, 1972), Leg 12 (Laughton and others, 1972), Leg 43 (Tucholke and others, 1979) Leg 44 (Benson and others, 1978), and Leg 76 (Sheridan and others, 1983). The excellent coring record of these operations has made detailed study of western North Atlantic Mesozoic paleontology and geology a reality. Summaries of the offshore drilling are given by Sheridan and Enos (1979), Poag (1978), Department of Energy, Mines, and Resources (1983) and Gradstein and Sheridan (1983a).

SAMPLE QUALITY AND REPOSITORIES

Industry drilling operations on the continental shelves and slopes primarily produce rotary-drilled rock-chip samples that are transported upward with the returning drilling mud, outside the drill string. Such ditch cuttings are routinely collected and bagged over drilling increments of 10 ft (3m). The main problem with these samples, other than highly variable recovery, is contamination. On the way up the hole, the chips mix with caving younger sediments already penetrated by the drillbit, so that biostratigraphic analysis is limited to last occurrences in time. The so-called assemblage-type biozones used in cutting samples are, in reality, interval zones and the lower zone limit is determined by the top of the next older assemblage of events (exits). As a result, zonations lose 50 to 75% of their time-resolving potential. Taxonomy in cuttings is based on typical specimens rather than populations of specimens, which creates problems with homeomorphs in time. Thus, evolution is next to impossible to address, and this takes away one of the paleontologic attractions of study of the passive margin in such samples.

Paleoecology based on cuttings is also a difficult and somewhat thankless undertaking, because the in situ assemblage is modified by cavings. Nevertheless, changing preservation with depth enables recognition of cavings and, where present, well casing prevents contamination by cavings.

There are often strings of side-wall cores of known depth available for each well. These may be cleaned of drilling-mud contamination. Unfortunately, the small core size (20 cm^3 or less) limits suitability of the cores for foraminifer and ostracod studies. Some wells do have conventional rotary cores, and several short upper Jurassic and Cretaceous intervals have thus been studied in more detail (e.g. Jansa and others, 1980; Doeven 1983; Scholle, 1977).

Technical details on the industry wells in Canadian waters are found in the Department of Energy, Mines, and Resources *Offshore Schedule of Wells* (1983). Most well data are made publicly available two years after completion of each well. Geological and technical data for each well are compiled in the Well History Reports which, together with the logs and all samples (including microfossil slides of foraminifers, ostracods, palynomorphs, and nannofossils) can be studied by appointment on the premises of the Department of Energy, Mines, and Resources at the Bedford Institute of Oceanography, P.O. Box 1006, Dartmouth, Nova Scotia, Canada B2Y 4A2.

Information on release of United States well information and study of samples can be obtained from: Director, Minerals

TABLE 1. LITERATURE ON MESOZOIC PALEONTOLOGY AND BIOSTRATIGRAPHY, WESTERN NORTH ATLANTIC OCEAN AND MARGIN

	FORAMINIFERS Smaller	FORAMINIFERS Larger	NANNOFOSSILS	DINOFLAGELLATES SPORES ACRITARCHS	RADIOLARIANS	CALPIONEL-LIDS
OCEANIC	Caron, 1972 Cita and Gardner, 1971 Gradstein 1978, 1983 Luterbacher, 1972 Olsson, 1977 McNulty, 1979		Bukry, 1972 Okada and Thierstein, 1979 Roth, 1978 Roth, 1983 Roth, Medd and Watkins, 1983 Schmidt, 1978 Thierstein, 1976 Thierstein and Okada, 1979 Wind, 1978 Wilcoxon, 1972	Habib, 1972, 1975, 1977, 1978, 1979 Habib and Knapp, 1982 Habib and Drugg, 1983	Baumgartner, 1983 Foreman, 1977	Lehmann, 1972 Remane, 1983
CONTINENTAL MARGIN	Ascoli, 1977 Dufaure et al., 1977 Exton and Gradstein (in press) Gradstein, 1977, 1978 Gradstein et al., 1975 Gradstein and Srivastava, 1980 Gradstein and Berggren, 1981 Hart, 1976 Jansa et al., 1980 Laughton, Berggren et al., 1972 Miller et al., 1982 Olsson, 1964	Hottinger, 1972 Sen Gupta and Grant, 1971 Schroeder and Cherchi, 1979	Doeven et al., 1982 Doeven, 1983 Laughton, Berggren et al., 1972	Aurisano and Habib, 1977 Barss et al., 1979 Bujak and Williams, 1977, 1978 Gradstein et al., 1975 Gradstein and Williams, 1976 Williams, 1975 Williams and Brideaux, 1975		Jansa et al., 1980

	APTYCHI, SMALL AMMONITES	OSTRACODS	PLANTONIC CRINOIDS	RUDISTS	CALCISPHER-ULIDS	SITE HISTORIES INCUDING BIO-CHRONOSTRATIGRAPHY
OCEANIC	Renz, 1972, 1978, 1979, 1983	Oertli, 1972, 1983 Swain, 1978	Hess, 1972		Bolli, 1978	Benson, Sheridan et al., 1978 Hollister, Ewing et al., 1972 Jansa et al., 1979 Laughton, Berrgren et al., 1972 Peterson, Edgar et al., 1970 Sheridan, Gradstein et al., 1983 Tucholke, Vogt et al., 1979
CONTINENTAL MARGIN		Ascoli, 1977 Bate, 1977 Jansa et al., 1980 Swain, 1980 Van Hinte et al., 1975		Perkins, 1979		Barss et al., 1979 Bhat et al., 1975 Gradstein, 1975 Gradstein et al., 1976 Jansa et al 1976, 1977 Jenkins et al., 1974 McWhae et, al., 1980 Ruffman and Van Hinte, 1973 Scholle et al., 1977, 1979, 1980, 1982 Srivastava (ed.) in press Upshaw et al., 1974 Williams et al., 1974

GSC

Management Service, U.S. Department of Interior, National Center, Reston, Virginia, U.S.A., 22092. Deep Basin cores obtained by D/V *Glomar Challenger* in the Deep Sea Drilling Project are archived at the Lamont-Doherty Geological Observatory, Palisades, New York, U.S.A.

WHO PUBLISHED WHAT AND WHERE

The published Mesozoic micropaleontological record of the western North Atlantic is spread over at least 75 titles (Table 1). For the purpose of orientation, I have split this scholarly edifice into two categories: Oceanic, and Continental Margin. This subdivision tends to be important biogeographically, not only for benthic, but also for pelagic microfossils. The pre-1970 literature is limited in size and scope and no efforts have been made to make a historically complete listing. Many single-well site reports with diverse paleontological information are listed in several categories. In the category of site histories, overlap with regional geology is strong. There may also be some omissions.

The listing in Table 1 shows that foraminifers, nannofossils, and palynomorphs are most frequently studied. This is a function of the long geological record of these groups, the relative abundance and good preservation of specimens in small samples from widely differing marine environments, and particularly of prolonged and rapid evolutionary diversification. Palynomorph (dinocysts and spores) zonations are more affected by provincialism than are those using foraminifers and nannofossils, but they correlate well in marginal-marine stratigraphic intervals.

DEVELOPMENT OF THE MICROFOSSIL RECORD IN THE OPEN ATLANTIC OCEAN

On the North Atlantic continental margin the marine Mesozoic sedimentary and microfossil record goes back as far as Early Jurassic, but the oldest, probably abyssal sediments sampled on oceanic crust are Callovian (late Middle Jurassic) in age. These dark shales were sampled at DSDP Site 534 (Plate 2) and dated by the *Stephanolithion hexum* nannofossil subzone, radiolarian zone A, and diagnostic dinoflagellates *Stephanelytron scarburghense* and *Energlynia indota* (Sheridan and others, 1983).

Africa and North America probably moved away from each other as early as Middle Jurassic time, widening the incipient southwest-northeast ocean passage from the Pacific into the Mediterranean Tethys (Plate 9). There is no reason to believe that the Callovian or younger midocean-ridge depth was anomalously shallow. Backtracking of drillsites in Jurassic ocean crust (Sites 99, 100, 105, 391, 534, and in Cretaceous crust (386 and 387; Plate 2) shows an initially rapid and later slow deepening from about 2.7 to 5 km. In Late Cretaceous time all the sites were in water roughly 4.5 km deep. Thus the Mesozoic oceanic microfossil record at these sites is labelled abyssal.

The oldest known abyssal lithologies are the above mentioned dark, varigated shales with radiolarian silts and turbiditic limestones of Callovian and Oxfordian age (see Fig. 1 of Jansa, this volume, and Plate 1B). Conformably overlying these layers are green-gray micritic to shaley limestones and reddish-brown, calcareous shales of the Cat Gap Formation (nomenclature after Jansa and others, 1979); these are dated Oxfordian-Tithonian by nannofossils, dinoflagellates, foraminifers, radiolarians and calpionellids. The CCD, which in the Middle Jurassic may have been rather shallow, probably was below 4 km in latest Jurassic time (Plate 1B). This may explain why even tiny aragonitic ammonite shells are locally preserved in the Cat Gap Formation.

Neocomian micritic nannofossil limestone and chalk, with some dark shale, clastic limestone, and carbonate sand form the Blake Bahama Formation; they provide an excellent nannofossil record (Schmidt, 1978) but are poor in foraminifers. There is also a good regional dinoflagellate stratigraphy (Habib, 1968, 1970, 1972, 1975, 1977, 1978). Sedimentation was strongly influenced by periodic changes (on the order of several thousands or hundreds of thousands of years) in either surface water fertility, deep circulation, and/or bioturbation or turbiditic influences, because hundreds of meters of limestone at Sites 534 and 391 in the Blake-Bahama Basin show an impressive alternation of laminated and non-laminated beds. The CCD probably resided near the seafloor (4–5 km), which accounts for the meager benthic foraminifer record.

The immediately overlying abyssal sediments are principally black and minor variegated shales of the Hatteras Formation. This lithology is generally thought to reflect a shallow CCD; it contains an increased proportion of carbonaceous, mostly terrigenous organic matter generated in a hot, wet climate. The dark shales show limited oxidizing capacity of bottom water and/or sediment (see Arthur, this volume). The shales contain the *Odontochitina operculata* to *Trithyrodinium suspectum* dinoflagellate zones (see Habib and Knapp, 1982; Jansa and others, 1979) and were deposited in Hauterivian/Barremian through Cenomanian time. Locally, there is a record of nannofossils, radiolarians, and planktonic foraminifers.

One of the few markedly condensed sequences in the western North Atlantic Ocean is between the *S. echinoideum* or *T. suspectum* dinoflagellate zones and the *Globotruncana stuartiformis* or *Globotruncanella mayaroensis* planktonic foraminifer zones (Plate 1B). The Cenomanian through Campanian or even younger strata are much condensed and could contain hiatuses. The exact stratigraphic extent of any hiatuses is unknown because in Sites 391, 534 and 105, and particularly in Sites 386 and 387, the record consists of cores with non-dated clays. This sediment forms the Plantagenet Formation, a multicoloured clay that is essentially devoid of carbonate, organic carbon, and siliceous microfossils. It is rich in authigenic minerals and undoubtedly accumulated slowly under highly oxidizing conditions below the CCD.

A review of the nature of depositional conditions in this interval is in Tucholke and others (1979), Gradstein and Sheridan (1983b), Robertson (1983), and Arthur and Dean (this volume).

Toward the close of Cretaceous time, the CCD dropped sharply (Plate 1B) to allow a relatively brief return to local, carbonate-rich sedimentation in Sites 385, 386, and 387, with preservation of upper Campanian-Maestrichtian globotruncanid planktonic foraminifers and nannofossils of the Tethyan province. In the western, deeper part of the basin, little carbonate was preserved; Site 391 contains no carbonate, but in Site 534 some poorly preserved chalk is present and is tentatively assigned a Maestrichtian age (M. Moullade, in Sheridan and others, 1983). In Site 105 the carbonate layer was eroded away during the Oligocene (Tucholke, 1979).

The temporal, taxonomic diversification of the Mesozoic oceanic microfossil record is schematically shown in Figure 1. This compilation is a crude one and merely serves as a guide to the following discussion. The microfossil record in the Jurassic North Atlantic Ocean typically consists of robustly built nannofossils; a patchily distributed, small-sized benthic foraminiferal fauna; a diverse dinoflagellate, acritarch and (redistributed) spore flora; locally diversified radiolarian assemblages that may be pyritized; rare aptychi; *Saccocoma*-type pelagic crinoids (Hess, 1972); and organic calcite spheres called calcispherulids (Bolli, 1978).

The calcispherulids are abundant in some Tithonian red shales.

Roth (1983) reports that in the Jurassic of Site 534, the relatively robust *Watznaueria* group of nannofossils is dominant and that in the upper section *Conusphaera, Polycostella* and small nannoconids (below 8 μm) also contribute to the nannofossil assemblage. Delicate genera such as *Corollithion, Stradnerlithus, Diadorhombus, Stephanolithion,* and *Zygodiscus* are much less common than in epicontinental seas. This forces specialists to develop local zonations for oceanic and epicontinental basins.

Preservation of calcareous nannofossils in Berriasian through Cenomanian strata is moderate. The *Watznaueria* group may be dominant, but *Rhagodiscus* and *Nannoconus* are also important contributors to the overall assemblage. The varicoloured sediments in the Upper Cretaceous section contain few specimens, and these could have been redeposited from the shallower parts of the basin, i.e. the slope and rise.

Particularly in Site 534, Jurassic-Valanginian radiolarians are locally common and display a high diversity (for example 50 to 80 Callovian-Oxfordian species). The best preserved tests are in pyrite. Fertility of the ocean surface waters may have been locally high, producing abundant radiolarians and coccolith-bearing algae, but undersaturation of bottom waters in silica hampered radiolarian preservation (Baumgartner, 1983). The Tethyan, Callovian-Valanginian zonation with 110 species in nine to fourteen unitary associations extends across the eastern and western North Atlantic.

In late Tithonian time, a dozen or so species of calpionellids developed and spread through the Tethys and into the Atlantic Ocean. Such a record is known from Mexico and Cuba and the Mediterranean Tethys. Detailed studies in the western Atlantic Ocean and in a few outer continental margin well sites by J. Remane (1983), and Jansa and others (1980), however, confirm the Tithonian-early Berriasian record, but have failed to disclose the late Berriasian to Valanginian zones. There is no sedimentary, magnetostratigraphic, nannofossil or dinoflagellate evidence for a sustained Berriasian-Valanginian hiatus in the ocean (e.g. Fig. 1, Plate 1B). The only reasonable answer at present is that this is a biogeographic peculiarity.

Renz (1972, 1978, 1979, 1983) reports that the Kimmeridgian through Valanginian aptychi faunule in the northwestern Atlantic Ocean, although sparse, may be compared to that in the Alpine-Mediterranean Province, and also to the Vinales Formation in Cuba. In Core 49 of Site 387 alone, 34 aptychi and 4 (Berriasian) ammonites were recovered in what is probably a condensed sequence. The calcitic tests of aptychi make them less vulnerable to deep-sea dissolution than the aragonitic ammonites to which they functionally belong.

At all western Atlantic Ocean sites, pollen grains, spores, acritarchs and dinoflagellates occur in the Jurassic and Lower Cretaceous (Cenomanian) strata. Supply of terrigenous or shelf-derived organic matter or deposition in faecal pellets brought this floral material into the deeper basin. Locally this organic biofacies is diversified and abundant in specimens (Habib, 1975; Habib and Drugg, 1983). There is an Early Cretaceous standard dinoflagellate zonation with eight partial or total ranges zones and three pollen zones. In particular the Hatteras Formation is dated using this zonation.

The abyssal, Middle and Late Jurassic benthic foraminiferal "fauna" of the Atlantic Ocean has a high species ($\pm$100) and high generic ($\pm$50) diversity but has patchy specimen representation and low communality at the species level (Gradstein, 1983; Luterbacher, 1972). The benthic microfauna consists principally of small agglutinated taxa in eight or so families, small nodosariids in a dozen or so genera, and variably represented epistominids, ophthalmidids, spirillinids, and turrilinids. Few benthic, calcareous taxa are uniquely abyssal, but test morphologies may differ slightly from the same taxa in epicontinental seas in Western Europe or Grand Banks. The abyssal, upper Cat Gap Formation (Plate 1B) is locally more rich in *Epistomina* (aragonitic), *Lenticulina, Ophthalmidium,* and *Neobulimina.*

A peculiar and little known microfauna occurs in the otherwise barren Upper Cretaceous red clays. It consists of small-sized agglutinated benthic foraminifers (Gradstein, 1978b; McNulty, 1979; Miller and others, 1982). This residual microfauna is cosmopolitan in the deep ocean, but in slope sites it occurs with many more taxa, larger specimen size, and better preservation. No stratigraphic-paleoecologic study of this abyssal and presumably indigenous faunule has been done to correlate and subdivide Atlantic red-clay sediments.

Jurassic planktonic foraminifera are virtually absent in the deep basin and are more typical for epicontinental seas. There is some evidence that the few tiny specimens of *Globuligerina* observed in DSDP Sites 534 and 105 were either transported down slope or deposited in faecal pellets (Gradstein, 1983; Luterbacher, 1972). Jurassic open-ocean surface waters probably were largely devoid of these organisms, which only proliferated since Aptian time. In DSDP Sites 105, 386 and 387, which were slightly shallower than 391 and 534, there is a patchy record of Albian-Cenomanian planktonic foraminifera, with *Ticinella, Planomalina, Rotalipora, Praeglobotruncana,* and *Schackoina* (Luterbacher, 1972; McNulty, 1979).

Ostracods are extremely rare in Sites 534 and 391 but less rare in Site 105; they consist largely of small, little ornamented types, deviating morphologically from the shallower water record elsewhere (Oertli, 1972; 1983; Swain, 1978).

DEVELOPMENT OF THE MICROFOSSIL RECORD IN THE OCEAN TO MARGIN TRANSITION ZONE

DSDP Sites 390 and 392 on the Blake Spur and Site 384 on the J-Anomaly Ridge, southeast of the Grand Banks (Plate 2, Fig. 2) take intermediate physiographic positions between the deep ocean and continental margin. They were in more open marine areas than adjacent shelves but were relatively shallow during much of their history compared to the abyssal sites. At these shallow sites there is a Lower Cretaceous, neritic, biogenic or bioclastic record, changing upward to a pelagic record of nannofossil ooze. Coeval Lower Cretaceous strata on the conti-

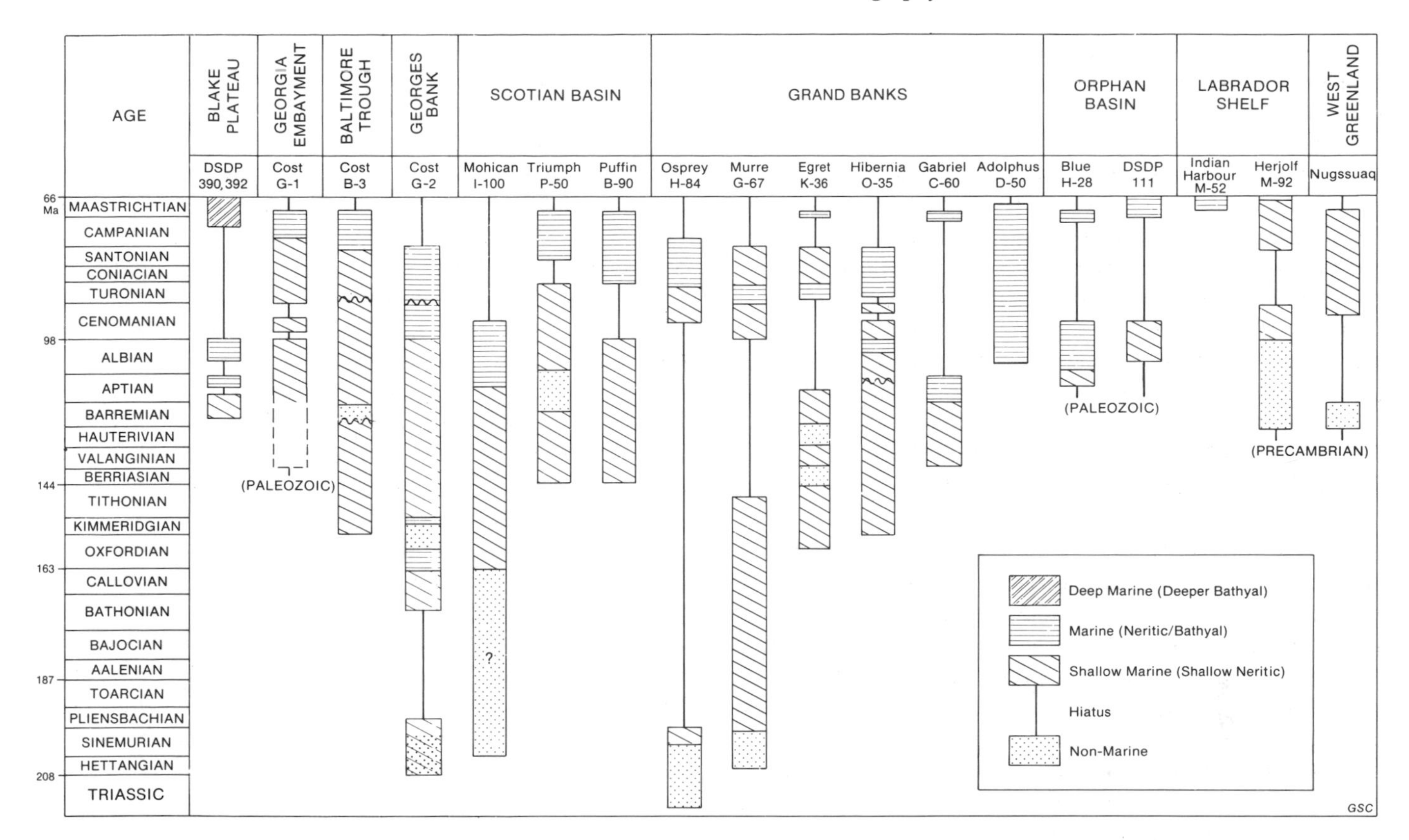

Figure 2. Stratigraphic distribution of Mesozoic sediments in western North Atlantic continental margin sites. Also shown is the sedimentary record in the circum-Labrador Sea. Sites were selected on the basis of representative biostratigraphic coverage.

nental margin are predominantly terrigenous clastics and in the deep ocean basin are hemipelagic to pelagic limestones and shales.

Sites 390 and 392 are on the upper edge of the Blake Escarpment, which is formed by a carbonate-bank facies of Jurassic(?) through Barremian age. This facies is the southwestern continuation of the extensive carbonate build-up, mainly in Jurassic time, along the eastern North American continental margin. Platform, ramp, and prograding types of carbonate build-up have been recognized along the outer Scotian shelf (Abenaki Limestones). Good descriptions of the biogenic and bioclastic facies can be found in Enos and Freeman (1978), Eliuk (1978) and Tucholke and others (1979). A warm, equitable climate, and a photic-zone regime with steady water temperatures of 15–20°C existed at least 10° farther north relative to the North American margin than at present.

At DSDP Sites 390 and 392, carbonate build-up on the Blake Spur drowned in Barremian time, although it may have persisted locally into the Late Cretaceous. The overlying oozes of Barremian, Aptian-Albian, and Maestrichtian age have an excellent and diversified nannofossil and planktonic and benthic foraminiferal record. Planktonic foraminifer zones recognized are *Globigerinelloides blowi, G. algerianus, Ticinella primula, T. breggiensis,* and *Globotruncanella mayaroensis* (Gradstein, 1978b; P. Marks, personal communication, 1982).

At Site 384 on the J-Anomaly Ridge and also on the Newfoundland Seamounts to the northeast (Plate 2), the Lower Cretaceous carbonate bank is on anomalously shallow ocean basement of Barremian-Aptian age. Unfortunately, core recovery in 123 m of basal bioclastic limestone at Site 384 was extremely poor, but principal features can be recognized. They include poor sorting (admixture of lime mud and fossil fragments), rudists and other pelecypods, gastropods, echinoderms, ostracods, larger foraminifers, bryozoans, solitary corals, and red algae. The contact with the underlying tholeitic basalt may be a weathered paleosol. Cores 16 and 20 recovered monopleurid and caprinid rudists fragments that correlate to the Albian of Texas and Mexico (Perkins, 1979). Rudists are typical of uppermost Jurassic and Cretaceous reef, bank, biostrome, or other shallow-water carbonate build-ups. At Site 384 there is no evidence of reef-type corals. Schroeder and Cherchi (1979) described a rich, shallow-marine, larger-benthic-foraminifer assemblage of orbitolinids in cores 16 and 20, with *Palorbitolina lenticularis* and *Paleodictyoconus arabicus* of Barremian to Aptian age. These are the oldest orbitolinids in the New World; they also occur on Flemish Cap further northward (Sen Gupta and Grant, 1971). In Albian-Cenomanian time, the J-Anomaly Ridge and probably the Newfoundland Seamounts subsided into deeper water, but hemipelagic to pelagic sediments did not accumulate until Maestrichtian time (Okada and Thierstein, 1979; McNulty, 1979).

On the J-Anomaly Ridge at Site 384 and in Sites 390 and 392, Blake Spur, the Cretaceous-Tertiary boundary beds have been preserved (Thierstein and Okada, 1979). This is in marked contrast to a substantial hiatus in the Maestrichtian-Paleocene record on the continental margin shelf and slope, including the Blake Spur (see below). The explanation for this gap on the margin may be erosion or non-deposition due to relative or eustatic sea-level fall near the Cretaceous/Tertiary boundary.

DEVELOPMENT OF THE MICROFOSSIL RECORD ALONG THE CONTINENTAL MARGIN

Although the subsidence and sedimentation history of Atlantic continental margin basins is more complex than that of the oceans, some simple observations stand out. Terrigenous clastic sediments predominate in the Mesozoic strata and the sequence is essentially shallow marine since Jurassic time. This observation holds true over the shelves of the Grand Banks, Nova Scotia, and the northeastern United States margin. The fossil continental-slope regime was beyond the present shelf edge. Subsidence and sedimentation apparently kept a delicate balance until relative sediment starvation brought much of the outer shelf into the bathyal realm in Late Cretaceous time (Fig. 2).

The formation nomenclature of the Mesozoic lithologic sequence off Canada is rather complex (see regional literature cited in Table 1). A formation may not express a distinct lithology, but is more an interval with certain types of rocks that can be traced seismically. Also, formations to some extent are treated as stage substitutes, as when, in a Grand Banks well, reference is made to Whale (Formation) equivalent strata, (Early–Middle Jurassic). It is sufficient to note here that marine Jurassic sediments consist of terrigenous clastics and some carbonates, whereas Lower Cretaceous sediments contain more sands, silts and shales. In Late Cretaceous time some chalks were deposited (Doeven, 1983). The latter are a response to widespread transgression, low hinterland relief and high pelagic carbonate productivity during warm and dry climatic conditions.

The normal salinity marine Mesozoic record goes back to late Sinemurian (approximately 200 m.y. ago), as correlated from Grand Banks beds in the *Involutina liassica* foraminifera zone to the *Frondicularia terquemi* foraminifer zone and the *Raricostatum* ammonite zone in Portugal (Exton and Gradstein, 1984). In both regions these beds immediately overlie marginal-marine dolomites, evaporites, and continental red beds. The younger part of the "continental" succession belongs in the spore interval zones *Classopollis meyerianus, Cypadopites subgranulosus,* and *Echinitosporites* cf *E. iliacoides* of Rhaetian through early Pliensbachian age (Barss and others, 1979; Bujak and Williams, 1977).

From late Sinemurian or early Pliensbachian time, shallow marine sedimentation on the Grand Banks continued until at least Aptian time, making the Jurassic Grand Banks record one of the most complete in North America. During this time the Grand Banks formed the northwest margin of the Tethys seaways, and larger foraminifers occur as far north as 50°, present latitude.

On the Scotian shelf and the United States continental shelf, continuous marine sedimentation began in Middle Jurassic time, with recognition of the palynomorph assemblage zones of *Mancodinium semitabulatum* and *Gonyaulacysta filapicata* (H. Cousminer, written communication, 1984; Bujak and Williams, 1977). In slightly younger marine beds occur the foraminifers *Epistomina regularis, E. mosquensis, E. coronata, E. stellicostata* and *Alveosepta jaccardi* (Ascoli, 1977; Scholle, 1980). The marine transgression may have been an expression of the Callovian eustatic sealevel rise postulated by Hallam (1975) from evidence of a general encroachment of the world seas over land, especially evident in the Osprey, Murre, and Egret wells (Fig. 2). Renewed subsidence and eustatic sea-level rise in middle to late Cretaceous time caused gradual transgression over the Grand Banks and also over old Paleozoic highs (with minor Jurassic cover) north of the Grand Banks (Orphan Knoll, DSDP Site 111; Texaco Blue H-28 well; Fig. 2). Away from the central Grand Banks, Early Cretaceous marine sedimentation continued without interruption. Several deltaic systems were built, as expressed in the Mississauga sands of the Orphan and Scotian basins (Jansa this volume). As far east as Flemish Pass, between Grand Banks and Flemish Cap (Plate 2), there is a 2-km-thick sequence of shallow marine shales and sands of Valanginian (*Lenticulina busnardoi* foraminifer assemblage) through Aptian age (*Conorotalites aptiensis–Lenticulina ouachensis* assemblages).

In the Gabriel, Blue, and DSDP Site 111 boreholes (Plate 2; Fig. 2) Turonian-Campanian sediments are missing, which compares to the condensed sequences or hiatuses in other transitional ("slope") sites and in the ocean basin, as discussed earlier. One explanation for the hiatus is deep-marine sediment starvation on relative highs, coupled with some erosion. Submarine erosion and submarine weathering may be concluded from a hardground at the base of the hiatus in DSDP Site 111 (Ruffman and Van Hinte, 1973) and in Dominion 0-23 (close to Blue H-28). Updip margin sites like Adolphus D-50, Hibernia 0-35, Murre G-67, and Egret K-36 (Fig. 2) have a depositional sequence with a deep neritic to bathyal *Globotruncana* foraminifer assemblage.

The Labrador Sea embayment did not open until later Cretaceous time, and then only in the south where marine magnetic anomalies 34-13 have been identified (Srivastava and Tapscott, this volume; Plate 3). This ocean passage remains largely unsampled at depth. Northern Labrador Sea opening may be slightly later, according to identification of marine magnetic anomalies 31-13 (Srivastava and Tapscott, this vol.). A mid-Cretaceous seaway probably extended northward here, connecting to the Arctic Ocean and/or Western Interior Sea (Williams and Stelck 1975; Gradstein and Srivastava 1980). Labrador Shelf deep drilling, east of the old landmass of the Precambrian shield, has recovered several hundreds of meters of shallow-marine Cen-

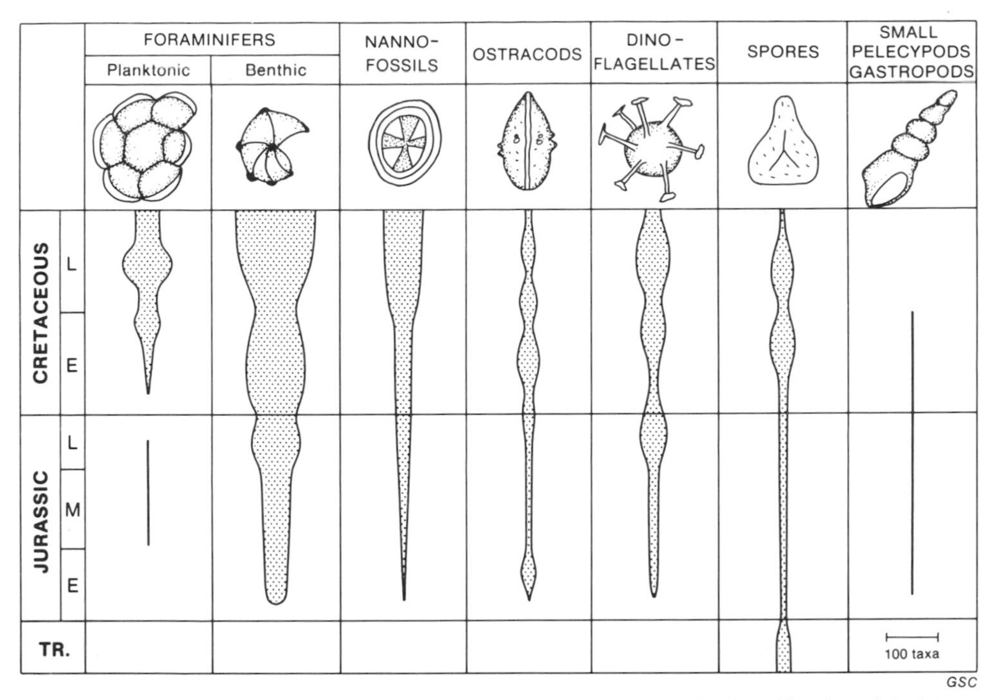

Figure 3. Schematic illustration of stratigraphic distribution and taxonomic diversification of the Mesozoic continental-margin microfossil record. On the Labrador Shelf the microfossil record is much impoverished and is post-Neocomian in age.

omanian and younger Cretaceous sediments (e.g. Herjolf well; Barss and others, 1979). Mid-Cretaceous (?) neritic sediments have been sampled on the southeastern Baffin Island shelf (Gradstein and Srivastava, 1980; MacLean and Falconer 1979). Around Nugssuaq, northeast of Davis Strait, Turonian-Danian shallow-marine strata under basalts have long been studied by Danish geologists (see Henderson and others, 1976, for review).

The Mesozoic microfossil record of the Canadian continental margin mainly harbours foraminifers, nannofossils, dinoflagellates, and spores (including pollen). Ostracods are locally common. Figure 3 gives a schematic impression of the variety and abundance of the preserved microflora and -fauna. From south to north this variety gradually decreases.

The Mesozoic foraminiferal record can be roughly subdivided into Jurassic to Neocomian and Barremian to Maestrichtian sections. The older record, which is transitional in Barremian throgh Albian strata, is definitely pre-actualistic; no comparable recent assemblages exist to aid with paleoecological interpretation. For example, the predominance of nodosariids and epistominids and local peaks of Jurassic globigerinids, make little sense when compared to any Recent shallow (or deep) marine environment. These families and their constituent genera are not predominant in post-Aptian rocks.

The Jurassic through Neocomian continental margin record includes many species of benthic foraminifera belonging in the genera *Lenticulina, Marginulina, Brizalina, Citharina, Frondicularia, Lingulina* (all nodosariids); *Garantella* (Middle Jurassic only); *Reinholdella, Epistomina* and *Conorboides* (all epistominids); *Trocholina, Patellina, Paalzowella, Eoguttulina, Spirrilina*; and some *Miliammina,* large *Ammobaculites* and *Ammomarginulina.* The latter three, and also *Trocholina, Paalzowella,* and the Jurassic-Early Cretaceous larger benthic foraminifera *Alveosepta* and *Choffatella,* are absent in deeper marine beds.

The uniquely Tethyan larger benthic foraminifer record of Jurassic through Early Cretaceous age extended as far north as the northern Grand Banks at present latitudes of 48°N (Sen Gupta and Grant, 1971; Gradstein, 1977). The Tethyan affinity contrasts with the modern record of larger benthic foraminifers that is confined to tropical and subtropical photic zones. The Tethyan record includes: *Orbitolina* spp, Barremian-Middle Cenomanian; *Choffatella decipiens,* (?) Barremian-early Aptian; *Palorbitolina lenticularis,* Barremian-early Aptian; *Everticyclammina virguliana,* late Oxfordian-Hauterivian; *Anchispirocyclina lusitanica,* Tithonian; *Feurtilia frequens,* Kimmeridgian-?Berriasian; *Rectocyclammina chouberti,* Kimmeridgian; *Kurnubia palastiniensis,* Oxfordian-Kimmeridgian; and *Alveosepta jaccardi,* late Oxfordian-early Kimmeridgian. A recent survey has confirmed the presence of these Tethyan shallow-neritic bottom dwellers in at least 44 well sites on the Grand Banks and Scotian Shelf (F. Thomas, personal communication, 1982).

Lower Cretaceous smaller benthic foraminifer assemblages are primarily composed of nodosariids, epistominids, *Trocholina,* and some species of *Gavelinella* (Hauterivian and younger), *Conorotalites* (Barremian and younger), and *Pleurostomella, Gyroidinoides,* and *Osangularia* (Aptian and younger). Among the smaller arenaceous taxa there are *Ammobaculites, Reophax, Ha-*

plophragmium, Marsonella, Uvigerinammina, and *Spiroplectinata.* The first three may occur with many specimens in virtually monotypic assemblages in shaley intervals of some wells; the depositional environment may have been brackish.

The Upper Cretaceous benthic foraminiferal assemblage includes taxa of *Neoflabellina, Kyphopyxa, Gavelinella, Lingulogavelinella, Bolivinoides, Gyroidinoides, "Pullenia," Globorotalites, Epistomina, Lenticulina* and varieties, *Gaudryina,* and *Ammobaculites.* The planktonic genera include many species of *Globotruncana, Globotruncanella, Rotalipora, Praeglobotruncana, Ticinella, Rugoglobigerina, Globigerinelloides, Hedbergella, Heterohelix, Pseudotextularia,* and *Sigalia.* Updip, closer to the (present) shoreline, well sites have much impoverished planktonic assemblages and progressively fewer single keeled forms, but rich benthic assemblages reflect the proximity to the Cretaceous shore line. This situation resembles a modern planktonic/benthic ratio in an ocean-margin transect.

Ascoli (1977) and Exton and Gradstein (1984) together report up to 100 species of Mesozoic ostracods. Kimmeridgian-Tithonian, Barremian-Aptian and Campanian strata in particular have numerous taxa but the number of specimens per sample is generally low compared to foraminifers. Many local morphological varieties of ostracods occur, but they have not been evaluated stratigraphically in any detail.

The Upper Cretaceous nannofossil record includes over 100 taxa and has strong affinities to other Tethyan regions as far north as the northern Grand Banks in the Adolphus well (Doeven, 1983).

Bujak and Williams (1977, 1978) give histograms of stratigraphic differentiation and species diversity of palynomorphs in Jurassic and Cretaceous margin well sites. Several hundred types of dinocysts and a hundred or more spore types occur. At least half have been found significant for stratigraphic correlation. The Middle-Late Jurassic and mid-Cretaceous transgressions caused dramatic increases from a few to several tens of dinocyst taxa per zone. Maximum diversities occurred in Callovian-Kimmeridgian, Albian, and Senonian time.

THE SEARCH FOR ZONATIONS

Zonations are empirical schemes that emphasize the temporal and spatial restriction of morphologically distinct fossil types. Good zonations have zonal units with well defined lower and upper limits, are easily recognizable in many sections, correlate well, and have been compared to other regional or extraregional zonations. Some evolutionary considerations are desirable. Chronostratigraphic meaning may have been verified in stratotype or other classical sections.

The biostratigraphic framework developed for Mesozoic strata in the western North Atlantic Ocean and margin is distinguished particularly by the nannofossil zonation of Doeven (1983) that used Late Cretaceous nannofossils in sidewall cores and conventional cores of margin wells, and by palynomorph zonations of Habib (1972, 1975) Habib and Knapp (1982) and Habib and Drugg (1983), based on deep sea cores. Both zonations are well defined and well illustrated. They have clear correlative power, and have been or are being compared to other regional and cosmopolitan zonations and to stratotype chronostratigraphy. Doeven's (1983) Upper Cretaceous nannofossil zonation yields the most detailed continental margin stratigraphy to date and Habib's zonation is widely correlated in the deep basin where carbonate dissolution is a problem. The nannofossil zonation has also been tested and reproduced quantitatively, using objective ranking techniques of first and last occurrences (Doeven and others, 1982).

Figures 4–8 summarize the state of the art in zonations and events schemes for the ocean basin and the continental margin. Details on each zonation may be found in the literature cited. Interval zones, using highest occurrences in well cutting samples, are marked with an asterisk (*).

Figure 4 shows the zonal framework for deep ocean sites based on nannofossils, dinoflagellates, foraminifers, radiolarians, and the combined events in pelagic crinoids, (*Saccocoma*) pelecypods ("filaments"), calpionellids, and aptychi. The radiolarian zonation incorporates the radiolarian events in DSDP Site 534 with those of the eastern Atlantic and Mediterranean Tethys. The methodology is quantitative, following programs developed by J. Guex and E. Davaud using overlap of stratigraphic ranges to create "associations unitaires" that resemble concurrent range zones (Baumgartner, 1983).

There is reasonable agreement for the Early Cretaceous between the stage assignments of the dinoflagellate and nannofossil stratigraphies, although there are boundary problems in Hauterivian and Aptian, possibly because of carbonate dissolution. In Jurassic strata there are more problems. The dinoflagellate chronostratigraphy may be half a stage or more younger than that based on nannofossils and sparse foraminifers. The adjustments of Oxfordian, Kimmeridgian, and Early Tithonian zonal boundaries needs more study (see Gradstein, 1983).

Figures 5–8 show the Jurassic and Cretaceous microfossil zonations in use for the continental-margin chronostratigraphy. In the foraminiferal record, the Jurassic *Reinholdella crebra* var. and *Epistomina stellicostata, E. uhligi* assemblages are beset with taxonomic problems and a complex record confused by well cavings. The Cretaceous benthic assemblages provide no firm zones but are successive assemblages based on cosmopolitan, low- to mid-latitude index fossils. Gradstein (1978a) and Doeven (1983) have studied interrelation of zones from various authors using different fossil assemblages, and zonal interrelations in Figure 6–8 follow their findings to date. Recently, Williamson (in press) has proposed a more definite zonation.

The marine record of the Labrador shelf is limited to the Upper Cretaceous and only the Maestrichtian has been found in more than one or two well sites. Zone markers differ from those to the south. It is interesting that, on the shelf, the *Abathomphalus mayaroensis* standard planktonic foraminifer zone of late Maestrichtian age has been recognized only north of the Grand Banks. Rocks of such age have been identified in a few wells on the

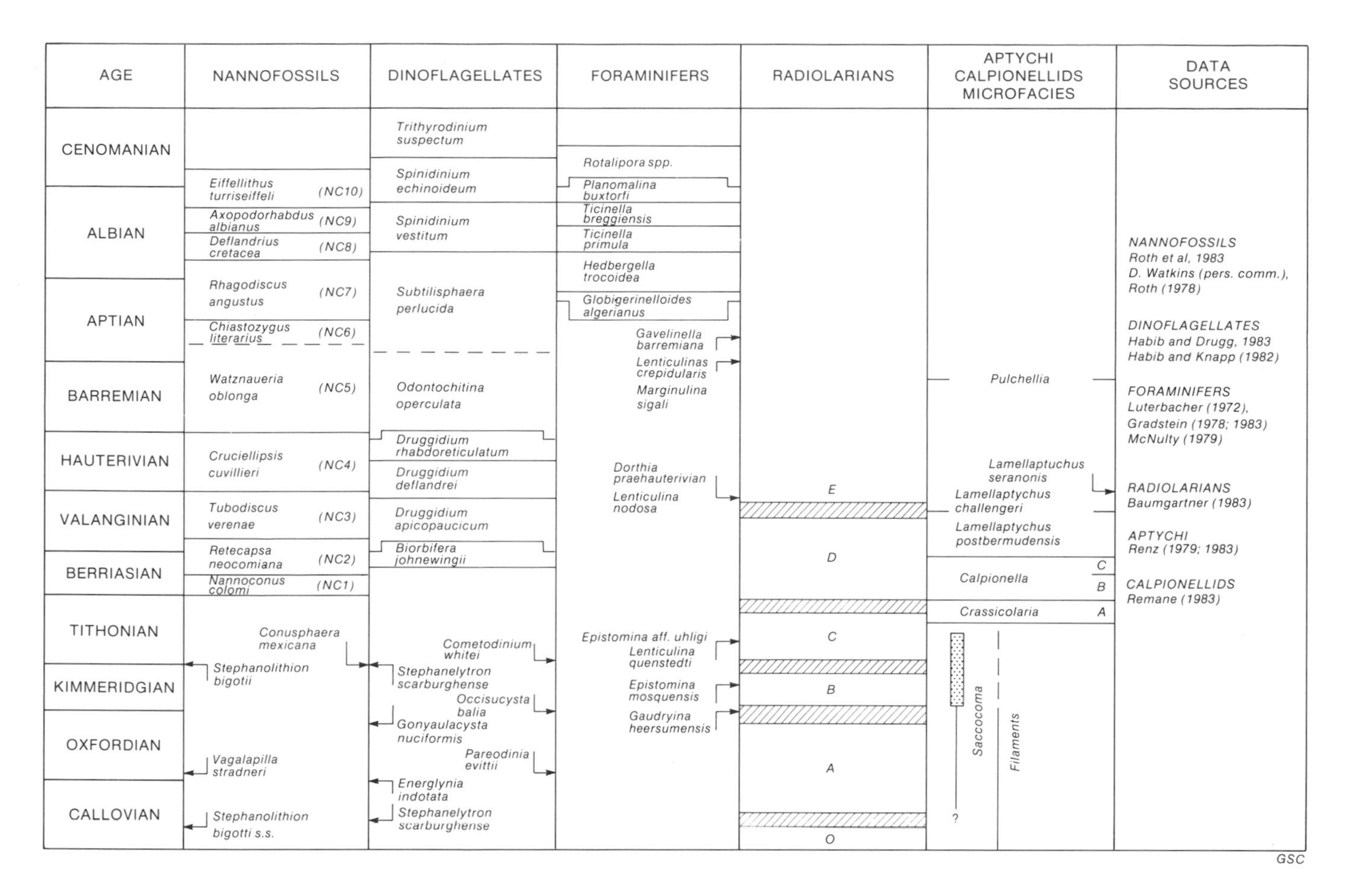

Figure 4. Mesozoic zonations, and appearance and disappearance events using nannofossils, dinoflagellates, foraminifera, radiolarians, aptychi, calpionellids, and microfacies in western North Atlantic ocean DSDP Sites 99, 100, 105, 384, 385, 386, 387, 391 and 534. The upper Campanian-Maestrichtian strata are zoned with standard nannofossil and foraminifer events.

Scotian and Grand Banks shelves using the coeval *Nephrolithus frequens* nannofossil zone. Elsewhere, the *Abathomphalus mayaroensis* zone is known from DSDP Site 390/392 on the Blake Plateau (P. Marks, personal communication, 1982) and Site 384 south of the Grand Banks (McNulty, 1979).

SYNTHESIS

The body of stratigraphic information cited for the western North Atlantic Ocean and margin has been integrated in Plate 1B. This chart expresses the observed stratigraphic relationship of the more cosmopolitan part of the Jurassic and Cretaceous microfossil zonations of Figures 4–8. Included are (1) circum-North Atlantic Mesozoic biostratigraphies using nannofossils, foraminifers, palynomorphs, radiolarians, and calpionellids, and (2) the Jurassic and Cretaceous ammonite zonation for low- to mid-latitudes. The chart also incorporates a stratigraphic recalibration of the geomagnetic reversal scale as discussed by Kent and Gradstein (this volume), the lithologic and seismic sequences of the deep ocean, and other major geological and climatic/oceanographic events.

I have used Hallam's (1975) standard Jurassic ammonite zonation. The Tithonian lower boundary is based on the appearance of *Gravesia,* which leads to the short (French) Kimmeridgian. Probably as a result of increasing geographic differentiation of morphologies and ranges, Upper Jurassic and particularly Cretaceous ammonite zonations become more complex and more latitudinally distinct (Hallam, 1975; Van Hinte, 1976a; Sigal, 1977; Robaszynski and others, 1979; and P. Marks, pers. communication, 1983).

There is a first-order correlation between the ranges of Lower Jurassic (Pliensbachian-Toarcian) foraminifers studied in Portugal and the ammonite zones. Moreover, the Lower Jurassic foraminiferal record of Portugal and the Grand Banks shows

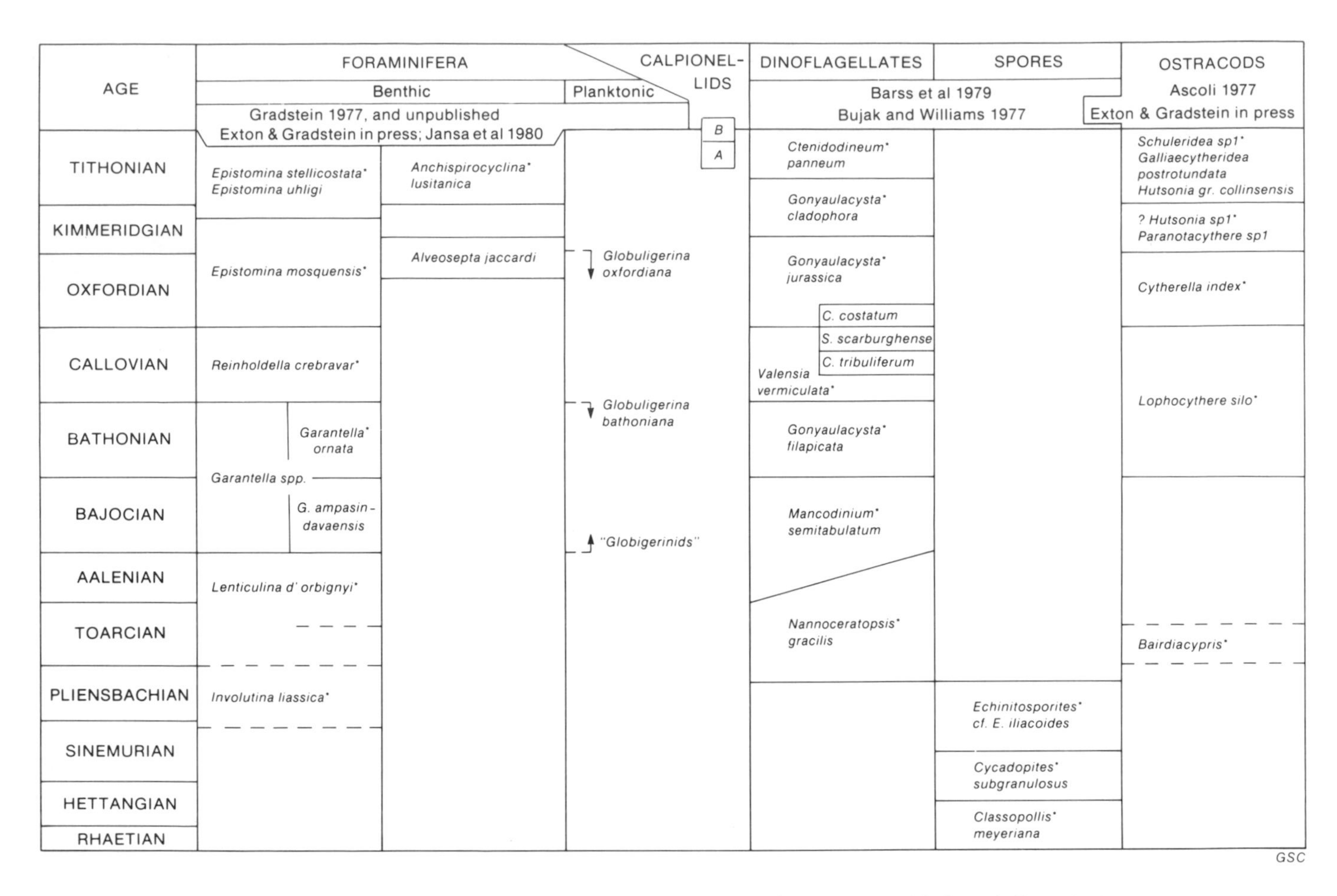

Figure 5. Jurassic zonations of Grand Banks and Scotian Shelf, based on benthic foraminifers, planktonic foraminifers, calpionellids, dinoflagellates, spores (including pollen), and ostracods. With few exceptions these are interval (assemblage) zones based on "tops."

good agreement (Exton and Gradstein, 1984; Copestake and Johnson, 1981). The benthic foraminifer *Garantella* events are recorded on the Grand Banks (Gradstein, 1977). The Upper Jurassic (Oxfordian-Tithonian) benthic foraminiferal events are as also known from the Canadian passive margin and to some extent from the Tethys in southern Europe (Jaffrezo, 1980) and North Africa. Several taxa, indicated with an asterisk (*) in Plate 1B occur only in the neritic biofacies. Increasing provincialism since Late Jurassic time precludes recognition of many of the events in higher latitudes, although along the western Atlantic strong Tethyan influences extended to the northern limit of the marine Jurassic and Early Cretaceous realm, at least as far north as present-day latitude 48°.

Middle Jurassic planktonic foraminiferal development is well recorded on the Grand Banks. The Jurassic record in Atlantic pelagic biofacies is sparse, to say the least, and confined to some specimens in Upper Jurassic (Oxfordian) strata (Gradstein, 1977; 1983). Until Barremian time, the planktonic foraminifera were mostly confined to epicontinental seas and shallow marginal basins.

The standardized Cretaceous planktonic zonation can be applied relatively easily in the northwestern Atlantic region, although the *Ticinella bejaouaensis* and *Globotruncana scutilla* zones are hard to recognize. This is because the zonal markers are not often encountered. The stratigraphic distribution of Cretaceous benthic foraminiferal events largely follows studies by Moullade (1966), Van Hinte (1976a), Ascoli (1977), Gradstein (1978b and unpublished), and Drushtchitz and Gorbatschik (1978). Good summaries of eastern Atlantic Cretaceous foraminiferal events are given by Butt (1982) and Sliter (1980).

Grand Banks age assignments using the Lower and lower Middle Jurassic micropaleontological zonation of organic-walled microfossils (Bujak and Williams, 1977) reasonably agree with those using foraminifers (see Gradstein, 1978a). The latter zonation has been compared with the ammonite succession shown in Plate 1B (Exton and Gradstein, 1984). For the younger Jurassic

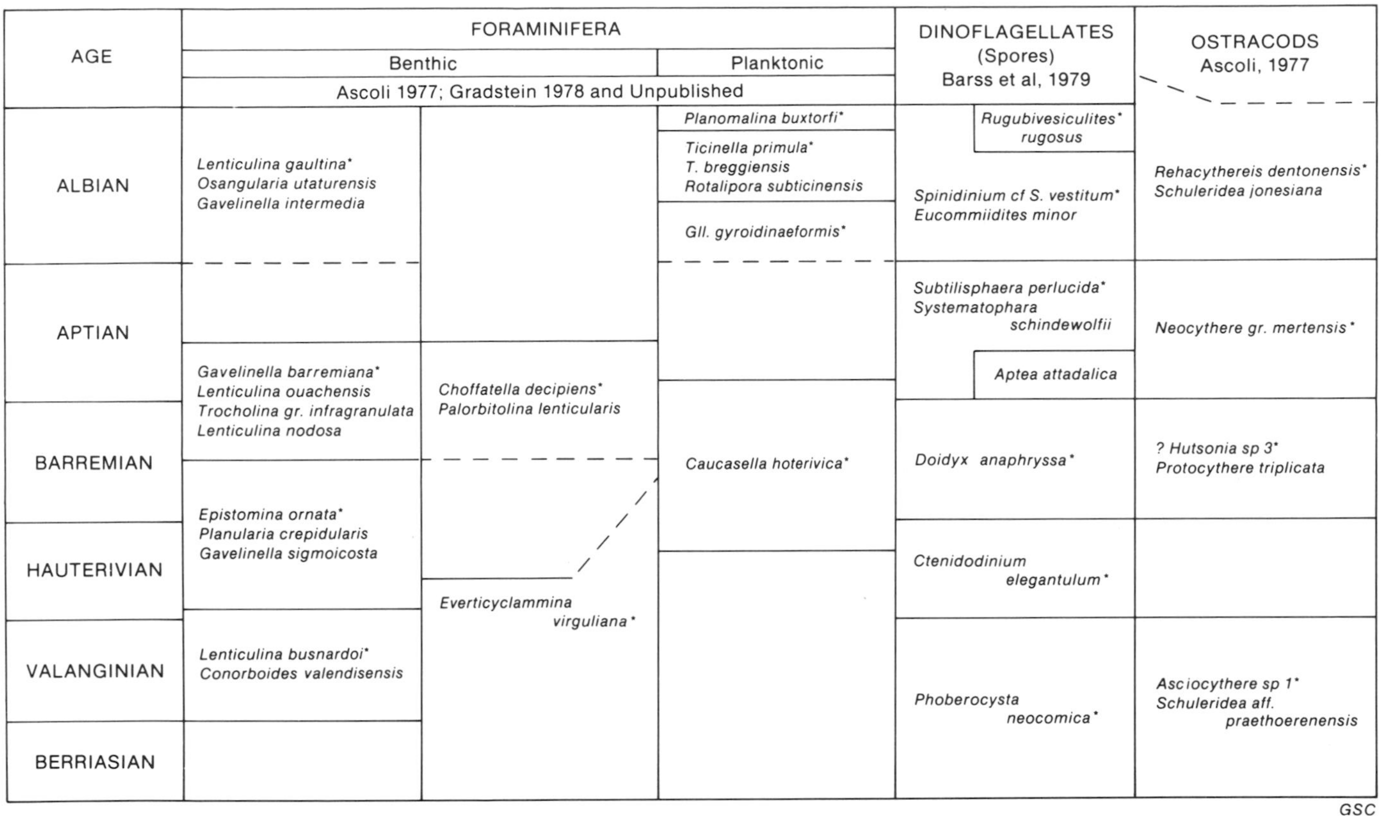

AGE	FORAMINIFERA — Benthic		FORAMINIFERA — Planktonic	DINOFLAGELLATES (Spores) Barss et al, 1979	OSTRACODS Ascoli, 1977
	Ascoli 1977; Gradstein 1978 and Unpublished				
ALBIAN	*Lenticulina gaultina**, *Osangularia utaturensis*, *Gavelinella intermedia*		*Planomalina buxtorfi**; *Ticinella primula**, *T. breggiensis*, *Rotalipora subticinensis*; *Gll. gyroidinaeformis**	*Rugubivesiculites** *rugosus*; *Spinidinium cf S. vestitum**, *Eucommiidites minor*	*Rehacythereis dentonensis**, *Schuleridea jonesiana*
APTIAN				*Subtilisphaera perlucida**, *Systematophara schindewolfii*; *Aptea attadalica*	*Neocythere gr. mertensis**
BARREMIAN	*Gavelinella barremiana**, *Lenticulina ouachensis*, *Trocholina gr. infragranulata*, *Lenticulina nodosa*	*Choffatella decipiens**, *Palorbitolina lenticularis*	*Caucasella hoterivica**	*Doidyx anaphryssa**	? *Hutsonia sp 3**, *Protocythere triplicata*
HAUTERIVIAN	*Epistomina ornata**, *Planularia crepidularis*, *Gavelinella sigmoicosta*	*Everticyclammina virguliana**		*Ctenidodinium elegantulum**	
VALANGINIAN	*Lenticulina busnardoi**, *Conorboides valendisensis*			*Phoberocysta neocomica**	*Asciocythere sp 1**, *Schuleridea aff. praethoerenensis*
BERRIASIAN					

GSC

Figure 6. Lower Cretaceous zonations of Grand Banks and Scotian Shelf, based on benthic and planktonic foraminifers, dinoflagellates and ostracods. With few exceptions these are interval (assemblage) zones based on "tops." For more details in foraminiferal zonation, see Williamson (in press).

and the Cretaceous stratigraphic section, the organic-microfossil biostratigraphy (dinoflagellates), as developed by D. Habib for the pelagic realm, is shown (Habib, 1977; Habib and Drugg, 1983). They correlate this biostratigraphy in terms of ammonite zonations and stratotype coverage, and chronostratigraphic application in the northwestern Atlantic ocean reasonably agrees with that of Roth and others (1983) and Roth (1978) who used nannofossils from the same deep sea drilling sections. However, there is disagreement on the position within the cores of the boundaries of Oxfordian, Kimmeridgian, early Tithonian, Hauterivian, and Aptian. The Jurassic nannofossil datums and their relation to ammonite zones (Plate 1B) follows a literature evaluation by Roth and others (1983) and D. Watkins (pers. communication, 1982).

The Upper Cretaceous nannofossil zonation is that developed on the northwestern Atlantic margin and adapted from Tethyan standard zonations (Doeven and others, 1982; Doeven, 1983). Doeven has tied it to the foraminiferal record of the Grand Banks. The calpionellid zonation is applicable to beds at the Jurassic-Cretaceous boundary and extends from the Alpine-Mediterranean realm into the western Atlantic (Remane, 1978; 1983; Jansa and others, 1980). The cephalopod (small ammonites and aptychi) correlation scheme in the pelagic biofacies (Renz, 1983) was developed in the western Atlantic. The Tethyan radiolarian zonation as extended in DSDP Site 534 (Blake Bahama Basin) is from Baumgartner (1983).

CONCLUSION

Nannofossils, planktonic and benthic foraminifers, and palynomorphs (spores, pollen and dinoflagellates) are most consistently present in the Jurassic and Cretaceous marine strata of the northwestern Atlantic deep basin and continental margin. Detailed nannofossil zonations have been developed for the Lower Cretaceous of the deep basin (Roth, 1978; 1983) and for the open marine Upper Cretaceous strata on the continental margin

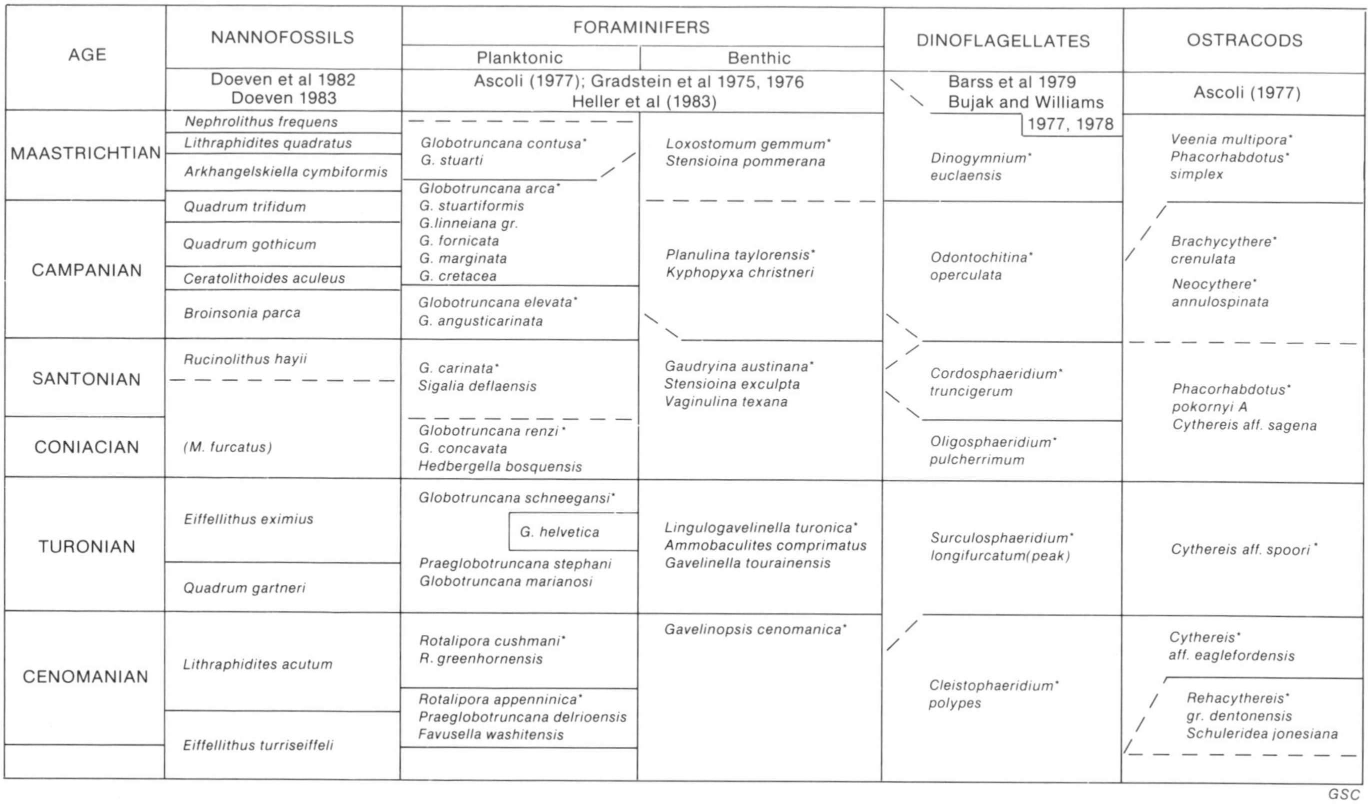

AGE	NANNOFOSSILS	FORAMINIFERS Planktonic	FORAMINIFERS Benthic	DINOFLAGELLATES	OSTRACODS
	Doeven et al 1982 Doeven 1983	Ascoli (1977); Gradstein et al 1975, 1976 Heller et al (1983)		Barss et al 1979 Bujak and Williams 1977, 1978	Ascoli (1977)
MAASTRICHTIAN	Nephrolithus frequens Lithraphidites quadratus Arkhangelskiella cymbiformis	Globotruncana contusa* G. stuarti	Loxostomum gemmum* Stensioina pommerana	Dinogymnium* euclaensis	Veenia multipora* Phacorhabdotus* simplex
CAMPANIAN	Quadrum trifidum Quadrum gothicum Ceratolithoides aculeus Broinsonia parca	Globotruncana arca* G. stuartiformis G.linneiana gr. G. fornicata G. marginata G. cretacea Globotruncana elevata* G. angusticarinata	Planulina taylorensis* Kyphopyxa christneri	Odontochitina* operculata	Brachycythere* crenulata Neocythere* annulospinata
SANTONIAN	Rucinolithus hayii	G. carinata* Sigalia deflaensis	Gaudryina austinana* Stensioina exculpta Vaginulina texana	Cordosphaeridium* truncigerum	Phacorhabdotus* pokornyi A Cythereis aff. sagena
CONIACIAN	(M. furcatus)	Globotruncana renzi* G. concavata Hedbergella bosquensis		Oligosphaeridium* pulcherrimum	
TURONIAN	Eiffellithus eximius Quadrum gartneri	Globotruncana schneegansi* G. helvetica Praeglobotruncana stephani Globotruncana marianosi	Lingulogavelinella turonica* Ammobaculites comprimatus Gavelinella tourainensis	Surculosphaeridium* longifurcatum(peak)	Cythereis aff. spoori*
CENOMANIAN	Lithraphidites acutum Eiffellithus turriseiffeli	Rotalipora cushmani* R. greenhornensis Rotalipora appenninica* Praeglobotruncana delrioensis Favusella washitensis	Gavelinopsis cenomanica*	Cleistophaeridium* polypes	Cythereis* aff. eaglefordensis Rehacythereis* gr. dentonensis Schuleridea jonesiana

GSC

Figure 7. Upper Cretaceous zonation of Grand Banks and Scotian Shelf, based on nannofossils, planktonic and benthic foraminifers, dinoflagellates and ostracods. The nannofossil zonation uses mainly first occurrences, the other zones are interval type assemblage zones based on "tops."

(Doeven, 1983). Twenty standard nannofossil zones span the 12 Cretaceous stages.

Many of the 24 standard planktonic foraminifer zones for the Aptian through Maestrichtian can be recognized in the region, although the much condensed Upper Cretaceous abyssal clays of the deep basin are virtually devoid of microfossils.

Benthic foraminiferal and ostracod events, assemblages, and assemblage zones of the Jurassic and Cretaceous strata on the continental margin provide stratigraphic resolution at approximately stage level. The Lower Jurassic regional zonation of the Grand Banks has been directly tied to the coeval foraminifer/ostracod and ammonite record in Portugal (Exton and Gradstein, 1984). In the deep basin the Mesozoic foraminifer and particularly the ostracod record is less diversified and patchy, which limits stratigraphic utility.

There is a well documented dinoflagellate/pollen zonation for the Lower Cretaceous limestone and shale sequences of the deep basin (see papers by D. Habib). Resolution is at or below stage level. The Jurassic and Cretaceous continental-margin sequence is regionally zoned by diversified dinoflagellate, pollen, and spore assemblages (Barss and others, 1979); resolution is approximately at stage level.

The Tethyan Upper Jurassic to Lower Cretaceous radiolarian and calpionellid zonations (Baumgartner, 1983; Remane, 1983) extend into the deep oceanic environment of the western Atlantic. Calpionellids also are of use in defining the Jurassic-Cretaceous boundary in outer continental-margin wells (Jansa and others, 1980).

Several outstanding problems provide stimulation and challenge for future research. In the deep basin one of the most pressing issues is to resolve discrepancies in Jurassic age assignments by using nannofossils, dinoflagellates, and foraminifers. Detailed comparative studies in easily accessible land sections in Portugal and Morocco will be helpful.

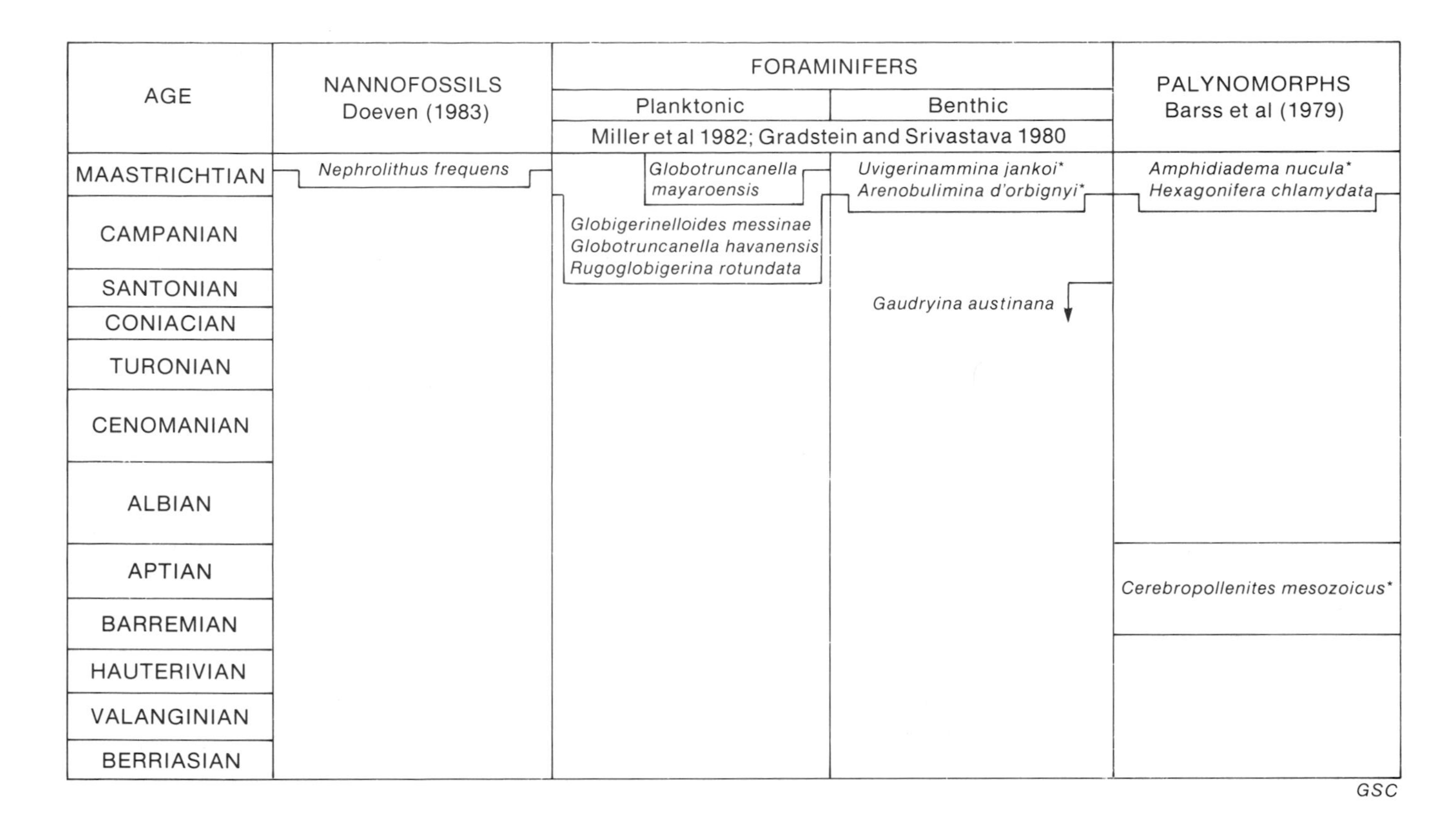

Figure 8. Cretaceous zonation of the Labrador Shelf, based on dinoflagellates, nannofossils and foraminifers.

Attention should be paid to reasons for differences in Hauterivian and Aptian age determinations using nannofossils and palynomorphs in the deep basin. A detailed study of arenaceous foraminifers in the abyssal and bathyal Cretaceous clays of several DSDP sites may help to zone and correlate these otherwise poorly fossiliferous strata.

On the Canadian continental margin the detailed Upper Cretaceous nannofossil zonation needs to be extended into older strata. This will benefit regional geology and help to provide a comprehensive deep-basin and continental-margin Jurassic and Cretaceous biostratigraphy of supra-regional value. The regional Mesozoic palynomorph zonations of the continental-margin basins will benefit from direct investigation of, and first order correlation to, conjugate Moroccan and Portuguese marine sections where Jurassic-Cretaceous ammonite and standard Cretaceous planktonic foraminifer stratigraphies have been applied in detail.

Jurassic and Cretaceous benthic foraminifer and ostracod assemblages of the continental margin should be better documented taxonomically. This will strengthen the biostratigraphic framework. Direct correlation to the conjugate margin record in Morocco and Portugal is very valuable for chronostratigraphic assignments, as is direct comparison to coeval assemblages in classical sections in Trinidad, Tunisia, and western Europe (Speeton clay).

The multiple biostratigraphic record of the continental margins is extensive, although sample quality is generally poor. Application of quantitative stratigraphic methods as used by Doeven and others (1982) and Baumgartner (1983) will objectively and efficiently reduce the large quantity of data to zonations. Careful comparison of the individual well records to the quantitative standard zonations will reveal the difference between "noise" and geological control on the microfossil distribution patterns in time and space. Ultimately such a quantitative approach will provide more detailed correlations, make the zonations better understood, and give non-specialists easier access to the information gathered.

REFERENCES CITED

Department of Energy, Mines, and Resources
1983 : Offshore Schedule of Wells: Ottawa, Canada.

Ascoli, P.
1977 : Foraminiferal and ostracod biostratigraphy of the Mesozoic–Cenozoic, Scotian Shelf, Atlantic Canada: in Proc. I. International Symposium on Benthonic Foraminifera, Halifax 1975; pt. B., Maritime Sediments; Special Publication no. 1, p. 653–771.

Aurisano, R., and Habib, D.
1977 : Upper Cretaceous dinoflagellate zonation of the subsurface Toms River section near Toms River, New Jersey: in Stratigraphic Micropaleontology of Atlantic Basin and Borderlands, ed. F. M. Swain; Elsevier, New York, p. 369–387.

Barss, M. S., Bujak, J. P., and Williams, G. L.
1979 : Palynological zonation and correlation of sixty-seven wells, eastern Canada: Geological Survey of Canada Paper 78–24, 118 p.

Bate, R. H.
1977 : Jurassic Ostracoda of the Atlantic Basin: in Stratigraphic Micropaleontology of Atlantic Basin and Borderlands, ed. F. M. Swain; Elsevier, New York, p. 231–245.

Baumgartner, P. O.
1983 : Summary of Middle Jurassic–Early Cretaceous radiolarian biostratigraphy of DSDP Site 534 (Blake–Bahama Basin) and correlation to Tethyan sections: in Initial Reports of the Deep Sea Drilling Project, v. 76, eds. R. E. Sheridan and F. M. Gradstein; U.S. Government Printing Office, Washington, D.C., p. 569–573.

Benson, W., and Sheridan, R. E., eds.
1978 : Initial Reports of the Deep Sea Drilling Project, v. 44: U.S. Government Printing Office, Washington, D.C., 1005 p.

Bhat, H., McMillan, N. J., Aubert, J., Porhault, B., and Surin, M.
1975 : North American and African drift—the record in Mesozoic coastal plain rocks, Nova Scotia and Morocco: in Canada's Continental Margins, eds. C. J. Yorath, E. R. Parker, D. J. Glass; Canadian Society of Petroleum Geologists, Memoir 4, p. 375–389.

Bolli, H. M.
1978 : Upper Jurassic Calcisphaerulidae from DSDP Leg 44, Hole 391C, Blake Bahama Basin, Western North Atlantic: in Initial Reports of the Deep Sea Drilling Project, v. 44, eds. W. E. Benson and R. E. Sheridan; U.S. Government Printing Office, Washington, D.C., p. 911–921.

Bujak, J. P., and Williams, G. L.
1977 : Jurassic palynostratigraphy of offshore eastern Canada: in Stratigraphic Micropaleontology of the Atlantic Basin and Borderlands, ed. E. M. Swain, Elsevier, New York, p. 321–340.

Bujak, J. P., and Williams, G. L.
1978 : Cretaceous palynostratigraphy of offshore southeastern Canada: Geological Survey of Canada Bulletin, v. 297, 19 p.

Burky, D.
1972 : Coccolith stratigraphy, Leg XI, Deep Sea Drilling Project: in Initial Reports of the Deep Sea Drilling Project, v. 11, eds. C. D. Hollister and J. I. Ewing; U.S. Government Printing Office, Washington, D.C., p. 475–483.

Butt, A.
1982 : Micropaleontological bathymetry of the Cretaceous of Western Morocco: Paleogeography, Paleoclimatology, and Paleoecology, v. 37, p. 235–275.

Caron, M.
1972 : Planktonic Foraminifera from the Upper Cretaceous of Site 398, Leg XI, Deep Sea Drilling Project: in Initial Reports of the Deep Sea Drilling Project, eds. C.D. Hollister and J.I. Ewing; U.S. Government Printing Office, Washington, D.C., p. 551–561.

Cita, M. B., and Gartner, S.
1971 : Deep Sea Upper Cretaceous from the western North Atlantic: Proceedings II Planktonic Conference, Rome 1970, vol. 1, p. 287–320.

Copestake, P., and Johnson, B.
1981 : Lower Jurassic (Hettangian–Toarcian) Foraminifera from the Mochras borehole, North Wales; systematic and biostratigraphical micropaleontology: Report of the Institute of Geological Sciences, p. 1–162.

Cushman, J.
1936 : Geology and paleontology of the Georges Bank Canyons; Part 4, Cretaceous and Late Tertiary Foraminifera: Geological Society of America Bulletin, v. 47, p. 413–440.

Doeven, P. H.
1983 : Cretaceous nannofossil stratigraphy and paleoecology of the northwestern Atlantic: Geological Survey of Canada Bulletin 356, 69 p.

Doeven, P. H., Gradstein, F. M., Jackson, A., Agterberg, F. P., and Nel, L. D.
1982 : A quantitative nannofossil range chart: Micropaleontology, v. 28, no. 1, p. 85–92.

Drushtichitz, V. V., and Gorbatschik, T. N.
1978 : Zonengliederung der Unteren Kreide der südlichen USSR nach Ammoniten und Foraminiferen: in Aspkte der Kreide Europas IUGS A, 6, p. 107–116.

Dufaure, Ph., McWhae, R., and Verdenius, J. G.
1977 : Tertiary and Upper Cretaceous in offshore Labrador boreholes: First stratigraphical results: in Proceedings, First International Symposium on Benthonic Foraminifera, Halifax 1975, Maritime Sediments, Special Publication 1B, p. 509–522.

Eliuk, L. S.
1978 : The Abenaki Formation, Nova Scotia Shelf, Canada—a depositional and diagenetic model for a Mesozoic carbonate platform: Bulletin of Canadian Petroleum Geology, vol. 26, p. 424–514.

Enos, P., and Freeman, T.
1978 : Shallow-water limestones from the Blake Nose, Sites 390 and 392; in Initial Reports of the Deep Sea Drilling Project, v. 44, eds. W. E. Benson, R. E. Sheridan; U.S. Government Printing Office, Washington, D.C., p. 413–463.

Exton, J., and Gradstein, F. M.
1984 : Early Jurassic stratigraphy and micropaleontology of the Grand Banks and Portugal; in Jurassic and Cretaceous Biochronology and Biogeography of North America, ed. G. E. Westermann; Geological Association of Canada, Special Paper 27, p. 13–30.

Foreman, H. P.
1977 : Mesozoic Radiolaria from the Atlantic basin and its borderlands; in Stratigraphic Micropaleontology of Atlantic Basin and Borderlands, ed. F. M. Swain; Elsevier, New York, p. 305–320.

Gradstein, F. M.
1975 : Biostratigraphy (foraminifera) and depositional environment of Amoco IOE Eider M–75, Grand Banks of Newfoundland: Geological Survey of Canada Report 334, 17 p.
1977 : Biostratigraphy and biogeography of Jurassic Grand Banks foraminifera; in Proceedings, First International Symposium on Benthonic Foraminifera, Halifax 1975, Part B, Maritime Sediments Special Publication, no. 1, p. 557–583.
1978a: Jurassic Grand banks Foraminifera: Journal of Foraminiferal Research, vol. 8, no. 2, p. 97–109.
1978b: Biostratigraphy of Lower Cretaceous Blake Nose and Blake–Bahama Basin Foraminifers, DSDP Leg 44, Western North Atlantic Ocean: in Initial Reports of the Deep Sea Drilling Project, v. 44, eds. W. E. Benson and R. E. Sheridan; U.S. Government Printing Office, Washington, D.C., p. 663–703.
1983 : Paleoecology and stratigraphy of Jurassic abyssal foraminifera in the Blake Bahama Basin, Deep Sea Drilling Project Site 534; in Initial Reports of the Deep Sea Drilling Project, v. 76, eds. F. M. Gradstein, R. E. Sheridan; U.S. Government Printing Office, Washington, D.C., p. 537–559.

Gradstein, F. M., and Berggren, W. A.

1981 : Flysch-type agglutinated foraminifera and the Maestrichtian to Paleogene history of the Labrador and North Seas: Marine Micropaleontology, v. 6, p. 211–268.

Gradstein, F. M., and Sheridan, R. E.

1983a: Introduction; in Initial Reports of the Deep Sea Drilling Project, v. 76, eds. R. E. Sheridan and F. M. Gradstein, U.S. Government Printing Office, Washington, D.C., p. 5–19.

1983b: On the Jurassic Atlantic Ocean and a synthesis of results of Deep Sea Drilling Project Leg 76; in Initial Reports of the Deep Sea Drilling Project, v. 76, eds. R. E. Sheridan and F. M. Gradstein; U.S. Government Printing Office, Washington, D.C., p.913–945.

Gradstein, F. M., and Srivastava, S. P.

1980 : Aspects of Cenozoic stratigraphy and paleoceanography of the Labrador Sea and Baffin Bay: Paleogeography, Palaeoclimatology, and Palaeoecology, v. 30, p. 261–295.

Gradstein, F. M., and Williams, G. L.

1976 : Biostratigraphy of the Labrador Shelf, pt. I: Geological Survey of Canada Report 349, 32 p.

Gradstein, F. M., Williams, G. L., Jenkins, W.A.M., and Ascoli, P.

1975 : Mesozoic and Cenozoic stratigraphy of the Atlantic continental margin, eastern Canada: Canadian Society of Petroleum Geologists, Memoir 4, p. 103–131.

Gradstein, F. M., Jenkins, W.A.M., and Williams, G. L.

1976 : Biostratigraphy and depositional history of Amoco Imp Skelly B-1 Egret K-36 Grand Banks, Newfoundland: Geological Survey of Canada Report 396, 22 p.

Habib, D.

1968 : Spores, pollen and microplankton from the Horizon Beta outcrop: Science, v. 162, p. 1480.

1970 : Middle Cretaceous palynomorph assemblages from days near the Horizon Beta deep-sea outcrop: Micropaleontology, v. 16, no. 3, p. 345–379.

1972 : Dinoflagellate stratigraphy, Leg XI, Deep Sea Drilling Project: in Initial Reports of the Deep Sea Drilling Project, v. 11, eds. C. D. Hollister and J. I. Ewing; U.S. Government Printing Office, Washington, D.C., p. 367–425.

1975 : Neocomian dinoflagellate zonation in the western North Atlantic: Micropaleontology, v. 21, no. 4, p. 373–392.

1977 : Comparison of Lower and Middle Cretaceous palynostratigraphy zonations in the western North Atlantic: in Micropaleontology of the Atlantic Basin and Borderlands, ed. F. M. Swain: Elsevier, New York, p. 341–367.

1978 : Palynostratigraphy of the Lower Cretaceous Section at DSDP Site 391, Blake-Bahama Basin, and its correlation in the North Atlantic: in Initial Reports of the Deep Sea Drilling Project, v. 44, eds. W. E. Benson and R. E. Sheridan; U.S. Government Printing Office, Washington, D.C., p. 887–899.

1979 : Cretaceous palynostratigraphy at Site 387, western Bermuda Rise: in Initial Reports of the Deep Sea Drilling Project, eds. B. E. Tucholke and P. R. Vogt; U.S. Government Printing Office, Washington, D.C., p. 585–591.

Habib, D., and Knapp, S. D.

1982 : Stratigraphic utility of Cretaceous small acritarchs: Micropaleontology 28(4), p. 335–371.

Habib, D., and Drugg, W. S.

1983 : Dinoflagellate age of Middle Jurassis-Early Cretaceous sediments in the Blake-Bahama Basin: in Initial Reports of the Deep Sea Drilling Project, v. 76, eds. R. E. Sheridan and F. M. Gradstein; U.S. Government Printing Office, Washington, D.C., p. 623–639.

Hallam, A.

1975 : Jurassic Environments: Cambridge University Press, 269 p.

Hart, M.B.

1976 : The mid-Cretaceous succession of Orphan Knoll (northwest Atlantic): micropaleontology and paleoceanographic implications: Canadian Journal of Earth Sciences, v. 13, no. 10, p. 1411–1421.

Heller, M., Gradstein, W. S., Gradstein, F. M., and Agterberg, F. P.

1983 : RASC Fortran IV computer program for ranking and scaling of biostratigraphic events: Geological Survey of Canada, Report 922, 54 p.

Henderson, G., Rosenkrantz, A., and Schiener, E. J.

1976 : Cretaceous-Tertiary sedimentary rocks of west Greenland: in Geology of Greenland, ed. A. Escher, Groenlands Geologiske Undersakning, Copenhagen, p. 341–362.

Hess, H.

1972 : Planktonic crinoids of Late Jurassic Age from Leg XI, Deep Sea Drilling Project: in Initial Reports of the Deep Sea Drilling Project, v. 11, eds. C.D. Hollister and J.I. Ewing; U.S. Government Printing Office, Washington, D.C., p. 631–645.

Hollister, C. D., and Ewing, J. I., eds.

1972 : Initial Reports of the Deep Sea Drilling Project, v. 11: U.S. Government Printing Office, Washington, D.C., 1077 p.

Hottinger, L.

1972 : Campanian larger foraminifera from Site 98, Leg 11 of the Deep Sea Drilling Project (Northwest Providence, Channel, Bahama Islands; in Initial Reports of the Deep Sea Drilling Project, v. 11, eds. C. D. Hollister and J. I. Ewing; U.S. Government Printing Office, Washington, D.C., p. 595–607.

Jaffrezo, M.

1980 : Les formations carbonatées des Corbières (France) du Dogger à L'Aptien. Micropaleontologie stratigraphique, biozonation, paléóécologie. Extension des resultats à la Mésogée; [Thèse doctorat] Université Pierre et Marie Curie; p. 1–165.

Jansa, L. F., Enos, P. Tucholke, B. E., Gradstein, F. M., and Sheridan, R. E.

1979 : Mesozoic-Cenozoic sedimentary formations of the North American Basin, western North Atlantic: in Deep Drilling Results in the Atlantic Ocean: Continental Margins and Paleoenvironment, eds. M. Talwani, W. Hay, W.B.F. Ryan, Maurice Ewing Series v. 3, American Geophysical Union, p. 1–58.

Jansa, L. F., Gradstein, F. M., Harris, I. M., Jenkins, W.A.M., and Williams, G. L.

1976 : Stratigraphy of the Amoco-10E Murre 6-67 Well, Grand Banks of Newfoundland: Geological Survey of Canada, Paper 75-30, 17 p.

Jansa, L. F., Gradstein F. M., Williams, G. L., and Jenkins, W.A.M.

1977 : Geology of the Amoco Imp Skelly A-1 Osprey H-84 Well, Grand Banks, Newfoundland: Geological Survey of Canada, Paper 77-21, 17 p.

Jansa, L. F., Remane, J., and Ascoli, P.

1980 : Calpionellid and foraminiferal-ostracod biostratigraphy at the Jurassic-Cretaceous boundary, offshore eastern Canada: Rivista Italiana di Paleontologia e Stratigrafia, v. 86, n. 1, p. 67–126.

Jenkins, W.A.M., Ascoli, P., Gradstein, F. M., Jansa, L. F., and Williams, G. L.

1974 : Stratigraphy of the Amoco-IOE A-1 Puffin B-90 well, Grand Banks of Newfoundland: Geological Survey of Canada Paper 74-61, 12 p.

JOIDES

1975 : Ocean drilling on the continental margin: Science 150, p. 709–716.

Laughton, A. S., and Berggren W. A., eds.

1972 : Initial Reports of the Deep Sea Drilling Project, v. 12: U.S. Government Printing Office, Washington, D.C., 1243 p.

Lehmann, R.

1972 : Microfossils in thin sections from the Mesozoic deposits of Leg XI: in Initial Reports of the Deep Sea Drilling Project, v. 11, eds. C. D. Hollister and J. I. Ewing; U.S. Government Printing Office, Washington, D.C., p. 659–667.

Loeblich, A. R., and Tappan, H.

1972 : Cretaceous planktonic Foraminifera, pt. 1, Cenomanian: Micropaleontology, v. 7, p. 257–304.

Luterbacher, H. P.

1972 : Foraminifera from the Lower Cretaceous and Upper Jurassic of the northwestern Atlantic: in Initial Reports of the Deep Sea Drilling Project,

v. 11, eds. C. D. Hollister and J. I. Ewing; U.S. Government Printing Office, Washington, D.C., p. 561–595.

MacLean, B., and Falconer, R.K.H.

1979 : Geological/geophysical studies in Baffin Bay and Scott Inlet–Buchan Gulf and Cape Dryer–Cumberland Sound areas of the Baffin Island shelf: Geological Survey of Canada, Paper 79–1B, p. 231–244.

McNulty, C. L.

1979 : Smaller Cretaceous foraminifers of Leg 43, Deep Sea Drilling Project: in Initial Reports of the Deep Sea Drilling Project, v. 43, eds. B. E. Tucholke and P. R. Vogt; U.S. Government Printing Office, Washington, D.C., p. 487–507.

McWhae, J.R.W., Elie, R., Laughton, K. C.,and Günther, P. R.

1980 : Stratigraphy and petroleum prospects of the Labrador Shelf: Bulletin of Canadian Petroleum Geology, v. 28, no. 4, p. 460–488.

Miller, K. G., Gradstein, F. M., and Berggren, W. A.

1982 : Late Cretaceous to Early Tertiary agglutinated benthic foraminifera in the Labrador Sea: Micropaleontology, v. 28, p. 1–30.

Moullade, M.

1966 : Etude stratigraphique et micropaleontologique du Cretacé inferieur de la "Fosse Vocontienne": Documents du Laboratorie Geologique de faculté des Sciences, v. 2, 369 p.

Oertli, H. J.

1972 : Jurassic ostracods of DSDP Leg XI (Sites 100 and 105)—preliminary account: in Initial Reports of the Deep Sea Drilling Project, v. 11, eds. C. D. Hollister and J. I. Ewing; U.S. Government Printing Office, Washington, D.C., p. 645–659.

1983 : Jurassic ostracoda of DSDP Leg 76, Hole 534A: in Initial Reports of the Deep Sea Drilling Project, v. 76, eds. R. E. Sheridan and F. M. Gradstein; U.S. Government Printing Office, Washington, D.C., p. 581–587.

Okada, H., and Thierstein, H. R.

1979 : Calcareous nannoplankton—Leg 43, Deep Sea Drilling Project: in Initial Reports of the Deep Sea Drilling Project, v. 43, eds. B. E. Tucholke and P. R. Vogt; U.S. Government Printing Office, Washington, D.C., p. 507–575.

Olsson, R. K.

1964 : Late Cretaceous planktonic foraminifera from New Jersey and Delaware: Micropaleontology, v. 10, p. 157–188.

1977 : Mesozoic Foraminifera—Western Atlantic: in Stratigraphic micropaleontology of Atlantic Basin and Borderlands, ed. F. M. Swain; Elsevier, New York, p. 205–231.

1978 : Summary of lithostratigraphy and biostratigraphy of Atlantic Coastal Plain (Northern Part): in Initial Reports of the Deep Sea Drilling Project, v. 44, eds. W. E. Benson and R. E. Sheridan; U.S. Government Printing Office, Washington, D.C., p. 939–941.

Perkins, B. R.

1979 : Rudists from DSDP Leg 43, Site 384: in Initial Reports of the Deep Sea Drilling Project, v. 43, eds. B. E. Tucholke and P. R. Vogt; U.S. Government Printing Office, Washington, D.C., p. 599–601.

Peterson, M.N.A., and Edgar, N. T., eds.

1970 : Initial Reports of the Deep Sea Drilling Project, v. 2, U.S. Government Printing Office, Washington, D.C., 490 p.

Poag, C. W.

1978 : Stratigraphy of the Atlantic continental shelf and slope of the United States: Annual Reviews of Earth and Planetary Sciences, v. 6, p. 251–280.

Remane, J.

1978 : Calponellids: in Introduction to Marine Micropaleontology, eds. B. U. Haq and A. Boersma, Elsevier, New York, p. 161–170.

1983 : Calpionellids and the Jurassic—Cretaceous boundary in Deep Sea Drilling Project Site 534, western North Atlantic: in Initial Reports of the Deep Sea Drilling Project, v. 76, eds. R. E. Sheridan and F. M. Gradstein; U.S. Government Printing Office, Washington, D.C., p. 561–569.

Renz, O.

1972 : Aptychi (Ammonoidea) from the Upper Jurassic and Lower Cretaceous of the Western North Atlantic (Site 105, Leg XI, Deep Sea Drilling Project): in Initial Reports of the Deep Sea Drilling Project, v. 11, eds. C. D. Hollister and J. I. Ewing; U.S. Government Printing Office, Washington, D.C., p. 607–631.

1978 : Aptychi (Ammonoidea) from the Early Cretaceous of the Blake–Bahama Basin, Leg 44, Hole 391C, DSDP: in Initial Reports of the Deep sea Drilling Project, v. 44, eds. W. E. Benson and R. E. Sheridan; U.S. Government Printing Office, Washington, D.C., p. 899–911.

1979 : Aptychi (Ammonoidea) and ammonites from the Lower Cretaceous of the Western Bermuda Rise, Leg 43, Site 387, DSDP: in Initial Reports of the Deep Sea Drilling Project, v. 43, eds. B. E. Tucholke and P. R. Vogt; U.S. Government Printing Office, Washington, D.C., p. 591–599.

1983 : Early Cretaceous Cephalopoda from the Blake–Bahama Basin (DSDP Leg 76, Site 534A) and their correlation in the Atlantic and south western Tethys: in Initial Reports of the Deep Sea Drilling Project, v. 76, eds. R. E. Sheridan and F. M. Gradstein; U.S. Government Printing Office, Washington, D.C., p. 639–645.

Robaszynski, F., Caron, M., and others

1979 : Atlas of Mid–Cretaceous planktonic foraminifera: Cah. Micropal, Part 1 (1979), p. 1–184; 2 (1979), p. 1–181.

Robertson, A.H.F.

1983 : Latest Cretaceous and Eocene palaeo–environments in the Blake–Bahama Basin, Western North Atlantic: in Initial Reports of the Deep Sea Drilling Project, v. 76, eds. R. E. Sheridan and F. M. Gradstein; U.S. Government Printing Office, Washington, D.C., p. 763–781.

Roth, P. H.

1978 : Cretaceous nannoplankton biostratigraphy and oceanography of the northwestern Atlantic Ocean: in Initial Report of the Deep Sea Drilling Project, v. 44, eds. W. E. Benson and R. E. Sheridan; U.S. Government Printing Office, Washington, D.C., p. 731–761.

1983 : Jurassic and Lower Cretaceous calcareous nannofossils in the Western Atlantic (Site 534): biostratigraphy, preservation and some observations on biogeography and paleoceanography: in Initial Reports of the Deep Sea Drilling Project, v. 76, eds. R. E. Sheridan and F. M. Gradstein; U.S. Government Printing Office, Washington, D.C., p. 587–623.

Roth, P. H., Medd, A. W., and Watkins, D. K.

1983 : Jurassic calcareous nannofossil zonation, an overview with new evidence from DSDP Site 534: in Initial Reports of the Deep Sea Drilling Project, v. 76, eds. R. E. Sheridan and F. M. Gradstein; U.S. Government Printing Office, Washington, D.C., p. 573–581.

Ruffman, A., and Van Hinte, J. E.

1973 : Orphan Knoll—A "chip" off the North American "Plate": in Earth Science Symposium Offshore Eastern Canada: Geological Survey of Canada, Paper 71–23, p. 407–449.

Ryan, W.B.F., Cita, M. B., Miller, E. L., Hanselman, D., Nesteroff, W. D., Hacker, B., and Nibbelink, M.

1978 : Bedrock geology in New England submarine canyons: Oceanologica Acta, v. 1, p. 233–254.

Saito, T., Burckle, L. H., and Ewing, M.

1966 : Lithology and paleontology of the reflective layer, Horizon A: Science, v. 154, p. 1173.

Schmidt, R. R.

1978 : Clacareous nannoplankton from the Western North Atlantic, DSDP Leg 44: in Initial Reports of the Deep Sea Drilling Project, v. 44, eds. W. E. Benson and R. E. Sheridan; U.S. Government Printing Office, Washington, D.C. p. 703–731.

Scholle, P.A., ed.

1977 : Geological studies on the COST no. B-2 well, United States mid–Atlantic outer continental shelf area: U.S. Geological Survey Circular 750, 71 p.

1979 : Geological studies of the COST GE-well, United States south Atlantic outer continental shelf area: U.S. Geological Survey Circular 800,

114 p.

1980 : Geological studies of the COST no B-3 well, United States mid-Atlantic continental shelf area: U.S. Geological Survey Circular 833, 132 p.

Scholle, P. A., and Wenkam, C. R., eds.

1982 : Geological studies of the COST no. G-1 and G-2 wells, United States north Atlantic outer continental shelf: U.S. Geological Survey Circular 861, 191 p.

Schroeder, R., and Cherchi, A.

1979 : Upper Barremian–lowermost Aptian orbitolinid foraminifers from the Grand Banks continental rise, northwestern Atlantic (DSDP Leg 43, Site 384): in Initial Reports of the Deep Sea Drilling Project, v. 43, eds. B. E. Tucholke and P. R. Vogt; U.S. Government Printing Office, Washington, D.C., p. 575–585.

Sen Gupta, B. K., and Grant, A. C.

1971 : *Orbitolina,* a Cretaceous larger foraminifer from Flemish Cap: Paleoceanographic implications: Science, v. 173, p. 934–936.

Sheridan, R. E., and Gradstein, F. M., eds.

1983 : Initial Reports of the Deep Sea Drilling Project, v. 76: U.S. Government Printing Office, Washington, D.C., 947 p.

Sheridan, R. E., and Osburn, W. L.

1975 : Marine geological and geophysical studies of the Florida–Blake Plateau–Bahama area: Canadian Society of Petroleum Geologists, Memoir 4, p. 9–33.

Sheridan, R. E., and Enos, P.

1979 : Stratigraphic evolution of the Blake Plateau after a decade of scientific drilling: in Deep Drilling Results in the Atlantic Ocean: Continental Margins and Paleoenvironment: eds. M. Talwani, W. Hay, and W.B.F. Ryan, Maurice Ewing Series v. 3, American Geophysical Union, p. 109–123.

Sigal, J.

1977 : Essai de zonation du Crétacé mediterranéen à l'aide des foraminiferes planctoniques: Geologie Mediterranéene, v. 2, p. 99–108.

Sliter, W. V.

1980 : Mesozoic foraminifers and deep-sea benthic environments from Deep Sea Drilling Project Sites 415 and 416, eastern North Atlantic: in Initial Reports of the Deep Sea Drilling Project, v. 50, eds. Y. Lancelot and E. L. Winterer; U.S. Government Printing Office, Washington, D.C., p. 353–427.

Srivastava, S. ed.

In press: Geological and Geophysical Atlas of the Labrador Sea: Geological Survey of Canada and Hydrographic Survey.

Sirvastava, S. P.

1978 : Evolution of the Labrador Sea and its bearing on the early evolution of the North Atlantic: Geophysical Journal of the Royal Astronomical Society, v. 52, p. 313–357.

Stephenson, L. W.

1936 : Geology and Paleontology of the Georges Bank Canyons; Part 2, Upper Cretaceous fossils from Georges Bank (including Species from Banquereau, Nova Scotia): Bulletin of the Geological Society of America, v. 47, p. 367–410.

Swain, F. M.

1978 : Notes on Cretaceous Ostracoda from DSDP Leg 44, Sites 390 and 392: in Initial Reports of the Deep Sea Drilling Project, v. 44, eds. W. E. Benson and R. E. Sheridan; U.S. Government Printing Office, Washington, D.C., p. 921–939.

1980 : Mesozoic Ostracoda in several COST Atlantic wells: in Microfossils from Recent and Fossil Shelf Seas, eds. J. S. Neale and M. D. Brasier; Horwood Ltd., p. 90–112.

Thierstein, H. R.

1976 : Mesozoic calcareous nannoplankton biostratigraphy of marine sediments: Marine Micropaleontology, v. 1, p. 325–362.

Thierstein, H. R., and Okada, H.

1979 : The Cretaceous–Tertiary boundary event in the North Atlantic: in Initial Reports of the Deep Sea Drilling Project, v. 43, eds. B. E. Tucholke and P. R. Vogt: U.S. Government Printing Office, Washington, D.C., p. 601–617.

Tucholke, B. E.

1979 : Relationships between acoustic stratigraphy and lithostratigraphy in the western North Atlantic basin: in Initial Reports of the Deep Sea Drilling Project, v. 43, eds. B. E. Tucholke and P. R. Vogt; U.S. Government Printing Office, Washington, D.C., p. 827–846.

Tucholke, B. E., and Vogt, P. R., eds.

1979 : Initial Reports of the Deep Sea Drilling Project, v. 43: U.S. Government Printing Office, Washington, D.C., 1115 p.

Upshaw, C. F., Armstrong, W. E., Creath, W. B., Kidson, E. J., and Sanderson, G. A.

1974 : Biostratigraphic framework of Grand Banks: American Association of Petroleum Geologists Bulletin, v. 58, no. 6, p. 1124–1132.

Van Hinte, J. E.

1976a: A Cretaceous time scale: American Association of Petroleum Geologists Bulletin, v. 60, p. 498–516.

1976b: A Jurassic time scale: American Association of Petroleum Geologists Bulletin, v. 60, p. 489–497.

Van Hinte, J. E., Adams, J.A.S., and Perry, D.

1975 : K/Ar age of Lower–Upper Cretaceous boundary at Orphan Knoll (Labrador Sea): Canadian Journal of Earth Sciences, v. 12, p. 1484–1491.

Wilcoxon, J. A.

1972 : Upper Jurassic–Lower Cretaceous calcareous nannoplankton from the Western North Atlantic basin: in Initial Reports of the Deep Sea Drilling Project, v. 11, eds. C. D. Hollister and J. I. Ewing; U.S. Government Printing Office, Washington, D.C., p. 427–459.

Williams, G. L.

1975 : Dinoflagellate and spore stratigraphy of the Mesozoic–Cenozoic, offshore eastern Canada: in Offshore Geology of Eastern Canada, Geological Survey of Canada, Paper 75–30(2), p. 107–161.

Williams, G. L., and Brideaux, W. W.

1975 : Palynological analyses of late Mesozoic–Cenozoic rocks of the Grand Banks of Newfoundland: Geological Survey of Canada, Bulletin 236, 162 p.

Williams, G. L., Jansa, L. F., Clark, D. F., and Ascoli, P.

1974 : Stratigraphy of the Shell Naskapi N-30 Well, Scotian Shelf, Eastern Canada: Geological Survey of Canada, Paper 74–50, 12 p.

Williams, G. D., and Steick, C. R.

1975 : Speculations on the Cretaceous paleogeography of North America: Geological Association of Canada Special Paper no. 13, p. 1–20.

Williamson, M. J.

In press: Quantitative biozonation of the Late Jurassic and Early Cretaceous of the Newfoundland Basin: Micropaleontology.

Wind, F. H.

1978 : Western North Atlantic Upper Jurassic calcareous nannofossil biostratigraphy: in Initial Reports of the Deep Sea Drilling Project, v. 44, eds. W. E. Benson and R. E. Sheridan: U.S. Government Printing Office, Washington, D.C., p. 761–775.

Wynne-Edwards, H. R.

1972 : The Grenville Province: in Variation in Tectonic Styles in Canada, eds. R. A. Price and R.J.W. Douglas; Geological Association of Canada Special Paper, no. 2, p. 263–334.

Manuscript Accepted by the Society November 12, 1984

ACKNOWLEDGMENTS

This study has benefited from the long-standing help and advice of many colleagues in North Atlantic stratigraphy and palep-oceanography. Several reviewers also took great pains to improve the text. I particularly thank P. Baumgartner, P. Doeven, D. Habib, L. Jansa, J. Ogg, R. Sheridan, B. Tucholke, D. Watkins, M. Williamson, and R. V. Wise.

Printed in U.S.A.

The Geology of North America
Vol. M, The Western North Atlantic Region
The Geological Society of America, 1986

Chapter 32

Paleogene biofacies of the western North Atlantic Ocean

Isabella Premoli-Silva
Department of Earth Sciences, University of Milan, Milan 20133, Italy
Anne Boersma
Microclimates, Box 404, RR 1, Stony Point, New York 10980

INTRODUCTION

This chapter presents an overview of the distributional patterns of the major calcareous and siliceous microfossil groups of the Paleogene in the western North Atlantic Ocean. From these patterns the evolution of open marine biofacies through space and time can be reconstructed. At first sight, this would seem to be an easy task, considering the large number of Deep Sea Drilling Project (DSDP) and commercial drill holes available, as well as the numerous piston cores which have retrieved pre-Quaternary sediments from the North Atlantic Basin in the last 25 years. However, two principal problems complicate detailed biofacies reconstructions. The first, and most frustrating, is the incompleteness of the Paleogene stratigraphic record from the North Atlantic. Existing deep-sea cores are complicated by extensive drilling gaps, large sedimentary hiatuses, and/or the absence of age-diagnostic fossils. The second problem is the difficulty of intercorrelation of the many different Paleogene biostratigraphic zonations and time-scales.

Since about 1969, biostratigraphic schemes based on the major calcareous and siliceous fossil groups have been extensively refined or newly constructed. Because different biostratigraphic schemes and taxonomies have been used by different researchers in earlier studies of North Atlantic cores, the majority of Paleogene calcareous DSDP sections and many of the other boreholes from the western North Atlantic have been restudied and rezoned during preparation of this chapter. The sections have all been recalibrated according to the time scale and zonation shown on Plate 1. As a result, much of the early Paleogene can now be accurately correlated and dated. However, for the Oligocene, ages still can only be estimated and the correlations of zones to time are not complete.

From analysis of the revised data, the picture of the distribution of marine biofacies through space and time, although incomplete, reveals some clear trends. These trends are outlined at the end of this chapter as a basis for future paleoceanographic interpretation and reconstruction.

DATA BASE

Selected Paleogene-age open marine biofacies summarized in this report are derived from sections drilled over the last 15 years by the Deep Sea Drilling Project up to and including Leg 82 (Figs. 1, 2; Plate 10, H–J). In addition, the early Joint Oceanographic Deep Earth Sampling Program (JOIDES) recovered six extensive Paleogene sections across the Blake Plateau (JOIDES, 1965). Continental margin drilling, aside from proprietary wells drilled by the petroleum industry, includes the COST B, COST G, and COST GE series, reported on by the United States Geological Survey (Poag, 1977, 1982; Valentine, 1977). Locations of all these samples are shown in Plate 2. Several non-proprietary wells from the United States offshore AMCOR series (Hathaway and others, 1979) and the Labrador offshore (Gradstein and others, 1975; Gradstein and Srivastava, 1980) are also discussed here. Continental-margin dredge samples recovered from canyons and on dives by the submersible *Alvin* of Woods Hole Oceanographic Institution are included, although they are spotty in their distribution (Gibson and others, 1986; Ryan and others, 1978; Freeman-Lynde and others, 1981).

More than 50 relatively short deep-sea cores containing Paleogene sediments also have been taken from the western North Atlantic on cruises of the research vessels *Chain* and R. V. *Knorr* of Woods Hole Oceanographic Institution, and the *Vema* and *Robert Conrad* of Lamont-Doherty Geological Observatory. They are not discussed in detail here, but are listed in Table 1.

DATA INTERPRETATION

The fact that no single, continuous Paleogene sequence has

Premoli-Silva, I., and Boersma, A., 1986, Paleogene biofacies of the western North Atlantic; *in* Vogt, P. R., and Tucholke, B. E., eds., The Geology of North America, Volume M, The Western North Atlantic Region: Geological Society of America.

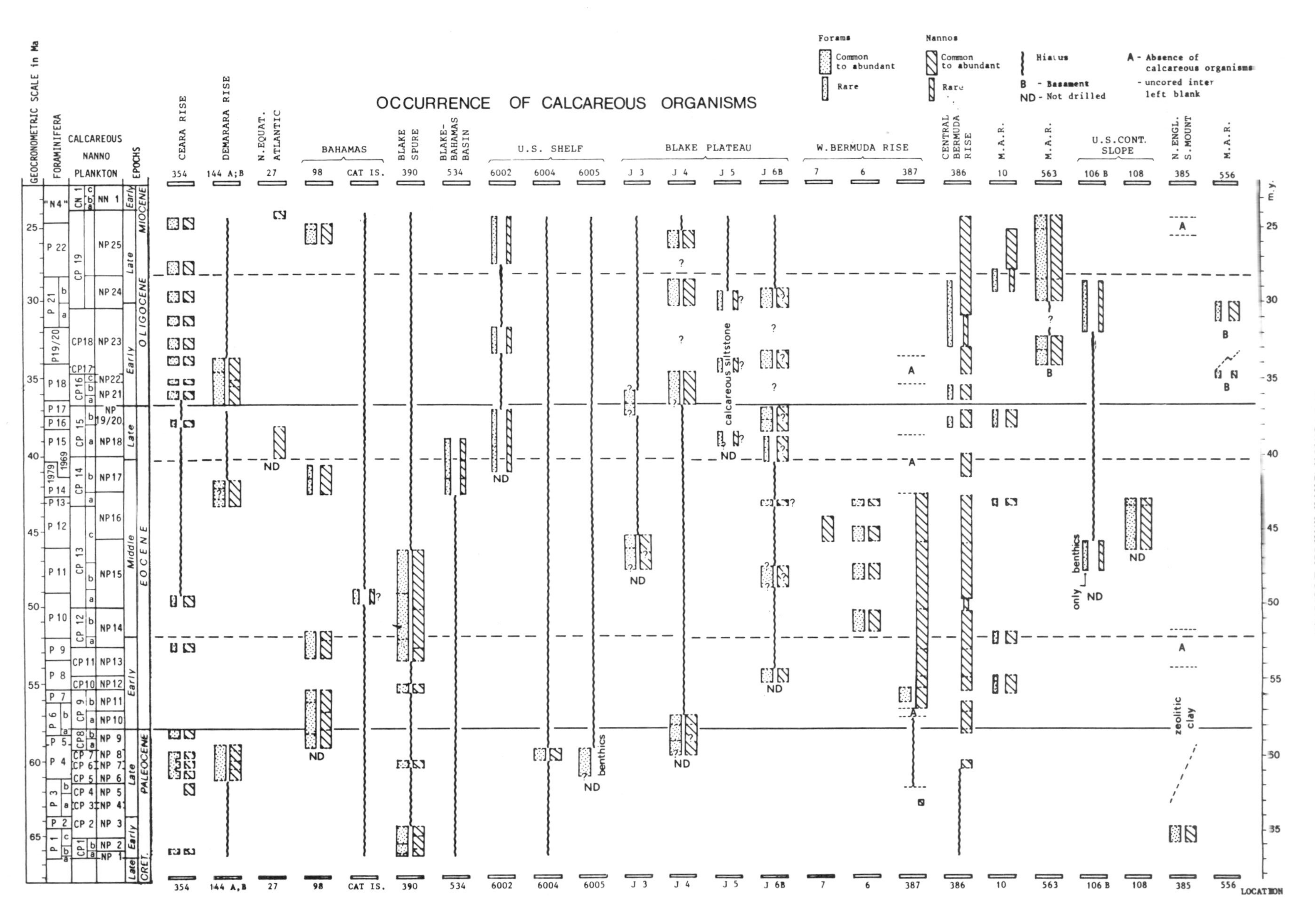
OCCURRENCE OF CALCAREOUS ORGANISMS
Forams
Common to abundant
Rare
Nannos
Common to abundant
Rare
Hiatus
B - Basement
ND - Not drilled
A - Absence of calcareous organisms
- uncored inter left blank
GEOCRONOMETRIC SCALE in Ma
FORAMINIFERA
CALCAREOUS NANNO PLANKTON
EPOCHS
CEARA RISE
DEMARARA RISE
N.EQUAT. ATLANTIC
BAHAMAS
CAT IS.
BLAKE SPURE
BLAKE-BAHAMAS BASIN
U.S. SHELF
BLAKE PLATEAU
W.BERMUDA RISE
CENTRAL BERMUDA RISE
M.A.R.
U.S.CONT. SLOPE
N.ENGL. S.MOUNT
calcareous siltstone
only benthics
benthics
zeolitic clay
MIOCENE
OLIGOCENE
EOCENE
PALEOCENE
CRET.
LOCATION

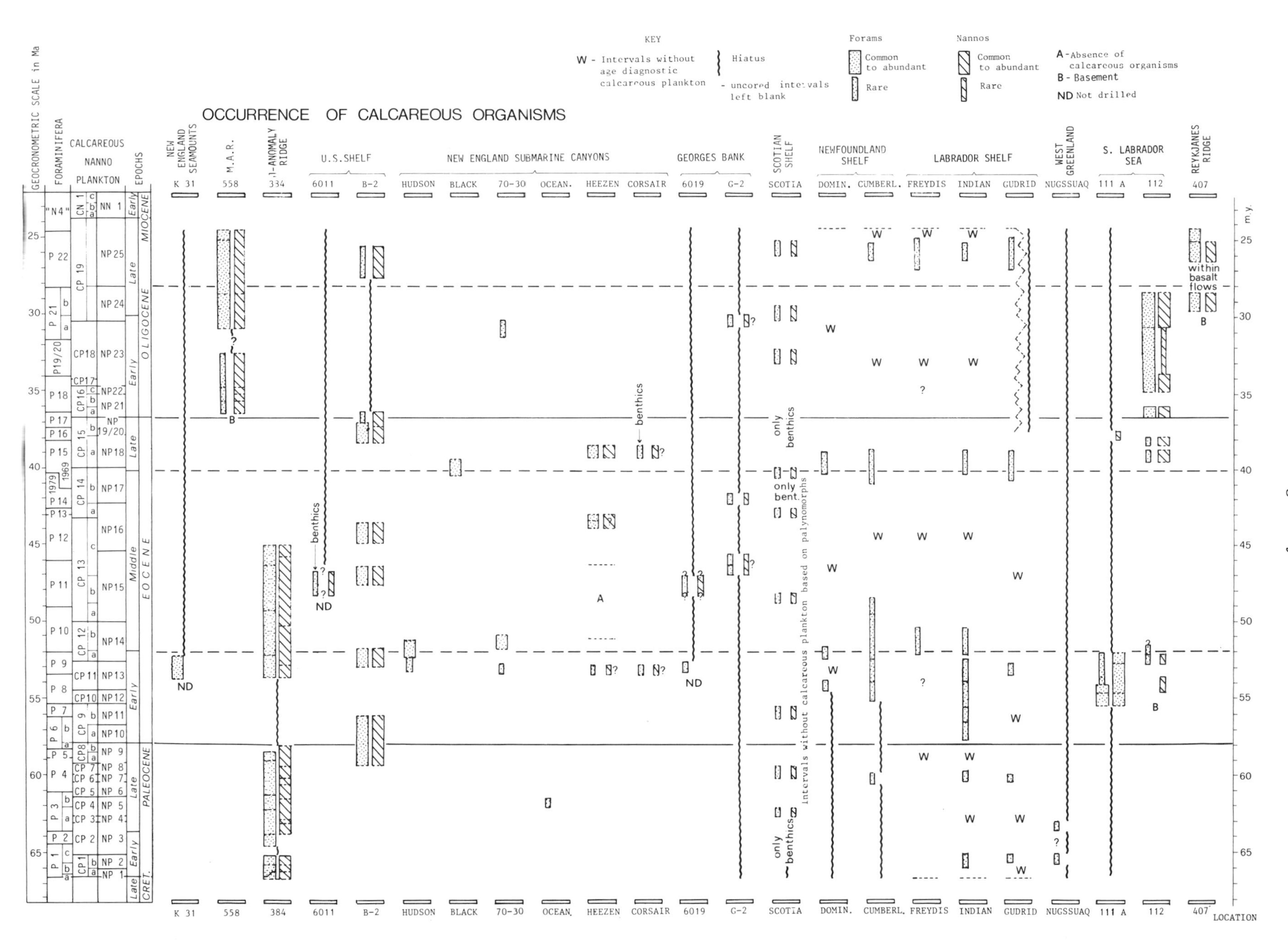

Figure 1 (this and previous page). See following page for explanation.

been recovered in any western North Atlantic open-marine site underscores the first major problem with interpretation of Paleogene biofacies of the North Atlantic. The most obvious gaps in the sedimentary record (Fig. 1) are extensive coring gaps in DSDP holes and USGS wells. These gaps resulted largely from the policy on early DSDP cruises to drill only selected portions of the sediment column on the way to basement. After DSDP Leg 39 a policy of continuous coring was adopted. In many oceanic areas the more complete sedimentary columns then recovered were, nevertheless, interrupted by numerous hiatuses resulting from erosion and/or non-deposition.

Dissolution of biogenic sediment in abyssal regions produces another sort of gap, the dissolution hiatus, that usually characterizes so-called 'telescoped' sedimentary sections. In these sediments one or several microfossil groups may be dissolved away. Aragonite-bearing fossils, such as the pteropods, marine molluscs, and a few genera of benthic foraminifera are the first to be dissolved, followed by the planktonic foraminifera, particularly those with high-Mg calcite in their tests. Near the calcite compensation depth (CCD), the level at which calcium carbonate dissolution equals accumulation, most planktonic foraminifera will also dissolve. Benthic foraminifera, because they are covered with protoplasm and often live as infauna within sediments, and nannofossils, which are often deposited within organic conglomerates such as fecal pellets, are preserved at greater seafloor depths than the planktonic foraminifera. Abyssal sediments thus may contain nannofossils in the fine fraction, but only benthic foraminifera in the coarse fraction. At the greatest depths and under upwelling zones where bottom water is undersaturated with respect to carbonate, siliceous organisms are the main contributors to the biogenic fraction of the sediment.

The exact depth and areas of calcareous and siliceous sediment dissolution or accumulation have varied through space and time because of the continual evolution of oceanic circulation patterns and related physico-chemical gradients. Although little borehole data exist to constrain the exact levels of the CCD during the Paleogene, the CCD can be located generally between 4,000 and 4,500 m in the western North Atlantic (Plate 1).

On shallow shelf areas further causes of stratigraphic gaps are the transgressive-regressive cycles of sea level. These produced erosional gaps and/or non-fossiliferous lithofacies during regressions. Such hiatuses are clearly expressed in Continental Offshore Stratigraphic Test (COST) and Atlantic Margin Coring Project (AMCOR) wells located on the United States continental margin.

Interpretation of biofacies is necessarily controlled by our ability to establish and compare coeval stratigraphic intervals on the basis of biozones. All microfossil biozonations have been significantly expanded, refined, and consequently improved over the past 30 years (inter alia Bolli, 1966; Blow, 1969, 1979; Berggren, 1969; Premoli-Silva and Bolli, 1973; Hardenbol and Berggren, 1978; Berggren and others, 1985; Bukry, 1973, 1975; Martini, 1970, 1971; Okada and Bukry, 1980; Riedel and Sanfilippo, 1978). Older papers with their sometimes more limited biostratigraphic coverage are difficult to compare with recent reports. These problems are exacerbated by the non-uniform coverage of microfossil groups in the various summary reports. Planktonic foraminifera and calcareous nannofossils have been surveyed more often, but a few studies exist on the siliceous, organic-walled, or benthic microfossils in the samples recovered to date.

In other areas of the ocean, paleomagnetism has provided an excellent chronostratigraphic tool and means of inter-site correlation. To date in the western North Atlantic, however, only the Paleogene section from DSDP Leg 82 has been measured. This leg was drilled on the Mid-Atlantic Ridge on oceanic basement known to be of early Oligocene age. The overlying sediment column at Sites 558 and 563 contains the record of all early Oligocene isochrons from magnetic anomalies 12 to 6 (Scientific Party Leg 82, 1982).

A significant complication to biostratigraphic zonations is the biogeographic imprint on biozonal index fossils (see Berggren and Olsson, this volume). Interpretation and correlation of low latitude sections (e.g., DSDP Sites 144 on the Demerara Rise and DSDP Site 98 in the Bahamas are facilitated by the fact that most Paleogene zonal index fossils are low latitude indigenes (Stainforth and others, 1975). In middle to high latitudes, however, biostratigraphic refinement and control are lost as the low

Figure 1 (previous two pages). Distribution of planktonic foraminifera and calcareous nannoplankton in the Paleogene sequences from the western North Atlantic. Time scale and biostratigraphic framework are after Berggren and others (1985; Plate 1). Locations (references) are: DSDP Site 354 (1–5); DSDP Site 144 (6–9); DSDP Site 27 (10–13); DSDP Sites 98, 105, 106B, 108 (14–16); Cat Island (17); DSDP Sites 417/418 (18, 19); DSDP Sites 384, 385, 386, 387 (31–33); DSDP Site 389 (20–23); DSDP Sites 8/8A, 9/9A, 10 (24, 25); DSDP Site 534 (26); AMCOR 6002, 6004, 6005, 6011, 6019 (27); JOIDES 3, 4, 5, 6B (28 and 54); DSDP Sites 6, 7 (29, 30); DSDP Sites 563, 556, 558 (53); Knorr 31 (54); COST B-2 (34, 35); COST G-2 (37); Hudson, Block, "70-30," Oceanographer canyons (52); Heezen and Corsair canyons (36); Commercial Bore-holes (38, 39); Nugssuaq (40, 41); DSDP Sites 111A, 112 (42–46); DSDP Site 407 (47–51). Reference Number: 1, Supko and others, 1977; 2, Boersma, 1977; 3, Perch-Nielsen, 1977a; 4, Perch-Nielsen, 1977b; 5, Bukry, 1977; 6, Hayes and others, 1972; 7, Beckmann, 1972; 8, Roth and Thierstein, 1972; 9, Petrushevskaya and Kozlova, 1972; 10, Bader and others, 1970; 11, Bolli, 1970; 12, Hay, 1970; 13, Riedel and Sanfilippo, 1970; 14, Hollister and others, 1972; 15, Luterbacher, 1972; 16, Wilcoxon, 1972; 17, Freeman-Lynde and others, 1981; 18, Donnelly and others, 1980; 19, Bukry, 1980; 20, Benson and others, 1978; 21, Gradstein and others, 1978; 22, Bukry, 1978a; 23, Weaver and Dinkelman, 1978; 24, Peterson and others, 1970; 25, Cita and others, 1970; 26, Sheridan and others, 1984; 27, Hathaway and others, 1979; 28, JOIDES, 1965; 29, Ewing and others, 1968; 30, Berggren and others, 1969; 31, Tucholke and others, 1979; 32, Okada and Thierstein, 1979; 33, Bukry, 1978b; 34, Poag, 1977; 35, Valentine, 1977; 36, Ryan and others, 1978; 37, Poag, 1982; 38, Gradstein and others, 1975; 39, Gradstein and Srivastava, 1980; 40, Hansen, 1970; 41, Boersma and Premoli-Silva, 1983; 42, Laughton and others, 1972; 43, Berggren, 1972; 44, Perch-Nielsen, 1972; 45, Bukry, 1972; 46, Benson, 1972; 47, Luyendyk and others, 1979; 48, Poore, 1979; 49, Steinmetz, 1979; 50, Martini, 1979; 51, Bukry, 1979; 52, Gibson and others, 1968; 53, Scientific Party Leg 82; 54, personal observations.

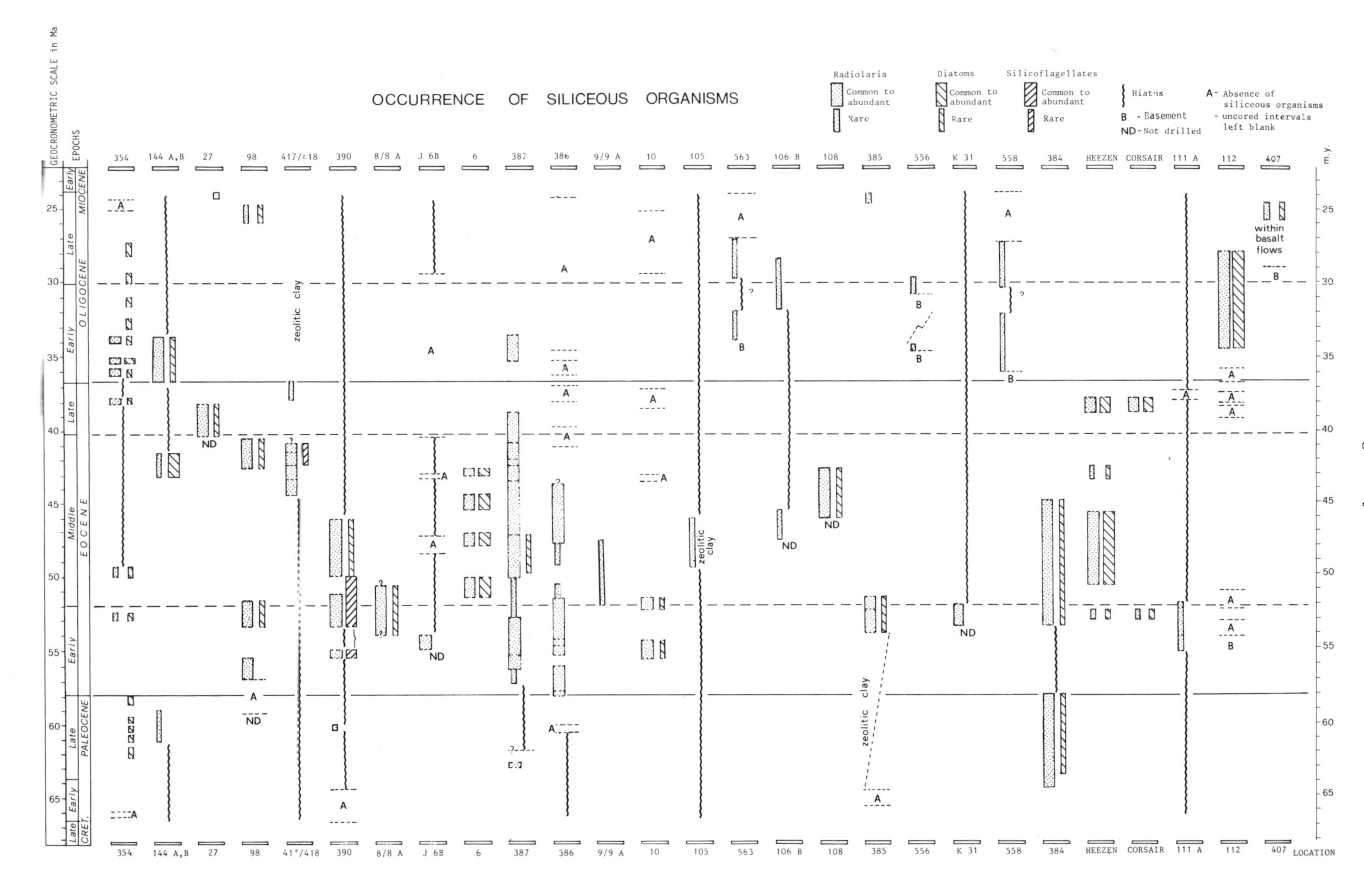

Figure 2. Distribution of siliceous organisms (radiolaria, diatoms, silicoflagellates) in the Paleogene sequences from the western North Atlantic. For time scale, biostratigraphic framework, and references see caption of figure 1.

TABLE 1. NORTH ATLANTIC PISTON SAMPLES OF PALEOCENE TO OLIGOCENE SEDIMENTS

Sample No.	Lat.	Long.	Water** Depth (m)	Sample Depth (cm)	CO_3%	Remarks
			North Atlantic Paleocene Samples			
A158-3	36° 42'N	67° 58'W	3790	167	11.6	
ET8, 67-11	32° 10'N	58° 58'W	3007	155	93.0	Some Mixing
MO D-76	44° 36'N	57° 42'W	58	1344*	13.0	Mid-Paleocene
MO I-59	44° 36'N	59° 38'W	92	917*	15.8	
RC9-5	32° 20'N	77° 37'W	543	122	46.6	
V18-18	05° 23'N	51° 03'W	152	520	66.9	Paleocene(?)
			North Atlantic Eocene Samples			
A150-1	32° 42'N	62° 30'W	1597	135	93.1	
A167-21	29° 49'N	76° 35'W	1454	365	75.1	late Eocene
AL706-6	20° 19'N	69° 03'W	2470	---	92.3	early(?) Eocene
AL780-2-1	41° 02'N	66° 21'W	1503	---	2.0	middle Eocene
AL783-10	41° 03'N	66° 22'W	1231	---	4.7	late Eocene
C10-1	33° 40'N	62° 30'W	1480	70	90.0	early Eocene
C22-5	32° 17'N	64° 38'W	1097	330	45.5	Eoc./Olig., some mixing
CH57-4PC	20° 03'N	66° 09'W	6618	281	26.4	late Eocene
C22-6	32° 16'N	64° 35'W	1509	185	26.6	late Eocene/Oligocene
ET-15480	35° 21'N	74° 50'W	1050-1025	---	72.2	middle Eocene
ET-15485	35° 07'N	75° 01'W	1200-1110	---	61.6	early Eocene
ET-15696	34° 14'N	75° 39'W	1550-1510	---	84.1	middle Eocene
ET-15717	33° 53'N	75° 44'W	1875-2100	---	91.3	middle Eocene; Eocene/Oligocene
ET-15725	34° 06'N	75° 38'W	2240-2170	---	89.7	late Eocene; Eocene/Oligocene
ET-17126	33° 57'N	75° 42'W	1788-1700	---	85.2	middle Eocene; Eocene/Oligocene
ET-17127	33° 56'N	75° 43'N	2050-1950	---	87.7	middle Eocene; Eocene/Oligocene

* Meters subbottom.
** Depth intervals are given for dredge samples.

latitude index species are ecologically eliminated. This problem is more acute with foraminifera and radiolarians than with the plants (nannofossils and diatoms).

Correlation of middle or high latitude zonations (Martini, 1979; Perch-Nielsen, 1972; Poore, 1979) is confounded by (1) the long stratigraphic ranges of many middle and high latitude-preferring species; (2) species migration through latitude that, according to the age refinement possible for the Paleogene, may occur over time intervals as great as 1 m.y.; (3) the current lack of independent means of correlation, such as paleomagnetics or absolute age dating; and (4) the profound, climatically induced circulation changes, particularly during the Eocene in the North Atlantic. These changes caused extensive geographic relocation of many fossil groups; many species were partitioned into ever-narrowing geographic areas rendering them less useful for either latitudinal or longitudinal biostratigraphic intercorrelations.

BIOSTRATIGRAPHY

Available data are compiled in Figures 1 and 2 and described in Table 2 to show: (1) the occurrences of major faunal and floral biota plotted against age; (2) sedimentary hiatuses; and (3) drilling gaps. Localities are arrayed roughly from lower to higher latitudes, shallower to deeper, and west to east. Where faunal and floral assemblages can be assigned to specific biozones, these zones are reported. In many instances, only occurrences and qualitative abundances can be recorded, particularly for the siliceous fossils which usually are described only from smear-slide analysis.

We have restudied the foraminifera in the Paleogene sections from many DSDP sites, including Sites 98, 106, 111, 112, 144, 354, 384, 386, 556, 558, 563; JOIDES boreholes J3, J4, J5, J6; the COST B2 well; commercial boreholes from West Greenland, the Nova Scotia shelf, and the Labrador margin; all U.S.G.S. dredges; *Alvin* dive samples listed in Table 1; and the giant gravity core KNR31-26GC. Foraminiferal zones have been recalibrated and correlated with nannofossil and/or radiolarian zonal schemes according to the zonations and time-scale in Plate 1.

BIOFACIES COMPONENTS

Biofacies included in this report consist of calcareous microfossils, including the calcareous planktonic and benthic foraminifera and the nannoplankton, and the siliceous microfossil groups, the radiolarians and diatoms. Foraminifera and radiolaria are protozoans, while nannoplankton and diatoms, classed as plants, are some of the ocean's most important primary producers. Because of the paucity of information on other microfossil groups, the organic-walled spores, pollen, dinoflagellates and hy-

TABLE 1. (CONTINUED)

North Atlantic Oligocene Samples

Sample No.	Lat.	Long.	Water** Depth (m)	Sample Depth (cm)	CO3%	Remarks
A156-2	29° 12'N	76° 49'W	2140	485	41.9	late Oligocene
A164-25	32° 13'N	64° 31'W	2953	357	77.6	Miocene
C10-11	32° 14'N	64° 32'W	2280	198	47.4	late Oligocene
C22-2	32° 11'N	64° 45'W	1939	88	41.9	Miocene; late Oligocene
C22-6	32° 16'N	64° 39'W	1509	50	31.5	early Oligocene
CH57-13PC	19° 52'N	68° 02'W	5384	494	11.2	late Oligocene
CH99-6PC	29° 18'N	36° 37'W	3936	382	93.2	Miocene; late Oligocene
ET8-67-13	32° 13'N	58° 58'W	1390	565	93.9	
ET-15481-I	35° 17'N	74° 52'W	1560-1490	---	55.7	early-middle Oligocene
ET-15481-II	35° 17'N	74° 52'W	1560-1490	---	37.9	late Oligocene
ET-17288	34° 30'N	75° 31'W	1945-1520	---	40.1	early-middle Oligocene
ET-19100	34° 39'N	75° 24'W	1740-1600	---	22.5	Oligocene/Miocene
GS6002-10/2	31c 09'N	80° 31'W	32	50	47.1	bottom at 79.2m (=Oligocene)
MO I-59	44° o9'N	59° 38'W	92	670*	10.3	early Oligocene
RC8-118	19° 12'N	65° 14'W	5620	180	18.6	Miocene; late Oligocene
RC10-277	24° 02'N	72° 51oW	5407	52	6.8	
V5-9	32° 13'N	64° 42'W	1710	160	59.2	Miocene; late Oligocene
V5-12	32° 15'N	64° 38'W	1926	180	45.4	Miocene; late Oligocene
V5-14	32° 20'N	64° 27'W	2365	168	37.6	
V5-26	32° 26'N	64° 28'W	1653	240	56.9	Miocene; late Oligocene
V15-204	29° o7'N	76° 47'W	2820	268	48.3	Miocene; late Oligocene
V21-243	24° 21'N	73° 01'W	5438	428	33.6	Miocene; late Oligocene
V24-15	24° 44'N	73° 45'W	5305	375	10.0	middle-late Oligocene
ET-17133	34° 11'N	75° 40'W	1565-1445	---	66.5	middle-late Eocene
ET-17145	34° 06'N	75° 36'W	2610-2380	---	54.7	middle Eocene
ET-17294	34° 48'N	75° 21'W	1776-1660	---	68.6	middle Eocene
ET-17295	34° 42'N	75° 26'W	1500-1345	---	81.1	late Eocene, some mixing
ET-17304	34° 51'N	75° 10'W	1460-1425	---	37.4	late Eocene
ET-19091	34° 06'N	75° 38'W	2340-2190	---	71.4	early-middle Eocene
ET-19093	34° 03'N	75° 42'W	1690-1575	---	55.6	early-middle Eocene
ET-31406A	22° 33'N	72° 20'W	3200	---	75.6	
G-13	40° 15'N	73° 59'W	13	220	19.5	early Eocene; Cretaceous
G-28	40° 15'N	73° 59'W	9	44	22.8	early Eocene
GS6002-20-2	31° 09'N	80° 31'W	32	100	71.3	bottom depth: 173.3m
GS6011-26-2	39° 43'N	73° 59'W	22	65	3.6	bottom depth: 229.7m
GS6019-8-1	41° 49'N	68° 17'W	174	100	86.3	bottom depth: 58.6m
KNR31 26CG	39° 21'N	67° 07'W	2726-2896	191	76.1	middle Eocene
MO I-59	44° 09'N	59° 38'W	92	716*	14.3	middle-late Eocene
RC8-2	11° 12'N	48° 05'W	4625	745	55.7	early-middle Eocene
RC10-27	22° 27'N	71° 36'W	5192	675	37.7	
RC10-279	24° 31'N	74° 26'W	4984	310	15.9	
RC11-247	24° 43'N	73° 47'W	5312	325	13.6	Paleocene(?)
RC16-30	21° 00'N	68° 23'W	5397	789	93.6	Eocene/Oligocene
V10-96	27° 52'N	54° 38'W	4939	620	51.9	late Eocene
V12-4	24° 17'N	53° 04'W	5009	670	71.8	middle Eocene
V15-195	27° 08'N	77° 21'W	1812	424	90.0	some mixing
V16-209	30° 00'N	51° 52'W	4673	540	93.2	early Eocene
V18-RD37	39° 40'N	65° 55'W	2496-2166	---	82.7	
V27-D19	52° 06'N	38° 48'W	3466	---	---	

* meters subbottom.
**Depth intervals are given for dredge samples.

Note: Ages are given for the bottom samples of each core unless a cm level downcore is speecified. Carbonate percentages were estimated to the nearest 0.1% for the sample depth (cm) listed in the adjacent column. Information on cores with prefixes KNR, AL, CH, or GS is available from Woods Hole Oceanographic Institution. All other listings can be located at Lamont Geological Observatory.

strichospheres, and the siliceous silicoflagellates, they cannot be discussed in any detail here.

Calcareous Biofacies

Foraminifera are omnivorous, approximately 10–400 μm sized protozoans, most of which precipitate external tests composed of calcium carbonate. A smaller number of benthic foraminifera encase themselves with detrital sediment grains cemented by an organic matrix. Planktonic species float in the water column from the near surface to several hundred meters water depth and are distributed from the Antarctic to the Arctic polar seas. Species' distributions are correlated with various oceanic physical and chemical variables such as the thermal gradient through depth and latitude, salinity, density, and the primary biotic factor, food availability. Benthic species inhabit

TABLE 2. TABULATION OF THE TIMES A PLANKTONIC FORAMINIFERAL ZONE WAS PENETRATED, RECOVERED, OR MISSED BECAUSE OF A HIATUS OR CORING GAP

Planktonic Zone	# Times Zone Penetrated	Percent Recovery	Hiatuses- % of section	# of Coring Gaps	Percent of Section-Coring Gaps
P1	28	8	28	10	35
P2	23	1	4	11	47
P3	15	33	6	9	60
P4	29	41	27	9	31
P5	26	23	42	9	34
P6	25	12	48	10	40
P7	27	29	37	9	33
P8	33	30	33	12	36
P9	34	61	29	3	8
P10	33	51	33	5	15
P11	31	45	29	8	25
P12	34	29	47	8	23
P13	35	25	45	10	28
P14	32	12	50	12	37
P15	40	30	40	12	30
P16	31	19	45	11	35
P17	34	29	44	9	26
P18	38	26	34	15	39
P19	34	11	44	15	44
P20	35	20	42	13	37
P21	40	32	35	13	32
P22	45	45	33	14	31

Note: Sources include all DSDP sites, AMCOR, COST, JOIDES, and commercial boreholes shown in Plate 1. The number of times a zone was penetrated was counted, as was the number of times the zone was absent due to coring gaps and hiatuses. The number of coring gaps, hiatuses, and the times the zone was actually recovered were then converted to percent of the times the zone was penetrated. Percentages are rounded to the nearest whole number.

sediments and/or marine plants from the marsh to the greatest depths of the oceans. At the two extremes of this depth range, only agglutinated species can flourish; the calcareous forms are excluded primarily by the high alkalinity which draws most carbonate into solution.

Paleogene-age carbonate sediment was distributed over half to two-thirds of the ocean bottom, at depths less than 4 to 4.5 km (Plate 10, H–J). Below the CCD the exposed carbonate of foraminiferal tests was dissolved; thinner-walled forms and juveniles dissolved first, and the thicker-walled forms remained as dissolution residues to depths of about 4,200 meters (Cita and others, 1970). The groups of Paleogene foraminifera that persist in a calcareous sediment altered by diagenesis, however, are not identical to those removed by dissolution; they are more related to the wall thickness. Besides the juveniles, all the smooth-, thin-walled group tend to disappear first and all together, while the largest forms were better preserved independently of their habitats, climatic or in the water column. (Boersma and Premoli-Silva, 1983).

Benthic species may survive at great depths, depending on the resistance of their organic coatings, the depth in the sediment at which they live (buffering them against dissolution), and carbonate-saturation levels at the sediment/water interface. After their deaths, however, dissolution proceeds rapidly (Douglas and Woodruff, 1981).

Calcareous nannoplankton, also called nannofossils, consist of a coccolithophorid, a sub-spherical organic mass covered with many calcareous platelets, or coccoliths. These micron-sized coccoliths are most often found in the sediment. Nannofossils float in the oceanic photic zone from approximately 60° North to 60° South latitude; their distributions are effectively confined between the Arctic and Antarctic Convergences. As primary producers their distributions are largely determined by nutrient and light availability. Secondary controlling factors include oceanic temperature, surface turbidity, vitamin concentrations, and predation pressure (Haq and Boersma, 1976).

Paleogene nannoplankton occur together with foraminifera in most calcareous sediments from the continental shelves to the abyss. At depths shallower than 1,500 meters the relative proportions of the two groups are controlled largely by current winnowing and redeposition (Shackleton, 1984). At seafloor depths greater than 4,000 meters nannofossils are concentrated primarily because of foraminiferal dissolution. The dissolution resistance of the nannofossils is currently attributed to their rapid delivery to the seafloor encased in fecal pellets of invertebrate predators (Honjo, this volume).

Siliceous Biofacies

The only common siliceous organisms included in this report are the 10->500 μm sized micro-omnivores, the radiolarians. These protozoans precipitate a siliceous latticework around their central, capsular protoplasmic mass. Although silica may be a bio-limiting element, radiolarian production is largely controlled by availability of nutrients and by live prey. Only in coastal and equatorial upwelling zones and in higher-latitude surface waters are conditions favorable for radiolarian blooms. Although radiolaria may occur elsewhere, their numbers in the sediment are usually diluted by the more abundant carbonate and terrigenous materials.

Large percentages of radiolarians in seafloor sediment characterize zones of high nutrient concentrations in surface waters, and/or abyssal areas in which most or all carbonate has been dissolved. In Paleogene-age sedimentary sections of the western North Atlantic, pure radiolarian biofacies are limited to the middle Eocene and to abyssal areas. Because there are few studies of radiolarians from this area, however, radiolarian presence could be determined only from shipboard smear-slide analysis done on DSDP core samples. Thus the abundances of radiolarians in the sediment are known only approximately.

Diatoms are microscopic plants (1 to 10 μm m in size) which secrete a siliceous external covering called a frustule. As primary producers, their distribution and abundance are regulated by the availability of bio-limiting nutrients and light. Like the radiolarians they proliferate in equatorial and coastal upwelling zones and in higher-latitude surface waters; they occur literally from pole to pole. Diatoms, however, are not commonly found in Paleogene sediments of the western North Atlantic. Like the radiolarians, they are preserved in abyssal facies and are concentrated where carbonate has been dissolved. For this report their presences have been determined only from smear-slide analysis of DSDP material.

A third siliceous group, the silicoflagellates, should occur in siliceous biofacies of this age. They have not yet, however, been studied from the sedimentary sections described in this report.

PALEOCENE

In general, the sedimentary record of the Paleocene is less continuous than that of the Eocene or Oligocene (Fig. 1). The only exception is found at DSDP Site 384 (south of the Grand Banks) where an almost complete Paleocene section was recovered. Otherwise the Paleocene record is spotty throughout latitude and depth. Also, the Cretaceous/Paleocene boundary is marked by hiatuses of varying duration except at DSDP Site 384 where an apparently continuous, little disturbed, and uniform nannofossil-ooze sequence was recovered.

Calcareous Biofacies

Lower Paleocene calcareous sediments are rare except at DSDP Site 384 on the J-Anomaly Ridge and at DSDP Site 390 on the Blake Spur (Plate 10,H). At Site 390 the entire foraminiferal Zone P1 and the correlative nannofossil Zones NP1–NP3 are recorded. At Site 384 the topmost part of Zone P1 (formerly called P1d, but now P1c, and the equivalent to NP3) is missing. This interval is preserved, however, at latitudes as high as Nugssuaq off West Greenland (Hansen, 1970) and along the Labrador shelf. In the deep basins calcareous sediments from Zone P1 (=NP2) have been recovered only at DSDP Site 385 at the foot of Vogel Seamount in the New England seamount chain.

At DSDP Site 384 all the calcareous planktonic zones from P2 (=NP3 partim) upward to the top of the Paleocene are preserved; in other locations the depositional intervals are very short compared to the duration of the hiatuses (see Table 2). Portions of foraminiferal Zone P3a (=NP4) are recorded in West Greenland and on the Nova Scotia shelf; the zone also occurs at abyssal DSDP Site 387, but most foraminifera are dissolved there. Portions of Zone P3b (=NP5) were recovered from the walls of Oceanographer Canyon and at DSDP Site 354 on the Ceara Rise in the equatorial North Atlantic. Similar lithofacies above and below the Paleocene coring gaps at Site 354 suggest that a more complete section may have been preserved, but was not cored.

The best represented part of the Paleocene in existing samples is foraminiferal Zone P4 (=NP7–NP8). The entire zone is recorded at DSDP Site 354 (Ceara Rise), Site 144 (equatorial Atlantic, Demerara Rise), and at Site 384 off the Grand Banks. Sediments containing nannoplankton Zones NP7 and NP8 were retrieved also at DSDP Sites 390 and 386 where the foraminifera were dissolved. Foraminifera and nannoplankton of this age were found in AMCOR 6004 on the Carolina Platform, on the Nova Scotian shelf, and along the Labrador shelf.

The uppermost Paleocene and the Paleocene/Eocene boundary are rarely represented in the western North Atlantic. Calcareous sediments straddling the boundary occur at DSDP Site 354 (Ceara Rise), although drilling gaps fragment this sequence. The boundary was cored at DSDP Site 98 in the Bahamas, at J4 on the Blake Plateau, and probably in the COST B2 well although the age data on this interval are less certain. Uppermost Paleocene Zone P5 (=NP9) is recorded in DSDP Site 384, but most of the lower Eocene, including its base, is missing at this site. The Paleocene/Eocene boundary is represented at the two abyssal, middle latitude DSDP Sites 386 and 385; at Site 386 planktonic foraminifera are absent and ages are based on calcareous nannoplankton and radiolarians. At Site 385 this boundary occurs within a sequence of zeolitic clays where both calcareous and siliceous fossils are uncommon.

In summary, Paleocene calcareous biofacies containing both nannofossils and foraminifera are recorded from the equator through the western North Atlantic, and episodically as far north as West Greenland. Foraminifera were preserved from shelf to abyssal depths (e.g., DSDP Site 386) where, however, they are strongly corroded. Nannofossils were preserved through all depths sampled, although in smaller numbers at sites near the CCD.

Siliceous Biofacies

The oldest known siliceous fossils of the North Atlantic Paleocene were radiolarians recovered at DSDP Site 384 off the Grand Banks (Fig. 2). These can be assigned to the early Paleocene equivalent of foraminiferal Zone P2. No Paleocene diatoms occur at this site. Beginning in Zone P4 and continuing through the upper Paleocene, radiolarians become geographically more widespread; in addition to DSDP Site 384, radiolarians have been found at DSDP Site 144 in the equatorial North Atlantic and in the Oceanic Formation on Barbados. At the latter two sites the radiolarians occur in a slightly recrystallized carbonate ooze. At the very top of the Paleocene some radiolarians also are found mixed with a nannofossil assemblage at abyssal DSDP Site 386.

Other Biofacies

Because they are occasionally transported to and preserved in the marine environment, the spores and pollen of land plants are often useful for biostratigraphic subdivision of marine sediments. A palynologic stratigraphy consisting of three assemblage zones has been used to subdivide the Paleocene on the southeastern Labrador and Newfoundland shelves (Barss and others, 1979). These zones, however, do not coincide in time with either the planktonic or benthic foraminiferal zones proposed in the same area, and they are not coeval with series boundaries (Gradstein and Berggren, 1981). In this same area benthic foraminifera have been surveyed, but they proved less useful for biostratigraphy (Dufaure and others, 1976). Only one benthic foraminiferal zone has been erected to represent the entire Paleocene.

To date, Paleocene benthic foraminifera have been reported only from the Labrador Sea and shelf (Dufaure and others, 1976; Berggren and Schnitker, 1983;) and from a few lower and middle latitude open-ocean DSDP sites (Tjalsma and Lohmann, 1983). The Labrador Sea faunas contained only agglutinated benthic foraminifera interpreted to have existed in compartmentalized, probably silled shelf basins which were flooded by fine-grained, carbon-rich clastics. Such an environment would support a rich benthic assemblage, but any calcareous material would be dissolved in the acidic sedimentary environment created by then presence of abundant organic matter (Gradstein and Berggren, 1981).

Summary

The known Paleocene record of calcareous fossils is patchy in both space and time because (1) there is only limited direct sampling; and (2) erosional hiatuses and intervals of carbonate dissolution punctuate the sedimentary record. Available samples demonstrate that lower Paleocene calcareous biofacies were preserved in the equatorial Atlantic, on the Blake Spur, and as far north as DSDP Site 384 off the Grand Banks. Although the basal zones are present at middle latitude sites, the important subzone P1c is absent at DSDP Site 384. This subzone represents a time of warm oceanic paleotemperatures throughout the Atlantic, when lower-latitude index species of planktonic foraminifera and nannoplankton extended their ranges into the higher middle latitudes (Haq and others, 1977) in warm, but apparently sluggish boundary currents (Boersma and Premoli-Silva, 1983).

The oldest Paleocene radiolarians were recovered in foraminiferal Zone P2 south of the Grand Banks (DSDP Site 384). By this time the ocean had recovered sufficiently from the Cretaceous/Tertiary boundary episode of very low primary production to support active carbon cycling and to increase primary productivity, at least among the radiolarians (Boersma and others, 1979).

Two important intervals (Zones P3a and P4 *in toto*) of the upper Paleocene are widely represented in the western North Atlantic and elsewhere (Gradstein and Agterberg, 1982). Zone P3a (=NP4) is preserved from the equator as far north as West Greenland (Hansen, 1970), and from shelf to abyssal depths (DSDP Site 387) where mainly nannofossils are preserved. A second period of oceanic warming occurred in this interval, indicated both in the oxygen isotopic record (Boersma and others, 1979) and by the incursion of lower-latitude plankton into higher latitudes (Gradstein and Srivastava, 1980). Only during this episode were truly high-latitude shelf faunas and carbonate sediments well preserved (Hofker, 1960; Boersma and Premoli-Silva, 1983).

Of all Paleocene zones, Zone P4 (=NP7–NP8) is best represented in marine sections worldwide. In the western North Atlantic, calcareous and (for the first time) siliceous sediments accumulated from the equator north to the Labrador shelf where Zone P4 is often the only zone of the Paleocene preserved. A major episode of open ocean and continental shelf productivity began at this time and continued through most of the late Paleocene. This is witnessed by (1) the maximal Tertiary excursion in the carbon isotope record (Boersma and others, 1979) implying a marked increase in primary production and carbon cycling; (2) increased biosiliceous productivity and preservation in equatorial regions (DSDP Site 144) and in the North Atlantic gyral system (DSDP Site 384); and (3) oceanwide increases in calcareous-sediment accumulation rates (Davies and others, 1977).

EOCENE

The Eocene stratigraphic record, although somewhat more complete than that of the Paleocene is patchy, and hiatuses are common (Table 2, Plate 10,I). Hiatuses are most common in the lower Eocene sections and on middle-latitude shelf edges and continental slopes.

Calcareous Biofacies

All lower Eocene calcareous planktonic foraminiferal zones are represented in high-latitude sites on the Nova Scotian shelf, on the Grand Banks, along the Labrador shelf, and offshore Newfoundland (Fig. 1). One complete lower Eocene section preserved in the Indian Harbour well on the Labrador shelf was dated by means of planktonic foraminifera only; sequences above and below, however, lacked age-diagnostic plankton.

In lower latitudes the lower Eocene is underrepresented because of numerous hiatuses and the non-recovery and coring gaps in DSDP sections. Similar lithofacies occur above and below many of the coring gaps, making it likely that nearly continuous calcareous sedimentation occurred at least on mid-oceanic topographic highs such as the Ceara Rise (DSDP Site 354), the Demerara Rise (DSDP Site 144), and in the Bahamas (DSDP Site 98).

Deep-basin, middle-latitude sites have a variable calcareous record. DSDP Site 385 through most of the Eocene was the locus of continuous deposition of zeolitic clays below the CCD, and it thus lacks age-diagnostic carbonate fossils. At DSDP Site 386, a deep-basin site farther south on the Bermuda Rise, a nearly complete lower and middle Eocene sequence (from Zone NP10 to NP16) can be identified from nannofossils, but foraminifera are missing.

Widespread preservation of calcareous fossils, particularly in middle-latitude, deeper water sites, makes the top of the lower Eocene and the boundary with the middle Eocene one of the best represented biostratigraphic intervals in the western North Atlantic. This entire interval was recovered in DSDP Site 384 south of the Grand Banks, on the Blake Spur at DSDP Site 390, and at abyssal DSDP Sites 386 and 387. Most other sections recovered only small portions of the lowermost middle Eocene (Zone P10).

Portions of the middle Eocene have been recovered at most locations. The record is most complete, although based primarily on nannofossils, at middle-latitude abyssal DSDP Sites 386 and 387, at DSDP Site 390 on the Blake Spur, and at DSDP Site 384. The section is least well represented at shallow-water sites (present depths <1000 m), such as DSDP Hole 111A on Orphan Knoll in the Labrador Sea, J4 on the Blake Plateau, and the AMCOR holes on the Carolina shelf. Zone P11 (=NP15) is most consistently found in the deeper sites, and short segments of the zone are present on the continental shelves.

The upper Eocene record is also patchy. In more than one-third of the 35 locations summarized in figure 1, upper Eocene sediments are missing. In 12 additional localities sediments of this age probably are present, but were not recovered. In the remaining areas the best records, although not continuous, are located at J6 on the Blake Plateau, in AMCOR 6002 in the Southeast Georgia Embayment, at abyssal DSDP Site 386 on the Bermuda Rise, and at DSDP Site 112 in the Labrador Sea.

Sequences considered continuous through the upper Eocene, and including both the lower and upper boundaries of this series, have been recovered in several commercial boreholes from the Nova Scotian and Labrador shelves as far north as 53°N. In the latter area, however, planktonic foraminifera and calcareous nannoplankton are confined to the lower part of the upper Eocene (foraminiferal Zone P15).

The Eocene/Oligocene boundary can be found in only three sites in the western North Atlantic. Because the actual Eocene/Oligocene boundary is detectable in only a few sequences worldwide, in most studies it is considered to be present if nannofossil Zone NP21 and foraminiferal Zone P17, both the Eocene and Oligocene parts, can be identified. By these criteria the Eocene/Oligocene transition is preserved in borehole J3 on the Blake Plateau, and it was identified on the basis of nannofossils in the COST B2 well on the continental shelf off New Jersey (Valentine, 1977). In these cases the boundary interval lies within a calcareous sequence deposited at upper bathyal depths. Although the boundary interval may be present in the commercial boreholes from the Nova Scotian shelf, in J5 on the Florida-Hatteras Slope, and in abyssal DSDP Site 417/418 on the southern Bermuda Rise, the absence of age-diagnostic fossils prevents its precise designation. At all other sites the boundary is marked by hiatuses or was not cored.

Siliceous Biofacies

Siliceous plankton became widespread for the first time during the Eocene (Fig. 2). They were distributed from the equatorial areas represented by DSDP Sites 354 and 144 (Petrushevskaya and Kozlova, 1972) as far north as the Labrador Sea (DSDP Hole 111A; Benson, 1972), and from the continental shelf (COST GE-1) to abyssal depths (DSDP Sites 386, 387, and 417/418; Bukry, 1980). Siliceous sediments were most widespread during the middle Eocene, where they were reported at 15 of the 28 sites surveyed. Radiolarians have been studied and zoned at only seven of these locations: DSDP Sites 10, 386, 387, 390, 417/418, and in Heezen and Corsair canyons. By contrast, significant occurrences of siliceous biofacies of upper Eocene age have been recognized at only four locations including DSDP Sites 27 (Riedel and Sanfilippo, 1970) and 387, and in Heezen and Corsair canyons.

During most of the Eocene a siliceous biofacies was common to dominant at abyssal DSDP sites 386 and 387, at deep-water Site 384 off the Grand Banks, and at DSDP Site 390 on the Blake Spur. Only at Site 387 did the biosiliceous sequence extend through the middle Eocene into the upper Eocene. After a hiatus of nearly 50 m.y., siliceous sedimentation was initiated at abyssal DSDP Site 417/418 at the top of the middle Eocene; a small portion of the uppermost upper Eocene was also recovered above a coring gap which encompassed most of the upper Eocene at this site.

Most upper lower to lower middle Eocene biosiliceous facies have been recrystallized to porcellanitic cherts, thus producing the seismic discontinuity originally named Horizon A (Ewing and Ewing, 1962) but recognized now as a diachronous reflector termed Horizon A^c in the northwest Atlantic (Tucholke and Mountain, 1979). Recrystallization has destroyed the original radiolarian assemblages, or at least altered the apparent species composition. The cherty sediments range from uppermost lower to middle Eocene (Zones P9/P10, =NP14 and NP15 *in toto*) in deep-basin DSDP Sites 6–10, 384–387, and possibly 106. Some siliceous fossils and chert of uppermost lower Eocene age are present in carbonate-rich sediments at DSDP Sites 98 (Bahamas) and 390 (Blake Spur); however seismic Horizon A^c has not been identified in these locations.

The oldest Eocene-age diatoms are recorded from the lower Eocene (=P7–P8) at DSDP Site 10 on the flank of the Mid-Atlantic Ridge. By the top of the lower Eocene, however, they occur in seven other localities, all located along the western margin of the western North Atlantic from the equatorial region (DSDP Site 354) as far north as the Grand Banks (DSDP Site 384). They are not, however, found in the fossiliferous intervals in the Labrador Sea (DSDP Site 112) during this time and are never abundant in abyssal sediments of Eocene age. In the middle Eocene, diatoms occur in several more locations, but they are most abundant in upper-middle latitudes off the Grand Banks at DSDP Site 384 and in Heezen canyon. Only a few upper Eocene sections contain diatoms; these fossils are recorded at DSDP Site 98 and in Heezen and Corsair canyons, but they are absent from the fossiliferous sediments of the Labrador Sea (DSDP Site 112 and Hole 111A).

Diatoms are not recorded from any sequence immediately below or at the Eocene/Oligocene boundary. Like the radiolarians, they are present immediately above the boundary only in the equatorial regions, represented by DSDP Site 144 on Demerara Rise and DSDP Site 354 on the Ceara Rise.

Other Fossil Groups

Macro- and microbenthos, in particular the benthic foraminifera, have been studied from West Greenland, the Canadian margin and the Labrador Sea (Gradstein and Berggren, 1981), the northeastern margin of the United States in the COST B2 well (Poag, 1977), the Blake Plateau (JOIDES, 1965), and open-ocean DSDP Sites 144 and 390 (Plate 10, I). In the Sable Island well on the Nova Scotian shelf, benthic foraminifera are present throughout the sequence but they are not age-diagnostic. This sequence has been assigned to the upper Eocene on the basis of the palynomorph zonation erected for this area by Barss and others (1979). Elsewhere on the Canadian margin a two-fold palynomorph zonation and three benthic foraminiferal assemblage zones can be used to subdivide these shallow-water sections (Gradstein and Berggren, 1981).

In the deep southern Labrador Sea, Eocene-age flysch-like benthic biofacies have been identified at DSDP Site 112. These assemblages terminate abruptly near the Eocene/Oligocene boundary and are replaced by deep-water calcareous benthic foraminifera. The change from an agglutinated to a calcareous benthic foraminiferal biofacies has been interpreted to reflect the replacement of old, corrosive bottom waters by vigorous bottom circulation in the Labrador Sea (Miller and others, 1982).

Eocene-age benthic foraminiferal index faunas, useful for the interpretation of approximate paleodepths of deposition, occur in JOIDES holes 1 and 2 (JOIDES, 1965) and in the slightly deeper hole J5 (personal observation). Holes J1 and 2, located on the present continental shelf off northern Florida (Plate 2) contain middle Eocene larger foraminifera in the shallower hole J1 and shelf-depth smaller benthic foraminifera in the slightly deeper hole J2. Hole J5, located on the upper slope, contains upper Eocene, upper-bathyal calcareous benthic foraminifera in a fine-grained calcareous clay that is also rich in small, smooth-walled planktonic foraminifera.

Tjalsma and Lohman (1983) studied deep water benthic foraminifera from DSDP Sites 390 and 144 as part of a general Atlantic Ocean survey. By studying hiatus distribution, and using the depth migrations of abyssal benthic biofacies to interpret implied changes in deep water circulation, they proposed increased deep-water circulation at the Paleocene/Eocene boundary and in the early Eocene. They found no correlation between fluctuations in calcareous benthic biofacies and the stable-isotopic record of deep-water paleotemperatures during the Eocene.

Summary

Despite numerous gaps in the sedimentary record, distribution patterns of Eocene-age calcareous and siliceous plankton reflect aspects both of the history of the CCD in the North Atlantic and of specific climatic and oceanographic events. The presence of a strong biosiliceous component beginning in Zone P6 time at abyssal DSDP Site 386 is a case in point. The subsequent, widespread development of the siliceous biofacies during the early Eocene in the western North Atlantic may indicate the incursion of high latitude, nutrient-rich waters emanating from the Greenland Sea immediately following shallow breaching of the Greenland-Scotland land bridge (Berggren and Hollister, 1974; see Plate 10, I).

From the late early to early middle Eocene conditions favorable to the accumulation of both calcareous and siliceous sediments developed throughout the western North Atlantic. A more complete sedimentary record accumulated, implying (1) a decrease in the corrosiveness of bottom water and possibly, an increase in circulation vigor, and (2) stimulation of siliceous and possibly general organic productivity from Paleocene through latest early Eocene time, which also suggest more vigorous surface circulation. These events have been correlated with the development of a marine connection between the Atlantic and the Norwegian-Greenland basins, and with a possible shallow-water connection to the Arctic Sea (Berggren and Hollister, 1974). There is no evidence, however, to prove that the Arctic Sea was particularly cold at this time (Boersma and others, 1985) nor to establish the density of Arctic water and thus its propensity to sink and flow through the North Atlantic basins.

Other factors which may have contributed to the development of multiple nutrient sources and hence to increased oceanic productivity include: (1) major sea level regressions that eroded shelf sediments and distributed more nutrients into the open ocean (Vail and others, 1979); (2) development of high-southern-latitude nutrient sources as circum-Antarctic waters into the South Atlantic circulated following deepening of the connection between Australia and Antarctica (Kennett, 1980); and (3) formation of deep-water channels from the South into the North Atlantic through equatorial fracture zones which experienced downward tectonic displacement during the Paleocene and early Eocene (Bonatti and others, 1977). The first appearance of several Indo-Pacific benthic foraminiferal species in the Bay of Biscay and the Gulf of Mexico at the early/middle Eocene boundary corroborates an increasing austral influence in the North Atlantic (Boersma, 1985).

The best represented interval of the middle Eocene corresponds to that part of Zone P11 correlative with nannofossil Zone NP15. Calcareous sediment was preserved from the deep basins to the continental shelves. Siliceous sediments were concentrated at abyssal sites and from the equator north along the western margin of the western North Atlantic. This was also the time of the first significant cooling of oceanic deep waters (Savin, 1977), represented by a temperature drop in the eastern North Atlantic (Vernaud-Grazzini and others, 1980). Development of steeper thermal gradients, both meridional and longitudinal, is indicated by stable isotopic studies of Atlantic foraminifera (Shackleton and Boersma, 1981; Corliss and others, 1984; Boersma and others, 1985). In the North Atlantic this thermal reorganization of the water column is indicated not by erosional hiatuses, but by increased biosiliceous productivity and sediment accumulation.

Evidence of more vigorous surface-water circulation in the main North Atlantic gyre may explain the widespread development of the biosiliceous facies extending from the equator northward along the course of the western boundary current system. The connection between increased surface circulation vigor and the more northern extent of siliceous fossils in the Gulf Stream was first established in studies of the Quaternary glacial cycles (L. Burckle, personal communication, 1981). For the middle Eocene, evidence is provided by the changing biogeographic distribution patterns of middle Eocene planktonic foraminifera, which in Zone P11 increased in diversity to their maximum value for the entire Paleocene; their provincialization also increased, both from north to south and east to west in the North Atlantic, and between the North and South Atlantic oceans. More importantly, foraminiferal species patterns suggest increased rates of delivery of warm waters across the higher middle latitude North Atlantic in a proto–North Atlantic Current which formed the northern arc of the main gyral system. Combined with information on diversity patterns, which suggest a southward squeezing of the North Atlantic gyre, the evidence indicates a more vigorous surface circulation (Boersma and others, 1985).

Due to the very large number of drilling gaps and hiatuses, sedimentary sections of late Eocene age are uncommon in the western North Atlantic (Table 2), rendering biofacies interpretation more difficult. Deposition of siliceous sediments seems to have been limited to abyssal areas and to the continental slope of the northeastern United States. Nannofossils, but not foraminifera occur in the few cored abyssal sites, indicating a relatively shallow CCD (<4500 meters, Plate 1). Calcareous facies deposited on the Labrador shelf in the early late Eocene (Zone P15) were rapidly replaced by detrital sediments bearing only benthic foraminifera and palynomorphs.

The Eocene/Oligocene boundary interval is represented at only a few sites in the western North Atlantic, including J3 on the Blake Plateau and the continental margin off New Jersey (COST B2). Lowermost Oligocene sediments are found in equatorial DSDP Sites 144 and 354, both of which are located on relatively deep topographic swells. Hiatuses are frequent in this interval, particularly along the continental margins where large parts of the Eocene record were removed by later bottom-current erosion (Tucholke and Mountain, this volume).

The extensive hiatuses in the western North Atlantic have been attributed to increased vigor of bottom-water circulation along the continental margin as seafloor spreading finally connected the North Atlantic with the Arctic (Berggren and Hollister, 1974; Tucholke and Miller, 1983). Subsequent stable-isotope evidence has demonstrated short-lived Antarctic glaciation just above the Eocene/Oligocene boundary (Matthews and Poore, 1980; Shackleton and others, 1984; discernible also in the records of Vergnaud-Grazzini and others, 1980), and correlated with the base of Anomaly 13 (Oberhänsli and Toumarkine, 1985), as well as the development of a proto-Antarctic Bottom Water which flowed from south to north through the western South and North Atlantic basins (McCoy and Zimmerman, 1977; Barker and others, 1977; Boersma, 1985). Hiatuses, accompanied by dissolution of carbonates, punctuate most abyssal and deep water sites throughout the Atlantic. We attribute the removal of section and carbonate dissolution in abyssal areas of the southwestern and equatorial North Atlantic to the intrusion of a proto-Antarctic Bottom Water derived from the southwestern South Atlantic during the Eocene/Oligocene transitional interval.

OLIGOCENE

The Oligocene stratigraphic record is very patchy due to a large number of hiatuses and coring gaps in the available boreholes (Figs. 1–2, Plate 10, D). The record is also difficult to interpret due to the nature of the biogeographic imprint on the plankton during this uniformly cool period. Biostratigraphic subdivision is very broad because of a lack of evolutionary events; in our opinion, time-stratigraphic and biostratigraphic correlations for the early Oligocene are not yet firmly established.

Calcareous Biofacies

The available record of the lower Oligocene is more complete than that of the upper Oligocene. In addition to the few sites where fossils were preserved through the Eocene/Oligocene boundary interval, part of lower Oligocene Zone P18 (=NP21 partim) is present on lower latitude topographic highs such as the Ceara Rise (DSDP Site 354) and the Demerara Rise (DSDP Site 144), at abyssal DSDP Site 386 and 417/418 on the Bermuda Rise, at middle latitude sites J3, J4, and possibly J6 on the Blake Plateau, and at DSDP Site 112 in the Labrador Sea (Fig. 1). Although not continuously cored, the record at DSDP Sites 354, 386, and 112 are considered to be relatively complete.

Hiatuses of varying duration interrupt nearly all the overlying sedimentary sequences beginning at the top of lower Oligocene foraminiferal Zone P20 (=basal NP23). Such gaps are more extensive on the continental shelves, for example in AMCOR 6002 in the Southeast Georgia Embayment, and in J5 located landward of the Blake Plateau and containing upper bathyal benthic foraminifera. Less obvious hiatuses occur in deep-water, telescoped sections that are punctuated by moderate carbonate dissolution which removed coarse fraction carbonate but preserved fine-fraction foraminifera and nannofossils (e.g., DSDP Sites 558 and 563).

The upper Oligocene sedimentary record corresponding to Zone P22 (=NP25) is better represented throughout latitude even though it is interrupted by drilling gaps and hiatuses, particularly in shallow-water (Fig. 1). Notably good sections occur at one continental shelf site in the Southeast Georgia Embayment (AMCOR 6002), at deep-basin DSDP Sites 10, 558, and 563, and at abyssal DSDP Site 386. Within Zone P22 calcareous biofacies are intercalated within basalt flows as far north as DSDP Site 407 located on the Reykjanes Ridge. The apparent northward extension of calcareous plankton and a slightly improved preservation of deep-basin carbonates is largely responsible for the better representation of the uppermost Oligocene section.

Siliceous Biofacies

During the early Oligocene, siliceous plankton consisting mainly of radiolarians were consistently present at equatorial Sites 354 and 144 (Petrushevskaya and Kozlova, 1972); deep-water Sites 106, 112, and 387; and ridge-flank Sites 556, 558, and 563 (fig. 2; Plate 10, J). By the middle Oligocene *T. tuberosa* Zone (equivalent to Zones P20 and NP23 partim), biosiliceous sedimentation decreased in equatorial regions (DSDP Site 354) but persisted at high/middle latitude DSDP Site 112 in the Labrador Sea (Benson, 1972). The only diatom-rich facies of the Oligocene also was preserved at DSDP Site 112 (beginning at the top of Zone P18) through the middle into the early late Oligocene. By the end of the Oligocene the extent of the siliceous biofacies was markedly diminished over all latitudes (Fig. 2).

Other Fossil Groups

Oligocene benthic foraminifera and palynomorphs have been studied from the margins of Canada (Gradstein and Agterberg, 1982). Benthic foraminifera also have been surveyed in the Labrador Sea (Miller and others, 1982; Berggren and Schnitker, 1983), in the central North Atlantic at DSDP Sites 558 and 563 (Boersma, 1985), in the COST B2 well off New Jersey (Poag, 1977), and in JOIDES cores from the Florida continental shelf (JOIDES, 1965). Like the Eocene, several benthic foraminiferal and palynomorph zones have been established for zonation of high-latitude, shallow-water sequences.

Sediments on the Mid-Atlantic Ridge flank south of the Azores (DSDP Sites 556, 558 and 563) are largely calcareous, contain moderate amounts of siliceous fossils in Zones P21b and early in Zone P22, and have experienced varying degrees of dissolution which removed parts of the section from the top of Zone P20 into Zone P21a. Benthic foraminiferal species distribution indicates the development of more corrosive seafloor environments during the P20 to P21a zonal interval. Preservation of calcareous material improved markedly during Zone P22, at which time the benthic indices for increased dissolution almost disappeared and increased numbers of low-latitude planktonic foraminifera migrated into the area (Boersma, 1985).

In the JOIDES drillholes, Oligocene benthic foraminifera are reported only in J1 (JOIDES, 1965). The benthics are accompanied for the first time by planktonic foraminifera which indicate a deepening of the water over this area of the Florida shelf sometime in the early Oligocene.

Summary

The Oligocene sedimentary record of the western North Atlantic is characterized by the low percentage of sections cored and recovered, as well as by a large number of drilling gaps and hiatuses (Table 2). Interpretation of biofacies is therefore rather speculative.

The Eocene/Oligocene boundary interval is represented in only a few shelf and relatively shallow plateau areas bounding the North Atlantic. In abyssal DSDP Site 386 the uppermost Eocene was not cored, but the basal Oligocene appears to be recognizable on the basis of nannofossils. Sedimentation resumed in equatorial areas of the Ceara Rise and Demerara Rise (DSDP Sites 354 and 144) immediately above the boundary. Poor representation of the basal (and later) Oligocene along the continental margin has been attributed to the extensive erosion by bottom currents of northern origin beginning near the Eocene/Oligocene boundary (Tucholke and Mountain, 1979). As discussed earlier, we believe that this erosion was caused by short-lived, vigorous bottom and deep-water currents of circum-Antarctic origin, as well as more rapid gyral circulation during a brief glacial episode at the very end of the Eocene. Hiatuses in equatorial and southwestern regions of the North Atlantic are some of the best evidence for increasing rates of proto-Atlantic bottom water generation (A. Gordon, personal communication, 1983), so the resumption of sedimentation in the basal Oligocene in equatorial regions (DSDP Sites 354 and 144) suggests decreased influence of this type of bottom-water, at least to the south (Boersma and others, 1985).

A second group of hiatuses on the shallow continental margin occurred high in Zone P 20 (=Zones NP23 and *T. tuberosa* partim), roughly coincident with (1) a contraction of the biosiliceous facies from equatorial and marginal areas to higher latitude areas, and (2) the first abundance of diatoms in the Labrador Sea. A major Oligocene sea level regression also occurred at this time, about 29–30 Ma (Vail and others, 1979). Evidence from stable-isotope studies (Savin, 1977) and reorganization of calcareous planktonic bioprovinces (Haq and others, 1977; Gradstein and Srivastava, 1980), suggest a southward contraction and invigoration of the main North Atlantic gyre during a second, transitory glacial episode.

Calcareous sediments of Zone P22 age have been recovered in numerous sites through latitude and depth, despite the lengthy hiatuses occurring in continental-margin sections of this age. Nannofossils were preserved at abyssal DSDP Site 386 located near 4,800 m paleodepth. This establishes the CCD in the western North Atlantic near 4,800 m during this time. Siliceous biofacies were rare in all but the highest latitude, open-marine site studied to date, DSDP Site 407 on the Reykjanes Ridge.

The most northward migration of subtropical foraminiferal species and most poleward extension of bioprovincial boundaries of the Oligocene occurred in Zone P22 (Haq and others, 1977). This faunal migration was accompanied by widespread calcareous sediment accumulation on the continental shelves and margins of Labrador and the northeastern United States, a slight decrease in the number of hiatuses, and global increases in sediment accumulation rates (Davies and others, 1979). Oceanwide warming of near-surface waters indicated by oxygen isotope analyses (Savin, 1977; Boersma and Shackleton, 1977; Vergnaud-Grazzini and others, 1980), and a shift to greater surface productivity indicated in the carbon isotope record (Shackleton and others, 1984) explain the biofacies patterns of the latest Oligocene.

BIOSTRATIGRAPHIC CONTROL OF AGE OF SEISMIC HORIZONS

To date, four regionally-extensive Paleogene-age seismic horizons have been identified in the western North Atlantic basin (Plate 1; Tucholke, 1979). In decreasing age they are Horizon A^c (early to middle Eocene), A^T (middle to late Eocene), A^U (early Oligocene), and A^V (late Oligocene). Distribution of these horizons throughout the North Atlantic is summarized by Tucholke and Mountain (this volume).

Horizon A^U, a regional unconformity along the deep continental margin, has been penetrated at DSDP Sites 4, 5, 8(?), 99, 100, 101, 105, 106, 391, and 534. The age of erosion is difficult to ascertain because (1) the unconformity is very large at some sites; for example, it extends from the Lower Cretaceous to the Pliocene at DSDP Site 99; and (2) spot coring failed to recover a continuous sedimentary section at most sites. The oldest sediments recovered above A^U at DSDP Site 5 are Oligocene Zone P21b in age (Cita and others, 1970). At DSDP Site 106 a general lower Miocene age for overlying sediments is based only on three recognizable nannofossil species (Bukry, 1972). The youngest sediments below the possible unconformity at DSDP Site 8 are unzonable and assigned to the lower to middle Eocene on the basis of the radiolarians present (Riedel and Sanfilippo, 1970). We consider these sediments to be correlative with the foraminiferal Zone P9–P10 transition at the lower to middle Eocene boundary. This roughly constrains the erosional episode to between Zones P9–P10 of the middle Eocene and Zone P21b of the upper Oligocene.

Horizon A^T, the top of a turbidite sequence lapping westward off the Bermuda Rise, ranges in age from Zone NP16 (= *P. triacantha*) at DSDP Site 386, to Zone NP17 (=*P. mitra*) at the more westerly DSDP Site 387, to middle Eocene (apparently equivalent to Zone NP14 = *P. striata striata*) at DSDP Site 8. There is some question regarding the age of the top of the cored interval at DSDP Site 8; it may have penetrated the *T. triacantha* Zone and would therefore be the same age as the top of the discontinuity at DSDP Site 386.

Horizon A^C, originally considered the world-wide synchronous Layer A (Ewing and Ewing, 1963), characteristically coincides with lower to middle Eocene cherts. The horizon actually varies in age from lower Eocene Zone P9 (=NP14b) at DSDP Site 384, to Zone P11 (=NP15b) at DSDP Site 387, with intermediate ages elsewhere (e.g., the R23 *P. cryptocephala* Zone =NP14 at DSDP Site 385). The age of the top of the chert and radiolarian-bearing layer may be time-transgressive from north to south, with the radiolarians disappearing earlier from the more northerly sites. Although DSDP Sites 386 and 387 appear to diverge from this pattern, the age assignment at Site 387 is not certain and may be an artifact of sediment recrystallization that was particularly destructive to the fossils at this site. This north-south pattern could represent (*inter alia*) the gradual, southerly contraction of radiolarian distribution with the first real cooling of the middle Eocene.

Horizon A^V marks the top of volcaniclastic turbidites surrounding Bermuda. Only a broad age range of Zones NP24–NP25 (upper middle to upper Oligocene) based on strongly dissolved calcareous biotas, can be assigned to these sediments at DSDP Site 386 (Bukry, 1978b).

SUMMARY AND CONCLUSIONS

We have summarized the available Deep Sea Drilling Project and JOIDES core samples, as well as United States Geological Survey and commercial borehole data to trace the time/space distributions of calcareous and siliceous biofacies through the Paleogene of the western North Atlantic. The following patterns emerge from this analysis.

The record of Paleogene-age calcareous and siliceous biofacies in the western North Atlantic currently is very patchy due to incomplete coring and the presence of numerous hiatuses (Fig. 1–2). As summarized in Table 2, recovery of planktonic foraminiferal zones averages only 25%, hiatuses constitute an average of nearly 40% of the recovered intervals, and drilled gaps account for the remaining 35%. To ascertain when accumulation of fossil biofacies was best represented, the number of times a penetrated zone was recovered can be compared with the times it was uncored or missing because of a hiatus. The best records of biotic sediment accumulation are found in the late Paleocene (foraminiferal Zone P4), the early to middle Eocene (foraminiferal Zones P9–P11), and the latest Oligocene (foraminiferal Zone P22). The upper Paleocene and Eocene sediments include both the calcareous and siliceous biofacies; the upper Oligocene biofacies are predominantly calcareous. The improved sediment preservation at these levels can be attributed to the decreased activity of bottom currents, increased sedimentation rates resulting from enhanced oceanic surface water productivity, chertification which armored at least some middle Eocene sediments against later erosion, and a possible deepening of the CCD which facilitated more widespread preservation of both biofacies.

Paleogene pelagic, calcareous biofacies are the most widespread throughout latitude and occur in all regions from the continental shelf edge to the abyssal depths of the Atlantic basin. At abyssal sites deeper than 4,000–4,500 m paleodepth, planktonic foraminifera are usually absent due to dissolution below the CCD; thus any calcareous biofacies present consist almost exclusively of the calcareous nannoplankton. An important exception occurs at DSDP Site 6, located at a present depth of 5,125 m on the western Bermuda Rise. Middle Eocene planktonic foraminifera at this site were transported to abyssal depths by turbidity currents and thus were buried before they could be dissolved (Ewing and others, 1968; personal observation).

We can use the co-occurrence of planktonic foraminifera and nannofossils in the calcareous biofacies to represent paleodepths above the CCD, nannofossils and/or nannofossils and radiolarians to represent paleodepths near the CCD, and samples containing only siliceous plankton to indicate paleodepths greater than the CCD. Available data then indicate a gradual drop in the CCD from near 4,000 m paleodepth to more than 4,500 m through the course of the Paleogene in middle-latitude abyssal areas of the western North Atlantic (Plate 1). Gaps in the stratigraphic record at some sites, the lack of age-diagnostic fossils at others, and the relatively small number of sites result in inadequate control for finer-scale delineation of the CCD in this region.

Planktonic foraminifera occurred in upper-middle to high latitude, shallow shelf sites in pulses during the early Paleocene (Zone P1c), late Paleocene (Zone P4), near the early/middle Eocene boundary (Zones P9–P10), in early late Eocene Zone P15, and in latest Oligocene Zone P22. Nannofossils have not yet been studied in these shallow-water areas. Preservation of carbonate on high-latitude shelves during Zones P4 and P9–10 is part of an overall global pattern of increased sediment accumulation through both depth and latitude. At both times there was also an increase in biosiliceous accumulation and eventually in induration of both the calcareous and siliceous biofacies. In the Eocene section only, this diagenesis produced a substantial amount of chert world wide. Increased productivity in oceanic surface waters at these times does not appear to be directly related to climate, however, because Zone P4 was a time of very warm oceanic temperatures, and Zones P9–10 marked the beginning of a worldwide decrease in surface water temperatures (Savin, 1977). It is more likely that tectonic breaching of circulation barriers allowed interchange of surface and intermediate waters with both the northern and southern circumpolar areas, thus stimulating oceanic productivity and enhancing preservation of biogenic sediment.

The occurrence and preservation of calcareous plankton generally in high-latitude regions during Zones Plc, mid-Paleocene Zone P3a, and late Oligocene Zone P22 are correlated with short-lived, climatic warming episodes indicated by the $\delta^{18}O$ record of paleotemperatures (Boersma and others, 1979). At such times subtropical calcareous plankton migrated as far north as West Greenland in the Paleocene, but they only reached the Labrador Sea in the Oligocene (Haq and others, 1977; Gradstein and Srivastava, 1980; Boersma and Premoli-Silva, 1983).

Siliceous organisms, primarily the radiolarians, were relatively rare in the Paleogene (Fig. 2). They first appeared during early Paleocene Zone P2 at DSDP Site 384 just south of the Grand Banks. During the Paleocene they very gradually increased in abundance and diversity, and began to spread over a large area of the Atlantic Ocean in the late Paleocene (Zone P4). By the very early Eocene the radiolarians were distributed from the Blake Plateau to the Mid-Atlantic Ridge, and as far north as Orphan Knoll in the southern Labrador Sea. In late early Eocene time biosiliceous sedimentation became widespread and continued so through the early middle Eocene. These sediments were later recrystallized to form the porcellanitic cherts (Horizon A^C) that occur through most of the western North Atlantic. By late Eocene time and continuing into the earliest Oligocene, siliceous organisms appear to have been most common at lower latitudes (Sites 354, 144), but rare at higher latitudes (Sites 111, 112 in the Labrador Sea). During the mid-Oligocene, siliceous organisms including the diatoms became common at DSDP Site 112 in the Labrador Sea. We have speculated that a glacial episode accompanied by increased bottom circulation was responsible for the increase in siliceous productivity at higher latitudes at this time. By the end of the Oligocene, however, the siliceous biofacies became geographically restricted and they are observed only at a few locations; these locations extended from near the equator to the highest latitude site included in this study, DSDP Site 407 in the Reykjanes Ridge.

REFERENCES

Bader, R. G., Gerard, R. D., and Hay, W. W., eds.,
1970 : Initial Reports of the Deep Sea Drilling Project, Volume 4: U.S. Government Printing Office, Washington, D.C., 753 p.

Barker, P., and Dalziel, I., eds.
1976 : Initial Reports of the Deep Sea Drilling Project, Volume 36: U.S. Government Printing Office, Washington, D.C., 1137 p.

Barss, M., Bujak, J., and Williams, G.
1979 : Palynological zonation and correlation of sixty-seven wells, eastern Canada: Geological Survey of Canada, Paper 78-24, p. 1–118.

Beckman, J. P.
1972 : The foraminifera and some associated microfossils of Sites 135 to 144; in Initial Reports of the Deep Sea Drilling Project, Volume 14, eds. D. E. Hayes and A. C. Pimm; U.S. Government Printing Office, Washington, D.C., p. 389–420.

Benson, R. N.
1972 : Radiolaria, Leg XII, Deep Sea Drilling Project; in Initial Reports of the Deep Sea Drilling Project, Volume 12. eds. A. Laughton and W. Berggren; U.S. Government Printing Office, Washington, D.C.,

p. 1085–1114.

Benson, W., and Sheridan, R. E., eds.

1978 : Initial Reports of the Deep Sea Drilling Project, Volume 44: U.S. Government Printing Office, Washington, D.C., 1005 p.

Berggren, W. A.

1969 : Paleogene biostratigraphy and planktonic foraminifera of northern Europe: Proceedings, First International Conference on Planktonic Microfossils, Geneva, 1967, D. J. Brill, Leiden, p. 121–160.

1972 : Cenozoic biostratigraphy and paleobiogeography of the North Atlantic: in Initial Reports of the Deep Sea Drilling Project, Volume 12, eds. A. Laughton and W. Berggren; U.S. Government Printing Office, Washington, D.C., p. 965–1002.

Berggren, W. and Gradstein, F.

1981 : Agglutinated benthic foraminiferal assemblages in the Paleogene of the central North Sea: their biostratigraphic and depositional environmental significance; in Petroleum Geology of the Continental Shelf of Northwestern Europe, eds. L. Y. Illington, G. D. Hobson; Institution of Petroleum, London, p. 282–295.

Berggren, W. and Hollister, C. D.

1974 : Paleogeography, paleobiogeography and the history of circulation of the Atlantic Ocean; in Symposium on Geologic History of the Oceans, ed. W. W. Hay; Society of Economic Paleontologists and Mineralogists, Special Publication 20, p. 120–186.

Berggren, W., Kent, D., and Flynn, J. J.

1985 : Paleogene geochronology and chronostratigraphy: in Geochronology and the Geologic Record, ed. N. J. Snelling, Geological Society of London, Special Paper, in press.

Berggren, W., Pessagno, G., and Bukry, D.

1969 : Biostratigraphy; in Initial Reports of the Deep Sea Drilling Project, Volume 1, eds. M. Ewing and J. C. Worzel; U.S. Government Printing Office, Washington, D.C., p. 594–623.

Berggren, W., and Schnitker, D.

1983 : Cenozoic marine environments in the North Atlantic and Norwegian-Greenland Sea; in Structure and development of the Greenland–Scotland Ridge, eds., M. Bott, S. Saxov, M. Talwani, and J. Thiede; Plenum Publishing Company, p. 495–548.

Blow, W. H.

1969 : Late Middle Eocene to Recent planktonic biostratigraphy: Proceedings, First International Conference Planktonic Microfossils, Geneva, 1967, E. F. Brill, Leiden, p. 199–422.

1979 : The Cainozoic Globigerinida: E. J. Brill, Leiden, 3 volumes, 1350 p.

Boersme, A.

1977 : Cenozoic Planktonic Foraminifera-DSDP Leg 39 (South Atlantic); in Initial Reports of Deep Sea Drilling Project, Volume 39, eds. P. R. Supko and K. Perch-Nielsen; U.S. Government Printing Office, Washington, D.C., p. 567–590.

Boersma, A.

1985 : Oligocene benthic foraminifera from North Atlantic DSDP Sites 558, 563, 549, and 366; examination of North and South Atlantic benthic faunas as paleo–watermass indices; in Initial Reports of the Deep Sea Drilling Project, Volume 82, eds. H. Bouga and S. Cande; U.S. Government Printing Office, Washington, D.C., p. 611–625.

Boersma, A. and Premoli-Silva, I.

1983 : Paleocene planktonic foraminiferal biogeography and the paleoceanography of the Atlantic Ocean: Micropaleontology, v. 29, p. 355–381.

Boersma, A., Premoli-Silva, I., and Shackleton, N.

1985 : Eocene planktonic foraminiferal biogeography and the paleooceanography of the Atlantic Ocean: Micropaleontology, in press.

Boersma, A., and Shackleton, N.

1977 : Oxygen and carbon isotope record through the Oligocene, DSDP Site 366, equatorial Atlantic; in Initial Reports of the Deep Sea Drilling Project, Volume 41, eds. E. Seibold, and Y. Lancelot; U.S. Government Printing Office, Washington, D.C., p. 957–962.

Boersma, A., Shackleton, N., Hall, M., and Given, Q.

1979 : Carbon and oxygen isotope records of the DSDP Site 384 (North Atlantic) and some Paleocene paleotemperatures and carbon isotope variations in the Atlantic Ocean: in Initial Reports of the Deep Sea Drilling Project, Volume 43, eds. B. Tucholke and P. Vogt; U.S. Government Printing Office, Washington, D.C., p. 695–717.

Bolli, H. M.

1966 : Zonation of Cretaceous to Pliocene marine sediments based on planktonic foraminifera: Asociacion Venezolana Geologia Mineira y Petroles Boletin Informativo, p. 3–32.

1970 : The foraminifera of Site 23-31, Leg 4; in Initial Reports of the Deep Sea Drilling Project, Volume 4, eds., R. Bader, R. D. Gerard and W. W. Hay; U.S. Government Printing Office, Washington, D.C., p. 577–644.

Bonatti, E., Sarnthein, M., Boersma, A., Gorini, M. and Honnorez, J.

1977 : Crustal emersion and subsidence at the Romanche Fracture Zone, equatorial Atlantic: Earth and Planetary Science Letters, v. 3, p. 75–89.

Bukry, D.

1972 : Further comments on Coccolith Stratigraphy, Leg XII, Deep Sea Drilling Project, Volume 12, eds. A. Laughton and W. Berggren; U.S. Government Printing Office, Washington, D.C., p. 1071–1084.

1973 : Low latitude coccolith biostratigraphic zonation: in Initial Reports of the Deep Sea Drilling Project, Volume 15, eds. T. Edgar and J. Saunders; U.S. Government Printing Office, Washington, D.C., p. 685–703.

1975 : Coccolith and silicoflagellate stratigraphy, northwestern Pacific Ocean, Deep Sea Drilling Project, Leg 32; in Initial Reports of the Deep Sea Drilling Project, Volume 32, eds. R. Larson and R. Moberly; U.S. Government Printing Office, Washington, D.C., p. 677–701.

1977 : Coccolith and silicoflagellate stratigraphy, South Atlantic Ocean: Deep Sea Drilling Project Leg 39; in Initial Reports of the Deep Sea Drilling Project, Volume 39, eds. P. Supko and K. Perch-Nielsen; U.S. Government Printing Office, Washington, D.C., p. 825–839.

1978a: Cenozoic silicoflagellate and coccolith stratigraphy, North-western Atlantic Ocean, Deep Sea Drilling Project, Leg 43; in Initial Reports of the Deep Sea Drilling Project, Volume 44, eds. R. Benson, and R. Sheridan; U.S. Government Printing Office, Washington, D.C., p. 775–806.

1978b: Cenozoic coccolith, silicoflagellate, and diatom stratigraphy, Deep Sea Drilling Project, Leg 44; in Initial Reports of the Deep Sea Drilling Project, Volume 44, eds. R. Benson, and R. Sheridan; U.S. Government Printing Office, Washington, D.C., p. 807–865.

1979 : Coccolith and silicoflagellate stratigraphy, northern Mid-Atlantic Ridge and Reykjanes Ridge, Deep Sea Drilling Project, Leg 49; in Initial Reports of the Deep Sea Drilling Project, Volume 49, eds. B. Luyendyk and J. Cann; U.S. Government Printing Office, Washington, D.C., p. 551–582.

1980 : Eocene diatoms and siliceous sponge spicules from north western Atlantic Ocean, DSDP Site 417 and 418; in Initial Reports of the Deep Sea Drilling Reports, Volume 51, 52, and 53, eds. T. Donnelly and J. Francheteau; U.S. Government Printing Office, Washington, D.C., p. 851–855.

Cita, M., Nigrini, C. and Gartner, R. S.

1970 : Biostratigraphy; in Initial Reports of the Deep Sea Drilling Project, Volume 2, eds. M. N. Peterson, and N. T. Edgar; U.S. Government Printing Office, Washington, D.C., p. 391–412.

Corliss, B., Aubry, M., Berggren, W., Fenner, J., Keigwin, L., and Keller, G.

1984 : The Eocene/Oligocene event in the deep sea: Science, v. 226, p. 806–801.

Davies, T. A., Hay, W., Southam, J., and Worsley, T.

1977 : Estimates of Cenozoic sedimentation rates: Science, v. 197, p. 53–55.

Douglas, R. and Woodruff, F.

1981 : Deep sea benthic foraminifera; in The Oceanic Lithosphere and The Sea, Volume 7, ed. C. Emiliani, Wiley Interscience, New York, p. 1233–1327.

Donnelly, T. and Francheteau, J., eds.

1980 : Initial Reports of the Deep Sea Drilling Project, Volumes 51, 52, and

53: U.S. Government Printing Office, Washington, D.C., 1613 p.

Dufaure, Ph., McWhae, R., and Verdenius, J. G.
1976 : Tertiary and upper Cretaceous in offshore Labrador boreholes; first stratigraphic results; in Proceedings International Symposium on Benthonic Foraminifera, Halifax, 1975; Maritime Sediments Special Publication 18, p. 509–522.

Ewing, J. I., and Ewing, M.
1962 : Reflection profiling in and around the Puerto Rico Trench; Journal of Geophysical Research, v. 67, p. 4729–4739.

Ewing, M. and Ewing, J. I.
1963 : Sediments at proposed LOCO drilling sites: Journal of Geophysical Research, v. 68, p. 251–256.

Ewing, M. and Worzel, T. M., eds.
1968 : Initial Reports of the Deep Sea Drilling Project, Volume 1: U.S. Government Printing Office, Washington, D.C., 672 p.

Freeman-Lynde, R. P., Cita, M., Jadoul, F., Miller, E., and Ryan, W.
1981 : Marine geology of the Bahama escarpment: Marine Geology, v. 44, p. 119–156.

Gibson, T., Hazel, J., and Mello, J. F.
1968 : Fossiliferous rocks from submarine canyons off the northeastern United States: U.S. Geological Survey Professional Paper 600-D, p. 222–230.

Gradstein, F. M., and Agterberg, F. P.
1985 : Model of Cenozoic foraminiferal stratigraphy-northwestern Atlantic margin: Proceedings, Quantitative Stratigraphic Correlation Symposium, 26th International Geological Congress, Paris, 1980, Wiley and Sons, London, p. 633–701.

Gradstein, F. and Berggren, W. A.
1981 : Flysch-type agglutinated foraminifera and the Maestrichtian to Paleogene history of the Labrador and North Seas: Marine Micropaleontology, v. 6, p. 211–268.

Gradstein, F., Bukry, D. and others
1978 : Biostratigraphic summary of DSDP Leg 44, western North Atlantic Ocean; Initial Reports of the Deep Sea Drilling Project, Volume 44, eds. W. Benson, R. Sheridan; U.S. Government Printing Office, Washington, D.C., p. 657–662.

Gradstein, F. M., and Srivastava, S. P.
1980 : Aspects of Cenozoic stratigraphy and paleooceanography of the Labrador Sea and Baffin Bay: Paleogeography, Paleoecology, Paleoclimatology, v. 30, p. 261–295.

Gradstein, F. M., Williams, G., Jenskins, W., and Ascoli, P.
1975 : Mesozoic and Cenozoic stratigraphy of the continental margin eastern Canada: Canada Continental Margins and Offshore Petroleum Exploration, Canadian Society Petroleum Geologists, Calgary, Memoir 4, p. 103–131.

Hansen, H. J.
1970 : Danian foraminifera from Nugssauq, West Greenland; Meddelelser om. Gronland, v. 193, 132 p.

Haq, B., and Boersma, A.
1976 : Introduction to Marine Micropaleontology: Elsevier, New York, 395 p.

Haq, B., Premoli-Silva, I., and Lohmann, P.
1977 : Calcareous planktonic paleogeographic evidence for major climatic fluctuations in the early Cenozoic Atlantic Ocean: Journal Geophysical Research, v. 92, p. 386–398.

Hardenbol, J. and Berggren, W. A.
1978 : A new Paleogene numerical time-scale; in Contribution to the Geologic Time Scale. eds. G. Cohee, M. Glaessner, and H. Hedburg, American Association of Petroleum Geologists, p. 213–234.

Hathaway, J., Poag, C. W., and others
1979 : U.S. Geological Survey core drilling on the Atlantic Shelf: Science, v. 206, no. 11, p. 515–527.

Hay, W. W.
1970 : Calcareous nannofossils from cores recovered on Leg 4: in Initial Reports of the Deep Sea Drilling Project, Volume 4; eds. R. G. Bader, R. D. Gerard and W. W. Hay; U.S. Government Printing Office, Washington, D.C., p. 455–502.

Hayes, D. E., and Pimm, A. C., eds.
1972 : Initial Reports of the Deep Sea Drilling Project, Volume 14: U.S. Government Printing Office, Washington, D.C., 1329 p.

Hollister, C. D. and Ewing, J. I., eds.
1972 : Initial Reports of the Deep Sea Drilling Project, Volume 11: U.S. Government Printing Office, Washington, D.C., 1077 p.

Hofker, J.
1960 : Planktonic foraminifera in the Danian of Denmark: Contributions Cushman Foundation Foraminiferal Research, v. 11, p. 73–86.

JOIDES
1965 : Ocean drilling on the continental margin: Science, v. 150, no. 3967, p. 709.

Kennett, J. P.
1980 : The development of planktonic biogeography in the southern Ocean during the Cenozoic: Marine Micropaleontology, v. 3, p. 301–345.

Laughton, A. G., and Berggren, W. A., eds.
1972 : Initial Reports of the Deep Sea Drilling Project, Volume 12: U.S. Government Printing Office, Washington, D.C., 1243 p.

Luterbacher, H. P.
1972 : Paleocene and Eocene planktonic foraminifera. Leg XI, Deep Sea Drilling Project: in Initial Reports of the Deep Sea Drilling Project, Volume 11; eds. C. D. Hollister, and J. I. Ewing; U.S. Government Printing Office, Washington, D.C., p. 547–550.

Luyendyk, B. and Cann, J., eds.
1979 : Initial Reports of the Deep Sea Drilling Project, Volume 49: U.S. Government Printing Office, Washington, D.C., 1020 p.

Martini, E.
1970 : Standard Paleogene calcareous nannoplankton zonation: Nature, v. 26, no. 5245, p. 560–561.
1971 : Standard Tertiary and Quaternary calcareous nannoplankton zonation: Proceedings, 2nd Planktonic Conference, Rome: ed. A. Farinacci, p. 739–785.
1979 : Calcareous nannoplankton and silicoflagellate biostratigraphy at Reykjanes Ridge, northeastern North Atlantic, DSDP Leg 49, Sites 407 and 409); in Initial Reports of the Deep Sea Drilling Project, Volume 49; eds. B. Luyendyk, and J. Cann; U.S. Government Printing Office, Washington, D.C., p. 533–550.

Matthews, R., and Poore, R.
1980 : Tertiary $^{18}0$ record and glacio-eustatic sea-level fluctuations: Geology, v. 8, no. 10, p. 501–504.

McCoy, F., and Zimmerman, H.
1977 : A history of sediment lithofacies in the South Atlantic Ocean: Initial Reports of the Deep Sea Drilling Project, Volume 39; eds. P. R. Supko and K. Perch-Nielsen; U.S. Government Printing Office, Washington, D.C., 1047–1081.

Miller, K. G., Gradstein, F. M., and Berggren, W. A.
1982 : Late Cretaceous to Early Tertiary agglutinated benthic foraminifera in the Labrador Sea: Micropaleontology, v. 28, p.1–30.

Miller, K., and Tucholke, B.
1983 : Development of Cenozoic abyssal circulation south of the Greenland-Scotland Ridge: Structure and Development of the Greenland-Scotland Ridge; ed. G. Saxov; Nato Advanced Research Institute, Plenum Press, Bressanone, p. 501–504.

Miller, K. G., Mountain, G. S., and Tucholke, B. E.
1985 : Oligocene glacio-eustasy and erosion on the margins of the North Atlantic: Geology, v. 13, p. 10–13.

Oberhansli, H. and Toumarkine, M.
1985 : The Paleogene oxygen and carbon isotope history of Sites 522, 523, and 524 from the central South Atlantic: in South Atlantic Paleoceanography, eds. K. Hsu, and H. J. Weissert, Cambridge University Press, Cambridge, p. 125–149.

Okada, H. and Bukry, D.
1980 : Supplementary modification and introduction of code numbers to the low-latitude coccolith biostratigraphic zonation: Marine Micropaleontology, v. 5, p. 321–325.
Okada, H. and Thierstein, H. R.
1979 : Calcareous nannoplankton, Leg 43, Deep Sea Drilling Project; in Initial Reports of the Deep Sea Drilling Project, Volume 43; eds. B. E. Tucholke and P. R. Vogt; U.S. Government Printing Office, Washington, D.C., p. 507–574.
Perch-Nielsen, K.
1972 : Remarks on Late Cretaceous to Pleistocene Coccoliths from the North Atlantic; in Initial Reports of the Deep Sea Drilling Project, Volume 12; eds. A. G. Laughton, W. A. Berggren; U.S. Government Printing Office, Washington, D.C., p. 1003–1070.
1977a: Albian to Pleistocene calcareous nannofossils from the western South Atlantic, DSDP Leg 39, in Initial Reports of the Deep Sea Drilling Project, Volume 39; in eds. P. R. Supko and K. Perch-Nielsen; U.S. Government Printing Office, Washington, D.C., p. 699–823.
1977b: Tertiary Silicoflagellates and other siliceous microfossils from the western South Atlantic: Deep Sea Drilling Project, Leg 39; in Initial Reports of the Deep Sea Drilling Project, Volume 39; eds. P. R. Supko and K. Perch-Nielsen; U.S. Government Printing Office, Washington, D.C., p. 863–866.
Peterson, M.N.A. and Edgar, N. T., eds.
1970 : Initial Reports of the Deep Sea Drilling Project, Volume 2: U.S. Government Printing Office, Washington, D.C., 491 p.
Petrushevskaya, M. G., and Kozlova, G. E.
1972 : Radiolaria: Leg 14, Deep Sea Drilling Project, in Initial Reports of the Deep Sea Drilling Project, Volume 14; eds. D. E. Hayes and A. C. Pimm; U.S. Government Printing Office, Washington, D.C., p. 495–648.
Poag, C. W.
1977 : Foraminiferal Biostratigraphy; in Geological Studies on the COST No. B-2 well, U.S. Mid-Atlantic Outer Continental Shelf Area, ed. P. A. Scholle, U.S. Geological Survey Circular, no. 750, p. 35–36.
1982 : Foraminiferal and seismic stratigraphic, paleoenvironments and depositional cycles in the Georges Bank Basin: in Geological Studies of the COST Nos. G-1 and G-2 wells, United States North Atlantic Outer Continental Shelf; eds. P. A. Scholle, and C. R. Wenkam, U.S. Geological Survey Circular no. 861; p. 43–91.
Poore, R. Z.
1979 : Oligocene through Quaternary planktonic foraminiferal biostratigraphy of the North Atlantic: DSDP Leg 49; in Initial Reports of the Deep Sea Drilling Project, Volume 49; eds. B. P. Luyendyk and J. C. Cann; U.S. Government Printing Office, Washington, D.C., p. 447–518.
Premoli-Silva, I. and Bolli, H. M.
1973 : Late Cretaceous to Eocene planktonic foraminifera and stratigraphy of the Leg 15 sites in the Caribbean Sea; in Initial Reports of the Deep Sea Drilling Project, Volume 15; eds. N. T. Edgar and J. B. Saunders; U.S. Government Printing Office, Washington, D.C., p. 499–548.
Reidel, W. R. and Sanfilippo, A.
1970 : Radiolaria, Leg 4, Deep Sea Drilling Project; in Initial Reports of the Deep Sea Drilling Project, Volume 4; eds. R. G. Bader and R. D. Gerard and W. W. Hay; U.S. Government Printing Office, Washington, D.C., p. 503–576.
1978 : Stratigraphy and evolution of tropical Cenozoic Radiolaria: Micropaleontology, v. 24, no. 1, p. 61–96.
Roth, R. H., and Thierstein, H.
1972 : Calcareous Nannoplankton: Leg XIV of the Deep Sea Drilling Project, in Initial Reports of the Deep Sea Drilling Project, Volume 14; eds. D. E. Hays and A. C. Pimm; U.S. Government Printing Office, Wahington, D.C., p. 421–486.
Ryan, W.B.F., Cita, M. B., Miller, E. L., Hanselman, D., Nesterhoff, W. D., Hecker, B., and Nibbelink, M.
1978 : Bedrock Geology in New England submarine canyons: Oceanologica Acta, v. 1, no. 1, p. 233–254.
Savin, S. M.
1977 : The history of the earth's surface temperature during the past one hundred million years: Annual Reviews of Earth and Planetary Sciences, v. 5, p. 319–344.
Scientific Party Leg 82
1982 : On Leg 82: Elements traced in Atlantic: Geotimes, July 1982, p. 21–23.
Shackleton, N. J. and Boersma, A.
1981 : The climate of the Eocene Ocean: Journal of the Geological Society of London, v. 138, p. 153–157.
Shackleton, N. J., and Members of the Shipboard Scientific Party
1984 : Accumulation rates in Leg 74 sediments: in Initial Reports of the Deep Sea Drilling Project, Volume 74; eds. T. C. Moore and P. D. Rabinowitz; U.S. Government Printing Office, Washington, D.C., p. 621–645.
Sheridan, R. E. and Gradstein, F. M., eds.
1984 : Initial Reports of the Deep Sea Drilling Project, Volume 76: U.S. Government Printing Office, Washington, D.C., 947 p.
Stainforth, R. M., Lamb, J. L., Luterbacher, H. P., Beard, J. H., and Jeffords, R. M.
1975 : Cenozoic planktonic foraminiferal zonation and characteristics of index species: University of Kansas Paleontology Contributions, Art. 62, 425 p.
Steinmetz, J. C.
1979 : Calcareous nannofossils from the North Atlantic Ocean, Leg 49, Deep Sea Drilling Project, in Initial Reports of the Deep Sea Drilling Project, Volume 49; eds. B. P. Luyendyk, and J. R. Cann; U.S. Government Printing Office, Washington, D.C., p. 519–532.
Supko, P. R., and Perch-Nielsen, K., eds.
1977 : Initial Reports of the Deep Sea Drilling Project, Volume 39: U.S. Government Printing Office, Washington, D.C., 1139 p.
Tjalsma, R. C., and Lohnman, G. P.
1983 : Paleocene-Eocene bathyal and abyssal benthic foraminifera from the Atlantic Ocean: Micropaleontology, Special Publication, no. 4, 90 p.
Tucholke, B. E.
1979 : Relationships between acoustic stratigraphy and lithostratigraphy in the western North Atlantic: in Initial Reports of the Deep Sea Drilling Project, Volume 43; eds. B. E. Tucholke, and P. R. Vogt; U.S. Government Printing Office, Washington, D.C., p. 827–846.
Tucholke, B. E. and Mountain, G. S.
1979 : Seismic stratigraphy, lithostratigraphy and paleosedimentation patterns in the North American Basin: in Deep Drilling Results in the Atlantic Ocean: Continental Margins and Paleoenvironments; eds. M. Talwani and W. W. Hay; American Geophysical Union, Washington, D.C., M. Ewing Series, v. 3, p. 58–86.
Tucholke, B. E., and Vogt, P. R.
1979a: Western North Atlantic: sedimentary evolution and aspects of tectonic history; in Initial Reports of the Deep Sea Drilling Project, Volume 43; eds. B. E. Tucholke and P. R. Vogt; U.S. Government Printing Office, Washington, D.C., p. 791–826.
Tucholke, B. and Vogt, P. R., eds.
1979b: Initial Reports of the Deep Sea Drilling Project, Volume 43: U.S. Government Printing Office, Washington, D.C., 1115 p.
Vail, P. R., Mitchum, R. M., Todd, R. G., Widimier, J. M., Thompson, S. III, Sangree, J. G., Bubb, J. N., and Hatlelid, W. G.
1979 : Seismic stratigraphy and global changes of sea level: in Seismic stratigraphy applications to hydrocarbon exploration, ed. C. E. Payton; American Association of Petroleum Geologists, Memoir 26, p. 49–250.
Valentine, P. C.
1977 : Nannofossil Biostratigraphy: in Geological Studies on the COST No. B-2 well, U.S. Mid-Atlantic Outer Continental Shelf Area; ed. P. A. Scholle; U.S. Geological Survey Circular, No. 750, p. 37–40.

Vergnaud-Grazzini, C., Pierre, C. and Letolle, R.
1980 : Paleoenvironment of the northwest Atlantic during the Cenozoic; Oxygen and carbon isotope analysis at the Deep Sea Drilling Project Sites 398, 400A and 401: Oceanologica Acta, v. 1, p. 381–390.

Weaver, F. M. and Dinkelman, M. G.
1978 : Cenozoic radiolarians from the Blake Plateau and the Blake-Bahama Basin, DSDP Leg 44, in Initial Reports of the Deep Sea Drilling Project, Volume 44; eds. W. E. Benson and R. E. Sheridan; U.S. Government Printing Office, Washington, D.C., p. 865–886.

Wilcoxon, J. A.
1972 : Calcareous nannoplankton ranges, Leg XI, Deep Sea Drilling Project; in Initial Reports of the Deep Sea Drilling Project, Volume 11; eds. C. D. Hollister and J. I. Ewing; U.S. Government Printing Office, Washington, D.C., p. 459–474.

Manuscript Accepted by the Society August 19, 1985

ACKNOWLEDGMENTS

We would like to thank B. Tucholke, A. R. Palmer and W. A. Berggren for their improvements to this chapter. We are also grateful to Paolo Malinverno, Silvia Aghib, Maurizio Orlando, and Elizabetta Erba for providing technical assistance.

Printed in U.S.A.

The Geology of North America
Vol. M, The Western North Atlantic Region
The Geological Society of America, 1986

Chapter 33

Neogene marine microfossil biofacies of the Western North Atlantic

C. Wylie Poag
U.S. Geological Survey, Woods Hole, Massachusetts 02543
Kenneth G. Miller
Lamont-Doherty Geological Observatory, Palisades, New York 10964

INTRODUCTION

The western North Atlantic Ocean was one of the first areas to be explored by the Deep Sea Drilling Project (DSDP; Leg 1, 1968). Subsequently, *Glomar Challenger* returned for 17 additional Legs (2, 4, 11, 12, 37, 39, 43, 44, 45, 49, 51, 76, 78, 82, 93, 94, 95; Fig. 1). In addition to deep-sea drilling, more than 200 gravity and piston cores, collected mostly by the Lamont-Doherty Geological Observatory, have sampled Neogene sediments in the western North Atlantic (Fig. 2). In the shallower water of the continental shelf, more than 100 boreholes have sampled the Neogene sedimentary section (Ascoli, 1976; Poag 1978, 1985, *in* Scholle, 1982; and Poag *in* Scholle and Wenkam, 1982; Hathaway and others, 1979; Barss and others, 1979; see Fig. 1). This paper summarizes microfossil biofacies data chiefly from DSDP Sites as published in the *Initial Reports* of that project. At the time of writing, detailed data from DSDP Legs 78, 82, 93, 94, and 95 had not yet been published. The *Initial Reports* data are supplemented by a few key published syntheses. Temporal and spatial composition and distribution of deep-sea zooplanktonic, phytoplanktonic, nektonic, and benthic biofacies (sediment assemblages) are briefly discussed and compared with data from North American continental margin sites (Fig. 1). We point out the relationships of the composition and distribution of biofacies to paleoclimates, paleoceanography, sea-level changes, and sedimentary processes. Using a narrow definition of the term Neogene, we limit our discussion to Miocene and Pliocene microfossils. For a dicussion of Quaternary microfossils, the reader should consult the DNAG Quaternary volume. The term biofacies has been defined in a number of different ways (e.g., Moore, 1949; Weller, 1958). We use it in a broad sense to refer to a distinctive assemblage of fossil organisms (Bates and Jackson, 1980, p. 66).

DATA BASE

The study area encompasses the western North Atlantic Ocean from 10 to 65°N latitude, bounded on the west by the coastline of North America and the Caribbean island arc and on the east by the axis of the Mid-Atlantic Ridge. It includes the Irminger, Labrador, Newfoundland, Scotian, Georges Bank, Blake-Bahama, and North American basins and the Carolina and Baltimore Canyon troughs.

Of 69 DSDP sites in this region, 49 yielded Neogene sediments (Fig. 1). However, the record of Neogene microfossil biofacies is quite incomplete because of discontinuous coring (especially during early cruises), generally poor core recovery, mechanical disturbance by rotary coring techniques, intervals of carbonate depletion, presence of many unconformities, and a propensity to drill on basement highs. At no single site has a continuous, undisturbed column of Neogene sediments been recovered. A similar incomplete Neogene record exists for more than 60 boreholes on the shelf and slope. Table 1 shows that upper Miocene facies are most frequently represented in the deep sea, whereas lower and middle Miocene facies are best represented on the margin; Pliocene facies have been sampled equally at margin and deep sea sites. The early Neogene record at many deep-sea sites has been interrupted by intervals of erosion and nondeposition caused by bottom currents (Tucholke and Mountain, 1979, this volume; Miller and Tucholke, 1983). The sedimentary record on the shallow continental margin has been modified in part by sea level changes (Poag and Schlee, 1984). The more complete middle Miocene section there may be the result of a high stand of sea level (Vail and Hardenbol, 1979; Poag, 1985).

Geographic distribution of DSDP Neogene sites is disproportionately biased toward mid latitudes (20°–40°N, 36 sites).

Poag, C. W., and Miller, K. G., 1986, Neogene marine microfossil biofacies of the Western North Atlantic; *in* Vogt, P. R., and Tucholke, B. E., eds., The Geology of North America, Volume M, The Western North Atlantic Region: Geological Society of America.

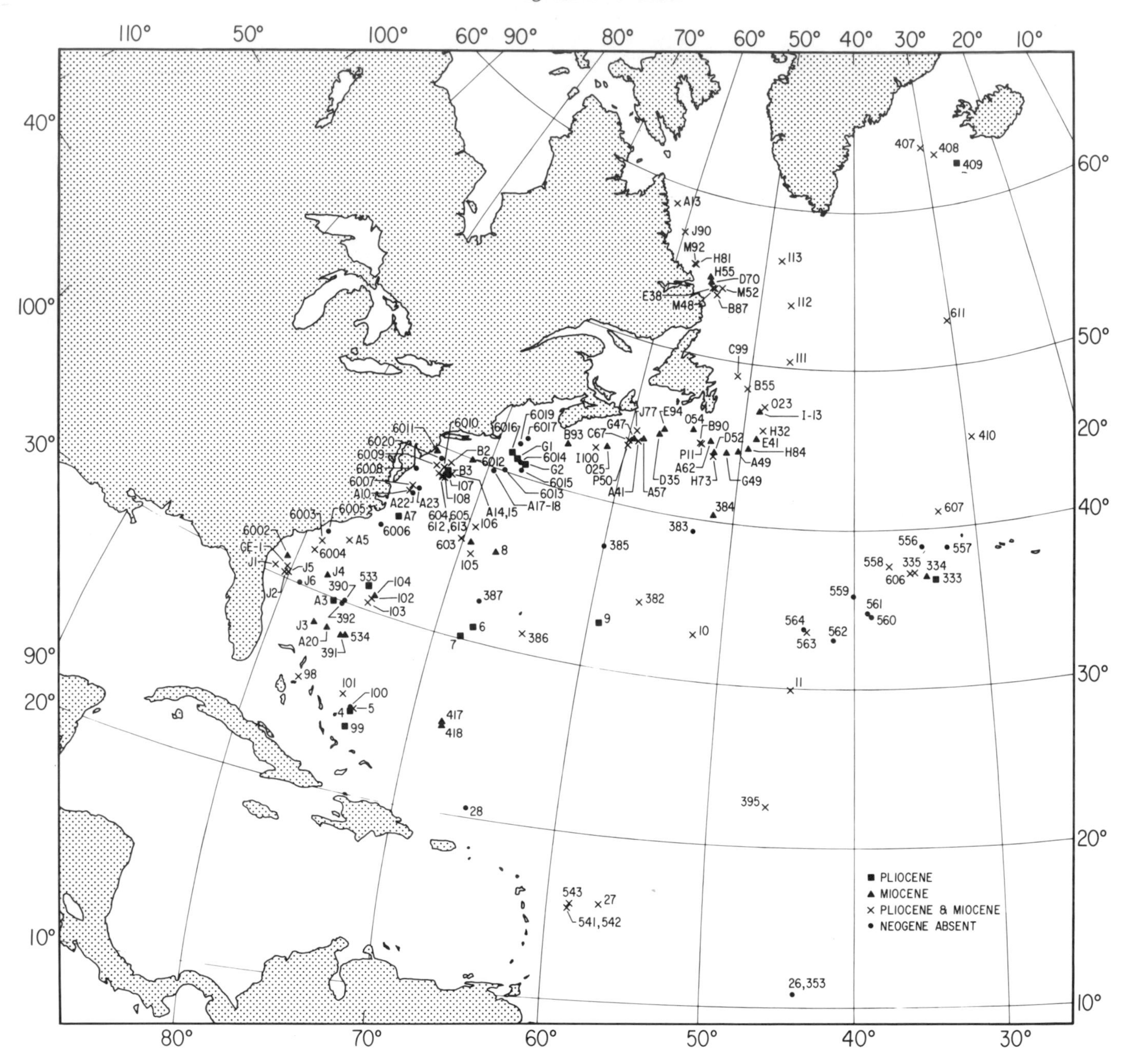

Figure 1. Location of DSDP and North American margin sites where Neogene sediments have been cored.

Nine sites lie above 40°N and four lie between 10° and 20°N; abyssal plain locations generally have been avoided. Planktonic foraminifera and calcareous nannofossils have been investigated in some detail at all sites where present, but other groups of microfossils have received little or no attention (Table 1).

ZOOPLANKTON

Calcareous Biofacies

The most thoroughly studied group of fossil marine zooplankton are the planktonic foraminifera, a diverse community of carbonate-shelled protistans that inhabit principally the upper 150 m of open oceanic waters (Boersma *in* Haq and Boersma, 1978; Fairbanks and others, 1980). Their great abundance, oceanwide distribution and sensitivity to water-mass properties, such as salinity, temperature, nutrients, and oxygen content, provide reliable means of estimating past environmental conditions, including climatic and oceanographic aspects. Their distinctive morphologic characteristics and relatively rapid rates of evolution have established them as a principal group for stratigraphic zona-

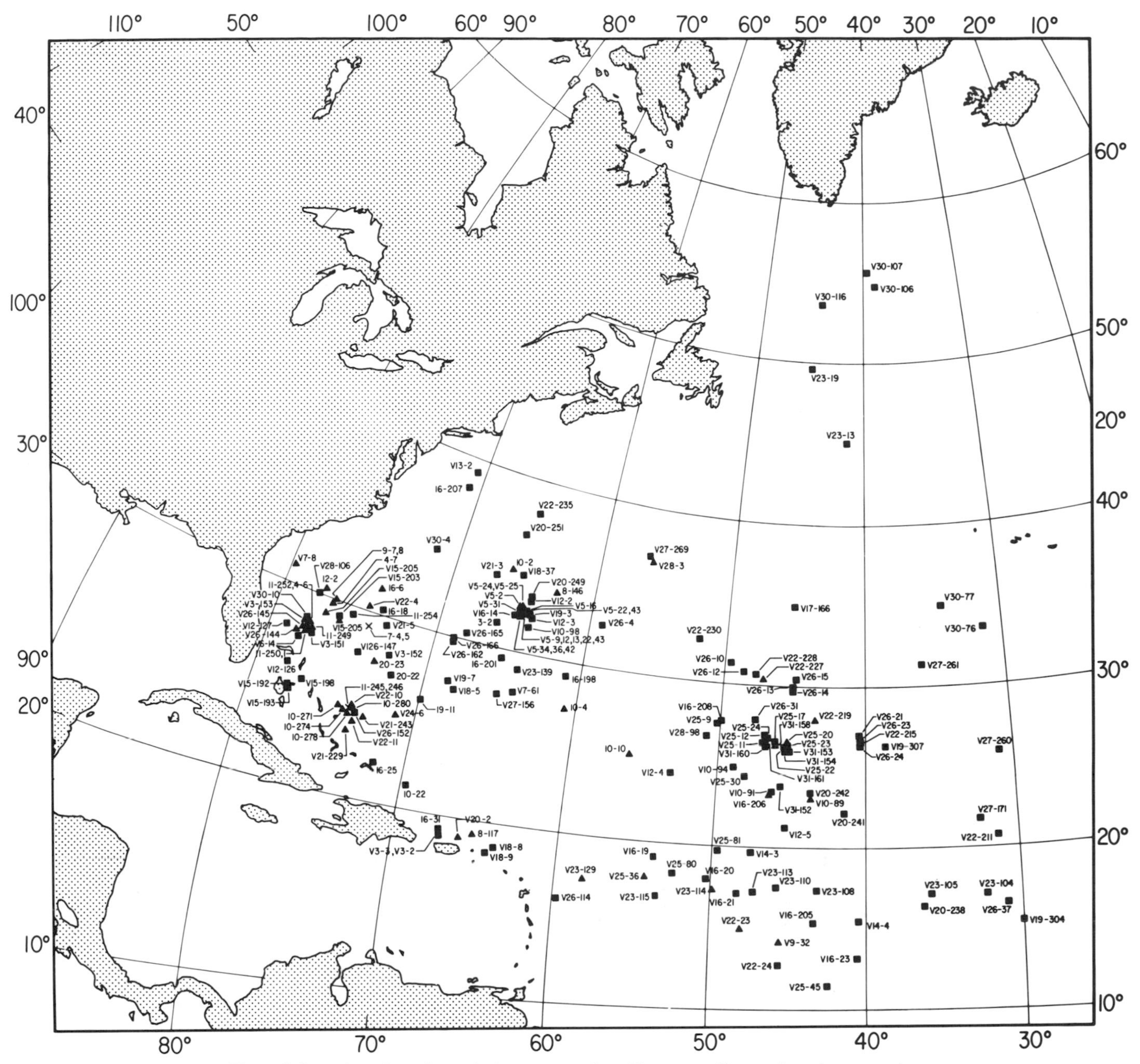

Figure 2. Location of gravity and piston cores where Neogene sediments have been cored.

tion of marine deposits and for fine-scale documentation of organic evolution (e.g., Malmgren and Kennett, 1980). Planktonic foraminifera are one of two chief groups analyzed since the beginning of oceanic microfossil studies (the other being calcareous nannoplankton). Planktonic foraminifera have been reported in varying detail from each Neogene unit reviewed here (Table 1).

Although planktonic foraminifera have been more extensively studied than most other microfossil groups, the number of sample sites is small relative to the size of the area under review. This makes biofacies synthesis extremely difficult. Nevertheless, two summaries have been published that emphasize paleobiogeographic aspects of Neogene planktonic foraminifera and provide general biofacies information. In the first summary, Thunell and Belyea (1982) subjected presence-absence data to Q-mode Varimax Factor Analysis, producing distribution maps of Neogene planktonic foraminiferal assemblages in the Atlantic Ocean. Their placement of assemblage boundaries within our review area is tenuous due to the sparsity of control points, so we have reinterpreted their data in a more conservative fashion (Fig. 3). Where possible, we have added qualitative data from margin

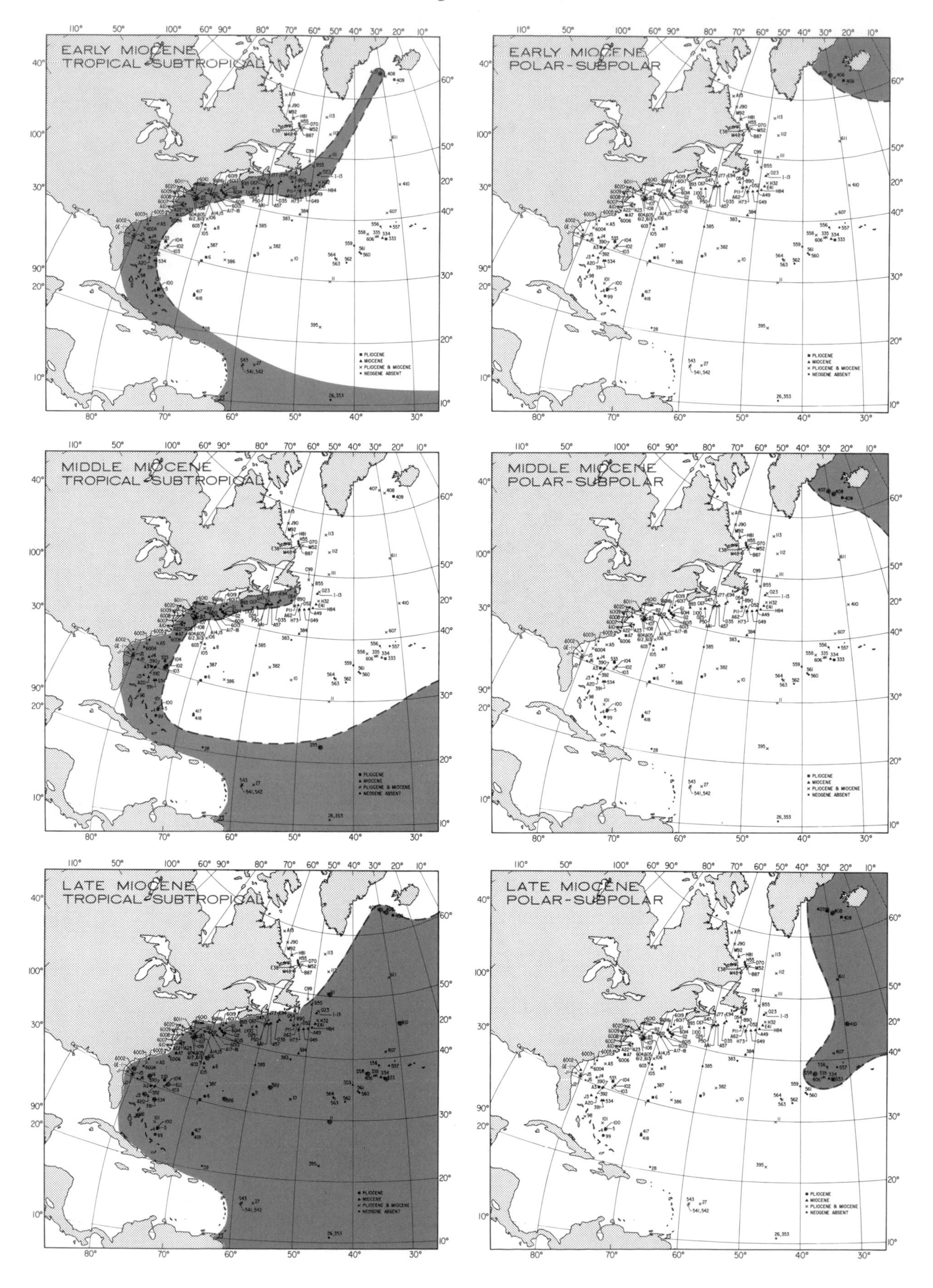
EARLY MIOCENE
TROPICAL-SUBTROPICAL
EARLY MIOCENE
POLAR-SUBPOLAR
MIDDLE MIOCENE
TROPICAL-SUBTROPICAL
MIDDLE MIOCENE
POLAR-SUBPOLAR
LATE MIOCENE
TROPICAL-SUBTROPICAL
LATE MIOCENE
POLAR-SUBPOLAR
PLIOCENE
MIOCENE
PLIOCENE & MIOCENE
NEOGENE ABSENT

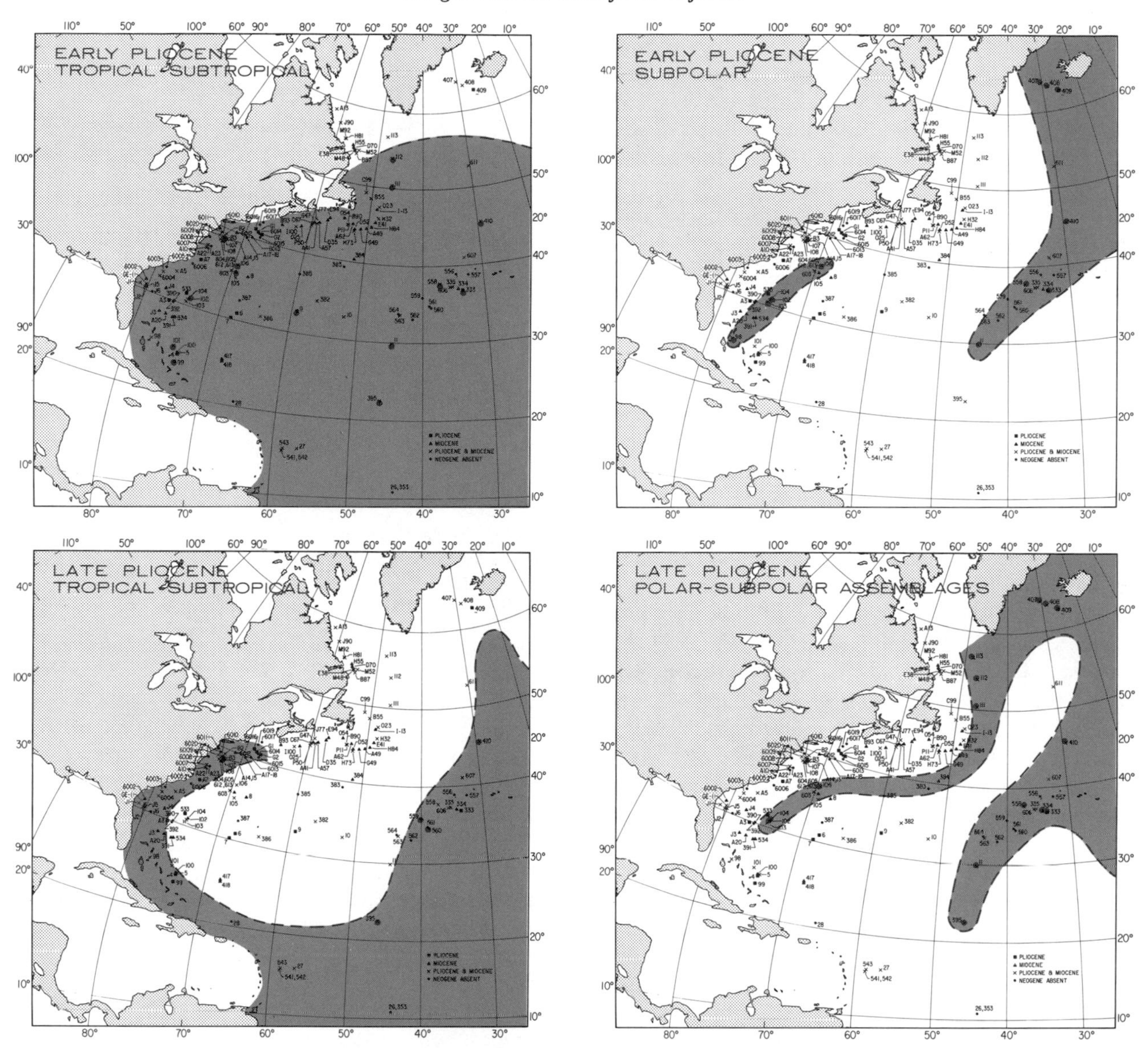

Figure 3 (this and facing page). Spatial distribution of planktonic foraminiferal biofacies in Neogene sediments (modified from Thunell and Belyea, 1982). Red circles are control points of Thunell and Belyea; light red pattern shows conservative estimate of biofacies distribution. A) Early Miocene tropical-subtropical foraminifera. B) Early Miocene polar-subpolar foraminifera. C) Middle Miocene tropical-subtropical foraminifera. D) Middle Miocene polar-subpolar foraminifera E) Late Miocene tropical-subtropical foraminifera H) Early Pliocene subpolar foraminifera I) Late Pliocene tropical-subtropical foraminifera J) Late Pliocene polar-subpolar foraminifera.

sites.

Thunell and Belyea (1982) mapped a widespread tropical-subtropical assemblage (we will call it a biofacies) in the western North Atlantic during the early Miocene. Predominant species are *Globigerinoides trilobus, Globorotalia peripheroronda,* and *Gl. siakensis.* However, they reported this biofacies from only one site in our review area (high-latitude Site 407; Fig. 3). Along the margin, though, similar biofacies are scattered sparsely from the Blake-Bahama basin to the Grand Banks (Poag, 1978; Poag *in* Scholle, 1980; Poag *in* Scholle and Wenkam, 1982; Ascoli, 1976; Gradstein and Srivastava, 1980; Gradstein and Agterberg, 1982). A transitional (temperate) biofacies characterized by *Globoquadrina dehiscens, Gq. altispira* and *Globigerinoides primordius* was extrapolated by Thunell and Belyea across the area of our review, but was not actually recorded at any site. A polar-subpolar biofacies (*Globigerina bulloides, G. glutinata,* and *Globorotalia miozea*) was also recorded by Thunell and Belyea at Site 407, accompanying the tropical-subtropical biofacies. The presence of both cold and warm water biofacies in the early Miocene at Site 407 indicates either that conditions changed drastically at this site sometime during this 7 m.y. interval, or that currents produced mixed biofacies. Thunell and Belyea do not provide sufficient information to resolve this matter.

A study by Berggren (1986) also notes widespread occurrences of rather uniform assemblages of planktonic foraminifera (chiefly globoquadrinids and catapsydracids), indicating little provincialization of this microfossil group during the early Miocene (see also Berggren and Olsson, this volume).

Middle Miocene tropical-subtropical biofacies (chiefly *Globigerinoides ruber* and *Globorotalia menardii*) were not as widespread (Thunell and Belyea 1982) in the open ocean, having been recorded at one, or possibly two, low-latitude sites in our study area (395 and 102?). But along the margin, tropical-subtropical biofacies (typified by *Globorotalia siakensis, Gl. peripheroronda,* and *Globigerinatella insueta*) are known from the tropics to the southern Grand Banks (Bartlett and Smith, 1970; Ascoli, 1976; Poag *in* Scholle, 1980; Poag and Hall, 1978; Gradstein and Srivastava, 1980; Gradstein and Agterberg, 1982; Poag, 1985).

Thunell and Belyea (1982) recognized a transitional middle Miocene biofacies (*Globigerina bulloides, G. nepenthes, Globigerinita glutinata, Globorotalia scitula,* and *Gl. miozea*) at two high-latitude sites (407, 408) and one mid-latitude site (102). A polar-subpolar biofacies (*Neogloboquadrina pachyderma,* both sinistral and dextral) was also recorded at 407 and 408. These authors recognized the first "true" Neogene polar biofacies in middle Miocene sediments and traced elements of it through eastern Atlantic sites to the northwest African margin. Such an expansion of polar associations is not recorded in our review area; nevertheless, increasing latitudinal differentiation of biofacies apparently took place in the middle Miocene.

Berggren (1986) also describes a developing trend toward provinciality among planktonic foraminifera of the middle Miocene. His analysis shows clearly the low-latitude development of the *Globorotalia fohsi* biofacies while the *Globorotalia miozea* biofacies evolved in mid- to high-latitudes.

During the late Miocene a tropical-subtropical biofacies (*Sphaeroidinellopsis seminulina, S. subdehiscens, Globoquadrina altispira, Orbulina universa, Globigerina nepenthes, Neogloboquadrina acostaensis*) expanded as far north as Sites 407 and 408 (Fig. 3; Thunell and Belyea, 1982). Fluctuating paleoceanographic conditions or current-induced mixing are indicated by polar-subpolar biofacies (*Neogloboquadrina pachyderma, Globigerina bulloides, Globigerinita glutinata*) that were recorded (along with tropical-subtropical assemblages) as far south as 37°N along the western flanks of the Mid-Atlantic Ridge. In fact, Poore (1981) interpreted the predominance of *Neogloboquadrina* at mid-latitude Site 410 (45°N) as an indication that subpolar conditions prevailed there throughout the late Miocene. Neogene biofacies that are transitional from tropical-subtropical to polar-subpolar, are also documented in upper Miocene sediments. Berggren (1986) shows that a temperate biofacies including *Globorotalia miozea, Gl. conoidea,* and *Gl. conomiozea* was rather widespread from 25° to 65°N. This biofacies has also been recorded from the COST B-3 well off New Jersey, and as far east as the flanks of the Mid-Atlantic Ridge (DSDP Sites 333, 334, 410). The general distribution pattern mimics that of Thunell and Belyea's transitional biofacies, and substantiates their recognition of increased latitudinal provincialization in mid- and high-latitudes. On the Canadian margin, the southward flowing, cold-water Labrador Current developed in the late Miocene, and that allowed polar-subpolar biofacies to occupy the shelf and slope of North America as far south as about 40°N. (Gradstein and Srivastava, 1980; see Fig. 3).

TABLE 1. TABULATION OF DSDP AND CONTINENTAL MARGIN SITES CONTAINING DATA REFERRED TO IN THIS PAPER

Section Studied	No. of DSDP Sites	No. of Continental Margin Sites
Pliocene	39	39
Miocene		
Upper	29	32
Middle	21	57
Lower	18	41
Microfossils Studied		
Planktonic forams	49	>60
Nannofossils	49	>60
Pteropods	0	0
Radiolarians	15	15
Diatoms	8	15
Silicoflagellates	6	3
Dinoflagellates	5	33
Ebridians	0	0
Benthic forams	14	14
Ostracodes	0	0
Sponge spicules	4	0
Ichthyoliths	4	1

According to Thunell and Belyea (1982), early Pliocene tropical-subtropical biofacies (*Globorotalia menardii, Gl. crassaformis, Globoquadrina altispira, Sphaeroidinellopsis seminulina, Sphaeroidinella dehiscens*) retreated somewhat to the south, failing to reach north of 55°N. (Fig. 3). A transitional biofacies (*Globigerina inflata, G. bulloides, Globorotalia scitula, Orbulina universa* and *Neogloboquadrina acostaensis*) was again well developed. Polar-subpolar biofacies (unchanged in species composition from late Miocene assemblages) spread southward as far as 25°N. and are recorded in the west central part of our review area for the first time in the Neogene. In Berggren's opinion (1986), the first truly Arctic biofacies developed near the end of the early Pliocene.

During the late Pliocene, tropical-subtropical biofacies (*Globigerinoides ruber, Globorotalia crassaformis, Gl. menardii, Gl. truncatulinoides, Orbulina universa, Pulleniatina obliquiloculata, Sphaeroidinella dehiscens*) were present along the eastern and western edges of the western Atlantic, but no central sites record them (Fig. 3; Thunell and Belyea, 1982). A transitional biofacies (*Globigerina inflata, G. bulloides, Globorotalia crassa-*

formis, Orbulina universa) was still well established in the deep sea, and polar-subpolar biofacies (composition as before, with the addition of *Globigerina inflata*) extended to nearly 20°N along the western flank of the Mid-Atlantic Ridge. Two occurrences of this biofacies near the western edge of the basin may be southward displacements caused by sediment transport in the deep, southward flowing Western Boundary Undercurrent.

A group of endemic planktonic foraminifera (*Globorotalia miocenica* group; Kaneps, 1970) evolved in the Atlantic during the late Pliocene. Their evolution attests to increased isolation from Pacific populations following closure of the Straits of Panama.

In summary, (Figs. 3, 4), Thunell and Belyea (1982) and Berggren (1986) agree that distinctive latitudinal climatic and oceanographic provinces, characterized by distinctive planktonic foraminiferal biofacies, developed in the middle Miocene, and became increasingly marked throughout the Neogene. Neogene biofacies provincialization culminated in the late Pliocene when polar-subpolar biofacies extended to low latitudes, reflecting increasingly steep latitudinal thermal gradients induced by climatic cooling (see Berggren and Olsson, this volume).

Siliceous Biofacies

Radiolarians are the best studied siliceous group of microfossils in the western North Atlantic. Members of this group of protistans live generally as single-celled individuals and build latticed, spongy, or perforate-plate skeletal structures of opaline silica (Kling *in* Haq and Boersma, 1978). Like their calcareous counterparts, the planktonic foraminifera, radiolarians inhabit chiefly the upper 150–300 m of the oceanic water column and their distribution is strongly influenced by circulation patterns and water-mass composition. Radiolarians provide a major means of biozonation, their skeletons replacing foraminifera in importance in such environments as upwelling zones and where deposition takes place below the calcite compensation depth (CCD). In the deep North Atlantic, radiolarians are relatively scarce in Neogene sediments because bottom waters appear to have been relatively depleted in silica (Casey and McMillen, 1977; Kling *in* Haq and Boersma, 1978). On the North American margin, however, radiolarians are abundant in areas where upwelling presumably took place (Poag, 1980, 1985; Palmer, 1983).

Up to now, radiolarians have not received much attention in the western North Atlantic (Table 1). This sparsity of data appears to be due largely to lack of study, and it probably does not reflect true distribution patterns. Radiolarians are present in at least one site during each Neogene interval, but they are best known in deep-sea high-latitude sites (407, 408) and near the Blake-Bahama Outer Ridge (391, 102–104). On the continental margin, radiolarians are especially prominent (and associated with abundant diatoms) in middle Miocene strata of the U.S. middle Atlantic states (Poag, 1985; Poag, *in* Scholle, 1980; Palmer, 1983; Fig. 5).

A second group of siliceous zooplankton, the silicoflagellates, has received minimal attention in the North Atlantic (Bukry *in* Warme and others, 1981). Silicoflagellates are minute (20–50 μm), unicellular flagellates having skeletons of hollow opaline spines or rings (Haq *in* Haq and Boersma 1978). Most species are thought to have been photosynthetic, which would limit their living distribution to the euphotic zone (upper 150 m). They constitute a relatively minor part of the zooplankton, and like radiolarians, are most abundant at high latitudes and below the CCD in more equatorial regions. The main research efforts with this group have focused on interpretations of paleoclimates, (paleoceanography), paleoproductivity, and biozonation (Bukry *in* Warme and others, 1981). However, low evolutionary rates reduce the relative effectiveness of silicoflagellates for biozonation (Haq *in* Haq and Boersma 1978). In the western North Atlantic, silicoflagellates have been reported at 6 DSDP sites and 3 margin sites (Table 1; Fig. 5). Wider distribution probably will be revealed by further study. This group has been reported from each stratigraphic interval considered here, but the best data come from mid- to high-latitude sites. The species diversity of silicoflagellates appears to have been higher during the Miocene than in the rest of the Cenozoic (Bukry *in* Warme and others 1981), thereby giving Miocene sediments the greatest potential for productive silicoflagellate research.

PHYTOPLANKTON

Calcareous Biofacies

The best known group of Neogene calcareous phytoplankton are the calcareous nannofossils; these are unicellular algae whose cells secrete a fossilizable skeleton of minute (1–15 μm) calcareous shields (Haq *in* Haq and Boersma 1978). These skeletal parts fall rapidly to the seafloor when incorporated in zooplankton fecal pellets (Honjo, 1976). The fecal pellets disintegrate among the sediments, releasing the nannofossils, which form the major constituents of many deep-sea oozes. Most nannofossils are believed to have been photosynthetic, a condition that would limit their living distribution to the euphotic zone (100–150 m maximum depth), where their distribution would be influenced by current patterns and water mass composition. They have been most abundant in high-latitude upwelling zones and narrow equatorial belts. They are best known as a means of precise biozonation, but also are important for interpreting evolution, productivity, paleoclimatology (paleoceanography), and paleobiogeography. In common with other calcareous fossils, they are subject to dissolution below the CCD and are especially susceptible to early diagenetic alteration and redistribution by currents.

Calcareous nannofossils, like planktonic foraminifera, have been recorded at every DSDP site where Neogene sediments were sampled, and they have been intensely studied (Table 1; Fig. 4). A few margin sites also have been thoroughly analyzed, but the published record is scant for that area (Valentine *in* Scholle, 1977; Valentine, 1979, and Valentine *in* Scholle and Wenkam, 1980). One broad synthesis of Neogene distribution patterns has

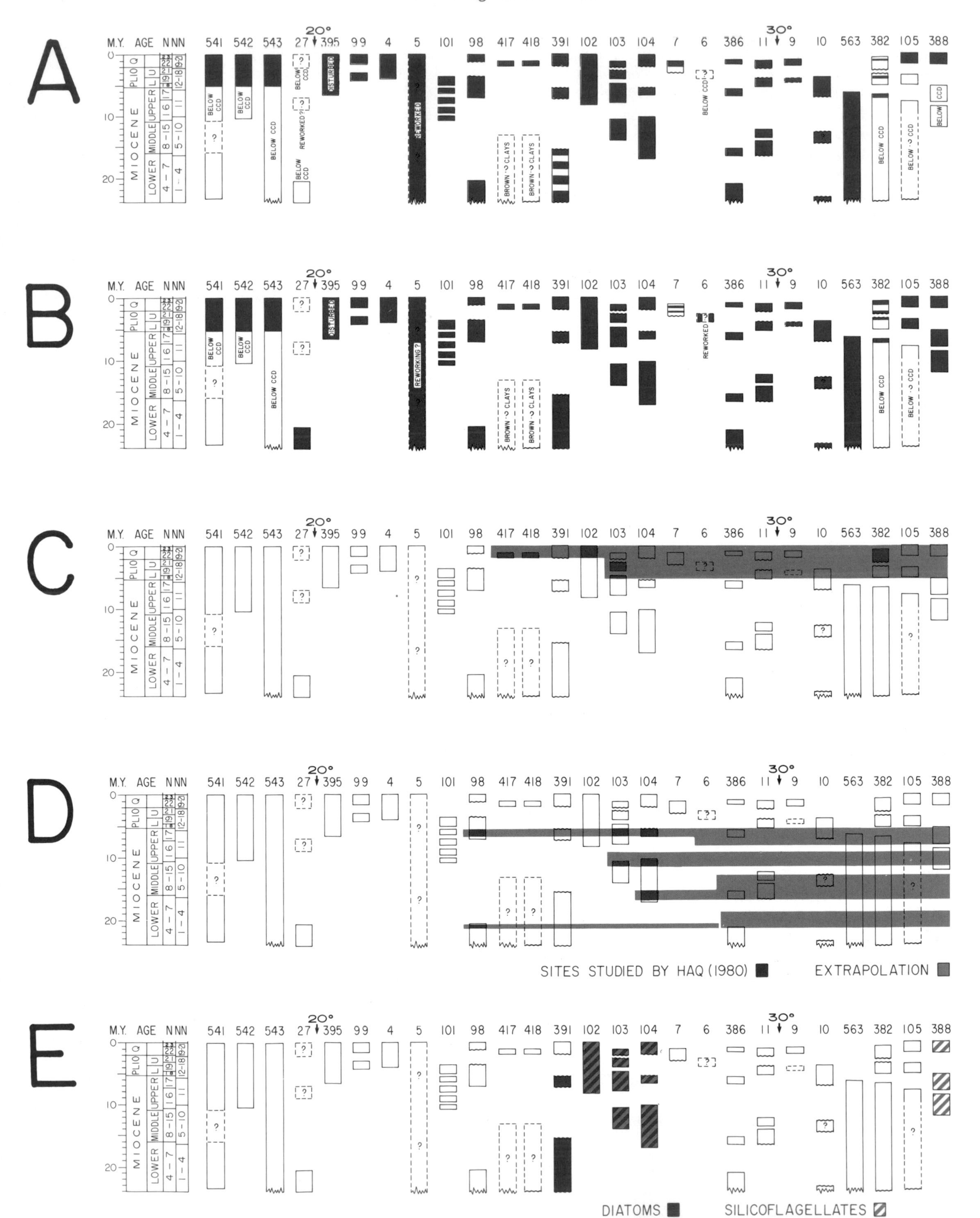
A
B
C
D
E
M.Y.
AGE
N
NN
PLIO
Q
MIOCENE
LOWER
MIDDLE
UPPER
20°
30°
541
542
543
27
395
99
4
5
101
98
417
418
391
102
103
104
7
6
386
11
9
10
563
382
105
388
BELOW CCD
DISTURBED
REWORKED
REWORKED?
REWORKING?
BROWN CLAYS
SITES STUDIED BY HAQ (1980)
EXTRAPOLATION
DIATOMS
SILICOFLAGELLATES

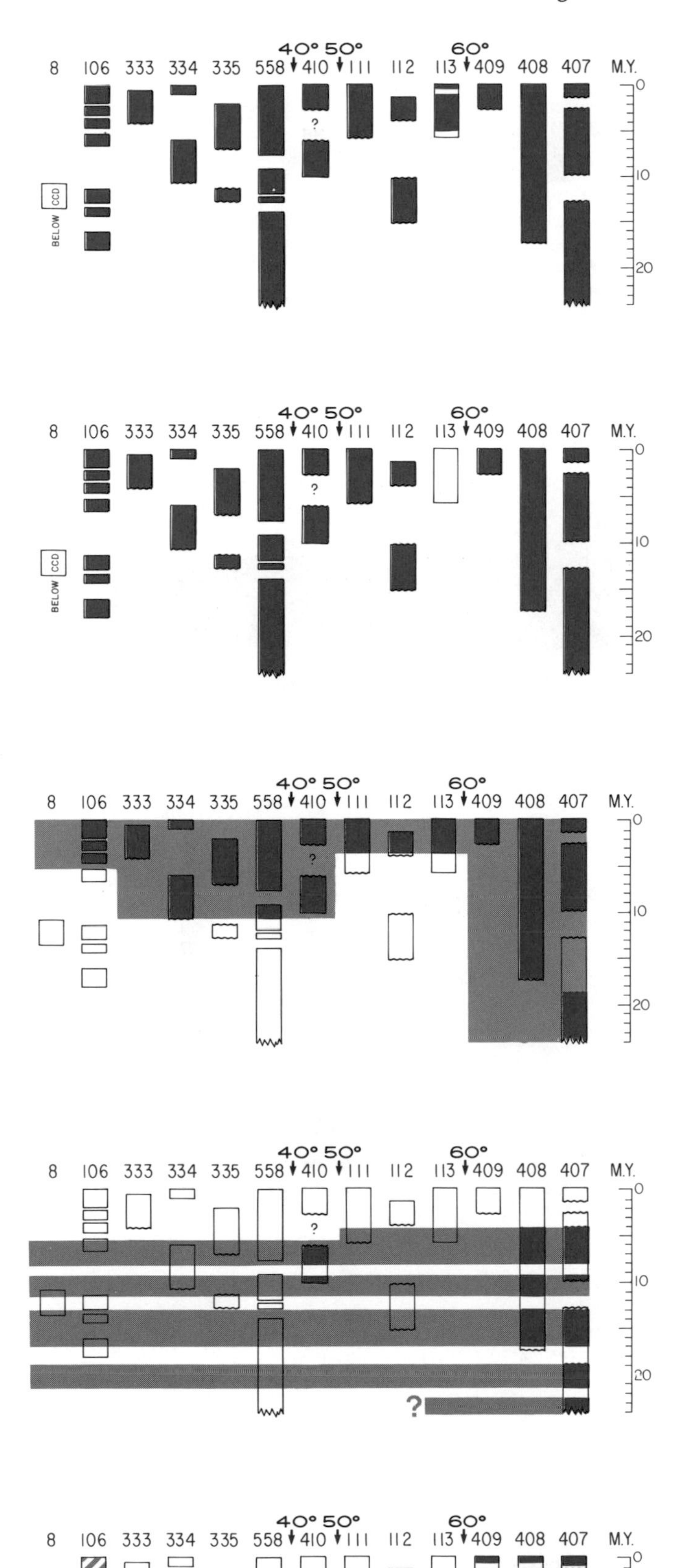

been published (Haq, 1980), based on Q-mode Varimax Factor Analysis of species census data. Haq found great similarities between the development of Neogene nannoplankton provinciality and that of planktonic foraminifera, although different sampling and analytical strategies may have produced some differences in detail. Rather uniform, widespread earliest Miocene biofacies, dominated by *Cyclicargolithus floridanus,* gave way to a latitudinal arrangement of four biofacies later in the early Miocene. The complexity of biofacies increased in mid-latitudes throughout the remaining Neogene. The warm, low-latitude biofacies was consistently characterized by *Discoaster* and *Sphenolithus.* The mid-latitudes of the western North Atlantic contained a *Cyclicargolithus floridanus* biofacies during the late early Miocene, a *Reticulofenestra pseudoumbilica–R. haqi* biofacies in the latest middle Miocene and a *Discoaster minutus* biofacies throughout the late Miocene. Cold high latitudes were continuously dominated by a *Coccolithus pelagicus* biofacies during the Miocene. Haq studied nine sites in the western North Atlantic: three high-latitude (Sites 407, 408, 418); five mid-latitude (98,102, 103, 4, 10), and one low-latitude (27). The high-latitude *C. pelagicus* biofacies expaned southward five times during the Neogene, reaching as far as Site 98 (Bahama Banks) during the middle of the early Miocene and during the latest Miocene (Fig. 4D). These southward expansions of early and late Miocene high-latitude (cold-water) nannofossil biofacies preceded the maximum southward expansion of northern planktonic foraminiferal biofacies, which took place in the late Pliocene (Thunell and Belyea, 1982; Berggren, 1986; Berggren and Olsson, this volume).

A second group of calcareous microfossils believed to have belonged to the phytoplankton are members of the extinct genus *Bolboforma* (Daniels and Spiegler, 1974). The small (150–75 μm diameter) unilocular or bilocular tests of these organisms are abundant in upper Eocene to lower Pliocene strata at several sites along the margins of the North Atlantic and in a few marine basins of northwestern Europe (Poag and Karowe, 1986). Only one site yielding *Bolboforma* has been documented outside the North Atlantic region (Bellingshausen Sea; Rögl and Hochuli, 1976), and the genus is not known from latitudes lower than 25 degrees. Although current knowledge of this group is rudimentary, a variety of distinctive morphologies exists among the sixteen known species, promising considerable biostratigraphic, paleogeographic, and paleoenvironmental utility in the near future.

Figure 4 (this and facing page). A) Reported temporal and spatial distribution of planktonic foraminiferal biofacies in DSDP sites; B) Nannofossil biofacies in DSDP sites; C) Polar-subpolar planktonic foraminiferal biofacies (modified from Thunell and Belyea, 1982); D) Polar-subpolar nannofossil biofacies (modified from Haq, 1980); E) Diatom and dinoflagellate biofacies at DSDP sites. All are sediment biofacies. N = Neogene planktonic foram biozones; NN = calcareous nannoplankton biozones. Blank columns indicate absence of the cited fossil groups.

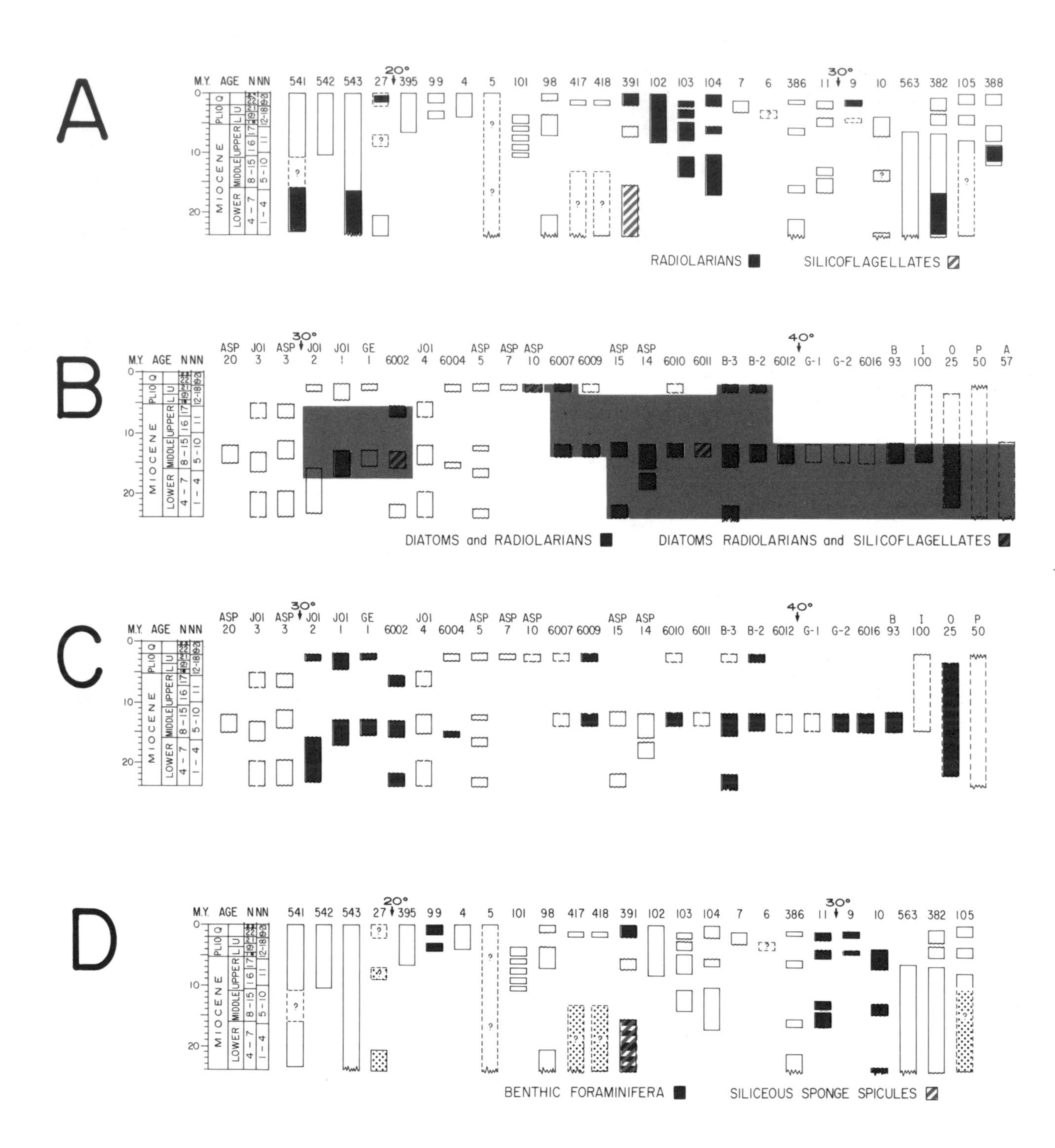

Figure 5 (this and facing page). A) Reported temporal and spatial distribution of radiolarian and silicoflagellate biofacies at DSDP sites; B) radiolarian, diatom, and silicoflagellate biofacies at margin sites; C) benthic foraminiferal biofacies at DSDP sites; D) benthic foraminiferal, siliceous sponge spicule, and ichthyolith biofacies at DSDP sites. All are sediment biofacies. Blank columns indicate absence of the cited fossil groups.

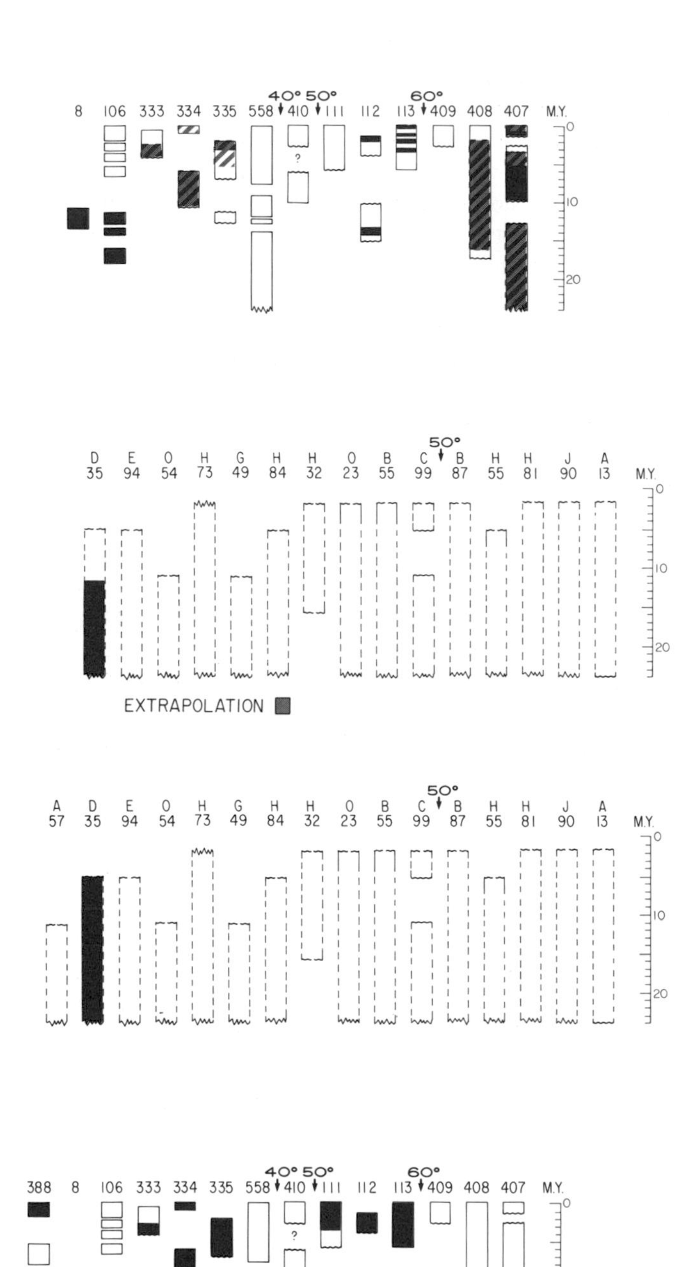

Siliceous Biofacies

Diatoms are the dominant group of Neogene siliceous marine phytoplankton. The silicified frustules of these photosynthetic, single-celled algae constitute a major part of all marine phytoplankton, and they are especially abundant in upwelling regions, notably at high latitudes. However, their study has lagged behind that of foraminifera, calcareous nannoplankton and radiolarians (Burkle *in* Haq and Boersma, 1978). While their paleoclimatic and biostratigraphic utility in the Pacific (Burkle, 1978; Donahue, 1970; Harper, 1977) and southern oceans (Weaver and Gombos *in* Warme and others, 1981) is well known, the published record of North Atlantic Neogene diatoms is sparse (Figs. 4, 5). Diatoms have especially important potential for the study of paleocurrents, particularly bottom currents, because the frustules are easily resuspended and transported. Diatoms, like radiolarians, are susceptible to dissolution by silica-depleted water masses such as those of the Neogene North Atlantic (Thiede and others, *in* Warme and others, 1981).

Planktonic diatoms have been noted at most North Atlantic sites where radiolarians were recorded (Table 1). The sparsity of control points is partly a product of limited research on this group of opaline phytoliths. Most Neogene intervals contain diatom records from at least two sites in the western North Atlantic. The deep-sea record is too sparse to reveal distribution trends, but along the margin, diatoms are particularly abundant in middle Miocene strata from Florida to New Jersey, (Fig. 5; *Melosira westii–Paralia sulcata–Thalassionema nitzschioides* biofacies) where they were associated with upwelling nutrient-rich waters (Abbott, 1980; Poag *in* Scholle, 1980).

Organic-Walled Biofacies

A potentially important group of organic-walled Neogene marine microfossils are the dinoflagellates. These small to relatively large (5 μm–2 mm) forms constitute the encysted stage of unicellular, photosynthetic algae (some of which also ingest food particles), which are coated with a resistant organic substance (sporopollenin) (Williams *in* Haq and Boersma, 1978). Living motile stages inhabit the euphotic zone of both marine and fresh waters, and therefore they are distributed according to current and water-mass patterns. They are very abundant among modern oceanic phytoplankton, being second only to diatoms. They have been useful in biozonation and in studies of paleoenvironments, paleobiogeography, and organic evolution. Their wide distribution provides a valuable means of correlating marginal marine strata with their open-marine equivalents.

Dinoflagellates have been recorded at most Canadian margin drilling sites (Barss and others, 1979), but from only 5 DSDP sites (Table 1; Fig. 4; all by Habib, 1972, Leg 11). Nevertheless, they are present throughout each Neogene section that has been searched for dinoflagellates; their apparent absence elsewhere is probably just a function of the meager research effort directed toward this group.

Nekton

Skeletal debris of fish (ichthyoliths), especially fish teeth, have been studied in some detail at four deep-sea sites (27, 105, 417, 418; Fig. 5; Table 1) where they have been useful in dating otherwise unfossiliferous brown zeolitic clays (Edgerton and others, 1977, Kaneps and others, 1981). Ichthyoliths have also been noted at the COST B-3 site off New Jersey (Poag *in* Scholle, 1980). Lower and middle Miocene strata along the western border of the Atlantic seem to be particularly enriched in these microfossils.

So far, the utility of ichthyoliths in the western North Atlantic has been limited to broad-scale biozonation (Edgerton and others, 1977; Gaemers, 1978), but they also may be useful for paleoenvironmental analysis (especially as indicators of upwelling; Diester-Haass, 1978) and for studies of organic evolution (especially the "ear bones" or otoliths; e.g., Gaemers, 1976).

BENTHOS

Calcareous Biofacies

The most useful and well studied group of Neogene marine benthic microfossils is benthic foraminifera (see Loeblich and Tappan, 1964). This group inhabits a complex variety of marine biotopes, ranging from estuaries to abyssal plains, and they thereby constitute excellent paleoenvironmental indicators (e.g., Bandy, 1964; Murray, 1973; Poag, 1981; Douglas and Woodruff *in* Emiliani, 1981). They also can be used for biozonation and analysis of sea-level changes, paleoclimates, and paleoceanographic conditions (e.g., Schnitker, 1980; Jenkins and Murray, 1981; Miller and others *in* Graciansky, Poag and others, 1985). In addition, there is a long record of evolutionary studies using benthic foraminifera (e.g., Galloway, 1933; Cushman, 1948). Thus it is somewhat surprising that Neogene deep-sea benthic foraminifera from the western North Atlantic have received little attention to date (partial records are available for 12 DSDP sites; Table 1; Fig. 5; see Berggren and Olsson, this volume). Continental-margin assemblages have been given preliminary analysis at 14 sites, but no detailed faunal studies have been published (Fig. 5). Evidence from the eastern North Atlantic (Berggren, 1972; Schnitker, 1979) suggests that a major evolutionary change in abyssal benthic foraminiferal assemblages took place in the middle Miocene, producing the modern association. However, not all authors agree with this assessment (e.g., Boltovskoy, 1980). At shallow continental margin sites, biofacies of Miocene and Pliocene benthic foraminifera are often distinctly different from those of the Oligocene and Eocene (Poag, 1980, 1985), due principally to major paleoenvironmental changes related to paleoceanography (e.g., upwelling in the Miocene) and sea-level fluctuations.

Siliceous Biofacies

Two groups of siliceous benthic microfossils have been reported in small numbers from North Atlantic Neogene sediments: 1) Within the planktonic diatom assemblages of the North American margin sites (Fig. 5), benthic diatoms are often present in considerable numbers (Abbott *in* Ray, 1980). These are photosynthetic organisms that lived within the photic zone and they are therefore useful for studies of paleoenvironments and sediment dispersal; 2) In the deep-sea, siliceous benthic biofacies composed of sponge spicules have been noted in four widely scattered lower and middle Miocene sections (Table 1; Fig. 5; Sites 8, 391, 407, 408), where they are associated with radiolarians. No interpretive studies of this group have been published, but they are potentially useful for biozonation, paleoenvironmental analysis, and evolutionary studies. Presumably, further research will reveal a much denser distribution of these poorly studied microfossils.

STABLE ISOTOPES AND BIOFACIES

An independent measure of Neogene climatic and oceanographic conditions is provided by analysis of oxygen and carbon isotopes from planktonic and benthic foraminiferal tests. Oxygen isotope variations are responses to changes in global ice volume (which influence sea level) and sea-water temperature (Shackleton and Opdyke, 1973). Carbon isotope variations reflect changes in the global ^{13}C budget and abyssal circulation (Shackleton, 1977; Broecker, 1982; Kroopnick and others, 1972). Thus analysis of oxygen and carbon isotope ratios can be used to infer changes in sea level, water temperature, and intensity of abyssal circulation, which are all important factors in determining the composition and distribution of marine biofacies. Until recently, inferences of such conditions in the western North Atlantic relied on extrapolation from Pacific DSDP sites or from shallow sites (<1 km paleodepth) in the eastern Atlantic (Blanc and others, 1980; Blanc and Duplessy, 1982). However, Miller and Fairbanks (1983) recently analyzed specimens of the benthic foraminiferal genus *Cibicidoides* from abyssal Site 563 (Fig. 1) on the western flank of the Mid-Atlantic Ridge (>2 km paleodepth). The $\delta^{18}O$ record at 563 (Fig. 6) mimics that of the Pacific (Savin and others, 1975; Shackleton and Kennett, 1975; Woodruff and others, 1981; Savin and others, 1981) and each reveals a middle Miocene increase of $\delta^{18}O$. Early workers suggested that this increase marked the first transition to glacial continental climates (Savin, 1977). This cooling, in turn, could have initiated North Atlantic plankton provincialization (Thunell and Belyea, 1982; Berggren, 1986). However, Miller and Fairbanks (1983), Keigwin and Keller (1984), and Miller and Thomas (1986) have noted that isotopic evidence suggests significant ice volume in the Oligocene.

The $\delta^{13}C$ record shows that: 1) in the early Miocene, Atlantic values were higher than Pacific values (Sites 289 and 77) by $\sim 0.6‰$. This difference is similar to the modern difference ($1.0‰$) between Pacific Deep Water and North Atlantic Deep Water and is taken as evidence that water like North Atlantic Deep Water was being produced during the early Miocene; 2) the offset between Atlantic and Pacific $\delta^{13}C$ records appears to have decreased somewhat near the boundary between the early and

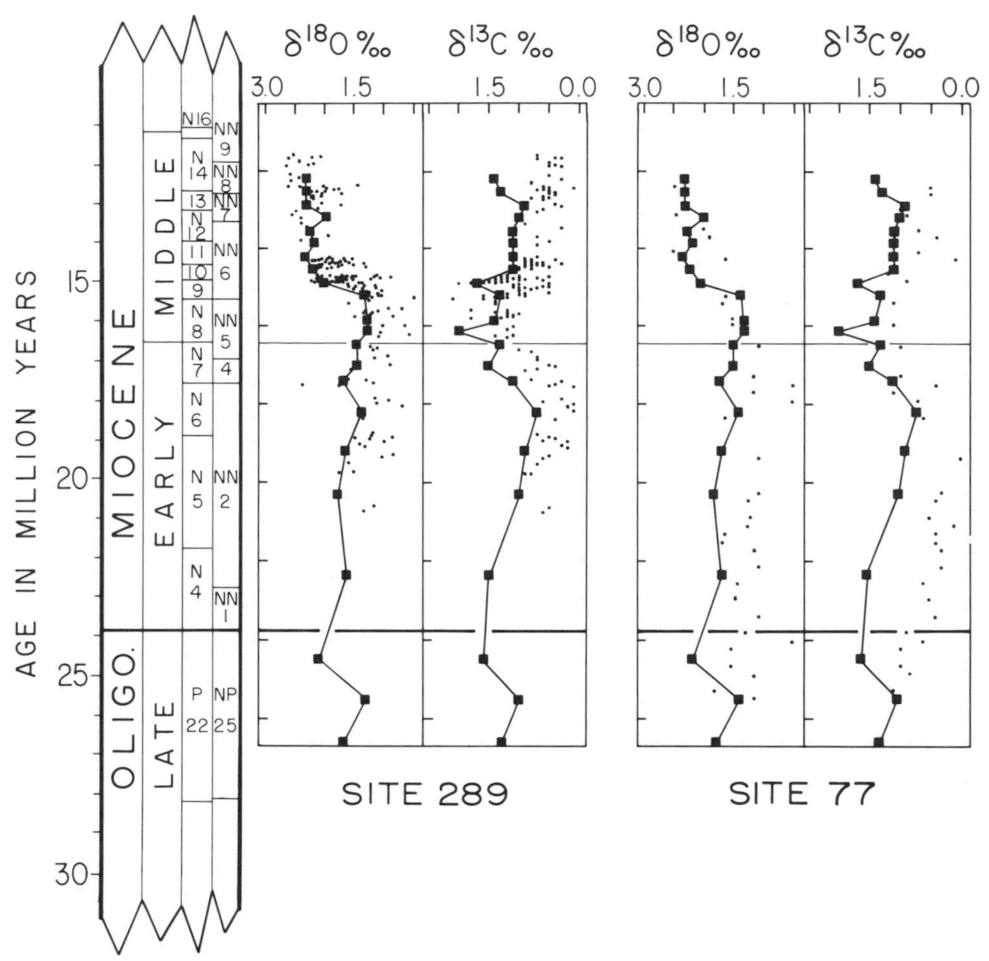

Figure 6. Comparison of benthic foraminiferal isotopic compositions (based on *Cibicidoides*) between Atlantic Site 563 (solid line connecting squares) and Pacific Sites 289 and 77 (solid circles). Note that Pacific sites are depleted in ^{13}C relative to the Atlantic sites (after Miller and Fairbanks, 1983).

middle Miocene (average difference 0.25 to 0.50‰). This decrease could be due to reduced bottom-water production in the North Atlantic. Such a decrease was postulated (Miller and Tucholke, 1983) on the basis of increased sedimentation rates beneath bottom boundary currents and widespread development of sediment drifts; 3) a late middle Miocene divergence of Pacific and Atlantic $\delta^{13}C$ values (~0.9‰ difference) may stem from increased North Atlantic bottom water production about 12 m.y. ago (see also Blanc and others, 1980; Blanc and Duplessy, 1983). Thus, carbon isotopic studies can provide a framework for interpreting bottom-water conditions that directly affect benthic biofacies.

The sparsity of late Miocene and Pliocene cores limits our understanding of the late Neogene isotopic record in the North Atlantic. The late Miocene $\delta^{13}C$ decrease (5.9 to 6.1 Ma), so well established in the Pacific (Keigwin and Bender, 1978; Vincent and others, 1979; Haq and others, 1980), has been tentatively identified only at DSDP Site 116 (Rockall Bank; Bender and Graham, 1981; Blanc and Duplessy, 1982). Furthermore, calibration of the known late Neogene isotopic record in the North Atlantic presently is hampered by problems of correlating high-latitude biozones (Site 116) with equatorial biozones at Site 397 (Shackleton and Cita, 1979) and 366 (Blanc and Duplessy, 1982), both off west Africa. This situation has led to conflicting inferences regarding the late Neogene Atlantic deep-water production (Bender and Graham, 1981; Blanc and Duplessy, 1982). A continuous, fossil-rich, abyssal, upper Miocene through Pliocene core is needed to resolve these problems.

INTERPRETATION OF BIOFACIES

When trying to interpret Neogene biofacies within the context of earth history, we face an exasperating problem. The biofacies sampled in sediments have been modified considerably by taphonomic processes (Lawrence, 1968), and they no longer represent their original compositions or distributions. On a broad scale, the distribution of carbonate, siliceous, and organic-walled sediment biofacies of the Holocene mimic their corresponding life biofacies; they consequently reflect latitudinal climatic belts and oceanographic properties such as water mass convergences and major current systems (Bé and Tolderlund *in* Funnell and Riedel, 1971; Lisitzin *in* Funnel and Riedel, 1971; Parker *in* Funnel and Riedel, 1971; Reidel, 1971; Williams *in* Funnell and Riedel, 1971; Ramsay *in* Ramsay, 1977; Kastner *in* Emiliani, 1981; Pisciatto *in* Warme and others, 1981). Thus, Haq (1980) and Thunell and Belyea (1982) interpreted the deep-sea Neogene biofacies patterns of calcareous nannoplankton and planktonic foraminifera, respectively, in terms of analogous Holocene climatic regimes and temperature gradients. Colder paleoclimates were presumed to have permitted equatorward migrations of high-latitude planktonic biofacies, and the steepest paleoclimatic gradients presumably produced the most complex planktonic biofacies patterns. But how reliable are these interpretations? Can one expect to achieve finer-scaled resolution of biofacies composition and boundaries? To help answer these questions, we can examine the pathway followed by the shells of microorganisms from a living community to a sedimentary deposit (see also Honjo, this volume).

Following Berger (1971), one can divide the historical development and modification of biofacies into three parts: 1) Life biofacies; 2) Death biofacies; and 3) Sediment biofacies. We sample only the third biofacies, but a satisfactory interpretaton of the samples depends upon how well one understands the conceptual relationships between all three biofacies.

We begin with a life biofacies (living community) whose productivity initially determines the diversity and quantity of skeletal parts eventually preserved. Heavy predation (especially among the plankton) immediately modifies the life biofacies, making a large proportion of the shells unavailable for sedimentation. The empty shells of the remaining community constitute a death biofacies. Further modification of the death biofacies takes place through shell dissolution both in the water column (plankton) and on the seafloor, by redistribution (erosion, suspension in currents), and by diagenesis. The final result is the sediment biofacies, which we can sample and analyze.

As an example of such an analysis, Thiede and others (*in* Warme and others, 1981) examined the components of Neogene sediment facies in the North Atlantic and found that since the early Miocene (ca. 20 Ma), biogenic carbonate has been the dominant constituent above 3.5 km water depth, but has been sparse or absent at greater depths, especially below 5.5 m, where red or brown clay predominates. There are important exceptions, however, where rapid mass transport has redistributed shallow-water carbonates to great depths. Some authors (e.g., Berger and von Rad, 1972) have reported a significant increase in carbonate dissolution at shallower depths (attributed to a rise in the carbonate compensation depth) in the late Miocene but neither Tucholke and Vogt (*in* Tucholke and others, 1979) nor Thiede and others (*in* Warme and others, 1981) confirmed this. These factors place important constraints upon the distribution of calcareous sediment biofacies and on our ability to interpret them accurately.

The sediments of the Neogene North Atlantic have been generally silica-poor because, like today, silica-enriched deep water was exported to other oceans in a lagoonal type of water exchange (Berger, 1970). Nevertheless, siliceous organisms presumably were abundant in some life biofacies. Heath (1974) estimated that around 96 percent of the original silica production from plankton is dissolved before it reaches the seafloor, and that only 2 percent is preserved following dissolution on the seafloor. Therefore, the Neogene siliceous sediment biofacies of the North Atlantic are poor representatives of the original life biofacies.

The organic-walled dinoflagellates present a slightly different problem. They are subject to predation in the life biofacies, but are resistant to chemical dissolution in the sediment biofacies. They are a minor part of the planktonic community and the encysted forms preserved as fossils represent only one stage in the total life cycle of a limited number of species. The sediment biofacies are, therefore, very biased samples of the life biofacies of this group.

The usual sparsity of fish teeth and bones compared to other microfossils limits their utility to very specialized conditions, in which they may be the only microfossil group represented. They are very resistant to dissolution and are preserved, sometimes in relatively rich concentrations, below the CCD and in regions of intense upwelling.

Another, often significant, modification of microfossil biofacies takes place in the vertical (stratigraphic) direction as a result of sediment mixing by burrowing organisms. Theoretical models (Berger and Heath, 1968) and field experiments (Glass, 1969; Ruddiman and Glover, 1972; Ruddiman, 1977) have described the vertical redistribution of particles. The field studies showed that post-depositional dispersal of "instantaneous" deposits, such as microtectites or volcanic glass shards, takes place upward and downward from a peak-abundance layer (representing the original deposition horizon). The dispersal zone averages 50–60 cm thickness and can represent 10,000 to 150,000 years depending upon the rate of sedimentation. Smaller particles are dispersed to greater distances than larger particles, suggesting by analogy that coccoliths or diatoms will be stratigraphically displaced over thicker intervals than foraminifera or radiolarians.

Processes of mass transport, such as turbidity currents and debris flows, redistribute sediments by addition (deposition) or subtraction (erosion) or both (see Embley and Jacobi, this volume). Turbidites (used in a broad sense; Kelts and Arthur *in* Warme and others, 1981) are not limited to bringing shallow-water (slope and shelf) sediments to the deep sea; they also redistribute pelagic deep-sea sediments, especially around the flanks of deep submarine highs. Kelts and Arthur (*in* Warme and others, 1981), who reviewed the distribution of turbidites cored during the first 54 Legs of the Deep Sea Drilling Project, found that at least 45 of 104 Neogene sites in the North Atlantic contain turbidites.

Bottom currents also contribute to modification of deep-sea biofacies by creating erosional or non-depositional gaps in the stratigraphic record, by size-sorting fossil assemblages, and by enhancing chemical dissolution due to prolonged exposure at the seafloor. Hopes for a continuous deep-sea stratigraphic record were quickly shattered by early results from the Deep Sea Drilling Project. Deep-sea unconformities were found to be widespread, and there is currently a lively debate as to their origins and significance (e.g., Moore and others, 1978; Vail and Hardenbol, 1979; Loutit and Kennett, 1981; Tucholke *in* Warme and others, 1981; Keller and Barron, 1983; Miller and Tucholke, 1983). Moore and others (*in* Warme and others, 1981) showed that 20–40 percent of the Miocene sections cored in the North Atlantic by the DSDP and 4–20 percent of the Pliocene sections were represented by hiatuses. Keller and Barron (1983) recognized seven apparently widespread Miocene deep-sea hiatuses; only two are clearly identified in the North Atlantic (Miller and Tucholke, 1983).

Life biofacies undergo similar alteration on continental margins, but additional complications can arise. For example, in the western North Atlantic, upwelling has produced zones of high productivity, especially notable from the Blake Plateau to New Jersey (Poag, 1985; Poag *in* Scholle, 1980; Gibson, 1967; Gibson *in* Ray, 1983; Abbott and Ernissee *in* Ray, 1983; Riggs, 1984); in this zone siliceous microfossils and ichthyoliths are more abundant at the expense of calcareous forms. Complicated coastal and along-slope offshore currents such as the Gulf Stream, Labrador Current, and Western Boundary Undercurrent have been particularly active in eroding wide swaths of the seafloor and redistributing sediment biofacies (Gradstein and Srivastava, 1980; Mountain and Tucholke *in* Poag, 1985). Deposition on the shelf and upper slope regions has been significantly modified by several Neogene sea-level falls, which exposed wide expanses of the shelf to intense subaerial and submarine erosion (Vail and others, 1977; Poag and Schlee, 1984). Shelf-edge submarine canyons formed conduits for mass gravity flows, which piled up thick layers of terrigenous debris at the base of the slope (Poag and others, 1985; Mountain and Tucholke *in* Poag, 1985).

In short, there is a significant information loss between formation of a life biofacies and an analysis of its equivalent sediment biofacies, especially in terms of species richness and specimen abundance. Any interpretation of geohistory based on sediment biofacies must take this information loss into account; therefore, broad-scale analyses of multiple biofacies patterns are likely to yield the most reliable conclusions.

CONCLUSIONS

We can conclude that the complex pathway from life to sediment biofacies reduces our ability to accurately construct fine-scale biozonations and to assess such properties as depositional environments, paleobiogeographic relationships and evolutionary lineages. However, we can improve our capabilities by focusing on those regions which contain optimal sedimentary records. For example, Ruddimann (1977) concluded that the Quaternary sediments of the North Atlantic are some of the least affected by carbonate dissolution, turbidite deposition, and severe bottom-current scour, and that sedimentation took place at a relatively rapid rate. Thus, the North Atlantic contains an optimal record of Quaternary deposition. We can reasonably extrapolate this optimal record to the Neogene, as shown by the recent coring carried out by DSDP Leg 94 (Scientific Party, Leg 94, unpublished data).

Regrettably, researchers have not taken full advantage of the availability of this rich data source. Biostratigraphic studies have received most attention up to now (Moore and Romine *in* Warme and others, 1981). Approximately 500 biostratigraphic papers were published between 1968 and 1978 using DSDP data (Riedel *in* Warme and others, 1981). Most of these have been qualitative studies of a single site or cruise-leg, published in the *Initial Reports* volumes. Moreover, the number of published DSDP biostratigraphic studies peaked ten years ago (1973-1974; Riedel *in* Warme and others, 1981).

Clearly there is a need for increased funding and intensified research aimed at relating the various deep-sea biofacies to each other and to their equivalents on the continental margins. On the basis of current knowledge, we can expect the Neogene biofacies to reveal important details of the paleoclimatic, paleoceanographic, and sedimentologic regimes that accompanied and, in part, stimulated the geological and biological evolution of the western North Atlantic basin. New research programs that integrate quantitative biofacies analyses (based on uniform taxonomy) with analysis of lithofacies, seismic facies, and stable isotopes are needed to provide a comprehensive geohistorical perspective of this region.

SELECTED BIBLIOGRAPHY

Abott, W. H.
1980 : Diatoms and stratigraphically significant silicoflagellates from the Atlantic Margin Coring Project and other Atlantic margin sites: Micropaleontology, v. 26, p. 49–80.

Ascoli, P.
1976 : Foraminiferal and ostracod biostratigraphy of the Mesozoic-Cenozoic Scotian Shelf, Atlantic Canada: Maritime Sediments Special Publication 1, pt. B., p. 653–677.

Bandy, O. L.
1964 : General correlation of foraminiferal structure with environment, in Approaches to Paleoecology, eds. J. Imbrie, and N. Newell, John Wiley and Sons, New York, p. 75–90.

Barss, M. S., Bujak, J. P., and Williams, G. L.
1979 : Palynological zonation and correlation of sixty-seven wells, eastern Canada: Geological Survey of Canada Paper 78-24, 118 p.

Bartlett, G. A., and Smith, L.
1971 : Mesozoic and Cenozoic history of the Grand Banks of Newfoundland: Canadian Journal of Earth Sciences, v. 8, p. 65–84.

Bates, R. L., and Jackson, J. A., eds.
1980 : Glossary of Geology, Second Edition: Falls Church, Virginia, American Geological Institution, 751 p.

Berger, W. H.
1970 : Biogenous deep-sea sediments: Fractionation by deep-sea circulation: Geological Society of America Bulletin, v. 81, p. 1385–1402.

1971 : Sedimentation of planktonic foraminifera: Marine Geology, v. 11, p. 325–358.
Berger, W. H., and Heath, G. R.
1968 : Vertical mixing in pelagic sediments: Journal of Marine Research, v. 26, p. 134–143.
Berger, W. H., and von Rad, U.
1972 : Cretaceous and Cenozoic sediments from the Atlantic Ocean: in Initial Reports of the Deep Sea Drilling Project, v. 14, eds. D. E. Hayes and A. C. Pimm; U.S. Government Printing Office, Washington, D.C., p. 787–794.
Berggren, W. A.
1972 : Cenozoic biostratigraphy and paleobiogeography of the North Atlantic: in Initial Reports of the Deep Sea Drilling Project, v. 12, eds. A. S. Laughton and W. A. Berggren; U.S. Government Printing Office, Washington, D.C., p. 965–1001.
1986 : Neogene planktonic foraminiferal biostratigraphy and biogeography; Atlantic, Mediterranean, and Indo-Pacific regions: in Pacific Neogene Datum planes: IGCP Project 114, ed. R. Tusuchi; University of Tokyo Press, Tokyo, in press).
Blanc, P. L., and Duplessy, J. C.
1982 : The deep-water circulation during the Neogene and the impact of the Messinian salinity crisis: Deep Sea Research, v. 29A, p. 1391–1414.
Blanc, P. L., Rabussier, D., Vergnaud-Grazzini, C., and Duplessy, J. C.
1980 : North Atlantic Deep Water formed by the later middle Miocene: Nature, v. 283, p. 553–555.
Boltovskoy, E.
1980 : On the benthonic bathyal-zone foraminifera as stratigraphic guide fossils: Journal of Foraminiferal Research, v. 10, p. 163–172.
Broecker, W. W.
1982 : Ocean chemistry during glacial time: Geochemica et Cosmochemica Acta, v. 46, p. 1689–1705.
Burkle, L. H.
1978 : Early Miocene to Pliocene diatom datum levels for the equatorial Pacific: Republic of Indonesia, Geological Research and Development Centre Special Publication No. 1, p. 25–44.
Casey, R. E., and McMillen, K. J.
1977 : Cenozoic radiolarians of the Atlantic Basin and margins: in Stratigraphic Micropaleontology of Atlantic Basin and Borderlands, ed. F. M. Swain; Elsevier, New York, p. 521–544.
Cushman, J. A.
1948 : Foraminifera their classification and economic use, 4th ed.: Cambridge, Massachusetts, Harvard University Press, 605 p.
Daniels, C. H. von, and Spiegler, D.
1974 : *Bolboforma* n. gen (Protozoa?)-eine neue stratigraphisch wichtig Gattung aus dem Oligozän/Miozän Nordwestdeutschlands: Paläontologische Zeitschrift, v. 48, p. 56–76.
Davies, T. A., Hay, W. W., Southam, J. R., and Worsley, T. R.
1977 : Estimates of Cenozoic oceanic sedimentation rates: Science, v. 197, p. 53–55.
de Graciansky, P. C., and Poag, C. W., eds.
1985 : Initial Reports of the Deep Sea Drilling Project, v. 80: U.S. Government Printing Office, Washington, D.C., 1260 p.
Diester-Haass, L.
1978 : Sediments as indicators of upwelling: in Upwelling Ecosystems, eds. R. Boje and M. Tomezak; Springer-Verlag, Heidelberg, p. 261–281.
Dohahue, J. G.
1970 : Diatoms as Quaternary biostratigraphic and paleoclimatic indicators in high latitudes of the Pacific Ocean [Ph.D. thesis]: New York, Columbia University, 230 p.
Edgerton, C. C., Doyle, P. S., and Reidel, W. R.
1977 : Ichthyolith age determinations of otherwise unfossiliferous Deep Sea Drilling Project Cores: Micropaleontology, v. 23, p. 194–205.
Emiliani, C.
1981 : The oceanic lithosphere: in ed. C. Emiliani, The Sea: Wiley-Interscience, New York, v. 7, 1783 p.
Fairbanks, R. G., Wiebe, P. H., and Bé, A.W.H.
1980 : Vertical distribution and isotopic composition of living planktonic foraminifera in the western North Atlantic: Science, v. 207, p. 61–63.
Funnell, B. M., and Riedel, W. R.
1971 : The micropaleontology of oceans: Cambridge University Press, Cambridge, 828 p.
Gaemers, P.A.M.
1976 : New concepts in the evolution of the Gadidae (Vertebrata, Pisces) based on their otoliths: Mededelingren Werkgroep Tertiaire en Kwartaire Geologie, v. 13, p. 3–32.
1978 : A biozonation based on Gadidae otoliths for the northwest European younger Cenozoic, with the description of some new species and genera: Mededelingren Werkgroep Tertiaire en Kwartaire Geologie, v. 15, p. 141–161.
Galloway, J. J.
1933 : A manual of Foraminifera; James Furman Kemp Memorial Series Publication 1: Principia Press, Bloomington, 483 p.
Gibson, T. G.
1967 : Stratigraphy and paleoenvironment of the phosphatic Miocene strata of North Carolina: Geological Society of America Bulletin, v. 78, p. 631–650.
Glass, B. P.
1969 : Reworking of deep-sea sediments as indicated by the vertical dispersion of the Australian and Ivory Coast microtectite horizons: Earth and Planetary Science Letters, v. 6, p. 409–415.
Gradstein, F. M., and Agterberg, F. P.
1982 : Models of Cenozoic foraminiferal stratigraphy: Northwestern Atlantic margin, in Quantitative Stratigraphic Correlation, eds. J. M. Cubitt and R. A. Reyment; John Wiley and Sons, London, p. 119–170.
Gradstein, F. M., and Srivastava, S. P.
1980 : Aspects of Cenozoic stratigraphy and paleoceanography of the Labrador Sea and Baffin Bay: Palaeogeography, Palaeoclimatology, Palaeoecology, v. 30, p. 261–295.
Gradstein, F. M., Williams, G. L., Jenkins, W.A.M., and Ascoli, P.
1975 : Mesozoic and Cenzoic stratigraphy of the Atlantic continental margin, eastern Canada: Canadian Society of Petroleum Geologists Memoir 4, p. 103–131.
Habib, D.
1972 : Dinoflagellate stratigraphy, Leg XI, Deep Sea Drilling Project: in Initial Reports of the Deep Sea Drilling Project, v. 11, eds. C. D. Hollister and J. I. Ewing; U.S. Government Printing Office, Washington, D.C., p. 367–426.
Haq, B. U.
1980 : Biogeographic history of Miocene calcareous nannoplankton and paleoceanography of the Atlantic Ocean: Micropaleontology, v. 23, p. 414–443.
Haq, B. U., and Boersma, A.
1978 : Introduction to Marine Micropaleontology: Elsevier, New York, 376 p.
Haq, B. U., Worsley, T. R., and 8 others
1980 : The late Miocene carbon-isotopic shift and the synchroneity of some phytoplanktonic biostratigraphic datums: Geology, v. 8, p. 427–431.
Harper, H. E.
1977 : Diatom stratigraphy of the Miocene/Pliocene boundary in marine strata of the circum North Pacific [Ph.D. thesis]: Cambridge, Massachusetts, Harvard University.
Hathaway, J. C., Poag, C. W., and 7 others
1979 : U.S. Geological Survey core drilling on the Atlantic Shelf: Science, v. 206, p. 515–527.
Hay, W. W.
1974 : Studies in paleo-oceanography: Society of Economic Paleontologists and Mineralogists Special Publication 20, 218 p.
Honjo, S.

1976 : Coccoliths; Production, transportation, and sedimentation: Marine Micropaleontology, v. 1, p. 65–79.

Hsü, K. J., Montadert, L., and 8 others
1977 : History of the Mediterranean salinity crisis: Nature, v. 267, p. 399–403.

Jenkins, D. G., and Murray, J. W.
1981 : Stratigraphic atlas of fossil foraminifera: The British Micropaleontological Society, Ellis Horwood, Ltd., Chichester, 310 p.

Kaneps, A. G.
1970 : Late Neogene biostratigraphy (planktonic foraminifera), biogeography, and dipositional history [Ph.D. thesis]: New York, Columbia University, 185 p.

Kaneps, A. G., Doyle, P. S., and Reidel, W. R.
1981 : Further ichthoyolith age determinations of otherwise unfossiliferous deep-sea cores: Micropaleontology, v. 27, p. 317–331.

Keigwin, L. D., and Bender, M.
1978 : A late Miocene change in the $\delta^{13}C$ of abyssal Pacific TCO_2: EOS Transactions of the American Geophysical Union, v. 59, p. 1118.

Keigwin, L. D., and Keller, G.
1984 : Middle Oligocene climatic change from equatorial Pacific DSDP Site 77: Geology, v. 12, p. 16–19.

Keller, G.
1981 : Miocene biochronology and paleoceanography of the North Pacific: Marine Micropaleontology, v. 6, p. 535–551.

Keller, G., and Barron, J. A.
1983 : Paleoceanographic implications of Miocene deep-sea hiatuses: Geological Society of America Bulletin, v. 94, p. 590–613.

Kroopnick, P., Weiss, R. F., and Craig, H.
1972 : Total CO_2 and dissolved oxygen $-^{18}O$ at Geosecs II in the North Atlantic: Earth and Planetary Science Letters, v. 16, p. 103–110.

Lawrence, D. R.
1968 : Taphonomy and information losses in fossil communities: Geological Society of America Bulletin, v. 79, p. 1315–1330.

Loeblich, A. R., Jr., and Tappan, H.
1964 : Foraminiferida: in ed. R. C. Moore; Treatise on invertebrate paleontology, Part C, Protista 2, Volume 2, Geological Society of America (and University of Kansas Press), Boulder, v. 1, p. C55–510; v. 2, p. C511–900.

Loutit, T. S., and Kennett, J. P.
1981 : Australasian Cenozoic sedimentary cycles, global sea level changes and the deep sea sedimentary record: Oceanologica Acta, Number Special, p. 45–63.

Malmgren, B. A., and Kennett, J. P.
1980 : Phyletic gradualism in a late Cenozoic planktonic foraminiferal lineage; DSDP Site 284, southwest Pacific: Paleobiology, v. 7, p. 230–240.

Miller, K. G., and Fairbanks, R. G.
1983 : Evidence for Oliogocene-middle Miocene abyssal circulation changes in the western North Atlantic: Nature, v. 306, p. 250–253.

Miller, K. G., and Tucholke, B. E.
1983 : Development of Cenozoic abyssal circulation south of the Greenland-Scotland Ridge: in Structure and Development of the Greenland-Scotland Ridge, eds. M. Bott, S. Saxov and others, Plenum Press, New York, p. 549–589.

Miller, K. G., and Thomas, E.
1986 : Late Eocene to Oligocene benthic foraminifera isotopic record, Site 574, equatorial Pacific: in Initial Reports of the Deep Sea Drilling Project, v. 85, eds. L. Mayer and F. Theyer; U.S. Government Printing Office, Washington, D.C., in press.

Moore, R. C.
1949 : Meaning of facies: Geological Society of America Memoir 29, p. 1–34.

Moore, T. C., Jr., van Andel, Tj. H., Sancetta, C., and Pisias, N.
1978 : Cenozoic hiatuses in pelagic sediments: Micropaleontology, v. 24, p. 113–138.

Murray, J. W.
1973 : Distribution and ecology of living benthic foraminiferids: Crane Russak, New York, 274 p.

Palmer, A. A.
1983 : Biostratigraphic and paleoenvironmental results from Neogene radiolarians, U.S. Mid-Atlantic Coastal Plain and continental margin: American Association of Petroleum Geologists Bulletin, v. 67, p. 528–529.

Poag, C. W.
1978 : Stratigraphy of the Atlantic Continental Shelf and Slope of the United States: Annual Review of Earth and Planetary Sciences, v. 6, p. 251–280.
1981 : Ecologic atlas of benthic foraminifera of the Gulf of Mexico: Hutchinson Ross, Stroudsburg, Pennsylvania, 175 p.
1984 : Neogene stratigraphy of the submerged U.S. Atlantic margin: Palaeogeography, Palaeoclimatology, Palaeoecology, v. 47, p. 103–127.
1985 : Geologic evolution of the United States Atlantic margin:New York, Van Nostrand Reinhold, 383 p.

Poag, C. W., and Hall, R. E.
1979 : Foraminiferal biostratigraphy, paleoecology, and sediment accumulation rates: in Geological Studies of the COST GE-1 well, United States South Atlantic Outer Continental Shelf Area, ed. P. A. Scholle;: U.S. Geological Survey Circular 800, p. 49–63.

Poag, C. W., and Schlee, J. S.
1984 : Depositional sequences and stratigraphic gaps on submerged United States Atlantic margin: in Interregional Unconformities and Hydrocarbon Accumulation, ed. J. S. Schlee; American Association of Petroleum Geologists Memoir 36, p. 165–182.

Poag, C. W., and Karowe, A. L.
1986 : Stratigraphic potential of *Bolboforme* significantly increased by new finds in the North Atlantic and South Pacific: Palaios, v. 1, p. 162–171.

Poore, R. Z.
1981 : Late Miocene biogeography and paleoclimatology of the central North Atlantic: Marine Micropaleontology, v. 6, p. 599–616.

Ramsay, A.T.S.
1977 : Oceanic micropaleontology: Academic Press, London, 1489 p.

Ray, C. E.
1983 : Geology and paleontology of the Lee Creek Mine, North Carolina, I: Smithsonian Contributions to Paleobiology, no. 53, 529 p.

Reidel, W. R.
1971 : Radiolarians from Atlantic deep-sea drilling: Proceedings of the Second Planktonic Conference, Roma, 1970, p. 194–205.

Riggs, S. R.
1984 : Paleoceanographic model of Neogene phosphorite deposition, U.S. Atlantic continental margin: Science, v. 223, p. 123–131.

Rögl, and Hochuli, P.
1976 : The occurrence of *Bolboforma,* a probably algal cyst, in the Antarctic Miocene of DSDP Leg 35: in Initial Reports of the Deep Sea Drilling Project, v. 35, eds. C. D. Hollister and C. Craddock; U.S. Government Printing Office, Washington, D.C., p. 713–719.

Ruddiman, W. F., and Glover, L. K.
1972 : Vertical mixing of ice-rafted volcanic ash in North Atlantic sediments: Geological Society of America Bulletin, v. 83, p. 2817–2836.

Sarnthein, M., Thiede, J., and 6 others
1982 : Atmospheric and oceanic circulation patterns off northwest Africa during the past 25 million years: in Geology of the Northwest African Continental Margin, eds. U. von Rod, K. Hinz, M. Sarnthein and E. Siebold; Springer-Verlag, Berlin, p. 584–604.

Savin, S. M.
1977 : The history of the earth's surface temperature during the past 100 million years: Annual Review of Earth and Planetary Sciences, v. 5, p. 319–355.

Savin, S. M., Douglas, R. G., and Stehli, F. G.
1975 : Tertiary marine paleotemperatures: Geological Society of America Bulletin, v. 86, p. 1499–1510.

Savin, S. M., Douglas, R. G., and 6 others

1981 : Miocene benthic foraminiferal isotope records; A synthesis: Marine Micropaleontology, v. 6, p. 423–450.
Schnitker, D.
1979 : Cenozoic deepwater benthic foraminifers, Bay of Biscay: in Initial Reports of the Deep Sea Drilling Project, v. 48, eds. L. Montadert and D. G. Roberts; U.S. Government Printing Office, Washington, D.C., p. 377–413.
1980 : Quaternary deep-sea benthic foraminifers and bottom water masses: Annual Review of Earth and Planetary Science, v. 8, p. 343–370.
Scholle, P. A.
1977 : Geologic studies of the COST No. B-2 well, U.S. Mid-Atlantic Outer Continental Shelf area: U.S. Geological Survey Circular 750, 71 p.
1980 : Geological studies of the COST No. B-3 well, U.S. Mid-Atlantic Continental Slope area: U.S. Geological Survey Circular 833, 132 p.
Scholle, P. A., and Wenkam, C. R.
1982 : Geological studies of the COST Nos. G-1 and G-2 wells, U.S. Outer Continental Shelf: U.S. Geological Survey Circular 861, 186 p.
Shackleton, N. J.
1977 : Carbon-13 in *Uvigerina;* Tropical rain forest history and the equatorial Pacific carbonate dissolution cycles: in The Fate of Fossil CO_2 in the Ocean, eds. N. R. Anderson and A. Malahoff; Plenum Press, New York, p. 401–427.
Shackleton, N. J., and Cita, M. B.
1979 : Oxygen and carbon isotope stratigraphy of benthic foraminifers at Site 397; Detailed history of climatic change during Late Neogene: in Initial Reports of the Deep Sea Drilling Project, v. 47, eds. U. von Rad and W.B.F. Ryan; U.S. Government Printing Office, Washington, D.C., p. 433–446.
Shackleton, N. J., and Kennett, J. P.
1975 : Paleotemperature history of the Cenozoic and the initiation of Antarctic glaciation; Oxygen and carbon isotope analyses in DSDP Sites 277, 279, and 281: in Initial Reports of the Deep Sea Drilling Project, v. 29, eds. J. P. Kennett and R. E. Houtz; U.S. Government Printing Office, Washington, D.C., p. 743–755.
Shackleton, N. J., and Opdyke, N. D.
1973 : Oxygen isotopic and plaeomagnetic stratigraphy of equatorial Pacific core V28-238; Oxygen-isotope temperatures and ice volumes on a 10^5 and 10^6 year scale: Quaternary Research, v. 3, p. 39–55.
Srinivasan, M. S., and Kennett, J. P.
1981 : Neogene planktonic foraminiferal biostratigraphy and evolution; Equatorial to subantarctic South Pacific: Marine Micropaleontology, v. 6, p. 499–533.
Thunell, R., and Belyea, P.
1982 : Neogene planktonic foraminiferal biogeography of the Atlantic Ocean: Micropaleontology, v. 28, p. 381–398.
Tucholke, B. E., and Mountain, G. S.
1979 : Seismic stratigraphy, lithostratigraphy, and paleosedimentation patterns in the North American Basin: in Deep Drilling Results in the Atlantic Ocean; Continental Margins and Paleoenvironments, eds. M. Talwani, W. W. Hay and W.B.F. Ryan; American Geophysical Union Maurice Ewing Series, v. 3, p. 58–86.
Tucholke, B. E. and Vogt, P. R., eds.
1979 : Initial Reports of the Deep Sea Drilling Project, v. 43: U.S. Government Printing Office, Washington, D.C., 1115 p.
Vail, P. R., and Hardenbol, J.
1979 : Sea-level changes during the Tertiary: Oceanus, v. 22, p. 71–79.
Vail, P. R., Mitchum, R. M., Jr., and Thompson, S., III
1977 : Seismic stratigraphy and global changes of sea level; Part 4; Global cycles of relative changes of sea level: in Seismic Stratigraphy; Applications to Hydrocarbon Exploration, ed. C. E. Payton; American Association of Petroleum Geologists Memoir 26, p. 83–98.
Valentine, P. C.
1979 : Calcareous nannofossil biostratigraphy and paleoenvironmental interpretation: in Geological Studies of the COST GE-1 well, U.S. South Atlantic Outer Continental Shelf Area, ed. P. A. Scholle; U.S. Geological Survey Circular 800, p. 64–70.
Vincent, E., Killingley, J. S., and Berger, W. H.
1979 : The magnetic epoch 6 carbon shift; A change in the oceans $^{13}C/^{12}C$ ratio 6.2 million years ago: Marine Micropaleontology, v. 5, p. 185–203.
Warme, J. E., Douglas, R. G., and Winterer, E. L., eds.
1981 : The Deep Sea Drilling Project; A decade of progress: Society of Economic Paleontologists and Mineralogists Special Publication No. 32, 564 p.
Weller, J. M.
1958 : Stratigraphic facies differentiation and nomenclature: American Association of Petroleum Geologists Bulletin, v. 42, p. 609–639.
Woodruff, F., Savin, S. M., and Douglas, R. G.
1981 : Miocene stable isotopic record; A detailed deep Pacific Ocean study and its paleoclimatic implications: Science, v. 212, p. 665–668.

Manuscript Accepted by the Society March 13, 1985

ACKNOWLEDGMENTS

We are grateful to Robert Thunell, Bilal Haq, and Brian Tucholke, whose perceptive criticisms led to significant improvement of the manuscript. K.G.M. was supported by a L-DGO post-doctoral fellowship, National Science Foundation contract OCE 83 10086, and by a consortium of oil companies. This is Lamont-Doherty Geological Observatory Contribution Number 4007.

Printed in U.S.A.

The Geology of North America
Vol. M, The Western North Atlantic Region
The Geological Society of America, 1986

Chapter 34

North Atlantic Mesozoic and Cenozoic paleobiogeography

W. A. Berggren
Woods Hole Oceanographic Institution, Woods Hole, Massachusetts 02543; Department of Geology, Brown University, Providence, Rhode Island 02912
R. K. Olsson
Department of Geological Sciences, Rutgers University, New Brunswick, New Jersey 08903

INTRODUCTION

A decade ago there was little information on the (quantitative) paleobiogeographic distribution of fossil calcareous plankton during the Mesozoic and Cenozoic (see early reviews by Berggren and Hollister, 1974; Berggren, 1978). The intervening years, however, have seen a concentrated effort towards delineating quantitatively based distribution patterns in both calcareous plankton and benthic foraminifera and relating these patterns to the evolution of surface and deep water masses in the Atlantic Ocean (*i. al.,* Haq and Lohmann, 1976; Haq and others, 1971; Tjalsma and Lohmann, 1983; Corliss, 1981; Miller, 1982; Miller and others, 1982; Nyong and Olsson, 1984; Roth and Bowdler, 1981; Thierstein, 1981).

Mesozoic paleobiogeography based on Atlantic microfossils is still in its infancy. Abyssal distributions of benthic foraminifera have been studied by Gradstein (1983), Nyong and Olsson (1984), and Sliter (1980). Numerous papers are available on the distribution of benthic foraminifera in shelf and epicontinental environments in Europe and North America but there is little information on the paleobiogeography of these in the North Atlantic basin. The record of Early and Middle Cretaceous planktonic foraminifera is scanty due to poor preservation because of an elevated calcite compensation depth (CCD). Late Cretaceous distribution of planktonic foraminifera is better known but quantitative studies are lacking. Calcareous nannofossil paleobiogeography is better known through quantitative studies by Doeven (1983), Roth and Bowdler (1981), and Thierstein (1981).

Summaries by Berggren (1978) and Thunell and Belyea (1982) on Atlantic Ocean Cenozoic planktonic foraminiferal biogeography and Berggren and Schnitker (1983) on North Atlantic Cenozoic benthic foraminiferal studies form the basis for the review presented in the second half of this paper. The summary presented in the section on Cenozoic biogeography is skewed towards a more substantive discussion of the benthic foraminifera in some instances (Paleogene). This is because calcareous plankton have been treated in greater detail in recent publications, whereas the field of deep-sea benthic foraminifera is a relatively new one and results are generally less familiar to readers.

MESOZOIC

Paleogeography and Paleocirculation

The formation of Atlantic Ocean crust probably began in late Bajocian to early Bathonian time (Klitgord and Schouten, this volume); deep oceanic basins soon formed in the area of Site 534 in the western North Atlantic and Site 416 in the eastern Atlantic (Plate 9). Surface water exchange was apparently unimpeded and stable between the Tethys in the east, the central Atlantic Ocean, and the Pacific on the west. The typical Upper Jurassic pelagic facies of *Saccocoma,* calcispherulids, and aptychi of the Tethys is now well known from the Deep Sea Drilling Project (DSDP) sites in the western and eastern central Atlantic (sites 100, 105, 367, 534), and it indicates free surface-water exchange of the expanding Atlantic Ocean with the Tethys (Jansa, this volume). The embryonic North Atlantic had a near-equatorial east-west position with wind-driven surface currents flowing westward from the Tethys through the Atlantic to the Pacific (Gradstein and Sheridan, 1983). During the Late Jurassic the North Atlantic may have widened enough for a return equatorial countercurrent to flow along its southeastern edge. Gradstein and Sheridan believe that this pattern persisted until the margin of the central Atlantic drifted north of 30° latitude. By the Early Cretaceous the clockwise North Atlantic gyre probably was well developed, and the incipient Gulf Stream Current was developed beneath northeast veering trade winds (Tucholke and McCoy, this volume).

Marine communication with the South Atlantic occurred by early Turonian time (Berggren and Hollister, 1977) when the North Atlantic had attained half its present width. The Grand

Berggren, W. A., and Olsson, R. K., 1986, North Atlantic Mesozoic and Cenozoic paleobiogeography; *in* Vogt, P. R., and Tucholke, B. E., eds., The Geology of North America, Volume M, The Western North Atlantic Region: Geological Society of America.

Banks/Portugal gateway to the north was epicontinental in character until Early Cretaceous time. Initial opening of the southern Labrador Sea began by Campanian-Maestrichtian time, and a marine passage with surface circulation towards the north probably linked the Atlantic and Arctic Oceans (see Plate 10, G; Gradstein and Srivastava, 1980). A surface connection via the region of the present Norwegian-Greenland Sea had not yet developed and Late Cretaceous paleoceanographic conditions of the North Atlantic were greatly influenced by circulation of warm surface currents from the equatorial region.

Backtracking of crustal depth at Site 534 (Sheridan, 1983) indicates that abyssal depths developed in the North Atlantic in Callovian time; so abyssal environments existed in the Atlantic almost from its inception. A deep connection with the South Atlantic followed in Campanian time, with the contemporaneous subsidence of the Rio Grande Rise and Walvis Ridge. There was no exchange of deep and intermediate waters with the Pacific and the Arctic.

Temperature gradients were lower in the Cretaceous than in the Cenozoic; this was apparently an acryogenic time with ice-free poles and sluggish surface currents (Roth and Bowdler, 1981). Circulation of deep and bottom waters were haline rather than thermohaline, apparently driven by excess evaporation in low latitudes (Brass and others, 1982). Black shales within the Hatteras Formation, and locally in the Blake-Bahama Formation and Callovian sediments (Plate 1) attest to periods of carbon-rich deposition in the Atlantic basin during Middle Jurassic, Early Cretaceous, and Middle Cretaceous times. Most of the organic matter is of terrestrial origin with lesser quantities from marine sources. The black shales generally are interpreted to indicate periods of widespread anoxia in the Atlantic basin (Tucholke and Vogt, 1979; Arthur, 1979; Summerhayes and Masran, 1983) although just how widespread these conditions were has been questioned (Habib, 1983; Waples, 1983). Well-preserved laminated sediments and lack of biogenic burrowing in many, but not all intervals suggest that marginal living conditions periodically existed on the sea bottom during the Jurassic and the Cretaceous.

The level of the CCD varied markedly during the Mesozoic and thereby affected the distribution of calcareous benthic organisms on the ocean floor. The CCD also affected preservation of the calcareous plankton record in the deeper parts of the Atlantic basin. Tucholke and Vogt (1979) have summarized the CCD history for the western North Atlantic (Plate 1). Deposition of calcareous claystones and limestones occurred near and above the CCD during the Late Jurassic into the Early Cretaceous. A sharp rise of the CCD to less than 2800 m occurred at the end of Neocomian time and was coincident with widespread black shale deposition. The CCD remained shallow until the Maestrichtian but the deep basin became oxygenated by late Cenomanian-Turonian time and the multicolored clays of the Plantagenet Formation were deposited. A drop of the CCD to more than 5400 m in late Maestrichtian to earliest Paleocene time resulted in basin-wide chalk deposition in the upper part of the Plantagenet Formation.

Benthic Foraminiferal Distribution

In the early stages of the North Atlantic opening (Callovian), a narrow but deep ocean basin was present (see Plate 9, B). The abyssal depths of this ocean were characterized by an agglutinated and calcareous assemblage of benthic foraminifera which had spread into the Atlantic basin from the Tethys (Gradstein, 1983). Very similar assemblages occurred in abyssal Jurassic settings at Indian ocean sites (Bartenstein, 1974). This Jurassic abyssal fauna continued into the Early Cretaceous and consisted of small-sized agglutinated taxa and small-sized nodosarids, as well as epistominids, ophthalminids, spirrilinids, and turrilinids. Stratigraphically persistent genera include: *Reophax, Bigenerina, Bathysiphon, Glomospira, Glomospirella, Lenticulina, Rhizammina, Psammosphaera, Dentalina, Lagena, Marginulina, Pseudonodosaria,* and *Trochammina.* The abyssal fauna is associated with radiolarians, calcispherulids, aptychi and *Saccocoma* (Gradstein, 1983). The agglutinated assemblage is composed of simple forms and has a strong communality, above the species level, with geologically younger deep-marine assemblages found below the CCD (Fig. 1).

The contrast between the deep oceanic assemblages and shallow marine-epicontinental assemblages is not well-known, but Gradstein (1983) notes that the shallow-marine environments contained fewer agglutinated families, lacked simple agglutinated genera, contained a greater diversity of epistominids, spirillinids, and Ammobaculites, and were characterized by calcareous species of *Miliammina, Trocholina, Patellina, Paalzowella, Citharina,* and *Tristix.* Diversity was higher in the abyssal assemblages than in the shallow-marine assemblages.

During the Late Jurassic and Early Cretaceous, a carbonate facies of reef, bank, and other buildups typical of shallow carbonate environments occurred along the western edge of the North Atlantic (Plate 9). This facies was drilled at Sites 390 and 392 on the Blake Spur and at Site 384 on the J-Anomaly Ridge. At Site 384, orbitolinids were recovered (Schroeder and Cherchi, 1979), and further north on the Flemish Cap orbitolinids also occur (Sen Gupta and Grant, 1971). Rudists also occur at Site 384. Thus along the shelf edge of the North Atlantic during the Late Jurassic and Early Cretaceous, a carbonate environment supported an assemblage of orbitolinids and other associated taxa.

The Albian to Cenomanian rise in sea level led to extensive flooding of the North Atlantic margins (Plate 1). Prior to the Albian, middle-depth waters were weakly stratified and were occupied by low-diversity assemblages of foraminifera (Sliter, 1980). The rise in sea level created broad epicontinental seas in Europe and wide continental shelves around the margins of the North Atlantic. Radiation of benthic foraminiferal species occurred in the expanded shelf and slope environments. Diversity increased significantly and many new genera appeared. Important new genera included *Arenobulimina, Clavulina, Globorotalites, Gyroidinoides, Eponides, Eouvigerina, Osangularia, Pleurostomella, Pyramidina, Pullenia, Quadrimorphina,* and *Valvulineria.* By early Turonian time, exchange of mid-depth waters

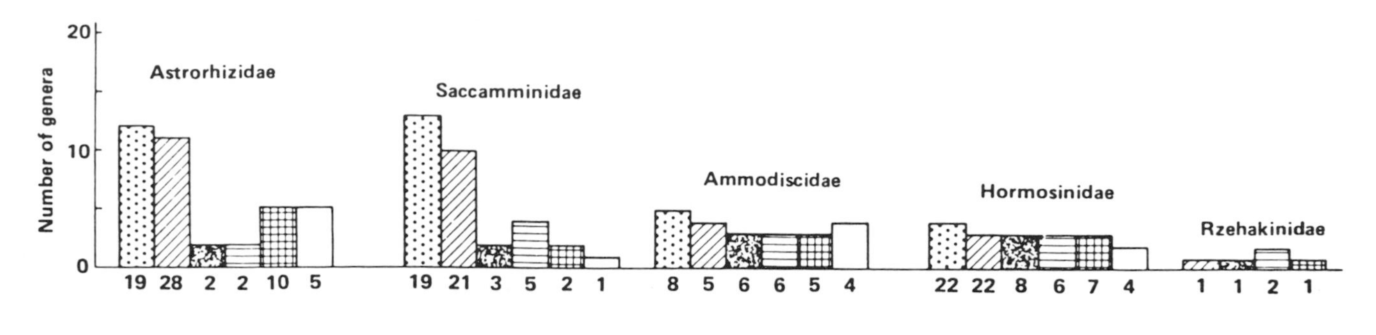

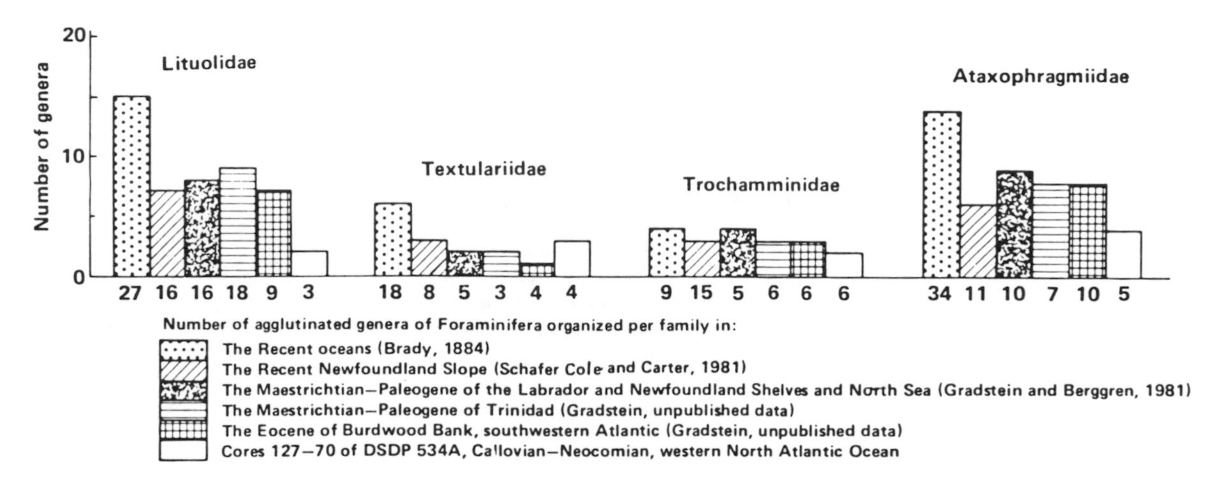

Figure 1. Comparison of North Atlantic Jurassic agglutinated deep-marine foraminifera with Recent, Maestrichtian-Paleogene, and Eocene assemblages. Family representation has remained similar since the Jurassic (after Gradstein, 1983).

with the South Atlantic allowed the spread of many species of these genera to the south. Abyssal connections allowed faunal interchange along the entire axis of the North and South Atlantic by Campanian time.

Little is known about the detailed biogeography of benthic foraminifera in the middle Cretaceous of the North Atlantic. The nodosariid- and epistominid-dominated assemblages of the Jurassic and Early Cretaceous had disappeared and were replaced by species of newly evolved genera. Important data on the composition of Middle Cretaceous abyssal and bathyal assemblages in deep-sea drilling sites in the central and North Atlantic Ocean are provided by Krasheninnikov and Pflaumann (1978), Mayne (1970), Sigal (1979), and Sliter (1976, 1977, 1980). Abyssal assemblages consisted almost exclusively of primitive agglutinated foraminifera similar to those of the Early Cretaceous and Late Cretaceous. Bathyal assemblages consisted predominantly of calcareous genera which include *Ellipsoidella, Pleurostomella, Stilostomella, Gyroidinoides, Osangularia,* and *Praebulimina.*

Middle Cretaceous shelf assemblages around the North Atlantic are well-known in Europe (Aubert and Bartenstein, 1976; Bartenstein and Brand, 1951; Gawor-Biedowa, 1972; Moullade, 1966, 1973; Neagu, 1965, 1968; Price, 1976, 1977), in Trinidad (Bartenstein and others, 1957, 1966; Bartenstein and Bolli, 1973), in the Scotian shelf (Ascoli, 1977), and in the middle Atlantic margin of North America (Gradstein, 1978; Olsson, 1977; Poag, 1980). The shelf assemblages, like the deep-sea assemblages, are only known in general distribution; biofacies distribution is not well understood. Important components of shelf assemblages included nodosarids, polymorphinids, *Gavelinella, Lingulogavelinella, Valvulineria, Epistomina, Buliminella,* and others.

With the development of shelf and slope foraminiferal assemblages, important speciation events took place in the genera *Arenobulimina* (Price, 1977), *Conorotalites* (Kaever, 1961), *Epistomina* (Ohm, 1967), *Gavelinella* (Malapris, 1965; Michael, 1966; Price, 1977), *Globorotalites* (Kaever, 1961), *Lingulogavelinella* (Malapris-Bizouard, 1967), and *Valvulineria* (Jannin, 1967). Species of these genera became important components of shelf assemblages. *Gavelinella,* in particular, underwent a radiation which resulted in the evolution of numerous species; these occur in a number of associations that suggest nearshore to bathyal ecologic niches.

In the western North Atlantic the distribution of Campanian to lower Maestrichtian benthic foraminifera is relatively uniform (Nyong and Olsson, 1984). Generic composition of common taxa include *Gavelinella, Dorothia, Conorbina, Osangularia, Gyroidinoides, Lenticulina, Praebulimina, Bolivina, Tritaxia, Valvulineria, Nuttallinella, Aragonia, Reusella, Pleurostomella, Nuttal-*

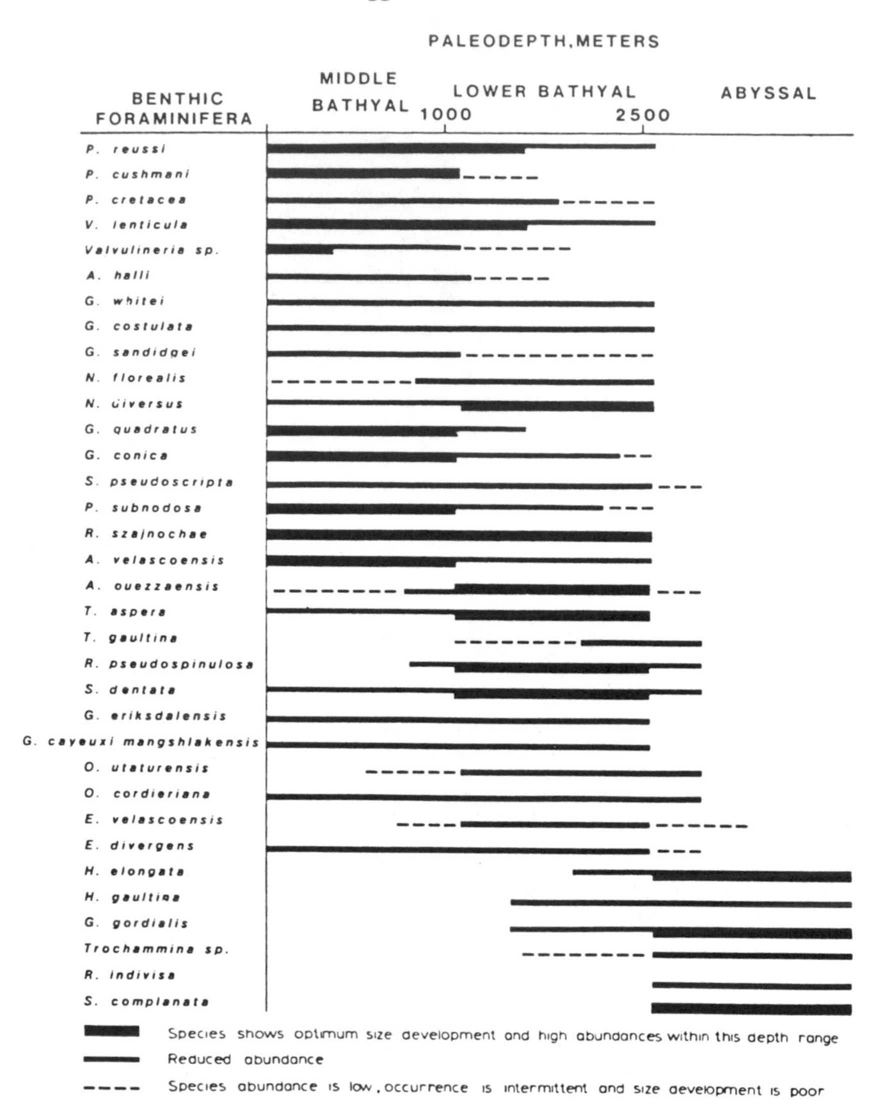

Figure 2. Model of bathyal to abyssal distribution of benthic foraminiferal species in the Campanian to lower Maestrichtian of the western North Atlantic (after Nyong and Olsson, 1984).

lides, Stilostomella, Pullenia, Quadrimorphina, Ellipsoidella, Globulina, Allomorphina, Berthelinella, Bandyella, and *Fissurina.* The species of these genera have been shown to extend over broad bathymetric ranges and in some cases to the level of the CCD in the South Atlantic, Pacific, and Indian oceans (McNulty, 1976, 1979; Sliter, 1976, 1977; Scheibnerova, 1977; Beckmann, 1978; Webb, 1973). Species of *Recurvoides, Glomospira, Haplophragmoides, Trochammina, Hyperammina, Rhabdammina, Rhizammina, Saccammina, Pelosina, Ammodiscus, Ammobaculites, Spiroplectammina,* and *Dendrophyra* are also present. Similar agglutinated assemblages were described from Upper Cretaceous to Lower Tertiary sections in the Labrador and North seas (Gradstein and Berggren, 1981; Miller and others, 1982), and they also are known from flysch deposits of the Carpathians and other ancient trench environments (Myatliuk, 1970; Geroch, 1966; Hanzlikova, 1972). Gradstein and Berggren (1981) designated this type of agglutinated assemblage as Type-A or flysch-type and differentiated them from Type-B (deep-sea) fauna of Krasheninnikov (1973, 1974).

Benthic foraminiferal changes within the Campanian to Maestrichtian deep-sea environment (Fig. 2) primarily involved shifts in species proportions rather than changes in the generic or species composition of the assemblages (Nyong and Olsson, 1984). Abundance was low but diversity was high. Mid-bathyal to abyssal biofacies strongly overlapped and most species had broad bathymetric ranges. This less stratified community structure suggests sluggish circulation of Atlantic bottom waters during Late Cretaceous time. Pronounced dominance of cassidulinids over nodosariids and buliminids was typical of lower bathyal assemblages. Thick-walled members of the cassidulinids were more common in the abyssal environment, but agglutinated taxa were dominant near and below the CCD.

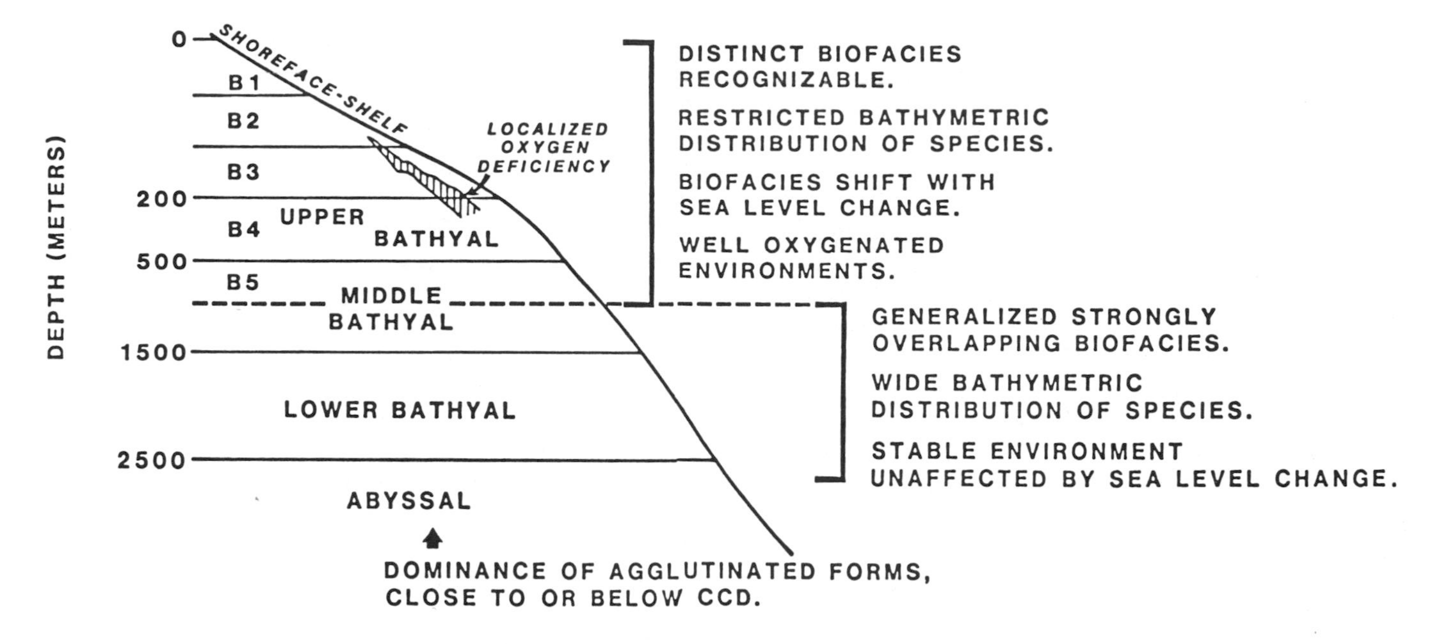

Figure 3. Paleobiogeographic model of the distribution of Campanian to lower Maestrichtian benthic foraminiferal biofacies in the western North Atlantic (after Nyong and Olsson, 1984). Biofacies are based on the distribution of foraminiferal assemblages in outcrops and wells in the coastal plain of New Jersey and Delaware, and their distribution in wells in the Baltimore Canyon Trough.

In contrast to the deep-sea assemblages, the shelf and upper-slope assemblages of the Campanian and Maestrichtian were organized into distinct biofacies and many species were restricted in their bathymetric distribution (Olsson and Nyong, 1984). The assemblages shifted back and forth along the shelf and replaced one another in time and space as environments of deposition changed in response to changes in sea level. The deep-sea assemblages were apparently little affected by sea level changes. A model for foraminiferal distribution in the Campanian and Maestrichtian in the western North Atlantic (Nyong and Olsson, 1984) is shown in Figure 3.

Distribution of Planktonic Foraminifera

North Atlantic Jurassic (Oxfordian) planktonic foraminifera are recorded in the Blake-Bahama Basin drill sites (Gradstein, 1983). Their first appearance in the Atlantic region, however, is known from the Grand Banks in epicontinental deposits of Bajocian age (Gradstein, 1978). These tiny (60 μm) globigerine forms are very rare or absent in pelagic facies, but they are sporadically abundant in Jurassic epicontinental and ocean margin environments (Gradstein, 1983). The two species reported from the Oxfordian of the Blake-Bahama Basin are *"Globigerina" helvetojurassica* and *Globuligerina oxfordiana.* Jurassic globigerinids disappeared from the Atlantic realm by Tithonian time.

Globigerinids did not reappear until Hauterivian time with the advent of *Caucasella hotervica* which, like the Jurassic globigerinids, occurs in ocean-margin facies. This was followed in the late Hauterivian with the appearance of the important genus *Hedbergella,* which had an open ocean habitat. *Hedbergella* was the stock for an evolutionary radiation which gave rise to a diversified assemblage of planktonic foraminifera by the end of Early Cretaceous time. The appearance of *Clavihedbergella, Globigerinelloides,* and *Schackoina* during the Barremian and Aptian were important evolutionary events which gave rise to numerous species that populated the Tethys and the widening North Atlantic Ocean. These Early Cretaceous planktonic foraminifera were generalized globigerine forms which speciated rapidly in the Aptian. More specialized forms with peripheral keels and supplementary apertures, as well as more heavily ornamented species, appeared in the Albian and Cenomanian. These included *Planomalina, Rotalipora, Praeglobotruncana, Ticinella,* and *Favusella.* Heterohelicids were rare. Many Tethyan species such as *Ticinella praeticinensis, T. roberti, Rotalipora appenninica, R. evoluta, R. cushmani, R. greenhornensis, R. ticinensis, Globigerinoides bentonensis,* and *Planomalina buxtorfi* have been reported from the western North Atlantic region as far north as the Grand Banks. Apparently, Tethyan influence was strong in the North Atlantic during Albian and Cenomanian time. Cooler northern sources of water were probably minor, but little data is available.

A faunal turnover occurred in the early Turonian with the elimination of the specialized genera from the Cenomanian and their replacement by the keeled *Marginotruncana* and the short-lived *Helvetotruncana.* Important globigerine forms which appeared and diversified during Turonian to Santonian time are

Whiteinella and *Archeoglobigerina*. Species which characterized these assemblages include *Marginotruncana imbricata, M. schneegansi, M. concavata, M. marginata, M. renzi, M. sigali, Whiteinella archeocretacea, W. baltica, Archeoglobigerina blowi,* and *A. cretacea.*

Another major radiation took place during the Campanian and Meastrichtian. Important speciation events occurred in the keeled *Globotruncana, Rugoglobigerina, Globigerinelloides,* and the heterohelicids. New heterohelicid genera that appeared include *Pseudoguembelina, Ventilabrella, Planoglobulina, Pseudotextularia,* and *Racemiguembelina. Abathomphalus* appeared in the latest Maestrichtian, and *A. mayaroensis* is the zonal marker for the uppermost Cretaceous. *Hedbergella monmouthensis* was relatively rare in the North Atlantic except in its northern parts, but it was an important species in the Maestrichtian because it gave rise to the globigerinids of the Cenozoic. The presence of Tethyan species such as *A. mayaroensis, Globotruncana contusa, G. stuarti, G. aegyptica, G. gansseri, Racemiguembelina fructicosa,* and *Pseudotextularia elegans* as far north as DSDP Site 111 on the Orphan Knoll indicates warm current transport of these species into the northern Atlantic.

Paleoceanography and Paleobiogeography

The western North Atlantic lay within the Tethyan realm during the Middle Jurassic to Early Cretaceous. The typical Tethyan pelagic facies of *Saccocoma,* calcispherulid, and aptychi as well as the filamentous facies of pelagic pelecypods occurred in the Late Jurassic. Tethyan species of benthic foraminifera such as *Lenticulina quenstedti, Alveosepta jaccardi, Pseudocyclammina virguliana, P. lituus,* and *Anchispirocyclina lusitanica* were present in the Grand Banks and the Scotian Shelf (Gradstein, 1977; Jansa and others, 1980; Williams and others, 1974). Nevertheless, the presence of species that also occurred in Alaska indicates that waters were cooler in this region. Jansa and others' (1980) study of the upper Tithonian and lower Berriasian calpionellid faunas, foraminifera, and ostracods suggests a northern, cold-water influence.

Early Cretaceous Tethyan species of benthic foraminifera occurring in the western Atlantic are *Lenticulina busnardoi, L. nodosa gibber, Marssonella praeoxycona, P. lituus, Choffatella decipiens,* and *Pseudocyclammina virguliana.* An increase in temperate species suggests somewhat stronger northern, cool-water influence than during the Jurassic. The northern borders of the Tethyan realm probably lay close to the margin of the western Atlantic Ocean (Plate 9D).

By Late Cretaceous time the Gulf Stream gyre dominated North Atlantic surface circulation with branches northward into the Labrador Sea and into the Arctic Ocean. Planktonic foraminifera were well established and abundant. The foraminiferal record is very spotty because much of the sea floor was below CCD during Middle and Late Cretaceous time. Only during the Campanian and Maestrichtian when the CCD deepened was a fair record preserved.

The biogeography of Campanian-Maestrichtian planktonic foraminifera in the North Atlantic shows distinct latitudinal trends. Globotruncanids were most diverse and abundant in low paleolatitudes (Davids, 1966); single-keeled globotruncanids were typical of low-latitude waters (Berggren and Hollister, 1977; Douglas and Sliter, 1966). Globotruncanid species also showed biogeographic patterns. *Globotruncana arca* was more abundant in the northern areas of the North Atlantic, whereas *Globotruncana aegyptiaca* (identified as *G. gagnebini*) was more abundant in the low-latitude regions (Davids, 1966). Gradstein and Srivastava (1980) noted that *Globotruncana arca* as well as *G. linneiana, G. fornicata, G. stuartiformis, Globotruncanella havanensis, Rugoglobigerina rugosa,* and *Globigerinelloides messinae* occurred in the Laborador Sea and as far north as Baffin Bay. They suggested that these were temperate species. Davids (1966) observed that *Hedbergella* increased in abundance in northern latitudes. This observation is borne out in DSDP Site 384 and especially in DSDP Site 111 (Plate 10G) where there is a marked increase in *Hedbergella.* In contrast, species of *Globotruncana* are less abundant at these sites. In Sweden and Denmark, *Hedbergella* was characteristic of Maestrichtian assemblages which also included *Globigerinelloides, Rugoglobigerina,* and *Heterohelix.*

Heterohelicids were very abundant in the Campanian-Maestrichtian planktonic assemblages of the North Atlantic, usually ranging to more than 40 percent of the assemblage (Davids, 1966). Greater diversity of heterohelicid genera and species occurred in low paleolatitudes. Characteristic genera include *Pseudoguembelina, Racemiguembelina, Planoglobulina, Pseudotextularia, Ventilabrella,* and *Heterohelix.* In the high paleolatitudes of the North Atlantic, heterohelicids were represented almost entirely by *Heterohelix.*

The general latitudinal trend of species distribution in the North Atlantic was similar to that of the South Atlantic Campanian-Maestrichtian where Sliter (1976) differentiated Tethyan, Transitional, and Austral planktonic foraminiferal assemblages. In contrast to the South Atlantic, polar waters probably did not greatly influence the North Atlantic because a deep-sea Atlantic connection did not open to the Arctic region until early Tertiary time. A Boreal province was probably not present in the North Atlantic (Berggren and Hollister, 1977). Nevertheless, the similarity of the Austral assemblages of Sliter to those of Denmark and Sweden suggests a Boreal influence, possibly by epicontinental flowage through the Labrador Sea and the Norwegian-Greenland region (Plate 10G).

Gradstein and Srivastava (1980) suggest that the northernmost occurrences (50°N latitude) of species of *Globotruncana* along the northwest Atlantic margin might serve to delimit the extent of Late Cretaceous tropical-subtropical (Tethyan) oceanic surface waters. However, the northern limit of *Globotruncana* might indicate a warm, seasonal extension of Tethyan species into an otherwise temperate or transitional province. The boundary between the transitional province and the tropical to subtropical water of the Tethyan realm can be interpreted to lie further south in the vicinity of DSDP Site 384 where a northward reduction in

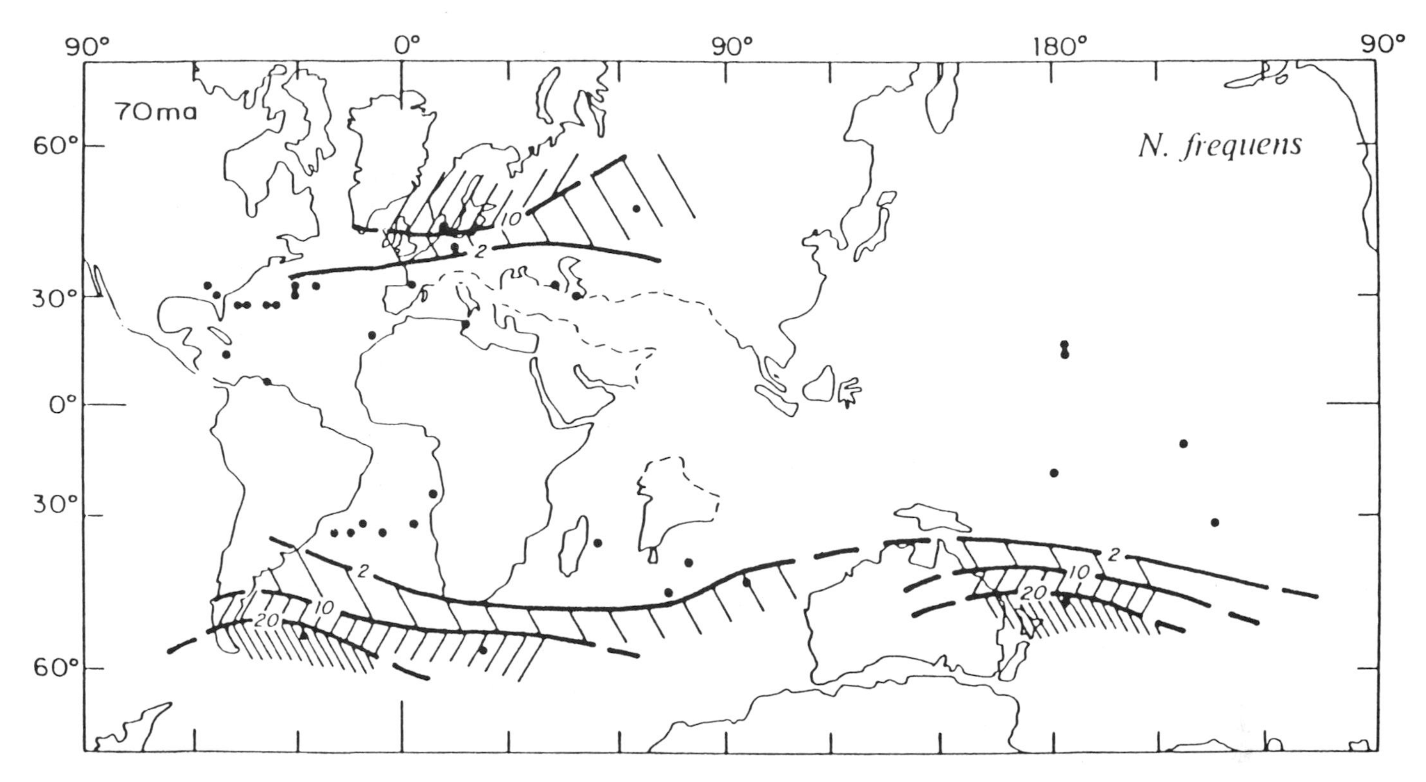

Figure 4. Average percent abundance of *Nephrolithus frequens* in latest Maestrichtian nannoplankton assemblages. Note strong boreal and austral affinities (after Thierstein, 1981, and modified to include data from Doeven, 1983).

globotruncanids is first noted (Plate 10G). Maestrichtian assemblages from the coastal plain of Delaware and New Jersey (Olsson, 1964, 1977) and the Newfoundland and Labrador shelves (Gradstein and Srivastava, 1980) are characterized by a general absence of Tethyan species and are distinctively temperate in composition. Although this might be due entirely to development of a shallow shelf environment which excluded for the most part incursions of Tethyan species, it might also indicate influence of cooler surface currents flowing south along the shelf margins of the western North Atlantic (Plate 10G).

The Middle Cretaceous nannoplankton biogeography of the Atlantic has been studied by Roth and Bowdler (1981). They distinguished between high-latitude assemblages, continental-margin assemblages, and oceanic assemblages. High-latitude assemblages were not identified in North Atlantic sites. Continental-margin assemblages are characterized by rare *Watznaueria barnesae* and common *Biscutum constans, Zygodiscus erectus, Z. elegans,* and *Z. diplogrammus.* Oceanic assemblages show abundant *W. barnesae, Parhabdolithus splendens,* and *P. asper.* Their study supports the concept of small surface-temperature gradients between low and high latitudes during the Middle Cretaceous and indicates a wide tropical belt extending from about 40°N to 40°S latitude. Minor latitudinal but strong east-west nannofloral gradients were present in the tropical belt.

In the Campanian and Maestrichtian, the nannofossil biogeography was strongly latitudinal (Thierstein, 1981). Species which dominate tropical and subtropical regions include *Watznaueria barnesae, Micula murus, Cylindralithus serratus,* and *Cretarhabdus surirellus. Nephrolithus frequens* showed strong boreal and austral affinities (Fig. 4). A sharp decrease in *Watznaueria barnesae* and a corresponding increase in *Micula "staurophora"* in the lower Maestrichtian of offshore wells, eastern Canada, indicates the onset of a cooling trend in the northwest Atlantic, a trend which is related to the opening of the Labrador Sea (Doeven, 1983). By latest Maestrichtian time, cool transitional waters, possibly influenced partly by boreal waters, existed along the northern borders of the North Atlantic. Circulation of tropical to subtropical waters were diverted further south (Plate 10G).

CENOZOIC

Paleogeography and Paleocirculation

Our understanding of the complex Cenozoic plate-tectonic history of the North Atlantic (Srivastava and Tapscott, this volume; Klitgord and Schouten, this volume) and North Sea has increased considerably as a result of several studies and summaries that have appeared over the past decade (Pitman and Talwani, 1972; Ziegler, 1975a, 1975b, 1979; Tucholke and Vogt, 1979; Eldholm and Thiede, 1980; Gradstein and Srivastava, 1980).

By Late Cretaceous time (anomaly 33 = 77–78 Ma) oceanic crust began to accrete between North America and Greenland. In the early Eocene (anomaly 24 = 55–56 Ma), a change from a two- to three-plate geometry occurred when spreading began along the incipient Reykjanes Ridge and in the Norwegian-Greenland Sea, resulting in the separation of Greenland from Rockall and Eurasia (Plate 10). During the Eocene, Greenland moved essentially in a NNW direction relative to Svalbard along a regional transform fault. Although there was no deep-water connection to the Arctic, the Greenland and Norwegian Basins were linked by deep water connections shortly after the initiation of spreading in the early Eocene. The possibility of further connections southward to the North Atlantic remains enigmatic and is discussed below.

During the late Eocene to earliest Oligocene (prior to anomaly 13 time) a change in relative motion between Greenland and Eurasia to WNW was associated with cessation of spreading in the Labrador Sea and a return to a two-plate geometry, with Greenland becoming a part of the North American plate. Separation (as opposed to translation) between Greenland and Svalbard began by early Oligocene time (anomaly 13 = 36 Ma) but a shallow-water connection of the Norwegian Sea to the Arctic may have existed prior to that time. Surface-water connection of the Atlantic with the Arctic probably existed through the Labrador Sea from Late Cretaceous time. During the Oligocene and Neogene the North Atlantic and Norwegian-Greenland Sea basins continued to widen. A deep-water connection of the Atlantic to the Arctic was probably established during the early Miocene.

The nascent Norwegian Sea was separated from the North Atlantic by a hot-spot generated, fairly flat-topped aseismic ridge (Greenland-Scotland Ridge). Subsidence and breaching of this barrier must have been a major event in the evolution of North Atlantic oceanography and in the exchange of terrestrial vertebrate faunas between Europe and North America. This was recognized by Berggren and Hollister (1974), who noted the temporal coincidence between the development of biogenic siliceous deposition in the North Atlantic (and, indeed the global, equatorial ocean) and the interruption in faunal continuity between European and North American terrestrial vertebrates about 49–50 Ma (McKenna, 1972, 1975).

Recent studies by Gradstein and Srivastava (1980) suggest that a sea strait linking the Arctic and Atlantic may have existed between Greenland and North America as early as Campanian-Maastrichtian time; judging from the far northward penetration of subtropical planktonic foraminifera, the surface circulation was directed towards the Arctic. McKenna (1980) also showed that the faunal evidence at Ellesmere Island indicates a warm but cyclic climate, with a weak equator-to-pole thermal gradient; this would be unlikely if poleward heat transfer during the winter was inhibited or blocked by land barriers or shallow shelves. Similarly, relatively warm waters from the North Atlantic must have penetrated into the Norwegian-Greenland Sea during the Eocene, because terrestrial flora indicate a warm, humid climate at that time on Spitsbergen.

What role did the Greenland-Scotland Ridge play as a possible barrier to the exchange? Nilsen (1978) and Thiede (1980) suggested that the main platform of the Iceland-Faeroe Ridge subsided below sea level during the early Miocene (*ca.* 24–14 Ma). Nilsen (1978) also suggested that the Greenland-Scotland Ridge (= Thulean land bridge) may have been broken or interrupted along its western connection with Greenland in the early Tertiary. Berggren and Schnitker (1983) suggested, on the other hand, that perhaps one should look in the other direction toward the Faeroe-Shetland Channel. If a micro-continent lies beneath the Paleocene lavas of the Faeroe Islands (Bott and others, 1971, 1974; Casten and Nielsen, 1975), the Faeroe-Shetland Channel could be a graben-like structure formed during the early Tertiary, Hebridean phase of volcanism (*ca.* 55 Ma; Brooks, 1980) and pre-drift tensional rifting. Subsidence of this area during the early Eocene could have formed a marine connection linking the North Atlantic with the Norwegian-Greenland Sea.

A brief interruption in this marine connection is required to allow a transitory early Eocene trans-Atlantic migration of terrestrial vertebrates (McKenna, 1972, 1975, 1983). This route could have been a subaerial or shallowly submerged "causeway" (Wyville-Thompson Ridge–Faeroe Bank) linking continental Europe with the Greenland-Scotland Ridge. This land bridge was terminated between 49 and 50 Ma (McKenna, 1973, 1975). Its transitory existence may be related to sea level changes. A rapid eustatic drop in sea level (TE1.2 of Vail and others, 1977) about 50 Ma could have temporarily exposed a shallow-water bridge that allowed the trans-Atlantic migration of terrestrial vertebrates. However, combined effects of subsidence and rising sea level would have quickly flooded the "land bridge" and reestablished a marine connection between the North Atlantic and Norwegian-Greenland Sea.

Marine connections of this Norwegian Sea with the Arctic are less certain. Prior to the early Oligocene (anomaly 13 = *ca.* 36 Ma), Svalbard and Greenland were essentially contiguous, although tectonically separated by a major transform fault. Separation accompanied by crustal thinning (Eldholm and Thiede, 1980) began by the early Oligocene with oceanic crust being generated no later than 32–34 Ma. It is possible that a shallow marine (epicontinental) connection existed between the Arctic and Barents Shelf from Jurassic time; this connection subsequently linked the Arctic-Barents sea with the Voring Plateau at least by early Eocene time (Thiede, 1980). This would vindicate the suggestion made by Berggren and Hollister (1971, 1974) that a deepening surface-water connection between the Arctic and North Atlantic was, in fact, initiated in the early Eocene, virtually concomitant with the initiation of seafloor spreading in the Norwegian Sea. Miller and Tucholke (1983) pointed out that the early Oligocene separation of Greenland and Svalbard could have allowed deeper, cooler Arctic water to flood the North Atlantic via passage(s) through the Greenland-Scotland Ridge, thus explaining dramatic early Oligocene erosion and changes in sedimentation patterns in the Atlantic.

Paleogene Calcareous Plankton Distribution. Latitudinal differentiation (provincialism) in both calcareous nannoplankton and planktonic foraminifera has existed essentially

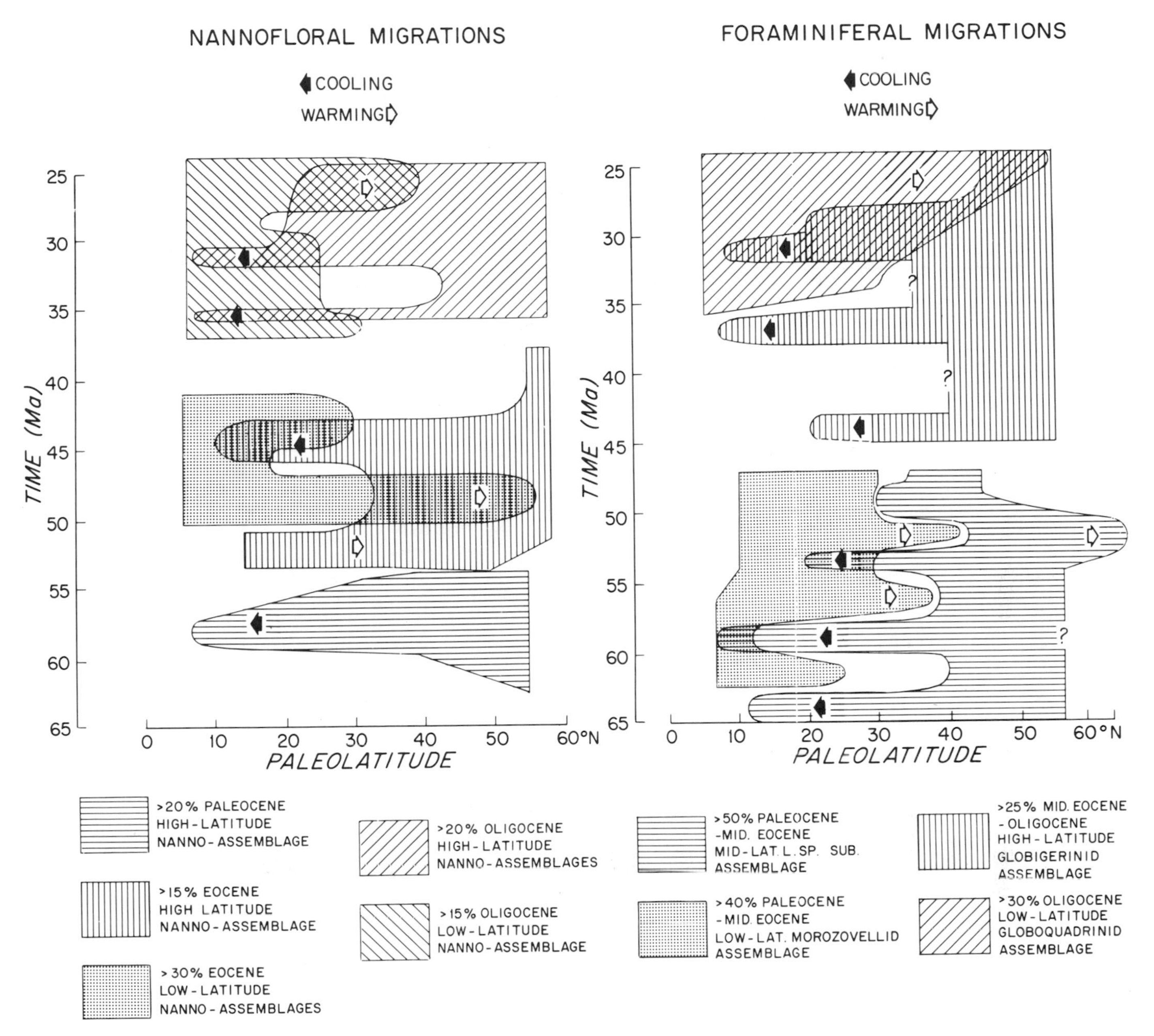

Figure 5. A summary of the major nannofloral and foraminiferal migrationary patterns through the early Cenozoic. Migrations toward higher latitudes are interpreted as being caused by climatic warming, and those toward lower latitudes, by climatic cooling. The patterns delineated enclose all samples which contain abundances greater than those indicated in the legend. Arrows in these areas indicate the directions of the major shifts of assemblages. Major and minor nannofloral assemblages having similar latitudinal preferences have been combined to obtain composite patterns in some cases (see text for details). (From Haq and Lohmann, 1976, Fig. 12).

throughout the Cenozoic Era. Only in the lowest Paleocene (basal Danian), following the massive extinction of planktonic organisms at the end of the Cretaceous, is a clear latitudinal differentiation not seen in either the low-diversity *Thoracosphaera - Braarudosphaera - Cyclagelosphaera* calcareous nannoplankton assemblage (Haq and others, 1971; Thierstein, 1981) or the correlative low-diversity *"Globigerina" eugubina* planktonic foraminiferal assemblage (Boersma and Premoli-Silva, 1983).

The repeated latitudinal migration (Figs. 5, 6, 7; Tables 1 and 2) of these microfloral and microfaunal assemblages has been interpreted in terms of climatic warming and cooling and concomitant displacement of surface currents (Haq and others, 1971). Marked warming episodes were recorded during the late Paleocene, early Eocene, and (less pronounced) late Oligocene, whereas four marked cooling episodes occurred during the middle Paleocene, middle Eocene, earliest Oligocene, and middle

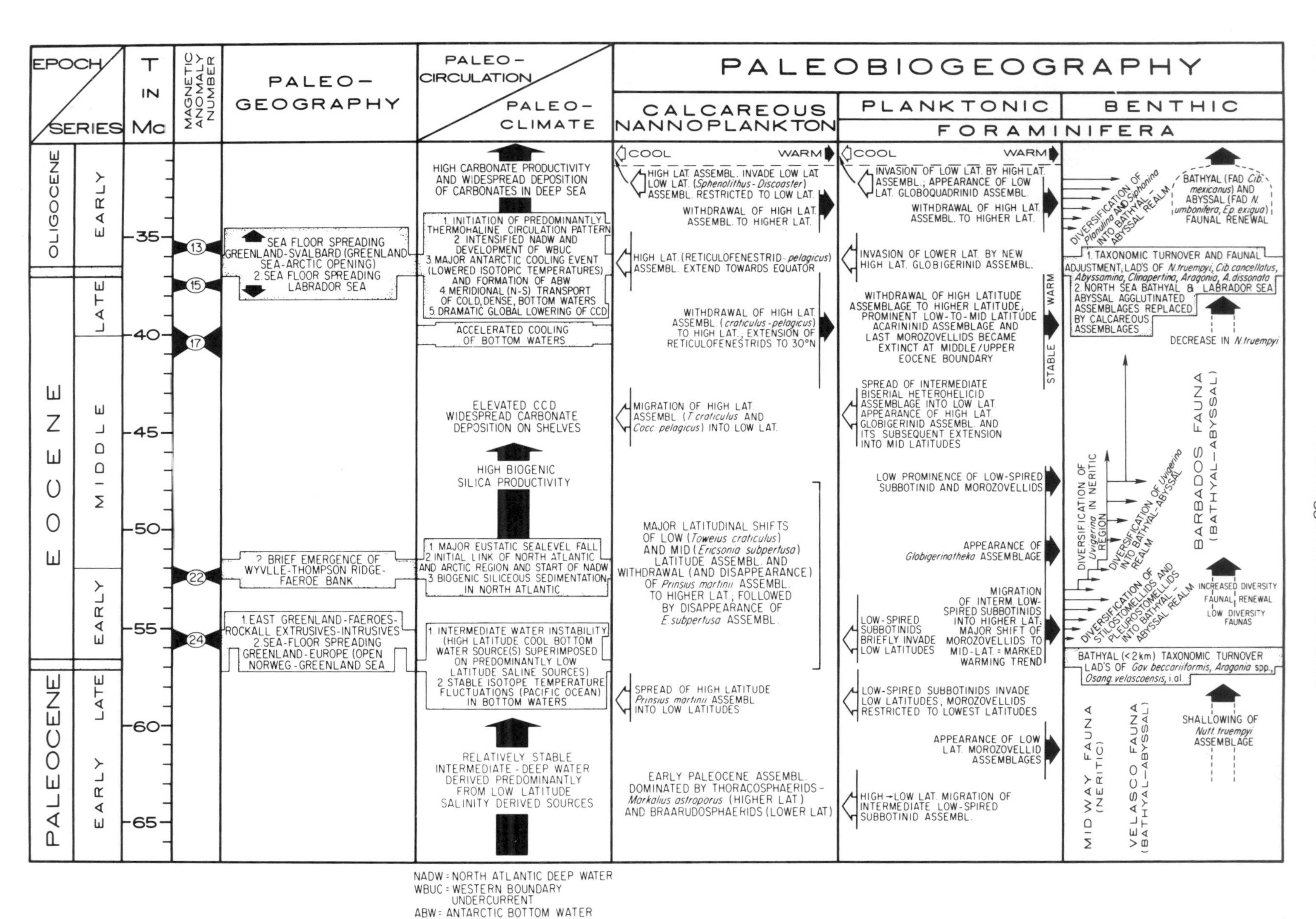

Figure 6. Paleogene event stratigraphy.

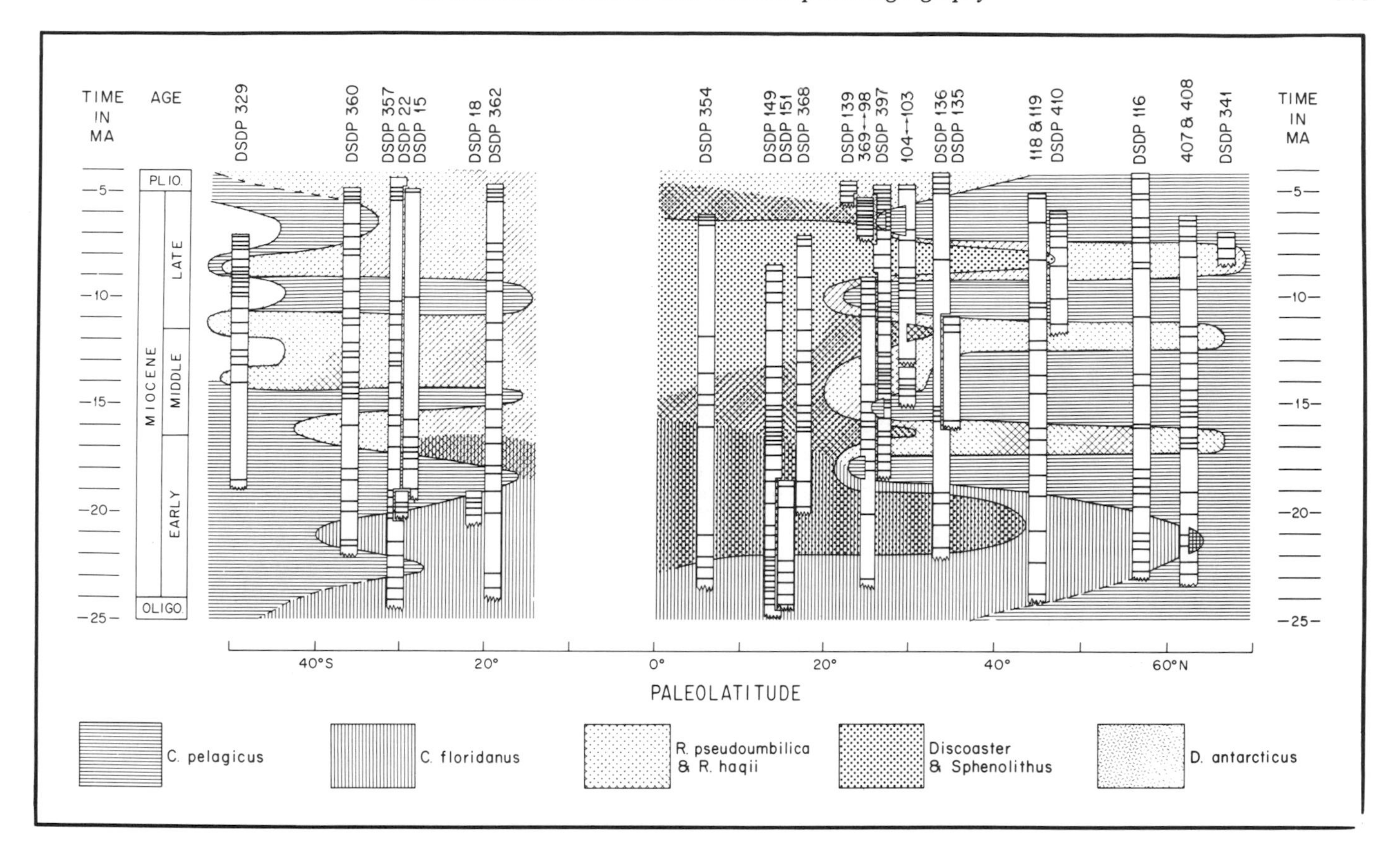

Figure 7. Summary of Miocene distribution patterns of 4 major assemblages and 1 minor assemblage in North and South Atlantic Ocean. To make migration patterns more discernable, the cosmopolitan and generally dominant *Dictyococcites minutus* assemblage has not been plotted. Migrations of assemblages characterizing low and mid-latitudes (*Discoaster-Sphenolithus* and *Reticulofenestra pseudoumbilica-R. haqii*) into higher latitudes is interpreted as a response to climatic warming; likewise extensions of an assemblage preferring high latitudes (*Coccolithus pelagicus* assemblage) into mid and low latitudes is interpreted as a response to cooling trends in climate. DSDP sites and their paleo-latitudinal positions are identified by numbers at the top of cores. Lines within core columns represent temporal levels of samples studied. Latitudinal shifts indicate generally cooler climates in earliest and latest Miocene and four warming and cooling cycles in the North and South Atlantic (and an additional minor warming episode in the South Atlantic). (From Haq, 1980, Fig. 12.)

Oligocene. During climatic optima in the early Paleocene (Zones Pld and P3a), poleward excursion of warm-water planktonic foraminiferal taxa occurred as far north as Greenland and the North Sea (Boersma and Premoli-Silva, 1983). These authors also suggest that the mid-Paleocene (Zone P3b) "cooling" event may, in fact, be an artifact of increased dissolution associated with increased productivity and higher surface-water values of carbon isotopes. Indeed, on the basis of a general correlation between oceanwide (as opposed to local) diversity of planktonic foraminiferal species and the carbon (rather than the oxygen) isotope record, Boersma and Premoli-Silva suggested that primary productivity and availability of nutrients exercised a controlling influence on the distribution of planktonic foraminifera.

Paleogene Benthic Foraminifera Distribution. Berggren and Aubert (1975) applied the terms "Midway-type" and "Velasco-type" to Paleocene benthic foraminiferal assemblages characteristic of neritic and bathyal-abyssal depth zones, respectively. The "Midway-type" fauna was differentiated, in turn, from a shallow-water (<50 m water depth), inner-middle-shelf assemblage (the Tethyan Carbonate Fauna) dominated by *Cibicides* spp., nonionids, discorbids, *Elphidium* spp., and larger benthic forms—*i.al. Nummulites, Operculina, Lockhartia, Rotalia, Ranikothalia.*

The Midway fauna is a rich, diverse association that is characterized by nodosariids, lenticulinids, dentalinids, marginulinids, vaginulinids, polymorphinids, anomalinids, gavelinellids, and *Cibicidoides,* and it is found primarily in clastic sediments. Faunal distribution is essentially cosmopolitan, and in the North Atlantic these assemblages have been recorded from upper Paleocene–lower Eocene sediments of the Rockall Plateau and as far north as Nugssuaq, West Greenland.

Velasco-type assemblages are characterized by robust gavelinellids (*G. beccariiformis, velascoensis*), *Nuttallides truempyi, Aragonia* spp., *Osangularia velascoensis,* gyroidinids, and several

TABLE 1. RELATIVE LATITUDINAL POSITIONS OF THE MAJOR NANNOPLANKTON ASSEMBLAGES (AFTER HAQ AND OTHERS, 1971, TABLE 1).

Latitudinal Preference	Assemblage	Time Interval When Dominant
Paleocene-early Eocene:		
Low	Toweius craticulus	middle Paleocene-early Eocene
Middle	Ericsonia subpertusa	early Paleocene-late Paleocene
High	Prinsius martinii	early Paleocene-late Paleocene
Eocene:		
Low to middle*	reticulofenestrid	early Eocene-late Eocene
	discoaster	early Eocene-middle Eocene
Middle	Reticulofenestra umbilica	early Eocene-middle Eocene
Middle to high*	Coccolithus pelagicus	early Eocene-middle Eocene
Oligocene:		
Low	sphenolith-discoaster	early Oligocene-late Oligocene
Cosmopolitan	Cyclicargolithus floridanus	early Oligocene-late Oligocene
Cosmopolitan	Dictyococcites hesslandii	early Oligocene-late Oligocene
High	reticulofenestrid	early Oligocene

*part of range emphasized

anomalinids. These assemblages have an essentially cosmopolitan distribution in the deep sea. Tjalsma and Lohmann (1983) provided detailed distribution data (quantitative and qualitative) on the various taxa of this assemblage. The Paleocene deep-sea fauna exhibits stratigraphic stability; faunal variation primarily is a result of gradual bathymetric restriction of the "shallower" *Gavelinella beccariiformis* assemblage and corresponding bathymetric expansion of the *Nuttallides truempyi* assemblage during the late Paleocene (Fig. 7).

A major benthic faunal turnover occurred near the Paleocene/Eocene boundary (Schnitker, 1979; Tjalsma and Lohmann, 1983). Following a gradual depth-dependent adjustment (depth restrictions vs. expansion) of various taxa, a rapid extinction event occurred (over 20 species), mainly in the "shallower" (2–3 km) *Gavelinella beccariiformis* assemblage which contained predominantly Cretaceous relict species. Basal and lower Eocene bathyal-abyssal assemblages are depauperate and characterized by low diversity; it took 2–3 m.y. for the deep-water benthic foraminifera to recover their normal vigor and develop to precrisis abundances and diversities. It appears that at least some of the new, predominantly deep-water species which appeared in the early Eocene evolved in the neritic realm and subsequently migrated to the greater depths of which they are characteristic. No corresponding crisis is seen in the neritic, "Midway-type" assemblages of the shelf regions. Following a rapid, relatively minor global eustatic lowering of sea level at the Paleocene/Eocene boundary, essentially similar faunas reestablished themselves in earliest Eocene time.

Deep-water benthic foraminifera gradually diversified during the early Eocene. This included the sequential appearance of several new taxa, *i. al. Cibicidoides grimsdalei, C. subspiratus, Alabamina dissonata,* and the development of species of *Stilostomella* and *Pleurostomella* as major components of bathyal and abyssal assemblages. The fauna appears to have remained relatively stable through the middle Eocene during a gradual lowering of deep-water temperatures. This Eocene fauna has been termed the "Barbados-type" fauna (Berggren and Schnitker, 1982) because of its documentation there 30 years ago (Beckmann, 1953).

Major late Eocene changes took place in the distribution patterns of deep-water benthic foraminifera following the relative middle Eocene stability. These changes were gradual and occurred over several million years, reflecting a change from predominantly haline to thermohaline circulation.

The major late Eocene faunal changes were the progressive restriction of mostly abyssal Eocene taxa to greater depths (>3 km), and the numerical reduction and ultimate extinction of several (predominantly) abyssal taxa (*Nuttallides truempyi, Abyssamina* spp., *Clinapertina* spp., *Aragonia* spp., and *Alabamina dissonata*). One of the more distinct bathyal extinctions was that of *Cibicidoides truncanus* (= *C. cancellatus* = *C parki*). Several species which are important components in the modern abyssal ocean appeared during the late Eocene–early Oligocene (*i.al., Nuttallides umbonifera, Epistominella exigua,* and *Eggerella bradyi*).

Oligocene abyssal faunas are dominated by stratigraphically

TABLE 2. RELATIVE LATITUDINAL POSITIONS OF PALEOGENE PLANKTONIC FORAMINIFERAL ASSEMBLAGES (AFTER HAQ AND OTHERS, 1971, TABLES 2 AND 3).

Latitudinal Preference	Assemblage	Time Interval When Dominant
High	globigerinid	late Eocene-early Oligocene
Middle* to high	biserial heterohelicid	late middle Eocene-early late Oligocene
Middle to low*	globoquadrinid	late early Oligocene-late Oligocene
Low	morozovellid	Paleocene-early Eocene
Low* to middle	*Acarinina spinuloinflata*	early and early middle Eocene
Middle* to low	low-spired subbotinid	Paleocene-early middle Eocene
Middle to high	*Globigerinatheka*	middle Eocene-late Eocene

*part of range emphasized

long-ranging and bathymetrically wide-ranging taxa, many of which survived the taxonomic turnover and faunal adjustment (depth displacement) at the end of the Eocene. During the Oligocene various long-ranging *Cibicidoides* taxa developed, the genus *Sphaeroidina* appeared, and the genera *Planulina* and *Siphonina* became common components of deep-sea assemblages. In the middle Oligocene, *Nuttallides umbonifera* became a numerically important component of abyssal faunas. Bathyal faunas exhibited many morphologic similarities with the Eocene bathyal *Lenticulina-Bulimina-Osangularia* assemblage (Tjalsma and Lohmann, 1983), although many of the constituent taxa changed.

Flysch-type agglutinated benthonic foraminiferal assemblages occur in the Paleocene and Eocene of the Canadian margin, the deep southern Labrador Sea, and the North Sea. These faunas terminate relatively abruptly near the Eocene/Oligocene boundary. On the Canadian margin and in the North Sea the distribution of these agglutinated assemblages is related to rapid deposition of carbon-rich, fine-grained clastics. Infilling of these basins during the late Eocene–Oligocene eliminated these faunas, and they were replaced by shallow-water (neritic), calcareous benthic assemblages. In the southern Labrador Sea (DSDP Site 112), agglutinated assemblages occurred at 3 km (paleodepth) and the faunal-replacement event coincided with the change in abyssal circulation already noted. The occurrence of agglutinated (flysch-type) foraminiferal assemblages appears to be related to certain hydrographic properties (low oxygen, low pH, high CO_2, corrosive older bottom waters). These conditions may develop in reducing substrates associated with high organic matter and poor circulation. Thus assemblages of (predominantly) agglutinated foraminifera are useful indicators of restricted circulation.

Paleoceanography. No major paleoceanographic changes have been documented in association with the deep-water benthic foraminiferal event at the Paleocene/Eocene boundary. It may be related to some event affecting intermediate and deep waters, although it is not clear at this time whether the Paleocene ocean was stratified into shallow, intermediate, and deep water masses. If the terminal Paleocene extinctions were a response to a "mid-water" event, then some surviving species responded by redeveloping in this new environment, while others avoided it by seeking greater depths. We explore the possible causes for this major event below in the light of recent data from the field of paleoceanography.

A revival and elaboration (Brass and others, 1982; Brass and others, 1983) of Chamberlin's salinity theory of bottom-water formation at low latitudes has led to a reconsideration of the history of deep- and bottom-water formation. High-latitude sources of cold bottom water probably were absent during the Cretaceous and early Cenozoic; thus, it has been suggested that low-latitude marginal seas with net evaporation could have been the source of warm, saline, dense bottom water (WSBW). Barron (1980) delineated the gradual, but marked, geographic repositioning of shallow seas from low latitudes (Cretaceous) to high latitudes (Cenozoic). Berger (1979) also suggested a change from haline to thermal ocean stratification during the Early Cenozoic; this is consistent with the view that the locus of deep-water formation changed from low to high latitudes (see also Kitchell and Clark, 1982, although their argument is somewhat tempered by Bukry's, 1984, revision of Arctic stratigraphy from Late Cretaceous to Early Eocene).

The predominance of hiatuses seen at 2.2–2.8 km and 3.8–5.2 km in the Late Paleocene–Early Eocene (Moore and others, 1981; Thiede, 1981; Ehrmann and Thiede, in press) suggests that injection of cool waters at intermediate and greater depths concomitant with the initial rifting and spreading of the NE Atlantic (Norwegian-Greenland Sea) might have occurred during Chron C24. The formation of (at least part of the) Early Eocene bottom waters in the North Atlantic may have taken place by sinking of relatively cooler, high-latitude surface waters

TABLE 3. MIOCENE CALCAREOUS NANNOPLANKTON VARIMAX FACTOR (BIOGEOGRAPHIC ASSEMBLAGE IN THE ATLANTIC OCEAN (DATA FROM HAQ, 1980).

ASSEMBLAGE	REMARKS
1. Cyclicargolithus floridanus	cosmopolitan in earliest Miocene (= Cyclococcolithus neogammation with D. hesslandi assemblage dominated the Oligocene.
2. Coccolithus pelagicus (sensu amplo)	dominant high northern latidude in most of the Miocene.
3. Discoaster-Sphenolithus	essentially 2:1 ratio between these two groups; important low latitude indicator in Miocene
4. Dictyococcites minutus	dominant Miocene assemblage; cosmopolitan, became numerically prominent in late early Miocene, but dominant in late Miocene.
5. Reticulofenestra pseudoumbilica - R. haqii	essentially 3:1 ratio between nominate taxa mid-latitude North Atlantic during late middle Miocene.

formed in the nascent Norwegian-Greenland Sea and their transport southwards via the longitudinal gateways of either (or both) the Faeroe-Shetland Channel or the Denmark Straits (Plate 1; Berggren and Schnitker, 1983; Miller and Tucholke, 1983; Miller, 1982). Evidence for increased North Atlantic bottom-water circulation in the Early Eocene is seen in the increase in biosiliceous and calcareous productivity, and enhanced preservation of the resulting sediments in the NW Atlantic and elsewhere (Berggren and Phillips, 1971; Berggren and Hollister, 1971, 1974). Berggren and Hollister (1974) developed a model which combined the following features by way of explanation: (a) increased productivity and preservation in response to an increase in silica and nutrient phosphorus resulting from ash alteration processes associated with extensive Early Eocene volcanism in the Caribbean (Gibson and Towe, 1971; contemporaneous volcanism in East Greenland and the Hebridian Province [Fitch and others, 1978] may also be expected to have played a significant role); and (b) initiation of "cold," deep water circulation in the North Atlantic. Concomitant formation of bottom waters from low-latitude, higher-salinity sources may have continued. The introduction of such multiple sources of intermediate and bottom water could have had a complicating effect on ocean geochemistry. Moore and others (1981; 1982), for example, noted relatively high variability in oxygen isotope measurements on early Paleogene benthic foraminifera. This suggested to them that multiple sources of deep and bottom waters were present, with a wide range of temperature and oxygen-isotopic compositions.

The data reviewed above suggest the following scenario for Late Cretaceous–Early Cenozoic circulation. (a) Late Cretaceous: formation of warm saline bottom waters (WSBW) from (predominantly) low-latitude sources. (b) Paleocene/Eocene boundary: initial formation of high-latitude North Atlantic intermediate waters by sinking of relatively cooler surface waters formed in the nascent Norwegian-Greenland Sea, and continued formation of bottom waters primarily from low-latitude, high-salinity sources. The resulting inhomogeneity (i.e., initial development of thermohaline stratification) may have caused disruption in the equilibrium-adaptation of bathyal and abyssal benthic foraminifera, thus causing massive extinctions. (c) Eocene: continued shift in the dominance of oceanic circulation from haline to thermohaline as a result of changing ocean-continent geometry. Thermohaline circulation became entirely dominant at the Eocene/Oligocene boundary with breaching of the connection between the Arctic and Norwegian-Greenland Seas (Miller and Tucholke, 1983). During the Eocene, deep-sea benthic foraminifera gradually became reestablished and adapted to the variable bottom-water conditions. As bottom-water sources became more homogeneous toward the late Eocene, albeit significantly colder, the deep-water benthic fauna responded in a series of orderly, sequential extinctions, appearances, opportunistic niche explorations, and faunal depth adjustments.

Low-latitude bottom-water temperatures decreased from about 15°C (Paleocene/Eocene boundary) to less than 5°C in the earliest Oligocene (Savin, 1977). This temperature decrease of some 10°C over a period of about 20 my had particularly marked accelerations near the Middle/Late Eocene boundary and at the Eocene/Oligocene boundary. It was a global phenomenon. The exact causes of the temperature decrease are uncertain but probably are linked to the gradual opening of gateways between the major oceans and high-latitude sources of cooler bottom waters (i.e., separation of Antarctica-Australia; subsidence of Tasman Ridge; breaching of Falkland-Antarctic connections; breaching of

TABLE 4. PLANKTONIC FORAMINIFERAL ASSEMBLAGES DERIVED FROM Q-MODE VARIMAX FACTOR ANALYSIS (FROM THUNELL AND BELYEA, 1982, TABLE 3).

Time Interval	Tropical-Subtropical Assemblage	Transitional Assemblage	Polar-Subpolar Assemblage
1. Quaternary	G. ruber G. crassaformis G. menardii G. truncatulinoides O. universa P. obliquiloculata S. dehiscens	G. inflata G. bulloides N. pachyderma (R)	N. pachyderma (L) N. pachyderma (R) G. bulloides
2. Late Pliocene	G. ruber G. sacculifer G. crassaformis O. universa P. obliquiloculata S. dehiscens	G. inflata G. bulloides G. crassaformis O. universa	N. pachyderma (L) N. pachyderma (R) G. bulloides G. glutinata
3. Early Pliocene	G. Menardi G. menardi G. altispira S. dehiscens S. seminulina	G. inflata G. inflata G. scitula O. universa N. acostaensis	N. pachyderma (L) N. pachyderma (R) G. glutinata G. bulloides
4. Late Miocene	S. semimulina S. subdehiscens G. altispira O. universa G. nepenthes N. acostaensis	G. dehiscens G. nepenthes S. subdehiscens	N. pachyderma (L) N. pachyderma (R) G. bulloides G. glutinata
5. Middle Miocene	G. ruber G. menardi	G. bulloides G. nepenthes G. glutinata G. scitula G. miozea	N. pachyderma (L) N. pachyderma (R)
6. Early Miocene	G. altispira G. trilobus G. peripheroronda G. siakensis G. venezuelana	G. dehiscens G. primordius G. altispira	G. bulloides G. glutinata G. miozea

the Rio Grande Rise; and, in the North Atlantic, linkage between the Arctic Ocean, the Norwegian-Greenland Sea, and the North Atlantic Ocean).

Neogene Calcareous Plankton Distribution. Latitudinal differentiation of calcareous plankton accelerated during the Neogene. Quantitative analysis shows five major assemblages (Table 3) in the Miocene, four of which exhibit distinct latitudinal shifts through time (Figs. 7 and 8; Haq, 1981). In the planktonic foraminifera, three "species associations" have been recognized in the Neogene and assigned (by Q-mode varimax factor analysis) to Tropical-Subtropical, Transitional, and Polar-Subpolar (Thunell and Belyea, 1982; Table 4). These papers and that of Berggren (1984) are the primary sources for the review of Neogene calcareous planktonic biogeography presented below.

Late Oligocene calcareous nannoplankton distribution patterns (characterized by cosmopolitan distribution of the *Cyclicargolithus floridanus* assemblage) continued into the earliest Miocene. Gradually a more complex pattern developed during the early Miocene with clear differentiation of the *Coccolithus pelagicus* assemblage in high latitudes (>45°N), the *Dictyococcites minutus* assemblage along the continental margins of the eastern North Atlantic (Bay of Biscay and N.W. Africa), and the *Discoaster-Sphenolithus* assemblage in the Caribbean and Gulf of Mexico.

Early Miocene planktonic foraminiferal distribution patterns exhibit a relatively marked uniformity over a large latitudinal extent. Tropical-Subtropical assemblages extended north almost to 65°N off eastern Greenland and to 40°N off the coast of NW Spain. This reflects the continued influence of the early Gulf Stream on North Atlantic paleobiogeography; it developed asymmetric paleobiogeographic distribution patterns on the western and eastern sides of the basin as a result of surface-water circulation patterns. The lack of high-latitude data precludes delineation of a distinct Polar-Subpolar assemblage or a northern boundary for the Transitional assemblage. Southward flow of a cool current along the eastern margin of the basin (to about 30°N) is suggested, however, by intermediate loadings (0–0.5) on "polar-subpolar" assemblages.

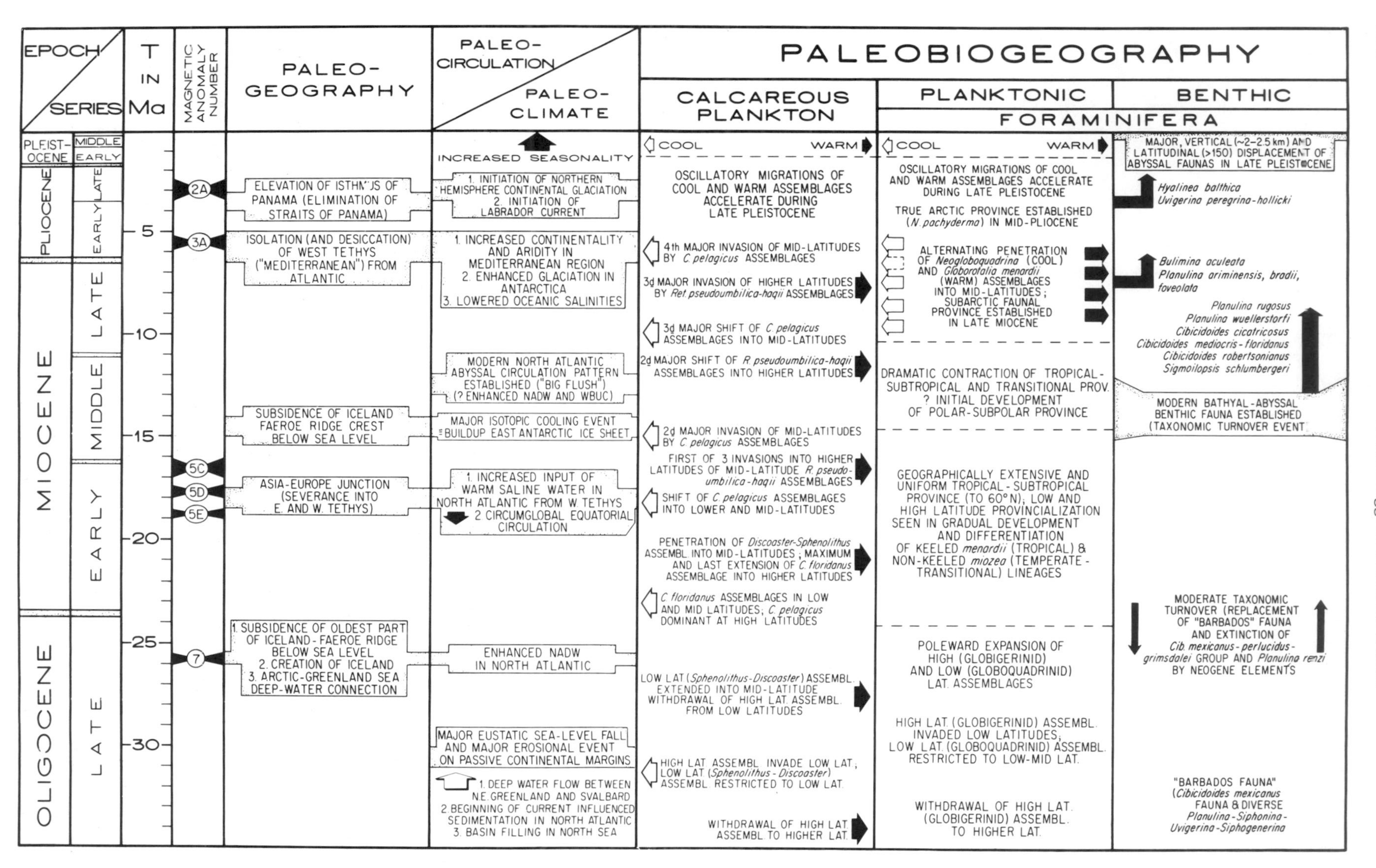

Figure 8. Neogene event stratigraphy.

Middle Miocene calcareous nannoplankton distribution patterns were like those of the late early Miocene except that the *D. minutus* assemblage spread to the Gulf of Mexico in the early middle Miocene, and the *Reticulofenestra pseudoumbilica-haqii* assemblage reached middle latitudes in the late early Miocene. The *Cyclicargolithus floridanus* assemblage was gradually restricted and ultimately eliminated during the middle Miocene, and it was replaced by the *Discoaster-Sphenolithus* assemblage.

Middle Miocene planktonic foraminiferal faunas exhibited a continuing trend towards latitudinal differentiation. There was a dramatic contraction of the Tropical-Subtropical province to within 30°–40°N of the equator (with maximum factor loadings [= >50 percent] restricted to within 15°N of the equator); a concomitant expansion of the Transitional Province from 25°N to 65°N occurred as a result of high-latitude cooling and southward extension of cool high-latitude water masses. The latitudinal differentiation was perhaps best exemplified by the essentially low versus mid-high latitude distribution patterns of the *Globorotalia fohsi* and *G. miozea* groups, respectively. The *fohsi* series has been recorded as far north as the eastern Canadian margin and the *miozea* group as far south as offshore of northwest Africa. This attests to the efficacy of western (Gulf Stream) and eastern (Canaries) boundary currents in biogeographic dispersal and in bringing groups with distinctly different environmental preferences into latitudinal juxtaposition.

The first true North Atlantic Polar-Subpolar assemblage is recognized in the Middle Miocene by Thunell and Belyea (1982) from the record of *Neogloboquadrina pachyderma* at DSDP Site 341 (Voring Plateau, *ca.* 68°N). However, neither the chronostratigraphic position nor the taxonomy of this planktonic taxon is firmly established. One of us (W.A.B.) recently examined samples from the interval in question of Site 341 and found a relatively diverse (5–6 species) fauna of *Neogloboquadrina* (ex. gr. *continuosa-acostaensis*), globigerinoids, and non-keeled globorotaliids. A typical early Pliocene sinistrally coiled *N. atlantica* fauna occurs above an unconformity with the lower upper Miocene, and there appears to be no evidence of a Polar fauna in the upper middle to lower upper Miocene at DSDP Site 341. Intermediate values of the Polar-Subpolar assemblage extend diagonally across the North Atlantic from approximately 65°N (western side) to 50°N (eastern side) and a southward extension of cool water is suggested to have occurred to the latitude of northwest Africa (~35°N).

The contraction of the Tropical-Subtropical Province and associated expansion of the Transitional Province is a direct expression of the mid-Miocene $\delta^{18}O$ increase seen in the isotopic record (Woodruff and others, 1981), and it is considered to represent major growth of the East Antarctic Ice Sheet.

During the late Miocene the *Dictyococcites minutus* assemblage spread throughout most of the North Atlantic Ocean, exhibiting a geographic distribution essentially similar to that of the *C. floridanus* assemblage during the early Miocene. At high latitudes the *pelagicus* assemblage remained dominant. In latest Miocene time, geographic distribution patterns remained essentially the same, although the *Discoaster-Sphenolithus* assemblage was reduced in dominance.

Late Miocene planktonic foraminiferal biogeography markedly departed from that of the middle Miocene. A distinct latitudinal differentiation occurred in North Atlantic faunas. Low-diversity subarctic assemblages consisted of *Globigerina bulloides,* (the apparently endemic) *Neogloboquadrina atlantica* (dextral), *N. acostaensis, Turborotalita quinqueloba,* orbulinids, and forms identified as *Neogloboquadrina pachyderma* (north of 60°N).

South of latitude 55°N these forms were joined by *Neogloboquadrina humerosa* and elements of the *Globorotalia tumida (merotumida-plesiotumida), G. menardii (G. pseudomiocenica),* and *G. conoidea-conomiozea* groups. *Globoquadrina dehiscens* has been reported in the late Miocene in the North Atlantic at sites ranging from 1800 to 4400 m; elsewhere it has an erratic upper stratigraphic limit which may be due to the shallow, essentially global, late Miocene excursion of the CCD (Ryan and others, 1974). Tropical faunas exhibited taxonomic and geographic instability during this time. The late Miocene was the time when the planktonic foraminiferal fauna developed an essentially modern aspect.

The late Miocene Tropical-Subtropical Province expanded considerably (to above 50°N) from its middle Miocene position and was particularly prominent in the central part of the North Atlantic. The Transitional assemblage also reached into higher latitudes (after its mid-Miocene restriction), extending from 50°N on the western, to 65°N on the eastern side of the basin. In fact the Transitional assemblage is the dominant late Miocene assemblage in the western North Atlantic between 25°N–45°N; the Tropical-Subtropical assemblage is of secondary importance there. The Transitional assemblage in this region of the western North Atlantic dominated the remainder of the Neogene to the present day. A Transitional fauna (dominated by *Globorotalia inflata*) characterizes the region around the Sargasso Sea today. Thus the Sargasso Sea may have existed as a distinct oceanographic entity at least since the late Miocene. Indeed, Berggren and Hollister (1974) and Olsson (this paper) argue for a Cretaceous subtropical gyre which suggests that the "Sargasso Sea" has been a distinct entity in the North Atlantic for 100 my; it may have existed even earlier, in the Late Jurassic (Tucholke and McCoy, this volume).

The late Miocene is marked by significant, multiple southerly excursions of the Polar-Subpolar assemblages into the North Atlantic, particularly in the central part of the basin. Intermediate factor loadings on the Polar-Subpolar assemblage as far south as 10°N off the coast of West Africa suggest that the eastern boundary current was episodically intensified during this interval.

Using numerical abundances of *Neogloboquadrina* (cool-water indicator) and *Globorotalia menardii* (warm-water indicator) at DSDP Sites 334 and 410, Poore (1981) delineated several, alternating equatorward and poleward migrations of surface-water masses with an essentially 1-my periodicity; cold events were identified at 9.5, 8.5, 7.5, and 5.5 Ma, with a subdued cold event at 6.5 Ma, and intervening warm events were identified at

9.8, and 7.6 Ma, with a questionable warm event at 10 Ma.

The Pliocene witnessed continued provincialization of planktonic foraminiferal faunas. Climatic cooling continued to compress essentially latitudinally-distributed faunal belts, and elevation of geographic barriers such as the Isthmus of Panama led to distinct, disjunct patterns of global biogeography. However, the tropical region continued to be the center of evolution and dispersal. Keeled globorotaliids and species of *Globigerinoides, Sphaeroidinellopsis,* and (locally) *Pulleniatina* were the dominant elements. The elevation of land areas in the area of the Isthmus of Panama in the early Pliocene resulted in the development of endemic Atlantic elements (*Globorotalia miocenica-exilis* group), the interruption in synchronous Atlantic-Indo-Pacific patterns of coiling changes in the genus *Pulleniatina,* and the temporary exclusion of the genus *Pulleniatina* from the Atlantic. *Pulleniatina* reappeared in the Atlantic Ocean once again in the late Pliocene (*ca.* 2.2 Ma), but subsequent coiling changes during the Pleistocene were out of phase with those in the Indo-Pacific.

Early Pliocene planktonic foraminiferal fauna north of latitude 60° were of low diversity, subarctic type (*i. al., Globigerina bulloides, Neogloboquadrina pachyderma, N. atlantica, N. acostaensis-humerosa* group, and *Globorotalia puncticulata*). Between latitudes 45° and 57°N, temperate assemblages included *Globorotalia margaritae,* the *Globorotalia conoidea-conomiozea* group, *Globigerina nepenthes,* and *Globigerinoides obliquus,* in addition to the forms listed above. A typical and relatively abundant form (in some instances comprising 50–90 percent of the total fauna) common to Pliocene faunas north of latitude 45° was the apparently endemic, sinistrally coiled *Neogloboquadrina atlantica.* A predominantly subtropical Pliocene faunal succession occurs at DSDP Site 111 in the Labrador Sea (lat. 50°N). It is unique in that it is a truly transitional type of fauna with tropical (*Sphaeroidinellopsis* spp., *Globorotalia menardii-tumida* group, *Globigerinoides* spp.) and temperate (*Globorotalia conoidea* group, *Neogloboquadrina atlantica-pachyderma*) elements occurring together. The northward extension of this fauna to latitude 50°N in the early Pliocene is attributed to the continued northerly path of the Gulf Stream, prior to mid-Pliocene inception of northern-hemisphere glaciation.

The greatest change in terms of biogeography during the early Pliocene was a significantly greater latitudinal extent (~30°N to 60°N) of the Transitional Province than during the late Miocene, indicating general climatic warming. The Polar-Subpolar assemblage was distributed as in the late Miocene except that the maximum loadings extend only to just south of Iceland (~65°N), 10° further north than in the late Miocene. Intermediate factor loadings indicate that a large tongue of cold water extended diagonally southwestward into low latitudes (~25°N) in the North Atlantic sometime during the early Pliocene.

The initiation of northern hemisphere glaciation over the interval of 3–2.4 Ma led to the formation of a true Arctic (Polar) faunal province and the development of the essentially monospecific *N. pachyderma* fauna characteristic of the present day (Berggren, 1972). A marked restriction of the Tropical-Subtropical assemblage occurred at the same time.

The transitional fauna continued its extensive geographic distribution. The sharply delineated, diagonal northern limit of this assemblage mirrors present-day conditions and attests to the long-term efficacy of the North Atlantic Drift (Gulf Stream system) in moderating northwestern European climate.

During the late Pliocene, moderate- to high-diversity faunas, including several tropical taxa, were present in the southern part of the Labrador Sea (DSDP Site 111), attesting to the periodic influence of the Gulf Stream. At a time coinciding approximately with the Pliocene/Pleistocene boundary (1.65 Ma), subpolar waters (reflected in faunal assemblages dominated by one species, *Neogloboquadrina pachyderma*) were established in the region of Orphan Knoll.

North Atlantic calcareous plankton biogeography during the Pleistocene is essentially a reflection of repeated water-mass migration across more than 20° of latitude (equivalent to oceanic surface water temperature oscillations of at least 12°C). It has been estimated that at least 11 such oscillations have occurred in the last 0.6 my and perhaps as many as 20 in the past 1.2 my. The southward penetration of Pleistocene polar waters (delineated by assemblages of sinistrally coiled *Neogloboquadrina pachyderma* and the absence of coccolithophorids) was about latitude 42°–45°N.

In the tropical North Atlantic (piston-core V16-205, lat. 15°N), early Pleistocene (Matuyama) cold episodes are characterized by the penetration of cold, high-latitude, dextrally coiled *Neogloboquadrina pachyderma,* whereas late Pleistocene (Brunhes) cold episodes are characterized by temperate-cool *Globorotalia inflata.* This suggests that maximum southward displacement of the Canaries Current occurred during the cold-water intervals of the early Pleistocene, whereas in the late Pleistocene a more moderate displacement of the Canaries Current brought temperate rather than subarctic waters into the tropical provinces.

The Tropical-Subtropical assemblage penetrated as far north as 60°N during Pleistocene interglacials. The Transitional fauna exhibit distribution patterns similar to those of the early and late Pliocene and they are most prominent along the margins of the North Atlantic basin between 20°N and 40°N. The Polar-Subpolar assemblage exhibits a striking NE–SW asymmetry, forming a distinct tongue-like extension from the Norwegian-Greenland Sea down along the western margin of the basin. This reflects the enhanced presence of the southward-flowing Labrador Current off the eastern coast of North America. The distribution of Pleistocene planktonic foraminiferal assemblages are seen to agree well with present-day plankton-tow and seafloor distribution patterns.

Neogene Benthic Foraminifera Distribution

Early Miocene deep-water benthic foraminiferal faunas (e.g., Hatton-Rockall Basin, Bay of Biscay) are essentially the same as those of the late Oligocene. Characteristic elements include *Oridorsalis umbonatus, Cibicidoides trincherasensis, Planulina renzi, Siphonina advena,* and *S. tenuicarinata.* Several typical Oligocene elements disappear near the Oligocene/Miocene boundary: *Cibicidoides mexicanus, C. perlucidus,* and *Planulina marialana.*

A major faunal-abundance change, from a *Siphonina*-dominated fauna to a *Cibicidoides*-dominated fauna, occurred just above the early/middle Miocene boundary in the Hatton-Rockall Basin, at a paleodepth of about 1 km. This change was coincident with the appearance of the modern deep-sea fauna (including such forms as *Planulina wuellerstorfi, Pyrgo murrhina, Laticarinina pauperata, Uvigerina peregrina, Sigmoilopsis schlumbergeri,* and *Anomalinoides cicatricosus*) in the middle Miocene (*ca.* 14–15 Ma). It appears to correspond closely with a major enrichment of $\delta^{18}O$ in benthic foraminifera (Fig. 8) which has been attributed to global change in the ^{18}O composition of seawater as a result of buildup of continental ice on Antarctica (Woodruff and others, 1981). The appearance of the new forms is associated with the disappearance of numerous taxa that had their initial appearance in the mid to late Paleogene (*i. al., Bulimina jarvisi, Buliminella grata, Hanzawaia cushmani*). The new fauna diversified and, by late middle Miocene time, the modern day fauna was well established.

Studies on the distribution of lower-bathyal and abyssal modern benthic foraminifera faunas of the North Atlantic have shown that the faunas conform essentially to the distribution of deep water masses in the North Atlantic (Streeter, 1973; Schnitker, 1974). Three assemblages have been noted; the shallowest (generally above 2.5 km) occurs on the lower part of the continental slope and the Mid-Atlantic Ridge and is characterized by species of *Uvigerina, Gyroidina,* and *Hoeglundina elegans*; a second, between 2.5 and 4 km, is dominated by *Epistominella exigua* and *Planulina wuellerstorfi*; the third assemblage, below 4 km, is strongly dominated by a single species, *Epistominella umbonifera.* North of 35°N, the *E. exigua* assemblage occurs over the entire basin below 3 km; the *E. umbonifera* assemblage occurs only south of 35°N.

These faunal assemblages experienced substantial geographic (>15° latitude) and bathymetric (up to 2–3 km) displacement during the Late Pleistocene. They responded to changes in bottom-water masses caused by glacial-interglacial fluctuations in the northern hemisphere. Composition and distribution of the bottom assemblages at the end of the last interglacial period, about 120,000 years ago, were almost the same as the present day.

The Neogene (last 24 my) has witnessed an inexorable and accelerating global climatic cooling. Causal mechanisms probably lie in a combination of continued changing ocean-continent geometry (northward translation of continents) and concomitant changing circulation patterns (latitudinal vs. longitudinal flows of surface and deep currents).

Neogene high-latitude, northern-hemisphere cooling is reflected in the tripartite division of calcareous plankton assemblages analogous to those of the present day, and their temporal and spatial fluctuation. Dramatic contractions of tropical-subtropical assemblages characterize the middle Miocene and middle Pliocene, reflecting the growth of the East Antarctic and northern high latitude ice caps respectively. These events are seen also in turnover in the benthic foraminifera faunas: in the middle Miocene in the form of the taxonomic turnover and development of the modern deep-sea fauna; in the mid-Pliocene in depth and geographic partitioning of deep-sea faunas.

Quo Vadim? The North Atlantic deep-sea fossil record provides a means of delineating major features of the biogeographic distribution of calcareous plankton and deep-sea benthic foraminifera during the Mesozoic and Cenozoic. However, a combined lack of adequate core coverage (particularly in the Mesozoic) and the temporal and spatial distribution of deep-sea stratigraphic hiatuses continue to hinder any detailed synopsis of the evolution of life in the sea. Seen against the background of the gradual change in ocean-continent geometry and attendant paleocirculation patterns, major changes in both faunal patterns (extinctions, turnovers, etc.) and stable isotopes are relatively abrupt (punctuational). Integration of data from the fields of lithostratigraphy, isotopic stratigraphy, magnetostratigraphy, and biostratigraphy is now providing the means for development of a *holistic* (Moore and Romine, 1981) or *systemic* (Berger and Vincent, 1981) approach to global chronostratigraphy. This is the conceptual framework within which paleoceanographers will be able to work in the years ahead as they seek to identify major threshold events in the evolution of the ocean and their relation to changing patterns of paleogeography, paleoclimate, and paleocirculation.

REFERENCES

Arthur, M. A.
1979 : North Atlantic Cretaceous black shales: The record at Site 398 and a brief comparison with other occurrences; in Initial Reports of the Deep Sea Drilling Project, v. 47, pt. 2, eds. J. C. Sibuet and W.B.F. Ryan; U.S. Government Printing Office, Washington, D.C., p. 719–752.

Ascoli, P.
1977 : Foraminiferal and ostracod biostratigraphy of the Mesozoic-Cenozoic, Scotian Shelf, Atlantic Canada; First International Symposium Benthonic Foraminifera, pt. B, Halifax, 1975; Maritime Sediments, Special Publication 1, p. 653–771.

Aubert, J., and Bartenstein, H.
1976 : *Lenticulina (L.) nodosa*—additional observations in the worldwide Lower Cretaceous; Bulletin Centre Recherches (Pau-SNPA), v. 10, p. 1–33.

Azema, J., and Jaffrezo, M.
1984 : Calpionellid stratigraphy in sediment across the Jurassic/cretaceous boundary in offshore Morocco (Deep Sea Drilling Project Leg 79) and their distribution in the North Atlantic Ocean; in Initial Reports of the Deep Sea Drilling Project, v 79, eds. K. Hinz and G. L. Winterer; U.S. Government Printing Office, Washington, D.C., p. 651–666.

Barron, E. J.
1980 : Paleogeography and climate: 180 million years to the present: [unpublished Ph.D. Dissertation]: University of Miami, Miami, Florida, 270 p.

Barron, E. J., and Washington, W. M.
1982a: Cretaceous climate: a comparison of atmospheric simulations with the geologic record: Palaeogeography, Palaeoclimatology, Palaeoecology, v. 40, p. 103–133.
1982b: The atmospheric circulation during warm geologic periods: is the equator-to-pole surface temperature the controlling factor? Geology, v. 10, p. 633–636.

Baron, E. J., Sloan, J. C., and Harrison, C.G.A.
1980 : Potential significance of land-sea distribution and surface albedo variations as a climatic forcing factor: 180 my to the present. Palaeogeography, Palaeoclimatology, Palaeoecology, v. 30, p. 17–40.

Barron, E. J., Thompson, S. L. and Schneider, S. H.
1981 : An ice-free Cretaceous? Results from climate model simulations. Science, v. 212, p. 501–508.

Bartenstein, H.
1974 : Upper Jurassic–Lower Cretaceous primitive arenaceous foraminifera from DSDP Sites 259 and 261, eastern Indian Ocean; in Initial Reports of the Deep Sea Drilling Project, v. 27, eds. J. J. Veevers and J. R. Heirtzler; U.S. Government Printing Office, Washington, D.C., p. 683–695.

Bartenstein, H., Bettenstaedt, F., and Bolli, H.
1957 : Die Foraminiferen der Unterkreide von Trinidad, B.W.I. 1 Teil: Cuche und Toco-Formation: Eclogae Geologicae Helvetiae, v. 50, p. 5–68.
1966 : Die Foraminiferen der Unterkreide von Trinidad, B.W.I. 2 Teil: Maridale Formation (Typlokalität): Eclogae Geological Helvetiae, v. 59, p. 129–177.

Bartenstein, H., and Bolli, H. M.
1973 : Die Foraminiferen der Unterkreide von Trinidad, W. I. Dritter Teil: Maridaleformation (Co-Typlokalität): Eclogae Geological Helvetiae, v. 66, p. 389–418.

Bartenstein, H., and Brand, E.
1951 : Micropalaontologische Untersuchungen zur Stratigraphie des nordwest-deutschen Valendis: Senckenbergiana Naturforschenden Gesellschaft Abhandlungen, v. 485, p. 239–336.

Beckmann, J. P.
1978 : Late Cretaceous smaller benthic foraminifera from Sites 363 and 364, DSDP Leg 40, southern Atlantic Ocean; in Initial Reports of the Deep Sea Drilling Project, v. 40, eds. H. M. Bolli and W.B.F. Ryan; U.S. Government Printing Office, Washington, D.C., p. 759–781.
1953 : Die Foraminiferen der Oceanic Formation (Eocaen-Oligocaenen) von Barbadoes: Eclogae Geological Helvetiae, v. 46, p. 301–412.

Berger, W. H.
1979 : Impact of deep-sea drilling on paleoceanography; in Deep Sea Drilling Results in the Atlantic Ocean, Continental Margins and Paleoenvironments, eds. M. Talwani, W. W. Hay, and W.B.F. Ryan; American Geophysical Union, Maurice Ewing Series, v. 3, Washington, D.C., p. 297–314.

Berger, W. H., and Vincent, E.
1981 : Chronostratigraphy and biostratigraphic correlation: exercises in systemic stratigraphy. Oceanologica Acta, 1981, Proceedings 26 International Geological Congress, Geological Oceans Symposium, Paris, July 7–17, 1980; p. 115–127.

Berger, W. H., Vincent, E., and Thierstein, H. P.
1981 : The deep sea record: major steps in Cenozoic ocean evolution; in The Deep Sea Drilling Project: a Decade of Progress, eds. J. E. Warme, R. G. Douglas and E. L. Winterer; Society of Economic Paleontologists and Mineralogist, Special Publication 32, p. 489–504.

Berggren, W. A.
1972 : Late Pliocene-Pleistocene glaciation; in Initial Reports of the Deep Sea Drilling Project, v. 12, eds. A. S. Laughton and W. A. Berggren; U.S. Government Printing Office, Washington, D.C., p. 965–1001.
1978 : Recent advances in Cenozoic planktonic foraminiferal biostratigraphy, biochronology, and biogeography: Atlantic Ocean: Micropaleontology, v. 24(4), p. 337–370.
1984 : Neogene planktonic foraminiferal biostratigraphy and biogeography: Atlantic, Mediterranean, and Indo-Pacific regions; in Pacific Neogene Datum Planes: Contributions to Biostratigraphy and Chronology, eds. N. Ikebe and R. Tsuchi; University of Tokyo Press, p. 111–161.

Berggren, W. A. and Aubert, J.
1975 : Paleocene benthonic foraminiferal biostratigraphy, paleobiogeography and paleoecology of Atlantic-Tethyan regions: Midway-type fauna: Palaeogeography, Palaeoclimatology, Palaeoecology, v. 18, p. 73–192.

Berggren, W. A., and Hollister, C. D.
1971 : Biostratigraphy and history of circulation of the North Atlantic: American Association of Petroleum Geologists, v. 55, p. 331.
1974 : Paleogeography, paleobiogeography and the history of circulation in the Atlantic Ocean; in Studies in Paleo-oceanography, ed. W. W. Hay; Society of Economic Paleontologists and Mineralogists, Special Publication 20, p. 126–186.
1977 : Plate tectonics and paleocirculation—Commotion in the ocean: Tectonophysics, v. 38, p. 11–48.

Berggren, W. A., and Phillips, J. D.
1971 : The influence of continental drift on the distribution of Cenozoic benthonic foraminifera in the Mediterranean and Caribbean, Gulf Coast regions; Symposium on Geology of Libya (Tripoli, 1969), Proceedings; Catholic Press, Beirut, p. 263–299.

Berggren, W. A. and Schnitker, D.
1983 : Cenozoic marine environments in the North Atlantic and Norwegian-Greenland Sea; in Structure and development of the Greenland-Scotland Ridge, NATO Conference Series IV: Marine Sciences, v. 8, eds. M.H.P. Bott, J. Saxov, M. Talwani, and J. Thiede; Plenum Press, New York, p. 495–548.

Boersma, A., and Premoli-Silva, I.
1983 : Paleocene planktonic foraminiferal biogeography and the paleoceanography of the Atlantic Ocean. Micropaleontology, v. 29(4), p. 355–381.

Bott, M.H.P., Browitt, C.W.A., and Stacey, A. P.
1971 : The deep structure of the Iceland-Faeroe Ridge: Marine Geophysical Research, v. 1, p. 328–351.

Bott, M.H.P., Sunderland, J., Smith, P. J., Casten, U., and Saxov, S.
1974 : Evidence for continental crust beneath the Faeroe Islands: Nature, v. 248, p. 202–204.

Brady, H. B.
1884 : Report on the Foraminifera dredged by H.M.S. Challenger, during the years 1873–1876: Reports of the Scientific Results of the Voyage of H.M.S. Challenger, v. 9 (Zoology), p. i-xx, 1–814, London.

Brass, G. W., Southern, J. R., and Peterson, W. H.
1982 : Warm saline bottom water in the ancient ocean: Nature, v. 296, p. 620–623.

Brass, G. W., Hay, W. W., Holser, W. J., Peterson, W. H., and others
1983 : Ocean circulation, plate tectonics and climate; National Research Council Report on Pre-Pleistocene Climate, eds. J. Crowell and W. Berger; Washington, D.C., p. 83–89.

Brooks, C. K.
1980 : Episodic volcanism, epeirogenesis and the formation of the North Atlantic Ocean: Palaeogeography, Palaeogeography, Palaeoclimatology, Palaeoecology, v. 30, p. 229–242.

Bukry, D.
1984 : Paleogene paleoceanography of the Arctic Ocean is constrained by the middle or late Eocene age of USGS core F1-422: evidence from silicoflagellates: Geology, v. 12, no. 4, p. 199–201.

Casten, U. and Nielsen, P. H.
1975 : Faeroe Islands—a microcontinental fragment?: Journal of Geophysics, v. 41, p. 357–366.

Corliss, B. H.
1981 : Deep-sea benthonic foraminiferal faunal turnover near the Eocene/Oligocene boundary: Marine Micropaleontology, v. 6, p. 367–384.

Davids, R. N.
1966 : A paleoecologic and paleo-biogeographic study of Maestrichtian planktonic foraminifera: [unpublished Ph.D. thesis]: Rutgers University, 241 p.

Doeven, P. H.
1983 : Cretaceous nannofossil stratigraphy and paleoecology of the Canadian Atlantic margin: Geological Survey of Canada, Bulletin 356, p. 1–69.

Douglas, R., and Sliter, W. V.
1966 : Regional distribution of some Cretaceous Rotaliporidae and Globotruncanidae (Foraminiferida) within North America: Tulane Studies in Geology, v. 4, p. 89–131.

Douglas, R. and Woodruff, F.
1981 : Deep sea benthic foraminifera. in The Oceanic Lithosphere, The Sea, vol. 7, ed. C. Emiliani; p. 1233–1327.

Ehrmann, W. U., and Thiede, J.
in press: History of Mesozoic and Cenozoic sediment fluxes to the North Atlantic Ocean: Contributions to Sedimentology.

Eldholm, O. and Thiede, J.
1980 : Cenozoic continental separation between Europe and Greenland: Palaeogeography, Palaeoclimatology, Palaeoecology, v. 30, p. 243–259.

Fitch, F. J., Hooker, P. J., Miller, J. A., and Brereton, N. K.
1978 : Glauconite dating of Palaeocene-Eocene rocks from East Kent and the time scales of Palaeogene volcanism in the North Atlantic region: Journal of the Geological Society of London, v. 135, p. 499–512.

Gawor-Biedowa, E.
1972 : The Albian, Cenomanian and Turonian foraminifera of Poland and their stratigraphic importance: Acta Paleontologica Polonica, v. 17, p. 1–155.

Geroch, S.
1966 : Lower Cretaceous small foraminifera of the Silesian series, Polish Carpathians: Rocznik Polskiego Towarzystwa Geologicznego, v. 36, p. 413–480.

Gibson, T. and Towe, K.
1971 : Eocene volcanism and the origin of Horizon A: Science, v. 172, p. 152–153.

Gradstein, F. M.
1977 : Biostratigraphy and biogeography of Jurassic Grand Banks foraminifera; in First International Symposium on Benthonic Foraminifera, Pt. B, Halifax, 1975; Maritime Sediments, Special Publication 1, p. 557–583.
1978 : Biostratigraphy of Lower Cretaceous Blake Nose and Blake-Bahama Basin foraminifera DSDP leg 44, Western North Atlantic Ocean; in Initial Reports of the Deep Sea Drilling Project, v. 44, eds. W. E. Benson and R. E. Sheridan; U.S. Government Printing Office, Washington, D.C., p. 663–701.
1983 : Paleocology and stratigraphy of Jurassic abyssal foraminifera in the Blake-Bahama Basin, Deep Sea Drilling Project Site 534; in Initial Report of the Sea Drilling Project, v. 76, eds. R. E. Sheridan and F. M. Gradstein; U.S. Government Printing Office, Washington, D.C., p. 537–559.

Gradstein, F. M., and Berggren, W. A.
1981 : Flysch-type agglutinated Foraminifera and the Maestrichtian to Paleocene history of the Labrador and North Seas: Marine Micropaleontology, v. 6, p. 211–268.

Gradstein, F. M., and Sheridan, R. E.
1983 : On the Jurassic Atlantic Ocean and a synthesis of results of Deep Sea Drilling Project Leg 76; in Initial Reports of the Deep Sea Drilling Project, v. 76, eds. R. E. Sheridan and R. M. Gradstein; U.S. Government Printing Office, Washington, D.C., p. 913–943.

Gradstein, F. M., and Srivastava, S. P.
1980 : Aspects of Cenozoic stratigraphy and paleoceanography of the Labrador Sea and Baffin Bay: Palaeogeography, Palaeoclimatology, Palaeoecology, v. 30, p. 261–295.

Habib, D.
1983 : Sedimentation-rate-dependent distribution of organic matter in the North Atlantic Jurassic-Cretaceous; in Initial Reports of the Deep Sea Drilling Project, v. 76, eds. R. E. Sheridan and F. M. Gradstein; U.S. Government Printing Office, Washington, D.C., p. 781–794.

Hanzlikova, E.
1972 : Carpathian Upper Cretaceous foraminifera of Moravia (Turonian-Maestrichtian): Czechoslovakia Ustredni Ustav Geologicky Rozpravy, v. 39, p. 1–159.

Haq, B. U.
1980 : Biogeographic history of Miocene calcareous nannoplankton and paleogeography of the Atlantic Ocean: Micropaleontology, v. 26, p. 414–443.
1981 : Paleogene paleoceanography: Early Cenozoic oceans revisited: in Oceanologica Acta, 1981, Proceedings 26 International Geological Congress, Geology of the Oceans Symposium, Paris, July 7–17, 1980; p. 71–82.

Haq, B. U., and Lohmann, G. P.
1976 : Early Cenozoic calcareous nannoplankton biogeography of the Atlantic Ocean. Marine Micropaleontology, v. 1, p. 111–194.

Haq, B. U., Premoli-Silva, I., and Lohmann, G. P.
1971 : Calcareous plankton biogeographic evidence for major climatic fluctuations in the early Cenozoic Atlantic Ocean: Journal of Geophysical Research, v. 82, p. 3861–2876.

Jannin, F.
1967 : Les "Valvulineria" de l'Albien de l'Aube: Revue de Micropaleontologie, v. 10, p. 153–178.

Jansa, L. F., Remane, J., and Ascoli, P.
1980 : Calpionellid and foraminiferal-ostracod biostratigraphy at the Jurassic-Cretaceous boundary, offshore eastern Canada: Rivista Italiana Paleontologia, v. 86, p. 67–126.

Kaever, M.
1961 : Morphologie, Taxionomie und Biostratigraphie con *Globorotalites* und *Conorotalites* (Kreide-Foram.): Geologie Jahrbuch, v. 78, p. 387–438.

Kitchell, J. A., and Clark, D. L.
1982 : Late Cretaceous–Paleogene paleogeography and paleocirculation: evidence of North Polar upwelling: Palaeogeography, Palaeoclimatology, Palaeoecology, v. 40, p. 135–165.

Krasheninnikov, V. A.
1973 : Cretaceous benthonic foraminifera, Leg 20, Deep Sea Drilling Project; in Initial Reports of the Deep Sea Drilling Project, v. 20, eds. B. C. Heezen and I. D. Macgregor; U.S. Government Printing Office, Washington, D.C., p. 205–219.
1974 : Upper Cretaceous benthonic agglutinated foraminifera, Deep Sea Drilling Project; in Initial Reports of the Deep Sea Drilling Project, v. 27, eds. J. J. Veevers and J. R. Heirtzler; U.S. Government Printing Office, Washington, D.C., p. 631–661.

Krasheninnikov, V. A., and Pflaumann, U.
1978 : Cretaceous agglutinated foraminifera of the Atlantic Ocean off West Africa (Leg 41, DSDP); in Initial Reports of the Deep Sea Drilling Project, v. 41, eds. Y. Lancelot and E. Seibold; U.S. Government Printing Office, Washington, D.C., p. 565–580.

Malapris, M.
1965 : Les Gavelinellidae et formes affines du gisement Albien de Courcelles (Aube): Revue de Micropaleontologie, v. 81, p. 131–150.

Malapris-Bizouard, M.
1967 : Les Lingulogavelinelles de l'Albien Inferieur et Moyen de L'Aube:

Revue de Micropaleontologie, v. 10, p. 128–150.
Mayne, W.
1970 : Lower Cretaceous foraminifera from Gorringe Bank, Eastern North Atlantic; in Initial Reports of the Deep Sea Drilling Project, v. 13, eds. W.B.F. Ryan and K. J. Hsu; U.S. Government Printing Office, Washington, D.C., p. 1075–1111.
McKenna, M. C.
1972 : Eocene final separation of the Eurasian and Greenland-North American land masses: Proceedings, 24th International Geological Congress, v. 7, p. 1201–1239.
1975 : Fossil mammals and early Eocene North Atlantic land continuity: Annals of the Missouri Botanical Gardens, v. 62, p. 335–353.
1980 : Eocene paleolatitudes, climate and mammals of Ellesmere Island: Palaeogeography, Palaeoclimatology, Palaeoecology, v. 30, p. 349–362.
1983 : Cenozoic paleogeography of North Atlantic land bridges: in Structure and development of the Greenland-Scotland Ridge, NATO Conference Series IV: Marine Sciences, v. 8, eds. M.H.P. Bott, S. Saxov, M. Talwani and J. Thiede; Plenum Press, New York, p. 351–399.
McNulty, C. L.
1976 : Cretaceous foraminiferal stratigraphy, DSDP Leg 33, Holes 315A, 316, and 317A; in Initial Reports of the Deep Sea Drilling Project, v. 33, eds. S. O. Schlanger and E. D. Jackson; U.S. Government Printing Office, Washington, D.C., p. 369–381.
McNulty, C. L.
1979 : Smaller Cretaceous foraminifera of Leg 43, Deep Sea Drilling Project; in Initial Reports of the Deep Sea Drilling Project, v. 43, eds. B. E. Tucholke and P. R. Vogt; U.S. Government Printing Office, Washington, D.C., p. 487–506.
Michael, E.
1966 : Die evolution der Gavelinelliden (Foram) in der NW-deutschen Unterkreide: Senckenbergiana Lethaea, v. 47, p. 411–459.
Miller, K. G.
1982 : Late Paleogene (Eocene to Oligocene) paleoceanography of the northern North Atlantic [unpublished Ph.D. Thesis]: Woods Hole Oceanographic Institution–Massachusetts Institute of Technology, 92 p.
Miller, K. G., Gradstein, F. M., and Berggren, W. A.
1982 : Late Cretaceous to early Tertiary agglutinated benthic foraminifera in the Labrador Sea: Micropaleontology, v. 28, p. 1–30.
Miller, K. G. and Tucholke, B. E.
1983 : Development of Cenozoic abyssal circulation south of the Greenland-Scotland Ridge: in Structure and Development of the Greenland-Scotland Ridge, Nato Conference Series IV; Marine Sciences, v. 8, eds. M.H.P. Bott, S. Saxov, M. Talwani, and J. Thiede; Plenum Press, New York, p. 549–589.
Moore, Jr., T. C., and Romine, K.
1981 : In search of biostratigraphic resolution: in The Deep Sea Drilling Project: a Decade of Progress, eds. J. E. Warme, R. G. Douglas and E. L. Winterer; Society of Economic Paleontologists and Mineralogists Special Publication 32, p. 317–334.
Moore, Jr., T. C., Pisias, N. G., and Keigwin, Jr., L. D.
1981 : Ocean basin and depth variability of oxygen isotopes in Cenozoic benthic foraminifera: Marine Micropaleontology, v. 6, p. 465–481.
1982 : Cenozoic variability of oxygen isotopes in benthic foraminifera: in Climate in Earth History (Studies in Geophysics), eds. W. H. Berger and J. C. Crowell; National Academy Press, Washington, D.C., p. 172–182.
Moore, Jr., T. C., van Andel, T. H., Sancetta, C., and Pisias, N. G.
1978 : Cenozoic hiatuses in pelagic sediments: Micropaleontology, v. 24, p. 113–138.
Moullade, M.
1966 : Etude stratigraphique et micropaleontologique du Cretace inferieur de la "Fosse Vocontinence": Documents, Laboratoire de Geologie, Faculte des Sciences, Lyon, No. 15, 369 p.
1973 : Zones de Foraminiferes du Cretace inferieur mesogeen: Comptes Rendus Hebdomadaires des Seances de l'Academie des Sciences, v. 278, ser. D., p. 1813–1816.
Myatliuk, E. V.
1970 : Foraminifery flishevikh otlozheniy vostochnykh karpat (Mel-paleogen): Vsesoyuznyy Neftyanoy Nanchno-Issledovatel'skiy Geologo razvedochnyy Institut, Trudy, v. 282, 360 p.
Neagu, T.
1965 : Albian foraminifera of the Rumanian Plains: Micropaleontlogy, v. 11, p. 1–38.
1968 : Biostratigraphy of Upper Cretaceous deposits in the southern Eastern Carpathians near Brasov: Micropaleontology, v. 14, p. 225–241.
Nilsen, T.
1978 : Lower Tertiary laterite on the Iceland-Faeroe Ridge and the Thulean Land Bridge: Nature, v. 274, p. 786–788.
Nyong, E. E., and Olsson, R. K.
1984 : A paleoslope model of Campanian to lower Maestrichtian foraminifera in the North American Basin and adjacent continental margin: Marine Micropaleontlogy, in press.
Ohm, U.
1967 : Zur Kenntnis der Gattungen *Reinholdella, Garantella,* und *Epistomina* (Foraminifera): Paleontographica, Abteil A, v. 127, p. 103–188.
Olsson, R. K.
1964 : Late Cretaceous planktonic foraminifera from New Jersey and Delaware: Micropaleontology, v. 10, p. 157–188.
1977 : Mesozoic Foraminifera-Western Atlantic: in Stratigraphic Micropaleontology of Atlantic basin and borderlands ed. F. M. Swain; Elsevier, New York, p. 205–221.
Olsson, R. K., and Nyong, E. E.
1984 : A paleoslope model for Campanian-lower maestrichtian Foraminifera of New Jersey and Delaware: Journal of Foraminiferal Research, v. 14, p. 50–68.
Pitman, W. C. III, and Talwani, M.
1972 : Sea floor spreading in the North Atlantic: Geological Society of America Bulletin, v. 83, p. 619–646.
Poag, C. W.
1980 : Foraminiferal stratigraphy, paleoenvironments and depositional cycles in the outer Baltimore Canyon trough: U.S. Geological Survey Circular 833, p. 44–65.
Poore, R. Z.
1981 : Late Miocene biogeography and paleoclimatology of the central North Atlantic: Marine Micropaleontology, v. 6, p. 599–616.
Price, R. J.
1976 : Paleoenvironmental interpretations in the Albian of western and southern Europe as shown by the distribution of selected foraminifera: Maritime Sediments, Special Publication 1, p. 625–648.
1977 : The evolutionary interpretation of the Foraminiferida *Arenobulimina, Gavelinella,* and *Hedbergella* in the Albian of north-west Europe: Palaeontology, v. 20, p. 503–527.
Roth, P. H., and Bowdler, J. L.
1981 : Middle Cretaceous calcareous nannoplankton biogeography and oceanography of the Atlantic Ocean; in The Deep Sea Drilling Project: A Decade of Progress, eds. J. E. Warme, R. G. Douglas, and E. L. Winterer, Society of Economic Paleontologists and Mineralogists, Special Publication 32, p. 517–546.
Ryan, W.B.F., Cita, M. B., Dreyfus Rowsan, M., Burckle, L. H. and Saito, T.
1974 : A paleomagnetic assignment of Neogene stage boundaries and the development of isochronous datum planes between the Mediterranean, the Pacific and Indian oceans in order to investigate the response of the world ocean to the Mediterranean "salinity crisis": Rivista Italiana Paleontologia, v. 80, no. 4, p. 631–688.
Sacco, P. A., and Olsson, R. K.
1980 : Upper Jurassic–Lower Cretaceous foraminifera in C.O.S.T. B-2 Well, Baltimore Canyon; in Abstracts with Programs, Geological Society of America, Northeastern Section Meeting, p. 80.
Savin, S. M.

1977 : The history of the earth's surface temperature during the past 100 million years: Annual Reviews of Earth and Planetary Sciences, v. 5, p. 319–344.

Schafer, C. T., Cole, F. E. and Carter, L.
1981 : Bathyal zone benthic foraminiferal genera off northeast Newfoundland: Journal of Foraminiferal Research, v. 11, no. 4, p. 296–313.

Scheibnerova, V.
1977 : Synthesis of the Cretaceous benthonic foraminifera recovered by the Deep Sea Drilling Project in the Indian Ocean; in Indian Ocean Geology and Biostratigraphy, ed. J. R. Heirtzler; American Geophysical Union, p. 585–597.

Schnitker, D.
1974 : West Atlantic abyssal circulation during the past 120,000 years: Nature, v. 248, p. 385–387.
1979 : Cenozoic deep water benthic foraminifera, Bay of Biscay, in Initial Reports of the Deep Sea Drilling Project, v. 48, eds. L. Montadert and D. G. Roberts; U.S. Government Printing Office, Washington, D.C., p. 377–413.

Schroeder, R., and Cherchi, A.
1979 : Upper Barremian–lowermost Aptian orbitlinid foraminifers from the Grand Banks continental rise, northwestern Atlantic (DSDP Leg 43, Site 384); in Initial Reports of the Deep Sea Drilling Project, v. 43, eds. B. E. Tucholke and P. R. Vogt; U.S. Government Printing Office, Washington, D.C., p. 575–584.

Sen Gupta, B. K., and Grant, A. C.
1971 : *Orbitolina,* a Cretaceous larger foraminifera, from Flemish Cap: paleoceanographic implications: Science, v. 173, p. 934–936.

Sheridan, R. E.
1983 : Phenomena of pulsation tectonics related to the breakup of the eastern North American continental margin: in Initial Reports of the Deep Sea Drilling Project, v. 76, eds. R. E. Sheridan and F. M. Gradstein; U.S. Government Printing Office, Washington, D.C., p. 897–912.

Sigal, J.
1979 : Chronostratigraphy and ecostratigraphy of Cretaceous formations; in Initial Reports of the Deep Sea Drilling Project, v. 47, part 2, eds. W.B.F. Ryan, and J. C. Sibuet; U.S. Government Printing Office, Washington, D.C., p. 287–326.

Sliter, W. V.
1976 : Cretaceous foraminifera from the southwestern Atlantic Ocean, Leg 36, Deep Sea Drilling Project; in Initial Reports of the Deep Sea Drilling Project, v. 36, eds. P. Barker and I.W.D. Dalziel; U.S. Government Printing Office, Washington, D.C., p. 519–573.
1977 : Cretaceous benthic foraminifera from the western South Atlantic, Leg 39, Deep Sea Drilling Project; in Initial Reports of the Deep Sea Drilling Project, v. 39, eds. P. R. Supko and K. Perch-Nielsen; U.S. Government Printing Office, Washington, D.C., p. 657–697.
1980 : Mesozoic foraminifera and Deep-sea benthic environments from Deep Sea Drilling Project Sites 415 and 416, eastern North Atlantic; in Initial Reports of the Deep Sea Drilling Project, v. 50, eds. Y. Lancelot and E. L. Winterer; U.S. Government Printing Office, Washington, D.C., p. 353–428.

Streeter, S. S.
1973 : Bottom water and benthonic foraminifera in the North Atlantic: Glacial-Interglacial contrasts. Quaternary Research, v. 3, p. 131 141.

Streeter, S. S. and Shackleton, N. J.
1979 : Paleocirculation of the deep North Atlantic: 150,000 year record of benthic foraminifera and oxygen-18: Science, v. 203, p. 168–170.

Summerhayes, C. P., and Masran, T. C.
1983 : Organic facies of Cretaceous and Jurassic sediments from Deep Sea Drilling Project Site 534 in the Blake-Bahama Basin, western North Atlantic; in Initial Reports of the Deep Sea Drilling Project, v. 76, eds. R. E. Sheridan and F. M. Gradstein; U.S. Government Printing Office, Washington, D.C., p. 469–480.

Thiede, J.
1980 : Palaeoceanography, margin stratigraphy and paleophysiology of the Tertiary North Atlantic and Norwegian-Greenland Seas: Philosophical Transactions Royal Society London,v . A295, p. 177–185.
1981 : Reworking in Upper Mesozoic and Cenozoic central Pacific deep sea sediments: Nature, v. 289, p. 667–670.

Thiede, J., Strand, J.-E., and Agdestein, T.
1981 : The distribution of major pelagic sediment components in the Mesozoic and Cenozoic North Atlantic Ocean; in The Deep Sea Drilling Project: a Decade of Progress, eds. J. E. Warme, R. G. Douglas and E. L. Winterer; Society of Economic Paleontologists and Mineralogists Special Publication 32, p. 67–90.

Thierstein, H. R.
1981 : Late Cretaceous nannoplankton and the change at the Cretaceous-Tertiary boundary; in The Deep Sea Drilling Project: a Decade of Progress, eds. J. E. Warme, R. G. Douglas and E. L. Winterer; Society of Economic Paleontologists and Mineralogists Special Publication 32, p. 355–394.

Thompson, S. L. and Barron, E. J.
1981 : Comparison of Cretaceous and present earth albedo: implications for the causes of paleoclimates: Journal of Geology, v. 89, p. 143–167.

Thunell, R. and Belyea, P.
1982 : Neogene planktonic foraminiferal biogeography of the Atlantic Ocean: Micropaleontology, v. 28, p. 381–398.

Tjalsma, R. C. and Lohmann, G. P.
1983 : Paleocene-Eocene bathyal and abyssal benthic foraminifera from the North Atlantic: Micropaleontology, Special Publication 4: 90 p.

Tucholke, B. E. and Vogt, P. R.
1979 : Western North Atlantic: Sedimentary evolution and aspects of tectonic history; in Initial Reports of the Deep Sea Drilling Project, v. 43, eds. B. E. Tucholke and P. R. Vogt; U.S. Government Printing Office, Washington, D.C., p. 791–825.

Vail, P. R., Mitchum, R. M., and Thompson II, S.
1977 : Seismic stratigraphy and global changes of sea level, part 4: global cycles of relative changes of sea level; Seismic Stratigraphy—applications to hydrocarbon exploration, ed. C. E. Payton, American Association of Petroleum Geologists Memoir, v. 26, p. 83–97.

Waples, D. W.
1983 : Reappraisal of anoxia and organic richness, with emphasis on Cretaceous of North Atlantic: American Association of Petroleum Geologists, Bulletin, v. 67, p. 963–978.

Webb, P. N.
1973 : Upper Cretaceous - Paleocene foraminifera from Site 208 (Lord Howe Rise, Tasman Sea); DSDP, Leg 21, in Initial Reports of the Deep Sea Drilling Project, v. 21, eds. R. E. Burns and J. E. Andrews; U.S. Government Printing Office, Washington, D.C., p. 541–573.

Williams, G. L., Jansa, L. F., Clark, D. F., and Ascoli, P.
1974 : Stratigraphy of the Shell Naskapi N-30 Well, Scotian Shelf, Eastern Canada: Geological Survey of Canada Paper 74–50, 12 p.

Woodruff, F., Savin, S. S., and Douglas, R. G.
1981 : Miocene stable isotope record: a detailed deep Pacific Ocean study and its paleoclimatic implications: Science, v. 212, p. 665–668.

Ziegler, P. A.
1975a: North Sea Basin history in the tectonic framework of North-Western Europe; in Petroleum Geology and the Continental Shelf of North-West Europe, ed. W. D. Wood; J. Wiley and Sons, New York, p. 131–151.
1975b: Geologic evolution of North sea and its tectonic framework: American Association of Petroleum Geologists Bulletin, v. 59, p. 1073–1097.
1979 : North-Western Europe: Tectonics and basin development: Geologie en Mijnbovw, v. 57, no. 4, p. 589–626.

ACKNOWLEDGMENTS

This paper has benefited considerably from the constructive criticism of F. M. Gradstein, R. M. Leckie, K. G. Miller, and B. E. Tucholke. Research by one of us (W.A.B.) was supported by a grant (OCE83-19052) from the National Science Foundation and by a consortium of oil companies. This is Woods Hole Oceanographic Institution Contribution No. 5977.

Printed in U.S.A.

The Geology of North America
Vol. M, The Western North Atlantic Region
The Geological Society of America, 1986

Chapter 35

Paleogeographic and paleobathymetric evolution of the North Atlantic Ocean

Brian E. Tucholke
Woods Hole Oceanographic Institution, Woods Hole, Massachusetts 02543
Floyd W. McCoy
Lamont-Doherty Geological Observatory of Columbia University, Palisades, New York 10964

INTRODUCTION

The principles of plate tectonics developed over the past 25 years provide a conceptual and predictive framework that allows relatively detailed reconstruction of the geologic history of ocean basins. Excluding extra-terrestrial effects, tectonics of the earth's crust is the primary mechanism controlling the evolution of the geologic record (Fig. 1). Secondary mechanisms related to tectonics through direct or indirect pathways create complicating and less easily predicted signals in the oceanic sedimentary record. In this paper we use plate-tectonic concepts to reconstruct the morphology and geologic conditions of the North Atlantic Ocean basin in twelve time slices from the Late Triassic (Atlantic closure) to the present. The specific time slices selected represent a balance between depiction of major tectonic/oceanographic "events" and portrayal of important, longer-term geologic episodes with relatively stable circulation and depositional patterns. Within each time-slice we have superimposed the major tectonic patterns, seamounts, oceanic circulation, and sedimentary lithofacies. While we have made this compilation as complete and coherent as possible within the time available, we point out that the synthesis is still a limited effort compared to the vast literature on the geology of the North Atlantic. Furthermore, because we have made the compilations at a small scale, many generalizations are necessary. Despite such deficiencies, these "first-generation" maps provide a good summary of our present knowledge of the geologic conditions in each time-slice.

Although this volume of The Geology of North America is primarily concerned with the western North Atlantic, we have found it necessary to consider here the entire North Atlantic basin and its marginal seas. From the time that the North Atlantic first began to have a "basinal" character in the Late Triassic, the western Atlantic region was affected by external environmental conditions that with time became increasingly more global in extent. Accordingly, we present an overview of North Atlantic evolution as a reasonable compromise between purely regional (western North Atlantic) and global viewpoints.

METHODS

Plate-kinematic reconstructions of continental positions and the associated tectonic patterns shown on the maps (Plates 9, 10) are primarily from the work of Klitgord and Schouten (this volume) and Srivastava and Tapscott (this volume). The reconstructions shown for the northern North Atlantic differ slightly from the final versions presented by Srivastava and Tapscott in this volume, principally by showing a looser fit of North America, Greenland, and Eurasia at closure (pre-magnetic anomaly M4). North America is held fixed at its modern latitude and longitude in all reconstructions. Rotated modern latitudes and longitudes depicted on adjacent plates provide a reference grid for geologic data on those plates. We have made no attempt to reconstruct paleolatitudes. However, it should be noted that the land masses depicted in the plates have drifted poleward by as much as 25°–30° since the Late Triassic (e.g., Barron and others, 1981; Parrish and Curtis, 1982); the changing latitudinal and longitudinal positions of the continents have effected changes in wind-stress and circulation patterns which *are* shown in the plates.

North of the major equatorial Atlantic fracture zones, the positions of the Mid-Atlantic Ridge axis and most transform-fault offsets are based on locations of identified magnetic anomalies (western Atlantic plus rotated eastern Atlantic). Notable exceptions are the northwestern part of the Labrador Sea, Baffin Bay, and the Greenland–Scotland Ridge. In the zone extending from Kings Trough to the Bay of Biscay the ridge-transform patterns are mostly inferred from data on the position of pole of relative plate rotation, and from geologic structure of the ocean crust.

Tectonic patterns are generalized in the Caribbean, the equatorial Atlantic, and the Tethys (Mediterranean). In these areas, magnetic lineations are absent or so poorly defined that they provide little constraint on the reconstructions. The Caribbean plate model is modified from that of Anderson and Schmidt (1983), with relative plate movements, thrusts, and other faults based largely on geological data (see Klitgord and Schouten, this volume). In the westernmost Mediterranean, tectonic patterns are

Tucholke, B. E. and McCoy, F. W., 1986, Paleogeographic and paleobathymetric evolution of the North Atlantic Ocean; *in* Vogt, P. R., and Tucholke, B. E., eds., The Geology of North America, Volume M, The Western North Atlantic Region: Geological Society of America.

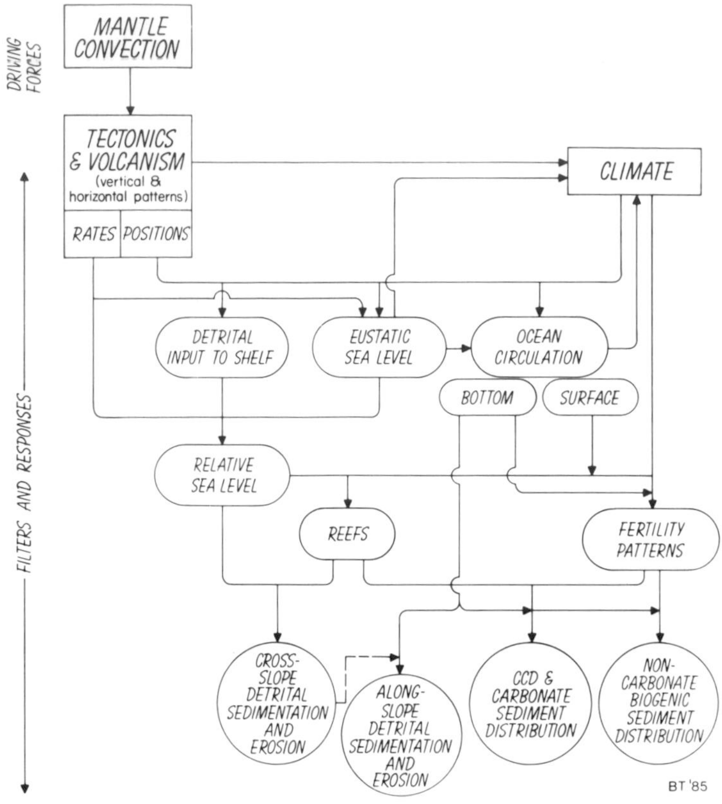

Figure 1. Whole-earth wiring diagram summarizing the hierarchy and interaction of the most direct controls on sedimentation and erosion in the North Atlantic basins. Numerous less direct controls and interactions are not included. Pathways for atmospheric dispersal of sediment (e.g., ash, eolian dust) involve an additional branch of tectonics/climate interaction that is not shown here.

adapted from Zeigler (1982), and Biju-Duval and others (1977).

Paleobathymetry was constructed from several kinds of information. In mid-ocean areas, we used the principle of increasing crustal depth with age (e.g., Sclater and others, 1971; Sclater and Wixon, this volume). We applied an empirical age-depth curve derived for the Atlantic (Fig. 2) to the crustal ages based on the magnetic-anomaly identifications of Klitgord and Schouten (this volume) and Srivastava and Tapscott (this volume). The paleomagnetic time-scale used is shown in Plate 1. The resulting paleo-isobaths were modified wherever depth anomalies were known to exist (e.g., Norwegian–Greenland Sea, Bermuda Rise, Azores Plateau) by using published studies of the anomalous localities as well as other available geological and geophysical data. In those areas away from mid-ocean ridge crests where significant sediment thicknesses existed during a given time slice, we examined selected seismic reflection profiles, identified the appropriate sediment thickness for that age from the seismic stratigraphy (e.g., Tucholke and Mountain, 1979; Mountain and Tucholke, 1985), and unloaded the crust to this sediment thickness to determine paleodepth. Spot depths from many Deep Sea Drilling Project (DSDP) sites backtracked along seafloor age-depth curves also were used. Corrections for sediment compaction were not made for either the reflection profiles or the drillsites. Continental-margin paleodepths were derived in a less quantitative fashion. These depths represent estimates based on geologic data (e.g., position of carbonate-bank edges, paleobathymetric estimates from microfossil assemblages in boreholes, and upper-margin seismic stratigraphic relationships) balanced against extrapolation of modern morphology backward in geologic time. On all the maps bathymetric trends (although usually not absolute depths) are based on observation of modern analogs. In tectonically complex areas such as the Caribbean and Tethys where the geologic evolution still is not well understood, we made no attempt to reconstruct paleodepths.

Surface circulation depicted in Plates 9 and 10 is based primarily on wind-stress patterns suggested by the atmospheric circulation models of Parrish and Curtis (1982). The circulation patterns also are consistent with paleobiogeographic information from North Atlantic planktonic fauna and flora as discussed in the Biofacies section of this book. Areas of upwelling shown on the maps apply more appropriately to nearshore, shelf, and shelf-break areas (Parrish and Curtis, 1982), but the upwelling labels have been placed farther offshore to avoid clutter on the maps. Deep circulation shown on the maps is based mainly on seismic- and rock-stratigraphic studies of Tucholke and Mountain (1979) and Miller and Tucholke (1983).

Sediment lithofacies were compiled from numerous sources (Table 1). In the main oceanic basin, lithofacies were based on sediments cored in DSDP boreholes, and these observations were then extrapolated throughout the basin in accordance with the level of the calcite compensation depth (CCD) for each time slice (Plate 1). Along continental margins and insular aprons, the lithofacies distributions are primarily from commercial and DSDP boreholes, outcrop data, and numerous summaries of Mesozoic-Cenozoic continental-margin sedimentary regimes for the Mesozoic and Cenozoic (Table 1). We also incorporated DSDP borehole results, together with data from numerous other cores and bottom samples, in the Holocene lithofacies distribution (Plate 10, L) by assuming that the uppermost sediment recovered in DSDP drilled or punched cores (upper Pleistocene to Holocene) was representative.

DSDP sites and other wells plotted on any particular time-slice penetrated that time-stratigraphic interval. However, they may not have recovered sediment from the interval if the section was spot cored or an unconformity was present. With the exception of the early Oligocene time-slice (Plate 10, J), we have made no attempt to evaluate the origin and distribution of hiatuses. Proper evaluation of the hiatuses inferred from DSDP boreholes will require careful study in a basinwide rock- and seismic-stratigraphic context and must consider the fact that there is strong bias in borehole siting. Most DSDP boreholes have been drilled in special-interest areas of thin sedimentary cover, on topographic highs, or on special geologic features along continental margins. We recognize this bias and have attempted to reduce it

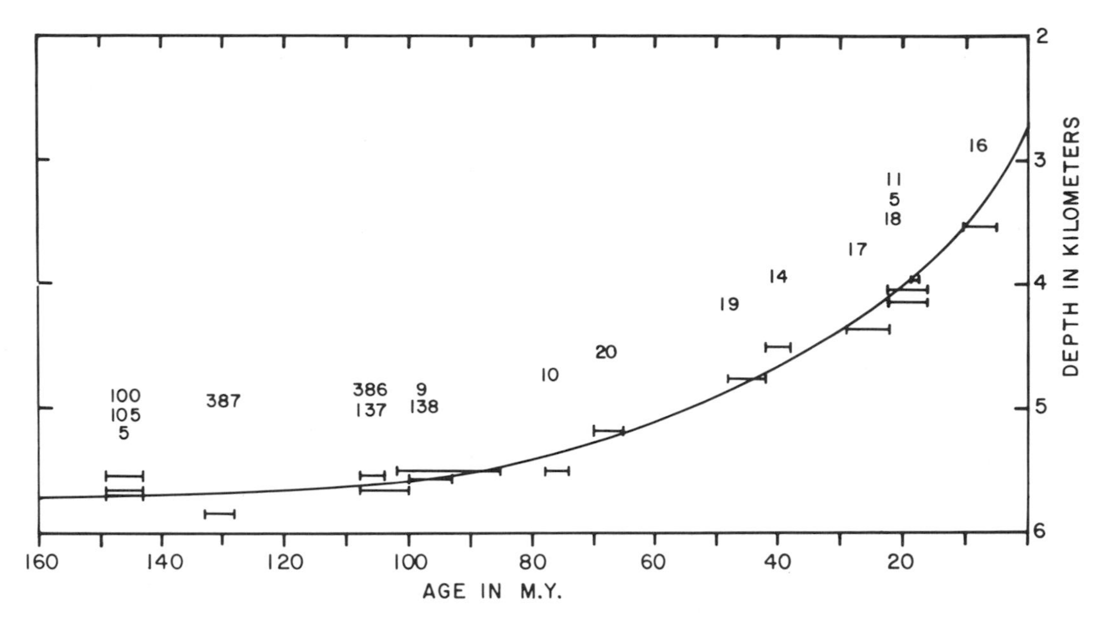

Figure 2. Empirical age-depth curve for Atlantic ocean crust based on DSDP borehole results (from Tucholke and Vogt, 1979a).

by mapping sediments in terms of only seven major lithofacies (Plate 9, A).

Many large areas have little or no data on lithofacies; either sampling has not been conducted or unconformities are present. The principal areas where data are sparse include offshore northeastern South America; the Caribbean basin; the central Gulf of Mexico; the continental shelves off Labrador, Baffin Island, Greenland, and Norway; in Rockall Trough; and on the northwest African continental shelf outside the Essaouira, Tarfaya-Aaiun, and Senegal basins. In these areas we have either omitted lithofacies patterns or extrapolated/interpolated them according to geologic relationships in nearby time and space.

Paleocoastlines are adapted primarily from data in Cook and Bally (1975), Jansa and Wade (1975), Biju-Duval and others (1977), Zeigler (1982), Uchupi and others (1984), and McCoy and Zimmerman (in prep.). However, appropriate data commonly are lacking or incomplete; in these circumstances paleocoastline positions were inferred (1) from modern coastlines, (2) by extrapolating coastlines between time-slices where data are available, or (3) through consideration of geologic data bearing on seaway formation. The occasionally convoluted coastline patterns on northwestern South America almost certainly represent complications resulting from displaced microplate terranes. Paleocoastline depiction in Central America and Mexico was not attempted.

In the following pages we discuss the morphologic evolution of the North Atlantic basin and note briefly the associated lithofacies and circulation patterns. The latter are discussed more fully in individual chapters in the Paleoceanography section of this book. Most localities discussed in the text are coded in Plates 9 and 10. Additional localities are labelled on the bathymetric chart of the modern Atlantic (Plate 2).

BASIN EVOLUTION

Seamounts, Banks, Plateaus, and Rises

Numerous seamounts, submarine banks, plateaus, and small rises occur across the floor of the North Atlantic Ocean. We have incorporated in our maps only those that are best studied or most distinctly defined (Plate 10, L). Because of the small scale of the maps, seamounts and other small elevations are mapped as black dots without paleobathymetric representation. Unfortunately, little is known about the age of most of these physiographic features. Igneous rocks dredged or otherwise sampled from their crests and slopes typically are either highly altered or significantly displaced from their original stratigraphic position. The few available radiometric ages thus must be viewed with circumspection. Many of the seamounts are flat-topped guyots having a sediment cap (Emery and Uchupi, 1984), but the ages of the sediments provide only minimum ages for the seamounts. In a few instances the stratigraphic position of the seamounts' volcaniclastic aprons can be determined in seismic reflection profiles with respect to other reflecting horizons of known age, thus constraining the timing of principal constructional episode(s) of the seamounts. The age data used to incorporate the seamounts on the paleogeologic maps are summarized below.

The oldest known seamounts in the North Atlantic occur at the western end of the New England seamount chain. Volcaniclastic aprons of two seamounts beneath the continental rise immediately adjacent to Georges Bank have a seismic-stratigraphic position indicating that they date to the Late Jurassic (~164–147 Ma; Mountain and Tucholke, 1985). They consequently are roughly contemporaneous with intermediate-age intrusives (~200–156 Ma) in the adjacent White Mountain igneous prov-

TABLE 1. PRINCIPAL SOURCES FOR DATA ON SEDIMENT FACIES*

Area	References
Caribbean and Gulf of Mexico	-Cook and Bally (1975), Khudoley and Meyerhoff (1971), Buffler and others (1984), Reeside (1957).
U.S. Atlantic Margin	-Cook and Bally (1975), Uchupi and others (1984), Mountain and Tucholke (1985).
Canadian Atlantic Margin	-Jansa and Wade (1975), King and MacLean (1975), Williams (1975), Hall (1981), McWhea (1981), Tucholke and Fry (1985).
Greenland Margin	-Buchardt (1981), Hall (1981), Henderson and others, (1981), Surlyk and others (1981).
Iceland and Greenland-Scotland Ridge	-Buchardt (1981), Hall (1981). [Holocene only: Gudmundsson and Saemundsson (1980)].
Eurasia	-Biju-Duval and others (1977), Rios (1978), Alverado (1980), Riveiro (1980), Buchardt (1981), Steel and others (1981), Zeigler (1982).
Northwestern Africa	-de Spengler and Delteil (1966), de Spengler and others (1966), Reyment (1966), Dillon and Sougy (1974), Biju-Duval and others (1977), Jansa and Wiedmann (1982), Ranke and others (1982), Wissmann (1982).
Atlantic Ocean Basins	-Tucholke (1979), Miller and Tucholke (1983). All appropriate Initial Reports of the Deep Sea Drilling Project. [Holocene only: Biscaye and others (1976), Rawson and Ryan (1978), Rona (1980), Simpkin and others (1981), McCoy (1983)].
Norwegian-Greenland Basin and Caribbean Basin	-All appropriate Initial Reports of the Deep Sea Drilling Project.

*References also include information on paleogeography, tectonics, and circulation. Numerous references of local significance are not cited here, in order to reduce the volume of the bibliography. Not all references have information for all time slices.

ince in New England, although both older (~230 Ma) and younger (125–100 Ma) series exist in this province (Foland and Faul, 1977). Duncan (1984) evaluated radiometric ages obtained from samples of the New England seamount chain farther to the east. He concluded that the seamounts formed as the North American plate passed over the New England hot spot between about 103 and 82 Ma in the Late Cretaceous, at a rate of 4.7 cm/yr (Plate 9, F). This hypothesis is consistent with age-progressive volcanism extending from the youngest magmatic series in New England (125 to 100 Ma), through the New England seamount chain, and into the Corner Seamounts, which are thought to date to about 75–70 Ma (Peterson and Edgar, 1970). The model also agrees with drilling results at DSDP Sites 385 (Vogel Seamount) and 382 (Nashville Seamount; Tucholke and Vogt, 1979b), but it does not explain the Upper Jurassic seamounts noted earlier at the western end of the New England chain. It is possible, as suggested by the nearby White Mountain ages, that oceanic crust in the area of the western New England seamounts has experienced more than one episode of magmatism.

A second, less well-defined chain of seamounts extends south of, and parallel to, the New England seamounts (Plate 9, F). Today, at its western end, it includes a large, unnamed seamount now buried beneath the upper continental rise adjacent to Hudson Canyon. Knauss Knoll, Caryn Seamount, Muir Seamount, and several other large peaks extend eastward to the northeasternmost Bermuda Rise and Sohm Abyssal Plain (Plate 2; Plate 10, L). These seamounts also may be related to hot-spot volcanism, but none are well dated. They are plotted on the paleogeologic maps to suggest formation at or near the crest of the spreading mid-ocean ridge.

Two "chains" of seamounts occur near the margin of the Grand Banks: (1) the Fogo Seamounts along the southern transform margin west of the J-Anomaly Ridge, and (2) the Newfoundland-Milne-Altair seamount chain east of Newfoundland (Plates 2, 5). The Fogo Seamounts include at least a dozen peaks that now are mostly buried beneath the continental rise. Many of these peaks follow fracture zone trends and they may have formed by a combination of volcanic and tectonic activity in the transform domain. Their age is unknown. However, seismic reflection profiles indicate that two of the Fogo seamounts are flat-topped and capped by possible reef carbonates (Uchupi and Austin, 1979). Reef carbonates of Barremian-Aptian age are known to be present on the adjacent J-Anomaly Ridge (DSDP Site 384, Tucholke and Vogt, 1979b); thus at least the two flat-topped Fogo seamounts also may date to the Early to middle Cretaceous. For the Newfoundland Seamounts, a middle Cretaceous age is inferred from a $^{40}Ar/^{39}Ar$ radiometric age-date of 98 Ma on trachyte from Scruncheon Seamount in the west (Sullivan and Keen, 1977) and a mid-Cretaceous limestone dredged from Shredder Seamount at the eastern end of the chain (Keen and others, 1977). Duncan (1984) suggested that the Newfoundland Seamounts formed as the North American plate passed over the Madeira hot-spot in the Cretaceous (Plate 9, F; 10, G); the Milne-Altair group of seamounts, although undated, is assumed to be part of this trend.

Several seamounts in the northern Rockall Trough, including Rosemary Bank, Anton Dohrn Seamount, and Hebrides Terrace Seamount, may date to the Late Cretaceous (Roberts and others, 1983; Plate 10, G). Thus we have assumed that they formed during the opening of Rockall Trough in the Cretaceous long-normal magnetic epoch (anomaly M-O to 34; Plate 9, E–F). However, it is quite possible that these and nearby seamounts, such as those at the western end of Rockall Plateau, were formed during the later "Thulean" (late Paleocene–early Eocene) volcanic episode that widely affected the northern North Atlantic (Plate 10, H).

An arcuate swath of seamounts and submarine banks extends from Kings Trough on the northeastern flank of the Mid-Atlantic Ridge northeastward into the Bay of Biscay (e.g., Coruna Seamount, North and South Charcot seamounts, and the Biscay and Cantabria seamounts). These probably developed during the latest Cretaceous and Tertiary along the Iberia-Europe plate boundary (Plate 10, G–J). The pole of relative plate rotation was nearby, just west of Iberia, during this period (Klitgord and Schouten, this volume) and the peaks probably owe their present-day arcuate trend to volcanism and/or uplift within a complex zone of extension, shear, and compression.

Several seamounts, not depicted in Plates 9 and 10, occur nearer the Iberian margin: Vasco da Gama, Vigo, and Porto seamounts are on the southern edge of Galicia Bank near DSDP Site 398 and consist of tilted, block-faulted fragments of continental crust that are capped by Upper Jurassic to Lower Cretaceous limestones (Dupeuble and others, 1976; Groupe Galice, 1979). To the southeast, another group of seamounts between Iberia and the Mid-Atlantic Ridge is presumably volcanic, but their ages are unknown. This group includes the large (100 km), caldera-shaped Tore Seamount (Plate 2) which occurs near the probable ocean-continent boundary (Tucholke and Ludwig, 1982). Seamounts and ridges extending west from Gibraltar towards the Madeira–Tore Rise owe their origin principally to volcanism and thrusting along the Iberia-Africa plate boundary during the Miocene Alpine orogeny (see Klitgord and Schouten, this volume).

Along the northwest African margin, several groups of seamounts are concentrated near the Madeira–Tore Rise, Canary Islands, Cape Verde Islands and Rise, and Sierra Leone Rise. The first three groups were constructed by volcanic activity primarily during the middle to late Miocene (e.g., Zbyszewski, 1971; Schmincke and von Rad, 1979; Duncan and Jackson, 1977). Ballard (1979) reported middle Eocene and Maestrichtian strata from the flank of Echo Bank in the Saharan seamount group south of the Canary Islands, suggesting that initial development of this seamount and the nearby Canary Islands ridge (McDougall and Schmincke, 1976) may have begun earlier. The age of volcanism that formed the seamounts on the Sierra Leone Rise and the adjacent, scattered seamounts to the north is unknown, but we assume that it dates to the Miocene (Plate 10, K).

Well seaward of central northwest Africa, a line of seamounts extends from the Great Meteor Seamount northwest onto the flanks of the Mid-Atlantic Ridge. Great Meteor Seamount is capped by Miocene pteropod chalk (Pratt, 1963) and thus may date to the Miocene. Volcanism along this trend has been attributed to the relative northwestward movement of the African plate over the New England hot spot during the mid-Tertiary; the mid-ocean ridge crest passed over the hot spot some time between about 70 and 38 Ma, thereby repositioning hot-spot volcanism from the North American plate to the African plate (e.g., Crough, 1981; Duncan, 1984). This model predicts that the age of these seamounts should become older toward the northwest, but this remains to be tested.

Volcanism forming the Azores Plateau and associated seamounts probably dates at least to the late Miocene (anomaly 5; Plate 10, K; Saemundson, this volume) because the present Azores triple junction and Iberia-Africa plate boundary had developed by this time (Klitgord and Schouten, this volume). Duncan (1984) suggests that volcanism above the Azores hot spot may have begun as early as 38 Ma (anomaly 13).

Evolution of Bathymetry, Coastlines, and Circulation

Late Triassic (Norian): Atlantic Closure (210 Ma; Plate 9, A). In the Late Triassic prior to formation of oceanic crust, the future North Atlantic margins were contained in a rift zone of shallow lacustrine and subaerial rift basins that were episodically flooded by sea water from the Tethys. These intermittent marine incursions from the northeast may have resulted in deposition of evaporites as far southwest as Cape Hatteras (Jansa, this volume). The evaporites are interbedded with clastic sands, silts, and shales. Depths below mean sea level in most of the rift basins probably were on the order of a few tens to a few hundred meters. About 180 Ma (Bajocian), some 30 Ma after the closure time-slice depicted in Plate 9, A, initial continental drift began between North America and Africa (Klitgord and Schouten, this volume), the first oceanic crust was emplaced, and a long narrow seaway extended southwest to Florida. Deposition of evaporites in local shallow basins such as the Carolina Trough and possibly in the Blake Plateau Basin and westernmost Bahamas region presumably resulted from restricted circulation into this seaway from the Tethys. Sands, silts, and shales eroded from surrounding rift mountains probably were marginally interfingered and interbedded with these evaporites.

Middle Jurassic (Late Bathonian—Early Callovian): Blake Spur Magnetic Anomaly (~170 Ma; Plate 9, B). The distribution of ammonite fauna (Westermann, 1980) indicates that by the time of the Blake Spur Magnetic Anomaly (late Bathonian–early Callovian), the Atlantic Ocean formed a continuous seaway from the Tethys to the Pacific. The exact position of the "Americas Seaway" between North America and South America is uncertain, but it presumably lay across what is now northern South America. Water depths across the juncture between North America, South America, and Africa probably were less than a kilometer, and deposition of evaporites may have occurred in bathymetrically isolated areas of the adjacent Blake Plateau Basin. Extended continental or "transitional" crust probably was present across this area (Dillon and others, 1979). A rift basin containing thinned continental and probably shallow oceanic crust formed in the Gulf of Mexico, and it became the site of accumulation of the massive Louann salt deposits. We infer that this basin was isolated from the Americas Seaway by rift mountains on the basin's southern flank (Plate 9, B).

Oceanic crust was present north of the present Blake Plateau and extended northeast to the transform margin between Africa and the Grand Banks. Oceanic crust also may have formed within small basins that were connected by transform offsets into the Tethys. The former Upper Triassic to Lower Jurassic Scotian-Moroccan salt basin probably was split shortly after seafloor spreading began about 180 Ma, forming two conjugate salt fronts. Between Georges Bank and the Americas Seaway fracture zone, the North Atlantic seafloor-spreading axis jumped eastward to the African margin about the time of the Blake Spur Magnetic Anomaly (Plate, 9, B); this stranded on the North American plate a wide segment of oceanic crust that has no correlative on the African plate (Vogt, 1973; Klitgord and Schouten, this volume). Significant thicknesses of sediment apparently had not accumulated in the central Atlantic basin at this time because there is no known structural or stratigraphic indication for this rift jump in seismic reflection profiles across either the pre- or post-jump positions of the spreading axis (Mountain and Tucholke, 1985). It would appear that marginal carbonate reefs and banks blocked significant detrital sediment input from the adjacent continents. If we assume that deep-basin sediment accumulation was minimal and that the oceanic crust followed the North Atlantic age-depth curve (Fig. 2), then a series of long, narrow basins deeper than 3 km probably flanked the pre-jump spreading axis (Plate 9, B).

For the Middle Jurassic, a general westerly circulation of surface water from the Tethys into the Pacific is suggested by models of wind-stress patterns (Parrish and Curtis, 1982). Limited gyral circulation may also have been present between North America and Africa. Some deep-water outflow from the Atlantic into the Tethys could have occurred through the "Gibraltar-Rif Seaway" (Jansa, this volume). Shallow or multiple-silled seafloor physiography leading to the Pacific and the Arctic ocean basins prevented any deep-water exchange with those oceans. However, a shallow-marine connection to the sub-Arctic is suggested by the occurrence of Middle Jurassic marine sediments on both Norway and east Greenland (Dalland, 1981; Surlyk and others, 1981).

Late Jurassic (Late Oxfordian): Anomaly M25 (157 Ma; Plate 9, C). By late Oxfordian time the oceanic basins of the North Atlantic were significantly wider but not markedly deeper than in the Middle Jurassic. The oldest crust near the North American continental margin was about 23 m.y. old, and would have subsided to a paleodepth slightly greater than 4 km according to the North Atlantic age-depth relationship (Fig. 2). However, more than 2 km of sediment had accumulated over much of this crust (Mountain and Tucholke, 1985), and maximum seafloor paleodepths probably were on the order of only 3 to 3.5 km. On the African plate, the southern part of the eastern North Atlantic basin contained oceanic crust that was only about 11–12 m.y. old because of the earlier (Blake Spur Magnetic Anomaly) ridge-jump; the oldest crust in the northern part of this basin was about 23 m.y. old. However, this relatively small age difference did not make a significant difference in the regional paleobathymetry of the eastern North Atlantic basin. The geometry of transform/spreading-ridge segments in the Gibraltar-Rif seaway is largely inferred, but small oceanic basins in this area could have been deeper than two kilometers. These basins probably did not attain depths predicted by the age-depth curve because they accumulated significant sediment fill from adjacent land masses. The depths of interbasin sills across the Atlantic-Tethys connection are unknown but probably also had depths greater than 2 km.

At the southern edge of the North Atlantic, the transform-fault zone between the Mid-Atlantic Ridge and the spreading center in the Gulf of Mexico cut across shallow continental and oceanic terrains that were responsible for the oceanographic isolation of the Gulf of Mexico. We presume that paleodepths in this zone were less than a kilometer. The main phase of salt deposition in the Gulf of Mexico ceased about late Oxfordian time, and the spreading center there began to split the salt deposits and form the separate salt fronts now observed in the subsurface along both the Yucatan and Mexican-American margins.

In the Bahamas region, a peculiar situation developed by this time, wherein very shallow crust accreted on the western side of the Mid-Atlantic Ridge and formed a substrate for reef growth, but no comparable feature formed in the conjugate area on the African plate (Plate 9, C). This asymmetry continued to develop into the middle Cretaceous. The plate-kinematic models on which these reconstructions are based require that oceanic crust underlies the Bahamas approximately east of Andros Island (see also Dietz and Holden, 1973). Morgan (1981) proposed that shallow oceanic crust beneath the Bahamas originated from volcanic activity as the North American plate moved over the Fernando de Noronha hot spot. Migration of the Mid-Atlantic Ridge spreading axis to the west of the hot spot (transfer of the hot spot to the South American plate) shortly following anomaly M-O (Aptian) would explain the abrupt eastward termination of the Bahamas platform at Navidad Bank (compare Plates 9, E and 9, F). This hypothesis currently provides the simplest explanation for the origin of the crust beneath the shallow carbonate banks of the southeast Bahamas.

Pelagic and hemipelagic lithofacies accumulated in the central Atlantic basin and contained a mixture of terrigenous and calcareous debris. Silts and shales were common along the continental margins and in the abyssal areas proximal to the margins. The North Atlantic calcite compensation depth (CCD) was near 3 km (Plate 1), so deeper basin deposits were carbonate-poor. Deep-sea drillsites in the western North Atlantic to date have recovered upper Oxfordian sediments deposited mostly at or below paleodepths of 3 km. The reddish and gray-green, variably calcareous shales in this stratigraphic interval have been defined as the Cat Gap Formation (Plate 1).

North Atlantic surface circulation in the Late Jurassic continued to include a westward, near-equatorial, circum-global flow from the Tethys through the Atlantic and Americas Seaway into the Pacific. With the widening of the Atlantic basins a steady clockwise (anticyclonic) gyral circulation probably was developed in surface waters of the central North Atlantic (Plate 9, C). Deep-water exchange between the Atlantic and Tethys may have allowed a relatively weak, contour-following bottom circulation.

Early Cretaceous (Late Valanginian): Anomaly M11 (133 Ma; Plate 9, D). By late in the Valanginian stage of the Early Cretaceous, paleodepths in both the western and eastern North Atlantic exceeded 4 km. Paleodepths in the Americas Seaway and the Caribbean are unconstrained and difficult to estimate, but it is likely that the developing Bahamas Platform continued to block a deep-water connection to the Atlantic. At the northern end of the Atlantic, estimates from the crustal age-depth relation suggest that a deep-water (>3 km) passage probably existed in the Gibraltar-Rif seaway. The Late Jurassic to Early Cretaceous Cimmerian cycle of rifting affected the entire trans-European platform (Sibuet and Ryan, 1979; Ziegler, 1982), and local basins there may have been deeper than a kilometer. A marine connection to the sub-Arctic (and possibly the Arctic) is suggested by the occurrence of Lower Cretaceous marine deposits in Norway, east Greenland, and Svalbard (Dalland, 1981; Surlyk and others, 1981; Birkelund and Perch-Nielsen, 1976).

Modelled wind-stress patterns (Parrish and Curtis, 1982) suggest that surface circulation was dominated by continued Tethys-to-Pacific, circum-global flow and that a well-established North Atlantic gyre was present. Currents in this gyre probably were westward-intensified, creating the beginnings of a Gulf Stream–like circulation along the eastern North American margin. Easterly surface flow from the northern Atlantic and possibly from the Arctic exited across the European platform into the Tethys via the Biscay and Polish seaways. A weak to moderate deep circulation in the central Atlantic is suggested by the presence of lens-shaped Neocomian carbonate deposits in the southwestern North Atlantic (Tucholke and Mountain, 1979). These deposits occur at the base of the steep Bahama Escarpment where bottom currents probably were topographically intensified; similarly intensified deep currents may have been present along the other steep, transform margins at the northern and southern ends of the central North Atlantic basin.

Along both the North American and African margins, wide, gently-sloping sediment prisms developed by deposition of terrigenous and shallow-water clastic debris that spilled off the continental shelves into the adjoining deep basins during a pronounced sea-level lowstand (Plate 1; see also Vail and others, 1984). Notable examples of such downslope and seaward transport of sediment occur off Baltimore Canyon Trough and northwestern Africa (Plate 9, D). The CCD was at a depth of approximately 4.5 km (Plate 1), so most pelagic sediments away from the continental margin were calcareous. White and gray limestones, locally with black marly interbeds, typify the pelagic and hemipelagic sediments recovered in deep-sea boreholes that penetrate this stratigraphic level. They constitute the Blake-Bahama Formation in the western North Atlantic (Plate 1). The dark interbeds appear to represent brief, episodic anoxic or dysoxic (low-oxygen) bottom-water conditions in the deep ocean basin (Tucholke and Vogt, 1979a); however, near the continental margins turbidites are common (Jansa and others, 1979) and they may have contributed to the formation of the dark interbeds by rapid influx and burial of carbon-rich sediments.

Middle Cretaceous (Early Aptian): Anomaly MO (118 Ma; Plate 9, E). The early Aptian marks a time of significant change in tectonic patterns and in paleobathymetry in and around the North Atlantic. For the first time, the Mid-Atlantic Ridge axis extended northward between the Grand Banks and Iberia and separated Flemish Cap from Galicia Bank. An incipient spreading axis probably continued to the north where it began to form the basin of Rockall Trough. A rift axis also extended eastward from a triple junction at the Mid-Atlantic Ridge and caused initial opening in the Bay of Biscay. In these areas where oceanic crust was just beginning to form, paleodepths were probably between 1 and 2 kilometers. The Mid-Atlantic Ridge southeast of the Grand Banks was exceptionally shallow, probably because it was above a hotspot (Tucholke and Ludwig, 1982). Parts of the ridge supported carbonate reefal banks as demonstrated at DSDP Site 384, and other nearby parts of the ridge probably were subaerial. Shortly following anomaly M-O, the spreading axis of this section of the Mid-Atlantic Ridge subsided rapidly to more normal elevations, and continued seafloor spreading divided the previously shallow crust into two subsidiary aseismic ridges, the J-Anomaly Ridge to the west and the Madeira–Tore Rise to the east (see Plates 9, F and 10, G–L). Paleodepths in the main parts of both the eastern and western basins of the North Atlantic were in excess of 4 km, but less than 5 km.

The Barremian through Cenomanian lithofacies of the Atlantic below about 2.5–3 km paleodepth characteristically consist of alternating black and gray-green shales that were deposited below the ambient CCD (Plate 1). The sedimentological and geochemical characteristics of these sediments indicate that they were deposited beneath anoxic and dysoxic bottom waters (Tucholke and Vogt, 1979a; Arthur and Dean, this volume), and it is likely that there was minimal renewal of bottom waters in the North Atlantic basins during the mid-Cretaceous. Barriers erected by compressional tectonics probably barred bottom-water exchange with extra-Atlantic areas such as the Caribbean (Klitgord and Schouten, this volume) and possibly the Tethys (Biju-Duval and others, 1977).

Atlantic surface circulation patterns followed those established in the Late Jurassic and Early Cretaceous. A narrow seaway may have extended south into the incipient rift zone that was developing between northeastern South America and central Africa. This seaway is inferred from the occurrence of possible upper Aptian marine sediments in the Ivory Coast marginal basin (de Spengler and Delteil, 1966). Marine transgression extending into this rift zone from the south to connect the North and South Atlantic Oceans did not occur until the late Albian (Premoli-Silva and Boersma, 1977; McCoy and Zimmerman, in prep.).

Late Cretaceous (Late Santonian): Anomaly 34 (84 Ma; Plate 9, F). A well-defined ridge-ridge-ridge triple junction between the Grand Banks and the Iberian Massif separated the North American, European, and Iberian-African plates in late Santonian time. The Labrador Sea basin began to form along the northwest arm of this triple junction (Srivastava and Tapscott, this volume). The eastward arm extended through the Bay of Biscay and Biscay Seaway where it was transformed into foredeep troughs along thrust zones of the northwest Tethys. Deep-oceanic basins (2–4 km) had formed in Rockall Trough, the Bay of Biscay, and between Iberia and the Grand Banks. The epicontinental sea across Europe connected the Arctic to the North Atlantic and Tethys, but the multiple-silled basins on continental crust in this region were too shallow to allow for deep-water exchange between the major oceanic basins.

With loading by the appropriate sediment cover taken into account, the western and eastern basins of the central North Atlantic had attained paleodepths of 5 km by Santonian time. The New England seamounts had been constructed across the western Atlantic (Duncan, 1984), separating the northern, Sohm basin from the southern, Hatteras basin. Seismic-stratigraphic relations indicate that the Sohm basin received large volumes of detrital sediment during the Late Jurassic and Cretaceous, so the basin was somewhat shallower than the Hatteras basin to the south. The principal lithofacies deposited in the North Atlantic basin during the Late Cretaceous consisted of carbonate-free, multicolored shales, defined as the Plantagenet Formation in the western North Atlantic. The CCD level was shallow, on the order of 2.5 km (Plate 1). The shallow CCD and implied low productivity in surface waters of the central Atlantic gyre probably reflect a warm, equitable climatic regime caused by the high sea levels and extensive transgressions of the Late Cretaceous.

The transition from deposition of black shales beneath anoxic/dysoxic bottom waters to deposition of red and multicolored shales beneath oxygenated bottom waters occurred very rapidly in the latest Cenomanian to early Turonian (Arthur and Dean, this volume). This dramatic change in lithofacies probably occurred in response to the establishment of deep-water exchange between the North and South Atlantic Oceans as the equatorial regions of South America and Africa separated (Tucholke and Vogt, 1979a). A deep-water connection between the North Atlantic and Tethys could have remained blocked because of compressional tectonics in the western Tethys. In the Caribbean, black shales continued to be deposited through Santonian time (Edgar and Saunders, 1973). This may reflect local basin isolation caused by uplift along circum-Caribbean compressional boundaries, or it could have been caused by locally strong oceanic upwelling and high productivity.

Surface circulation in the North Atlantic remained similar to that of the Early Cretaceous with a continually expanding central North Atlantic gyre. However, surface-water flow through the Caribbean into the Pacific probably increased because South Atlantic surface waters now contributed to the flow. Some surface-water exchange between the Tethys and South Atlantic had occurred via the Trans-African seaway across Africa since the late Cenomanian, but by the late Santonian this flow was blocked in the south by folding in the Benue Trough (Reyment, 1980).

Latest Cretaceous (Late Maestrichtian): Anomaly 30 (69 Ma; Plate 10, G). The tectonic patterns of the Santonian persisted into the Maestrichtian. With continued spreading between North America and Greenland, the Labrador Sea–Baffin Bay seaway developed and possibly connected north to the Arctic. Distribution patterns of microfauna indicate the northward incursion of temperate, western Atlantic surface waters as far north as Baffin Island (Gradstein and Srivastava, 1980). Igneous activity between Rockall Bank and the Shetland Plateau marked the beginning of the extensive "Thulean" volcanism during the early Tertiary. In the Caribbean, a long spreading center reaching toward the Pacific was abandoned, and relative motion between North and South America diminished (Klitgord and Schouten, this volume).

Continued separation between South America and Africa allowed increased exchange of deep and surface waters between the North and South Atlantic. Overall paleocirculation patterns remained the same as during the Santonian, except that a generally counterclockwise (cyclonic) surface-water gyre began to form in the northern North Atlantic between eastern Canada, Greenland, and Europe (Plate 10, G). Shallow-water circulation between the Tethys and the South Atlantic occurred through the Trans-african seaway during most of the Campanian and Maestrichtian, but it may have been very limited during the sea-level regression of the latst Maestrichtian (Reyment, 1980).

Extensive deposition of calcareous oozes and marls occurred in the late Maestrichtian, forming the Crescent Peaks member of the Plantaganet Formation in the western North Atlantic Ocean. This depositional event was caused by a relatively short-lived deepening of the CCD to more than 5 km (Plate 1), and even the deepest basins accumulated carbonate-rich sediments. The CCD rebounded to shallow levels very near the Cretaceous-Tertiary boundary (Tucholke and Vogt, 1979a).

Early Tertiary (Late Paleocene): Anomaly 25 (59 Ma; Plate 10, H). The late Paleocene was a time of extensive "Thulean" volcanic and intrusive igneous activity extending from Davis Strait through east Greenland, across the Greenland–Scotland Ridge and Wyville Thomson Ridge to the western British Isles (Hall, 1981). Igneous activity peaked during latest-stage rifting and initial drift between Greenland and Rockall Bank/northern Europe. Oceanic crust began to form in this rift by anomaly-24 time, creating a triple junction south of Greenland (Talwani and Eldholm, 1977; Srivastava and Tapscott, this volume). The paleogeography around the North Atlantic is reasonably well documented from the distribution of terrestrial vertebrate fauna (McKenna, 1983), and it correlates with the tectonic evolution of the region as follows. Volcanism along the "Thulean line" produced a continuous Thulean land bridge from England through Greenland and across Davis Strait to Canada. In combination with uplift of the Artois Arch across the English Channel, and a probable link to Iberia across the Biscay Seaway (related to Pyrenean orogenesis), a continuous dry-land connection existed from southern Europe to North America. A more northerly De Geer land route between North America and Scandinavia also existed, apparently because of uplift at the Yermak triple junction off northern Greenland. The Norwegian-Greenland marine basin was isolated from the North Atlantic, and it connected to the Tethys and the world ocean only through the Polish Seaway. Baffin Bay was similarly isolated from the Atlantic by the temporary land bridge across Davis Strait.

By this time the Labrador Sea had become a well developed oceanic basin with off-ridge depths of between 3 and 4 kilometers. Depth distribution in the central North Atlantic began to resemble modern hypsometric relationships with large seafloor basins deeper than 5 km. On the Mid-Atlantic Ridge in the equatorial Atlantic, separation of the shallow, paired aseismic ridges, Ceara Rise and Sierra Leone Rise, was occurring after their formation at the ridge crest earlier in the Paleocene. The North Atlantic CCD was 3.5 to 4 km deep (Plate 1), so extensive deep-basin areas were again carpeted with pelagic clays of the Plantagenet Formation. Sizable tracts of calcareous sediments accumulated above the CCD along the African and South American margins, along the Mid-Atlantic Ridge, and in the Yucatan-Florida-Bahamas region off southeastern North America (Plate 10, H).

Surface circulation patterns in the central Atlantic continued to be dominated by a large anticyclonic gyre north of a westward-flowing Tethys-Atlantic-Pacific surface current. The cyclonic surface gyre of the northern North Atlantic became better developed as this basin widened and because peripheral seaways to extra-Atlantic areas were blocked by land bridges. Convergent plate boundaries around the Caribbean probably continued to block significant deep-water connections between the Atlantic and the Pacific. Seafloor physiography in the equatorial Atlantic posed no barriers to free deep-water exchange between the North and South Atlantic Oceans (McCoy and Zimmerman, 1977). The Trans-African seaway allowed surface circulation between the Tethys and the South Atlantic Ocean (Reyment, 1980).

Early Tertiary (Early Middle Eocene): Anomaly 21 (49.5 Ma; Plate 10, I). By early middle Eocene time, changing patterns of volcanism, in combination with continued basin enlargement and associated tectonic events, had modified the paleogeography of the northern North Atlantic (Hall, 1981; McKenna, 1983; Nunns, 1983). The Thulean land bridge was severed both at Davis Strait and between Greenland and the British Isles. Volcanism had ceased in Davis Strait and the slowly subsiding crust was tectonized in a major shear zone (Ungava transform complex) that formed a shallow seaway between Baffin Bay and the Labrador Sea. "Thulean" volcanism continued along east Greenland and on portions of the Greenland-Scotland Ridge. Parts of this ridge subsided below sea level, but the ridge continued to form a significant barrier that isolated the deep (2 km) Norwegian-Greenland Sea from both Rockall Trough and the developing Irminger Basin between Greenland and Rockall Plateau. The De Geer land bridge continued to connect North America to Fennoscandia via northern Greenland and Svalbard. With subsidence of the Artois Arch, the land bridge between the British Isles and Europe was disconnected and circulation through the English Channel was renewed. The Norwegian-Greenland Sea was now connected to the North Atlantic in two locations and to the Tethys via the Polish Seaway, but it remained isolated from the Arctic.

The Iberia-Africa plate rotated counterclockwise with respect to Europe, about a pole of relative plate rotation just west of Iberia (Klitgord and Schouten, this volume). This led to the initial opening of Kings Trough and continued a complex shear/thrust geometry along the northern Iberian margin (Pyrenean Orogeny). In the Caribbean, compression occurred across most of the basin's encircling plate boundaries and the basin probably was bathymetrically isolated from both the deep Atlantic and Pacific.

The CCD during the early middle Eocene remained near 3.5 to 4 km depth in the North Atlantic, and pelagic shales were deposited over large areas. Unlike pelagic deposits in any previous or succeeding epochs, however, basin shales contained a significant biosiliceous component, particularly in the western North Atlantic. The deep western North Atlantic basin also received large volumes of turbidites derived from the North American continental margin; these were locally rich in calcareous and biosiliceous debris (Tucholke and Mountain, this volume). The oozes and cherts of the Bermuda Rise Formation are formed by both these pelagic and allochthonous hemipelagic components (Plate 1). Biosiliceous material accumulated because of high productivity by silica-secreting plankton in surface waters. The highest productivity apparently followed the path of circum-equatorial flow across the central Atlantic and was concentrated around the large, anticyclonic gyre centered on the western Atlantic. Similar associations with the circum-equatorial circulation have been observed in middle Eocene sediments of the Pacific and Indian Oceans (Berggren and Hollister, 1974). Additional zones of high biosiliceous productivity related to upwelling were present off eastern North America, off northwest and equatorial Africa, around Rockall Plateau, and in the Caribbean.

In addition to the main Atlantic circulation, latitudinally elongated cyclonic gyres probably dominated surface circulation patterns in the semi-enclosed basins of Baffin Bay and the Norwegian-Greenland Sea. A generally cyclonic gyre was present in the adjacent northern North Atlantic basin. Bottom circulation patterns are difficult to assess, but deep-water exchange of Atlantic bottom water with the Tethys and the Caribbean probably was constrained in narrow passageways and by tectonic sills. Free exchange of bottom water with the South Atlantic continued.

Early Tertiary (Earliest Oligocene): Anomaly 13 (35.5 Ma; Plate 10, J). Perhaps the most significant paleoceanographic event of the Tertiary in the North Atlantic was the development of strong abyssal circulation during the earliest Oligocene (Tucholke and Mountain, 1979; Miller and Tucholke, 1983). Climatic cooling had commenced by the late middle Eocene (Plate 1), and with physiographic changes caused by evolving plate boundaries in the sub-Arctic, the circulation of the early Oligocene ocean was drastically modified. Tectonic shear between Svalbard and Greenland appears to have changed to extensional opening by the time of anomaly 13 (Talwani and Eldholm, 1977), and a deep passage developed between the Arctic Ocean and Norwegian-Greenland Sea. Cool Arctic water flowed through this passage and south into the North Atlantic over the Greenland-Scotland Ridge at its southeast end (Wyville-Thomson Ridge), and/or possibly at its northeast end (Denmark Strait). This formed a deep, contour-following flow that caused extensive erosion along the margins of basins in the northern and western North Atlantic. Extensive erosion along the North American margin removed thick sequences of sediments and strongly modified the morphology of the continental rise. Erosion cut down to Lower Cretaceous stratigraphic levels off the southeastern United States (Tucholke and Mountain, this volume). The average slope of the continental margin was thereby steepened, and this led to additional slope defacement caused by slumping and related sediment mass movements. Similar, but apparently less intense and "patchier" erosion also occurred along the north

African and European continental margins (Arthur and others, 1979). This difference in intensity of bottom circulation, compared to the western Atlantic, is substantiated by differences in carbon-isotope composition of benthic foraminifera between the eastern and western North Atlantic (Miller and others, in press). The bottom water in the eastern North Atlantic basins may have included components of Arctic-derived water and cool bottom water that flowed from high southern latitudes into the western North Atlantic and through equatorial fracture zones into the eastern North Atlantic (McCoy and Zimmerman, 1977).

Significant changes in morphology of the North Atlantic seafloor were also caused by changing tectonic patterns. In the Labrador Sea, seafloor spreading ceased some time prior to anomaly 13 (Srivastava and Tapscott, this volume) and the average basin depth peripheral to the defunct spreading ridge was on the order of 4 kilometers. A marine connection through the Nares Strait to the Arctic probably had not been established yet, and only this portion of the Early Tertiary De Geer land bridge survived (McKenna, 1983). In the west central North Atlantic Ocean, the Bermuda Rise was uplifted in the middle to late Eocene; by late Eocene time the Bermuda pedestal had reached sea level and was being subaerially eroded (Tucholke and Vogt, 1979a). In the Caribbean region, underthrusting along the northern plate boundary changed to largely strike-slip motion in the middle Eocene; concomitant underthrusting at the eastern boundary of the Caribbean formed the Lesser Antilles volcanic belt by early Oligocene time. A small component of underthrusting at the northeastern edge of the Caribbean probably caused initial formation of the present Puerto Rico Trench by late Eocene time (Tucholke and Ewing, 1976).

The trans-Atlantic surface equatorial current was not interrupted by tectonic barriers around the Caribbean, and the North Atlantic surface circulation continued to be a principally two-gyre system. With a CCD level between 4 and 4.5 km (Plate 1), pelagic clays mixed with hemipelagic debris eroded from the continental margins formed the dominant lithofacies in the central parts of the abyssal basins. Biosiliceous sediments were deposited locally throughout the North Atlantic, but they had less regional significance than in the early to middle Eocene (Plate 10, I). This hemipelagic-pelagic facies forms the lower part of the Blake Ridge Formation in the western North Atlantic; it interfingers with calcareous sediments that were deposited above the contemporary CCD level, as for example on the Mid-Atlantic Ridge. Where the seafloor was not being eroded, hemipelagic calcareous facies were widespread along both the North American and Africa continental margins. Clastic components were more important in the northern North Atlantic, in the deeper basins of the Greenland-Norwegian Sea, and in current-winnowed deposits in the shallow seaways across Western Europe.

Late Tertiary (Late Miocene): Anomaly 5 (9.5 Ma; Plate 10, K). By the late Miocene, the overall morphology of the North Atlantic seafloor had changed extensively from that of the Oligocene in response to two factors: the Alpine orogeny, and changes in intensity of abyssal circulation. The effects of both were particularly pronounced along the continental margins.

In the eastern North Atlantic the Alpine orogeny included a shift of the Africa-Eurasia plate boundary from the north to the south side of the Iberian peninsula (Klitgord and Schouten, this volume). At the intersection of this new plate boundary with the Mid-Atlantic Ridge the Azores triple-junction formed a volcanic plateau. Collision between Iberia and Africa created Gorringe Ridge extending from Iberia toward the Madeira-Tore Rise, the Atlas Mountains across North Africa, and associated uplifts both along the north African margin and in the adjacent ocean basin. Within about 4 million years after the late Miocene time slice depicted in Plate 10, K, this convergence closed the Strait of Gibraltar and, with only a shallow sill into the Red Sea left as a remnant of the Tethyan connection to the east, led to dessication of the Mediterranean basin during the Messinian (Ryan and Hsü, 1973).

Extensive igneous activity accompanied the Alpine orogeny off northwest Africa. It built volcanic edifices on and around the Madeira-Tore Rise, constructed the Canary Islands, formed the Cape Verde Rise and islands, and also probably affected the Sierra Leone Rise. This tectonic and igneous activity resulted in a significant reduction in average depth of the eastern Atlantic basin (compare Plates 10, J and 10, K). Current-controlled sedimentation on Oligocene erosional surfaces throughout much of the Miocene added to the physiographic changes of the eastern continental margin (Arthur and others, 1979) and the adjacent abyssal basin. Sediment drifts such as the Madeira Drift on the Madeira-Tore Rise formed locally.

The northern and western North Atlantic continental margins became ornamented through the deposition of large sediment drifts beneath a moderated flow of the deep boundary current system (Tucholke and Mountain, this volume). The deep contour-following circulation along both the North American and African continental margins contained water from Arctic and Antarctic sources, with the latter entering the eastern Atlantic basin through deep equatorial fracture zones.

The Atlantic surface circulation continued in a two-gyre pattern, and the northern, cyclonic gyre had large excursions into the Norwegian-Greenland and Labrador Seas. Circum-global flow from the Tethys to the Pacific, however, had been broken briefly in the early and middle Miocene, and permanently in the late Miocene by uplift in the eastern Tethys (Rögl and Steininger, 1983), and by progressive restriction at the Panama isthmus through the late Miocene (Zimmerman, 1982). Changing faunal affinities in both planktonic and benthic foraminifera suggest that the Atlantic-Pacific connection through Panama was largely closed by the late Pliocene (~3 Ma; Keigwin, 1982). This containment of the Atlantic surface circulation probably led to intensification of Gulf Stream flow along the eastern margin of North America.

In the northern North Atlantic, continued spreading between Greenland and Svalbard widened this passageway to the Arctic and enhanced the exchange of surface and deep water between the Arctic and the Norwegian-Greenland Sea. The only remaining marine connection from the Atlantic and Tethys to the Arctic was across the subsiding Greenland-Scotland Ridge. The Nares Strait probably remained closed, thereby isolating Baffin Bay and the Labrador Sea from the Arctic (McKenna, 1983). Within the Labrador Sea, the extinct Paleogene spreading center had been buried by sediment.

Calcareous lithofacies became more widespread in the late Miocene, reflecting an increase in the depth of the CCD to near 5 km (Plate 1), coupled with generally decreased basin depths in the eastern Atlantic. Hemipelagic shales deposited from bottom currents were prevalent in the western Atlantic. Both lithofacies constitute the upper part of the Blake Ridge Formation. Mass movement of shallow-water carbonates from the Bahamas and Blake Plateau formed the Great Abaco Member of this formation in the Blake-Bahama Basin.

Holocene: O Ma (Plate 10, L). Conditions in the modern North Atlantic reflect a relatively uniform continuation of the tectonic, bathymetric, and circulation patterns that evolved during the Neogene. However, in terms of lithofacies, the Holocene Atlantic is not typical of these trends because of the effects of intense Pleistocene climatic and sea-level cycles. On the continental shelves, sea-level lowstands reworked sediments and left behind coarse-grained residual deposits; at high latitudes, glaciers scoured the shelves, exposed older sediments, and deposited glacial debris on the continental margins both directly and from drifting ice. Deep basins contain an imprint of glacial material in the form of ice-rafted debris; the primary effect, however, on pelagic sedimentation has been the varying production and preservation of biogenic carbonate, resulting in pronounced glacial-interglacial cycles of carbonate content. The deep-basin facies in Plate 10, L are based on average carbonate contents for the upper Pleistocene and Holocene sediments.

The lithofacies, bathymetry, and circulation patterns in the modern North Atlantic are mapped at approximately the same level of detail as in the previous time slices. By comparing the Holocene panel (Plate 10, L) with the larger-scale foldouts (e.g., Plate 2), the reader can better understand the level of detail depicted in the earlier time slices. This gives an indication of the current status of our paleofacies, paleobathymetry, and paleocirculation information.

CONCLUSIONS

We have presented a summary of important elements in the evolution of the North atlantic, but we emphasize that it is only a broad overview. The maps in Plates 9 and 10 contain a host of tectonic, geomorphic, and geologic subtleties that cannot be dealt with in the limited space available here. Even so, these maps are based only upon a part of available geologic literature that pertains to the evolution of the Atlantic basin and its margins. More detailed analysis, coupled with increasingly refined plate-kinematic models that are becoming available for this region, will make it possible to place many puzzling geologic relationships into a coherent regional context and thereby to better understand their significance. The Caribbean and Mediterranean remain the most complex marginal areas of the North Atlantic, and much remains to be learned from judiciously selected deep-sea drillsites there and from field studies on land. In the main North Atlantic basin, two outstanding problems are evident. First, the origin, age, and evolution of both individual seamounts, and seamount groups and chains, are poorly understood. Careful study of seamount age and structure will yield important information about tectonic relationships and plate driving mechanisms. Second, most DSDP drillsites very incompletely represent the oceanography history of the North Atlantic because coring gaps and hiatuses are present, and because many boreholes are sited in anomalous areas. The hiatuses and site locations result from the tendency to drill thin sedimentary sections in order to penetrate the older stratigraphic sections, and to drill in interesting (complex) areas. However, if we want to establish standards for oceanic basin evolution, it will be necessary to drill into and continuously core from the most uniform, complete, and fossiliferous stratigraphic sections available. Good reference or type sections are vital. The new Ocean Drilling Program holds promise for satisfying this need.

REFERENCES

Alvarado, M.
1980 : Géologie des Pays Européens: comité National Francais de Géologie, 26th Congrès Géologie International, p. 1–54.

Anderson, T. H., and Schmidt, V. A.
1983 : The evolution of Middle America and the Gulf of Mexico–Caribbean Sea during Mesozoic time: Geological Society of America Bulletin, v. 94, p. 941–966.

Arthur, M. A., von Rad, U., Cornford, C., McCoy, F., and Sarnthein, M.
1979 : Evolution and sedimentary history of the Cape Bojador continental margin, northwestern Africa: in Initial Reports of the Deep Sea Drilling Project, v. 47, (Part 1), eds. U. von Rad and W.B.F. Ryan; U.S. Government Printing Office, Washington, D.C., p. 773–816.

Ballard, J. A.
1979 : Geology of a stable intraplate region: the Cape Verde/Canary Basin: [Unpublished Ph.D. Thesis], University of North Carolina, 197 p.

Barron, E. J., Harrison, C.G.A., Sloan, J. L., and Hay, W. W.
1981 : Paleogeography, 180 million years ago to the present: Eclogae Geologicae Helvetiae, v. 74, p. 443–470.

Berggren, W. A., and Hollister, C. D.
1974 : Paleogeography, paleobiogeography, and the history of circulation in the Atlantic Ocean; in Studies in Paleo-oceanography, ed. W. W. Hay; Society of Economic Paleontologists and Mineralogists Special Publication no. 20, p. 126–186.

Biju-Duval, B., Dercourt, J., and Le Pichon, X.
1977 : From the Tethys Ocean to the Mediterranean Seas: A plate tectonic model of the evolution of the western Alpine system; in International Symposium on the Structural History of the Mediterranean Basins, Split, Yugoslavia, 25–29 October 1976, eds. B. Biju-Duval and L. Montadert; Editions Technip, Paris, p. 143–164.

Birkelund, T., and Perch-Nielsen, K.

1976 : Late Paleozoic–Mesozoic evolution of central East Greenland; in Geology of Greenland, eds. A. Escher and W. S. Watt; Geological Survey of Greenland, Copenhagen, p. 305–339.

Biscaye, P. E., Kolla, V., and Turekian, K. K.
1976 : Distribution of calcium carbonate in surface sediments of the Atlantic Ocean: Journal of Geophysical Research, v. 81, p. 2595–2603.

Buchardt, B.
1981 : Tertiary deposits of the Norwegian-Greenland Sea region (Svalbard, northeast and east Greenland, Iceland, the Faeroe Islands and the Norwegian-Greenland Sea) and their correlation to northwest Europe: in Geology of the North Atlantic Borderlands, eds. J. W. Kerr and A. J. Fergusson; Canadian Society of Petroleum Geologists, Memoir 7, p. 585–610.

Buffler, R. T., Locker, S. D., Bryant, W. R., Hall, S. A., and Pilger, Jr., R. H.
1984 : Gulf of Mexico, Atlas 6, Ocean Margin Drilling Program Regional Atlas Series: Marine Science International, Woods Hole.

Cook, D. A., and Bally, A. W., eds.
1975 : Stratigraphic Atlas of North and Central America: Princeton University Press, Princeton, 272 p.

Crough, S. T.
1981 : Mesozoic hot spot epeirogeny in eastern North America: Geology, v. 9, p. 2–6.

Dalland, A.
1981 : Mesozoic sedimentary succession at Andøy, northern Norway and relation to structural development of the North Atlantic area; in Geology of the North Atlantic Borderlands, eds. J. W. Kerr, and A. J. Fergusson; Canadian Society of Petroleum Geologists, Memoir 7, p. 563–584.

de Spengler, A., Castelain, J., Cauvin, J., and Leroy, M.
1966 : Le bassin Secondaire-Tertiare du Sénégal; in Sedimentary Basins of the African Coasts: Part 1. Atlantic Coast, ed. D. Reyre; Association Service Geologique Afrique, Paris, p. 80–94.

de Spengler, A., and Delteil, J. R.
1966 : Le bassin Secondaire-Tertiare de Cote d'Ivoire; in Sedimentary Basins of the African Coasts: Part 1. Atlantic Coast, ed. D. Reyre; Association Service Geologique Afrique, Paris, p. 99–113.

Dietz, R. S., and Holden, J. C.
1973 : Geotectonic evolution and subsidence of Bahama platform: reply: Geological Society of America Bulletin, v. 84, p. 3477–3482.

Dillon, W. P., and Sougy, J.M.A.
1974 : Geology of west Africa and Canary and Cape Verde Islands; in The Ocean Basins and Margins, v. 2, The North Atlantic, eds., A.E.M. Nairn, and F. G. Stehli; Plenum, New York, p. 315–390.

Dillon, W. P., Paull, C. K., Dahl, A. G., and Patterson, W. C.
1979 : Structure of the continental margin near the COST No. GE-1 drill site from a common depth-point seismic-reflection profile: in Geological Studies of the COST GE-1 Well, United States South Atlantic Outer Continental Shelf Area, ed. P. A. Scholle; U.S. Geological Survey Circular 800, p. 97–114.

Duncan, R. A.
1984 : Age progressive volcanism in the New England Seamounts and the opening of the central Atlantic Ocean: Journal of Geophysical Research, v. 89, p. 9980–9990.

Duncan, R. A., and Jackson, E. D.
1977 : Geochronology of basaltic rocks recovered by DSDP Leg 41, eastern Atlantic Ocean; in Initial Reports of the Deep Sea Drilling Project, v. 41, eds. Y. Lancelot and E. Seibold; U.S. Government Printing Office, Washington, D.C., p. 1113–1118.

Dupeuble, P.-A., Réhault, J. P., Auxietre, J. L., Dunand, J. P., and Pastouret, L.
1976 : Résultats de dragages et essai de stratigraphie des bancs de Galice, et des montagnes de Porto et de Vigo (Marge Occidentale Ibérique): Marine Geology, v. 22, p. M37–M49.

Edgar, N. T., and Saunders, J. B., eds.
1973 : Initial Reports of the Deep Sea Drilling Project, v. 15: U.S. Government Printing Office, Washington, D.C., 1137 p.

Emery, K. O., and Uchupi, E.
1984 : The Geology of the Atlantic Ocean: Springer-Verlag, New York, 1050 p.

Foland, K. A., and Faul, H.
1977 : Ages of the White Mountain intrusives—New Hampshire, Vermont and Maine, USA: American Journal of Science, v. 277, p. 888–904.

Gradstein, F. M., and Srivastava, S. P.
1980 : Aspects of Cenozoic stratigraphy and paleoceanography of the Labrador Sea and Baffin Bay: Palaeogeography, Palaeoclimatology, Palaeoecology, v. 30, p. 261–295.

Groupe Galice
1979 : The continental margin off Galicia Bank and Portugal: acoustical stratigraphy, dredge stratigraphy, and structural evolution; in Initial Reports of the Deep Sea Drilling Project, v. 47 (Part 2), eds. J.-C. Sibuet and W.B.F. Ryan; U.S. Government Printing Office, Washington, D.C., p. 633–662.

Gudmundsson, G., and Saemundsson, K.
1980 : Statistical analysis of damaging earthquakes and volcanic eruptions in Iceland from 1550–1978: Journal of Geophysics, v. 47, p. 99–109.

Hall, J. M.
1981 : The Thulean volcanic line; in Geology of the North Atlantic Borderlands, eds. J. W. Kerr and A. J. Fergusson; Canadian Society of Petroleum Geologists, Memoir 7, p. 231–244.

Henderson, G., Schiener, E. J., Risum, J. B., Croxton, C. A., and Anderson, B. B.
1981 : The west Greenland basin; in Geology of the North Atlantic Borderlands, eds. J. W. Kerr and A. J. Fergusson; Canadian Society of Petroleum Geologists, Memoir 7, p. 399–428.

Jansa, L. F., Enos, P., Tucholke, B. E., Gradstein, F., and Sheridan, R. E.
1979 : Mesozoic-Cenozoic sedimentary formations of the North American Basin; western North Atlantic; in Deep Drilling Results in the Atlantic Ocean: Continental Margins and Paleoenvironment, eds. M. Talwani, W. Hay, and W.B.F. Ryan; American Geophysical Union, Washington, D.C., M. Ewing Series 3, p. 1–57.

Jansa, L. F., and Wade, J. A.
1975 : Geology of the continental margin off Nova Scotia and Newfoundland; in Offshore Geology of Eastern Canada, v. 2, Regional Geology, eds. W.J.M. van der Linden and J. A. Wade; Geological Survey of Canada, Paper 74-30, p. 51–105.

Jansa, L. F., and Wiedmann, J. W.
1982 : Mesozoic-Cenozoic development of the eastern North American and northwest African continental margins: A comparison; in Geology of the Northwest African Continental Margin, eds. U. von Rad, K. Hinz, M. Sarnthein, and E. Seibold; Springer-Verlag, New York, p. 215–269.

Keen, C. E., Hall, B. R., and Sullivan, K. D.
1977 : Mesozoic evolution of the Newfoundland basin: Earth and Planetary Science Letters, v. 37, p. 307–320.

Keigwin, L. D., Jr.
1982 : Neogene planktonic foraminifers from Deep Sea Drilling Project Sites 502 and 503; in Initial Reports of the Deep Sea Drilling Project, v. 68, eds. W. L. Prell, and J. V. Gardner; U.S. Government Printing Office, Washington, D.C., p. 269–288.

Khudoley, K. M., and Meyerhoff, A. A.
1971 : Paleogeography and Geological History of Greater Antilles:, Geological Society of America, Memoir 129, 199 p.

King, L. H., and MacLean, B.
1975 : Stratigraphic interpretation of the central Grand Banks from high-resolution seismic reflection profiles; in Offshore Geology of Eastern Canada, Volume 2, Regional Geology, eds. W.J.M. van der Linden, and J. A. Wade; Geological Survey of Canada, Paper 74-30, p. 175–185.

McCoy, F. W.
1983 : Seafloor sediment; in Geologic Map of the Circum-Pacific Region Northeast Quadrant (1:10,000,000), ed. K. J. Drummond; Circum-Pacific Map Project of the Circum-Pacific Council for Energy and Mineral Resources, American Association of Petroleum Geologists.

McCoy, F. W., and Zimmerman, H. B.
1977 : A history of sediment lithofacies in the South Atlantic Ocean; in Initial Reports of the Deep Sea Drilling Project, v. 39, eds. P. Supko, and K. Perch-Nielsen; U.S. Government Printing Office, Washington, D.C., p. 1047–1079.
in prep: Paleogeography, paleosedimentation and paleoceanography of the Mesozoic-Cenozoic Atlantic Ocean: American Association of Petroleum Geologists Special Publications.
McDougall, I., and Schmincke, H. U.
1976 : Geochronology of Gran Canaria, Canary Islands; age of shield building volcanism and other magmatic phases: Bulletin Volcanologique, v. 40, p. 1–21.
McKenna, M. C.
1983 : Cenozoic paleogeography of North Atlantic land bridges; in Structure and Development of the Greenland-Scotland Ridge, eds. M.H.P. Bott, S. Saxov, M. Talwani, and J. Thiede; Plenum, New York, p. 351–399.
McWhae, J.R.H.
1981 : Structure and spreading history of the northwestern Atlantic region from the Scotian Shelf to Baffin Bay; in Geology of the North Atlantic Borderlands, eds. J. W. Kerr and A. J. Fergusson; Canadian Society of Petroleum Geologists, Memoir 7, p. 299–332.
Miller, K. G., and Tucholke, B. E.
1983 : Development of Cenozoic abyssal circulation south of the Greenland-Scotland Ridge; in Structure and Development of the Greenland-Scotland Ridge, eds. M.H.P. Bott, S. Saxov, M. Talwani, and J. Thiede; Plenum, New York, p. 549–589.
Miller, K. G., Fairbanks, R. G., and Thomas, E.
in press: Benthic foraminiferal carbon isotopic records and the development of abyssal circulation in the eastern North Atlantic; in Initial Reports of the Deep Sea Drilling Project, v. 94, eds., R. Kidd and W. F. Ruddiman; U.S. Government Printing Office, Washington, D.C.
Morgan, W. J.
1981 : Hotspot tracks and the opening of the Atlantic and Indian Oceans; in The Sea, v. 7, ed. C. Emiliani; Wiley, New York, p. 443–488.
Mountain, G. S., and Tucholke, B. E.
1985 : Mesozoic and Cenozoic geology of the U.S. Atlantic continental slope and rise; in Geologic Evolution of the United States Atlantic Margin, ed. C. W. Poag; Van Nostrand Reinhold, Stroudsburg, p. 293–341.
Nunns, A. G.
1983 : Plate tectonic evolution of the Greenland-Scotland Ridge and surrounding regions; in Structure and Origin of the Greenland-Scotland Ridge, eds. M.H.P. Bott, S. Saxov, M. Talwani, and J. Thiede; Plenum, New York, p. 11–30.
Parrish, J. T., and Curtis, R. L.
1982 : Atmospheric circulation, upwelling and organic-rich rocks in the Mesozoic and Cenozoic eras: Palaeogeography, Palaeoclimatology, Palaeoecology, v. 40, p. 31–66.
Peterson, M.N.A., and Edgar, N. T., eds.
1970 : Initial Reports of the Deep Sea Drilling Project, v. 2: U.S. Government Printing Office, Washington, D.C., 491 p.
Pratt, R. M.
1963 : Great Meteor Seamount: Deep-Sea Research, v. 10, p. 17–25.
Premoli-Silva, I., and Boersma, A.
1977 : Cretaceous planktonic foraminifers—DSDP Leg 39 (South Atlantic); in Initial Reports of the Deep Sea Drilling Project, v. 39, eds. P. Supko and K. Perch-Nielsen; U.S. Government Printing Office, Washington, D.C., p. 615–641.
Ranke, U., von Rad, U., and Wissmann, G.
1982 : Stratigraphy, facies and tectonic development of the on- and off-shore Aaiun-Tarfaya Basin—a review; in Geology of the Northwest African Continental Margin, eds. U. von Rad, K. Hinz, M. Sarnthein, and E. Seibold; Springer-Verlag, New York, p. 86–105.
Rawson, M. D., and Ryan, W.B.F.
1978 : Ocean Floor Sediment and Polymetallic Nodules (Map; 1:23,230,300); U.S. Department of State.
Reeside, J. B., Jr.
1957 : Paleoecology of the Cretaceous Seas of the Western Interior of the United States: Geological Society of America, Memoir 67, p. 505–542.
Reyment, R. A.
1966 : Sedimentary sequence of the Nigerian coastal basin; in Sedimentary Basins of the African Coasts: Part 1, Atlantic Coast, ed. D. Reyre; Association Service Geologique Afrique, Paris, p. 115–141.
1980 : Biogeography of the Saharan Cretaceous and Paleocene epicontinental transgressions: Cretaceous Research, v. 1, p. 299–327.
Reyre, D.
1966 : Particularités géologiques des bassins cotiers de l'Ouest Africain; in Sedimentary Basins of the African Coasts: Part 1. Atlantic Coast, ed. D. Reyre; Association Service Geologique Afrique, Paris, p. 253–302.
Ribeiro, A.
1980 : Portugal; Synthèse de l'évolution paléogéographique et tectonique, Géologie des Pays Européens: Comité National Francais de Géologie, 26th Congrès Géologie International, p. 89–149.
Rios, J. M.
1978 : The Mediterranean coast of Spain and the Alboran Sea; in The Ocean Basins and Margins, v. 4B, The Western Mediterranean, eds. A.E.M. Nairn, W. H. Kanes, and F. G. Stehli; Plenum, New York, p. 1–65.
Roberts, D. G., Bott, M.H.P., and Uruski, C.
1983 : Structure and origin of the Wyville-Thomson Ridge; in Structure and Origin of the Greenland-Scotland Ridge, eds. M.H.P. Bott, S. Saxov, M. Talwani and J. Thiede; Plenum, New York, p. 133–158.
Rögl, F., and Steininger, F. F.
1983 : Vom zerfall der Tethys zu Mediterran und Paratethys: Annalen des Naturhistorischen Museums in Wien, v. 85A, p. 135–163.
Rona, P. A.
1980 : The Central North Atlantic Ocean Basin and Continental Margins: Geology, Geophysics, Geochemistry and Resources, including the Trans-Atlantic Geotraverse (TAG): NOAA Atlas 3, U.S. Department of Commerce, 99 p.
Ryan, W.B.F., and Hsu, K. J., eds.
1973 : Initial Reports of the Deep Sea Drilling Project, v. 13, U.S. Government Printing Office, Washington, D.C., 1447 p.
Schminke, H. U., and von Rad, U.
1979 : Neogene evolution of Canary Island volcanism inferred from ash layers and volcaniclastic sandstones of DSDP Site 397 (Leg 47A); in Initial Reports of the Deep Sea Drilling Project, v. 47, (Part 1), eds. U. von Rad and W.B.F. Ryan; U.S. Government Printing Office, Washington, D.C., p. 703–725.
Sclater, J. G., Anderson, R. N., and Bell, M. L.
1971 : Elevation of ridges and evolution of the central eastern Pacific: Journal of Geophysical Research, v. 76, p. 7888–7915.
Sibuet, J.-C., and Ryan, W.B.F.
1979 : Site 398: Evolution of the west Iberian passive continental margin in the framework of the early evolution of the North Atlantic Ocean; in Initial Reports of the Deep Sea Drilling Project, v. 47, (Part 2), eds. J.-C. Sibuet and W.B.F. Ryan; U.S. Government Printing Office, Washington, D.C., p. 761–775.
Simpkin, T., Siebert, L., McClelland, L., Bridge, D., Newhall, C., and Latter, J. H.
1981 : Volcanoes of the World: Hutchinson Ross, Stroudsburg, 232 p.
Steel, R. J., Dalland, A., Kalgraff, K., and Larsen, V.
1981 : The central Tertiary basin of Spitsbergen: sedimentary development of a sheared-margin basin; in Geology of the North Atlantic Borderlands, eds. J. W. Kerr and A. J. Fergusson; Canadian Society of Petroleum Geologists, Memoir 7, p. 647–664.
Sullivan, K. D., and Keen, C. E.
1977 : Newfoundland Seamounts; in Volcanic Regimes of Canada, eds. W.R.A. Baragar, L. C. Coleman, and J. M. Hull; Geological Association of Canada, Special Paper 16, p. 461–476.

Surlyk, F., Clemmensen, L. B., and Larsen, H. C.
1981 : Post-Paleozoic evolution of the east Greenland continental margin; in Geology of the North Atlantic Borderlands, eds. J. W. Kerr and A. J. Fergusson; Canadian Society of Petroleum Geologists, Memoir 7, p. 611–645.

Talwani, M., and Eldholm, O.
1977 : Evolution of the Norwegian-Greenland Sea: Geological Society of America Bulletin, v. 88, p. 969–999.

Tucholke, B. E.
1979 : Relationships between acoustic stratigraphy and lithostratigraphy in the western North Atlantic; in Initial Reports of the Deep Sea Drilling Project, v. 43, eds. B. Tucholke and P. R. Vogt; U.S. Government Printing Office, Washington, D.C., p. 827–846.

Tucholke, B. E., and Ewing, J. I.
1976 : Bathymetry and sediment geometry of the Greater Antilles Outer Ridge and vicinity: Reply: Geological Society of America Bulletin, v. 87, p. 1371–1374.

Tucholke, B. E., and Fry, V. A.
1985 : Basement structure and sediment distribution in the Northwest Atlantic Ocean: American Association of Petroleum Geologists Bulletin, v. 69, p. 2077–2097.

Tucholke, B. E., and Ludwig, W. J.
1982 : Structure and origin of the J Anomaly Ridge, western North Atlantic Ocean: Journal of Geophysical Research, v. 87, p. 9389–9407.

Tucholke, B. E., and Mountain, G. S.
1979 : Seismic stratigraphy, lithostratigraphy and paleosedimentation patterns in the North American basin; in Deep Drilling Results in the Atlantic Ocean: Continental Margins and Paleoenvironment, eds. M. Talwani, W. Hay, and W.B.F. Ryan; American Geophysical Union, Washington, D.C., M. Ewing Series, v. 3, p. 58–86.

Tucholke, B. E., and Vogt, P. R.
1979a: Western North Atlantic: sedimentary evolution and aspects of tectonic history; in Initial Reports of the Deep Sea Drilling Project, v. 43, eds. B. E. Tucholke and P. R. Vogt; U.S. Government Printing Office, Washington, D.C., p. 791–825.

Tucholke, B. E., and Vogt, P. R., eds.
1979b: Initial Reports of the Deep Sea Drilling Project, v. 43: U.S. Government Printing Office, Washington, D.C., 1115 p.

Uchupi, E., and Austin, J. A., Jr.
1979 : The stratigraphy and structure of the Laurentian Cone region: Canadian Journal of Earth Sciences, v. 16, p. 1726–1752.

Uchupi, E., Sancetta, C., Eusden, J. D., Jr., Bolmer, S. T., Jr., McConnell, R. L., and Lambiase, J. J.
1984 : Appendix (Lithofacies Maps) d–h; Eastern North American Continental Margin and Adjacent Ocean Floor: Ocean Margin Drilling Program Atlas, v. 1–5, Marine Science International, Woods Hole.

Vail, P. R., Hardenbol, J., and Todd, R. G.
1984 : Jurassic unconformities, chronostratigraphy, and sea-level changes from seismic stratigraphy and biostratigraphy; in Interregional Unconformities and Hydrocarbon Accumulation, ed. J. S. Schlee; American Association of Petroleum Geologists, Memoir 36, p. 129–144.

Vogt, P. R.
1973 : Early events in the opening of the North Atlantic; in Implications of Continental Drift to the Earth Sciences, v. 2, eds. D. H. Tarling and S. K. Runcorn; Academic Press, London, p. 693–712.

Westermann, G.E.G.
1980 : Ammonite biochronology and biogeography of the circum-Pacific Middle Jurassic; in Systematics Association Special Volume no. 18, "The Ammonoidea," eds., M. R. House and J. R. Senior; Academic Press, London, p. 459–498.

Williams, G. L.
1975 : Dinoflagellate and spore stratigraphy of the Mesozoic-Cenozoic, offshore eastern Canada; in Offshore Geology of Eastern Canada, v. 2, Regional Geology, eds. W.J.M. van der Linden and J. A. Wade; Geological Survey of Canada, Paper 74-30, p. 107–161.

Wissmann, G.
1982 : Stratigraphy and structural features of the continental margin basin of Senegal and Mauritania; in Geology of the Northwest African Continental Margin, eds. U. von Rad, K. Hinz, M. Sarnthein, and E. Seibold; Springer-Verlag, New York, p. 160–181.

Zbyszewski, G.
1971 : Recontiecimento geologico da parte occidental da ilha da Madeira: Academia das Ciências de Lisboa, Memórias, v. 15, p. 7–23.

Ziegler, P. A.
1982 : Geological Atlas of Western and Central Europe: Elsevier, Amsterdam, 130 p.

Zimmerman, H. B.
1982 : Lithologic stratigraphy and clay mineralogy of the western Caribbean and eastern equatorial Pacific, Leg 68, Deep Sea Drilling Project; in Initial Reports of the Deep Sea Drilling Project, v. 68, eds. W. L. Prell, and J. V. Gardner; U.S. Government Printing Office, Washington, D.C., p. 383–395.

MANUSCRIPT ACCEPTED BY THE SOCIETY JULY 1, 1985

ACKNOWLEDGMENTS

This synthesis depended in large part on our research that has been supported over the past ten years by the Office of Naval Research and the National Science Foundation. In this respect, we especially acknowledge ONR Contracts N00014-79-C-0071 and N00014-82-C-0019, and NSF Grant OCE79-09382 to Woods Hole Oceanographic Institution, and NSF Grant OCE76-02038 to Lamont-Doherty Geological Observatory. B. Tucholke also acknowledges support of a Mellon Senior Study Award at WHOI during preparation of this paper. We thank the scientists listed as contributors on Plates 9 and 10 for their contributions to these reconstructions. We also thank E. Uchupi for his general help in locating published data pertinent to the syntheses. The manuscript was reviewed by S. Cande, D. A. Johnson, M. Leckie, K. G. Miller, and E. Uchupi. Contribution No. 5980 of Woods Hole Oceanographic Institution and Contribution No. 3965 of Lamont-Doherty Geological Observatory.

Printed in U.S.A.

The Geology of North America
Vol. M, The Western North Atlantic Region
The Geological Society of America, 1986

Chapter 36

Paleoceanography and evolution of the North Atlantic Ocean basin during the Jurassic

Lubomir F. Jansa
Geological Survey of Canada, Atlantic Geoscience Center, Bedford Institute of Oceanography, Dartmouth, Nova Scotia B2Y 4A2, Canada

INTRODUCTION

Jurassic paleoceanographic conditions in the Atlantic Ocean can be reconstructed from information derived from the preserved sediments and the biogeography of the fossil fauna, from reconstructed basin paleogeographies (shape, size, depth, and positions of landmasses and seaways), and from indications of paleotopography, paleobathymetry, and paleoclimate. An extensive literature exists about Mesozoic sediments and biota of the Tethyan realm from Europe and parts of northwestern Africa. Such data, supplemented by my own field studies in central and southern Europe and Morocco, and by studies of oil exploratory wells on the eastern North American margin, were major sources of the Jurassic lithofacies data summarized in Plate 9. Additional data comes from Continental Offshore Stratigraphic Test (COST) wells drilled on the eastern North Atlantic U.S. shelf and summarized in U.S. Geological Survey reports. However, the most important information is derived from the Deep Sea Drilling Project (DSDP), which drilled numerous deep holes in the deep North Atlantic basin and its continental margins.

Greater insight into the early evolution of the North Atlantic basin was provided particularly by DSDP Leg 79, which drilled four holes on the deep-water Moroccan continental margin, at water depths of 3151 to 3992 m (Sites 554 to 547, Hinz and others, 1984; Plates 2, 9). These holes provide the only information about the Lower Jurassic in the central North Atlantic basin, and are used extensively in this chapter on the assumption that basin development was similar on both sides of the Atlantic. This assumption is supported by the similarity of conjugate shelf areas of the Mesozoic North Atlantic (Bhat and others, 1975; Jansa and Wade, 1975; Jansa and Wiedmann, 1982) and of the eastern and western central North Atlantic Upper Jurassic deep-sea deposits (Jansa and others, 1978; Fig. 1).

Jurassic sedimentation in the North Atlantic extended northward from the central North Atlantic into an area separating the Grand Banks and Iberian Peninsula (Dupeublé and others, 1976). It covered parts of the Grand Banks (Jansa and Wade, 1975) and extended north into the Celtic Sea, Western Approaches, and Bay of Biscay Rift (Ziegler, 1981). Because the aim of this chapter is to document the development of the oceanic basin, the history of the adjacent shallow epeiric seas is excluded from the discussion below. The Middle to Late Jurassic ocean that extended from the Gulf of Mexico into central and northern South America is not discussed here either, but it is covered in other DNAG volumes.

In the North American Basin, Jurassic sedimentary rocks were recovered mainly from five DSDP sites: 99, 100, 105, 391 and 534 (Fig. 1). Stratigraphic studies of these sites demonstrate that basin-wide oceanographic changes are reflected by sediment composition so that major lithofacies can be correlated from site to site (Lancelot and others, 1972; Bernoulli, 1972). Jansa and others (1979) reviewed the post-Middle Jurassic lithologic and seismic-stratigraphic correlations in the North American Basin and placed the lithofacies into a formal stratigraphic framework. More recently, Site 534 was drilled, extending our knowledge of Middle Jurassic events (Fig. 1; Ogg and others, 1983).

Mesozoic circulation in the Atlantic basin has been discussed in detail by Berggren and Hollister (1974) and Berger (1981) on the basis of the marine microfauna distribution, but the authors were influenced by results of experimental modeling of the paleocirculation (Luyendyk and others, 1972). Proposed paleocirculation patterns presented in this contribution differ from those of Berggren and Hollister (1974), partly because more data are available for paleogeographic reconstructions and lithofacies distribution, but mainly because we include climatic variability, which was not considered by Luyendyk and others (1972).

The paleocirculation and lithofacies data presented here are summarized in Plate 9. The data are plotted on plate reconstruction maps of Klitgord and others (this volume) and Srivastava and Tapscott (this volume). At the request of the editors, the reference list is limited only to the most important literature related directly to the discussion of paleoceanography.

Jansa, L. F., 1986, Paleoceanography and evolution of the North Atlantic Ocean basin during the Jurassic; *in* Vogt, P. R., and Tucholke, B. E., eds., The Geology of North America, Volume M, the Western North Atlantic Region: Geological Society of America.

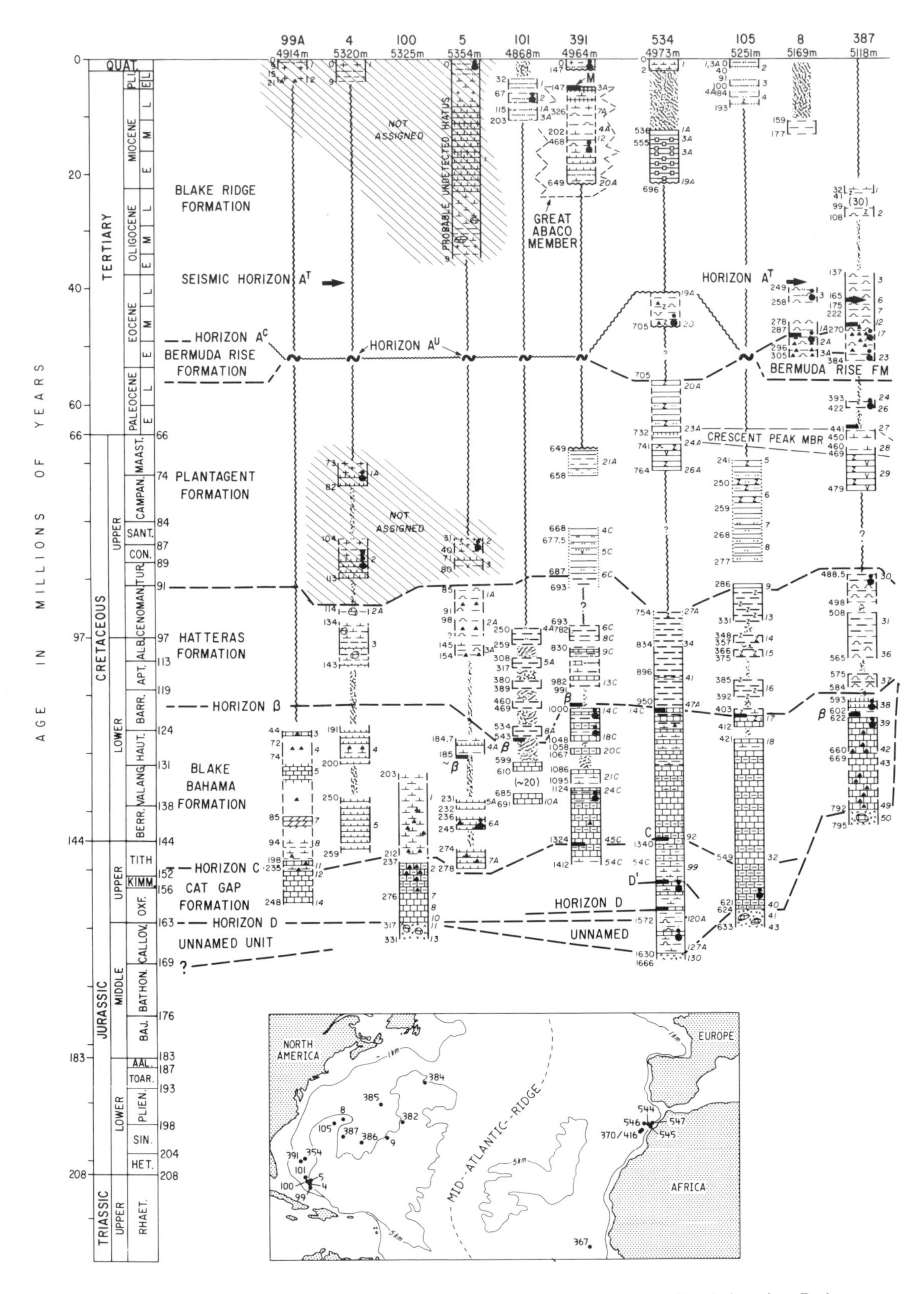

Figure 1. Correlation chart of DSDP Sites that encountered Jurassic rocks in the North American Basin. The chart shows schematic lithology of cores, sub-bottom depth, age, formation names, and important seismic markers (updated from Tucholke, 1979 and Jansa and others, 1979).

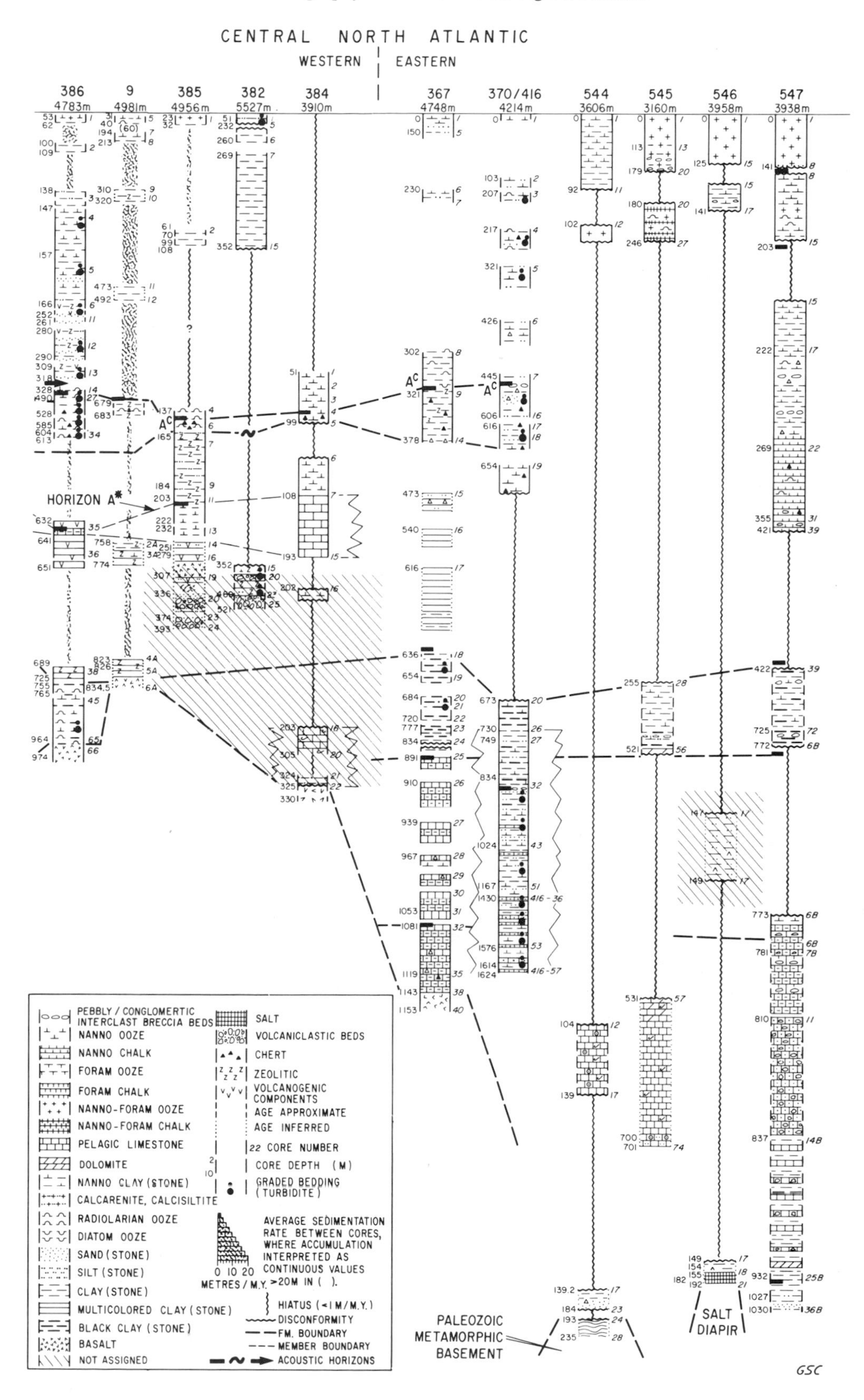

Figure 1 (*Continued*).

TRIASSIC

Basin Evolution

It is generally agreed that the first marine incursion into the developing rifts of the North Atlantic basin probably occurred during the Late Triassic and led to the precipitation of evaporites. Evaporites of Late Triassic age underlie the Grand Banks and extend southward to the Scotian shelf and to Georges Bank (Plate 9A). The evaporites are mainly halite and form sequences 1 to 2 km thick (McIvor, 1972; Jansa and Wade, 1975).

Palynological dating of evaporites on the Canadian margin indicates two distinct intervals of evaporite deposition. The older sequence (Carnian-Norian) contains more than 2000 m of predominantly halite and is located at the eastern Grand Banks (Jansa and others 1980a). Evaporites underlying the shelf seaward of the basement hinge zone of Nova Scotia and the southern Grand Banks are younger, being of Rhaetian-Sinemurian/Hettangian age (Barss and others, 1978; Jansa and Wade, 1975).

Diapirs beneath the continental slope off northeastern North America have not been sampled, but a diapir in the conjugate evaporite belt off Morocco was sampled at DSDP Site 546 (Plate 9A; Fig. 1). This site was drilled in 3958 m of water, with salt being reached at 155 m subbottom (Hinz and others, 1984). The evaporite sequence consists mainly of halite with minor clays, anhydrite, carnallite, and sylvite (Fig. 2a). The evaporites are overlain by several meters of unfossiliferous, reddish sandy mudstone. Geochemical and mineralogical studies indicate that the evaporites are of dominantly marine origin and were deposited from shallow brine (Holser and others, 1984).

Spores embedded in the salt provide an age similar to the Scotian shelf evaporites (Rhaetian-Hettangian; E. Davies personal communication, 1983). Apparently contradictory results were derived from study of sulfur isotopes in the halite sequence. According to Holser and others (1984), the isotope ratios are more compatible with a Permian to Scythian age. However, the Triassic and Late Permian marine deposits are unknown on eastern and western central North Atlantic margins, which suggests that palynological dating of the salt is likely to be correct, and it is used here in basin reconstruction. Drilling results off Morocco indicate that the salt sequence overlies continental basement. A similar conclusion was suggested earlier from reflection seismic profiles studies offshore from eastern Canada by Jansa and Wade (1975), and by Lehner and DeRuiter (1977) off western Africa.

Physiography and Paleoceanography

Extensive geophysical studies along both central North Atlantic continental margins allowed us to map the salt basin. Closing the North Atlantic on the seaward outlines of the diapiric salt off Nova Scotia and Morocco (Plate 9A) shows a good fit (Hinz and others, 1984) and supports a model involving original salt deposition in a single basin, which later was broken and drifted apart.

According to this reconstruction, the central intracratonic rift system that became the site of evaporite deposition was about 350 km wide and 900 km in length. The southern limit was probably near the line connecting the prolongation of the Anti-Atlas and Long Island. The chemistry of the evaporites suggests that the water over the rift floor was shallow. One to two km of evaporites were deposited along the margins of the central rift, and more may have been deposited in the center of the basin. Therefore, the basin subsidence and accumulation rates (uncorrected for compaction) were approximately 200 m/m.y. or greater.

Distribution of evaporitic basins and lithofacies for western Europe (Ziegler, 1981; Plate 9A) suggests that the main connection with Tethys was northward, either through the Bay of Biscay Rift (Jansa and others, 1980a) or through western Europe; the future "Iberia-Rif" seaway did not yet exist or was very shallow.

The paleo-latitude of the basin ranged from 10° to 25°N (Van der Voo and French, 1974, Scotese and others, 1981). The climate, as witnessed by the red beds and evaporites, was warm and arid. Atmospheric circulation was probably weak and variable, since the southern part of the basin was equatorial. The northern part of the basin may have been influenced by trade winds (Parrish and Curtis, 1982) which, together with density gradients in the basin, are assumed to be the dominant driving force for surface currents. Inflow of fresher, lighter water from the northeast, excess evaporation in the basin, and wind stress are likely to have resulted in surface flow to the southwest approximately parallel to the long axis of the basin. Reddish-brown clay beds intercalated within the salt indicate oxygenated bottom conditions; thus thermohaline vertical convection and mixing probably resulted from density differences caused by evaporation of surface waters. The circulation regime was comparable to a narrow estuary with complete vertical mixing. Alternation of halite beds and unfossiliferous beds of reddish-brown silty shales implies periodic inflow of fresh sea water into the basin, as well cyclic climatic conditions.

EARLY JURASSIC

Basin Evolution

Evaporite deposition continued into earliest Jurassic time (Hettangian/Sinemurian; Barss and others, 1979; Jansa and Wade, 1975) from the Grand Banks to the Baltimore Canyon Trough. In the latter area an Early Jurassic radiometric age of 183 ± 12 Ma was obtained for the salt (Houston Oil and Minerals #1-Well History Report). Termination of evaporite deposition (evaporites wedged into red beds) in marginal basins of Portugal (Zbyszewski, 1965) and Morocco (Salvan, 1974) generally is assumed to be Late Triassic from its stratal position. However, this is not paleontologically confirmed and the termination could have occurred in the Early Jurassic.

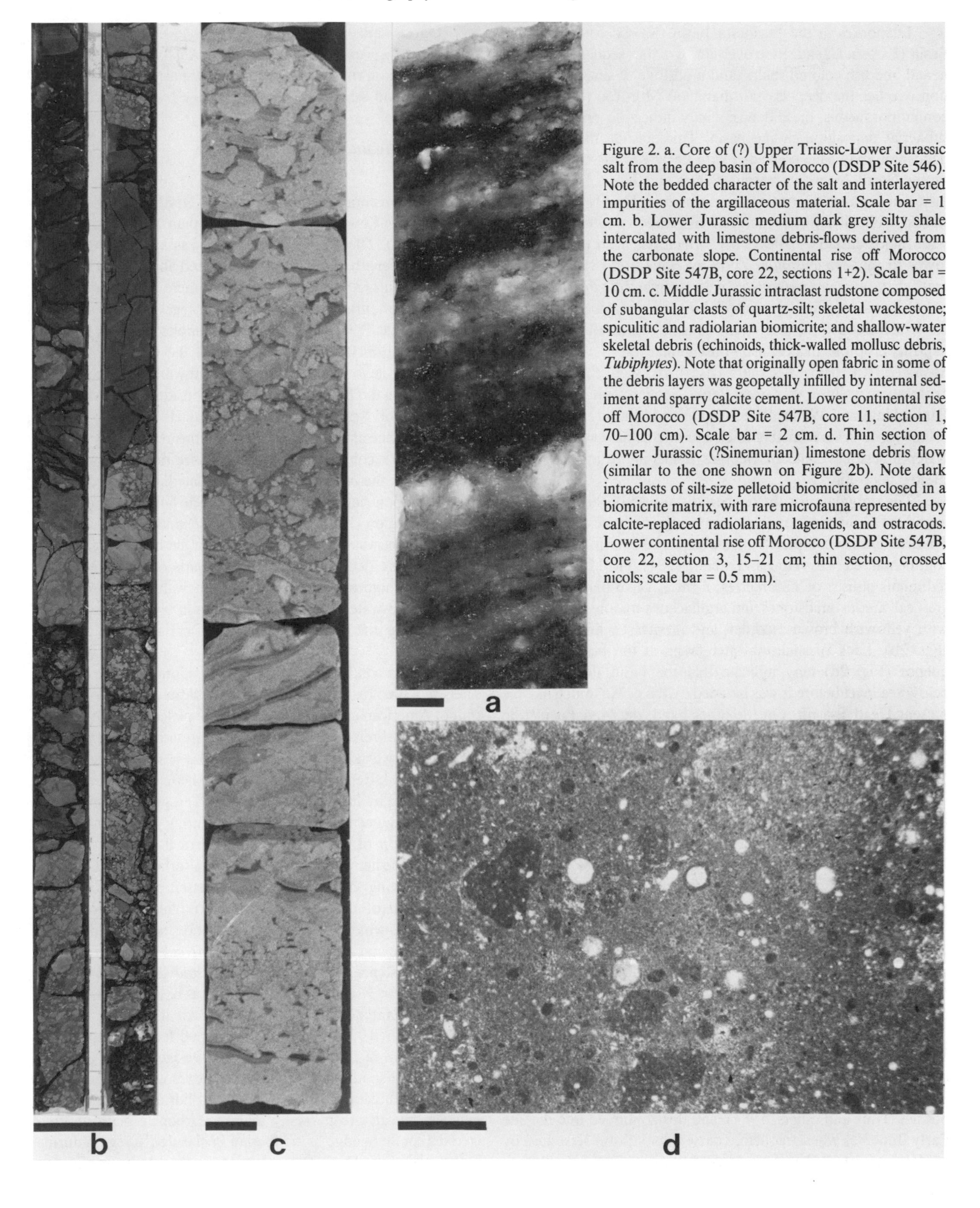

Figure 2. a. Core of (?) Upper Triassic-Lower Jurassic salt from the deep basin of Morocco (DSDP Site 546). Note the bedded character of the salt and interlayered impurities of the argillaceous material. Scale bar = 1 cm. b. Lower Jurassic medium dark grey silty shale intercalated with limestone debris-flows derived from the carbonate slope. Continental rise off Morocco (DSDP Site 547B, core 22, sections 1+2). Scale bar = 10 cm. c. Middle Jurassic intraclast rudstone composed of subangular clasts of quartz-silt; skeletal wackestone; spiculitic and radiolarian biomicrite; and shallow-water skeletal debris (echinoids, thick-walled mollusc debris, *Tubiphytes*). Note that originally open fabric in some of the debris layers was geopetally infilled by internal sediment and sparry calcite cement. Lower continental rise off Morocco (DSDP Site 547B, core 11, section 1, 70–100 cm). Scale bar = 2 cm. d. Thin section of Lower Jurassic (?Sinemurian) limestone debris flow (similar to the one shown on Figure 2b). Note dark intraclasts of silt-size pelletoid biomicrite enclosed in a biomicrite matrix, with rare microfauna represented by calcite-replaced radiolarians, lagenids, and ostracods. Lower continental rise off Morocco (DSDP Site 547B, core 22, section 3, 15–21 cm; thin section, crossed nicols; scale bar = 0.5 mm).

Evaporites in the Essaouira Basin (Morocco) and Scotian Basin (Eastern Canada) are overlain by a thin sequence of continental, reddish colored shales, and mudstone of similar composition overlies the deep-sea salt diapir off Morocco (Site 546). If contemporaneous, these deposits may indicate a brief period of subaerial deposition in the basin. This episode of continental deposition may have coincided with Early Kimmerian tectonism on the North American margin (Schlee and Jansa, 1981).

The Jurassic depositional cycle in the central North Atlantic started with Sinemurian marine transgression contemporaneous with major eustatic sea-level rise, which has been placed in the lower to mid-*semicostatum* Zone by Hallam (1981). Brownish grey, shallow-water carbonates with stromatolitic dolomites and peloid limestones were deposited along the margins of the epeiric sea which flooded the area of the Triassic central rift basin. Such deposits crop out in Portugal (Lusentanian Basin, Algarve) and Morocco (Rif chains, Essaouira Basin), and were encountered in exploratory wells on the Grand Banks, Scotian shelf, and Georges Bank (Jansa and Wade, 1975; Amato and Simonis, 1980). Beds of anhydrite several meters thick are intercalated with shallow-water limestones beneath Georges Bank, indicating locally restricted deposition.

The only data for the deeper, distal part of the basin come from DSDP Site 547, on the lower continental rise off Morocco. There an almost complete Jurassic sequence overlies the undated continental red beds (Hinz and others, 1984). The Lower Jurassic sediments consist of 120 metres of thick, olive black and olive grey calcareous mudstones and argillaceous micrites intercalated with yellowish brown, nodular, and intraclastic limestone beds (Fig. 2b). Lack of shallow-water fauna at the base of the sequence (Fig. 2d) may indicate that the basin floor subsided below sea level before it was invaded by the ocean, much like the present Dead Sea rift. On the other hand, the faunal depletion also could be due to higher salinity of sea water because of partial dissolution of underlying evaporite beds, or it may reflect low diversity of Lower Jurassic pelagic fauna.

A significant change of lithology occurred during the Pliensbachian on the continental margins. The principally dolomite deposits were replaced by more argillaceous facies (sometimes bituminous) and by limestones that are frequently biomicrites enclosing chert nodules. There was an accompanying increase in richness of the biota, represented by molluscs, echinoids, foraminifera, ostracoda, belemnites, ammonites, and radiolarians. This change suggests a deeper depositional environment and the establishment of normal salinity. The deepening corresponds to the eustatic sea-level rise placed by Hallam (1978) in the *margaritatus* Zone. This sea-level rise had little effect on proximal basin deposition, but, on the basis of the section at Site 547, displaced limestone debris became more frequent as a carbonate slope formed on the Moroccan margin. During the Pliensbachian, thin beds of carbon-rich claystones were also deposited (Hinz and others, 1984), and this continued into the late Early Jurassic, when frequent, coarse debris flows separated by hardgrounds marked an early Middle Jurassic eustatic drop of sea level (Jansa and others, 1984). This sea-level drop is also observed in Morocco (Ambroggi, 1963), Algarve (Rocha, 1977), and on the eastern North American margin (Jansa and Wade, 1975; Bhat and others, 1975).

Physiography and Paleoceanography

The maximum depth of the central North Atlantic basin at the end of the Liassic was probably less than 1000 m (Hinz and others, 1984). The width of the basin was approximately 500 km, and the length of the basin is estimated about 1700 km. The southern limit of the basin is uncertain; there is no evidence for marine Lower Jurassic deposits in the Senegal Basin or southern Florida and the Blake-Bahama Basin (Templeton, 1971; Klitgord and others, 1984). The diapiric salt off the Carolinas and off Senegal may have originated during southward advancement of the sea during the Early Jurassic and is considered to be younger than the salt off Nova Scotia and Morocco. However, no data are available for the age of the evaporites in this region.

The dark color of Lower Jurassic sediments off Morocco indicates that the basin was poorly ventilated. Occasional occurrence of carbon-rich layers in the basin (Site 547) and in outcrops of the Rif region (Wildi and others, 1977) and Lusetanian Basin (personal observation) also point to periodic oxygen deficiency. However, rare occurrence of reddish marls off-shore Morocco indicate periodic oxygenation of bottom waters, thus requiring episodic deep-water overturn, or changes in bottom-water circulation during the early stages of the central North Atlantic development.

Climatic conditions during the Early Jurassic were progressively modified by the developing ocean basin and the northward shift of American and African continental plates. Climate became temperate and relatively humid at the northern end of the basin, where marine shales were deposited (Jansa and Wade, 1975; Ziegler, 1981). Reconstruction of atmospheric circulation for the Pliensbachian by Parrish and Curtis (1982) suggests easterly trade winds over the central North Atlantic during "northern winter" and formation of a weak low pressure cell during "northern summer." The net wind drag on the ocean surface would result in currents flowing to the southwest. Increased evaporation at the southern end of the basin may have resulted in a combined thermohaline-wind driven surface circulation, with the flow being anticlockwise and strongest on the western side of the basin.

The existence of a northern connection via the Grand Banks to the northern North Atlantic (Celtic Sea–Bay of Biscay region) can be substantiated by the distribution of the Lower Jurassic facies (Ziegler, 1981) and their brachiopod fauna (Ager, 1974). The extension of Lower Jurassic dolomite facies from the Subbetic Zone of southern Spain to the Algarves and the occurrence of similar lithofacies in the Rif area (Wildi and others, 1977) indicate that an exceptionally shallow carbonate platform covered the southern edge of the Iberian Peninsula-Rif region during Sinemurian time. The platform was too shallow to be an effective seaway connecting central North Atlantic and the western Tethys.

That marine incursions during Early Jurassic came into the central North Atlantic from the north and not from the east is well documented by the fauna. As pointed out by Lancelot and Winterer (1980), there is a striking difference between faunas from both the Lusetanian Basin (Portugal) and the Essaouira Basin (Morocco), where the ammonites and brachiopods are clearly subboreal (Mouterde and Ruget, 1975; Ager, 1974), and faunas from the central High and Middle Atlas, where they are definitely Tethyan (Choubert and Faure-Muret, 1962; Faugeres and Mouterde, 1979). The onset of Pliensbachian time marks the first mixing of the Tethyan and subboreal faunas in the Lusetanian basin and the Rif. Thus, unrestricted flow of the Tethyan water into the central North Atlantic was initiated either as a result of subsidence of the region between the southern Iberian Peninsula and northern Africa, or as a consequence of sea-level rise. The presence of Pliensbachian cherty limestones in the Algarve and Rif areas indicates either higher fertility or "upwelling" in this region, thus providing evidence for intermediate or bottom-water flow through the "Iberia-Rif" seaway.

Interpretation of the geologic data suggests that the Early Jurassic circulation was comparable to that of a deep estuary, with vertical stratification in which the bottom and intermediate water circulation was sluggish. However, the overlying Pliensbachian and younger sediments contain a larger variety of biota which is thought to be the result of increased biological productivity and nutrients in the surface water.

MIDDLE JURASSIC

Basin Evolution

On the continental margins off eastern Canada, northeastern United States, and Morocco, the Lower Jurassic carbonates are overlain by lower Middle Jurassic continental clastics (McIver, 1972; Amgroggi, 1963). Shoaling also occurred on the Grand Banks (Jansa and Wade, 1975), in Georges Bank Basin (Amato and Simonis, 1980), and in southern Portugal (Rocha, 1977), documenting a major regressive period in the central North Atlantic during Aalenian(?)-Bajocian time. Carbonate deposition resumed in the Scotian Basin and southern Georges Bank during Bathonian time (Barss and others, 1978), but on the central Grand Banks and Algarve the rise of sea level led to the deposition of neritic shales (Jansa and Wade, 1975; Rocha, 1977).

The next major rise of sea level, noticeable particularly on the shelf areas, occurred during Callovian time when deeper neritic shales were deposited above shallow-water carbonates. This transgression reached the eastern Essaouira Basin, where oolitic and oncolitic carbonates overlie continental red beds (Ambroggi, 1963).

Data for the deep central North Atlantic basin come from two areas. Offshore Morocco (Site 547, Plate 9B) a condensed Middle Jurassic sequence at the foot of the continental margin is represented by sandy limestones and coarse, proximal, limestone debris flows (Hinz and others, 1984; Fig. 2c). A similar situation was found by Wildi and others (1977) in the Rif area, where Middle Jurassic pelagic carbonates and marls are highly condensed and only a few meters thick. In the Blake-Bahama Basin, the basal 63 m of strata in Site 534, dated as Callovian, are predominantly dark greenish gray, silty calcareous claystones intercalated with carbon-rich claystone, light grey limestone turbidites (Figs. 3a, c), and radiolarian siltstone beds (Fig. 3b; Ogg and others, 1983). Lithologically, the Callovian silty claystones of the Blake-Bahama Basin resemble the dark, silty, calcareous claystones of the "Terres Noires" that outcrop in southeastern France (deGraciansky and others, 1980).

The tectonic instability that is a characteristic feature of the circum-central North Atlantic Middle Jurassic is well documented in northwestern Europe, where parts of the Northwest European Graben were uplifted and eroded (Ziegler, 1981), as were the outer edges of the eastern North American and northwest African continents (Jansa and Wade, 1975; Jansa and Wiedmann, 1982). The tectonism is referred to as the "Mid-Cimmerian Phase" by Ziegler (1981). It was accompanied by widespread volcanic activity in the North Sea, the Celtic Sea, southern Spain, the Tarfaya-Aauin Basin and, to a limited degree, along the eastern North American margin (Dixon and others, 1981; Jansa and Wiedmann, 1982). Because drilling at Site 534 confirmed the presence of Callovian oceanic basement (Sheridan and others, 1983), the tectonism and volcanism at the margin is interpreted as marking the period of final separation of the North American and African continental plates, which is dated as Aalenian-early Bajocian (180–187 Ma).

Physiography and Paleoceanography

Although a seaway could have been open intermittently between the North Atlantic and Pacific during the late Early Jurassic (Hallam, 1983), communication became well established in Middle Jurassic, as indicated by distribution of ammonites (Westermann, 1981). The exact location of this seaway is unresolved, however. The southern boundary of the Middle Jurassic central North Atlantic basin probably extended south of Florida (the Cuba transform fault(?); Klitgord and others, 1984; Plate 9B). The size of the oceanic basin increased to approximately 600 × 3000 km, with the basin axis oriented NE–SW. The depth of the sea floor in Callovian time off northwestern Africa was estimated by Wildi (1981) as 2000 m. However Sheridan and others (1983), by applying the subsidence calculation of Parsons and Sclater (1977) for theoretical ocean-crust subsidence, and assuming that the elevation of the mid-ocean ridge was the same as in present time, estimated the depth of the Callovian Blake-Bahama Basin to be 3000 m.

The opening of the central North Atlantic basin to the Pacific changed the paleocirculation pattern in the basin from a semirestricted gulf or "estuary" to a hydrodynamically open "channel" system (Plate 9B). This opening probably allowed in-

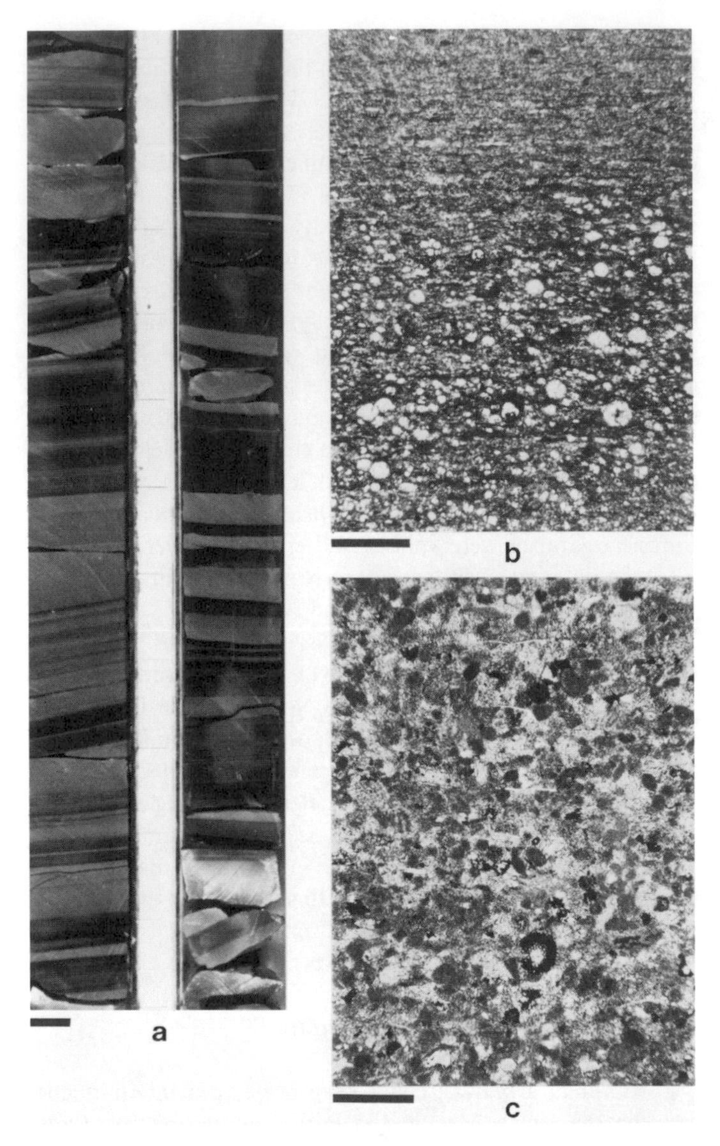

Figure 3. a. Carbonate turbidites (light grey) intercalated within Callovian greenish-black radiolarian mudstones in the Blake-Bahama Basin (DSDP Site 534A, core 22, sections 1+2). Fabric and composition of main lithologies are shown in Figures 3b and 3c. Scale bar = 2 cm. b. Radiolarian mudstone; tests of radiolarians are reprecipitated silica. The tests were concentrated into thin laminae by weak? bottom currents (DSDP Site 534A, core 126, section 2, 97–98 cm; thin section, crossed nicols; scale bar = 0.5 mm). c. Limestone turbidite similar to that shown in Figure 3a. Note the peletoid packestone texture; peloids are silt-size, rounded to subangular, and well sorted. The additional components are sparry-calcite- and pyrite-replaced radiolarian tests, rare foraminifera, and short filaments (pelagic bivalves) cemented by ferroan sparry calcite. Callovian, Blake-Bahama Basin (DSDP Site 534A, core 124, section 4, 22 cm, thin section, parallel nicols, scale bar = 0.5 mm).

creased bottom-current velocities and thus led to widespread nondeposition, erosion, and/or intermittent deposition in the regions of seaways such as the Rif area (Wildi and others, 1977). The increased vigor of the bottom circulation in the basin may be reflected in the prominent change from a dominantly medium gray color of basinal deposits to reddish brown in the Middle Jurassic off Morocco (Site 547; Jansa and others, 1984). The occurrence of Early Callovian(?) reddish colored marls in the Blake-Bahama Basin (Ogg and others, 1983) is compatible with the episodic presence of oxygen-rich bottom waters in the western central North Atlantic. However, the progression of reddish marls into dark claystones during the Callovian in the Blake-Bahama Basin, suggests that oxygen-poor bottom conditions became dominant in the southern portion of the central North Atlantic at this time. These conditions extended to the Gulf of Mexico as shown by occurrence of brown colored, carbon-rich limestones of the lower Smackover Formation at the Gulf Coast and by the lithologically similar Tepexic Formation in Mexico, (Cantu Chapa, 1971). These deposits provide firm evidence of a low-oxygen bottom-water regime and sluggish circulation during the late Callovian in the southwestern part of the central North Atlantic basin and in the Gulf of Mexico. Such interpretations appear to contradict evidence for the well-oxygenated bottom sediments in the northeastern central North Atlantic (Jansa and others, 1984). Three different explanations can be suggested: 1) the connection to the Pacific either was not open during Callovian time or it was very restricted and thus did not demonstrably influence bottom circulation in the southern part of the basin. This hypothesis is not supported by paleontologic evidence (Hallam, 1977; Westermann, 1981), because the distribution of ammonites indicates marine connections between the Andes, Mexico, and Eurafrica toward the end of early Callovian time; 2) the circulation in the basin was asymmetrical; or 3) the seaway probably opened along one of the transform fault zones south of Yucatan, with the Gulf of Mexico being a semiclosed embayment as suggested by Olson and Leyden (1973). An implication of the latter hypothesis, preferred by the author, is that the Louan salt was deposited from North Atlantic and not from Pacific water.

The surface-water circulation pattern in the central North Atlantic basin during the early Middle Jurassic is assumed to be anticlockwise (Plate 9B) because the wind stress direction is presumed to have remained about the same as during Early Jurassic time. The outflow of the "deep" bottom or intermediate water to the Mediterranean Tethys took place through the opening between Spain and north Africa, as indicated by development of highly condensed Middle Jurassic sections in the external zone of the Rif (Wildi and others, 1977). Occurrence of Toarcian ammonitic shales in Algarve (Rocha, 1977) and the evidence from the Rif (Wildi, 1981) suggest that this seaway was more than several hundred meters deep. In this respect, the plate-kinematic reconstruction in Plate 9B may show too restrictive a fit between Iberia and Africa. Another connection between Tethys and the central North Atlantic was via shallow, epeiric seas covering the Grand Banks, the Celtic Sea, Western Approaches, and western Europe (Plate 9B). Such a connection most likely generated southward-oriented colder, fresher, surface-water flow along the eastern North American margin. Hence, the circulation in the basin was probably in part thermohaline and in part wind driven.

Additional information about the circulation regime is provided by the relative abundance and distribution of radiolarians.

They are almost the only microfossils in Callovian sediments in the Blake-Bahama Basin (Ogg and others, 1983; Fig. 3b); they are similarly dominant in eastern North Atlantic Middle Jurassic sediments (Jansa and others, 1978), in correlative sequences in some of Rifean nappes of the Beni M'hamed unit (Wildi and others, 1977), and in the Tethyan Realm and parts of the Middle East and California (Jenkyns and Winterer, 1982). This widespread, explosive radiation of radiolaria during Middle Jurrasic time in both the western Tethys Ocean and the eastern Pacific Ocean strongly suggests that the cause was not a local event, but a worldwide oceanic phenomenon.

Several major oceanographic events, including the opening of the seaways between the western Tethys and Pacific oceans and between Tethys and the Arctic (via Greenland Sea; Arkell, 1956), increased vigor of bottom circulation. Possible upwellings, intensive volcanic activity in northwestern and southern Europe, initiation of extensive volcanism with formation of incipient oceanic crust in the central North Atlantic and Mediterranean, and high levels of transfer of silica and aluminosilicate minerals from continental areas to ocean basins, all occurred during the Middle Jurassic. The cumulative effect of all these processes could have contributed to the increase of silica and other nutrients in the oceans and thus have provided favorable conditions for radiolarian blooms. However, I believe that it was mainly a change in the circulation pattern that resulted in upwellings and increased surface productivity.

LATE JURASSIC

Basin Evolution

The Late Jurassic evolution of the North Atlantic Basin is much better constrained by the deep-sea data than were the Middle Jurassic events. The oceanic character of the basin is well attested to by the evidence of drilled basaltic sequences that are overlain by pelagic carbonates punctuated by turbidites near the continental margins. The turbidites are of dominantly carbonate composition where they fronted extensive carbonate platforms and banks (Scotian Basin), and they are terrigenous down-dip from major clastic depocenters (Baltimore Canyon Trough).

Because of sea-floor spreading, the basin width during Late Jurassic time was increasing and the basin physiography was progressively changing. In the North American Basin, this resulted in progressively younger pelagic and turbidite deposits onlapping eastward onto the oceanic basement (Tucholke and Mountain, 1979).

The Upper Jurassic sequence in the North American Basin belongs to the Cat Gap Formation (Jansa and others, 1979) which is 230 m thick at Site 534 (Ogg and others, 1983) and consists of Oxfordian to lower/upper Tithonian pelagic carbonate, marl, and limestone turbidites (Fig. 1). This Cat Gap Formation was informally subdivided into three members: basal variegated calcareous claystone, greenish-grey marl, and a topmost reddish-brown marl and clayey limestone.

The variegated claystone member, penetrated only in Site 534, consists of reddish, greenish and blackish colored calcareous claystones with minor beds of silt-size peloid packstone turbidites. The characteristic feature of these claystones is their lack of microfossils and low carbonate content, which resulted either from deposition near the CCD or more probably from a high input of terrigenous clay from the continent.

The variegated claystone grades upward into greenish-grey marl with thick limestone turbidite beds. These turbidites are composed of a) silt-size peloid packstones derived from the hemipelagic carbonate slope; and b) fine- to medium-grained peloid-skeletal grainstones composed of debris of shallow shelf biota such as thick-walled mollusc shells and echinoid and bryozoan debris, mixed with peloid grains, oolites, algal oncolites, and micrite intraclasts (Fig. 4b).

The youngest member consists of pelagic reddish brown marls and clayey nodular limestones (Fig. 4c) largely made up of material transported from the shelf. These rock units are identical to those that were encountered at DSDP Sites 99, 100, 105, 534, and 391 in the North American Basin and at Sites 367 and 547 in the eastern central North Atlantic (Jansa and others, 1979; Lancelot and others, 1978; Jansa and others, 1984; Plate 9C). The radiolarians are important constituents of the Oxfordian limestones only in the Cape Verde Basin (Site 367, Fig. 4a). Due to the oxidized nature of the latter sediments, the organic-matter content in this member is less than 0.1 percent.

The lithologic boundary between the dominant rock types of the Cat Gap Formation and the overlying Blake-Bahama Formation (Fig. 1) varies from sharp to transitional over an interval of 1 to 45 m thick. The transition zone consists of alternating reddish-brown marls and limestones with greenish grey and light grey bioturbated marly chalks and pelagic limestones (Fig. 4d). The age of the reddish brown member of the Cat Gap Formation is late Oxfordian to late Tithonian (Jansa and others, 1979). The sedimentation rate at drillsites located farther from the margin is 8 m/m.y. and it increases closer to the margin (Sites 391, 534) to 14–16 m/m.y. The increase in sedimentation rates reflects increased frequency and thickness of turbidite beds.

The top of the Cat Gap Formation is approximately correlative with seismic horizon C in the Blake-Bahama Basin (Benson and others, 1978; Jansa and others, 1979) and Horizon J_1 elsewhere in the western North Atlantic (Plate 1B; Fig. 1).

Physiography and Paleoceanography

During latest Jurassic time the North Atlantic basin was about 1300–1500 km wide and 3300 km long. The basin was connected to the western Tethys by the 'Iberia-Rif' seaway and by epicontinental seas that covered the Grand Banks, the western Iberian Peninsula, and western Europe (Plate 9C). According to ammonite evidence, (Hallam, 1977), the seaway to the Pacific was also open via the region west of the Merida Andes, Venezuela, and east of the Yucatan Peninsula. At this time Atlantic sea water also covered most of the Gulf of Mexico area. The depth of

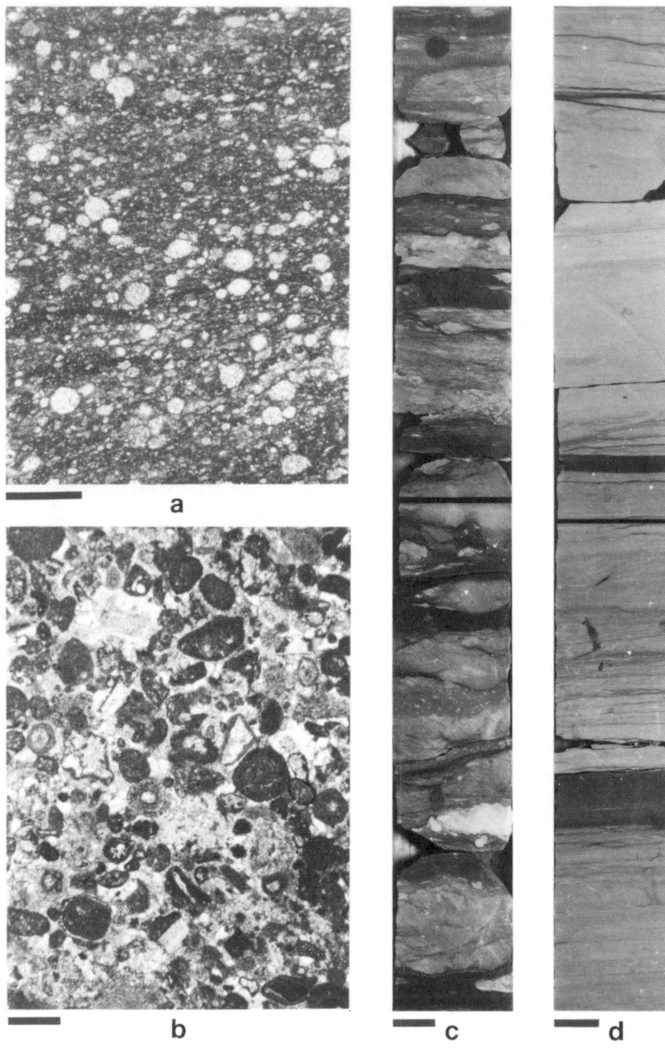

Figure 4. a. Silica-replaced tests of radiolarians in Oxfordian, reddish colored, clayey, nodular limestone in Cape Verde Basin. The thin whispy lines marking concentration of clay minerals probably originated by diagenetic dissolution (DSDP Site 367, core 37, section 1, 144 cm, thin section, crossed nicols, scale bar = 0.5 mm). b. Limestone turbidite composed mainly of a shallow-water, intensively micritized skeletal debris (echinoderms, bryozoans, belemnite and algal fragments, foraminifera tests), concentric and radial oolites, oncoliths, coated grains, and micrite intraclasts. Grains are cemented by slightly ferroan sparry calcite; echinoids show syntaxial overgrowth. Kimmeridgian, Blake-Bahama Basin (DSDP Site 534A, core 106, section 1, 138 cm; thin section, crossed nicols; scale bar = 0.5 mm). c. Reddish brown clayey, slightly nodular limestone typical of the Cat Gap Formation in the North Atlantic basin. The carbonate nodules are separated by irregular clayey laminae. Good preservation of nannofossils in the clayey laminae indicate that the clay concentration is mostly a depositional feature. Kimmeridgian, Cape Verde Basin (DSDP Site 367, core 35, section 2, 110-140 cm). Scale bar = 2 cm. d. Light grey nannofossil-rich pelagic limestone of the Blake-Bahama Formation. Note the more regularly bedded character of the limestone when compared to the Upper Jurassic Cat Gap Formation (Figure 4c). Intercalated are thin beds of dark grey and greenish marl. Minor bioturbation is observed in more clayey beds. Microstylolites are common. Late Tithonian-Early Berriasian, Blake-Bahama Basin (DSDP Site 391C, core 44, section 4, 50–100 cm). Scale bar = 2 cm.

the Blake-Bahama Basin at the end of the Jurassic was about 3600 m according to Sheridan and others (1983), and it is assumed that the central North Atlantic was similarly deep at this time.

Interpretation of paleoceanography, as derived mainly from the distribution of lithofacies in the North Atlantic basin and its margins (Plate 9C) indicates that the Late Jurassic paleocirculation pattern probably had some similarities to the present circulation in the North Atlantic. An important factor for the Late Jurassic paleoceanography was change of the latitudinal position of the central North Atlantic. According to Scotese (1979) Cape Hatteras was approximately at 15°N and the North Atlantic basin extended from the equator to 25°N. Paleoclimatic reconstruction of the Late Jurassic by Parrish and Curtis (1982) shows atmospheric anticyclones located off California and the Middle East and an atmospheric low centered over the central North American continent during the "northern summer." Such an atmospheric pattern would have resulted in southwest blowing trade winds over the southern central North Atlantic, with the southernmost part of the basin located in the doldrum belt. The wind stress would have generated southward surface flow, and a gyre with a cyclonic (anticlockwise) flow may have formed in the southern part of the basin. In the northern part of the basin, wind torque could have caused strong current transport to the north. It is suggested that the surface current flow was northward near Cape Hatteras (15°N in Late Jurassic time), and this led to the formation of a northern gyre, with a clockwise flow (Plate 9C).

Inspection of the lithofacies distribution in Europe (Plate 9C) shows that carbonate deposition was dominant in southwestern Europe, in contrast to deposition of clastics and carbon-rich shales in the northwestern Europe and the North Sea (Ziegler, 1981). A similar facies contrast existed between Upper Jurassic sequences on the Grand Banks and in western Portugal. Such a paleogeographic configuration clearly suggests a climatic difference. This either could have been related to the presence of two water masses or could be a reflection of paleolatitude. In the latter case, the paleo-equator position would have to be parallel to the direction derived from the Galula magnetic pole (Briden, 1961). If two water masses were present, then it is likely that the warm North Atlantic water-mass circulated in a pattern similar to the present Gulf Stream, passing northeastward along the Iberian Peninsula, warming the climate of southwestern Europe and permitting carbonate deposition. At the same time, cooler and lower-density water flow was directed southward from the Arctic (Callomon and others, 1972) along the coast of eastern Greenland, like the present day East Greenland Current (Plate 9C). The presence of such flow is supported by mixing of Tethyan and subboreal faunas on the Grand Banks (Jansa and others, 1980b) and by occurrence of some subboreal faunal elements on the Scotian Shelf (Ascoli, personal communication), which suggest that the Scotian shelf area was also a region of mixing between the proto-East Greenland Current and warmer southern waters.

Occurrence of calpionellids in Algarve (Duran Delga and Rey, 1982), in drill-sites west of Portugal, in the Bay of Biscay, and on the eastern Grand Banks (Jansa and others, 1980b) is an indication of divergence of the flow past the "Iberia-Rif" seaway. Part of the Mediterranean Tethys surface water outflow experienced a northward deflection along the southwestern edge of Iberian margin because of the Coriolis force. Obviously, the strength of this current, as well as that of the postulated incipient proto-Gulf Stream, is unknown. The major outflow from "the Iberia-Rif" seaway would have been directed southwestward, as indicated by experimental studies (Luyendyk and others, 1972).

No data exist to document the characteristics and position of the seaway connecting the central North Atlantic and the Pacific. On the basis of the modeling experiments by Luyendyk and others (1972), Berggren and Hollister (1974) suggested that surface Atlantic water flowed into the Pacific Ocean. Such a pattern is accepted here with an assumption that this seaway was never wide and deep, even though there are no supporting data at this time. During Late Jurassic time, it appears that the Gulf of Mexico was connected to the southern central North Atlantic by a broad seaway between Florida and Cuba that allowed free water interchange. Existence of such a marine connection is supported by the occurrence of Upper Jurassic coral bioherms in southwestern Arkansas and Florida (Baria and others, 1982). These coral bioherms are faunally similar to the coeval coral-sponge bioherms that were present off eastern Canada (Jansa and others, 1982) and they would have required seawater of normal salinity.

The presence of a highly condensed Upper Jurassic sedimentary sequence in the Rif area (Wildi and others, 1977) and on the Mazagan platform slope (Hinz and others, 1984) could be evidence that, like the Middle Jurassic situation, the deep part of the basin off the Rif region was an area that experienced bottom- or intermediate-water outflow to the east.

The properties of the basinal sediments in the North American Basin are compatible with the presence of weak currents, but they do not indicate the presence of strong contour or bottom currents (Bernoulli, 1972; Jansa and others, 1978). However, the presence of oxidized bottom sediments in the central North Atlantic does indicate permanent mixing.

CCD and Late Jurassic Shift in Carbonate Productivity

The apparent shallow depth of the CCD in the Early Jurassic and part of the Middle Jurassic is attributed to low productivity of carbonate-secreting organisms in the oceanic basin, which resulted in carbonate undersaturation of deep ocean water. What then was the cause of the widespread deposition of pelagic carbonates and the related deepening of the CCD during the Late Jurassic? No major paleoceanographic change is known from the North Atlantic at this time. A hypothesis involving a deepening of the CCD by the internal mechanism of basin development, when the carbonate-producing factory shifted from the shelf to the open oceanic basin, is discussed.

During Early and Middle Jurassic time, extensive carbonate sedimentation occurred on the circum-North Atlantic continental margins (Jansa, 1981; Jansa and Wiedmann, 1982; Schlee and others, 1979; Dillon and others, 1979). In contrast, during Late Jurassic time the terrigenous clastics began to displace the carbonate deposits from the inner shelf. In some of the more rapidly subsiding marginal basins the terrigenous clastics even became the dominant deposits. For example, over 1300 m of clastics accumulated during Late Jurassic time in the Scotian Basin and on Georges Bank (Jansa and Wade, 1975; Schlee and Jansa, 1981), clastic deposits in excess of 2000 m are known from the Baltimore Canyon Trough, and 1500 to 2500 m of terrigenous deposits accumulated in the northern Lusetanian Basin (Zbyszewski, 1965). I have no data from the Senegal Basin area, but a similar trend is expected to exist along the western African margin (Lehner and DeRuiter, 1977). This progressive offshore displacement of carbonate deposition culminated during Early Cretaceous time, when clastics smothered carbonate deposition on the Atlantic shelves and overran the shelf-edge carbonate banks. The destruction of the ecological niche for the carbonate-secreting benthic organisms on the shelf may have increased the availability of calcium carbonate for calcareous plankton and led to a progressive shift of the carbonate production factory from the shelf into adjacent oceanic basin. This would have resulted in increased carbonate saturation of North Atlantic waters and progressive deepening of the CCD in the basins.

CONCLUSIONS

A shallow Upper Triassic evaporitic basin marks the initiation of the North Atlantic basin according to evidence from salt sequences encountered in Deep Sea Drilling Project Leg 79. The geochemistry of the halite, which includes small amounts of potash salts, favours a shallow-marine origin for the evaporites. Connection with the western Tethys at this time is speculative. Based on Late Triassic paleogeography, the sea connection was probably northward through the Bay of Biscay Rift and/or Western Europe. The incipient North Atlantic basin was about 350 km wide and 900 km long and perhaps not more than few hundred meters deep. The wind driven surface currents approximately paralleled the long axis of the basin. Excessive evaporation toward the southern end of the basin contributed to the southward flow and the thermohaline circulation was superimposed on the wind driven circulation. Oxidized basin sediments indicate complete vertical mixing. The basin was later split apart by seafloor spreading.

Early in Jurassic time (Sinemurian), an epeiric basin was established as a result of continuing basin subsidence and a eustatic sea-level rise which affected the area previously occupied by Upper Triassic evaporites. Normal salinity in the northern part of the basin was achieved during the Pliensbachian. Fine-grained radiolarian mudstones and minor carbonates dominated the basin's axial deposition, and limestone debris flows and turbidites accumulated at the toes of marginal carbonate platforms encircling the basin. Evaporite deposition probably continued at the southern, shallowest, end of the basin. The average annual wind stress on the ocean surface was by trade winds blowing from the east and provided a major driving force for weak anticlockwise surface currents. The Early Jurassic Atlantic connection to the western Tethys was through the "Iberia-Rif" seaway, and to northwestern Europe via shelf seas flooding the Grand Banks, western Portugal, the Celtic Sea, and the Western Approaches. A more humid, and probably slightly colder, climate over northern Europe initiated weak, thermohaline surface flow along the northeastern North American coast. The salinity difference between the northern and southern part of the basin produced weak salinity-driven, deep circulation in the basin. Lower Jurassic sediments on the North Atlantic margins and in the basin confirm the establishment of a vertically stratified system in the proto-Atlantic and suggest that overall circulation was sluggish.

The Middle Jurassic basin physiography, chemistry, and circulation were strongly influenced by the final separation of continental plates during Aalenian-Bajocian time (approximately 180 Ma, Plate 1B), by the initiation of the seaway to the Pacific via central America, by the opening of the seaway between Tethys and the Arctic, by a major regression in the central North Atlantic during Bajocian time, and by volcanism in northwestern Europe. All these changes, including large fluvial influxes of alumino-silicates into the basin, increased the levels of silica and other nutrients in surface waters; combined with upwellings, this produced extensive blooms of radiolaria, and they became the "dominant" upper Middle Jurassic microfauna in the basin. The opening of the southern seaways increased bottom- or intermediate-water flow; in the "Iberia-Rif" seaway this led to depositional hiatuses and condensed sequences.

The presence of carbon-rich Callovian shales in the Blake-Bahama Basin and Gulf of Mexico may indicate that bottom circulation and mixing did not improve in the southern end of the basin. The wind-driven surface circulation, deduced from the paleoclimatic restoration of the North Atlantic (Parrish and Curtis, 1982), was similar in direction to that during the Early Jurassic. Nevertheless, during late Middle Jurassic time, the whole central North Atlantic ocean system changed from a semiclosed basin to a partially open system. The increased width of the basin then allowed development of an asymetrical circulation pattern in the basin.

During Late Jurassic time, the North American Basin was rimmed by carbonate platforms along its entire length. Coral-stromatoporoid bioherms were scattered near the paleoshelf edges from the Scotian shelf to the Gulf of Mexico. The debris from these carbonate shelves was transported into the basin by turbidity currents and accumulated at the toe of the margin, particularly during major regressive intervals (early Oxfordian, Kimmeridgian). Deposition was dominated by marls and calcareous shales, with clayey limestones dominating during Tithonian time. The upward increase of carbonate in the basinal sediments is interpreted to be a result of the progressive deepening of the CCD, related to the displacement of carbonate production from the shelf to the oceanic basin at the end of Jurassic time. The mean sedimentation rate in the basin at this time was 8 m/m.y. and toward the margins increased to 16 m/m.y.

The Late Jurassic central North Atlantic surface paleocirculation was principally wind driven (with superimposed thermohaline circulation), leading to the development of a clockwise gyre in the northern part of the basin and to an anti-clockwise gyre in the southern part of the basin. A proto-"Gulf Stream" and a proto-East Greenland Current could also have been initiated at this time. In the "Iberia-Rif" seaway, the inflow of Tethyan surface water to the Atlantic probably proceeded along the northern side of the seaway and the outflow of denser North Atlantic water moved at a deeper level on the southern side of the seaway. Deflection of the inflowing Tethyan water by the Coriolis force produced some northward surface flow along the Iberian peninsula which merged with the proto-"Gulf Stream." Oxygenated bottom sediments in the basin indicate circulating bottom waters, but occurrence of only minor current features in the sediments suggests low-velocity flows.

This reconstructed history of the North American Basin evolution and paleoceanography has practical implications for hydrocarbon exploration of the continental margins. Lower and Middle Jurassic times were more favorable periods for preservation of organic matter along the deep continental margins than was the Late Jurassic period when most bottom sediments in the basin were oxidized and organic matter destroyed. Inferred paleocirculation patterns also allow one to deduce possible locations of more fertile surface waters and less oxygenated bottom waters, such as the Blake-Bahama Basin during the Middle Jurassic.

SELECTED BIBLIOGRAPHY

Ager, D. V.
1974 : The western High Atlas of Morocco and their significance in the history of the North Atlantic: Proceedings of the Geologists Association, v. 85, p. 23–41.

Ambroggi, R.
1963 : Etude geologique du versant meridional du Haut Atlas occidental et de la Plaine du Souss: Notes et Memoires du Service Geologique du Maroc, No. 157, 321 pp.

Arkell, W. J.
1956 : Jurassic Geology of the World: Hafner Publishing Company, New York, 806 pp.

Amato, R. V., and Simonis, E. K., eds.
1980 : Geologic and operational summary, COST G-2 well, Georges Bank area, North Atlantic OCS: U.S. Geological Survey Open File Report, 80-269, 116 pp.

Baria, L. R., Stoudt, D. L., Harris, P. M., and Crevello, P. D.
1982 : Late Jurassic reefs of the Smackover Formation—U.S. Gulf Coast: American Association of Petroleum Geologists Bulletin, v. 66, p. 1449–1482.

Barss, M. S., Bujak, J. P., and Williams, G. L.
1978 : Palynological zonation and correlation of sixty seven wells, eastern Canada: Geological Survey of Canada, Paper 78-24, 118 pp.

Benson, W. E., and Sheridan, R. E., eds.
1978 : Initial Reports of the Deep Sea Drilling Project, v. 45: U.S. Govern-

ment Printing Office, Washington, D.C., 1005 pp.
Berger, W. H.
1981 : Paleoceanography: the deep-sea record: in The Sea, v. 7, ed. C. Emiliani; John Wiley and Sons, New York, p. 1347–1519.
Berggren, W. A., and Hollister, C. D.
1974 : Paleogeography, paleobiogeography and the history of paleocirculation in the Atlantic Ocean: in Studies in Paleooceanography, ed. W. W. Hay; Society of Economic Paleontologists and Mineralogists Special Publication, no. 2, p. 126–186.
Bernoulli, D.
1972 : North Atlantic and Mediterranean Mesozoic facies: a comparison: in Initial Reports of the Deep Sea Drilling Project, v. 11, eds. C. D. Hollister and J. I. Ewing; U.S. Government Printing Office, Washington, D.C., p. 801–871.
Bhat, H., McMillan, N. J., Aubert, J., Porthault, B., and Surin, B.
1975 : North American and African drift—The record in Mesozoic coastal plain rocks, Nova Scotia and Morocco: in Canada's Continental Margins and Offshore Petroleum Exploration, eds. C. J. Yorath, E. R. Parker and D. J. Glass; Canadian Society of Petroleum Geologists, Memoir 4, p. 375–389.
Callomon, J., Donovan, D. T., and Trumpy, R.
1972 : An annotated map of the Permian and Mesozoic formations of east Greenland: Meddelelser om Gronland, Kobenhaven, v. 168, 35 p.
Cantu Chapa, A.
1971 : La serie Huasteca (Jurassico Medico-Superior) del centro este de Mexico: Revista del Instituto Mexicano del Petroleo, April 1971, p. 17–40.
Choubert, G., and Faure-Muret, A.
1962 : Evolution du domaine atlasique marocain depuis les temps paleozoiques: Mémoires hors serie Société Géologique de France, t. 1: p. 447–527.
de Graciansky, P. C., Lemoine, M., Arnaud-Vanneau, A., Arnaud, H., Beaudoin, B., Bourbon, M., Chenet, P. Y., Elmi, S., and Ferry, S.
1980 : European continental margin of the Mesozoic Tethys in the western Alps: 26th International Geological Congress, Excursions 27 C, 114 p.
Dillon, W. P., Paul, C. K., Buffler, R. T., and Fail, J-P.
1979 : Structure and development of the Southeast Georgia Embayment and northern Blake Plateau, Preliminary analysis: in Geological and Geophysical Investigations of Continental Margins, eds. J. S. Watkins, L. Montadert and P. W. Dickerson; American Association of Petroleum Geologists Memoir, v. 29, p. 27–41.
Dixon, J. E., Fitton, J. G., and Frost, R.T.C.
1981 : The tectonic significance of post-Carboniferous igneous activity in the North Sea Basin: Petroleum Geology of Continental Shelf of North-West Europe: Institute of Petroleum, London, p. 121–137.
Dupeuble, P. A., Réhault, J. P., Auxiètre, J. L., Dunand, J. P., and Pastouret, L.
1976 : Résultats de dragages et essai de stratigraphie des Bancs de Galice, et des montagnes de Porto et de Vigo (marge occidentale Ibérique): Marine Geology, 22, p. 37–49.
Durand Delga, M., and Rey, M.
1982 : Decouvertes de calpionelles dans le Jurassique terminal et le Cretace de l'Algarve (Portugal): Compte Rendu Academie Science de Paris, v. 295, p. 237–242.
Faugères, J. C., and Mouterde, R.
1979 : Paléobiogéographie et paléogéographie aux confins atlantico-mésogéens: Données fournies par le Lias sudrifain (Maroc): 7ième Réunion annualle des Sciences de la Terre, Lyon, 183 p.
Hallam, A.
1977 : Biogeographic evidence bearing on the creation of Atlantic seaways in the Jurassic: in Paleontology and Plate Tectonics, ed. R. M. West; Milwaukee Public Museum Special Publication in Biology and Geology, no. 2, p. 23–39.
1978 : Eustatic cycles in the Jurassic: Paleogeography, Paleoclimatology, and Paleoecology, v. 1123, p. 1–32.
1981 : A revised sea-level curve for the early Jurassic: Journal of Geological Society of London, v. 138, p. 735–743.
1983 : Early and Mid-Jurassic molluscan biogeography and the establishment of the central Atlantic seaway: Paleogeography, Paleoclimatology, and Paleoecology, v. 43, p. 181–193.
Hinz, K. and Winterer, E. L., eds.
1984 : Initial Reports of the Deep Sea Drilling Project, v. 79: U.S. Government Printing Office, Washington, D.C., 934 p.
Holser, W. T., Saltzman, E. S., and Brookins, D. G.
1984 : Geochemistry and petrology of evaporites cored from a deep-sea diapir at Site 546, DSDP Leg 79, Offshore Morocco: in Initial Reports of the Deep Sea Drilling Project, v. 79, eds. K. Hinz and E. L. Winterer; U.S. Government Printing Office, Washington, D.C., p. 509–542.
Jansa, L. F.
1981 : Mesozoic carbonate platforms and banks of the eastern North American margin: Marine Geology, v. 44, p. 97–117.
Jansa, L. F., Bujak, J. P., and Williams, G. L.
1980a: Upper Triassic salt deposits of the western North Atlantic: Canadian Journal of Earth Sciences, v. 17, p. 547–559.
Jansa, L. F., Enos, P., Tucholke, B. E., Gradstein, F. M., and Sheridan, R. E.
1979 : Mesozoic-Cenozoic sedimentary formations of the North American Basin; western North Atlantic: in Deep Drilling Results in the Atlantic Ocean: Continental Margins and Paleoenvironment, eds. M. Talwani, W. Hay and W.B.F. Ryan; Maurice Ewing Series v. 3: American Geophysical Union, p. 1–57.
Jansa, L. F., Gardner, J. V., and Dean, W. E.
1978 : Mesozoic sequences of the central North Atlantic: in Initial Reports of the Deep Sea Drilling Project, v. 41, eds. Y. Lancelot and E. Seibold; U.S. Government Printing Office, Washington, D.C., p. 991–1031.
Jansa, L. F., Remane, J., and Ascoli, P.
1980b: Calpionellid and foraminiferal-ostracod biostratigraphy at the Jurassic-Cretaceous boundary, offshore eastern Canada: Rivista Italiana di Paleontologia e Stratigrafia, v. 86, p. 67–126.
Jansa, L. F., and Wade, J. A.
1975 : Geology of the continental margin off Nova Scotia and Newfoundland: in Offshore Geology of Eastern Canada, 2. Regional Geology, eds. W.J.M. Van der Linden and J. A. Wade; Geological Survey of Canada, Paper 74-30, p. 51–106.
Jansa, L. F., and Wiedmann, J.
1982 : Mesozoic-Cenozoic development of the eastern North American and northwest African continental margins: a comparison: in Geology of the Northwest African Continental Margin, eds. U. von Rad, K. Hinz, M. Sarnthein and E. Seibold; Springer-Verlag, Berlin, p. 215–269.
Jansa, L. F., Steiger, T. H., and Bradshaw, M. J.
1984 : Mesozoic carbonate deposition on the outer continental margin off Morocco: in Initial Reports of the Deep Sea Drilling Project, v. 79: U.S. Government Printing Office, Washington, D.C., p. 857–891.
Jansa, L. F., Termier, G., and Termier, H.
1982 : Les biohermes a algues, spongiaires et coraux des séries carbonatées de la flexure bordière du "paleoshelf" au large du Canada oriental: Revue de Micropaleontologie, v. 25: 181–219.
Jenkyns, H. C., and Winterer, E. L.
1982 : Paleoceanography of Mesozoic ribbon radiolarites: Earth and Planetary Science Letters, v. 60, p. 351–375.
Klitgord, K. D., Popenoe, P., and Schouten, H.
1984 : Florida: a Jurassic transform plate boundary: Journal of Geophysical Research, v. 89, p. 7753–7772.
Lancelot, Y., Hathaway, J. C., and Hollister, Ch.D.
1972 : Lithology of sediments from the western North Atlantic, Leg XI, Deep Sea Drilling Project: in Initial Reports of the Deep Sea Drilling Project, v. 11, eds. C. D. Hollister and J. I. Ewing: U.S. Government Printing Office, Washington, D.C., p. 901–950.
Lancelot, Y., and Seibold, E., eds.
1978 : Initial Reports of the Deep Sea Drilling Project, v. 41: Washington,

D.C., U.S. Government Printing Office, p. 1259.
Lancelot, Y., and Winterer, E. L.
1980 : Evolution of the Moroccan oceanic basin and adjacent continental margin—a synthesis: in Initial Reports of the Deep Sea Drilling Project, v. 50, eds. Y. Lancelot and E. L. Winterer: U.S. Government Printing Office, Washington, D.C., p. 801–821.
Lehner, P., and DeRuiter, P.A.C.
1977 : Structural history of Atlantic margin of Africa: American Association of Petroleum Geologists Bulletin, v. 61, p. 961–981.
Luyendyk, B. P., Forsyth, D., and Phillips, J. D.
1972 : Experimental approach to the paleocirculation of the oceanic surface waters: Geological Society of America Bulletin, v. 83, p. 2649–2664.
McIver, N. L.
1972 : Cenozoic and Mesozoic stratigraphy of the Nova Scotia shelf: Canadian Journal of Earth Sciences, v. 9, p. 54–70.
Mouterde, R., and Ruget, Ch.
1975 : Esquisse de la paléogéographie du Jurassic inférieur et moyen au Portugal: Bulletin Société géologique du France, v. 7, p. 779–786.
Ogg, J. G., Robertson, A.H.F., and Jansa, L. F.
1983 : Jurassic sedimentation history of Site 534 (western North Atlantic) and of the Atlantic-Tehys Seaway: in Initial Reports of the Deep Sea Drilling Project, v. 76, eds. R. S. Sheridan and F. M. Gradstein; U.S. Government Printing Office, Washington, D.C., p. 829–884.
Olson, W. S., and Leyden, R. J.
1973 : North Atlantic rifting in relation to Permian-Triassic salt deposition: in The Permian and Triassic System and their Mutual Boundary, eds. A. Logan and L. V. Hills; Canadian Society of Petroleum Geologists, Memoir 2, p. 720–732.
Parrish, J. T., and Curtis, R. L.
1982 : Atmospheric circulation, upwelling, and organic-rich rocks in the Mesozoic and Cenozoic Eras: Paleogeography, Paleoclimatology, and Paleoecology, v. 40, p. 31–66.
Parsons, B., and Sclater, J. G.
1977 : An analysis of the variation of ocean floor bathymetry and heat flow with age: Journal of Geophysical Research, v. 82, p. 802–827.
Rocha, B. R.
1977 : Estudo estratigráfico e paleontólogico do Jurássico do Algarve ocidental: Universidade Nova De Lisboa, Ciencias da Terra 2, 178 pp.
Salvan, H. M.
1974 : Les séries salifères du Trias marocain; caràcteres généraux et posibilités d'interprétation: Bulletin de la Society Géologique de France, v. 16, p. 724–731.
Schlee, J. S., Dillon, W. P., and Grow, J. A.
1979 : Structure of the continental slope off the Eastern United States: Society of Economic Paleontologists and Mineralogists, Special Publication 27, p. 95–117.
Schlee, J. S., and Jansa, L. F.
1981 : The paleoenvironment and development of the eastern North American continental margin: Oceanologica Acta, 4: Special issue—Geology of Continental margins, p. 71–80.
Scotese, C. R.
1979 : Phanerozoic continental drift base maps: in Paleogeographic Reconstructions: State of the Art, eds. R. K. Bamback and C. R. Scotese;. Southeastern Section: Geological Society of America, Short Course.
Shell Oil Company, Exploration Department
1975 : Stratigraphic Atlas of North and Central America: eds. T. D. Cook and A. W. Bally; Princeton University Press, Princeton, 272 p.
Sheridan, R. E., and Gradstein, F. M., ed.
1983 : Initial Reports of the Deep Sea Drilling Project, v. 76, U.S. Government Printing Office, Washington, D.C., 947 p.
Templeton, R.S.M.
1971 : The geology of the continental margin between Dakar and Cape Palmas: in The Geology of the East Atlantic Margin, ed. F. M. Delany; Great Britain Institute of Geological Sciences Report, 70/16, p. 47–60.
Tucholke, B. E.
1979 : Relationships between acoustic stratigraphy and lithostratigraphy in the western North Atlantic Basin; in Initial Reports of the Deep Sea Drilling Project, v. 43, eds. B. E. Tucholke and P. R. Vogt: U.S. Government Printing Office, Washington, D.C., p. 827–846.
Tucholke, B. E., and Mountain, G. S.
1979 : Seismic stratigraphy, lithostratigraphy and paleosedimentation patterns in the North American Basin: in Deep Drilling Results in the Atlantic Ocean: Continental Margins and Paleoenvironment, eds. M. Talwani, W. Hay, and W.B.F. Ryan; American Geophysical Union, Maurice Ewing Series, v. 3, p. 58–86.
Van der Voo, R., and French, R. B.
1974 : Apparent polar wandering for the Atlantic-bordering continents: Late Carboniferous to Eocene: Earth-Science Reviews, v. 10, p. 99–119.
Westermann, G.E.G.
1981 : Ammonite biochronology and biogeography of the circum-Pacific Middle Jurassic: in Systematics Association Special Volume 18, "The Ammonoidea," ed. M. R. House and J. R. Senior; Academic Press, London, p. 459–498.
Wildi, W.
1981 : Le Ferrysch: cône de sédimentation détritique en eau profonde à la bordure nord-ouest de l'Afrique au Jurassique moyen à supérieur (Rif externe, Maroc): Eclogae Geologicae Helvetiae, v. 74, p. 481–527.
Wildi, P., Nold, M., and Uttinger, J.
1977 : La Dorsale calcaire entre Tetouan et Assifane (Rif interne, Maroc): Eclogae Geologicae Helvetiae, v. 70, p. 371–415.
Zbyszewski, G.
1965 : Noticia explicativa da Folha 30-D Alenquer: Serviços Geológicos de Portugal, Lisboa, p. 104.
Ziegler, P. A.
1981 : Evolution of sedimentary basins in north-west Europe: in Petroleum Geology of the Continental Shelf of North-West Europe, eds. L. V. Illing and G. D. Hobson; Institute of Petroleum, London, p. 3–39.

Manuscript Accepted by the Society January 18, 1985

ACKNOWLEDGMENTS

The author is grateful to W. A. Berggren, B. Tucholke, K. Klitgord, E. L. Winterer, and S. Bell for constructive comments improving the manuscript. The help of many colleagues in France, Spain, Portugal, Italy, and Morocco who assisted in the author's field studies is also acknowledged.

The Geology of North America
Vol. M, The Western North Atlantic Region
The Geological Society of America, 1986

Chapter 37

Cretaceous paleoceanography of the western North Atlantic Ocean

Michael A. Arthur
Graduate School of Oceanography, University of Rhode Island, Narragansett, Rhode Island 02882
Walter E. Dean
U.S. Geological Survey, Denver Federal Center, Denver, Colorado 80225

INTRODUCTION

In this paper we summarize available information on the Cretaceous lithostratigraphy and paleoceanography of the western North Atlantic. The data and some of our interpretations draw in large part on papers published in the Deep Sea Drilling Project (DSDP) volumes. We have attempted to cite relevant references when possible, but space limitations make it difficult to give proper credit to all sources; we apologize for any omissions.

Organic carbon (Corg) and carbonate ($CaCO_3$) analyses were tabulated for each site from papers in the DSDP Initial Report volumes and other published works (e.g., Summerhayes, 1981). Corg, $CaCO_3$, and non-$CaCO_3$ mass accumulation rates (MARS) were calculated using core by core averages of component percentages for the more continuously cored sites; core averages for wet bulk density and porosity (from DSDP data files); biostratigraphies of de Graciansky and others (1982), Roth and Bowdler (1981), and Cool (1982); and the time scales of the Decade of North American Geology (Palmer, 1983; Kent and Gradstein, this volume) or Harland and others (1982; see Plate 1).

Backtracked paleodepths for western North Atlantic DSDP Sites from Tucholke and Vogt (1979) with the revised stratigraphy of de Graciansky and others (1982) were used in plotting Corg and $CaCO_3$ in Figures 2, 3, 4 and 5 (see also Thierstein, 1979).

LITHOLOGIC CHARACTERISTICS

Facies Trends

The Cretaceous section in the deep North Atlantic Ocean basin is characterized by variations in color, and concentrations of $CaCO_3$ and Corg that reflect variations in productivity, $CaCO_3$ dissolution rates, and redox conditions in the water column and/or sediments. The Neocomian to mid-Aptian sequence consists mainly of pelagic carbonate facies, but most of the later Cretaceous is characterized by cyclic interbeds of red, green, and black claystone or shale (Plate 1). Interbeds of Tithonian to Neocomian laminated, dark-olive to black marlstone within bioturbated, white to light-gray limestone were recovered at DSDP Sites 101, 105, 387, 391 and 534 in the western North Atlantic (Fig. 1). The laminated dark marlstone beds contain 2%–5% Corg. The Tithonian through Neocomian carbonate unit in the western North Atlantic basin was named the Blake-Bahama Formation by Jansa and others (1979). This is the temporal and lithologic equivalent to the Maiolica Formation that crops out in the Tethyan regions of the Mediterranean (Bernoulli, 1972; Jansa and others, 1979); it is also similar to Neocomian carbonate facies in the eastern N. Atlantic and Gulf of Mexico.

The end of deposition of the Lower Cretaceous Blake-Bahama Formation apparently was caused by a sudden rise in the carbonate compensation depth (CCD) (Fig. 2; Arthur, 1979a; Thierstein, 1979; Tucholke and Vogt, 1979). At most DSDP sites in the North Atlantic, the Blake-Bahama Formation is overlain by the middle Cretaceous (Aptian-Albian to Turonian) black-shale facies of the Hatteras Formation (Jansa and others, 1979); but at several sites the two are separated by a thin (<10 m) unit of interbedded, oxidized red and green claystones of middle to late Aptian age (Plate 1; Jansa and others, 1978). Jansa and others (1979) considered the red and green claystone unit to be a subfacies of the black-shale facies. This oxidized unit also occurs in sequences of the same age in the Tethyan region (e.g., Arthur and Premoli-Silva, 1982) and in the eastern North Atlantic (Jansa and others, 1978, Dean and Gardner, 1982).

The middle Cretaceous black-shale facies represents the main interval of high Corg contents in the Atlantic, and at most DSDP sites it consists of interbedded green and black clay-rich lithologies (see reviews by Arthur, 1979a; Tucholke and Vogt, 1979; Thierstein, 1979; Arthur and Natland, 1979; Tissot and others, 1979, 1980; Summerhayes, 1981, 1985; Dean and Gardner, 1982; Weissert, 1981; Arthur, Dean, and Stow, 1984). Sediments of this facies have been recovered at eleven DSDP sites in the western North Atlantic (Fig. 1).

Concentrations of Corg in the black beds in the western

Arthur, M. A., and Dean, W. E., 1986, Cretaceous paleoceanography of the western North Atlantic Ocean; *in* Vogt, P. R., and Tucholke, B. E., eds., The Geology of North America, Volume M, the Western North Atlantic Region: Geological Society of America.

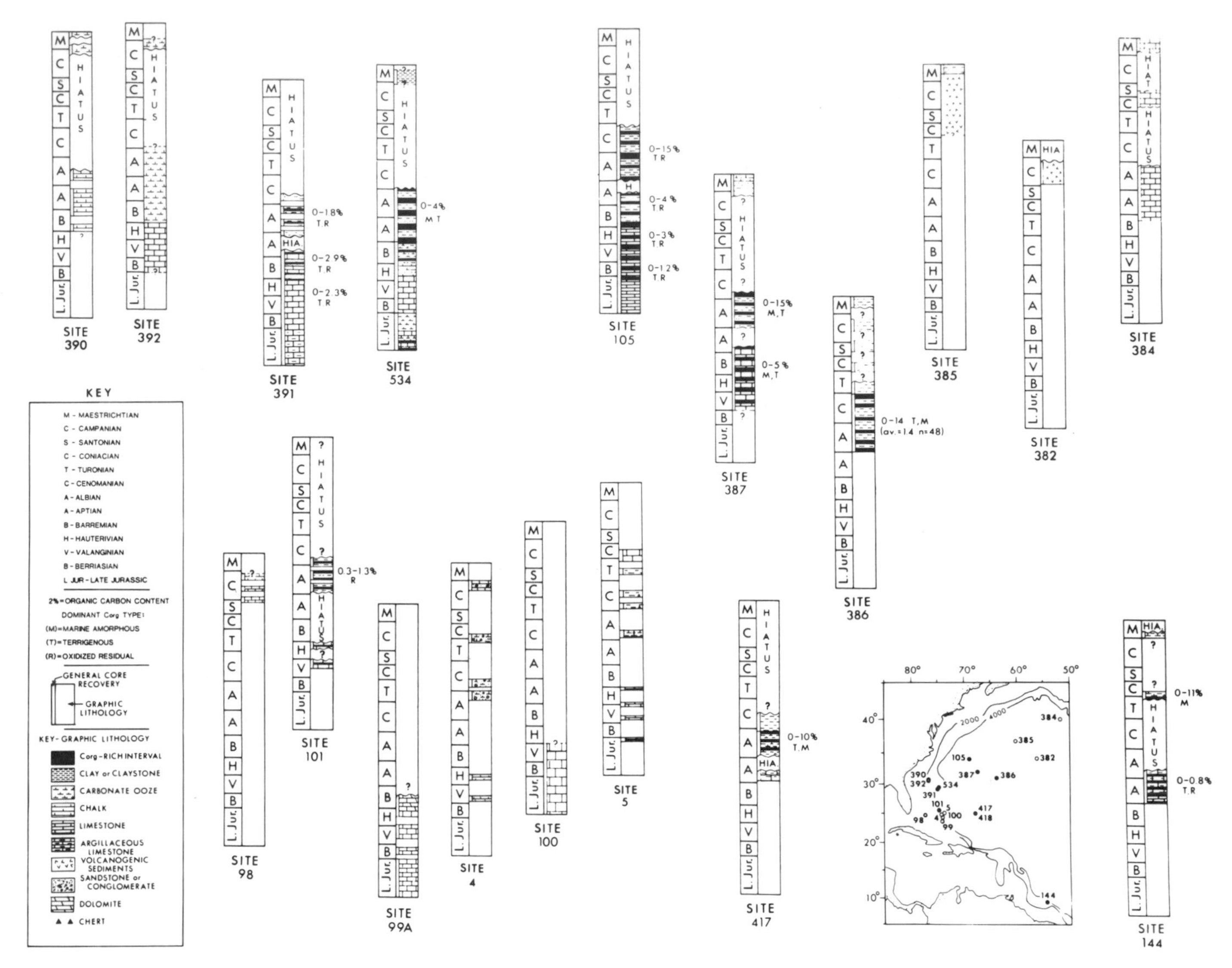

Figure 1. Lithologic columns for western North Atlantic DSDP sites—Cretaceous interval only. See inset for locations, and key for explanation of symbols.

North Atlantic range from <1% to 15% (Figs. 1, 3, 4) and usually are more than 1% (Fig. 4). In comparison, Corg concentrations in black shales in the eastern North Atlantic are as high as 37% and usually are more than 5% (Summerhayes, 1981; Dean and Gardner, 1982). True black-colored shale or claystone usually amounts to less than half of the "black shale" facies. Despite differences in sources of organic matter (see discussion below), the dominant appearance of the black-shale facies in all Atlantic basins is a cyclic interbedding of dark, commonly Corg-rich strata with bioturbated, lighter-colored Corg-poor strata.

The black-shale facies is overlain by an Upper Cretaceous to Lower Tertiary red and green claystone facies (Plate 1) termed the Plantagenet Formation by Jansa and others (1979) and recovered at DSDP Sites 7, 9, 105, 382, 386, 391, and 534 in the North American basin. This facies represents a return to more oxidizing conditions in the sediments, presumably in response to a slower rate of supply of organic detritus and (or) more effective oxygenation of bottom waters during the Late Cretaceous.

The lithologic facies described above can easily be correlated over most of the North American basin and even across the Mid-Atlantic Ridge into the basins of the eastern North Atlantic (Jansa and others, 1978, 1979). On the North American continental margin the lateral equivalents of the Neocomian carbonates of the Blake-Bahama Formation consist of deltaic sequences north of Cape Hatteras, and shallow-water carbonates farther south. In the basin the contact between the Neocomian carbonates and the overlying claystones and shales of the Hatteras Formation corresponds to a very strong acoustic reflector, Horizon β, that can be traced over great distances (Tucholke and Mountain, 1979). Benson, Sheridan, and others (1978) correlated the base of the Blake-Bahama Formation with an acoustic reflector (Horizon C), but correlation of lithology with acoustic charac-

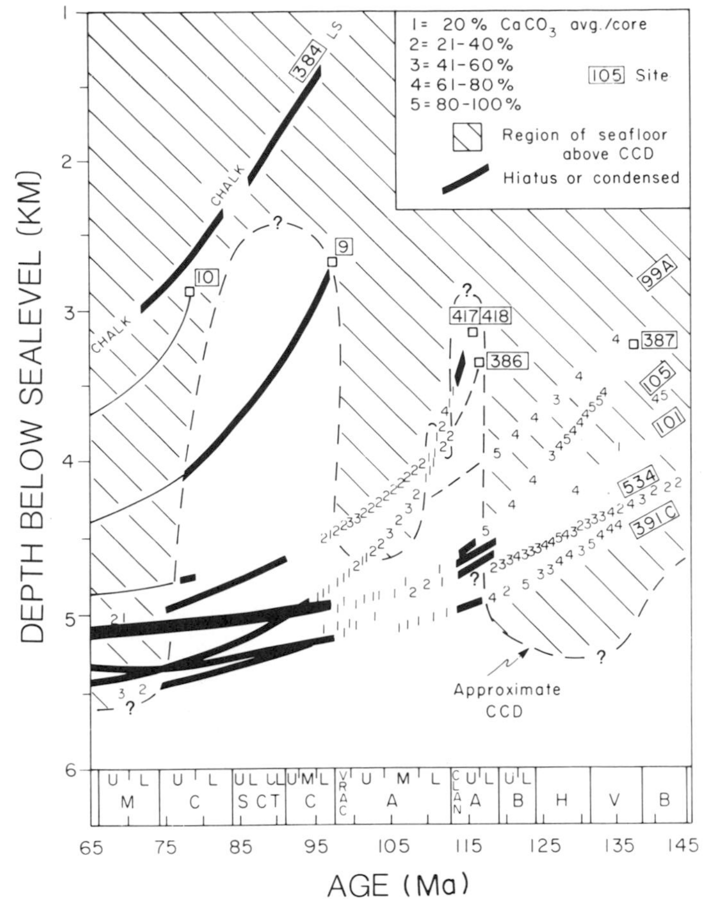

Figure 2. Backtracking curves of seafloor paleodepth versus age (Sclater and others, 1977; Tucholke and Vogt, 1979) for selected western North Atlantic DSDP sites. Average $CaCO_3$ concentrations per core are shown by code number (see key). *Shaded region* represents area of seafloor above carbonate compensation depth (CCD) as inferred from the average $CaCO_3$ concentrations. *Heavy solid lines* represent hiatuses or extremely slow sediment accumulation. Cretaceous stages in time scale represented by initials.

ter below the strong reflector β may be questionable (Tucholke and Mountain, 1979).

Lithologic variations within the Hatteras Formation and differences between the Hatteras and Plantagenet formations are mainly changes in color in response to differences in Corg content and type, and minor differences in $CaCO_3$ and biogenic silica content. Because the physical properties of the two formations are rather uniform there are few, if any, distinctive seismic reflectors within these clay-rich units. Consequently, it is difficult to determine thickness variations of the black-shale facies seismically, although the Hatteras Formation appears to thicken toward the North American continental margin (Tucholke and Mountain, 1979). On the continental shelf and slope, the black-shale facies passes laterally into transgressive, marginal-marine clastic sequences including extensive coal deposits of Aptian-Albian age. The multicolored claystones of the Plantagenet Formation have no well-defined lateral equivalents on the North American continental margin; coeval strata consist mainly of calcareous glauconitic claystones (Jansa and others, 1979). A brief, sharp depression of the CCD during the late Maestrichtian resulted in the accumulation of marls within the slowly deposited, predominantly clay-rich facies of the Plantagenet Formation. These marls correspond to seismic Horizon A*, which can be correlated over most of the North American basin but gradually fades in intensity toward deeper paleodepths (Tucholke and Vogt, 1979; Tucholke and Mountain, 1979).

Accumulation Rates

$CaCO_3$ and non-$CaCO_3$ mass accumulation rates (MARS) for continuously cored intervals of western North Atlantic DSDP sites are illustrated in Figure 5, and Corg MARS are shown in Figure 6. Uncertainties in these rates are probably fairly large.

Figure 5 shows that time-scale differences notwithstanding, major changes in $CaCO_3$ and detrital-mineral MARS occurred between the Valanginian-Hauterivian and the Aptian-Albian. Valanginian-Hauterivian $CaCO_3$ MARS are high, whereas the MARS of non-$CaCO_3$ material are low. Barremian through late Cretaceous MARS of $CaCO_3$ are very low at all sites, but there does seem to be some relationship between higher MARS of $CaCO_3$ and shallower paleodepths (e.g., DSDP Site 386). However, some of the carbonate material in the Albian-Cenomanian at Sites 386, 417, and 418, for example, may have been redeposited from the flanks of nearby basement pinnacles having as much as 1.8 km relief above the surrounding seafloor (Tucholke and Vogt, 1979). Nannofossil-rich carbonate in Site 386 occurs in discrete graded beds with sharp basal contacts in association with other lithologies that are presumed to have been redeposited (e.g., McCave, 1979b).

There is relatively little difference in the noncarbonate MARS between the Valanginian-Hauterivian and Barremian-Albian intervals, although late Aptian–early Albian non-$CaCO_3$ MARS may be slightly higher. It appears that the more landward sites (391, 101, 105) have higher noncarbonate MARS than those sites at the ridge crest (417, 418). An exception is Site 386, which is characterized by more rapid MARS, probably because of redeposition from surrounding paleohighs. The Late Cretaceous MARS at all sites are uniformly low. Much of the terrigenous material was probably of eolian derivation (Lever and McCave, 1984).

Corg MARS are variable (Fig. 6), but Hauterivian values are apparently the highest of the Cretaceous in the western North Atlantic basin. Middle to late Aptian and possibly earliest Albian Corg MARS are also high, with relatively low values in the Albian to early Cenomanian. A peak in Corg MARS in the middle to late Cenomanian does not show up in the core-averaged data in Figure 6.

CRETACEOUS PALEOENVIRONMENTS

Variations in the carbonate compensation depth (CCD), the

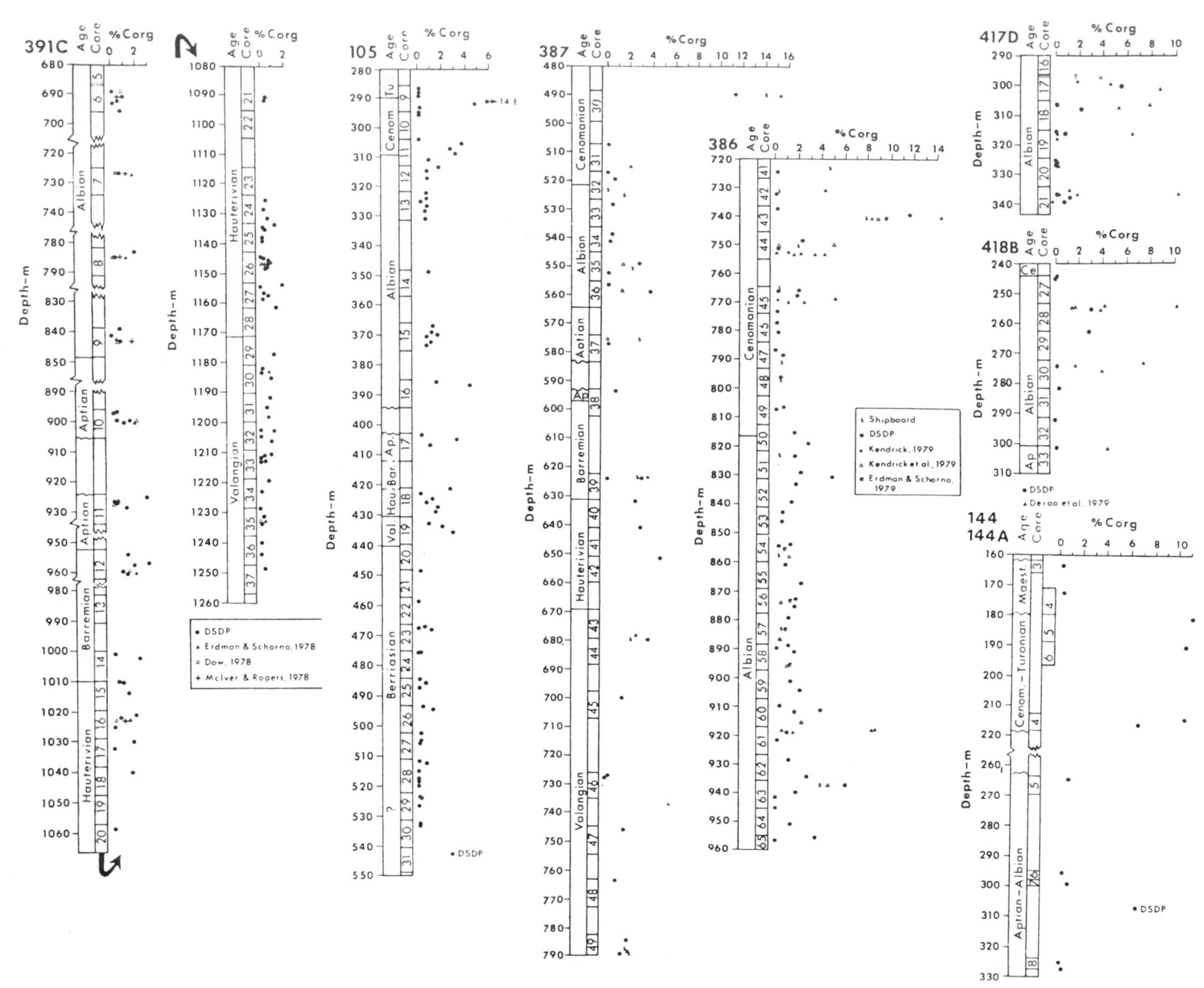

Figure 3. Organic carbon (Corg) contents in Cretaceous strata at selected western North Atlantic DSDP sites. Panel is arranged by approximate paleodepth (left-deep, right-shallow). Sources of data are shown by symbols for each site. Age, core number, and sub-bottom depth also are shown.

depth below which little or no $CaCO_3$ accumulated in sediments on the seafloor (Berger and Winterer, 1974), apparently were substantial in the Cretaceous western North Atlantic (Plate 1; Fig. 2). Although the backtracking curves do not completely define the entire Cretaceous seafloor depth range for the western North Atlantic basin, the following patterns of CCD variation have been suggested by Tucholke and Vogt (1979), Thierstein (1979), and Chenet and Francheteau (1980). The CCD was deeper than 4 km during the Early Cretaceous until about mid-Aptian time. In the mid-Aptian, the CCD rose rapidly to possibly as shallow as about 2 km during the late-Aptian to early Albian. The CCD dropped to depths near 4 km in the Albian–early Cenomanian (Fig. 2), rose again in the middle Cenomanian, remained relatively shallow during most of the Late Cretaceous, and finally dropped in latest Campanian through late Maestrichtian to a depth of more than 4.5 km. The overall pattern shown in Figure 2 corresponds in general to those suggested by Tucholke and Vogt (1979) for the western North Atlantic.

Times of high CCD level coincided with hiatuses or extremely reduced rates of sedimentation (Fig. 2). A hiatus occurred in the mid-Aptian (de Graciansky and others, 1982), and hiatuses may be present in the Upper Cretaceous at most DSDP Sites. In general, rates of sedimentation were extremely slow for the red and multicolored claystones of the Plantagenet Formation (<1 m./m.y.).

Black Shale Deposition

Detailed examination of middle Cretaceous black shales re-

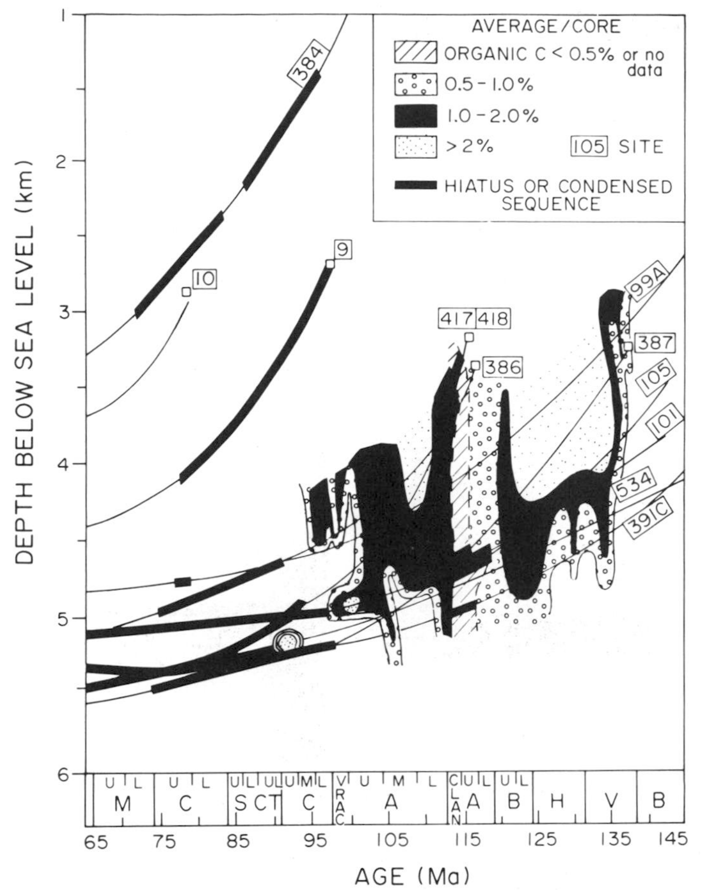

Figure 4. Backtracking curves of seafloor paleodepth versus age (Sclater and others, 1977; Tucholke and Vogt, 1979) for selected western North Atlantic DSDP sites. Average organic carbon (Corg) contents per core are shown by code number (see key). *Shaded areas* show depth-time patterns of changing concentrations of Corg. Intervals in *heavy solid lines* represent hiatuses or extremely slow sediment accumulation. Cretaceous stages in time scale represented by initials.

veals characteristics that cannot be explained by simple local models. These characteristics include: time stratigraphic coincidence (middle Cretaceous) in the Atlantic, Tethys, and on isolated plateaus and seamounts in the Pacific; black-shale occurrence in a variety of different environments including large, open-ocean basins and restricted basins in deep and shallow water; and variations in amount and type of organic matter that commonly occurs even within the same lithologic sequence. In general, sea levels were much higher during the middle Cretaceous than today, and this produced much larger areas of warm, shallow, highly productive marginal and epicontinental seas. Global climates were warm and equable, and rainfall along Tethyan margins probably was high (e.g., Barron and Washington, 1982), which aided the rapid expansion of terrestrial vegetation. There were, therefore, abundant sources of both autochthonous-marine and terrestrial organic matter. Because of the warmer global climate, surface and bottom waters of the ocean were warmer and probably more saline (e.g., Brass and others, 1982) with a relatively small difference between surface and bottom-water temperatures (Savin, 1977; Fischer and Arthur, 1977; Brass and others, 1982; Wilde and Berry, 1982). Relatively warm, salty, shelf waters that formed in evaporative settings may have sunk and formed the dominant deep-water mass (Brass and others, 1982; Wilde and Berry, 1982; Saltzman and Barron, 1982). It is not clear what the relative rates of vertical mixing would be under such conditions, although Brass and others (1982) argued that mixing rates could be as high or higher than those of today, and Barron and Washington (1982) suggested that oceanic circulation on the whole was not "sluggish" as is commonly assumed. In any case, bottom-water circulation was sufficient to supply some oxygen to maintain oxidizing conditions in the deep basins of the Pacific and Indian Oceans. However, rates of oxygen supply were low enough that depletion of dissolved oxygen apparently occurred periodically over a broad range of mid-water depths in areas of relatively high organic productivity; in these areas, dissolved-oxygen contents in intermediate- and deep-water masses were already low due to higher temperature and salinity. During the middle Cretaceous, much of the world ocean below a few hundred meters may have been delicately poised in terms of oxygen supply and consumption rates. Therefore, relatively small changes in the flux of organic matter or the rate of deep-water supply at any one place may have caused anoxia or near-anoxia within midwater oxygen-minimum zones and possibly throughout much of the bottom-water mass; this was particularly true in more tectonically restricted basins such as the North and South Atlantic.

An expanded and intensified oxygen-minimum zone during much of the Early and Middle Cretaceous would explain (1) the excellent preservation of organic carbon, and (2) the increase in accumulation rate of organic carbon over much larger areas in continental slope environments, on flanks of high-standing features on the seafloor (e.g., Schlanger and Jenkyns, 1976; Fischer and Arthur, 1977), and in deep-sea settings influenced by redeposition (Dean and Gardner, 1982; Arthur, Dean, and Stow, 1984). Oxygen-minimum zones probably were most expanded and intense in areas of pronounced upwelling and high organic productivity, such as off northwest Africa (e.g., Berger and von Rad, 1972; Thierstein, 1979; Arthur and Natland, 1979; Summerhayes, 1985). This would explain the higher overall concentrations, accumulation rates, and more marine character of organic matter in middle Cretaceous strata from the eastern North Atlantic basin, relative to that in the western North Atlantic (Tissot and others, 1979; Tissot and others, 1980; de Graciansky and others, 1982). The oxygen-minimum zone may have periodically expanded to the deepest parts of the seafloor so that the entire basin was anoxic.

Fluctuations in sea level and climate probably were the dominant causes for variations in amount of organic matter supplied to and preserved in the basin. The rapid rate of supply of terrigenous organic matter to a basin that was already poorly oxygenated (but not necessarily anoxic) also enhanced oxygen deficits in some basins, and caused increased preservation of marine organic matter.

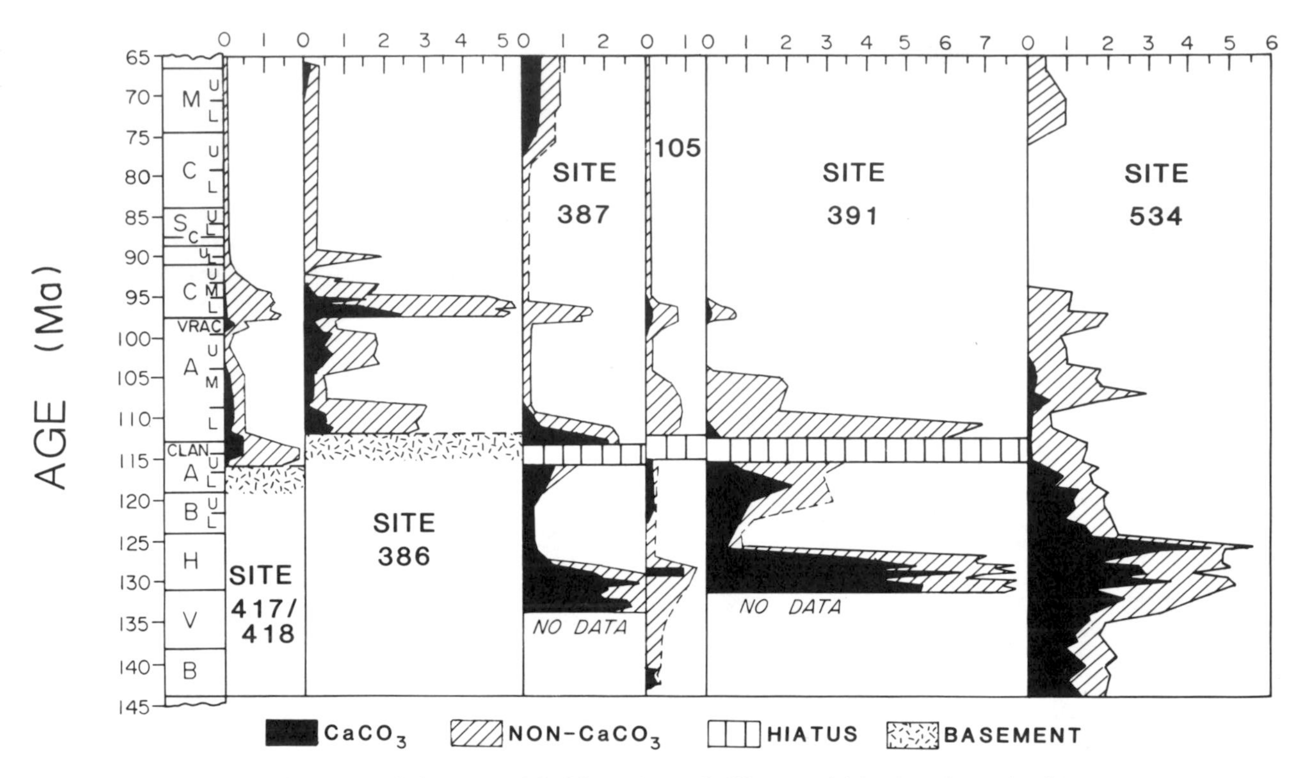

Figure 5. Mass accumulation rates of $CaCO_3$ and non-$CaCO_3$ material for those intervals of western North Atlantic DSDP sites which have adequate recovery, sampling, and biostratigraphy. Time scale used is from Plate 1. Cretaceous stages in time scale represented by initials.

In addition to the more global factors outlined above, the evolution of relatively Corg-rich strata in the western North Atlantic basin was dependent on a variety of regional or local processes. The Hatteras Formation at deeper paleodepths and areas closer to the continental margin (Sites 105, 387, 391, 534) is characterized by lower average Corg contents but relatively higher Corg accumulation rates than at shallower sites (386, 417/418) that were at or near the Aptian-Albian ridgecrest (Figs. 3, 4 and 6). The Corg at the deeper sites appears to be predominantly mixtures of terrestrial organic matter and highly degraded marine autochthonous Corg (Summerhayes, 1981, 1985; Herbin and Deroo, 1982); organic matter in Corg-rich beds at sites higher on the flanks of the mid-Cretaceous ridgecrest is largely autochthonous marine Corg (Fig. 7). Some Corg-rich beds at Sites 417/418 also have high hydrogen indices, indicating a marine origin and relatively good preservation (e.g., Herbin and Deroo, 1982; Fig. 7); these may represent mono-specific dinoflagellate blooms (Hochuli and Kelts, 1980) with enhanced preservation under periodic marginally oxic to anoxic conditions.

Much of the Corg in the Hatteras Formation in western North Atlantic DSDP sites may have been redeposited from shallower paleodepths. The relationship between terrigenous turbidites and Corg-rich strata (with terrestrial Corg) has been noted at Sites 391 and 534 (e.g., Summerhayes, 1981; de Graciansky and others, 1982; Habib, 1983; Summerhayes and Masran; 1983; Arthur, Dean, Bottjer, and Scholle, 1984; and Robertson and Bliefnick, 1983), suggesting redeposition of terrigenous Corg from the adjacent, humid North American margin. Many of the Corg-rich beds at Site 386 also may have been redeposited from somewhat shallower paleodepths of local basement highs along the ridgecrest (e.g., Tucholke and Vogt, 1979; McCave, 1979b). Corg-rich beds are commonly associated with apparently winnowed or redeposited, thin radiolarites having sharp basal contacts, and they could represent sediment originally preserved within an expanded and intensified oxygen-minimum zone which periodically sloughed off steep basement slopes.

Corg contents reach a maximum at many sites some time between the middle Cenomanian and earliest Turonian (Fig. 3). This appears to be a manifestation of the global "oceanic anoxic" event (e.g., Schlanger and Jenkyns, 1976; Schlanger and others, 1985; Arthur and others, 1985). This brief but widespread period of Corg enrichment in ocean sediments may have been due to (1) a sudden burst of productivity as the result of more rapid deep-water turnover rates as the deep-water connection opened between the North and South Atlantic oceans (Tucholke and Vogt, 1979; Summerhayes, 1985), or (2) greater production of warm,

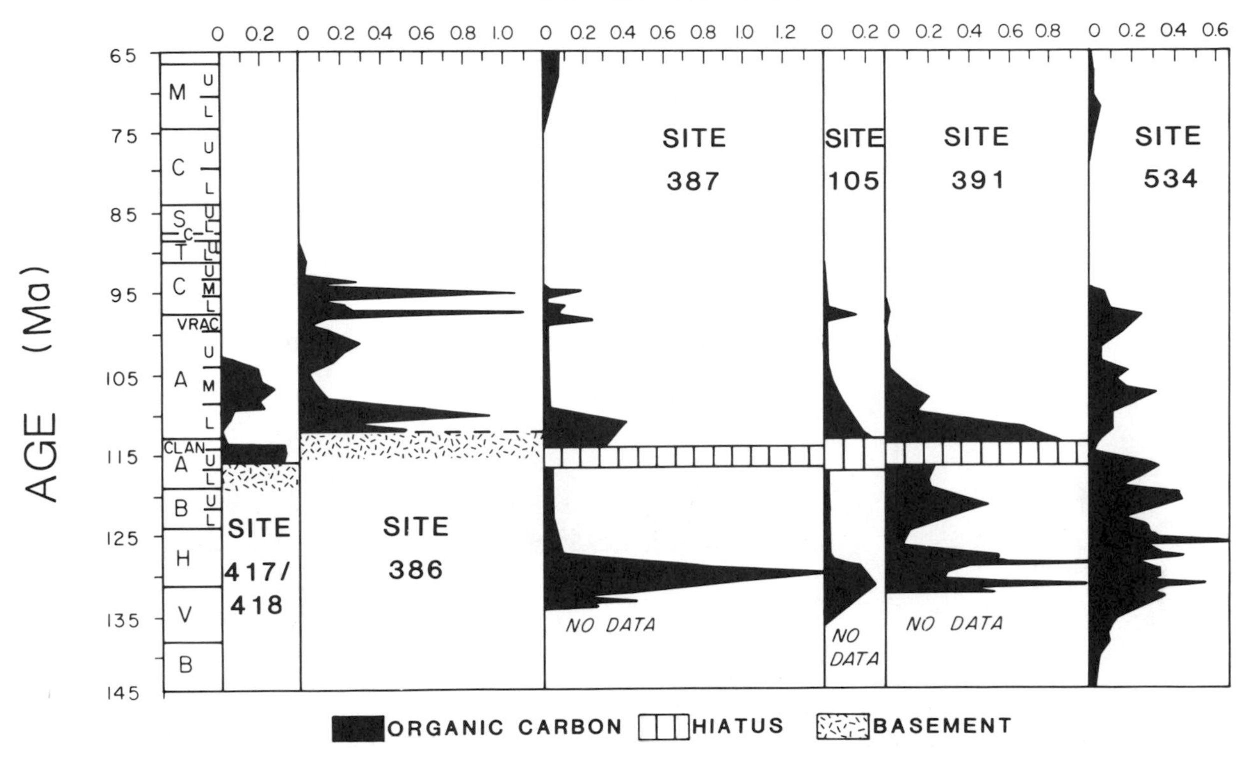

Figure 6. Organic-carbon (Corg) mass accumulation rates for selected western North Atlantic DSDP sites, calculated using DNAG time scale (Plate 1). Cretaceous stages in time scale represented by initials.

saline deep water as the result of the Late Cenomanian–Early Turonian global transgression (Arthur and others, 1985).

Cretaceous Corg preservation, both in the Blake-Bahama and the Hatteras formations, does not appear to be simply a function of sedimentation rate as proposed by Habib (1979; 1983) and Ibach (1982), following Muller and Suess (1979) for the Holocene (Fig. 8). There is an overall increase in Corg MARS with increasing sedimentation rate, but with substantial scatter. The Corg MARS also are significantly higher than predicted, even taking compaction into account. The highest Corg MARS, however, do occur within alternating laminated and bioturbated limestones of the rapidly deposited Blake-Bahama Formation, and high sedimentation rates may have been a factor in the apparently enhanced Corg MARS there. The Corg in the Blake-Bahama Formation has low hydrogen indices (Fig. 7; Herbin and Deroo, 1982), and it is therefore a mixture of terrestrial and relatively poorly preserved marine Corg (Summerhayes, 1981).

Paleofertility of the Cretaceous Western North Atlantic Basin

Definitive indicators of paleofertility are not readily available in ancient marine sediments. High Corg and biogenic silica contents are commonly cited as indicators of increased fertility, but the possibility of enhanced preservation under an anoxic and silica-saturated water column, or overall high sedimentation rates, are additional factors that must be considered. A number of authors have argued for relatively low fertility in the western North Atlantic during deposition of the Hatteras Formation (e.g., Fischer and Arthur, 1977; Berger, 1979; Roth, 1978; Thierstein, 1979; Arthur and Kelts, 1979; Roth and Bowdler, 1981; Cool, 1982). These authors supply various reasons for the low fertility including (1) enhanced stability of the water column, (2) low rates of upwelling, and (3) low overall nutrient content of the oceans due to rapid burial of nutrients in massive marine evaporites or in flooded epicontinental seas. Enhanced organic-carbon contents in deeper basins are most likely due to: (1) preservation of marine Corg under oxygen-deficient deep water masses; (2) possibly higher sedimentation rates in some cases (e.g., Ibach, 1982; Habib, 1983); (3) increased supply of marine organic matter by redeposition from shallow-water sites of enhanced preservation (e.g., Dean and Gardner, 1982; Arthur, Dean, and Stow, 1984); and/or (4) terrestrial organic-matter influx carried by surface currents (e.g., Hochuli and Kelts, 1980), bottom nepheloid layers (Summerhayes, 1981), or redeposition from humid shelf or deltaic regions (e.g., Habib, 1979, 1983; Arthur,

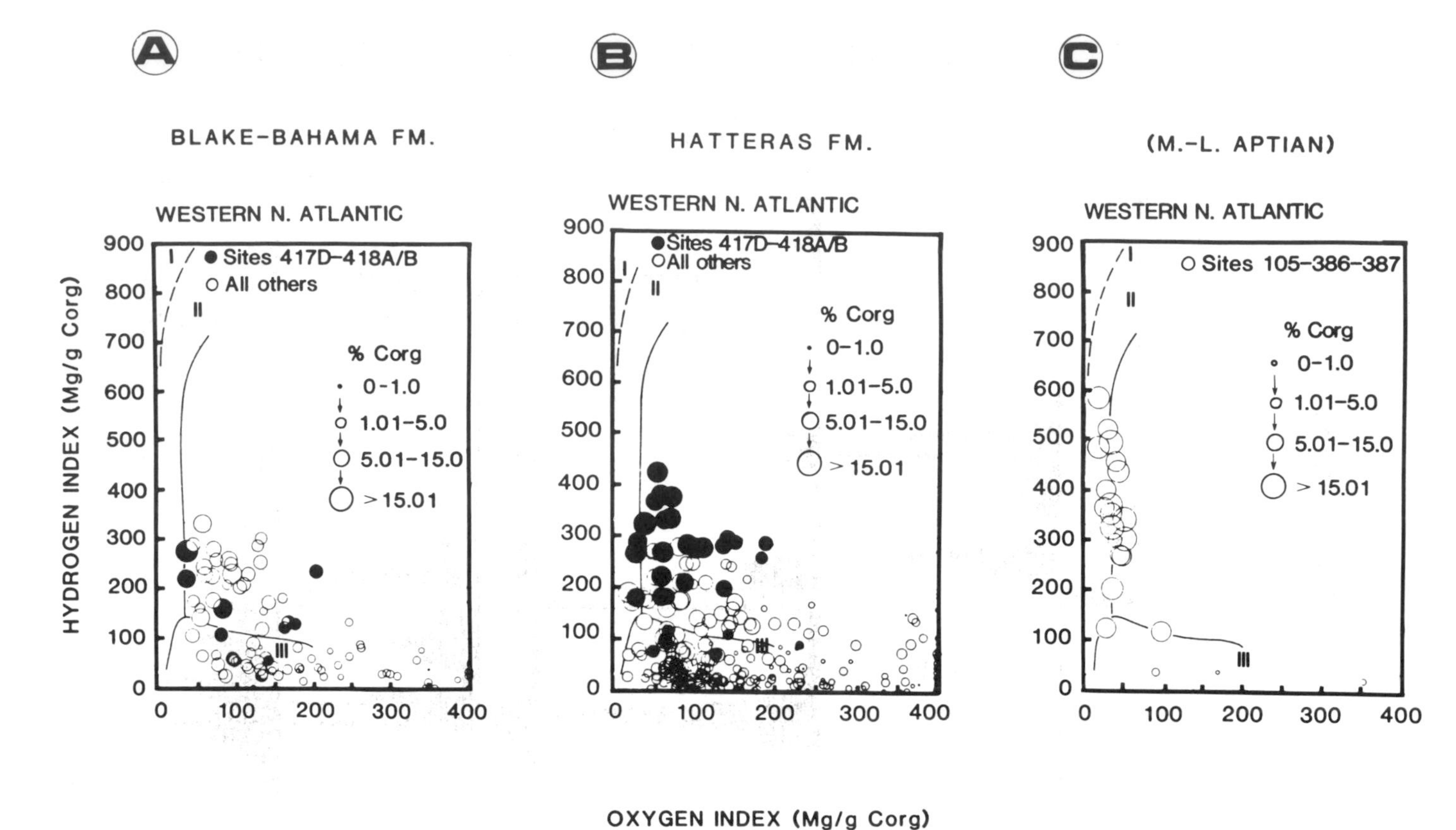

Figure 7. Hydrogen and oxygen indices for organic matter from western North Atlantic DSDP sites (modified from Herbin and Deroo, 1982). Symbols are shown in key. Curves marked *I, II,* and *III* refer to thermal maturation pathways of type I, II, and II Kerogen (Tissot and others, 1980).

Dean, and Stow, 1984; de Graciansky and others, 1979; Robertson and Bliefnick, 1983).

Organic-carbon accumulation rates (Fig. 8) also may indicate rates of primary productivity (e.g., Muller and Suess, 1979). The Cretaceous Corg MARS generally indicate that primary productivity in the western North Atlantic was low relative to recent rates if we assume that preservation was enhanced under periodically anoxic conditions, and we use the method outlined by Muller and Suess (1979) to estimate Holocene rates of primary productivity. Bralower and Thierstein (1985) came to a similar conclusion on the basis of MARS data. However, we suggest that their inferences from the Muller and Suess (1979) technique may not be entirely applicable for the following reasons: (1) the Cretaceous seafloor depths at most of the western North Atlantic drillsites are deeper than most localities in the Holocene data set; (2) redeposition of Corg may have been important; (3) initial enhanced preservation may have occurred in shallower, oxygen-deficient regions; (4) the seafloor at the ultimate site of deposition may not have been anoxic; and (5) periodic oxidation (see Cyclic Sedimentation) may have led to reduced preservation of Corg. Therefore, the overall slower sedimentation rates, with oxygenation and later bioturbation of many Corg-rich layers, could have degraded much of the Corg and could lead to biased MARS results. Such relationships require much further study before conclusions on paleoproductivity can be drawn from the organic-carbon content of the Cretaceous sediments.

Biogenic silica occurs in both the upper part of the Blake-Bahama Formation and in the Hatteras Formation at several different DSDP Sites. It takes the form of disseminated, poorly preserved radiolarian tests that are commonly pyritized or replaced by chalcedony. Minor chert and silicified limestone was found in the Hauterivian of Site 387, and numerous radiolarian sand/silt beds are present in DSDP Sites 386 and 417/418. There is the distinct possibility, however, that the radiolarian-rich beds at these sites are largely redeposited (e.g., McCave, 1979b). There is also a distinct possibility that the beds result from intense "blooms" in the surface water.

In modern environments, the accumulation of biogenic opal on the seafloor generally occurs where the flux or production rate exceeds rates of dissolution in an entirely undersaturated water column (e.g., Heath, 1974; Berger, 1976). Therefore, rates of biogenic silica accumulation may be a good indication of paleoproductivity. We calculated average biogenic silica accumulation rates for the Hauterivian-Cenomanian interval of DSDP Sites 105, 391, and 387, and the Aptian-Cenomanian interval at DSDP Sites 386 and 417/418. For these estimates we calculated both excess silica accumulation rates (Leinen, 1979: excess Si is greater than an average Si/Al ratio of 5:1) using both available

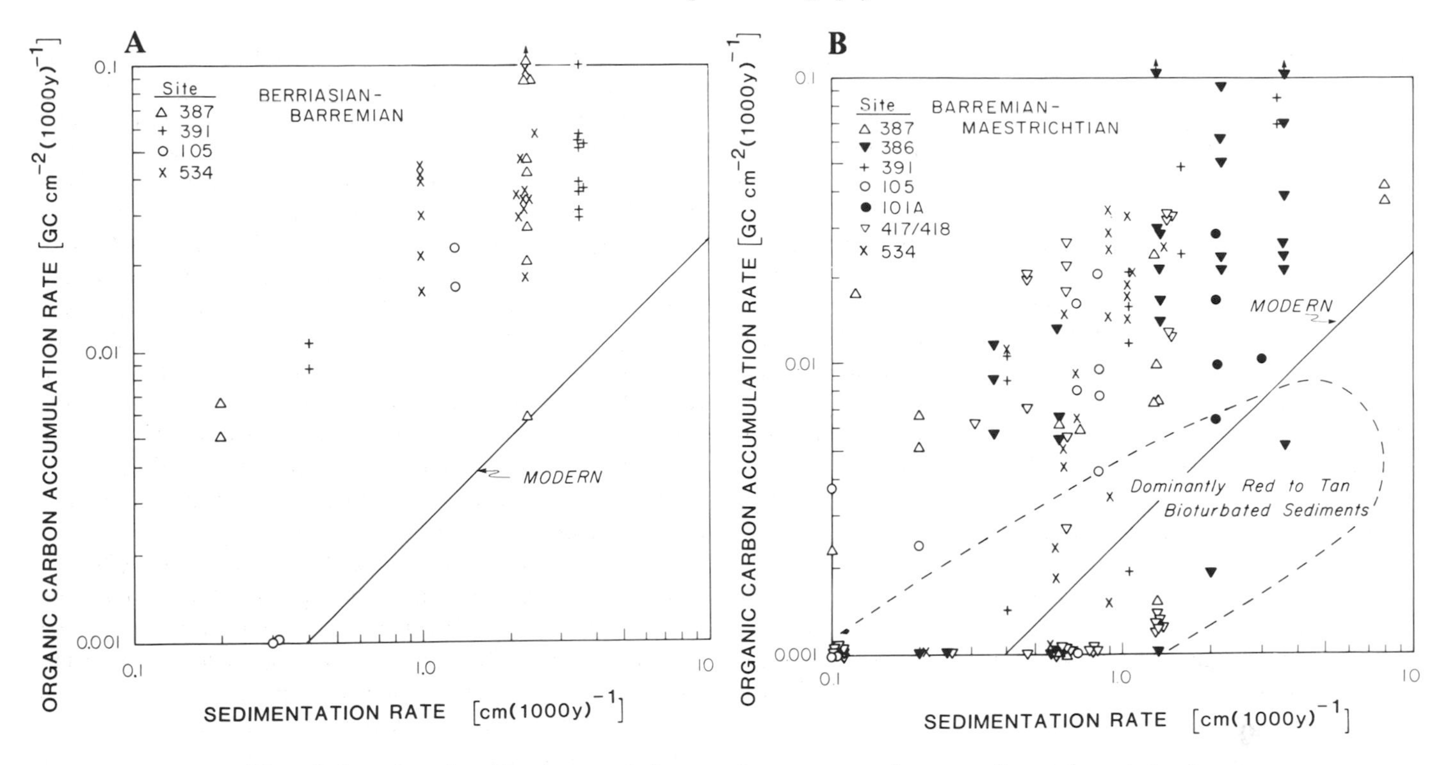

Figure 8. Organic carbon (Corg) accumulation rates (average per core) *versus* sedimentation rate for *A*, Berriasian through upper Barremian; and *B*, upper Barremian through Maestrichtian strata at western North Atlantic DSDP sites. All values were calculated using the DNAG time scale (Plate 1). Note the larger scatter in data, suggesting that sedimentation rate alone does not control Corg preservation. The diagonal line represents the best fit to data from modern environments (Muller and Suess, 1979), with no correction for compaction.

geochemical data (Murdmaa and others, 1978; Murdmaa and others, 1979; Donnelly, 1980; Debrabant and Chamley, 1982) and using estimated radiolarian percentages from smear slides. Both methods have their problems but they roughly agree for each site. The values average 120 $gSiO_2/cm^2/my$ at DSDP Sites 105, 391 and 387; 330 $gSiO_2/cm^2/my$ at Site 386; and 280 $gSiO_2/cm^2my$ at Sites 417/418. These averages are probably maximum values, except that discrete beds may have much higher, "instantaneous" accumulation rates. The averages, however, are much lower than modern accumulation rates for highly productive regions (about 3000–50,000 $gSiO_2/cm^2/my$; DeMaster, 1981), and they are well below maximum accumulation rates under the equatorial oceanic divergence in the Pacific during the Late Cenozoic (about 850 $g/cm^2/m.y.$; Leinen, 1979; DeMaster, 1981). Therefore, the average rates of biogenic silica accumulation do not appear to indicate high biological productivity in the western North Atlantic basin during the Early to mid-Cretaceous. However, the rates of accumulation are somewhat higher than those in most low-productivity regions of the world ocean today. The Cretaceous rates may be somewhat elevated because of redeposition from local highs, particularly at Sites 386 and 417/418, and (or) because of enhanced preservation under more silica-saturated deep-water masses. The possibility remains, however, that individual radiolarian-rich beds represent periodic surface-water blooms.

Cyclic Sedimentation

Cyclicity is a pronounced characteristic of nearly all Cretaceous pelagic sequences in the Atlantic. For example, intercalated carbonate-rich and carbonate-poor beds are particularly common in environments that were close to the CCD and where there was an abundant supply of clay. Additionally, in anoxic or marginally oxic depositional environments, intervals of oxygen depletion apparently were periodic. Such periodic enrichment of Corg occurs in the relatively clay-rich intervals in much of the Blake-Bahama Formation. In the deeper parts of the western North Atlantic, the middle to Upper Cretaceous sequences are predominantly clay-rich, and they contain color and(or) carbon-content cycles. Fluctuations in rate of supply of organic matter produced cyclic variations in redox conditions within the sediments and (or) within the bottom-water mass; this in turn produced cyclic interbeds of green/red or green/black clay-rich lithologies. Individual couplets of interbedded black/green, or red/green lithologies that dominate the middle Cretaceous sequences are mainly several decimeters thick, but they can range from several centimeters to as much as 80 cm. Periodicities of these cycles at most sites range from 20,000 to 100,000 years, with averages of about 40,000 to 50,000 years (Dean and others, 1978; Arthur and Fischer, 1977; Arthur, 1979a; McCave, 1979a; Arthur, Dean, Bottjer, and Scholle, 1984). It is difficult to define the periodicities more accurately because of a lack of high-resolution biostratigraphy, imprecise time scales, bioturbation, and diagenetic overprints. It is also likely that some of the cyclicity in deeper sites may be due to episodic redeposition from shallower regions of the seafloor and thus may not be strictly periodic. The cyclicity typical of most Cretaceous pelagic strata is most likely the result of variations in one or more climate-related forcing mechanisms, probably related in turn to cycles in solar-terrestrial orbital parameters (Milankovitch cycles). Arthur, Dean, Bottjer, and Scholle (1984) suggested that periodic changes in insolation, evaporation, wind stress, and (or) rainfall in a wide variety of environments caused changes in input of terrigenous detritus, water mass stratification, surface productivity, deep-water oxygen content, and rates of carbonate dissolution. The importance of different climate-related forcing mechanisms may have varied between depositional environments and between paleolatitudes, but the overall periodicities are the same, and the manifestations in lithologies are similar regardless of the mechanisms that caused the cycles. The cyclicity of supply and (or) preservation of organic matter and $CaCO_3$ had a pronounced influence on subsequent diagenetic events and on geochemical partitioning in adjacent beds.

CRETACEOUS OCEANIC CIRCULATION AND CHEMISTRY

Deposition of white to gray bioburbated nannofossil chalks alternating with finely laminated Corg-rich marls or clayey limestones (Blake-Bahama Formation) occurred at all depths in the western North Atlantic basin during most of the Late Tithonian to early Aptian. The CCD appears to have been generally deep during this interval, in a continuation of conditions which prevailed throughout the Late Jurassic. Deposition rates of the limestones were very high. The deposition of the Blake-Bahama Formation occurred during a time when the North Atlantic was a relatively narrow, possibly tectonically restricted basin; global eustatic sea level was low, but reef growth along the North American margin may have limited input of much terrigenous material. Corg deposition apparently was not important during the Early Cretaceous until the late Valaginian–early Hauterivian. The highest Corg MARS appear in Hauterivian strata along with high $CaCO_3$ MARS. Tucholke and Mountain (1979) noted lensoid depositional units within the Blake-Bahama Formation in seismic profiles across the Cat Gap region. They suggested that bottom currents were intensified along the steep Bahama Banks and that current-influenced sediment accumulation may have occurred in such areas. The unusually high $CaCO_3$ and Corg MARS in Hauterivian strata at several DSDP sites may be related to such current-influenced sediment deposition. It is also possible that the fine laminations that characterize much of the Blake-Bahama Formation may be current laminations (contourites) rather than laminations caused by very brief, periodic anoxia at the seafloor.

Clay mineral assemblages (e.g., Debrabant and Chamley, 1982) reflect a dominantly terrigenous source at most drill sites during the Early Cretaceous, with a gradual transition to dominantly volcanically derived smectitic clays at all sites through the Aptian-Albian. This change in clay mineralogy suggests a general decrease in terrestrial input, correlating with an increase in width of the Atlantic and with transgression and flooding of marginal seas beginning in the Aptian-Albian (e.g., Vail and others, 1977). A similar trend is reflected in changes of organic-matter type at some sites, namely decreasing influence of terrigenous Corg in Aptian-Albian strata (e.g., Hochuli and Kelts, 1980; Herbin and Deroo, 1982). MARS of $CaCO_3$, Corg, and non-$CaCO_3$ material progressively decreased from Neocomian through the Late Cretaceous time. These decreased MARS are probably also a result of the reduction of terrigenous influence during the overall Middle to Late Cretaceous transgression. In addition, the decrease in $CaCO_3$ MARS may relate to increased burial of $CaCO_3$ in shallow seas, with a consequent CCD rise to shallow levels during the Aptian through the Late Cretaceous.

A mid-Aptian increase in $CaCO_3$ and non-$CaCO_3$ MARS followed by a significant hiatus or a period of slow deposition (Fig. 5) may be a manifestation of a brief mid-Aptian regression (Vail and others, 1977) followed by a rapid sea level rise. Arthur and Kelts (1979) linked the sudden and pronounced mid-Aptian CCD rise and subsequent hiatus to the extraction of large volumes of evaporites and to related ocean-chemical changes in the tectonically restricted northern South Atlantic during the mid-Aptian regression. A brief oxygenation event associated with local red-clay deposition occurred in the North Atlantic at that time (e.g., Cool, 1982). Reconnection of the northern South Atlantic to the rest of the world ocean during the late Aptian may have occurred partly as the result of renewed global transgression. Possible formation of saline deep water resulting from this connection, and a brief increase in Corg production, may have led to maximum preservation of Corg (e.g. Herbin and Deroo, 1982) as evidenced by the Corg MARS maximum in the late Aptian–earliest Albian.

Tucholke and Mountain (1979) have shown that the thickness of the Hatteras Formation is relatively uniform over much of the western North Atlantic basin, but that there are several depositional lobes near the continental margin. Thus supply of terrigenous sediment and Corg at point-sources from the adjacent margin through turbidity currents is likely. For example, terrigenous turbidites have been noted in the Blake-Bahama Formation and in part of the Hatteras Formation at Sites 391 and 534 (e.g., Habib, 1983).

Preservation and MARS of Corg and $CaCO_3$ in the western North Atlantic basin declined through mid-Albian and Cenomanian time during a period of overall sea level regression. Deposition of cyclic green and black (locally red-brown) clays dominated. Surface biologic productivity was apparently low (e.g., Roth, 1978; Thierstein, 1979; Bralower and Thierstein, 1985) and deeper water masses were supplied with sufficient oxygen to allow oxidation of most Corg, except for the refractory terrigenous material. Periodically, however, productivity may have been higher (e.g., Hochuli and Kelts, 1980; McCave, 1979a); or dissolved oxygen in deep-water masses was reduced enough to allow enhanced Corg preservation in discrete beds. Periodic redeposition of Corg-rich sediment from shallow basement pinnacles within an expanded and intensified oxygen-minimum zone to deeper-water sites by downslope transport also may have been important. In this regard, Tucholke and Mountain (1979) note locally thick black-shale accumulations in topographic lows. Increased preservation of Corg during the earliest mid-Cenomanian to earliest Turonian may reflect a short-term, global burst in organic productivity due to increased deep-water turnover rates that were associated with a major Cenomanian-Turonian transgression.

Deposition of carbonate-free multicolored or red clays, characterized by low bulk and Corg MARS, occurred over most of the western North Atlantic basin after late Cenomanian time. Exceptions occurred at paleodepths above a very shallow CCD (~2800 m), where pelagic carbonates were deposited (e.g., Sites 384 and 390). The shallow CCD and low sediment accumulation rates may have been due to overall low fertility in the western North Atlantic and to high rates of $CaCO_3$ burial and trapping of terrigenous detritus in extensive shelf seas (e.g., Hays and Pitman, 1973). Metal enrichments in the Upper Cretaceous multicolored clays (e.g., Murdmaa and others, 1978; Murdmaa and others, 1979) have been ascribed to the influence of volcanic emanations (Lancelot and others, 1972). On the basis of the enrichment patterns, however, Arthur (1979b) argued that the metals could have been derived by expulsion of metal-rich, anoxic pore waters from the underlying black shales into the highly oxygenated bottom waters that prevailed in the western North Atlantic during the Late Cretaceous.

A strong depression of the CCD to more than 5000 m depth occurred in the western North Atlantic during the middle to late Maestrichtian (Tucholke and Vogt, 1979). This may have been a global event (Thierstein, 1979), linked to a gradual withdrawal of seas from the continents and reduced shelf-basin fractionation of carbonate supply.

It is clear that changing sea levels, subsidence of the North American continental margin, and the gradual evolution of the North Atlantic from a tectonically restricted basin to a broad, open-ocean basin all had a pronounced influence on Cretaceous paleoceanography and sedimentary processes. Better understanding of the paleoceanography and sedimentary evolution, however, awaits further deep drilling, adequate biostratigraphic control, and integration on a global scale.

The major remaining paleoceanographic problems in the western North Atlantic largely revolve around the origin of the Corg and biogenic silica-enrichments in Lower to Middle Cretaceous rocks. There remains much debate as to the relative role of productivity *versus* anoxia in the preservation of the organic matter, and the importance of redeposition to the lithologic heterogeneity of the Blake-Bahama and Hatteras formations. Additional geochemical and mineralogical studies of the Upper Cretaceous multicolored claystones of the Plantagenet Formation may lead to some interesting insights into the origin of metal enrichments in marine sediments. These slowly accumulated clays also may provide a long-term record of extraterrestrial debris fluxes. Finally, careful study of the intriguing limestone /marlstone cycles of the Blake-Bahama Formation will be important for evaluating the role of episodic, downslope sediment transport *versus* periodic, oceanographic fluctuations in controlling the depositional record.

REFERENCES

Arthur, M. A.
1979a: North Atlantic Cretaceous black shales: the record at Site 398 and a brief comparison with other occurrences; in Initial Reports of the Deep Sea Drilling Project, v. 47, Part 2, eds. J.-C. Sibuet and W.B.F. Ryan; U.S. Government Printing Office, Washington, D.C., p. 719–751.
1979b: Origin of Upper Cretaceous multicolored claystones of the Western Atlantic; in Initial Reports of Deep Sea Drilling Project, v. 43, eds. B. E. Tucholke and P. R. Vogt; U.S. Government Printing Office, Washington, D.C., p. 417–420.

Arthur, M. A., Dean, W. E., and Stow, D.A.V.
1984 : Models for the deposition of Mesozoic-Cenozoic fine-grained organic-carbon-rich sediment in the deep sea; in Fine-Grained Sediments: Processes and Products, eds. D.A.V. Stow, and D. Piper; Geological Society of London, Special Publication 15, p. 527–559.

Arthur, M. A., Dean, W. E., Bottjer, D., and Scholle, P. A.
1984 : Rhythmic bedding in Mesozoic-Cenozoic pelagic carbonate sequences: the primary and diagenetic origin of Milankovitch-like cycles; in Milankovitch and Climate, Part 1. ed. A. L. Berger; Riedel Publishing Company, Holland, p. 191–222.

Arthur, M. A., and Fischer, A. G.
1977 : Paleocene magnetic stratigraphy at Gubbio, Italy: lithostratigraphy and sedimentology: Geological Society of America Bulletin, v. 88, p. 367–371.

Arthur, M. A., and Kelts, K. R.
1979 : Evaporites, black shales and perturbations of ocean chemistry and fertility: Geological Society of America, Abstracts with Program, v. 11, Annual Meeting, San Diego, p. 381.

Arthur, M. A., and Natland, J. H.
1979 : Carbonaceous sediments in the North and South Atlantic: the role of salinity in stable stratification of early Cretaceous basins; in Deep Drilling Results in the Atlantic Ocean: Continental Margins and Paleoenvironment, eds. M. Talwani, W. W. Hay, and W.B.F. Ryan; American Geophysical Union, Maurice Ewing Series, v. 3, p. 297–344.

Arthur, M. A., and Premoli-Silva, I.
1982 : Development of widespread organic carbon-rich strata in the Mediterranean Tethys; in Nature and origin of Cretaceous carbon-rich facies, eds. S. O. Schlanger and M. B. Cita; Academic Press, London, p. 7–54.

Arthur, M. A., Schlanger, S. O., and Jenkyns, H. C.
1986 : The Cenomanian-Turonian Oceanic Anoxic Event, II. Paleoceanographic controls on organic matter production and preservation; in Marine Petroleum Source Rocks, eds. J. Brooks and A. Fleet, Geological Society of London, Special Publication, in press.

Barron, E. J., and Washington, W. M.
1982 : Cretaceous climate: a comparison of atmospheric simulations with the geologic record: Palaeogeography, Palaeoclimatology, Palaeoecol-

ogy, v. 40, p. 103–133.

Benson, W. E., Sheridan, R. E., eds.
1978 : Initial Reports of the Deep Sea Drilling Project, v. 44: U.S. Government Printing Office, Washington, D.C., 1005 p.

Berger, W. H.
1976 : Biogenous deep-sea sediments: production preservation and interpretation; in Chemical Oceanography, b. 5, 2nd Edition, eds. J. P. Riley, and R. Chester; Academic Press, London, p. 265–388.

Berger, W. H.
1979 : Impact of Deep Sea Drilling on paleoceanography; in Results of Deep Drilling in the Atlantic Ocean: Continental Margins and Paleoenvironment, eds. M. Talwani, W. W. Hay, and W.B.F. Ryan; American Geophysical Union, Maurice Ewing Series, v. 3, p. 297–344.

Berger, W. H., and von Rad, U.
1972 : Cretaceous and Cenozoic sediments from the Atlantic Ocean; in Initial Reports of the Deep Sea Drilling Project, v. 14, eds. D. E. Hayes and A. C. Pimm; U.S. Government Printing Office, Washington, D.C., p. 787–953.

Berger, W. H., and Winterer, E. L.
1974 : Plate stratigraphy and the fluctuating carbonate line, in Pelagic sediments on land and under the sea, eds. K. J. Hsü and H. C. Jenkyns; Special Publication International Association of Sedimentology, v. 1, p. 11–48.

Bernoulli, D.
1972 : North Atlantic and Mediterranean Mesozoic facies: a comparison; in Initial Reports of the Deep Sea Drilling Project, v. 11, eds. G. D. Hollister and J. I. Ewing; U.S. Government Printing Office, Washington, D.C., p. 801–872.

Bralower, T. J., and Thierstein, H. R.
1985 : Organic carbon and metal accumulation rates in Holocene and mid-Cretaceous marine sediments: paleoceanographic significance; in Marine Petroleum Source Rocks, eds. J. Brooks, and A. Fleet; Geological Society of London, Special Publication, in press.

Brass, G. W., Southam, J. R., and Peterson, W. H.
1982 : Warm saline bottom water in the ancient ocean: Nature, v. 296, p. 620–623.

Chenet, P. Y., and Francheteau, J.
1980 : Bathymetric reconstruction method: application to the central Atlantic basin between 10°N and 40°N; Initial Reports of the Deep Sea Drilling Project, v. 51–53, Pt. 2, eds. T. W. Donnelly and J. Francheteau; U.S. Government Printing Office, Washington, D.C., p. 1501–1513.

Cool, T. E.
1982 : Sedimentological evidence concerning the paleoceanography of the Cretaceous western North Atlantic Ocean: Palaeogeography, Palaeoclimatology, Palaeoecology, v. 39, p. 1–35.

Dean, W. E., and Gardner, J. V.
1982 : Origin and geochemistry of redox cycles of Jurassic to Eocene Age, Cape Verde Basin (DSDP Site 367), continental margin of northwest Africa; in Nature and origin of Cretaceous carbon-rich facies, eds. S. O. Schlanger, and M. B. Cita; Academic Press, London, p. 55–78.

Dean, W. E., Gardner, J. V., Jansa, L. F., Cepek, P., and Seibold, E.
1978 : Cyclic sedimentation along the continental margin of northwest Africa; in Initial Reports of the Deep Sea Drilling Project, v. 41, eds. Y. Lancelot and E. Seibold; U.S. Government Printing Office, Washington, D.C., p. 965–986.

Debrabant, P., and Chamley, H.
1982 : Influences océaniques et continentales dans les prémier s dépots de l'Atlantique Nord: Bulletin Societe Geologique de France, v. 24, p. 473–486.

de Graciansky, P. C., Auffret, G. A., Dupeuble, P., Montadert, L., and Müller, C.
1979 : Interpretation of depositional environments of the Aptian/Albian black shales on the north margin of the Bay of Biscay (DSDP Sites 400 and 402); in Initial Reports of the Deep Sea Drilling Project, v. 48, eds. L. Montadert and D. G. Roberts; U.S. Government Printing Office, Washington, D.C., p. 877–907.

de Graciansky, P. C., Brosse, E., Deroo, G., Herbin, J.-P., Montadert, L., Müller, C., Sigal, J. and Schaaf, A.
1982 : Les formations d'age Cretace de l'Atlantique Nord et leur matiere organique: paleogeographie et milieux de depot: Revue de l'Institut Francais du Petrole, v. 37, p. 275–337.

deMaster, D. J.
1981 : The supply and accumulation of silica in the marine environment: Geochimica et Cosmochimica Acta, v. 45, p. 1715–1732.

Donnelly, T. W.
1980 : Chemistry of sediments of the western Atlantic: Site 417 compared with Sites 9, 105, 386 and 387; in Initial Reports of the Deep Sea Drilling Project, v. 51–53, Pt. 2, eds. T. W. Donnelly and J. Francheteau; U.S. Government Printing Office, Washington, D.C., p. 1515–1523.

Fischer, A. G., and Arthur, M. A.
1977 : Secular variations in the pelagic realm; in Deepwater carbonate environments, eds. H. E. Cook and P. Enos, Society of Economic Paleontologists and Minerologists Special Publication 25, p. 19–50.

Habib, D.
1979 : Sedimentary origin of North Atlantic Cretaceous palynofacies; in Deep drilling results in the Atlantic Ocean: continental margins and Paleoenvironments, eds. M. Talwani, W. W. Hay, and W.B.F. Ryan, American Geophysical Union, Maurice Ewing Series, Washington, D.C., v. 3, p. 420–437.

Habib, D.
1983 : Sedimentation-rate-dependent distribution of organic matter in the North Atlantic Jurassic-Cretaceous; in Initial Reports of the Deep Sea Drilling Project, v. 76, eds. R. E. Sheridan and F. M. Gradstein; U.S. Government Printing Office, Washington, D.C., p. 781–794.

Harland, W. B., Cox, A. U., Llewellyn, P. G., Pickton, C.A.G., Smith, A. G., and Walters, R.
1982 : A geologic time scale: Cambridge University Press, Cambridge, 131 p.

Hays, J. D., and Pitman, W. C. III.
1973 : Lithospheric plate motion, sea level changes and climatic and ecological consequences: Nature, v. 246, p. 18–22.

Heath, G. R.
1974 : Dissolved silica and deep-sea sediments, in Studies in paleoceanography, ed. W. W. Hay; Society of Economic Paleontologists and Minerologists Special Publication 20, p. 77–93.

Herbin, J. P., and Deroo, G.
1982 : Sedimentologie de la matiére organique dans les formations du Mesozoique de l'Atlantique Nord: Bulletin Societe Geologique de France, v. 24, p. 497–510.

Hochuli, D., and Kelts, K.
1980 : Palynology of middle Cretaceous black clay facies from Deep Sea Drilling Project Sites 417 and 418 of the western North Atlantic; in Initial Reports of Deep Sea Drilling Project, v. 51-53, Part 2, eds. T. Donnelly and J. Francheteau; U.S. Government Printing Office, Washington, D.C., p. 897–936.

Ibach, L.E.J.
1982 : Relationship between sedimentation rate and total organic carbon in sediments: American Association of Petroleum Geologists, Bulletin, v. 66, p. 170–188.

Jansa, L. F., Enos, P., Tucholke, B. E., Gradstein, F. M., and Sheridan, R. E.
1979 : Mesozoic-Cenozoic sedimentary formations of the North Atlantic basin; western North Atlantic; in Deep drilling results in the Atlantic Ocean: continental margins and paleoenvironment, eds. M. Talwani, W. W. Hay, and W.B.F. Ryan; American Geophysical Union, Maurice

Ewing Series, v. 3, p. 1–57.

Jansa, L. F., Gardner, J. V., and Dean, W. E.
1978 : Mesozoic sequences of the central North Atlantic; in Initial Reports of the Deep Sea Drilling Project, v. 41, eds. Y. Lancelot and E. Seibold; U.S. Government Printing Office, Washington, D.C., p. 991–1031.

Lancelot, Y., Hathaway, J. C., and Hollister, D. C.
1972 : Lithology of sediments from the western North Atlantic, Leg 11, Deep Sea Drilling Project; in Initial Reports of the Deep Sea Drilling Project, v. 11, eds. C. D. Hollister and J. I. Ewing; U.S. Government Printing Office, Washington, D.C., p. 901–950.

Leinen, M.
1979 : Biogenic silica accumulation in the central equatorial Pacific and its implications for Cenozoic paleoceanography: Geological Society of America Bulletin, v. 90, p. 801–803.

Lever, A., and McCave, I. N.
1984 : Eolian components in Cretaceous and Tertiary North Atlantic sediments: Sedimentology, in press.

McCave, I. N.
1979a: Depositional features of organic black and green mudstones at DSDP Sites 386 and 387, western North Atlantic; in Initial Reports of the Deep Sea Drilling Project, v. 43, eds. B. E. Tucholke and P. R. Vogt; U.S. Government Printing Office, Washington, D.C., p. 411–416.
1979b: Diagnosis of turbidites at Sites 386 and 387 by particle-counter size analysis of the silt (2–4 μm) fraction; in Initial Reports of the Deep Sea Drilling Project, v. 43, eds. B. Tucholke and P. R. Vogt; U.S. Government Printing Office, Washington, D.C., p. 395–405.

Müller, P. S., and Suess, E.
1979 : Productivity, sedimentation rate, and sedimentary organic carbon content in the oceans: Deep-Sea Research, v. 26, p. 1347–1362.

Murdmaa, I. O., Gordeev, V. V., Bazilevskaya, E. S., and Emelyanov, E. M.
1978 : Inorganic geochemistry of the Leg 44 sediments; in Initial Reports of Deep Sea Drilling Project, v. 44, eds. W. E. Benson and R. E. Sheridan; U.S. Government Printing Office, Washington, D.C., p. 575–582.

Murdmaa, I. O., Gordeev, V. V., Emelyanov, E. M., and Bazilevskaya, E. S.
1979 : Inorganic geochemistry of Leg 43 sediments; in Initial Reports of Deep Sea Drilling Project, v. 43, eds. B. E. Tucholke and P. R. Vogt; U.S. Government Printing Office, Washington, D.C., p. 675–694.

Palmer, A. R.
1983 : The DNAG time scale: Geology, v. 11, p. 503–504.

Robertson, A.H.F., and Bliefnick, D. M.
1983 : Sedimentology and origin of Lower Cretaceous pelagic carbonates and redeposited clastics, Blake-Bahama Formation, Deep Sea Drilling Project Site 534, western equatorial Atlantic; in Initial Reports of the Deep Sea Drilling Project, v. 74, eds. R. E. Sheridan and F. M. Gradstein; U.S. Government Printing Office, Washington, D.C., p. 795–828.

Roth, P. H.
1978 : Cretaceous nannoplankton biostratigraphy and oceanography of the northwestern Atlantic Ocean; in Initial Reports of the Deep Sea Drilling Project, v. 44, eds. W. E. Benson and R. E. Sheridan; U.S. Government Printing Office, Washington, D.C., p. 731–752.

Roth, P. H., and Bowdler, J. L.
1981 : Middle Cretaceous nannoplankton biogeography and oceanography of the Atlantic Ocean; in Deep-sea drilling: a decade of progress, eds. J. E. Warme, R. G. Douglas, and E. L. Winterer; Society of Economic Paleontologists and Mineralogists, Special Publication 32, p. 517–546.

Saltzman, E. S., and Barron, E. J.
1982 : Deep circulation in the Late Cretaceous: oxygen isotope paleotemperatures from *Inoceramus* remains in D.S.D.P. cores: Palaeogeography, Palaeoclimatology, Palaeoecology, v. 40, p. 167–181.

Savin, S. M.
1977 : History of the earth's surface temperature during the past 100 million years: Annual Review of Earth and Planetary; Sciences, v. 5, p. 319–355.

Schlanger, S. O., Arthur, M. A., Jenkyns, H. C., and Scholle, P. A.
1986 : The Cenomanian-Turonian oceanic anoxic event, I., Stratigraphy and distribution of organic carbon-rich beds and the marine $S^{13}C$ excursion; in Marine Petroleum Source Rocks, eds. J. Brooks and A. Fleet; Geological Society of London, Special Publication, in press.

Schlanger, S. O., and Jenkyns, H. C.
1976 : Cretaceous oceanic anoxic events—causes and consequences: Geologie en Mijnbouw, v. 55, p. 179–184.

Sclater, J. G., Hellinger, S., and Tapscott, C.
1977 : The paleobathymetry of the Atlantic Ocean from the Jurassic to the present: Journal of Geology, v. 85, p. 509–552.

Summerhayes, C. P.
1981 : Organic facies of Middle Cretaceous black shales in deep North Atlantic: American Association of Petroleum Geologists Bulletin, v. 65, p. 2364–2380.

Summerhayes, C. P.
1986 : Organic rich Cretaceous sediments form the North Atlantic; in Marine Petroleum Source Rocks, eds. J. Brooks, and A. Fleet; Geological Society of London, Special Publication, in press.

Summerhayes, C. P., and Masran, P.
1983 : Organic facies of Cretaceous and Jurassic sediments from DSDP Site 534 in the Blake Bahama Basin, western North Atlantic; in Initial Reports of the Deep Sea Drilling Project, v. 76, eds. R. E. Sheridan and F. M. Gradstein; U.S. Government Printing Office, Washington, D.C., p. 469–480.

Thierstein, H. R.
1979 : Paleooceanographic implications of organic carbon and carbonate distribution in Mesozoic deep sea sediments; in Deep drilling results in the Atlantic Ocean: continental margins and paleoenvironment, eds. M. Talwani, W. W. Hay, and W.B.F. Ryan; American Geophysical Union, Maurice Ewing Series, v. 3, p. 249–274.

Tissot, B., Demaison, G., Masson, P., Delteil, J. R., and Combaz, A.
1980 : Paleoenvironment and petroleum potential of middle Cretaceous black shales in Atlantic basins: American Association of Petroleum Geologists Bulletin, v. 64, p. 2051–2063.

Tissot, B., Deroo, G., and Herbin, J. P.
1979 : Organic matter in Cretaceous sediments of the North Atlantic: contributions to sedimentology and paleogeography; in Deep Drilling in the Atlantic Ocean: continental margins and paleoenvironments, eds. M. Talwani, W. W. Hay, and W.B.F. Ryan; American Geophysical Union, Maurice Ewing Series, v. 3, p. 362–374.

Tucholke, B. E., and Mountain, G. S.
1979 : Seismic stratigraphy, lithostratigraphy and paleosedimentation patterns in the North American Basin; in Deep Drilling Results in the Atlantic Ocean: Continental margins and paleoenvironments, eds. M. Talwani, W. W. Hay, and W.B.F. Ryan; American Geophysical Union, Maurice Ewing Series, v. 3, p. 58–86.

Tucholke, B. E., and Vogt, P. R.
1979 : Western North Atlantic: sedimentary evolution and aspects of tectonic history; in Initial Reports of the Deep Sea Drilling Project, v. 43, eds. B. E. Tucholke and P. R. Vogt; U.S. Government Printing Office, Washington, D.C., p. 791–825.

Vail, P. R., Mitchum, R. J., Jr., and Thompson, S.
1977 : Seismic stratigraphy and global changes of sea-level, Part 4: Global cycles of relative change of sea-level: American Association of Petroleum Geologists, Memoir 26, p. 83–97.

van Hinte, J. E.
1976 : A Cretaceous time scale: Bulletin of the American Association of Petroleum Geologists, v. 60, p. 498–516.

Weissert, H.
1981 : The environment of deposition of black shales in the Early Cretaceous: an ongoing controversy; in The Deep Sea Drilling Project: a

decade of progress, eds. J. E. Warme, R. G. Douglas, and E. L. Winterer; Society of Economic Paleontologists and Minerologists, Special Publication 32, p. 547–560.

Wilde, P., and Berry, W.B.N.
1982 : Progressive ventilation of the oceans—potential for return to anoxic conditions in the post-Paleozoic; in Nature and origin of Cretaceous carbon-rich facies, eds. S. O. Schlanger and M. B. Cita; Academic Press, London, p. 209–224.

MANUSCRIPT ACCEPTED BY THE SOCIETY JULY 15, 1985

The Geology of North America
Vol. M, The Western North Atlantic Region
The Geological Society of America, 1986

Chapter 38

Tertiary paleoceanography of the western North Atlantic Ocean

Brian E. Tucholke
Woods Hole Oceanographic Institution, Woods Hole, Massachusetts 02543
Gregory S. Mountain
Lamont-Doherty Geological Observatory of Columbia University, Palisades, New York 10964

INTRODUCTION

The Tertiary paleoceanography of the North Atlantic Ocean is known in greater detail than that of earlier periods because the upper part of the geologic record is more accessible to geologic sampling and geophysical observation. The Tertiary record also appears more complex, partially because of the greater data density. Nonetheless, there are large segments of this record about which we know relatively little. In particular, the Paleocene and Oligocene sedimentary records are not well represented on the shallow continental margin because erosion during sea-level lowstands removed much of the section. Equivalent records in the deep basin are poorly known because low sediment accumulation rates or erosion by abyssal currents resulted in preservation of abnormally thin sections. Many of these sections also are poorly sampled in existing Deep Sea Drilling Project (DSDP) drillsites.

In contrast to earlier periods, the Tertiary North Atlantic Ocean basin was strongly affected by the action of abyssal contour-following currents, particularly following Eocene time. The currents intermittently caused extensive seafloor erosion along the continental margins and they decorated the basin with large sediment drifts. The development and patterns of surface currents, and to some extent deep currents, are dealt with from a paleobiogeographic viewpoint by Berggren and Olsson (this volume). In this paper, we summarize the Tertiary paleoceanography based on other information from the geologic record, principally seismic stratigraphic interpretations that are correlated with rock stratigraphy at DSDP and other boreholes. These data are interpreted within the framework of the plate-tectonic evolution of the North Atlantic Ocean basins. Although we concentrate on the paleoceanographic development of the western North Atlantic, we also discuss "external" factors that affected the geologic record in the western basins. Important aspects of the geologic and oceanographic development of the North Atlantic are summarized in time series in Figure 1, and in time-slice maps for the late Maestrichtian, late Paleocene, early middle Eocene, early Oligocene, and late Miocene in Plate 10. Locations discussed are given in Plate 2.

PALEOCENE

For a period of several million years in the late Maestrichtian, the calcite compensation depth (CCD) in the North Atlantic was very deep, near 5.3 km (Plate 1). The CCD rebounded to shallower levels across the Cretaceous/Tertiary boundary and probably was near 3.5 to 4.5 km depth during most of the Paleocene. The temporary depression of the CCD resulted in widespread deposition of a carbonate layer in the deep basin (Plate 10, G). This layer was sandwiched between low-carbonate shales and it forms the Crescent Peaks Member of the Plantagenet Formation (Jansa and others, 1979). The upper surface of the carbonates is recognized as seismic Horizon A* (Tucholke and Mountain, 1979). Within the continental rise, Horizon A* is often difficult to identify and trace. Beneath the lower rise, this may be due to strong dilution of the carbonates by terrigenous debris; on the upper rise (above the ambient CCD), the carbonates are only part of a much thicker calcareous sequence and they often lack a unique seismic signature.

Along the continental shelf of eastern North America a widespread unconformity occurs between Cretaceous and Tertiary strata, so the carbonates forming Horizon A* are usually absent; much of the Paleocene sedimentary record is missing as well (Poag and Hall, 1979; Poag, 1980; Olsson and Wise, 1985). The unconformity can be traced readily in seismic reflection profiles across Baltimore Canyon trough (Schlee, 1981) and Georges Bank basin (Austin and others, 1980). Paleocene lowstands of eustatic sea-level that followed the globally elevated late Cretaceous sea levels are one possible explanation for this erosion (Fig. 1; Vail and others, 1977; Poag, 1980). However, Paleocene marine sediments that are coeval with offshore hiatuses are present in the U.S. Atlantic coastal plain, so it is possible that surface currents reaching to the sea floor were responsible for much of the erosion on the middle and outer continental shelf (Olsson and Wise, 1985).

Paleocene carbonate-rich sediments accumulated across northern Yucatan, Florida, the Blake Plateau, and the Bahama

Tucholke, B. E., and Mountain, G. S., 1986, Tertiary paleoceanography of the western North Atlantic Ocean; *in* Vogt, P. R., and Tucholke, B. E., eds., The Geology of North America, Volume M, The Western North Atlantic Region: Geological Society of America.

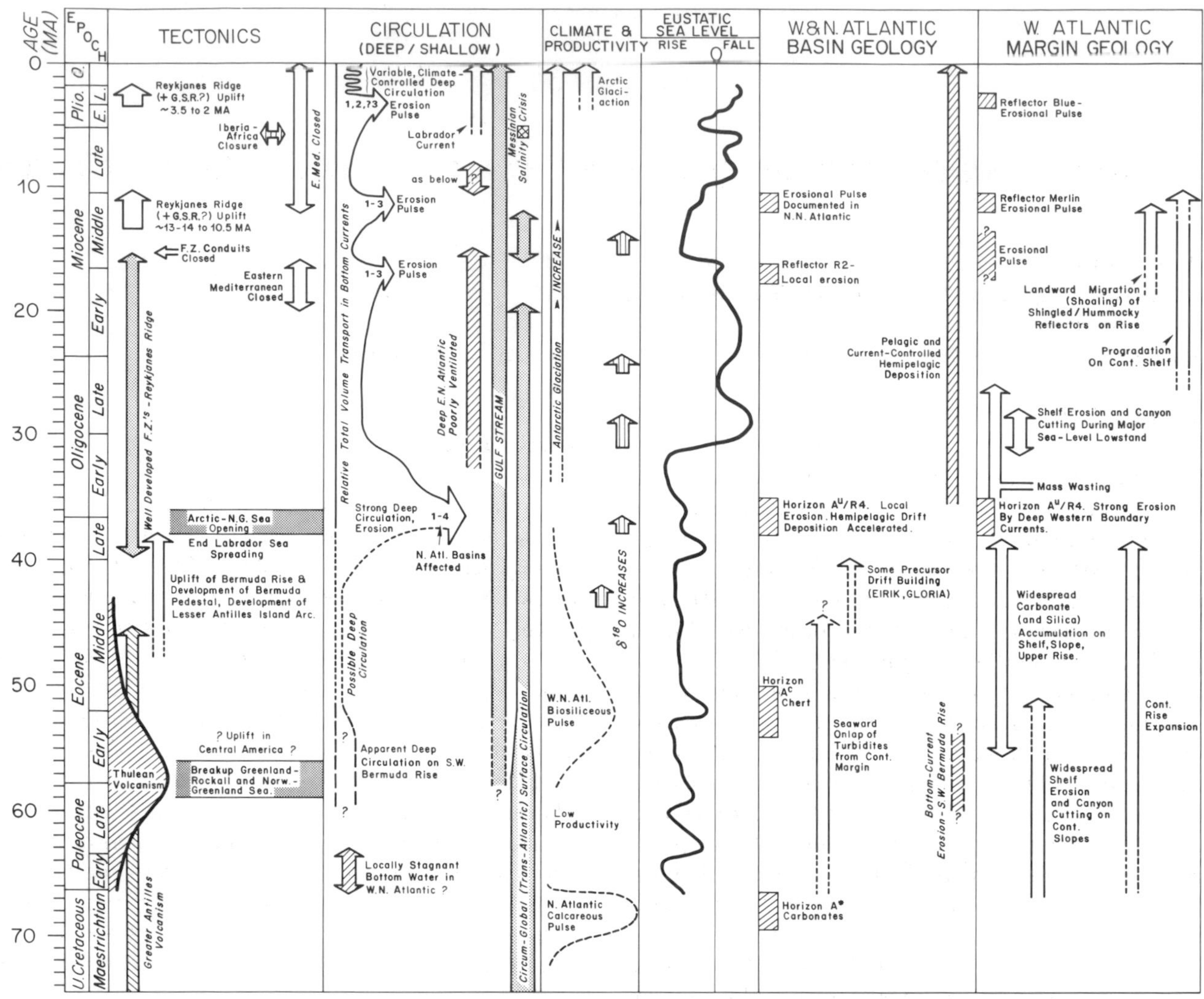

Figure 1. Simplified summary of geologic and paleoceanographic events affecting the development of the Cenozoic sedimentary record in the North Atlantic Ocean. Compiled from numerous published and unpublished data; principal references are cited in the text. Eustatic sea-level curve adapted from Vail and Hardenbol (1979) using data of Miller and others (1985b). Under "Circulation" the basins affected are 1) western North Atlantic, 2) northwestern Atlantic (Labrador/Irminger basins), 3) northeastern Atlantic north of Charlie-Gibbs F.Z., and 4) eastern North Atlantic south of Charlie-Gibbs F.Z.

Banks; on isolated topographic highs such as the J-Anomaly Ridge south of the Grand Banks (Site 384); and along the crest of the Mid-Atlantic Ridge (Plate 10, H). Carbonates increasingly diluted by terrigenous detritus probably were deposited along the North American continental margin northward from Cape Hatteras, but there is little control by samples. Sediments have been recovered from the Paleocene continental slope only at DSDP Site 605 (Fig. 2; Plate 10, H) and from a few seafloor outcrops. They consist of lower Paleocene silty glauconitic marls capped by lower to upper Paleocene clayey nannofossil limestones (van Hinte and Wise, 1986).

Seismic reflection profiles from the western Atlantic continental margin show Paleocene cutting of channels and small canyons on the continental slope (Fig. 3), and lower Paleogene sediment isopachs suggest correspondent upbuilding of the continental rise (Mountain and Tucholke, 1985). Seismically laminated sediments characteristic of turbidites immediately overlie Cretaceous strata beneath the lower continental rise, Hatteras Abyssal Plain, and western-most Bermuda Rise (Tucholke and Mountain, 1979). The laminated sediments above Horizon A* also filled topographic lows and smoothed previously irregular sea-floor along their eastern margin. These features indicate seaward progradation of an abyssal plain onto the western fringes of what is now the Bermuda Rise. The development of this regime of downslope sedimentation probably was concentrated in areas where Cretaceous fans had developed previously, namely seaward of Georges Bank and Baltimore Canyon trough, and at the mouths of Northeast Providence Channel and Great Abaco Canyon (Plate 2; Mountain and Tucholke, 1985). However, the locations of these depocenters is partly speculative because Olig-

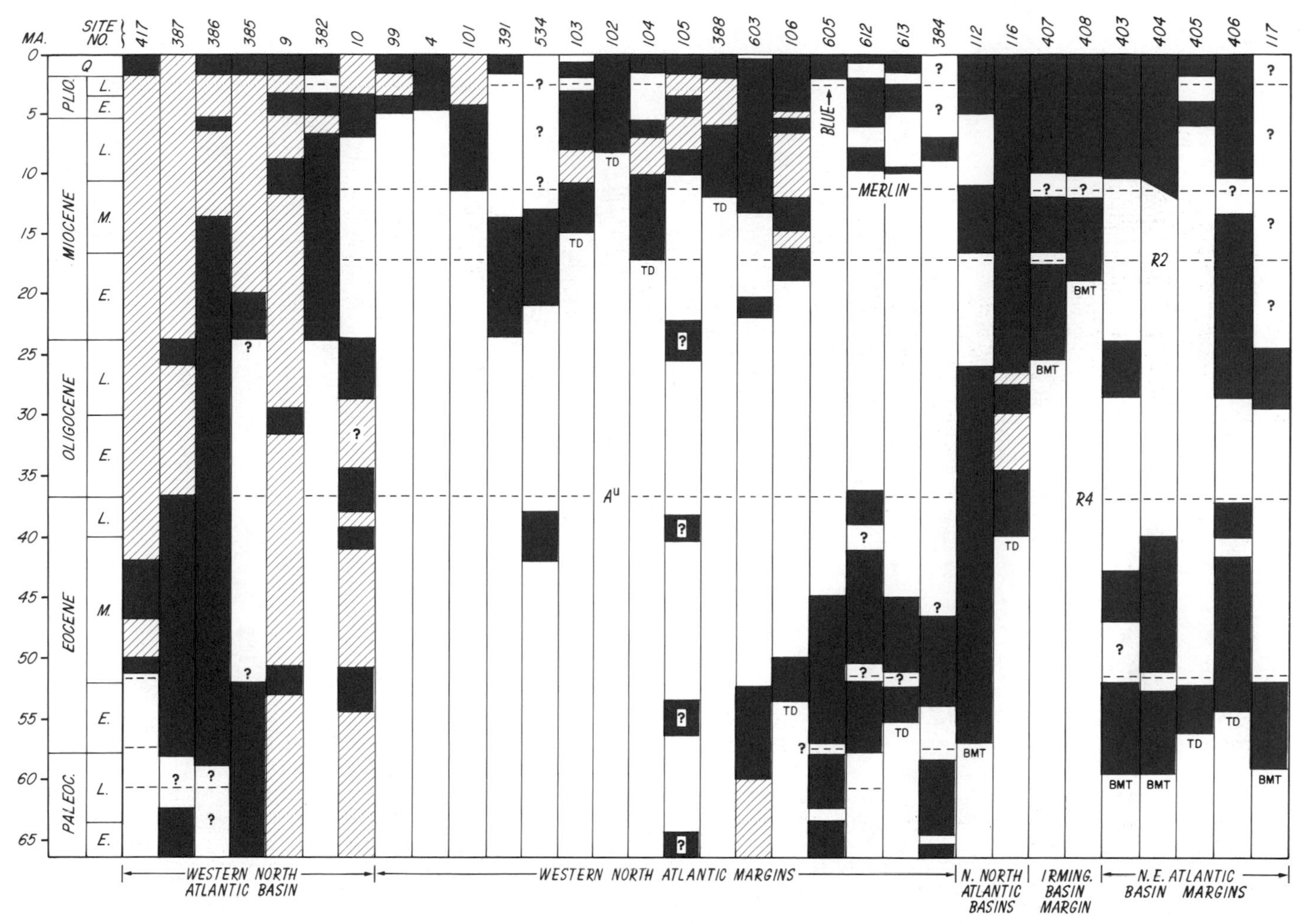

Figure 2. Cenozoic sedimentary record in selected DSDP drillsites in the western and northern North Atlantic. Solid red - recovered section; hachured - uncored or undated, but inferred continuous sedimentation; blank - hiatus, except below total depth (TD) of hole or where hole bottomed in basement (BMT). Significant unconformities of regional extent are indicated by dashed lines; color code shows known correlation to labelled seismic reflection boundaries. Core record based on Tucholke (1979), Kaneps and others (1981), Miller and Tucholke (1983), Sheridan and Gradstein (1983), Poag (1985), van Hinte and Wise (1986), and Poag and Watts (1986). For drillsite locations, see Plates 2 and 10.

ocene erosion by bottom currents removed much of the Paleocene sedimentary section on the continental rise, especially south of Cape Hatteras (see Fig. 4 and Early Oligocene, below).

Very few core samples exist to define the Paleocene depositional environment in deep water (Fig. 2). At DSDP Site 387 in the west central basin (Plate 10, H), the entire Paleocene section is less than 60 m thick. Lower Paleocene sediments recovered from this site were deposited below the CCD at paleodepths of about 5.5 km (Tucholke and Vogt, 1979). The section contains alternating gray-green and black shale layers, with the latter intermittently enriched in organic carbon (up to 1.3%); these sediments are very similar to the mid-Cretaceous "black shales" of the North Atlantic (Arthur and Dean, this volume). The occurrence of this lithofacies could indicate that the deepest part of the western North Atlantic basin was intermittently anoxic during the early Paleocene. However, the sediments also contain fine-grained, cryptically graded turbidites (McCave, 1979), and the reducing conditions responsible for the black layers may have occurred only within the pore waters of rapidly deposited beds. Because of their fine grain size it is likely that these were distal turbidites derived from sources along the U.S. Atlantic continental margin.

Outside the areas influenced by this downslope sediment movement, the deep Paleocene basin accumulated fine-grained pelagic shales (Plate 10, H). Total accumulation of Paleocene sediments in such areas ranged from less than 11 m at DSDP Site 386 in the center of the western North Atlantic basin, to less than 47 m of fine-grained zeolitic shales at Site 385 on the flank of Vogel Seamount (Fig. 2). At Site 105 near the continental margin, the seafloor formed a topographic swell that probably was bypassed by turbidity currents. Biostratigraphic study of fish debris in fine-grained pelagic shales there (Kaneps and others, 1981)

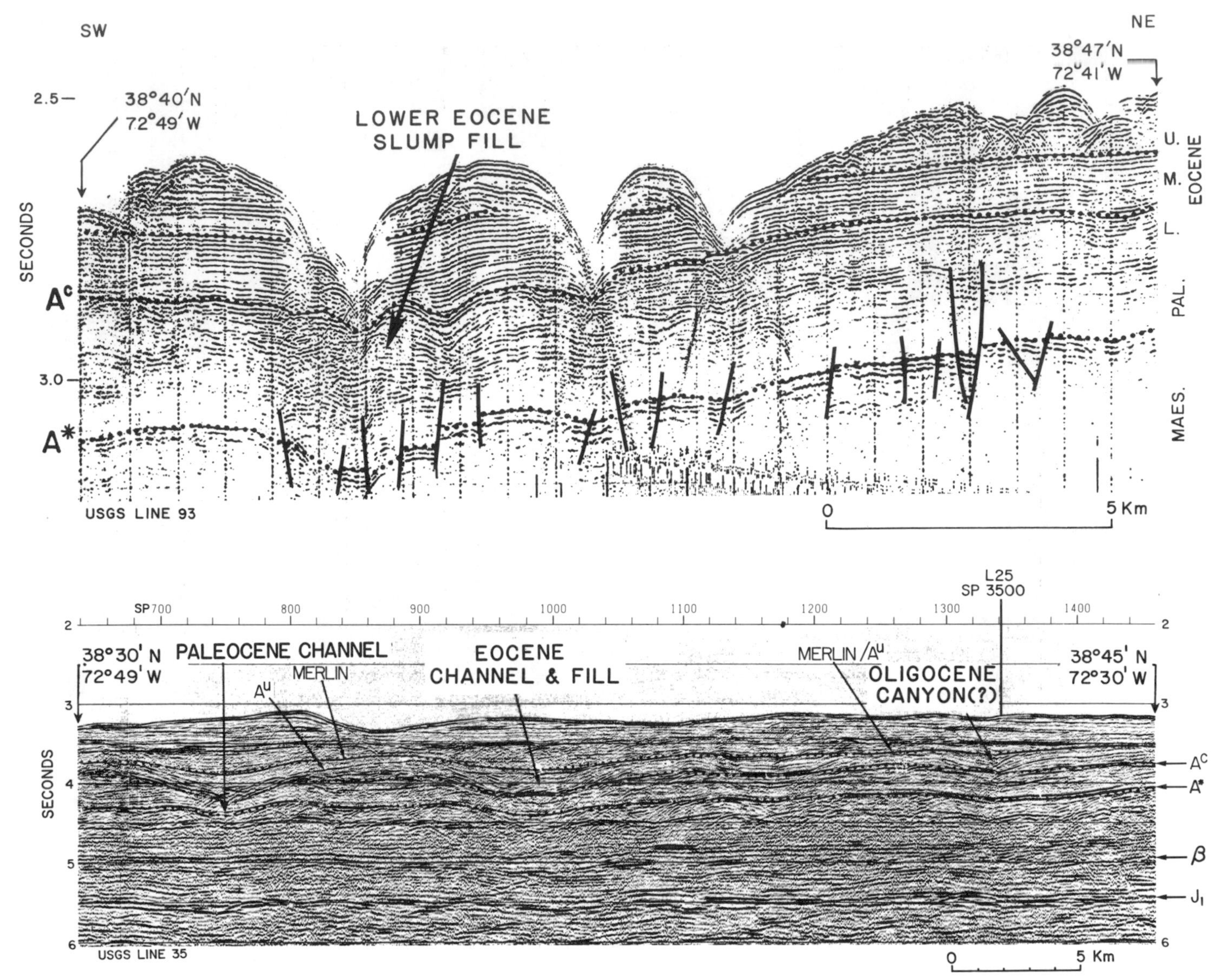

Figure 3. Seismic reflection profiles along the U.S. Atlantic continental slope seaward of Baltimore Canyon trough. Top - Segment of USGS line 93 (Paleogene-to-present continental slope) landward of lower profile. Note Paleocene to lower Eocene channels or small canyons, faulting of Cretaceous to lower Eocene sediments, and coherently laminated and draped middle and to upper Eocene sedimentary section. Horizon A^U and the record of all subsequent events have been removed in this area by Pleistocene cross-slope erosion that formed the large modern seafloor canyons. Bottom - Segment of USGS line 35 (uppermost continental rise; Paleogene continental slope) showing Paleocene and lower Eocene channels together with Eocene channel fill. The small Oligocene(?) canyon cut through Horizon A^U, and subsequently was planed off by bottom-current erosion in the late middle Miocene (reflector Merlin). Erosion at Merlin has re-excavated Horizon A^U along right half of profile. Note shallow late Pliocene-Quaternary channels at sea floor.

suggests that some of these sediments are of Paleocene age. The widespread occurrence of pelagic shales and thin stratigraphic sections in the deep basin suggest that biologic productivity was low in the central-ocean surface waters of the Paleocene North Atlantic Ocean (Plate 10, H); such central-gyre locations in the modern oceans normally exhibit low fertility.

EOCENE

In contrast to the Paleocene sedimentary record, the Eocene is well represented in cores, dredges, and drill samples from both the western North Atlantic continental margin and the deep basin. By Eocene time, shelf erosion and fore-shelf sedimentation

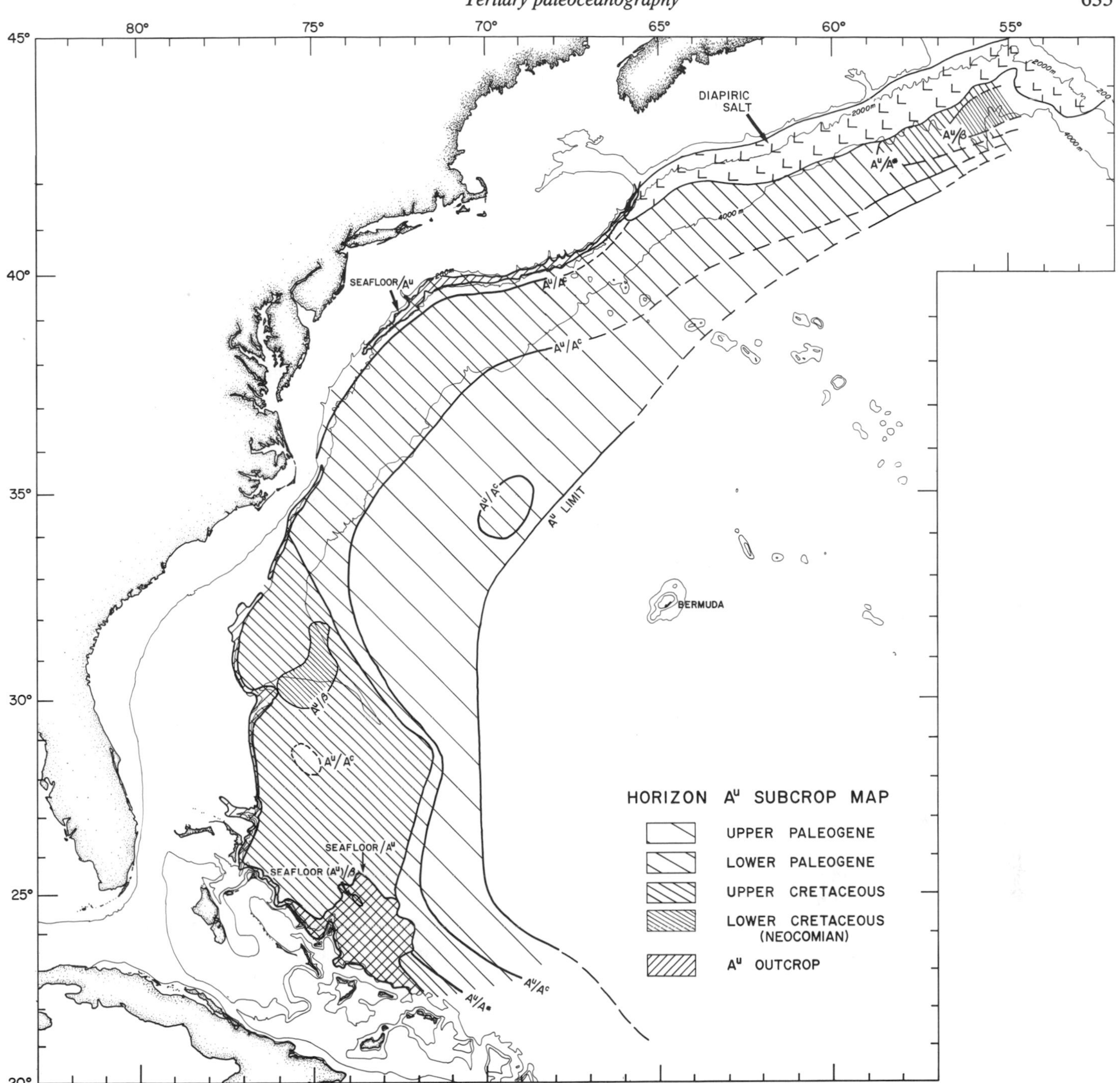

Figure 4. Subcrop geology at Horizon A^U, modified from Mountain and Tucholke (1985).

had reduced the physiographic breaks in slope that normally are used to define the continental shelf, slope, and rise. The boundaries between these provinces were blurred and a sedimentary ramp with seaward-increasing gradient extended from the shoreline to the upper continental rise.

Thick sequences of Eocene chalk and limestone occur beneath both the U.S. Atlantic coastal plain and the present continental shelf (Plate 10, I; e.g. Rhodehamel, 1977). Eocene glauconite beds and calcareous clays are present in the coastal plain from Virginia to New Jersey; farther south and extending into the eastern Gulf coast region the Eocene sediments are largely carbonate rocks (Gibson, 1970). The upper lower to lower middle Eocene carbonates along the U.S. Atlantic shelf and coastal plain are rich in silica. The silica is of biogenic and volcanic origin; it commonly is diagenetically altered and consists principally of cristobalite, montmorillonite, and clinoptilolite (Gibson, 1970).

Non-volcanic terrigeneous detritus is uncommon in these carbonates. On the U.S. shelf, Eocene carbonates sampled in COST wells were deposited in relatively deep settings, at outer-shelf to upper slope-depths (Poag and Hall, 1979; Poag, 1980). The borehole results and seismic-stratigraphic relations show that these sequences are relatively thick; where the section was interrupted by Eocene erosion, the unconformities occur primarily at stage boundaries. To the north along the Nova Scotian, Grand Banks, and Labrador shelf, the Eocene section is primarily nearshore sandstones and offshore mudstones and shales that locally are calcareous (Plate 10, I; Hardy, 1975; Jansa and Wade, 1975; Srivastava and others, 1981).

The area of the Eocene continental "slope" accumulated thick sequences of chalk and marl, and the sediments are exceptionally rich in biosiliceous debris in the lower middle Eocene section (Plate 10, I). Unconformities are developed at the lower/middle and middle/upper Eocene stage boundaries (e.g., DSDP Sites 612, 613), and they apparently were eroded during sea-level lowstands (Figs. 1, 2). The lower Eocene continental slope was cut by small canyons and channels (below Horizon A^c, Fig. 3), none of which rival their Quaternary counterparts in size. Most of these channels occur above older, Paleocene channels and represent continued growth or redevelopment of the earlier features. Slumping and other mass movements probably played an important role in the evolution of the channels. Slope instability would have been accentuated by sediment loading as rapid accumulation of biosiliceous chalk occurred beneath the fertile Eocene surface waters. In addition, the Eocene channels seaward of Baltimore Canyon trough lie across and downslope from a zone of faults that can be traced down into Upper and middle Cretaceous strata draping over a Mesozoic carbonate structure (Fig. 3, top). It appears that the faulting was active into early Eocene time, possibly in response to porosity reduction in the underlying carbonates. The Mesozoic carbonates occur along the entire U.S. and Nova Scotian Atlantic margin (Plate 9, C; Mountain and Tucholke, 1985), so similar faulting and slumping could have occurred along much of the western edge of the ocean basin. Early Eocene headward erosion of many slope canyons and channels probably was facilitated by these instabilities, and their growth also may have been augmented by downslope sediment transport during the eustatic sea-level lowstand at the early/middle Eocene boundary.

In middle Eocene time, however, most canyons appear to have become inactive. Seismic reflection profiles suggest that many canyons were filled with slump debris (Fig. 3), and a middle Eocene slump has been drilled in an inter-canyon area at DSDP Site 613 (Plate 10, I; Poag, 1985). Middle and upper Eocene seismic sequences along the continental slope commonly are uniformly laminated and draped over both filled and unfilled canyons (Fig. 3, top). From existing data it is uncertain whether any significant slope erosion or canyon development accompanied the sea-level lowstand at the middle/late Eocene boundary, or whether the unconformity encompassing this boundary was largely a result of subsequent, Oligocene erosion (Fig. 2). However, the uniformly developed seismic sequences suggest that there was little mass movement of sediment.

The deep margin of the western Atlantic basin contains abundant fine-grained lower to middle Eocene turbidites (Tucholke and Mountain, 1979). The turbidites were deposited as far east as the present central Bermuda Rise, up to 1,200 km from the edge of the continental shelf (the topographic swell forming the Bermuda Rise had not yet formed at this time). The turbidites are characterized by a high biogenic-opal content that includes radiolarians, diatoms, and sponge spicules. High carbonate content occurs locally in the turbidites (Plate 10, I). This silica-rich sedimentary sequence is defined as the Bermuda Rise Formation (Plate 1; Jansa and others, 1979).

Eocene downslope sedimentation along the outer U.S. Atlantic margin probably followed patterns established during Paleocene sea-level lowstands when turbidity currents first reached the area of the present Bermuda Rise. However, it is likely that the turbidites deposited in the deep basin were derived mainly from the continental rise near or below the ambient CCD at 4 km (Plate 1). If they had originated in shallower water (e.g. continental "slope") they should have consistently higher carbonate contents than are observed. Consequently, the sediment mass movements noted above on the continental slope appear to have been rather localized phenomena, although they may have been important in loading the continental rise and thus triggering deeper-water slides, debris flows, and turbidity currents.

Uplift forming the Bermuda Rise began in the middle Eocene, and it was accompanied by development of the Bermuda pedestal. By early late Eocene time the volcanic pedestal had reached sea level and it was being subaerially eroded and shedding volcaniclastic turbidites to the adjacent basin; the erosion continued into the middle Oligocene (Tucholke and Vogt, 1979). On the Bermuda Rise the widespread deposition of turbidites derived from the continental margin ceased in the late middle Eocene as the Rise was uplifted (Tucholke and Mountain, 1979). The top of the turbidite sequence is marked by a distinct seismic reflector, Horizon A^T. This reflector may be diachronous (Premoli-Silva and Boersma, this volume), becoming younger toward the west as the rise was uplifted, but presently available stratigraphic control in DSDP drillsites is not adequate to evaluate this possibility fully. West of the Bermuda Rise, there is no stratigraphic control to determine whether or not the cessation of turbidite deposition was part of a general, margin-wide phenomenon caused by changes in the continental-margin source area. However, as already noted, there is some evidence that the continental slope became more stable in the middle and late Eocene.

Outside the areas influenced by turbidity currents, the deep western Atlantic basin accumulated clays throughout the Eocene. The CCD probably was near 4 km (Plate 1), and calcareous sediments were deposited at shallower depths along the Mid-Atlantic Ridge (Plate 10, I). Like the sediments along the continental margin, these shales and carbonates contain a significant biosiliceous component in the upper lower to lower middle Eocene section. The biogenic silica is most abundant in the western

North Atlantic basin (Plate 10, I), and recrystallization of the silica has formed widespread beds of porcelanitic chert (Reich and von Rad, 1979). The top of the cherts forms the seismic reflector, Horizon A^c (Tucholke and Mountain, 1979), which generally correlates with the upper boundary of the Bermuda Rise Formation throughout all facies in the deep basin.

The origin of the silica-rich sediments in the Bermuda Rise Formation and in time-equivalent facies along the shallow continental margin has received considerable attention in the literature, but it still is not clearly understood. Gibson and Towe (1971) noted that volcanogenic components are common in, and generally restricted to, sediments of this age along the continental margin and that these components appear to become less common northward along the margin. They postulated that the volcanic material was derived from the Caribbean and dispersed northward by winds, the Antilles Current, and the Gulf Stream. The volcanic detritus presumably would have increased dissolved silica and nutrient phosphorous in the water column, and it therefore may have stimulated surface-water productivity of siliceous organisms and enhanced preservation of their tests in the seafloor sediments. This idea was corroborated by Mattson and Pessagno (1971), who pointed out that widespread early and middle Eocene volcanism occurred in the Caribbean. Recent deep-sea drilling at Sites 605 and 613 on the uppermost continental rise (Eocene continental slope) off New Jersey recovered a middle Eocene ash bed (Poag, 1985), and the most likely source for this ash was the Caribbean. However, extensive Caribbean volcanism also occurred during the Late Cretaceous and Paleocene (Mattson, 1984), so stratigraphic restriction of the most silica-rich sediments to the Eocene cannot be explained solely by temporal patterns of volcanism.

The timing of development of the Gulf Stream may help explain the stratigraphic distribution of biogenic silica. Pinet and Popenoe (1985) interpreted the seismic stratigraphy of the Blake Plateau and found that development of strong Gulf Stream flow began in the late Paleocene to early Eocene; prior to that time a flow exited eastward from the Gulf of Mexico through the Suwanee Strait across northern Florida. We are not certain what initiated the Gulf Stream flow, but it is possible that uplift associated with collision tectonics between Cuba and the Bahamas and/or in the region of central America (Mattson, 1984) restricted Atlantic-to-Pacific circumglobal flow and thus diverted surface water into the western Atlantic boundary currents. A combination of surface circulation patterns and Caribbean volcanism (providing silica and bio-limiting nutrients) explains several observations about the early to middle Eocene biosiliceous episode: 1) silica-rich sediments are most concentrated in the central-Atlantic gyre centered over the western basins, 2) the onset of the episode coincides with development of strong Gulf Stream flow, and 3) biosiliceous sedimentation in the western Atlantic basin generally declined after the middle Eocene. The last observation correlates with a sharp reduction in circum-Caribbean volcanism as plate-kinematic patterns shifted (Fig. 1); generally north-south underthrusting along the northern and southern Caribbean plate boundaries was replaced by westward underthrusting to form the Lesser Antilles (Mattson, 1984).

There is limited evidence for abyssal circulation during the Eocene (Plate 10, I). Sediment accumulation rates were unusually low in the upper Paleocene and some of the lower Eocene sections cored at DSDP sites in the western North Atlantic. The low rates are consistent with the fine grain size and general lack of biogenic components in these pelagic shales, so it is possible that the sedimentary sections are mostly continuous. However, even minor erosion can create significant hiatuses in such slowly accumulating sediments. Several hiatuses are thought to be present in this section and others may exist undetected (Fig. 2). Marked sediment thickness variations occur locally in the Horizon A^*-A^c interval (Maestrichtian-middle Eocene) on the present southern Bermuda Rise and they probably were caused by current-controlled deposition (Tucholke and Mountain, 1979). Recent work in this area has identified an angular seismic unconformity that can be traced to DSDP Hole 417A and correlated to an unconformity between the Upper Cretaceous and middle Eocene sections (Mountain and others, 1985). The erosion appears to be current-controlled, but its exact timing and the origin of the deep currents are still poorly understood.

In the Labrador Sea, the apparently Eocene sections at the base of two sediment drifts are unusually thick and locally unconformable to the underlying basement topography; both of these sequences could have been deposited from abyssal currents. The anomalous thickness variations are observed below reflector R4 in profile J, Plate 6 (Gloria Drift) and below a possibly correlative reflector beneath Eirik Ridge (profile I, Plate 6). Because abyssal currents tend to bear to the right in the northern hemisphere and to follow bathymetric contours, the likely source area for bottom currents over the Eirik and Gloria drifts in the Eocene was the nascent Norwegian-Greenland Sea (Plate 10, I; see Plates 2 and 10, K for drift locations).

Berggren and Hollister (1974) invoked the development of abyssal circulation and correlative upwelling to help explain the early to middle Eocene episode of biosiliceous productivity discussed earlier. However, it is not clear that the deep circulation would have had any significant effect on upwelling or productivity. Furthermore, abyssal currents show their best development *after* the middle Eocene (see below). Therefore deep circulation is not in itself a sufficient explanation for the enhanced biosiliceous sedimentation which occurred in early and middle Eocene time.

EARLY OLIGOCENE

Like the Paleocene, the Oligocene of the western North Atlantic is not well represented in available drillhole and outcrop samples, especially near the continental margins. This results from the facts that 1) the continental shelf was deeply eroded during sea-level lowstands (Fig. 1), and 2) the continental slope and rise were excavated by abyssal contour-following currents, as discussed below.

At the beginning of the Oligocene global temperatures became significantly cooler and substantial continental ice caps probably were present (Miller and Fairbanks, 1985). A cooling trend extending from middle Eocene to early Oligocene time is well documented from oxygen-isotope, planktonic faunal and floral, and terrestrial floral records (see summaries by Savin, 1977, and Miller and Tucholke, 1983). In the North Atlantic a marked shift from agglutinated to calcareous benthic foraminifera occurred between the middle Eocene and early Oligocene, but bottom-water temperatures dropped most sharply in the earliest Oligocene (Miller and others, 1985a). The temperature drop at the Eocene/Oligocene boundary is interpreted to have been caused by development of northern sources of bottom water for the North Atlantic basin (Miller and Tucholke, 1983).

The seismic- and rock-stratigraphic records of the North Atlantic corroborate this interpretation. In the northern North Atlantic a widespread seismic reflector termed R4 (Roberts, 1975) correlates in time with the $\delta^{18}O$ increase near the Eocene/Oligocene boundary. Along the margins of the basins this reflector is an unconformity that appears to have been eroded by contour-following bottom currents (Plate 10, J). Away from the margins, where the abyssal currents were less intense, reflector R4 marks the base of a sedimentary sequence characterized by obvious indicators of current-controlled deposition. These indicators include lensing, migrating sediment waves, thick unconformable sequences of non-laminated sediments, and locally chaotic reflectors that possibly represent erosional rills (Miller and Tucholke, 1983). Such features are best developed in the sediment drifts of the northern North Atlantic (e.g. Plate 6, Profiles I, J).

Along the continental margins of the western North Atlantic, the Eocene/Oligocene boundary is contained within a long-ranging hiatus that is observed beneath the entire continental slope and rise (Fig. 2). The correlative seismic boundary is Horizon A^U, a seismic unconformity that clearly cuts into older reflecting horizons (Plate 6, Profiles N-P). Because the unconformity is distributed along the margin of the basin, there is little doubt that the widespread erosion was accomplished by contour-following bottom currents (Plate 10, J). The erosion reached its deepest stratigraphic level (~Neocomian) seaward of Laurentian Channel, beneath the present Blake Outer Ridge, and along the northwestern edge of the Bahama Banks (Fig. 4). We are uncertain why the stratigraphic depth of erosion was less between Georges Bank and Cape Hatteras, but three factors probably were important. First, southerly flow of the abyssal currents in this region was opposed by northerly flow at the base of the overlying Gulf Stream, thus reducing the intensity of the abyssal currents. Second, a wide, thick prism of continental rise sediments had accumulated along this part of the margin in Cretaceous to Eocene time (Mountain and Tucholke, 1985). Consequently much thicker sedimentary sequences had to be eroded to reach a given stratigraphic level. Finally, the same thick accumulation of rise sediments caused the seafloor gradients between Cape Hatteras and Georges Bank to be lower than those to either the north or the south; the abyssal current thus covered a much wider area of seafloor over its depth range, and for a given volume transport the current intensity would have been reduced.

Local seafloor erosion of comparable age also occurred along the margins of the eastern North Atlantic (Plate 10, J). However, this erosion was much more restricted in scope and the general intensity of abyssal currents in the eastern basins must have been significantly lower than that in the western North Atlantic.

The distribution of unconformities near the Eocene/Oligocene boundary in the North Atlantic indicates a northerly source for the bottom water (Plate 10, J). As already noted, there are faunal, isotopic, and seismic indications for gradually developing deep circulation in the Eocene, even though the erosion that occurred near the Eocene/Oligocene boundary has the appearance of a geologically sudden event. This abruptness may be more apparent than real, because the sedimentary record of any precursor phenomena was eroded away in most areas drilled to date. In any case, the development of the abyssal currents was rapid enough and intense enough that it is inviting to link the establishment of strong deep circulation to a tectonic cause. The opening of a deep connection between the North Atlantic and the Arctic Ocean is the most plausible explanation to date (Miller and Tucholke, 1983). Plate-kinematic models indicate that relative motion between Spitsbergen and Greenland changed from strike-slip to extension about the time of magnetic anomaly 13 (early Oligocene; e.g. Srivastava and Tapscott, this volume). This is thought to have allowed cool, deep Arctic water to enter the Norwegian-Greenland Sea and, by flowing over the Greenland-Scotland Ridge, to flood the floor of the North Atlantic Ocean (Plate 10, J).

There are two significant uncertainties about this tectonic link. First, finite-difference poles of plate rotation have been calculated on the "golden spike" of anomaly 13 purely as a matter of expedience, so it is uncertain how close to anomaly 13 the actual change to extension between Greenland and Spitsbergen occurred. It is possible, for example, that the extension began in the latter part of the Eocene and that progressive widening and deepening of this tectonic gap continued into the early Oligocene. The slowly increasing volume transport of deep and bottom water through the gap could explain the middle and late Eocene shifts in isotopic and species composition of benthic foraminifera noted earlier. Second, there still remain questions about the locations and depths of marine connections across the Greenland-Scotland Ridge. Terrestrial and marine faunal and floral patterns indicate that such connections existed between the North Atlantic and Norwegian-Greenland Sea by middle to late Eocene time (McKenna, 1983; also summarized by Miller and Tucholke, 1983), although the exact locations of the passages are unclear. Subsidence models of the Greenland-Scotland Ridge are not well constrained because the age, crustal structure, and origin of various parts of the ridge are poorly known. Model estimates for the first appearance of marine passages range from late Eocene (Vogt, 1972) to late Oligocene (Talwani and Udintsev, 1976). Following the arguments presented by Miller and Tucholke (1983), we suggest that early Oligocene marine connections existed at the southeastern end of the Greenland-Scotland Ridge, and possibly also across the northwestern end of the Ridge near Greenland (Plate 10, J).

Although the principal source of bottom water is thought to have been the Arctic, via the Norwegian-Greenland Sea, it is also possible that some bottom water was formed in the Labrador Sea and Baffin Bay (Miller and Tucholke, 1983). Modern formation of deep and intermediate North Atlantic water occurs in the Norwegian-Greenland and Labrador seas during winter convection of high-salinity water that is advected from subtropical areas (Worthington, 1976). During the late Eocene and early Oligocene, the combined conditions of cooling global climate and northward advection of saline water by the Gulf Stream (Plate 10, J) could have contributed to similar high-latitude production of deep and bottom water.

There is an apparent coincidence in time between the first global development of a deep, cold thermohaline circulation ("psychrosphere"; Kennett and Shackleton, 1976) and the initiation of strong, widespread, deep circulation in the North Atlantic. Johnson (1983) pointed out that modern production of Antarctic Bottom Water near Antarctica is linked by a "reinforcing teleconnection" to the production of North Atlantic Deep Water in the Norwegian-Greenland and Labrador Seas. The high-salinity North Atlantic Deep Water reaching circum-Antarctic areas such as the Weddell Sea mixes with cold ambient waters to form Antarctic Bottom Water, which in turn sinks and floods the sea floor in most of the world's ocean basins. It is possible that the same process operated at the Eocene/Oligocene boundary and that the development of the psychrosphere is therefore a direct result of the tectonic evolution of the northern North Atlantic (Johnson, 1983; Tucholke and Miller, 1983). If the volume transport and southward penetration of Arctic-derived bottom water increased gradually (over several million years), rather than suddenly, in response to tectonic enlargement of the Greenland-Spitsbergen passage, then the volume production of the early Antarctic-derived bottom water probably also would have increased gradually. Furthermore, if the Antarctic-derived bottom water was produced principally in the Atlantic sector of the circum-Antarctic as it is today, then basins near the Antarctic source (e.g. Cape Basin and Indian Ocean) would have witnessed the development of the psychrosphere earlier than remote basins (e.g. North Pacific Ocean). Thus it may be possible to test the concept of North Atlantic control on development of the psychrosphere by careful time-stratigraphic analysis of changes in the isotopic composition of benthic foraminifera with respect to their locations along the probable dispersal paths of the Arctic and Antarctic-derived bottom waters.

In the abyssal North Atlantic outside the path of the deep boundary current, the Oligocene seafloor accumulated calcareous oozes above the ambient CCD (~4.5 km) and low-carbonate clays below that level (Plate 10, J). The clays included sediment eroded by currents from the continental margin and advected to the interior of the western North Atlantic basin. The total area of seafloor eroded by bottom currents between the Greenland - Scotland Ridge and 5°N was about 4.5 to 5×10^6 km^2 (Plate 10, J). Average depth of erosion is difficult to estimate, but available seismic and borehole data suggest values of a few meters to several hundred meters. At an average erosion depth of 100 m, up to 0.5×10^6 km^2 of sediment were removed from the basin margins. The abyssal boundary current probably exported much of this material from the North Atlantic to depositional sites in the South Atlantic. The Argentine Basin is one possible locus of deposition; it contains large volumes of acoustically non-laminated, current-deposited beds, including a significant section of rapidly deposited Oligocene sediments (~50 m/my; Supko and Perch-Nielsen, 1977).

The oldest sediments thus far cored above Horizon A^U in DSDP sites along the North American continental margin are lower Miocene at DSDP Sites 104, 106, 391, 534, and 603 (Fig. 2; Plate 10, K). However, these sites did not sample the oldest beds, which are present above the unconformity in other locations, notably beneath the central continental rise off New Jersey and beneath the central and western Blake Outer Ridge. Seismic-stratigraphic studies show that more than 2 km of lower Miocene and older sediments overlie Horizon A^U in these areas (Fig. 5). In reflection profiles across the western Blake Outer Ridge and locally beneath the upper continental rise the oldest beds above Horizon A^U appear to be chaotically deposited sediments. These deposits are thought to have been emplaced by sediment mass movements that occurred when bottom currents eroded and oversteepened the continental margin; consequently the sequences may be as old as lower Oligocene (Mountain and Tucholke, 1985). Farther seaward, the oldest beds above Horizon A^U commonly are seismically non-laminated to well laminated and they may have been deposited from turbidity currents and/or contour currents, possibly under plane-bed flow conditions. We have no direct information on the age of these beds, but we can make an estimate. For example, DSDP Site 104 demonstrated that middle Miocene sediments accumulated at rates up to 190 m/m.y. on the central Blake Outer Ridge (Plate 10, K); if we apply this rate to the ~2 km, undrilled sedimentary sequence between the bottom of the borehole and Horizon A^U, an early late Oligocene age (ca. 28 Ma) is suggested for beds immediately above the unconformity.

The continental shelves of the North Atlantic were subjected to widespread erosion during a major sea-level lowstand late in the early Oligocene (Fig. 1; Miller and others, 1985b). On the U.S. Atlantic shelf, this is manifested in a hiatus that spans most of the Oligocene (Poag, 1978). The subaerial and wave-base erosion of the shelf delivered large quantities of sediment to the western Atlantic basin; however, it appears that most of this sediment bypassed the continental slope because there are modern widespread exposures of Eocene, Miocene, and Pliocene sediments on the slope but virtually no exposures of Oligocene strata (Weed and others, 1974; Tucholke and Raymond, in prep.). In places, this bypassing occurred in turbidity currents and other mass wasting processes, both on the continental slope and upper

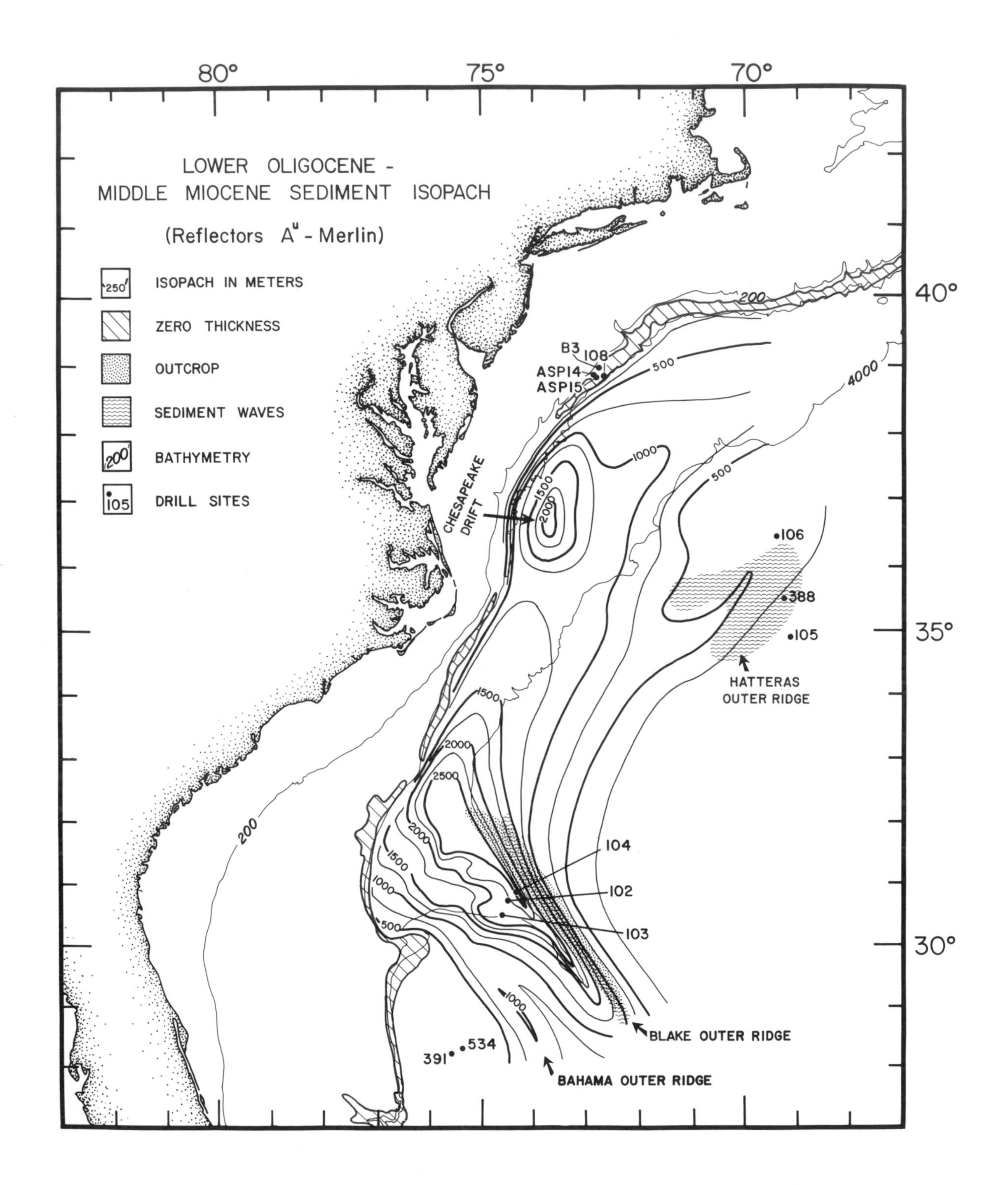

Figure 5. Thickness, in meters, of lower(?) Oligocene to upper middle Miocene sediments along the U.S. Atlantic continental margin; adapted from Mountain and Tucholke (1985). Modern outcrops seaward of the continental slope and distribution of sediment waves within this stratigraphic interval are shown. In this figure and Figures 7 and 8, zones of zero thickness along the continental slope are generalized from Tucholke and Raymond (in prep.) and slope outcrops are not otherwise indicated; these zones locally may have several tens of meters of younger overburden.

rise. For example, well developed submarine canyons probably cut by turbidity currents during the late early Oligocene are present beneath the western end of the Blake Outer Ridge (see Figs. 3 and 4 of Dillon and others, this volume). In other areas, such as the continental slope off New Jersey, sediment bypass apparently occurred with little erosion of submarine canyons; this, however, is difficult to document because later erosion by bottom currents and turbidity currents removed or overprinted most Oligocene features (Fig. 3).

Erosion of the continental shelf and slope related to the mid-Oligocene sea-level lowstand created an unconformity that re-excavated Horizon A^U on the continental slope. Thus, Horizon A^U and the shelf unconformity can be traced more or less continuously as one seismic discontinuity, even though the onshelf and offshelf unconformities were created by different processes and at different times. Taken together, the early Oligocene erosion by abyssal currents, the subsequent mass wasting, and the mid-Oligocene erosion related to the sea-level lowstand cut several kilometers into the pre-existing continental slope, which had maintained a nearly constant position from late Jurassic time. Subsequent erosion of the slope and the form of later deposition on the outer shelf (both discussed below) caused this "retreat" of the continental slope to continue into the Neogene. Compared to its Cretaceous to early Paleogene position, the setback of the shelf-edge now varies from about 5 to 30 km, with the largest retreat having occurred in the vicinity of Baltimore Canyon Trough (Schlee and others, 1979).

LATE OLIGOCENE-LATE PLIOCENE

Lithofacies

A significant change in the style of sedimentation in the western North Atlantic, in both deep and shallow water, occurred after the early Oligocene. The North Atlantic had become a more complex system because global climate had cooled significantly, glaciation was occurring, high-latitude sources of bottom water were in full production, deep-water pathways were open to abyssal currents, and sea-level fluctuations were large (Fig. 1).

On the U.S. Atlantic continental shelf, lower Paleogene calcareous deposits gave way to more siliciclastic sedimentation in the late Oligocene. South of Cape Hatteras the core of the Gulf Stream migrated seaward and landward with changes in eustatic sea level, causing significant erosion across the Blake Plateau (Pinet and Popenoe, 1985). Preserved strata were typically silty clays, occasionally phosphatic, and enriched in biosiliceous or calcareous fossils. North of Cape Hatteras and continuing along the Canadian margin, thicker shelf deposits accumulated outside the influence of the Gulf Stream. Glauconitic sands and biosiliceous debris were common in this area, but calcareous fossils were rarely preserved. The contrast with older Paleogene deposits is clearly expressed in the COST B2 and B3 wells in Baltimore Canyon trough, where upper Oligocene sands and silty claystones rest unconformably on upper Eocene chalks and calcareous mudstones (Poag, 1980).

Upper Oligocene sediments in the deep western and northern North Atlantic basins are not well sampled. The existing record, principally from DSDP Sites 9, 10, 386, and 387 (western N. Atlantic) and 112, 116, 117, 403 and 406 (northern North Atlantic), suggests a depth-dependent distribution of shales, marls and calcareous oozes like that of the early Oligocene (Plate 10, J). The gray-green and brown shales of Oligocene and younger ages in the western Atlantic basin have been defined as the Blake Ridge Formation by Jansa and others (1979; Plate 1).

Relatively little is known about lower and middle Miocene sediments beneath the continental shelf, slope, and rise of eastern North America. Samples recovered in COST, AMCOR, ASP, and DSDP wells along the U.S. Atlantic shelf and slope are largely terrigenous sands, silts, and shales that are locally glauconitic, diatomaceous, or calcareous (Hathaway and others, 1979; Poag, 1980; 1985). Middle Miocene sediments recovered from DSDP Sites 103 and 104 on the Blake Outer Ridge and Sites 106 and 603 on the lower continental rise are gray-green mudstones with variable contents of biosiliceous debris, and at Site 106 the sediments contain abundant quartz and feldspar (Jansa and others, 1979). Lower to middle Miocene pollen assemblages in the Blake Outer Ridge sediments suggest a proximal, temperate source area (Hollister and Ewing, 1972).

Eustatic sea level was generally rising during the early and middle Miocene (Fig. 1), and seismic reflection profiles show that a delta prograded seaward across the continental shelf off southern New Jersey at this time, reaching its maximum seaward extent in the late Miocene (Schlee, 1981). Progradation presumed to be of a similar age has been observed along the Nova Scotian continental shelf (Swift, 1985). Some of these sigmoidal beds reached the shelf edge (e.g. off the mouth of Chesapeake Bay), thus physically building or amplifying the physiographic expression of the continental slope. The modern result is that Miocene exposures are relatively widespread along the upper part of the U.S. Atlantic continental slope (Tucholke and Raymond, in prep.). Seabed sampling and seismic reflection profiles show that much of the lower slope has no Miocene sediments; these areas were bypassed by the shelf-derived sediments, or else the sediments were later removed by mass movements or contour-following bottom currents (Poag, 1985; van Hinte and Wise, 1986).

Upper Miocene sediments have been cored at numerous sites along the western Atlantic continental rise (Fig. 2; Plate 10, K). These are mostly gray-green mudstones with a variable content of terrigenous silt and sand, carbonate content less than about 20%, and a notable rarity of siliceous microfossils.

The Blake-Bahama Basin accumulated an unusual sequence of intraclastic chalks during the Miocene (Plate 10, K). These have been cored at DSDP Sites 391 and 534, and they form the Great Abaco Member of the Blake Ridge Formation (Jansa and others, 1979; Sheridan and Gradstein, 1983). The chalks contain both pelagic (including biosiliceous) components and abundant shallow-water debris derived from the adjacent carbonate banks.

They were deposited by massive debris flows and turbidity currents that probably originated in nearby Great Abaco Canyon and Northeast Providence Channel.

Miocene sediments in the deep western North Atlantic basin are mostly gray-green and brownish shales. The level of the CCD during the early and middle Miocene is not well known, but probably was between 3.5 and 4.5 km (Plate 1); the CCD during late Miocene time deepened to 4.5-5 km. Calcareous oozes accumulated above this level in areas not influenced by significant input of terrestrial detritus (e.g. Yucatan, Blake, and Bahama platforms; Mid-Atlantic Ridge; Plate 10, K).

Pliocene sedimentation was generally similar to that of the late Miocene. Pliocene sediments of the continental shelf are poorly sampled but consist mainly of silty shales and sands in existing cores. The Pliocene CCD was near 5 km, and gray-green and brown shales and silty shales are characteristic of the deep basin below this depth (Plate 1; Jansa and others, 1979).

Depositional Patterns

Following the latest Eocene to early Oligocene episode of strong sea-floor erosion, the intensity of North Atlantic abyssal boundary currents appears to have decreased. This is deduced from the fact that in the northern North Atlantic, upper Oligocene sediments overlie the reflector R4 unconformity at several DSDP sites in near-margin locations (Miller and Tucholke, 1983); the boundary currents should have been most erosive along the steeper slopes in such areas. As noted earlier, possibly lower Oligocene hummocky sediments and upper Oligocene laminated sediments occur above the Horizon A^U unconformity beneath the upper and central U.S. Atlantic continental rise.

Current-controlled deposition was important throughout the late Oligocene and Neogene in the western North Atlantic basin. Deposition seaward of the U.S. Atlantic shelf was concentrated on the central and upper continental rise, and on the Blake Outer Ridge (Fig. 5). For the Oligocene to middle Miocene, along-margin elongation of sediment isopachs clearly demonstrates the importance of contour-following bottom currents in controlling depositional patterns. There were two principal depocenters, one off Baltimore Canyon trough (Chesapeake Drift) and a second forming the Blake-Bahama Outer Ridge system. The thick sediments that accumulated in the Chesapeake Drift suggest significant sediment input from the adjacent margin; the prograding shelf delta noted earlier probably was an important source of sediments in this area. The Blake Ridge appears to have accreted in the zone of flow crossover between the Gulf Stream and the southward-flowing abyssal boundary current (Bryan, 1970). Much of this sediment could have been derived from Gulf Stream erosion of the adjacent Blake Plateau (Pinet and Popenoe, 1985), thereby explaining the proximal, temperate source of sediments recovered in DSDP boreholes on the outer ridge (Hollister and Ewing, 1972). Thus the continental shelf and Blake Plateau supplied much of the sediment deposited on the continental rise, but deep currents were important in reworking and redistributing this detritus.

Additional evidence for current control on deposition along the continental rise is found in the seismic facies of the Oligocene to middle Miocene section. Migrating sediment waves were formed along the north flank of the Blake Outer Ridge and on the juvenile Hatteras Outer Ridge (Fig. 5). Landward of these bedforms, hyperbolic and shingled reflectors are widespread and they suggest intermittent erosion and differential deposition, respectively (Mountain and Tucholke, 1985). At least one seismic discontinuity, showing local truncation and onlap of reflectors, interrupts this depositional sequence (bold arrow, Fig. 6; see also Tucholke and Laine, 1982). This suggests that brief erosional pulses of bottom currents occurred in the late Oligocene to middle Miocene, but the ages of most such events presently are not well constrained along the North American continental margin.

Upper Oligocene to middle Miocene sedimentary sequences drilled at DSDP sites in the northern North Atlantic, especially near basin margins, are commonly abbreviated or interrupted by long-ranging hiatuses (Fig. 2). The correlative seismic reflection sequences show contorted reflectors, seismic discontinuities, bed lensing, and local sediment waves, all of which suggest current-controlled erosion and deposition. A widespread reflector termed R2 in the northeastern Atlantic (Rockall Trough, South Iceland Basin) dates to the late early Miocene and in many areas appears to correlate with a current-eroded unconformity (Figs. 1 and 2; Miller and Tucholke, 1983). Counterpart unconformities may exist in the Labrador Sea (Site 112, Fig. 2) and along the U.S. Atlantic margin (e.g. bold arrow, Fig. 6). A second, widespread erosional pulse that is documented in DSDP boreholes occurred during the late middle Miocene (Fig. 2). This erosion affected both the northeastern Atlantic and Laborador-Irminger Sea basins, and it correlates in time with erosion on the U.S. Atlantic margin (reflector Merlin, Fig. 1).

Beginning in the middle Miocene, indicators of current-controlled deposition in the sedimentary record changed significantly, from rather irregular patterns of deposition and erosion to remarkably coherent sediment waves and seismically well laminated sedimentary beds (Fig. 6). Along most of the U.S. Atlantic margin this change occurred immediately above reflector Merlin ($\sim$12 Ma; Mountain and Tucholke, 1985), but locally in both the western and northern North Atlantic it began somewhat earlier, about 18-17 Ma (above reflector R2; Miller and Tucholke, 1983). In the western North Atlantic the sediment waves became best developed in the deeper parts of the basin ($\lesssim$4 km), notably on the Hatteras and Bahama outer ridges (Figs. 6 and 7). Rapid accumulation of acoustically non-laminated or weakly laminated sediments having apparently well developed sediment waves also began on the Caicos and Greater Antilles outer ridges (Plate 6, Profile P). Margin-parallel elongation of sediment isopachs indicates that currents also controlled depositional patterns on the shallower part of the continental rise, although the Chesapeake Drift depocenter shifted seaward relative to its late Paleogene-early Neogene position (compare Figs. 5 and 7).

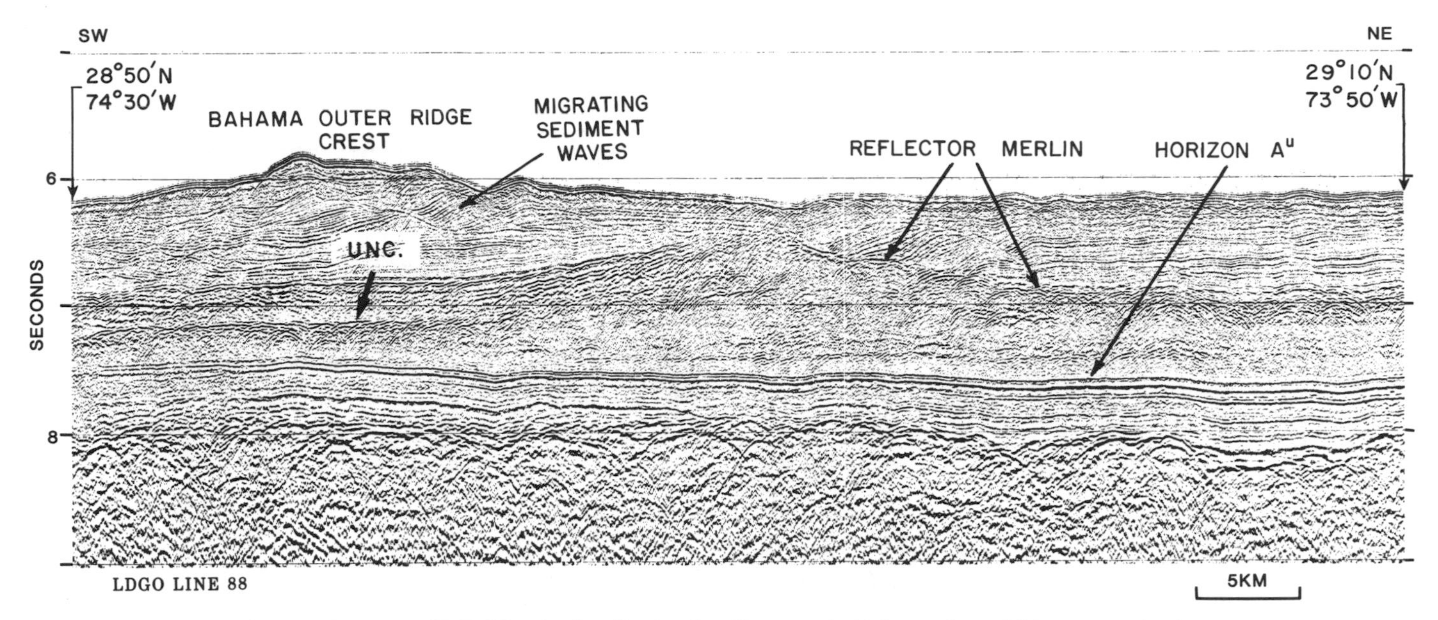

Figure 6. Segment of L-DGO seismic reflection line 88 across the Bahama Outer Ridge. Note seismic discontinuity (arrow), interpreted as an erosional unconformity, between Horizon A^U and reflector Merlin. This horizon may correlate with late early Miocene erosion (~18-17 Ma) in the northern North Atlantic. Well developed sediment waves forming the Bahama Outer Ridge post-date reflector Merlin (upper middle Miocene, ~12-11 Ma).

The same depositional patterns continued to develop along the western Atlantic margin until the late Pliocene, about 3-2 Ma, when another pulse of erosion by bottom currents occurred (Figs. 1, 2). This event is manifested seismically by a reflector termed Blue by Mountain and Tucholke (1985). Correlative unconformities in the northern North Atlantic are not well documented.

OVERVIEW OF CURRENT-CONTROLLED SEDIMENTATION

The summary presented above briefly outlines our current knowledge of how and when contour-following bottom currents affected the sedimentary record. Our understanding of what this record means in terms of specific current patterns and intensities, and how these relate to tectonic, sea-level, and climatic events, is considerably less advanced. However, by integrating all these factors and considering potential cause-and-effect relations, we can attempt a more complete interpretive framework as outlined below.

There was a significant reduction in bottom-current speeds following the early Oligocene erosional episode (Fig. 1), but the cause of the reduction is unclear. It may represent attainment of a new equilibrium in abyssal hydrography following the introduction of cold bottom water from the Arctic. If the introduction of this bottom water was prolonged by continued widening of the Greenland-Spitsbergen tectonic gap, and if the bottom-water effects were as globally significant as we have suggested, then it probably took at least several million years before the deep circulation achieved a relatively stable configuration. A further reduction in current speed could have been caused by the major mid-Oligocene sea-level lowstand (Fig. 1), which presumably would have reduced the volume transport of bottom water across the Greenland-Scotland Ridge.

During the early Oligocene episode of maximum volume transport, bottom water eroded the seafloor along the margins of the western, northeastern and locally the east-central North Atlantic basins (Plate 10, J; Fig. 2). When the volume transport subsequently was reduced, most of the bottom water probably was funnelled to the western Atlantic basins along two pathways: 1) direct overflow into the western Atlantic at the northwest end of the Greenland-Scotland Ridge, and 2) overflow across the southeast part of the Ridge; this flow entered the northeast Atlantic but quickly exited westward through Charlie-Gibbs Fracture Zone and numerous other fracture-zone conduits in the Reykjanes Ridge. This circulation pattern would have resulted in reduced ventilation of the eastern Atlantic compared to the western Atlantic, as is suggested by studies of carbon isotopes (Fig. 1; Miller and others, 1986).

As sea level rose and the Greenland-Scotland Ridge subsided in late Oligocene to early Miocene time, volume transport of bottom water across the Ridge presumably increased. By about 17 Ma (late early Miocene) this culminated in local seafloor erosion in areas where the boundary current was active, both in the northeast Atlantic north of Charlie-Gibbs Fracture Zone and in the western North Atlantic. Erosion in the western North Atlantic may have been short-lived, however (e.g. DSDP sites 603, 112, Fig. 2). It is known from detailed magnetic studies (Vogt and Avery, 1974; Voppel and Rudloff, 1980) that the numerous transform faults transecting the Reykjanes Ridge disappeared from ocean crust generated after about 15 Ma (midway

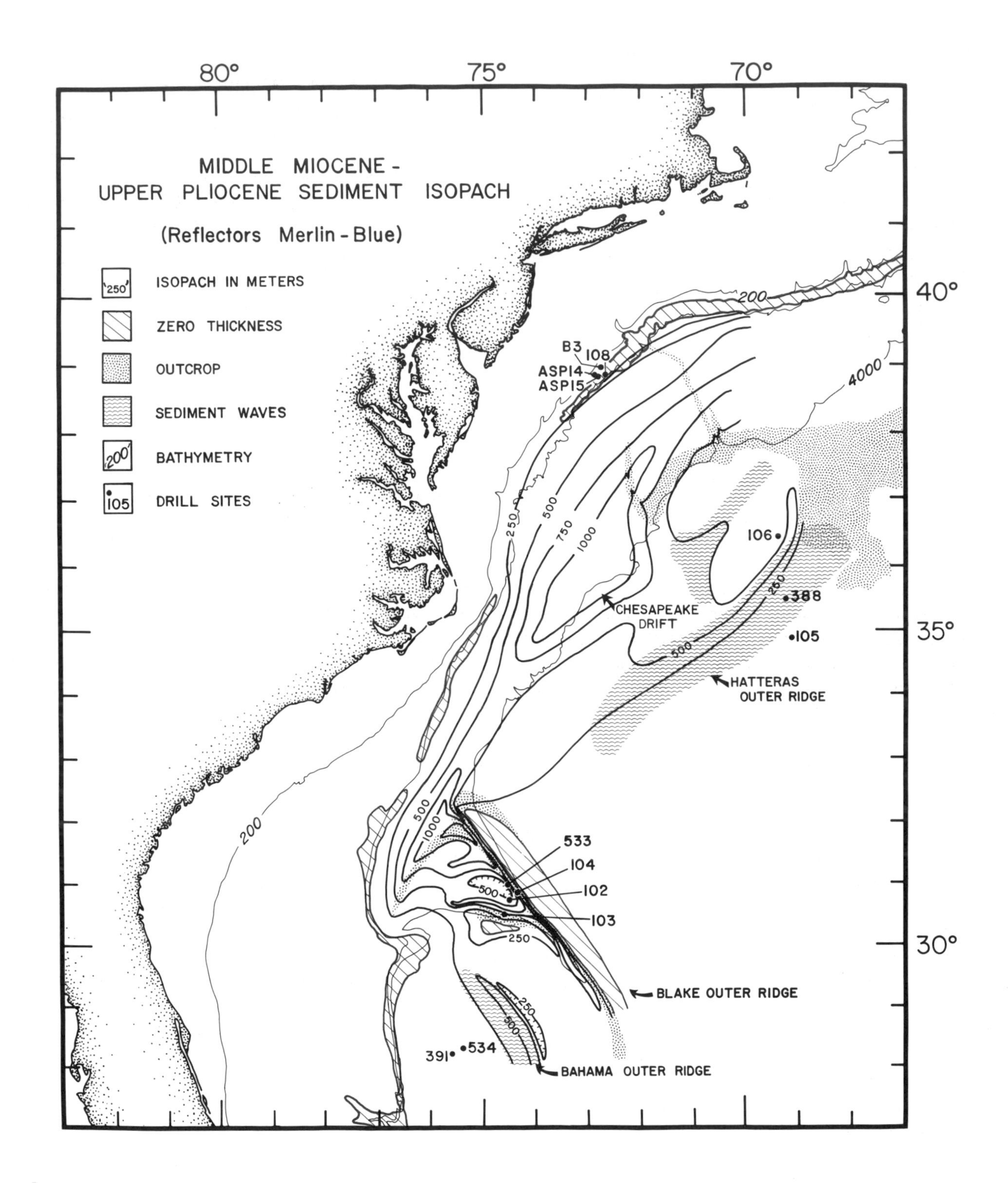

Figure 7. Thickness, in meters, of upper middle Micoene to upper Pliocene sediments along the U.S. Atlantic continental margin; adapted from Mountain and Tucholke (1985). Note that sediment waves are best developed at large seafloor depths (4-5 km) on the Hatteras and Bahama Outer Ridges (see Fig. 6).

between magnetic anomalies 6 and 5), when seafloor spreading became organized along a more continuous ridge axis. We suggest that the loss of these fracture-zone conduits caused a reduction in the volume transport of bottom water across the Reykjanes Ridge to the western basin, thereby increasing the volume transport to the eastern North Atlantic south of Charlie-Gibbs Fracture Zone. This interpretation agrees with carbon-isotope studies (Miller and others, 1986) which imply that the eastern North Atlantic once again became well ventilated by about 15 Ma (Fig. 1).

In the middle Miocene, volume transport across the Greenland-Scotland Ridge continued to increase as the Ridge subsided. The distribution of current-produced seismic facies along the U.S. Atlantic margin provides some indirect evidence for this assertion. The zone of hummocky and shingled seismic reflections migrated up the continental rise during the middle Miocene (Mountain and Tucholke, 1985). Seaward of this zone, reflectors are relatively flat with some local lensing; on the lower rise, seaward progradation of sediments also occurred above Horizon A^U. We interpret these features to mean that the thickness of the deep boundary current increased during the middle Miocene. The irregular bedforms probably were produced at the shallow edge of the current where intermittent erosion and deposition occurred. Farther seaward on the continental rise, currents probably were strong enough to create a generally erosional regime. However net deposition occurred, including progradation eastward onto the Horizon A^U unconformity, because the rate of sediment input from the continental margin exceeded the rate of erosion. By about 12-11 Ma (late middle Miocene), volume transport of bottom water had increased to the point where a pulse of widespread seafloor erosion, particularly important on the steeper slopes, occurred throughout the northeastern and western North Atlantic (Figs. 1, 2).

Why this pulse of seafloor erosion terminated is uncertain. One possibility is that volume transport of bottom water across the Greenland-Scotland Ridge was reduced because of uplift of the ridge. Seismic reflection profiles across Reykjanes Ridge (e.g. Plate 5, Profile A; Talwani and others, 1971) consistently show that between about 14-13 Ma and 10.5 Ma, the ridge-crest elevation increased by 0.8 to 1.0 km (see also Vogt, this volume, Ch. 24). This was presumably a thermal event that generated uplift in proto-Iceland and in adjacent parts of the Greenland-Scotland Ridge. Reduced volume transport across the ridge also may have been caused by eustatic sea-level lowering beginning about 11.5 Ma (Zone N13, Vail and others, 1977; Fig. 1); the magnitude of this effect, compared to that of ridge uplift, is difficult to evaluate. Reduced abyssal flow following ~11 Ma is indicated not only by re-established deposition above the current-eroded unconformity (Fig. 2), but also by carbon-isotope measurements which suggest that both the eastern and western North Atlantic again became less well ventilated (Fig. 1; Miller and others, 1986, Figs. 4, 5b).

Between about 10.5 Ma and 3-2 Ma the volume transport across the Greenland-Scotland Ridge to both the northeastern and western Atlantic basins once again increased in response to Ridge subsidence (Fig. 1). The minor episode of decreased ventilation in the North Atlantic ended by about 8 Ma (Miller and others, 1986), probably in response to this increased flow. At about 3-2 Ma, local seafloor erosion on the U.S. Atlantic continental rise again occurred, forming the seismic unconformity identified in seismic reflection profiles as reflector Blue (Mountain and Tucholke, 1985). There is some evidence in DSDP boreholes for a similar upper Pliocene unconformity in the northeastern North Atlantic, but this is not yet documented in seismic stratigraphic studies. As in the middle Miocene, an abrupt end to sea-floor erosion correlates with presumably reduced volume transport of bottom water caused by uplift of the Greenland-Scotland Ridge and by lowered sea level (Fig. 1). Elevation of the Reykjanes Ridge crest by about 0.8 km, taken as an indicator of possible uplift of the Greenland-Scotland Ridge, occurred between about 3.5 Ma and 2.0 Ma (Plate 5, Profile A; see also Talwani and others, 1971).

We have interpreted the pulses of strong abyssal circulation at about 37-33 Ma, 18-17 Ma, 12-11 Ma, and 3-2 Ma (Fig. 1) principally from seismic- and rock-stratigraphic data. Miller and Fairbanks (1985) used benthic foraminiferal $\delta^{13}C$ differences between the Atlantic and Pacific to infer abyssal Atlantic circulation changes of the Oligocene and Miocene. Although Miller and Fairbanks caution that their results are subject to change with better stratigraphic control, it is noteworthy that three of their four peak circulation episodes occur at the ages noted above. Their fourth episode, about 24-23 Ma (Oligocene/Miocene boundary), is poorly controlled, and available seismic- and rock-stratigraphic evidence is inadequate to evaluate the reality of the event.

The evolution of North Atlantic abyssal currents appears to show relatively direct correlations with known aspects of the tectonic evolution of the northern North Atlantic, and it therefore appears that tectonics may have been the primary factor in controlling patterns of bottom-water circulation. In contrast, possible cause-and-effect correlations with global climate changes are mostly uncertain, with the possible exception of the glacioeustatic sea-level changes already noted. Even where such correlations can be made, it is usually unclear whether the circulation changes were a cause or a result of climate changes, or whether both were controlled by tectonics.

By way of example, we can consider several significant climate changes that did occur during the Tertiary (Fig. 1; see also Berggren and Olsson, this volume). Global cooling during the middle Eocene to early Oligocene correlates with the establishment of deep, north-south circulation in the Atlantic and other ocean basins, but any cause-and-effect relations are unclear. The major sea-level lowstand in the latest early Oligocene appears to have been a glacio-eustatic phenomenon (Miller and others, 1985b); we currently do not have enough seismic- and rock-stratigraphic data in the deep basins to interpret whether this event had any connection with changes in abyssal circulation. A global $\delta^{18}O$ increase in the middle Miocene (~15-14 Ma) has been interpreted to represent development of widespread glaciation in Antarctica (e.g. Shackleton and Kennett, 1975) with a concomitant decrease in bottom-water temperatures (Miller and others, 1986). This cooling event appears to have had no immediate effect on the deep circulation of the North Atlantic, although it may have contributed to generally increased production of bottom water there (Fig. 1). Perhaps the most direct correlation that presently can be made between climate and abyssal circulation is the event at 3-2 Ma. Ice-rafted sediment cored at DSDP Site 552A on Rockall Plateau shows that major northern hemisphere ice-growth occurred by 2.5-2.4 Ma (Shackleton and others, 1984); associated cooling and sinking of high-latitude surface waters may have stimulated the deep North Atlantic circulation and caused the observed late Pliocene erosion of the western North Atlantic seafloor.

LATE PLIOCENE-HOLOCENE

The Quaternary geology of the North American plate, including the western North Atlantic, is dealt with in detail by a separate DNAG volume. However, it is pertinent to summarize here some geologic aspects of the "glacial" western North Atlantic.

The late Pliocene development of large glacial ice caps in Greenland, northern North America, and Eurasia had a profound effect on sedimentation patterns in the North Atlantic. Downslope sedimentary processes including turbidity currents, slumps, slides, and debris flows became important along the continental margins as large volumes of sediment were delivered to the continental slopes and rises by glacial ice and by rivers crossing the continental shelves during sea-level lowstands. Laine (1980) has estimated that more than 1×10^6 km^3 of terrestrial sediment was delivered in this manner to the western North Atlantic basins during the past 2.8 m.y.; somewhat over half of this volume was deposited by turbidity currents in the deepest parts of the basins to form the modern abyssal plains (Plate 2). Consequently, there was a major shift from continental-margin to deep-basin depocenters in the late Pliocene and Quaternary.

The locations of depocenters within the provinces of the western Atlantic continental margins also shifted. For example, on the U.S. Atlantic margin, sediments transported by downslope processes accumulated in a sedimentary basin on the lower continental rise landward of the Hatteras Outer Ridge, and on the gently dipping landward margin of Chesapeake Drift (Fig. 8). Southeast of Cape Hatteras the seafloor gradient was steeper and most terrigenous sediment bypassed this area to be deposited on the Hatteras Abyssal Plain. Major cross-rise conduits such as Hudson and Wilmington channels were cut by turbidity currents during the late Pliocene and Quaternary (Fig. 8). Whether the feeder canyons across the continental slope were also cut at this time, or whether pre-existing canyons simply were excavated more deeply is uncertain. Some large canyons, notably Oceanographer Canyon on the southern margin of Georges Bank, are known to be as old as Cretaceous, and they have experienced several cycles of filling and re-excavation (Ryan and others, 1978). In contrast, numerous smaller canyons along the continental slope are probably Pleistocene features, and even now they are undergoing headward erosion (Fig. 3; Twichell and Roberts, 1982).

Despite the strong influence of cross-slope sedimentary processes following the onset of northern-hemisphere glaciation, a clear signature of bottom-current control on sedimentation continued to be present. This is reflected in margin-parallel elongation of upper Pliocene-Quaternary sediment isopachs on the upper continental rise seaward of New Jersey and on the Hatteras Outer Ridge (Fig. 8). Outside the areas affected by direct input of terrestrial sediments, the upper Pliocene-Quaternary depocenters remained similar to those controlled by bottom currents at earlier times (e.g. the Blake Outer Ridge; compare Figs. 7 and 8). In addition, Quaternary bottom currents eroded or prevented significant deposition in areas that also had been affected previously by strong bottom-current erosion. Consequently, Miocene and Pliocene sediments now crop out along the seaward edge of Chesapeake Drift, across the northern margin of Hatteras Outer Ridge, and on the northern flank of the Blake Outer Ridge (Fig. 8). Quaternary fluctuations in bottom-current intensity still are not fully understood, but it appears that during at least some full-glacial stages, volume production of North Atlantic bottom water was reduced due to extensive sea-ice cover in sub-Arctic areas (e.g. Streeter and Shackleton, 1979; Mix and Fairbanks, 1985). Sedimentological studies and observations of sediment distribution patterns show that the Quaternary sedimentary record of the continental margin has been produced by the dynamic interaction of along-slope (bottom-current) and cross-slope (e.g. turbidity-current) processes (McCave and Tucholke, this volume).

CONCLUSIONS

The Tertiary paleoceanography of the western North Atlantic Ocean was divided into two very different kinds of regimes, approximately on either side of the Eocene/Oligocene boundary. Before late Eocene to early Oligocene time there was only limited abyssal circulation, and the basin sediments primarily record the influence of downslope sedimentation processes and varying productivity patterns in surface waters.

We now have a general perception of how the western Atlantic evolved in Paleocene and Eocene time, but many questions remain. For example, the paleoceanographic conditions leading to the late early to early middle Eocene episode of biosiliceous productivity in surface waters still are not well constrained, although many samples are now available from drill holes, piston cores, dredges, and subaerial outcrops. A rigorous quantitative evaluation of surface-water productivity patterns across the entire North Atlantic (and Pacific) for this time period would help significantly to clarify the paleoceanographic controls on this phenomenon. Another major uncertainty is in exactly how the continental slopes of the western North Atlantic developed in response to sea-level transgressions and regressions and to the correlative variations in sediment supply. How important was mass wasting to submarine canyon development and degradation

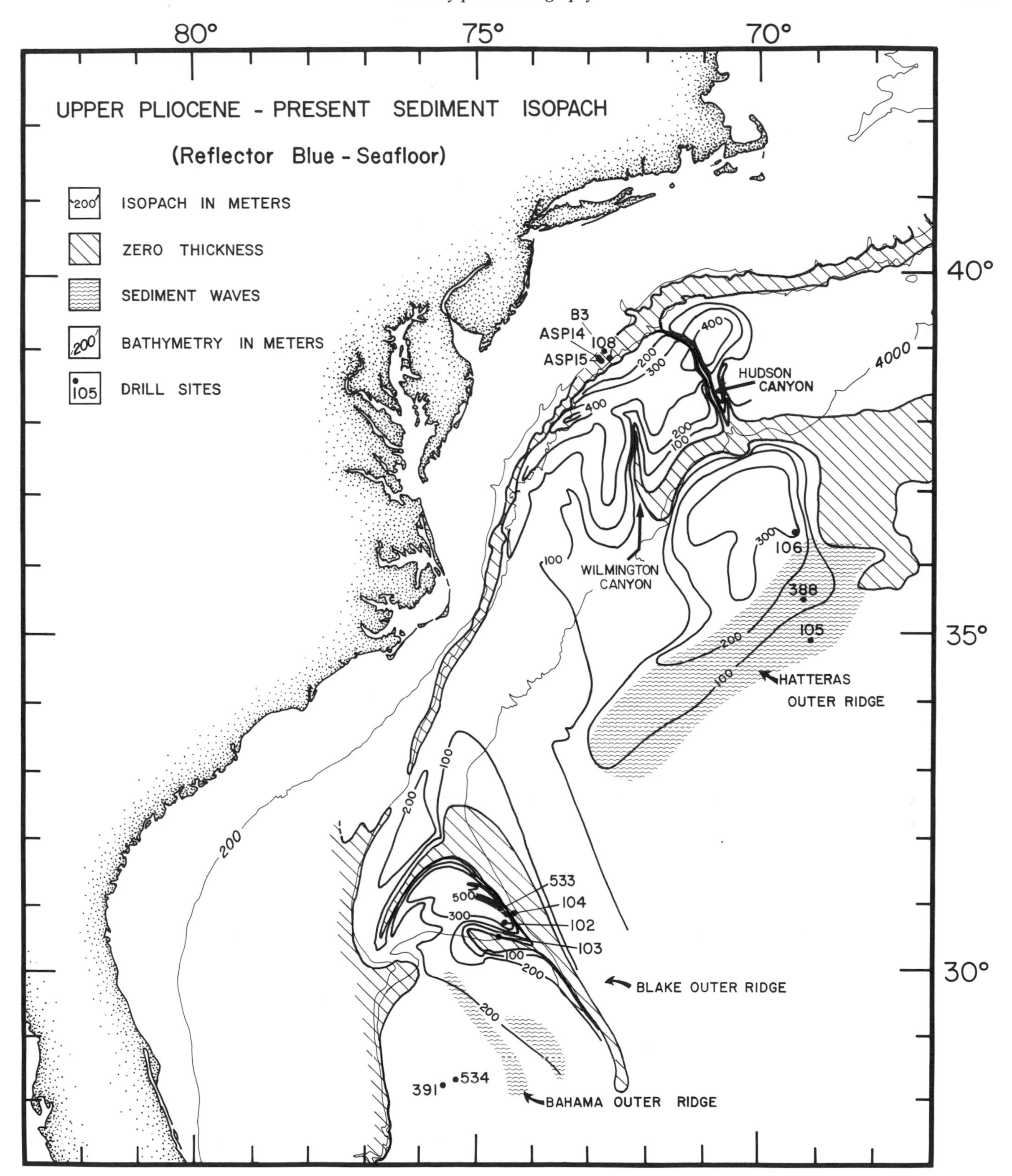

Figure 8. Thickness, in meters, of uppermost Pliocene and Quaternary sediments along the U.S. Atlantic continental margin; modified from Mountain and Tucholke (1985). Areas of zero thickness are steeper-gradient sea floor that was eroded by abyssal currents during the late Pliocene (reflector Blue) and intermittently during the Quaternary, probably during interglacial periods.

in areas where the slope was loaded by sediments? When did slope bypass and canyon erosion occur along the seaward extension of fluvial drainage systems? What was the timing of these processes relative to fluctuations in relative sea level along the margins? A third topic needing more detailed study is the relative importance of bottom currents in early Paleogene time. As already noted, there are local indications of erosion and current-controlled deposition by bottom currents in the early Eocene and possibly late Paleocene. Although these current effects were not nearly as well developed nor as widespread as later current-controlled sedimentation and erosion, their detailed documentation will be valuable in clarifying how sources of bottom-water developed for the North Atlantic.

By the early Oligocene, and continuing to the present, the deep circulation of the North Atlantic played a principal role in molding the sedimentary record. Most of the major sedimentary features of the deep North Atlantic seafloor developed during this period. We can propose some reasonable scenarios for evolution of the abyssal current system, but they still are incomplete and need to be tested. Throughout the western and northern North Atlantic detailed study of high-resolution seismic reflection profiles and correlative sediment samples, including but not restricted to DSDP cores, is needed. A careful reassessment of biostratigraphy and the application of other stratigraphic methods must be accomplished to establish a detailed time-stratigraphic framework. This is critical to the understanding of cause and effect relationships between events that are separated by only a few million years or less. Only after this time framework is more precisely defined will we be able to apply the complete spectrum of geologic data including tectonics, seismic stratigraphy, stable-isotope stratigraphy, sedimentology, and paleobiogeography of benthic fauna for paleoceanographic analysis. The result will be not only a more complete comprehension of the paleoceanographic evolution of the North Atlantic, but also a better understanding of how North Atlantic events were related to, and possibly even helped to control, the paleoceanographic development of the global oceans.

REFERENCES

Austin, J. A., Jr., Uchupi, E. Shaughnessy, D. R., III, and Ballard, R. D.
1980 : Geology of New England passive margin: American Association of Petroleum Geologists Bulletin, v. 64, p. 501–526.

Berggren, W. A., and Hollister, C. D.
1974 : Paleogeography, paleobiogeography, and the history of circulation in the Atlantic ocean; in Studies in Paleo-oceanography, ed. W. W. Hay; Society of Economic Petrologists and Mineralogists Special Publication no. 20, p. 126–186.

Bryan, G. M.
1970 : Hydrodynamic model of the Blake Outer Ridge: Journal of Geophysical Research, v. 75, p. 4530–4537.

Gibson, T. G.
1970 : Late Mesozoic-Cenozoic tectonic aspects of the Atlantic coastal margin: Geological Society of America Bulletin, v. 81, p. 1813–1822.

Gibson, T. G., and Towe, K. M.
1971 : Eocene volcanism and the origin of Horizon A: Science, v. 172, p. 152–154.

Hardy, I. A.
1975 : Lithostratigraphy of the Banquereau Formation on the Scotian Shelf; in Offshore Geology of Eastern Canada, v. 2, eds., W.J.M. van der Linden and J. A. Wade; Geological Survey of Canada Paper 74-30, p. 163–174.

Hathaway, J. C., Poag, C. W., Valentine, P. C., Miller, R. E., Schultz, D. M., Manheim, F. T., Kohout, F. A., Bothner, M. H., and Sangrey, D. A.
1979 : U.S. Geological Survey core drilling on the Atlantic continental shelf: Science, v. 206, p. 515–527.

Hollister, C. D., and Ewing, J. I., eds.
1972 : Initial Reports of the Deep Sea Drilling Project, v. 11: Washington, D.C., U.S. Government Printing Office, 1077 p.

Jansa, L. F., and Wade, J. A.
1975 : Geology of the continental margin off Nova Scotia and Newfoundland; in Offshore Geology of Eastern Canada, v. 2, eds. W.J.M. van der Linden and J. A. Wade; Geological Survey of Canada Paper 74-30, p. 51–105.

Jansa, L. F., Enos, P., Tucholke, B. E., Gradstein, F. M., and Sheridan, R. E.
1979 : Mesozoic-Cenozoic sedimentary formations of the North American Basin; Western North Atlantic; in Deep Drilling Results in the Atlantic Ocean: Continental Margins and Paleoenvironment, eds. M. Talwani, W. W. Hay and W.B.F. Ryan; American Geophysical Union Maurice Ewing Series, v. 3, p. 1–57.

Johnson, D. A.
1983 : Paleocirculation of the southwestern Atlantic; in Initial Reports of the Deep Sea Drilling Project, v. 72, eds., P. F. Barker, R. L. Carlson and D. A. Johnson; U.S. Government Printing Office, Washington, D.C., p. 977–994.

Kaneps, A. G., Doyle, P. S., and Riedel, W. R.
1981 : Further icthyolith age determinations of otherwise unfossiliferous Deep Sea Drilling cores: Micropaleontology, v. 27, p. 317–331.

Kennett, J. P., and Shackleton, N. J.
1976 : Oxygen isotopic evidence for the development of the psychrosphere 38 Myr ago: Nature, v. 260, p. 513–515.

Laine, E. P.
1980 : New evidence from beneath the western North Atlantic for the depth of glacial erosion in Greenland and North America: Quaternary Research, v. 14, p. 188–198.

Mattson, P. H.
1984 : Caribbean structural breaks and plate movements; in The Caribbean-South American Plate Boundary and Regional Tectonics, eds. W. E. Bonini, R. B. Hargraves and R. Shagam; Geological Society of America Memoir 162, p. 131–152.

Mattson, P. H. and Pessagno, E. A.
1971 : Caribbean Eocene volcanism and the extent of Horizon A: Science, v. 174, p. 138–139.

McCave, I. N.
1979 : Diagnosis of turbidites at Sites 386 and 387 by particle-counter size analysis of the silt (2-40μm) fraction; in Initial Reports of the Deep Sea Drilling Project, v. 43, eds. B. E. Tucholke and P. R. Vogt; U.S. Government Printing Office, Washington, D.C., p. 395–405.

McKenna, M. C.
1983 : Cenozoic paleogeography of North Atlantic land bridges: in Structure and Development of the Greenland-Scotland Ridge, eds. M.H.P. Bott, S. Saxov, M. Talwani and J. Thiede; Plenum Press, New York, p. 351–399.

Miller, K. G., and Fairbanks, R. G.

1985 : Oligocene to Miocene carbon isotopic cycles and abyssal circulation changes; in The Carbon Cycle and Atmospheric CO_2: Natural Variations Archean to Present, eds. E. T. Sundquist and W. S. Broecker; American Geophysical Union Geophysical Monograph 32, p. 469–486.

Miller, K. G., and Tucholke, B. E.
1983 : Development of Cenozoic abyssal circulation south of the Greenland-Scotland Ridge; in Structure and Development of the Greenland-Scotland Ridge, eds. M.H.P. Bott, S. Saxov, M. Talwani and J. Thiede; Plenum Press, New York, p. 549–589.

Miller, K. G., Curry, W. B., and Ostermann, D. R.
1985a: Late Paleogene (Eocene to Oligocene) benthic foraminiferal oceanography of the Goban Spur region, Deep Sea Drilling Project Leg 80; in Initial Reports of the Deep Sea Drilling Project, v. 80, eds. P. C. de Graciansky and C. W. Poag; U.S. Government Printing Office, Washington, D.C., p. 505–538.

Miller, K. G., Mountain, G. S., and Tucholke, B. E.
1985b: Oligocene glacio-eustasy and erosion on the margins of the North Atlantic: Geology, v. 13, p. 10–13.

Miller, K. G., Fairbanks, R. G., and Thomas, E.
1986 : Benthic foraminiferal carbon isotopic record and the development of abyssal circulation in the eastern North Atlantic; in Initial Reports of the Deep Sea Drilling Project, v. 94, eds. R. Kidd and W. F. Ruddiman; U.S. Government Printing Office, Washington, D.C. (in press).

Mix, A. C., and Fairbanks, R. G.
1985: North Atlantic surface-ocean control of Pleistocene deep-ocean circulation: Earth and Planetary Science Letters, v. 73, p. 231–243.

Mountain, G. S., and Tucholke, B. E.
1985 : Mesozoic and Cenozoic geology of the U.S. Atlantic continental slope and rise; in Geologic Evolution of the United States Atlantic Margin, ed. C. W. Poag; Van Nostrand Reinhold, New York, p. 293–341.

Mountain, G. S., Driscoll, N. W., and Miller, K. G.
1985 : Cenozoic seismic stratigraphy of the southwest Bermuda Rise: Geological Society of America Abstracts with Programs, v. 17, p. 670.

Olsson, R. K., and Wise, S. W., Jr.
1985 : The east coast sequential unconformity (ECSU): Geological Society of America Abstracts with Programs, v. 17, p. 681–682.

Pinet, P. R., and Popenoe, P.
1985 : A scenario of Mesozoic-Cenozoic ocean circulation over the Blake Plateau and its environs: Geological Society of America Bulletin, v. 96, p. 618–626.

Poag, C. W.
1978 : Stratigraphy of the Atlantic continental shelf and slope of the United States: Annual Review of Earth and Planetary Science, v. 6, p. 251–280.
1980 : Foraminiferal stratigraphy, paleoenvironments, and depositional cycles in the outer Baltimore Canyon Trough; in Geological Studies of the COST No. B-3 Well, U.S. Mid-Atlantic Continental Slope Area, ed. P. A. Scholle; U.S. Geological Survey Circular 833, p. 44–66.
1985 : Cenozoic and Upper Cretaceous sedimentary facies and depositional systems of the New Jersey slope and rise; in Geological Evolution of the United States Atlantic Margin, ed. C. W. Poag; Van Nostrand Reinhold, New York, p. 343–365.

Poag, C. W., and Hall, R.
1979 : Foraminiferal biostratigraphy, paleoecology and sediment accumulation rates; in Geological Studies of the COST No. GE-1 Well, United States South Atlantic Continental Shelf Area, ed. P. A. Scholle; U.S. Geological Survey Circular 800, p. 49–63.

Poag, C. W., and Watts, A. B., eds.
1986 : Initial Reports of the Deep Sea Drilling Project, v. 95: U.S. Government Printing Office, Washington, D.C., (in press).

Reich, V., and von Rad, U.
1979 : Eocene porcelanites and Early Cretaceous cherts from the western North Atlantic Basin; in Initial Reports of the Deep Sea Drilling Project, v. 43, eds. B. E. Tucholke and P. R. Vogt; U.S. Government Printing Office, Washington, D.C., p. 437–455.

Rhodehamel, E. C.
1977 : Lithologic descriptions; in Geologic Studies on the COST No. B-2 Well; U.S. Mid-Atlantic Outer Continental Shelf Area, ed. P. A. Scholle; U.S. Geological Survey Circular 750, p. 15–22.

Roberts, D. G.
1975 : Marine geology of the Rockall Plateau and Trough: Philosophical Transactions of the Royal Society of London, Series A, v. 278, p. 447–509.

Ryan, W.B.F., Cita, M. B., Miller, E. L., Hanselman, D., Nesterhoff, W. D., Hecker, B., and Nibbelink, M.
1978 : New England Submarine Canyons: Oceanologica Acta, v. 1, p. 233–254.

Savin, S. M.
1977 : The history of the earth's surface temperature during the last 100 million years: Annual Review of Earth and Planetary Science, v. 5, p. 319–355.

Schlee, J.
1981 : Seismic stratigraphy of Baltimore Canyon Trough: American Association of Petroleum Geologists Bulletin, v. 65, p. 26–53.

Schlee, J. S., Dillon, W. P., and Grow, J. A.
1979 : Structure of the continental slope off the eastern United States; in Geology of Continental Slopes, eds. L. J. Doyle and O. H. Pilkey; Society of Economic Paleontologists and Mineralogists Special Publication 27, p. 95–117.

Shackleton, N. J., and Kennett, J. P.
1975 : Paleotemperature history of the Cenozoic and the initiation of Antarctic glaciation: oxygen and carbon isotopic analyses in DSDP Sites 277, 279, and 281; in Initial Reports of the Deep Sea Drilling Project, v. 29, eds. J. P. Kennett and R. E. Houtz; U.S. Government Printing Office, Washington, D.C., p. 743–755.

Shackleton, N. J., and 16 others
1984 : Oxygen isotope calibration of the onset of ice-rafting and history of glaciation in the North Atlantic region: Nature, v. 307, p. 620–623.

Sheridan, R. E., and Gradstein, F. M., eds.
1983 : Initial Reports of the Deep Sea Drilling Project, v. 76: United States Government Printing Office, Washington, D.C., 974 p.

Srivastava, S. P., Falconer, R.K.H., and MacLean, B.
1981 : Labrador Sea, Davis Strait, Baffin Bay: geology and geophysics–a review; in Geology of the North Atlantic Borderlands, eds. J. W. Kerr and A. J. Fergusson; Canadian Society of Petroleum Geologists Memoir 7, p. 333–398.

Streeter, S. S., and Shackleton, N. J.
1979 : Paleocirculation of the deep North Atlantic: 150,000-year record of benthic foraminifera and oxygen-18: Science, v. 203, p. 168–171.

Supko, P. R., and Perch-Nielsen, K., eds.
1977 : Initial Reports of the Deep Sea Drilling Project, v. 39: U.S. Government Printing Office, Washington, D.C., 1139 p.

Swift, S.A.
1985 : Cenozoic geology of the continental slope and rise off western Nova Scotia [Ph.D. thesis]: Woods Hole Oceanographic Institution–Massachusetts Institute of Technology Joint Program in Oceanography, W.H.O.I. Report 85-34, 188 p.

Talwani, M., and Udintsev, G., eds.
1976 : Initial Reports of the Deep Sea Drilling Project, v. 38: U.S. Government Printing Office, Washington, D.C., 1256 p.

Talwani, M., Windisch, C. C., and Langseth, M. G.
1971 : Reykjanes Ridge crest: a detailed geophysical study: Journal of Geophysical Research, v. 76, p. 473–517.

Tucholke, B. E.
1979 : Relationships between acoustic stratigraphy and lithostratigraphy in the Western North Atlantic Basin; in Initial Reports of the Deep Sea Drilling Project, v. 43, eds. B. E. Tucholke, and P. R. Vogt; U.S. Government Printing Office, Washington, D.C., p. 827–846.

Tucholke, B. E., and Miller, K. G.
1983 : Late Paleogene abyssal circulation in the North Atlantic: American Association of Petroleum Geologists Bulletin, v. 67, p. 559.

Tucholke, B. E., and Mountain, G. S.
1979 : Seismic stratigraphy, lithostratigraphy, and paleosedimentation patterns in the North American Basin; in Deep Drilling Results in the Atlantic Ocean: Continental Margins and Paleoenvironment, eds. M. Talwani, W. Hay and W.B.F. Ryan; American Geophysical Union Maurice Ewing Series, v. 3, p. 58–86.
Tucholke, B. E., and Laine, E. P.
1982 : Neogene and Quaternary development of the lower continental rise off the central U.S. east coast; in Studies in Continental Margin Geology, eds. J. S. Watkins and C. L. Drake; American Association of Petroleum Geologists Memoir 34, p. 295–305.
Tucholke, B. E., and Vogt, P. R.
1979 : Western North Atlantic: sedimentary evolution and aspects of tectonic history; in Initial Reports of the Deep Sea Drilling Project, v. 43, eds. B. E. Tucholke and P. R. Vogt; U.S. Government Printing Office, p. 791–825.
Twichell, D. C., and Roberts, D. G.
1982 : Morphology, distribution, and development of submarine canyons on the United States Atlantic continental slope between Hudson and Baltimore canyons: Geology, v. 10, p. 408–412.
Vail, P. R., Mitchum, R. M., Jr., Todd, R. G., Widmier, J. M., Thompson, S., III, Sangree, J. B., Bubb, J. N., and Hatlelid, W. G.
1977 : Seismic stratigraphy and global changes of sea level; in Seismic Stratigraphy—Application to Hydrocarbon Exploration, ed. C. E. Payton; American Association of Petroleum Geologists Memoir 26, p. 49–212.
Vail, P. R., and Hardenbol, J.
1979 : Sea-level changes during the Tertiary: Oceanus, v. 22, no. 3, p.71–79.
van Hinte, J., and Wise, S. W., Jr., eds.
1986 : Initial Reports of the Deep Sea Drilling Project, v. 93: U.S. Government Printing Office, Washington, D.C., (in press).
Vogt, P. R.
1972 : The Faeroe-Iceland-Greenland aseismic ridge and the Western Boundary Undercurrent: Nature, v. 238, p. 79–81.
Vogt, P. R., and Avery, D. E.
1974 : Detailed magnetic surveys in the northeast Atlantic and Labrador Sea: Journal of Geophysical Research, v. 79, p. 363–389.
Voppel, D., and Rudloff, R.
1980 : On the evolution of the Reykjanes Ridge south of 60°N between 40 and 12 million years before present: Journal of Geophysics, v. 47, p. 61–66.
Weed, E.G.A., Minard, J. P., Perry, W. J., Jr., Rhodehamel, E. C., and Robbins, E. I.
1974 : Generalized pre-Pleistocene geologic map of the northern United States continental margin: U.S. Geological Survey Miscellaneous Investigations, Map I-861.
Worthington, L. V.
1976 : On the North Atlantic circulation: Johns Hopkins Oceanographic Studies No. 6, Johns Hopkins University Press, Baltimore, 110 p.

MANUSCRIPT ACCEPTED BY THE SOCIETY MARCH 31, 1986
WOODS HOLE OCEANOGRAPHIC INSTITUTION CONTRIBUTION NO. 6125
LAMONT-DOHERTY GEOLOGICAL OBSERVATORY CONTRIBUTION NO. 3988

ACKNOWLEDGMENTS

This paper was prepared with the support of a Mellon Senior Study Award to B. Tucholke and National Science Foundation Grant OCE83-10086 to G. Mountain at Lamont-Doherty Geological Observatory. The work is based in part on research supported by the National Science Foundation (Grants OCE76-02038 and OCE79-09382) and by the Office of Naval Research (Contracts N00014-79-C-0071 and N00014-80-C-0098). The manuscript was typed by P. Foster. We thank K. G. Miller for stimulating discussions about the Cenozoic paleoceanography of the North Atlantic. The manuscript was critically reviewed by K. G. Miller, C. W. Poag, E. Uchupi, and S. W. Wise, Jr., although any sins of (c)ommission are purely ours.

Printed in U.S.A.

The Geology of North America
Vol. M, The Western North Atlantic Region
The Geological Society of America, 1986

Chapter 39

Space systems as marine geologic sensors

R. J. Anderle
Naval Surface Weapons Center, Dahlgren, Virginia 22448

INTRODUCTION

Space systems that provide measurements applicable to marine geologic studies can be divided into two categories: *ground-based* observations of artificial and natural satellites and stars, and *satellite-borne* sensors that make visual, infrared, and radio-frequency measurements of the ground and ocean. The ground-based measurements provide information on plate motion and intraplate deformation, as well as long-wavelength (>1000 km) information about the earth's gravity field useful in tests of hypotheses of plate motion. The satellite-borne observations provide information on short-wavelength (<100 km) variations in the gravity field that correlate with seamounts and other geologic features, observations of terrestrial lineations and other surface features, limited information about ocean depth in shallow areas, and direct observations of sedimentary plumes and plankton in surface water. Satellite-borne instruments also yield indirect data relating to sedimentary processes and plankton productivity and dispersal through observation of ocean currents and their meanders, rings, ocean surface temperature, icebergs and sea ice, surface waves and internal waves that affect these processes. Satellites also provide a valuable service in collecting and relaying data from remote fixed stations and from drifting buoys and also determine the position of buoys so that current patterns can be determined.

GROUND-BASED OBSERVATIONS

Long-wavelength features of the earth's gravity field are determined from laser range or from radio frequency Doppler observations of artificial satellites. Doppler observations are an indirect form of range measurements, since the maximum rate of change of the Doppler measurements within a pass varies inversely with the range between the transmitter and receiver at the time of closest approach of the satellite to the station. In the case of either range or Doppler observations, the gravity field is derived in a massive, linearized, least-squares adjustment of thousands of parameters that best fit the observations. The parameters include not only coefficients of a harmonic series assumed to represent the gravity field, but also the positions of the observing stations, the constants of integration for each span of observations for each satellite, atmospheric drag and solar radiation force-scaling parameters, tidal parameters, and various instrument biases. Plate and crustal motions can also be deduced from range or Doppler measurements on satellites, as well as from laser range observations to lunar reflectors, short and medium base-line electronic phase difference measurements of artificial satellites, and very long base-line interferometer (VLBI) observations of quasars (Rogers and others, 1978). Laser range observations, which have centimeter precision for the best stations, are more accurate than the radio frequency Doppler measurements (Anderle, 1976). However, the higher cost, weather limitations, and restricted geographic distribution of the laser equipment result in nearly equivalent overall performance of laser and Doppler systems for gravity field applications. An overriding factor currently in favor of the laser systems for gravity applications is that the most complete Doppler-derived gravity fields are not generally available to the scientific community. The use of passive laser reflectors on the satellites, rather than active frequency transmitters, also favors the laser systems, although several Doppler satellites have been in virtually continuous operation for over 15 years. Laser observations made a significant contribution in the definition of the longer wave portion of the Goddard Earth Model (GEM) 10 (Lerch and others, 1979). The errors in individual coefficients in this spherical harmonic expansion of the gravity field are pessimistically estimated to range from 10% to 80% for coefficients of degrees 10 to 20 (corresponding to wavelengths of 4000 to 2000 km) by comparison with older solutions (Khan, 1983), and perhaps optimistically estimated to range from 13% to 40% for the same wavelengths from internal tests.

The space systems that offer the most promise for the accurate determination of plate motions and intraplate deformations are the lunar laser and the VLBI. The greater accuracy is possible because the large distances to the target or source limit the effects of certain model errors on base-line determinations. To date, base stations have been limited to the major plates and in many in-

Present address: General Electric Company, Systems Division, Box 8555-43A30, Philadelphia, Pennsylvania 19101.

Anderle, R. J., 1986, Space systems as marine geologic sensors, *in* Vogt, P. R., and Tucholke, B. E., eds., The Geology of North America, Volume M, The Western North Atlantic Region: Geological Society of America.

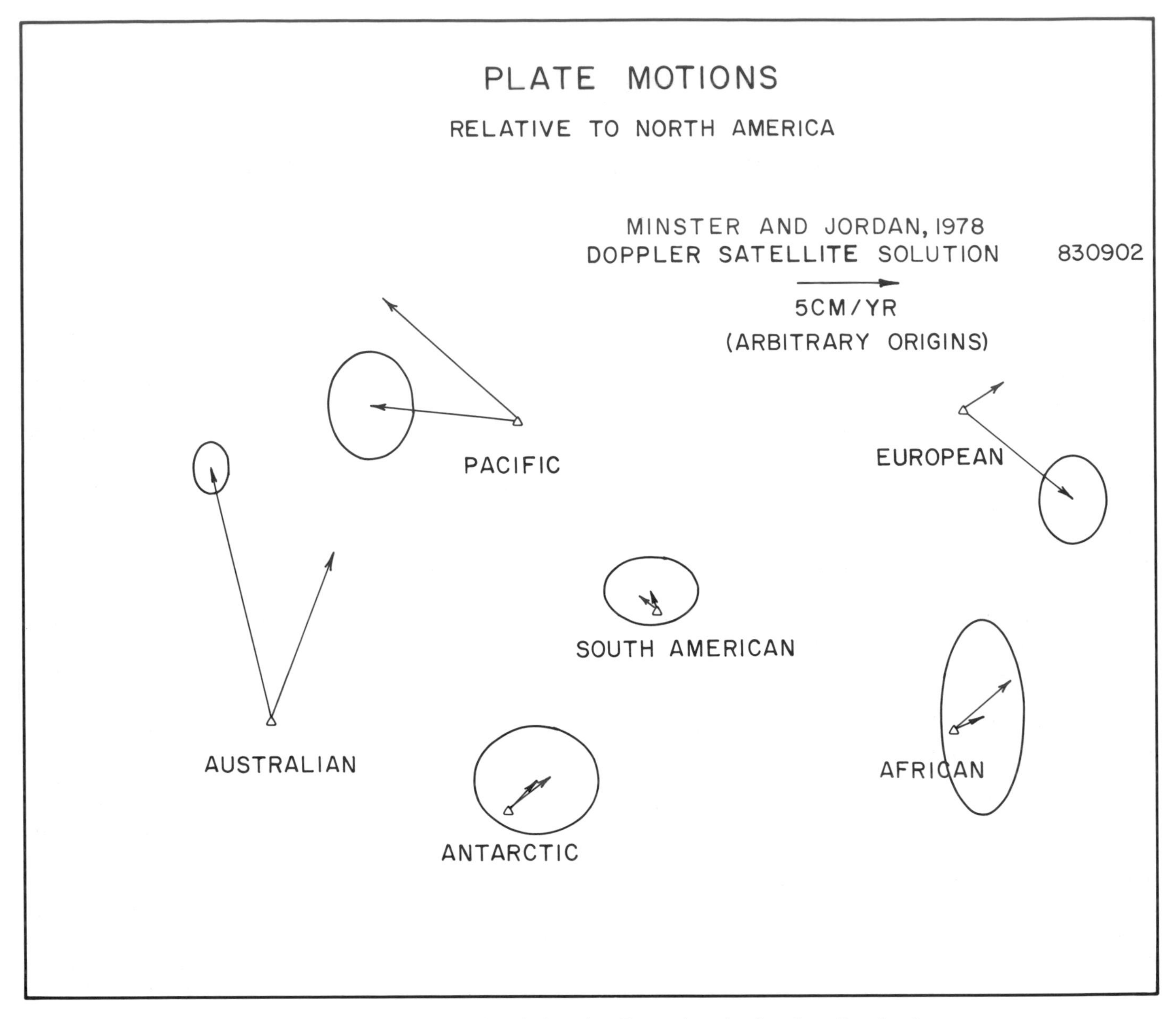

Figure 1. Plate motions relative to North America. Rates of motion based on Doppler data were computed from 50% of the observations made on one satellite at 20 sites on 9 plates over a 9 year period. Standard errors of the average geological rates at these sites are negligible compared to the error ellipses shown for the Dopper results.

stances are only available for geophysical applications a small percentage of the time. A few transportable lunar lasers and VLBI are available or under construction which will extend this capability. For example, it would be possible to locate mobile equipment on both sides of the Mid-Atlantic Ridge in Iceland and also in the Azores so that internal deformation within the North American Plate and within the Eurasian Plate can be separated from the plate motions determined from the fixed VLBI sites on the North American and European continents. No lunar laser results are available on plate motion because only one such station has been in operation for a sufficiently long time. Model errors in the satellite orbit are troublesome in applying laser or electronic Doppler observations of artificial satellites to plate motion studies; however, Figure 1 shows that even the less accurate Doppler system can produce interesting plate motion results because of the wealth of data produced by the system (Anderle and Malyevac, 1983). It will be interesting to obtain corresponding data from one of the other space systems to explain the factor of 2 difference in the magnitude of the Doppler and geologically determined plate motions. The latter represent the average motion over the past few million years (Minster and Jordan, 1978). At this time it is not known whether the discrepancy is due to the more recent measurements over a short period at a different geographic location by the Doppler system and therefore perhaps indicative of a recent acceleration in plate motion or whether it is due to biases in the Doppler system. Certainly the "geologic" rates could not be in error by so much in view of the 2 to 3 mm/yr standard error of these rates at the plate boundaries. Mobile VLBI and artificial satellite observations have been used in California to measure plate and crustal motion in the vicinity of the San Andreas fault which forms the principal boundary between the Pacific and North American plates. The laser data also showed motions significantly larger (50%) than those inferred from geologic records (Smith and others, 1979).

Transportable lunar lasers will also be used for short baseline measurements as well as for long base lines. However, the most exciting prospect for the determination of motions over short distances (e.g., across active plate boundaries) is by electronic measurements of satellites in the NAVSTAR Global Positioning System (GPS). Since these satellites are 20,000 km above the earth, errors of even a few meters in the satellite orbit would be acceptable in the precise determination of base lines shorter than 100 km.

Portable, all-weather GPS receivers could not only monitor plate separation at the boundaries in Iceland and the Azores but, if sufficiently accurate orbits are available, could also be deployed in regions where logistics are a problem, such as the plate boundaries between Spitsbergen or Jan Mayen and the East Coast of Greenland.

Three types of portable or mobile equipment that are under development to exploit the GPS signals have the capability of providing measurements of the relative position of two or more sites to an accuracy of about 1 cm after a few hours of measurement. Equipment conceived by Counselman (Counselman and Steinbrecher, 1982) and equipment developed by MacDoran (MacDoran and others, 1982) can determine relative site positions without detailed knowledge of the structure of the coded signals transmitted by the GPS satellites. Equipment that has been developed by Texas Instrument Corporation for an Interagency Coordinating Committee (Sims, 1982) accomplishes the same function as well as others. Measurements over short and intermediate distances with GPS receivers, coupled with long distance measurements with VLBI and mobile VLBI equipment will provide a considerable amount of data on current plate motions and internal plate deformations.

SATELLITE-BORNE SENSORS

Data from a wide variety of satellite-borne instruments are available for geologic applications. Stewart (1981) listed some 18 instruments that have provided or are providing data from 1 or more of 29 spacecraft. This list can be extended to include magnetometers, which are of secondary interest to the oceanographic applications he addressed, the LANDSAT-4 satellite which was launched after the publication of his paper, and other spacecraft launched to extend the missions he described. Three of the instruments listed are active radio-frequency transmitters: the altimeter, the synthetic aperture radar (SAR), and the scatterometer.

The radar altimeter, which was carried as an experiment aboard Skylab and produced measurements over the oceans from the GEOS-3 and SEASAT-1 spacecraft, is one of the most productive satellite sensors from the viewpoint of the geology of the North Atlantic (see Vogt, this volume, Ch. 14).

The design of the altimeters varies somewhat depending on the accuracy goal and power available for the mission. The SEASAT altimeter had a precision of 10 cm for 1-second averaging times for significant wave heights of 1 to 20 m (Townsend, 1980). The effective footprint of the raw signals, which is limited by the width of the transmitted pulse, is 1.6 km for a smooth ocean. The effect of ocean roughness is averaged out due to the size of the footprint, although a bias correction is made according to the wave height, which is inferred from the sharpness of the rise in power of the signal reflected from the surface. There are small errors in the corrections made for ionospheric and tropospheric effects on the signal, but the major error source is in the uncertainty in the position of the satellite. The error in the height component of satellite position is about 1 m, although enforcement of consistency in the altimeter measurements to a mean ocean surface yields a relative accuracy of 20 to 30 cm (Tapley and Born, 1980). Fortunately, the principal error in the satellite orbit has a wavelength of several thousand kilometers. Therefore, deviations in the ocean surface of 10 cm due to ocean currents or sea-floor topography are detectable in the altimetry signal owing to their shorter wavelength. Wavelengths shorter than 50 km are generally not detectable because the geophysical or oceanographic effects at shorter wavelengths are not large enough to be extracted from the noise in the altimeter measurements. The noise in the GEOS-3 altimeter was about three times larger than that of the SEASAT altimeter, whereas the altimeter in the proposed TOPEX (topographic experiment) satellite is four or five times better than that of SEASAT. The SEASAT satellite had a power failure after about 4 months in orbit, so that observations were limited to a grid of about 1°. Although the altimeter aboard the GEOS-3 satellite operated for about 3½ years, the quantity of data did not greatly exceed that subsequently acquired by SEASAT because of the GEOS-3 power limitations and its requirement for real-time readout of altimetry data. Very dense data are available from the GEOS satellite over the western North Atlantic because of the availability of ground telemetry stations in that area.

The GEOS-3 and SEASAT-1 altimetry measurements have been used to map the height of the mean ocean surface with respect to the reference ellipsoid at intervals of 1° or better, and the measurements have been converted to gravity anomalies by Rapp (1983) and others (see also frontispiece). Figure 2 shows the intermediate wavelength geoid anomalies that are revealed when the long-wavelength effects are removed from the sea surface derived from altimetric data. Mass anomalies, principally due to oceanic basement topography, cause measurable variations in the height of the mean ocean surface and are evident in the altimetric data as shown by the comparisons between altimetry, bathymetry, and gravimetry (Vogt, this volume,

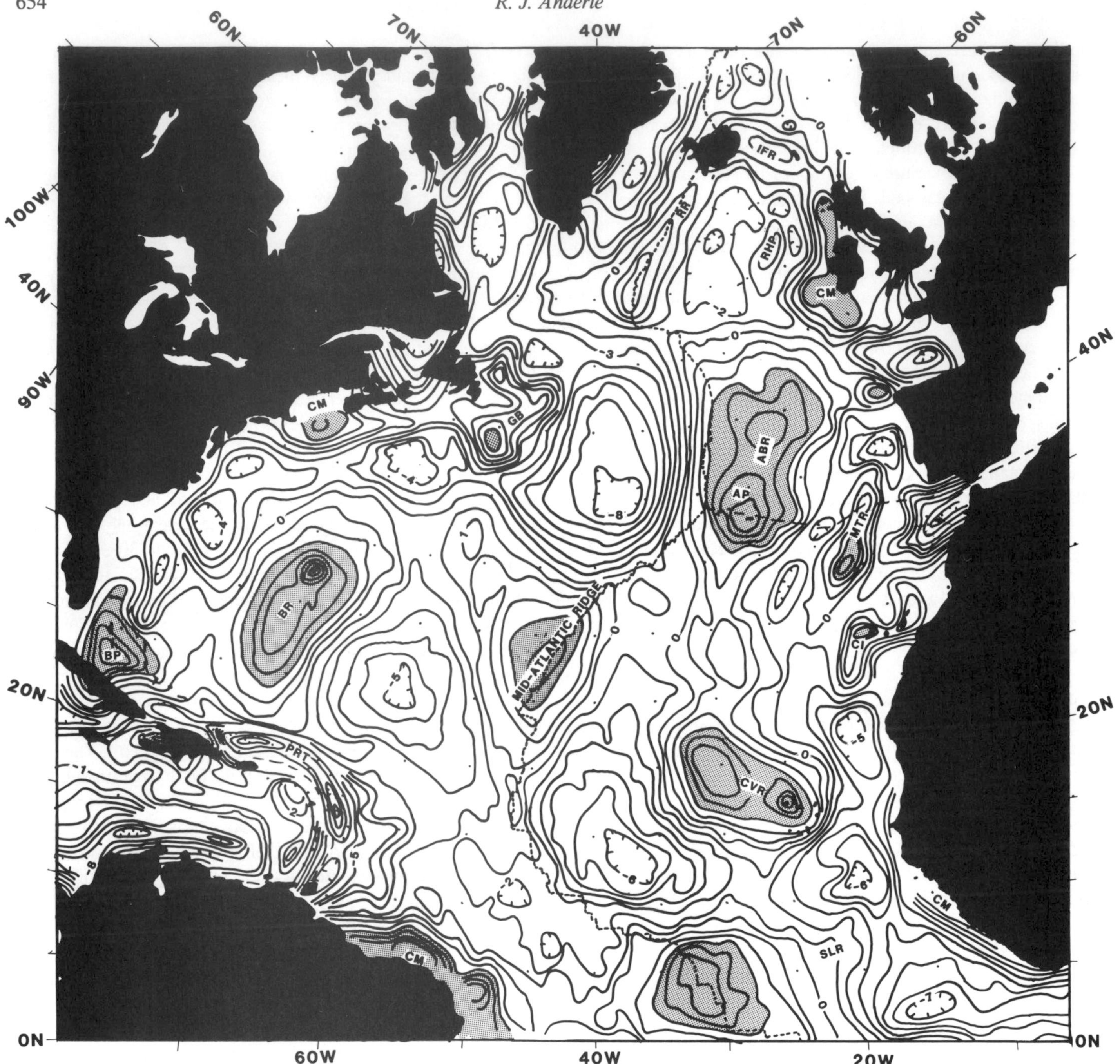

Figure 2. Intermediate wavelength geoid anomaly contours (altimetric minus GEM-9) plotted on 40°W transverse Mercator projection. Contour interval 1 m. Geoid anomalies correlating with topographic features include: BP, Bahamas Platform; BR, Bermuda Rise; GB, Grand Banks; RR, Reykjanes Ridge; IFR, Iceland-Faeroe Ridge; RHP, Rockall-Hatton Plateau; ABR, Azores-Biscay Rise; AP, Azores Platform; MTR, Madeira-Tore Rise; CI, Canary Islands; CVR, Cape Verde Rise; SLR, Sierra Leone Rise; CM, Continental Margins; PRT, Puerto Rico Trench. [Contoured and interpreted by Peter Vogt from the differences of the GEM-9 geoid derived from Lerch et al. (1979) from the altimetric geoid provided by Marsh and Martin (1982).]

Ch. 14). The mean ocean surface is often treated as the geoid to a first approximation because altimetric measurements alone cannot distinguish between surface variations due to gravity disturbances and those due to such effects as a steady-state current. However, removal of a gravimetrically determined geoid from the observed altimetry yields the dynamic sea-surface topography from which the global ocean circulation can be calculated. Encouraging results have been obtained by Cheney and Marsh (1982) with the use of a long-wavelength geoid determined by laser observations of satellites. However, improvements in the satellite-derived geoid achievable in the future are required to advance the state of knowledge of global circulations available from terrestrial observations. But even without a precise geoid, measurements of variations in the surface with time allow the detection and mapping through time of convergence zones and currents, including such features as meanders and the warm- and cold-water rings that may be generated from cut-off meanders. Such features have already been mapped along the Gulf Stream (Cheney and Marsh, 1981). For such studies it is important to combine altimetry with SAR and infrared (IR) data from the ocean surface. For example, winds may push the surface location of a front, as seen in the IR, many tens of kilometers from its average deep location revealed by altimetric data (Cheney, 1981; Hayes, 1981). Knowledge of the behavior of convergence zones, currents, and rings is important in the interpretation of the geologic significance of sediment accumulation and distribution of biofacies (Ch. 25 to 34, this volume). For example, warm rings are one mechanism for increasing the abundance of warm-water microfossil assemblages in otherwise colder biofacies.

The SAR is another active satellite sensor that has returned extremely valuable and, in some instances, unexpected geologic information. The SAR has not only provided detailed land topographic maps with clearly discernible faults and other lineaments (Foster and Hall, 1981) but has also given indications of shallow-water bathymetry (Beal and others, 1981) and even evidence of ancient river beds now buried under sand (McCauley and others, 1982). Of more direct interest to the geology of the North Atlantic is the capability of the SAR to detect traveling and internal waves, currents, and rings (Gower, 1981) which bear on the dispersal of sediments and marine life. Figure 3 shows the Gulf Stream in a SEASAT-1 SAR image in August 1978.

The third satellite-borne radio-frequency transmitter of interest to geologists is the scatterometer. The scatterometer, for example the SEASAT-A Satellite Scatterometer (SASS), provides accurate measurements of wind speed and direction. Such information is not only useful in correcting and interpreting measurements by other instruments such as the altimeter and the SAR, but also provides important information on global weather patterns for use in determining air-sea interactions and developing ocean circulation models that must form the basis for understanding present-day sedimentary processes on the sea bed.

Among the passive satellite-borne sensors, the visible light return beam vidicon (RBV) and visible to infrared multispectral scanners (MSS) on the early LANDSAT satellites have been used to map faults and other geologic features on land (Cochrane and Browne, 1981), while the more advanced MSS and thematic mapper aboard the LANDSAT-4 satellite provide more detailed resolution of geologic features as a result of their smaller pixel size and additional frequency bands. Because the western North Atlantic crust is almost entirely submerged, such studies are limited to areas such as Iceland (Vogt and Tucholke, this volume), the Azores, Bermuda, and a few shallow banks.

A wide variety of passive instruments is also available for geologic applications in water-covered areas. The ocean color scanner (Khorran, 1981) and the Coastal Zone Color Scanner (CZCS) (Mueller and LaViolette, 1981) have provided information on the density of chlorophyll, solids, and turbidity in surface waters. Such data bear on studies discussed e.g. in Honjo (this

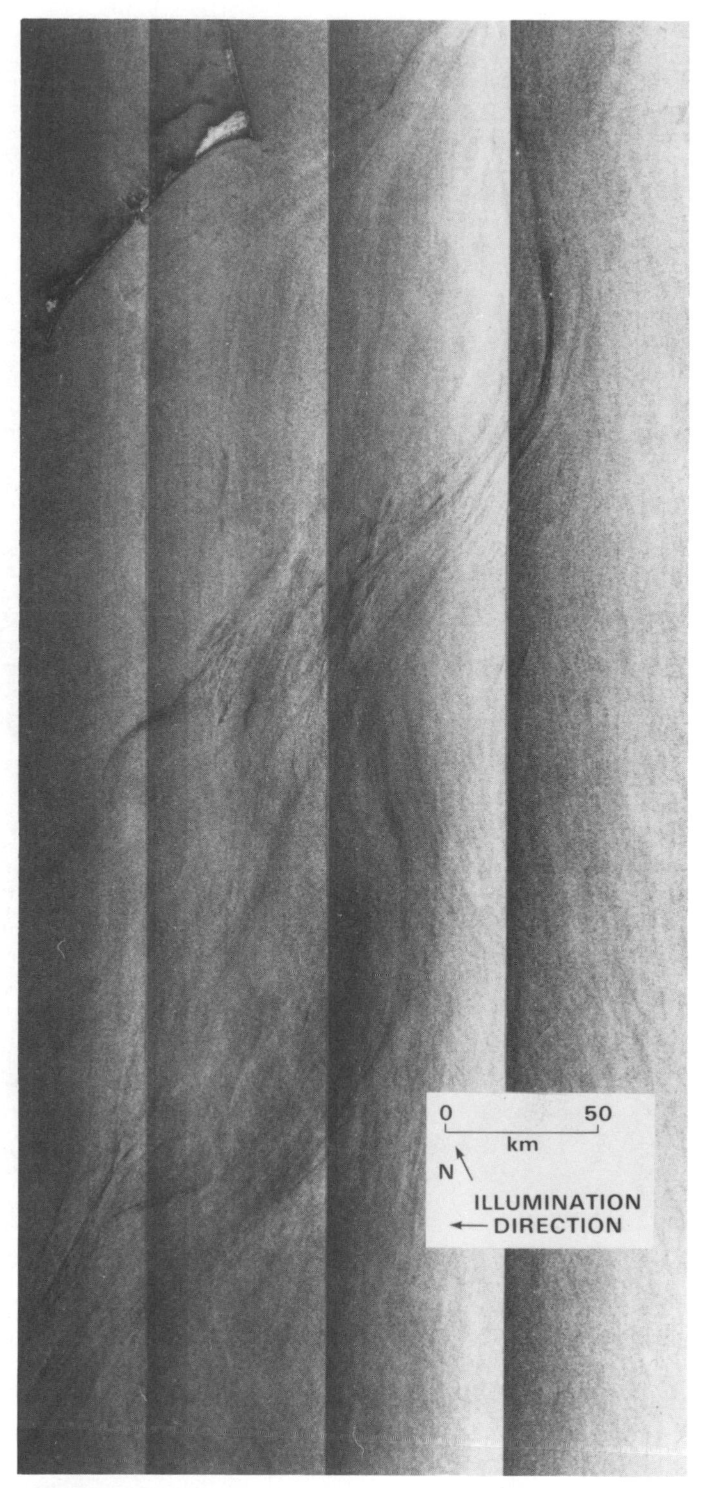

Figure 3. SEASAT SAR image of Gulf Stream. Cape Hatteras is in upper left.

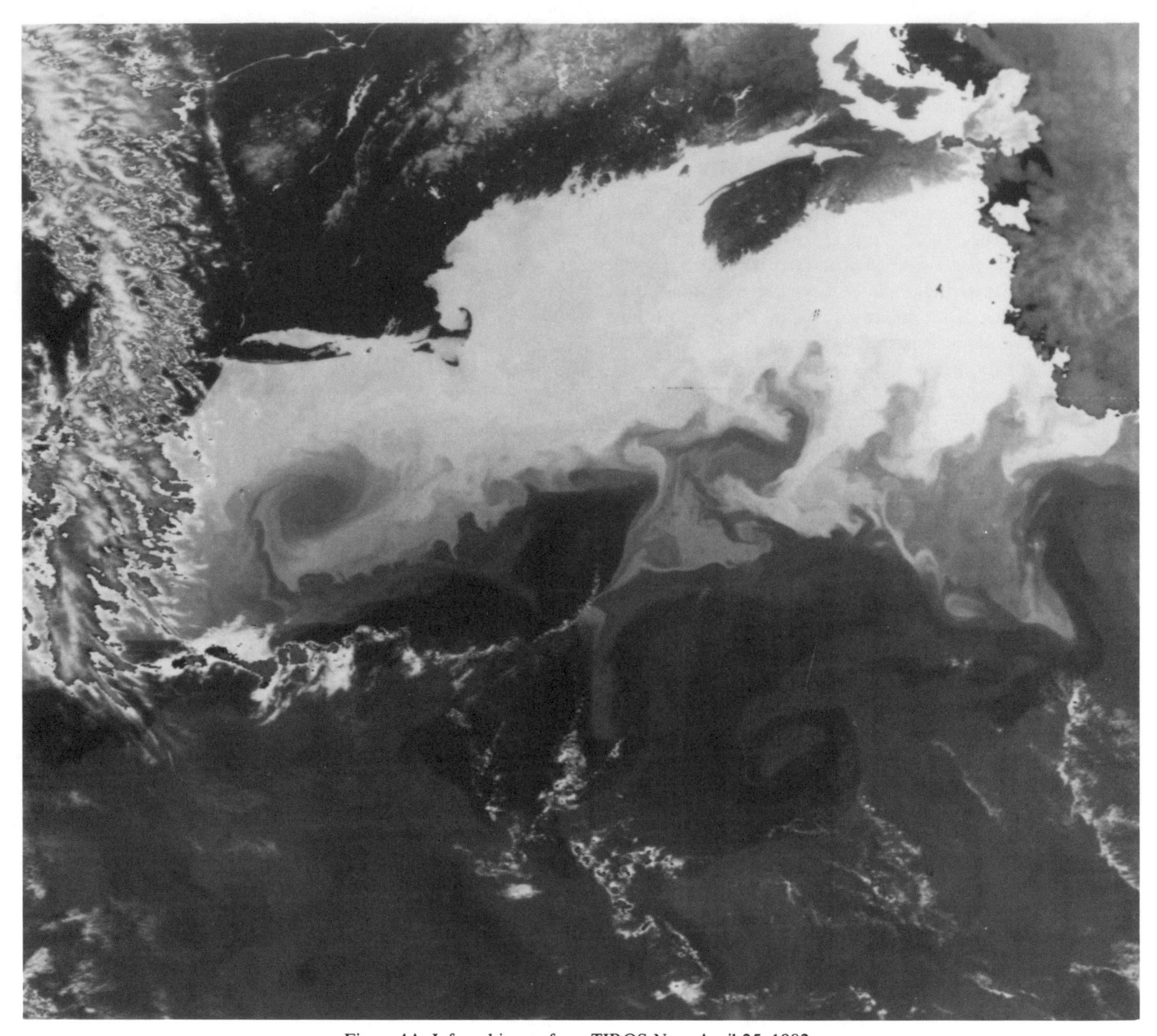

Figure 4A. Infrared image from TIROS-N on April 25, 1982.

volume). The CZCS along with microwave instruments, such as the Scanning Microwave Multichannel Radiometer (SMMR) aboard SEASAT-1 (Njoku and others, 1980) and the Very High Resolution Radiometer (VHRR) aboard TIROS and NOAA satellites (Parmenter, 1977), give detailed maps of currents and warm-water rings which are also necessary in such analyses. The microwave instruments detect the currents and rings from the temperature differentials; in Figure 4B an oceanographic analysis of radiometer data is superimposed on a copy of the radiometer image shown in Figure 4A. The SMMR also provides water-vapor information and wind data (Lipes, 1982) needed to calibrate other instruments such as the altimeter.

The last satellite sensor to be mentioned here is the magnetometer. A number of satellites have carried magnetometers, but the most extensive mapping of the earth's magnetic field was carried out by the MAGSAT satellite which was launched into a near polar orbit (96.76° inclination) in October 1979 and mapped the vector field until its reentry in June 1980. A good correlation found between the measured magnetic field and the geologic features is found in some parts of the world (Frey, 1982). Vogt (this volume) shows the magnetic anomalies for the North Atlantic region. The lower limit of the spatial resolution of the magnetic field derived from MAGSAT data is 250 km (Sailor et al., 1982), which precludes "seeing" anomalies due to specific seamounts or linear anomalies caused by sea-floor spreading and geomagnetic reversals. However, comparison between upward

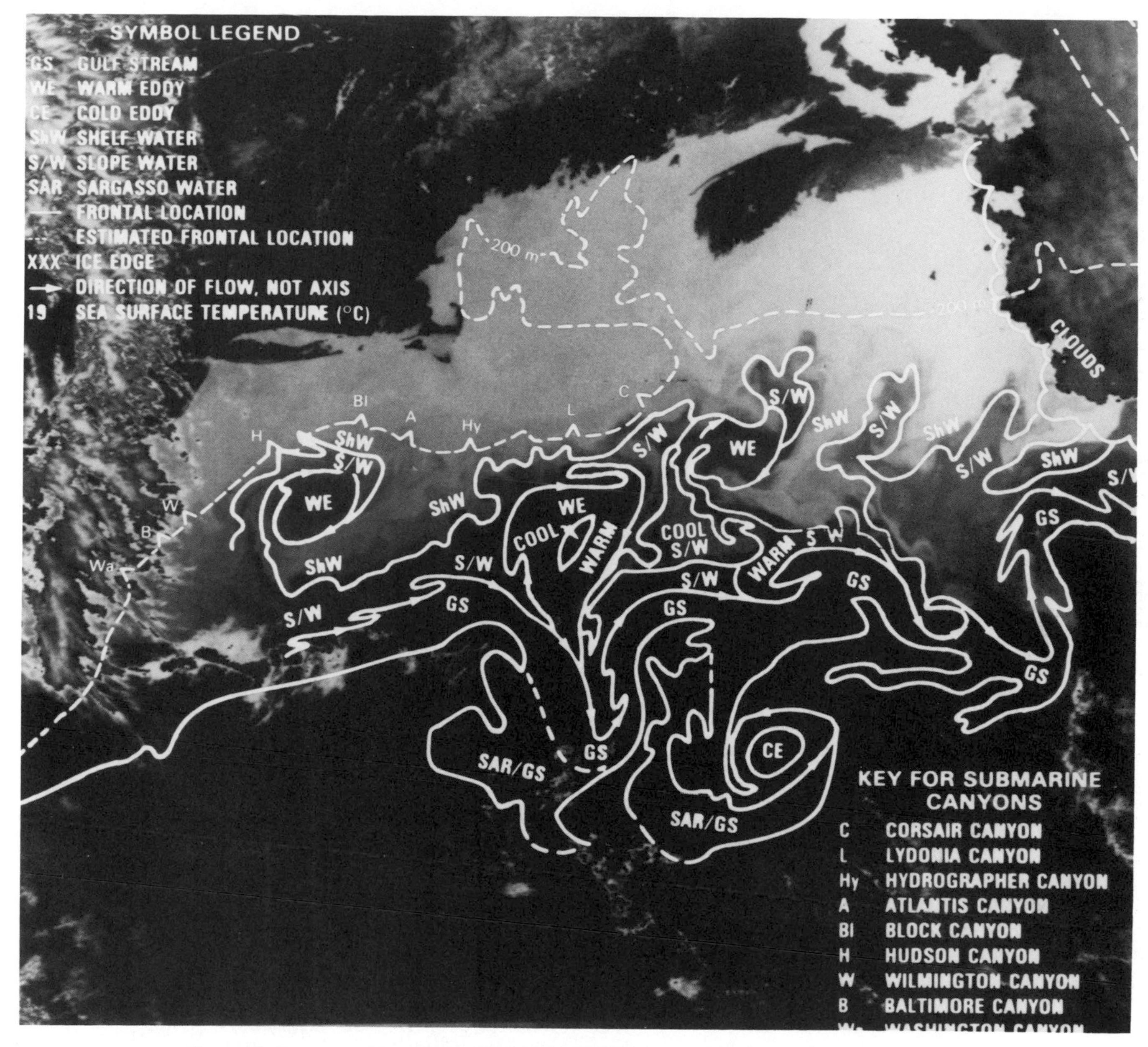

Figure 4B. Oceanographic analysis of TIROS-N AVHRR image of 25 April 1982. (After Jennifer W. Clark, National Earth Satellite Service.)

continued ship data in the Pacific resembles MAGSAT-measured anomalies, demonstrating that the latter are "real" (LaBrecque and Cande, 1984). Such long-wavelength anomalies may be useful in estimating the average magnetization structure for entire provinces, such as island arcs, seamount chains, or oceanic plateaus.

THE PROMISE OF SPACE

Table 1 shows spacecraft that have carried the satellite-borne sensors discussed above as well as many of the planned and proposed programs. Improvements in the information obtained are continuing to be made in a variety of ways. In some instances, such as the altimeter, higher precision is obtained by redesign of the sensor. In others, such as the radiometers, additional frequency bands are added to assist in differentiating among signals produced by various types of surface features. For most sensors, advanced digital processing allows extraction of more information. Satellite on-board processors reduce the amount of ground processing required in the case of altimetry data, and data storage or satellite relay allows acquisition of global data for systems that were previously limited to observations within view of a telemetry readout station. LANDSAT-4 was the first satellite equipped with a GPS-to-satellite receiver and navigation processor, so that satellite position was available in real-time when sensor data were acquired.

The latter half of the 1980s marks the beginning of major

TABLE 1. SPACE BORNE SENSORS PERTINENT TO GEOLOGY

Satellite	Launch	Radar altimeter	Synthetic-Aperture Radar	Scatter-ometer	Photographic, Vidicon	Infrared	Radiometer	Magnet-ometer
NIMBUS 1	1964				AVCS	HRIR		
NIMBUS 2	1966				AVCS	HRIR		
NIMBUS 3	1969					HRIR		
NIMBUS 5	1972						NEMS,ESMR	
NIMBUS 6	1975						ESMR	
NIMBUS 7	1978				CZCS		SMMR	
Skylab	1972	ALT		SCAT				
LANDSAT 1	1972				MSS,RBV			
LANDSAT 2	1975				MSS,RBV			
LANDSAT 3	1978				MSS,RBV			
LANDSAT 4	1982				MSS,TM	MSS,TM		
GEOS-3	1978	ALT						
HCMM-1	1978					HCMR		
SEASAT-1	1978	ALT	SAR	SCAT			SMMR	
ITOS-1,NOAA-1	1970				AVCS	SR		
NOAA 2,3,4,5	1972-76					VHRR,SR		
TIROS 14(TIROS-N)	1978					AVHRR		
NOAA-6	1979					AVHRR		
SMS 1,2	1974-78							
GOES 1,2,3						VISRR		
DMSP(F1,F2,F3)	1976-78					OLS		
OGO 2,4,6	1965-71							SCALAR
Columbia Shuttle	1981		SIR-A					
MAGSAT	1979							VECTOR
STS-17 Shuttle	Planned (84)		SIR-B					
GEOSAT	Planned (84)	ALT						
LANDSAT-5	Planned							
DMSP	Planned (84)						SSMI	
ERS-1	Planned (87)	ALT	AMI	AMI				
SPOT	Planned (84)				HRV			
GRM	Proposed (87/88)							VECTOR
TOPEX	Proposed (87/88)	ALT					TMR	
MOS-1	Proposed (85)				MSS,AVHRR	MSS,AVHRR	2-Frequency	
RADARSAT/FIREX	Proposed (90)		SAR	SCAT				
POSEIDON	Proposed	ALT						
ERS-1(Japan)	Proposed (90)		SAR				Visible & Near IR	
NROSS	Proposed (88)	ALT		SCAT			4 Channel	
						7 Channel	2-Frequency	
Shuttle (s)	Proposed		MRSE	MRSE			MRSE	

Satellite Acronyms

DMSP	Defense Materological Satellite Program
ERS	European Space Agency Remote Sensing Satellite
FIREX	Free-Flying Imaging Radar
GOES	Geostationary Observational Environment Satellite
GRM	Geopotential Research Mission
HCMM	Heat Capacity Mapping Mission
ITOS	Improved TIROS Operational Satellite
MOS	Maritime Observation Satellite
NROSS	Naval Remote Ocean Sensing Satellite
OGO	Orbiting Geophysical Observatory
SMS	Synchronous Meterological Satellite
SPOT	System Probatoire d'Observation de la Terre (Earth Observation Test System)
TIROS	Television and Infrared Observational Satellite

Sensor Acronyms

ALT	Altimeter
AMI	Active Microwave Instrumentation
AVCS	Advanced Vidicon Camera Systems
AVHRR	Advanced Very High Resolution Radiometer
CZCS	Coastal Zone Color Scanner
ESMR	Electrically Scanned Microwave Radiometer
HCMR	Heat Capacity Mapping Radiometer
HRIR	High Resolution Infrared Radiometer
HRV	High Resolution Visible Imaging Instruments
MRSE	Microwave Remote Sensing Experiment
MSS	Multispectral Scanner
NEMS	NIMBUS-E Microwave Scatterometer
OLS	Optical Line Scanner
RBV	Return Beam Vidicon
SAR	Synthetic-Aperture Radar
SCAMS	Scanning Microwave Spectrometer
SCAT	Scatterometer
SIR	Shuttle Imaging Radar
SMMR	Scanning Multi-frequency Microwave Radiometer
SR	Scanning Radiometer
TM	Thematic Mapper
TMR	TOPEX Microwave Radiometer
VHRR	Very High Resolution Radiometer
VISSR	Visible and Infrared Spin Scan Radiometer

Note: Table entries largely taken from Stewart (1981) and personal correspondence with Stewart.

initiatives by governments other than the United States in the acquisition of satellite-borne sensor data applicable to geologic investigations. The European Space Agency is moving rapidly ahead with plans for the launch of the ice and ocean monitoring satellite, ERS-1; the French Government is developing the SPOT land-mapping satellite and considering the Poseidon oceanographic satellite; the Canadian Government is actively considering RADARSAT; and the Japanese Government is deeply involved in plans for the MOS (Maritime Observation Satellite) and an earth resources satellite. These programs, as well as continuing and new U.S. programs, will ensure a continuous source of increasingly useful satellite-borne sensor data.

Table 1 does not show developments planned for ground observations of satellites (to obtain data on the earth's gravity field and on plate motions) or satellite-borne instrumentation for measuring the gravity field. The GRM (Geopotential Research Mission) program includes a magnetometer and a system for measuring the relative velocity between two spacecraft which are kept in a low-altitude "drag free" orbit by a system that senses and compensates for external nonconservative forces. The low altitude (170 km) will reduce measurement errors and provide better spatial resolution of magnetic field data; in addition, the measured variations in velocity can be used to deduce gravity data to 1-mgal accuracy and the geoid to 3-cm accuracy at wavelengths as short as 100 km (Kaula, 1983). A geoid to this accuracy is a necessary complement to satellite altimetric data (e.g., TOPEX) for use in global ocean circulation studies. A cryogenic gravity gradiometer is being developed which could be flown as a follow-on program in the 1990s for further refinement of the gravity field (Paik, 1981). Such an instrument would measure spatial derivatives of gravity to an unprecedented accuracy of 10^{-4}E (10^{-13}cm s^{-2}/cm), which is believed to be about two orders of magnitude better than the accuracy needed to achieve the gravity and geoid accuracies cited above.

REFERENCES

Anderle, R. J.
1976: Error model for geodetic positions derived from Doppler satellite observations, Bulletin Geodesique, v. 50, no. 1, p. 43–77.

Anderle, R J., and Malyevac, C. A.
1983: Current plate motions based on Doppler satellite observations; Geophysical Research Letters, v. 10, no. 1, p. 67–70.

Beal, R. C., DeLeonibus, P. S., and Katz, I., eds.,
1981: A search for cold water rings; *in* Spaceborne Synthetic Aperture Radar for Oceanography, eds. R. C. Beal, P. S. DeLeonibus, and I. Katz; The Johns Hopkins University Press, Baltimore, Maryland, p. 163.

Cheney, R. E.
1981: A search for cold water rings; *in* Spaceborne Synthetic Aperture Radar for Oceanography, eds. R. C. Beal, P. S. DeLeonibus, and I. Katz; The Johns Hopkins University Press, Baltimore, Maryland, p. 163.

Cheney, R. E., and Marsh, J. G.
1981: Seasat altimeter observations of dynamic topography in the Gulf Stream region; Journal of Geophysical Research, v. 86, no. C1, p. 473–483.
1982: Global ocean circulation from satellite altimetry; EOS, v. 63, no. 45, p. 997.

Cochrane, G. R., and Browne, G. H.
1981: Geomorphic mapping from Landsat-3 Return Beam Vidicon (RBV) imagery: Photogrammetric Engineering and Remote Sensing; Journal of the American Society of Photogrammetry, v. XLVII, no. 8, p. 1205–1213.

Counselman III, C. C., and Steinbrecher, D. H.
1982: The Macrometer™: A compact radio interferometry terminal for geodesy with GPS; *in* Proceedings of the Third International Geodetic Symposium on Satellite Doppler Positioning, Volume 2; New Mexico State University, p. 1165–1172.

Foster, J. L., and Hall, D. K.
1981: Multisensor analysis of hydrologic features with emphasis on the Seasat SAR: Photogrammetric Engineering and Remote Sensing; Journal of the American Society of Photogrammetry, v. XLVII, no. 5, p. 655–664.

Frey, H.
1982: MAGSAT scalar anomalies and major tectonic boundaries in Asia; Geophysical Research Letters, v. 9, no. 5, p. 299–302.

Gower, J.F.R., ed.
1981: Oceanography from Space; Plenum Press, New York, New York, 978 p.

Hayes, R. M.
1981: Detection of the Gulf Stream; *in* Spaceborne Synthetic Aperture Radar for Oceanography, eds. R. C. Beal, P. S. DeLeonibus, and I. Katz; The Johns Hopkins University Press, Baltimore, Maryland, p. 146–160.

Kaula, W. M.
1983: Inference of variations in the gravity field from satellite-to-satellite range-rate; Journal of Geophysical Research, v. 88, no. B10, p. 8345–8349.

Khan, M. A.
1983: Accuracy of earth's gravity field models; Physics of the Earth and Planetary Interiors, v. 31, no. 3, p. 231–240.

Khorran, S.
1981: Use of ocean color scanner data in water quality mapping: Photogrammetric Engineering and Remote Sensing; Journal of the American Society of Photogrammetry, v. XLVII, no. 5, p. 667–676.

LaBrecque, J. L., and Cande, S. C.
1984 : Intermediate wavelength magnetic anomalies over the Central Pacific; Journal of Geophysical Research, B, v. 89, p. 11,124–11,134.

Lerch, F. J., Klosko, S. M., Laubscher, R. E., and Wagner, C. A.
1979: Gravity model improvement using GEOS 3 (GEM 9 and 10); Journal of Geophysical Research, v. 84, no. B8, p. 3897–3915.

Lipes, R. G.
1982: Description of Seasat radiometer status and results; Journal of Geophysical Research, v. 87, no. C5, p. 3385–3395.

MacDoran, P. F., Spitzmesser, D. J., and Buennagel, L. A.
1982: SERIES: Satellite emission range inferred earth surveying; *in* Proceedings of the Third International Geodetic Symposium on Satellite Doppler Positioning, Volume 2; New Mexico State University, p. 1143–1155.

Marsh, J. G., and Martin, T. V.
1982: The SEASAT altimeter mean sea surface model; Journal of Geophysical Research, v. 87, no. C5, p. 3269–3280.

McCauley, J. F., Schaber, G. G., Breed, C. S., Grolier, M. J., Haynes, C. V., Issawi, B., Elachi, C., and Blom, R.
1982: Subsurface valleys and geoarcheology of the Eastern Sahara revealed by shuttle radar; Science, v. 218, no. 1004-1020, p. 295–312.

Minster, J. B., and Jordan, T. H.
1978: Present day plate motions; Journal of Geophysical Research, v. 83, no. B11, p. 5331–5354.

Mueller, J. L., and LaViolette, P. E.
1981: Color and temperature signatures of ocean fronts observed with the Nimbus-7 CZCS; *in* Oceanography from Space, Plenum Press, New York, New York, p. 295–312.

Njoku, E., Stacey, J. M., and Barath, F. T.
1980: The Seasat Scanning Multichannel Microwave Radiometer (SMMR): Instrument description and performance; Journal of Oceanic Engineering, v. OE-5, no. 2, p. 125–137.

Paik, H. J.
1981: Superconducting tensor gravity gradiometer; Bulletin Geodesique, v. 55, no. 4, p. 370–381.

Parmenter, F. C.
1977: An overview of oceanic features and air-sea interaction processes as viewed from the NOAA operational satellites; *in* A Collection of Technical Papers; American Institute of Aeronautics and Astronautics, Paper 77-1569, p. 62–69.

Rapp, R. H.
1983: The determination of geoid undulations and gravity anomalies from SEASAT altimeter data; Journal of Geophysical Research, v. 88, no. C3, p. 1552–1562.

Rogers, A.E.E., Knights, C. A., Hinteregger, H. F., Whitney, A. R., Counselman III, C. C., Shapiro, I. I., and Gourevitch, S. A.
1978: Geodesy by radio interferometry: Determination of a 1.24-km base

line vector with ~5 mm repeatability; Journal of Geophysical Research, v. 83, no. B1, p. 325–334.

Sailor, R. V., Lazarewicz, A. R., and Brammer, R. F.
1982: Spatial resolution and repeatability of MAGSAT crustal anomaly data over the Indian Ocean; Geophysical Research Letters, v. 9, no. 4, p. 289–292.

Sims, M. L.
1982: GPS Geodetic Receiver System; in Proceedings of the Third International Geodetic Symposium on Satellite Doppler Positioning, Volume 2; New Mexico State University, p. 1103–1121.

Smith, D. E., Kolenkiewicz, R., Dunn, P. J., and Torrence, M. H.
1979: The measurement of fault motion by satellite laser ranging; Tectonophysics, v. 52, p. 59–67.

Stewart, R. H.
1981: Satellite oceanography: The instruments; Oceanus, v. 24, no. 3, p. 66–74.

Tapley, B. D., and Born, G. H.
1980: The Seasat precision orbit determination experiment; Journal of the Astronautical Sciences, v. XXVIII, no. 4, p. 315–326.

Townsend, W. F.
1980: An initial assessment of the performance achieved by the Seasat-1 radar altimeter; Oceanic Engineering, v. OE5, no. 2, p. 80–92.

MANUSCRIPT ACCEPTED BY THE SOCIETY JANUARY 21, 1984

ACKNOWLEDGMENTS

Special data tapes or photographic material used in preparing this chapter were provided by Richard Rapp of Ohio State University, James Marsh of Goddard Space Flight Center, Jennifer Wartha-Clark of the National Environmental Satellite Data and Information Service, Customer Services of EROS Data Center, and the European Space Agency through the Deutsche Forschungs-und Versuchsanstalt fur Lüft-und Raumfahrt.

I appreciate the valuable suggestions and corrections provided by reviewers Bernard Chovitz, A. R. Palmer, Mark Settle, and especially Peter Vogt.

Note added to proof, November, 1985

Dramatic advances were made in 1985 in the application of space techniques to geophysics. The stability of the North American plate and a component of the relative motion of the North American and European plates were measured to an accuracy of better than 1 cm/year while a special VLBI campaign has acquired data in the relative motion of the Pacific and North American plates. Lageos laser determinations of the relative motions of several major plates have been refined to an accuracy of better than 1 cm/year. The motions generally conform to those derived from geologic data.

References:

Carter, W. E., Robertson, D. S., and MacKay, J. R.
1985 : Geodetic Radio Interferometric Surveying: Application and Results, Journal of Geophysical Research, v. 90, p. 4577–4587.

Clark, Thomas A., Ryan, J. W., and Gordon, D.
1985 : Geodesy by interferometry: Recent Measurements of North American to Pacific Plate Motions: EOS, Transactions, American Geophysical Union, v. 66, p. 848.

Smith, David E., Christodoulidis, Demosthenes C., Kolenkiewicz, Ronald, Dunn, Peter J., Torrance, Mark H., and Kosko, Steven
1985 : "Global Plate Motion Results from Satellite Laser Ranging": EOS, Transactions American Geophysical Union, v. 66, p. 848.

The Geology of North America
Vol. M, The Western North Atlantic Region
The Geological Society of America, 1986

Chapter 40

Resource potential of the western North Atlantic Basin

William P. Dillon and Frank T. Manheim
U.S. Geological Survey, Woods Hole, Massachusetts 02543
Lubomir F. Jansa
Geological Survey of Canada, Bedford Institute of Oceanography, Halifax, Nova Scotia B27 4A2, Canada
Gudmundur Pálmason
ORKUSTOFNUN, National Energy Authority, Reykjavik, Iceland
Brian E. Tucholke
Woods Hole Oceanographic Institution, Woods Hole, Massachusetts 02543
Richard S. Landrum
Union Oil Company of California, 4615 Southwest Freeway, Houston, Texas 77027

INTRODUCTION

The principal geologic resources of the western North Atlantic Basin probably include petroleum, ferromanganese nodules and crusts, phosphorite, mid-ocean-ridge sulfides, and geothermal energy. Of these, only geothermal energy has been developed, at present, and some of the others never may be developed. In our opinion, the most valuable resource of the western North Atlantic ultimately will be petroleum, once techniques of drilling and completing wells in deep water are perfected.

PETROLEUM

We here consider the petroleum resources only of the off-shelf portion of the western North Atlantic Ocean. Very little information is available for this region; off the eastern United States, only four petroleum exploration holes have been drilled in one restricted area seaward of the shelf, off the Baltimore Canyon trough. However, by interpreting seismic reflection profiles and stratigraphic data from the Deep Sea Drilling Project (DSDP) and other wells on the adjacent slope and shelf, we can evaluate the geologic conditions that existed during development of the basin and that might lead to petroleum accumulations.

The well-known factors that lead to oil and gas accumulations are availability of source beds, adequate maturation, and the presence of reservoir beds and seals configured to create a trap. The western boundary of the area considered in this paper, the present slope-rise break, is one that has developed from the interplay of sedimentation and erosion at the continental margin; these processes are affected by variations in margin subsidence, sediment input, oceanic circulation, sea level, and other factors. Thus the slope-rise break has migrated over time and is locally underlain by slope and shelf deposits, as well as deep-basin facies. These changes in depositional environments may well have caused juxtaposition of source and reservoir beds with effective seals.

Several papers have been written on the hydrocarbon prospects of the oceanic basins (Jansa and MacQueen, 1978; Roberts and Caston, 1975; Dow, 1979; Hedberg, 1976; McIver, 1975; Schott and others, 1975; Byramjee and others, 1975) with greatly differing opinions expressed because of the lack of constraints by data. Any conclusions about hydrocarbon prospects of the deep continental margin remain speculative, therefore, even though more data have become available as a result of Deep Sea Drilling Project studies.

Source Beds

Large enough quantities of sediments rich in organic carbon probably are present in the deep sea to allow some strata to be considered source beds for petroleum (Benson and Sheridan, 1978; Arthur and Schlanger, 1979; Tucholke and Vogt, 1979; Habib, 1982). Organic carbon-rich deposits have been drilled only in the deep basin, but we can trace updip equivalents in seismic reflection profiles and these may form adequate source rocks.

Both supply and preservation of organic matter are critical. Analysis of 7,300 samples from DSDP cores shows that the distribution of organic carbon in deep-sea deposits is skewed towards low values; the average is about 0.3%, with 0.1% being

Dillon, W. P., Manheim, F. T., Jansa, L. F., Pálmason, G., Tucholke, B. E., and Landrum, R. S., 1986, Resource potential of the western North Atlantic Basin; *in* Vogt, P. R., and Tucholke, B. E., eds., The Geology of North America, Volume M, The Western North Atlantic Region: Geological Society of America.

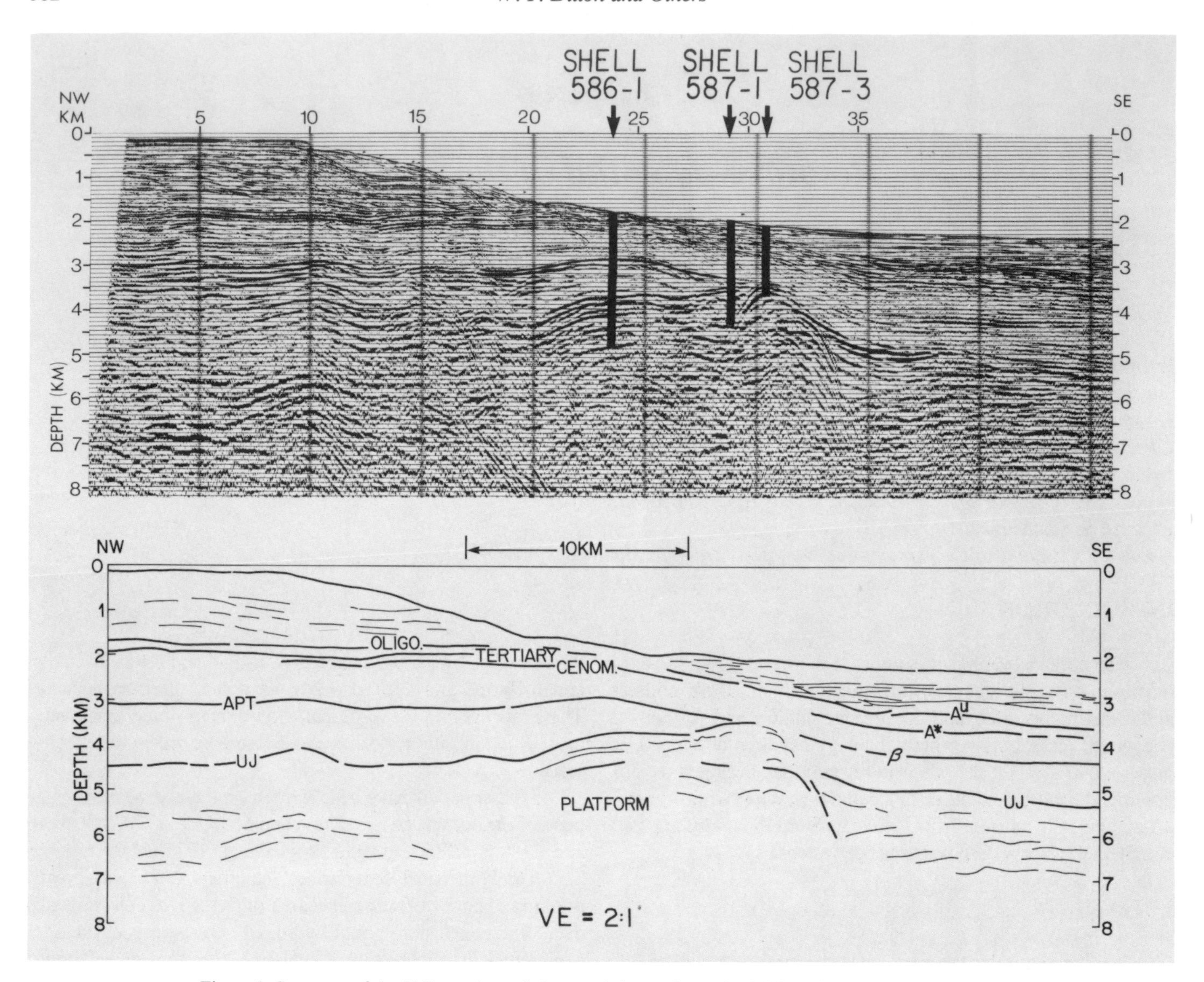

Figure 1. Structure of the U.S. continental slope and rise as shown by USGS seismic line 6, located off Delaware Bay, New Jersey, eastern United States (Schlee and others, 1979). This depth-converted line shows apparent closure on carbonate platform edge. Four Shell Oil Co. deep-water test wells, a few kilometers to the north, penetrated a Jurassic-Lower Cretaceous reef complex; three of the wells are projected onto the profile. One of the wells on this structure set a world's water-depth record for petroleum exploration holes at 2,119 m (6,952 ft). Some of the irregularities in reflectors between km 10 and 20 may result from growth faulting (John F. Karlo, Shell Oil Company, written communication, 1985). Reflecting horizons are UJ - Upper Jurassic, β - top of Neocomian, A* - Maestrichtian, and A^u - Oligocene unconformity. Location is shown in Figure 5.

most common (McIver, 1975). The type of organic matter varies from hydrogen-rich, amorphous organic matter of probable marine origin to thermally inert and vitrinite-rich material derived from terrigenous sources (Kendrick and others, 1979). Preservation of the organic matter is controlled by oceanographic conditions such as bottom circulation rates, degree of oxygenation of bottom waters, deposition rates, supply rate of terrestrial organic matter from the continents, and productivity in the surface water. Changes in these factors create a stratigraphic sequence with highly variable quantities and types of organic carbon (Jansa and others, 1979; Tissot and others, 1979, 1980; Tucholke, 1979; Sheridan and others, 1982; Jansa, this volume).

Drilling at abyssal depths off Morocco indicates that bottom waters in the Early Jurassic North Atlantic Basin were oxygen depleted; claystone that is locally rich in organic carbon (up to 6.7%) was deposited at this time (Hinz and others, 1982). Middle Jurassic siltstones in the Blake-Bahama Basin apparently were deposited in an oxygen-depleted environment and they con-

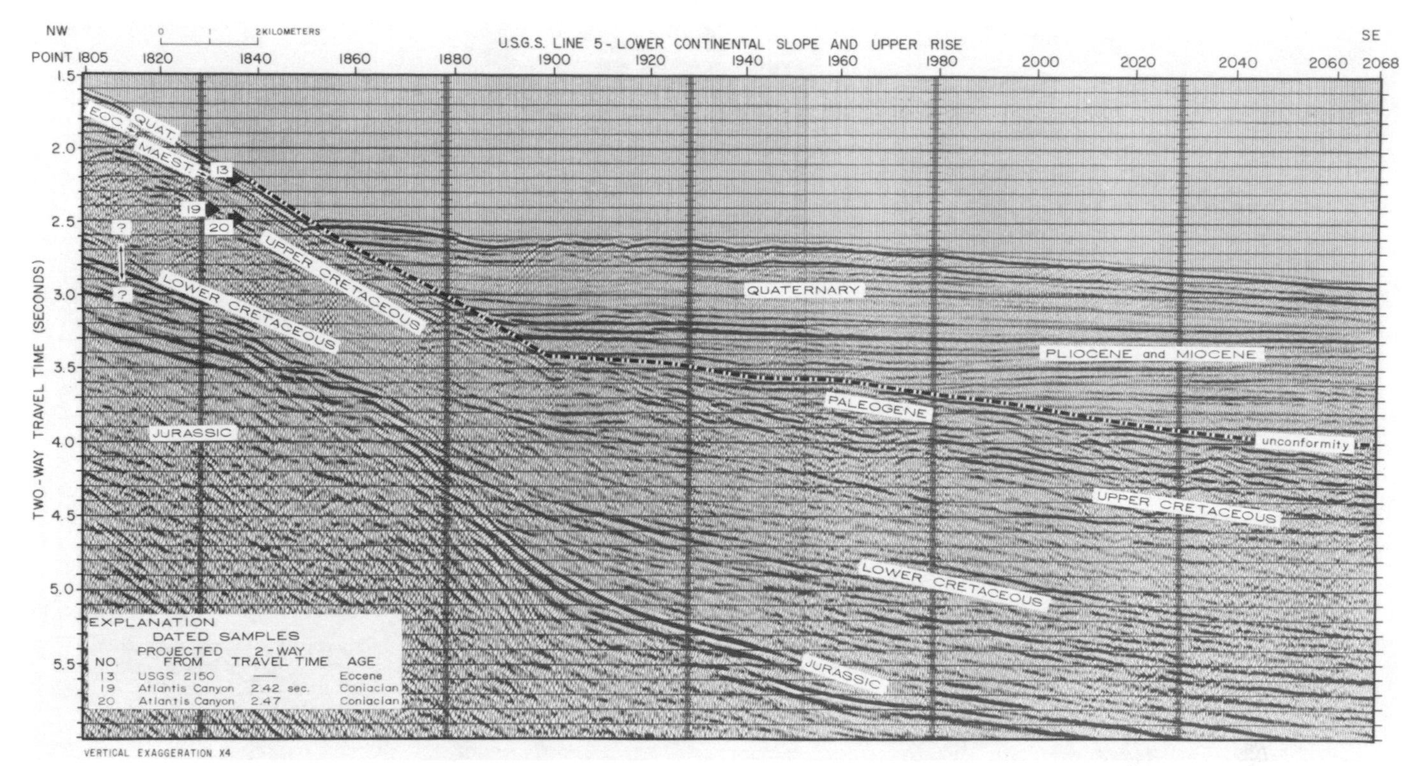

Figure 2. Stratigraphy of the lower continental slope and upper continental rise as shown on USGS seismic line 5, off Martha's Vineyard Island, Massachusetts, eastern United States (Valentine, 1981). Updip pinchouts of Paleogene strata beneath the unconformity and of Lower Cretaceous beds against the Jurassic shelf-edge structure (near shotpoint 1880) could lead to petroleum-trapping conditions. Location is shown in Figure 5.

tain both marine and terrestrial organic matter in quantities of 0.4% to 3.1%; where they were drilled, their maturation level is close to but below that required to generate oil (Herbin and others, 1983). Near the Cenomanian-Turonian boundary, at the top of the Hatteras Formation in the western North Atlantic, organic carbon values of as much as 17% are recorded; this organic matter is largely pyrolisable matter of marine origin. In contrast, older strata in the Hatteras Formation (Barremian-Cenomanian) typically have lower carbon values, and the carbon is dominantly of terrestrial origin. For example, in the Blake-Bahama Basin the black shales of the Hatteras Formation have organic carbon contents of 0.4 to 3%, but the carbon is mainly inert terrigenous material (Tissot and others, 1979; Herbin and others, 1983). Beds rich in reactive marine carbon, such as those near the top of the Hatteras Formation, are potential source rocks, provided that they extend landward beneath the thick wedge of continental rise deposits, where thermal maturation could have occurred. Analysis of drilling results correlated to seismic reflection studies indicate that these beds do indeed extend beneath the upper continental rise to the foot of the paleocontinental slope (Figs. 1, 2) (Mountain and Tucholke, 1985).

Maturation

The most restrictive factor with regard to petroleum generation in deep-sea sediments probably is not availability of organic matter, but rather maturation. Maturation requires relatively high temperatures, which generally are reached only with considerable depth of burial. The maturation temperatures for liquid petroleum lie within the range of 50°C to 150°C (Tissot and others, 1974). The burial depth required depends on the geothermal gradient, which, on the eastern North American margin, is a relatively low 2.3°C/100 m (Robbins, 1977), although slightly higher gradients (2.5–3.5°C/100 m) are found locally on the deeper margin (Jansa and MacQueen, 1978). The deep sea drilling in the North Atlantic at Sites 534, 391, 398 showed that the oil window was not quite reached in those wells at 1,600 m depth (Kendrick and others, 1979; Herbin and others, 1983). The requisite sediment thicknesses are present in the basin only near the continental margins, where sediments shed by the continents have built up to thicknesses as great as 8 km (Plate 6) and as much as 2-3 km of sediment overlie Lower Cretaceous beds there (Mountain and Tucholke, 1985). Thermal modeling for the United States east coast margin suggests that sediments buried beneath the continental rise are mature for oil and, in places, for gas (Sawyer and others, 1982).

Reservoir Beds and Seals

Many carbonate strata are interpreted to be present beneath

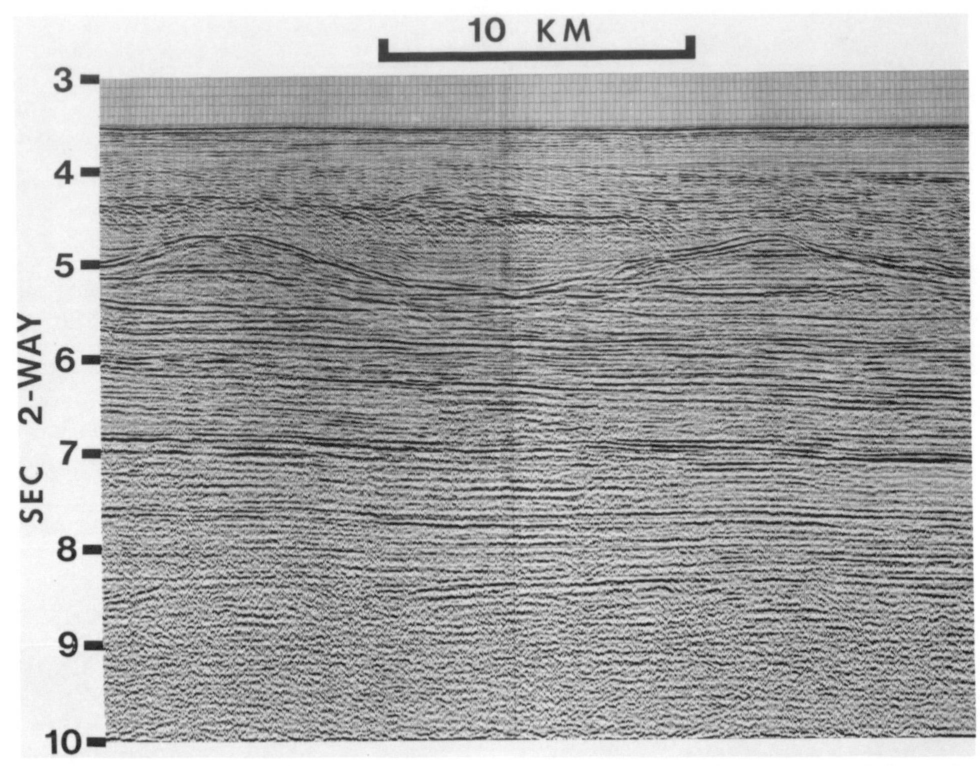

Figure 3. Multichannel seismic reflection profile across part of a buried, Paleogene (middle Oligocene?) submarine canyon system within the continental rise off Georgia, eastern United States (USGS line BT4; Dillon and Paull, 1978). Location is shown in Figure 5. A structure contour map of part of this region is shown in Figure 4.

the upper continental rise (Schlee and others, 1979) and they provide ample opportunity to develop reservoir beds in locations where the sedimentary section is thick enough to generate hydrocarbons. Under the very uppermost part of the present continental rise are Jurassic-Cretaceous reefs, carbonate platforms, and bank-edge carbonate shoals that were once at sea level (Figs. 1, 2) (Schlee and others, 1979; Jansa, 1981; Mountain and Tucholke, 1985). These carbonate structures, now far below sea level, could have experienced fresh-water leaching and dolomitization during Late Jurassic and Early Cretaceous sea-level lowstands, processes that would generate significant porosity. Carbonate talus deposits and debris flows that accumulated on the seaward slope of this carbonate margin could be important petroleum reservoirs (Poag, 1982), although initial porosity probably was occluded by carbonate cement, as has been noted in cores drilled off the Mazagan Plateau (Hinz and others, 1982). Thus, formation of secondary porosity would be necessary, as in the case of the Poza Rica trend, Mexico (Enos, 1974). Because these fore-reef deposits always have been at considerable depths (rise and lower slope), solution porosity created by phreatic leaching probably would not have formed. Significant fracture porosity would not be anticipated in this tectonically quiet region, although historical earthquake activity (Zoback and others, this volume) suggests that some fracturing might have occurred.

Coarse turbidite and debris-flow deposits, forming channel-filling wedges, might be a very important reservoir type on the continental margins. Numerous canyons and channels have been recognized in seismic reflection profiles (Figs. 3, 4) (Mountain and Tucholke, 1985; Schlee and others, 1985), and channel-fill deposits have been identified in drill samples (Poag, 1985).

The necessary seals for reservoir rocks like those described above easily could be formed by the many fine-grained deposits that are characteristic of the continental rise and deep oceanic basin.

We conclude that source beds, maturation, reservoir beds, and seals probably all are present in the region of the western North Atlantic basin, but that these factors probably would combine to produce significant petroleum traps only near the continental margin. The remaining question is, what sort of traps are likely to be present? We will consider three types of possible traps: stratigraphic traps, structural traps, and traps created by geochemical processes that form gas hydrates.

Stratigraphic Traps

Along the continental margin of eastern North America, where there is essentially no tectonic activity but probably large sedimentologic changes from the continental shelf to the abyssal plains, stratigraphic traps probably are far more common than

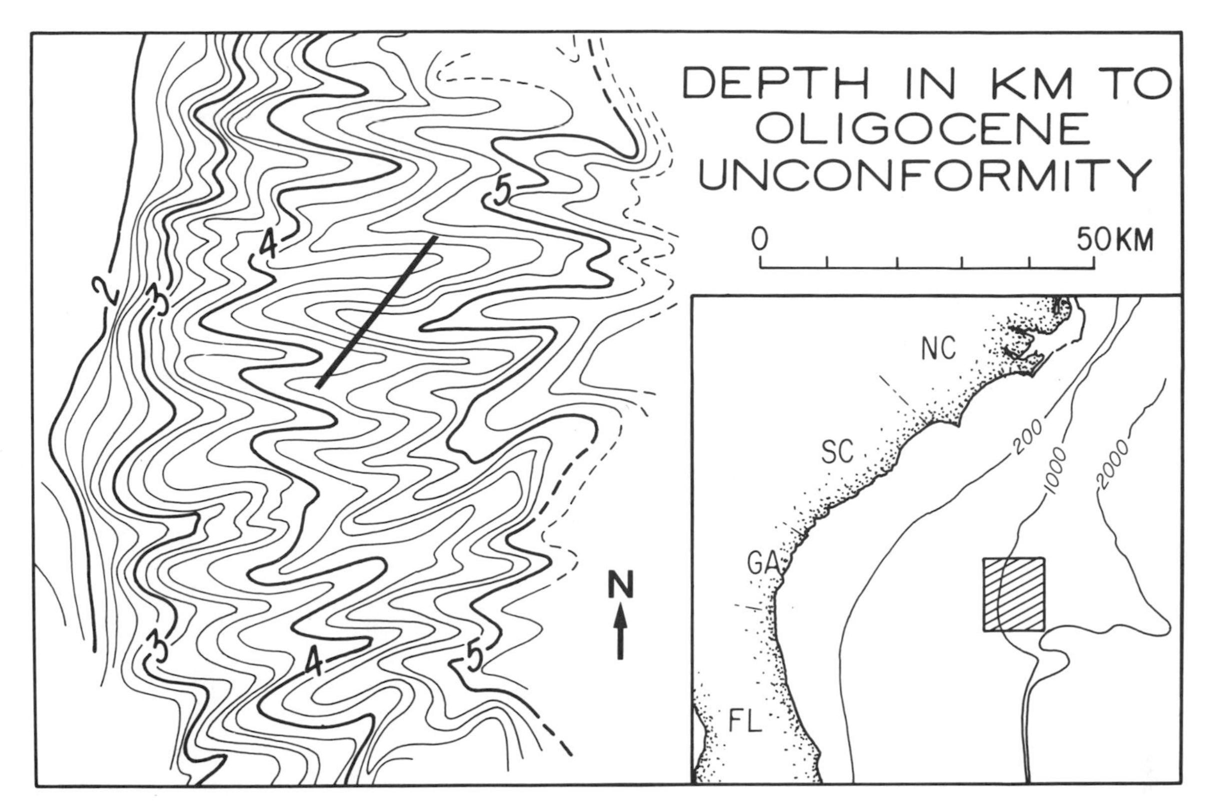

Figure 4. Structure contours, in kilometers below sea level, on Paleogene unconformity formed by submarine-canyon erosion. Heavy line shows location of seismic profile shown in Figure 3.

structural traps. The most obvious traps observed in seismic profiles across the continental rise are features that initially formed the edge of the continental shelf, although they are now beneath the continental rise. An example off New Jersey is shown in Figure 1 (Schlee and others, 1979), where the old Jurassic-Lower Cretaceous carbonate platform edge resides well seaward of the present shelf break. This profile is depth-converted, and demonstrates significant apparent closure. Four Shell exploration wells were drilled on this structure to test the carbonate bank-edge structure or reef, as well as back-reef and fore-reef facies. All wells were declared dry. Two of the wells set drilling records for water depth when they were drilled, and one remains the world's deepest-water offshore petroleum test well at 2,119 m water depth. In other areas of the continental slope, the carbonate reefs or banks have so little sediment cover that they are unlikely to be sealed, but they may have shed fore-reef detritus to form reservoirs that are sealed by younger deep-sea clays. Such situations are suggested by Mattick and others (1978) and Poag (1982).

Simple pinchouts of deep-sea sands may form traps along the margin. Two examples of such structures are suggested (Fig. 2) by termination of Lower Cretaceous reflection events against the Jurassic shelf-edge bank, and termination of Paleogene reflections against the superjacent unconformity (Valentine, 1981). The Lower Cretaceous pinchouts occur in rocks that are the updip equivalent of the Hatteras Formation of approximately Barremian-Cenomanian age, which is the most prominent example of an organic-rich unit in the basin. If sand layers are present here, near the old continental shelf edge, then traps may have been formed.

Some lateral, as well as vertical, changes in permeability are required in order to form a stratigraphic trap, of course. A variety of geological processes could create these changes, but perhaps the most apparent in the western North Atlantic Basin are the erosional-depositional processes that have created the many irregular unconformities that are observed in the rise deposits. The "unconformity" in Figure 2 is part of the well-known Horizon A^u, eroded principally by bottom currents that flowed parallel to depth contours beginning in latest Eocene to early Oligocene time (Tucholke, 1979; Tucholke and Vogt, 1979). More local erosion of channels across the continental margin at various times also is common. One such well-developed channel sequence lies buried within the continental rise off the southeastern United States (Figs. 3, 4). Two kinds of traps may have developed. Coarse strata in intercanyon ridges could be sealed along the canyon walls by fine channel-filling. Alternatively, after these channels were cut, it is possible that they were filled or partially filled by coarse-grained deposits. Such seaward-dipping, tongue-like, porous and permeable pods could be sealed by younger fine-grained deposits to form petroleum traps (Amato and Simonis, 1981). Similar channel-fill deposits also have been described for continental slopes off New Jersey (Poag, 1985) and Georges Bank (Schlee and others, 1985).

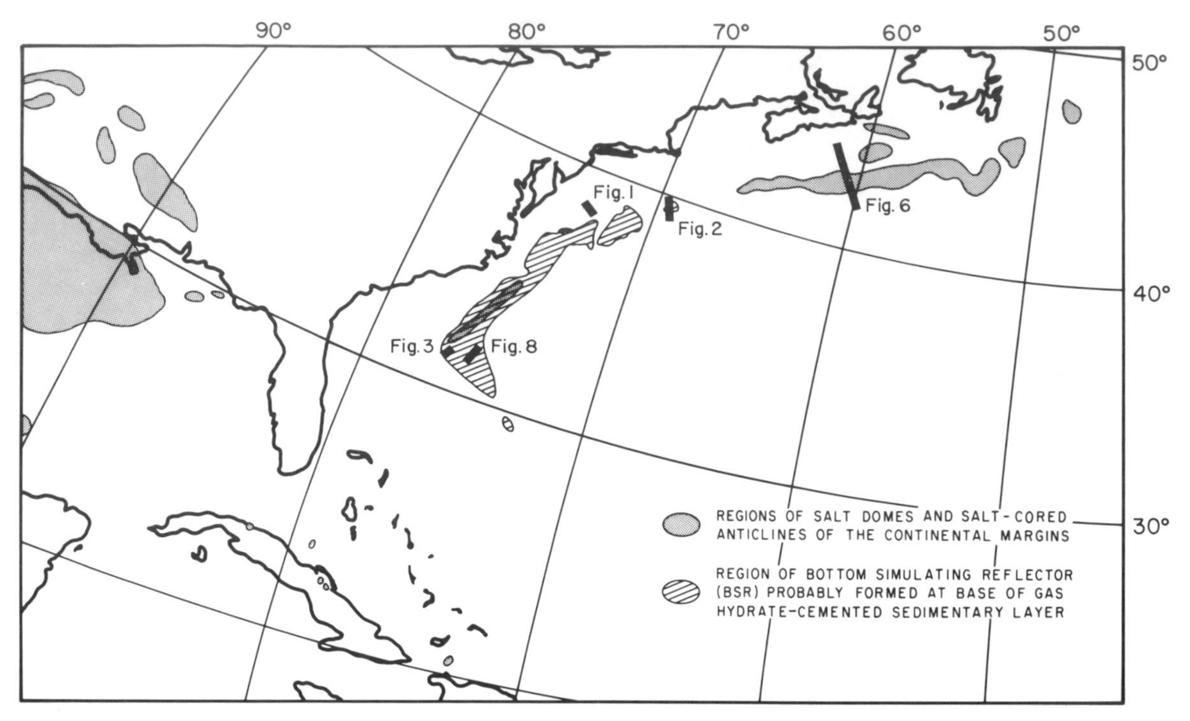

Figure 5. Reported salt diapirs, and bottom simulating reflectors (BSR's) resulting from gas hydrate (clathrate) formation in the western North Atlantic region. Salt diapirs reported in King (1969); Meyerhoff and Hatten (1974); Jansa and Wade (1975); Uchupi and Austin (1979); Collins (1980); Grow (1980); Martin (1980); Dillon and others (1982); Popenoe and others (1982). Gas hydrates mapped by Tucholke and others (1977); Shipley and others (1979); Dillon and others (1980); Paull and Dillon (1981); Popenoe and others (1982). Locations of seismic reflection profiles are shown by heavy bars.

Structural Traps

The western boundary of the North Atlantic Basin is formed by trailing-edge (passive type) continental margins, except for the eastern and northern boundaries of the Caribbean plate. Passive continental margins are initiated by stretching and rifting of the lithosphere, but during the subsequent drift episode, which is the marine phase of continental margin accretion (probably the main hydrocarbon-related phase), relatively few tectonic structures are formed. Notable tectonic structures formed in passive margins during the drift phase are growth faults and salt domes. Growth faults are normal faults that move during accumulation of strata by a gravity-driven sliding to seaward of a major block of strata above a concave upward fault. Such faults are scattered along the continental margin (Mattick and others, 1978). In areas of high sedimentation rates, this type of growth fault can generate anticlines in the form of so-called rollover structures, in which strata in the upper, gliding block subside against the curved fault surface. Folds also can form by crumpling of strata in the forward part of the gliding block. Growth faults also form due to salt migration, where removal of salt below a block of strata allows it to subside. Along the margin of the western Atlantic Basin, a growth fault system 300 km long has been reported just west of the salt dome belt off the southeastern United States (Fig. 5). Presumably, the faulting was caused by salt flow into that belt of salt structures (Dillon and others, 1982).

Reported salt diapirs along eastern North America (Fig. 5) are located principally in two zones that extend to slope depths, one off Nova Scotia and Newfoundland at the outer edge of the Scotian Basin (Jansa and Wade, 1975) and another off North Carolina seaward of the Carolina Trough (Dillon and others, 1982). These groups of domes clearly are related to the basins that lie to landward. Salt is a common deposit of early-stage rift basins, and is common around the North Atlantic (Aymé, 1965; Pautot and others, 1970; Burke, 1975; Jansa and Wade, 1975; Kinsman, 1975; Evans, 1978; Uchupi and Austin, 1979; Jansa and others, 1980; Manspeizer, 1982; Dillon and others, 1982). Salt commonly is squeezed seaward from its site of initial deposition and then rises in diapirs (Lehner, 1969; Humphris, 1979). Salt diapirs off eastern North America are observed as single domes or groups of domes (Fig. 6). Many of these actively rising domes deform the sea floor, as indicated in seismic profiles and side-scan sonar. Warping of strata on the crest and flanks of salt domes and termination of the strata against the domes commonly forms traps. Traps can also result from associated faulting. The east coast domes, both off Canada (Fig. 6) and off the Carolina Trough commonly seem to reach almost to the sea floor, so that there is extensive faulting of shallow strata above the domes due

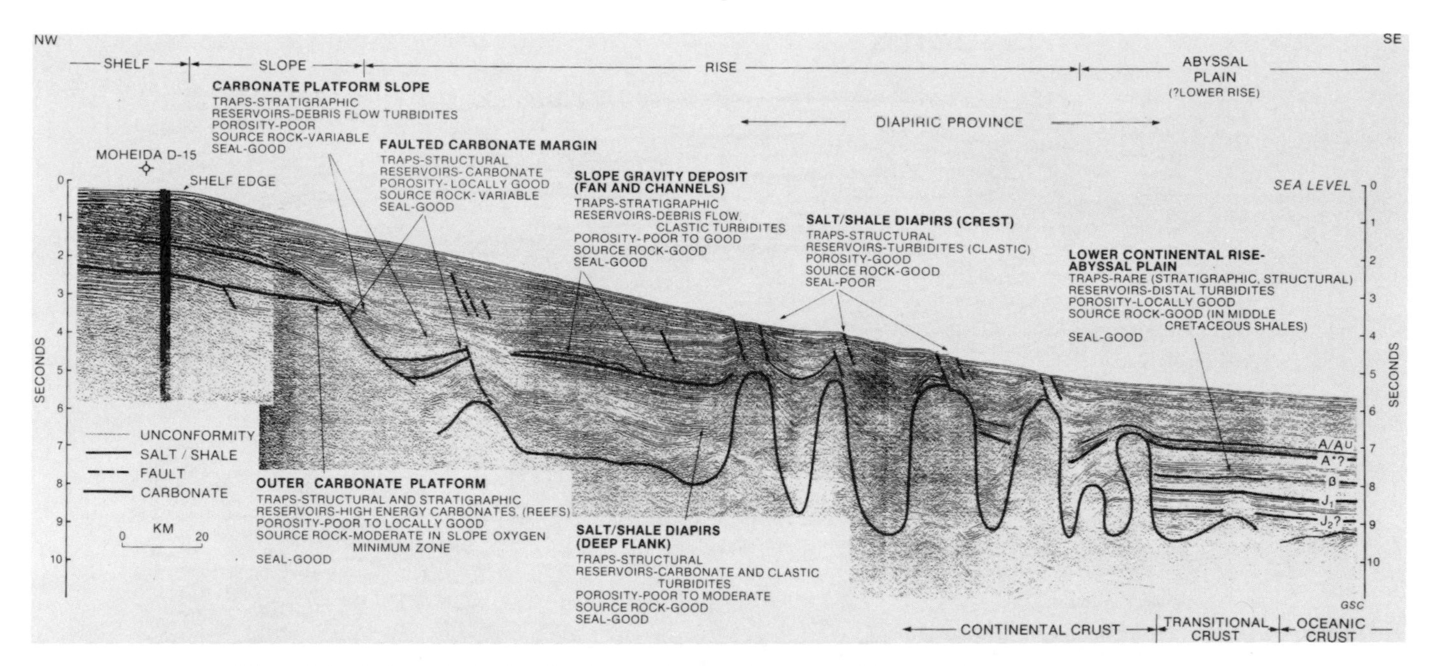

Figure 6. Multichannel seismic reflection profile extending from the Nova Scotian shelf southeast across the Scotian salt ridge to the Sohm Abyssal Plain. Petroleum potential of various geologic features is summarized. Correlations of reflectors to time-rock units are given in Plate 1. Location is shown in Figure 5.

to salt movement and solution collapse. This shallow faulting may have destroyed petroleum traps over the domes.

Gas Traps Formed by Geochemical Processes (Gas-Hydrate Traps)

Gas hydrates, also known as clathrates, are ice-like crystalline solids that form as a cage-like structure of water molecules surrounding a gas molecule (Davidson, 1983). The gas, in nature, can be CO_2 or H_2S, but most commonly is methane (CH_4) or several other low-molecular-weight hydrocarbons. Gas hydrates are stable above the stability temperatures of solid water (ice) at the elevated pressures present at seafloor depth (Sloan and Parrish, 1983). For example, Figure 7A indicates that at 0.3 km water depth, methane hydrate is stable at a few degrees C, whereas at 3 km it is stable up to temperatures exceeding 20°C. Consider what happens with a sea floor at 2 km and a normal temperature profile, as indicated in Figure 7B. At depths where the temperature curve falls below the phase boundary, gas hydrate will be stable and it is likely to be present within the sediment where biogenic generation of methane from organic matter can produce enough gas to create significant amounts of gas hydrate or where gas might be supplied by leakage from deeper strata. Gas hydrates will persist down to the depth where the phase boundary curve is intersected by the geothermal temperature curve. Below that depth, gas hydrate would be unstable, and any gas would be dissolved in pore water or present as free gas.

Because the geothermal gradient is fairly constant over sizeable sea-floor areas, the temperature at which gas hydrate becomes unstable will occur at an approximately constant subbottom depth; thus a uniformly thick layer of sediment just below the sea floor is capable of having its pore space filled with solid gas hydrate. Cementation of the sediment by filling the pores with gas hydrate increases its seismic velocity (Stoll and others, 1971; Stoll, 1974). Thus the surface layer of sea-floor sediment, if it contains gas hydrate, has a higher seismic velocity

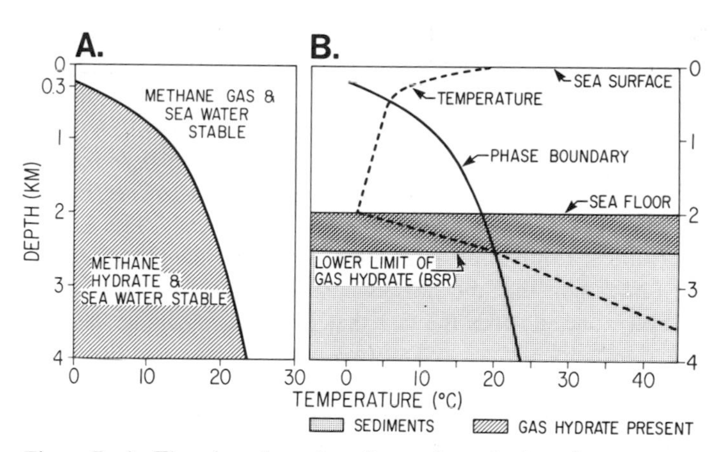

Figure 7. A. The phase boundary for methane hydrate in sea water, adapted from Claypool and Kaplan (1974). B. Relationship of phase boundary to temperature/depth in sea water and sediments using values typical of those off the southeastern United States. The phase boundary is the same as that shown in A. In this example the sea floor is placed at 2 km water depth. BSR indicates position of the bottom-simulating reflector that is observed in seismic profiles at the base of the zone of gas-hydrate formation.

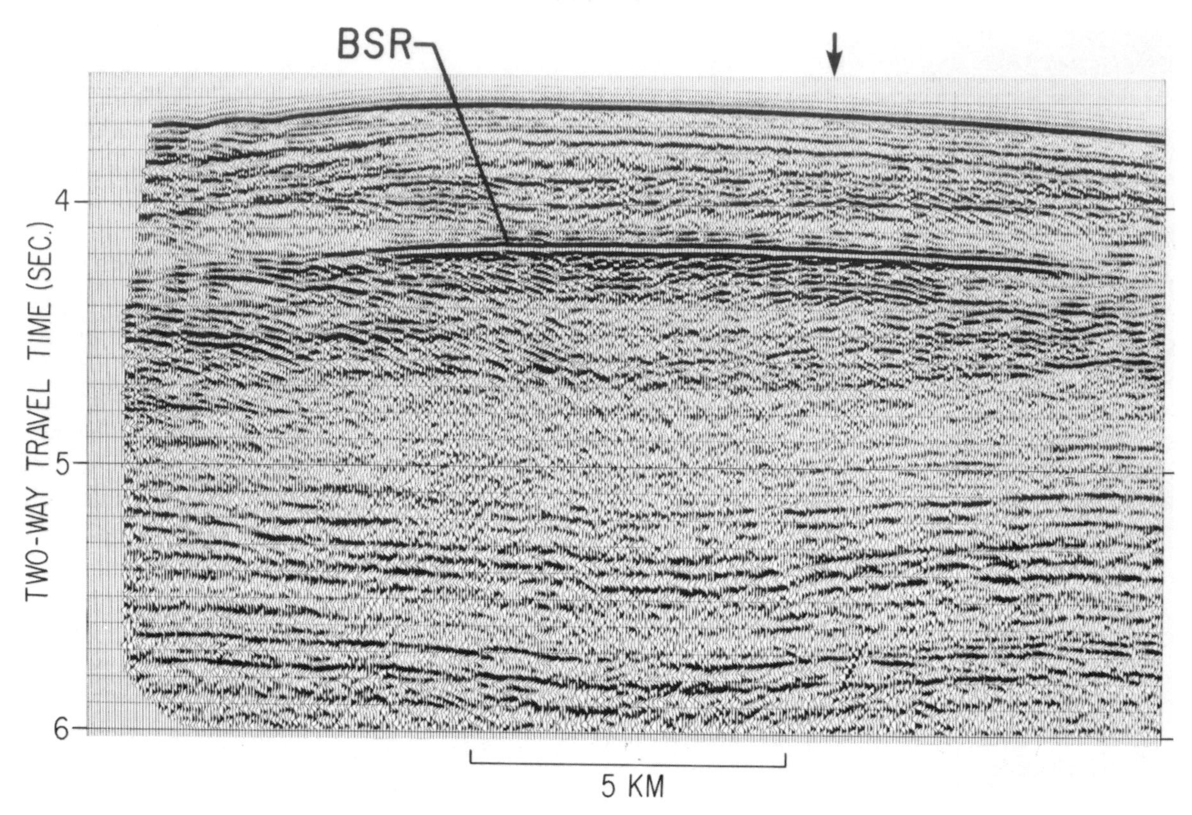

Figure 8. Bottom-simulating reflector (BSR) in a multichannel seismic reflection profile across the Blake Outer Ridge off the southeastern United States (USGS line TD2; Dillon and Paull, 1983). Location is shown in Figure 5. The BSR is considered to form at the base of a zone of gas-hydrate-cemented sediment; in this case the hydrate zone at the ridge crest may trap free gas. The very strongly reflective portion of the BSR may represent the region where gas is trapped beneath the hydrate-filled sediment. This relationship would account for the strength of the reflection there (free gas would cause an unusually low velocity and, therefore, cause a large reflectivity contrast at the velocity inversion). Gas trapped beneath the strongly reflective part of the BSR, resulting in a low-velocity region, also could account for the apparent pull-down of strata visible beneath that region, most notably at 5-6 seconds depth.

than underlying sediment below the phase boundary, and the acoustic impedance change at this velocity interface would produce a seismic reflection at the base of the cemented zone. A reflection event called a bottom simulating reflector (BSR), that nearly parallels the sea floor in seismic profiles, has been considered to result from gas hydrate formation in the sediments (Fig. 8) (Tucholke and others, 1977; Shipley and others, 1979; Dillon and others, 1980; Paull and Dillon, 1981). Evidence supporting the conclusion that the BSR is created at the base of the gas-hydrate zone includes the fact that the BSR occurs at the predicted level for the gas hydrate/gas phase boundary at the water depths and temperature gradients observed (Tucholke and others, 1977; Kvenvolden and Barnard, 1982; Kvenvolden, 1983). Furthermore, analysis of the seismic velocity structure (Dillon and Paull, 1983) shows that velocities are high above the BSR, as would be predicted from laboratory measurements (Stoll and others, 1971), and below the BSR, velocities are extremely low (less than the velocity of water). The presence of velocities less than water velocity requires that free gas be present below the BSR and shows that gas hydrate-cemented sediment can act as a seal to form a gas trap. Evidence for a decrease in seismic velocity at the BSR also has been observed as a phase reversal in the seismic wavelet returned from that level (Shipley and others, 1979).

Bottom simulating reflections have been mapped by several workers off eastern North America (Fig. 5). Undoubtedly, this map is incomplete because the BSRs often have been viewed as artifacts and ignored by seismic interpreters.

Although it ultimately may be possible to produce gas from sub-seabed gas hydrates (Holder and others, 1983), the introduction of heat into sediment to break down the hydrate, or of other chemicals to act as antifreeze, presents considerable technical difficulties. Sediments saturated with gas hydrate may be far more important as seals for gas traps (Tucholke and others, 1977; Dillon and others, 1980). Diagrams of several possible gas traps having gas hydrate seals are shown in Figure 9. The simplest case (Fig. 9, upper left) occurs when the sea floor forms a dome. In such a case, the gas hydrate layer (paralleling the sea floor topography) will also form a dome and can trap gas. Velocity analyses at such a location (at arrow in Fig. 8) indicate that free gas is trapped beneath a gas hydrate seal, in this case beneath the Blake Ridge off the southeastern United States. Other possible types of gas hydrate-sealed traps that have been interpreted from seismic profiles include situations where permeable beds, interlayered with impermeable beds, dip opposite to the sea floor slope and are sealed at their updip ends by gas hydrate (Fig. 9, right). Another type of gas hydrate-sealed trap is produced where a gas hydrate layer crosses a salt diapir (Fig. 9, lower left). Salt diffusing from the diapir will act as antifreeze for the hydrate, and the diapir will conduct heat upward more effectively than surrounding sediments because salt has a higher thermal conductivity. Both factors inhibit gas hydrate formation above the diapir and raise the phase boundary, thus producing a dome in the base of the gas hydrate layer. Such traps generally may be small and non-commercial, but such shallow gas traps could be a hazard to drilling if they are not recognized.

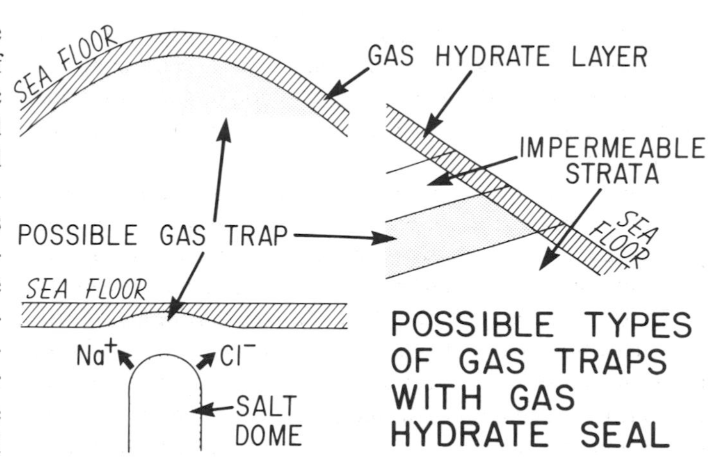

Figure 9. Schematic diagrams of situations in which a layer of gas-hydrate-cemented sediment can form gas traps. Compare situation at the upper left with that in Figure 8.

Economic Considerations

Given the great water depths that we are considering here (<2,000 m), any discussion of potential petroleum resources should take into account the economic viability of their present and possible future recovery. For this example we will consider only liquid hydrocarbons. The importance of this hypothetical example is not in the exact numbers used, but in the general magnitude of the problems, the type of analysis required, and the size of accumulation necessary for commercial production. Due to space limitations, this scenario is necessarily simplistic. The area that we will take to be prospective is approximately the size of one federal lease block (2,000 hectares). A reservoir productivity is chosen that is conservatively appropriate, based on experience with carbonate reservoirs, and a rather thick net pay zone will be assumed.

Prospective area —	2,000 hectares
Reservoir productivity —	1,250 bbls/hect m
Net pay —	200 m
Reserves —	500,000,000 bbls
Development parameters	
Platforms —	2 at $500,000,000 each
Wells —	80 at $5,000,000 each
Infrastructure (includes transportation, storage, shore facilities)	$500,000,000

In this example, development expenditure approaches $2,000,000,000. One may argue specifics, but this is probably a realistic figure based on current costs that are experienced in moderate water depths and that are extrapolated into the deep-ocean realm (>2,000 m). Furthermore, lease acquisition costs, royalty payments, and the cost of borrowed money must be considered.

With 80 available well slots, a figure of 70 productive development wells is chosen as reasonable. A daily oil production rate of 1,000 bbls/day/well is estimated. Some deep-sea reservoirs may not be capable of sustaining this rate, while others, such as reef carbonates with secondary porosity, may produce at higher rates. Total production in this example would average 70,000 bbls of oil/day. A reasonable annual production rate, allowing for down time, would be 23,000,000 barrels. Balancing the geological and other risks, lead time for initial and peak production, cost of borrowed money, royalties, taxes, and so forth, against probable income at various oil price levels, it can be seen that price is critical in determining economic viability. At higher future prices of $50/bbl or more, the rate of return may be worth the large risk, whereas at today's prices these resources cannot be developed. Good reservoir-rock porosity and permeability, thick pay section, and a high ratio of pay thickness to areal extent would be required in order to minimize development costs while maximizing production rates to achieve a reasonable rate of return.

METALLIFEROUS AND PHOSPHORITE DEPOSITS

Excluding seaward extensions of continental metalliferous deposits and continental-shelf deposits, metalliferous deposits in the North Atlantic basin can be divided into three categories: ferromanganese nodules, ferromanganese crusts, and hydrothermal deposits. In addition, phosphorite deposits occur on the Blake Plateau.

Ferromanganese Nodules

Reported locations of nodules and crusts are indicated on the bathymetric map at the back of this volume (Plate 2). The paucity of data permits only general conclusions about nodule distribution and composition, but the general pattern of metal deposits can be presented in a schematic cross section (Fig. 10). Abyssal nodules, which tend to be enriched in Mn (Cronan, 1975), occur primarily on low marginal slopes of the North Atlantic basins on the sides away from the continents, where input of terrigenous material is small. Another major concentration of nodules and pavements occurs on the Blake Plateau off the coast of the southeastern United States. These deposits consist of ferromanganese oxides developed on phosphorite nuclei, and they differ significantly from the deep-ocean nodules in form and in environment of origin.

The Blake Plateau has been known as a site of phosphorite and ferromanganese occurrence since John Murray's (1885) report on samples from cruises of the Coast Survey vessel *Blake.* In the 1960s, extensive manganese pavement and nodule deposits were discovered in the area (Pratt and McFarlin, 1966). These were shown by Manheim and other (1980, 1982) to be formed on a lag rubble of mid-Tertiary phosphorites, partially by replacement of the phosphorite and partially by layer-by-layer accretion of ferromanganese oxides. These authigenic deposits form very close to the North American continent because the seabed scouring effect of the Gulf Stream has prevented sediment accumulation. The Gulf Stream has been active in this region since early Tertiary time (Tucholke and Mountain, this volume). The area covered by these deposits exceeds 14,000 km^2 and most of it is underlain by dense phosphorite pavement, on the basis of recent multi-frequency acoustic measurements and other techniques (Manheim and others, 1982).

Although the Blake Plateau nodules have metal compositions that are less valuable than those of nodules in some Pacific areas, certain factors have produced interest in mining the Blake Plateau area. These include location of the deposits in relatively shallow water adjacent to the U.S. coast, large potential tonnages of deposits (hundreds of millions of tons), presence of the deposits within the U.S. Exclusive Economic Zone, and the possibility of using the nodules directly as a catalyst material in petroleum refining. The use of the nodule material as a catalyst requires

TABLE 1. COMPARISON OF PRINCIPAL METAL CONCENTRATIONS IN NODULES FROM THE ATLANTIC AND PACIFIC OCEANS*

	PACIFIC			ATLANTIC		
	McKelvey and others (1983)		Cronan (1980)	McKelvey and others (1983)		Cronan (1980)
Component	Mean	Standard Deviation	Mean	Mean	Standard Deviation	Mean
Mn	20.1	6.9	19.8	13.3	6.2	15.8
Fe	11.4	5.3	12.0	17.0	5.0	15.4
Ni	.76	.43	.63	.32	.23	.33
Cu	.54	.40	.30	.13	.10	.12
Co	.27	.19	.33	.27	.20	.32

*More samples are available from the Pacific sites (about 1780) than for Atlantic locations (less than 300) (Frazer and Fisk, 1980). The Pacific data of McKelvey and others (1983) are higher in Cu and Ni than those of Cronan (1980) because the former contain a larger number of samples from the prime abyssal nodule area near the Clarion-Clipperton Fracture Zone (SE of the Hawaiian Islands). Average depth of the Pacific samples is 4379 m (SD = 1195 m), and of the Atlantic samples is 3614 m SD = 1513 m). Mean values are in percent of oven-dried material.

relatively minor processing and could be followed by recovery of metals from spent catalyst. It is pointed out by Michael Cruickshank (written communication, 1983) that V and Ni content of the catalyst would be further heightened by additions of these metals from petroleum feedstocks. These considerations may have led to the attempt by Reynolds International, in 1978, to obtain leases to mine the region. The lease request was cancelled in 1982 at a time of low mineral prices.

Taken as a whole (Table 1), Atlantic nodules tend to have lower Mn, Ni, and Cu values than Pacific nodules (but note that far more Pacific samples are available—see footnote Table 1). Furthermore, only about 5% of Atlantic nodules exceed 1% nickel plus copper (on the basis of 180 samples), whereas, worldwide, more than 30% of nodules analyzed exceed a total of 1% of those two elements (on the basis of 1,236 samples). Co values of Atlantic and Pacific nodules are rather similar, whereas Atlantic nod-

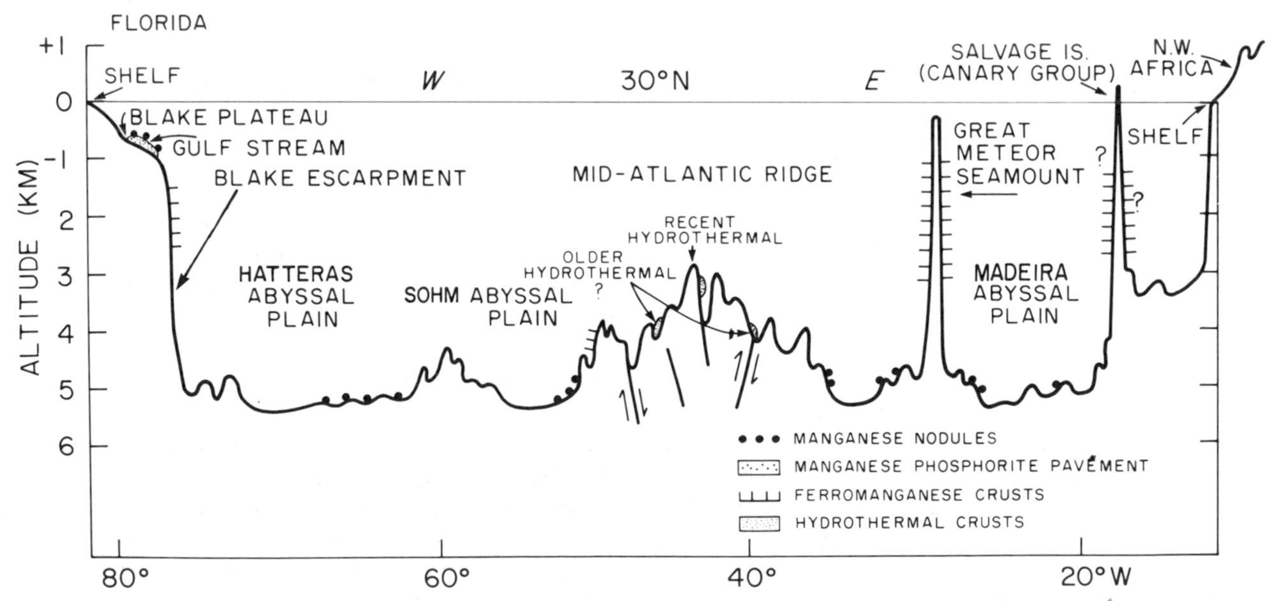

Figure 10. Semi-schematic cross section of the North Atlantic Basin at 30° N latitude showing distribution of manganese nodules, pavements, and crusts, and hydrothermal deposits on the sea floor of the North Atlantic. Vertical relief is greatly exaggerated. The section is a modified version of sketches in Emelyanov (1982).

ules studied to date tend to have higher Fe values as well as larger amounts of certain minor constituents such as V and As (Manheim and others, 1982; and unpublished data). These relationships are linked to the greater supply of terrigenous, iron-rich detritus to the North Atlantic and to lesser contributions from decomposed planktonic debris, as compared to Pacific nodules (Cronan, 1975; Ewing and others, 1973; Heath, 1981; Emelyanov, 1982). V and As have been previously noted to be correlated to iron content in deep ocean nodules (Calvert and Price, 1977). Mineralogically, Atlantic nodules are dominated by vernadite (hydrous MnO_2), and amorphous oxides (Cronan, 1975). Todorokite (another form of poorly crystalline, hydrous MnO_2) is present principally in Blake Plateau pavements and in some abyssal nodules. Mn, Ni, Cu, Zn, Mg, Ca, K, and Na are enriched in the todorokite nodules with respect to vernadite nodules. The latter tend to be enriched in Fe, Co, V, P, and As (Manheim and others, 1980).

Cobalt-rich Ferromanganese Crusts

Ferromanganese crusts on exposed rock surfaces in the ocean (Plate 2) have long been known (Murray and Renard, 1891). However, only recently has the high cobalt content of many of these crusts been reported, particularly from the Mid-Pacific Seamounts (Cronan, 1977; Frazer and Fisk, 1980), and their possibly significant economic potential noted (Halbach and others, 1982; Halbach, 1982; Halbach and Manheim, 1986). The Atlantic Ocean traverse (Fig. 10) schematically indicates types of locations where crusts that are potentially Co-rich have been observed. The crusts grow very slowly and, thus, are found only in areas free of detrital sedimentation, commonly at relatively shallow water depths (less than 3,000 m). The New England and Corner seamounts are known to bear crusts as thick as 14 cm on basaltic and phosphatic substrates (Heirtzler and others, 1977). Thin, but extensive crusts also are developed on exposed Cretaceous carbonates along the Blake Escarpment (Dillon and others, 1985; Dillon and others, 1986).

Enhanced cobalt and nickel concentration (>1%) has created much interest in possible mining of Pacific crusts (Commeau and others, 1984). Comparable data from Atlantic sites are sparse, although values of approximately 1% Co have been reported for crusts from Kelvin Seamount in the New England seamount chain, and an average concentration of 0.54% Co was measured in seven crust samples from isolated North Atlantic seamounts (Cronan, 1975).

Phosphorites

The phosphorite deposits of the Blake Plateau are a part of the massive phosphogenic development in the middle Tertiary along the U.S. Atlantic margin south of Cape Hatteras (Manheim and others, 1980). Lag deposits of primary phosphorite remain after winnowing and scouring away of Oligocene through middle Miocene sediments by the Gulf Stream. These deposits have been recemented to form continuous pavements that cover a total area of about 20,000 km^2. The thickness of the deposits is only known at the outer feather-edge (a few cm), but boulders greater than one meter thick have been observed from submersibles. Water depths range from <400 m to >1,000 m. Only a minimum tonnage of about 2 billion tons has been estimated because samplers have been unable to penetrate through the pavements.

The phosphorite resembles other marine phosphorites, consisting primarily of carbonate fluorapatite with subsidiary dolomite, glauconite, quartz, and calcium carbonate. No primary phosphatization is presently occurring on the Blake Plateau. Uranium content of the cobbles, plates and pavements is lower than normal in-place primary phosphorite due to oxidative leaching by Gulf Stream waters.

Although the deposits are very large, their economic potential is lowered by their greater water depth and distance from shore compared to the shelf deposits off North Carolina and Georgia (Riggs, 1984), and perhaps partly by their impregnation with iron and manganese oxides, discussed previously.

Hydrothermal Deposits

Studies by Rona and collaborators (Rona, 1980; Rona and Lowell, 1980) in the 1970s revealed the presence of hydrothermal deposits on the Mid-Atlantic Ridge, particularly in the Trans-Atlantic Geotraverse (TAG) geothermal field (Edmond, this volume). As was expected from the fact that the Mid-Atlantic Ridge is a slow-spreading, ocean-opening center, initial discoveries consisted of thin oxidized crusts having chemical composition similar to hydrothermal oxides from more northerly (FAMOUS) Mid-Atlantic sites (Toth, 1980) and from Pacific sites. The typical pattern of hydrothermal oxide deposits was observed: extreme separation of Mn and Fe (i.e. where Mn is high, Fe is low and vice versa), low Al content in Mn oxides, and relative depletion in valuable trace elements. More recently, relatively pure manganese oxide deposits, locally up to 3 m or more in thickness, have been reported for the TAG area (Rona and others, 1984). Rona (1985) reported recovery of massive Cu-Fe-Zn sulfides, sulfates and layered Fe-Mn oxides, as well as photographic observations of hydrothermal vents ("black smokers"). This discovery suggests that well developed hydrothermal systems and associated deposits are not restricted to fast-spreading ridges, as in the Pacific.

No deposits of significant value or tonnage have yet been reported from, or predicted for the Mid-Atlantic Ridge. Even hypothetical rich deposits may be unlikely to quality as economically valuable, considering the availability of resources on land, poor long-term market conditions for the metals (Cu, Zn, and Ag) currently known from polymetallic sulfide deposits, and the practical and political problems of mining on the sea floor in the middle of the Atlantic Ocean. Nevertheless, the unexpected discoveries regarding metalliferous deposition on the sea floor in the past 8 years, as well as uncertainty in predicting future changes in political and market developments, suggest that it may be wise to leave open the question of future economic potential.

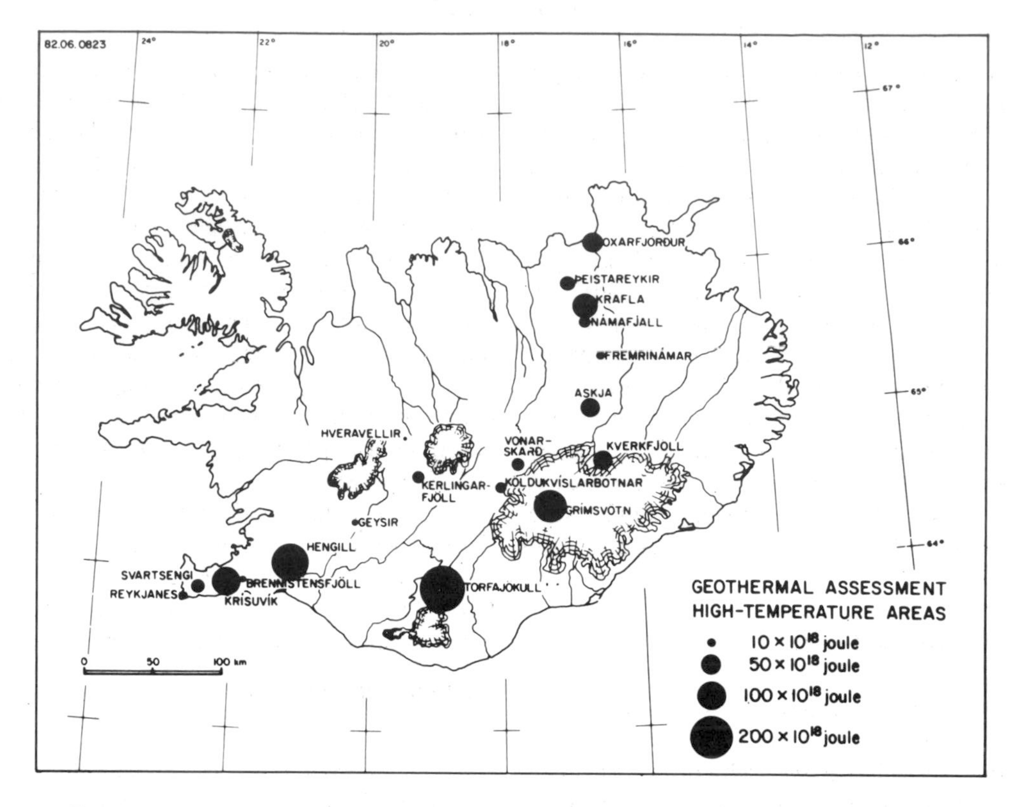

Figure 11. Geothermal assessment of the total stored heat for high-temperature fields in Iceland.

GEOTHERMAL RESOURCES

Iceland

Geothermal resources are relatively abundant in Iceland, due to the country's location astride the Mid-Atlantic Ridge (Plate 2). Commercial exploitation of geothermal energy began just over 50 years ago when the first geothermal heating was installed in a few houses in Reykjavik. At the present time, geothermal energy supplies about 80% of the space-heating requirements of Iceland, at an average cost which is about 18% of the cost of oil heating. Other uses of geothermal energy include generation of electricity, industrial drying, greenhouse farming, aquaculture, and heating of swimming pools (see Fig. 2 of Edmond, this volume). Today, geothermal energy provides about one-third of the total energy sold to users, and its thus plays a major role in the energy economy of the country. In spite of this, only a small fraction of the estimated resource is being used.

The strongest geothermal activity occurs in the so-called neovolcanic zone (high-temperature fields), while lower-grade activity appears on the flanks of this zone (low-temperature fields) (Bodvarsson, 1961; Edmond, this volume). About 150 wells have been drilled to depths of more than 1,000 meters in the last 25 years. Reservoir temperatures range from less than 100°C to about 350°C. Salinity ranges from some 200 ppm total dissolved solids to about 32,000 ppm. The physical state of the reservoir fluid ranges from single-phase, liquid-dominated, to a two-phase boiling state, with dry steam conditions developing around wells in some of the two-phase reservoirs.

An assessment of the geothermal potential of the high-temperature fields, based on stored-heat calculations, recently was made by the Geothermal Division of Orkustofnun (Pálmason, 1981). The area of the fields was estimated on the basis of surface manifestations and geophysical surveys where they were available. The reservoir thickness from which production can be obtained is assumed to be 3 km. The stored-heat assessment, together with the location of the high-temperature fields, is given in Figure 11. Assuming a geothermal recovery factor of 20% and reasonable estimates for the accessibility of the various areas, the total available heat from the high-temperature fields is estimated to be 10^{20} Joule, or, when converted to electricity, 175,000 MWyrs. On the basis of geological considerations, it seems possible that an additional potential of 2-3 times the above values may lie hidden within the neovolcanic zone without direct surface expression. The assessment makes no predictions of the rate at which the energy can be withdrawn, but a considerable body of production data is gradually accumulating (Pálmason and others, 1983).

Azores

Geothermal activity occurs on four of the nine main islands in the Azores group astride the Mid-Atlantic Ridge. The main hot springs are on Sao Miguel Island. A few deep wells have been drilled in the Ribeira Grande area on Sao Miguel Island, and one of them is connected to a 3-MW portable power plant (Mete and Rivera-Rodriguez, 1982). The Ribeira Grande reservoir is liquid-dominated with a reservoir temperature of 220°–230°C. As far as is known, no assessment has been made of the total geothermal potential of the Azores. A summary of the hot-spring activity is given by Waring (1965).

CONCLUSIONS

The State of Resources of the Deep Basin

We have concentrated, here, on the resources of the deep ocean basin, primarily seaward of the slope/rise break, because other volumes of the DNAG series will describe resources closer to shore. Offshore resources certainly are valuable, but, as is shown clearly in the history of offshore petroleum exploration, economic factors militate that nearshore and shallow-water sites will be developed first. In the case of petroleum, now that many of the continental shelves of the world have been studied, a search for the most geologically and economically favorable off-shelf traps is underway. Oil and gas are the most important mineral resources of the offshore region, and this will probably continue to be true as the search for these resources extends into the deep sea, seaward of the continental slope. Petroleum is likely to exist in the thick-sediment regions of the continental rise. Its development will depend primarily on economic factors.

Perhaps the next two potentially most valuable marine mineral resources are sand and gravel, and phosphorite, but these are most abundant and also most easily mined in the continental shelves and, therefore, probably will not be mined in the deep ocean for the foreseeable future. Conversely, ferromanganese crusts and nodules generally are deep-basin resources, not present on continental margins of the western North Atlantic except on the Blake Plateau. The value of these deposits probably is dependent on (1) a need for some of the rarer elements in the deposits, such as cobalt, nickel, or copper, or (2) dual usage of the material as catalyst as well as metal source. Their development depends on the state of the world metals market and on political/strategic considerations with regard to such materials as cobalt, which have a restricted worldwide supply.

In general, hydrothermal deposits of metal sulfides seem to be best developed at fast-spreading ridge crests, as in the Pacific. However, recent discovery of such sulfides on the Mid-Atlantic Ridge suggest that they may occur more commonly in the North Atlantic than was previously anticipated. The competitiveness of mining such deposits, located at mid-ocean sites, against on-land mining seems very problematical, especially considering the long-term depressed state of on-land metals mining. The discovery of rich deposits of valuable metals at sea-floor sites could change this conclusion. Although such deposits would not be anticipated with our present information, our present state of knowledge is in a very early stage of development.

The lands around the western North Atlantic basin with the largest potential geothermal energy resources (Iceland and the Azores) are also those with, perhaps, the least likelihood of developing fossil-fuel resources. Far more geothermal energy probably is available and could be of huge benefit to the countries where its development is feasible.

New Work

What research should be targeted as most valuable to development of the resources in the western North Atlantic Basin? First, even though economic considerations may prevent development of petroleum resources in deep water (2,000-5,000 m) in the near future, we should make an effort to learn the nature and extent of such resources so that we can plan rationally for the energy future of the world. Seismic data collection over the continental rise and drilling in the thin sediment areas of the deep basins have given us preliminary information. Next we need to drill in the thick sedimentary prism of the continental rise to learn whether source beds, reservoirs, and seals exist, whether maturation has occurred, as has been predicted, and whether possible traps really hold petroleum. Furthermore, we need to study, with the drill and other methods (e.g., seismic, heat flow), some of the proposed unconventional petroleum traps, especially gas-hydrate-sealed traps, and we need to consider the possibility of recovery of gas from gas hydrate. Also, for the purposes of understanding world energy resources, more research on geothermal sources should be mounted.

A far greater effort should be made to sample ferromanganese crusts and hydrothermal deposits, to analyze for the rarer elements and to relate the variation in valuable metals to geochemical conditions. Although this is a poorly understood subject and such research is needed worldwide, the Atlantic is marked by an exceptional dearth of information on this topic.

REFERENCES

Amato, R. V., and Simonis, E. K.
1981 : Deep-water petroleum prospects, U.S. Atlantic continental margin [abs.]: American Association of Petroleum Geologists Bulletin, v. 65, p. 1658.

Arthur, M. A., and Schlanger, S. O.
1979 : Cretaceous "oceanic anoxic events" as causal factors in development of reef-reservoired giant oil fields: American Association of Petroleum Geologists Bulletin, v. 63, p. 870–885.

Aymé, J.-M.
1965 : The Senegal salt basin; in Salt Basins Around Africa, The Institute of Petroleum: Amsterdam, Elsevier, p. 83–90.

Benson, W. E., Sheridan, R. E., eds.
1978 : Initial Reports of the Deep Sea Drilling Project, v. 44: U.S. Government Printing Office, Washington, D.C., 1005 p.

Bodvarsson, G.
1961 : Physical characteristics of natural heat resources in Iceland: Reykjavik, Jokull, v. 11, p. 29–38.

Burke, K.
1975 : Atlantic evaporites formed by evaporation of water spilled from Pacific, Tethyan, and southern oceans: Geology, v. 3, p. 613–616.

Byramjee, R. S., Mugniot, J. F., and Biju Duval, B.
1975 : Petroleum potential of deep-water areas of the Mediterranean and

Caribbean Seas: Ninth World Petroleum Congress Proceedings, v. 2, p. 299–312.
Calvert, S. E., and Price, N. B.
1977 : Geochemical variation in ferromanganese nodules and associated sediments from the Pacific: Marine Chemistry, v. 5, p. 43–75.
Claypool, G. E., and Kaplan, I. R.
1974 : The origin and distribution of methane in marine sediments: in Natural Gases in Marine Sediments; ed. I. R. Kaplan; Plenum Press, New York, p. 99–139.
Collins, S. E.
1980 : Jurassic Cotton Valley and Smackover reservoir trends, east Texas, north Louisiana, and south Arkansas: American Association of Petroleum Geologists Bulletin, v. 64, p. 1004–1013.
Commeau, R. F., Clark, A., Johnson, C., Manheim, F. T., Aruscavage, P. J., and Lane, C. M.
1984 : Ferromanganese crust resources in the Pacific and Atlantic Oceans; in Conference Record, Oceans: '84: Washington, D.C., Marine Technology Society, and Piscatawy, New Jersey, Institute of Electronic and Electrical Engineers, Ocean Engineering Society, p. 421–430.
Cronan, D. S.
1975 : Manganese nodules and other ferromanganese deposits from the Atlantic Ocean: Journal of Geophysical Research, v. 80, p. 3831–3837.
1977 : Deep-sea nodules: distribution and geochemistry; in Marine Manganese Deposits, ed. G. P. Glasby; Elsevier, Amsterdam, p. 11–44.
1980 : Underwater Minerals: Academic Press, London, 318 p.
Davidson, D. W.
1983 : Gas hydrates as clathrate ices: in Natural Gas Hydrates: Properties, Occurrences, and Recovery, ed. J. L. Cox; Butterworth, Boston, p. 1–15.
Dillon, W. P., and Paull, C. K.
1978 : Interpretation of multichannel seismic-reflection profiles of the Atlantic continental margin off the coasts of South Carolina and Georgia: U.S. Geological Survey Miscellaneous Field Studies Map MF-936.
1983 : Marine gas hydrates, II: Geophysical evidence: in Natural Gas Hydrates: Properties, Occurrences, and Recovery: Butterworth, Boston, p. 73–90.
Dillon, W. P., Grow, J. A., and Paull, C. K.
1980 : Unconventional gas hydrate seals may trap gas off southeast U.S.: Oil and Gas Journal, v. 78, no. 1, p. 124, 126, 129, 130.
Dillon, W. P., Popenoe, P., Grow, J. A., Klitgord, K. D., Swift, B. A., Paull, C. K., C. K., and Cashman, K. V.
1982 : Growth faulting and salt diapirism: Their relationship and control in the Carolina Trough, eastern North America: in Studies in Continental Margin Geology, eds. J. S. Watkins and C. L. Drake; American Association of Petroleum Geologists Memoir No. 34, p. 21–46.
Dillon, W. P., Paull, C. K., and Gilbert, L. E.
1985 : History of the Atlantic continental margin off Florida: The Blake Plateau Basin: in Geologic evolution of the United States Atlantic Continental Margin, ed. C. W. Poag; Van Nostrand Reinhold, New York, p. 189–215.
Dillon, W. P., Valentine, P. C., and Paull, C. K.
1986 : Geology of the Blake Escarpment: NOAA Symposium Volume on Undersea Research (in press).
Dow, W. G.
1979 : Petroleum source beds on continental slopes and rises: in Geological and Geophysical Investigations of Continental Margins, eds. J. S. Watkins, L. Montadert, and P. W. Dickerson; American Association of Petroleum Geologists Memoir 29, p. 423–442.
Emelyanov, E. M.
1982 : Sedimentogene basseine Attanticheskogo Okeana (Sediment genesis in basins of the Atlantic Ocean): Moscow, USSR, Izdat. Nauka, 190 p.
Enos, P.
1974 : Reefs, platforms, and basins of middle Cretaceous in northeast Mexico: American Association of Petroleum Geologists Bulletin, v. 58, p. 800–809.
Ewing, M., Carpenter, G., Windisch, C., and Ewing, J.
1973 : Sediment distribution in the oceans. The Atlantic: Geological Society of America Bulletin 84, p. 71–88.
Evans, R.
1978 : Origin and significance of evaporites in basins around Atlantic margin: American Association of Petroleum Geologists Bulletin, v. 62, p. 223–234.
Frazer, J. Z., and Fisk, M. B.
1980 : Availability of copper, nickel, cobalt, and manganese from ocean ferromanganese nodules (III): Scripps Institute of Oceanography Reference Series 80-16, 116 pp.
Grow, J. A.
1980 : Deep structure and evolution of the Baltimore Canyon Trough in the vicinity of the COST No. B-3 well: in Geological Studies of the COST No. B-3 Well, United States Mid-Atlantic Continental Slope Area, ed. P. A. Scholle; U.S. Geological Survey Circular 833, p. 117–132.
Habib, D.
1982 : Sedimentation of black clay organic facies in a Mesozoic oxic North Atlantic: Third North American Paleontological Convention, Proceedings, v. 1, p. 217–220.
Halbach, P.
1982 : Co-rich ferromanganese seamount deposits of the Central Pacific Basins: in Marine Mineral Deposits–New Research Results and Economic Prospects, eds. P. Halbach and P. Winter; Essen, Marine Rohstoffe und Meerestechnik, Bd 6, Verlag Gluckauf, p. 60–85.
Halbach, P., and Manheim, F. T.
1986 : Cobalt and other metal potential of ferromanganese crusts on seamounts of the central Pacific basins: Marine Mining (in press).
Halbach, P., Manheim, F. T., and Otten, P.
1982 : Co-rich ferromanganese deposits in the marginal seamount regions of the central Pacific Basin–Results of the Midpac '81: Erzmetall. v. 35, p. 447–453.
Heath, G. R.
1981 : Ferromanganese nodules of the deep sea: Economic Geology, 75th Anniversary Volume, p. 736–765.
Hedberg, H. D.
1976 : Ocean boundaries and petroleum resources: Science, v. 191, p. 1009–1018.
Heirtzler, J. R., Taylor, P. T., Ballard, R. D., and Houghton, R. L.
1977 : A visit to the New England Seamounts: American Scientist, v. 65, p. 466–472.
Herbin, J. P., Deroo, G., and Roucaché, J.
1983 : Organic geochemistry in the Mesozoic and Cenozoic formations of Site 534, Leg 76, Blake-Bahama Basin, and comparison with Site 391, Leg 44: in Initial Reports of the Deep Sea Drilling Project, v. 76, eds. R. E. Sheridan and F. M. Gradstein; U.S. Government Printing Office, Washington, D.C., p. 481–493.
Hinz, K., and thirteen others
1982 : Preliminary results of DSDP Leg 79 seaward of the Mazagan Plateau off central Morocco: in Geology of the Northwest African Continental Margin, eds. U. von Rad, K. Hinz, M. Sarnthein, and E. Seibold; Springer-Verlag, New York, p. 23–33.
Holder, G. D., Angert, P. F., and Pereira, V.
1983 : Implications of gas hydrates associated with gas reserves: in Natural Gas Hydrates: Properties, Occurrences, and Recovery, ed. J. L. Cox, Butterworth, Boston, p. 91–114.
Hollister, C. D., Ewing, J. I., eds.
1972 : Initial Reports of the Deep Sea Drilling Project, v. 12: U.S. Government Printing Office, Washington, D.C., 1077 p.
Humphris, C. C., Jr.
1979 : Salt movement on Continental Slope, northern Gulf of Mexico: American Association of Petroleum Geologists Bulletin, v. 63, p. 782–798.
Jansa, L. F.

1981 : Mesozoic carbonate platforms and banks of the eastern North American margin: Marine Geology, v. 44, p. 97–117.
Jansa, L. F., and MacQueen, R. W.
1978 : Stratigraphy and hydrocarbon potential of the central North Atlantic basin: Geoscience Canada, v. 5, no. 4, p. 176–183.
Jansa, L. F., and Wade, J. A.
1975 : Geology of the continental margin off Nova Scotia and Newfoundland: in Offshore Geology of Eastern Canada, v. 2, Regional Geology, eds. W.J.M. vander Linden and J. A. Wade; Geological Survey of Canada Paper 74-30, p. 51–105.
Jansa, L. F., Enos, P., Tucholke, B. E., Gradstein, F. M., and Sheridan, R. E.
1979 : Mesozoic-Cenozoic sedimentary formations of the North American Basin: Western North Atlantic: in Deep Drilling Results in the Atlantic Ocean: Continental Margins and Paleoenvironment, eds. M. Talwani, W. Hay, and W.B.F. Ryan; American Geophysical Union, Maurice Ewing Series, v. 3, p. 1–57.
Jansa, L. F., Bujak, J. P., and Williams, G. L.
1980 : Upper Triassic salt deposits of the western North Atlantic: Canadian Journal of Earth Sciences, v. 17, p. 547–559.
Kendrick, J. W., Hood, A., and Castaño, J. R.
1979 : Petroleum-generating potential of Cretaceous sediments from Leg 43, Deep Sea Drilling Project: in Initial Reports of the Deep Sea Drilling Project, v. 43, eds. B. E. Tucholke and P. R. Vogt; U.S. Government Printing Office, Washington, D.C., p. 663–668.
King, P. B.
1969 : Tectonic map of North America: U.S. Geological Survey, 2 sheets, scale 1:5,000,000.
Kinsman, D.J.J.
1975 : Salt floors to geosynclines: Nature, v. 255, p. 375–378.
Kvenvolden, K. A.
1983 : Marine gas hydrates–I: Geochemical evidence: in Natural Gas Hydrates: Properties, Occurrences, and Recovery, ed. J. L. Cox; Butterworth, Boston, p. 63–72.
Kvenvolden, K. A., and Barnard, L. A.
1982 : Hydrates of natural gas in continental margins: in Studies in Continental Margin Geology, eds. J. S. Watkins and C. L. Drake; American Association of Petroleum Geologists Memoir No. 34, p. 631–640.
Lehner, P.
1969 : Salt tectonics and Pleistocene stratigraphy on Continental Slope of northern Gulf of Mexico: American Association of Petroleum Geologists Bulletin, v. 53, p. 2431–2479.
Manheim, F. T., Pratt, R. M., and McFarlin, P. F.
1980 : Composition and origin of phosphorite deposits of the Blake Plateau: Society of Economic Paleontologists and Mineralogists Special Publication 29, p. 117–137.
Manheim, F. T., Popenoe, P., Siapno, W. D., and Lane, C. M.
1982 : Manganese-phosphorite deposits of the Blake Plateau: in Marine Mineral Deposits, eds. P. Halbach and P. Winter; Marine Rohstoffe und Meerestechnik, v. 6, p. 9–44.
Manspeizer, W.
1982 : Dynamics of the Atlantic passive margin: A Triassic-Liassic record of tectonic climatic interactions: Geological Society of America Abstracts with Programs, v. 14, p. 555.
Martin, R. G.
1980 : Distribution of salt structures in the Gulf of Mexico: Map and descriptive text: U.S. Geological Survey Miscellaneous Field Studies Map MF-1213.
Mattick, R. E., Girard, O. W., Jr., Scholle, P. A., and Grow, J. A.
1978 : Petroleum potential of the U.S. Atlantic slope, rise, and abyssal plain: American Association of Petroleum Geologists Bulletin, v. 62, p. 592–608.
McIver, R. D.
1975 : Hydrocarbon occurrences from JOIDES Deep Sea Drilling Project: Ninth World Petroleum Congress Proceedings, v. 2, p. 269–280.
McKelvey, V. E., Wright, N. A., and Bowen, R. W.
1983 : Analysis of the world distribution of metal-rich subsea manganese nodules: U.S. Geological Survey Circular 886, 53 p.
Mete, L., and Rivera-Rodriguez, J.
1982 : Reservoir characterization of the Ribeira Grande (Azores) field: Stanford, California, Proceedings, Eighth Workshop, Geothermal Reservoir Engineering, p. 73–89.
Meyerhoff, A. A., and Hatten, C. W.
1974 : Bahamas salient of North America: Tectonic framework, stratigraphy, and petroleum potential: American Association of Petroleum Geologists Bulletin, v. 58, no. 6, p. 1201–1239.
Mountain, G. S., and Tucholke, B. E.
1985 : Mesozoic and Cenozoic Geology of the U.S. Atlantic Continental Slope and Rise: in Geologic Evolution of the United States Atlantic Continental Margin, ed. C. W. Poag; Van Nostrand Reinhold, New York, p. 293–341.
Murray, J.
1885 : Report on the specimens of bottom deposits collected by the U.S. Coast Survey steamer *Blake,* 1877–1880: Bulletin of the Museum of Comparative Zoology, v. 12, p. 37–61.
Murray, J., and Renard, A. F.
1891 : Report on deep-sea deposits, Report of Scientific Results, Voyage of *H.M.S. Challenger:* Deep Sea Deposits, p. 1–525.
Pálmason, G.
1981 : An assessment of the geothermal resources in Iceland: Reykjavik, Proceedings, Energy Conference, June 1981, p. 121–137 (in Icelandic).
Pálmason, G., Stefansson, V., Thorhallsson, S., and Thorsteinsson, T.
1983 : Geothermal field developments in Iceland: Stanford, California, Proceedings, Ninth Workshop, Geothermal Reservoir Engineering, p. 37–52.
Paull, C. K., and Dillon, W. P.
1981 : Appearance and distribution of the gas hydrate reflection in the Blake Ridge region, offshore southeastern United States: U.S. Geological Survey Miscellaneous Field Studies Map MF-1252.
Pautot, G., Auzende, J. M., and Le Pichon, X.
1970 : Continuous deep-sea salt layer along North Atlantic margins related to early phase of rifting: Nature, v. 227, p. 351–354.
Poag, C. W.
1982 : Stratigraphic reference section for Georges Bank basin-depositional model for New England passive margin: American Association of Petroleum Geologists Bulletin, v. 66, p. 1021–1041.
1985 : Cenozoic and Upper Cretaceous sedimentary facies and depositional systems of the New Jersey slope and rise: in Geologic Evolution of the United States Atlantic Continental Margin, ed. C. W. Poag; Van Nostrand Reinhold, New York, p. 343–365.
Popenoe, P., Coward, E. L., and Cashman, K. V.
1982 : A regional assessment of potential environmental hazards to and limitations on petroleum development of the southeastern United States Atlantic Continental Shelf, Slope, and Rise, offshore North Carolina: U.S. Geological Survey Open-File Report 82-136, 67 p., 1 plate.
Pratt, R. M., and McFarlin, P. F.
1966 : Manganese pavements on the Blake Plateau: Science, v. 151, p. 1080–1082.
Riggs, S. R.
1984 : Paleoceanographic model of Neogene phosphorite deposition, U.S. Atlantic continental margin: Science, v. 223, p. 123–131.
Robbins, E. I.
1977 : Geothermal gradients: in Scholle, P. A., ed., Geological studies on the COST B-2 well: U.S. mid-Atlantic outer continental shelf area: U.S. Geological Survey Circular 750, p. 44–45.
Roberts, D. G., and Caston, V. D.
1975 : Petroleum potential of the deep Atlantic Ocean: Ninth World Petroleum Congress Proceedings, v. 2, p. 281–298.
Rona, P. A.

1980 : Hydrothermal deposits of the Mid-Atlantic Ridge crest (latitude 26°N): in Geology and Geochemistry of Manganese, eds. I. M. Varentsov and G. Grasselly; Stuttgart, Schweitzerbart'sche Verlag, v. 3, p. 195–210.

1985 : Black smokers on the Mid-Atlantic Ridge: EOS Transactions American Geophysical Union, v. 66, p. 682–683.

Rona, P. A., and Lowell, R. P., eds.

1980 : Sea floor spreading centers, hydrothermal systems: Benchmark Papers in Geology 56: Dowden, Hutchinson and Ross, Inc., Stroudsburg, Pennsylvania, 425 p.

Rona, P. A., Thompson, G., Mottl, M. J., Jenkins, D., Mollette, M., von Damm, K., and Edmond, J. M.

1984 : Hydrothermal activity at the TAG hydrothermal field: Journal of Geophysical Research, v. 89, p. 11365–11378.

Sawyer, D. S., Toksoz, M. N., Sclater, J. G., and Swift, B. A.

1982 : Thermal evolution of the Baltimore Canyon Trough and Georges Bank Basin: in Studies in Continental Margin Geology, eds. J. S. Watkins and C. L. Drake; American Association of Petroleum Geologists Memoir No. 34, p. 743–762.

Schlee, J. S., Dillon, W. P., and Grow, J. A.

1979 : Structure of the continental slope off the eastern United States: in Geology of Continental Slopes, eds. L. J. Doyle and O. H. Pilkey; Society of Economic Paleontologists and Mineralogists Special Publication No. 27, p. 95–117.

Schlee, J. S., Poag, C. W., and Hinz, K.

1985 : Seismic stratigraphy of the continental slope and rise seaward of Georges Bank: in Geologic Evolution of the United States Atlantic Continental Margin, ed. C. W. Poag; Van Nostrand Reinhold, New York, p. 265–292.

Schott, W., Branson, J. C., and Turpie, A.

1975 : Petroleum potential of the deep-water regions of the Indian Ocean: Ninth World Petroleum Congress Proceedings, v. 2, p. 319–335.

Sheridan, R. E., and 25 others

1982 : Early history of the Atlantic Ocean and gas hydrates on the Blake Outer Ridge; Results of the Deep Sea Drilling Project Leg 76: Geological Society of America Bulletin, v. 93, p. 876–885.

Shipley, T. H., Houston, M. H., Buffler, R. T., Shaub, F. J., McMillen, K. J., Ladd, J. W., and Worzel, J. L.

1979 : Seismic evidence for widespread possible gas hydrate horizons on continental slopes and rises: American Association of Petroleum Geologists Bulletin, v. 63, p. 2204–2213.

Sloan, E. D., and Parrish, W. R.

1983 : Gas hydrate phase equilibrium: in Natural Gas Hydrates: Properties, Occurrences, and Recovery, ed. J. L. Cox; Butterworth, Boston, p. 17–34.

Stoll, R. D.

1974 : Effects of gas hydrates in sediments: in Natural gases in marine sediments, ed. I. R. Kaplan; Plenum Press, New York, p. 235–248.

Stoll, R. D., Ewing, J., and Bryan, G. M.

1971 : Anomalous wave velocities in sediments containing gas hydrates: Journal of Geophysical Research, v. 76, no. 8, p. 2090–2094.

Tissot, B. P., Durand, B., Espitalie, J., and Combaz, A.

1974 : Influence of the nature and diagenesis of organic matter in formation of petroleum: American Association of Petroleum Geologists Bulletin, v. 58, p. 499–506.

Tissot, B. P., Deroo, G., and Herbin, J. P.

1979 : Organic matter in Cretaceous sediments of the North Atlantic; contribution to sedimentology and paleontology: in Deep Drilling in the Atlantic Ocean, Continental Margins and Paleoenvironment, eds. M. Talwani, W. Hay, and W.B.F. Ryan; American Geophysical Union, Maurice Ewing Series 3, p. 362–374.

Tissot, B. P., Demaison, G., Masson, P., Delteil, D. R., and Combaz, A.

1980 : Paleoenvironment and petroleum potential of middle Cretaceous black shales in Atlantic basins: American Association of Petroleum Geologists Bulletin, v. 64, p. 2051–2063.

Toth, J. R.

1980 : Deposition of submarine crusts rich in manganese and iron: Geological Society of America Bulletin, pt. I, v. 91, p. 44–54.

Tucholke, B. E.

1979 : Relationship between acoustic stratigraphy and lithostratigraphy in the western North Atlantic Basin: in Initial Reports of the Deep Sea Drilling Project, v. 43, eds. B. E. Tucholke and P. R. Vogt; U.S. Government Printing Office, Washington, D.C., p. 837–846.

Tucholke, B. E., and Vogt, P. R.

1979 : Western North Atlantic sedimentary evolution and aspects of tectonic history: in Initial Reports of the Deep Sea Drilling Project, v. 43, eds. B. E. Tucholke and P. R. Vogt; U.S. Government Printing Office, Washington, D.C., p. 791–825.

Uchupi, E., and Austin, J. A.

1979 : The geologic history of the passive margin off New England and the Canadian Maritime Provinces: Tectonophysics, v. 59, p. 53–69.

Valentine, P. C.

1981 : Continental margin stratigraphy along the U.S. Geological Survey seismic line 5—Long Island Platform and western Georges Bank Basin: U.S. Geological Survey Miscellaneous Field Studies Map MF-857.

Waring, G. A.

1965 : Thermal springs of the United States and other countries of the world—a summary: Washington, D.C., U.S. Geological Survey Professional Paper 492, 303 pp.

MANUSCRIPT ACCEPTED BY THE SOCIETY JANUARY 2, 1986
WOODS HOLE OCEANOGRAPHIC INSTITUTION CONTRIBUTION NO. 6099.

ACKNOWLEDGMENTS

Preparation of this paper was partially supported by a Mellon Senior Study Award at Woods Hole Oceanographic Institution (BET). We are indebted to J. F. Karlo, S. Snelson, G. W. Moore, M. M. Ball, and M. H. Bothner for review of the manuscript, although the opinions expressed are solely the responsibility of the authors.

NOTE ADDED IN PROOF:

Since this chapter was written, hydrothermal metal sulfides have been drilled on the Mid-Atlantic Ridge at Ocean Drilling Project (ODP) Site 649 (Leg 106 Scientific Party, 1986, Drilling succeeds on bare rock: Geotimes, v. 31, no. 5, p. 10–12). The site (23°22.08′N, 44°57.00′W) is about 350 km south of the TAG area and 25 km south of the Kane Fracture Zone. The vent field that had been recognized in earlier bottom photographs was surveyed by television and side-scan sonar. The survey disclosed a large black smoker, two other large chimneys that probably are active, and many inactive chimneys in an area of at least 40,000 m^2. Ten shallow holes were drilled, the first ever drilled in an ocean-floor hydrothermal area. Hydrothermal deposits were at least 13 m thick at the visibly active vent. The deposits include sulfides of iron, copper and zinc in massive lenses, and disseminated fine grains, both within a matrix of talc or chlorite and possibly sulfate.

Printed in U.S.A.

The Geology of North America
Vol. M, The Western North Atlantic Region
The Geological Society of America, 1986

Chapter 41

The juridical ocean basin

John A. Knauss
Graduate School of Oceanography, University of Rhode Island, Kingston, Rhode Island 02881

INTRODUCTION

Since World War II, claims of national jurisdiction over areas of the ocean and its resources have been moving seaward in an irregular and often confusing way. The trend has been aptly described as "creeping jurisdiction." A strong case can be made for the proposition that the fisheries and mineral resources of the ocean have been the primary driving force for these claims.

Before 1958

"Every nation is free to travel to every other nation and to trade with it," wrote Hugo Grotius, often called the father of international law, in *Mare Liberum* published in 1609. He viewed the sea as *res communis,* a common possession for all and the private property of none. His views slowly gained international acceptance, in part because it was indeed in the general interest to encourage common use of the ocean for trade, and there was little in the ocean that could be subject to private property rights. Even fish were sufficiently abundant, given the crude methods of catching, so that whatever regulations were necessary could be controlled by local agreements.

The Grotius view was generally unchallenged for more than three centuries, a few international fisheries and whaling agreements notwithstanding. The first significant challenge came in 1945 when President Truman declared U.S. jurisdiction "with respect to the natural resources of the subsoil and seabed of [its] continental shelf" and implied that other coastal nations should do likewise. The reason, of course, was oil. Although "under water" oil wells had been drilled from piers or slant drilled from land since the turn of the century, geological exploration strongly indicated that many oil bearing structures in the Gulf of Mexico extended further offshore. Unlike the fishermen whose tradition has been to assume that the fish are a common property resource available to anyone who can catch them, the oil industry tradition was one of well-marked claims and jurisdiction; there had to be some legal entity with whom the industry could register its claims. The Truman Proclamation of 1945 provided official recognition by the United States that beyond the territorial sea at least some parts of the seabed were subject to private development. The first, truly offshore oil well was drilled 12 miles from shore in the Gulf of Mexico in 22 feet of water in 1947.

The Truman Proclamation on the Continental Shelf was carefully worded and its claims were narrowly prescribed; for example, the character of the water above the continental shelf was not affected and remained as high seas. If that was all that was done and if other nations had followed the U.S. suggestion, Grotius might have been bruised, but not badly wounded. However, President Truman issued a second proclamation at the same time; this one on fisheries. It read in part: "[T]he Government of the United States regards it as proper to establish conservation zones in those areas of the high seas contiguous to the coast of the United States wherein fisheries activities . . . shall be subject to regulation and control of the United States." The United States did not establish such fisheries zones immediately, but others did. Peru, with a narrow continental shelf and a deep sea trench off its shore, also possesses what at one time was the world's largest fishery. In 1945 and 1946, Mexico, Argentina, and Panama claimed jurisdiction over the resources of the Continental Shelf, and in 1947, Peru laid claim to its offshore continental shelf and to the adjacent sea out to 200 miles offshore. These nations were not as careful as the United States in limiting the nature of the conservation and resource zones. In August 1952, Chile, Ecuador, and Peru joined in the Declaration of Santiago to claim "sole sovereignty and jurisdiction" over the sea, sea floor, and subsoil of a zone extending *not less* than 200 nautical miles from the coast." In effect, the Declaration of Santiago established a 200-mile territorial sea for the west coast nations of South America.

The First and Second UN Law of the Sea Conferences

The first UN Conference on the Law of the Sea (LOS) was convened in 1958. Many saw as its goal and greatest achievement the codification of customary international law. Although there were a few regional treaties dealing with fisheries and a series of bilateral agreements between neighboring coastal nations, as well as a body of admiralty law on the rights and responsibilities of ships on the high seas, there were no treaty agreements relating to the different juridical zones of the ocean such as the high seas or territorial seas prior to the first UN Conference on the Law of the

Knauss, J., 1986, The juridical ocean basin; *in* Vogt, P. R., and Tucholke, B. E., eds., The Geology of North America, Volume M, The Western North Atlantic Region: Geological Society of America.

Sea. The 1958 Conference produced four conventions, three of which—on the territorial sea, high seas, and fisheries—did indeed codify customary international law, and the fourth was built on Truman's Continental Shelf Proclamation. At the same time it highlighted the disagreements. For example, the fisheries convention never did receive widespread adherence and there was no agreement on the breadth of the territorial sea. Thus, although the 1958 LOS Convention on the Territorial Sea and the Contiguous Zone carefully spells out the mixture of coastal nation and flag nation rights, it fails to prescribe a breadth to the territorial sea. A second UN Conference in 1960 failed by one vote to achieve the necessary two-thirds vote on a six-nautical-mile-wide territorial sea compromise between those who wanted three and those who wanted a wider territorial sea.

From a geological perspective, the most interesting of the four conventions of 1958 was the Convention on the Continental Shelf. The shelf was given a juridical, rather than a geological definition. It was defined as "the seabed and subsoil of the submarine areas adjacent to the coast [including coasts of islands] . . . but outside the area of the territorial sea . . . to a depth of 200 meters or beyond that limit to where the depth of the superjacent waters admit to the exploitation of the natural resources of the said area." The coastal nation can exercise sovereign rights over the continental shelf for purposes of exploring and exploiting its natural resources but the waters above the continental shelf remain as high seas. The permission of the coastal nation is also required for any research "concerning the continental shelf and undertaken there."

The outer edge of the Continental Shelf (when capitalized in this article Continental Shelf means a juridically defined shelf as distinguished from a geologically defined continental shelf) is determined by technology. As technology and economics make exploration and exploitation possible in ever deeper waters, the depth (and distance from shore) of the Continental Shelf increases. A hotly debated issue among international lawyers has been the limits, if any, to this definition. If commercial exploitation is occurring at 400 meters in one part of the world, does this mean that the outer edge of the Continental Shelf is 400 meters all over the world? There is a substantial body of opinion that believes this to be the case. Is there any limit to how far from shore the outer edge of the Continental Shelf can extend? Again there is a substantial body of opinion that believes the phrase in the definition "adjacent to the coast" places some outer limit. Neither issue has been specifically tested in court and both questions may be moot because of the recent agreement at the third UN Conference on Law of the Sea on a new definition for the outer edge of the Continental Shelf.

The Law of the Sea Convention of 1982

Some believe the third UN Conference on Law of the Sea was called because the US and the USSR wanted to resolve the question of the width of the territorial sea. Others believe it stemmed directly from a speech by Ambassador Pardo of Malta to the General Assembly of the United Nations in 1967 on the potential wealth from the manganese nodules on the deep sea floor far beyond national jurisdiction and the need to establish a legal regime to manage this resource. There is probably truth in both claims. What three years of preparatory work by the UN Seabed Committee did make clear is that the third UN Conference on LOS was going to have a dynamics of its own and that it was not going to be possible to contain the conference to one or two issues. All LOS issues, including those that many thought had been resolved by the four conventions of 1958, were to be reopened, reexamined and changed if found wanting. The Law of the Sea Conference began in Caracas in 1974 and labored for eight years. The resulting treaty has 320 articles in fifteen chapters and an additional eight annexes. The treaty was approved in April 1982, and was open for signatures in December 1982. At this writing a number of major maritime nations have not signed the Convention, including the United Kingdom, the Federal Republic of Germany, Italy, Peru, Ecuador, Argentina, and the United States. However, only the United States has publicly renounced the Convention. Its effectiveness will be strongly conditioned by how many nations stand with the United States. In balance the United States found the provisions of the Convention acceptable except for seabed mining. In the words of President Reagan, "Our review recognizes . . . that the deep seabed mining part of the Convention does not meet United States objectives."

JURIDICAL DIVISIONS OF THE OCEAN

The Law of the Sea Convention probably has sufficient support to enter into force. However, no nation is bound by a treaty to which it is not a party. What follows is a brief description of the juridical divisions of the ocean with particular emphasis on its provisions for exploration and development of its geological resource. A number of the provisions of the Law of the Sea Convention are either already embodied in the 1958 conventions or are widely perceived as being a part of customary international law. Their status will not change, even if the Law of the Sea Convention is not widely adopted.

Internal Waters

The water area landward of the baselines from which one determines the breadth of the territorial sea are internal waters. The coastal nation exercises sovereignty over these waters and the land beneath it as it does over its lands.

Territorial Sea

Coastal state sovereignty extends beyond its land and internal waters to an adjacent belt of ocean, not to exceed 12 nautical miles, described as the territorial sea. Sovereignty is complete except ships, including warships, have the right of innocent passage. The key word is passage. Innocent passage does not include stopping or anchoring (except in distress), fishing, research or survey activities, or "any other activity not having a direct bearing on passage." In addition, submarines or other underwater vehicles are required to navigate on the surface and show their flag.

Archipelagic Waters

Island nations such as Indonesia or the Philippines are "archipelagic nations," but Portugal cannot claim archipelagic status for the Azores or the United States for Hawaii. Archipelagic nations may draw straight base lines joining the outermost islands and drying reefs to enclose their nation provided that (a) the ratio of the area of the water to the area of the land of the enclosed area does not exceed nine to one and (b) the length of the individual base lines shall not exceed 100 nautical miles except that three percent can be up to 125 nautical miles (Figure 1). For purposes of computing the land to water ratio, a land area may include waters within fringing reefs of islands and atolls, "including that part of a steep-sided oceanic plateau which is enclosed or nearly enclosed by a chain of limestone islands and drying reefs lying on the perimeter of the plateau." For an archipelagic nation, these archipelagic base lines are used to determine the 12-mile territorial sea that extends seaward from the archipelagic waters.

Archipelagic nations exercise the same sovereignty over their archipelagic waters that other nations exercise over their territorial seas, with one exception. Sea lanes must be established through their archipelagoes to allow for free passage of ships and aircraft of all kinds over, on, or under the water.

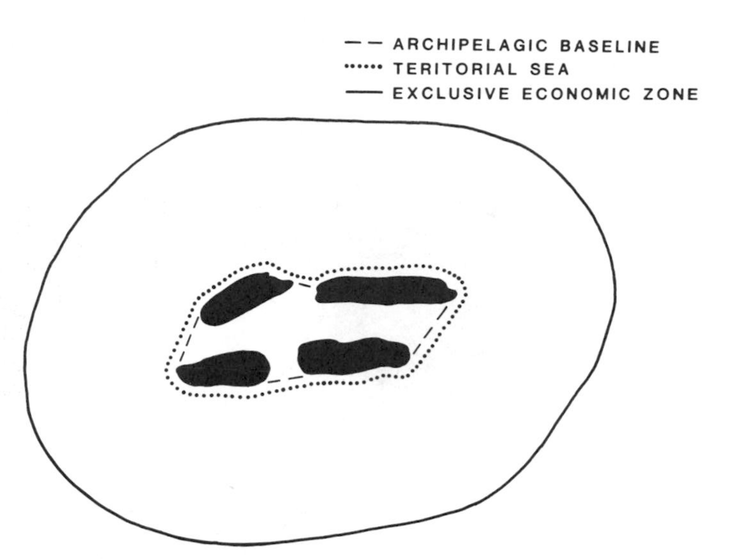

Figure 1. Archipelagic water is water internal to baselines that join the outermost points of the outermost islands and drying reefs, provided the baselines do not exceed 100 nautical miles (3 percent of the baselines can be 125 nautical miles) and that the ratio of land to water enclosed by the baselines is between 1 to 1 and 1 to 9. Seaward of the archipelagic waters is the 12-mile territorial sea and the 200-mile exclusive economic zone.

Straits Used for International Navigation

These are straits that connect one area of the high seas or exclusive economic zone with another area of the high seas or exclusive economic zone. When the LOS Convention moved to a 12-mile territorial sea a number of international straits less than 24 miles wide suddenly lost their high sea status; among them are the straits of Gibraltar and Malacca. In agreeing to a 12-mile territorial sea, the LOS Convention also established a special designation of "international straits" where ships and aircraft of all kinds are guaranteed free transit on, over, and beneath the sea. However, for such activities as survey, research, exploration, and exploitation of resources, these straits are part of the territorial sea of the contiguous nation.

Exclusive Economic Zone

The 1958 LOS Convention granted the coastal nation control over the resources of its adjacent seabed: the new LOS treaty extends control of resources to the adjacent water column. This new juridical area, the exclusive economic zone (EEZ), extends to 200 nautical miles beyond the base lines of the territorial sea; thus, if a nation establishes a 12-mile territorial sea, the EEZ would be 188 miles wide. In the EEZ the coastal nation exercises sovereign rights over the exploitation, management, and conservation of all resources of the water column and seabed, both living and nonliving, as well as the production of energy from waters, currents, or winds. The exclusive economic zone is a resource zone. The coastal nation exercises considerable control over all matters affecting the resources of the area including marine scientific research, which may or may not be directed toward resource exploration, as well as the protection and preservation of the marine environment. For navigational purposes, both commercial and military, the EEZ is the high seas (although the Convention carefully refrains from making that statement explicit) except that the coastal nation can enforce certain internationally agreed upon pollution regulations on commercial ships in its EEZ that it cannot enforce for ships on the high seas.

A considerable part of the world's oceans are included in the exclusive economic zone, including the entire Mediterranean and Caribbean. Except for Antarctica there is a 200-mile EEZ extending off of every coast and surrounding every habitable island (Figure 2). It has been estimated that the combined areas of territorial sea, archipelagic waters, and exclusive economic zone amount to about 32 percent of the ocean. The United States EEZ covers 2.2 million square nautical miles, or about 80 percent of the land area of the lower forty-eight states.

The Continental Shelf

To the extent that the exclusive economic zone gives the coastal nations control of seabed resources within the EEZ, the Continental Shelf jurisdiction out to 200 miles is redundant. But because the Continental Shelf juridical zone was defined in the 1958 Convention and because the outer limit of the Continental Shelf may extend beyond 200 miles, it is treated separately in the

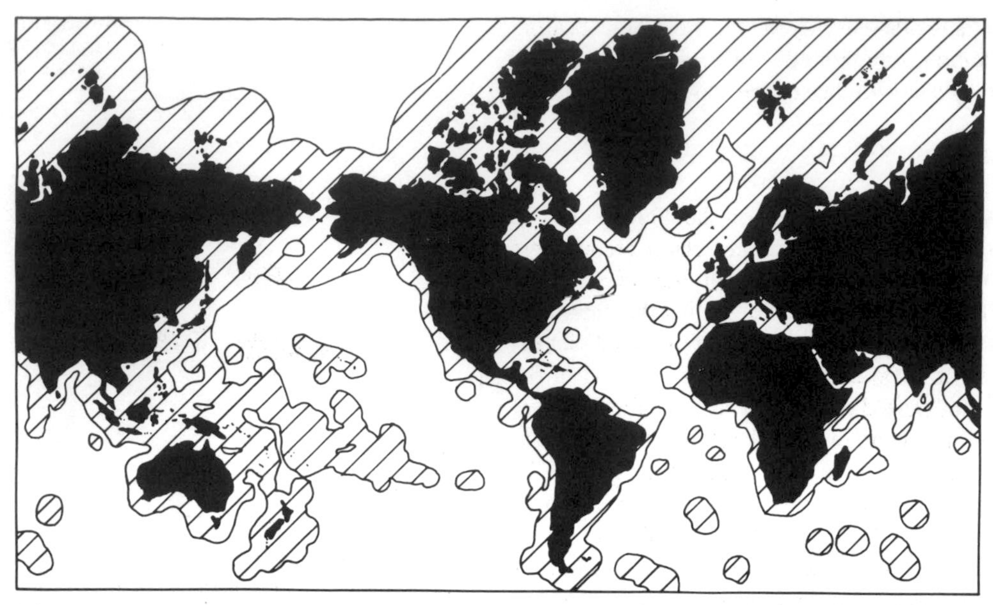

Figure 2. About 32 percent of the world's oceans are covered by territorial seas, archipelagic waters, and exclusive economic zones. The Mercator projection makes the percent covered look even larger.

new LOS treaty. Its landward edge is the outer limit of the territorial sea. Its seaward edge is the "natural prolongation of [a nation's] land territory to the outer edge of the continental margin, or to a distance of 200 nautical miles from the base lines from which the breadth of the territorial sea is measured where the outer edge of the continental margin does not extend to that distance." Thus for all coastal nations with relatively narrow continental margins, the outer edge of the Continental Shelf is well defined; it is 200 miles. For defining the outer edge of the Continental Shelf for broad margin nations the Convention substitutes a very confusing attempt at a geological definition for the exploitability definition of the 1958 Convention (see next section).

In both the 1958 Convention and new LOS treaty the coastal nation has absolute jurisdiction over the exploration and exploitation of resources of the seabed, both living and nonliving. In both, the coastal nations exercise jurisdiction over marine scientific research.

The Area

That part of the seabed beyond the limits of national jurisdiction, i.e., beyond a coastal nation's Continental Shelf is defined as "the Area" in the LOS treaty and "the Area and its resources are the common heritage of mankind." No nation can "claim or exercise sovereignty or sovereign rights over any part of the Area or its resources." "All rights to the resources of the Area are vested in mankind as a whole on whose behalf the [International Sea-Bed] Authority shall act."

The stated reasons for the United States' refusal to sign the Law of the Sea Convention were the following: relative power of nations or groups of nations to control the policy of the Authority; production controls levied on those nations who wish to mine; the mandatory requirement that in certain circumstances mining technology must be sold to other nations and the Authority; and concern about how this section of the treaty might be amended.

High Seas

The water column above the Area and above the Continental Shelf beyond 200 miles is the high seas, which are open to all nations, coastal or landlocked. All nations enjoy the rights of freedom of navigation, overflight, freedom to fish (with some qualifications), and freedom to conduct marine scientific research, construct artificial islands, and lay submarine cables and pipe lines. Each nation is responsible for determining that its nationals exercise freedom of the high seas in accordance with accepted international law and the provisions of the treaty.

Islands

Islands can have a territorial sea, a continental shelf, and an exclusive economic zone. However, *rocks* that "cannot sustain human habitation or economic life of [their] own" are entitled to a territorial sea, but not to an exclusive economic zone or a continental shelf. By this definition, St. Paul's Island near the equator in the Atlantic is presumably a rock.

Antarctica

The jurisdiction of the Antarctic continent and the waters surrounding it are controlled by the Antarctic Treaty (in force as of 1961), to which all nations laying claim to parts of the Antarctic Continent are parties, as well as a number of nations that do not have claims. The treaty is explicitly neutral concerning national claims. In the absence of internationally recognized claims to national jurisdiction over the continent there are, consequently, no recognized national claims to a territorial sea, exclusive economic zone, or Continental Shelf adjacent to the Antarctic. The waters surrounding Antarctica are high seas.

The treaty is subject to reopening in 1991. In the meantime there has been growing interest in the exploration and exploitation of the natural resources of the area. In 1982, agreement was reached on a Convention on the Conservation of Antarctica Marine Living Resources. The Convention covers an area from the Antarctic continent northward to a line that runs between 45 and 60°S and roughly coincides with the Antarctic Convergence.

Significant fishing is already taking place around Antarctica. Mineral exploitation in and around Antarctica is probably some years off; however, the eleventh Antarctic Treaty consultative meeting in 1981 recommended that "a regime on Antarctic mineral resources should be concluded as a matter of urgency" and that such a regime should apply not only to the Antarctic Continent but to "its adjacent offshore areas but without encroachment on the deep seabed." Negotiations on this regime began in 1982.

RULES FOR DETERMINING JURISDICTIONAL BOUNDARIES

Determining boundaries between jurisdictions can be difficult. In this section we briefly discuss three types of problems: base lines, boundaries between adjacent or opposite nations, and the determination of the outer edge of the Continental Shelf.

Base Lines

There is relatively little difference in the guidelines given to a nation by the 1958 Convention or by the new LOS treaty in determining those base lines that prescribe the perimeter of a nation, inside of which are its internal waters. The guidelines use the following definitions: the normal base line is the low water line along the coast; along deeply indented or island fringed coasts straight base lines may be drawn, but these "must not depart to any appreciable extent from the general direction of the coast" (a straight base line may be drawn across the mouth of a river); where there are bays, straight base lines should not exceed 24 nautical miles.

Even the simplest rules can be subject to differing interpretations. Consider the case of *United States* v. *Maine, Massachusetts, Rhode Island and New York.* The four states wish to extend the base line as far offshore as possible and thus extend the internal waters under their jurisdiction. The federal government wishes to limit this area. A key to the argument is whether Long Island is an island or part of the mainland, which in turn is determined in part by whether the East River is truly a river, that is, a branch of the Hudson with Manhattan Island lying between the branches, or whether the East River is a strait that separates Long Island Sound from the Atlantic (Figure 3a). If Long Island is an extension of the mainland, then several new base line options are available which, in the words of both the 1958 Convention and the new LOS treaty, "do not depart in any appreciable extent from the general direction of the coast." If it is an island then the existing base line more closely conforms to the LOS guidelines (Figure 3b). (Note added in proof: The United States Supreme Court in February 1985 opted for a line approximating the option running north-south in Figure 3b.)

With the possible exception of the passing reference to "historic bays" in Article 10, and which is nowhere defined in the Convention, the most ambiguous guideline for determining base lines in both the LOS treaty and the 1958 Convention reads in part, "account may be taken, in determining particular base lines, of economic interests peculiar to the region concerned, the reality and importance of which are clearly evidenced by long usage." This language in turn was derived from the Anglo-Norwegian Fisheries case decided by the International Court of Justice (ICJ) in 1951 that allowed Norway to connect the outermost rocks and islands in determining its baselines (Figure 4). Among the reasons given by the ICJ was that "certain economic interests peculiar to a region, the reality and importance of which are clearly evidenced by long usage" should not be overlooked.

The coastal nation is responsible for drawing its own baselines and publicizing them. If other nations object they can ask for a ruling by the International Court of Justice (ICJ). However, as demonstrated by the ICJ decision in the case of Norway, it would appear that nations have considerable flexibility in extending their internal waters seaward.

Delimitation Between Adjacent or Opposite Nations

Boundary disputes are to be settled by agreement by the parties involved. The early drafts of the LOS Convention read, "Such an agreement shall be in accordance with equitable principles, employing the median or equidistance line, where appropriate, and taking account of all circumstances prevailing in the area concerned." However ambiguous the above guidelines may seem, the final version is less useful, and makes reference only to achieving an equitable solution on the basis of international law.

One important reason why the equidistant or median line concept is not predominant is a 1969 decision of the ICJ that resolved claims of Denmark, West Germany, and the Netherlands to their share of North Sea Continental Shelf. Figure 5 shows the division that was determined by the ICJ as well as that based on the equidistant rule. Among its geographical, geological, and proportionality arguments the Court found it unacceptable, "that a State should enjoy Continental Shelf rights considerably different than its neighbors merely because in one case the coastline is roughly convex in form and in the other it is markedly concave, although those coastlines are comparable in length."

Including equitable principles with equidistance criteria for resolving boundary disputes allows for a significant increase in the degrees of freedom that can be applied to such discussions, as can be seen in a recent Canada-US boundary dispute. A simple equidistant line divides fisheries rich (and perhaps oil rich)

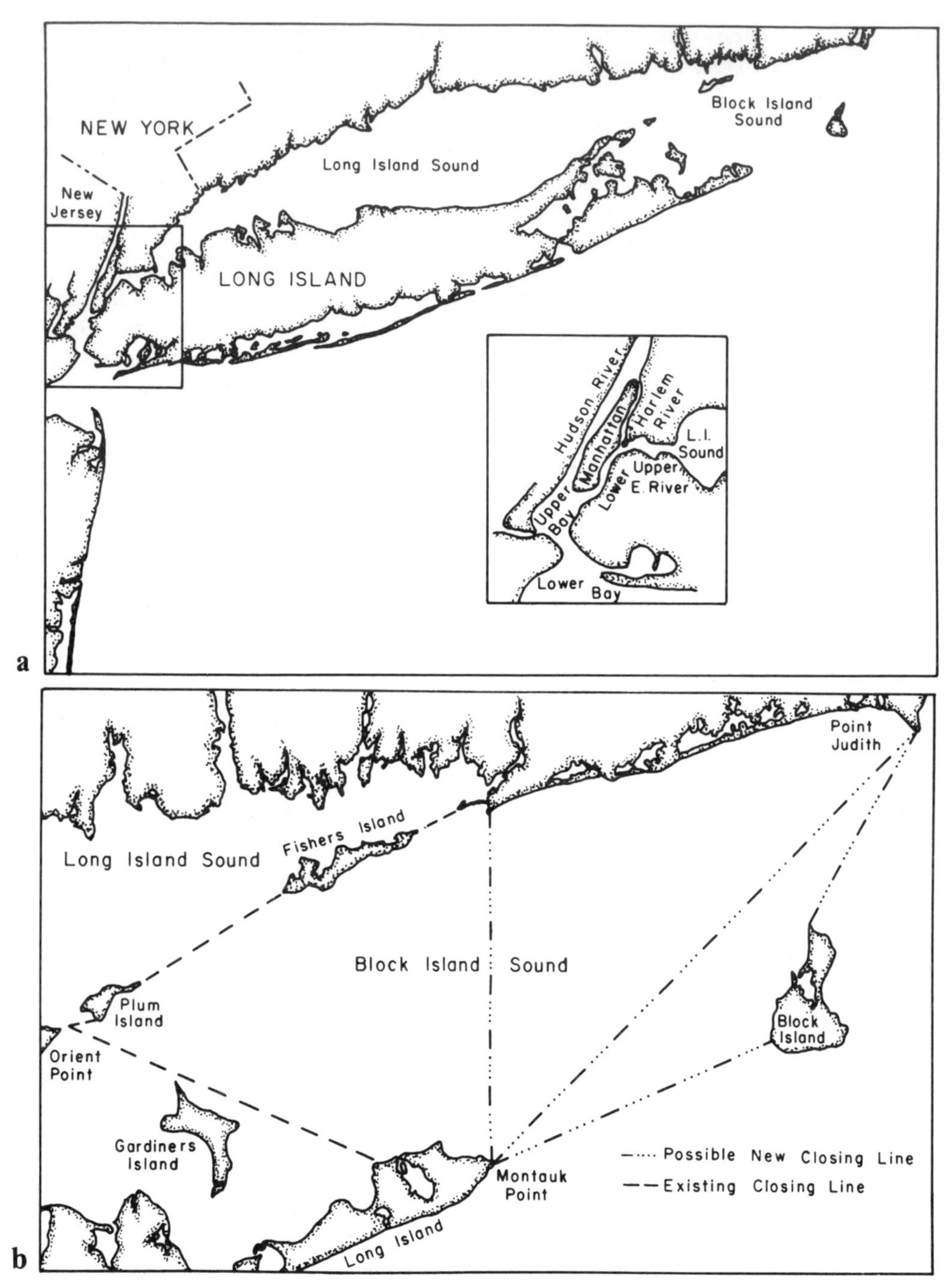

Figure 3. An example of baseline ambiguity. If Long Island is an island, then the existing closing lines are probably the best interpretation of treaty guidelines for the drawing of baselines. However, if Long Island is considered part of the mainland, then other options are possible which in accordance with Article 7, "do not depart in any appreciable extent from the general direction of the coast." (Redrawn from Figures 1 and 2 of Swanson, R. L., C. A. Parker, M. C. Meyer, and M. C. Champ, 1982, Is the East River New York a River or Long Island an Island?, NOAA Technical Report NOS 93, U.S. Dept. of Commerce, National Oceanic and Atmospheric Administration, Washington, D.C., 23 p.)

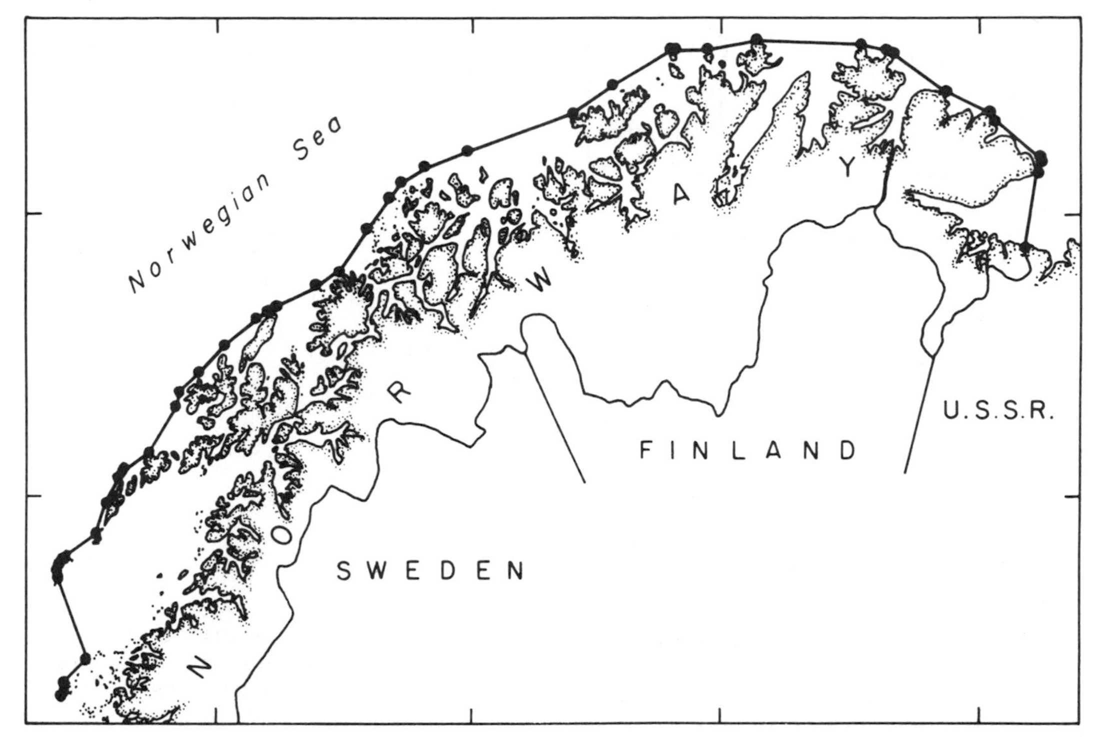

Figure 4. The International Court of Justice (ICJ) decreed in 1935 that Norway could use straight baselines to connect the outermost fringe of islands that generally parallel the highly indented Norwegian coast. The language in several articles in the LOS Convention follow the precedent of this case.

Georges Bank in such a way as to give about one-third to Canada and two-thirds to the United States (Figure 6). The United States claimed that for ecological, geological, and historical reasons, all of Georges Bank belongs to the United States and that the boundary should be determined in large part by the Northeast Channel. Canada, on the other hand, argued that since the shoreline, except for Cape Cod, runs to the Southwest and since Cape Cod is little more than a migrating sand spit, it should be ignored. Canada's proposal, which ignores Cape Cod, might be called an equitable equidistance line. At this writing the case is before the International Court of Justice. (Note added in proof: The ICJ division was basically an equidistant line with a small concession made to the United States in recognition of the proportionality of the length of coastline. [International Court of Justice, Year 1984, 12 October 1984. Case concerning Delimitation of the Maritime Boundary in the Gulf of Maine Area: Canada/United States of America.])

Outer Edge of the Continental Shelf

Perhaps even more ambiguous than the LOS treaty guidelines for base lines and delimitations of boundaries between adjacent and opposite nations, is the definition of the Continental Shelf as given by Article 76.

1. The continental shelf of a coastal State comprises the sea-bed and subsoil of the submarine areas that extend beyond its territorial sea throughout the natural prolongation of its land territory to the outer edge of the continental margin, or to a distance of 200 nautical miles from the baselines from which the breadth of the territorial sea is measured where the outer edge of the continental margin does not extend up to that distance.

2. The continental shelf of a coastal State shall not extend beyond the limits provided for in paragraphs 4 to 6.

3. The continental margin comprises the submerged prolongation of the land mass of the coastal State, and consists of the sea-bed and subsoil of the shelf, the slope and the rise. It does not include the deep ocean floor with its oceanic ridges or the subsoil thereof.

4. (a) For the purposes of this Convention, the coastal State shall establish the outer edge of the continental margin wherever the margin extends beyond 200 nautical miles from the baselines from which the breadth of the territorial sea is measured, by either:

(i) a line delineated in accordance with paragraph 7 by reference to the outermost fixed points at each of which the thickness of sedimentary rocks is at least 1 percent of the shortest distance from such point to the foot of the continental slope; or

(ii) a line delineated in accordance with paragraph 7 by reference to fixed points not more than 60 nautical miles from the foot of the continental slope.

(b) In the absence of evidence to the contrary, the foot of the continental slope shall be determined as the point of maximum change in the gradient at its base.

5. The fixed points comprising the line of the outer limits of the continental shelf on the sea-bed, drawn in accordance with paragraph 4 (a) (i) and (ii), either shall not exceed 350 nautical miles from the baselines from which the breadth of the territorial sea is measured or shall not exceed 100 nautical miles from the 2,500 metre isobath, which is a line connecting the depth of 2,500 metres.

6. Notwithstanding the provisions of paragraph 5, on submarine ridges, the outer limits of the continental shelf shall not exceed 350 nautical miles from the baselines from which the breadth of the territorial sea is measured. This paragraph does not apply to submarine elevations

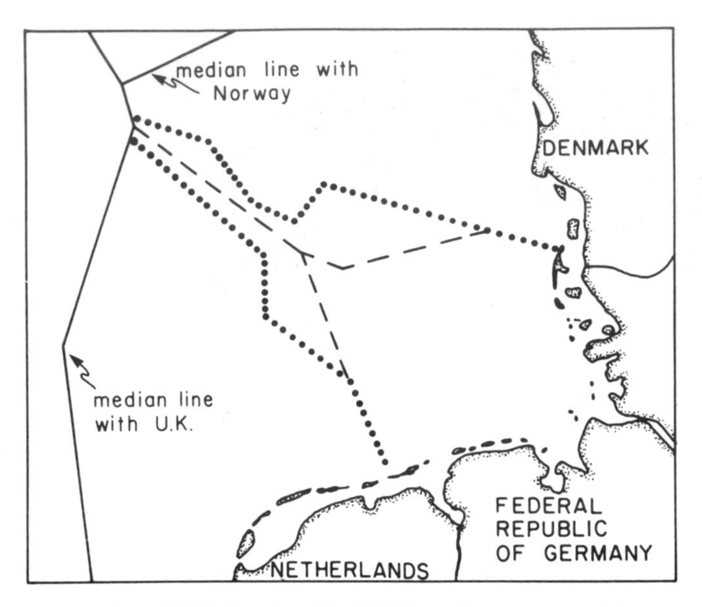

Figure 5. In resolving claims of Denmark, the Federal Republic of Germany, and Netherlands to the Continental Shelf, the ICJ ruled that the median line took unfair advantage of the FRG's concave coastline. The ICJ therefore gave Germany a significantly larger share of the Shelf.

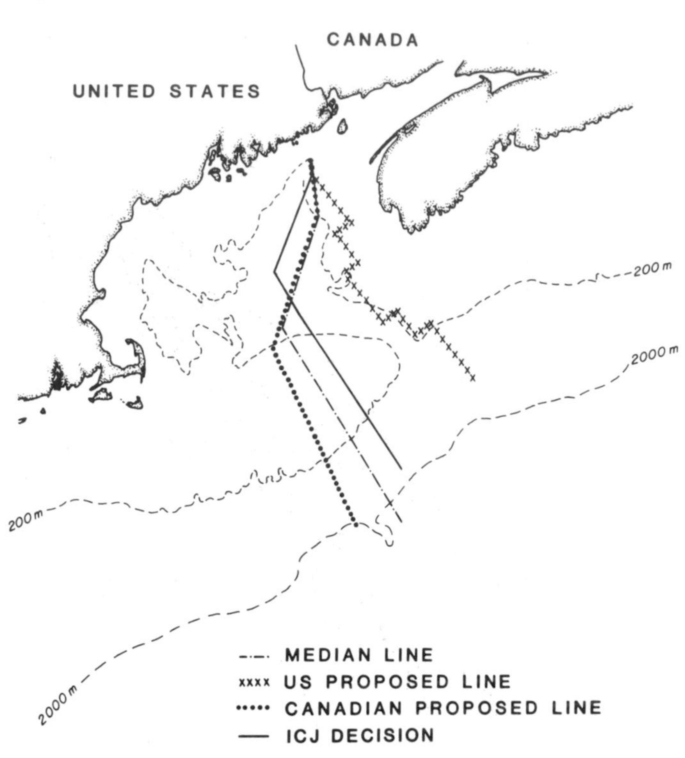

Figure 6. The median line between the United States and Canada cuts across Georges Bank. The United States claimed before the ICJ that the division should generally follow the Northeast Channel and that all of Georges Bank belonged to the United States. Canada's counterclaim was that Cape Cod should be ignored in determining a more equitable median line. The boundary decreed by the ICJ in its decision of October, 1984 is also shown.

that are natural components of the continental margin, such as its plateaux, rises, caps, banks and spurs.

7. The coastal State shall delineate the outer limits of its continental shelf, where that shelf extends beyond 200 nautical miles from the baselines from which the breadth of the territorial sea is measured, by straight lines not exceeding 60 nautical miles in length, connecting fixed points, defined by co-ordinates of latitude and longitude.

8. Information on the limits of the continental shelf beyond 200 nautical miles from the baselines from which the breadth of the territorial sea is measured shall be submitted by the coastal State to the Commission on the Limits of the Continental Shelf set up under Annex II on the basis of equitable geographical representation, establishment of the outer limits of their continental shelf. The limits of the shelf established by a coastal State on the basis of these recommendations shall be final and binding.

9. The coastal State shall deposit with the Secretary-General of the United Nations charts and relevant information, including geodetic data, permanently describing the outer limits of its continental shelf. The Secretary-General shall give the publicity thereto.

10. The provisions of this article are without prejudice to the question of delimitation of the continental shelf between States with opposite or adjacent coasts.

There was general agreement during the LOS treaty negotiations that there should be a fixed breadth to the edge of the Continental Shelf, as distinguished from exploitability criteria of the 1958 Convention that allowed the outer edge of the shelf to move deeper and farther offshore as technology advanced. Implicit in all discussions was an understanding that the new definition of the Continental Shelf should include all expected hydrocarbon deposits. Interpreting Article 76 should provide gainful employment for a number of marine geologists and lawyers for some time to come.

For those parts of the world where the outer edge of the margin does not extend more than 200 miles offshore the definition is straightforward; the outer edge of the Continental Shelf is 200 miles offshore. Where the margin extends beyond 200 miles, the outer edge of the Continental Shelf is determined by either the thickness of the sediments (paragraph 4(a)(i)), or is a point 60 nautical miles seaward of the foot of the slope (paragraph 4(a)(ii)) or is a point 100 nautical miles seaward of the 2500 meter isobath (paragraph 5). However, in all cases the outer edge of the Continental Shelf cannot extend more than 350 nautical miles from the base lines unless there is a submarine elevation such as a plateau or rise that is a natural prolongation of the land mass (paragraph 6), in which case the Continental Shelf can extend 100 nautical miles beyond the 2500 meter isobath even if that means extending the edge of the Continental Shelf seaward 350 nautical miles. Although many delegations to the LOS treaty negotiations had geologists to advise them, it is no exaggeration to say that no party to the negotiations had worked through the consequence of the various interpretations to Article 76, in large part because the necessary data are not available.

Paragraph 4(a)(i) is particularly intriguing. After the foot of the slope is determined, one then measures the thickness of sediment seaward of the slope. At 100 km from the foot of the slope the sediment must be at least 1 km thick; at 200 km it must be at least 2 km thick and so forth. The edge of the Continental Shelf is that point at which the sediment layer thins to a thickness of less than one percent of the distance from the foot of the continental slope.

Article 76 raises a number of intriguing questions. What significance should be attached to *continental*? Does Article 76 preclude Iceland from claiming a Continental Shelf of more than 200 miles under paragraph 3, or can Iceland extend its jurisdiction to 350 miles over that part of the mid-Atlantic ridge less 2,500 meters deep, or can it claim, under paragraph 6, jurisdiction over the entire ridge south of Iceland until it runs into the counterclaim of the Azores? Is the New Zealand claim to the New Zealand plateau determined by sediment thickness as in paragraph 4 or natural prolongation under paragraph 6?

Some observers believe that coastal nations will not rush to file claims since once a boundary is determined, it is presumably not open for renegotiation on the basis of newer information. Until such time as the coastal nation has reason to believe these regions are of some economic importance there is little to gain by submitting its proposed boundaries to the Commission on the Limits of the Continental Shelf, a 21-person elected board of experts. Although the coastal nation makes the final decision on its Continental Shelf boundary, it is "on the basis of these recommendations" of the Commission.

IMPACT OF TREATIES ON OCEAN USE

The areas of the ocean over which nations claim varying degrees of national jurisdiction are increasing. Whether or not the United States and a number of other major maritime nations remain outside the Law of the Sea Convention, some form of the 200-mile Exclusive Economic Zone will become part of customary international law. For example, the United States claimed an Exclusive Economic Zone in the spring of 1983 by Presidential Proclamation, even though it said it would not be party to the complete treaty. What may become hazy in time, if the Convention does not gain wide adherence, are the carefully negotiated differences between the mix of national and international rights that distinguish the territorial sea, the exclusive economic zone, the archipelagic waters, and the Continental Shelf beyond 200 miles. Given the trends of the past 40 years one might assume these areas will come under increasingly national jurisdiction with fewer recognized international rights to these areas. What follows is a brief description of how a number of practices of interest to geologists are being affected by the changing law of the sea.

Marine Scientific Research

At one extreme the coastal nation exercises complete discretion concerning marine scientific research in internal waters, archipelagic waters and the territorial sea. At the other, all nations are free to conduct marine scientific research on the high seas and in the deep seabed area. There is a mixed regime in the exclusive economic zone and the Continental Shelf.

"Marine scientific research in the exclusive economic zone and on the continental shelf shall be conducted with the consent of the coastal State," according to Article 246, which goes on to say "Coastal States shall, in normal circumstances, grant their consent to marine scientific research projects by other States or competent international organizations in their exclusive economic zone or on their continental shelf." But, finally "Coastal States may however, in their discretion withhold their consent . . . if that project: (a) is of direct significance for the exploration and exploitation of natural resources, whether living or non-living; (b) involves drilling into the continental shelf, the use of explosives or the introduction of harmful substances in the marine environment; (c) involves construction, operation or use of artificial islands, installations, and structures. . . .Permission can also be denied if the request contains information about the research program which is inaccurate or if the researching State or competent international organization has outstanding obligations to the coastal State from a prior research project."

The regulations are spelled out in more detail in the 1982 treaty than in the 1958 Convention. For example, under the new LOS Convention the researching nation is obligated to describe the proposed research in some detail, submit a request through official channels six months in advance of the planned starting date, allow coastal nation participation if possible, share data and samples to the extent possible, and if requested, provide the coastal nation with a preliminary report and an assessment of the results. The researching nation is obliged to make the results of the work internationally available unless the research is "of direct significance for the exploration and exploitation of natural resources" as determined by the coastal nation. The coastal nation can inform the research group at the time permission is granted of whatever rules it will use to control publication of such results.

The coastal nation must respond to the request for permission to do research within four months of the request. Thus a researching group will have at least two months notice of whether its request has been granted. If the coastal nation fails to respond within four months of the request, the Convention says the researching group can go forward with its work under implied consent.

The regulations for marine scientific research under the treaty are the same for the exclusive economic zone and the Continental Shelf, with one significant exception. For the Continental Shelf beyond 200 miles, the coastal nation may not withhold its permission for research under Article 246 unless the research is planned for an area the coastal nation had previously publicly designated as an area in which "exploitation or detailed exploratory operations focused on those areas are occurring or will occur within a reasonable period of time." In other words, until such time as a coastal nation is preparing detailed exploration of its Continental Shelf more than 200 miles offshore it cannot refuse permission to conduct marine scientific research on that portion of its shelf.

Oil and Mineral Exploration and Exploitation

On its Continental Shelf, the coastal nation exercises "sovereign rights for the purpose of exploring it and exploiting its natural resources, . . . no one may undertake these activities without the express consent of the coastal state." The regulations governing the coastal nation's exercise of its sovereignty are established by national legislation. In the United States the controlling legislation for Continental Shelf resources beyond three miles is the Outer Continental Shelf (OCS) Lands Act. To the extent this legislation does not adequately address such resources as polymetallic sulphides or Blake Plateau manganese crusts, the OCS Lands Act will require amendment or new legislation will be required.

In the LOS Convention the deep seabed beyond the edge of the Continental Shelf is called the Area, and as previously noted, its "resources are the common heritage of mankind . . . all rights in the resources of the Area are vested in mankind as a whole on whose behalf the [International Sea-Bed] Authority shall act." The treaty defines resources as "all solid, liquid or gaseous mineral resources *in situ* in the Area or beneath the sea-bed including polymetallic nodules."

The deep seabed mining section of the Convention and its two annexes read as much like the kind of business contract one might sign when buying a condominium as it does an international treaty that has been hailed as a constitution for the ocean. Although the Convention gives the Authority jurisdiction over all resources of the seabed, in fact, the details of the Convention speak to only one resource, manganese nodules. Those charged with negotiations were faced with an extraordinarily difficult task. On the one hand the industrial nations had to negotiate for a nonexistent industry. On the other hand, the developing nations were concerned, and probably with some reason, that if they put aside agreement on the regime until a viable marine mining industry was in place they would be in a weaker bargaining position. In the end the United States concluded that the final treaty provisions made it impossible for any but a state-subsidized industry to mine the deep seabed for manganese nodules.

Other mineral resources of commercial value may some day be found in the Area, and some mineral deposits such as polymetallic sulphides may be found in both the Area and on the Continental Shelf. According to the LOS Convention those found landward of the edge of the Continental Shelf are to be exploited by rules established by the coastal nation, while those in the Area are to be exploited by rules established by the Authority. Presumably, nations such as the United States that are not party to the treaty can exploit the resources of the Area independent of the Authority.

Military Uses of the Ocean

The Law of the Sea treaty states that the high seas and the Area shall be reserved for peaceful purposes and Article 240 of the treaty says "marine scientific research shall be conducted exclusively for peaceful purposes." Nowhere in the 1982 Convention is marine scientific research defined, and there is considerable precedence over the last 35 years that interprets the movement of military ships and submarines through the ocean as "for peaceful purposes." Such ships are free to roam the seas and engage in maneuvers and other peaceful activities anywhere in the ocean beyond the territorial sea. Thus a military vessel may conduct a detailed topographic survey on a foreign nation's Continental Shelf under the navigational provisions of the LOS Convention. It need not ask permission to do the work as long as it maintains that the survey is not marine scientific research, and the coastal nation has no recourse under the treaty to question the issue. In essence, therefore, military ships are free to conduct what most scientists might define as marine research anywhere in the ocean outside internal waters, although the territorial sea provisions of the Convention may limit the marine research in territorial and archipelagic waters to those activities that can reasonably be done while passing through the area and not stopping or anchoring.

Dumping and Seabed Emplacement

Of special interest is the proposal to emplace high level nuclear wastes from commercial and military reactors 30 to 100 meters below the surface in low energy, low tectonic activity sites in the deep North Pacific and Atlantic. Although nations such as the United States may have a number of geological options, nations such as the United Kingdom, Japan, and the Netherlands may not. Currently a number of nations are exploring the deep seabed option under a coordinated research program.

Since the deep seabed emplacement program is still in its research phase, the juridical issue has received limited attention. The proposed disposal areas are generally far from shore and not on a national Continental Shelf. Deep seabed emplacement is not addressed explicitly in either the Law of the Sea treaty or in the Convention on the Prevention of Marine Pollution by Dumping Wastes and Other Matter, usually referred to as the London Dumping Convention. The issue was never debated during the negotiations at the Law of the Sea Conference. However, some might argue the broad powers granted the International Sea-Bed Authority in the Area over resources, protection of marine environment including disposal of waste, and protection of human life could be interpreted to include jurisdiction over deep seabed emplacement. The juridical issue of the London Dumping Convention revolves around the definition of dumping, which is defined as "any deliberate disposal at sea." Dumping of high-level radioactive waste is forbidden under the London Dumping Convention.

A number of observers believe that although a strong case can be made for excluding the International Sea-Bed Authority from jurisdiction over deep seabed emplacement, a new annex at least will be required to the London Dumping Convention. This matter was discussed at the 1983 consultative meeting of the signatories to the London Dumping Convention, and one can assume that it will remain an agenda item for future meetings. Except as it may affect the issue of deep seabed emplacement, the London Dumping Convention does not directly affect any geological activities in the ocean. The jurisdiction of the London Dumping Convention is the dumping of waste material in all regions of the ocean beyond internal waters. However, dumping must be deliberate. Dumping does not include disposal at sea of material "incidental to, or derived from the normal operations of vessels, aircraft and platforms" or the "placement of matter for a purpose other than mere disposal thereof, provided such placement is not contrary to the aims of this convention." The latter exclusion would seem to cover scientific research.

CONCLUSION

National claims and international conventions have radically changed the juridical ocean. Prior to World War II, the oceans and its resources were *res communis,* a common property for all and the private property of none. National claims were mostly limited to a three-mile territorial sea. The 1982 Law of the Sea treaty: (1) gives archipelagic nations almost complete sovereignty over the water between its islands; (2) provides each nation and island with a 12-mile territorial sea over which it has complete sovereignty except for the right of foreign ships to innocent passage; (3) provides all coastal nations and habitable islands with sovereign rights to explore and exploit the resources and to control marine scientific research out to 200 nautical miles, thereby creating a new juridical regime, called an Exclusive Economic Zone; (4) provides the coastal nation with sovereign rights to explore and exploit the resources of the seabed to 350 miles (and sometimes further), if its Continental Shelf (roughly equivalent to the edge of the continental margin) extends that far; and (5) establishes an International Sea-Bed Authority to develop the rules and regulations by which the resources of the deep seabed are to be exploited.

It seems likely that the 1982 Law of the Sea Convention will gain the necessary 60 ratifications to enter into force, but for the near term at least it will fail to gain the adherence of a number of major maritime powers including the United States. The official United States position at the time of the treaty vote was that the United States could accept all parts of the proposed Convention except those dealing with the provision of the deep seabed beyond national jurisdiction. Most believe that the other parts of the Convention will be widely accepted in a general way, that is, varying degrees of national jurisdiction will be extended out to 200 miles. For example, in 1983 the United States claimed an Exclusive Economic Zone that closely tracks the EEZ provisions of the LOS Convention. However, if the United States and a number of other major maritime nations do not ratify the treaty, then the details of the closely negotiated text for these new juridical zones may not gain sufficient force to become widely accepted as customary international law. As long as man finds new ocean uses and resources, it is unlikely the clock will turn back. Even with a widely accepted Law of the Sea Convention in place, many believe that the creeping jurisdiction of slowly extending national jurisdiction will continue. Without a widely accepted treaty, the creep rate may be accelerated.

SELECTED BIBLIOGRAPHY

Background

O'Connell, D. P.
1982 : The International Law of the Sea: Vol. 1, Clarendon Press, Oxford, U.K., p. 633.
Hollick, Ann L.
1981 : U.S. Foreign Policy and the Law of the Sea: Princeton University Press, Princeton, N.J., p. 496.

Source Material

Oda, S.
1972 : The International Law of Ocean Development: Basic Documents, A. W. Sijthoff, Leiden, Netherlands, p. 519.
1962 : International Legal Materials, Vol. I-XXII, American Society of International Law.
(The Oda volume is a source of marine treaties, ICJ decisions, UN resolutions, etc., dealing with the oceans prior to 1970. The International Legal Materials series includes all significant UN and international legal materials on all subjects since about 1961. The 1982 UN Convention on the LOS is in 1982, Vol. XXI, No. 6, p. 1245–1354.)

Juridical Divisions

Shalowitz, A. L.
1962 : Shore and Sea Boundaries: 2 vols., U.S. Government Printing Office, Washington, D.C., p. 419 and 749.
Alexander, L.
1983 : Baseline Delimitations and Maritime Boundaries: Virginia Journal of International Law, v. 23, p. 503–5037.
Adede, A. O.
1979 : Toward the Formulation of the Rules of Delimitation of Sea Boundaries Between States with Adjacent or Opposite Coasts: Virginia Journal of International Law, v. 19, p. 207–255.
Hodgson, R. O., and Alexander, L.
1972 : Towards an Objective Analysis of Special Circumstances: Bays, Rivers, Coastal and Oceanic Archipelagos, and Atolls: Law of the Sea Institute Occasional Paper No. 13, Law of the Sea Institute, University of Hawaii, p. 54.

Manuscript Accepted by the Society August 31, 1984

ACKNOWLEDGMENT

I wish to acknowledge the assistance of Lewis Alexander who over the years has introduced me to many nuances of the ever changing juridical ocean basin.

Printed in U.S.A.

Index

Glossary of Acronyms

AABW or ABW, Antarctic Bottom Water
AEM, Airborne Electromagnetic Bathymetric Mapping (System name)
AFRC, African plate
AGU, American Geophysical Union
AM0-2, AM1-2, AM2-2, AM3-2, Absolute motion models by Minster and Jordan, 1978
AMAR, American Mid-Atlantic Ridge (Project name)
AMC, Axial magma chamber (generic feature name)
AMCOR, Atlantic Margin Coring Project
ANGUS, Acoustically Navigated Geologic Underwater System
AP, Abyssal Plain
BFP, Best fitting poles (plate rotation, Minster and Jordan, 1978)
BHST, Bathymetric Hazard Survey Test (Project name)
BO'SUN, Shallow-water swath-mapping system (name), now called Hydrochart or Bathymetric Swath Survey System (BS^3)
BS^3, Bathymetric Swath Survey System (See BO'SUN)
BSMA, Blake Spur Magnetic Anomaly
BSR, Bottom-simulating reflector
CCD, Calcite compensation depth (or carbonate compensation depth)
CGFZ, Charlie Gibbs Fracture Zone
CH-S, Cooling half-space (lithospheric thermal model)
CHUR, Chondritic meteorites uniform reservoir
CMQZ, Cretaceous Magnetic Quiet Zone
CNEXO, Centre National pour L'Exploitation des Oceans (now called IFREMER—Institut Francais pour l'Exploitation de la Mere)
CORG, Organic Carbon
COST, Continental Offshore Stratigraphic Test (drilling program)
CP, Cooling plate (lithospheric thermal model)
CRM, Chemical remanent magnetization
CZCS, Coastal Zone Color Scanner (satellite system)
D, Bulk solid/melt partition coefficient
DBDB, Digital Bathymetric Data Base (succeeded SYNBAPS)
DEEPTOW, Deep-towed sidescan sonar system (Scripps Inst. of Oceanography)
DMA, Defense Mapping Agency
DNAG, Decade of North American Geology
DSDP, Deep Sea Drilling Project
EEZ, Exclusive Economic Zone
EM, Electromagnetic
ECMA, East Coast Magnetic Anomaly
EPR, East Pacific Rise
ERTS, Earth Resources Technology Satellite
ESP, Expanding spread profile (seismic method)
EURA, Eurasian plate
FAMOUS, French-American Mid-Ocean Undersea Study (program)
FETI, Titanium ferrobasalt
FHAZ, Fracture high amplitude zone (magnetic anomaly province)
FNRS, Submersible (bathyscaphe) name
FZ, Fracture zone, generally used interchangeably with transform fault
FZA, Fracture Zone A, FAMOUS area (specific feature)
FZB, Fracture Zone B, FAMOUS area (specific feature)
GEBCO, General Bathymetric Chart of the Oceans
GEM, Goddard Earth Model (gravity field model)
GEOS, Geodetic Satellite (name)
GEOSAT, Geodetic Satellite (name)
GLORIA, Geologic Long Range Inclined Asdic; Asdic is British equivalent of "sonar" and stood for Allied Submarine Detection Investigation Committee
GPS, Global Positioning System (navigation satellites)
GRM, Geopotential Research Mission (satellites)
GSA, Geological Society of America
GSC, Galapagos Spreading Center
GSR, Greenland-Scotland Ridge
H-Zone, High-amplitude magnetic anomaly zone
HEBBLE, High Energy Benthic Boundary Layer Experiment (Project name)
HFU, Heat flow unit ($1 HFU = 1 \mu cal/cm^2/sec$)
HREE, Heavy rare earth elements
ICJ, International Court of Justice
IMQZ, Inner Magnetic Quiet Zone
IOS, Institute of Oceanographic Sciences (U.K.)
IPOD, International Phase of Ocean Drilling (Project)
IR, Infrared
IRDP, Iceland Research Drilling Project (drill site)
J-anomaly, Unique linear high amplitude magnetic anomaly in western North Atlantic
JAR, J-Anomaly Ridge (specific basement ridge)
JMQZ, Jurassic Magnetic Quiet Zone
JOIDES, Joint Oceanographic Institutions for Deep Earth Sampling
KFZ, Kane Fracture Zone
LAA, Large amorphous aggregates (marine snow)
LADLE, Lesser Antilles Deep Lithosphere Experiment (project)
LAGEOS, Laser Geodynamics Satellite (name)
LANDSAT, Earth surface imaging satellite (name)
LASE, Large Aperture Seismic Experiment (Project)
LFC, Large Format Camera System used on Space Shuttle to photograph earth's surface
LIBEC, Light-Behind Camera (system name)
LILE, Large ion lithophile element
LOS, Law of the Sea (convention)
LREE, Light rare earth elements
M-series, Mesozoic geomagnetic reversal sequence (also called Keathley Sequence)
MAGNET, Aeromagnetic project name (U.S. Naval Oceanographic Office)
MAGSAT, Aeromagnetic Satellite (name)
MAR, Mid-Atlantic Ridge
MARS, Mass accumulation rates
MOR, Mid-oceanic ridge
MORB, Mid-ocean ridge basalt
MPB, Magnetic plate boundary
MSS, Multi-Spectral Scanner (system)
NAVSTAR (GPS), Navigation satellite system (Global Positioning System)
NAVOCEANO, U.S. Naval Oceanographic Office
NAZC, Nazca Plate
NOAA, National Oceanic and Atmospheric Administration (agency name)
NOAM, North American Plate
NOO, U.S. Naval Oceanographic Office
NORDA, Naval Ocean Research and Development Activity (laboratory name)
NR-1, Submersible name
NRL, Naval Research Laboratory
NRM, Natural remanent magnetization
NSF, National Science Foundation
NUVEL-2, Relative plate motion models (DeMets and others, 1986)
OBS, Ocean bottom seismometer (generic feature name)
OCR, Overconsolidation ratio (sediment property)
OCS, Outer Continental Shelf (Lands Act)
OGO, Orbiting Geophysical Observatory (satellite name)
OMD, Ocean Margin Drilling (project name)
ONR, Office of Naval Research
OSC, Overlapping spreading center (generic feature name)
OTTER, Oceanographer Transform Tectonic Exploration Research (specific project)
PCFC, Pacific Plate
PDR, Precision Depth Recorder (system name)

PMEL, Pacific Marine Environmental Laboratory (NOAA lab)
PTDZ, Principal Transform Displacement Zone (generic tectonic province name)
RADARSAT, Satellite (name)
RBV, Return Beam Vidicon (imaging system used on LANDSAT satellite)
REE, Radio-echosounding (method)
RM1, RM2, Relate (plate) motion models (Minster and others, 1974)
ROV, Remotely operated (underwater) vehicle (generic system name)
RT, Ridge transform (intersection) (generic feature name)
RTR, Ridge-Transform-Ridge (plate boundary type)
SAR, Synthetic Aperture Radar (generic method)
SASS, Sonar Array Sounding System
SASS, SEASAT Satellite Scatterometer (specific system name)
SCUBA, Self-contained underwater breathing apparatus (generic system name)
SEABEAM, Multibeam bathymetric swath-mapping system (trade name)
SEAMARC, Sea Mapping and Remote Characterization (trade name) (side-scan sonar system)
SEASAT, Geodetic satellite (name)
SEIR, Southeast Indian Ridge
SENR, Southeast Newfoundland Ridge
SFS, Sea floor spreading
SIVGP, Smithsonian Institution Volcanic Glass Project
SLAR, Side-looking airborne radar (generic method name)
SLR, Satellite laser ranging (generic method name)
SOAM, South American plate
SMMR, Scanning Microwave Multichannel Radiometer (Satellite system)
SWH, Significant Wave Height (quantity)
SYNBAPS, Synthetic Bathymetric Profiling System (gridded data base)
TAG, Trans-Atlantic Geotraverse (program)
TFZ, Transform Fault Zone (generic tectonic province name)
TM, Thematic Mapper (system on LANDSAT satellites)
TOPEX, Satellite (name)
TPT, Trans-polar trend (arrangement of plate rotation poles for Atlantic plate pairs)
TRM, Thermoremanent magnetization
TTZ, Transform Tectonized Zone
URI, University of Rhode Island
USGS, U.S. Geological Survey
VD, Vertical deflection (parameter)
VHRR, Very High Resolution Radiometer (satellite system)
VLBI, Very Long Base Line Interferometry (positioning system using distant astronomic sources)
VRM, Viscous remanent magnetization
WBUC, Western Boundary Undercurrent
WHOI, Woods Hole Oceanographic Institution
WSBW, Warm Saline Bottom Water
WWSSN, Worldwide Standardized Seismograph Network

Typeset by WESType Publishing Services, Inc., Boulder, Colorado
Printed in U.S.A. by Malloy Lithographing, Inc., Ann Arbor, Michigan

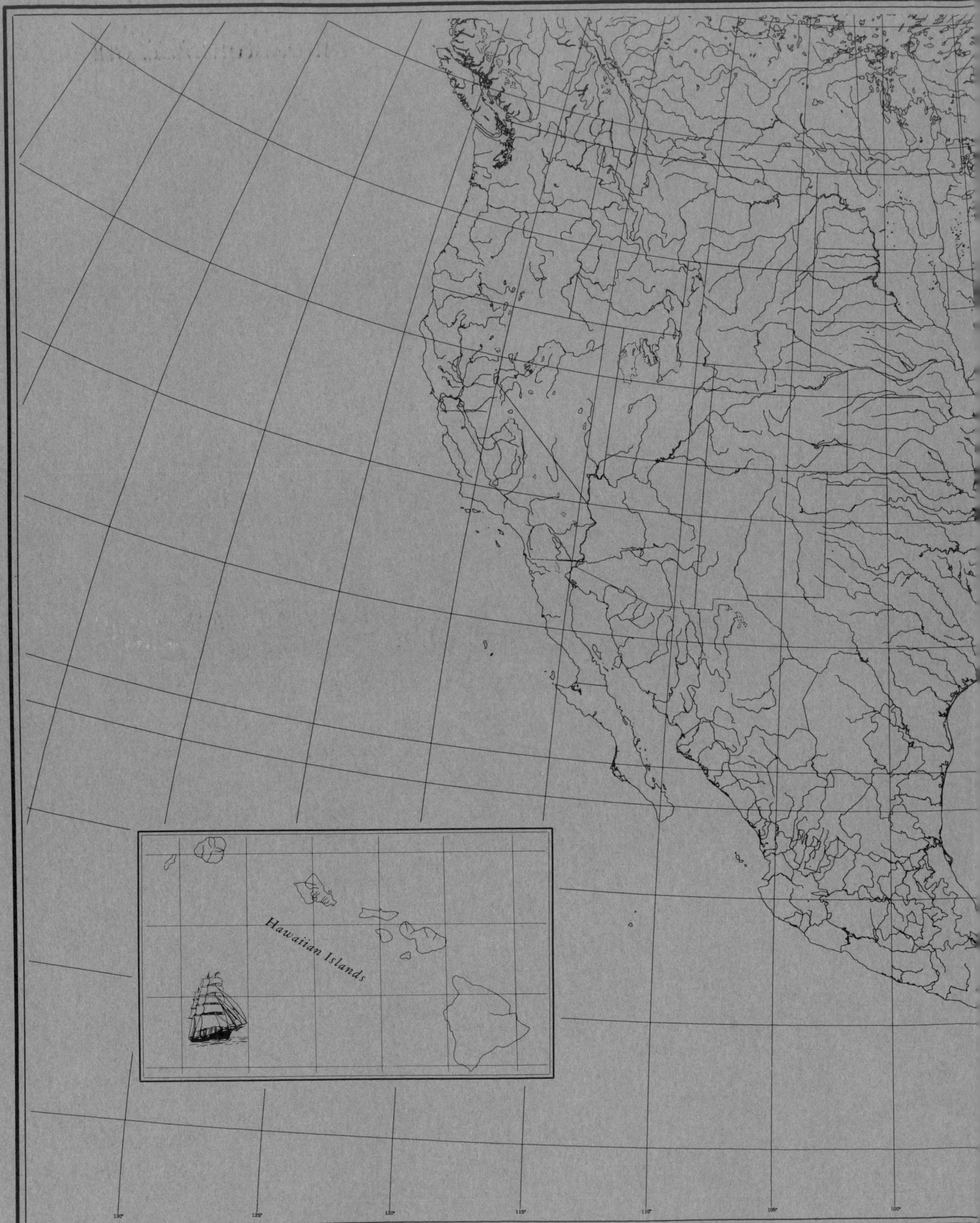

Hawaiian Islands

130° 125° 120° 115° 110° 105° 100°